RÖMPP
ENCYCLOPEDIA

RÖMPP ENCYCLOPEDIA

Natural Products

Editors

Prof. Dr. Wolfgang Steglich
Dr. Burkhard Fugmann
Dr. Susanne Lang-Fugmann

Authors

Prof. Dr. Günter Adam
Prof. Dr. Heidrun Anke
Prof. Dr. Wilhelm Boland
Dr. Martina Breiling
Dr. Jens Donath
Prof. Dr. Wittko Francke
Dr. Burkhard Fugmann
Dr. Fritz Hansske
Prof. Dr. Thomas Hartmann
Prof. Dr. Shin-Ichi Hatanaka
Prof. Dr. Gerhard Höfle
Dr. Rudolf Hopp
Dr. Udo Huber
Dr. Siegfried Huneck
Dr. Dirk Kusch
Dr. Susanne Lang-Fugmann
Dr. Heinrich F. Moeschler
Prof. Dr. Franz-Peter Montforts
Prof. Dr. Martin G. Peter
Dipf.-Chem. Jörn Piel
Dr. Hartmut Schick
Dr. Willibald Schliemann
Dr. Jürgen Schmidt
Prof. Dr. Klaus Schreiber
Prof. Dr. Horst-Robert Schütte
Dr. Stefan Schulz
Dipl.-Chem. Volker Stanjek
Dr. Bert Steffan
Prof. Dr. Wolfgang Steglich
Prof. Dr. Joachim Stöckigt
Prof. Dr. Dieter Strack
Dr. Horst Surburg
Prof. Dr. Joachim Thiem
Dr. Nikolaus Weber
Prof. Dr. Klaus Weinges
Dr. Peter Werkhoff
Dr. Ludger Witte
Prof. Dr. Axel Zeeck
Prof. Dr. Hans D. Zinsmeister

Georg Thieme Verlag
Stuttgart · New York

Die Deutsche Bibliothek – CIP-Einheitsaufnahme

Römpp encyclopedia natural products / ed. Wolfgang Steglich … Authors Günter Adam … [Transl. by R. Dunmur]. – Stuttgart ; New York : Thieme, 2000
Dt. Ausg. u. d. T.: Römpp-Lexikon Naturstoffe
ISBN 3-13-117711-X
ISBN 0-86577-988-0

Illustrated:
Hanne Haeusler
Kornelia Wagenblast

Cover design:
Schaaf + Killinger, Stuttgart

This book is an authorized and revised translation of the German edition published and copyrighted 1997 by Georg Thieme Verlag, Stuttgart, Germany.
Title of the German edition:
Römpp Lexikon Naturstoffe

Translated by Dr. R. Dunmur, Stuttgart

Rüdigerstraße 14, D-70469 Stuttgart
Printed in Germany by Konrad Triltsch, Print und digitale Medien GmbH, D-97199 Ochsenfurt-Hohestadt

Printed on acid-free paper, bleached without the use of chlorine by Arjowiggins Deutschland GmbH, Dettingen / Teck

ISBN 3-13-117711-X (GTV)
ISBN 0-86577-988-0 (TNY) 1 2 3 4 5 6

Preface

The *Roempp Encyclopedia Natural Products* intends to provide the reader with rapid information on organic natural products and phenomena connected with them. This book is directed to chemical scientists, biologists, pharmacists as well as to interested laymen who wish to learn about e.g. the lacrymator in onions or the typical flavor compounds of beetroot. Teaching staff at high schools and universities will be motivated to incorporate the fascinating realm of natural products into their classes.

The *Roempp Encyclopedia Natural Products* focusses on so-called *Secondary Metabolites*, compounds which occur in certain organisms only, often serving beneficial functions for the species. They are separated conceptually from the so-called *Primary Metabolites* which are essential for maintenance of cellular processes. The latter are ubiquitous in practically all organisms and provide the building blocks for the biosynthesis of secondary metabolites. Examples of secondary metabolites include toxins and defence compounds, which plants and animals employ against enemies or competitors; pheromones and flower pigments provide insects with essential information. Nevertheless, the biological function of many secondary metabolites remains unclear, e.g. it remains an open question why the fly agaric produces the red pigments in its cap, for which it is known. Secondary metabolites show remarkable structural diversity. Even structurally very "unusual" compounds like isocyanides, nitro compounds and halogenated aromatic ring systems, which were formerly thought to be of synthetic origin only, can be found among natural products. This encyclopedia is a rich source of such substances, which are usually not included in textbooks. This knowledge deficit has certainly contributed to the widespread opinion, that "natural" chemicals are benign and "synthetic" ones are hazardous.

The editors of this encyclopedia had the difficult task of choosing a representative selection from the more than 170,000 known secondary metabolites. The original first German edition of this book, published in 1997, has been completely revised, updated and expanded by the editors with the addition of new entries. The literature has been covered until the end of 1999. To keep the size of this book within reasonable limits, only primary metabolites that are related to secondary compounds have been mentioned. This applies especially to carbohydrates and nucleosides/nucleotides. Even macromolecular compounds and proteins are mentioned marginally only. Basic concepts and definitions that appear in standard textbooks are not covered as full entries. The main emphasis of this encyclopedia is on those types of microbial, plant or animal secondary metabolites which show interesting biological activities or are responsible for conspicuous properties of the organisms like color and smell. Our intention was to provide insight into the vast structural diversity of secondary metabolites which is becoming more and more important for the discovery of lead structures in pharmaceutical and crop protection research. Chemical ecology was another important focus. For certain phylogenetic groups of organisms, characteristic secondary metabolites are mentioned with respect to chemotaxonomic aspects. They can be found as specific entries under the compound names. Collective names like actinomycetes, toadstools, insect attractants or macrolides also facilitate searching. The numerous cross references are designed to make the information readily accessible. Furthermore, the *Roempp Encyclopedia Natural Products* contains an appendix with an extensive index of Latin species names and a molecular formula index.

The selection of compounds remains subjective and there are certainly important entries that are missing and should be considered for a subsequent edition of this book. The editors would very much appreciate your suggestions.

Due to the restricted size of this book, the names of authors have been omitted in the literature references. The coverage of total syntheses is restricted to references to the literature. Special attention was given to provide correct stereochemistry and concise texts.

Leverkusen, Ratingen, Munich,
March 2000

B. Fugmann
S. Lang-Fugmann
W. Steglich

Preface

Although our ability to synthesize molecules has progressed dramatically over the course of the past several decades, organic synthesis is still in its infancy compared to the dazzling variety and complexity of molecules which nature can so deftly prepare. Taking the same number and types of constituent atoms, nature combines them with seemingly limitless variation, creating in the process molecular architectures which not even the most fanciful of chemists could be expected to conjure on their own. More significantly, nature's ingenious collection of molecular designs exhibits a dazzling array of biological properties, affording opportunities to probe important biochemical processes as well as to develop therapeutic agents to treat some of the most serious ailments afflicting mankind. With new molecular constructs isolated and characterized daily, the constantly enriched library of natural products provides a vibrant engine that will undoubtedly continue to drive forward our quest for understanding and improving the world around us.
As one pages through the *Roempp Encyclopedia Natural Products*, these concepts come to life through a well selected sample of nature's library of molecular diversity. More than a simple catalogue listing of structures, the editors provide thorough and meticulously researched information on the biology, biochemistry, and pharmacology of important natural products isolated from a myriad of organisms from all corners of the world. Whether searching for a historically important natural product such as the poison strychnine or a leading cancer therapeutic such as Taxol™, scientists and laymen alike are certain to find answers to their questions as well as key references, should their interest be piqued. This volume represents a fine addition to the reference literature, and hopefully it will be followed by a continuing series which will provide not only a detailed collection of important information, but also a source of inspiration for new generations of chemists and biologists.

K. C. Nicolaou
The Scripps Research Institute and
University of California, San Diego
9 June 2000

Preface

Natural products, whether their sources are animals, plants or microorganisms, are sources of inspiration not only for organic chemists, but also for physical and biological chemists and scientists in a variety of other fields. They are the motivating power that drives creative thought processes in various scientific disciplines.
The compounds included in this encyclopedia were selected from an enormous number, approximately 170,000, of natural products by 39 recognized German experts from the various areas of natural products chemistry. The compounds were chosen in a well-balanced way from the most significant classes of natural products such as antibiotics, alkaloids, pheromones, flavonoids, and ionophores. Secondary metabolites are given priority, but important primary metabolites are also described, though fewer in number.
The stereochemical illustration of the structure of natural products seems to be the most effective way of drawing out creativity from scientists and even artists. In addition to the structure, various biological and physical properties are concisely described; the profiles of the compounds from discovery to application are also soundly arranged in this book. Useful synthetic derivatives and biosynthesis of natural products have been added. Thus the depth of coverage mirrors the breadth of natural products research.
One of the principal features of this book is that it serves as a reference to some important general terms related to natural products such as alkaloids, carotinoids, glycoproteins, and snake venoms. Moreover, historical coverage and overview texts, even headings like "screening", can be found, indicating that this book is much more than a mere dictionary of natural products.
The original and review articles cited in this encyclopedia are not only related to the discovery of the compounds, but are also suitable for their comprehensive understanding.
The highly acclaimed original first German edition of this book was published in 1997. It was completely revised and updated by the editors and translated into English as an international version.
I believe that this encyclopedia is one of the most valuable books that people interested in natural products chemistry should have on hand, regardless of whether they are experts or newcomers in the field.

Satoshi Ōmura, Ph. D.
(June, 2000)

Frequently cited books and periodicals

ApSimon **1**	ApSimon (ed.), The Total Synthesis of Natural Products, Vol. 1–9, New York: Wiley and Sons 1973–1992
Arctander	Arctander, Perfume and Flavor Materials of Natural Origin, Elisabeth, N. J.: published by the author 1960
Asahina-Shibata	Asahina & Shibata, Chemistry of Lichen Substances, Tokyo: Japan Soc. for the Promotion of Science 1954
Atta-ur-Rahman **15**	Atta-ur-Rahman (ed.), Studies in Natural Product Chemistry, 20 Vol., Amsterdam: Elsevier since 1988
Bauer et al. (2.)	Bauer, Garbe & Surburg, Common Fragrance and Flavor Materials, 2. edition, Weinheim: VCH Verlagsges. 1990
Bedoukian (3.)	Bedoukian, Perfumery and Flavoring Synthetics, 3. edition, Wheaton: Allured 1986
Beilstein E IV **7**	Beilsteins Handbook of Organic Chemistry, 4. edition, Berlin: Springer since 1918, analog E III/IV 17 for the 3./4. and EV 17/11 for the 5., since 1994 Beilstein Informationssysteme GmbH
Betina	Betina (ed.), Mycotoxins, Amsterdam: Elsevier 1984
Barton-Nakanishi **1**	Barton & Nakanishi, Comprehensive Natural Products Chemistry, Vol. 1–9, Amsterdam: Elsevier Pergamon 1999
C. I. **1**	Colour Index, 3. edition, 4. revision, 9 Vol. and Supplements, Bradford: Society of Dyers and Colourists 1971–1992
Cole-Cox	Cole & Cox, Handbook of Toxic Fungal Metabolites, New York: Academic Press 1981
Culberson	Culberson, Chemical and Botanical Guide to Lichen Products, Chapel Hill: The University of North Carolina 1969
Collins-Ferrier	Collins & Ferrier, Monosaccharides, their Chemistry and Roles in Natural Products, Chichester: J. Wiley & Sons 1995
Czygan (2.)	Czygan (ed.), Pigments in Plants, 2. edition, Berlin: Akademie-Verl. 1980
Dictionary of Terpenoids **2**	Conolly & Hill, Dictionary of Terpenoids, 3 Vol. (Vol. 1 Mono- and Sesquiterpenoids, Vol. 2 Di- and higher Terpenoids, Vol. 3 indexes), London: Chapman & Hall 1991
Dolphin **I**	Dolphin (ed.), The Porphyrins, 7 Vol., New York: Academic Press 1978, 1979
Dolphin B_{12}, **I**	Dolphin (ed.), Vitamin B_{12}, Vol. I, II, New York: Wiley 1982
Elsevier **14**	Elsevier's Encyclopedia of Organic Chemistry, Series III: Carboisocyclic Condensed Compounds (Vol. 12, 13, 14 and Supplement), Amsterdam: Elsevier 1940–1954; Berlin: Springer 1954–1969
Fenaroli (2.) **1**	Fenaroli's Handbook of Flavor Ingredients, 2. edition, Cleveland: CRC Press 1975
Florey **6**	Florey (ed.), Analytical Profiles of Drug Substances, 23 Vol., New York: Academic Press 1972–1992
Gill-Steglich	Gill & Steglich, Pigments of Fungi (Macromycetes), in Progress in the Chemistry of Organic Natural Compounds (ed. Zechmeister), Vol. 51, Wien-New York: Springer 1987
Goodwin I (2.)	Goodwin (ed.), Chemistry and Biochemistry of Plant Pigments, 2. edition, Vol. 1, London: Academic Press 1976
Goodwin II	Goodwin (ed.), Plant Pigments, London: Academic Press 1988
Hager (5.) **7b**	Hagers Handbuch der Pharmazeutischen Praxis (ed. Bruchhausen et al.), 5. edition, 10 Vol., Berlin: Springer 1991–1999
Handbook of Terpenoids **I**	Dev & Narula, Handbook of Terpenoids, Monoterpenoids, Vol. I, II, Boca Raton: CRC Press, Inc. 1982
Harborne (1988)	Harborne (ed.), The Flavonoids, part 1 & 2, London: Chapman & Hall 1975; 1. Supplement: Advances in Research, published 1982; 2. Supple-

	ment: Advances in Research since 1980, published 1988; 3. Supplement: Advances in Research since 1986, published 1994
Hegnauer **I**	Hegnauer, Chemotaxonomie der Pflanzen, X Vol., Basel: Birkhäuser 1962, 1986
Isler	Isler et al. (eds.), Carotenoids, Basel: Birkhäuser 1971
Jordan	Jordan (ed.), Biosynthesis of Tetrapyrroles, Amsterdam: Elsevier 1991
Karrer, Nr. 100	Karrer et al., Konstitution und Vorkommen der organischen Pflanzenstoffe (exklusive Alkaloide), Basel: Birkhäuser 1958, 1977 (Supplement 1), 1981 (Supplement 2/1), 1985 (Supplement 2/2)
Kirk-Othmer (4.) **17**	Kirk & Othmer (eds.), Encyclopedia of Chemical Technology, 26 Vol., 4. edition, New York: Interscience since 1992
Kleemann-Engel	Kleemann & Engel, Pharmaceutical Substances, 3. edition, Stuttgart: Thieme 1999
Lindberg **3**	Lindberg (ed.), Strategies and Tactics in Organic Synthesis, 3 Vol., New York: Academic Press 1984, 1989, 1991
Luckner (3.)	Luckner, Secondary Metabolism in Microorgnisms, Plants, and Animals, 3. edition, Berlin: Springer 1990
Maarse	Maarse (ed.), Volatile Compounds in Food and Beverages, New York: Marcel Dekker 1991
Manske **11**	eds.: Manske & Holmes, Vol. 1–4; Manske, Vol. 5–16; Manske & Rodrigo, Vol. 17; Rodrigo, Vol. 18–20; Brossi, Vol. 21–40; Brossi & Cordell, Vol. 41; Cordell, Vol. 42–44; Cordell & Brossi, Vol. 45, The Alkaloids, Chemistry and Pharmacology, 53 Vol. up to 2000, New York: Academic Press since 1950
Martindale (29.)	Martindale, The Extra Pharmacopoeia (Reynolds, ed.), 29. edition, London: the Pharmaceutical Press 1989; analog Martindale 30. edition of 1993
Merck-Index (12.)	The Merck-Index, An Encyclopedia of Chemicals, Drugs and Biologicals, 12. edition, Whitehouse Station, N. Y.: Merck & Co., Inc. 1996
Mothes et al.	Mothes, Schütte & Luckner (eds.), Biochemistry of Alkaloids, Berlin : VEB Verl. der Wissenschaften 1985 ; Weinheim : Verl. Chemie 1985
Müller-Lamparsky	Müller & Lamparsky (eds.), Perfumes, Art, Science, Technology, London: Elsevier Appl. Sci. 1991
Negwer (6.), No. 100	Negwer, Organic-Chemical Drugs and their Synonyms, 6. edition, Berlin: Akademie-Verl. 1987; New York: VCH Publishers 1987; analog 7. edition of 1994
Nicolaou	Nicolaou & Sorensen, Classics in Total Synthesis, Weinheim: VCH Verlagsges. 1996
Ohloff	Ohloff, Riechstoffe u. Geruchssinn, Berlin: Springer 1990
Pelletier **1**	Pelletier (ed.), Alkaloids, Chemical and Biological Perspectives, Vol. 1–14, New York: Wiley 1983; Oxford: Pergamon 1994
Pfander (2.)	Pfander (ed.), Key to Carotenoids, 2. edition, Basel: Birkhäuser 1987
Phillipson et al.	Phillipson, Roberts & Zenk (eds.), The Chemistry and Biology of Isoquinoline Alkaloids, Berlin: Springer 1985
Phytochemical Dictionary	Harborne & Baxter, Phytochemical Dictionary, London: Taylor & Francis 1993
Rowe	Rowe (ed.), Natural Products of Woody Plants II, Berlin: Springer 1989
Sax (8.)	Lewis (ed.), Sax's Dangerous Properties of Industrial Materials, 8. edition, 3 Vol., New York: Van Nostrand Reinhold 1992
Scheuer I **1**	Scheuer (ed.), Marine Natural Products, Vol. 1–5, New York: Academic Press 1978–1983
Scheuer II **1**	Scheuer (ed.), Bioorganic Marine Chemistry, 6 Vol., Berlin: Springer 1987 –1992
Shamma	Shamma, Isoquinoline Alkaloids, Chemistry and Pharmacology, New York: Academic Press 1972
Shamma-Moniot	Shamma & Moniot, Isoquinoline Alkaloids Research 1972–1977, New York: Plenum Press 1978
Smith	Smith (ed.), Porphyrins and Metalloporphyrins, Amsterdam: Elsevier 1975
Snell-Ettre **18**	Snell & Hilton (eds.), from Vol. 8 Snell & Ettre (eds.), Encyclopedia of Industrial chemical Analysis, 20 Vol., New York: Interscience 1966–1975
Stryer 1995	Stryer, Biochemistry, New York: Freeman 1995
Thomson **2**	Thomson, Naturally Occurring Quinones, Vol. 1–4, London: Butterworth 1957; London: Academic Press 1971; London: Chapman & Hall 1987

TNO list (6.) TNO list (6.) Suppl. 1	Maarse & Visscher (eds.), Volatile Compounds in Food – Qualitative and Quantitative Data, 6. edition, Zeist: TNO 1989–1994 [Suppl. 1 (1990); 2 (1991); 3 (1992); 4 (1993); 5 (1994)]
Turner **1**	Turner or Turner & Aldrige, Fungal Metabolites, Vol. 1 & 2, London: Academic Press 1971, 1983
Ullmann (5.) **A12**	Ullmann's Encyclopedia of Industrial Chemistry, 5. edition, 36 Vol., Weinheim: Wiley-VCH since 1985
Waterman **8**	Waterman (ed.), Methods in Plant Biochemistry, Vol. 8, Alkaloids and Sulfur Compounds, London: Academic Press 1993
Zechmeister **35**	Zechmeister (ed.), Progress in the Chemistry of Organic Natural Products, 79 Vol. up to 2000, Berlin: Springer since 1938
Zeelen	Zeelen, Medicinal Chemistry of Steroids, Amsterdam: Elsevier 1990

List of abbreviations

a = year
$[\alpha]_D$ = specific optical rotation
abbr. = abbreviation
abs. = absolute
bp. = boiling point
CAS = Chemical Abstracts Service
cf. = confer
cryst. = crystals
decomp. = decomposes, decomposition
D. = density
d = day
EC = enzyme commission
ED_{50} = effective dose (effective for 50% of tested subjects)
edn. = edition
eds. = editors
e.g. = exempli gratia (for example)
ε = molar extinction coefficient
f., ff. = the (next) following page, the following pages
fp. = flash point
gen. = general
h = hour
HS = harmonized commodity
i.e. = id est (that is)
i.m. = intramuscular
i.p. = intraperitoneal
i.v. = intravenous
LC = lethal concentration
LD_{50} = lethal dose (lethal for 50% of tested subjects)
Lit. = Literature
λ_{max} = absorption maximum
MIC. = minimal inhibitory concentration
mol. = molecular
Mol. = molecule
mp. = melting point
M_R = relative molecular mass
p. = page
pK_a = negative logarithm of equilibrium constant K_a
p.o. = per os (orally)
ppb = parts per billion = 10^{-9}
ppm = parts per million = 10^{-6}
ppt = parts per trillion = 10^{-12}
® = trade mark
s.c. = subcutaneous
sp. = species
subl. = sublimates, sublimation
Suppl. = Supplement
var. = variant
vol. = volume
* = treated as keyword in this work
°C = Celsius degrees

A

Aaptamine.

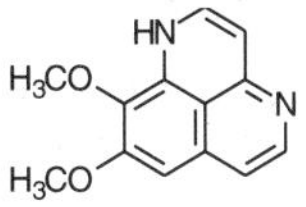

$C_{13}H_{12}N_2O_2$, M_R 228.25, mp. 110–113 °C. The brilliant green A. with a 1*H*-benzo[*de*]-[1,6]naphthyridine skeleton was isolated from the Pacific sponge *Aaptos aaptos*. A. blocks α-adrenoceptors.

Lit.: Heterocycles **49**, 543 (1998); **50**, 543–559 (1999). ▪ J. Chem. Soc., Perkin Trans. 1 **1987**, 173. – *Synthesis:* Heterocycles **48**, 2089 (1998) (synthesis: Aaptosamine) ▪ J. Org. Chem. **52**, 616 (1987) ▪ Synthesis **1996**, 1199 ▪ Tetrahedron **43**, 4803 (1987) ▪ Tetrahedron Lett. **23**, 5555 (1982); **31**, 569 (1990); **34**, 4683 (1993). – *[CAS 85547-22-4]*

Abbeymycin see anthramycins.

Abhexon(e) see hydroxyfuranones.

Abieta-7,13-diene. $C_{20}H_{32}$, M_R 272.47, $[\alpha]_D$ +127° ($CHCl_3$), from roots of *Solidago missouriensis* and *Helichrysum chionosphaerum*. A. is formed by metal ion-dependent cyclization of geranylgeranyl diphosphate as intermediate in the biosynthesis of *abietic acid.

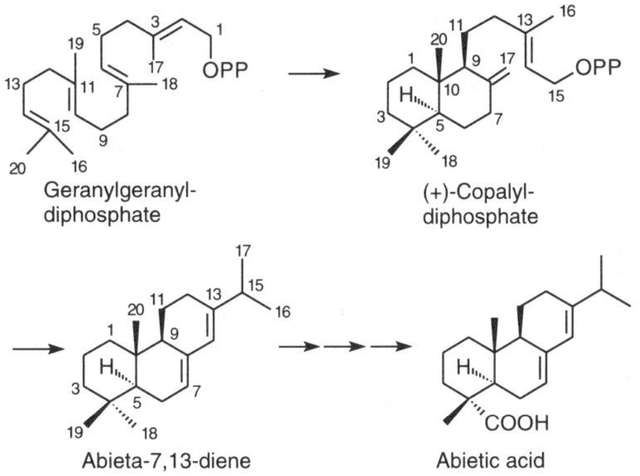

Figure: Biosynthesis of Abietadiene and Abietic acid.

Lit.: Arch. Biochem. Biophys. **313**, 139–149 (1994) ▪ Beilstein E III **5**, 1310. – *[CAS 42895-82-9]*

Abietanes. Structural type of diterpenes, e. g., *abieta-7,13-diene, the most important representative is *abietic acid.

Abietic acid. Formula see abieta-7,13-diene. $C_{20}H_{30}O_2$, M_R 302.46. A resin acid belonging to the diterpenes, monoclinic plates, mp. 172–175 °C, $[\alpha]_D^{15}$ –102° (C_2H_5OH). Occurs above all in *Pinus* and *Abies* species; the most important component of colophony from which it can be separated by distillation. A. is a component of the defensive secretion of the trees against insects and infections by microorganisms. The biosynthesis from tricyclic *abieta-7,13-diene (figure see above) involves sequential oxidation of the C-18 methyl group to a carboxy function.

Use: To form esters, varnish components, soaps, metal soaps (resin soaps from abietates), additive in lactic and butyric acid fermentations.

Lit.: Arch. Biochem. Biophys. **308**, 258–266 (1994) ▪ Beilstein E IV **9**, 2175–2178 ▪ Karrer, No. 1952 ▪ Merck Index (12.), No. 3 ▪ Org. Synth. Coll. **4**, 1–4 (1963) ▪ Phytochemistry **19**, 2655 (1980) ▪ Sax (8.), AAC 500. – *[HS 3806 90; CAS 514-10-3]*

$\Delta^{8,13}$-Abietic acid see palustric acid.

Abietin see coniferin.

Abietospiran [(23*S*,25*R*)-17,23-epoxy-23-hydroxy-3α-methoxy-(5α)-9β,19-cyclolanostane-26-carboxylic acid lactone].

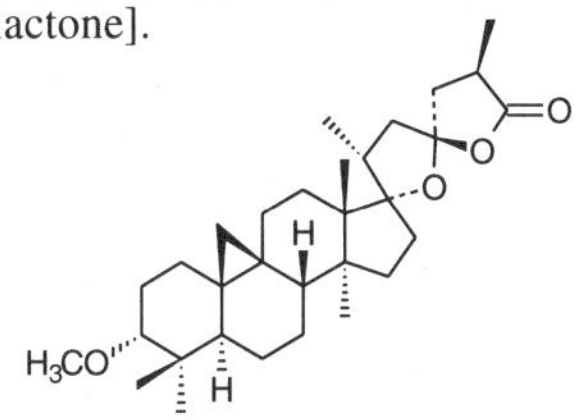

$C_{31}H_{48}O_4$, M_R 484.72, colorless cryst., mp. 219–221 °C, $[\alpha]_D$ –16.8° ($CHCl_3$). *triterpene of the bark of silver fir (*Abies alba*, Pinaceae), responsible for the gray-white appearance of the tree. For isolation see *Lit.*[1].

Lit.: [1] Angew. Chem. Int. Ed. Engl. **18**, 698 (1979).
gen.: Beilstein E V **19/5**, 641. – *[CAS 71648-15-2]*

Abrine [(*S*)-2-methylamino-3-(3-indolyl)propanoic acid, *N*-methyl-L-tryptophan]. Not to be confused with the toxic agglutinin abrin. Formula, see tryptophan. $C_{12}H_{14}N_2O_2$, M_R 218.26, mp. 296 °C (decomp.), $[\alpha]_D^{21}$ +47° (0.5 m HCl); monohydrochloride, mp. 222 °C. Isolated from seeds of *Abrus precatorius* and *Gastrolobium callistachys* (Fabaceae). A powerful antigen.

Lit.: Beilstein E V **22/14**, 40 f. – *[HS 2933 90; CAS 526-31-8]*

Abscisic acid [abscisin II, ABA; histor. name dormin; (2*Z*,4*E*)-5-((*S*)-1-hydroxy-2,6,6-trimethyl-4-oxo-2-cyclohexenyl)-3-methyl-2,4-pentadienoic acid].

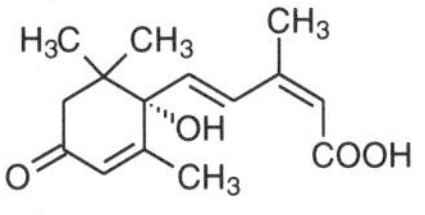

1

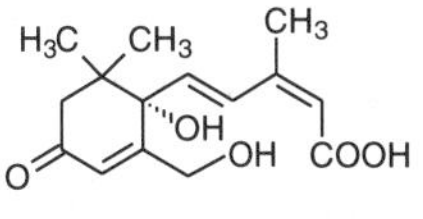

3

A. is isolated from potatoes, avocado pears, cabbage, rose leaves, and numerous trees [1]. A. is a widely distributed sesquiterpene with the skeleton of the *ionones. For the most important data of *A.* (**1**), *A. β-D-glucopyranoside* (**2**), and *nigellic acid* (**3**) see table.

Table: Abscisic acid and derivatives.

	molecular formula	M_R	mp. [°C]	$[\alpha]_D$	CAS
1	$C_{15}H_{20}O_4$	264.32	160–161	+430°	21293-29-8
2	$C_{21}H_{30}O_9$	426.46	114	+180° (C_2H_5OH, 17 °C)	21414-42-6
3	$C_{15}H_{20}O_5$	280.32	185–187		91897-25-5

Biological activity: A. acts as a phytohormone for defoliation, inhibition of flowering, and induces hibernation-like states [2]. It is thus an antagonist of the *plant growth substances. A. plays a role in plant signal transduction [3], gene expression, and in environmental stress like pest infestation [4]. A. is used to regulate the ripening process of fruit. Several syntheses have been reported for A.[5–8]. The biosynthesis involves metabolization of *all-trans*-*violaxanthin [9,10]. Detection by chromatographic methods and monoclonal immunoassays.

Lit.: [1] Nature (London) **205**, 1269 (1966); **210**, 627, 742 (1966). [2] Plant Growth Regul. **11**, 225–238 (1992); Plant Physiol. **108**, 573 (1995). [3] Adv. Bot. Res. **19**, 103–187 (1993). [4] Davies & Jones (eds.), Environmental Plant Biology: Abscisic Acid. Physiology and Biochemistry, p. 125–135, 189–199, Oxford (UK): BIOS Sci. Publ. 1991. [5] Helv. Chim. Acta **61**, 2616 (1978); **72**, 361 (1989). [6] J. Org. Chem. **51**, 253 (1986). [7] Aust. J. Chem. **45**, 179 (1992). [8] Tetrahedron **48**, 8229 (1992); Trends Org. Chem. **4**, 371 (1993). [9] Phytochemistry **29**, 3473 (1990); **31**, 2649 (1992). [10] Methods Plant Biochem. **9**, 381–402 (1993).
gen.: Ullmann (5.) **A 20**, 421.

Absinthin.

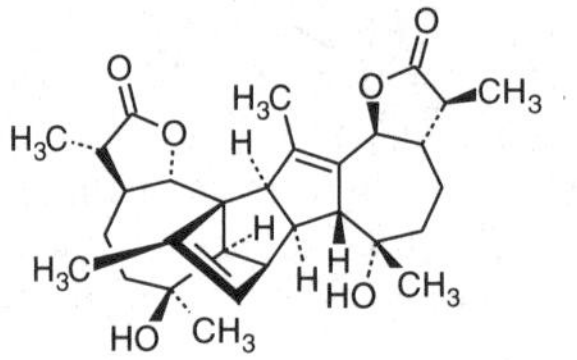

$C_{30}H_{40}O_6$, M_R 496.64, orange-yellow needles, mp. 182–183 °C (decomp.; from CH_3OH), $[\alpha]_D$ +180° ($CHCl_3$). A constituent of *Artemisia absinthium* and *A. sieversiana* (Asteraceae) with a very bitter taste; it was used in bitter aperitifs until early in the 20th century („absinthe“). Later on, it was banned in some countries due to the toxic side effects of thujone (see thujan-3-ones) and *santonin, also present in Artemisia oil, from which absinthe was prepared. Recently, absinthe, a green liqueur, is becoming fashionable again [1].

Lit.: [1] Chem. Eng. News **77** (12), 56 (1999, Mar. 22).
gen.: Phytochemistry **24**, 1009–1016 (1985) ▪ Tetrahedron Lett. **21**, 3191 (1980); **22**, 2269 (1981) (isolation). – *[CAS 1362-42-1]*

Aburamycin B see chromomycins.

Acanthifolicin see okadaic acid.

Acanthostral.

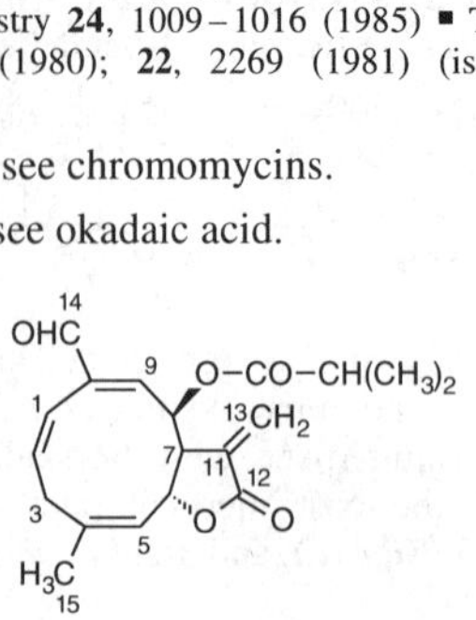

$C_{19}H_{22}O_5$, M_R 330.37, oil, $[\alpha]_D^{20}$ –30° (CH_3OH). Antineoplastically active *cis,cis,cis*-*germacranolide from the Paraguayan medicinal plant *Acanthospermum australe* ("Tapecue").

Lit.: Tetrahedron Lett. **37**, 1455 (1996). – *[CAS 174720-94-6]*

Acarbose (Glucobay®, Precose®, Prandase®). Formula see glycosidase inhibitors. $C_{25}H_{43}NO_{18}$, M_R 645.61; amorphous solid, $[\alpha]_D^{20}$ +171.3°. Pseudotetrasaccharide from the class of carbaoligosaccharide antibiotics, isolated from cultures of *Actinoplanes* strains. A. is a very effective inhibitor of intestinal α-D-glucosidases and saccharidase, it reduces glucose absorption in the gastrointestinal tract and is recommended as a drug for diet-supported diabetes therapy.

Lit.: Angew. Chem. Int. Ed. Engl. **20**, 744–761 (1981) ▪ Aust. J. Chem. **50**, 193–228 (1997) (synthesis) ▪ Carbohydr. Res. **128**, 235–268 (1984) (Isolation & structure elucidation) ▪ Creutzfeldt (ed.), Acarbose for the Treatment of Diabetes mellitus, Berlin: Springer 1988 ▪ Hager (5.) **7**, 1 ff. ▪ J. Antibiot. **40**, 855 ff. (1987) (biosynthesis) ▪ J. Chem. Soc., Chem. Commun. **1988**, 605 f. (synthesis) ▪ Nachr. Chem. Tech. Lab. **42**, 1119 (1994) ▪ Truscheit et al., Progress in Clinical Biochemistry and Medicine, p. 17–100, Berlin: Springer 1988. – *[HS 2940 00; CAS 56180-94-0]*

α-Acaridial see mites.

ACC. Abbreviation for *1-aminocyclopropanecarboxylic acid.

Acetaldehyde see alkanals.

Acetic acid benzyl ester see benzyl acetate.

Acetic acid esters, acetates see fruit esters.

Acetogenins. Originally a class name for natural products that could be considered as being made up of C_2 building blocks (acetate). Biosynthesis of A. starts from acetyl- and malonyl-CoA and proceeds on multienzyme complexes (polyketide synthases). The A. are a subgroup of *polyketides, which additionally may carry propionate and/or butyrate building blocks. A. are often cyclic compounds containing phenol and hydroxyquinone structures. *Examples:* *actinorhodin, *tetracyclines.

Lit.: O'Hagan, The Polyketide Metabolites, Chichester: Ellis Harwood 1991.

Acetoin (3-hydroxy-2-butanone).

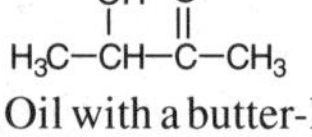

$C_4H_8O_2$, M_R 88.11. Oil with a butter-like odor; D. 0.997, bp. 148 °C, mp. 15 °C, fp. 50 °C; miscible with water, soluble in ethanol, poorly soluble in ether and petroleum ether. LD_{50} (rat orally) >5 g/kg. Both enantiomers are widely distributed [1], often as by-products of microbial processes. Important among others in *butter flavor [1–2 ppm (–)-A., ca. 78% ee], yoghurt, *cheese, asparagus (see vegetable flavors), *sherry and *wine flavor (white burgundies, Chardonnay wines) but also the reason for bad taste in fruit juices and beer. Production by fermentation from molasses or glucose.

Lit.: [1] TNO list (6.), Suppl. 5, p. 312.
gen.: Arctander, Nr. 43 ▪ Beilstein E IV **1**, 3991 f. ▪ Fenaroli (2.) **2**, 6. – *[HS 2914 49; CAS 52217-02-4 ((±)-A.); 53584-56-8 ((–)-(R)-A.); 78183-56-9 ((+)-(S)-A.)]*

Acetomycin.

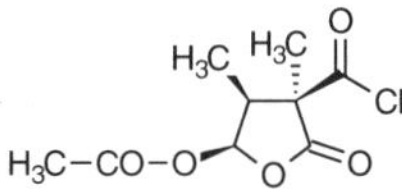

$C_{10}H_{14}O_5$, M_R 214.22, cryst., mp. 115–116 °C, $[\alpha]_D$ –157° (C_2H_5OH). γ-Lactone isolated from *Streptomyces ramulosus*. A. is active against Gram-positive bacteria, some fungi and protozoa. A. inhibits *in vitro* the growth of some tumor cell types but is, however inactivated *in vivo* by esterases. The biogenesis starts from 2 molecules acetate and 1 molecule glycerol, later 1 acetyl unit and 1 CH_3 group (from methionine) are added.

Lit.: Beilstein E III/IV **18**, 1135 ▪ Chem. Pharm. Bull. **47**, 517 (1999) (synthesis (±)-A.) ▪ J. Antibiot. **38**, 1684–1690 (1985); **40**, 73–80 (1987) (activity) ▪ J. Org. Chem. **57**, 3789–3798 (1992) ▪ Tetrahedron: Asymmetry **9**, 1215 (1998) (synthesis (–)-A.) ▪ Tetrahedron Lett. **34**, 2669–2672 (1993) (synthesis). – *[CAS 510-18-9]*

Acetosyringone (4-hydroxy-3,5-dimethoxyacetophenone, acetosyringenone).

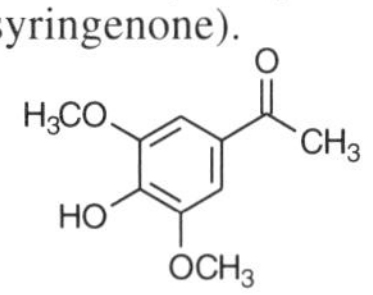

$C_{10}H_{12}O_4$, M_R 196.20, mp. 123–124 °C, colorless crystals. A. occurs in waste water from wood pulping and is isolated from many types of wood. The 4-*O*-β-D-glucopyranoside, {*glucoacetosyringone*, $C_{16}H_{22}O_9$, M_R 358.35, mp. 218–219 °C, $[\alpha]_D$ –38.1° (pyridine)} is a constituent of *Ranzania japonica* [1] (Berberidaceae). uv_{max} 302 nm ($E^{1\%}_{1\,cm}$ 557); in alkali: uv_{max} 362 nm. A. is a plant stress hormone, stimulates gene expression in *Agrobacterium tumefaciens*, and also acts as an insect attractant.

Lit.: [1] Planta Med. **47**, 253–254 (1987).
gen.: Beilstein E IV **8**, 2735 ▪ Luckner (3.), p. 398–400. – *[HS 2914 50; CAS 2478-38-8 (A.); 86402-41-7 (gluco-A.)]*

Acetylcholine [(2-acetoxyethyl)trimethylammonium hydroxide].

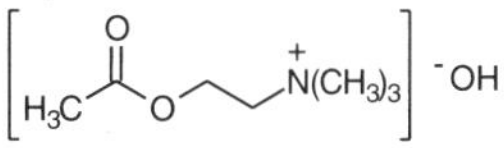

$C_7H_{17}NO_3$, M_R 163.22. The cation of A. is widely distributed as a neurotransmitter in animals and also occurs in plants, e. g., *ergot and *Capsella bursa-pastoris* (Brassicaceae). A. is easily hydrolysed by alkali. A. chloride, $C_7H_{16}ClNO_2$, M_R 181.66, hygroscopic crystalline powder, mp. 149–152 °C.

Toxicology: LD_{50} 2500 mg/kg (rat p.o.), 250 mg/kg (rat s.c.), 22 mg/kg^{-1} (rat i.v.).

Activity: A distinction is made between the muscarinic and the nicotinic activity according to the agonists muscarine (see fly agaric constituents) and *nicotine. The muscarinic effects comprise peripheral vessel dilatation, reduction of blood pressure, and a slowing of heart rate. *Atropine as a muscarinic antagonist suppresses these effects. Higher doses of A. lead to an increase in blood pressure (nicotinic effect).

Biosynthesis: A. is formed from *choline and *acetyl-CoA. A. deficiency of the enzyme choline *O*-acetyltransferase (EC 2.3.1.6) in the region of the cerebral cortex is being discussed as a possible cause of Alzheimer's disease [1].

Detection: After alkaline hydrolysis, enzymatically by means of [^{32}P]ATP and choline kinase (EC 2.7.1.32) [2].

Lit.: [1] Biochem. Soc. Trans. **17**, 76–79 (1989). [2] Bergmeyer, Methods of Enzymatic Analysis (3.), vol. 8, p. 462–473, Weinheim: Verl. Chemie 1985.
gen.: Acta Crystallogr., Sect. B **31**, 1581–1586 (1975) (crystal structure) ▪ Angew. Chem. Int. Ed. Engl. **23**, 195 (1984) ▪ Beilstein E IV **4**, 1446–1447 ▪ Stryer 1995, p. 292–297. – *Toxicology:* Sax (8.), ABO 000 (A. chloride), CMF 250 (A. acetate). – *[HS 2923 90; CAS 51-84-3 (A. cation); 60-31-1 (A. chloride)]*

Acetyl-CoA (*S*-acetyl-coenzyme A, "activated acetic acid").

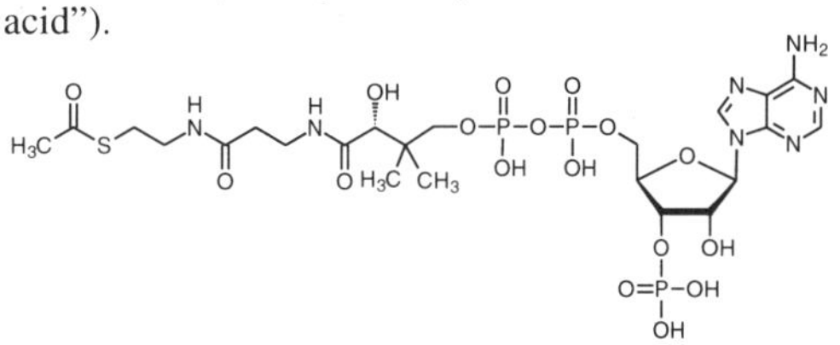

$C_{23}H_{38}N_7O_{17}P_3S$, M_R 809.57. A. plays a central part in metabolism as an acetyl group transfer reagent. In aerobic organisms the acetyl group is formed by degradation of carbohydrates, amino acids, and fats, in anaerobic organisms [1] by carbonylation of methyl groups bonded to *tetrahydrofolic acid.

Lit.: [1] Trends. Biochem. Sci. **11**, 14–18 (1986).
gen.: Beilstein E V **26/16**, 431 ▪ Nachr. Chem. Tech. Lab. **36**, 993–997 (1988). – *[CAS 72-89-9]*

***N*-Acetyldopamine** see *N*-acylcatecholamines.

***N*-Acetylneuraminic acid** see neuraminic acid.

***N*-Acetylnoradrenaline** see *N*-acylcatecholamines.

2-Acetylpyridine see cocoa flavor.

2-Acetyl-1-pyrroline [1-(3,4-dihydro-2*H*-pyrrol-5-yl)-ethanone].

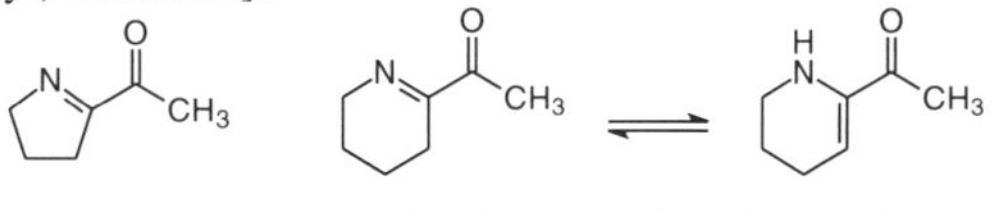

2-Acetyl-Δ¹- and -Δ²-tetrahydropyridine

C_6H_9NO, M_R 111.14. Rusty liquid with popcorn-like odor, bp. 26–28 °C (10 Pa); rapid darkening of color at 20 °C, limited storage life at –20 °C, 1% in pentane; olfactory threshold in water 0.1 ppb [1]. Typical *bread and *rice flavor [1]. A. is formed from yeast constituents on heating L-*ornithine and 2-oxopropanal. The bread aroma *2-acetyl-Δ¹(Δ²)-tetrahydropyridine* [$C_7H_{11}NO$, M_R 125.17, bp. 33–34 °C (0.12 kPa), olfactory threshold 1.6 ppb] is formed analogously from L-*proline and 2-oxopropanal.

Lit.: [1] J. Agric. Food Chem. **36**, 1006 (1988). – *[CAS 85213-22-5 (A.); 25343-57-1, 97073-22-8 (2-acetyl-Δ¹- & Δ²-tetrahydropyridine)]*

16-*O*-Acetylsamandarine see salamander steroid alkaloids.

2-Acetyl-Δ¹(Δ²)-tetrahydropyridine see 2-acetyl-1-pyrroline.

2-Acetylthiazol(in)e see meat flavor.

Acivicin [(*S*)-amino-((*S*)-3-chloro-4,5-dihydro-5-isoxazolyl)-acetic acid)].

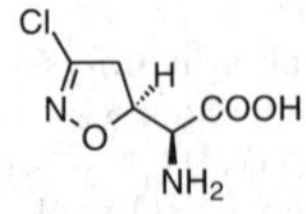

(αS, 5S)

$C_5H_7ClN_2O_3$, M_R 178.58, mp. 228–230 °C, $[\alpha]_D^{20}$ +135° (H_2O). Isoxazoline antibiotic from cultures of *Streptomyces sviceus* [1]. A. exhibits cytostatic activity [2] as well as herbicidal effects [3]. The biosynthesis proceeds from L-*ornithine [4].

Lit.: [1] Tetrahedron Lett. **1973**, 2549; Antimicrob. Agents Chemother. **7**, 807 (1975). [2] Cancer Chemother. Rep. I **58**, 935 (1974). [3] Europ. Pat. A. EP 640 286 (1. 3. 1995). [4] J. Am. Chem. Soc. **114**, 10166 (1992).

gen.: J. Org. Chem. **53**, 4074 (1988) ▪ Tetrahedron **47**, 6079 (1991) (synthesis). – *[CAS 42228-92-2]*

Aclarubicin (aclacinomycin A).

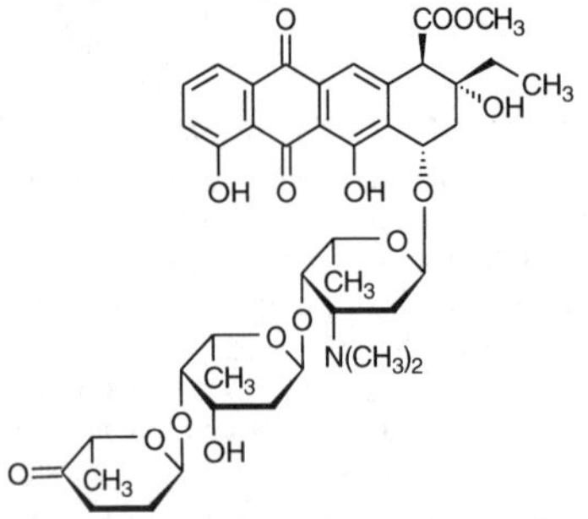

$C_{42}H_{53}NO_{15}$, M_R 811.88, yellow powder, mp. 129–135 °C, $[\alpha]_D$ +29° ($CHCl_3$). A pigment belonging to the group of *anthracyclines isolated from cultures of *Streptomyces galilaeus*. On acid hydrolysis A. furnishes aclavinone and the deoxysugars L-rhodosamine, 2,6-dideoxy-L-*lyxo*-hexose, and L-cinerulose. A. is used clinically as an antitumor agent and is less cardiotoxic than *adriamycin.

Lit.: Hager (5.) **7**, 60 ff. ▪ J. Antibiot. **28**, 830 (1975); **32**, 801 (1979); **34**, 331, 916 (1981); **36**, 1458 (1983) (synthesis) – *[HS 2941 90; CAS 57576-44-0]*

Aconine (jesaconine). Formula see aconitine. $C_{25}H_{41}NO_9$, M_R 499.60, amorphous powder, mp. 132 °C, $[\alpha]_D$ +23° (H_2O), soluble in water, insoluble in ether. A diterpene alkaloid from the blue monk's hood (*Aconitum napellus*), bitter-tasting, poisonous, antipyretic agent. A. is formed from *aconitine by heating in water at 170 °C.

Lit.: Beilstein E V **21/6**, 310 ▪ Hager (5.) **3**, 15–17; **4**, 14; 65–80 ▪ Merck Index (12.), No. 117 ▪ Sax (8.), ADG 500. – *[HS 3004 40, 2939 90; CAS 509-20-6]*

Aconitic acid (propene-1,2,3-tricarboxylic acid).

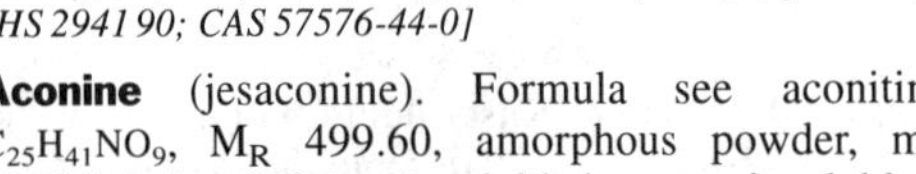

$C_6H_6O_6$, M_R 174.11. The *Z* form, mp. 125 °C, which is widely distributed in animal organisms as an intermediate in the citric acid cycle, readily forms an anhydride with mp. 74 °C. The corresponding *E* form obtained through dehydration of *citric acid with sulfuric acid [1] forms plates (from water) with mp. 194–195 °C (decomp.), pK_{a1} 2.8, pK_{a2} 4.46 (25 °C). LD_{50} 180 mg/kg (mouse i. v.). It is present in monk's hood (*Aconitum napellus*, see aconitine) in large amounts, and found in asarabacca (*Asarum europaeum*), in sugar beet, in horsetail plant, in types of grain, and in beetroot.

Use: For synthesis of unsaturated polyesters and to stabilize edible fats and oils; for determination of tert. amines [2].

Lit.: [1] Org. Synth. (Coll. Vol.) **2**, 12–14 (1943). [2] Pure Appl. Chem. **56**, 468–477 (1984).

gen.: Beilstein E IV **2**, 2405 ▪ Karrer, No. 876 ▪ Stryer 1995, p. 510, 513 f. – *Toxicology:* Sax (8.), ADH 000. – *[HS 2917 19; CAS 499-12-7; 585-84-2 (Z); 4023-65-8 (E)]*

Aconitine.

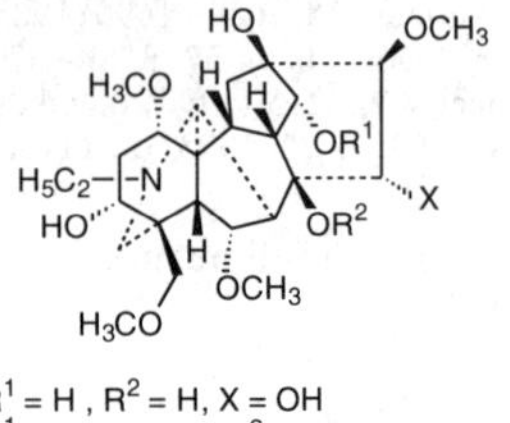

R^1 = H , R^2 = H, X = OH : Aconine
R^1 = CO–C_6H_5 , R^2 = CO–CH_3 , X = OH : Aconitine

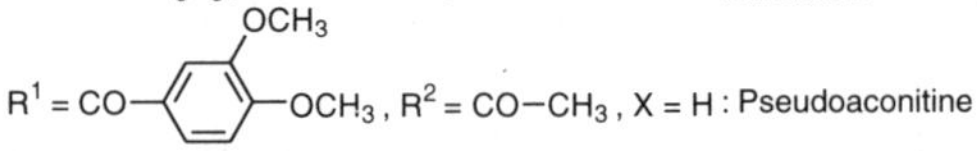

$C_{34}H_{47}NO_{11}$, M_R 645.75, highly poisonous hexagonal plates, mp. 204 °C, $[\alpha]_D$ +19° ($CHCl_3$), soluble in ether, practically insoluble in water. A. is extracted from leaves of the blue monk's hood (*Aconitum napellus*, Ranunculaceae) and other *Aconitum* species with alcohol. Other monk's hood species contain related A.; these are always acetic acid and benzoic acid esters of diterpene alkaloids (aconines) substituted with hydroxy and methyl groups. The east Indian *Aconitum ferox* contains *pseudoaconitine* {$C_{36}H_{51}NO_{12}$, M_R 689.80, mp. 211–213 °C, $[\alpha]_D$ +24° ($CHCl_3$)}; this is one of the strongest plant toxins and was used for poisoned arrows. As little as 0.0006 mg A. changes the cardiac movements of frogs in a characteristic way. It causes cardiac arrhythmia. The lethal dose for man appears to be 1–2 mg. In drugs a maximum of 0.2 mg may be administered at one time. If A. is applied in an ointment to the skin, irritation occurs (burning sensation, tingling, itching), followed by paralysis of nerve endings. In some countries allopathic and homeopathic A. preparations are used for neuralgia and fibrile diseases; in Germany use is declining. LD_{50} 0.166 mg/kg i. v., 0.328 mg/kg i. p.

Lit.: Beilstein E V **21/6**, 310 f. (A.), 309 (pseudoaconitine) ▪ Hager (5.) **3**, 72; **5**, 608 ▪ Manske **4**, 275–330; **17**, 1–103 ▪ Merck Index (12.), No. 120 ▪ Sax (8.), ADH 500, ADH 750 ▪ Zechmeister **16**, 26–89. – *[HS 2939 90; CAS 302-27-2 (A.); 127-29-7 (pseudoaconitine)]*

Acorenone see calamus oil.

Acridones. A group of yellow, weakly-basic alkaloids derived from acridine. A. occur exclusively in the Rutaceae (in bark, wood, leaves, roots e. g., *Euodia xanthoxyloides*) especially in more than 20 species of the subfamilies Rutoideae, Toddalioideae, and Aurantioideae. Most of the naturally occurring A. (>100) are methylated at N and carry oxygen functions at positions 1 and 3. These OH groups may be free, alkylated (methylated, isoprenylated) or incorporated in furan or pyran rings. *Rutacridone* ($C_{19}H_{17}NO_3$, M_R 307.35, mp. 161–162 °C) occurs in the roots of *Ruta graveolens*, from which a series of other A. is derived. Formation of A. has also been demonstrated in tissue cultures of *R. graveolens*. *Acronycine* (acronine, $C_{20}H_{19}NO_3$, M_R 321.38, mp. 175–176 °C) has cytostatic activity.

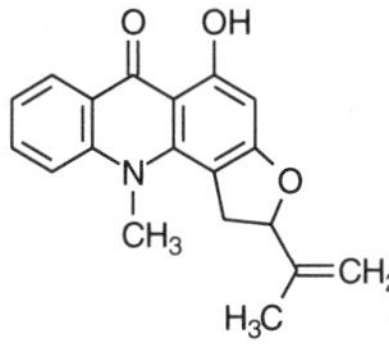

Rutacridone

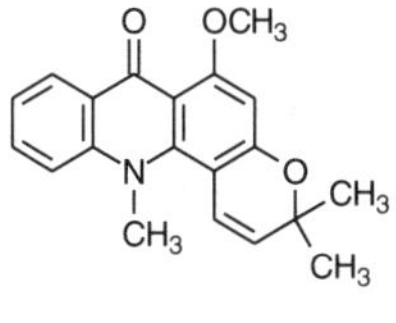

Acronycine

In addition to the structural types shown there are also homo-A., bis-A., and acrimarines (composed of A. and coumarin parts).
The preferred occurrence of A. in roots and their later storage in periderm is indicative of a possible protective function of A. They have a broad antimicrobial activity.
Biosynthesis: From *N*-methylanthraniloyl-CoA and 3 mol. malonyl-CoA [1]; the acridone synthase exhibits a high sequence homology with a synthase of the *chalcones [2].
Lit.: [1] FEBS Lett. **187**, 311 (1985); J. Chem. Soc., Chem. Commun. **1982**, 1263. [2] Z. Naturforsch. C **49**, 26–32 (1994).
gen.: Beilstein E V **27/14**, 219, 220 ▪ Mothes et al., p. 322–328 ▪ Nat. Prod. Rep. **10**, 99–108 (1993) (synthesis) ▪ Pharmazie **43**, 815–826 (1988); **46**, 745 (1991) ▪ Phytochemistry **18**, 161 (1979). – *[CAS 17948-33-3 (rutacridone); 7008-42-6 (acronycine)]*

Acromelic acids (acromelinic acids).

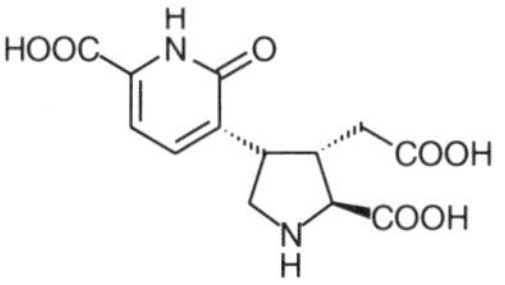

Acromelic acid A

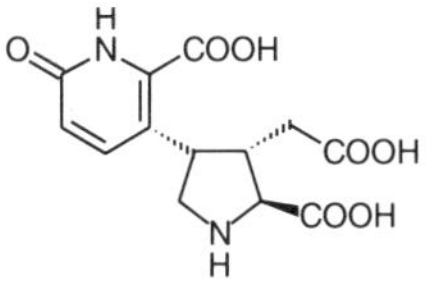

Acromelic acid B

The Japanese toadstool *Clitocybe acromelalga* (Dokusa-sako, Basidiomycetes) is one of the most intensively investigated fungi on account of its toxicity. After consumption of the toadstool severe pain occurs accompanied by red edemas on the hands and feet which last for about one month. In the search for the responsible toxins the strongly neuroexcitator active, isomeric *A. A* [$C_{13}H_{14}N_2O_7$, M_R 310.26, cryst., mp. >310 °C, $[\alpha]_D$ +27.8° (H_2O)] and *A. B* [powder, $[\alpha]_D$ +50.1° (H_2O)] were isolated [1]. The biosynthesis of A. probably starts from L-DOPA by ring cleavage and condensation of the product with L-glutamic acid. In addition to numerous other amino acids [2] the toxic pyridine nucleoside *clitidine was isolated from *C. acromelalga*.
Lit.: [1] J. Am. Chem. Soc. **110**, 4807, 6926 (1988); Tetrahedron Lett. **31**, 3901 (1990); Chem. Lett. **1993**, 21; Tetrahedron **49**, 2427 (1993). [2] Z. Naturforsch. C **49**, 707 (1994) (review).
gen.: Br. J. Pharmacol. **104**, 873 (1991) (activity). – *Synthesis:* Heterocycles **29**, 1473 (1989); **40**, 1009 (1995) ▪ Tetrahedron **54**, 7465 (1998) ((+)-A.) ▪ Tetrahedron Lett. **34,** 331 (1993); **39**, 707 (1998). – *[CAS 86630-09-3 (A. A); 86630-10-6 (A. B)]*

Acronycine see acridone.

Actinidine.

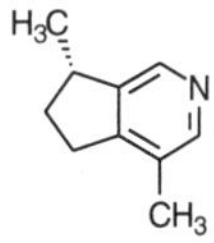

$C_{10}H_{13}N$, M_R 147.22, oil, bp. 100–103 °C (1 kPa), $[\alpha]_D^{11}$ –7.2° ($CHCl_3$), a monoterpene alkaloid (*iridoids) from the east Asian creeper *Actinidia polygama* [1] and other Actinidiaceae, Bignoniaceae, and Valerianaceae. It is a component of the defensive secretion of black beetles [2], ants [3], grasshoppers [4], and flies [5]. Only the (*S*) enantiomer occurs in nature [1]. A. exhibits antimicrobial properties [6] and a deterrent effect against birds and arthropods [2,7], while cats are strongly attracted [1].
Synthesis: There are various syntheses of the racemate and the unnatural (*R*) enantiomer [8], but only three for the (*S*) enantiomer [9]. The biosynthesis presumably proceeds from *iridodial.
Lit.: [1] Bull. Chem. Soc. Jpn. **32**, 315–316 (1959); **33**, 712–713 (1960). [2] Biochem. Syst. Ecol. **21**, 143–162 (1993). [3] J. Chem. Ecol. **3**, 241–344 (1977). [4] J. Entomol. Sci. **21**, 97–101 (1986). [5] Biochem. Syst. Ecol. **20**, 107–111 (1992). [6] Oecologia **82**, 446–449 (1990). [7] J. Appl. Entomol. **113**, 128–137 (1992). [8] Tetrahedron Lett. **29**, 6113–6114 (1988). [9] Heterocycles **31**, 1727–1731 (1990).
gen.: Beilstein E V **20/6**, 394 f. – *[HS 2939 90; CAS 524-03-8]*

Actinobolin see bactobolin.

Actinomycetes (from Greek: aktín = ray, mykes = fungus). Prokaryotic organisms of the order Actinomycetales. The earlier classification of organisms to this order on the basis on morphological-biochemical features has been mostly confirmed by sequence determination of the 16S-ribosomal RNA. The A. are divided into 7 taxonomic main groups: Actinobacteria, Nocardioformi, Actinoplanetes, Thermomonospora, Maduromycetes, Streptomycetes and a group with multilocular sporangia. Important criteria for the classification of A. are, in addition to morphology, especially "chemotaxonomical markers" such as the chemical composition of the cell wall, the cell wall sugars, and the phospholipids. All A. have an unusually high proportion of DNA bases G+C (55–78 mol%) for a bacterium.
The A. are a group of morphologically heterogeneous, Gram-positive bacteria (cell diameter: 0.5–2.0 μm). Colonies formed of A. have above all the ability to form mycelia with a fungus-like appearance.
A. in general have a strictly aerobic metabolism. Only in the group of Actinobacteria, anaerobic and facultative anaerobic types are also known. The natural location of A. is the aerobic region of the soil but they also occur in sediments of fresh or salt water or as endosymbionts of plants and insects. In the soil A. degrade especially complex organic materials as well as xenobiotics. Various A., especially their thermophilic types, can reproduce greatly on agricultural products (e.g., hay, grain, cotton), in compost, or in the water of humidifiers. The spores spread through the air so that, in closed rooms, they can cause allergies or infections in man (e.g., farmer's lung). There are some plant- and human-pathogenic A., those of risk group 3, however, are only known from the genus *Mycobacterium*.

Because of their ability to form secondary metabolites, A. have acquired major significance for biotechnology. In a special anabolic metabolism and sometimes in simple nutrient media A. can form unusual, low-molecular-weight (M_R <2000) natural products exhibiting a wide range of structures and which in many cases inhibit the growth of pathogens or the functions of selected enzymes with a high efficacy. A., especially those from the family *Streptomycetes, are producers of clinically important *antibiotics (e.g. *erythromycin, *streptomycin, *tetracyclines). Over 60% of the about 20000 microbial secondary metabolites described to date originate from the Actinomycetes. The reason why the secondary metabolism is so pronounced in just this order is the subject of numerous hypotheses. The genetics of the Streptomycetes in particular is undergoing intensive research.

Lit.: Annu. Rev. Microbiol. **48**, 257–289 (1994) ▪ Balows et al. (ed.), The Prokaryotes, 2nd. edn., vol. 1 & 2, Berlin: Springer 1992 ▪ Goodfellow et al. (eds.), Actinomycetes in Biotechnology, London: Academic Press 1988 ▪ Holt et al. (eds.), Bergey's Manual of Determinative Bacteriology, 9th. edn., Baltimore: Williams & Wilkins 1994 ▪ Holt et al. (eds.), Bergey's Manual of Systematic Bacteriology, vol. 4, Baltimore: Williams & Wilkins 1989 ▪ Piepersberg & Zeeck, Sekundärstoffwechsel, in Präve et al. (eds.), Handbuch der Biotechnologie, 4th. edn. p. 141–177, München: Oldenbourg 1994.

Actinomycins.

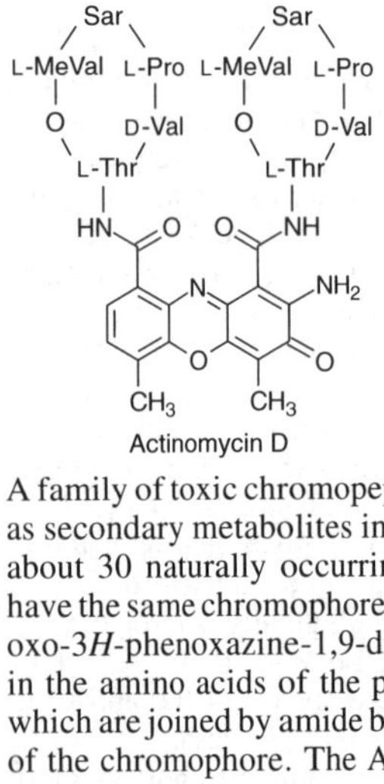

Sar = Sarcosine = *N*-Methyl-glycine
MeVal = *N*-Methyl-L-valine

Actinomycin D

A family of toxic chromopeptides, which occur widely as secondary metabolites in the *Streptomycetes. The about 30 naturally occurring, orange-red colored A. have the same chromophore (2-amino-4,6-dimethyl-3-oxo-3*H*-phenoxazine-1,9-dicarboxylic acid) but differ in the amino acids of the pentapeptide lactone rings, which are joined by amide bonds to the carboxy groups of the chromophore. The A. were first discovered by Waksman in 1940. The A. exhibit high antibacterial and cytostatic activities. They inhibit DNA-dependent RNA-synthesis and DNA replication. For this the chromophore part selectively inserts itself between two guanine/cytosine base pairs of the DNA double helix (intercalation), while the two peptide lactone rings are fixed in the small helix grooves. A. thus hinder the reading of the information stored in the DNA. The biosynthesis and directed biosynthesis of A. have been studied.

A. D (A. C_1, *dactinomycin*): $C_{62}H_{86}N_{12}O_{16}$, M_R 1255.44, red rhomb. cryst., mp. 246–247 °C, $[\alpha]_D$ −315° (CH_3OH). This is the only A. used therapeutically to treat malignant tumors; however, it exhibits considerable side effects.

Lit.: Angew. Chem. Int. Ed. Engl. **37**, 2381 (1998) (crystal structures: A. D, A. Z_3). – *Reviews:* Angew. Chem. Int. Ed. Engl. **14**, 375 (1975) ▪ Annu. Rev. Microbiol. **43**, 173–206 (1989) ▪ Hager (5.) **7**, 1169 ▪ Keller, in: Vining & Stuttard (eds.), Genetic and Biochemistry of Antibiotic Production, p. 173–196, Boston: Butterworth-Heinemann 1995 ▪ Mauger, in: Sammes (ed.), Topics in Antibiotic Chemistry, vol. 5, p. 225–306, Chichester: Ellis Harwood 1980 ▪ Sengupta, in: Foye (ed.), Cancer Chemotherapeutic Agents, p. 261–292, Washington: Am. Chem. Soc. 1995 ▪ Zechmeister **18**, 1–54. – *[CAS 56-76-0 (A. D)]*

Actinorhodin.

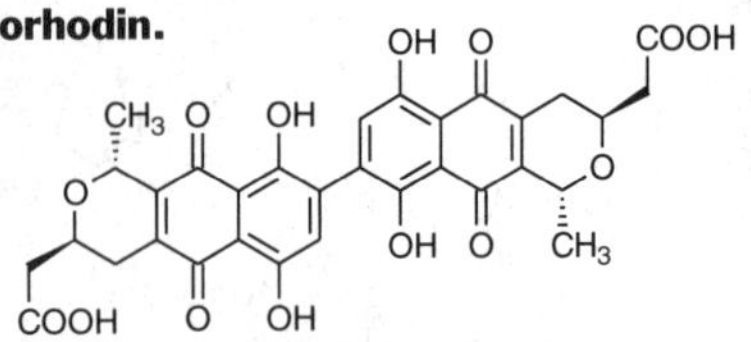

$C_{32}H_{26}O_{14}$, M_R 634.55, mp. 270 °C, $[\alpha]_D$ +117°, red needles, insoluble in water, poorly soluble in organic solvents. A. was first isolated in 1947 by Brockmann from *Streptomyces coelicolor* and is the first described member of the *benzoisochromanequinone antibiotics. A. is a pH indicator (color change at pH 8.5 from red to blue) and can be cleaved into two identical parts by diazomethane which are isolated as indazolequinones. A. has antimicrobial and antiviral (HIV) activities. The two halves are built up by a polyketide synthase of type II from acetate units (see acetogenins, polyketides) and undergo dimerization in a later step of the biosynthesis. The methods of *streptomycetes genetics were developed by Hopwood on *S. coelicolor* A3(2). Today metabolites related to A. can be constructed biosynthetically by manipulation of the known biosynthesis genes.

Lit.: Hopwood et al., in: Vining & Stuttard (eds.), Genetics and Biochemistry of Antibiotic Production, p. 65–102, Boston: Butterworth-Heinemann 1995 (genetics) ▪ J. Chem. Soc., Perkin Trans. 1 **1997**, 1399 (synthesis: racemic monomeric unit) ▪ J. Org. Chem. **46**, 455 (1981) (biosynthesis) ▪ Justus Liebigs Ann. Chem. **1983**, 510–512 ▪ Nature (London) **375**, 549–554 (1995). – *[HS 2941 90; CAS 1397-77-9]*

Actinospectracin see spectinomycin.

***N*-Acylcatecholamines.**

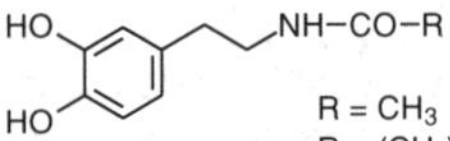

R = CH_3 : *N*-Acetyldopamine (**1**)
R = $(CH_2)_2$—NH_2: *N*-β-Alanyldopamine (**2**)

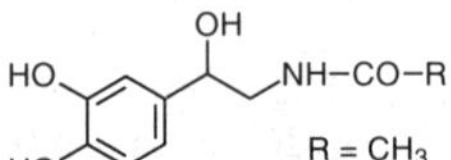

R = CH_3 : *N*-Acetylnoradrenaline (**3**)
R = $(CH_2)_2$—NH_2: *N*-β-Alanylnoradrenaline (**4**)

Table: *N*-Acylcatecholamines.

no.	molecular formula	M_R	mp. [°C]	CAS
1	$C_{10}H_{13}NO_3$	195.22	78–80	2494-12-4
2	$C_{11}H_{16}N_2O_3$	224.26		54653-62-2
3	$C_{10}H_{13}NO_4$	211.22	77–79	(±): 34649-26-8 (*R*)-(–): 30959-88-7 67083-59-4
4	$C_{11}H_{16}N_2O_4$	240.26		(*R*): 68299-73-0

N-Acetyldopamine {*N*-[2-(3,4-dihydroxyphenyl)ethyl]-acetamide} and *N-β-alanyldopamine* (3-amino-*N*-[2-(3,4-dihydroxyphenyl)ethyl]propanamide) occur in insects where they serve to form the sclerotized cuticle required for the exoskeleton. They are transformed by enzymatic oxidation to the corresponding *o*-quinones and *p*-quinonemethides which cross-link structural proteins with each other and with *chitin. The biosynthesis of *N*-acetyldopamine starts from *dopamine and *acetyl-CoA by way of arylamine-*N*-acyltransferase (EC 2.3.1.5). In contrast to the biosynthesis of (*R*)-*noradrenaline, the *p*-quinonemethides, generated non-selectively by enzymatic oxidation from *N*-acetyldopamine or *N-β*-alanyldopamine, undergo addition of water to furnish the *racemic* side-chain hydroxylated derivatives *N-acetylnoradrenaline* (*N*-[2-(3,4-dihydroxyphenyl)-2-hydroxyethyl]acetamide) or, respectively *N-β-alanylnoradrenaline* (3-amino-*N*-[2-(3,4-dihydroxyphenyl)-2-hydroxyethyl]propanamide).

Lit.: Adv. Insect Physiol. **15**, 317–473 (1980) ▪ Science **217**, 364–366 (1982) ▪ Tetrahedron Asymmetry **6**, 839–842 (1995) ▪ Z. Naturforsch. C **33**, 498–503 (1978) (synthesis) ▪ see also melatonin.

Acylphloroglucinols. *Phloroglucinol derivatives also known as phloroglucides. They contain up to six phloroglucinol residues linked to each other by methylene bridges. To date more than 50 such compounds are known and can be classified as the acylphloroglucinol, acylfilicinic acid, or 6-propyl-2*H*-pyran-2,4(3*H*)-dione types. In general they are rather stable to acids but very sensitive to bases. They occur in numerous fern species, especially of the family Polypodiaceae (=polypody ferns) and serve as chemotaxonomic markers [1,2]. Their occurrence in some families of monocotyledonous and dicotyledonous angiosperms (e.g., Liliaceae, Rosaceae, Euphorbiaceae, Lamiaceae, and Asteraceae) has been confirmed. Mixtures of acylphloroglucinols from various fern species (=*Dryopteris* sp.) are known as filicin or *filixic acids.

Lit.: [1] Stud. Geobotanica **1**, 275 (1980). [2] Hegnauer **VII**, 437. *gen.:* Ann. Bot. Fennici **20**, 407 (1983) ▪ Zechmeister **54**, 4.

Adaline.

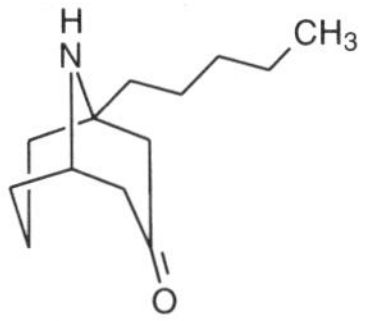

$C_{13}H_{23}NO$, M_R 209.33, $[\alpha]_D^{25}$ −13° ($CHCl_3$), cryst., mp. 204–205 °C (hydrochloride). Bicyclic *alkaloid from ladybird beetles of the genus *Adalia*. It is biosynthetically related to *coccinelline, the major alkaloid of ladybird beetles. Homologues of A. have been identified in *Euphorbia* plants.

Lit.: J. Org. Chem. **57**, 4211–4214 (1992) ▪ Manske **31**, 193–315. – *[CAS 41267-60-1]*

Adenine see purines.

***S*-Adenosylmethionine** [*S*-(5′-deoxyadenosine-5′-yl)-L-methionine, abbr.: SAM].

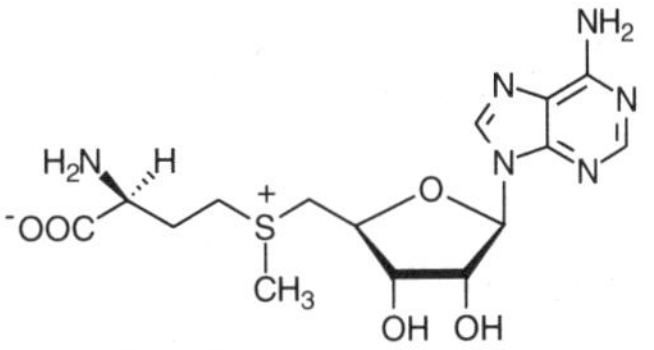

$C_{15}H_{22}N_6O_5S$, M_R 398.44, an important methyl group donor in metabolism, e.g., in the biosynthesis of methoxybenzene derivatives and methyl esters, as well as the propylamine units of *spermidine and *spermine. SAM is transformed to *S*-Adenosylhomocysteine in the process. The biosynthesis starts from homocysteine by transfer of the methyl groups from N^5-methyltetrahydrofolate.

Lit.: Annu. Rev. Biochem. **44**, 435–451 (1975) ▪ Eur. J. Biochem. **58**, 31–41 (1975) (UV, CD) ▪ J. Am. Chem. Soc. **103**, 6015–6019 (1981) (^1H NMR) ▪ Merck Index (12.), No. 155 ▪ Stryer 1995, p. 721–723. – *[CAS 29908-03-0]*

Adiposine see glycosidase inhibitors.

Adrenal hormones see aldosterone and corticosteroids.

(*R*)-Adrenaline {L-adrenaline, (*R*)-epinephrine, 4-[1-hydroxy-2-(methylamino)ethyl]-1,2-benzenediol}.

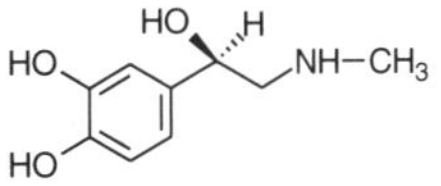

*β-Phenylethylamine alkaloid, $C_9H_{13}NO_3$, M_R 183.21. cryst., mp. 211–212 °C, on rapid heating mp. 216 °C (decomp.), $[\alpha]_D^{20}$ −53° (1 m HCl), soluble in water and alcohol, insoluble in ether, acetone, oils. Easily oxidized, especially by atmospheric oxygen in alkali, with formation of adrenochrome and higher molecular brown pigments (see melanins). (*R*)-A. was isolated from adrenal glands in 1901 and was the first hormone to be obtained in crystalline form. It has also been found in callus tissue of *Portulaca grandiflora* (Centrospermae, Portulacaceae) (configuration not determined). The synthesis of *rac.* A. carried out in 1904 by Stolz in laboratories of the Farbwerke Hoechst was the first ever synthesis of a hormone.

Toxicology: LD_{50} 4 mg/kg (mouse i.p.). A. can cause a contact dermatitis.

Activity: Like the chemically and physiologically related (*R*)-*noradrenaline (*R*)-A., as an *adrenal hormone*, increases the degradation of glycogen in the liver and of fat in adipose tissue as well as the oxidative metabolism in muscle. As *neutrotransmitter* of the adrenergic nerve system (*R*)-A. increases heart rate as a sympathicomimetic, constricts blood vessels of the skin, mucous membranes, and abdominal viscera, and dilates vessels of the skeletal musculature and liver. The relaxation of smooth musculature in the intestine or bronchi effected by (*R*)-A. leads to a reduction of peristalsis (intestinal movements) or to dilatation of the bronchi. (*S*)-A. is about 12 times less active than (*R*)-adrenaline.

Biosynthesis: In the adrenal cortex from L-tyrosine → L-DOPA → dopamine → (*R*)-noradrenaline → (*R*)-adrenaline.

Use: On account of its adrenergic action in the form of salts (e.g. hydrogen tartrate) (*R*)-A. is contained in some parenterally administered sympathicomimetics, broncholytics, or antiasthmatics. (*R*)-A. hydrogen tartrate is also used for hyperallergic reactions, especially after insect bites, in cardiac arrest, and unconsciousness.

Lit.: Beilstein EIV **13**, 2927 ▪ J. Chem. Soc. Faraday Trans. 2 **77**, 227–233 (1981) (^{13}C NMR) ▪ Stryer 1995, p. 340–347, 594f., 600 ▪ Ullmann (5.) **A 2**, 455. – *Toxicology:* Sax (8.), AES 000 [(*R*)-A. hydrogen tartrate], AES 250 [(*S*)-A.], EBB 500 (*rac.* A.). – *[HS 2937 99; CAS 51-43-4 (R); 150-05-0 (S); 329-65-7 (±); 51-42-3 ((R)-A. hydrogen tartrate)]*

Adriamycin (doxorubicin, 14-hydroxydaunomycin).

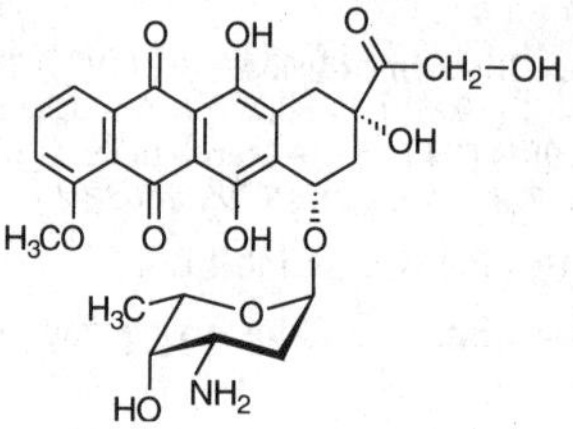

$C_{27}H_{29}NO_{11}$, M_R 543.53, hydrochloride: orange-red needles, mp. 204–205 °C, $[\alpha]_D$ +248°. A. is a cytotoxic pigment from *Streptomyces peucetius* belonging to the group of *anthracyclines. Acid hydrolysis of A. affords L-daunosamine and adriamycinone. A. is used in chemotherapy for tumor diseases, it exerts its activity by binding to DNA (intercalation) and inhibition of topoisomerase II. Its use is limited because of cardiotoxic side effects. A. also has an antiviral action (HIV-1).

Lit.: Arcamone, Doxorubicin, New York: Academic Press 1981 ▪ Beilstein E V **18/10**, 352 ▪ Hager (5.) **7**, 1431–1435 ▪ J. Antibiot. **40**, 396 (1987) ▪ J. Med. Chem. **29**, 1225 (1986) (synthesis). – *[HS 2941 90; CAS 23214-92-8]*

Aequorin. Photoprotein from lamp jellyfish (medusa, *Aequorea*-species) with a molecular mass of ca. 19 500. Work-up of 2 t of the organisms gave 125 mg aequorin. In the presence of traces of Ca^{2+} or Sr^{2+} ions A. emits light (*bioluminescence). This reaction is used to analyse Ca^{2+} ions in biological systems using A. produced by gene technology. The Ca^{2+} binding sites in protein are the SH groups of 3 molecules cysteine and 1 molecule histidine; *coelenterazine was identified as the functional chromophore. Light emission occurs from a phenolate monoanion of coelenteramide [1].

Lit.: [1] J. Chem. Soc., Chem. Commun. **1994**, 165 ff.
gen.: Biochemistry **26**, 1326 (1987) ▪ Methods Enzymol. **83**, 124 (1986) ▪ Proc. Natl. Acad. Sci. USA **83**, 8107 (1986) ▪ Scheuer I **3**, 179 ff.

Aerobactin. $C_{22}H_{36}N_4O_{13}$, M_R 564.55, powder, $[\alpha]_D^{22}$ –10.8° (H_2O). A. is formed in cultures of *Aerobacter aerogenes*, *Escherichia coli*, and *Klebsiella pneumoniae* under iron-deficit conditions. It belongs to the family of citrate hydroxamates *siderophores. Related compounds are *schizokinen* ($C_{16}H_{28}N_4O_9$, M_R 420.42) from *Bacillus megaterium*, *arthrobactin* ($C_{20}H_{36}N_4O_9$, M_R 476.53) from *Arthrobacter* sp., and *nannochelin A* ($C_{38}H_{48}N_4O_{13}$, M_R 768.82) from a *Nannocystis* species. The siderophores are excreted under iron-deficit conditions, after complexation with iron(III) ions they are

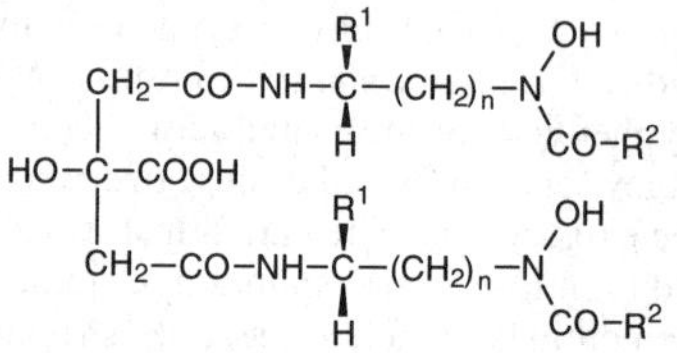

	n	R^1	R^2
Aerobactin	4	COOH	CH_3
Schizokinen	2	H	CH_3
Arthrobactin	4	H	CH_3
Nannochelin A	4	$COOCH_3$	$CH=CH-C_6H_5$

taken up again by the producers. During reduction the appreciably more weakly bound iron(II) ions are liberated and the ligand can be recycled.

Lit.: Biochim. Biophys. Acta **192**, 175 (1969) ▪ Bioorg. Chem. **17**, 13 (1989) ▪ J. Antibiot. **45**, 147 (1992). – *Synthesis:* J. Am. Chem. Soc. **104**, 3096 (1982) ▪ J. Org. Chem. **57**, 7140 (1992). – *[CAS 26198-65-2 (A.); 35418-52-1 (schizokinen); 39007-57-3 (arthrobactin); 133705-25-6 (nannochelin A)]*

Aerocyanidin.

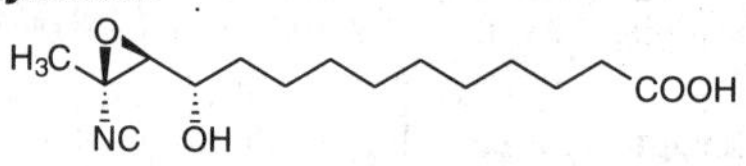

relative configuration

$C_{15}H_{25}NO_4$, M_R 283.37, cryst., mp. 63.5–66.5 °C, $[\alpha]_D$ –20° (CH_3OH). An unusual isonitrile derivative from *Chromobacterium violaceum* with high activity against Gram-positive bacteria. Easily evolves HCN and is transformed thereby through the epoxycyanohydrin (Payne rearrangement) to the α,β-epoxyketone. This explains the toxicity *per os*, which on a molar basis corresponds to that of NaCN.

Lit.: J. Antibiot. **41**, 454 (1988). – *[CAS 113701-99-8]*

Aerosporin see streptothricins.

Aerothionin.

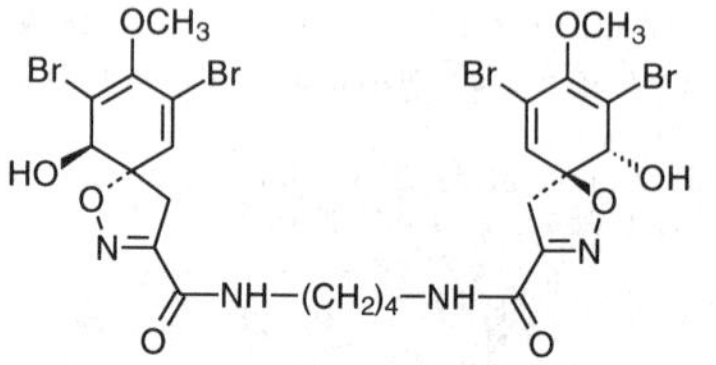

Aerothionin

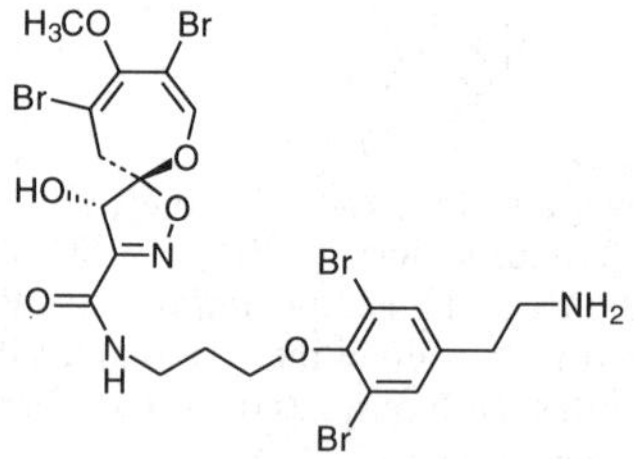

Psammaplysin A

Sponges of the order *Verongida* contain numerous metabolites having a spiroisoxazoline system, which can be derived from bromotyrosine. A.[1] from *Aplysina fistularis*, $C_{24}H_{26}Br_4N_4O_8$, M_R 818.11, mp. 134–137 °C, $[\alpha]_D$ +252°; *psammaplysin A* [2] from *Psammaplysilla purpurea* with a 1,6-dioxa-2-aza[4,6]undecane system:

$C_{21}H_{23}Br_4N_3O_6$, M_R 733.05, $[\alpha]_D$ −65.2°. Biogenetically psammaplysin may be derived from bromotyrosine by way of benzene oxide-oxepin intermediates.

Lit.: [1] J. Chem. Soc., Chem. Commun. **1970**, 752; J. Chem. Soc. Perkin Trans. 1 **1972**, 18; J. Org. Chem. **63**, 5581 (1998); Tetrahedron Lett. **22**, 39 (1981). – *Synthesis:* Tetrahedron Lett. **24**, 3351 (1983); Bull. Chem. Soc. Jpn. **58**, 3453 (1985). [2] J. Am. Chem. Soc. **107**, 2916 (1985); J. Nat. Prod. **55**, 822 (1992). *gen.:* Beilstein E V **27/16**, 123. – *[CAS 28714-26-3 (A.); 85819-66-5 (psammaplysin A)]*

A-factor [(3*R*)-3-hydroxymethyl-2-(6-methylheptanoyl)-4-butanolide].

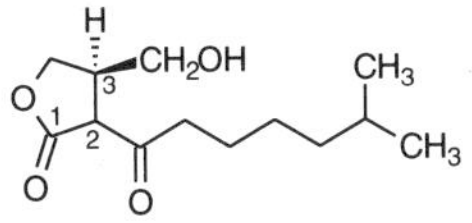

$C_{13}H_{22}O_4$, M_R 242.31, wax-like solid, $[\alpha]_D$ −13.1° ($CHCl_3$). It was recognized by Khokhlov in 1973 as a signal substance in *Streptomyces griseus* that induces not only spore formation/air mycelium formation but also *streptomycin production; also known as an *autoregulator*. The action starts at a concentration of $<10^{-9}$ M by occupation of specific receptors (binding proteins), thus A-f. is comparable to the hormones in higher organisms. The promotors of cell processes are controlled at the gene level in *S. griseus*. The signal system has been investigated by genetic methods and underlines the special position of *streptomycetes as producers of secondary metabolites. Stimulation of protein phosphorylation has also been observed. A-f. also influences *anthracycline and *nosiheptide production but, according to current knowledge, is not a universal regulator. There are other butanolides (e.g., from *S. virginiae*) with similar functions, homologues with shorter chains but without any function, and also substances from streptomycetes with completely different structures (e.g., *pamamycins, *hormaomycin), but with similar actions to A-f. A-f. undergoes easy epimerization at C-2.

Lit.: J. Antibiot. **35**, 349–358 (1982) ▪ Z. Allg. Mikrobiol. **13**, 647–655 (1973). – *Biosynthesis:* J. Am. Chem. Soc. **114**, 663–668 (1992). – *Synthesis:* Tetrahedron **38**, 2919 ff. (1982); **39**, 3107 ff. (1983). – *Review:* Annu. Rev. Microbiol. **46**, 377–398 (1992) ▪ Gene **115**, 159–165 (1992); **146**, 47–52 (1994). – *[CAS 51311-41-2]*

Aflastatin A.

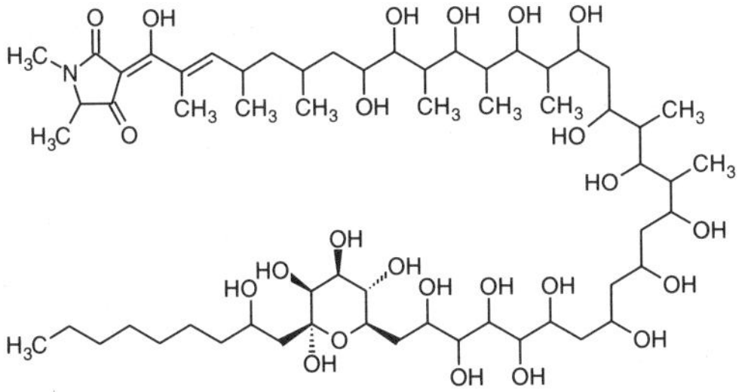

$C_{62}H_{115}NO_{24}$, M_R 1258.57, powder, $[\alpha]_D^{19}$ −2.6° (DMSO). A. A is the first known strong inhibitor of aflatoxin biosynthesis, also effective for the treatment of infections with *Candida* and *Trichophyton* sp. and *Staphylococcus aureus*.

Lit.: Chem. Eng. News **74** (35), 31 (1996, Aug. 26) ▪ J. Am. Chem. Soc. **118**, 7855 (1996). – *Structure:* J. Antibiot. **50**, 111 (1997) ▪ J. Org. Chem. **65**, 438 (2000) (abs. configuration). – *[CAS 179729-59-0]*

Aflatoxins. Group of metabolic products from the mold fungi *Aspergillus flavus*, *A. parasiticus* and others, which attack in particular nuts, corn, and grain flour. A. are among the most dangerous fungal toxins (*mycotoxins) and strongest liver carcinogens [1,2]. They became known as the cause of the mass turkey death in 1960 in Great Britain when 100000 turkeys were affected [3].

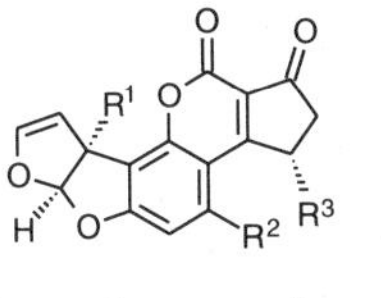

R^1 = H, R^2 = OCH_3, R^3 = H : A. B_1
R^1 = OH, R^2 = OCH_3, R^3 = H : A. M_1
R^1 = H, R^2 = OH, R^3 = H : A. P_1
R^1 = H, R^2 = OCH_3, R^3 = OH : A. Q_1

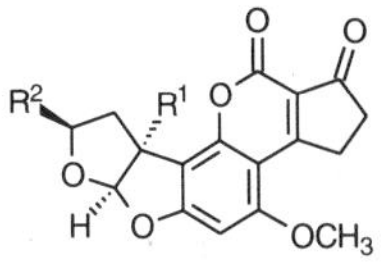

R^1 = H, R^2 = H : A. B_2
R^1 = OH, R^2 = H : A. M_2
R^1 = H, R^2 = OH : A. B_{2a}

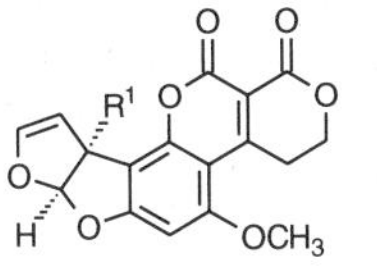

R^1 = H : A. G_1
R^1 = OH : A. GM_1

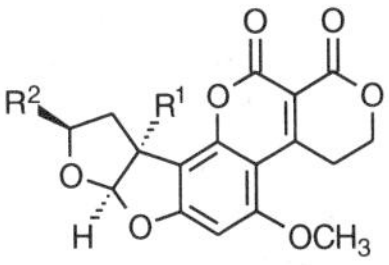

R^1 = H, R^2 = H : A. G_2
R^1 = OH, R^2 = H : A. GM_2
R^1 = H, R^2 = OH : A. G_{2a}

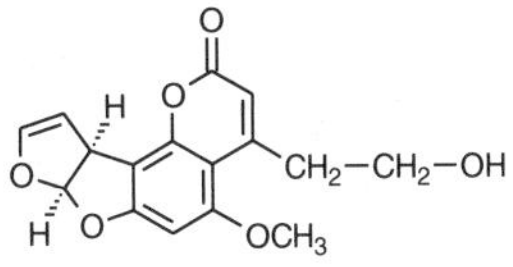

A. B_3

Table: Aflatoxins.

Aflatoxin	molecular formula	M_R	mp. [°C]	optical rotation $[\alpha]_D$	CAS
B_1	$C_{17}H_{12}O_6$	312.28	268–69	–562° ($CHCl_3$)	1162-65-8
B_2	$C_{17}H_{14}O_6$	314.29	305	–492° ($CHCl_3$)	7220-81-7
B_{2a}	$C_{17}H_{14}O_7$	330.29	217		17878-54-5
P_1	$C_{16}H_{10}O_6$	298.25	>320	–574° (CH_3OH)	32215-02-4
B_3	$C_{16}H_{14}O_6$	302.28	217		23315-33-5
G_1	$C_{17}H_{12}O_7$	328.28	247–50	–556° ($CHCl_3$)	1165-39-5
G_2	$C_{17}H_{14}O_7$	330.29	237–40	–473° ($CHCl_3$)	7241-98-7
G_{2a}	$C_{17}H_{14}O_8$	346.29	190		20421-10-7
M_1	$C_{17}H_{12}O_7$	328.28	299	–280° (DMF)	6795-23-9
M_2	$C_{17}H_{14}O_7$	330.29	293		6885-57-0
GM_1	$C_{17}H_{12}O_8$	344.28	276		23532-00-5
GM_2	$C_{17}H_{14}O_8$	346.29	270–72		
Q_1	$C_{17}H_{12}O_7$	328.28	266		52819-96-2

A. flavus forms the A. B_1 and B_2; *A. parasiticus* additionally G_1 and G_2 (B = blue and G = green [4] for the flu-

orescence in UV light). Other A. are animal or microbial metabolic products of these four (e. g. M_1, M_2, GM_1, GM_2, P_1, Q_1). Compounds of the M series occur especially in milk and dairy products. All A. consist of a dihydrofurofuran or tetrahydrofurofuran ring system annelated to a substituted coumarin system.
For detection in the pg-ng range, antibodies for ELISA tests are available. A. B_1 is the most frequently occurring and most highly toxic and carcinogenic aflatoxin. The toxicity of the A. depends both on the structure and on the individual precondition of the respective organism or cell type. The liver is the primary target of attack but kidney damage also occurs. A. B_1 is metabolically activated by microsomal enzymes whereupon a reactive epoxide is formed. This can bind both to DNA and to chromosomal proteins. The mutagenic and carcinogenic actions are explained by a covalent bonding to N-7 of a guanosine of DNA with subsequent further reactions which so cause a mutation of gene p53, a tumor suppressor gene. In hepatocytes in the presence of A. B_1 a G→T transformation occurs which leads to exchange of arginine by serine in coded protein. This G→T transformation is also found in liver tumors[5]. A. B_1 and hepatitis B virus act synergistically in the liver[6]. Trout and ducklings react particularly sensitively to A. B_1 (LD_{50} 18 µg/50 g body weight). The biosynthetic steps including the participating enzymes and some of the genes have been elucidated[6].

Lit.: [1]Nature (London) **267**, 863 (1977). [2]Proc. Natl. Acad. Sci. USA **80**, 2695–2698 (1983). [3]Endeavour **1963**, 75–79. [4]J. Am. Chem. Soc. **85**, 1705 (1963); **87**, 882 (1965). [5]Nature (London) **350**, 427–428, 429–432 (1991). [6]Handbook of Applied Mycology, vol. 5, Mycotoxins in Ecological Systems, chapters 10 & 11, New York: Marcel Dekker 1992; Appl. Environ. Microbiol. **59**, 3273–3279 (1993); Chem. Rev. **97**, 2537–2555 (1997); Lancet **339**, 943–964 (1992); Barton-Nakanishi **1**, 443–472.
gen.: Angew. Chem. Int. Ed. Engl. **23**, 493 (1984) ▪ Betina, chapter 7 ▪ Cole-Cox, p. 1–66. – *Biosynthesis:* Microbiol. Rev. **52**, 274–295 (1988) ▪ Biochemistry **30**, 4343–4350 (1991) ▪ Appl. Environ. Microbiol. **59**, 3564–3571 (1993). – *Carcinogenesis:* Proc. Natl. Acad. Sci USA **83**, 9418–9422 (1986); **90**, 8586–8590 (1993) ▪ J. Biol. Chem. **264**, 12226–12231 (1989). – *Epidemiology:* Annu. Rev. Phytopathol. **25**, 249–270 (1987). – *Synthesis:* J. Org. Chem. **51**, 1006 (1986) ▪ Sax (8.), AET-AEW 500 ▪ Synlett **1997**, 665 (A. B_2) ▪ Synthesis **1988**, 760 ▪ Tetrahedron Lett. **40**, 8513 (1999) (A. M_2). – *[HS 2932 90]*

Aflatrem (α,α-dimethylallylpaspalinin).

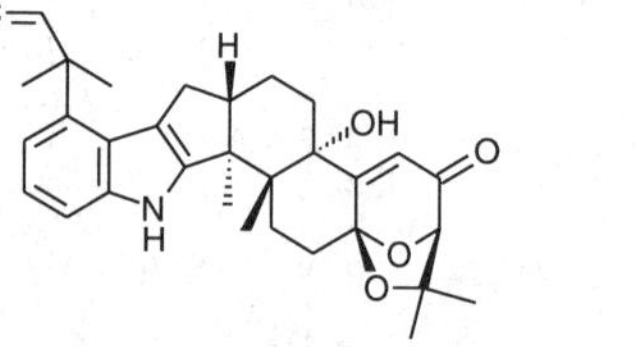

$C_{32}H_{39}NO_4$, M_R 501.67, needles, mp. 222–224 °C, component of *Aspergillus flavus*-*sclerotia and mycelia, belongs to the class of tremorgenic toxins which cause trembling and cramps in animals (see also paspalin under paspalitrems). In sclerotia A. has been attributed a protective function against animal attack.

Lit.: Cole-Cox, p. 410–413 ▪ Science **144**, 177 (1964) ▪ Tetrahedron Lett. **1980**, 239. – *[CAS 70553-75-2]*

Africanol.

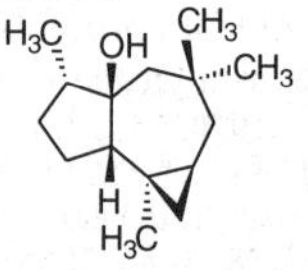

Africanol

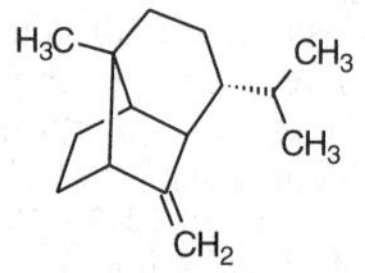

Sinularene

Substances with unusual carbon skeletons have been isolated from soft corals of the family Alcyonacea and the closely related Gorgonacea: e.g., the sesquiterpenes *A.*[1] from *Lemnalia africana* ($C_{15}H_{26}O$, M_R 222.37, mp. 58–60 °C, $[\alpha]_D$ +59.5°) and *sinularene*[2] from *Sinularia mayi* ($C_{15}H_{24}$, M_R 204.36, oil, $[\alpha]_D$ –142°).

Lit.: [1]J. Org. Chem. **58**, 3557 (1993). [2]Can. J. Chem. **65**, 144 (1987); Tetrahedron Lett. **1977**, 2395; Scheuer I **2**, 247 (review: terpenoids from coelenterates).
gen.: Synlett **1998**, 621 (synthesis isoafricanol) ▪ see capnellene, sarcophytol A. – *[CAS 53823-07-7 (A.); 64845-75-6 (sinularene)]*

Agar(-agar). A gel-forming heteropolysaccharide from the cell wall of numerous red algae (commercial production from *Gelidium* species). A. is a mixture of the gelating *agarose (up to 70%) and the non-gelating *agaropectin* [β-1,3-linked D-*galactose units (some of which are esterified in 6 position by sulfuric acid), anhydrosugar 3,6-anhydrogalactose and corresponding uronic acids; proportion up to 30%]. M_R 110 000–160 000, colorless and tasteless, insoluble in cold, soluble in hot water. A solid gel is formed even in a 1% solution that melts between 80 and 100 °C and resolidifies at ca. 45 °C.

Lit.: Hager (5.) **1**, 702 ff. ▪ Ullmann (5.) **A 11**, 503, 570 f.; **A 15**, 185; **A25**, 45 f. – *[HS 1302 31; CAS 9002-18-0]*

Agaricone.

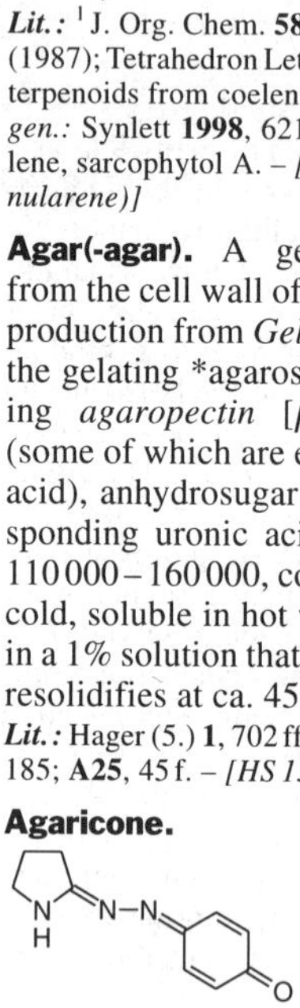

Agaricone

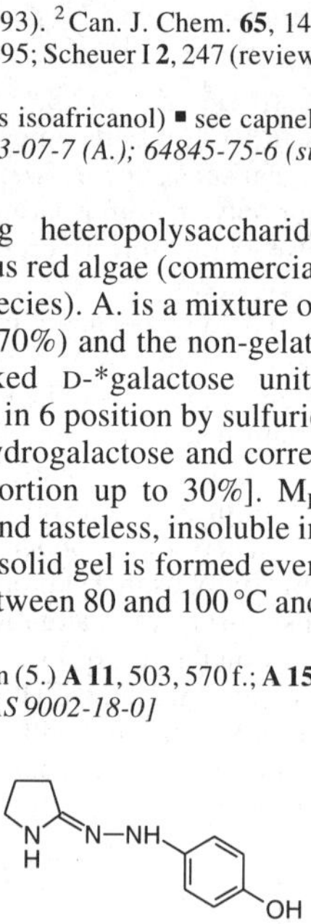

Leukoagaricone

$C_{10}H_{11}N_3O$, M_R 189.22, red cryst., mp. 115 °C (decomp.). Merocyanin pigment from *Agaricus xanthoderma* (Agaricales). A. exists in the toadstool as the colorless *leukoagaricone* ($C_{10}H_{13}N_3O$, M_R 191.23; hydrochloride: mp. 224 °C), which upon injury to the fruitbody is transformed to the orange-yellow, water-soluble A. by the action of atmospheric oxygen and oxidases (see also xanthodermin).

Lit.: Angew. Chem. Int. Ed. Engl. **24**, 1063 (1985) ▪ Zechmeister **51**, 238–241. – *[CAS 99280-73-6 (A.); 99280-75-8 (leukoagaricone)]*

Agaricus formazan, agaricus hydrazones see 4-(hydroxymethyl)benzenediazonium ion.

Agaridin see 4-diazo-2,5-cyclohexadien-1-one.

Agaridoxin see agaritine.

Agaritine (L-glutaminic acid 5-{*N*′-[4-(hydroxymethyl)-phenyl]hydrazide}). A compound from the cultivated mushroom *Agaricus bisporus* and related species[1].

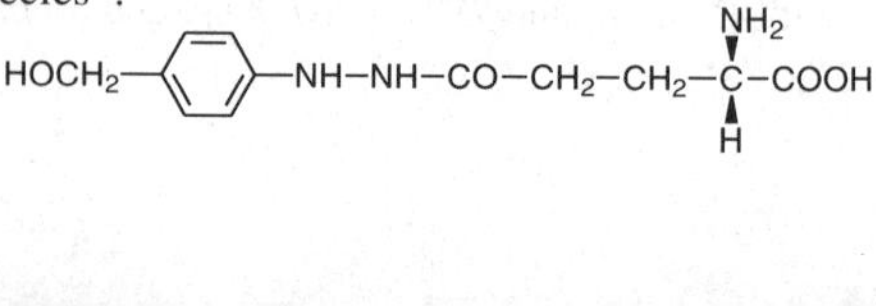

$C_{12}H_{17}N_3O_4$, M_R 267.28, mp. 205–209 °C (decomp.), $[\alpha]_D^{25}$ +7° (H_2O), for synthesis see *Lit.*[2]. A. is a procarcinogen[3]. Other L-glutamine compounds in *Agaricus* species are *N^5-(4-hydroxyphenyl)-L-glutamine* [$C_{11}H_{14}N_2O_4$, M_R 238.24, mp. 225–226 °C, $[\alpha]_D$ +42.5° (0.1 N NaOH), uv_{max} 440 nm], which is present in the spores of *A. bisporus* up to 1–2% of the dry weight as well as the *ortho*-quinone *N^5-(3,4-dioxo-1,5-cyclohexadienyl)-L-glutamine* ($C_{11}H_{12}N_2O_5$, M_R 252.23, uv_{max} 440 nm) generated therefrom by tyrosine oxidase which can be isolated as the acetyl or benzoyl derivative. *A. campestris* contains *N^5-(3,4-dihydroxyphenyl)-L-glutamine* (agaridoxin, $C_{11}H_{14}N_2O_5$, M_R 254.24, mp. 218–221 °C), which turns blue on spotting with iron(III) chloride. The aromatic ring system of A. and related fungal constituents are formed biosynthetically from *shikimic acid via 4-hydroxybenzoic acid[4]. The *ortho*-quinone is transformed with cleavage of pyroglutamic acid into *2-hydroxy-4-imino-2,5-cyclohexadienone.

Lit.: [1] J. Biol. Chem. **239**, 2267–2273 (1964). [2] J. Am. Chem. Soc. **83**, 3333 f. (1961); Helv. Chim. Acta **70**, 1261 (1987). [3] J. Agric. Food Chem. **30**, 521–525 (1982); Helv. Chim. Acta **70**, 1261–1267 (1987). [4] J. Biol. Chem. **261**, 13203–13209 (1986).
gen.: J. Org. Chem. **41**, 1603–1606 (1976); **45**, 3540 (1980) ▪ Phytochemistry **50**, 555 (1999) ▪ Zechmeister **51**, 236 (review). – *[CAS 2757-90-6]*

Agarofurans. *Sesquiterpene alkaloids and esters from Celastraceae. Most are polyacylated polyol derivatives of dihydro-β-agarofuran. Examples for some naturally occurring agarofurans:

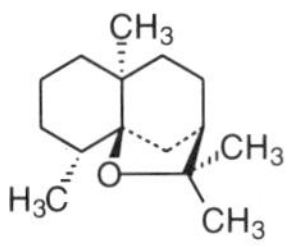

Dihydro-β-agarofuran

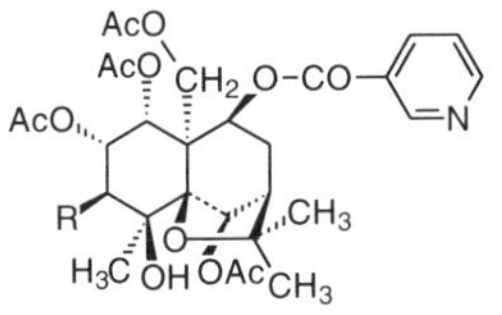

R = H : Maytine
R = OH : Maytoline
Ac = CO–CH₃

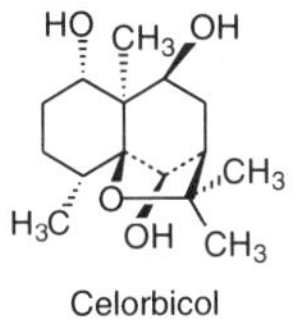

Celorbicol

Table: Data of Agarofurans.

	Maytine	Maytoline	Celorbicol
molecular formula	$C_{29}H_{37}NO_{12}$	$C_{29}H_{37}NO_{13}$	$C_{15}H_{26}O_4$
M_R	591.61	607.61	270.37
mp. [°C]			222
optical rotation		$[\alpha]_D^{25}$ +0.3° ($CHCl_3$)	$[\alpha]_D^{26}$ 24° ($CHCl_3$)
CAS	31146-56-2	31146-55-1	59812-41-8

A. often show biological activities. Thus, some have insecticidal activity while others are cytostatically active.

Lit.: Chin. Chem. Lett. **8**, 491 (1997) ▪ Manske **16**, 215 ▪ Phytochemistry **17**, 1821 (1978) ▪ Tetrahedron **43**, 5557 (1987).

Agaropectin see agar(-agar).

Agarose.

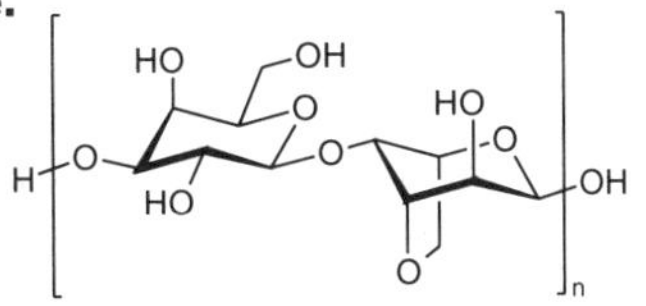

Gel forming β-D-/α-L-galactan, present to ca. 70% in *agar(-agar) from which it is isolated. Built up of α-1-3′-polymer-linked agarobiose (formula of the repeating unit). This consists of D-galactopyranose, which is β-1,4-glycosidically linked with 3,6-anhydro-L-galactopyranose. Depending on the type of the natural product differing amounts of methylated, sulfated, and with pyruvic acid acetalized hydroxy groups are found along the polysaccharide chain[1]. A. (commercial product e. g. Sepharose) is important for preparation of materials for gel and affinity chromatography[2].

Lit.: [1] The Polysaccharides, vol. 2, chapter 4, p. 217, New York: Academic Press 1983. [2] Methods Enzymol. **22**, 345–378 (1971).
gen.: Hager (5.) **2**, 243 ff. ▪ Ullmann (5.) **B3**, 11–18 f. – *[HS 3913 90; CAS 9012-36-6]*

Agathadiol see labdanes.

Agelastatin A see oroidin.

Agelastatins. Cytotoxic bromopyrroles from the sponge *Agelas dendromorpha*. The main compound is *A. A*, $C_{12}H_{13}BrN_4O_3$, M_R 341.16. A. A exhibits potent insecticial activity, but is also very toxic for brine shrimp.

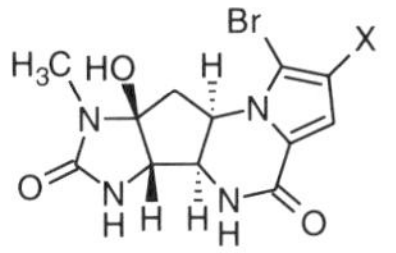

X = H : A. A
X = Br : A. B

Lit.: Helv. Chim. Acta **77**, 1895 (1994); **79**, 727 (1996) ▪ J. Chem. Soc., Chem. Commun. **1993**, 1305 ▪ J. Nat. Prod. **61**, 158 (1998) ▪ J. Am. Chem. Soc. **121**, 9574 (1999) ▪ J. Org. Chem. **63**, 7594 (1998). – *[CAS 152406-28-5 (A. A)]*

Ageratochromene see precocenes.

Aggregation pheromones see pheromones.

Aglaiastatin.

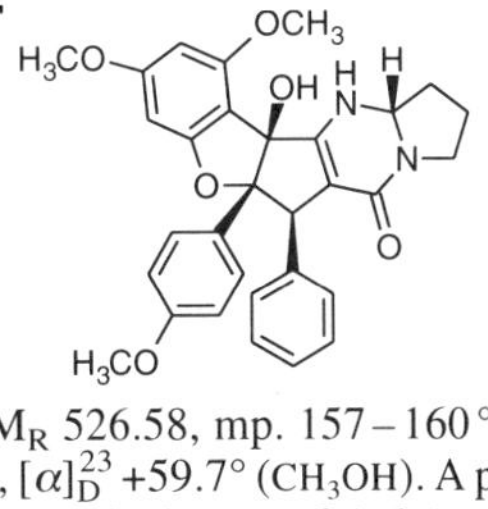

$C_{31}H_{30}N_2O_6$, M_R 526.58, mp. 157–160 °C (cryst. with 1.5 mol H_2O), $[\alpha]_D^{23}$ +59.7° (CH_3OH). A protein synthesis inhibitor from the leaves of *Aglaia odorata* (Meliaceae). A. is structurally related to *rocaglamide.

Lit.: J. Chem. Soc., Chem. Commun. **1994**, 773; **1998**, 1097 (synthesis) ▪ J. Nat. Prod. **59**, 650 (1996). – *[CAS 176785-78-7]*

Agmatine [(4-aminobutyl)guanidine].

$HN{=}C(NH_2){-}NH{-}(CH_2)_4{-}NH_2$

$C_5H_{14}N_4$, M_R 130.19, mp. 101.5–103 °C. A. is a guanidine derivative formed by decarboxylation of L-*arginine. It is an intermediate in the, in plants especially widely distributed, biosynthesis of *putrescine from arginine. Natural sources include ergot, the Asteraceae *Ambrosia artemisifolia*, the sea anemone *Anthopleura japonica* and herring semen.

Lit.: Beilstein E IV **4**, 1291 ▪ Merck-Index (12.), No. 186 ▪ Phytochemistry **10**, 731 (1971) (biosynthesis). – *[HS 2925 20; CAS 306-60-5]*

Agrobactin.

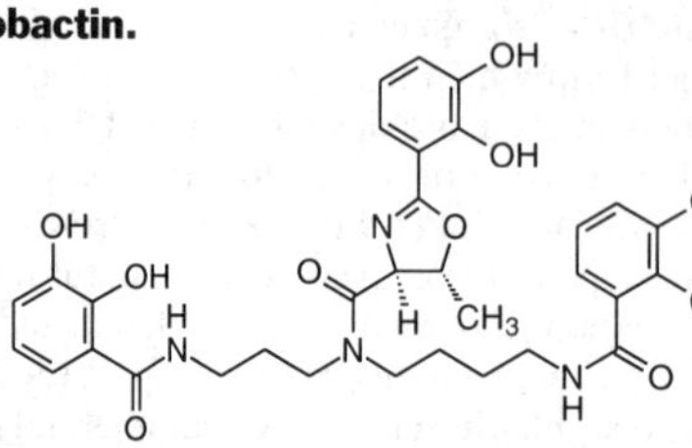

$C_{32}H_{36}N_4O_{10}$, M_R 636.66, mp. 108–112 °C (decomp.), a *siderophore of the catechol type from *Agrobacterium tumefaciens*. A. is built up from a threonyl-*spermidine and three 2,3-dihydroxybenzoyl units. The closely related *parabactin* ($C_{32}H_{36}N_4O_9$, M_R 620.66) formed by *Paracoccus denitrificans* lacks a phenolic hydroxy group, *vibriobactin* ($C_{35}H_{39}N_5O_{11}$, M_R 705.72) from *Vibrio cholerae* carries two dihydroxyphenyloxazolinecarboxylic acid units on bis(3-aminopropyl)-amine. Biosynthesis and excretion of the compound are induced by iron deficiency. After loading with iron the complex is degraded on the cell surface and the iron taken up by the organism.

Lit.: Biochem. J. **146**, 191–204 (1975) ▪ J. Am. Chem. Soc. **102**, 7715–7718 (1980) ▪ J. Biol. Chem. **254**, 1860–1865 (1979) ▪ J. Chem. Soc., Perkin Trans. 1 **1984**, 183–187 ▪ J. Org. Chem. **50**, 2780–2782 (1985). – *[CAS 70393-50-9 (A.); 74149-70-5 (parabactin); 88217-23-6 (vibriobactin)]*

Agrocin 84.

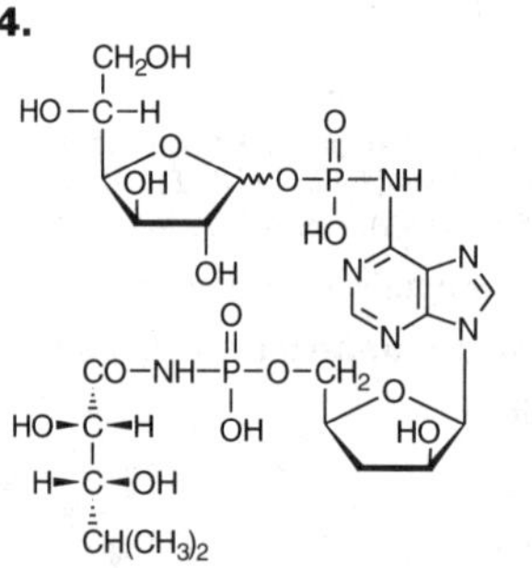

$C_{22}H_{36}N_6O_{16}P_2$, M_R 702.51. A nucleoside antibiotic, adenosine derivative substituted in the pentose part and on the amino group with phosphoramidate groups, it is isolated from *Agrobacterium radiobacter*. A. shows bacteriocidal action against *Agrobacterium tumefaciens* by inhibition of DNA synthesis.

Lit.: Clare, in Vining & Stuttard (eds.), Genetics and Biochemistry of Antibiotic Production, p. 619–632, Boston: Butterworth-Heinemann 1995 ▪ J. Antibiot. **41**, 1711–1739 (1988) ▪ Nature (London) **265**, 379, 697 (1977). – *[HS 2941 90; CAS 59111-78-3]*

Agroclavine (8,9-didehydro-6,8-dimethylergoline). $C_{16}H_{18}N_2$, M_R 238.33, $[\alpha]_D$ –155° ($CHCl_3$); needles, mp. 208–209 °C (decomp.), soluble in pyridine, chloroform, alcohol; poorly soluble in water. A. is a non-peptide ergot alkaloid of the ergoline type, it occurs as the first complete tetracyclic ergoline in ergot biosynthesis [1]. A. occurs not only in ergot (see ergot) but is also synthesized by *Penicillium* species and higher plants such as non-European bindweeds (Convolvulaceae).

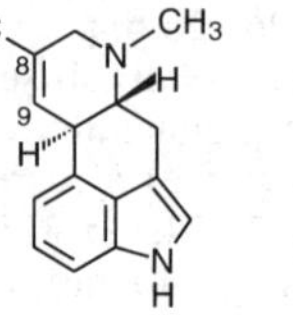

Lit.: [1] J. Am. Chem. Soc. **90**, 6500 (1968).

gen.: Biochem. Physiol. Pflanz. **172**, 531 (1978) (biosynthesis). – *Synthesis:* Chem. Pharm. Bull. **34**, 948 (1986) ▪ Heterocycles **45**, 1263 (1997) ((–)-A.) ▪ Tetrahedron Lett. **27**, 3469 (1986). – *[HS 2939 60; CAS 548-42-5]*

Aizumycin see bicyclomycin.

Ajmalicine (raubasine, δ-yohimbine, tetrahydroserpentine).

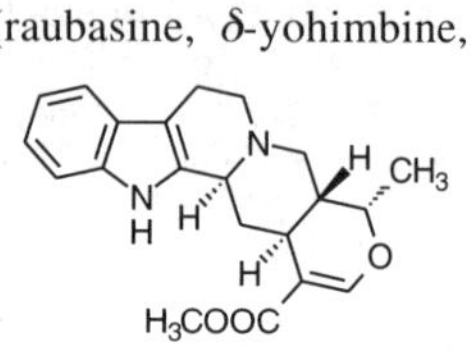

$C_{21}H_{24}N_2O_3$, M_R 352.43, $[\alpha]_D$ –66° ($CHCl_3$), mp. 261–263 °C; occurs in *Rauvolfia* and *Catharanthus* (syn. *Vinca*) species [1], in *Corynanthe johimbe* (Rubiaceae) and in many related plants; *monoterpenoid indole alkaloid, with 4 centers of chirality; eight isomers are known in the plant kingdom. LD_{50} in mice <400 mg/kg, has deterrent activity against herbivores, and antimicrobial activity against Gram-negative bacteria. The biosynthesis was clarified through isolation of the most important enzymes from *Catharanthus* and *Rauvolfia* cell cultures [2].

Use & activity: A. is used as an antihypertensive, sedative, and tranquilizer; as antihypertensive agent especially in combination with reserpine: it exerts a blood pressure lowering effect by widening the blood vessels (vasodilatator), the arterial blood pressure is lowered; see also Rauvolfia alkaloids.

Lit.: [1] Taylor, in Farnsworth (ed.), The Catharanthus Alkaloids, New York: Marcel Dekker 1975. [2] Phytochemistry **18**, 965 (1979).

gen.: Beilstein E III/IV **27**, 7927 f. – *Biosynthesis:* Manske **7**, 1, 59; **8**, 693–723. – *Synthesis:* Heterocycles **24**, 2117 (1986) ▪ J. Am. Chem. Soc. **115**, 807 f. (1993); **117**, 9139–9150 (1995) ▪ J. Chem. Res. (S) **8**, 190 (1983) ▪ J. Org. Chem. **60**, 3236–3242 (1995) ▪ Tetrahedron Lett. **26**, 865 (1985). – *Activity:* Hager (5.) **3**, 30 ff.; **6**, 361–380; **9**, 495. – *[HS 2939 90; CAS 483-04-5]*

Ajmaline (rauvolfine, tachmaline, cardiorhythmine).

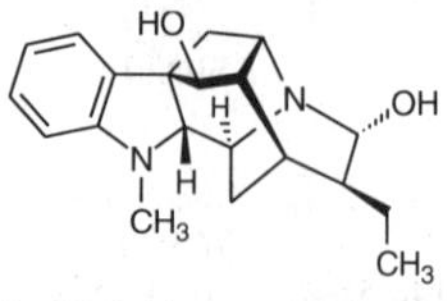

$C_{20}H_{26}N_2O_2$, M_R 326.44, mp. 205–207 °C, $[\alpha]_D$ +144° ($CHCl_3$), 158–160 °C (anhydrous). A. is a *Rauvolfia alkaloid and occurs in the roots of various *Rauvolfia* species, especially the Indian *R. serpentina* and the Af-

rican *R. vomitoria* as well as in cell and tissue cultures of these plants. A. is readily soluble in methanol, chloroform.
Use & activity: In medicine as an antihypertensive, tranquilizer, especially as anti-arrhythmic agent; the latter activity corresponds to that of the *Cinchona alkaloid *quinidine by blockade of sodium channels; inhibition of glucose uptake in mitochondria of cardiac muscle tissue. A. is poorly resorbable and has a low bioavailability.
Biosynthesis: A. is formed from tryptamine and the monoterpene secologanin (see secoiridoids) via *strictosidine, catalysed by ca. 10 known enzymes (see also figure under monoterpenoid indole alkaloids).
Lit.: Arzneim.-Forsch. **24**, 874 (1974) ▪ Beilstein EV **23/13**, 349 f. ▪ Hager (5.) **6**, 362–380; **3**, 32 ff.; **7**, 87; **9**, 495 ▪ J. Org. Chem. **63**, 4166 (1998) (synthesis (+)-A.) ▪ Manske **8**, 785, 789; **11**, 41–72 ▪ Zechmeister **43**, 267–346. – *[HS 2939 90; CAS 4360-12-7 (R); 4360-12-15 (S)]*

Ajoene see allium.

Ajugarins.

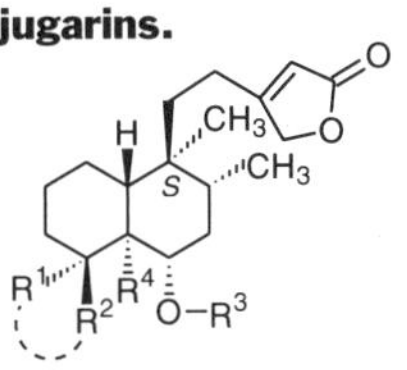

R^1	R^2	R^3	R^4	
—O—CH$_2$—		CO—CH$_3$	CH$_2$—O—CO—CH$_3$	A. I
—O—CH$_2$—		H	CH$_2$—O—CO—CH$_3$	A. II
OH	CH$_2$OH	CO—CH$_3$	CH$_2$—O—CO—CH$_3$	A. III
COOCH$_3$	H	CO—CH$_3$	CH$_3$	A. IV
—O—CH$_2$—		CO—CH$_3$	CH$_3$	A. V

Table: Data of Ajugarins.

	molecular formula	M_R	mp. [°C]	CAS
A. I	$C_{24}H_{34}O_7$	434.53	155–157	62640-05-5
A. II	$C_{22}H_{32}O_6$	392.49	188–189	62640-06-6
A. III	$C_{24}H_{36}O_8$	452.54	243–245	62640-07-7
A. IV	$C_{23}H_{34}O_6$	406.52	119–120.5	82225-47-6
A. V	$C_{22}H_{32}O_5$	376.49	217–218	82231-14-9

The A. are clerodane diterpenes, isolated from various Lamiaceae (*Ajuga remota, A. ciliata, A. decumbens, Scutellaria galericulata, Teucrium* sp.). A distinction is made between the A. I–V [1,2] as well as the structurally related *ajugamarins* from *A. nipponensis* [3].
Biological activity: Extracts of *A. remota* have antifeedant activity against several insects [4]. The behavior changing effects have also been investigated by electrophysiology [5]. Structure-activity relationships have been deduced for derivatives of A. and related compounds [6,7]. The A. are used as model compounds for mechanistic investigations (A. I) and as extracts for plant protection [8].
Lit.: [1] J. Chem. Soc., Chem. Commun. **1982**, 618. [2] Chem. Lett. **1983**, 223. [3] Chem. Pharm. Bull. **37**, 354, 988 (1989). [4] Abstracts Brighton Crop Prot. Conf. Pests Dis. (3.), p. 1041–1046 (1988). [5] Entomol. Exp. Appl. **46**, 267–274 (1988). [6] Experientia **39**, 403 (1983). [7] Phytochemistry **28**, 1069 (1989). [8] Greenhalgh & Roberts (eds.), Pestic. Sci. Biotechnol., Proc. Int. Congr. Pestic. Chem. (6.), p. 125 ff., Oxford, UK: Blackwell 1987.
gen.: J. Org. Chem. **57**, 862 (1992) ▪ Merck-Index (12.), No. 197 ▪ Recl. Trav. Chim. Pays-Bas **106**, 1–18 (1987) ▪ Phytochemistry **29**, 1793 (1990). – *Pharmacology:* East Afr. Med. J. **71**, 587 ff. (1994). – *Synthesis:* J. Chem. Soc., Chem. Commun. **1983**, 503 ▪ Phytochemistry **37**, 147–157 (1994) ▪ Tetrahedron **42**, 6519–6534 (1986) ▪ Tetrahedron Lett. **23**, 1751 (1982).

Akimycin see streptothricins.

Aknadinine see 1-nitroaknadinine.

Akuammicine, Akuammidine s. Alstonia alkaloids.

Alamethicin see ionophores.

Alangiside see ipecac alkaloids.

Alanine (2-aminopropanoic acid, abbreviation: Ala or A).

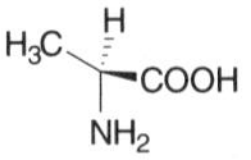

(2*S*)-, L-form

$C_3H_7NO_2$, M_R 89.09, mp. 297 °C (decomp.); $[\alpha]_D^{25}$ +2.8° (H_2O), +9.55° (5 m HCl), pK_a 2.35 and 9.87, pI 6.00. monohydrochloride, mp. 204 °C, $[\alpha]_D^{25}$ +8.5° (H_2O).
The (2*S*) form is a component of almost all proteins. The average content is 9.0% [1]. Genet. Code: GCU, GCC, GCA and GCG. Silk fibroin has a particularly high content of L-A. (25%). The free form is ubiquitous. Since A is mainly formed from pyruvic acid with the aid of aminotransferase it is not an essential amino acid. Its metabolization also proceeds through pyruvic acid. A. plays a central role in nitrogen metabolism.
The (*2R*) form (D-) is a component of the peptidoglycans in bacterial cell walls and also a constituent of the cell walls of higher plants. *γ-glutamyl-D-alanine* is found in the shoots of young peas (*Pisum sativum*) [2], mp. 295–297 °C (decomp.), $[\alpha]_D^{25}$ –1.8° (H_2O).
Lit.: [1] Biochem. Biophys. Res. Commun. **78**, 1018–1024 (1977). [2] Biochim. Biophys. Acta **304**, 363–366 (1973).
gen.: Beilstein EIV **4**, 2480–2485 ▪ Davies, Amino Acids and Peptides, p. 11, London: Chapman and Hall 1985 ▪ Merck-Index (12.), No. 205. – *[HS 2922 49; CAS 56-41-7 (L-A.); 338-69-2 (D-A.)]*

***β*-Alanine** (3-aminopropanoic acid).
$H_2N–CH_2–CH_2–COOH$, $C_3H_7NO_2$, M_R 89.09, mp. 197–198 °C (decomp.); monohydrochloride, mp. 122.5 °C, pK_a 3.60 and 10.19, pI 6.90. Component of *pantothenic acid, coenzyme A (see acetyl-CoA), *anserine, *carnosine, and of bacterial cell walls. The free form is ubiquitous in low concentrations in plants and fungi. *β*-A. is formed from *aspartic acid by decarboxylation.
Lit.: Beilstein EIV **4**, 2526–2528 ▪ Merck-Index (12.), No. 206. – *[HS 2922 49; CAS 107-95-9]*

2-(Alanin-3-yl)clavam see clavams.

L-Alanosine [(*S*)-2-amino-3-(hydroxynitrosoamino)-propanoic acid].

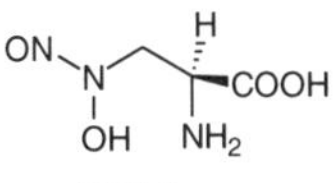

(2*S*)-form

$C_3H_7N_3O_4$, M_R 149.11, mp. 190 °C (decomp.), $[\alpha]_D$ +8.0° (1 m HCl), –46.0° (0.1 m NaOH), pK_a 4.8 and 8.6; LD_{50} (mouse i. v.) 300 mg/kg. Antibiotic from cultures of *Streptomyces alanosinicus*. A. inhibits the biosynthesis of adenine. It is undergoing preclinical development for tumor therapy [1].

Lit.: [1] Drug News, R & D Focus (23.9) **1996**, 3.
gen.: Biological activity: Nature (London) **211**, 1198 f. (1973). – *Isolation & structure:* Farmaco Ed. Sci. **21**, 269–277 (1966). – *Synthesis:* Bull. Chem. Soc. Jpn. **46**, 1847–1850 (1973). – *[HS 2928 00; CAS 16931-22-9; 5854-93-3 (L); 5854-95-5 (DL)]*

Alantolactone see helenin.

***N*-β-Alanyldopamine, *N*-β-alanylnoradrenaline** see *N*-acylcatecholamines.

Alarm pheromone see pheromones.

Alarm substances (Alarm pheromones). Name for substances that cause increased alertness and preparation for defense in the target organism. A. s. are signal substances, e.g. *pheromones, which are secreted by insects in times of danger and induce among the same species either defensive measures (e.g. biting) or fleeing behavior depending on aggressivity. A. s. exhibit various chemical structures and occur especially in social insects. Sites of formation of A. s. in *ants are the mandible glands, poison glands, Dufour's glands, and anal glands. Alkylpyrazines [1] (which represent a trace pheromone in some species) act on some species of ants while others respond to hexanal (see alkanals) and *hexanol* [2] ($C_6H_{14}O$, M_R 102.18), *dendrolasin, various ketones, or *dimethyl trisulfide* ($C_2H_6S_3$, M_R 126.25) as alarm substances. *Undecane, which was identified in ants very early, is widely distributed among ants [3,4]. In the alpine ant *Formica lugubris* it acts simultaneously as a sexual pheromone [5]. *trans-β*-*Farnesene acts as alarm substance for a several of greenfly species [6]. *Isopentenyl acetate* ($C_7H_{12}O_2$, M_R 128.17) is released when a honeybee stings and stimulates other bees to attack [7]. In some fresh-water swarming fish, e.g. minnows, a substance is released in the water when a single fish is injured which causes other fish of the same species to flee; see also pheromones, defensive secretions.

Lit.: [1] Science **182**, 501 (1973). [2] Nature (London) **285**, 230 (1975). [3] Ber. Dtsch. Chem. Ges. **25**, 1489 (1892). [4] Blum, Chemical Defenses of Arthropods, New York: Academic Press 1981. [5] Naturwissenschaften **93**, 30–34 (1993). [6] Science **177**, 1121 (1972). [7] Nature (London) **195**, 1018–1020 (1962). – *[CAS 111-27-3 (hexanol); 1191-16-8 (isopentyl acetate)]*

Albizziine [(*S*)-2-amino-3-ureidopropanoic acid].

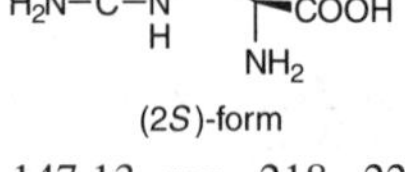

(2*S*)-form

$C_4H_9N_3O_3$, M_R 147.13, mp. 218–220 °C (decomp.), $[\alpha]_D^{25}$ –66.2° (H_2O), $[\alpha]_D^{24}$ –22.2° (1 N HCl). Non-proteinogenic amino acid from the seeds of *Albizia julibrissin* and other Mimosaceae species.

Lit.: Angew. Chem. Int. Ed. Engl. **30**, 712 (1991) ▪ Chem. Pharm. Bull. **23**, 2669 (1975) ▪ Synth. Commun. **27**, 2345 (1997) (synthesis D-A.). – *[HS 2924 10; CAS 1483-07-4]*

Alboinone.

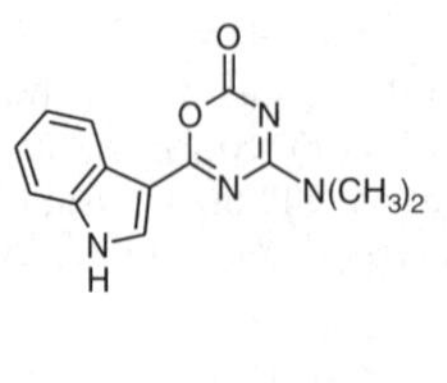

$C_{13}H_{12}N_4O_2$, M_R 256.26. Alkaloid from the ascidian *Dendrodoa grossularia*. A. is the first naturally occurring oxadiazinone alkaloid.

Lit.: Tetrahedron **53**, 2055 (1997). – *[CAS 188547-40-2]*

Albomycins. Iron-containing antibiotics belonging to the sideromycins from various *streptomycetes strains (e.g. *Streptomyces subtropicus*). The iron-free form is a peptide composed of 3 molecules N^δ-acetyl-N^δ-hydroxy-L-ornithine, 1 mol. L-serine, and an unusual amino acid, with a variable pyrimidine part and an *N*-glycosidically bound thiofuranose. The iron(III) complexes of the hydroxamate*siderophores have a red color and are highly active against bacteria.

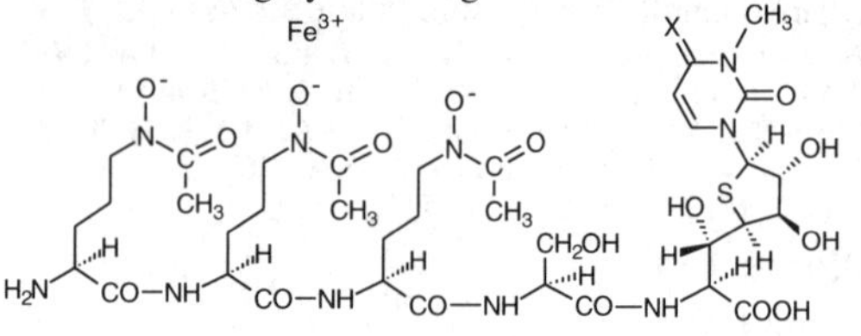

X = O : A. δ_1
X = N–CO–NH$_2$: A. δ_2
X = NH : A. ε

Table: Data of Albomycins.

A.	molecular formula	M_R	CAS
δ_1	$C_{36}H_{55}FeN_{10}O_{18}S$	1003.80	12676-11-8
δ_2	$C_{37}H_{57}FeN_{12}O_{18}S$	1045.84	34755-52-7
ε	$C_{36}H_{56}FeN_{11}O_{17}S$	1002.82	12676-10-7

Lit.: Angew. Chem. Int. Ed. Engl. **21**, 527, Suppl. 1322 (1982) ▪ Justus Liebigs Ann. Chem. **1984**, 1399 f. ▪ Pure Appl. Chem **28**, 603–636 (1971). – *[HS 2941 90]*

Alchornine see imidazole alkaloids.

Alder buckthorn (*Rhamnus frangula*). A member of the Rhamnaceae endemic to Europe, north west Asia, and north Africa whose bark is used to prepare extracts with laxative activity and for mordant dyeing of wool (sandy brown to red-brown) on account of its contents of *emodin, *frangulins, *chrysophanol, *physcion, and other *anthranoids.

Lit.: Hager (5.) **1**, 587 ff.; **6**, 397–410.

Aldosterone (aldocortene, electrocortine, oxocorticosterone, Reichstein substance X).

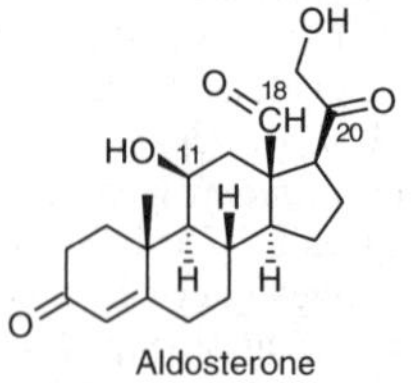

Aldosterone

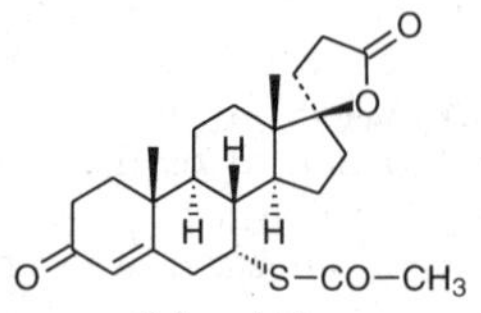

Spironolactone

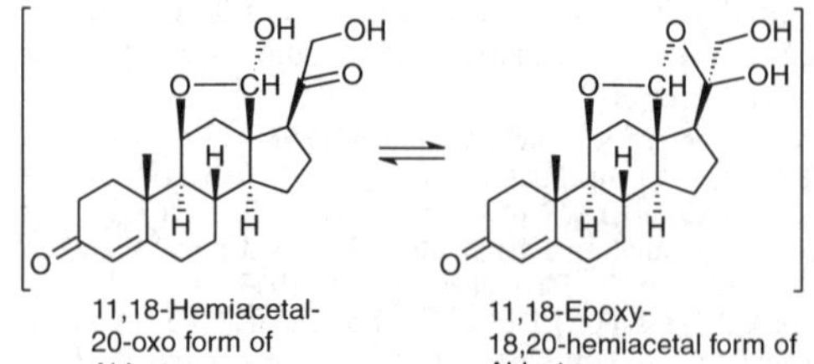

11,18-Hemiacetal-20-oxo form of Aldosterone

11,18-Epoxy-18,20-hemiacetal form of Aldosterone

International generic name for 11β,21-dihydroxy-3,20-dioxopregn-4-en-18-al, $C_{21}H_{28}O_5$, M_R 360.45. The *corticosteroid, mp. 164–169 °C (anhydrous), 108–112 °C (hydrate), $[\alpha]_D$ +161° ($CHCl_3$), isolated in 1953 from adrenal cortex extracts, exists in solution as an equilibrium mixture of two hemiacetal forms (see structures, ratio ca. 3:4), with the less hindered (18*S*,20*S*) stereoisomer predominating. A. was isolated from adrenal cortex in a yield of merely $5 \cdot 10^{-6}$% but is readily accessible by a four-stage synthesis from corticosterone, see *Lit.*[1]. As a typical mineralocorticoid A. is responsible not only for sodium, chloride, hydrogen carbonate ions, and water retention in the kidneys but also for release of potassium ions into urine. Thus A. regulates the water and electrolyte households. An elevated serum level of A. results in high blood pressure. Use of A. antagonists such as, e.g., *spironolactone* which acts as a diuretic, is recommended as treatment. A. is also employed in therapy for Addison's disease.

Lit.: [1] J. Am. Chem. Soc. **83**, 4083–4089 (1961).
gen.: Beilstein E IV **8**, 3491 ▪ Hager (5.) **7**, 98 ▪ Kleemann-Engel, p. 1742 (spironolactone) ▪ Müller, Regulation of Aldosterone Biosynthesis, Berlin: Springer 1988 ▪ Zeelen, p. 129–142. – *[HS 293729; CAS 52-39-1 (A.); 52-01-7 (spironolactone)]*

Aleuriaxanthin.

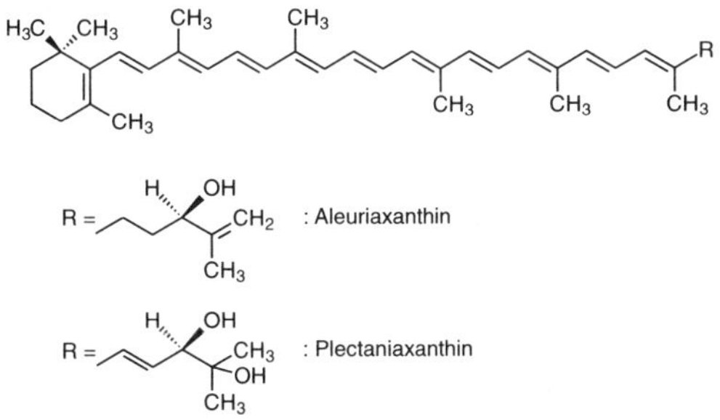

$C_{40}H_{56}O$, M_R 552.88, mp. 122–122.5 °C, opt. active pigment of the orange-red cup fungus *Aleuria aurantia* (Ascomycetes), where it occurs together with β- and γ-carotene. The abs. configuration of A. was determined by synthesis. *Plectaniaxanthin* ($C_{40}H_{56}O_2$, M_R 568.88, cryst., mp. 172–173 °C) is responsible for the brilliant red color of the fruitbody of the scarlet elf cup (*Sarcoscypha coccinea*, Ascomycetes), which can be found in mountain forests during the spring thaw. Both compounds occur in the fungi, in part combined with long-chain fatty acids.

Lit.: Phytochemistry **6**, 995 (1967); **21**, 2087 (1982) ▪ Zechmeister **51**, 193–207 (review). – *[CAS 51599-07-6 (A.); 51599-07-6 (plectaniaxanthin)]*

Aleuritic acid see hydroxy fatty acids.

Alexine.

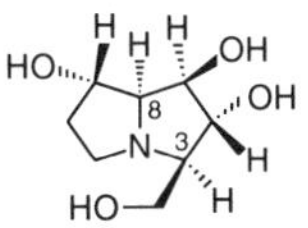

$C_8H_{15}NO_4$, M_R 189.21, mp. 162–163 °C, $[\alpha]_D$ +40° (H_2O). A. is a polyhydroxylated pyrrolizidine alkaloid (see polyhydroxy alkaloids) from the tree *Alexa leiopetala*, a south American Fabaceae. The 8-epimer *australine* and the 3,8-diepimer occur together with the isomeric polyhydroxylated indolizidine alkaloid *castanospermine in the tree-like Fabaceae *Castanospermum australe* growing in Australia. A. is, like castanospermine and *swainsonine, a glycosidase inhibitor. It shows particularly strong activity against amyloglucosidase of fungi of the genus *Aspergillus*. Australine acts as a specific inhibitor of exo-1,4α-glucosidase.

Lit.: J. Am. Chem. Soc. **120**, 7359 (1998); **121**, 3046 (1999) (synthesis (+)-australine) ▪ Proc. Phytochem. Soc. Eur. **33**, 271–282 (1992) ▪ Rec. Adv. Phytochem. **23**, 395–427 (1989). – *[HS 293990; CAS 116174-63-1 (A.); 118396-02-4 (australine); 119065-82-6 (3,8-diepimer)]*

Algae. Name for numerous species of lower plants composed of single cells or cell aggregates. Algae are subdivided into the strains:

– Section Cyanophyta (=Cyanobacteria, previously: blue algae) in the class Cyanophyceae. These are unicellular microorganisms, related to the bacteria, with blue or red pigments in the plasma (phycocyanin or phycoerythrin) which if necessary are capable of photosynthesis and nitrogen fixation, see also majusculamides (*Lyngbya majuscula*), tantazoles, anatoxins;

– Section Rhodophyta (red algae), the color is caused by the accessory pigment phycoerythrin which is localized in the chloroplast, see teurilene (*Laurencia* spp.), obtusallene I;

– Section Heterokontophyta, classes: Chrysophyceae (gold algae), Xanthophyceae, Eustigmatophyceae, Bacillariophyceae (=diatoms, with SiO_2 skeleton), e.g. *Nitzschia pungens* (see domoic acid), Raphidophyceae, Dictyochophyceae, Phaeophyceae (brown algae), e.g., bladder kelps are mostly anchored to a solid base and can form colonies more than 100 m long (sea weed). Their color is due to fucoxanthin, further components are e.g., ectocarpene, dictyodial (*Ectocarpus siliculosus*), acetylsanadaol (*Pachydiction* sp.), and *okamurallene;

– Section Haptophyta (=Prymnesiophyta), only one class: Haptophyceae, e.g. *Chrysochromulina polylepis*, which led to an enormous toxic blooming in the south coastal seas of Norway in 1988;

– Section Dinophyta with the single class Dinophyceae, usually gold-brown in color (*peridinin); e.g., *Gymnodinium breve*; constituents see gonyaulin, brevetoxins, maitotoxin, okadaic acid, saxitoxin;

– Section Chlorophyta (green algae), here the green chlorophyll is not concealed by other pigments; however, the algae do certainly contain many carotinoids. They are generally uni- to multicellular organisms living in fresh or salt water (e.g., *Chlorella vulgaris*). Photosynthesizing micro-A. as so-called phytoplankton, serves as food for many aquatic animals, e.g., the toothless whales. The A. are mostly autotrophic and in part form unusual compounds, see caulerpin (*Caulerpa taxifolia*); classes: Prasinophyceae, Chlorophyceae, Ulvophyceae, Cladophorophyceae, Bryopsidophyceae, etc.

In fresh, moist condition brown algae contain 75–90% water, in the dry state 20–35% *alginic acid, 4–12% mannitol, 9–12% laminarin, 7–12% cellulose, 16–23% other carbohydrates, 5–15% protein, 1–2% fat, also salts, vitamins, toxins and organic compounds with in part very unusual structures that may also contain Br, Cl, and I. Many algal components with antibacterial activity are known. Some A. are surprisingly specialized, e. g., an A. living in the Dead Sea produces glycerol (up to 85% of its dry weight!), others enrich uranium from sea water. Lichens represent a symbiotic form of algae with fungi.

Lit.: Abstract source: Algae Abstracts, New York: Plenum (since 1973).

Algal pheromones. Group of lipophilic unsaturated C_{11}-hydrocarbons and epoxides excreted from the female sex cells of marine brown algae to attract male spermatozoids [1]. In highly developed species [e. g., redware (*Laminariales*) and spiney kelp (*Desmarestiales*)] these compounds additionally act as synchronizing release factors, by first effecting the release of male sex cells from their gametangia. The *Fucales* (e. g., bladder kelp) in contrast use the C_8-hydrocarbon, *fucoserratene* [(1,3*Z*,5*E*)-octa-1,3,5-triene], as pheromone. In *Dictyopteris* species C_{11}-hydrocarbons are continuously released from the thallus into sea water in small amounts and then presumably take on ecologically important functions in the fight for living space [2]. Interference with the signalling systems of neighboring species [3] and antifeedant activities towards sea urchins and fish have been demonstrated [4].

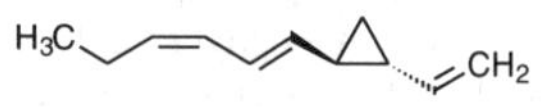

Hormosirene

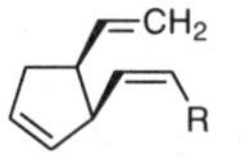

R = C_2H_5 : Multifidene
R = CH=CH_2: Viridiene

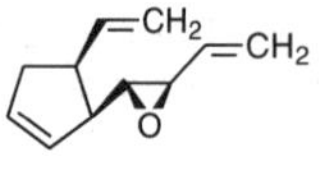

Caudoxirene

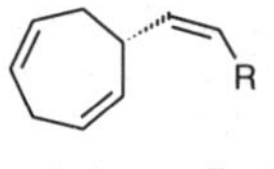

R = C_2H_5 : Ectocarpene
R = CH=CH_2: Desmarestene

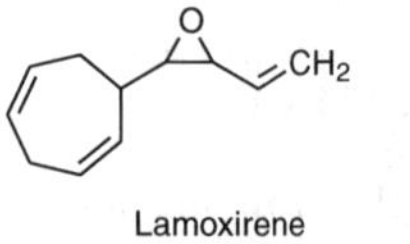

Lamoxirene

Finavarrene

Table: Data of Algal pheromones.

pheromone	molecular formula	M_R	optical activity	CAS
Caudoxirene	$C_{11}H_{14}O$	162.23	$[\alpha]_{25}^{578}$ +238.3° (CH_2Cl_2)	117415-46-0
Desmarestene	$C_{11}H_{14}$	146.23	$[\alpha]_D^{22}$ +168° (CH_2Cl_2)	83013-90-5
Ectocarpene	$C_{11}H_{16}$	148.25	$[\alpha]_{578}^{20}$ +17.4° (CH_2Cl_2)	33156-93-3
Finavarrene	$C_{11}H_{16}$	148.25		29837-19-2
Hormosirene	$C_{11}H_{16}$	148.25	$[\alpha]_D^{21}$ −43° ($CHCl_3$)	29837-20-5
Lamoxirene	$C_{11}H_{14}O$	162.23		92675-19-9
Multifidene	$C_{11}H_{16}$	148.25	$[\alpha]_{20}^{578}$ +261° (CH_2Cl_2)	52886-04-1
Viridiene	$C_{11}H_{14}$	146.23	$[\alpha]_{25}^{578}$ +228° (pentane)	83013-89-2
Fucoserratene	C_8H_{12}	108.18		40087-61-4

The A. are formed from highly unsaturated C_{20}-fatty acids by oxidative processes. The biosynthesis of cycloheptadienes (e. g., ectocarpene) proceeds via a spontaneous *Cope*-rearrangement of a primarily formed *cis*-disubstituted cyclopropane. Higher plants synthesize ectocarpene by oxidative decarboxylation of dodeca-3,6,9-trienoic acid. Some of the C_{11}-hydrocarbons are also found in *diatoms* and higher plants but their biological function is still unknown.

Lit.: [1] Angew. Chem. Int. Ed. Engl. **21**, 643 (1982); Biol. Bull. **170**, 145 (1986); Bot. Acta **101**, 149 (1988); Plant Cell Environ. **16**, 891 (1993); Proc. Natl. Acad. Sci. USA **92**, 37 (1995). [2] Tetrahedron Lett. **28**, 307 (1987). [3] Ecology **71**, 776 (1990). [4] Marine Ecol. **48**, 185 (1988).

gen.: Biosynthesis: Angew. Chem. Int. Ed. Engl. **31**, 1246 (1992); **34**, 1602 (1995) ▪ Eur. J. Biochem. **191**, 453 (1990) ▪ Proc. Natl. Acad. Sci. USA **92**, 31–43 (1995) ▪ Tetrahedron **54**, 11 033 (1998) (fuscoserratene). – *Synthesis:* Angew. Chem. Int. Ed. Engl. **38**, 546 (1999) (multifidene) ▪ J. Chem. Soc., Chem. Commun. **1993**, 551 ▪ Helv. Chim. Acta **78**, 447 (1995) ▪ Scheuer I **1**, 98 ff. ▪ Scheuer II **2**, 107 ff. ▪ Tetrahedron **51**, 7927 (1995); **53**, 145 (1997) ((+)-multifidene); **53**, 14 651 (1997) ((±)-lamoxirene) ▪ Tetrahedron Lett. **36**, 8745 (1995).

Algal toxins see shellfish poisons.

Algin. Common term for the polysaccharide constituents of brown algae (see algae and alginic acid).

Alginic acid.

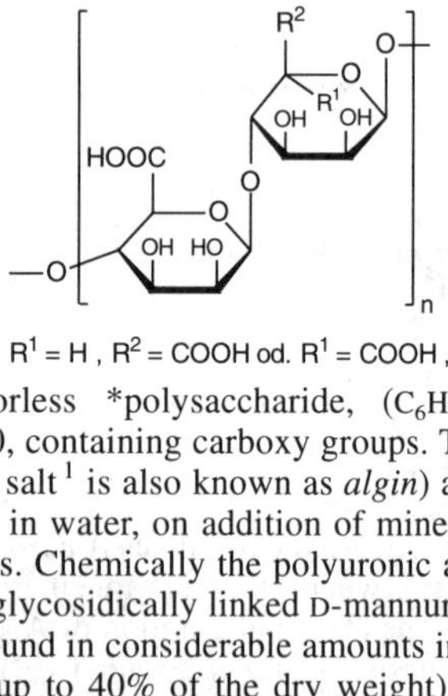

R^1 = H , R^2 = COOH od. R^1 = COOH , R^2 = H

A colorless *polysaccharide, $(C_6H_8O_6)_n$, M_R ca. 200 000, containing carboxy groups. The alkali metal (the Na salt [1] is also known as *algin*) and Mg salts are soluble in water, on addition of mineral acid A. precipitates. Chemically the polyuronic acid A. consists of 1,4-glycosidically linked D-mannuronic acid units. A. is found in considerable amounts in marine brown algae (up to 40% of the dry weight) as matrix substance. For separation, see *Lit.* [2] and for use, see *Lit.* [3]

Lit.: [1] Merck-Index (12.), No. 240, 241. [2] Kirk-Othmer (3.) **17**, 768–774. [3] Kirk-Othmer (3.) **12**, 45; Martindale (31.), p. 1535.

gen.: Environ. Sci. Res. **44**, 209 (1992) (biosynthesis) ▪ Hager (5.) **1**, 181 ff.; **5**, 200 ff.; 740 f.; **7**, 109 ff. ▪ Ullmann (5.) **A 11**, 503, 563; **A25**, 34–40 – *[HS 3913 10; CAS 9005-38-3 (A. Na salt); 9005-32-7 (A.)]*

Alizarin(e) (1,2-dihydroxyanthraquinone, C. I. 58 000).

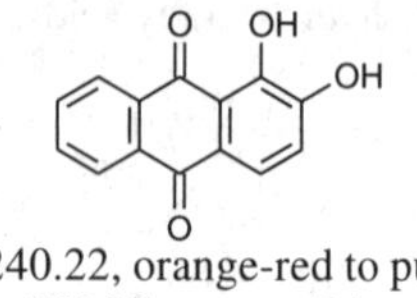

$C_{14}H_8O_4$, M_R 240.22, orange-red to purple-red needles or prisms, mp. 290 °C, uv_{max} 434 nm (C_2H_5OH); very poorly soluble in boiling water, soluble in alcohols, ethers, aromatic hydrocarbons, and glacial acetic acid. A. occurs in rhizomes of *madder (*Rubia tinctorum*,

Rubiaceae) in the form of the 2-*O*-*β*-primveroside (*ruberythric acid) from which it is liberated on drying the rootstock, constituting together with *purpurin and other di- and trihydroxyanthraquinone glycosides the colored components.

Use: With some metal salts (Al, Fe, Cr) A. forms brilliantly colored complexes known as *madder lakes* or *rubiates*. In the process of *Turkey red dying*, already known in antiquity and developed further in 13th and 14th centuries especially in Turkey, threads or textiles of cotton, wool, or silk are treated with rancid olive oil, followed by a slurry of chalk and alum, treatment with steam, and finally dyed with madder. The constitution of Ca-Al alizarinate (Turkey red) was elucidated by X-ray crystallography [1]. Since A. binds to calcifying tissue it is used in histological examination of bone formation; further, A. serves as an acid-base indicator and is used for photometric determination of trivalent metal cations as well as F^- and SO_4^{2-}.

Synthesis: First synthesis in 1868 was technically not realizable; in 1869 Graebe, Liebermann and Caro found an access to synthetic A. by sulfonation of anthraquinone and subsequent melting with alkali; after its market introduction in 1871 the synthetic product quickly displaced the natural substance.

The biosynthesis [2] starts from *isochorismic acid, 2-oxoglutaric acid, and mevalonic acid lactone, see anthranoids.

Lit.: [1] Chem. Ber. **127**, 1185–1190 (1994). [2] Phytochemistry **12**, 1669 (1973); Planta Med., Suppl. **214** (1975); Nat. Prod. Rep. **10**, 233–263 (1993); Recent Adv. Phytochem. **20**, 243–261 (1986).
gen.: Beilstein E IV **8**, 3256 ▪ J. Org. Chem. **52**, 5685 (1987) (synthesis) ▪ Merck-Index (12.), No. 247 ▪ Sigma-Aldrich Library of Stains, Dyes and Indications, p. 75 ▪ Ullmann (4.) **7**, 603, 639–642 ▪ Zechmeister **49**, 79–149. – *[HS 291469; CAS 72-48-0]*

Alkaloidal dyes. Representatives of colored alkaloids that are suitable for dying wool either directly or after treatment with alum. These include *berberine hydroxide* {$C_{20}H_{19}NO_5$, M_R 353.37, mp. 145 °C ($CHCl_3$-Solvate), C. I. 75160)} isolated from root bark of barberry (*Berberis vulgaris*, Berberidaceae) and *harmalol* [harmala red, $C_{12}H_{12}N_2O$, M_R 200.24, mp. 100–105 °C (trihydrate)], a red substance easily oxidized in air, isolated from seeds of the xerophyte *Peganum harmala* (Zygophyllaceae).

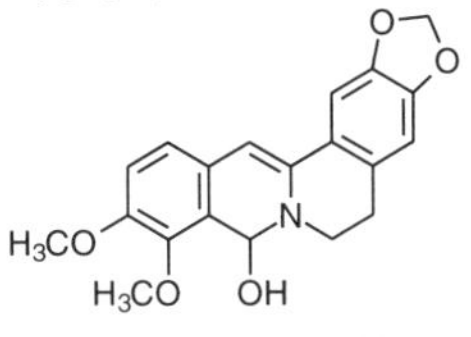

Berberine hydroxide

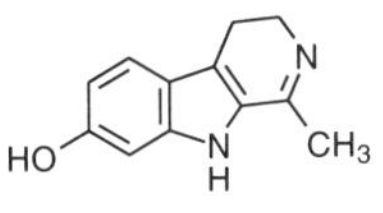

Harmalol

Lit.: Biochem. Soc. Trans. **20**, 3595 (1992) ▪ Collect. Czech. Chem. Commun. **50**, 2299–2309 (1985) ▪ J. Nat. Prod. **40**, 384 (1977) ▪ Phytochemistry **19**, 2009 (1980) – *[HS 243990; CAS 117-74-8 (berberine hydroxide); 525-57-5 (harmalol)]*

Alkaloids. General name for basic natural products occurring mostly in plants containing one or more basic nitrogen atoms, mostly in heterocyclic rings. A. often show pronounced pharmacological activities. The N atoms usually originate from amino acids.

There is no uniform classification for the A. In the literature divisions according to origin (*examples:* Aconitum, *Amaryllidaceae, *Aspidosperma, *cactus, *Catharanthus, *Cephalotaxus, *Cinchona, coca, *Corydalis, curare, *Dendrobates, *ergot, *Erythrina, *Iboga, *Lycopodium, Maytenus, *opium, *Rauvolfia, Senecio, *Strychnos, *tobacco, *Vinca alkaloids, *salamander, *Solanum, *Veratrum steroid alkaloids) in addition to divisions according to chemical structure (*examples:* *aporphine, benzylisoquinoline, *bisbenzylisoquinoline, berberine, carboline, *diterpene, *imidazole, *indole, *indolizidine, *isoquinoline, lupinane, macrocyclic, morphine, *peptide, *β-phenylethylamine, *piperidine, purine, pyridine, pyrrolidine, *pyrrolizidine, *quinoline, *quinolizidine, quinuclidine, spermine, spermidine, *steroid, terpene, *tropane, tropolone alkaloids) are used.

The individual A. or their parent N-bases often derive their names from the natural product (*example:* tropane = parent base of *atropine and of *cocaine) or from the plants in which the A. occur (*example:* solanine, *vincamine), and sometimes from their physiological properties, like, e. g. *morphine (Greek.: Morpheus = god of sleep) and emetine (emetic, see ipecac alkaloids).

Most A. are colorless (yellow colored: e. g., chelidonine), as free bases they are readily soluble in water, even more soluble in alcohols, ethers, and chloroform. With acids they often form nice crystalline salts that are well soluble in water; because of their solubility the latter are used preferentially in medicine (e. g., morphinum hydrochloricum, pilocarpine hydrochloride). Up to the end of 1957 2233 A. from 3761 plant species of 156 families had been isolated, in 1976 the numbers had increased to 5000 A. from 7000 plant species, and in 1989 over 10 000 alkaloids were known. In 1995 the number of known A. was estimated to be ca. 15 000. As early as in the 19th century, structure investigations on A. led to the development of useful degradation methods which provided a major improvement to the arsenal of organic chemistry. Investigations on biosynthesis of A. started much later.

Biosynthesis: In contrast to steroids and terpenes, there is no equivalent to the isoprene rule for alkaloids as a useful aid for structure determination and biogenetic investigations. It is generally accepted today that A. are formed in cyclization, condensation, and dimerization reactions from amino acids and biogenic amines with biogenic aldehydes and ketones.

Occurrence: Since most A. are nerve poisons they cannot form in larger amounts in the animal world, except in skin glands where they are excreted for specific purposes as *neurotoxins (e. g. the *amphibian venoms samandarine and other salamander alkaloids, *tetrodotoxin, *batrachotoxins, bufotenine and other *toad poisons, *glomerine, *serotonine, *histamine, tyramine and other biogenic amines). Most A. occur in plants as by-products of amino acid metabolism where they are mostly stored in peripheral parts of the plant (leaves, roots, barks, fruits) and much less frequently in the wood. A. occur only rarely as free bases in plants and are mostly found as salts with oxalic, acetic, lactic, and citric acids, etc. Some higher plant families are particularly rich in A., such as, e. g. periwinkle (Apocynaceae), poppy (Papaveraceae), and nightshade (Solanaceae), in contrast Compositae (Asteraceae), conifers, and lower plants (ferns, mosses, algae) are usually free of A. It is conspicuous that terpene- and resin-rich plants are usually low in A. Related plant families usually have similar alkaloid patterns.

Detection, analysis: A. precipitation reagents that effect formation of poorly water soluble salts are, e. g., Dragendorff's, Fröhde's, and Mayer's reagents, picric acid, picrolonic acid, dipicrylamine, phosphomolybdic acid, Reinecke's salt, and sodium tetraphenylborate. Special color reactions are used for some groups of A., e. g., Vitali's reaction to detect tropane A., the thalleioquine reaction for *Cinchona* A., the murexide reaction for purines.

Production/extraction: Isolation and separation of A. from natural sources are based on the solubility of the free bases in lipophilic (ether, chloroform) and of the salts in hydrophilic solvents (water, ethanol). Many A. can also by synthesized but the development of partial and total syntheses served mainly for structure elucidation and only very few procedures have practical significance.

Use: On account of their activities, especially on the central nervous system, many A. have a long history of pharmaceutical use, e. g., the *morphinan alkaloids (as analgesic and narcotic agents), ergot, quinine, tropane, *Rauvolfia* A., *Vinca* A. from evergreen species as cytostatic agents. Among the A. there are not only addictive drugs but also strong poisons: heroin (*morphine and *morphinan alkaloids), lysergic acid diethylamide, *mescaline, *cocaine and, resp. *strychnine, *batrachotoxins, *tetrodotoxin, *saxitoxin, *aconitine etc.

Historical: The term A. was introduced by the apothecary C. F. W. Meissner in 1819 for "alkali-like" plant substances. The development of A. chemistry is illustrated by some dates and names: morphine (1803 or 1816 Sertürner), strychnine (1818 Pelletier and Caventou), solanine (1820 Desfosses), caffeine (1820 Runge), quinine (1820 Pelletier and Caventou), nicotine (1828 Posselt and Reimann), atropine (1831 Mein), codeine (1832 Robiquet), theobromine (1842 Woskresensky), cocaine (1860 Wöhler), ephedrine (1887 Nagai), scopolamine (1881 Ladenburg and 1888 E. Schmidt), mescaline (1896 Heffter). The first A. synthesis was realized in 1886 by Ladenburg (coniine). In the 20th century A. chemistry is especially associated with the names Willstätter, Woodward, Schöpf, Robinson, Wieland, Karrer, Späth, Wiesner, and Hesse.

Lit.: ApSimon (ed.), The Total Synthesis of Natural Products, vols. 1 – 10, New York: Wiley 1971 – 1997 ▪ Blum, The Toxic Action of Marine and Terrestrial Alkaloids, Fort Collins: Alaken Inc. 1995 ▪ Blum, Chemistry and Toxicology of Diverse Classes of Alkaloids, Fort Collins: Alaken Inc. 1996 ▪ Helv. Chim. Acta **75**, 647 – 688 (1992) (Review) ▪ Manske (53 vols. up to 2000) ▪ Mothes, Schütte, and Luckner, Biochemistry of Alkaloids, Weinheim: Verl. Chemie 1985 ▪ Pelletier, Alkaloids: Biological Perspectives, vols. 1 – 6, New York: Wiley 1983 – 1988; vols. 7 – 9, Oxford: Pergamon 1992 – 1995 ▪ Rodd's Chem. Carbon Compds. (2.) **4B** (1997) (complete volume) ▪ Pharmazie **52**, 546 (1997) (history) ▪ Saxton (ed. 1971 – 1975) and Grundon (ed. 1976 – 1983), The Alkaloids, Specialist Periodical Reports, 13 vols. up to 1983, London: Royal Soc. Chem. 1971 – 1983 ▪ Southon and Buckingham (eds.), Dictionary of Alkaloids, London: Chapman & Hall 1989 ▪ Ullmann (5.) **A 1**, 353 – 407 ▪ Waterman (ed.), Alkaloids and Sulphur Compounds, London: Academic Press 1993. – *[HS 2939 10 – 2939 90]*

Alkanals.

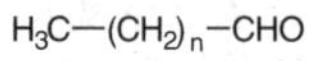
n-Alkanals

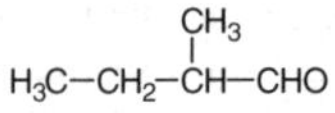
2-Methylbutanal

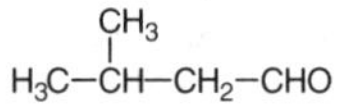
3-Methylbutanal

n-A. up to C_{14} from the fatty acid catabolism [1] and lower, in part methyl-branched A. from the Strecker degradation [1] are widely distributed in small amounts in nature and are used as *aroma and *flavor compounds. Apart from acetaldehyde (LD_{50} 1.9 g/kg) the LD_{50} values (rat oral) are between 2.3 g/kg (C_{12}) and >5 g.

Lower A. up to C_6 are especially of interest as aroma substances. *Acetaldehyde*, e. g., is responsible for a fresh fruity taste [2] and is also present in many alcoholic drinks; it is also an important raw material for synthesis. *2-Methylbutanal* when diluted has a fruity cocoa-like smell and is present, e. g., in aromas of cocoa, tee, and peanuts. *3-Methylbutanal* has a pungent smell and stimulates coughing, when diluted it has a green-fruity smell and is important in *bread, *cocoa and tomato aroma. *Hexanal* irritates the eyes and airways, when diluted it has a fatty-green smell like unripe fruit; olfactory threshold in water: 5 ppb [3]. The biosynthesis [4] from linoleic acid by lipoxygenases and lyases starts after the destruction of the plants cellular integrity, e. g., in apples, cucumbers, tomatoes, tea leaves, and grapes. Hexanal is mainly used in fruit aromas.

Table: Data of some Alkanals.

	n	molecular formula	M_R	bp. [°C]	D_4^{20}	flash point [°C]	CAS
Acetaldehyde	0	C_2H_4O	44.05	21	0.78	–38	75-07-0
2-Methylbutanal	–	$C_5H_{10}O$	86.13	90 – 92	0.804	4	96-17-3
3-Methylbutanal	–	$C_5H_{10}O$	86.13	92 – 93	0.785	– 1	590-86-3
Hexanal	4	$C_6H_{12}O$	100.16	128	0.814	35	66-25-1
Octanal	6	$C_8H_{16}O$	128.21	171	0.821	76	124-13-0
Nonanal	7	$C_9H_{18}O$	142.24	191	0.826	82	124-19-6
Decanal	8	$C_{10}H_{20}O$	156.27	209	0.827	86	112-31-2
Undecanal	9	$C_{11}H_{22}O$	170.30	223	0.827	103	112-44-7
Dodecanal	10	$C_{12}H_{24}O$	184.32	237	0.832	–	112-54-9

The A. C_8–C_{12} occur in many essential oils, especially in *citrus oils, *coriander, *caraway (C_8 and C_9) and *rose oil (C_8–C_{10}), as well as in traces in many aromas [5]. Their odors are described as orange-like (C_8), fatty-flowery, citrus-like (C_9), fatty orange-like (C_{10}), flowery-fatty, rose-like (C_{11}), and flowery-waxy, when highly diluted violet-like (C_{12}). Apart from citrus aromas C_8–C_{12}-A. are mainly used in perfumery for fragrances with an emphasis on aldehyde odors, e. g., of the Chanel No. 5 type.

Lit.: [1] Müller-Lamparsky, p. 101 ff. [2] Maarse, p. 309–322. [3] Ohloff, p. 61. [4] Müller-Lamparsky, p. 101 ff. [5] TNO list (6.), Suppl. 5, p. 341, 353, 390, 433, 444, 533.

gen.: Bauer et al. (2.), p. 10–12 ▪ Bedoukian (3.), p. 19–25 – *[HS 2912 12 (acetaldehyde); 2912 19]*

Alkannin (C. I. 75530).

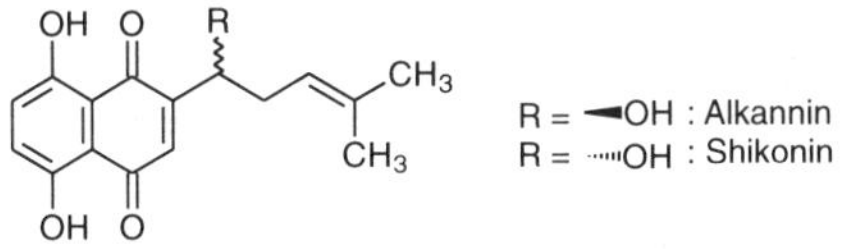

$C_{16}H_{16}O_5$, M_R 288.30, red-brown needles, mp. 149 °C, uv_{max} 546 nm (CH_3OH), $[\alpha]_D^{20}$ –157° (benzene). Dying naphthazarine derivative from the roots of alkanet (*Alkanna tinctoria*, Boraginaceae; up to 30 cm high bushy plant) grown in Turkey and south Europe. A. also occurs in the form of *O*-1′-esters of acetic, 2-methyl-2-butenoic, and 3-methyl-2-butenoic acids in the roots of *Alkanna nobilis*. A. is used to color foods and cosmetics (not in Germany) and is used to treat stasis ulcers on account of its antimycotic activity. The biosynthesis starts from 4-hydroxybenzoic acid and geranyl diphosphate [1]. The (*S*)-form of A. is *shikonin.

Lit.: [1] Luckner (3.) p. 402.

gen.: Angew. Chem. Int. Ed. Engl. **37**, 839 (1998) (synthesis); **38**, 270 (1999) (review) ▪ Hager (5.) **1**, 701; **4**, 175 ff. ▪ Merck-Index (12.), No. 253 ▪ Tetrahedron Lett. **38**, 7263 (1997) (synthesis (–)-A.) – *[HS 3203 00; CAS 517-88-4]*

Alkanolides (hydroxyalkanoic acid lactones). *γ*- and *δ*-lactones usually occur together in flavors [1], as well as in flower extracts, e. g. *tuberose absolute. In *fruit flavors such as apricot, mango, nectarine, and peach 4-A. are more frequent, while in *butter and *coconut flavor the *δ*-lactones predominate. The lactones C_8–C_{12}

4-Alkanolides (γ-Lactones, 5-Alkyl-dihydro-2(3*H*)-furanones)

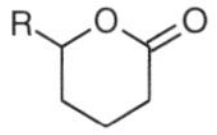

5-Alkanolides (δ-Lactones, 6-Alkyl-tetrahydro-2*H*-pyran-2-ones)

are of interest for use as *aroma and *flavor compounds, and in perfumery especially the strongly smelling *γ*-lactones: thresholds in water ca. 7–60 ppm (C_9); *δ*-lactones ca. 100 (C_{10})–400 ppm (C_8) [2]. The (*R*)-configuration dominates in the natural 4-A., whereas the (*S*)-form is also found in the 5-A., e. g., (*S*)-5-decanolide in raspberries [3]. There are only slight sensory differences between the (*R*)- and (*S*)-isomers [4]. For biosynthesis from fatty acids and production using microorganisms, see *Lit.* [5]. For synthesis see *Lit.* [6]. Pure 4- and 5-A. are colorless liquids with coconut-like (especially C_8 and C_9), fruity peach-like (C_{10} and C_{11}) or fruity-fatty odors (C_{12}) [4]. The 5-A. have somewhat weaker, cream-like odors. For characteristic data of some important 4- and 5-alkanolides see table, $[\alpha]_D$ values are only given for the (*R*)-isomers (measured in methanol). LD_{50} value (rat oral) for 4-octanolide: 4400 mg/kg, for all other examples >5000 mg/kg.

Lit.: [1] TNO list (6.), Suppl. 4, p. 375, 386, 460, 471, 553. [2] J. Agric. Food Chem. **36**, 551 (1988); Maarse, p. 502 f. [3] J. Chromatogr. **481**, 363–367 (1989). [4] J. Agric. Food Chem. **37**, 413–418 (1989) [5] Hopp, Mori (eds.), Recent Developments in Flavor and Fragrance Chemistry, p. 291–304, Weinheim: VCH 1993. [6] Bedoukian (3.), p. 256–266.

gen.: Bauer et al. (2.), p. 117 ff. ▪ Beilstein E V **17/9**, 68 ff. – *[HS 2932 29]*

Alkenals. A. are formed from fatty acids by autooxidation or enzyme action [1] and are widely distributed in *essential oils and aromas [2]. The C_6–C_{13}-A. (see also hexenals) have sensory effects even in low concentrations. Odors of (*E*)-2-A. resemble those of the *alkanals but are stronger and less fatty; C_{10}: citrus, orange; C_{12}: orange, mandarine. Clearly different are the odors of the (*Z*)-A. (see table, p. 20). C_7: green, fatty cream; C_9: cucumber, melon; C_{10}: citrus, flowery [3].

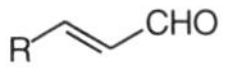

(*E*)-2-A.

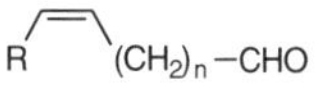

(*Z*)-A.

Table: Data of Alkanolides.

Alkanolide	R	molecular formula	M_R	bp. [°C] (kPa)	D.	$[\alpha]_D^{20}$	CAS
4-Octanolide	C_4H_9	$C_8H_{14}O_2$	142.20	116–117 (1.3)	0.977	+56.2° (*R*)	104-50-7 107797-24-0 (*R*)
4-Nonanolide	C_5H_{11}	$C_9H_{16}O_2$	156.22	136 (1.7)	0.968	+51.8° (*R*)	104-61-0 63357-96-0 (*R*)
4-Decanolide	C_6H_{13}	$C_{10}H_{18}O_2$	170.25	156 (2.3)	0.952	+48.5° (*R*)	706-14-9 107797-26-2 (*R*)
5-Decanolide	C_5H_{11}	$C_{10}H_{18}O_2$	170.25	122 (0.2)	$D_4^{27.5}$ 0.954	+60° (*R*)	705-86-2 2825-91-4 (*R*) 59285-67-5 (*S*)
4-Undecanolide	C_7H_{15}	$C_{11}H_{20}O_2$	184.28	167–169 (2.0)	0.944	+45.3° (*R*)	104-67-6 74568-06-2 (*R*)
4-Dodecanolide	C_8H_{17}	$C_{12}H_{22}O_2$	198.31	130 (0.07) (mp. 17–18 °C)	0.934	+42.2° (*R*)	2305-05-1 69830-91-7 (*R*)
5-Dodecanolide	C_7H_{15}	$C_{12}H_{22}O_2$	198.31	124 (0.01)	0.942	+46° (*R*)	713-95-1 29587-89-1 (*R*)

Table: Data of sensorially important Alkenals.

	R	n	molecular formula	M_R	bp. [°C, (kPa)]	D.	T_{GC}* [pmol/ stimulus]	CAS
(*Z*)-4-Heptenal	$H_3C–CH_2$	2	$C_7H_{12}O$	112.17	41 (1.3)	0.850	1.5	6728-31-0
(*E*)-2-Octenal	$H_3C–(CH_2)_4$	–	$C_8H_{14}O$	126.20	174	0.854	100	2548-87-0
(*E*)-2-Nonenal	$H_3C–(CH_2)_5$	–	$C_9H_{16}O$	140.23	189	0.847	0.14	18829-56-6
(*Z*)-3-Nonenal	$H_3C–(CH_2)_4$	1	$C_9H_{16}O$	140.23			0.1	31823-43-5
(*Z*)-6-Nonenal	$H_3C–CH_2$	4	$C_9H_{16}O$	140.23	72 (1.3)	0.851	0.05	2277-19-2
(*E*)-2-Decenal	$H_3C–(CH_2)_6$	–	$C_{10}H_{18}O$	154.25	118 (2.9)	0.849	25	3913-81-3
(*Z*)-4-Decenal	$H_3C–(CH_2)_4$	2	$C_{10}H_{18}O$	154.25	75 (0.8)	0.848	1	21662-09-9
(*E*)-2-Dodecenal	$H_3C–(CH_2)_8$	–	$C_{12}H_{22}O$	180.29	126 (1.3)	0.854	2	4826-62-4

* T_{GC} = perceptibility threshold, determined by gas chromatography[4]; 1 pmol/stimulus corresponds approximately to an odour limit of 0.5 ppb in water.

Occurrence: (*E*)-2-A. C_8–C_{13} in *citrus oils, especially bitter orange, C_8 also in guava and ginger aromas, C_9 in *bread, cucumber, carrot (see vegetable flavors) and *rice flavor, C_{10} in *coriander oil, *butter, chicken and guava aroma, C_{12} in coriander oil, peanut and *meat flavor. (*Z*)-4-Heptenal is found, among others, in *butter, seafood and *tea flavor, (*Z*)-3- and (*Z*)-6-nonenal in cucumber, melon; and fish aroma, and (*Z*)-4-decenal in *calamus oil and *Citrus junos* oil[2,3].

Use: A. have many applications in food flavoring and perfumery. LD_{50} values (rat oral) ≥5 g/kg.

Lit.: [1] Müller-Lamparsky, p. 101–126. [2] TNO list (6.), Suppl. 5, p. 345, 355, 385, 436, 450. [3] Perfum. Flavor. **12** (5), 31–43 (1987).

gen.: Arctander, No. 842, 1110, 1505, 2356. – *Synthesis:* Bauer et al. (2.), p. 13 ▪ Bedoukian (3.), p 29–33. – *[HS 2912 19]*

Alkoxylipids see ether lipids.

Allamandin.

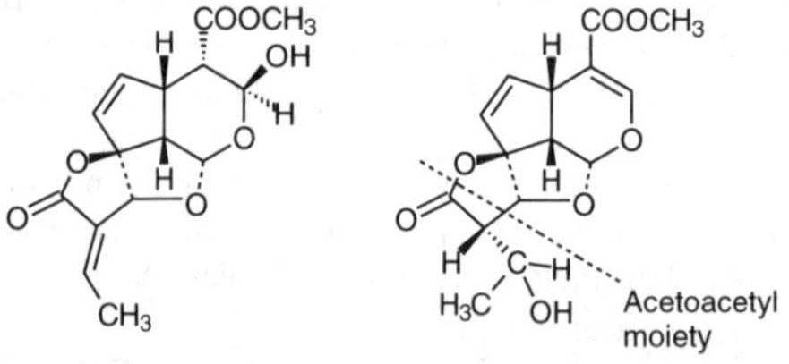

Allamandin Allamandicin

A. and its isomer allamandicin have the *iridoid skeleton to which acetoacetic acid has been added in an alkylation reaction. *A.*, $C_{15}H_{16}O_7$, M_R 308.29, mp. 212–215 °C, $[\alpha]_D$ +15° (CH_3OH). *Allamandicin*, mp. 117–118 °C, $[\alpha]_D$ +293° ($CHCl_3$).

Occurrence: Both compounds occur in the periwinkle family (Apocynaceae), e. g. in *Allamanda cathartica*[1] and in most *Plumeria* species[2]. A. exhibits *in vivo* activity against P-388 leukemia in mice and *in vitro* activity against human carcinomas of the nasopharynx (epipharynx) (KB)[3].

Lit.: [1] J. Org. Chem. **39**, 2477 (1974). [2] Hegnauer **III**, 150. [3] Cancer Chemother. Rep. **25**, 1 (1962).

gen.: J. Nat. Prod. **43**, 649 (1980). – *[HS 2938 90; CAS 51820-82-7 (A.); 51838-83-6 (allamandicin)]*

Allantoic acid (diureidoacetic acid).

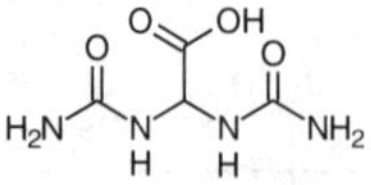

$C_4H_8N_4O_4$, M_R 176.13, mp. 173 °C (decomp.), sinters at 168 °C, needles. A. occurs in numerous plants, especially the Fabaceae. A. is formed in plants and osseous fish through hydrolytic ring opening of *allantoin by allantoinase (EC 3.5.2.5).

Lit.: Acta Crystallogr., Sect. B **24**, 1686–1692 (1968) (crystal structure) ▪ Beilstein EIV **3**, 1492 ▪ Luckner (3.), p. 437–439. – *[HS 2924 10; CAS 99-16-1]*

Allantoin (5-ureidohydantoin).

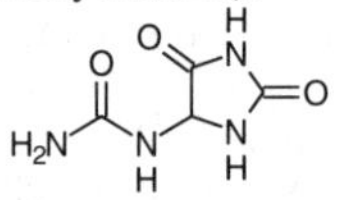

$C_4H_6N_4O_3$, M_R 158.12. A. is widely distributed as a degradation product from purine metabolism. The (*R*)-form has mp. 226–229 °C (decomp.), $[\alpha]_D^{20}$ –92.4° (H_2O), pK_a 8.5 (25 °C), and is moderately soluble in water. The (*S*)-form is isolated from *Platanus orientalis*. The racemate forms monoclinic prisms with mp. 238–240 °C (decomp.); it is less soluble in water than the enantiomer. A. is used for wound healing and as an antipsoriatic agent.

Biosynthesis: A. is formed in microorganisms, plants, and animals, except primates, by oxidative degradation of uric acid by urate oxidase (uricase; EC 1.7.3.3).

Lit.: Beilstein E V **25/15**, 338 ▪ J. Org. Chem. **42**, 3132–3140 (1977) (^{13}C NMR) ▪ Karrer, No. 2569 ▪ Luckner (3.) p. 437–439. – *[HS 2924 29; CAS 97-59-6 (A.); 7303-80-2 (R); 5377-33-3 (rac.)]*

Allelochemicals. General term for signal substances acting between different species of organisms; see also allomones, kairomones, synomones, semiochemicals.

Allelopathy (from Greek: allelo = each other and pathe = action). The term was first used by the plant physiologist Hans Molisch in 1937[1] to describe chemical interactions between plants. He included not only disadvantageous but also advantageous interactions. Although the term is occasionally extended to include biochemical interactions between plants and microorganisms[2], today the term A. is mostly limited to interactions between higher plants. A. is generally observed as an inhibition of germination and growth or special forms of soil exhaustion. Responsible for these effects are substances generally known as allelopathic substances or phytotoxic compounds. They are mostly lower-molecular-weight natural products, i.e., typical secondary plant products. Most of the compounds currently identified with certainty as allelopathic substances are volatile *terpenes or simple phenolic compounds. A typical example of A. is the observation that under and in the proximity of some trees (e.g. walnut, pine) other plants are inhibited in their germination and growth. As early as 1925 Massey[3], observed that tomato and alfalfa plants developed poorly and mostly died within a radius of more than 25 m around the trunk of a walnut tree. Later, *juglone was identified as the responsible phytotoxic substance. Juglone occurs only in the leaves and fruits of the tree and is washed out of dead tissue by rain. In the soil juglone completely inhibits the germination of, e.g., lettuce seeds at a concentration of as little as 0.002%. Allelopathic interactions have been described in many ecosystems, especially where there is strong competition for the available water in the soil[4].

Lit.: [1] Molisch, Der Einfluß einer Pflanze auf die andere – Allelopathie, Jena: Fischer 1937. [2] Rice, Allelopathy, 2nd edn., New York: Academic Press 1984. [3] Phytopathology **15**, 773–784 (1925). [4] Rec. Adv. Phytochem. **3**, 106–121 (1970). *gen.:* Alkaloids **43**, 1–104 (1993) ■ Harborne, Introduction to Ecological Biochemistry (3.), London: Academic Press 1994.

Alliacol(ide). *Alliacolide*, $C_{15}H_{22}O_4$, M_R 266.34, cryst., mp. 192–194 °C, $[\alpha]_D$ –35° ($CHCl_3$). Sesquiterpenoid from cultures of the garlic mushroom (*Marasmius alliaceus*, Basidiomycetes) where it occurs with a series of hydroxy derivatives and the deoxy compound alliacide[1]. *Alliacol A* {$C_{15}H_{20}O_4$, M_R 264.32, cryst., mp. 156–158 °C, $[\alpha]_D$ +10° ($CHCl_3$)} and the isomeric *alliacol B* {cryst., mp. 95 °C or 121–123 °C, $[\alpha]_D$ –5.5° ($CHCl_3$)} are weak antibiotics with cytotoxic activity against ascites cells of the Ehrlich carcinoma[2].

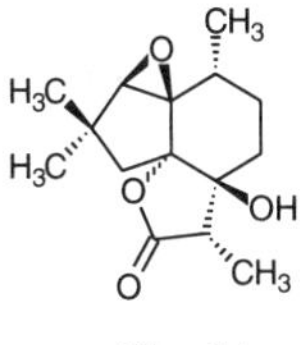

Alliacolide

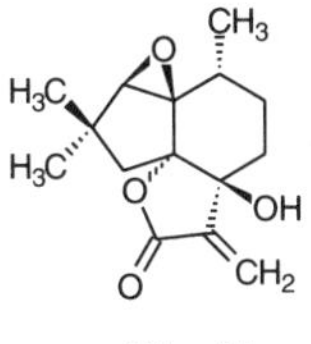

Alliacol A

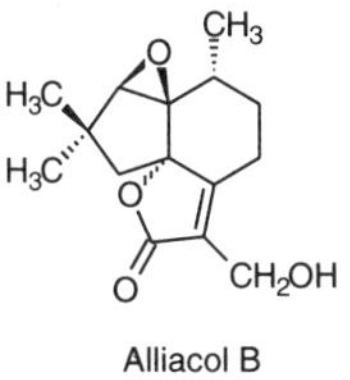

Alliacol B

Lit.: [1] J. Chem. Soc., Perkin Trans. 1 **1981**, 1790, 1794; **1982**, 2787; **1985**, 2749; **1988**, 1107 (synthesis). [2] J. Antibiot. **34**, 1271 (1981); J. Org. Chem. **60**, 4822–4833 (1995). – *[CAS 66389-08-0 (A.); 79232-29-4 (alliacol A); 79232-33-0 (alliacol B)]*

Allicin, alliin see Allium.

Allium. A genus of the Liliaceae widely distributed in the northern hemisphere of which ca. 450 species exhibit the typical leek odor especially when injured. E.g., onion (*Allium cepa*), Welsh onion (*A. fistulosum*), shallot (*A. ascalonicum*), leek (*A. ampeloprasum* var. *porrum*), garlic (*A. sativum*), wild garlic (*A. ursinum*), chive (*A. schoenoprasum*) and Chinese chive (*A. tuberosum*) are cultivated as crop plants; according to popular belief, the bulbs of alpine leek (*A. victorialis*, wild alraun) shall make invulnerable. Together with a relatively large number of soluble carbohydrates, the storage organs also contain high levels of *ascorbic acid and *thiamin(e). Secondary metabolites and sulfur-containing amino acids make up 1–5% of the dry weight of *Allium* plants.

The organosulfur chemistry[1] of the genus A. is characterized by a large number of unusual types of compounds, many of which are highly reactive and exhibit manifold physiological activities. A. species also accumulate selenoamino acids and metabolize them to secondary metabolites analogous to those of the sulfur-amino acids[2]. The biosynthesis of sulfur-containing components proceeds from sulfate through cysteine and γ-glutamylcysteine to glutathione. The SH groups of these two peptides are alkylated by methacrylic acid (from valine) or, in the case of glutathione, also methylated. Hydrolysis of the amide bond to glycine and decarboxylation of the *S*-2-carboxymethyl group leads to dipeptides with *S*-propenyl – or, after reduction, with *S*-propylcysteine residues. The thioethers are transformed to *S*-oxides by an oxidase, γ-glutamyltransferase finally liberates the four key compounds: *S*-2-propenylcysteine *S*-oxide [*alliin*[3], $C_6H_{11}NO_3S$, M_R 177.22, mp. 163 °C (decomp.), $[\alpha]_D^{21}$ +67.7° (H_2O)], *S*-propylcysteine *S*-oxide, *S*-(*E*)-1-propenylcysteine *S*-oxide, and *S*-methylcysteine *S*-oxide. Differing concentrations of these compounds in the various A. species determine their typical taste and odor.

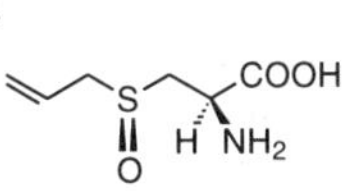

Alliin (**1**)

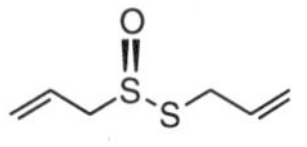

Allicin (**2**)

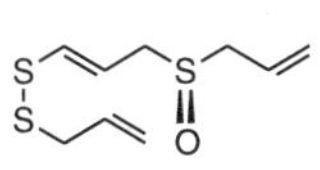

(*E*)-Ajoene (**3**)

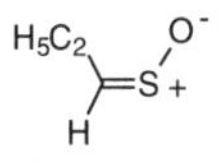

Propanthial *S*-oxide (**4**)

Allinase, occurring in A. species and activated upon injury, cleaves cysteine *S*-oxides to ammonia, pyruvate, and sulfenic acids (R–S–OH). These react rapidly with elimination of water to give thiosulfinates (R–SO–S–R). *Allicin*[4] [2-propene-1-thiosulfinic acid *S*-allyl ester, $C_6H_{10}OS_2$, M_R 162.28, pale yellow liquid, slightly toxic and irritating; LD_{50} (mouse i.v.) 60 mg/kg] can be extracted from chopped garlic with ethanol as a liquid with a pleasant, garlic-like odor. It decomposes in aqueous media preferentially to diallyl disulfide (R–S–S–R) and (*E*)- and (*Z*)-*ajoene*[5] (allyl-[3-(2-propenylsulfinyl)-1-propenyl] disulfide, $C_9H_{14}OS_3$, M_R 234.41). On heating in oil vinyldithiins and various sulfides are formed. For practical reasons a distinction is made between primary and secondary volatile compounds (isolated as essential oils and used, e.g., in foods) and secondary, non-volatile compounds (in solution). The oils of the various A. species differ in the pattern of the groups R in the thiosulfinates (C_2-, C_4-, C_6-thiosulfinates) and the polysulfides derived therefrom and can thus be characterized analytically. The tear inducing principle (lacrimal factor, LF) of onion,

(*Z*)-*propanethial S-oxide*[6] (C_3H_6OS, M_R 90.14) is also formed through the action of allinase and its coenzyme pyridoxal 5′-phosphate on the substrate (*E*)-(+)-*S*-(1-propenyl)-L-cysteine *S*-oxide. LF is rapidly hydrolysed to form the cyclic diethyl-*cis*- and -*trans*-1,2,4-trithiolanes in a multistep process. In contrast, spontaneous condensation of LF leads, among others, with elimination of propanal to diethyl-1,2,3-trithiolane 1,1-dioxide . It is assumed that the various A. species can, in principle, produce all of the large number of compounds derivable from the four cysteine *S*-oxides but that the reactions proceed at very different rates. Thus, in garlic, thiosulfinates containing exclusively propenyl groups are produced ten times more rapidly than those that also contain methyl groups.

Pharmacology: Garlic and – in higher doses – wild garlic were historically used to fight epidemics. Allicin, even at a dilution of 1 : 10^5, inhibits the growth of Gram-positive and Gram-negative bacteria. Onion extracts have a weaker effect than garlic. Allicin is also active against yeast and fungi, garlic oil has recently been used in plant protection. Garlic's antimicrobial activity can be traced to enzyme inhibition: Allicin blocks the action of cysteine proteinases and alcohol dehydrogenases by reacting with thiol groups of the enzymes[7]. The demonstrated cancer-inhibiting action of A. components has been linked with the neutralization of nitrosamines. The inhibiting effects of garlic and onion on the aggregation of human blood platelets have been attributed to various component substances. The γ-glutamyl-*S*-alkylcysteines affect blood pressure regulation, ajoenes and dithiines lower the serum lipid content. Onion extracts have antiasthmatic effects. Ajoenes are linked with many physiological effects. A large proportion of the activities ascribed to A. components can be classified as spontaneous reactions with SH groups of enzymes. Phenomena like the breath odor occurring after consumption of A. species confirm the continuation of the degradation chain of cysteine *S*-oxides, as well as that of the analogous selenium compounds, in the human body. Garlic preparations in low doses are considered to be harmless; however, concentrates may have toxic effects in the airways, on the skin, and in the gastrointestinal tract.

Lit.: [1] Angew. Chem. Int. Ed. Engl. **31**, 1135 (1992). [2] J. Agric. Food Chem. **43**, 1751, 1754 (1995). [3] Beilstein E IV **4**, 3147; Bioscience Biotechnol., Biochem. **58**, 108 (1994); J. Agric. Food Chem. **42**, 146 (1994); J. Nat. Prod. **56**, 864 (1993); Phytochemistry **33**, 107 (1993). [4] ACS Symp. Ser. **534**, 306–330 (1993) (Human Medicinal Agents from Plants); Beilstein E IV **4**, 7; J. Org. Chem. **59**, 3227 (1994); Pure Appl. Chem. **65**, 625 (1993); Sax (8.), AFS 250. [5] J. Am. Chem. Soc. **108**, 7045 (1986). [6] J. Agric. Food Chem. **19**, 269 (1971); J. Am. Chem. Soc. **118**, 7492 (1996); J. Org. Chem. **53**, 2026 (1988). [7] Antimicrob. Agents Chemother. **41**, 2286 (1997).

gen.: Gazz. Chim. Ital. **127**, 523 (1997) (review) ▪ J. Agric. Food Chem. **45**, 4414 (1997) (synthesis) ▪ Med. Res. Rev. **16**, 111–124 (1996). – *[CAS 556-27-4 (1); 539-86-6 (2); 92284-99-6 (E form of 3); 92285-00-2 (Z form of 3); 32157-29-2 (4)]*

Allixin.

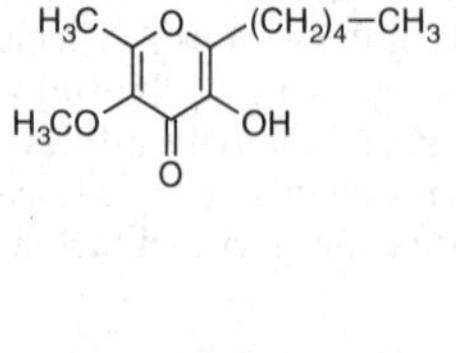

$C_{12}H_{18}O_4$, M_R 226.27, needles, mp. 80–81 °C. The antitumour agent A. is formed by garlic (*Allium sativum*) under stress conditions.

Lit.: Chem. Pharm. Bull. **37**, 1656 (1989). – *Synthesis:* Tetrahedron Lett. **38**, 7761 (1997); **39**, 2339 (1998). – *[CAS 125263-70-9]*

Allocryptopine see protopine alkaloids.

Allomones. Term derived from Greek allos = the other, in context: another, foreign and hormòn = set in motion, for (mostly low-molecular-weight) signal substances that act between organisms of different types (inter-species) and provide an advantage for the producer. Typical A. are *antibiotics, *repellents, *defensive secretions and *attractants, if, e.g., the latter serves to attract prey. A. that are biologically active towards other types of organisms and concomitantly show a behavior-modifying action to other members of the same species are called *pheromones. Bola spiders produce the sexual attractant of certain female butterflies in order to attract the corresponding males[1]. Biogenic signal substances that (merely) benefit the receiver are known as *kairomones. A. that are advantageous for both sender and receiver are known as *synomones; see also semiochemicals.

Lit.: [1] Science **236**, 964 (1987).

gen.: Agosta, Bombardier Beetles and Fever Trees: Chemical Warfare and Signals in Animals and Plants, New York: Addison-Wesley 1995 ▪ Harborne, Introduction to Ecological Biochemistry (3.), London: Academic Press 1994 ▪ Nordlund, Jones & Lewis, Semiochemicals, their Role in Pest Control, New York: Wiley 1981 ▪ Schlee, Ökologische Biochemie, Berlin: Springer 1986.

Allosamidin.

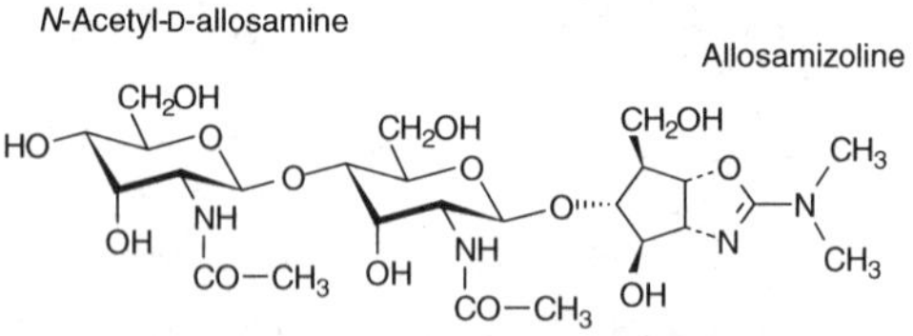

$C_{25}H_{42}N_4O_{14}$, M_R 622.63, $[\alpha]_D$ −24.8°. Pseudotrisaccharide from *Streptomyces* sp. made up from 2 mol. *N*-acetyl-D-allosamine and allosamizoline, a bicyclic cyclopentanoid aminocyclitol, by β-glycosidic attachment. A. specifically inhibits chitinase of the silk worm.

Lit.: Agric. Biol. Chem. **52**, 1615 (1988) (structure) ▪ Carbohydr. Res. **294**, 29 (1996) (synthesis) ▪ Curr. Opin. Ther. Pat. **1992**, 1645 ff. ▪ J. Antibiot. **40**, 296–300 (1987); **46**, 1582–1588 (1993) (biosynthesis) ▪ J. Am. Chem. Soc. **118**, 9526–9538 (1996) (synthesis) ▪ J. Chem. Soc., Perkin Trans. 1 **1992**, 1649; **1994**, 3411–3421 ▪ J. Org. Chem. **64**, 1782 (1999) (synthesis) ▪ Tetrahedron Lett. **27**, 2475 (1986); **33**, 793, 7565 (1992) (synthesis). – *[CAS 103782-08-7]*

Allylglycine see 2-amino-4-pentenoic acid.

Allyl isothiocyanate see mustard oils.

Aloe-emodin [1,8-dihydroxy-3-(hydroxymethyl)-anthraquinone].

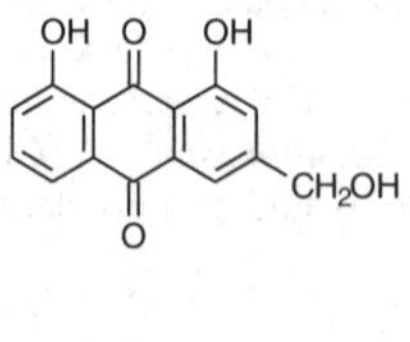

$C_{15}H_{10}O_5$, M_R 270.24, orange needles, mp. 223–225 °C, uv_{max} 430 nm (C_2H_5OH), soluble in ether, benzene, and hot ethanol. A. is named after the aloe, the concentrated leaf juices of various *Aloe* species (*A. ferox, A. barbadensis*, Liliaceae) and occurs additionally in free form and as the corresponding anthrone glycosides in *Rhamnus*, *Rheum*, *Rumex*, and *Cassia* species. In addition to laxative action it also exhibits an antileukemic effect and is used as starting material for the synthesis of *anthracycline antibiotics.

Lit.: Ann. Pharm. Fr. **40**, 357–363 (1982) ▪ J. Org. Chem. **50**, 139 (1985). – *[CAS 481-72-1]*

Alpinigenine see rhoeadine alkaloids.

Alprostadil see prostaglandins.

Alstonia alkaloids. A group of over 250 differing monomeric and dimeric *monoterpenoid indole alkaloids from the genus *Alstonia* (Apocynaceae) having aspidospermidine, corynanthean, heteroyohimbane, and related indole skeletons. The A. a. are isolated from leaves, roots, and bark of wild *Alstonia* species (bushes, often fairly large evergreen trees) with the exception of *A. scholaris* which can be cultivated.
The tree *A. scholaris* (as already suggested by the species name "lignum scholare" – it was used in the past to make blackboards) contains over 18 alkaloids of which, among others, those listed in the table are typical[1].

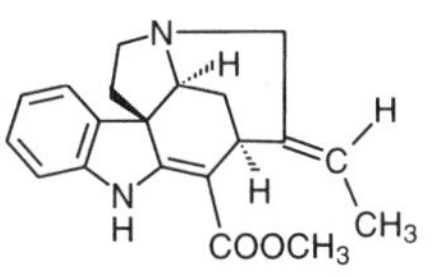

Akuammicine (**1**)

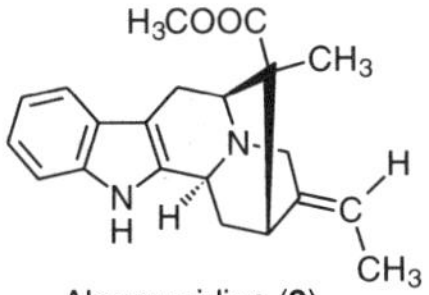

Akuammidine (**2**)

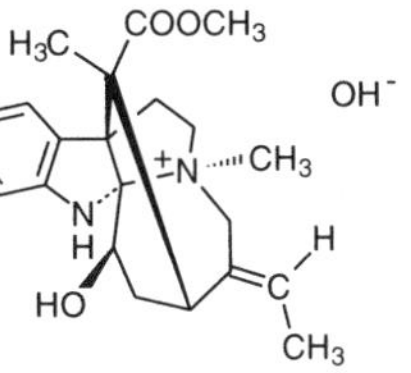

Echitamine (**3**)

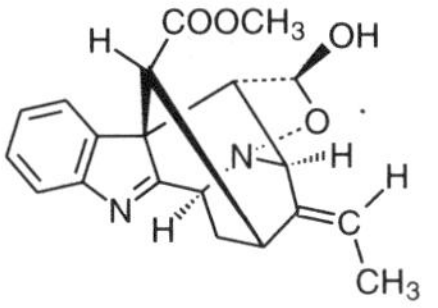

Nareline (**4**)

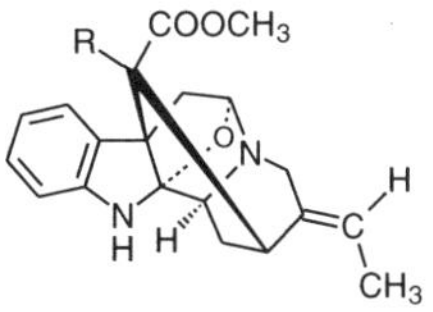

R = CHO : Picralinal (**5**)
R = H : Picrinine (**6**)

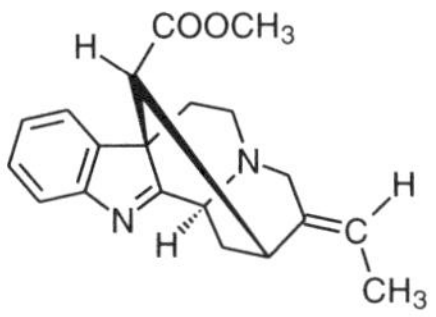

Strictamine (**7**)

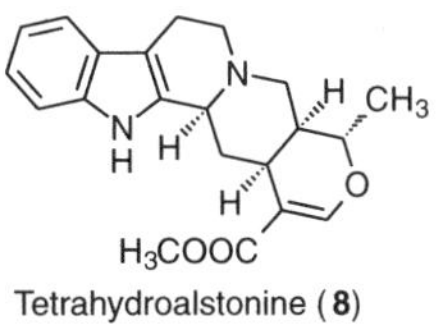

Tetrahydroalstonine (**8**)

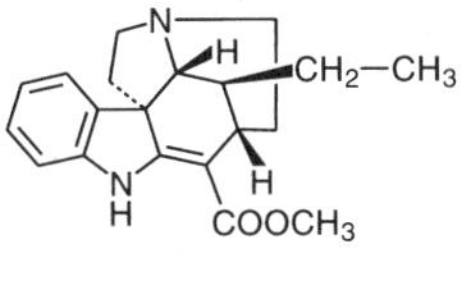

Tubotaiwine (**9**)

Toxicology: Strictamine has antidepressive activity (LD_{50} mouse 162 mg/kg) and inhibits monoamine oxidase.

Table: Data of Alstonia alkaloids.

no.	molecular formula	M_R	mp. [°C]	$[\alpha]_D$	CAS
1	$C_{20}H_{22}N_2O_2$	322.41	182	–745° (C_2H_5OH)	639-43-0
2	$C_{21}H_{24}N_2O_3$	352.43	234	+24° (CH_3OH)	639-36-1
3	$C_{22}H_{30}N_2O_4$	386.49	206 (hydroxide)	–29° (hydroxide) C_2H_5OH)	6871-44-9
4	$C_{20}H_{20}N_2O_4$	352.39	275 (decomp.)	–71° ($CHCl_3$/ CH_3OH)	63950-46-9
5	$C_{21}H_{22}N_2O_4$	366.42	180	–180° ($CHCl_3$)	20045-06-1
6	$C_{20}H_{22}N_2O_3$	338.41	225	–47° ($CHCl_3$)	4684-32-6
7	$C_{20}H_{22}N_2O_2$	322.41	112	+103°	6475-05-4
8	$C_{21}H_{24}N_2O_3$	352.43	231	–107° ($CHCl_3$)	6474-90-4
9	$C_{20}H_{24}N_2O_2$	324.42	amorphous	+611° ($CHCl_3$)	6711-69-9

Use: Alstonia extracts are used in traditional medicine for treatment of malaria, febrile attacks, constipation, chronic diarrhea, and dysentery.

Lit.: [1] Pelletier **1**, 321 (1983); Planta Med. **30**, 86 (1976).
gen.: Fitoterapia **61**, 225 (1990) ▪ J. Am. Chem. Soc. **118**, 9804 (1996) (synthesis (±)-akuammicine) ▪ J. Nat. Prod. **55**, 1323 (1992) ▪ Manske **8**, 159; **12**, 207 (pharmacological/biological activity); **14**, 157 ▪ Phytother. Res. **6**, 121 (1992). – *[HS 2939 90]*

Alternariol.

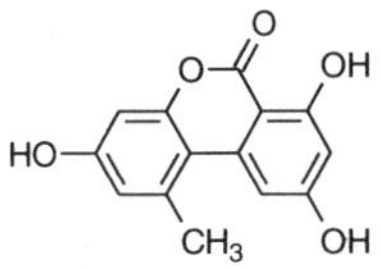

$C_{14}H_{10}O_5$, M_R 258.23, mp. 350 °C, (decomp.). Phytotoxic, cytotoxic, and antibacterial metabolite of *Alternaria dauci*, *A. tenuis*, and *A. cucumerina*.

Lit.: Cole-Cox, p. 615–618 ▪ Sax (8.), AGW 476. – *Biosynthesis:* J. Chem. Soc., Chem. Commun. **1982**, 1011. – *Synthesis:* J. Chem. Soc., Chem. Commun. **1986**, 15. – *[CAS 641-38-3]*

Altohyrtins see spongistatins.

Amanitins. The group of A., also known as *amatoxins*, consists of eight defined compounds. Together with the *phallotoxins they represent the strongly toxic poisons of the deadly amanita or death cup (*Amanita phalloides*). They are also responsible for the toxic activity of *A. verna*, *Galerina marginata*, and other *Galerina* species. They are cyclic octapeptides with a sulfur-containing bridge. Th. Wieland made major contributions to the structure elucidation of A. In addition to the amino acid (4*R*)-4,5-dihydroxy-L-isoleucine, discovered for the first time in the A., the toxins contain the following building blocks (for explanation of the symbols, see under amino acids) : Asx, Cys, Gly, Hypro, Ile, Trp; the A. exist as the sulfoxides (see table, p. 24). The toxicity is very high, hence even 1 µg of γ-A. or 2.5 µg of α-A. are sufficient to kill mice. The poison is not destroyed by drying or boiling or by

Table: Data of Amanitins.

name	molecular formula	M_R	mp. [°C]	LD_{50}	K_i [M]	CAS
α-Amanitin	$C_{39}H_{54}N_{10}O_{14}S$	918.98	254–255 (decomp.)	0.3	0.5×10^{-8}	23109-05-9
β-Amanitin	$C_{39}H_{53}N_9O_{15}S$	919.97	300	0.5	0.5×10^{-8}	21150-22-1
γ-Amanitin	$C_{39}H_{54}N_{10}O_{13}S$	902.98		0.2	0.5×10^{-8}	21150-23-2
ε-Amanitin	$C_{39}H_{53}N_9O_{14}S$	903.97		0.3	0.5×10^{-8}	21705-02-2
Amanullin	$C_{39}H_{54}N_{10}O_{12}S$	886.98		>20	10^{-8}	21803-57-6
Amanullinic acid	$C_{39}H_{53}N_9O_{13}S$	887.97		>20	–	54532-45-5
Proamanullin	$C_{39}H_{54}N_{10}O_{11}S$	870.98		>20	5×10^{-5}	54532-46-6
Amanin	$C_{39}H_{53}N_9O_{14}S$	903.97		0.5	0.5×10^{-8}	21150-21-0

(LD_{50} values in mg/kg white mouse i.p., K_i for 50% inhibition of RNA polymerase B)

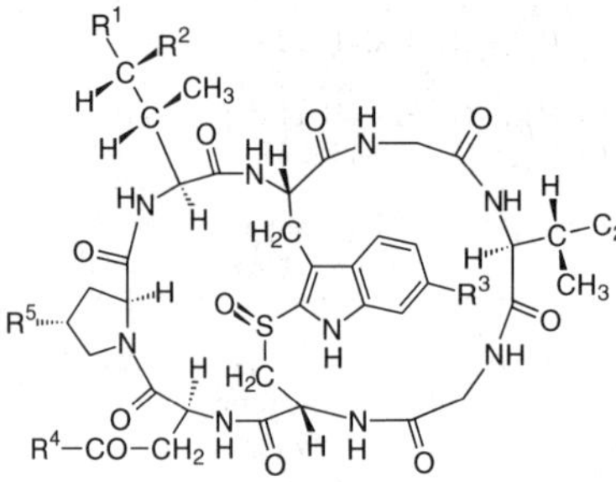

	R^1	R^2	R^3	R^4	R^5
α-Amanitin	CH_2OH	OH	OH	NH_2	OH
β-Amanitin	CH_2OH	OH	OH	OH	OH
γ-Amantitin	CH_3	OH	OH	NH_2	OH
ε-Amanitin	CH_3	OH	OH	OH	OH
Amanullin	CH_3	H	OH	NH_2	OH
Amanullinic acid	CH_3	H	OH	OH	OH
Proamanullin	CH_3	H	OH	NH_2	H
	CH_2OH	OH	H	OH	OH

proteases in the digestive tract; its action is attributed to allosteric blockage of mRNA synthesis by complexation with RNA polymerase in the cell nucleus, by which the entire enzyme/protein synthesis in the liver is completely stopped. For details of the mechanism of action of the A., see *Lit.*[1]. A characteristic of the toxic action of A. is its delayed onset; even with lethal doses death does not occur until 15 hours after consumption of the poison. The death cup fungus also contains the cyclic decapeptide *antamanide with antitoxic action against phalloidin[2]. According to *Lit.*[3] *cytochrome c, and according to other sources *lipoic acid are effective as antidotes.

Lit.: [1] Faulstich, Kommerel & Wieland, Amanita Toxins and Poisoning, p. 59–78, Baden-Baden: G. Witzstrock 1980; Wieland, Peptides of Poisonous Amanita Mushrooms, Berlin: Springer 1986. [2] Chem Pharm. Bull. **36**, 3196 (1988); **37**, 1684 (1989). [3] Schweiz. Med. Wochenschr. **108**, 185 (1978).
gen.: Drug Metab., Drug Interact. **6**, 265 (1988) ▪ Experientia **47**, 1186 (1991) (review) ▪ J. Am. Chem. Soc. **111**, 4791 (1989) (crystal structure) ▪ J. Chromatogr. **563**, 299 (1991); **598**, 227 (1992) (HPLC) ▪ Magn. Reson. Chem. **27**, 173 (1989) (NMR, conformation) ▪ Manske **40**, 189 (review) ▪ Sax (8.), AHI 625 ▪ Turner **1** & **2** ▪ Wieland, Peptides of Poisonous Amanita Mushrooms, Berlin: Springer 1986.

Amanullin, amanullinic acid see amanitins.

Amara see bitter principles.

Amarantin (2′-*O*-glucuronosylbetanin).

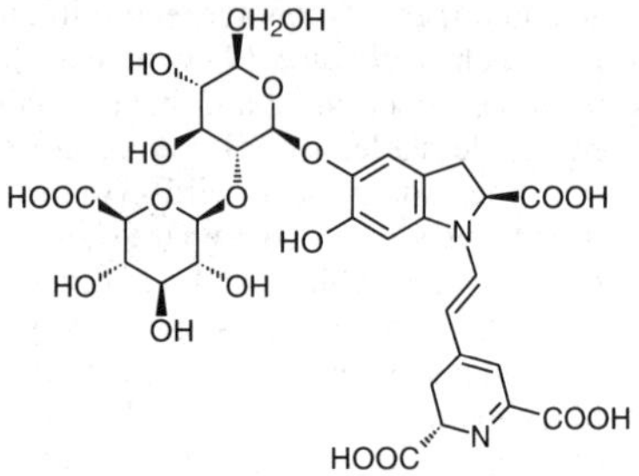

$C_{30}H_{34}N_2O_{19}$, M_R 726.60, dark red cryst., uv_{max} 536 nm. A. belongs to the group of betacyanins (see betalains) and occurs especially in *Amaranthus* (Amaranthaceae) and *Chenopodium* species (Chenopodiaceae).

Lit.: Hager (5.) **5**, 552 ▪ Manske **39**, 1–62. – *[CAS 15167-84-7]*

Amarin see cucurbitacins.

Amarogentin.

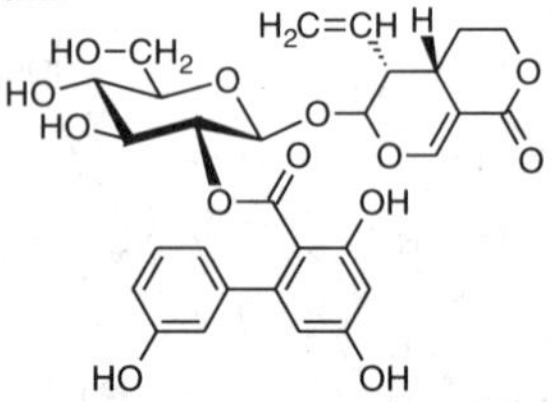

$C_{29}H_{30}O_{13}$, M_R 586.54, mp. 229–230 °C, $[\alpha]_D^{20}$ –117° (CH_3OH). A very bitter glycoside from *Gentiana* and *Swertia* spp., see also gentiopicroside. A. is the most *bitter principle currently known. It still has a detectable bitter taste at a dilution of 1 : 60 000 000.

Lit.: J. Nat. Prod. **59**, 27 (1996) ▪ Merck Index (12.), No. 393. – *[CAS 21018-84-8]*

Amaryllidaceae alkaloids. A. a. occur especially in the family Amaryllidaceae and are soluble in acetone, ethanol, and chloroform (see table, p. 26).

Biosynthesis[1]**:** The various structures can formally be derived from the same precursor, norbelladine, simple or multiple couplings of which give the galanthamine type (*p-o* coupling), the lycorine type (*o-p* coupling), and the haemanthamine type (*p-p* coupling). Norbelladine consists of tyrosine (for the C_6–C_2 unit and nitrogen) and phenylalanine (for the C_6–C_1 unit).

Use: Lycorine is the most widely distributed A. a., it is highly toxic and responsible for *Narcissus* poison-

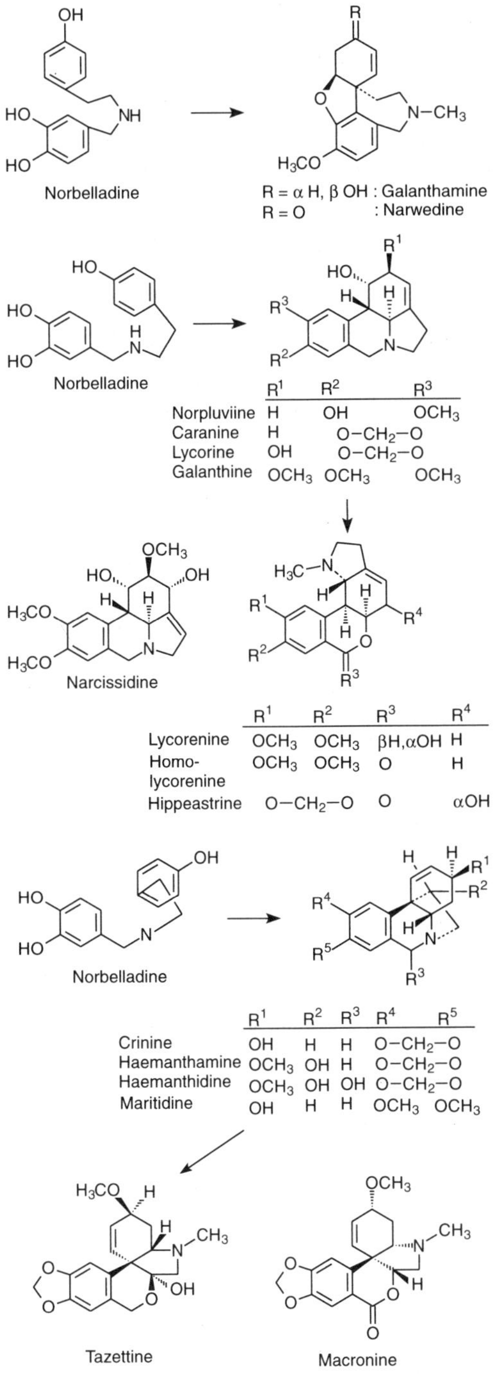

Figure: Biosynthesis of Amaryllidaceae alkaloids.

ings, LD_{50} (dog i. v.) 41 mg/kg. Lycorine and related compounds have antiviral and antineoplastic activities. Galanthamine is a powerful analgesic as well as a cholinesterase inhibitor, toxicity, LD_{50} (mouse s.c.) 11 mg/kg. Tazettine LD_{50} (dog i. v.) 71 mg/kg. Galanthamine was approved in Japan in 1996 for use in therapy for Alzheimer's disease and is in an advanced state of clinical development in the USA. G. was registered in Europe for treatment of mild forms of Alzheimer's disease[2].

Lit.: [1]Mothes et al., p. 241–246. [2]Manufacturing Chemist, No. 2 (1999).

gen.: Acta Crystallogr. Sect. C **54**, 1653 (1998) (tazettine, X-ray structure) ▪ Beilstein EV **27/9**, 277 (galanthamine); EIV **27**, 6463 (lycorine); 6418 (crinine) ▪ Drugs of Today **33**, 251, 259, 265, 273 (1997) (review, galanthamine) ▪ Hager (5.) **3**, 748–750; **5**, 213–218 ▪ J. Org. Chem. **63**, 3607 (1998) (crinine) ▪ Kirk-Othmer (4.) **1**, 1053–1066 ▪ Manske **15**, 83–164; **30**, 251–376; **51**, 324–424 ▪ Nat. Prod. Rep. **6**, 79–84 (1989); **9**, 183–191 (1992); **10**, 291–299 (1993); **11**, 329–332 (1994) ▪ Rodd's Chem. Carbon Compds. (2.) **4B**, 165–249 (1997) ▪ Sax (8.), AHI 635, GBA 000 ▪ Ullmann (5.) **A 1**, 382. – *Synthesis:* J. Heterocycl. Chem. **32**, 195 (1995) (galanthamine) ▪ J. Am. Chem. Soc. **106**, 6431 ff. (1984); **110**, 8250 ff. (1988); **118**, 6210 (1996); **120**, 3664 (1998) [(±)-tazettine] ▪ J. Chem. Soc., Perkin Trans. 1 **1996**, 571–580; **1999**, 1251 (oxomaritidine) ▪ J. Org. Chem. **49**, 157–163 (1984); **58**, 4662–4672 (1993); **63**, 3607 [(±)-crinine], 6625 [(±)-galanthamine etc.], 7586 [(+)-crinamine etc.] (1998) ▪ Tetrahedron Lett. **38**, 7931 (1997) [(±)-narwedine]; **39**, 2087 (1998) (galanthamine, large scale). – *[HS 2939 90]*

Amatoxins see amanitins.

Amavadin. Three species of *Amanita* mushrooms: *A. muscaria*, the "fly agaric", *A. regalis*, and *A. velatipes* concentrate vanadium to levels up to 400 times those typically found in plants. The vanadium is equally distributed between stem, skin and cap. Responsible for the accumulation is a blue, vanadium-containing compound, amavadin, first characterised from *A. muscaria*. It is a 1:2 complex with the dianionic ligand derived from (*S,S*)-2,2′-(hydroxyimino)dipropanoic acid ((*S,S*)-hidpaH_3).

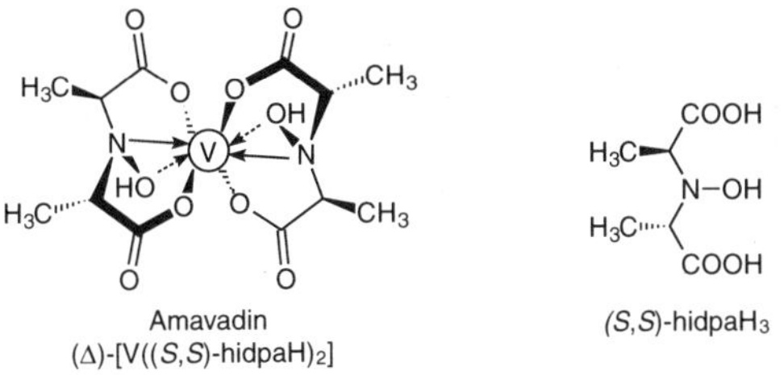

Amavadin (Δ)-[V((*S,S*)-hidpaH)$_2$]   (*S,S*)-hidpaH_3

The vanadium center of A. exists as an eight-coordinate, non-oxo complex, first postulated in 1995[1] and confirmed in 1999[2]. It was crystallized as the salt $[Ca(H_2O)_5][(\Delta)\text{-}V((S,S)\text{-hidpa})_2] \cdot 2\,H_2O$. The ions are linked in the form of infinite chains.

V.: $C_{12}H_{18}N_2O_{10}V$, M_R 401.22, no mp., optically active, stable in dilute acids and aqueous NH_3, soluble in methanol.

Lit.: [1]Met. Ions Biol. Syst. **31**, 407–421 (1995). [2]Angew. Chem. Int. Ed. Engl. **38**, 795–797 (1999). – *[CAS 12705-99-6]*

Amber (succinite). General term for fossilized terpenoid resins from the Tertiary period formed from resinous secretions of coniferous trees. They owe their hardness and poor solubility to the evaporation of volatile *terpenes as well as to autoxidation and polymerization of resin acids and alcohols. The resin alcohols are mostly esterified with succinic acid. – *[HS 2530 90]*

Ambergris. Name given to a secretion from the digestive tract of the sperm whale that, after the death of the

Table: Data and occurrence of Amaryllidaceae alkaloids.

name	molecular formula	M_R	mp. [°C]	$[\alpha]_D$	occurrence	CAS
Galanthamine type						
Galanthamine	$C_{17}H_{21}NO_3$	287.36	127–129	–122° (C_2H_5OH)	*Galanthus voronovii*	357-70-0
Narwedine	$C_{17}H_{19}NO_3$	285.34	189–192	+405° ($CHCl_3$)	*Narcissus cyclamineus, Galanthus nivalis*	510-77-0
Lycorine type						
Caranine	$C_{16}H_{17}NO_3$	271.32	178–180	–197° ($CHCl_3$)	*Amaryllis belladonna*	477-12-3
Galanthine	$C_{18}H_{23}NO_4$	317.38	132–134	–85° ($CHCl_3$)	*Galanthus voronovii*	517-78-2
Hippeastrine	$C_{17}H_{17}NO_5$	315.33	214–215	+160° ($CHCl_3$)	*Hippeastrum vittatum*	477-17-8
Homolycorenine	$C_{18}H_{21}NO_4$	315.37	175	+85° (C_2H_5OH)	*Narcissus poeticus*	477-20-3
Lycorenine	$C_{18}H_{23}NO_4$	317.38	202	+125° (C_2H_5OH)	*Lycoris radiata*	477-19-0
Lycorine	$C_{16}H_{17}NO_4$	287.32	280–281	–120° (C_2H_5OH)	*Lycoris radiata*	476-28-8
Narcissidine	$C_{18}H_{23}NO_5$	333.38	218–219	–32° ($CHCl_3$)	*Narcissus poeticus*	27857-07-4
Norpluviine	$C_{16}H_{19}NO_3$	273.33	239–241 (decomp.)	–232° (CH_3OH)	*Lycoris radiata, Pancratium maritimum*	517-99-7
Haemanthamine type						
Crinine	$C_{16}H_{17}NO_3$	271.32	207–208	+26° ($CHCl_3$)	*Hippeastrum vittatum Crinum moorei*	510-67-8
Haemanthamine	$C_{17}H_{19}NO_4$	301.34	203–203.5	+19.7° (CH_3OH)	*Haemanthus puniceus*	466-75-1
Haemanthidine	$C_{17}H_{19}NO_5$	317.34	189–190	–41° ($CHCl_3$)	*Haemanthus puniceus*	466-73-9
Macronine	$C_{18}H_{19}NO_5$	329.35	203–205	+413° ($CHCl_3$)	*Crinum macrantherum,*	2124-70-1
Maritidine	$C_{17}H_{21}NO_3$	287.36	253–256	+25.1° (C_2H_5OH)	*Pancratium maritimum, Hippeastrum ananuca, Narcissus tazetta*	22331-07-3
Tazettine	$C_{18}H_{21}NO_5$	331.37	210–211	+150° ($CHCl_3$)	*Narcissus tazetta*	507-79-9

animal, floats on water on account of its density (ca. 0.9). A. forms a gray to black, often white marbleized, wax-like mass; it is insoluble in water and poorly soluble in alcohol, mp. ca. 60 °C. It should not be confused with spermaceti or cetaceum also often named as white ambergris or amber. The so-called gray A. smells like moss-covered forest soil, tobacco, and sandalwood. The main components of A. are the triterpene alcohol *ambrein* [$C_{30}H_{52}O$, M_R 428.74, mp. 83 °C, bp. 210 °C (30 Pa)], metabolites of ambrein, and varying amounts of (+)-*epi*-coprosterol[1]. The fragrant component amounts to less than 0.5 wt-% of A.; it is probably composed of oxidation products of ambreins including, among many others, dihydro-γ-ionone which has a strong odor of A.

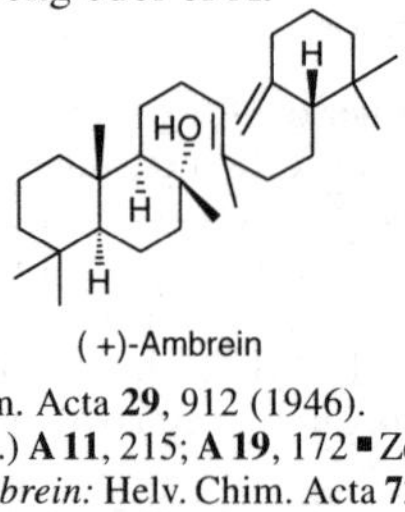

(+)-Ambrein

Lit.: [1] Helv. Chim. Acta **29**, 912 (1946).
gen.: Ullmann (5.) **A 11**, 215; **A 19**, 172 ▪ Zechmeister **12**, 186. – *Synthesis of ambrein:* Helv. Chim. Acta **72**, 996 (1989) ▪ Justus Liebigs Ann. Chem. **1990**, 361 ▪ Tetrahedron **53**, 3527 (1997). – *[HS 0510 00; CAS 473-03-0 (ambrein)]*

Ambiguine isonitrile D see hapalindoles.

(–)-Ambra oxide see clary sage oil.

Ambrein see Ambergris.

Ambrette seed oil. Pasty yellow mass, often also called ambrette butter or concrete, consisting mostly of palmitic acid. The actual oil (also known as ambrette absolute) is obtained by treatment with alkali or precipitation of the palmitic acid with alcohol. The odor is flowery-sweet, musk-like with a brandy-like note, the taste is unique, fresh-aromatic.
Production: By steam distillation from seeds of *Abelmoschus moschatus* (Malvaceae), a hibiscus species growing in the tropics. *Origin:* India, South America.
Composition[1]***:*** The main component is *farnesyl acetate* (50–60%) ($C_{17}H_{28}O_2$, M_R 264.41), accompanied by some **farnesol* (3–5%). The typical odor and taste determining component is **ambrettolide*.
Use: A. s. o. is one of the most expensive raw materials for perfumes. It is thus used in small amounts in very expensive perfume oils only. Small amounts are also used to improve the aromatic character of some liquors.
Lit.: [1] Perfum. Flavor. **3** (4), 54 (1978); **16** (5), 80 (1991); Flavour Fragr. J. **7**, 65 (1992).
gen.: Arctander, p. 58 ▪ Bauer et al. (2.), p. 134. – *Toxicology:* Food Cosmet. Toxicol. **13**, 705 (1975). – *[CAS 8015-62-1 (A. seed oil); 4128-17-0 (farnesyl acetate)]*

Ambrettolide [(*Z*)-7-hexadecen-16-olide].

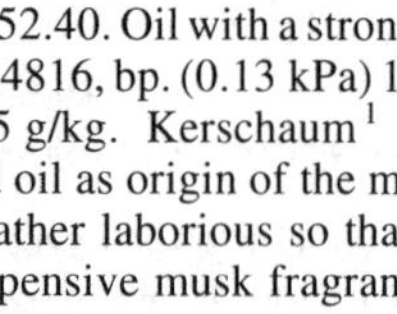

$C_{16}H_{28}O_2$, M_R 252.40. Oil with a strong musk-like odor, D. 0.958, n_D^{20} 1.4816, bp. (0.13 kPa) 154–156 °C, LD_{50} (rat orally) >5 g/kg. Kerschaum[1] identified A. in *ambrette seed oil as origin of the musk-like odor. Its synthesis[2] is rather laborious so that A. still belongs to the more expensive musk fragrances. Its main use is in fine perfumery.
Lit.: [1] Ber. Dtsch. Chem. Ges. **60B**, 902 (1927). [2] Theimer (ed.), Fragrance Chemistry, p. 469 ff., San Diego: Academic Press 1982; Synlett **1997**, 600 (thia-analogue).
gen.: Bedoukian (3.), p. 303 ff. ▪ Beilstein E V **17/9**, 283 ▪ Fenaroli (2.) **2**, 242 ▪ Ohloff, p. 198 f. – *[HS 2932 29; CAS 123-69-3]*

Ambrox.

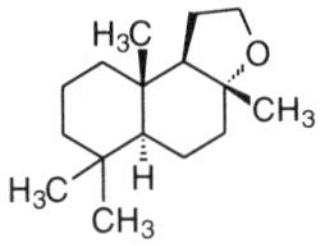

Sesquiterpenoid natural product of the drimane type, $C_{16}H_{28}O$, M_R 236.38, mp. 75–76 °C, used for a "woody" fine odor in perfumes.

Lit.: Can. J. Chem. **75**, 1136 (1997) (synthesis) ▪ Helv. Chim. Acta **72**, 996 (1989) ▪ J. Org. Chem. **61**, 2215 (1996) ▪ Tetrahedron **49**, 6251 (1993); **50**, 10095 (1994); **51**, 8333 (1995); **55**, 8567 (1999) (synthesis derivatives) ▪ Tetrahedron: Asymmetry **7**, 1695–1704 (1996); **9**, 1789 (1998) (synthesis) ▪ Tetrahedron Lett. **34**, 629 (1993). – *[CAS 6790-58-5]*

Ambruticins.

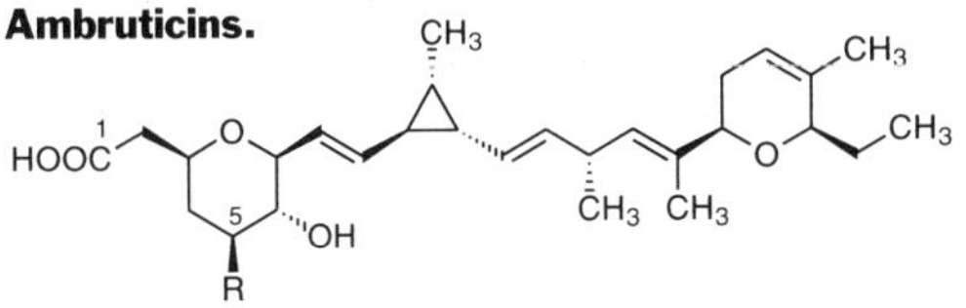

R = OH : A.S
R = $N(CH_3)_2$: A.VS -3

Antifungal *polyether antibiotics from myxobacteria, characterized by a 1,2,3-trisubstituted cyclopropane building block and variable substituents at C-5. Thus, A. F and S from *Polyangium cellulosum* have a 5α- or 5β-hydroxy group, the A. VS from *Sorangium cellulosum* have differently substituted 5β-amino groups, e.g., a dimethylamino group in the main component A. VS-3 ($C_{30}H_{47}NO_5$, M_R 501.7, $[\alpha]_D^{22}$ +90.5°). A. S ($C_{28}H_{42}O_6$, M_R 474.3, $[\alpha]_D^{22}$ +42°) is active against yeasts and pathogenic fungi in the range 0.01 – 1 µg/mL *in vitro*. Although also active *in vivo*, the slightly toxic A. [LD_{50} (mouse i.v.) 315 mg/kg, (mouse p.o.) >1 g/kg] have not been developed for therapeutic use. Their chemistry has, however, been investigated thoroughly, including total syntheses.

Lit.: J. Am. Chem. Soc. **112**, 9645 (1990) ▪ Justus Liebigs Ann. Chem. **1991**, 941 ▪ Kuhn & Fiedler (eds.), The Biosynthetic Potential of the Myxobakteria, p. 61–78, Tübingen: Attempto 1995 ▪ Tetrahedron **49**, 8015 (1993). – *[CAS 58857-02-6 (A. S); 135074-48-5 (A. VS-3)]*

Amentoflavones. *Biflavonoids, formed by C,C-linkage (oxidative coupling) of the A ring of an *apigenin molecule with the B ring of a second apigenin molecule, e.g. 3′,8-biapigenin (amentoflavone) from the leaves of *Metasequoia glyptostroboides* (Taxodiaceae), $C_{30}H_{18}O_{10}$, M_R 538.47, yellow cryst., mp. >300 °C.

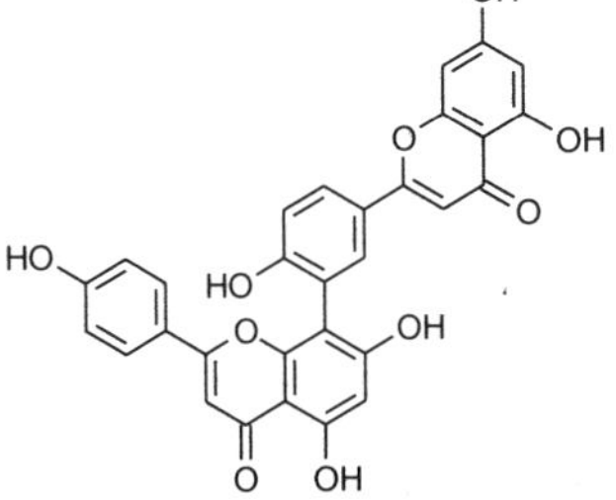

3′,8-Biapigenin

In addition to its existence in unsubstituted form, 3′,8-biapigenin often occurs in form of various glucosides and methyl ethers.

Lit.: Harborne (1975), p. 692 ff.; (1982), p. 505 ff.; (1988), p. 99 ff. – *[CAS 1617-53-4]*

Americanin A.

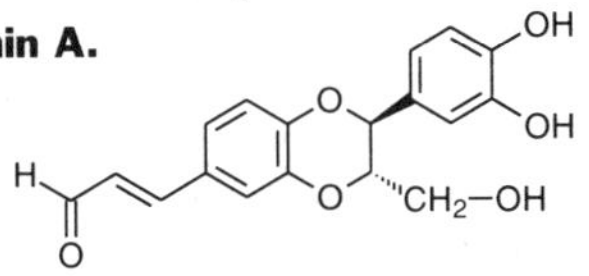

(2*S*)-*trans*

$C_{18}H_{16}O_6$, M_R 328.32, yellow plates (from methanol), mp. 246–247 °C, $[\alpha]_D^{17}$ +23.7°; *lignan from *Phytolacca americana*. A. is also isolated as the racemate. The allylic alcohol, *americanol A*, isolated from the same plant exhibits neurotrophic activity [1]. The biosynthesis presumably proceeds through oxidative dimerization of 3,4-dihydroxycinnamyl alcohol and dehydrogenation of the allylic alcohol. For americanin B, see lignans.

Lit.: [1] Chem. Pharm. Bull. **40**, 252–254 (1992).

gen.: Beilstein E V **19/6**, 616 ▪ J. Chem. Res. (S) **1998**, 436 (synthesis americanol) ▪ Justus Liebigs Ann. Chem. **1986**, 647–654 ▪ Tetrahedron Lett. **1978**, 3239–3242. – *[CAS 69506-79-2]*

Amicoumacins.

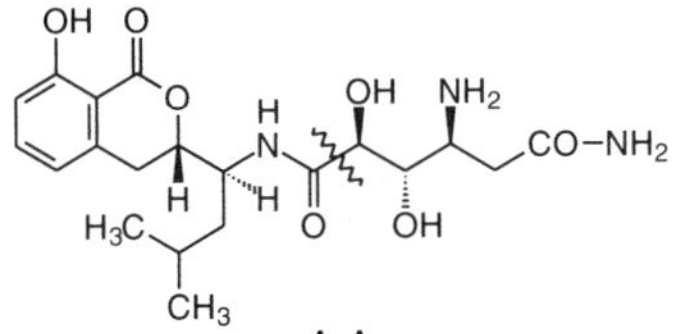

A. A

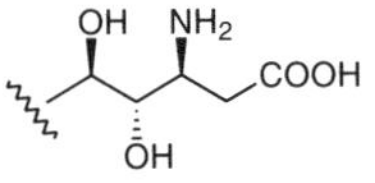

A. B

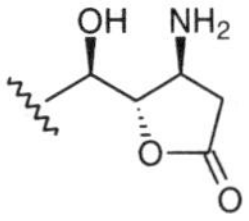

A. C

Table: Data of Amicoumacins.

	A	B	C
molecular formula	$C_{20}H_{29}N_3O_7$	$C_{20}H_{28}N_2O_8$	$C_{20}H_{26}N_2O_7$
M_R	423.47	424.45	406.44
mp. [°C]	132–135	137–145	131–133
$[\alpha]_D^{23}$ (CH_3OH)	–97.2°	–106.1°	-81.6°
CAS	78654-44-1	82768-33-0	77682-31-6

A family of 8-hydroxycoumarin *antibiotics formed by *Bacillus pumilus* BN-103. A. A exhibits good activity against Gram-positive bacteria as well as *Salmonella* and *Shigella* species. Also anti-inflammatory and protective effects against stress-induced stomach ulcers were observed in a rat model. The acute LD_{50} amounts to 132 mg/kg (mouse p.o.). A. are used as acaricides.

Lit.: Agric. Biol. Chem. **46**, 1255, 2659 (1982) ▪ J. Antibiot. **34**, 611 (1981) ▪ Tetrahedron **40**, 2519 (1984).

2-Aminoacetophenone see fruit flavors (grape aroma).

Amino acids (aminocarboxylic acids).

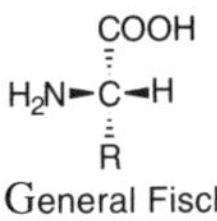

General Fischer projection of the L-form.

General term for carboxylic acids containing one or more amino or, respectively, basic groups. Naturally occurring A. are mostly α-A., i. e., carboxy and amino groups are linked through a common α-carbon atom. In relation to this carbon atom there are two optically active stereoisomers, L- and D-, with the exception of glycine. The majority of the naturally occurring A. belongs to the L-series. A. are components of all proteins in which they are linked by peptide bonds, –CO–NH–. There are 20 proteinogenic A.: 1) those with nonpolar side chains: *glycine (Gly, G), L-*alanine (Ala, A), L-*L-valine (Val, V), L-*leucine (Leu, L), L-*isoleucine (Ile, I), L-*L-proline (Pro, P), L-*tryptophan (Trp, W), L-*phenylalanine (Phe, F), L-*methionine (Met, M). 2) those with polar side chains: *L-serine (Ser, S), *L-tyrosine (Tyr, Y), *L-threonine (Thr, T), L-*cysteine (Cys, C), L-*asparagine (Asp, N), L-*glutamine (Gln, Q).
3) those with negatively charged side chains (acidic A.): L-*aspartic acid (Asp, D), L-*glutamic acid (Glu, E).
4) those with positively charged side chains (basic A.): L-*lysine (Lys, K), L-*arginine (Arg, R), L-*histidine (His, H).
Apart from these proteinogenic A, there are numerous so-called non-proteinogenic A. in the free form. These include intermediate products in the biosyntheses of the proteinogenic A. and are widely distributed in plants and fungi.
Occurrence: A. are components of the peptidoglycan in bacterial cell walls, peptide hormones (e. g. insulin, oxytocin, β-endorphin, etc.), *mycotoxins (*gramicidin A, *valinomycin, *amanitins, etc.), and γ-glutamyl compounds. Peptidoglycan and mycotoxins also contain the D-forms of amino acids. *4-Hydroxyproline, *5-hydroxylysine, and other non-proteinogenic A. are post-translationally formed in intact peptide chains.
Gen. Properties: A. do not have a defined mp., but decompose at higher temperatures. In the solid state or in the physiological pH range A. exist as "internal salts" ("zwitterions," "betaines"), the carboxy and amino groups are ionized and, depending on pH value of the solution, they act as acids or bases (ampholytes). A. have minimal solubilities in the non-polar state (isoelectric points, pI). The large differences in polarity, acidity, basicity, aromaticity, size, conformation, or, respectively, ability to form hydrogen bridging bonds, are responsible for the differing properties of proteins, especially their catalytic activities.
Biochemistry: Each proteinogenic A. is coded with 2 to 4 base triplets in DNA. According to this ordering principle, the A. are bound to their specific tRNAs, transported to ribosomes, and incorporated there into the growing peptide chains. A. that cannot be synthesized by mammals and must be taken up with food are called *"essential amino acids"*. For humans these are Arg, His, Ile, Leu, Lys, Met, Thr, Trp, Phe, and Val. Plants and microorganisms can synthesize all proteinogenic amino acids. Glu has a central role in A. biosynthesis. In plants and microorganisms Glu and Gln are first formed in the glutamate synthase cycle. Some processes of A. biosynthesis are directly coupled with the citric acid cycle and are thus closely connected with the metabolism of carbohydrates and fatty acids.
Activity: Many neurotransmitters, e. g., Gly, Glu, GABA (see 4-aminobutanoic acid), *histamine, *dopamine, *noradrenaline, *(*R*)-adrenaline, *serotonin are known among the A. and their derivatives. A. are precursors of many other cell components.
Use & synthesis: Essential A. are used as food additives. They are produced by fermentation or synthetic means (Strecker, Erlenmeyer, or malonic ester synthesis) and separated by column chromatography. As the esters, racemates can be separated into the enantiomers by acylases.
Detection: E.g., by color reactions with ninhydrin (mostly violet).
Lit.: Amino acids, Pept., Proteins **29**, 1–125 (1998) (review) ▪ Barrett (ed.), Chemistry and Biochemistry of the Amino Acids, London: Chapman & Hall 1985 ▪ Barton-Nakanishi **4**, 1–12 (review) ▪ Barrett & Elmore, Amino Acids and Peptides, Cambridge: Univ. Press 1998 ▪ Davies (ed.), Amino Acids and Peptides, London: Chapman & Hall 1985 ▪ Greenstein & Winitz, Chemistry of the Amino Acids, New York: Wiley 1961 ▪ Nat. Prod. Rep. **16**, 485–498 (1999) ▪ Rosenthal, Plant Non Protein Amino and Imino Acids, New York: Academic Press 1982 ▪ Ullmann (5.) **A 2**, 57–85 ▪ Zechmeister **59**, 1–140 (A. from fungi). – *Synthesis:* Adv. Asymm. Synth. **1**, 45–94 (1995) ▪ Amino acids, Pept., Proteins **27**, 1–101 (1996) ▪ Spec. Chem. **19**, 10 (1999) (large scale synthesis).

2-Aminoadipic acid (2-aminohexanedioic acid, homoglutamic acid).

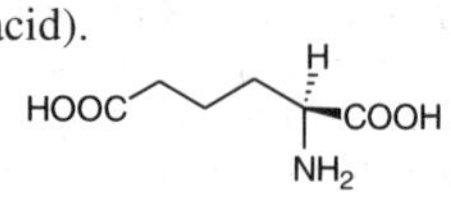

(2*S*)-form

$C_6H_{11}NO_4$, M_R 161.16, mp. 195 °C (decomp.), $[\alpha]_D^{22}$ +23° (6 m HCl). Non-proteinogenic amino acid, the free (2*S*) form is widely distributed in small amounts in plants and fungi. It is isolated from pea seedlings (*Pisum sativum*). (2*R*)-Form: component of the *cephalosporins C and N. Monohydrate, mp. 205–207 °C (decomp.), $[\alpha]_D^{20}$ –25° (5 m HCl). A. is an intermediate product of lysine catabolism.
Lit.: *Synthesis:* J. Org. Chem. **59**, 3676 (1994) ▪ Synth. Commun. **18**, 1707 (1988) ▪ Synthesis **1994**, 601. – *[HS 2922 49; CAS 1118-90-7 (+)-(S); 7620-28-2 (–)-(R)]*

2-Aminobenzoic acid see anthranilic acid.

4-Aminobenzoic acid (*p*-aminobenzoic acid, vitamin H′, PABA).

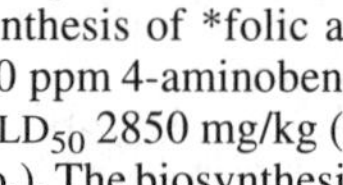

$C_7H_7NO_2$, M_R 137.14, prisms, mp. 188–188.5 °C, D. 1.374, pK_{a1} 2.41, pK_{a2} 4.85 (25 °C), soluble in water. A. is easily oxidized, e. g., with Fe(III) salts; a slight yellow to red discoloration occurs on exposure to air or light. Bacteria require A. as an essential growth substance for the synthesis of *folic acid. Baker's yeast contains ca. 5–10 ppm 4-aminobenzoic acid. A. is not toxic to rodents: LD_{50} 2850 mg/kg (mouse p. o.), 1830 mg/kg (rabbit p. o.). The biosynthesis starts from *chorismic acid and glutamate [1]. A. can be determined in urine by thin layer densitometry [2].
Use: As starting material for the synthesis of azo dyes, folic acid, local anesthetics, and sun protection agents.
Lit.: [1] Nat. Prod. Rep. **10**, 233–263 (1993); **11**, 173–203 (1994). [2] Int. Lab. **1976**, No. 1, 39–41.

gen.: Beilstein E IV **14/2**, 1126 ▪ J. Chromatogr. **351**, 532 (1986) (use) ▪ Luckner (3.), p. 326–327 ▪ Photodermatology **3**, 6 (1986) (review). – *Toxicology:* Sax (8.), AIH 600. – *[HS 2922 49; CAS 150-13-0]*

4-Aminobutanoic acid (γ-aminobutyric acid, γ-A., GABA). $H_2N-(CH_2)_3-COOH$, $C_4H_9NO_2$, M_R 103.12, mp. 203 °C (decomp.), pK_a 4.03 and 10.56; monohydrochloride, mp. 135–136 °C. A non-proteinogenic amino acid neurotransmitter acting by blockage of ganglions. 4-A. was discovered almost simultaneously in beets (*Beta vulgaris*), yeasts, and brain. It is widely distributed in plants and fungi. The biosynthesis starts from L-Glu through glutamate decarboxylase (EC 4.1.1.15)[1] and also from spermine[2].

Lit.: [1] J. Chem. Soc., Perkin Trans. 1 **1982**, 455–459. [2] Phytochemistry **17**, 550 f. (1978).
gen.: Beilstein E IV **4**, 2600 ff. ▪ Iversen, in Lipton et al. (eds.), Psychopharmacology: A Generation of Progress, p. 25–38, New York: Raven 1978. – *[HS 2922 49; CAS 56-12-2]*

2-Amino-4-chloro-4-pentenoic acid see 2-amino-4-pentenoic acid.

1-Aminocyclopropanecarboxylic acid (ACC).

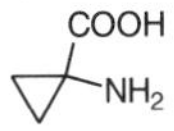

$C_4H_7NO_2$, M_R 101.11, mp. 229–231 °C. A non-proteinogenic amino acid first known as a synthetic product and later isolated from pears and apples. ACC is formed from methionine via *S-adenosylmethionine with the help of ACC synthase (EC 4.4.1.14) and cleaved by ACC oxidase to the multifunctional *plant growth substance ethylene which plays key roles in various plant physiological processes such as ripening of fruit, aging, germination, and response to stress.

Lit.: Acc. Chem. Res. **32**, 711 (1999) (biosynthesis ethylene) ▪ Annu. Rev. Plant Physiol. **35**, 155–189 (1984) ▪ ACS Symp. Ser. **551**, chap. 31 (1994) ▪ Beilstein E IV **14**, 973 ▪ Synth. Commun. **28**, 3159 (1998) ▪ Tetrahedron Lett. **38**, 1677 (1997). – *[HS 2922 49; CAS 22059-21-8]*

(*S*)-2-Amino-3-cyclopropylpropanoic acid.

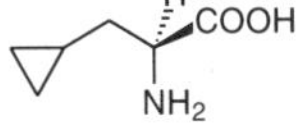

$C_6H_{11}NO_2$, M_R 129.16, mp. 240 °C (decomp.), $[\alpha]_D$ –12.4° (H_2O). A non-proteinogenic amino acid in the fruit bodies of *Amanita virgineoides*[1]. It inhibits the growth of *Escherichia coli*[2] as well as spore germination of *Pyricularia oryzae*[1] (mold as cause of mildew on rice).

Lit.: [1] Chem. Lett. **1986**, 511–512. [2] J. Am. Chem. Soc. **77**, 6675–6677 (1955). – *[CAS 102735-53-5]*

Aminodeoxy sugars see amino sugars.

2-Amino-6-diazo-5-oxohexanoic acid (6-diazo-5-oxonorleucine).

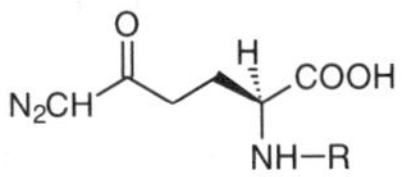

R = H : 2-Amino-6-diazo-5-oxohexanoic acid
R = CO–CH3 : Duazomycin

(*S*)-2-A. was discovered in cultures of *Streptomyces ambofaciens* in 1956 as a diazo compound with antitumor activity but is too toxic for clinical use. [$C_6H_9N_3O_3$, M_R 171.16, yellow cryst., mp. 142–150 °C, $[\alpha]_D$ +21° (H_2O)]. The *N*-acetyl derivative *duazomycin* [$C_8H_{11}N_3O_4$, M_R 213.19, yellow powder) and peptides of 2-A. are also active against tumors.

Lit.: J. Am. Chem. Soc. **78**, 3075 (1956); **80**, 3941 (1958) ▪ J. Org. Chem. **48**, 741 (1983) ▪ Sax (8.), DCQ 400. – *[CAS 157-03-9 (2-A.); 2508-89-6 (duazomycin)]*

Aminoglycosides see aminosugars.

(*R*)-2-Aminoheptanedioic acid see 2-aminopimelic acid.

2-Amino-4,5-hexadienoic acid.

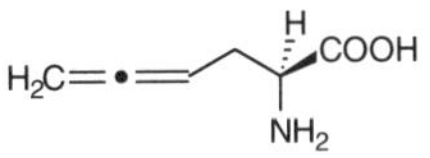

(2*S*)-form

$C_6H_9NO_2$, M_R 127.14, mp. 174 °C (decomp.), $[\alpha]_D^{22}$ –52.1° (H_2O). Non-proteinogenic amino acid from the fruit bodies of *Amanita* spp.: *A. smithiana*[1], *A. solitaria*[1], *A. pseudoporphyria*[2], *A. neoovoidea*[3], *A. abrupta*[4]. The racemate has hypoglycemic activity and causes respiratory depression and hypothermia[5].

Lit.: [1] Tetrahedron Lett. **1968**, 6283 f. [2] Trans. Mycol. Soc. Jpn. **26**, 61–68 (1985). [3] Sci. Pap. Coll. Gen. Educ. Univ. Tokyo **30**, 147–150 (1980). [4] Toxicology **38**, 161–173 (1986); Phytochemistry **26**, 565 f. (1987). [5] J. Nat. Prod. **36**, 169–173 (1973).
gen.: J. Chem. Soc., Chem. Commun. **1984**, 1284 (synthesis) ▪ Turner **2**, 387. – *[HS 2922 49; CAS 52521-29-6 (±)]*

2-Amino-4-hexynoic acid.

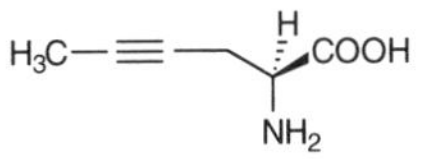

(2*S*)-form

$C_6H_9NO_2$, M_R 127.14, mp. >199 °C (decomp.), $[\alpha]_D$ –54.4° (H_2O). Non-proteinogenic amino acid in the toadstools *Tricholomopsis rutilans*[1], *Amanita pseudoporphyria*[2] and *A. miculifera*. Ninhydrin reaction: brown. It inhibits L-methionine-*S*-adenosyltransferase[3].

Lit.: [1] Phytochemistry **11**, 3327–3329 (1972). [2] Trans. Mycol. Soc. Jpn. **26**, 61–68 (1985). [3] Mol. Pharmacol. **10**, 293-304 (1974).
gen.: J. Nat. Prod. **36**, 169 (1973) ▪ Turner **2**, 387 ▪ Zechmeister **59**, 32 f. – *[HS 2922 49; CAS 29834-76-2]*

δ-Aminolevulinic acid see 5-amino-4-oxovaleric acid.

5-Amino-4-oxovaleric acid (5-amino-4-oxopentanoic acid; δ-aminolevulinic acid, δ-ALA). $H_2N-CH_2-CO-CH_2-CH_2-COOH$, $C_5H_9NO_3$, M_R 131.13. cryst., mp. 118–119 °C, as hydrochloride (M_R 167.59) mp. 156–158 °C (decomp.). δ-ALA is an intermediate in the biosynthesis of the *porphyrins, *hydroporphyrins, and *corrins, it is formed in animals, yeasts, and some bacteria from succinyl-coenzyme A and glycine via decarboxylation under catalytic action of δ-aminolevulinate synthase (EC 2.3.1.37) and pyridoxal phosphate. In green plants it is formed in several enzymatic steps from *glutamic acid with participation of a transfer ribonucleic acid as cofactor[1]. The glutamic acid pathway is also possible in animals in addition to the

classic formation of δ-ALA[2]. δ-ALA is transformed further to *porphobilinogen.
Increased excretion of δ-ALA in urine can be an indication for lead poisoning since, as a result of inhibition of the specific δ-ALA dehydratase (porphobilinogen synthase, EC 4.2.1.24), no condensation to porphobilinogen can occur. δ-ALA can, however, also be degraded to 4,5-dioxopentanoic acid. A. is used in the U. S. A. for the treatment of actinic keratosis of the face and scalp in photodynamic therapy (Levulan®).
Lit.: [1]Trends Biochem. Sci. **13**, 139–143 (1988). [2]Angew. Chem. Int. Ed. Engl. **23**, 998 (1984).
gen.: Beilstein E IV **4**, 3265 ▪ Jordan, p. 4–19, 155, 158 ▪ Synthesis **1999**, 568 (synthesis). – *[HS 2922 50; CAS 106-60-5]*

6-Aminopenicillanic acid (6-APA).

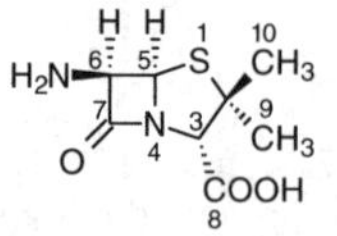

$C_8H_{12}N_2O_3S$, M_R 216.26, mp. 208–209 °C (cryst. from dil. HCl), $[\alpha]_D$ +273° (0.1 M HCl). Bicyclic skeleton of all *penicillin antibiotics. A. is produced industrially by chemical or enzymatic hydrolysis of penicillin G or V. Currently ca. 7000 t penicillin G are used per year of which over 90% originate from *Escherichia coli* immobilized on polymer carriers. It is released with penicillin amidase. A. is the precursor for all semisynthetic penicillins that arise through acylation of A. at the amino group. By chemical or enzymatic ring expansion, derivatives of 7-aminocephalosporanic acid (parent compound of the *cephalosporins) are also accessible from A.
Lit.: BioEngineering **8**, 60–77 (1992) ▪ Kirk-Othmer (4.) **3**, 129 ▪ Nature (London) **283**, 257 (1959) ▪ Tetrahedron Lett. **21**, 2291 (1980). – *[HS 2934 90; CAS 551-16-6]*

2-Amino-4-pentenoic acid (allylglycine).

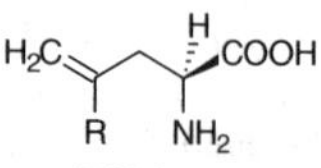

(2S)-form

R = H : 2-Amino-4-pentenoic acid
R = Cl : 2-Amino-4-chloro-4-pentenoic acid

$C_5H_9NO_2$, M_R 115.13, mp. ca. 196 °C, $[\alpha]_D^{22}$ −33.7° (H_2O). Non-proteinogenic amino acid in the fruit bodies of *Amanita pseudoporphyria*[1] and *A. abrupta*[2]. Synthetic A. is long known as an antimetabolite against *Escherichia coli* and *Saccharomyces cerevisiae* and as an inhibitor of glutamate decarboxylase (EC 4.1.1.15). Ninhydrin reaction: brown. A. inhibits protein synthesis. The 4-chloro derivative of A. [*2-amino-4-chloro-4-pentenoic acid*, (2S)-form: $C_5H_8ClNO_2$, M_R 149.58] is a major constituent of the free amino acids in fruit bodies of *Amanita pseudoporphyria*[1]. It has antibacterial activity.
Lit.: [1]Trans. Mycol. Soc. Jpn. **26**, 61–68 (1985). [2]Phytochemistry **26**, 565 f. (1987). – *[CAS 16338-48-0 (A.); 55528-30-8 (chloro derivative)]*

(S)-2-Amino-4-pentynoic acid (propargylglycine).

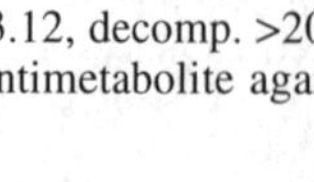

$C_5H_7NO_2$, M_R 113.12, decomp. >200 °C, $[\alpha]_D^{22}$ −32.6° (H_2O). A. is an antimetabolite against *Bacillus subtilis*[1] produced by *Streptomyces* sp. and occurs together with allylglycine (see 2-amino-4-pentenoic acid) in fruit bodies of *Amanita* spp.. Ninhydrin reaction: brown. It inactivates various B_6-enzymes, e. g., cystathionine γ-synthase (EC 4.4.1.1)[2], methionine γ-lyase (EC 4.4.1.11)[3], and L-alanine aminotransferase (EC 2.6.1.2)[4].
Lit.: [1]J. Antibiot. **24**, 239–244 (1971). [2]J. Am. Chem. Soc. **95**, 6124 f. (1973). [3]Biochemistry **20**, 4325–4333 (1981). [4]J. Biol. Chem. **255**, 3487–3491 (1980). – *[HS 2922 49; CAS 23235-01-0]*

2-Aminophenol (questiomycin B).

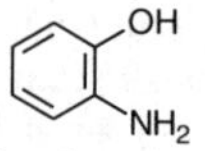

C_6H_7NO, M_R 109.13, rhomb. cryst., D. 1.328, mp. 174 °C, subl. 153 °C (1.5 k Pa), pK_{a1} 4.78, pK_{a2} 9.97, it is readily oxidizable to brown products. The hydrochloride, mp. 207 °C, is stable to oxidation but sensitive to light. A. is isolated from *Streptomyces* sp.; LD_{50} (rat p. o.) 1300 mg/kg, (mouse p. o.) 1250 mg/kg, (cat s. c.) 37 mg/kg. A. exhibits tuberculostatic and antibacterial activities, causes skin and eye irritations; the vapors are toxic when inhaled. The biosynthesis proceeds via the shikimic acid pathway.
Use: Starting material for the production of pharmaceuticals, hair dyes, and other industrial dyes.
Lit.: Beilstein E III/IV **13/2**, 805 ▪ Merck-Index (12.), No. 481 ▪ Ullmann (5.) **A 2**, 99–105; **A17**, 446. – *Toxicology:* Sax (8.), ALT 000. – *[HS 2922 29; CAS 95-55-6]*

2-Aminopimelic acid [(*R*)-2-aminoheptanedioic acid].

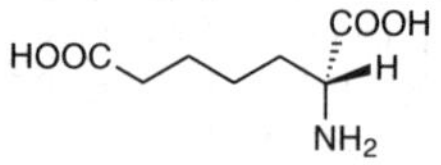

$C_7H_{13}NO_4$, M_R 175.18, mp. 219–220 °C (decomp.), $[\alpha]_D^{25}$ −21.0° (5 m HCl), $[\alpha]_D^{26}$ −45.5° (1 m HCl). Non-proteinogenic amino acid in spleenworts (finger ferns) *Asplenium* spp.[1–3]. 2-A. and the *trans*-3,4-dehydro derivative as (2*R*)-form occur in *A. unilaterale*[2]. 2-A. in *A. wilfordii* is partially racemized[3]. In contrast, 4-Hydroxy-2-aminopimelic acid always occurs in the (2*S*)-form in both species[3,4]. For chemotaxonomy see *Lit.*[5].
Lit.: [1]Karrer, No. 2380. [2]Phytochemistry **22**, 2735 ff. (1983). [3]Phytochemistry **24**, 2291–2294 (1985); Karrer, No. 2381. [4]Phytochemistry **18**, 1505–1509 (1979). [5]Bot. Mag. Tokyo **101**, 353–372 (1988).
gen.: Beilstein E IV **4**, 3080. – *[CAS 3721-85-5]*

3-Aminoproline (3-amino-2-pyrrolidinecarboxylic acid).

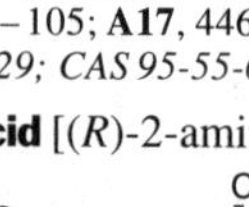

(2S, 3R)-, *cis*-L-form

$C_5H_{10}N_2O_2$, M_R 130.15, hydrochloride, mp. 215 °C (decomp.), $[\alpha]_D$ +23.0° (5 m HCl), +5.8° (H_2O)[1]. Non-proteinogenic amino acid from the morel *Morchella esculenta* and related species. It is isolated both as the γ-glutamyl compound[2] and as the free amino acid[3] from cultivated mycelia of *M. esculenta*. Ninhydrin reaction: light blue.

Biosynthesis: From L-proline[4], the mechanism is unknown.

Lit.: [1] Phytochemistry **8**, 1305 – 1308 (1969). [2] Trans. Mycol. Soc. Jpn. **24**, 191 – 195 (1983). [3] Appl. Environ. Microbiol. **38**, 1018 – 1019 (1979). [4] Sci. Pap. Coll. Gen. Educ. Univ. Tokyo **26**, 33 – 38 (1976).
gen.: Tetrahedron **51**, 5169 (1995) (synthesis). – *[CAS 24279-08-1 (2S,3R-form); 24279-22-9 (hydrochloride)]*

3-Aminospirostanes see Solanum steroid alkaloids.

Amino sugars (aminodeoxy sugars). Name for *monosaccharides, in which at least one or even more arbitrary hydroxy groups have been replaced by amino groups (exception: the anomeric hydroxy function; in this case the compounds are glycosylamines). A. have been successfully synthesized by numerous classical (Strecker synthesis, aminolytic epoxide opening; S_N2 reactions with amines, azides; red. of oximes, nitro compounds, azido sugars; retro-Amadori rearrangement, etc.) and enzymatic routes (e.g. epimerization of other A.; construction with aldolases). *Aminoglycosides* form the base for the important class of aminoglycoside antibiotics.

Lit.: Jeanloz & Balazs, The Amino Sugars (3 vols.), New York: Academic Press 1965 – 1969 ▪ Zechmeister **20**, 200 – 270.

Ammajin see marmesin.

Ammodendrine (1′-Acetyl-1′,4′,5′,6′-tetrahydroanabasine).

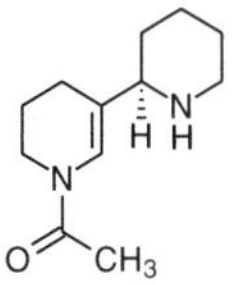

(+)-(*R*)-form

$C_{12}H_{20}N_2O$, M_R 208.30, mp. 43 – 46 °C, $[\alpha]_D$ +15° (C_2H_5OH); racemate, mp. 50 – 60 °C. A. is a *piperidine alkaloid, occurring together with *quinolizidine alkaloids in many Fabaceae genera (e.g., *Ammodendron, Baptisia, Cytisus, Lupinus*)[1]. Its structure is closely related to that of *anabasine.

Lit.: [1] Waterman **8**, 197 – 239.
gen.: Beilstein E V **23/5**, 282 f. ▪ Pelletier **2**, 105 – 148; **3**, 26 f. – *[HS 2939 90; CAS 27542-15-0; 494-15-5 ((+)-form); 20824-32-2 (racemate)]*

Amoorastatin see meliatoxins.

Amphibian venoms. General term for *toxins serving exclusively for defence and protection against microorganisms secreted from skin glands by frogs, toads, and amphibians. Among the A. are some of the strongest known venoms (*batrachotoxins and *tetrodotoxin). Their chemical composition does not follow any uniform principle; they are e.g., steroids, peptides, biogenic amines, and alkaloids of which the alkaloids represent the largest class with about 300 compounds identified to date; see frog toxins, toad poisons, salamander steroid alkaloids.

Lit.: Annu. Rev. Biochem. **59**, 395 – 414 (1990) ▪ Manske **43**, 185 – 288.

Amphibine A see cyclopeptide alkaloids.

Amphidinolides. A family of marine macrolides with potent antineoplastic activity. A. are produced by symbiontic microalgae *Amphidinium* sp., dinoflagellates, harvested from the inner tissue of the host Okinawan flatworms *Amphiscolops* sp.. A. show cytotoxic and antineoplastic activity, e.g., *A. A*: $C_{31}H_{46}O_7$, M_R 530.69, needles, mp. 130 – 133 °C, $[\alpha]_D^{24}$ +46° ($CHCl_3$).

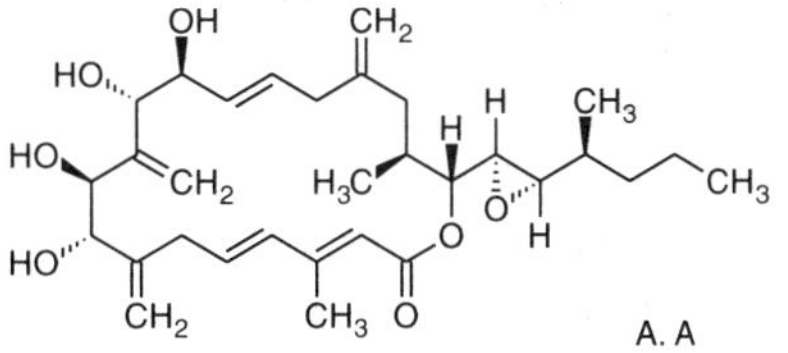

A. A

Lit.: Heterocycles **44**, 543 – 572 (1997) (review) ▪ J. Chem. Soc., Chem. Commun. **1994**, 1455 (isolation) ▪ J. Org. Chem. **58**, 2645 (1993) ▪ Tetrahedron **53**, 7827 (1997) (biosynthesis) ▪ Tetrahedron Lett. **27**, 5755 (1986) (isolation). – *[CAS 106463-75-6 (A. A)]*

Amphikuemine.

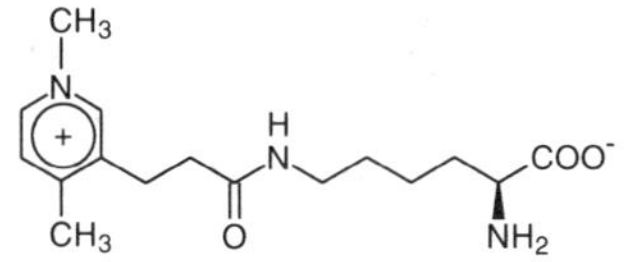

$C_{16}H_{25}N_3O_3$, M_R 307.39. A., which is produced together with similar substances by the sea anemone *Radianthus kuekenthali*, serves for mutual recognition and orientation in the symbiosis between the anemone and the anemone fish *Amphiprion perideraion*.

Lit.: Heterocycles **30**, 247 (1990) (synthesis) ▪ Science **234**, 585 (1986). – *[CAS 105870-54-0]*

Amphotericin B.

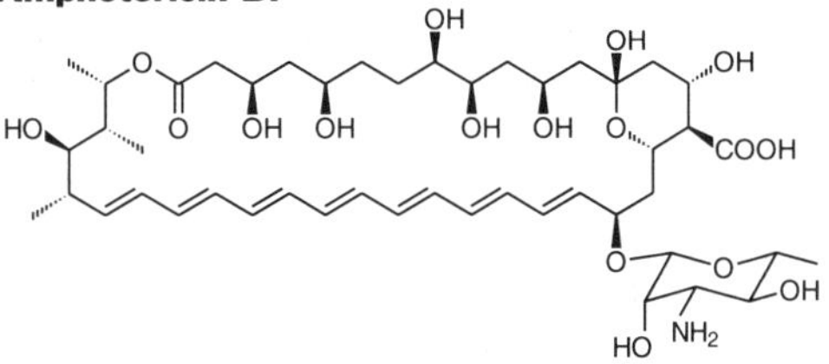

$C_{47}H_{73}NO_{17}$, M_R 924.09, dark yellow cryst., mp. above 170 °C (decomp.), $[\alpha]_D$ +333° (DMF), a polyene*macrolide of the heptaene series with a 38-membered lactone ring and *O*-glycosidically linked amino sugar (D-mycosamine) produced by *Streptomyces nodosus*. A. B is amphoteric and contains lipophilic and hydrophilic regions in the molecule. It exerts its effects in interactions with steroid-containing membranes. A. B has antifungal action and is used clinically in the treatment of mycosis. On systemic administration considerable side effects occur, particularly serious is the nephrotoxicity of A. B. Liposome formulations improve its tolerability. In cases of mycoses occurring through impairment of the immune system (e.g. in AIDS or after organ transplantations) A. B is often the only effective agent.

Lit.: Antimicrob. Agents Chemother. **38**, 1604 (1994) ▪ J. Antibiot. **38**, 1699 (1985); **49**, 1232 (1996) (absolute configuration A. A) ▪ Martindale (30.), p. 315 ▪ Omura (ed.), Macrolide Antibiotics, p. 351 – 493, Orlando: Academic Press 1984. – *Synthesis:* J. Am. Chem. Soc. **110**, 4672, 4685, 4696 (1988) ▪ Nicolaou, p. 421 – 450 ▪ Sax (8.), AOC 500. – *[CAS 1397-89-3]*

3′-AMP monobutyl ester see B factor.

Amsonia alkaloids. Group of *monoterpenoid indole alkaloids from various species of the genus *Amsonia*

(Apocynaceae), e. g., *A. angustifolia*, *A. tabernaemontana*, and *A. elliptica*. To date more than 40 alkaloids (e.g. *eburnamonine, *tabersonine, *yohimbine) have been identified.
Lit.: Pelletier **1**, 322. – *[HS 2939 90]*

Amurensine see pavine and isopavine alkaloids.

Amurine see morphinan alkaloids.

Amygdalin see cyanogenic glycosides.

Amylopectin. Main component of *starch (figure see there) amounting to ca. 70–80% (remainder: *amylose). It is a branched, high molecular weight, polydisperse glucopyranosyl polymer, M_R 10^6–$2 \cdot 10^7$. Cluster-like accumulations of α-1,4-glucan chains with 20–30 glucose units are bound through α-1,6-branches to an α-1,4-glucan backbone which has a structure similar to that of amylose. A. is insoluble in cold water but swells in hot water and forms a paste which solidifies on cooling. As a result of the α-1,6-glucosidic bonding it is rather resistant to attack by the enzyme β-amylase which can only split it into the so-called boundary dextrins. In contrast α-amylase degrades A. to maltose, glucose, and isomaltose.
Lit.: Kirk-Othmer (3.) **21**, 492–507 ▪ Prog. Biotechnol. **1985**, 45–54, 55–60 ▪ Starch: Properties and Potential, Gulliard (ed.), New York: Wiley 1987 ▪ Ullmann (5.) **A 12**, 462; **A 25**, 2–20 ▪ Whistler & BeMiller (eds.), Industrial Gums (3.), p. 579 ff., San Diego: Academic Press 1993 ▪ see also amylose, starch. – *[HS 3913 90; CAS 9037-22-3]*

Amylose. A component (20–30%) of *starch surrounded by *amylopectin. A. is a linear α-1,4-glucan, M_R 50 000–200 000 (see figure at starch). Crystalline A. occurs in various polymorphic forms (A, B, C, and V-A.), that differ in conformation and crystal packing. A. is soluble in water and gives the characteristic blue color with iodine-potassium iodide solution (Lugol's solution) (formation of inclusion compounds, traces of iodide ions are necessary for occurrence of the blue color, formation of I_5^- ions: $I-I\cdots I^-\cdots I-I$). Because of its predominately unbranched structure, A. can be degraded to oligosaccharides both by α- and by β-amylase. The screw-like (helical) conformation also allows the formation of inclusion compounds with alcohols.
Lit.: ACS Symp. Ser. **141**, 459–482 (1980) ▪ Kirk-Othmer (3.) **21**, 492–507 ▪ New Food Ind. **24**, 82 (1982) ▪ Starch: Properties and Potential, Gulliard (ed.), New York: Wiley 1987 ▪ Ullmann (5.) **A 12**, 462; **A 25**, 7. – *[HS 3913 90; CAS 9005-82-7]*

Amylostatins see glycosidase inhibitors.

Amyrins.

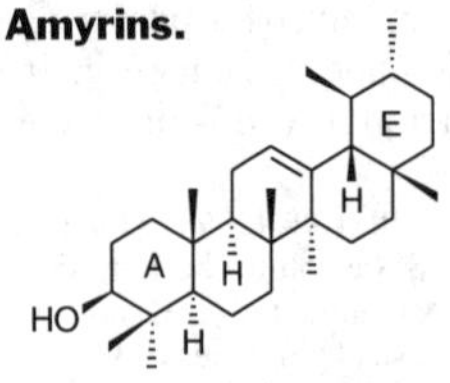

α-Amyrin

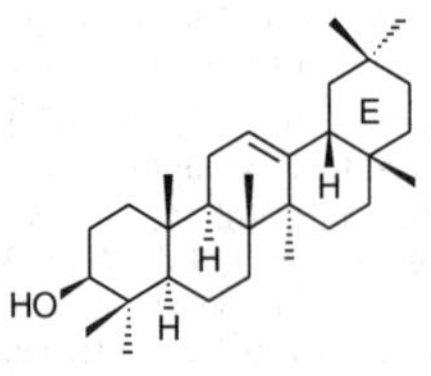

β-Amyrin

α-A. (urs-12-en-3β-ol), mp. 186 °C (needles), $[\alpha]_D$ +91.6° (benzene) and β-A. (olean-12-en-3β-ol), mp. 197.5 °C (needles), $[\alpha]_D$ +99.8° (benzene), $C_{30}H_{50}O$, M_R 426.73, are both optically active pentacyclic *triterpenes that occur in the latex of rubber trees (as acetates). They are simple unsaturated alcohols with 4 angular methyl groups and 2 geminal methyl groups on ring A. δ-A. [olean-13(18)-en-3β-ol; mp. 212.5 °C] is a double bond isomer of β-Amyrin. The saturated, substituted picene skeleton of α-A. is known as *ursane, that of β-A. as *oleanane (see triterpenes).
Lit.: Beilstein E IV **6**, 4191 (α-A.), 4195 (β-A.), 4196 (δ-A.) ▪ Karrer, No. 2011 ▪ Merck-Index (12.), No. 659, 660 ▪ Phytochemistry **23**, 407 (1984); **27**, 3579 (1988) (δ-A.). – *Synthesis:* J. Am. Chem. Soc. **115**, 8871 (1993) ▪ Tetrahedron **41**, 2513 (1985) (β-A.). – *[CAS 638-95-9 (α-A.); 559-70-6 (β-A.)]*

Anabasine [nicotimine, 3-((*S*)-2-piperidinyl)-pyridine].

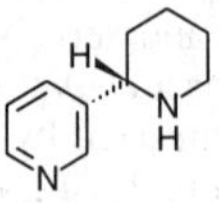

(-)-form

$C_{10}H_{14}N_2$, M_R 162.23, D. 1.046, mp. 9 °C, bp. 270–272 °C, $[\alpha]_D$ −82.2°. A. is a *tobacco alkaloid from many *Nicotiana* species (Solanaceae). Its name is derived from the plant *Anabasis aphylla* (Chenopodiaceae) in which it occurs as the main alkaloid; in addition it is found in several other plants from various families (including Fabaceae, Berberidaceae) and in marine ribbon worms; it has also been identified as a venom component in north American ants. It exhibits toxicity similar to that of *nicotine and is used as an insecticide.
Lit.: Beilstein E V **23/6**, 96 f. ▪ J. Chem. Soc., Chem. Commun. **1975**, 9 (biosynthesis) ▪ Manske **26**, 122 ▪ Sax (8.), p. 249 ▪ Ullmann (5.) **A 1**, 360; **A 14**, 271. – *Synthesis:* J. Org. Chem. **45**, 1515 (1980); **54**, 4261 ff. (1989). – *[HS 2939 90; CAS 494-52-0]*

Anacardic acid [(*Z,Z*)-6-(8,11-pentadecadienyl)salicylic acid].

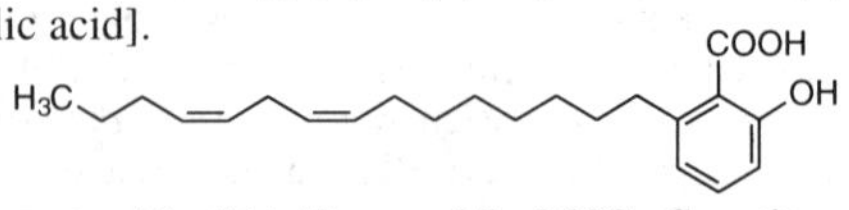

$C_{22}H_{32}O_3$, M_R 344.49, mp. 25–26 °C. Constituent of the shell of the cashew nut (*Anacardium occidentale*); it is well soluble in ethanol. A. has antineoplastic, nematicidal, molluscicidal, and weak antibacterial activities.
Lit.: Chem. Soc. Rev. **8**, 499 (1979) (review) ▪ J. Agric. Food Chem. **34**, 979 (1986) ▪ Justus Liebigs Ann. Chem. **1995**, 2009–2020 (synthesis). – *[CAS 25377-74-6]*

Anagyrine.

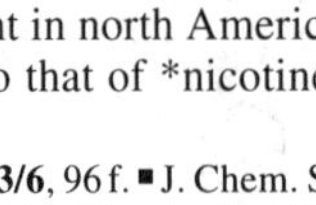

$C_{15}H_{20}N_2O$, M_R 244.34, mp. 295–297 °C (hydrochloride), $[\alpha]_D$ −165° (C_2H_5OH). A. is a tetracyclic *quinolizidine alkaloid with a *sparteine skeleton of the α-pyridone type found in many genera of the Fabaceae (e. g. *Anagyris*, *Baptisia*, *Lupinus*)[1]. In pharmacology A. is known for its cardiotonic action that induces tachycardia, it also causes malformations in calves[2,3]. The 11-epimer is called *thermopsine*.
Lit.: [1]Waterman **8**, 197–239. [2]J. Toxicol. Environ. Health **1**, 887–898 (1976). [3]Pelletier **2**, 105–148.
gen.: Beilstein E V **24/3**, 410 f. ▪ Merck-Index (12.), No. 666. – *[HS 2939 90; CAS 486-89-5; 34389-11-2 (racemate); 486-90-8 (thermopsine)]*

Anandamide (arachidonic acid 2-hydroxyethylamide).

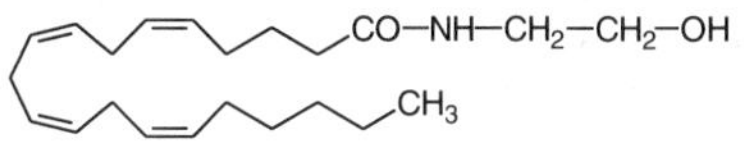

$C_{22}H_{37}NO_2$, M_R 347.54, colorless liquid. The name is derived from the Sanskrit word "Ananda", which means something like a state of blissfulness. A. was first isolated from pig brains where it serves as endogenous ligand for *cannabinoid receptors ("endogenous cannabinoid"). A. has also been isolated from cocoa products and is held responsible for psychoactive effects [1]. In mice A. acts similarly to Δ^9-*tetrahydrocannabinol (Δ^9-THC), the psychoactive principle of *hashish. The psychotropic action of A. is attributed to the binding on these receptors which lie in the membranes of synapses [2]. Recently, A. has been identified as a neurotransmitter, involved in regulating movement in animals [3]. Some further fatty acids are also isolated from brain tissue, e. g. palmitic acid and docosatetraenoic acid ethanolamides. A. is deactivated in the brain by an amidase which converts the compound in the presence of water to *arachidonic acid and ethanolamine.

Lit.: [1] Nature (London) **382**, 677 (1996); Naturwiss. Rundsch. **49**, 481 (1996). [2] J. Med. Chem. **40**, 3617 (1997). [3] Nat. Neurosci. **2**, 358 (1999).

gen.: Drugs of Today **32**, 275–285 (1996) (review) ▪ J. Biol. Chem. **240**, 1019 (1965); **269**, 22937 (1994) ▪ J. Med. Chem. **36**, 3032 (1992) ▪ Nature (London) **376**, 591 (1995) ▪ Science **258**, 1946 (1992). – *[CAS 94421-68-8]*

Anandimycin see gilvocarcins.

Anastrephin.

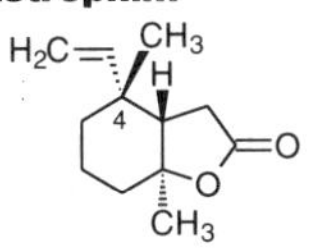

Anastrephin

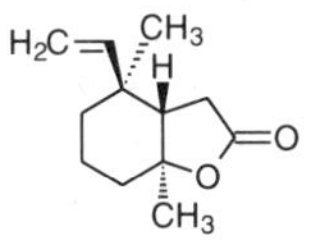

Epianastrephin

$C_{12}H_{18}O_2$, M_R 194.27, $[\alpha]_D^{23.5}$ –50.4° (hexane), cryst., mp. 94–95 °C. As a mixture with *epianastrephin*, the 4-epimer of A., and the monocyclic isomer *suspensolide, A. forms the *pheromone system of tropical fruit flies of the genus *Anastrepha*. In *A. suspensa* (–)-A. and its epimer are present in low enantiomeric excess only.

Lit.: ApSimon **9**, 360–362 (synthesis) ▪ Tetrahedron Lett. **24**, 2611–2614 (1983). – *[CAS 77670-94-1 (A.); 77670-93-0 (epianastrephin)]*

Anatoxins.

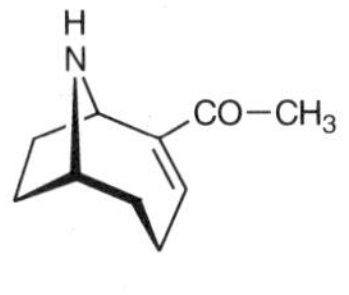

Anatoxin a

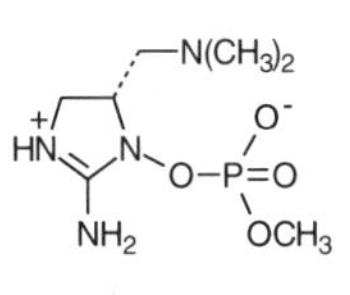

Anatoxin a(s)

Toxins formed by various strains of the cyanobacterium *Anabaena flos-aquae* ("blue algae") which often cause animal poisonings.

A. a {1-(9-azabicyclo[4.2.1]non-2-en-2-yl)ethanone} [1]: $C_{10}H_{15}NO$, M_R 165.24, oil, occurs in nature as the (1*R*)-(+)-form. Mp. of the *N*-acetyl derivative 117–118 °C; an alkaloid from *A. flos-aquae*. The neurotoxin kills mice within 2 to 5 min and is thus known as very fast death factor (VFDF), LD_{min} (mouse i. p.) 0.25 mg/kg. The activity is due to blockage of postsynaptic depolarization in muscle cells.

A. a(s) {(*S*)-2-amino-4,5-dihydro-1-[(hydroxymethoxyphosphinyl)oxy]-*N,N*-dimethyl-1*H*-imidazole-5-methanamine}: $C_7H_{17}N_4O_4P$, M_R 252.21; decomposes in basic solutions and slowly on storage at –20 °C. It is a highly active neurotoxin [LD_{50} (mouse) 20–40 µg/kg] that inhibits choline esterase [2]. The structure and absolute configuration have been elucidated by spectroscopic methods and by comparison with a synthetic model compound.

Lit.: [1] Beilstein E V **21/7**, 269; Angew. Chem. Int. Ed. Engl. **38**, 1985 (1999); Bioorg. Med. Chem. Lett. **7**, 2867 (1997); J. Am. Chem. Soc. **121**, 3057 (1999); J. Chem. Soc., Chem. Commun. **1995**, 831, 1461; J. Org. Chem. **55**, 5025–5033 (1990); Acta Chem. Scand., Ser. B **41**, 180 (1987); **43**, 917 (1989); J. Am. Chem. Soc. **107**, 8066 (1985); **111**, 8021 (1989); Tetrahedron: Asymmetry **3**, 1263–1270 (1992); Tetrahedron Lett. **36**, 8867 (1995) (absolute configuration of A. a); Tetrahedron **52**, 6025–6061, 11637 (1996) (review); **39**, 2133 (1998) (synthesis). [2] Toxicon **26**, 750 (1988); Gazz. Chim. Ital. **123**, 329 (1993) [review on A.-a(s)] – *[CAS 64285-06-9 (A. a); 103170-78-1 (A. a(s))]*

Ancistrocladus alkaloids see naphthyl(tetrahydro)isoquinoline alkaloids.

Ancistrodial.

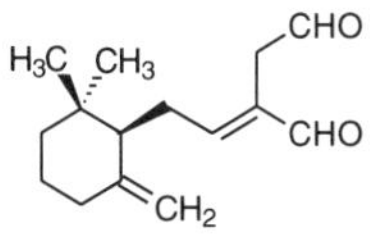

$C_{15}H_{22}O_2$, M_R 234.34, soluble in dichloromethane and chloroform. It is the active principle in the defensive secretions of the west African termite species *Ancistrotermes cavithorax* [1]. A. is only produced by the small soldier ants; the defensive secretions of the large soldier ants do not contain any ancistrodial. A. is active against the termites' main predator, ants of the species *Megaponera foetens* [2]. The formula shows the stereochemistry of synthetic (*R*)-(–)-A. The stereochemistry of the natural product is still unknown.

Lit.: [1] J. Chem. Ecol. **13**, 1171 (1987). [2] Oecologia **32**, 101 (1977).

gen.: Synthesis: Acta Chem. Scand. **46**, 625 (1992); Eur. J. Org. Chem. **1998**, 2851 [(–)-A.]; Tetrahedron: Asymmetry **7**, 3009 (1996). – *[CAS 68398-28-7]*

Andrimid.

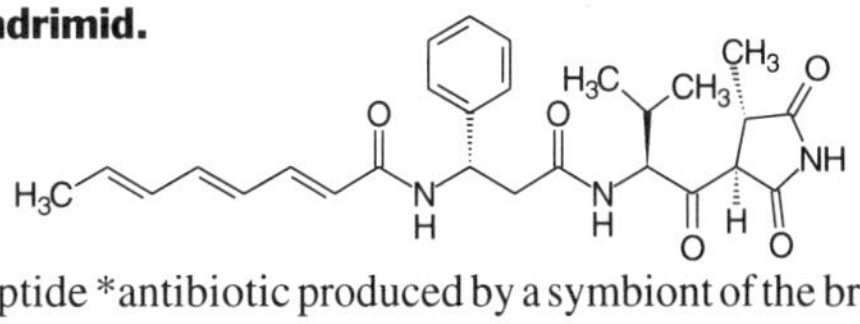

Peptide *antibiotic produced by a symbiont of the brown grasshopper (*Nilaparvata lugens*). $C_{27}H_{33}N_3O_5$, M_R 479.58, mp. 172–173.5 °C. The symbiont, an *Enterobacter* species, has been isolated from the egg mass, from which it can be passed on to the young insects. A. has a highly specific activity against the rice pathogen *Xanthomonas campestris* pv. *oryzae* (MIC 0.1 µg/mL), but is not or only weakly active against other Gram-positive and Gram-negative bacteria. Unusual is the β-ketoamide structure in the succinimide residue, similar to that occurring also in, e. g. pepstanone, an inhibitor of acid peptidase. Analogs with different configuration or lacking methyl groups were found to be very poorly active.

Lit.: J. Am. Chem. Soc. **109**, 4409 (1987) ▪ Pestic. Sci. **27**, 117 (1987). – *Synthesis:* J. Chem. Soc., Chem. Commun. **1989**, 299 ▪ Tetrahedron Lett. **32**, 4393 (1991). – *[HS 2941 90; CAS 108868-95-7]*

Androcymbine see phenethyltetrahydroisoquinoline alkaloids.

Androgens. The sex hormones of the human male are *testosterone* and *androstanolone* (stanolone, 4,5α-dihydrotestosterone). Both belong to the A. and contain the C_{19} carbon skeleton of 5α-androstane (see steroids). According to chemical nomenclature rules they are named as 17β-hydroxyandrost-4-en-3-one (testosterone) and 17β-hydroxy-5α-androstan-3-one (androstanolone). The naturally occurring A also include other androstane derivatives, albeit with weaker androgenic activities, all of which are biosynthetic intermediates or metabolites of the actual hormones, e. g., *androstenolone* (prasterone, 3β-hydroxyandrost-5-en-17-one), *androstenedione* (androst-4-ene-3,17-dione), and *androsterone* (3α-hydroxy-5α-androstan-17-one).

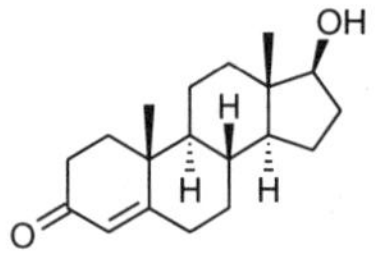

Testosterone

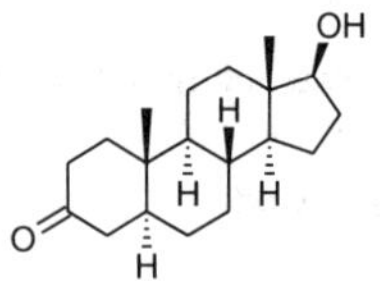

Androstanolone

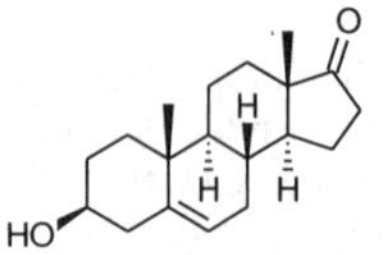

Androstenolone

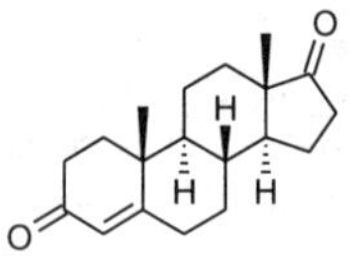

Androstendione

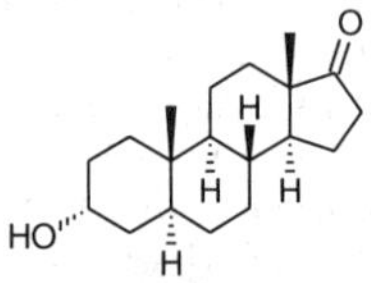

Androsterone

Figure 1: Structures of the Androgens.

Table: Data of Androgens.

Steroid	molecular formula	M_R	mp. [°C]	$[\alpha]_D$	CAS
Testosterone	$C_{19}H_{28}O_2$	288.43	155	+109° (C_2H_5OH)	58-22-0
Androstanolone	$C_{19}H_{30}O_2$	290.45	181	+32.4° (C_2H_5OH)	521-18-6
Androstenolone	$C_{19}H_{28}O_2$	288.43	152–153	-3° ($CHCl_3$)	53-43-0
Androstendione	$C_{19}H_{26}O_2$	286.41	173–174	+194° ($CHCl_3$)	63-05-8
Androsterone	$C_{19}H_{30}O_2$	290.45	184–186	+94.6° (C_2H_5OH)	53-41-8

Testosterone and androstanolone are jointly responsible for the development and maintenance of the male sexual characteristics although their functions differ. During embryogenesis of the male androstanolone is responsible for development of the external sex organs and the urogenital region. Testosterone on the other hand is important for the prenatal development of those portions of the urogenital tract arising from the Wolffian duct, such as epididymis, spermatic duct, and seminal vesicle. It has also been shown that androstanolone participates in the feedback regulation of testosterone production and also plays a role in the postnatal development of epididymis, seminal vesicle, prostate, and penis. After puberty testosterone induces male sexual behavior, development of potency and libido, increase of muscle mass, spermatogenesis, as well as enlargement of larynx, vocal cords, penis, and scrotum while androstanolone is responsible for the increase in facial and body hair, the occurrence of acne, as well as the later loss of hair and increase in prostate size [1].

Biosynthesis & metabolism: Testosterone is formed principally in the interstitial cells of testes tissue (3–10 mg/d) and to a lesser extent in men and women also in the adrenal cortex (ca. 0.5 mg/d). The dominating biosynthetic pathway starts from 3β-hydroxypregn-5-en-20-one (product of biological degradation of the cholesterol side chain) and proceeds via its 17α-hydroxy derivative and androstenolone to testosterone, from which androstanolone is formed by steroid 5α-reductase. A further pathway proceeds from the same precursor via progesterone (see progestogens), its 17α-hydroxy compound and androstenedione also to testosterone.

Testosterone circulating in the blood is mainly metabolized in the liver to androstenedione which is converted via intermediates into androsterone and its 5β-isomer. These are then excreted in urine as glucuronides or sulfate conjugates.

Occurrence other than in man: The spectrum of A. in vertebrates is, in principle, qualitatively the same as that in humans although there may be quantitative differences and differences with regard to biosynthetic and metabolic details. A. have also occasionally been isolated from plants as metabolites of steroidal secondary substances, e. g., testosterone and androstenedione from pollen of Scottish pine *Pinus sylvestris* (Pinaceae) and *rubrosterone* (2β,3β,14α-trihydroxy-5β-androst-7-ene-6,17-dione), a biological degradation product of an *ecdysteroid from *Achyranthes rubrofusca* (Amaranthaceae).

Physiological activity: For the functional significance of A., especially testosterone and androstanolone, see above. By means of the *cock's comb test* (capon test) the following relative androgenic activities were determined (µg-equivalent of an international unit = capon unit): testosterone (15 µg), androstanolone (20 µg), androsterone (100 µg), androstenedione (120 µg), and androstenolone (200 µg). Some A. are considered to have antidepressive, psychotonic properties; testosterone and androstanolone also increases in activity, aggressivity, tension, and performance on account of their anabolic action.

Use: A. are used clinically for the treatment of A. deficiency syndromes. Orally administered testosterone, however, is rapidly metabolized in intestinal walls and liver; so it is better to administer testosterone 17-*O*-esters dissolved in oils by i. m. injection. The undesired

deactivation of testosterone by dehydrogenation to the 17-oxo compound can be mostly prevented by 17α-substitution, thus, e. g., *methyltestosterone* can also be administered orally.
The stimulating effect of A. on the development of muscle tissue and performance (*anabolic* action) can be of interest for certain diseases as well as in animal fattening, body building, and sports (doping). For this a separation of the androgenic and anabolic actions would be useful but is in fact not possible. To a certain extent, however, a dissociation of testosterone- and androstanolone-typical (anabolic) actions does seem possible. Numerous orally administered "anabolic agents" (substituted androstanolone derivatives and 19-norsteroids) having been developed according to this concept and are in use, e. g., *oxymetholone* or *norethandrolone*.
Production: Testosterone and its derivatives as well as the unnatural A. and anabolic agents are industrially produced currently either by total synthesis or from androst-4-ene-3,17-dione, obtained by microbial side chain degradation from *sterols.
Lit.: [1] Endocrinology **125**, 2434 (1989).
gen.: Bohl & Duax (eds.), Molecular Structure and Biological Activity of Steroids, Boca Raton: CRC Press 1982 ▪ Fieser & Fieser, Steroide, p. 551–587, Weinheim: Verl. Chemie 1961 ▪ Zeelen, p. 177–195. – *[HS 2937 99; CAS 19466-41-2 (rubrosterone); 58-18-4 (methyltestosterone); 434-07-1 (oxymetholone); 52-78-8 (norethandrolone)]*

Andromedol, andromedotoxin see grayanotoxins.

Androstane see steroids.

Androstanolone, androstenedione, androstenolone, androsterone see androgens.

5α-Androst-16-en-3α-ol, 5α-androst-16-en-3-one see steroid odorants.

Anemonin.

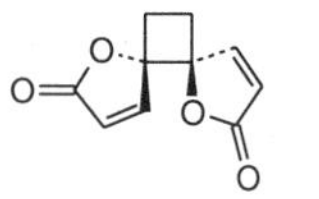

(±)-Anemonin

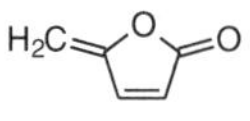

Protoanemonin

A. (*trans*-1,7-dioxadispiro[4.0.4.2]dodeca-3,9-diene-2,8-dione), $C_{10}H_8O_4$, M_R 192.17, needles or prisms, mp. 157–158 °C, steam-distillable. Component of crowfoot plants (Ranunculaceae) like *Ranunculus thora*, *Anemone pulsatilla*, *A. altaica*, *A. alpina*, and *Clematis recta*. A. is the dimer of the steam-distillable, easily polymerizing *proto-A.* [isomycin, 5-methylene-2(5*H*)-furanone], $C_5H_4O_2$, M_R 96.09, pale yellow oil, bp. 45 °C (195 Pa) from which A. forms on drying. Both substances exhibit antibacterial, antipyretic, sedative [1] and insecticidal [2] activities against *Drosophila melanogaster* (fruit fly), with proto-A. being more active than A. Proto-A. is also a strong irritant of mucous membranes. LD_{50} (mouse i. p.) 150 mg/kg.
Lit.: [1] J. Ethopharmacol. **24**, 185–191 (1988). [2] Indian J. Exp. Biol. **31**, 85–86 (1993).
gen.: An. Quim., Ser. C **85**, 5–8 (1989) (synthesis) ▪ Beilstein E V **17/9**, 371 (proto-A.); **19/5**, 101 f. (A.) ▪ Bull. Chem. Soc. Jpn. **55**, 1584–1587 (1982) (biosynthesis); **65**, 2366–2370 (1992) (synthesis) ▪ Merck-Index (12.), No. 681 – *[CAS 508-44-1 (A.); 108-28-1 (proto-A.)]*

Anethole [1-methoxy-4-(1-propenyl)benzene].

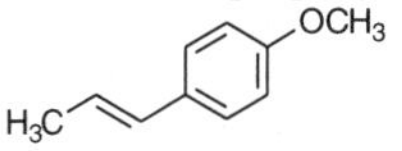

$C_{10}H_{12}O$, M_R 148.20. (*E*)-A. forms crystals with anise-like smell and sweet taste, D. 0.9883, mp. 22 °C, bp. 81 °C (306.6 Pa), n_D^{20} 1.5615. The (*Z*)-isomer melts at –22 °C, D. 0.9878, bp. 79 °C (306.6 Pa), n_D^{20} 1.5546. A. occurs in many essential oils, especially anise oil (80–90%) from which it can be obtained by freezing it out. (*Z*)-A.: LD_{50} (rat i. p.) 93 mg/kg or (rat p. o.) 150 mg/kg; (*E*)-A.: LD_{50} (rat i. p.) 900 mg/kg or (rat p. o.) 2090 mg/kg.
Activity: A. is an attractant (see semiochemicals) [1] for the corn root pests *Diabrotica undecimpunctata* and *D. virgifera* widely distributed in South and West USA. Biosynthesis from L-phenylalanine → *(*E*)-*p*-coumaric acid.
Use: Natural and synthetic A. are used in the production of liquors (anisette etc.), cosmetic products, anisaldehyde, flavor substances and in color photography (sensitizer).
Lit.: [1] J. Chem. Ecol. **13**, 959–975 (1987).
gen.: Beilstein E IV **6**, 3796 ▪ Hager (5.) **5**, 159, 516 ▪ Synth. Commun. **10**, 225–231 (1980) (synthesis) ▪ Ullmann (5.) **A 11**, 195 f. – *Toxicology:* Sax (8.), PMQ 750, PMR 000. – *[HS 2909 30; CAS 104-46-1 (A.); 4180-23-8 (E); 25679-28-1 (Z)]*

Angelica acid esters [(*Z*)-2-methyl-2-butenoic acid esters] see chamomile oils.

Angelica root/seed oil. *Root oil:* yellow to brown oil. *Seed oil:* yellowish oil. Greenish, spicy, slightly musk-like, earthy-peppery odor; peculiar sweet-aromatic, slightly bitter taste.
Production: By steam distillation from the roots or seeds of *Angelica archangelica*. Main areas of cultivation are Germany, France, Belgium, and Holland.
Composition [1]***:*** Main components are monoterpene hydrocarbons such as *α-*pinene* (ca. 20%), *α-phellandrene* (see *p*-menthadienes, ca. 20%), 3-**carene* (ca. 10%), *limonene* (see *p*-menthadienes, ca. 10%), and *β-phellandrene* (ca. 20%). The macrocyclic lactone **15-pentadecanolide* (ca. 1%) is responsible for the typical musk-like odor.
Use: On account of the laborious separations (poor yields, long distillation times) A. belongs to the more expensive natural sources of crude materials for perfumery; A. is thus used only in low doses, e. g., to prepare perfumes with a masculine touch and to improve the aromatic character of liquors.
Lit.: [1] Perfum. Flavor. **7** (2), 35 (1982); **14** (4), 41 (1989); J. Essent. Oil Res. **3**, 229 (1991); **5**, 447 (1993); J. Agric. Food Chem. **42**, 1979, 2235 (1994).
gen.: Arctander, p. 63, 65 ▪ Bauer et al. (2.), p. 135 – *Toxicology:* Food Cosmet. Toxicol. **12**, 821 (1974); **13**, 713 (1975). – *[HS 3301 29; CAS 8015-64-3]*

Angelicin see furocoumarins.

Angucyclins.

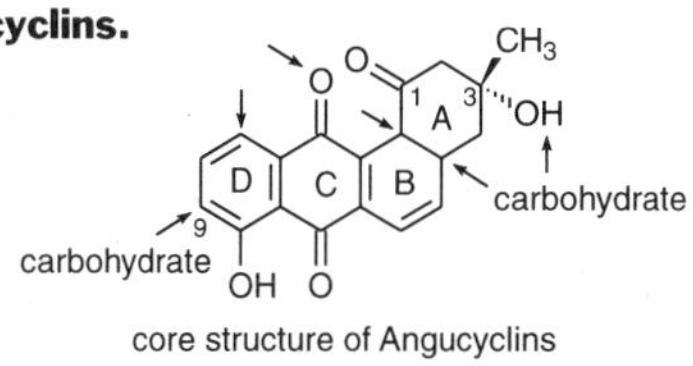

core structure of Angucyclins

Term for *O*-glycosides which on hydrolysis give an angularly (Latin.: angularis = corner) anellated, tetracyclic quinone or quinone-like derivative, an *angucyclinone*. The unspecific names benz[*a*]anthracene derivatives or *isotetracenones* are also used. The basic skeleton, a *polyketide, can be modified in various positions (see arrows in the structure). In many cases the first deoxy sugar in position C-9 is C-glycosidically bound (see aquayamycin). The A. are mainly produced by *actinomycetes, over 100 metabolites of this type are known, e. g., *saquayamycins, *urdamycins, *vineomycins. The biological activity of A. and angucyclinones is not restricted to a specific type of action. There are reports of antitumor (*in vitro* against various cell lines), enzyme inhibiting (e. g., tyrosine and prolylhydroxylases), inhibition of blood coagulation, antibacterial (Gram-positive), and antiviral (including HIV) activities. The biosynthesis has been elucidated for the example of the *urdamycins [1].

Lit.: [1] J. Antibiot. **42**, 1151–1157 (1989); Angew. Chem. Int. Ed. Engl. **29**, 1051–1053 (1990).
gen.: J. Antibiot. **39**, 1657–1669 (1986) ▪ J. Org. Chem. **58**, 2447–2551 (1993) ▪ Nat. Prod. Rep. **9**, 103–138 (1992) ▪ Tetrahedron Lett. **34**, 3769 (1993). – *Reviews on synthesis and biosynthesis:* Isr. J. Chem. **37**, 3–22 (1997); Top. Curr. Chem. **188**, 127–195 (1997).

Anguibactin.

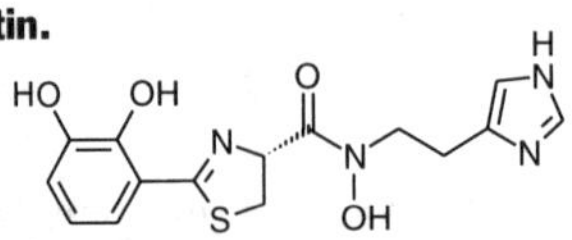

The *siderophore A. ($C_{15}H_{16}N_4O_4S$, M_R 348.38) has been isolated from cultures of the bacterium *Vibrio anguillarum* which is pathogenic for fish. It is composed of 2,3-dihydroxybenzoic acid, *N*-hydroxyhistamine, and cysteine. The affinity of the compound for iron(III) is so high that the microorganisms must be cultivated in a low-iron medium and the metabolites separated by chromatography on deionized material.

Lit.: J. Am. Chem. Soc. **111**, 292 (1989) ▪ J. Bacteriol **167**, 57 (1986). – *[CAS 117308-63-1]*

Anguidin see scirpenols.

Angustifoline.

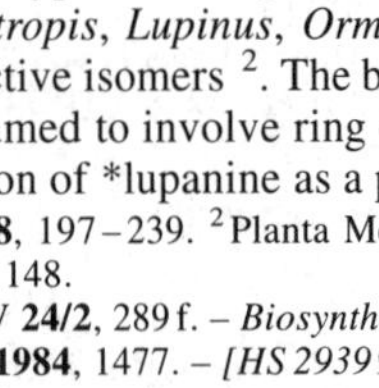

(-)-form

$C_{14}H_{22}N_2O$, M_R 234.34, mp. 80.5–81 °C, $[\alpha]_D$ −7.5° (+5.2°) (C_2H_5OH). A tricyclic *quinolizidine alkaloid of the sparteine type from 4 genera of the Fabaceae (*Cytisus, Diplotropis, Lupinus, Ormosia*) [1]. A. exists as 2 optically active isomers [2]. The biosynthesis of A. in plants is assumed to involve ring opening and side chain degradation of *lupanine as a precursor [3].

Lit.: [1] Waterman **8**, 197–239. [2] Planta Med. **59**, 289 (1993). [3] Pelletier **2**, 105–148.
gen.: Beilstein E V **24/2**, 289 f. – *Biosynthesis:* J. Chem. Soc., Chem. Commun. **1984**, 1477. – *[HS 2939 90; CAS 550-43-6]*

Anhaline see hordenine.

Anhalonium alkaloids. A group of *isoquinoline alkaloids from the Mexican cactus species *Anhalonium (Lophophora) williamsii* (Peyotl) and *A. lewinii,* with contents of up to 5%. A. a. are soluble in water, ethanol, chloroform, and form nicely crystallizing salts with acids. The most important A. a. is anhalonine. Carnegine and pellotine are isolated as racemates.

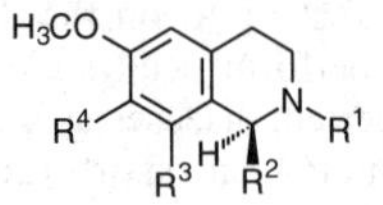

	R^1	R^2	R^3	R^4
Anhalamine (1)	H	H	OH	OCH_3
Anhalonidine (2)	H	CH_3	OH	OCH_3
Anhalonine (3)	H	CH_3	O–CH_2–O	
Carnegine (4)	CH_3	CH_3	H	OCH_3
Lophophorine (5)	CH_3	CH_3	O–CH_2–O	
Pellotine (6)	CH_3	CH_3	OH	OCH_3
Salsolidine (7)	H	CH_3	H	OCH_3

Table: Structure and data of Anhalonium alkaloids.

no.	molecular formula	M_R	mp./bp. [°C]	$[\alpha]_D$	CAS
1	$C_{11}H_{15}NO_3$	209.25	189–191		643-60-7
2	$C_{12}H_{17}NO_3$	223.27	160–161	−21.2°	3851-33-0
3	$C_{12}H_{15}NO_3$	221.26	85–86	−56.3° ($CHCl_3$)	519-04-0
4	$C_{13}H_{19}NO_2$	221.30	262–263/ 170 (133 Pa)	−24.9°	490-53-9
5	$C_{13}H_{17}NO_3$	235.28	oil, 140–145 (2.6 Pa)	−47.3° ($CHCl_3$)	17627-78-0
6	$C_{13}H_{19}NO_3$	237.30	100.5	−11.1° (C_2H_5OH)	83-14-7
7	$C_{12}H_{17}NO_2$	207.27	53–53.5/ 140 (133 Pa)	−24.9°	38520-68-2

Salsolidine occurs in *Salsola richteri* (Chenopodiaceae). It exhibits antihypertensive activity. Lophophorine (*N*-methylanhalonine) is the most poisonous member of this group; LD_{50} (rabbit i. v.) 15–20 mg/kg, physiological action, see mescaline. Anhalonidine is also highly toxic.

Biosynthesis: from tyrosine and dopa.

Lit.: Hager (5.) **5**, 707–712 ▪ Kirk-Othmer (4.) **1**, 1053–1066 ▪ Manske **4**, 8–14; **21**, 255–327; **31**, 1–28; **41**, 1–40 ▪ Merck-Index (12.), No. 691–693 ▪ Mothes et al., p. 195–197 (biosynthesis) ▪ Phillipson et al., p. 47–61 ▪ Shamma, p. 1–43 ▪ Shamma-Moniot, p. 1–26 ▪ Ullmann (5.) **A 1**, 368 (1985). – *Synthesis:* J. Chem. Soc., Perkin Trans. 1 **1992**, 309–310 ▪ J. Org. Chem. **55**, 1086–1093 (1990) ▪ Synthesis **1998**, 162 [(–)-(*S*)-salsolidine] ▪ Tetrahedron Lett. **29**, 6949–6950 (1988). – *[HS 2939 90]*

15′,20′-Anhydrovinblastine.

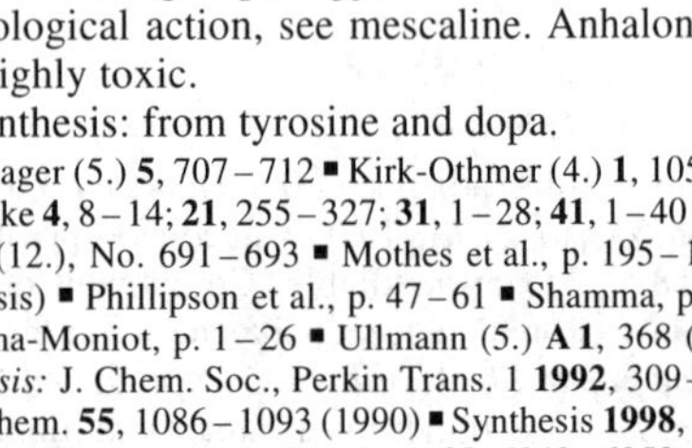

$C_{46}H_{56}N_4O_8$, M_R 792.97, highly unstable *monoterpenoid indole alkaloids of the group of dimeric *Catharanthus alkaloids (bisindole alkaloids) isolated from

freshly extract leaves [1,2] or shoot cultures of *Catharanthus roseus* (Apocynaceae). In solution A. is easily oxidized by air to other dimeric alkaloids, e. g., *vinblastine, leurosine, leurosidine, catharine. A. may be a key biosynthetic intermediate for many bisindoles of the vinblastine type [3] and be formed biogenetically by enzymatic coupling of *catharanthine and *vindoline.
Lit.: [1] Phytochemistry **27**, 1713 (1988). [2] J. Am. Chem. Soc. **100**, 6253 (1978). [3] J. Chem. Soc., Chem. Commun. **1979**, 582. *gen.:* Manske **37**, 11. – *[HS 2939 90; CAS 38390-45-3]*

Anigorufone, anigozanthin see phenalenones.

Animal toxins (zootoxins). Collective name for poisons secreted by animals, mostly from specific venom glands, for catching prey or for defence and which can have painful or even fatal effects on humans. They occur in almost all classes of animals. Animal poisons find applications especially in neurochemistry and as therapeutics for neuralgic diseases as well as in the form of *arrow poisons. *Examples* of animal toxins are fish poisons, *spider venoms, *scorpion venoms, *snake venoms, *amphibian venoms, toxins of cnidaria (jelly fish, sea anemones), insects (*ants, bees, and wasps, see bee venom, wasp venom), mollusks and *echinodermata (see echinoderm toxins). Some animal toxins are taken up with food, e. g., the *shellfish poisons originate from algae. Both the groups of animals producing toxins and the chemical structures of the produced toxins are highly diverse. They include peptides and proteins, but also alkaloids, steroids, terpenes, heterocyclic compounds, saponins, biogenic amines, and aliphatic compounds. According to their modes of action they can be classified as nerve, cardiac, muscle, or blood poisons as well as hallucinogenic agents.
Lit.: Tu, Handbook of Natural Toxins, Vol. 2, 3 & 5, New York: Dekker 1982 – 1991.

***p*-Anisaldehyde** (4-methoxybenzaldehyde).

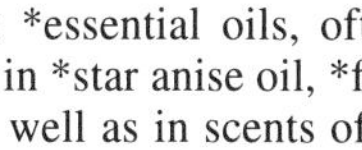

$C_8H_8O_2$, M_R 136.15. Colorless to yellowish oil with an odor like hawthorn and mimosa flowers, poorly soluble in water, readily soluble in ethanol and other organic solvents, D. 1.123, mp. +2 °C, bp. 248 °C; oxidized at air, sensitive to light and alkalis, LD_{50} (rat oral) 3210 mg/kg.
Occurrence: In *essential oils, often together with *anethole, e. g., in *star anise oil, *fennel oil, and Tahiti *vanilla, as well as in scents of acacia and wallflower blossoms. Aroma substance of *Russula laurocerasi* (Basidiomycetes).
Use: In flower perfumes and aromas, luster additive in galvanization, and as intermediate for pharmaceuticals and UV-absorbers (*p*-methoxycinnamates) [1].
Lit.: [1] Ullmann (5.) **A 24**, 233.
gen.: Bauer et al. (2.), p. 104 ■ Bedoukian (3.), p. 46 f. ■ Fenaroli (2.), **2**, 336. – *Synthesis:* DE.P 2 848 397 (1978), BASF ■ EP.A 330036 (1989), Bayer AG. – *[HS 2912 49; CAS 123-11-5]*

Anisatin. (–)-*A.*, $C_{15}H_{20}O_8$, M_R 328.31, mp. 227 – 228 °C (H_2O), $[\alpha]_D^{20}$ – 28° (dioxan) and *neoanisatin*, $C_{15}H_{20}O_7$, M_R 312.31, mp. 237 – 238° (H_2O), $[\alpha]_D^{25}$ – 25° (dioxan) are sesquiterpenoid spasmodic toxins

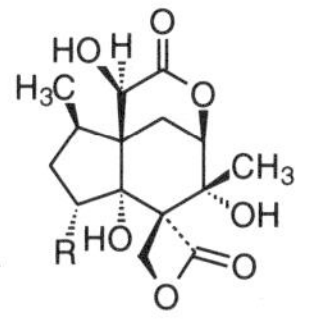

R = OH : Anisatin
R = H : Neoanisatin

from the Japanese star anise (*Illicium anisatum*, Japanese name Shikimi). They are antagonists of the neurotransmitter GABA (*4-aminobutanoic acid). With an LD_{100} (mouse i. p.) of 1 mg/kg they are among the strongest known plant toxins.
Lit.: Aust. J. Chem. **41**, 1071 (1988) ■ Beilstein EV **19/7**, 92 ■ J. Am. Chem. Soc. **112**, 9001 (1990) (anisatin) ■ Chem. Lett. **1991**, 639 (synth. of neoanisatin) ■ Synlett **1997**, 1387 (synthesis of anisatin). – *[CAS 5230-87-5 (A.); 15589-82-9 (neoanisatin)]*

Anisomycin (flagicidin).

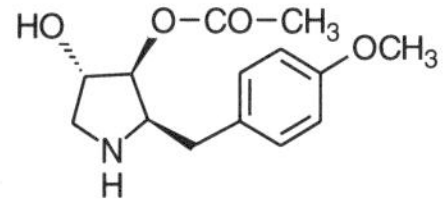

$C_{14}H_{19}NO_4$, M_R 265.31, needles, mp. 144 – 145 °C, $[\alpha]_D$ – 30° (CH_3OH). Isolated from *Streptomyces griseolus*. A. has antibacterial and antifungal properties and is active against tumors and protozoa. It inhibits protein biosynthesis.
Lit.: Heterocycles **41**, 2423 (1995) ■ J. Antibiot. **46**, 1300 (1993) (derivatives) ■ Jiménez & Vázques, in Hahn (ed.), Antibiotics, vol. V-2, p. 1 – 19, Berlin: Springer 1979. – *Synthesis:* Bull. Kocran Chem. Soc. **20**, 321 (1999) ■ Heterocycles **41**, 2423 (1995) ■ J. Org. Chem. **57**, 1316 – 1320 (1992) ■ Synthesis **1999**, 1473 ■ Tetrahedron: Asymmetry **6**, 2907 (1995) ■ Tetrahedron Lett. **40**, 8223 (1999). – *[HS 2941 90; CAS 22862-76-6]*

Annastatin see bestatin.

Annatto see bixin.

Annonaceous acetogenins (Annonins). A class of compounds from Annonaceae (tropical bushes and trees growing mainly in central and South America). The Annonaceae are one of the largest families of magnolia-like plants with 120 genera and ca. 2000 species, including *Annona*, *Rollinia*, *Uvaria* and *Goniothalamus*. The class of mono- and ditetrahydrofuran fatty acid γ-lactones, known since 1982, now encompasses over 90 representatives (examples, see table, p. 38). Three different structural types are distinguished according to the number and positions of the tetrahydrofuran rings: mono- and neighboring or distant ditetrahydrofurans. The A. exhibit pronounced cytotoxic (especially against P338, L 1210, and KB cells) and insecticidal properties. Because of their toxicity, however, they are not used in plant protection [1]. Fruits of some Annonaceae, including Cherimoya, are marketed. A. are derivatives of higher, unsaturated fatty acids.
Activity: The A. inhibit electron transport in mitochondria by decreasing the activity of the NADH-ubiquinone reductase (complex I) [2] and they inhibit plasma membrane NADH-oxidase.
Lit.: [1] Can. J. Chem. **72**, 287 – 293, 1533 – 1536 (1994); Helv. Chim. Acta **76**, 2433 – 2444 (1993); Heterocycles **41**, 229 – 236 (1995); Nat. Prod. Rep. **12**, 9 f. (1995); J. Chem. Soc., Perkin Trans. 1 **1994**, 1975 – 1981; J. Med. Chem. **37**, 1971 – 1976 (1994); J. Nat. Prod. **56**, 1095 – 1100, 1688 – 1694 (1993); **57**,

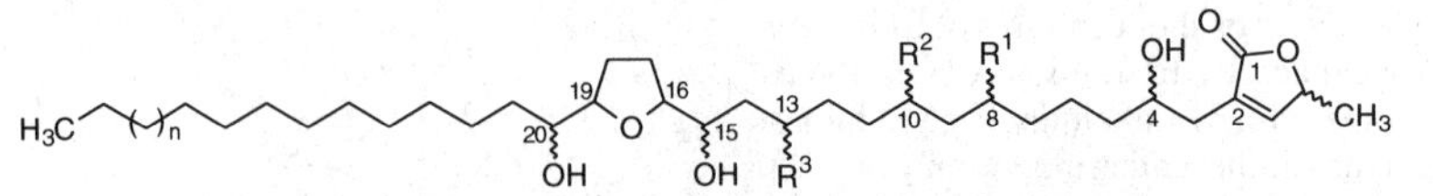

n = 1; $R^1 = R^3 = OH, R^2 = H$: Annomonicin
n = 1; $R^1 = R^3 = H, R^2 = OH$: Annonacin
n = 4; $R^1 = R^2 = R^3 = H$;(15*R*,20 *R*): Annonastatin

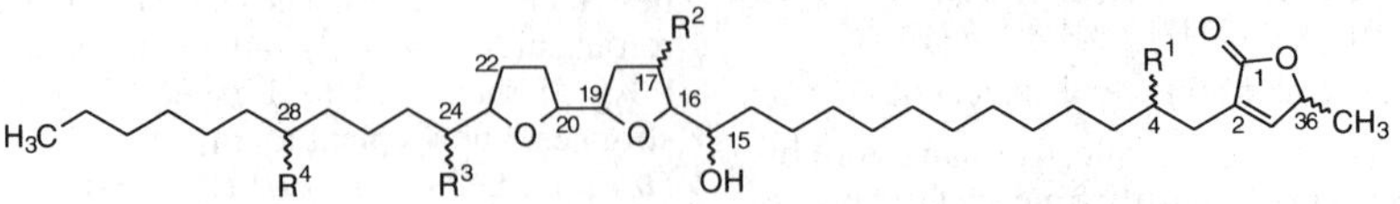

$R^1 = R^2 = H, R^3 = R^4 = OH$; (15*R*,16*R*,19*R*,20*R*,23*R*,24*S*,28*S*,36*S*) : Annonin I
$R^1 = R^3 = H, R^2 = R^4 = OH$; (15*R*,16*S*,19*R*,20*R*,23*R*,24*S*,28*S*,36*S*) : Annonin XVI
$R^1 = R^3 = OH, R^2 = R^4 = H$: Asimicin
$R^1 = R^2 = R^4 = H, R^3 = O{-}CO{-}CH_3$;(15*S*,16*S*,19*S*,20*S*,23*S*,24*R*,36*S*) :Uvaricin

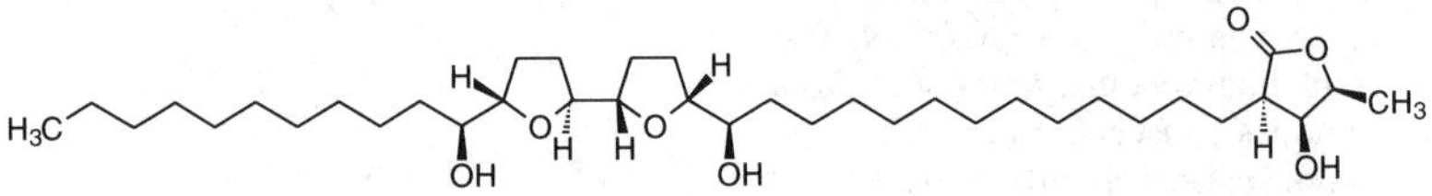

Laherradurin

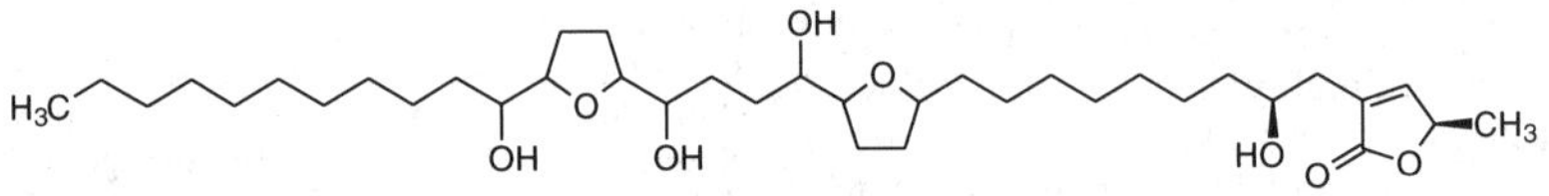

Bullatalicin

Table: Data and occurrence of Annonaceous acetogenins.

compound	molecular formula	M_R	mp. [°C]	optical rotation	occurrence	CAS
Annomonicin	$C_{35}H_{64}O_8$	612.89	45–48	$[\alpha]_D^{20}$ +4° (CH_3OH)	*Annona montana*	128741-22-0
Annonacin	$C_{35}H_{64}O_7$	596.89	wax	–	*Annona densicoma, Goniothalamus giganteus*	111035-65-5
Annonastatin	$C_{38}H_{70}O_6$	622.97	amorphous	$[\alpha]_D^{25}$ +15° (CH_2Cl_2)	*Annona squamosa*	129138-50-7
Annonin I (Squamocin)	$C_{37}H_{66}O_7$	622.93	–	$[\alpha]_D^{22}$ +14.9° (CH_3OH)	*Annona squamosa, Uvaria narum*	120298-30-8
Asimicin[a]	$C_{37}H_{66}O_7$	622.93			*Annona squamosa,*	102989-24-2
Rolliniastatin 1			wax 81–83	$[\alpha]_D$ +25.2° (CH_2Cl_2)	*Asimina triloba, Rollinia mucosa*	111056-97-4
Rolliniastatin 2			wax 73–76	$[\alpha]_D^{27}$ +5.3° ($CHCl_3$)		121917-13-3
Bullatacin			cryst. 69–70	$[\alpha]_D^{25}$ +13° ($CHCl_3$)		123123-32-0
Uvaricin[2]	$C_{39}H_{68}O_7$	648.96	wax ~25°	$[\alpha]_D^{25}$ +11.3° (CH_3OH)	*Uvaria acuminata*	82064-83-3
Laherradurin	$C_{37}H_{68}O_7$	624.94	85–86		*Annona cherimola*	125276-75-7
Bullatalicin	$C_{37}H_{66}O_8$	638.93	amorphous 120–121	$[\alpha]_D$ +13.25° (C_2H_5OH)	*Annona bullata*	125882-64-6

[a] mixture of several stereoisomers, out of which 3 have been determined.

486–493, 911–915 (1994); J. Org. Chem. **59**, 1598f., 3472–3479 (1994); Phytochemistry **33**, 1065–1073 (1993); **35**, 1325–1329 (1994); Planta Med. **56**, 312–319 (1990); Synthesis **1995**, 1447–1464; Tetrahedron **50**, 8479–8490 (1994); Tetrahedron Lett. **34**, 5043–5046 (1993); **35**, 157–160; 2517–2520 (1994); **37**, 7001 (1996). [2]J. Org. Chem. **47**, 3151–3153 (1982); **63**, 5863 (1998).

gen.: Acc. Chem. Res. **28**, 359–365 (1995) ▪ J. Nat. Prod. **53**, 237–278 (1990); **62**, 504–540 (1999) (review) ▪ Phytochemistry **42**, 253–271 (1996); **48**, 1087–1117 (1998) ▪ Zechmeister **70**, 81–305. – *Absolute configuration:* Atta-ur-Rahman **17**, 251–282. – *Biological activity:* J. Med. Chem. **40**, 2102 (1997). – *Pharmacology:* Biochem. J. **301**, 161–167 (1994) ▪ Cancer Chemother. Pharmacol. **34**, 166–170 (1994) ▪ Eur. J.

Biochem. **219**, 691–698 (1994). – *Synthesis reviews:* Atta-ur-Rahman **18**, 193–228 ▪ Isr. J. Chem. **31**, 97 (1997) ▪ J. Nat. Prod. **62**, 504 (1999) ▪ J. Org. Chem. **62**, 5989, 5996 (1997); **63**, 5863 (1998) (uvaricin) ▪ Nat. Prod. Rep. **13**, 275–306 (1996) ▪ Phytochemistry **42**, 253–271 (1996); **48**, 1087–1117 (1998).

Annonalide see pimaranes.

Annonins see Annonaceous acetogenins.

Annotinine.

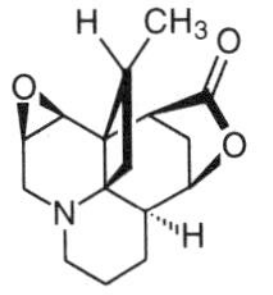

Annotinine

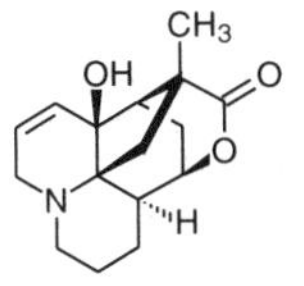

Annotine

A., $C_{16}H_{21}NO_3$, M_R 275.35, mp. 232 °C, $[\alpha]_D$ −5.3° ($CHCl_3$) and *annotine*, $C_{16}H_{21}NO_3$, M_R 275.35, mp. 174.5–176 °C, $[\alpha]_D^{21}$ −114° ($CHCl_3$) are the main alkaloids of *Lycopodium annotinum* (forest club moss).

Lit.: Beilstein E V **27/14**, 45 (annotine) ▪ Manske **10**, 305 ff.; **14**, 347 ff. – *Absolute configuration:* Tetrahedron Lett. **1961**, 187; **1969**, 1307. – *Structure:* Can. J. Chem. **67**, 1765 (1989) (annotine) ▪ J. Am Chem. Soc. **78**, 2867 (1956) ▪ Tetrahedron **4**, 87 (1956). – *Synthesis:* Can. J. Chem. **47**, 433 (1969). – *[CAS 559-49-9 (annotinine); 5096-59-3 (annotine)]*

Anonaine see aporphine alkaloids.

Ansamitocins see maytansinoids.

Ansamycins. Term coined by Prelog from Latin ansa = handle for antibiotically active, macrocyclic natural products with an aromatic/quinonoid structural unit bridged by an aliphatic chain. A. are macrolactams, they are classified according to the planar structural unit (benzoquinone/hydroquinone, naphthoquinone) and the type and length of the aliphatic bridge. Aminobenzoic acids are the biosynthetic precursors of the planar unit. *Examples:* geldanamycin, *rifamycins, *ansatrienins, *naphthomycins.

Lit.: Angew. Chem. Int. Ed. Engl. **28**, 146 (1989) ▪ Hager (5.) **9**, 515 ▪ Helv. Chim. Acta **56**, 2279–2287 (1973) ▪ Zechmeister **33**, 231–307.

Ansatrienins (mycotrienins).

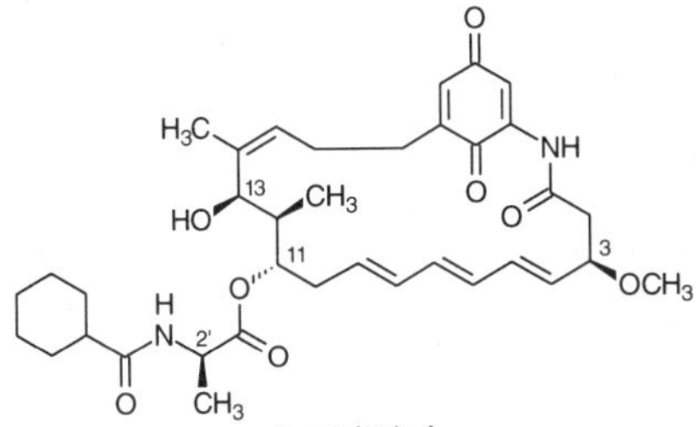

Ansatrienin A

Antifungally active antibiotics of the *ansamycin type from *streptomycetes (e.g., *Streptomyces collinus*). *A. A:* $C_{36}H_{48}N_2O_8$, M_R 636.79, yellow powder, decomp. above 170 °C, $[\alpha]_D$ +110° ($CHCl_3$). *A. B* (hydroquinone of A. A): $C_{36}H_{50}N_2O_8$, M_R 638.80, mp. 168 °C, $[\alpha]_D$ +334° ($CHCl_3$). The ansa skeleton is acylated with *N*-cyclohexylcarbonyl-D-alanine at an OH group. The A. differ in their side chains. The related trienomycins contain a phenol ring instead of a benzoquinone.

Lit.: J. Am. Chem. Soc. **120**, 4123 (1998) [synthesis (+)-mycotrienin] ▪ J. Antibiot. **35**, 1460–1474 (1982); **36**, 187–189 (1983); **38**, 799 (1985) ▪ J. Org. Chem. **62**, 8290 (1997) (synthesis) ▪ Tetrahedron Lett. **23**, 59–62 (1982); **32**, 841–844 (1991). – *Biosynthesis:* Angew. Chem. Int. Ed. Engl. **28**, 146–178 (1989) ▪ J. Am. Chem. Soc. **115**, 5254–5266 (1993) ▪ Pure Appl. Chem. **61**, 485 (1989). – *[HS 2941 90; CAS 80111-47-3 (A. A); 80111-48-4 (A. B)]*

Anserine (*N*-*β*-alanyl-3-methyl-L-histidine, 3-ethylcarnosine).

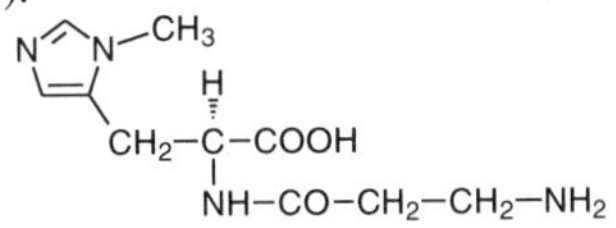

$C_{10}H_{16}N_4O_3$, M_R 240.26, mp. 240–242 °C (decomp.), hygroscopic, $[\alpha]_D^{23}$ +11.4° (H_2O), pK_a 2.64, 7.04 and 9.49. Dipeptide from N^3-methyl-L-histidine and *β*-alanine in muscles of geese, other birds, and crocodiles.

Lit.: Adv. Enzyme Regul. **30**, 175 (1990) (review) ▪ Beilstein E V **25/16**, 430 ▪ J. Chromatogr. **432**, 315 (1988) ▪ Greenstein & Winitz, Chemistry of the Amino Acids, vol. 2, p. 1521 f., New York: Wiley 1961. – *[CAS 584-85-0 (L-form)]*

Antamanide.

L-Ala—L-Phe—L-Phe—L-Pro—L-Pro
L-Pro—L-Pro—L-Val—L-Phe—L-Phe

Antitoxic cyclodecapeptide from the death cup (*Amanita phalloides*), antidote for phalloidin (see phallotoxins), $C_{64}H_{78}N_{10}O_{10}$, M_R 1147.38, mp. 172 °C, $[\alpha]_D$ −180° (CH_3OH).

Lit.: An. Quim., Ser. C **80**, 118–122 (1984) (synthesis) ▪ Chem. Pharm. Bull. **36**, 3196 (1988); **37**, 1684 (1989) (synthesis) ▪ CRC Crit. Rev. Biochem. **5**, 185–260 (1978) (activity) ▪ Justus Liebigs Ann. Chem. **1989**, 903, 913 ▪ J. Am. Chem. Soc. **112**, 2908 (1990) (structure) ▪ see also amanitins. – *[CAS 16898-32-1]*

Antheraxanthin (5,6-epoxyzeaxanthin, 5,6-epoxy-5,6-dihydro-*β*,*β*-carotene-3,3′-diol).

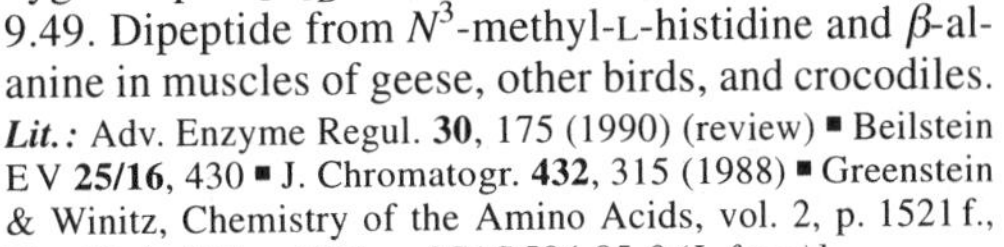

$C_{40}H_{56}O_3$, M_R 584.88, mp. 197 °C (9*E*-isomer) and 108 °C (9*Z*-isomer). Both isomers belong to the xanthophylls and occur in *Lilium* species (Liliaceae) as well as in *Capsicum annuum* (Solanaceae). The *E*-isomer has been isolated from *Euonymus europaeus* (Celastraceae), *C. annuum*, and others.

Lit.: Straub, Key to Carotinoids (2.), p. 231, Basel: Birkhäuser 1987 ▪ Pfander (2.), p. 97. – *[CAS 640-03-9 (all-E); 68831-78-7 (9Z)]*

Antheridiol. Steroid lactone formed by the aquatic fungus *Achlya bisexualis* (*A. heterosexualis*) and isolated in 1971 as the first plant sex hormone. A. is secreted by female mycelia and induces formation and development of male sex organs, so-called antheridia. In addition, A. induces the antheridia to form a second type of sex hormone, the oogoniols, which in turn promote development of oogonia, the egg-bearing gametes, in female cells. Fertilization then occurs by direct contact of an antheridium with an oogonium [1]. The oogoniols are a mixture of various esters in which the 3*β*-hydroxy group of the *sterol derivative *oogoniol* is esterified by acetic, propanoic, or 2-methylpropanoic acid. Both A. and oogoniol have the 24-ethylcholestane skeleton.

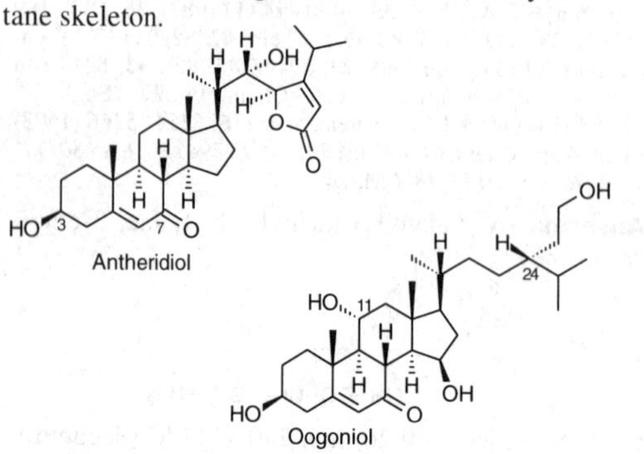

Antheridiol [(22*S*,23*R*)-3β,22,23-trihydroxy-7-oxo-stigmasta-5,24(28)-diene-29-carboxylic acid-23-lactone], $C_{29}H_{42}O_5$, M_R 470.65, mp. 250–255 °C. *Oogoniol* (3β,11α,15β,29-tetrahydroxyporiferast-5-en-7-one), $C_{29}H_{48}O_5$, M_R 476.70, mp. 165–167 °C (from CH_3OH) or 181–183 °C (from C_2H_5OH).

Lit.: [1] J. Biol. Chem. **259**, 15324 (1984).
gen.: J. Org. Chem. **39**, 669 (1974) (synthesis) ▪ Zeelen, p. 274–275. – *[CAS 22263-79-2 (A.); 66786-17-2 (oogoniol)]*

Anthocerodiazonin.

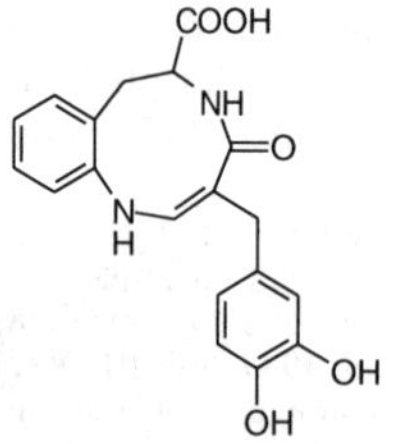

$C_{19}H_{18}N_2O_5$, M_R 354.36, colorless oil, $[\alpha]_D^{20}$ +7.3°, well soluble in methanol. An *alkaloid formed by the moss *Anthoceros agrestis* (*in vitro* cultures on agar nutrient medium and in suspension cultures).

Lit.: Phytochemistry **37**, 899 (1994). – *[HS 2938 90; CAS 159903-66-9]*

Anthocyanins (Marquart, 1835: Greek anthos = flower and kyanos = blue color). A. belong to the group of *flavonoids and occur ubiquitously in flowers and fruits as well as often in shoot axes and leaves of Angiosperms (exceptions see betalains in nine families of the Caryophyllales). A. give colors from blue through various shades to red. The content of A. is often low but can amount to up to a third of the dry weight, e.g., in blue-black pansies (*Viola tricolor*, Violaceae).
The A. are glycosides, the actual chromophores of which are their aglycones, the anthocyanidins. A common feature of all known anthocyanidins is the skeleton of a C-4′-hydroxylated 2-phenylchromene, existing primarily as a flavylium cation. Secondary structures such as quinoid bases, and carbinol or chalcone pseudobases occur in aqueous solution in dependence on pH value.
In weakly acid or neutral aqueous solutions the carbinol and chalcone pseudobases exist in a ratio of 4 : 1 [1]. A. are derived from three basic structures: *pelargonidin, *cyanidin, and *delphinidin, which differ in substitution on the B ring.

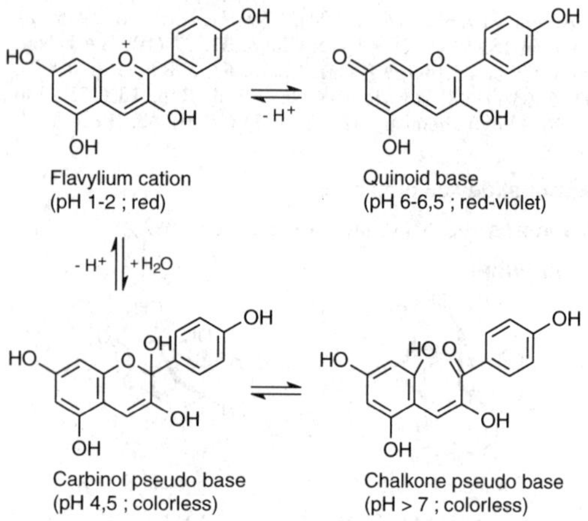

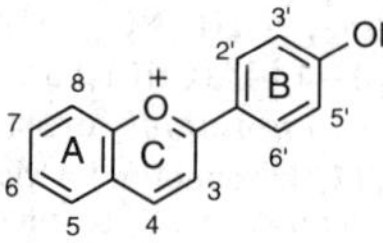

Table: Substitution pattern of Anthocyanidins.

Anthocyanidin	substitution pattern C-3	C-5	C-6	C-7	C-3′	C-5′
basic structures						
*Pelargonidin	OH	OH	H	OH	H	H
*Cyanidin	OH	OH	H	OH	OH	H
*Delphinidin	OH	OH	H	OH	OH	OH
methoxylated structures						
*Peonidin	OH	OH	H	OH	OCH_3	H
*Petunidin	OH	OH	H	OH	OCH_3	OH
*Malvidin	OH	OH	H	OH	OCH_3	OCH_3
3-deoxy structures						
*Apigeninidin	H	OH	H	OH	H	H
Luteolinidin	H	OH	H	OH	OH	H
Tricetinidin	H	OH	H	OH	OH	OH

The variety of the up to 300 reported A. arises through multiple degrees of glycosidation with various hexoses and pentoses as well as from additional acylation with aliphatic and aromatic acids. Further structural differences result from simple conjugation, e.g., glycosidation at C-3 of the pyran ring with formation of monosides and further glycosidation, e.g., at C-5 of the A ring with formation of 3,5-diglycosides (glucosides, galactosides, rhamnosides, arabinosides, and xylosides). Complex aromatic polyacyl conjugates, with long side chains can form folded structures in dependence on the nature and position of the acyl group (hydroxycinnamic acid). A well-known example is "heavenly blue anthocyanin" (HBA) from *Ipomoea tricolor* (Convolvulaceae). HBA bears a branched group (1,2-glycosidic) at C-3 of peonidins with 5 glucose and 3 caffeic acid units as well as a glucose group at C-5. The caffeic acid groups have ester bonds at C-6 of the glucose-units and a glycosidic bond with one of the phenolic hydroxy groups.

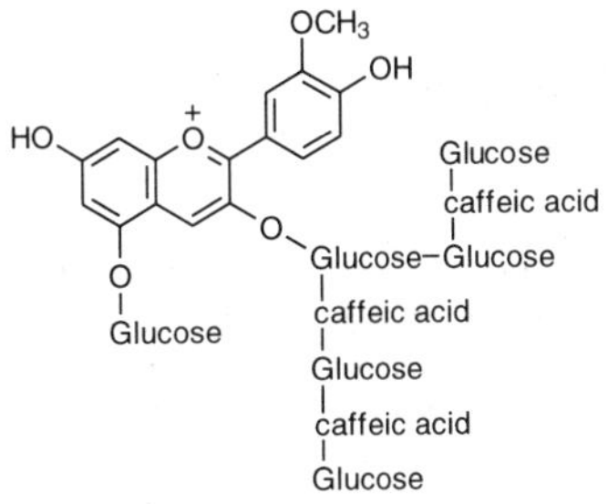

"Heavenly Blue Anthocyanin" (HBA)

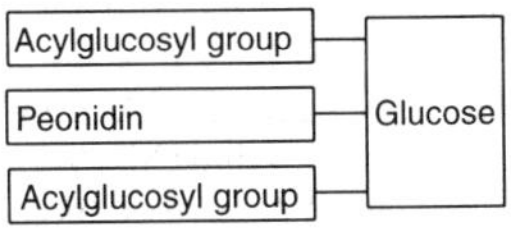

model of "Sandwich Stacking"

The folded structure of HBA results in a so-called "sandwich stacking"[2] in which the caffeic acids groups are arranged in parallel to peonidin on account of hydrophobic interactions. This is the reason for the phenomenon of the intramolecular copigmentation[3], reflected in a marked deepening of the color (bathochromic shift of light absorption). Furthermore, this stacking protects the anthocyanidin from tautomerization and hydration at the average pH values between 4 and 6 in the plant vacuoles. At pH values above 3 the colorless quinoid base forms and can lead to pyran ring opening through addition of water at C-2. The pH-dependent color behavior of the A. is a conspicuous indication for tautomerism of the anthocyanidin structure. In acidic aqueous solution, isolated A. exhibit a red to reddish violet color which turns to a blue to bluish-green color upon addition of weak alkali. Stabilization of the anthocyanidin primary structure and protection from water addition in A. that are not protected by intramolecular stacking is achieved either by self-association or by intermolecular stacking (intermolecular copigmentation) with other phenylpropane derivatives, such as, e.g., flavones or hydroxycinnamic acid conjugates.

Use: A. play a major role in food coloring (E 163). The problem of their instability could be solved by use of the considerably more stable polyacylated A.

Biosynthesis: The biosynthesis of anthocyanidins proceeds through proanthocyanidins (*leucoanthocyanidins). The enzymes participating in the transformation of the proanthocyanidins to the A. have not yet been identified. However, it is assumed that hydroxylation at C-2, catalyzed by a dioxygenase, and subsequent dehydratase reactions give the flavylium structure. The hydroxylation reaction has been confirmed by molecular genetic studies. The glycosidations of anthocyanidins are catalyzed by specific nucleotide sugar-dependent glycosyltransferases. In glucosidation reactions UDP-glucose serves as glycosyl donor. In acylation reactions coenzyme A thioesters of aliphatic and aromatic acids are accepted. In esterification reactions with hydroxycinnamic acids the corresponding 1-*O*-acylglucosides can serve as acyl donors[4], as has been demonstrated for numerous other acylation reactions[5,6].

Lit.: [1]Can. J. Chem. **68**, 775 (1990). [2]Zechmeister **52**, 113. [3]Biochem. J. **25**, 1687 (1931). [4]Planta **186**, 582 (1992). [5]Harborne & Lea (eds.), Methods in Plant Biochemistry, vol. 9, p. 45–97, London: Academic Press 1993. [6]Bot. Acta **105**, 146 (1992).

gen.: Harborne (1994), p. 1–22, 499–535, 565–588 ▪ J. Chem. Soc., Perkin Trans. 1 **1996**, 735 (A. in wine) ▪ Markakis (ed.), Anthocyanins as Food Colors, New York: Academic Press 1982 (use as food colorants) ▪ Zechmeister **52**, 113–158.

Anthracyclines.

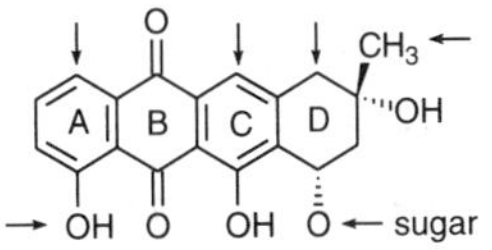

core structure of Anthracyclines

Name for highly active antibiotic and cytostatic *O*-glycosides from which, upon hydrolysis, the linear anellated tetracyclic anthraquinone derivative, *anthracyclinone*, is formed. The basic skeleton, a *polyketide, can be changed in several positions (see arrows in the formula). A. are only formed by *actinomycetes, about 200 representatives have been isolated and described, including the clinically used antitumor agents *adriamycin, *daunorubicin, and *aclarubicin. The activity of A. results from their ability to insert the planar chromophore part (rings A, B, and C) between two base pairs of the DNA double helix (intercalation) while the D ring and the sugar units hold the molecule in this position. The consequences are an inhibition of DNA and RNA polymerases as well as topoisomerase II.

Lit.: Angew. Chem. Int. Ed. Engl. **25**, 790 (1986) (synthesis) ▪ El Khadem (ed.), Anthracycline Antibiotics, New York: Academic Press 1982 ▪ Hutchinson, in: Vining & Stuttard (eds.), Genetics and Biochemistry of Antibiotic Production, p. 331–357, Boston: Butterworth-Heinemann 1995 ▪ Lown (ed.), Anthracycline and Anthracenedione-based Anticancer Agents, Amsterdam: Elsevier 1988 ▪ Priebe (ed.), Anthracycline Antibiotics, ACS Sympos. Ser. **574**, Washington: ACS 1995 ▪ Zechmeister **21**, 121–182. – *[HS 2941 30; 2941 90]*

Anthracyclinone see rhodomycins.

Anthramycins.

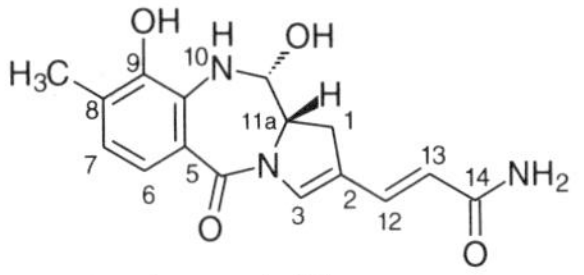

Anthramycin (**1**)

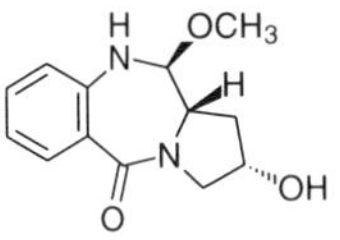

Abbeymycin (**2**)

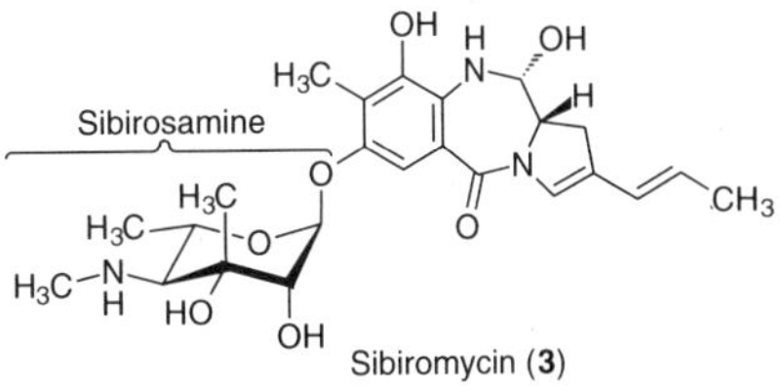

Sibiromycin (**3**)

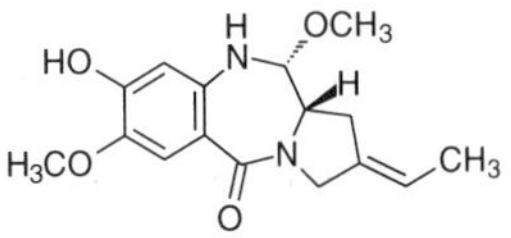

Tomaymycin (**4**)

Table: Data of Anthramycins.

no.	molecular formula	M_R	mp. [°C]	$[\alpha]_D$	CAS
1	$C_{16}H_{17}N_3O_4$	315.33	188–194	+930° (DMF)	4803-27-4
2	$C_{13}H_{16}N_2O_3$	248.26	142–144	+303° (H_2O)	108073-64-9
3	$C_{24}H_{33}N_3O_7$	475.50	>120 (decomp.)	+525° (DMF)	12684-33-2
4	$C_{16}H_{20}N_2O_4$	304.32	145–146	+423° (pyridine)	35050-55-6

Pyrrolo [2,1-*c*] [1,4] benzodiazepine antibiotics with antitumor activity from *Streptomyces* species. The main representative of the group is *anthramycin:* pale yellow prisms, soluble in hot methanol and water, epimerizes in solution, active against Gram-positive bacteria and tumors, DNA complexing activity. A. are produced by *Streptomyces refuineus* and *S. spadicogriseus*, for isolation, see *Lit.*[1], biosynthesis, see *Lit.*[2], synthesis, see *Lit.*[3], total synthesis, see *Lit.*[4]; for *abbeymicin*, see *Lit.*[5], *tomaymycin*, see *Lit.*[6]

Lit.: [1] J. Am. Chem. Soc. **87**, 5791 (1968). [2] Tetrahedron Lett. **1976**, 1419. [3] J. Chem. Soc., Chem. Commun. **1982**, 741. [4] J. Am. Chem. Soc. **111**, 5417–5424 (1989). [5] J. Antibiot. **40**, 145 (1987). [6] J. Am. Chem. Soc. **110**, 2992 (1988).

gen.: Foye, Cancer Chemotherapeutic Agents, ACS Professional Reference Book, Washington DC 1995 ▪ J. Antibiot. **30**, 349 (1997) ▪ J. Org. Chem. **53**, 482–487 (1988) ▪ Pharm. Res. **1984**, 52. – *[HS 2941 90]*

Anthranilic acid (2-aminobenzoic acid).

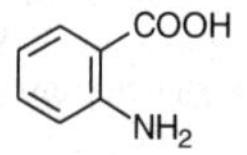

$C_7H_7NO_2$, M_R 137.14. Colorless to pale yellow, blue-fluorescing, sweet-tasting plates, D. 1.412, mp. 146–147 °C, pK_{a1} 1.97, pK_{a2} 4.79 (25 °C), sublimes without decomposition; soluble in water, ethanol, ether, chloroform, hot benzene. On stronger heating it decomposes to aniline and CO_2. Several crystal modifications are known. LD_{50} (rat p.o.) 4450 mg/kg, (mouse i.p.) 2500 mg/kg. A. is an intermediate in the biosynthesis of *tryptophan and occurs in bacteria and fungi. Biosynthesis from *chorismic acid and glutamate.

Use: A. currently serves as important starting material for the synthesis of A. esters, important as aroma substances, indigo, as well as pharmaceuticals and plant protection agents (acarizides).

Lit.: Acta Crystallogr., Sect. B **33**, 3205 f. (1977) (crystal structure) ▪ Beilstein E IV **4**, 1004 ▪ Nat. Prod. Rep. **11**, 173–203 (1994) ▪ Ullmann (5.) **A 3**, 565. – *Toxicology:* Sax (8.), API 500 ▪ see also 4-aminobenzoic acid. – *[HS 2922 49]*

Anthranoids. Term applied to all naturally occurring compounds derived from anthracene[1]; these include (in free or glycosidic form) *preanthraquinones, anthrones (*chrysarobin) and bianthrones (*sennosides), emodines (1,8-dihydroxyanthraquinones, *chrysophanol, *aloe-emodin, *rhein, *emodin, *endocrocin, *physcion, *frangulins) and anthraquinones in the strict sense (*tectoquinone, *alizarin(e), *purpurin, *islandicin, *morindone, *carminic, *kermesic and *laccaic acids).

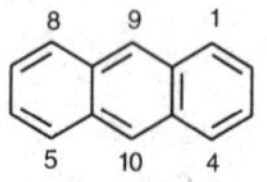

Anthracene

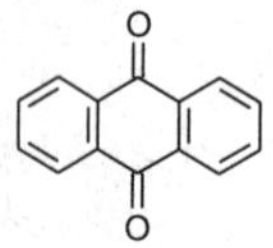

Anthraquinone

On account of the presence of the chromogenic tricyclic ring system in part with quinone structure as well as auxochromic hydroxy groups the A. are yellow, orange, or red to violet pigments. A. are isolated from higher fungi, especially from certain *Cortinarius* and *Dermocybe* species (*Dermocybe pigments) as well as from *Penicillium* and *Aspergillus* species and lichens, on the other hand from higher plants and in few cases from insects. Naturally occurring A. can be divided into two groups according to their biosynthetic routes. The first group contains all A. formed biosynthetically from acetate by the polyketide route. *Polyketides (*acetogenins) in general show a regular pattern of oxygen-substituted and oxygen-free C atoms. The biosynthesis begins with the starter molecule *acetyl-CoA which reacts with 7 mol. malonyl-CoA with loss of the free carboxy group. The particular spatial arrangements of the primarily formed polyoxo acids initially give rise to preanthraquinones such as *atrochrysone and anthrone derivatives by aldol condensation reactions, which then undergo further modifying reactions[2,3]. Acetogenic A., which occur as glycosides, are particularly frequent in plants of the families Rhamnaceae (*Rhamnus* spp.), Polygonaceae (*Rheum* spp.), and Fabaceae (*Cassia* spp.). They are used as phytopharmaceuticals in form of extracts or teas on account of their laxative actions[4,5].

A. of the second group are formed from *shikimic acid, 2-oxoglutaric acid, and mevalonic acid lactone. Isochorismic acid, formed from shikimic acid via chorismic acid, first reacts with 4-hydroxybutyryl-2′-thiamine diphosphate ("activated" succinyl semialdehyde, from 2-oxoglutaric acid) to give 2-succinylbenzoic acid, which is activated by formation of the CoA ester on the succinyl unit. This reactive intermediate reacts with the dimethyl allyl pyrophosphate formed from mevalonic acid lactone through double cyclization to give the tricyclic system of the A. which, in turn, is further modified by, e.g., oxidation reactions[6,7].

Members of the second group occur frequently in roots, stems, bark, and wood of plants of the family (Rubiaceae, *madder) and related families, e.g. Bignoniaceae, Verbenaceae, Liliaceae, and Scrophulariaceae. Extracts of madder root are used for kidney stone problems on account of the Ca^{2+} and Mg^{2+} complexing properties of their active ingredients[8]. Furthermore, hypotensive[9] and antileukemic[10,11] activities of Rubiaceae anthraquinones have been reported. Important A. formed by insects are *carminic acid, *kermesic acid, and *laccaic acids, which all have, like alizarin and related compounds from madder root, historical significance as dyes.

Numerous plant cell cultures that produce A., sometimes in considerable amounts, have been described[12].

Lit.: [1] Turner **1**, 157–166; Turner **2**, 140–147. [2] Luckner (3.), p. 165, 177. [3] J. Chem Soc., Chem. Commun. **1969**, 210. [4] Banks et al., in: Avery (ed.), Drug Treatment (2.), p. 712, Syd-

ney-New York: ADIS-Press 1980. [5]Fingl, in Goodman et al. (eds.), Pharmacological Basis of Therapeutics (6.), p. 1007, New York: MacMillan Publ. Co., Inc. 1980. [6]Zechmeister **49**, 79–149. [7]Luckner (3.), p. 330. [8]Pharmazie **29**, 478 (1974). [9]J. Inst. Chem., Calcutta **54**, 22 (1982). [10]J. Nat. Prod. **45**, 206 (1982). [11]Phytochemistry **23**, 1723 (1984). [12]Suzuki & Matsumoto, in: Bajaj (eds.), Biotechnology in Agriculture and Forestry, vol. 4, Medicinal and Aromatic Plants I, p. 237–250, Berlin: Springer 1988.
gen.: Naturwissenschaften **58**, 585 (1971); **65**, 439 (1978).

Anthraquinones (from fungi). Over 40 monomeric A. derivatives are known from ascomycetes and imperfect fungi; and there are several dimeric compounds (e.g., *luteoskyrin, *rugulosin, *skyrin). Many A. have also been isolated from plants, lichens, basidiomycetes, bacteria (streptomycetes), and insects (*anthranoids). Some A. occur both in fungi and in lichens, e.g. *chrysophanol, *emodin, erythroglaucin, and *physcion.

Lit.: Thomson **3**, chap. 3 ▪ Turner **1**, 156–166; **2**, 140–154.

Anthricin see podophyllotoxin.

Antibiotics. Low molecular weight secondary metabolites that are produced by microorganisms and, in low concentrations, inhibit the growth of or kill other microorganisms. The term was introduced by Waksman (1941) (biotikos, Greek = of life). Producers of A. are especially bacteria (e.g., *actinomycetes, ca. 65%) and fungi (e.g., *Aspergillus*, *Penicillium*, *Acremonium*, ca. 25%). The significance of the formation of A. for the producer is uncertain; A. possibly serve to suppress competing organisms. The broader definition of A. also includes semisynthetic derivatives and those produced by biotransformation as well as antibiotically active substances from plants (e.g., anacardic acids) and animals (e.g., *magainins, defensins). Also the *phytoalexins formed by microbial infections in plants are classified as antibiotics. Since the discovery of staphylococci inhibition (*penicillins from *Penicillium notatum*) by A. Fleming (1929), an intensive screening in the industrialized countries after World War 2 led to a large number of new antibiotics. Up to 1996, ca. 9000 A. of microbial origin had been described. In addition, there are ca. 4000 natural products from plants, algae, lichens, marine organisms, and animals with antibiotic properties. The need for new A. will increase further on account of the increasing number of resistant pathogens (e.g., *Mycobacterium tuberculosis, Staphylococcus aureus*) and new indications (e.g., *Helicobacter pylori* and stomach ulcers, *Chlamydia pneumoniae* and atherosclerosis) (at present ca. 300/year). Of the new A. ca. 35–40% are derivatives of previously described antibiotics. In commercial and clinical use are ca. 130 A. produced by fermentation and ca. 50 semisynthetic substances.
Classification: Classification systems are highly arbitrary according to spectrum of action, mechanism of action, producer strain, biosynthetic origin, or chemical structure. *Structural* distinctions are made as: carbohydrate A. (e.g., aminoglycosides like *kanamycins), macrolide A. (e.g., lactones like *erythromycin A), quinone A. (e.g., *anthracyclines), amino acid derived A. (e.g., *β-lactam antibiotics such as penicillin, *cephalosporins or peptides like defensins), N-heterocyclic A. (e.g., nucleosides like *nikkomycins), O-heterocyclic A. (e.g., polyethers like *ionomycin), alicyclic A. (e.g., steroids like viridins), *ansamycins (e.g., *rifamycins), aromatic A. (e.g., *chloramphenicol), aliphatic A. (e.g., *fosfomycin). According to their *spectrum of activity* distinctions are made between the less selective broad-spectrum A. (e.g., chloramphenicol) and the narrow-spectrum A. that are effective solely against few groups of microorganisms. In addition to their use in human medicine (e.g. as antimicrobial principles in chemotherapy), A. are also used in plant protection, food conservation, and animal breeding. Because of the risk of rapid development of resistance the simultaneous use of an A. for human therapy and as animal food additive or in veterinary medicine is no longer allowed and strictly supervised by the regulatory authorities.
Mechanism of action: A. reversibly inhibit the growth of microorganisms (e.g., bacteriostatics, fungistatics) or kill them (e.g., bactericides, fungicides). The most important cell molecular targets of A. are cell wall biosynthesis (e.g., penicillins, cephalosporins, vancomycins, cycloserine), cytoplasm membranes (polyene and polypeptide A.), replication (e.g., novobiocin, mitomycin C), transcription (e.g., actinomycin, rifampicin), translation (e.g., tetracyclines, chloramphenicol, streptomycin, erythromycin, lincomycin, puromycin), respiratory metabolism by decoupling oxidative phosphorylation (e.g., antimycin, valinomycin, *strobilurins), or the transport of iron (*siderophores). Because of their ubiquitous mechanisms of action, many A. are extremely toxic to humans. But also the clinically used A. can have detrimental and in individual cases lethal effects, especially in long-term therapy and high doses (e.g., allergic reactions up to rare anaphylactic shock, kidney-liver damage, neurotoxic reactions, hearing impairments, osteological complications). A further problem of A. therapy is the *development of resistance*. In addition to multiple resistance, cross resistance frequently occurs, i.e., acquired resistance against one A. is accompanied by simultaneous resistance against one or more other A. with the same mechanism of action without the pathogen ever being in contact with the original A. The most frequent mechanisms of resistance are uptake resistance (impairment of transmembrane transport), surrogate enzyme formation (an A.-resistant enzyme takes over the blocked reaction), deactivation (A. is deactivated enzymatically), action resistance (mutated enzyme or ribosomal protein is A.-resistant), over production (overproduced target enzyme can no longer be inhibited or deactivated by the A.). The genetic information for A. resistance is localized on plasmids and is often rapidly transferred to daughter populations on replication, which, especially in hospitals can lead to the much feared resistant pathogens (hospitalism). One factor responsible for the increasing resistance is the wide use of A. (e.g., for mild diseases and for prophylaxis) in earlier years and in countries with liberal prescription practice. The discovery of many A. and their use in biochemistry and molecular biology have contributed much to the elucidation of specific molecular cell functions. The simplest detection of antibiotic activity is the agar diffu-

sion test (inhibition zone test). Production of most natural A. is by batch fermentation (50–300 m^3), in some cases in semicontinuous processes. Concerning β-lactams, semisynthetic compounds are used almost exclusively in therapy. The world-wide production is well over 100 000 t/year with a market value of over 10 billion US \$. The most widely marketed are the cephalosporins, ampicillin, tetracyclines, and quinolone antibiotics. New A. can be obtained by synthetic modification, mutasynthesis, protoplast fusion, and using genetically engineered actinomycetes and filamentous fungi (*hybrid antibiotics).

Lit.: Bycroft, Dictionary of Antibiotics and Related Substances, London: Chapman & Hall 1988 ▪ Chem. Rev. **95**, 1859 (1995) ▪ Curr. Opin. Biotechnol. **4**, 531 (1993) ▪ J. Antibiotics **41**, 1711 (1988) ▪ Mol. Microbiol. **16**, 385 (1995) ▪ Nat. Prod. Rep. **5**, 47 (1988) ▪ Rehm & Reed, Biotechnology, vol. 4, Weinheim: Verl. Chemie 1986 ▪ Science **264**, 373 (1994) ▪ Vandamme, Biotechnology of Industrial Antibiotics, New York: Marcel Dekker 1984.

Anticapsin see chlorotetaine.

Antifreeze proteins. The freezing point of blood serum of some fish species living in the polar regions (e.g., *Trematomus borchgrevinki, Dissostichus mawsoni, Boreogadus saida*) is about –2 °C and thus markedly lower than those of other fish species (–0.6 to –0.8 °C). "Antifreeze glycoproteins" are responsible for these low values. The sequences of the currently investigated proteins are strictly periodic:

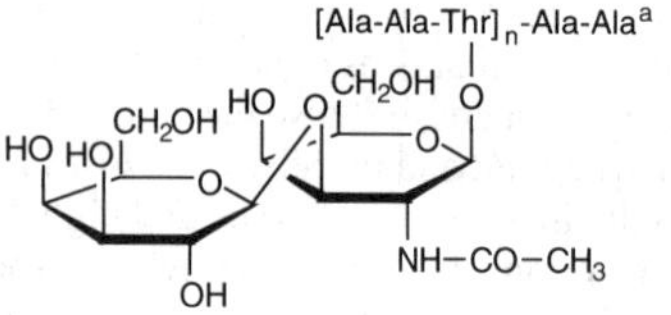

[a] 1 or 2 Alanine residues occupy the C-terminal position

The molecular masses are in the range 10 500 to 27 000. The conformation is mainly stretched with some α-helical regions. An X-ray crystallographic analysis of the A. p. from the winter flounder revealed the ice-binding region in the protein helix [1]. A. p. have also been found in some insects and hibernating plants. A.p. are studied for application in cancer cryosurgery.[2]

Lit.: [1] Nature (London) **375**, 427 (1995); Pat. WO 9728,260 (7.8.1997). [2] Drug News, R & D Focus, 20.10.1997, p. 7. *gen.:* Chem. Rev. **96**, 601–617 (1996) (review) ▪ Cryo-biology **29**, 69–79 (1992) ▪ Food Technol. **47**, 82–90 (1993) ▪ J. Am. Chem. Soc. **121**, 941 (1999) (activity).

Antimycins. Group of closely related antibiotics isolated in 1947 on the basis of their activities against phytopathogenic fungi (e.g., *Piricularia oryzae*) from

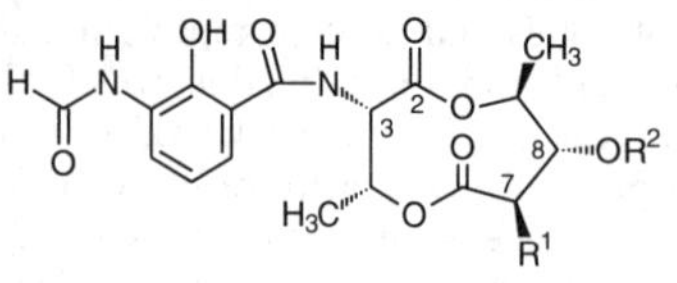

R^1	R^2	
$(CH_2)_5–CH_3$	$CO–C_4H_9$	: A. A_1
$(CH_2)_5–CH_3$	$CO–C_3H_7$	: A. A_2
$(CH_2)_3–CH_3$	$CO–C_4H_9$	: A. A_3
$(CH_2)_3–CH_3$	$CO–C_3H_7$	: A. A_4

Table: Main components of the Antimycin A complex.

Antimycin	molecular formula	M_R	CAS
A_1	$C_{28}H_{40}N_2O_9$	548.63	642-15-9
A_2	$C_{27}H_{38}N_2O_9$	534.61	27220-57-1
A_3	$C_{26}H_{36}N_2O_9$	520.58	522-70-3
A_4	$C_{25}H_{34}N_2O_9$	506.55	27220-59-3

streptomycete cultures. The so-called A. A complex was later separated into numerous components by HPLC on reversed phase silica gel. The composition of the A. A complex from varying streptomycetes varies considerably, it mostly contains the A. A_1 to A_4.

The A. consist of a 9-membered, asymmetrical bislactone with 3-*N*-formylaminosalicylic acid being linked to the threonine unit by an amide bond. The A. differ in the alkyl chain at C-7, which is preferentially even-numbered and unbranched, and in the 8-*O*-acyl group which is preferentially derived from a short-chain, methyl-branched carboxylic acid. It has been shown by HPLC/NMR methods that, e.g., *A. A_1* [mp. 149–150 °C, $[\alpha]_D$ +76° ($CHCl_3$), colorless crystals, insoluble in water, soluble in ethanol, acetone, chloroform] is isomeric in the 8-*O*-acyl group [A. A_{1a}: R^2=–CO–CH(CH_3)–C_2H_5, A. A_{1b}: R^2=–CO–CH_2–CH(CH_3)$_2$]. The same is true for *A. A_3* [*blastmycin*, colorless crystals, mp. 174–175 °C, $[\alpha]_D$ +79.4° ($CHCl_3$)]. A. selectively inhibit the respiratory chain by binding to the cytochrome bc_1 complex. Because of this property A. are among the most important tools for molecular biological studies on the respiratory chain. The A. are strong fish poisons (LD_{50} <1 ng/mL, use in the fishery industry), have antifungal, insecticidal, nematocidal, and antiviral (e.g., against AIDS viruses) activities. Their toxicity to higher organisms [LD_{50} (mouse i.v.) 0.9 mg/kg] is putatively attributed to other mechanisms of action.

Lit.: *Structure:* J. Antibiot. **25**, 373 ff. (1972); **29**, 804–808 (1976); **46**, 241–246 (1993) ▪ Magn. Reson. Chem. **25**, 1078 ff. (1987). – *Synthesis:* Tetrahedron **48**, 7059–7070 (1992) ▪ Tetrahedron Lett. **24**, 2657 ff. (1983). – *Separation:* Anal. Chem. **61**, 404–408 (1989) ▪ J. Chromatogr. **447**, 65–79 (1988); **464**, 453–458 (1988). – *Activity:* Biochem. Biophys. Acta **1185**, 271–278 (1994); **1229**, 149–154 (1995) ▪ J. Biol. Chem. **265**, 11409ff. (1990) ▪ Methods Enzymol. **126**, 253–271 (1986).

Antirrhinoside.

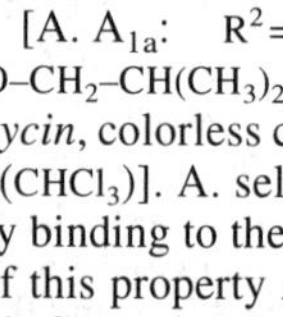

Antirrhinoside

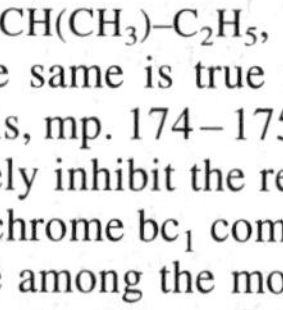

Procumbide

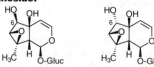

A. and its C-6 epimer *procumbide* are *iridoid glucosides with the sum formula $C_{15}H_{22}O_{10}$, M_R 362.33. – A., amorphous, $[\alpha]_D$ –79° (dioxane); procumbide, crystalline, mp. 212 °C, $[\alpha]_D$ –81° (C_2H_5OH).

Occurrence: A. in twisting snapdragon (*Antirrhinum tortuosum*, Scrophulariaceae), procumbide together with harpagide and harpagoside in the roots of the South African rampion (*Harpagophytum procumbens*, Pedaliaceae).

Lit.: J. Nat. Prod. **43**, 649, No. 27, 28 (1980) ▪ Justus Liebigs Ann. Chem. **1985**, 1063 ▪ Phytochemistry **31**, 795 (1992); **33**, 1087 (1993); **39**, 549 (1995) (biosynthesis). – *[HS 2938 90; CAS 20770-65-4 (A.); 20486-27-5 (procumbide)]*

Ants (Formicidae). Colony-forming insects with over 6000 species worldwide; especially numerous in the tropics and with ca. 200 species in Europe. A. have a finely developed orientation sense, sun compass, scent marks, landmarks. Various *pheromones serve for communication, especially for trail-marking and for giving alarm. A series of poisons serve for defence or hunting purposes, are often species-specific and are applied with the help of poison systems at the end of the abdomen or in the biting apparatus. The odor and defence substances of ants are mostly low-molecular-weight compounds such as ketones, di- and trisulfides and monoterpenes, e.g., *2-heptanone* ($C_7H_{14}O$, M_R 114.19), (*S*)-*4-methyl-3-heptanone, *manicone, *6-methyl-5-hepten-2-one*, ($C_8H_{14}O$, M_R 126.20), *iridomyrmecin, *iridodial, dolichodial (see teucrium lactones), *dendrolasin, *citrals, *citronellals, *citronellols, limonene (see *p*-menthadienes) and others. Alkylpyrazines are found in some tropical ants and are used as *trail pheromones*. Because their action leads to necrosis, toxins of the fire ant are particularly feared; they consist of 2-methyl-6-alkylpiperidines or 2,5-dialkylpyrrolidines. Some ants use *formic acid as poison and defensive weapon. Leaf-cutting ants which cultivate a specific fungus in their nests as source of food, secrete the *plant growth substances *phenylacetic acid* ($C_8H_8O_2$, M_R 136.15), *3-indolylacetic acid, and 3-hydroxydecanoic acid (*myrmicacin), in order to keep their cultured fungi free of other fungi and bacteria.

Lit.: Hölldobler & Wilson, The Ants, Berlin: Springer 1990. – *[HS 2939 90; CAS 110-43-0 (heptanone); 103-82-2 (phenylacetic acid)]*

Apamin.

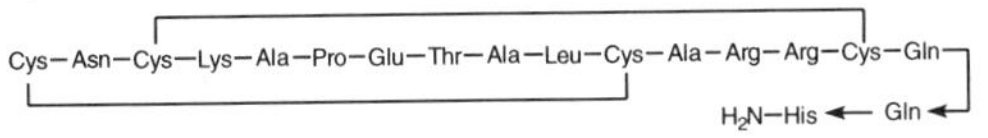

$C_{79}H_{131}N_{31}O_{24}S_4$, M_R 2027.34, LD_{50} (mouse i.v.) 4 mg/kg. Polypeptide of 18 amino acid residues and 2 cyclic disulfide bridges. It amounts to ca. 2% in the dry substance of *bee venom and represents the smallest neurotoxic polypeptide. A. selectively attacks the CNS, but does not play a significant role in the poisoning of warm-blooded animals on account of its low concentration.

Lit.: Bioorg. Khim. **7**, 5–15, 165–178 (1981) ▪ Eaker & Wadström, Natural Toxins, p. 481–486, New York: Pergamon Press 1980 ▪ Eur. J. Biochem. **196**, 639–645 (1991). – *Synthesis:* Int. J. Peptide Res. **11**, 238 (1978). – *[CAS 24345-16-2]*

Aphidicolin

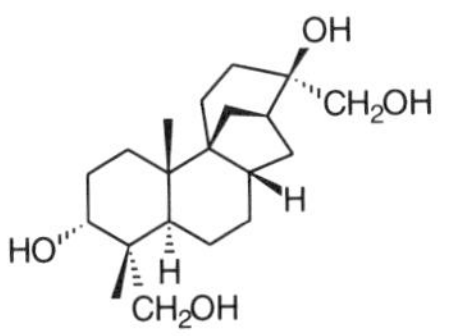

Aphidicolin

$C_{20}H_{34}O_4$, M_R 338.49, mp. 227–233 °C, $[\alpha]_D$ +12° (CH_3OH); diterpenoid with antibiotic and antiviral activity from *Cephalosporium aphidicola* and other lower fungi. A. is a specific inhibitor of DNA polymerase. The biosynthesis proceeds from labda-8(17),13-dien-15-yl diphosphate. A. is used in biochemistry to investigate cell differentiation. A. antagonizes the cytotoxic action of *camptothecin.

Lit.: Helv. Chim. Acta **71**, 872–875 (1988) ▪ J. Am. Chem. Soc. **109**, 1597 ff. (1987) ▪ J. Chem. Soc., Perkin Trans. 1 **1984**, 2751; **1985**, 2705 ▪ J. Org. Chem. **53**, 4929 (1988) ▪ Merck-Index (12.), No. 769 ▪ Tetrahedron **55**, 5641 (synthesis-review), 7541 (1999) (biosynthesis) ▪ Tetrahedron Lett. **36**, 5379 (1995). – *[HS 2941 90; CAS 38966-21-1]*

Apigenin (4′,5,7-trihydroxyflavone). Formula see flavones, $C_{15}H_{10}O_5$, M_R 270.24, yellow needles, mp. 352 °C, uv_{max} 380 nm (log ε 4.36); moderately soluble in hot alcohol, insoluble in water. A *flavonoid belonging to the class of *flavones that is widely distributed as aglycone and in numerous glycosides, e.g. in *apiin.

Lit.: Beilstein EV **18/4** 575 ▪ Hager (5.) **4**, 449 f. ▪ Ullmann (5.) **A 12**, 586. – *[HS 2932 90; CAS 520-36-5]*

Apigeninidin (4′,5,7-trihydroxyflavylium, apigenidin). Formula see anthocyanins, $C_{15}H_{11}O_4^+$, M_R 255.25, red prisms, mp. (chloride) >350 °C. A. belongs to the deoxyanthocyanidins. It occurs as the 5-glucoside ("gesnerin") in *Gesneria fulgens* (Gesneriaceae) and as the 7-glucoside and 5,7-diglucoside in the fern *Blechnum procerum* (Blechnaceae).

Lit.: Tetrahedron **37**, 1481 (1981). – *[CAS 1151-98-0]*

Apiin [7-(2-*O*-D-apio-*β*-D-furanosyl-*β*-D-glucopyranosyloxy)-4′,5-dihydroxyflavone].

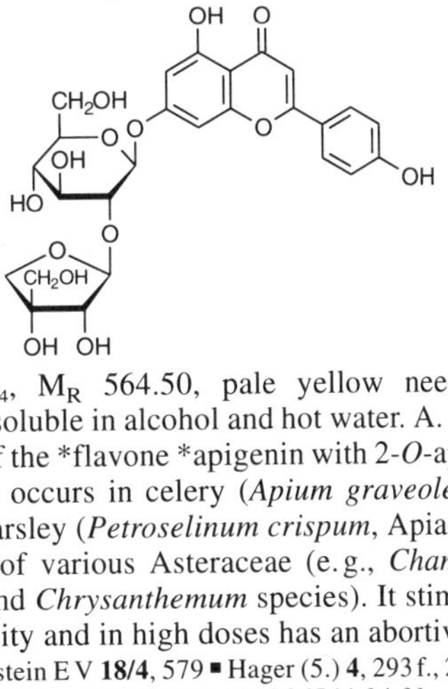

$C_{26}H_{28}O_{14}$, M_R 564.50, pale yellow needles, mp. 236 °C, soluble in alcohol and hot water. A. is the glycoside of the *flavone *apigenin with 2-*O*-apiosylglucose. A. occurs in celery (*Apium graveolens*, Apiaceae), parsley (*Petroselinum crispum*, Apiaceae), and flowers of various Asteraceae (e.g., *Chamaemelum nobile* and *Chrysanthemum* species). It stimulates renal activity and in high doses has an abortive action.

Lit.: Beilstein EV **18/4**, 579 ▪ Hager (5.) **4**, 293 f., 297 f. ▪ Karrer, No. 1449. – *[HS 2938 90; CAS 26544-34-3]*

Apiol (5-allyl-4,7-dimethoxy-1,3-benzodioxole, parsley camphor).

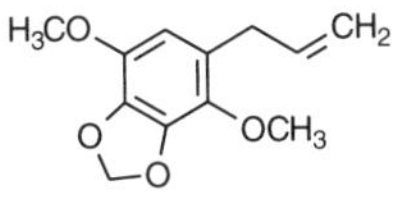

$C_{12}H_{14}O_4$, M_R 222.24. needles smelling mildly of parsley, mp. 30 °C, bp. 294 °C. A. is obtained from parsley oil. LDL_0 (mouse s.c.) 1000 mg/kg. For isolation of A. from *Sassafras albidum* see *Lit.*[1] (see also dill apiol). A. has abortive and diuretic activities and exhibits a synergistic action with insecticides[2]. Biosynthesis from cinnamoyl-CoA.

Lit.: [1] Phytochemistry **15**, 1773–1775 (1976). [2] J. Agric. Food. Chem. **22**, 658–664 (1974).
gen.: Beilstein EV **19/3**, 307 ▪ Karrer, No. 246 ▪ Luckner (3.), p. 387. – *Toxicology:* Sax (8.), AGE 500. – *[HS 2932 90; CAS 523-80-8]*

Aplasmomycins.

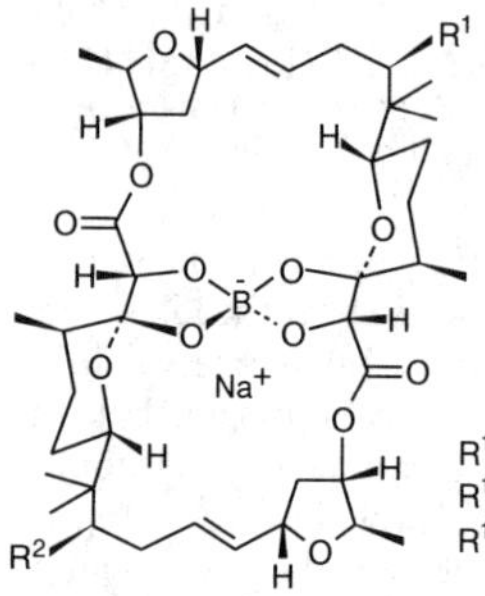

$R^1 = R^2 = OH$: A. A
$R^1 = OH$, $R^2 = O{-}CO{-}CH_3$: A. B
$R^1 = R^2 = O{-}CO{-}CH_3$: A. C

Table: Data of Aplasmomycins.

A.	molecular formula	M_R	$[\alpha]_D$	CAS
A*	$C_{40}H_{60}BNaO_{14}$	798.71	+225°	61230-25-9
B	$C_{42}H_{62}BNaO_{15}$	840.75	+188°	68193-20-4
C	$C_{44}H_{64}BNaO_{16}$	882.78	+134°	68193-21-5

* mp. 283–285 °C, needles

Family of boron-containing macrocyclic dilactones (macrodiolides), produced by *Streptomyces griseus* (from marine sediments). A. are active against Gram-positive bacteria and plasmodia. They belong to the small group of boron-containing ionophores (*boromycin, *tartrolones, borophycin).
Lit.: Angew. Chem. Int. Ed. Engl. **28**, 146 (1989) (biosynthesis) ▪ J. Antibiot. **29**, 1019 (1976); **30**, 714 (1977); **33**, 1316 (1980) ▪ Merck-Index (12.), No. 779 ▪ Tetrahedron **46**, 3469–3488 (1990) (synthesis); **47**, 3511–3520 (1991) (structure).

Aplysi... From sea snails of the family Aplysiidae (lumpfish) a series of structurally widely differing substances has been isolated, e. g. the highly poisonous aplysiatoxin [1,2] (from *Stylocheilus longicauda*), a tumor promoter; the cytostatically active aplysistatin [3] (inhibitor of murine lymphocytic leukemia); aplysianin [4], a glycoprotein of molecular mass ca. 250000 with potent antitumor activity; the violet, ink-like defence substance aplysioviolin [5]; further aplysinopsin [6] from marine sponges (*Fascaplysinopsis reticulata*) of the Great Barrier Reef with antitumor activity; aplysin [7] from *Aplysia kurodai*, aplysin 20 [8], aplysinal [9], aplysinol [10], aplysterol [11] and the aplykurodins [12] A and B.
Lit.: [1] Acc. Chem. Res. **10**, 33 (1977); Pure Appl. Chem. **41**, 1 (1975); Adv. Cancer Res. **49**, 223 (1987). [2] J. Am. Chem. Soc. **110**, 5768 (1988) (total synthesis). [3] J. Am. Chem. Soc. **99**, 262 (1977); Tetrahedron Lett. **21**, 2787 (1980); **23**, 4643 (1982) (Synth.). [4] Cancer Res. **47**, 5649 (1987). [5] Dolphin **VI**, 521 (1979); Naturwissenschaften **53**, 613 (1966). [6] Tetrahedron Lett. **1977**, 61. [7] J. Org. Chem. **45**, 3989 (1980). [8] Bull. Chem. Soc. Jpn. **44**, 2560 (1971). [9] Phytochemistry **16**, 1062 (1977). [10] Tetrahedron Lett. **1976**, 4219. [11] J. Chem. Soc., Chem. Commun. **1973**, 825; Tetrahedron Lett. **1978**, 4373. [12] Tetrahedron Lett. **27**, 1153 (1986).
gen.: Beilstein EV **17/2**, 112; **17/4**, 306; **17/5**, 721; **17/10**, 254; **17/11**, 88; **19/4**, 200; **19/7**, 153 ▪ Bull. Chem. Soc. Jpn. **70**, 1479 (1997) (review) ▪ Scheuer I **1**, 74; **4**, 18, 73, 123.

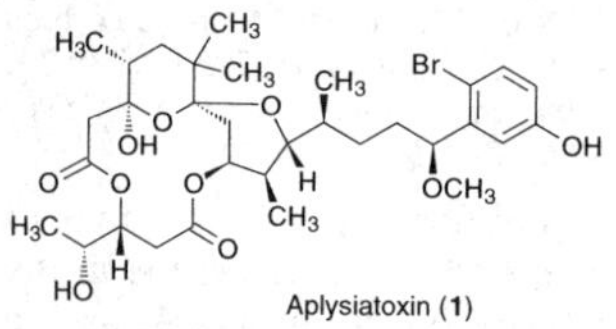

Aplysiatoxin (**1**)

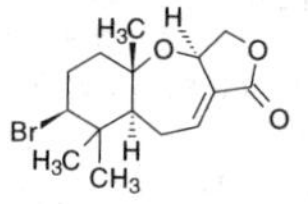

Aplysistatin (**2**)

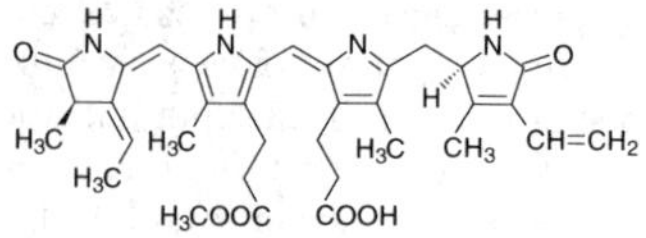

Aplysioviolin (**3**)

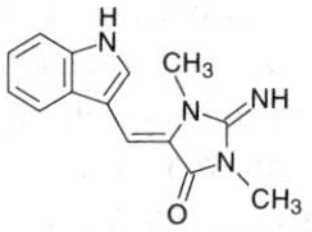

Aplysinopsin (**4**)

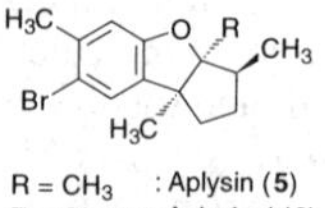

R = CH_3 : Aplysin (**5**)
R = CH_2OH: Aplysinol (**6**)
R = CHO : Aplysinal (**7**)

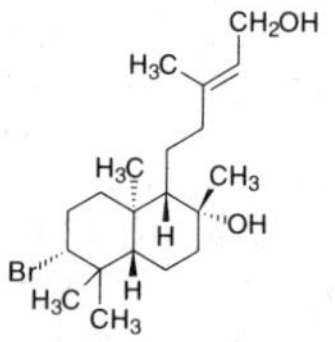

Aplysin 20 (**8**)

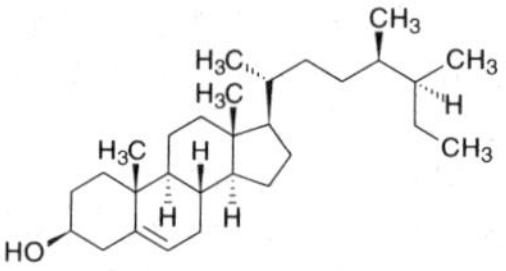

Aplysterol (**9**)

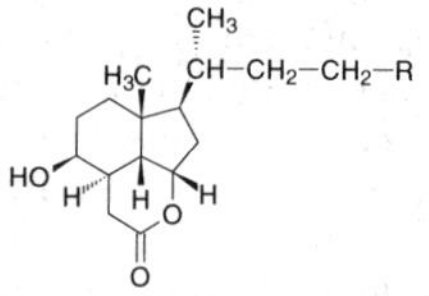

R = $CH_2{-}CH(CH_3)_2$: Aplykurodin A (**10**)
R = $CH{=}C(CH_3)_2$: Aplykurodin B (**11**)

Table: Data of Aplysi... substances.

no.	molecular formula	M_R	mp. [°C]	$[\alpha]_D$	CAS
1	$C_{32}H_{47}BrO_{10}$	671.62			52659-57-1
2	$C_{15}H_{21}BrO_3$	329.23	173–175	$[\alpha]_D^{20}$ –375° (CH_3OH)	62003-89-8
3	$C_{34}H_{40}N_4O_6$	600.71	310–315 green cryst.		15265-71-1
4	$C_{14}H_{14}N_4O$	254.29	232–233 yellow needles		63153-56-0
5	$C_{15}H_{19}BrO$	295.22	85–86	$[\alpha]_D^{27}$ –85° ($CHCl_3$)	6790-63-2
6	$C_{15}H_{19}BrO_2$	311.22	158–160	$[\alpha]_D^{19}$ –56° ($CHCl_3$)	6790-64-3
7	$C_{15}H_{17}BrO_2$	309.20			64052-99-9
8	$C_{20}H_{35}BrO_2$	387.40	146–147	$[\alpha]_D^{15}$ –78° (CH_3OH)	17941-22-9
9	$C_{29}H_{50}O$	414.72	135–136	$[\alpha]_D^{20}$ –25° ($CHCl_3$)	38636-49-6
10	$C_{20}H_{34}O_3$	322.49	138	$[\alpha]_D^{20}$ –44° ($CHCl_3$)	101691-10-5
11	$C_{20}H_{32}O_3$	320.47	130	$[\alpha]_D^{20}$ –36° ($CHCl_3$)	101691-09-2

Apoatropine.

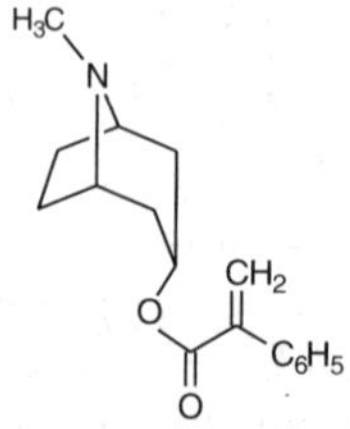

$C_{17}H_{21}NO_2$, M_R 271.36, mp. 60–62 °C. A *tropane alkaloid [LD_{50} (mouse p.o.) 160 mg/kg]. A. is formed readily under acidic conditions by dehydrating *hyoscyamine and *atropine. A. has been detected in various Solanaceae genera [e.g., *Atropa belladonna* (deadly nightshade), *Datura* (thorn apple), *Duboisia, Hyoscyamus* (henbane), and *Mandragora* (mandrake)].

Lit.: Beilstein E V **21/1**, 228 ▪ Manske **44**, 1–114 ▪ Merck-Index (12.), No. 780 ▪ Sax (8.), p. 280. – *[HS 2939 90; CAS 500-55-0]*

Apocynaceae alkaloids. Group of alkaloids from the plant families Apocynaceae (periwinkle family), Loganiaceae (poison nuts), and Rubiaceae, which are among the most important families for the occurrence of *indole alkaloids.

Lit.: Alkaloids (London) **1977**, 268 (review) ▪ Biochem. Syst. Ecol. **12**, 159 (1984) ▪ Nat. Prod. Rep. **3**, 443 (1986) (review). – *[HS 2939 90]*

Apocynaceae steroid alkaloids. This term is used for the *steroid alkaloids occurring in genera of the Apocynaceae. These alkaloids are occasionally also found in the Didymelaceae as well as in Buxaceae genera together with the actual *Buxus steroid alkaloids. The longest known and most important A. s. a., *conessine*, occurs in the Indian bush *Holarrhena antidysenterica* ("kurchee" or "kurchi"), so that these alkaloids are often collected together under the general term *Holarrhena* or *kurchee alkaloids*.

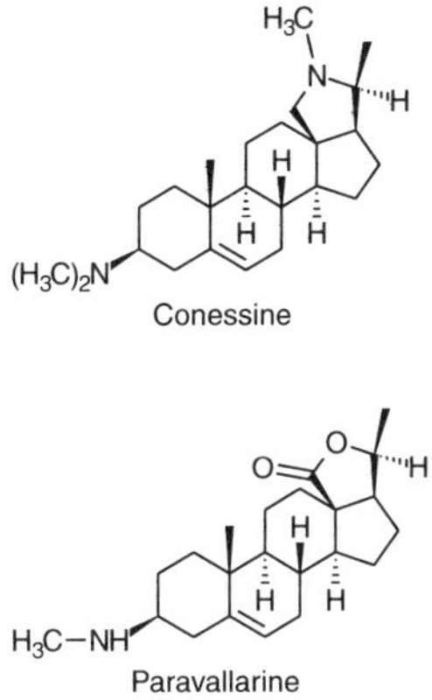
Conessine

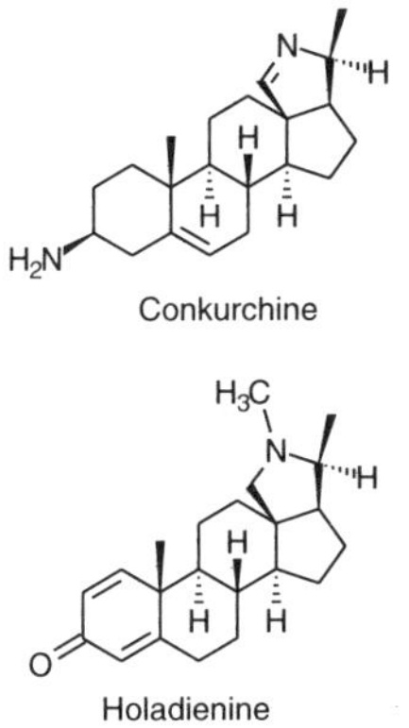
Conkurchine

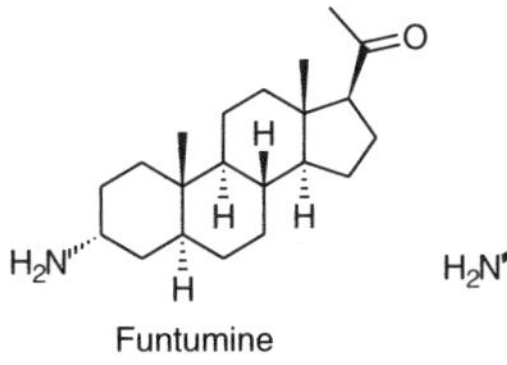
Paravallarine

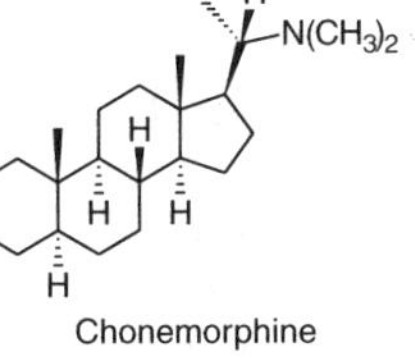
Holadienine

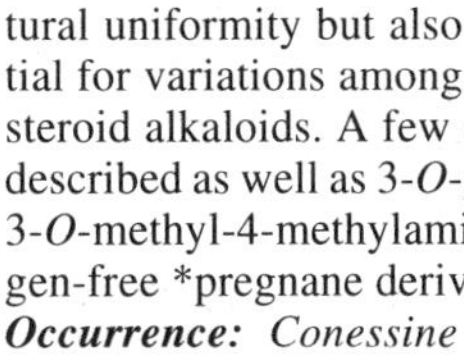
Funtumine

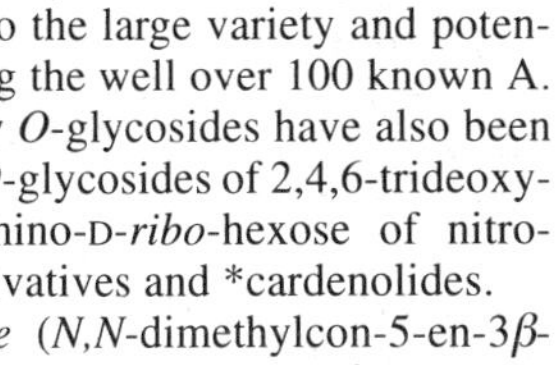
Chonemorphine

The 6 formulae shown emphasize not only the structural uniformity but also the large variety and potential for variations among the well over 100 known A. steroid alkaloids. A few *O*-glycosides have also been described as well as 3-*O*-glycosides of 2,4,6-trideoxy-3-*O*-methyl-4-methylamino-D-*ribo*-hexose of nitrogen-free *pregnane derivatives and *cardenolides.

Occurrence: *Conessine* (*N,N*-dimethylcon-5-en-3β-amine) is isolated from *Holarrhena antidysenterica* and many other *H.* species as well as *Wrightia tomentosa* and *Funtumia elastica*; *conkurchine* (*N*-demethylcona-5,18(*N*)-dien-3β-amine) from *H. antidysenterica* and other *H.* species as well as *W. tomentosa, F. elastica*, and *Malouetia arborea*; *paravallarine* [(20*S*)-20-hydroxy-3β-(methylamino)pregn-5-en-18-carboxylic acid-20-lactone] from *Paravallaris microphylla*; *holadienine* (cona-1,4-dien-3-one) from *H. floribunda* and *Didymeles cf. madagascariensis* (Didymelaceae); *funtumine* (3α-amino-5α-pregnan-20-one) from *F. latifolia, H. febrifuga*, and *H. congolensis*; and finally, *chonemorphine* [(20*S*)-20-(dimethylamino)-5α-pregnan-3β-amine] from *Chonemorpha macrophylla, C. penangensis, M. bequaertiana, Dictyophleba lucida*, and *F. latifolia*.

Table: Data of Apocynaceae steroid alkaloids.

Alkaloid CAS	molecular formula	M_R	mp. [°C]	$[\alpha]_D$
Conessine 546-06-5	$C_{24}H_{40}N_2$	356.60	123–124	–1.9° ($CHCl_3$)
Conkurchine 3792-62-9	$C_{21}H_{32}N_2$	312.50	151	–51° ($CHCl_3$)
Paravallarine 510-31-6	$C_{22}H_{33}NO_2$	343.51	181	–52° ($CHCl_3$)
Holadienine 2841-61-4	$C_{22}H_{31}NO$	325.49	109 (sublimation)	+80° ($CHCl_3$)
Funtumine 474-45-3	$C_{21}H_{35}NO$	317.52	126	+95° ($CHCl_3$)
Chonemorphine 4282-07-9	$C_{23}H_{42}N_2$	346.60	144–146	+25° ($CHCl_3$)

Biosynthesis: The biosynthesis of A. s. a. proceeds from *cholesterol via *pregnane derivatives (e.g., 3β-hydroxypregn-5-en-20-one and progesterone (see progestogens), which have also been isolated as metabolites together with A. s. a. from *H. floribunda* [1].

Biological activity: Conessine has a narcotic effect on frogs but not on mammals. It has local anesthetic, insecticidal, and antibacterial properties. The conessine-rich bark of the kurchee tree is used in India for the treatment of dysentery. Other A. s. a. exhibit hypotensive, vasodilating, hypocholesterolemic, antigonadotropic, corticotropic, and antiphlogistic activities. Some A. s. a. are used as starting materials for the synthesis of 18-oxygenated steroids (e.g., *aldosterone).

Lit.: [1] Mothes et al., S. 363–384.

gen.: Beilstein E V **22/10**, 215 f. ▪ Goutarel, Les Alcaloides Steroidiques des Apocynacees, Hermann: Paris 1964 ▪ J. Am. Chem. Soc. **118**, 9876 (1996) [synthesis (+)-conessine] ▪ Manske **7**, 319–342; **9**, 305–426; **32**, 79–239; **43**, 1–118. – *[HS 2939 90]*

Apomorphine see aporphine alkaloids.

Aporphine alkaloids.

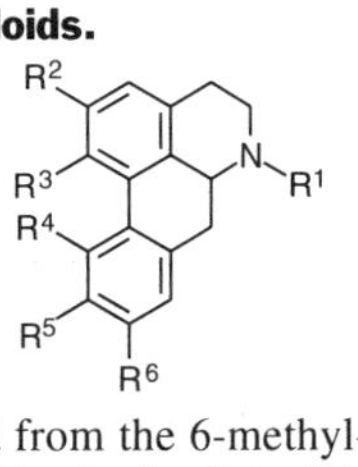

A. a. are derived from the 6-methyl-5,6,6a,7-tetrahydro-4*H*-dibenzo[*de,g*]quinoline skeleton (aporphine) and occur in various plant families (e.g., Annonaceae,

Table: Structures and data of Aporphine alkaloids.

name (configuration)	R^1	R^2	R^3	R^4	R^5	R^6	molecular formula	M_R	mp. [°C]	$[\alpha]_D$	CAS
Anonaine (*R*)	H	O–CH_2–O		H	H	H	$C_{17}H_{15}NO_2$	265.31	122–123	–68° (C_2H_5OH)	1862-41-5
Apomorphine (*R*)	CH_3	H	H	OH	OH	H	$C_{17}H_{17}NO_2$	267.33	195		58-00-4
Boldine (*S*)	CH_3	OH	OCH_3	H	OCH_3	OH	$C_{19}H_{21}NO_4$	327.38	161–163	+111° (C_2H_5OH)	476-70-0
Bulbocapnine (*S*)	CH_3	O–CH_2–O		OH	OCH_3	H	$C_{19}H_{19}NO_4$	325.36	201–203	+232° (C_2H_5OH)	298-45-3
Corydine (*S*)	CH_3	OCH_3	OH	OCH_3	OCH_3	H	$C_{20}H_{23}NO_4$	341.41	149	+204° (C_2H_5OH)	476-69-7
Corytuberine (*S*)	CH_3	OCH_3	OH	OH	OCH_3	H	$C_{19}H_{21}NO_4$	327.38	240	+286° (C_2H_5OH)	517-56-6
Dicentrine (*S*)	CH_3	O–CH_2–O		H	OCH_3	OCH_3	$C_{20}H_{21}NO_4$	339.39	168–169	+57° (C_2H_5OH)	517-66-8
Glaucine (*S*)	CH_3	OCH_3	OCH_3	H	OCH_3	OCH_3	$C_{21}H_{25}NO_4$	355.43	120–121	+116° (C_2H_5OH)	475-81-0
Isoboldine (*S*)	CH_3	OCH_3	OH	H	OCH_3	OH	$C_{19}H_{21}NO_4$	327.38	178–180	+54° (C_2H_5OH)	5164-93-2
Isocorydine (*S*)	CH_3	OCH_3	OCH_3	OH	OCH_3	H	$C_{20}H_{23}NO_4$	341.41	185	+215° ($CHCl_3$)	475-67-2
Isothebaine (*S*)	CH_3	OCH_3	OH	OCH_3	H	H	$C_{19}H_{21}NO_3$	311.38	164–166	+281° ($CHCl_3$)	568-21-8
Magnoflorine (*S*)	$(CH_3)_2$	OCH_3	OH	OH	OCH_3	H	$C_{20}H_{24}NO_4^+$	342.41			2141-09-5
Nornuciferine (*R*)	H	OCH_3	OCH_3	H	H	H	$C_{18}H_{19}NO_2$	281.35	128–129	–145° (C_2H_5OH)	4846-19-9
Nuciferine (*R*)	CH_3	OCH_3	OCH_3	H	H	H	$C_{19}H_{21}NO_2$	295.38	165.5	–164° (C_2H_5OH)	475-83-2
Roemerine (*R*)	CH_3	O–CH_2–O		H	H	H	$C_{18}H_{17}NO_2$	279.34	102–103	–97°	548-08-3

Aristolochiaceae, Euphorbiaceae, Fabaceae, Liliaceae, Magnoliaceae, Menispermaceae, Papaveraceae, Ranunculaceae). Anonaine is found mainly in *Annona reticulata* (Annonaceae), bulbocapnine in *Corydalis cava* (Fumariaceae), isothebaine especially in *Papaver orientale* and *P. bracteatum*, magnoflorine in *Magnolia obovata* (Magnoliaceae) and *Aquilegia hybrida* (Ranunculaceae), and roemerine in *Roemeria refracta* (Papaveraceae) and *Nelumbo nucifera* (Nelumbonaceae). A. a. are soluble in chloroform, alcohol, and ether, poorly soluble in water.

Activity & use: *Apomorphine* [1] is the second strongest (after *vomicine) known vomitive (emetic), it acts on the vomiting center of the CNS. Apomorphine is in clinical development for male erectile dysfunction (sublingual formulation). Higher doses (20 mg) damage the respiratory center. LD_{50} (mouse i. p.) 160 mg/kg; induces contact dermatitis. *Glaucine* and *dicentrine* exhibit a broad spectrum of pharmacological activities. They exert antithrombotic, analgesic, anti-inflammatory, and fungicidal actions, induce spasms and unconsciousness, as well as respiratory paralysis [2,3]. As antitussives they show effects similar to codeine. *Corytuberine* induces tonic convulsions and stimulates secretion of saliva and tears. It slows the pulse rate by vagus activity and increases blood pressure during convulsions. *Bulbocapnine* induces cataleptic states and acts as a sedative. It is sometimes used in treatment of Parkinson's disease and saltation (chorea). It causes hypokinesia up to elimination of the involuntary and reflex movements (bulbocapnine rigidity). *Magnoflorine* has weak curare-like and hypotensive activity. *Nuciferine* is an inhibitor of adenylate cyclase and has a strong sedative effect. *Boldine* is used as a diuretic [4]. Low doses of *isothebaine* increase muscular tonus in intestinal and uterine tissue. *Isocorydine* (artabotrine, luteanine) exhibits sedative, cholinergic activity, LD_{50} (rat i. p.) 10.9 mg/kg.

Biosynthesis: A. a. are formed from phenolic tetrahydrobenzylisoquinolines either by direct oxidative coupling or via dienones (proaporphines). The direct oxidative coupling can be either *ortho-para* (to isoboldine type) or *ortho-ortho* (to bulbocapnine type) [5,6].

Lit.: [1] Eur. J. Pharmacol. **219**, 67–74 (1992). [2] Planta Med. **57**, 406 ff. (1991). [3] Biochem. Pharmacol. **43**, 323–329 (1992). [4] Planta Med. **57**, 519–522 (1991). [5] Mothes et al., p. 207–215. [6] Phytochemistry **32**, 897–903 (1993); J. Nat. Prod. **57**, 1033 ff. (1994).

gen.: Beilstein EV **21/5**, 333 (apomorphine) ▪ Florey **20**, 121–171 ▪ Hager (5.) **4**, 1, 721 f.; **2**, 144; **4**, 480–497, 835–848, 1013–1027, 1155–1158; **5**, 110–115, 745–750; **7**, 277 ff., 506 f. ▪ Manske **4**, 119–145; **9**, 2–37; **17**, 385–544; **24**, 153–251; **36**, 72–90; **53**, 58–119 ▪ Pelletier **5**, 133–270, 271–673 ▪ Phillipson et al., p. 146–170 ▪ Sax (8.), AQP 250, AQP 500, DNZ 100, HLT 000, TDI 475, TDI 500, TDI 600 ▪ Shamma, p. 195–228 ▪ Shamma-Moniot, p. 123–156 ▪ Ullmann (5.) **A 1**, 373. – *Synthesis:* Chem. Pharm. Bull. **34**, 1946 ff. (1986) ▪ Tetrahedron **51**, 5341 (1995). – *[HS 2939 10, 2939 90]*

Apovincamine.

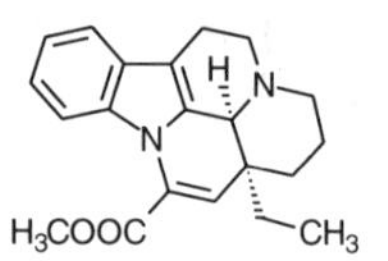

$C_{21}H_{24}N_2O_2$, M_R 336.43, cryst., mp. 160–161 °C, $[\alpha]_D$ +118° ($CHCl_3$), soluble in chloroform; an *indole alkaloid from *Tabernaemontana rigida* (Apocynaceae, periwinkle family) formed by dehydrating *vincamine. A. acts a vasodilator and thus lowers blood pressure.

Lit.: Neurobiol. Aging **11**, 39 (1990) (pharmacology) ▪ J. Org. Chem. **33**, 1055 (1968) ▪ Tetrahedron **39**, 3737 (1983). – *[HS 2939 90; CAS 4880-92-6]*

Apple aroma see fruit flavors.

Apricot aroma see fruit flavors.

Aquayamycin.

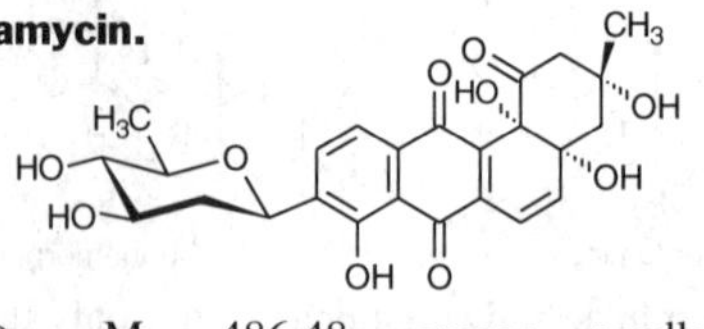

$C_{25}H_{26}O_{10}$, M_R 486.48, orange needles, mp. 200–203 °C, $[\alpha]_D$ +160°. Angucyclinone antibiotic from *Streptomyces misawanensis* with a quinonoid polyketide part and a C-glycosidically bound 2,6-dideoxy sugar. A. is the hydrolysis product of various *angucyclines (e. g., *vineomycins, *urdamycins, *saquayamycins), in which A. is O-glycosidically bound to deoxy sugars. A. has weak antibacterial and cytotoxic activities, it inhibits some hydrolases and monooxygenases. For biosynthesis, see *Lit.* [1].

Lit.: [1] J. Antibiot. **42**, 1151–1157 (1989); J. Org. Chem. **58**, 2547–2551 (1993).
gen.: Beilstein EV **18/5**, 715 ▪ Chem. Pharm. Bull. **32**, 4350–4359 (1984) ▪ J. Antibiot. **39**, 1657–1669 (1986); **42**, 1151–1157 (1989) ▪ Nat. Prod. Rep. **9**, 103–137 (1992) ▪ Tetrahedron **26**, 5171–5190 (1970). – *[HS 2941 90; CAS 26055-63-0]*

Ara A see mycalamides.

Arabinose.

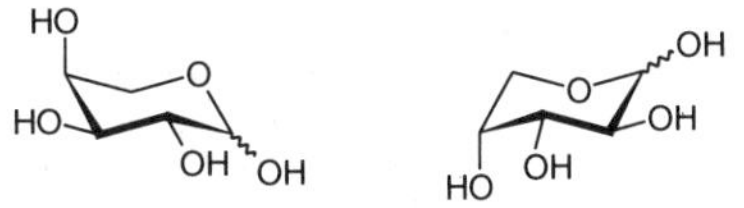

L- D-

Arabinopyranose

$C_5H_{10}O_5$, M_R 150.13. D-A.: mp. 159–160 °C, $[\alpha]_D$ –175 → –105° (H_2O); L-A.: mp. 160 °C, $[\alpha]_D$ +190 → +105° (H_2O). Monosaccharide of the aldopentose series. D-A. occurs only rarely in nature in bacterial polysaccharides. L-A. is frequently found in nature in plants, gums , hemicelluloses, pectins, bacterial polysaccharides, in the form of glycosides, glycoconjugates, and heteropolysaccharides. L-A. is isolated from dried beet pulp, Aloe, and gums [1].
Lit.: [1] Methods Carbohydr. Chem. **1**, 76 (1962).
gen.: Beilstein EIV **1**, 4215 ff. ▪ Collins & Ferrier, Monosaccharides, p. 47, 53, New York: Wiley 1995 ▪ Karrer, No. 583 f. ▪ Kirk-Othmer (4.) **4**, 913. – *[HS 2940 00; CAS 147-81-9; 10323-20-3 (D-A.); 5328-37-0 (L-A.)]*

Arabitol (arabinitol).

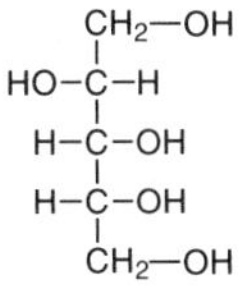

D-form

$C_5H_{12}O_5$, M_R 152.15, water-soluble, sweet-tasting, colorless prisms, mp. 103 °C, $[\alpha]_D$ +8° (H_2O) (D-form) or wart-like crystals, mp. 102 °C, $[\alpha]_D$ –5° (H_2O) (L-form). A. belongs to the group of pentitols or pentahydroxy alcohols and can be prepared by reduction of *arabinose or lyxose. D-A. is frequently found in lichens and fungi.
Lit.: J. Am. Chem. Soc. **104**, 1109 (1982) (synthesis) ▪ Karrer, No. 147 ▪ Kirk-Othmer (3.) **1**, 754 ff. ▪ Merck-Index (12.), No. 800. – *[CAS 488-82-4 (D(+)); 7643-75-6 (L(–))]*

Arach(id)ic acid (eicosanoic acid).
$H_3C–(CH_2)_{18}–COOH$, $C_{20}H_{40}O_2$, M_R 312.54, mp. 75.5–76.5 °C, bp. 203–205 °C (133 Pa), $n_D^{85.6}$ 1.4307; soluble in organic solvents, insoluble in water. A saturated *fatty acid with 20 C atoms. A. occurs in small amounts in the triglycerides of peanut oil (1.5%; *Arachis hypogaea*), olive seed oil (3%), and *rapeseed oil (0.5–1%) as well as cocoa butter (1.1%). The seed fat of *Nephelium lappaceum* ("rambutan tallow") and other Sapindaceae contain up to 35% arachic acid.
Lit.: Beilstein EIV **2**, 1275 ▪ Karrer, No. 707. – *[HS 2915 90; CAS 506-30-9]*

Arachidonic acid [(*all-Z*)-5,8,11,14-eicosatetraenoic acid].

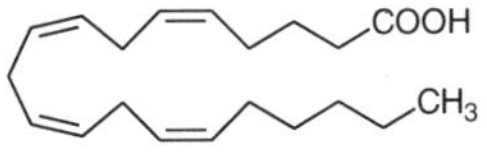

$C_{20}H_{32}O_2$, M_R 304.47, mp. –49.5 °C, bp. 163 °C (133 Pa). A biochemically important essential *fatty acid, precursor of various *eicosanoids. In animals, A. is formed in several steps from *linoleic acid by introduction of 2 double bonds and chain extension by 2 carbon atoms: linoleic acid → γ-*linolenic acid → *dihomo-γ-linolenic acid ↔ arachidonic acid. In nature A. occurs in animal lipids, especially in phospholipids of cell membranes, from which phospholipase A_2 releases A. on stimulation by hormones or mediators. A. serves as starting material for the biosynthesis of numerous physiologically active oxygen derivatives (especially hydroxy-, hydroperoxy-, and epoxy compounds) known as the *eicosanoids. Their separation into different groups in the so-called A. cascade is shown in the figure [1]:

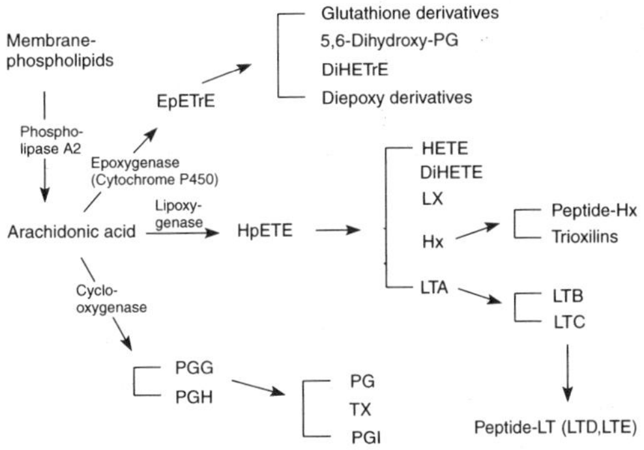

Figure: Arachidonic acid cascade.
EpETrE=epoxyeicosatrienoic acids; (Di)HETE=(di)hydroxyeicosatetraenoic acids; HpETE=hydroperoxyeicosatetraenoic acids; DiHETrE=dihydroxyeicosatrienoic acids; LX=lipoxins; PGI=prostacyclins; TX=thromboxanes; LT=leukotrienes; PG=prostaglandins; Hx=hepoxilins; PGG, PGH=prostaglandin endoperoxides

Large proportions of glycerol esters of A. are sometimes found in the lipids of lower plants (algae, mosses, and ferns) and microorganisms.
Lit.: [1] Annu. Rev. Biochem. **55**, 69 (1986); J. Neurochem. **55**, 1 (1990).
gen.: Beilstein EIV **2**, 1802 ▪ Biotechnol. Appl. Biochem. **15**, 1 (1992) ▪ Dunn et al. (eds.), Arachidonic Acid Metabolism, Auckland: Adis Press Ltd. 1987 ▪ Free Radical Biol. Med. **7**, 409 (1989) ▪ Prog. Lipid Res. **27**, 271 (1988); **33**, 329 (1994) ▪ see also eicosanoids and linolenic acid. – *[HS 2916 19; CAS 506-32-1]*

Arachidonic acid 2-hydroxyethylamide [arachidonic acid ethanolamide] see anandamide.

Araguspongine D see xestospongines.

Arborescin [(+)-arborescin].

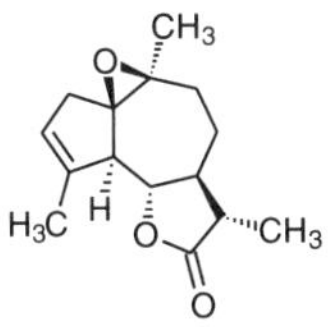

$C_{15}H_{20}O_3$, M_R 248.32, cryst., mp. 140–142 °C, $[\alpha]_D^{20}$ +64° ($CHCl_3$); soluble in acetone and DMF. A *sesquiterpene lactone from *Artemisia arborescens* (Asteraceae) and *Matricaria globifera*. A. was used in ancient Greece and Arabia as a contraceptive [1].

Lit.: [1] Tetrahedron Lett. **1963**, 1127.
gen.: Beilstein E V **19/4**, 278 ▪ J. Org. Chem. **47**, 3903 (1982) (synthesis). – *[CAS 6831-14-7]*

Arbutin [(4-hydroxyphenyl)-*β*-D-glucopyranoside].

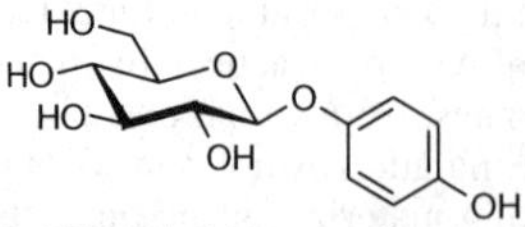

$C_{12}H_{16}O_7$, M_R 272.25. Bitter-tasting needles that exist in an unstable (mp. 165 °C) and a stable form (mp. 200 °C), $[\alpha]_D^{18}$ −60.3° (H_2O). TLD_0 (rat p. o.) 13.6 g/kg. Cleavage to hydroquinone and glucose occurs on reaction with emulsin or dilute acid. The black color of many dead plants can be explained by oxidation of the hydroquinone. A. occurs in pear leaves (*Pyrus communis*, Rosaceae) and is widely distributed in leaves of Ericacae, e. g., in bearberry species (*Arctostaphylos uva-ursi*), as well as in cranberries, heather, and azalea. The leaves of *Saxifraga crassifolia* are especially rich in A.

Activity: Hydroquinone released by cleavage of A. has antimicrobial properties.

Use: As color stabilizer in the photographic industry; previously as a diuretic and for disinfection of the urinary tract.

Lit.: Beilstein E V **17/7**, 110 ▪ Hager (5.) **4**, 327 ff.; 497 f.; **7**, 291 ff. ▪ Helv. Chim. Acta. **75**, 2009 ff. (1992) (synthesis in plant cell cultures) ▪ Karrer, No. 204. – *Toxicology:* Sax (8.), HIH 100. – *[HS 2938 90; CAS 497-76-7]*

Arcaine [1,1′-(1,4-butanediyl)bisguanidine].

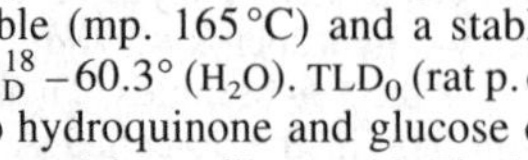

$C_6H_{16}N_6$, M_R 172.23, mp. (sulfate) 291 °C (decomp.). A. is a bisguanidino derivative that has been found in some fungi, plants, and marine organisms. A. reduces the blood sugar level in warm-blooded organisms.

Lit.: Annu. Rev. Plant Physiol. **36**, 117 (1985) ▪ Phytochemistry **10**, 731 (1971) (biosynthesis). – *[CAS 544-05-8]*

Archaebacteria (Archea, from Greek: archaios = ancient; bakterion = rod). Name for a group of prokaryotes which, according to recent phylogenetic considerations, represent a third distinct ancient kingdom besides the remaining prokaryotes (eubacteria including cyanobacteria) and the eukaryotes. The name already indicates that the A. possess characteristics that were probably adapted to the conditions prevailing during the early history of life on the Earth. They represent a primary form of living organisms that are not derived from bacteria.

Important distinguishing features between eubacteria and A. are: the cell wall structures of A. are heterogeneous (pseudomurein, proteins, glycoproteins, heteropolysaccharides), all A. lack muramic acid, a typical building block of the eubacterial cell wall. Thus, various antibiotics (e. g. penicillin, chloramphenicol) do not exhibit any inhibitory action. The cytoplasm membranes contain glycerol ethers with C_{20}- and C_{40}-isoprenoids instead of fatty acid glycerol esters. In addition to special biosynthetic pathways and unusual coenzymes, the main differences lie in the structure of the DNA-dependent RNA polymerases, the tRNA, and the nucleic acid sequence of the ribosomal nucleic acids; the 16S rRNA data were thus used as the basis for a molecular taxonomy. A. are gaining increasing practical significance in biotechnology, e. g. to obtain thermostable enzymes, to extract ores directly from their deposits, or for the isolation of secondary metabolites.

Lit.: Eur. J. Biochem. **173**, 473 (1988) ▪ Forum Mikrobiol. **10**, 209 (1987); **11**, 82 (1988) ▪ J. Mol. Evol. **11**, 245 (1978) ▪ J. Org. Chem. **64**, 3139 (1999) (glycolipids) ▪ Kandler & Zillig, Achaebacteria '85, Stuttgart: G. Fischer 1986 ▪ Kates, Kushner & Matheson, The Biochemistry of Archea (Archaebacteria), Amsterdam: Elsevier 1993 ▪ Nature (London) **356**, 148 (1992) ▪ Syst. Appl. Microbiol. **7**, 278 (1986).

Arcyria pigments (bisindolylmaleimides and indolo[2,3-*a*]carbazoles). The tiny fruit bodies of the slime molds *Arcyria denudata*, *A. nutans*, and related myxomycetes contain indole pigments of the bisindolylmaleimide type. Parent compounds are the red *arcyriarubins* A, B, and C. Oxidative cyclization in the 2,2′-position furnishes the pale yellow and poorly soluble *arcyriaflavins* A, B, and C (table 1). Indolocarbazoles of this type are also known as active principles from streptomycetes (*rebeccamycin).

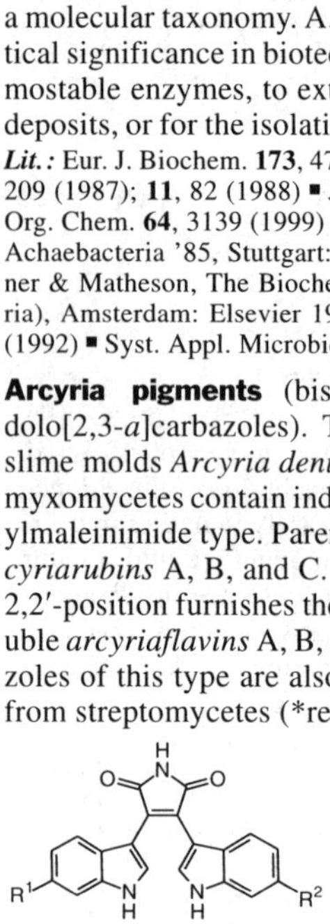

$R^1 = R^2 = H$: Arcyriarubin A
$R^1 = OH, R^2 = H$: Arcyriarubin B
$R^1 = R^2 = OH$: Arcyriarubin C

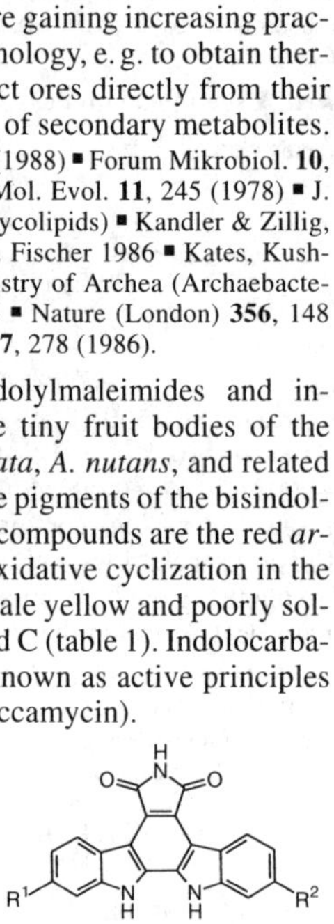

$R^1 = R^2 = H$: Arcyriaflavin A
$R^1 = OH, R^2 = H$: Arcyriaflavin B
$R^1 = R^2 = OH$: Arcyriaflavin C

Table 1: Data of Arcyriaflavins and Arcyriarubins.

	molecular formula	M_R	mp. [°C]	CAS
Arcyria-rubin A	$C_{20}H_{13}N_3O_2$	327.34	161 red cryst.	119139-23-0
Arcyria-rubin B	$C_{20}H_{13}N_3O_3$	343.34	154–155	73697-62-8
Arcyria-rubin C	$C_{20}H_{13}N_3O_4$	359.34	205–206 red cryst.	73697-63-9
Arcyria-flavin A	$C_{20}H_{11}N_3O_2$	325.33	>250 pale yellow	118458-54-1
Arcyria-flavin B	$C_{20}H_{11}N_3O_3$	341.33	>350 pale yellow	73697-64-0
Arcyria-flavin C	$C_{20}H_{11}N_3O_4$	357.33	>350 pale yellow	73697-65-1

Cyclization of the arcyriarubins in the 2,4′-position gives rise to the blue *arcyriacyanins* (table 2).

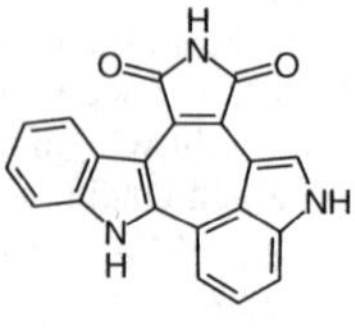

Arcyriacyanin A

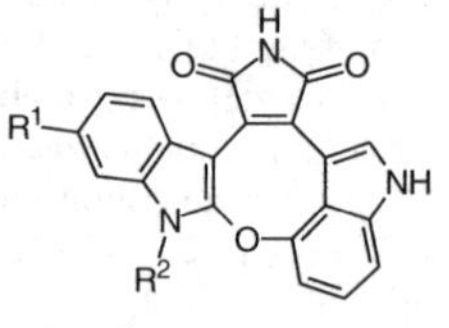

$R^1 = R^2 = H$: Arcyroxocin A
$R^1 = OH, R^2 = H$: Arcyroxocin B
$R^1 = H, R^2 = OH$
: N-Hydroxyarcyroxocin A

Table 2: Data of Arcyriacyanin A and Arcyroxocins.

	molecular formula	M_R	mp. [°C]
Arcyria-cyanin A	$C_{20}H_{11}N_3O_2$	325.33	350 (bluegreen microcryst.)
Arcyroxocin A	$C_{20}H_{11}N_3O_3$	341.33	>300 (decomp.) (red microcryst.)
Arcyroxocin B	$C_{20}H_{11}N_3O_4$	357.33	>235 (decomp.) (red violet flaces)
N-Hydroxy-arcyroxocin A	$C_{20}H_{11}N_3O_4$	357.33	268–270 (red cryst.)

In *arcyroxocins* A and B (table 2) the two indole units are linked by an oxygen atom. *N-Hydroxyarcyroxocin A* was previously considered to be an oxepin derivative ("arcyroxepin B"). It turns violet on contact with alkali. The arcyriarubins, arcyriacyanins, and arcyroxocins also occur in the sporangia in the form of their colorless dihydro derivatives.

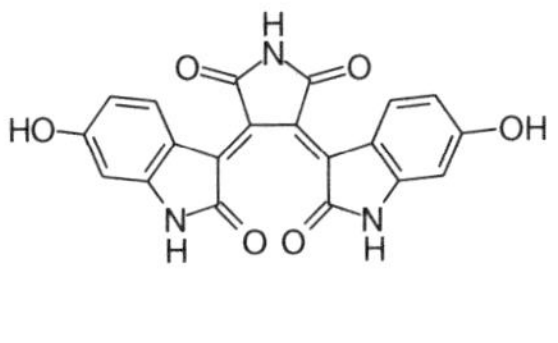

Arcyriaverdin C

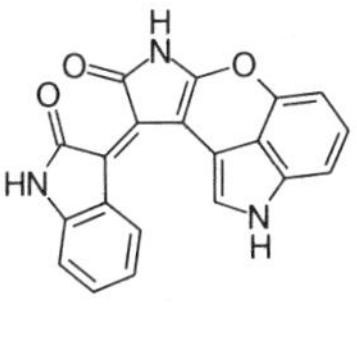

Arcyroxindole A

Biogenetically related metabolites from fruit bodies of *Arcyria* are *arcyriaverdin C* ($C_{20}H_{11}N_3O_6$, M_R 389.32) and *arcyroxindole A* {$C_{20}H_{11}N_3O_3$, M_R 341.33, orange cryst., mp. >240 °C (decomp.), $[\alpha]_D$ −28°} which is chiral on account of its helical structure.

Lit.: Angew. Chem. Int. Ed. Engl. **19**, 459 (1980) ▪ Atta-ur-Rahman **12**, 365 ff. (review) ▪ Ciba Found. Symp. **154**, 56, 1990 (isolation, constitution) ▪ Pure Appl. Chem. **61**, 281 (1989). – *Synthesis:* Chem. Eur. J. **3**, 70 (1997) (arcyriacyanin A) ▪ Chem. Pharm. Bull. **46**, 889 (1998); **48**, 81 (2000) (arcyriacyanin A) ▪ Tetrahedron **44**, 2887 (1988); **48**, 8869–8880 (1992) ▪ Tetrahedron Lett. **36**, 2689 (1995); **37**, 4483 (1996). – *Review:* Zechmeister **51**, 216 f.

Areca alkaloids. General term for dehydro-*piperidine alkaloids of the betel (areca) nut, the seeds of the betel palm *Areca catechu* (Palmae, Areceae). Pyridine alkaloids are isolated from the hemispherical, ca. 3-cm large nuts of the betel palm cultivated in India and east Asia.

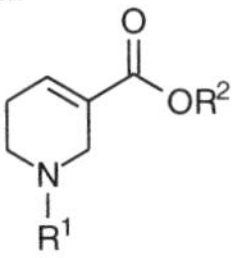

$R^1 = R^2 = CH_3$: Arecoline
$R^1 = CH_3$, $R^2 = H$: Arecaidine
$R^1 = H$, $R^2 = CH_3$: Guvacoline
$R^1 = R^2 = H$: Guvacine

Table: Data of Areca alkaloids.

	molecular formula	M_R	mp. (bp.) [°C]	CAS
Arecoline	$C_8H_{13}NO_2$	155.20	(209)	63-75-2
Arecaidine	$C_7H_{11}NO_2$	141.17	232 (decomp.)	499-04-7
Guvacoline	$C_7H_{11}NO_2$	141.17	27	495-19-2
Guvacine	$C_6H_9NO_2$	127.14	293–295 (decomp.)	498-96-4

The main alkaloid is *arecoline*, a potent parasympathomimetic with muscarine-like (see fly agaric constituents) and *pilocarpine-like activities. It promotes saliva and sweat secretion and stimulates intestinal activity. Today the use of areca preparations is limited to veterinary medicine as anthelmintics. The "betel chewers" of south east and east Africa, estimated to about 200–400 million, exploit the stimulating and relaxing effects of the betel nut. The so-called betel plugs consist of powdered Areca seeds, some leaves of the aromatic betel pepper (*Piper betel*), other aromatic additives, and lime. On chewing the betel plugs, the arecoline is hydrolysed under the weak alkaline conditions (lime!) to *arecaidine*. Thereby the parasympathomimetic action is mostly lost while the stimulating action is retained. The increased incidence of oral cavity carcinomas among betel chewers is presumably attributable to the alkylating action of arecaidine.

Lit.: Beilstein E V **22/1**, 322 f. ▪ Biochem. J. **122**, 503 (1971) (metabolism) ▪ Hager (5.) **3**, 88 ff.; **7**, 292 ▪ Manske **26**, 144 ▪ Pelletier **3**, 10–12 ▪ Sax (8.), p. 285 f., 425

Arenemycin see pentalenolactones.

Argemonine see pavine and isopavine alkaloids.

L-Arginine [(*S*)-2-Amino-5-guanidinopentanoic acid, abbreviation Arg or R].

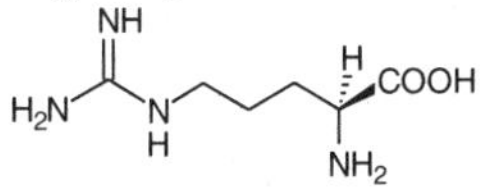

(2*S*)-, L-form

$C_6H_{14}N_4O_2$, M_R 174.20; dihydrate, mp. 244 °C (decomp.), $[\alpha]_D^{25}$ +12.5° (H_2O), +27.6° (5 m HCl); monohydrochloride, mp. 235 °C (decomp.); pK_a 1.82, 8.99 (NH_3) and 12.48 (guanidine), pI 11.5. Strongly alkaline aqueous solutions absorb CO_2 from the air. A. is a proteinogenic amino acid occurring in almost all proteins. Genetic code: CGU, CGC, CGA, CGG. The average content in proteins amounts to 4.7% [1]. A. was first obtained as the silver salt from the hydrolysate of horn shavings. It is also found in free form in plants and fungi where A. serves as a reserve substance for nitrogen, especially in shoots and storage cells.

Detection: Monosubstituted guanidino compound, with diacetyl-α-naphthol (Sakaguchi reaction) [2].

Biosynthesis & metabolism: Intermediate product in the urea cycle. Formed from argininosuccinate by argininosuccinate lyase (EC 4.3.2.1). Since A. is cleaved by arginase (EC 3.5.3.1) to urea and *L-ornithine it serves as a direct precursor of urea. In the human growth period, most A. is used for urea production and it is thus considered to be an essential amino acid in this period. A. plays a central role in the biosynthesis of guanidine derivatives. Through N^{ω}-oxidation and the action of NO synthase, the tissue hormone *nitrogen monoxide (NO), important for the function of smooth muscles, is formed. For regulation of biosynthesis in plants, see *Lit.* [3].

Lit.: [1] Biochem. Biophys. Res. Commun. **78**, 1018–1024 (1977). [2] Biochem. J. **5**, 25, 133 (1925). [3] Phytochemistry **27**, 1571–1574 (1988).

gen.: Beilstein E IV **4**, 2648 ▪ Merck-Index (12.), No. 817. – *[HS 2925 20; CAS 74-79-3]*

Argiopinins.

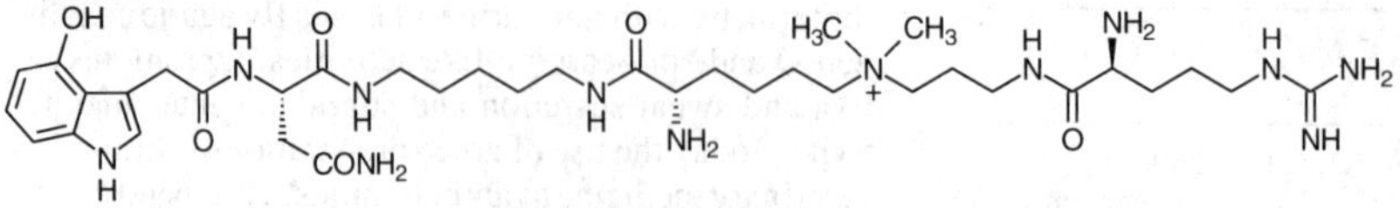

Argiopinin I

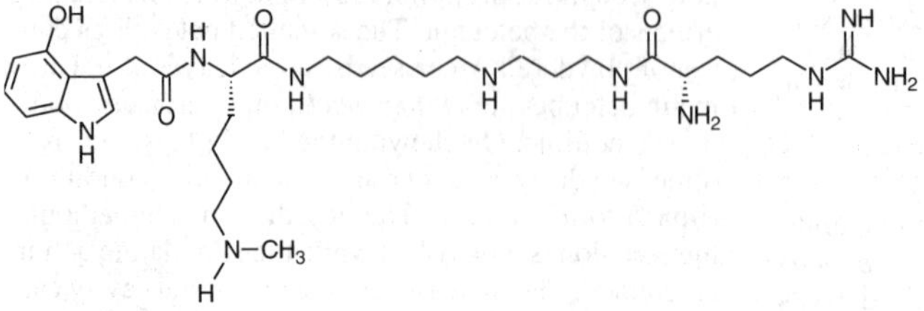

Argiopinin IV

Neurotoxic *polyamines from the venom of *Argiope* spiders [1]. A terminal amino group of the polyamine chain is substituted with an N^2-(4-hydroxy-3-indolylacetyl)-L-asparaginyl or N^2-(4-hydroxy-3-indolylacetyl)-N^6-methyl-L-lysyl group (in pseudoargiopinins the 4-hydroxy group is absent) while the other carries an L-arginyl residue. The differences are in the structure of the polyamine chain. A. block glutamate receptors and the associated Ca ion channels [2].

Table: Data of Argiopinins.

Argiopinin	molecular formula	M_R	CAS
A. I	$C_{36}H_{63}N_{12}O_6^+$	759.97 (ion)	117233-41-7
A. II	$C_{35}H_{60}N_{12}O_6$	744.94	117233-42-8
A. III (Argiotoxin-659)	$C_{31}H_{53}N_{11}O_5$	659.83	111944-83-3
A. IV	$C_{31}H_{54}N_{10}O_4$	630.83	117233-43-9
A. V	$C_{32}H_{56}N_{10}O_4$	644.86	117233-44-0
Pseudo-A. I	$C_{36}H_{63}N_{12}O_5^+$	743.97 (ion)	117255-12-6
Pseudo-A. II	$C_{35}H_{60}N_{12}O_5$	728.94	117233-45-1
Pseudo-A. III	$C_{19}H_{27}N_5O_3$	373.46	117233-46-2

Lit.: [1] Toxicon **27**, 541–549 (1989). [2] Pharmacol. Ther. **52**, 245–268 (1991).
gen.: J. Chem. Ecol. **19**, 2411–2451 (1993).

Argiotoxins. Acylpolyamines from the venom of various *Argiope* spiders (Araneidae, round web spiders) [1,2]. The numbers after the names give the molecular masses of the respective toxins. Like all other currently known acylpolyamines A. block the Ca^{2+}-dependent glutamate receptor non-competitively.

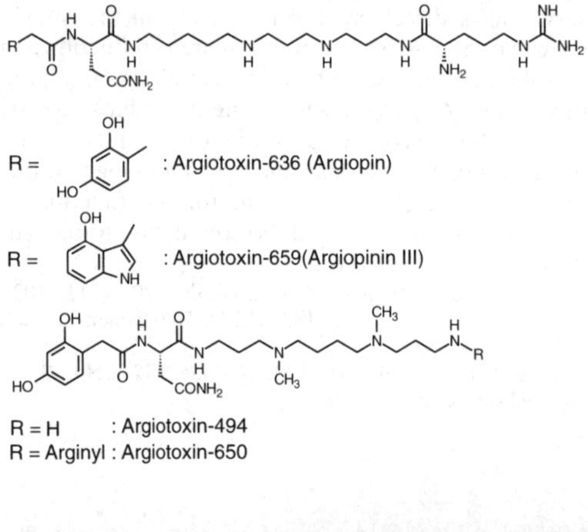

Table: Data of Argiotoxins.

Argiotoxin	molecular formula	M_R	CAS
A.-494	$C_{24}H_{42}N_6O_5$	494.63	
A.-636 (Argiopin)	$C_{29}H_{52}N_{10}O_6$	636.80	105029-41-2
A.-650	$C_{30}H_{54}N_{10}O_6$	650.82	
A.-659 s. Argiopinin III.			

Lit.: [1] Brain Res. **448**, 30–39 (1988). [2] Bioorg. Khim. **12**, 1121–1124 (1989).
gen.: J. Chem. Ecol. **19**, 2411–2451 (1993) ▪ Pharmacol. Ther. **52**, 245–268 (1991). – *Synthesis:* Biochem. Biophys. Res. Commun. **148**, 678–683 (1987) ▪ Bioorg. Khim. **14**, 704–706 (1988) ▪ Tetrahedron Lett. **28**, 6015–6018 (1988); **29**, 6223–6226 (1989) ▪ Tetrahedron **49**, 5777–5790 (1993).

Argopsin.

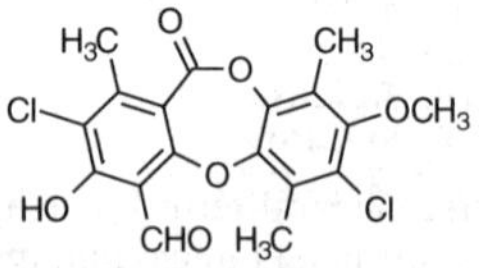

$C_{18}H_{14}Cl_2O_6$, M_R 397.21, prisms, mp. 220–221 °C; orange-red color on reaction with *p*-phenylenediamine. A *depsidone from the lichen *Argopsis friesiana*.

Lit.: Phytochemistry **14**, 1625–1628 (1975). – *[CAS 52809-10-6]*

Aristeromycin see neplanocins.

Aristolactone.

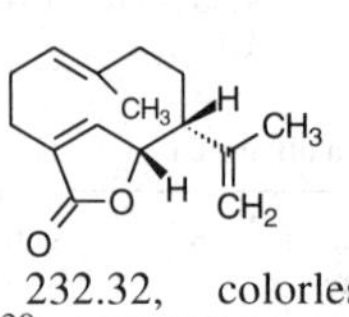

$C_{15}H_{20}O_2$, M_R 232.32, colorless cryst., mp. 110–111 °C, $[\alpha]_D^{20}$ +145.8° ($CHCl_3$); soluble in chloroform, ethanol, and THF. An α,β-unsaturated *sesquiterpene lactone with the germacrene skeleton (see germacrenes) from *Aristolochia reticulata* and *A. serpentaria* [1]. The absolute configuration has been determined both by X-ray diffraction [2] and enantiospecific synthesis [3].

Lit.: [1] J. Pharm. Pharmacol. **6**, 1005 (1954); J. Chem. Soc. **4**, 3289 (1959). [2] J. Chem. Res. (*S*) **1982**, 403. [3] J. Am. Chem. Soc. **110**, 2925 (1988).
gen.: Beilstein E V **17/10**, 91. – *[CAS 6790-85-8]*

Aristolochic acids. Various *Aristolochia* species (Aristolochiaceae) contain derivatives of 10-nitrophenanthrene-1-carboxylic acid, the so-called aristolochic acids.

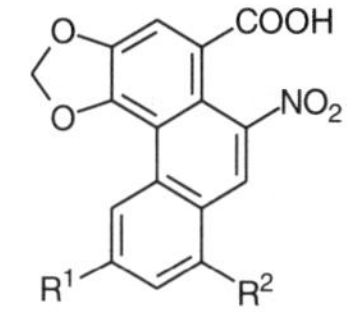

	R^1	R^2
A. I	H	OCH_3
A. C	OH	H
A. D	OH	OCH_3
A. IV	OCH_3	OCH_3
Aristoloside	*O*-β-D-Gluc	OCH_3

Table: Data of Aristolochic acids and derivatives.

compound	molecular formula	M_R	mp. [°C]	CAS
A. I	$C_{17}H_{11}NO_7$	341.28	281–286 (decomp.) yellow, bitter tasting needles	313-67-7
A. C	$C_{16}H_9NO_7$	327.25	needles · $5H_2O$: 280 (decomp.)	4849-90-5
A. D	$C_{17}H_{11}NO_8$	357.28	254–259 wine red cryst.	17413-38-6
A. IV	$C_{18}H_{13}NO_8$	371.30	268–270 (decomp.) deep-red prisms	15918-62-4
Aristoloside	$C_{23}H_{21}NO_{13}$	519.42	193–196 orange prisms $[\alpha]_D$ −69.5° (CH_3OH)	84014-70-0

A. I, isolated for the first time in 1851 from the roots of birthwort (*Aristolochia clematitis*), is widely distributed in the Aristolochiaceae and also occurs together with other nitrophenanthrenecarboxylic acids such as A. C in the Chinese drug Fang-chi (*Asarum canadense*). Also widely distributed are A. D and A. IV. Aristoloside is the *β*-D-glucoside of A. D and is isolated from *Aristolochia manshuriensis*.
A. I is characterized by cytotoxic activities and an inhibitory effect on platelet aggregation. It increases the phagocytosis activity of leukocytes, thus explaining the use of birthwort as a wound-healing agent by the ancient Egyptians and Romans. The use of A. preparations for wound-healing and in acne preparations was forbidden in Germany in 1982 because of its suspected cancerogenic action. A. I is known since 1993 to be the cause of a kidney disease (Chinese herbal medicine nephropathy), which results from confusion of the Chinese herbal medicine *Stephania tetrandra* with *Aristolochia fang-chi* (use in slimming preparations). Consumption of the extract leads to carcinomas and renal failure requiring transplantation.
The biosynthesis of A. involves oxidative degradation of aporphine precursors. They are taken up with food by caterpillars living on *Aristolochia* sp., stored, and later protect the butterfly (e.g., *Pachlioptera aristolochiae*) from potential enemies.

Lit.: Beilstein E V **19/7**, 611 f. ▪ Biochim. Biophys. Acta **1001**, 1 (1989) (pharmacology) ▪ Can. J. Chem. **47**, 481 (1969) (biosynthesis) ▪ J. Nat. Prod. **45**, 657 (1982) (review) ▪ Phytochemistry **21**, 1759 (1982) (aristoloside); **36**, 1063 (1994) ▪ Sax (8.), AQY 125, AQY 250.

Aristotelia alkaloids. *Indole alkaloids from *Aristotelia* spp. (Elaeocarpaceae), a family of approximately 50 compounds, e.g. *aristoteline* [$C_{20}H_{26}N_2$, M_R 294.43, mp. 164 °C, $[\alpha]_D^{20}$ +16° CH_3OH)] with hypotensive properties and *peduncularine* [$C_{20}H_{24}N_2$, M_R 292.42, mp. 155–157 °C, $[\alpha]_D^{19}$ −24° (CH_3OH)].

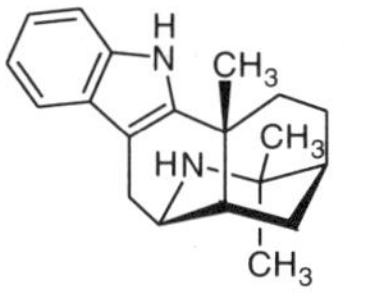

Aristoteline

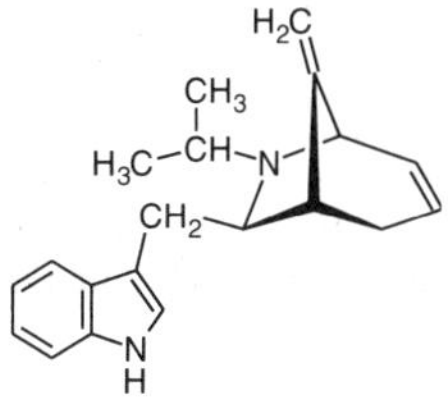

Peduncularine

Lit.: Chimia **45**, 329–341 (1991) ▪ Indian J. Chem., Sect B **38**, 789 (1999) ▪ J. Am. Chem. Soc. **111**, 2588 (1989) (peduncularine) ▪ J. Org. Chem. **57**, 5848 (1992); **58**, 564 (1993) (synthesis) ▪ Manske **24**, 113–152 ▪ Tetrahedron: Asymmetry **3**, 1197 (1992) (synthesis). – *[CAS 57103-59-0 (aristoteline); 34964-75-5 (peduncularine)]*

Aristoteline see Aristotelia alkaloids.

Armillaria sesquiterpenoids.

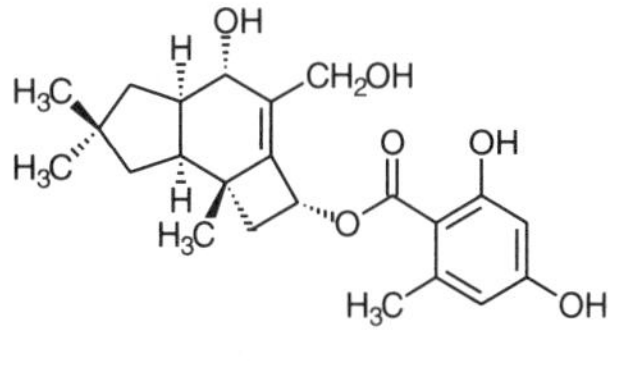

Armillyl orsellinate

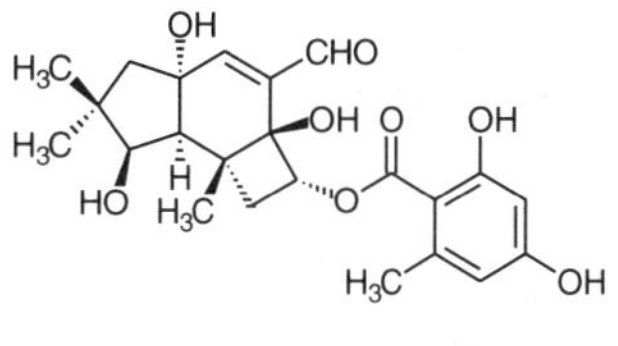

Melledonal A

Numerous orsellinic acid esters of hydroxylated *protoilludanes, active against Gram-positive bacteria and fungi, are isolated from cultures and mycelia extracts of the phytopathogenic honey mushrooms (*Armillaria mellea*, Basidiomycetes) and *A. novae-zelandiae*. *Armillyl orsellinate* {$C_{23}H_{30}O_6$, M_R 402.49, cryst., mp. 91–93 °C, $[\alpha]_D$ −126.9° (CH_3OH)} and *melledonal A* {$C_{23}H_{28}O_8$, M_R 432.47, cryst., mp. 136–137 °C, $[\alpha]_D$ +195.3° (CH_3OH)} are shown as examples. Structural variations can occur in the orsellinic acid part (chlorosubstitution, *O*-methylation, hydroxylation) and in the protoilludane skeleton.

Lit.: Chinese Chem. Lett. **1**, 173 (1990) ▪ Gazz. Chim. Ital. **118**, 517, 523 (1988) ▪ J. Chem. Soc., Chem. Commun. **1982**, 135; **1984**, 22 ▪ J. Chem. Soc., Perkin Trans. 1 **1988**, 301, 503 ▪ J. Nat. Prod. **48**, 10 (1985); **49**, 111 (1986) ▪ Phytochemistry **25**, 471 (1986); **26**, 3075 (1987); **29**, 179, 2569 (1990); **31**, 2047 (1992); **44**, 1473 (1997) (new aryl esters) ▪ Planta Med. **50**, 288 (1984); **55**, 479, 564 (1989); **57**, 578 (1991) ▪ Proc. Phytochem. Soc. of Europe **27**, 19–32 (1987) (review) ▪ Tetrahedron Lett. **23**, 2515 (1982); **26**, 5343 (1985). – *[CAS 82105-51-9 (armillyl orsellinate); 103847-15-0 (melledonal A)]*

Armoise (artemisia) oil. Light yellow to yellow oil with a fresh, green-herby, camphor-like, somewhat bitter-sweet odor.
Production: By steam distillation of the wild mugwort species *Artemisia herba-alba* from Morocco.
Composition [1]***:*** *A. herba-alba* exists as numerous chemotypes that are, however, limited to specific regions. Main components of the classical A., the so-called "Marrakech" type, are α- and *β-*thujan-3-ones* (35 and 5%) and **camphor* (40%); other types contain up to 60% α-thujanone.
Use: Exclusive use in perfume production, mainly for the so-called Chypre types of perfume (fresh-tart with masculine note).
Lit.: [1] Perfum. Flavor. **6** (1), 37 (1981); **7** (5), 44 (1982); **14** (3), 71 (1989); **19** (5), 94 (1994).
gen.: Arctander, p. 73 ▪ Bauer et al. (2.), p. 137 – *Toxicology:* Food Cosmet. Toxicol. **13**, 719 (1975). – *[HS 3301 29; CAS 8022-37-5]*

Aroma chemicals, fragrance chemicals. Name for natural or synthetic organic compounds used, on account of the usually pleasant odors, in the manufacture of perfumes and perfumed products. The *flavor compounds are distinguished from aroma chemicals by their occurrence and use in foods [1–5]. The term *odorant chemical* applies generally to compounds that can be detected using the sense of smell [6]. The term *fragrance chemicals* is used, except for aroma chemicals, also for odor complexes, e. g., the absolutes and *essential oils used in the perfume industry. For the physiology of the sense of smell, see *Lit.* [7,8].
Historical (see *Lit.* [9]): The use of aroma chemicals for religious and personal purposes has a long history and is mentioned in the oldest documents of the Assyrians and Egyptians [8]; however, the production of uniform aroma chemicals such as *anethole, *cedrol, *citral, *eugenol, *geraniol and menthol (see 3-*p*-menthanols) by crystallization or distillation from essential oils only began in the 18th and 19th centuries. The extensive development of organic chemistry in the 19th century finally enabled the preparation of synthetic aroma chemicals such as *coumarin, *ionones, or vanillin (see 3,4-dihydroxybenzaldehydes).
Structure & aroma: Since only little is known about the molecular structure of the appropriate receptors, empirically determined relationships between odor and molecular structure still provide the basis for the specific development of aroma chemicals [10,11]. The most important factors are molecular shape and size (vapor pressure) and the nature and position of polar, so-called "osmophoric" groups such as carbonyl, ester, and hydroxy groups or double bonds. Rigid molecules such as, e. g., *camphor, exhibit only one specific odor type while flexible molecules, especially those with differing molecular parts, have complex odors since they can apparently influence several different types of receptors.
Classification: A rough classification into the following groups of compounds can be made (approximate economic value in %): terpenoids (40%), brenzcatechol derivatives (17%), phenol derivatives (13%), other aromatic compounds (20%), aliphatic, alicyclic, and heterocyclic compounds (10%).
Qualitative and quantitative evaluations of an odor furnish odor profiles [11] for classification [12] that are amenable to electronic data processing. The Haarmann & Reimer odor cycle according to U. Harder [13] provides a good foundation as does the further developed odor landscape according to J.-N. Jaubert [14]. For the odor classes, see table.

Table: The most important classes of perfume scents.

class	olfactory character	typical examples
1 flowery	flower scents (rose, jasmine, etc.)	*Geraniol, *Ionones
2 citrus	like citrus fruits	*Citral
3 green	grass, green leaves	3-*Hexen-1-ol
4 herblike	green (spice) herbs	*Thujan-3-ones
4a minty	peppermint, spearmint	Menthone *(–)-Carvone
4b camphor-like	celluloid, earthlike	*Camphor
5 aldehyde	fatty, waxlike, like ozone, watery	*Alkanals and *Alkenals C_8–C_{13}
6 fruity	fruit fragrances	*fruit esters, *Alkanolides
7 sweet-balsamy	like resin, chocolate	Vanillin (see 3,4-Dihydroxy-benzaldehydes)
8 spicy	spice fragrances	*Eugenol
9 woody	e.g., cedar, sandalwood	*Santalol, *Cedrol
10 sensual	musk, ambergris, civet	*Muscone, Ambra oxide, *civet

Production: Even today many aroma chemicals are still isolated from essential oils, others are prepared semisynthetically from the components of the oils or from other suitable organic compounds. The proportion of natural and nature identical aroma chemicals used in the perfume industry is ca. 70%. The aroma chemicals not occurring in nature are often structural analogues of natural products that are difficult to synthesize, e. g., the sandalwood aroma chemicals prepared from campholene aldehyde or *camphene and *guaiacol as substitutes for the expensive *sandalwood oil or *santalols.
Use: In perfumes and for perfuming products for body hygiene, washing and cleaning products, for scenting rooms and for improving the odors of technical products. The world market for aroma chemicals and *flavor compounds amounted to 9.7 billion US-$ [15] in 1994.
Lit.: [1] Arctander. [2] Bauer et al. (2.). [3] Bedoukian (3.). [4] Ohloff. [5] Hopp & Mori (eds.), Recent Developments in Flavor and Fragrance Chemistry, Weinheim: VCH Verlagsges. 1993. [6] Theimer (ed.), Fragrance Chemistry, p. 1–25, San Diego: Academic Press 1982; Ohloff, p. 1–9; Müller-Lamparsky, p. 127–149, 481–498. [7] Toller & Dodd (ed.), Fragrance: The Psychology and Biology of Perfume, p. 37–50, 63–90, 142–160, London: Elsevier 1992. [8] Ohloff, Irdische Düfte – himmlische Lust. Eine Kulturgeschichte der Duftstoffe, p. 10–69, Basel: Birkhäuser 1992. [9] Bauer et al. (2.), p. 1–2; Bedoukian (3.), p. XI–XV; Ohloff, p. VII–X. [10] Ohloff, p. 11–56; Müller-Lamparsky, p. 287–330; Theimer (ed.), Fra-

grance Chemistry, p. 77–122, San Diego: Academic Press 1982. [11] Hopp & Mori (ed.), Recent Developments in Flavor and Fragrance Chemistry, p. 29–67, Weinheim: VCH Verlagsges. 1993; ACS Symp. Ser. **705** (1998). [12] ACS Symp. Ser. **674** (1997); Müller-Lamparsky, p. 3–12, 251–286, 333–335; Theimer (ed.), Fragrance Chemistry, p. 28–76, San Diego: Academic Press 1982. [13] H & R Contact, Nr. 23, 18–27 (6/1979) [14] Parfums Cosmet. Arômes **77**, 53–56 (1987); **78**, 71–82 (1987). [15] Chem. Ind. (London) **1996**, 170 ff.

Aroma compounds see flavor compounds.

Arrow poisons. Toxins used by various native tribes in tropical or subtropical regions to poison arrowheads for use in hunting or warfare. They are mostly plant toxins, e.g., *aconitine or the Brazilian Tike-Uba toxin, but also *snake, *spider, *scorpion, and *amphibian venoms as well as toxins from caterpillars and beetles, e.g., the African bushman poison *diamphotoxin* (polypeptide, M_R ca. 60000) from pupae of *Diamphidia nigro-ornata*, toxins from marine animals and algae, etc., were used. The plant toxins used by African tribes include extracts from *Strophanthus*, *Adenium*, *Acokanthera*, *Calotropis*, and *Strychnos* species. Animal toxins were used preferentially by South and Central American tribes, especially the highly toxic secretions of colored frogs ("arrow poison frogs", see toad poisons, amphibian venoms, frog toxins, batrachotoxins, histrionicotoxins, gephyrotoxin, pumiliotoxins) of the species *Phyllobates* and *Dendrobates*. Curare and other components of *Strychnos* species were used as plant poisons. In East Asia the milk juice of the Upas tree and bark extracts of *Lophopetalum toxicum* were used as A. p. The chemical structures of the A. p. vary widely, cardiac glycosides on the basis of *cardenolides and *bufadienolides, *steroid alkaloids, *diterpene alkaloids, *strychnine, and *curare are widespread. The effects differ accordingly: cardiac arrest, paralysis, hypoglycemia, convulsions, etc. have been described.

Lit.: J. Ethnopharmacol. **25**, 1–41 (1989) ▪ Lewin, Die Pfeilgifte (2.), Hildesheim: Gerstenberg 1984 ▪ Med. Monatsschr. Pharm. **16**, 101–107 (1993) ▪ Naturwiss. Rundsch. **47**, 379–385 (1994) (arrow poison frogs) ▪ Neuwinger, Afrikanische Arzneipflanzen u. Jagdgifte, Stuttgart: Wiss. Verlagsges. 1994 ▪ Toxicon **27**, 1351–1366 (1989); **28**, 435–444 (1990). – *[CAS 87915-42-2 (diamphotoxin)]*

Arsenic in Natural Products.

$(H_3C)_3\overset{+}{As}–CH_2–COO^-$

Arsenobetaine

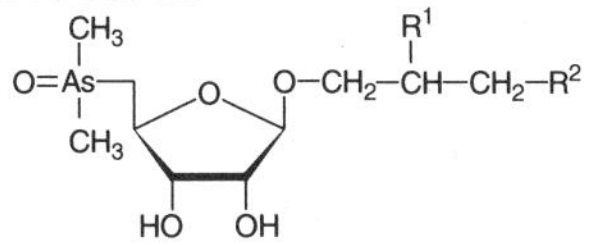

$[(H_3C)_4\overset{+}{As}]\ Cl^-$

Tetramethylarsonium chloride

1, R^1 = OH, R^2 = OSO_3H

2, R^1 = OH, R^2 = SO_3H

3, R^1 = OH, R^2 = $O–P(=O)(O^-)–O–CH_2–CH(OH)–CH_2–OH$

$(H_3C)_3As{=}O$

Trimethylarsine oxide

4, R^1 = NH_2, R^2 = SO_3H

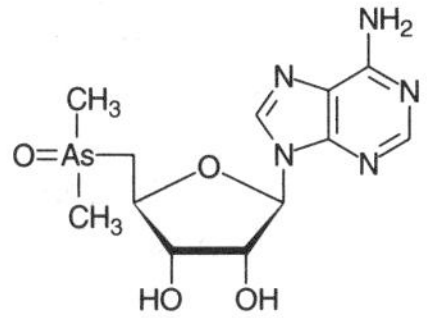

5

Various arsenic containing natural products have been isolated together with inorganic arsenates from marine organisms. The most widely distributed compounds in marine species seem to be *arsenobetaine* ($C_5H_{11}AsO_2$, M_R 178.06, cryst., mp. 204–210 °C), *tetramethylarsonium chloride* ($C_4H_{12}AsCl$, M_R 170.51), and *trimethylarsine oxide* (C_3H_9AsO, M_R 136.03) [1]. Arsenobetaine is weakly toxic and is excreted unchanged after consumption by marine species. The various *dimethylarsinoyl-* and *trimethylarsonio-β-D-ribosides*, which occur especially in brown algae, have more complicated structures. Thus, compound **1** accounts for 50% of the total arsenic content (10 µg/g fresh weight) of *Hizikia fusiforme* ("Hijiki", Fucales), a much esteemed food in Japan [2]. **1** can also be isolated from the kidneys of the giant clam *Tridacna maxima* (1 g As/kg dry weight) [3]. Compounds **2** and **3** occur in *H. fusiforme* and the brown kelp (*Ecklonia radiata*) [2,4], **4** occurs only in *H. fusiforme* [1]. The occurrence of the nucleoside **5** in the kidneys of *T. maxima* has biosynthetic significance [1,5].

Lit.: [1] Nat. Prod. Rep. **10**, 421–428 (1993) (review). [2] J. Chem. Soc., Perkin Trans. 1 **1987**, 577. [3] J. Chem. Soc., Perkin Trans. 1 **1982**, 2989. [4] J. Chem. Soc., Perkin Trans. 1 **1983**, 2375. [5] J. Chem. Soc., Perkin Trans. 1 **1991**, 928; **1992**, 1349. – *[CAS 64436-13-1 (arsenobetaine); 5814-22-2 (tetramethylarsonium chloride); 4964-14-1 (trimethylarsine oxide)]*

Artabotrine (isocorydine) see aporphine alkaloids.

Artabsin see proazulenes.

Artemether see qinghaosu.

Artemisia ketone (3,3,6-trimethyl-1,5-heptadien-4-one).

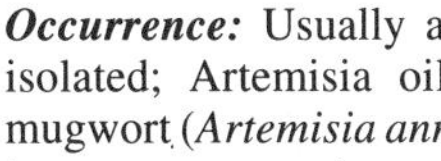

Artemisia ketone

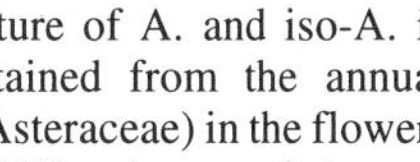

Isoartemisia ketone

$C_{10}H_{16}O$, M_R 152.24, oil, bp. 182–183 °C. A. and its isomer iso-A. (2,5,5-trimethyl-1,6-heptadien-4-one) are *monoterpene ketones for which, as yet, no practical uses have been found.

Occurrence: Usually a mixture of A. and iso-A. is isolated; Artemisia oil, obtained from the annual mugwort (*Artemisia annua*, Asteraceae) in the flowering season contains a ca. 30% mixture of A. and iso-A. 65% mixtures occur in the essential oil of lavender cotton (*Santolina chamaecyparissus*, Asteraceae).

Biosynthesis: A. is one of the few monoterpenes that is not formed by the known biosynthetic routes for *terpenes since there are two methyl groups on C-3 of A. A. prepared biosynthetically from [2-^{14}C]-mevalonic acid contains 80% of the radioactivity at C-9/C-10 and only 13% at C-7/C-8. This difference in activity means that the C_5 precursors of the two halves of A. undergo different enzymatic reactions [1], i.e., a head-to-tail reaction of the C_5 precursors does not occur.

Lit.: [1] Chem. Lett. **1972**, 533.

gen.: Beilstein E II **1**, 813 ▪ Experientia **34**, 1121 (1978) ▪ Helv. Chim. Acta **70**, 1745 (1987) (synthesis) ▪ Karrer, No. 436. – *[CAS 546-49-6 (A.); 512-37-8 (iso-A.)]*

Artemisin (8-hydroxysantonin).

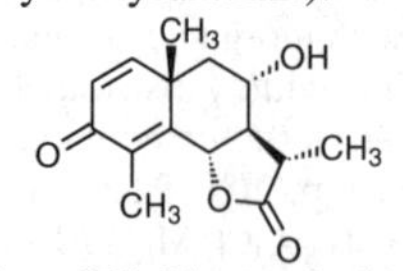

$C_{15}H_{18}O_4$, M_R 262.31, colorless cryst., mp. 202–203 °C, $[\alpha]_D^{23}$ –84.9° (C_2H_5OH); soluble in ethyl acetate, insoluble in petroleum ether. A *sesquiterpene from various *Artemisia* species (Asteraceae) (cf. santonin) with a bitter taste. Since A. can be prepared enantioselectively [1], it finds frequent use as starting material for natural product synthesis[2].

Lit.: [1] Tetrahedron Lett. **1966**, 2615; Bull. Chem. Soc. Jpn. **42**, 3366 (1969). [2] Tetrahedron **45**, 5925 (1989); J. Org. Chem. **57**, 3610 (1992); Tetrahedron Lett. **33**, 5253 (1992).
gen.: Beilstein E V **18/3**, 304. – *[CAS 481-05-0]*

Artemisinin see qinghaosu.

Arthogalin.

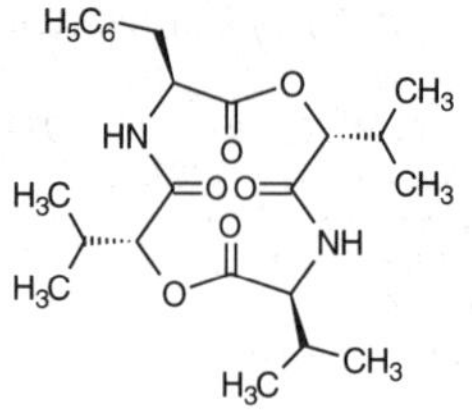

$C_{24}H_{34}N_2O_6$, M_R 446.54, thin needles with mp. 285–286 °C. The first cyclic depsipeptide from a lichen (*Arthothelium galapagoense*, crustaceous lichen endemic to the Galapagos Islands). On acid hydrolysis A. gives L-phenylalanine, L-valine, and D-α-hydroxyisovaleric acid.

Lit.: Z. Naturforsch. B **50**, 1101–1103 (1995).

Arthonin [(–)-*N*-benzoyl-L-isoleucine-*N'*-benzoyl-L-valinyl ester].

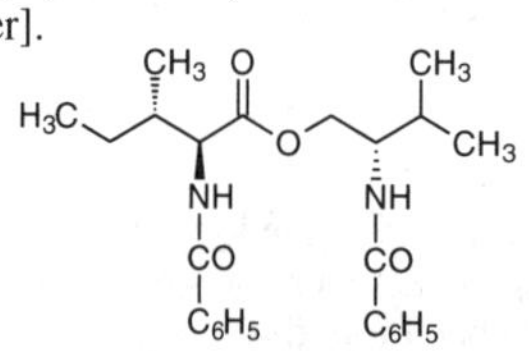

$C_{25}H_{32}N_2O_4$, M_R 424.54, needles, mp. 165–167 °C, $[\alpha]_D^{20}$ –8.8°; soluble in methanol. Amino acid-amino alcohol-ester (L-isoleucine – L-valinol) from the crustaceous lichen *Arthonia endlicheri*.

Lit.: Tetrahedron: Asymmetry **4**, 303–311 (1993). – *[CAS 149182-40-1]*

Arthrobacilins. Glycolipid *antibiotic complex formed by an *Arthrobacter* sp. No. 2967 and stored in the cells. A. represent cyclic trimeric esters of long-chain 3-(β-D-galactopyranosyloxy)-fatty acids which,

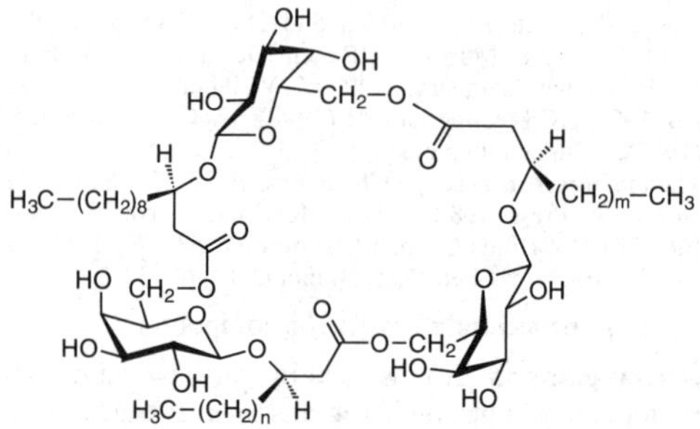

m = 8 n = 8 : A
m = 8 n = 10 : B
m = 10 n = 10 : C

with their 27-membered ring structures (3-fold symmetry axis) and the 9 hydroxy groups, resemble the *cyclodextrins. However, as a result of their hydrophobic alkyl side chains they are soluble in organic solvents. A. weakly inhibit the growth of HeLa cells (IC_{50} 18 µg/mL).

Lit.: Tetrahedron Lett. **33**, 2705 (1992).

Table: Data of Arthrobacilins.

	A	B	C
molecular formula	$C_{54}H_{96}O_{21}$	$C_{56}H_{100}O_{21}$	$C_{58}H_{104}O_{21}$
M_R	1081.34	1109.40	1137.45
$[\alpha]_D^{22}$ ($CHCl_3$)	–31°	–26°	–28°
CAS	142547-11-3	142547-12-4	142547-13-5

Arthrobactin see aerobactin.

Arthrographol see asperfuran.

Arthrosporone s. triquinanes.

Asarones.

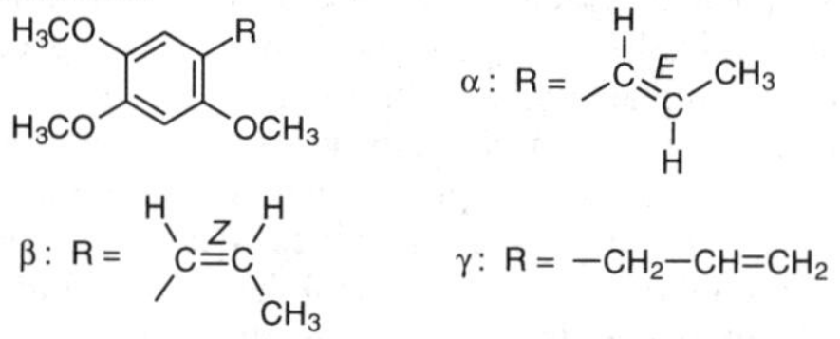

$C_{12}H_{16}O_3$, M_R 208.26.

1. α-A. [asarin, 1,2,4-trimethoxy-5-((*E*)-1-propenyl)benzene] needles, mp. 62–63 °C, bp. 296 °C. A. occurs in asarabacca (*Asarum europaeum*, Aristolochiaceae) and is the main active principle of *calamus oil; it has a cholesterol-lowering action, but also has skin- and mucous membrane-irritating, nauseating, abortive, and narcotic properties; it can lead to respiratory paralysis; LD_{50} (mouse p. o.) 418 mg/kg. It is also suspected to have cancerogenic activity.

2. β-A. [1,2,4-trimethoxy-5-((*Z*)-1-propenyl)benzene] yellow oil, D. 1.082, bp. 163 °C (1.6 kPa), also occurs in calamus oil as well as *Acorus* sp. and acts as a chemosterilant for insects [1].

3. γ-A. (isoasarone, sekishone, 1-allyl-2,4,5-trimethoxybenzene), oil, mp. 25 °C, bp. 145–147 °C (267 Pa), occurs in *Caesulia axillaris*, *Acorus calamus*, and *Aniba hostmannia* and exhibits antifeedant effects on insects.

Lit.: [1] Nature (London) **270**, 512 (1977); Z. Naturforsch. B **35**, 1449 f. (1980).
gen.: Angew. Chem. Int. Ed. Engl. **22**, 623, Suppl. 887 (1983) (synthesis) ▪ Beilstein E IV **6**, 7476 ▪ Hager (5.) **3**, 100 ff.; **4**, 377–395 ▪ J. Nat. Prod. **44**, 668 f. (1981) (^{13}C NMR). – *Toxicology:* Sax (8.), IHX 400. – *[CAS 2883-98-9 (α); 5273-86-9 (β); 5353-15-1 (γ)]*

Ascaridole (1,4-epidioxy-*p*-menth-2-ene). $C_{10}H_{16}O_2$, M_R 168.24, pale greenish-yellow, clear liquid, mp.

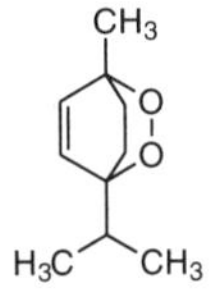

3 °C, bp. 75 °C (900 Pa), D. 1.01, racemic A. is a *monoterpene peroxide. It explodes on warming to 130 °C and on treatment with acids.

Occurrence: A. is the main component (45–70%) of oleum chenopodii (*oil of chenopodium, wormseed oil), an oil obtained by steam distillation of withered, fruit-bearing twigs of the fragrant goosefoot (*Chenopodium ambrosioides*, Chenopodiaceae).

Use: Anthelmintic, active against eelworms and hookworms. Excessive doses lead to severe, often lethal poisonings so that A. has been replaced by synthetic anthelmintics with specific activities.

Lit.: Beilstein E V **19/1**, 319 ▪ J. Org. Chem. **48**, 4918 (1983) (synthesis) ▪ Karrer, No. 571 ▪ Merck-Index (12.), No. 864 ▪ Sax (8.), ARM 500 ▪ Ullmann (5.) **A 19**, 205. – *[HS 2909 60; CAS 512-85-6]*

Ascididemnin see shermilamines.

Ascochitin.

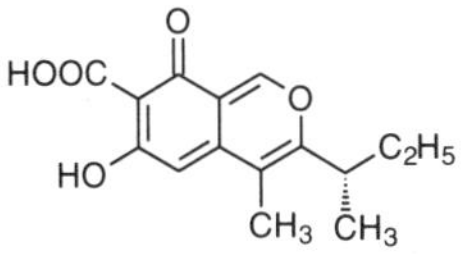

$C_{15}H_{16}O_5$, M_R 276.29, yellow cryst., mp. 196–198 °C, $[\alpha]_D^{25}$ –86° ($CHCl_3$). Antibacterial, antifungal, phytotoxic, and cytotoxic metabolite of the ascomycetes *Ascochyta pisi* and *A. fabae*. The biological activities are similar to those of *citrinin.

Lit.: J. Chem. Soc. (C) **1971**, 3557 (synthesis) ▪ J. Chem. Soc. Perkin Trans. 1 **1980**, 675, 2549 (biosynthesis) ▪ Mycol. Res. **98**, 1069 (1994) (production). – *[CAS 3615-05-2]*

Ascochlorin.

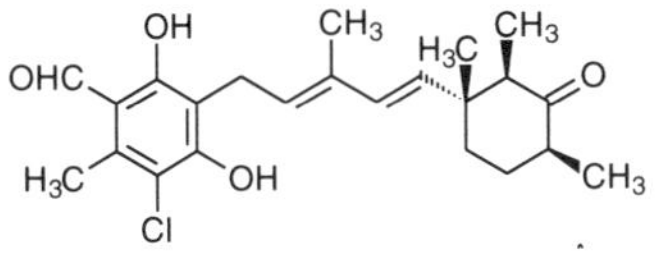

$C_{23}H_{29}ClO_4$, M_R 404.93, needles, mp. 172–173 °C, $[\alpha]_D^{25}$ –31° (CH_3OH), metabolite of *Ascochyta vicae*, *Nectria coccinea*, and other fungi with antibiotic and antiviral properties.

Lit.: Tetrahedron **25**, 1323 (1969) (isolation); **42**, 2635 (1986) (synthesis). – *[CAS 26166-39-2]*

Ascomycin (Immunomycin, FK-520, FR-900520).

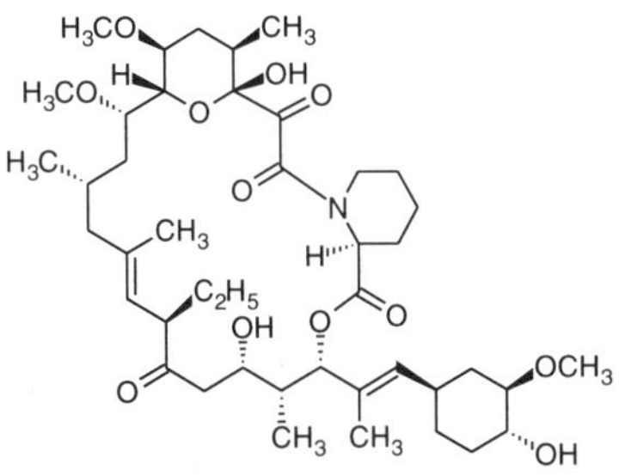

A macrolide antibiotic, related to *FK-506 (ethyl group instead of allyl group), $C_{43}H_{69}NO_{12}$, M_R 792.01, plates, mp. 163–165 °C, $[\alpha]_D^{23}$ –84° ($CHCl_3$). A. is isolated from cultures of *Streptomyces hygroscopicus* and *S. tsukubaensis*. It shows immunosuppressive properties.

Lit.: Bioorg. Med. Chem. Lett. **6**, 2193 (1996) (synthesis) ▪ J. Antibiotics **45**, 126–128 (1992); **47**, 806 (1994) (isolation); **50**, 701 (1997) (biosynthesis) ▪ Tetrahedron **49**, 8771 (1993) (structure). – *[CAS 104987-12-4]*

Ascorbic acid (vitamin C).

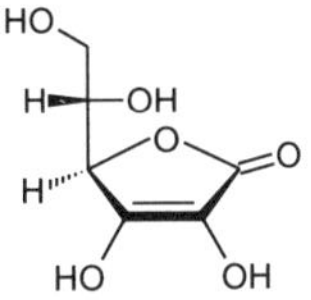

$C_6H_8O_6$, M_R 176.13. Pleasantly tart tasting cryst., D. 1.65, mp. 192 °C (decomp.). Natural L-A. is dextrorotatory, $[\alpha]_D^{25}$ +24° (H_2O), $[\alpha]_D^{18}$ +49° (CH_3OH); pK_{a1} 4.04, pK_{a2} 11.34 (25 °C/0.1 m KNO_3). It is widely distributed in all higher plants and animals, citrus fruits and many berries are particularly rich in L-A. It has a relatively strong reducing potential, ε_0' 0.127 V (at pH 5), is sensitive to light and alkali and easily undergoes autoxidation, especially in the presence of traces of heavy metals (Cu, Fe). Dry A. is stable as are its alkali salts and esters. LD_{50} (rat p.o.) 11900 mg/kg, (mouse p.o.) 3367 mg/kg, TLD_0 (human i.v.) 2300 mg/kg.

Activity: L-A. has anti-scurvy activity. The use of A. as prophylaxis against colds and cancer is questionable while positive effects on wound-healing and the immune system have been demonstrated. Unnatural D-(–)-A. {$[\alpha]_D^{18}$ –23° (H_2O), $[\alpha]_D^{18}$ –48° (CH_3OH)} has practically no biological activity. The biosynthesis proceeds through D-glucose → D-glucuronic acid → L-gulonic acid → L-gulonic acid γ-lactone → 2-oxo-L-gulonic acid γ-lactone.

Use: As antioxidant for technical and food technological purposes and in vitamin preparations.

Lit.: Acta Crystallogr., Sect. B **25**, 2214–2223 (1969) (crystal structure) ▪ Adv. Appl. Microbiol. **12**, 11–33 (1979) (biotechn. production) ▪ Adv. Carbohydr. Chem. **37**, 7–77, 79–155 (1980) ▪ Anal. Biochem. **123**, 389 (1982) (detection) ▪ Beilstein E V **18/5**, 26 ▪ Chem. Ind. (London) **1993**, 280 ▪ Hager (5.) **7**, 299 ff.; **4**, 261 f.; **6**, 992 f. ▪ J. Chem. Soc., Perkin Trans. 1 **1998**, 3141 (synthesis) ▪ Kirk-Othmer (4.) **25**, 17–47 (review) ▪ Machlin (ed.), Handbook of Vitamins, p. 195–232, New York: Dekker 1991 ▪ Nature (London) **393**, 365 (1998) (biosynthesis) ▪ Phytochemistry **52**, 193–210 (1999) (biosynthesis) ▪ Tetrahedron **33**, 1587 ff. (1977) (^{13}C NMR) ▪ Ullmann (5.) **A 9**, 448; **A 11**, 563, 568; **A 27**, 547–607 (review) ▪ Vitamin C: its Chemistry and Biochemistry, London: R. Chem. Soc. 1991 ▪ U.S. Pat. 5,998,634 (7. 12. 1999) (synthesis). 1. – *Toxicology:* Sax (8.), ARN 000. – *[HS 2936 27; CAS 50-81-7 (A.); 10504-35-5 (D-(–))]*

Asiaticoside. Saponin-like *triterpene glycoside {$C_{48}H_{78}O_{19}$, M_R 959.14, mp. 235–238 °C, $[\alpha]_D$ –14° (C_2H_5OH)} from the leaves of the Asian marsh pennywort, *Centella asiatica* (*Hydrocotyle asiatica*, Apiaceae). A. shows antibiotic activity against Gram-positive bacteria and mycobacteria and has a wound-healing effect.

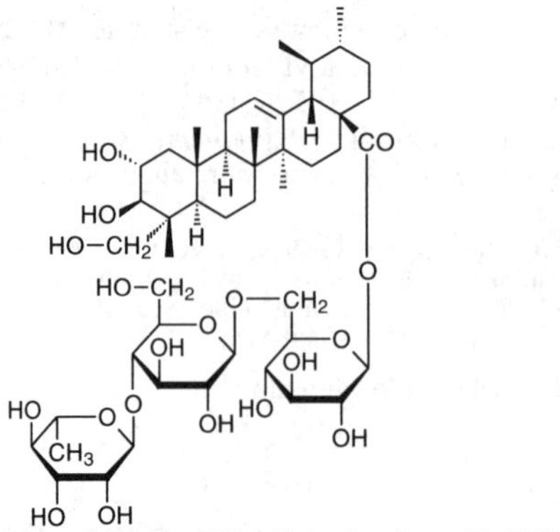

Lit.: Merck-Index (12.), No. 869 ▪ Phytochemistry **37**, 1131 (1994). – *[HS 2938 90; CAS 16830-15-2]*

Asimicin see annonaceous acetogenins.

ASP see domoic acid.

Asparagamine A.

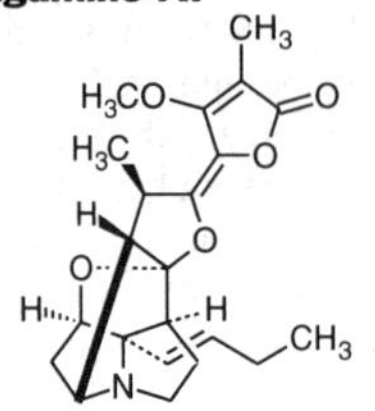

(relative configuration)

An alkaloid with cage-like structure from roots of *Asparagus racemosus* (Liliaceae). $C_{22}H_{27}NO_5$, M_R 385.45, prisms, mp. 180 °C, $[\alpha]_D$ +202.5° (CH_3OH). A. A shows anti-oxytocin and antineoplastic activities.
Lit.: J. Chem. Soc., Perkin Trans. 1 **1995**, 391. – *[CAS 156798-15-1]*

Asparagine [2-aminosuccinic acid 4-amide, abbreviation: Asn, Asp(NH_2), or N].

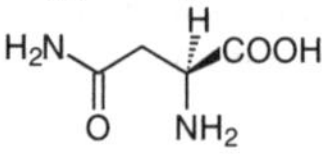

(2*S*)-, L-form

$C_4H_8N_2O_3$, M_R 132.12, monohydrate, mp. 226–227 °C (decomp.). $[\alpha]_D^{25}$ –5.6° (H_2O), +33.2° (3 m HCl); pK_a 2.1 and 8.84, pI 5.41. Proteinogenic amino acid that is widely distributed in plants in the free form. The genetic code in protein synthesis is AAU and AAC. The average content in proteins is 4.4% [1]. A. was isolated as the first natural amino acid from the juice of asparagus (*Asparagus officinalis*) and later from various plants such as shoots of vetch, lupines, soy beans, etc. Free Asn sometimes serves as a storage form for nitrogen in plants. Some glycoproteins contain N^2-glycosides of Asn at the end of the side chain. Asn is formed from L-aspartic acid, NH_3 and ATP by A. synthetase (EC 6.3.1.1.) and is classified as a non-essential amino acid. Asn is hydrolysed by asparaginase (EC 3.5.1.1.).
Isoasparagine (α-Asn, 2-aminosuccinic acid 1-amide). (2*S*)-form, mp. ca. 190 °C (decomp.), $[\alpha]_D^{20}$ +14.9° (0.1 m HCl), pI 5.50. A non-proteinogenic amino acid in *Chara corallina* (Charophyceae) and related species [2].
N^4-(2-Hydroxyethyl)-L-asparagine. (2*S*)-Form, mp. 199–200 °C, $[\alpha]_D^{20}$ +2.9° (H_2O). Non-proteinogenic amino acid from he squirting cucumber (*Ecballium elaterium*) and the red bryony (*Bryonia dioica*).
Lit.: [1] Biochem. Biophys. Res. Commun. **78**, 1018–1024 (1977). [2] Phytochemistry **22**, 2313 f. (1983); Merck-Index (12.), No. 872.
gen.: Beilstein EIV **4**, 3004–3006 ▪ Greenstein & Winitz, Chemistry of the Amino Acids, vol. 3, p. 1856–1865, New York: Wiley 1961 ▪ Karrer, No. 2369 ▪ Phytochemistry **27**, 663 (1988) (review). – *[HS 2924 10; CAS 70-47-3 (A.); 28057-52-5 (iso-A.); 7175-33-9 (N^4-(2-hydroxyethyl)-L-A.)]*

Asparagus flavor see vegetable flavors.

Aspartic acid (2-Aminosuccinic acid, abbreviation: Asp or D).

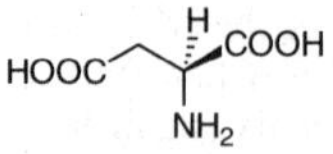

(2*S*)-, L-form

$C_4H_7NO_4$, M_R 133.10, mp. 269–271 °C (decomp.), $[\alpha]_D^{25}$ +5.05° (H_2O), +25.4° (5 N HCl); pK_a 1.99, 9.90 and 3.90 (β-COOH), pI 2.77. Proteinogenic amino acid, widely distributed as protein building block (average 5.5% [1]) and in the free form. Free A. is found in higher concentrations in sugar beet and sugar beet molasses; and is easily accessible by enzymatic hydrolysis of asparagine.
Biosynthesis & metabolism: Asp is formed from oxaloacetic acid by aspartate aminotransferase (EC 2.6.1.1) and serves as starting material in the biosyntheses of threonine, methionine, and lysine. The first step is catalysed by aspartate kinase (EC 2.7.24) which only occurs in plants and microorganisms. This enzyme exists as 3 isozymes in *Escherichia coli* and exhibits a typical example of feedback regulation. Asp plays a central role in the biosyntheses of *pyrimidines and *purines. In the urea cycle Asp condenses with *citrulline to argininosuccinate, a stimulating neurotransmitter [2].
L-Aspartyl-L-phenylalanine methyl ester (Aspartam®) is an artificial sweetener that is almost 200 times sweeter than saccharose.
Lit.: [1] Biochem. Biophys. Res. Commun. **78**, 1018–1024 (1977). [2] Physiol. Pharmacol. **54**, 70 (1976).
gen.: Beilstein EIV **4**, 2998–3001 ▪ Collingridge (ed.), The NMDA Receptor, Oxford: Univ. Press 1994 ▪ Collect. Czech. Chem. Commun. **50**, 2122 (1985) (manufacture) ▪ Greenstein & Winitz, Chemistry of the Amino Acids, vol. 3, p. 1856–1878, New York: Wiley 1961 ▪ Merck-Index (12.), No. 875. – *[HS 2922 49; CAS 56-84-8]*

Asperdiol [(1*R*,2*S*,3*E*,7*R*,8*R*,11*E*)-7,8-epoxy-3,11,15-cembratriene-2,18-diol].

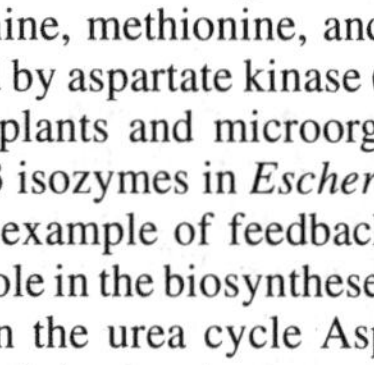

$C_{20}H_{32}O_3$, M_R 320.47, needles, mp. 109–110 °C, $[\alpha]_D^{20}$ –87° ($CHCl_3$). Diterpene from the gorgonian *Eunicea asperula* with antitumor activity, it belongs to the *cembranoids.
Lit.: Beilstein E V **17/5**, 242 ▪ J. Am. Chem. Soc. **108**, 6389 ff. (1986) ▪ J. Org. Chem. **51**, 858–872 (1986); **52**, 1803–1810 (1987) ▪ Scheuer I 2, 189 f. – *[CAS 64180-67-2]*

Asperentin see cladosporin.

Asperfuran (arthrographol).

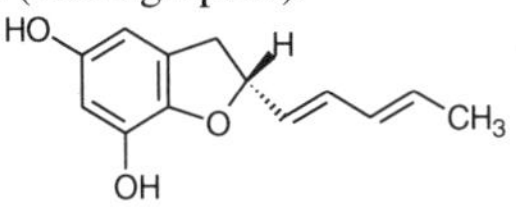

$C_{13}H_{14}O_3$, M_R 218.25, cryst., mp. 130 °C, $[\alpha]_D$ –20.9° (acetone). Antifungal dihydrobenzofuran derivative from culture media of *Aspergillus oryzae* and *Arthrographis pinicola*.

Lit.: Can. J. Microbiol. **36**, 77, 83 (1990) ▪ J. Antibiot. **43**, 648 (1990). – *Synthesis* (*rac*-A.): Can. J. Chem. **69**, 1909 (1991) ▪ Chem. Lett. **1993**, 1683 ▪ Heterocycles **43**, 665 (1996). – *[CAS 129277-10-7]*

Aspergillic acid.

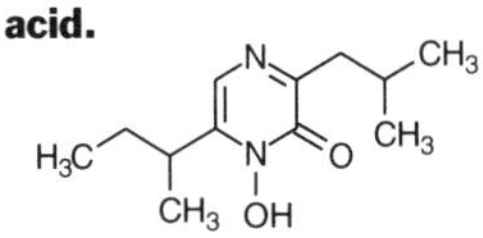

$C_{12}H_{20}N_2O_2$, M_R 224.30, pale yellow needles, mp. 97–99 °C, $[\alpha]_D^{18}$ +13.3°; red complex with Fe^{3+}, green Cu(II) salts. Toxic, antibacterial metabolite from *Aspergillus flavus*, *A. sojae*, and *A. quercinus*, LD_{50} (mouse p.o.) 4–5 mg/kg. A. is formed from leucine and isoleucine. Other amino acids can be built-in by specific feeding during the fermentation, e.g., neoaspergillic acid (2 mol. leucine).

Lit.: Biochem. J. **96**, 533–538 (1965) (biosynthesis) ▪ Can. J. Biochem. **51**, 1311–1315 (1973) ▪ Cole-Cox, p. 794 ▪ Gottlieb & Shaw (eds.), Antibiotics II, p. 43–51, Heidelberg: Springer 1967 ▪ Laskin & Lechevalier (eds.), Handbook of Microbiology, vol. 5, p. 623–626, Boca Raton: CRC Press 1984 ▪ Manske **29**, 185 (1986) (review) ▪ Sax (8.), ARO 000. – *[CAS 490-02-8]*

Aspergillin see gliotoxin.

Asperlicin.

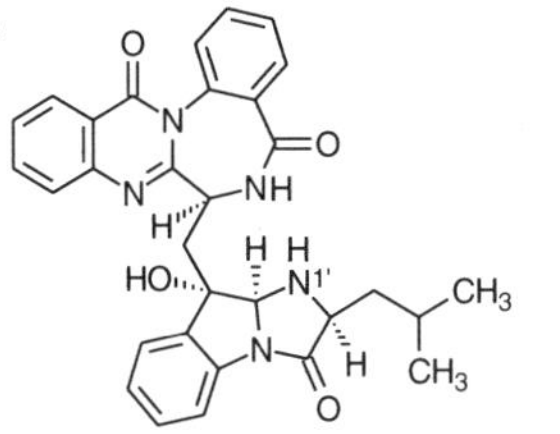

$C_{31}H_{29}N_5O_4$, M_R 535.60, cryst., mp. 211–213 °C, $[\alpha]_D$ –185.3° (CH_3OH), parent compound of the asperlicins produced by *Aspergillus alliaceus*[1]. A. has a benzodiazepine skeleton biosynthetically formed from 2 molecules anthranilic acid and L-tryptophan (with attached L-leucine). Derivatives of A. are obtained by precursor-directed biosynthesis[2]. *A. B* (1′-hydroxyasperlicin, $C_{31}H_{29}N_5O_5$, M_R 551.60) is an effective antagonist of cholecystokinins (CCK) and acts 7 times stronger than asperlicin. CCK (= pancreozymin) is a gastrointestinal peptide hormone that stimulates enzyme secretion of the pancreas and stomach and induces contractions of the gallbladder. A. and its derivatives have potential applications in the treatment of stomach ulcers and pancreatic tumors[3].

Lit.: [1]Science **230**, 177 (1985). [2]J. Antibiot. **41**, 878 ff., 882–891 (1988). [3]Proc. Natl. Acad. Sci. USA **83**, 4918, 4923 (1986); Merck-Index (12.), No. 878.

gen.: J. Antibiot. **47**, 599 ff. (1994) (NMR) ▪ J. Am. Chem. Soc. **120**, 6417 (1998) (synthesis) ▪ J. Ind. Microbiol. **4**, 97 (1989) (review) ▪ J. Med. Chem. **29**, 1941 (1986) ▪ J. Org. Chem. **52**, 1644 ff. (1987) (synthesis). – *[CAS 93413-04-8 (A.); 93413-08-2 (A. B)]*

Asperpentin see siccayne.

Asperuloside. A. and the compounds listed in the table are *iridoid glucosides with an intramolecular lactone ring or an acid function. Some of the acids may be formed during work-up.

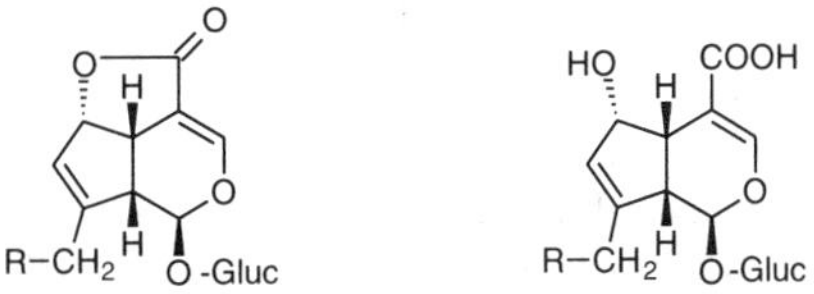

R = O–CO–CH3 : Asperuloside (**1**)
R = OH : Deacetylasperuloside (**2**)
R = O–CO–S–CH3 : Paederoside (**3**)

R = O–CO–CH3 : Asperulosidic acid (**4**)
R = OH : Deacetylasperulosidic acid (**5**)
R = O–CO–S–CH3 : Paederosidic acid (**6**)

Table: Data of Asperuloside and related compounds.

no.	molecular formula	M_R	mp. [°C]	$[\alpha]_D$	CAS
1	$C_{18}H_{22}O_{11}$	414.37	131–132	–200° (H_2O)	14259-45-1
2	$C_{16}H_{20}O_{10}$	372.33	118–120	–88° (C_2H_5OH)	18843-01-1
3	$C_{18}H_{22}O_{11}S$	446.43	122–123	–195° (CH_3OH)	20547-45-9
4	$C_{18}H_{24}O_{12}$	432.38	127–131	+8.6° (CH_3OH)	25368-11-0
5	$C_{16}H_{22}O_{11}$	390.34	138–145	+33° (H_2O)	14259-55-3
6	$C_{18}H_{24}O_{12}S$	464.44	124–129	+54° (CH_3OH)	18842-98-3

A. and the listed compounds occur mainly in Rubiaceae[1] but have also been detected in other plant families. A. is isolated from fresh woodruff (*Asperula odorata*) and the bark of the New Zealand plants *Coprosma robusta* (1.2%), *C. tenuifolia* (2.9%), *C. repens* (1.7%), and *C. arborea* (0.6%) (all Rubiaceae). Paederoside and paederosidic acids occur in the Japanese plant *Paederia scandens* (Rubiaceae)[2]. In the paederoside compounds the acetate group of A. is replaced by *S*-methylthiocarbonate; these are as yet the only known sulfur-containing iridoid glucosides and appear to be the main cause of the unpleasant smell of *Paederia* species.

Lit.: [1]Hegnauer **VI**, 132. [2]Tetrahedron Lett. **1968**, 683.

gen.: Beilstein E V **18/7**, 565 f.; **19/6**, 201 f. ▪ Bull. Chem. Soc. Jpn. **66**, 2646 (1993) (synthesis) ▪ J. Nat. Prod. **43**, 649 (1980) ▪ Merck-Index (12.), No. 879 ▪ Phytochemistry **25**, 2515 (1986) (biosynthesis). – *[HS 2938 90]*

Aspicilin.

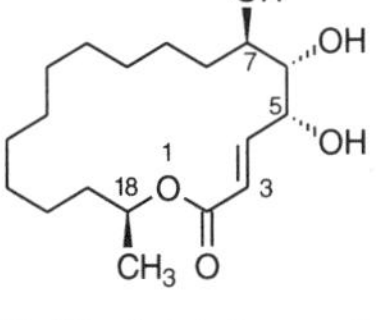

$C_{18}H_{32}O_5$, M_R 328.45, platelets, mp. 153–154 °C, $[\alpha]_D^{20}$ +32°, soluble in ether and benzene. Macrocyclic lactone from the crustaceous lichens *Aspicilia gibbosa* and *A. caesiocinerea* and target of various syntheses [1].

Lit.: [1] Helv. Chim. Acta **71**, 1719–1794 (1988); **72**, 1753–1786 (1989); J. Org. Chem. **59**, 949 ff. (1994); **62**, 377 (1997); **63**, 7505 (1998) [(+)-A.]; Justus Liebigs Ann. Chem. **1995**, 1185–1191; Org. Lett. **2**, 123 (2000); Tetrahedron: Asymmetry **2**, 801–819 (1991) [(−)-A.]; Tetrahedron Lett. **28**, 2409 ff. (1987) [(−)-A.]; **32**, 7397 ff. (1991); **36**, 2607 ff. (1995); **38**, 8883 (1997); **39**, 5597 (1998) [(+)-A.]; **40**, 161 (1999).
gen.: Angew. Chem. Int. Ed. Engl. **24**, 987 (1985) ▪ Beilstein EV **18/4**, 258 ▪ Tetrahedron Lett. **36**, 2607 (1995). – *[CAS 52461-05-9]*

Aspidosperma alkaloids. A group of over 100 monomeric and dimeric *monoterpenoid indole alkaloids with the aspidospermidine or the related eburnamenine basic skeleton that can be divided into further subgroups (structure types) [1,2] occurring in the genus *Aspidosperma* (Apocynaceae) with over 40 species. A. a. are also found in many other plant genera such as *Alstonia*, *Catharanthus*, *Hazunta*, *Pandaca*, *Tabernaemontana*, and *Vinca* and can be isolated from the respective cell culture systems. Important members of this group are: aspidospermine, *eburnamonine, *tabersonine, vincadifformine, *vincamine, and *vindoline.

Lit.: [1] J. Nat. Prod. **37**, 219 (1974). [2] Phillipson et al., S. 11.
gen.: Alkaloids: Chem. Biol. Perspectives **3**, 275 (1985) ▪ Chem. Heterocycl. Compd. **25**, 331 (1983) ▪ J. Org. Chem. **62**, 6855 (1997) (aspidospermidine) ▪ Manske **17**, 200; **50**, 343–376; **51**, 2–198 ▪ Pelletier **9** ▪ Stud. Nat. Prod. Chem. **19 E**, 89–116 (1997). – *[HS 2939 90]*

Aspochalasins.

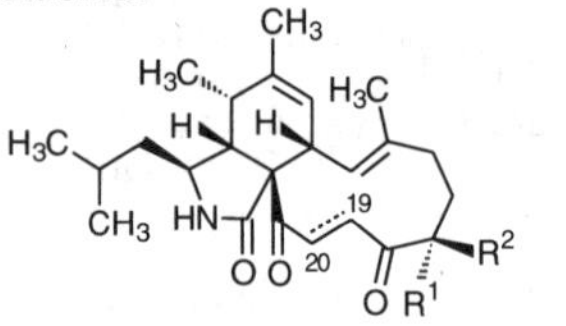

A. A: $\overset{20}{-CH_2}-\overset{19}{CH_2}-$ $R^1, R^2 = O$

A. B: $\overset{20}{-CH}\overset{E\,19}{=CH}-$ $R^1 = OH, R^2 = H$

The A. A u. B are antibacterial and antifungal metabolites from *Aspergillus microcysticus* while A. C, D and E have no antimicrobial activity. They belong to the class of *cytochalasins. The are formed biogenetically from leucine and a polyketide chain. *A. A:* $C_{24}H_{33}NO_4$, M_R 399.53, light yellow powder, $[\alpha]_D^{25}$ −20°.

Lit.: Helv. Chim. Acta **62**, 1501–1524 (1979); **65**, 1426 (1982) ▪ J. Am. Chem. Soc. **111**, 8284 (1989) (synthesis) ▪ J. Antibiotics **46**, 679 (1993) (A. E). – *[CAS 72363-48-5 (A. A)]*

Astaxanthin [3,3′-dihydroxy-*β,β*-carotene-4,4′-dione].

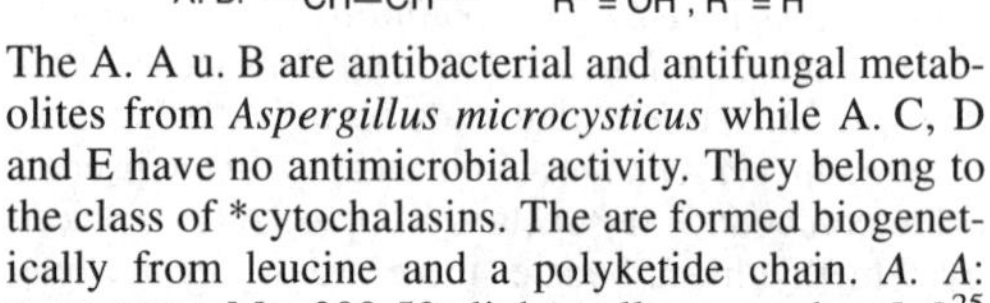

$C_{40}H_{52}O_4$, M_R 596.85, violet plates, mp. 215–216 °C (decomp.). A. is a *carotinoid belonging to the group of xanthophylls that is widely distributed in the animal kingdom as a red pigment, especially in crabs and starfish (Echinodermata). A. also occurs in bird feathers, foot and leg skin of flamingos and other birds – a result of eating crabs. Native A. exists either in the free form, as esters (e. g. dipalmitate), or as chromoprotein. In lobster shell (*Astacus*) A. exists in a protein complex, a blue-black pigment, from which A. is liberated on boiling and oxidized to the, also red, tetraketone astacene (astacin). A. is used to color farm-raised salmon. It is produced in genetically engineered yeast cell cultures.

Lit.: Acta Chem. Scand., Ser. B **28**, 730 (1974) ▪ Beilstein E IV **8**, 3318 ▪ Britten, Liaaen-Jensen & Pfander (eds.), Carotenoids, Basel: Birkhäuser 1995 ▪ Helv. Chim. Acta **64**, 2405, 2447 (1981) ▪ J. Nat. Prod. **58**, 1929 (1995) (glucoside) ▪ Merck-Index (12.), No. 890 ▪ Phytochemistry **15**, 1003, 1009 (1976). – *[HS 3203 00; CAS 472-61-7]*

Asteltoxin.

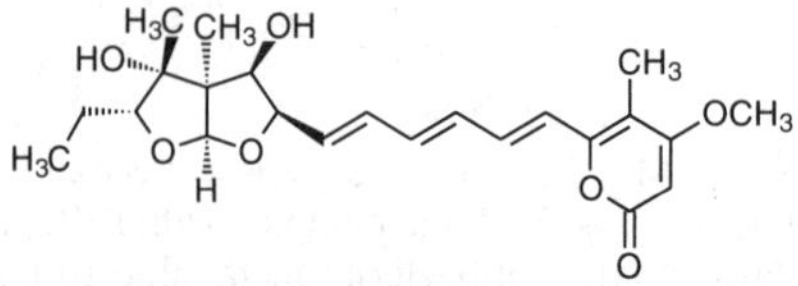

$C_{23}H_{30}O_7$, M_R 418.49, pale yellow needles, mp. 130–132 °C, $[\alpha]_D^{20}$ +20° (CH_3OH), metabolite from *Aspergillus stellatus* and *Emericella variecolor* with weak antibacterial activity. A. inhibits energy production of cells (mitochondrial ATPase). The mechanism of action is similar to that of the *citreoviridins.

Lit.: J. Chem. Soc., Chem. Commun. **1979**, 441; **1984**, 977 (biosynthesis) ▪ J. Chem. Soc., Perkin Trans. 1 **1981**, 1298 ▪ Nicolaou, p. 317 ▪ Tetrahedron **46**, 2353 (1990) (synthesis) ▪ Tetrahedron Lett. **27**, 2575 (1986) (absolute configuration); **36**, 3095 (1995) (synthesis). – *[CAS 79663-49-3]*

Asterin see chrysanthemin.

Asterosaponins.

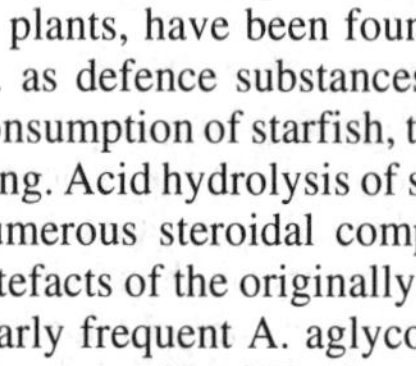

Saponins from *echinodermata, the only animal species in which saponins, otherwise widely distributed in plants, have been found. Starfish (Asteroidea) use A. as defence substances (see also imbricatine). On consumption of starfish, the A. cause nausea and vomiting. Acid hydrolysis of starfish saponins gives rise to numerous steroidal compounds, some of which are artefacts of the originally present aglycones. A particularly frequent A. aglycone is the pregnane *asterone* ($C_{21}H_{32}O_3$, M_R 332.48, cryst., mp. 196–198 °C, $[\alpha]_D$ +98.4°) which readily isomerizes. *Asterogenol* ($C_{21}H_{34}O_3$, M_R 334.50, cryst., mp. 263–264 °C) has been isolated from the European starfish *Asteria rubens*. In addition, C_{27}-steroids, some of which are epoxidized in the side chain, have also been found. A. with sulfate groups and oligosaccharide units are also known.

Lit.: Comp. Biochem. Physiol. B **64**, 25 (1979). – *Review:* Pharmazie **42**, 2 (1987) ▪ Scheuer I **5**, 287–389; II **2**, 60–86 ▪ see also echinoderm saponins.– *[CAS 37717-02-5 (asterone); 75921-90-3 (asterogenol)]*

Astringenin see hydroxystilbenes.

Atisane see diterpenes.

Atisine.

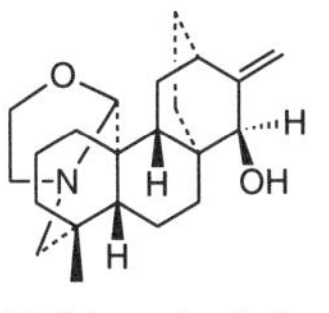

$C_{22}H_{33}NO_{20}$, M_R 343.51, resin, $[\alpha]_D$ –21° (C_2H_5OH); diterpene alkaloid from *Aconitum heterophyllum*, *A. gigans*, and *A. anthara*.

Lit.: Bull. Soc. Chim. Belg. **103**, 67 (1994) ■ J. Am. Chem. Soc. **112**, 1164 (1990). – *[HS 2939 90; CAS 466-43-3]*

Atractyligenin.

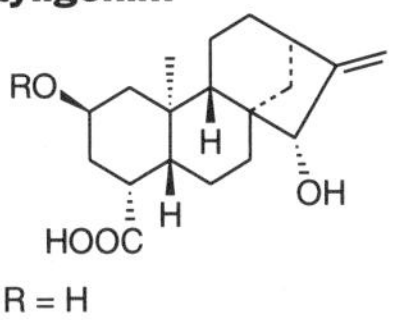

R = H : Atractyligenin
R = 2-(3-Methylbutyryl)-3,4-di-*O*-sulfo-β-D-glucopyranosyl : Atractyloside

$C_{19}H_{28}O_4$, M_R 320.43, mp. 189 °C, $[\alpha]_D$ –146° (C_2H_5OH). Diterpene belonging to the norkauranes isolated from *Atractylis gummifera*.

Lit.: Beilstein E III **10**, 1869 ■ Karrer, No. 6051 ■ Phytochemistry **27**, 1532 (1988); **37**, 353 (1994). – *[CAS 10391-47-6]*

Atractyloside. $C_{30}H_{46}O_{16}S_2$, M_R 726.81, formula, see atractyligenin. A. occurs in *Atractylis gummifera* and is highly toxic with a *strychnine-like mechanism of action; LD_{50} (rat i. m.) 431 mg/kg.

Lit.: Merck-Index (12.), No. 899 ■ Phytochemistry **27**, 1532 (1988). – *[CAS 17754-44-8]*

Atranorin.

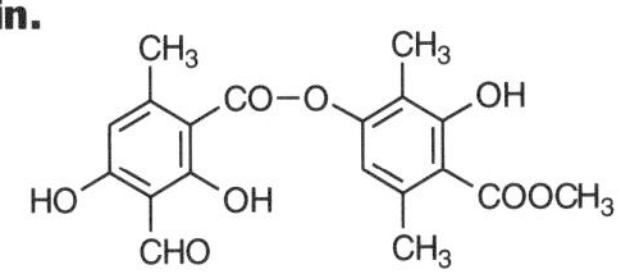

$C_{19}H_{18}O_8$, M_R 374.35, prisms, mp. 195 °C, yellow color with KOH and *p*-phenylenediamine, soluble in benzene and chloroform. A. is a *depside with antibiotic properties occurring in many lichens (e. g., in *Hypogymnia physodes* and *Lecanora atra*). The biosynthesis proceeds via acetyl-CoA and *orsellinic acid.

Lit.: Acta Crystallogr., Sect. B **38**, 3126 (1982) ■ Beilstein E IV **10**, 3872 ■ Karrer, No. 1043 ■ Tetrahedron **21**, 3531–3536 (1965). – *[CAS 479-20-9]*

Atrochrysone.

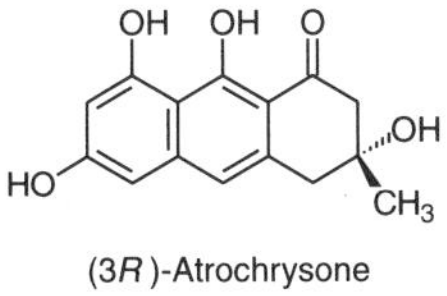

(3*R*)-Atrochrysone

$C_{15}H_{14}O_5$, M_R 274.27, greenish-yellow needles, mp. 215–218 °C, $[\alpha]_{546}$ –8° (CH_3OH). The (3*R*)-compound has been isolated from the toadstools *Cortinarius atrovirens* (Basidiomycetes) and *C. odoratus*, the (3*S*)-compound from an Australian *Dermocybe* species. A. is the key compound in the biosynthesis of many polyketidic *anthraquinones and *preanthraquinones.

Lit.: J. Nat. Prod. **55**, 372 (1992) (biosynthesis) ■ Phytochemistry **28**, 2647 (1989) ■ Sydowia **37**, 284 (1984). – *Reviews:* Zechmeister **51**, 149 ■ Nat. Prod. Rep. **11**, 79 (1984). – *[HS 2914 69; CAS 124903-85-1]*

Atromentin.

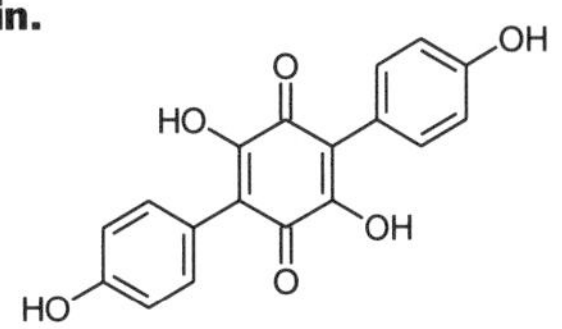

$C_{18}H_{12}O_6$, M_R 324.29, bronze-colored plates, mp. >300 °C. The *terphenylquinone A. is found in the brown outer skin of the toadstool *Paxillus atrotomentosus* (Basidiomycetes). In the fruit body of the fungus A. exists in the form of the colorless *leucomentin which easily decomposes into A. and the insecticidal *osmundalactone. Biosynthetically, A. is formed by condensation of 2 molecules 4-hydroxyphenylpyruvic acid and occurs in the form of esters of various acids in fungi (*aurantiacin, *flavomentins, *spiromentins). The constitution of A. was elucidated in the 1920's by F. Kögl.

Lit.: Beilstein E IV **8**, 3699 ■ Thomson **2**, 158 ff. ■ Zechmeister **51**, 13 ff. – *[CAS 519-67-5]*

Atropic acid s. tropic acid.

Atropine [(±)-hyoscyamine, tropine tropate].

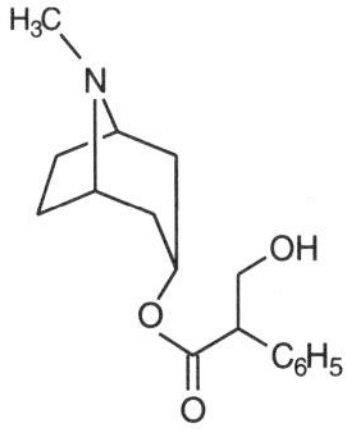

$C_{17}H_{23}NO_3$, M_R 289.37, mp. 114–116 °C, prisms. Like the structurally similar *scopolamine, A. belongs to the *tropane alkaloids and is the racemate of *hyoscyamine; it is formed from (–)-hyoscyamine during isolation of the latter from the natural sources deadly nightshade (*Atropa belladonna*), henbane species (e. g., *Hyoscyamus niger*), thorn apple species (e. g., *Datura stramonium*), and mandrake (*Mandragora officinalis*) by treatment with alkali. A. is a potent poison (lethal dose for humans ca. 100 mg). A., named after the Greek goddess of fate Atropos, was used as a pharmaceutical in antiquity on account of its particular effects, e. g., as a mydriatic [deadly nightshade extract applied dropwise to the eyes caused dilatation of the pupils for several days, an effect especially sought after by ladies in the middle ages, "flashing eyes" (Bella donna!)]. The pharmacological action of A. is based on blockade of the muscarinic acetylcholine receptors of the parasympathetic nervous system as a competitive antagonist (parasympatholytic). In addition to its characteristic mydriatic action, A. also has a soothing effect on spasms of the gastrointestinal tract and bronchi. A. acts at first stimulating and then paralysing on the CNS. Like other tropane alkaloids it favorably influences the symptoms of Parkinson's disease. A. has a local anesthetic action and is given as an antidote for poisonings with parasympathomimetic compounds such as nerve gases and organophosphorus insecticides [1] or *pilocarpine. The highest permissible single dose of A. sulfate is 0.5–1 mg. A. also finds use in veterinary medicine.

Lit.: [1]Nat. Prod. Rep. **11**, 443–450 (1994).
gen.: Beilstein E V **21/1**, 236 ff. ▪ Hager (5.) **3**, 112 ff.; **4**, 424 f., 1140 f.; **5**, 462 f.; **7**, 315 ▪ Kirk-Othmer (4.) **1**, 1047 f.; **12**, 388 ▪ Manske **9**, 269–303 (synthesis); **44**, 1–114, 115–187 ▪ Merck-Index (12.), No. 907 ▪ Sax (8.), p. 311 ▪ Ullmann (5.) **A 1**, 360 f. – *Pharmacology:* Nat. Prod. Rep. **8**, 606 f. (1990); **10**, 202 (1991); **11**, 445 f. (1994). – *[HS 2939 90; CAS 51-55-8]*

Atrovenetin (deoxynorherqueinone).

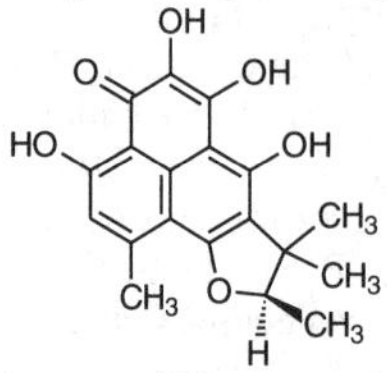

(*R*)

$C_{19}H_{18}O_6$, M_R 342.35, brownish yellow plates or prisms, mp. 295 °C, $[\alpha]_D^{20}$ +116°, *antibiotic isolated from the mycelium of *Penicillium atrovenetum* and *P. herquei* belonging to the class of *phenalenones.

Lit.: *Biosynthesis:* Pure Appl. Chem. **34**, 515 (1973). – *Synthesis:* J. Org. Chem. **51**, 4813 (1986). – *[HS 2941 90; CAS 2582-86-7]*

Atrovirin.

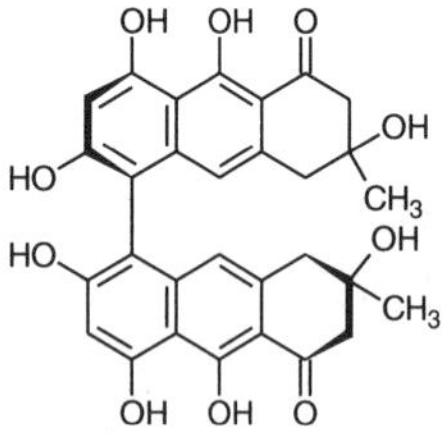

Atrovirin B

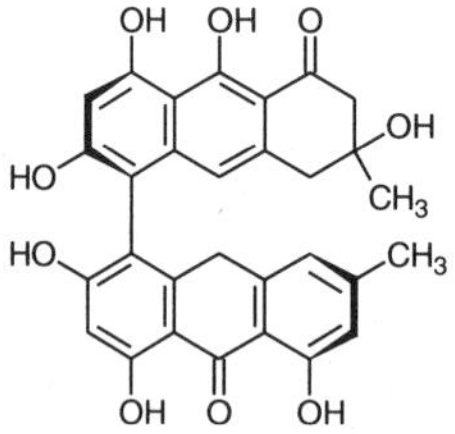

Austrovenetin

A. B: $C_{30}H_{26}O_{10}$, M_R 546.53, green-yellow powder, mp. 216–218 °C, $[\alpha]_{546}$ +350° (CH_3OH). Dimeric *preanthraquinone from the black-green toadstool *Cortinarius atrovirens* (Basidiomycetes) and the Australian agaric *Dermocybe icterinoides* (different stereochemistry at the sp^3-centers!). The absolute configuration at the biphenyl linkage was established by comparison with synthetic models according to the Horeau method. On dehydration and oxidation A. furnishes (+)-(*S*)-*skyrin which occurs together with other pigments of the A. type in fungi. *Austrovenetin* {$C_{30}H_{24}O_9$, M_R 528.52, green powder, mp. >340 °C, $[\alpha]_D$ +567° (CH_3OH)}, (+)-(*S*)-skyrin and a series of related compounds with 5,5′ linkages of the preanthraquinone halves have been isolated from the olive green Australian *Dermocybe austroveneta*. During isolation austrovenetin is easily converted to protohypericin and *hypericin.

Lit.: J. Nat. Prod. **51**, 1251 (1988) ▪ Nat. Prod. Rep. **11**, 81 f. (1994) ▪ Phytochemistry **30**, 951 (1991); **37**, 1679 (1994) ▪ Steglich, in: Atta-ur Rahman & Le Quesne (eds.), Natural Products Chemistry III, p. 305–315, Berlin: Springer 1988 ▪ Zechmeister **51**, 151 f. – *[CAS 133632-58-3 (A. B); 133632-57-2 (austrovenetin)]*

Attractants. General term for single or combined signal substances (*semiochemicals) which have a specific attracting effect on certain organisms: *insect attractants and *pheromones, sexual pheromones of lower organisms (e.g., gamones) or higher animals (*aroma chemicals). To measure the efficacy of natural or synthetic A. 1 *attractant unit* is defined as that amount of A. (contained in 1 mL of petroleum ether solution) which induces a change in behavior in 50% of the target organisms when a glass rod moistened with the test solution is placed near the organisms. For determination of an A. unit for algae (gamones), see *Lit.*[1].

Lit.: [1]Angew. Chem. Int. Ed. Engl. **21**, 643–653 (1982); Jacobs & Renner, Biologie u. Ökologie der Insekten, Stuttgart: Fischer 1988.

Aucubin (rhinanthin, aucuboside). A. belongs to the *iridoid glucosides with a C_9 aglycone. It occurs in nature in the free form and as esters with aromatic carboxylic acids (benzoic and 4-hydroxybenzoic acid). Apart from *catalpol, A. is the most frequently occurring iridoid glucoside and can be isolated on a 100g scale. Its absolute configuration was determined by correlative connection with that of catalpol[1].

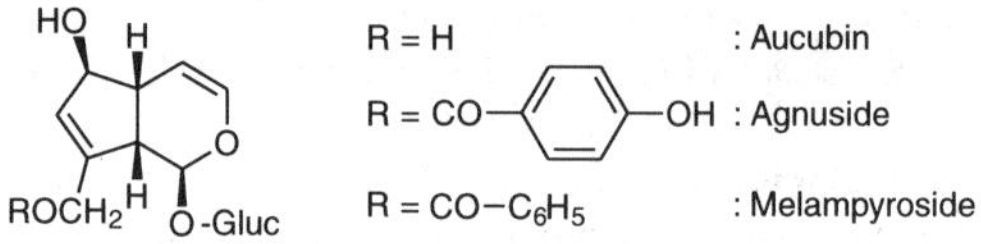

Table: Aucubin and derivatives.

name	Aucubin[2]	Agnuside[3]	Melampyroside[4]
molecular formula	$C_{15}H_{22}O_9$	$C_{22}H_{26}O_{11}$	$C_{22}H_{26}O_{10}$
M_R	346.33	466.44	450.44
mp. [°C]	180	145–146	amorphous
$[\alpha]_D$	–162° (H_2O)	–91.5° (C_2H_5OH)	–50° (acetone)
CAS	479-98-1	11027-63-7	55785-60-9

Occurrence: A. has been detected in more than 80 different plants, especially in the family Scrophulariaceae. It owes its name to its occurrence (2–3%) in *Aucuba japonica*, a plant endemic to Japan (Cornaceae). In Europe A. occurs in the widely distributed ribwort (*Plantago lanceolata*, Plantaginaceae) in larger amounts (1.1%).

Lit.: [1]Justus Liebigs Ann. Chem. **1990**, 715. [2]Bobbitt & Segebarth, in Taylor & Battersby (eds.), Cyclopentanoid Terpene Derivatives, chapter 1, New York: Dekker 1969. [3]Arch. Pharm. (Weinheim, Ger.) **293**, 556 (1960). [4]Tetrahedron **30**, 4049 (1974).
gen.: Hager (5.) **6**, 221 f., 224 ff., 385 f., 1116 f., 1183 ff. ▪ J. Nat. Prod. **43**, 649, No. 32, 35, 36 (1980) ▪ Karrer, No. 1791. – *[HS 2938 90]*

Aurachins. Group of isoprenoid quinoline alkaloids from *Stigmatella aurantiaca* (Myxobacteria)[1] having a remarkable similarity to the pseudanes, e.g., 2-heptyl-4-quinolinol *N*-oxide from pseudomonads. Although A. C and D are 4-quinolones carrying a farnesyl group in the 3 position, during the course of the biosynthesis of A. A and B the farnesyl groups and the oxygen substituent are exchanged by an as yet unknown mechanism. The A. inhibit the growth of Gram-positive bacteria in the range 0.15–5 mg/mL, and

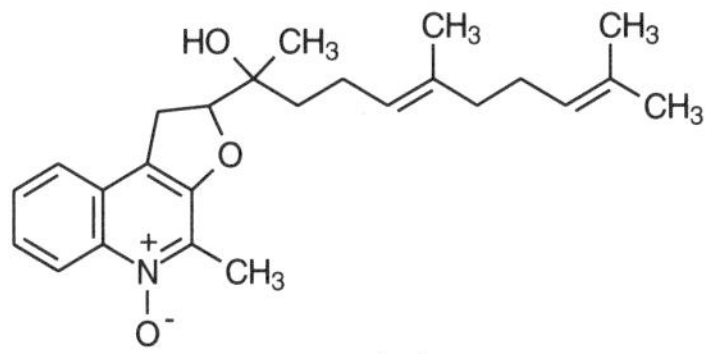

A. A

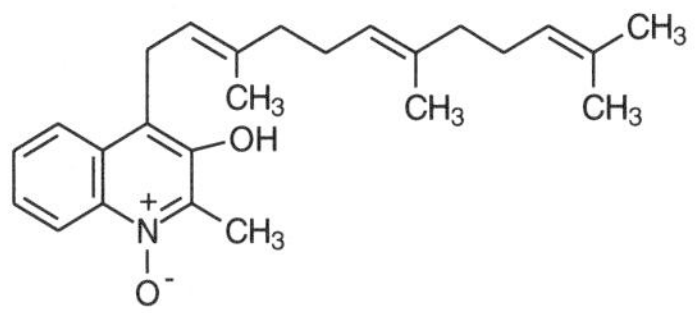

A. B

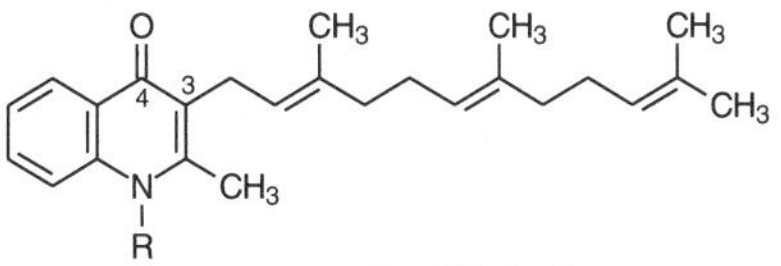

R = OH : A. C
R = H : A. D

Table: Data of Aurachins.

A.	molecular formula	M_R	mp. [°C]	$[\alpha]_D$	CAS
A	$C_{25}H_{33}NO_3$	395.54	111–112	−49.2°	108354-15-0
B	$C_{25}H_{33}NO_2$	379.54	93–94	–	108354-12-7
C	$C_{25}H_{33}NO_2$	379.54	124–125	–	108354-14-9
D	$C_{25}H_{33}NO$	363.54	165–168	–	108354-13-8

weakly inhibit the growth of some yeasts and fungi. Like the structurally related HQNO [see 2-heptyl-1-hydroxy-4(*1H*)-quinolinone] A. inhibit NADH oxidation in the eukaryotic respiratory chain. A. C and D, in particular, are potent inhibitors of photosynthesis; thus A. C preferentially inhibits the cytochrome b_6/f complex and A. D the photosystem II [2].

Lit.: [1] J. Antibiot. **40**, 258 (1987). [2] Z. Naturforsch. **450**, 322 (1990).

Aurantiacin.

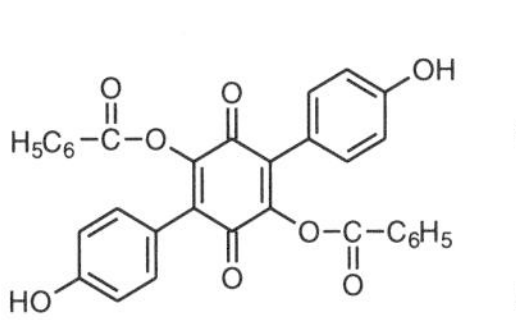

Aurantiacin

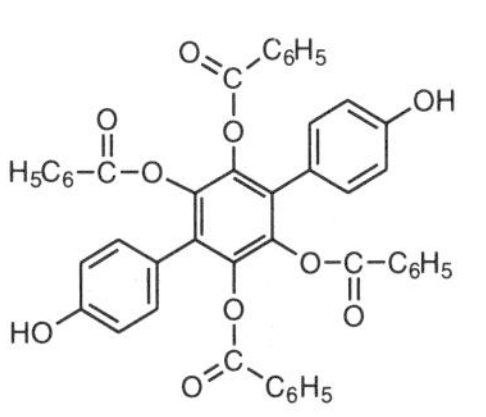

Dihydroaurantiacin dibenzoate

$C_{32}H_{20}O_8$, M_R 532.51, dark red needles, mp. 285–295 °C. A. is the orange-red pigment of some *Hydnellum* species, e. g., *H. aurantiacum* (Basidomycetes). It is accompanied by the colorless dihydroaurantiacin dibenzoate ($C_{46}H_{30}O_{10}$, M_R 742.74, cryst., mp. 305–307 °C). Alkaline saponification gives rise to benzoic acid and *atromentin.

Lit.: Zechmeister **51**, 25 f. – *[CAS 548-32-3]*

Aurantosides see tetramic acid.

Aureobasidins (basifungins). Group of about 20 (found so far) cyclic depsipeptide antibiotics from cultures of the yeast *Aureobasidium pullulans* [1]. The main component, A. A exhibits good fungicidal activity *in vitro* and *in vivo* against human pathogenic fungi, e. g., *Candida albicans*, *Cryptococcus* sp., *Histoplasma* sp., and was thus undergoing clinical testing as an antimycotic [2]. It acts as an inhibitor of inositolphosphorylceramide (IPC)-synthase and therewith on the constitution of phosphoinositol containing sphingolipids (biosynthesis of fungal cell walls) [3].

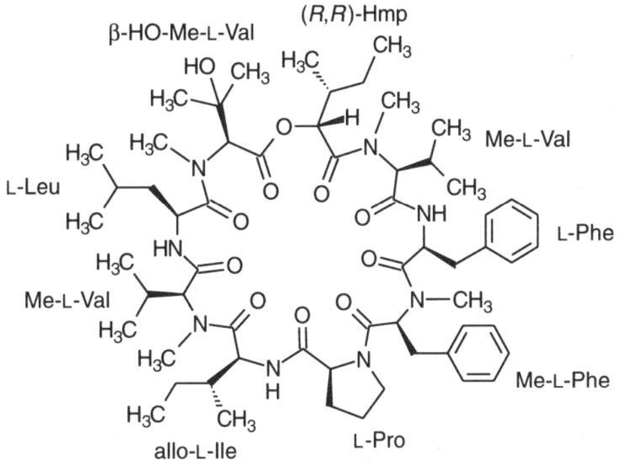

A. A: $C_{60}H_{92}N_8O_{11}$, M_R 1101.44, mp. 155–157 °C (rods), 128–130 °C (dihydrate), $[\alpha]_D^{20}$ −247° (CH_3OH), LD_{50} (mouse p. o.) 230 mg/kg, (mouse i. v.) >1000 mg/kg. The total synthesis of the compound has been reported [4] but the synthetic procedure is too laborious to compete with the fermentation route.

Lit.: [1] J. Antibiot. (Tokyo) **44**, 919–924, 925–933, 1187–1198, 1199–1207 (1991); **46**, 1347–1354, 1414 (1993); **48**, 525 ff. (1995); J. Chem. Soc., Chem. Commun. **1992**, 1231 ff.; J. Org. Chem. **59**, 570–578 (1994). [2] J. Antibiot. (Tokyo) **46**, 1414–1420 (1993). [3] J. Biol. Chem. **1997**, 9809. [4] Chem. Lett. **1993**, 1873 ff.; Tetrahedron **52**, 4327–4346 (1996); Tetrahedron Lett. **37**, 5661 (1996).

gen.: J. Antibiot. **49**, 676 (1996) (biosynthesis); **51**, 353 (biological activity), 359 (1998) (synthesis). – *[CAS 127785-64-2 (A. A); 127939-17-7 (A. B)]*

Aureolic acid (mithramycin, plicamycin).

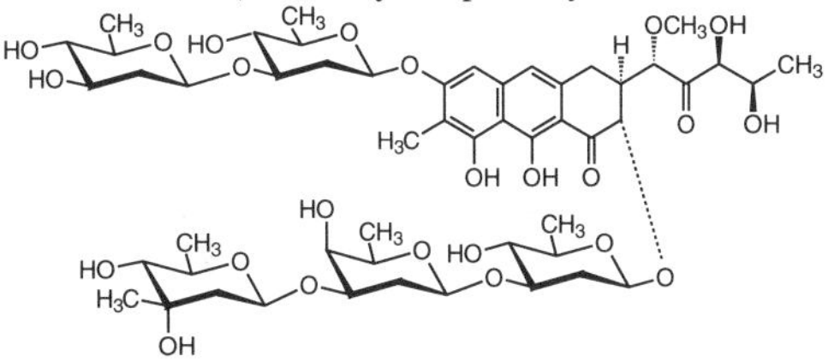

$C_{52}H_{76}O_{24}$, M_R 1085.16, yellow cryst., mp. 180–183 °C. A. belongs to the substance class of the *chromomycins which exhibit anti-tumor activity. It is isolated from *Streptomyces argillaceus* and *S. tanashiensis*. It cannot be used clinically on account of its toxicity.

Lit.: Angew. Chem. Int. Ed. Engl. **22**, 58 (1983) ▪ J. Antibiot. **43**, 1543–1552 (1990) ▪ J. Nat. Prod. **62**, 119 (1999) (structure). – *[HS 2941 90; CAS 18378-89-7]*

Aureomycin see tetracyclines.

Aureothin (mycolutein).

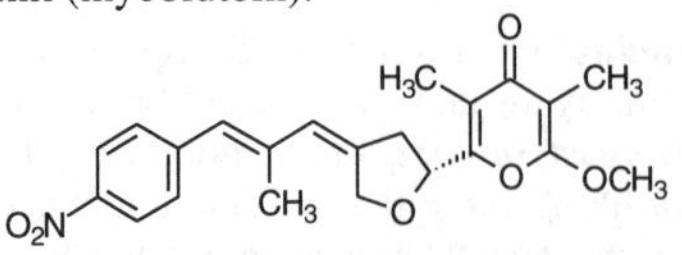

$C_{22}H_{23}NO_6$, M_R 397.43, yellow prisms, mp. 158 °C, $[\alpha]_D^{25}$ +54° ($CHCl_3$). Antibiotic isolated from various *streptomycetes (e.g., *Streptomyces thioluteus*). A. exhibits antitumor activity, and has antifungal, anti-HIV, insecticidal, and herbicidal actions.

Lit.: Chem. Pharm. Bull. **23**, 569 (1975) ▪ Chem. Lett. **1987**, 1381 (synthesis) ▪ J. Antibiot. **29**, 236 (1976) ▪ Tetrahedron **30**, 459 (1974) (biosynthesis) ▪ Tetrahedron Lett. **33**, 521 (1992). – *[HS 2941 90; CAS 2825-00-5]*

Aureothricin see holomycin.

Aurodox see efrotomycin.

Aurones (2-benzylidene-3-benzofuranones). A small group of golden-yellow *flavonoids, biosynthetically derived directly from *chalcones. The actual enzymatic reaction has not yet been elucidated.

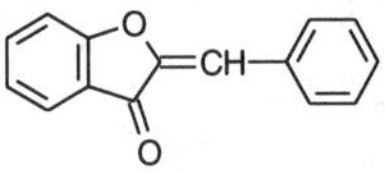

The A. differ in their hydroxylation pattern. They occur as glycosides especially in the petals of members of the Asteraceae, but have also been isolated from Cyperaceae, Ericaceae, Fabaceae, and Rubiaceae.

Lit.: Harborne (1988), p. 340–342; (1994), p. 399–491.

Austin.

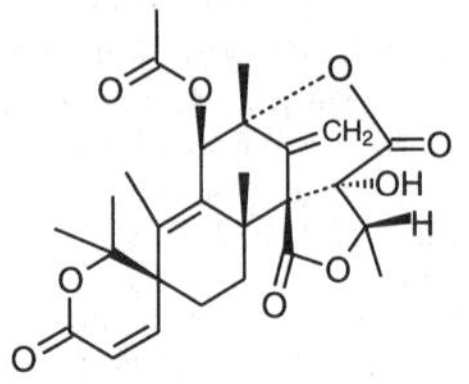

$C_{27}H_{32}O_9$, M_R 500.51, colorless cryst., mp. 298–300 °C, mycotoxin, metabolite of *Aspergillus ustus*. LD_{50} (chick p.o.) 350–375 mg/kg.

Lit.: J. Am. Chem. Soc. **98**, 6748–67 (1976) ▪ J. Chem. Soc., Chem. Commun. **1986**, 214 ▪ J. Chem. Soc., Perkin Trans. 1 **1989**, 807 (biosynthesis). – *[CAS 61103-89-7]*

Austocystins. The A. A to I are antimicrobially active and mutagenic metabolites of *Aspergillus ustus*. They are structurally related to *sterigmatin* from *Aspergillus versicolor*. Biogenetically they are derived

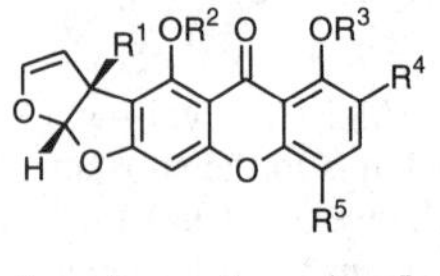

	R^1	R^2	R^3	R^4	R^5
Sterigmatin	H	H	H	H	H
A. A	H	CH_3	CH_3	Cl	H
A. B	H	H	H	H	$CH_2–CH_2–C(CH_3)_2–OH$
A. C	H	CH_3	H	H	$CH_2–CH_2–C(CH_3)_2–OH$
A. D	OH	H	H	H	$CH_2–CH_2–C(CH_3)_2–OH$
A. E	OH	CH_3	H	H	$CH_2–CH_2–C(CH_3)_2–OH$
A. F	OH	H	H	H	H
A. G	OH	H	CH_3	Cl	H
A. H	OH	H	H	H	$CH_2–CH=C(CH_3)_2$
A. I	OH	CH_3	H	H	H

Table: Data of Austocystins.

compound	molecular formula	M_R	mp. [°C]	CAS
Sterigmatin	$C_{17}H_{10}O_6$	310.26		55256-49-0
A. A	$C_{19}H_{13}ClO_6$	372.76	204–205	55256-58-1
A. B	$C_{22}H_{20}O_7$	396.40	172–173	55256-57-0
A. C	$C_{23}H_{22}O_7$	410.42	168–170	55256-55-8
A. D	$C_{22}H_{20}O_8$	412.40	114–116	55256-53-6
A. E	$C_{23}H_{22}O_8$	426.42		55256-54-7
A. F	$C_{17}H_{10}O_7$	326.26	230–233	55256-56-9
A. G	$C_{18}H_{11}ClO_7$	374.73	168–170	58775-49-8
A. H	$C_{22}H_{18}O_7$	394.38	141–143	58775-50-1
A. I	$C_{18}H_{12}O_7$	340.29		58775-51-2

from versicolorin A, an intermediate in aflatoxin biosynthesis.

Lit.: Betina, chap. 7 ▪ J. Chem. Soc., Perkin Trans. 1 **1974**, 2250–2256; **1983**, 1745 f. ▪ Rodricks, Hesseltine & Mehlman (eds.), Mycotoxins in Human and Animal Health, p. 419–467, Park Forest South: Pathotox Publ. 1977 ▪ Sax (8.), ARU 250 ▪ Stud. Org. Chem. **26**, 387 (1986).

Australine see alexine.

Austrocorticin.

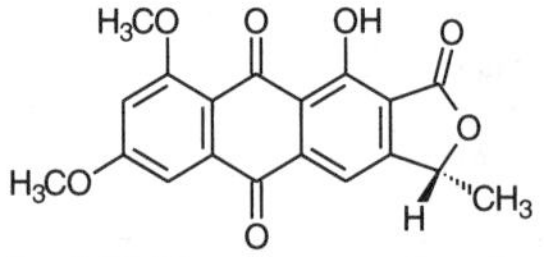

$C_{19}H_{14}O_7$, M_R 354.32, orange-colored needles, mp. 246–250 °C, $[\alpha]_D$ +59° ($CHCl_3$). Main pigment of the brilliant orange-colored Australian agaric *Dermocybe* WAT 19353. As shown by feeding experiments, the biosynthesis of A. proceeds from a propionate starter unit and 7 mol. malonyl-CoA, an as yet unique combination in the basidiomycetes.

Lit.: Aust. J. Chem. **47**, 1363 (1994) [synthesis of (*S*)-(–)-A.] ▪ J. Chem. Soc., Perkin Trans. 1 **1990**, 1159. – *[CAS 118214-58-7]*

Austrocortilutein.

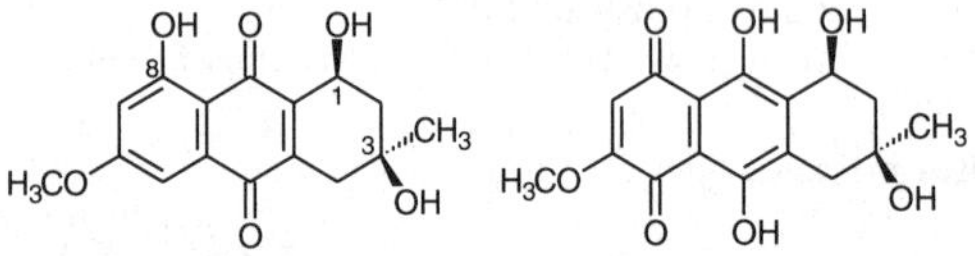

(1*S*,3*S*)-Austrocortilutein (1*S*,3*S*)-Austrocortirubin

$C_{16}H_{16}O_6$, M_R 304.30, orange-yellow needles, mp. 183–185 °C, $[\alpha]_D$ +62° (C_2H_5OH). A. occurs together with *austrocortirubin* {$C_{16}H_{16}O_7$, M_R 320.30, dark red needles, mp. 192–195 °C, $[\alpha]_D$ +109° (C_2H_5OH)} in the splendid red Australian agaric *Dermocybe splendens*. The compounds exist predominately as the (1*S*,3*S*)-stereoisomers and in the form of 8-*O*-*β*-D-gentiobiosides. Interestingly, all four possible stereoisomers of A. can be isolated from closely related *Dermocybe* species. Compounds of this type exhibit antibiotic activity against bacteria and fungi. (1*S*,3*S*)-A. exerts selective cytotoxicity against melanoma cells and is being tested *in vivo* against skin cancer.

Lit.: Aust. J. Chem. **44**, 1427, 1447 (1991) (synthesis of rac. A.) ▪ J. Chem. Soc., Perkin Trans. 1 **1990**, 1583 ▪ Nat. Prod. Lett. **1**, 187 (1992) ▪ Nat. Prod. Rep. **11**, 75 (1994) (review). – *[CAS 97400-70-9 (A.); 97400-69-6 (austrocortirubin)]*

Austrovenetin see atrovirin.

Autumnaline see phenethyltetrahydroisoquinoline alkaloids.

Auxins. Term for naturally occurring or synthetic *plant growth substances which resemble the natural plant hormone auxin (*3-indolylacetic acid) in their activities. In low doses A. have growth stimulating activities and at higher doses herbicidal effects.
Lit.: Ullmann (4.) **24**, 53 f.; (5.) **A 20**, 415, 417.

Avarol.

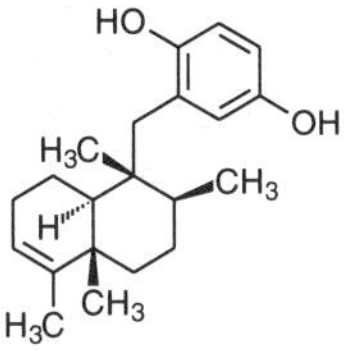

Avarol

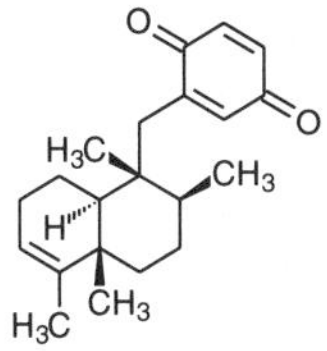

Avarone

A. ($C_{21}H_{30}O_2$, M_R 314.47, cryst., mp. 148–150 °C, $[\alpha]_D$ +6.1°) and *avarone* ($C_{21}H_{28}O_2$, M_R 312.45, oil, optically inactive) are components of the marine sponge *Dysidea avara* with interesting antileukemic and cytostatic activities; they do not affect healthy cells. Both substances inhibit the replication of AIDS viruses.
Lit.: Isolation: J. Chem. Soc., Perkin Trans. 1 **1976**, 1408 ▪ Tetrahedron Lett. **1974**, 3401. – *Pharmacology:* Biochem. Arch. **3**, 275 (1987) ▪ Cancer Res. **47**, 6565 (1987) ▪ Cell. Biochem. Funct. **6**, 123 (1988) ▪ J. Natl. Cancer Inst. **78**, 663 (1987). – *Synthesis:* Angew. Chem. Int. Ed. Engl. **33**, 853 (1994); **38**, 3089 (1999) ▪ J. Chem. Soc., Chem. Commun. **1996**, 2717 ▪ J. Org. Chem. **47**, 1727 (1982); **61**, 8775 (1996) ▪ J. Serb. Chem. Soc. **53**, 229–249 (1988). – *[CAS 55303-98-5 (avarol); 55303-99-6 (avarone)]*

Avenaciolide.

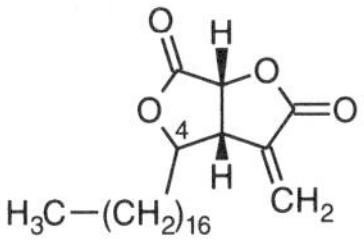

Avenaciolide

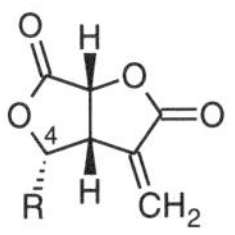

R = $(CH_2)_{16}-CH_3$: Isoavenaciolide
R = C_2H_5 : Ethisolide

Bislactone isolated from fermentation broths of *Aspergillus avenaceus* and other *Aspergillus* and *Penicillium* species[1]. *A.* [$C_{15}H_{22}O_4$, M_R 266.34, cryst., double mp. 49–50° and 54–56 °C, $[\alpha]_D$ –41.6° (C_2H_5OH)], inhibits fungal spore germination, shows antibacterial activity and inhibits glutamate transport in rat liver mitochondria. The highly functionalised structure has led to numerous total syntheses that are notable in their strategic diversity. The 4-epimer *isoavenacolide* [$C_{15}H_{22}O_4$, M_R 266.34, needles, mp. 129–130 °C, $[\alpha]_D$ –167.2° (C_2H_5OH)], is an antifungal co-metabolite of A. from *A. avenaceus*[2]. *Ethisolide* [$C_9H_{10}O_4$, M_R 182.18, needles, mp. 122–123 °C, $[\alpha]_D$ –214° (C_2H_5OH)] is produced by *Penicillium* ssp. and exhibits similar antifungal activities[3]. A. is biosynthetically related to *canadensolide[4].
Lit.: [1]J. Am. Chem. Soc. **97**, 3870 (1975); Tetrahedron Lett. **1975**, 3657. – *Biological activity:* Biochim. Biophys. Acta **325**, 375 (1973). – *Syntheses:* J. Org. Chem. **57**, 2228 (1992) (and literature cited therein); **61**, 8448 (1996); **64**, 8311 (1999); Tetrahedron Lett. **35**, 5841 (1994). [2]J. Chem. Soc. C **1971**, 2431. [3]Tetrahedron Lett. **1975**, 3657, **1977**, 2865. [4]*Biosynthesis:* Turner **2**, 378. – *[CAS 20223-76-1 (A.); 33644-09-6 (isoavenaciolide); 33644-10-9 (ethisolide)]*

Avenalumin.

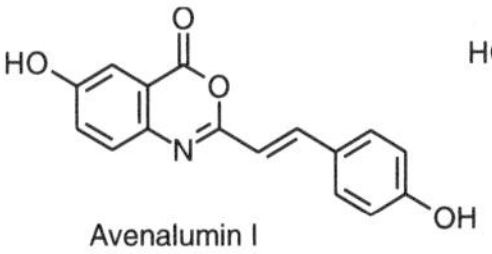

Avenalumin I

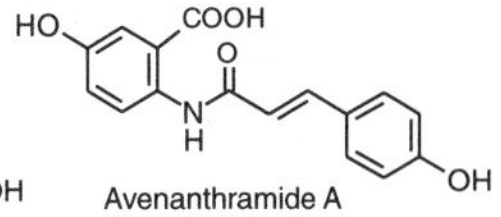

Avenanthramide A

$C_{16}H_{11}NO_4$, M_R 281.27 or $C_{16}H_{13}NO_5$, M_R 299.28, amorphous powder. Main phytoalexin isolated in the form of the acetate from leaves of oat (*Avena sativa*) after infection with the rust fungus *Puccinia coronata*[1]. According to *Lit.*[2] the natural product exists in the form of the open chain amide *avenanthramide A*.
Lit.: [1]Physiol. Plant Pathol. **19**, 217–226 (1981). [2]J. Agric. Food Chem. **37**, 60 (1989); Tetrahedron Lett. **31**, 2647–2648 (1990). – *[CAS 78164-38-2 (A.); 108605-70-5 (avenanthramide A)]*

Avenanthramide A see avenalumin.

Avermectins.

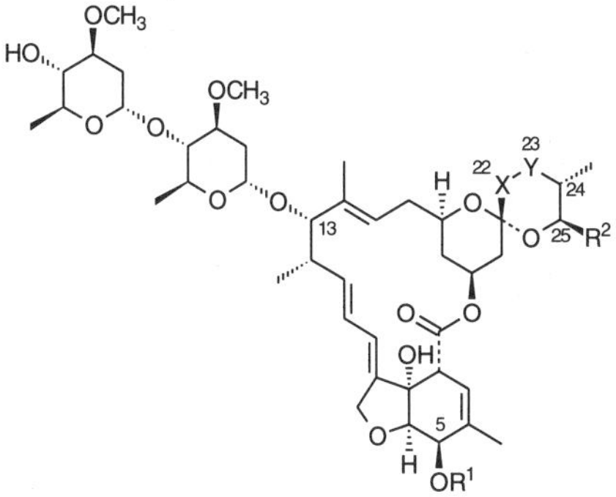

R^1	R^2	X–Y	
CH_3	$CH(CH_3)-C_2H_5$	CH=CH	A. A 1a
CH_3	$CH(CH_3)_2$	CH=CH	A. A 1b
CH_3	$CH(CH_3)-C_2H_5$	$CH_2-CH(OH)$	A. A 2a
CH_3	$CH(CH_3)_2$	$CH_2-CH(OH)$	A. A 2b
H	$CH(CH_3)-C_2H_5$	CH=CH	A. B 1a
H	$CH(CH_3)_2$	CH=CH	A. B 1b
H	$CH(CH_3)-C_2H_5$	$CH_2-CH(OH)$	A. B 2a
H	$CH(CH_3)_2$	$CH_2-CH(OH)$	A. B 2b

Group of macrolide *antibiotics from *Streptomyces avermitilis* with broad anthelmintic and antiarthropodal spectra of action in extremely low doses (10 µg/kg)[1]. The A. are disaccharide derivatives of pentacyclic, 16-membered lactones, all are optically active. In spite of their macrolide structures, they do not inhibit protein biosynthesis. They have no antibacterial or antifungal activity. The A. are divided into two main groups: group A with a methoxy group in position 5 and B with a hydroxy group (see formula). The *milbemycins, which do not have an α-L-oleandrosyl-α-L-oleandrosyl group at C-13 are closely related to the A. (for chemistry of the A. and milbemycins, see *Lit.*[2,3]) and are used in veterinary medicine.
Mechanism of action: The A. act on GABA-controlled chloride ion channels in the CNS of invertebrate animals[4,5]. For the problem of resistance to A., see *Lit.*[6,7].
Use: The A. derivative *ivermectin is used in veterinary medicine[8] and as an antiparasitic agent in human

Table: Data of Avermectins.

	molecular formula	M_R	$[\alpha]_D^{27}$ ($CHCl_3$)	CAS
A. A 1a	$C_{49}H_{74}O_{14}$	887.12	+68.5°	65195-51-9
A. A 1b	$C_{48}H_{72}O_{14}$	873.09		65195-52-0
A. A 2a	$C_{49}H_{76}O_{15}$	905.13	+48.8°	65195-53-1
A. A 2b	$C_{48}H_{74}O_{15}$	891.11		65195-54-2
A. B 1a	$C_{48}H_{72}O_{14}$	873.09	+55.7°	65195-55-3
A. B 1b	$C_{47}H_{70}O_{14}$	859.06		65195-56-4
A. B 2a	$C_{48}H_{74}O_{15}$	891.11	+38.3°	65195-57-5
A. B 2b	$C_{47}H_{72}O_{15}$	877.08		65195-58-6

medicine[9] [active against the buffalo gnat that carries the pathogen of filariasis (river blindness), with which about 20–40 million people are afflicted].
Synthesis & biosynthesis: For synthesis, see *Lit.*[10–15]. For biosynthesis and genetic regulation, see *Lit.*[16,17]. The use of genetically engineered strains of *Streptomyces avermitilis*[18] and bioconversion[19] are gaining significance for the preparation of new A. derivatives.
Detection: Quantitative determination of A. in animal food, forms of administration, and biological samples by HPLC[20].

Lit.: [1] Antimicrob. Agents Chemother. **15**, 361 & 638 (1979). [2] Nat. Prod. Rep. **3**, 87–121 (1986). [3] ACS Symp. Ser. **551** (Natural and Engineered Pest Management Agents) 54–73 (1994). [4] J. Neurochem. **64**, 2354 ff. (1995). [5] Life Sci. **56**, 757–765 (1995). [6] ACS Symp. Ser. **591** (Molecular Action of Insecticides on Ion Channels), 284–292 (1995). [7] Annu. Rev. Entomol. **40**, 1–30 (1995). [8] Annu. Rev. Entomol. **36**, 91–117 (1991). [9] J. Med. Chem. **23**, 1134 (1980). [10] Synform **6**, 179–196 (1988). [11] ApSimon **9**, 613–644; J. Am. Chem. Soc. **117**, 1908–1939 (1995); J. Org. Chem. **63**, 2591 (1998). [12] Synform **8**, 285–312 (1990). [13] Chem. Soc. Rev. **20**, 211–269, 271–339 (1991). [14] Rev. Stud. Nat. Prod. Chem. **12**, 3–33 (1993). [15] Spec. Publi.-R. Soc. Chem. **147** (Advances in the Chemistry of Insect Control III), 16–26 (1994). [16] Biotechnol. Ser. **28** (Genetics and Biochemistry of Antibiotic Production), 421–442 (1995); Chem. Rev. **97**, 2591–2609 (1997). [17] Ann. N. Y. Acad. Sci. **721** (Recombinant DNA Technology II), 123–132 (1994). [18] J. Antibiot. **48**, 92 ff., 549–562 (1995). [19] US P. 93-123181 (1993), Merck Co., Inc., USA, Inv.: Arison. [20] Vet. Parasitol. **48**, 59–66 (1993).
gen.: ACS Symp. Ser. **362** (Impact Chem. Biotechnol.), 242–255 (1988); **443** (Synth. Chem. Agrochem. 2), 422–435 (1991); **524** (Pest Control with Enhanced Environmental Safety), 169–182 (1993) ▪ Biol.-Chem. Interface **1999**, 257–269 ▪ Godfrey (ed.), Agrochemicals from Natural Products, p. 215–255, New York: Dekker 1995 ▪ Lindberg **3**, 237 ff. ▪ Lukacs & Ohno (eds.), Recent Prog. Chem. Synth. Antibiot., p. 65–102, Berlin: Springer 1990. – *Pharmacology:* Annu. Rev. Pharmacol. Toxicol. **32**, 537–553 (1992). – *[HS 2941 90]*

Averufin

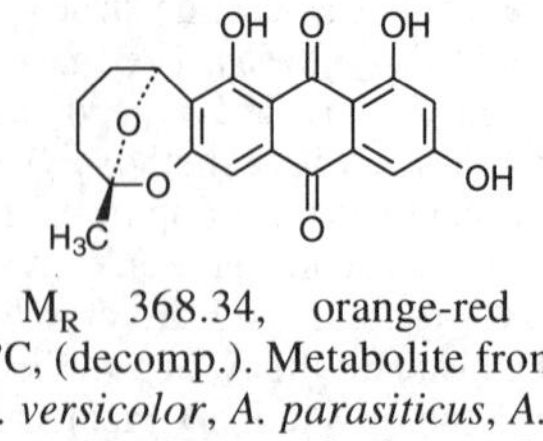

$C_{20}H_{16}O_7$, M_R 368.34, orange-red cryst., mp. 280–282 °C, (decomp.). Metabolite from *Aspergillus* species (*A. versicolor, A. parasiticus, A. ustus, A. nidulans, A. multicolor*), and phytopathogenic fungi (*Cercospora smilacis, Dothistroma pini*), and *Monocillium nordinii*. A. is an intermediate in aflatoxin biosynthesis and shows weak mutagenic activity in the Ames test.

Lit.: App. Environ. Microbiol. **57**, 1340–1345 (1991); **59**, 2486–2492 (1993) (biosynthesis) ▪ Beilstein EV **19/7**, 73 ▪ Cole-Cox, p. 107 ff. ▪ J. Am. Chem. Soc. **107**, 270 (1985) (abs. configuration) ▪ J. Chem. Soc., Chem. Commun. **1996**, 301 f. (biosynthesis) ▪ J. Org. Chem. **50**, 5533 (1985) (synthesis) ▪ Z. Naturforsch. B **40**, 319 (1985) (isolation). – *[CAS 14016-29-6]*

Avidin. Tetrameric glycoprotein (M_R 66000), occurring in the albumen of bird and amphibian eggs and in the genital tracts of all animals. Denaturization occurs on boiling and irradiation.
Each subunit consists of 128 amino acids with alanine at the N-terminal, glutamic acid at the C-terminal end and a carbohydrate unit on the asparaginyl residue in position 17. The extremely strong and selective affinity of A. for *biotin (dissociation constant = 10^{-15} M) is the reason for possible deficiency symptoms after longer-term consumption of raw egg white (avitaminosis). This interaction with biotin is one of the strongest non-covalent molecular interactions known today. The complex is stable at extreme pH values, in the presence of denaturizing agents, of proteolytic enzymes, and of organic solvents. When biotin is bound to macromolecules it retains its affinity for A. For this reason A. is used in combination with biotin for immobilization of ligands or reactants, for example, in immunoassays and other bioanalytical detection systems, in affinity chromatography, and in sequence determinations of deoxyribonucleic acids on solid phase. One unit A. is defined as the amount of A. that binds 1 mg D-biotin at pH 8.9. The histone-like properties (pI = 10.0) can lead to high non-specific binding which can then exclude the use of A.

Lit.: Adv. Protein Chem. **29**, 85 (1975) ▪ Biochem. J. **118**, 67, 71 (1970) ▪ Biochim. Biophys. Acta **264**, 165 (1972) ▪ J. Biol. Chem. **246**, 698 (1971) ▪ Merck-Index (12.), No. 920 ▪ Methods Biochem. Anal. **26**, 1 (1980). – *[CAS 110539-97-4]*

Avilamycins.

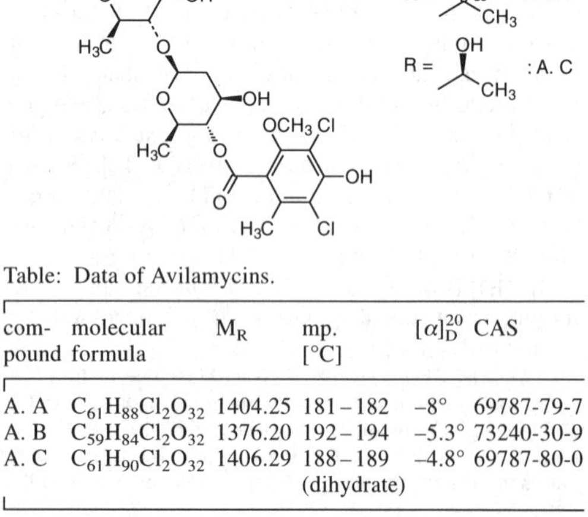

Table: Data of Avilamycins.

compound	molecular formula	M_R	mp. [°C]	$[\alpha]_D^{20}$	CAS
A. A	$C_{61}H_{88}Cl_2O_{32}$	1404.25	181–182	–8°	69787-79-7
A. B	$C_{59}H_{84}Cl_2O_{32}$	1376.20	192–194	–5.3°	73240-30-9
A. C	$C_{61}H_{90}Cl_2O_{32}$	1406.29	188–189 (dihydrate)	–4.8°	69787-80-0

*Antibiotic complex with 16 components formed by *Streptomyces viridochromogenes*. The A. belong to the *orthosomycin* class of antibiotics which are characterized by an *oligosaccharide structure and two orthoester groups. The main component is *A. C* (fine platelets). Other components with known structures are *A. A* (needles) and the crystalline *A. B* (45-*O*-acetyl-45-*O*-deisobutyryl-A. A), also known as *curamycin A*. A. are active against Gram-positive pathogens, they inhibit the binding of aminoacyl-tRNA to the *30S* subunit of bacterial ribosomes. A. have been tested as food antibiotics in animal breeding.

Lit.: Carbohydr. Res. **248**, 107 (1993) ▪ Heterocycles **15**, 1621 (1981) ▪ J. Antibiot. **39**, 877 (1986) ▪ Lukacs, Recent Progress in the Chemical Synthesis of Antibiotics and Related Microbial Products, p. 475–550, Berlin: Springer 1993 ▪ Tetrahedron **35**, 1207 (1979); **43**, 4133 f. (1987) ▪ Tetrahedron Lett. **28**, 1105 f. (1987) (synthesis). – *[HS 2941 90]*

Axisonitriles see isocyanides.

Azadirachtin(s).

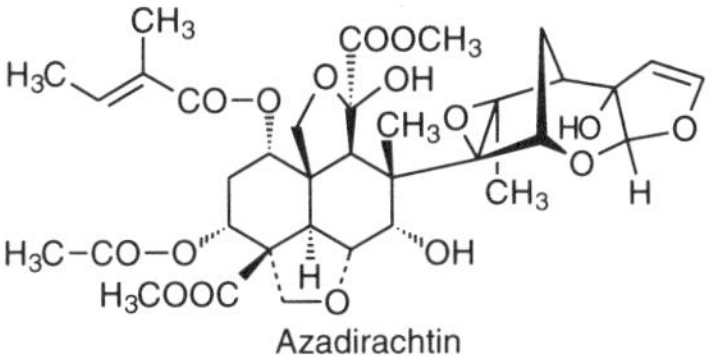

Azadirachtin

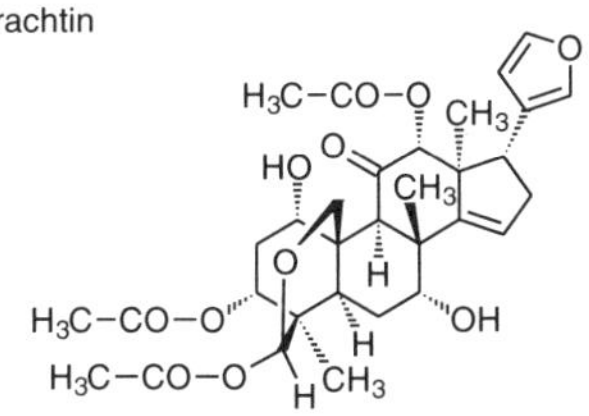

Azadirachtanin

$C_{35}H_{44}O_{16}$, M_R 720.72, crystalline powder, mp. 155–158 °C, $[\alpha]_D$ –53° ($CHCl_3$), unstable in solution. Like *azadirachtanin* ($C_{32}H_{40}O_{11}$, M_R 600.66, cryst., mp. 225–226 °C) A. is a highly oxidized triterpene [1] (*limonoids) isolated from the Indian tree *Azadirachta indica* (*neem tree). A. is a very effective, systemically active feeding deterrent for insects. It has an ecdysone-like action and causes growth disorders in the larval stage of the creatures [2]. It is non toxic for mammals [3].

Lit.: [1] Tetrahedron Lett. **26**, 6435 (1985). [2] Heterocycles **23**, 2321 (1985). [3] Z. Naturforsch. C **42**, 4 (1987).

gen.: Nat. Prod. Rep. **10**, 109–157 (1993) (review) ▪ Schmutterer, The Neem Tree-Source of Unique Natural Products for Integrated Pest Management, Medicine, Industry and other Purposes, Weinheim: VCH Verlagsges. 1995 ▪ Zechmeister **78**, 47–150 (review). – *Biological activity:* ACS Symp. Ser. **387**, 150–156 (1989); **595**, 108–133 (1995) ▪ Brighton Crop. Prot. Conf.-Pests Dis. **1**, 53–58 (1994) ▪ Experienta **51**, 831 ff. (1995). – *Synthesis:* Nat. Prod. Rep. **10**, 109–157 (1993) ▪ Pure Appl. Chem. **66**, 2099 ff. (1994) ▪ Tetrahedron **50**, 11553–11568 (1994); **51** 6591–6604 (1995). – *[CAS 11141-17-6 (A.); 98798-58-4 (azadirachtanin)]*

Azalomycin B see elaiophylin.

Azalomycin M see nigericin.

Azaserine (*O*-diazoacetylserine).

$N_2{=}CH{-}CO{-}O{-}CH_2{-}CH(NH_2){-}COOH$

$C_5H_7N_3O_4$, M_R 173.13. Light yellow to greenish cryst., mp. 153–155 °C (decomp.), $[\alpha]_D$ +9.7° (2 N HCl). A. is an antibiotic, isolated in optically active L-form, from *Streptomyces* species, the racemic form can be synthesized. As antagonist of glutamine it has mutagenic, bactericidal, fungicidal, and cytostatic activities.

Lit.: Beilstein E IV **4**, 3124 ▪ IARC Monogr. **10**, 73–77 (1976) ▪ J. Antibiot. **40**, 1657 (1987) (isolation) ▪ Sax (8.), ASA 500 ▪ Snell-Hilton **5**, 477, 599 ff. – *[HS 2941 90; CAS 115-02-6 (L)]*

Azaspiracid.

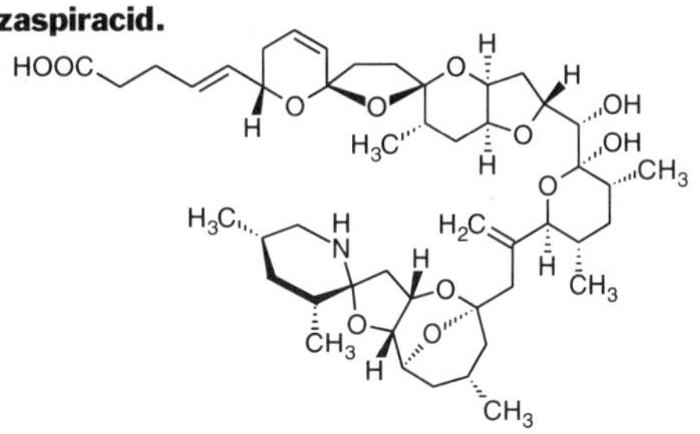

Marine toxin from mussels (*Mytilus edulis*) cultivated in Ireland, $C_{47}H_{71}NO_{12}$, M_R 842.07, colorless, amorphous solid, $[\alpha]_D^{20}$ –21° (CH_3OH), LD_{50} (mouse i.p.) 0.2 mg/kg. People who became ill in 1995 after consumption of mussels presented with symptoms very similar to those of diarrhetic shellfish poisoning (DSP): nausea, vomiting, severe diarrhea and stomach cramps.

Lit.: J. Am. Chem. Soc. **120**, 9967 (1998). – *[CAS 214899-21-5]*

Azepinostatin see balanol.

Azetidine-2-carboxylic acid.

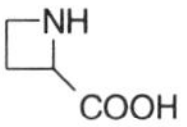

$C_4H_7NO_2$, M_R 101.11, cryst., decomposition without melting at >200 °C. A. is the only naturally occurring azetidine [(*S*)-(–)-form]. It is present in roots and leaves of various lily species, also in the seed of some Fabaceae and in fruit bodies *Clavaria miyabeana*. A. has an inhibitory effect on the germination of seeds in which it does not exist and is also an antagonist of *L-proline [1].

Lit.: [1] Nature (London) **200**, 148 (1963).

gen.: Beilstein E V **22/1**, 15 f. ▪ Chem. Rev. **79**, 331 (1979) ▪ Sax (8.), ASC 500 ▪ Tetrahedron **44**, 637 (1988) (synthesis). – *[CAS 2133-34-8]*

Azicemicins.

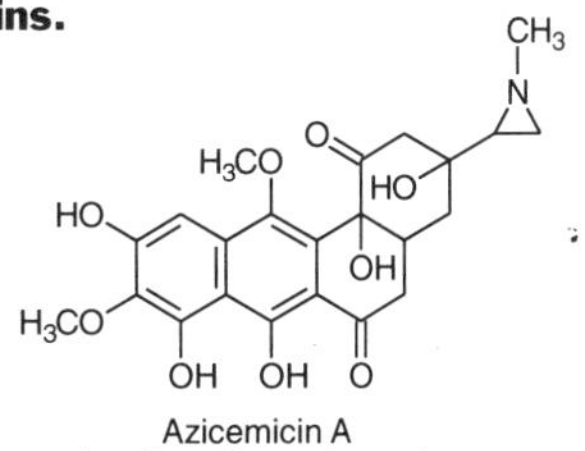

Azicemicin A
(configuration unknown)

A. A: $C_{23}H_{25}NO_9$, M_R 459.45, yellow powder, $[\alpha]_D$ –185° (CH_3OH). The aziridine derivative A. A isolated from an *Amycolatopsis* strain exhibits, like the corresponding *N*-demethyl compound *A. B* {$C_{22}H_{23}NO_9$, M_R 445.43, yellow powder, $[\alpha]_D$ –121° (CH_3OH)}, a moderate inhibitory action towards Gram-positive bacteria and mycobacteria. A. are non-toxic in mice.

Lit.: J. Antibiot. **46**, 1772 (1993); **48**, 217, 1148 (1995). – *[CAS 154163-93-6 (A. A)]*

Azinomycin B see carzinophilin.

Azirinomycin (3-methyl-2*H*-azirine-2-carboxylic acid).

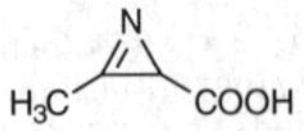

$C_4H_5NO_2$, M_R 99.09, pale yellow, unstable oil. An antibiotically active azirine derivative from cultures of *Streptomyces aureus*. A. is active *in vitro* against Gram-positive and Gram-negative bacteria.

Lit.: Chem. Lett. **1976**, 1063 (synthesis) ▪ J. Antibiot. **24**, 42, 48 (1971). – *[CAS 31772-89-1]*

Azotobactin D see siderophores.

Azotochelin [N^2,N^6-bis(2,3-dihydroxybenzoyl)-L-lysine].

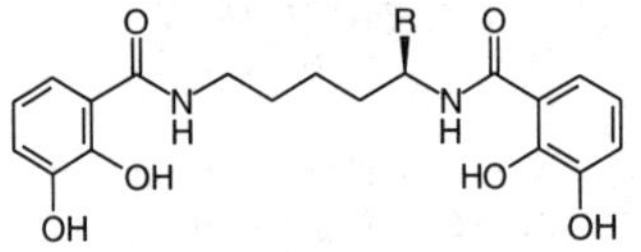

R = COOH : Azotochelin
R = CH_2OH : Myxochelin A

$C_{20}H_{22}N_2O_8$, M_R 418.40. A., isolated from *Azotobacter vinelandii* (nitrogen fixing bacterium) under iron deficit conditions, belongs to the *siderophores of the catechol type, mp. 82–87 °C. A. allows the uptake of iron essential for life. In closely related *myxochelin A* ($C_{20}H_{24}N_2O_7$, M_R 404.42) from *Angiococcus disciformis* the carboxy group of the lysine residue is reduced to a CH_2OH group.

Lit.: J. Antibiot. **42**, 14 (1989) ▪ J. Bacteriol. **159**, 341–347 (1984). – *[CAS 23369-85-9 (A.); 120243-02-9 (myxochelin)]*

Azoxybacilin.

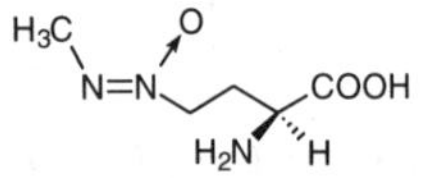

$C_5H_{11}N_3O_3$, M_R 161.16, needles, mp. 203–205 °C, $[\alpha]_D$ +9.4° (H_2O). A strongly antifungal azoxy compound from cultures of *Bacillus cereus*. A. inhibits sulfur fixation in methionine biosynthesis.

Lit.: Chem. Pharm. Bull. **42**, 1703 (1994) (synthesis) ▪ J. Antibiot. **47**, 833, 909 (1994). – *[CAS 157998-96-4]*

Azoxybenzene-4,4′-dicarboxylic acid.

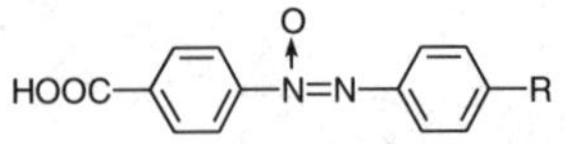

R = COOH : Azoxybenzene-4,4′-dicarboxylic acid
R = CH_2OH : 4′-(Hydroxymethyl)azoxybenzene-4-carboxylic acid

$C_{14}H_{10}N_2O_5$, M_R 286.24, yellow amorphous powder, mp. >360 °C. A. is isolated, together with *4′-(hydroxymethyl)azoxybenzene-4-carboxylic acid* [$C_{14}H_{12}N_2O_4$, M_R 272.26, yellow powder, mp. 251 °C (decomp.)] from cultures of *Entomophthora virulenta* (Zygomycetes), a fungal parasite on insects. 4′-(Hydroxymethyl)azoxybenzene-4-carboxylic acid is responsible for the insecticidal activity of the culture extract.

Lit.: J. Chem. Soc., Perkin Trans. 1 **1978**, 171; Perkin Trans. 2 **1995**, 1679 ▪ Tetrahedron **51**, 11305 (1995) (synthesis). – *[CAS 582-69-4 (A.); 66641-30-3 (4′-(hydroxymethyl)azoxybenzene-4-carboxylic acid)]*

Azoxy compounds (natural). Aliphatic A. c. are produced by streptomycetes (*elaiomycin, *jietacin A, *maniwamycins, *valanimycin), the amino acid *azoxybacilin by bacteria. The azoxy cyanide and carboxamide *calvatic acid and *lyophyllin are metabolites of basidiomycetes. The glycosides of the carcinogenic methylazoxymethanol (macrozamin, see cycasin) which occur in nuts of palm ferns (Zamiaceae, Cycadaceae) play a role as risk factors in human nutrition. Insecticidal derivatives of azoxybenzenes have been isolated from a zygomycete parasite of insects (*azoxybenzene-4,4′-dicarboxylic acid). The as yet only known natural azo compound is 4,4′-dihydroxyazobenzene [1] which has been isolated from the extract of *Agaricus xanthoderma* (Basidiomycetes) but it probably represents an artefact (*4-diazo-2,5-cyclohexadien-1-one). Various natural A. c. show antifungal activities and are strong carcinogens.

Lit.: [1] Z. Naturforsch. C **39**, 1027 (1984).

Azulenes (Spanish azul = blue).

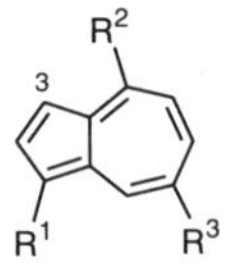

$R^1 = R^2 = R^3 = H$	: Azulene (**1**)
$R^1 = R^2 = CH_3$, $R^3 = H$	: 1,4-Dimethylazulene (**2**)
$R^1 = COOCH_3$, $R^2 = CH_3$, $R^3 = H$	: Methyl-4-methylazulene-1-carboxylate (**3**)
$R^1 = R^2 = CH_3$, $R^3 = C_2H_5$	: Chamazulene (**4**)
$R^1 = R^2 = CH_3$, $R^3 = CH(CH_3)_2$	: Guajazulene (**5**)

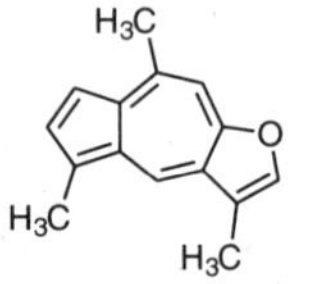

Linderazulene (**6**)

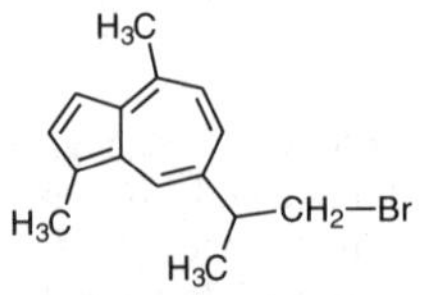

Ehuazulene (**7**)

General name for the blue to violet aromatic hydrocarbons derived from bicyclo[5.3.0]decapentaene (*azulene*, $C_{10}H_8$, M_R 128.17, mp. 99 °C, blue needles).

A. occur in the blue oil bodies of various liverworts (*mosses), e.g., *Calypogeia azurea* (also in cell culture) contains *1,4-dimethylazulene* ($C_{12}H_{12}$, M_R 156.23, deep blue prisms, mp. 23 °C) and *methyl 4-methylazulene-1-carboxylate* ($C_{13}H_{12}O_2$, M_R 200.24, violet cryst., mp. 48–49 °C) [1]. The latter has also been isolated from *Helichrysum acuminatum* (Asteraceae) [2]. Sea fans (Paramuriceidae) produce *linderazulene* ($C_{15}H_{14}O$, M_R 210.28, shining violet-black platelets) and halogenated guajazulenes such as *2-bromoguajazulene* ($C_{15}H_{17}Br$, M_R 277.20) and the side-chain brominated *ehuazulene* [$C_{15}H_{17}Br$, M_R 277.20, oil, $[\alpha]_D$ +2° (hexane)] [3]. Also the deep blue milk of the agaric *Lactarius indigo* owes its color to an azulene derivative, see Lactarius pigments.

Most A. isolated from plants are artefacts formed during work-up, e. g., by steam distillation of *essential oils, from certain colorless sesquiterpenes, the *proazulenes. Especially rich in A. [e. g., *chamazulene* (dimethulene), $C_{14}H_{16}$, M_R 184.28, blue oil, bp. 161 °C (1.2 kPa)] are the essential oils of mugwort and chamomile. Furthermore, *chamomile oils contain *guajazulene* [azulone, $C_{15}H_{18}$, M_R 198.31, blue-violet platelets, mp. 31 °C, bp. 167–168 °C (1.2 kPa), uv_{max} 662 nm (ether)], which can also be obtained by partial synthesis from guajol of pockwood oil. The therapeutic relevance of the anti-inflammatory and antiallergic effects of plant A. is controversial[4].

Lit.: [1] Chem. Ber. **99**, 2669 (1966); Phytochemistry **31**, 1667, 1671 (1992). [2] Phytochemistry **26**, 803 (1987). [3] Experientia **37**, 680 (1981); **43**, 624 (1987); Tetrahedron Lett. **25**, 587 (1984). [4] Arch. Pharm. (Weinheim, Ger.) **330**, 7 (1997).
gen.: Angew. Chem. Int. Ed. Engl. **12**, 793 (1973) ▪ Bull. Chem. Soc. Jpn. **66**, 892 (1993) (synthesis azulene) ▪ Elsevier **12A**, 420–430 ▪ Ginsburg, Non-benzenoid Aromatic Hydrocarbons, p. 171–337, New York: Interscience 1959 ▪ J. Am. Chem. Soc. **101**, 251 (1979) (synthesis guaiazulene) ▪ Prog. Inorg. Chem. **11**, 53–98 (1970) ▪ Zechmeister **19**, 32–119. – *[HS 2902 19 (**4, 5**); CAS 275-51-4 (**1**); 1127-69-1 (**2**); 10527-06-7 (**3**); 529-05-5 (**4**); 489-84-9 (**5**); 489-79-2 (**6**); 90052-62-3 (**7**); 90052-61-2 (2-bromoguajazulene)]*

B

Baccharane see triterpenes.

Baccharinoids. Group of (more than 30) strongly cytotoxic, macrocyclic *tric(h)othecene derivatives from South American bushes of the genus *Baccharis* (Asteraceae). The compounds are plant transformation products of the *roridins which are formed by fungi (*Myrothecium* species) in the root region and are taken up by the plants. *Baccharis megapotamica* forms B. from roridins and *B. cordifolia* forms miotoxins. Roridins are also isolated from these plants. Poisoning of grazing animals is not uncommon since the toxins occur not only in the roots but also in aboveground parts of the plants. In calves 2 g fresh young leaves per kg body weight or 0.5 g of flowering plants lead to death.

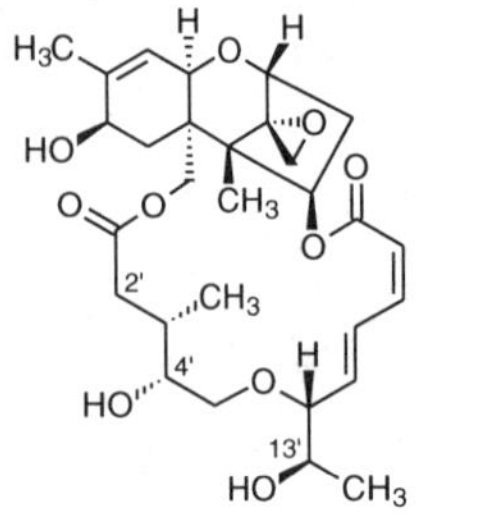

B. B_1 = Baccharol

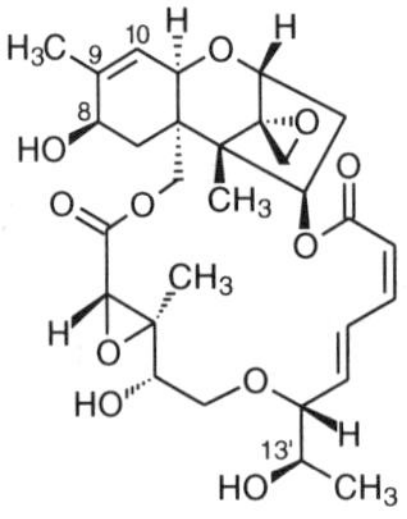

B. B_4 = Baccharinol

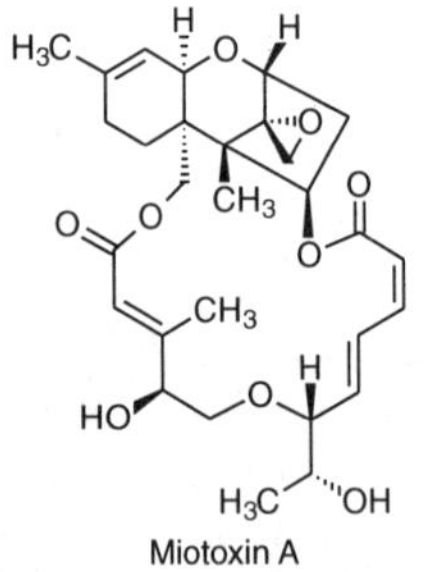

Miotoxin A

Table: Data of Baccharinoids and Miotoxins.

compound	molecular formula	M_R	mp. [°C]	CAS
Baccharinoids:				
B. B1	$C_{29}H_{40}O_{10}$	548.63	162–164	71695-69-7
B. B4	$C_{29}H_{38}O_{11}$	562.61	259–263	63783-94-8
Miotoxins:				
M. A	$C_{29}H_{38}O_9$	530.62		90790-03-7
M. B	$C_{29}H_{38}O_9$	530.62		93633-90-0

Lit.: Betina, chap. 10 ▪ J. Nat. Prod. **50**, 815 (1987); **51**, 736–744 (1988) ▪ J. Org. Chem. **52**, 45 (1987) ▪ Justus Liebigs Ann. Chem. **1984**, 1746; **1985**, 633 ▪ Phytochemistry **30**, 789–797 (1991) ▪ Toxicon **23**, 731–745 (1985) ▪ Z. Naturforsch. C **39**, 212 (1984).

Bacillus thuringiensis. A bacterial species which, upon destruction of the sporangium, releases an endotoxin (B. t. toxin)[1] that kills caterpillars in very low concentrations. B. t. is used as biological pesticide (insecticide) under different trade names. In humans and mammals it seems to be safe (no signs of toxic or infectious effects). The B. t. genes have been expressed by genetic engineering in crop plants[2,3], resulting in pest resistance of these transgenic plants. On the other hand, pests can aquire resistance against B. t.[4]. For mechanism of action see *Lit.*[5–9]. Pollen from corn, genetically engineered to produce B. t. toxin (which protects the plant against European corn borers) kills certain butterfly larvae in lab tests[10].

Lit.: [1] Pestic. Sci. **45**, 95–105 (1995). [2] ACS Symp. Ser. **524**, 267–280 (1993). [3] Agric. Ecosyst. Environ. **49**, 85–93 (1994). [4] Fla. Entomol. **78**, 414–443 (1995). [5] Annu. Rev. Entomol. **37**, 615–636 (1992). [6] Arch. Insect Biochem. Physiol. **22**, 357–371 (1993). [7] Mol. Microbiol. **7**, 489 (1993). [8] Adv. Insect Physiol. **24**, 275–308 (1994). [9] ACS Symp. Ser. **591**, 308–319 (1995). [10] Nature (London) **399**, 214 (1999).
gen.: ACS Symp. Ser. **524**, 258–266 (1993) ▪ Curr. Opin. Biotechnol. **6**, 305–312 (1995) ▪ Entwistle et al. (eds.), Bacillus thuringiensis, An Environmental Biopesticide: Theory and Practice, Chichester, New York: Wiley 1993 ▪ Genet. Eng. (N. Y.) **17**, 99–117 (1995) ▪ Mol. Microbiol. **18**, 1–12 (1995).

Bacilysin see chlorotetaine.

Bacimethrin.

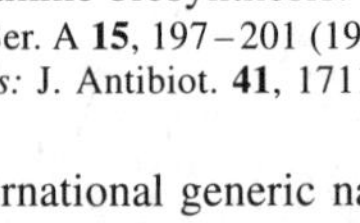

$C_6H_9N_3O_2$, M_R 155.16, crystals (from methanol), mp. 175 °C. *Thiamine antagonist isolated from culture filtrates of *Bacillus megaterium* and *Streptomyces albus*. B. is assumed to inhibit phosphorylation of the pyrimidine part in thiamine biosynthesis.

Lit.: J. Antibiot., Ser. A **15**, 197–201 (1962); **40**, 1431–1439 (1987). – *Synthesis:* J. Antibiot. **41**, 1711 (1988) (review). – *[CAS 3690-12-8]*

Bacitracin. International generic name for a peptide antibiotic complex with broad activity against Gram-positive pathogens, especially staphylococci, penicillin- and sulfonamide-resistant streptococci; and some Gram-negative cocci and bacteria. B. is isolated from *Bacillus subtilis* and *B. licheniformis*. Commercial B. is a mixture with B. A as main component (70%). B. A has the following constitution:

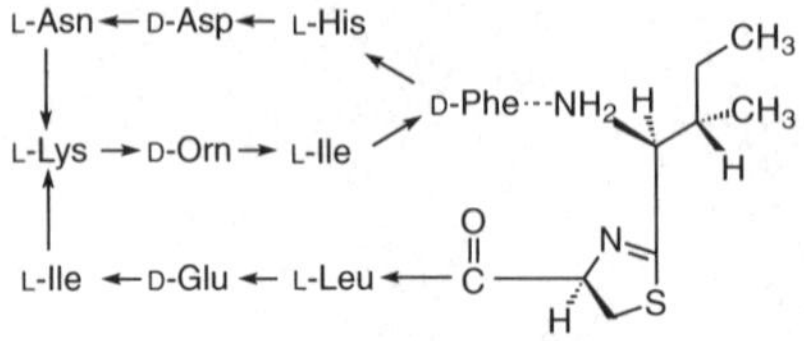

$C_{66}H_{103}N_{17}O_{16}S$, M_R 1422.71, $[\alpha]_D^{23}$ +5° (0.02 M HCl). hygroscopic powder with a very bitter taste, readily soluble in water, slightly soluble in alcohol, insoluble in ether, chloroform, and acetone. At acidic pH of 4–5 B. is relatively stable, but it is unstable in alkaline solution. The intact thiazoline ring is essential for the antibiotic activity.

Use: B. is used especially in topical application (wound infections, burns, skin grafts). On oral administration B. is practically not resorbed. B. is widely used in animal breeding as animal food additive.

Mechanism of action: In cell wall synthesis B. acts as an inhibitor of the biosynthesis of murein.

Lit.: Drugs Pharm. Sci. **22**, 665 (1984) ▪ J. Antibiot. **45**, 1325 (1992) ▪ J. Org. Chem. **61**, 3983 (1996) (synthesis of B. A) ▪ Kirk-Othmer (4.) **2**, 895; **3**, 268, 273 ▪ Martindale (30.), p. 121 ▪ Merck-Index (12.), No. 965 ▪ PCT Pat. Appl. WO 9747,313 (1997) ▪ Sax (8.), BAC 250. – *[HS 2941 90; CAS 1405-87-4 (B. complex); 22601-59-8 (B. A)]*

Bacteriochlorins. The structural type of the B. (7,8,17,18-tetrahydroporphyrin) is derived from *porphin by formal reduction of two peripheral double bonds (C7 and C17) in oppositely positioned pyrrole rings of the porphyrin basic skeleton (formula, see porphyrins). The name is based on the *bacteriochlorophylls, of which some possess the basic skeleton of B. with a central magnesium(II) ion. A further representative of this structural type is *tolyporphin A.

Lit.: Chem. Rev. **94**, 327 (1994) ▪ Dolphin **II**, 6, 62–65 ▪ Progr. Heterocycl. Chem. **10**, 1–24 (1998) (synthesis) ▪ Smith, p. 643.

Bacteriochlorophylls.

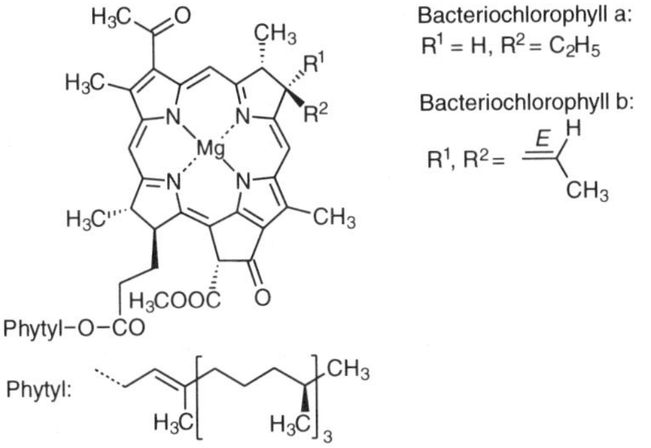

Main pigments of phototrophic bacteria of the orders Chlorobinea (green bacteria) and Rhodospirillinea (purple bacteria). B. a and b have the actual *bacteriochlorin skeleton while B. c–e have the *chlorin skeleton on which the *chlorophylls of higher plants and cyanobacteria are based; they are also very similar to chlorophylls in other respects. Elucidation of the detailed spatial construction of the crystallized bacterial photosynthesis system from membranes of *Rhodopseudomonas viridis* was realized in 1984 by Deisenhofer, Huber, and Michel by X-ray structural analysis for which they received the Nobel Prize for Chemistry in 1988 [1].

Lit.: [1] Angew. Chem. Int. Ed. Engl. **28**, 848–891 (1989).
gen.: Bioorg. Med. Chem. Lett. **9**, 1631 (1999) (synthesis) ▪ Chimia **41**, 277–292 (1987) ▪ Dolphin **II**, 306–312; **IV**, 216–227 ▪ Nat. Prod. Rep. **6**, 171 (1989) ▪ New Compr. Biochem. **19**, 237 (1991) ▪ Smith, p. 465 f. – *[HS 3203 00; CAS 17499-98-8 (B. a); 53199-29-4 (B. b)]*

(32*R*,33*R*,34*S*)-Bacteriohopane-32,33,34,35-tetraol see hopanoids.

Bactobolin.

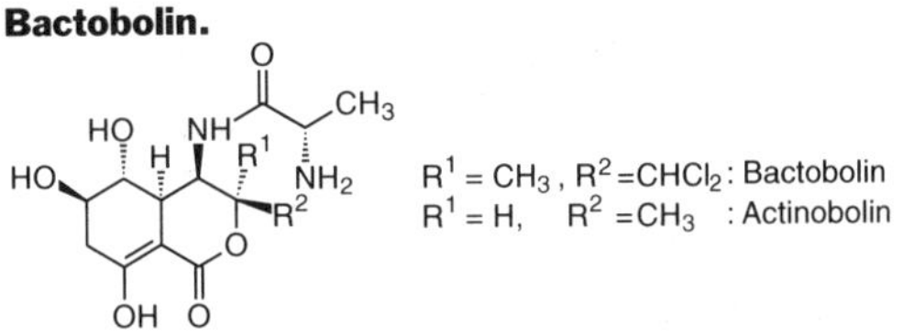

$C_{14}H_{20}Cl_2N_2O_6$, M_R 383.23, mp. 214 °C (decomp.; hydrochloride), $[\alpha]_D^{22}$ (hydrochloride) –9.2° (H_2O). Broad-spectrum *antibiotic from *Pseudomonas yoshitomensis*, active against Gram-positive and -negative bacteria with an unusual hexahydroisocoumarin structure. Three hydroxy groups and an alanyl side chain render B. so polar that it is only well soluble in polar solvents like water and methanol. A close structural relative is *actinobolin* from *Streptomyces griseoviridis*, $C_{13}H_{20}N_2O_6$, M_R 300.31, $[\alpha]_D^{22}$ (sulfate) +54.5° (H_2O). Actinobolin acts against caries (periodontitis mice, rats); LD_{50} (mouse i.p.) 2.8 g/kg. B. has pronounced activity against experimental tumors but is not mutagenic. In spite of its laborious isolation it is produced on a large scale by fermentation and its total synthesis is the subject of intensive effort.

Lit.: J. Am. Chem. Soc. **92**, 4933–4942 (1970); **112**, 3475–3482 (1990) (synthesis) ▪ J. Chem. Soc., Perkin Trans. 1 **1990**, 731–738 (actinobolin) ▪ Merck-Index (12.), No. 139. – *[HS 2941 90; CAS 72615-20-4 (B.); 24397-89-5 (actinobolin)]*

Badiones.

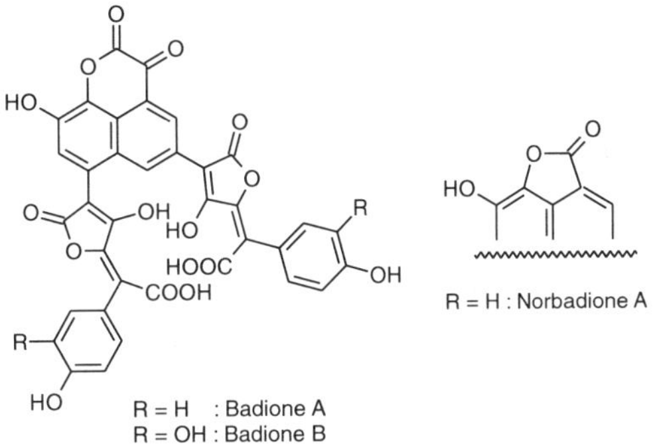

Chocolate brown pigments from the cap skin of some mushrooms such as the bay bolete (*Xerocomus badius*), the with bolete (*Boletus erythropus*), and related *Boletus* species (Basidiomycetes) [1]. They are formed biosynthetically by oxidative dimerization of *pulvinic acids. When the white flesh under the brown cap skin of *Xerocomus badius* is treated with *xerocomic acid, *B. A* ($C_{36}H_{18}O_{16}$, M_R 706.53, black-brown, metallic shining crystals, decomp. >250 °C) is formed. In *Boletus erythropus B. B* ($C_{36}H_{18}O_{18}$, M_R 738.53) is formed analogously from *variegatic acid. Both B. form stable salts with potassium ions. This ability to form complexes with alkali metal ions explains the high radioactivity of *Xerocomus badius* after emission of ^{137}Cs during the Chernobyl reactor incident [2]. *Xerocomus badius* and, especially, *Pisolithus tinctorius* (Gasteromycetes) [3], a pioneering fungus on barren soil, contain the potassium salts of *norbadione A* ($C_{35}H_{18}O_{15}$, M_R 678.52, red cryst., mp. >300 °C).

Lit.: [1] Angew. Chem. Int. Ed. Engl. **23**, 445 (1984). [2] Angew. Chem. Int. Ed. Engl. **28**, 453 (1989). [3] Phytochemistry **24**, 1351 (1985).

gen.: Angew. Chem. Int. Ed. Engl. **24**, 711 (1985). – *Review:* Nat. Prod. Rep. **11**, 72 (1994) ▪ Zechmeister **51**, 55. – *[HS 3203 00; CAS 90295-66-2 (B. A); 97486-15-2 (B. B); 90295-68-4 (norbadione A)]*

Baeocystine see psilocybine.

Bafilomycins. Group of 16-membered macrocyclic lactones produced by *actinomycetes (e.g., *Streptomyces griseus*). Most thoroughly investigated are *B. A*$_1$ [$C_{35}H_{58}O_9$, M_R 622.84, mp. 103–106 °C (decomp.)] and. *B*$_1$ [*setamycin*, $C_{44}H_{65}NO_{13}$, M_R 815.00, mp. 134–135 °C, $[\alpha]_D$ +18.3° (CH_3OH)]; they have a side chain which has folded to form a cyclic semiketal (group name: *pleco-*macrolides*). B. A_2 and B. B_2 are the methyl ketals at C-19 and are considered, like B. D and B. E (ketone in place of the semiketal), as artifacts.

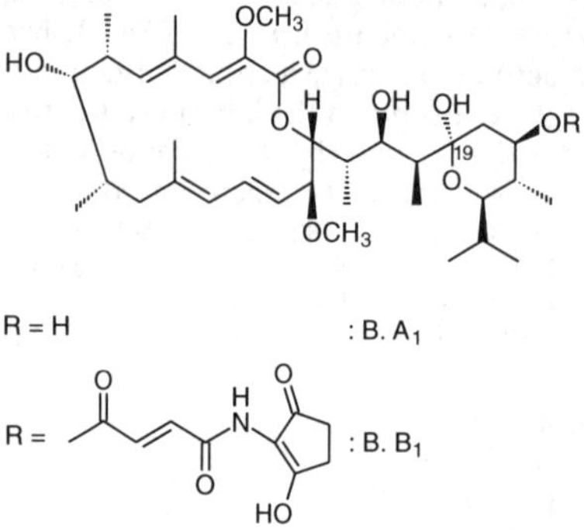

Activity: B. inhibit the growth of Gram-positive bacteria, protozoa, fungi, and yeasts; insecticidal, nematocidal, and immunosuppressive properties have also been reported. Development of the B. has so far been prevented by their general toxicity. B. A_1, in particular, is a specifically acting inhibitor of vacuolar H^+-ATPase [1] of the V type (K_i = 0.5 nM), P type ATPases are less affected and those of the F type are not inhibited.

Lit.: [1] J. Med. Chem. **41**, 1883 (1998).

gen.: Angew. Chem. Int. Ed. Engl. **38**, 1652 (1999) [synthesis (–)-B. A_1] ▪ Helv. Chim. Acta **68**, 83–94 (1985) ▪ J. Antibiot. **37**, 110, 970 (1984); **41**, 250 (1988) ▪ J. Chem. Soc., Perkin Trans. 2 **1990**, 717 ▪ J. Org. Chem. **62**, 3271 (1997) [synthesis (–)-B. A_1] ▪ Tetrahedron Lett. **28**, 5565 (1987); **37**, 1069, 1073 (1996) (synthesis). – *Activity:* Biochemistry **15**, 3902–3906 (1993). – *[CAS 88899-55-2 (B. A$_1$); 88899-56-3 (B. B$_1$)]*

Baikiain (1,2,3,6-tetrahydro-2-pyridinecarboxylic acid).

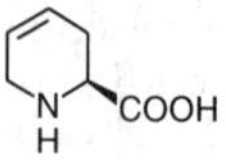

(2*S*)-, L-form

$C_6H_9NO_2$, M_R 127.14, mp. 274 °C (decomp.), $[\alpha]_D^{20}$ –288° (H_2O); hydrochloride: mp. 264 °C, $[\alpha]_D^{20}$ –90.1° (H_2O). Non-proteinogenic amino acid first isolated from Rhodesian teak wood *Baikiaea plurijuga* (Fabaceae) and later also from *Caesalpinia tinctoria* and red algae. B. inhibits the activity of *glutamic acid as neurotransmitter [1].

(2S,3R)-3-Hydroxybaikiain [2] {$C_6H_9NO_3$, M_R 143.14, mp. 300–302 °C, $[\alpha]_D^{20}$ –332.7° (H_2O)} occurs in high concentrations in the fruit bodies of the poisonous toadstool *Russula subnigricans*.

Lit.: [1] J. Neurobiol. **10**, 355 (1959). [2] Chem. Pharm. Bull. **35**, 3482–3486 (1987).

gen.: Beilstein E V **22/1**, 319 ▪ Greenstein & Winitz, Chemistry of the Amino Acids, vol. 3, p. 2534–2537, New York: Wiley 1961 ▪ Karrer, No. 2397 ▪ Tetrahedron: Asymmetry **8**, 1855 (1997) [synthesis (–)-B.]. – *[CAS 31456-71-0 (L-B.)]*

Balanol (Azepinostatin).

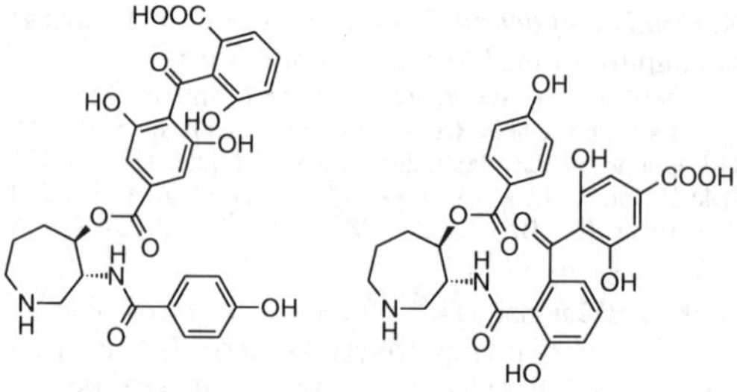

Balanol Ophiocordin

$C_{28}H_{26}N_2O_{10}$, M_R 551.17, mp. 180 °C (decomp.), $[\alpha]_D^{23}$ –111.0° (CH_3OH). Hexahydroazepine derivative from the hyphomycete *Verticillium balanoides*.

Activity [1]: Inhibitor of protein kinase C in the lower nanomolar range, more potent than *staurosporine and thus of major pharmaceutical interest, e.g., for cancer therapy. Longer known is a regioisomer of B., the antifungally active *ophiocordin* [2] {$[\alpha]_D^{20}$ +70° (CH_3OH)}, from the fungus *Cordyceps ophioglossoides*. Because of its interesting activity and modular structure, B. has been one of the first natural products to become a target for combinatorial synthesis [3].

Lit.: [1] J. Chem. Soc., Perkin Trans. 1 **1995**, 2355–2362. [2] Chem. Ber. **113**, 2221 (1980). [3] J. Med. Chem. **40**, 226–235 (1997); PCT Pat. Appl. WO 9729091 (1997).

gen.: Bioorg. Med. Chem. Lett. **5**, 2151, 2155 (1995); **6**, 973–978 (1996) ▪ Chem. Eur. J. **1** 454–466, 467–494 (1995) (synthesis, review) ▪ Chimia **50**, 530 (1996) [synthesis (–)-B.] ▪ J. Am. Chem. Soc. **115**, 6452, 10 468 (1993) (isolation); **116**, 8402 (1994) ▪ J. Antibiot. **47**, 639 (1994) ▪ J. Org. Chem. **59**, 5147 (1994); **61**, 4572 (1996); **63**, 4397 (1998) ▪ Synlett **1997**, 580 (synthesis) ▪ Tetrahedron **51**, 6061–6070 (1995); **53**, 4857, 17 177 (1997) ▪ Tetrahedron: Asymmetry **10**, 4521 (1999) (synthesis) ▪ Tetrahedron Lett. **37**, 8439 (1996); **38**, 1693 (1997). – *[CAS 63590-19-2; 147397-03-3; 158931-29-4]*

Balata. A *gutta-percha-like tough, hard, poorly elastic polyisoprene (structure see under natural rubber) obtained from the resin-rich milk of wild-growing tropical trees (*Mimusops balata*, Sapotaceae) in Venezuela and Brazil.

Lit.: Kirk-Othmer (3.) **20**, 488 ▪ Staudinger et al., Das wissenschaftliche Werk von Hermann Staudinger, vol. 1: Arbeiten über Isopren, Kautschuk u. Balata, Heidelberg: Hüthig 1969 ▪ Ullmann (4.) **13**, 588. – *[HS 4001 30]*

Bambermycin see moenomycin.

Banana aroma see fruit flavors.

Baptifoline (13-hydroxyanagyrine).

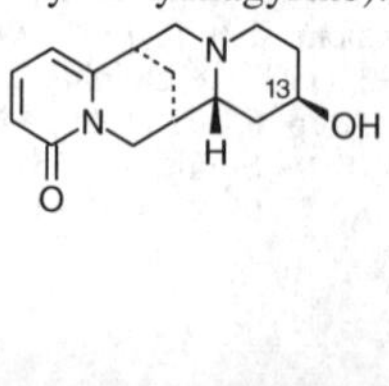

$C_{15}H_{20}N_2O_2$, M_R 260.34, mp. 210 °C, $[\alpha]_D$ –147.7° (C_2H_5OH). Tetracyclic *quinolizidine alkaloid of the *sparteine type with an α-pyridone structure from several genera of the Fabaceae (e. g., *Baptisia, Caulophyllum, Sophora*). In analogy to *13-hydroxylupanine, esters of B. occasionally occur, and generally as the acetates. The 13-epimer *epi-baptifoline* (mp. 215 °C, $[\alpha]_D^{21}$ –138.9°) is often found together with B.[1].

Lit.: [1] Waterman **8**, 197–239.
gen.: Beilstein E V **25/2**, 65 f. ▪ Pelletier **2**, 105–148. – *[HS 2939 90; CAS 732-50-3]*

Baptitoxine see cytisine.

Barbamide.

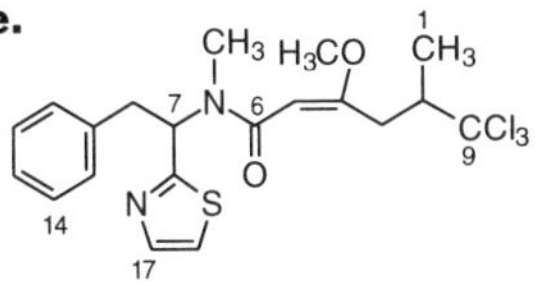

$C_{20}H_{23}Cl_3N_2O_2S$, M_R 461.96, yellow oil. Metabolite of the Caribbean cyanobacterium *Lyngbya majuscula* with molluscicidal activity.

Lit.: J. Nat. Prod. **59**, 427 (1996). – *Biosynthesis:* Angew. Chem. Int. Ed. Engl. **38**, 1209 (1999) ▪ J. Am. Chem. Soc. **120**, 7131 (1998).

Barbatanes, α-barbatene see gymnomitranes.

Bark beetles. Insects related to the *weevils that bore into the wood and bark of trees and often cause extreme economic damage to forests. Many species live in symbiosis with fungi; the elm bark beetle *Scolytus multistriatus* transmits the feared elm tree disease caused by the fungus *Ceratocystis ulmi*. Some B. species are controlled by the application of synthetic *pheromones in trap devices, that contain combinations of pheromones with synthetic insecticides; see also brevicomin, chalcogran, conophthorin, frontalin, ipsdienol, lineatin, multistriatin, pityol, sulcatol, seudenol.

Lit.: J. Chem. Ecol. **21**, 1043–1063 (1995).

Barrigenol A_1 see oleanane.

Barringtogenic acid see medicagenic acid.

Bartramiaflavone.

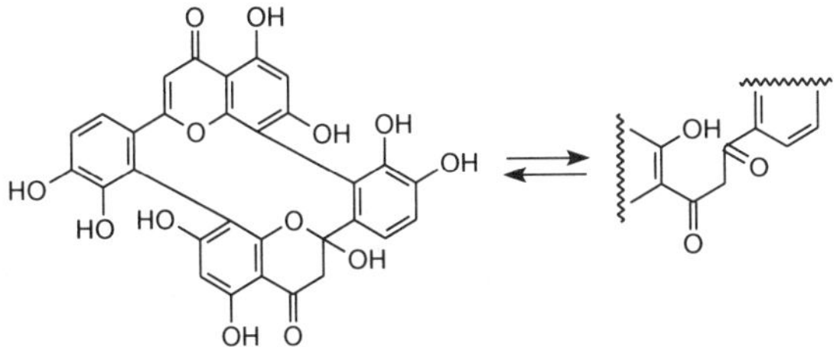

cyclic form oxo form

$C_{30}H_{18}O_{13}$, M_R 586.47, amorphous powder, soluble in 80% acetone, DMF, DMSO, insoluble in H_2O. Macrocyclic *biflavonoid, consisting of two cyclo/oxo tautomers, in equilibrium with each other. B. is found in the moss species *Bartramia pomiformis* and *B. halleriana*. On heating (250 °C) or boiling with 20% H_2SO_4 B. is converted by dehydration to *anhydrobartramiaflavone* ($C_{30}H_{16}O_{12}$, M_R 568.45), a doubly linked biluteolin.

Lit.: Phytochemistry **30**, 1653 (1991) ▪ Z. Naturforsch. C **48**, 531 (1993). – *[CAS 135117-91-8 (cyclo form); 135117-93-0 (oxo form)]*

Basidalin.

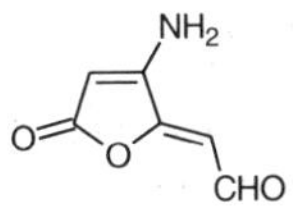

$C_6H_5NO_3$, M_R 139.11, mp. 142–149 °C. Lactone antibiotic from cultures of *Leucoagaricus naucina* (Basidiomycetes). B. exhibits antitumor activity.

Lit.: J. Antibiot. **36**, 448 (1983) ▪ Tetrahedron **50**, 7783 (1994) (synthesis). – *[CAS 82501-56-2]*

Basifungins see aureobasidins.

Bastadins.

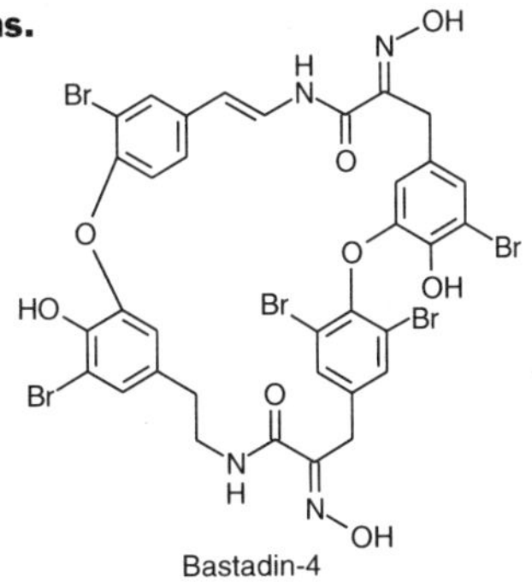

Bastadin-4

Open chain or macrocyclic oximes (more than 30 described) from the south Pacific sponge *Ianthella basta*. They are formed from 4 bromotyrosine units by single or double oxidative phenol coupling. B. act on the calcium channels of skeletal muscle [1]. An example is *B.-4* [$C_{34}H_{25}Br_5N_4O_8$, M_R 1017.11, yellow needles, mp. 250 °C (decomp.)].

Lit.: [1] J. Biol. Chem. **269**, 23236–23249 (1994); **272**, 15687–15696 (1997).
gen.: Aust. J. Chem. **34**, 765 (1981) ▪ J. Nat. Prod. **53**, 1441 (1991); **56**, 782 (1993); **59**, 1121 (1996) (isolation) ▪ J. Org. Chem. **63**, 4269 (1998) (synthesis B. 2, 3, 6) ▪ Tetrahedron Lett. **40**, 7023, 7027 (1999) (synthesis B. 12). – *[CAS 75513-48-3 (B.-1); 79067-76-8 (B.-4)]*

Batrachotoxins. Highly poisonous *steroid alkaloids (see table) isolated from the skin of the Colombian frogs *Phyllobates aurotaenia, P. terribilis, P. bicolor, P. vittatus*, and *P. lugubris*. B. (as skin preparations) are used as arrow poisons.

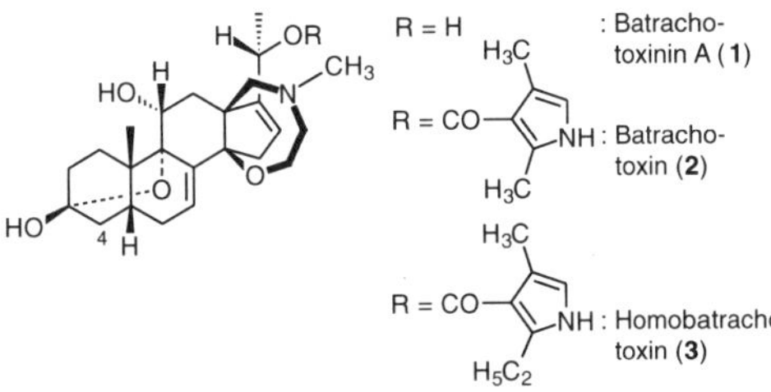

Table: Data of Batrachotoxins.

no.	molecular formula	M_R	$[\alpha]_D$ (CH_3OH)	LD_{50} (µg/kg; mouse s.c.)	CAS
1*	$C_{24}H_{35}NO_5$	417.55	–42°	1000	19457-37-5
2	$C_{31}H_{42}N_2O_6$	538.68	–5°→–10°	2	23509-16-2
3	$C_{32}H_{44}N_2O_6$	552.71		2–3	23509-17-3

* mp. 160–162 °C

The physiological action is based on depolarization of neurons and muscle cells by specific blockade of sodium ion transport through plasma membrane channels whereby muscular and finally respiratory paralysis results [1]. Antagonist of B. is *tetrodotoxin.

Lit.: [1] Bioorg. Med. Chem. Lett. **7**, 181 (1997).

gen.: Helv. Chim. Acta **56**, 139 (1973) (synthesis) ▪ J. Am. Chem. Soc. **120**, 6627 (1998) (synthesis B. A) ▪ Manske **43**, 185–288 ▪ Science **172**, 995 (1971) ▪ Zechmeister **41**, 206–340. – *Toxicology:* Clin. Toxicol. **4**, 331 (1971) ▪ J. Gen. Physiol. **88**, 841 (1987) ▪ Sax (8.), BAR 750.

Batyl alcohol (1-*O*-octadecyl-*sn*-glycerol).

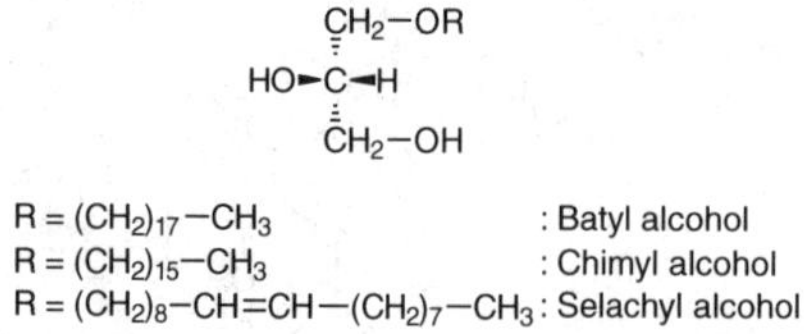

$C_{21}H_{44}O_3$, M_R 344.58, mp. 70.5–71.5 °C, $[\alpha]_D^{22}$ –11.8° (diacetate). B. exists in animal tissue either esterified with two fatty acid units as a neutral *ether lipid or with one fatty acid unit and phosphoric acid or phosphoric acid ester group as a polar *ether lipid (alkoxylipid). B. as well as *chimyl alcohol* {1-*O*-hexadecyl-*sn*-glycerol, $C_{19}H_{40}O_3$, M_R 316.52, mp. 64.5–65.5 °C, $[\alpha]_D$ –12.8° (diacetate)} and *selachyl alcohol* {1-*O*-[(*Z*)-9-octadecenyl]glycerol, $C_{21}H_{42}O_3$, M_R 342.56, $[\alpha]_D^{22}$ –15.2° (isopropylidene derivative)} are the main components of the alkylglycerols of animal lipids. B. and other alkylglycerols are biologically active compounds with, e. g., bacteriostatic, fungicidal, and immunostimulating properties .

Lit.: Braquet, Mangold & Vargaftig (eds.), Biologically Active Ether Lipids, p. 48–57, Basel: Karger 1988. – *[HS 290949; CAS 10567-22-3 (B.); 506-03-6 (chimyl alcohol); 34043-91-9 (selachyl alcohol)]*

Batzelladines. Alkaloids with unusual polycyclic structures from the caribbean *Batzella* sponges. B. induce $P56^{ICK}$-CD4-dissociation (cell cycle).

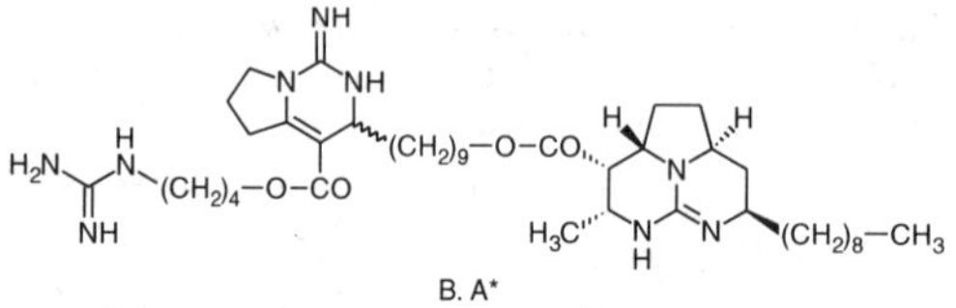

B. A*

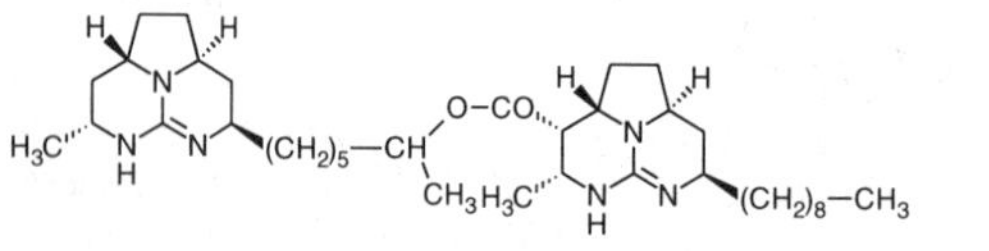

B. F*

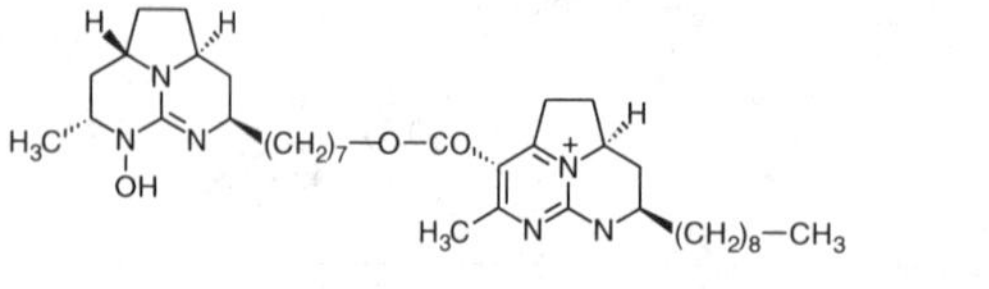

B. H*

* relative configurations shown.

Table: Data of Batzelladines (major metabolites).

B.	molecular formula	M_R	mp.	$[\alpha]_D^{25}$ (CH₃OH)	CAS
A	$C_{42}H_{73}N_9O_4$	768.09	powder	+ 8.9°	147664-18-4
F	$C_{37}H_{64}N_6O_2$	624.94	gum	+ 19.4°	188112-82-5
H	$C_{35}H_{57}N_6O_2^+$	609.87	oil (formate)	+ 33.7° (formate)	188112-85-8

Lit.: Eur. J. Org. Chem. **1999**, 1173–1183 (synthesis) ▪ J. Org. Chem. **60**, 1182 (1995); **62**, 1814 (1997) (isolation); **64**, 1512 (1999) (B. A–G) ▪ Tetrahedron **54**, 9481 (1998) (synthesis) ▪ Tetrahedron Lett. **37**, 6977 (1996); **39**, 5697 (1998) (synthesis).

Batzellines.

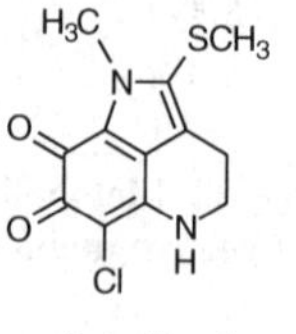

Batzelline A
N-Demethyl B. A = B. B
De(methylthio) B. A = B. C

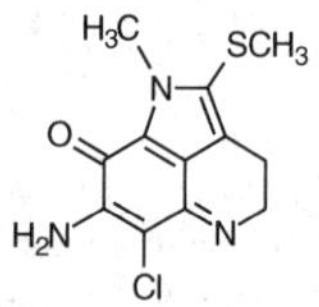

Isobatzelline A

The bioactive pyrrolo[4,3,2-*de*]quinoline alkaloids B. are isolated from the Caribbean sponge *Batzella* sp., living at a depth of about 120 m. *B. A* [1]: $C_{12}H_{11}ClN_2O_2S$, M_R 282.74, black prisms, mp. 205 °C; *isobatzelline A* [2]: $C_{12}H_{12}ClN_3OS$, M_R 281.77, brown solid, active against P388 leukemia cells and *Candida albicans*. They may be biosynthetic precursors of the *discorhabdins.

Lit.: [1] Tetrahedron Lett. **30**, 2517 (1989). [2] J. Org. Chem. **55**, 4964 (1990).

gen.: Chem. Lett. **1991**, 1785 (synthesis) ▪ Eur. J. Org. Chem. **1999**, 1173 (synthesis B. A, B) ▪ Heterocycles **41**, 1905 (1995) ▪ J. Org. Chem. **62**, 568 (1997) (synthesis B. C) ▪ Tetrahedron **50**, 2017 (1994). – *[CAS 123064-89-1 (B. A); 133401-01-1 (iso-B. A); 123064-90-4 (B. B); 123064-91-5 (B. C)]*

Beer flavor. Primary odor and taste substances from malt and hops determine the type of beer: bitter substances (see humulone) and aroma substances of hops characterize Pilsener beer, a relatively high content of *Furaneol® provides the caramel note of dark beer. The most important aroma substances in light full beer are 3-methylbutanol, *2-phenylethanol, *4-vinylguaiacol, 2-methylbutanoic acid, isovaleric acid, Furaneol®, and butanoic acid. In white (wheat) beer the content of 4-vinylguaiacol is significantly higher. Many aroma substances are present in concentrations in the range of the perception threshold or lower but can contribute additively to B. f., e. g. the *fruit esters (ethyl butanoate, ethyl hexanoate, etc.) or *acetoin and *2,3-butanedione. In higher concentrations the latter gives an off flavor (bad smell) as do *phenylacetaldehyde, (*E*)-2-nonenal (see alkenals), and 3-mercapto-3-methylbutyl formate (see coffee flavor) in old beer.

Lit.: Maarse, p. 581–616, 716–719.

Bees (Apidae). Insects belonging to the Hymenoptera that feed mainly on pollen and nectar of flowering plants. In addition to the bumblebee, the honeybee (*Apis mellifica*) deserves special attention. The *beeswax used to build the honeycomb is secreted from wax glands on the underside of the abdomen of 10–20-day-old female workers. A two-piece sting equipped with a barb at the end of the abdomen injects the *bee venom produced in two venom glands and stored in a venom bladder into attacking enemies. On being stung, the enemy is marked with an alarm *pheromone which prompts members of the same species to attack; the main component of this signal is *3-methylbutyl acetate* ($C_7H_{14}O_2$, M_R 130.19). Honeybees form colonies lasting several years, having a strict social order. Female workers hatch from the germinated eggs or – on special nourishment of the young female larvae with so-called *royal jelly – queen bees. These secrete a pheromone from their mandible glands, the main component of which is *trans-9-oxo-2-decenoic acid* (see queen substance).

Lit.: Hermann (ed.), Social Insects, Vol. III, p. 361–423 (stingless bees), New York: Academic Press 1982 ▪ Jacobs & Renner, Biologie u. Ökologie der Insekten (3.), Heidelberg: Spektrum 1998. – *[CAS 123-92-2 (3-methylbutyl acetate).]*

Beeswax. Kneadable excretion product (*waxes) from glands of the honey-*bee from which honeycomb is built. Honeycomb, freed from honey by centrifugation is melted, the melt (mp. 61–68 °C) separated from solid impurities, and the yellow (cera flava), brown, or red crude wax allowed to solidify. Treatment of the crude wax with oxidation agents result in complete bleaching (cera alba). B. consists of a mixture of complex wax esters (ca. 70%), normal *fatty acids, and *hydroxyfatty acids (13–14%) as well as hydrocarbons (10–14%). The wax esters contain mainly *1-triacontanol as alcohol component which is esterified especially with palmitic acid and *cerotic acid. One of the main hydroxyfatty acid wax esters contained in B. is ceryl hydroxypalmitate (8–9%). 14-Hydroxypalmitic as well as 16-hydroxy- and 17-hydroxystearic acids are also constituents of B.; free wax acids present include lignoceric, cerotic, and *melissic acid. For compositions and purity control, see *Lit.*[1], for the typical odor of B., see *Lit.*[2].

Lit.: [1]Hager (4.) **7b**, 499–504. [2]J. Sci. Food Agric. **28**, 511–518 (1977).
gen.: Hamilton (ed.), Waxes: Chemistry, Molecular Biology and Functions, p. 1–90, Dundee: The Oily Press 1995 ▪ Kolattukudy (ed.), Chemistry and Biochemistry of Natural Waxes, Amsterdam: Elsevier 1976 ▪ Naturwissenschaften **77**, 34 f. (1990) ▪ Ullmann (5.) **A 24**, 222. – *[HS 1521 90; CAS 8006-40-4 (white); 8012-89-3 (yellow)]*

Beet root flavor see vegetable flavors.

Beet roots, red beets. The cultured form of *Beta vulgaris* (Chenopodiaceae) has been used in Europe as a biennial vegetable plant since the 13th century. They are rich in oxalic acid (>300 mg/kg, in leaves >900 mg/kg), nitrate (200 mg/kg), and sodium (>80 mg/kg). The dark red color is due to the content of betanin (see betanidin) and other *betalains and the typical odor to *geosmin. The pressed juice is permitted for use as a food colorant.

Lit.: Franke, Nutzpflanzenkunde, Stuttgart: Thieme 1989. – *[HS 0706 90]*

Bee venom. Secretion from the venom bladder of the honeybee (*Apis mellifica*). Yellowish, amorphous powder, easily soluble in water, typical composition: 55% *melittin, 13% phospholipase A, 3% hyaluronidase, and 2% *apamin. LD_{50} (mouse i. v.) 6 mg/kg. Further components are *histamine, the so-called *mast cell degranulating peptide (MCD peptide) and other low molecular substances. Fatal accidents after bee stings are not unusual and are mostly due to the anaphylactic action of B. (allergic shock). B. preparations are used in the treatment of rheumatic diseases, multiple sclerosis (MS), muscle and joint pain, neuralgias, and muscle sprains.

Lit.: Endeavour **12**, 60 (1988). – *[HS 0510 00]*

Behenic acid (docosanoic acid). $H_3C-(CH_2)_{20}-COOH$, $C_{22}H_{44}O_2$, M_R 340.59, mp. 80–81.5 °C. Saturated *fatty acid with 22 C atoms; name derived from the behen (or ben) nut (= seed of *Moringa oleifera*, Moringaceae) which contains about 6% B. as glycerol ester. Small amounts of B. also occur as esters in peanut (3%) and rapeseed oils (0.5%) as well as in *carnauba wax and other waxes. Seed fat of *Lophira alata* (Ochnaceae) contains B. in larger amounts (15–30%). B. is easily produced by hydrogenation of *erucic acid and other unsaturated C_{22} fatty acids.

Lit.: Beilstein E IV **2**, 1290 ▪ Hilditch & Williams, The Chemical Constitution of Natural Fats, p. 304 ff., 583 ff., London: Chapman & Hall 1964 ▪ Ullmann (5.) **A 10**, 247. – *[HS 2915 90; CAS 112-85-6]*

Benanomicins see pradimicins.

Bengamides.

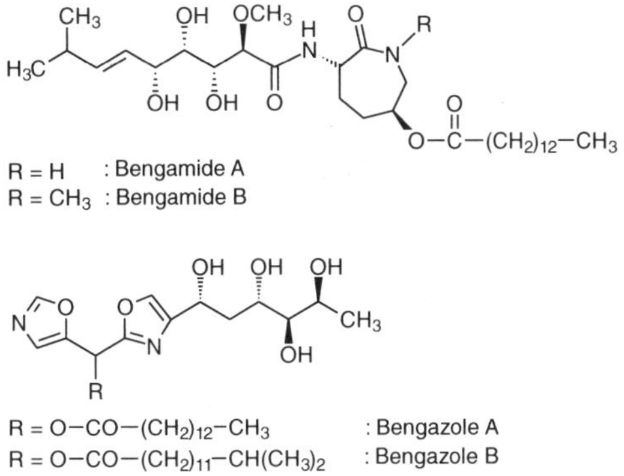

R = H : Bengamide A
R = CH_3 : Bengamide B

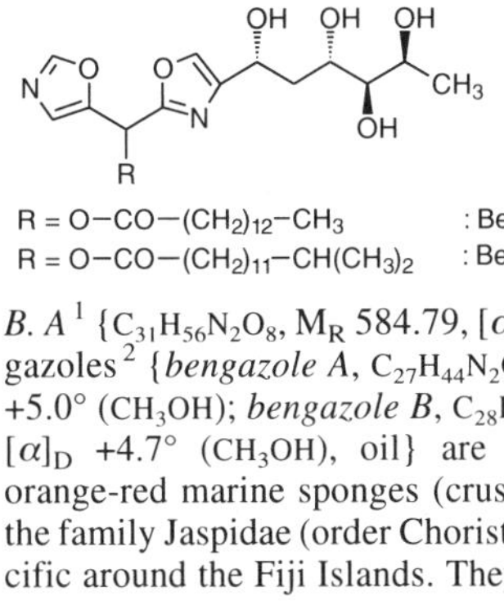

R = O–CO–$(CH_2)_{12}$–CH_3 : Bengazole A
R = O–CO–$(CH_2)_{11}$–$CH(CH_3)_2$: Bengazole B

B. A[1] {$C_{31}H_{56}N_2O_8$, M_R 584.79, $[\alpha]_D$ +30.3°} and bengazoles[2] {*bengazole A*, $C_{27}H_{44}N_2O_8$, M_R 524.65, $[\alpha]_D$ +5.0° (CH_3OH); *bengazole B*, $C_{28}H_{46}N_2O_8$, M_R 538.68, $[\alpha]_D$ +4.7° (CH_3OH), oil} are components of the orange-red marine sponges (crustaceous sponges) of the family Jaspidae (order Choristida) found in the Pacific around the Fiji Islands. They exhibit anthelmintic activity against the nematode *Nippostrongylus braziliensis*. B. A and *B. B* {$C_{32}H_{58}N_2O_8$, M_R 598.82, $[\alpha]_D$ +34.6° (CH_3OH)} have antimicrobial activity against *Streptococcus pyogenes* (MIC 3.9 and 1.9 µg/mL).

Lit.: [1]J. Org. Chem. **51**, 4494 (1986). [2]J. Am. Chem. Soc. **110**, 1598 (1988); **111**, 647 (1989); Chem. Rev. **93**, 1718 (1993); J. Org. Chem **61**, 4073 (1996) (bengazoles C-G); **64**, 4995 (1999) (synthesis).
gen.: J. Nat. Prod. **60**, 814 (1997) ▪ see discodermin A, jaspamide. – *[CAS 104947-68-4 (B. A); 104947-69-5 (B. B); 112549-08-3 (bengazole A); 112549-09-4 (bengazole B)]*

Bengazoles see bengamides.

Benzaldehyde. H_5C_6–CHO, C_7H_6O, M_R 106.12. Oily liquid with a bitter almond-like smell and strong refractive power, bp. 179 °C, flash point 64 °C, poorly soluble in water, fully miscible with ethanol and ether, LD_{50} (rat p. o.) 1300 mg/kg.
Occurrence: In small amounts in many *flavor compounds, *essential oils, and flower aromas. The main ingredient of bitter almond oil B. is formed by hydrolysis of the glycoside amygdalin. For analytical distinction from synthetic B., see *Lit.*[1].
Use: As flavor and aroma substance, as intermediate for *aroma chemicals, e. g., *cinnamaldehyde, dyes, and pharmaceuticals.
Lit.: [1]Tetrahedron Lett. **22**, 3525 (1981); Parfum. Cosmet. Arom. **94**, 95 (1990).
gen.: Bauer et al. (2.), p. 78 ▪ Beilstein E IV **7**, 5057 ▪ Merck-Index (12.), No. 1085. – *[HS 2912 21; CAS 100-52-7]*

Benzene-1,2,4-triol (1,2,4-trihydroxybenzene).

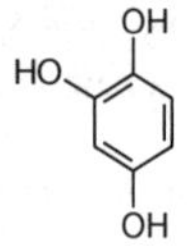

Benzene-1,2,4-triol

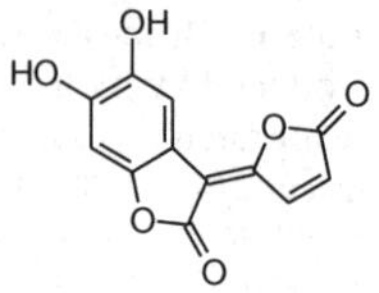

Gomphilactone

$C_6H_6O_3$, M_R 126.11, cryst., mp. 140.5 °C (subl.). Typical metabolite of mushrooms of the genus *Gomphidius* (Basidiomycetes) which causes the characteristic reddening and subsequent blackening of bruises in *G. glutinosus* and *G. maculatus*[1]. The easily oxidizable, antibacterially active B. also occurs in the marine sponge *Axinella polycapella*[2]. In *Gomphidius* species, oxidative dimerization of B. gives the red *gomphilactone* ($C_{12}H_6O_6$, M_R 246.18, dark-red rods, mp. 192–195 °C) which is structurally similar to bovilactone-4.4[1].
Lit.: [1]Z. Naturforsch. C **29**, 446 (1974); **36**, 488 (1981). [2]Experientia **37**, 13 (1981).
gen.: Beilstein E IV **6**, 7338 ▪ Tetrahedron **50**, 6377 (1994) (synthesis). – *[CAS 5333-73-3 (B.); 78570-66-8 (gomphilactone)]*

Benzoic acid. H_5C_6–COOH, $C_7H_6O_2$, M_R 122.12. Colorless, shining plates, mp. 122 °C, bp. 249 °C, sublimes at 100 °C with formation of vapors that strongly irritate mucous membranes, can be steam distilled; poorly soluble in cold water. B. occurs in free form in benzoin gum and in *defensive secretions of water beetles (*Dytiscus* sp.); in *Malus pumila* (common apple) B. acts as a *phytoalexin. B. sublimed from resin was observed in 1556 by Nostradamus, its structure was determined in 1832 by Liebig and Wöhler. Esters of B. (benzoates) are widely distributed in essential oils. *Benzyl benzoate*, $C_{14}H_{12}O_2$, M_R 212.25, plates, mp. 21 °C, bp. 316–317 °C, is contained in *peru balsam, in *cinnamon leaf (bark) oil, and in *ylang-ylang oil.
Toxicology: LD_{50} (rat p. o.) 2530 mg/kg, LDL_0 (human p. o.) 500 mg/kg, TDL_0 (human skin, topical) 6 mg/kg.
Activity: B. is a mild irritant of the skin, mucous membranes, and eyes and acts as an expectorant. Benzyl benzoate is an insect repellent and has weak action against mites (acaricide) and against mycosis pedis (athlete's foot). Its biosynthesis involves oxidative degradation of *cinnamic acid.
Use: B. is registered as a conservation agent for acidic foods[1]. On account of its antiseptic properties B. is also used as an additive in certain pharmaceutical preparations. The benzoates of lower alcohols are used in perfumery and as aroma substances.
Lit.: [1]Food Sci. **54**, 650 (1989).
gen.: Beilstein E IV **9**, 273–276 ▪ Hager (5.) **7**, 47, 429 ff., 439 ff. ▪ Luckner (3.), p. 397 f. ▪ Merck-Index (12.), No. 1122 ▪ Ullmann (5.) **A 3**, 555–569. – *[HS 2916 31; CAS 65-85-0 (B.); 120-51-4 (benzyl benzoate)]*

Benzoisochromanequinones.

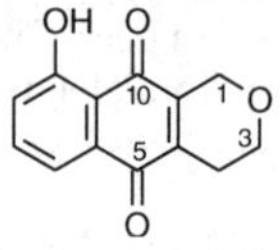

Name of colored hydroxynaphthoquinone derivatives with a tricyclic *polyketide skeleton. For substitution on the dihydropyran ring there are two stereochemical series (A=*actinorhodins, G=*granaticins). B. are typical metabolites of *streptomycetes, over 50 representatives are known which exhibit differing biological activities, including antibacterial, antifungal, antineoplastic, antiviral (also against HIV) actions.
Lit.: Angew. Chem. Int. Ed. Engl. **28**, 146–178 (1989) (biosynthesis) ▪ Justus Liebigs Ann. Chem. **1974**, 1100–1125; **1987**, 751–758 ▪ Thomson **3**, 270–321 (review).

Benzomalvins. Benzodiazepinone alkaloids from *Penicillium* spp.. They inhibit *substance P, e.g., *B. B*: $C_{24}H_{17}N_3O_2$, M_R 379.41, mp. 260 °C, $[\alpha]_D$ +158° (CH_3OH). B. are structurally related to *asperlicin.

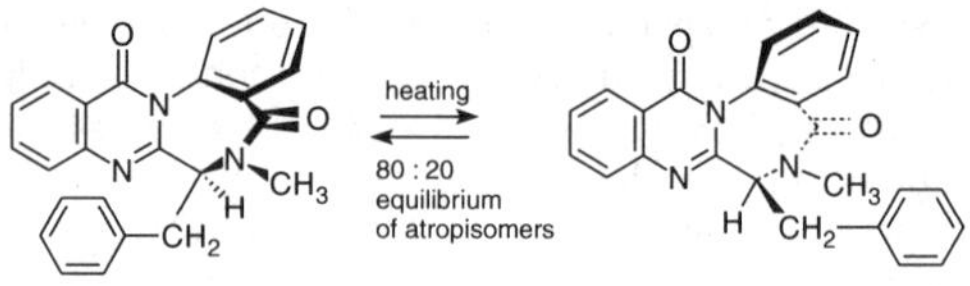

B. A B. D

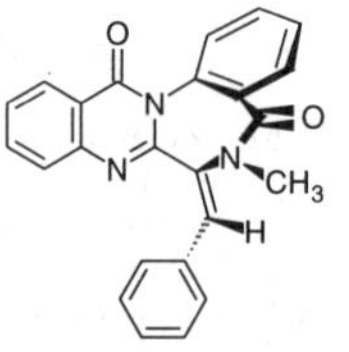

B. B
(dehydrobenzomalvin A)

Lit.: Chem. Lett. **1997**, 869 (synthesis) ▪ J. Antibiot. **47**, 515 (1994) ▪ J. Nat. Prod. **58**, 1575 (1995) (isolation). – *[CAS 157047-97-7 (B. B); 157-047-96-6 (B. A)]*

Benzo[c]phenanthridine alkaloids. Group of alkaloids with the benzophenanthridine skeleton. They occur in various genera of the Papaveraceae (including *Argemone*, *Bocconia*, *Chelidonium*, *Eschscholtzia*, *Glaucium*, *Macleaya*, *Meconopsis*, *Papaver*, *Sanguinaria*) and Fumariaceae (including *Corydalis*, *Dicentra*, *Fumaria*) and Rutaceae (*Fagara*, *Zanthoxylum*). The B. a. are insoluble in water, soluble in alcohol, ether, and chloroform.
Use: *Chelerythrine* (**1**) is the most active alkaloid from *Chelidonium majus* (swallowwort), topically it has an irritating effect, taken orally, it causes vomiting, gas-

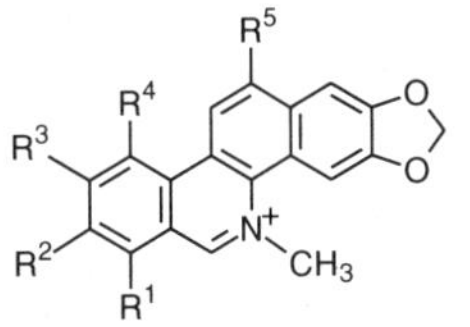

	R^1	R^2	R^3	R^4	R^5
Chelerythrine (**1**)	OCH_3	OCH_3	H	H	H
Macarpine (**2**)	O–CH_2–O		H	OCH_3	OCH_3
Sanguinarine (**3**)	O–CH_2–O		H	H	H
Nitidine (**4**)	H	OCH_3	OCH_3	H	H

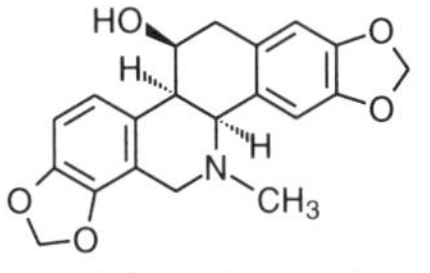

(+)Chelidonine (**5**)

Table: Data of Benzophenanthridine alkaloids.

name	molecular formula	M_R	mp. [°C]	$[\alpha]_D$	CAS
1	$C_{21}H_{18}NO_4^+$	348.38	213 (chloride)		34316-15-9
2	$C_{22}H_{18}NO_6^+$	392.39			23594-80-1
3	$C_{20}H_{14}NO_4^+$	332.34			2447-54-3
4	$C_{21}H_{18}NO_4^+$	348.38			6872-57-7
5	$C_{20}H_{19}NO_5$	353.37	135–136	+115° (C_2H_5OH)	476-32-4

troenteritis, and massive diarrhea, in larger doses it has a central paralyzing effect and causes death through respiratory paralysis. It is recommended as an antimicrobial and anti-inflammatory agent against infections of the oral cavity. (**1**) is an inhibitor of rat liver aminotransferase, a potent cytotoxic agent and decoupler of oxidative phosphorylation. *Chelidonine* (**5**) has a relaxing effect on smooth musculature and is recommended for the alliviation of gastrointestinal pain and asthmatic spasms. It is a weak mitotic poison which is the reason for the wart-removing property of swallowwort. *Nitidine* (**4**) is a highly toxic antitumor agent and an inhibitor of reverse transcriptase in oncongenic viruses. *Sanguinarine* (**3**) has antimicrobial and anti-inflammatory activity. It is a potent cytotoxic agent and inhibitor of alanine aminotransferase. It causes a temporary change in intraoccular pressure and may be implicated in the development of glaucoma; LD_{50} (rat i.p.) 18 mg/kg.
The biosynthesis of *macarpine* (**2**) proceeds in 12 enzymatic steps from (*S*)-reticuline. The key step is the oxidative transformation of protopine via 6-hydroxyprotopine to dihydrosanguinarine[1,2].

Lit.: [1] Pure Appl. Chem. **66**, 2023–2028 (1994). [2] Zenk in: Goldwin et al. (eds.), Organic Reactivity Physical and Biological Aspects, Publ. 148, p. 89–109, Cambridge: Royal Society of Chemistry 1995.
gen.: Adv. Heterocycl. Chem. **67**, 345–389 (1997) (synthesis review) ▪ Beilstein EIV **27**, 6874 ff. (chelidonine, sanguinarine) ▪ Chem. Lett. **1986**, 739 ▪ Chem. Pharm. Bull. **44**, 1634 (1996) (synthesis chelerythrine) ▪ Hager (5.) **2**, 519; **3**, 264, 1054 ff.; **4**, 835–848 ▪ Heterocycles **50**, 627 (1999) (review) ▪ Ind. J. Chem. Sect. B **36**, 679 (1997) ▪ J. Chem. Soc., Perkin Trans. 1 **1996**, 1647 (synthesis) ▪ Kirk-Othmer (4.) **1**, 1053–1066 ▪ Manske **25**, 178–188; **26**, 185–240; **49**, 273 ▪ Phytochemistry **29**, 1113–1122 (1990); **31**, 2713–2717 (1992) ▪ Sax (8.), CCS 650, CDL 000 ▪ Shamma, p. 317–343 ▪ Shamma-Moniot, p. 271–292 ▪ Tetrahedron **54**, 9875–9894 (1998) (synthesis) ▪ Ullmann (5.) **A 1**, 380. – *[HS 2939 90]*

Benzoquinones. Derivatives of *p*- or *o*-benzoquinones belonging to the group of quinone pigments that are widely distributed in nature. *p*-B. itself was recognized in 1838 by Woskresensky (student of Liebig) as the oxidation product of *quinic acid and named quinoyl; the name quinone was introduced by Berzelius. The chromophore of B. consists of two carbonyl groups which are in conjugation with two double bonds.

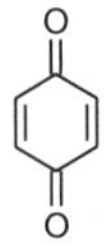

p-Benzoquinone

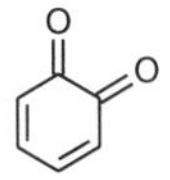

o-Benzoquinone

p-B. and B. derivatives are constituents of animal venoms (*bombardier beetle, *defensive secretions, e. g., *ethyl-1,4-benzoquinone, *methyl-1,4-benzoquinone). More than 100 different B. have been isolated from higher plants (e. g., *primin, embelin, rapanone, *perezone) and from lower plants, higher fungi and their culture filtrates (e. g., *helicobasidin, *pleurotin, *boviquinones, *spinulosin). The *plastoquinones occur ubiquitously in small amounts in all green plants and participate in photosynthesis and electron transfer within the respiratory chain.
Biosyntheses of the B. are numerous and at least 6 different routes leading to B. are known[1]. Thus, 2-methoxybenzoquinone and the glucoside of benzohydroquinone *arbutin are formed by β-oxidation of cinammic acid or its derivatives[2]. On the other hand the B. nucleus of the plastoquinones arises from *homogentisic acid while the lipophilic side chain is built up from mevalonic acid[3]. Structurally very similar B. can be formed from completely different precursors (acetate or *p*-hydroxybenzoate) in the same fungal organism[3].

Lit.: [1] Czygan (2.), p. 356. [2] Z. Pflanzenphysiol. **59**, 439 (1968). [3] Luckner (3.), p. 415.
gen.: Hager (5.) **3**, 163 ff. ▪ Thomson, Naturally Occurring Quinones, p. 1 ff., London: Chapman and Hall 1986.

Benzotetronic acid see 4-hydroxycoumarin.

Benzyl acetate (acetic acid benzyl ester). H_5C_6–CH_2–O–CO–CH_3, $C_9H_{10}O_2$, M_R 150.18. Liquid with fruity, jasmine-like odor, D. 1.055, bp. 215 °C, flash point 100 °C; LD_{50} (rat p. o.) 2.49 g/kg. B. occurs, e. g., in jasmine oil (up to 65%), *Gardenia* oil, and *ylang-ylang oil.
Use: Mainly as fragrance substance for jasmine-like odors, in small amounts as solvent for printing inks. One of the most widely used flavor substances (ca. 4000 t/year world consumption).

Lit.: Bauer et al. (2.), p. 88 ▪ Bedoukian (3.), p. 65 ff. ▪ Food Cosmet. Toxicol. **11**, 875 (1973) ▪ Merck-Index (12.), No. 1158. – *[HS 2915 39; CAS 140-11-4]*

Benzyl alcohol (phenylmethanol). H_5C_6–CH_2–OH, C_7H_8O, M_R 108.14. Liquid with a weak aromatic odor, mp. −15 °C, bp. 205 °C, flash point 94 °C. LD_{50} (rat

p. o.) 1230 mg/kg, (mouse i. p.) 650 mg/kg. B. constitutes about 6% of jasmine flower oil and also occurs in both the free form and as esters in many other essential oils, e. g., wallflower oil, clove oil, tuberose oil, peru balsam (Peruvian balsam), balsam of tolu, and *storax. Some plants contain glycosides of B. B. has also been isolated from the cuticle of insects.

Biosynthesis: β-Hydroxydihydrocinammic acid → benzaldehyde (retroaldol cleavage) → benzyl alcohol.

Lit.: Beilstein EIV **6**, 2222 ▪ Hager (5.) **1**, 148 ff. (use in dermatitics); **6**, 1144 f.; **7**, 438 ff. ▪ Karrer, No. 248 ▪ Luckner (3.), p. 400 ▪ Merck-Index (12.), No. 1159 ▪ Ullmann (5.) **A 4**, 1–8. – *[HS 2906 21; CAS 100-51-6]*

Benzyl benzoate see benzoic acid.

Benzyl cinnamate see peru balsam.

Benzyl(tetrahydro)isoquinoline alkaloids. B. a. occur especially in various Annonaceae, Lauraceae, Rhamnaceae, Ranunculaceae, Papaveraceae, and Fabaceae. They are soluble in ethanol, chloroform, ether and poorly soluble in water. *Laudanidine* is toxic, it acts as a convulsive and respiratory stimulant, *laudanosine* is a tetanic toxin, *orientaline* is the biosynthetic precursor of isothebaine. *Protosinomenine* occurs especially in *Erythrina lithosperma* (Fabaceae). *Papaverine* [1-(3,4-dimethoxybenzyl)-6,7-dimethoxyisoquinoline] acts as a parasympathicolyticum relaxing smooth musculature, it promotes cerebral perfusion (vasodilatation)[1]. It is an antiasthmatic and oral antispasmodic agent used in the treatment of gastrointestinal spasms; LD_{50} (mouse i. v.) 25 mg/kg. *Reticuline* is an important precursor of protoberberines, morphinans, aporphines, spirobenzylisoquinolines, and rhoeadine alkaloids.

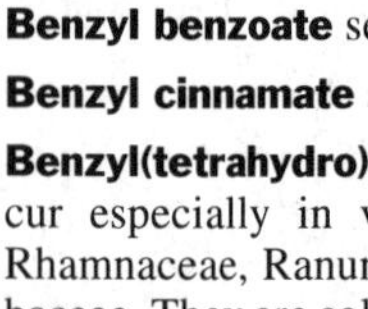

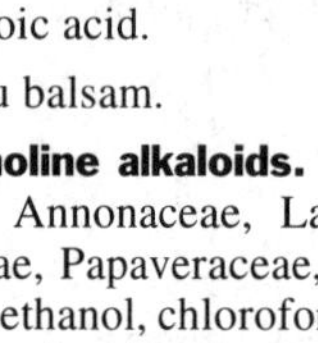

Papaverine (7)

	R¹	R²	R³	R⁴	R⁵
(*S*) Coclaurine (1)	H	CH_3	H	H	OH
(*R*) Laudanidine (2)	CH_3	CH_3	CH_3	OH	OCH_3
(*S*) Laudanosine (3)	CH_3	CH_3	CH_3	OCH_3	OCH_3
(*R*) Orientaline (4)	CH_3	CH_3	H	OCH_3	OH
(*S*) Protosinomenine (5)	CH_3	H	CH_3	OH	OCH_3
(*S*) Reticuline (6)	CH_3	CH_3	H	OH	OCH_3

Table: Data of Benzyl(tetrahydro)isoquinoline alkaloids.

no.	molecular formula	M_R	mp. [°C]	$[\alpha]_D$	CAS
1	$C_{17}H_{19}NO_3$	285.34	217–218	+47° (C_2H_5OH)	486-39-5
2	$C_{20}H_{25}NO_4$	343.42	181–182	+134° (CH_3OH)	3122-95-0
3	$C_{21}H_{27}NO_4$	357.45	89	+103° (C_2H_5OH)	2688-77-9
4	$C_{19}H_{23}NO_4$	329.40	128–130 ($HClO_4$)		27003-74-3
5	$C_{19}H_{23}NO_4$	329.40			30883-59-1
6	$C_{19}H_{23}NO_4$	329.40		+132° (CH_3OH)	485-19-8
7	$C_{20}H_{21}NO_4$	339.39	147–148		58-74-2

Biosynthesis: Starting from dopa via dopamine and reaction with 4-hydroxyphenylacetaldehyde[2]. Detection by thin layer chromatography.

Lit.: [1] Eur. J. Pharmacol. **196**, 183–187 (1991). [2] Justus Liebigs Ann. Chem. **1990**, 550.

gen.: Beilstein E V **21/6**, 46 ff. (laudanosine, reticuline), 182 f. (papaverine) ▪ Florey **17**, 367–447 ▪ Hager (5.) **2**, 256–301, 519, 912 ▪ Manske **17**, 385–544; **31**, 1–28 ▪ Phillipson et al., p. 126–141 ▪ Sax (8.), PAH 000, PAH 250, SLX 500 ▪ Shamma, p. 45–89 ▪ Shamma-Moniot, p. 27–55 ▪ Ullmann (5.) **A 1**, 369 – *Synthesis:* Kleemann-Engel, p. 1443 f. (papaverine) ▪ Pharmazie **43**, 313 f. (1988) ▪ Tetrahedron **53**, 16 327 (1997) (laudanosine) ▪ Tetrahedron: Asymmetry **7**, 2711 (1996) ▪ Tetrahedron Lett. **32**, 2995 f. (1991). – *[HS 2939 10, 2939 90]*

Berbamine see bisbenzylisoquinoline alkaloids.

Berberine see protoberberine alkaloids.

Berberine hydroxide see alkaloidal dyes.

β-Bergamotene.

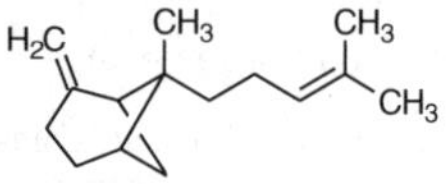

$C_{15}H_{24}$, M_R 204.36, oil, bp. 120–130 °C (133 Pa), $[\alpha]_D$ +35.8° ($CHCl_3$). *Sesquiterpene from *Aspergillus fumigatus*[1] and *Valeriana wallichii*, occurs in nature only in the *trans* form. Biosynthesis from farnesyl pyrophosphate[2]; β-B. is the biosynthetic precursor of the antibiotics *ovalicin and *fumagillin. For synthesis, see *Lit.*[3]; absolute configuration, see *Lit.*[4].

Lit.: [1] Tetrahedron Lett. **1976**, 4625. [2] J. Am. Chem. Soc. **111**, 1152 (1989). [3] Tetrahedron Lett. **26**, 3535 (1985); J. Org. Chem. **52**, 4508 (1988). [4] J. Am. Chem. Soc. **112**, 1285 (1990). – *[CAS 6895-56-3]*

Bergamot oil. Greenish-yellow to green oil; its odor is fresh, sparkling, sweet-fruity, its taste sweet-fruity, refreshing.

Production: By mechanical compression of the peel (*citrus oils) of unripe fruits of bergamot, *Citrus aurantium* subsp. *bergamia*. *Origin:* southern Italy (Calabria) and Ivory Coast.

Composition[1]***:*** Main constituents are *limonene* (see *p*-menthadienes) (30–45%), *linalyl acetate* (see linalool) (20–40%), and **linalool* (5–20%).

Use: Mainly for the production of perfumes, especially Eau de Colognes and fresh toilet waters. Pressed B. contains up to ca. 1% of the *furocoumarin *bergapten which causes phototoxic reactions (so-called "berlock" or "berloque" (perfume) dermatitis). Thus mild rectification of B. is recommended before its use in perfume oils, since bergapten remains in the distillation residue. B. is also used in small amounts to aromatize confectionery and bakery products, often in combination with other citrus oils; well known is the aromatization of tea ("Earl Grey" tea).

Lit.: [1] Perfum. Flavor. **12** (2), 68 (1987); **13** (2), 67 (1988); **16** (5), 75 (1991); **19** (6), 29, 57 (1994); Flavour Fragr. J. **10**, 33 (1995).

gen.: Arctander, p. 91 ▪ Bauer et al. (2.), p. 146 ▪ ISO 3520 (1980). – *Toxicology:* Food Cosmet. Toxicol. **11**, 1031, 1035 (1973). – *[HS 3301 11; CAS 8007-75-8; 68648-33-9 (furocoumarin-free)]*

Bergamottin see bergapten.

Bergapten (4-methoxy-7*H*-furo[3,2-*g*][1]benzopyran-7-one).

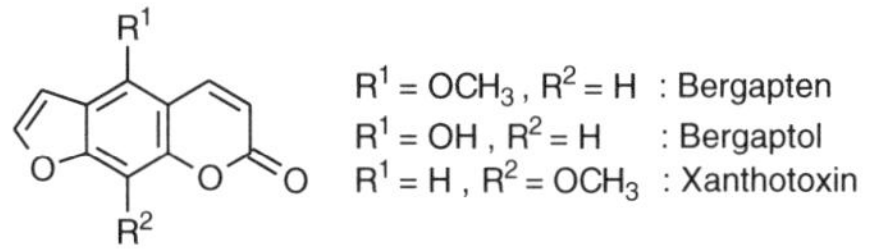

$C_{12}H_8O_4$, M_R 216.19, needles, mp. 188 °C (subl.). B. is found in numerous plants, first (1891) described in *bergamot oil. B. occurs especially in the Rutaceae and Apiaceae. B. is the methyl ether of *Bergaptol* ($C_{11}H_6O_4$, M_R 202.17, needles, mp. 277–278 °C), also a component of bergamot oil; the geranyl ether is known as *Bergamottin*. *Xanthotoxin* (meloxin, 9-methoxy-7*H*-furo[3,2-*g*][1]benzopyran-7-one), $C_{12}H_8O_4$, M_R 216.19, needles or prisms, mp. 148 °C, bitter taste with a tingling aftertaste, occurs in the African tree *Zanthoxylum senegalense (Fagara xanthoxyloides)*, in bishop's weed *Ammi majus*, goutweed, and other Apiaceae, in parsnip roots it acts as a *phytoalexin, it is poisonous for amphibians and reptiles.
Activity: The former practice in dermatology of using B. and xanthotoxin in browning suntan creams is contraindicated by their phototoxicity (see furocoumarins). *Cytochrome P-450 is deactivated by reactive xanthotoxin metabolites which covalently bind to the apoprotein [1]; Xanthotoxin inhibits the degradation of *caffein(e) in blood [2]. Recently, B. has been introduced into the therapy of severe forms of psoriasis [3].
Toxicology: B. is assumed to be carcinogenic. Xanthotoxin is cytotoxic for cultured human lymphocytes at a concentration of 100 µM. The biosynthesis proceeds from *umbelliferone and mevalonic acid → *marmesin → bergapten.

Lit.: [1] Biochem. Pharmacol. **38**, 1647–1655 (1989); J. Pharmacol. Exp. Ther. **254**, 720–731 (1990). [2] Clin. Pharmacol. Ther. **42**, 621–626 (1987). [3] Drug News **1996** (17.6.), 7.
gen.: Beilstein EV **19/6**, 4 f. ▪ Hager (5.) **5**, 431–440; 436 f.; 664–670; **6**, 506–521 ▪ J. Heterocycl. Chem. **17**, 985 ff. (1980) (synthesis) ▪ Karrer, No. 1369 ▪ Luckner (3.) p. 388–391, 471 ▪ Nat. Prod. Rep. **12**, 477–505 (1995) ▪ Phytochemistry **12**, 1657–1667 (1973) (biosynthesis) ▪ Synlett **1995**, 573 (synthesis). – *Pharmacology:* Br. J. Dermatol. **112**, 469–473 (1985). – *Toxicology:* Sax (8.), MFN275, XDJ000. – *[HS 2932 29; CAS 484-20-8 (B.); 486-60-2 (bergaptol); 7380-40-7 (bergamottin); 298-81-7 (xanthotoxin)]*

Berkheyaradulene see isocomene.

Bestatin {annastatin, ubenimex, *N*-[(2*S*,3*R*)-3-amino-2-hydroxy-4-phenylbutyryl]-L-leucine}.

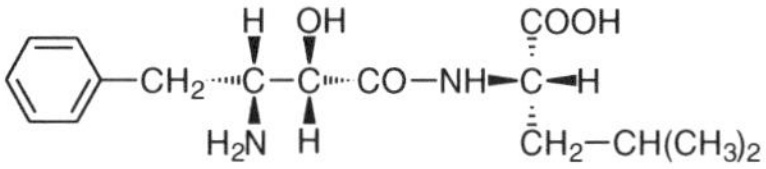

$C_{16}H_{24}N_2O_4$, M_R 308.38, needles, mp. 233–236 °C, $[\alpha]_D^{20}$ –15.5° (1 M HCl). Peptide antibiotic from *Streptomyces olivoreticuli* with antitumor activity, it is a specific inhibitor of aminopeptidase B, a leucine aminopeptidase. B. activates macrophages and T-lymphocytes and *in vitro* increases not only blastogenesis of lymphocytes but also the manifestation of delayed hypersensitivity. It is used clinically in cancer therapy.

Lit.: Annu. Rev. Microbiol. **36**, 75 (1982) ▪ J. Org. Chem. **54**, 4235 (1989); **55**, 2232 (1990) ▪ Merck-Index (12.), No. 9973 ▪ Tetrahedron Lett. **33**, 6803 (1992) ▪ Ullmann (5.) **A 2**, 531. – *[CAS 58970-76-6]*

Betacyans see betalains.

Betalains. Group of water-soluble, nitrogen-containing flower and fruit pigments which occur in nine families of the plant order Caryophyllales and in some higher fungi, e. g., in the cap skin of the fly agaric (*Amanita muscaria*). The occurrence of B. in members of the Caryophyllales is of high chemotaxomic significance. B. are iminium derivatives of *betalamic acid. The biosynthesis proceeds from tyrosine through dihydroxyphenylalanine (DOPA) which, on the one hand cyclizes to cyclo-DOPA and, on the other hand, recyclizes to betalamic acid after extradiol cleavage (dioxygenase activity). To the B. belong the red-violet *betacyans* (the glycosides and acylglucosides of *betanidin; e. g., betanin) and the yellow *betaxanthins* (condensation products from betalamic acid and proteinogenic and non-proteinogenic amino acids or biogenic amines; e. g., *indicaxanthin). The betacyans celosianin, *gomphrenin, and lampranthin as well as the betaxanthins *miraxanthin, muscaaurin, *portulacaxanthin, and *vulgaxanthin are widely distributed in the Caryophyllales.

Lit.: Bot. Mag. **101**, 175 (1988) ▪ Endeavour **10**, 31 (1986) ▪ Manske **39**, 1–62 ▪ Methods Plant Biochem. **8**, 421–450 (1993).

Betalamic acid {(*S*)-1,2,3,4-tetrahydro-4-[(*E*)-oxoethylidene]-2,6-pyridinedicarboxylic acid}.

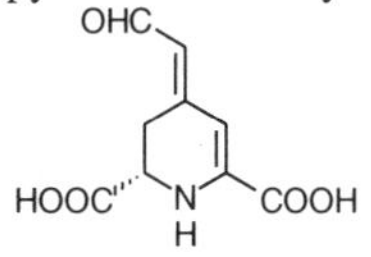

$C_9H_9NO_5$, M_R 211.17, uv_{max} 428 nm. B. is the central building block of all *betalains, but also occurs in the free form, especially in members of the Aizoaceae (*Mesembryanthemum*). B. can be obtained from betaxanthins by base-catalyzed hydrolysis.

Lit.: Helv. Chim. Acta **67**, 1547 (1984) ▪ Manske **39**, 1–62. – *[CAS 18766-66-0]*

Betanidin.

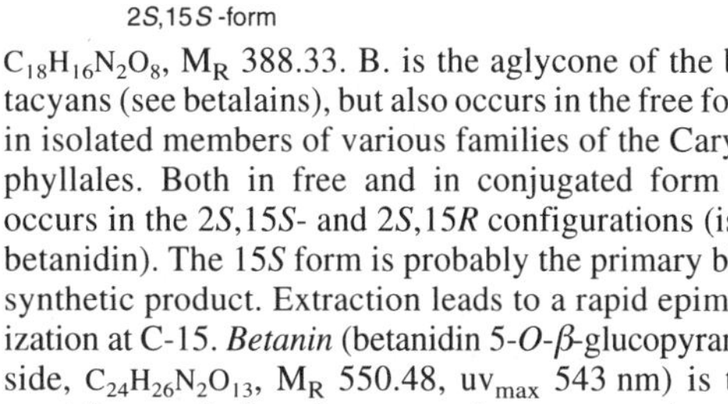

$C_{18}H_{16}N_2O_8$, M_R 388.33. B. is the aglycone of the betacyans (see betalains), but also occurs in the free form in isolated members of various families of the Caryophyllales. Both in free and in conjugated form B. occurs in the 2*S*,15*S*- and 2*S*,15*R* configurations (isobetanidin). The 15*S* form is probably the primary biosynthetic product. Extraction leads to a rapid epimerization at C-15. *Betanin* (betanidin 5-*O*-*β*-glucopyranoside, $C_{24}H_{26}N_2O_{13}$, M_R 550.48, uv_{max} 543 nm) is the most frequently found compound among members of the Caryophyllales containing betacyans. Betanin is the pigment responsible for the color of *beet roots and is used as a coloring substance in foods.

Lit.: Karrer, No. 4402 ▪ Manske **39**, 1–62. – *[HS 2925 20 (B.); 3203 00 (betanin); CAS 2181-76-2 (B.); 7659-95-2 (betanin); 4934-32-1 (iso-B.)]*

Betaxanthins see betalains.

Betel nut see Areca alkaloids.

Betulafolienetriol see ginseng.

Betulin [lup-20(29)-ene-3*β*,28-diol, trochol, betulol].

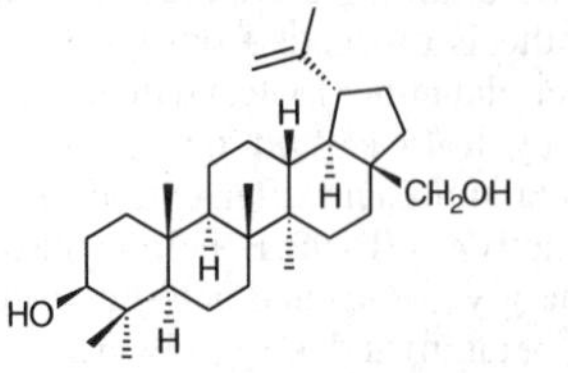

$C_{30}H_{50}O_2$, M_R 442.73, mp. 251–252 °C, $[\alpha]_D$ +20° (pyridine), soluble in acetic acid, less soluble in water, poorly soluble in organic solvents.
Occurrence: In the outer parts of birch bark (*Betula alba*, Betulaceae) in amounts of up to 24% dry weight and in other barks together with *saponins and sesquiterpene alcohols such as betulenol. B. is a triterpene alcohol with the lupane skeleton (cf. figure under triterpenes) and was one of the first natural products to be investigated (first mentioned in 1788 [1]).
Lit.: [1] Phytochemistry **28**, 2229-2242 (1989).
gen.: Beilstein E IV **6**, 6534 ▪ Karrer, No. 2024 ▪ J. Nat. Prod. **51**, 229 (1988) ▪ Justus Liebigs Anm. Chem. **1991**, 1245 (synthesis) ▪ Phytochemistry **28**, 2229 (1989) (review). – *[CAS 473-98-3]*

Betulinic acid [3*β*-hydroxy-20(29)-lupene-28-carboxylic acid].

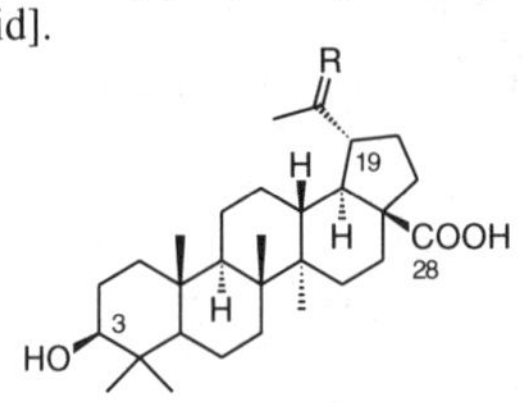

R = CH_2 : Betulinic acid
R = O : Platanic acid

$C_{30}H_{48}O_3$, M_R 456.7, needles (from CH_3OH), mp. 290–293 °C, $[\alpha]_D^{25}$ +7.5° (pyridine). A *triterpene from the leaves of *Syzygium claviflorum* (Myrtaceae) with significant activity against HIV viruses and human melanoma cells (inhibits phosphokinase C). Several patent applications cover compounds with analogous structures [1]. *S. claviflorum* also contains the related *platanic acid*, $C_{29}H_{46}O_4$, M_R 458.7, needles (from $CH_3OH/CHCl_3$), mp. 279–282 °C, $[\alpha]_D$ –38° (pyridine).
Lit.: [1] Bioorg. Med. Chem. Lett. **8**, 1707 (1998).
gen.: J. Med. Chem. **39**, 1016, 1056, 1069 (1996) ▪ J. Nat. Prod. **57**, 243–247 (1995) ▪ Nat. Med. **1**, 1046 (1995) ▪ Synth. Commun. **27**, 1607 (1997) (synthesis). – *[CAS 472-15-1 (B.)]*

Beyerane see diterpenes.

B-factor [adenosine-3′-(monobutyl phosphate), 3′-AMP-monobutyl ester]. $C_{14}H_{22}N_5O_7P$, M_R 403.34, Na salt: powder, mp. 182–186 °C. Substance isolated from yeast extracts that induce in *Nocardia* sp. *rifamycin B production in the concentration range of 2–30 ng/ml. The regulation of rifamycin biosynthesis occurs prior to the incorporation of 3-aminohydroxy-

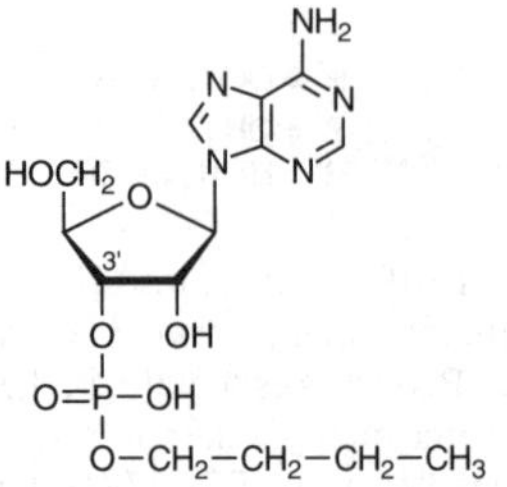

benzoic acid, the C_7N starter unit. Chemical modifications of B-f. showed that the *n*-octyl ester is the most effective, but also that 3′-GMP esters are active. Substances with B-f. activity have also been detected in *Nocardia* sp., they are presumably formed from cAMP by alcoholysis.
Lit.: Annu. Rev Microbiol. **46**, 377–398 (1992) (review) ▪ J. Antibiot. **37**, 1587–1595 (1984); **41**, 360–365 (1998); **43**, 321 ff. (1990). – *[CAS 52278-63-4]*

Biacetyl see 2,3-butanedione.

Bialaphos (phosphinothricylalanylalanine).

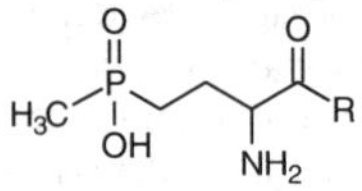

R = OH : Phosphinothricin
R = Ala–Ala–OH : Bialaphos

$C_{11}H_{22}N_3O_6P$, M_R 323.29, amphoteric powder, mp. 159–161 °C, $[\alpha]_D$ –34° (H_2O), tripeptide antibiotic produced by *Streptomyces viridochromogenes* and *S. hygroscopicus* with antibacterial, antifungal, and herbicidal activity. LD_{50} (mouse p.o.) 500 mg/kg. B. consists of 2 molecules L-alanine and the unusual amino acid phosphinothricin (PT). *Phosphinothricin* [$C_5H_{12}NO_4P$, M_R 181.13, mp. 241–243 °C (*S*-form)] was the first amino acid with a phosphinic acid group to be described. The non-ribosomal biosynthesis of B. (13 steps) has also been well elucidated at the genetic level. PT is formed from phosphoenol pyruvate (C-3/C-4) and acetate (C-1/C-2), the methyl group at phosphorus is introduced into the intact tripeptide from methionine or methylcobalamine.
The alanylalanine residue of B. is responsible for its transport into the cell where the active principle PT is formed by a peptidase. PT inhibits glutamine synthetase (types I and II). B. and synthetic PT (Basta®, DL-form of the monoammonium salt of PT) are used in agriculture as herbicides, both are readily degradable. Resistance against B. and PT is achieved by *N*-acetylation of PT. The resistance gene from producing organisms which induces the formation of a specific acetyltransferase has been cloned in crop plants.
Lit.: Bull. Chem. Soc. Jpn. **61**, 3705 (1988) (synthesis) ▪ Helv. Chim. Acta **55**, 224–239 (1972). – *Biosynthesis:* Barton-Nakanishi **1**, 865–880 ▪ Gene **115**, 127–132 (1992) ▪ J. Antibiot. **38**, 687 (1985); **52**, 925 (1999) ▪ Thompson & Seto, in Vining & Stuttard (eds.), Genetics and Biochemistry of Antibiotic Production, p. 197–222, Boston: Butterworth-Heinemann 1995 (review). – *Toxicology* (PT): Food Chem. Toxicol. **32**, 461–470 (1994). – *[HS 2941 90; CAS 35597-43-4 (B.); 35597-44-5 (PT, S-form); 77182-82-2 (Basta)]*

Bicyclomycin (aizumycin). $C_{12}H_{18}N_2O_7$, M_R 302.28, cryst., mp. 187–189 °C (decomp.), $[\alpha]_D$ +63.5° (CH_3OH), water-soluble, bicyclic diketopiperazine antibiotic produced by *Streptomyces sapporonensis* and *S. aizunensis* with *in vitro* and *in vivo* activity

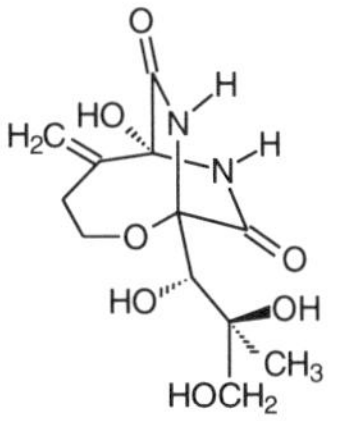

against Gram-negative bacteria, LD_{50} (mouse p. o., i. p. or s.c.) >4 g/kg. It is used in veterinary medicine for diarrhea (calves, pigs) and also acts as a growth promoter in chickens and pigs (Bicozamycin®). B. binds irreversibly to B. binding proteins (BBP) from *Escherichia coli* and thus influences the degradation of peptidoglycans (cell wall) in the region of the cross-linking peptide bond from diaminopimelic acid. Recent data with structural analogues of B. show, that they reversibly bind to Rho transcription termination factor and inhibit transcription termination *in vitro*. The biosynthesis starts from leucine and isoleucine. The additional oxygen atoms in the molecule are introduced with the help of cytochrome-dependent oxidases. Semisynthetically produced B. derivatives do not possess improved activities, total syntheses of B. are also known. B. is produced industrially by fermentation.

Lit.: Chem. Rev. **88**, 511–540 (1988) (review) ▪ J. Antibiot. **25**, 569, 576, 582 (1972) ▪ J. Org. Chem **53**, 6035 (1988) (synthesis); **57** 5523 (1992); **61**, 7750, 7756, 7764 (1996) (activity, synthesis) ▪ Tetrahedron Lett. **37**, 6935 (1996) (biosynthesis). – *[HS 2941 90; CAS 38129-37-2]*

Biflavonoids. B. were originally defined as various dimers of the *flavone *apigenin [1] (e.g., *amentoflavone and cupressoflavone) and were distinguished from other B. (e.g., the *proanthocyanidins). Now a series of other B. structures are known that can be assigned to the various classes of flavonoids, e.g., the *isoflavones, *aurones or *chalcones. In addition to the frequent C–C-linkages of monomers, C–O–C bonds are also known. Their occurrence is limited to members of the gymnosperms (universal) and certain families of the angiosperms. Triflavonoids built up from the flavone *luteolin have also been described [2].

Lit.: [1] Harborne, Comparative Biochemistry of the Flavonoids, London: Academic Press 1967. [2] Z. Naturforsch. C **47**, 527 (1992).
gen.: Harborne (1975, 1988, 1994).

Bilberry flavor see fruit flavors.

Bile acids. A group of physiologically important C_{24}-steroid carboxylic acids which, mainly as amides with *taurine or *glycine conjugated in the form of salts, represent the digestion-promoting constituents of bile. They are formed in the liver of humans and higher vertebrates, stored in the gallbladder, and excreted into the duodenum. The B. are the main components of bile dry substance. They possess dispersing, micelle-forming properties and thus facilitates absorption of fat in the small intestine (see table p. 82).
Most B. are derived from 5β-cholane-24-carboxylic acid by introduction of α-hydroxy groups at C-3, -6, -7, and -12. *Cholic acid, deoxycholic acid,* and *chenodeoxycholic acid* are the main B. in human bile

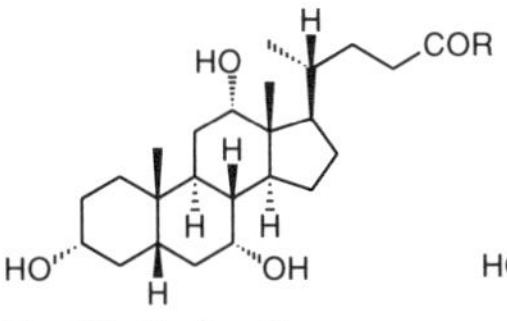

R = OH : Cholic acid
R = NH–CH_2–CH_2–SO_3H : Taurocholic acid
R = NH–CH_2–COOH : Glycocholic acid

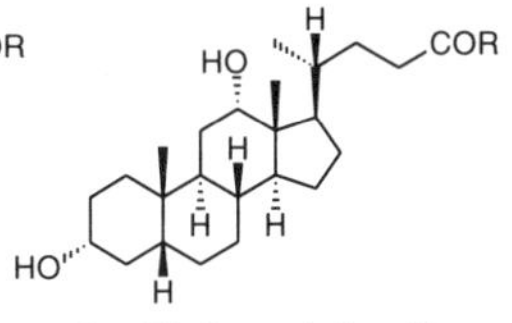

R = OH : Deoxycholic acid
R = NH–CH_2–CH_2–SO_3H : Taurodeoxycholic acid
R = NH–CH_2–COOH : Glycodeoxycholic acid

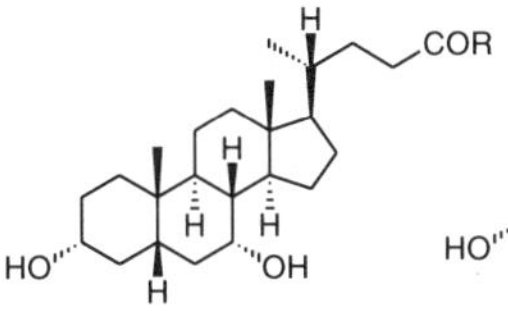

R = OH : Chenodeoxycholic acid
R = NH–CH_2–CH_2–SO_3H : Taurochenodeoxycholic acid
R = NH–CH_2–COOH : Glycochenodeoxycholic acid

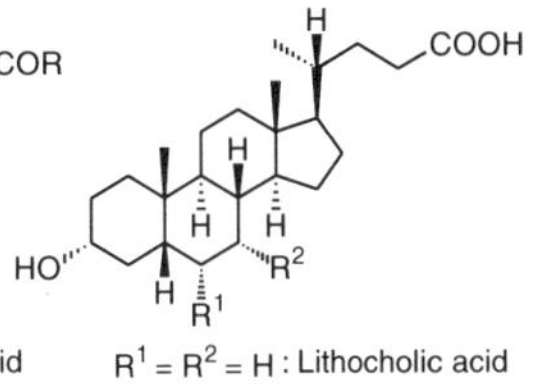

$R^1 = R^2 = H$: Lithocholic acid
$R^1 = R^2 = OH$: Hyocholic acid
$R^1 = OH$, $R^2 = H$: Hyodeoxycholic acid

(ratio ca. 1:2:2). Bile also contains small amounts of *lithocholic acid* and its 7β-hydroxy derivative (*ursodeoxycholic acid*). It shows antiviral activity and activity in biliary cirrhosis. The B. themselves are absorbed to ca. 90% (20–30 g/d) in the gastrointestinal tract, conjugated in the liver, and returned to the bile so that a high concentration of B. (ca. 30 g/L bile) is maintained. The spectrum of B. in higher vertebrates varies widely. Thus, e. g., *hyocholic acid* has been isolated from rat and pig bile; while the main component of pig bile is *hyodeoxycholic acid.* The bile of lower vertebrates such as frogs, toads, and sharks do not contain B., but rather bile alcohols in the form of their sulfates. A typical example is the hexahydroxylated *sterol *scymnol* (5β-cholestan-3α,7α,12α,24,26,27-hexol) from shark bile.

Biosynthesis: The primary B. are synthesized in the liver from *cholesterol by a complicated, multi-step reaction sequence. The 7α-hydroxy group is introduced first while the 12-hydroxy group is added later to a further intermediate with subsequent formation of both chenodeoxycholic acid and cholic acid. Deoxycholic acid is not synthesized in the liver but rather in the intestines by 7α-dehydroxylation of cholic acid by intestinal bacteria.

Use: A low conversion rate of *cholesterol to B. in the liver leads to an oversaturation of bile with cholesterol and can result in the formation of cholesterol gallstones. Oral administration of a large amount of chenodeoxycholic acid can dissolve small gallstones. However, the 7α-dehydroxylation of the administered exogenous chenodeoxycholic acid by intestinal bacteria can result in the formation of high concentrations of toxic lithocholic acid (risk of development of liver and bile cirrhosis).

Lit.: Compr. Supramol. Chem. **1**, 777 (1996) (use) ▪ Danielsson & Sjövall (eds.), Sterols and Bile Acids, Amsterdam: Elsevier 1985 ▪ Kleemann-Engel, p. 1980 f. (activity) ▪ Nachr. Chem. Tech. Lab. **43**, 1047–1055 (1995) ▪ Zeelen, p. 243–249. – *[HS 2918 19; CAS 128-13-2 (ursodeoxycholic acid); 6785-34-8 (scymnol)]*

Table: Data of Bile acids.

Bile acid	molecular formula	M_R	mp. [°C]	$[\alpha]_D$	CAS
Bile acids:					
Cholic acid	$C_{24}H_{40}O_5$	408.58	197	+37° (C_2H_5OH)	81-25-4
Deoxycholic acid	$C_{24}H_{40}O_4$	392.58	187–189	+53° ($CHCl_3$)	83-44-3
Chenodeoxycholic acid	$C_{24}H_{40}O_4$	392.58	119 and 145–146	+11° (C_2H_5OH)	474-25-9
Lithocholic acid	$C_{24}H_{40}O_3$	376.58	186	+33.7° (C_2H_5OH)	434-13-9
Hyocholic acid	$C_{24}H_{40}O_5$	408.58	188–189	+ 5.5° (C_2H_5OH)	547-75-1
Hyodeoxycholic acid	$C_{24}H_{40}O_4$	392.58	197–198	+ 8° (CH_3OH)	83-49-8
Taurocholic acids:					
Taurocholic acid (Choloyltaurine)	$C_{26}H_{45}NO_7S$	515.71	125 (decomp.)	+38.8° (C_2H_5OH/H_2O)	81-24-3
Taurodeoxycholic acid (Deoxycholoyltaurine)	$C_{26}H_{45}NO_6S$	499.71		+33° (H_2O)	516-50-7
Taurochenodeoxycholic acid (Chenodeoxycholoyltaurine)	$C_{26}H_{45}NO_6S$	499.71	182–184 (Na salt)	+ 9° (C_2H_5OH, Na salt)	516-35-8
Glycocholic acids:					
Glycocholic acid (Choloylglycine)	$C_{26}H_{43}NO_6$	465.63	140–142 and 154–155 (decomp.)	+27.8° (C_2H_5OH)	475-31-0
Glycodeoxycholic acid (Deoxycholoylglycine)	$C_{26}H_{43}NO_5$	449.63	187–188	+53° ($CHCl_3$)	360-65-6
Glycochenodeoxycholic acid (Chenodeoxycholoylglycine)	$C_{26}H_{43}NO_5$	449.63	181–182	+ 9.2° (C_2H_5OH)	640-79-0

Bile pigments. General term for *bilirubin and its precursors and degradation products. Bilirubin is formed in bile by oxidative degradation of *heme and its derivatives. Further bacterial degradation in the intestines leads to other B. p. such as *stercobilin, *stercobilinogen, *urobilin, and *urobilinogen. *Biliverdin is an intermediate in the formation of bilirubin from heme. Tetrapyrrole pigments with the general name *phycobilins occurring in the bile, proteins of algae and plants belong to the bile pigments. Also included in the group of B. p. are the products of *chlorophyll degradation, the *luciferins of dinoflagellates, and *substance F.

Lit.: Angew. Chem. Int. Ed. Engl. **16**, 470f. (1977) ▪ Falk, The Chemistry of Linear Oligopyrroles and Bile Pigments, p. 46ff. Wien: Springer 1989.

Biliproteins. Chromoproteins that contain *phycobilins and other *bile pigments such as phytochrombilins as functional chromophores; see also phycobilins and phytochromes.

Bilirubin.

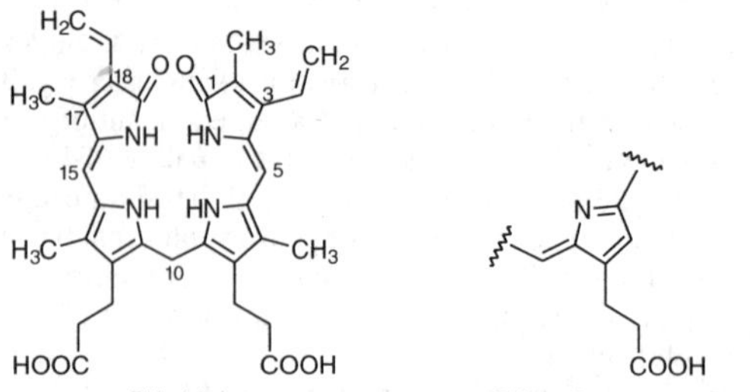

Bilirubin Biliverdin

$C_{33}H_{36}N_4O_6$, M_R 584.67. Light to dark red crystals that turn black without melting on heating. Insoluble in water, easily soluble in benzene, chloroform, chlorobenzene, carbon disulfide, acids, and bases, moderately soluble in alcohols and ethers. Structurally, the substitution pattern is analogous to that of *protoporphyrin IX; a semi-systematic name for B. is biladiene-*ac* IXα, for nomenclature, see bile pigments.

Occurrence & physiology: Degradation product of blood pigment in bile (see bile pigments; the name is derived from Latin: bilis = bile and ruber = red), occurs to a higher extent in blood in hepatitis (jaundice). Gallstones contain a calcium salt of bilirubin. B. are normally formed in the spleen by oxidative cleavage of the porphyrin ring of *heme originating mainly from erythrocytes *h(a)emoglobin and hydrogenation of the green intermediate *biliverdin. The oxidative ring opening requires three equivalents of oxygen. The first intermediate is heme oxidized at position 5, other intermediates have not yet been characterized. The carbon atom 5 of heme is lost as carbon monoxide. The color changes observed in hematoma (bruises) are also a result of this degradation. B. is delivered from the gallbladder to the gut as the diglucuronide. Bacteria afford further metabolization of B. in the intestines; first the vinyl groups and then the 4,5- and 15,16-double bonds on the A,D lactam rings are reduced with formation of *urobilinogen. Further reduction of the 2,3- and 17,18-double bonds in the lactam rings furnishes *stercobilinogen, which is reoxidized by dehydrogenation at C-10 to *stercobilin during isolation. If the level of B. in the blood increases too strongly through disease or in new born infants, brain damage may occur. In most cases a phototherapy usually helps to lower the B. level. Besides L-*ascorbic acid, B. is another natural antioxidant that traps hydroperoxy radicals.

Lit.: Beilstein E V **26/15**, 523 ▪ Falk, The Chemistry of Linear Oligopyrroles and Bile Pigments, p. 40–59, Wien: Springer 1989 ▪ Merck-Index (12.), No. 1263. – *[HS 2933 90; CAS 635-65-4]*

Biliverdin (dehydrobilirubin, bilatriene IXα). Formula see under bilirubin. $C_{33}H_{34}N_4O_6$, M_R 582.66. Dark green (name from Latin: bilis = bile and Italian/Spanish: verde = green) plates or prisms that decompose at 300 °C, soluble in methanol, ether, chloroform, benzene, and alkalis. B. occurs as an intermediate in the degradation of *heme to *bilirubin to an increased extent in the serum of patients with liver diseases and can be determined for diagnostic purposes. Amphibians and birds only have B. and no bilirubin as *bile pigment. The deep blue color of the emu egg is caused by B. For nomenclature, see bile pigments.

Lit.: Beilstein E V **26/15**, 527 ▪ Falk, The Chemistry of Linear Oligopyrroles and Bile Pigments, p. 46–58, 382, Wien: Springer 1989 ▪ Merck-Index (12.), No. 1264. – *[HS 2933 90; CAS 114-25-0]*

Bilobalide (bilobalide A).

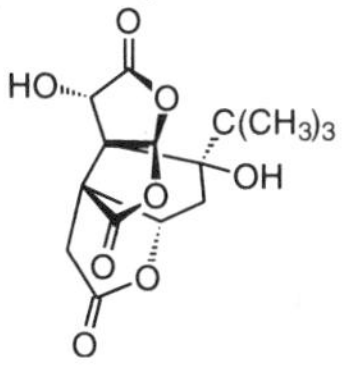

$C_{15}H_{18}O_8$, M_R 326.30, mp. >300 °C, $[\alpha]_D^{20}$ −64° (acetone), soluble in organic solvents. *Sesquiterpene from the ginkgo tree (*Ginkgo biloba*). It is a C_{15} trilactone, similar to the ginkgolides which are hexacyclic *diterpenes and, like them, possesses the as yet only naturally occurring *tert*-butyl groups. For synthesis, see *Lit.*[1].

Lit.: [1] J. Am. Chem. Soc. **109**, 7534 (1987); **114**, 5445 f. (1992); **115**, 3146–3155 (1993); Tetrahedron Lett. **29**, 3423 (1988). *gen.:* Justus Liebigs Ann. Chem. **1987**, 1079 (crystal structure) ▪ Phytother. Res. **2**, 1–24 (1988) (review) ▪ Tang & Eisenbrand, Chinese Drugs of Plant Origin, p. 557 f., Berlin: Springer 1992. – *[CAS 33570-04-6]*

Biochanin A see isoflavones.

Biogenesis. In the widest sense, B. means the formation of substances and supramolecular structures in biological systems. Although this can, in principle, also include degradation products (e. g., biogenic substances in sediments and fossils), the term is also used frequently as an equivalent for biosynthesis, i.e., for the enzymatic formation of natural products from simpler precursors. Labelled compounds are very useful in investigations of B. pathways (radioactive as well as stable isotopes: ^2H, ^3H, ^{13}C, ^{32}P, ^{15}N). Specific *in vitro* biosyntheses with the help of enzymes, cell cultures (fermentations) or with use of complex organisms are one of the fields of biotechnology. B. must be distinguished from the formation of chemical compounds in prebiotic evolution, in geological processes, and in chemical synthesis in the laboratory. These are also known as abiotic processes.

Lit.: Herbert, The Biosynthesis of Secondary Metabolites (2.), London: Chapman & Hall 1989 ▪ Luckner ▪ Mann, Secondary Metabolism (2.), Oxford: Univ. Press 1987.

Biogenic amines see polyamines and monoamines.

Bioluminescence. Name for the part of chemiluminescence involving the formation of so-called "cold light" in bacteria, flagellates (marine phosphorescence), fungi, sponges, jellyfish, worms, crabs, and fish as well as in insects. The emission of light quanta results from a prior stimulation by an enzymatically regulated oxidation. The phenomenon has been most intensively investigated in glow-worms (fireflies), in which *luciferins, magnesium ions, ATP, and oxygen must be present for the reaction to start. The enzyme *luciferase catalyzes the oxidation of luciferin with emission of a photon. The B. system of the ostracod *Cypridina* is more simple: *Cypridina*-luciferin, -luciferase, and oxygen. It is assumed that the various *luciferins (figure see there) are oxidized to α-peroxylactones (1,2-dioxetan-3-ones) which emit light quanta on decomposition. The emitted light may be circularly polarized. Jellyfish use *aequorin and calcium ions to generate B.; the mechanisms in luminous bacteria have not been completely clarified. It is however clear that a certain concentration of a specific "autoinducer" must be reached to induce *de novo* synthesis of the enzyme apparatus required for B. *Lampteroflavin and panal seem to be involved in the luminescence of various fungi. The light source can be localized in special light cells, in light tissue, or in light organs. Light glands which secrete luminescent mucus are also known. The biological function of B. is often unknown. It may serve as recognition for members of the same species, to attract a sex partner, but also to deter enemies or to search for and attract prey. B. reactions are used in luminometry to determine very small amounts of ATP, O_2, NAD(NADP) or Ca^{2+} ions.

Lit.: Chimia **49**, 45 (1995) ▪ De Luca u. McElroy (eds.), Bioluminescence and Chemiluminescence, N. Y.: Academic Press 1981 ▪ Nachr. Chem. Tech. Lab. **33**, 785 (1985).

Biopterin [2-amino-6-(1,2-dihydroxypropyl)-pteridin-4-ol].

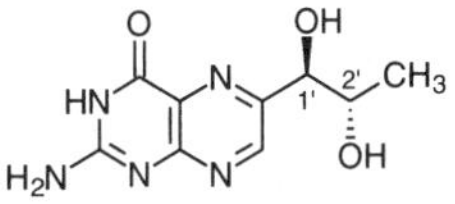

$C_9H_{11}N_5O_3$, M_R 237.22. The (1′*R*,2′*S*) form (L-*erythro* form) forms pale yellow cryst., mp. 250–280 °C (decomp.), $[\alpha]_D^{20}$ −66° (0.1 m HCl), pK_{a1} 2.23, pK_{a2} 7.89. In alkaline solution B. shows blue fluorescence. It is widely distributed in microorganisms, insects (e. g. in *royal jelly of queen bees), algae, amphibians, and mammals and is also found in urine. In metabolism tetrahydro-B. acts as a cofactor for phenylalanine 4-monooxygenase (EC 1.14.16.1). B. is a growth factor for insects. The (1′*R*,2′*R*) form (D-*threo* form) known as *dictyopterin* melts at >300 °C and has pK_{a1} 2.20, pK_{a2} 7.92. It occurs in the slime mold *Dictyostelium discoideum*. The biosynthesis starts from guanosine triphosphate.

Lit.: Beilstein E V **26/18**, 422 ▪ Chem. Rev. **97**, 1755–1792 (1997) (synthesis review) ▪ Eur. J. Biochem. **187**, 665–669 (1990) (dictyopterin; CD, MS) ▪ J. Org. Chem. **61**, 8698 (1996) (total synthesis L-B.) ▪ Justus Liebigs Ann. Chem. **1989**, 963–967, 1267 ff. (synthesis) ▪ Karrer, No. 2572 ▪ Science **213**, 1129 ff. (1981). – *[CAS 22150-76-1 (1R,2S); 13019-52-8 (1R,2R)]*

Bioquinones see plastoquinones, tocopherols, ubiquinones, and vitamin K_1.

Biosynthesis see biogenesis.

Biotin (vitamin B_7, vitamin H, coenzyme R).

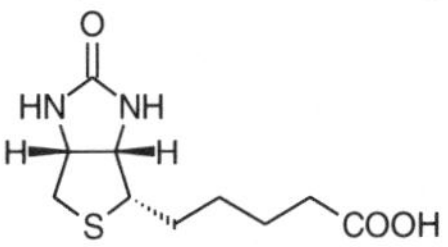

$C_{10}H_{16}N_2O_3S$, M_R 244.31. needles, mp. 232–233 °C, $[\alpha]_D^{20}$ +91° (0.1 M NaOH). It is stable to air and heat, neutral solutions are stable up to 100 °C. TDL_0 (rat s. c.) 100 mg/kg. B is teratogenic in animal experiments. B. occurs in all living cells and is especially abundant in yeasts, eggs, liver, kidneys, pancreas, and milk. It has also been isolated from various higher plants, e.g., maize sprouts. Bound to the ε-amino group of a specific lysine residue of proteins (amide bond) it acts as cofactor for many carboxylases. B. belongs to the vitamins (previously called vitamin H). Deficiency symptoms manifesting as seborrhea, dermatitis, loss of appetite, muscular pain, tiredness, and nervous disorders rarely occur because B. is synthesized by human intestinal flora. Vitamin deficiency may, however, occur after excessive consumption of raw egg white on account of the high affinity binding of B. to the protein *avidin (use in biochemical analysis).

Biosynthesis: L-Alanine + pimelyl-CoA → 7-oxo-8-aminopelargonic acid → 7,8-diaminopelargonic acid → dethiobiotin. The sulfur atom originates from L-methionine.

Lit.: Beilstein E III/IV **27**, 7979–7982 ▪ Biochemistry **5**, 713 f. (1966) (abs. configuration) ▪ Luckner (3.), p. 273 ▪ Stryer 1995, p. 572 f.; 614 f. – *Synthesis:* Angew. Chem. Int. Ed. Engl. **34**, 2391 (1995) ▪ Chem. Rev. **97**, 1755–1792 (1997) (review) ▪ J. Org. Chem. **60**, 321–330 (1995) ▪ Tetrahedron **46**, 1057–1062, 7667–7676 (1990) ▪ Tetrahedron Lett. **34**, 4365 (1993). – *Toxicology:* Sax (8.), VSU100. – *[HS 2936 29; CAS 58-85-5]*

(Z)-(1S,2R,4S)-α-Bisabolene-1,2-epoxide see bugs.

Bisabolenes. $C_{15}H_{24}$, M_R 204.36. Mixture of isomeric *sesquiterpenes in various plant oils, e.g. *bergamot and *citrus oil, which find use as fragrances.

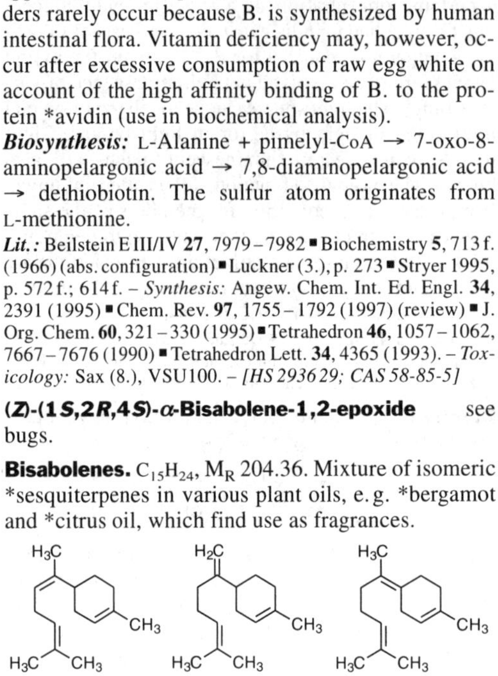

α β γ

Table: Data of Bisabolenes.

		bp. [°C]	$[\alpha]_D^{20}$ (C_2H_5OH)	CAS
α-B.		156 (1.6 kPa)		17627-44-0
	(R)-(E)-		+54.3°	70286-31-6
	(S)-(E)-		–38.9°	70286-32-7
	(R)-(Z)-		–12.0°	70286-33-8
	(S)-(Z)-		+8.0° [1]	58845-44-6
β-B.		129–130 (1.4 kPa)		489-79-6
	(R)-	colourless oil	+74°	
	(S)-		–84.4° [2]	495-61-4
γ-B.				495-62-5
	(E)-	oil		53585-13-0
	(Z)-[3]			13062-00-5

Lit.: [1] Helv. Chim. Acta **62**, 369 (1979). [2] J. Am. Chem. Soc. **94**, 4298 (1972). [3] J. Org. Chem. **41**, 697 (1976).

gen.: Beilstein E IV **5**, 1174 ▪ Can. J. Chem. **62**, 2079 (1984) (biosynthesis) ▪ Synth. Commun. **25**, 2909–2921 (1995) ▪ Tetrahedron **39**, 883 (1983) (synthesis) ▪ Ullmann (4.) **22**, 545. – *[HS 2902 19]*

(–)-α-Bisabolol {(2S)-6-methyl-2-[(1S)-4-methyl-3-cyclohexenyl]-5-hepten-2-ol; levomenol}.

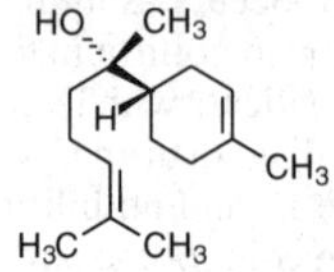

$C_{15}H_{26}O$, M_R 222.37, viscous, weakly smelling liquid, bp. 153 °C (1.6 kPa), $[\alpha]_D^{20}$ –60.2° (neat); soluble in alcohols. (–)-(α)-B. occurs in *chamomile oils as well as the buds of some poplar species. Because of its antiphlogistic and spasmolytic action it is used in medicine and in cosmetics (fixative). For synthesis, see *Lit.*[1].

Lit.: [1] Aust. J. Chem. **42**, 2035, 2041 (1989); J. Org. Chem. **58**, 5528 (1993); Tetrahedron Lett. **34**, 4939 (1993).

gen.: Beilstein E IV **6**, 416 ▪ Ullmann (5.) **A 11**, 221; **A 26**, 215. – *[CAS 23089-26-1]*

Bisbenzylisoquinoline alkaloids.

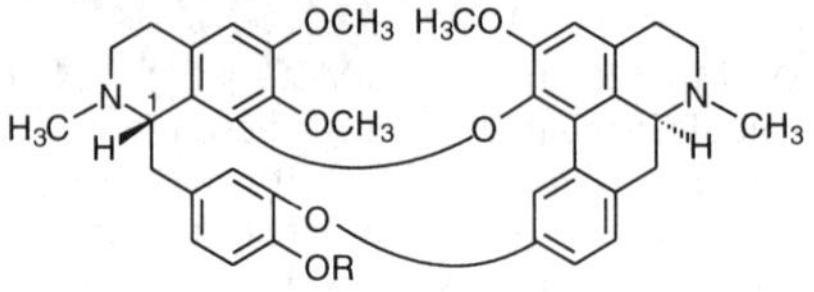

R = CH3 : Tetrandrine
R = CH3 ; 1αH : Isotetrandrine
R = H ; 1αH : Berbamine

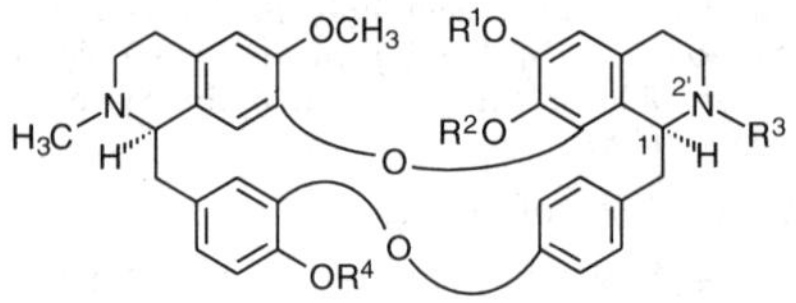

	R1	R2	R3	R4
Cepharanthine	CH2		CH3	CH3
Oxyacanthine	CH3	CH3	CH3	H
Epistephanine	CH3	CH3	*	CH3

*1', 2'-C=N-double bond

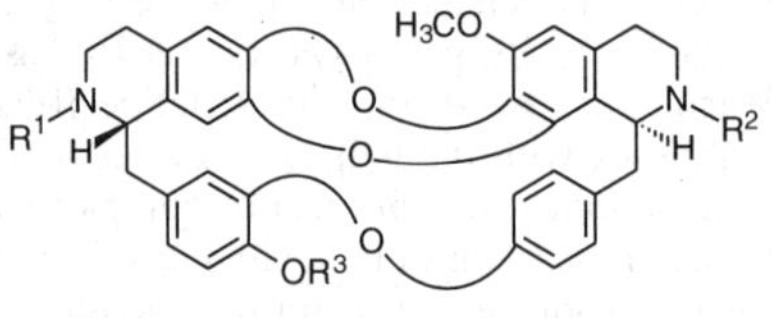

	R1	R2	R3
Cocsoline	H	CH3	H
Trilobine	CH3	H	CH3
Isotrilobine	CH3	CH3	CH3

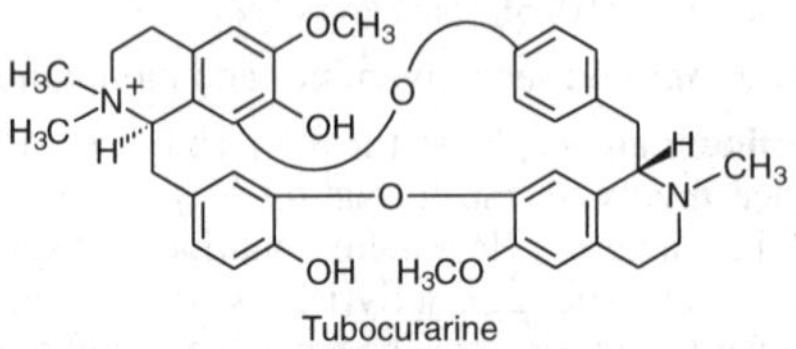

Tubocurarine

The B. a. represent the largest group of the *isoquinoline alkaloids. They consist of two benzyltetrahydro-

Table: Data and occurrence of Bisbenzylisoquinoline alkaloids.

name	molecular formula	M_R	mp. [°C]	$[\alpha]_D$ ($CHCl_3$)	occurrence	CAS
Berbamine	$C_{37}H_{40}N_2O_6$	608.73	155–157	+109°	*Atherosperma moschatum* and *Berberis* species	478-61-5
Cepharanthine	$C_{37}H_{38}N_2O_6$	606.72	145–155	+277°	*Stephania cepharantha, S. erecta, S. sasakii, S. epigeae*	481-49-2
Cocsoline	$C_{34}H_{32}N_2O_5$	548.64	197–199	+204°	*Cocculus pendulus, C. leaebe*	54352-70-4
Epistephanine	$C_{37}H_{38}N_2O_6$	606.72	203–204	+183.5°	*Stephania* species	549-08-6
Isotetrandrine	$C_{38}H_{42}N_2O_6$	622.76	182–183	+150.7°	*Berberis* species, *Cyclea barbata, Mahonia* species, *Stephania* species	477-57-6
Isotrilobine	$C_{36}H_{36}N_2O_5$	576.69	217–218	+325°	*Anisocycla grandidieri, Cocculus* species, *S. hernandifolia*	26195-62-0
Oxyacanthine	$C_{37}H_{40}N_2O_6$	608.73	212–214	+285.6°	*Berberis* species, *Mahonia* species, *Thalictrum lucidum*	548-40-3
Tetrandrine	$C_{38}H_{42}N_2O_6$	622.76	218	+241.38°	*Cocculus-* and *Stephania* species (e. g., *S. tetrandra)*	518-34-3
Trilobine	$C_{35}H_{34}N_2O_5$	562.67	235–239	+305°	*Anisocycla grandidieri, Cocculus* species	6138-73-4
Tubocurarine [1]	$C_{37}H_{41}N_2O_6^+$	610.75			*Chondrodendron tomentosum*	57-95-4

isoquinoline units, linked by 1, 2, or rarely 3 ether bridges and often also by C–C bonds. More than 225 various B. a. have been isolated from 14 different plant families, especially from Berberidaceae, Menispermaceae, Monimiaceae, and Ranunculaceae (see table). In the Menispermaceae 21 genera produce no less than 112 different B. alkaloids. Besides bisbenzylisoquinolines with 2 benzylisoquinoline units there also exist combinations of one benzylisoquinoline and one aporphine or proaporphine unit. They are soluble in ethanol, ether, and chloroform, poorly soluble in water.
Biosynthesis: B. a. are formed by oxidative dimerization of 2 molecules of benzylisoquinolines, e. g., coclaurine or *N*-methylcoclaurine [2,3].
Use [4]***:*** The B. a. are generally toxic and have a curare-like action. Cycleanine and tetrandrine as well as isotetrandrine are reported to have anti-inflammatory and anesthetic properties, isochondodendrine antispasmodic and sedative effects, while oxyacanthine is a sympathicolytic agent, an adrenaline antagonist, and a vasodilator, it is the hypotensive principle of *Berberis vulgaris*. Only one bond appears to be necessary for the observed tumor-inhibiting activity. Dauricine is toxic with an LD_{50} (mouse i. p.) 6 mg/kg, tubocurarine, a powerful curare-like toxin is the oldest known muscle relaxant [5,6]. It is used as a diagnostic tool, e. g., of Myasthenia gravis, and by South American Indians as an arrow poison (see curare). The dextrorotatory isomer is 50 times more active than the laevorotatory isomer.
Lit.: [1] Beilstein E III/IV **27**, 8727; Manske **16**, 363; Merck-Index (12.), No. 9939; Adv. Biochem. Psychopharmacol. **21**, 67–80 (1980). [2] Mothes et al., p. 204–207. [3] Phytochemistry **27**, 2557–2565 (1988). [4] J. Pharm. Pharmacol. **43**, 589 ff. (1991). [5] Br. J. Pharmacol. **103**, 1607–1613 (1991). [6] J. Nat. Prod. **54**, 645–749 (1991); **60**, 934–953 (1997) (reviews).
gen.: Florey **7**, 477–500 ▪ Hager (5.) **4**, 852–858 ▪ J. Nat. Prod. **54**, 645–749 (1991); **60**, 934–953 (1997) (reviews) ▪ J. Org. Chem. **64**, 2381 (1999) (synthesis trilobin) ▪ Kirk-Othmer (4.) **1**, 1053–1066 ▪ Manske **25**, 163–178; **30**, 1–222 ▪ Nat. Prod. Rep. **3**, 477–488 (1986) ▪ Sax (8.), CCX 550, TDX 380, TDX 830, TDX 835, TEH 250, TNY 750, TNZ 000, TOA 000 ▪ Shamma, p. 115–152 ▪ Shamma-Moniot, p. 71–102 ▪ Ullmann (5.) **A 1**, 370. – *[HS 2939 90]*

***N,N'*-Bis(2,6-dimethylphenyl)-2,2,4,4-tetramethyl-1,3-cyclobutanediimine.**

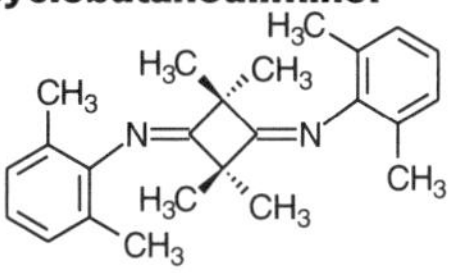

$C_{24}H_{30}N_2$, M_R 346.52, cryst., mp. 169–170 °C. Unusual cyclobutane derivative from the giant reed (*Arundo donax*, Poaceae). The compound, whose structure has been confirmed by X-ray crystallography does not fit in any known biosynthetic scheme.
Lit.: Phytochemistry **34**, 1277 (1993). – *[CAS 52199-12-9]*

Bisindole alkaloids. *Alkaloids formally containing two (hydrogenated) indole units occurring, e. g., in Apocynaceae, micoorganisms, or marine animals.
Lit.: Manske **47**, 173–226.

Bisindolylmaleimides see arcyria pigments.

Bis(2-methyl-3-furyl) disulfide see 2-methyl-3-furanthiol.

Bisnoryangonin see styrylpyrones.

Bitter principles (amaroids). Name for compounds of plant or fungal origin exhibiting a very high degree of bitterness at very low concentrations. There are no uniform structural features for the interaction of B. p. with the receptors of the tongue. Not only *glycosides [e. g., from gentian (amarogentin), blessed thistle, buckbean, and centaury] but also isoprenoids [e. g., from hops (humulone, lupulone), vermouth (*absinth), angelica, dandelion, hemlock, citrus fruits (*limonin, nomilin, *naringin), fungi [crystallopicrin (see iridals), *pistillarin, etc.] act as B. p. Also *santonin, *lactucin, *cascarillin, *quassin(oids), *cynaropicrin and *columbin belong to this category. In a large proportion of the terpenoid B. p. the lactone ring responsible for the bitter taste is formed during isolation (artefacts). Bitter tastes are also known for amino acids and peptides [1], alkaloids, e. g., *strychnine, *vomicine and *quinine. Quinine hydrochloride is used as standard substance for classification of the bitter value. The most bitter substance currently known is **amarogen-*

tin, which still has a detectable bitter taste at a dilution of 1:60 000 000.

Use: B. p. are used in medicine as a stomachic or gastric tonic in cases of loss of appetite and dyspeptic complaints. Other uses are in the production of alcoholic drinks (bitters, various aperitifs) and soft drinks (tonic water, bitter lemon, bitter orange, etc.). The formation of B. p. plays an important role in the production of beer and cheese.

Lit.: [1] Nachr. Chem. Tech. Lab. **25**, 442 (1977).
gen.: Food Rev. Int. **1**, 271–354 (1985) ▪ Parfums, Cosmet., Aromes **1**, 16, 53 (1975) ▪ Zechmeister **17**, 124–182; **26**, 190–244; **30**, 101; **44**, 101; **47**, 221.

Bixin.

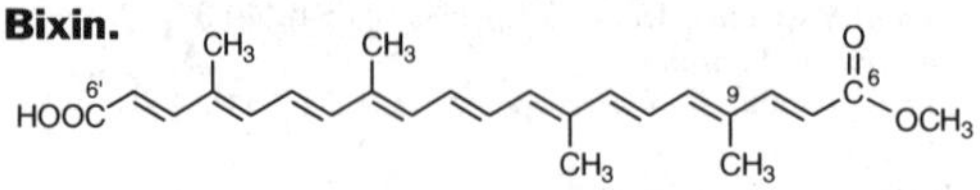

(*all-E*)-form (β-Bixin, Isobixin)

$C_{25}H_{30}O_4$, M_R 394.51, orange to purple-colored plates, mp. 217 °C. The more labile (9*Z*) form (α-bixin, bixin) of the *carotinoid B. is contained in ripe seeds of the tropical annatto tree (roucou) *Bixa orellana* (Bixaceae). The crude concentrate is known as *annatto* (orlean) and is used to color foods (E 160b), e.g., margarines and cheeses as well as cosmetics.

Lit.: Beilstein E IV **2**, 2355 ▪ Britten et al., Carotenoids, Basel: Birkhäuser 1995 ▪ Helv. Chim. Acta **82**, 696 (1999) (synthesis) ▪ Merck-Index (12.), No. 1347 ▪ Phytochemistry **46**, 1379 (1997) (carotinoids from annatto) ▪ Zechmeister **18**, 320–330. – *[HS 3203 00; CAS 6983-79-5 (α-B.); 39937-23-0 (β-B.)]*

Bizelesin see CC-1065.

Black oak see dye plants.

Blasticidins.

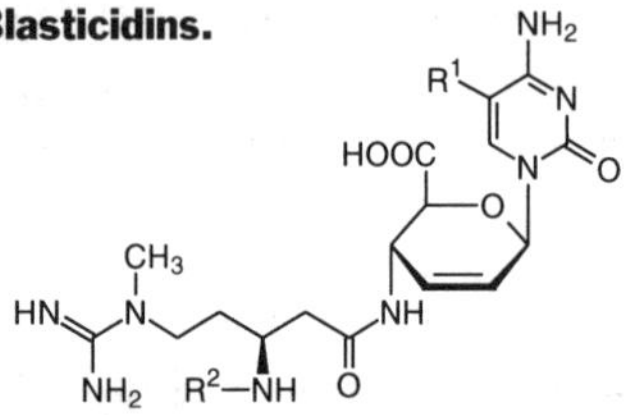

$R^1 = R^2 = H$: B.S
$R^1 = CH_2OH$, $R^2 = CO{-}CH(NH_2){-}CH_2{-}CH(CH_3)_2$: Rodaplutin

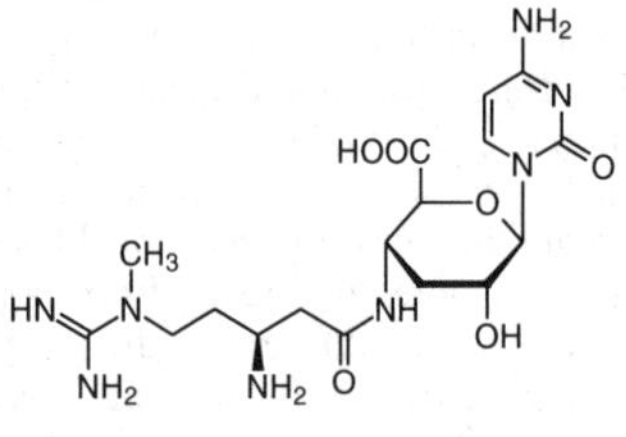

B.H

The most important members of a series of *nucleoside antibiotics from various *Streptomyces* species are *B. H*[1] [$C_{17}H_{28}N_8O_6$, M_R 440.46, needles, mp. 230–235 °C (dihydrochloride, decomp.)] and *B. S*[2] {*cytovirin*, $C_{17}H_{26}N_8O_5$, M_R 422.44, needles, mp. 235 °C (decomp.), $[\alpha]_D^{20}$ +108.4°, pK_{a1} 2.4, pK_{a2} 4.6, pK_{a3} 8.0, pK_{s4} >12.5, LD_{50} (rat p.o.) 16 mg/kg, (rat s.c.) 3100 mg/kg}. B. S is composed of an unusual *pyrimidine nucleoside (*cytosinin*) in which the cytosine is *N*-glycosidically linked to a 2,3-unsaturated 4-aminouronic acid. The amino group is peptidically linked with a β-amino acid (N^δ-methyl-L-β-arginine = *blastidic acid*). B. S is highly toxic to mammals and its use in plant protection is limited by its very high phytotoxity. B. S specifically inhibits protein biosynthesis by inhibiting peptidyltransferase. To combat mildew on rice plants (caused by *Pyricularia oryzae*) a commercial preparation is sprayed which contains B. S as its benzylaminobenzenesulfonate salt. *Rodaplutin* {$C_{24}H_{39}N_9O_7$, M_R 565.63, amorphous powder, mp. >230 °C (decomp.), $[\alpha]_D^{20}$ +58.2° (H_2O)} from cultures of *Nocardioides albus* and *Streptomyces* sp. exhibits acaricidal and insecticidal as well as weak antimicrobial activities.

Lit.: [1] Tetrahedron **32**, 1493 (1976); J. Antibiot. **30**, 1019 (1977). [2] Tetrahedron **56**, 693 (2000) (biosynthesis); Tetrahedron Lett. **27**, 3815 (1986); J. Antibiot. **42**, 470 (1989).
gen.: Tetrahedron Lett. **38**, 7399 (1997) (B. A). – *[CAS 61461-71-0 (B. H); 2079-00-7 (B. S); 108351-49-1 (rodaplutin)]*

Blastmycin see antimycins.

Blattaria see cockroaches.

Blattellastanosides see cockroaches.

Bleomycins.

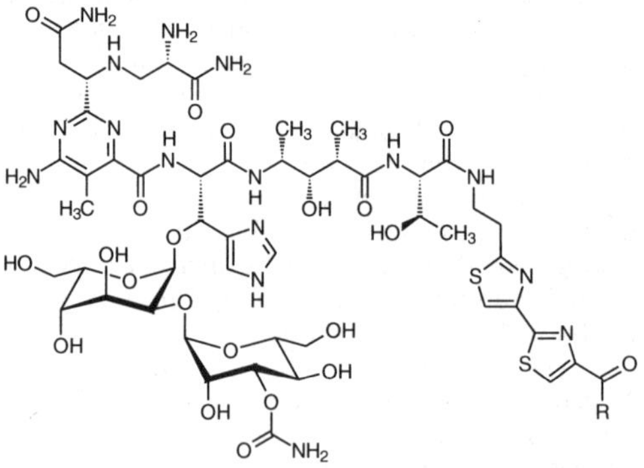

$R = NH{-}(CH_2)_3{-}\overset{+}{S}(CH_3)_2$: B. A_2
$R = NH{-}(CH_2)_4{-}NH{-}C(=NH){-}\overset{+}{N}H_3$: B. B_2

Glycopeptide antibiotics from *Streptomyces verticillus* discovered in 1966. In natural substance mixtures (9 components) *B. A_2* ($C_{55}H_{84}N_{17}O_{21}S_3^+$, M_R 1415.57) predominates. *B. B_2* is the most important minor component ($C_{55}H_{85}N_{20}O_{21}S_2^+$, M_R 1425.52, ion). Salts of B. are soluble in water. Structure elucidations were accomplished by classical peptidochemical methods and completed in 1982 with total synthesis. The B. contain an unusual hexapeptide, bound to a disaccharide (of D-gulose and D-mannose); they differ in the amide side-chain on the bisthiazole. Various transition metal ions form colored complexes with B. B. have antibacterial activities and are active against some rapidly growing types of cancer. They are used clinically, e.g., for lymph, skin, or testicular cancers. The action is based on a base-specific binding to DNA, B. cause oxidative single strand and double strand cleavage: They take up Fe^{2+} in their metal-binding region. The complex activates molecular oxygen, which induces radical cleavage by expulsion of an H atom from C-4 of deoxyribose. There are numerous synthetic and semisynthetic B. derivatives prepared for therapeutic and mechanistic studies. Phleomycins and tallysomycins also belong to the class of bleomycins.

Lit.: Acc. Chem. Res. **29**, 322 (1996) (review) ▪ Hager (5.) **7**, 501–506 ▪ Hecht, in Foye (ed.), Cancer Chemotherapeutic Agents, p. 369–388, Washington: ACS 1995 ▪ J. Antibiot. **44**, 357–365 (1991). – *Review:* Angew. Chem. Int. Ed. Engl. **38**, 448–476 (1999). – *Synthesis:* Angew. Chem. Int. Ed. Engl. **32**, 273 (1993) ▪ Eur. J. Med. Chem. **2**, 543–560 (1995) ▪ J. Am. Chem. Soc. **116**, 5647 (1994); **120**, 11285 (1998) (B. A_2) ▪ Ohno & Otsuka, in Lukacs &Ohno (eds.) Recent Progress in the Chemical Synthesis of Antibiotics, p. 387–414, Berlin: Springer 1990 ▪ Tetrahedron **48**, 1193 (1992) ▪ Ullmann (5.) **A 2**, 498f., 526f.; **A5**, 18. – *Toxicology:* Sax (8.), BLY 250. – *Mechanism of action:* Angew. Chem. Int. Ed. Engl. **34**, 746–770 (1995) ▪ Chem. Rev. **98**, 1153 (1998) ▪ Exp. Opin. Ther. Patents **6**, 893–899 (1996) ▪ J. Am. Chem. Soc. **112**, 3997 (1990); **117**, 3883 (1995). – *[CAS 11116-31-7 (B. A_2)]*

Blepharismins.

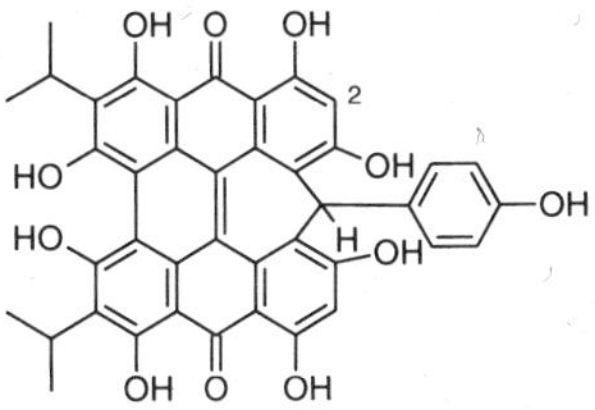

Blepharismin BL-3 (= B. C)

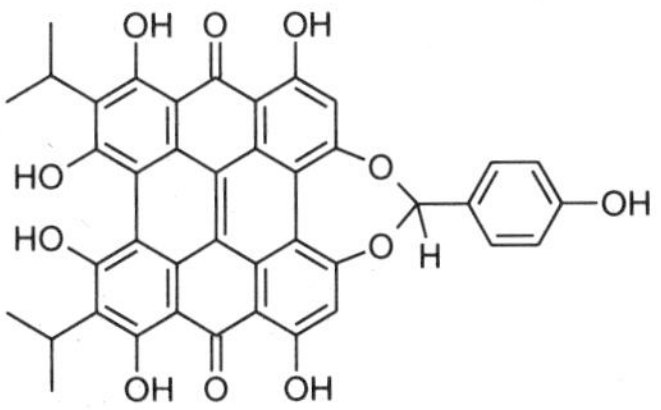

Oxyblepharismin C

$C_{41}H_{30}O_{11}$, M_R 698.68, red powder.

Blepharismin (BL-3) is the main component of a pigment complex from the ciliate *Blepharisma japonicum*[1]. The other B. differ by substitution of one (BL-2, BL-4) or both isopropyl groups (BL-1) against ethyl and may carry an additional methyl group at C-2 (BL-4, BL-5). Interestingly the 4-hydroxybenzylidene methine proton occurs at $\delta = 7.00$ in the ^{1}H NMR spectrum. The red BL-3 ($\lambda_{max} = 576$ nm, color due to fluorescence emission) yields the weakly fluorescent blue BL-3 ($\lambda_{max} = 590$ nm), upon irradiation with red light. On irradiation with daylight (λ_{max} 400–600 nm) the B. undergo an interesting transformation into the corresponding oxyblepharismins[2].

B. japonicum is capable of both light intensity and color-sensory perception, a process in which the B. act as a unique visual sensory system[3]. The pigments induce a step-up photophobic response (ciliary reversal) followed by swimming acceleration (photo-kinetic response)[4]. For a similar photodetector molecule from ciliates see stentorin.

Lit.: [1] J. Am. Chem. Soc. **119**, 5762, 9588 (1997); Tetrahedron Lett. **38**, 7411 (1997). [2] Tetrahedron Lett. **39**, 4003 (1998). [3] FEMS Microbiol. Lett. **155**, 67 (1997). [4] Microbios **99**, 89 (1999).

gen.: Angew. Chem. Int. Ed. Engl. **38**, 3117 (1999) (review). – *[CAS 129898-49-3 (BL-3); 209669-05-6 (oxyblepharismin C)]*

Boar odorant see steroid odorants.

Boforsin see forskolin.

Bolegrevilol see suillin.

Boletus mushroom pigments. Boletes and related mushrooms (Boletales) are characterized by their rich diversity of colors. Lead pigments are the yellow hydroxypulvinic acids such as *variegatic acid and *xerocomic acid which are responsible for the blue color acquired by many boletes when their fruit bodies are injured. Oxidation of *pulvinic acid can furnish red pigments such as variegatorubin or brown cap pigments of the *badione type. Interestingly, the cornflower boletus (*Gyroporus cyanescens*) and the gasteromycete *Chamonixia caespitosa* have further systems for forming blue colors based on oxidation of the cyclopentanediones *gyrocyanin and *chamonixin. The yellow *grevillins typical for the genus *Suillus* are biogenetically closely related. While the above-mentioned pigments are biosynthetically derived from tyrosine, polyprenylated 2,5-dihydroxybenzoquinones such as *boviquinone-4 and *tridentoquinone arise from 4-hydroxybenzoic acid. Examples of individual pigments are *retipolide A and the brilliant yellow chromoalkaloid *curtisine A.

Lit.: Zechmeister **51**, 4ff.

Bombardier beetles. Beetles of the genus *Brachinus* and related species. They possess an abdominal defensive gland with which they can shoot explosion-like a hot secretion at an enemy. For this a 25% H_2O_2 solution and a 10% solution of hydroquinone are injected into a "combustion chamber". The catalases and polyphenol oxidases (temperature optimum of the enzymes 70–80 °C) present in the chamber effect an explosive reaction to water and *benzoquinone or *methyl-1,4-benzoquinone and the up to 70 °C hot mixture is shot specifically at the enemy.

Lit.: Angew. Chem. Int. Ed. Engl. **9**, 1–8 (1970) ▪ J. Insect. Physiol. **16**, 749–789 (1970).

Bombesin.

5-Oxo—Pro—Gln—Arg—Leu—Gly—Asn—Gln—Trp
|
H_2N—Met—Leu—His—Gly—Val—Ala

$C_{71}H_{110}N_{24}O_{18}S$, M_R 1619.86, mp. 185 °C (hydrochloride, decomp.). A tetradecapeptide from the skin of toads of the genus *Bombina*. B. stimulates the secretion of gastric and pancreatic mucous membranes and exhibits hyperglycemic, hypertensive, and antidiuretic properties. High concentrations are found in the cells of human lung carcinomas[1].

Lit.: [1] Science **241**, 1246–1248 (1981).

gen.: Am. J. Respir. Cell Mol. Biol. **3**, 189f. (1990) ▪ Handb. Psychopharmakol. **16**, 363ff. (1983) ▪ J. Biol. Chem. **265**, 7709 (1990) (review) ▪ Med. Res. Rev. **15**, 389–417 (1995) (bombesin receptors). – *[CAS 31362-50-2]*

Bombykol [(10*E*,12*Z*)-10,12-hexadecadien-1-ol].

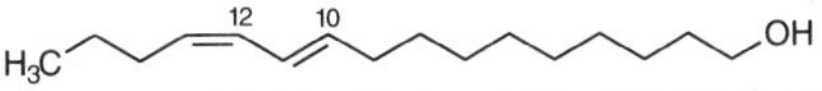

$C_{16}H_{30}O$, M_R 238.41, oil, bp. 119–120 °C (1.33 Pa). Sex pheromone of the female silk worm (*Bombyx mori*); for humans odorless. For silk worm males still detectable at a dilution of 10^{-16}–10^{-18} g/mL[1]. B. was the first *pheromone to be identified; it was isolated by A. Butenandt from the scent glands of over 500000 female insects[2]. The corresponding aldehyde (*bombykal*, $C_{16}H_{28}O$, M_R 236.40) and the (*E*,*E*)-stereoisomer are also contained in the female's scent glands[3].

Lit.: [1]Z. Naturforsch. B **14**, 283 f. (1959). [2]Naturwissenschaften **57**, 23–28 (1970). [3]Angew. Chem. Int. Ed. Engl. **17**, 60 (1978).

gen.: ApSimon **9**, 59 ▪ Beilstein EIV **1**, 2295 ▪ Biosci. Biotechnol. Biochem. **57**, 2144 ff. (1993); **59**, 560 ff. (1992); **60**, 369 ff. (1996) ▪ Hummel & Miller (eds.), Techniques in Pheromone Research, p. 1–44, Berlin: Springer 1984 ▪ Zool. Sci. **13**, 79–87 (1996). – *[CAS 765-17-3 (B.); 63024-98-6 (bombykal); 965-19-5 ((E,E)-stereoisomer)]*

Bonellin.

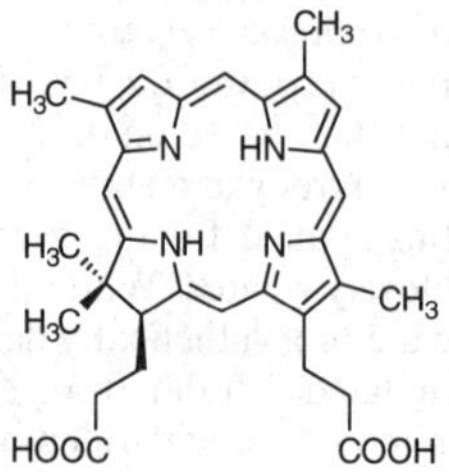

$C_{31}H_{34}N_4O_4$, M_R 526.64. Green-black crystals that decompose at ca. 300 °C. uv$_{max}$ (6% HCl) 402, 636 nm. Readily soluble in methanol, dichloromethane, moderately soluble in ether. B. belongs to the structural class of the *chlorins.

Occurrence: B. is the gender differentiating green pigment of *Bonellia viridis*, an echiuroid worm widely distributed in the Mediterranean Sea. It is obtained by methanol extraction.

Lit.: Chem. Rev. **94**, 327 (1994) ▪ J. Chem. Soc., Chem. Commun. **1983**, 1237 ▪ Justus Liebigs Ann. Chem. **1990**, 415; **1991**, 709–725. – *[CAS 62888-19-1]*

Bongkrekic acid.

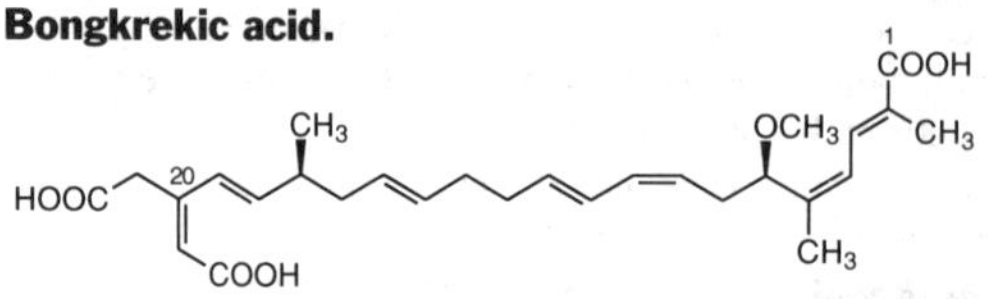

$C_{28}H_{38}O_7$, M_R 486.61, colorless, amorphous powder, melting range 50–60 °C, $[\alpha]_D^{22}$ +165° (aqueous $NaHCO_3$ solution), acid-labile toxin from *Pseudomonas cocovenenans*, that occurs in partially defatted coconuts. The name is derived from "bongkrek", an Indonesian coconut recipe fermented with *Rhizopus oryzae* but which is highly poisonous when infiltrated with *P. cocovenenans* on storage. Symptoms such as headache, vertigo, and spasms start within a few hours of consumption, coma and death occur after 1–2 days after sufficiently high doses. *Isobongkrekic acid*[1], mp. 190 °C (decomp.), is formed from B. on treatment with alkali, but is also formed by an as yet unidentified microorganism. It differs from B. by the (20*E*)-configuration. B. acts as an inhibitor for the transport of ADP through the inner mitochondrial membrane [LD_{50} for B. (mouse i. v.) 1.4 mg/kg].

Lit.: [1]J. Org. Chem. **53**, 4883 (1988).

gen.: Beilstein EIV **3**, 1285 ▪ Tetrahedron **29**, 1541 (1973). – *Pharmacology:* Biochem. Biophys. Res. Commun. **39**, 363 (1970) ▪ J. Biol. Chem. **245**, 1319 (1970). – *Synthesis:* J. Am. Chem. Soc. **106**, 462 (1984). – *[CAS 11076-19-0 (B.); 60132-21-0 (iso-B.)]*

Borneols (1,7,7-trimethylbicyclo[2.2.1]heptan-2-ol, bornan-2-ol).

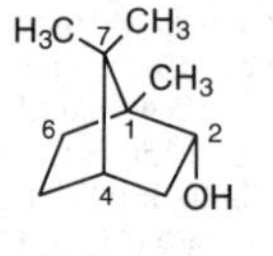

(-)-Borneol

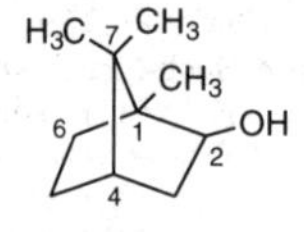

(+)-Isoborneol

$C_{10}H_{18}O$, M_R 154.25, cryst. from petroleum ether, bicyclic *monoterpene alcohols with bornane structure that occur in nature in four optically active and two racemic forms: (+)-(1*R*)- and (–)-(1*S*)-*endo* form [(+)- and (–)-B.], mp. 208–209 °C, $[\alpha]_D$ ±37° to ±38° (C_2H_5OH); (+)-(1*S*)- and (–)-(1*R*)-*exo* form [(+)- and (–)-isoborneol], mp. 212–214 °C, $[\alpha]_D$ ±33° to ±34°; (±)-B., mp. 212 °C; (±)-iso-B., mp. 210–215 °C.

Occurrence: (+)- and (–)-B. in free and ester form have been detected in more than 150 plant families, most frequently in the Pinaceae. Crystalline (+)-B. occurs in the hollow spaces in the wood of *Dryobalanops aromatica* (a tree of the Dipterocarpaceae endemic in Sumatra and Borneo). (+)- and (–)-iso-B. also occur occasionally.

Use: Both B. and iso-B. as well as their acetylated derivatives are used in the production of bath salts, soaps, and sprays with pine needle odors.

Lit.: Beilstein EIV **6**, 281 ▪ Karrer, No. 312, 313 ▪ Merck-Index (12.), No. 1366 ▪ Ullmann (5.) **A 11**, 171. – *[HS 2906 19; CAS 464-43-7 ((+)-B.); 464-45-9 ((–)-B.); 16725-71-6 ((+)-iso-B.); 10334-13-1 ((–)-iso-B.)]*

Bornyl acetate see juniper berry oil and fir and pine needle oils.

Boromycin.

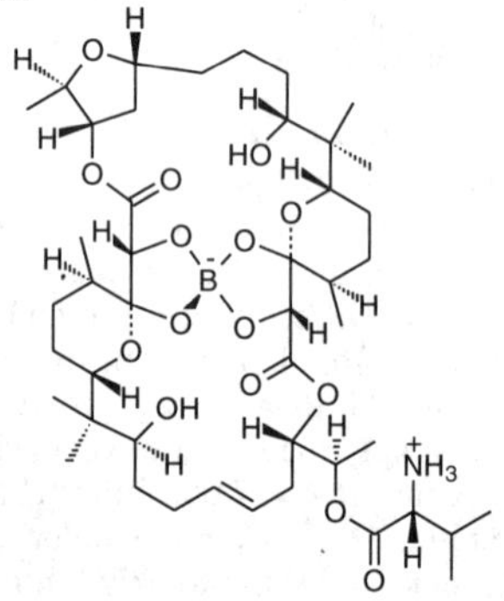

$C_{45}H_{74}BNO_{15}$, M_R 879.89, cryst., mp. 223–228 °C, soluble in organic solvents, $[\alpha]_D$ +63.5° ($CHCl_3$). B. is isolated from the culture broth of *Streptomyces antibioticus*[1] and was the first known natural product containing boron. The biosynthesis from D-valine as well as polyketide units has been investigated[2]. B. is active against Gram-positive bacteria and exhibits coccidiostatic properties.

Lit.: [1]Helv. Chim Acta **50**, 1533–1539 (1967). [2]J. Org. Chem. **46**, 2661–2665 (1981).

gen.: Helv. Chim. Acta **54**, 1709–1713 (1971) ▪ J. Antibiot. **38**, 1444 ff. (1985) ▪ J. Org. Chem. **46**, 2661 (1981) ▪ Tetrahedron Lett. **25**, 3671–3674 (1984). – *[CAS 34524-20-4]*

Boschnaloside. B. belongs to the *iridoid glucosides with a C_{10} aglycone possessing an aldehyde function. $C_{16}H_{24}O_8$, M_R 344.36, amorphous, $[\alpha]_D$ –104° (CH_3OH).

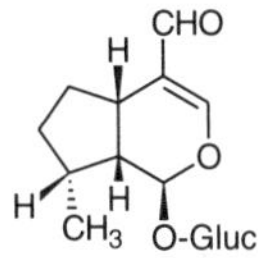

Occurrence: B. was first isolated from *Boschniakia rossica* (Orobanchaceae)[1]. It was later found in the tropical American endemic plant *Leucocarpus perfoliatus* (Scrophulariaceae)[2]. B. exhibits, like nepetalactone (see iridodial) and the *valepotriates a strongly attractive action on cats and other cat-like predatory animals[3].

Lit.: [1]Inouye, in Wagner & Hörhammer (eds.), Pharmacognosy and Phytochemistry, p. 290, New York: Springer 1971. [2]Helv. Chim. Acta **62**, 2708 (1979). [3]Hegnauer **V**, 252.
gen.: Chem. Pharm. Bull. **28**, 1730 (1980) ▪ Phytochemistry **20**, 2717 (1981). – *[HS 2938 90; CAS 72963-55-4]*

Boseiomycins see streptothricins.

Boswellic acids. Constituents of frankincense (*Boswellia carterii, B. frereana*) and Olibanum oil.

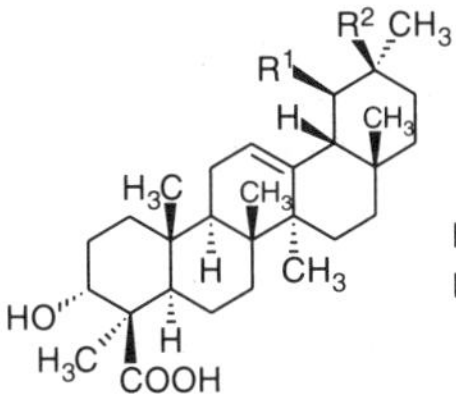

α-B. {$C_{30}H_{48}O_3$, M_R 456.70, cryst., mp. 289 °C, $[\alpha]_D^{20}$ +114.5° ($CHCl_3$)} or *β-B.* {cryst., mp. 228–232 °C (238–240 °C), $[\alpha]_D$ +119° ($CHCl_3$)}, found in frankincense and hops may be responsible for the antitremor activity of frankincense.

Lit.: Nature (London) **390**, 667 (1997) ▪ Phytochemistry **39**, 453 (1995). – *[CAS 631-69-6 (β-B.); 471-66-9 (α-B.)]*

Botox® see botulism toxin.

Botrydial.

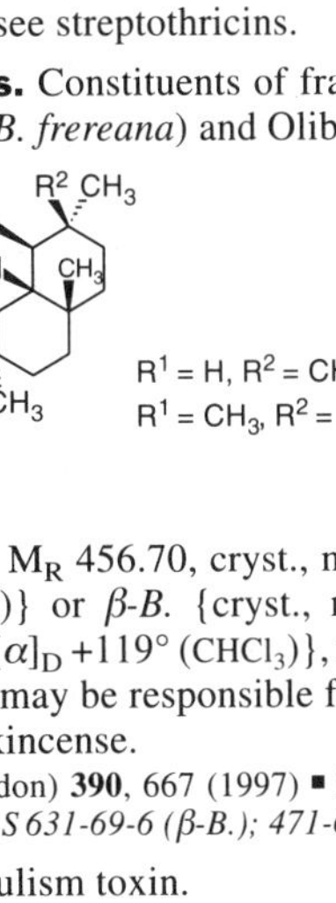

$C_{17}H_{26}O_5$, M_R 310.39, cryst., mp. 108–110 °C, $[\alpha]_D^{20}$ +34° ($CHCl_3$), soluble in benzene, chloroform, and methanol. Metabolite of the grey mold fungus *Botrytis cinerea*. It has antibiotic and fungistatic properties[1]. B. is chemically unstable, and very sensitive to light and heat. Biosynthesis from mevalonic acid[2].

Lit.: [1]Chem. Ber. **107**, 1720 (1974). [2]J. Chem. Soc., Perkin Trans. 1 **1982**, 2187.
gen.: J. Chem. Soc., Chem. Commun. **20**, 924 (1979) ▪ Phythopathol. Z. **63**, 193 (1968) ▪ Tetrahedron Lett. **26**, 5433 (1985). – *[CAS 54986-75-3]*

Botulism toxin, botulin (botulinum toxin). Botulism is today a very infrequently occurring form of food poisoning by B., a metabolite of *Clostridium botulinum*. This rod-like bacterium develops anaerobically in conserved meat products, occasionally also in conserved vegetable or fish products as a result of an inadequate addition of salt or (in case of fish) of inadequate acidification. B. consists of a single polypeptide chain with a molecular mass of ca. 150 000. 7 types of B. are known, which differ immunologically and in their molecular masses (from ~90 000–180 000). On account of the extreme toxicity [LD_{50} (mouse) 0.00003 μg/kg, i.e., 1 g of toxin would probably be sufficient to kill several million people] the toxin has been considered in the context of biological warfare; however fortunately the lability of the toxin renders it unsuitable for such uses (it is denaturized by 30 min heating at 80 °C). Occasional cases of B. poisoning are managed by injection of polyvalent B. antitoxin. Usually B. is carried by the blood circulation to the motor end-plates of peripheral transverse striated muscles. There, it inhibits the liberation of acetylcholine from the nerve endings. Thus muscles do not receive signals to contract and flaccid paralysis occurs. The toxic symptoms usually occur after 12–40 h (first headache and abdominal pains, later swallowing, speech, and visual impairments). In the absence of treatment death occurs after 3–6 d in over 50% of the cases through paralysis of the respiratory system or cardiac arrest. Besides man, animals are also afflicted by botulism. Thus bovine botulism in South Africa causes high losses in cattle. In recent years botulism was assumed to be the cause of mass deaths of birds in shallow waters when oxygen deficiency occurred in long periods of high temperature: aqueous plants died and rapid reproduction of *C. botulinum* took place.
B. A (Botox®, Dysport®) has an M_R of ca. 90 000 and is used in extremely small doses in ophthalmology to treat idiopathic blepharospasm (squinting) and other spasms such as hemifacial spasm, facial paralysis, TMJ (temporomandibular joint disorder), writer's cramp, extreme sweating, etc.[1]. Recent clinical data show, that B. is effective in the treatment of essential hand tremor and upper limb spasticity after a stroke, as well as in certain cases of headache/migraine.

Lit.: [1]J. Neurology, Neurosurgery and Psychiatry **64**, 751 (1998).
gen.: Ullmann (5.) **A 17**, 147. – *[CAS 107231-12-9]*

Bougainvilleins. B. belong to the betacyans (see betalains). They are the 5-*O*- and 6-*O*-sophorosides of *betanidin and their acyl derivatives (esterified with hydroxycinammic acid). They occur in *Bougainvillea* species (Nyctaginaceae) and are responsible for the intense colors of the bracts. The most complex B. structure is that isolated from *B. glabra*, namely 6-*O*-(6-*O*-caffeoyl-6′-*O*-*p*-coumaroyl-2′-*O*-sophorosyl-*β*-sophoroside)[1].

Lit.: [1]Phytochemistry **37**, 761 (1994).
gen.: Manske **39**, 15, 60.

Bourbonal see 3,4-dihydroxybenzaldehydes.

Boviquinones.

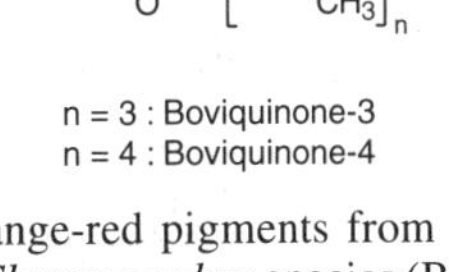

Group of orange-red pigments from *Suillus bovinus* and various *Chroogomphus* species (Basidiomycetes).

The pigments consist of a 2,5-dihydroxy-1,4-benzoquinone nucleus substituted in the 3 position by a polyprenyl side chain. Depending on the number of isoprene units in this side chain one speaks of *B.-3* (n=3=farnesyl-: $C_{21}H_{28}O_4$, M_R 344.45, orange-yellow cryst., mp. 98–100 °C) and *B.-4* (n=4=geranylgeranyl-: $C_{26}H_{36}O_4$, M_R 412.57, yellow cryst., mp. 84–85 °C). With alkali the B. furnish dark red anions (typical color reaction of fungal tissue with ammonia vapor). Dimers can be formed by direct 6,6′-linkage, bridging through a 6,6′-methylene group or condensation to bovilactones. For ansaquinones related to B.-4, see tridentoquinone.

Lit.: Tetrahedron Lett. **39**, 5167 (1998) (biosynthesis) ▪ Zechmeister **51**, 98 ff. (review). – *[HS 3203 00; CAS 34198-83-9 (B.-3); 28129-52-4 (B.-4)]*

Bovolide see tobacco flavor.

Boxazomycins. Benzoxazoles from *Pseudonocardia* sp. with activity against Gram-positive bacteria and anaerobic organisms, including resistant strains with clinical relevance. *B. A*: $C_{14}H_{12}N_4O_5$, M_R 316.27, yellow needles, mp. >275 °C (decomp.); *B. B*: $C_{14}H_{12}N_4O_4$, M_R 300.27, yellow powder, mp. >270 °C (decomp.).

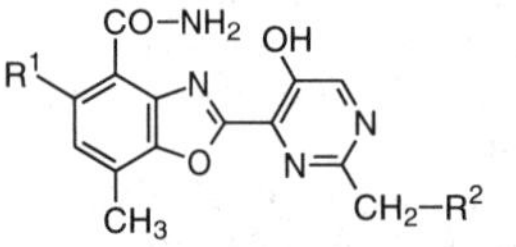

R^1 = OH, R^2 = OH : B. A
R^1 = OH, R^2 = H : B. B
R^1 = H, R^2 = OH : B. C

Lit.: J. Am. Chem. Soc. **110**, 2954 (1988) ▪ Tetrahedron Lett. **36**, 7213 (1995). – *[CAS 107021-64-7 (B. A); 107021-65-8 (B. B); 107021-66-9 (B. C)]*

Bracken (*Pteridium aquilinum*). Cosmopolitan fern occurring everywhere except for polar and extreme drought regions and in temperate South America. The entire plant is toxic and cancerogenic.

Components: Numerous *pterosins and pterosides [1,2] as well as the mutagenic *ptaquiloside* (aquilide A) (see pterosins) have been isolated from the rhizomes and leaves; the latter is responsible for the phytotoxicity and cancerogenicity. Leaves and rhizomes contain an acidic mucus, aquilinan (contains galactose, fucose, xylose, mannose, glucuronic acid, and arabinose), as well as many organic acids and free, proteinogenic and non-proteinogenic amino acids. B. contains prunasin and *cyanogenic glycosides. Numerous polyphenols such as cinammic acid glycosides [3] and their sulfate esters, flavonols, and phytoecdysone are found in all organs.

Use: In spite of their toxicity and cancerogenicity, the starch-rich rhizomes were once the main food of the New Zealand Maoris. In Japan the young leaves of the variety *latiusculum* ("Warabi") are a favored vegetable. Leaves are used in a tea for rheumatism [4], leafstalk bases are used in Portugal as a tapeworm remedy [5]. Leaf fronds are also used as a gritting material.

Lit.: [1] Chem. Pharm. Bull **30**, 3640 (1982). [2] Tetrahedron Lett. **24**, 4117, 5371 (1983). [3] Phytochemistry **31**, 3489 (1992). [4] Pharm. Acta Helv. **8**, 138 (1933). [5] Ann. Fac. Port. **11**, 5 (1951).

gen.: Bot. J. Linn. Soc. **73** (1976) ▪ Hager (5.) **6**, 295, 305 ▪ Kunkel, Plants for Human Consumption, p. 297, Koenigstein: Koeltz Scientific Books 1984 ▪ Science **159**, 1472 (1968). – *[HS 1211 90]*

Brain lipids see oleamide.

Brassicasterol see sterols.

Brassinin.

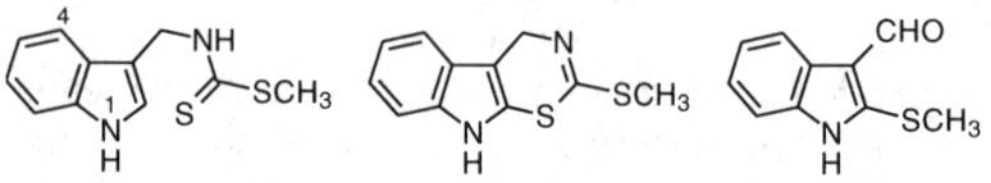

Brassinin (**1**) Cyclobrassinin (**2**) Brassicanal A (**3**)

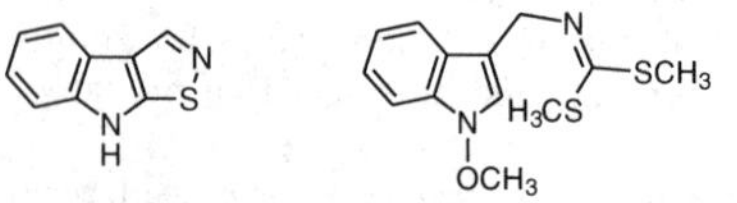

Brassilexin (**4**) Methoxybrassenin A (**5**)

Cruciferae (Brassicaceae), e. g., cabbage, rape, radish, protect themselves against microbial attack by production of unusual, sulfur-containing indole derivatives. Thus, on inoculation of *Brassica oleracea* var. *capitata* (white cabbage) with *Pseudomonas cichorii* the *phytoalexins *brassinin* ($C_{11}H_{12}N_2S_2$, M_R 236.35, mp. 132–133 °C) [1], *1-methoxybrassinin* (methoxybrassinin, $C_{12}H_{14}N_2OS_2$, M_R 266.38, oil) [2], *4-methoxybrassinin* ($C_{12}H_{14}N_2OS_2$, M_R 266.38, amorphous) [3], *methoxybrassenin A* ($C_{13}H_{16}N_2OS_2$, M_R 280.40, amorphous) [4] and cyclic compounds such as *cyclobrassinin* ($C_{11}H_{10}N_2S_2$, M_R 234.33, mp. 136–137 °C) [1,5] and *spirobrassinin* (see oxindole alkaloids) are formed [3,6]. In *B. campestris* subsp. *pekinensis* (Chinese cabbage) besides B. and other phytoalexins *brassicanal A* ($C_{10}H_9NOS$, M_R 191.25, mp. 210–213 °C) [7] is formed. 4-Methoxybrassinin and brassicanal A inhibit the germination of conidia of *Bipolaris leersiae* and other fungi. Spirobrassinin is also a phytoalexin of radish, *Raphanus sativus* var. *hortensis*. [6] From the leaves of *Brassica juncea*, a mustard species that is resistant to the fungus *Leptosphaeria maculans*, cyclobrassinin [9] and its sulfoxide as well as the unusual isothiazole derivative *brassilexin* ($C_9H_6N_2S$, M_R 174.22, mp. 164–167 °C) [8] have been isolated. B. and the other Brassicaceae phytoalexins are formed biosynthetically from the mustard oil glucosides glucobrassicin [9] and neoglucobrassicin (see glucosinolates) [10]. After cleavage of glucose and rearrangement, the corresponding 3-isothiocyanatomethyl indoles could be formed and have therefore been discussed as precursors [11].

Lit.: [1] J. Chem. Soc., Chem. Commun. **1986**, 1077; Heterocycles **36**, 2783 (1993); Tetrahedron **54**, 3549 (1998) (synthesis). [2] Heterocycles **31**, 1605 (1990); **33**, 77 (1992). [3] Phytochemistry **29**, 1499 (1990). [4] Phytochemistry **30**, 3921 (1991). [5] Phytochemistry **29**, 1087 (1990). [6] Chem. Lett. **1987**, 1631. [7] Chem. Lett. **1990**, 209; Heterocycles **34**, 1877 (1992) (synthesis). [8] Tetrahedron Lett. **29**, 6447 (1988); J. Chem. Soc., Perkin Trans. 1 **1990**, 2856; Synthesis **1990**, 214 (synthesis); Bioorg. Med. Chem. Lett. **8**, 3037 (1998). [9] Experientia **21**, 520 (1965); Phytochemistry **9**, 1629 (1970); **29**, 1315 (1990); Tetrahedron Lett. **31**, 1417 (1990); J. Chem. Res. **1992** (*M*), 1669; (*S*), 207. [10] Acta Chem. Scand. **16**, 1378 (1962). [11] J. Am. Chem. Soc. **116**, 6650 (1994); Z. Naturforsch. **49**, 281 (1994).

gen.: Synlett **1997**, 289 (synthesis). – *[CAS 105748-59-2 (1); 105748-58-1 (2); 113866-44-7 (3); 119752-76-0 (4); 105748-60-5 (1-methoxy-B.); 129602-03-5 (4-methoxy-B.)]*

Brassinosteroids. B. represent a new class of steroidal plant hormones ubiquitously distributed in the plant kingdom. Since the characterization of the first member of this group of active principles, *brassinolide*, from rape pollen (*Brassica napus*)[1], intensive investigations have led to the isolation and structure elucidation of over 30 different B. from numerous plant species.

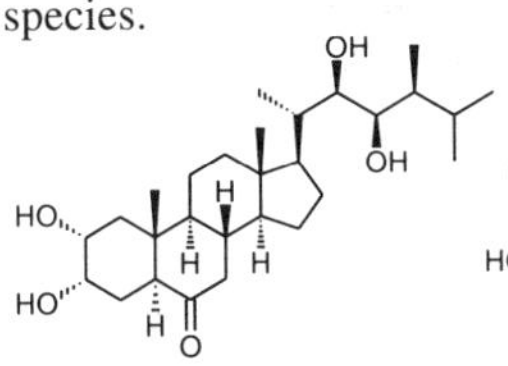

Castasterone

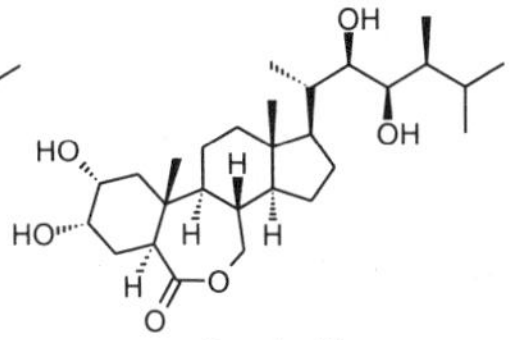

Brassinolide

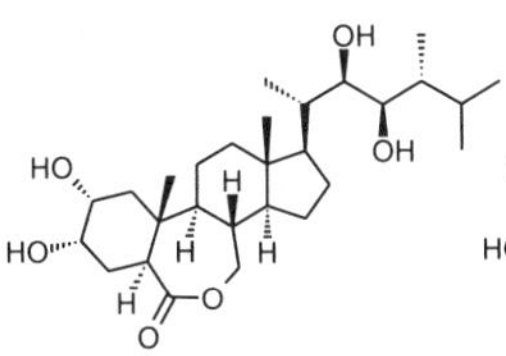

24-Epibrassinolide

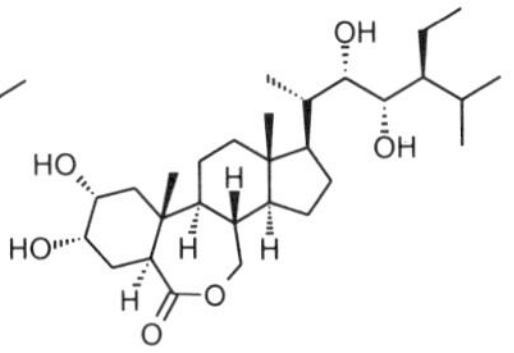

(22*S*,23*S*)-Homobrassinolide

Some of the most important naturally occurring members are *brassinolide* {$C_{28}H_{48}O_6$, M_R 480.69, mp. 274–275 °C, $[\alpha]_D$ +41.9° ($CHCl_3/CH_3OH$, 9 : 1)}, *24-epibrassinolide* {$C_{28}H_{48}O_6$, M_R 480.69, mp. 256–258 °C, $[\alpha]_D$ +30° ($CHCl_3/CH_3OH$, 9 : 1)}, and *castasterone* {$C_{28}H_{48}O_5$, M_R 464.69, mp. 252–255 °C, $[\alpha]_D$ –4° ($CHCl_3$)}.
Particularly rich in B. are pollens (for *Brassica napus* ca. 100 µg/kg). B., especially brassinolide and 24-epibrassinolide, have unusually high phytohormonal activity. B. effect a general promotion of growth in plants (e. g., in roots), stimulate photosynthesis and protein biosynthesis, modulate the mode of action of other phytohormones, and delay senescence processes. B. exhibit antiecdysteroid effects on insects. For practical uses of B. the so-called "anti-stress" action on plants is of major interest and has been detected in field experiments under unfavorable cultivation conditions (e. g., drought, cold, nutrient deficiency, pest infestation).

Lit.: [1]Nature (London) **281**, 216f. (1979); Tetrahedron **55**, 8341 (1999).
gen.: Annu. Rev. Plant Physiol. Plant Mol. Biol. **39**, 23–52 (1988) ▪ Atta-ur-Rahman **16**, 321–364; **18**, 495–550 ▪ Biosci., Biotechnol. Biochem. **61**, 757 (1997) (biosynthesis review) ▪ Bohl & Duax (eds.), Molecular Structure and Biological Activity of Steroids, p. 317–339, Boca Raton: CRC Press 1992 ▪ Cell **90**, 825 ff. (1997) (biosynthesis review) ▪ Chem. Nat. Compd. **33**, 389–416 (1997) (synthesis review) ▪ Cutler, Advances in the Use of Brassinosteroids (ACS Symp. Ser. **551**, p. 85–102), Washington: Am. Chem. Soc. 1994 ▪ Cutler, Yokota & Adam (eds.), Brassinosteroids, Chemistry, Bioactivity, and Applications (ACS Symp. Ser. **474**), Washington: Am. Chem. Soc. 1991 ▪ Ebing (ed.), Chemistry of Plant Protection, p. 103–139, Heidelberg: Springer 1991 ▪ J. Chem. Res. (S) **1997**, 360 (synthesis), 418 (biosynthesis) ▪ J. Chem. Soc., Perkin Trans. 1 **1996**, 295–302 (synthesis of brassinolide) ▪ Magn. Reson. Chem. **35**, 629 (1997) (conformation) ▪ Phytochemistry **25**, 1787–1799 (1986) ▪ Pure Appl. Chem. **62**, 1319 ff. (1990) ▪ Sembdner, Schneider & Schreiber (eds.), Methoden zur Pflanzenhormonanalyse, Berlin: Springer 1988 ▪ Stud. Nat. Prod. Chem. **19 E**, 245–287 (1997) (synthesis) ▪ Tetrahedron **52**, 2435–2448 (1996) ▪ Zechmeister **78**, 1–46. – *[CAS 72962-43-7 (brassinolide); 78821-43-9 (24-epibrassinolide); 80736-41-0 (castasterone)]*

Brazilin.

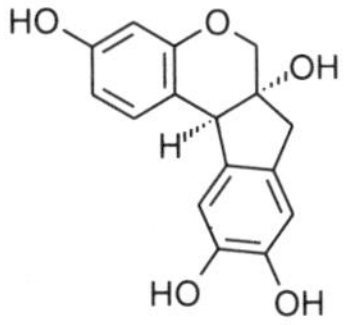

Brazilin
(relative configuration)

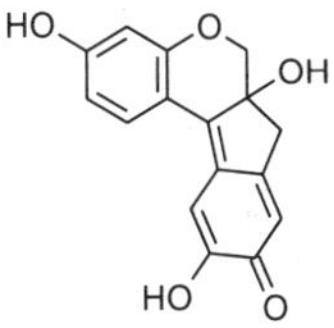

Brazilein

$C_{16}H_{14}O_5$, M_R 286.28, pale yellow needles, mp. 250 °C. B. is together with *brazilein* ($C_{16}H_{12}O_5$, M_R 284.27) the main colored component of the redwoods [genuine Brazilian wood (*Caesalpinia brasiliensis*), sappanwood (*C. sappan*), Nicaragua wood (*C. echinata*), Jamaica redwood (*Haematoxylum brasiletto*), pernambucco wood (*C. crista*), all Fabaceae], whose heart woods are used in wool and cotton dying. B. was first isolated in 1808 by Chevreul from brazilwood. From brazilwood the South American state Brazil (Terra de Brazile) got its name.

Lit.: Beilstein E V **19/4**, 278 ▪ Bull. Soc. Chim. Fr. **1975**, 1770 (synthesis) ▪ Karrer, No. 1699 ▪ Melliand Textilberichte **1997**, 733 ▪ Phytochemistry **46**, 177 (1997) (NMR brazilein) ▪ Rowe, p. 847. – *[HS 3203 00; CAS 474-07-7 (B.); 600-76-0 (brazilein)]*

Brazil wax see carnauba wax.

Brazzein. Polypeptide of 54 amino acids, M_R 6473, from the west African tree *Pentadiplandra brazzeana* (Tiliaceae). In a 2% solution it is 2000 times sweeter than cane sugar and shows good thermal stability (no loss of activity after 4 h at 80 °C), compare with monellin, thaumatin. B. can also be obtained by synthesis (solid phase method)[1] or biotechnologically in genetically engineered *Escherichia coli* cultures[2].

Lit.: [1]Biopolymers **39**, 95–101 (1996). [2]Pat. WO 9531 547 (publ. 23.11.1995).
gen.: FEBS Lett. **355**, 106 ff. (1994).

Bread flavor. *White bread:* During the baking process the impact substances *2-acetyl-1-pyrroline and 2-acetyl-Δ^1(Δ^2)-tetrahydropyridine form in the crust, on storage their concentrations decrease rapidly. Precursors are metabolites of yeast. Also of importance are (*E*)-2-nonenal (see alkenals), 3-methylbutanal (see alkanals), *2,3-butanedione, and *methional. In bread crumbs, degradation products of linolic acid such as (*E*)- and (*Z*)-2-nonenal and (*E,E*)-*2,4-decadienal predominate.
Rye bread: In comparison to white bread the crust only contains small amounts of 2-acetyl-1-pyrroline, but instead more 3-methylbutanal, methional, 2,3-butanedione, 2-ethyl-3,5-dimethylpyrazine (see pyrazines), *Furaneol®, *phenylacetaldehyde and *4-vinylguaiacol.

Lit.: J. Agric. Food Chem **44**, 1515 (1996) ▪ Maarse, p. 41–77 ▪ Ohloff.

Brefeldins.

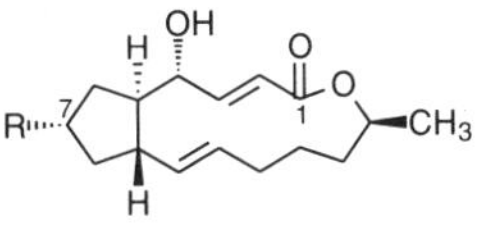

R = OH : B. A
R = H : B. C

Group of unusual *macrolide antibiotics obtained, e. g., from *Penicillium brefeldianum* with a broad spectrum of biological activities (antifungal, antiviral, nematicidal, cytotoxic, fish toxic effects, inhibitor of wheat germination). The most important compound is *B. A* {$C_{16}H_{24}O_4$, M_R 280.37, prisms, mp. 204–205 °C, $[\alpha]_D$ +96° (CH_3OH)}, a macrolactone with a 13-membered ring and annelated cylopentane ring; slightly reminiscent of the *prostaglandins. B. A is an inhibitor of nucleic acid and protein biosynthesis, it also exhibits immunological effects that might be of interest for transplantations. The 7-OH-group of B. A can be epimerized or glycosylated with β-D-galactose. *B. C* differs from B. A by the lack of the OH-group on the cyclopentane ring ($C_{16}H_{24}O_3$, M_R 264.36, cryst., mp. 161 °C).

Lit.: Beilstein E V **18/3**, 49 (B. A); 238 (B. C) ▪ Cole-Cox, p. 844 (review) ▪ J. Chem. Soc., Perkin Trans. 1 **1982**, 2387 ▪ J. Nat. Prod. **57**, 541 ff. (1994). – *Biosynthesis:* J. Am. Chem. Soc. **108**, 2448 (1986). – *Synthesis:* J. Am. Chem. Soc. **108**, 284 (1986); **110**, 5198 (1988) ▪ J. Org. Chem **58**, 4555 (1993); **60**, 6670 (1995); **62**, 4552 (1997) [(+)-B. A] ▪ Tetrahedron Lett. **30**, 4845 (1989); **39**, 5377 (1998). – *Activity:* Nature (London) **339**, 223 (1989); **346**, 63 (1990). – *[CAS 20350-15-6 (B. A); 73899-78-2 (B. C)]*

Brevetoxins (BTX).

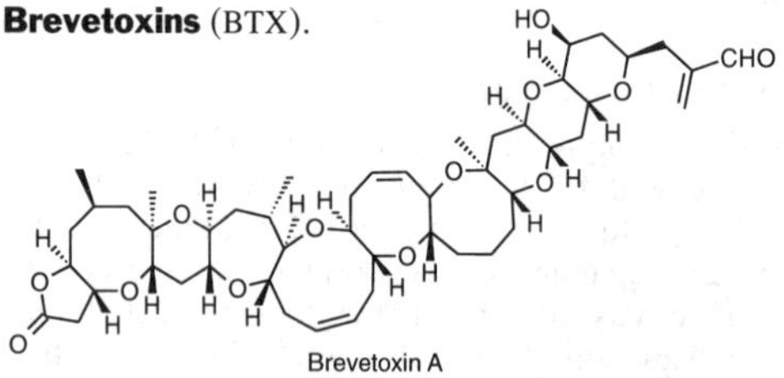

Brevetoxin A

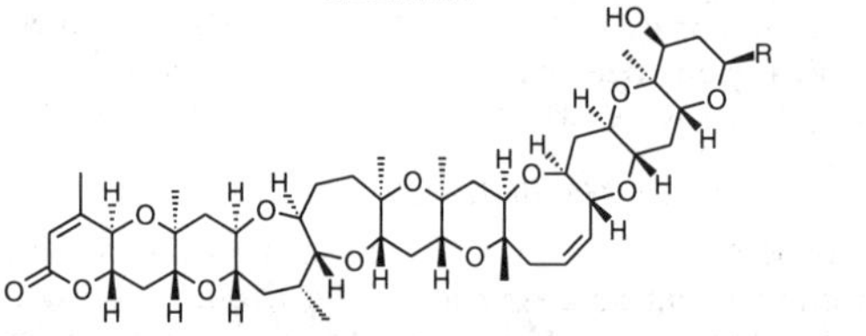

R = CH_2–C(=CH_2)–CHO : Brevetoxin B R = CH_2–C(=O)–CH_2–Cl : Brevetoxin C

The B. are linear polycyclic neurotoxins with unusual structures from the dinoflagellate *Ptychodiscus brevis* (*Gymnodinium breve*), a unicellular alga which occurs in the Gulf of Mexico and along the coasts of Florida (massive occurrence in the "*red tide"). B. are responsible for extensive death of fish (e. g., death of humpback whales near Cape Cod, Mass. 1987) and mollusk poisoning as well as fish poisoning in man after consumption of contaminated animals; in this respect they are comparable with *saxitoxins and gonyautoxins from dinoflagellates but, in contrast to these, they are not soluble in water. B. were first obtained in crystalline form in 1981 from cultures of algae[1]. *B. B* [BTX-B, GB-2, $C_{50}H_{70}O_{14}$, M_R 895.10, cryst., mp. 295–297 °C, LD_{100} (zebra fish) 16 ng/mL water] is contained in ca. 10-fold higher concentration than *B. A* [BTX-A, GB-1, $C_{49}H_{70}O_{13}$, M_R 867.09, fine prisms, double mp. 197–199 °C and 218–220 °C, LD_{100} (zebra fish) 3 ng/mL water] and *B. C* [$C_{49}H_{69}ClO_{14}$, M_R 917.53, LD_{100} (zebra fish) 60 ng/mL water], see also yessotoxin and ciguatoxin.
B. are activators of potential-dependent sodium channels in cell membranes; their activity is counteracted by *tetrodotoxin[2]. It is assumed that the conformational changes of B. have direct influence on the opening mechanism of the ion channels[3].

Lit.: [1] J. Am. Chem. Soc. **108**, 514 (1986). [2] J. Biol. Chem. **479**, 204 (1986); FEBS Lett. **219**, 355 (1987); J. Org. Chem. **53**, 4156 (1988) (structure). [3] Tetrahedron Lett. **33**, 525 (1992); J. Am. Chem. Soc **108**, 7855 (1986); **109**, 2184 (1987) (biosynthesis).
gen.: Chem. Rev. **93**, 1897 (1993) ▪ New J. Chem. **11**, 753 (1987) ▪ Tetrahedron **54**, 735 (1998) (isolation B. B2) ▪ Tetrahedron Lett. **36**, 8995 (1995) (structure) ▪ Toxicon **26**, 97 (1988). – *Synthesis:* Angew. Chem. Int. Ed. Engl. **35**, 588–607 (1996) ▪ Chem.-Eur. J. **5**, 599, 618, 628, 646 (1999) (B. A) ▪ Chemtracts **11**, 1041–1052 (1998) (B. A) ▪ J. Am. Chem. Soc. **117**, 1171, 1173, 10227–10263 (1995) ▪ Nature (London) **392**, 264–269 (1998) (B. A) ▪ Nicolaou, p. 731–786. – *[CAS 98112-41-5 (B. A); 79580-28-2 (B. B); 82983-92-4 (B. C)]*

Brevicomin (7-ethyl-5-methyl-6,8-dioxabicyclo-[3.2.1]octane).

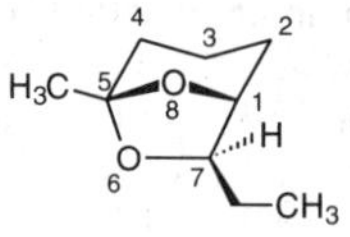

(+)-(1*R*)-7*exo*-form

$C_9H_{16}O_2$, M_R 156.22, liquid with a flowery odor, bp. 95–100 °C (14.67 kPa), n_D^{20} 1.4372. Aggregation pheromone of *bark beetles of the genera *Dendroctonus*, *Dryocoetes*, and *Leperisinus*. B. occurs as *exo-* and *endo*-isomers in nature. In some species the substance is expressed by females, in others by males and acts as an attractant to both sexes. The insects apparently produce different enantiomer ratios in order to increase the information content of the volatile signal. Males of the six-dentated bark beetle *Dendroctonus ponderosae* secrete (1*R*)-7-*exo*-B. of high optical purity, $[\alpha]_D^{26}$ +84.1° (diethyl ether). The biosynthesis of B. appears to proceed through 6-nonen-2-one ($C_9H_{16}O$, M_R 140.23)[1]. *endo*-B. has also been detected in the fragrance of the orchid *Ophrys speculum*. B. belongs to the same class of bicyclic acetals as *frontalin and *multistriatin. *3,4-Didehydrobrevicomin* (7-ethyl-5-methyl-6,8-dioxabicyclo[3.2.1]oct-3-ene, $C_9H_{14}O_2$, M_R 154.21) is a *pheromone of the house mouse *Mus musculus*[2].

Lit.: [1] J. Chem. Ecol. **18**, 1389–1404 (1992). [2] Experientia **40**, 217 ff. (1984).
gen.: Zechmeister **37**, 1–190 (review). – *Synthesis:* ApSimon **4**, 150 ff.; **9**, 394–424 ▪ Chem. Lett. **1989**, 1755 f. [(+)-B.] ▪ J. Chem. Soc., Chem Commun. **1996**, 1477 [(+)-B.] ▪ J. Chem. Soc., Perkin Trans. 1 **1990**, 1375–1382 ▪ J. Org. Chem. **64**, 2524 (1999) ▪ Justus Liebigs Ann. Chem. **1990**, 123 f. ▪ Org. Lett. **1**, 1827 (1999) ▪ Tetrahedron: Asymmetry **8**, 779–800 (1997) [(–)-B.] ▪ Tetrahedron Lett. **36**, 6253 (1995). – *[CAS 64313-75-3 (B.); 88525-42-2 (3,4-didehydro-B.); 20290-99-7 ((+)-exo-B.)]*

Brevioxime.

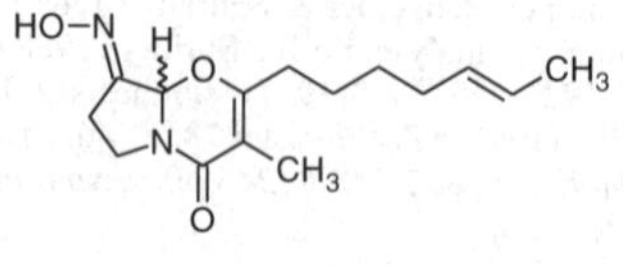

(natural product: (-)-form)

A *juvenile hormone biosynthesis inhibitor, isolated from fermentation broths of *Penicillium brevicompactum*: $C_{15}H_{22}N_2O_3$, M_R 278.35, $[\alpha]_D^{23}$ –39° ($CHCl_3$).
Lit.: Eur. J. Org. Chem. **1999**, 221–226 (synthesis) ▪ J. Org. Chem. **62**, 8544 (1997) ▪ Chem. Commun. **1999**, 2251. – *[CAS 198773-28-3]*

Breynins.

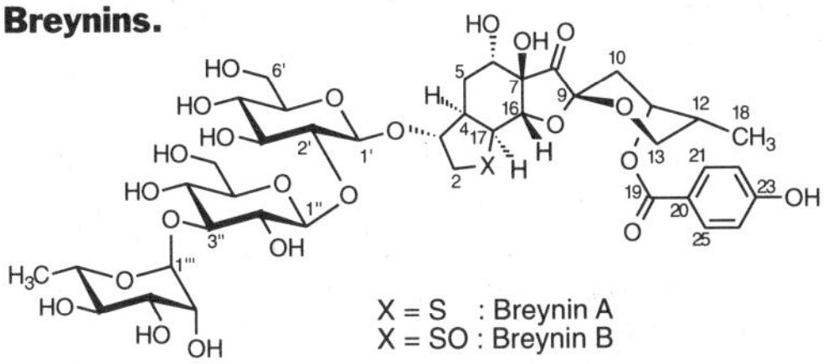

B. A: $C_{40}H_{56}O_{23}S$, M_R 936.93, mp. 195–197 °C.
B. B: $C_{40}H_{56}O_{24}S$, M_R 952.93, mp. 208–210 °C.
Sulfur-containing spiroketal glycosides with significant hypocholesteremic activity. They are isolated from the wood of the Taiwanese shrub *Breynia officinalis* (Euphorbiaceae).
Lit.: Chem. Pharm. Bull. **24**, 114, 169 (1976) ▪ IRCS Med. Sci. **14**, 905 (1986) ▪ J. Am. Chem. Soc. **112**, 4552 (1990); **114**, 9419 (1992) ▪ J. Org. Chem. **57**, 5115 (1992) ▪ Tetrahedron Lett. **14**, 2439 (1973); **34**, 4281 (1993); **40**, 9 (1999) (synthesis of aglycone Breynolide). – *[HS 2938 90; CAS 50863-74-6 (B. A); 58857-03-7 (B. B)]*

Briaranes, briarein A see erythrolides.

Broussonetine G.

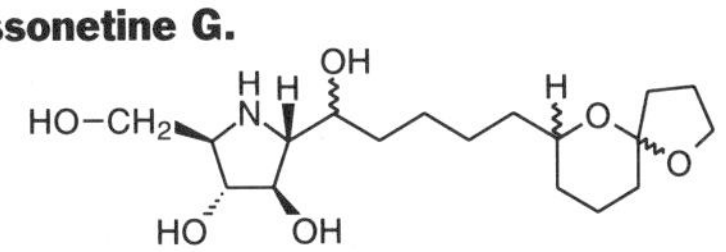

$C_{18}H_{33}NO_6$, M_R 359.46, oil, $[\alpha]_D$ +17.5° (CH_3OH). A potent inhibitor (IC_{50}<50 nM) of β-glucosidase, β-galactosidase, and β-mannosidase, isolated from the mulberry *Broussonetia kazinoki* (Moraceae). It is a lead structure for therapeutics against diabetes, cancer, and AIDS.
Lit.: Chem. Pharm. Bull. **46**, 1048 (1998). – *[CAS 198953-21-8]*

Browniine.

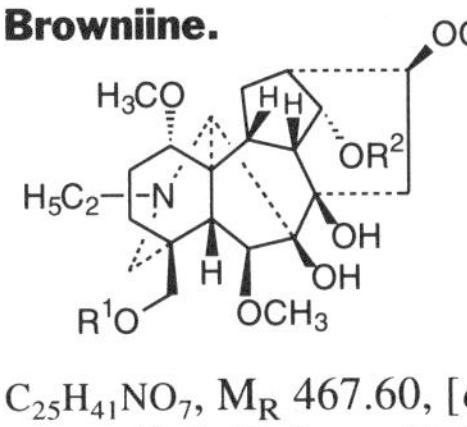

$C_{25}H_{41}NO_7$, M_R 467.60, $[\alpha]_D^{25}$ +40.2° (C_2H_5OH). Diterpene alkaloid from *Delphinium brownii* and other *Delphinium* species (Ranunculaceae). *Lycotonine*, $C_{25}H_{41}NO_7$, M_R 467.60, mp. 151–153 °C, $[\alpha]_D^{20}$ +50° (C_2H_5OH), is an alkaloid from *D. barbeyi*, *Inula royleana* (Asteraceae), *Aconitum lycoctonum*, and many other *Delphinium* and *Aconitum* species.
Lit.: Manske **12**, 2–206 (B.); **34**, 95–179 (lycoctonine). – *[HS 2939 90; CAS 5140-42-1 (B.); 26000-17-9 (lycoctonine)]*

Bruceantin.

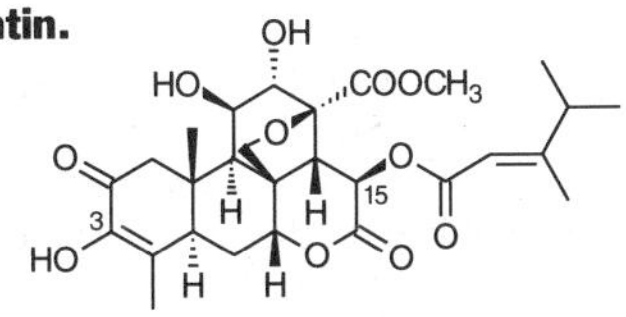

$C_{28}H_{36}O_{11}$, M_R 548.59, cryst., mp. 225–226 °C, $[\alpha]_D^{25}$ –27.7° (pyridine). Quassinoid with antileukemic activity from the tree *Brucea antidysenterica* (Simaroubaceae). LD_{50} (mouse i. v.) male: 1.95 mg/kg, female: 2.58 mg/kg.
Lit.: Beilstein E V **19/7**, 134 ▪ J. Am. Chem. Soc. **115**, 5841 (1993) ▪ Merck-Index (12.), No. 1475 ▪ Sax (8.), BOL 500 ▪ Pharmacol. Ther. **37**, 425 (1988) (activity) ▪ Zechmeister **47**, 221–264. – *[CAS 41451-75-6]*

Brucine. Formula see under strychnine. $C_{23}H_{26}N_2O_4$, M_R 394.47, mp. 178 °C (anhydrous), 105 °C (as hydrate), $[\alpha]_D$ –127° ($CHCl_3$), soluble in chloroform and ethanol. *Indole alkaloid (indoline type) from *Strychnos nux-vomica* ("poison nut") and many other *Strychnos* species (Loganiaceae). It occurs especially in the seeds together with *strychnine. Like strychnine, B. is highly toxic, lethal dose for man ~200 mg [LD_{50} (rat p. o.) 1 mg/kg], and has central stimulating activity.
Use: For photometric detection of ClO_3^-, NO_3^-, NO_2^-. As an optically active base it is used for racemate resolution of racemic acids. B. has an extremely bitter taste (bitter value 1:220 000) and serves as standard for bitter value determinations.
Lit.: Beilstein E IV **27**, 7876 ▪ Hager (5.) **6**, 817–845 ▪ Manske **1**, 375–500; **8**, 591–671; **34**, 211; **36**, 1 ▪ Martindale (30.), p. 1229 ▪ Sax (8.), No. BOL 750 ▪ Tetrahedron **24**, 4573 (1968). – *[HS 2939 90; CAS 357-57-3]*

Bryamycin see thiostrepton.

Bryoflavone.

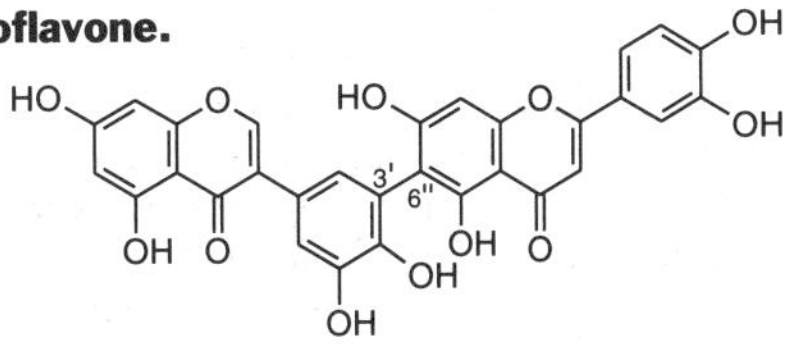

$C_{30}H_{18}O_{12}$, M_R 570.47, crystalline, yellow powder, soluble in 80% acetone, DMF, DMSO, insoluble in H_2O, mp. 250–252 °C (decomp.). *Biflavonoid isolated from the moss *Bryum capillare*, composed of one flavone and one isoflavone unit. It represents the prototype for a new class of biflavonoids which has so far not been found in any other plant family.
Lit.: Roth & Rupp (eds.), Roth Collection of Natural Products Data, Weinheim: VCH Verlagsges. 1995 ▪ Z. Naturforsch. C **48**, 863 (1987). – *[CAS 111200-22-7]*

Bryophytes, constituents of (constituents of *mosses). Bryophytes produce numerous secondary metabolites that have not yet been found in other plants. They are mostly phenolic compounds and terpenes, there are considerable differences between the individual classes of mosses. *Hornworts* (Anthocerotae) are characterized by the presence of lignans, *liverworts* (Hepaticae) principally by terpenes, especially *sesquiterpenes that are often enantiomers of the compounds found in higher plants and bibenzyl derivatives. *Mosses* (Musci) contain biflavonoids (see flavonoids), sphagnorubins (*peat mosses*) and *coumarin derivatives. Alkaloids are rare and have as yet only been found in liverworts.
Lit.: Heterocycles **46**, 795–848 (1997) (review) ▪ Zechmeister **65**, 1–575 (review).

Bryostatins.

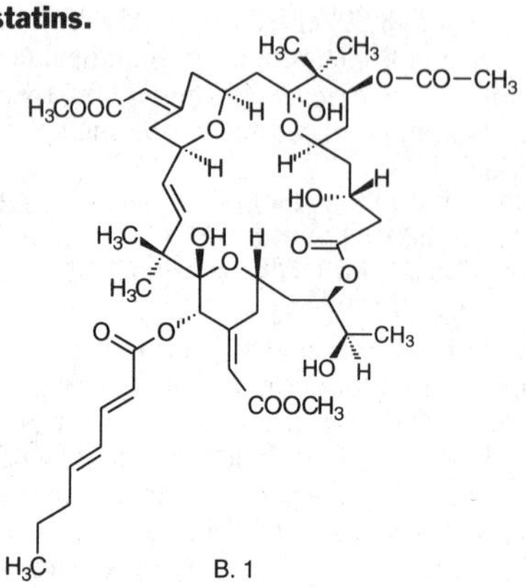

B. 1

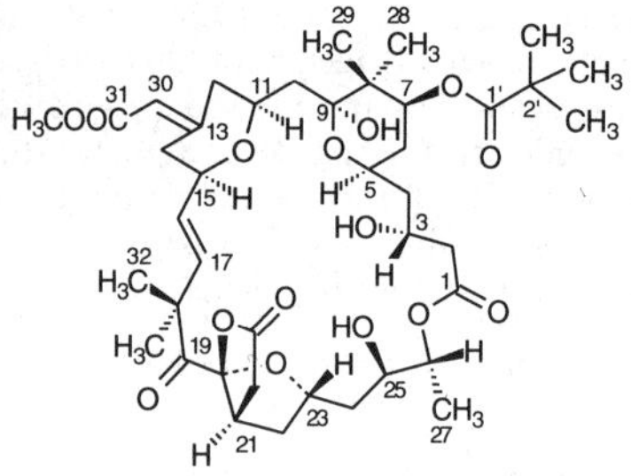

Neristatin 1

The B.[1] from the marine bryozoan *Bugula neritina* exhibit interesting biological properties: activation of protein kinase C in picomolar concentrations, stimulation of hematopoietic precursor cells, activation of T-cells, cytotoxicity against murine P 388-lymphocytic leukemia (52–96% increase in survival rate at a dose of 10–70 μg/kg)[1,2]. *B. 1*: $C_{47}H_{68}O_{17}$, M_R 905.04, mp. 230–235 °C, $[\alpha]_D$ +34.1° (CH_3OH). B. 1 is being clinically tested against metastatic tumors in phase II trials. To date 17 B. have been described, all exhibiting cytostatic activity. For synthesis, see *Lit.*[3]. In addition, 8 mg of *Neristatin 1* have been obtained from 1000 kg *B. neritina*[4] [$C_{41}H_{60}O_{15}$, M_R 792.92, mp. 214–216 °C, $[\alpha]_D$ +98° (CH_2Cl_2)]. The first synthesis of B. 2 has been published in 1998 (70 steps)[5]. The first synthesis of B. 1 had a yield of $5\times10^{-3}\%$ [3,4]. One gram B. 1 from *Bryozoa* cultures represents a value of ca. 18.000 US $.

Lit.: [1] J. Am. Chem. Soc. **104**, 6846 (1982); J. Org. Chem. **52**, 2854 (1987); **56**, 1337 (1991); Tetrahedron **41**, 985 (1985); **47**, 360 (1991); **52**, 2369 (1996); Tetrahedron Lett. **34**, 4981 (1993). [2] J. Nat. Prod. **54**, 1265 (1991) (review). [3] J. Am. Chem Soc. **112**, 7407 (1990); **121**, 7540 (1999); J. Chem. Soc., Chem. Commun. **1996**, 1931; Med. Res. Rev. **19**, 388–407 (1999); Tetrahedron: Asymmetry **7**, 2889 (1996); Tetrahedron Lett. **37**, 7695 (1996). [4] J. Am. Chem. Soc. **113**, 6693 (1991); **120**, 4534 (1998) (analogues). [5] Angew. Chem. Int. Ed. Engl. **37**, 2354 (1998)

gen.: Chem. Rev. **95**, 2041 (1995) ▪ Proc. Natl. Acad. Sci. USA **81**, 917 (1995) ▪ Tetrahedron Lett. **37**, 8305 (1996) (biosynthesis). – *Review:* Nachr. Chem. Tech. Lab. **47**, 432–438 (1999). – *[CAS 83314-01-6 (B. 1); 135004-30-7 (neristatin 1)]*

Bryozoan alkaloids. A group of *indole alkaloids from marine bryozoans. There are 4000 bryozan species widely distributed in the oceans (down to a depth of 8000 m) and waters worldwide. B. a. are brominated tryptamine derivatives that can be prenylated in many positions (N- and C-prenylation), the amino group of tryptamine is often cyclized to give a further pyrrolidine ring (the pyrrolidino-indoline system is formed). A typical example is *flustramine B.

Lit.: Acta Chem. Scand., Ser. B **39**, 517 (1985) ▪ Scheuer I **5**, 269 ▪ Stud. Nat. Prod. Chem. **17**, 73–112 (1995) (review) ▪ Zechmeister **57**, 153 ff.

BTX see brevetoxins.

Buchu leaf oil. Dark yellow to brownish oil with a fresh, sweet-minty odor resembling somewhat that of black currants, with slightly bitter taste.
Production: By steam distillation from the leaves of the Buchu trees growing in South Africa, which include various species of the genus *Barosma* (Rutaceae), e. g., *B. betulina.*
Composition[1]*:* The main components are *limonene* (see *p*-menthadienes) (ca. 10%), *menthone* and *isomenthone* (see *p*-menthanones) (together 40%), *pulegone* (see *p*-menthenones) (ca. 10%), *diosphenol* ($C_{10}H_{16}O_2$, M_R 168.24) and *ψ-diosphenol* ($C_{10}H_{16}O_2$, M_R 168.24) (together 20%). The typical cassis odor is due to *(+)-trans-8-mercapto-p-menthan-3-one* ($C_{10}H_{18}OS$, M_R 186.31) which is present at an amount of 0.2% only.

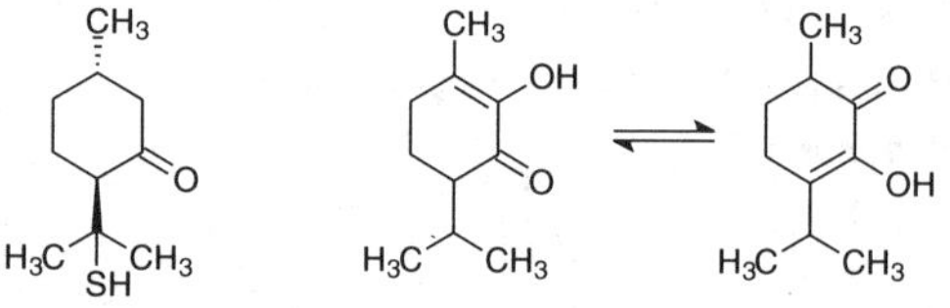

(+)-*trans*-8-Mercapto-*p*-menthan-3-one — Diosphenol — ψ-Diosphenol

Buccocamphor

Use: In very small amounts only for the production of perfumes; mainly for the aromatization of foods (fruit aromas).

Lit.: [1] Bull. Chem. Soc. Jpn. **49**, 2292 (1976); Perfum. Flavor. **1** (2), 76 (1976).

gen.: Arctander, p. 107 ▪ Bauer et al. (2.), p. 139 ▪ Chem. Lett. **1980**, 779 ▪ Müller-Lamparsky, p. 535. – *[HS 3301 29; CAS 68650-46-4 (B.); 490-03-9 (diosphenol); 54783-36-7 (ψ-diosphenol); 35117-85-2 ((+)-trans-8-mercapto-p-menthan-3-one)]*

Buckthorn. General term for various bushes belonging to the Rhamnaceae the barks of which are used to prepare laxatives on account of their contents of *anthraquinone glycosides. The unripe berries contain high concentrations of *flavonoids (*quercetin, rhamnetin, *kaempferol, rhamnazin), hence they give a yellow color with a good washing resistance to mordant wool. In a strict botanical sense, B. is *Rhamnus cathartica.*

Lit.: Hager (5.) **6**, 393–396.

Bufadienolides (5*β*-bufa-20,22-dienolides). General name for compounds with the steroid basic skeleton shown in the formulae with *cis*-linkage of rings A and B (5*βH*) and C and D (14*βH*) as well as 17*β*-oriented *α*-*pyrone group. The (20*R*)-20,21,22,23-tetrahydro compound is known as 5*β*-bufanolide. B. are related to the *cardenolides in their chemistry and biological activities. Compounds with this basic structure occur

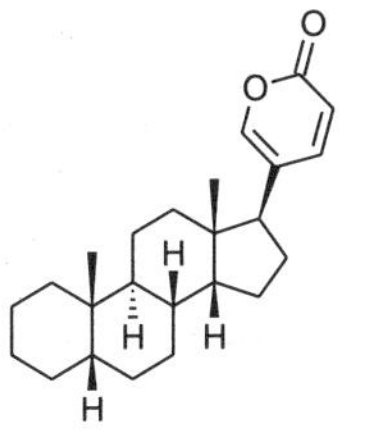

5β-Bufa-20,22-dienolide

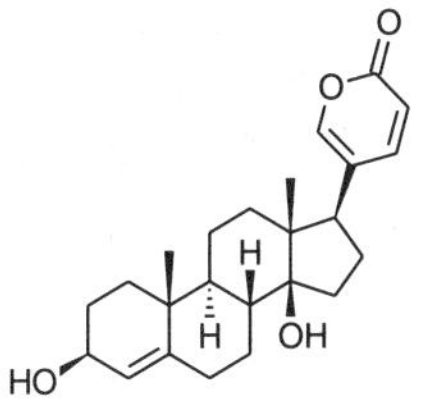

Scillarenin (**1**)

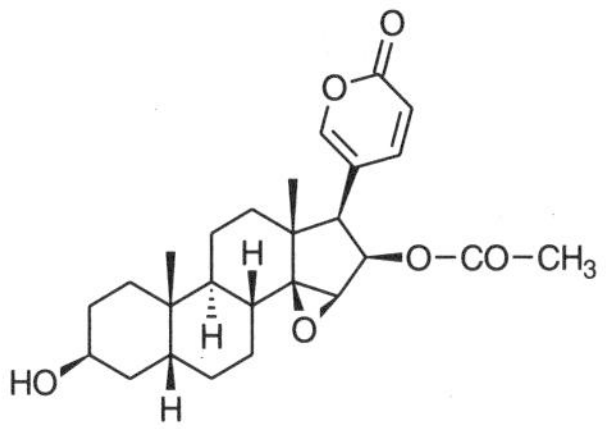

Cinobufagin (**2**)

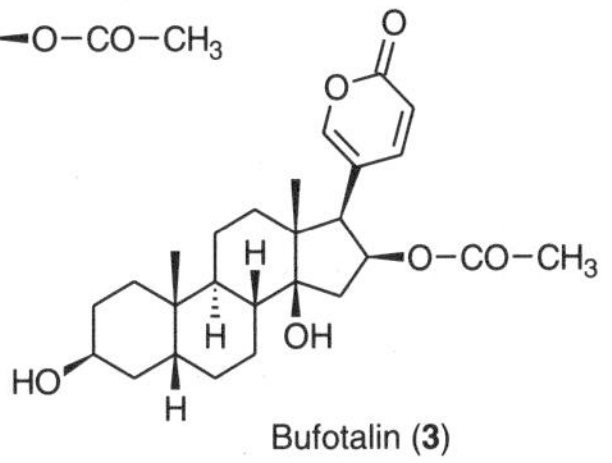

Bufotalin (**3**)

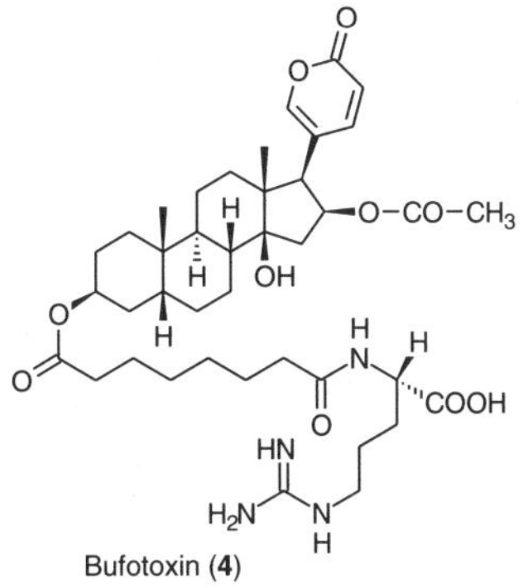

Bufotoxin (**4**)

Table: Data of Bufadienolides.

no.	molecular formula	M_R	mp. [°C]	$[\alpha]_D$	CAS
1	$C_{24}H_{32}O_4$	384.52	232–238	−16.8° (CH_3OH)	465-22-5
2	$C_{26}H_{34}O_6$	442.55	213–215		470-37-1
3	$C_{26}H_{36}O_6$	444.57	223 (decomp.)	+5.4° ($CHCl_3$)	471-95-4
4	$C_{40}H_{60}N_4O_{10}$	756.94	204–205	+3.9° (CH_3OH)	464-81-3

in the plant families Hyacinthaceae (*Scilla*, *Urginea* and *Bowiea* species) and Ranunculaceae (*Helleborus* species) as well as in the skin glands (especially parotid glands) of toads (Bufonidae), from which they derive their name. The B. are toxic and, like cardenolide glycosides, have cardiac activities.

The *Scilla* glycosides include, e. g., *proscillaridin*, an L-rhamnoside of scillarenin, which can be obtained by enzymatic cleavage of the glycoside: $C_{30}H_{42}O_8$, M_R 530.66, mp. 219–222 °C, $[\alpha]_D$ −91.5° (CH_3OH). Proscillaridin can be isolated from the sea onion (*Urginea maritima*). Cinobufagin is a major B. of the east Asian toad *Bufo asiaticus*, gland preparations of which ("Ch'an Su") are used in traditional Chinese medicine to treat hydrops. The European toad *B. vulgaris* contains mainly bufotalin and bufatoxin.

Lit.: Beilstein E V **18/4**, 443; **19/6**, 60 ▪ Hager (5.) **3**, 222 ff. ▪ Nat. Prod. Rep. **15**, 397–413 (1998) (review) ▪ Progr. Nucl. Magn. Res. Spectr. **19**, 131–181 (1987) ▪ Phytochemistry **48**, 1–29 (1998) (review) ▪ Ullmann (5.) **A 5**, 277 ▪ Sax (8.), BOM 650–BON 000, DPG 109 ▪ Zeelen, p. 290. – *[CAS 466-06-8 (proscillaridin)]*

Bufotenidine, bufotenine, bufoviridine see indolylalkylamines.

Bugs (not usually used technically). Insects of the suborder Heteroptera, ca. 40 000 species, of the order Hemiptera (Rhyncota, ca. 90 000 species). The B. are subdivided into the aquatic B. (Hydrocorisae) and the terrestrial B. (Geocorisae). All B. are plant- or bloodsuckers with a flattened body. B. much feared as carriers of diseases are, e. g., the South American winged *Triatoma* that carries Chagas' disease and the European bedbug *Cimex lectularius*. B. often possess glands from which they can emit *defensive secretions.

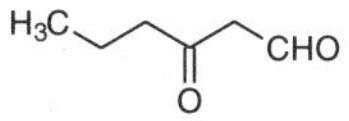

3-Oxohexanal (**1**)

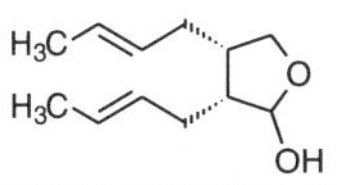

(3*R*,4*S*)-3,4-Bis((*E*)-2-butenyl)-tetrahydro-2-furanol (**2**)

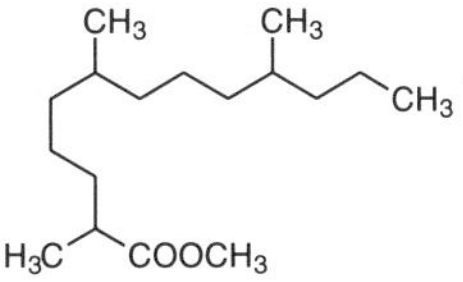

Methyl 2,6,10-trimethyltridecanoate (**3**)

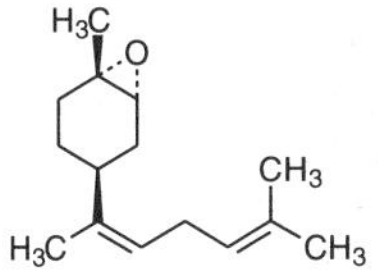

(*Z*)-(1*S*,2*R*,4*S*)-α-Bisabolene 1,2-epoxide (**4**)

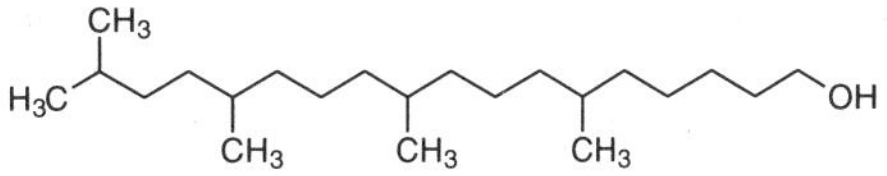

6,10,13-Trimethyl-1-tetradecanol (**5**)

These secretions are often short-chain aldehydes such as *(E)-2-hexenal* [see hexenals] or more highly oxygenated, low-molecular weight carbonyl compounds such as *3-oxohexanal* ($C_6H_{10}O_2$, M_R 114.14) or *(E)-4-oxo-2-hexenal* ($C_6H_8O_2$, M_R 112.13) and their analogues with 8 or 10 carbon atoms[1]. *Pheromones of B. include simple esters of *(E)-2-hexen-1-ol* [see hexen-1-ols][2], for *Biprorulus bibax* the hemiacetal *(3R,4S)-3,4-bis((E)-2-butenyl)tetrahydro-2-furanol*[3] ($C_{12}H_{20}O_2$, M_R 196.29) has been identified which may be considered as the dimerization product of two C_6 compounds. Terpenes such as *(Z)-(1S,2R,4S)-α-bisabolene 1,2-epoxide*[4] ($C_{15}H_{24}O$, M_R 220.35), the sexual pheromone of *Nezara viridula* have also been identified. The biosyntheses of *6,10,13-trimethyl-1-tetradecanol* ($C_{17}H_{36}O$, M_R 256.47)[5] and *methyl 2,6,10-trimethyltridecanoate* ($C_{17}H_{34}O_2$, M_R 270.46)[6], the pheromones of *Stiretrus anchorago* and *Euschistus heros*, are assumed to involve mevalonate and/or propionate units.

Lit.: [1] Annu. Rev. Entomol. **33**, 211–238 (1988). [2] J. Chem. Ecol. **21**, 973–985 (1995). [3] Justus Liebigs Ann. Chem. **1993**, 1287–1294. [4] Naturwissenschaften **76**, 173 ff. (1989); Justus

Liebigs Ann. Chem. **1994**, 785–789. [5]J. Chem. Ecol. **15**, 1717–1728 (1989); Justus Liebigs Ann. Chem. **1991**, 783–788. [6]J. Chem. Ecol. **20**, 1103–1111 (1994); Justus Liebigs Ann. Chem. **1994**, 1153–1160. – *[CAS 2492-43-5 ((E)-4-oxo-2-hexenal); 90131-02-5 (4); 100508-45-0 (5)]*

Bulbocapnine see aporphine alkaloids.

Bulgarein.

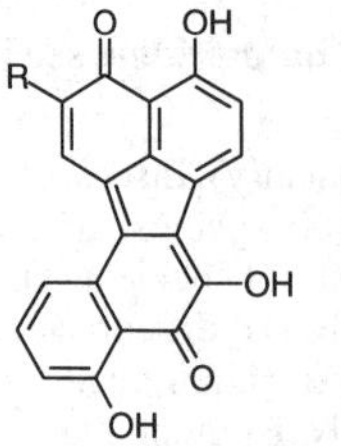

$C_{20}H_{10}O_5$, M_R 330.30, purple colored needles, mp. >300 °C, soluble in ethanol to give a red solution which turns to deep blue on dilution. Pigment from the black-brown fruit bodies of the ascomycete *Bulgaria inquinans*, which is often found on freshly felled oak trunks. In addition, the fungus contains *bulgarhodin* ($C_{20}H_{10}O_6$, M_R 346.30, purple colored needles, mp. >300 °C, insoluble in most organic solvents). The *Bulgaria* pigments are probably formed by oxidation of *1,8-naphthalenediol, as supported by their co-occurrence with 4,9-dihydroxyperylene-3,10-quinone. They are currently the only known natural products with a benzo[*j*]fluoranthene skeleton, a widely distributed carcinogenic substance that is formed in combustion processes etc.

Lit.: J. Chem. Soc., Perkin Trans. 1 **1976**, 2149. – *[CAS 62417-80-5 (B.); 62417-79-5 (bulgarhodin)]*

Bullatacin, bullatalicin see annonaceous acetogenins.

Bungarotoxins. Toxins of the very poisonous southeast Asian snake *Bungarus multicinctus* (striped krait, Elapidae). Crude extract of venom: LD_{50} (mouse s. c.) 0.019 to 0.33 mg/kg. The postsynaptic *neurotoxin α-B. is a polypeptide (M_R ca. 8000) of 74 amino acids and 5 disulfide bridges exhibiting *curare-like activity. β-B. contains different polypeptides and is a presynaptic neurotoxin.

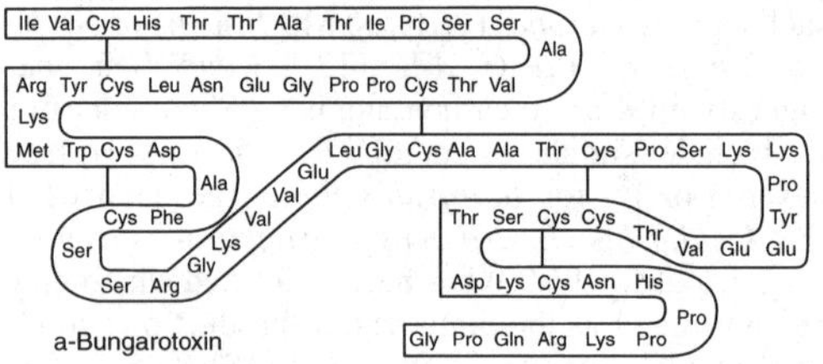

a-Bungarotoxin

Use: In experimental biochemistry to investigate neuromuscular processes.

Lit.: Biochemistry **27**, 2775 (1988) ▪ Tu, Handbook of Natural Toxins, vol. 5, New York: Dekker 1991. – *[CAS 32709-28-2]*

2,3-Butanedione (Biacetyl). $H_3C–CO–CO–CH_3$, $C_4H_6O_2$, M_R 86.09, bp. 88 °C, mp. –3 °C, flash point 10 °C, greenish-yellow liquid with an acrid odor, butter-like odor in high dilution; soluble in water, ethanol, and ether. LD_{50} (rat p. o.) >3 g/kg.

Occurrence: trace component in many essential oils and foods. Important, e. g., in *butter, *cheese, *whisky, *bread, *coffee flavor and black currants (see fruit flavors). It is formed as by-product in the biosyntheses of *leucine and *L-valine and in the metabolism of lactic bacteria.

Lit.: Arctander, No. 872 ▪ Beilstein E IV **1**, 3644 ▪ Fenaroli (2.) **2**, 123 ▪ Ullmann (5.) **A 15**, 90. – *[HS 2191 9; CAS 431-03-8]*

Butanoic acid esters (butanoates) see fruit esters.

Butein (2′,3,4,4′-tetrahydroxychalcone).

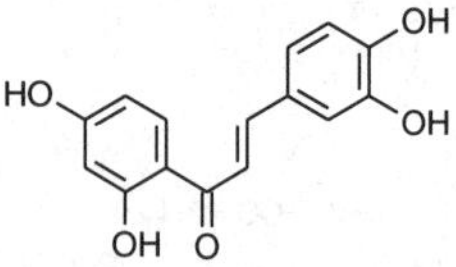

$C_{15}H_{12}O_5$, M_R 272.26, orange-yellow needles, mp. 213–215 °C. B. and its 4′-*O*-glucoside (*coreopsin*) are the most common tetrahydroxychalcones (*chalcones) in nature. Free B. often occurs in wood and bark of various trees, e. g., *Acacia* (Fabaceae) or *Rhus* species (sumac, tanner's sumac, Anacardiaceae). Coreopsin occurs in various members of the Asteraceae, e. g., in *Viguiera* species.

Lit.: Harborne (1975), p. 459; (1994), p. 394. – *[CAS 487-52-5]*

Butenolide(s). 1. Trivial name for *5-(acetylamino)-2(5H)-furanone*, $C_6H_7NO_3$, M_R 141.13, mp. 116–118.5 °C, $[\alpha]_D$ 0° ($CHCl_3$), a *mycotoxin from *Fusarium equiseti*, *F. graminearum*, and *F. tricinctum*. B. has antibacterial, antifungal, and skin irritating properties. Longer exposure leads to skin necrosis; known in cattle as "fescue foot disease" after consumption of infected fescue grass (*Festuca arundinacea*). LD_{50} (mouse i. p.) 43.6 mg/kg.

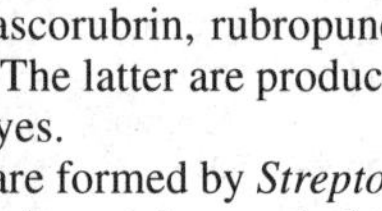

2. General name for compounds containing a dihydrofuran-2-one group, e. g., *variegatic acid and other pigments from Basidiomycetes, *patulin, *penicillic acid, *tetronic acid, which occur in fungi and lichens, *cardenolides and *ascorbic acid from plants as well as pigments (monascorubrin, rubropunctatin) from *Monascus* species. The latter are produced biotechnologically as food dyes.

3. Virginiae B. are formed by *Streptomyces* species as *antibiotic-inducing and sporulation-triggering factors.

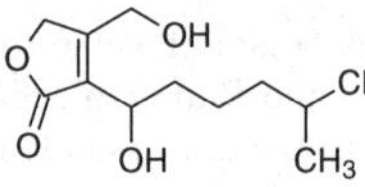
Virginiae butenolide A

Virginiae B. A, $C_{12}H_{22}O_4$, M_R 230.30.

Lit. (to 1): Cole-Cox, p. 898–901 ▪ J. Org. Chem. **38**, 815 (1973) ▪ Phytochemistry **7**, 139–146 (1968). – *(to 2):* Turner **1**, 138, 284–287, 358; **2** 16–23, 111 ff., 369 ff., 470 f., 485, 530 ▪ Zechmeister **35**, 133–198. – *Monascus pigments:* Bio-Engineering **3**, 18–27 (1987). – *(to 3):* J. Antibiotics **41**, 1828 (1988); **42**, 1873 (1989) ▪ J. Am. Chem. Soc. **112**, 898 (1990); **114**, 663 (1992). – *[CAS 16275-44-8 (1.); 88169-49-7 (3.)]*

Butter flavor. Aroma-active components in *creamery butter* are the 5-*alkanolides C_8, C_{10}, and C_{12}, free *fatty acids C_{10} and C_{12}, as well as trace components, e. g., (Z)-4-heptenal (see alkenals) and *skatole. *Sour cream butter* additionally contains metabolic products of the starter cultures such as *acetoin, *2,3-butanedione, *2,3-pentanedione* ($C_5H_8O_2$, M_R 100.12), and acetic acid. Rancid butter contains elevated concentrations (>50 ppm) of the fatty acids C_4, C_6, and C_8, liberated by lipases.

Lit.: TNO list (6.), Suppl. 5, p. 101–108. – *[CAS 600-14-6 (2,3-pentanedione)]*

Butterflies (moths, Lepidoptera). An insect order with more than 100000 species, wing span between 2 and 300 mm; sucking mouthparts to take up flower juices. Caterpillars of large butterflies have abdominal prolegs with hooks, those of smaller butterflies prolegs with crochets. The latter are frequently designated as moths. Male B. have extraordinarily sensitive and selective sensory organs in their antennae to detect the sexual pheromones produced by the females, these pheromones originate mainly from the fatty acid pool. Like *bombykol they consist of unbranched, simply or doubly unsaturated, even-numbered chains with a functional group at one end (alcohols or their acetates, *tetradecenyl acetate, aldehydes, less frequently fatty acid alkyl esters) and 10–18 carbon atoms. (Homo)conjugated polyenes and *epoxy polyenes derived from fatty acids, branched hydrocarbons such as *5,9-dimethylheptadecane*[1] ($C_{19}H_{40}$, M_R 268.53), ketones such as *(Z)-6-heneicosen-11-one*[2] ($C_{21}H_{40}O$, M_R 308.55), secondary alcohols[3] or epoxides like *disparlure are less common. For comprehensive reviews on the sexual pheromones of B., see *Lit.*[4]. In some species the pheromones are also produced by the males, e. g., *danaidal and *hepialone.

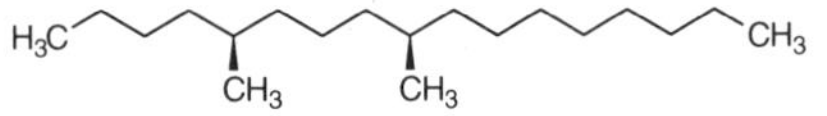

(5*S*, 9*S*)-5,9-Dimethylheptadecane

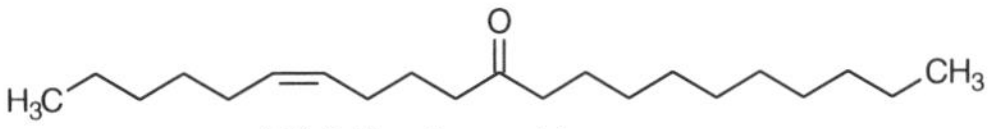

(*Z*)-6-Heneicosen-11-one

Lit.: [1]Naturwissenschaften **74**, 143 f. (1987). [2]J. Org. Chem. **43**, 2361 (1978); ApSimon **4**, 60 ff.; **9**, 165–175. [3]J. Chem. Ecol. **21**, 13–28, 29–44 (1995). [4]Morgan u. Mondava, CRC Handbook of Natural Pesticides, vol. IV, Pheromones, p. 35–94, Boca Raton: CRC Press 1988; Mayer & McLaughlin, Handbook of Insect Pheromones and Sex Attractants, Boca Raton: CRC Press 1991; Arn, Toth, & Priesner, List of Sex Pheromones of Lepidoptera and Related Attractants (2.), Montfavet: Int. Org. for Biological Control 1992. – *[CAS 108195-53-5 (5,9-dimethylheptadecane); 54844-65-4 ((Z)-6-heneicosen-11-one)]*

Butyl isothiocyanate see mustard oils.

3-Butylphthalide see celery seed oil.

Buxus steroid alkaloids. *Steroid alkaloids with 9β,19-cyclopregnane or 9(10→19)-*abeo*-pregnane structure, but with 1 to 3 additional methyl groups at C-4 and/or C-14 as well as amino functions at C-3 and/or C-20. They are isolated from numerous species of the Buxaceae genus *Buxus*. To date about 200 B. s. are known. In order to achieve transparency and especially to reduce the number of incomprehensible trivial names for new compounds it was proposed[1] to accept a name only for new basic structures and to designate the respective methyl substitution pattern at the 3- or 20-amino groups by letters as shown in the accompanying table 1.

Table 1: Identification of the methyl group substitution pattern at N(3) and N(20) of Buxus steroid alkaloids.

	N(3)		N(20)			N(3)		N(20)	
A	CH_3	CH_3	CH_3	CH_3	I	H	H	H	H
B	CH_3	CH_3	H	CH_3	K	CH_3	CH_3	–	–
C	H	CH_3	CH_3	CH_3	L	–	–	CH_3	CH_3
D	H	CH_3	H	CH_3	M	H	CH_3	–	–
E	CH_3	CH_3	H	H	N	–	–	H	CH_3
F	H	H	CH_3	CH_3	O	H	H	–	–
G	H	CH_3	H	H	P	–	–	H	H
H	H	H	H	CH_3					

In the following discussion – with simultaneous demonstration of the mentioned principle – some important structural types are presented. Thus, oxygen functions at C-16 and the 4-methyl group are frequently present, in the case of *buxaquamarine K* even in the form of a tetrahydrooxazine ring, the formation of which involves the 4β-methyl and an *N*-methyl group. Acylations of various amino and hydroxy groups are not uncommon (example: *N-benzoylbaleabuxidienine F*). *Cyclobuxine D* possesses a 4-methylene group instead of a 4-methyl group.

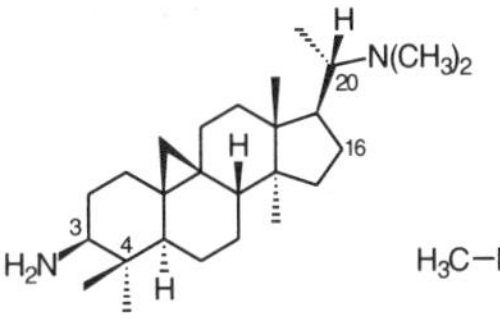

Cycloprotobuxine F

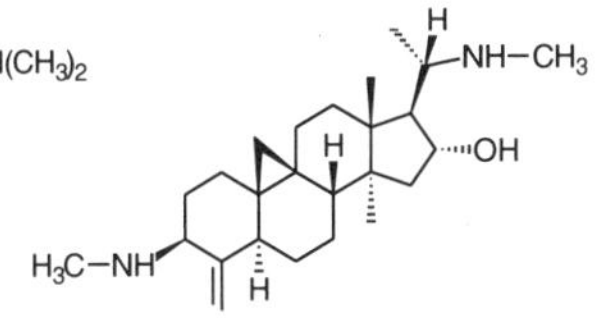

Cyclobuxine D

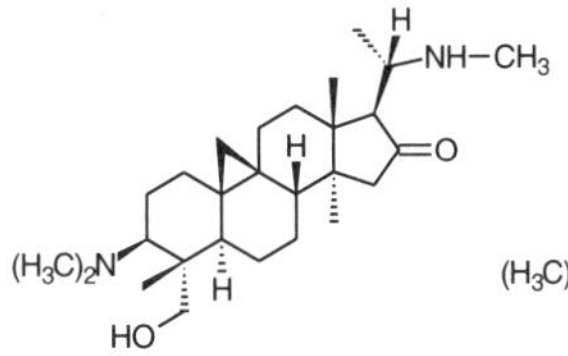

Buxazidine B

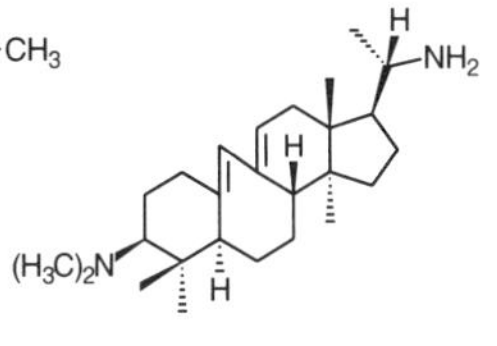

Buxamine E

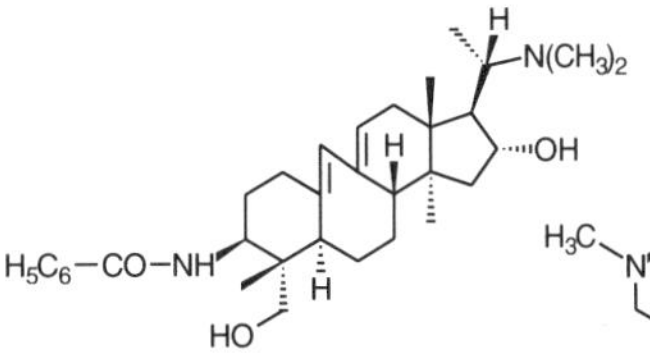

N-Benzoylbaleabuxidienine F

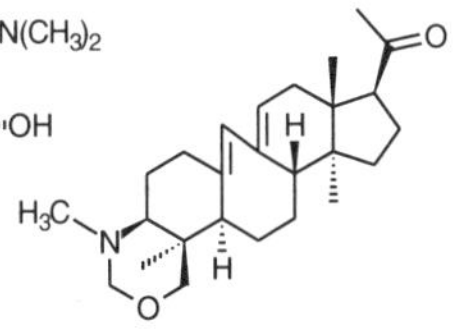

Buxaquamarine K

Biosynthesis: For the B. s. biosynthesis does not start from *cholesterol, but from the latter's biosynthetic precursor *cycloartenol, which already exhibits the 9β,19-cyclo structure present in most of these alkaloids. Degradation of the 17β-side chain to give 4,4,14-trimethyl-9β,19-cyclopregnane is realized analogously to that of cholesterol. This genesis of B. s. from cycloartenol is supported by the results of isotope labelling experiments. Compounds in which the reactive

Table 2: Data of selected Buxus steroid alkaloids.

Alkaloid	molecular formula	M_R	mp. [°C]	$[\alpha]_D$	occurrence	CAS
Cycloprotobuxine F	$C_{26}H_{46}N_2$	386.66	163	+42° ($CHCl_3$)	*Buxus madagascarica*	36151-05-0
Cyclobuxine D	$C_{25}H_{42}N_2O$	386.62	245–247	+98° ($CHCl_3$)	*B. sempervirens*, *B.* species	2241-90-9
Buxazidine B	$C_{27}H_{46}N_2O_2$	430.67	234–236	−31° ($CHCl_3$)	*B. sempervirens*	15208-57-8
Buxamine E	$C_{26}H_{44}N_2$	384.65	amorphous	+32° ($CHCl_3$)	*B. sempervirens*, *B.* species	14317-17-0
N-Benzoylbaleabuxidienine F	$C_{33}H_{48}N_2O_3$	520.76	291	−36° ($CHCl_3$)	*B. balearica*	14155-72-7
Buxaquamarine K	$C_{26}H_{39}NO_2$	397.60	amorphous	+24° ($CHCl_3$)	*B. papilosa*	98776-20-6

11-oxo-9β,19-cyclopregnane system is present, e.g., buxarine F, are assumed to be biogenetic precursors of those alkaloids that are characterized by the 9(10→19)-*abeo*-pregnane structure with its conjugated transoid diene system (e.g. *buxamine E*). However, nothing is yet known about the special biosynthetic mechanism of this rearrangement.

Biological activity: For some B.s., e.g., cyclobuxine D, anti-inflammatory and hypotensive activities have been observed.

Lit.: [1] Tetrahedron Lett. **1964**, 2895.
gen.: Manske **9**, 305–426; **14**, 1–82; **32**, 79–239 ▪ Mothes et al., p. 363–384 (biosynthesis) ▪ Phytochemistry **29**, 1293 (1990) ▪ Tetrahedron **42**, 5747 (1986). – *[HS 2938 90; CAS 11011-80-6 (buxarin F)]*

Byssochlamic acid.

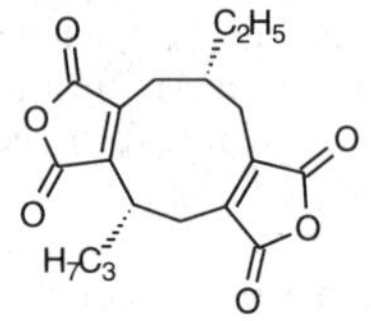

$C_{18}H_{20}O_6$, M_R 332.35, mp. 163.5 °C, $[\alpha]_D$ +108° ($CHCl_3$), prisms. *Mycotoxin from *Byssochlamys fulva* (*Paecilomyces variotii*) and *Byssochlamys nivea*, two ascomycetes, that often contaminate fruit juices and conserved fruits since their ascospores are thermally resistant and can be killed only if heated to over 85 °C.

Lit.: Beilstein E V **19/5**, 406 ▪ Helv. Chim. Acta **64**, 2791 (1981) ▪ J. Am. Chem. Soc. **114**, 9673 (1992) (synthesis) ▪ Zechmeister **25**, 131–149. – *[CAS 743-51-1]*

C

Cabagin see *S*-methyl-L-methionine sulfonium chloride.

Cabenegrins (cabenigrins). *Phytoalexins with the *pterocarpan skeleton from the Brazilian plant "*Cabeca de negra*" (*Annona coriacea*, Amazon region), used by the native population for treatment of snake and spider bites. The components C. A-1 and C. A-2 prevent (at 2–3 mg/kg in mice) from an otherwise lethal course of the poisoning. *C. A-1*: $C_{21}H_{20}O_6$, M_R 368.37, mp. 167–168 °C.

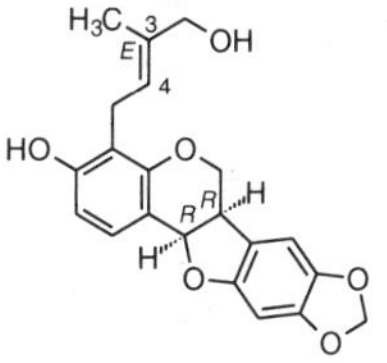

Cabenegrin A-1

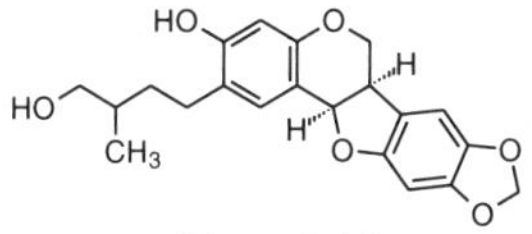

Cabenegrin A-2

Lit.: ACS Symp. Ser. **551**, 14 (1994) ▪ Chem. Ind. (London) **1984**, 303 ▪ Tetrahedron **55**, 9283–9296 (1999) (synthesis, abs. configuration) ▪ Tetrahedron Lett. **23**, 3859 (1982) (synthesis). – *[CAS 84297-59-6 (C. A-1); 84297-60-9 (C. A-2)]*

Cactus alkaloids. About 2000 cactus species (stem succulents) of tropical and subtropical desert areas are known, some of which contain hallucinogenic principles, e. g., *Lophophora* species indigenous to Mexico; together with other hallucinogenic cacti the latter are known as *peyote (see Lophophora alkaloids). They contain ca. 5% alkaloids such as *mescaline, anhalonine, lophophorine, pellotine, etc., mainly *isoquinoline alkaloids, and *phenylethylamine alkaloids.
Lit.: J. Chem. Ecol. **13**, 2069–2081 (1987) ▪ J. Nat. Prod. **49**, 735 (1986) ▪ Phytochemistry **22**, 1263, 2101 (1983).

Cadaveric poisons see ptomaines.

Cadaverine (1,5-pentanediamine). $H_2N\text{–}(CH_2)_5\text{–}NH_2$, $C_5H_{14}N_2$, M_R 102.18, mp. 9 °C, bp. 178–180 °C. C. is a *polyamine, formed by bacteria through decarboxylation of *lysine in the gastrointestinal system or on putrefaction of cadavers, contributing to the odor of putrefaction. It is the biosynthetic precursor of some alkaloids; see also quinolizidine alkaloids, Lycopodium alkaloids, piperidine alkaloids.
Lit.: Beilstein E IV **4**, 1310 ▪ Merck-Index (12.), No. 1645 ▪ Ullmann (5.) **A 2**, 27. – *[HS 2921 29; CAS 462-94-2]*

Cadinane see cadinene and termites.

Cadinene. $C_{15}H_{24}$, M_R 204.36. Of the possible isomers of C. (eight are known to date) which differ by the position of the double bond and by ring linkages *β*-C. (colorless oil with a pleasant odor, bp. 275 °C, $[\alpha]_D^{20}$ –251°) is the most important. It is the most widely dis-

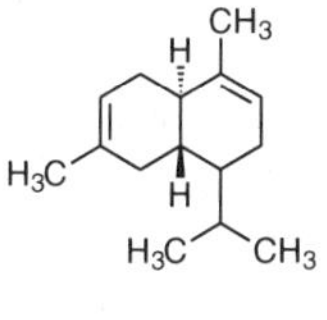

(-)-β-Cadinene

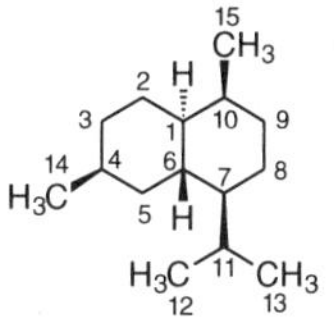

Cadinane skeleton

tributed *sesquiterpene in the plant kingdom and the major component of juniper tar oil which is used in ointments for skin rashes, medicinal tar soaps, and hair preparations.
Lit.: Beilstein E IV **5**, 1183 ▪ Indian J. Chem., Sect. B **21**, 145 (1982) (synthesis) ▪ J. Chem. Soc., Perkin Trans. 1 **1997**, 2065 [biosynthesis (–)-*γ*-C.] ▪ Justus Liebigs Ann. Chem. **1986**, 87 ▪ Tetrahedron **39**, 883 (1983). – *[HS 2902 19; CAS 523-47-7 (β-C.)]*

Caerulomycin A [4-methoxy-2,2′-bipyridine-6-carbaldehyde-(*E*)-oxime].

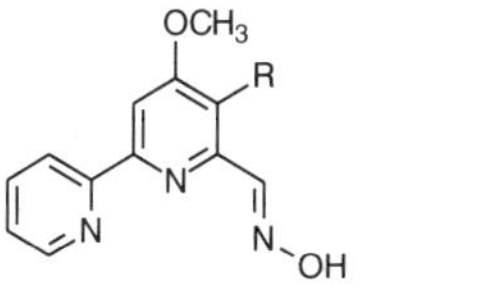

R = H : Caerulomycin A
R = SCH_3 : Collismycin A
CHO for oxime: Caerulomycin E

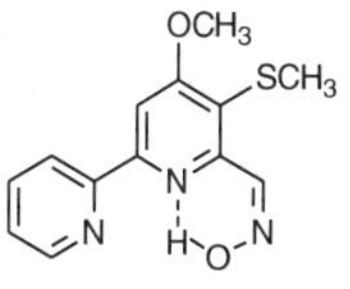

Collismycin B

$C_{12}H_{11}N_3O_2$, M_R 229.24, cryst., mp. 169–170 °C. An antibiotically active 2,2′-dipyridyl derivative from cultures of *Streptomyces caeruleus* and *Nocardiopsis cirriefficiens*[1]. C. is active against yeasts, fungi, and *Entamoeba histolytica*. The related streptomycetes metabolites *collismycin A* [$C_{13}H_{13}N_3O_2S$, M_R 275.33, powder, mp. 170–172 °C (decomp.)] and *B* [mp. 148–150 °C (decomp.)] contain a methylthio group and are the isomeric (*E*)- and (*Z*)-oximes[2]. They inhibit the binding of dexamethason to glucocorticoid receptors with IC_{50} values of 1.5 and 1.0×10^{-5} M, respectively.
Lit.: [1] Can. J. Chem. **45**, 1215 (1967) (structure); **47**, 165 (1969) (synthesis); **66**, 191 (1988) (biosynthesis) [2] J. Antibiot. **47**, 1072 (1994).
gen.: Beilstein E V **25/2**, 135; E V **25/3**, 134 ▪ J. Org. Chem. **63**, 2892 (1998) (synthesis). – *[CAS 21802-37-9 (C.); 158792-24-6 (collismycin A); 158792-25-7 (collismycin B)]*

Cafestol.

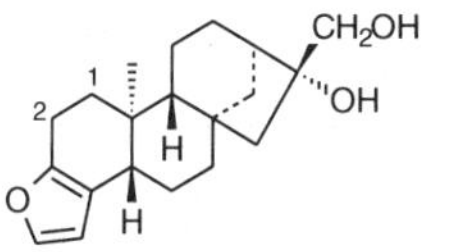

$C_{20}H_{28}O_3$, M_R 316.44, mp. 160–162 °C (decomp.), $[\alpha]_D$ –101° ($CHCl_3$). A diterpene of the *kaurane type from coffee bean oil. C. also occurs in boiled coffee and raises serum cholesterol levels in humans. This effect is due to the down-regulation of cholesterol 7α-hydroxylase and sterol 27-hydroxylase in hepatocytes, resulting in bile acid synthesis suppression [1].

Lit.: [1] Arterioscler., Thromb., Vasc. Biol. **17**, 3064–3070 (1997).

gen.: Beilstein E V **17/5**, 357 ▪ Hager (5.) **4**, 931 f., 938 ▪ J. Am. Chem. Soc. **109**, 4717 f. (1987) ▪ Karrer, No. 1965a ▪ Merck-Index (12.), No. 1672 ▪ Tetrahedron Lett. **28**, 5403 (1987) (synthesis). – *[CAS 469-83-0]*

Caffeic acid [(*E*)-3,4-dihydroxycinnamic acid, (*E*)-3-(3,4-dihydroxyphenyl)-2-propenoic acid].

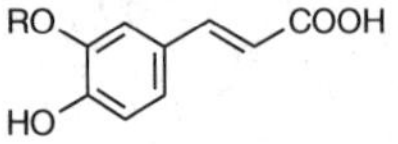

R = H : Caffeic acid
R = CH_3 : Ferulic acid

$C_9H_8O_4$, M_R 180.16, yellow cryst., mp. 223–225 °C (decomp.), pK_{a1} 4.62, pK_{a2} 9.07 (25 °C), forms as monohydrate upon crystallization from dilute aqueous solutions. C. is widely distributed in plants in the free form as well as in ester (see, e. g., chlorogenic acid, cynarin(e)) and glycoside forms. The derivatives are formed by acylation with caffeoyl-CoA. The (*Z*)-form is known as iso-C.; it occurs in olive and peanut oils and in other plant products. *Ferulic acid* (4-hydroxy-3-methoxycinnamic acid, C. 3-methyl ether), $C_{10}H_{10}O_4$, M_R 194.19: (*E*)-form, orthorhomb. needles, mp. 174 °C, is widely distributed in plants, e. g., in beets, grain crops, and giant fennel *Ferula asa-foetida*. The (*Z*)-form of ferulic acid, uv_{max} (C_2H_5OH): 316 nm, is a yellow oil and also occurs as a natural product [1]. *Vinyl caffeate* ($C_{11}H_{10}O_4$, M_R 206.20, mp. 133–140 °C) occurs as an antioxidant in *Perilla frutescens* (Lamiaceae) [2].

Activity: On account of its easy oxidation to the *o*-quinone (*caffeoquinone* $C_9H_6O_4$, M_R 178.14) it is an allergen (see also chlorogenic acid).

Use: C.: As reagent for detection of iron(III). Ferulic acid: as conservation agent for foods.

Biosynthesis: From L-*phenylalanine → *cinnamic acid → *(*E*)-*p*-coumaric acid → C. → ferulic acid.

Lit.: [1] ACS Symp. Ser. **506**, 180 (1992); Phytochemistry **4**, 383–399 (1965). [2] Biosci., Biotechnol. Biochem. **60**, 1115 (1996).

gen.: ApSimon **4**, 533 ▪ Arch. Mikrobiol. **154**, 206 (1990) ▪ IARC Monogr. **56**, 115 (1993) (review toxicology) ▪ Luckner (3.), p. 383–386 ▪ Nat. Prod. Rep. **12**, 101–133 (1995) ▪ Zechmeister **35**, 83–85. – *[HS 291829, 291890; CAS 501-16-6 (C.); 4361-87-9 (iso-C.); 1135-24-6 (ferulic acid); 537-98-4 (E-form); 1014-83-1 (Z-form); 15416-77-0 (caffeoquinone)]*

Caffeine, coffein(e) (theine, methyltheobromine, guaranine, 1,3,7-trimethylxanthin).

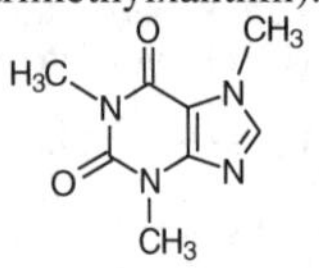

$C_8H_{10}N_4O_2$, M_R 194.19, odorless, bitter tasting cryst., D. 1.23, mp. 238 °C (sublimes above 178 °C), well soluble in water and chloroform, moderately soluble in alcohols. C. is a plant alkaloid belonging to the *purines and forms salts with acids that are readily soluble in water.

Occurrence: C. is found as the salt of *chlorogenic acid in coffee beans (1–1.5%), in dried black tea (up to 5%; this C. in tea was also previously called *theine*), in Brazil or Paraguay tea (0.3–1.5%), in cola (ca. 1.5%), and in guarana paste (up to 6.5%); cocoa kernels also contain C. (ca. 0.2%).

Physiological activity: C. has a stimulating effect on the CNS because it delays the transformation of cAMP into AMP through inhibition of phosphodiesterase. Moderate amounts of C. stimulate cardiac activity, metabolism, and respiration, blood pressure and body temperature increase, blood vessels in the brain dilate somewhat while the intestines contract; the diuretic action of coffee is well known. The improved perfusion in the cerebrum results, in the conscious regions, in elimination of tiredness, temporary improvement of performance, and improvements in mood. As an antagonist of adenosine C. can eliminate the inhibitory action of adenosine on the liberation of neurotransmitters. Furthermore, C. has a marked lipolytic effect in adipose tissue (increase of free fatty acids in serum). Chronic abuse can lead to a mild form of addiction which, upon withdrawal, can afford symptoms such as headache. Higher doses of C. (above about 300 mg) cause hand tremor, blood congestion in the head, pressure in the heart region; the lethal dose for humans is between 5 and 30 g. C. has a half-life in the organism of 3–5 h. In the liver C. is primarily demethylated, oxidized, and acetylated. The main metabolites are paraxanthine (80%), *theobromine (11%), and *theophylline (4%). People with hypertension, coronary artery disease, and nervous disorders should avoid C.-containing drinks. Extraction with, e.g., supercritical carbon dioxide, is used as a method of decaffeination.

Use: Stimulant in coffee and tea as well as C.-containing soft drinks which must have a C. content of between 65 mg and 250 mg/L; Cola beverages contain ca. 160 mg/L and coffee ca. 500 mg/L. C. is used in medicine for treatment of cardiac insufficiency, neuralgia, headache, asthma attacks, hayfever, nicotine, morphine, and alcohol poisonings.

Lit.: Angew. Chem. Int. Ed. Engl. **17**, 702–709 (1978) (decaffeination) ▪ Beilstein E III/IV **26**, 2338 ▪ Chem. Ber. **126**, 1955 (1993) (synthesis) ▪ Chem. Eng. World **33**, 110 (1998) (synthesis) ▪ Hager (5.) **4**, 940 ff.; **7**, 1073–1078 ▪ IARC Monographs **51**, 291 (1991) (review toxicology) ▪ J. Biol. Chem. **1962**, 1941 (biosynthesis) ▪ Karrer, No. 2565 ▪ Kihlmann, Caffeine and Chromosomes, Amsterdam: Elsevier 1977 ▪ Med. Monatsh. Pharm. **14**, 271 (1991); **15**, 258 (1992) ▪ Merck-Index (12.), No. 1674. – *[HS 293930; CAS 58-08-2]*

Caffeoquinone see caffeic acid.

Calamus oil (sweet flag oil). Yellow to brown, rather viscous oil with a heavy, warm-spicy, sweet, root-like odor and a warm-spicy, bitter taste.

Production: By steam distillation from the roots of calamus (*Acorus calamus*); *origin*: eastern Europe.

Composition [1]***:*** C. consists mostly of sesquiterpenoid compounds. Typical components are *shyobunone* ($C_{15}H_{24}O$, M_R 220.35) and its isomers (together ca. 8–10%), as well as *preisocalamendiol* ($C_{15}H_{24}O$, M_R 220.35) (8%) and *acorenone* ($C_{15}H_{24}O$, M_R 220.35) (11%). The characteristic component of calamus is *cis-asarone* (β-asarone, see asarones) present to about

10% in oils of European origin and up to over 80% in Indian and east Asian oils. The unique odor is mainly due to trace components such as *(Z,Z)-4,7-decadienal* ($C_{10}H_{16}O$, M_R 152.24).

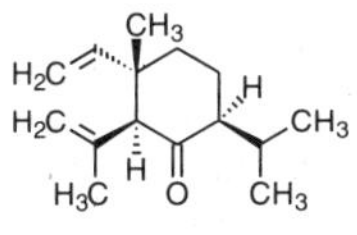
Shyobunone

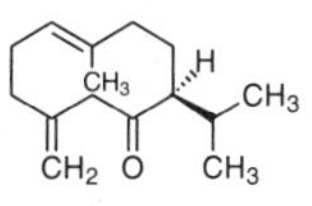
Preisocalamendiol

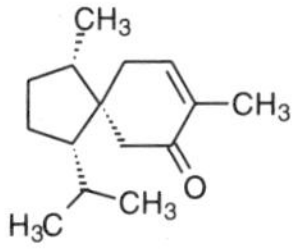
Acorenone

Use: In low doses in the production of perfumes with herby-spicy notes; to improve the aromatic nature of alcoholic beverages such as liquors and bitters, etc. On account of the cancerogenic activity of *cis*-asarone the use of calamus oil is subject to legal restrictions (content <1 ppm).

Lit.: [1]Perfum. Flavor. **5** (7), 49 (1980); **8** (4), 63 (1983); **11** (3), 52 (1986). [2]Brunke, Progress in Essential Oil Research, p. 215, Berlin: de Gruyter 1986.
gen.: Arctander, p. 111 ▪ Bauer et al. (2.), p. 140. – *Toxicology:* Food Cosmet. Toxicol. **15**, 623 (1977). – *Synthesis:* J. Chem. Soc., Chem. Commun. **1990**, 1778 ▪ J. Chem. Soc., Perkin Trans. 1 **1992**, 229 (acorenone) ▪ Tetrahedron **49**, 4761 (1993) (shyobunone). – *[HS 3301 29; CAS 8015-79-0 (C.); 21698-44-2 (shyobunone); 25645-19-6 (preisocalamendiol); 5956-05-8 (acorenone); 22644-09-3 (4,7-decadienal)]*

Calanolides. Pyranocoumarins from *Calophyllum* spp., especially *Calophyllum lanigerum* (Guttiferae). *C. A* [NSC 675 451, $C_{22}H_{26}O_5$, M_R 370.44, oil, $[\alpha]_D$ +60° ($CHCl_3$)] from *C. lanigerum* is a non-nucleosidic HIV-1 specific reverse transcriptase inhibitor currently under development as an AIDS chemotherapeutic. Further known C.: B, C, D, E_1, E_2, F.

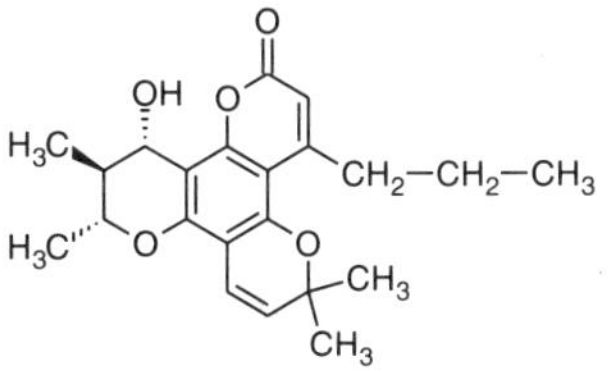
C. A

Lit.: Chin. Chem. Lett. **9**, 433 (1998) (synthesis) ▪ J. Am. Chem. Soc. **120**, 9074, 9400 (1998) (synthesis) ▪ J. Med. Chem. **39**, 1303–1313 (1996) (synthesis) ▪ J. Nat. Prod. **59**, 754 (1996) (isolation); **61**, 1252 (1998) ▪ Modern Drug Discovery (Nov./Dec.) **1999**, 31–36 (review) ▪ Synth. Commun. **26**, 4005–4021 (1996) ▪ Tetrahedron Lett. **36**, 5475 (1995) (synthesis) ▪ Tetrahedron: Asymmetry **7**, 3315–3326 (1996). – *[CAS 142632-32-4 (C. A)]*

Calbistrin A.

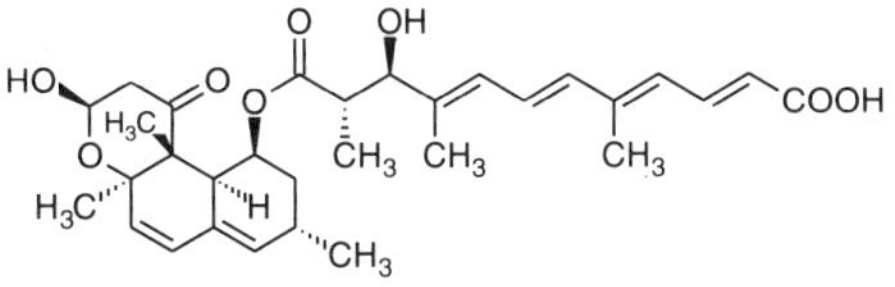

$C_{31}H_{40}O_8$, M_R 540.64, mp. 133 °C, $[\alpha]_D$ +69° ($CHCl_3$). A metabolite from cultures of *Penicillium restrictum* with multiple biological activities. C. A is an antifungal and cholesterol lowering agent. It promotes the production of nerve growth factor (NGF).

Lit.: J. Antibiot. **46**, 34, 39 (1993) (isolation) ▪ J. Nat. Prod. **56**, 1779 (1993) ▪ Tetrahedron Lett. **38**, 583 (1997) (synthesis). – *[CAS 147384-55-2]*

Calcidiol, Calcifediol, Calciferols see calcitriol.

Calcimycin (A 23187).

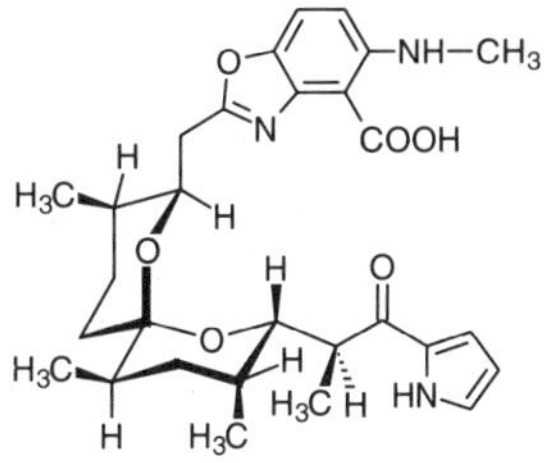

$C_{29}H_{37}N_3O_6$, M_R 523.63, mp. 181–182 °C, $[\alpha]_D$ +362° ($CHCl_3$); readily soluble in chloroform, dimethyl sulfoxide, ethyl acetate, ethanol, methanol, sparingly soluble in water. Antibiotic with weak *in vitro* activity against Gram-positive and -negative bacteria and fungi; LD_{50} (rat) 9.2 mg/kg. It is produced by *Streptomyces chartreusis*. On account of its ionophoric properties of forming stable complexes with divalent cations it is often used for the transmembrane transport and to increase the intracellular concentration of these ions. Its ion specificity can be described as follows:

$$Mn^{2+} > Ca^{2+} > Mg^{2+} > Sr^{2+} > Ba^{2+} \approx Li^+ > Na^+ > K^+$$

As fluorescent reagent it is used to examine protein hydrophobicity. It is a decoupler of oxidative phosphorylation and inhibits ATPase activity in mitochondria. C. reinforces the action of *N*-methylaspartate on hippocampus neurons (rat).
In its biosynthesis the 2-pyrrolecarbonyl unit originates from proline, the spiroketal from acetate/propanoate, and the substituted benzene ring from shikimic acid.

Lit.: Annu. Rev. Biochem. **45**, 501 (1976) ▪ ApSimon **9**, 584–600 ▪ Brain Res. **540**, 322 (1991) ▪ Eur. J. Pharmacol. **257**, 275 (1994) (activity) ▪ J. Am. Chem. Soc. **109**, 7553 (1987); **113**, 5337 (1991) (synthesis) ▪ J. Org. Chem. **45**, 3537 (1989); **54**, 3347 (1989) (synthesis) ▪ Merck-Index (12.), No. 1678 ▪ Synform **8**, 38 (1990) (review) ▪ Tetrahedron **39**, 1255 (1983) (biosynthesis); **42**, 6465 (1986) ▪ Zechmeister **58**, 1–81 (biosynthesis). – *[HS 2941 90; CAS 52665-69-7]*

Calciol see calcitriol.

Calcitic acid see calcitriol.

Calcitriol (1*α*,25-dihydroxycholecalciferol, 1*α*,25-dihydroxyvitamin D_3). A *steroid hormone with the systematic chemical name (1*S*,3*R*,5*Z*,7*E*)-9,10-secocholesta-5,7,10(19)-triene-1,3,25-triol. It is an endogenous metabolite of *cholecalciferol* (Calciol), the so-called antirachitic *vitamin D_3, and thus the actual natural, biologically active form of this "vitamin".
Biosynthesis: Starting compound for the biosynthesis of C. is 7,8-didehydrocholesterol (*provitamin D_3*, procalciol, procholecalciferol, see sterols) which, in turn, is formed in the liver from *cholesterol and occurs abundantly in the skin of humans and animals. UV irradiation of the skin furnishes (6*Z*)-tacalciol (*previtamin D_3*, precalciol, precholecalciferol, tachysterol) which undergoes thermal isomerization to calciol (cholecalciferol, vitamin D_3). Calciol is oxidized in the liver to *calcidiol* (*25-hydroxycholecalciferol*, cholecalcifediol, calcifediol). Calcidiol bound to protein (*α*-globulin) is the actual prohormone (storage form of the hormone).

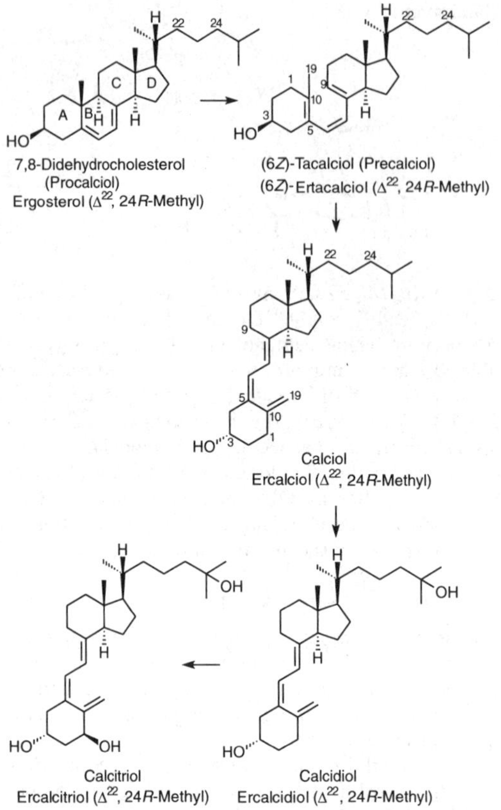

Figure 1: Formation of calcitriol (or ercalcitriol) from 7,8-didehydrocholesterol (or ergosterol).

By an analogous route from ergosterol (see sterols) *ercalciol* (*ergocalciferol*, *vitamin D_2) is formed, which is further transformed to *ercalcitriol* (*1α,25-dihydroxyergocalciferol*, 1α,25-dihydroxyvitamin D_2). Ercalcitriol exhibits the same antirachitic activity as calcitriol in humans and animals. Also, $\Delta^{5,7}$-unsaturated sterols have been tested as starting materials for the synthesis of structurally modified calciols, e. g., 22,23-dihydroergosterol, 7,8-didehydrositosterol, -stigmasterol, and -campesterol (see sterols and vitamin D). However, these calciol analogues exhibit low or no activity.

Occurrence: Calciol (cholecalciferol) has been detected in all investigated vertebrates, sometimes together with small amounts of ercalciol, probably formed from ergosterol consumed with food. Fish liver oils (cod-liver oil) are especially rich in calciol (e. g., halibut and tuna 5 μg/100 g). Egg yolk contains 2 μg/100 g, most other foods markedly less (calf liver, e. g., 0.25 μg/100 g). Surprisingly, calciol and its metabolites are occasionally found in higher plants, especially in the so-called calcinogenous plants which, when fed to animals, can cause symptoms similar to hypervitaminosis D, e. g., calciol in yellow oats *Trisetum flavescens* and in alfalfa *Medicago sativa* (in the latter 0.1 μg/100 g dry weight) as well as calciol and the 3-*O*-glucosides of calcidiol or calcitriol in *Solanum malacoxylon* [1,2].

Metabolism: Calcidiol is hydroxylated to C. in the kidneys.

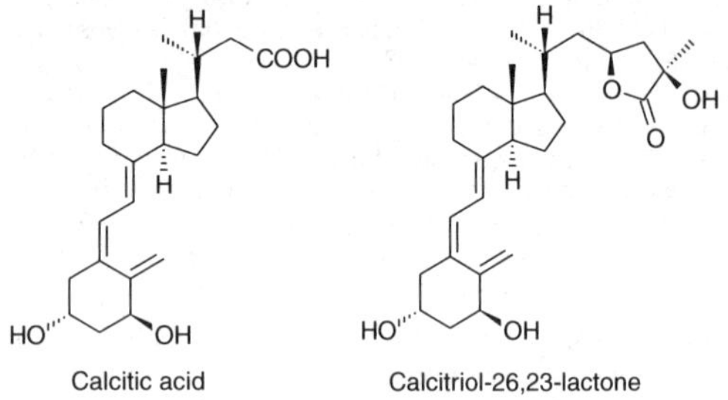

Figure 2: Metabolites of calcitriol.

The main metabolic pathway of C. proceeds through 23- and 26-hydroxylation to 23*S*,26-dihydroxycalcitriol and further oxidation to *calcitriol-26,23-lactone*. In total over 50 metabolites of calciol and ercalciol have been isolated and identified to date. Most of these secondary products have no or a very reduced antirachitic activity but do possess other interesting physiological properties the relevance of which is being investigated.

Physiological activity & therapeutic use: Under normal physiological conditions humans and mammals can produce sufficient amounts of calciol. The daily need for humans is about 2.5 μg for adults or 10 μg for children. The main action of C. in humans and vertebrates is to control gastrointestinal calcium and phosphate ion resorption and to promote mineralization of bones. The result of calciol deficiency is the occurrence of rachitis (rickets, English disease, Glisson's disease) in infants and small children. In adults demineralization of the skeleton leads to bone remodelling

Table: Data of Calcitriol, Ercalcitriol and their precursors.

Steroid	molecular formula	M_R	mp. [°C]	$[\alpha]_D$	CAS
7,8-Didehydrocholesterol → Calcitriol:					
(6*Z*)-Tacalciol	$C_{27}H_{44}O$	384.65	oil		1173-13-3
Calciol	$C_{27}H_{44}O$	384.65	87–88	+84.8° (acetone)	67-97-0
Calcidiol	$C_{27}H_{44}O_2$	400.65	oil		19356-17-3
Calcitriol	$C_{27}H_{44}O_3$	416.64	118–119	+47.9° (C_2H_5OH)	32222-06-3
Ergosterol → Ercalcitriol:					
(6*Z*)-Ertacalciol	$C_{28}H_{44}O$	396.66	oil	+43° (benzene)	21307-05-1
Ercalciol	$C_{28}H_{44}O$	396.66	114.5–117	+102.5° (C_2H_5OH)	50-14-6
Ercalcidiol	$C_{28}H_{44}O_2$	412.66	96–97	+56.8° (C_2H_5OH)	21343-40-8
Ercalcitriol	$C_{28}H_{44}O_3$	428.66	169–170	+47.2° (C_2H_5OH)	60133-18-8

and softening (osteomalacia). Cause of calciol deficiency is often insufficient exposure to sunlight and thus reduced formation of calciol in the skin. Prophylaxis for rachitis is exposure to sunlight or UV radiation and intake of calciol or ercalciol. On overdosing calcium is mobilized from bones (hypercalcemia) and deposited especially in the kidneys and blood vessels, leading to numerous symptoms of diseases and finally impairment of renal function. Recent investigations suggest that C. is possibly involved in cell differentiation processes, e. g., of human leukocytes and myelocytic leukemia cells, as well as of melanoma cells, and in epidermal keratinositis in mice [3]. C. and synthetic analogs have successfully been used in therapy for psoriasis. Steroid inhibitors of calcitriol biosynthesis have been developed. Growth regulating activities of calciol (or its metabolites) have also been detected in higher plants, e. g., for formation of adventive roots [2].

Lit.: [1] Science **194**, 853 ff. (1976). [2] Physiol. Plant. **74**, 391–396 (1988). [3] Med. Res. Rev. **7**, 333 (1987); Bohl & Duax (eds.), Molecular Structure and Biological Activity of Steroids, p. 293–316, Boca Raton: CRC Press 1992.
gen.: Adv. Drug Res. **28**, 269–312 (1996) (review of pharmacology) ▪ Annu. Rep. Med. Chem. **19**, 179 (1984) ▪ Annu. Rev. Biochem. **52**, 411 (1983) ▪ Fieser & Fieser, Steroids, New York: Reinhold 1959 ▪ Norman et al. (eds.), Vitamin D, Berlin: de Gruyter 1982, 1985, & 1988 ▪ Pure Appl. Chem. **64**, 1809 (1992) (synthesis of calcitriol) ▪ Zechmeister **39**, 63–121 ▪ Zeelen, p. 251–263. – *Synthesis review:* Chem. Rev. **95**, 1877–1952 (1995) ▪ Process Chem. Pharm. Ind. **1999**, 73–89 ▪ Synthesis **1994**, 1383–1398. – *[HS 2936 29]*

Calcium 4-*O*-β-D-glucopyranosyl-*cis-p*-coumarate see leaf-movement factors.

Caldariellaquinone.

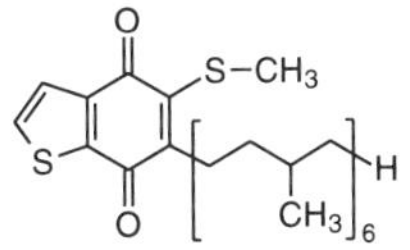

$C_{39}H_{66}O_2S_2$, M_R 631.07, orange-red oil, $[\alpha]_D$ −5° ($CHCl_3$), component of the extremely thermophilic and acidophilic *archaebacterium *Caldariella acidophila*. C. is obtained from the lipid fraction by chromatography on silica gel.

Lit.: J. Bacteriol. **171**, 6610 (1989) ▪ J. Chem. Soc., Perkin Trans. 1 **1977**, 653; **1990**, 2346 ff. (biosynthesis) ▪ J. Chem. Soc., Chem. Commun. **1986**, 733 ▪ J. Org. Chem. **48**, 4312 (1983) (synthesis). – *[CAS 63693-26-5]*

Calendula oil. Extract of the fat oil from flowers of marigold (*Calendula officinalis*, Asteraceae) for pharmaceutical and cosmetic uses, e. g., anti-inflammatory ointments. C. for technical uses is also obtained from marigold seeds. The seed oil contains large amounts (ca. 60%) of esterified *calend(ul)ic acid* [(8*E*,10*E*,12*Z*)-octadecatrienoic acid, $C_{18}H_{30}O_2$, M_R 278.44, mp. 40–40.5 °C]. The conjugated double bonds of calendulic acid are readily oxidized and polymerized so that C. can also serve as starting material for paints and varnishes.

Lit.: Beilstein E IV **2**, 1791 f. (calendulic acid) ▪ Murphy (ed.), Designer Oil Crops, p. 254 ff., Weinheim: VCH Verlagsges. 1994 ▪ Nachr. Chem. Tech. Lab. **47**, 1325 (1999) (biosynthesis). – *[HS 3301 29; CAS 5204-87-5 (calendulic acid)]*

Calend(ul)ic acid see Calendula oil.

Calicheamicins.

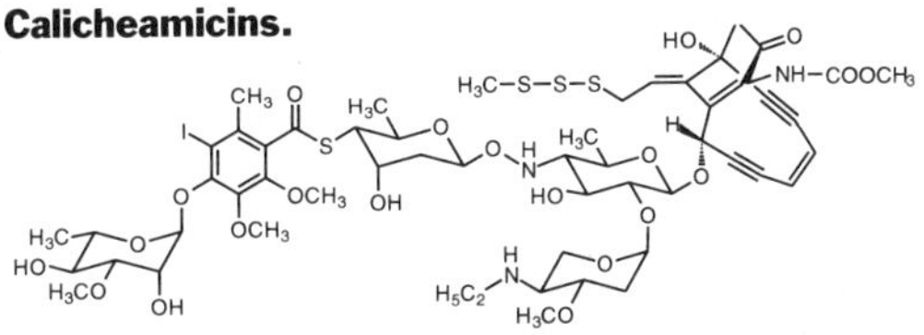

Series of closely related *enediyne antibiotics produced by *Micromonospora echinospora* subsp. *calichensis*. The main component is *C.* γ_1^I {$C_{55}H_{74}IN_3O_{21}S_4$, M_R 1368.36, powder, $[\alpha]_D$ −103° (C_2H_5OH)}. The other C. differ in having bromine in place of iodine, different *N*-alkyl residues or lacking carbohydrate groups. C. are active against Gram-positive bacteria in the subpicogram range. Together with the structurally related *esperamycins they are the most potent known antitumor antibiotics in some animal models (4000 times more active than doxorubicin, see adriamycin), but are too toxic for practical use. The necessary dose for antitumor activity is merely 0.5–1.5 µg/kg body weight. C., linked to anti-CD33 monoclonal antibody, is tested in phase III of clinical trials against acute myeloid leukemia (AML). Like other enediynes C. also effect cleavage of the DNA double helix with retention of the single strands. The first total synthesis of *C.* γ_1^I was realized in 1992 by K.C. Nicolaou's group.

Lit.: *Review:* Acc. Chem. Res. **24**, 235 (1991) ▪ Angew. Chem. Int. Ed. Engl. **30**, 1387–1416 (1991). – *Structure:* J. Am. Chem. Soc. **110**, 631, 6890 (1988); **114**, 985–997, 10082 (1992). – *Synthesis:* Angew. Chem. Int. Ed. Engl. **32**, 1377–1385 (1993); **33**, 855, 858 (1994) ▪ J. Am. Chem. Soc. **115**, 7593–7635 (1993); **118**, 4904 (1996); **119**, 6739 (1997); **120**, 3518, 10332, 10350 (1998) [(±)-calicheamicinone] ▪ Synform **8**, 83–112 (1990) ▪ Tetrahedron **55**, 3277 (1999) (x-ray). – *Activity:* Angew. Chem. Int. Ed. Engl. **33**, 183 (1994); **35**, 2797–2801 (1996) ▪ J. Am. Chem. Soc. **112**, 4554 (1990); **116**, 3197 (1994); **117**, 5750, 8074 (1995) ▪ Proc. Natl. Acad. Sci. U. S. A. **89**, 4608 (1992) ▪ Science **240**, 1198 (1988). – *[CAS 108212-75-5 (C.* γ_1^I*)]*

Callicarpone.

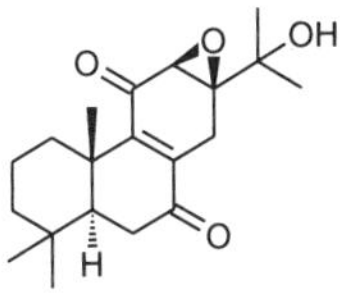

$C_{20}H_{28}O_4$, M_R 332.44, mp. 111–112 °C, yellow needles, $[\alpha]_D^{23}$ −188° ($CHCl_3$), a diterpene with *abietane structure from *Callicarpa candicans*; fish toxin.

Lit.: Agric. Biol. Chem. **31**, 498 (1967) ▪ Karrer, No. 6006. – *[CAS 5938-11-4]*

Callimorphine.

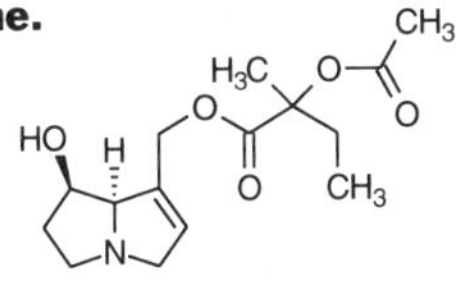

$C_{15}H_{23}NO_5$, M_R 297.35. A *pyrrolizidine alkaloid found in some butterfly species of the family of the tiger moths (Arctiidae). The caterpillars feed on pyrrolizidine-containing plants and store the alkaloids in their body [1,2]. During pupation C. is formed in the moths by transesterification of a part of the stored pyrrolizidine alkaloids with an acid produced by the insect[3,4]. On account of the high alkaloid content, which can amount to several percent of the dry weight[3,5], they are protected against predators. The moths often have a bright color pattern as warning signal[6].

Lit.: [1] Tetrahedron Lett. **21**, 1383 f. (1980). [2] J. Chem. Ecol. **16**, 543 (1990). [3] Z. Naturforsch. C **45**, 1185 (1990). [4] Biochem. Syst. Ecol. **18**, 549 (1990). [5] Biol. J. Linn. Soc. **12**, 305 (1979). [6] J. Chem. Ecol. **16**, 165 (1990).
gen.: J. Nat. Prod. **52**, 360 (1989) ▪ Pelletier **9**, 155–233. – *[CAS 74991-73-4]*

Callistephin (*pelargonidine-3-*O*-*β*-glucopyranoside, 3-*β*-D-glucopyranosyloxy-4′,5,7-trihydroxyflavylium). $C_{21}H_{21}O_{10}^+$, M_R 433.39, uv_{max} 515 nm (ethanol + HCl), dark brown-red needles with bronze. C. belongs to the *anthocyanins and occurs in the flowers of the purple-red aster (*Callistephus chinensis*, Asteraceae) and carnation (*Dianthus caryophyllus*, Caryophyllaceae) as well as in fruits of strawberry species (*Fragaria*, Rosaceae).

Lit.: Acta Chem. Scand. **46**, 872 (1992) ▪ Karrer, No. 1706. – *[CAS 18466-51-8]*

Calonectrin see scirpenols.

Calvatic acid [4-(cyano-*NNO*-azoxy)benzoic acid].

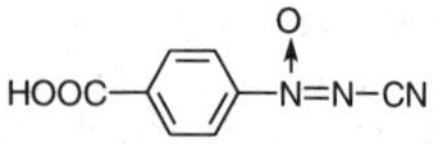

$C_8H_5N_3O_3$, M_R 191.15, light yellow powder, mp. 111–112 °C. Azoxycyanide with antitumor activity from cultures of the puffballs *Calvatia lilacina*, *C. craniiformis*, and *Lycoperdon pyriforme*. C. has antifungal and antibacterial properties and exhibits an intraperitoneal LD_{50} of 125 mg/kg in mice.

Lit.: Gazz. Chim. Ital. **106**, 1107 (1976) ▪ J. Antibiot. **28**, 87 (1975); **39**, 864 (1986) (synthesis) ▪ Tetrahedron Lett. **1974**, 3431. – *[CAS 54723-08-9]*

Calycanthine.

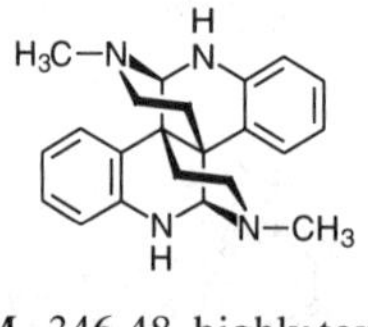

(+)-form

$C_{22}H_{26}N_4$, M_R 346.48, highly toxic crystals, mp. 245 °C (vacuum), readily soluble in organic solvents, poorly soluble in water. (+)-C. occurs in plants of the genus *Calycanthus* (Calycanthaceae) and in the form of its optical antipode in skin secretions of the South American arrow poison frog. (+)-form from *Calycanthus floridus*, $[\alpha]_D$ +684° (C_2H_5OH); (−)-form from *Phyllobates terribilis*, $[\alpha]_D^{25}$ −570° (CH_3OH).

Lit.: Beilstein E V **26/12**, 42 ▪ J. Am. Chem. Soc. **118**, 8166 (1996) (synthesis *meso*-C.) ▪ Phytochemistry **31**, 317 (1992) ▪ Tetrahedron **23**, 4131 (1967). – *[CAS 595-05-1 (+); 85548-42-1 (−)]*

Calyculins. Strongly antitumor active metabolites from the Japanese marine sponge *Discodermia calyx*. Phosphatases 1 and 2A have been identified as the molecular sites of action. IC_{50} 0.5–2 nM (PP1); 0.1–1 nM (PP2A). With regard to PP1 C. A is about 30–200 times more effective than *okadaic acid. Acidic, alkaline, and protein-tyrosine phosphatases are not affected. C. is cell permeable and can be used to study Ca^{2+}-activated K^+ channels in smooth muscle cells. Use of C. enables a correlation between fibroblast differentiation and the degree of protein phosphorylation to be made. Insecticidal properties have been found for C. E and F. C. are well soluble in DMSO and ethanol.

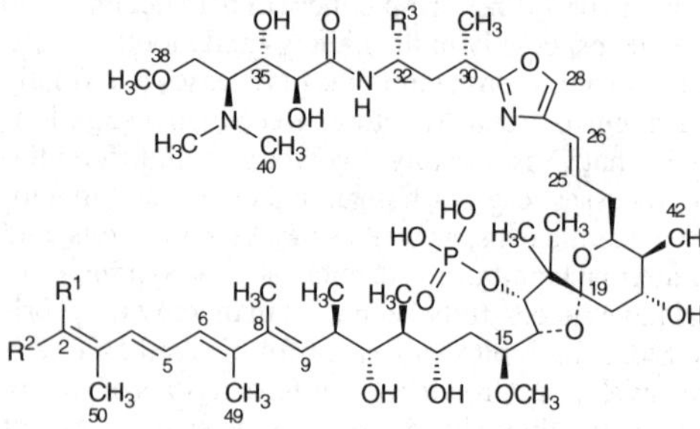

R¹	R²	R³	
CN	H	H	C. A
H	CN	H	C. B
CN	H	CH_3	C. C
H	CN	CH_3	C. D
(6Z)-Isomer of C. A			C. E
(6Z)-Isomer of C. B			C. F
$CONH_2$	H	H	Calyculinamide A[1]
H	$CONH_2$	H	Calyculinamide B[1]

Table: Data of Calyculins.

	molecular formula	M_R	mp. [°C]	$[\alpha]_D^{23}$ (C_2H_5OH)	CAS
C. A	$C_{50}H_{81}N_4O_{15}P$	1009.18	247–249	−60°	101932-71-2
C. B	$C_{50}H_{81}N_4O_{15}P$	1009.18	amorphous	−61°	107537-44-0
C. C	$C_{51}H_{83}N_4O_{15}P$	1023.21	amorphous	−65°	107537-45-1
C. D	$C_{51}H_{83}N_4O_{15}P$	1023.21	amorphous	−41°	107447-09-6
C. E	$C_{50}H_{81}N_4O_{15}P$	1009.18	137–140	−83°	133445-05-3
C. F	$C_{50}H_{81}N_4O_{15}P$	1009.18	152–155	−33°	133445-06-4

Lit.: [1] J. Org. Chem. **62**, 2636, 2640 (1997).
gen.: Agric. Biol. Chem. **55**, 2765 (1991) ▪ Angew. Chem. Int. Ed. Engl. **35**, 673 ff. (1994) ▪ Biochem. Biophys. Res. Commun. **159**, 871 (1989); **176**, 288 (1991) ▪ Bioorg. Med. Chem. Lett. **9**, 717 (1999) (structure) ▪ Biotechnology **8**, 732 (1990) ▪ Cancer Res. **50**, 3521 (1990) ▪ FEBS Lett **314**, 149 (1992) ▪ Helv. Chim. Acta **75**, 1593–1603 (1992) ▪ J. Am. Chem. Soc. **114**, 9434–9453 (1992); **120**, 12435 (1998) (synthesis C. C); **121**, 10468–10486 (1999) ▪ J. Chem. Soc., Chem. Commun. **1992**, 1236–1242; **1996**, 871 ▪ J. Org. Chem. **61**, 6139–6161 (1996); **63**, 7596 (1998) (total synthesis C. A, C. B) ▪ J. Pharmacol. Exp. Ther. **250**, 388 (1989) ▪ J. Toxicol. Sci. **14**, 326 (1989) ▪ Spec. Publ. Royal Soc. Chem. **119**, 117–134 (1993) ▪ Tetrahedron Lett. **32**, 5605, 5983 (1991) (abs. configuration).

Calystegines. General name for a group of polyhydroxylated *tropane alkaloids[1].

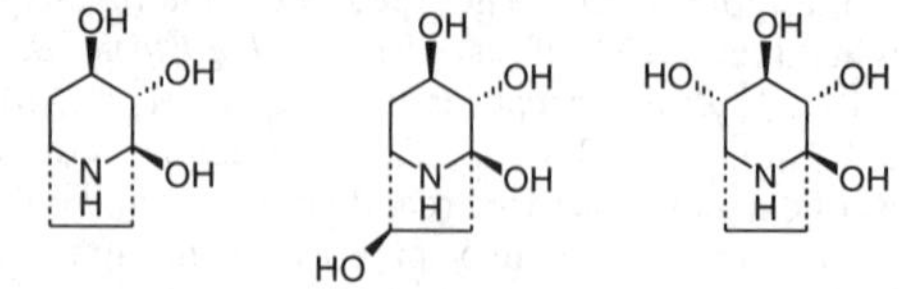

Calystegine A_3 Calystegine B_1 (relative configuration) Calystegine B_2

The C. were first detected in *Calystegia sepium* (Convolvulaceae)[1], and have since been found in several species of the Solanaceae[2]. Their occurrence in tro-

Table: Data of Calystegines.

	molecular formula	M_R	CAS
C. A_3	$C_7H_{13}NO_3$	159.19	131580-36-4
C. B_1	$C_7H_{13}NO_4$	175.18	127414-86-2
C. B_2	$C_7H_{13}NO_4$	175.18	127414-85-1

pane alkaloid-forming species such as *Atropa* and *Hyoscyamus* suggests that they are biogenetically derived from pseudotropine occurring in these species. Like other *polyhydroxy alkaloids, C. are glycosidase inhibitors. The tetrahydroxylated C. B in particular are good α- and β-glucosidase inhibitors ($K_i=4.3 \cdot 10^{-6}$ and $7 \cdot 10^{-6}$), see *Lit.*[3]. To date, the C. A_3, A_5, B_1, B_2, B_3, C_1, C_2 and B_4 (a trehalase inhibitor from *Scopolia japonica*)[4] are known.

Lit.: [1] Phytochemistry **29**, 2125 ff. (1990). [2] Plant Cell, Tissue Org. Cult. **38**, 235–240 (1994). [3] Arch. Biochem. Biophys. **304**, 81–88 (1993). [4] Carbohydr. Res. **293**, 195 (1996).
gen.: Carbohydr. Res. **304**, 173 (1997) ▪ Phytochem. Analysis **4**, 193–204 (1993) ▪ Phytochemistry **45**, 425 (1997) (occurrence) ▪ Synlett **1998**, 316 (synthesis C. B_2) ▪ Tetrahedron **52**, 15137 (1996) (synthesis).

Calysterol (23,24-(*R*)-ethylidene-cholesta-5,23-dien-3β-ol).

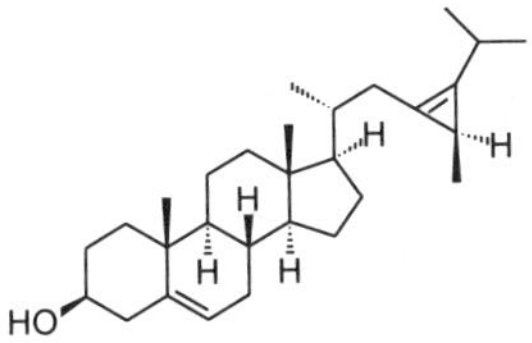

Sterol from the Mediterranean sponge *Calyx nicaensis* containing the unusually reactive cyclopropene structure. $C_{29}H_{46}O$, M_R 410.68, mp. 114–116 °C, $[\alpha]_D^{20}$ −29.3° ($CHCl_3$).

Lit.: J. Am. Chem. Soc. **105**, 4407 (1983); **117**, 1849 (1995) ▪ J. Chem. Soc., Chem. Commun. **1987**, 1441 (biosynthesis) ▪ Tetrahedron **31**, 1715 (1975). – *[CAS 57331-04-1]*

Camelliagenin C see oleanane.

Camoensine.

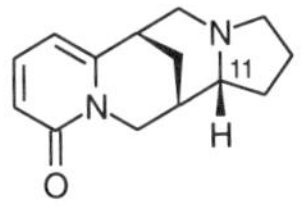

$C_{14}H_{18}N_2O$, M_R 230.31, non-crystalline, $[\alpha]_D^{20}$ −108° ($CHCl_3$), *quinolizidine alkaloid from the Fabaceae genera *Camoensia* and *Melolobium*[1]. It provides the name for the C.-type which, in contrast to the mostly six-membered ring system of the quinolizidine alkaloids, contains a five-membered ring. The tetrahydro compound is called *camoensidine* {$C_{14}H_{22}N_2O$, M_R 234.34, $[\alpha]_D^{20}$ −67° (C_2H_5OH)}, the 11-epimer *leontidine*, {$[\alpha]_D$ −180° (C_2H_5OH)}. Leontidine is also found is some species of the Berberidaceae genus *Leontice*[1].

Lit.: [1] Waterman **8**, 197–239.
gen.: Beilstein E V **24/3**, 403; E V **24/2**, 292 ▪ Pelletier **2**, 105–148. – *[HS 2939 90; CAS 58845-83-3 (C.); 58845-84-4 (camoensidine); 35721-27-8 (leontidine)]*

Campachy wood (logwood) see haematoxylin.

Campestane, Campesterol see sterols.

Camphene (2,2-dimethyl-3-methylenbicyclo[2.2.1]-heptane).

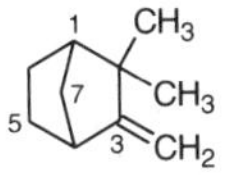

$C_{10}H_{16}$, M_R 136.24, crystalline, bicyclic *monoterpene with isocamphane structure occurring in nature in two optically active forms and a racemate: (1*R*,4*S*)(+)- and (1*S*,4*R*)(−)-C., mp. 50–51 °C, $[\alpha]_D$ ±119° (benzene); (±)-C., mp. 47 °C. Optically active C. racemizes readily, thus various values for optical rotation are found in the literature. C. occurs in many essential oils, e.g., Ceylon, *citronella, *turpentine, and *bergamot oil.

Use: In the perfume industry and for production of camphechlor (toxaphen), a chlorinated C. that was used as an insecticide. Since it accumulates in the body fat of mammals its use as crop protection agent is no longer permitted in Germany.

Lit.: Beilstein E IV **5**, 461 ▪ Chem. Ber. **111**, 2527 (1978) (synthesis) ▪ Karrer, No. 65 ▪ Merck-Index (12.), No. 1777. – *[HS 2902 19; CAS 5794-03-6 ((+)-C); 5794-04-7 ((−)-C.)]*

Camphor (1,7,7-trimethylbicyclo[2.2.1]heptan-2-one, bornan-2-one).

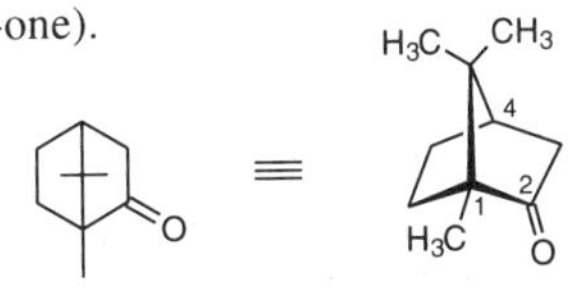

(+)-Camphor

$C_{10}H_{16}O$, M_R 152.24, crystalline bicyclic *monoterpene with bornane structure occurring in nature in two optically active forms and a racemate: (1*R*,4*R*)(+)-C. (Japanese camphor), (1*S*,4*S*)(−)-C. (matricaria camphor), mp. 178–179 °C, $[\alpha]_D$ ±40° (95% C_2H_5OH); (±)-C., mp. 179 °C. (+)-C. is the stereochemical reference substance for many monoterpenes because its absolute configuration was chemically correlated with glucose by Freudenberg[1].

Occurrence: In essential oils of many plants (in particular Lauraceae, Lamiaceae, and Asteraceae). (+)-C. in pure form was isolated in the second half of the 18th century from the camphor tree (*Cinnamomum camphora*, Lauraceae). (±)-C. is produced from α-*pinene.

Use: Mainly as a softener of celluloid. Since C. sublimes at 20 °C it is used against moths in textiles. C. is a component of many popular ointments for use against rheumatic pain, neuralgias, and inflammations, whereas its internal use as analeptic and secretogogue is declining.

Lit.: [1] Justus Liebigs Ann. Chem. **587**, 213 (1954); **594**, 76 (1955).
gen.: Beilstein E IV **7**, 213 ff. ▪ Helv. Chim. Acta **76**, 607 (1993) (review) ▪ Karrer, No. 562 ▪ Merck-Index (12.), No. 1779 ▪ Ullmann (5.) **A 11**, 173. – *[HS 2914 21; CAS 464-49-3 ((+)-C.); 464-48-2 ((−)-C.)]*

Camptothecin.

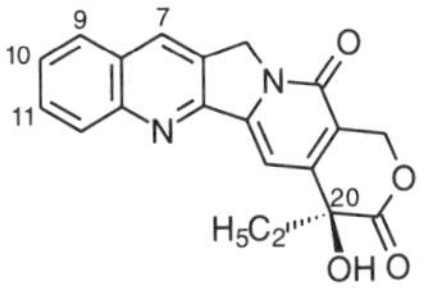

$C_{20}H_{16}N_2O_4$, M_R 348.36, light yellow needles (CH_3OH), mp. 264–267 °C (decomp., also given: mp. 275–277 °C or 287–288 °C), $[\alpha]_D$ +31° ($CHCl_3/CH_3OH$, 8:2), very poorly soluble in water. C. is isolated from the wood and bark of the endemic Chinese tree *Camptotheca acuminata* (Nyssaceae), as well as from *Ervatamia heyneana* (Apocynaceae) and *Ophiorrhiza mungos* (Rubiaceae). C. belongs to the *quinoline alkaloids. Numerous syntheses have been described[1].

Use & activity: C. exhibits cytotoxic and cancer inhibiting action[2,3], caused by inhibition of topoisomerase I and DNA strand cleavage, but it is too toxic for use in tumor therapy. Many derivatives of C. with lower toxicity and much improved water solubility are already marketed or are undergoing clinical development for cancer therapy, e. g., 9-amino-camptothecin, DX8951 or lurtotecan; all are inhibitors of topoisomerase I, e. g., irinotecan (campotesin®, camptosar®, topotecin®)[4] or topotecan[5]. Irinotecan was authorized for use against various tumors in 1994 in Japan and in 1995/96 in Europe, topotecan was authorized for use in the U. S. A. 1996 as Hycamtin®. C. also shows insecticidal and growth-inhibiting properties in plants. Antiviral activity through inhibition of DNA replication has also been reported[6].

Biosynthesis: C. is formed from tryptamine and the monoterpene loganin by way of *strictosidine and strictosamide[7,8]. Numerous naturally occurring and cytotoxic derivatives of C. are known, especially derivatives hydroxylated and methoxylated in the 7-, 9-, 10-, or 11-positions.

Lit.: [1] J. Synth. Org. Chem. **57**, 181–195 (1999); Merck-Index (12.), No. 1783; Nachr. Chem. Tech. Lab. **43**, 686–692 (1995); Tetrahedron **52**, 11385 (1996); **53**, 11049–11060 (1997); Tetrahedron Lett. **39**, 6745 (1998); **40**, 3817, 8211 (1999). [2] J. Biol. Chem. **260**, 14873 (1986). [3] Manske **21**, 101–138. [4] Drugs of Today **34**, 777–803 (1998) (review). [5] Drug News, R & D Focus, 24. 2. 1997; 2. 3. 1998; Kleemann-Engel, p. 1916. [6] Kinghorn (ed.), ACS Symp. Ser. **534**, 149–169, Human Med. Agents From Plants (1993); Potmesil & Pinedo (eds.), Camptothecins: New Anticancer Agents, Boca Raton: CRC Press 1995; Cancer Res. **53**, 2823–2829 (1993); J. Med. Chem. **33**, 972 (1990); **36**, 2689–2700 (1993); **38**, 395–401 (1995); J. Org. Chem. **58**, 611–617 (1993); Med. Res. Rev. **18**, 299–314 (1998) (therapy with irinotecan, topotecan). [7] Tetrahedron **37**, 1047 (1981). [8] J. Am. Chem. Soc. **101**, 3358 (1979); **114**, 10971 (1992).

gen.: J. Nat. Cancer Inst. **85**, 271–288 (1993) ▪ Manske **50**, 509–536 ▪ Tang & Eisenbrand (eds.), Chinese Drugs of Plant Origin, p. 239–261, Berlin: Springer 1992. – *Synthesis:* Angew. Chem. Int. Ed. Engl. **35**, 1692 (1996) ▪ Bioorg. Med. Chem. Lett. **7**, 847 (1997) ▪ J. Am. Chem. Soc. **114**, 10971 (1992) ▪ J. Chem. Soc., Perkin Trans. 1 **1998**, 389 ▪ Nachr. Chem. Tech. Lab. **43**, 686–692 (1995) ▪ Tetrahedron **53**, 11049 (1997) ▪ Tetrahedron: Asymmetry **9**, 2285 (1998) ▪ Tetrahedron Lett. **35**, 3613 ff., 5331 ff. (1994); **37**, 5679, 5683 (1996); **41**, 859 (2000) (analogues). – *[HS 2939 29; CAS 7689-03-4]*

Canadensolide.

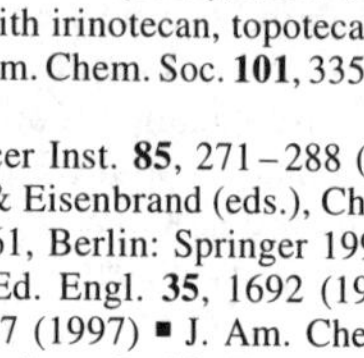

$C_{11}H_{14}O_4$, M_R 210.23, mp. 46–47.5 °C, $[\alpha]_D$ −141° (−168.9°).

C. is produced by *Penicillium canadense*, *Aspergillus flavus* and other species of these genera. C. exhibits the germination of fungi and is biosynthetically related to *avenaciolide. C. has been the target of several total syntheses.

Lit.: J. Chem. Soc., Chem. Commun. **1971**, 1561 ▪ Tetrahedron Lett. **1968**, 727; **1978**, 3233. – *Syntheses:* Tetrahedron **49**, 1211 (1993) (and literature cited therein). – *Biosynthesis:* Turner **2**, 376–377. – *[CAS 20421-31-2]*

Canadine see protoberberine alkaloids.

Cananga oil see ylang-ylang oil.

Canarione.

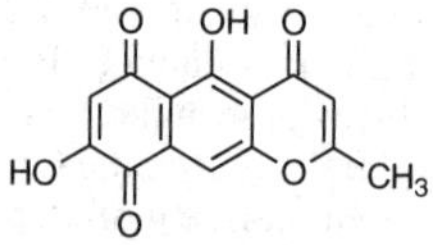

$C_{14}H_8O_6$, M_R 272.21, orange platelets that decompose above 250 °C; with KOH they give a dirty-green color and with $Mg(OAc)_2$ a red-violet color, soluble in acetone. A naphthoquinone from the beard lichen *Usnea canariensis* indigenous to the Canary Islands.

Lit.: Phytochemistry **16**, 121–123 (1977). – *[HS 2914 69; CAS 63681-86-1]*

Canavanine (*O*-guanidino-L-homoserine).

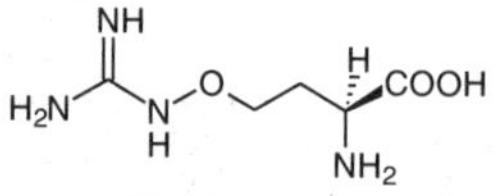

(2*S*)-, L-form

$C_5H_{12}N_4O_3$, M_R 176.18, mp. 184 °C, $[\alpha]_D^{20}$ +7.9° (H_2O); sulfate, mp. 172 °C (decomp.), $[\alpha]_D^{17}$ +19.4° (H_2O).

Occurrence: Non-proteinogenic amino acid in the sword or jack bean (*Canavalia ensiformis*) and in alfalfa (*Medicago sativa*). Widely distributed in seeds of Papilionoidae species.

Biosynthesis & metabolism: C. is probably formed by reactions similar to the urea cycle and cleaved by arginase (EC 3.5.3.1) to canaline (*O*-amino-L-homoserine) and urea. Canaline can convert to L-homoserine.

Toxicity: On account of its structural similarity to L-*arginine, C. is considered to be its antimetabolite and causes poisoning symptoms in mammals[1] and plants[2]. C. inhibits biosynthesis, uptake, and transport of arginine[3]. It can be bound to tRNA by arginyl-tRNA synthetase and be incorporated in place of arginine in proteins[4]. This changes the tertiary and quaternary structure of the protein[5]. C. exhibits diuretic activity[6].

Lit.: [1] Pharmazie **17**, 621 (1962). [2] Proc. Soc. Exp. Biol. Med. **88**, 79 (1955). [3] Cold Spring Harbor Symp. Quant. Biol. **26**, 183 (1960); Naturwissenschaften **50**, 179 (1963). [4] J. Mol. Biol. **14**, 474–489 (1967); J. Biol. Chem. **242**, 5490–5494 (1967). [5] J. Biol. Chem. **244**, 3810–3817 (1969). [6] Nature (London) **283**, 872 (1980).

gen.: Beilstein E IV **4**, 3188 f. ▪ Bell, in: Harborne (ed.), Biochem. Aspects of Plant-Animal Coevolution, p. 143, New York: Academic Press 1978 ▪ Greenstein & Winitz, Chemistry of the Amino Acids, vol. 3, p. 2622–2628, New York: Wiley 1961 ▪ Karrer, No. 2393 ▪ Merck-Index (12.), No. 1786 ▪ Phytochemistry **30**, 1055 (1991) (activity) ▪ Q. Rev. Biol. **52**, 155 (1977) ▪ Rosenthal, Plant Nonprotein Amino and Imino Acids, p. 95–113, New York: Academic Press 1982. – *Chemotaxonomy:* Biochem. Syst. Ecol. **6**, 201 (1978). – *[HS 2925 20; CAS 543-38-4]*

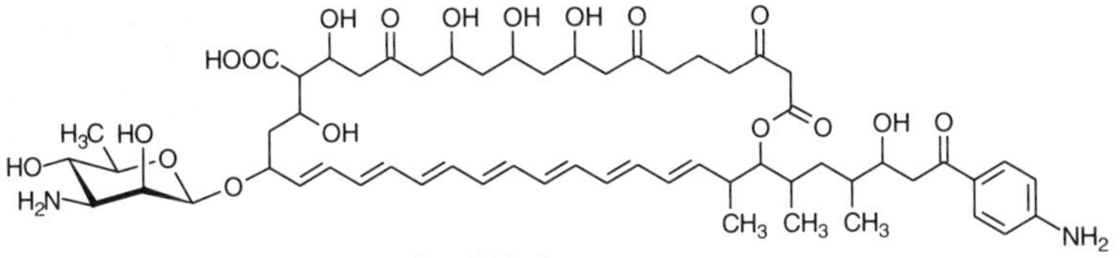

Candicidin D

Candicidin (levorin A). A heptaene *macrolide antibiotic, mixture of four components isolated from *streptomycetes (e.g., *Streptomyces griseus*) and active against yeasts and fungi. The main component *C. D* (levorin A_2)[1], $C_{59}H_{84}N_2O_{18}$, M_R 1109.32, yellow needles or rosettes, insoluble in water, alcohols, hydrocarbons, soluble in DMSO, DMF and lower aliphatic carboxylic acids, $[\alpha]_D$ +175° (pyridine), mp. 150 °C (decomp.). The biosynthesis has been investigated to the epigenetic level[2] (especially phosphate regulation), formation of the aglycone from 4-aminobenzoic acid as starter unit, acetyl- and propanoate units, β-glycosidic linkage to D-mycosamine. C. is purported to have a favorable effect on prostatic hypertrophy.

Lit.: [1] Tetrahedron Lett. **1979**, 1791–1794. [2] J. Antibiot. **25**, 1116–1121 (1972); Trends Biotechnol. **8**, 184–189 (1990); Gene **79**, 47–58 (1989).

gen.: Lipids **16**, 423 (1981) ■ Martindale (30.), p. 319 ■ Mol. Pharmacol. **24**, 270 (1983). – *[HS 2941 90; CAS 39372-30-0 (C. D)]*

Candidin.

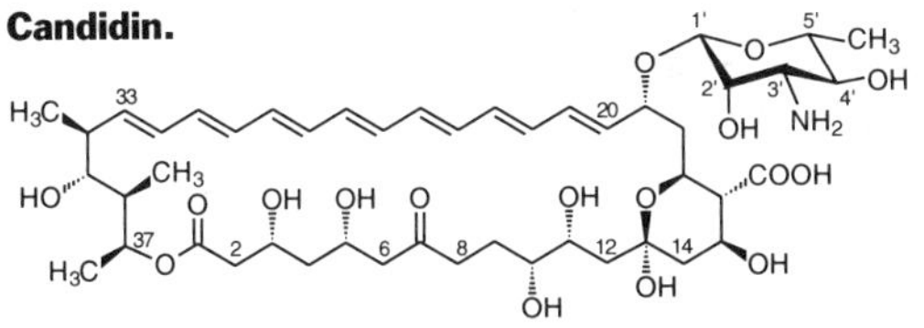

A heptaene *macrolide antibiotic isolated from *Streptomyces viridoflavus* and active against *Candida albicans* and dermatophytes, $C_{47}H_{71}NO_{17}$, M_R 922.08, golden-yellow needles, mp. 180 °C (decomp.), optically dextrorotatory; insoluble in water and organic solvents except pyridine, DMF, and glacial acetic acid. D-Mycosamine is β-glycosidically bound at C-19. C. is similar to *amphotericin B. Unfortunately the name candidin has also been given to two completely different natural products, one an indigo-like dye, the other a chromone dye.

Lit.: J. Antibiot. **36**, 1415 ff. (1983); **46**, 1598–1604 (1993). – *[HS 2941 90; CAS 1405-90-9]*

Canescine see reserpine.

Cane sugar see saccharose.

Cannabinoids. General term for compounds from *Cannabis* species and synthetic derivatives of the compounds. Characteristic for the C., apart from their psychotropic activity, is a series of other pharmacological actions. As narcotic drugs are used the female flowering tips (*marihuana) of the Indian hemp (*Cannabis sativa*) or the resin obtained therefrom (*hashish). Among others, the resin contains dibenzopyran derivatives. Main constituents are *cannabidiol*, antiepileptic and hypnotic effects, *cannabinol* and *tetrahydrocannabinols* (THC); of the two isomers Δ^9-THC (known previously as Δ^1-THC) and Δ^8-THC (with 8,9-double

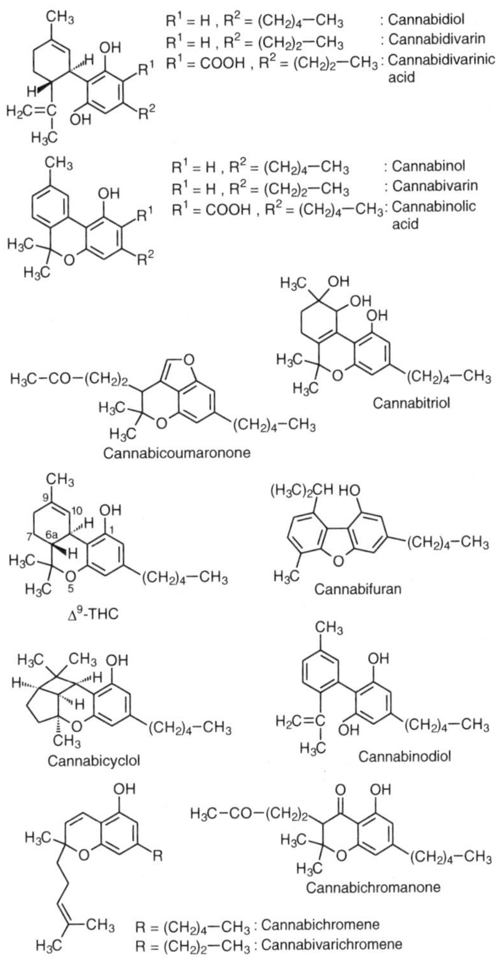

bond, previously known as $\Delta^{1(6)}$-THC) the illustrated Δ^9-THC [for short nabilon (racemic form) or dronabinol] is the physiologically more active form. Dronabinol was available as an antiemetic in USA under the free name Marinol®. Responsible for the pharmacological effects are C. receptors in the brain, of these the THC-binding receptor was cloned and sequenced at the beginning of the 1990's. Endogenous cannabinoids, such as *anandamide, are known to play a role in modulating the pain pathway, synthetic C. have various pharmacological effects[1]. The mentioned compounds as well as those listed in the table are insoluble in water and soluble in organic solvents.

For uses, see tetrahydrocannabinols. The C. can be detected in blood, urine, and saliva with the help of dansyl chloride/fluorimetry, GC, TLC, MS, or HPLC.

Table: Data of Cannabinoids.

name	molecular formula	M_R	mp. (bp.) [°C]	CAS
Cannabinol	$C_{21}H_{26}O_2$	310.43	76–77	521-35-7
Cannabidiol	$C_{21}H_{30}O_2$	314.47	66–67	13956-29-1
Cannabitriol	$C_{21}H_{30}O_4$	346.47	171–173	11003-36-4
Cannabinolic acid	$C_{22}H_{26}O_4$	354.45		2808-39-1
Δ^9 (Δ^1)-Tetra-hydrocanna-binol*	$C_{21}H_{30}O_2$	314.47	(155–157/ 6.67 Pa)	1972-08-3
Δ^8-Tetrahy-drocanna-binol*	$C_{21}H_{30}O_2$	314.47	(175–178/ 13.3 Pa)	5957-75-5
Cannabi-nodiol	$C_{21}H_{26}O_2$	310.43		39624-81-2
Cannabi-varin	$C_{19}H_{22}O_2$	282.37		33745-21-0
Cannabidi-varin	$C_{19}H_{26}O_2$	286.41		24274-48-4
Cannabidi-varinic acid	$C_{20}H_{26}O_4$	330.41	102–105	31932-13-5 (*R,R*) 112420-11-8 (*S,S*)
Cannabifuran	$C_{21}H_{26}O_2$	310.43		56154-58-6
Cannabi-chromanone	$C_{20}H_{28}O_4$	332.43		56154-57-5
Cannabi-chromene	$C_{21}H_{30}O_2$	314.47		20675-51-8
Cannabi-coumaronone	$C_{21}H_{28}O_3$	328.45	amorphous	70474-97-4
Cannabi-varichromene*	$C_{19}H_{26}O_2$	286.41		41408-19-9
Cannabi-cyclol	$C_{21}H_{30}O_2$	314.47	146–147	21366-63-2

*=hallucinogenic

Lit.: [1] J. Neurosci. **18**, 451 (1998); Scrip Magazine Dez. 1997, 22–26.
gen.: Bruneton, Pharmacognosy, Phytochemistry, Medicinal Plants, p. 371, Paris: Lavoisier Publ. 1995 ▪ Hager (5.) **4**, 640–660; **3**, 1155f., 1174f. ▪ Phytother. Res. **3**, 219–231 (1989) ▪ Zechmeister **25**, 175–213. – *Synthesis:* Agurell (ed.), Cannabinoids: Chem., Pharmacol., Ther. Aspects, Orlando: Academic Press 1984 ▪ ApSimon **4**, 185–262 ▪ Bioorg. Med. Chem. **6**, 2383 (1998) ▪ J. Chem. Soc., Perkin Trans. 1 **1992**, 605 ▪ Progr. Med. Chem. **24**, 159–207 (1987) (THC). – *Toxicology:* Sax (8.), CBD 599–CBD 760; TCM 000, TCM 250. – *Activity:* Curr. Med. Chem. **6**, 685–773 (1999) ▪ Kleiber & Kovar, Auswirkungen des Cannabiskonsums, Stuttgart: Wiss. Verlagsges. 1998 ▪ Lancet **365**, 61 (1993) ▪ Mechoulam, Cannabinoids as Therapeutic Agents, Boca Raton: CRC Press 1986 ▪ Modern Drug Discovery (11./12.) **1999**, 39–46 ▪ Pharmacol. Rev. **38**, 1, 151 (1986) ▪ Pharm. Biochem. Behav. **40**, 695 (1991) ▪ Planta Medica **57**, S 60–67 (Suppl. 1) (1991) ▪ Stud. Nat. Prod. Chem. **19 E**, 185–244 (1997).

Cannabis see hashish.

Cantharidin.

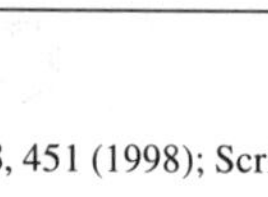

$C_{10}H_{12}O_4$, M_R 196.20, orthorhombic platelets, mp. 218 °C, sublimes above 84 °C, insoluble in water, poorly soluble in organic solvents except acetone and chloroform, readily soluble in acids and alkalis. C. is a toxic component (ca. 10–50 mg are lethal for humans) of the hemolymph (0.4–1%) of oil beetles (Meloidae, also in Oedemeridae), such as, e.g., *Lytta vesicatoria* ("spanish fly"). These beetles used to be dried to prepare C.-containing extracts for medical use (vesicant, aphrodisiac). C. exhibits insecticidal and antifeedant activities but is used by the beetles only as a pheromone [1]. The biosynthesis starts from farnesol [2].
Lit.: [1] Z. Naturforsch. C **47**, 290–299 (1992). [2] McCormick & Carrel, in Prestwich & Blomquist, Pheromone Biochemistry, p. 307–350, Orlando: Academic Press 1987. – *[CAS 56-25-7]*

Canthaxanthin (β,β'-carotene-4,4′-dione).

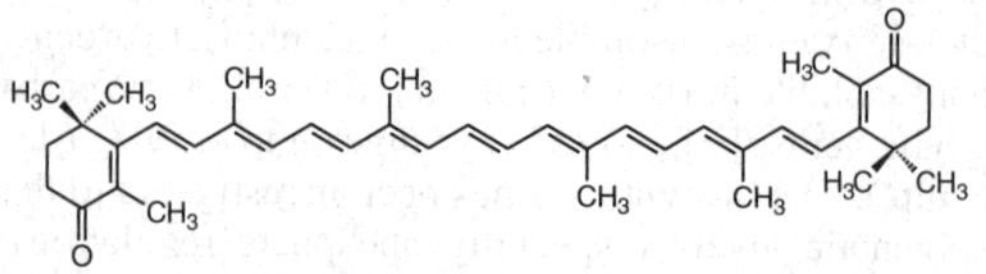

$C_{40}H_{52}O_2$, M_R 564.85, purple crystals, mp. 218 °C. The pink color of the north American chanterelle *Cantharellus cinnabarinus* (Basidiomycetes) is due to its content of canthaxanthin. The edible yellow fungus, much favored in Europe, *C. cibarius* on the other hand contains mainly *β-carotene and related *carotinoids. C. is used to color foods.
Lit.: Beilstein E IV **7**, 2680 ▪ Straub et al., Key to Carotenoids (2.), p. 129, 380, Basel: Birkhäuser 1987 ▪ Zechmeister **51**, 195 ff. – *[CAS 514-78-3]*

Canthin-6-one (6*H*-indolo[3,2,1-*de*][1,5]naphthyridin-6-one).

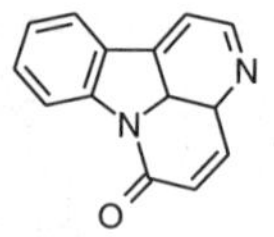

$C_{14}H_8N_2O$, M_R 220.23, pale yellow needles, mp. 162.5–163.5 °C. Parent compound for a group of alkaloids from Rutaceae and Simaroubaceae as well as the fungus *Boletus curtisii* (*curtisine A). C. also cooccurs with its *N*-oxide ($C_{14}H_8N_2O_2$, M_R 236.23, mp. 244–245 °C) in the wood and bark of the tree of heaven (*Ailanthus altissima*, Simaroubaceae). C. exhibits antimicrobial properties and, like the *N*-oxide, has cytotoxic activity against guinea pig keratinocytes.
Lit.: Chem. Pharm. Bull. **29**, 390 (1981) ▪ J. Nat. Prod. **46**, 374 (1983) ▪ Planta Med. **46**, 187 (1982). – *Synthesis:* Chem. Ber. **109**, 705 (1976) ▪ Heterocycles **3**, 7 (1975); **14**, 975 (1980) ▪ J. Am. Chem. Soc. **105**, 907 (1983). – *[CAS 479-43-6 (C.); 60755-87-5 (C. N-oxide)]*

Capnellene [$\Delta^{9(12)}$-capnellene].

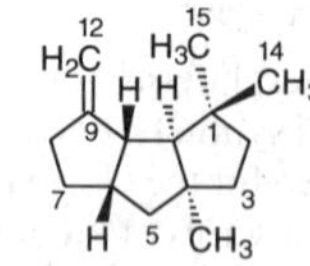

$C_{15}H_{24}$, M_R 204.36, oil, $[\alpha]_D^{20}$ –145° ($CHCl_3$). The tricyclic sesquiterpene hydrocarbon C. (triquinane, decahydrocyclopenta[*a*]pentalene skeleton) is found in the soft coral *Capnella imbricata*. C. inhibits the growth of microorganisms and prevents the attachment of larvae to the coral.
Lit.: *Structure:* Tetrahedron Lett. **22**, 4389, 4393 (1981). – *Synthesis:* J. Am. Chem. Soc. **118**, 7108 (1996) ▪ J. Chem. Soc.,

Perkin Trans. 1 **1992**, 865 ▪ J. Org. Chem. **63**, 4011 (1998) ▪ Tetrahedron **48**, 4559 (1992); **50**, 403, 655 (1994) ▪ Tetrahedron Lett. **38**, 2911 (1997). – *[CAS 68349-51-9]*

Capsaicin [(*E*)-*N*-(4-hydroxy-3-methoxybenzyl)-8-methyl-6-nonene amide].

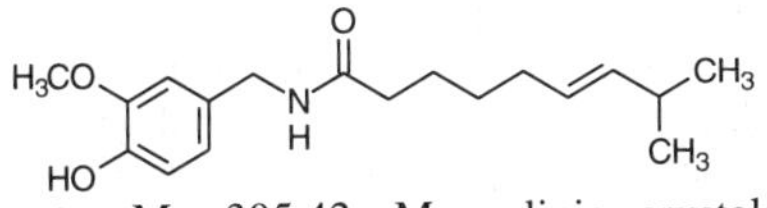

$C_{18}H_{27}NO_3$, M_R 305.42. Monoclinic crystals, mp. 64–65 °C, bp. 210–220 °C (1.3 Pa), barely soluble in water, readily soluble in most organic solvents. C. is responsible for the hot taste of peppers, chilli, and other *Capsicum* species (even at a dilution of $1:10^5$) in which it is present at up to 0.3–0.5%. For extraction and purification, see *Lit.*[1]. C. is a strong irritant.
Toxicology: LD_{50} (mouse p.o.) 47 mg/kg. On i.v. and i.p. administration C. is highly toxic.
Activity: At low doses C. increases hydrochloric acid secretion in the stomach. A chronic overdosing of the spice causes chronic gastritis, kidney and liver damage. On mucous membranes even a small amount of C. causes sensations of burning, heat, and pain; on longer exposure on skin necroses and ulcers arise[2].
Biosynthesis: L-Phenylalanine →→ vanillylamine (rearrangement with retention of the nitrogen atom) and subsequent acylation.
Use: In alcoholic solutions for topical administration for chilblains, rheumatism, etc. The desensitizing activity on neurons is utilized in neurobiological research. As e.g. Dolenon®, C. is applied in therapy of pain and diabetic polyneuropathy.
Lit.: [1] J. Agric. Food Chem. **25**, 1419f. (1977). [2] J. Clin. Toxicol. **25**, 591–601 (1987); Phytother. Res. **2**, 175 (1988).
gen.: Arch. Int. Pharmacodyn. Ther. **303**, 147–166 (1990) (pharmacology) ▪ Beilstein E IV **13**, 2588 ▪ Biosci. Biotechnol. Biochem. **56**, 946–948 (1992) (synthesis) ▪ Chem. Eng. News (4.3.) **1996**, 30f. (pharmacological action) ▪ J. Chromatogr. **144**, 149ff. (1977) (detection) ▪ J. Org. Chem. **53**, 1064–1071 (1988) (review); **54**, 3477f. (1989) (synthesis) ▪ Karrer, No. 1019 ▪ Luckner (3.), p. 366ff. – *Toxicology:* Sax (8.), CBF750. – *[HS 2939 90; CAS 404-86-4]*

Capsanthin [(3*R*,3′*S*,5′*R*)-3,3′-dihydroxy-*β*,*κ*-caroten-6′-one].

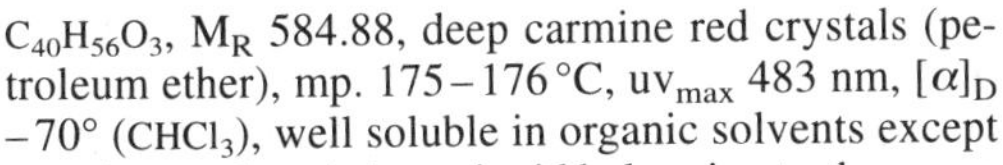

$C_{40}H_{56}O_3$, M_R 584.88, deep carmine red crystals (petroleum ether), mp. 175–176 °C, uv_{max} 483 nm, $[\alpha]_D$ –70° ($CHCl_3$), well soluble in organic solvents except petroleum ether. A *carotinoid belonging to the group of xanthophylls which occurs as red pigment in peppers (*Capsicum annuum*, Solanaceae). C. is authorized in the EU for use in coloring foods as well as in cosmetics.
Lit.: Britton, Liaan-Jensen & Pfander (eds.), Carotenoids, Basel: Birkhäuser 1995 ▪ J. Am. Diet. Assoc. **93**, 284–296, 318–323 (1993) ▪ J. Chem. Soc., Perkin Trans. 1 **1983**, 1465 ▪ J. Chromatograph. **757**, 89 (1997) (HPLC separation from capsorubin) ▪ Karrer, No. 1858 ▪ Packer (ed.) Carotinoids, San Diego: Academic Press 1993 ▪ Tetrahedron Lett. **27**, 2535 (1986). – *[HS 3203 00; CAS 465-42-9]*

Capsidiol.

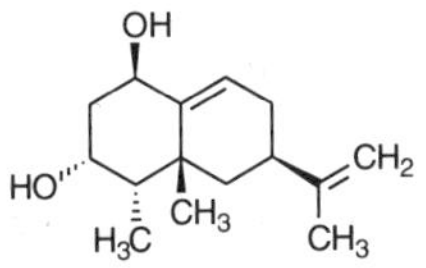

$C_{15}H_{24}O_2$, M_R 236.35, cryst., mp. 152–153 °C, $[\alpha]_D^{22}$ +21° ($CHCl_3$). Fungicidal *sesquiterpene from tobacco plants and pepper (*Capsicum annuum*). It is liberated in stress situations caused by fungal attack[1]. Well soluble in ethanol and methanol, less soluble in ether and chloroform, insoluble in water. For biosynthesis, see *Lit.*[2].
Lit.: [1] Phytopathol. Z. **74**, 141 (1972). [2] Can. J. Chem. **58**, 1894 (1980); Biosci., Biotechnol., Biochem. **58**, 305 (1994); Phytochemistry **28**, 775 (1989). – *[CAS 37208-05-2]*

Capsorubin [(3*S*,3′*S*,5*R*,5′*R*)-3,3′-dihydroxy-*κ*,*κ*-carotene-6,6′-dione].

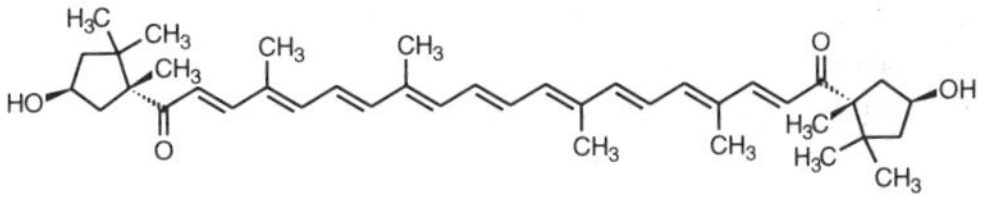

$C_{40}H_{56}O_4$, M_R 600.88, violet-red needles, mp. 201 °C, uv_{max} 468, 503, 541 nm, well soluble in ethanol and acetone. A *carotinoid belonging to the group of xanthophylls that occurs together with *capsanthin in the red fruit skin of pepper (*Capsicum annuum*, Solanaceae) as well as in lilies. C. is authorized for use in the EU for coloring foods and cosmetics.
Lit.: Beilstein E IV **8**, 3304 ▪ Helv. Chim. Acta **66**, 1939 (1983) (synthesis) ▪ Karrer, No. 1859 ▪ Packer (eds.), Carotinoids, San Diego: Academic Press 1993 ▪ Pure Appl. Chem. **35**, 113 (1973). – *[HS 3203 00; CAS 470-38-2]*

Caranine see Amaryllidaceae alkaloids.

Caraway oil. Colorless to yellowish oil, readily undergoes partial resinification on exposure to air. It has a typical strong spicy odor and taste of caraway. It is produced by steam distillation of caraway (*Carum carvi*).
Composition[1]***:*** The major components, also mainly responsible for the typical taste and odor, are (+)-**carvone* (50–60%) and (+)-*limonene* (see *p*-menthadienes) (30–45%); the two components together constitute ca. 95% of the oil. Minor components include carvone derivatives such as *1,6-dihydrocarvone* (see *p*-menthenones).
Use: For improving the aromatic character of foods and alcoholic beverages; in small amounts in the perfume industry (masculine notes); in medicine as carminatives and balneotherapeutic agents.
Lit.: [1] Perfum. Flavor. **13** (2), 68 (1988); **14** (2), 45 (1989); **17** (1), 54 (1992).
gen.: Arctander, p. 124 ▪ Bauer et al. (2.), p. 141 ▪ ISO 8896 (1987). – *Toxicology:* Food Cosmet. Toxicol. **11**, 1051 (1973). – *[HS 3301 29; CAS 8000-42-8]*

Carbapenems.

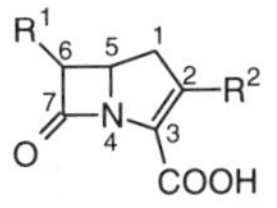

A group of **β*-lactam antibiotics, in which the *β*-lactam ring is condensed to a 2,3-dihydropyrrole ring. The name C. is formally derived from the penem skeleton of 2,3-didehydro*penicillin, which can formally be transformed to a C. by replacement of the sulfur atom by a CH_2 group. In 1979 the first member of the C. group, *thienamycin, was isolated from *Streptomyces cattleya*. Further natural C. are the epithienamycins, asparenomycins, pluracidomycins, and olivanic acids. In addition to their antimicrobial activity, many C. have a competitive inhibiting action with regard to the *β*-lactamases responsible to a major extent for penicillin resistance.

Lit.: Angew. Chem. Int. Ed. Engl. **24**, 180–202 (1985) ▪ Butterworth et al., The New *β*-Lactam-Antibiotics, in Mizrahi & van Wezel (eds.), Advances in Biochemical Processes 1, p. 252–292, New York: Alan R. Liss 1983 ▪ J. Am. Chem. Soc. **100**, 6491 (1978). – *[CAS 83200-96-8]*

Carbazomycins. The carbazole alkaloids C. A–H are isolated from cultures of *Streptoverticillium ehimense* (syn. *S. abikoense*), e.g. *C. A:* $C_{16}H_{17}NO_2$, M_R 255.31, yellow needles, mp. 143–146 °C. C. are antibiotics and interesting synthetic targets.

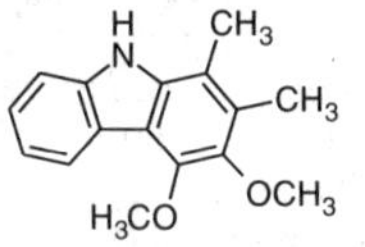

C. A

Lit.: J. Antibiot. **43**, 1623 (1990) (biosynthesis) ▪ J. Chem. Soc., Perkin Trans. 1 **1997**, 349 (synthesis, C. C, D); **1998**, 173 (synthesis C. G, H) ▪ Tetrahedron **52**, 3029 (1996) ▪ Tetrahedron Lett. **38**, 4051 (1997); **40**, 6915 (1999) (synthesis, C. A, B). – *[CAS 75139-39-8 (C. A)]*

Carbohydrates. General name for the very widely distributed natural products of the types polyhydroxyaldehydes (aldoses) and polyhydroxyketones (ketoses) as well as higher molecular compounds that can be converted to these by hydrolysis.

Monosaccharides usually have a chain length of five (pentoses) or six (hexoses) carbon atoms. M. with more (heptoses, octoses, etc.) or less C atoms (tetroses) are relatively rare. The most important and widely distributed M. are: D-*glucose, D-*galactose, D-*mannose, D-*fructose, L-*arabinose, D-*D-xylose, D-*ribose, and D-deoxyribose. In *oligosaccharides* 2–10 monosaccharide molecules are linked together through elimination of water to larger molecules which can be considered as *glycosides (full acetals)[1]. The simplest oligosaccharides are the *disaccharides*, of which three have major significance and occur in the free form, namely *sucrose (cane sugar, beet sugar), *lactose (milk sugar), and maltose (malt sugar). Disaccharides are often components of glycosides (*example:* *gentiobiose of amygdalin). For nomenclature of C. see *Lit.*[1]. For glycosides and oligosaccharides, see also *Lit.*[2], glycosides, and oligosaccharides.

The C. have many functions in nature, as skeletal substances (cellulose in wood), storage substances (starch, saccharose, and other sugars), components of nucleic acids, glycolipids, glycoproteins, glycosphingosides, etc.

For *methods of formation* of oligo- and polysaccharides, see *Lit.*[3]. A considerable proportion of the consumed C. is often indigestible for humans (*cellulose, dextran, lichenin), while herbivores and especially ruminants can digest cellulose and degrade it to acetic, propanoic, and butanoic acids. On account of man's inability to use or digest them the alginates *carrageenan, pectic acid, tragacanth, *agar-agar, carob tree seed flour, etc. are often used in foods not only as thickeners but also as fillers of low nutritional value. In addition, C. are important raw materials for industrial products such as thickening agents and emulsifiers, auxiliary substances for paper and textiles, etc. C. are components of industrial glues[4]. As components of proteins and lipids C. are also of major significance in biochemical regulatory processes such as, e.g., the immune response, now included in the special field "glycerobiochemistry"[5]. Numerous C. with unusual structures are components of *antibiotics[6]. C. are valuable building blocks and raw materials for organic synthesis[7].

Lit.: [1]Pure Appl. Chem. **68**, 1919–2008 (1996); Carbohydr. Res. **187**, 165–171 (1989). [2]Angew. Chem. Int. Ed. Engl. **34**, 1432 (1995); Angew. Chem. Int. Ed. Engl. **34**, 1 (1995); Levy & Tang, The Chemistry of C-Glycosides, Oxford: Elsevier 1995; Contemp. Org. Synth. **3**, 173–200 (1996); Tetrahedron **52**, 1095–1121 (1996); **54**, 9913–9959 (1998); 11317–11362 (1998). [3]Prep. Carbohydr. Chem. **1997**, 1–636; J. Am. Chem. Soc. **121**, 734 (1999). [4]ACS Symp. Ser. **385**, 271–288 (1989); Carbohydr. Res. **189**, 103–112 (1989). [5]Chem. Eng. News (30.9) **1996**, 36–40; J. Am. Chem. Soc. **120**, 11567 (1998); Protein Science **8**, 410 (1999); Nat. Prod. Rep. **14**, 99–110 (1997); Rev. Chem. Biochem.: Top. Curr. Chem. **188**, 1–84 (1997); Townsend & Hotchkiss, Techniques in Glycobiology, New York: Marcel Dekker 1997; Chem. Soc. Rev. **26**, 463 (1997); Allen & Kisailus, Glycoconjugates, New York: Marcel Dekker 1998. [6]Krohn et al., Antibiotics and Antiviral Compounds, p. 327–402, Weinheim: VCH Verlagsges. 1993. [7]Carbohydrates as Organic Raw Materials, Vol. I: Ed. Lichtenthaler 1991; Vol. II: Ed. Descotes 1993; Vol. III: Ed. van Bekkum et al. 1996, Weinheim: VCH Publ.

gen.: ACS Symp. Ser. **386**: Trends in Synth. Carbohydr. Chem. (1989) ▪ Angew. Chem. Int. Ed. Engl. **38**, 2300–2324 (C. mimics) ▪ Barton-Nakanishi **3**, 1–826 ▪ Chapleur, Carbohydrate Mimics, Weinheim: Wiley-VCH 1998 ▪ Collins-Ferrier ▪ David, The Molecular and Supramolecular Chemistry of Carbohydrates, New York: Oxford Univ. Press 1998 ▪ Ernst, Hart & Sinaÿ, Carbohydrates in Chemistry and Biology (4 vol.), Weinheim: Wiley-VCH 2000 ▪ Ferrier (ed.), Carbohydrate Chemistry (SPR Reports No. **27**), London: Royal Soc. Chem. 1995 ▪ Lehmann, Carbohydrates – Structure and Biology, Stuttgart: Thieme 1998 ▪ Lindhorst, Essentials of Carbohydrate, Chemistry and Biochemistry, Weinheim: Wiley VCH 2000. – *Synthesis:* Bols, Carbohydrates Building Blocks, Chichester: Wiley 1996 ▪ Györgydeak & Pelyvas, Monosaccharide Sugars, San Diego: Academic Press 1997 ▪ Prep. Carbohydr. Chem. **1997**, 1–636. – *Analytics:* Capillary Electrophor. (2.), 273–362 (1998) ▪ Carbohydr. Chem. **1998**, 448–502.

***β*-Carboline** (norharmane, 2-carboline, 9*H*-pyrido-[3,4-*b*]indole).

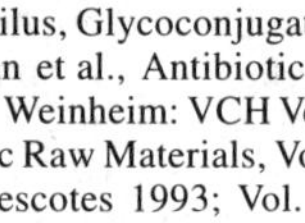

$C_{11}H_8N_2$, M_R 168.20, mp. 198.5 °C (needles), blue fluorescence in dilute acidic solution. *β*-C. is isolated from plants (*Catharanthus roseus*, *Lolium perenne*, *Festuca* species) and microorganisms (streptomycetes and *Nocardia* species).

Activity: Increases the benzo[*a*]pyrene induced mutagenicity and acts as an enzyme inhibitor. Structurally and biologically interesting *β*-C. alkaloids have also been isolated from marine sources. Two groups of these alkaloids are the *eudistomines and the *manzamines.

Lit.: Org. Prep. Proced. Int. **28**, 1–64 (1996) ▪ Prog. Drug Res. **29**, 415 (1985). – *[HS 2939 90; CAS 244-63-3]*

β-Carboline-1-propanoic acid see infractines.

β-Carbolines see harmans.

Carbomycin.

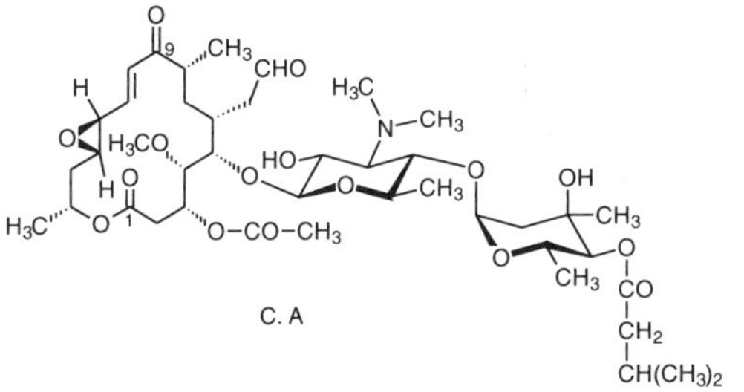

International free name for C. A (magnamycin, deltamycin A_4). C. A is a weakly basic 16-membered *macrolide antibiotic from *Streptomyces halstedii* with the sugars β-D-mycaminose (middle) and α-L-mycarose (right). *C. A:* $C_{42}H_{67}NO_{16}$, M_R 841.99, colorless, fusiform crystals, mp. 210–214 °C, $[\alpha]_D$ −54° (CH_3OH), poorly soluble in water and petroleum ether, soluble in alcohol and chloroform, broad activity against Gram-positive bacteria. Apart from the illustrated C. A there is also a *C. B* ($C_{42}H_{67}NO_{15}$, M_R 825.99) containing a double bond instead of the epoxy ring. The biosynthesis genes of C. were investigated and used through transfer to other *streptomycetes for obtaining hybrid antibiotics.

Use: Against Gram-positive bacteria, rickettsiae, and many viruses, also in veterinary medicine. C. acts on the peptidyltransferase center in bacterial cells by blocking the formation of peptide bonds.

Lit.: Angew. Chem. Int. Ed. Engl. **10**, 236–248 (1971) ▪ Heterocycles **31**, 5 (1990) ▪ J. Am. Chem. Soc. **103**, 1222 ff. (1981) ▪ J. Antibiot. **32**, 878 (1979) ▪ Omura (ed.), Macrolide Antibiotics, New York: Academic Press 1984 ▪ Zechmeister **30**, 336–359. – *Genetics:* Gene **85**, 293–301 (1989) ▪ Katz & Donadio, Genetics and Biochemistry of Antibiotic Production, p. 385–420, Boston: Butterworth-Heinemann 1995. – *Synthesis:* Tetrahedron **41**, 3569–3624 (1985). – *[HS 2941 90; CAS 4564-87-8 (C. A); 21238-30-2 (C. B)]*

Carbon monoxide. CO, M_R 28.01, mp. −213 °C, bp. −190 °C. Gaseous, toxic substance formed in trace amounts by algae and higher plants. CO binds to *hemoglobin (Hb) ca. 250 times more strongly than O_2 (formation of "carboxyhemoglobin", CbOHb); about 3% of human Hb exists as CbOHb.

In the human body CO is formed in the catabolism of *heme after cleavage of globin (protein part) under catalysis by heme oxygenase (EC 1.14.99.3) with the help of O_2 and NADPH. The α-methylene group of heme is quantitatively oxidized to CO. This is the only endogenous source of CO in the human organism. The green tetrapyrrole *biliverdin IXa is formed as further reaction product.

In neurons of the olfactory system (nerves of smell) and in the cerebellum there are relatively high concentrations of the constitutive heme oxygenase-2. There appears to be a locally limited region of CO biosynthe-

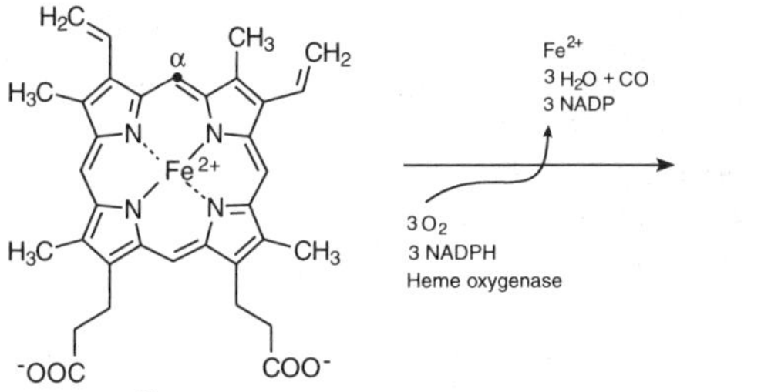

Heme

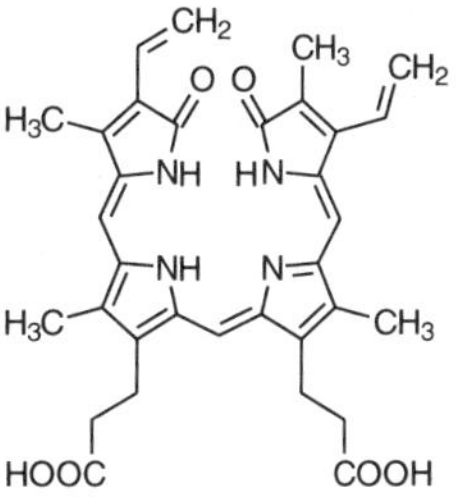

Biliverdin IXa

sis which correlates with the concentration of the intracellular messenger cGMP. Participation of CO in the sense of smell is feasible. An inducible heme oxygenase in the peripheral tissue (e.g., smooth muscle cells) is known. Zinc *protoporphyrin-IX, a selective inhibitor of this enzyme, suppresses the endogenous formation of CO. CO is apparently a physiological regulator of the synthesis of cGMP by means of guanylyl cyclase.

Lit.: J. Biol. Chem. **263**, 3348 (1988) ▪ Nachr. Chem. Tech. Lab. **41**, 339 (1993) ▪ Science **259**, 381 (1993) ▪ Trends Pharmacol. Sci. **12**, 185 (1991). – *[HS 2811 29; CAS 630-08-0]*

Cardenolides [5β-card-20(22)-enolides].

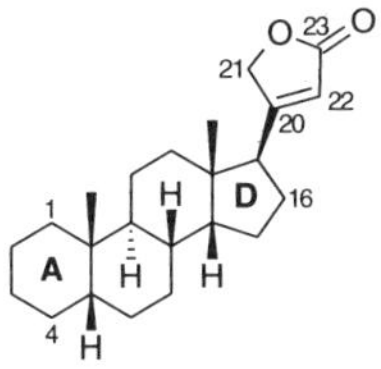

5β-Card-20(22)-enolide

Name for a group of compounds having the steroid skeleton shown in the figure with *cis*-linkage of rings A and B (5β*H*) and C and D (14β*H*) as well as a 17β-positioned, 20(22)-unsaturated γ-lactone group. The (20*R*)-20,22-dihydro compound is known as 5β-cardanolide. The C. are related to the *bufadienolides in their chemical and biological activities. C. are isolated from numerous plant species of the Liliaceae, Ranunculaceae, Asclepiadaceae, Apocynaceae, and Scrophulariaceae. Particularly important and characteristic members are digitoxigenin (see Digitalis glycosides) and strophanthidin (see strophanthins). The C. occurring as glycosides have cardiac activities. They exhibit positive ionotropic action, increase systolic contraction forces of the heart, reduce the heart rate, and thus improve cardiac efficiency. C. can also be taken up with food by insects and used as defensive substances [1].

Lit.: [1] Naturwiss. Rundsch. **20**, 499–511 (1967); J. Chem. Ecol. **12**, 1171 (1986).

gen.: Fieser & Fieser, Steroids, New York: Reinhold 1959 ▪ Hager (5.) **3**, 469 ff.; **5**, 645 ff. ▪ Planta Med. **64**, 491 (1998) (biosynthesis) ▪ Ullmann (5.) **A 5**, 271–278 ▪ Zechmeister **13**, 137–231 ▪ Zeelen, p. 287–299. – *[HS 2938 90]*

Cardiac glycosides see Digitalis glycosides and strophanthins.

Cardinalins. **Dermocybe* pigments from the New Zealand toadstool *Dermocybe cardinalis*. C. are the first pyranonaphthoquinones to be found in Higher fungi. Quinones based on 3,4-dihydro-1*H*-naphtho[2,3-*c*]pyran that bear carbon substituents at C-1 and C-3 in the heterocyclic ring form an important class of biologically active natural products that includes the *nanaomycins and griseusins, deoxyfrenolicin, *kalafungin, *granaticin and the *actinorhodins. The simplest examples, in which both of the carbon substituents are methyl groups, include the eleutherins, protoaphins and the ventiloquinones. To date, quinones of this type seem to be restricted to plants (particularly those belonging to Rhamnaceae), soil bacteria (especially various *Streptomyces*) and insects. Currently known are the C. 1–6, e.g. *C. 1*: $C_{32}H_{36}O_{10}$, M_R 580.62, cryst., mp. 119–124 °C, $[\alpha]_D$ +3.4° ($CHCl_3$) and *C. 2*: $C_{32}H_{34}O_{10}$, M_R 578.61, yellow needles, mp. 17–175 °C, $[\alpha]_D$ –25.9° ($CHCl_3$).

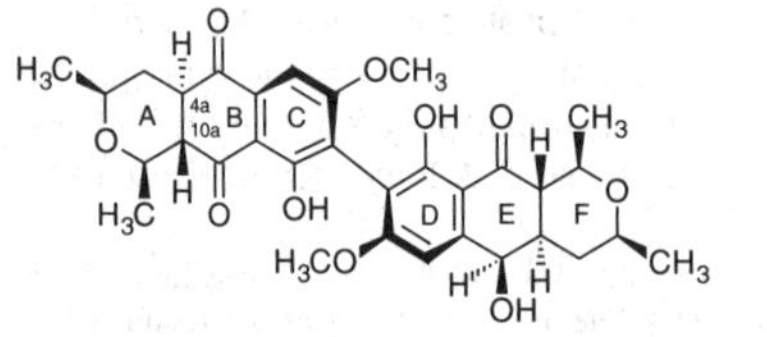

(relative configuration)

C.1; (4a,10a-dihydro-C.1 = C.2)

Lit.: J. Chem. Soc., Perkin Trans. 1 **1997**, 919–925. – *[CAS 160669-40-9 (C. 1); 160700-43-6 (C. 2)]*

Cardol [5-(8,11-pentadecadienyl)resorcinol].

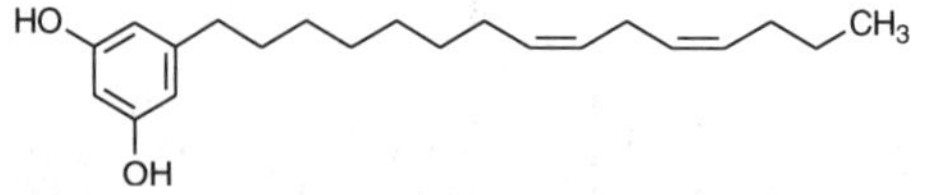

$C_{21}H_{32}O_2$, M_R 316.48. Strongly skin irritating yellowish oil. The name "cardol" is sometimes also used for a mixture of phenols substituted with different unsaturated and saturated alkyl groups (see also urushiol(s)). C. occurs in the resin of Anacardiaceae, e.g., in fruit peels of *Anacardium occidentale* (cashew nut), that are used in Java to poison animals; death occurs after 1–2 d.

Lit.: Hager (5.) **4**, 254 ff.; **6**, 636 ▪ J. Am. Chem. Soc. **117**, 12683 (1995) (isolation) ▪ J. Org. Chem. **62**, 2332 (1997) (synthesis) ▪ Lewin, Die Pfeilgifte, p. 95, Hildesheim: Gerstenberg 1984 ▪ Lipids **21**, 241 (1986). – *[CAS 25702-11-8]*

Carenes (3,7,7-trimethylbicyclo[4.1.0]hept-2- and -3-enes; 2- and 3-carenes).

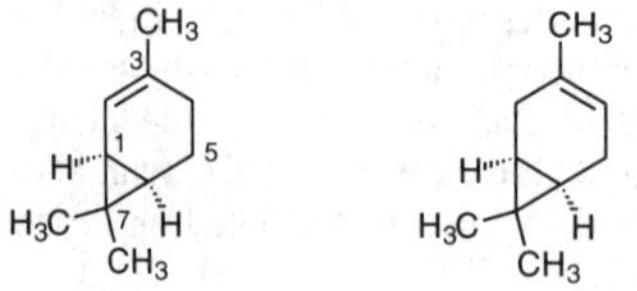

(+)-2-Carene (+)-3-Carene

$C_{10}H_{16}$, M_R 136.24, oil, bicyclic *monoterpenes with carane structure. 2-C. occurs only as the (1*S*,6*R*)(+)-form while 3-C. exists in nature in both enantiomeric forms. 2-C., bp. 46 °C (1.1 kPa) $[\alpha]_D$ +97.7° (neat). (1*S*,6*R*)(+)- and (1*R*,6*S*)(–)-3-C., bp. 45 °C (1.1 kPa), $[\alpha]_D$ ±17.3° (neat). C. readily undergo autoxidation. (+)-2-C. constitutes up to 24% of the essential oil of a sweet grass (*Andropogon himalayensis*, Poaceae) that is widely distributed in the mountains of India, (+)-3-C. up to 60% of that of the tropical pine *Pinus longifolia* and (–)-3-C. up to 19% in the common forest pine (*Pinus sylvestris*, Pinaceae). Ozonization products of C. are used in the perfume industry.

Lit.: Karrer, No. 56, 58 ▪ Merck-Index (12.), No. 1885 ▪ Naturwissenschaften **79**, 416 (1992) ▪ Ullmann (5.) **A 11**, 166, 169. – *Synthesis:* Gazz. Chim. Ital. **107**, 433 (1977) ▪ J. Chem. Soc., Perkin Trans. 1, **1978**, 1370 ▪ J. Org. Chem. **52**, 1493 (1987). – *[HS 2902 19; CAS 4497-92-1 ((+)-2-C.); 498-15-7 ((+)-3-C.); 13466-78-9 ((–)-3-C.)]*

Carminic acid (7-*C*-glucopyranosyl-3,5,6,8-tetrahydroxy-1-methyl-anthraquinone-2-carboxylic acid, E 120, C. I. natural red 4, C. I. 75470).

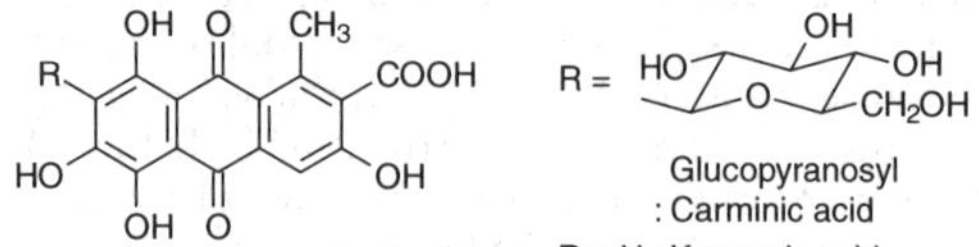

Glucopyranosyl : Carminic acid

R = H : Kermesic acid

$C_{22}H_{20}O_{13}$, M_R 492.39, red prisms, mp. 136 °C (decomp.), uv_{max} 491 nm (CH_3OH), soluble in hot water and ethanol. Solutions of C. have indicator properties (pH 4.8 yellow, pH 6.2 violet). C. is the main component of the dye *cochineal. Glucose-free C. is called *kermesic acid. The coccids use C. as a defence agent against ants. Caterpillars that feed on these insects also acquire protection against ants by consumption of C.[1].

Lit.: [1] Science **208**, 1039 (1980).

gen.: Beilstein E III/IV **18**, 6697 ▪ C. I. **4**, 4632 ▪ J. Chem. Res. (S) **1998**, 546 (synthesis) ▪ J. Chem. Soc., Perkin Trans. 1 **1997**, 575 (synthesis) ▪ Merck-Index (12.), No. 1891. – *[HS 3203 00; CAS 1260-17-9]*

Carnauba wax (Brazil wax). A plant wax obtained from the leaf surface of the endemic Brazilian carnauba palm (*Copernicia prunifera*). C. contains ca. 85% wax esters, especially those with a chain length of 48–64 C atoms, in addition 2–3% each of long-chain alcohols and diols, free fatty acids, and saturated hydrocarbons. The fatty acids of C. mostly have 20–30 C atoms [so-called wax acids, e.g. *behenic, *cerotic, lignoceric acid, *melissic acid], ω-hydroxyfatty acids with 20–30 C atoms, e.g., 22-hydroxydocosanoic acid and aromatic carboxylic acids (e.g., *cinnamic acid) esterified with wax alcohols (30–34 C atoms) and long-chain α,ω-diols (22–34 C atoms), e.g., 1,22-docosanediol, 1,24-tetracosanediol, and 1,26-hexacosanediol.

Use: C. are used as separating agents in bakery and confectionery products, as coatings for citrus fruits, glazing for coffee beans, and as a raw material for the chewing mass of chewing gum. They are also used in polishes for shoes and furniture and in cosmetic and pharmaceutical preparations.

Lit.: Hamilton (ed.), Waxes: Chemistry, Molecular Biology and Functions, p. 257–310, Dundee: The Oily Press 1995 ▪ Ullmann (5.) **A 9**, 571 ▪ see also beeswax. – *[HS 1521 10; CAS 8015-86-9]*

Carnegine see Anhalonium alkaloids.

Carnitine [((*R*)-3-carboxy-2-hydroxypropyl)trimethylammonium betaine, vitamin B_T].

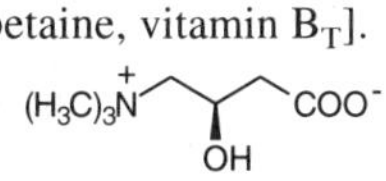

(3*R*)-, L-form

$C_7H_{15}NO_3$, M_R 161.20, hygroscopic crystals, mp. 212 °C (decomp.), $[\alpha]_D^{22}$ −23.5° (H_2O); monohydrochloride, mp. 137–139 °C (decomp.), $[\alpha]_D^{22}$ −22.4° (H_2O). Characteristic component of striated muscles and liver. C. plays a decisive role in fatty acid metabolism by transporting acyl groups through the mitochondrial inner membrane. The acyl group is transferred from acyl-CoA to C. and the thus formed *O*-acylcarnitine is transported through the membrane into the mitochondrial matrix, where acylcarnitine reacts through retroacylation with free CoA to acyl-CoA and C., which is transported back into the cytosol.

Lit.: Beilstein E IV **4**, 3185 f. ▪ Frenkel & McGarry (eds.), Carnitine, Biosynthesis, Metabolism and Functions, New York: Academic Press 1980 ▪ Hager (5.) **7**, 712 ▪ Merck-Index (12.), No. 1898. – *Synthesis:* Helv. Chim. Acta **70**, 2058–2064 (1987) ▪ Tetrahedron: Asymmetry **8**, 2663 (1997). – *[HS 2923 90; CAS 541-15-1]*

Carnosine (*N*-*β*-alanyl-L-histidine).

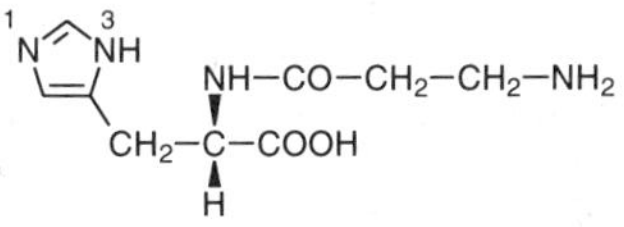

$C_9H_{14}N_4O_3$, M_R 226.24, mp. 246–250 °C (decomp.), $[\alpha]_D^{20}$ +24.1° (H_2O), pK_a 2.64, 6.83 and 9.51; monohydrochloride, mp. 245 °C (decomp.). A non-proteinogenic amino acid in muscles that was first isolated from beef extracts. In muscle tissue of humans and many animals C. represents the largest proportion of all amino acids with the exception of *carnitine. Muscles of other animals, e. g., doves and geese, contain in contrast 3-methylcarnosine (see anserine). Activities in the olfactory sense and in wound healing have been discussed.

Lit.: Adv. Enzyme Regul. **30**, 175 (1990) ▪ Beilstein E V **25/16**, 428 f. ▪ Int. J. Biochem. **22**, 129 (1990) (review) ▪ Merck-Index (12.), No. 1899. – *[HS 2933 29; CAS 305-84-0]*

Carnosol (picrosalvin).

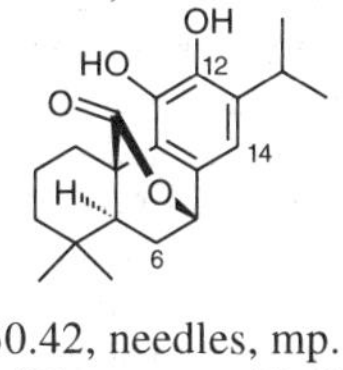

$C_{20}H_{26}O_4$, M_R 330.42, needles, mp. 221–226 °C, $[\alpha]_D$ −66° (C_2H_5OH). Diterpene with *abietane structure; bitter principle of *Salvia carnosa* and other *Salvia* species as well as *Rosmarinus officinalis*.

Lit.: Beilstein E V **18/3**, 481 ▪ Karrer, No. 6098 ▪ Phytochemistry **25**, 269 (1986). – *[CAS 5957-80-2]*

***β*-Carotene** (*β,β*-carotene, provitamin A). $C_{40}H_{56}$, M_R 536.88, dark-red to violet prisms (C_6H_6/CH_3OH), mp.

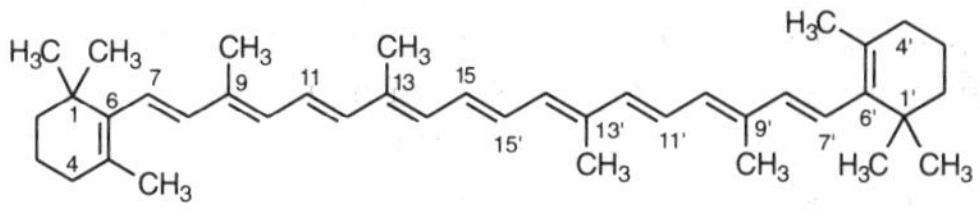

183 °C; uv_{max} 455, 520 nm. *β*-C. is the most abundant carotene in the animal and plant kingdoms. It was first isolated by Wackenroder in 1826 from carrots (*Daucus carota*, Apiaceae). Willstätter determined the molecular formula in 1907 and Karrer the constitutional formula in the early 1930's. *β*-C. is the provitamin A, which is oxidatively cleaved in animals to yield 2 molecules of retinal and subsequently reduced to retinol (vitamin A) (see pigments of vision).

Use: *β*-C. is used as an antioxidant, as an vitamin A precursor in medical preparations, as food and cosmetic dye, and fodder additive.

Lit.: Angew. Chem. Int. Ed. Engl. **16**, 423–429 (1977) (synthesis) ▪ Britton, Liaan-Jansen & Pfander (eds.), Carotenoids, Basel: Birkhäuser 1995 ▪ Hager (5.) **4**, 85 ff., Monographie **A11**, **7**, 715 ▪ Int. J. Vit. Nutr. Res. **63**, 93–121 (1993) ▪ J. Am. Chem. Soc. **117**, 2747 (1995) (synthesis) ▪ J. Chem. Soc., Perkin Trans. 1 **1997**, 269 (biosynthesis) ▪ Karrer, No. 1821 ▪ Merck-Index (12.), No. 1902 ▪ Ullmann (5.) **A 11**, 508, 572. – *[HS 2936 10; CAS 7235-40-7]*

Carotenes. $C_{40}H_{56}$, M_R 536.88. Group of 11- to 12-fold unsaturated tetraterpenes belonging to the *carotinoids. Although carotene was originally thought to be a homogeneous natural product, Kuhn and Lederer in 1931 separated it into the three isomers *α*-carotene (red cryst., mp. 187 °C), **β*-carotene, and *γ*-carotene (violet prisms, mp. 178 °C). *α*-C. and *γ*-C. differ from *β*-C. in the structure of one of the two terminal groups.

terminal group symbol:

α-Carotene (β,ε-Carotene):
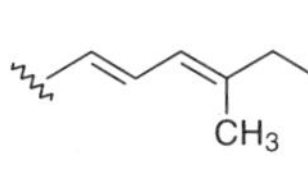

ε

β -Carotene (β,β-Carotene):
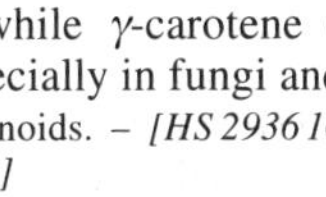

β

γ-Carotene (β,ψ-Carotene):
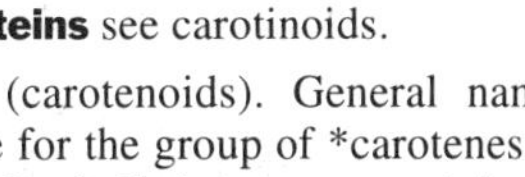

ψ

α-Carotene is found in small amounts together with *β*-carotene; while *γ*-carotene occurs only in trace amounts, especially in fungi and bacteria.

Lit.: see carotinoids. – *[HS 2936 10; CAS 432-70-2 (α-C.); 472-93-5 (γ-C.)]*

Carotenoproteins see carotinoids.

Carotinoids (carotenoids). General name derived from carotene for the group of *carotenes (hydrocarbons) and xanthophylls (oxygen-containing carotenes) and other carotene derivatives with a basic skeleton made up of eight isoprene units (tetraterpenes). The colors of the C. (mostly yellow, orange, or red) result from the polyene structures with (usually) 11–12 conjugated double bonds. More than 600 natural C. have been described to date. The C_{40} basic skeleton is not only substituted by hydroxy or oxo groups; there are also apo-, nor-, or seco-C. (shortened chain, ring opening), and retro-C. (double bond shifts) derivatives. Although there is a strict IUPAC nomenclature [1] system for C., many specialists continue to use the trivial names – which often end with ...xanthine, see, e.g., anthera-, citrana-, crypto-, mactra-, rhodo-, and zeaxanthin. A correlation of trivial and systematic names is given in *Lit.*[1,2]. In nature, C. often exist as *carotenoproteins*, i.e., they are non-covalently associated with proteins to give water-soluble complexes and are thus protected from external influences (cf. chromoproteins).

Occurrence: In *higher plants* the C. occur in the plastids. In the thylakoid membrane of chloroplasts only *β-carotene, *lutein, *violaxanthin, and *neoxanthin occur. The color becomes apparent in autumn leaves when the *chlorophyll has been degraded and the chloroplasts transformed to chromoplasts. Similar processes occur during the ripening of fruit. Although the yellow to red colors of flower petals are mostly due to *flavonoids, C. do occasionally occur as flower pigments, e.g., *crocetin and *violaxanthin (in pansies, Arnica, and daffodil flowers). Further information on the occurrence of C. is given in *Lit.*[3]. The C. of the orange to brilliant red-colored *algae* are localized in chromatophores. A pattern of evolution can be deduced from their occurrence patterns. C. are also constituents of higher fungi [4]. Cyanobacteria (blue algae) also contain C., especially β-carotene. A series of C. with unusual structures are found in *bacteria*. The C. in *animals* are mostly transformation products of carotinoids from plant foods. They are responsible for the yellow to red colors of many bird feathers (e.g., flamingos) as well as the red colors of lobster shell or salmon (*astaxanthin), other crustaceans and the ladybird beetle. C. or their transformation products also occur in butter, blood plasma, eye pigments (*retinal). For their occurrence in marine organisms, see *Lit.*[5].

Physiology: In plants, C. function as light filters and are involved in the energy transport of photosynthesis [6]. In flower petals and fruits C. are relevant as attracting colors for animals. In animals, their function as provitamin is important, therefore, C. are precursors of retinal and thus of significance for the vision process [7] (see pigments of vision). Many aroma substances may be considered as degradation products of the C.[8].

Use: On account of their toxicological harmlessness, many C. are used to color foods (margarine, butter, cheese, fruit juices), in cosmetics, and as fodder additives (see *Lit.*[9]). C. are used in medicinal preparations as vitamin A precursors. Their antioxidant activity is also employed in the prophylaxis and therapy for certain types of cancer [10].

Biosynthesis: The biosynthesis of the C. proceeds in chloroplasts or chromoplasts in analogy to other isoprenoids by way of 3-isopentenyl diphosphate ("active isoprene"). The colorless 15-*cis*-phytoene (C_{40}) is formed from 2 molecules of geranylgeranyl diphosphate, further transformations furnish, via *all-trans-ζ*-carotene (colored), *all-trans*-lycopene. Subsequent cyclization reactions allow the formation of other carotinoids. Xanthophylls are formed from C. by incorporation of oxygen [11].

Lit.: [1] Carotinoids Photosynth. **1993**, 1–15; Pure Appl. Chem. **41**, 405 (1975). [2] Bauernfeind, Carotenoids as Colorants and Vitamin A Precursors, New York: Academic Press 1981. [3] Packer (ed.), The Carotenoids, San Diego: Academic Press 1993. [4] Zechmeister **51**, 193 f. (1987). [5] Scheuer I **2**, 2–75; Monogr. Oceanogr. Methodol. **10**, 578–594 (1977) (UV-spectra). [6] Physiol. Plants **69**, 561 (1987). [7] New Food Ind. **30**, 30 (1988). [8] Zechmeister **35**, 431–527; ACS Symp. Ser. **317**, 157 (1986). [9] Crit. Rev. Food Sci. Nutr. **18**, 59–97 (1982). [10] Pure Appl. Chem. **57**, 717 (1985); Prog. Clin. Biol. Res. **259**, 177–200 (1988). [11] Surgeon et al. (eds.), Biosynthesis of Isoprenoid Compounds, vol. 2, p. 1–122, New York: Wiley 1983.
gen.: Anal. Chem. **68**, 299–304A (1996) (analysis) ▪ Chem. Ind. (London) **1993**, 79 ▪ Goodwin, The Biochemistry of the Carotenoids, New York: Methuen 1984 ▪ Goodwin, The Biochemistry of the Carotenoids (2.), vol. 1, London: Chapman & Hall 1980 ▪ Pfander (2.) ▪ Ullmann (5.) **A 11**, 308, 572. – *Synthesis:* Pure Appl. Chem. **51**, 435–675, 857–886 (1979); **57**, 735 (1985); **69**, 2027, 2047–2060 (1997) ▪ Stud. Nat. Prod. Chem. **20**, 561–612 (1998). – *Series:* Britton, Liaan-Jansen & Pfander (eds.), Carotenoids, vols. 1, 2, Basel: Birkhäuser 1995, 1996. – *[HS 3204 19]*

Carpaine.

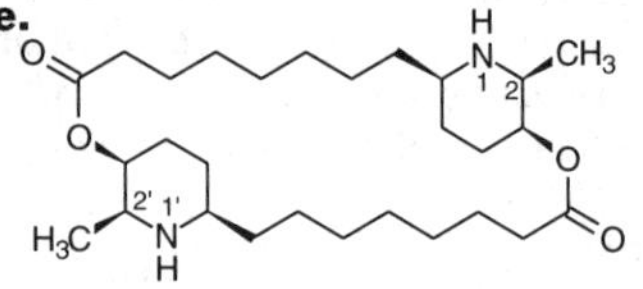

$C_{28}H_{50}N_2O_4$, M_R 478.72, mp. 121 °C, $[\alpha]_D$ +21.65° (C_2H_5OH), monoclinic prisms. C., a macrocyclic symmetrical dilactide, is an alkaloid from the leaves of the papaya tree (*Carica papaya*) which causes bradycardia (slowing of heart rate). It is an effective amoebicide and exhibits (*in vitro*) anti-tumor activity even at low concentrations. Derivatives of C. from *C. papaya* are the 2-epimer (*pseudocarpaine*), the 1,2-didehydro compound (*dehydrocarpaine I*), and 1,1′,2,2′-tetradehydrocarpaine (*dehydrocarpaine II*).

Lit.: Beilstein E III/IV **27**, 8781 ▪ Manske **26**, 96 ▪ Pelletier **3**, 66–68 ▪ Res. Commun. Chem. Pathol. Pharmacol. **22**, 277–289 (1978) (pharmacology). – *Synthesis:* Heterocycles **14**, 169 (1980 ▪ Tetrahedron **31**, 1047 (1975) ▪ Tetrahedron Lett. **1979**, 3391. – *[HS 2939 90; CAS 3463-92-1 (C.); 3760-91-6 (pseudo-C.); 72362-02-8 (didehydro-C.); 72362-03-9 (dehydro-C.)]*

Carrageenan (carrageenin). A gel-forming extract obtained from north Atlantic red algae (*Chondrus crispus* and *Gigartina stellata*) belonging to the Florideae, named after the Irish coastal town Carragheen, with a structure similar to that of *agar (and like agar also known as Florideae starch). The algae collected on the coasts of Ireland, northern France, and USA are also known as Irish moss or charag(h)een. The heteropolysaccharide C. precipitated from the hot water extract of the algae is a colorless to sand-colored powder with M_R 100 000–800 000 and a sulfate content of ca. 25%; it is readily soluble in warm water and forms a thixotropic gel on cooling even when the water content amounts to 95–98%. The firmness of the gel results from the double helix structure of C.[1]. In carrageenan three main components are distinguished:

Iota(ι)-C.: β-D-galactose-4-sulfate (3-*O*-glycosylated) and 3,6-anhydro-α-D-galactose-2-sulfate (4-*O*-glycosylated) alternately linked, water soluble and gel-forming.

Kappa(κ)-C.: β-D-galactose-4-sulfate (3-*O*-glycosylated) and 3,6-anhydro-α-D-galactose (4-*O*-glycosy-

lated) alternately linked (in agar: 3,6-anhydro-L-galactose), gel-forming.
Lambda(λ)-C.: β-D-galactose-2-sulfate (30% without sulfate residue; 3-*O*-glycosylated) and α-D-galactose-2,6-disulfate (4-*O*-glycosylated) alternately linked, not gel-forming, readily soluble in cold water.

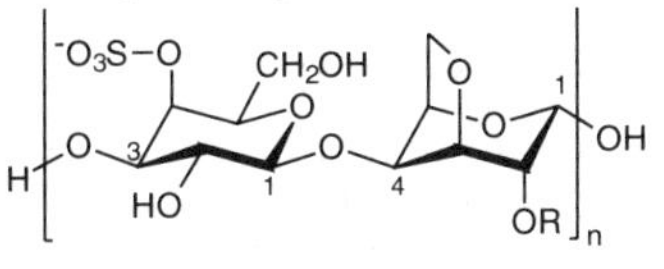

R = SO_3^- : ι-C.   R = H : κ-C.

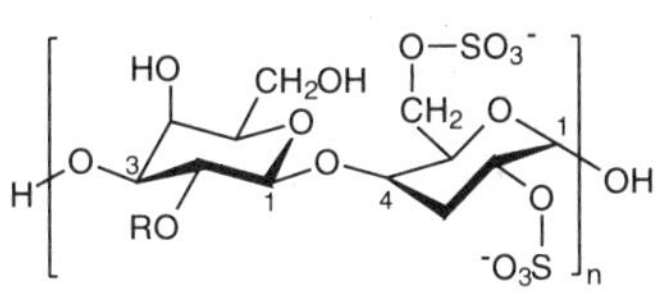

R = 30% H , 70% SO_3^- : λ-C.

The nature of the accompanying cations (K, NH_4, Na, Mg, Ca) also influences the solubilities of the carrageenans.
Use: As gelating agent and stabilizer in foods, binding agent and emulsifier in cosmetics.
Lit.: [1] Angew. Chem. Int. Ed. Engl. **16**, 214–224 (1977).
gen.: Bioact. Mol. **2**, 121 (1987) ▪ Bot. Mar. **27**, 189 (1984) ▪ Food Hydrocolloids **2**, 73–113 (1983) ▪ Hager (5.) **2**, 901; **4**, 859 ff. ▪ Pathol. Biol. **27**, 615 (1979) ▪ Ullmann (5.) **A 11**, 503, 570 f. ▪ Whistler & BeMiller (eds.), Industrials Gums, p. 145 ff., San Diego: Academic Press 1993. – *[HS 1302 39; CAS 9000-07-1 (C.); 11114-20-8 (χ-C.)]*

Carrot flavor see vegetable flavors.

Carthamin (safflor red, safflower red).

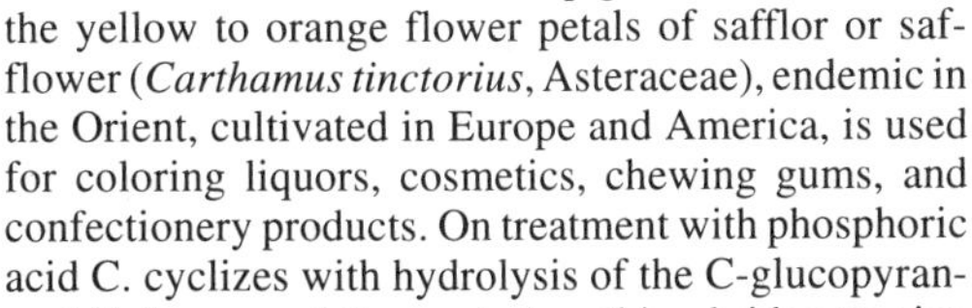

$C_{43}H_{42}O_{22}$, M_R 910.79, red needles, mp. 228–230 °C, poorly soluble in water, soluble in ethanol and dilute alkali carbonate solution. The pigment isolated from the yellow to orange flower petals of safflor or safflower (*Carthamus tinctorius*, Asteraceae), endemic in the Orient, cultivated in Europe and America, is used for coloring liquors, cosmetics, chewing gums, and confectionery products. On treatment with phosphoric acid C. cyclizes with hydrolysis of the C-glucopyranosyl linkages and the central methine bridge to give carthamidine, a flavone derivative.
Lit.: Chem. Lett. **1979**, 201; **1986**, 495; **1996**, 833 (absolute configuration) ▪ C. I. **4**, 4625 ▪ J. Chromatogr. **438**, 61 (1988) ▪ Tetrahedron Lett. **23**, 5163 (1982). – *[HS 3203 00; CAS 36338-96-2]*

Carvacrol see cymenols.

Carvone (*p*-mentha-6,8-dien-2-one).

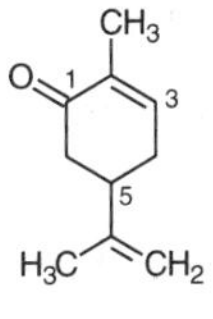

$C_{10}H_{14}O$, M_R 150.22, thin oil, monocyclic *monoterpene ketone. Enantiomeric and racemic forms occur in nature: (4*S*)(+)- and (4*R*)(–)-C., bp. 230–231 °C, $[\alpha]_D$ ±62° (neat.). (±)-C., bp. 231 °C. Example for the differing physiological properties of enantiomers: (+)-C. has the typical odor of caraway while (–)-C. has a mint-like odor.
Occurrence: (+)-C. up to 85% in *caraway oil (*Carum carvi*, Umbelliferae) and 60% in *dill oil from the fruits of *Anethum graveolens* (Apiaceae). (–)-C. up to 72% in *spearmint oil from *Mentha spicata* (Lamiaceae). (±)-C. up to 2% in ginger-grass oil from *Cymbopogon martinii* (Poaceae). *C. martinii* is distributed throughout India and occurs in two forms which differ widely in their chemistry but are very difficult to distinguish by morphology. The form *motia* furnishes *palmarosa oil containing 75–95% *geraniol while the form *sofia* provides the ginger-grass oil [1]. C. is used in liquors, cosmetics, and soaps.
Lit.: [1] Hegnauer **II**, 187.
gen.: Beilstein E IV **7**, 315 ▪ Karrer, No. 557 ▪ Ullmann (5.) **A 11**, 172. – *Synthesis:* Indian J. Chem. **4**, 275 (1966) ▪ Justus Liebigs Ann. Chem. **1993**, 403, 1133. – *[HS 2914 29; CAS 2244-16-8 ((+)-C.); 6485-40-1 ((–)-C.); 22327-39-5 (racemate)]*

Carvotanacetone see *p*-menthenones.

Caryophyllenes.

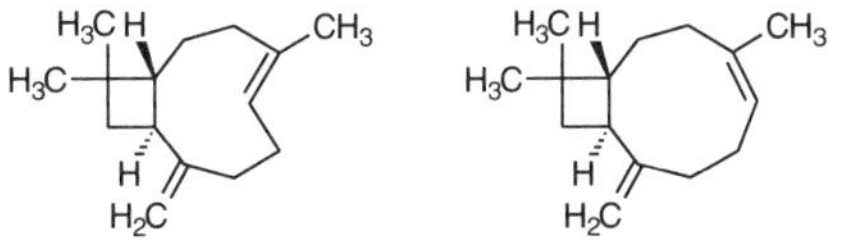

β   γ

$C_{15}H_{24}$, M_R 204.36. *Sesquiterpenes isolated from essential oils (e.g., clove oil, *Caryophylli flos*). A distinction is made between α-C. (see humulene), β-C., and γ-C. The β-isomer is a colorless liquid [bp. 129–130 °C (13 hPa), $[\alpha]_D^{20}$ –9.2° (neat)] with a clove- to turpentine-like odor. It is also found in larvae of *Parides arcas*. The γ-isomer (isocaryophyllene) [colorless oil, bp. 125 °C (19 hPa), $[\alpha]_D$ –24.1°] has the *Z*-configuration at the ring double bond. C. are used as aroma substances, e.g., in chewing gum and as fixatives.[1].
Lit.: [1] Food Cosmet. Toxicol. **12**, Suppl., 841 (1974).
gen.: Beilstein E IV **5**, 1182 ▪ Helv. Chim. Acta **80**, 1980 (1997) (synthesis) ▪ Karrer, No. 1929 ▪ Nat. Prod. Rep. **15**, 187–204 (1998) (review) ▪ Phytochemistry **51**, 873 (1999) ▪ Q. Rev. Chem. Soc. **21**, 331 (1967) ▪ Ullmann (5.) **A 11**, 166. – *[HS 2902 19; CAS 87-44-5 (β-C); 118-65-0 (γ-C.)]*

Carzinophilin (azinomycin B).

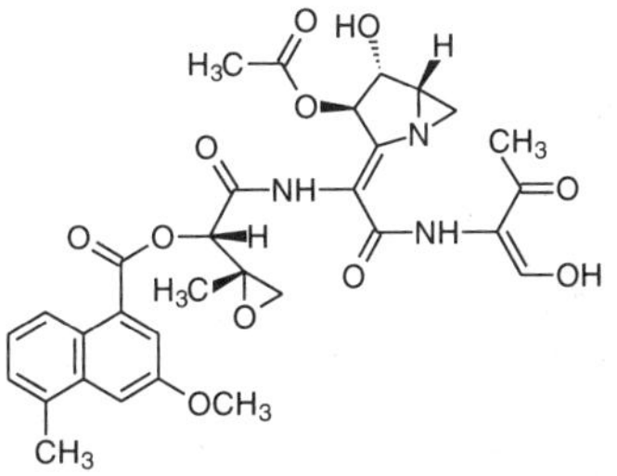

$C_{31}H_{33}N_3O_{11}$, M_R 623.62, colorless needles, mp. 217–222 °C, $[\alpha]_D$ +57.8° ($CHCl_3$). Antitumor antibiotic from *Streptomyces sahachiroi* and *S. griseofuscus*.

C. forms strand linkages in the DNA double helix through double alkylation between guanine and another purine base. The reaction starts with a nucleophilic attack of guanine on the reactive 2-methylene-1-azabicyclo[3.1.0]hexane system. C. also exhibits antibacterial activity.

Lit.: J. Antibiot. **39**, 1527–1532 (1986) ▪ Tetrahedron Lett. **32**, 3807 (1991). – *Synthesis:* Chem. Pharm. Bull. **42**, 285 (1994) ▪ Heterocycles **47**, 59 (1998) ▪ J. Org. Chem. **58**, 7848 ff. (1993); **63**, 5738 (1998) ▪ Tetrahedron Lett. **35**, 2207 ff., 9405–9412 (1994). – *Mechanism of action:* J. Am. Chem. Soc. **114**, 3144 f. (1992) ▪ Tetrahedron Lett. **33**, 3711 ff. (1992). – *[CAS 106486-76-4]*

Casbene.

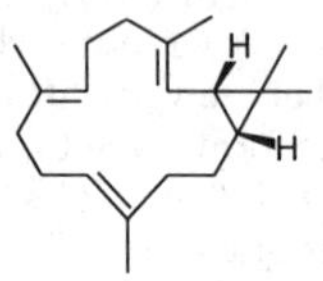

(relative configuration)

$C_{20}H_{32}$, M_R 272.47, diterpene with casbane structure from seedlings of *Ricinus communis* with antifungal properties.

Lit.: Arch. Biochem. Biophys. **276**, 270–277 (1990) ▪ Proc. Natl. Acad. Sci. USA **91**, 8497–8501 (1994) ▪ Tetrahedron Lett. **39**, 2033 (1998) (enzymatic synthesis). – *[CAS 24286-51-9]*

Cascarillin.

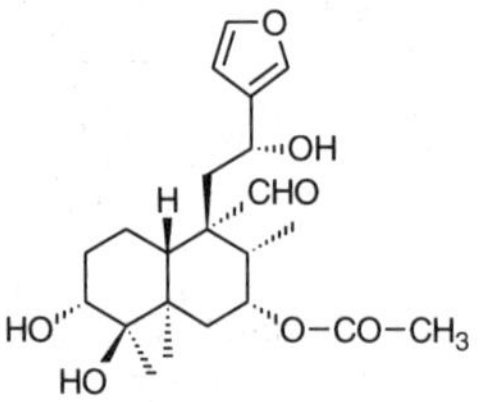

$C_{22}H_{32}O_7$, M_R 408.49, prisms, mp. 203.5 °C. Diterpene with *clerodane structure from *Croton eleuteria.*

Lit.: Beilstein E IV **18**, 3072 ▪ Karrer, No. 5987 ▪ Merck-Index (12.), No. 1933. – *[CAS 10118-56-6]*

Cassaine see Erythrophleum alkaloids.

Cassia oil. Red-brown oil, strong spicy, cinnamon-like odor, warm, aromatic, sweet-spicy taste.

Production: By steam distillation from leaves, twigs, young branches, and bark of the Chinese cinnamon tree (*Cinnamomum aromaticum*, synonym *C. cassia*, Lauraceae), also known as Chinese *cinnamon leaf (bark) oil.

Composition[1]: Main components are **cinnamaldehyde* (ca. 75%), *cinnamyl acetate* ($C_{11}H_{12}O_2$, M_R 176.22) (5%), and *2-methoxycinnamaldehyde* ($C_{10}H_{10}O_2$, M_R 162.19) (12%), by the occurrence of which C. differs from cinnamon oil. Important minor components of C. are **benzaldehyde, salicylaldehyde,* ($C_7H_6O_2$, M_R 122.12), and **coumarin.*

Use: In small amounts in perfume production for spicy, oriental nuances; mainly for aromatization of foods (bakery, confectionery products, refreshing drinks, liquors).

Lit.: [1] Perfum. Flavor. **19** (4), 33 (1994); Dev. Food Sci. **34**, 411–425 (1994).

gen.: Arctander, p. 132 ▪ Bauer et al. (2.), p. 145 ▪ ISO 3216 (1974). – *Toxicology:* Food Cosmet. Toxicol. **13**, 109 (1975). – *[HS 3301 29; CAS 8007-80-5 (C.); 21040-45-9 (cinnamyl acetate); 1504-44-1 (2-methoxycinnamaldehyde); 90-02-8 (salicylaldehyde)]*

Castanogenol, castanopsol see oleanane.

Castanospermine.

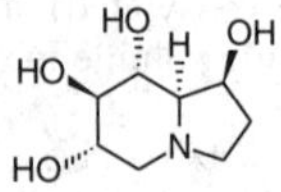

$C_8H_{15}NO_4$, M_R 189.21, large cubic cryst., mp. 212–215 °C (decomp.), $[\alpha]_D^{25}$ +79.7° (H_2O). Toxic alkaloid from the Australian leguminous tree *Castanospermum australe*. It is active against AIDS viruses in that it changes the surface glycoproteins of HIV (glycoprotein processing) in such a way that the virus cannot attach to the host cells (T4 lymphocytes) and reproduce. C. also exhibits activity against cancer cells and Herpes viruses. C. inhibits α- and β-glucosidases, but has a less specific activity than the structurally related *swainsonine. Furthermore, C. effects the accumulation of highly mannosylated glycoproteins such as, e.g., *concanavalin A. *6-Epi-C.* ($C_8H_{15}NO_4$, M_R 189.21) also occurs in small amounts together with C. and is a potent inhibitor of amyloglucosidases (exo-1,4-α-glucosidases).

Lit.: Arch. Biochem. Biophys. **221**, 593 (1983); **230**, 668 (1984) ▪ Biochem. Pharmacol. **36**, 2381 (1987) ▪ J. Am. Chem. Soc. **121**, 3046 (1999) (synthesis) ▪ J. Chem. Soc., Chem. Commun. **1998**, 1353 ▪ J. Org. Chem. **52**, 5492 (1987); **58**, 52–61, 7096 (1993); **61**, 6762 (1996) (synthesis) ▪ Merck-Index (12.), No. 1944 ▪ Nature (London) **330**, 74 (1987) ▪ Proc. Natl. Acad. Sci. USA **84**, 8120 (1987) ▪ Tetrahedron **50**, 2131 (1994) ▪ Tetrahedron Lett. **25**, 165 (1984); **30**, 705 (1989) ▪ Trends Biochem. Sci. **9**, 32 (1984). – *[CAS 79831-76-8 (C.); 107244-34-8 (epi-C.)]*

Castasterone see brassinosteroids.

Castoramine.

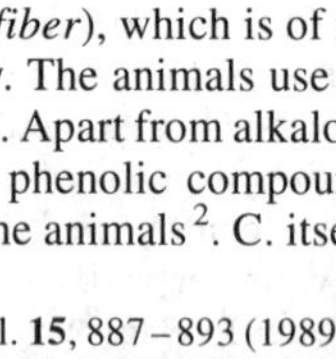

$C_{15}H_{23}NO_2$, M_R 249.35, mp. 64.5–65.6 °C, $[\alpha]_D^{25}$ –80° ($CHCl_3$). C. occurs in the castoreum, a glandular sac in beavers (*Castor fiber*), which is of importance for the perfume industry. The animals use the glands for territorial marking[1]. Apart from alkaloids like C. the castoreum contains phenolic compounds that influence the behavior of the animals[2]. C. itself does not appear to play a role[3].

Lit.: [1] J. Chem. Ecol. **15**, 887–893 (1989). [2] J. Chem. Ecol. **19**, 1491–1500 (1993); **20**, 3063–3081 (1994). [3] Am. Midl. Nat. **120**, 144–149 (1988). – *[CAS 6874-86-8]*

Castor oil (ricinus oil). The oil obtained from seeds of the castor-oil plant or castor beans (*Ricinus communis*, Euphorbiaceae), mp. –10 °C. It contains about 80–85% *ricinoleic acid as glycerol ester together with oleic (7%), linolic (3%), palmitic (2%), and stearic (1%) acids. The *Ricinus* seeds contain the highly poisonous constituents *ricin, a mixture of proteins, and *ricinine, a pyridone alkaloid. The presence of these substances leaves the oilcake remaining after pressing of C. unsuitable for use as animal fodder.

Use: Mainly for technical purposes as lubricants, in the production of paints, varnishes, and plastics; raw material for the production of cosmetics, in medicine as a laxative.

Lit.: Murphy (ed.), Designer Oil Crops, p. 81 f., 181 f., Weinheim: VCH Verlagsges. 1994 ▪ Ullmann (5.) **A 10**, 233. – *[HS 1515 30; CAS 8001-79-4]*

Catalpol. C. is an *iridoid glucoside with a very stable epoxide ring. $C_{15}H_{22}O_{10}$, M_R 362.33, mp. 202–205 °C, $[\alpha]_D$ −102° (95% C_2H_5OH). The absolute configuration of C. has been correlated to that of leonuride by chemical means[1].

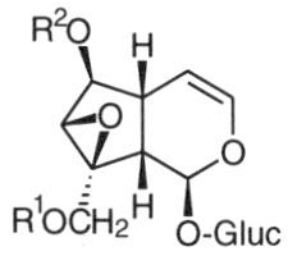

Occurrence: C. occurs mainly as esters of various aromatic carboxylic acids. The esters form at the primary OH- (R^1) group, e.g., *scutellarioside* I {globularin, R^1=cinnamoyl, $C_{24}H_{28}O_{11}$, M_R 492.48, $[\alpha]_D$ −73° (C_2H_5OH)} or at the secondary OH- (R^2) group, e.g., *catalposide* {R^2=4-hydroxybenzoyl, $C_{22}H_{26}O_{12}$, M_R 482.44, mp. 215–217 °C, $[\alpha]_D$ −184° (CH_3OH)} but also at the OH group at C-6 of glucose, e.g., *picroside* I {cinnamoyl at C-6 of glucose, $C_{24}H_{28}O_{11}$, M_R 492.48, $[\alpha]_D$ −82° (CH_3OH)}.

C. is isolated from the plant *Catalpa bignonioides* (Bignoniaceae) endemic in North America and East Asia[2] and is besides *aucubin the most commonly occurring iridoid glucoside. Today ca. 1 kg C. can be obtained from the alkaline soda extract of 5 kg of dried *Picrorhiza kurrooa* (Scrophulariaceae) widely distributed in northern India and Nepal[3].

Lit.: [1] Justus Liebigs Ann. Chem. **1985**, 1063. [2] Am. Chem. J. **10**, 328 (1888). [3] Justus Liebigs Ann. Chem. **1986**, 46.

gen.: Bobbitt & Segebarth, in Taylor & Battersby (eds.), Cyclopentanoid Terpene Derivatives, chap. 1, New York: Dekker ▪ Hager (5.) **6**, 384 ff. ▪ J. Nat. Prod. **43**, 649, No. 37–50 (1980); **48**, 957 (1985) ▪ Karrer, No. 5654 ▪ Phytochemistry **35**, 1187 (1994) (biosynthesis). – *Synthesis:* Justus Liebigs Ann. Chem. **1991**, 893. – *[HS 2938 90; CAS 2415-24-9 (C.); 1399-49-1 (scutellarioside I); 6736-85-2 (catalposide); 27409-30-9 (picroside I)]*

Catechins (catechols). Group of crystalline colorless compounds that can be considered as hydrogenated *flavones or anthocyanidins. The C. form the parent substances of a series of natural oligo- or polymeric tanning agents (non-hydrolysable C.-tanning agents that are also known as condensed proanthocyanidins), e.g., in tea. They occur together with other phenols in many types of fruit and are involved in the brown coloration of bruised and cut surfaces (e.g., of apples) catalyzed by phenol oxidases. For example, *catechin:* $C_{15}H_{14}O_6$, M_R 290.27. (2*R*,3*S*)-form, crystalline powder, can form a tetrahydrate, $[\alpha]_D$ +17°, mp. 175–177 °C; (±)-C. with 3 mol of water of crystallization, mp. 212–216 °C, moderately soluble in cold water, well soluble in hot water, glacial acetic acid, ethanol, and acetone. Dark green coloration with $FeCl_3$ solution. The epimeric *epicatechin* [(2*R*,3*R*), cryst., mp. 242 °C] is leavorotatory, $[\alpha]_D$ −68° (C_2H_5OH).

Catechin is widely distributed in plants and amounts to 2–10% of catechu[1]. Extracts from *dye plants with

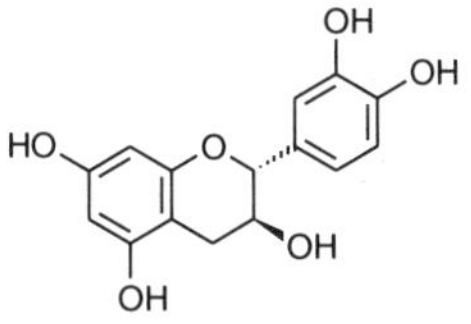

Catechin

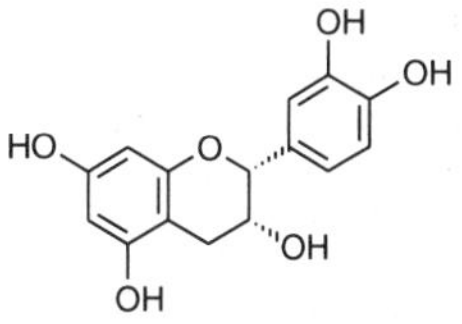

Epicatechin

a high proportion of condensed tanning agents of the C.-type are classified under the name *catechi*. For epimerization of catechin in alkaline medium, see *Lit.*[1]. Catechin is used for dying, staining, and tanning.

Lit.: [1] Helv. Chim. Acta **60**, 1665 (1977).

gen.: Beilstein E V **17/8**, 447 f. ▪ J. Chem. Soc., Perkin Trans. 1 **1977**, 1637 (biosynthesis) ▪ Karrer, No. 1762, 1763 ▪ Sharma (ed.), Chemistry and Technology of Catechin and Catechu Manufacturing, Dehradun, Indien: International Book Distributors 1985 ▪ Tetrahedron **28**, 2819 1972 (isolation) ▪ Tetrahedron Lett. **38**, 3089 (1997) (synthesis) ▪ Zechmeister **27**, 158–260. – *[HS 2932 90; CAS 154-23-4 ((+)-catechin); 490-46-0 (epicatechin)]*

Catecholamines. Name for the group of biogenic amines derived from L-*DOPA that are of major significance as neurotransmitters and neurohormones. See under: L-adrenaline, dopamine, L-noradrenaline, see also *N*-acylcatecholamines.

Catharanthine.

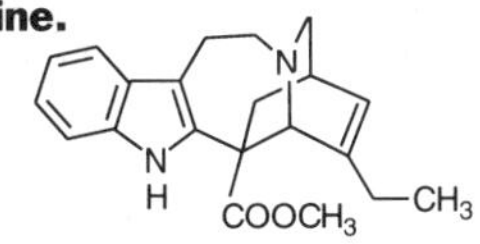

$C_{21}H_{24}N_2O_2$, M_R 336.43, crystalline, mp. 126–128 °C ($CHCl_3$), 61–63 °C (racemate), $[\alpha]_D$ +29.8°. C. is a major alkaloid of *Catharanthus* species. It is the biosynthetic precursor of dimeric *Catharanthus (Vinca) alkaloids. C. exhibits moderate blood sugar lowering and diuretic activities.

Lit.: Beilstein E V **25/5**, 227. – *Biosynthesis:* J. Chem. Soc., Chem. Commun. **1966**, 810, 890. – *Pharmacology:* Hager (5.) **3**, 259; **6**, 890 ▪ J. Ethnopharmacol. **2**, 119–127 (1980); **5**, 1–71 (1982). – *Synthesis:* Chem. Pharm. Bull. **30**, 4052 (1982) ▪ J. Org. Chem. **50**, 3236 (1985) ▪ Org. Lett. **1**, 973 (1999) ▪ Tetrahedron **46**, 1711 (1990) ▪ see also Catharanthus alkaloids. – *[HS 2939 90; CAS 2468-21-5]*

Catharanthus alkaloids. Alkaloids (*monoterpenoid indole alkaloids) from the tropical and subtropical weed (shrub) *Catharanthus roseus* (Madagascar evergreen, previously *Vinca rosea*) as well as other species. C. a. are often falsely named as Vinca alkaloids. They occur in 7 species of the genus *Catharanthus* (Apocynaceae). *C. roseus* is of commercial importance, to date over 100 structurally different alkaloids have been isolated. Phytochemically, *C. roseus* is the most thoroughly investigated alkaloid-containing plant, cell and tissue cultures have been developed for the production of the alkaloids. The main alkaloids of *Catharanthus* plants are *catharanthine and *vindoline. Undifferentiated cell cultures synthesize over 40 C. a.[1], including catharanthine, but not vindoline.

Use & activity: The commercially most important C. a. are the dimeric or bisindole bases *vinblastine and *vincristine, formed by coupling of catharanthine and vindoline. The dimeric C. a. are obtained commercially by cultivation of *C. roseus* (in Israel, Pakistan, Texas,

Hungary) and have been used successfully in therapy of leukemias and Hodgkin's disease.

Lit.: [1]Zechmeister **55**, 89.

gen.: Taylor, in Farnsworth (ed.), The Catharanthus Alkaloids, New York: Marcel Dekker 1975. – *Biosynthesis:* Atta-ur-Rahman & Basha, Biosynthesis of Indole Alkaloids, Oxford: Clarendon Press 1983 ▪ Manske **49**, 222–290 ▪ Stud. Org. Chim. **26**, 497 (1986). – *Cultivation:* Biotechnol. Lett. **8**, 863 (1986) ▪ Enzyme Microb. Technol. **9**, 466 (1987) ▪ Planta Med. **53**, 479 (1987). – *Pharmacology & therapeutic use:* Hager (5.) **3**, 259f. ▪ Manske **37**, 205. – *Synthesis:* Hasan, in Atta-ur-Rahman (ed.), Natural Product Chemistry, p. 121–133, Berlin: Springer 1986. – *[HS 293990]*

Cathenamine (dehydroajmalicine).

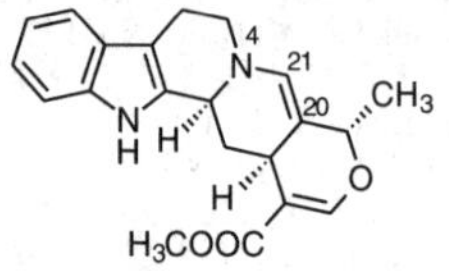

$C_{21}H_{22}N_2O_3$, M_R 350.42, amorphous, $[\alpha]_D$ −52° ($CHCl_3$). An unstable, reactive alkaloid, existing as $\Delta^{4(21)}$-immonium salts, isolated from *Guettarda eximia* [1,2] (Rubiaceae) and synthesized enzymatically with enzymes from *Catharanthus roseus* cell suspension cultures [3,4]. C. plays a key role in the biosynthesis of *monoterpenoid indole alkaloids , e.g., the *heteroyohimbine alkaloids such as *ajmalicine, 19-epi-ajmalicine, and tetrahydroalstonine (see Alstonia alkaloids).

Lit.: [1]Tetrahedron Lett. **1977**, 1889. [2]J. Chem. Soc., Chem. Commun. **1979**, 1015. [3]J. Chem. Soc., Chem. Commun. **1977**, 164. [4]Phytochemistry, **18**, 965 (1979).

gen.: Synthesis: Heterocycles **45**, 779 (1997) ▪ J. Am. Chem. Soc. **110**, 5925 (1988). – *[HS 293990; CAS 63661-74-5]*

Cathinone see kat and phenylethylamine alkaloids.

Caudoxirene see algal pheromones.

Caulerpenyne.

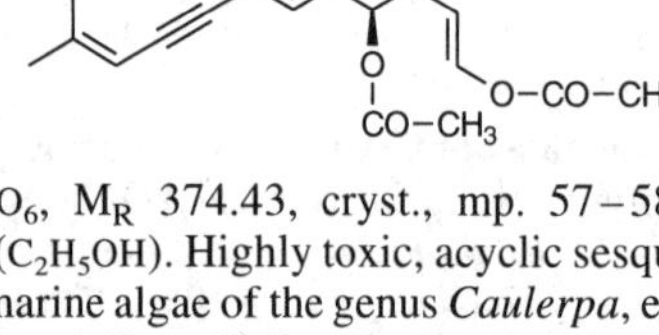

$C_{21}H_{26}O_6$, M_R 374.43, cryst., mp. 57–58 °C, $[\alpha]_D^{20}$ +7.1° (C_2H_5OH). Highly toxic, acyclic sesquiterpenoid from marine algae of the genus *Caulerpa*, e.g., *C. prolifera* [1] and *C. taxifolia*. The latter, although actually endemic in tropical waters, has recently been brought into the Mediterranean Sea near Monaco. The toxic C., which is produced by the algae in greater amounts in the colder water of the Mediterranean Sea than in its original home waters [2] is causing problems for the fishing industry.

Lit.: [1]Tetrahedron Lett. **1978**, 3593. [2]Helv. Chim. Acta **75**, 689 (1992).

gen.: Helv. Chim. Acta **76**, 855 (1993). – *[CAS 70000-22-5]*

Caulerpin.

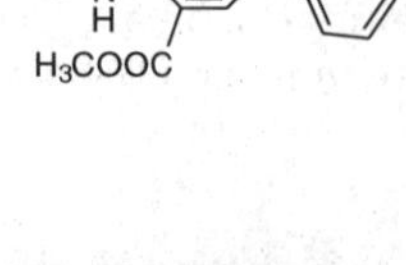

Green algae of the genus *Caulerpa* are much favored in the Pacific region as a food. In the Mediterranean Sea, however, *Caulerpa* is a source of considerable ecological concern since algae were released from an aquarium in Monaco and have displaced the endemic algal species. *C. racemosa* contains an interesting cyclooctatetraene system, *C.*: $C_{24}H_{18}N_2O_4$, M_R 398.42, red prisms, mp. 317 °C.

Lit.: J. Chem. Res. (S) **1978**, 126 ▪ J. Struct. Chem. **13**, 472–476 (1994) (X-ray crystal structure) ▪ Phytochemistry **26**, 619f. (1987); **30**, 3041 (1991). – *[CAS 26612-48-6]*

CBD_2 see tric(h)othecenes.

CC-1065 (rachelmycin).

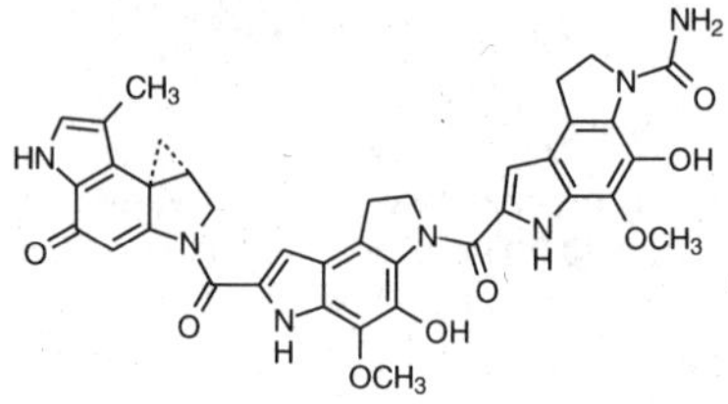

CC-1065

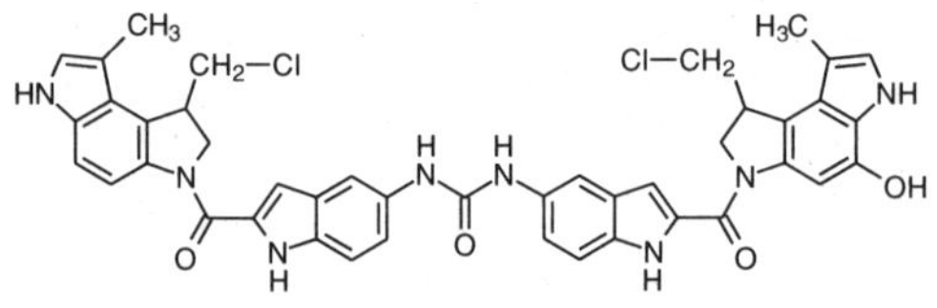

Bizelesin

$C_{37}H_{33}N_7O_8$, M_R 703.71, yellow cryst., $[\alpha]_D$ +97° (DMF), soluble in DMF, DMSO, and pyridine.

An extremely cytotoxic antitumor antibiotic produced by *Streptomyces zelensis* with three partially hydrogenated pyrrolo[3,2-*e*]indole units rotated to the right along the longitudinal axis. The unit with the anellated cyclopropane ring is the pharmacophoric group, the other two units effect the strong DNA binding. CC-1065 binds in the small groove of AT-rich, double-stranded DNA by way of alkylation of *N*-3 of adenine with opening of the cyclopropane ring. In cell cultures 1 ng/mL is sufficient to inhibit the growth of various types of tumors. In animal experiments it was found to be markedly more active than *adriamycin (400-fold) or *actinomycin D (10-fold). It exhibits delayed hepatotoxicity; decreases in the undesired toxicity [LD_{50} (mouse i.v.) 9 µg/kg] with retention of the antitumor activity have now been realized by derivatization [1] (e.g., bizelesin [2], which is 10000 times more potent than CC-1065, carzelesin, or adozelesin). The unnatural enantiomer of CC-1065 acts like the natural product. The biosynthesis starts from tyrosine, serine, and methionine. Particular safety precautions are needed when isolating CC-1065 on account of its toxicity. CC-1065 also exhibits antibacterial and antifungal activities. *Duocarmycin A is related to CC-1065; see also gilvusmycin.

Lit.: [1]Curr. Pharm. Des. **4**, 249–276 (1998). [2]J. Am. Chem. Soc. **115**, 5929–5933 (1993); Sem. Oncol. **24**, 219–240 (1997).

gen.: Structure: J. Am. Chem. Soc. **103**, 7629–7635 (1981) ▪ J. Antibiot. **34**, 1119–1125 (1981). – *Synthesis:* Angew. Chem. Int. Ed. Engl. **34**, 1366ff. (1995) ▪ Chem. Rev. **97**, 787–828 (1997) ▪ Chem.-Eur. J. **4**, 1554 (1998) ▪ Heterocycles **45**, 2303

(1997) ▪ J. Am. Chem. Soc. **110**, 1321 ff. (1988); **116**, 5523 (1994) ▪ J. Org. Chem. **57**, 1277 (1992); **61**, 4894 (1996) (derivatives); **62**, 5849 (1997); **64**, 5241 (1999) ▪ Synthesis **1999**, 1505 ▪ Synth. Commun. **25**, 1725–1739 (1995). – *Review:* Adv. Heterocycl. Nat. Prod. Synth. **2**, 1–188 (1992) ▪ Angew. Chem. Int. Ed. Engl. **35**, 1438–1474 (1996) ▪ Foye (ed.), Cancer Chemotherapeutic Agents, p. 326, Washington: ACS 1995 ▪ J. Antibiot. **39**, 319–334 (1986). – *Activity:* Acc. Chem. Res. **32**, 1043 (1999) ▪ Chem. Eng. News (17.11.) **1997**, 37 ▪ J. Am. Chem. Soc. **119**, 4977, 4987 (1997) ▪ J. Med. Chem. **35**, 1773–1782 (1992). – *[HS 2941 90; CAS 69866-21-3]*

Cedar camphor see cedrol.

Cedarwood oil. Depending on the origin, the following types of cedarwood oil used in the perfume and fragrance industries are distinguished:
A) ***American cedarwood oils:*** Virginia C. is a somewhat viscous, yellowish oil with a soft, woody, sweet and balsamy odor of pencils. Texas C. is a viscous, reddish to brownish oil with an odor similar to that of the Virginia oil. Part of the content of cedrol often crystallizes from the oil.
Production: Both oils are produced by steam distillation from the wood of various *Juniperus* species (Cupressaceae): Virginia oil from the "Southern red cedar", *J. virginiana*, growing mainly in the south east of the USA, Texas oil from species growing mainly in Texas such as, e.g., *J. mexicana* and *J. deppeana*. In the past the Virginia oil was a by-product from the processing of cedar wood for wardrobes, crates, and pencils. Today, the major part of the estimated annual production of ca. 1500 t is distilled in Texas.
B) ***Chinese cedarwood oil:*** A clear, mobile, mostly yellowish oil with a typical odor of cedarwood but with a much more smoky and fatty note than the American oils.
Production: By steam distillation from the wood of the Chinese red pine *Chamaecyparis funebris* (Cupressaceae). The annual production amounts to ca. 200–300 t.
Composition [1]***:*** The main components of all mentioned oils are the sesquiterpene compounds *α-*cedrene* (ca. 10–25%), **thujopsene* (ca. 20–35%), and **cedrol* (ca. 20–40%, in Chinese oil only 10–15%).
Use: C. find wide use in the perfume industry; rectification of the oils furnishes starting materials for further substances used in perfumes, e.g., cedrol methyl ether [2] and cedryl acetate [3] from cedrol, acetylcedrene [4] from the mixture of cedrene/thujopsene, and cedrene epoxide [5] from α-cedrene.
Lit.: [1] Perfum. Flavor. **5** (3), 63 (1980); **16** (5), 79 (1991). [2] Bauer et al. (2.), p. 49. [3] Bauer et al. (2.), p. 60. [4] Bauer et al. (2), p. 57. [5] Ohloff, p. 171.
gen.: Arctander, p. 144 f. ▪ Bauer et al. (2.), p. 143 ▪ ISO 4724 (1984), 4725 (1986), 9843 (1991) ▪ Ohloff, p. 170. – *Toxicology:* Food Cosmet. Toxicol. **12**, 845 (1974); **14**, 711 (1976). – *[HS 3301 29; CAS 8000-27-9]*

Cedrene.

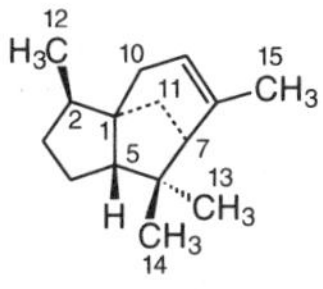

α-Cedrene

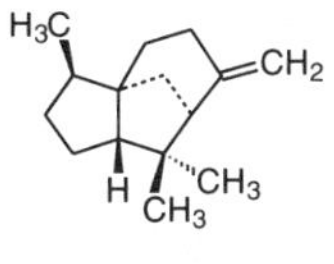

β-Cedrene

(–)-*α*-C., $C_{15}H_{24}$, M_R 204.36, oil, bp. 124–126 °C (16 hPa), $[\alpha]_D^{20}$ –91.3° (neat), and *(–)-thujopsene each constitute about a quarter of *cedarwood oil [1]. Accompanying component (ca. 15%) is *β*-C., oil, bp. 120 °C (6.7 hPa), $[\alpha]_D^{25}$ +9.7°. Friedel-Crafts acylation of the oil produces a complex mixture with a woody smell, that finds wide use in the perfume industry as a fixative [2]. Treatment of *α*-C. with peracids leads to an epoxide which, besides its cedarwood-like basic odor, has a nuance resembling that of *patchouli oil. The 1,7-epimer of *α*-C. is found in cypress oils as *α-funebrene* [3].
Lit.: [1] Beilstein E III **5**, 1095; Herba Hung. **24**, 27 (1985). [2] Ohloff, p. 170 ff. [3] J. Nat. Prod. **47**, 924 (1984).
gen.: Tetrahedron Lett. **24**, 2125 (1983) (synthesis); **37**, 4421 (1996) (synthesis *δ*-2-C.); **39**, 7713 (1998) (synthesis *α*-C.). – *[CAS 469-61-4 (α-C.); 546-28-1 (β-C.); 50894-66-1 (α-F.)]*

Cedrol (*α*-cedrol, cedar camphor, cypress camphor).

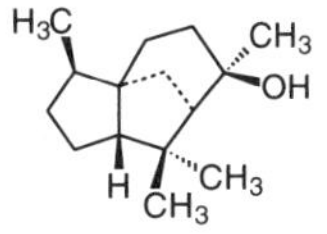

$C_{15}H_{26}O$, M_R 222.37, mp. 86–87 °C, $[\alpha]_D^{20}$ +10.5° ($CHCl_3$). C. constitutes the actual fragrance substance of *cedarwood oil [1]; it is often found in cypress oils, sometimes in juniper species (red cedar). C. is the main component in the oil of *Juniperus chinensis*. To increase the valuable woody, balmy notes, the alcoholic components of cedarwood oil are esterified [2] (e.g., *O*-acetyl-C. = cedryl acetate). C. and *cedrene-containing distillates serve as fixatives for soap perfumes, pine scents, and for perfumes with a woody note in general.
Lit.: [1] J. Nat. Prod. **47**, 924 (1984); Anal. Chem. **60**, 472 (1988). [2] Ohloff, p. 170 ff.
gen.: Phytochemistry **41**, 1361 (1996) (isolation) ▪ Tetrahedron **37**, 4401 (1981) (synthesis). – *[CAS 77-53-2]*

Celenamides.

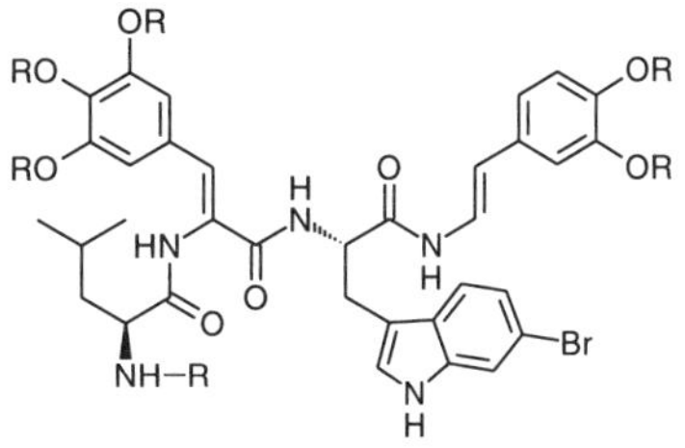

R = H : Celenamide A
R = CO—CH_3 : Hexaacetylcelenamide A

The C. A–D are isolated from the yellow borer sponge *Cliona celata* [1]. The major part of the sponge is found in rocks in which it has chemically etched a system of fine tubes and passages comparable to the thread-like mycelia of fungi; only the deep yellow in- and outflow openings are visible. The unstable tetrapeptide derivatives contain two dehydroamino acid residues and have been characterized in the form of their peracetyl derivatives: e.g. *hexaacetylcelenamide A*, $C_{46}H_{48}BrN_5O_{14}$, M_R 974.82, $[\alpha]_D$ +40°. They are related with integerrine [2] from *Ceanothus integerrimus* (Rhamnaceae) and with the *tunichromes from the tunicate *Ascidia nigra* [3].

Lit.: [1] J. Org. Chem. **45**, 3687 (1980); Can. J. Chem. **58**, 2121 (1980). [2] Tetrahedron Lett. **1968** 1311. [3] J. Am. Chem. Soc. **110**, 6162 (1988).
gen.: Angew. Chem. Int. Ed. Engl. **23**, 991 (1984) (synthesis). – *[CAS 74144-98-2 (C. A)]*

Celery seed oil. Colorless to brown-yellow oil with a very adherent, warm, spicy-sweet, odor of celery and a somewhat burning, aromatic, spicy-sweet taste typical of celery.
Production: By steam distillation from the seeds (fruits) of celery, *Apium graveolens*.
Composition [1]***:*** Main components are (+)-*limonene* (see *p*-menthadienes) (70–80%) and (+)-*β*-**selinene* (ca. 10%). Mainly responsible for the typical organoleptic impressions are *3-butylphthalide* [ca. 3%, $C_{12}H_{14}O_2$, M_R 190.24, bp. 177–178 °C (15 hPa)] and *sedanenolide* (ca. 5%, $C_{12}H_{16}O_2$, M_R 192.26).

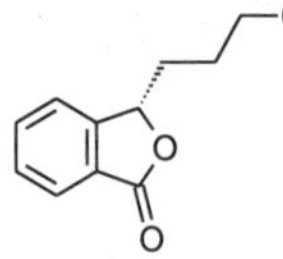
3-Butylphthalide

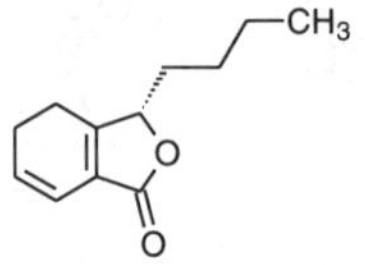
Sedanenolide

Use: The oil finds widespread use in the perfume industry albeit only in very low doses; it imparts a natural warmth and brilliance to many perfume compositions. It is also used to improve the aromatic nature of soups, salads, and meat products.
Lit.: [1] Perfum. Flavor. **14** (5), 52 (1989); **15** (5), 57 (1990); **20** (1), 52 (1995); Dev. Food Sci. **34**, 329 (1994).
gen.: Arctander, p. 149 ▪ Bauer et al. (2.), S. 144 ▪ ISO 3760 (1979) ▪ Perfum. Flavor. **15** (3), 55 (1990). – *Toxicology:* Food Cosmet. Toxicol. **12**, 849 (1974). – *[HS 3301 29; CAS 8015-90-5 (S.); 6066-49-5 (3-butylphthalide); 62006-39-7 (sedanenolide)]*

Cellobiose see cellulose.

Cellulose.

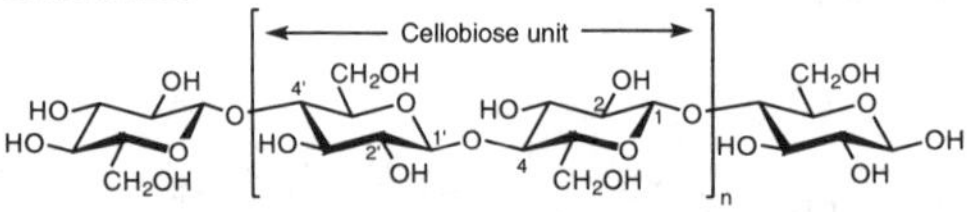

C. is the isotactic *β*-1,4′-polyacetal of cellobiose $(C_6H_{10}O_5)_n$. The average degree of polymerization (DP) lies between 600 (C. from *Acetobacter*) and 15 000 (raw cotton). Native C. forms pseudomonoclinic unit cells made up of a central glucan strand with four further glucan strands at the edges. C. adopts a slightly helical form on account of intramolecular hydrogen bridges, in which in particular the hydroxy groups at C-3 and C-6, intermolecular hydrogen bridges as well as the pyranose ring oxygen atom participate. This hinders free rotation and causes relatively high rigidity which provides the basis for C.'s function as a skeletal substance.
Physical properties: C. is relatively hygroscopic, it absorbs 8–14% of water. C. is insoluble in water or dilute acids; in alkaline solutions it swells strongly and the short-chain units with DP < 200 (*α*-C.) dissolve. C. dissolves in zinc chloride solution or copper salt solutions containing ammonia (Schweizer's reagent, Cuoxam) or ethylenediamine (Cuen); C. does not melt, flame point > 290 °C. C. is colored blue-violet by chlorine and zinc iodide, D. 1.52–1.59. C. is degraded by cellulases that occur in bacteria, fungi, insects, and mollusks.
Occurrence: C. is the most abundant organic compound. It is a component of almost all plant cell walls, including green algae, diatoms, fungi; tunicates and bacteria (e. g., *Acetobacter xylinum* and *A. aceti*) also form cellulose. C. biosynthesized by *A. xylinum* is anchored in the so-called "terminal complexes" in the bacterial membrane and arranged linearly along the longitudinal axis of the rod-shaped bacteria. Fibers up to 1 m in length are formed (in cotton on the other hand only a few cm). Through cell division a C.-network is finally formed – without polyoses and lignin – that can be recovered easily. Unbound *Acetobacter*-C. has a very high capacity for binding water (700-fold dry weight) and usually excellent firmness. Algal C. exists in large crystallites and is used to produce microcrystalline C.
Production: The most important sources of crude C. are cotton and bast plants: flax, ramie, jute, hemp (at present still very small amounts) for the textile industry and wood for the paper and pulp industry, in smaller amounts from bagasse. The isolation of C. from wood needs a series of processes in order to separate C. from lignin and polyoses.
Biosynthesis: In 1996, the biosynthetic gene for C. in higher plants was identified [1].
Lit.: [1] Proc. Natl. Acad. Sci. USA **93**, 12637 (1996).
gen.: ACS Symp. Ser. **340**, 1–310 (1987) ▪ Barton-Nakanishi **3**, 529–598 (review) ▪ Fengel & Wegener, Wood Chemistry, Ultrastructure, Reactions, Berlin: de Gruyter 1989 ▪ Kennedy et al., Cellulose and its Derivatives, Chichester: Horwood 1985 ▪ Kirk-Othmer (4.) **5**, 476–497; **6**, 1023; **7**, 292; **8**, 451 ▪ Klemm et al. (eds.), Comprehensive Cellulose Chemistry (2 Vol.: Vol. 1: Fundamentals and Analytical Methods; Vol. 2: Derivatization of Cellulose), Weinheim: Wiley-VCH 1998 ▪ Methods Biotechnol. **10**, 57–69 (1999) (enzymatic synthesis) ▪ Nature (London) **311**, 165 (1984) (biosynthesis) ▪ Nevell et al., Cellulose Chemistry, Chichester: Horwood 1985 ▪ Tappi J. **72** (3), 169 (1989) ▪ Ullmann (5.) **A 5**, 377–418 ▪ Wilke et al., Chem. Technol. Rev. 218: Enzymatic Hydrolysis of Cellulose, Park Ridge: Noyes Data Corp. 1983 ▪ Wood, Methods Enzymol. 160/A, San Diego: Academic Press 1988 ▪ Young et al., Cellulose, New York: Wiley 1986. – *Journal:* Cellulose (Roberts, ed.), London: Chapman & Hall (since 1994). – *[HS 3912 11, 3912 90; CAS 9004-34-6 (C.); 528-50-7 (cellobiose)]*

Celorbicol see agarofurans.

Cemadotin see Dolastatins.

Cembranoids. Group of macrocyclic *diterpenes (unsaturated cyclotetradecane derivatives) found in tobacco, insects, soft corals (especially in the Great Barrier Reef), and pine trees. Activities against tumors and psoriasis have been reported. *Cembrenene* ($C_{20}H_{30}$, M_R 270.46, oil) exemplifies C. structures:

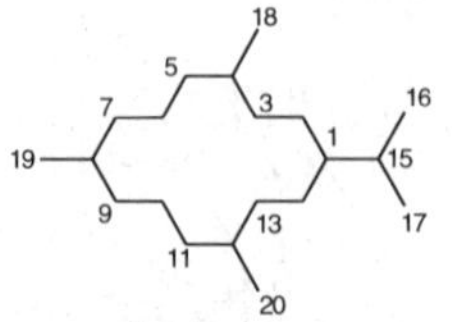
Cembrane

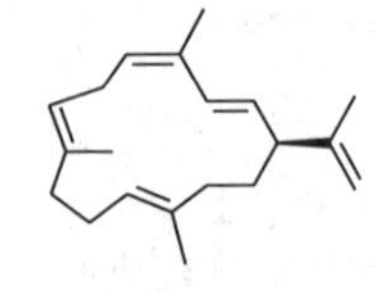
Cembrenene

Lit.: Atta-ur-Rahman **10**, 3–42 (1992) ▪ Chem. Pharm. Bull. **36**, 3780–3786 (1988) ▪ Chem. Rev. **88**, 719–732 (1988) ▪

Heterocycles **49**, 531–556 (1998) (review) ▪ J. Chem. Soc., Perkin Trans. 1 **1996**, 57 (synthesis) ▪ J. Org. Chem. **53**, 1616–1623 (1988) ▪ Tetrahedron **45**, 103 (1989); **49**, 4975–4992 (1993) (C. from tobacco) ▪ Tetrahedron: Asymmetry **10**, 1877 (1999) (synthesis sinulariol) ▪ Tetrahedron Lett. **33**, 1225 (1992) ▪ Zechmeister **36**, 285–387; **59**, 141–320; **60**, 1–141. – *[CAS 79296-91-6 (cembrenene)]*

Cembranolides. Name for *cembranoids containing a lactone ring, e.g., 3,7,11,15(17)-cembratetraen-16,2-olide, $C_{20}H_{28}O_2$, M_R 300.44, oil, $[\alpha]_D$ −29.0° ($CHCl_3$) isolated from the soft coral *Sinularia mayi*.

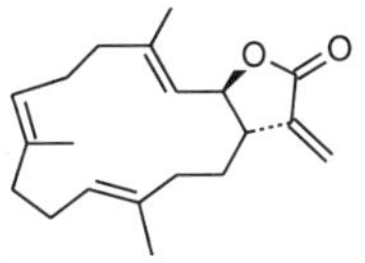

Lit.: Chem. Lett. **1982**, 277 ▪ Heterocycles **36**, 1957 (1993); **37**, 679 (1994).

Cembrenes. Group of monocyclic diterpenes belonging to the *cembranoids consisting of *cembrene* [$C_{20}H_{32}$, M_R 272.47, cryst., mp. 58–59 °C, $[\alpha]_D^{23}$ +238° ($CHCl_3$)], *C. A* [*neocembrene*, colorless oil, bp. 150–152 °C (0.12 kPa), $[\alpha]_D$ −19.7° ($CHCl_3$)], and *C. C* [(1*E*,3*E*,7*E*,11*E*-cembratetraene, $C_{20}H_{32}$, M_R 272.47, oil)].

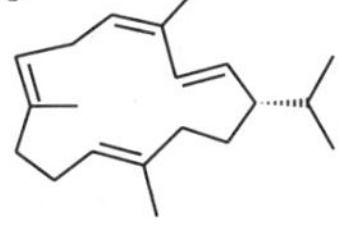

(+)-Cembrene

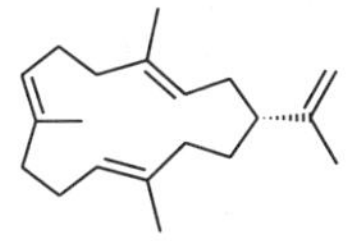

Cembrene A

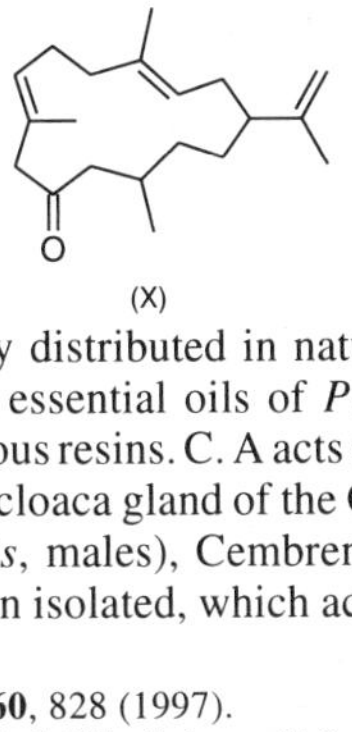

(X)

The C. are widely distributed in nature. They occur, among others, in essential oils of *Pinus* species, tobacco, and in various resins. C. A acts as a termite pheromone. From the cloaca gland of the Chinese alligator (*Alligator sinensis*, males), Cembrene A and the ketone (X) have been isolated, which act as pheromones on females [1].

Lit.: [1] J. Nat. Prod. **60**, 828 (1997).

gen.: Beilstein E IV **5**, 1497 ▪ Scheuer II **2**, 264 ▪ Synlett **1992**, 577. – *Synthesis:* Bull. Soc. Chim. Belg. **104**, 69 ff. (1995) (synthesis C. A) ▪ Chem. Lett. **1994**, 741 ▪ Indian J. Chem., Sect. B **24**, 918–922 (1985) ▪ J. Chem. Soc., Perkin Trans. 1 **1996**, 57 ▪ Synthesis **1996**, 736 ▪ Tetrahedron: Asymmetry **7**, 2851 (1996). – *[CAS 1898-13-1 (C.); 31570-39-5 (C. A); 64363-64-0 (C. C)]*

Cephaeline see ipecac alkaloids.

Cephalonic acid see ophiobolins.

Cephalosporins. Group name for an important group of *β-lactam antibiotics based on the parent skeleton of 7β-aminocephalosporanic acid (3-acetoxymethyl-7β-amino-3-cephem-4-carboxylic acid).

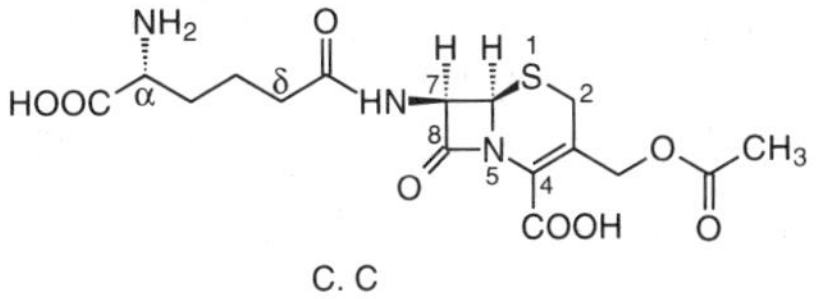

C. C

Natural C. are β-lactam-dihydrothiazine derivatives with (*R*)-2-aminoadipic acid as N^7-acyl residue. The most important member is *C. C* ($C_{16}H_{21}N_3O_8S$, M_R 415.42, as Na salt soluble in water, hardly soluble in ethanol, ether), first detected in 1953 in culture filtrates of the fungus *Acremonium chrysogenum*. C. act, like the closely related *penicillins, as inhibitors of bacterial cell wall synthesis in growing microorganisms. Only semisynthetic C. are used in therapy which, apart from oral efficacy and excellent β-lactamase stability as broad-spectrum antibiotics, also exhibit activity against Gram-negative pathogens. Besides production of the C. skeleton by fermentation, chemical processes for ring expansion of penicillins to C. have found wide technical applications.

Lit.: Angew. Chem. Int. Ed. Engl. **24**, 180–202 (1985) ▪ Kirk-Othmer (4.) **3**, 28 ▪ Morin & Gorman, Chemistry and Biology of β-Lactam Antibiotics, vol. 1, Penicillins and Cephalosporins, New York: Academic Press 1982 ▪ Page (ed.), The Chemistry of β-Lactams, London: Blackie Academic and Professional 1992 ▪ Reuben & Wittkoff, Pharmaceutical Chemicals in Perspective, New York: Wiley 1989.

Cephalostatins.

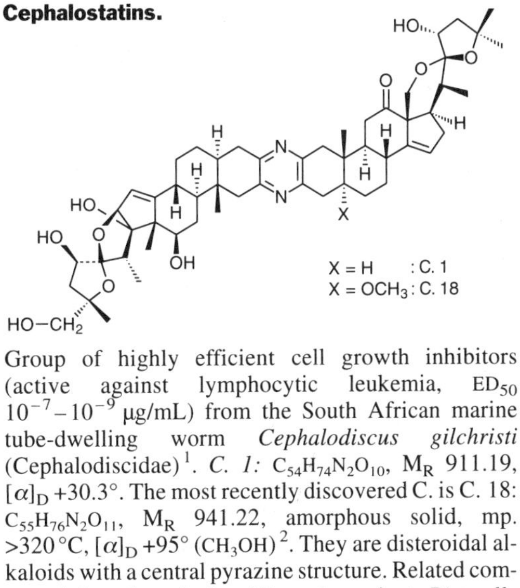

X = H : C. 1
X = OCH_3 : C. 18

Group of highly efficient cell growth inhibitors (active against lymphocytic leukemia, ED_{50} 10^{-7}–10^{-9} μg/mL) from the South African marine tube-dwelling worm *Cephalodiscus gilchristi* (Cephalodiscidae) [1]. *C. 1:* $C_{54}H_{74}N_2O_{10}$, M_R 911.19, $[\alpha]_D$ +30.3°. The most recently discovered C. is C. 18: $C_{55}H_{76}N_2O_{11}$, M_R 941.22, amorphous solid, mp. >320 °C, $[\alpha]_D$ +95° (CH_3OH) [2]. They are disteroidal alkaloids with a central pyrazine structure. Related compounds have been isolated from the tunicate *Ritterella tokioka* [3].

Lit.: [1] J. Am. Chem. Soc. **110**, 2006 (1988); **117**, 10157 (1995); J. Org. Chem. **57**, 429 (1992). [2] J. Nat. Prod. **61**, 955 (1998). [3] J. Org. Chem. **60**, 608 (1995).

gen.: Angew. Chem. Int. Ed. Engl. **35**, 611–615 (1996) (review) ▪ Bioorg. Med. Chem. Lett. **5**, 2027 (1995) ▪ Can. J. Chem. **67**, 1509 (1989) ▪ Eur. J. Org. Chem. **1998**, 2811–2831 (synthesis) ▪ J. Am. Chem. Soc. **117**, 10157 f. (1995); **118**, 10672 (1996); **120**, 692–707 (1998); **121**, 2071 (1999) (synthesis) ▪ J. Chem. Soc., Chem. Commun. **1988**, 865, 1440 ▪ J. Chem. Soc., Perkin Trans. 1 **1993**, 2865 ▪ Tetrahedron Lett. **40**, 4655 (1999) (synthesis). – *[CAS 112088-56-9 (C. 1).]*

Cephalotaxin(on)e see Cephalotaxus alkaloids.

Cephalotaxus alkaloids. Conifers of the East Asian endemic genus *Cephalotaxus* (Cephalotaxaceae) contain a group of alkaloids with pentacyclic structures. The most important alkaloid is cephalotaxine which occurs together with some derivatives and various esters. In *Cephalotaxus harringtoniana* alkaloids with the cephalotaxin structure occur concomitantly with others having the homoerythrinane skeleton such as schelhammericine and schelhammeridine, the main alkaloids of *Schelhammera pedunculata* (Liliaceae). Harringtonine and homoharringtonine exhibit antitumor activity, they are more active than *vincristine against murine leukemia cells. They act as inhibitors of protein and DNA biosynthesis and are used clinically against acute myelocytic leukemia. The esters are inhibitors of eukaryotic protein synthesis.

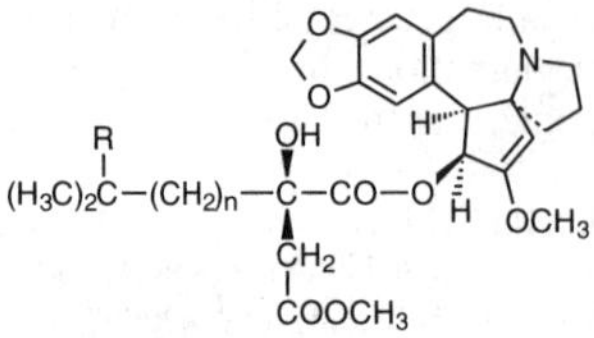

R = OH, n = 2 : Harringtonine (**1**)
R = OH, n = 3 : Homoharringtonine (**2**)
R = H, n = 2 : Deoxyharringtonine (**3**)

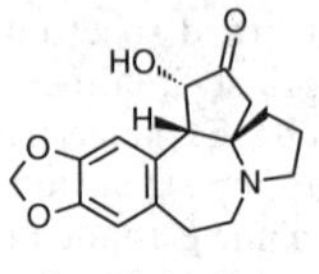

Demethylcephalotaxine (**4**)

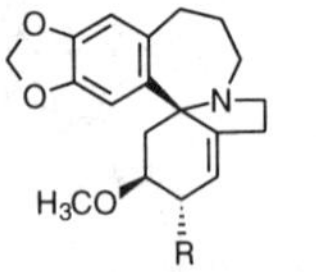

R = H : Schelhammericine (**5**)
R = OH : Schelhammerine (**6**)

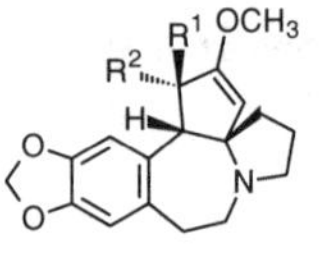

R^1 = H, R^2 = OH: Cephalotaxine (**7**)
R^1, R^2 = O : Cephalotaxinone (**8**)

Table: Data of Cephalotaxus alkaloids.

no.	molecular formula	M_R	mp. [°C]	$[\alpha]_D$ ($CHCl_3$)	CAS
1	$C_{28}H_{37}NO_9$	531.6	73–75	–104.6°	26833-85-2
2	$C_{29}H_{39}NO_9$	545.63	144–146	–119°	26833-87-4
3	$C_{28}H_{37}NO_8$	515.6		–125.4°	36804-95-2
4	$C_{17}H_{19}NO_4$	301.34	109–111	–110°	39707-71-6
5	$C_{19}H_{23}NO_3$	313.4	76–77	+122°	21030-78-4
6	$C_{19}H_{23}NO_4$	329.4	173–174	+186°	21030-71-7
7	$C_{18}H_{21}NO_4$	315.37	132–133	–204°	24316-19-6
8	$C_{18}H_{19}NO_4$	313.35	198–200	–146°	38750-57-1

Biosynthesis: The concomitant occurrence of C. a. and homoerythrina alkaloids indicates a common biosynthesis. Both types of alkaloids are derived from 1-phenethylisoquinoline with (see phenethyltetrahydroisoquinoline alkaloids) ring A originating from tyrosine and ring B from phenylalanine. The acid components, e.g., harringtonic, isoharringtonic, and deoxyharringtonic acids are derived from leucine.

Lit.: Anticancer Res. **11**, 1879 (1991) ▪ Biochim. Biophys. Acta **1129**, 177–182 (1992) ▪ Manske **23**, 157–226; **25**, 57–69; **51**, 199–270 ▪ Nat. Prod. Rep. **6**, 55–66 (1989); **10**, 464 (1993) ▪ Pelletier **5**, 639–690; **7**, 1–41 ▪ Sax (8.), HGI 575, IKS 500 ▪ Tang, Eisenbrand (eds.), Chinese Drugs of Plant Origin, p. 281 f., Berlin: Springer 1992 ▪ Tetrahedron Lett. **37**, 7053 (1996) ▪ Ullmann (5.) **A 1**, 379. – *Biosynthesis:* Rec. Adv. Phytochem. **13**, 55–84 (1979) ▪ J. Am. Chem. Soc. **102**, 1099–1111 (1980). – *Synthesis:* Angew. Chem. Int. Ed. Engl. **36**, 1124 (1997) ▪ Chem. Pharm. Bull. **44**, 500 (1996); **47**, 983 (1999) ▪ J. Am. Chem. Soc. **112**, 9601–9613 (1990); **121**, 10264 (1999) ▪ J. Org. Chem. **55**, 831–838 (1990); **60**, 115–119 (1995); **61**, 7335–7347 (1996) ▪ Tetrahedron Lett. **38**, 4347 (1997). – *[HS 2939 90]*

Cephamycins (7α-methoxycephalosporins). Group of *β-lactam antibiotics that differ from the *cephalosporins by a 7α-methoxy group.

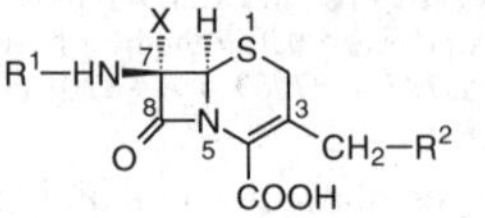

X = H : Cephalosporins
X = OCH_3 : Cephamycins

$R^1 = CO{-}(CH_2)_3{-}\overset{(R)}{C}H(NH_2){-}COOH$

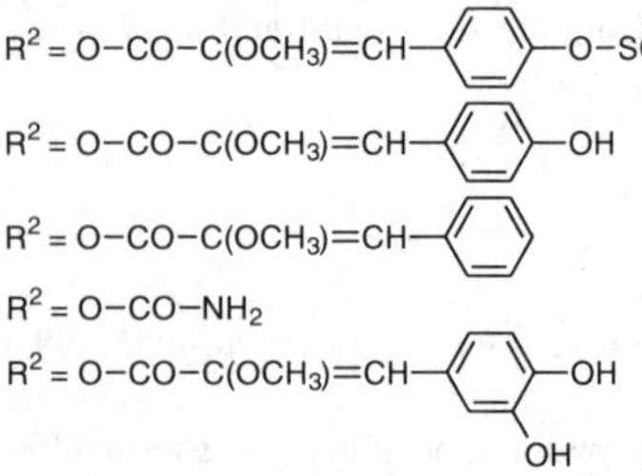

: C. A
: C. B
: Deoxy-C. B
: C. C
: C 2801X

The C. isolated from *Streptomyces* and *Nocardia* culture broths are active against Gram-positive and Gram-negative bacteria and exhibit a higher stability towards β-lactamases in comparison with the *cephalosporins. The biosynthesis of the C. proceeds similarly to that of the cephalosporins and *penicillins. The additional oxygen atom on the β-lactam ring originates from atmospheric oxygen and the *O*-methyl group from **S*-adenosylmethionine. Starting from the naturally occurring C. (C. A–C, *deoxycephamycin B, antibiotic C 2801X*) a series of semisynthetic derivatives with improved properties has been prepared, e. g., cefotaxime, cefoxitin, and cefotetan.

Lit.: Heterocycles **8**, 719 (1977) (review) ▪ J. Antibiot. **29**, 113–120 (1976). – *[HS 2941 90]*

Cepharanthine see bisbenzylisoquinoline alkaloids.

Ceramides.

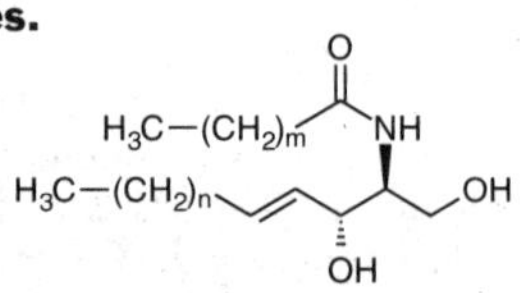

ceramides

m	*N*-acyl residues	aminodiol backbones	n
12	*N*-myristoyl-	-hexadecasphingosine	10
14	*N*-palmitoyl-	-sphingosine*	12
16	*N*-stearoyl-*	-eicosasphingosine	14
18	*N*-arachidoyl-	-docosasphingosine	16
20	*N*-behenoyl-		
22	*N*-lignoceroyl-		
24	*N*-ceroyl-		

* major constituents.

C., e.g., *N*-stearoylsphingosine ("ceramide"), $C_{36}H_{71}NO_3$, M_R 565.95, oil, play a central role in sphingolipid metabolism and have an equally important function as second messengers in the sphingolipid-dependent pathway of cell signal transmission. Increase in ceramide levels may inhibit cell growth as a cellular defense/repair mechanism and initiate apoptosis, etc. C. and diacylglycerols (DAG) may define opposing pathways of cell regulation [1]. Moreover, *sphingosine, the major constituent of C., displays potent inhibiting properties on protein kinase C (PKC) both *in vivo* and *in vitro* and thus plays a pivotal role in cell recognition, cell growth modulation and signal transmission. Glycosphingolipids derived from C. have also been shown to mediate cell recognition events.

Lit.: [1] Angew. Chem. Int. Ed. **39**, 1440 (2000); J. Biol. Chem. **269**, 3125 (1994).
gen.: Justus Liebigs Ann. Chem. **1996**, 2079 (synthesis). – *[CAS 104404-17-3 (C.); 2304-81-6 (N-stearoylsphingosine)]*

Ceratiopyrones.

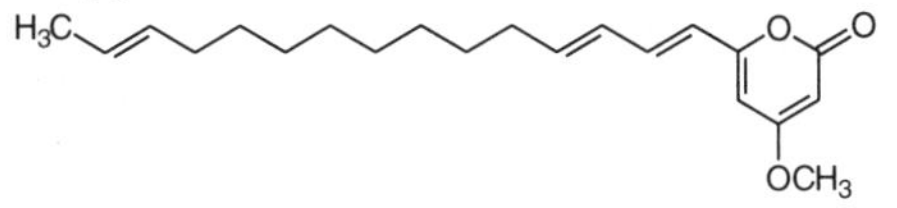

Ceratiopyrone A

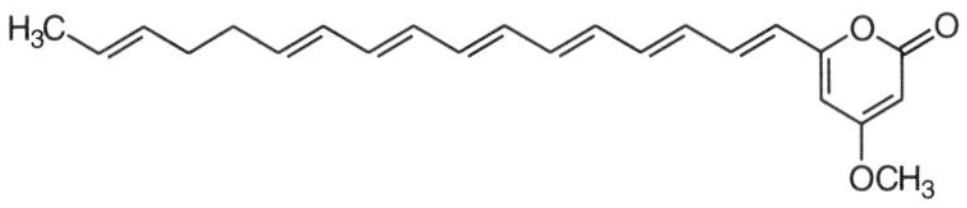

Ceratioflavin A

Unsaturated 6-alkylpyrones from the colorless to bright yellow plasmodia of *Ceratiomyxa fruticulosa* (Myxomycetes), a slime mold distributed from the tropics to the Arctic regions. The main constituent is *ceratiopyrone A* ($C_{21}H_{30}O_3$, M_R 330.47, solid, mp. 73–77 °C), and is accompanied by analogues with two and four double bonds in the C_{15}- and C_{17} side chains as well as the polyene pigment *ceratioflavin A* ($C_{23}H_{26}O_3$, M_R 350.46, yellow oil).

Lit.: Justus Liebigs Ann. Chem. **1995**, 81. – *[CAS 160700-45-8 (C. A); 160669-46-5 (C. B); 160669-49-8 (ceratioflavin A)]*

Cercosporamide.

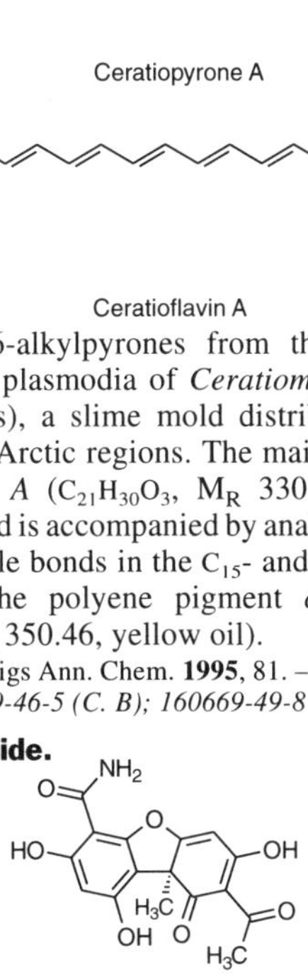

$C_{16}H_{13}NO_7$, M_R 331.28, cryst., mp. 188–189 °C, $[\alpha]_D$ –26°. *Usnic acid derivative from the fungus *Cercosporidium henningsii*. Phytotoxin.

Lit.: J. Org. Chem. **56**, 909 f. (1991). – *[CAS 131436-22-1]*

Cerebrodienes. Cranial lipids structurally related to *sphingosine which are of significance for the induction of sleep, e. g., *C. A*, $C_{18}H_{35}NO$, M_R 281.48, solid.

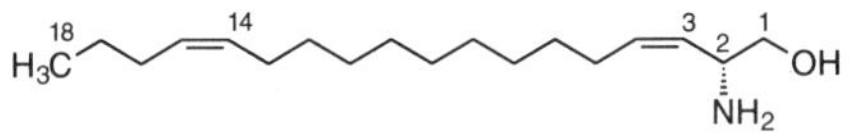

C.A

Lit.: Proc. Natl. Acad. Sci. USA **91**, 9505 (1994).

Cerebrose see galactose.

Cerebrosides.

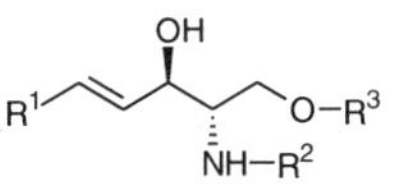

R^1 = long chain alkyl residue
R^2 = fatty acid residue
R^3 = β-D-Galactopyranosyl or β-D-Glucopyranosyl

C. is the general name for a group of glycolipids made up of *sphingosine, a fatty acid, and a monosaccharide (usually galactose). They always occur as mixtures and are important components of fats, especially in brain and nerve tissue. In C. from plant sources, there may also be multiply unsaturated, branched alkyl groups R^1, as well as α-hydroxy-fatty acids as acyl components. The C. have antifungal properties.

Lit.: Angew. Chem. Int. Ed. Engl. **24**, 65 f. (1985) ▪ J. Antibiot. **41**, 469–480 (1988) (^1H, ^{13}C-NMR, MS) ▪ Luckner (3.), p. 150 ff. ▪ Tetrahedron **41**, 2369–2386 (1985) (synthesis).

Cerniltone see DIBOA.

Cernuine.

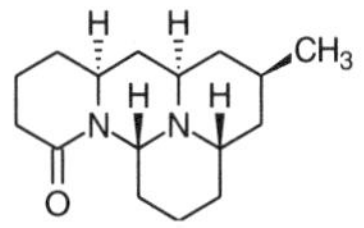

$C_{16}H_{26}N_2O$, M_R 262.40, mp. 103–104 °C, $[\alpha]_D$ –20.5° (CH_3OH). Alkaloid from some Lycopodiaceae species (*clubmosses, *Lycopodium cernuum* and *L. carolianum*). It is weakly toxic. The biosynthesis starts from lysine with cadaverine, 1-piperideine, and *pelletierine as intermediates [1].

Lit.: [1] Can. J. Chem. **49**, 3352 (1971); Pelletier **3**, 185–240.
gen.: Beilstein E V **24/2**, 306. – *[HS 2939 90; CAS 6880-84-8]*

Cerotic acid (ceric aid, cerinic acid, hexacosanoic acid). $H_3C–(CH_2)_{24}–COOH$, $C_{26}H_{52}O_2$, M_R 396.70, mp. 87.7–88.5 °C. Component of *bee and other insect waxes, wool wax, *carnauba and other palm leaf and grass waxes. C. also occurs in the wax of tubercle bacilli (*Mycobacterium tuberculosis*). For the role of C. in the metabolic disease adrenoleukodystrophia, see erucic acid.

Lit.: Beilstein E IV **2**, 1310 ▪ Ullmann (5.) **A 10**, 247. – *[HS 2915 90; CAS 506-46-7]*

Cerulenin (helicocerin).

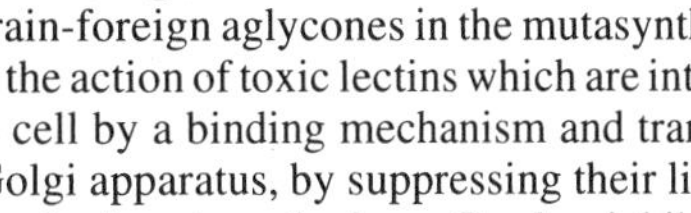

$C_{12}H_{17}NO_3$, M_R 223.27, needles, mp. 93 °C, $[\alpha]_D$ +63° (CH_3OH), antifungal *antibiotic produced by *Cephalosporium caerulens* that inhibits fatty acid and steroid biosynthesis. It is also used to inhibit *polyketide biosynthesis, e. g., of *macrolides, in order to glycosylate strain-foreign aglycones in the mutasynthesis. C. inhibits the action of toxic lectins which are introduced into the cell by a binding mechanism and transported to the Golgi apparatus, by suppressing their liberation in the endoplasmic reticulum. C. also inhibits HIV, interaction with the viral proteases has been demonstrated. The open-chain form is in equilibrium with a cyclic semiaminal form.

Lit.: Chem. Pharm. Bull. **40**, 2945, 2954 (1992) ▪ Methods Enzymol. **72**, 520 (1981) (review). – *Synthesis:* J. Med. Chem. **42**, 4932 (1999) (analogues) ▪ J. Org. Chem. **58**, 6779 (1993); **61**, 6121 (1996); **62**, 636 (1997) ▪ Tetrahedron Lett. **32**, 6771 (1991) ▪ Yoda, in Lukacs (ed.), Recent Progress in the Chemical Synthesis of Antibiotics, vol. 2, p. 939–970, Berlin: Springer 1993. – *[CAS 17397-89-6]*

Cevadine, Cevane see Veratrum steroid alkaloids.

Ceveratrum alkaloids see Veratrum steroid alkaloids.

Chaconine see Solanum steroid alkaloid glycosides.

Chaenomeloidin see salicyl alcohol.

Chaetoglobosins. Ch. A–G and J are metabolic products from *Chaetomium cochliodes*, *Ch. globosum*, and other *Chaetomium* species (Ascomycetes) [1,2], Ch. K,

L, and M from *Diplodia macrospora* (Sphaeropsidales)[3]. Ch. C is also formed by *Penicillium aurantiovirens*[4] (on spoiled pecan nuts). The Ch. are *mycotoxins related to the *cytochalasins and are all cytotoxic, some also have antimicrobial (Ch. A) and phytotoxic (Ch. K) activities. They are formed biosynthetically from tryptophan and a *polyketide chain (nonaketides)[5]. Some C. induce the formation of multinuclear cells in mammalian cell cultures.

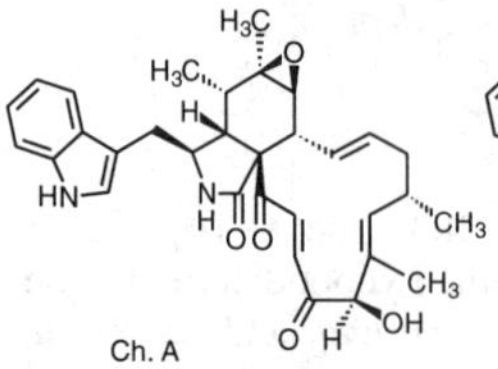

Ch. A

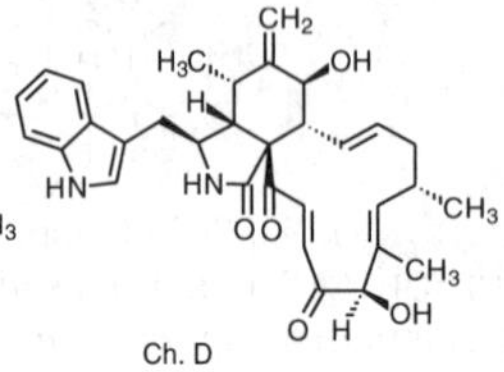

Ch. D

Table: Data of selected Chaetoglobosins.

compound	molecular formula	M_R	mp. [°C]	CAS
Ch. A	$C_{32}H_{36}N_2O_5$	528.65	188	50335-03-0
Ch. C	$C_{32}H_{36}N_2O_5$	528.65	259–261	50645-76-6
Ch. D	$C_{32}H_{36}N_2O_5$	528.65	216	55945-73-8
Ch. K	$C_{34}H_{40}N_2O_5$	556.70	235–240	72509-61-6

Lit.: [1] Tetrahedron Lett. **1973**, 2109; **1976**, 1349; **1977**, 2771 ff. [2] Can. J. Microbiol. **25**, 170 (1979); Chem. Pharm. Bull. **30**, 1609–1638 (1982). [3] J. Agric. Food Chem. **28**, 139 (1980); Helv. Chim. Acta **65**, 1543 (1982); **71**, 1881 (1988). [4] Tetrahedron Lett. **1976**, 1355. [5] Helv. Chim. Acta **64**, 2056 (1981); Chem. Pharm. Bull. **31**, 490 (1983).
gen.: Betina, chap 13 ▪ Cole-Cox, p. 304–333 ▪ J. Chem. Soc., Perkin Trans. 1 **1992**, 2949, 2955 ▪ Sax (8.), CDG 750.

Chaetomidin see oosporein.

Chalciporone.

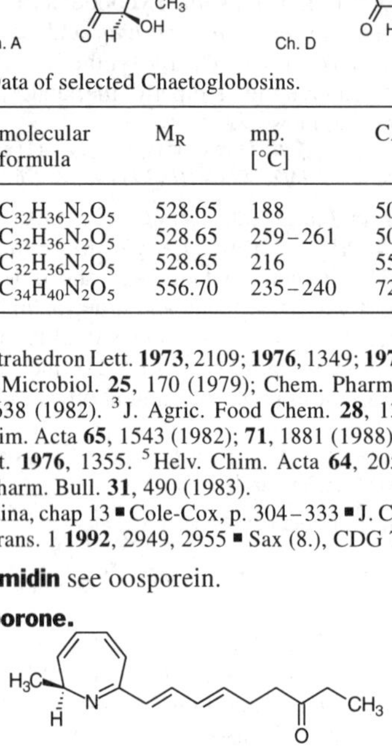

Chalciporone

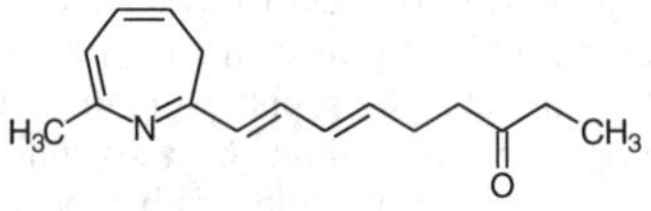

Isochalciporone

$C_{16}H_{21}NO$, M_R 243.35, yellowish oil, $[\alpha]_D$ −452° (ether). Alkaloid from *Chalciporus piperatus* (Basidiomycetes) that is responsible for the burning taste of this mushroom. Besides C. and some related 2*H*-azepines the corresponding 3*H*-azepines (e.g., *isochalciporone*, light yellow oil) have been isolated; the latter are formed from the 2*H*-isomers by sigmatropic [1,5]-H-shifts, are achiral, and have a mild taste. The mushroom is used as a spice.
Lit.: Tetrahedron **43**, 1075 (1987); **52**, 10883 (1996). – *[CAS 112448-74-5 (C.); 112448-72-3 (iso-C.)]*

Chalcogran (2-ethyl-1,6-dioxaspiro[4.4]nonane).

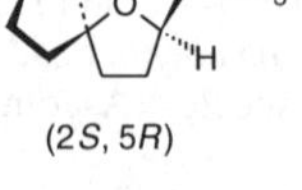

(2*S*, 5*R*)

$C_9H_{16}O_2$, M_R 156.22, bp. 73 °C (4.0 kPa); (2*S*,5*R*)-isomer $[\alpha]_D^{25}$ −100° (hexane); (2*S*,5*S*)-isomer $[\alpha]_D^{25}$ +96° (hexane). Aggregation pheromone of bark beetles of the genus *Pityogenes*. Males of *P. chalcographus* excrete this *spiroacetal as a mixture of the (2*S*,5*S*)- and (2*S*,5*R*)-isomers to attract males and females of the same species to colonize a host tree. Only the (2*S*,5*R*)-isomer has an attracting action which is dramatically increased by simultaneously excreted, extremely small amounts of *methyl (2E,4Z)-2,4-decadienoate* ($C_{11}H_{18}O_2$, M_R 182.26). Isomers of C. are the pheromones *conophthorin and *olean, the sexual attractant of the olive fruit fly *Dacus oleae*. C. is currently the economically most important *pheromone with a market value in 1995 of ca. 20 Mio. DM for its use against forest pests.
Lit.: ApSimon **8**, 562; **9**, 446–451 (synthesis) ▪ Carbohydr. Res. **261**, 231 (1994) (synthesis) ▪ J. Chem. Ecol. **15**, 685–695 (1989). – *[CAS 38401-84-2 (C.); 1191-03-3 (methyl (2E,4Z)-2,4-decadienoate)]*

Chalcones.

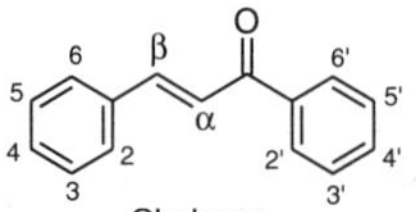

Chalcone

C. belong to the *flavonoids and accumulate as glycosides in few plant families only. The most abundant C. in nature is 2′,3,4,4′-tetrahydroxychalcone (*butein). 2′,4,4′-Trihydroxychalcone (*isoliquiritigenin*, $C_{15}H_{12}O_4$, M_R 256.26, mp. 200–204 °C), occurs in the form of four different glycosides in the flowers of common gorse (*Ulex europaeus*, Fabaceae). The 2′,3,4,4′,5-pentahydroxychalcone *robtein* ($C_{15}H_{12}O_6$, M_R 288.26, mp. >345 °C) occurs in the heartwood of *Robinia pseudoacacia* (Fabaceae). The 2′,4,4′,6′-tetrahydroxychalcone is the central precursor for the biosynthesis of the most important classes of flavonoids.
Lit.: Ullmann (5.) **A 13**, 614. – *[CAS 961-29-5 (isoliquiritigenin); 2679-65-4 (robtein)]*

(−)-Chamaecynone see eudesmanes.

Chamazulene see azulenes.

Chamigrenes.

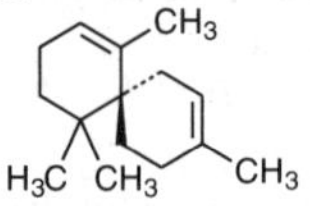

α

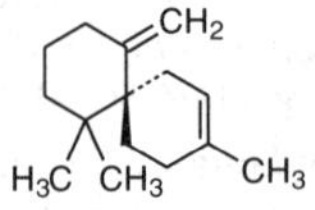

β

$C_{15}H_{24}$, M_R 204.36. Spirobicyclic *sesquiterpenes with the chamigrane skeleton. Halogenated or oxidized derivatives of *α-C.*[1] {oil, $[\alpha]_D$ −14.5° ($CHCl_3$)} occur in brown algae of the genus *Laurencia*. *β-C.*[2] {oil, bp. 110–113 °C (17.3 · 10^3 Pa), $[\alpha]_D^{15}$ −52.7° ($CHCl_3$)} is found in false cypress species, e.g., *Chamaecyparis taiwanensis*.
Lit.: [1] Helv. Chim. Acta **60**, 515 (1977); Tetrahedron **52**, 13181 (1996) (synthesis). [2] J. Org. Chem. **49**, 1001 (1984); Pure Appl. Chem. **58**, 395 (1986).
gen.: Phytochemistry **27**, 1761 (1988) ▪ Scheuer I **1**, 138; **2**, 331; **5**, 164. – *[CAS 19912-83-5 (α-C.); 18431-82-8 (β-C.)]*

Chamomile oils. Products known as C. are obtained from two different chamomile species.

1. ***Blue chamomile oil*** (German C.): Blue oil that quickly turns green to brown under the action of light and air and finally changes to a brown viscous mass. It has a sweet-aromatic, herby odor with a fresh fruity tone of cocoa and an aromatic bitter taste.
Production: By steam distillation from chamomile flowers [1].
Composition [2]***:*** The blue color of the oil is caused by *chamazulene* (see azulenes) formed under the distillation conditions from the colorless precursor *matricin present in commercial oils (from Hungary and Egypt) to ca. 1–5%. Further characteristic components are **bisabolol* and derivatives (ca. 25–50%) and the shown *dienediyne spiroether* (2–4%; $C_{13}H_{12}O_2$, M_R 200.24). Another major component is *trans-β-*farnesene* (25–50%).

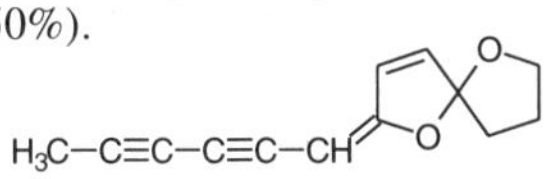

Dienediyne spiroether

Use: In medicine as a carminative, anti-inflammatory, and spasmolytic agent; in very small amount in the perfume industry and to improve the aromatic character of alcoholic beverages (liquors).
2. ***Roman chamomile oil:*** Yellow to light greenish blue oil with a fresh, sweet herb-like odor with nuances of tea.
Production: By steam distillation of the flowers of the Roman chamomile *Chamaemelum nobile*.
Composition [2]***:*** Typical constituents and main components are esters of *angelic acid* [(*Z*)-2-methyl-2-butenoic acid], e.g., *isobutyl angelicate* ($C_9H_{16}O_2$, M_R 156.22), *methallyl angelicate* ($C_9H_{14}O_2$, M_R 154.21), *2-* and *3-methylbutyl angelicate* ($C_{10}H_{18}O_2$, M_R 170.25).
Use: In small amounts to add aromatic character to alcoholic beverages (liquors) and in the perfume industry.
Lit.: [1] Perfum. Flavor. **12** (1), 35 (1987). [2] Perfum. Flavor. **15** (4), 63 (1990); **17** (5), 145 (1992); **18** (4), 71 (1993).
gen.: Arctander, p. 154, 157 ▪ Bauer et al. (2.), p. 144, 145. – *Toxicology:* Food Cosmet. Toxicol. **12**, 851, 853 (1974). – *[HS 3301 29; CAS 8002-66-2 (C.); 16863-61-9 (dienediyne spiroether); 7779-81-9 (isobutyl angelicate); 61692-78-2 (methylallyl angelicate); 61692-77-1 & 10482-55-0 (2- and 3-methylbutyl angelicate)]*

Chamonixin.

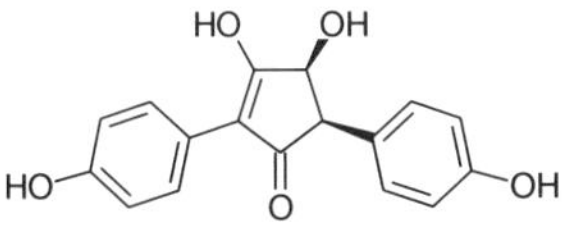

(–)-C.

$C_{17}H_{14}O_5$, M_R 298.30, colorless cryst., monohydrate, decomp. >140 °C. Occurs in the optical dextrorotatory form $[\alpha]_D$ +31.9° (C_2H_5OH) in the gasteromycete *Chamonixia caespitosa* (Basidiomycetes) and is responsible for the intense blue color of bruises on the white sporophore. The (–)-enantiomer is isolated from *Gyrodon lividus* and *Paxillus involutus* (Basidiomycetes); see also gyrocyanin and involutin.
Lit.: Z. Naturforsch. C **32**, 46 (1977); **35**, 824 (1980) ▪ Zechmeister **51**, 63–70. – *[CAS 62501-37-5]*

Charamin (2,6-dihydroxy-4-azoniaspiro[3.3]heptane).

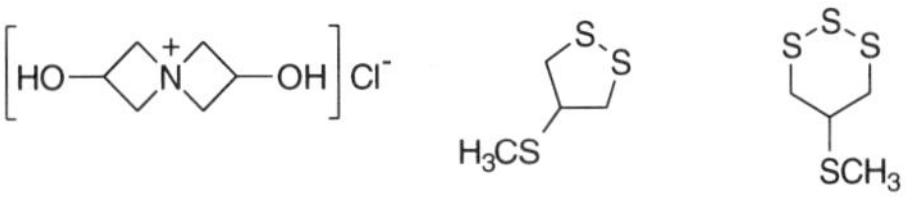

Charamin 4-Methylthio-1,2-dithiolane 5-Methylthio-1,2,3-trithiane

$C_6H_{12}ClNO_2$, M_R 165.62. Antibacterially active, quaternary ammonium compound of unknown stereochemistry from the water horsetail *Chara globularis* growing in freshwater [1]. This alga which often dominates in the ecological system owes its unpleasant smell to sulfur compounds such as the insecticide *4-methylthio-1,2-dithiolane* ($C_4H_8S_3$, M_R 152.29, oil) and the herbicide *5-methylthio-1,2,3-trithiane* ($C_4H_8S_4$, M_R 184.35) [2].
Lit.: [1] J. Org. Chem. **52**, 694 (1987). [2] Phytochemistry **19**, 1228 (1980) (isolation); Experientia **40**, 186 (1984) (pharmacology); Tetrahedron **38**, 2425 (1982); Agric. Biol. Chem. **54**, 1719 (1990) (synthesis of dithiolane). – *[CAS 106500-22-5 (C.); 75679-69-5 (4-methylthio-1,2-dithiolane); 75679-70-8 (5-methylthio-1,2,3-trithiane)]*

Chartarin see chartreusin.

Chartelline A.

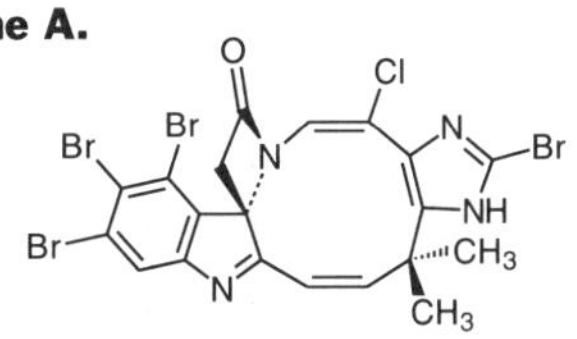

$C_{20}H_{13}Br_4ClN_4O$, M_R 680.42, mp. 214–216 °C, $[\alpha]_D^{20}$ –421° (C_2H_5OH). Unusual polyhalogenated alkaloid with a *β*-lactam unit from the marine bryozoan *Chartella papyracea*.
Lit.: J. Am. Chem. Soc. **107**, 4542 (1985) ▪ J. Org. Chem. **52**, 4709 (1987). – *[CAS 96845-55-5]*

Chartreusin (lambdamycin).

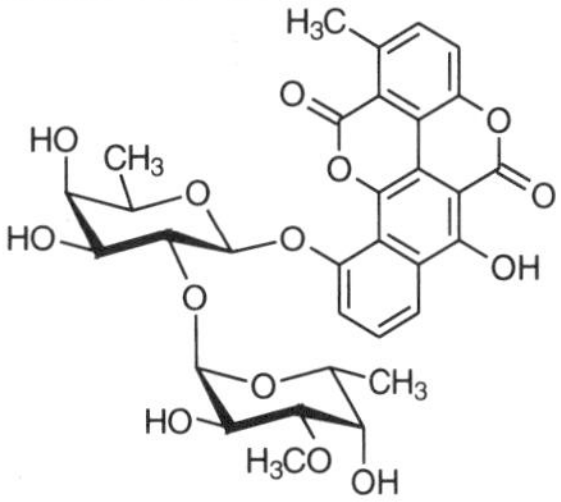

$C_{32}H_{32}O_{14}$, M_R 640.60, yellow plates, mp. 184–186 °C, $[\alpha]_D^{25}$ +127.5° (pyridine), weak acid. *Antibiotic from the African *Streptomyces chartreusis* and other strains from north American soils. C. consists of a *disaccharide and the aglycone *chartarin*. The aglycone is biosynthetically built up of eleven acetate units. C. is active against Gram-positive bacteria, mycobacteria, and tumor cells, see also ellagic acid.
Lit.: Can. J. Chem. **55**, 2450 (1977) (biosynthesis) ▪ Cancer Research **37**, 1666 (1977) ▪ J. Antibiot. **45**, 875 (1992) (derivatives) ▪ J. Org. Chem. **45**, 4071 (1980) (synthesis); **52**, 996 (1987) ▪ Prog. Med. Chem. **19**, 247 (1982) (review) ▪ Sax (8.), CDK 250 ▪ Z. Allg. Mikrobiol. **29**, 362 (1977). – *[HS 2941 90; CAS 6377-18-0]*

Charybdotoxin. $C_{176}H_{277}N_{57}O_{55}S_7$, M_R 4295.95. A high affinity neurotoxin isolated from the scorpion *Leiurus quinquestriatus hebraeus*. This presynaptic K^+ chan-

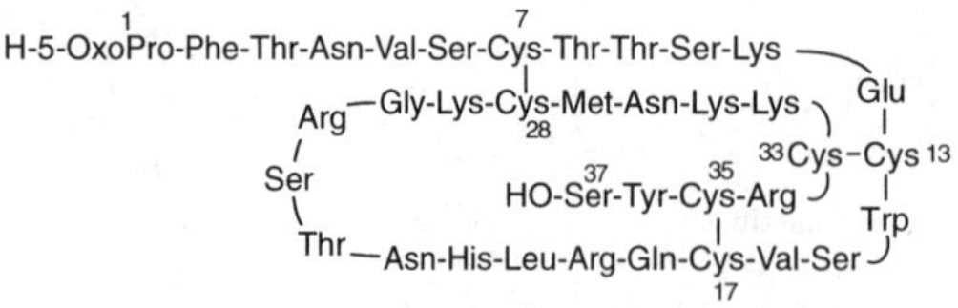

nel antagonist increases neuromuscular transmission via liberation of acetylcholine at the neuromuscular branching points. C. consists of 37 amino acids and is tricyclic through S–S-bridges at the following cysteines: 7→28, 13→33 and 17→35.

Lit.: Br. J. Pharmacol. **108**, 214 (1993) ▪ Eur. J. Biochem. **217**, 157 (1993) (synthesis) ▪ J. Biol. Chem. **265**, 18745 (1990) ▪ J. Membr. Biol. **109**, 95 (1988) ▪ Nature (London) **313**, 316 (1985) ▪ Pharm. Ther. **46**, 137 (1990) ▪ Proc. Natl. Acad. Sci. U.S.A. **85**, 3329 (1988). – *[CAS 95751-30-7]*

Chatancin.

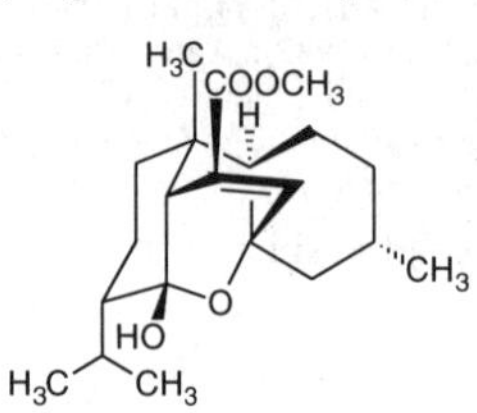

$C_{21}H_{32}O_4$, M_R 348.48, needles, mp. 106–108 °C (CH_3OH); $[\alpha]_D$ +10.5° ($CHCl_3$). A PAF-antagonist from a species of the soft corals *Sarcophyton.*

Lit.: Angew. Chem. Int. Ed. Engl. **37**, 2226 (1998) ▪ J. Org. Chem. **55**, 5803 (1990). – *[CAS 129895-81-4]*

Chaulmoogric acid [13-(2-cyclopentenyl)tridecanoic acid].

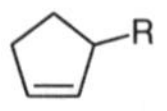

R = $(CH_2)_{12}$–COOH : Chaulmoogric acid (*S*)
R = $(CH_2)_{10}$–COOH : Hydnocarpic acid (configuration unknown)
R = $(CH_2)_6$–CH=CH–$(CH_2)_4$–COOH: Gorlic acid (*S*)

$C_{18}H_{32}O_2$, M_R 280.45, mp. 69 °C, bp. 247–248 °C (2.7 kPa), $[\alpha]_D^{25}$ +61.7°; soluble in organic solvents, insoluble in water. C. belongs to the group of cyclopentenyl-fatty acids, the glycerol esters of which are typical components of seed oils of endemic tropical plants of the family Flacourtiaceae. C. and other cyclopentenyl-fatty acids, e. g., *hydnocarpic acid* {11-(2-cyclopentenyl)undecanoic acid, $C_{16}H_{28}O_2$, M_R 252.40, mp. 58–59 °C, $[\alpha]_D^{25}$ +68°} and *gorlic acid* {$C_{18}H_{30}O_2$, M_R 278.44, mp. 6 °C, bp. 232.5 °C (1.33 kPa), $[\alpha]_D^{25}$ +60.7°} are mainly obtained from the seed fat of *Hydnocarpus kurzii* (chaulmoogra oil), *H. wightiana* (maratti oil), and *Caloncoba (Oncoba) echinata*, which all contain 80–90% cyclopentenyl-fatty acids. For biosynthesis see *Lit.*[1]. Cyclopentenyl-fatty acids are also found in the lipids of Passifloraceae and Turneraceae[2,3]. C. and other cyclopentenyl-fatty acids exhibit antibacterial activities, especially against mycobacteria (leprosy).

Lit.: [1]Biochemistry **13**, 2241 (1974); Mangold & Spener, in: Stumpf & Conn (eds.), The Biochemistry of Plants, vol. 4, p. 647, New York: Academic Press 1980. [2]J. Am. Chem. Soc. **92**, 6378 (1970). [3]Phytochemistry **19**, 1863 (1980).
gen.: Phytochemistry **19**, 1685 (1980) ▪ Ullmann (5.) **A 10**, 233. – *[HS 2916 20; CAS 29106-32-9 (C.); 459-67-6 (hydnocarpic acid); 502-31-8 (gorlic acid)]*

Chavicine see piperine.

Chavicol see estragole.

Cheese flavor. C. f. is formed from milk fat, milk protein, lactose during the maturation of cheese mainly through enzymatic and microbial processes. Quantitative and, sometimes, qualitative differences are responsible for the diversity of cheese flavors. Typical aroma substances are the free C_4–C_{12} fatty acids, C_7, C_9, and C_{11} 2-alkanones (also in Roquefort cheese), the butter aroma substances *acetoin, *2,3-butanedione, and 5-*alkanolides, *(–)-(*R*)-1-octen-3-ol (fungus note in Camembert), 4-alkanolides and alkylpyrazines with nut-like nuances, *indole, *skatole, and phenols with "stable-like" odors, as well as numerous sulfur compounds such as *methional, methyl mercaptan (moldy, coal-like), dimethyl sulfide and dialkyl polysulfides with, in part, onion- and garlic-like nuances. *Furaneol® and homofuraneol (see hydroxyfuranones) are responsible for the sweetish odor of Emmental cheese.

Lit.: TNO list (6.), p. 335–386; Suppl. 5, p. 32–98.

Cheilanthifoline see protoberberine alkaloids.

Chelerythrine, chelidonine see benzo[*c*]phenanthridine alkaloids.

Chenodeoxycholic acid see bile acids.

Cherimoline (4*H*-pyrano[3,4-*c*]quinolin-4-one).

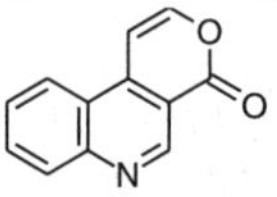

$C_{12}H_7NO_2$, M_R 197.19, white powder, mp. 203–205 °C. C. is a *quinoline alkaloid from the stems of *Annona cherimolia.*

Lit.: Tetrahedron Lett. **38**, 6247 f. (1997). – *[CAS 196958-70-0]*

Cherry flavor see fruit flavors.

Chicoric acid see tartaric acid.

Chilocorine A, B see coccinelline.

Chimonanthine see pyrroloindole alkaloids.

Chimyl alcohol see batyl alcohol.

China alkaloids see Cinchona alkaloids.

Chitin (from Greek chitōn = tunic).

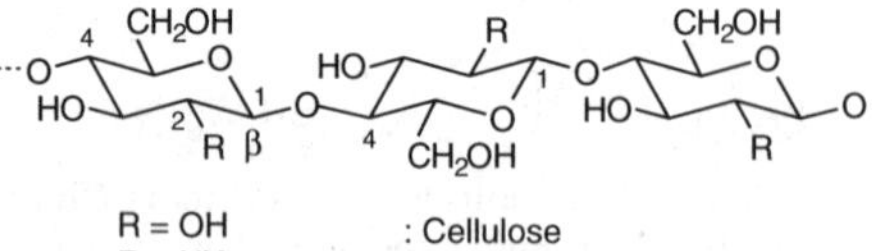

R = OH : Cellulose
R = NH_2 : Chitosan
R = NH–CO–CH_3 : Chitin

*Amino sugar-containing polysaccharide with the general formula $(C_8H_{13}NO_5)_x$, M_R ca. 400 000, isolated especially from animals. C. consists of chains of *β*-1,4-glycosidically linked *N*-acetyl-D-glucosamine (NAG) residues. C. can be considered as a derivative of *cellulose (with 2-acetamido instead of 2-hydroxy groups). In various invertebrate organisms C. has a supporting function similar to that of cellulose in plants. The shells (exoskeletons) of arthropods are made of C., it is also

found in the scleroproteins (structure proteins) of mollusks, in the cell walls of algae, yeasts, fungi, and lichens. Rather pure C. occurs in the wings of may bugs and lobster shells. C. has a highly ordered, crystalline structure and occurs in three polymorphous forms: α-, β-, and χ-chitin. In α-C. the chains have an antiparallel orientation. α-C. occurs where particular hardness is required. In contrast, β-C. and χ-C. are found where a combination of flexibility and toughness is needed.
Strong acids cleave C. into D-*glucosamine (chitosamine) and acetic acid. On decomposition with alkalis, acetates and the weakly basic deacetylated, partially depolymerized, and crystallizable *chitosan* are formed; the latter can form gels and films. The *chitinases* (EC 3.2.1.14) found in snail stomachs, some mold fungi, and bacteria can – like some lysozymes – dissolve C. The thus formed chitobiose, the disaccharide of β-1,4-linked NAG, is then cleaved to monomers by chitobiase (EC 3.2.1.30). The formation of C. is catalyzed by the enzyme chitin synthase (EC 2.4.1.16). The activity of, e.g., the insecticide diflubenzurone is based on inhibition of this enzyme.
Use: Regenerated C. fibers as degradable – by endogenous lysozyme – wound dressing. Chitosan as auxiliary agent in paper and dying, binding agent for fleece substances, glue in the leather industry, for sausage skins, dialysis membranes, contact lenses, and as chelation and flocculation agents (e.g., for waste water processing).
Lit.: Beilstein EIV **18**, 7535; EV **18/11**, 147 (chitobiose) ▪ Muzzarelli, Chitin, Oxford: Pergamon 1977 ▪ New Sci. **129**, 46 (1991) ▪ Polysaccharides **1998**, 569–594 (technical use of C.) ▪ Roberts, Chitin Chemistry, London: MacMillan 1992 ▪ Tetrahedron Lett. **38**, 2111 (1997) (synthesis chitobiose) ▪ Ullmann (5.) **A 6**, 231 f. – *Journal:* Adv. Chitin Sci. (from 1996). – *[HS 3913 90; CAS 1398-61-4 (C.); 9012-76-4 (chitosan)]*

Chloramphenicol.

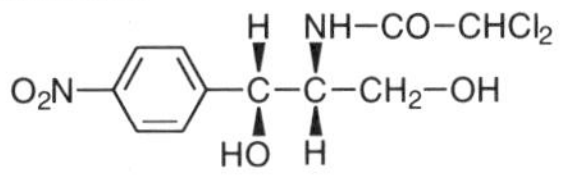

International generic name for (–)-D-*threo*-2-(dichloroacetamido)-1-(4-nitrophenyl)-1,3-propanediol, $C_{11}H_{12}Cl_2N_2O_5$, M_R 323.13. Bitter tasting crystals, mp. 149–152 °C, $[\alpha]_D$ +18.6° (C_2H_5OH). C., discovered in 1947 in culture broths of *Streptomyces*, contains nitro and dichloroacetyl groups – a most unusual combination for natural products. C. was the first synthetically prepared, so-called broad-spectrum antibiotic. It is active against rickettsiae, certain viruses, Gram-positive and -negative bacteria and cocci, actinomycetes, spirochetes, and leptospira; it is mainly used therapeutically for parrot fever, typhoid fever, paratyphoid fever, spotted fever, and meningitis. C. acts by inhibiting protein biosynthesis in the pathogen[1]; for details on significant chromosome damage, see *Lit.*[2]. The resistance of various bacteria against C. arises through acetylation of the C. molecule with the help of endogenous bacterial enzymes. For the history of its discovery, elucidation of its constitution, and technical synthesis, see *Lit.*[3].
Lit.: [1] Mol. Pharmacol. **2**, 158 (1966). [2] Naturwiss. Rundsch. **22**, 214 f. (1969). [3] Synth. Commun. **27**, 1857 (1997); Tetrahedron Lett. **39**, 8503 (1998).

gen.: Chem. Ind. (London) **1993**, 249 (absolute configuration) ▪ Florey **4**, 47–90, 517 ▪ Hager (5.) **7**, 847 ff. ▪ IARC Monogr. **10**, 85–98 ▪ J. Chem. Soc., Chem. Commun. **1992**, 859 ▪ Kirk-Othmer (4.) **2**, 961 ▪ Pongs, Antibiotics 5/1, p. 26–42, Berlin: Springer 1979 ▪ Sax (8.), CDP 250–CDP 725 ▪ Snell-Hilton **5**, 493–507 ▪ Ullmann (5.) **A 2**, 535. – *[HS 2941 40; CAS 56-75-7]*

Chlorins.

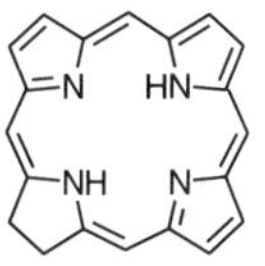

The structural type of C. (2,3-dihydroporphyrins) is formally derived from that of porphyrin by reduction of a peripheral double bond in a pyrrole ring of the porphyrin skeleton. The name is based on that of *chlorophyll which possesses the C. skeleton with a central magnesium(II) ion. Other substances with the C. skeleton are, in addition to the chlorophylls and their degradation products, some *bacteriochlorophylls, *bonellin, *chlorophyllone, cyclopheophorbide, *factor I, *heme d, and *tunichlorin.
Lit.: Chem. Rev. **94**, 327 (1994) ▪ Chimia **41**, 277 (1987) ▪ Dolphin **II** 8 ff., 30; **III**, 415–428 ▪ Smith, p. 158–170, 217 ▪ Tetrahedron **31**, 367 (1975). – *Synthesis:* Angew. Chem. Int. Ed. Engl. **31**, 1592 ff. (1992) ▪ J. Chem. Soc. Perkin Trans. 1 **1988**, 3119–3131. – *[HS 3203 00]*

Chloroatranorin.

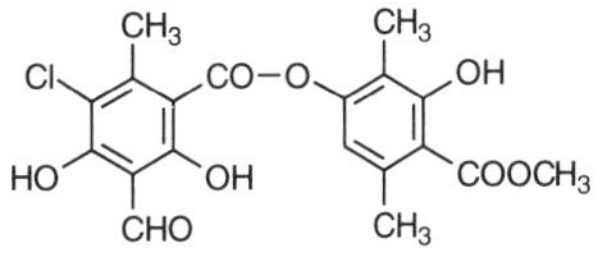

$C_{19}H_{17}ClO_8$, M_R 408.79, prisms, mp. 208–209 °C. C. is a *depside that often occurs together with *atranorin, the biogenetic precursor of C., in lichens (e.g., *Anaptychia neoleucomelaena*).
Lit.: Asahina-Shibata, p. 96 ff. ▪ Aust. J. Chem. **46**, 301 (1993) ▪ Karrer, No. 1052. – *[CAS 479-16-3]*

Chlorocruoroporphyrin (Spirographis porphyrin, photoprotoporphyrin).

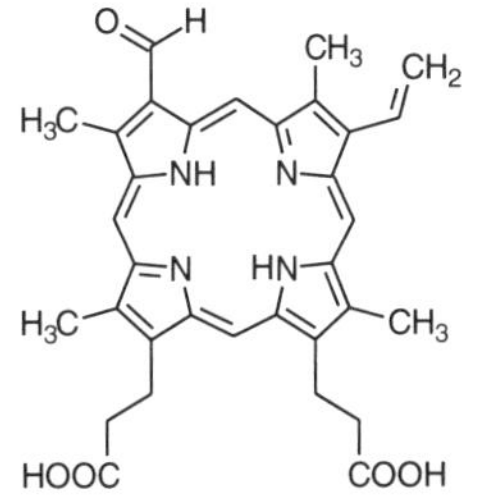

$C_{33}H_{32}N_4O_5$, M_R 564.64. Red-black crystals, mp. (methyl ester) 276–278 °C. uv_{max} (methyl ester in $CHCl_3$) 420, 519.5, 559, 584, 642 nm. C. is the porphyrin derived from *chlorocruoroheme* (Spirographis heme; $C_{33}H_{30}FeN_4O_5$, M_R 618.47) by removal of the central Fe(II) ion. The *heme serves as oxygen-transporting prosthetic group of the blood pigment chlorocruorin of some polychete worms.
Lit.: Dolphin **I**, 304 f. – *Synthesis:* Tetrahedron Lett. **31**, 3779 ff. (1966). – *[CAS 24869-67-8]*

Chlorogenic acid [(1*S*)-3β-(3,4-dihydroxycinnamoyloxy)-1α,4α,5α-trihydroxycyclohexanecarboxylic acid, 5-*O*-caffeoylquinic acid)].

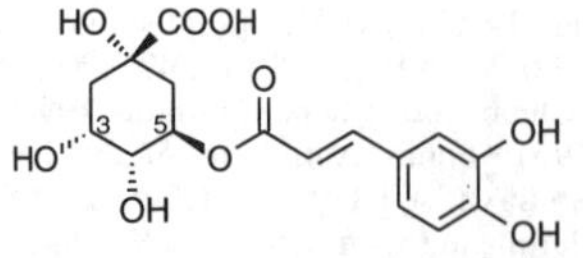

$C_{16}H_{18}O_9$, M_R 354.31. As hemihydrate mp. 208–210 °C (decomp.), $[\alpha]_D^{16}$ −35.2° (H_2O). TDL_0 40 mg/kg (rat i.p.). C. is an experimental teratogenic agent. C. occurs as mixture with isomeric caffeoyl esters of quinic acid in numerous plants, e.g., *Chrozophora*, *Cinchona*, *Scabiosa*, *Valeriana*, *Senecio*, and *Hypericum* species and was originally isolated from Liberian coffee in which it is present in about 6.5 to 8%. C. forms colored complexes with many metal ions, e.g., Fe(III), that are typical for catechols (1,2-benzenediols).

Activity: C. is a *phytoalexin and exhibits activity against phytopathogenic viruses and fungi. The postulate that C. belongs to the stimulating components of coffee is contradicted by the fact that no corresponding effects are observed upon consumption of many C.-containing foods. The biosynthesis proceeds through *(*E*)-*p*-c(o)umaric acid → 5-*O*-*p*-coumaroylquinic acid → chlorogenic acid.

Lit.: Beilstein EIV **10**, 2259 ▪ Chem. Ind. (London) **1995**, 836 ff. ▪ Chem. Pharm. Bull. **34**, 1419 (1986); **35**, 2133 (1987) ▪ Karrer, No. 990 ▪ Luckner (3.), p. 385 f. ▪ Phytochemistry **15**, 703 ff. (1976) (biosynthesis) ▪ Zechmeister **35**, 80 ff. – *Toxicology:* Sax (8.), CHK175. – *[HS 291829; CAS 327-97-9]*

Chloroh(a)emin see h(a)emin.

2-Chloro-4-nitrophenol see drosophilin A.

Chlorophyllone a.

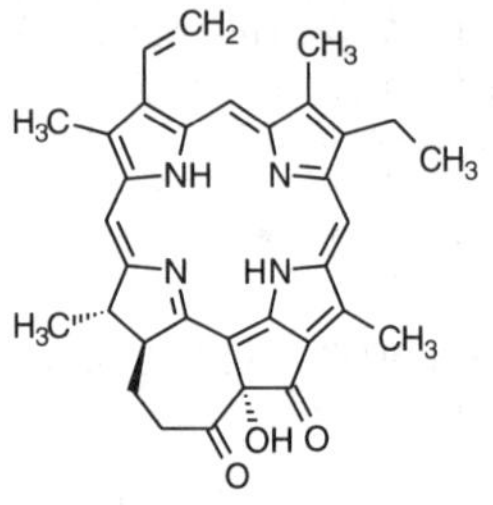

$C_{33}H_{32}N_4O_3$, M_R 532.64; uv_{max} (CH_3OH) 408, 503, 534, 608 and 665 nm. Green pigment. C.a belongs to the structural class of the *chlorins, while the substitution pattern is suggestive of an origin from *chlorophyll a.

Occurrence: C.a is the green pigment of *Ruditapes philippinarum*, a Japanese fresh-water mussel and serves as an antioxidant. C.a is structurally very closely related to cyclopheophorbide. Through opening of the cyclopentenone ring in C.a the natural metabolites of C.a, chlorophyllonic acid and chlorophyllone lactone are formed.

Lit.: ACS Symp. Ser. **547**, 164 (1994) ▪ Chem. Nat. Symp. Papers **32**, 57 (1990) ▪ Chem. Rev. **94**, 327 (1994) ▪ J. Org. Chem. **61**, 2501 (1996) ▪ Tetrahedron Lett. **31**, 1165 (1990). – *[CAS 127266-93-7]*

Chlorophylls (from Greek chloros = yellow-green and phyllon = leaf). The name coined by Pelletier and Caventou for the leaf pigment of higher plants and green algae responsible for their green color and which makes photosynthesis possible [1,2,3]. With the exception of C.c and e, which are based on the porphyrin skeleton, all C. have the *chlorin skeleton. In plants C. is found in the chloroplasts. The C. of higher plants and green algae is not a uniform substance but a mixture of C.a and b, which was separated chromatographically for the first time by Tswett. C.a is blue-green, C.b yellow-green. Green leaves contain about three times as much C.a as C.b, which is an oxidation product of C.a. C.c, d, and e have been isolated from marine algae. The *bacteriochlorophylls c–d are structurally more closely related to C. than to bacteriochlorophylls a and b.

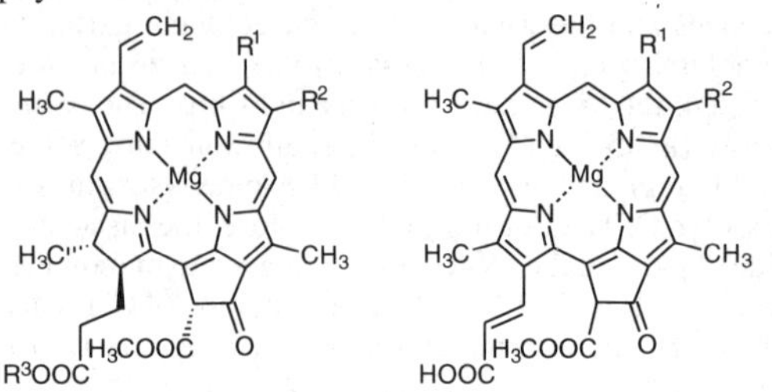

Chlorophyll a	Chlorophyll b	Chlorophyll c
$R^1 = CH_3$	$R^1 = CHO$	c_1 $R^1 = CH_3$, $R^2 = C_2H_5$
$R^2 = C_2H_5$	$R^2 = C_2H_5$	c_2 $R^1 = CH_3$, $R^2 = CH{=}CH_2$
R^3 = Phytyl	R^3 = Phytyl	c_3 $R^1 = COOCH_3$, $R^2 = CH{=}CH_2$

Phytyl: 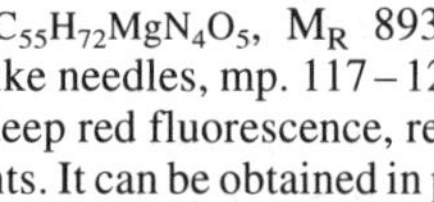

Structure: Up to 1997 ca. 50 C. protein complexes have been described which are designated, among other systems, with letters and numbers according to the nature of the C. and the wavelength of the absorbed light, e.g., a685, b650, c705, P700.

C. a, $C_{55}H_{72}MgN_4O_5$, M_R 893.51; forms blue-black, wax-like needles, mp. 117–120 °C, soluble in alcohol with deep red fluorescence, readily soluble in organic solvents. It can be obtained in pure form from the blue-green alga *Anacystis nidulans*.

C. b, $C_{55}H_{70}MgN_4O_6$, M_R 907.49; dark green platelets, mp. 120–130 °C, solubility similar to that of C.a. It is used mainly by algae for photosynthesis.

C. c_1, $C_{35}H_{30}MgN_4O_5$, M_R 610.95; red-black, hexagonal crystals., strictly speaking it is a chlorophyllide. Chlorophyllides are defined as the water-soluble C. structures with a free 17-propanoic acid residue formed by loss of *phytol or other long-chain alcohols. C. c_2 is the divinyl derivative of C. c_1, C. c_3 is an oxidized derivative of C. c_2.

Biosynthesis: Labelling experiments with radioactive isotopes have shown that in plants starting from succinyl-coenzyme A and glycine and/or glutamic acid (through *5-amino-4-oxovaleric acid and *porphobilinogen) *protoporphyrins and related compounds are formed and converted to the water-soluble chlorophyllide a by incorporation of Mg, hydrogenation, oxidation, and cyclopentane ring closure [4]. The enzyme chlorophyllase is involved in the esterification of the latter species with *phytol (to C.a); in place of the phytyl group the bacteriochlorophylls c–e contain farnesyl or geranylgeranyl groups. The biosynthesis of C. is inhibited by lead ions. Iron(III) (as component of *ferredoxin) is necessary for formation of C. and for photosynthesis. Most higher plants require light for the formation of C. (otherwise turning yellow), however, many algae and cryptogams can form C. in total darkness. The deep sea dragon fish *Malocosteus niger* has retinal pigments derived from C. that absorb the far-red wavelengths and then pass the light signal on to the visual pigments[5]. The autumn colors of leaves observed in temperate climates[6] are due to the degradation of C. in the leaves after which the colors of the other leaf pigments, e.g., the *carotinoids, become dominant (most of the contained magnesium is resorbed by the plants). Each year ca. 10^9 tons of C. are degraded worldwide[7]. Identified degradation products include the so-called *metabolite RP 14* (**1**, $C_{35}H_{41}KN_4O_{10}$, M_R 716.83, pale yellow powder) and the red *chlorophyll a catabolite* **2** ($C_{34}H_{38}N_4O_5$, M_R 582.36, reddish crystals, mp. 128–129 °C) in barley seedling or, respectively, in the unicellular alga *Chlorella protothecoides*[8]. The central Mg atom can also be removed from C. and chlorophyllides by chemical means to furnish pheophytins and pheophorbides. By means of hydrolysis and Mg exchange the natural C. are converted to the chlorophyllines as water-soluble and light-resistant pigments.

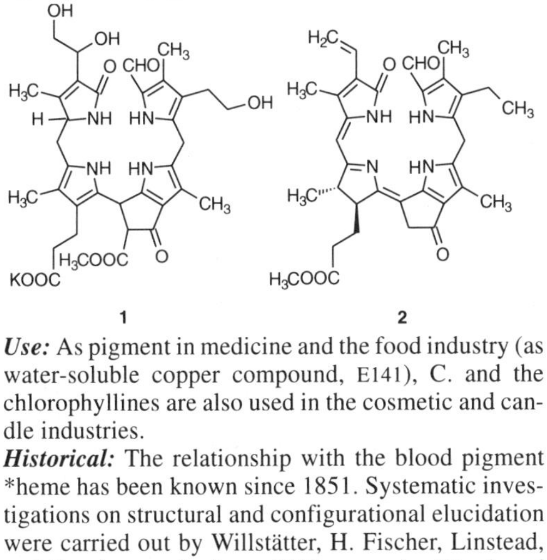

1 2

Use: As pigment in medicine and the food industry (as water-soluble copper compound, E141), C. and the chlorophyllines are also used in the cosmetic and candle industries.

Historical: The relationship with the blood pigment *heme has been known since 1851. Systematic investigations on structural and configurational elucidation were carried out by Willstätter, H. Fischer, Linstead, and Brockmann Jr. from 1906 onwards. Woodward achieved the total synthesis in 1960[1,9].

Lit.: [1]Dolphin **II**, 288–326. [2]Nature (London) **361**, 326 (1993). [3]Angew. Chem. Int. Ed. Engl. **28**, 848–869 (1989). [4]Nat. Prod. Rep. **4**, 441–469 (1987); J. Nat. Prod. **51**, 629–642 (1988); Phytochemistry **46**, 1151–1167 (1997). [5]Nature (London) **393**, 423 (1998). [6]Acc. Chem. Res. **32**, 35–43 (1999); Angew. Chem. Int. Ed. Engl. **36**, 401–404 (1997); Helv. Chim. Acta **80**, 1355 (1997). [7]J. Photochem. Photobiol. B.: Biology **32**, 141–151 (1996). [8]Angew. Chem. Int. Ed. Engl. **30**, 1315 (1991); FEBS Lett. **293**, 131 (1991). [9]Tetrahedron **46**, 7599 (1990).

gen.: Beilstein E V **26/15**, 505 f., 537 f. ▪ Hager (5.) **7**, 881 ▪ Karrer, No. 2507, 4368, 4369 ▪ Scheer, Chlorophylls, Boca Raton: CRC Press 1991 ▪ Ullmann (4.) **11**, 126 ff. ▪ Zechmeister **26**, 284–355 ▪ see also bacteriochlorophylls, porphyrins and other entries mentioned in the text. – *[HS 3203 00; CAS 479-61-8 (a); 519-62-0 (b); 18901-56-9 (c); 519-63-1 (d)]*

Chlorotetaine [*N*-L-alanyl-3-((*S*)-3-chloro-4-oxo-2-cyclohexen-1-yl)-L-alanine].

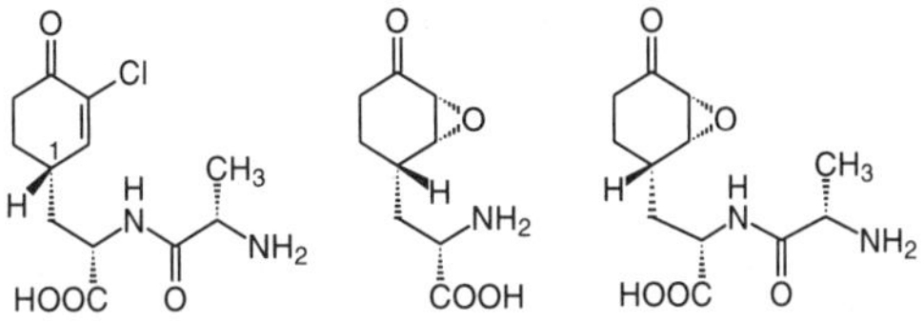

Chlorotetaine Anticapsin Bacilysin

A peptide antibiotic from *Bacillus subtilis*. $C_{12}H_{17}ClN_2O_4$, M_R 288.73, stable in aqueous solution at pH 5 and 20 °C, the biological activity decreases markedly in strongly acidic or basic media, especially on warming, on account of an intramolecular 1,4-addition of the amide to the enone system furnishing a 6-oxo-octahydroindole. C. inhibits glucosamine 6-phosphate synthetase. It has antifungal activity and, in high concentrations, is active against Gram-positive and -negative bacteria. C. can enter the cell through the dipeptide transport system and probably then impairs cell wall synthesis.

Structurally close relatives of C. are *bacilysin* [$C_{12}H_{18}N_2O_5$, M_R 270.29, amorphous, $[\alpha]_D^{20}$ +103° (H_2O)] and *anticapsin* [$C_9H_{13}NO_4$, M_R 199.21, mp. 150 °C (decomp.), $[\alpha]_D^{25}$ +125° (H_2O)], structurally similar is *lascivol from the agaric *Tricholoma lascivum*. Bacilysin is produced by *B. subtilis* and *B. pumilus* and, like C., is a glucosamine 6-phosphate synthetase inhibitor. It acts against Gram-positive and -negative bacteria, and thus kills growing staphylococci. The biosynthesis of bacilysin proceeds from shikimic acid and alanine[1]. Anticapsin from *Streptomyces griseoplanus* is active against some Gram-positive organisms.

For isolation of C., see *Lit.*[2], for stereoselective synthesis, see *Lit.*[3,4].

Lit.: [1]Biochem. **118**, 557–569 (1985); J. Gen. Microbiol. **94**, 37–53 (1976). [2]Angew. Chem. Int. Ed. Engl. **27**, 1733 f. (1988); **30**, 1685 (1991). [3]Krohn et al. (eds.), Antibiotics and Antiviral Compounds, p. 215–228, Weinheim: VCH Verlagsges. 1993. [4]J. Chem. Soc., Chem. Commun. **1993**, 1332; J. Org. Chem. **59**, 2748–2761 (1994); Tetrahedron **51**, 5193–5206 (1995) (anticapsin). – *[HS 2941 90; CAS 117985-03-2 (C.); 29393-20-2 (bacilysin); 28978-07-6 (anticapsin)]*

Chlorotetracycline see tetracyclines.

Chlorothricin.

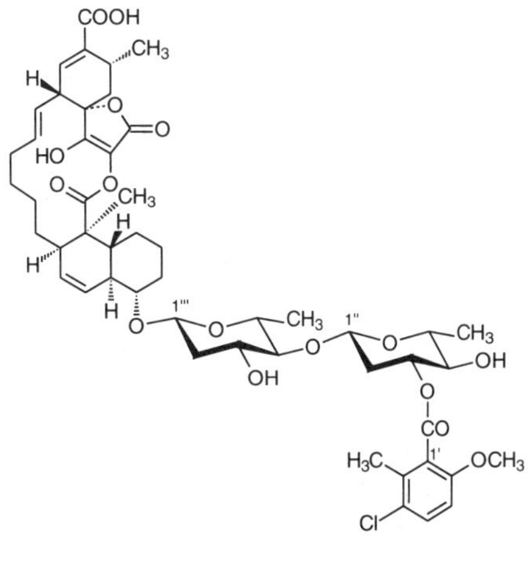

$C_{50}H_{63}ClO_{16}$, M_R 955.49, crystalline, mp. 206–207 °C, *antibiotic against Gram-positive bacteria from *Streptomyces antibioticus*. C. occurs as a mixture with dechlorothricin. Both compounds represent diprotic acids. C. acts as an inhibitor of pyruvate carboxylase and malate dehydrogenase. The aglycone is (–)-chlorotricolide [1].

Lit.: [1] J. Am. Chem. Soc. **120**, 7411–7419 (1998); J. Org. Chem. **63**, 5473–5482 (1998).
gen.: Helv. Chim. Acta **52**, 127 (1969); **55**, 2071 (1972) ▪ J. Antibiot. **39**, 1123 (1986) (biosynthesis); **45**, 207 (1992) (derivatives) ▪ J. Org. Chem. **48**, 4370 (1983) (synthesis). – *[HS 2941 90; CAS 34707-92-1]*

Cholane see steroids.

Cholecalcifediol, Cholecalciferol see calcitriol.

Cholera toxin. The toxin of the cholera pathogen *Vibrio cholerae* is a lipid- and carbohydrate-free protein with a molecular mass of ca. 82 000. Like *diphtheria toxin, it consists of an A subunit and 5 or 6 B subunits. Subunit A has a molecular mass of ca. 28 000 (240 amino acid residues), subunit B a molecular mass of 11 600 (103 amino acid residues). The B subunits are responsible for the binding of C. to the cell surface of intestinal mucous membranes and facilitate the transport of the actual toxic principle, the A subunit, through the cell membrane. After proteolytic and reductive cleavage (S–S bridges) the A subunit effects a continuous activation of adenylate cyclase, resulting in an unrestricted synthesis of cyclo-AMP. The cyclo-AMP level regulates, among others, the excretion of water and salts. Because of this overstimulation of adenylate cyclase more than ten times the normal amount of water is liberated in the small intestine and its subsequent resorption is not possible; this is the reason for the typical loss of water and electrolytes in cholera and results in complete dehydration of the body with lethal results. An epidemic in south America as late as 1991 caused several thousand deaths.

Lit.: Kuwahara et al. (eds.), New Perspectives in Clinical Microbiology, vol. 6, Advances in Research on Cholera, Boston: Martinus Nijhoff 1983, 1985 ▪ Nachr. Chem. Tech. Lab. **31**, 779 (1983) ▪ Sax (8.), CMC 800. – *Mechanism of action:* Chem. Eng. News (1.3.) **1999**, 46 ff. ▪ Nature Struct. Biol. **6**, 134 (1999) ▪ Science **274**, 1859 (1996).

Cholestane see steroids and sterols.

Cholesterol (cholesterin, cholest-5-en-3β-ol). $C_{27}H_{46}O$, M_R 386.66, mp. 148.5 °C, $[\alpha]_D$ –39.5° ($CHCl_3$). C. is the most important zoosterol (see sterols) and is found in humans and all animals. The brain (ca. 10% of the dry weight), adrenals, egg yolk, and wool fat are particularly rich in C.; blood usually contains 0.15–0.25%, the heart ca. 2% of dry weight. It also occurs in small amounts in plants (especially in chloroplasts) and has also been found in bacteria.

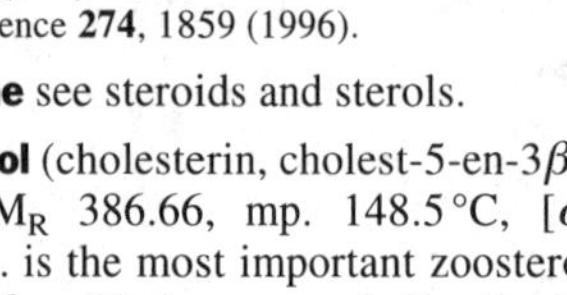

Biosynthesis: The biosynthesis follows the pathway described under *steroids, in humans, animals, and fungi via *lanosterol, in plants via *cycloartenol as the primary squalene-2,3-epoxide cyclization product. Insects, including arthropods, are not capable of steroid biosynthesis and must therefore take up the C. necessary for biosynthesis of *ecdysteroids either with food or produce it by conversion of the phytosterols (see sterols) present in plant foods. The main site of synthesis in human and animal organisms is the liver but C. is also formed in the adrenal cortex, skin, intestine, testicles, and aorta. C. and its esters (see cholesteryl esters) are transported in blood in the form of lipoproteins.

Biological importance: In humans and animals C. is the biogenetic precursor of all *steroid hormones, *bile acids (this is the main degradation route for C., in humans ca. 1.0–1.5 g/d), and other steroidal metabolites. In plants C. or its precursors (e. g., *cycloartenol) are biosynthetically converted to the typical phytosterols, *steroid sapogenins, *steroid alkaloids, *cardenolides, and plant *bufadienolides. C. and the cholesteryl esters belong to the lipids and are, besides phospho- and glycolipids, important components of biomembranes, especially the plasma membranes of eukaryotes where they regulate fluidity. In higher animal organisms, C. also has importance as skin protecting substance, swelling regulator, nerve isolator, etc. Through false eating habits as well as certain enzyme and receptor defects, pathologically high serum levels of C. may occur (hypercholesterolemia) which are considered to be partly responsible for development of arteriosclerosis (for therapy see compactin, mevinolin). C. is also the main component of some gallstones (see bile acids).

Lit.: Beilstein E IV **6**, 4000 ▪ Chem. Pharm. Bull. **45**, 944 (1997) (biosynthesis) ▪ Danielsson & Sjövall (eds.), Sterols and Bile Acids, Amsterdam: Elsevier 1985 ▪ Magn. Reson. Chem. **34**, 137 (1996) (NMR) ▪ Martindale (31.), p. 1409 ▪ Patterson & Nes (eds.), Physiology and Biochemistry of Steroids, Champaign, Ill.: Am. Oil Chem. Soc. 1991 ▪ Stryer 1995, p. 691–703 ▪ Ullmann (5.) **A 13**, 112 ▪ Zeelen, p. 43–71. – *[CAS 57-88-5]*

Cholesteryl esters (cholesterol esters). As a monofunctional secondary alcohol *cholesterol is able to form esters. As lipids these esters are very important for the construction of cell membranes (see cholesterol). Esters of C. with aliphatic or aromatic carboxylic acids are also important as liquid crystals because of their special properties (*cholesteric phases*).

Lit.: Beilstein E IV **6**, 4007; **9**, 320 ▪ Kelker & Hatz, Handbook of Liquid Crystals, Weinheim: Verl. Chemie 1980 ▪ Kontakte (Merck) **1971**, No. 1, p. 3 ff. ▪ Stryer 1995, p. 689.

Cholic acid see bile acids.

Choline [(2-hydroxyethyl)trimethylammonium].

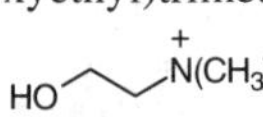

$C_5H_{14}NO^+$, M_R 104.17 (ion); the hydroxide, $C_5H_{15}NO_2$, M_R 121.18, is a highly hygroscopic syrup that crystallizes only with difficulty, is very soluble in water and alcohols but not soluble in ether. It has a strongly basic reaction and absorbs carbon dioxide. C. is weakly toxic, LD_{50} 400 mg/kg (rat i. p.), 3400 mg/kg (rat, hydrochloride p. o.), 6640 mg/kg (cat p. o.). The chloride, $C_5H_{14}ClNO$, M_R 139.63, forms hygroscopic crystals. C. occurs both in the free form and bound in lipids (e. g., sphingomyelines and lecithines) in bile, brain, egg yolk

as well as in hops, *Belladonna*, *Strophanthus*, and in numerous seeds, leaves, and stalks of plants. In the nerve system C. is formed by enzymatic hydrolysis of *acetylcholine. The *O*-(3,4-dimethoxy)benzoate is known as the alkaloid hesperalin from *Hesperis matronalis* (Brassicaceae).
Activity: C. has vasodilator and antihypertensive activities, regulates intestinal movements, and is supposed to reduce fat deposition in the liver (lipotrophic action). The biosynthesis starts from L-serine.
Use: In medicaments for arteriosclerosis.
Lit.: Annu. Rev. Nutr. **1**, 95–121 (1981) (review) ▪ Beilstein EIV **4**, 1443f. ▪ Hager (5.) **7**, 925–930 ▪ Kirk-Othmer (4.) **6**, 199 ▪ Luckner (3.), p. 270ff. ▪ Negwer (6.), No. 368 ▪ Ullmann (5.) **A7**, 39ff.; **A9**, 452. – *Toxicology:* Sax (8.), CMF 000, CMF 750. – *[HS 2923 10; CAS 62-49-7 (C. cation); 123-41-1 (hydroxide); 67-48-1 (chloride)]*

Chondramides.

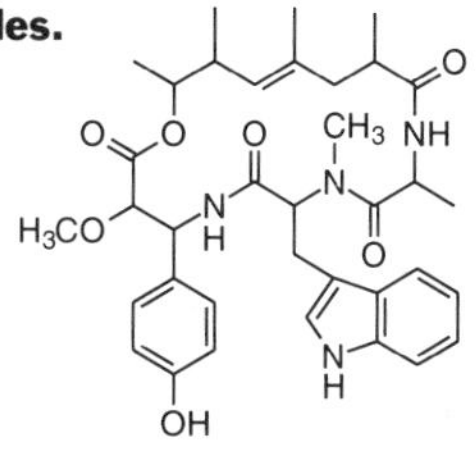

C. A

Cytostatic and antifungal depsipeptides from *Chondromyces crocatus* (Myxobacteria). They are structurally related with *jaspamide from sponges. *C. A:* $C_{36}H_{46}N_4O_7$, M_R 646.75, glass, soluble in methanol, acetone, insoluble in hexane. *Cytostatic mechanism of action:* accumulation of cell nuclei and stabilization of actin occur, the result is death of the cell.
Lit.: J. Antibiot. **48**, 1262 (1995) ▪ Justus Liebigs Ann. Chem. **1996**, 285–290. – *[CAS 172430-60-3 (C. A); 172430-63-6 (C. D)]*

Chondrillin see plakinic acids.

Chondrine [(1*S-trans*)-thiomorpholine-3-carboxylic acid 1-oxide].

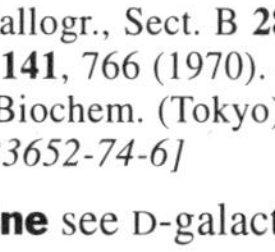

(1*S*,3*R*)

$C_5H_9NO_3S$, M_R 163.19, cryst., mp. 255 °C, $[\alpha]_D$ +20.91°. An unusual amino acid from red algae of the genus *Chondria* and brown algae of the genus *Undaria* or *Zonaria*.
Lit.: Acta Crystallogr., Sect. B **28**, 2789 (1972) ▪ Arch. Biochem. Biophys. **141**, 766 (1970). – *Synthesis:* Chem. Ber. **98**, 781 (1965) ▪ J. Biochem. (Tokyo) **54**, 222 (1963) ▪ Scheuer I **3**, 122. – *[CAS 23652-74-6]*

Chondrosamine see D-galactosamine.

Chonemorphine see Apocynaceae steroid alkaloids.

Chorismic acid [(3*R*)-*trans*-3-(1-carboxyvinyloxy)-4-hydroxy-1,5-cyclohexadiene-1-carboxylic acid].

$C_{10}H_{10}O_6$, M_R 226.19. Crystallizes as a monohydrate, mp. 148–149 °C, $[\alpha]_D$ –295.5° (H_2O). In bacteria, fungi, and higher plants C. is an important intermediate in the biosynthesis of aromatic natural products via the *shikimic acid pathway. The biosynthesis proceeds via shikimic acid → shikimic acid 3-phosphate → 5-*O*-(1-carboxyvinyl)-shikimic acid 3-phosphate → chorismic acid.
Lit.: J. Am. Chem. Soc. **91**, 5893f. (1969) (biosynthesis); **112**, 8907ff. (1990) (synthesis) ▪ J. Org. Chem. **64**, 4935 (1999) (synthesis) ▪ Luckner (3.), p. 323–329, 363 ▪ Nat. Prod. Rep. **11**, 173–203 (1994) ▪ Synthesis **1993**, 179 (review). – *[CAS 617-12-9]*

Chromomycins.

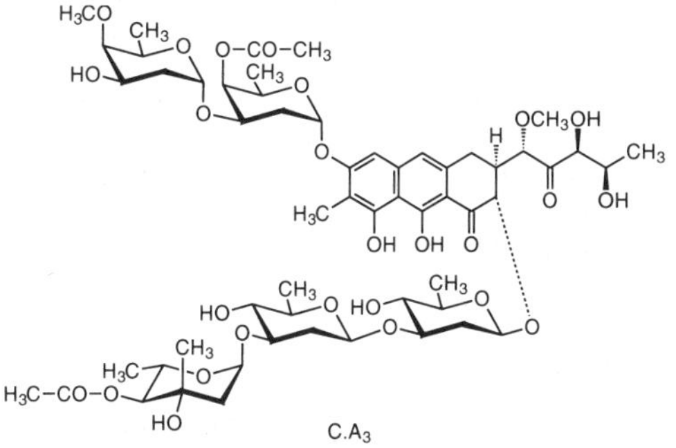

C.A_3

Group of *antibiotics (C. A–C) obtained from *Streptomyces griseus* and *S. aburaviensis*. The main component, *C. A_3 (aburamycin B)*, $C_{57}H_{82}O_{26}$, M_R 1183.26, has antitumor and antimicrobial activity; LD_{50} (mouse p.o.) 1.85 mg/kg. Biosynthetic investigations have shown that C. A_3 from *S. griseus* is formed from acetate, methionine, and glucose.
Lit.: Adv. Pharmacol. Chemother. **12**, 1 (1975) ▪ Aszalos (ed.), Antitumor Compounds of Natural Origin, vol. 1, p. 191–235, Boca Raton: CRC Press 1981 ▪ J. Antibiot. **43**, 110ff., 883–889 (1990). – *[CAS 7059-24-7 (C. A_3)]*

Chromoproteins. Name for proteins that appear colored on account of the presence of prosthetic groups with pigment nature and/or functionally important metal ions. Members of the physiologically important class of C. are *h(a)emoglobin, *cytochrome, flavin enzymes, *rhodopsin, *phytochrome and protein complexes of *chlorophyll.

Chrysanthemin (asterin, *cyanidin-3-*O*-*β*-glucopyranoside). $C_{21}H_{21}O_{11}^+$, M_R 449.39, as chloride metallic shining, red-brown platelets or prisms, mp. 205 °C (decomp.). C. belongs to the *anthocyanins and occurs, e.g., as pigment in asters (*Aster*, Asteraceae), chrysanthemums (*Chrysanthemum*, Asteraceae), elderberries (*Sambucus*, Caprifoliaceae), blackberries (*Rubus*, Rosaceae), bilberries (*Vaccinium*, Ericaceae), blood oranges (*Citrus*, Rutaceae), peaches (*Prunus persica*, Rosaceae), or in the red autumn leaves of some maple species (*Acer*, Aceraceae).
Lit.: Beilstein EV **17/8**, 475 ▪ Karrer, No. 1713 ▪ Phytochemistry **15**, 1395 (1976). – *[HS 3203 00; CAS 7084-24-4]*

Chrysanthemumic acid see pyrethrum.

Chrysanthenol see pinenols.

Chrysanthenone see pinenones.

Chrysarobin (chrysophanolanthrone, 1,8-dihydroxy-3-methylanthrone).

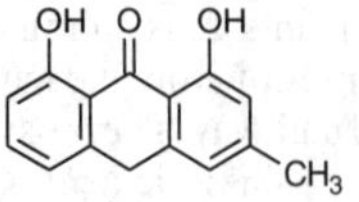

$C_{15}H_{12}O_3$, M_R 240.26, yellow needles, mp. 206–208 °C, soluble in alcohol, chloroform, or fats. C. is the main component of a mixture of various anthrones and anthranols known as chrysarobinum obtained by benzene extraction from araroba or goa powder (yellow-brown powder from cavities in the heartwood of the 20–30 m high tree *Andira araroba*, Fabaceae, endemic to Brazil and for long cultivated in India) which, after dying on wool, gives a dark violet color. C. is also isolated from *Cassia* and *Rumex* species and from *Ferreirea spectabilis* (Fabaceae); it is also formed by *Penicillium islandicum*.

Biosynthesis: An octaketide formed in the polyketide pathway (*acetyl-CoA and 7 molecules of malonyl-CoA), see anthranoids.

Lit.: Hager (5.) **1**, 575; **7**, 937 ▪ Karrer, No. 1256 ▪ Luckner (3.), p. 176 ▪ Merck-Index (12.), No. 2312. – *[CAS 491-58-7]*

Chrysin (5,7-dihydroxyflavone). Formula, see flavones, $C_{15}H_{10}O_4$, M_R 254.24, pale yellow crystals, mp. 285–286 °C, insoluble in water, soluble in alkalis and ethanol; used as complexation reagent for Cu. C. occurs together with its 7-methyl ether in poplar buds (*Populus nigra*, Salicaceae), in the heartwood of various *Pinus* species (Pinaceae) and the sweet cherry (*Prunus avium*, Rosaceae).

Lit.: Beilstein E V **18/4**, 76 ▪ Karrer, No. 1438 ff. ▪ Merck-Index (12.), No. 2316. – *[CAS 480-40-0]*

Chrysobactin see siderophores.

Chrysophanol (chrysophanic acid, 1,8-dihydroxy-3-methylanthraquinone).

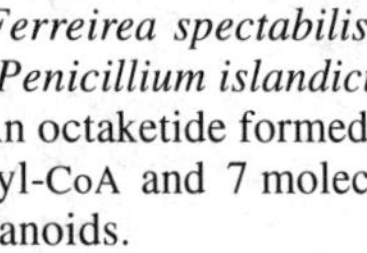

$C_{15}H_{10}O_4$, M_R 254.24, golden-yellow platelets, mp. 200–201 °C, uv_{max} 429 nm (C_2H_5OH). C. occurs in the free form, as the 1-*O*-*β*-D-glucoside (chrysophanein), or in the form of dianthrones (dichrysophanoldianthrone, palmidine D, rheidine B) in the roots of rhubarb and sorrel, in the form of glycosides also in the bark of buckthorn, in buckthorn berries, and in senna leaves and fruits. The drug prepared from these plant starting materials is used as a laxative. For biosynthesis, see *Lit.*[1].

Lit.: [1] Phytochemistry **25**, 103 (1986).
gen.: Hager (5.) **4**, 719 f.; **6**, 392 ff. ▪ Karrer, No. 1254 ▪ Luckner (3.), p. 176 ▪ Merck-Index (12.), No. 2318 ▪ Synthesis **1994**, 255. – *[HS 291469; CAS 481-74-3]*

Chrysophanolanthrone see chrysarobin.

Chrysorrhealactone. The milkcaps *Lactarius chrysorrheus* and *L. scrobiculatus* (Basidiomycetes) contain a white latex which turns sulfur yellow when the fruitbody is injured and takes on a burning, bitter taste. Responsible for this is the enzymatic transformation

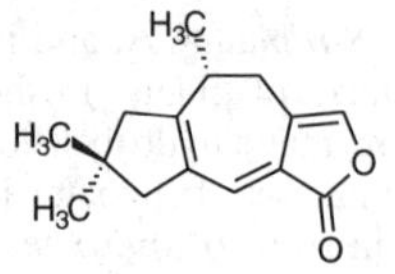

Chrysorrhealactone

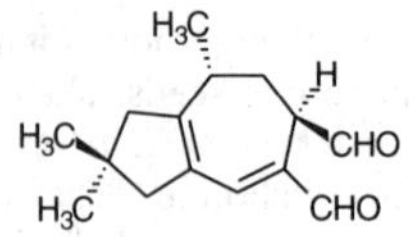

Chrysorrheadial

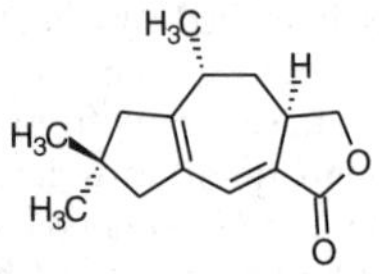

Lactaroscrobiculide A

of 6-oxostearoyl- or *stearoylvelutinal to yellow *C.* {$C_{15}H_{18}O_2$, M_R 230.31, unstable oil, $[\alpha]_D$ −30.4° (CH_2Cl_2)}, the burning tasting *chrysorrheadial* {$C_{15}H_{20}O_2$, M_R 230.32, oil, $[\alpha]_D$ +60.2° (CH_2Cl_2)}, bitter *lactaroscrobiculide A* {$C_{15}H_{20}O_2$, M_R 230.32, cryst., mp. 86–88 °C, $[\alpha]_D$ +466° (CH_2Cl_2)} and other lactaranes. Many of the lactaranes and secolactarane derivatives isolated earlier from *L. scrobiculatus* and other milkcaps are artefacts of the wok-up procedures.

Lit.: Tetrahedron **49**, 1489 (1993). – *[CAS 147396-19-8 (C.); 147396-20-1 (chrysorrheadial); 59476-60-7 (lactaroscrobiculide A)]*

Chuangxinmycin (chuanghsinmycin).

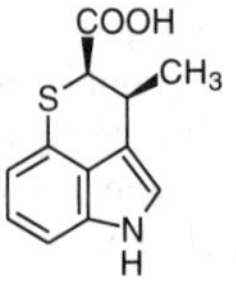

$C_{12}H_{11}NO_2S$, M_R 233.29, mp. 186–189 °C, $[\alpha]_D^{20}$ −29° (C_2H_5OH). C. is a broad-spectrum antibiotic isolated from *Actinoplanes jinanensis*. In China C. is used to treat *Escherichia coli* infections, the activity seems to be based on inhibition of tryptophan biosynthesis.

Lit.: Chem. Pharm. Bull. **42**, 271–276 (1994) (synthesis) ▪ J. Chem. Soc., Perkin Trans. 1 **1992**, 323 ff. ▪ Sci. Sinica Engl. Ed. **20**, 106 (1977) ▪ Tetrahedron: Asymmetry **8**, 2295 (1997) (synthesis) ▪ Tetrahedron Lett. **34**, 489 (1993) (synthesis); **38**, 1805 (1997) [synthesis (±)-C.]. – *[CAS 63339-68-4]*

Chymostatins. Known are the C. A, B, and C isolated from actinomycetes which have inhibitory effects on proteases such as chymotrypsin, papain, and most serine/cysteine endoproteases.

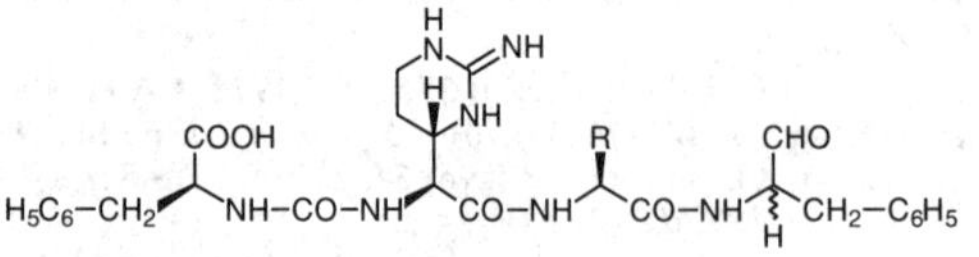

R = CH_2–$CH(CH_3)_2$: Chymostatin A
R = $CH(CH_3)_2$: Chymostatin B
R = $CH(CH_3)$–CH_2–CH_3 : Chymostatin C

Table: Data of Chymostatins.

name	molecular formula	M_R	CAS
C. A	$C_{31}H_{41}N_7O_6$	607.71	51759-76-3
C. B	$C_{30}H_{39}N_7O_6$	593.68	51759-77-4
C. C	$C_{31}H_{41}N_7O_6$	607.71	51759-78-5

Lit.: J. Antibiot. **23**, 425 (1970); **26**, 625 (1973) ▪ Methods Enzymol. **45**, 678 (1976) ▪ Prog. Ind. Microbiol. **27**, 403 f. (1989) ▪ Umezawa & Aoyagi, in Medical and Biological Aspects, p. 3–15, Berlin: Springer 1983.

Cibaric acid.

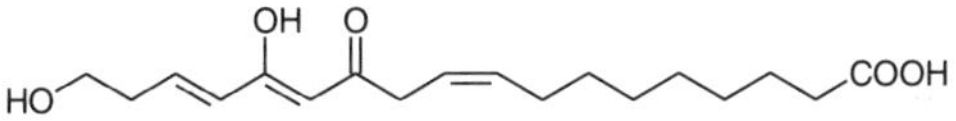

$C_{18}H_{28}O_5$, M_R 324.42. Oleic acid derivative from injured fruit bodies of the chanterelle (*Cantharellus cibarius*, Basidiomycetes). The corresponding (*E*)-10-hydroxy-8-decenoic acid was isolated from *C. tubaeformis*.

Lit.: Acta Chem. Scand., Ser. B **46**, 301 (1992) ▪ J. Org. Chem. **56**, 1233 (1991). – *[CAS 130523-93-2]*

Cicu(to)toxin (8,10,12-heptadecatriene-4,6-diyne-1,14-diol).

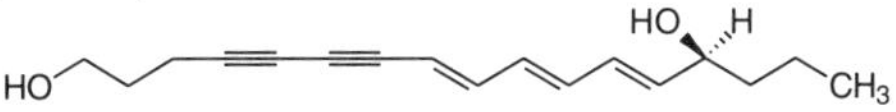

$C_{17}H_{22}O_2$, M_R 258.36, (*all-E*)-(–)-form: amorphous powder, mp. 54 °C, $[\alpha]_D^{15}$ –14.5° (C_2H_5OH), UV (C_2H_5OH): λ_{max} 242, 252, 318.5, 335.5. Toxic component of the roots of the water hemlock (*Cicuta virosa*, Apiaceae) and *C. maculata*. (*all-E*)-(±)-form, cryst., mp. 67 °C. C. inhibits electron transport in the respiratory chain and photosynthesis. It also exhibits antileukemic action and acts as a centrally attacking spasmogenic toxin. LD_{50} (mouse i. p.) 9.2 mg/kg; LDL_0 (cat p. o.) 7 mg/kg.

Lit.: Beilstein EIV **1**, 2744 ▪ J. Nat. Prod. **49**, 1117–1121 (1986) (^1H NMR, MS, IR) ▪ Karrer, No. 138 ▪ Merck-Index (12.), No. 2328 ▪ Planta Med. **61**, 439 (1995) ▪ Tetrahedron **55**, 12087–12098 (1999) (abs. configuration). – *Toxicology:* Sax (8.), CMN 000 ▪ Vet. Hum. Toxicol. **29**, 240 f. (1987). – *[CAS 82905-09-7 (C.); 505-75-9 (all-E)-(–)]*

Ciguatera toxin. The term ciguatera is used to describe poisoning caused by eating various tropical and subtropical reef fish usually considered to be edible, e. g., *Lutjanus bohar* (red snapper), *Epinephelus fuscoguttatus* (reef cod), *Variola louti* (reef perch). About 20 000 cases of poisonings with low mortality rate are reported each year. Ciguatera poisonings are especially frequent in the Pacific region. Nausea, vomiting, diarrhea, headache, restlessness, and muscular paralysis are observed as symptoms. Neurological symptoms often last for months, e. g., the dry ice effect: inflicted subjects who immerse their hands in cold water experience a scalding feeling.

The various effects that have been described are ultimately the result of a direct action on excitable membranes: the toxins increase the Na^+ permeability of Na channels. Two groups of compounds responsible for the poisoning are distinguished: *ciguatoxin (CTX) and *maitotoxin and compounds derived from them. Both groups are produced by the epiphytic dinoflagellate *Gambierdiscus toxicus*, passed on to herbivorous fish, and finally from there to carnivorous fish. However, only the less polar and less toxic precursor CTX3C has been isolated from *G. toxicus*, this is converted to the highly toxic ciguatoxin in the course of the food chain in fish [1].

Lit.: [1] Tetrahedron Lett. **34**, 1975 (1993).

gen.: Biol. Bull. (Woods Hole, Mass.) **172**, 137 (1987) ▪ FEBS Lett. **219**, 355 (1987) ▪ J. Pharmacol. Exp. Ther. **235**, 783 (1985) ▪ Med. Mol. Pharm. **14**, 206–216 (1991) ▪ Naturwiss. Rundsch. **43**, 120 (1990); **45**, 481 f. (1992).

Ciguatoxin (CTX 1) (formula see below). For structural elucidation 0.35 mg of C. ($C_{60}H_{86}O_{19}$, M_R 1111.33) were isolated from 125 kg intestines of the moray *Gymnothorax javanicus* (primary origin: *Gambierdiscus toxicus*). CTX acts as an agonist of voltage-sensitive sodium channels (VSSC). The lethal dose is 0.35 µg/kg (mouse). Thus, the toxin is about 100 times more potent than *tetrodotoxin.

Lit.: Biosci. Biotechnol. Biochem. **60**, 2103 (1996) (isolation of C.-4 A). – *Synthetic studies:* Angew. Chem. Int. Ed. Engl. **37**, 965 (1998) ▪ Synthesis **1999**, 1431. – *Pharmacology:* J. Biol. Chem. **259**, 8353 (1984) ▪ Med. Mol. Pharm. **14**, 206–210 (1991) ▪ Toxicon **22**, 169 (1984). – *Structure:* Chem. Rev. **93**, 1897 (1993) ▪ J. Am. Chem. Soc. **111**, 8929 (1989); **112**, 4380 (1990); **119**, 11325 (1997); **120**, 5914 (1998) (absolute configuration) ▪ Tetrahedron **53**, 3057 (1997) (absolute configuration) ▪ Tetrahedron Lett. **33**, 525 (1992); **34**, 1975 (1993). – *[CAS 11050-21-8]*

Ciliaric acid see trachylobanes.

Ciliatine.

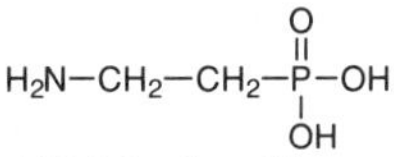

$C_2H_8NO_3P$, M_R 125.06, rhombic crystals or needles, mp. 295–297 °C (decomp.). C. was the first natural product to be discovered containing a C–P bond. C. occurs in the form of glyceryl esters in the protozoa *Tetrahymena pyriformis*, sea anemones, bovine brain, and mycobacteria.

Lit.: Synthesis **1989**, 52 ▪ Tetrahedron **45**, 2557 (1989) ▪ see also phosphonothrixin. – *[CAS 2041-14-7]*

Cilofungin see echinocandin B.

Cinachyrolides s. spongistatins.

Cinchona alkaloids. *Indole alkaloids (cinchonamine group) and *quinoline alkaloids (cinchonine group), only the latter group is of therapeutic significance. C. a. are isolated from *Cinchona bark* (*Jesuit's bark; Cinchonae cortex*). The name is derived from the Indian word quina (bark). The drug consists of the dried tube

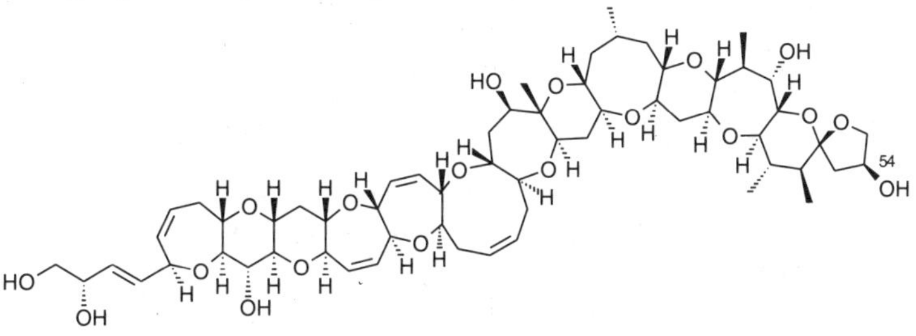

Ciguatoxin

or half-tube-like pieces of trunk and branch bark of *Cinchona* species such as *C. pubescens* (syn. *C. succirubra*) and *C. calisaya* (syn. *C. ledgeriana*, Rubiaceae) that are endemic in south America and are cultivated in south east Asia and Africa (Zaire). The bark is harvested from 6–8 year-old trees. The red ("succirubra" = red juicy), officinal bark is used for the production of cinchona dry extract according to the Swiss pharmacopoeia or cinchona tincture according to the German pharmacopoeia DAB 10; it contains between 4 and 12% alkaloids. The non-officinal "yellow bark" of *C. calisaya* is used for the technical isolation of C. a. In Zaire, Indonesia, and Malaysia in particular ca. 5000–10000 t of bark are processed annually to yield ca. 300–500 t of the main alkaloids *quinine and *quinidine [1], which comprise 80% of the bark's alkaloid content. Further main alkaloids are *cinchonine and *cinchonidine. *Cinchonaminal is of biogenetic interest as a possible intermediate. In the quinoline bases the quinoline ring system is linked with the quinuclidine ring system by a hydroxymethylene bridge (1-azabicyclo-[2.2.2]octane) and carries a vinyl group in the 3-position. The quinuclidine nitrogen atom is protonated with formation of salts under physiological conditions.
Minor alkaloids are the so-called epibases (9-hydroxy epimers) as well as 10,11-dihydroalkaloids. Quinine and quinovine are responsible for the bitter taste of the drug. C. a. exist in the drug as salts of *quinic acid and other plant acids. Attempts to produce C. a. in cell suspension cultures of *Cinchona* were as yet unsuccessful, although differentiated plant material such as root and shoot cultures do form quinoline alkaloids.
Biosynthesis: Knowledge about the biosynthesis of C. a. is insufficient and is based almost exclusively on speculations derived from feeding experiments with radiolabelled precursors [2]; the transformation from indole- to quinoline-type and the precursor role of *tryptophan and secologanin, as well as that of *strictosidine are confirmed [3]. The cinchoninone: NADPH oxidoreductases I and II from cell cultures of *C. calisaya* are as yet the only known enzymes that may be of biogenetic significance for the formation of C. a. [4].
Use: The pure alkaloids are used as antimalarials, with activity decreasing in the order quinine, quinidine, cinchonine, cinchonidine. They find further use as tonic and as bitter stomachic. Cinchona tincture containing additional essential oil (orange oil, cinnamon) and bitter substances (gentian root) stimulates appetite and promotes secretion of gastric juices.

Lit.: [1] Manske **34**, 331–398. [2] Mothes et al., p. 272–313. [3] Hager (5.) **4**, 873; J. Chem. Soc. Perkin Trans. 1 **1979**, 2308. [4] Phytochemistry **26**, 393 (1987).
gen.: Curr. Res. Med. Aromat. Plants **9**, 34–56 (1987) ▪ Hager (5.) **1**, 577 ff.; **4**, 877–882 ▪ Saxton (ed.), The Monoterpenoid Indole Alkaloids, Suppl. vol. 25, Heterocyclic Compounds, p. 647, Chichester: John Wiley & Sons 1994 – *Synthesis:* Chem. Heterocycl. Compd. **25**, 729–752 (1983) ▪ J. Chem. Soc., Perkin Trans. 1 **1987**, 1053–1058. – *[HS 2939 29]*

Cinchona bark see Cinchona alkaloids.

Cinchonaminal. $C_{19}H_{22}N_2O$, M_R 294.40, an alkaloid discussed as intermediate in the biosynthesis of *cinchona alkaloids (*quinine, *quinidine). C. still belongs

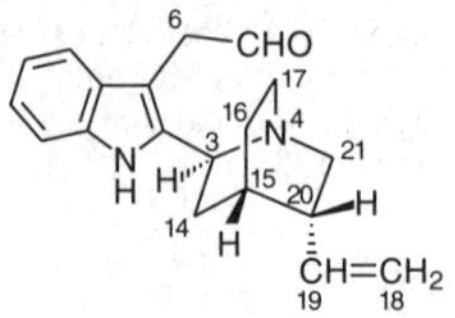

to the indole type, already carries the intact quinuclidine ring of the Cinchona alkaloids but does not yet belong to the *quinoline alkaloids, to which the Cinchona alkaloids must be attributed. C. may serve as a biogenetic precursor of *cinchonamine, to which it can be transformed by a reductase (reduction of the aldehyde group at C-6).

Lit.: Hager (5.) **4**, 872 f. ▪ Luckner (3.), p. 358 ▪ Mothes et al., p. 293. – *[HS 2939 90; CAS 29560-34-7]*

Cinchonamine.

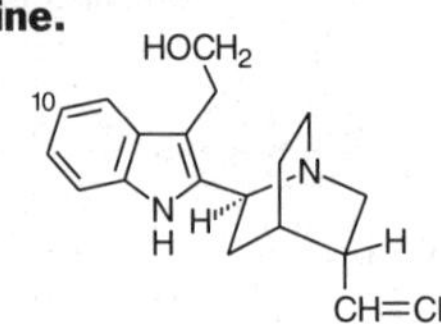

$C_{19}H_{24}N_2O$, M_R 296.41; prisms from methanol, mp. 185–186 °C (also 194 °C), $[\alpha]_D$ +128° (C_2H_5OH); insoluble in water, readily soluble in benzene, chloroform. C. is an alkaloid of the Cinchona group but also belongs to the indole type. C. was first isolated over one hundred years ago from the bark of *Remijia purdieana* (Rubiaceae) [1] and later also found in *Cinchona pubescens* [2]. *10-Methoxycinchonamine* ($C_{20}H_{26}N_2O_2$, M_R 326.44) has been found in the leaves of *C. calisaya* (synonym *C. ledgeriana*).
Biosynthesis: Formally C. could represent a biosynthetic intermediate of the *Cinchona alkaloids of the quinoline series, however, the respective *in vivo* investigations could not clarify this.
C. has no antimalarial activity. For stereochemistry and synthesis of C., see *Lit.* [3,4].

Lit.: [1] C. R. Acad. Sci. **93**, 593 (1881). [2] Planta Med. **50**, 17 (1984). [3] J. Am. Chem. Soc. **80**, 3484 (1958). [4] Helv. Chim. Acta **59**, 2268 (1976).
gen.: Hager (5.) **4**, 872 f. ▪ Merck-Index (12.), No. 2344. – *[HS 2939 90; CAS 482-28-0]*

Cinchonidine [(8*S*,9*R*)-cinchonan-9-ol]. Formula, see quinine. $C_{19}H_{22}N_2O$, M_R 294.40; plates or prisms, mp. 210 °C, $[\alpha]_D$ −110° (C_2H_5OH); well soluble in alcohol and chloroform. C. is an alkaloid of the Cinchona group (*Cinchona alkaloid), a *quinoline alkaloid. C. is isolated from many *Cinchona* species (*C. tucujensis*, *C. pubescens*) and also occurs in *Remijia* species (Rubiaceae) and in some members of the olive family such as *Olea europaea* (olive tree) and *Ligustrum vulgare* (common privet). C. is obtained commercially from cinchona bark.
Biosynthesis: *In vivo* experiments with the labeled ketone of C. (= cinchonidinone) revealed its incorporation into the Cinchona alkaloids, especially in non-methoxylated members such as C. and cinchonine [1]. For enzymatic studies, see. *Lit.* [2], for synthetic investigations, see *Lit.* [3]. The uses of C. are similar to those of *quinine.

Lit.: [1] J. Chem. Soc., Chem. Commun. **1971**, 31. [2] Phytochemistry **26**, 393 (1987). [3] J. Chem. Soc., Chem. Commun. **1986**, 573.

gen.: Angew. Chem. Int. Ed. Engl. **35**, 2143 (1996) ■ Beilstein EV **23/12**, 406f. ■ Hager (5.) **4**, 877; **5**, 939 ■ J. Org. Chem. **39**, 2413 (1974) ■ Manske **34**, 331 ■ Tetrahedron **23**, 3253 (1967). – *[HS 2939 29; CAS 485-71-2]*

Cinchonine [(9*S*)-cinchonan-9-ol]. Formula, see quinine. $C_{19}H_{22}N_2O$, M_R 294.40; needles or prisms from ether or alcohol, mp. ca. 265 °C (sublimation starts at 220 °C). C. is practically insoluble in water, well soluble in alcohol, $[\alpha]_D$ +229° (C_2H_5OH). C. belongs to the *Cinchona alkaloids, is the most important minor alkaloid of *quinine, and occurs in many *Cinchona* species (Rubiaceae).
Activity: C. is active against malaria but markedly less effective than quinine. C. can also be used in racemate separation and for precipitation of Bi, Cd, Ge, Mo, and W.
Lit.: Acta Crystallogr., Sect. B **35**, 440 (1979) ■ Beilstein EV **23/12**, 406 ■ Hager (5.) **4**, 877; **5**, 939 ■ J. Chem. Soc., Chem. Commun. **1971**, 31 ■ Manske **34**, 331. – *[HS 2939 29; CAS 118-10-5]*

Cineoles {1,4-C.=1-methyl-4-(1-methylethyl)-7-oxabicyclo[2.2.1]heptane, 1,4-epoxy-*p*-menthane; 1,8-C.= 1,3,3-trimethyl-2-oxabicyclo[2.2.2]octane, 1,8-epoxy-*p*-menthane}.

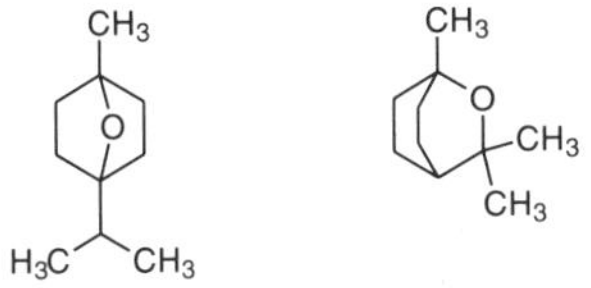

1,4-Cineole 1,8-Cineole

$C_{10}H_{18}O$, M_R 154.25, colorless oils with camphor-like odor, bicyclic *monoterpene epoxides. 1,4-C., bp. 173–174 °C; 1,8-C., bp. 176–177 °C. 1,4-C. occurs in the essential oil of "Fructus Cubebae" [a drug obtained from fully grown but unripe fruits of the Indonesian endemic plant *Piper cubeba* (Piperaceae)] and 1,8-C. occurs to 40–60% in *eucalyptus oils (*Eucalyptus globulus*, Myrtaceae). C. are used as expectorant for bronchial catarrh and as aroma substances in the perfume industry.
Lit.: Beilstein EV **17/1**, 273 ■ Karrer, No. 569f. – *Synthesis:* Aust. J. Chem. **50**, 35 (1997) ■ Chem. Lett. **1979**, 373. – *[HS 2932 90 (1,8-C.); CAS 470-67-7 (1,4-C.); 470-82-6 (1,8-C.)]*

Cinerins.

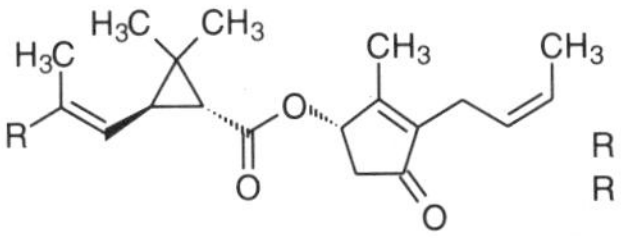

R = CH3 : Cinerin I
R = COOCH3: Cinerin II

Toxic, insecticidal components of *Pyrethrum* species (cf. pyrethrum), especially *Chrysanthemum cinerariaefolium. C. I* {$C_{20}H_{28}O_3$, M_R 316.42, bp. 137 °C (10.66 Pa), $[\alpha]_D^{20}$ –22° (hexane)} is a viscous oil, extremely sensitive to air; *C. II* {$C_{21}H_{28}O_5$, M_R 360.43, bp. 184 °C (1.33 Pa), $[\alpha]_D^{20}$ +16° (isooctane)} is also very sensitive to air. C. are used in admixture with other Pyrethrum components as insecticides.
Lit.: Crombie, Int. Congr. Ser. – Exerpta Med. **832**, 3–25 (1988) (biosynthesis). ■ J. Chem. Soc., Chem. Commun. **1972**, 1276 ■ J. High Resolut. Chromatogr. **14**, 48ff. (1991) (analysis) ■ Ullmann (5.) **A 14**, 273 ■ Zechmeister **19**, 120–164. – *Synthesis:* J. Agric. Food Chem. **31**, 151–156 (1983) ■ Pestic. Sci. **11**, 129–133 (1980) – *Toxicology:* Hayes et al. (eds.), Handbook of Pesticide Toxicology, p. 585f., London: Academic Press 1991 ■ Sax (8.), p. 1130f. ■ Zechmeister **19**, 120–164. – *[HS 2916 20; CAS 25402-06-6 (C. I); 121-20-0 (C. II)]*

Cinerolone see pyrethrum.

Cinnabarin [polystictin, 2-amino-9-(hydroxymethyl)-3-oxo-3*H*-phenoxazine-1-carboxylic acid].

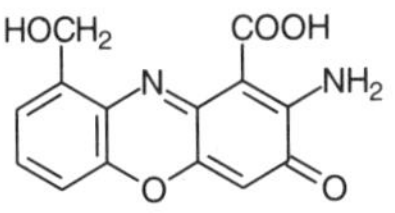

Cinnabarin

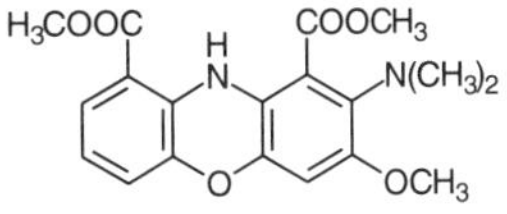

Pycnosanguin

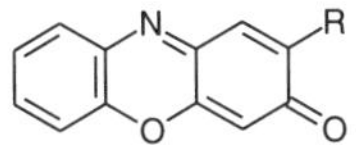

R = H : Phenoxazin-3-one
R = SCH3 : 2-(Methylthio)phenoxazin-3-one

$C_{14}H_{10}N_2O_5$, M_R 286.24, red needles, mp. 320 °C (decomp.). Antibiotically active phenoxazinone pigment from the world-wide distributed cinnabar polypore (*Pycnoporus cinnabarinus = Trametes cinnabarina*, Basidiomycetes). Like other phenoxazinones C. is formed by oxidative dimerization of 3-hydroxyanthranilic acid. *Pycnosanguin* ($C_{19}H_{20}N_2O_6$, M_R 372.38, yellow needles, mp. 246–249 °C) is a dihydro derivative. *Phenoxazin-3-one* ($C_{12}H_7NO_2$, M_R 197.19) and its 2-methylthio derivative ($C_{13}H_9NO_2S$, M_R 243.28) are isolated from cultures of the St George's mushroom (*Calocybe gambosa*, Agaricales) [1].
Lit.: [1] Arch. Pharm. (Weinheim, Ger.) **321**, 363 (1988); **324**, 3 (1991).
gen.: Zechmeister **51**, 208f. (review). – *[CAS 146-90-7 (C.), 133056-31-2 (pycnosanguin); 1916-63-8 (phenoxazin-3-one)]*

Cinnalutein.

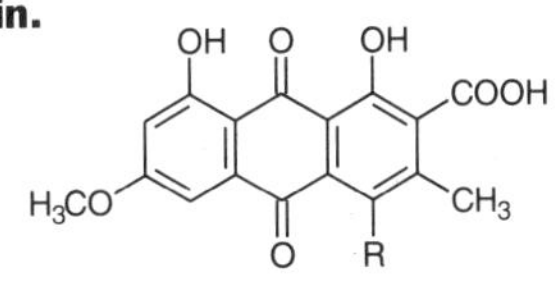

R = H : Cinnalutein
R = OH : Cinnarubin

The anthraquinonecarboxylic acids *C.* [$C_{17}H_{12}O_7$, M_R 328.28, yellow needles, mp. 277 °C (decomp.)] and *cinnarubin* [$C_{17}H_{12}O_8$, M_R 344.28, red needles, mp. >290 °C (decomp.)] occur together with *physcion, fallacinol, and other neutral anthraquinones in *Dermocybe cinnabarina* (Basidiomycetes) and a group of related agarics. The pigments exist in the fruit bodies mainly as *O*-glycosides.
Lit.: Zechmeister **51**, 131ff. (review). – *[CAS 26687-55-8 (C.), 39012-25-4 (cinnarubin)]*

Cinnamaldehyde (3-phenyl-2-propenal). H_5C_6–CH=CH–CHO, C_9H_8O, M_R 132.16; yellowish liquid with a strong odor of cinnamon. Naturally occurring and synthetic C. consist of more than 98% *trans*-C.: bp. 253 °C, mp. –7.5 °C. Soluble in organic solvents; LD_{50} (rat p.o.) 2.22 g/kg, oxidized in air to *cinnamic acid and sensitive towards alkali. C. occurs in *cinnamon leaf (bark) oil (65–75%) and *cassia oil (ca. 75%).

Use: For artificial cinnamon oil, in flavors and aromas, and to a limited extent in perfumes [on account of its sensitizing properties only in combination with equal amounts of (+)-limonene (see *p*-menthadienes) or *eugenol], in addition as an intermediate in the production of cinnamyl alcohol and subsequent products. World-wide requirements: ca. 4000 t per year.
Lit.: Bauer et al. (2.), p. 82 ▪ Bedoukian (3.), p. 98–105 ▪ Beilstein E IV **7**, 984 ff. ▪ Food Cosmet. Toxicol. **17**, 253 (1979) ▪ Synthesis **1994**, 369 (synthesis). – *[HS 2912 29; CAS 104-55-2 (Z.); 14371-10-9 (trans-Z.)]*

Cinnamates see cinnamic acid.

Cinnamic acid (3-phenyl-2-propenoic acid).

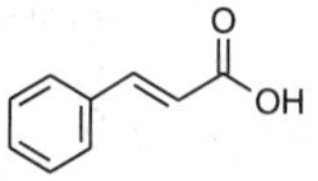

$C_9H_8O_2$, M_R 148.16; (*E*)-form, needles or prisms, mp. 133 °C, bp. 300 °C, pK_a 4.46 (25 °C), steam distillable. The acid occurs either in the free form of esterified in *storax (oil), *peru balsam, and other essential oils. The most important simple derivatives of (*E*)-C. a. occurring as natural products are: a) the *methyl ester* $C_{10}H_{10}O_2$, M_R 162.20, mp. 36 °C, bp. 263 °C, bp. 132 °C (2.0 kPa); fruity odor of strawberries. – b) The *ethyl ester*, $C_{11}H_{12}O_2$, M_R 176.21, mp. 12 °C, bp. 159 °C (3.2 kPa). – c) The *cinnamyl ester* (styracin), $C_{18}H_{16}O_2$, M_R 264.32, needles or prisms, mp. 44 °C. – d) *Cinnamic acid amide*, C_9H_9NO, M_R 147.18, needles, mp. 147 °C, isolated from the fern *Cornopteris decurrentialata*. The (*Z*)-form of C. a. exists in three different crystalline forms, mp. 68, 58, and 42 °C; the highest melting form is known as α-C. a., the other two as β-C.; pK_a 3.85 (25 °C). Irradiation of C. a. leads to *truxillic and truxinic acids.
Use: Simple esters of (*E*)-C. a. are used as odor and taste additives in perfumes and foods. Esters of C. a. and *p*-methoxy-C. a. are used as sun blockers.
Biosynth.: By elimination of ammonia from L-*phenylalanine, catalyzed by L-phenylalanine ammonia lyase (PAL, EC 4.3.1.5).
Lit.: Beilstein E IV **9**, 2001–2014 ▪ Chem. Pharm. Bull **67**, 1475 (1994) ▪ J. Org. Chem. **58**, 6364 (1993) (synthesis) ▪ Luckner (3.), p. 383–389 ▪ Zechmeister **54**, 12–16, 108–112. – *Toxicology:* Sax (8.), MIO 500. – *[HS 2916 39; CAS 621-82-9 (Z.); 140-10-3 (E-form); 102-94-3 (Z-form); 1754-62-7 (a); 4192-77-2 (b); 122-69-0 (c); 40918-97-6 (c, E,E-form); 22031-64-7 (d)]*

Cinnamon leaf (bark) oil. Bark and leaves of the cinnamon tree give essential oils with different compositions.
Production: By steam distillation from the leaves or bark of the Sri Lankan cinnamon tree [1] *Cinnamomum verum*.
1. *Cinnamon leaf oil:* Warm, spicy odor of cinnamon and cloves; warm, spicy taste.
Composition[2]*:* The main component is **eugenol* (ca. 70%); typical accompanying compounds are **linalool* (2%), **cinnamaldehyde* (2%), **safrole* (2%), *eugenyl acetate* (4-allyl-2-methoxyphenyl acetate, $C_{12}H_{14}O_3$, M_R 206.24) (2%), and *benzyl benzoate* (see benzoic acid) (4%).
Use: In the perfume industry for spicy-oriental notes; for the production of eugenol with a high-quality odor; to improve the aromatic nature of foods such as confectionery, refreshing beverages, liquors, etc.
2) *Cinnamon (bark) oil:* Stronger warmer, spicy-sweet odor of cinnamon; sweet, spicy taste.
Composition[2]*:* The main component is **cinnamaldehyde* (ca. 75%); typical minor components are **linalool* (3%), **eugenol* (2%), and *cinnamyl acetate* (see cassia oil) (5%).
Use: In the perfume industry for warm, oriental notes; to improve the aromatic nature of foods such as puddings, confectionery, bakery products, liquors, and refreshing beverages (cola-lemonades); in medicine in carminatives and in oral and pharyngeal therapeutics on account of its antimicrobial activity.
Lit.: [1] Perfum. Flavor. **2** (3), 53 (1977); **3** (4), 55 (1978); **19** (2), 69 (1994). [2] Perfum. Flavor. **3** (4), 55 (1978); **19** (3), 59 (1994); Dev. Food Sci. **34**, 411–425 (1994).
gen.: Arctander, p. 163, 167 ▪ Bauer et al. (2.), p. 145, 146 ▪ ISO 3524 (1977). – *Toxicology:* Food Cosmet. Toxicol. **13**, 111, 749 (1975). – *[HS 3301 29; CAS 8015-96-1 (1.); 8015-91-6 (2.); 93-28-7 (eugenyl acetate)]*

Cinnamyl acetate see cassia oil.

Cinnamyl alcohol (cinnamic alcohol, 3-phenyl-2-propen-1-ol). H_5C_6–CH=CH–CH_2OH, $C_9H_{10}O$, M_R 134.17, (*E*)-form: needles with a sweet-balsamy, hyacinth-like odor, mp. 33 °C, bp. 258 °C, bp. 142–145 °C (1.86 kPa), readily soluble in alcohol and ether, it is slowly oxidized by atmospheric oxygen under warmer conditions and on exposure to light. It occurs mostly in the form of cinnamic acid esters (see cinnamic acid) in *storax (oil), *peru balsam, cinnamon leaf, narcissus, and hyacinth oils.
Use: (*E*)-Z. is used in differing qualities in the perfume industry: mp. 33 °C (highly pure); mp. 28 °C ("prime", specification of the Essential Oil Association); mp. 20 °C, crude product from styrax. Simple esters of (*E*)-C. a. such as cinnamyl formate, acetate, propionate, or benzoate are used as fixatives, they usually have the same pleasant odor as the alcohol.
Lit.: Beilstein E IV **6**, 3799 ▪ J. Chem. Soc., Perkin Trans. 1 **1990**, 2775–2790 (1990) (synthesis) ▪ J. Org. Chem. **57**, 2599 (1992) (synthesis) ▪ Luckner (3.), p. 396 ▪ Org. Prep. Proceed. Int. **17**, 251 ff. (1985) (synthesis) ▪ Snell-Ettre **10**, 76–84. – *[HS 2906 29; CAS 104-54-1 (C.); 4407-36-7 (E); 4510-34-3 (Z)]*

Cinobufagin see bufadienolides.

Cispentacin [(1*R*)-*cis*-2-aminocyclopentanecarboxylic acid].

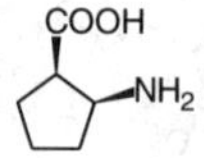

$C_6H_{11}NO_2$, M_R 129.16. *Antibiotic from *Bacillus cereus* and *Streptomyces setonii*. Prisms, mp. 195–196 °C (decomp.), $[\alpha]_D^{25}$ –10.7° (H_2O). In a model of systemic candidiasis (mouse p.o.), C. has antimycotic activity and is nontoxic at 1 g/kg. The synthetically prepared stereoisomers [1] are biologically inactive. For synthesis, see *Lit.*[2].
Lit.: [1] J. Antibiot. **43**, 1–7, 513–518 (1990). [2] Chem. Pharm. Bull. **41**, 1012–1018 (1993); J. Chem. Soc., Chem. Commun. **1991**, 2276; J. Chem. Soc., Perkin Trans. 1 **1994**, 1411; Synlett **1993**, 461; Tetrahedron: Asymmetry **7**, 3565 (1996).
gen.: J. Antibiot. **42**, 1749–1762 (1989). – *[HS 2941 90; CAS 122672-46-2]*

Cistus oil see labdanum (-absolute, -oil, -resinoid).

Citrals [(*E*)- and (*Z*)-3,7-dimethylocta-2,6-dienal]. Formula, see geraniol. $C_{10}H_{16}O$, M_R 152.23, oils with a lemon-like odor. Unsaturated acyclic *monoterpene aldehydes: (*E*)-C. (C. a, *geranial*), bp. 77 °C (1.8 hPa); (*Z*)-C. (C. b, *neral*), bp. 76 °C (2.3 hPa). Commercial C. is a mixture of *E/Z*-isomers. East Indian *lemongrass oil (*Cymbopogon flexuosus*, Poaceae) contains up to 45–50% (*E*)-C. and 28–30% (*Z*)-C. The essential oil of the leaves of *Backhousia citriodora* (Myrtaceae), a plant growing only in tropical and subtropical areas, contains up to 90–97% C. as a mixture. Synthesis by oxidation of *geraniol or nerol. C. is used in the flavor and fragrance industry as starting material for *β*-*ionone.

Lit.: Beilstein E IV **1**, 3569 f. ▪ Karrer, No. 371 ▪ Merck-Index (12.), No. 2383 ▪ Ullmann (5.) **A 11**, 159. – *[HS 2912 19; CAS 5392-40-5 (C.); 141-27-5 ((E)-C.); 106-26-3 ((Z)-C.)]*

Citranaxanthin (5′,6′-dihydro-5′-apo-18′-nor-*β*,*ψ*-caroten-6′-one).

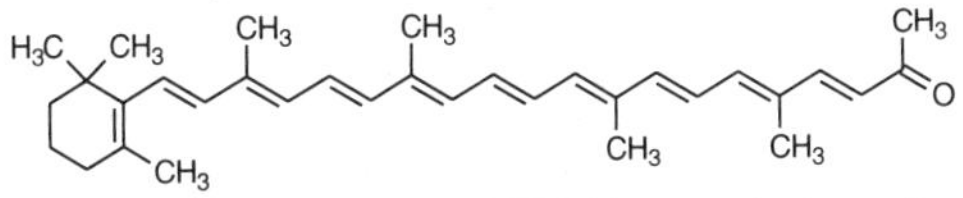

$C_{33}H_{44}O$, M_R 456.71, mp. 156 °C (decomp.). A xanthophyll (see also carotinoids) occurring naturally in citrus fruits but which is also accessible by Wittig syntheses [1]. C. is used as a fodder additive for fattened poultry to improve pigmentation of egg yolk and skin.

Lit.: [1] Angew. Chem. Int. Ed. Engl. **16**, 423–429 (1977).
gen.: see carotinoids. – *[HS 2936 10; CAS 3604-90-8]*

Citreoviridins.

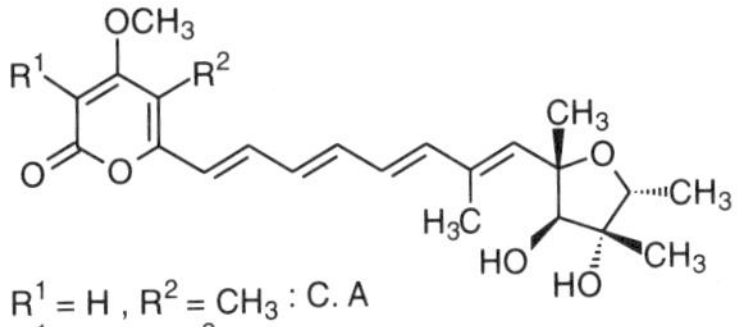

R^1 = H , R^2 = CH_3 : C. A
R^1 = CH_3 , R^2 = H : C. C
R^1 = R^2 = CH_3 : C. D

Neurotoxic *mycotoxins from *Penicillium citreoviride* (= *P. toxicarium*), *P. charlesii*, *P. citrinum*, *P. ochrosalmoneum*, and *Aspergillus terreus*. The most important compound is *C. A*, $C_{23}H_{30}O_6$, M_R 402.49, dark yellow cryst., mp. 107–111 °C, $[\alpha]_D$ –68.9°. C. are formed via the *polyketide pathway (nonaketides), 5 methyl groups originate from methionine. C. A occurs in mold-infested rice (east Asia), consumption of which leads to the symptoms of an acute cardiac beriberi, often with death by respiratory arrest. In Europe C. A can occur in sausage products through colonization by a C. A-producing *Penicillium* strain during uncontrolled maturation and storage. Mode of action: inhibition of mitochondrial ATPase, i. e. ATP synthesis.

Lit.: Appl. Environ. Microbiol. **54**, 1096 ff. (1988) (occurrence) ▪ Betina (ed.), Mycotoxins, p. 262 ff., Amsterdam: Elsevier 1984 ▪ Cole-Cox, p. 876 ▪ J. Biol. Chem. **256**, 557 (1981) (mode of action) ▪ J. Chem. Soc., Chem. Commun. **1985**, 1531 ▪ J. Chem. Soc., Perkin Trans. 1 **1982**, 2175 ▪ J. Korean Chem. Soc. **40**, 674 (1996) (crystal structure) ▪ J. Org. Chem. **56**, 7174 (1991) (synthesis) ▪ Phytochemistry **11**, 3215 (1972) (biosynthesis) ▪ Sax (8.), CMS 500 ▪ Tetrahedron Lett. **26**, 231 (1985); **29**, 711 (1988). – *[CAS 25425-12-1 (C. A); 74145-78-1 (C. C); 74145-79-2 (C. D)]*

Citric acid (2-hydroxy-1,2,3-propanetricarboxylic acid).

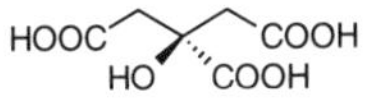

$C_6H_8O_7$, M_R 192.13. Rhombic crystals with a pleasant tart taste, as monohydrate D. 1.542, mp. 100 °C; anhydrous mp. 153 °C, pK_{a1} 3.13, pK_{a2} 4.78, pK_{a3} 6.43 (25 °C). On heating to above 175 °C decomposition occurs with formation of *citraconic acid anhydride* (methylmaleic acid anhydride, $C_5H_4O_3$, M_R 112.08), action of sulfuric acid leads to *aconitic acid. C. occurs in all organisms as an intermediate of the citric acid cycle; in adults ca. 2000 g C. are formed daily. C. is one of the most widely distributed plant acids and occurs in high concentration in lemon juice (5–7%). Technical production proceeds by aerobic fermentation with yeasts, e. g. *Candida* sp. (e. g., *Jarrowia lipolytica*) as well as with *Aspergillus niger.*

Use: As acidifying agent in beverages and other foodstuffs as well as in anticoagulant and drug preparations (effervescent tablets). For complexation of heavy metal ions in galvanizing industry and in waste water. Esters of C. are used as softening agents.

Lit.: Beilstein E IV **3**, 1272–1275 ▪ Biotechnol. Adv. **13**, 209 (1995) (manufacture) ▪ Karrer, No. 877 ▪ Kirk-Othmer (4.) **6**, 354 ▪ Ullmann (5.) **A 7**, 103–108. – *[HS 2918 14; CAS 77-92-9]*

Citrinin.

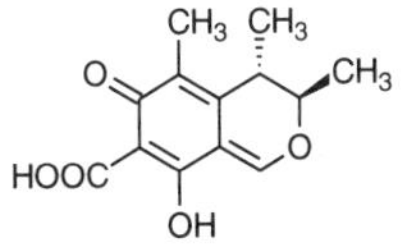

$C_{13}H_{14}O_5$, M_R 250.25, lemon yellow crystals, mp. 179 °C, $[\alpha]_D$ –37.4° (C_2H_5OH), reacts with NH_3 → deep red. *Mycotoxin formed by many *Penicillium* and *Aspergillus* species. C. was discovered in 1931 and described as an antibiotic in 1941. Its numerous biological activities (including phytotoxic, mutagenic, teratogenic, and cancerogenic effects) have been intensively investigated [1]. In all experimental animals the kidneys are attacked primarily. C. has also been isolated from *Crotalaria crispata*, an Australian Fabaceae and *Pythium ultimum*, a phytopathogenic oomycete. Large amounts of C. are formed by *Penicillium citrinum*, *P. expansum* (= *P. griseofulvum*), *P. roqueforti*, and *Aspergillus terreus* not only on rice but also on other cereal crops. C. can also be formed in bread by fungi and thus lead to mycotoxicosis [2]. The biosynthesis proceeds via the *polyketide pathway (pentaketide), two methyl branchings and the carboxy group originate from methionine.

Lit.: [1] Betina, chap. 9. [2] Getreide, Mehl, Brot **31**, 265–270 (1977); Tetrahedron **39**, 3583 (1983) (biosynthesis).
gen.: Beilstein E V **18/9**, 61 ▪ Cole-Cox, p. 824 ▪ J. Chem. Soc., Perkin Trans. 1 **1981**, 2577, 2594; **1986**, 2101; **1987**, 2743 ▪ Justus Liebigs Ann. Chem. **1995**, 885 ff. ((–)-C., synthesis) ▪ Sax (8.), CMS 775 ▪ Zechmeister **24**, 289. – *[HS 2932 90; CAS 518-75-2]*

Citronellals (3,7-dimethyl-6- and -7-octenal). $C_{10}H_{18}O$, M_R 154.25, colorless liquids with melissa-like odor. Acyclic *monoterpene aldehydes occurring in nature as two optically active forms and a racemate: (3*R*)(+)- and (3*S*)(–)-*β*-C., bp. 206–208 °C, $[\alpha]_D$ ±11.5–14.2°; (±)-C., bp. 208–209 °C. Natural C. consist mainly or

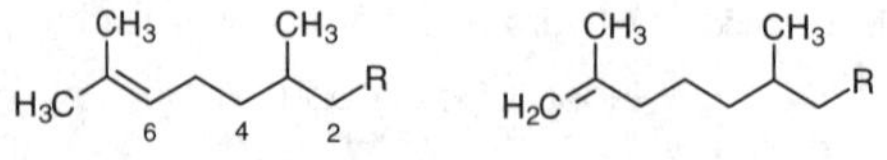

R = CHO : (β-)-Citronellal α-Citronellal
R = CH_2OH : (β-)-Citronellol α-Citronellol

completely of the *β*-form. C. are very sensitive to atmospheric oxygen, acids, and bases and polymerize easily.
Occurrence: C. occur mainly in tropical plants belonging to the Myrtaceae that are rarely found in Europe: (3*R*)(+)-*β*-C. in the essential oil of *Baeckea citriodora* up to 80%, (3*S*)(−)-*β*-C. up to 80% in the essential oil of *Backhousia citriodora*. (±)-C. in the essential oil of *Eucalyptus citriodora* up to 70%.
Use: C. find use in perfume and soap industries and for the preparation of menthol, *citronellol, hydroxycitronellal, and isopulegol (see *p*-menthenols). Dog collars impregnated with C. are purported to calm dogs and induce them to bark less [1].
Lit.: [1] Chem. Eng. News (20. 5.) **1996**, 64.
gen.: Beilstein EIV **1**, 3515 ▪ J. Org. Chem. **26**, 3072 (1961) (synthesis) ▪ Karrer, No. 366, 367 ▪ Ullmann (5.) **A 11**, 161. – *[HS 2912 19; CAS 141-26-4 (α-C.); 2385-77-5 ((3R)(+)-β-C.); 5949-05-3 ((3S)(−)-β-C.); 26489-02-1 (racemate)]*

Citronella oil. C. is obtained from two tropical grass species of the genus *Cymbopogon* (Poaceae).
A. *Citronella oil, type Ceylon (Sri Lanka):* light yellow to brownish oil with a herbal fresh, grass-like, camphor-like odor produced by steam distillation of the aerial parts of the citronella grass sort "lenabatu", *Cymbopogon nardus*. Main components are *limonene* (see *p*-menthadienes) (ca. 10%), **citronellal* (ca. 10%), **borneol* (ca. 5%), **citronellol* (ca. 5%), **geraniol* (ca. 20%), and *isoeugenol methyl ether* (*isoeugenol; ca. 10%) [1].
Use: As perfume in household articles (soaps, cleaners, washing powders).
B. *Citronella oil, type Java:* light yellow to light brown oil with a mildly sweet, flowery, rosy-citrus odor; prepared by steam distillation of the aerial parts of the citronella grass variety "mahapengiri", *Cymbopogon winterianus*. Java citronella grass is mainly cultivated in China, Indonesia, Taiwan, and south America. Main components are (+)-**citronellal* (30–40%), (+)-**citronellol* (10–15%), and **geraniol* (20–30%) [1].
Use: Today mainly as perfume for household articles like soaps, cleaners, washing powders; medicinal use in balneotherapeutics ("Indian melissa oil"). C. used to be rectified in Java on a technical scale for production of (+)-citronellal as well as citronellol and geraniol, which were both used as aroma substances.
Lit.: [1] Perfum. Flavor. **19** (2), 29 (1994).
gen.: Arctander, p. 170 ▪ Bauer et al. (2), p. 151 ▪ ISO 3849 (1981), 3848 (1976). – *Toxicology:* Food Cosmet. Toxicol. **11**, 1067 (1973). – *[HS 3301 29; CAS 8000-29-1]*

Citronellols (3,7-dimethyloct-6- and -7-en-1-ol). $C_{10}H_{20}O$, M_R 156.27, formula, see citronellals. Acyclic *monoterpene alcohols, of which the *β*-forms of both enantiomers occur in nature: (3*R*)(+)-*β*- and (3*S*)(−)-*β*-C., bp. 224–224.5 °C, $[\alpha]_D$ ±5° (neat). Both have strong, rose-like odors.
Occurrence: (3*R*)(+)-*β*-C. in Java citronella oil (*Cymbopogon winterianus*) up to 54% and in glandular secretions of alligators [1] (initially known as "yacarol"). (3*S*)(−)-*β*-C. in Turkish *rose oil together with *geraniol up to 50%.
Use: Perfume and cosmetic industry. C. serve as starting materials for the production of citronellyl esters: formates (fruity, flowery), acetates (bergamot, lavender), propionates (roses, lily of the valley), isobutyrates (sweet, fruity), isovalerates (herby), tiglates (fungus-like).
Lit.: [1] Chem. Ber. **70**, 37 (1937).
gen.: Beilstein EIV **1**, 2188 ▪ Karrer, No. 111 ▪ Merck-Index (12.), No. 2391 ▪ Org. Synth. **72**, 74 (1995) (synthesis) ▪ Ullmann (5.) **A 11**, 157. – *[HS 2905 22; CAS 6812-78-8 ((3S)(−)-α-C.); 1117-61-9 ((3R)(+)-β-C.); 7540-51-4 ((3S)(−)-β-C.); 26489-01-0 ((±)-β-C.)]*

Citronellyl formate see geranium oil.

Citrovorum factor see folinic acid.

Citrulline (2-amino-5-ureidovaleric acid).

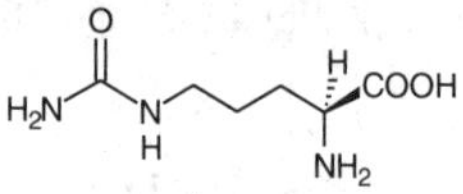

(2*S*)-form

$C_6H_{13}N_3O_3$, M_R 175.19, mp. 222 °C, $[\alpha]_D^{25}$ +7.5° (H_2O), +42.4° (1 M HCl), pK_a 2.43 and 9.41, pI 5.92; monohydrochloride mp. 185 °C (decomp.), $[\alpha]_D^{25}$ +17.9° (H_2O). Widely distributed non-proteinogenic amino acid and intermediate product in the urea cycle. C. was first isolated from the juice of water melons (*Citrullus vulgaris*, Cucurbitaceae) and then from alder (*Alnus incana, A. glutinosa*). C. is formed in the urea cycle from carbamoyl phosphate and *L-ornithine with ornithine carbamoyltransferase (EC 2.1.3.3).
Lit.: Angew. Chem. Int. Ed. Engl. **30**, 712 (1991) (synthesis) ▪ Beilstein EIV **4**, 2647 ▪ Karrer, No. 2391 ▪ Merck-Index (12.), No. 2392. – *[HS 2924 10; CAS 372-75-8]*

Citrusin C see eugenol.

Citrusin F see (di)hydrocaffeic acid.

Citrus oils. *Essential oil obtained from the peel of citrus fruits (bergamot, grapefruit, lime, mandarin, orange, lemon). Since the oil is stored in small bubbles in the outer peel (albedo) it can be isolated by mechanical processes. C. are thus known as "pressed" or "cold-pressed" oils.
C. consist mainly of monoterpene hydrocarbons, above all *limonene* (see *p*-menthadienes) (exception: *bergamot oil, that only contains ca. 40%). Since the important odor and taste components are thus only present in relatively low concentrations, so-called concentrates of C. are produced, especially for use as aromatizing agents, by removing a large portion of the unwanted non-polar terpene hydrocarbons, thereby the content of the polar components is increased. This can be achieved technically by distillation, distribution, extraction, or adsorption processes.
Since C. cannot be distilled they also contain small amounts (1–5%) of non-volatile components including pigments, waxes, coumarins, etc. The latter class of compounds also includes the so-called *furocoumarins, which are present in some C. at such concentrations that the application of these oils on the skin (e.g., as perfume) may lead to phototoxic reactions. These

oils can thus only be used in cosmetic products in low, harmless doses or must previously be rectified so that the poorly volatile furocoumarins are removed.

Lit.: Arctander, p. 38. – *Coumarins/furocoumarins:* Flavour Fragr. J. **7**, 129 (1992) ▪ J. Chromatogr. **672**, 177 (1994) ▪ J. Essent. Oil Res. **1**, 139 (1989) ▪ Perfum. Flavor. **7** (3), 57 (1982). – *Constituents:* Perfum. Flavor. **16** (2), 17 (1991) ▪ see also bergamot, lime, mandarin, orange, lemon oil.

Civet. C. is a secretion of the civet cat obtained by pressing out ("curetage") the anal sac of the animal; the following species of civet cat are known: *Civettictis civetta* (African civet cat), *Viverra zibetha* (Asian civet cat), and *Viverricula indica* (Chinese civet cat). The animals are bred in captivity for production of the secretion. The main producing country is Ethiopia with ca. 2 t/year. Freshly obtained C. is a light to yellow, cream-like mass which becomes increasingly darker and harder on exposure to the air.
For use in perfume oils either an alcoholic solution (C. tincture) is prepared or crude C. is extracted with a suitable solvent such as ethanol or acetone. The product (C. absolute) is a viscous, gray to gray-brown mass with a strong animal, musk-like, slightly fecal odor.
Composition [1]***:*** C. absolute consists of more than 50% fatty acids. The main component responsible for the odor is the macrocyclic ketone **civet(t)one*, with a content of ca. 3–5%.
Use: As a component of perfumes but, on account of its high price, only in very small amounts in very expensive perfumes.
Lit.: [1] Ohloff, p. 199.
gen.: Arctander, p. 173 ▪ Dev. Food Sci. **18**, 587 (1988) ▪ Parfum. Cosmet. Arom. **63**, 65 (1985); **90**, 79 (1990). – *Toxicology:* Food Cosmet. Toxicol. **12**, 863 (1974). – *[HS 0510 00; CAS 68991-27-5]*

Civet(t)one [(*Z*)-9-cycloheptadecen-1-one].

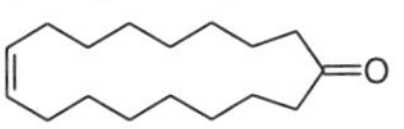

$C_{17}H_{30}O$, M_R 250.42. Crystals with a sweet animalish, musk-like odor; mp. 31–32 °C, bp. 103 °C (6.7 Pa), soluble in alcohol and other organic solvents.
Occurrence: In *civet (ca. 3%) together with other macrocyclic ketones [1]. Recognized as the component responsible for the civet odor in 1915 by Sack, structure elucidation in 1926 by Ruzicka [2], for synthesis, see *Lit.* [3,4]. C. is used in fine perfumes for a fixating musk-like odor.
Lit.: [1] Ohloff, p. 199. [2] Helv. Chim. Acta **9**, 230 (1926). [3] Helv. Chim. Acta **62**, 2661 (1979). [4] Seifen Fette Öle Wachse **115**, 538 ff. (1989); J. Am. Oil Chem. Soc. **71**, 911 ff. (1994); An. Quinn. **78**, 304 (1982).
gen.: Bedoukian (3.), p. 303–322 ▪ Beilstein E III **7**, 524 ▪ Theimer (ed.), Fragrance Chemistry, p. 433–494, San Diego: Academic Press 1982. – *[HS 2914 29; CAS 542-46-1 (Z); 1502-37-0 (E)]*

Cladosporin (asperentin).

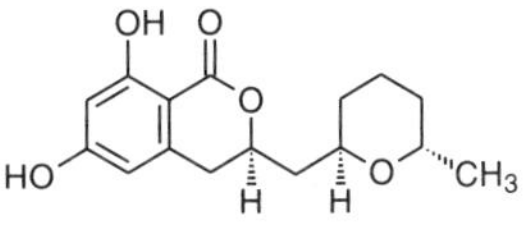

$C_{16}H_{20}O_5$, M_R 292.33, cryst., mp. 188–189 °C, $[\alpha]_D$ –24.8° (C_2H_5OH). An isocoumarin derivative isolated from culture broths of *Cladosporium cladosporioides* and *Aspergillus* species (e. g., *A. flavus*, *A. repens*); its biosynthesis proceeds through the *polyketide pathway from eight acetate units. C. has a selective action against certain Gram-positive bacteria by inhibiting the uptake of uracil and leucine. It also has insecticidal, phytotoxic, and antitumor activities.
Lit.: Arch. Microbiol. **116**, 253–257 (1978) ▪ Dev. Food. Sci. **8**, 457 (1984) (review) ▪ J. Agric. Food Chem. **29**, 853 ff. (1981) ▪ J. Am. Chem. Soc. **111**, 3382–3390 (1989) (biosynthesis) ▪ J. Antibiot.. **24**, 747–755 (1971); **32**, 952–958 (1979) ▪ J. Nat. Prod. **56**, 1397–1401 (1993); **57**, 640–643 (1994). – *[CAS 35818-31-6]*

Clary sage oil. Light yellow oil with a sweet-herby, lavender-bergamot-like odor with hints of ambergris and tobacco and a bitter aromatic taste.
Production: By steam distillation from the flowering shrub of clary (*Salvia sclarea*, Lamiaceae). *Origin*: Mediterranean region (e. g., France) and parts of the former USSR.
Composition [1]***:*** Main components are *linalyl acetate* (see linalool) (45–70%) and **linalool* (10–20%). Depending on the distillation conditions it also contains small amounts (up to a few %) of the diterpene alcohol (–)-*sclareol* (see labdanes). These components constitute the main part (ca. 70%) of the solvent extract of clary ("clary concrete"). Larger areas in USA, Israel, and parts of the former USSR are used for the cultivation of C. to produce sclareol. (–)-Sclareol is the starting material for the valuable ambergris odor substance (–)-8,12-epoxy-13,14,15,16-tetranorlabdane ("*ambroxide*", "ambrox", $C_{16}H_{28}O$, M_R 236.40); The tricyclic lactone (+)-*norambreinolide* ("sclareolide", $C_{16}H_{26}O_2$, M_R 250.38) is usually an intermediate.

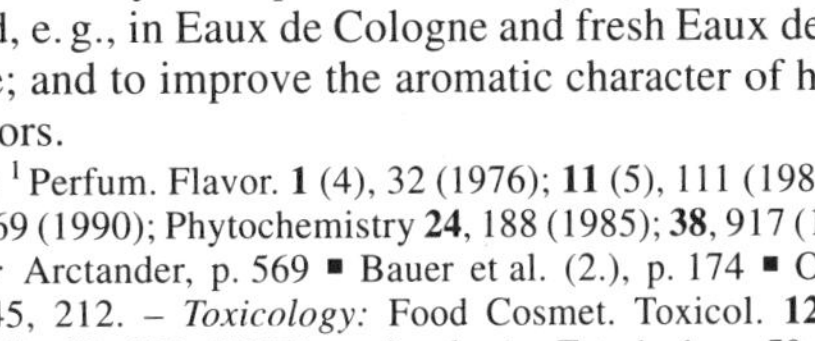

(+)-Norambreinolide (-)-Ambroxide

Use: Mainly in the perfume industry where it is widely used, e. g., in Eaux de Cologne and fresh Eaux de Toilette; and to improve the aromatic character of herbal liquors.
Lit.: [1] Perfum. Flavor. **1** (4), 32 (1976); **11** (5), 111 (1986); **15** (4), 69 (1990); Phytochemistry **24**, 188 (1985); **38**, 917 (1995).
gen.: Arctander, p. 569 ▪ Bauer et al. (2.), p. 174 ▪ Ohloff, p. 145, 212. – *Toxicology:* Food Cosmet. Toxicol. **12**, 865 (1974); **20**, 823 (1982). – *Synthesis:* Tetrahedron **50**, 6653 (1994). – *[HS 3301 29; CAS 8016-63-5 (oil); 6790-58-5 ((–)-ambroxide); 564-20-5 ((+)-norambreinolide)]*

Clathridine. C. [1] exists in the brilliant yellow Mediterranean sponge *Clathrina clathrus* (Calcispongiae) as a 2:1 Zn complex; $C_{16}H_{15}N_5O_4$, M_R 341.33, cryst., mp. 260–262 °C (decomp.), it has antimycotic activity against *Candida albicans* and *Saccharomyces cerevisiae*. *Leucetta microrhaphis* contains the related *(9E)-clathridine-9-(2-sulfoethylimine)* [2], yellow needles, $C_{18}H_{20}N_6O_6S$, M_R 449.12, mp. 275–277 °C (decomp.).

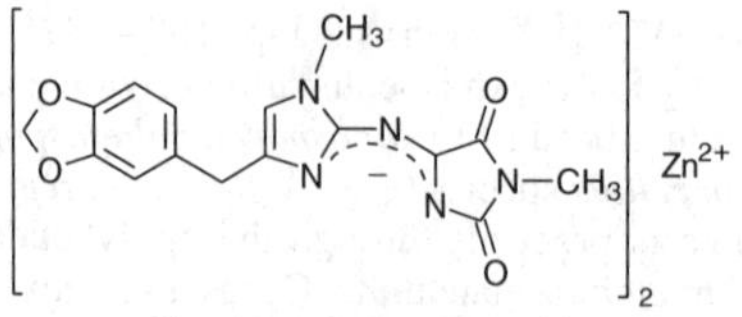

Clathridine zinc complex

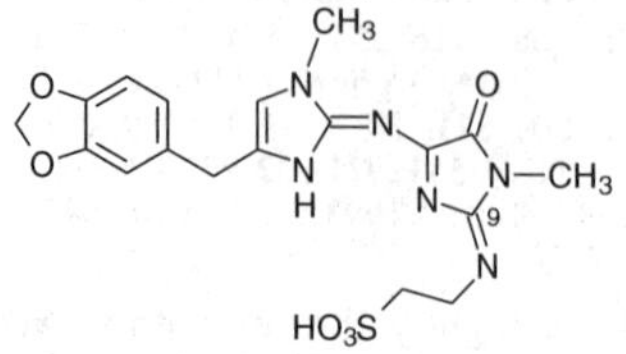

(9*E*)-Clathridine-9-(2-sulfoethylimine)

Lit.: [1] Tetrahedron **45**, 3873 (1989); **46**, 4387 (1990); **49**, 329 (1993). [2] J. Org. Chem. **57**, 2176 (1992). – *[CAS 122759-55-1 (C.); 122780-90-9 (C.-Zn complex)]*

Clausenamide [(±)-3-hydroxy-5-(hydroxyphenylmethyl)-1-methyl-4-phenyl-2-pyrrolidinone].

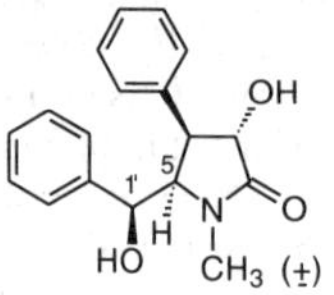

Clausenamide

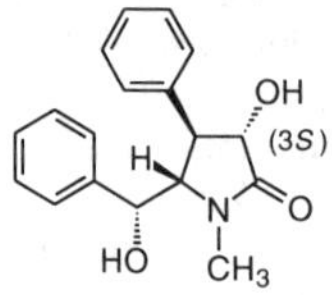

Neoclausenamide

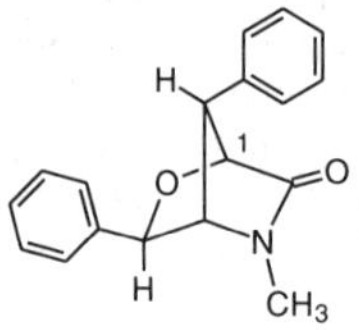

Cycloclausenamide

$C_{18}H_{19}NO_3$, M_R 297.35, needles, mp. 239–240 °C, a γ-lactam from the leaves of the Chinese fruit tree *Clausena lansium* (Rutaceae). In contrast to the stereoisomeric 1′,5-diepimer *neo-C.*, mp. 205–206 °C, C. exists in nature as a racemate. Dehydration of neo-C. affords the bicyclic *cyclo-C.* (ζ-C.[1]), $C_{18}H_{17}NO_2$, M_R 279.34, prisms, mp. 164–166 °C, $[\alpha]_D^{24.5}$ –40° (CH_3OH), which has also been isolated from *C. lansium*.

Activity: C. is a radical scavenger and induces *cytochrome P_{450}; it has hepatoprotective properties and is used in traditional Chinese medicine for acute and chronic hepatitis.

Lit.: [1] Chin. Chem. Lett. **2**, 291 (1991).
gen.: Chin. Chem. Lett. **5**, 267 f. (1994) (synthesis) ▪ Chin. Sci. Bull. **34**, 957–959 (1989) ▪ Indian J. Chem. Sect. B **37**, 422 (1998) (lansamides) ▪ J. Chem. Soc., Chem. Commun. **1998**, 1159 (synthesis) ▪ J. Org. Chem. **52**, 4352–4356 (1987) (synthesis) ▪ Phytochemistry **27**, 445–450 (1988) ▪ Synlett **1991**, 343 f. (synthesis). – *[CAS 103541-15-7 (C.); 114528-82-4 (neo-C.); 103541-16-8 (cyclo-C.)]*

Clavamine.

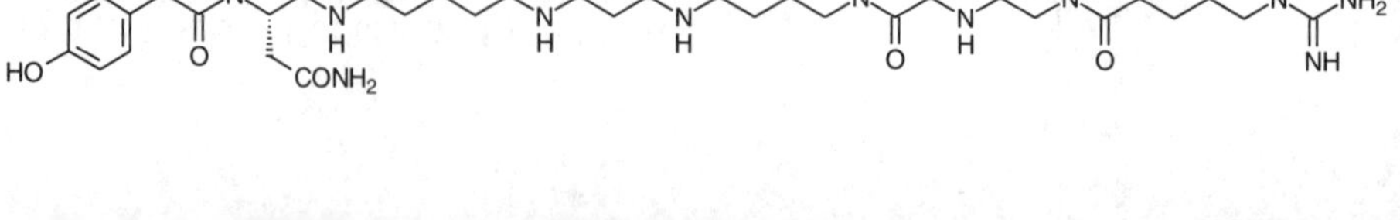

$C_{35}H_{60}N_{12}O_9$, M_R 792.94, $[\alpha]_D^{15}$ –12.6° (H_2O). An (acyl)*polyamine from the tropical round-web spider *Nephila clavata* (joro spider) in which it occurs together with other acylpolyamines (*JSTX, NPTX). C. has a strong insecticidal activity[1].

Lit.: [1] Biog. Amines **7**, 375–384 (1990)
gen.: J. Chem. Ecol. **19**, 2411–2451 (1993) ▪ Pharmacol. Ther. **52**, 245–268 (1991). – *Synthesis:* Tetrahedron **46**, 3819–3822 (1990). – *[CAS 129121-68-2]*

Clavams.

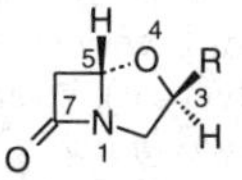

R = COOH : Clavam-2-carboxylic acid
R = CH₂–CH(NH₂)–COOH : 2-Alanylclavam
R = CH₂–CH₂–OH : Hydroxyethylclavam

R = 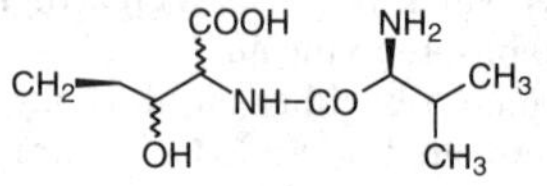 : Valclavam

Table: Data of Clavams.

compound	molecular formula	M_R	occurrence	CAS
Clavam-2-carboxylic acid[1]	$C_6H_7NO_4$	157.13	*Streptomyces clavuligerus*	71657-61-9
2-(Alan-3-yl)-clavam[2]	$C_8H_{12}N_2O_4$	200.19	*S. clavuligerus*	74758-63-7
2-(2-Hydroxyethyl)clavam[3]	$C_7H_{11}NO_3$	157.17	*S. antibioticus*	79416-52-7

The C. are *β-lactam antibiotics of the oxapenam type (4-oxa-1-azabicyclo[3.2.0]heptan-7-one), in contrast to the penams the sulfur atom is replaced by oxygen. The C. have antifungal properties and are isolated from culture broths of *Streptomyces* species. The first C. to be discovered was clavam-2-carboxylic acid; important C. are also the clavamycins from *S. hygroscopicus*[4].

Lit.: [1] J. Chem. Soc., Chem. Commun. **1979**, 282. [2] J. Antibiot. **36**, 208–212, 213 ff., 217–225 (1983). [3] Tetrahedron Lett. **22**, 2539 (1981). [4] J. Antibiot. **39**, 510 (1986).
gen.: Biosynthesis: J. Am. Chem. Soc. **114**, 2762 f. (1992). – *Synthesis:* J. Org. Chem **50**, 3457 (1985) ▪ Tetrahedron **43**, 2467 (1987); **47**, 6079–6111 (1991) ▪ Tetrahedron Lett. **29**, 1609 (1988); **32**, 1979 (1991); **34**, 5645 (1993) ▪ see also clavulanic acid. – *[HS 2941 90]*

Clavicipitic acid.

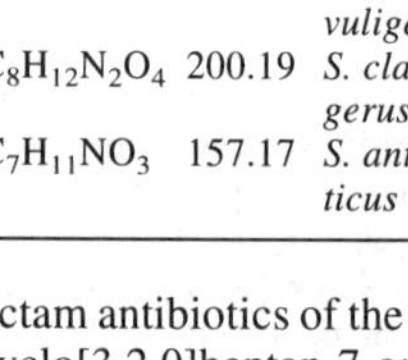

$C_{16}H_{18}N_2O_2$, M_R 270.33, plates, mp. 264 °C (decomp.). Metabolite of *Claviceps* spp. including *C. fusiformis*, cf. lysergic acid and ergot alkaloids. The biosynthesis

proceeds from *tryptophan via *N*-(3-methyl-2-butenyl)tryptophan[1].
Lit.: [1] Phytochemistry **14**, 735–737 (1975).
gen.: Beilstein E V **25/5**, 76f. ▪ J. Org. Chem. **60**, 1486 (1995) (synthesis) ▪ Tetrahedron **53**, 51 (1997) [synthesis (±)-C.]; **55**, 10989 (1999) ▪ Turner **1**, 319; **2**, 403. – *[CAS 33062-26-9]*

Clavicoronic acid.

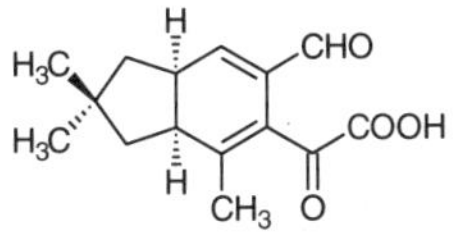

$C_{15}H_{18}O_4$, M_R 262.31, powder, mp. 75–77 °C, $[\alpha]_D$ +42.7° (CH_3CN). Biologically active sesquiterpenoid of the seco-sterpurene type (see sterpuranes) from mycelium cultures of *Artomyces pyxidata* (= *Clavicorona pyxidata*, Basidiomycetes). C. is an inhibitor of various RNA-dependent DNA-polymerases (reverse transcriptases) and acts against vesicular stomatitis viruses.
Lit.: J. Antibiot. **45**, 29 (1992). – *[CAS 139748-98-4]*

Clavilactones.

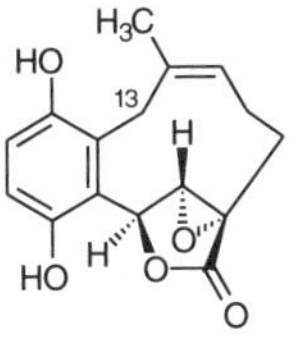

Clavilactone A

Antibiotically active cyclic geranylhydroquinones from cultures of *Clitocybe clavipes* (Basidiomycetes). In addition to *C. A* {$C_{16}H_{16}O_5$, M_R 288.30, yellow cryst., mp. 176 °C, $[\alpha]_D$ +81° (CH_3OH)} the corresponding quinone *C. B*, {$C_{16}H_{14}O_5$, M_R 286.28, yellow cryst., mp. 76–79 °C, $[\alpha]_D$ –55° ($CHCl_3$)} and the 13*β*-hydroxy derivative *C. C* {$C_{16}H_{16}O_6$, M_R 304.30, yellow cryst., mp. 178–182 °C, $[\alpha]_D$ +110° (CH_3OH)} have been isolated. The C. inhibit germination of cress seeds. They are biogenetically related with the *flavidulols.
Lit.: J. Chem. Soc., Perkin Trans. 1 **1994**, 2165. – *[CAS 158681-46-0 (C. A); 158681-47-1 (C. B); 158681-48-2 (C. C)]*

Clavulanic acid.

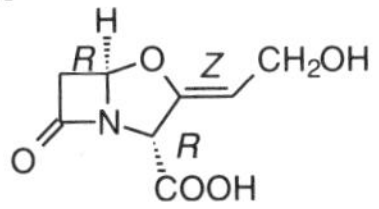

$C_8H_9NO_5$, M_R 199.16. A *β-lactam antibiotic with an oxapenem skeleton (3-methylidene-4-oxa-1-azabicyclo[3.2.0]-heptan-7-one) formed by *Streptomyces clavuligerus*, *S. jumonjinensis*, and *S. katsurahamanus*. C., an unstable oil, has broad but low activity against Gram-positive and Gram-negative bacteria. C. is commercially used as a *β*-lactamase inhibitor. The inhibition is irreversible and leads in combination with *β*-lactamase-sensitive *penicillins and *cephalosporins to a marked increase in the activities of these antibiotics, even against pathogens that are resistant to *β*-lactam antibiotics. C. is used in combination with amoxycillin in a commercial product (Augmentin®).
Lit.: Antimicrob. Agents Chemother. **11**, 852 (1977) ▪ Drugs **22**, 337 (1981) ▪ J. Antibiot. **29**, 668 (1976) ▪ J. Chem. Soc., Chem. Commun. **1976**, 226; **1990**, 617; **1997**, 1025 (biosynthesis) ▪ J. Chem. Soc., Perkin Trans. 1 **1984**, 635–650; **1990**, 21–33, 1513–1520 (biosynthesis) ▪ Merck-Index (12.), No. 2402. – *[HS 2941 90; CAS 58001-44-8]*

Cleistanthane see diterpenes.

Clementeins. *Sesquiterpene lactones with the rare oxetane structure (e. g., *taxol) from *Centaurea canariensis* (syn. *C. clementei*), e. g., *C. A*: $C_{21}H_{26}O_7$, M_R 390.41, mp. 193–195 °C.

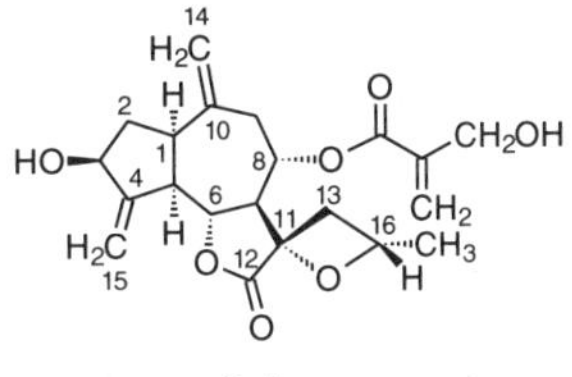

C. A
16-Epimer : C. B
4,15-Dihydro-16-Epimer : C. C

Lit.: Tetrahedron **42**, 3611–3622 (1986); **49**, 2499–2508 (1993). – *[CAS 86939-93-7 (C. A); 106621-84-5 (C. B); 106533-05-5 (C. C)]*

Clerodanes. The C. are a group of bicyclic diterpenoids with over 650 members, they occur in higher plants (especially Lamiaceae), microorganisms, and marine organisms.

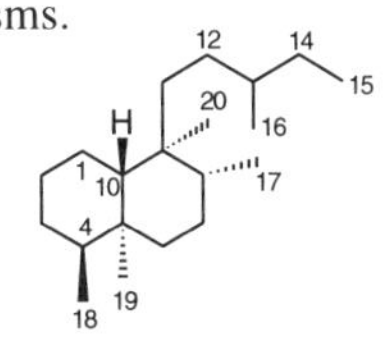

Their biosynthesis proceeds through geranylgeranyl diphosphate, cyclization and rearrangement of the resulting labdadienyl cation. Antibiotic, antitumor, and insect antifeedant properties have been described.
Lit.: J. Indian Chem. Soc. **75**, 552 (1998) (isolation) ▪ Nat. Prod. Rep. **9**, 243–287 (1992) ▪ Synlett **1998**, 912 (synthesis) ▪ Tetrahedron **44**, 6607–6622 (1988) ▪ Zechmeister **63**, 107–196.

Clerodin (3-deoxycaryoptinol).

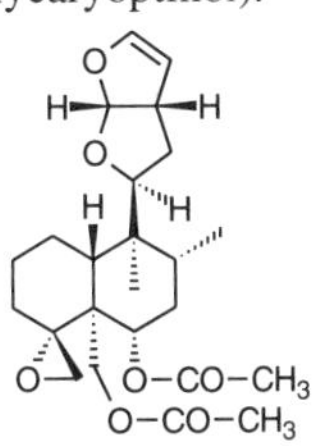

$C_{24}H_{34}O_7$, M_R 434.53, mp. 161–162 °C, $[\alpha]_D^{30}$ –37.6° (C_2H_5OH). Diterpene of the *clerodane type from leaves of *Clerodendron infortunatum* (Verbenaceae), exhibits anthelmintic properties.
Lit.: Zechmeister **63**, 107–196. – *[CAS 464-71-1]*

Clinimycin. Synonym for oxytetracycline, see tetracyclines.

Clithioneine [*S*-((2*S*,3*S*)-2-hydroxy-3-amino-3-carboxypropyl)ergothioneine].

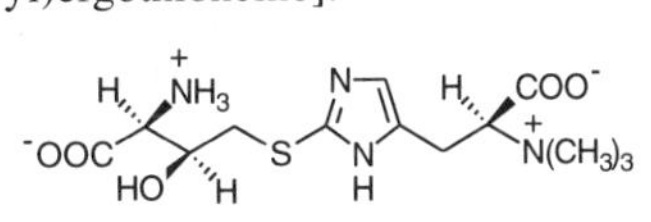

$C_{13}H_{22}N_4O_5S$, M_R 347.41, $[\alpha]_D^{23}$ +44.2° (H_2O). Amino acid betaine from the toadstool *Clitocybe acromelalga*.
Lit.: Phytochemistry **23**, 1003–1006 (1984) ▪ Tetrahedron Lett. **22**, 1617–1618 (1981). – *[CAS 78873-51-5]*

Clitidine.

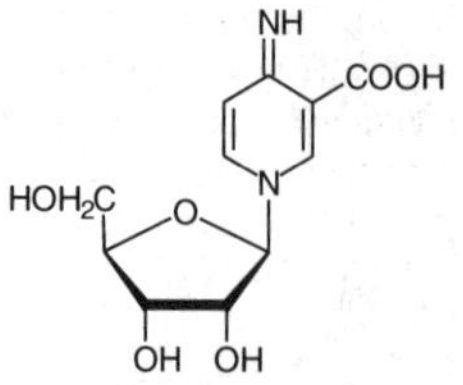

$C_{11}H_{14}N_2O_6$, M_R 270.24, cryst., mp. 189–191 °C (monohydrate), $[\alpha]_D$ −50.6° (H_2O). A highly toxic pyridine nucleoside from the Japanese toadstool Dokusasako (*Clitocybe acromelalga*, Basidiomycetes). The LD_{50} (mouse) is less than 50 µg/kg. C. acts as an NAD analogue. The corresponding nucleotide is also present in the fungus.
Lit.: Chem. Lett. **1977**, 1449 ▪ Heterocycles **47**, 661 (1998) (tautomerism) ▪ J. Chem. Soc., Perkin Trans. 1 **1976**, 2077 ▪ Phytochemistry **35**, 897 (1994) ▪ Tetrahedron **38**, 3281 (1982) ▪ Tetrahedron Lett. **1977**, 481. – *[CAS 63592-84-7]*

Clitocine.

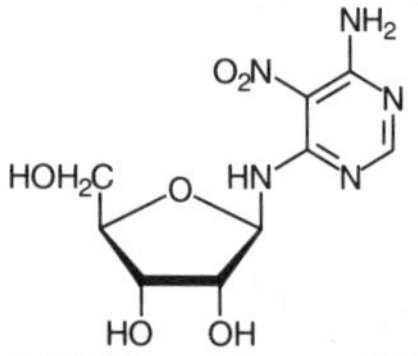

$C_9H_{13}N_5O_6$, M_R 287.23, cryst., mp. 228–231 °C. Insecticidal nucleoside from the tawny funnel cap (*Clitocybe inversa*, Basidiomycetes). The nitro group on the pyrimidine ring is an unusual feature; it is presumably formed by oxidative cleavage from the imidazole ring of adenine.
Lit.: Acta Cryst. C, Cryst. Struct. Commun. **44**, 1076 (1988) ▪ J. Antibiot. **41**, 1711 (1988) (review) ▪ J. Med. Chem. **31**, 786 (1988) ▪ Tetrahedron Lett. **27**, 4277 (1986). – *Synthesis:* J. Chem. Soc., Chem. Commun. **1988**, 195 ▪ Tetrahedron Lett. **31**, 279 (1990). – *[CAS 105798-74-1]*

Clove oils. C. are mainly produced by steam distillation of the various parts of the clove tree *Syzygium aromaticum* (Myrtaceae) predominantly cultivated in Madagascar and Indonesia. Clove flower oil has the highest organoleptic value, while clove leaf oil is the quantitatively most important oil. The world-wide annual production of C. amounts to 2000 to 3000 t.
1. ***Clove flower oil:*** Yellow to brownish, slightly viscous oil with a warm, sweet-spicy, mildly fresh-fruity odor and a warm-spicy, mildly burning taste. It is obtained from dried flower buds (cloves).
Composition [1]***:*** Main component, also principally responsible for the odor and taste, is **eugenol*. The high content (5–10%) of *eugenyl acetate* (see cinnamon leaf (bark) oil) distinguishes the flower oil from the oils obtained from other parts of the plant (see below). A further major constituent is **caryophyllene* (ca. 10%).
Use: In the perfume industry for sweet-spicy and flowery compositions; however, the main use is to improve the aromatic character of foods, clove flower oil is widely used for this purpose in, e.g., bakery products, sauces and spice mixtures. In addition to the oil the oleoresin, obtained by various extraction procedures, is also used for aromatic flavors. Clove flower oil is also used medically in carminatives and products for use in the mouth and throat.
2. ***Clove leaf oil:*** The crude oil is a dark brown to lilac brown, somewhat viscous liquid with a medical-phenolic, rough, burnt taste. For use in perfumes or flavors the crude oil is usually rectified to furnish a clear, yellow to light brown mobile liquid with a less burnt, sweet-spicy taste which is more dry-woody than the flower oil. It is produced from the leaves.
Composition [1]***:*** Like the flower oil, the leaf oil contains **eugenol* (ca. 85%) and **caryophyllene* (ca. 10% or more) as major components, but hardly any *eugenyl acetate*.
Use: Mainly in the perfume industry for herby-spicy compositions. The main portion of clove leaf oil is distilled to furnish pure eugenol which is also used as a component of perfumes or is transformed into other fragrance compounds (e. g., *isoeugenol). Caryophyllene obtained as a by-product is also employed as a precursor of fragrance substances.
3. ***Clove stem oil:*** Yellow to light brown oil with a sweet-spicy, somewhat woody odor, very similar to the flower oil but without the fresh-fruity note of the latter. It is obtained from dried flower stems.
Composition [1]***:*** The stem oil has a very similar composition to the flower oil but has appreciably less eugenyl acetate (mostly 5% or less). It is used similarly to the flower oil but gives products of lower quality.
Lit.: [1] Perfum. Flavor. **12** (4), 69 (1987); **13** (6), 57 (1988); **16** (4), 49 (1991); **19** (5), 92 (1994); Dev. Food Sci. **34**, 483–500 (1994).
gen.: Arctander, p. 179, 182, 184 ▪ Bauer et al. (2.), p. 150 ▪ ISO 3141 (1986), 3142 (1974), 3143 (1975). – *Toxicology:* Food Cosmet. Toxicol. **13**, 761, 765 (1975); **16**, 695 (1978). – *[HS 3301 29; CAS 8000-34-8 (1.); 8015-97-2 (2.); 8015-98-3 (3.)]*

Clubmosses (Lycopodiatae or Lycopodiophytina). A class of the *Pteridophyta* (*pteridophytes; ferns) with three recent orders: Lycopodiales (*lycopods*), Selaginellales (*selaginella*), and Isoetales (*quillworts*), with one family each. The exclusively herbaceous plants have small or narrow to scale-like leaves (=microphylla). They were particularly highly developed in the Carboniferous period (see pteridophytes; ferns (Filicatae)).
Constituents: Remarkably, various *Isoetes* species exhibit the Crassulacean acid metabolism [1]. Secondary plant substances appear to be represented exclusively by *flavones. The *lycopods* have similar flavonoid patterns to the previous group but differ through the occurrence of additional polyphenols, triterpenes with onocerane and serratane skeletons, and alkaloids (cf. Lycopodium alkaloids). In contrast to the quillworts and lycopods, the *selaginella* contain trehalose (2–17%) instead of saccharose as the main sugar. Many *Selaginella* species form the trisaccharide selaginose (2-α-glucoside of trehalose) which is unknown in other plants. Their flavonoid pattern is conspicuous by the occurrence of biflavones (*amentoflavone, *hinokiflavone, sotetsuflavone, and others).

Lit.: [1] Nature (London) **304**, 310 (1983); Oecologia **58**, 57 (1983).
gen.: Biochem. Syst. Ecol. **6**, 99 (1978) ▪ Hegnauer **I**, 223; **VII**, 403 ▪ Phytochemistry **4**, 57 (1965); **6**, 663 (1967); **19**, 803 (1980). – *[HS 1211 90]*

Clupa(no)donic acid [(*all-Z*)-4,8,12,15,19-docosapentaenoic acid].

$C_{22}H_{34}O_2$, M_R 330.51. Pale yellow liquid, D. 0.937, mp. −78 °C, bp. 331 °C. C. is a docosapentaenoic acid (DPA) of the ω3-series with an unusual distribution of the C=C double bonds; it was first isolated from the Japanese sardine (*Clupanodon melanostica*), the oil of which contains ca. 8% C. The DPA of the ω3-series normally formed in animals is (*all-Z*)-7,10,13,16,19-DPA built from *eicosapentaenoic acid by chain lengthening. The corresponding fatty acid of the ω6-series is (*all-Z*)-4,7,10,13,16-DPA, formed as metabolic product from *linoleic acid. Many fish and fish liver oils contain glycerol esters of C. in varying amounts, e. g. sardine oil 2.8%, mackerel oil 0.6%, anchovy oil 1.2%, and herring oil 0.4% as well as cod-liver oil 10%.
Lit.: Beilstein E III **2**, 1528 ▪ Biochim. Biophys. Acta **409**, 304 (1975) ▪ Kinsella, Seafoods in Human Health and Desease, p. 241–260, New York: Dekker 1987. – *[CAS 2548-85-8 (C.); 2234-74-4 ((all-Z)-7,10,13,16,19-DPA); 2313-14-6 ((all-Z)-4,7,10,13,16-DPA)]*

Cnicin.

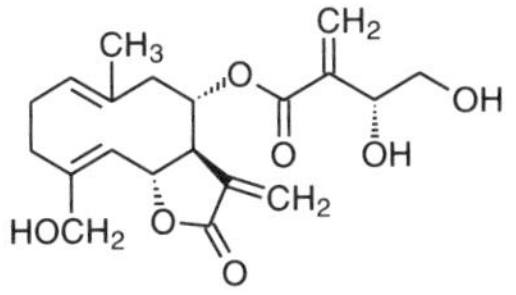

$C_{20}H_{26}O_7$, M_R 378.42, cryst., mp. 143 °C, $[\alpha]_D^{20}$ +158° (C_2H_5OH); soluble in ethanol, poorly soluble in water. Bitter substance of common benedict (*Cnicus benedictus*, Asteraceae). C. is obtained by ethanol extraction of the dried and powdered aerial parts of the plant. It has antibiotic activity against trichomonads. It has the *germacrane type structure.
Lit.: Beilstein E V **19/10**, 508 ▪ Justus Liebigs Ann. Chem. **1997**, 527 ▪ Merck-Index (12.), No. 2486 ▪ Nat. Prod. Lett. **5**, 47 (1994). – *[CAS 24394-09-0]*

Coagulin see snake venoms.

Cobalamins. The semi-systematic name (abbr.: Cbl) recommended by IUPAC/IUB for derivatives of *vitamin B_{12} [see also corrin(s)].

Cobratoxins (cobrotoxins, cobra venoms). The cobras or poisonous vipers (Elapidae) are the most widely distributed family of poisonous snakes and occur in all parts of the world except Europe. Particularly dangerous species are the African spitting cobra (*Naja nigricollis*), ringed cobra (*Haemachatus haemachatus*), black mamba (*Dendroaspis polylepis*), the Asian cobra (*Naja naja*), king cobra (*Naja bungarus=Ophiophagus hannah*) as well as various kraits (*Bungarus fasciatus* and *B. caeruleus*). The venoms consist of polypeptides with 60–74 amino acids and 4–5 disulfide bridges. They have neuro- and cardiotoxic activities and comprise up to 70% of the content of the venom. The raw venoms also contain enzymes, especially phospholipases, endopeptidases, exopeptidases, and proteinases that are also highly poisonous. Typical symptoms for an Asian cobra bite are at first severe pain, followed by swelling, blood blisters, and necrosis (onset of digestion). The wounds heal very poorly, the healing process requires several months and often surgical treatment. General symptoms are vomiting, unconsciousness, disorders of vision and speech, convulsions and paralysis; see also snake venoms.
Lit.: J. Protein Chem. **6**, 365–373 (1987); **8**, 575–583 (1989) ▪ Tu, Handbook of Natural Toxins, Vol. 5, p. 3–470, New York: M. Dekker 1991.

Cobyrinic acid see vitamin B_{12}.

Coca. Dried leaves of the 1–2 m high coca shrub (*Erythroxylum coca*, Erythroxylaceae) endemic in Bolivia, Peru, Columbia, Brazil, and Java, which contains *tropane alkaloids (*cocaine and other ecgonine derivatives) in addition to *hygrine and *cuscohygrine. South American shrubs contain ca. 0.5–1% alkaloids, mainly cocaine, while species cultivated in Java contain mainly cinnamoylecgonine derivatives (1.5–2.5%). For use and cocaine addiction, see cocaine.
Lit.: Hager (5.) **5**, 88 ff. ▪ Wolters, Drogen Pfeilgift u. Indianermedizin, p. 82–99, Greifenberg: Urs Freud 1994. – *[HS 1211 90]*

Coca acids see truxillic and truxinic acids.

Coca alkaloids see tropane alkaloids.

Cocaine [methyl (1*R*)-3-*exo*-benzoyloxy-2-*exo*-tropanecarboxylate, benzoylecgonine methyl ester].

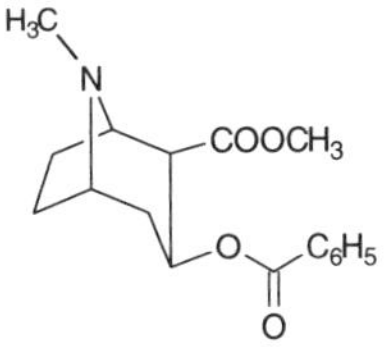

$C_{17}H_{21}NO_4$, M_R 303.36, mp. 98 °C, $[\alpha]_D$ −30° (CH_3OH), racemate: mp. 79–80 °C, bitter tasting monoclinic crystals, readily soluble in organic solvents, moderately soluble in water with alkaline reaction. C. forms water-soluble, crystalline salts [hydrochloride: $C_{17}H_{22}ClNO_4$, M_R 339.82, mp. 195 °C, $[\alpha]_D^{20}$ −71.9° (water), slight bitter taste, very well soluble in water, well soluble in alcohol and chloroform, insoluble in ether, sensitive to heat and light].
The poisonous C.[1] [LD_{50} (rat i. v.) 17.5 mg/kg] is the main alkaloid of the *coca shrub (*Erythroxylum coca*) (see also tropane alkaloids). C. has a local anesthetic effect and was first used in surgery in 1884; on account of its disadvantages – easy decomposition during sterilization and addictive effects – it is merely used clinically today. The habitual chewing of *coca leaves mixed with lime as is still practiced especially by inhabitants of the Andean mountains in South America is known as *Cocaism*. On chewing, the cocaine extracted into the weakly alkaline medium is partly saponified to the non-addictive *ecgonine. A characteristic feature of C. is the performance increasing and appetite inhibiting effect. The addictive consumption of coca paste or of pure C. by inhalation, smoking, or parenteral administration (injection) is known as *co-*

cainism (cocaine addiction). Frequent use leads to a psychic dependence but a physical habituation does not occur so that discontinuation of the drug use does not lead to withdrawal symptoms. On consumption C. effects a short-term hyperstimulation of the sympathetic nervous system characterized by euphoria, feeling of power and, especially, liveliness. In USA cocaine hydrochloride (street name: coke, snow) and the free base C. (street name: crack)[2] are the most frequently consumed, illegal drugs. New developments in the fight against cocaine addiction are the development of catalytic antibodies against C. as well as a vaccine based on C.-antibodies[3] and the discovery of 4-iodo-cocaine as the currently most potent C.-antagonist[4]. New investigations have revealed a connection between cocaine addiction and the increased liberation of certain endogenous analgesics (dynorphin, endorphins) in the brain[5]. Cocaine abuse alters gene expression[6]. C. also has an appreciable activity against insect larvae[7].

Lit.: [1]J. Med. Chem. **35**, 2178 (1992). [2]Chem. Eng. News (3.11.) **1997**, 9. [3]J. Am. Chem. Soc. **118**, 5881 (1996); Nature (London) **378**, 725 (1995). [4]Chem. Ind. (London) **1995**, 295. [5]Modern Drug Discovery (11./12.) **1999**, 24–31; Mol. Brain Res. **38**, 71 (1996). [6]Chem. Eng. News (2.12.) **1996**, 19. [7]Proc. Nat. Acad. Sci. USA **90**, 9645 (1993).

gen.: Beilstein EV **22/5**, 54f. ▪ Hager (5.) **3**, 333ff.; **7**, 1060–1070 ▪ Kirk-Othmer (4.) **1**, 1048f.; **9**, 723; **11**, 918 ▪ Manske **44**, 1–114 ▪ Ullmann (5.) **A1**, 360f. ▪ Wolters, Drogen Pfeilgift u. Indianermedizin, p. 82–99, Greifenberg: Urs Freud 1994. – *Biosynthesis:* Manske **33**, 50–51 ▪ Planta Med. **56**, 339–352 (1990). – *Pharmacology:* Manske **44**, 98–99 ▪ Nat. Prod. Rep. **11**, 446f. (1994). – *Reviews:* Florey **15**, 151–231 (1986) – *Synthesis:* J. Chem. Soc., Chem. Commun. **1985**, 233 ▪ J. Org. Chem. **63**, 4069–4078 (1998) (EPC synthesis). – *[HS 2939 90; CAS 50-36-2 (C.); 21206-60-0 (racemate)]*

Coccinelline [(3a*β*,6a*α*,9a*β*)-dodecahydro-2*α*-methylpyrido[2,1,6-*de*]quinolizine *Nβ*-oxide].

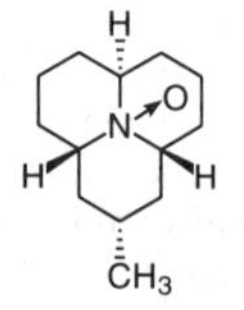

Coccinelline

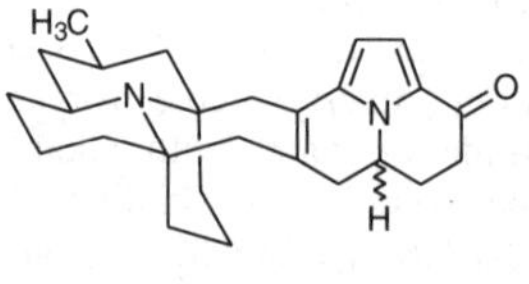

Chilocorine A

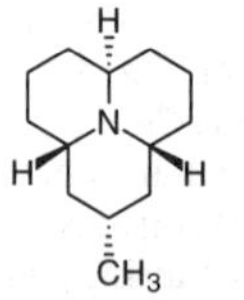

Precoccinelline

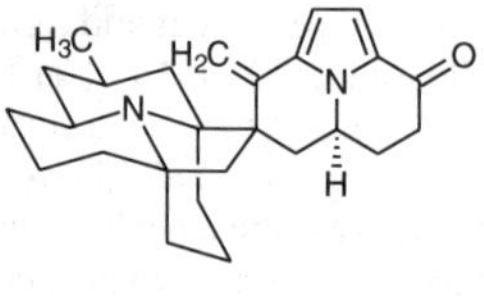

Chilocorine B

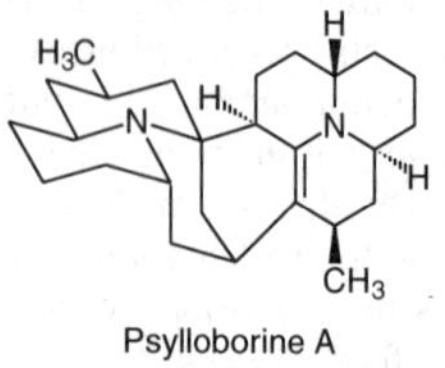

Psylloborine A

$C_{13}H_{23}NO$, M_R 209.33, cryst., mp. 204 °C. Defence alkaloid (*allomone) of the ladybird beetle (Coccinellidae). C. possesses five stereocenters. Precursor of C. in the ladybird is *precoccinelline* [(3a*β*,6a*α*,9a*β*)-dodecahydro-2*α*-methylpyrido[2,1,6-*de*]quinolizine, $C_{13}H_{23}N$, M_R 193.33] of which diastereomers have also been identified. Polycyclic "dimers" of C. are also known, e.g., *chilocorine A* and *chilocorine B* ($C_{26}H_{34}N_2O$, M_R 390.57), and dimers with a new heptacyclic ring system: e.g. *psylloborine A*[1]. Like *epilachnene also found in the Coccinellidae, the whole group of alkaloids is probably formed from unsaturated fatty acids.

Lit.: [1]Tetrahedron **54**, 12243 (1998).

gen.: Beilstein EV **20/5**, 155 ▪ Chem. Rev. **96**, 1005–1122 (1996) (activity) ▪ Heterocycles **7**, 685–707 (1977) ▪ Proc. Natl. Acad. Sci. USA **91**, 12790–12793 (1994) ▪ Tetrahedron **49**, 9333–9342 (1993); **51**, 8711–8718 (1995). – *[CAS 34290-97-6 (C.); 38211-56-2 (pre-C.)]*

Cocculidine, Cocculine see Erythrina alkaloids and picrotoxin.

Cochineal. Dye mixture from the dried, female scale insect (*Dactylopius coccus*, Coccidae, Homoptera). From 1 kg of the insect ca. 50 g C. are obtained in the form of bright red lumps which can be ground; partly soluble in hot water, readily soluble in alkaline solution. The main components are *carminic acid and its shining red mordants with Al and Ca.

Use: C. preparations are used in histology for staining. In addition, C. is used to color foods as well as pharmaceutical and cosmetic products, inks, and artist's paints. C. was introduced to Europe from Central America in the 16th century as a dye for textiles and carpets.

Lit.: Endeavour **2**, 85–92 (1978) ▪ Kirk-Othmer (4.) **2**, 801; **6**, 895, 930; **7**, 587; **8**, 787; **10**, 874; **11**, 812 ▪ Schweppe et al., in Feller (ed.), Artists Pigments, vol. 1, p. 255–283, Cambridge: University Press 1986 ▪ Ullmann (5.) **A 11**, 572; **A 18**, 211. – *[HS 3203 00; CAS 1260-17-9]*

Cochineals see scales.

Cochliobolin A see ophiobolins.

Cockroaches (Blattaria). An evolutionarily very old family of insects belonging to the order Blattaria. The insects are mostly active at night and are characterized by a flattened body, long, fiber-like antennae, chewing mouthparts, and typical legs. Many species of C. are pests in food storage and carriers of various diseases such as, e.g. salmonella, intestinal bacteria, pyogenic bacteria, worm eggs, etc. In Central Europe the German C. (*Blattella germanica*) is widely distributed together with the Oriental C. or kitchen C. (*Blatta orientalis*).

Derivatives of the sesquiterpene germacrene have been identified as the sexual pheromone of females of the American C. *Periplaneta americana* (see also periplanones). The structures of the *pheromones of *Cryptocercus punctulatus* have been elucidated: *(4R,5R,6S,7E,9E)-4,6,8-trimethylundeca-7,9-dien-5-ol* ($C_{14}H_{26}O$, M_R 210.36)[1] and *5-[(2R,4R)-2,4-dimethylheptyl]-3-methyl-2H-pyran-2-one* (supella pyrone, $C_{15}H_{24}O_2$, M_R 236.35), also identified as the sexual pheromone of females of *Supella longipalpa*[2]; they are probably formed from propionate units and are closely related to pheromones such as *invictolide, *lardolure, *matsuone, *serricorole and *stegobinone.

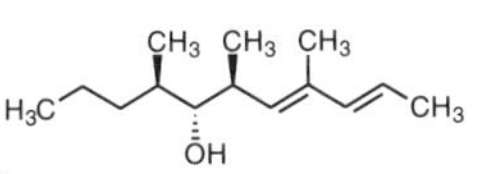

(4R,5R,6S,7E,9E)-4,6,8-Trimethyl-undeca-7,9-dien-5-ol (**1**)

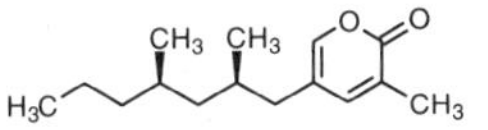

Supella pyrone (**2**)

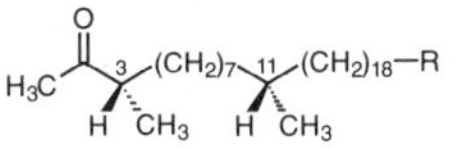

R = H : (3S,11S)-3,11-Dimethyl-2-nonacosanone (**3**)
R = OH : (3S,11S)-29-Hydroxy-3,11-dimethyl-2-nonacosanone (**4**)

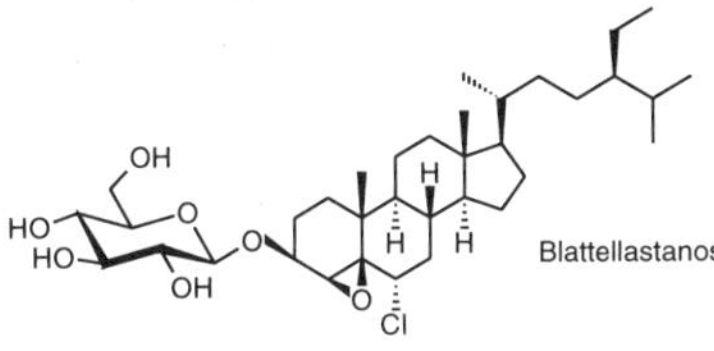

Blattellastanoside A (**5**)

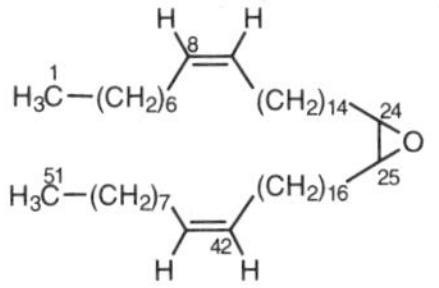

(8Z,42Z)-24,25-Epoxy-8,42-henpentacontadiene (**6**)

Pheromones of *Blattella germanica* are also known: *(3S,11S)-3,11-dimethyl-2-nonacosanone* ($C_{31}H_{62}O$, M_R 450.83) and *(3S,11S)-29-hydroxy-3,11-dimethyl-2-nonacosanone* ($C_{31}H_{62}O_2$, M_R 466.83) are produced by the females[3]. Interestingly the non-natural stereoisomers exhibit the same biological activities[4]. Similar branched ketones have been found in desert locusts. The chlorine-containing steroid glycosides *blattellastanoside A* ($C_{35}H_{59}ClO_7$, M_R 627.30) and *blattellastanoside B* ($C_{35}H_{61}ClO_7$, M_R 629.32), in which the epoxide groups are reduced to 5-hydroxy functions have been reported as aggregation pheromones[5]. *(8Z,42Z)-24,25-Epoxy-8,42-henpentacontadiene* ($C_{51}H_{98}O$, M_R 727.34) has been identified as a nymph recognition signal of *Nauphoeta cinerea*[6]. Some C. species use hydrocarbons as pheromones[7].

Lit.: [1]Justus Liebigs Ann. Chem. **1990**, 2149–1255. [2]Proc. Natl. Acad. Sci. (USA) **92**, 1033ff. (1995). [3]J. Chem. Ecol. **2**, 449–455 (1976); **5**, 289–297 (1979). [4]Mem. Coll. Agric. Kyoto Univ. **1983**, No. 122, 1. [5]Justus Liebigs Ann. Chem. **1993**, 665–670; Tetrahedron Lett. **34**, 6059 (1993). [6]Justus Liebigs Ann. Chem. **1994**, 695–700. [7]Justus Liebigs Ann. Chem. **1997**, 815.

gen.: Synthesis: ApSimon **4**, 135 ff.; **9**, 185 ff. ▪ Tetrahedron **46**, 4473–4486 (1990) ▪ Tetrahedron Lett. **39**, 3541 (1998) (blattelastanoside B). – *[CAS 135557-66-3 (1); 69274-90-4 (3); 60789-53-9 (4); 149864-63-1 (5); 151397-98-7 (blattellastanoside B)]*

Coclaurine see benzyl(tetrahydro)isoquinoline alkaloids.

Cocoa flavor. Some of the over 500 known volatile components of C. f. are already present in raw cocoa, but most are formed after drying and roasting (at 110–130 °C), mainly by Maillard or Strecker reactions from amino acids, peptides, and sugars resulting from anaerobic fermentation. C. f. is not determined by one *impact compound but is rather a composition of various aromas: caramel-like (*maltol, *Furaneol®, and *2-hydroxy-3-methyl-2-cyclopenten-1-one), flowery (*linalool, *2-phenylethanol, *phenylacetaldehyde), cocoa/chocolate-like [2- and 3-methylbutanals (see alkanals), *5-methyl-2-phenyl-2-hexenal* ($C_{13}H_{16}O$, M_R 188.27)], roasted [*pyrazines, *2-acetylpyridine* (C_7H_7NO, M_R 121.14)], smoky-clove-like (including *guaiacol, *eugenol), and nutty [e.g., *4-methyl-5-vinylthiazole* (C_6H_7NS, M_R 125.19)]. The bitter taste is mostly due to *theobromine and diketopiperazines. The latter are formed by thermal degradation of proteins.

Lit.: Food Rev. Int. **5**, 317–414 (1989) ▪ Maarse, p. 617–669 ▪ TNO list (6.), Suppl. 5, p. 137–151. – *[CAS 21834-92-4 (5-methyl-2-phenyl-2-hexenal); 1122-62-9 (2-acetylpyridine); 1759-28-0 (4-methyl-5-vinylthiazole)]*

Coconut flavor. Key compounds in C. f. are the C_8–C_{12} and C_{14} 5-*alkanolides, with 5-decanolide as the main component. The 4-nonanolide (see alkanolides) used in commercial C. f. cannot be detected in natural C. f. The sweet flavor of roasted coconuts is due to the additional formation of *maltol and *2-hydroxy-3-methyl-2-cyclopenten-1-one.

Lit.: Maarse, p. 675 f. ▪ TNO list (6.), p. 754 f.

Cocsoline see bisbenzylisoquinoline alkaloids.

Codeine (methylmorphine, morphine 3-methyl ether). Data and formula, see morphinan alkaloids. With acids C. forms salts that are readily soluble in water, of these codeine phosphate is most frequently used in medicine. C. belongs to the opium alkaloids (Greek Kodeia = poppyhead) and is present to 0.3–3% in poppy juice. On account of the soothing effects on the coughing center C. salts are used in antitussive agents. In contrast to morphine, C. has practically no analgesic effect. LD_{50} (rat s.c.) 500 mg/kg, (mouse p.o.) 250 mg/kg, (mouse i.p.) 200 mg/kg, (rabbit i.v.) 34 mg/kg.

Lit.: Beilstein E V **27/9**, 319 ff. ▪ Hager (5.) **7**, 1067–1073 ▪ J. Org. Chem. **64**, 7871–7884 (1999) (synthesis) ▪ Manske **2**, 171–189; **6**, 219–245; **45**, 128–232 ▪ Martindale (30.), p. 1069 ▪ Sax (8.), CNF 500 ▪ Tetrahedron **39**, 2393 (1983) ▪ see also morphinan alkaloids. – *[HS 2939 10]*

Codeinone see morphinan alkaloids.

Codlemone [(8E,10E)-8,10-dodecadien-1-ol].

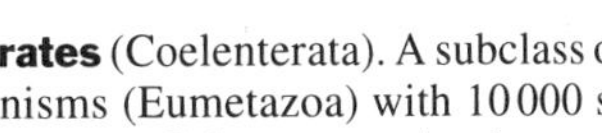

$C_{12}H_{22}O$, M_R 182.31, cryst., mp. 29–30 °C, bp. (1.33 Pa) 80–85 °C. Sexual pheromone of the apple or codling moth *Laspeyresia pomonella* (pest in apple orchards), see also farnesene.

Lit.: J. Org. Chem. **47**, 4801 (1982); **51**, 4934 (1986) ▪ Tetrahedron Lett. **24**, 1247 (1983); **33**, 3643 (1992). – *[CAS 33956-49-9]*

Coelenterates (Coelenterata). A subclass of multicellular organisms (Eumetazoa) with 10000 species living in the oceans. C. have very simple structures. The body usually consists of only 2 epithelia (epidermis and flagellated gastrodermis, formed from ectodermis or endodermis). In between there is a supporting substance (mesogloea) in which cells are present.

1) Strain: Cnidaria.

a) Class: Hydrazoa; Cnidaria with simple, firmly anchored small polyps or as free-swimming hydromedusae (e.g., *Aequorea aequorea*, blue phosphorescent,

see also aequorin); siphonopheres or tubular jellyfish (e. g., *Physalia physalis*, "Portugese man-of-war", with up to 30 m long tentacles, highly toxic, deaths have been reported from the Western Atlantic);
b) Class: Cubozoa (stinging jellyfish, e. g., *Carybdea marsupialis*, causes severe burns);
c) Class: Scyphozoa (umbrella jellyfish, e. g., *Aurelia aurita*, *Pelagia noctiluca*, lamp jellyfish, medusa glowing pink-violet);
d) Class: Anthozoa [animal flowers or coral polyps, e. g., *Cerianthus membranaceus*, cylinder anemone, *Anemonia sulcata*, *Palythoa toxica* (produces the most potent non-proteinogenic marine toxin *palytoxin), *Alcyonium palmatum*, dead-man's-fingers, *Madrepora oculata*, white coral, *Corallium rubrum*, precious coral; sea feather: *Pennatula phosphorea*, salmon-pink, phosphorescent, *Lituaria australasia*, see lituarines].
2) Strain: Ctenophora (comb jellyfish, ctenophores). Mostly pelagarous living, very fine and fragile metazoa, sometimes violet shiny, on stimulation blue-green to blue fluorescent, e. g., *Cestus veneris*, girdle of Venus.
Lit.: Meier & White, Handbook of Clinical Toxicology of Animal Venoms and Poisons, Boca Raton: CRC Press 1995.

Coelenterazine. $C_{26}H_{21}N_3O_3$, M_R 423.47. Prosthetic group with the imidazopyrazine structure of the photoprotein *aequorin from bioluminescent jelly fish species of the genus *Aequorea*. A blue luminescence is observed in the presence of traces of Ca^{2+} ions. Natural aequorin does not need oxygen for bioluminescence, thus it is assumed to exist in the form of a peroxide.

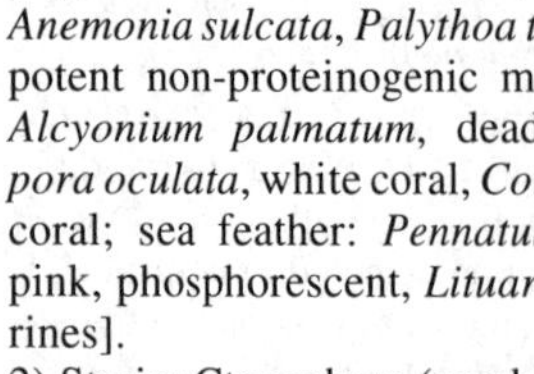

Coelenterazine

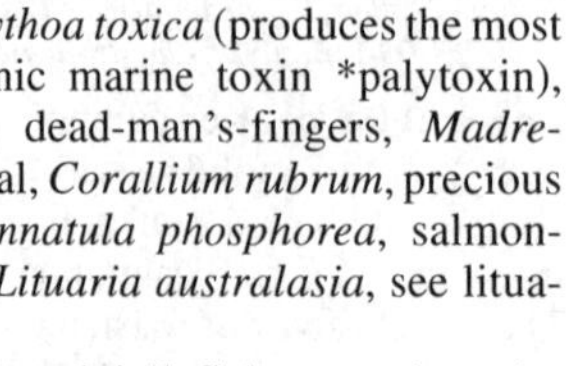

Coelenteramide

This peroxide decomposes in the presence of Ca^{2+} ions with emission of light to CO_2 and *coelenteramide*, $C_{25}H_{21}N_3O_3$, M_R 411.46; for mechanism, see *Lit.*[1]. Functioning aequorin can be regenerated *in vitro* from the apoprotein, C., and oxygen[2]. Synthetic analogues have been prepared by elaborated synthesis of C. and studied for their luminescence mechanism[3].
Lit.: [1] Chem. Lett. **1998**, 95 f.; Proc. Natl. Acad. Sci. USA **75**, 2611 (1978). [2] Nature (London) **256**, 236 (1975); Tetrahedron Lett. **33**, 1303 (1992). [3] J. Chem. Soc., Chem. Commun. **1997**, 323; Tetrahedron **53**, 12903 (1997).
gen.: Anal. Biochem. **240**, 308 (1996) (synthesis) ▪ Bull. Chem. Soc. Jpn. **63**, 3132 (1990) ▪ Chem. Lett. **1975**, 141; **1979**, 249 ▪ Scheuer I **3**, 179–222. – *[CAS 55779-48-1 (C.); 50611-86-4 (coelenteramide)]*

Coenzyme B$_{12}$. The physiologically active form of *vitamin B$_{12}$.

Coenzyme F 430 see factor F 430.

Coenzyme Q see ubiquinones.

Coenzymes (obsolete: coferments). Name for low-molecular-weight organic components of enzymes that do not belong to the regular peptide chains but participate in the catalysis, especially of group transfer and redox reactions. They are also designated as *cofactors* although this general term also includes metal ions (e. g., Zn^{2+} in the case of carboxypeptidase A). C. can also be bound to proteins non-covalently as *cosubstrates* or can be bound very firmly or even covalently as a *prosthetic group*. Many C. or their precursors are vitamins.
Lit.: Stryer 1995, p. 451–455 ▪ Voet & Voet, Biochemistry, 2nd ed., New York: Wiley 1995.

Cofactors see coenzymes.

Coffee flavor. Green coffee beans contain only a few of the around 800 known volatile components of coffee flavor, e. g., alkylmethoxypyrazines (see pyrazines). Most of the *flavor compounds are first formed on roasting, especially from *sucrose, free amino acids (e. g., *cysteine, *L-serine, *L-threonine), and *chlorogenic acid. Furan derivatives and *pyrazines dominate qualitatively and quantitatively, but larger numbers of ketones and diketones, pyrroles, pyridines, thiophenes, thiazoles, oxazoles, and phenol are also present. Particularly dominating aroma substances include[1,2]: the sulfur compounds *2-furylmethanethiol, *3-mercapto-3-methylbutyl formate* ($C_6H_{12}O_2S$, M_R 148.22), *2-methyl-3-furanthiol, *methional, and *kahweofuran*[3] (C_7H_8OS, M_R 140.20), 2-isopropyl-3-methoxy-, 2-isobutyl-3-methoxy-, 2-ethyl-3,5-dimethyl-, and 2,3-diethyl-5-methylpyrazines, β-*damascenone, *hydroxyfuranones (including *Furaneol®, sotolone, abhexone), *maltol, and *2-hydroxy-3-methyl-2-cyclopenten-1-one, as well as *4-vinylguaiacol and *4-ethylguaiacol* ($C_9H_{12}O_2$, M_R 152.19). Also of organoleptic significance are *cafestol* [$C_{20}H_{28}O_3$, M_R 316.44, cryst., mp. 160–162 °C, $[\alpha]_D$ –101° ($CHCl_3$)] and 1,2-dehydrocafestol (kahweol)[4].

Kahweofuran

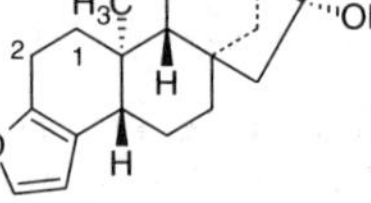

3-Mercapto-3-methylbutyl formate

4-Ethylguaiacol

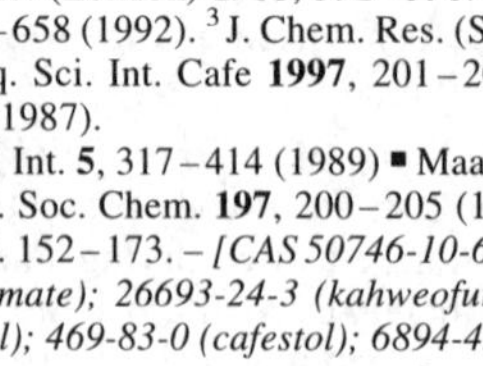

Cafestol

Lit.: [1] Chem. Ind. (London) **1988**, 592–596. [2] J. Agric. Food Chem. **40**, 655–658 (1992). [3] J. Chem. Res. (S) **1998**, 74 (synthesis). [4] Colloq. Sci. Int. Cafe **1997**, 201–204; Tetrahedron Lett. **28**, 5403 (1987).
gen.: Food Rev. Int. **5**, 317–414 (1989) ▪ Maarse, p. 617–669 ▪ Spec. Publ. R. Soc. Chem. **197**, 200–205 (1996) ▪ TNO list (6.), Suppl. 5, p. 152–173. – *[CAS 50746-10-6 (3-mercapto-3-methylbutyl formate); 26693-24-3 (kahweofuran); 2785-89-9 (4-ethylguaiacol); 469-83-0 (cafestol); 6894-43-5 (kahweol)]*

Cognac lactone.

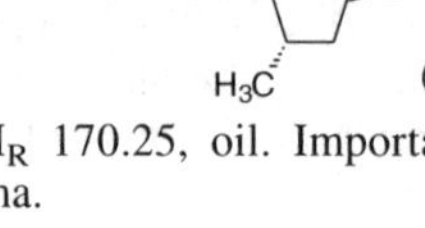

(+)-C.

$C_{10}H_{18}O_2$, M_R 170.25, oil. Important component of cognac aroma.

Lit.: Chem. Pharm. Bull. **40**, 2579 (1992) ▪ Heterocycles **36**, 1017 (1993) ▪ J. Heterocycl. Chem. **35**, 485 (1998) (synthesis). – *[CAS 114485-30-2]*

Colchicine see phenethyltetrahydroisoquinoline alkaloids.

Coleons.

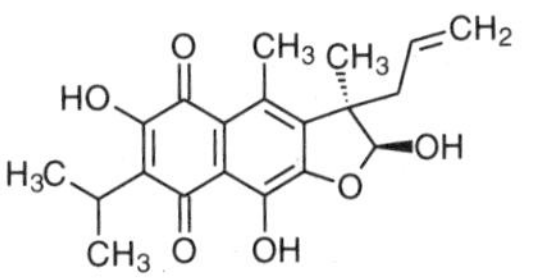

Coleon A
(absolute configuration)

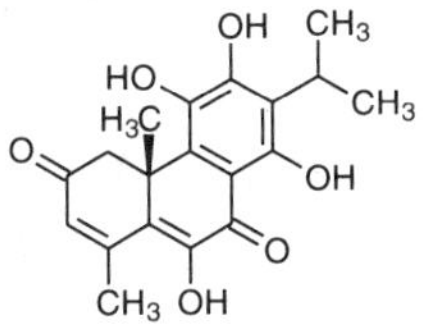

Coleon B

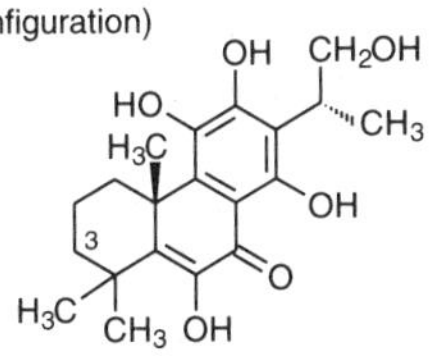

Coleon C

Group of quinonoid diterpenoids (C. A–O) from leaf glands and leaves of African *Coleus* species (Lamiaceae). C. are mostly tricyclic (hydrophenanthrene type) crystalline compounds with yellow to red colors. To date about 20 different compounds (coleons A–Z) have been isolated and investigated, e.g., *C. A* [$C_{20}H_{22}O_6$, M_R 358.39, mp. 136–136.5 °C, $[\alpha]_D^{23}$ +100° (C_2H_5OH)], *C. B* [$C_{19}H_{20}O_6$, M_R 344.36, $[\alpha]_D$ +130° (C_2H_5OH)] deep orange pigment from leaves of *Coleus igniarius*, and *C. C* [$C_{20}H_{26}O_6$, M_R 362.42, mp. 210 °C, $[\alpha]_D^{25}$ +27° ($CHCl_3$)] from *C. aquaticus*.

Lit.: Bull. Chem. Soc. Jpn. **55**, 1168 (1982) ▪ Chem. Pharm. Bull. **39**, 3041 (1991) ▪ Contact Dermatitis **19**, 217 f. (1988) ▪ Helv. Chim. Acta **66**, 429 (1983); **67**, 201 (1984); **71**, 577–587, 1638 (1988) (synthesis) ▪ Phytochemistry **22**, 2005 (1983); **38**, 195 (1995) (absolute configuration). – *[CAS 1984-44-7 (C. A); 20710-79-6 (C. B); 35298-85-2 (C. C)]*

Colforsin see forskolin.

Colistins see polymyxins.

Collinomycin see rubromycins.

Collismycin A (B) see caerulomycin A.

Collybolide.

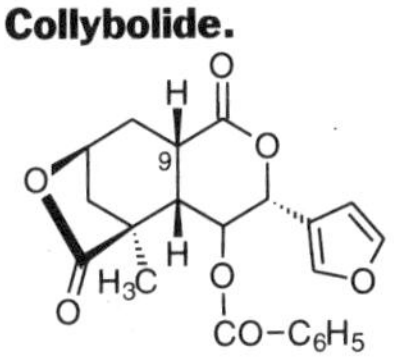

Collybolide

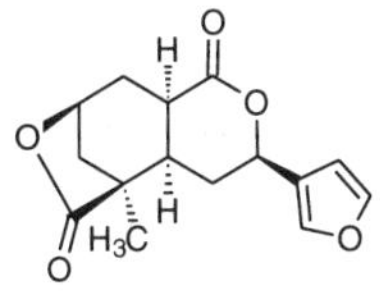

Deoxycollybolidol

(relative configuration)

$C_{22}H_{20}O_7$, M_R 396.40, cryst., mp. 210 °C, $[\alpha]_D$ +17° ($CHCl_3$). Unusual sesquiterpenoid benzoate from fruit bodies of the bitter agaric *Collybia maculata* (basidiomycetes). Occurs together with the 9-epimer, *isocollybolide* {cryst., mp. 193 °C, $[\alpha]_D$ –17° ($CHCl_3$)}. *Deoxycollybolidol* {$C_{15}H_{16}O_5$, M_R 276.29, cryst., mp. 189–190 °C, $[\alpha]_D$ +21° ($CHCl_3$)} is isolated from *C. peronata*.

Lit.: Acta Crystallogr., Sect. C **28**, 331 (1972) ▪ Beilstein E V **19/10**, 525 f. ▪ Phytochemistry **25**, 2661 (1986) ▪ Tetrahedron **30**, 1327 (1974). – *[CAS 33340-30-6 (C.); 106794-13-2 (deoxy-C.); 31199-75-4 (iso-C.)]*

Color reactions of mushrooms. The fruit bodies of macromycetes often show characteristic color changes at injury or bruising sites; in addition mycologists have described numerous chemical spot reactions. The responsible chromogens are summarized in the table.

Lit.: Zechmeister **51**, 249 ff.

Table: Color reactions of mushrooms.

mushroom	cause/reagent	color change	chromogene
Boletes	bruising	yellow → deep blue	*Variegatic acid, *Xerocomic acid
Gyroporus cyanescens, *Chamonixia*, *Gyrodon lividus*	bruising	pale yellow → azure	*Gyrocyanin, *Chamonixin
Paxillus involutus	bruising	colorless → brown	*Involutin
Strobilomyces floccopus, *Hygrocybe conica* and other blackening mushrooms	bruising	colorless → red → black	DOPA (*3,4-Dihydroxy-L-phenylalanine)
Russula nigricans	bruising	colorless → orange → black	L-*Tyrosine
Gomphidius	bruising	colorless → red → black	*Benzene-1,2,4-triol
Yellow Stainer (*Agaricus xanthoderma*)	bruising	colorless → chrome yellow	*Xanthodermin, *Agaricone
Inocybe, *Psilocybe*, *Panaeolus*	bruising	colorless → blue	*Psilocine
Hapalopilus nidulans	alkali	brown → purple violet	*Polyporic acid
Paxillus atrotomentosus	alkali	brown → purple violet	*Atromentin
Anthracophyllum	alkali	brown → green	*Cycloleucomelone
Thelephoraceae	alkali	→ blue	*Thelephoric acid
Cortinarius, *Dermocybe*	alkali	yellow, orange → red, violet	*Preanthraquinones, *Dermocybe pigments
Fomes fomentarius	alkali	brown → bloodred	*Fomentariol
Albatrellus ovinus	alkali	colorless → yellow	*Scutigeral
Suillus bovinus, *Chroogomphus*	alkali	orange → red	*Boviquinones
Giant club, coral fungi, *Gomphus*	$FeCl_3$	colorless → green	*Pistillarin
Lyophyllum connatum	$FeCl_3$	colorless → purple blue	*Connatin
Inonotus hispidus, *Gymnopilus*	$FeCl_3$	yellow → green	*Hispidin
Gloeophyllum odoratum, *G. sepiarium*	$FeCl_3$	yellowish brown → green	*Trametin
Numerous yellowing *Agaricus* species	aniline/HNO_3	orange red (Schäffer's cross reaction)	Agaricus hydrazones

Columbamine see protoberberine alkaloids.

Columbin.

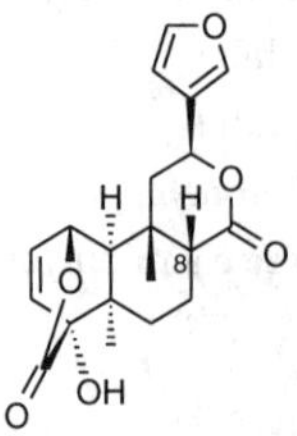

$C_{20}H_{22}O_6$, M_R 358.39, very bitter needles, insoluble in water, soluble in acetone, mp. 195–196 °C, $[\alpha]_D$ +52.7° (pyridine). Diterpene with *clerodane structure from *Colombo* root (*Jateorhiza palmata*), and *Tinospora cordifolia*.

Lit.: Beilstein EV **19/10**, 546 ▪ Hager (5.) **5**, 557 ▪ Indian J. Chem., Sect. B **29**, 521 (1990); **35**, 630 (1996) ▪ Karrer, No. 1146a ▪ Merck-Index (12.), No. 2558. – *[CAS 546-97-4]*

Combinatorial biosynthesis see hybrid antibiotics.

Combretastatins. Antineoplastic compounds from the South African tree *Combretum caffrum* (Combretaceae) which has many uses in traditional medicine. *C. caffrum* contains stilbenes, bibenzyls, phenanthrenes, and diaryl ethers such as the cytotoxic C. D2 which are grouped together under the name C.

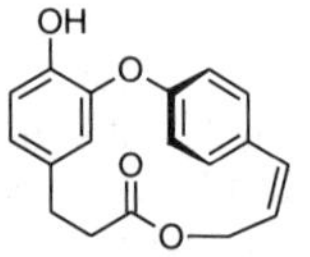

C.D2

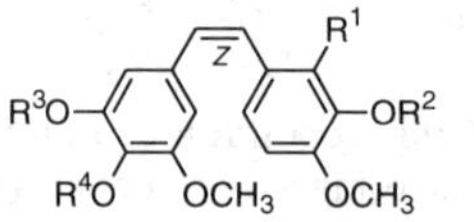

	R^1	R^2	R^3	R^4
C.A1	OH	H	CH_3	CH_3
C.A2	H	H	—CH_2—	
C.A4	H	H	CH_3	CH_3

Table: Data of Combretastatins.

C.	molecular formula	M_R	mp. [°C]	CAS
A1	$C_{18}H_{20}O_6$	332.35	114	109971-63-3
A2	$C_{17}H_{16}O_5$	300.31	oil	111394-44-6
A4	$C_{18}H_{20}O_5$	316.35	oil	117048-59-6
D2	$C_{18}H_{16}O_4$	296.32	150	126191-23-9

The stilbenes, especially C. A4, inhibit the polymerization of tubulin and the binding of colchicin to tubulin. C. A1, A2, and A4 have good activities against human tumor cells[1]. The semisynthetic C. A4 is in phase II clinical trials against advanced solid tumors.

Lit.: [1] J. Med. Chem. **38**, 1666, 2994 (1995).

gen.: Carbohydr. Res. **301**, 95–109 (1997) (synthesis C. A) ▪ Chem. Eur. J. **4**, 33–43 (1998) (synthesis C. D) ▪ J. Med. Chem. **41**, 3022 (1998) (synthetic analogues) ▪ J. Nat. Prod. **50**, 119, 1189 (1987); **57**, 1136 (1994) ▪ J. Org. Chem. **55**, 2797 (1990) (C. C1) ▪ Justus Liebigs Ann. Chem. **1992**, 399 ▪ Pharm. Res. **8**, 776 (1991) (pharmacology) ▪ Synthesis **1999**, 1656 (synthesis C. A4) ▪ Tetrahedron Lett. **33**, 4213 (1992); **36**, 9369 (1995) (C. D2)

Commersonine see Solanum steroid alkaloid glycosides.

Compactin [ML-236 B, mevastatin].

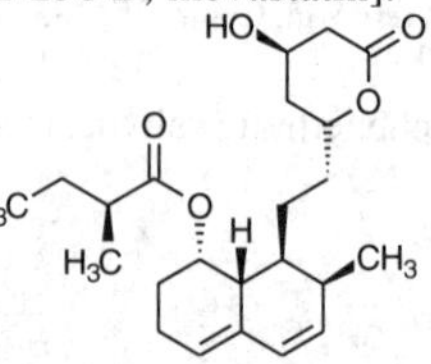

$C_{23}H_{34}O_5$, M_R 390.52, cryst., mp. 152 °C, $[\alpha]_D^{29}$ +221.2° (CH_2Cl_2); soluble in methanol and acetone. Metabolite of *Penicillium brevi-compactum*[1]. C. is a pharmaceutically important inhibitor of cholesterol biosynthesis; it competitively binds to 3-hydroxy-3-methylglutaryl-coenzyme-A reductase[2], see also mevinolin. For biosynthesis, see *Lit.*[3]; for asymmetric synthesis, see *Lit.*[4].

Lit.: [1] J. Chem. Soc., Perkin Trans. 1 **1976**, 1165. [2] Martindale (30.), p. 991; Trends Biochem. Sci. **6**, 10 (1981); J. Med. Chem. **28**, 401 (1985). [3] J. Antibiot. (Tokio) **38**, 444 (1985); J. Chem. Soc., Perkin Trans. 1 **1996**, 2357. [4] Atta-ur-Rahman **11**, 335–377; J. Chem. Soc., Perkin Trans. 1 **1994**, 2417; **1995**, 777–783; J. Am. Chem. Soc. **105**, 1043 (1983); **110**, 6914 (1988); **112**, 3018–3028 (1990); J. Org. Chem. **51**, 2487 (1986); Synlett **1997**, 1111; Tetrahedron **42**, 4909–4951 (1986); Stud. Nat. Prod. Chem. **11**, 335 (1992) (review).

gen.: Beilstein EV **18/3**, 145. – *[HS 2932 29; CAS 73573-88-3]*

Complicatic acid.

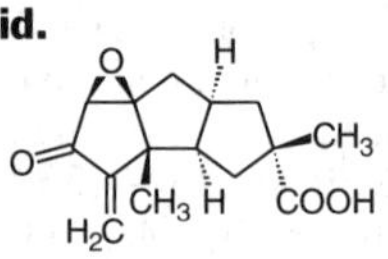

$C_{15}H_{18}O_4$, M_R 262.31, oil, $[\alpha]_D$ –79°. Sesquiterpenoid of the *hirsutane type from cultures of the fungus *Stereum complicatum* (Basidiomycetes); see also hirsutic acid.

Lit.: Chem. Pharm. Bull. **38**, 3230 (1990) (synthesis) ▪ J. Org. Chem. **51**, 2742 (1986) (synthesis) ▪ Phytochemistry **12**, 2717 (1973) (isolation) ▪ Tetrahedron **37**, 2202 (1981) (review). – *[CAS 51741-93-6]*

Concanamycins (folimycins).

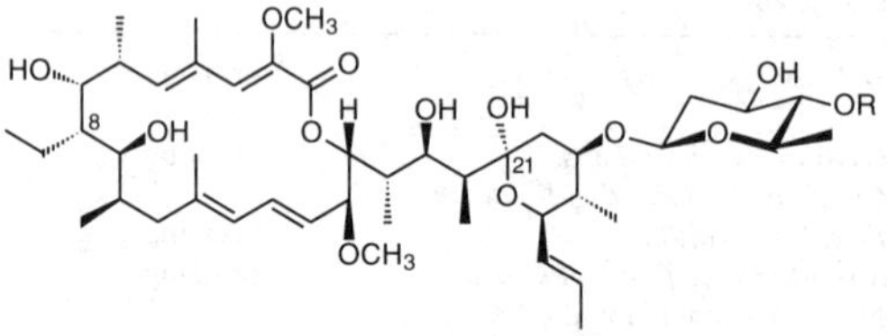

R = CO–NH_2 : C. A
R = H : C. C

Family of 18-membered macrolactones having a side-chain that folds to a cyclic hemiketal (group name: *pleco-macrolides*, see macrolides) and are glycosylated with 2-deoxy-D-rhamnose. C. A–G are isolated from *streptomycetes (e.g., *Streptomyces diastatochromogenes*); *C. A* {$C_{46}H_{75}NO_{14}$, M_R 866.09, cryst., mp. 162–163.5 °C, $[\alpha]_D$ –21.7° (CH_3OH)} and *C. C* ($C_{45}H_{74}O_{13}$, M_R 823.07, mp. 153–155 °C) are formed predominately. The aglycone *concanolide A* (= *C. F*: $C_{39}H_{64}O_{10}$, M_R 692.93) is formed on the *polyketide pathway from 4 acetate, 7 propionate, and 1 butyrate units, in addition, 2 C_2-units are incorporated in the chain which originate from glycerol. C. B and

C. D have a methyl group at C-8 in place of the ethyl group.
Activity: The C. exhibit antifungal, antiviral, immunosuppressive, and cytotoxic activities. The C. became excellent aids in biochemistry after it was recognized that they are specific inhibitors of ATPases of the V-type (C. A: $K_i = 0.02$ nM) and in this respect are even superior to the related *bafilomycin A_1. Since V-type ATPases are active in the plasma membranes of osteoclasts inhibition of these enzymes has been discussed as a possibility for treatment of osteoporosis. ATPases of the P-type are inhibited in the micromolar range while no effect has been observed on F-type ATPases. Semi-synthetic, more stable 21-deoxyconcanamycins are promising candidates for further development.

Lit.: Acta Crystallogr. Sect. C **48**, 1519 (1992) ▪ Folia Microbiol. **36**, 99 (1991) ▪ J. Antibiot. **37**, 1333, 1738 (1984) ▪ Justus Liebigs Ann. Chem. **1994**, 305–312 (biosynthesis) ▪ Tetrahedron Lett. **39**, 6003, 6007 (1998) [synthesis (+)-C. F]. – *Derivatives:* J. Org. Chem. **58**, 5487–5492 (1993). – *Activity:* Biochemistry **15**, 3902–3906 (1993) ▪ Biol. Pharm. Bull. **19**, 297 ff. (1996) ▪ J. Antibiot. **45**, 1108–1116 (1992); **48**, 488–494 (1995). – *[CAS 80890-47-7 (C. A); 81552-34-3 (C. C)]*

Concanavalin A (Con A). A carbohydrate-free seed protein of the jack bean (*Canavalia ensiformis*) belonging to the group of the lectins (phytohemagglutinins), a metalloprotein made up of 4 identical subunits, each with M_R 26 000, consisting of 238 amino acid residues, one Ca^{2+} and one Mn^{2+} center, and a binding site for carbohydrates [1]. The biosynthesis of C. A proceeds through rearrangement of the amino acid chain (transpeptidization) of a precursor protein in such a way that the original amino terminal ends up in the middle of the sequence [2]. Other plant lectins show sequence similarities with this precursor. C. A is able to bind specifically to the carbohydrate receptors of membrane glycoproteins of cells (e. g. erythrocytes, lymphocytes) [3] and thus to agglutinate the cells. It is a mitogen for T-lymphocytes (T-cells), since it induces proliferation of T-cells in cultures to which macrophages have been added by stimulating the expression of interleukin 2. Cytotoxic T-cells which normally have antigen-specific activity, are induced by C. A to unspecific killing of cells. This is probably mediated by agglutination of the cytotoxic T-cells in the target cells. C. A antibody conjugates also stimulate B-cells.
Use: C. A can be used as stimulant in lymphocyte cultures, for the detection of D-glucopyranosyl- and D-mannopyranosyl-rich oligosaccharides on account of its affinity for these carbohydrates, and in affinity chromatography.

Lit.: [1] J. Biol. Chem. **250**, 1490–1547 (1975). [2] J. Cell Biol. **102**, 1284–1297 (1986). [3] Annu. Rev. Biochem. **56**, 24–28 (1987).
gen.: Annu. Rev. Plant Physiol. **36**, 209 (1985) ▪ Biochemistry **29**, 3599 (1990) ▪ Franz (ed.), Advances in Lectin Research, vol. 2, Berlin: Springer 1989 ▪ Methods Enzymol. **150**, 29 (1987) ▪ Sharon, Lectins, N. Y.: Chapman and Hall 1989 ▪ Trends Biochem. Sci. **13**, 60 (1988). – *[CAS 11028-71-0]*

Condurangin see conduritols.

Conduritols. $C_6H_{10}O_4$, M_R 146.14. 1,2,3,4-Cyclohexenetetrols from the bark of the vine *Marsdenia condurango* (Asclepiadaceae). The 6 theoretically possible diastereomers have been named with the letters A–F. In nature only two of these, *C. A* (1α,2β,3β,4α-meso-form, tetragonal prisms, mp. 142–143 °C) and *C. B* (racemic 1α,2β,3α,4β-form) have been found. C. act as inhibitors of glucosidases.

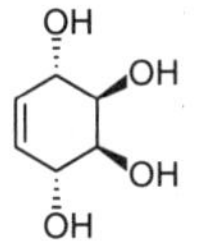

C. A

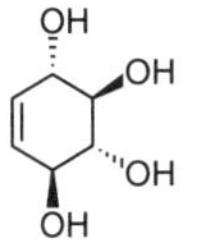

C. B (racemic)

The powdered bark of *M. condurango* is used as a bitter stomachic due to its content of glycosides (pregnane derivatives), that is called "condurangin".

Lit.: Chem. Commun. **1999**, 1985 ▪ Karrer, No. 277 ▪ Pure Appl. Chem. **69**, 97 (1997) ▪ Tetrahedron **46**, 3715 (1990) ▪ Tetrahedron: Asymmetry **10**, 3273 (1999). – *[CAS 526-87-4 (C. A); 25348-64-5 (C. B); 1401-98-5 (condurangin)]*

Conessine see Apocynaceae steroid alkaloids.

Conhydrine, γ-Coniceine see hemlock alkaloids.

Conidendrin (α-conidendrin, tsugalactone, tsugaresinol).

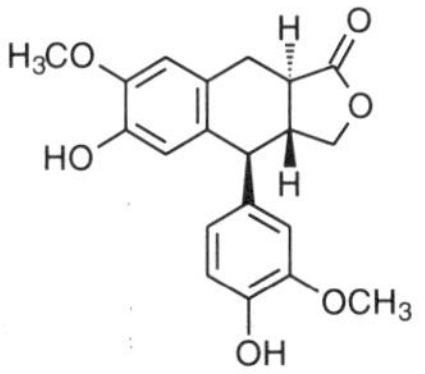

$C_{20}H_{20}O_6$, M_R 356.38, mp. 254–255 °C, $[\alpha]_D^{20}$ –54.5° (acetone). C. is a *lignan occurring in pine resin, hemlock fir resin, and other *Tsuga* species as well as in *Trachelospermum asiaticum*, *Picea* species, and in *Abies amabilis*. It can be isolated from the waste water of pine cellulose production. Demethylation furnishes conidendrols. C. is used as an antioxidant in cotton oil and peanut oil, etc.

Lit.: Beilstein E V **18/5**, 352 ▪ Chem. Lett. **1983**, 1543–1546. – *Synthesis:* Tetrahedron **42**, 2005–2016 (1986); **48**, 687–694 (1992) ▪ Tetrahedron Lett. **30**, 4371 (1989). – *[CAS 518-55-8]*

Coniferin {[2-methoxy-4-(3-hydroxy-1-propenyl)-phenyl]-β-D-glucopyranoside, abietin, coniferoside}.

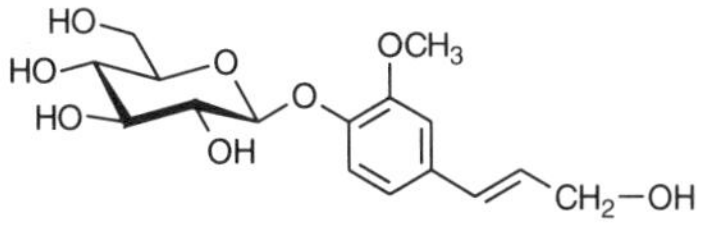

$C_{16}H_{22}O_8$, M_R 342.35. Weakly bitter tasting needles, as dihydrate mp. 186 °C, $[\alpha]_D^{20}$ –66.9° (aqueous pyridine). C. is the main glycoside of conifers and is isolated from, e. g., *Larix decidua* (syn. *L. europaea*), *Fraxinus quadrangulata*, and *Scorzonera hispanica*; it also occurs in asparagus, black salsify, sugar beet, and other plants. C. represents the storage and transport form of *coniferyl alcohol that is needed for the biosyntheses of *lignin and numerous other *phytoalexins.

Lit.: Merck-Index (12.), No. 2567 ▪ Phytochemistry **39**, 409 (1995); **45**, 47 (1997) ▪ Synthesis **1998**, 157 (synthesis) ▪ see also coniferyl alcohol. – *[CAS 531-29-3]*

Coniferyl alcohol [4-(3-hydroxy-1-propenyl)-2-methoxyphenol].

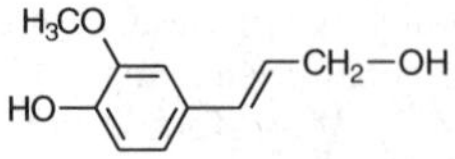

$C_{10}H_{12}O_3$, M_R 180.20. Prisms, mp. 73–74 °C, bp. 163–165 °C (400 Pa); readily soluble in ether, soluble in alcohol and alkali, practically insoluble in water. C. is easily polymerized by dilute acids and oxidizing agents and forms amorphous gums. C. is a monomeric precursor of *lignin. It is formed by digestion of lignin in cellulose production or from *coniferin by treatment with emulsin.

Lit.: Beilstein E IV **6**, 7477 ▪ J. Agric. Food Chem. **40**, 1108 (1992) ▪ Luckner (3.) p. 387 ▪ Synthesis **1994**, 369 (synthesis). – *[CAS 458-35-5]*

Coniine see hemlock alkaloids.

Conkurchine see Apocynaceae steroid alkaloids.

Connatin [N^5-(dimethylcarbamoyl)-N^5-hydroxy-L-ornithine].

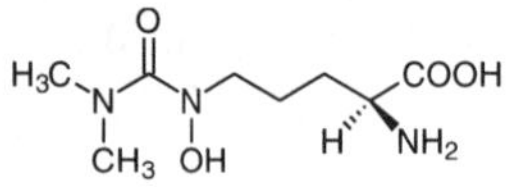

Connatin

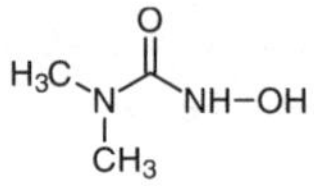

N'-Hydroxy-*N,N*-dimethylurea

$C_8H_{17}N_3O_4$, M_R 219.24, cryst., mp. 184–188 °C (decomp.), $[\alpha]_D$ +25.5° (H_2O), soluble in water and methanol. C. occurs together with *lyophyllin and *N'*-hydroxy-*N,N*-dimethylurea in *Lyophyllum connatum* (Agaricales). The biosynthesis starts from *citrulline by methylation and subsequent *N*-hydroxylation. C. is the biosynthetic precursor of *N'-hydroxy-N,N-dimethylurea* ($C_3H_8N_2O_2$, M_R 104.11, mp. 107–109 °C) which, together with C., is responsible for the blue color reaction of the fruit bodies of *L. connatum* with iron(III) chloride.

Lit.: Angew. Chem. Int. Ed. Engl. **23**, 72 (1984). – *[CAS 88245-12-9 (C.); 52253-32-4 (N'-hydroxy-N,N-dimethylurea)]*

Conophthorin [(5*S*,7*S*)-7-methyl-1,6-dioxaspiro-[4.5]decane).

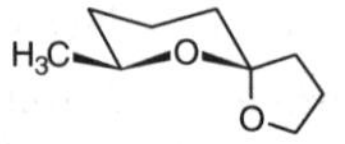

$C_9H_{16}O_2$, M_R 156.22, liquid with a flower-like odor, bp. 77 °C (6.53 kPa). Isolated from the sting apparatus of wasps (*Paravespula* and *Dolichovespula*) identified as a *spiroacetal. C. is also produced by bark beetles of the genera *Conophthorus*, *Cryphalus*, and *Leperisinus* and used by them as *pheromone to deter members of the same species (Alarm pheromone) in order to prevent local overpopulations. In *Conophthorus* species C. has the (5*S*,7*S*)-configuration, $[\alpha]_D^{22}$ −78.3° (pentane). C. has also been detected in the bark of pine trees, cork oak, and the odor of orchids; see also chalcograne, oleane.

Lit.: ApSimon **9**, 455–459 (synthesis) ▪ J. Chem. Ecol **21**, 143–197, 169–185 (1995) ▪ Naturwissenschaften **66**, 618–619 (1979). – *[CAS 77715-03-8]*

Conotoxin KK-0 see King Kong peptide.

Conotoxins (cone toxins). Peptide toxins from subtropical and tropical marine cone snails (shells) (Conidae). The snails sting to trap prey and as defence against predators with harpoon-like denticles (length ca. 7 mm) loaded with the neurotoxin. Each of the more than 500 cone snail species has a unique characteristic mixture of up to 100 various C., peptides consisting of at least 10–30 amino acids and biosynthetically originating from precursors with 60–90 amino acid units (see figure).

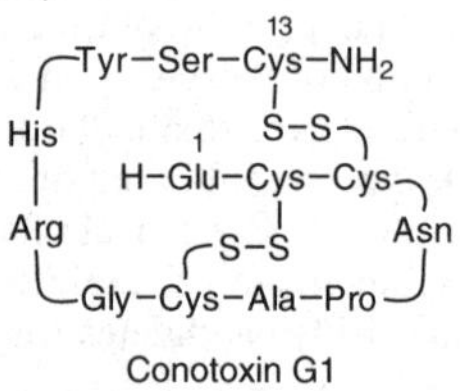

Conotoxin G1

The presumably tens of thousands of C. which are limited in their conformational flexibility by disulfide bridges and other structural features may be compared with the peptide libraries prepared (by combinatorial synthesis) for, e. g., pharmaceutical screening. Synthetic conopeptide libraries are prepared to target NMDA receptor subtypes, neuronal and neuromuscular subtypes of the nicotinic acetylcholine receptor, voltage-gated sodium channels in muscle and neuronal tissue, and voltage-sensitive potassium channels [1]. *Conus purpurascens* uses a cocktail of slow-acting and rapidly-acting conotoxins to catch prey [2]. For humans the sting of a cone snail is very painful and is followed by a progressive numbness which soon affects the entire body. In severe cases muscular paralysis occurs. Human deaths as a result of heart failure are mainly attributed to *Conus geographus*, *C. textile*, and *C. tulipa*. ω-C. from the venom of *Conus magus* is a promising new drug for the treatment of severe neuropathic pain. Its analgetic effect is due to a highly selective modulation of N-type neuronal calcium channels. Chronic administration of ω-C. (Ziconotide®) does not elicit tolerance or lead to addiction (phase III of clinical development). Peptidomimetic analogues of ω-C. have been prepared [3].

Lit.: [1] Chem. Rev. **93**, 1923–1936 (1993); Drug News, R & D Focus, 27. 4. 1998, p. 6; Science **257**, 257–263 (1990); Trends Biotechnol. **13**, 422–426 (1995). [2] Nature (London) **381**, 148 (1996). [3] Innovation Perspect. Solid Phase Synth. Comb. Libr., Collect. Pap., Int. Symp. 4th 1995 (publ. 1996), 539–542; Tetrahedron Lett. **39**, 7619 (1998).

gen.: Biochemistry **26**, 2086 (1987); **27**, 6256 (1988) ▪ J. Am. Chem. Soc. **115**, 10492 (1993) (synthesis of α-C.) ▪ J. Biol. Chem. **258**, 12247 (1983); **262**, 15821 (1987) ▪ J. Med. Chem. **42**, 2364 (1999) (structure α-C.) ▪ J. Toxicol., Toxin Rev. **4**, 107–132 (1985). – *Review:* Baldomero et al., in Tu (ed.), Marine Toxins and Venoms, p. 327–352, New York: Dekker 1988. – *[CAS 76862-65-2 (C. G1)]*

Contortin.

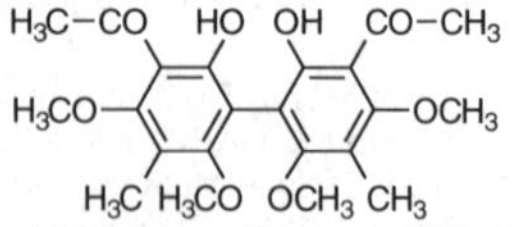

$C_{22}H_{26}O_8$, M_R 418.44, needles, mp. 129–130 °C. As yet the only known biphenyl derivative from lichens (*Psoroma contortum*).

Lit.: Aust. J. Chem. **37**, 1531–1538 (1984). – *[CAS 91925-83-6]*

Contraceptives see estrogens and progestogens.

Convallatoxin, Convalloside see strophanthins.

Copacamphene see sativene.

Coprine [N^5-(1-hydroxycyclopropyl)-L-glutamine].

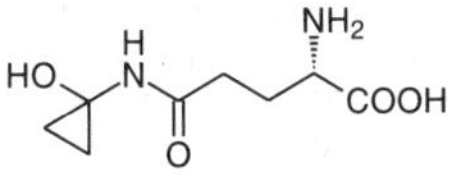

$C_8H_{14}N_2O_4$, M_R 202.21, cryst., mp. 197–199 °C, $[\alpha]_D$ +7.6° (H_2O). Amino acid from the inky cap (*Coprinus atramentarius*, Basidiomycetes). C. is responsible for the toxic action of this fungus when alcohol is consumed at the same time. C. acts in such a way that the ability of the liver to degrade alcohol is lost temporarily (inhibition of acetaldehyde dehydrogenase by the transition state analogue 1-aminocyclopropanol formed by hydrolysis in gastric juices [1,2]) so that an elevated acetaldehyde blood level results ("antabuse alcohol reaction", acetaldehyde poisoning). Signs for this are severe headache and intense feeling of indisposition up to circulatory collapse. Repeated oral administration of C. to rats resulted in damage to testicular tissue so that consumption of this fungus is not recommended [3].

Lit.: [1] Acta Pharmacol. Toxicol. **40**, 476 (1977). [2] J. Pharm. Sci. **65**, 1774 (1976); J. Chem. Soc., Perkin Trans. 1 **1977**, 684; Helv. Chim. Acta **34**, 1078 (1992) (synthesis). [3] Toxicology **12**, 89 (1979).

gen.: Bresinsky & Besl, Giftpilze, p. 119ff., Stuttgart: Wissenschaftliche Verlagsges. 1985 ▪ J. Nat. Prod. **38**, 489 (1975) ▪ Turner **2**, 386 ▪ Zechmeister **39**, 236, 240, 262. – *[CAS 58919-61-2]*

Coproporphyrins.

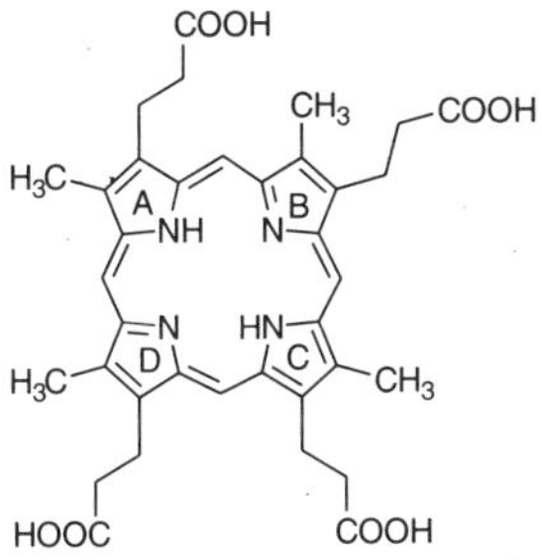

Coproporphyrin III

C. I (tetramethyl ester of C. III): $C_{40}H_{46}N_4O_8$, M_R 710.83, uv_{max} ($CHCl_3$) 400, 498, 532, 566, 594, 621 nm. Prism-shaped crystals, mp. 192 °C.

Occurrence: C. III, shown in the formula, is one of the four constitutional isomers C. I–IV. The reduced form of C. III, copro-*porphyrinogen III, occurs as an intermediate in the biosynthesis of the red blood pigment *heme and is itself formed in the course of the biosynthesis by decarboxylation of uroporphyrinogen III. C. III, in which the methyl groups and propanoic acid side chains of the pyrrole rings A–C are arranged in series and exchanged in ring D, is formed during isolation through oxidation by atmospheric oxygen. It occurs in small amounts in feces (Greek: copros= feces) and urine. The occurrence of larger amounts of C. III and its biologically abnormal constitutional isomer C. I, with a serial arrangement of methyl groups and propanoic acid side chains, is an indication for a metabolic defect (coproporphyria) of genetic origin or caused by chronic poisoning. The constitutional isomers C. II and C. IV do not occur in nature.

Lit.: Dolphin **IV**, 19–20; **VI** 46, 249 ▪ Jordan, p. 67–84 ▪ Org. Prep. Proced. Int. **27**, 224 (1995) ▪ Tetrahedron **46**, 7483 (1990) (synthesis). – *[CAS 531-14-6, 25767-20-8 (C. I); 14643-66-4 (C. III)]*

Coprostanol see sterols.

Coptisine see protoberberine alkaloids.

Corallistin A.

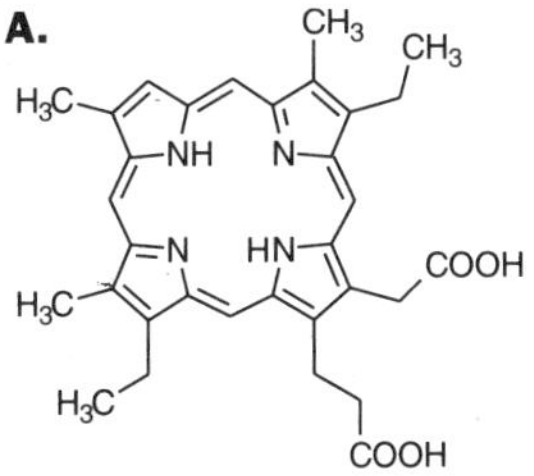

$C_{32}H_{34}N_4O_4$, M_R 538.65; uv_{max} (dimethyl ester in $CHCl_3$) 400, 498, 538, 565, and 618 nm. C. A is one of the few naturally occurring, metal-free *porphyrins.

Occurrence: C. A is isolated from the New Caledonian marine sponge *Corallistes* sp. [1] which contains further porphyrins designated as C. B–E [2] in addition to the main component C. A. Their substitution patterns differ slightly from that of C. A.

Lit.: [1] Helv. Chim. Acta **72**, 1451 (1989). [2] Helv. Chim. Acta **76**, 1489 (1993). – *[CAS 125398-29-0]*

Corallopyronin A see myxopyronins.

Cordycepin (3′-deoxyadenosine).

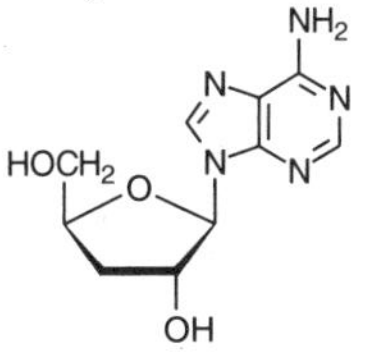

$C_{10}H_{13}N_5O_3$, M_R 251.24, needles, mp. 225–226 °C, $[\alpha]_D^{20}$ −47° (H_2O), well soluble in water. Metabolite from culture broths of *Cordyceps*, *Isaria*, *Emericella*, and *Aspergillus* species. C. exhibits strong cytotoxic activity by inhibition of RNA biosynthesis. Because of this property C. is used for investigations on messenger-RNA transcription. C. was the first nucleoside antibiotic to be described (1951) and was isolated from *Cordyceps militaris*. C. selectively inhibits the polyadenylation of hnRNA in HeLa cells at 50 µg/mL. Furthermore not only 45S-rRNA biosynthesis but also adenylate cyclase are inhibited.

Lit.: Angew. Chem. Int. Ed. Engl. **34**, 350 (1995) ▪ Beilstein EV **26/16**, 300f. ▪ J. Am. Chem. Soc. **103**, 6739 (1981) ▪ Merck-Index (12.), No. 2593 ▪ Phytochemistry **31**, 1409 (1992) ▪ Tetrahedron Lett. **26**, 4295 (1985). – *[CAS 73-03-0]*

Coreopsin see butein.

Coreximine see protoberberine alkaloids.

Coriamyrtin.

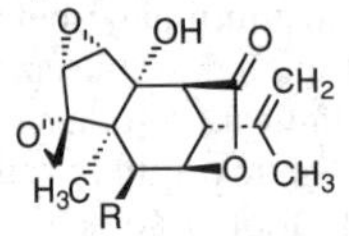

R = H :Coriamyrtin
R = OH : Tutin

$C_{15}H_{18}O_5$, M_R 278.30, cryst., mp. 229–230 °C, $[\alpha]_D^{14}$ +79.0°. *Sesquiterpene from *Coriaria myrtifolia, C. japonica*, and the parasitic mistle *Loranthus parasiticus*, the aqueous extract of which is used in south west China as shock therapy for schizophrenia. The activity is attributed to C. and *tutin* ($C_{15}H_{18}O_6$, M_R 294.30, mp. 209–210 °C, $[\alpha]_D^{20}$ +9.3°) and comparable with that of an electroshock or an overdose of insulin. Tutin is named after the Maori name for *C. ruscifolia* ("tutu"), a New Zealand tree which has caused major loss of lives among the settlers. The 4β-hydroxy derivative *mellitoxin* {$C_{15}H_{18}O_7$, M_R 310.30, mp. 225–240 °C (decomp.), $[\alpha]_D$ +31.9°} is isolated from poisonous honey from bees that have taken up excretions (mildew) from lice living on tutu.

Lit.: Aust. J. Chem. **42**, 1881 (1989) ▪ Beilstein EV **19/10**, 439f. ▪ Merck-Index (12.), No. 2594 ▪ Synform **3**, 188 (1985) ▪ Tetrahedron **42**, 5551 (1986) (synthesis tutin). – *[CAS 2571-86-0 (C.); 2571-22-4 (tutin); 3484-46-6 (mellitoxin)]*

Coriander oil. Colorless to light yellowish oil, sweet-spicy, aromatic odor with flowery, basalmy-woody undertones and a sweet, spicy, warm, slightly burning taste.

Production: By steam distillation from the fruits of coriander (*Coriandrum sativum*).

Composition[1]***:*** The main component is (+)-**linalool* (coriandrol) (60–80%). Important for the typical odor are small amounts of saturated and unsaturated aliphatic aldehydes. These aldehydes are the main components of the herbage oil, principally *2-decenal* and *2-dodecenal* (see alkenals).

Use: In the perfume industry (modern Eaux de Toilette, masculine notes); in the production of spice mixtures to improve the aromatic character of foods, and in medical carminatives.

Lit.: [1] Perfum. Flavor. **16** (1), 49; (6), 52 (1991); **19** (1), 42 (1994).
gen.: Arctander, p. 191 ▪ Bauer et al. (2.), p. 150 ▪ ISO 3516 (1980). – *Toxicology:* Food Cosmet. Toxicol. **11**, 1077 (1973). – *[HS 3301 29; CAS 8008-52-4]*

Coriolin.

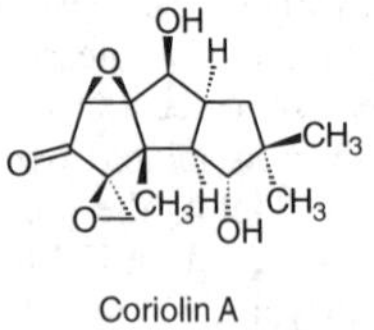

Coriolin A

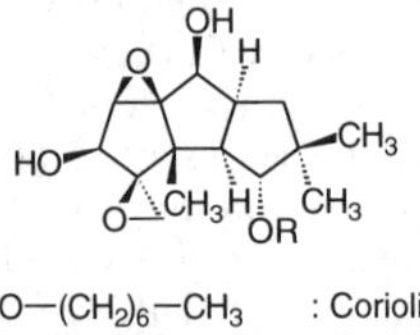

R = CO–$(CH_2)_6$–CH_3 : Coriolin B
R = CO–CH(OH)–$(CH_2)_5$–CH_3 : Coriolin C

C. A: $C_{15}H_{20}O_5$, M_R 280.32, cryst., mp. 175 °C, $[\alpha]_D$ –20.7° ($CHCl_3$). Sesquiterpenoid of the *hirsutane type from cultures of the wood-rotting fungus *Coriolus consors* (Polyporaceae) with antibiotic activity against Gram-positive bacteria, trichomonades, and Yoshida sarcoma cells. The fungus also produces the esters *C. B* ($C_{23}H_{34}O_6$, M_R 406.52, cryst., mp. 215–216 °C) and *C. C* ($C_{23}H_{34}O_7$, M_R 422.52). Diketocoriolin B, obtained by oxidation of C. B., exhibits a stronger antitumor activity than the starting compound and, in contrast, is not an immunosuppressive agent.

Lit.: Chem. Commun. **1999**, 2519 (synthesis) ▪ J. Am. Chem. Soc. **108**, 4149 (1986) [(–)-C.] ▪ J. Org. Chem. **52**, 4647 (1987) (*rac.*-C.); **64**, 2648 (1999) (synthesis) ▪ Justus Liebigs Ann. Chem. **1993**, 1133; **1994**, 99 [(–)-C.] ▪ Mulzer et al. (eds.), Organic Synthesis Highlights, p. 323–334, Weinheim: VCH Verlagsges. 1991 (synthesis, review) ▪ Tetrahedron **37**, 2202 (1981) ▪ Tetrahedron Lett. **38**, 465 (1997) (synthesis) ▪ Turner **2**, 244–247 (review). – *[CAS 33404-85-2 (C. A); 33400-89-4 (C. B); 33400-90-7 (C. C)]*

Coronamic acid, Coronafacic acid see coronatine.

Coronaridine see iboga alkaloids.

Coronatine.

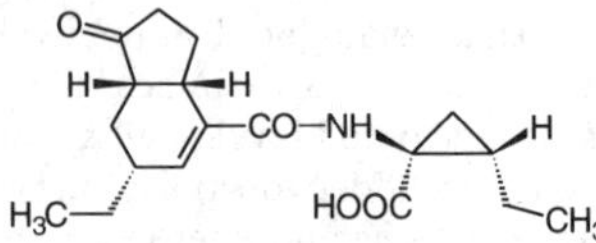

$C_{18}H_{25}NO_4$, M_R 319.40, mp. 151–153 °C, $[\alpha]_D^{20}$ +68.4° (CH_3OH). Phytotoxic metabolite from *Pseudomonas coronafaciens* and *P. glycinea* that causes systemic chlorosis and in potatoes hypertrophic tuber growth. C. induces the biosynthesis of volatile compounds in plants that attract enemies of herbivores. Chemical synthesis proceeds like the biosynthesis through the amino acid building block (*coronamic acid*, biogenetically from L-*isoleucine via L-*allo*-isoleucine) and a bicyclic carboxylic acid (*coronafacic acid*, biogenetically from acetate, butyrate, and pyruvate).

Lit.: Biosci., Biotechnol. Biochem. **61**, 752 (1997) (synthesis) ▪ Can. J. Chem. **72**, 86–99 (1994) ▪ J. Am. Chem. Soc. **102**, 2463, 6353 (1980); **108**, 4681 (1986) ▪ Phytochemistry **17**, 2028 (1978) ▪ Tetrahedron **53**, 9509–9524 (1997) (synthesis) ▪ Tetrahedron Lett. **269**, 365 (1979). – *[CAS 62251-96-1 (C.); 63393-56-6 (coronamic acid); 62251-98-3 (coronafacic acid)]*

Corphin(s).

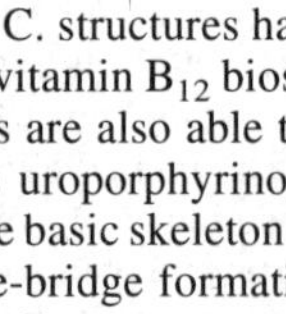

Corphin

Name coined by Eschenmoser for a heterocyclic ring system[1,2] with a structure between that of *corrin and that of *porphine. C. structures have been detected as intermediates in *vitamin B_{12} biosynthesis[3]. The participating enzymes are also able to convert unnatural substrates such as uroporphyrinogen I *in vitro* to C. compounds[2,3]. The basic skeleton of *factor F 430, involved in methane-bridge formation in bacteria, also belongs to the corphins.

Lit.: [1] Helv. Chim. Acta **65**, 600–610 (1982). [2] Chem. Rev. **94**, 327 (1994). [3] Angew. Chem. Int. Ed. Engl. **32**, 1223 (1993).

Corrin(s).

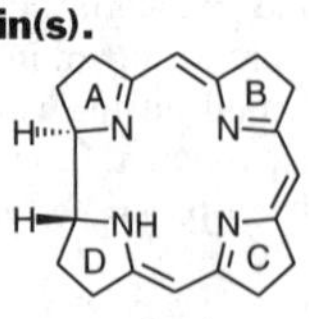

Corrin

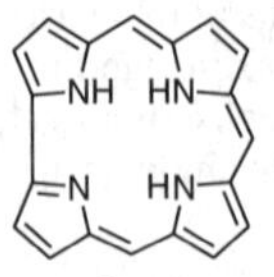

Corrol

$C_{19}H_{22}N_4$, M_R 306.41. Name derived from "core" for the basic skeleton upon which the *cobalamins (see also vitamin B_{12}) and in general the corrinoids are based [1,2]. The 4 partially hydrogenated pyrrole rings of C. are designated clockwise with the letters A–D. A characteristic structure element that distinguishes the C. from the *porphyrins is the direct linkage of the rings A and D. The completely unsaturated system derived from C. is called *corrol* ($C_{19}H_{14}N_4$, M_R 298.35, octadehydrocorrin, see formula). Partially unsaturated C. derivatives between corrin and corrol are also known [2]. Formal introduction of a methine group between C-1 and C-19 in corrol leads to porphyrin, in the same way *corphins, *isobacteriochlorins, *bacteriochlorins, and *chlorins can formally be derived from C. and its partially unsaturated derivatives. Compounds with the basic skeleton of C. are also known as corrinoids.

Lit.: [1] Pure Appl. Chem. **48**, 495–502 (1976). [2] Angew. Chem. Int. Ed. Engl. **27**, 5–39 (1988).
gen.: Dolphin, **I**, 19–21, 357–364; **II**, 328–391 ▪ Dolphin B_{12} **I**, 116–118; **II**, 5–7, 263–287. – *[CAS 262-76-0 (C.); 26444-09-7 (corrol)]*

Cortexolone, Cortexone see corticosteroids.

Corticocrocin see corticrocin.

Corticosteroids (corticoids, cortins). Physiologically important group of *steroid hormones formed in the adrenal cortex. They are structurally derived from the parent hydrocarbon pregnane (see steroids), possess an α,β-unsaturated 3-oxo group, a second oxo group at C-20, and further oxygen functions in positions 11, 17, 18, and/or 21. Over 30 natural C. are currently known. Of these 7 are biologically active including, besides *aldosterone (Reichstein's substance X), *cortisone* (cortone, Reichstein's substance Fa), *hydrocortisone* (cortisol, Reichstein's substance M), *cortexolone* (cortodoxone, 17α-hydroxydeoxycorticosterone, Reichstein's substance S), *11-dehydrocorticosterone* (Kendall's substance A), *corticosterone* (Reichstein's substance H), and *cortexone* (deoxycorticosterone, deoxycortone, Reichstein's substance Q).

Activity: C. are divided into two groups according to their activity profile as *mineralocorticoids* (mineralocorticosteroids) and *glucocorticoids* (glucocorticosteroids). The former regulate the electrolyte equilibrium in the body and promote especially sodium ion retention and potassium ion excretion. The most important, most active mineralo-C. is aldosterone. However, the gluco-C. corticosterone and hydrocortisone also have a high affinity for the mineralocorticoid receptors and thus exhibit corresponding activities.

The most important gluco-C. hormone is *hydrocortisone* which is excreted by the adrenal cortex into the blood as response to physical and psychic stress (the normal average rate of formation in humans amounts to about 20 mg/d; for comparison corticosterone ca. 3 mg and aldosterone 0.3 mg). Hydrocortisone promotes the conversion of proteins and lactate into energy-rich carbohydrates (gluconeogenesis), storage of the latter as *glycogen, and stimulates the formation of storage (depot) fats (lipids). In addition, hydrocortisone prevents or reduces tissue reactions to inflammatory processes and has immunosuppressive activity.

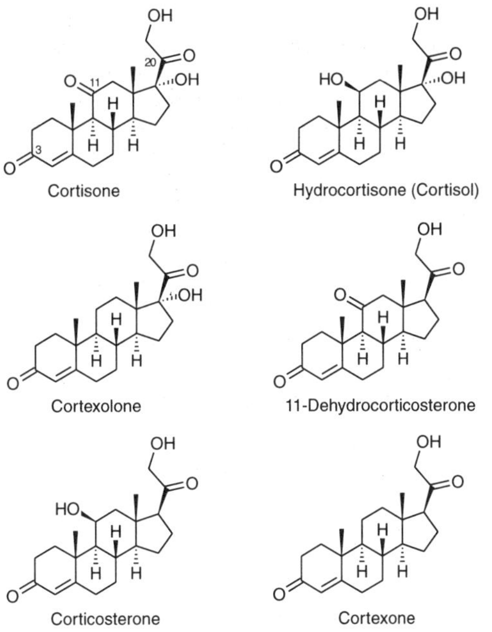

Figure: Naturally occurring corticosteroids.

Table: Data of Corticosteroids.

Corticosteroid CAS	molecular formula	M_R	mp. [°C]	$[\alpha]_D$
Cortisone 53-06-5	$C_{21}H_{28}O_5$	360.45	230–231	+209° ($CHCl_3$)
Hydrocortisone 50-23-7	$C_{21}H_{30}O_5$	362.47	220 (decomp.)	+167° ($CHCl_3$)
Cortexolone 152-58-9	$C_{21}H_{30}O_4$	346.47	212.8–216.8	+132° (dioxane)
11-Dehydro-corticosterone 72-23-1	$C_{21}H_{28}O_4$	344.45	183–183.5	+239° (dioxane)
Corticosterone 50-22-6	$C_{21}H_{30}O_4$	346.47	180–182	+223° (C_2H_5OH)
Cortexone 64-85-7	$C_{21}H_{30}O_3$	330.47	141–142	+178° (C_2H_5OH)

Biosynthesis & metabolism: Hydrocortisone is biosynthesized from 3β-hydroxypregn-5-en-20-one via 17α-hydroxy- and 17α,21-dihydroxyprogesterone (see pregnane derivatives). The other gluco-C. and aldosterone are derived from intermediates of this biosynthetic pathway. Hydrocortisone in the blood circulation is rapidly metabolized. Thus, dehydration of the 11β-hydroxy group gives cortisone. The final metabolites of hydrocortisone excreted by the kidneys as glucuronides are mainly 5α- and 5β-pregnane-3α,11β,17α,20,21-pentols.

Medicinal use: C. are used in physiological doses for substitution therapy in cases of functional disorders of the adrenals. Gluco-C. are used in pharmaceutical doses especially on account of their anti-inflammatory, anti-rheumatic, and immunosuppressive activities for, e.g., rheumatoid arthritis or allergic problems. For external local administrations in cases of skin diseases or asthma where a systemic distribution is not wanted, special, highly-active synthetic C. derivatives have been developed and used with great success.

Lit.: Bohl & Duax (eds.), Molecular Structure and Biological Activity of Steroids, Boca Raton: CRC Press 1992 ▪ Fieser & Fieser, Steroids, New York: Reinhold 1959 ▪ Hager (5.) **8**, 473 f.; **9**, 328 f. ▪ Lee & Fitzgerald (eds.), Progress in Research and Clinical Applications of Corticosteroids, Philadelphia: Heyden 1982 ▪ Ullmann (5.) **A 13**, 135–154 ▪ Zeelen, p. 143–175. – *[HS 293721]*

Corticrocin [(*all-E*)-2,4,6,8,10,12-tetradecahexaenedioic acid)].

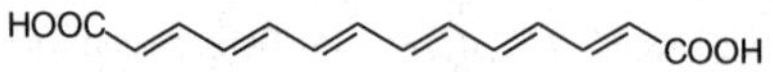

$C_{14}H_{14}O_4$, M_R 246.26, yellow cryst., mp. 270 °C (subl.), 317 °C (decomp.), very poorly soluble in all organic solvents. Isolated from the yellow-orange rhizomorphs of the fungus *Piloderma croceum* (= *Corticium croceum*) that occurs on wood and in the acidic soil of conifer forests.

Lit.: Acta Chem. Scand. **2**, 209 (1948) ▪ Beilstein E IV **2**, 2349 ▪ J. Chem. Soc. **1954**, 3217 (synthesis) ▪ Macromolecules **18**, 2088 (1985) ▪ Z. Naturforsch. C **53**, 4 (1998). – *[CAS 505-53-3]*

Cortins see corticosteroids.

Cortisalin.

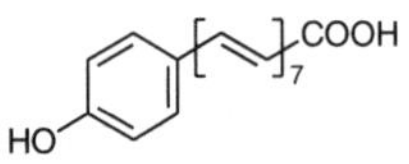

Cortisalin

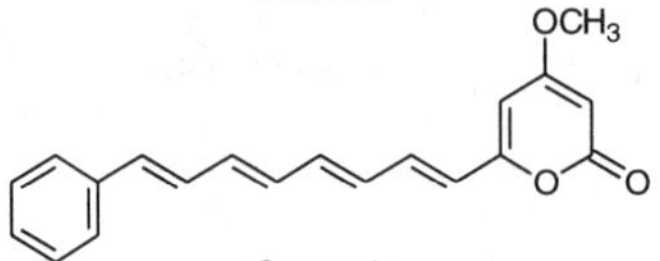

Geogenin

$C_{21}H_{20}O_3$, M_R 320.39, red-violet cryst., mp. >290 °C (decomp.), very poorly soluble in organic solvents. A polyenecarboxylic acid from the brilliant red fruit bodies of *Cytidia salicina* (Basidiomycetes) growing on willow trees. A structural relative is the sulfur-yellow pyrone derivative *geogenin* ($C_{20}H_{18}O_3$, M_R 306.36, yellow needles, mp. 236–240 °C) from the yellow fungus *Hydnellum geogenium*.

Lit.: Zechmeister **51**, 97–98 (review).

Cortisol, Cortisone see corticosteroids.

Corydaline see protoberberine alkaloids.

Corydalis alkaloids. General name for a series of *isoquinoline alkaloids with aporphine, protoberberine, or benzo[*c*]phenanthridine skeletons from the roots of the yellow fumitory and some other *Corydalis* species (see aporphine, protoberberine, and benzo[*c*]phenanthridine alkaloids). Some of the C.a. have narcotic, muscle paralyzing activities and the powdered root stocks of the C.a.-containing plants had traditional use as anthelmintic and menstruation-promoting agents.

Lit.: Hager (5.) **4**, 1013–1027. – *[HS 293990]*

Corydine see aporphine alkaloids.

Corynantheine alkaloids (Corynanthé alkaloids).

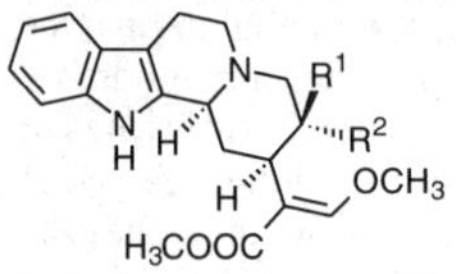

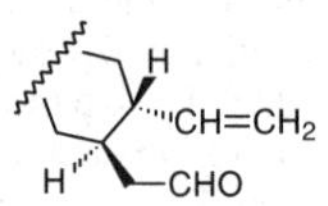

$R^1 = H$, $R^2 = CH{=}CH_2$: Corynantheine Corynantheal

$R^1 = C_2H_5$, $R^2 = H$: Corynantheidine

C. a. (*indole alkaloids) are structurally related to the *heteroyohimbine alkaloids. Between 1981 and 1995 more than 400 new structures from both groups were reported. C. a. are found especially in *Corynanthé*, *Strychnos*, *Pseudocinchona*, *Rauvolfia*, and *Rubiaceae* species. Important examples of this group of alkaloids are *corynantheine* ($C_{22}H_{26}N_2O_3$, M_R 366.46) which acts as a sympatholytic agent; *corynantheal* ($C_{19}H_{22}N_2O$, M_R 294.40) which is of biosynthetic interest, in particular as a precursor of *Cinchona alkaloids, and *corynantheidine* ($C_{22}H_{28}N_2O_3$, M_R 368.48, yellow cryst., mp. 117 °C); although similar in name corynanthine is not a coryantheine alkaloid but rather a *yohimbine alkaloid.

Lit.: Hager (5.) **4**, 1030 f. ▪ Lounasmaa & Tovanen, in: Saxton (eds.), The Monoterpenoid Indole Alkaloids, Suppl. vol. 25, Heterocyclic Compounds, p. 57, Chichester: John Wiley 1994 ▪ Manske **27**, 131 ▪ Nat. Prod. Rep. **1**, 21 (1984); **2**, 49 (1985); **3**, 353 (1986); **4**, 591 (1987); **6**, 1, 433 (1989); **7**, 191 (1990); **8**, 251 (1991); **9**, 393 (1992). – *[HS 293990; CAS 18904-54-6 (corynantheine); 572-78-1 (corynantheal); 23407-35-4 (corynantheidine)]*

Corynanthine see yohimbine.

Corynetoxins see tunicamycins.

Corytuberine see aporphine alkaloids.

Cosmene [(3*E*,5*E*)-2,6-dimethylocta-1,3,5,7-tetraene].

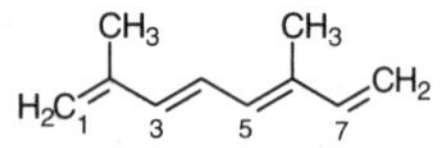

$C_{10}H_{14}$, M_R 134.22, bp. ca. 30 °C (30 Pa). Conjugated unsaturated acyclic *monoterpene widely distributed in essential oils, e. g., with a content of 20–40% in the essential oil of *Cosmos bipinnatus* (Asteraceae).

Lit.: Beilstein E IV **1**, 1128 ▪ Karrer, No. 33. – *[CAS 460-01-5]*

Costunolide see germacranolides.

Cotinine [(*S*)-1-methyl-5-(3-pyridyl)-2-pyrrolidinone].

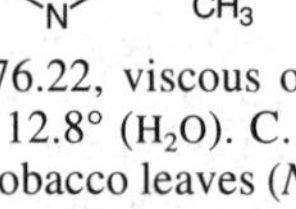

$C_{10}H_{12}N_2O$, M_R 176.22, viscous oil, bp. 210–211 °C (0.8 kPa), $[\alpha]_D^{20}$ –12.8° (H_2O). C. is a *tobacco alkaloid occurring in tobacco leaves (*Nicotiana tabacum*), in *Duboisia hopwoodii*, and in tobacco smoke. C. is the main metabolite of *nicotine. It is probably also formed as an artefact in mammalian organisms by autoxidation of nicotine.

Lit.: Beilstein E V **24/2**, 504 f. ▪ Manske **26**, 131–133 ▪ Merck-Index (12.), No. 2619 ▪ Pelletier **1**, 85–152; **3**, 30. – *[HS 293990; CAS 486-56-6]*

Cotton. The name cotton (C.) covers not only the seed of the yellow-flowering (*gossypetin) cotton plant [*Gossypium* (Malvaceae)] which has been cultivated in tropical and subtropical regions for over 5000 years, but also the textile fibers obtained therefrom (abbreviation: Cf., cotton fibers). Cf. today represent about 35% of the world market for textile fibers. After ripening the burst fruits contain fine, white seed hairs up to 5 cm in length, in addition the seeds are covered with a short-fibered wool. The long hairs consist to over 90% of *cellulose with average M_R 320 000 and an average degree of polymerization of 10 000 – 14 000 (purified 500 – 3000). The fiber stem, the so-called secondary wall, contains only about 5% non-cellulose material; in contrast the thin outer skin or primary wall consists to about 90% of cotton wax and *pectins and only about 10% cellulose. Raw Cf. can be bleached and dyed; mercerization (treatment with cold sodium hydroxide solution) gives luster, strength, and receptiveness to dyeing. Working with raw C. results in a risk for diseases of the lungs and the respiratory tract (*byssinosis* = cotton-mill fever; allergic bronchitis in cotton workers), thus a maximum working place concentration for C. dust of 1.5 mg/m^3 has been set in Germany [1]. Cotton wool and cellulose (from lint) are also made from C., the seeds furnish cottonseed oil which contains *gossypol. C. is extremely sensitive to pests – a yearly loss of about 50% of the world harvest through animal pests and diseases is estimated – and special crop protection agents (mostly synthetic pyrethroids, see pyrethrum) must be used [2]. The world cotton harvest in 1994 amounted to 18.4 million t.

Lit.: [1] Br. J. Ind. Med. **44**, 577 (1987); Wakelyn & Jacobs (eds.), Proc. of the 11th Cotton Dust Res. Conf., Cotton Dust, vol. 12, New Orleans, Jan. 1988, Memphis: Natl. Cotton Council 1988. [2] Annu. Rev. Entomol. **22**, 451 – 482 (1977).
gen.: Ullmann (5.) **A 5**, 375 f. – *Colors:* Textilveredlung **22**, 19 (1987) ▪ Wool Sci Rev. **62**, (1986). – *Mercerization:* J. Soc. Dyers Colour. **103**, 342 (1987) ▪ Text. Dyer Printer **21**, 25 (1988).

Cotylenin F.

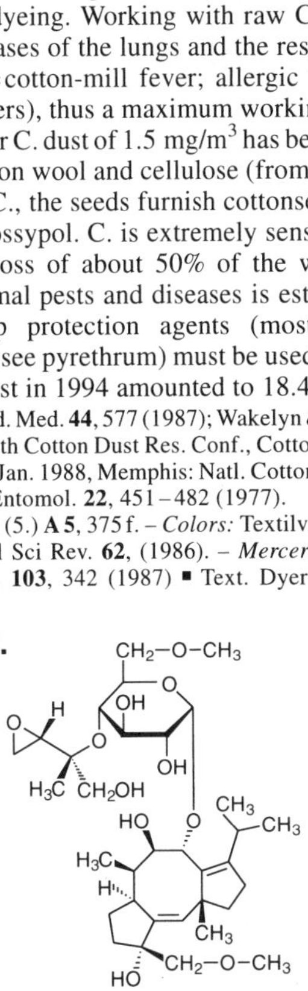

$C_{33}H_{54}O_{11}$, M_R 626.79, mp. 178 – 179 °C, $[\alpha]_D$ +16.4° (CH_3OH), like cotylenins A – J: a diterpene with the *fusicoccin H structure isolated from cultures of *Cladosporium* species; it exhibits plant growth regulating properties.

Lit.: Agric. Biol. Chem. **43**, 385 (1979) ▪ Biosci., Biotechnol. Biochem. **62**, 1815 (1998) (x-ray structure C. A). – *[CAS 58045-03-7]*

(*E*)-*p*-Coumaric acid [(*E*)-4-hydroxycinnamic acid].

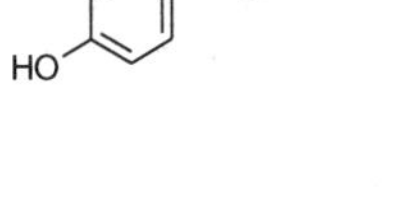

$C_9H_8O_3$, M_R 164.16. needles, mp. 210 – 213 °C, crystallizes from aqueous solution as monohydrate. TDL_0 50 mg/kg (mouse p.o.). On exposure to light dihydroxytruxillic acid (see also truxillic and truxinic acids) is formed. *p*-C. is widely distributed in plants, e.g., in the bark of *Prunus serotina* as well as in *Trifolium pratense* and in *Daviesia latifolia*. Numerous 4-*O*-glycosides are known, e.g., the glucoside of the methyl ester known as *linocinnamarin* [$C_{16}H_{20}O_8$, M_R 340.33, mp. 167 °C, $[\alpha]_D^{27}$ –73° (CH_3OH)] isolated from *Linum usitatissimum*, as well as esters of higher fatty alcohols. The biosynthesis involves hydroxylation of *cinnamic acid.

Lit.: Beilstein E IV **10**, 1005 ▪ Karrer, No. 951, 953, 955 ▪ Luckner (3.), p. 388 – 390. – *Toxicology:* Sax (8.), CNU 825. – *[CAS 501-98-4]*

***o*-Coumaric acid** see coumarinic acid.

***p*-Coumaric acid 4-*O*-(hydrogen sulfate)** [(*E*)-4-sulfooxycinnamic acid].

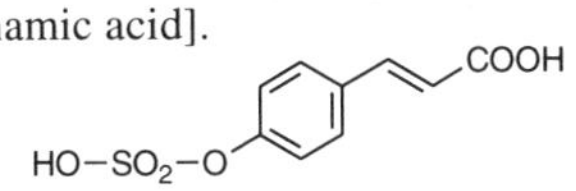

$C_9H_8O_6S$, M_R 244.22. The hydrogen sulfate of 4-hydroxycinnamic acid and related compounds inhibit the growth of marine bacteria and balanids on the eelgrass *Zostera marina*. The compounds are nontoxic for animals and are easily hydrolyzed by acids. They may be suitable for the control of biofouling in water environments.

Lit.: Phytochemistry **34**, 401 (1993). – *[HS 291829; CAS 151481-49-1]*

Coumarin (2*H*-1-benzopyran-2-one).

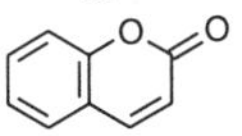

$C_9H_6O_2$, M_R 146.15. Rhombic prisms with a burning taste and pleasant, spicy odor of vanilla, mp. 70 °C, bp. 297 – 299 °C, light sensitive (formation of dimers). C. is derived from 2*H*-chromene and represents the basic skeleton of a series natural compounds such as, e.g., *umbelliferone and *furocoumarins (see also dihydroxycoumarins) [1], some of which have photoallergenic and irritating properties. C. occurs, sometimes in the form of the glucoside of *coumarinic acid, in the flowers and leaves of many grass and clover species, in sweet clover (*Melilotus* species), in woodruff (*Galium odoratum*), in lavender oil, and in the tonca bean (*Dipteryx odorata*).

Toxicology: Larger doses of C. cause severe headache, vomiting, vertigo, hypersomnia, even higher doses result in central paralysis and respiratory arrest in coma. LD_{50} 293 mg/kg (rat p.o.), 202 mg/kg (guinea pig p.o.). In addition liver and kidney damage occurs [2]. C. is carcinogenic in animal experiments.

Activity: C. and some of its derivatives act as inhibitors in plants (see plant germination inhibitors).

Biosynthesis: By spontaneous lactonization of *coumarinic acid. The glucoside of coumarinic acid occurring as a precursor is cleaved, e.g., when the plant is injured. The aromas of freshly cut grass, woodruff, and hay contain coumarin as characteristic component. For detection, see *Lit.* [3].

Use: As optical brightener and dye for lasers. As fixative in perfumes and for odorizing technical products. For coumarinoic natural products see *Lit.*[4].

Lit.: [1] Nat. Prod. Rep. **6**, 591–624 (1989); **12**, 477–505 (1995); Zechmeister **58**, 83–295 (review). [2] Food Cosmet. Toxicol. **12**, 385–405 (1974); **17**, 277 (1979). [3] Int. Flavours Food Additives **9**, 223–228 (1978). [4] *Reviews:* Barton-Nakanishi **1**, 623–638; Chem. Nat. Compd. (engl. transl.) **34**, 345 (1999); Nat. Prod. Rep. **14**, 465–475 (1997); Zechmeister **72**, 1–120.
gen.: Atta-ur-Rahman **18**, 971–1080 (^{13}C NMR coumarins) ▪ Beilstein E V **17/10**, 143 ▪ Hager (5.) **4**, 898 f.; **7**, 1112 ff. ▪ J. Chem. Soc., Perkin Trans. 1 **1987**, 1753–1756 (synthesis) ▪ Karrer, No. 1318 ▪ Ullmann (5.) **A 11**, 208–209 ▪ Zechmeister **35**, 199–430. – *Toxicology:* Sax (8.), CNV 000. – *[HS 2932 21; CAS 91-64-5]*

Coumarinic acid [(*Z*)-2-hydroxycinnamic acid].

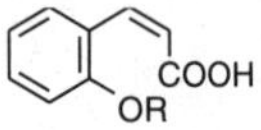

R = H : Coumarinic acid
R = β-D-Glucopyranosyl : Coumarinic acid glucoside

$C_9H_8O_3$, M_R 164.16. C. is formed from *o-coumarinic acid* [(*E*)-2-hydroxycinnamic acid, $C_9H_8O_3$, M_R 164.16, needles, mp. 207–208 °C (subl.), pK_a 4.61] by photoisomerization. The glucoside, also known as *melilotoside* {$C_{15}H_{18}O_8$, mp. 240–241 °C, $[\alpha]_D$ −60.9° (50% aqueous C_2H_5OH)} is isolated from *Melilotus alba, Hierochloe odorata, Dipteryx odorata*, and *Trigonella* species and is an intermediate in the biosynthesis of *coumarin from which free C. is formed by spontaneous lactonization. The biosynthesis of coumarinic acid proceeds through hydroxylation of *cinnamic acid.

Lit.: Beilstein E IV **10**, 999 f. ▪ Karrer, No. 949 ▪ Luckner (3.) p. 388–390 ▪ Nachr. Chem. Tech. Lab. **26**, 206–209 (1987) ▪ Zechmeister **35**, 73–132. – *[HS 2916 39; CAS 495-79-4 (C.); 614-60-8 (o-coumarinic acid); 618-67-7 (melilotoside)]*

Coumestrol (3,9-dihydroxy-6*H*-benzofuro[3,2-*c*][1]-benzopyran-6-one, 3,9-dihydroxycoumestan).

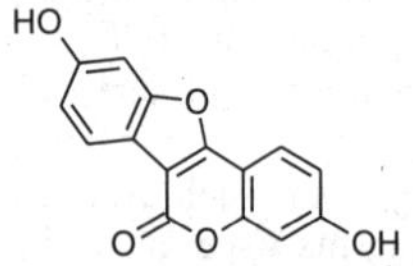

Coumestrol

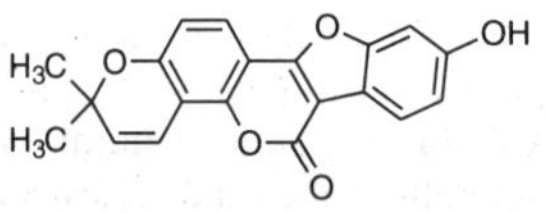

Plicadin

$C_{15}H_8O_5$, M_R 268.23. cryst., mp. 385 °C, sublimes at 325 °C, practically insoluble in acidic or neutral solvents and petroleum ether, sparingly soluble in alkalis, chloroform, ether, and methanol. Acidic to neutral aqueous solutions show blue fluorescence, alkali solutions a yellow-green fluorescence. TDL_0 35 g/kg (mouse p. o.). C. is a coumestan derivative. It occurs mainly in Fabaceae, subfamily Papilionaceae, and has been isolated from, e. g., *Medicago* species, *Glycine max, Astragalus sinicus, Centrosema pubescens, Dolichos biflorus, Melilotus alba, Phaseolus* species, *Psoralea corylifolia, Trifolium* species, *Trigonella corniculata*, and *Vigna unguiculata* as well as from *Spinacea oleracea* (Chenopodiaceae) and *Brassica oleracea* (Brassicaceae). The 3-*O*-*β*-D-glucopyranoside is known as *coumestrin*. Related to C. is *Plicadin* ($C_{20}H_{14}O_5$, M_R 334.33, needles, mp. 127 °C), isolated from *Psoralea plicata*. The total synthesis of plicadin has been published[1]. C. shows estrogenic activity and is used as an experimental carcinogen.

Biosynthesis: C. is formed by oxidative ring closure from the corresponding *isoflavone.

Lit.: [1] Angew. Chem. Int. Ed. Engl. **38**, 1435 (1999).
gen.: Beilstein E V **19/6**, 405 ▪ Luckner (3.), p. 413–414 ▪ Phytochemistry **30**, 2800 (1991) (isolation) ▪ Synth. Commun. **18**, 157–166 (1988) (synthesis of dimethyl ether) ▪ Zechmeister **43**, 1–266. – *[CAS 479-13-0 (C.); 137551-37-2 (plicadin)]*

CP-263,114. The closely related structures CP-263,114 and CP-225,917 were isolated from a *Phoma* species (phytopathogenic fungus). They are potent inhibitors of Ras farnesyl transferase (antitumor target) and squalene synthase (antifungal target) and provide a challenging task for synthetic chemists; e.g., *CP-263,114*: $C_{31}H_{36}O_9$, M_R 552.61, amorphous solid, $[\alpha]_D^{25}$ −11° (CH_2Cl_2). *Mechanism of action:* The CP compounds act as tetraanion mimics and interfere with the biosynthesis of certain organophosphates.

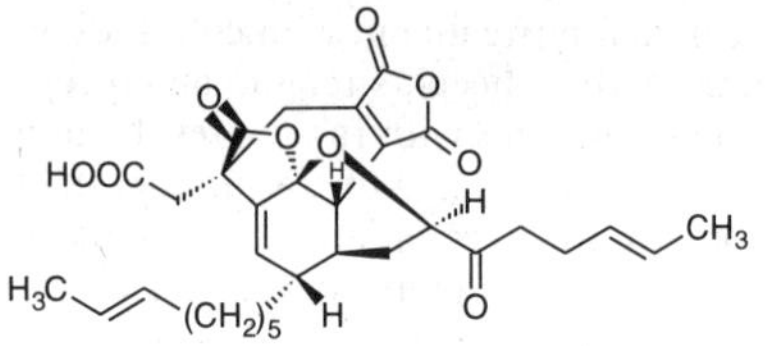

CP-263,114

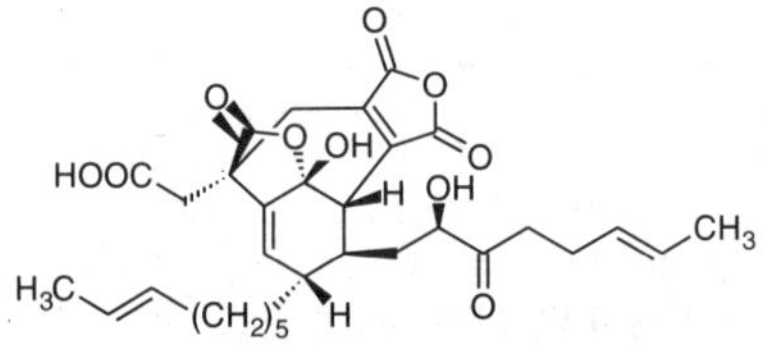

CP-225,917

Lit.: Angew. Chem. Int. Ed. Engl. **38**, 549, 1485, 1669, 1676, 3197 (1999) (synthesis); **39**, 1415–1421, 1829 (2000) (review, abs. configuration) ▪ Chem. Eng. News **77**, (23), 8 (June 7, 1999) ▪ J. Am. Chem. Soc. **119**, 1594 (1997); **121**, 890 (1999); **122**, 420 (2000) (biosynthesis) ▪ J. Antibiot. **50**, 1 (1997) (isolation) ▪ J. Org. Chem. **65**, 337 (2000) (synthesis) ▪ Org. Lett. **1**, 63 (1999). – *[CAS 186700-09-4 (CP-263,114); 166527-60-2 (CP-225,917)]*

Crambescidin see ptilomycalin.

Creatine (*N*-amidinosarcosine).

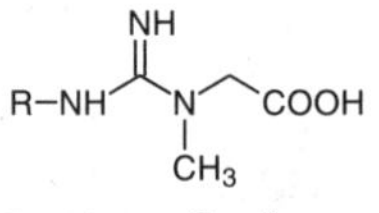

R = H : Creatine
R = PO_3^{2-} : Creatine phosphate

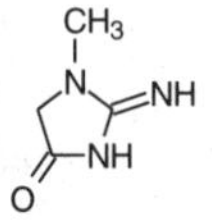

Creatinine

$C_4H_9N_3O_2$, M_R 131.13, the monohydrate in the form of monoclinic prisms looses water at 100 °C, mp. 300 °C, soluble in hot water, practically insoluble in alcohol and ether. In aqueous solution *creatinine* [$C_4H_7N_3O$, M_R 113.12, orthorhombic prisms, mp. 305 °C (decomp.), usually forms well crystalline salts with acids] is formed. C. is found in the muscle tissue of vertebrates (0.05–0.4%), and in smaller amounts in brain and blood; the name is derived from the Greek: kreas = flesh. It is formed in the organism by transamidination from L-arginine to glycine furnishing guanidinoacetic acid and subsequent methylation by *S-adenosylmethionine (through *guanidinoacetate methyltransferase*, EC 2.1.1.2). C. occurs in fresh muscle as *creatine phosphate* ($C_4H_{10}N_3O_5P$, M_R 211.12) and plays an important role as energy storage compound (so-called phosphagen). In the working muscle creatine phosphate and adenosine 5′-diphosphate form adenosine 5′-triphosphate (ATP) and C. under the action of the enzyme *creatinine kinase* ; the reverse reaction proceeds in resting muscle. Since the phosphate transfer potential of creatine phosphate is about 13 J/mol higher than that of ATP the chemical equilibrium lies more on the side of the latter. In some invertebrates L-arginine phosphate exists as phosphagen in place of creatine phosphate. C. is absorbed in the kidneys, thus its occurrence in urine is an indication for an impairment of renal function. The excreted form of C. is creatinine; the amount of creatinine excreted is proportional to muscle mass and remains more or less constant in individual subjects. The degradation of C. to urea and *sarcosine is catalysed by *creatinase* (creatine amidinohydrolase, EC 3.5.3.3) present in microorganisms. For enzymatic determination of C. and creatinine, see *Lit.*[1]. C. is considered to be an appetite stimulating component of beef and meat extracts.

Lit.: [1] Bergmeyer, Methods of Enzymatic Analysis (3.), vol. 8, p. 488–507, Weinheim: Verl. Chemie 1985.
gen.: Annu. Rev. Biochem. **54**, 831–862 (1985) ▪ Beilstein E IV **4**, 2426; E V **25/14**, 80 ▪ J. Heterocycl. Chem. **9**, 203 (1972) (synthesis of creatinine) ▪ Science **269**, 671 (1995) (analysis, diagnostics of creatinine). – *[HS 2925 20, 2933 29, 2929 90; CAS 57-00-1 (C.); 60-27-5 (creatinine); 67-07-2 (creatine phosphate)]*

Cremeomycin.

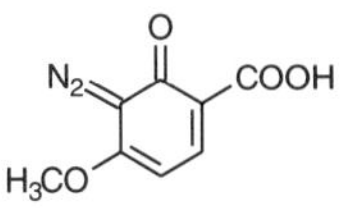

$C_8H_6N_2O_4$, M_R 194.15. Diazo compound with antitumor activity from *Streptomyces cremeus*. It inhibits the growth of murine lymphocytic leukemia L-1210 cells (yellow powder, mp. 142–143 °C, IC_{50} = 1.5 µg/mL).
Lit.: J. Antibiot. **48**, 516 (1995). – *[CAS 11050-22-9]*

Crepenynic acid see polyynes.

***p*-Cresol methyl ether** see ylang-ylang oil.

Cribrochalinamine oxide A.

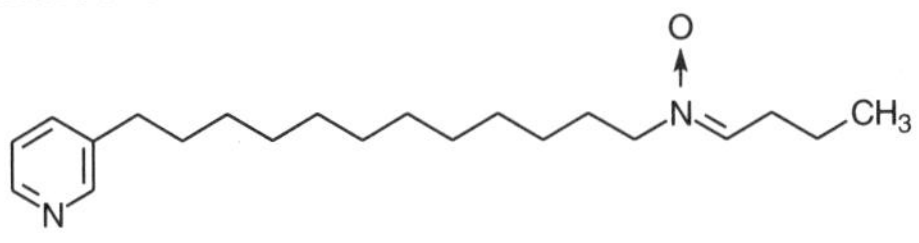

$C_{21}H_{36}N_2O$, M_R 332.53. Nitrone from the Pacific sponge *Cribrochalina* sp. with antifungal activity.
Lit.: Tetrahedron Lett. **34**, 5953 (1993). – *[CAS 152273-70-6]*

Crinine see Amaryllidaceae alkaloids.

Crinipellins.

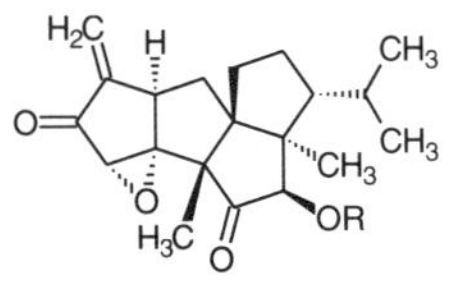

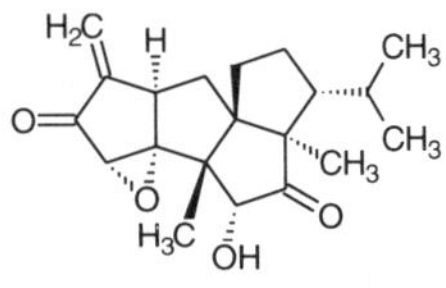

R = H : Crinipellin A
R = CO–CH$_3$: *O*-Acetylcrinipellin A
Crinipellin B

As yet the only known natural compounds with a tetraquinane skeleton, isolated from cultures of the agaric *Crinipellis stipitaria* (Basidiomycetes). The diterpenoids *C. A* {$C_{20}H_{26}O_4$, M_R 330.42, cryst., mp. 148 °C, $[\alpha]_D$ –168° ($CHCl_3$)} and *C. B* {cryst., mp. 150–151 °C, $[\alpha]_D$ –118.5° ($CHCl_3$)} differ only in the position of the α-hydroxyketone group. In addition, *O-acetyl-C. A* {$C_{22}H_{28}O_5$, M_R 372.46, cryst., mp. 107 °C, $[\alpha]_D$ –136° (ethyl acetate)} and the biologically inactive compounds dihydro-C. B and tetrahydro-C. A have been isolated. The C. and *O*-acetyl-C. A have antimicrobial and cytotoxic activities.

Lit.: Angew. Chem. Int. Ed. Engl. **24**, 118 (1985) ▪ Forum Mikrobiol. **11**, 21 (1988) ▪ J. Antibiot. **32**, 130 (1979) ▪ J. Chem. Soc., Perkin Trans. 1 **1991**, 693 (synthesis of C. A) ▪ J. Org. Chem. **58** 11 ff. (1993) (synthesis of C. B) ▪ Synthesis **1998**, Spec. Issue, 590–602 (synthesis C. B) ▪ Tetrahedron Lett. **39**, 8589 (1998) (C. B). – *[CAS 97294-60-5 (C. A); 97294-61-6 (C. B); 97315-00-9 (O-acetyl-C. A)]*

Cripowellines. The *Amaryllidaceae alkaloids C. A and B from bulbs of *Crinum powellii* represent a novel structural type. They are potent insecticides, comparable with the pyrethroids (see pyrethrum): *C. A*: amorphous solid, $C_{25}H_{31}NO_{12}$, M_R 537.51, $[\alpha]_D^{20}$ –44° (CH_3OH); *C. B*: amorphous solid, $C_{25}H_{33}NO_{11}$, M_R 523.53, $[\alpha]_D^{20}$ –64.1° (CH_3OH).

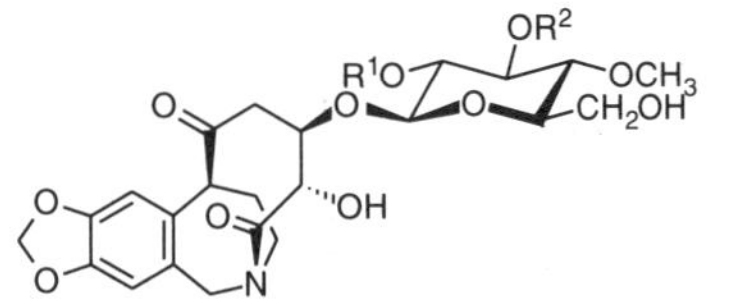

R^1, R^2 = —CH$_2$–O–CH$_2$— : C. A
R^1 = CH$_3$, R^2 = CH$_3$: C. B

Lit.: Tetrahedron Lett. **39**, 1737 (1998). – *[CAS 196814-62-7 (C. A); 196812-55-2 (C. B)]*

Crispolide see tansy.

Cristatic acid.

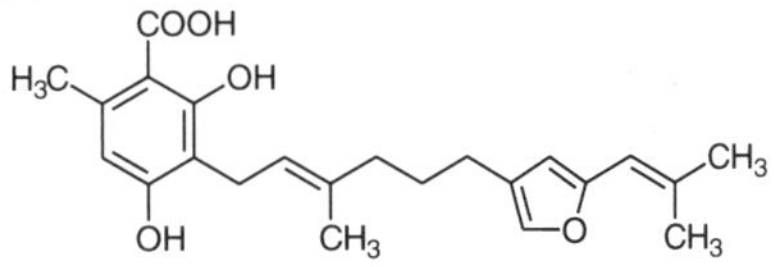

$C_{23}H_{28}O_5$, M_R 384.47, cryst., mp. 104 °C. C. is a modified farnesylphenol from fruit bodies of the green fungus *Albatrellus cristatus* (Basidiomycetes) with antibacterial, cytotoxic, and hemolytic activities.
Lit.: Justus Liebigs Ann. Chem. **1981**, 2099 ▪ Tetrahedron **44**, 41 (1988) (synthesis). – *[CAS 80557-13-7]*

Crocetin (8,8′-diapocarotene-8,8′-dicarboxylic acid).

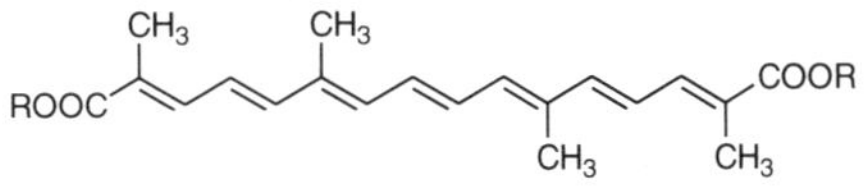

R = H : Crocetin
R = β-Gentiobiosyl : Crocin

$C_{20}H_{24}O_4$, M_R 328.41, brick-red, rhombic crystals, mp. 285–287 °C (*all-E*-form, α-C.), soluble in pyridine and other organic bases, uv_{max} 464, 436, 411 nm (pyridine). The di-β-gentiobiosyl ester of C. is called *crocin* [$C_{44}H_{64}O_{24}$, M_R 976.97, brown-red needles, mp. 180–190 °C (decomp.), soluble in hot water, poorly soluble in organic solvents] and occurs in *Crocus* (Iridaceae) and *Gardenia* (Rubiaceae) species, it is the coloring principle of saffron (*Crocus sativus*, content 24–27%). The apo-*carotinoid C. has a photosensitizing action, increases oxygen diffusion in plasma, and influences the formation of bilirubin. The monomethyl ester is *β-C.* ($C_{21}H_{26}O_4$, M_R 342.44, mp. 218 °C), the dimethyl ester is *γ-C.* ($C_{22}H_{28}O_4$, M_R 356.46, mp. 222–225 °C).

Lit.: Beilstein E V **17/8**, 112 ▪ Chem. Ber. **110**, 3582 (1977) ▪ Helv. Chim. Acta **58**, 1608 (1975); **62**, 1944 (1979) ▪ Karrer, No. 1862, 1864 ▪ Merck-Index (12.), No. 2657 ▪ Phytochemistry **21**, 1039 (1982) (biosynthesis). – *[CAS 27876-94-4 (C.); 42553-65-1 (crocin)]*

Crocin (gardenin). Formula, see under crocetin. $C_{44}H_{64}O_{24}$, M_R 976.98, brown-red needles, mp. 180–190 °C (decomp.), soluble in hot water, poorly soluble in organic solvents, uv$_{max}$ 464, 434 nm (CH_3OH). C. is the di-*β*-gentiobiosyl ester of *α*-*crocetin and occurs in *Crocus* (Iridaceae) and *Gardenia* (Rubiaceae) species and as the colored principle in saffron (*Crocus sativus*, content 24–27%).

Lit.: Beilstein E V **17/8**, 112 ▪ Helv. Chim. Acta **58**, 1608 (1975) ▪ Karrer, No. 1864 ▪ Merck Index (12.), No. 2657. – *[CAS 42553-65-1]*

Crotamine. Component of rattlesnake venom (Crotalinae) that destroys muscle fibers. This myotoxin is composed of 42 amino acids and has three disulfide bridges, isoelectric point at 10.3.

Lit.: Pharmacol. Ther. **48**, 223 (1990). – *[CAS 58740-15-1]*

Crotocin see tric(h)othecenes.

Croton oil. Fat, weakly rancid smelling oil with a burning taste from the seeds of the tree *Croton tiglium* (purging croton, Euphorbiaceae). Brown-yellow, highly viscous, toxic liquid, poorly soluble in alcohol, well soluble in glacial acetic acid, carbon disulfide, oils, petroleum ether, ether, and chloroform. C. contains, among others, the glycerol esters of stearic, palmitic, myristic, lauric, and tiglic acids. C. is not only a drastic laxative but also has a strong local irritating effect and especially acts as a cocarcinogenic agent; the active principles for the latter effect have been identified as phorbol esters (for details, see tigliane).

Lit.: Hager (4.) **4**, 346–348 ▪ Merck-Index (12.), No. 2665 ▪ Sax (8.), COC 250 ▪ Zechmeister **31**, 377–467.

Crotonosine see proaporphine alkaloids.

Crotoxin. Main component of the *snake venoms of rattlesnakes (Crotalinae). It is a complex of a basic phospholipase A_2 (C. B, M_R 13500) with an acidic protein (C. A, M_R 10000) which transports the phospholipase to its site of action, the presynaptic membrane of the neuromuscular end-plates. Poisoning leads to local pain and necrosis. Systemic sequelae are tiredness, collapse, and shock through to death. In addition to the neurotoxicity, C. also has hemolytic action.

Lit.: Hawgood & Bon, in Tu (ed.), Handbook of Natural Toxins, vol. 5, p. 53–84, New York: Dekker 1991 ▪ Pharmacol. Ther. **48**, 223 (1990). – *[CAS 9007-40-3]*

Crotsparin(in)e see proaporphine alkaloids.

Crustecdysone see ecdysteroids.

Crustinic acid. $C_{24}H_{20}O_{11}$, M_R 484.42, needles, mp. 178–180 °C (decomp.). Tridepside from the lichen

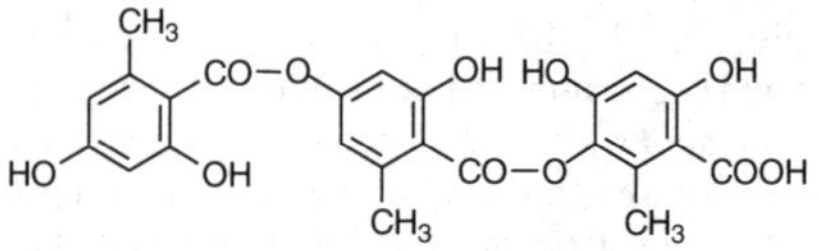

Umbilicaria crustulosa; on hydrolysis with concentrated sulfuric acid it furnishes 2 molecules *orsellinic acid and 1 molecule 5-hydroxyorsellinic acid.

Lit.: Phytochemistry **32**, 475–477 (1993).

Crustulinol diester see fasciculic acid.

Cryptaustoline see dibenzopyrrocoline alkaloids.

Cryptolepine (5-methyl-5*H*-quindoline).

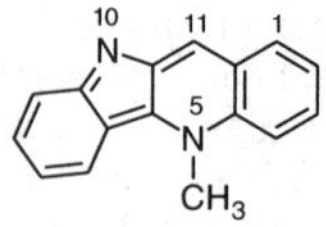

$C_{16}H_{12}N_2$, M_R 232.28, deep-purple needles, mp. 175–178 °C; an alkaloid with antihyperglycemic activity from *Cryptolepis* species, e.g., *Cryptolepis sanguinolenta* (Periplocaceae). The indoloquinoline alkaloid C. is a rare example of a natural product whose synthesis was reported prior to its isolation from nature. C. was synthesized in 1906 by Fichter and Boehringer [1] for use as a possible dye while its isolation from *Cryptolepis triangularis* was reported by Clinquart [2] 23 years later. Cryptolepine-containing plants have been used by indigenous peoples of Africa as a dye [3], and extracts obtained from various *Cryptolepis* spp. have a number of reported and current ethnomedical uses. Cryptolepine and its hydrochloride salt possess numerous biological properties, including antimalarial activity. C. was investigated as a lead structure for antidiabetic research.

Lit.: [1] Chem. Ber. **39**, 3932–3942 (1906). [2] Bull. Acad. R. Med. Belg. **9**, 627–635 (1929). [3] Kew Bull. **1926**, 225–238.
gen.: Isolation: Planta Med. **62**, 22 (1996). – *Activity:* J. Nat. Prod. **60**, 688 (1997). – *Synthesis:* J. Med. Chem. **41**, 2754–2764 (1998) ▪ J. Nat. Prod. **62**, 976 (1999) ▪ Tetrahedron Lett. **37**, 4283 (1996). – *[CAS 480-26-2]*

Cryptopine see protopine alkaloids.

Cryptopleurine see Tylophora alkaloids.

Cryptoporic acids.

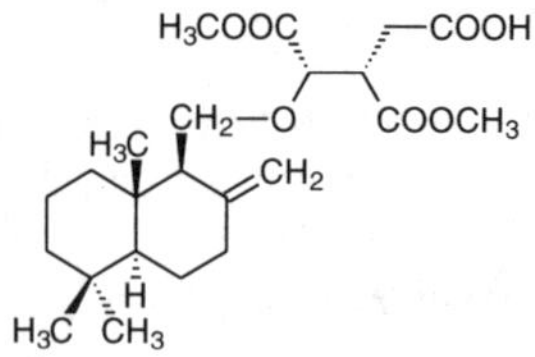

Cryptoporic acid A

Bitter sesquiterpenoids (C. A–G) of the *drimane type from the Japanese fungus *Cryptoporus volvatus* (Aphyllophorales) which grows especially on dead wood after forest fires. In addition to constituents with a sesquiterpene unit such as *C. A* {$C_{23}H_{36}O_7$, M_R 424.53, oil, $[\alpha]_D$ +45.3° ($CHCl_3$)} the macrocyclic dilactone *C. D* {$C_{44}H_{64}O_{14}$, M_R 816.98, powder, mp. 238–241 °C, $[\alpha]_D$ +39.9° ($CHCl_3$)} and further compounds with two drimane units also occur. From 40 kg fresh fungi 700 g

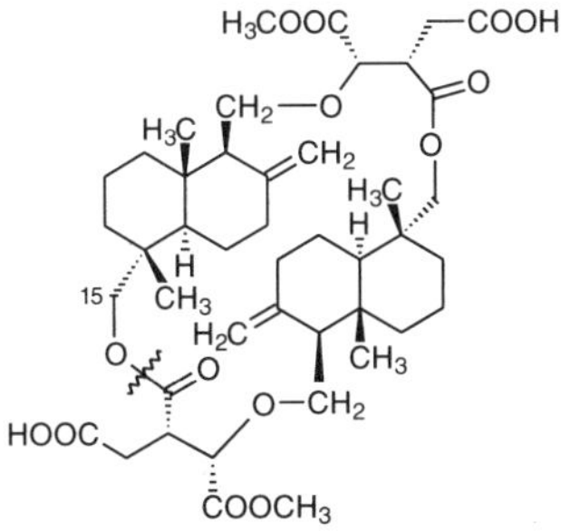

Cryptoporic acid D

C. E were isolated [$C_{45}H_{68}O_{15}$, M_R 849.02, mp. 85–88 °C (structure: open chain at C-15, methyl ester)].

Lit.: Heterocycles **47**, 1067–1110 (1998) (review) ▪ Phytochemistry **30**, 1555 (1991); **31**, 579 (1992); **32**, 891 (1993); **33**, 1055 (1994) ▪ Tetrahedron Lett. **28**, 6303 (1987). – *[CAS 113592-87-3 (C. A); 119979-95-2 (C. D)]*

Cryptosporiopsin.

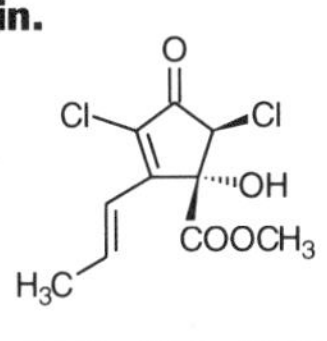

(+)-Cryptosporiopsin

$C_{10}H_{10}Cl_2O_4$, M_R 265.09, cryst., mp. 133–137 °C, $[\alpha]_D$ +129° ($CHCl_3$). Antifungal metabolite from *Cryptosporiopsis* sp. and *Sporormia affinis* (Deuteromycetes), finds use to protect pinewood against the harmful fungus *Lenzites trabea*. Interestingly, the (–)-enantiomer is produced by *Phialophora asteris* [1]. C. and some related cyclopentane derivatives probably arise, like *terrein by oxidative ring contraction of an aromatic polyketide precursor [2].

Lit.: [1] J. Am. Chem. Soc. **95**, 3000 (1973); Experientia **32**, 331 (1976); J. Chem. Soc., Perkin Trans. 1 **1983**, 2595 (synthesis). [2] Turner **2**, 94 ff. – *[CAS 25707-30-6 (+); 21652-66-4 (–)]*

Cryptostyline I see 1-phenyltetrahydroisoquinoline alkaloids.

Cryptowoline see dibenzopyrrocoline alkaloids.

Cryptoxanthin [*β*-cryptoxanthin, (3*R*)-*β*,*β*-caroten-3-ol].

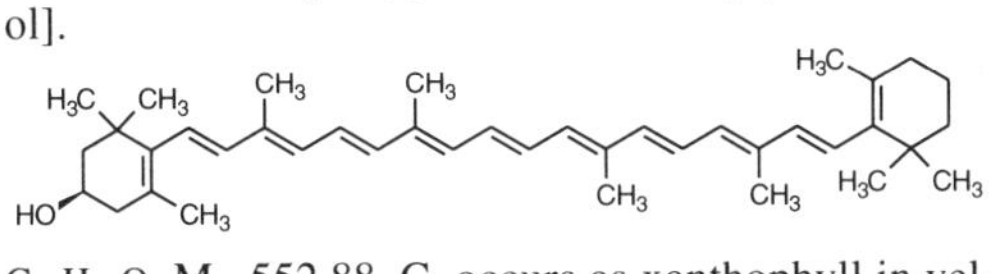

$C_{40}H_{56}O$, M_R 552.88. C. occurs as xanthophyll in yellow corn (*Zea mays*, Poaceae), egg yolk, butter, pumpkins (*Cucurbita* species, Cucurbitaceae), and in the fruit of ground cherry (*Physalis* species, Solanaceae). The *all-E*-C. forms red crystals, mp. 158–159 °C (racemate) or 169 °C (natural product), uv_{max} 452, 480 nm, well soluble in chloroform, poorly soluble in methanol and ethanol.

Lit.: Beilstein E IV **6**, 5111 ▪ Karrer, No. 1837 ▪ see also carotinoids. – *[CAS 472-70-8]*

Crystallopicrin see iridals.

CTX 1 see ciguatoxin.

Cubebin see lignans.

Cucujolides. Group of unsaturated macrocyclic lactones with 12 or 14 carbon atoms identified as *pheromones of the saw-toothed grain beetles of the genera *Cryptolestes* and *Oryzaephilus*. Some of these compounds, which occur in differing mixtures in the different species, are chiral and species-specific. Enantiomer ratios have also been observed, see also ferrulactone II.

Lit.: ApSimon **9**, 265–273, 364–365 (synthesis) ▪ J. Chem. Ecol. **13**, 1525–1542, 1543–1554 (1987).

Cucumber flavor see vegetable flavors.

Cucumins see triquinanes.

Cucurbitacins. Large group (>40) of tetracyclic *triterpenes, mostly occurring in glycosidic form (exceptions: C. A, C. C, and C. F) as poisonous bitter substances in cucumber and pumpkin plants (Cucurbitaceae) as well as in some cruciferous plants (Brassicaceae); they have also been isolated from an agaric (*hebevinosides).

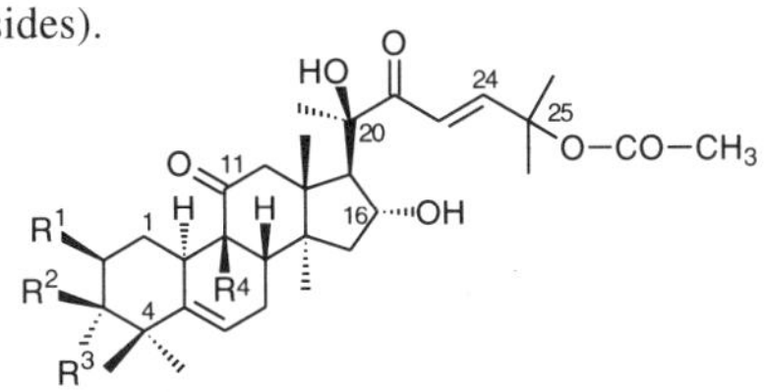

	R^1	R^2	R^3	R^4	
C. A	OH	O		CH_2OH	
C. B	OH	O		CH_3	
C. C	H	OH	H	CH_2OH	
C. E	OH	O		CH_3	1,2-Didehydro

Table: Data of Cucurbitacins.

C.	molecular formula	M_R	mp. [°C]	$[\alpha]_D$	CAS
C. A	$C_{32}H_{46}O_9$	574.71	207–208	+97.3° (C_2H_5OH)	6040-19-3
C. B	$C_{32}H_{46}O_8$	558.71	184–186	+87.5° (C_2H_5OH)	6199-67-3
C. C	$C_{32}H_{48}O_8$	560.73	207.5	+95.2° (C_2H_5OH)	5988-76-1
C. E	$C_{32}H_{44}O_8$	556.70	232–233	-60° ($CHCl_3$)	18444-66-1

The C. are unsaturated lanostane derivatives with hydroxy and oxo substituents that may be acetylated. In contrast to *lanosterol, an angular methyl group is present at C-9 and not at C-10. An 11-oxo group is characteristic for the C. To date 18 different C. have been isolated. The most abundant members are *C. B* [amarine, LD_{50} (mouse p.o.) 5 mg/kg], *C. C*, and *C. E* [*α*-elaterine, hexagonal plates, LD_{50} (mouse p.o.) 340 mg/kg]. The C. are responsible, e.g., for the taste in cucumbers that have turned bitter (gustatory threshold 10^{-6} Mol/L) and are strong laxatives. The wild cucumber was feared for this reason in ancient times. Cultivated cucumbers (*Cucumis sativus*) that have a bitter taste should not be eaten. In addition, the C. exhibit diuretic, blood-pressure lowering, and antirheumatic activities. C. are also considered to be natural defence substances of the plants against harmful pests. All C. are highly toxic. C. E can be used for "attract and kill" strategy in crop protection, because it attracts corn rootworms, a severe insect pest[1].

Lit.: [1]Chem. Eng. News (4.5.) **1998**, 40.

gen.: J. Chem. Ecol. **12**, 1109–1124 (1986) ▪ Karrer, No. 3923–3928 ▪ Tetrahedron Lett. **33**, 6755 (1992) ▪ Zechmeister **29**, 307–362.

Cularine alkaloids. A small group of benzylisoquinoline alkaloids that have an intramolecular ether bond between the A ring of the isoquinoline unit and the 1-benzyl group (oxepine ring). C. a. occur especially in Fumariaceae and in particular in *Corydalis, Sarcocapnos, Ceratocapnos,* and *Dicentra* species. They exhibit anesthetic and hypotensive activities[1,2].

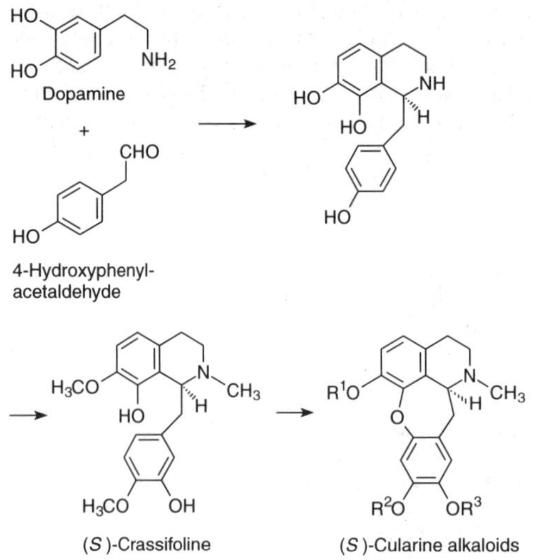

	R^1	R^2	R^3
Cularine	CH_3	CH_3	CH_3
Cularidine	H	CH_3	CH_3
Cularicine	H	CH_2	
Culacorine	H	CH_3	H

Figure: Biosynthesis of Cularine alkaloids.

Table: Data of Cularine alkaloids.

name	molecular formula	M_R	mp. [°C]	$[\alpha]_D$ (CH_3OH)	CAS
Culacorine	$C_{18}H_{19}NO_4$	313.35	250	+188°	16209-79-3
Cularicine	$C_{18}H_{17}NO_4$	311.34	185	+295°	2271-08-1
Cularidine	$C_{19}H_{21}NO_4$	327.38	157	+292°	5140-50-1
Cularine	$C_{20}H_{23}NO_4$	341.41	115	+285°	479-39-0

The biosynthesis of C. a. proceeds through oxidative phenol coupling[3,4], see figure.

Lit.: [1]Eur. J. Pharmacol. **196**, 183–187 (1991). [2]Nat. Prod. Rep. **10**, 453 (1993). [3]J. Chem. Soc., Chem. Commun. **1993**, 73 ff.; Mothes et al., p. 225 f.; Phytochemistry **25**, 111 ff. (1986). [4]Justus Liebigs Ann. Chem. **1993**, 557.

gen.: Manske **29**, 287–325 ▪ Phillipson et al., p. 102–125 ▪ Shamma, p. 165–168 ▪ Shamma-Moniot, p. 107–111 ▪ Ullmann (5.) **A 1**, 372. – *[HS 2939 90]*

Cumalin see pyrones.

Cuparenone (α-cuparenone). $C_{15}H_{20}O$, M_R 216.32, cryst., mp. 52–53 °C, $[\alpha]_D^{30}$ +177° ($CHCl_3$). A constit-

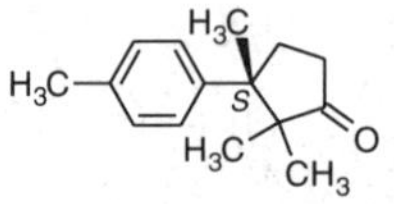

uent of the wood of "Mayur pankhi" [*Thuja occidentalis, T. occidentalis* var. *compacta* (Cupressaceae) and of the moss *Mannia fragrans*].

Lit.: Can. J. Chem. **75**, 621 (1997) (synthesis) ▪ J. Chem. Soc., Chem. Commun. **1997**, 1167 ▪ J. Org. Chem. **64**, 8728 (1999) (cuparenes) ▪ Tetrahedron **53**, 3167 (1997); **55**, 13417 (1999) (synthesis) ▪ Tetrahedron Lett. **38**, 4069 (1997). – *[CAS 16196-32-0]*

Cuprimyxin see iodinin.

Curacin A.

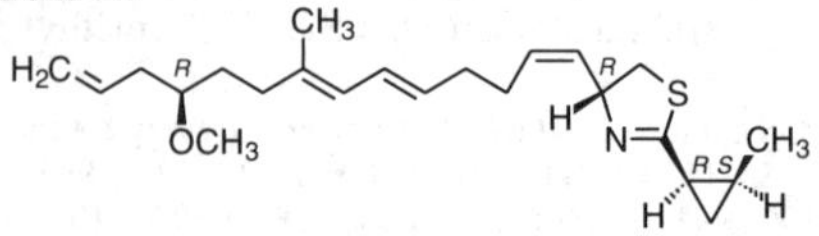

Antineoplastically active lipid from the marine cyanobacterium *Lyngbya majuscula.* $C_{23}H_{35}NOS$, M_R 373.60, oil, $[\alpha]_D^{20}$ +86° ($CHCl_3$). C. A inhibits tubulin polymerization (binds to the colchicine binding site of tubulin). IC_{50} (various tumor cell lines): 1–10 nmol/L. C. B is a stereoisomer of C. A. C. A has been a target of intense synthetic efforts.

Lit.: Bioorg. Med. Chem. Lett. **7**, 2657 (1997) (analogues) ▪ J. Am. Chem. Soc. **117**, 5612 (1995); **119**, 103 (1997) ▪ J. Chem. Soc., Perkin Trans. 1 **1999**, 2455 ▪ J. Nat. Prod. **58**, 1961 (1995) ▪ J. Org. Chem. **61**, 6556 (1996) ▪ Phytochemistry **49**, 2387 (1998) (isolation C. D) ▪ Tetrahedron **52**, 14543 (1996); **53**, 11087 (1997) ▪ Tetrahedron Lett. **36**, 1189 (1995); **37**, 953, 1795, 1799, 4397, 7167 (1996); **39**, 2861 (1998). – *[CAS 155233-30-0 (C. A); 157319-51-2 (C. B)]*

Curamycin A see avilamycin.

Curare. General name for arrow poisons produced by the Indians in tropical South America from the barks of many *Strychnos* species, especially *S. toxifera.* Other plants such as *Chondrodendron tomentosum* are used for the same purpose. For further names for the toxin, see *Lit.*[1]. Of the 40 alkaloids contained in C., only those with two quaternary nitrogen atoms show the typical toxic action. *C*-Curarine I (see toxiferines) and *C*-dihydrotoxiferine are the most active compounds. Depending on the compositions, distinctions are made between:

Calabash-C.: Thick syrupy mass, packed in gourds, the alkaloids present are mostly of the *strychnine type, rarely of the *yohimbine type. Individual compounds are *C*-*toxiferine I, *C-dihydrotoxiferine, C-curarine I,* and *C-calebassine*[2] (see toxiferines) (*C* stands for calabash).

Tubo-C.: Paste-like or hard dark mass, marketed in bamboo cans, and *pot-C.:* dry, brown-black extract, packed in small pots. Main alkaloids of this preparations are *tubocurarine* (see bisbenzylisoquinoline alkaloids) and *curarine.*

Use: C. is used in hunting, most frequently with blowguns[1].

Mechanism of action & physiological activity: The toxiferines and tubocurarine are neuromuscular blocking alkaloids and bind as antagonists to the nicotinic receptors. If C. enters the blood circulation extremely small amounts (in frogs as little as 10 μg) are sufficient to paralyze the neuromuscular end-plates in striated, voluntarily moveable muscles, so that the legs and arms, head, body and thorax successively become incapable of movement; death occurs rapidly through respiratory paralysis; cardiac muscles are not affected by the paralysis. The meat of animals killed with C. is edible since C. taken up from the gastrointestinal tract is toxic at relatively high doses only. In humans ca. 50–120 mg p.o. are poisonous. Washing of arrow wounds with potassium permanganate solution is recommended as remedy since this destroys the toxin oxidatively. Acetylcholine esterase inhibitors such as neostigmine can be administered as an antidote.

Historical[1]***:*** Isolated findings suggest a very long history for use of the poison, information about it first reached Europe after the circumnavigations by Magellan (1522), Sir Walter Raleigh (1596), and Herrera (1601). Barrère reported in 1741 that the toxin originated from a liana while Humboldt and Bonpland around 1800 were the first Europeans to witness a curare production[1]. The components of C. became known through the investigations of H. O. Wieland and especially P. Karrer who isolated over 30 different alkaloids alone from calabash-curare.

Lit.: [1] Lewin, Die Pfeilgifte, p. 413 f., Hildesheim: Gerstenberg 1984. [2] Helv. Chim. Acta **65**, 2587–2597 (1982).
gen.: Chem. Ber. **110**, 449–462 (1977) ▪ Conseiller et al., Curares and Curarisation, Amsterdam: Excerpta Med. 1980 ▪ Hager (5.) **4**, 854; **6**, 818–830, 842 ▪ Ullmann (5.) **A 1**, 370 ff. ▪ Sax (8.), COF 750, COF 825. – *[CAS 8063-06-7]*

Curcumenes.

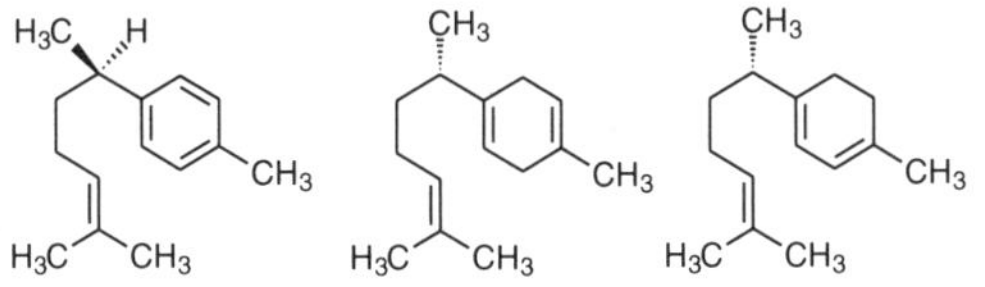

α β γ

A widely distributed group of *sesquiterpenes in the plant kingdom, e.g., from *Curcuma aromatica*, a ginger plant, and *Nectandra elaiophora* (Lauraceae) (see table below).

Lit.: Beilstein E IV **5**, 1173, 1465 ▪ J. Org. Chem. **40**, 3306 (1975); **62**, 5219 (1997) [(+)-α-C., synthesis] ▪ Karrer, No. 42, 1870, 1871 ▪ Tetrahedron **21**, 2593 (1965); **44**, 4757 (1988).

Curcuminoids. Group of substituted bis(hydroxycinnamoyl)methane pigments occurring in the rhizomes of *Curcuma* species (e.g., turmeric, *Curcuma domestica, C. zedoaria,, C. xanthorrhiza,* Zingiberaceae) and representing the main color-giving components. They include *curcumin* [bis(feruloyl)methane], *demethoxycurcumin* [feruloyl-(4-coumaroyl)methane], and *bisdemethoxycurcumin* [bis(4-coumaroyl)methane].

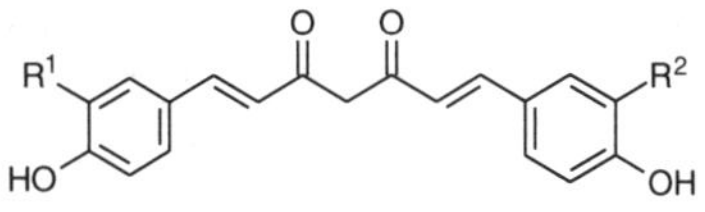

$R^1 = R^2 = OCH_3$: Curcumin
$R^1 = OCH_3$, $R^2 = H$: Demethoxycurcumin
$R^1 = R^2 = H$: Bisdemethoxycurcumin

Table: Data of Curcuminoids.

compound	molecular formula	M_R	mp. [°C]	CAS
Curcumin (C. I.75300)	$C_{21}H_{20}O_6$	368.39	183 (orange-yellow prisms)	458-37-7
Demethoxy-curcumin	$C_{20}H_{18}O_5$	338.36	168	22608-11-3
Bisdemethoxy-curcumin	$C_{19}H_{16}O_4$	308.33	224 (monohydrate)	22608-12-4

The pigment mixture from turmeric is used to dye textiles (yellow to brown, cotton, wool, and silk) and to color certain foodstuffs. Curcumin is an important constituent of Indian curry spice. Curcumin inhibits lipid peroxidation and has antibacterial activity[1]. It inhibits activation of transcription factor *NFκB* (relevant for signal transduction, septic shock, inflammatory diseases)[2].

Lit.: [1] Chem. Ind. (London) **1994**, 289. [2] Pat. WO 9709877.
gen.: Acta Chem. Scand., Ser. B **36**, 475 (1982) ▪ Hager (5.) **4**, 1084–1102 ▪ J. Chem. Soc., Perkin Trans. 1 **1973**, 2379 (biosynthesis) ▪ Phytochemistry **33**, 501–502 (1993) ▪ Ullmann (5.) **A 11**, 572.

Curtisine A.

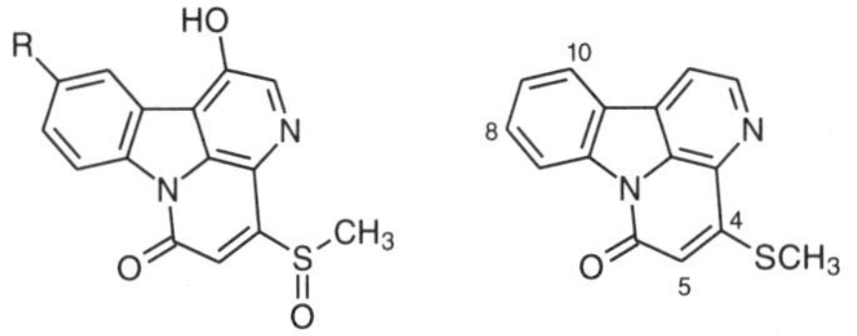

R = OH : Curtisine A
R = H : 9-Deoxycurtisine A

4-(Methylthio)canthin-6-one

$C_{15}H_{10}N_2O_4S$, M_R 314.32, light yellow needles, mp. >340 °C, $[\alpha]_D$ –149° (CH_3OH). C. and *9-deoxycurtisine A* [$C_{15}H_{10}N_2O_3S$, M_R 298.32, pale yellow solid, optically active] are responsible for the brilliant yellow

Table: Data of Curcumenes.

		molecular formula	M_R	bp. [°C] (Pa)	optical activity	CAS
α-C.		$C_{15}H_{22}$	202.34	137 ($2.26 \cdot 10^3$)		
	(*R*)-(–)-			(oil)	$[\alpha]_{546}$ –41.5°	4176-17-4
	(*S*)-(+)-				$[\alpha]_D^{20}$ +36.2° (undiluted)	4176-06-1
β-C.		$C_{15}H_{24}$	204.36	98–100 ($2.93 \cdot 10^2$)		451-56-9
	(*S*)-(+)			(oil)	$[\alpha]_D^{20}$ +26.7° (undiluted)	
γ-C.		$C_{15}H_{24}$	204.36	94 ($3.99 \cdot 10^2$)		28976-63-3 (*R*)
	(*S*)-(+)			(oil)	$[\alpha]_D^{20}$ +31.8°	

color of the north American *Boletus curtisii* (Basidiomycetes). Solutions of these compounds show an intense yellow-green fluorescence. C. occurs in the mushroom together with *canthin-6-one, *4-(methylthio)canthin-6-one* ($C_{15}H_{10}N_2OS$, M_R 266.32, colorless crystals, mp. 257–258 °C), and its 5-, 8-, and 10-methylthio-substituted isomers. Interestingly, 4-(methylthio)canthin-6-one also occurs in the Australian mangrove *Pentaceras australis* (Rutaceae)[1].

Lit.: [1] Aust. J. Res., Ser. A **5**, 387(1952).
gen.: Schering Lectures **24**, 15 (1994).

Curvularin.

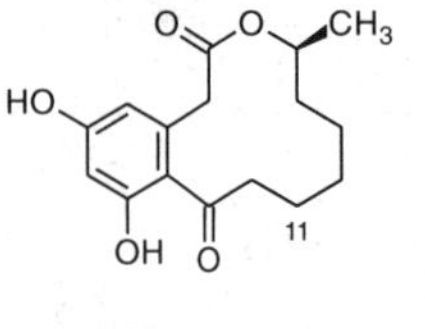

Curvularin

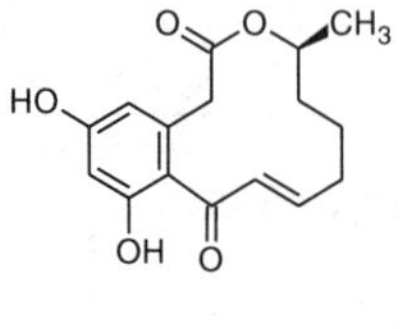

(*E*)-α,β-Dehydrocurvularin

$C_{16}H_{20}O_5$, M_R 292.33, mp. 206 °C, $[\alpha]_D$ −36.3°. Phytotoxic macrolide from *Curvularia* spp., *Penicillium gilmanii*, *P. citreo-viride*, *Alternaria cinerariae* and *A. macrospora*[1]. *11β-hydroxycurvularin* [$C_{16}H_{20}O_6$, M_R 308.33, mp. 138–140 °C, $[\alpha]_D$ −15.1° (C_2H_5OH)] from *Alternaria tomato*, *Aspergillus aureofulgens* and *Penicillium roseopurpureum* suppresses the sporulation in the producing organisms[2]. C. and 11β-hydroxy-C. act as spindle poisons and effectively inhibit cell division[3]. *(E)-α,β-dehydrocurvularin* [$C_{16}H_{18}O_5$, M_R 290.32, amorphous powder, $[\alpha]_D$ +7.3° (C_2H_5OH)] exhibits similar biological activities[4]. C. has been the target of several total syntheses[5].

Lit.: [1] J. Chem. Soc. **1959**, 3146; Aust. J. Chem. **21**, 783 (1968); J. Chem. Soc. C **1971**, 3069; Z. Naturforsch. C **36**, 1081 (1981). [2] Agric. Biol. Chem. **40**, 1663 (1976). [3] Agric. Biol. Chem. **52**, 3119 (1988); Aust. J. Chem. **46**, 571 (1993). [4] J. Chem. Soc. C **1967**, 947; J. Nat. Prod. **48**, 139 (1985). [5] Justus Liebigs Ann. Chem. **1997**, 1979; Tetrahedron **49**, 1211 (1993) (and literature cited therein).
gen.: Biosynthesis: J. Am. Chem. Soc. **111**, 3391 (1989) ▪ Heterocycles **32**, 307 (1991) ▪ Tetrahedron **54**, 15937. – *[CAS 20421-31-2 (C.); 60821-04-7 (11β-hydroxycurvularin); 21178-57-4 (α,β-dehydrocurvularin)]*

Cuscohygrine [cuskhygrine, bellardine, 1,3-bis(1-methyl-2-pyrrolidinyl)-2-propanone].

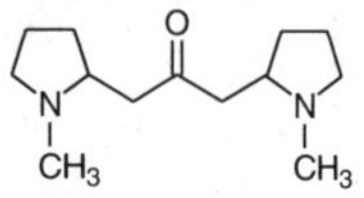

$C_{13}H_{24}N_2O$, M_R 224.35, oil, bp. 118–125 °C (0.27 kPa), crystals with 3.5 mol H_2O, mp. 40–41 °C. Pyrrolidine alkaloid occurring in many plants including coca shrub (*Erythroxylum coca*), in the roots of deadly nightshade (*Atropa belladonna*), in thorn apple species (*Datura* sp.), and in henbane (*Hyoscyamus* sp.). Many plants containing C. found use in the traditional medicine of many countries as sedatives or narcotics. Natural C. is either the *meso*-form or a mixture of the *meso*-form with the racemate. Like other alkaloids with similar structures, C. undergoes racemization readily and the optically active form is not known.

Lit.: Beilstein E V **24/1**, 458 ▪ Hager (5.) **4**, 424 f. ▪ Manske **1**, 91–106; **6**, 31–34; **27**, 272–274, 294–296 ▪ Phytochemistry **19**, 2351 (1980); **20**, 1403 (1981). – *Biosynthesis:* Phytochemistry **22**, 699 (1983) ▪ Planta Med. **56**, 339–352 (1990). – *[HS 2939 90; CAS 454-14-8 (meso-C.)]*

Cuticular hydrocarbons. Components of the hydrophobic layer on the surface of plants (cuticula) and animals (skin). For insects C. h. appear to be important not only for protection (against dehydration, infection, etc.) but also for use in chemical communication. Even in closely related species, different but specific hydrocarbon patterns can be observed. In colony-forming insects the C. h. may even be colony-specific (contribution to "hive odor"). These patterns (mixtures) generally consist of series of bis-homologous, odd-numbered alkanes, alkenes, and alkadienes (decarboxylation products of fatty acids) with 17–39 carbon atoms. This pattern is often accompanied by branched hydrocarbons in which the methylene groups in the chain are regularly separated by an odd number of methyl groups. The branching sites are the result of incorporation of propanoate or methylmalonate units in the chain. Some hydrocarbons of this type are known to be sex *pheromones of insects; see also flies.

Lit.: Arch. Insect Biochem. Physiol. **6**, 227–265 (1987) ▪ Stanley-Samuelson & Nelson (eds.), Insect Lipids: Chemistry, Biochemistry and Biology, Lincoln: University of Nebraska Press 1993.

Cutin. Structural component of the outer lipophilic protective layer (cuticle) of the aerial parts of plants, especially leaves. *Suberin serves similar functions in roots and bark. C. is a natural polyester, formed enzymatically from *hydroxyfatty acids with 16 and 18 C atoms. ω-Hydroxy- and dihydroxyfatty acids, e. g., 10,16-dihydroxypalmitic acid, as well as epoxy- and oxofatty acids, and α,ω-dicarboxylic acids are the main components of cutin. *Cutinases* (C.-cleaving enzymes) occur especially in pollen and in plant-pathogenic fungi, e. g., *Fusarium solani* (white rot in potatoes).

Lit.: Annu. Rev. Plant Physiol. **32**, 539 (1981) ▪ Holloway in Mangold (ed.), Handbook of Chromatography Lipids, vol. 1, p. 321–345, Boca Raton: CRC Press 1984 ▪ Kolattukudy, in Stumpf & Conn (eds.), The Biochemistry of Plants, vol. 4, p. 571–645; vol. 9, p. 291–314, New York: Academic Press 1980.

Cyanidin (3,3′,4′,5,7-pentahydroxyflavylium).

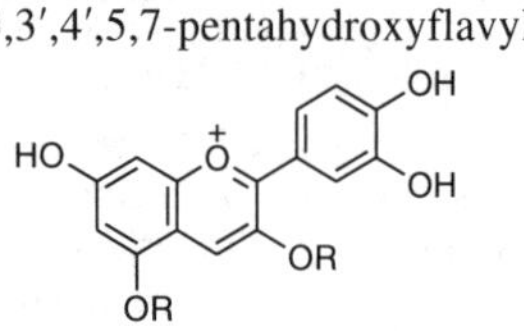

R = H : Cyanidin
R = β-D-Glucopyranosyl : Cyanin

$C_{15}H_{11}O_6^+$, M_R 287.25, as chloride ($C_{15}H_{11}ClO_6$, M_R 322.70) metallic brown-red needles (monohydrate), mp. >300 °C (decomp.), uv_{max} 535 nm, readily soluble in methanol, ethanol, and amyl alcohol (violet solution); see also cyanin.

Lit.: Beilstein E V **17/8**, 474 ▪ Karrer, No. 1712 ▪ Merck-Index (12.), No. 2757. – *[CAS 528-58-5]*

Cyanin (*cyanidin-3,5-di-*O*-β-glucopyranoside). $C_{27}H_{31}O_{16}^+$, M_R 611.53, as chloride green-yellow prismatic needles (aqueous HCl), mp. 203–204 °C, uv_{max} 522 nm (CH_3OH+HCl), $[\alpha]_D$ −258° (aqueous HCl). C.

is a frequently occurring *anthocyanin and is found, above all, in the flowers of cornflower (*Centaurea cyanus*, Asteraceae), red roses (*Rosa*, Rosaceae), greater knapweed (*Centaurea scabiosa*, Asteraceae), and purple foxglove (*Digitalis purpurea*, Scrophulariaceae).

Lit.: Beilstein E V **17/8**, 476 ▪ Karrer, No. 1715. – *[HS 3203 00; CAS 2611-67-8]*

3-Cyano-L-alanine [(*S*)-2-amino-3-cyanopropanoic acid].

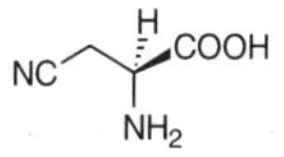

$C_4H_6N_2O_2$, M_R 114.10, mp. 218 °C, $[\alpha]_D^{26}$ −2.9° (H_2O). First isolated as a neurotoxic non-proteinogenic amino acid from the seeds of fodder vetch (*Vicia sativa*)[1] which had been contaminated with meadow pea (*Lathyrus sativus*) (a plant causing lathyrism) during harvesting. It was later found in many other *Vicia* spp.[2] as well as in the poisonous toadstool *Clitocybe acromelalga* (Tricholomataceae)[3].

Toxicity: Causes neurolathyrism[4] and cystathionuria by inhibition of cystathionine γ-lyase (EC 4.4.1.1)[5].

Biosynthesis & metabolism: From serine or cysteine and cyanide[6] catalyzed by 3-cyanoalanine synthase (EC 4.4.1.9)[7]. 3-Cyanoalanine hydratase (EC 4.2.1.65) hydrates C. to asparagine[8].

Lit.: [1] J. Biol. Chem **237**, 733–735 (1962). [2] Biochem. J. **97**, 104–111 (1965). [3] Phytochemistry **33**, 53–55 (1993). [4] J. Am. Chem. Soc. **85**, 3311–3312 (1963). [5] Biochem. Pharmacol. **16**, 2299–2308 (1967). [6] Nature (London) **206**, 110–112 (1965); **208**, 1206–1208 (1965). [7] J. Biol. Chem. **244**, 2632–2640 (1969). [8] Arch. Biochem. Biophys. **152**, 62–69 (1972).

gen.: J. Am. Chem. Soc. **110**, 2237 (1988) (synthesis) ▪ Luckner (3.), p. 248 (biosynthesis) ▪ Phytochemistry **33**, 53 f. (1993) ▪ Rosenthal, Plant Nonprotein Amino and Imino Acids, p. 60–68, New York: Academic Press 1982. – *[CAS 6232-19-5]*

Cyanocobalamine see vitamin B_{12}.

Cyanogenic glycosides (hydrogen cyanide glycosides, nitrilosides).

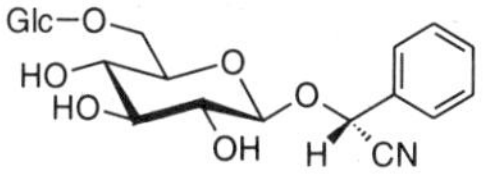

(*R*)-(-)-Form: Amygdalin

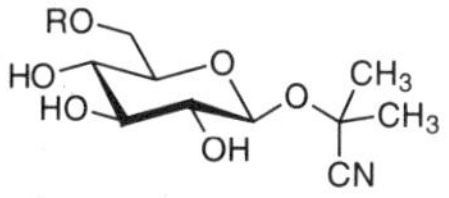

R = H : Linamarin
R = Glc : Linustatin

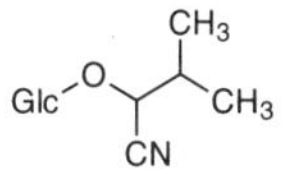

(*S*)-(-)-Form : Heterodendrin
(*R*)-(+)-Form: Epiheterodendrin

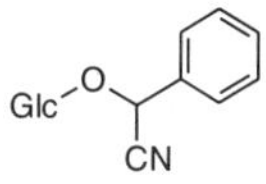

(*S*)-(-)-Form : Sambunigrin
(*R*)-(-)-Form : Prunasin
(±)-Form : Prulaurasin

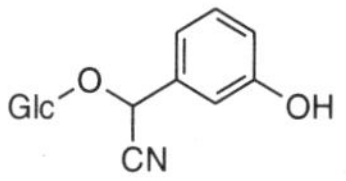

(*R*)-(-)-Form : Holocalin
(*S*)-(-)-Form : Zierin

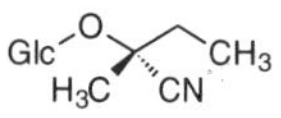

(*R*)-(-)-Form : Lotaustralin

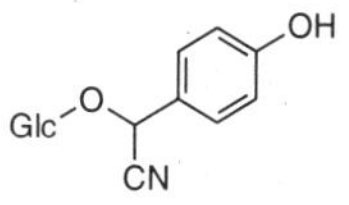

(*R*)-(-)-Form : Taxiphyllin
(*S*)-Form : Dhurrin

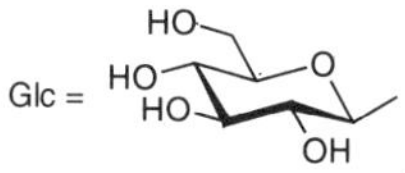

Table: Data and occurrence of Cyanogenic glycosides.

Cyanogenic glycoside	molecular formula	M_R	mp. [°C]	optical activity	occurrence	CAS
Amygdalin	$C_{20}H_{27}NO_{11}$	457.43	214 (trihydrate)	$[\alpha]_D$ −40.6° (H_2O)	widely distributed, see text	29883-15-6
Linustatin	$C_{16}H_{27}NO_{11}$	409.39	123–123.5	$[\alpha]_D^{25}$ −37° (H_2O)	*Passiflora pendens*	72229-40-4
Heterodendrin	$C_{11}H_{19}NO_6$	261.27	106–107 (tetraacetate)	$[\alpha]_D^{25}$ −45° (CH_3OH)	*Acacia sieberiana*, *Heterodendron oleaefolium*	66465-22-3
Epiheterodendrin	$C_{11}H_{19}NO_6$	261.27	139–140 (tetraacetate)	$[\alpha]_D^{25}$ +11.5° (CH_3OH)		57103-47-6
Sambunigrin	$C_{14}H_{17}NO_6$	295.29	151–152	$[\alpha]_D$ −76.3° (ethyl acetate)	*Sambucus nigra*	99-19-4
Prunasin	$C_{14}H_{17}NO_6$	295.29	147–148	$[\alpha]_D$ −29.6° (H_2O)	*Prunus* spp., *Cotoneaster* spp.	99-18-3
Prulaurasin	$C_{14}H_{17}NO_6$	295.29	122–123	$[\alpha]_D$ −53° (H_2O)		138-53-4
Holocalin	$C_{14}H_{17}NO_7$	311.29	154–155	$[\alpha]_D^{20}$ −59.1° (C_2H_5OH)	*Holocalyx balansae*	41753-54-2
Zierin	$C_{14}H_{17}NO_7$	311.29	156	$[\alpha]_D^{20.3}$ −29.5°	*Zieria laevigata*	645-02-3
Lotaustralin	$C_{11}H_{19}NO_6$	261.27	139	$[\alpha]_D^{20}$ −26.4° (H_2O)	*Lotus australis*, *Trifolium repens*	534-67-8
Taxiphyllin	$C_{14}H_{17}NO_7$	311.29	168–169	$[\alpha]_D^{20}$ −66° (C_2H_5OH)	*Taxus* spp., *Bambusa* spp., *Sorghum bicolor*, *Macadamia ternifolia*	21401-21-8
Dhurrin	$C_{14}H_{17}NO_7$	311.29	165 (monohydrate)		*Sorghum bicolor*	499-20-7
Linamarin (Phaseolunatin)	$C_{10}H_{17}NO_6$	247.25	143–144	$[\alpha]_D$ −28.5° (H_2O)	*Linum usitatissimum*, *Manihot* spp., *Phaseolus lunatus*, *Trifolium repens*	554-35-8

A widely distributed group of cyanohydrin glycosides in plants; the best known example is amygdalin (mandelonitrile gentiobioside, glucoprunasin), crystalline powder; like most aromatic C. G. it occurs mainly in the seeds of Rosaceae (stone fruits) such as cherries, peaches, apricots, plums, apples, and bitter almonds (see Table, p. 163).
For detection and for removal of linamarin from foods (*Phaseolus*, *Cassava*) and animal fodder (flaxseed) by enzymatic cleavage and extraction processes, see *Lit.*[1]. Linamarin and lotaustralin occur in the hemolymph of caterpillars of the butterfly *Zygaena trifolii*[2].
Activity: Enzymatic cleavage of C. G. by β-glucosidases (EC 3.2.1.21, e. g., emulsin or linamarase) affords *cyanohydrins which release hydrogen cyanide, thus consumption of the corresponding foodstuffs can lead to poisoning. Dhurrin inhibits the β-D-glucosidase from corn with $K_i = 1$ mM.
Toxicology: LD_{50} (mouse p. o.) 880 mg/kg (amygdalin).
Lit.: [1] Food Chem. **48**, 99–101, 263–269 (1993); J. Am. Oil Chem. Soc. **71**, 603–607 (1994); J. Sci. Food Agric. **66**, 31–33 (1994). [2] Insect Biochem. Mol. Biol. **24**, 161–165 (1994).
gen.: ACS Symp. Ser. **533**, 170–204 (1992) ▪ Barton-Nakanishi **1**, 881–896 (biosynthesis) ▪ Biosci., Biotechnol. Biochem. **62**, 453 (1998) ▪ Nat. Prod. Rep. **12**, 101–133 (1995) ▪ Phytochemistry **33**, 847–850 (1993); **34**, 433–466 (1993) ▪ Planta Med. **57**, S1–S9 (1991) ▪ Zechmeister **28**, 74–108. – *Toxicology:* Mikrooekol. Ther. **15**, 257–260 (1985) ▪ Sax (8.), AOD 500, GFC 100, IHO 700, MAP 250, MLC 750. – *[HS 2938 90]*

Cyanohydrins (natural). C. occur in the form of glycosides (*cyanogenic glycosides) or more rarely esterified with fatty acids (oleic acid and C_{20}-acids) as cyanolipids in plants. Hydrogen cyanide is liberated on enzymatic cleavage and protects the plants from attack by predators. Cyanogenic glycosides are also taken up with food by caterpillars of *Zygaena* (burnet moth) and *Heliconius* species and stored as defence substances[1]. Millipedes (Myriapoda) of the order Polydesmida use mandelic acid nitrile as defence substance which is decomposed to hydrogen cyanide and benzaldehyde in a defence gland[2]. Unusually stable C. have been isolated from marine sponges of the genus *Laxosuberites*[3].
Lit.: [1] Spencer (ed.), Chemical Mediation of Coevolution, p. 167–240, San Diego: Academic Press 1988. [2] Exp. Pharmakol. **48**, 41 (1978). [3] J. Org. Chem. **45**, 4980 (1980).

Cyanopsin. Fish *pigment of vision from 3-dehydroretinal (oxidized vitamin A_2) and opsin of the retinal uvula. The replacement of retinal by 3-dehydroretinal leads to a red-shifted absorption of visual pigment and is considered as an adaptation to the changed spectral distribution of incident light underwater.
Lit.: Merck-Index (12.), No. 2766 ▪ see also pigments of vision.

Cyasterone see ecdysteroids.

Cyathanes. Numerous diterpenoids with the cyathane skeleton have been isolated from cultures of various *Cyathus* species (birdnest fungi, basidiomycetes). Examples are the widely antibiotically active *cyathatriol* ($C_{20}H_{32}O_3$, M_R 320.47, cryst., mp. 172–173 °C), *cyathin A_3* {$C_{20}H_{30}O_3$, M_R 318.46, powder, mp. 148–150 °C, $[\alpha]_D$ –160° (CH_3OH)}, and *allocyathin*

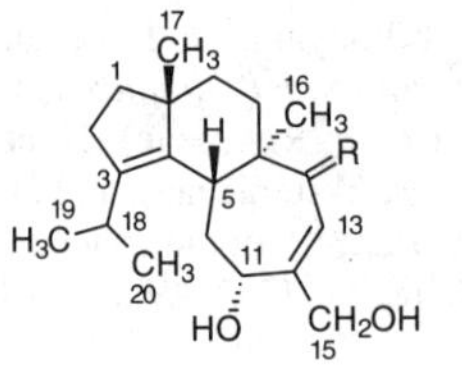

R = ····H , ◄OH : Cyathatriol
R = O : Cyathin A_3

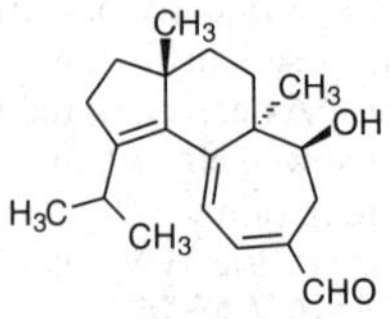

Allocyathin B_2

B_2 {$C_{20}H_{28}O_2$, M_R 300.44, yellow syrup, $[\alpha]_D$ +87.1° (CH_3OH)}. Biogenetic close relatives are *herical as well as *sarcodonin and the *striatals. C. from the monkey head mushroom (*Hericium erinaceum*) show anti-ischemic activities (improvement of cerebral ischemia in rat model).
Lit.: Beilstein E V **17/5**, 306. – *Reviews:* Phytochemistry **22**, 699 (1983) (biosynthesis) ▪ Rev. Latinoam. Quim. **9**, 177 (1978) ▪ Tetrahedron **37**, 2199 (1981) ▪ Turner **2**, 288–291. – *Synthesis:* J. Am. Chem. Soc. **118**, 7644 (1996) ▪ J. Org. Chem. **63**, 4732 (1998). – *[CAS 70116-99-3 (cyathatriol); 38598-35-5 (cyathin A_3); 70117-00-9 (allocyathin B_2)]*

Cycasin [(methyl-*ONN*-azoxy)methyl-β-D-glucopyranoside].

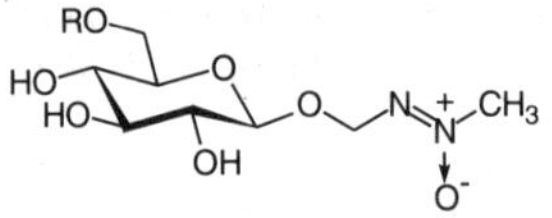

R = H : Cycasin

R = 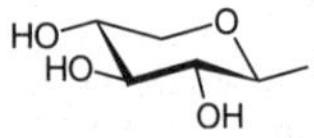 : Macrozamin

$C_8H_{16}N_2O_7$, M_R 252.22, needles, mp. 154 °C (decomp.), (144–145 °C, decomp.); $[\alpha]_D^{18}$ –44° (H_2O); alkaloid from the seeds of fern palms *Cycas circinalis* and *C. revoluta*. LD_{50} (rat) 562 mg/kg. The 6-*O*-β-D-xylopyranosyl derivative *macrozamin*, $C_{13}H_{24}N_2O_{11}$, M_R 384.34, mp. 199–200 °C, $[\alpha]_D^{16}$ –70° (H_2O) occurs in *Macrozamia riedlei*, *Bowenia serrulata*, *C. media*, and *Encephalartos* spp. (Zamiaceae, Cycadaceae). Higher oligoglycosides of methylazoxymethanol are known as neocycasins. C. also occur as defence substances in the butterflies *Eumaeus atala florida* and *Seirarctia echo* that feed on Cycadaceae[1]. Enzymatic cleavage of the glycosides affords the carcinogenic aglycone, *(methyl-ONN-azoxy)methanol* [$C_2H_6N_2O_2$, M_R 90.08, bp. 51 °C (78 Pa)] which is also present in the seeds of *C. circinalis* and has significance in food toxicology.
Lit.: [1] J. Chem. Ecol. **15**, 113–1146 (1989); Phytochemistry **25**, 1853–1854 (1986); **31**, 1955–1957 (1992).
gen.: Aust. J. Chem. **48**, 1059 (1995) (structure) ▪ Beilstein E V **17/7**, 149 ▪ Bioact. Mol. **2**, 3–24 (1987) (review) ▪ Environ. Exp. Bot. **30**, 429–434 (1990) ▪ Matsumoto, Handbook Natural Occurring Food Toxicants, p. 43–61, Boca Raton: CRC Press 1983. – *Pharmacology:* Biochim. Biophys. Acta **1193**, 151–154 (1994) ▪ Mutat. Res. **228**, 1–50 (1990) ▪ Neurology **42**, 1336–1340 (1992). – *Toxicology:* Sax (8.), COU 000, HMG 000, MGS 750. – *[HS 2939 90; CAS 14901-08-7 (C.); 6327-93-1 (macrozamin); 590-96-5 ((methyl-ONN-azoxy)-methanol)]*

Cycloalliin [(1*S*,3*R*,5*S*)-5-methyl-3-thiomorpholinecarboxylic acid *S*-oxide].

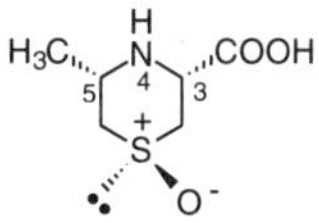

$C_6H_{11}NO_3S$, M_R 177.22, $[\alpha]_D^{20}$ −17.4° (H_2O). Non-proteinogenic amino acid in onions (*Allium cepa*) [1]. The biosynthesis proceeds through cyclization of *S*-(1-propenyl)-L-cysteine *S*-oxide [1].

Lit.: [1] Acta Chem. Scand. **13**, 623–628 (1959); **19**, 2257–2259 (1965).
gen.: Acta Crystallgr., Sect. B **28**, 2615 (1972) (absolute configuration) ▪ Beilstein E III/IV **27**, 3960 ▪ Luckner (3.), p. 279. – *[CAS 455-41-4]*

Cycloartenol.

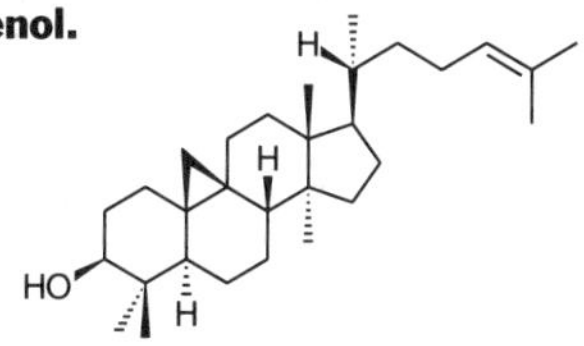

$C_{30}H_{50}O$, M_R 426.73, mp. 215 °C (anhydrous), 99 °C (hydrate), $[\alpha]_D$ +54° ($CHCl_3$). Tetracyclic triterpene alcohol with an extra cyclopropane ring. C. is isolated technically e. g. from the fruits of *Artocarpus integrifolius*, the poison nut (*Strychnos nux-vomica*), leaves of the potato plant (*Solanum tuberosum*), and the latex of Euphorbiaceae. It is present, at least in traces, in all green photosynthetically active plants since it is the first detectable cyclization product of 2,3-epoxysqualene in the course of the biosynthesis of *cholesterol (and thus all other plant steroids) in plants.

Lit.: Annu. Rev. Plant Physiol. **26**, 209–236 (1975) ▪ Beilstein E IV **6**, 4202 ▪ Helv. Chim. Acta **72**, 1–13 (1989) (structure) ▪ J. Chem. Soc., Perkin Trans. 1 **1989**, 261 (biosynthesis) ▪ Zeelen, S. 43–71. – *[CAS 469-38-5]*

Cyclobrassinin see brassinin.

Cyclobuxine D see Buxus steroid alkaloids.

Cyclochlorotine.

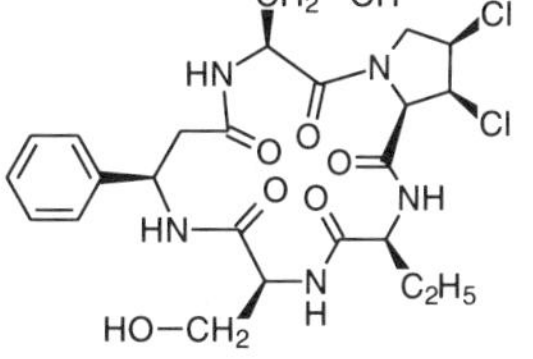

$C_{24}H_{31}Cl_2N_5O_7$, M_R 572.45, needles (CH_3OH) mp. 255 °C, $[\alpha]_D$ −92.9° (CH_3OH). Cyclic peptide built up from 2 mol L-serine and 1 mol each of L-3-amino-3-phenylpropanoic acid, L-2-aminobutanoic acid, and 3α,4α-dichloro-L-proline. It is a *mycotoxin from *Penicillium islandicum* cultures and a hepatotoxin that accelerates the degradation of glycogen so that glycogen granules in the liver disappear. The organ then appears pale and anemic. LD_{50} (mouse s. c.) 0.47 mg/kg.

Lit.: Chem. Lett. **1973**, 1319 ▪ Cole-Cox, p. 708–711 ▪ Sax (8.), COW 750. – *[CAS 12663-46-6]*

Cyclodextrins (cycloamyloses, cycloglucans; after their discoverer also Schardinger dextrins). Cyclic dextrins are formed on degradation of *starch by *Bacillus macerans* or *B. circulans* under the action of cyclodextrin glycosyltransferase. The C. consist of 6, 7, or 8 α1-4′-linked glucose units (α-, β-, or γ-C.). The figure shows α-C. ($C_{36}H_{60}O_{30}$, M_R 934.43).

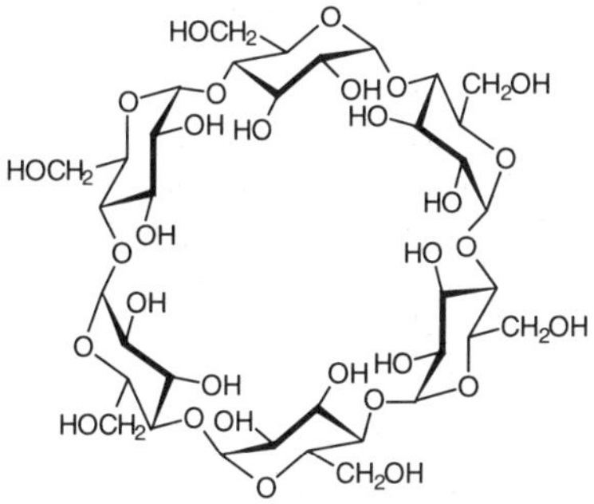

These *cyclohexa-(-hepta-, -octa-)amyloses* are arranged in the crystal lattice of C. in such a way that open, intramolecular channels are formed in which guest molecules can be enclosed in varying amounts up to saturation ("molecular encapsulation"), e. g., gases, alcohols, or hydrocarbons. α-C. also forms a blue-colored inclusion compound with iodine in which the iodine atoms are arranged like a string-of-pearls in the channels. On account of this property C. are used in the production of food, cosmetics, pharmaceuticals, and pesticides as well as for solid phase extractions and for use as high performance separation phases for enantiomeric and diastereomeric mixtures.

Lit.: Adv. Carbohydr. Chem. Biochem. **46**, 205–249 (1988) ▪ Adv. Photochem. **21**, 1–133 (1996) (photochemistry) ▪ Angew. Chem. Int. Ed. Engl. **19**, 344–362 (1980) ▪ Bender & Komiyama, Cyclodextrin Chemistry, Berlin: Springer 1978 ▪ Chem. Eur. J. **1**, 33–55 (1995) (literature review) ▪ Duchêne, New Trends in Cyclodextrins and Derivatives, Paris: Ed. de la Santé 1991 ▪ Duchêne, Cyclodextrins and their Industrial uses, Paris: Ed. de la Santé 1987 ▪ Exp. Opin. Ther. Patents **9**, 1697–1717 (1999) ▪ Kirk-Othmer (4.) **14**, 129 ff. ▪ Szejtli, Cyclodextrins and their Inclusion Complexes, Budapest: Akademiai Kiado 1982 ▪ Ullmann (5.) **A 14**, 121–126. – *[CAS 10016-20-3 (α-C.); 7585-39-9 (β-C.); 17465-86-0 (γ-C.)]*

Cycloleucomelone.

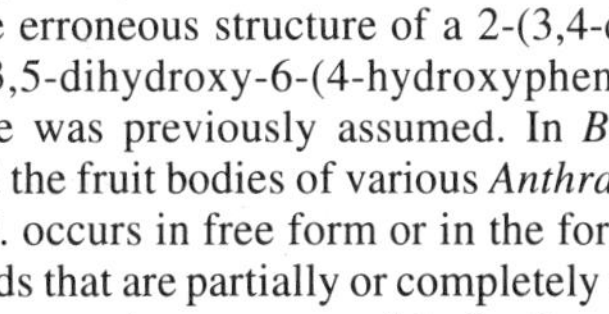

$C_{18}H_{10}O_7$, M_R 338.27, brown plates or powder, mp. 320 °C (decomp.). Component of the fruit bodies of *Boletopsis leucomelaena* (=*Polyporus leucomelas*, Basidiomycetes). C. is identical to "leucomelone" for which the erroneous structure of a 2-(3,4-dihydroxyphenyl)-3,5-dihydroxy-6-(4-hydroxyphenyl)-*p*-benzoquinone was previously assumed. In *B. leucomelaena* and the fruit bodies of various *Anthracophyllum* species C. occurs in free form or in the form of leuco compounds that are partially or completely acetylated. These compounds are responsible for the green color reaction of fruit bodies on treatment with alkalis. Polyacetylated leucocycloleucomelones are effective inhibitors of 5-lipoxygenase.

Lit.: Chem. Pharm. Bull. **40**, 3194 (1992) (activity) ▪ Z. Naturforsch. B **42**, 1349, 1354 (1987) ▪ Zechmeister **51**, 27–30 (review). – *[CAS 112209-48-0]*

Cyclopamine see Veratrum steroid alkaloids.

Cyclopenin(e).

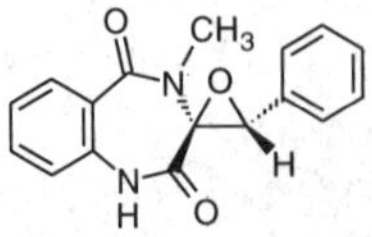

$C_{17}H_{14}N_2O_3$, M_R 294.30, needles, mp. 183–184 °C, $[\alpha]_D^{20}$ –291° (CH_3OH). A metabolite of *Penicillium cyclopium* and *P. corymbiferum* with growth-inhibiting activity on plants.

Lit.: J. Chem. Soc., Perkin Trans. 1 **1976**, 1564 (biosynthesis) ▪ J. Org. Chem. **42**, 3650 (1977); **47**, 2456 (1982) (synthesis of (±)-C.). – *[CAS 20007-87-8]*

Cyclopentenyl fatty acids see chaulmoogric acid.

Cyclopeptide alkaloids. *Peptide alkaloids with complex structures occurring in tree- or shrub-like members of the Rhamnaceae family, later found also in other plant families (Sterculiaceae, Pandaceae, Rubiaceae, Urticaceae, Hymenocardiaceae, and Celastraceae). For most of the well over 100 representatives a common structural pattern can be recognized (see figure).

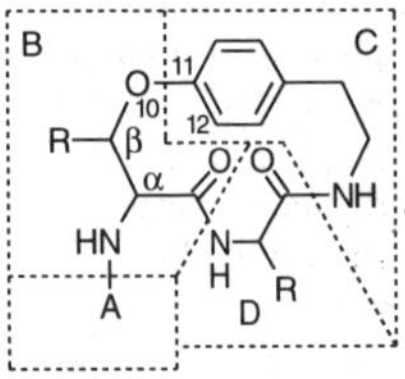

Figure: Structural features of Cyclopeptide alkaloids.
A = basic amino acid; B = β-hydroxyamino acid, C = tyramine derivative, D = ring-linked amino acid. In the so-called type 4 (4 building blocks) A is an *N,N*-dimethylamino acid, in type 5 (5 building blocks) a further amino acid is inserted between A and B. Many C. a. form a 14-membered ring system; in the 13-membered basic skeleton O-12 is linked to C-12.

As typical examples of C. a. *amphibine A* (type 4) and *zizyphine A* (type 5) from *Zizyphus* species as well as *frangulanine* (type 4) occurring in small amounts in bark and leaves of buckthorn (*Rhamnus frangula*) are illustrated:

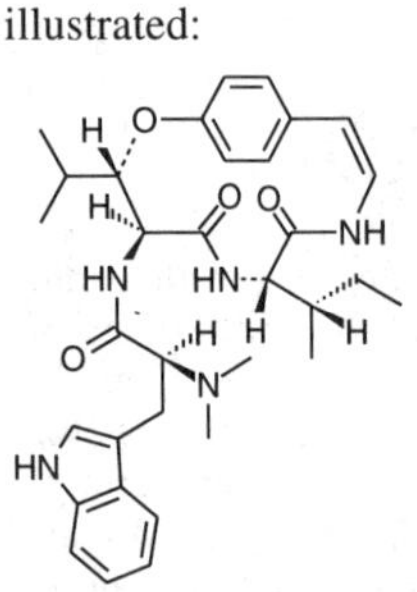

Amphibine A

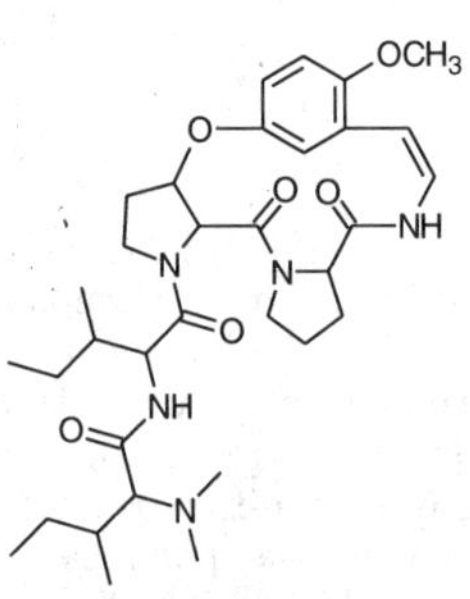

Zizyphine A

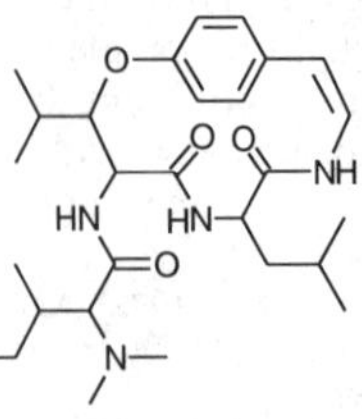

Frangulanine

Table: Data of Cyclopeptide alkaloids.

	Amphibine A	Frangulanine	Zizyphine A
molecular formula	$C_{33}H_{43}N_5O_4$	$C_{28}H_{44}N_4O_4$	$C_{33}H_{49}N_5O_6$
M_R	573.74	500.68	611.78
mp. [°C]	237–239	276–279	124–126
$[\alpha]_D$	–310° (CH_3OH)	–288° ($CHCl_3$)	–411° ($CHCl_3$)
CAS	36535-97-4	25350-22-5	51059-42-8

Little is known on the biological activities of C. a. because it is difficult to obtain adequate amounts of the pure substances. Apart from weak antibiotic activities, membrane effects (ionophoric) on mitochondria have been reported and an inhibition of phosphorylation in isolated spinach chloroplasts has been demonstrated.

Lit.: Hegnauer **VI**, 65–68; **IX**, 353–358 ▪ Manske **15**, 165–205 ▪ Nachr. Chem. Techn. Lab. **37**, 1034–43 (1989) (review) ▪ Nat. Prod. Rev. **14**, 75 (1997) (review) ▪ Pelletier **3**, 113–168 ▪ Tetrahedron Lett. **39**, 283 (1998) (synthesis zizyphine); **40**, 83, 455 (1999) ▪ Zechmeister **75**, 1–179 (review). – *[HS 2939 90]*

Cyclophellitol.

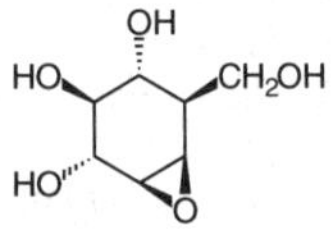

$C_7H_{12}O_5$, M_R 176.17, platelets, mp. 149–151 °C, $[\alpha]_D$ +103° (H_2O). β-Glucosidase inhibitor from wood-rotting fungi of the genus *Phellinus* (Aphyllophorales).

Lit.: Arch. Biochem. Biophys. **297**, 362 (1992) (biological activity) ▪ J. Antibiot. **43**, 49, 1579 (1990). – *Synthesis:* Bull. Chem. Soc. Jpn. **66**, 3760 (1993) ▪ Carbohydr. Res. **222**, 189 (1991) ▪ Chem. Lett. **1994**, 37 ▪ J. Antibiot. **44**, 456 (1991) ▪ J. Chem. Soc., Perkin Trans. 1 **1994**, 2017 ▪ J. Org. Chem. **62**, 3360 (1997); **63**, 7920 (1998) ▪ Synlett **1991**, 831. – *[CAS 126661-83-4]*

Cyclophostin.

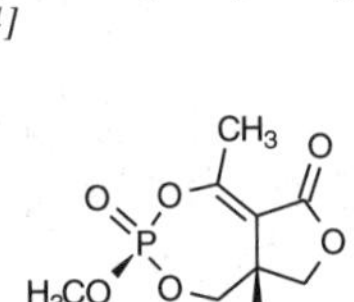

$C_8H_{11}O_6P$, M_R 234.15, cryst., mp. 113–114 °C, $[\alpha]_D$ –7.56° (C_2H_5OH). Insecticide from cultures of *Streptomyces lavendulae* [1]. Structurally related with an insecticidal cyclic organophosphate from *S. antibioticus* [2]. C. is one of the most effective inhibitors of acetylcholine esterase of common house flies (I_{50} = 7.6×10^{-10} M).

Lit.: [1] J. Antibiot. **46**, 1315 (1993). [2] Experientia **43**, 1235 (1987). – *[CAS 144773-26-2]*

Cyclopiazonic acid (α-cyclopiazonic acid).

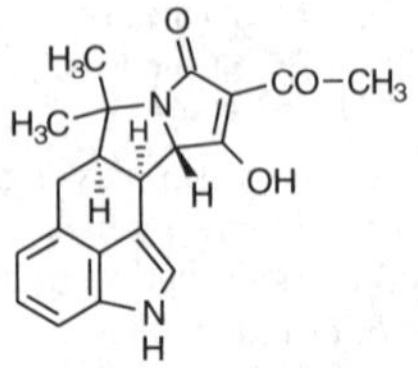

(relative configuration)

$C_{20}H_{20}N_2O_3$, M_R 336.39, cryst., mp. 245–246 °C. *Mycotoxin from many *Penicillium* species (e. g., *P. cyclopium, P. camemberti, P. viridicatum*) and *Aspergillus* species (e. g., *A. flavus, A. oryzae, A. versicolor*) that attack beans, corn flour, wheat, and sausage products. *P. camemberti* and other fungal strains that are used in cheese production are free of the toxin! C. exhibits neurotoxic activity [LD_{50} (rat p. o.) 36–63 mg/kg; (rat i. p.) 2.3–3.8 mg/kg].

Lit.: Aust. J. Chem. **45**, 99–107 (1992) ▪ Betina, p. 417–420 ▪ Cole-Cox, p. 497–500 ▪ J. Biol. Chem. **272**, 2794 (1997) (pharmacology) ▪ J. Chem. Soc., Chem. Commun. **1982**, 1367 ▪ J. Pharmacol. Exp. Ther. **283**, 286 (1997) (pharmacology) ▪ Phytochemistry **10**, 351 (1971) (biosynthesis) ▪ Reiß (ed.), Mycotoxine in Lebensmitteln, p. 297–341, Stuttgart: Fischer 1981 ▪ Sax (8.), CQD 000 ▪ Steyn & Vleggar (eds.), Mycotoxins and Phycotoxins, p. 239–250, Amsterdam: Elsevier 1986. – *[CAS 18172-33-3 (C.); 83136-88-3 (other enol tautomer)]*

Cycloposine see Veratrum steroid alkaloids.

Cycloserine.

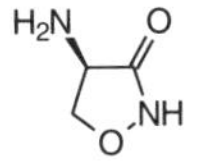

Generic name for (+)-4-amino-3-isoxazolidinone, $C_3H_6N_2O_2$, M_R 102.09, mp. 155–156 °C, $[\alpha]_D^{23}$ +116° (H_2O), readily soluble in water, poorly soluble in alcohols; stable in alkaline aqueous solutions. The *antibiotic obtained from cultures of *streptomycetes (e. g., *Streptomyces orchidaceus*) acts as an antagonist of D-*alanine required for the construction of murein for bacterial cell walls. LD_{50} (mouse p. o.) 5290 mg/kg. C. inhibits alanine racemase (EC 5.1.1.1) and D-alanyl-D-alanine synthetase (EC 6.3.2.4) of bacteria.

Use: As broad-spectrum antibiotic, especially against mycobacteria (tuberculosis), rickettsiae, protozoa, and infections of the urinary tract. It finds only limited clinical applications on account of side effects.

Lit.: Anal. Profiles Drug Subst. **18**, 567 (1989) ▪ Beilstein E V **27/20**, 3 ▪ Martindale (30.), p. 156 ▪ Met. Ions Biol. Syst. **19**, 295 (1985). – *[HS 2941 90; CAS 68-41-7]*

Cyclosporins.

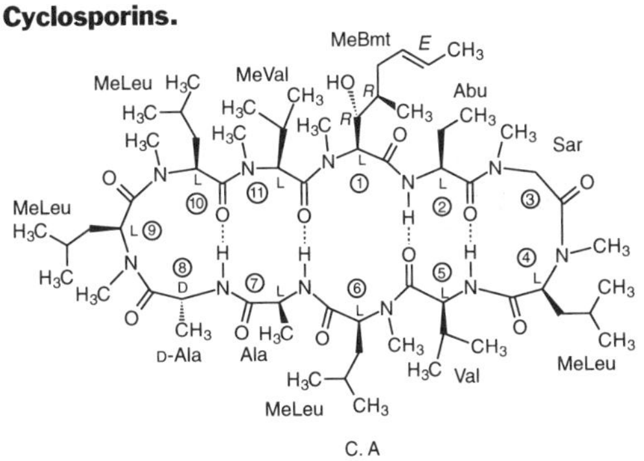

C. A

Abu = L-α-Aminobutyric acid
MeBmt = (4*R*)-4-((*E*)-2-Butenyl)-*N*,4-dimethyl-L-threonine
MeLeu = *N*-Methyl-L-leucine
MeVal = *N*-Methyl-L-valine
Sar = *N*-Methylglycine (*Sarcosine)

A group of cyclic oligopeptide *antibiotics (C. A to Z) discovered in 1972 from lower fungi (e.g., *Trichoderma polysporum*) with remarkable biological activities. *C. A*: $C_{62}H_{111}N_{11}O_{12}$, M_R 1202.63, needles, mp. 148–151 °C, $[\alpha]_D^{20}$ –244° ($CHCl_3$), LD_{50} (rat p. o.) 1489 mg/kg; poorly soluble in chloroform, consists of 11 *amino acids, some of which are *N*-methylated, linked to form a ring. The minor components exhibit close structural and in some cases also activity relationships to C. A. In these the amino acids in almost all positions are chemically modified or exchanged. The building blocks MeBmt, α-aminobutyric acid, *sarcosine, and *N*-methylvaline in the positions 1, 2, 3, and 11 are essential for immunological activity.

Activity: C. A has strong immunosuppressive activity (commercially available as Sandimmun®) and is used in transplantation medicine to suppress the immune response [1]. The activity is directed at the T-helper cells that play a significant role in the cellular immune response to transplanted tissue. The absorbed C. A is bound to serum proteins and biotransformed to hydroxylated derivatives by blood serum.

The occurrence of numerous secondary metabolites and the unusual structural elements L-α-aminobutyric acid, D-alanine, and the C_9-amino acid (2*S*,3*R*,4*R*,6*E*)-2-amino-3-hydroxy-4-methyl-6-octenoic acid [(4*R*)-4-((*E*)-2-butenyl)-4-methyl-L-threonine = Bmt] indicate that this peptide is not produced ribosomally but rather post-translationally by a multi-enzyme complex [2]. Additionally offered "foreign" amino acids are incorporated, e. g., further cyclosporins are obtained in directed biosyntheses by replacement of the α-aminobutyric acid in position 2 by D- or L-allylglycine, of D-alanine in position 8 by D-serine or 3-fluoroalanine, and of butenylmethylthreonine in position 1 by L-β-cyclohexylalanine [3].

The biogenesis of the amino acids in position 2-11 in C. A proceeds through the usual biosynthetic routes with subsequent *N*-methylation by *S-adenosylmethionine-dependent methyltransferases. The amino acid in position 1 is built up on the *polyketide pathway and subsequently methylated at C-4 [4]. It has been demonstrated that MeBmt is the "starter" of the non-ribosomal peptide synthesis by means of a mutant strain of *Tolypocladium inflatum* [5]. Recent clinical data show, that C. application in transplantation medicine significantly increases the risk of cancer [6].

Lit.: [1] Lancet **2**, 1323, 1327 (1978); Chem. Eng. News (15. 4.) **1996**, 30 f.; Krohn et al., Antibiotics and Antiviral Compounds, p. 229–240, Weinheim: VCH Verlagsges. 1993; Immunology Today **13**, 136–142 (1992); Naturwiss. Rundsch. **43**, 541 (1990) (C. A). [2] Phytochemistry **23**, 549 (1984); Eur. J. Appl. Microbiol. Biotechnol. **14**, 237 (1982). [3] J. Antibiot. **42**, 591–597 (1989). [4] Kahan (ed.), Cyclosporins, London: Grune & Stratton **1987**. [5] J. Antibiot. **43**, 707–712 (1990). [6] Nachr. Chem. Tech. Lab. **47**, 439 (1999).

gen.: Barton-Nakanishi **1**, 533–556 (review biosynthesis) ▪ Drugs **43**, 440 (1992); **45**, 953 (1993) (review) ▪ Helv. Chim. Acta **78**, 355 (1995) (structure) ▪ Martindale (30.), p. 1882 (pharmacology) ▪ Tejane, Cyclosporin in the Therapy of Renal Disease, Basel: Karger 1995 (review). – *[CAS 59865-13-3 (C. A)]*

Cyclotenes see 2-hydroxy-3-methyl-2-cyclopenten-1-one.

Cyclotheonamides see theonellamides.

Cyclothialidine.

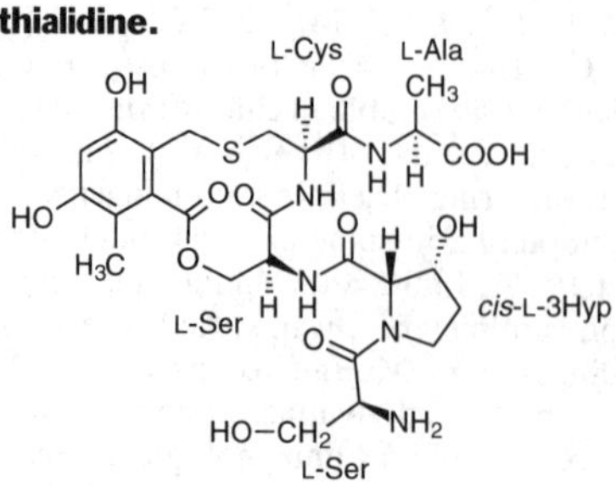

C.

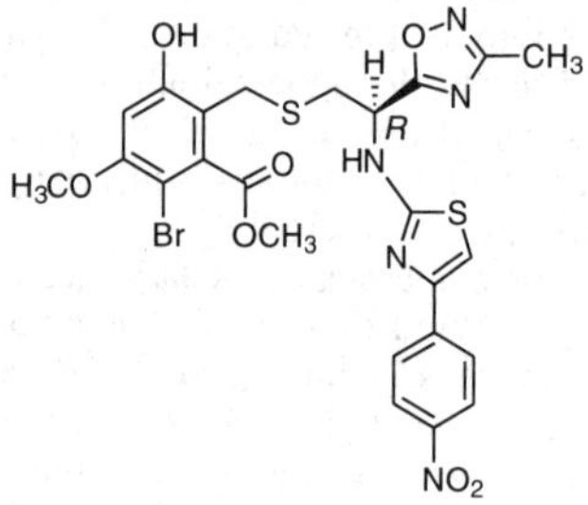

RO-61-6653

$C_{26}H_{35}N_5O_{12}S$, M_R 641.65, yellow powder, $[\alpha]_D$ –13.2° (H_2O). A topoisomerase II (DNA gyrase) inhibitor from cultures of *Streptomyces filipinensis*. C. is a novel antibacterial lead structure. Significant improvement of activity has been found with the analogue RO-61-6653 of seco-cyclothialidine.

Lit.: Antimicrob. Agents Chemother. **40**, 473 (1996) (activity) ▪ Helv. Chim. Acta **79**, 2219–2234 (1996) (synthesis) ▪ J. Antibiot. **47**, 32, 37 (1994) (activity); **50**, 402–411 (1997) (synthetic analogues). – *[CAS 147214-63-9 (C.); 173153-97-4 (RO-61-6653)]*

Cylindrospermopsin.

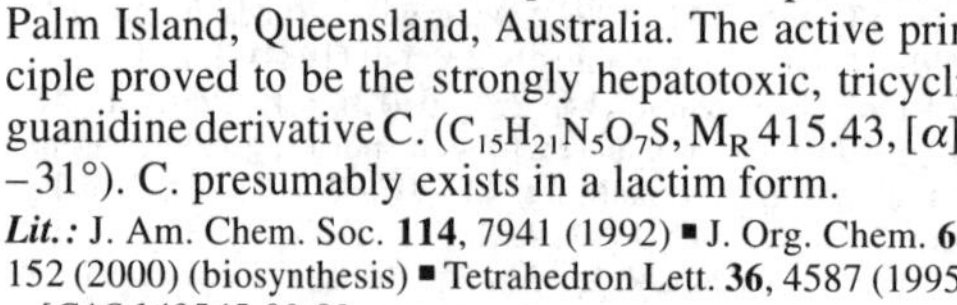

The cyanobacterium *Cylindrospermopsis raciborskii*, which is normally nontoxic, was found to be responsible for the outbreak of a hepatoenteritis epidemic on Palm Island, Queensland, Australia. The active principle proved to be the strongly hepatotoxic, tricyclic guanidine derivative C. ($C_{15}H_{21}N_5O_7S$, M_R 415.43, $[\alpha]_D$ –31°). C. presumably exists in a lactim form.

Lit.: J. Am. Chem. Soc. **114**, 7941 (1992) ▪ J. Org. Chem. **65**, 152 (2000) (biosynthesis) ▪ Tetrahedron Lett. **36**, 4587 (1995). – *[CAS 143545-90-8]*

Cymarin see strophanthins.

Cymenes [cymol(e)s].

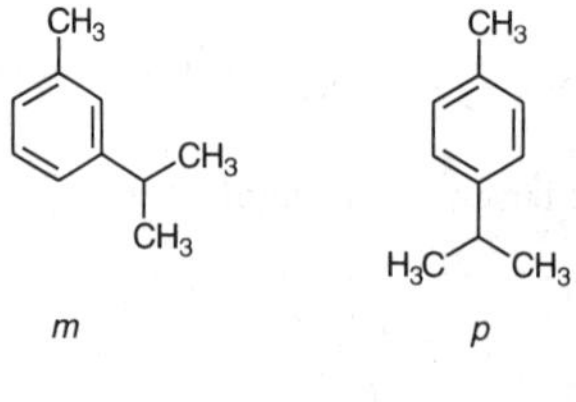

m *p*

$C_{10}H_{14}$, M_R 134.22, oil; *m*-C. bp. 176–177 °C, *p*-C. bp. 186–188 °C. Aromatic monocyclic *monoterpenes, the *o*-C. isomer does not occur naturally. In higher concentrations the vapors irritate the eyes and airways, LD_{50} (rat p. o.) 4.7 g/kg.

Occurrence: *m*-C. in the essential oils of leaves and fruits of the black currant(*Ribes nigrum*, Saxifragaceae). *p*-C. is widely distributed in the plant kingdom and is isolated from, e. g., *turpentine, cypress oil, *cinnamon, *eucalyptus, *thyme oil and many others. C. is used as an aroma substance in cosmetics.

Lit.: Beilstein E IV **5**, 1060 ▪ Karrer, No. 39, 5620 ▪ Merck-Index (12.), No. 2832. – *[HS 2902 90; CAS 535-77-3 (m-C.); 99-87-6 (p-C.)]*

Cymenols.

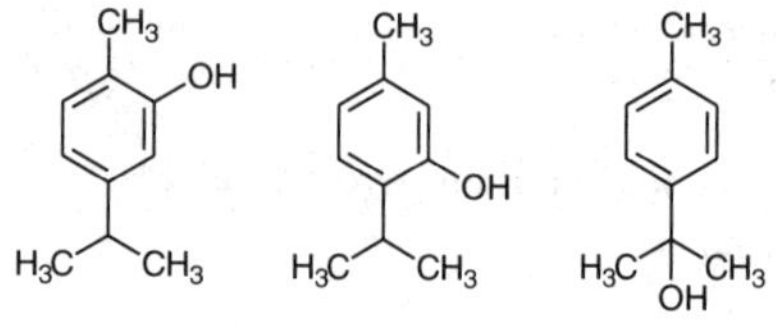

p-Cymen-2-ol (*Carvacrol*) *p*-Cymen-3-ol (*Thymol*) *p*-Cymen-8-ol

$C_{10}H_{14}O$, M_R 150.22; *p*-cymen-2-ol (*carvacrol*), liquid, bp. 238 °C; *p*-cymen-3-ol (*thymol*), cryst., mp. 50–51 °C; *p*-cymen-8-ol, liquid, bp. 111–112 °C (16 mbar). Aromatic *monoterpene phenols or alcohols.

Occurrence: In numerous essential oils: cymen-2-ol especially in *origanum (oregano) oil (*Coridothymus capitatus*, Lamiaceae) up to 80%, in savory and *thyme oil up to 70%; cymen-3-ol especially in essential oil of seeds of *Orthodon angustifolium* (Lamiaceae) up to 72%; cymen-8-ol from the frass of the old house borer beetle *Hylotrupes bajulus* (Cerambycidae)[1].

Use: Carvacrol as disinfectant and solvent; thymol as stabilizer (0.01%) of halothanes and trichloroethene. Mixtures of thymol with methyl *N*-methylanthranilate in appropriate ratios have a typical mandarin odor[2].

Lit.: [1] Tetrahedron Lett. **1975**, 3585. [2] Helv. Chim. Acta **46**, 1480 (1963).

gen.: Beilstein E IV **6**, 3334 ▪ Karrer, No. 174, 176, 5642. – *[HS 2907 19; CAS 499-75-2 (p-cymen-2-ol); 89-83-8 (p-cymen-3-ol); 1197-01-9 (p-cymen-8-ol)]*

Cymol(e)s see cymenes.

Cynarin(e) (1,3-di-*O*-caffeoylquinic acid).

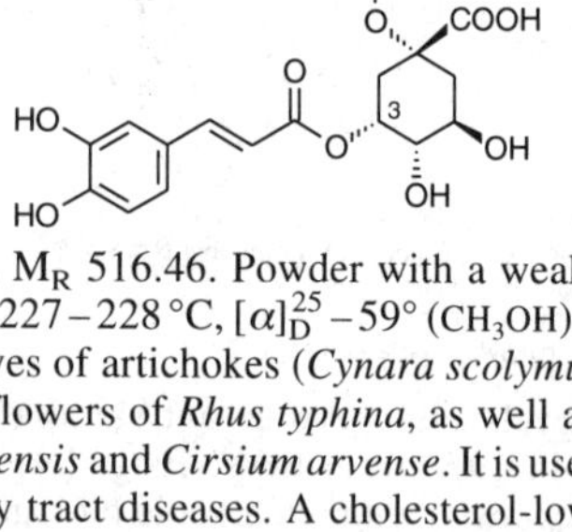

$C_{25}H_{24}O_{12}$, M_R 516.46. Powder with a weak sweetish taste, mp. 227–228 °C, $[\alpha]_D^{25}$ –59° (CH_3OH). C. occurs in the leaves of artichokes (*Cynara scolymus*, Asteraceae), in flowers of *Rhus typhina*, as well as in *Senecio nemorensis* and *Cirsium arvense*. It is used for liver and biliary tract diseases. A cholesterol-lowering activity has also been reported. The biosynthesis pro-

ceeds through acylation of *quinic acid with caffeoyl-CoA (see caffeic acid).
Lit.: Beilstein EIV **10**, 2261 ▪ Hager (5.) **7**, 1149 ff. ▪ J. Agric. Food Chem. **32**, 538–540 (1984) (^{1}H NMR) ▪ Luckner (3.), p. 385–386. – *[HS 2818 29; CAS 30964-13-7]*

Cynaropicrin.

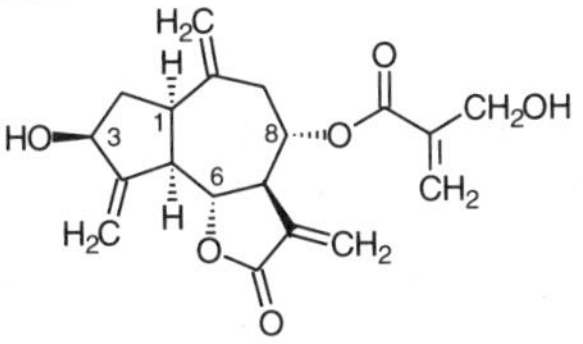

$C_{19}H_{22}O_6$, M_R 346.38, non-crystalline, $[\alpha]_D^{20}$ +108.6°. C. is an example of the plant guaianolides. These occur particularly frequently in the genera *Saussurea* (Himalayas) and *Cousinia* (Turkey) which are especially widely distributed in Asia; other sources are *Centaurea* species (thistles). C. occurs especially in artichokes (*Cynara scolymus*). The individual compounds differ in the acyl group at C-8 or by modifications at C-3 (e.g., 3-ketone, *dehydrocynaropicrin*, $C_{19}H_{20}O_6$, M_R 344.36, mp. 126 °C, $[\alpha]_D^{20}$ +60°; often also as C-3 glycosides). Like many other unsaturated *sesquiterpene lactones of this type (see also lactucin) C. also shows antitumor activity against certain cancer cell lines (ED_{50} 5 μg/mL for He-La cells)[2]. Inhibition of the growth of germinating garden lettuce, comparable to that of *auxins, by C. has been reported[3].
Lit.: [1] J. Chem. Soc., Chem. Commun. **1972**, 386; Phytochemistry **24**, 2013 (1985); **30**, 3810 (1991); Collect. Czech. Chem. Commun. **56**, 1106 (1991); **59**, 1175 (1994). [2] Planta Med. **33**, 356 (1978). [3] ACS Symp. Ser. **268**, 83 ff. (1985). – *[CAS 35730-78-0 (C.); 35821-02-4 (dehydro-C.)]*

Cynodine, Cynometrine see imidazole alkaloids.

Cyperones.

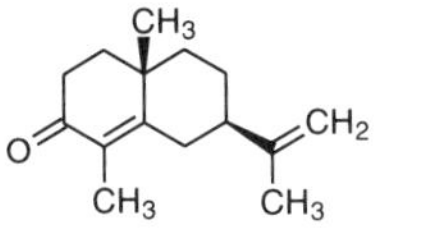

(+)-α-C.

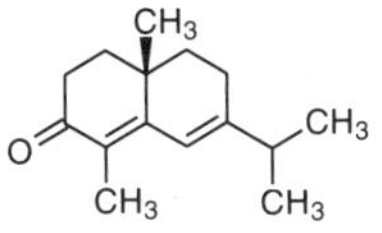

(+)-β-C.

Sesquiterpenes from *Cyperus rotundus* and *C. scariosus* oils: *(+)-α-C.* (eudesma-4,11-dien-3-one): $C_{15}H_{22}O$, M_R 218.33, bp. 177 °C, $[\alpha]_D$ +138° ($CHCl_3$); *(+)-β-C.* (eudesma-4,6-dien-3-one): oil, $C_{15}H_{22}O$, $[\alpha]_D$ +67.3° ($CHCl_3$).
Lit.: Helv. Chim. Acta **77**, 1707–1720 (1994) (synthesis β-C.) ▪ J. Chem. Res. (S) **1996**, 477 (synthesis α-C.); **1998**, 650 (synthesis β-S.) ▪ Tetrahedron **52**, 10507–10518 (1996) (synthesis β-C.) ▪ Tetrahedron: Asymmetry **7**, 2607 (1996). – *[CAS 473-08-5 ((+)-α-C.); 23665-63-6 ((+)-β-C.)]*

Cypress camphor see cedrol.

Cystathionine [2L,2′L-cystathionine, *S*-((*R*)-2-amino-2-carboxyethyl)-L-homocysteine].

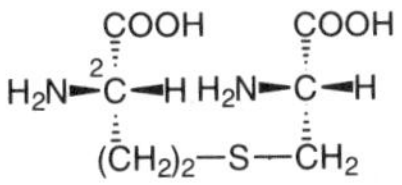

$C_7H_{14}N_2O_4S$, M_R 222.26, mp. 312 °C, $[\alpha]_D^{20}$ +23.1° (1 m HCl). Widely distributed non-proteinogenic amino acid, first isolated from vetch (*Astragalus pectinalis*).

Biosynthesis: Formed from *methionine and *L-serine through *S-adenosylmethionine, *S*-adenosylhomocysteine, and homocysteine[1] ($C_4H_9NO_2S$, M_R 135.18, mp. 233 °C).
Lit.: [1] Beilstein EIV **4**, 3189.
gen.: Luckner (3.), p. 281 ▪ Merck-Index (12.), No. 2847. – *[CAS 56-88-2 (C.); 6027-13-0 (homocysteine)]*

Cysteic acid (3-sulfoalanine, 2-amino-3-sulfopropanoic acid).

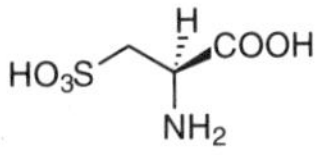

(2*R*)-, L-form

$C_3H_7NO_5S$, M_R 169.15, $[\alpha]_D^{20}$ +8.66° (H_2O). C. occurs in sheep's wool that has been exposed to sunlight.
Lit.: Anal. Biochem. **220**, 249–256 (1994) ▪ Beilstein EIV **4**, 3296 ▪ J. Chromatogr. **354**, 482–485 (1986). – *[CAS 13100-82-8]*

Cysteine (2-amino-3-mercaptopropanoic acid, abbr.: Cys or C).

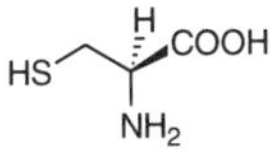

(2*R*)-, L-form

$C_3H_7NO_2S$, M_R 121.15, mp. 220–240 °C (decomp.), $[\alpha]_D^{25}$ −16.5° (H_2O), +6.5° (5 m HCl); hydrochloride, mp. 175–178 °C (decomp.), pK_a 1.92, 10.78 and 8.33 (SH). Generally classified as a non-essential amino acid, average content in proteins 2.8%. C. is easily oxidized in neutral or alkaline aqueous solutions to *cystine with a disulfide bridge, stronger oxidizing conditions lead to *cysteic acid. In proteins C. participates in the construction and stabilization of specific three-dimensional structures through formation of disulfide bridges. In enzymes, the mercapto group of C. is often responsible for their functions. These enzymes are inhibited by metal ions that bind to the SH group.
Numerous naturally occurring derivatives of C. are known, e.g., **S*-methyl-L-cysteine from pole beans, cabbage, and beets, *S*-propyl-C. and *S*-(1-propenyl)-C. from onions, and *S*-allyl-C. from garlic. Human urine contains various derivatives of Cys.
Biosynthesis & metabolism: In plants from *O*-acetylserine; in animal tissue, on the other hand, by cleavage of *cystathionine formed from **S*-adenosylmethionine. Since *methionine is an essential amino acid, Cys may also be considered as essential. C. can be converted by several routes to pyruvic acid and is metabolized in the citric acid cycle. Cys is a component of *glutathione.
Lit.: Annu. Rev. Biochem. **52**, 187–222 (1983) ▪ Beilstein EIV **4**, 3144 f. ▪ Merck-Index (12.), No. 2850. – *Synthesis:* Angew. Chem. Int. Ed. Engl. **20**, 647–667 (1981). – *[HS 2930 90; CAS 52-90-4 (L-form)]*

Cystine [3,3′-dithiobis(2-aminopropanoic acid), $(Cys)_2$].

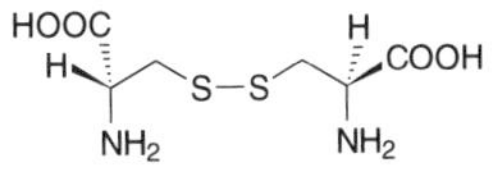

(2*R*, 2′*R*)-, L-form

$C_6H_{12}N_2O_4S_2$, M_R 240.39, mp. 260–261 °C (decomp.), $[\alpha]_D^{25}$ −232° (5 m HCl), pK_a 1.65, 2.26, 7.85, and 9.85.

C. was first isolated from urinary tract stones and can be obtained from the hydrolysates of keratin-rich proteins [1] such as horse hair (content 8%). C. is reductively cleaved by 2-mercaptoethanol or 1,4-dithiothreitol (Cleland's reagent) and oxidized by peroxyformic acid to cysteic acid.

Lit.: [1] J. Biol. Chem. **118**, 101 (1937).
gen.: Beilstein E IV **4**, 3155 f. ▪ Merck-Index (12.), No. 2851. – *[HS 2930 90; CAS 56-89-3 (L-form)]*

Cystodytin A see shermilamines.

Cytisine (baptitoxine, ulexine).

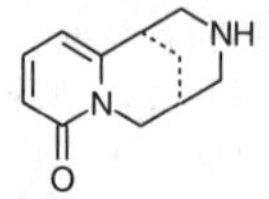

$C_{11}H_{14}N_2O$, M_R 190.24, mp. 155 °C (subl.), bp. 218 °C (260 Pa), $[\alpha]_D^{17}$ −119° (H_2O). Toxic *quinolizidine alkaloid of the α-pyridone type from many Fabaceae genera (e.g., *Anagyris, Baptisia, Laburnum*) [1] including golden-chain (laburnum) (*Laburnum anagyroides*) and common gorse (*Ulex europaeus*). C. is mostly accompanied by *N-methylcytisine* {$C_{12}H_{16}N_2O$, M_R 204.27, mp. 137 °C, $[\alpha]_D$ −221.6° (H_2O)}. C. acts like *nicotine as a ganglion blocker. It acts on the CNS (on the vomiting, vasomotor, and respiratory centers) first as a stimulant and then paralytically. After lethal doses death occurs through respiratory arrest. Teratogenic and uterotonic activities of C. have been reported [2]. The nicotine-like action explains why heavy smokers can smoke laburnum leaves without experiencing the toxic symptoms that occur in those not accustomed to nicotine. The C.-content is largest in golden-chain which, in Germany, is the plant most frequently responsible for poisoning among children. 3–4 fruits or 15–20 seeds can be lethal for a child. Among animals the toxicity of C. varies widely: LD_{50} (cat) 3 mg/kg, (dog) 4 mg/kg, (mouse) 100 mg/kg, (goat) 110 mg/kg (all values p.o.); Snails are not sensitive to C. or to *strychnine. C. has found occasional medicinal use as an antiemetic and antitussive agent, in the former USSR it was also used in low doses as a respiratory stimulant.

Lit.: [1] Waterman **8**, 197–239. [2] Pelletier **2**, 105–148. [3] Br. J. Pharmacol. **35**, 161 (1969).
gen.: Beilstein E V **24/2**, 535 f. ▪ Hager (5.) **3**, 382; **4**, 461; **5**, 624 ▪ Manske **3**, 143–156; **7**, 268–272 ▪ Phytochemistry **16**, 1460 (1977); **21**, 1470, 2385 (1982) (synthesis) ▪ Sax (8.), S. 1014 ▪ Ullmann (5.) **A 1**, 365. – *[HS 2939 90; CAS 485-35-8 (C.); 486-86-2 (N-methyl-C.)]*

Cytochalasins (cytochalasans). Term derived from Greek: Kytos = cell and chalasis = relaxation, diminution, for a group of *mycotoxins. C. are produced by a number of phytopathogenic fungi, e.g., the imperfect fungi *Curvularia lunata, Drechslera dematioidea, Phoma* sp., and related mold fungi or by ascomycetes such as *Gnomonia, Hypoxylon, Xylaria*, and *Daldinia* sp. (Xylariaceae). A. C. is also isolated from the basidiomycete *Coriolus vernicipes*. Structurally related to the C. are the *aspochalasins, *chaetoglobosins, *zygosporins, and phomins, the latter are in part identical with the C. Biogenetically the C. are composed of an amino acid (mostly Phe) and a polyketide chain (octa- or nonaketide), additional methyl branchings are derived from methionine.

The C. are of interest on account of their diverse biological activities. They reversibly inhibit cytoplasm division [cytokinesis, but not nuclear division (mitosis)] and can thus lead to multi-nuclear giant cells or at higher concentrations to expulsion of the nucleus from the cell (denucleation). In mammalian cells they inhibit movement processes that are associated with the aggregation of microtubuli and the formation of actin filaments (e.g., formation of blood clots by blood platelets, formation of contractile rings during cytokinesis, formation of microvilli). Whereas colchicine, *vinblastine, and *taxol attack tubulin, C. inhibits the attachment of actin molecules at the "barbed" end of actin filaments and leads to their depolymerization. C. inhibit phagocytosis, pinocytosis of macrophages, the movement of fibroblasts and amoebae, but not muscle contractions. Further biological activities include antibiotic properties as well as inhibition of glucose transport and secretion of thyroid and growth hormones. C. A and 18-deoxy-C. H inhibit HIV-1 protease. Of greatest importance are C. A and B. A novel structural type represent the phomacines [1].

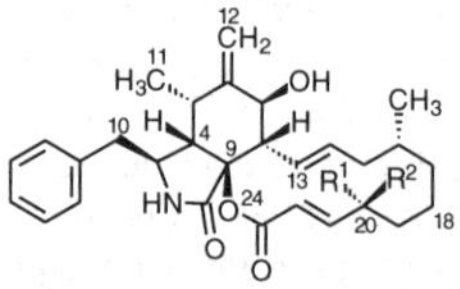

R^1 = OH, R^2 = H : Cytochalasin B
R^1, R^2 = O : Cytochalasin A

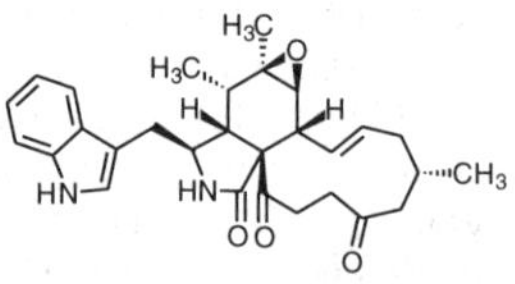

Cytochalasin G

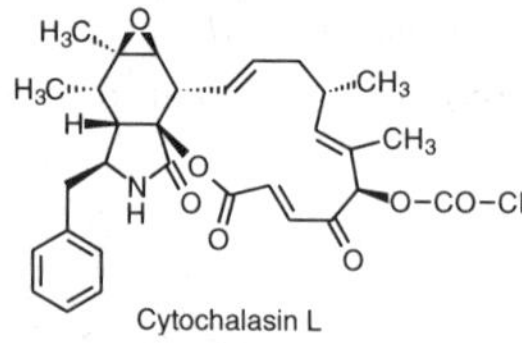

Cytochalasin L

Table: Data of Cytochalasins.

C.	molecular formula	M_R	mp. [°C]	$[\alpha]_D$	CAS
A	$C_{29}H_{35}NO_5$	477.60	185–187	+92° (C_2H_5OH)	14110-64-6
B	$C_{29}H_{37}NO_5$	479.62	218–221	+83° (CH_3OH)	14930-96-2
G	$C_{29}H_{34}N_2O_4$	474.60	255–257	−99° (CH_3OH)	70852-29-8
L	$C_{32}H_{37}NO_7$	547.65		−165° (C_2H_5OH)	79637-87-9

Lit.: [1] J. Org. Chem. **62**, 2148 (1997).
gen.: Agric. Biol. Chem. **51**, 2625–2629 (1987); **53**, 1699 (1989) (rosellichalasin); **55**, 1899 f. (1991) (pyrichalasin H) ▪ Atta-ur-Rahman **13**, 107–153 (1993) ▪ Betina, chap. 13 ▪ Chem. Pharm. Bull. **35**, 902 (1987) (isolation of C. O, N, P) ▪ Cole-Cox, p. 264–343 ▪ J. Antibiot. **45**, 671–691 (1992) (18-deoxy-C. H.) ▪ J. Am. Chem. Soc. **112**, 4351 (1990) (general synthesis) ▪ J. Chem. Soc., Perkin Trans. 1 **1989**, 57–65 (synthesis, isolation of C. N, O-1, P-1, Q, R), 489–497, 499, 507 (synthesis of C. H), 519, 525 (synthesis of C. G); **1991**, 1411–1417; **1999**, 3269, 3285 ▪ Pendse (ed.), Recent Advances in Cytochalasans, London: Chapman & Hall 1986 (reviews) ▪ Phytochemistry **30**, 3945 (1991); **41**, 821 (1996) ▪ Pure Appl.

Chem. **65**, 1309–1318 (1993) (biosynthesis) ▪ Sax (8.), CQM 125, CQM 250, PAM 775, ZVS 000 ▪ Tetrahedron **45**, 2323–2335, 2417–2429 (1989) (synthesis); **50**, 5615–5620 (1994); **53**, 6485 (1997) (new C.).

Cytochromes. Group of widely distributed heme proteins acting as redox catalysts in the respiratory chain and photosynthesis, as electron transfer agents, and in numerous metabolic processes to activate oxygen. The redox reactions are based on the change of valency of iron bound in *heme according to the equation $Fe^{3+} + e^- \rightleftarrows Fe^{2+}$. More than 30 C. are known and are designated by letter and number indices or by the wavelength of the most important absorption band. The redox potentials of the individual C. (see table) are influenced by the protein content.

Table: Standard redox potentials of Cytochromes.

Cytochrome	E_0 (V)
e^-+C. $c^{(3+)} \rightleftarrows$ C. $c^{(2+)}$	+0.22
e^-+C. $b^{(3+)} \rightleftarrows$ C. $b^{(2+)}$	+0.07
e^-+C. $a^{(3+)} \rightleftarrows$ C. $a^{(2+)}$	+0.29
cf. $NAD^+ + H_2 \rightleftarrows$ NADH	–0.32

Cytochrome a together with C. a_3 forms the so-called complex IV (ferrocytochrome c: dioxygen oxidoreductase, EC 1.9.3.1) of the respiratory chain. It is a lipid-containing enzyme localized in the inner mitochondrial membrane.

Cytochrome b is a dimeric membrane protein of the inner mitochondrial membrane, M_R 60000, with one heme group per subunit[1]. In the respiratory chain it transfers electrons from *ubiquinone to C. c. Also known are a C. b_2 from yeast, a C. b_5 from mitochondria, a C. b_6 from chloroplasts[2], and a C. b_1 as well as a C. b_{562} from *E. coli*.

Over 100 amino acid sequences are known for *cytochromes c* of the respiratory chain[3]. Comparisons show that C. c has changed its structure relatively slightly through mutations over 2 billion years, this must be interpreted as an indication of the major significance of the function of C. c for living organisms. Human C. c has M_R 12400 and a chain length of 104 amino acid residues. In all C. c heme is bound through thioether groups to the cysteine residues 14 and 17. X-ray crystallographic analyses have shown that C. c possesses a recognition site for C. oxidase in the lysine residues 72 and 73. Cyanide, carbon monoxide, and oxygen do not react with the prosthetic (heme) group of the native protein. Further C. c's are: C. c_1 which together with C. b forms the complex III of the respiratory chain and C. c_{551} which is produced by *Pseudomonas* spp. Numerous C. c occur in prokaryotes but differ from those of eukaryotes in that they do not react with C. oxidase.

The *cytochromes* P_{450} are widely distributed redox catalysts and effect the hydroxylation of xenobiotic substances, these processes can be considered both as detoxification reactions as well as toxification reactions for activation of certain carcinogenic agents, e.g., polycondensed aromatics[4]. The hydroxylation observed in the biosynthesis of some steroids[5] (including *ecdysteroids) is an insertion reaction of oxygen in a carbon-hydrogen bond which occurs under participation of C. P_{450} as a mixed functional oxygenase (monooxygenase). C. P_{450} also catalyze the epoxidation of double bonds, e.g., in the biosynthesis of *juvenile hormones in insects. They are assigned to the family of C. b and occur in mammals in the mitochondria of the adrenal glands and in the endoplasmatic reticulum of the liver. They consist of complexes of 16 subunits, M_R 850000. For substrate specificity, see *Lit.*[6], for models of the catalytic center, see *Lit.*[7], for structural relationships within the P_{450} superfamily, see *Lit.*[8], for function in *arachidonic acid metabolism, see *Lit.*[9]; for discovery of the C. by McMunn in 1884, see *Lit.*[10].

Use: C. c is used in pharmaceutical preparations for hypoxemia, against poisoning by an overdose of sleeping pills and carbon monoxide poisonings, etc.

Lit.: [1] Trends Biochem. Sci. **9**, 209 f. (1984). [2] Trends Biochem. Sci. **10**, 125–129 (1985). [3] Trends Biochem. Sci. **8**, 316–320 (1983). [4] Ortiz de Montellano, Cytochrome P_{450}: Structure, Mechanism, and Biochemistry, New York: Plenum 1986; Guengerich, Mammalian Cytochromes P_{450}, 2 vols., Boca Raton: CRC Press 1987. [5] Trends Biochem. Sci. **9**, 393–396 (1984). [6] Nature (London) **339**, 632–634 (1989). [7] Nachr. Chem. Tech. Lab. **36**, 890–895 (1988). [8] Annu. Rev. Biochem. **56**, 945–993 (1987). [9] Annu. Rev. Biochem. **55**, 69–102 (1986). [10] Trends Biochem. Sci. **9**, 364–366 (1984).

gen.: Stryer 1995, p. 534, 537–544, 662–670, 703 f. – *For C.* P_{450}: Nat. Prod. Rep. **8**, 527–551 (1991) ▪ Phytochemistry **43**, 1–21 (1996). – *For C. c:* Acta Crystallogr., Sect. D **50**, 687–694 (1994) ▪ Biochim. Biophys. Acta **1058**, 71 (1991) (NMR) ▪ Moore & Pettigrew, Cytochrome C, Berlin: Springer 1990 ▪ Scott & Mank (eds.), Cytochrome C, a multidisciplinary approach, Sansolito (CA): University Science Books 1996. – *For C* c_{551}: FEBS Lett. **70**, 180–184 (1976) (^{1}H NMR); **261**, 196–198 (1990) (sequencing and cloning). – *[CAS 9035-34-1 (C. a); 9035-37-4 (C. b); 9035-39-6 (C. b_5); 9035-40-9 (C. b_6); 9035-38-5 (C. b_1); 9007-43-6 (C. c); 9035-42-1 (C. c_1); 9048-77-5 (C. c_{551}); 9035-51-2 (C. P_{450})]*

Cytohemin (hemin a).

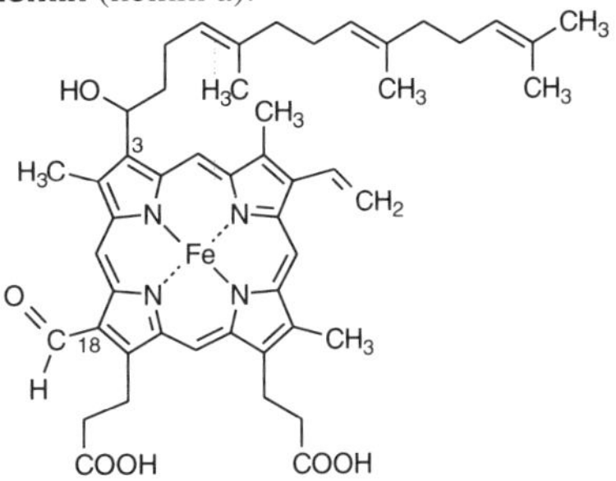

$C_{49}H_{56}ClFeN_4O_6$, M_R 888.31; uv_{max} (pyridine) 427, 585–587 nm. Green *h(a)emin, prosthetic group of cytochrome oxidase. C. differs from hemin by the lipophilic side chain in position 3 and the formyl group in position 18. A close relative of C. is the *heme of an oxidase from *Escherichia coli* in which the formyl group in position 18 is replaced by a methyl group[1]. Removal of the central iron atom leads to *cytoporphyrin*[2] (porphyrin a): $C_{49}H_{58}N_4O_6$, M_R 799.02.

Lit.: [1] J. Am. Chem. Soc. **114**, 1182 (1992). [2] Beilstein E III/IV **26**, 3338 f.; Dolphin **II**, 309. – *[CAS 19554-22-4]*

Cytokinins. Term derived from cyto and Greek kinein = movement for a group of *plant growth substances that are predominately formed in young roots, stimulate cell division, and in some case together with

*gibberellins regulate seed germination, development and differentiation processes during fruit formation and ripening, budding, etc. C. are the class of plant growth substances that delay senescence with the widest spectrum of activity. C. counteract the apical dominance of *auxins and promote the development of side shoots. Examples of C. that can be considered in the broadest sense as plant hormones (phytohormones, see plant growth substances) are *kinetin, *zeatin, N^6-isopentenyladenine, and related natural and synthetic adenine derivatives. The C. were previously distinguished as phytokins from the "actual" kinins (plasma kinins).
Use: One commercial application is, e. g., the combination of gibberellins A_4 u. A_7 with the synthetic C. N^6-benzyladenine which improves the development of fruit in certain apple species [1].

Lit.: [1] ACS Symp. Ser. **557**, 1 – 14, Bioregulators for Crop Protection and Pest Control (1994).
gen.: Horgan & Jeffcoat (eds.), Cytokinins: Plant Hormones in Search of a Role, Bristol: The British Plant Regulator Group 1987 ▪ Kaminek, Mok & Zazimalova (eds.), Physiology and Biochemistry of Cytokinins in Plants, Den Haag: SPB Acad. Publ. 1992 ▪ Mohr & Schopfer, Plant Physiology, Berlin: Springer 1995 ▪ Nickell, Plant Growth Regulators – Agricultural Uses, Berlin: Springer 1982 ▪ Ullmann (5.) **A 20**, 421.

Cytoporphyrin see cytohemin.

Cytosine see pyrimidines.

Cytovirin see blasticidins.

D

Dactinomycin see actinomycins.

DAHP see 3-deoxy-D-*arabino*-2-heptulosonic acid 7-phosphate.

Daidzein see isoflavones.

Daidzin (7-β-D-glucopyranosyloxy-4′-hydroxyisoflavone). $C_{21}H_{20}O_9$, M_R 416.38, cryst., mp. 233–235 °C, $[\alpha]_D^{20}$ −36.4° (0.02 m KOH). *Isoflavone-glycoside from soybeans, *Phaseolus lobata* and other Fabaceae, 7-*O*-β-D-glucopyranoside of daidzein (for structure, see isoflavones). D. reduces the alcohol uptake of alcohol-addicted experimental animals [1].

Lit.: [1] Alcohol **11**, 279 (1994); Physiol. Behav. **57**, 1155 (1995).
gen.: Chem. Eng. News (4. 12.) **1995**, 72. – *[CAS 552-66-9]*

Damascenine see 3-hydroxyanthranilic acid.

β-Damascenone.

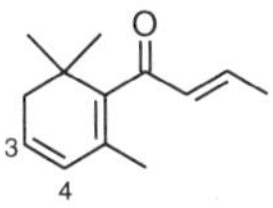

$C_{13}H_{18}O$, M_R 190.29, oil, bp. (1.7 kPa) 116–118 °C. Component of Bulgarian *rose oil, obtained from *Rosa damascena*. β-D. is only present to 0.05% in this oil but is mainly responsible for the rose-like fragrance. β-D. is an essential component of perfumes and gives them fresh and brilliant notes.

Lit.: Can. J. Chem. **70**, 2094 (1992) ▪ Helv. Chim. Acta **69**, 228 (1986); **71**, 1587 (1988) ▪ J. Org. Chem. **55**, 1106 (1990). – *[CAS 23726-93-4]*

Damascones [4,8-, 5,8-, 5(13),8-, 3,8-, and 2,8-megastigmadien-7-one].

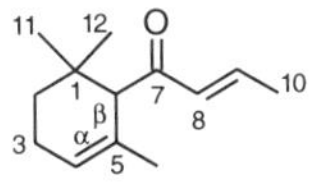

α-Damascone

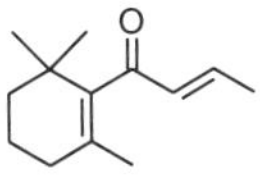

β-Damascone

$C_{13}H_{20}O$, M_R 192.30, isomers of *ionone which differ only in the position of the ring double bond (α-, β-, γ-, δ-, and ε-damascone). The name "damascones" or "rose ketones" is derived from their occurrence in the essential oil of damask rose (*Rosa damascena*) [1]. *α-D.*: bp. 77–80 °C (0.13 Pa), $[\alpha]_D^{20}$ +487° ($CHCl_3$); *β-D.*: bp. 52 °C (0.13 Pa). D. possess the common olfactory feature of a narcotic-herby odor of exotic flowers with an undertone resembling black currents. Although α- and β-D. in *rose oil have a combined content of merely ca. 0.15%, they determine the basic odor of the oil (olfactory threshold: 0.009 ppb). D. occur in other flower oils [2] as well as in tea, certain tobacco, and fruit aromas, e. g., apples and raspberries [3]. α- and β-D. are indispensable components of modern perfumes (*Passion, Cool Water*). β-D. can be synthesized from β-ionone by carbonyl transposition [4]. Natural D. are formed by oxidative degradation of the corresponding carotinoid precursors [5].

Lit.: [1] Helv. Chim. Acta **53**, 541 (1970); Ohloff, p. 120, 152. [2] Helv. Chim. Acta **55**, 1866 (1972). [3] Perfum. & Flavor **3**, 11 (1978). [4] Helv. Chim. Acta **56**, 310 (1973); J. Am. Chem. Soc. **110**, 6909 (1988); J. Org. Chem. **55**, 1106 (1990); Tetrahedron **49**, 1871–1878 (1993). [5] Tetrahedron Lett. **31**, 6521 (1990). – *[CAS 24720-09-0 (α-D.); 23726-91-2 (β-D.)]*

Dammarenes.

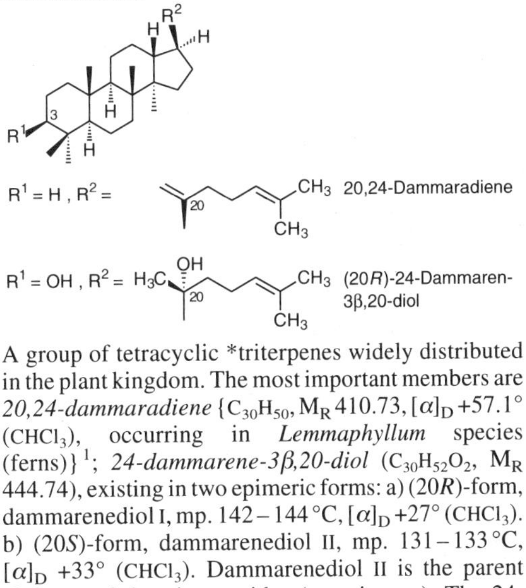

A group of tetracyclic *triterpenes widely distributed in the plant kingdom. The most important members are *20,24-dammaradiene* {$C_{30}H_{50}$, M_R 410.73, $[\alpha]_D$ +57.1° ($CHCl_3$), occurring in *Lemmaphyllum* species (ferns)} [1]; *24-dammarene-3β,20-diol* ($C_{30}H_{52}O_2$, M_R 444.74), existing in two epimeric forms: a) (20*R*)-form, dammarenediol I, mp. 142–144 °C, $[\alpha]_D$ +27° ($CHCl_3$). b) (20*S*)-form, dammarenediol II, mp. 131–133 °C, $[\alpha]_D$ +33° ($CHCl_3$). Dammarenediol II is the parent skeleton of the ginsenosides (see ginseng). The 24-dammarene-3,20-diols occur in *dammar resin and in ginseng roots [2]. Also important are *24-dammarene-3,12,20-triol*, $C_{30}H_{52}O_3$, M_R 460.74, as (3α,12β,20*S*) form "*betulafolienetriol*" and as (3β,12β,20*S*)-form "*protopanaxadiol*", occurring in ginseng roots [3]. The D. usually exist as the glycosidically bound *saponins, see also triterpenes and ginseng.

Lit.: [1] Chem. Pharm. Bull. **31**, 2530 (1983). [2] J. Chem. Soc. **1956**, 2196 (isolation); Chem. Pharm. Bull. **22**, 1213 (1974) (synthesis). [3] Chem. Pharm. Bull. **24**, 400 (1976); **30**, 2380, 2393 (1982); Phytochemistry **22**, 1473 (1983).
gen.: Chem. Pharm. Bull. **33**, 3176 (1985) (isolation) ▪ J. Am. Chem. Soc. **118**, 8765 (1996) (synthesis) ▪ Pharm. Unserer Zeit **16**, 164 (1987) (biosynthesis) ▪ Phytochemistry **21**, 2420 (1982). – *[CAS 8774-88-6 (20,24-dammaradiene); 19132-83-3 (24-dammarene-3,20-diol); 14351-28-1 ((3β,20R)-dammarenediol I); 14351-29-2 ((3β,20S)-dammarenediol II); 7755-01-3 (betulafolienetriol); 6892-79-1 (protopanaxadiol)]*

Dammar resin [dam(m)ar]. Pale yellow, transparent drop-like or irregularly shaped pieces of resin with a weakly aromatic odor from the south east Asian pitch tree (*Shorea wiesneri*, Dipterocarpaceae) and other *Shorea* or *Hopea* species. D. has varying hardnesses depending on its origin.
Use: As binder in paints, for embedding microscopic preparations, for plasters (can cause irritations), glue for theatrical wigs, in photography etc.[1]. Most D. originates from Sumatra where the natives collect it from scratched tree trunks. A solution of D. in chloroform or xylene is used to preserve microscopic preparations. Triterpenoids from D. (see dammarenes) exhibit antiviral properties[2].
Lit.: [1] Parfüm. Kosmet. **57**, 248 (1976). [2] J. Nat. Prod. **50**, 706 (1987).
gen.: Janistyn (3.) **1**, 224 ▪ Pharm. Unserer Zeit **16**, 169 (1987) ▪ Ullmann (5.) **A 23**, 77. – *[HS 1301 90; CAS 9000-16-2]*

Danaidal.

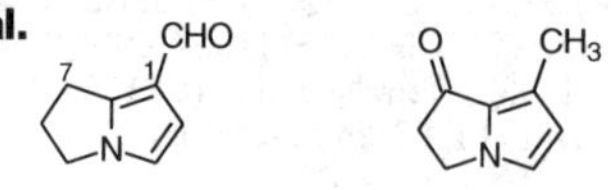

Danaidal Danaidone

C_8H_9NO, M_R 135.17, mp. 59–60 °C. Insect pheromone of males of some butterfly species (Arctiidae and Danaidae). The butterflies take up *pyrrolizidine alkaloids from plants as larvae and/or as fully developed insects[1]. The *pheromone D. and other analogous substances are formed from the *necine(s) part of the alkaloids. Members of this group include the 7*R*-hydroxy derivative *hydroxydanaidal* {$C_8H_9NO_2$, M_R 151.17, mp. 55 °C (subl.), $[\alpha]_D^{25}$ –140° ($CHCl_3$)} and *danaidone* (C_8H_9NO, M_R 135.17, mp. 74–75 °C). Further compounds often also occur in the pheromone glands so that species-specific odors result. The amount of the pheromone apparently informs the female butterfly about the pyrrolizidine content of the male[2], which is in part transferred to fertilized eggs during copulation and serves as protection from predators[3].
Lit.: [1] Naturwissenschaften **73**, 17 (1986); J. Chem. Ecol. **16**, 165 (1990); Biol. Unserer Zeit **25**, 8–17 (1995). [2] J. Chem. Ecol. **16**, 543 (1990). [3] Proc. Natl. Acad. Sci. USA **85**, 5992 (1988).
gen.: ApSimon **9**, 478 ▪ Beilstein E V **21/7**, 434; **21/12**, 158 ▪ Pelletier **9**, 155–233. – *Synthesis:* Justus Liebigs Ann. Chem. **1986**, 1645 ▪ Quim. Nova **6**, 74 (1983). – *[CAS 27628-46-2 (D.); 28379-58-0 (hydroxy-D.); 6064-85-3 (danaidone)]*

Dandelion. A widely distributed, yellow-flowering, milk juice(latex)-producing plant [*Taraxacum officinale* (Asteraceae)] occurring in diverse forms and related with chicory, lettuce, and black salsify; its healing properties were known in antiquity. The pharmaceutically used roots and leaves of plants harvested before flowering contain sesquiterpene *bitter principles (taraxin), triterpenes such as *taraxasterol, and *taraxerol as well as *pectins, *choline, *inulin (especially in the roots), etc. L. has a stimulating effect on bile and renal function, is thus often used as a component of blood-cleaning teas, and the freshly collected herbage is eaten as a salad.
Lit.: Fintelmann et al., Phytotherapie Manual, p. 92, Stuttgart: Hippokrates 1989 ▪ Taraxaci radix cum herba (Monograph 228/05.12.84), Köln: Bundesanzeiger-Verlagsges. 1984. – *[HS 0709 90]*

Daphnetin (7,8-dihydroxycoumarin).

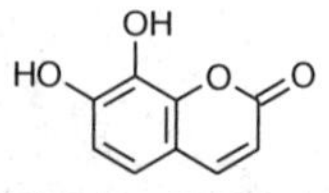

$C_9H_6O_4$, M_R 178.14. Yellowish needles, mp. 253–255 °C (decomp.), sublimes, is soluble in hot water, hot alcohol, and alkalis (with yellow color). D. can be isolated from daphne (*Daphne*) species and other plants. The 7-*O*-*β*-D-glucopyranoside *daphnin* {$C_{15}H_{16}O_9$, M_R 340.29, mp. 223–224 °C (dihydrate), $[\alpha]_D^{22}$ –114.7° (CH_3OH)} occurs in *Daphne* species and in *Thymelaea hirsuta* (Thymelaeaceae). D. inhibits the germination of seeds and the growth of wheat. It exhibits cytotoxic activity. The biosynthesis proceeds via *(*E*)-*p*-coumaric acid → *umbelliferone → daphnetin.
Lit.: Beilstein E V **18/3**, 211 ▪ Z. Naturforsch. C **41**, 247–252 (1986) (biosynthesis). – *[HS 2932 29; CAS 486-35-1 (D.); 486-55-5 (daphnin)]*

Daphnetoxin.

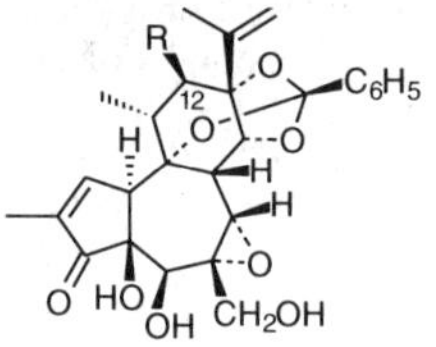

R = H : Daphnetoxin
R = Cinnamoyloxy : Gnidicin

$C_{27}H_{30}O_8$, M_R 482.53, cryst., mp. 194–196 °C, $[\alpha]_D$ +63°, a very poisonous diterpenoid from daphne species (*Daphne* sp.). It can cause contact dermatitis in humans (blisters on skin and mucosa), LD_{50} (mouse p.o.) 275 mg/kg, (mouse i.p.) 1100 µg/kg.
Lit.: Hager (5.) **3**, 388 f. ▪ J. Am. Chem. Soc. **92**, 1070 f. (1970) ▪ Sax (8.), DAB 850 ▪ Schmidt, in Evans (ed.), Naturally Occurring Phorbol Esters, p. 217–243, Boca Raton: CRC Press 1986 ▪ Zechmeister **44**, 73–100. – *[CAS 28164-88-7]*

Daphnin see daphnetin.

Daphniphyllum alkaloids.

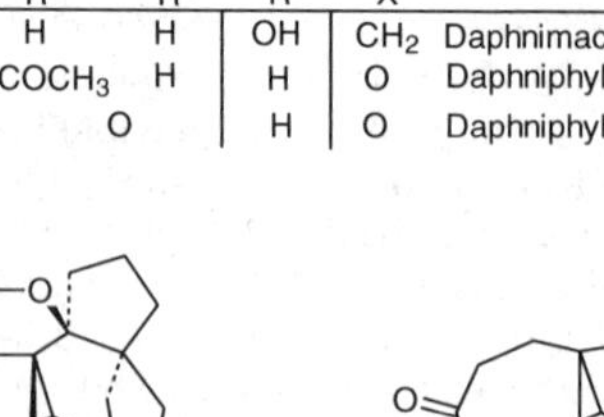

R¹	R²	R³	R⁴	R⁵	X	
O		H	H	OH	CH_2	Daphnimacropine (1)
O		$OCOCH_3$	H	H	O	Daphniphylline (2)
H	OH	O		H	O	Daphniphyllidine (3)

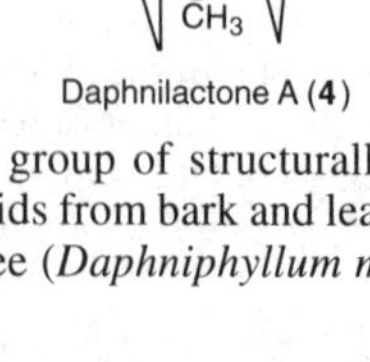

Daphnilactone A (**4**)

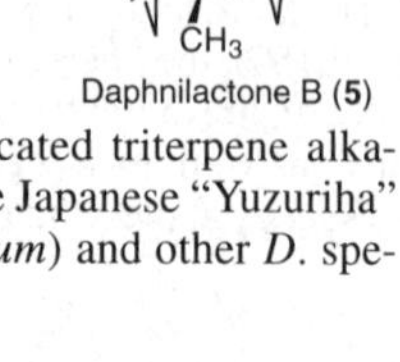

Daphnilactone B (**5**)

A group of structurally complicated triterpene alkaloids from bark and leaves of the Japanese "Yuzuriha" tree (*Daphniphyllum macropodum*) and other *D.* species. These unusual trees develop a completely new generation of leaves before they lose the old leaves. The most important members are *daphniphylline* {$C_{32}H_{49}NO_5$, M_R 527.74, hydrochloride: needles, mp. 238–240 °C, $[\alpha]_D^{20}$ +44° ($CHCl_3$)}, *daphniphyllidine* ($C_{30}H_{47}NO_4$, M_R 485.71, amorphous) as well as *daphnilactone A* ($C_{23}H_{35}NO_2$, M_R 357.54, needles, mp. 195 °C) and *daphnilactone B* ($C_{22}H_{31}NO_2$, M_R 341.49, plates, mp. 92–94 °C). The biosynthesis proceeds through squalene [1].

Lit.: [1] Tetrahedron Lett. **1973**, 799.
gen.: J. Am. Chem. Soc. **110**, 8734 (1988); **111**, 1530 (1989) ▪ J. Chem. Res. S **1993**, 262 ▪ J. Org. Chem. **57**, 2531–2594 (1992); **60**, 1120 (1995) ▪ Manske **15**, 41; **29**, 265–286 ▪ Nachr. Chem. Tech. Lab. **37**, 916 (1989). – *Synthesis, review:* Angew. Chem. Int. Ed. Engl. **31**, 665–681 (1992) ▪ Pure Appl. Chem. **62**, 1911–1920 (1990). – *[CAS 15007-67-7 (2); 50764-62-0 (3); 38210-98-9 (4); 38826-56-1 (5)]*

Darlucins.

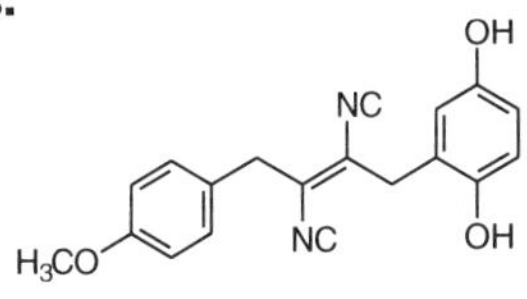

Darlucin A

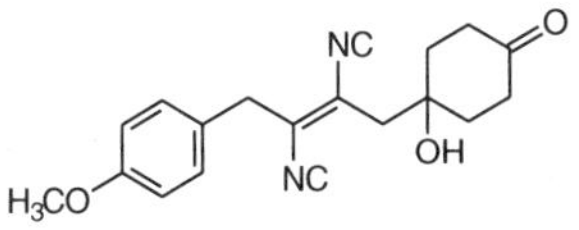

Darlucin B

Antibiotically active diisocyanides of the *xanthocillin type from cultures of the deuteromycete *Sphaerellopsis filum* (*Darluca filum*). *D. A* ($C_{19}H_{16}N_2O_3$, M_R 320.35, oil) and *D. B* ($C_{19}H_{20}N_2O_3$, M_R 324.38, oil) are the first alkenes with isocyano groups in the 1,2-positions. The configuration of the double bond has not yet been elucidated.

Lit.: J. Antibiot. **48**, 36 (1995). – *[CAS 162341-15-3 (D. A); 162341-16-4 (D. B)]*

Darnel, rye-grass. Common name for *Lolium temulentum* (bearded darnel). Another European sweet grass of the genus *Lolium* (wall barley) is *L. perenne*, perennial rye-grass. *L. temulentum* occurs in moist grain fields, waysides, and waste lands and is toxic for grazing animals. The symptoms such as nervous disorders, muscular twitching, and unsteady gait, correspond to those of poisonings by tremorigenic *mycotoxins (*penitrems, *aflatrem, *territrems, *paspalitrems). Similar poisonings are known as rye-grass staggers (*L. perenne*), paspalum staggers, and Bermuda grass staggers. The toxicity with which the grasses protect themselves from herbivorous insects, is attributed to bacterial toxins (corynetoxins) and mycotoxins. Endophytic fungi which are able to produce the respective toxins have been isolated from a series of grasses.

Lit.: Annu. Rev. Entomol. **39**, 401 (1994) ▪ Aust. J. Agric. Res. **36**, 35–42 (1985) (corynetoxins) ▪ N. Z. Vet. J. **29**, 185 (1981) ▪ Steyn & Vleggaar (eds.), Mycotoxins and Phycotoxins, p. 501–511, Amsterdam: Elsevier 1986 ▪ Zechmeister **48**, 1–80.

Daturadiol see oleanane.

Daunorubicin (daunomycin, rubidomycin, leukaemomycin, rubomycin C).

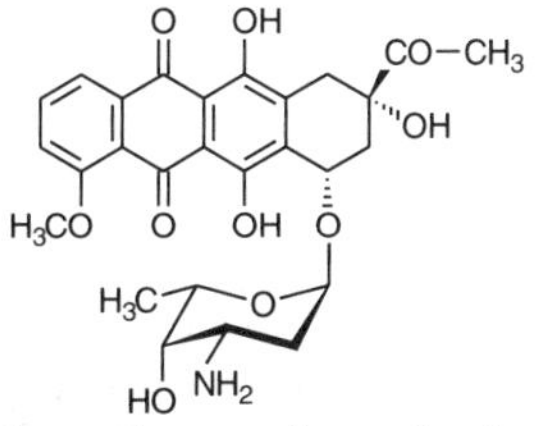

International generic name for an *anthracycline antibiotic from *streptomycetes (e. g., *Streptomyces peucetius*). $C_{27}H_{29}NO_{10}$, M_R 527.53, hydrochloride: red needles, decomp. >177 °C. Acidic hydrolysis of D. furnishes L-*daunosamine* and the aglycone *daunomycinone* {$C_{21}H_{18}O_8$, M_R 398.37, mp. 213–214 °C, $[\alpha]_D$ +193° (dioxane)}.

D. was the first anthracycline to be used clinically in chemotherapy for malignant tumors, however it has strong cardiotoxic side effects. The mechanism of action is complex: intercalation into DNA, inhibition of topoisomerase II, chelating agent for metal ions such as Fe^{3+} and Cu^{2+}, participation in redox reactions, effects on membrane functions. D. also shows anti-HIV activity.

Lit.: Beilstein E V **18/10**, 349 ▪ J. Am. Chem. Soc. **86**, 5334 (1964) ▪ Tetrahedron Lett. **1968**, 3353. – *Reviews:* Arcamone, Doxorubicin, New York: Academic Press 1981 ▪ Foye (ed.), Cancer Chemotherapeutic Agents, p. 210–217, Washington: ACS 1995 ▪ Hutchinson, in Genetics and Biochemistry of Antibiotic Production, p. 331–357, Boston: Butterworth-Heinemann 1995 ▪ Priebe (ed.), Anthracycline Antibiotics, ACS Symposium Ser. **574**, Washington: ACS 1995. – *[HS 2941 30; CAS 20830-81-3 (D.); 21794-55-8 (daunomycinone)]*

Davana oil. Brown viscous oil with a warm sweet, tea-like odor, strongly resembling that of dried fruits.

Production: By steam distillation of the herb *Artemisia pallens* (Asteraceae) growing in southern India.

Composition [1]***:*** D. consists mainly of sesquiterpenoid compounds. The main component (ca. 50%) is (+)-*davanone* {$C_{15}H_{24}O_2$, M_R 236.35, $[\alpha]_D^{21}$ +77.7° (neat)}.

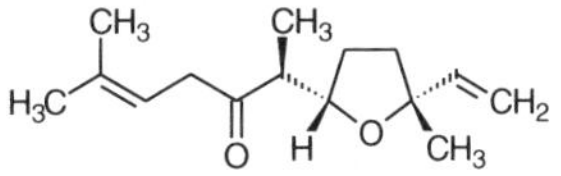

(+)-Davanone

Use: Used in only small amounts in fine perfumes on account of its high price; the main use is the aromatization of foods.

Lit.: [1] Perfum. Flavor. **3** (2), 45 (1978); **13** (3), 50 (1988); **20** (1), 54 (1995).
gen.: Bauer et al. (2.), p. 153 ▪ Tetrahedron **55**, 617 (1999) [synthesis (±)-davanone]. – *Toxicology:* Food Cosmet. Toxicol. **14**, 737 (1976). – *[HS 3301 90; CAS 8016-03-3 (D.); 20482-11-5 (davanone)]*

Deadly nightshade, belladonna see atropine.

Debromoaplysiatoxin see majusculamides.

2,4-Decadienals. $C_{10}H_{16}O$, M_R 152.24. (*E,E*)-2,4-D.: Liquid with a fatty aldehyde-like odor, bp. 110 °C (0.9 kPa), D. 0.871, LD_{50} (rat p.o.) >5 g/kg. Occur-

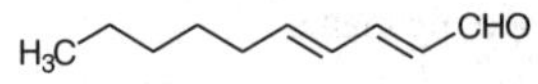

(*E,E*)-2,4-D.

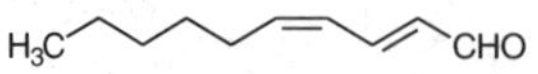

(*E,Z*)-2,4-D.

rence [mostly together with (*E,Z*)-2,4-D.] in fat-containing foods like fish, meat, and dairy products, as well as in *citrus oils, in which 2,4-D. makes a major contribution to the odor on account of its low olfactory threshold of 0.07 ppb [1]. Both 2,4-D. are formed by autoxidation of linoleic acid [2]. (*E,E*)-2,4-D. is produced in amounts of some 100 kg worldwide and is prepared by the Hoaglin synthesis [3] from 2-octenal and vinyl ether. It is used in meat and citrus aromas. The structurally related (*Z,Z*)-4,7-decadienal is an important trace constituent of *calamus oil.

Lit.: [1] Perfum. Flavor. **16** (1), 14 (1991). [2] Belitz-Grosch (4.), p. 188. [3] J. Am. Chem. Soc. **80**, 3069 (1958).
gen.: Bedoukian (3.), p. 26 ff. ▪ Fenaroli (2.) **2**, 117 ▪ J. Chem. Soc., Perkin Trans. 1 **1988**, 2061 (synthesis) ▪ TNO list (6.), Suppl. 5, p. 340 – *[CAS 25152-84-5 (E,E); 25152-83-4 (E,Z)]*

Decaline see Lythraceae alkaloids.

Decanal see alkanals.

4- and 5-Decanolide see alkanolides.

Decarestrictin D see tuckolide.

Decarestrictins.

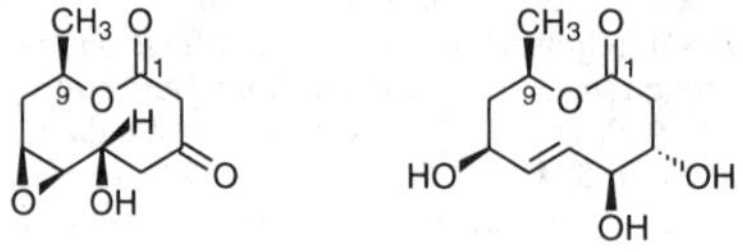

D. B D. D

Family of ten-membered ring lactones (decan-9-olides) produced by *Penicillium* strains (e.g., *P. simplicissimum*) with a methyl group at C-9. 15 D. are known, including the main components B and D with a fermentation yield of 500–600 mg/L. *D. B:* $C_{10}H_{14}O_5$, M_R 214.23, oil, $[\alpha]_D$ –49° (CH_3OH), soluble in organic solvents, insoluble in water or hexane. *D. D:* $C_{10}H_{16}O_5$, M_R 216.24, cryst., mp. 116 °C, $[\alpha]_D$ –62° (CH_3OH). The D. differ in their oxygenation patterns at C-3 to C-8 as well as by the presence of an (*E*)-double bond at C-4 or C-5. The biosynthesis proceeds through a common pentaketide precursor that is transformed by enzymatic steps (e.g., by monooxygenases, incorporation of oxygen) and non-enzymatic reactions (e.g., acid-catalyzed addition of water or allyl rearrangement) [1]. In particular, D. D and M inhibit the *de novo* synthesis of cholesterol in HEP-G2 cell cultures and regulate the lipid level in rats. D. D is also isolated from the sclerotia of *Polyporus tuberaster*. The fungus ("Indian bread") is used against rheumatism in traditional medicine.

Lit.: [1] J. Chem. Soc., Perkin Trans. 1 **1993**, 495–500.
gen.: Biosci. Biotechnol. Biochem. **59**, 1657 (1995) ▪ J. Antibiot. **45**, 56–73, 1176–1181 (1992); **46**, 1372–1380 (1993) ▪ J. Chem. Soc., Perkin Trans. 1 **1993**, 495–500 ▪ J. Nat. Prod. **55**, 649–653 (1992) ▪ J. Org. Chem. **63**, 2332 (1998) (synthesis D. L) ▪ Tetrahedron Lett. **39**, 4421 (1998) (synthesis D. D); **41**, 1199 (2000) (D. C_2). – *[CAS 127393-91-3 (D. B); 127393-89-9 (D. D)]*

Decenals see alkenals.

Decinine, Decodine see Lythraceae alkaloids.

Decorticasine see lolines.

Defensive secretions. Secretions emitted especially by animals to drive off predators or to overcome prey. The D. s. of insects have been investigated in detail. They are mostly low-molecular weight substances that trigger an escape reaction in the predator by an olfactory stimulus or act by direct contact on the body surface (see hydroquinone, bombardier beetles). For example, many bug species use unpleasantly-smelling unsaturated aldehydes such as (*E*)-2-*hexenal to repel predators [1]. The skin hairs of pupa of the Mexican bean beetle *Epilachna varivestis* have drops of azamacrolides [2] (see epilachnene) at their ends. Contact causes ants to immediately turn away from the pupa and to clean their mouthparts. In some cases the activity of the compounds is increased by solvents. Thus, some predatory rove beetles (Staphylinidae) use *methyl-1,4-benzoquinone in a solution of hydrocarbons, lactones, or aliphatic isopropyl esters. Only these solutions of the quinone have an effect while the pure substance or the solvents alone have no activity [3]. Some D. s. are also of industrial relevance such as, e.g., the anthraquinone *C*-glucoside *carminic acid, formed by scales apparently for defence against predators [4]. The structures of the compounds used as D. s. vary widely. In addition to simple aliphatic compounds, they include terpenes, alkaloids, aromatic compounds, steroids, glycosides, isoxazolines, nitriles, biogenic amines, and amino acids. The substances used are often only known from the insects that use them, e.g., the defensive alkaloids of ants and ladybirds (*coccinelline) or the *pyridopyrazines of the springtail. Little is known about a possible participation of microorganisms in the synthesis of the compounds, see also cantharidin, pederine, polyzonimine, monomorines, coccinelline, glomerine, alarm substances, allomones.

Lit.: [1] Biochem. Syst. Ecol. **18**, 369–376 (1990). [2] Proc. Natl. Acad. Sci. USA **90**, 5204–5208 (1994). [3] Entomol. Gener. **15**, 275–292 (1991). [4] Science **208**, 1039 ff. (1979).
gen.: Agosta, Bombardier Beetles and Fever Trees, New York: Addison Wesley 1996 ▪ Annu. Rev. Entomol. **32**, 17–48, 381–413 (1987) ▪ Biochem. Syst. Ecol. **21**, 143–162 (1993) ▪ Blum, Chemical Defenses of Arthropods, New York: Academic Press 1981 ▪ Pasteels et al., in Jolivet et al., Biology of Chrysomelidae, Amsterdam: Kluywer 1988 ▪ Proc. Natl. Acad. Sci. USA **92**, 14–18 (1995).

Deferrithiocin see ferrithiocin.

Deguelin see tephrosin.

Dehydroajmalicine see cathenamine.

Dehydrobilirubin see biliverdin.

11-Dehydrocorticosterone see corticosteroids.

Dehydrodecodine see Lythraceae alkaloids.

Dehydrogeosmin.

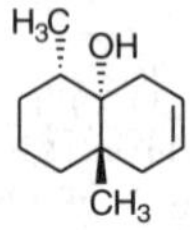

$C_{12}H_{20}O$, M_R 180.29, colorless liquid, $[\alpha]_D^{20}$ +27.6° (CH_2Cl_2). D. is a flower aroma constituent of various

Cactaceae[1] [e.g., *Rebutia marsoneri*, *R. knupperiana*, *Mammillaria sphaerica* (syn.: *Dolichothele*) and others.]. It has the same absolute configuration as natural (–)-*geosmin and possesses a similar odor. The biosynthesis proceeds from *farnesol[2]. For synthesis, see *Lit.*[3].

Lit.: [1] Helv. Chim. Acta **73**, 133 (1990). [2] Helv. Chim. Acta **76**, 2547 (1993). [3] Helv. Chim. Acta **76**, 1949 (1993). – *[CAS 126370-54-5]*

Dehydromatricaria ester see polyynes.

3-Dehydroquinic acid [(1*R*)-(1α,3β,4α)-1,3,4-trihydroxy-3-oxocyclohexanecarboxylic acid].

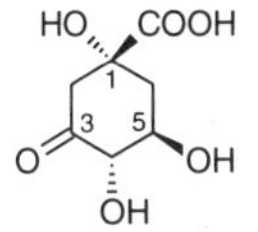

$C_7H_{10}O_6$, M_R 190.15, mp. 140–142 °C, $[\alpha]_D^{28}$ –82.4° (C_2H_5OH). Isolated from culture filtrates of *Escherichia coli*. Intermediate in the biosynthesis of *quinic acid.

Lit.: Beilstein E IV **10**, 4006 ▪ Biochemistry **28**, 7560–7572, 7572–7582 (1989) ▪ J. Am. Chem. Soc. **111**, 2299 f. (1989) ▪ Nat. Prod. Rep. **11**, 173–203 (1994). – *[CAS 10534-44-8]*

3-Dehydroshikimic acid [(4*S*)-*trans*-4,5-dihydroxy-3-oxo-1-cyclohexenecarboxylic acid].

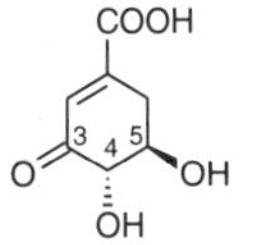

(4*S*, 5*R*)-form

$C_7H_8O_5$, M_R 172.14, needles, mp. 146–147 °C, $[\alpha]_D^{28}$ –57° (C_2H_5OH). D. is isolated as the direct biosynthetic precursor of *shikimic acid from culture filtrates of *Escherichia coli* and is widely distributed in nature. The biosynthesis proceeds through cleavage of water from *3-dehydroquinic acid.

Lit.: Beilstein E IV **10**, 3855 ▪ J. Org. Chem. **50**, 5897 f. (1985) (synthesis) ▪ Luckner (3.), p. 323–326 ▪ Nat. Prod. Rep. **11**, 173–203 (1994) ▪ see also shikimic acid. – *[CAS 2922-42-1]*

Dehydrovomifoliol see honey flavor.

Dekamycin III see neomycin.

Delicial see Lactarius pigments.

Delphin (*delphinidin 3,5-di-*O*-β-glucopyranoside). $C_{27}H_{31}O_{17}^+$, M_R 627.53, as chloride ($C_{27}H_{31}ClO_{17}$, M_R 662.99) bronze-shining crystals, mp. 202–203 °C (decomp.), uv_{max} 534 nm. D. belongs to the *anthocyanins and occurs in flowers of *Salvia patens* (Lamiaceae) and *Verbena hybrida* (Verbenaceae) as well as in red grapes (*Vitis vinifera*, Vitaceae); D. is also a pigment component of pomegranate (*Punica granatum*, Punicaceae).

Lit.: Karrer, No. 1733 ▪ Merck-Index (12.), No. 2931 ▪ Phytochemistry **16**, 591 (1977); **36**, 613 (1994). – *[CAS 17670-06-3]*

Delphinidin (3,3′,4′,5,5′,7-hexahydroxyflavylium). Formula, see anthocyanins, $C_{15}H_{11}O_7^+$, M_R 303.25, as chloride ($C_{15}H_{11}ClO_7$, M_R 338.70) chocolate-brown prisms of needles with metallic luster (from 5% aq. HCl), mp. >350 °C, uv_{max} 544 nm, well soluble in alcohols and ethyl acetate; see also anthocyanins.

Lit.: Beilstein E V **17/18**, 514 ▪ Karrer, No. 1727–1735 ▪ Schweppe, p. 398 ▪ Tetrahedron **39**, 3005 (1983). – *[CAS 528-53-0]*

Deltamycin A₄ see carbomycin.

Delvomycin see kirromycin.

Demecolcine see phenethyltetrahydroisoquinoline alkaloids.

6-Demethyltetracycline see tetracyclines.

Demissidine, Demissine see Solanum steroid alkaloid glycosides.

Dendrobates alkaloids. Mostly toxic alkaloids present in cutaneous gland secretions of South American colored poison frogs of the Dendrobatidae; they are used by the Naonama, Cuna, and Choco Indians of Columbia as arrow poisons. The structures of the alkaloids vary widely[1]. To date more than 20 different classes have been identified, the main representatives include the *batrachotoxins, *pumiliotoxins, *histrionicotoxins and *epibatidine which are discussed under the individual entries. Several hundred minor alkaloids have been described. The various D. a. are designated by their molecular mass and a subsequent letter assigned in the order of their identification[1]. Thus, e. g., histrionicotoxin is D. a. 283A; see also frog toxins. The close similarity between alkaloids isolated from the ant *Solenopsis azteca* and those from the strawberry poison frog *Dendrobates pumilio* supports the hypothesis that alkaloids in the frog's skin are of dietary origin[2].

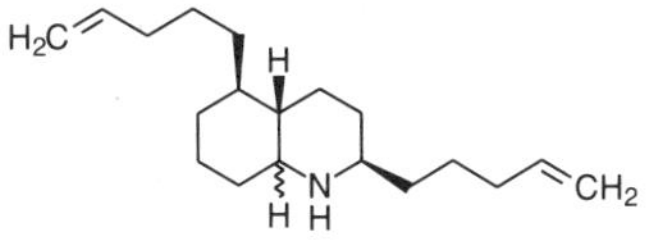

Ant alkaloid

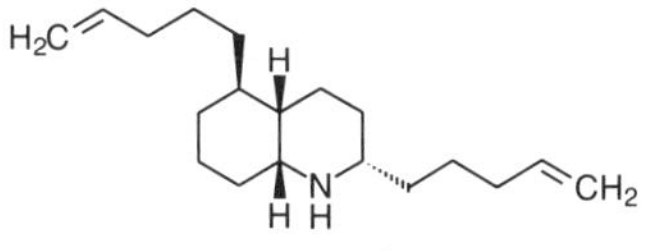

Dendrobates poison

Lit.: [1] Manske **43**, 185–288. [2] J. Nat. Prod. **62**, 5 (1999).
gen.: J. Nat. Prod. **49**, 265 (1986); **60**, 2 (1997) ▪ Stud. Nat. Prod. Chem. **19 E**, 3–88 (1997) (synthesis review) ▪ Tetrahedron **43**, 643–652 (1987) ▪ Zechmeister **41**, 206–340.

Dendrobium alkaloids. Alkaloids occurring in the orchid genus *Dendrobium*. Similar to the *Nuphar alkaloids, these are "pseudoalkaloids" whose C-skeletons originate from terpene metabolism (sesquiterpenes). About 15 alkaloids are known, they were first isolated from *Dendrobium nobile* and later also from other species; a typical representative is *dendrobine* {$C_{16}H_{25}NO_2$, M_R 263.38, mp. 134–136 °C, $[\alpha]_D^{14}$ –48.4° (CH_3OH)}.

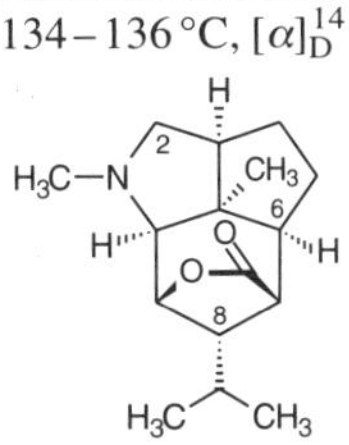

Dendrobine

In China and Japan the dried stems of endemic *Dendrobium* species are used as antipyretic and tonic drug, called "Chin-Shi-Hu". *D. nobile* is generally named as the principle species although other species are also used in the drug.

Lit.: Chem. Lett. **1993**, 213 ▪ J. Am. Chem. Soc. **114**, 4089 (1992); **119**, 4130 (1997) (synthesis); **121**, 6072 (1999) (synthesis dendrobin) ▪ J. Org. Chem. **57**, 3519 (1992); **59**, 5633 (1994) (synthesis) ▪ Zechmeister **38**, 47. – *[CAS 2115-91-5 (dendrobine)]*

Dendrodoine see grossularines.

Dendrolasin [3-(4,8-dimethylnona-3,7-dienyl)furan].

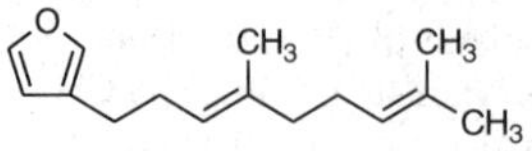

$C_{15}H_{22}O$, M_R 218.34, oil, bp. 148–150 °C (2.13 kPa). Alarm and defence substance of the ant species *Lasius (Dendrolasius) fuliginosus*[1]. D. also occurs in the leaves of *Torreya nucifera*[2] (Taxaceae) as well as in marine organisms. For synthesis, see *Lit.*[3].

Lit.: [1]Tetrahedron **1**, 177 (1957). [2]Bull. Chem. Soc. Jpn. **38**, 381 (1965). [3]J. Chem. Soc., Chem. Commun. **1995**, 2135; J. Org. Chem. **45**, 5225 (1980); **48**, 5183, 5356 (1983); Bull. Chem. Soc. Jpn. **61**, 4029–4035 (1988); Tetrahedron **53**, 8513 (1997).
gen.: Beilstein E V **17/1**, 660 ▪ Helv. Chim. Acta **52**, 15 (1969) (biosynthesis) ▪ J. Nat. Prod. **46**, 481 (1983) (review); **50**, 482 (1987) (isolation). – *[CAS 23262-34-2]*

Dendrotoxin.

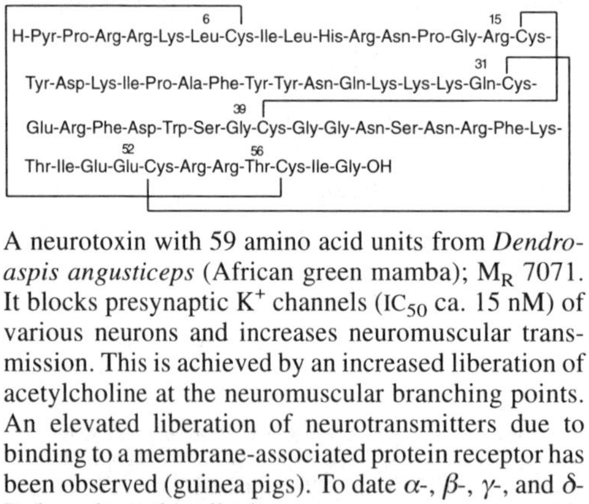

A neurotoxin with 59 amino acid units from *Dendroaspis angusticeps* (African green mamba); M_R 7071. It blocks presynaptic K^+ channels (IC_{50} ca. 15 nM) of various neurons and increases neuromuscular transmission. This is achieved by an increased liberation of acetylcholine at the neuromuscular branching points. An elevated liberation of neurotransmitters due to binding to a membrane-associated protein receptor has been observed (guinea pigs). To date α-, β-, γ-, and δ-D. have been described.

Lit.: Biochemistry **31**, 12297 (1992) ▪ Eur. J. Pharmacol. **209**, 87 (1991) ▪ FEBS Lett. **255**, 159 (1989) ▪ J. Med. Chem. **39**, 2141–2155 (1996) (mechanism, structure) ▪ J. Physiol. **420**, 365 (1990) ▪ Mol. Pharmacol. **34**, 152 (1988) ▪ Neuroscience **40**, 29 (1991) ▪ Pharmacology **77**, 153 (1982) ▪ Proc. Natl. Acad. Sci. USA **38**, 493 (1986) ▪ Protein Pept. Lett. **1**, 239–245 (1995) (review). – *[CAS 74811-93-1]*

Denticulatins.

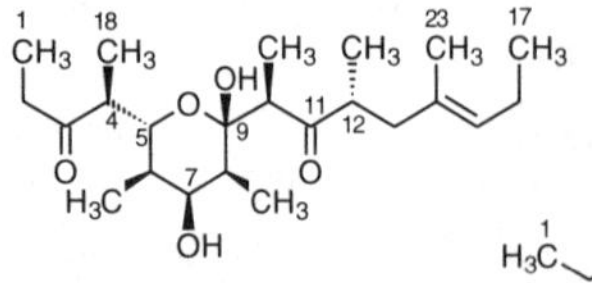

D. A

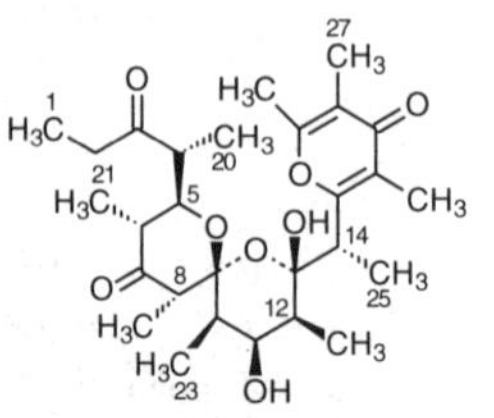

Siphonarin A
(relative configuration)

Snails of the genus *Siphonaria* live as lung-breathing mollusks in tidal zones of the oceans. They contain secondary metabolites biosynthetically formed by condensation of polypropionate units [1], e.g., *D. A*[2] from *S. denticulata*, $C_{23}H_{40}O_5$, M_R 396.57, oil, $[\alpha]_D$ −30.7° and *siphonarin A*[3] from *S. zelandica* and *S. atra*, $C_{28}H_{42}O_8$, M_R 506.64, cryst., mp. 164–166 °C, $[\alpha]_D$ +21.7°.

Lit.: [1]J. Chem. Soc., Chem. Commun. **1988**, 1061. [2]J. Am. Chem. Soc. **105**, 7413 (1983). [3]J. Am. Chem. Soc. **106**, 6748 (1984).
gen.: Tetrahedron **53**, 9169–9202 (1997) (synthesis review). – *[CAS 87697-98-1 (D. A); 92125-62-2 (siphonarin A)]*

Denudatin B see kadsurenone.

3′-Deoxyadenosine see cordycepin.

Deoxyanthocyanins. Glycosides of deoxyanthocyanidins. The deoxyanthocyanidins differ from the anthocyanidins (see anthocyanins) by the absence of a hydroxy group at C-3 of the heterocyclic ring. D. are rare flower and leaf pigments occurring especially in the representatives of the subfamily Gesnerioideae of the Gesneriaceae.

Lit.: Harborne and Baxter, Handbook of the Natural Flavonoids, New York: Wiley 1999.

Deoxycholic acid see bile acids.

Deoxycollybolidol see collybolide.

Deoxycorticosterone see corticosteroids.

Deoxyharringtonine see Cephalotaxus alkaloids.

3-Deoxy-D-*arabino*-2-heptulosonic acid 7-phosphate (DAHP).

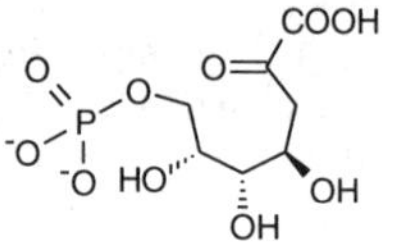

$C_7H_{11}O_{10}P$, M_R 286.13. DAHP is an intermediate in the biosynthesis of *shikimic acid and is formed by enzymatic addition of *phosphoenolpyruvic acid to D-*erythrose 4-phosphate by means of phospho-2-dehydro-3-deoxyheptanoate aldolase (DAHP synthase, EC 4.1.2.15).

Deoxymannojirimycin (1,5-dideoxy-1,5-imino-D-mannitol).

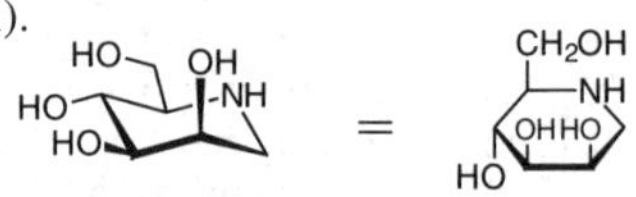

$C_6H_{13}NO_4$, M_R 163.17, mp. 188 °C, $[\alpha]_D^{20}$ −26.7° (CH_3OH). A piperidine alkaloid, mannosidase inhibitor from the seeds of *Lonchocarpus sericeus* and *L. costaricensis* (Fabaceae) as well as from cultures of *Streptomyces lavendulae*.

Lit.: Angew. Chem. Int. Ed. Engl. **27**, 716 (1988) ▪ Carbohydrate Res. **254**, 25 (1994) ▪ J. Antibiot. **42**, 1302 (1989) ▪ J. Chem. Soc., Chem. Commun. **1999**, 41 (synthesis) ▪ Synlett **1997**, 533 ▪ Tetrahedron Lett. **33**, 189 (1992); **34**, 3613 (1993); **37**, 1461 (1996); **38**, 7977 (1997). – *[CAS 84444-90-6]*

Deoxynivalenol (vomitoxin).

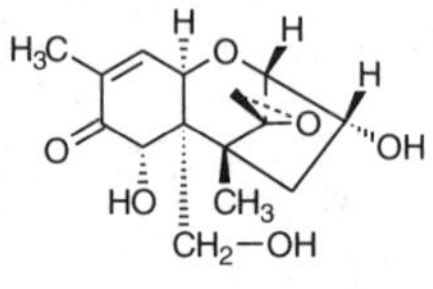

$C_{15}H_{20}O_6$, M_R 296.32, needles, mp. 151–153 °C, $[\alpha]_D$ +6.35° (C_2H_5OH), D., also known as DON, is a *mycotoxin of the *tric(h)othecenes group formed by *Fusarium* species [LD_{50} (mouse i. p.) 70 mg/kg]. D. occurs world-wide, in Europe especially in corn and wheat infected with *Fusarium* species (e. g., *F. culmorum* and *F. graminearum*). Some countries, e. g. Canada, have set a maximum permissible quantity in wheat for food production at 2 mg/kg, for baby food 1 mg/kg. D. is very rapidly metabolized in animals so that there is no carry-over into meat products. Intestinal bacteria open the epoxide ring, leading to rapid detoxification.

Lit.: Appl. Environ. Microbiol. **57**, 672–677 (1991) (occurrence); **58**, 3857–3863 (1992) (bacterial transformation) ▪ Chelkowski (ed.), Fusarium – Mycotoxins, Taxonomy and Pathogenicity, p. 441–472, Amsterdam: Elsevier 1989 (occurrence) ▪ Cole-Cox, p. 202 ▪ J. Biol. Chem. **271**, 27353 ▪ J. Chem. Soc., Perkin Trans. 1 **1989**, 887 (synthesis) ▪ Sax (8.), VTF 500 ▪ Tetrahedron **45**, 2277 (1989) (biosynthesis). – *[CAS 51481-10-8]*

Deoxynojirimycin (1,5-dideoxy-1,5-imino-D-glucitol).

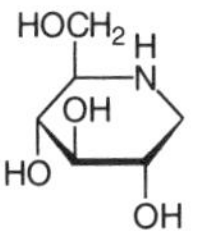

$C_6H_{13}NO_4$, M_R 163.17, mp. 192–195 °C, $[\alpha]_D^{20}$ +47.5° (H_2O). D. is a glucose-like piperidine derivative from the bark of the mulberry tree (*Morus*). It can also be obtained by fermentation from *Streptomyces* species. On account of their ability to inhibit intestinal α-glucosidases, D. derivatives have been developed as oral antidiabetic drugs (e. g., Miglitol®, Emiglitate®). D. derivatives also exhibit antiviral activity against HIV (inhibition of the glycosylation of viral proteins).

Lit.: Beilstein E V **21/6**, 8 ▪ Merck-Index (12.), No. 2952. – *Biosynthesis:* Tetrahedron **48**, 6285 ff. (1992); **49**, 6707–6717 (1993). – *Synthesis:* Angew. Chem. Int. Ed. Engl. **33**, 2320 ff. (1994) ▪ Heterocycles **29**, 1469 (1989); **46**, 637 (1997) ▪ J. Am. Chem. Soc. **111**, 3924 ff. (1989) ▪ Justus Liebigs Ann. Chem. **1989**, 423–428 ▪ Synthesis **1999**, 571 ▪ Tetrahedron Lett. **39**, 7173 (1998). – *[CAS 19130-96-2]*

Deoxynorherqueinone see atrovenetin.

2-Deoxy-D-ribose (D-*erythro*-2-deoxypentose).

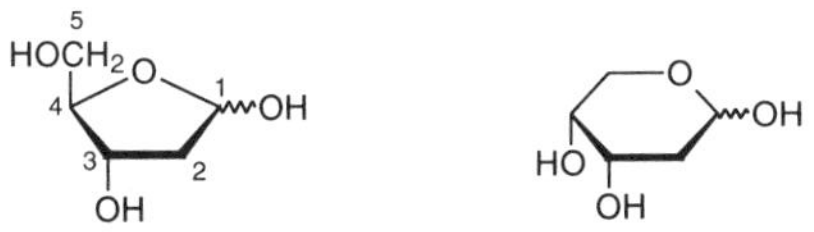

$C_5H_{10}O_4$, M_R 134.13, soluble in water and pyridine, poorly soluble in alcohol. In aqueous solution at 31 °C 2-D. exists to 40% as the α-pyranose, to 35% as the β-pyranose form, to 13% as the α-furanose, and to 12% as the β-furanose form. 2-D. is widely distributed in all animal and plant cell nuclei as the carbohydrate building block of deoxyribonucleic acids (DNA) or, respectively, the cell nucleus nucleosides (deoxynucleosides). As a typical *deoxy sugar, 2-D. has one hydroxy group less than the corresponding, fully hydroxylated sugar, D-ribose. 2-D. – in contrast to the other sugars – gives a color reaction with fuchsin/sulfurous acid and thus allows the detection of DNA (Feulgen's color reaction).

Lit.: Beilstein E IV **1**, 4181 ff. ▪ Carbohydr. Res. **123**, 320 (1983) ▪ J. Org. Chem. **44**, 2472 (1979) (synthesis) ▪ Nucleosides, Nucleotides **18**, 2357 (1999) (synthesis). – *[HS 2940 00; CAS 533-67-5]*

16-Deoxysarcophin see sarcophytol A.

15-Deoxyspergualin {(*S*)-*N*-[4-(3-aminopropylamino)butyl]-2-(7-guanidinoheptanoyl)-2-hydroxyacetamide}. Formula see spergualin; $C_{17}H_{37}N_7O_3 \cdot 3HCl \cdot 2H_2O$ (syrup), M_R 387.53. 15-D. is a derivative of the bacterial metabolite *spergualin from *Bacillus laterosporus*. 15-D. is characterized by a particularly pronounced antitumor and immunosuppressive activity and is used clinically in kidney transplantations. Further fields of application include autoimmune diseases (e. g., multiple sclerosis) and general organ transplantations. 15-D. acts at a specific point in T cell activation. As potential sites of molecular attack the heat-shock proteins HSP 70 and/or HSP 90 appear to be of special relevance for the immunosuppressive activity.

Lit.: Ann. N. Y. Acad. Sci. **696**, 123 (1993) ▪ Biochemistry **33**, 2561 (1994) ▪ Drugs Fut. **20**, 196 (1995) ▪ J. Antibiot. **35**, 1665 (1982); **44**, 1033 (1991) ▪ Science **258**, 484 (1992). – *[CAS 89149-10-0]*

Deoxy sugars. General term for monosaccharides in which one or more hydroxy groups have been replaced by hydrogen. Important examples are *2-deoxy-D-ribose, a building block of deoxyribonucleic acids, *fucose, and *L-rhamnose, the common sugar building blocks of polysaccharides, glycoproteins, and plant glycosides.

Lit.: Carbohydr. Chem. **16**, 122 (1985); **17**, 120 (1985) ▪ Collins (ed.), Carbohydrates, p. 691–696, London: Chapman & Hall 1987 ▪ Collins-Ferrier, p. 206–217 ▪ Kennedy, Carbohydrate Chemistry, Oxford: Clarendon Press 1988 ▪ Nat. Prod. Rep. **16**, 283 (1999) (biosynthesis in bacteria) ▪ Pure Appl. Chem. **60**, 1655 (1988).

Dephostatin.

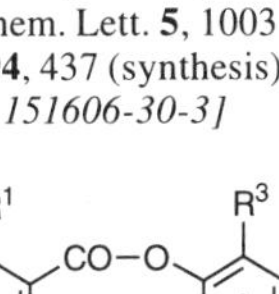

$C_7H_8N_2O_3$, M_R 168.15. The nitrosamine derivative D. produced by a streptomycete strain inhibits protein tyrosine phosphatase (PTPase) with an IC_{50} of 7.7 μM.

Lit.: Bioorg. Med. Chem. Lett. **5**, 1003 (1995) ▪ J. Chem. Soc., Chem. Commun. **1994**, 437 (synthesis) ▪ J. Antibiot. **46**, 1342, 1716 (1993). – *[CAS 151606-30-3]*

Depsides.

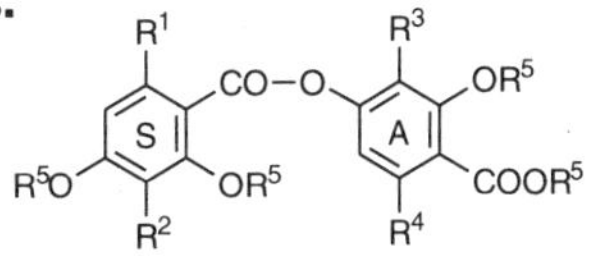

D. are condensation products of aromatic hydroxycarboxylic acids in which the carboxy group of the acidic molecule (S) is esterified by the phenolic hydroxy group of the 2nd acidic molecule (A, alcohol). Depending on the number of hydroxycarboxylic acids in the molecule distinction is made between di- and tridepsides and on the linkage between *para*- and *meta*-depsides. D. occur mainly in lichens (*lichen substances) where the substituents in the above formula can be: R^1, R^4=CH_3, *n*-C_3H_7, *n*-C_5H_{11}; R^2, R^3=H, CH_3, CHO,

CH_2OH; $R^5 = H, CH_3$. Most D. are didepsides, such as, e.g., *atranorin, *chloroatranorin, *gyrophoric acid, and *lecanoric acid. In some D. (e.g., alectorialic acid and barbatolic acid) the acid part (S) is linked with a benzylic hydroxy group of the A part. The structures of about 100 D. isolated from lichens have been elucidated by hydrolysis and alcoholysis to the corresponding carboxylic acids or esters (S part) and to the phenols (A part), respectively.
Activity: Many D. have antibiotic activity and, at concentrations of 10^{-3} molar inhibit plant growth. The physiological role of D. in lichens is mostly unknown; it is assumed that (1) they protect the algae from intensive light on account of their strong UV absorptions, (2) they protect the lichens from attack by microorganisms by means of their antibiotic activities, (3) they influence the permeability of the cell membranes of the symbiotic algae, (4) they dissolve metal ions necessary for growth from the substrate by way of their ability to form metal complexes, or (5) as hydrophobic substances they protect the medulla of the lichens from complete wetting by water so that gas exchange and thus also photosynthesis in the algae is possible even under rainy conditions.
Biosynthesis: The aromatic hydroxycarboxylic acid is first formed from polyacetates and then esterified enzymatically. Besides in lichens, D. also occur in tanning substances, e.g., in coffee or savory; a well-known representative is digallic acid (see tannins).
Lit.: Culberson ▪ Karrer, No. 1023–1053 ▪ Zechmeister **41**, 1–46; **45**, 103–234.

Depsidones.

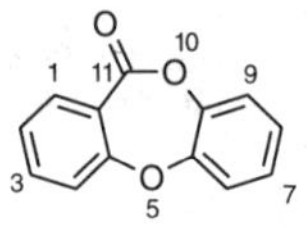

With 66 members (in 1994) the D. are the second largest group of aromatic *lichen substances, and have the basic skeleton of 11*H*-dibenzo[*b,e*][1,4]dioxepin-11-one (formula). *Examples:* *argopsin, *diploicin, eriodermin, *fumarprotocetraric acid, *lobaric acid, *norstictic acid, *pannarin, *physodic acid, *salazinic acid, and *stictic acid. The UV spectrum of a typical D., *psoromic acid, shows the following maxima: 211 (log ε 4.64), 240 (4.58), s 271 (4.18), and 317 nm (3.59) (CH_3OH). The biosynthesis of lichen D. probably proceeds through the corresponding *depsides, this is supported by the simultaneous occurrence of the coupling partners olivetoric acd – physodic acid in the same genus (*Parmelia*). Some D. have also been found in fungi of the genera *Aspergillus*, *Emericella*, and *Preussia*.
Lit.: Asahina-Shibata, p. 112–150 ▪ J. Antibiot. **45**, 1195–1201 (1992) ▪ J. Chem. Soc., Perkin Trans. 1 **1978**, 395–400; **1988**, 2611–2614 ▪ J. Nat. Prod. **54**, 213–217 (1991) ▪ Karrer, No. 1054–1070a ▪ Zechmeister **29**, 209–306; **45**, 103–234 (1984). – *[CAS 3580-77-6 (depsidone)]*

Depsones.

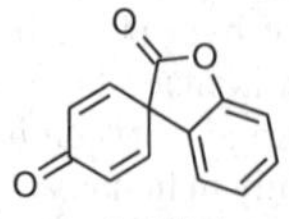

A small group of aromatic *lichen substances [*picrolichenic acid, subpicrolichenic acid, and superpicrolichenic acid]. The biosynthesis presumably proceeds via the *depsides by phenol oxidation.
Lit.: Culberson, p. 140 f.

Depudecin [(2*R*,3*S*,4*S*,7*S*,8*S*,9*R*)-3,4;7,8-diepoxy-5,10-undecadiene-2,9-diol].

$C_{11}H_{16}O_4$, M_R 212.25, oil, $[\alpha]_D^{24}$ −35.8° (CH_3OH). A phytotoxin from cultures of *Alternaria brassicicola* and *Nimbya scirpicola*. D. effects the conversion of tumor cells (fibroblasts) transformed by ras-oncogenes back to normal cells. D. changes the structure of the cytoskeleton.
Lit.: Biosci., Biotechnol., Biochem. **58**, 565 (1994) ▪ Chem. & Biol. **2**, 517–525 (1995) ▪ J. Antibiot. **45**, 879 (1992). – *[CAS 139508-73-9]*

Dermadin see isocyanides.

Dermocanarin I.

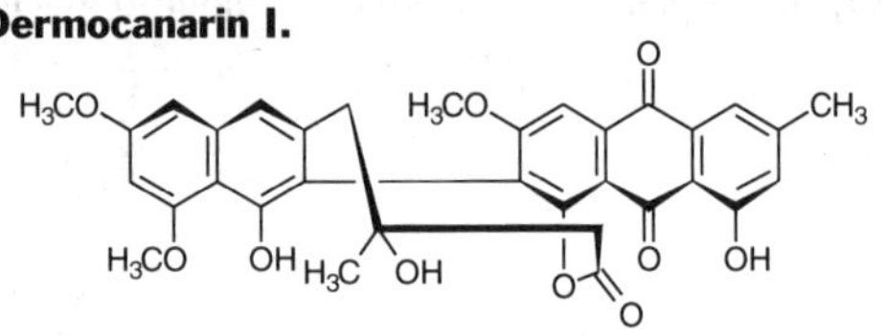

$C_{33}H_{28}O_{10}$, M_R 584.58, yellow powder, mp. 215–218 °C, $[\alpha]_D$ +27° ($CHCl_3$). A member of a group of unusual dimeric *anthraquinone pigments from the mycelium of the Tasmanian agaric *Dermocybe canaria* (Basidiomycetes). Some *preanthraquinones of the same type occur in *Cortinarius sinapicolor* and Australian *Dermocybe* species.
Lit.: J. Chem. Soc., Perkin Trans. 1 **1998**, 3431 (biosynthesis) ▪ Nat. Prod. Rep. **11**, 75, 83 (1994) ▪ Tetrahedron Lett. **31**, 3505 (1990). – *[CAS 129656-55-9]*

Dermochrysone.

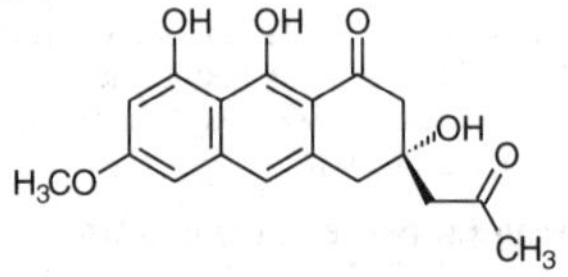

Dermochrysone

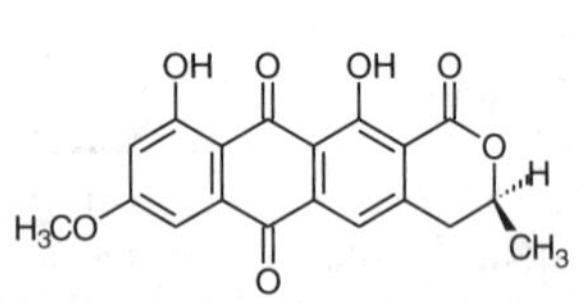

Dermolactone

$C_{18}H_{18}O_6$, M_R 330.34, green prisms, mp. 189–192 °C, $[\alpha]_D$ +28° ($CHCl_3$). D. is a *preanthraquinone of the rare nonaketide type from the Australian agaric *Dermocybe sanguinea* and is accompanied by the anthraquinone *dermolactone* {$C_{19}H_{14}O_7$, M_R 354.32, orange needles, mp. 255–257 °C, $[\alpha]_D$ +169.3° ($CHCl_3$)} and related compounds of the nonaketide type. Dermolactone exists in the fungus in partially racemized form ($[\alpha]_D$ +45.9°).
Lit.: J. Chem. Soc., Perkin Trans. 1 **1990**, 2585 f.; **1995** 1215–1223 (synthesis of dermolactone) ▪ Nat. Prod. Rep. **11**,

84f. (1994) ▪ Tetrahedron Lett. **34**, 3155 (1993). – *[CAS 130968-95-5 (D.); 130968-97-7 (dermolactone)]*

Dermocybe pigments.

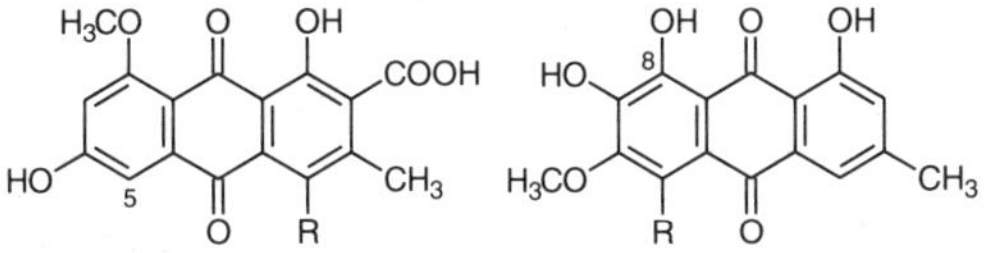

R = H : Dermolutein
R = OH : Dermorubin

R = H : Dermoglaucin
R = OH : Dermocybin

A chemotaxonomically important type of colored octaketides of the *anthraquinone type from agarics of the genus *Cortinarius*, subgenus *Dermocybe*. Especially widely distributed are the anthraquinonecarboxylic acids *dermolutein* [$C_{17}H_{12}O_7$, M_R 328.28, orange cryst., mp. 270 °C (decomp.)] and *dermorubin* [$C_{17}H_{12}O_8$, M_R 344.28, red cryst., mp. >300 °C (decomp.)], biosynthetically derived from *endocrocin. The 5-chloro derivatives of dermolutein and dermorubin also occur in small amounts in *Dermocybe*. The neutral anthraquinones *physcione, *dermoglaucin* ($C_{16}H_{12}O_6$, M_R 300.27, brick-red cryst., mp. 236 °C) and *dermocybin* ($C_{16}H_{12}O_7$, M_R 316.27, permanganate-colored cryst., mp. 227 °C) are formed biosynthetically by hydroxylation and *O*-methylation of *emodin. They exist in the fruitbodies mainly as the 8-*O*-glucosides. D. were among the first more exactly described pigments of higher fungi and are used especially in Scandinavian countries to dye wool. The D. from the northern and southern hemispheres reveal interesting structural differences (*austrocorticin, *austrocortilutein, *dermocanarin I, *dermochrysone).

Lit.: Nat. Prod. Rep. **11**, 74f. (1994) ▪ Zechmeister **51**, 125–143 (reviews). – *[CAS 26071-13-6 (dermolutein); 26071-14-7 (dermorubin); 7213-59-4 (dermoglaucin); 7229-69-8 (dermocybin)]*

Deserpidine see reserpine.

Desertomycins.

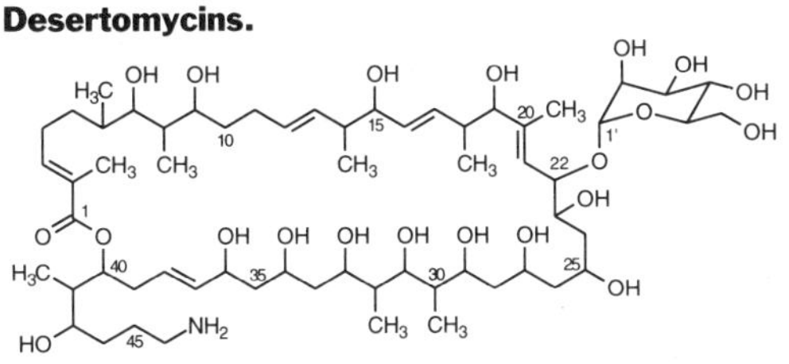

D. A

Macrolactones produced by *Streptomyces flavofungini* besides the polyene *macrolide flavofungin. *D. A:* $C_{61}H_{109}NO_{21}$, M_R 1192.53, mp. 189–190 °C. The stereochemistry of the 42-membered lactone ring is not known. The biosynthesis of D. A follows the *polyketide pathway with *4-aminobutanoic acid (from ornithine) as starter unit. The oxygen atom at C-22 which carries the α-D-mannopyranosyl unit originates from atmospheric oxygen. D. A was the first member of a family of unusual, antifungal macrolactones [e.g., *oasomycins, azalomycins (see nigericin), monazomycins] whose starter units are derived from ornithine or arginine (group name: marginolactones). D. A is active against phytopathogenic fungi and yeasts and is used to isolate human pathogenic fungi for diagnostic purpose. It is proposed that the biological activity is based on ionophoric properties.

Lit.: Acta Microbiol. Hung. **33**, 271–283 (1986) ▪ Can. J. Microbiol. **36**, 609–616 (1990) ▪ J. Am. Chem. Soc. **108**, 8056 (1986) ▪ J. Antibiot. **45**, 1016ff., 1987 (1992) ▪ J. Chem. Soc., Perkin Trans. 1 **1993**, 2525–2531 ▪ J. Org. Chem. **59**, 6986–6993 (1994). – *[HS 2941 90; CAS 12728-25-5 (D.); 121820-50-6 (D. A)]*

Desmarestene see algal pheromones.

Desoxy... see deoxy...

Destruxins. Cyclic depsipeptides from *Metarrhizium anisopliae* and *Oospora destructor* (Hyphomycetes) with insecticidal activity[1]. *Destruxin B* is a host-specific phytotoxin from *Alternaria brassicae*[2]. The D. are optically active, mostly crystalline compounds. For the most important D., see the table.

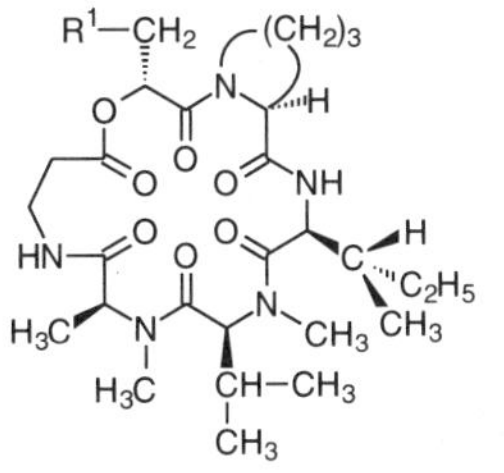

	R^1
D. A	$CH{=}CH_2$
D. B	$CH(CH_3)_2$
D. C	$CH(CH_3){-}CH_2OH$
D. D	$CH(CH_3){-}COOH$
D. E	oxiranyl
D. F	$CH(OH){-}CH_3$

Table: Data of Destruxins.

name	molecular formula	M_R	mp. [°C]	CAS
D. A	$C_{29}H_{47}N_5O_7$	577.72	126–129	6686-70-0
D. B	$C_{30}H_{51}N_5O_7$	593.76	238	2503-26-6
D. C	$C_{30}H_{51}N_5O_8$	609.76	219–220	27482-49-1
D. D	$C_{30}H_{49}N_5O_9$	623.75	oil	27482-50-4
D. E	$C_{29}H_{47}N_5O_8$	593.72	172–173	76689-14-0
D. F	$C_{29}H_{49}N_5O_8$	595.74		148440-85-1

There is no commercial use for the individual substances. The D. have supporting effects in the parasitization of insects by *Metarrhizium anisopliae*[3]. For biosynthesis and incorporation experiments of, e.g., ^{13}C labelled acetate, see *Lit.*[4], for detection, see *Lit.*[5]. D. B reversibly suppresses the expression of hepatitis B viral surface antigen (HBsAg) gene in human hepatoma cells and may have potential for development as a specific anti-HBV drug[6].

Lit.: [1] Phytochemistry **20**, 715 (1981); J. Am. Chem. Soc. **106**, 2388 (1984); Adv. Cell Cult. **4**, (1985); C. R. Acad. Sci. Ser. C **303**, 641 (1986); **304**, 229 (1987); J. Liq. Chrom. **12**, 383–395 (1989); J. Chem. Soc., Perkin Trans. 1 **1989**, 2347–2357; J. Nat. Prod. **56**, 643–647 (1993). [2] Plant Pathol. **41**, 55–63 (1993). [3] J. Insect. Physiol. **35**, 97–105 (1989). [4] Phytochemistry **33**, 1403ff. (1993). [5] Biol. Mass Spectrom. **21**, 33–42 (1992); J. Chromatogr. **830**, 115–125 (1999). [6] Antiviral Res. **34**, 137 (1997).

gen.: Pharmacology: Biochim. Biophys. Acta **1126**, 41–48 (1992) ▪ Insect. Biochem. Mol. Biol. **23**, 43ff. (1993) ▪ Phytochemistry **49**, 1815 (1998) (isolation D. Ed1) ▪ Toxicon **28**, 1249–1254 (1990). – *Synthesis:* Tetrahedron Lett. **38**, 339 (1997) (D. B).

Deterrol see Lactarius pigments.

Deuteroporphyrin see porphyrins.

Deutzioside (mentzeloside)

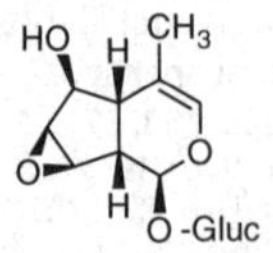

An *iridoid glucoside with an epoxide ring and in which C-10 is absent. $C_{15}H_{22}O_9$, M_R 346.33, mp. 266–267 °C, $[\alpha]_D$ −101° (H_2O). Reductive cleavage of the epoxide ring affords *deutziol* {$C_{15}H_{24}O_9$, M_R 348.35, mp. 108–111 °C, $[\alpha]_D$ −150° (CH_3OH).

Occurrence: D. was isolated from the plant *Mentzelia decapetala* (Loasaceae) cultivated in a botanical garden [1] from which the name "mentzeloside" was derived. Only after deutziol had been isolated from the ornamental shrub *Deutzia scabra* (Saxifragaceae) [2] and the chemical relationships between the two natural products recognized was the name changed to deutzioside.

Lit.: [1] Can. J. Chem. **51**, 760 (1973). [2] Gazz. Chim. Ital. **104**, 17 (1974); **106**, 57 (1976).
gen.: J. Nat Prod. **43**, No. 5, 6, 649 (1980) ▪ Phytochemistry **25**, 2515 (1986) (biosynthesis). – *[CAS 41753-47-3 (D.), 60352-72-9 (deutziol)]*

Dextrose see D-glucose.

Dhurrin see cyanogenic glycosides.

Diaminopimelic acid (2,6-diaminoheptanedioic acid).

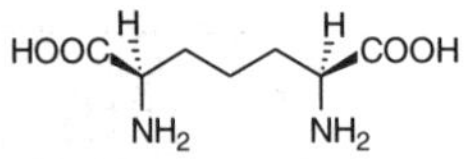

(2*S*, 6*R*)-, *meso*-form

$C_7H_{14}N_2O_4$, M_R 190.20, mp. 313–315 °C (decomp.); hydrochloride, mp. 264–265 °C (decomp.); (2*S*,6*S*)-, L form, mp. 310–312 °C (decomp.), $[\alpha]_D^{16}$ +30.4° (6 m HCl), pK_a of *meso*- and L-form: 1.80, 2.20, 8.80, 9.90; pI 5.50. Biosynthetic precursor of *lysine in plants and bacteria.

Occurrence: The *meso*-form and occasionally its chiral isomers occur in the peptidoglycans of algal and bacterial cell walls. The *meso*-form is, in peptide-bound form, also a component of bacterial spores. The content amounts to 0.02 to 2.0% of the dry weight of the bacteria. D. is widely distributed.

Biosynthesis: 2,3,4,5-Tetrahydropyridine, formed from aspartic acid and pyruvic acid, and succinyl-CoA are converted by *N*-succinyl-2-amino-6-oxopimelic acid synthase to *N*-succinyl-(2*S*,6*S*)-2,6-diaminopimelic acid. Succinic acid is cleaved by succinyldiaminopimelic acid desuccinylase (EC 3.5.1.18.) to furnish the (2*S*,6*S*) form of D. The *meso*-form is generated by D. epimerase.

Lit.: Beilstein EIV **4**, 3081 ▪ Chem. Pharm. Bull. **43**, 535 (1995) (synthesis) ▪ J. Bacteriol. **178**, 6496 (1996) (biosynthesis) ▪ J. Org. Chem. **57**, 6519–6527 (1992). – *[CAS 922-54-3 (meso); 14289-34-0 (L-form)]*

Diamphotoxin see arrow poisons.

Dianthalexin (7-hydroxy-2-phenyl-4*H*-3,1-benzoxazin-4-one).

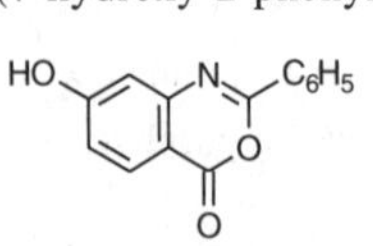

$C_{14}H_9NO_3$, M_R 239.23, cryst., mp. 229–231 °C. Alkaloid and *phytoalexin of the garden carnation (*Dianthus caryophyllus*) (Caryophyllaceae); it is formed on infection with the fungus *Fusarium oxysporum* f. sp. *dianthi*; see also avenalumin.

Lit.: J. Heterocycl. Chem. **25**, 715–718 (1988); **28**, 2075 (1991) (synthesis) ▪ J. Phytopathol. **126**, 281–292 (1989) ▪ Luckner (3.), p. 333, 471 ▪ Phytochemistry **31**, 3761–3767 (1992) ▪ Phytopathology **81**, 728–734 (1991) ▪ Tetrahedron Lett. **31**, 2647 f. (1990) (synthesis). – *[CAS 85915-62-4]*

Diazo compounds (natural). Various types of D. c. occur in nature. Thus, derivatives of diazoacetic acid (*azaserine) and diazoketones (*2-amino-6-diazo-5-oxohexanoic acid) are known as metabolic products of streptomycetes with anti-tumor activity. Quinonediazide (*4-diazo-2,5-cyclohexadien-1-one) and phenyldiazonium salts [*4-(hydroxymethyl)benzenediazonium ion] have been detected in the fruitbodies of the mushroom genus *Agaricus*. Further biologically active D. c. are *cremeomycin, *4-diazo-3-methoxy-2,5-cyclohexadien-1-one, and *kinamycins.

4-Diazo-2,5-cyclohexadien-1-one (4-hydroxybenzenediazonium betaine).

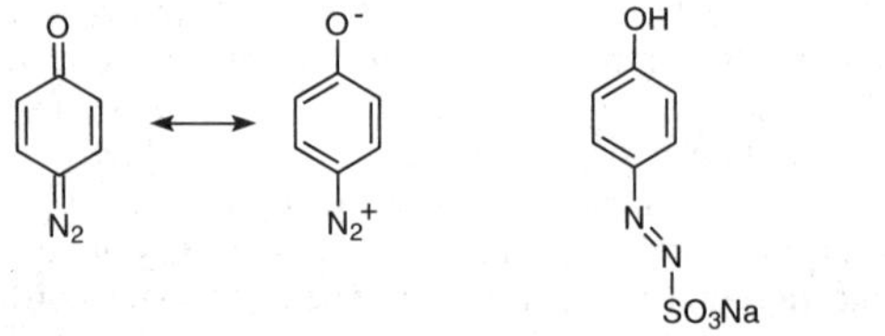

4-Diazo-2,5-cyclohexadien-1-one — Sodium 4-hydroxyphenyldiazenesulfonate (agaridin)

$C_6H_4N_2O$, M_R 120.11, orange solid, explodes at 92 °C. 4-D. occurs in the fruit bodies of *Agaricus xanthoderma* (Basidiomycetes) and is probably biosynthetically closely related to *xanthodermin. On decomposition 4-D. releases phenol, hydroquinone, and 4,4′-biphenyldiol. On addition of sodium hydrogen sulfite a stable diazenesulfonate (*"agaridin"*, $C_6H_5N_2NaO_4S$, M_R 224.17) is formed which, like 4-D., is a highly active antibiotic and inhibits various tumor cell lines. 4-D. can be detected in methanol extracts of the fungus by azo-coupling with 2-naphthol or resorcinol.

Lit.: Angew. Chem. Int. Ed. Chem. **24**, 1063 (1985) ▪ Tetrahedron Lett. **27**, 559 (1986) ▪ Zechmeister **51**, 238–242. – *[CAS 932-97-8 (4-D.); 99280-77-0 (agaridin)]*

4-Diazo-3-methoxy-2,5-cyclohexadien-1-one.

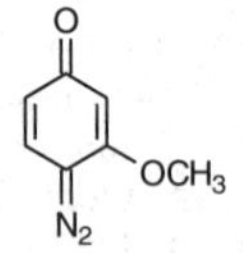

$C_7H_6N_2O_2$, M_R 150.14, yellow, photolabile crystals; diazo compound from cultures of *Penicillium funiculosum*. D. shows good activity against aerobic bacteria (MIC 0.05–0.4 µg/mL). LD_{50} (mouse i. p.) <17 mg/kg.

Lit.: J. Antibiot. **39**, 1054 (1986). – *[CAS 105114-23-6]*

Diazonamides. The halogenated cyclopeptides D. A and B with unusual CC bridges are isolated from the tunicate *Diazona chinensis*; they have strong *in vitro* cytotoxic activities against the colon tumor HCT-16

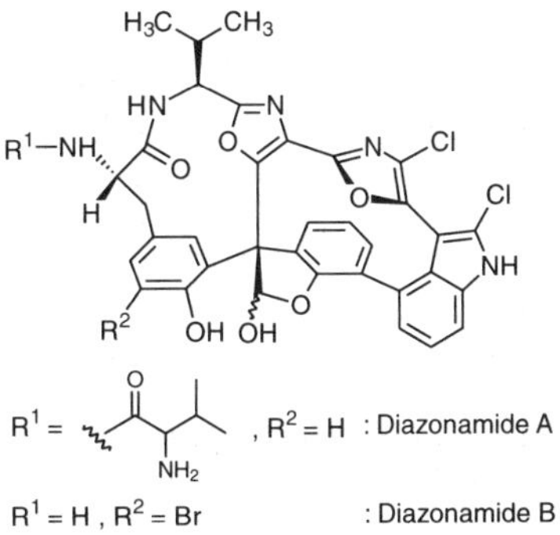

and B 16 murine melanoma (IC_{50} each <15 ng/mL); *D. A:* $C_{40}H_{36}Cl_2N_6O_7$, M_R 783.67, $[\alpha]_D$ −217.3°; *D. B:* $C_{35}H_{26}BrCl_2N_5O_6$, M_R 763.43.

Lit.: Angew. Chem. Int. Ed. Engl. **39**, 937 (2000). ▪ J. Am. Chem. Soc. **113**, 2303 (1991) ▪ J. Chem. Soc., Perkin Trans. 1 **1997**, 2413 (synthesis). – *[CAS 131727-01-0 (D. A); 131703-15-6 (D. B)]*

6-Diazo-5-oxonorleucine see 2-amino-6-diazo-5-oxohexanoic acid.

Dibenz[*d,f*]azonine alkaloids see morphinan alkaloids.

Dibenzo[*b,g*]pyrrocoline alkaloids (dibenzo[*b,g*]indolizine alkaloids, indolo[2,1-*a*]isoquinoline alkaloids). Only two naturally occurring D.a. are known, *cryptaustoline* {$C_{20}H_{24}NO_4^+$, M_R 342.41, mp. 214 °C (iodide), $[\alpha]_D$ −151° (C_2H_5OH)} and *cryptowoline* {$C_{19}H_{20}NO_4^+$, M_R 326.37, mp. 245–246 °C (iodide), $[\alpha]_D$ −186° (C_2H_5OH)} which had already been synthesized before their detection in nature. They occur in the Australian endemic species *Cryptocarya bowiei* (Lauraceae).

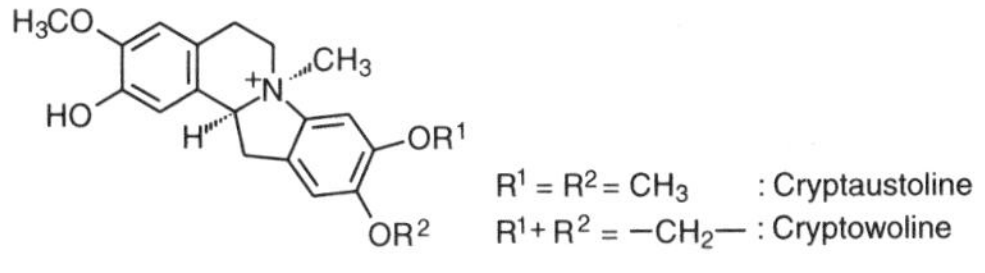

Lit.: Beilstein EIV **27**, 6482 ▪ Can. J. Chem. **67**, 947 (1989) (isolation) ▪ Chem. Pharm. Bull. **37**, 1682 (1989) (synthesis) ▪ J. Am. Chem. Soc. **114**, 8483 (1992) (absolute configuration) ▪ Ullmann (5.) **A1**, 380. – *[HS 293990; CAS 26754-21-2 (cryptaustoline); 15583-51-4 (cryptowoline)]*

DIBOA [2,4-dihydroxy-2*H*-1,4-benzoxazin-3-(4*H*)-one].

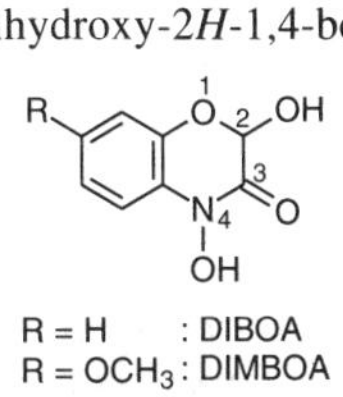

$C_8H_7NO_4$, M_R 181.14, solid. Occurs together with the methoxy derivative DIMBOA [$C_9H_9NO_5$, M_R 211.17, pink needles, mp. 156–157 °C (decomp.)] free and as glucoside in various grain species (wheat, rye, corn, etc.) and wild grasses, where it protects the plants from attack by insects (aphids), fungi, and bacteria. Activity against benign prostatic hyperplasia, chronic prostatitis, and cell lines from prostatic carcinomas has been demonstrated for DIBOA from a rye pollen extract (Cernilton)[1].

Lit.: [1] J. Med. Chem. **38**, 735 (1995).
gen.: Crop Sci. **10**, 1496 (1991) ▪ J. Chem. Ecol. **18**, 469, 945 (1992) (activity) ▪ Phytochemistry **27**, 3349 (1988); **36**, 893 (1994) (biosynthesis); **43**, 551 (1996) (review). – *[CAS 17359-54-5 (DIBOA); 15893-52-4 (DIMBOA)]*

Dibromphakellin, Dibromsceptrin see oroidin.

1,3-Di-*O*-caffeoylquinic acid see cynarin(e).

Dicentrine see aporphine alkaloids.

2,6-Dichlorophenol see ticks.

Dichroine B see febrifugine.

Dicoumarol [3,3′-methylenbis-(4-hydroxycoumarin)].

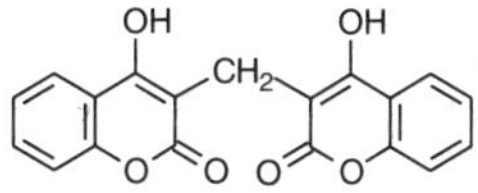

$C_{19}H_{12}O_6$, M_R 336.30, slightly bitter-tasting crystals with a pleasant odor, mp. 287–293 °C, soluble in benzene, chloroform, aqueous and organic bases. D. was isolated in 1938 from fermented alfalfa (*Melilotus alba*) in which it is formed from *coumarin; it also occurs in *Anthoxanthum* (Gramineae). D. is the cause of a cattle disease prevalent in USA and Canada since the 1920's. The afflicted animals show a severe tendency for bleeding usually leading to death.

Activity: D. inhibits vitamin K epoxide reductase and thus the synthesis of the blood clotting factor prothrombin. D. acts as an anticoagulant (vitamin K antagonist). D. and its analogues (e.g., warfarin, phenprocoumon) are used clinically for thrombosis therapy. It causes tissue and skin hemorrhages in rodents and thus finds use as a rat poison. For the mechanism of action, see *Lit.*[1]. LD_{50} (rat p.o.) 541.6 mg/kg.

Lit.: [1] J. Am. Chem. Soc. **103**, 3910–3915 (1981); Nat. Prod. Rep. **11**, 591–606 (1994).
gen.: Beilstein E V **19/6**, 682 ▪ Hager (5.) **7**, 1270ff. (monograph) ▪ J. Chem. Educ. **61**, 87f. (1984) (synthesis) ▪ J. Nutr. **117**, 1325 (1987) (review) ▪ Phytochemistry **27**, 1933–1941 (1988) (review) ▪ Stryer 1995, p. 255. – *Toxicology:* Sax (8.), BJZ 000. – *[HS 293229; CAS 66-76-2]*

Dictyopterin see biopterin.

Dicurone see D-glucurono-6,3-lactone.

7,8-Didehydrocholesterol see calcitriol and sterols.

3,4-Didehydroretinal (dehydroretinal, vitamin A_2 aldehyde).

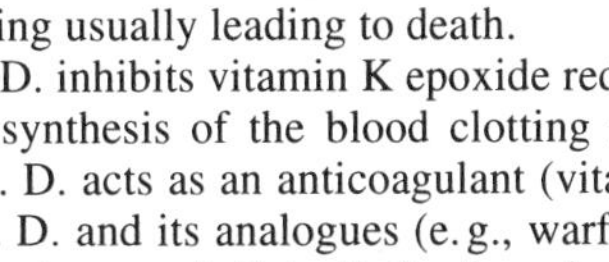

$C_{20}H_{26}O$, M_R 282.43, orange-red cryst., mp. 77–78 °C. A *carotinoid which together with *retinal plays a key role as visual pigment in the sight processes in freshwater fish by reacting with opsin (see pigments of vision) to afford *porphyropsin and *cyanopsin. The compounds, as well as the corresponding alcohol *3,4-didehydroretinol* (vitamin A_2), $C_{20}H_{28}O$, M_R 284.44, golden-yellow oil, occurs in fish oils. The alcohol exhibits 40% of the biological activity of vitamin A_1 (retinal).

Lit.: Bioorg. Med. Chem. Lett. **6**, 2049 (1996) (synthesis) ▪ Tetrahedron Lett. **32**, 4115–4118 (1991); **34**, 319 (1993); **35**,

1209 (1994) (synthesis) ▪ Tetrahedron **50**, 3389 (1994) ▪ see also retinal. – *[CAS 472-87-7 (3,4-D.); 79-80-1 (3,4-didehydroretinol)]*

Didemnimides. *Indole alkaloids with potent antifeedant activity on fish from the Caribbean mangrove ascidian *Didemnum conchyliatum*, e.g. *D. A*: $C_{15}H_{10}N_4O_2$, M_R 278.27, orange needles, mp. 234 °C.

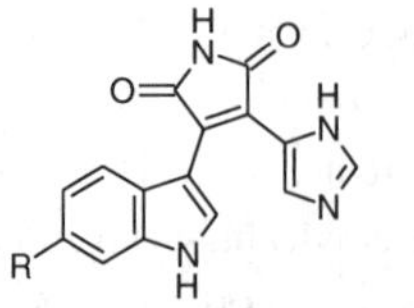

Lit.: J. Org. Chem. **62**, 1486 (1997) ▪ Tetrahedron Lett. **39**, 9629 (1998) (synthesis). – *[CAS 186143-93-1 (D. A); 186144-09-2 (D. B)]*

Didemnins.

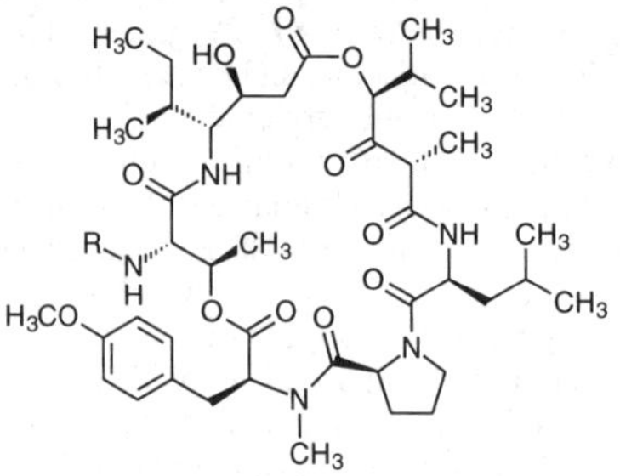

The D. are cyclodepsipeptides from Caribbean *tunicates (e.g., *Trididemnum solidum* or *Aplidium albicans*) with immunosuppressive, antiviral, and antitumor activities. The particularly active D. B[1] inhibits the DNA viruses *Herpes simplex* types I and II *in vitro* at a concentration of 0.05 μM and impedes *Varicella* viruses as well. Very good results have also been obtained in animal models for lethal virus infections such as Rift Valley fever, yellow fever, sandfly fever with survival rates of 90% at 0.25 mg/kg. D. B is active against P 388, murine leukemia, and B 16 melanoma. D. stop cell growth by inhibiting protein biosynthesis through blockade of the GTP-binding receptor EF-1α[3]. D. B is undergoing clinical testing as anti-cancer drug. The D. are related to the *roseotoxins. *Examples*: *D. A* ($C_{49}H_{78}N_6O_{12}$, M_R 943.19), *D. B* ($C_{57}H_{89}N_7O_{15}$, M_R 1112.37), and *D. C* ($C_{52}H_{82}N_6O_{14}$, M_R 1015.25). The terpenoids *didemnaketals* A and B can be obtained from *Didemnum* sp., these compounds inhibit HIV-1 protease (IC_{50} 2 and 10 μM)[2]. Structure-activity relationships have been studied with natural and synthetic D.[4].

Lit.: [1]Helv. Chim. Acta **73**, 25–47 (1990); J. Nat. Prod. **51**, 1 (1988); Proc. Natl. Acad. Sci. USA **85**, 4118 (1988) (structure). [2]J. Am. Chem. Soc. **113**, 6321 (1991). [3]J. Biol. Chem. **269**, 15411 (1994) (D. B receptor). – *Pharmacology:* Antiviral Res. **3**, 269 (1983); Cancer Res. **44**, 1796 (1984); J. Org. Chem. **54**, 617 (1989). – *Synthesis:* Tetrahedron Lett. **30**, 3053 (1989); Synthesis **1988**, 475; **1989**, 832; J. Am. Chem. Soc. **109**, 6846 (1987); **111**, 669 (1989); **112**, 7659–7672 (1990); Tetrahedron **44**, 3489 (1988). [4]J. Med. Chem. **39**, 2819–2834 (1996); Bioorg. Med. Chem. Lett. **8**, 3653 (1998); Tetrahedron **55**, 313–334 (1999).

gen.: Atta-ur-Rahman **10**, 201–302 ▪ Attaway & Zaborsky (eds.), Marine Biotechnology, vol. 1, p. 312 ff., New York: Plenum 1993 ▪ Chem. Rev. **93**, 1771 (1993) ▪ J. Am. Chem. Soc. **117**, 8885 (1995) (new D.) ▪ J. Peptide Res. **54**, 146–161 (1999) (synthesis, activity) ▪ Nachr. Chem. Tech. Lab. **37**, 1040 (1989) ▪ Scheuer II **1**, 101, 120, 150; **3**, 31 ▪ Tetrahedron **45**, 181 (1989) ▪ Zechmeister **49**, 151. – *[CAS 77327-04-9 (D. A); 77327-05-0 (D. B); 77327-06-1 (D. C)]*

Differentiation factors. Name for low-molecular-weight substances that trigger cell differentiation in microorganisms (e.g., formation of aerial mycelia, formation of spores) and/or induce formation of secondary products. Often, they are intracellular signalling substances (autoregulators), e.g., the *pamamycins and *hormaomycin. Chemical signals between organisms of different species have also been observed (e.g., *B-factor, homoserine lactones).

Lit.: Annu. Rev. Microbiol. **46**, 377–398 (1992) ▪ Gene **115**, 159–165 (1992) ▪ J. Bacteriol. **176**, 269–275 (1994).

Digallic acid see gallic acid and tannins.

Digenic acid see domoic acid.

Digipurpurin see Digitalis glycosides.

Digitalis glycosides. Highly toxic substances used in cardiac therapy from the Scrophulariaceae *Digitalis purpurea* (purple foxglove), *D. lanata*, *D. lutea* (lesser yellow foxglove), and *D. grandiflora* (syn. *D. ambigua*). A large number of compounds has been isolated to date. These include those shown in the figure; the genuine tetraosides (primary glycosides) of 3-hydroxylated *cardenolides are made up from *glucose and *digitoxose and are acetylated in the carbohydrate part, e.g., digitoxigenin, gitoxigenin, and digoxigenin. These can, on work-up of the plant material, be readily transformed to the therapeutically important triosides (secondary glycosides) by enzymatic cleavage of the terminal glucose and acetic acid.

In addition to those listed in the table on p. 185 and further glycosides with cardiac activity {such as *digitalin*, a 3-*O*-[β-D-glucopyranosyl-(1→4)-D-digitaloside] of gitoxigenin}, other steroid glycosides, albeit without cardiac activity, have been isolated from *Digitalis* species. To be mentioned in this context are the digitanol glycosides such as *digipurpurin*, a 3-*O*-glycoside of 3β,12α,14-trihydroxy-14β-pregn-5-en-20-one (digipurpurogenin) corresponding to purpurea glycosides, an intermediate of cardenolide glycoside biosynthesis, and a series of *steroid saponins, such as digitonin.

Use: In contemporary cardiac therapy, especially for the long-term treatment of chronic myocardial insufficiency, arrhythmias, and valvular defects, the better dosable, pure crystalline secondary glycosides, mostly digitoxin or semi-synthetic derivatives (e.g. partially acetylated compounds), are used almost exclusively instead of *Digitalis* leaf powders, extracts or glycoside mixtures. On account of their high toxicity, exact dosing of the D. g. and adjustment and close monitoring of the patient are necessary. High doses of digitoxin cause cardiac paralysis and death. Many animals have a markedly lower sensitivity than humans. Borderline doses (i. v.): LD_0 (man) 12 μg/kg, (rat) 50 μg/kg, (cat) 62 μg/kg, (mouse) 200 μg/kg, and (pigeon) 670 μg/kg. D. g. show cumulative action. The half-life of the

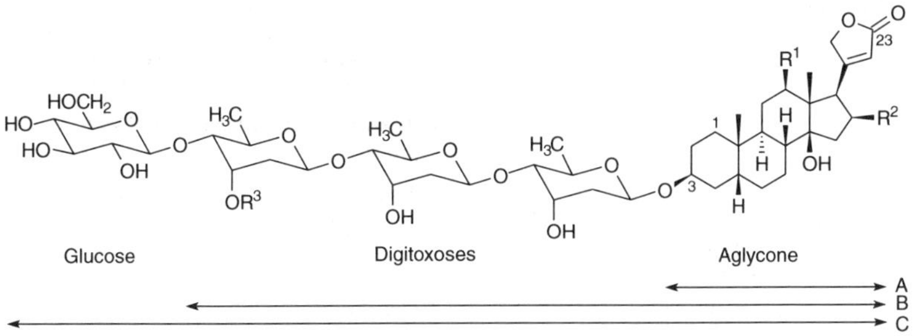

part of formula	name	R^1	R^2	R^3
part A (aglycone)	Digitoxigenin	H	H	-
	Gitoxigenin	H	OH	-
	Digoxigenin	OH	H	-
part B (triosides, secondary glycosides)	Digitoxin	H	H	H
	Gitoxin	H	OH	H
	Digoxin	OH	H	H
part C (tetraosides, primary glycosides)	Purpurea glycoside A	H	H	H
	Purpurea glycoside B	H	OH	H
		OH	H	H
acetylated primary glycosides	Lanatoside A	H	H	$CO-CH_3$
	Lanatoside B	H	OH	$CO-CH_3$
	Lanatoside C	OH	H	$CO-CH_3$

Table: Data of Digitalis glycosides and aglyca.

Glycoside	molecular formula	M_R	mp. [°C]	$[\alpha]_D$	CAS
Digitoxigenin	$C_{23}H_{34}O_4$	374.52	253	+19.1° (CH_3OH)	143-62-4
Gitoxigenin	$C_{23}H_{34}O_5$	390.52	220–225	+32.6° (CH_3OH)	545-26-6
Digoxigenin	$C_{23}H_{34}O_5$	390.52	220	+23° (CH_3OH)	1672-46-4
Digitoxin	$C_{41}H_{64}O_{13}$	764.95	256–257	+4.8° (dioxane)	71-63-6
Gitoxin	$C_{41}H_{64}O_{14}$	780.95	282–285 (decomp.)	+5° (pyridine)	4562-36-1
Digoxin	$C_{41}H_{64}O_{14}$	780.95	265 (decomp.)	+13.3° (pyridine)	20830-75-5
Purpurea glycoside A	$C_{47}H_{74}O_{18}$	927.09	270–280 (decomp.)	+12° (C_2H_5OH)	19855-40-4
Purpurea glycoside B	$C_{47}H_{74}O_{19}$	943.09	240	+15.5° (C_2H_5OH)	19855-39-1
Lanatoside A	$C_{49}H_{76}O_{19}$	969.13	245–248 (decomp.)	+2.5° (pyridine)	17575-20-1
Lanatoside B	$C_{49}H_{76}O_{20}$	985.13	245–248 (decomp.)	+11° (pyridine)	17575-21-2
Lanatoside C	$C_{49}H_{76}O_{20}$	985.13	245–248 (decomp.)	+16° (pyridine)	17575-22-3

plasma concentration after i.v. administration (in humans) varies between 32–34 h (α-acetyldigoxin) and 144–192 h (digitoxin). The cardiac active steroids influence Na^+/K^+ exchange in myocardial membranes and act through control of Na^+/K^+-ATPase.

Lit.: Adv. Drug Res. **19**, 313–562 (1990) (activity) ▪ Beilstein E V **18/3**, 348, 354; **18/4**, 388; **19/3**, 338–342, 603 ▪ Bodem & Dengler, Cardiac Glycosides, Berlin: Springer 1978 ▪ Erdmann et al. (eds), Cardiac Glycosides, p. 1785–1985, Darmstadt: Steinkopff 1986 ▪ Hager (5.) **3**, 468 ff., 725–745; **4**, 1063 ff., 1168–1190; **7**, 1297 ff. ▪ J. Am. Chem. Soc. **118**, 10660 (1996) (synthesis digitoxigenin) ▪ Merck-Index (12.), No. 3196–3210 ▪ Pharm. Unserer Zeit **7**, 33-45 (1978); **16**, 81 (1987) ▪ Sax (8.), DKL 200–DKL 875, GEU 000–GEW 000, LAT 000–LAU 400 ▪ Zechmeister **69**, 71–156 (review). – *[HS 2938 90]*

Digitogenin see steroid sapogenins.

Digitonin see steroid saponins.

Digitoxigenin, Digitoxin see Digitalis glycosides.

Digitoxose (2,6-dideoxy-D-*ribo*-hexose). $C_6H_{12}O_4$, M_R 148.16, mp. 108 °C, $[\alpha]_D^{20}$ +46.4° (H_2O), +37° (CH_3OH). D. is a *deoxy sugar that is liberated from *Digitalis glycosides on mild hydrolysis.

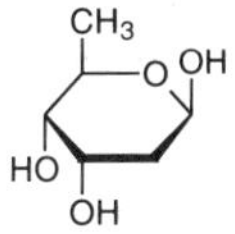

β-pyranose form

Lit.: Beilstein E IV **1**, 4191 ▪ Karrer, No. 602 ▪ Merck-Index (12.), No. 3207 ▪ Synlett **1991**, 750 ff. – *[HS 2940 00; CAS 527-52-6]*

Digoxigenin, Digoxin see Digitalis glycosides.

Dihomo-γ-linolenic acid [(*all-Z*)-8,11,14-eicosatrienoic acid].

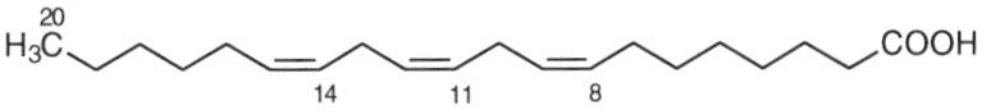

$C_{20}H_{34}O_2$, M_R 306.49. D. is formed in the animal organism by chain-extension of γ-*linolenic acid by two C-atoms, it is transformed to *arachidonic acid by introduction of a 5,6-double bond. In animal cells prostaglandins of the PGE_1 type are formed from D.

Lit.: see eicosanoids. – *[CAS 1783-84-2]*

Dihydroactinidiolide [5,6,7,7a-tetrahydro-4,4,7a-trimethylbenzofuran-2(4*H*)-one].

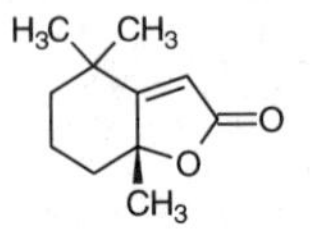
(*R*)-(-)-Dihydroactinidiolide

$C_{11}H_{16}O_2$, M_R 180.25, mp. 69–71 °C, $[\alpha]_D^{20}$ –119.9° ($CHCl_3$). A norisoprenoid[1] with hay-like odor arising from *carotinoid degradation; the (*R*)-form occurs in the aromas of tea, tobacco, currants, passion fruit, tomatoes, etc.[2].

Lit.: [1] J. Agric. Food Chem. **38**, 237–243 (1990). [2] TNO list (6.), Suppl. 5, p. 272.
gen.: Beilstein E V **17/9**, 515 ▪ Chem. Lett. **1995**, 53 ▪ Justus Liebigs Ann. Chem. **1993**, 77 ▪ Tetrahedron Lett. **32**, 4871 (1991) (synthesis). – *[CAS 15356-74-8 (D.); 17092-92-1 (R-form)]*

Dihydrocaffeic acid [hydrocaffeic acid, 3-(3,4-dihydroxyphenyl)-propanoic acid, 3,4-dihydroxydihydrocinnamic acid]

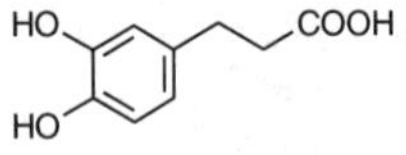

$C_9H_{10}O_4$, M_R 182.18, plates, mp. 139 °C, pK_{a1} 4.56, pK_{a2} 9.36, pK_{a3} 11.6 (30 °C in 0.1 M $NaClO_4$), occurs in the free form in spores of *Lycopodium clavatum* as well as in higher plants. *Lemon oil (from *Citrus limon*) contains the 4-*O*-(6-*O*-*β*-D-glucopyranosyl-*α*-D-glucopyranoside) of the methyl ester {*citrusin F*, $C_{22}H_{32}O_{14}$, M_R 520.49, $[\alpha]_D^{22}$ –4.68° (CH_3OH)}. The methyl ester of 4-*O*-*β*-D-glucopyranosyl-3-*O*-methyl-D., *citrusin E* {$C_{17}H_{24}O_9$, M_R 372.37, mp. 108 °C, $[\alpha]_D^{17}$ –35.7° (CH_3OH)} exhibits hypotensive activity in animal experiments[1] and is an inhibitor of phenylalanine ammonia lyase (PAL, EC 4.3.1.5).

Lit.: [1] Agric. Biol. Chem. **55**, 647–650 (1991).
gen.: Appl. Environ. Microbiol. **54**, 712–717 (1988) (metabolism) ▪ Indian J. Exp. Biol. **31**, 847–877 (1993) ▪ J. Am. Chem. Soc. **115**, 7752–7760 (1993) ▪ J. Chromatogr. **655**, 119–125 (1993) ▪ J. Org. Chem. **59**, 264 f. (1994) (oxide.) ▪ Plant Cell, Tissue Organ Cult. **13**, 15–26 (1988) (biosynthesis) ▪ Synth. Commun. **25**, 3187 (1995) (synthesis). – *[CAS 1078-61-1 (D.); 134860-04-1 (citrusin F)]*

Dihydrocarveol see *p*-menthenols.

Dihydrocarvone see *p*-menthenones.

Dihydroconiferyl alcohol.

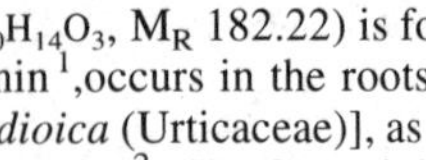

D. (R=CH_3, $C_{10}H_{14}O_3$, M_R 182.22) is formed on degradation of *lignin[1], occurs in the roots of the stinging nettle [*Urtica dioica* (Urticaceae)], as well as in grape juice and raspberries[2]. The free triol [4-(3-hydroxypropyl)-1,2-benzenediol (R = H), $C_9H_{12}O_3$, M_R 168.19] is a constituent of *Onopordon corymbosum* (Cynareae, Asteraceae) and *Hymenoxys linearifolia* (syn. *Tetraneuris linearifolia*, Heliantheae, Asteraceae) as well as the Asian yew *Taxus baccata*.

Lit.: [1] Cell. Chem. Technol. **23**, 579–583 (1989); Phytochemistry **33**, 1395–1401 (1993). [2] J. Agric. Food Chem. **35**, 519–524 (1987); **39**, 173 ff. (1991).
gen.: Flavour Fragrance J. **8**, 215–220 (1993) ▪ Phytochemistry **28**, 3477–3482 (1989); **33**, 1489 ff. (1993); **36**, 1031 ff. (1994) ▪ Plant Cell, Tissue Organ Cult. **13**, 15–26 (1988) (biosynthesis). – *[CAS 2305-13-7 (D.); 46118-02-9 (free triol)]*

5,6-Dihydro-5*β*,6*α*-dihydroxyzeaxanthin see zeaxanthin.

Dihydro-4*H*-1,3,5-dithiazines (1,3,5-dithiazinanes).

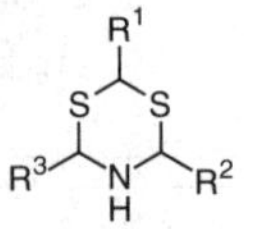
1

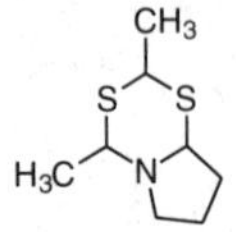
2,4-Dimethyl-tetrahydro-pyrrolo[2,1-*d*]-1,3,5-dithiazine
2

D. of the types **1** and **2** are highly active secondary *flavor compounds which are formed from aldehydes, hydrogen sulfide, and ammonia (**1**) or 1-pyrroline (**2**) when food is heated, i.e., from degradation products of lipids, *amino acids and *thiamin. D. occur especially in yeast extracts and *meat flavor as well as in aromas of peanuts, cocoa, and seafoods[1]. Widely distributed is *thialdine*[2,3] (**1**: R^1,R^2,R^3=CH_3), $C_6H_{13}NS_2$, M_R 163.30, cryst. rusty and meat-like odor, mp. 46 °C. (C_2–C_4)-Alkyldimethyl- and dialkylmethyl-D.[1,4] exhibit interesting odor notes such as peanut, egg, cocoa, or fried onions, as does compound **2**[5] ($C_8H_{15}NS_2$, M_R 189.33, mp. 40.5 °C), which occurs in yeast extracts, pork, and seafood aromas.

Lit.: [1] Food Rev. Int. **8**, 391–442 (1992). [2] Ann. Chem. Pharm. **61**, 2 (1847). [3] J. Agric. Food Chem. **20**, 177 (1972). [4] EU. P. 186026 (1985), Haarmann & Reimer. [5] J. Agric. Food Chem. **39**, 1123 ff. (1991). – *[CAS 638-17-5 (thialdine); 116505-60-3 (2)]*

(*S*)-2,3-Dihydrofarnesol see terrestrol.

3,4-Dihydroxybenzaldehydes. Group of phenolic aldehydes that often occur as methyl ethers in essential oils and are intermediates in the biosynthesis of some *alkaloids.

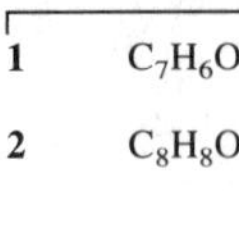

R^1	R^2	
H	H	: Protocatechualdehyde (**1**)
H	CH_3	: Vanillin (**2**)
βDGlc	CH_3	: Glucovanillin (**3**)
CH_3	CH_3	: Veratrumaldehyde (**4**)
C_2H_5	CH_3	: Ethylvanillin (**5**)

Table: Data of 3,4-Dihydroxybenzaldehydes and derivatives.

no.	molecular formula	M_R	mp. (b.p.) [°C]	CAS
1	$C_7H_6O_3$	138.12	153–154 (decomp.)	139-85-5
2	$C_8H_8O_3$	152.15	81 (285, CO_2-atmosphere) (170 [2 kPa])	121-33-5
3	$C_{14}H_{18}O_8$	314.29	188–192	494-08-6
4	$C_9H_{10}O_3$	166.18	44 (58) (285, 172–175 [2.3 kPa])	120-14-9
5	$C_{10}H_{12}O_3$	180.20	64–65	120-25-2

Protocatechualdehyde (3,4-dihydroxybenzaldehyde): Flat, pale yellow cryst., pK_{a1} 7.27, pK_{a2} 11.4 (25 °C).

Vanillin (4-hydroxy-3-methoxybenzaldehyde): colorless, light-sensitive prisms or needles with an odor of vanilla which are slowly oxidized in moist air to vanillic acid (see dihydroxybenzoic acids), D. 1.056, D_4^{20} 1.056, pK_{a1} 7.40 (25 °C). Occurs to 1.5–4% in *vanilla pods, also in *storax, *clove oils, in flowers of black salsify, potatoes, spiraea, and in various foods such as milk, wine, rice wine, etc. Also the smell of old, yellowed paper with high wood content is due to vanillin. Vanillin is also produced by male bugs of *Eurygaster integriceps* as an *insect attractant[1]. *Glucovanillin* (vanilloside, avenein): needles, $[\alpha]_D^{20}$ −88.7° (H_2O), occurs in oak and in wheat *Triticum repens*. *Veratrum aldehyde* (methylvanillin): needles, odor like vanillin, occurs in the oils of *Cymbopogon javanensis*, *Eryngium poterium*, and other plants. *Ethylvanillin* (4-ethoxy-3-methoxybenzaldehyde): prisms, contained in the wood hyacinth *Platanthera bifolia* and in styrax. The synthetic 3-ethoxy-4-hydroxybenzaldehyde (*bourbonal*), which is important as a fragrance and odor substance, is known as ethylvanillin.

Use: Protocatechualdehyde is used in the form of a 0.1% aqueous solution for photometric determination of molybdenum. 3,4-D., vanillin, and veratrum aldehyde are intermediates in the synthesis of L-*DOPA and numerous other natural products[2]. Vanillin and ethylvanillin are used widely as a spice in chocolate, confectionery products, liquors, bakery products, and other sweet foods as well as in the production of vanillin sugar (not vanilla sugar) instead of the expensive natural vanillin. Small amounts are used in cosmetic products (deodorants, perfumes, etc.) and to improve the taste of pharmaceutical preparations. The vanillin content of the wood of wine barrels contributes to the aromatization of wine[3]. Vanillin is a component of Günzburg's reagent to determine HCl in gastric juices and contained in staining reagents for thin layer chromatography of amino acids, steroids, phenols, essential oils, it is used as reference substance in microanalysis and also as a glazing agent in galvanization and serves in 1% solution for the gravimetric analysis of zirconium.

Biosynthesis: By oxidative degradation of *caffeic acid and methylation with **S*-adenosylmethionine.

Lit.: [1]Naturwissenschaften **62**, 348 (1975). [2]J. Chem. Soc., Perkin Trans. 1 **1984**, 1539–1545; J. Org. Chem. **48**, 3653–3657 (1983); **53**, 1064–1071 (1988). [3]Weinwirtsch. Tech. **1990** No. 2, 23–29.
gen.: Analyst (London) **110**, 1283–1287 (1985) ▪ J. Am. Chem. Soc. **120**, 10545 (1998) (short synthesis vanillin) ▪ Kirk-Othmer (3.) **10**, 485; **23**, 704–717; **15**, 161 f.; **21**, 185 ▪ Luckner (3.), p. 368, 400, 474 ▪ Negwer (6.), No. 865 ▪ Perfum. Flavor. **15**, 45, 50, 52 (1990). – *[HS 2912 41, 2912 49]*

Dihydroxybenzoic acids.

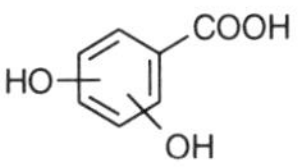

$C_7H_6O_4$, M_R 154.12. The isomeric D. are crystalline and some have reducing activity. D. are often present in natural products in the form of glycosides and/or methyl ethers.

2,3-D. (mp. 206 °C, pK_a 2.91) occurs e. g. in *Centaurium erythraea* and *Gentiana lutea* and is also produced by *Rhizobium*. The diacetate (*dipyrocetyl*, $C_{11}H_{10}O_6$, M_R 238.20, mp. 160 °C) has anti-inflammatory and antirheumatic activity.

2,4-D. (*β-resorcylic acid*, mp. 216–221 °C, pK_a 3.22) is used in the production of dyes, drugs, and cosmetics. The methyl ester of 2-hydroxy-4-methoxybenzoic acid (*primrose camphor*, $C_9H_{10}O_4$, M_R 182.18, mp. 49 °C) as well as its 2-*O*-glycoside *primeverin occur in primrose species.

2,5-D. see gentisic acid.

2,6-D. (*γ-resorcylic acid*, mp. 150 to 170 °C, pK_a 1.08), is present in the form of 2-hydroxy-6-methoxybenzoic acid ($C_8H_8O_4$, M_R 168.15, mp. 135 °C) as a natural product in *Gloriosa superba* and *Colchicum* species. The 4-bromoanilide of 2,6-D. [$C_{13}H_{10}BrNO_3$, M_R 308.13, mp. 221–230 °C (*resorantel*)] is used as an anthelmintic in veterinary medicine.

3,4-D. (mp. 199 °C, pK_a 4.49) can be extracted from *lignin and occurs in free form in many higher plants. Furthermore, many derivatives (esters and ethers) of 3,4-D. are known as natural products. The 4-*O*-β-D-glucoside is a constituent of the hard outer shell of the egg capsule (ootheca) of cockroaches. 4-Hydroxy-3-methoxybenzoic acid (*vanillic acid*, $C_8H_8O_4$, M_R 168.15, mp. 211 °C) is used as preservation and disinfecting agent.

3,5-D. (*α-resorcylic acid*) is formed oxidatively from *orsellinic acid.

Lit.: J. Med. Chem. **32**, 257 (1989) (derivatives, ^1H NMR) ▪ Luckner (3.), p. 397–402 ▪ Planta Med. **30**, 174 (1974) (glucosides) ▪ Zechmeister **54**, 1–312. – *Toxicology:* Sax (8.), HJB 500 (2,3-D.), HOE 600 (2,4-D.), GCU 000 (2,5-D.), HJF 000 u. VFT 000 (3,4-D.). – *[CAS 303-38-8 (2,3-D.); 486-79-3 (dipyrocetyl); 89-86-1 (2,4-D.); 5446-02-6 (primrose camphor); 303-07-1 (2,6-D.); 3147-64-6 (2,6-D.); 99-50-3 (3,4-D.); 121-34-6 (vanillic acid); 99-10-5 (3,5-D.)]*

1α,25-Dihydroxy(chole)calciferol see calcitriol.

(*E*)-3,4-Dihydroxycinnamic acid see caffeic acid.

6,7-Dihydroxycoumarins. A widely distributed group of derivatives of *coumarin.

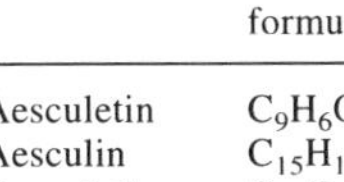

R¹	R²	
H	H	: Aesculetin
βDGlcp	H	: Aesculin
CH_3	H	: Scopoletin
CH_3	βDGlcp	: Scopolin
H	CH_3	: Isoscopoletin
βDGlcp	CH_3	: Magnolioside
CH_3	CH_3	: Scoparone

Table: Data of 6,7-Dihydroxycumarins.

	molecular formula	M_R	mp. [°C]	CAS
Aesculetin	$C_9H_6O_4$	178.14	270	305-01-1
Aesculin	$C_{15}H_{16}O_9$	340.29	204–206	531-75-9
Scopoletin	$C_{10}H_8O_4$	192.17	202–204	92-61-5
Scopolin	$C_{16}H_{18}O_9$	354.31	217–219	531-44-2
Isoscopoletin	$C_{10}H_8O_4$	192.17	185	776-86-3
Magnolioside	$C_{16}H_{18}O_9$	354.31	227	20186-29-2
Scoparone	$C_{11}H_{10}O_4$	206.20	144	120-08-1

Aesculetin (6,7-dihydroxy-2*H*-1-benzopyran-2-one) occurs in horse chestnut, *Aesculus hippocastanum*, as well as in deadly nightshade *Atropa*, thorn apple *Da-*

tura, foxglove *Digitalis*, some ferns and other plants. It has also been isolated from infected sweet potatoes. UV (ethanol): λ_{max} 352 nm. – *Aesculin* [6-β-D-glucopyranosyloxy-7-hydroxycoumarin, esculin, bitter-tasting needles, with 1.5 mol H_2O, $[\alpha]_D$ −78.4° (50% aqueous dioxane)] occurs in the bark of horse chestnut and has also been isolated from ash trees *Fraxinus ornus* and *F. japonica*, as well as from flowers of *Solanum pinnatisectum* and other plants. The blue fluorescence described by Krais in 1929 led to the discovery of optical brighteners for textile fibers. – *Scopoletin* (needles or prisms) occurs besides *scopolamine free or as its 7-*O*-β-D-glucoside (*scopolin*, murrayin) in the roots of *Scopolia* and *Atropa belladonna* (deadly nightshade), as well as in *Convolvulus scammonia*, tobacco plants, oat shoots, oleander bark, or *Gelsemium* roots. Numerous other glycosides are known, the carbohydrate unit being partially acetylated in some cases. – *Isoscopoletin* (6-hydroxy-7-methoxycoumarin, yellow crystals) occurs together with the 6-*O*-β-D-glucopyranoside (*magnolioside*), $[\alpha]_D$ −28°, in *Artemisia* species and magnolia. – *Scoparone* (scoparin, 6,7-dimethoxycoumarin, needles) occurs in *Artemisia* species and citrus oils.

Activity: Aesculetin exhibits antifungal activity. Aesculin reduces the permeability of blood capillaries. Scopoletin acts as an inhibitor of plant growth and is contained in the Chinese drug "Toki" from *Angelica acutiloba*. Scoparone is mutagenic.

Use: Like other coumarins as light- and sun-protection agent. Drugs to treat varicose veins, sports injuries, hemorrhoids, and thrombosis often contain horse chestnut extracts; the vascular sealing effect is attributed not only to aesculetin but also to the saponin *escin. Aesculin is used in UV filters.

The biosynthesis proceeds from *o*-coumaric acid glucoside.

Lit.: Arch. Pharm. **328**, 737 (1995) (scopoletin, technical synthesis) ▪ Chem. Pharm. Bull. **34**, 4012–4017 (1986) ▪ Heterocycles **41**, 1979, 2889 (1995) (synthesis of aesculetin) ▪ Indian J. Chem., Sect. B **26**, 574 f. (1987) (synthesis) ▪ J. Biochem. Biophys. Methods **18**, 297–307 (1989) ▪ J. Indian Chem. Soc. **67**, 785 (1990) ▪ Luckner (3.) p. 388 ff., 471 ▪ Nat. Prod. Rep. **12**, 477–505 (1995) ▪ Org. Magn. Reson. **7**, 339 ff. (1975) (^{13}C NMR) ▪ Phytochemistry **11**, 657–662 (1972) (biosynthesis); **23**, 467 f. (1984); **31**, 717 ff. (1992) ▪ Ullmann (5.) **A 18**, 155 f.; 164 ff. – *[HS 2338 90]*

3,9-Dihydroxycoumestan see coumestrol.

1α,25-Dihydroxyergocalciferol see calcitriol.

1α,15-Dihydroxymarasmene see mniopetals.

4,9-Dihydroxy-3,10-perylendione.

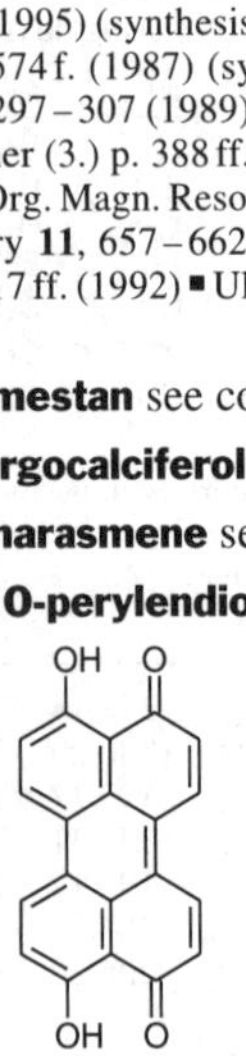

$C_{20}H_{10}O_4$, M_R 314.30, dark red needles, mp. 350 °C (decomp.). 4,9-D. is isolated from the brown-black, wood-like fruit bodies of the tree fungus *Daldinia concentrica* and other Ascomycetes. It is probably formed by oxidative coupling of two molecules *1,8-naphthalenediol.

Lit.: Zechmeister **51**, 117 f. (review). – *[CAS 10190-97-3]*

α-3,4-Dihydroxyphenylalanine (DOPA).

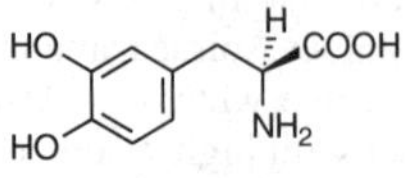

(2*S*)-, L-form

$C_9H_{11}NO_4$, M_R 197.19, mp. 276–278 °C (decomp.), $[\alpha]_D^{13}$ −13.1° (1 M HCl). Non-proteinogenic amino acid in some plants and fungi, including *Vicia faba* (broad bean, horsebean, Windsor bean), *Mucuna* (syn. *Stizolobium* spp.), and the genera *Baptisia*, *Lupinus*, *Vicia* [1,2].

Metabolism: 3,4-D. is formed from tyrosine by means of tyrosine 3-monooxygenase (EC 1.14.16.2.) and tetrahydrobiopterin. DOPA is a biosynthetic precursor of *dopamine, *noradrenaline, and *adrenaline. It is used for therapy of Parkinson's disease (dopamine deficiency). Oxidation and subsequent polymerization lead to the formation of *melanins. Extradiol cleavage of the benzene ring gives rise to *betalamic acid, *stizolobic acid, stizolobinic acid, muscaflavin, etc.[3] 3,4-D. is toxic for the development of certain insects and for the reproduction of the duckweed *Lemna minor*[4].

Lit.: [1] J. Med. Chem. **14**, 463 (1971). [2] Nature (London) **229**, 136 f. (1971). [3] Tetrahedron **35**, 2843–2853 (1979). [4] Environ. Exp. Bot. **21**, 221–230 (1981).

gen.: Beilstein E IV **14**, 2492 ▪ Greenstein & Winitz, Chemistry of the Amino Acids, vol. 3, p. 2713–2723, New York: J. Wiley 1961 ▪ Rosenthal, Plant Nonprotein Amino and Imino Acids p. 143 f., New York: Academic Press 1982 ▪ Ullmann (5.) **A 2**, 60, 87; **A18**, 187, 230; **A19**, 4, 5, 9. – *[HS 2922 50; CAS 59-92-7]*

β-3,4-Dihydroxyphenylalanine (β-DOPA).

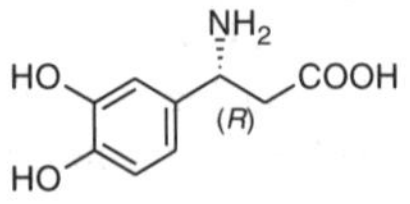

$C_9H_{11}NO_4$, M_R 197.19, mp. 194 °C, $[\alpha]_D^{23}$ −12.1° (6 n HCl).

Isolated as (*R*)-enantiomer from the toadstool *Cortinarius violaceus*. Forms a violet complex with iron(III) ions that is responsible for the color of the fungus. Various heterocycles can be synthesized with D.

Lit.: Angew. Chem. Int. Ed. Engl. **37**, 3292–3295 (1998).

2-(3,4-Dihydroxyphenyl)ethanol see echinacoside.

3-(3,4-Dihydroxyphenyl)lactic acid [3-(3,4-dihydroxyphenyl)-2-hydroxypropanoic acid, Dan Shen Su].

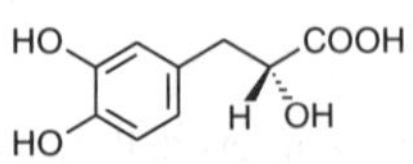

(*R*)-, D-form

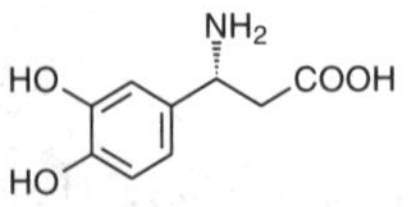

(*R*)

$C_9H_{10}O_5$, M_R 198.18, (*R*)-form: prisms, mp. 84–87 °C, $[\alpha]_D^{18}$ +10.8° (CH_3OH), occurs with the 4′-*O*-β-D-glucopyranoside, $C_{15}H_{20}O_{10}$, M_R 360.32, hygroscopic powder, $[\alpha]_D^{25}$ −41.0° (CH_3OH) in *Coptis japonica* (Ra-

nunculaceae), *Clinopodium umbrosum* (Lamiaceae) and in the roots of *Salvia miltiorrhiza* (Chinese drug Danshen). The (*R*)-form acts as coronary vasodilatator and is a hepatoprotective agent, it inhibits platelet aggregation in blood, cholesterol biosynthesis, and thromboxane A_2 synthase.
Lit.: Chem. Pharm. Bull. **33**, 527 (1985) ▪ Nat. Prod. Rep. **10**, 233–263 (1993) ▪ Phytochemistry **30**, 2877–2881 (1991) ▪ Plant Cell, Tissue Organ Cult. **13**, 15–26 (1988) (metabolism). – *[CAS 23028-17-3 (D.); 76822-21-4 (R-form); 96552-85-1 (glucopyranoside)]*

(2*R*)-3β,4α-Dihydroxy-2α,5β-pyrrolidinedimethanol (DMDP).

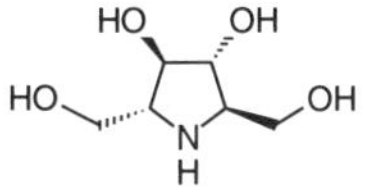

$C_6H_{13}NO_4$, M_R 163.17. D. is a *polyhydroxy alkaloid occurring in various Fabaceae (*Derris elliptica*, *Lonchocarpus* sp.) and, like *alexine, *castanospermine, *deoxynojirimycin and *swainsonine, acts as a glucosidase inhibitor. It is a nitrogen-containing structural analogue of β-D-fructofuranose. DMDP has an antifeedant effect in insects. The moth *Urania fulgens* has overcome this inhibiting activity. Its larvae are adapted to the high DMDP content of the fodder plant and store the substance. The moths themselves are brightly colored (warning signal) and presumably protected from predators by their DMDP content. Similar relationships have been observed for other natural products, especially the *pyrrolizidine alkaloids. DMDP exerts antiviral properties.
Lit.: Angew. Chem. Int. Ed. Engl. **33**, 1242 (1994) (synthesis) ▪ Phytochemistry **46**, 255 (1997) (isolation) ▪ Proc. Phytochem. Soc. Eur. **33**, 271–282 (1992) ▪ Rec. Adv. Phytochem. **23**, 395–427 (1989) ▪ Tetrahedron Lett. **26**, 1469–1472 (1985) (synthesis). – *[CAS 59920-31-9]*

2,3-Dihydroxysuccinic acid see tartaric acid.

1α,25-Dihydroxyvitamin D$_3$ see calcitriol.

3,5-Diiodo-L-thyronine, 3,5-diiodo-L-tyrosine see iodoamino acids.

Diisocyanoadociane.

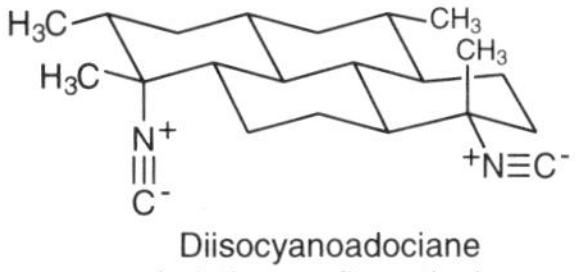

Diisocyanoadociane
(relative configuration)

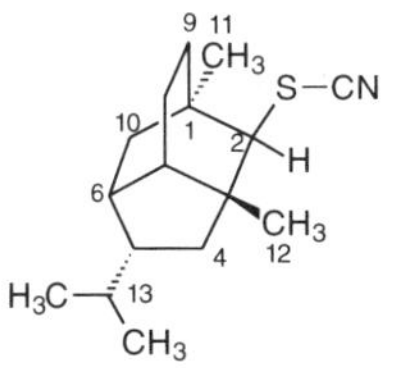

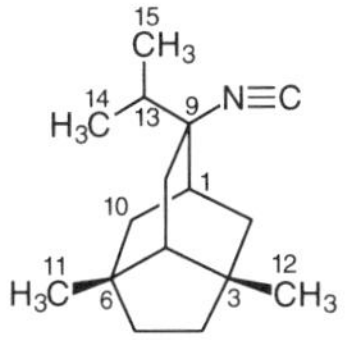

2-Thiocyanatopupukeanane 9-Isocyanoneopupukeanane
(relative configuration)

Marine sponges contain numerous isonitriles and thiocyanates. The biosynthetic precursors of the isocyanide groups are the corresponding isothiocyanates. *D.*[1] from *Adocia* sp., $C_{22}H_{32}N_2$, M_R 324.51, $[\alpha]_D$ +45.8°. *9-Isocyanoneopupukeanane*[2] from *Ciocalypta* sp., colorless oil, $C_{16}H_{25}N$, M_R 231.38, $[\alpha]_D$ +33° (CH_2Cl_2). A thiocyanate with a corresponding skeleton was isolated from *Axinyssa aplysinoides* (*Halichondria*): (*1R,2R,3R,5R,6S,7S*)-*2-thiocyanatopupukeanane*[3], $C_{16}H_{25}NS$, M_R 263.45, $[\alpha]_D$ +5.8° ($CHCl_3$).
Lit.: [1] J. Am. Chem. Soc. **98**, 4010 (1976); Tetrahedron Lett. **21**, 315 (1980). – *Biosynthesis:* J. Chem. Soc., Chem. Commun. **1986**, 35. [2] J. Org. Chem. **54**, 2095 (1989); **61**. 3259–3267 (1996); J. Nat. Prod. **59**, 710–716 (1996). – *Biosynthesis:* Scheuer II **6**, 2092; Tetrahedron Lett. **40**, 3909 (1999). – *Synthesis:* J. Org. Chem. **54**, 3820 (1989); Scheuer II **6**, 139. [3] J. Org. Chem. **57**, 3191 (1992). – *[CAS 60197-58-2 (D.); 119323-93-2 (9-isocyanoneopupukeanane)]*

Dill apiol (6-allyl-4,5-dimethoxy-1,3-benzodioxole).

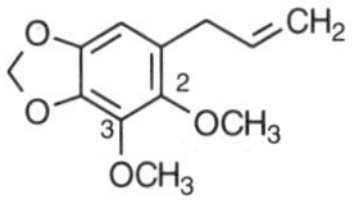

$C_{12}H_{14}O_4$, M_R 222.24, mp. 29.5 °C, n_D^{25} 1.5278. Constituent of Japanese and Indian (*Anethum sowa*) and European (*A. graveolens*) dill oils. For isolation from the Brazilian medicinal plant *Heckeria umbellata* (Piperaceae), see *Lit.*[1] (compare apiole). D. shows synergistic activity with pyrethroids (see pyrethrum).
Lit.: [1] Helv. Chim. Acta **61**, 2273 f. (1978).
gen.: Beilstein E V **19/3**, 307 ▪ Chem. Ber. **102**, 2663–2676 (1969) (synthesis) ▪ Merck-Index (12.), No. 776. – *[HS 2932 90; CAS 484-31-1]*

(+)-Dill ether see dill (weed) oil.

Dill (weed) oil. Pale yellow oil, with a typical sweet-herby, peppery, somewhat caraway-like odor and pleasant, warm-herby, peppery taste with an anise-like undertone.
Production: By steam distillation form dill weed (*Anethum graveolens*) before the fruit are ripe (seeds).
Composition[1]***:*** Major components are (+)-*limonene* (see *p*-menthadienes) (30–40%) and (+)-**carvone* (30–40%); mainly responsible for the organoleptic impressions are (+)-*α-phellandrene* (see *p*-menthadienes) (10–20%) and (+)-*dill ether* ($C_{10}H_{16}O$, M_R 152.24).

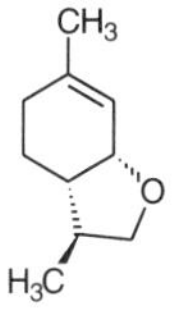

(+)-Dill ether

The oil of ripe dill seeds consists mainly of (+)-limonene (up to 40%) and (+)-carvone (up to 60%). The so-called Indian dill oil distilled from the seeds of *A. sowa* contains carvone and limonene together with larger amounts of *dill apiole. The organoleptic properties of both oils which are of less significance than the weed oil strongly resemble those of *caraway oil.
Use: Small amounts are used in the perfume industry to create herby-heady nuances, the main use is aromatization of foods (pickles, sauces, fish products, etc.).
Lit.: [1] Perfum. Flavor. **10** (6), 29 (1985); **16** (1), 52 (1991); **19** (5), 90 (1994); Food Chem. **43**, 337 (1992).
gen.: Arctander, p. 216 ▪ Bauer et al. (2.), p. 153 ▪ Gildemeister **6**, 450 ▪ H & R, p. 162. – *Toxicology:* Food Cosmet. Toxi-

col. **14**, 747 (1976). – *[HS 330129; CAS 8006-75-5 (D.); 74410-10-9 ((+)-dill ether)]*

Dimethylammonium salt see leaf-movement factors.

Dimethylarsinoyl-β-D-ribosides see arsenic in natural products.

5,9-Dimethylheptadecane see butterflies.

15,19(17,21)-Dimethylheptatriacontane see flies.

5-[(2*R*,4*R*)-2,4-Dimethylheptyl]-3-methyl-2*H*-pyran-2-one see cockroaches.

2,4-Dimethylindole see lascivol.

(3*S*,11*S*)-3,11-Dimethyl-2-nonacosanone see cockroaches.

(*E*)-4,8-Dimethylnona-1,3,7-triene see homoterpenes.

1,7-Dimethylnonyl propanoate see leaf beetles.

(*Z*)-3,7-Dimethyl-2,7-octadienyl propanoate see scales.

(6*R*,12*R*)-6,12-Dimethyl-2-pentadecanone see leaf beetles.

Dimethyl trisulfide see alarm substances.

Dimethylxanthines see theobromine and theophylline.

Dinactin see nonactin.

Dinophysistoxin-1 see okadaic acid.

Dinorargemonine see pavine and isopavine alkaloids.

Dioncophylline C, Dioncopeltatine A see naphthyl(tetrahydro)isoquinoline alkaloids.

Dioscin see steroid saponins.

Dioscorine.

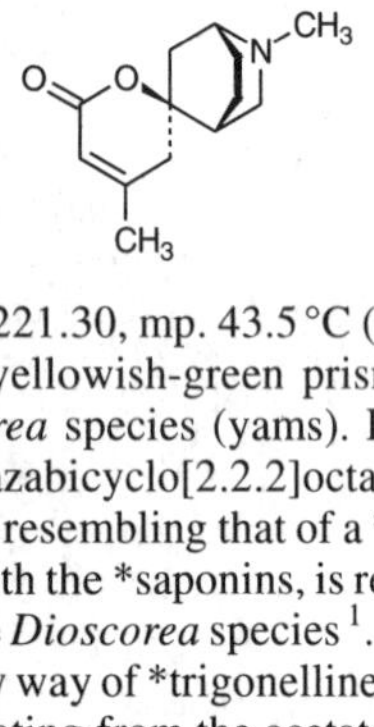

$C_{13}H_{19}NO_2$, M_R 221.30, mp. 43.5 °C (34–35 °C), $[\alpha]_D^{18}$ −35° ($CHCl_3$); yellowish-green prisms. D. occurs in various *Dioscorea* species (yams). It is a convulsive toxin with a 2-azabicyclo[2.2.2]octane ("isoquinuclidine") structure resembling that of a *tropane alkaloid and, together with the *saponins, is responsible for the toxicity of some *Dioscorea* species [1]. It is formed from nicotinic acid by way of *trigonelline by coupling with a C_7-unit originating from the acetate metabolism [2].

Lit.: [1] Merck-Index (12.), No. 3347; Ullmann (5.) **A 22**, 305; Hegnauer **II**, 133–152; **VII**, 609–617. [2] J. Am. Chem. Soc. **109**, 6179 (1989); Phytochemistry **28**, 3325 (1989). – *[HS 293990; CAS 3329-91-7]*

Diosgenin see steroid sapogenins.

Diosphenols see buchu leaf oil.

Dioxapyrrolomycin see pyrrolomycins.

1,2-Dioxolanes see plakinic acids.

Dioxybrassinin see oxindole alkaloids.

Dipentene see *p*-menthadienes.

Diphenyl *N*-cyclooctylphosphoramidate see PB toxin.

Diphtheria toxin. Protein, consisting of the subunits A (M_R ca. 24000) and B (M_R ca. 38000) linked by disulfide bridges, that is formed by the diphtheria bacterium (*Corynebacterium diphtheriae*) as an *exotoxin. Only diphtheria bacteria that have been attacked by a specific bacteriophage β, that possesses the gene for D. t. are able to synthesize D. t. Damage to the host cell occurs by binding of the less stable fragment B of D. t. to a certain ganglioside on the cell surface so that the more stable fragment A can penetrate into the cell. Thus, the elongation factor EF-2 of translation is ADP-ribosylated at the unusual amino acid diphthamide and deactivated. Protein biosynthesis at the 80 S-ribosomes is blocked [1]. It is probable that a single A-fragment is sufficient to kill a host cell within 24 h. Massive release of the toxin into the blood stream can lead to collapse or a lethally-ending necrosis of heart and liver tissue. D. t. was discovered in 1890 by E. von Behring and S. Kitasato. LD_{50} (mouse i.p.) 0.3 µg/kg; (hamster i.p.) 6.5 µg/kg. The lethal dose for rabbits, guinea pigs, and also human beings amounts to merely 0.1 µg/kg. Another compound, exotoxin A, formed on infections caused by *Pseudomonas aeruginosa* has an action similar to that of D. t.

Lit.: [1] Annu. Rev. Biochem. **46**, 69 (1977); FEBS Lett. **103**, 253 (1979); Nachr. Chem. Tech. Lab. **29**, 66, 212 (1981). *gen.:* Harvard Lect. **76**, 45–73 (1980–1981) ▪ Trends Biochem. Sci. (Pers. Ed.) **12**, 28 (1987).

Dipicolinic acid (2,6-pyridinedicarboxylic acid).

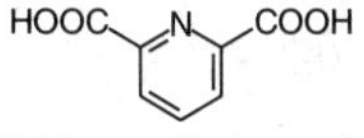

$C_7H_5NO_4$, M_R 167.12, needles, mp. (anhydrous) 252 °C (decomp.), soluble in alkalis, poorly soluble in glacial acetic acid. D. is formed in bacterial endospores and is probably necessary for the development of the thermoresistance in certain bacteria.

Lit.: Beilstein E V **22/4**, 128. – *[HS 293339; CAS 499-83-2]*

Diploicin.

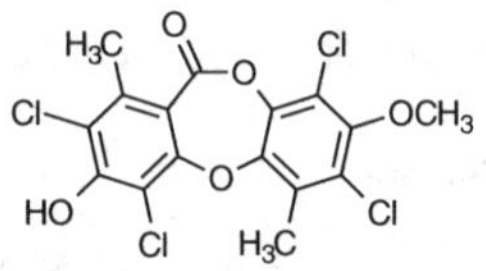

$C_{16}H_{10}Cl_4O_5$, M_R 424.06, needles, mp. 232 °C. A *depsidone with 4 chlorine atoms in the molecule; occurs in the crustaceous lichen *Diploicia canescens*.

Lit.: Aust. J. Chem. **43**, 197, 419 (1990) ▪ J. Chem. Soc., Perkin Trans. 1 **1976**, 147–151; **1981**, 849, 855 (synthesis) ▪ Karrer, No. 1068. – *[CAS 527-93-5]*

Diprionol (3,7-dimethyl-2-pentadecanol).

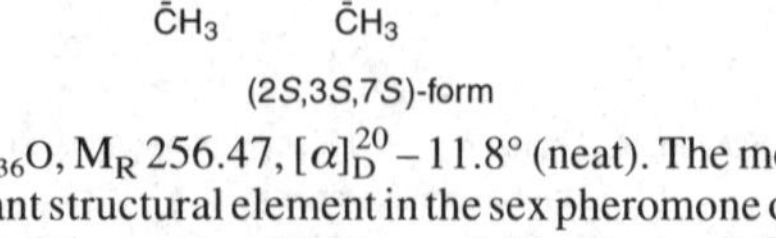

(2*S*,3*S*,7*S*)-form

$C_{17}H_{36}O$, M_R 256.47, $[\alpha]_D^{20}$ −11.8° (neat). The most important structural element in the sex pheromone of sawflies of the genera *Diprion* and *Neodiprion*. Mixtures of diastereoisomers of the acetate or propanoate form the species-specific *pheromones of these insects.

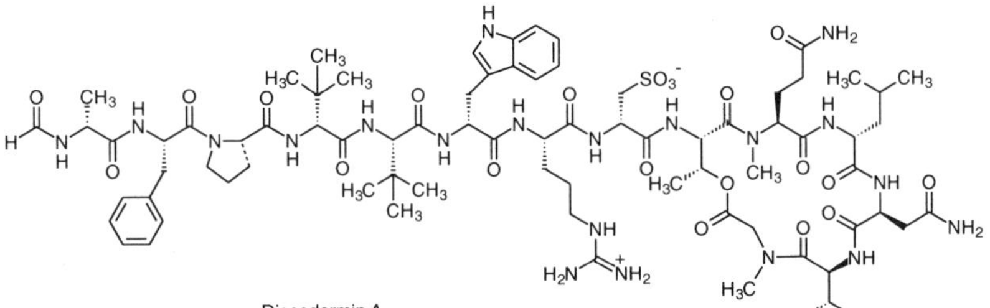

Discodermin A

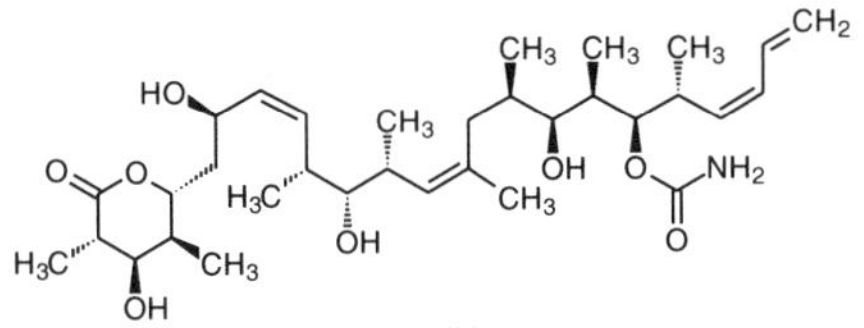

Discodermolide

Lit.: ApSimon **9**, 122–128 (synthesis) ▪ J. Chem. Ecol. **14**, 1131–1144 (1988) ▪ J. Insect Physiol. **41**, 395–401 (1995) ▪ Science **192**, 51 ff. (1976). – *[CAS 59056-73-4 (diastereomeric mixture); 67253-83-2 (2S,3S,7S-form)]*

Disaccharides. Term for compounds consisting of two glycosidically linked *monosaccharide molecules (e.g., D-*glucose, D-*fructose). The most important D. are cellobiose (see cellulose), maltose (malt sugar), *lactose (milk sugar), and *sucrose (can sugar). Further D. include *gentiobiose, melibiose, *trehalose, *turanose. The D. occur either in free form (saccharose), as components of oligo- and polysaccharides (cellobiose), or are glycosidically linked to plant pigments and other plant components (to aglycones such as anthocyanidins). D. can be cleaved by acids or enzymes (disaccharidases such as galactosidases, glucosidases, β-fructofuranosidases) to *monosaccharides.
Lit.: Collins-Ferrier, p. 462–469.

Discodermide see tetramic acid.

Discodermin A. *D. A*[1] (formula see above), $C_{77}H_{116}N_{20}O_{22}S$, M_R 1705.96, mp. 226–227 °C, $[\alpha]_D$ −6.3° (CH_3OH), is a tetradecapeptide from the sponge *Discodermia kiiensis* (Lithistida) that inhibits the growth of *Bacillus subtilis* and *Proteus mirabilis* (IC_{50} 3.0 and 1.6 µg/mL) as well as phospholipase A_2 (IC_{50} 3.5–7.0×10^{-7} M)[2]. The polyhydroxylactone *(+)-discodermolide*[3], $C_{33}H_{55}NO_8$, M_R 593.80, mp. 115–116 °C, $[\alpha]_D$ +7.2° (CH_3OH), is isolated from *D. dissoluta*. It inhibits *in vitro* the proliferation of murine P388 leukemia cells (IC_{50} 0.5 µg/mL) by binding to microtubuli even more strongly than taxol[4] and shows immunosuppressive action.
Lit.: [1]Tetrahedron Lett. **25**, 5165 (1984); **26**, 855 (1985); **35**, 8251 (1994); Tetrahedron **50**, 13409 (1994); J. Nat. Prod. **48**, 236 (1985). [2]New. J. Chem. **14** 721 (1990). [3]Angew. Chem. Int. Ed. Engl. **39**, 377 (2000); J. Org. Chem. **55**, 4912 (1990); **62**, 6098 (1997); **63**, 7885 (1998); Ann. N. Y. Acad. Sci. **94**, 696 (1993); J. Am. Chem. Soc. **115**, 12621 (1993); **117**, 12011 (1995); **118**, 9509, 11054–11080 (1996) (synthesis); Org. Lett. **1**, 1823 (1999) (multi-gram synthesis). [4]Biochemistry **35**, 243 (1996). – *[CAS 94552-47-3 (D.); 127943-53-7 (discodermolide)]*

Discodermolide see discodermin A.

Discorhabdins. Spiro-heterocyclic compounds from marine sponges of widely differing origin (New Zealand: *Latrunculia brevis*; Fiji: *Zyzzya* sp., Japan: *Prianos melanos*). These topoisomerase II inhibitors contain the unusual pyrrolo[4,3,2-*de*][1,7]phenanthroline ring system with a spiro-cyclohexadienone or -cyclohexenone system attached. Of the five D. A–E, only D. D shows antitumor activity. The D. are red or green compounds with mp.'s (hydrochlorides) of over 360 °C. D. are, in part, identical to the prianosines. For data see table.

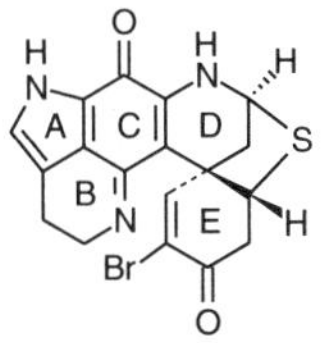

Discorhabdin A
(Prianosin A)

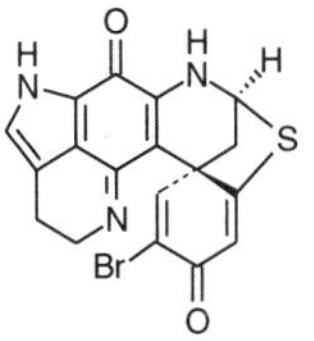

Discorhabdin B

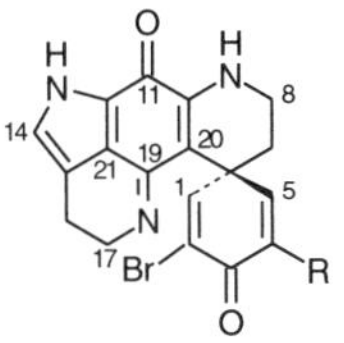

R = Br : Discorhabdin C
R = H : Discorhabdin E

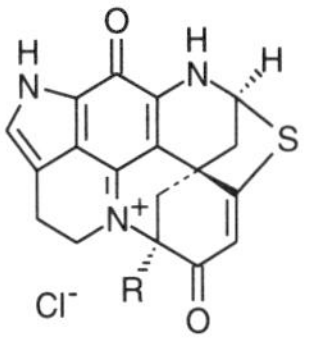

R = H : Discorhabdin D
(Prianosin D)

Table: Data of Discorhabdins.

D.	molecular formula	M_R	CAS
D. A	$C_{18}H_{14}BrN_3O_2S$	416.29	112515-41-0 118490-52-1 (HCl)
D. B	$C_{18}H_{12}BrN_3O_2S$	414.28	115439-61-7 118490-51-0 (HCl)
D. C	$C_{18}H_{13}Br_2N_3O_2$	463.13	105372-81-4 105372-82-5 (HCl)
D. D	$C_{18}H_{14}ClN_3O_2S$	371.84	115384-95-7 (cation) 115459-47-7 (HCl)
D. E	$C_{18}H_{14}BrN_3O_2$	384.23	159308-95-9 159308-96-0 (TFA)

Lit.: J. Am. Chem. Soc. **114**, 2175 (1992); **115**, 1632 (1993) ▪ J. Nat. Prod. **62**, 636 (1999) (isolation) ▪ J. Org. Chem. **51**, 5476 (1986); **53**, 4127 (1988); **59**, 8233 (1994); **60**, 1800 (1995); **64**,

16 (1999) (synthesis) ▪ Synthesis **1999**, 885–897 (synthesis) ▪ Tetrahedron **44**, 1727 (1988) ▪ Tetrahedron Lett. **32**, 4151 (1991).

Disparlure (*cis*-7,8-epoxy-2-methyloctadecane).

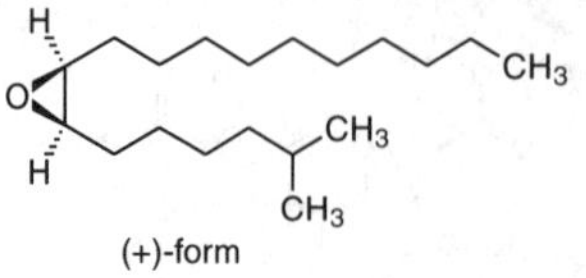

(+)-form

$C_{19}H_{38}O$, M_R 282.51, odorless liquid. Sexual attractant (*pheromone) of the female gypsy moth, *Lymantria dispar*, and *L. monacha*. The enantiomeric composition of the natural product varies, the (7*R*,8*S*)-enantiomer {$[\alpha]_D^{27}$ +0.9° (CCl_4)}, however, plays a major role for both species [1]. The epoxide ring of D. is opened by enzymes in the antennae of the male gypsy moth to give the (7*R*,8*R*)-diol [2]. Recent results have shown that the unbranched analogue of D. ("nordisparlure") occurs in *Lymantria* and plays an important role as a further pheromone component.

Lit.: [1] Naturwissenschaften **63**, 582 f. (1976); **70**, 466 f. (1983). [2] Pure Appl. Chem **61**, 551 ff. (1989).

gen.: ApSimon **4**, 146–152; **9**, 85–94 (synthesis) ▪ Beilstein EV **17/1**, 170 ▪ J. Chem. Soc., Perkin Trans. 1 **1990**, 639 ▪ J. Org. Chem. **64**, 2152, 3719 (1999) [synthesis (+)- and (–)-D.] ▪ Tetrahedron: Asymmetry **8**, 375 (1997) [synthesis (–)-D.] ▪ Zh. Org. Khim. **34**, 1509 (1998) (synthesis). – *[CAS 29804-22-6]*

Distamycins. Pyrrole antibiotics from *Streptomyces distallicus*, used topically to treat herpes simplex infections, e.g., *D. A* (stallimycin): $C_{22}H_{27}N_9O_4$, M_R 481.51, mp. 154–156 °C; used as the hydrochloride (Herperal®).

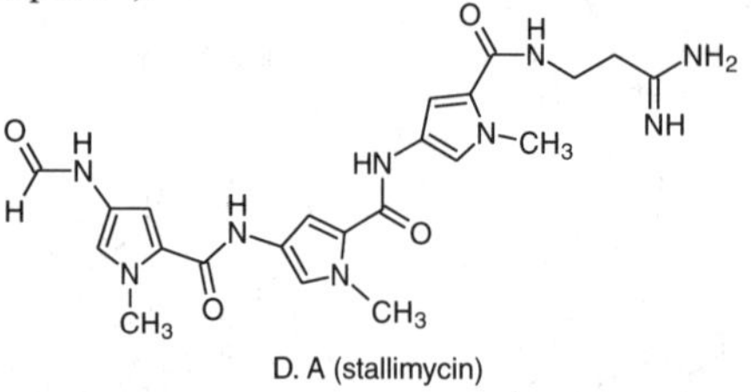

D. A (stallimycin)

D. are undergoing preclinical testing against carcinomas caused by viruses (adenoviruses, leukemia, papilloma, etc.) [1].

Lit.: [1] Pat. WO 97/28,123; 98/21,202.

gen.: J. Med. Chem. **26**, 1042 (1983) ▪ J. Org. Chem. **50**, 3774 (1985); **65**, 1102 (2000) (analogues) ▪ Martindale (30.), p. 555 ▪ Sax (8.), DXG 600, SLI 300. – *[CAS 636-47-5 (D. A)]*

Diterpene alkaloids. General term for terpene alkaloids whose N-containing skeleton is formed from a diterpene (C_{20}) precursor. D. a. are highly oxidized and complex natural products. They can be divided into two large groups, the highly toxic, norditerpenoid alkaloids based on a hexacyclic C_{19} skeleton and the diterpenoid alkaloids based on a C_{20} skeleton. Examples of the C_{19}-alkaloids include the aconitine type (*aconitine, delphinine) and the lycoctonine type (lycoctonine, *browniine) while examples of the C_{20}-alkaloids include the atisine type (*atisine, hetidine, *hetisine) and the veatchine type (*veatchine, *napelline). D. a. occur mainly in plants of the genera *Aconitum* (monk's hood) and *Delphinium* (larkspur) but have also been found in *Garrya*, *Inula*, *Erythrophleum* (cassaine see Erythrophleum alkaloids), *Anopteris*, and *Spiraea* species.

Lit.: Alkaloids (London) **7**, 247–267 (1977); **8**, 219–245 (1978); **9**, 221–237 (1979); **10**, 211–226 (1981); **11**, 203–224 (1981); **12**, 248–274 (1982); **13**, 281–308 (1983) ▪ Alkaloids (Wiley) **2**, 205–462 (1984) ▪ Heterocycles **27**, 1813, 2467 (1988); **28**, 107, 205 (1989) ▪ J. Nat. Prod. **43**, 41–71 (1980) ▪ Manske **12**, 1–206; **17**, 1–103; **18**, 99–216; **42**, 152–247 ▪ Mothes et al., p. 361 ▪ Nat. Prod. Rep. **3**, 451–464 (1986); **10**, 471–486 (1993) **16**, 619–635 (1999) ▪ Pelletier **1**, 155–210 ▪ Pure Appl. Chem. **69**, 119 (1997).

Diterpenes (diterpenoids). Natural products containing 20 carbon atoms built up of 4 isoprene units belonging to the group of the terpenes. The name is applied not only to the hydrocarbons but also to their derivatives which are very widely distributed in nature. Their structures vary widely.

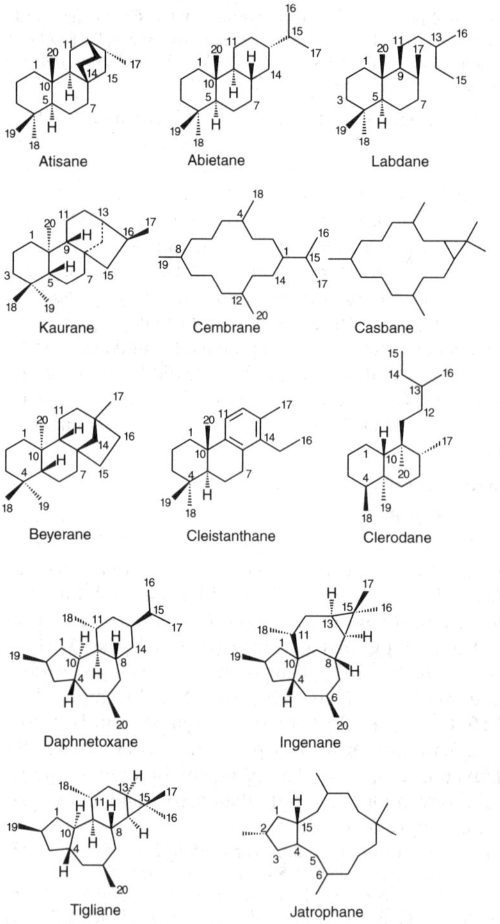

Figure: Carbon skeletons of various cyclic diterpenes.

D. occur in the higher boiling fractions of *essential oils and resins. Examples of the open-chain compounds include, e. g., *geranylgeraniol, *crocetin, and *phytol, the latter is a component of vitamins E and K. Most D., however, are bi- and tricyclic compounds, predominately perhydronaphthalene or phenanthrene derivatives, e.g., the resin acid *abietic acid. The tetracyclic phorbol, a *tigliane derivative is found in the seeds of *Croton tiglium*. Phorbol esters were subject of numerous investigations on account of their cocarcinogenic and proinflammatory actions. The tetracyclic *taxol is commercially available as an anti- cancer drug (Paclitaxel). The *coleons and *forskolin are tricyclic, the *quassin(oids) are tetracyclic, *cafestol is pentacyclic, and the *ginkgolides are hexacyclic diterpenes; often cyclic ethers and lactones contribute to the number of rings. Numerous monocyclic D. are also known, such as vitamin A or the *cembrenes. Important diterpene acids are the *gibberellins, which can act as growth hormones in higher plants; many plant pigments also are diterpenes. Numerous *diterpene alkaloids are derived from D. All D. are derived form the C_{20}-isoprenoid intermediate geranylgeranyl diphosphate which is converted into derivatives of the various skeletal types by the action of D. synthases (cyclases).

Lit.: ApSimon **8**, 1–243 (1992) (synthesis) ▪ Atta-ur-Rahman **12**, 233–274 (1993) ▪ Evans, Naturally Occurring Phorbol Esters, p. 1–31, 139–170, Boca Raton: CRC 1986 ▪ Karrer, No. 1934–1965a, 3657–3712 ▪ Luckner (3.), p. 169–202 ▪ Nat. Prod. Rep. **6**, 347-358 (1989); **9**, 1–16 (1992); **10**, 159–174 (1993); **13**, 59–71 (1996); **16**, 209 (1999) ▪ Pure Appl. Chem. **49**, 1423 (1977) ▪ Scheuer II **1**, 125–130 ▪ Terpenoids Steroids **10**, 106–134 (1981); **11**, 91–109 (1982); **12**, 186–206 (1983) ▪ West, Biosynthesis of Diterpenes, in Porter & Spurgeon (eds.), Biosynthesis of Isoprenoid Compds., vol. 1, p. 375–411, New York: Wiley 1981 ▪ Zechmeister **44**, 1–100; **46**, 77–149.

1,3,5-Dithiazinanes see dihydro-4*H*-1,3,5-dithiazines.

1,2-Dithiines see thiarubrins.

3*H*-1,2-Dithiole, 1,2-Dithiolane-4-carboxylic acid (methyl ester) see vegetable flavors (asparagus).

Djenkolic acid (*S,S'*-methylenebis-L-cysteine).

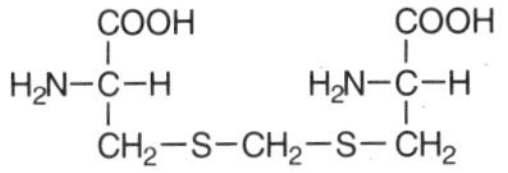

$C_7H_{14}N_2O_4S_2$, M_R 254.32, mp. 300–350 °C (decomp.), $[\alpha]_D^{20.5}$ –65.0° (1 m HCl); monohydrochloride, mp. 300–350 °C (decomp.). Non proteinogenic amino acid from *Pithecellobium lobatum* and other Fabaceae; it occurs in *P. bubalium* (4%) and many species of *Acacia*. *N*-Acetyldjenkolic acid is found in *Mimosa acanthocarpa*. When taken orally D. can form crystals in the kidneys and cause severe pain. For its action on the nervous system, see *Lit.*[1]. The synthesis proceeds from L-Cys and formaldehyde in strongly acidic solution.

Lit.: [1] J. Neurosci. **14**, 3881–3897 (1994).
gen.: Beilstein E III **4**, 1591 ▪ Karrer, No. 2423 ▪ Phytochemistry **23**, 265–270 (1984). – *[HS 2930 90; CAS 498-59-9]*

DMDP. Abbreviation for *(2*R*)-3*β*,4*α*-dihydroxy-2*α*,5*β*-pyrrolidinedimethanol.

Docosanedioic acid (phellogenic acid) see Japan wax.

Docosanoic acid see behenic acid.

(*all-Z*)-4,8,12,15,19-Docosapentaenoic acid see clupa(no)donic acid.

(*Z*)-13-Docosenoic acid see erucic acid.

Dodecanal see alkanals.

4- and 5-Dodecanolide see alkanolides.

(3*Z*,6*Z*,8*E*)-3,6,8-Dodecatrien-1-ol see termites.

(*E*)-2-Dodecenal see alkenals.

(*Z*)-3-Dodecen-1-ol see weevils.

(*Z*)-7-Dodecenyl acetate see elephant pheromone.

(*Z*)-7-Dodecenyl acetate see tetradecenyl acetate and elephant pheromone.

Dolabellanes.

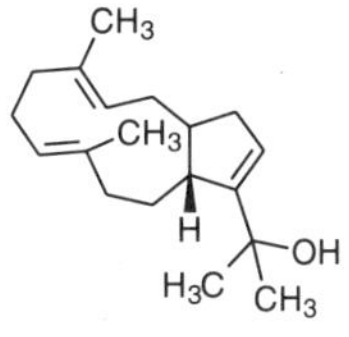

Palominol

*Diterpenes of the bicyclic D. group were first isolated from the herbivorous lumpfish *Dolabella californica*[1]. Since then similar metabolites have been found in brown algae of the family Dictyotaceae on which *D. californica* feed. Finally, D. were also detected in sea fans. For example, *palominol* [(1*S*,3*E*,7*E*,11*R*)-dolabella-3,7,12-trien-18-ol][2]: $C_{20}H_{32}O$, M_R 288.47, mp. 52–53 °C, $[\alpha]_D^{27}$ –33° ($CHCl_3$), was isolated from *Eunicea laciniata*.

Lit.: [1] J. Am. Chem. Soc. **98**, 4664 (1976); J. Org. Chem. **42**, 3157 (1977). [2] J. Org. Chem. **56**, 3392 (1991).
gen.: J. Am. Chem. Soc. **118**, 1229 f. (1996) (synthesis, corrected absolute configuration) ▪ Tetrahedron **54**, 11683–11729 (1998) (review). – *[CAS 126222-05-7 (palominol)]*

Dolastatins.

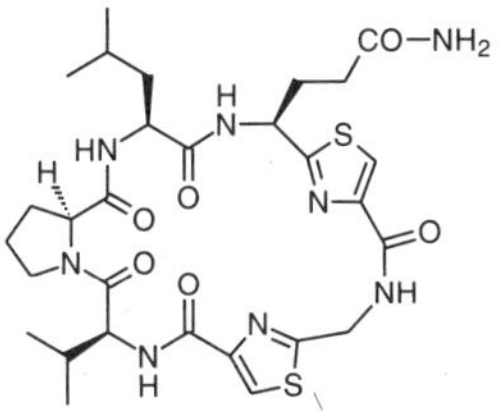

D. 3

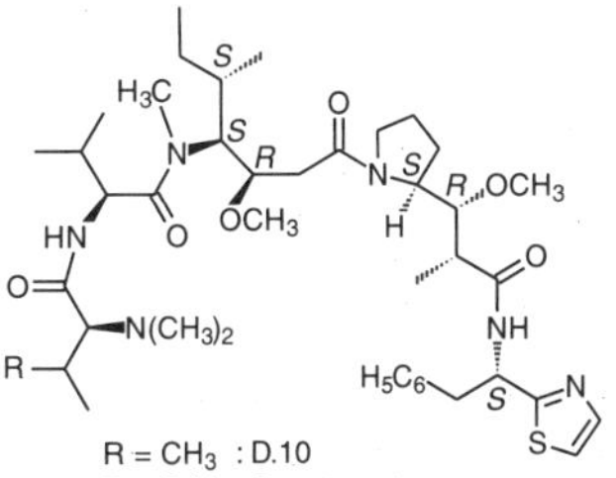

R = CH_3 : D.10
R = C_2H_5 : Symplostatin

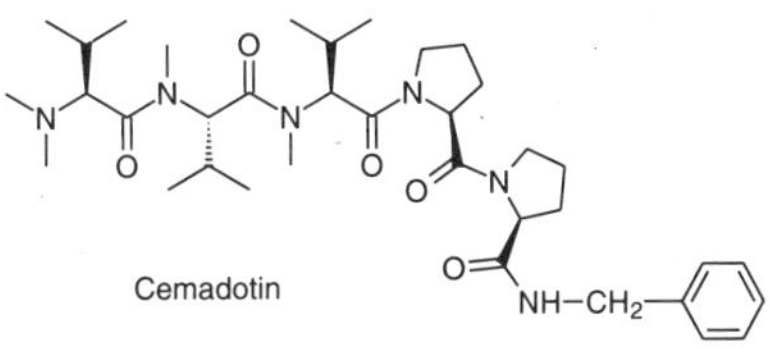

Cemadotin

Mostly cyclic pseudopeptides from the lumpfish *Dolabella auricularia* (living in the Indian Ocean) with remarkable antitumor effects. The most important compounds are *D. 3*[1] {(cyclo[Val-Pro-Leu-(gln)Thz-(gly)Thz]}: $C_{29}H_{40}N_8O_6S_2$, M_R 660.81, mp. 133–137 °C and *D. 10*[2]: $C_{42}H_{68}N_6O_6S$, M_R 785.10, $[\alpha]_D$ –68°, colorless, amorphous, D. 11, D. 14, D. 15[3–5], and symplostatin[6], $C_{43}H_{71}N_6O_6S$, M_R 799.5, $[\alpha]_D^{20}$ –45° (CH_3OH). The open-chain pseudo-pentapeptide D. 10 is one of the most active antineoplastic substances for human melanoma cells (in clinical tests 67% cure rate at a dose of 26 µg/kg)[2]. D. 10 is being clinically tested in lung cancer patients[7] (phase I/II). It inhibits the aggregation of microtubuli by binding to the tubulin *β*-subunit, the tubulin-dependent GTP hydrolysis, and the binding of *vincristine to tubulin[8]. Most advanced is the clinical development of the D. 15-analogue cemadotin[9] (LU-103793), Phase II, IC_{50} values in various tumor cell lines 0.1–1 nM, $C_{35}H_{56}N_6O_5$, M_R 640.84.

Lit.: [1]J. Am. Chem. Soc. **104**, 905 (1982); J. Org. Chem. **50**, 2654 (1985); **51**, 4580, 4586 (1986); Tetrahedron **42**, 2695 (1986); Angew. Chem. Int. Ed. Engl. **23**, 725 (1984); Synthesis **1987**, 233, 236. [2]Anticancer Drug Res. **13**, 243–277 (1998); Chemtracts: Org. Chem. **3**, 54 (1990); J. Am. Chem. Soc. **109**, 6883 (1987); **111**, 5463 (1989); J. Chem. Soc., Perkin Trans. 1 **1996**, 853, 859; Tetrahedron **49**, 1913–24 (1993) (synthesis). [3]Scheuer II **1**, 118. [4]J. Am. Chem. Soc. **119**, 2111 (1997); J. Org. Chem. **55**, 2989 (1990); **64**, 405 (1999). [5]J. Org. Chem. **54**, 6005 (1989); J. Am. Chem. Soc. **113**, 6692 (1991) (synthesis). [6]J. Nat. Prod. **61**, 1075 (1998); **62**, 655 (1999). [7]Drug News (29.4.) **1996**, 6; Antimicrob. Agents Chemother. **42**, 2961 (1998) (antifungal activity). [8]Biochem. Pharmacol. **40**, 1859 (1990); **45**, 1503 (1993); J. Med. Chem. **41**, 1524 (1998). [9]Cancer Res. **55**, 3085 (1995); J. Nat. Prod. **62**, 655 (1999).

gen.: Bioorg. Med. Chem. Lett. **7**, 827 (1997) (D. 18) ▪ Curr. Pharm. Res. **5**, 139–162 (1999) (review) ▪ Heterocycles **47**, 491 (1998) (D. 17) ▪ J. Am. Chem. Soc. **111**, 5015 (1989); **118**, 1874 (1996) (D. H) ▪ Nachr. Chem. Tech. Lab. **37**, 1040 (1989) ▪ J. Org. Chem. **59**, 2935 (1994) (synthesis); **61**, 6340 (1996) (D. G); **64**, 405 (1999) (D. 15) ▪ Synthesis **1996**, 719 ▪ Tetrahedron **49**, 9151–9170 (1993); **53**, 8149 (1997); **55**, 12301 (1999) (D. I) ▪ Tetrahedron Lett. **36**, 5057, 5059 (1995) (D. E); **37**, 7299 (1996) (D. G) ▪ Zechmeister **70**, 1–79 (review). – *[CAS 80387-90-2 (D. 3); 110417-88-4 (D. 10); 159776-69-9 (cemadotin)]*

Dolichodial see Teucrium lactones.

Dolichols. Group of polyisoprenoid alcohols (polyisoprenes) with an *α*-saturated isoprene unit and a terminal primary alcohol group. The number of isoprene units varies in higher eukaryotes between 14 and 20. A D. with 20 isoprene units ($C_{100}H_{164}O$, M_R 1382.40, viscous, colorless oil, mp. –10 °C, CAS [2067-66-5]) was first isolated from kidney tissue[1]. D. occur in animal and plant organisms as free D. or esterified with long-chain *fatty acids or, respectively, phosphoric acid (dolichyl phosphates)[2].

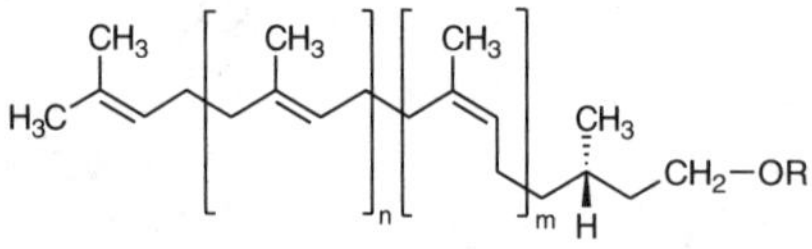

n = 2, m = 10-16

R = H : Dolichols
R = PO_3H_2 : Dolichyl phosphates

D. and their esters are present in practically all membranes of eukaryotic cells (except those of mitochondria and plastids) and function in the biosynthesis of glycoproteins and glycolipids as carriers of and transport molecules for oligosaccharides. D. are also components of the lipid membranes of nerve cells. An elevated content is a sign for pathological changes, also in connection with Alzheimer's disease[3]. D. are formed from isoprene units[4].

Lit.: [1]Biochem. J. **88**, 470 (1963). [2]Biochem. J. **251**, 1–9 (1988); Trends Biochem. Sci. **12**, 443 ff. (1987). [3]Lancet **2**, 99 (1982); Chem. Scr. **27**, 79–84 (1987); Excerpta Med. Int. Congress Ser. **782**, 389–411 (1988). [4]Biochim. Biophys. Acta **143**, 448–532 (1994); New Compr. Biochem. **299**, 127–143 (1995) (biosynthesis).

gen.: Biochem. Cell Biol. **70**, 382 ff. (1992) (function in human tissue) ▪ Chem. Phys. Lipids **51**, 159 (1989) (synthesis) ▪ Dolichols and Dolichyl Derivatives, in Sherma, Mukherjee, & Weber (eds.), Handbook of Chromatography – Analysis of Lipids, Boca Raton: CRC Press 1993 (analysis) ▪ Dolichols, Polyprenols and Derivatives, Warschau: Techgen 1995 ▪ J. Plant Physiol. 143, 448–452 (1994) (occurrence in plants) ▪ Tetrahedron Lett. **24**, 5103 (1983) (synthesis).

Dominicalure.

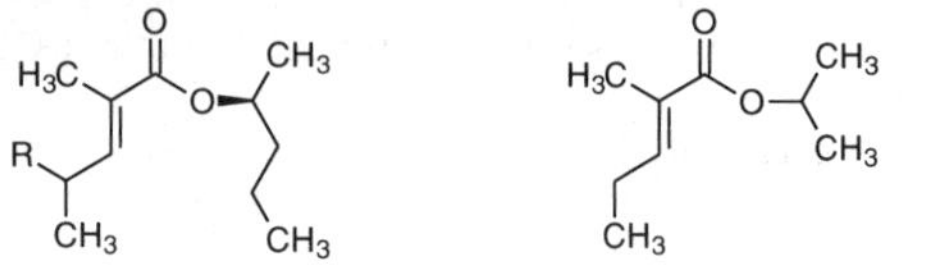

R = H : D. I
R = CH_3 : D. II

Trunc-call 1

Aggregation pheromone of the male grain beetle *Rhyzoptera dominica*[1]. It consists of the components *D. I* [((*S*)-1-methylbutyl) (2*E*)-2-methyl-2-pentenoate]: $C_{11}H_{20}O_2$, M_R 184.28, $[\alpha]_D^{19}$ +19.2° (diethyl ether) and *D. II* [((*S*)-1-methylbutyl) (2*E*)-2,4-dimethyl-2-pentenoate]: $C_{12}H_{22}O_2$, M_R 198.31, $[\alpha]_D^{19}$ +10.9° (diethyl ether). The related isopropyl (2*E*)-2-methyl-2-pentenoate is called "trunc-call 1"; together with isopropyl (2*E*,4*E*)-2,4-dimethyl-2,4-heptadienoate ("trunc-call 2") it forms the sexual pheromone of the closely related grain beetle *Prostephanus truncatus*[2].

Lit.: [1]J. Chem. Ecol. **7**, 759–780 (1981). [2]J. Chem. Ecol. **17**, 789–803 (1991).

gen.: ApSimon **9**, 99 f. (synthesis) ▪ J. Chem. Ecol. **22**, 673–680 (1996) (synthesis). – *[CAS 80510-15-2 (D. I); 80510-16-3 (D. II)]*

Domoic acid.

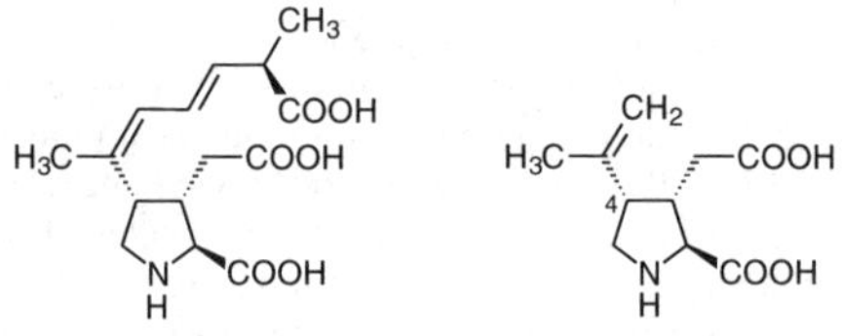

Domoic acid

Kainic acid
(Digenic acid)

After consumption of mussels cases of poisoning with the clinical picture of ASP (amnesic shellfish poisoning) have been described (compare shellfish poisons). Responsible are the proline derivative *D.*[1] ($C_{15}H_{21}NO_6$, M_R 311.33, $[\alpha]_D$ –109.6°) and *kainic acid*[2] {$C_{10}H_{15}NO_4$, M_R 213.23, cryst., mp. 253–254 °C (hydrate, decomp.), $[\alpha]_D$ –14.8°} which cause severe damage to the brain because they act as excitatory glutamate antagonists. D. was first isolated as the anthelmintic principle from the red alga *Chondria armata* and was subsequently found in the diatoms *Nitzschia pungens* and *Pseudonitzschia australis*. Mussels take up the diatoms with their food and accumulate the toxins. Kainic acid occurs in the red algae *Digenea simplex* and *Centroceras clavulatum*. The name kainic acid is derived from "kaininso", the Japanese name for the alga. The 4-epimer of kainic acid, *α-allo-kainic acid* (mp. 238–242 °C) is also contained in the algae. Kainic acid analogues (kainoids) are the subject of intensive pharmacological research[3].

Lit.: [1] Pure Appl. Chem. **61**, 513 (1989); J. Phycol. **28**, 439 (1992); J. Chem. Soc., Chem. Commun. **1992**, 714 (biosynthesis); Can. J. Chem. **68**, 22 (1990) (derivatives). [2] Heterocycles **44**, 129 (1997); J. Am. Chem. Soc. **121**, 11139 (1999); J. Chem. Soc., Perkin Trans. 1 **1992**, 553; J. Chem. Soc., Chem. Commun. **1993**, 125; **1999**, 245; **2000**, 317; J. Org. Chem. **61**, 7116 (1996) (synthesis); **62**, 1896 (1997); McGeer et al. (eds.), Kainic Acid as a Tool in Neurobiology, New York: Raven Press 1978; Pure Appl. Chem. **70**, 259 (1998); Synlett **1996**, 60 ff., 95; **1997**, 275; **1998**, 507; Tetrahedron **51**, 4195–4212 (1995); **53**, 5233 (1997); **55**, 6153 (1999); Tetrahedron Lett. **34**, 3435 (1993); **36**, 9309 (1995); **38**, 857 (1997); **39**, 2171 (1998). [3] Adv. Nitrogen Heterocycles **3**, 159–218 (1998) (review); Synlett **1996**, 95; Tetrahedron **52**, 1931–1941 (1996). – *Pharmacology:* Ciba Found. Symp. **126**, 186–203 (1987) (review). – *Synthesis:* Tetrahedron **27**, 3979 (1989); J. Chem. Soc., Perkin Trans. 1 **1992**, 553; J. Org. Chem. **57**, 6279 (1992); J. Org. Chem. **61**, 5418–5424 (1996) ((–)-α-K.).

gen.: Beilstein E V **22/4**, 371 ▪ Chem. Eng. News (3.1.) **2000**, 14 f. ▪ Chem. Rev **93**, 1897 (1993) ▪ Merck-Index (12.), No. 3475 (D.), 5289 (kainic acid) ▪ Scheuer I **3**, 105 ff. ▪ Scheuer II **6**, 3. – *[CAS 14277-97-5 (D.); 487-79-6 (kainic acid); 4071-39-0 (α-allo-kainic acid)]*

Donaxine see gramine.

DOPA. Abbreviation for *α-3,4-dihydroxyphenylalanine.

(*R*)-*β*-DOPA see *β*-3,4-dihydroxyphenylalanine.

Dopamine [hydroxytyramine, 3,4-dihydroxyphenethylamine, 4-(2-aminoethyl)pyrocatechol].

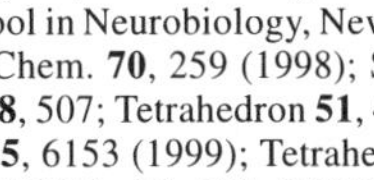

$C_8H_{11}NO_2$, M_R 153.18. The free base forms coarse prisms, rapidly discoloring on contact with air; as hydrochloride [mp. 241 °C (decomp.)] stable and soluble in water, insoluble in ether. It is a neurotransmitter which, among others, controls the motor system of the central nervous system. The symptoms observed in Parkinson's disease are caused primarily by a breakdown of dopaminergic neurons in the substantia nigra of the brain, generating a deficiency of D. Since D. cannot pass the blood-brain barrier while L-*DOPA can, the latter is used therapeutically in combination with DOPA decarboxylase inhibitors for the therapy of Parkinson's disease, usually with good results. D. occurs in various higher plants such as broom (*Cytisus scoparius*) and bananas (*Musa sapientum*) as well as in algae (*Monostroma fuscum*). In plants D. is the biogenetic precursor of numerous alkaloids [*adrenaline, *mescaline; see also benzyl(tetrahydro)isoquinoline alkaloids] and in insects of *N*-acetyldopamine (see *N*-acylcatecholamines).

Activity: D. constricts the vessels of the skin and musculature, increases the contractile force of the heart, and dilates the vessels of the kidneys and mesenterium. Furthermore, it inhibits the expression of the neurohormone prolactine.

Use: As sympathicomimetic, for treatment of cardiogenic shock conditions, and to increase arterial blood pressure.

The biosynthesis proceeds through decarboxylation of L-*DOPA by DOPA decarboxylase (EC 4.1.1.28).

Lit.: Beilstein E IV **13**, 2603 ▪ Florey **11**, 257 (review) ▪ Hager (5.) **7**, 1421 ff. ▪ Luckner (3.), p. 363–380 ▪ Negwer (6.), No. 946 ▪ Recent Progr. Horm. Res. **42**, 251 (1986) (review) ▪ Ullmann (5.) **A 13**, 107; **A19**, 1–15. – *Toxicology:* Sax (8.), DYC 400, DYC 600. – *Pharmacology:* Pharmacotherapy **6**, 304 (1986). – *[HS 2922 29; CAS 51-61-6]*

Dopastin {(*E*)-[(*S*)-2-(Hydroxynitrosoamino)-3-methylbutyl]-2-butenamide}.

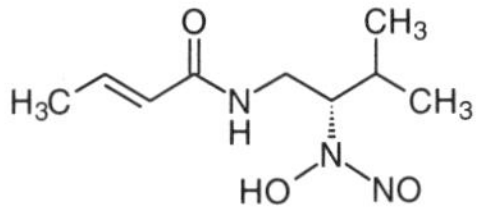

$C_9H_{17}N_3O_3$, M_R 215.25, needles, mp. 116–119 °C, pK_a 5.1, $[\alpha]_D^{21}$ –246° (C_2H_5OH). D. is isolated from cultures of a *Pseudomonas* strain and is an inhibitor of dopamine *β*-hydroxylase; it exhibits hypotensive activity and inhibits the germination of barley.

Lit.: Agric. Biol. Chem. **38**, 2093–2111 (1974) ▪ Pure Appl. Chem. **33**, 129 (1973). – *[CAS 37134-80-8]*

Dormin see abscisic acid.

Doxorubicin see adriamycin.

Drimanes.

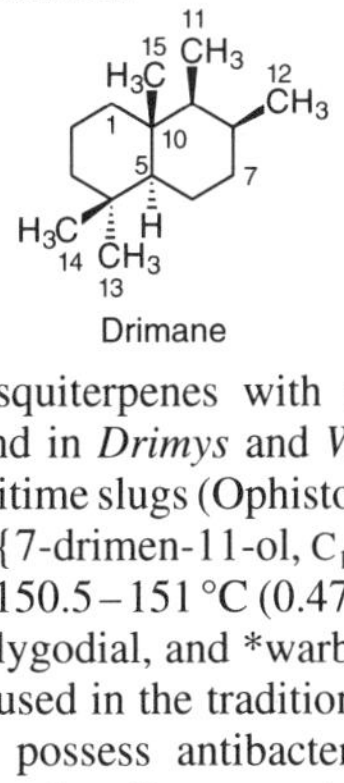

Drimane

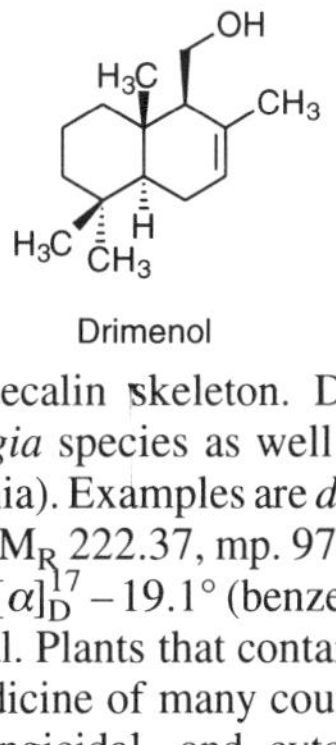

Drimenol

*Sesquiterpenes with *trans*-decalin skeleton. D. are found in *Drimys* and *Warburgia* species as well as in maritime slugs (Ophistobranchia). Examples are *drimenol* {7-drimen-11-ol, $C_{15}H_{26}O$, M_R 222.37, mp. 97.8 °C, bp. 150.5–151 °C (0.47 kPa), $[\alpha]_D^{17}$ –19.1° (benzene)}, *polygodial, and *warburganal. Plants that contain D., are used in the traditional medicine of many countries and possess antibacterial, fungicidal, and cytotoxic properties. On account of their spicy taste they are used as a condiment. More complex D. are obtained from cultures of basidiomycetes (*marasmenes, *mniopetals). The biosynthesis of D. proceeds from *farnesol[1].

Lit.: [1] Nat. Prod. Rep. **8**, 309 (1991).

gen.: J. Am. Chem. Soc. **101**, 4398, 4400 (1979) ▪ J. Chem. Res. (S) **1996**, 108 f. ▪ Nat. Prod. Rep. **8**, 319 (1991) (synthesis) ▪ Tetrahedron **45**, 1567–1576 (1989); **55**, 1561 (1999) (synthesis) ▪ Tetrahedron Lett. **27**, 61 (1986). – *[CAS 5951-58-6 (drimane); 468-68-8 (drimenol)]*

Dronabinol see cannabinoids.

Drosophilin A (2,3,5,6-tetrachloro-4-methoxyphenol).

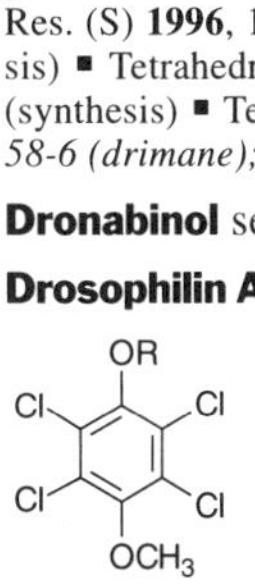

R = H : Drosophilin A
R = CH_3: D.-A methyl ether

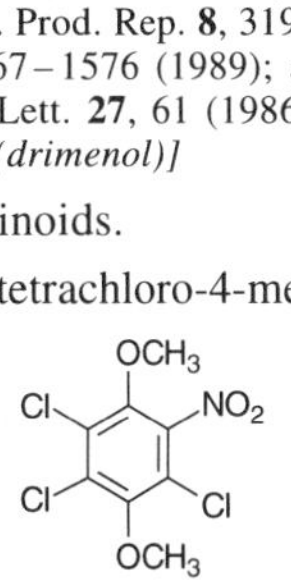

1,2,4-Trichloro-3,6-dimethoxy-5-nitrobenzene

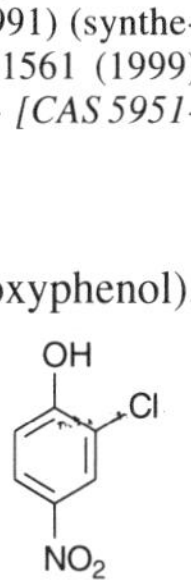

2-Chloro-4-nitrophenol

$C_7H_4Cl_4O_2$, M_R 261.92, cryst., mp. 116–117 °C. The antibiotically active D. A was isolated from cultures of *Psathyrella subatrata* (syn. *Drosophila subatrata*, Basidiomycetes)[1]. Related, polychlorinated hydroquinone derivatives are the corresponding *methyl ether* ($C_8H_6Cl_4O_2$, M_R 275.95, needles, mp. 164 °C)[2] from the wood rotting fungus *Fomes fastuosus* and *1,2,4-trichloro-3,6-dimethoxy-5-nitrobenzene* ($C_8H_6Cl_3NO_4$, M_R 286.50, cryst., mp. 115–116 °C) from *Fomes robiniae*[3]. The subterranean gasteromycete *Stephanospora caroticolor* contains *2-chloro-4-nitrophenol* [$C_6H_4ClNO_3$, M_R 173.56, needles, mp. 111 °C, pK_a 3.74 (25 °C)], the potassium/sodium salt of which is responsible for the orange color of the fruit body and was previously used in antifungal skin ointments[4].

Lit.: [1] Beilstein E IV **6**, 5775; Phytochemistry **23**, 2392 (1984). [2] Tetrahedron Lett. **1966**, 1229. [3] Tetrahedron Lett. **1972**, 211. [4] Steglich et al., unpublished; J. Pharm. Pharmacol. **18**, 41 (1945). – *[CAS 484-67-3 (D.); 944-78-5 (D. methyl ether); 35282-83-8 (1,2,4-trichloro-3,6-dimethoxy-5-nitrobenzene); 619-08-9 (2-chloro-4-nitrophenol)]*

Drosopterin see pteridines.

DSP toxins see shellfish poisons.

Dulcitol see galactitol.

Dunawithagenin see withanolides.

Duocarmycins (pyrindamycins).

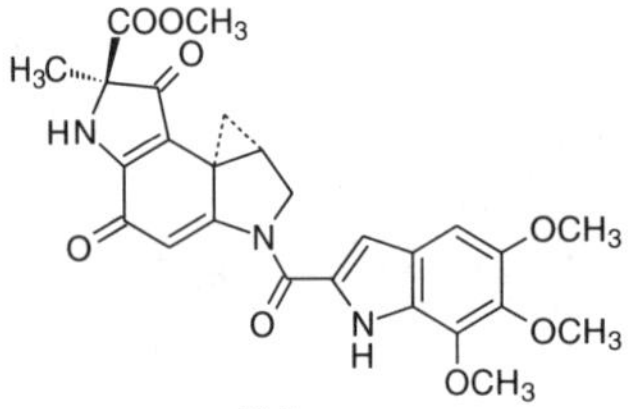

D.A

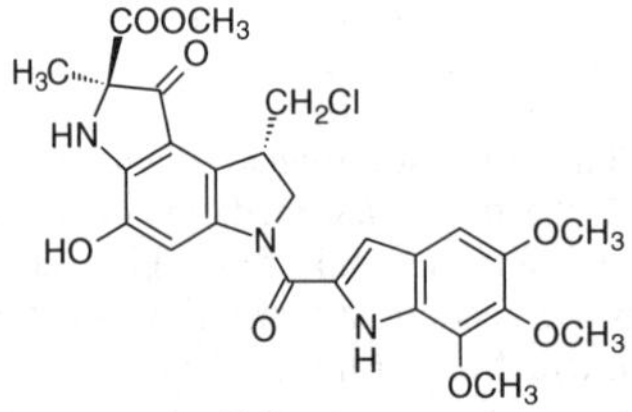

D.C₂
(Pyrindamycin A)

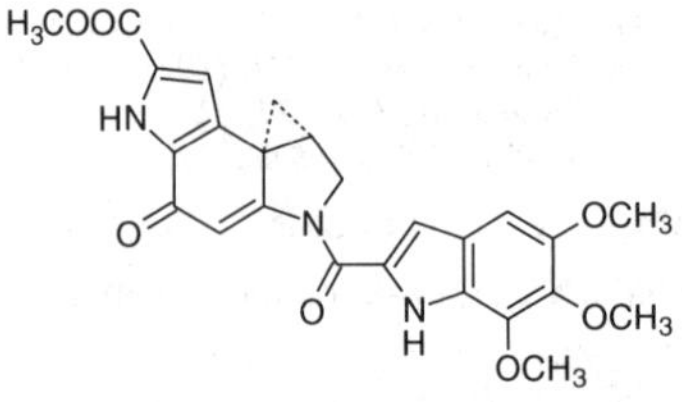

D. SA

Highly active antitumor antibiotics from cultures of streptomycetes. The most important members are *D. A*, $C_{26}H_{25}N_3O_8$, M_R 507.50, yellow powder, mp. 148 °C, $[\alpha]_D^{22}$ +282° (CH_3OH), *D. C₂*, $C_{26}H_{26}ClN_3O_8$, M_R 543.96, yellow needles, mp. 256 °C (237 °C, decomp.), $[\alpha]_D^{22}$ –51° (CH_3OH) and D. SA[1], $C_{25}H_{23}N_3O_7$, M_R 477.47, yellow powder, $[\alpha]_D^{24}$ +180° (CH_3OH).

Mechanism of action: The D. bind to AT-rich sequences of DNA, affording selective alkylation of N3-atom of adenine (A) at 3′-terminus of 3 or more connected AT-basepairs in DNA. CC-1065 acts similarly. *Distamycin-addition can shift alkylation to guanine[2].

Lit.: [1] Bioorg. Med. Chem. Lett. **6**, 2207 (1996); Chem. Pharm. Bull. **43**, 378, 1064 (1995); **44**, 1631 (1996); **46**, 400, 559 (1998); Tetrahedron Lett. **38**, 7207, 7209 (1997). [2] Angew. Chem. Int. Ed. Engl. **38**, 650–653 (1999); J. Am. Chem. Soc. **119**, 4977, 4987 (1997).

gen.: Angew. Chem. Int. Ed. Engl. **35**, 1438–1474 (1996) (activity) ▪ Bioorg. Med. Chem. Lett. **6**, 1215, 2147 (1996) (derivative) ▪ Chem. Heterocycl. Compd. **34**, 1386–1405 (1998) ▪ Chem. Pharm. Bull. **43**, 378, 1064, 1530 (1995); **44**, 67–79 (1996) ▪ Chem. Rev. **97**, 787–828 (1997) (review) ▪ J. Am. Chem. Soc. **114**, 10056 (1992); **115**, 9025 (1993); **118**, 2301 (1996) (synthesis); **119**, 311–325, 4977 (1997) (synthesis D. A, analogues) ▪ J. Antibiot. **44**, 1045 (1991) ▪ J. Med. Chem. **40**, 972 (1997) ▪ J. Org. Chem. **61**, 4894 (1996); **62**, 8868 (1997) (derivatives); **64**, 5946 (1999) (synthesis analogues) ▪ Pure Appl. Chem. **66**, 2255 (1994) (review) ▪ Tetrahedron **50**, 2793, 2809 (1994) (synthesis). – *[CAS 118292-34-5 (D. A); 118292-36-7 (D. C₂); 130288-24-3 (D. SA)]*

Dwarf pine oil see fir and pine needle oils.

Dye plants. Term for all wild-growing or cultivated plants whose flowers, fruits, seeds, leaves, stems, woods, roots, and barks contain natural pigments or pigment precursors that are used for dyeing wool, cotton, silk, leather, paper, wood, foods, and cosmetics or serve as pigments of paints. See table for the important dye plants; see also madder (see table, p. 197).

Lit.: [1] Rowe, p. 848. [2] Angew. Bot. **67**, 128 (1993). [3] Franke, Nutzpflanzenkunde (5.), Stuttgart: Thieme 1992.

gen.: Ullmann (4.) **11**, 99–134.

Dyer's madder see madder.

Dyer's greenweed, dyer's knotweed see dye plants.

Dyer's sawwort, dyer's woad, dyer's weld see dye plants.

Dyes see natural dyes.

Dye-woods. Heartwood of various, mostly tropical, plant species from which pigments or pigment precursors are extracted and used for dyeing wool, leather, paper, etc., e. g., redwoods (Brazil wood, sappan wood, Nicaragua wood, Jamaica redwood, Pernambuco wood, see brazilin), logwood (see haematoxylin), red sandalwood (see santalins), yellowwood (dyer's mulberry, see dye plants).

Lit.: Rowe, p. 843 ff.. – *[HS 1404 10]*

Dynemicins. Cytostatically active *enediyne antibiotics with the ability to form on *in vivo* activation benzenoid diradicals that damage deoxyribonucleic acids. D. were model compounds for the synthesis of highly active structurally analogous compounds , e. g., golfomycin[1], see also calicheamicins, esperamycins, neocarzinostatins. *D. A*[2] is the parent compound: $C_{30}H_{19}NO_9$, M_R 537.48, violet powder, mp. 208–210 °C (decomp.), $[\alpha]_D^{24}$ +270° (DMF) from *Micromonospora chersina* and *M. globosa* cultures. Further D. such as *D. L*[3] {$C_{30}H_{22}ClNO_9$, M_R 575.96, blue, amorphous powder, mp. 222–225 °C, $[\alpha]_D^{27}$ –820°

Table: Dye plants.

plant	occurrence	dyes	color	use
black oak (*Quercus velutina*, Fagaceae)	eastern North America	Quercetin-3-*O*-L-rhamnoside	yellow	utilization of the inner bark (quercitron bark, C. I. Natural Yellow 9) for cotton prints, dyeing of wool and paper; discovered by Bancroft 1797
dyer's greenweed (*Genista tinctoria*, Fabaceae)	UK, central and southern Europe	*Luteolin, *Genistein	yellow	flowers and leaves employed for wool dyeing by ancient Greeks and Romans
dyer's knotweed (*Polygonum tinctorium*, Polygonaceae)	indigenous to East Asia, southern China, and Japan; cultivated in the Caucasus, France, and Germany	*Indican	blue	leaves are used to dye wool
old fustic [1] (*Chlorophora tinctoria*, Moraceae)	Central America including Antilles and tropical South America	*Morin, *Maclurin; *Kaempferol, 1,3,6,7-Tetrahydroxy-xanthone	yellow	the heartwood (yellowwood) was exported to Europe for wool dyeing
sawwort (*Serratula tinctoria*, Asteraceae)	Europe and Asia	*Apiin, *Kaempferol	yellow	use of the epigeous parts for wool dyeing
dyer's woad (*Isatis tinctoria*, Brassicaceae)	indigenous to Central Asia and the Near East; cultivated in East Asia, India, North Africa, and Europe	*Isatan B (Indigo precursor)	blue	leaves utilized for colourfast wool dyeing, most important dye plant in the Middle Ages; nowadays cultivated again in eastern Germany
dyer's weld [2] (*Reseda luteola*, Resedaceae)	indigenous to Morocco and the Mediterranean, nowadays also in Central and South Europe	*Luteolin, Isorhamnetin, Kaempferol glycosides	yellow	the plant and its seeds were used even in the antiquity for dyeing wool and silk
safflower [3] (*Carthamus tinctorius* Asteraceae)	Egypt and Near East	*Carthamin	yellow orange red	utilization of the blossom leaves to dye silk, cotton, and leather; the plant was also grown in ancient Egypt 4000 years ago

$(CH_3OH)\}$, *D. M* $\{C_{29}H_{23}NO_9$, M_R 529.50, blue powder, mp. 238–240 °C, $[\alpha]_D^{27}$ −2460° $(CH_3OH)\}$ with cytotoxic and antibacterial activity, are presumably formed from D. A in cultures. The activity features of D. stimulated the development of synthetic approaches to the natural product itself and of analogs thereof[4].

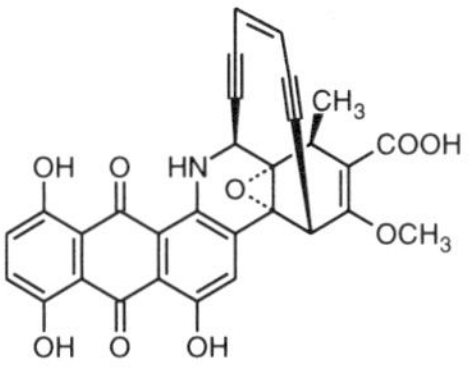

Dynemicin A

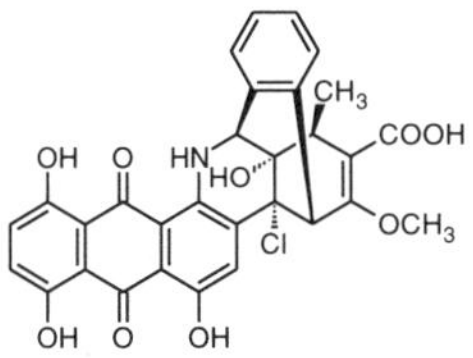

Dynemicin L

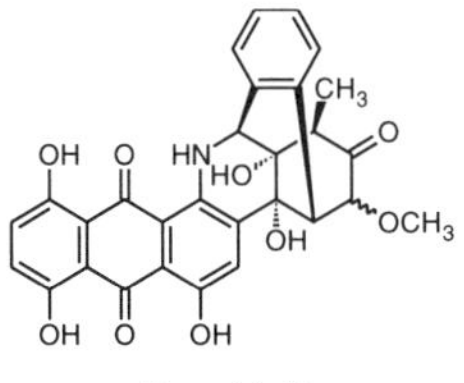

Dynemicin M

Lit.: [1] Angew. Chem. Int. Ed. Engl. **29**, 1064 (1990). [2] *Isolation:* J. Antibiot. **92**, 1449 (1989). – *Biosynthesis:* J. Am. Chem. Soc. **114**, 4107 (1992). – *Activity, structure:* J. Am. Chem. Soc. **112**, 3715, 4040 (1990); Science **256**, 1172 (1992); Tetrahedron Lett. **31**, 1521 (1990). – *Review:* Angew. Chem. Int. Ed. Engl. **30**, 1387 (1991). – *Synthesis:* Angew. Chem. Int. Ed. Engl. **34**, 1721 (1995); Angew. Chem. Int. Ed. Engl. **33**, 781 f. (1994); Chem. Biol. **2**, 33 (1995); J. Am. Chem. Soc. **118**, 9509–9525 (1996); **119**, 6072–6094 (1997). [3] J. Antibiot. **44**, 1300 (1991). [4] Eur. J. Org. Chem. **1999**, 1–13. – *[CAS 124412-57-3 (D. A); 127032-74-0 (D. L); 127003-54-7 (D.M)]*

Dysidazirine.

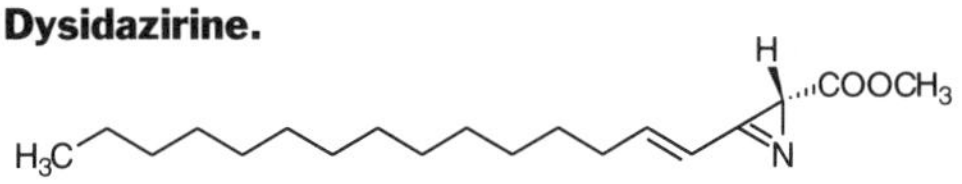

$C_{19}H_{33}NO_2$, M_R 307.48, low-melting solid; $[\alpha]_D$ −165° (CH_3OH). Azirine derivative from the Pacific Ocean sponge *Dysidea fragilis*. D. has cytotoxic activity against L1210 leukemia cells and inhibits the growth of Gram-negative bacteria and yeasts. The natural compound exists in partially racemized form.

Lit.: J. Am. Chem. Soc. **117**, 3651 (1995) [synthesis, (*R*)-(–)-D.] ▪ J. Org. Chem. **53**, 2103 (1988). – *[CAS 113507-74-7]*

Dysidiolide.

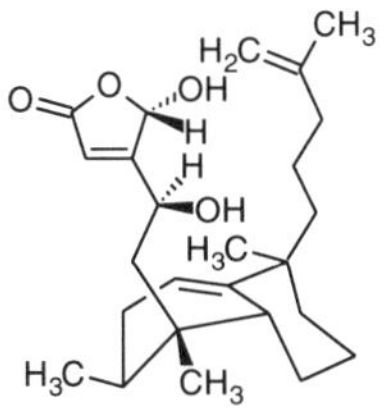

$C_{25}H_{38}O_4$, M_R 402.57, cryst., mp. 186–187 °C, $[\alpha]_D^{24}$ −11.1° $(CH_2Cl_2/CH_3OH, 1:1)$. Protein tyrosine phosphatase cdc 25 A-inhibitor (antimitotic activity by arresting the cell cycle, the cells become unable to divide) from the Caribbean sponge *Dysidea etheria*. D. is a *sesterterpenoid, it is being tested in various cancer models, e. g., lung carcinoma and leukemia.

Lit.: *Isolation:* J. Am. Chem. Soc. **118**, 8759 (1996). – *Synthesis:* Bioorg. Med. Chem. Lett. **9**, 2537 (1999) ▪ Chemtracts **12**, 512 (1999) ▪ J. Am. Chem. Soc. **119**, 12425 (1997); **120**, 1615 (1998) ▪ J. Org. Chem. **63**, 228 (1998) ▪ Pat WO 9940079 (12.08.1999) ▪ Tetrahedron Lett. **39**, 3995 (1998); **41**, 911 (2000). – *[CAS 182136-94-3]*

Dysoxysulfone (2,4,5,7,9-pentathiadecane 2,2,9,9-tetraoxide).

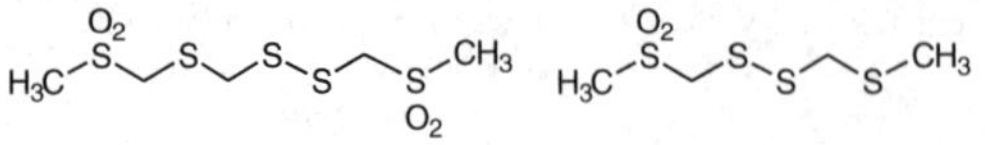

Dysoxysulfone

2,4,5,7-Tetrathiaoctane 2,2-dioxide

$C_5H_{12}O_4S_5$, M_R 296.45, cryst., mp. 107–108 °C. Active principle of tea prepared from the leaves of *Dysoxylum richii* (Meliaceae) used against pain in the traditional medicine of Fiji [1]. A related sulfur derivative, *2,4,5,7-tetrathiaoctane 2,2-dioxide* ($C_4H_{10}O_2S_4$, M_R 218.36), occurs in *Tulbaghia violacea* (Liliaceae) used in South Africa as a remedy for intestinal worms. For further aliphatic polysulfur compounds, see under lenthionine.

Lit.: [1] Tetrahedron Lett. **30**, 4919 (1989).
gen.: J. Org. Chem. **59**, 2273 (1994) (synthesis) ▪ Planta Med. **58**, 295 (1992). – *[CAS 125292-92-4 (D.); 143113-67-1 (2,4,5,7-tetrathiaoctane 2,2-dioxide)]*

E

Ebelactones. Enzyme inhibitors with β-lactone structure from cultures of *Streptomyces aburaviensis. E. A* {$C_{20}H_{34}O_4$, M_R 338.49, needles, mp. 86 °C, $[\alpha]_D^{26}$ −221° (CH_3OH)} and *E. B* {$C_{21}H_{36}O_4$, M_R 352.51, needles, mp. 77 °C, $[\alpha]_D^{26}$ −203° (CH_3OH)} inhibit membrane-bound esterases, lipases, and aminopeptidases of animal cells as well as cutinases of fungal cells (antifungal activity).

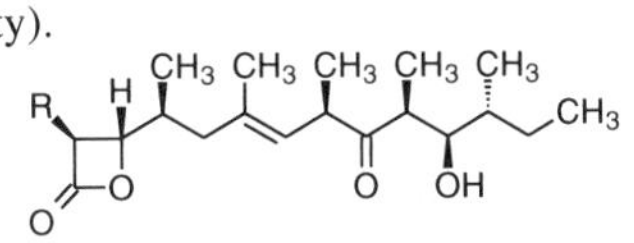

R = CH_3 : E.A
R = C_2H_5 : E.B

Lit.: J. Antibiot. **33**, 1594 (1980) (isolation); **35**, 1495, 1670 (1982) (biosynthesis) ▪ J. Org. Chem. **60**, 3288–3300 (1995) (synthesis). – *[CAS 76808-16-7 (E. A); 76808-15-6 (E. B)]*

Eburicoic acid. Formula, see eburicol; $C_{31}H_{50}O_3$, M_R 470.74, needles, mp. 293 °C, $[\alpha]_D$ +34° (pyridine). Widely distributed C_{31}-triterpene carboxylic acid with eburicane skeleton from wood-rotting fungi such as *Fomes officinalis, Lentinus lepideus, Gloeophyllum* sp., *Polyporus* sp. and other Aphyllophorales [1]. E. is formed biosynthetically from *eburicol that occurs in many fungi.

Lit.: [1] J. Chem. Soc. C **1967**, 2002.
gen.: Beilstein EIV **10**, 1172 ▪ Chem. Pharm. Bull. **22**, 877 (1974) ▪ Turner **1** 260; **2**, 323 (distribution). – *[CAS 560-66-7]*

Eburicol [24-methylenelanost-8-en-3β-ol, eburica-8,24(28)-dien-3β-ol].

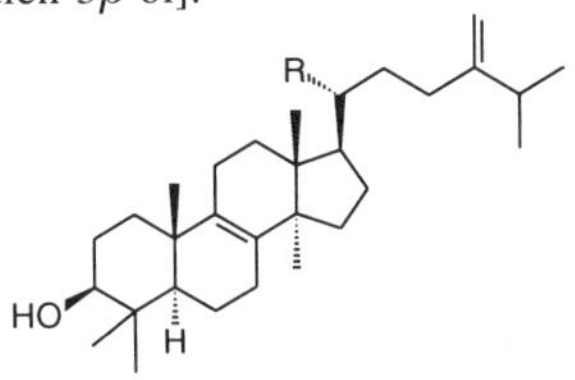

R = CH_3 : Eburicol
R = COOH : Eburicoic acid

$C_{31}H_{52}O$, M_R 440.75, cryst., mp. 158–159 °C, $[\alpha]_D$ +66° ($CHCl_3$). Biosynthetic precursor of numerous tetracyclic triterpenoids in fungi. Widely distributed in imperfect fungi, yeasts, basidiomycetes, and other fungi.

Lit.: C. R. Acad. Sci., Ser. D **276**, 205 (1973) (biosynthesis) ▪ J. Chem. Soc., Chem. Commun. **1966**, 595 ▪ Phytochemistry **37**, 201 (1994) (E.-3-ketone) ▪ Turner **2**, 325 (distribution). – *[CAS 6890-88-6]*

Eburnamonine [huntericine, eburnamenin-14(15*H*)-one]. $C_{19}H_{22}N_2O$, M_R 294.40, cryst., mp. 183 °C, $[\alpha]_D$ +89° ($CHCl_3$). An *indole alkaloid from many genera of the family Apocynaceae, e.g., *Amsonia, Hunteria,*

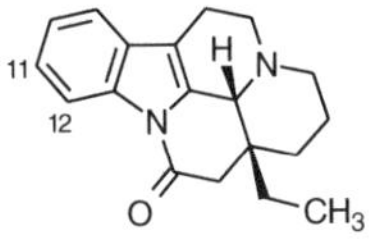

(+)-form

Vinca species. It occurs in the (−) form {*vincamone*, mp. 173–174 °C, $[\alpha]_D$ −102° ($CHCl_3$)} in *Vinca minor*, and also as racemate (*vincanorine*, mp. 203–204 °C).
Activity: Vasodilative, hypotensive effects, as well as bronchoconstrictory activity. On chronic administration of (−)-E. to rats increased enzyme activities (mitochondrial cytochrome oxidase) were measured in rat brains.
Biosynthesis: The biogenetic origin of E. is mostly unknown. Since E. belongs to the group of Aspidosperma alkaloids, *strictosidine is the first intermediate which reacts further through rearrangement of the monoterpene part by a still unknown mechanism. The alkaloids preakuammicine, stemmadenine, secodine derivatives, and *tabersonine may participate in the formation of E., as they attribute to the biosynthesis of *vincamine.

Lit.: Beilstein EV **24/4**, 192 ▪ Manske **8**, 253–259; **11**, 108–110; **42**, 1–116 ▪ Szántay, in Saxton (ed.), Monoterpenoid Indole Alkaloids, Suppl. Vol. 25, part. 4 to Heterocyclic Compounds, p. 478 ff., 733–736, Chichester: Wiley & Sons 1994. – *Pharmacology:* Arzneim.-Forsch. **29**, 1094 (1979) ▪ Hager (5.) **6**, 1128. – *Synthesis:* Chem. Pharm. Bull. **30**, 1521 (1982); **31**, 1191 (1983) ▪ J. Org. Chem. **53**, 1953 (1988); **59**, 7197 (1994); **62**, 6855 (1997); **64**, 7586 (1999) ▪ Tetrahedron **42**, 3215 (1986); **43**, 493 (1987); **50**, 9487–9494 (1994) ▪ Tetrahedron Lett. **30**, 873 (1989); **41**, 587 (2000). – *[HS 2939 90; CAS 474-00-0 ((+)-E.); 4880-88-0 ((−)-E.); 2580-88-3 (racemate)]*

Ecdysteroids. A characteristic group of steroid hormones that trigger the developmentally very important skin sloughing processes in insects and crustaceans (see insect hormones); the compounds are thus also known as *sloughing* or *cocoon hormones.*
The first E. to be discovered in 1954 by Butenandt and Karlson was *ecdysone* isolated from larvae of the silk worm (*Bombyx mori*) (yield: merely 4.5×10^{-6}%). Since then, ecdysone and the almost twice as active *ecdysterone* have been detected in numerous insect species. Mollusks, annelids, and nematodes appear to require E. for the regulation of various vital processes. *Ponasterone* A was isolated from the crab *Carcinus maenas*. Surprisingly, E. have also been found in many plant species, often in high concentrations, e. g., α-ecdysone and ecdysterone in fern species, yew trees, verbena, and foxtail, ecdysterone in roots of *Achyranthes fauriei*, and ponasterone A in leaves of the conifer *Podocarpus nakaii. Makisterones* A and C as well as *cyasterone* also belong to these so-called *phytoecdy-*

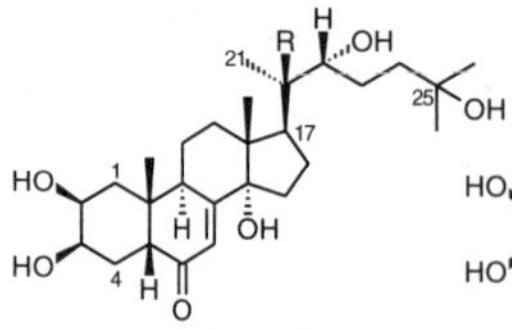

R = H : (α-)Ecdysone (**1**)
R = OH : Ecdysterone (β-Ecdysone, Crustecdysone) (**2**)

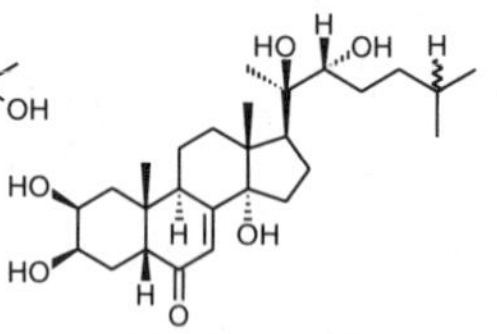

Ponasterone A (**3**)

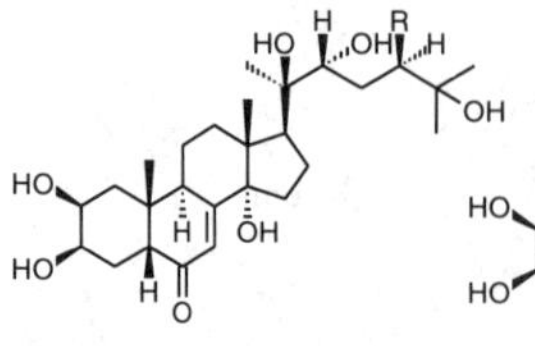

R = CH_3 : Makisterone A (**4**)
R = CH_2—CH_3: Makisterone C (Lemmasterone, Podecdysone) (**5**)

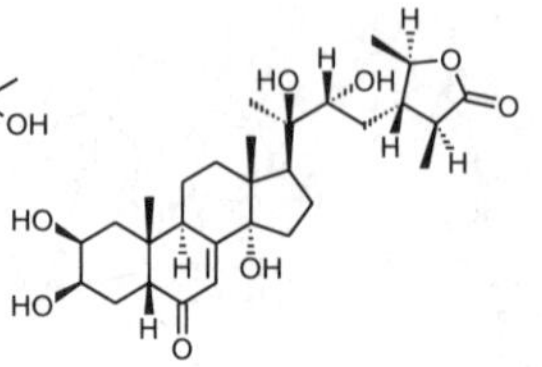

Cyasterone (**6**)

Table: Data of Ecdysteroids.

no.	molecular formula	M_R	mp. [°C]	$[\alpha]_D$	CAS
1	$C_{27}H_{44}O_6$	464.64	242	+64.7° (C_2H_5OH)	3604-87-3
2	$C_{27}H_{44}O_7$	480.64	237.5–239.5 (243)	+61.8° ($CHCl_3$)	5289-74-7
3	$C_{27}H_{44}O_6$	464.64	259–260	+90° (CH_3OH)	13408-56-5
4	$C_{28}H_{46}O_7$	494.67	286–287	+83.3° (CH_3OH)	20137-14-8
5	$C_{29}H_{48}O_7$	508.70	263–265 (decomp.)		19974-41-5
6	$C_{29}H_{44}O_8$	520.66	164–166	+64.5° (pyridine)	17086-76-9

steroids. The C_{28}-E. makisterone was found to be the sloughing hormone of the honey bee *Apis mellifera*. In general the concentrations in plants are up to 1000-fold higher than those in insects [1]. To date over 30 different E. and about the same number of conjugated E. (e.g., glycosides such as *ponasteroside* A) have been isolated from invertebrates. The number of known phytoecdysteroids exceeds 70. Their protective action for plants is presumed to involve the prevention of accumulation of larger insect populations [2]. The E. are formed biosynthetically from *cholesterol or, respectively, phytosterols (see sterols). However, these sterols or the phyto-E. must be taken up by the insects and other arthropods as essential nutritional components since themselves they have no capacity for steroid biosynthesis.

Lit.: [1] Phytochemistry **32**, 1361 (1993); Zechmeister **28**, 256. [2] J. Nat. Prod. **38**, 195 (1975).

gen.: Chem. Nat. Compd. **33**, 506–526 (1997) (review) ▪ Danielsson & Sjövall (eds.), Sterols and Bile Acids, p. 199–230, Amsterdam: Elsevier 1985 ▪ Koolman (eds.), Ecdysone, Stuttgart: Thieme 1989 ▪ Tetrahedron Lett. **38**, 2697, 2701 (1997) (biosynthesis) ▪ Zeelen, p. 269–274.

Ecgonine [(1*R*)-3*exo*-hydroxytropane-2*exo*-carboxylic acid]. $C_9H_{15}NO_3$, M_R 185.22, cryst., mp. (as hydrate) 205 °C (anhydrous 198 °C), $[\alpha]_D$ −45.4° (H_2O). E. is the parent compound of many *tropane alkaloids of the Erythroxylaceae. It is mostly esterified with ar-

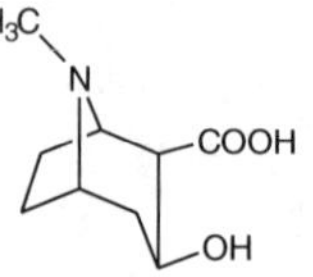

omatic acids and occurs in free form in some Erythroxylaceae only [1]. E. was first obtained in 1923 by Willstätter through hydrolysis of *cocaine. E. is highly toxic, the racemate is accessible through synthesis.

Lit.: [1] Manske **44**, 1–114.

gen.: Beilstein EV **22/5**, 53 ▪ J. Am. Chem. Soc. **101**, 2435 (1979) (synthesis) ▪ Ullmann (5.) **A1**, 361. – *Biosynthesis:* Planta Med. **56**, 339–352 (1990) ▪ Manske **44**, 142. – *[HS 2939 90; CAS 481-37-8]*

Echinacoside.

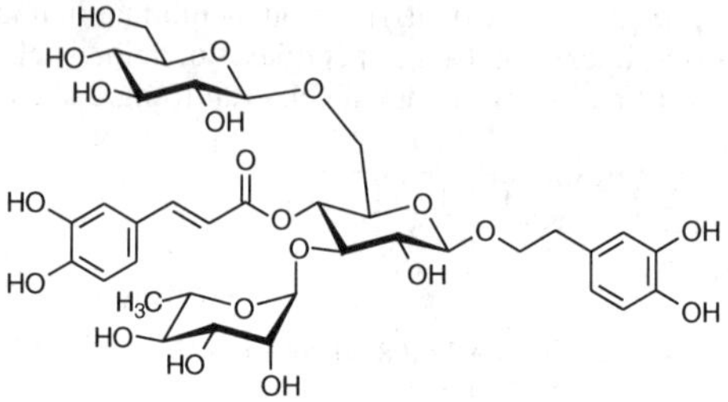

$C_{35}H_{46}O_{20}$, M_R 786.74, needles, mp. >200 °C (decomp.), $[\alpha]_D^{20}$ −56.6° (H_2O). E. is a glycoside of *2-(3,4-dihydroxyphenyl)ethanol* [$C_8H_{10}O_3$, M_R 154.17, bp. 170–175 °C (3 Pa)] with 2 mol D-*glucose and 1 mol *L-rhamnose, with one glucose unit being esterified by *caffeic acid. E. occurs in the roots of *Echinacea angustifolia* and *E. purpurea* (Asteraceae).

Activity: E. has antibiotic and anti-inflammatory properties. Preparations (especially tinctures and the pressed juice) were already in use at the end of the 19th century.

Use: Internally, E. is used in urology as well as an anti-influenza, anti-inflammatory, and mode-changing agent; externally to treat wounds. Most applications are based on the immunostimulating activity of E.

Lit.: Beilstein EIV **17**, 3629 ▪ Chem. Pharm. Bull. **32**, 3009 (1984) ▪ Hager (5.) **6**, 389 f. ▪ Phytochemistry **26**, 1981 ff. (1987); **27**, 2787–2794 (1988) ▪ Z. Naturforsch. C **37**, 351 ff. (1982); **40**, 585 ff. (1985) (NMR). – *[CAS 82854-37-3 (E.); 10597-60-1 (2-(3,4-dihydroxyphenyl)ethanol)]*

Echinatine see indicine.

Echinocandin B.

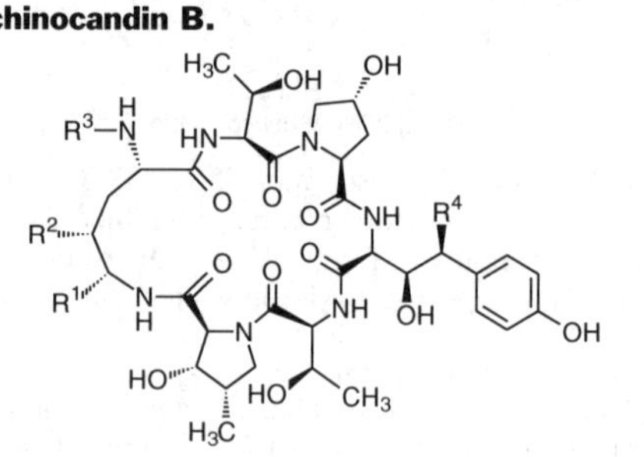

R^1	R^2	R^3	R^4	
OH	OH	Linoleoyl	OH	E. B
OH	OH	Linoleoyl	H	E. C
H	H	Linoleoyl	H	E. D
OH	OH	4-Octyloxybenzoyl	OH	Cilofungin

[Linoleoyl = (*Z,Z*)-9,12-Octadecadienoyl]

$C_{52}H_{81}N_7O_{16}$, M_R 1060.25, amorphous powder, mp. 160–163 °C, $[\alpha]_D^{20}$ −48° (CH_3OH). E. B is a cyclic hexapeptide and belongs to the neutral lipopeptide*antibiotics with antifungal properties (anti-*Candida in vitro* and *in vivo*)[1]. It is isolated from cultures of *Aspergillus rugulosus*. Besides E. B, E. C[2] and E. D[3] have also been described. An important derivative of E. B is *cilofungin*, with good antimycotic action against pathogenic yeasts, due to the inhibition of (1,3)-β-D-glucan synthase[4]. Semisynthetic echinocandins have been developed clinically as systemic antimycotic agents[5].

Lit.: [1]Diagn. Microbiol. Infect. Dis. **12**, 1–4 (1989); FEBS Lett. **173**, 134–138 (1984). [2]Chem. Abstr. **105**, 173030 (1986). [3]Chemtracts: Org. Chem. **1**, 148–151 (1988). [4]Antimicrob. Agents Chemother. **32**, 1331, 1901 ff. (1988); Eur. J. Clin. Microbiol. Infect. Dis. **7**, 77, 432 (1988); J. Antibiot. **42**, 389–397 (1989) (synthesis); J. Antimicrob. Chemother. **22**, 891–897 (1988); Mycoses **31**, 330 (1988). [5]Exp. Opin. Ther. Patents **5**, 771–786 (1995); J. Antibiotics **50**, 562 (1997); J. Org. Chem. **64**, 2411 (1999).
gen.: Helv. Chim. Acta **57**, 2459 (1974); **62**, 1252 (1979) ▪ J. Antibiot. **42**, 389–397 (1989) ▪ Tetrahedron **49**, 6195–6222 (1993) ▪ Tetrahedron Lett. **1976**, 4147. – *[CAS 54651-05-7 (E. B); 71018-12-7 (E. C); 71018-13-8 (E. D); 79404-91-4 (cilofungin)]*

Echinochromes. Group of hydroxy-1,4-naphthoquinone derivatives that occur in the eggs and perivisceral fluid of sea urchins (Echinoidae), e.g., *E. A* (2-ethyl-3,5,6,7,8-pentahydroxy-1,4-naphthoquinone, $C_{12}H_{10}O_7$, M_R 266.21, dark red needles, mp. 223 °C).

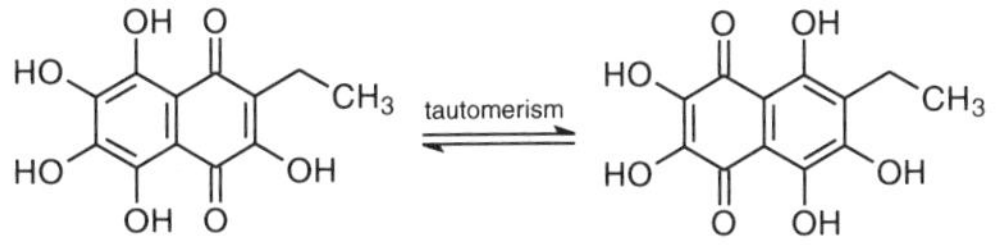

Echinochrome A

E. A represents the prosthetic group of a high-molecular-weight complex that acts as an activating and agglutinating agent for spermatosomes.

Lit.: Exp. Cell Res. **134**, 65–72 (1981) ▪ J. Org. Chem. **31**, 3645 (1966) (synthesis) ▪ Kardiologiya **31**, 79 (1991) ▪ Scheuer I **5**, 82 (biosynthesis). – *[CAS 517-82-8]*

Echinodermata. General term for spiny marine organisms (echinoderms). In mature state mostly five-armed, radially symmetric bilateria, sometimes with calcareous skeleton in the subdermis from which the spines protrude above the surface of the body. 1. Substrain: Pelmatozoa, class: Crinoidea (feather stars, sea lilies); 2. substrain: Eleutherozoa, class: Holothuroidea (sea cucumbers); class: Echinoidea (sea urchins); class: Asteroidea (starfish); class: Ophiuroidea (brittle stars). Many of the E. secrete toxins whose active principles are steroid glycosides (starfish, Asteroidea) or triterpene glycosides (sea cucumbers, Holothuroidea, e.g., *holothurins).

Echinoderm saponins. Asteroids (starfish) and holothuroids (sea cucumbers) secrete saponins for defense. The E. s. are responsible for the general toxicity of these creatures, they have cytotoxic, hemolytic, ichthyotoxic, and microbicidal activities. As yet E. s. have not been detected in sea urchins and feather stars. Starfish contain only steroid saponins while sea urchins contain only triterpene saponins, the *holothurins. The so-called *asterosaponins have a sulfate group in the 3 position, e.g. thornasteroside A from *Acanthaster planci*. Cyclic saponins such as *sepositoside A* {$C_{45}H_{70}O_{19}$, M_R 915.04, sodium salt: amorphous, $[\alpha]_D$ −68.5° (H_2O)} are typical for the starfish species *Echinaster*. For a comprehensive review of the various E. s., see *Lit.*[1].

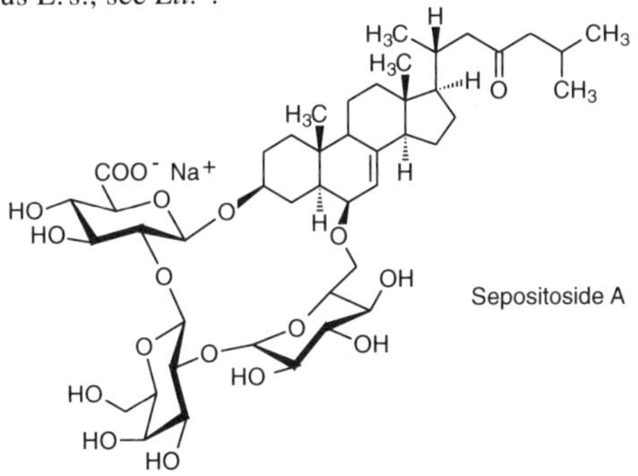

Sepositoside A

Lit.: [1]Hostettmann (ed.), Biologically Active Natural Products, p. 153–165, Oxford: Clarendon Press 1987.
gen.: de Couet et al., Gefährliche Meerestiere, Hamburg: Jahr 1981 ▪ Gazz-Chim. Ital. **126**, 667 (1996) ▪ J. Chem. Soc., Perkin Trans. 1 **1981**, 1855 ▪ J. Nat. Prod. **53**, 1000, 1225 (1990); **54**, 1254 (1991) ▪ J. Org. Chem. **63**, 4438 (1998) (synthesis) ▪ Justus Liebigs Ann. Chem. **1991**, 595 ▪ Zechmeister **62**, 75–340. – *[HS 2938 90; CAS 79154-52-2 (septositoside A)]*

Echinoderm toxins. Among the *echinodermata, the stalked grasping organ (pedicellaria) of isolated species of sea urchins and starfish may also serve as a poison weapon. The sea urchin *Toxopneustes pileolus* is much feared in East Asian waters. Other species of sea urchins also possess poisonous spines (e.g., *Diadema setosum*). The chemical composition of the venoms is mostly unknown. The cuvierian organs extruded from the cloaca of sea cucumbers when disturbed from sticky fibers in water may also secrete a venom, e.g., the *holothurins; see also echinoderm saponins.

Lit.: de Couet et al., Gefährliche Meerestiere, Hamburg: Jahr 1981.

Echinodol.

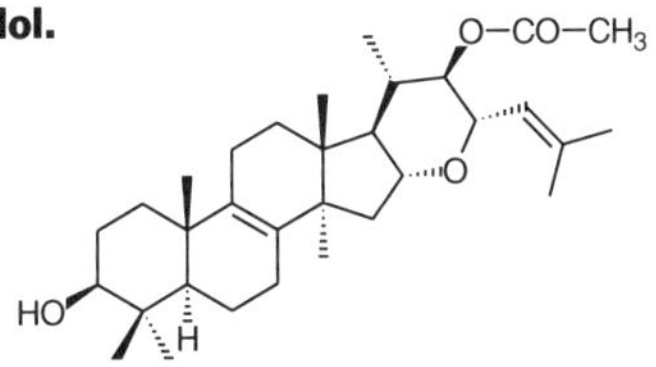

$C_{32}H_{50}O_4$, M_R 498.75, cryst., mp. 236–238 °C, $[\alpha]_D$ +48°. A *triterpene from the brown-red fruit bodies of the Indian paint fungus *Echinodontium tinctorium* (Aphyllophorales). The Japanese *E. tsugicola* contains the corresponding ketone *echinodone*, $C_{32}H_{48}O_4$, M_R 496.73, needles, mp. 225–227 °C, $[\alpha]_D$ +67° ($CHCl_3$).

Lit.: Beilstein E V **17/5**, 325; **18/2**, 98 ▪ Chem. Pharm. Bull. **20**, 1993 (1972). – *[CAS 10178-38-8 (E.); 10178-41-3 (echinodone)]*

Ecteinascidins. The antineoplastically active E., tris-tetrahydroisoquinoline alkaloids, are isolated from the Caribbean mangrove ascidia *Ecteinascidia turbinata*, e.g., *Et 743* ($C_{39}H_{43}N_3O_{11}S$, M_R 761.84), active against

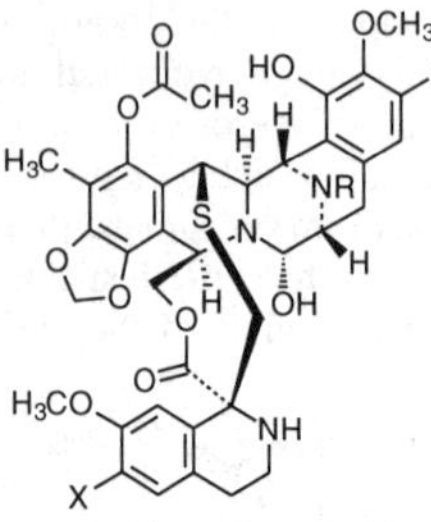

R = CH_3, X = OH: Et 743

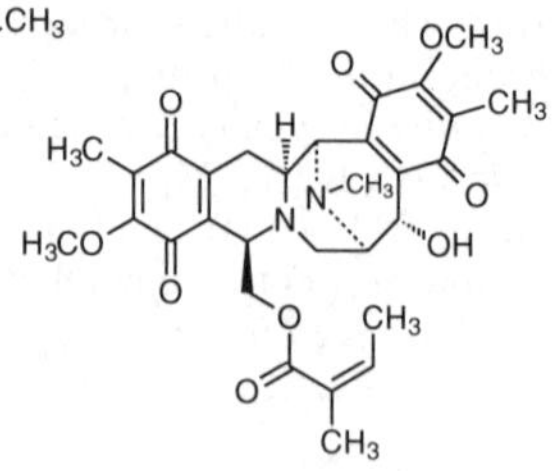

Renieramycin A

L1210 cells (IC_{50} 0.5 ng/mL) and P388-murine leukemia. The E. are closely related to the renieramycins from the sponges *Reniera* sp., e. g., *renieramycin A*, $C_{30}H_{34}N_2O_9$, M_R 566.61, $[\alpha]_D$ −36.3° (CH_3OH). Et 743 is in clinical phase II of development against tumors of lung, bone, colon, breast, ovarials, kidneys and against melanomas and soft tissue sarkomas[1].

Lit.: [1] R & D Focus, Drug News, 22.2.1999, p. 7; J. Med. Chem. **42**, 2493 (1999); Proc. Natl. Acad. Sci. USA **96**, 3496 (1999).

gen.: Ind. J. Chem. B **37**, 1258 (1998) (renieramycins) ▪ J. Am. Chem. Soc. **104**, 265 (1982); **118**, 9017 (biosynthesis, absolute configuration), 9202 (1996) (synthesis E. 743); **119**, 5475 (1997) (NMR of DNA adduct) ▪ J. Chem. Soc., Perkin Trans. 1 **1997**, 53–69 (synthesis) ▪ J. Org. Chem. **54**, 5822 (1989); **55**, 4508, 4512 (1990) ▪ Nachr. Chem. Tech. Lab. **41**, 6 (1993) ▪ Org. Lett. **1**, 75 (1999) (activity) ▪ Proc. Natl. Acad. Sci. USA **89**, 11456 (1992) ▪ Synlett **1999**, 1103 ▪ Tetrahedron Lett. **33**, 3721 (1992). – *[CAS 114899-77-3 (Et 743); 79644-60-1 (renieramycin A)]*

Ectocarpene see algal pheromones.

Edeines.

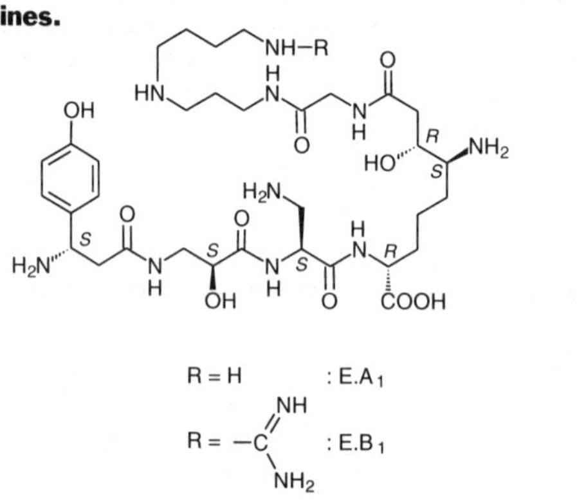

A group of linear pentapeptide alkaloids from *Bacillus brevis*. *E. A_1* ($C_{33}H_{58}N_{10}O_{10}$, M_R 754.88) and *E. B_1* ($C_{34}H_{60}N_{12}O_{10}$, M_R 796.92) contain, besides glycine, only non-proteinogenic amino acids and are conjugated with polyamines such as spermidine (A_1) or *N*-amidinospermidine (B_1). The E. are active against Gram-positive and -negative bacteria, fungi, and yeasts by inhibiting both DNA replication and protein biosynthesis at the 30S-ribosomal subunit. They are too toxic for medicinal uses.

Lit.: Int. J. Pept. Protein. Res. **26**, 279 (1985) ▪ J. Antibiot. **36**, 1001, 1239 (1983) ▪ Methods Enzymol. **94**, 441 (1983). – *[CAS 27656-72-0 (E. A_1); 27656-73-1 (E. B_1)]*

EDRF see nitrogen monoxide.

Efrotomycin. *E. A_1* ($C_{59}H_{88}N_2O_{20}$, M_R 1145.35) is a pale yellow polyene *antibiotic[1,2] from *Streptomyces* species which, like *goldinomycin* {aurodox, $C_{44}H_{62}N_2O_{12}$, M_R 810.98, $[\alpha]_D$ −82.8° (C_2H_5OH), isolated from *S. goldiniensis*, formula see below}, belongs to the group of *kirromycin/elfamycin-like antibiotics. Mechanism of action: inhibition of protein biosynthesis by binding to elongation factor Tu[3,4] (formula see below).

Use: To promote growth in pigs[5,6]. No cross resistance with antibiotics, applications in therapy for human and animal infections[7]. For synthesis, see *Lit.*[8], for biosynthesis *Lit.*[9], and for detection *Lit.*[10].

Lit.: [1] J. Antibiot. (Tokyo) **38**, 1691–1698 (1985). [2] Diagn. Microbiol. Infect. Dis. **6**, 49–52 (1987). [3] Biochem. J. **283**, 649–652 (1992). [4] J. Gen. Microbiol. **139**, 769–774 (1993). [5] J. Anim. Sci. **65**, 877–880 (1987). [6] Fed. Regist. **57**, 38441 f. (1992). [7] Methods Find. Exp. Clin. Pharmacol. **11**, 697–701 (1989). [8] J. Am. Chem. Soc. **107**, 1691–1698 (1985); J. Chem. Soc., Chem. Commun. **1985**, 1016 ff. [9] J. Antibiot. **42**, 944–951 (1989); J. Ind. Microbiol. **8**, 265–271 (1991); Bioprocess Eng. **7**, 257–263 (1992). [10] Biol. Mass Spectrom. **21**, 51–59 (1992).

gen.: J. Antibiot. **42**, 1453–1459, 1610–1618 (1989); **45**, 1697 ff. (1992) (pharmacology). – *[HS 2941 90; CAS 56592-32-6 (E. A_1); 12704-90-4 (goldinomycin)]*

EGCG. Abbreviation for *epigallocatechin 3-gallate.

Egomaketone [1-(3-furanyl)-4-methyl-3-penten-1-one].

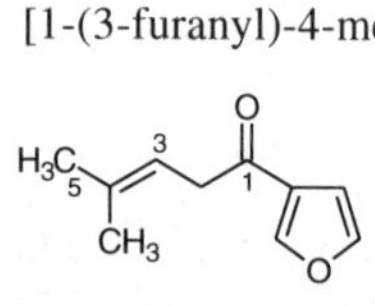

$C_{10}H_{12}O_2$, M_R 164.20, bp. 122–126 °C (2 kPa), 233 °C (76.4 kPa). A *monoterpene with an oxo group and fu-

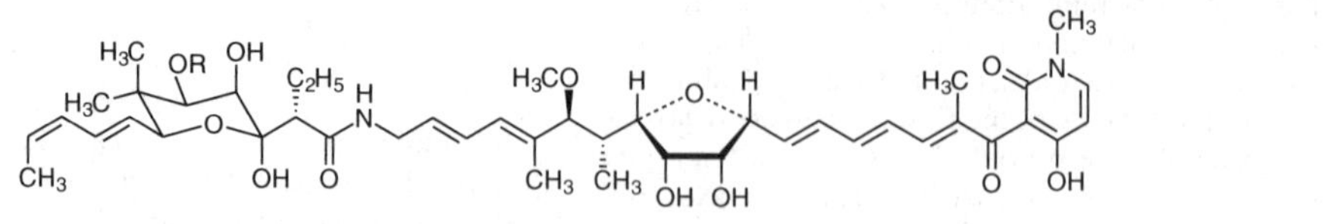

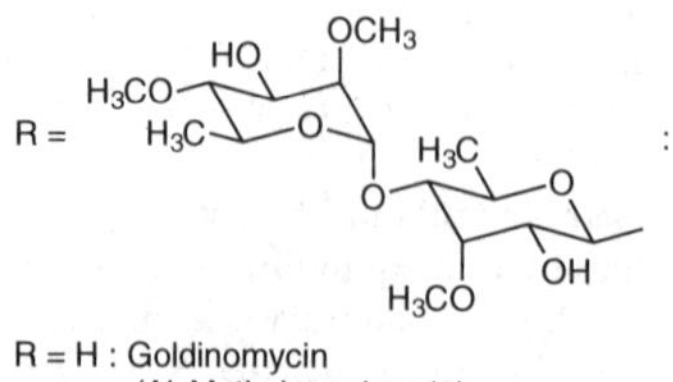

: Efrotomycin A_1 (Efrotomycin)

R = H : Goldinomycin (*N*-Methyl-mocimycin)

ran ring. The analogue hydrogenated at the C-3/4 double bond is *perilla ketone*, $C_{10}H_{14}O_2$, M_R 166.22, bp. 72–73 °C (300 Pa). E. and perilla ketone are suspected of causing severe pulmonary edemas and emphysemas in grazing animals. LD_{50} (mouse i.p.) 2.5–6 mg/kg; (rat i.p.) 10 mg/kg. E. occurs in Japanese "perilla oil" *Perilla frutescens* (Lamiaceae) in up to 70% [1]. For synthesis, see *Lit.*[2].

Lit.: [1] Chem. Ind. (London) **1962**, 1618. [2] J. Nat. Prod. **58**, 1955 (1995) (egomaketone); J. Org. Chem. **44**, 2807 (1979); Synthesis **1991**, 242; Tetrahedron Lett. **33**, 5245 (1992) (perilla ketone).
gen.: Karrer, No. 5649. – *[CAS 59204-74-9 (E.); 553-84-4 (perilla ketone)]*

Eicosanoids (icosanoids). General term for oxygen derivatives of unsaturated fatty acids with 20 C atoms formed especially from *arachidonic and *eicosapentaenoic acid by the action of oxygenases, e.g., cyclooxygenase and lipoxygenase (see also oxylipins). Cyclic (e.g., *prostaglandins and *thromboxanes) and acyclic (e.g., *leukotrienes and *lipoxins) compounds are formed via biosynthetic intermediates. These compounds are physiologically active mediators that transmit receptor controlled signals in animal cells. An increased formation of E. has been detected in some, especially chronic, diseases, e.g., joint rheumatism, inflammatory gastrointestinal diseases, asthma, and arteriosclerosis.

Lit.: Adv. Lipid Res. **23**, 169, 199 (1989) ▪ Angew. Chem. Int. Ed. Engl. **30**, 1100 (1990) ▪ Annu. Rev. Nutr. **11**, 41 (1991) ▪ Barton-Nakanishi **1**, 159–272 (review) ▪ Fat Sci. Technol. **96**, 7 (1994) ▪ Folco, Samuelsson, Maclouf, Velo (eds.), Eicosanoids from Biotechnology to Therapeutic Applications. NATO ASI Ser. A: Life Sciences, vol. 283, New York: Plenum 1996 ▪ Rev. Physiol. Biochem. Pharmacol. **121**, 1 (1992). – *Journal:* Eicosanoids, Berlin: Springer (since 1988).

Eicosapentaenoic acid [EPA, (*all-Z*)-5,8,11,14,17-(e)icosapentaenoic acid].

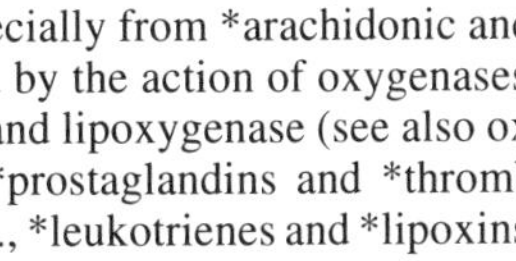

$C_{20}H_{30}O_2$, M_R 302.46, n_D^{20} 1.4986, oil. Highly unsaturated essential *fatty acid (polyene fatty acid) formed especially in algae (e.g., red algae) and enriched by way of the food chain in fish oil where it exists as the glycerol ester. E. is the biochemical precursor of a series of *prostacyclins, *thromboxanes, and *leukotrienes (see also eicosanoids). It is used as dietetic and as drug for treatment of coronary artery diseases, arterial thrombosis, and rheumatoid arthritis.

Lit.: Annu. Rev. Nutr. **10**, 149 (1990) ▪ Chow (ed.), Fatty Acids in Foods and their Health Implications, p. 446 ff., 746 ff., New York: Marcel Dekker 1992 ▪ Fat Sci. Technol. **94**, 506 (1992) ▪ J. Org. Chem. **60**, 6627 (1995) (synthesis) ▪ Merck-Index (12.), No. 3572 ▪ New Engl. J. Med. **318**, 549–557 (1988) ▪ Phytochemistry **43**, 63 (1996) (isolation) ▪ Prog. Food Nutr. Sci. **12**, 111–150 (1988) ▪ Stansby (ed.), Fish oils in nutrition, p. 26 ff., New York: Van Nostrand Reinhold 1990. – *[CAS 10417-94-4]*

(*all-Z*)-5,8,11,14-Eicosatetraenoic acid see arachidonic acid.

(*all-Z*)-8,11,14-Eicosatrienoic acid see dihomo-γ-linolenic acid.

(*Z*)-11-Eicosen-1-ol acetate see flies.

Elaeocarpus alkaloids.

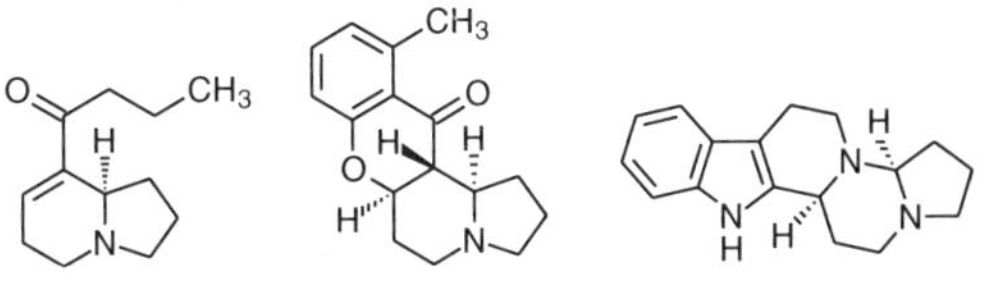

Elaeokanine A Elaeocarpine Elaeocarpidine (relative configuration)

Table: Data of Elaeocarpus alkaloids.

	Elaeo-kanine A [1]	Elaeo-carpine [2,3]	Elaeo-carpidine [4]
molecular formula	$C_{12}H_{19}NO$	$C_{17}H_{19}NO_2$	$C_{18}H_{21}N_3$
M_R	193.29	257.33	267.37
mp. [°C]	163–165 (picrate)	81–82	229–230
$[\alpha]_D$	+13° (CHCl$_3$)	+206° (CHCl$_3$)	±0° (CHCl$_3$)
CAS	33023-01-7	30891-90-8	20069-07-2

*Alkaloids with indolizidine or quinolizidine structure occurring only in a few species of the genus *Elaeocarpus* (Elaeocarpaceae). Two types of E. are known, C_{12}-alkaloids such as *elaeokanine A* and C_{16}-alkaloids such as *elaeocarpine*. The E.a. are probably formed biogenetically by coupling of Δ^1-pyrroline or Δ^1-pyrrolidine, already carrying a C_3 chain at the N-atom, with a *polyketide of the appropriate chain length. The indole alkaloid *elaeocarpidine* can be derived similarly by coupling of Δ^1-pyrroline with tryptamine and a C_3 unit.

Lit.: [1] J. Org. Chem. **55**, 292 (1990); Nat. Prod. Rep. **8**, 560 ff. (1991); **11**, 26 (1994); Manske **28**, 210–218; **44**, 221 ff. [2] Tetrahedron Lett. **21**, 1373 (1980). [3] Tetrahedron **29**, 1285 (1973). [4] Synth. Commun. **17**, 377 (1987).
gen.: Experientia **46**, 22 f. (1990) (biosynthesis) ▪ Mothes et al., p. 117 ff. ▪ Pelletier **3**, 241–273 ▪ Tetrahedron **54**, 1153–1168 (1998) (synthesis) ▪ Tetrahedron: Asymmetry **3**, 535 (1992). – *[HS 2939 90]*

Elaiolide see elaiophylin.

Elaiomycin [4-methoxy-3-(1-octenyl-*ONN*-azoxy)-2-butanol].

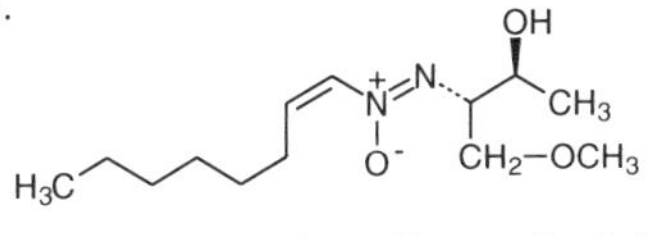

$C_{13}H_{26}N_2O_3$, M_R 258.36, yellow oil, $[\alpha]_D$ +38.4° (C_2H_5OH). Azoxy compound with tuberculostatic activity from cultures of *Streptomyces hepaticus*. E. is an experimental carcinogen and exhibits in mice a subcutaneous LD_{50} of 63 mg/kg.

Lit.: Beilstein E IV **4**, 3376 ▪ J. Am. Chem. Soc. **99**, 1643 (1977) (synthesis); **104**, 339 (1982); **106**, 5764 (1984) (biosynthesis) ▪ Merck-Index (12.), No. 3575 ▪ Sax (8.), EAG 000 (toxicology). – *[CAS 23315-05-1]*

Elaiophylin (azalomycin B, salbomycin).

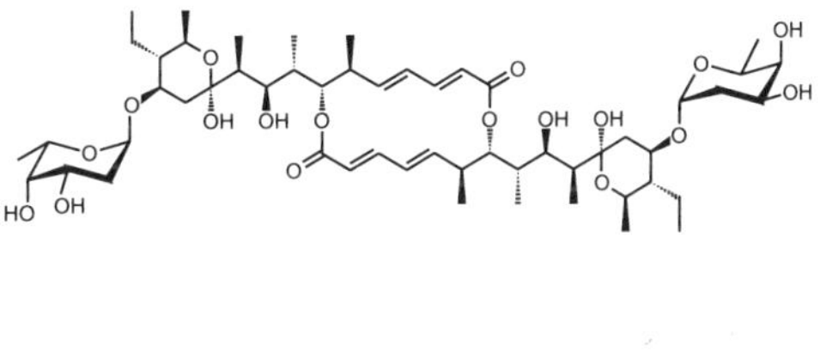

$C_{54}H_{88}O_{18}$, M_R 1025.28, colorless needles, mp. 210–212 °C (decomp.), $[\alpha]_D$ −48° (CH_3OH), C_2-symmetrical macrodiolide with the side chain folded to a cyclic hemiketal (group name: *plecomacrolides*, see macrolides) and glycosylated with 2-deoxy-L-fucose. E. is formed by various *streptomycetes (e. g., *Streptomyces melanosporofacien*), the C-atom skeleton of the aglycone (*elaiolide*, $C_{42}H_{68}O_{12}$, M_R 764.99, oil) is formed by the *polyketide pathway. E. is active against Gram-positive bacteria and protozoa. E. derivatives exhibit anthelmintic and growth-promoting activity in ruminants.

Lit.: Helv. Chim. Acta **64**, 407–424 (1981); **65**, 262–267 (1982) ▪ J. Org. Chem. **57**, 4030 (1992) (biosynthesis). – *Derivatives:* J. Antibiot. **43**, 1431 (1990) ▪ J. Org. Chem. **58**, 5487 (1993). – *Synthesis:* Chem. Pharm. Bull. **38**, 2435 (1990) ▪ J. Org. Chem. **56**, 6530 (1991); **62**, 454 (1997) (elaiolide) ▪ Synform **4**, 289–304 (1986) (review) ▪ Tetrahedron Lett. **38**, 695 (1997) (core structure). – *[CAS 37318-06-2]*

α-Elaterin. Synonym for *cucurbitacins.

Elatol [(+)-elatol].

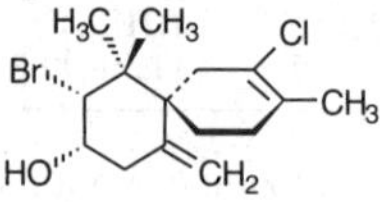

$C_{15}H_{22}BrClO$, M_R 333.70, oil, $[\alpha]_D^{25}$ +75.4° ($CHCl_3$); sesquiterpene with chamigrane skeleton from the red algae *Laurencia elata* (Australia)[1] and *L. obtusa* (Canary Islands, Jamaica)[2]. (−)-E. is found in some Caribbean algae[3]. E. has cytotoxic properties.

Lit.: [1]Tetrahedron Lett. **1974**, 3487. [2]Tetrahedron Lett. **1976**, 3051. [3]Phytochemistry **26**, 1053 (1987).
gen.: J. Nat. Prod. **51**, 1302 (1988); **55**, 1561 (1992). – *[CAS 55303-97-4]*

Elemenes. *Sesquiterpenes from elemi, Java citronella oil, *Kadsura japonica* and *Dysoxylon frazeranum*.

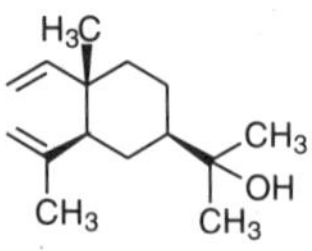

Elemol

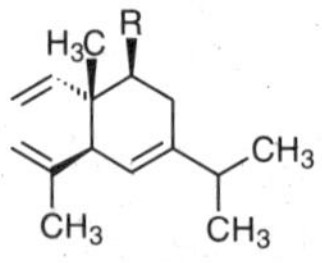

R = H : δ-Elemene
R = OH : δ-Elemen-9-ol

Table: Data of Elemenes.

	Elemol	δ-Elemene	δ-Elemen-9-ol
molecular formula	$C_{15}H_{26}O$	$C_{15}H_{24}$	$C_{15}H_{24}O$
M_R	222.37	204.36	220.35
mp./(bp.) [°C]	53 (cryst.)	(107/1.33 kPa) (oil)	oil
optical activity	$[\alpha]_D^{20}$ −5.82° ($CHCl_3$)	$[\alpha]_D$ −2.88° ($CHCl_3$)	$[\alpha]_D^{24}$ −13.1° (CH_3OH)
CAS	639-99-6	20307-84-0	20482-29-5

Particularly important E. are elemol[1], δ-elemene[2], and δ-elemen-9-ol[3]. Elemol-containing fractions of elemi oil serve as fixatives with a peppery-balsamic odor.

Lit.: [1]Tetrahedron **20**, 2647 (1964). [2]Aust. J. Chem. **17**, 1270 (1964). [3]Tetrahedron Lett. **1968**, 2899; **30**, 685 (1989).
gen.: ApSimon **2**, 265–276 ▪ J. Am. Chem. Soc. **119**, 7165 (1997) (synthesis α-E.) ▪ J. Ind. Chem. Soc. **75**, 601 (1998) (synthesis α-E.) ▪ Scheuer I **1**, 163; **2**, 280.

Elemicin see safrole.

Elemi oil/resin. E. oil is a colorless to light yellow oil, E. resin a golden-yellow to light brown, viscous mass. ***Production:*** The oil by steam distillation, the resin by solvent extraction (e. g., with toluene) from Manila elemi resins, an exudate from the tree species *Canarium luzonicum* (Burseraceae) growing in the Philippines. The odor is fresh, herby-peppery, citrus-like with a woody-balsamic after odor. The odor of the resin is less intense than that of the oil.
Composition[1]***:*** The resin consists of ca. 20–30% essential oil and the rest as resinous components. The oil contains ca. 15–25% *(+)-α-phellandrene*, 50–65% *(+)-limonene* (see *p*-menthadienes), 7–14% *(−)-elemol* (see elemenes), and ca. 1–5% *elemicin* (see safrole).
Use: Mainly in the production of perfumes; the oil for perfumes with a fresh-heady note, the resin on account of its fixative properties.

Lit.: [1]Perfum. Flavor. **5** (1), 58 (1980); **9** (3), 38 (1984); Planta Med. **1986**, 305; Flavour Fragr. J. **8**, 35 (1993).
gen.: Arctander, p. 221 ▪ Bauer et al. (2.), p. 153 ▪ Gildemeister **5**, 673 ▪ H & R, p. 164. – *Toxicology:* Food Cosmet. Toxicol. **14**, 755 (1976). – *[HS 3301 29; CAS 8023-89-0 (oil); 9000-74-2 (resin)]*

Elemol see elemenes.

Eleo... see elaeo...

Elephant pheromone. Female elephants, e.g., the Asian elephant (*Elephas maximus*), use *(Z)-7-dodecenyl acetate* ($C_{14}H_{26}O_2$, M_R 226.36, oil) as a pheromone to indicate their readiness for mating. The same compound is also used by more than 126 species of insects, especially by lepidoptera (butterflies), in their pheromone mixtures for attracting males[1].

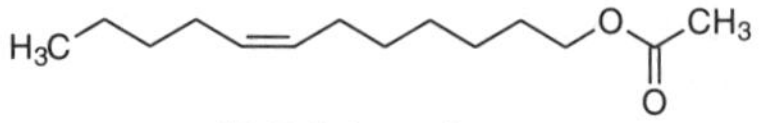

(*Z*)-7-dodecenyl acetate

Lit.: [1]Nature (London) **379**, 684 (1998). – *[CAS 14959-86-5]*

Eleuthosides. Marine natural products from soft corals (*Eleutherobia albiflora*) which exhibit strong cytotoxic and anticancer properties with a *taxol®-like mechanism of action (tubulin polymerization and microtubule stabilization)[1]. Most active is *eleutherobin*, first isolated in 1994: $C_{35}H_{48}N_2O_{10}$, M_R 656.76, $[\alpha]_D^{20}$ −67° (CH_3OH), oil. Closely related are E. A and B from

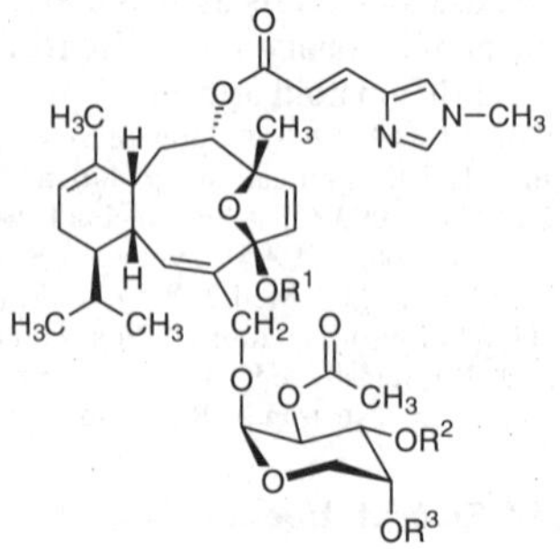

R[1] = CH₃, R[2] = H, R[3] = H : eleutherobin
R[1] = H, R[2] = CO–CH₃, R[3] = H : E. A
R[1] = H, R[2] = H, R[3] = CO–CH₃ : E. B

Eleutherobia aurea. Also related are the *sarcodictyins and valdivone A:

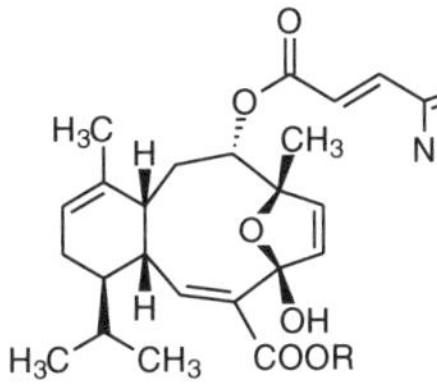

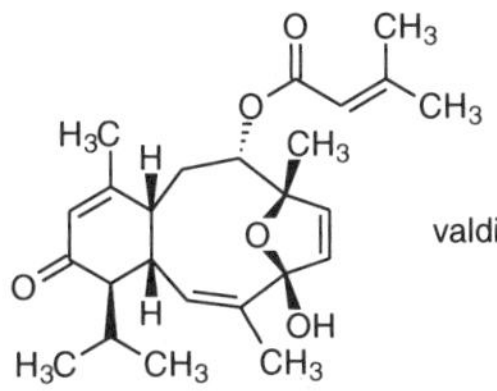

Lit.: [1] Cancer Res. **58**, 1111 (1998); J. Am. Chem. Soc. **119**, 8744 (1997).
gen.: Angew. Chem. Int. Ed. Engl. **36**, 2520 (1997); **37**, 789 (1998) ▪ Chem. Eng. News **75** (47), 64 (Nov. 24, 1997) ▪ J. Am. Chem. Soc. **119**, 11 353 (1997); **120**, 8674–8680 (1998); **121**, 6563 (1999) (synthesis eleutherobin and E.) ▪ J. Nat. Prod. **59**, 873 (1996) (isolation E.) ▪ Nach. Chem. Tech. Lab. **47**, 1228–1233 (1999) (review). – *[CAS 174545-76-7 (eleutherobin); 180692-76-6 (E. A); 180692-77-7 (E. B)]*

Elfamycins see kirromycin.

Elicitors see phytoalexins.

Ellagic acid (2,3,7,8-tetrahydroxy[1]benzopyrano-[5,4,3-*cde*][1]benzopyran-5,10-dione).

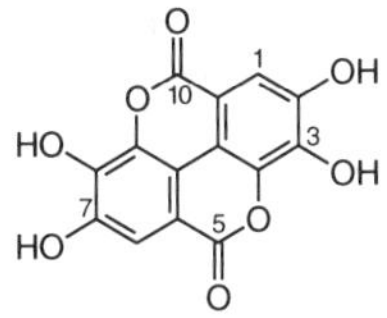

$C_{14}H_6O_8$, M_R 302.20, needles, mp. >360 °C. E. is widely distributed in free form and as glycosides and/or methyl ethers in dicotyledons in the plant kingdom, especially in gallnuts, leaf gall, also in leaves of Eucalyptus species (*Eucalyptus* spp.), oak trees (e.g., European oak, *Quercus robur*), and spurges (*Euphorbia*). E. is a component of many *tanning agents, ellagitannins. The methyl derivatives *nasutin B* [(2,3,7-tri-*O*-methyl-E.), $C_{17}H_{12}O_8$, M_R 344.28, mp. 288–290 °C (decomp.)] and *nasutin C* [(2,7-di-*O*-methyl-E.), $C_{16}H_{10}O_8$, M_R 330.25, mp. 336–338 °C (decomp.)] occur in the hemolymph of the termite *Nasutitermes exitiosus* and also in plants, *Sonneratia apetala, Tamarix gallica, T. nilotica* (Tamaricaceae), etc. 3,8-Dideoxy-E. ($C_{14}H_6O_6$, M_R 270.20, yellow cryst., subl. with partial decomposition at 300 °C), also known as *nasutin A*, also occurs in the hemolymph of *N. exitiosus* and has been isolated from castoreum (dried scent glands of the Canadian beaver).
Activity: E. inhibits the mutagenicity of benzpyrene and other aromatic hydrocarbons [1], as well as aflatoxins, nitropyrene, and nitrosourea derivatives. E. and E.-derivatives inhibit *E. coli* DNA gyrase supercoiling with approximately the same potency as nalidixic acid (antibacterial activity) [2]. Ellagitannins from tea and redwine are attributed cancer-prophylactic properties.
Use: As mordant, intestinal astringent, hemostyptic, antioxidant in pork and fish products [3]; ellagitannins and E. contribute to the taste of brandy and cognac that have matured in oak barrels [4].
For synthesis, see *Lit.* [5,6]. The biosynthesis proceeds through oxidative dimerization of gallic acid and lactone formation.
Lit.: [1] ACS Symp. Ser. **546**, 294–302 (1994); Proc. Natl. Acad. Sci. USA **79**, 5513–5517 (1982). [2] Bioorg. Med. Chem. Lett. **8**, 97 (1998). [3] J. Agric. Food Chem. **40**, 17–21 (1992); J. Food Lipids **1**, 69–78 (1993); J. Food Sci. **58**, 318 ff. (1993). [4] Food Sci. Technol. (London) **25**, 350 ff. (1992); J. Agric. Food Chem. **41**, 1872–1879 (1993); J. Prakt. Chem. **341**, 159 (1999). [5] Wein-Wiss. **49**, 83 ff. (1994). [6] Phytochemistry **36**, 1253–1260 (1994).
gen.: Beilstein E V **19/7**, 108 ▪ Chem. Pharm. Bull. **45**, 1751 (1997) ▪ Karrer, No. 1143 ▪ Merck-Index (12.), No. 3588 ▪ Phytochemistry **26**, **29**, 251–256 (1990) (glycosides); **36**, 793–798 (1994). – *Pharmacology:* Carcinogenesis (London) **14**, 1321 ff. (1993) ▪ Mutat. Res. **308**, 191–203 (1994); **322**, 97–110 (1994) ▪ Nutr. Cancer **18**, 181–189 (1992). – *[HS 2932 29; CAS 476-66-4 (E.); 5145-53-9 (nasutin B); 3374-77-4 (nasutin C); 71540-38-5 (nasutin A)]*

Ellipticine (5,11-dimethyl-6*H*-pyrido[4,3-*b*]carbazole).

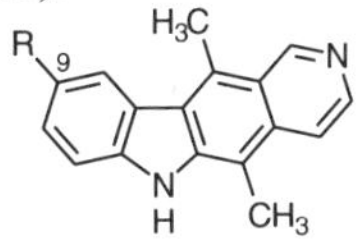

$C_{17}H_{14}N_2$, M_R 246.31, bright yellow needles (CH_3OH), mp. 311–315 °C (decomp.). E. is an *indole alkaloid from Ranunculaceae and Apocynaceae, e.g., *Aspidosperma olivaceum*, it also occurs in *in vitro* callus cultures of *Ochrosia elliptica*. E. exists in nature together with its 9-methoxy and/or 9-hydroxy derivatives. E. has mutagenic and clastogenic activity (chromosome breakage) [1] and is toxic, LD_{50} (mouse p.o.) 200 mg/kg, (i.v.) 20 mg/kg. Putative mechanisms for the cytotoxic action include: DNA-intercalation and inhibition of DNA-dependent RNA-polymerase, formation of complexes with DNA/topoisomerase and resultant DNA-double strand breakages. Endogenous formation of 9-hydroxy-E. involves cytochrome P 450-dependent enzymes. The advantage of E. over other cytostatic agents is its low cardiotoxicity [2], as yet, however, only one synthetic preparation is on the market (Celyptium) [3]. Elliptinium and Datelliptium have been used for treatment of advanced breast cancer. For the numerous syntheses, see *Lit.* [4]; details of the biosynthesis are still unknown.
Lit.: [1] Environ. Mutagenesis **9**, 161 (1987); Exp. Opin. Ther. Patents **6**, 1285–1294 (1996). [2] Arch. Biochem. Biophys. **259**, 1–14 (1987). [3] Sainsbury, in Wilman (ed.), The Chemistry of Antitumor Agents, p. 410–435, Glasgow: Blackie 1990; J. Org. Chem. **57**, 5878–5899 (1992). [4] Heterocycles **53**, 11 (2000); Tetrahedron **56**, 193–207 (2000); Tetrahedron Lett. **30**, 297 (1989); **31**, 1081 (1990).
gen.: Beilstein E V **23/9**, 417 ▪ Manske **25**, 116; **39**, 239–352 ▪ Sax (8.), EAI 850, HKH 000 ▪ Synlett **1998**, 157. – *[HS 2939 90; CAS 519-23-3]*

Elliptone see rotenoids.

Elloramycins see tetracenomycins.

Elsinochromes. Red perylenequinone pigments from cultures of the ascomycetes *Elsinoë annonae* and *Sphaceloma randii*, e.g., *E. A* (phycarone, $C_{30}H_{24}O_{10}$, M_R 544.51, dark red cryst., mp. 255 °C) and *E. D* ($C_{30}H_{26}O_{10}$, M_R 546.53, orange cryst., mp. 159–161 °C).

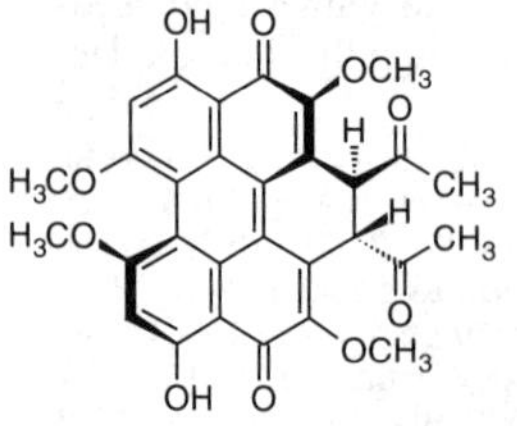

E.A

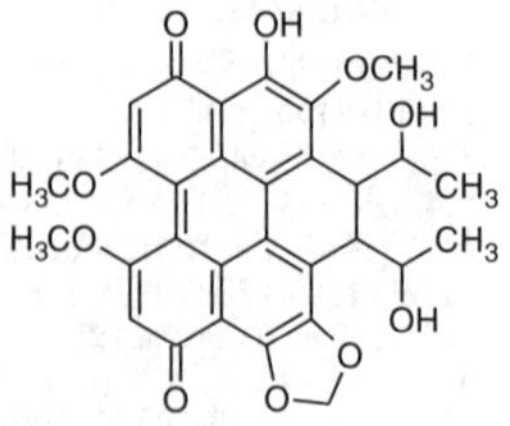

E.D

Like *hypericin the E. exhibit photodynamic activity and inhibit protein kinase C. The E. exist in solution as tautomeric mixtures, some E. can be separated into diastereomeric pairs [1]. On account of their helical chirality, some E. exhibit very high optical rotation.

Lit.: [1] Gazz. Chim. Ital. **123**, 131–136 (1993).
gen.: Acta Crystallogr., Sect. C **45**, 628 (1989); **46**, 267 (1990) (structure) ▪ Can. J. Chem. **59**, 422 (1981) (biosynthesis). – *[CAS 24568-67-0 (E. A); 32500-05-3 (E. D)]*

Emerin.

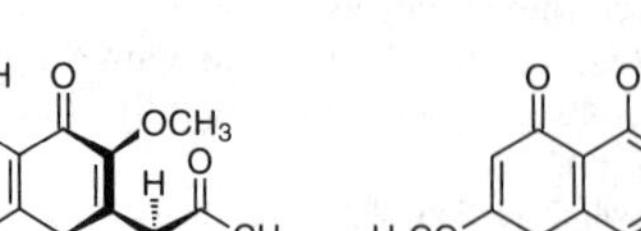

$C_{20}H_{16}N_2O_2$, M_R 316.36, greenish-yellow cryst., mp. 225–226 °C. Dinitrile from cultures of *Aspergillus nidulans* with structural similarity to diisonitriles of the *xanthocillin type.

Lit.: Agric. Biol. Chem. **39**, 2181 (1975). – *[CAS 40581-18-8]*

Emestrin (mycotoxin EQ-1).

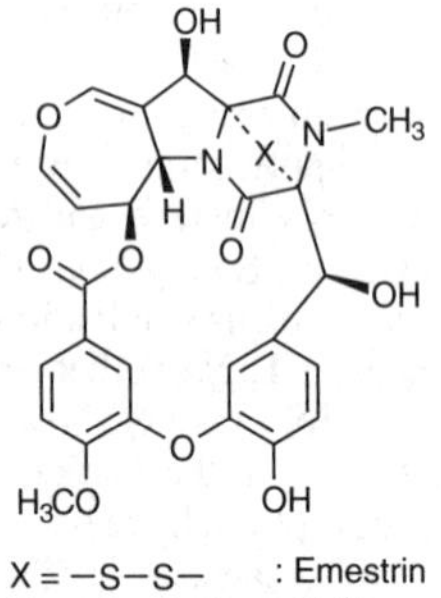

X = –S–S– : Emestrin
X = –S–S–S– : E. B

$C_{27}H_{22}N_2O_{10}S_2$, M_R 598.60, cryst., mp. 233–236 °C (decomp.), $[\alpha]_D$ +184° ($CHCl_3$). E. is isolated from cultures of ascomycetes of the genus *Emericella*. The epidithiodiketopiperazine E. exhibits a strong antifungal activity. The analogous epitrithiodiketopiperazine *E. B* [$C_{27}H_{22}N_2O_{10}S_3$, M_R 630.66, cryst., mp. 230–238 °C (decomp.)] is isolated from *Emericella striata* [1].

Lit.: [1] Chem. Pharm. Bull. **35**, 3460 (1987).
gen.: J. Chem. Soc., Perkin Trans. 1 **1986**, 109 ▪ Phytochemistry **46**, 123 (1997) (NMR). – *[CAS 97816-62-1 (E.); 107395-35-7 (E. B)]*

Emetine see ipecac alkaloids.

Emodin (1,3,8-trihydroxy-6-methylanthraquinone).

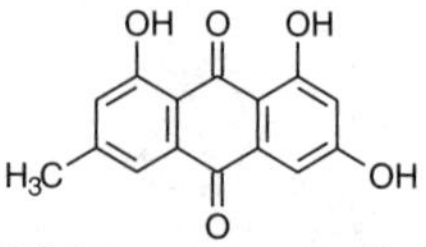

$C_{15}H_{10}O_5$, M_R 270.24, orange or yellowish-brown needles, mp. 259–260 °C, uv_{max} 436 (lg ε 3.94, C_2H_5OH); insoluble in water, soluble in organic solvents. E. is a widely distributed pigment not only in plants but also in higher fungi [1] and lichen. E. occurs in the free form, in the form of dianthrones, and glycosidically bound in roots of rhubarb and sorrel as well as in buckthorn bark (*frangulins) and is responsible for the laxative action of plant drugs prepared therefrom.

Biosynthesis: In the polyketide pathway (*acetyl-CoA and 7 molecules malonyl-CoA) an initial, non-aromatic precursor is formed, *endocrocin, from which E. is formed by decarboxylation [2,3]; E. represents the starting product for *chrysophanol, the *Dermocybe pigments, and the E.-dimer, *hypericin.

Lit.: [1] Zechmeister **51**, 125–135. [2] Acta Chem. Scand. **12**, 1211 (1958). [3] J. Chem. Soc., Chem. Commun. **1972**, 102.
gen.: Beilstein E IV **8**, 3575 ▪ Hager (5.) **5**, 145 f.; **6**, 392 ff. ▪ Merck-Index (12.), No. 3602. – *[HS 2914 69; CAS 518-82-1]*

Endiandric acids.

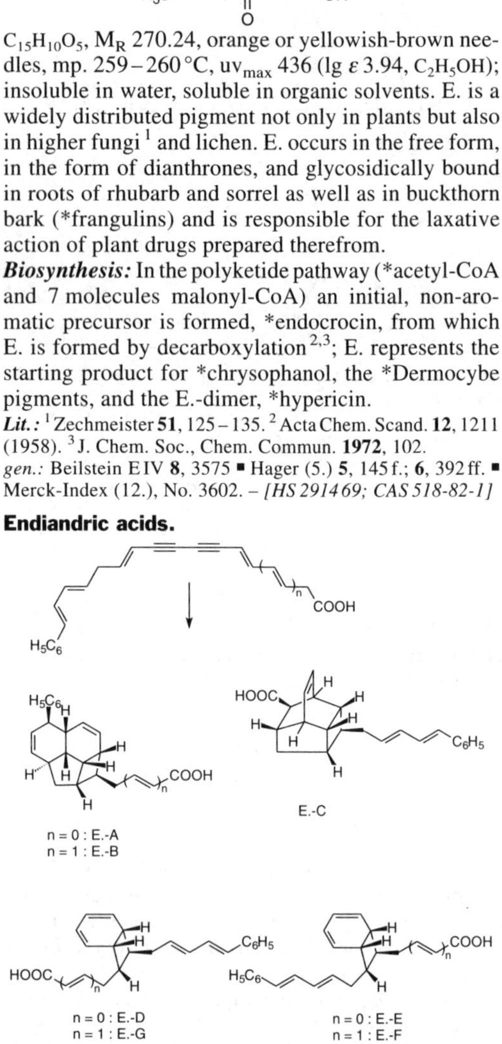

Group of polycyclic compounds from the leaves of *Endiandra introrsa* (Lauraceae), a large tree of the Australian rain forests, belonging to the Lauraceae (laurels).

Table: Data of Endiandric acids.

(±)-E.	molecular formula	M_R	mp. [°C]	CAS
A*	$C_{21}H_{22}O_2$	306.40	147–149 rods	74591-03-0
B	$C_{23}H_{24}O_2$	332.44	163–165 rosettes	76060-33-8
C	$C_{23}H_{24}O_2$	332.44	125–132	76060-34-1
D	$C_{21}H_{22}O_2$	306.40		82679-68-3
E	$C_{21}H_{22}O_2$	306.40		88825-08-5
F	$C_{23}H_{24}O_2$	332.44		82808-36-4
G	$C_{23}H_{24}O_2$	332.44		82863-35-2

* pK_s 5.1

The E. possess several asymmetric centers. In nature they occur as mixtures of stereoisomers. For synthesis, see *Lit.*[1]. The biosynthesis proceeds non-enzymatically from ω-phenylpolyenynoic acids by pericyclic reactions (endiandrin cascade)[2].

Lit.: [1] J. Am. Chem. Soc. **104**, 5555–5562 (1982); Kocovský et al., Synthesis of Natural Products II, p. 269–273, Boca Raton: CRC Press 1986; Lindberg **1**, 155–174; Nicolaou, p. 265; Synlett **1997**, 493. [2] Angew. Chem. Int. Ed. Engl. **35**, 289–291 (1996).

gen.: Aust. J. Chem. **34**, 1655 (1981) (E.-A); **35**, 557–565 (1982) (E.-B); **35**, 567–579 (1982) (E.-C); **35**, 2247–2256 (1982) (E.-D); **36**, 627–632 (1983) ▪ Merck-Index (12.), No. 3611.

Endocrocin (1,6,8-trihydroxy-3-methylanthraquinone-2-carboxylic acid).

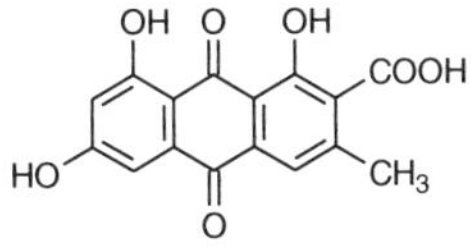

$C_{16}H_{10}O_7$, M_R 314.25, orange plates, mp. 340 °C (decomp.), uv_{max} 442 nm, pigment from *Dermocybe sanguinea* (Basidiomycetes), *D. cinnamomeolutea*, *D. cinnabarina*, *Aspergillus amstelodami*, and the ascomycete *Nephromopsis endocrocea* as well as lichens (*Cetraria* species) and *ergot. E., an octaketide formed in the polyketide pathway, is the direct precursor of the 6- and 8-methyl ethers *cinnalutein and dermolutein (*Dermocybe pigments).

Lit.: J. Chem Soc., Perkin Trans. 1 **1990**, 1159–1167 ▪ Luckner (3.), p. 177 ▪ Tetrahedron **49**, 771 (1993) ▪ Zechmeister **51**, 129. – *[CAS 481-70-9]*

Endorphins. General term derived from endogenous (= originating from within) and *morphine for analgesically active peptides from the brain of mammals and man. The E. are biogenetically closely related to the *enkephalins and dynorphins. Like these compounds, the E. block opiate receptors in the brain and spinal cord, thus preventing transmission of pain signals. The therapeutic use is hindered by the fact that they cannot pass the blood-brain barrier and, like synthetic opiates, cause addiction.

Lit.: J. Chem. Soc., Perkin Trans. 1 **1983**, 65 ▪ Kirk-Othmer (4.) **17**, 858–881 ▪ Merck-Index (12.), No. 3613 ▪ Prog. Med. Chem. **17**, 1 (1980) ▪ Science **220**, 721 (1983) ▪ Trends Pharmacol. Sci. **5**, 137 (1984) (review). – *[CAS 61214-51-5 (β-endorphin, human)]*

Endothelium-Derived Relaxing Factor see nitrogen monoxide.

Endotoxins. Name for bacterial toxins which, in contrast to the *exotoxins, are not secreted by living bacteria but are released from the cells by autolysis (e.g. in the intestines). In the case of classic E. they consist of the thermostable lipopolysaccharide (LPS-) fraction of the cell membrane anchored on the outer membranes of Gram-negative bacteria. The LPS consists of *lipid A, the chain polysaccharide, and the O-specific chain; lipid A is responsible for the toxic action of LPS. E. are found in all Enterobacteriaceae, e.g., *Salmonella* (typhus), *Shigella* (dysentery), and many other Gram-negative pathogens. In the host organism E. stimulate mediators (cytokines) of the immune system. One of the most important subsequent reactions is the production of interleukin I and tumor necrosis factor (TNF) by macrophages. This induces an increased synthesis of prostaglandin E_2 in endothelial cells of the hypothalamus resulting in an increase in the temperature of the organism. In humans merely 1–2 μg/kg i.v. LPS induce fever. Besides fever, E. promote numerous other pathophysiological effects. On liberation of larger amounts of E., an irreversible E. shock may occur. In contrast, orally taken LPS does not cause any symptoms even in higher concentrations.

In the wider sense, protein toxins localized in the cytoplasm of Gram-negative bacteria are also designated as E., for example, toxins of *Vibrio cholerae* (*cholera toxin), the bacteriocins (e.g., colicins) or the polypeptides γ-E. formed by *Bacillus thuringiensis* which represent up to 30% of the cell weight and are used in plant protection as antifeedants against insect pests.

Lit.: Agarwal, Bacterial Endotoxins and Host Response, Amsterdam: Elsevier 1980.

Enediyne antibiotics. Family of DNA-cleaving secondary metabolites (e.g., *calicheamicins, *esperamycins, *dynemicins from *actinomycetes and *namenamicin from the marine ascidian *Polysyncraton lithostrotum*). The antitumor activity against leukemia and solid tumors occurs at a concentration about 1000-fold lower than that of *adriamycin. After a starting step the conjugated multiple bond system (enediyne) undergoes a cycloaromatization (Bergman cyclization) with formation of radicals (E. often have a blue color) which effect cleavage of DNA strands (see formula). The labile chromophore is stabilized by a protein in some cases (e.g., *neocarzinostatins, kedarcidin, macromomycin). C-1027 is undergoing clinical development[1]. The most recently discovered E. is N 1999 A_2[2].

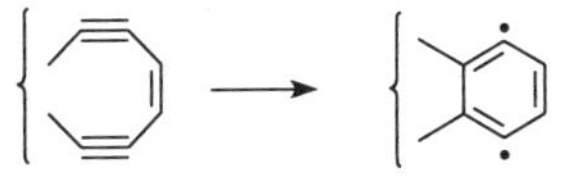

Lit.: [1] Chem. Eng. News (13.3.) **2000**, 47–50. [2] Tetrahedron Lett. **39**, 6495 (1998).

gen.: Angew. Chem. Int. Ed. Engl. **30**, 1387–1416 (1991); **32**, 842–845 (1993) ▪ Antimicrob. Agents Chemother. **44**, 382 (2000) (synthesis) ▪ Barton-Nakanishi **1**, 557–572 (biosynthesis) ▪ Borders & Doyle (eds.), Enediyne Antibiotics as Antitumor Agents, New York: Marcel Dekker 1995 ▪ Contemp. Org. Synth. **3**, 41–63, 93–124 (1996) (synthesis) ▪ J. Am. Chem. Soc. **121**, 9881 (1999); **122**, 939, 1556 (2000) (biosynthesis) ▪ J. Chem. Soc., Chem. Commun. **1998**, 483 f. (mechanism) ▪ J. Med. Chem. **39**, 2103–2117 (1996) (review) ▪ J. Org. Chem. **61**, 16 (1996) (review) ▪ Krohn et al., Antibiotics and Antiviral Compounds, p. 279–326, Weinheim: VCH Verlagsges. 1993 ▪ Lukacs, Recent Progress in the Chemical Synthesis of Antibiotics and Related Products, p. 213–330, Berlin: Springer 1993 ▪ Nature (London) **361**, 21 (1993) ▪ Hirama, Neocarcinostatin 1997, 47–82 (review neocarcinostatin) ▪ Proc. Natl. Acad. Sci. USA **90**, 5881–5888 (1993) ▪ Pure Appl. Chem. **65**, 1271–80 (1993); **69**, 525 (1997) (mechanism) ▪ Science **256**, 1172–78 (1992) ▪ Stang & Diederich (eds.), Modern Acetylene Chemistry, p. 203–283, Weinheim VCH Verlagsges. 1995 ▪ Tetrahedron **52**, 6453–6518 (1996). – *[HS 2941 90]*

Enkephalins. Term derived from the Greek: enképhalos = brain for certain oligopeptides with analgesic activity (brain peptides, neuropeptides) which were discovered in human brain tissue in 1975. Distinctions are drawn between Met[5]-and Leu[5]-E.:

H-Tyr-Gly-Gly-Phe-X-OH X = Met or Leu

Met-E. or, respectively, Leu-E. are identical with the first 5 amino acids of the later discovered polypeptidic *endorphins or, respectively, the dynorphins. Met-E. is formed from β-lipotropin (amino acids 61–65).
E. and endorphins, like *morphine and opiates, have a blocking effect on specific opiate receptors and prevent – as so-called endogenous opiates – transmission of pain signals. Numerous analogues of E. have been synthesized as potential analgesics.

Lit.: Chem. Pharm. Bull. **35**, 1044 (1987) ▪ Kirk-Othmer (4.) **17**, 858–881 ▪ Life Sci. **31**, 2249 (1982) ▪ Nature (London) **258**, 277 (1975); **295**, 202–208, 663 ff. (1982) ▪ Synth. Commun. **21**, 459 (1991). – *[CAS 58569-55-4 (Met-E.); 58822-25-6 (Leu-E.)]*

Enmein.

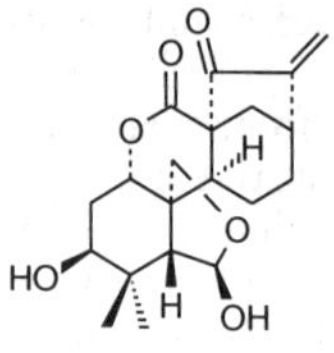

Enmein

$C_{20}H_{26}O_6$, M_R 362.42, mp. 297–299 °C decomp., $[\alpha]_D$ –131.3° (pyridine), a diterpene of the kaurane type; bitter principle from *Rabdosia trichocarpa* and *R. japonica*; shows antitumor and anti-inflammatory activity against Ehrlich ascites tumors and inhibits insect growth. E. can be transformed into numerous other diterpenes such as *abietanes, Aconitum alkaloids, *gibberellins. The biosynthesis proceeds from 16-kaurene.

Lit.: Agric. Biol. Chem. **43**, 71 (1979) ▪ Karrer, No. 3712 ▪ Zechmeister **46**, 77–157. – *[CAS 3776-39-4]*

Enniatins.

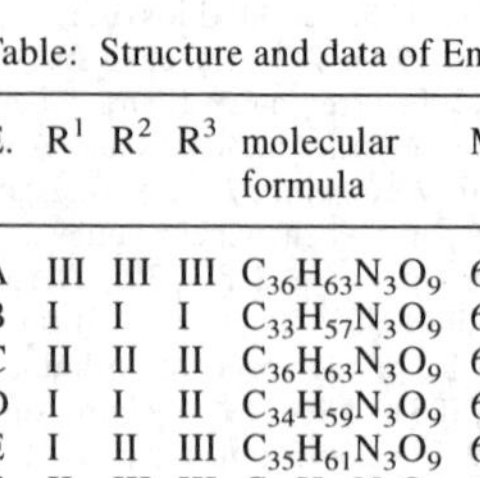

R^1, R^2, R^3 =
I : $CH(CH_3)_2$
II : $CH_2{-}CH(CH_3)_2$
III : $CH(CH_3){-}C_2H_5$ (*S*)

Table: Structure and data of Enniatins.

E.	R^1	R^2	R^3	molecular formula	M_R	$[\alpha]_D$ ($CHCl_3$)	CAS
A	III	III	III	$C_{36}H_{63}N_3O_9$	681.91	-90°	2503-13-1
B	I	I	I	$C_{33}H_{57}N_3O_9$	639.83	–107.9°	917-13-5
C	II	II	II	$C_{36}H_{63}N_3O_9$	681.91	–83°	19893-23-3
D	I	I	II	$C_{34}H_{59}N_3O_9$	653.42	–63°	19893-21-1
E	I	II	III	$C_{35}H_{61}N_3O_9$	667.44	–77°	144470-22-4
F	II	III	III	$C_{36}H_{63}N_3O_9$	681.91	–70°	144446-20-8

Ionophoric cyclic depsipeptides isolated from *Fusarium* strains (to date 12 described), alternately built up of 3 residues each of D-α-hydroxyisovaleric acid and *N*-methyl-L-amino acids. The E. exhibit insecticidal and antifungal activity. They form crypates with K^+ ions and thus accomplish their selective transport through biological membranes [1]. Since this activity is not limited to microbial membranes, their high toxicity prevents the therapeutic use of the E.; E. D to F inhibit acyl-CoA: cholesterol acyltransferase (ACAT). The E. synthetases of *Fusarium lateritium* and *F. sambucinum* have been characterized [2].

Lit.: [1] Chem. Unserer Zeit **7**, 120 (1973). [2] J. Antibiot. **45**, 1273–1277 (1992).
gen.: Appl. Microbiol. Biotechnol. **20**, 83 (1984) ▪ Can. J. Chem. **70**, 1281 (1992); **71**, 1362 (1993) ▪ J. Antibiot. **45**, 1207–1215 (1992) ▪ Met. Ions Biol. Syst. **19**, 139 (1985).

Enterobactin (enterochelin).

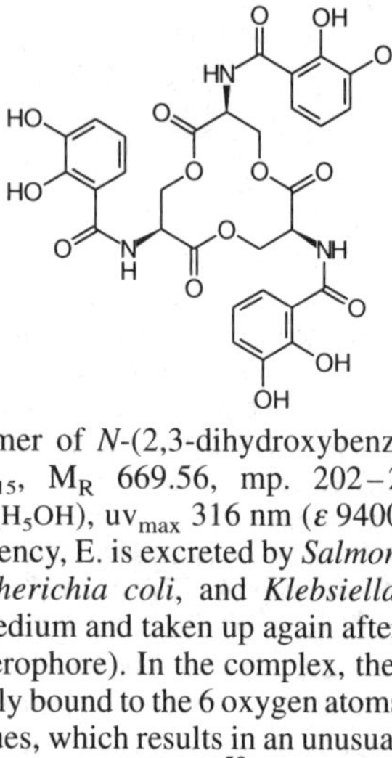

Cyclic trimer of *N*-(2,3-dihydroxybenzoyl)-L-serine, $C_{30}H_{27}N_3O_{15}$, M_R 669.56, mp. 202–203 °C, $[\alpha]_D^{25}$ +7.40° (C_2H_5OH), uv_{max} 316 nm (ε 9400). In cases of iron deficiency, E. is excreted by *Salmonella typhimurium*, *Escherichia coli*, and *Klebsiella pneumoniae* into the medium and taken up again after loading with iron (*siderophore). In the complex, the iron (Fe^{3+}) is octahedrally bound to the 6 oxygen atoms of the 3 catechol residues, which results in an unusually large complexation constant of $K = 10^{52}$. On protonation, the neutral complex $(Fe\ Ent_3)^0$ is formed from which the iron is liberated by the organism under reduction to Fe^{2+}.

Lit.: J. Am. Chem. Soc. **105**, 4617, 4623 (1983); **114**, 661 (1992) ▪ J. Chem. Soc., Chem. Commun. **1983**, 846 ▪ Tetrahedron Lett. **38**, 749 (1997). – *[CAS 28384-96-5]*

Enterocin see vulgamycin.

Enterolactone see lignans.

Enteromycin.

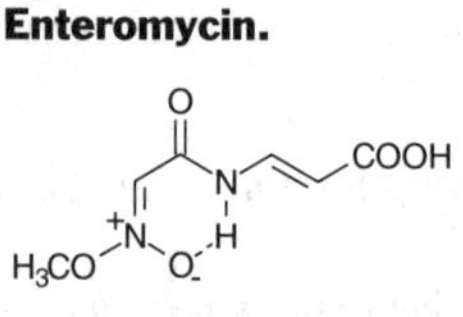

Enteromycin

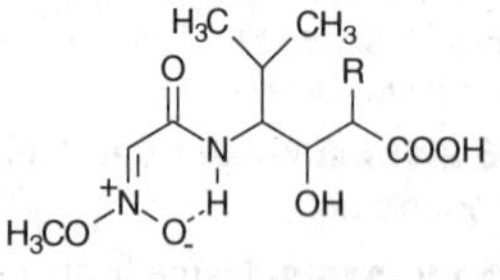

R = CH_3 : Nitracidomycin A
R = C_2H_5 : Nitracidomycin B

$C_6H_8N_2O_5$, M_R 188.14, needles or prisms, mp. 172 °C (decomp.). An *O*-methylnitronic acid derivative from cultures of *Streptomyces albireticuli* with activity against Gram-positive and -negative bacteria [1]. The *nitracidomycins* A {$C_{11}H_{20}N_2O_6$, M_R 276.29, $[\alpha]_D$ –31.9° ($CHCl_3$)} and B ($C_{12}H_{22}N_2O_6$, M_R 290.32) produced by *S. viridochromogenes* are of the same structural type [2].

Lit.: [1] Bull. Chem. Soc. Jpn. **34**, 1419, 1631 (1961); J. Org. Chem. **30**, 2330 (1965); Tetrahedron **21**, 267 (1965). [2] J. Antibiot. **42**, 329, 332 (1989). – *[HS 2941 90; CAS 3552-16-7 (E.); 103527-97-5 (nitracidomycin A); 103527-98-6 (nitracidomycin B)]*

Enterotoxins. Toxic proteins formed by bacteria with molecular masses in the range from 27000 to 30000 which are usually excreted into the medium (*exotoxins). E. can be taken up with contaminated food or be formed by the bacteria colonizing the intestinal walls. Finally, the bacteria can penetrate the intestinal walls and then start to excrete the E. Some E. are thermally very stable and survive when food is boiled. E. from *Salmonella* and *Staphylococcus* species are the most frequent causes of food poisoning. Shortly after uptake, the symptoms of nausea, vomiting, diarrhea, and circulatory complaints occur. Deaths are rare and occur only when the subject is already in a weakened state. The sites of attack by E. vary, e. g., at intestinal epithelial cells or in the vegetative nervous system. For the production of antitoxins, E. are obtained by lysis of bacterial cells or from cell-free culture filtrates. E. have been detected, e. g., in the following bacterial species: *Bacillus cereus, Clostridium perfringens, Escherichia coli, Vibrio cholerae, Staphylococcus aureus,* and *Streptococcus faecalis.*

Lit.: Nature (London) **339**, 221 (1989).

Ephedra alkaloids. Alkaloids from the shrubby joint fir (*Ephedra*, Ephedraceae) occurring in warm-temperate regions of the earth. The main alkaloid is *ephedrine. Many of the constituents of *Ephedra* (Ma-Huang drug in Chinese traditional medicine) are hypotensive.

Lit.: Tang & Eisenbrand (eds.), Chinese Drugs of Plant Origin, p. 481–490, Berlin: Springer 1992. – *[HS 2939 90]*

Ephedrine. Formula, data, and occurrence, see *β*-phenylethylamine alkaloids. E. is an oral sympathicomimetic with a weaker but longer-lasting action than adrenaline. Apart from *Ephedra* species, E. is also found in monk's hood, yew trees, and other plants. It has hypertensive, cardiotonic, bronchial dilating, and appetite reducing activities, and is thus used in drugs for hypotension, chronic bronchitis, asthma attacks, and as a mucosal decongestant for colds, as well as in appetite reducing agents. On repeated use, however the activity can decrease (tachyphylaxia). The (1*S*,2*R*)-enantiomer exhibits about 1/3 of the pharmacological activity of the natural (1*R*,2*S*)-form. The two diastereomers of E. and its synthetic enantiomer are known as pseudoephedrines.

Lit.: Beilstein E IV **13**, 1879 ▪ Florey **8**, 489–507; **15**, 233–281 ▪ Hager (5.) **5**, 46 ff.; **8**, 39 ff; **9**, 439 ff. ▪ Merck-Index (12.), No. 3645 ▪ see phenylethylamine alkaloids. – *[HS 2939 40]*

Epibatidine {(1*R*)-*exo*-2-(6-chloro-3-pyridyl)-7-azabicyclo[2.2.1]heptane}. $C_{11}H_{13}ClN_2$, M_R 208.69, solid, mp. 63 °C, $[\alpha]_D$ −5° ($CHCl_3$); hydrochloride: cryst.,

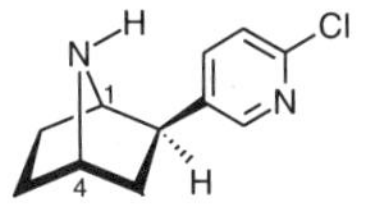

mp. 217 °C, $[\alpha]_D$ −42.9° (CH_3OH). Unusual chloropyridine alkaloid from the skin of the Ecuadorian arrow poison frog *Epipedobates tricolor* (1 mg from 750 frogs)[1]. The absolute configuration was demonstrated by comparison with the synthetic compound[2]. E. is an extremely potent *nicotine receptor agonist and exhibits noteworthy analgesic effects. Thus, in the hot-plate analgesic test and by induction of the tail-bristling reaction in mice it proved to be 200–500 times more active than *morphine, but does not bind to opioid receptors. The activity is thus not antagonized by naloxone[1]. Caution is required in handling E.: even scratching it off from TLC plates can cause irritation of nasal mucosa and spasms. On account of its unusual structure and interesting pharmacological activities E., together with the *epothilones, is currently one of the most frequently synthesized natural products[2,3]. E. is a lead structure for the synthesis of novel analgesics[4]. The E.-analogue ABT-594 is under clinical development (phase III) as analgesic with a new mechanism of action[5].

Lit.: [1]Bioorg. Med. Chem. Lett. **6**, 279 (1996); J. Am. Chem. Soc. **114**, 3475 (1992); Science **261**, 1117 (1993). [2]J. Org. Chem. **59**, 1771 (1994); **63**, 760, 8397 (1998); J. Chem. Soc., Perkin Trans. 1 **1998**, 3689; J. Prakt. Chem. **337**, 167–174 (1995). [3]J. Med. Chem. **40**, 2293 (1997); **41**, 674 (1998); J. Org. Chem. **58**, 5600 (1993); **61**, 4600, 7189 (1996); **64**, 836 (1999); J. Chem. Soc., Chem. Commun. **1993**, 1216; **1994**, 1775; **1997**, 1857; **1998**, 2251; J. Chem. Soc. Perkin Trans. 1 **2000**, 329–343; Synthesis **1998**, 1335; Tetrahedron **52**, 11053 (1996); **53**, 17177 (1997); Tetrahedron Lett. **34**, 3251, 4477, 7493 (1993); **35**, 3171, 5417, 7439, 9297 (1994); **37**, 1463, 7485 (1996); **39**, 1023, 2059, 4513 (1998); **40**, 557 (1999); Synlett. **1994**, 343; **1997**, 589. [4]Chem. Pharm. Bull. **43**, 901 (1995); **47**, 1501 (1999); J. Org. Chem. **64**, 8968 (1999); J. Prakt. Chem. **341**, 506 (1999). [5]Bioorg. Med. Chem. **6**, 2103 (1998); J. Med. Chem. **41**, 407 (1998); Pharm. Unserer Zeit **27**, 289 f. (1998). – *[CAS 140111-52-0]*

Epidermin (staphyloccin). $C_{98}H_{141}N_{25}O_{23}S_4$, M_R 2165.59, together with *gallidermin* {[Leu[6]]epidermin, $C_{98}H_{141}N_{25}O_{23}S_4$, M_R 2165.59, basic amphiphilic polypeptide} and *Pep 5* ($C_{153}H_{256}N_{48}O_{39}S_3$, M_R 3488.19) member of the *lantibiotics*. These are polycyclic peptide antibiotics that contain thioether amino acids such as *lanthionine and *α,β*-unsaturated amino acids. E. is produced by *Staphylococcus epidermidis* originating from isolates of human skin. Gallidermin is produced by *S. gallinarum* isolated from cockscombs. Both are active against Gram-positive pathogens and against the

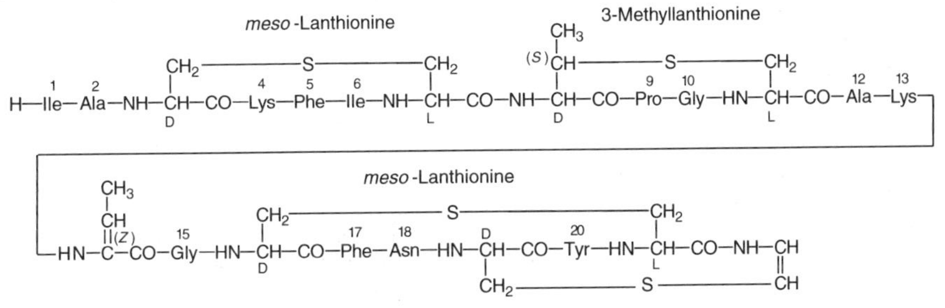

pathogen responsible for acne disease, *Propionibacterium acnes* (MIC 0.25 µg/mL), as well as streptococci, e. g., *Streptococcus pyogenes* (MIC 1 µg/mL). Gallidermin has higher activity than epidermin. Since the producing organisms are found on human skin it is assumed that they exert a protective effect against pathogens. Only little is known about the site and mechanism of action. The bactericidal activity can probably be attributed to a depolarization of the energetically excited bacterial cytoplasmic membrane and induction of autolysis.

Biosynthesis: E., like gallidermin and Pep 5, belongs to the few antibiotics that are formed from ribosomally synthesized precursors by posttranslational enzymatic modification.

Lit.: Angew. Chem., Int. Ed. Engl. **28**, 616–619 (1989) (Pep 5); **35**, 2104–2107 (1996) (biosynthesis) ▪ Angew. Chem. Int. Ed. Engl. **24**, 1051 (1985) (epidermin); **30**, 1051 (1991) ▪ Eur. J. Biochem. **177**, 53–59 (1988) (gallidermin). – *[HS 2941 90; CAS 99165-17-0 (E.); 117978-77-5 (gallidermin); 84931-86-2 (Pep 5)]*

Epiderstatin.

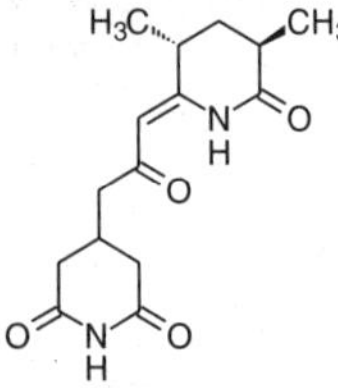

$C_{15}H_{20}N_2O_4$, M_R 292.33, powder, mp. 185–187 °C, $[\alpha]_D^{21}$ +5.3° (CH_3OH). E. is a glutarimide antibiotic from *Streptomyces pulveraceus* and an inhibitor of mitosis.

Lit.: J. Antibiot. **42**, 1599, 1607 (1989); **45**, 1963 (1992); **48**, 1176 (1995) (synthesis) ▪ Nat. Prod. Lett. **1**, 149 (1992) ▪ Tetrahedron Lett. **33**, 309 (1992). – *[CAS 126602-16-2]*

Epigallocatechin 3-gallate (EGCG; Teatannin II).

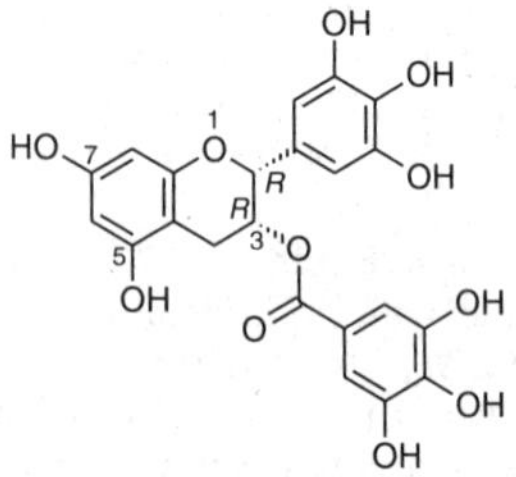

One cup of green tea contains up to 150 mg of EGCG. This polyphenol [$C_{22}H_{18}O_{11}$, M_R 458.37, cryst., mp. 217 °C and 245–247 °C, $[\alpha]_D$ –179° (C_2H_5OH)] from *Thea sinensis* (syn. *Camellia sinensis*), *Mallotus japonicus*, etc. occurs in green, but not in black tea and has been found to be responsible for the anticancer property of green tea[1]. It also shows anti-HIV activity[2]. E. inhibits the formation of thymine dimers in mouse epidermal DNA, thus protecting mice from mutagenic UV-induced lesions associated with the induction of skin cancer.

Lit.: [1]Nature (London) **387**, 561 (1997). [2]Biochemistry **29**, 2841 (1990).

gen.: Chem. Eng. News (12.4.) **1999**, 51 ▪ J. Am. Chem. Soc. **121**, 12073 (1999) ▪ Phytochemistry **27**, 579 (1988); **31**, 2117 (1992) ▪ Prev. Med. **21**, 503 (1992). – *[CAS 989-51-5]*

Epilachnene.

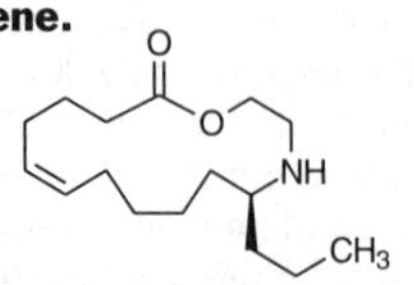

(*S*)-E.

The larva of the Mexican bean seed beetle *Epilachna varivestis* excretes a protective secretion from its glandular hair. The active principle is the azamacrolide E.[1]: $C_{16}H_{29}NO_2$, M_R 267.41, oil. The biosynthesis proceeds from oleic acid and L-serine[2]. For absolute configuration (*S*) see *Lit.*[3].

Lit.: [1]Proc. Natl. Acad. Sci. USA **90**, 5204–5208 (1993). [2]Proc. Natl. Acad. Sci. USA **91**, 12790–12793 (1994); Tetrahedron **55**, 955 (1999). [3]Tetrahedron Lett. **37**, 8613 (1996); **38**, 2787 (1997). – *[CAS 147363-82-4]*

22,26-Epimino(epoxy)cholestane see Solanum steroid alkaloids.

(*R*)-Epinephrine see (*R*)-adrenaline.

Epistephanine see bisbenzylisoquinoline alkaloids.

Epothilones.

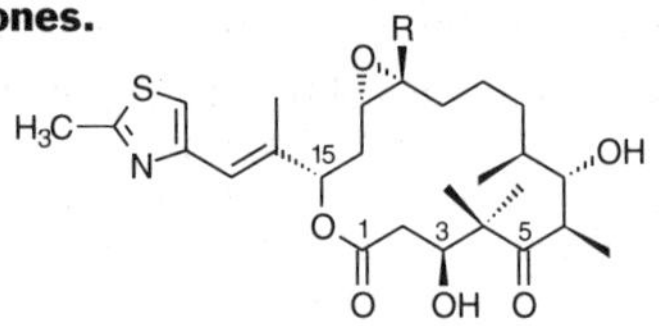

R = H : E.A
R = CH_3 : E.B

Homologous 16-membered macrolides from *Sorangium cellulosum* (Myxobacteria) with antifungal and cytostatic activity, *E. A*[1]: $C_{26}H_{39}NO_6S$, M_R 493.66, $[\alpha]_D^{21}$ –47.1° (CH_3OH), mp. 95 °C; *E. B*[2]: $C_{27}H_{41}NO_6S$, M_R 507.69, $[\alpha]_D^{21}$ –35.0° (CH_3OH), mp. 93–94 °C. E. inhibit growth of a broad spectrum of tumor cell lines with emphasis on breast and large bowel cell lines, IC_{50} (mouse fibroblast line L929) 2 ng/mL. The activity, like that of *taxol®, is based on an inhibition of cell division by stabilization of the microtubuli and apoptosis; on account of their higher affinity, the E. displace taxol from the microtubuli[3]. First syntheses of analogues have been described[4]. The discovery of E. has triggered extremely high synthetic efforts, reflected by the literature below. For sequencing of E. genes, see *Lit.*[5].

Lit.: [1]J. Am. Chem. Soc. **119**, 7960–7973, 7974–7991 (1997); Proc. Natl. Acad. Sci. U. S. A. **95**, 14603 (1998) (antibody catalysis). [2]Angew. Chem. Int. Ed. Engl. **36**, 757, 2097 (1997); **37**, 2675 (1998); Chem. Eur. J. **5**, 2483, 2492 (1999); J. Chem. Soc., Chem. Commun. **1997**, 2343; **1998**, 1597; **1999**, 519; J. Org. Chem. **62**, 454 (1997); **64**, 684, 7224 (1999); Synlett **1998**, 861. [3]Angew. Chem. Int. Ed. Engl. **36**, 2093–2096 (1997); **37**, 2014–2045 (1998) (review). [4]Angew. Chem. Int. Ed. Engl. **37**, 81, 84 (1998); **39**, 209 (2000); Chem. Biol. **5**, 365 (1998); Chem. Eur. J. **3**, 1957–1970, 1971–1986 (1997); J. Am. Chem. Soc. **119**, 2733 (1997); Synthesis **1999**, 1469; Tetrahedron **54**, 7127–7166 (1998); **55**, 8199 (1999). [5]Chem. Biol. **7**, 97 (2000).

gen.: Angew. Chem. Int. Ed. Engl. **35**, 2801 (1996) (synthesis E. A); **36**, 166, 523, 525 (1997) (synthesis E. A, B); **38**, 1971 (1999); **39**, 209, 581 (2000) ▪ Bioorg. Med. Chem. Lett. **6**, 893–898 (1996) (wrong structure) ▪ Cancer Res. **55**, 2325–2333 (1995) ▪ Chemtracts **11**, 678–696 (1998) (review) ▪ J. Am. Chem. Soc. **119**, 7960, 7974, 10073–10092 (1997) ▪ J. Antibiot. **49**, 560–563 (1996) ▪ Nature (London) **387**, 268 (1997); **390**, 100 (1997) ▪ Tetrahedron Lett. **39**, 8633 (1998) (synthesis). – *[CAS 152044-53-6 (E. A); 152044-54-7 (E. B)]*

(2S,3S)-2,3-Epoxyneral see mites.

Epoxy polyenes. Group of sexual attractants of, especially, the female looper (Geometridae), but also of other butterflies such as, e. g., tiger moths. The compounds are optically active and contain, besides an epoxy ring, 1–3 homoconjugated double bonds in an unbranched carbon chain with 17, 19, 21, or 23 carbon atoms. The E. are formed biogenetically from unsaturated fatty acids through decarboxylation and regio- as well as enantioselective oxidation. The absolute configuration of the natural product is decisive for its biological activity [1,2]. The corresponding non-oxygenated polyenes often occur concomitantly as less active, accompanying *pheromones. For some species, however, these unsaturated hydrocarbons are the sole sexual attractant. Higher oxygenated E. are found in brown algae but their function is unknown [3].

Lit.: [1] J. Chem. Ecol. **19**, 2721–2735 (1993). [2] Z. Naturforsch. C **49**, 516–521 (1994). [3] Nat. Prod. Rep. **9**, 323–364 (1992).
gen.: ApSimon **9**, 95–99 (synthesis).

Eptastatin see pravastatin.

Equilenin, Equilin see estrogens.

Equisetum alkaloids see horse tail.

Erabutoxins. A peptide complex, three components, E. A, E. B and E. C have been isolated. Each consists of a single chain of 62 amino acid residues with 4 intramolecular disulfide bridges. They are isolated from the venom of the sea snake *Laticauda semifasciata*.

Lit.: J. Biol. Chem. **264**, 9239 (1989) ▪ Sax (8.), ECX 000. – *[CAS 59536-69-5]*

Erbstatin [*N*-((*E*)-2,5-dihydroxystyryl)formamide].

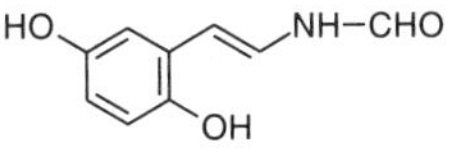

$C_9H_9NO_3$, M_R 179.18, yellow cryst., mp. 78–82 °C. E. is isolated from *Streptomyces* sp.[1]. E. is a highly active, competitive inhibitor of tyrosine specific protein kinases [2] and has, on account of its resultant antineoplastic activity, significance for pharmaceutical research [3].

Lit.: [1] J. Antibiot. **39**, 170 ff., 314 (1986). [2] J. Antibiot. **40**, 1209 f., 1471 ff. (1987); Biochem. Int. **15**, 989–995 (1987). [3] Biochem. Pharmacol. **48**, 549 (1994); Jpn. J. Cancer Res. **78**, 329 (1987).
gen.: Synthesis: Chem. Pharm. Bull. **37**, 2214 (1989) ▪ J. Antibiot. **40**, 1207 f. (1987); **46**, 785 (1993) ▪ J. Org. Chem. **52**, 2945 (1987) ▪ Synth. Commun. **18**, 1483–1489 (1988) ▪ Synthesis **1988**, 970 ff.; **1990**, 568 f. ▪ Tetrahedron Lett. **27**, 5799 (1986); **28**, 2217 (1987). – *[CAS 100827-28-9]*

Ercalcidiol, Ercalciol, Ercalcitriol see calcitriol.

Eremofortins see PR toxin.

Eremophilanes. *Sesquiterpenes with a 7-isopropyl-1,8*a*-dimethyldecalin skeleton that are often found in cornflowers (Asteraceae).

On the basis of the stereochemistry, two types can be distinguished. In compounds of the eremophilane type all three side-chains are in the *β*-position. *Examples: eremophilone* from the Australian tree *Eremophila mitchellii* [1]. In compounds of the valencane type the two methyl groups have *α*-orientation. *Examples:* (+)-*nootkatone* from the cypress *Chamaecyparis nootkatensis* and the essential oil of grapefruit (*Citrus paradisi*) [2]. Nootkatone is used to aromatize beverages. For synthesis, see *Lit.* [3]. Further important E. include E. lactones [e. g., (+)-*eremophilenolide* [4]] and the furanoeremophilanoids [e. g., *petasalbin* (ligularol) [5]].

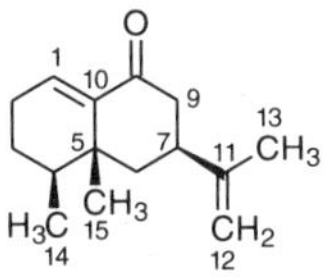

Eremophilone (**1**)

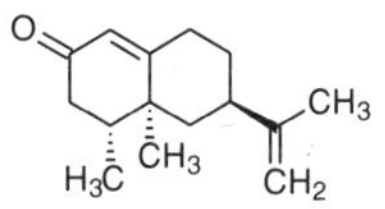

(+)-Nootkatone (**2**)

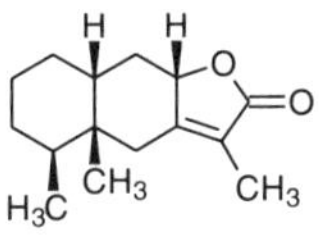

(+)-Eremophilenolide (**3**)

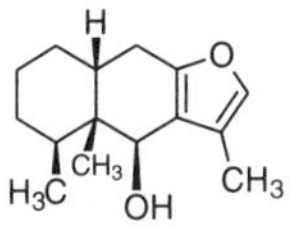

Petasalbin (**4**)

Table: Data of Eremophilanes.

	molecular formula	M_R	mp. (bp.) [°C]	optical rotation	CAS
1	$C_{15}H_{22}O$	218.34	41–42 (171/ 15 hPa)	$[\alpha]_{546}$ −207° (CH_3OH)	562-23-2
2	$C_{15}H_{22}O$	218.34	36–37	$[\alpha]_D$ +195.5° ($CHCl_3$)	4674-50-4
3	$C_{15}H_{22}O_2$	234.34	124	$[\alpha]_D$ +202° ($CHCl_3$)	4871-90-3
4	$C_{15}H_{22}O_2$	234.34	81–82	$[\alpha]_D^{24}$ −11.8° ($CHCl_3$)	4176-11-8

Lit.: [1] Phytochemistry **45**, 195 (1997); Tetrahedron Lett. **1977**, 1081. [2] Tetrahedron Lett. **1965**, 4779; J. Org. Chem. **36**, 594 (1971). [3] Bull. Chem. Soc. Jpn. **55**, 887 (1982); J. Org. Chem. **47**, 4622 (1982); **50**, 3615 (1985). [4] Beilstein E V **17/9**, 639. [5] J. Am. Chem. Soc. **103**, 4611 (1981) (synthesis); Bull. Chem. Soc. Jpn. **46**, 2840 (1973).
gen.: Angew. Chem. Int. Ed. Engl. **12**, 793–806 (1973) ▪ ApSimon **2**, 361–380; **5**, 180 (synthesis) ▪ Pure Appl. Chem. **21**, 263 (1970) ▪ Zechmeister **34**, 81–186.

Ergocalciferol see calcitriol.

Ergochromes (secalonic acids, previously: ergochrysins). The dark-green *sclerotia (*ergot) of the ergot fungus (*Claviceps purpurea*) growing on rye contain a mixture of pigments designated as E. However, they are also formed by other fungi, e. g., *Aspergillus*, *Penicillium* species, and *Phoma terrestris* and occur in lichens. All E. are dimers of seven different xanthone monomers linked in the 2,2′-positions. For systematic reasons, the E. are named after the monomers from which they are built up, e. g., E. BE = *secalonic acid F* {$C_{32}H_{30}O_{14}$, M_R 638.58, yellow needles, mp. 218–221 °C, $[\alpha]_D^{20}$ +202° (pyridine)}; E. AC = *ergochrysin A* [$C_{31}H_{28}O_{14}$, M_R 624.55, yellow cryst., mp. 285 °C (decomp.)] or E. CC = *ergoflavin* {$C_{30}H_{26}O_{14}$, M_R 610.53, yellow needles, mp. >350 °C (decomp.), $[\alpha]_D^{20}$ +105° (pyridine)}. Biosynthetically E. are formed from octaketides through ring opening and recyclization of anthraquinone intermediates such as *emodin [1]. Secalonic acid has cytostatic properties [2].

Ergochrome	alternative name
AA	Secalonic acid A
BB	Secalonic acid B
AB	Secalonic acid C
CC	Ergoflavin
AC	Ergochrysin A
BC	Ergochrysin B
AD	–
BD	–
CD	–
DD	–
EE	Secalonic acid D
GG	Secalonic acid E
BE	Secalonic acid F
AG	Secalonic acid G
CF	Ergoxanthin

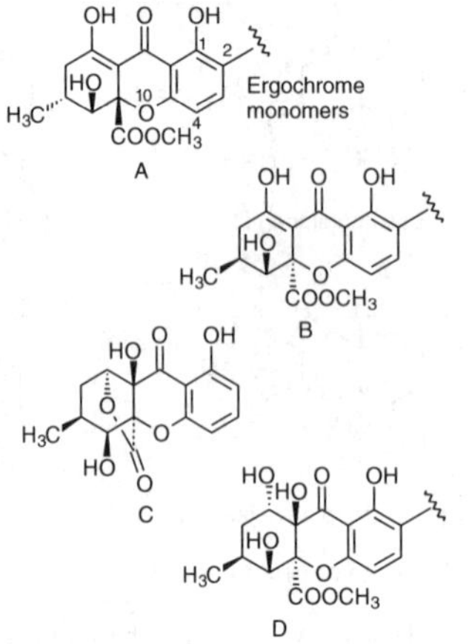

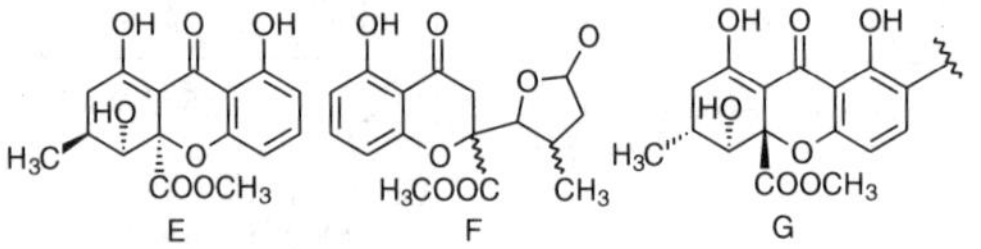

Figure: Ergochromes.

Lit.: [1] Angew. Chem. Int. Ed. Engl. **19**, 460 f. (1980); Franck, in Steyn, Biosynthesis of Mycotoxins: Study of Secondary Metabolism, p. 157–191, New York: Academic Press 1980. [2] Drugs Exptl. Clin. Res. **13**, 339–344 (1987)
gen.: Cole-Cox, S. 646–669 ▪ Hager (5.) **3**, 534; **4**, 913 f. ▪ J. Org. Chem. **42**, 352 (1977) ▪ Pure Appl. Chem. **54**, 2220 (1982) ▪ Sax (8.), EDA 500 ▪ Tetrahedron Lett. **19**, 1379 (1978); **20**, 4633 (1979) (isolation); **21**, 1185 (1980) ▪ Turner **2**, 160–164. – *[CAS 35287-69-5 (secalonic acid D); 60687-07-2 (secalonic acid F); 3419-11-2 (ergochrysin A); 3101-51-7 (ergoflavin)]*

Ergochrysins see ergochromes.

Ergocornine see ergot alkaloids.

α(β)-Ergocryptine, ergometrine, ergosine, ergostine see ergot alkaloids.

Ergoflavin see ergochromes.

Ergostane, Ergosterol (ergosterol) see sterols.

Ergot. Dried *sclerotia of the ergot fungus (*Claviceps purpurea*, Hypocreaceae); sclerotia are persisting, winter-surviving forms. E. contains up to 1% of its weight of *ergot alkaloids. These alkaloids are produced by large-scale cultivation of rye-ergot (especially in Canada, but also in Spain, France, and Germany). The cultivation is strictly controlled (drug laws). Extraction of E. furnishes the pharmaceutically important ergot alkaloids. E. also contains polyketide pigments (*ergochromes, secalonic acids) as well as ergosterol and ricinoleic acid, the later serves as an indicator for the detection of ergot contaminations in flour and other grain products. Contamination of rye flour with E. in the Middle Ages often led to serious mass poisonings by the contained ergot alkaloids. Besides *C. purpurea* other *Claviceps* species also form alkaloid-containing sclerotia (see table).
Lit.: Hager (5.) **4**, 911–925 ▪ Lyon Pharm. **37**, 323–331 (1986) ▪ Merck-Index (12.), No. 3702 ▪ Modern Drug Discovery March/April **1999**, 20–31 (pharmacology) ▪ Planta Med. **46**, 131–144 (1982) ▪ Sax (8.), EDB 500 ▪ Ullmann (5.) **A 1**, 385.

Ergot alkaloids (secale alkaloids). The *sclerotia of the ergot fungus (*E* ergot, from old French: argot = cock's spur) contain numerous alkaloids. The E. a. are responsible for the so-called *ergotism*, a disease frequently occurring in the Middle Ages through consumption of flour or bread contaminated by ergot. Symptoms of the acute form: nausea, headache, spasms, uterine contractions. *Convulsive ergotism* is due to damage to the nervous system while epidemic gangrene (St. Anthony's fire) is mainly due to vascular constrictions leading even to death of the affected extremity, the afflicted body parts become dark red to black in color, lose feeling, die, and are discarded by the body without loss of blood. The various forms result from differing toxin contents and composition of the sclerotia. A single dose of 8–10 g sclerotia can be lethal for an adult human.

Historical: The Assyrians mentioned the toxicity of ergot about 600 BC. The first proven report originated in the year 857 AD (Xanten am Rhein), in 922 the disease is supposed to be responsible for about 40000 deaths in France and Spain. In 1676 ergot was identified as the cause of St. Anthony's fire and the mills subsequently were controlled. In 1853 the fungus was recognized for the first time as the actual pathogen by Tulasne (French mycologist).

Pharmacological activity: On account of their toxicity, E. a. soon became of interest to pharmacologists. The first report of their use to induce birth was published in 1808 in the USA but this indication was given

Table: Sclerotia-forming *Claviceps* species and their alkaloids.

species	host	sclerotia	contents (%)	alkaloids	poisoning
C. purpurea	rye, barley, durum wheat, spelt, pasture grasses and others	5–40 mm long and 2–6 mm broad	~1	Peptide alkaloids Ergometrines	yes
C. paspali	*Paspalum dilatum*	2–4 mm roundish	0.003	Lysergamides Paspaline Elymoclavine	yes
C. fusiformis	millet (*Panicum*)	2–4 mm roundish	0.3	Agroclavines Chanoclavine Elymoclavine	yes
C. gigantea	maize	30–35 mm 5–8 mm	0.03	Festuclavine Pyroclavine Chanoclavine	yes
Sphacelia sorgi	sorghum	2 mm 2–4 mm	0.5	Pyroclavine Festuclavine Dihydroergosine	

up again in 1824, it was thereafter used only to stop postnatal hemorrhage. They also find use, mainly as the 9,10-dihydro derivatives, as sympathicolytics (alpha-receptor blockers) in the therapy of hypertension (rare) and migraine. For further uses, see *Lit.*[1]. Ergot peptide alkaloids are also formed by endophytic fungi (e.g., *Acremonium coenophialum*) in pasture grasses and cause poisoning of grazing animals (see darnel, ryegrass). The fungus *Epichloe typhina*, related to *Claviceps*, forms lysergic acid amide. Alkaloids of the clavine type are found in *Balansia obtecta*, *Geotrichum candidum*, various *Aspergillus*- and *Penicillium* species and the zygomycetes *Mucor hiemalis* and *Rhizopus nigricans*. For further occurrence of E.a., see *Lit.*[2].

Structures: The E.a. belong to three groups. The first contains alkaloids of the clavine type, the other two are alkaloids derived from *lysergic acid. The names of the alkaloids carry endings with ...ine (see table); the corresponding isomeric derivatives of *isolysergic acid* (epimeric at C-8) are pharmacologically inactive and their names end with...inine (e.g., ergotaminine). Chemically related with the E.a. is the hallucinogenic *lysergic acid diethylamide* [LSD, see figure with $-CO-N(C_2H_5)_2$ at C-8]. A simply constructed E.a. is the amide of lysergic acid with 2-amino-1-propanol, international free name *ergometrine* (also known as ergobasine, ergonovine, with $CO-NH-CH(CH_3)-CH_2OH$ at C-8), $C_{19}H_{23}N_3O_2$, M_R 325.41, mp. 159–162 °C. The complexly built group of E.a. consists of lysergic acid tripeptides which all contain proline (see figure) linked with α-hydroxyvaline [$R^1=CH(CH_3)_2$, ergotoxine group] or α-hydroxyalanine ($R^1=CH_3$, ergotamine group) and other amino acids (Phe, Leu, Val, see R^2).

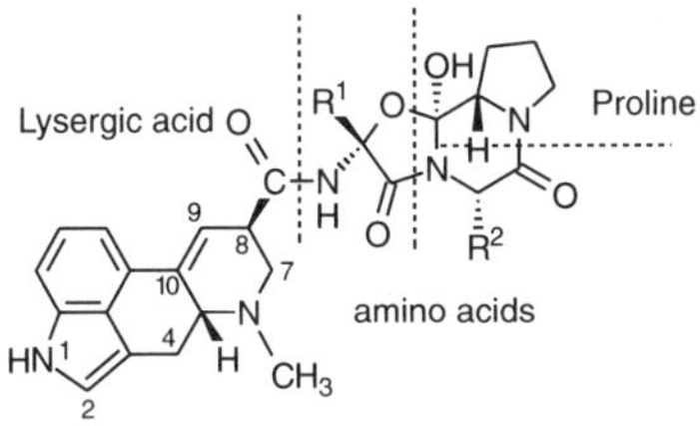

The biosynthesis of E.a. proceeds from tryptophan, mevalonic acid, and the CH_3-group of methionine and is understood in the key steps[3]. Although all E.a. are also accessible by synthesis (through lysergic acid), they are mainly obtained from submersion cultures of *Claviceps* species or from field cultures, derivatives are mostly obtained by semisynthetic routes or by specific fermentations[4].

Lit.: [1]Chem. Eng. News **55**, No. 49, 5 (1977). [2]Handbook of Experimental Pharmacology **49**, 29 (1978); Helv. Chim. Acta **58**, 2484–2500 (1975); Appl. Environ. Microbiol. **58**, 857 (1992). [3]Tetrahedron **32**, 873 (1976); Adv. Biotechnol. Processes **3**, 197–239 (1984); Chim. Ind. (Milan) **66**, 335 (1984). [4]Trends Biotechnol. **2**, 166–172 (1984); Current Microbiol. **15**, 97–101 (1987); Biotechnology **4**, 569–609 (1986); Appl. Environ. Microbiol. **59**, 2029–2033 (1993); J. Nat. Prod. **55**, 424 (1992); Top. Curr. Chem. **186**, 45–64 (1997).
gen.: Manske **15**, 1–40; **38**, 1–156; **50**, 172–218 ▪ Sax (8.), EDA 600–EDC 575 ▪ Ullmann (5.) **A 1**, 385–387. – *Isolation:* J. Agric. Food. Chem. **29**, 653 (1981); **31**, 655 (1983) ▪ J. Nat. Prod. **47**, 970 (1984); **52**, 506 (1989); **56**, 489 (1993). – *Review:* Med. Aromat. Plants Ind. Profiles **6**, 201–228 (1999) ▪ Reháĉek & Sajdl (eds.), Ergot Alkaloids: Chemistry, Biological Effects, Biotechnology, Amsterdam: Elsevier 1990. – *Synthesis:* Heterocycles **45**, 1263 (1997) ▪ J. Chem. Soc., Perkin Trans. 1 **1990**, 707–713 ▪ Nachr. Chem. Tech. Lab. **38**, 246–250 (1990). – *Activity:* Hager (5.) **3**, 525–550; **4**, 911–920; **8**, 56–70. – *[HS 293960]*

Table: Structures and data of Ergot alkaloids.

	R^1	R^2	molecular formula	M_R	mp. [°C]	CAS
Ergotamine	CH_3	$CH_2-C_6H_5$	$C_{33}H_{35}N_5O_5$	581.67	213–214 (decomp.)	113-15-5
Ergosine	CH_3	$CH_2-CH(CH_3)_2$	$C_{30}H_{37}N_5O_5$	547.65	228 (decomp.)	561-94-4
Ergostine[3]	C_2H_5	$CH_2-C_6H_5$	$C_{34}H_{37}N_5O_5$	595.70	204–208 (decomp.)	2854-38-8
Ergocristine	$CH(CH_3)_2$	$CH_2-C_6H_5$	$C_{35}H_{39}N_5O_5$	609.73	175 (decomp.)	511-08-0
α-Ergocryptine	$CH(CH_3)_2$	$CH_2-CH(CH_3)_2$	$C_{32}H_{41}N_5O_5$	575.71	214 (decomp.)	511-09-1
β-Ergocryptine	$CH(CH_3)_2$	$CH(CH_3)-C_2H_5$ (*S*)	$C_{32}H_{41}N_5O_5$	575.71	173	20315-46-2
Ergocornine	$CH(CH_3)_2$	$CH(CH_3)_2$	$C_{31}H_{39}N_5O_5$	561.68	182–184 (decomp.)	564-36-3

Ergotamine see ergot alkaloids.

Ergothioneine (2-mercapto-N^α,N^α,N^α-trimethyl-L-histidinium betaine).

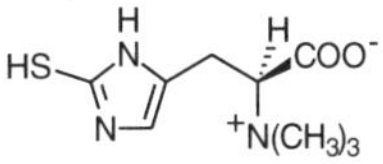

$C_9H_{15}N_3O_2S$, M_R 229.30, needles, dihydrate, mp. 256–257 °C (decomp.), $[\alpha]_D^{20}$ +116° (H_2O).

Occurrence: First found in *sclerotia of the fungus *Claviceps purpurea* (*ergot). E. also occurs in blood, seeds, and in various mammalian tissues, especially in the liver and kidneys, as well as in the king crab (*Limulus polyphemus*). It chelates divalent metal ions and serves as a transport vehicle. The biosynthesis proceeds from histidine.

Lit.: Beilstein E V **25/17**, 205 ▪ Biochem. J. **211**, 605 (1983) ▪ J. Org. Chem. **60**, 6295 (1995) (synthesis (L)(+)-E.) ▪ Med. Hypotheses **18**, 351-370 (1985). – *[CAS 497-30-3]*

Ericaceae diterpenes see grayanotoxins.

Erinacin A see herical.

Eritadenine [(*R,R*)-6-amino-α,β-dihydroxy-9*H*-purine-9-butanoic acid].

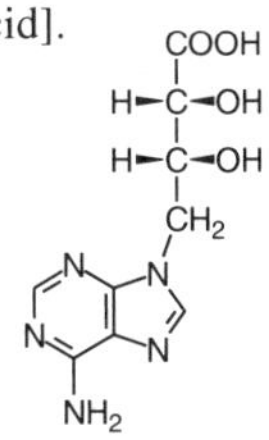

$C_9H_{11}N_5O_4$, M_R 253.22, needles, mp. 261–263 °C, $[\alpha]_D$ +16° (1 N HCl), +50° (0.1 N NaOH). Adenine derivative from the Japanese edible mushroom Shiitake (*Lentinus edodes*). E. exhibits cholesterol-lowering and antiviral activities.

Lit.: Coll. Czech. Chem. Commun. **47**, 1392 (1982) ▪ Experientia **29**, 271 (1973) ▪ J. Med. Chem. **17**, 846 (1974) ▪ J. Org. Chem. **38**, 2887 (1973) ▪ Tetrahedron **28**, 899 (1972) ▪ Tetrahedron Lett. **1970**, 1359. – *[CAS 23918-98-1]*

(6Z)-Ertacalciol see calcitriol.

Erucic acid [(*Z*)-13-docosenoic acid].

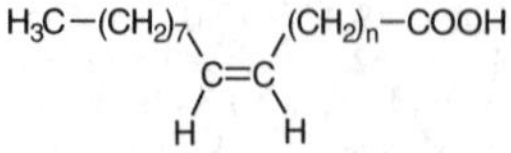

n = 11 : Erucic acid
n = 13 : Nervonic acid

$C_{22}H_{42}O_2$, M_R 338.57, mp. 33.5 °C, bp. 225 °C (1.33 kPa); soluble in organic solvents, insoluble in water. Singly unsaturated *fatty acid with 22 C-atoms that occurs as the glycerol ester in seed oils of crucifers, e. g., rape, mustard, and cabbage species, as well as in nasturtium plants, e. g. Indian cress (*Tropaeolum majus*). For the contents of E. in edible oils and fats as well as the nutritional-physiological effects of E., see rapeseed oil. In the seed oil of nasturtiums E. occurs together with some other very long chain monoene fatty acids, e. g. *nervonic acid* [(*Z*)-15-tetracosenoic acid, $C_{24}H_{46}O_2$, M_R 366.63, mp. 44–45 °C]. Recently a synthetic, E.-rich oil (Lorenzo's oil, named after the first patient to be treated with this oil) was used to treat the metabolic disease adrenoleukodystrophy (ALD). Because of a functional disorder of the peroxisomes *cerotic acid-rich membrane lipids (e. g., sphingomyelins) are formed in the nerve tissue of ALD patients which finally leads to the destruction of myelin. In ALD patients the biosynthesis and thus the accumulation of cerotic acid and other saturated, very long-chain fatty acids (VLCFA) is suppressed by uptake of E. and other long-chain monoene fatty acids with food. E. itself does not have a harmful effect on membrane lipids of myelin.

Lit.: J. Chem. Soc., Perkin Trans. 1 **1993**, 1183 ▪ Lipid Technol. **1994** (1), 10 ▪ Medical Hypotheses **42**, 237 (1994) ▪ N. Engl. J. Med. **329**, 745 (1993) ▪ Science **268**, 1506 (1995) ▪ see also rapeseed oil. – *[HS 2916 19; CAS 112-86-7 (E.); 506-37-6 (nervonic acid)]*

Erucic acid amide see oleamide.

Erucin, erysolin see mustard oils.

Erysodine, erysonine, erysopine, erysotrine, erysovine, erythraline, erythratine see Erythrina alkaloids.

Erythrin.

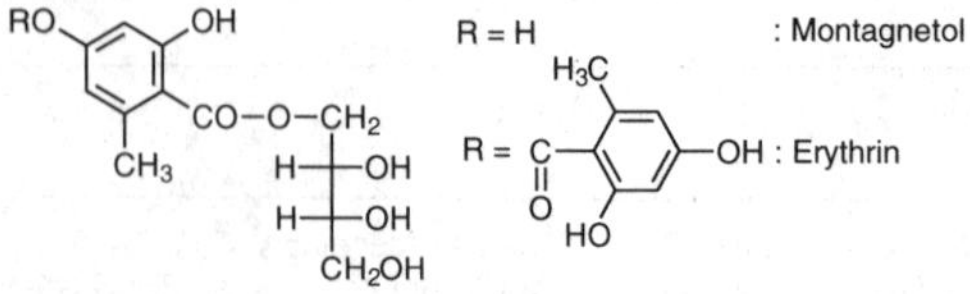

$C_{20}H_{22}O_{10}$, M_R 422.39, needles, mp. 156–157 °C, $[\alpha]_D^{24}$ +8°. Ester of *lecanoric acid with erythritol. E. occurs, e. g., in the lichen family Roccellaceae and furnishes the bitter-tasting *montagnetol* ($C_{12}H_{16}O_7$, M_R 272.25, mp. 156–157 °C; identical with pikroerythrin) after cleavage of the *depside bond.

Lit.: Asahina-Shibata, p. 59–61 ▪ Aust. J. Chem. **41**, 1789 (1988) (synthesis E.) ▪ Karrer, No. 1025. – *[CAS 480-57-9]*

Erythrina alkaloids. Group of alkaloids from *Erythrina* species (Fabaceae) growing in tropical and subtropical regions, e. g., *E. crista-galli.* They are spirocyclic *isoquinoline alkaloids and are found in all parts of the plants. Three main types are distinguished, the dienoid, enoid, and lactone alkaloids: erythrinadienes, erythrinenes, and oxaerythrinanones. E. a. exhibit curare-like activities, as well as sedative, hypotensive, CNS-depressive, laxative, and diuretic properties. For synthesis, see *Lit.*[1].

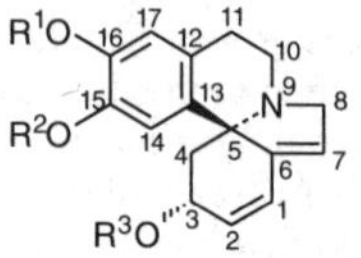

	R1	R2	R3
Erysodine (1)	H	CH_3	CH_3
Erysonine (2)	H	CH_3	H
Erysopine (3)	H	H	CH_3
Erysovine (4)	CH_3	H	CH_3
Erythraline (5)	CH_2		CH_3
Erysotrine (6)	CH_3	CH_3	CH_3

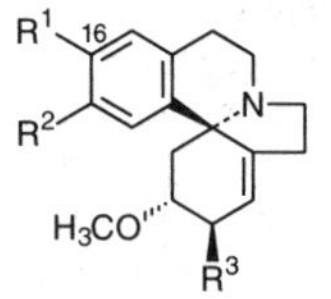

	R1	R2	R3
Erythratine (7)	O–CH2–O		OH
Cocculidine (8)	H	OCH_3	H
Cocculine (9)	H	OH	H

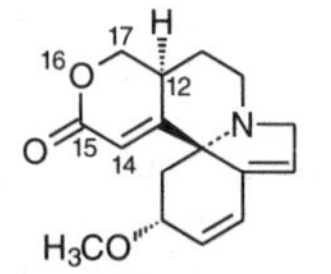

α-Erythroidine (10)

Table: Data of Erythrina alkaloids.

no.	molecular formula	M_R	mp. [°C]	$[\alpha]_D$	CAS
1	$C_{18}H_{21}NO_3$	299.37	204–205	+239° ($CHCl_3$)	7290-03-1
2	$C_{17}H_{19}NO_3$	285.34	241–243	+289° (0.5% HCl)	7290-05-3
3	$C_{17}H_{19}NO_3$	285.34	241–242	+265.2° (C_2H_5OH/ glycerol)	545-68-6
4	$C_{18}H_{21}NO_3$	299.37	178–179.5	+252° (C_2H_5OH)	466-72-8
5	$C_{18}H_{19}NO_3$	297.35	120	+228° (C_2H_5OH)	466-77-3
6	$C_{19}H_{23}NO_3$	313.40	159–161 (picrate)	+142° (picrate/ C_2H_5OH)	27740-43-8
7	$C_{18}H_{21}NO_4$	315.37	170	+145.5° (C_2H_5OH)	5550-20-9
8	$C_{18}H_{23}NO_2$	285.39	86–87	+250.9° ($CHCl_3$)	27675-40-7
9	$C_{17}H_{21}NO_2$	271.36	217–218	+271.1° (CH_3OH)	27675-39-4
10	$C_{16}H_{19}NO_3$	273.33	58–60	+136° (H_2O)	466-80-8

On the basis of recent investigations[2] the previously assumed biosynthetic pathway[3] to E. a. must be revised.

Lit.: [1]Chem. Ber. **112**, 1329–1347 (1979); Chem. Pharm. Bull. **46**, 906 (1998); Heterocycles **48**, 1599 (1998); J. Nat. Prod. **54**,

329–363 (1991); Stud. Nat. Prod. Chem. **3**, 455–493 (1989). [2]Phytochemistry **52**, 373–382 (1999). [3]Mothes et al., p. 220–223.
gen.: Beilstein E V **21/5**, 185 f., 633 f.; **21/6**, 25 f., 81 ff.; **21/13**, 313, 510 ff., 538 ff., 547, 563 ▪ Eur. J. Org. Chem. **1999**, 909 (synthesis) ▪ Manske **18**, 1–98; **35**, 177–214 ▪ Nat. Prod. Rep. **1**, 371 (1984); **3**, 555–564 (1986); **10**, 463 f. (1993) ▪ Phillipson et al., p. 62–78 ▪ Phytochemistry **39**, 677 (1995) ▪ Sax (8.), EDE 000, EDE 500, EDF 000, EDG 500, EDH 000, PIE 500 ▪ Ullmann (5.) **A 1**, 379. – *[HS 2939 90]*

Erythrolides.

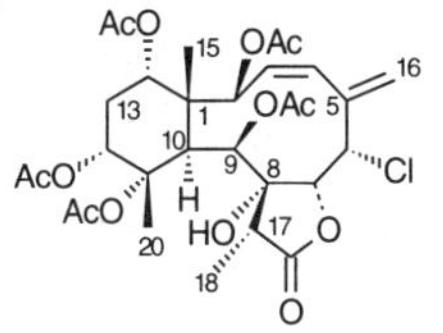

Briarein A

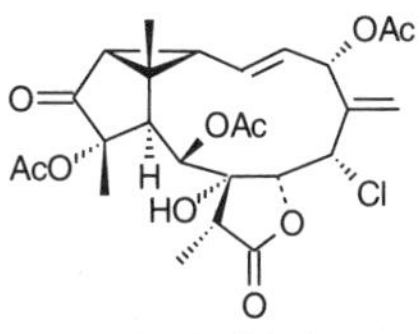

Erythrolide A

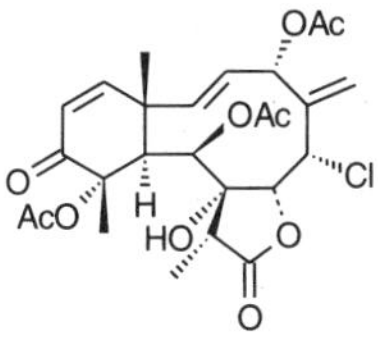

Erythrolide B

Ac = CO−CH_3

Numerous *diterpenes with the briarane skeleton are isolated from sea fans, sea feathers, and soft corals, e.g. *briarein A* [1], a chlorinated diterpene lactone from the gorgonian *Briareum asbestinum* ($C_{30}H_{39}ClO_{13}$, M_R 643.08) or *E. A* [2] ($C_{26}H_{31}ClO_{10}$, M_R 538.98) from the sea fan *Erythropodium caribaeorum*, which is formed by a di-π-methane rearrangement from *E. B* ($C_{26}H_{31}ClO_{10}$, M_R 538.98). The compounds have anti-inflammatory activities.

Lit.: [1]J. Nat. Prod. **59**, 15 (1996). [2]J. Am. Chem. Soc. **106**, 5026 (1984); J. Nat. Prod. **56**, 1051 (1993); J. Org. Chem. **56**, 2344 (1991).
gen.: Scheuer I **2**, 173 ▪ Scheuer II **3**, 85. – *[CAS 89999-14-4 (E. A); 89999-15-5 (E. B); 62681-06-5 (briarein A)]*

Erythromycin.

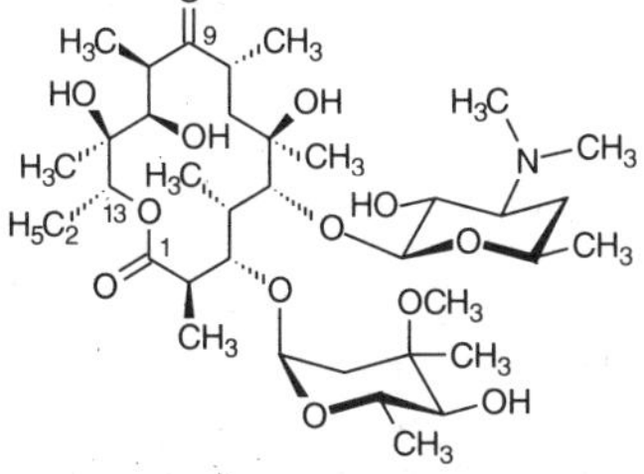

International free name for the main component (E. A, *erythrocin*) of the mixtures of 14-membered *macrolide antibiotics formed by *Saccharopolyspora erythraea* (previously: *Streptomyces erythraeus*). The sugar building blocks of E. A are α-L-cladinose, the 3-*O*-methyl ether of L-*mycarose, and the amino sugar β-D-desosamine. The sugars are necessary for the antibacterial activity. A 9,12-hemiketal is readily formed in the lactone ring.

E. A: $C_{37}H_{67}NO_{13}$, M_R 733.94, mp. 135–140 °C and 190–193 °C, $[\alpha]_D$ −78° (C_2H_5OH), well soluble in ethanol, chloroform, acetone, and acetonitrile. The biosynthesis of E. has been investigated in detail down to the genetic level. The first stable intermediate built up from 7 propionate units on the polyketide synthase is 6-deoxy-*erythronolide B, later in the biosynthesis it is hydroxylated at C-6 and C-12, glycosylated and methylated. E. derivatives, e.g. troleandomycin (see oleandomycin), also *hybrid antibiotics, are accessible through genetic manipulations. The first total synthesis was achieved by Woodward in 1981. E. B is 12-deoxyerythromycin A. E. C and D contain L-mycarose in place of L-cladinose.

E. shows low toxicity and finds wide clinical use against Gram-positive bacteria, some Gram-negative bacteria, and against mycoplasma (causes of severe airway infections). It is frequently used against pathogens that have become insensitive to *β-lactam and aminoglycoside antibiotics. E. acts as an inhibitor of bacterial protein synthesis by binding to the 50S-subunit of ribosomes, especially the L15 protein, and stimulating the liberation of peptidyl-tRNA from ribosomes during translocation. Resistance against E. occurs by changes in the binding sites on the ribosome.

Lit.: J. Chem. Soc., Perkin Trans. 1 **1985**, 2599–2603 ▪ J. Chem. Res. (S) **1998**, 727 (structure). – *Biosynthesis:* Angew. Chem. Int. Ed. Engl. **30**, 1302 (1991) ▪ Barton-Nakanishi **1**, 495–532 ▪ Chem. Rev. **97**, 2611–2629 (1997). – *Derivatives:* Drugs **44**, 117 (1992). – *Genetics:* Annu. Rev. Microbiol. **47**, 875–912 (1993); **49**, 201–238 (1995) ▪ Chem. Biol. **2**, 583–589 (1995) ▪ J. Antibiot. **48**, 647–651 (1995) ▪ Science **268**, 1487 ff. (1995). – *Resistance:* J. Antibiot. **43**, 977–984 (1990). – *Review:* Ōmura (ed.), Macrolide Antibiotics, New York: Academic Press 1984. – *Synthesis:* Angew. Chem. Int. Ed. Engl. **30**, 1452 ff. (1991) ▪ J. Am. Chem. Soc. **103**, 3210–3217 (1981); **119**, 3193 (1997) (E. B) ▪ Lukacs (ed.), Recent Progress in the Chemical Synthesis of Antibiotics and Related Microbial Products, p. 121–140, Berlin: Springer 1993 ▪ Tetrahedron **41**, 3569–3624 (1985). – *[HS 2941 50; CAS 114-07-8 (E. A)]*

Erythronolide B (12-deoxyerythronolide A).

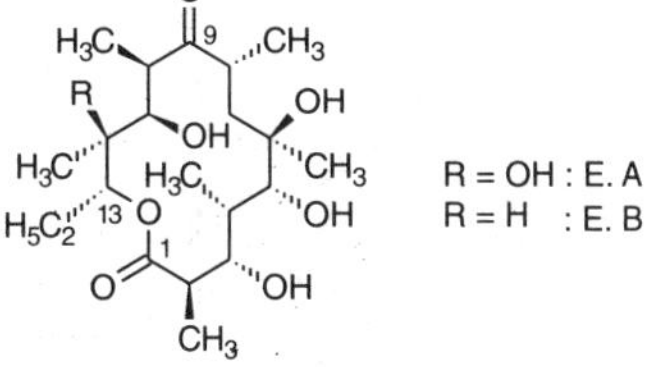

$C_{21}H_{38}O_7$, M_R 402.53, cryst., mp. 223–225 °C, $[\alpha]_D$ −66.5° (CH_3OH). E. B is the aglycone of the *macrolide antibiotic *erythromycin B and a stable intermediate in the biosynthesis of 6-deoxyerythronolide B ($C_{21}H_{38}O_6$, M_R 386.53, mp. 147–149 °C) and erythromycin A in *Saccharopolyspora erythraea*. Because of the absence of a sugar unit it has only weak antibacterial activity.

Lit.: Angew. Chem. Int. Ed. Engl. **30**, 1302 (1991) (biosynthesis) ▪ Beilstein E V **18/5**, 409. – *Synthesis:* Angew. Chem. Int. Ed. Engl. **30**, 1452 (1991) ▪ J. Am. Chem. Soc. **111**, 7634 ff. (1989) ▪ Nicolaou, p. 167. – *[CAS 3225-82-9 (E. B); 15797-36-1 (6-deoxyerythronolide B)]*

Erythrophleum alkaloids. Group of diterpene alkaloids occurring in *Erythrophleum* species (Fabaceae). They are derivatives (esters and amides) of cassaic acid. Data for important E. a. are shown in the table.

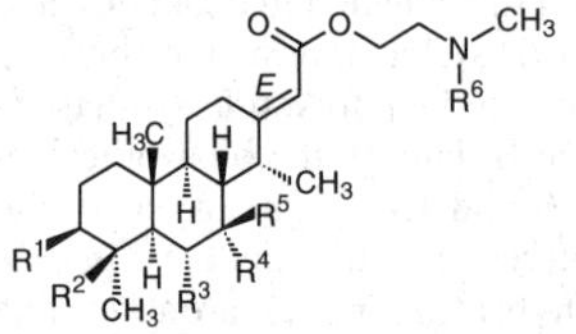

R^1	R^2	R^3	R^4	R^5	R^6	
OH	CH_3	H	O		CH_3	Cassaine (**1**)
OH	CH_3	H	H	OH	CH_3	Cassaidine (**2**)
H	$COOCH_3$	H	O		CH_3	Cassamine (**3**)
3-Hydroxy-isovaleryloxy	CH_3	H	O		CH_3	Coumingine (**4**)
H	$COOCH_3$	H	H	OH	H	Erythrophleine (**5**)
OH	$COOCH_3$	H	O		CH_3	Erythrophlamine (**6**)
H	$COOCH_3$	OH	O		CH_3	Erythrophleguine (**7**)

Table: Data of Erythrophleum alkaloids.

no.	molecular formula	M_R	mp. [°C]	$[\alpha]_D$ (C_2H_5OH)	CAS
1	$C_{24}H_{39}NO_4$	405.58	142.5	−103°	468-76-8
2	$C_{24}H_{41}NO_4$	407.59	139.5	−98°	26296-41-3
3	$C_{25}H_{39}NO_5$	433.59	86−87	−56°	471-71-6
4	$C_{29}H_{47}NO_6$	505.70	142	−70°	26241-81-6
5	$C_{24}H_{39}NO_5$	421.58	115	−18.5°	36150-73-9
6	$C_{25}H_{39}NO_6$	449.59	149−151	−62.5°	511-00-2
7	$C_{25}H_{39}NO_6$	449.59	77−78	−38°	4829-28-1

E. a. are toxic on account of their digitalis-like action. Poisonings can be lethal due to cardiac arrest. Some E. a. have local anesthetic properties.

Lit.: Aust. J. Chem. **27**, 179 (1974) ▪ Manske **4**, 265−273; **10**, 287−303; **25**, 5.

Erythropterin.

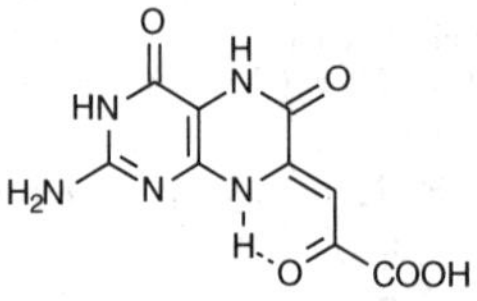

$C_9H_7N_5O_5$, M_R 265.19, red hydrated crystals. Red pigment of butterfly wings; also occurs in *Mycobacterium lacticula* and other bacteria.

Lit.: Beilstein EIII/IV **26**, 4061; EV **26/18**, 495 ▪ Comp. Biochem. Physiol. B.: Biochem. Mol. Biol. **101**, 115−133 (1992); **108**, 79−94 (1994) ▪ J. Heterocycl. Chem. **29**, 583 (1992) (review). – *[CAS 7449-03-8]*

D-Erythrose 4-phosphate.

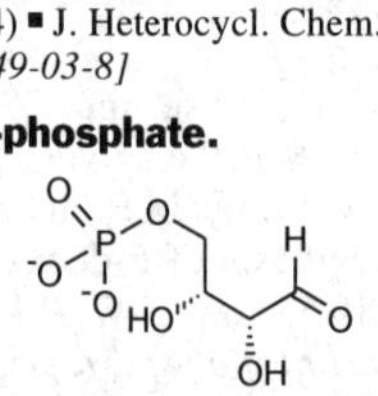

$C_4H_7O_7P$, M_R 198.07 (dianion). Occurs in trace amounts in muscles. E. is an important intermediate in the Embden-Meyerhoff-Parnas pathway for glycolysis and a precursor in the biosynthesis of *3-deoxy-D-*arabino*-2-heptulosonic acid 7-phosphate (DAHP).

Eschscholtzine see pavine and isopavine alkaloids.

Escin. A mixture of more than 30 different *saponins isolable from (horse) chestnuts (*Aesculus hippocastanum*, Hippocastanaceae) consisting mainly (up to 20%) of one glycoside {$C_{55}H_{86}O_{24}$, M_R 1131.27, mp. 224−225 °C, $[\alpha]_D^{27}$ −24° (CH_3OH)} of protoaescigenin (see oleanane) with 2 mol. glucose and 1 mol. glucuronic acid and esterified with acetic acid and tiglic acid or angelic acid. E. is used for treatment of hemorrhoids and varicose veins; it also exhibits cancerostatic properties.

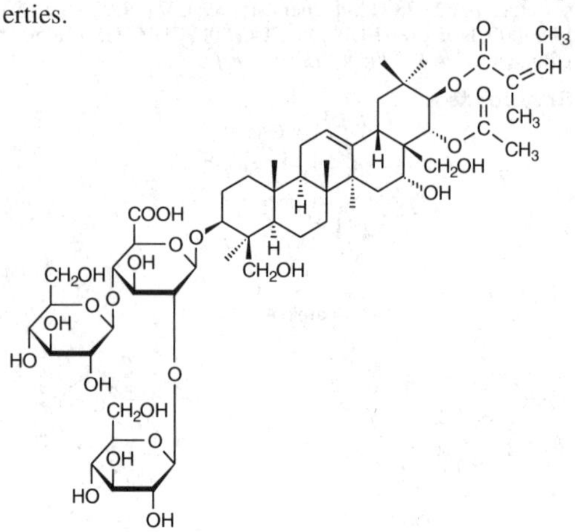

Lit.: Chem. Pharm. Bull. **32**, 3378 (1984); **33**, 1043 (1985); **42**, 1357 (1994); **44**, 1454 (1996) (structure, activity) ▪ Zechmeister **30**, 527−532. – *[HS 2938 90; CAS 6805-41-0]*

Esculetin, esculin see 6,7-dihydroxycoumarins.

Esculin see 6,7-dihydroxycoumarins.

Eserine see physostigmine.

Esperamycins (esperamicins).

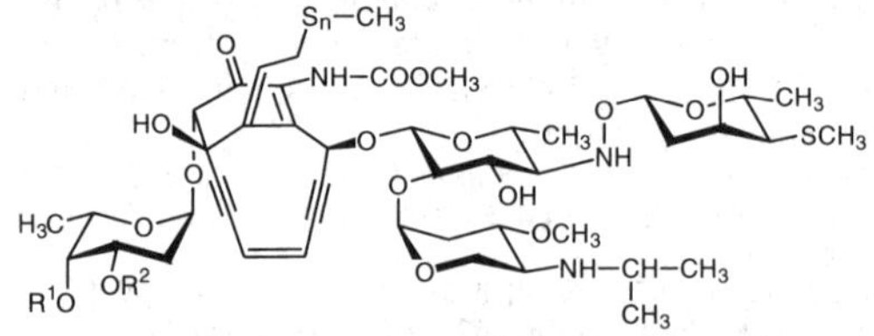

Aroyl =

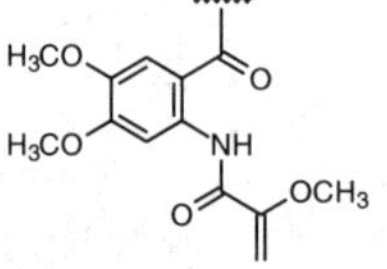

	n	R^1	R^2
E. A_1	3	H	Aroyl
E. P	4	H	Aroyl
E. A_2	3	Aroyl	H

Group of naturally occurring, DNA-cleaving *enediyne antibiotics from cultures of *Actinomadura verrucospora*. Examples are: *E. A_1*, $C_{59}H_{80}N_4O_{22}S_4$, M_R 1325.56, cryst., mp. 156−158 °C, $[\alpha]_D^{24}$ −207° ($CHCl_3$), *E. P*, $C_{59}H_{80}N_4O_{22}S_5$, M_R 1357.63, $[\alpha]_D$ −227° (CH_3OH) and *E. A_2*, $C_{59}H_{80}N_4O_{22}S_4$, M_R 1325.56, cryst., mp. 147−149 °C, $[\alpha]_D^{27}$ −179.4° ($CHCl_3$). E. show a very strong action as broad-spectrum antibiotics and belong, together with the *calicheamicins, to the most effective known cytostatic agents.

Lit.: Angew. Chem. Int. Ed. Engl. **30**, 1387−1416 (1991) ▪ J. Am. Chem. Soc. **115**, 12340 (1993) (biosynthesis) ▪ J. Antibiot. **38**, 1605 (1985); **48**, 1497 (1995) (biosynthesis) ▪ Proc. Natl. Acad. Sci. USA **86**, 7672 (1989). – *Synthesis:* J. Am. Chem. Soc. **114**, 2560−2567 (1992) ▪ J. Org. Chem. **55**, 1709 ff. (1990). – *[CAS 99674-26-7 (E. A_1); 157078-54-1 (E. P); 99674-27-8 (E. A_2)]*

Essential oils. General term for concentrates obtained from plants and used mainly as natural starting materials in the perfume and food industries, consisting of more or less volatile components, such as, e. g., genuine essential oils, *citrus oils, absolues, resinoids, etc. The term is often used for the volatile components contained in the plants. In the stricter sense, however, essential oils are the volatile components obtained by steam distillation from plant raw materials.
Production/extraction: Steam is passed though the plant material or the latter is boiled in water. On cooling the vapors, the essential oil separate from the aqueous phase of the condensate. Sometimes various parts of one plant species yield qualitatively different essential oils, e. g., the leaves of the cinnamon tree furnish an oil rich in *eugenol, while the oil of the bark contains *cinnamaldehyde as the main component. For a review of the various forms of apparatus used for the technical-scale isolation of essential oils, see *Lit.*[1]. For commercially used plants the yields of essential oils are usually in the range of 0.5 to 5%.
Composition: Genuine essential oils consist exclusively of volatile components with boiling points mainly between 150 and 300 °C. They contain predominantly hydrocarbons or monofunctional compounds such as aldehydes, alcohols, esters, ethers, and ketones. Parent compounds are mono- and sesquiterpenes, phenylpropane derivatives, and longer-chain aliphatic compounds. Accordingly, essential oils are relative non-polar mixtures, i. e., they are soluble in most organic solvents. Often the organoleptic properties are not determined by the main components but by minor and trace compounds such as, e. g., 1,3,5-undecatrienes and pyrazines in *galbanum oil. In many of the commercially important oils, the number of identified components exceeds 100. Very many of the constituents are chiral, frequently one isomer predominates or is exclusively present, e. g., (–)-menthol in *peppermint oils or (–)-linalyl acetate in *lavender oil.
Analysis/quality control/investigation[2,3]***:*** For commercially used oils, quality control tests (guidelines, see ISO 4720) generally involve determination of density, refractive index, optical rotation, and solubility. The qualitative and quantitative composition is analyzed by gas chromatography (GC).
Use (For special uses, see the individual entries): Financially about half of the production of essential oils is used in the food industry as aromatization agents. The rest is principally used in the perfume industry, either directly as components or as starting material for the production of other component. Only a relatively small portion is used in pharmaceutical preparations. Recently, essential oils have gained some popularity for the so-called aroma (fragrance) therapy.
Lit.: [1] Perfum. Flavor. **9** (2), 93 (1984). [2] Perfum. Flavor. **15** (1), 1 (1990). [3] Sandra & Bicchi, Capillary Gas Chromatography in Essential Oil Analysis, Heidelberg: Hüthig 1987.
gen.: Arctander ▪ Bauer et al. (2.) ▪ Ohloff. – *Journals:* Perfumer & Flavorist; J. Essent. Oil Res. (both Carol Stream, IL, USA: Allured Publishing Corp.) ▪ Flavour Fragr. J. (Chichester: Wiley).

Esterastin.

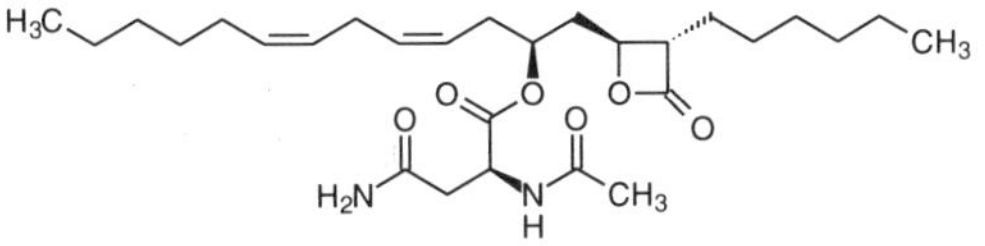

$C_{28}H_{46}N_2O_6$, M_R 506.67, mp. 99–100 °C, $[\alpha]_D^{23}$ +11° ($CHCl_3$), isolated from a *Streptomyces lavendulae* strain. E. blocks pancreatic lipase; cf. lipstatin.
Lit.: J. Antibiot. **31**, 639, 797 (1978). – *[CAS 67655-93-0]*

Estradiol see estrogens.

Estragole [1-methoxy-4-(2-propenyl)benzene].

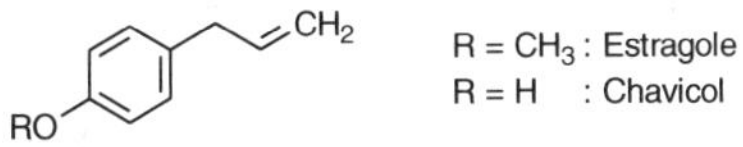

$C_{10}H_{12}O$, M_R 148.20; oil, bp. 216 °C, D^{15} 0.976, n_D^{27} 1.5230, forms an azeotrope with water, has an anise-like odor. LD_{50} (mouse p. o.) 1250 mg/kg. Main component (60–75%) of the essential oil of tarragon, *Artemisia dracunculus*, and occurs in many other essential oils, e. g., anise, star anise, bay, fennel, pine, turpentine oil. *Chavicol* (4-allylphenol), $C_9H_{10}O$, M_R 134.18, oil, mp. 16 °C, bp. 237 °C, D. 1.0175, n_D^{18} 1.5441, pK_a 10.23 (25 °C), occurs together with terpenes in the volatile components of betel oil and together with E. in *sweet basil oil. Glycosides of chavicol are also widely distributed.
Activity: E. and chavicol are attractants for the beetles *Diabrotica undecimpunctata* and *D. virgifera*[1] occurring in the South and West USA where they attack corn roots.
Use: E. is used for perfumes, to aromatize food, and as an aroma substance in the production of food and liquors.
Lit.: [1] J. Chem. Ecol. **13**, 959–975 (1987).
gen.: Beilstein E IV **6**, 3817 ▪ Food Chem. Toxicol. **28**, 537 (1990) (toxicology) ▪ Helv. Chim. Acta **61**, 401–429 (1978) (synthesis) ▪ J. Org. Chem. **60**, 1856 (1995) (synthesis) ▪ Karrer, No. 165, 166 ▪ Sax (8.), AFW 750 ▪ Synthesis **1983**, 701 ff. (synthesis) ▪ Xenobiotica **17**, 1223 (1987) (metabolism). – *[HS 2909 30; CAS 140-67-0 (E.); 501-92-8 (chavicol)]*

Estranes see steroids.

Estriol see estrogens.

Estrogens (oestrogens). The naturally occurring E. include *estradiol, estrone, estriol, equilin,* and *equilenin.* Estradiol and estrone are, together with progesterone (see progestogens), the most important female sex hormones. The chemical structures of the E. are derived from the C_{18}-hydrocarbon parent *estrane*, thus, they are 19-norsteroids, have an aromatic A-ring with a phenolic 3-hydroxy group and an additional oxygen function at C-17. *Estrone, estradiol,* and *estriol* are isolated from the urine of pregnant women; *equilin* and *equilenin* from the urine of pregnant horses. Surprisingly, estrone is also found in traces in plants, for example, in date seeds (*Phoenix dactylifera*), apples (*Pirus malus*), pomegranates (*Punica granatum*), and together with estradiol in apricots (*Prunus armeniaca*), estriol is found in willow catkins (*Salix* species). Equilenin was the first natural *steroid to be totally synthesized in 1939.

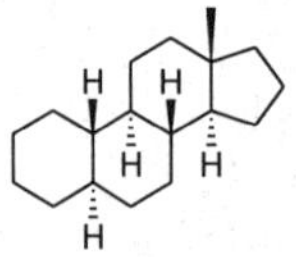

5α-Estrane

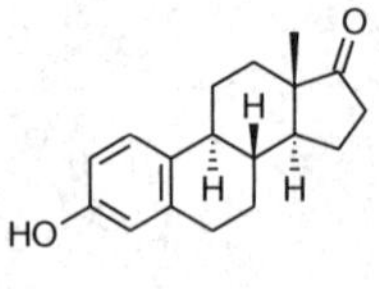

Estrone

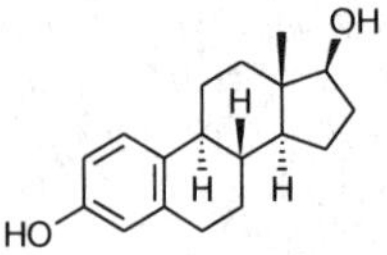

Estradiol

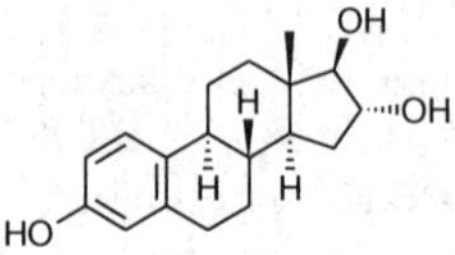

Estriol

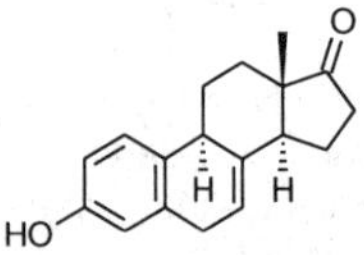

Equilin

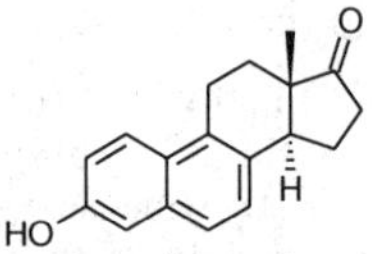

Equilenin

Table: Data of Estrogens.

Estrogen	molecular formula	M_R	mp. [°C]	$[\alpha]_D$	CAS
Estrone	$C_{18}H_{22}O_2$	270.37	261–264	+165° ($CHCl_3$)	53-16-7
Estradiol	$C_{18}H_{24}O_2$	272.39	178	+78° (C_2H_5OH)	50-28-2
Estriol	$C_{18}H_{24}O_3$	288.39	277–279	+61° (dioxane)	50-27-1
Equilin	$C_{18}H_{20}O_2$	268.36	238–240	+308° ($CHCl_3$)	474-86-2
Equilenin	$C_{18}H_{18}O_2$	266.34	258–259	+87° (dioxane)	517-09-9

Metabolism: E. are formed in the Graafian follicles of the ovaries and in the corpus luteum, and during pregnancy also in the placenta; they also occur in small amounts in the male organism. Estrone is principally biosynthesized from androst-4-ene-3,17-dione, the 17-dehydro product of testosterone, by an aromatase. Estradiol is metabolized in the liver to 2-methoxyestrone via estrone. Estriol is also formed by side reactions. Both metabolites are excreted by the kidneys. Equilenin and equilin are also metabolic excretion products of estrogens.

Activity: In the female organism, E. effect the normal course of the menstrual cycle. E. are responsible for the proliferation of uterine mucosa, growth of the mammary glands, and the development of the secondary sexual characteristics, in combination with the gestagen progesterone and other factors. Determination of the biological activity of E. is done with the help of the Allen-Doisy test on castrated female mice or rats. Estradiol is the most active E. [ED_{50} (rat s.c.) 0.1 µg]. The other mentioned E. exhibit the following ED_{50} values: estrone 0.8 µg, estriol 10 µg, equilin 10 µg, and equilenin 17 µg. Non-steroidal E. are, e.g. some longer known stilbene derivatives such as *diethylstilbestrol*. Non-steroidal plant constituents with E.-like activity are also known (phytoestrogens), such as the *isoflavone genistein, the naphthochromene *miroestrol, and the *pterocarpan *coumestrol.

Use: E. are used therapeutically for, e.g., menstrual disorders, climactic complaints, and in combination with gestagens to inhibit ovulation (oral contraceptives). Among others, the synthetic E. *ethinylestradiol* and *mestranol* are very effective for the latter purpose.

Lit.: Acta Crystallogr., Sect. C **55**, 425 (1999) (structure equilin) ▪ Bohl & Duax (eds.), Molecular Structure and Biological Activity of Steroids, Boca Raton: CRC Press 1992 ▪ J. Am. Chem. Soc. **121**, 8237 (1999) (synthesis estradiol) ▪ Nicolaou, p. 153–166 (synthesis) ▪ Tetrahedron Lett. **40**, 907 (1999) (synthesis equilenin) ▪ Ullmann (5.) **A 13**, 119–123 ▪ Zeelen, p. 201–234. – *[HS 293792; CAS 56-53-1 (diethylstilbestrol); 57-63-6 (ethinylestradiol); 72-33-3 (mestranol)]*

Estrone see estrogens.

Ether lipids (alkoxylipids). Name for a group of lipids containing one or more long-chain alkyl or 1-alkenyl groups linked by ether bridges. The most widely distributed are the ether glycerolipids in which the alkyl or 1-alkenyl groups are bound to the oxygen atoms of glycerol (see batyl alcohol). 1-Alkenyl ether compounds of glycerol are better known as *plasmalogens*. E. of glycerol typically occur in animals and archaebacteria. The liver oils of some cartilaginous fish (sharks, rays or skates), e.g., dogfish (*Squalus acanthias*) and *Chimaera monstrosa* contain 45–70% E. E. with highly unsaturated alkyl chains, so-called *fecapentaenes, have been isolated from human feces; they have been connected with the development of large intestine cancer. Synthetic E. are used as cytostatic agents in cancer chemotherapy, see also platelet activating factor.

Lit.: Baumann (ed.), Ether Lipids in Oncology, Champaign: Am. Oil Chem. Soc. 1987 ▪ Braquet, Mangold & Vargaftig (eds.), Biologically Active Ether Lipids, Basel: Karger 1988 ▪ J. Clin. Biochem. Nutr. **14**, 71 (1993) ▪ Kates, Kushner & Matheson (eds.), Biochemistry of Archaea (Archaebacteria), Amsterdam: Elsevier Sci. Publ. 1993 ▪ Mangold & Paltauf (eds.), Ether Lipids: Biochemical and Biomedical Aspects, New York: Academic Press 1983.

Ethisolide see avenaciolide.

Ethyl-1,4-benzoquinone (ethyl-*p*-benzoquinone).

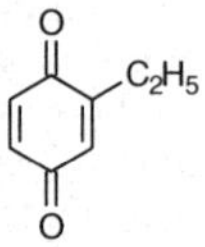

$C_8H_8O_2$, M_R 136.15, yellow cryst., mp. 37–38 °C. Component of the *defensive secretions of many insects (especially beetles) and millipedes. In the beetle *Eloides longicollis* E. and *methyl-1,4-benzoquinone are formed via the acetate metabolism, while the also present 1,4-benzoquinone is formed from aromatic amino acids [1].

Lit.: [1] J. Am. Chem. Soc. **88**, 1590 ff. (1966).
gen.: Blum, Chemical Defenses of Arthropods, p. 181–195, New York: Academic Press 1981 ▪ Biochem. Syst. Ecol. **21**, 143–162 (1993). – *[CAS 4754-26-1]*

Ethyl (2*E*,4*Z*)-2,4-decadienoate see fruit esters.

2-Ethyl-3,5(6)-dimethylpyrazine see pyrazines and pheromones.

5-Ethyl-2,4-dimethylthiazole see meat flavor.

Ethylene see 1-aminocyclopropanecarboxylic acid and plant growth substances.

4-Ethylguajacol see coffee flavor.

Ethyl hexanoate see fruit esters.

Ethyl 3-hydroxy-3-methylbutanoate see fruit flavors (bilberry flavor).

(*E*)-4-Ethylidene-L-glutaminic acid see 4-methyleneglutamic acid.

Ethyl 3-mercaptopropanoate see fruit flavors (grape flavor).

3-Ethyl-5-methyl-1,2,4-trithiolane see vegetable flavors (leek).

Ethyl octanoate see fruit esters.

4-Ethylphenol see wine flavor.

Ethyl propanoate see fruit esters.

Ethylvanillin see 3,4-dihydroxybenzaldehydes.

Etioline see Solanum steroid alkaloids.

Eucalyptus oils. E. mostly occur as by-products in the treatment of eucalyptus trees cultivated for the production of wood, paper, cellulose, charcoal, etc. The total area world-wide used for the cultivation of eucalyptus plants is estimated as 6,000,000 ha [1]. With over 600 species, originally endemic to Australia and Tasmania, *Eucalyptus* is now grown in all warmer regions of the world. The commercially important *essential oils obtained from *Eucalyptus* plants are:
1. ***Cineole-rich eucalyptus oil:*** A readily mobile, almost colorless to light yellow oil with a characteristic fresh, camphor-like odor and a content of 80–85% 1,8-cineole.
Production: By steam distillation of leaves of *Eucalyptus globulus* (Spain, Portugal, China, Chile), *E. smithii* (South Africa), *E. polybractea* and *E. radiata* (Australia). The world-wide annual production is probably between 2000 and 3000 t.
Composition [1,2]***:*** Besides the main component *1,8-*cineole* minor components are *α-*pinenes*, *p-cymol* (see cymenes), *limonene* (see *p*-menthadienes), and *trans-pinocarveol* (see pinenols).
Use: For perfuming household articles, for technical perfuming; for aromatizing confectionery ("eucalyptus sweets") and oral hygiene products; in medicine for rhinologic, balneotherapy, mouth and throat, bronchological, expectorant, and antirheumatic ointment uses; as starting material for pure 1,8-cineole.
2. ***Eucalyptus citriodora oil:*** Almost colorless to light yellowish-green oil with a strong, fresh, citrus-, rose-like odor.
Production: By steam distillation of the leaves of *Eucalyptus citriodora*. Main areas of cultivation are Brazil, China, and India. Annual production world-wide somewhat more than 800 t.
Composition [3]***:*** Main component is racemic **citronellal* (>70%). Minor components are **citronellol*, *isopulegol*, and *neo-isopulegol* (see *p*-menthenols).
Use: For perfuming household articles, as starting material for the isolation of pure citronellal, which is also used in perfume industry or for production of other fragrance substances by chemical reactions.

Lit.: [1] Boland, Brophy & House, Eucalyptus Leaf Oils, Melbourne/Sidney: Inkata Press 1991. [2] Perfum. Flavor. **11** (6), 39 (1986); **15** (6), 58 (1990); **18** (3), 72 (1993); **19** (6), 61 (1994). [3] Perfum. Flavor. **4** (1) 49 (1979); **11** (4) 74 (1986); **13** (1), 46 (1988).
gen.: Arctander, p. 227 ▪ Bauer et al. (2.), p. 153, 154 ▪ ISO 770 (1980), 3044 (1974). – *Toxicology:* Food Chem. Toxicol. **26**, 323 (1988) ▪ Food Cosmet. Toxicol. **13**, 107 (1975). – *[HS 3301 29; CAS 8000-48-4 (1.); 85203-56-1 (2.)]*

Eucarvone (2,6,6-trimethyl-2,4-cycloheptadien-1-one).

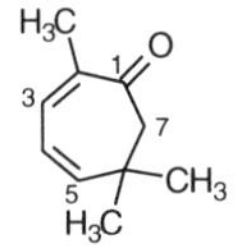

$C_{10}H_{14}O$, M_R 150.22, oil, bp. 82–84 °C (800 Pa). Monocyclic 7-membered *monoterpene, occurring in the essential oils of the rhizomes of the Korean and Chinese plants "Xi-Xin" (*Asarum sieboldii*, *A. heterotropoides*, Aristolochiaceae) [1]. The content of E. in the Chinese commercial drug varies widely, from traces to the main component. For synthesis, see *Lit.* [2].
Lit.: [1] Hegnauer **III**, 188 f. [2] Tetrahedron Lett. **1981**, 645.
gen.: Karrer, No. 568. – *[CAS 503-93-5]*

Eucommiol.

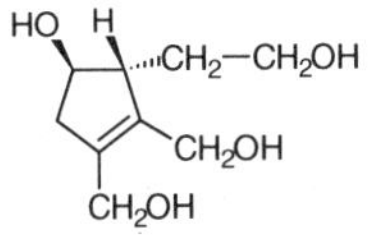

$C_9H_{16}O_4$, M_R 188.22, oil, bp. 200 °C (20 Pa), $[\alpha]_D$ −30.5° (CH_3OH). E. is a cyclopentanoid natural product formed by reduction of a C_9-*iridoid and occurs in the leaves of *Eucommia ulmoides* (sole species of the Eucommiaceae) [1]. *E. ulmoides* is endemic to China, but is cultivated in many countries [2]; in China the extract is a valued drug (antirheumatic, hypotensive) [3].
Lit.: [1] Tetrahedron **30**, 4117 (1974). [2] Hegnauer **IV**, 100. [3] Pharmazie **13**, 436 (1958).
gen.: Chem. Pharm. Bull. **45**, 1094 (1997) ▪ J. Nat. Prod. **43**, 649, No. 173 (1980). – *[CAS 55930-44-4]*

Eudesmanes (selinanes). Bicyclic *sesquiterpenes with 7-isopropyl-1,4a-dimethyldecalin skeleton.

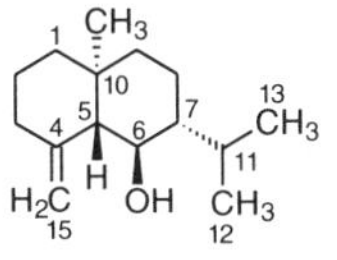

(-)-Laevojunenol (**1**)

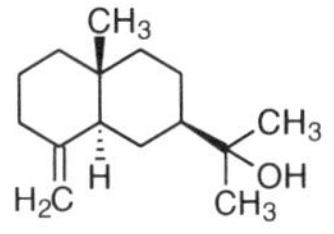

(+)-(β)-Eudesmol (**2**)

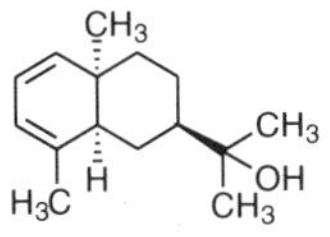

(+)-Occidentalol (**3**)

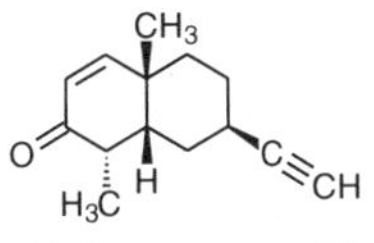

(-)-Chamaecynone (**4**)

Four types can be distinguished on the basis of the ring linkages, their biogenesis proceeds via the corresponding conformers of the *germacrane cation. Two of them are *trans*-decalins with the basic structures of (−)-eudesmane, e.g., (−)-*laevojunenol* [1], and (+)-eudesmane, e.g. (+)-(*β*)-*eudesmol* [2]. The epi-eudesmanes, e.g., (+)-*occidentalol* [3], and the chamaecynanes (nor-

Table: Data of Eudesmanes.

	molecular formula	M_R	mp. [°C]	optical activity	CAS
1	$C_{15}H_{26}O$	222.37	65 cryst.	$[\alpha]_D^{20}$ −57°	30951-17-8
2	$C_{15}H_{26}O$	222.37	76 cryst.	$[\alpha]_D^{20}$ +68.3° ($CHCl_3$)	473-15-4
3	$C_{15}H_{24}O$	220.35	97.5–98 cryst.	$[\alpha]_D^{20}$ +363.2° ($CHCl_3$)	473-17-6
4	$C_{14}H_{18}O$	202.30	92 cryst.	$[\alpha]_D$ −93.3° (CH_3OH)	10208-54-5

eudesmanes with an ethynyl group), e.g., *chamaecynone*[3], possess the basic skeleton of *cis*-decalin. Eudesmols are major components of *eucalyptus oils. They are valued in perfumery as fixatives. Eudesmyl acetate serves as a substitute for *linalool and bergamot.

Lit.: [1]Tetrahedron **48**, 851 (1992); **49**, 4761 (1993). [2]J. Org. Chem. **48**, 4380 (1983); **50**, 1359 (1985); **53**, 2555 (1988); Phytochemistry **28**, 1909 (1989); Tetrahedron **50**, 4755 (1994). [3]Tetrahedron **49**, 4761 (1993).

Eudesmanolides. *Sesquiterpene lactones with the *eudesmane basic skeleton mainly from Asteraceae. On the basis of the ring closure of the γ-lactone group two forms are distinguished: the *santanolides* in which the lactone oxygen is linked to carbon atom 6 (e.g., *santonin, *artemisin) and the *alantolides* (e.g., *helenin[1]) in which linkage occurs at carbon atom 8.

Lit.: [1]Beilstein E V **17/10**, 97; Tetrahedron **33**, 617 (1977).
gen.: Planta Med. **14**, Suppl., 97 (1966).

Eudesmene see selinenes.

Eudesmol see eudesmanes and geranium oil.

Eudistomins.

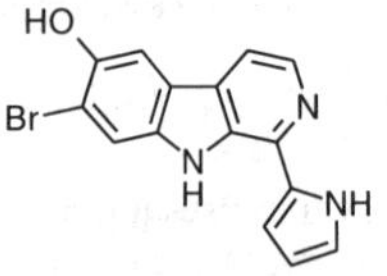

Eudistomin A

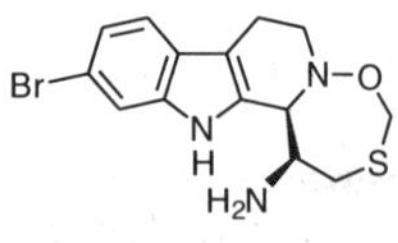

Eudistomin K

The E. A–V are β-carbolines from the tunicate *Eudistoma olivaceum.* They are cytotoxic against murine leukemia cells L1210 (E. B: IC_{50} 3.4 µg/mL) and L51784 (E. B: IC_{50} 3.1 µg/mL), calmodulin antagonists (E. A: $IC_{50}=3\times10^{-5}$ M), and active against Herpes simplex virus type 1. *E. A*, $C_{15}H_{10}BrN_3O$, M_R 328.17, yellow oil; *E. K*, $C_{14}H_{16}BrN_3OS$, M_R 354.26, yellow oil, $[\alpha]_D$ −102° (CH_3OH). The eudistomidins from *E. glaucus* are structurally related to the E.[1].

Lit.: [1]J. Org. Chem. **55**, 3666–3670 (1990).
gen.: Chem. Rev. **93**, 1771 (1993) (review) ▪ J. Am. Chem. Soc. **109**, 3378–3387 (1987) ▪ J. Nat. Prod. **61**, 959 (1998) (E. V) ▪ J. Org. Chem. **56**, 5369 (1991) ▪ Nat. Prod. Lett. **9**, 7 (1996). – *Pharm. Review:* Exp. Opinion Ther. Patents **6**, 201 ff. (1996). – *Synthesis:* Heterocycles **29**, 2507 (1989); **30**, 229 (1990); **49**, 499 (1998) (review) ▪ Synth. Commun. **25**, 3373–3379 (1995) ▪ Tetrahedron Lett. **30**, 2099, 3605, 7117 (1989); **36**, 7085 (1995); **37**, 9353 (1996). – *[CAS 88704-36-3 (E. A); 88704-52-3 (E. K)]*

Eudistone A see shermilamines.

Eugenol (4-allyl-2-methoxyphenol).

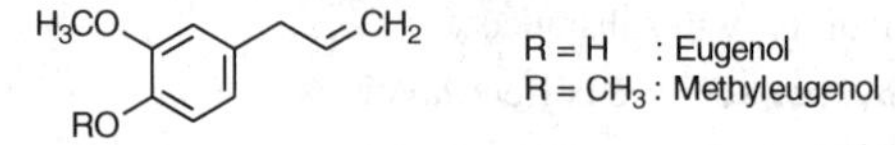

$C_{10}H_{12}O_2$, M_R 164.20, oil, bp. 253 °C, mp. −9 °C, D^{20} 1.0664, n^{20} 1.5410. E. is the main component of the flower odor of *Eugenia* and *Cassia* species as well as a component of numerous essential oils: *clove oils (80%), allspice and allspice leaf oils (60–90%), bay oil (60%), cinnamon bark oil (4–10%), and *sweet basil oil, it smells of carnations. *Methyleugenol*: $C_{11}H_{14}O_2$, M_R 178.23, bp. 255 °C, is a component of numerous essential oils. The acetate (*aceteugenol*), $C_{12}H_{14}O_3$, M_R 206.24, mp. 30–31 °C, bp. 280–281 °C, occurs in *Eugenia caryophyllata* and in *Cinnamomum zeylanicum.* E. palmitate (*oryzarol*), $C_{26}H_{42}O_3$, M_R 402.62, is isolated from rice (*Oryza sativa*). E. *O*-β-D-glucopyranoside (*citrusin C*), $C_{16}H_{22}O_7$, M_R 326.35, mp. 130–131 °C, $[\alpha]_D^{22}$ −50.7° (CH_3OH), is a component of *Perilla frutescens.*

Toxicology: In experimental studies at high concentrations and long exposure times E. exhibits carcinogenic, mutagenic, and skin-irritating properties.

Use: E. is used in dentistry as an anesthetic and as an additive to fillings (together with E. ethers, E. benzoate, E. cinnamate, E. acetate, etc.). E. and methyleugenol are important fragrance and aroma substances. E. is an insect attractant.

Lit.: Beilstein E IV **6**, 6337 ▪ J. Chem. Soc., Chem. Commun. **1975**, 879 f. (biosynthesis) ▪ J. Econ. Entomol. **82**, 123–129 (1989) ▪ Karrer, No. 185 ▪ Phytochemistry **31**, 3265 ff. (1992) (citrusin C) ▪ Synthesis **1983**, 701 (synthesis) ▪ Ullmann (5.) **A 8**, 558. – *Toxicology:* Sax (8.), EQR500. – *[HS 2909 50; CAS 97-53-0 (E.); 93-15-2 (Methyleugenol); 93-28-7 (aceteugenol); 18604-50-7 (citrusin C)]*

Eugenyl acetate see cinnamon leaf (bark) oil.

Euphohelionone see lathyranes.

Euphol see triterpenes.

Euphorbiaceae and Thymelaeaceae poisons. The discovery of skin-irritating and concomitant tumor-promoting (cocarcinogenic) activities in Euphorbiaceae and Thymelaeaceae prompted research into the isolation and structural elucidation of their components. The irritating and cocarcinogenic principles were found to be derivatives of the *diterpene hydrocarbons **tigliane*, ingenane, and *14,15-seco-tigliane*[1], such as, e.g., esters of phorbols (see tigliane), *ingenol, and *daphnetoxin as well as some of their deoxy- and hydroxy products. 10–12 berries of mezereon can be lethal for children. A prerequisite for the biological activity is esterification at C_{12} and/or C_{13} (see formula of phorbol under tigliane) by mostly short- and long-chain fatty acids or aromatic acids. These compounds are strong skin irritants, lead to formation of blisters, necrosis, and on contact with the eyes to blindness under certain circumstances. They are among the most active of the currently known cocarcinogens, and influence cell growth and proliferation by activation of protein kinase C. Thus, for example mouse epidermal cells proceed directly from the G_1-phase to DNS synthesis and mitosis. The tigliane skeleton is formed

biosynthetically by cyclization of a *casbene derivative.
Some species of Euphorbiaceae and Thymelaeaceae are grown as ornamental plants and many are cultivated for commercial purposes. Whole plants and plant parts are used as traditional and veterinary drugs or Euphorbiaceae are added to animal fodder. Furthermore, some Euphorbiaceae are of technological importance, e. g., for obtaining fats and oils as well as rubber and starch.

Lit.: [1] Naturwissenschaften **65**, 640 (1978).
gen.: Arch. Toxikol. **44**, 279 (1980) ▪ Experientia **37**, 681 (1981) ▪ Synlett **1994**, 325 ▪ Zechmeister **44**, 1.

Eusiderin A see lignans.

Eustomorusside, eustomoside see secoiridoids.

Eutypin(e) see siccayne.

Euxanthinic acid, euxanthone see Indian yellow.

Evernic acid ethyl ester see oak moss absolute.

Everninomicin see ziracin.

Evodin see limonin.

Exaltolide® see 15-pentadecanolide.

Exotoxins. Bacterial toxins which, in contrast to *endotoxins, are secreted by living, mostly Gram-positive bacteria (*Staphylococcus aureus* E.; *Cornyebacterium diphtheriae* see diphtheria toxin; *Clostridium botulinum* see botulism toxin; *Clostridium tetani*). E. are mostly proteins (molecular mass between 24000 and 1300000), some of which have enzyme functions (e. g., phospholipase, DNase, protease); *β*-exotoxins are nucleosides. Even at low concentrations E. exhibit specific, in some cases extremely toxic, activities for the respective host cell, but can be deactivated usually by heat or chemical modification (e. g., with formaldehyde) without loss of the antigen properties (toxoids). Inactivated E. are thus used for active immunizations (diphtheria, tetanus).

Extremozymes (extremoenzymes). Enzymes from microorganisms living under extreme conditions: in hot springs (temperature >90 °C to well over 100 °C at high water pressures), saturated salt concentrations, pH-values >12 and <2, and temperatures below 0 °C. On account of particular structural properties, they are stable and able to function under such unfavorable conditions (compare with archaebacteria). E. have gained major industrial significance as biocatalysts in biochemistry and synthesis. Thus, the development of molecular biology sequencing methods and the polymerase chain reaction (PCR) essential for DNA (gene) analysis would not have been possible without the E.

Lit.: Review: ACS Symp. Ser. **498** (1992) ▪ Adv. Biochem. Eng. Biotechnol. **61** (1998) ▪ Biotechnol. **13**, 662–668 (1995) ▪ Chem. Eng. News (18.12.) **1995**, 32–42; (14.10.) **1996**, 31 ff. ▪ Chem. Ind. (London) **1995**, 689–693 ▪ Chemistry (ACS), Oct. **1998**, p. 15–20 ▪ J. Biotechnol. **33**, 1–14 (1994) ▪ World J. Microbiol. Biotechnol. **11**, 85–94 (1995).

F

F-acids (furan fatty acids, furanoid fatty acids). The F-a., first isolated in 1966 from *Exocarpus* seed oil and then in 1974 from fish oils, are multiply substituted furan derivatives of fatty acids.

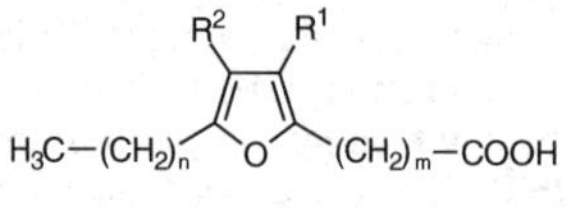

R^1,R^2 = H,CH_3; m = 2,4,6-12,14; n = 2-6

The content of esterified F-a. in cod-liver oil and other fish oils is ca. 1%. The F-a. in fish oils and mammalian lipids originate mainly from consumed food, e. g. from algae [1]. The seed oil of *Exocarpus cupressiformis* (Santalaceae) contains *5-hexyl-2-furanoctanoic acid* ($C_{18}H_{30}O_3$, M_R 294.43, mp. 20–21 °C) as the glycerol ester. Esters of F-a. occur in small amounts in plant oils [2], e. g., soya, wheatgerm, rape, and corngerm oils as well as in the membrane lipids of animal and plant cells; cholesterol esters of F-a., in particular, are found in blood. An intermediate in the biosynthesis of F-a. is the 13-hydroperoxide of linoleic acid, which is transformed to F-a. by a specific 13-lipoxygenase. Degradation products of F-a., the so-called urofuran acids, are found in mammalian urine [3]. Saturated F-a., so-called tetrahydrofuran (THF) fatty acids have also been found in animals and plants [3,4], e. g., *(6R,9S,10S)-6,9-epoxy-10-hydroxyhexadecanoic acid* ($C_{16}H_{30}O_4$, M_R 286.41, mp. 43.5 °C, $[\alpha]_D^{18}$ +108°) in wool fat (see also annonaceous acetogenins). THF fatty acids are formed via epoxy-hydroxy compounds as intermediates or by transfer of hydrogen to F-a.[3]. After injury to plant tissue F-a. and their unstable ene-dione degradation products are probably formed as defence compounds (*phytoalexins) against phytopathogenic fungi [1].

Lit.: [1] J. Lipid Mediators **7**, 199 (1993); Lipids **23**, 1032 (1988). [2] Fat Sci. Technol. **93**, 249 (1991). [3] Justus Liebigs Ann. Chem. **1985**, 1168; **1986**, 127; **1989**, 449; **1993**, 251 ff. [4] Aust. J. Chem. **33**, 891 (1980); **43**, 895 (1990).
gen.: Nachr. Chem. Tech. Lab. **35**, 1240 (1987) ▪ Tetrahedron Lett. **39**, 333 (1998) (synthesis F-a. F_S).

Factor I (2-methyl-2,3-dihydrouroporphyrin III).

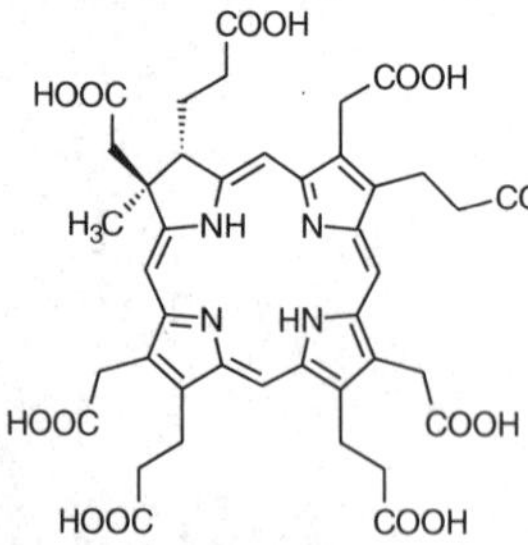

$C_{41}H_{42}N_4O_{16}$, M_R 846.80, uv$_{max}$ (octamethyl ester in MeOAc) 644, 615, 592, 498, 489, 393 nm. F. I belongs to the structural class of the *chlorins.

Occurrence: F. I is formed as oxidation product of *precorrin 1 on isolation from *vitamin B_{12}-producing bacteria, e. g., *Propionibacterium shermanii* and *Clostridium tetanomorphum*. Precorrin 1 is the tetrahydro form of F. I and has been identified as the biosynthetic intermediate directly following uroporphyrinogen III (see uroporphyrins) in the biosynthetic pathway leading to vitamin B_{12}.

Synthesis & biosynthesis: In connection with biosynthetic studies on vitamin B_{12} Battersby realized the total synthesis of F. I. The 2-methyl group on ring A of F. I originates from **S*-adenosylmethionine, and the carbon skeleton from uroporphyrinogen III, the key building block in porphyrin biosynthesis.

Lit.: Chem. Rev. **94**, 327 (1994) ▪ Dolphin B12 **I**, 130 f. ▪ J. Biol. Chem. **266**, 23840 (1991) (biosynthesis) ▪ J. Chem. Soc., Chem. Commun. **1982**, 455; **1985**, 583; **1989**, 428 ▪ J. Chem. Soc., Perkin Trans. 1 **1988**, 1577 ▪ Jordan, p. 108 ▪ Z. Physiol. Chem. **358**, 339 (1978). – *[CAS 73528-45-7]*

Factor F 430.

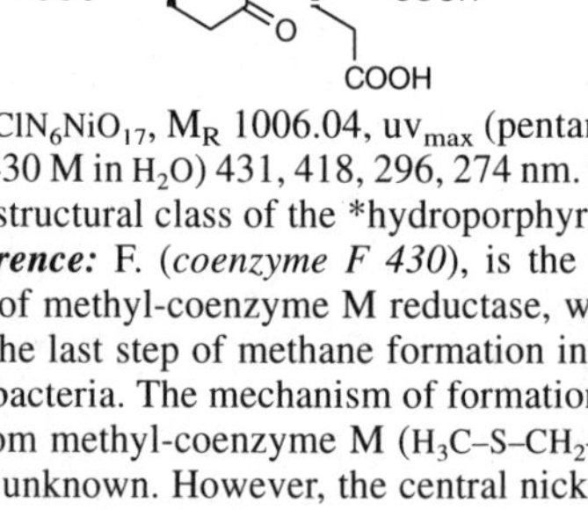

$C_{42}H_{51}ClN_6NiO_{17}$, M_R 1006.04, uv$_{max}$ (pentamethyl ester, F 430 M in H_2O) 431, 418, 296, 274 nm. F. belongs to the structural class of the *hydroporphyrins.

Occurrence: F. (*coenzyme F 430*), is the prosthetic group of methyl-coenzyme M reductase, which catalyzes the last step of methane formation in methanogenic bacteria. The mechanism of formation of methane from methyl-coenzyme M ($H_3C–S–CH_2–CH_2–SO_3^-$) is still unknown. However, the central nickel atom in F. most certainly participates in the reduction.

Biosynthesis: F. is formed from *5-amino-4-oxovaleric acid via the normal biosynthetic route for porphinoid compounds. The 2- and 7-methyl groups originate from **S*-adenosylmethionine.

Lit.: Chem. Rev. **94**, 327 (1994) ▪ Chimia **41**, 277 (1987); **48**, 50 (1994) ▪ J. Am. Chem. Soc. **119**, 733 (1997) (synthesis) ▪ Jordan, p. 143–147 ▪ New Compr. Biochem. **19**, 139 (1991) (review). – *[CAS 73145-13-8 (cation)]*

FAD. Abbreviation for flavine adenine dinucleotide (see riboflavin(e)).

Fagopyrin see hypericin.

Falcarinol (1,9-heptadecadiene-4,6-diyn-3-ol).

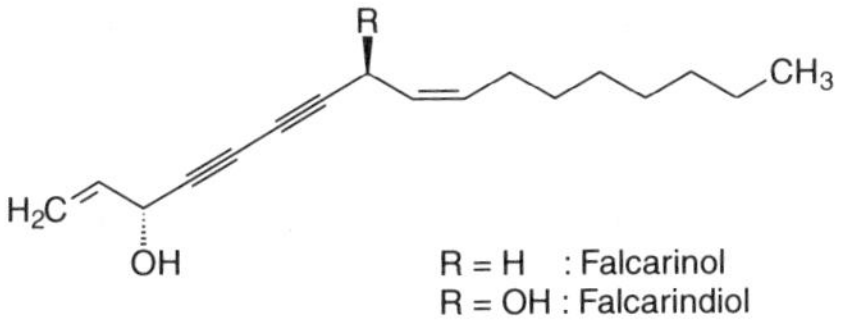

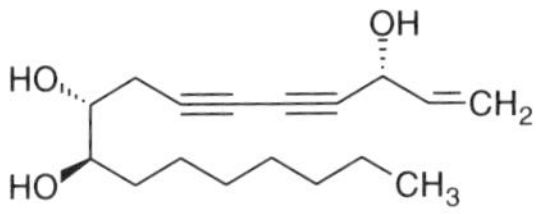

Panaxytriol

$C_{17}H_{24}O$, M_R 244.38, (*R*)-(*Z*)-(−)-form, F.: oil, bp. 115 °C (2.6 Pa), $[\alpha]_D^{20}$ −22.5° (ether). For the absolute configuration of F. and the (+)-enantiomer *panaxynol*, see *Lit.*[1]. The enantiomers are constituents of *Falcaria vulgaris*, ginseng roots (*Panax ginseng*), the Chinese drug Fangfeng (roots of *Saposhnikovia divaricata*), English ivy (*Hedera helix*), vegetable species of the Apiaceae, e.g., carrots (*Daucus carota*), celery and parsley as well as *Schefflera* species (Araliaceae). The corresponding diol, *falcarindiol*, $C_{17}H_{24}O_2$, M_R 260.38, occurs, among others, in the Chinese drug Shuifangfeng (roots of *Libanotis laticalycina*) and carrots. *Panaxytriol* [falcarintriol, $C_{17}H_{26}O_3$, M_R 279.38, needles, $[\alpha]_D^{20}$ −21.2° ($CHCl_3$)] does also occur in ginseng roots and is a potent tumor inhibitor[2]. The total synthesis is described in *Lit.*[3]. F. is formed by the plants under conditions of stress and can trigger an allergic contact dermatitis in humans.

Activity: F. exhibits antibacterial[4] and cytotoxic (antitumor)[5] activities and has as well antifungal[6] and analgesic properties, it also inhibits the aggregation of blood platelets[7].

Lit.: [1] Tetrahedron Lett. **35**, 993 f. (1994). [2] Chem. Pharm. Bull. **39**, 521 (1991); **45**, 1114 (1997). [3] Synlett **1998**, 737. [4] Herba Pol. **38**, 137–140 (1992). [5] BioMed. Chem. Lett. **3**, 581–584 (1993); Chem. Pharm. Bull. **41**, 549–552 (1993); J. Med. Chem. **35**, 3007–3011 (1992); J. Nat. Prod. **55**, 667–671 (1992). [6] J. Sci. Food Agric. **63**, 313–317 (1983). [7] Fitoterapia **64**, 179 (1993).

gen.: Bull. Chem. Soc. Jpn. **62**, 2977–2980 (1989) ▪ Contact Dermatitis **17**, 1–9 (1987) (toxicology) ▪ J. Nat. Prod. **59**, 748–753 (1996) (isolation); **62**, 626 (1999) (synthesis falcarindiol) ▪ Phytochemistry **30**, 3327–3333, 4151 f., 4189 f. (1991); **31**, 3621 ff. (1993) ▪ Tetrahedron Lett. **40**, 2181 (1999) (absolute configuration). – *[CAS 21852-80-2 ((R)-(Z)); 81203-57-8 ((S)-(Z)); 55297-87-5 ((3R,8S,9Z) (falcarindiol)); 87005-03-6 (panaxytriol)]*

Falconensins. Azaphilone derivatives (F. A–H) from the ascomycete fungus from Venezuelan soil *Emericella falconensis*, e.g., *F. A*: $C_{23}H_{24}Cl_2O_7$, M_R 483.34, pale yellow needles with blue fluorescence, mp. 84 °C, $[\alpha]_D^{20}$ +238° (CH_3OH), related to *mitorubrin.

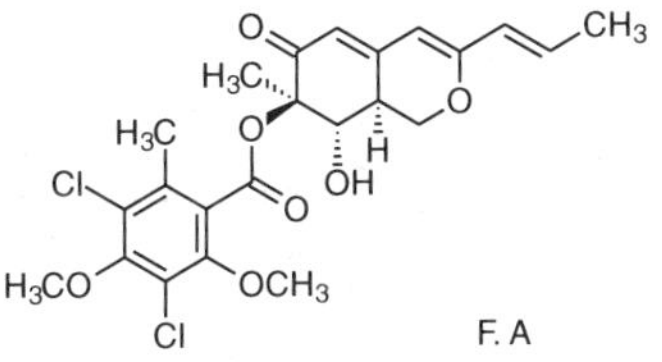

F. A

Lit.: Chem. Pharm. Bull. **40**, 3142–3144 (1992); **44**, 2213–2217 (1996). – *[CAS 147614-33-3 (F. A)]*

Faranal [(3*S*,4*R*,6*E*,10*Z*)-3,4,7,11-tetramethyl-6,10-tridecadienal].

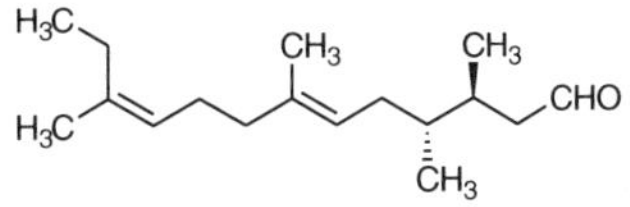

$C_{17}H_{30}O$, M_R 250.42, oil, bp. 60–70 °C $[\alpha]_D^{23}$ +16.2° (hexane). Trail-marking *pheromone of the Pharaoh ant (*Monomorium pharaonis*) with an efficacy of 1 pg/cm trail.

Lit.: Tetrahedron Lett. **1977**, 2617–2620 (isolation); **22**, 461–464 (1981). – *Synthesis:* ApSimon **9**, 148–152 ▪ Justus Liebigs Ann. Chem. **1995**, 2089 ff. ▪ Tetrahedron **49**, 1025–1042 (1993). – *[CAS 65395-77-9]*

Farnesene.

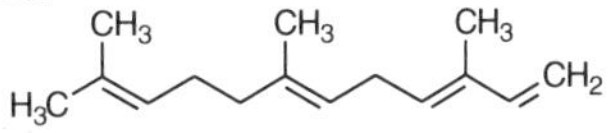

α

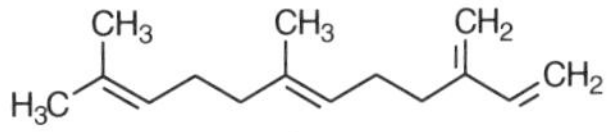

β

$C_{15}H_{24}$, M_R 204.36. F. exists in nature as *α*-F. [colorless liquid, bp. 98–102 °C (665 Pa), n_D^{25} 1.4790] and *β*-F. [colorless liquid, bp. 95–107 °C (665 Pa), D_{18} 0.8363, n_D^{20} 1.4995]. *α*-F. is a component of the glandular secretion of the ant species *Aphaenogaster longiceps*[1]. It acts as an attractant in larvae of the apple moth (*Laspeyresia pomonella*)[2] (see also codlemone). *β*-F. was first detected in hop oil[3]. It is an extremely effective alarm pheromone of aphids. Both structural isomers as well as mixtures of their *E*- and *Z*-isomers occur in many essential oils (e.g., of apples and citrus fruits[4]).

Lit.: [1] Experientia **43**, 345 (1987); J. Chem. Ecol. **14**, 825 (1988). [2] ApSimon **9**, 273 f.; Nature (London) **239**, 170 (1972). [3] J. Chem. Ecol. **15**, 2061 (1989); Synth. Commun. **20**, 423 (1990). [4] Aust. J. Chem. **23**, 2101 (1970); J. Agric. Food Chem. **28**, 680 (1980).

gen.: Beilstein E IV **1**, 1133 ▪ Bioorg. Med. Chem. **4**, 283 (1996) (synthesis *α*-F.) ▪ J. Org. Chem. **60**, 6211 (1995) (synthesis *α*-F.) ▪ Merck-Index (12.), No. 3976, 3977. – *[CAS 502-61-4 (α-F.); 18794-84-8 (β-F.)]*

Farnesol (3,7,11-trimethyl-2,6,10-dodecatrien-1-ol).

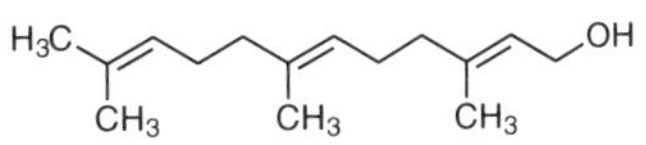

E/*E*-form

$C_{15}H_{26}O$, M_R 222.37, colorless oil, bp. 160 °C ($17.9 \cdot 10^3$ Pa); soluble in ether and THF. It occurs in musk kernels, lime-tree blossoms, *Acacia* species and in essential oils. On account of its intense odor of lily of the valley it is used in the perfume and soap industries. F. acts as a *pheromone for bumblebees. Furthermore, F. diphosphate is the starting material for the biosynthesis of all sesquiterpenes (e.g. *dehydrogeosmin, *drimanes etc.). It is sensitive to air, light, and heat and is oxidized to farnesal, farnesenic acid, etc.

Lit.: Beilstein E IV **1**, 2335 ▪ Bioorg. Med. Chem. **2**, 631 (1994) (synthesis) ▪ Helv. Chim. Acta **59**, 1233 (1976) (synthesis) ▪ J. Chem. Ecol. **14**, 1153 (1988) (isolation) ▪ Merck-Index (12.), No. 3978 ▪ Phytochemistry **29**, 3469 (1990); **36**, 1203 (1994) (biosynthesis). – *[HS 2905 22; CAS 4602-84-0]*

Farnesyl acetate see ambrette seed oil.

Fascaplysin(e) see oxindole alkaloids.

Fasciculic acids (fasciculols).

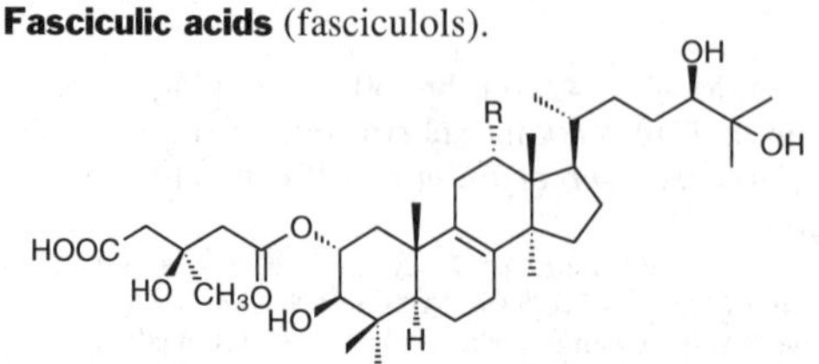

R = H : Fasciculic acid A
R = OH : Fasciculic acid B

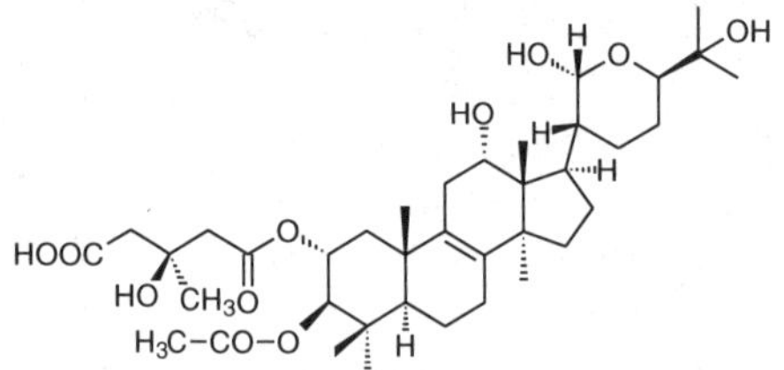

Crustulinol-diester

The F. are terpenoids of the lanostane type isolated from fruit bodies of the sulfur tuft (*Hypholoma fasciculare*, Basidiomycetes)[1]. *F. A* {$C_{36}H_{60}O_8$, M_R 620.87, needles, mp. 177–179 °C, $[\alpha]_D$ +8.3° (CH_3OH)} and a series of closely related compounds act as specific calmodulin antagonists, with *F. B* {$C_{36}H_{60}O_9$, M_R 636.87, powder, mp. 98–103 °C, $[\alpha]_D$ +19.3° (CH_3OH)} being particularly active. Related compounds, such as *3β-acetyl-2α-((S)-3-hydroxy-3-methylglutaryl)-crustulinol* {$C_{38}H_{60}O_{11}$, M_R 692.89, cryst., mp. 204–205 °C, $[\alpha]_D$ –10.8° (CH_3OH)} which is cytotoxic and causes paralysis in mice after intraperitoneal administration, occur in *Hebeloma crustuliniforme*, *H. sinapizans*, and *H. spoliatum*[2].

Lit.: [1] Chem. Pharm. Bull. **33**, 3821 (1985); **37**, 3247 (1989); **41**, 1738 (1993); Phytochemistry **20**, 1867 (1981); **27**, 1439 (1988); **31**, 4355 (1992). [2] Tetrahedron Lett. **24**, 1635 (1983); Chem. Pharm. Bull. **40**, 869 (1992). – *[CAS 126906-00-1 (F. A); 126882-55-1 (F. B); 86894-26-0 (crustulinol diester)]*

Fasciculins.

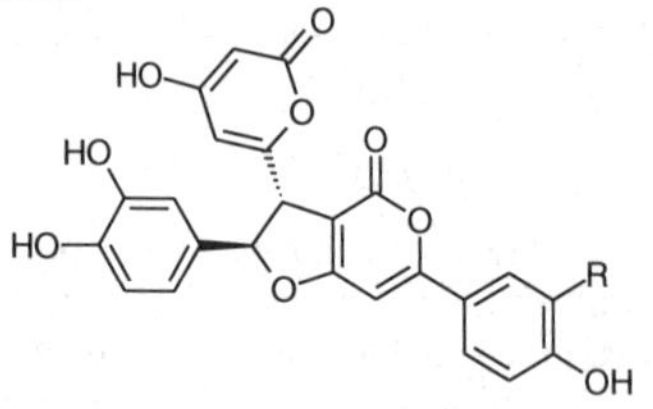

R = H : Fasciculin A
R = OH : Fasciculin B

F. A ($C_{24}H_{16}O_9$, M_R 448.39; tetramethyl ether: pale yellow cryst., mp. 110–112 °C, optically active) and *F. B* {$C_{24}H_{16}O_{10}$, M_R 464.39; pentamethyl ether: ivory-colored needles, mp. 140–145 °C, $[\alpha]_{578}$ +3.7° (CH_3OH)} occur in the fruit bodies of the sulfur tuft (*Hypholoma fasciculare*) as well as other *Hypholoma* and *Pholiota* species. The F. exhibit blue fluorescence under UV light and are biosynthetically closely related to the yellow *hypholomines.

Lit.: Chem. Ber. **110**, 1047 (1977) ▪ Tetrahedron Lett. **38**, 363 (1997) ▪ Zechmeister **51**, 92–96 (review). – *[CAS 62350-93-0 (F. A); 62350-95-2 (F. B)]*

Fats and oils. The differentiation between fats (solid) and oils (liquid) is based on the physical state at room temperature. The physical properties are determined by the chain length and the number of *cis*-C=C double bonds in the fatty acid parts of the triglycerides (triacylglycerols). Longer chains and saturated fatty acids lead to higher melting points while shorter chains and unsaturated fatty acids result in lower melting points. Chemically F. and o. are composed of *triglycerides and accompanying fatty substances, e.g., *sterols of animal or plant origin, *tocopherols, *carotinoids, and phenolic compounds. They are stored as energy-rich compounds in oil seeds or adipose tissues of plants and animals, respectively ("storage fat"). These energy reserves are mobilized if required in plants (e.g., seed germination) and animals (e.g., food deficits).

Composition: The most important plant F. consist almost exclusively of *fatty acids with unbranched chains and an even number of C atoms. Animal F. contain appreciable amounts of fatty acids with singly branched chains and an odd number of C atoms [*iso*- = terminal isopropyl group $-CH(CH_3)_2$, *anteiso*-fatty acids = terminal *sec*-butyl group $-CH(CH_3)C_2H_5$]. In addition, the double bonds of the unsaturated fatty acids in most plant oils have the *cis*-configuration, whereas animal F. often also contain appreciable quantities of *trans*-fatty acids which, e.g., in ruminants especially, originate from the lipids of bacteria of the rumen. The saturated fatty acids esterified with glycerol occurring most frequently in F. are lauric, myristic, palmitic, and stearic acids, as well as the singly unsaturated palmitoleic and oleic acids, the doubly unsaturated linoleic acid, and the triply unsaturated linolenic acid. Unusual fatty acids such as *ricinoleic acid in *Ricinus communis*, *vernolic acid in *Vernonia* sp., calendulic acid in *Calendula officinalis*, *petroselinic acid in *Petroselinum hortense* occur in the seed oils of some plants.

Biosynthesis: F. and o. are formed and stored in the tissue of seeds (e.g., rape, sunflower, soya) or fruits (e.g., olives, palm trees) in the form of minute oil droplets. The droplets occur in the cytoplasm; they are enclosed by a unilamellar membrane of phospholipids and proteins, so-called oleosins. *Triglycerides, the main components of F. and o., are formed by the Kennedy biosynthetic pathway [1].

Lit.: [1] Prog. Lipid Res. **32**, 247 (1993).
gen.: Bockisch, Nahrungsfette u. -öle, Stuttgart: Ulmer 1993 ▪ Gunston, Harwood & Padley, The Lipid Handbook, 2nd edn., London: Chapman & Hall 1994 ▪ Ullmann (5.) **A 10**, 173–243. – *[HS 1519 11–1519 30, 2915 11, 2915 21, 2915 60, 2915 70, 2915 90]*

Fatty acids. General term for carboxylic acids occurring as glycerol esters in the triglycerides of *fats and oils and other lipids (e.g., membrane lipids, phospholipids). F. are principally classified according to their chain lengths (short, medium, long, and very long-chain F.), to the degree of unsaturation (saturated and unsaturated F.), to the type (double or triple bond) and number (mono-, di-, polyene-F.), as well as geometry [(*Z*)- or *cis*- and (*E*)- or *trans*-F.] and position (F. with double bonds in differing positions; isolene- and conjuene-F.) of the multiple bonds, to the branching (e.g., *iso*- and *anteiso*-F., see fats and oils), or the presence of other functional groups in the molecule, e.g., hydroxy (*hydroxy fatty acids), epoxy (see vernolic acid), keto groups and heterocyclic (e.g., *annonaceous acetogenins, *F-acids) and carbocyclic rings, e.g., cyclopropane, cyclopropene (see malvalic acid), cyclopentene rings (see chaulmoogric acid). Essential (necessary for life) F. are not synthesized in animal organisms and must therefore be taken up as food. The cells of higher animals in particular are not able to produce F. with a double bond in the $\omega3$- or $\omega6$-positions. Thus, e.g., *linoleic acid, and *linolenic acid as well as the F. derived therefrom are essential for these organisms.
Occurrence: F. occur in nature in unbound form (so-called free F.) and in chemically bound forms, especially as esters of glycerol (e.g., *triglycerides), glycero-3-phosphate and its derivatives (glycerophospholipids). Derivatives of F.-amides, e.g., sphingomyelines and cerebrosides, are also known. Most widely distributed are palmitic (C_{16}), stearic (C_{18}), oleic (C_{18}), and linoleic acids (C_{18}), occurring as glycerol esters in many plant oils and animal fats. In addition, lauric (C_{12}) and myristic acids (C_{14}) as components of coconut and *palm kernel oil as well as *erucic acid (C_{22}) as the main component of rapeseed oil from older cultivations are widely distributed. Furthermore, F. esters of other natural products such as sterols, sugars and glycosides, aliphatic alcohols are also known.
Biosynthesis: The biosynthesis of long-chain, saturated F. from acetyl- or malonyl-coenzyme A as well as reduced nicotinamide adenine dinucleotide phosphate (NADPH) is catalyzed by fatty acid synthase (EC 2.3.1.85, 2.3.1.86), a multi-enzyme complex. The growing F.-chain is bound as a thioester to one component, the acyl carrier protein (ACP or ACP-SH) and is so transported from one catalytic enzyme center of the enzyme complex to the next.
Production & use: F. are mainly produced by saponification (alkaline hydrolysis) of natural fats. F. and their derivatives are mainly used in the production of soaps, detergents, lubricants, greases, textile auxiliaries, epoxide and alkyde resins, plastics, paints, softeners, cosmetics, and pharmaceuticals. F. methyl esters are also used as a fuel ("biodiesel").
Lit.: Chem. Ind. (London) **1994**, 131–134, 360 (F. and nutrition) ▪ Prog. Lipid Res. **32**, 151–194 (1993) (synthesis of rare F. with unusual structures) ▪ Recent Res. Dev. Lipids **2** (part 2), 371–380 (1998) (review methoxy-F.) ▪ Spec. Publ., Royal Soc. Chem. **180**, 34–56 (1996) (unusual structures) ▪ Ullmann (5.) **A 10**, 245–276 ▪ see also fats and oils.

Fawcettimine see Lycopodium alkaloids.

FDM A₁ see flavomannins.

Febrifugine {dichroine B, 3-[3-((2*S*)-*trans*-3-hydroxy-2-piperidinyl)-2-oxopropyl]-4(3*H*)-quinazolinone}.

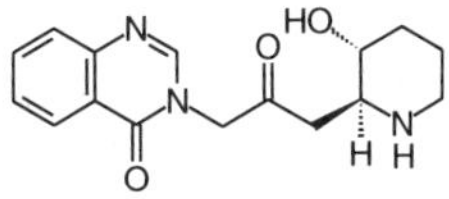

$C_{16}H_{19}N_3O_3$, M_R 301.35, needles, mp. 139–140 °C or 154–156 °C (dimorphous), $[\alpha]_D^{25}$ +6° ($CHCl_3$), $[\alpha]_D^{25}$ +28° (C_2H_5OH). A toxic alkaloid from leaves and roots of the Saxifragaceae *Dichroa febrifuga* (Chinese drug Ch'ang Shan) and *Hydrangea* species (Hydrangeaceae). LD_{50} (mouse p.o.) 2–3 mg/kg.
Activity: F. is 100 times stronger than *quinine against malaria pathogens but is too toxic for clinical use. In addition F. exhibits antipyretic, emetic, and antineoplastic properties and acts as a coccidiostatic agent[1].
Lit.: [1] Tang & Eisenbrand, Chinese Drugs of Plant Origin, p. 455 ff., Berlin: Springer 1992.
gen.: Beilstein E III/IV **24**, 376 ▪ Manske **29**, 129 ▪ Zechmeister **46**, 172 ff. – *Synthesis:* Chem. Pharm. Bull. **47**, 905 (1999) ▪ Heterocycles **51**, 1869 (1999) ▪ J. Med. Chem. **42**, 3163 (1999) (analogues) ▪ J. Org. Chem. **64**, 6833 (1999) ▪ Synthesis **1999**, 1814 ▪ Tetrahedron Lett. **37**, 3255 (1996); **40**, 2175 (1998). – *[HS 2939 90; CAS 24159-07-7 ((2S)-(trans)-(+))]*

Fecapentaenes. The F. are strongly mutagenic metabolites formed by bacteria in human intestines (among others, *Bacteroides*) which are presumed to promote the occurrence of stomach cancer. They are isolated from excretion products (feces). The feces of about 3% of the population of North America exhibit mutagenic activity (Ames test). The F. are mainly *trans*-conjugated pentaenes linked as vinyl ethers with glycerol. They are very unstable, air- and acid-sensitive compounds. With other lipids the F. form micelles; when the lipids are removed by progressive enrichment steps the F. are subjected more and more to oxygen and acids so that decomposition occurs rapidly. Biological studies are mainly performed with the synthetic racemates. The enantiomers have comparable mutagenic activities.

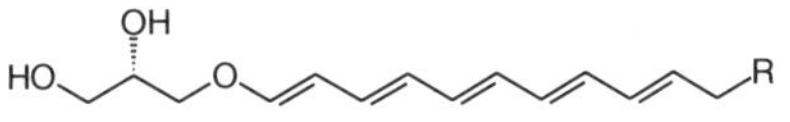

R = CH_3 : F.-12
R = C_3H_7 : F.-14

Examples: F.-12 [(*S*)-3((*all-E*)-1,3,5,7,9-dodecapentaenyloxy)-1,2-propanediol], $C_{15}H_{22}O_3$, M_R 250.34, oil (main metabolite) and F.-14 [(*S*)-3((*all-E*)-1,3,5,7,9-tetradecapentaenyloxy)-1,2-propanediol], $C_{17}H_{26}O_3$, M_R 278.39, oil. Other F. with longer chain lengths probably also occur in low concentrations.
Lit.: J. Nat. Prod. **48**, 622–630 (1985) ▪ Justus Liebigs Ann. Chem. **1988**, 449–454 ▪ Naturwissenschaften **69**, 557 (1982) ▪ Science **225**, 521 (1984). – *Biosynthesis:* Prog. Biochem. Pharmacol. **22**, 35–47 (1988) ▪ Prog. Clin. Biol. Res. **206**, 199–211 (1986). – *Synthesis:* J. Nat. Prod. **50**, 75–83 (1987); **51**, 176–179 (1988) ▪ J. Chem. Soc., Chem. Commun. **1984**, 349. – *Activity:* Cancer Lett. **39**, 287–296 (1988) ▪ Chem. Res. Toxikol. **3**, 391 (1990) ▪ Mutat. Res. **206**, 3–9 (1988); **208**, 9–15 (1988); **259**, 387 (1991). – *[CAS 91423-46-0 (F.-12); 91379-15-6 (F.-14)]*

Fellutamides see fungi.

Fenchenes {α-F. = 7,7-dimethyl-2-methylenebicyclo[2.2.1]heptane; β-F. = (1*R*,4*R*)-(+)-2,2-dimethyl-5-methylenebicyclo[2.2.1]heptane}.

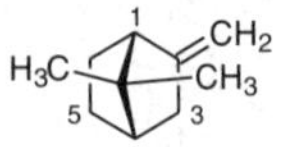

(+)-α-F.

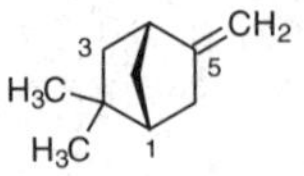

(+)-β-F.

$C_{10}H_{16}$, M_R 136.24, oil. Isomeric bicyclic *monoterpenes: (1*S*,4*R*)-(−)-*α-F.*, bp. 155–158 °C, $[\alpha]_D$ −32° (neat); *β-F.*, bp. 153 °C, [α]D +84.9° (neat). α-F. occurs in the essential oil of leaves of the western red cedar (*Thuja plicata*, Cupressaceae)[1] and the medicinal valerian (*Valeriana officinalis*, Valerianaceae)[2], while β-F. occurs in the fruits of common caraway (*Carum carvi*, Apiaceae)[3].

Lit.: [1] Phytochemistry **1**, 195 (1962). [2] Collect. Czech. Chem. Commun. **31**, 1113 (1966). [3] Planta Med. **30**, 93 (1976).
gen.: Beilstein EIV **5**, 465 f. ▪ Bull. Chim. Soc. Fr. **1976**, 1583 (synthesis) ▪ Karrer, No. 5621. – *[CAS 471-84-1 (α-F.); 497-32-5; 33404-67-0 ((+)-β-F.)]*

Fenchol (fenchyl alcohol, 1,3,3-trimethylbicyclo[2.2.1]heptan-2-ol).

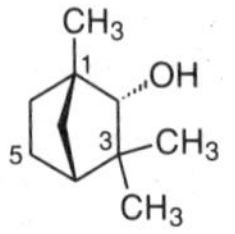

(-)-α-Fenchol (1*S*,2*S*,4*R*)
(-)-(1*S*)-*endo*-Fenchol

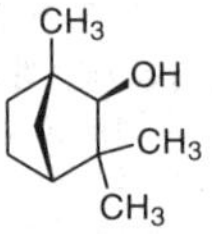

(-)-β-Fenchol (1*S*,2*R*,4*R*)
(-)-(1*S*)-*exo*-Fenchol

$C_{10}H_{18}O$, M_R 154.25. A bicyclic *monoterpene with the fenchane structure that was first detected in the turpentine oil of the pine *Pinus palustris*. Various stereoisomers and their mixtures occur in numerous plant oils, e. g., (−)-β-F. (oil, $[\alpha]_D^{20}$ −31°), in the essential oil of Lawson's cypress (*Chamaecyparis lawsoniana*, Cupressaceae) or (+)-α-F. (cryst., mp. 47–47.5 °C, $[\alpha]_D^{20}$ +12.5°) from the marsh resin obtained from peat bogs (= kauri resin of the tree *Agathis australis*, Araucariaceae). Fenchyl acetate is used in perfumes with a coniferous nuance.

Lit.: ApSimon **4**, 547 f. ▪ Beilstein EIV **6**, 278 ▪ Karrer, No. 310. – *[CAS 470-08-6 ((−)-β-F.); 512-13-0 ((+)-α-F.)]*

Fenchone (1,3,3-trimethylbicyclo[2.2.1]heptan-2-one).

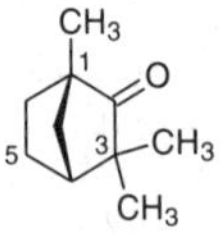

(+)-form

$C_{10}H_{16}O$, M_R 152.24, colorless oil with a camphor-like odor. A bicyclic *monoterpene ketone (fenchane structure), isomeric with *camphor (bornane structure). Both enantiomers occur in nature: (1*S*,4*R*)(+)- and (1*R*,4*S*)(−)-F., bp. 193 °C, $[\alpha]_D$ ±66.9° (C_2H_5OH); (±)-F. (synthetic), bp. 192–194 °C.

Occurrence: (+)-F. in *fennel oil from the seeds of *Foeniculum vulgare* (Apiaceae), content 10–15% (a fennel species cultivated in central Europe); (−)-F. in the essential oil of the western white cedar (*Thuja occidentalis*, Cupressaceae), content 8%. For synthesis, see *Lit.*[1,2]. F. is used as a carminative and to modify odors and tastes.

Lit.: [1] J. Chem. Soc., Chem. Commun. **1971**, 395. [2] Justus Liebigs Ann. Chem. **1981**, 2093.
gen.: Beilstein EIV **7**, 212 ▪ Karrer, No. 561 ▪ Merck-Index (12.), No. 4008 ▪ Ullmann (5.) **A 11**, 173. – *[HS 2914 29; CAS 7787-20-4 ((−)-F.); 4695-62-9 ((+)-F.); 18492-37-0 ((±)-F.)]*

Fennel oil. Two types of fennel oil are used in the perfume and aroma industries.

1. ***Bitter fennel oil:*** Colorless to yellowish oil[1] with a typical camphor-like aromatic, sweet-spicy odor and a slightly bitter, camphor-like, sweet taste.

Production: By steam distillation of the aerial parts of bitter fennel, *Foeniculum vulgare* ssp. *piperitum*, harvested when the fruits (seeds) just begin to ripen. Origin, e. g., Mediterranean region.

Composition[2]***:*** The main components are, e. g., *α-*pinene* (ca. 5%), *α-phellandrene* (10–20%), *limonene* (see *p*-menthadienes) (ca. 20%), **fenchone* (10–20%), and **anethole* (30–40%). As the fruits ripen the oil contains less terpene hydrocarbons and more anethole and fenchone.

Use: Mainly to improve the aromatic character of foods (confectionery, bakery products, marinades), liquors, and oral hygiene products; medicinal uses in carminatives, urological/diuretic agents, mouth and throat preparations, bronchological agents and expectorants.

2. ***Sweet fennel oil:*** Colorless to yellowish oil with a clean, sweet-aromatic odor and a warm spicy, aromatic sweet taste, strongly resembling anise[1].

Production: By steam distillation of the dried fruits (seeds) of the sweet fennel, *F. vulgare*. In the past years sweet fennel has mainly been cultivated in Tasmania.

Composition[2]***:*** The main component is *anethole* (usually more than 80%). In addition, small amounts of **estragole* are usually found; the fenchone content is merely 1–2%.

Use: Similar to star anise oil, mainly to improve the aromatic character of liquors and similar alcoholic beverages.

Lit.: [1] Fenaroli (2.) **1**, 350/1. [2] Perfum. Flavor. **9** (1), 59 (1984); **14** (2), 47 (1989); **17** (2), 44 (1992); **19** (1), 31 (1994).
gen.: Arctander, p. 239 ▪ Bauer et al. (2.), p. 155. – *Toxicology:* Food Cosmet. Toxicol. **17**, 529 (1976). – *[HS 3301 29; CAS 8006-84-6]*

Fernanes (*D*: *C*-friedo-*B'*:*A'*-neogammaceranes).

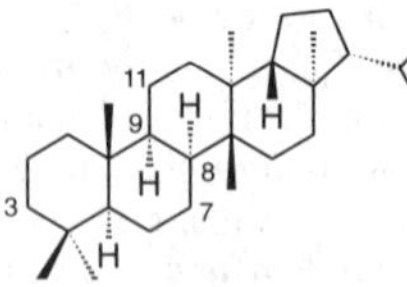

Fernane skeleton

Name for a group of *triterpenes with structures derived from the skeleton of fernane (see formula). They are widely distributed in the Filicatae (*pteridophytes; ferns) where more than 20 different derivatives are known [*fern-9(11)-ene*, $C_{30}H_{50}$, M_R 410.73; *fern-9(11)-en-3β-ol*, $C_{30}H_{50}O$, M_R 426.73; *fern-9(11)-en-3-one*,

$C_{30}H_{48}O$, M_R 424.71; *ferna-7,9(11)-diene*, $C_{30}H_{48}$, M_R 408.71 and others]. They occur in lichens [1], naked-seed plants [2], mosses [3], and angiosperms [4,5].

Lit.: [1] Culberson, Culberson & Johnson, Second Supplement to Chemical and Botanical Guide to Lichen Products, St. Louis: Am. Bryol. Lichenol. Soc. Missouri Bot. Garden 1977. [2] Rev. Latinoam. Quim. **5**, 163 (1974). [3] Phytochemistry **15**, 1178 (1976). [4] Phytochemistry **9**, 2137 (1970). [5] Phytochemistry **17**, 154 (1978).

gen.: Chem. Pharm. Bull. **42**, 39 (1994); **43**, 1849 (1995) (fernenes) ▪ Hegnauer **VII**, 445; **X**, 592 ▪ Zechmeister **54**, 77. – *[CAS 1615-99-2 (fern-9(11)-ene); 4966-00-1 (fern-9(11)-en-3β-ol); 6090-29-5 (fern-9(11)-en-3-one); 2318-80-1 (ferna-7,9(11)-diene)]*

Ferredoxins. Name for a group of non-hem iron proteins of animal, plant, or microbial origin that usually contain equal numbers of iron and labile sulfide ions; 2, 4, 6, or 8 of each. The *2-iron-F.* [96–98 amino acid residues, 4–6 of which are cysteine (Cys)] contain the four-membered rings shown in the Figure; these are bound by two cysteine residues per iron atom to the protein. The *4-iron-F.* and the *8-iron-F.* (55 amino acid residues) possess one or two cubic clusters (see figure).

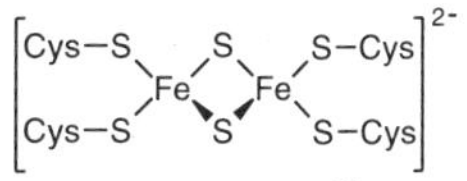

oxidized Fe_2S_2—Cys_4-Cluster

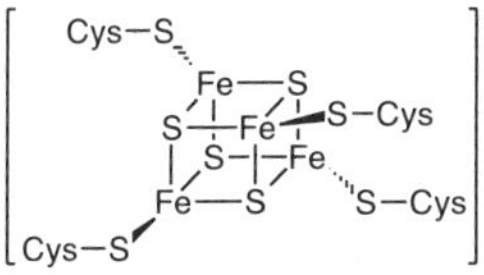

oxidized Fe_4S_4—Cys_4-Cluster

Iron and sulfur can be extracted from F.; the resulting apoferredoxin is reactivated by iron(II) salts and sulfides. The synthesis of the iron-free protein has been achieved by the Merrifield technique. On account of their properties as redox systems ($Fe^{3+}+e^- \rightleftarrows Fe^{2+}$) the F. effect electron transport between enzyme systems but do not exhibit any enzymatic activity. They transport electrons in the respiratory chain, in photosynthesis, and in nitrogen fixation. The iron-sulfur protein P_{439} of the Fe_4S_4-type (M_R 11 600) plays a role in photosynthesis. Conclusions can be drawn about the evolutionary histories of plants from the similarities and differences in the amino acid sequences. For the evolutionary history of F. in photosynthesis, see *Lit.* [1]. F. with Fe_3S_3- and Fe_3S_4-clusters also occur in bacteria [2].

Lit.: [1] Trends Biochem. Sci. **13**, 30–33 (1988). [2] FEMS Microbiol. Rev. **54**, 155–176 (1988); Trends Biochem. Sci. **13**, 369 f. (1988).

gen.: ACS Symp. Ser. **372**, 258 (1987) ▪ Adv. Inorg. Chem. **33**, 39–67 (1989) ▪ FASEB J. **4**, 2483–2491 (1990) (review) ▪ Mol. Struct. Biol. **6**, 79 (1985) (review) ▪ Struct. Bonding (Berlin) **72**, 113 (1990) (^{1}H NMR, review) ▪ Stryer 1990, p. 418 f., 549 ff., 602, 818.

Ferrichromes. Iron(III) complexes of the trihydroxamate-type isolated from fungi and possessing growth-promoting activities. F. and F. A were first discovered in 1952 in culture filtrates of *Ustilago sphaerogena*. *F.:* yellow cryst., mp. 248–251 °C, $[\alpha]_D$ +304° (H_2O), soluble in water and methanol. The iron-free compounds (*siderophores) are cyclic hexapeptides, their biosynthesis proceeds non-ribosomally with the help of peptide synthetases. Desferri-ferrichrome contains the characteristic tripeptide sequence of N^5-acyl-N^5-

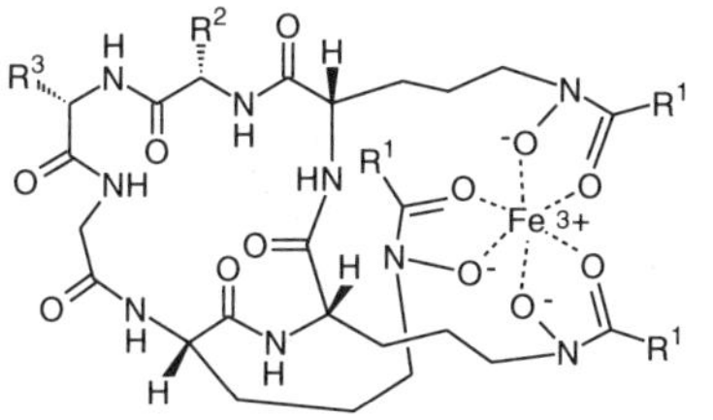

R^1	R^2	R^3	
CH_3	H	H	Ferrichrome
CH_3	H	CH_2OH	Ferricrocin
CH_3	H	CH_3	Ferrichrome C
CH_3	CH_2OH	CH_2OH	Ferrichrysin
$CH{=}C(CH_3){-}CH_2{-}CH_2OH$ (*cis*)	CH_2OH	CH_2OH	Ferrirhodin
$CH{=}C(CH_3){-}CH_2{-}COOH$ (*trans*)	CH_2OH	CH_2OH	Ferrichrome A
$CH{=}C(CH_3){-}CH_2{-}CH_2OH$ (*trans*)	CH_2OH	CH_2OH	Ferrirubin

Table: Data of Ferrichromes.

compound	molecular formula	M_R	CAS
F.	$C_{27}H_{42}FeN_9O_{12}$	740.53	15630-64-5
F. A	$C_{41}H_{58}FeN_9O_{20}$	1052.81	15258-80-7
F. C	$C_{28}H_{44}FeN_9O_{12}$	754.56	56665-78-2

hydroxy-L-ornithine and the tripeptide from glycine. The small amino acids can be varied (exchange of glycine for serine or alanine) as well as the *N*-acyl groups (e. g., acetyl in F. for *trans-β*-methylglutaconoyl in F. A). The hydroxamate groups of the 3 ornithine side chains form stable, crystallizable, octahedral complexes (K_f ~1030) with Fe^{3+}. The F. are recognized by specific membrane transport systems and actively taken up in the membrane. The release of iron occurs at the inner side of the membrane by reduction to Fe^{2+}, the ligand is passed out through the membrane and is then available for the next transport process.

Iron(III) trihydroxamates related to the F. are ferricrocin, ferrichrysin, ferrirhodin and ferrirubin (see formula). Other iron-transporting compounds from fungi are fusigen, caprogen, rhodotorulic acid, and, with regard to type, the new polycarboxylate siderophores (e. g., rhizoferrin).

Lit.: J. Am. Chem. Soc. **102**, 4224–4231 (1980) (structure) ▪ Proc. Natl. Acad. Sci. USA **90**, 903–907 (1993) (genetics) ▪ Topics Curr. Chem. **123**, 49–102 (1984) ▪ Winkelmann (ed.), Handbook of Microbial Iron Chelates, Boca Raton: CRC Press 1991 ▪ Winkelmann & Drechsel, in Rehm & Reed (eds.), Biotechnology, 2nd edn., vol. 7, chap. 7, Weinheim: VCH Verlagsges. 1996 ▪ Winkelmann & Winge (eds.), Metal Ions in Fungi, New York: Dekker 1994 ▪ Zechmeister **22**, 279–322.

Ferrimycins. Iron(III) trihydroxamates formed by *Streptomyces griseoflavus* possessing antibiotic properties. *F. A_1:* $C_{41}H_{65}FeN_{10}O_{14}$, M_R 977.87. The iron-complexing part of F. A_1 corresponds to ferrioxamine B, the free amino group of which forms an amide bond to an antibiotically active group. The activity of F. against Gram-positive bacteria can be eliminated by *ferrioxamines; a transport antagonism is assumed to be responsible.

Lit.: Winkelmann (ed.), Handbook of Microbial Iron Chelates, Boca Raton: CRC Press 1991 ▪ Winkelmann & Drechsel, in Rehm & Reed (eds.), Biotechnology, 2nd edn., vol. 7, chap. 7,

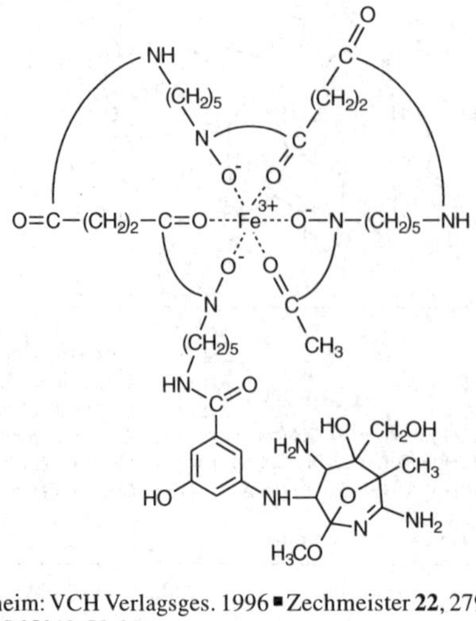

F. A_1

Weinheim: VCH Verlagsges. 1996 ▪ Zechmeister **22**, 279–322. – *[CAS 15319-50-3]*

Ferrioxamines.

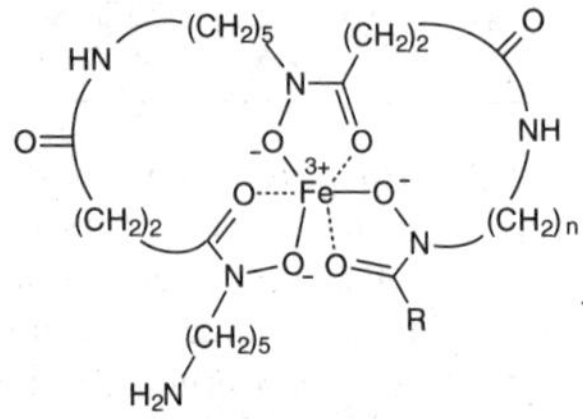

n	R	
4	CH_3	: A_1
5	CH_3	: B
5	CH_2–CH_2–COOH	: G

Iron(III) complexes (components A–H) of the trihydroxamate-type isolated from *actinomycetes (e.g., *Streptomyces pilosus*) with growth regulating properties. *F. B:* $C_{25}H_{45}FeN_6O_8$, M_R 613.51, red-brown, hygroscopic solid (hydrochloride), uv_{max} 428 nm, soluble in water and methanol. The iron-free compound (*siderophore) of F. B is an open-chain trihydroxamic acid (*desferrioxamine B*, $C_{25}H_{48}N_6O_8$, M_R 576.69) which on acid hydrolysis affords acetic acid, succinic acid, and *N*-hydroxy-1,5-pentanediamine in the ratio 1:2:3. Other *desferrioxamines* contain instead *N*-hydroxy-1,4-butanediamine in the amide bond (e.g., component A_1) or a further molecule of succinic acid in place of acetic acid (component G). The desferrioxamines, recently discovered also in pseudomonads and enterobacteria, form stable, octahedral 1:1 complexes (K_f ~1030) in the presence of Fe^{3+}. Desferrioxamine B (Desferal®) is produced by large-scale fermentation and used as a salt (mesylate) to treat iron-storage diseases in humans.

Lit.: Gross, Aumiller & Gelzer (eds.), Desferrioxamin, Geschichte, Stellenwert, Perspektiven, München: MMV Medizin 1992 ▪ J. Am. Chem. Soc. **114**, 2224–2230 (1992) ▪ further Lit. see ferrimycins. – *[CAS 14710-31-7 (F. A1); 14836-73-8 (F. B); 29825-05-6 (F. G); 70-51-9 (desferrioxamine)]*

Ferrirubin see siderophores.

Ferrithiocin. Iron complex with 2 molecules of *desferrithiocin* [(*S*)-4,5-dihydro-2-(3-hydroxy-2-pyridinyl)-4-methyl-4-thiazolinecarboxylic acid].

$C_{20}H_{16}FeN_4O_6S_2$, M_R 532.38, red-brown needles, mp. 160 °C (decomp.), $[\alpha]_D$ –578° (H_2O). A secondary metabolite from *Streptomyces antibioticus*. The producing organism excretes the iron-free form – the *side-

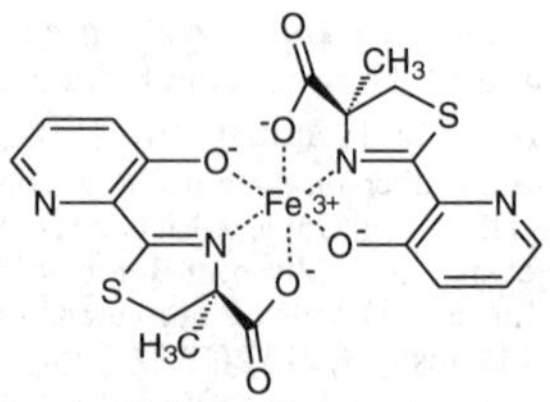

rophore desferrithiocin – into the culture filtrate. The iron complex ferrithiocin is first formed on addition of iron(III) salts. The formation constant K_f of F. is $10^{29.6}$ M^{-1}. The chemical conversion of F. into desferrithiocin is possible. In some bacteria F. effects a potentiation of the activity of *cephalosporins but it does not have antibiotic properties.

Desferrithiocin {$C_{10}H_{10}N_2O_3S$, M_R 238.27, mp. 154 °C (decomp.), sulfur-yellow needles, $[\alpha]_D$ +30.1° (CH_3OH)}. Besides iron, desferrithiocin can also bind Al, Zn, and Mg but not as strongly as iron. Thus desferrithiocin has an iron-specific activity. As a consequence of its iron complexation, desferrithiocin inhibits the growth of *Escherichia coli* K_{12} and is a potential oral drug for use in the treatment of diseases caused by an iron excess. Desferrithiocin functions *in vitro* and *in vivo* as an iron chelator but also, like F., exerts toxic side effects.

Lit.: Br. J. Haematol. **81**, 424–431 (1992) ▪ J. Chem. Soc., Chem. Commun. **1990**, 1194ff. ▪ J. Med. Chem. **34**, 2072–2078 (1991); **37**, 1411–1417, 2889–2895 (1994) ▪ Tetrahedron **49**, 5359 (1993) (synthesis). – *[CAS 76045-30-2]*

Ferruginine.

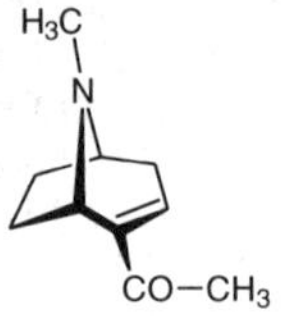

$C_{10}H_{15}NO$, M_R 165.23, amorphous powder, $[\alpha]_D$ +37° ($CHCl_3$). An alkaloid from *Darlingia ferruginea* (Proteaceae).

Lit.: J. Org. Chem. **62**, 6704 (1997). – *[CAS 73069-63-3]*

Ferrulactone II (3-dodecen-11-olide).

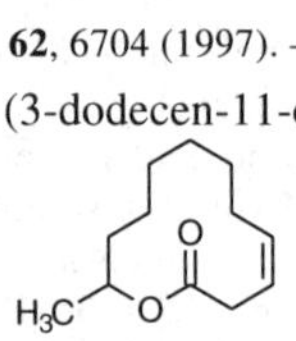

(±)-Ferrulactone II

$C_{12}H_{20}O_2$, M_R 196.29. A *pheromone belonging to the group of *cucujolides of male grain beetles. While *Cryptolestes ferrugineus* produces the (*S*)-enantiomer, *Oryzaephilus mercator* uses the (*R*)-enantiomer as an attractant [1].

Lit.: [1] J. Chem. Ecol. **13**, 1543–1560 (1987).

gen.: ApSimon **9**, 265 (synthesis) ▪ Chem. Nat. Compd. (engl. transl.) **32**, 909, 912 (1997) (synthesis F. I) ▪ Justus Liebigs Ann. Chem. **1992**, 1011 ▪ Tetrahedron: Asymmetry **6**, 2219 (1995) (synthesis). – *[CAS 87583-38-8 ((R)-F.); 86578-99-6 ((S)-F.); 87583-37-7 ((±)-F.)]*

Ferulenol.

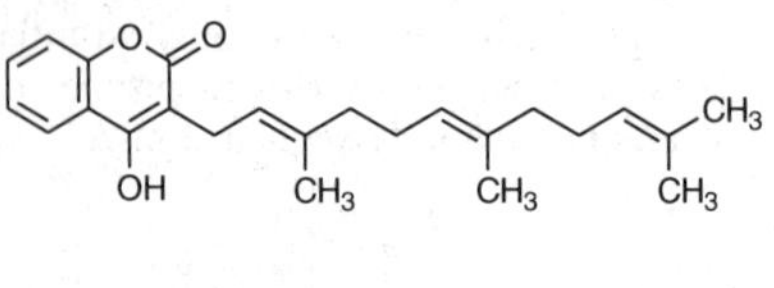

$C_{24}H_{30}O_3$, M_R 366.50, colorless cryst., mp. 61 °C. 3-Farnesyl-4-hydroxycoumarin from *Ferula communis*, an Umbelliferous species occurring in the Mediterranean region [1], has significance for homeopathic medicine. Its high toxicity for domestic animals is a frequent cause of problems in the animal breeding industry [2].

Lit.: [1] Boll. Soc. Ital. Biol. Sper. **16**, 544 (1941); Tetrahedron Lett. **1964**, 2783. [2] Phytochemistry **26**, 253 (1987).
gen.: Beilstein E V **18/2**, 245 ▪ Phytochemistry **26**, 1613 (1987). – *[CAS 6805-34-1]*

Ferulic acid see caffeic acid.

Fervenulin {6,8-dimethylpyrimido[5,4-*e*]-1,2,4-triazine-5,7(6*H*,8*H*)-dione}.

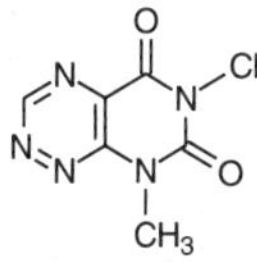

$C_7H_7N_5O_2$, M_R 193.16, yellow cryst., mp. 178–179 °C, soluble in organic solvents. An antibiotic from *Streptomyces fervens* and *S. rubrireticuli*; F. exhibits antitumor properties and is a weak antimycotic agent also having activity against trichomonads.

Lit.: Beilstein E V **26/19**, 134 ▪ J. Heterocycl. Chem. **33**, 949 (1996) (synthesis) ▪ J. Org. Chem. **43**, 469 (1978). – *[CAS 483-57-8]*

Festucine see lolines.

Ficellomycin.

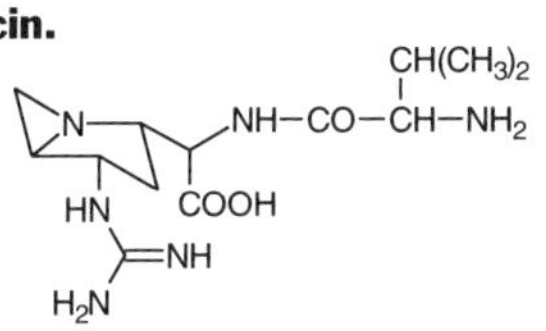

$C_{13}H_{24}N_6O_3$, M_R 312.37, amorphous solid, $[\alpha]D$ +39° (H_2O). Dipeptide antibiotic from cultures of *Streptomyces ficellus*. F. contains the 1-azabicyclo[3.1.0]hexane unit otherwise unknown in natural products and is active against Gram-positive bacteria, possibly through alkylation of the DNA.

Lit.: Biochemistry **16**, 3406 (1977) ▪ J. Antibiot. **29**, 1001 (1976); **42**, 357 (1989). – *[HS 2941 90; CAS 59458-27-4]*

Ficine.

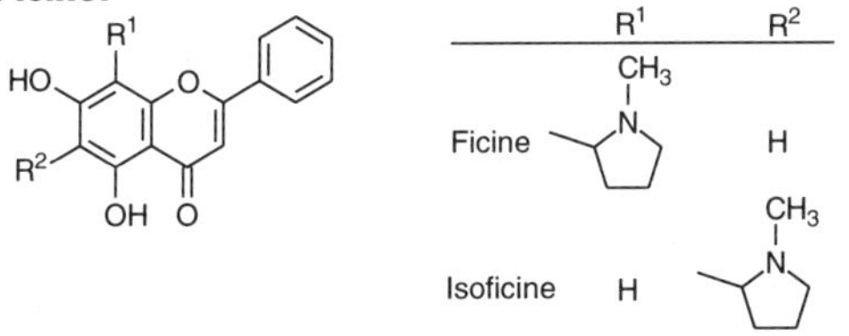

Pyrrolidine alkaloid, $C_{20}H_{19}NO_4$, M_R 337.38, cryst., mp. 235 °C, originating from *Ficus pantoniana* (Moraceae). F. as well as the isomeric compound *isoficine*, mp. 168 °C, also from *F. pantoniana*, were the first alkaloids with a flavonoid partial structure to be discovered.

Lit.: Tetrahedron Lett. **1965**, 1987. – *[CAS 2520-36-7 (F.); 2255-61-1 (isoficine)]*

Filbert flavor (hazelnut flavor). The most important of the over 300 known components of F. are methyl-alkenones and -alkanones, together with alkylpyrazines (see pyrazines), *alkanals, and *alkenals. (+)-(2*E*,5*S*)-5-Methyl-2-hepten-4-one (see filbertone) has been recognized as the *impact compound in F. [1].

Lit.: [1] Angew. Chem. Int. Ed. Engl. **28**, 1022 f. (1989).
gen.: Maarse, p. 676 ff. ▪ TNO list (6.), p. 714–719.

Filbertone (5-methyl-2-hepten-4-one).

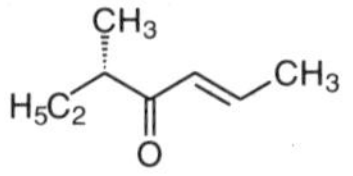

(+)-(2*E*,5*S*)-F.

$C_8H_{14}O$, M_R 126.20. Liquid, dilute solutions have a hazelnut-like odor, bp. 72 °C (2 kPa). Olfactory threshold in water 5 ppt [1]. It is considered as an *impact compound of *filbert flavor, especially the (+)-(*E*,5*S*)-isomer [2].

Lit.: [1] DE. P. 35 25 604 (1983), Haarmann & Reimer. [2] Angew. Chem. Int. Ed. Engl. **28**, 1022 (1989).
gen.: J. Braz. Chem. Soc. **9**, 583 (1998) (synthesis). – *[HS 2914 19; CAS 102322-83-8 ((+)-E); 103070-07-1 ((+)-Z); 122440-59-9 (2E,5S)]*

Filicin see filixic acids.

Filifolone (4,7,7-trimethylbicyclo[3.2.0]hept-3-en-6-one).

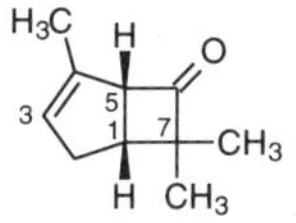

(+)-form

$C_{10}H_{14}O$, M_R 150.22, oil. Bicyclic *monoterpene with unusual structure. (1*R*,5*R*)(+)- and (1*S*,5*S*) (–)-F., bp. 86–88 °C (2 kPa), $[\alpha]_D$ ±307° (neat).

Occurrence: (–)-F. in the essential oil of *Artemisia filifolia* (Asteraceae), a sand sage from Arizona, content up to 11%; (+)-F. in the essential oil of the Australian plant *Zieria smithii* (Rutaceae) [1]. For synthesis, see *Lit.* [2–5].

Lit.: [1] Chem. Commun. **1967**, 1037. [2] J. Am. Chem. Soc. **85**, 3525 (1963). [3] Tetrahedron Lett. **1980**, 691. [4] Helv. Chim. Acta **67**, 1854 (1984). [5] J. Org. Chem. **63**, 2389 (1998). – *[CAS 4613-37-0]*

Filipin. Antibiotic complex produced by *Streptomyces filipinensis* and *S. durhamensis* consisting of the main components F. I–IV. F. IV is an isomer of F. III. A structural relative of *amphotericin B.

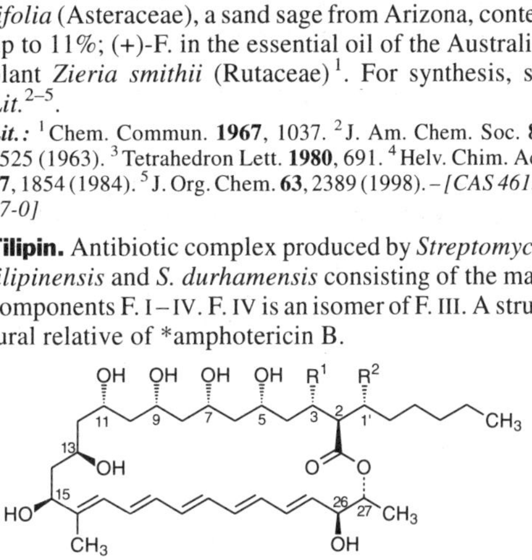

Table: Data of Filipin components.

component	R^1	R^2	molecular formula	M_R	CAS
F. I	H	H	$C_{35}H_{58}O_9$	622.84	38723-93-2
F. II	OH	H	$C_{35}H_{58}O_{10}$	638.84	38620-77-8
F. III	OH	OH	$C_{35}H_{58}O_{11}$	654.84	480-49-9

F. exhibits predominately antifungal, but also cytotoxic and antiviral activities, it acts on the membrane

building block of sensitive eukaryotes, thereby changing the membrane permeability. This activity is related to the 28-membered pentaene macrolactone structure. In the past F. was used as crop protection agent.
Lit.: Angew. Chem. Int. Ed. Engl. **34**, 1227–1230 (1995) ▪ J. Antibiot. **23**, 414, 603 (1970); **35**, 988–996 (1982); **42**, 322 ff. (1989) ▪ J. Org. Chem. **61**, 4219–4231 (1996) ▪ Magn. Reson. Chem. **35**, 538 (1997) (NMR F. III). – *Review:* Holz, in: Antibiotics 5, vol. 2, p. 313–340, New York: Springer 1979 ▪ Ōmura (ed.), Macrolide Antibiotics, New York: Academic Press 1984.

Filixic acids (filicin).

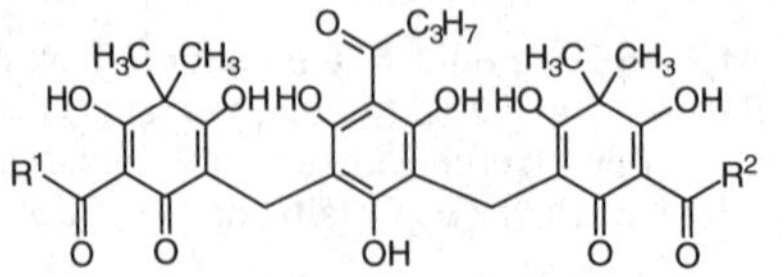

$R^1 = R^2 = C_3H_7$: F.BBB
$R^1 = C_2H_5$, $R^2 = C_3H_7$: F.PBB
$R^1 = R^2 = C_2H_5$: F.PBP

Table: Main components of Filixic acids.

	molecular formula	M_R	mp. [°C]	CAS
F. BBB	$C_{36}H_{44}O_{12}$	668.74	168–174 [1]	4482-83-1
F. PBB	$C_{35}H_{42}O_{12}$	654.71	184–186	49582-09-4
F. PBP	$C_{34}H_{40}O_{12}$	640.68	192–194 [2]	51005-85-7

Phenolic natural products (acylphloroglucinols) from the male fern *Dryopteris filix-mas*. The natural product is a mixture of six homologues (R^1, R^2=methyl, ethyl, propyl) with three main components, see table. ***Use:*** F. exhibit anthelmintic activity. Ether extracts or solutions of the active principles in a fatty oil in combination with the subsequent administration of a laxative are used as a remedy for tapeworms. As a result of the low therapeutic range and the difficulties in standardization, F. have been mostly replaced by synthetic agents (they are still used occasionally in veterinary medicine).
Lit.: [1] Phytochemistry **11**, 1850 (1972). [2] Phytochemistry **12**, 1493 (1973); Planta Med. **28**, 144 (1975).
gen.: Ann. Bot. Fenn. **20**, 407 (1983) ▪ Helv. Chim. Acta **77**, 1985 (1994) (ms) ▪ Schröder et al., Arzneimittelchemie, vol. III, p. 215, Stuttgart: Thieme 1976 ▪ Zechmeister **54**, 10.

Finavarrene see algal pheromones.

Fir and pine needle oils. General term for the *essential oils, obtained by steam distillation from needles (branch tips, young shoots) of various Pinaceae species of the genera *Pinus*, *Abies*, *Picea*, and *Tsuga*. They mostly possess a fresh, resiny odor and consist mainly of monoterpene hydrocarbons such as **pinenes*, *phellandrenes* (see *p*-menthadienes), **camphene*, **myrcene*, *3-*carene*, and *limonene* (see *p*-menthadienes). The component mainly responsible for the odor is (–)-*bornyl acetate* ($C_{12}H_{20}O_2$, M_R 196.29) which can be present, as in Siberian pine needle oil, to more than 30%. The oils are used in the production of perfumes for men, for perfuming household articles like cleaners, bath products, sauna oils, and in pharmaceutical preparations such as anti-rheumatic ointments, massage oils, balneotherapeutics, bronchial and rhinological agents. They readily form peroxides which have sensitizing properties. Therefore, oils with a peroxide content of more than 10 mmol peroxide/L should not be used. The most frequently used oils are:
1. *Pine needle oil:* Thin, colorless to yellowish oil, produced from the needles of *Pinus sylvestris* and *P. nigra*.
2. *Dwarf pine oil:* Thin, colorless oil with a fresh, balmy odor, produced by distillation from the needles of various subspecies of the dwarf pine *P. mugo*.
3. *Siberian fir needle oil:* Colorless to light yellow oil with a strict, fresh-spicy odor, produced from the needles of fir tree *Abies sibirica* growing in east Asia.
Lit.: Arctander, p. 249, 539, 540 ▪ Bauer et al. (2.), p. 172 ▪ Perfum. Flavor. **16** (2), 64 (1991). – *Toxicology:* Food Cosmet. Toxicol. **13**, 450 (1975); **14**, 843, 845 (1976). – *[HS 3301 29; CAS 8023-99-2 (1.); 8000-26-8 (2.); 8021-29-2 (3.); 5655-61-8 ((–)-bornyl acetate)]*

Fischerellins. Photosystem-II-inhibiting *allelochemicals from the cyanobacterium *Fischerella muscicola*. F. exhibit antifungal and herbicidal activities; e.g., *F. A*: $C_{26}H_{36}N_2O_2$, M_R 408.58, amorphous powder.

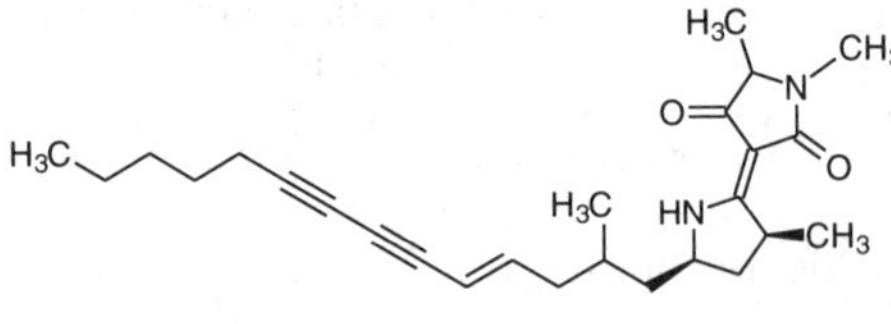

F. A
(relative configuration)

Lit.: Tetrahedron Lett. **37**, 6539 (1996). – *[CAS 182227-56-1 (F. A)]*

12-*epi*-Fischer indole G-isonitrile see indole alkaloids from cyanobacteria.

Fisetin (3,3′,4′,7-tetrahydroxyflavone, 5-deoxyquercetin). Formula see flavones. $C_{15}H_{10}O_6$, M_R 286.24, yellow needles, mp. 330 °C (decomp.), insoluble in water, soluble in alcohol. A *flavonoid belonging to the *flavones; pigment from the smoke tree (*Cotinus coggyria*, syn. *Rhus cotinus*, Anacardiaceae). F. is weakly toxic [LD_{50} (mouse i. v.) 180 mg/kg] and is assumed to be mutagenic [1]. F. reduces the toxicity of aflatoxins [2].
Lit.: [1] Sax (8.), p. 1706; Mutat. Res. **66**, 223 (1979); Biochem. Soc. Trans. **5**, 1489 (1977). [2] J. Environ. Pathol. Toxicol. **2**, 1021 (1979).
gen.: Beilstein E V **18/5**, 291 ▪ Hager (5.) **4**, 27, 30 f. ▪ Merck-Index (12.), No. 4129 ▪ Phytochemistry **17**, 827 (1978) ▪ see also flavonoids. – *[HS 2932 90; CAS 528-48-3]*

FK-409 (4-ethyl-2-hydroxyimino-5-nitro-3-hexenoic acid amide).

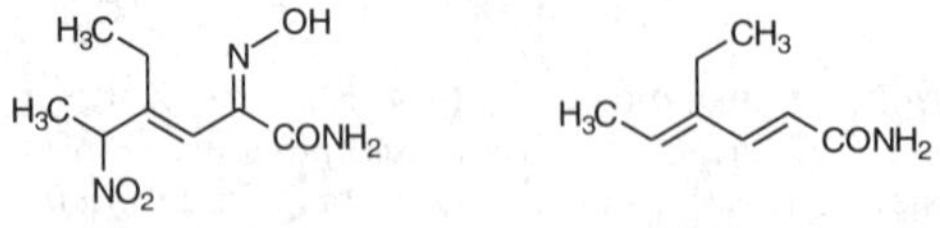

FK-409 FR-900411

$C_8H_{13}N_3O_4$, M_R 215.21, mp. 140 °C, optically inactive, isolated from cultures of *Streptomyces griseosporeus* after acidification (pH 3) prior to extraction. It is also obtained semisynthetically from *FR-900411* (4-ethyl-2,4-hexadienoic acid amide: $C_8H_{13}NO$, M_R 139.20, mp. 57–58 °C) produced by the strain through reduction with nitrite at pH 3. FK-409 liberates NO (*nitrogen monoxide) *in vivo*, it has vasodilatory properties and inhibits blood platelet aggregation. It may be useful clinically for treatment of angina pectoris.

Lit.: Cardiovasc. Drugs Rev. **12**, 2–15 (1994) ▪ Chem. Pharm. Bull. **37**, 2864 ff. (1989) ▪ Eur. J. Pharmacol. **257**, 123 (1994); **272**, 39 (1995) ▪ J. Antibiot. **42**, 1578–1583, 1584–1588, 1589–1592 (1989). – *[CAS 92454-60-9]*

FK-506, tsukubaenolide (tacrolimus, fujimycin).

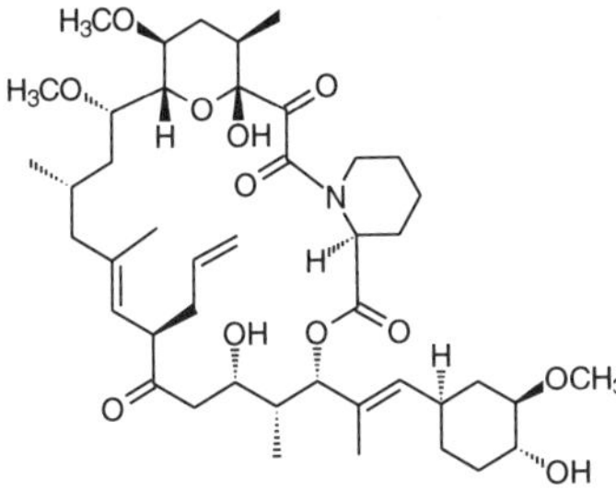

$C_{44}H_{69}NO_{12}$, M_R 804.04, prisms, mp. 127–129 °C, $[\alpha]_D$ –84.4° ($CHCl_3$), a macrolactam lactone produced by *Streptomyces tsukubaensis*, in which a long-chain hydroxycarboxylic acid is cyclized with L-*pipecolic acid as bridging unit. FK-506 is structurally related to *rapamycin and has been prepared synthetically. It exhibits immunosuppressive activity by suppression of cell-mediated and humoral immune responses. It has been available in Japan since 1993, in Europe and USA since 1996 under the name Prograf® (*tacrolimus*) for use in transplantation medicine to suppress rejection reactions [1]. FK-506 is also effective in the treatment of autoimmune diseases, e.g., multiple sclerosis, psoriasis, or rheumatoid arthritis.

Activity: Like *cyclosporin A (CsA) and rapamycin, FK-506 binds to cytoplasmic receptors, the so-called *immunophilins*, and prevents the signal transduction leading to activation of T lymphocytes. FK-506 and CsA inhibit the production of interleukin 2 by lymphocytes. FK-506 is effective at doses 10–100 times lower than those required with CsA.

The FK-506 binding protein (*FKBP*) isolated from calf thymus and human spleen has a molecular mass of ca. 12 000 and shows the activity of a peptidylprolyl *cis/trans*-isomerase (rotamase; EC 5.2.1.8), which is inhibited by binding of FK-506. However, this is probably not the reason for the immune response; inhibition of the phosphatase activity of calcineurin by the FK-506/FKBP complex is assumed to be responsible, this impairs the formation and/or activation of the transcription factor (NF-AT) for IL 2 gene expression.

Lit.: [1] Am. J. Gastroenterol. **90**, 771 (1995); Exp. Opin. Ther. Patents **8**, 1109–1124 (1998); Kleemann-Engel, p. 1804 ff.; Nature Medicine **3**, 421 (1997); New Drug Commentary **23**, No. 4, 68 f. (1996).

gen.: Annu. Rep. Med. Chem. **26**, 211–220 (1991); **29**, 347 (1994) ▪ Drugs **46**, 746 (1993) ▪ J. Am. Chem. Soc. **113**, 1409 ff. (1991) ▪ Nature (London) **341**, 755–769 (1989) ▪ Science **251**, 283–287 (1991) ▪ Trends Pharmacol. Sci. **12**, 218–223 (1991). – *Synthesis:* J. Org. Chem. **61**, 6856–6872 (1996) ▪ Lindberg **3**, 417–494 ▪ Lukacs, Recent Progress in the Chemical Synthesis of Antibiotics, p. 141–212, Berlin: Springer 1993 ▪ Nicolaou, p. 600 ff. ▪ Synlett **1994**, 381–392 ▪ Tetrahedron **53**, 13221–13256, 13257–13284 (1997). – *Mechanism of action:* Ann. Pharmacother. **28**, 501 (1994) ▪ Angew. Chem. Int. Ed. Engl. **31**, 384–400 (1992); **33**, 1437–1452 (1994) ▪ Drug. Metab. Dispos. **23**, 28 (1995) ▪ Immunol. Today **13**, 136 (1992). – *FK-BP:* Nature (London) **351**, 248 ff. (1991) ▪ Science **252**, 836–842 (1991). – *[CAS 104987-11-3]*

FK-520 see ascomycin.

Flagicidin see anisomycin.

Flavidulols.

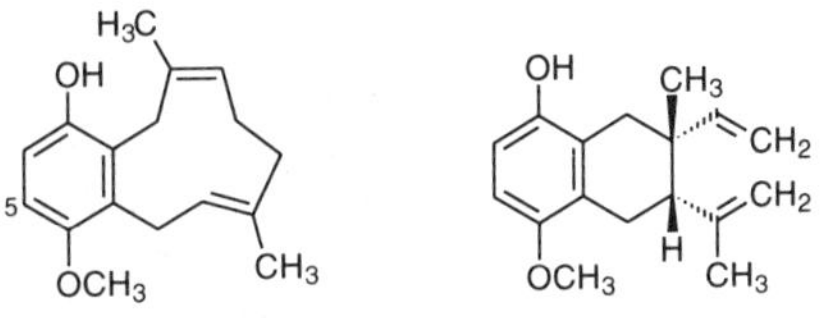

Flavidulol A Flavidulol B (±)

Cyclic geranylhydroquinones with antibacterial, antifungal, and immunosuppressive activities from the Japanese milkcap *Lactarius flavidula* (Basidiomycetes). Besides the main component *F. A* (1.83 g from 1.5 kg fungi, $C_{17}H_{22}O_2$, M_R 258.36, oil), the stearoyl ester *F. D* ($C_{35}H_{56}O_3$, M_R 524.83, wax, mp. 40–41 °C) and the 5,5′-dehydrodimer *F. C* ($C_{34}H_{42}O_4$, M_R 514.71, cryst., mp. 185–186 °C) also occur in the fungus. The optically inactive F. B is probably formed by Cope rearrangement of F. A.

Closely related are the *clavilactones, the cordiachromes [1] from *Cordia millenii* (Boraginaceae), and wigandol [2] from *Wigandia kunthii* (Hydrophyllaceae).

Lit.: [1] J. Chem. Soc., Perkin Trans. 1 **1973**, 1352. [2] Phytochemistry **19**, 2202 (1980).

gen.: Chem. Pharm. Bull. **36**, 2366 (1988); **41**, 654, 2032 (1993). – *[CAS 117568-32-8 (F. A); 117568-33-9 (F. B); 156980-40-4 (F. D); 117568-34-0 (F. C)]*

Flavinantine see morphinane alkaloids.

Flavine adenine dinucleotide, flavine mononucleotide see riboflavin(e).

Flavins.

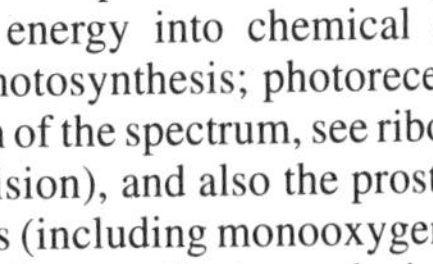

Trivial name for all low-molecular-weight compounds containing the skeleton of 7,8-dimethylisoalloxazine (flavin). The F. are, among others, components of the reaction centers of proteins that catalyze the conversion of light energy into chemical energy (photosystems for photosynthesis; photoreceptor activity in the blue region of the spectrum, see riboflavin; see also pigments of vision), and also the prosthetic groups of redox enzymes (including monooxygenases of the respiratory chain), see flavine adenine dinucleotide (FAD), flavine mononucleotide (FMN) under *riboflavin.

Lit.: Angew. Chem. Int. Ed. Engl. **27**, 333–343 (1988) ▪ Biotechnol. Genet. Eng. Rev. **5**, 297–318 (1987) ▪ Chem. Biochem. Flavoenzymes **1**, 1–193 (1991) ▪ Flavins and Flavoproteins, Proc. of the 9th Int. Symp., Edmondson & McCormick (eds.), Berlin: De Gruyter 1987 ▪ Flavins and Flavoproteins, 8th Int. Symp., Bray et al. (eds.), Berlin: De Gruyter 1984 ▪ Free Radical. Biol. Med. **3**, 215–230 (1987) ▪ J. Photochem. Photobiol. B **2**, 143 (1988) ▪ Synth. Commun. **25**, 2315 (1995) (synthesis).

Flavomannins.

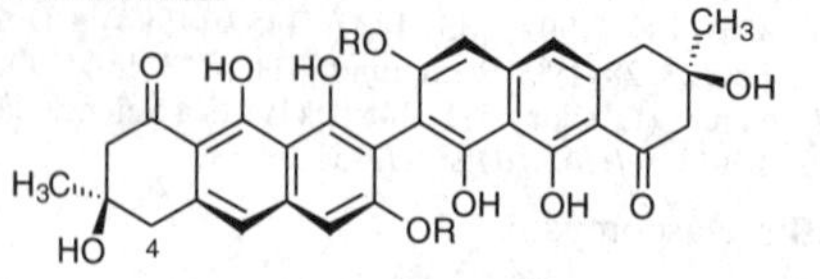

R = H : Flavomannin A
R = CH_3 : Flavomannin-6,6'-di-*O*-methyl ether A_1 (FDM A_1)

A group of dimeric *preanthraquinones from *Penicillium* sp. (Hyphomycetes)[1] as well as the fruit bodies of basidiomycetes of the genera *Cortinarius* (subgenus *Phlegmacium*), *Dermocybe*, and *Tricholoma*, in which they are responsible for the yellow, orange, and green colors[2]. The 7,7′-linkage of the preanthraquinone units is characteristic for the F. and distinguishes them from the *phlegmacins. The parent compound *F. A* {$C_{30}H_{26}O_{10}$, M_R 546.53, yellow plates (monohydrate), mp. 224–226 °C (decomp.), [α]D – 1400° (C_2H_5OH)} is isolated from *Penicillium wortmannii* and *Cortinarius odoratus*. In most of the naturally occurring F. derivatives from gill fungi some of the phenolic OH groups are methylated, in addition the 4- and 4′-positions may by hydroxylated or oxidized to keto groups. Elimination of water to the monoanthrone or oxidation to the anthraquinone is frequently observed. The most common F. derivative is the *6,6′-di-O-methyl ether* [″*FDM*″, $C_{32}H_{30}O_{10}$, M_R 574.58, yellow cryst., mp. 245–250 °C (decomp.)], occurring in the two stereoisomeric forms A_1 and B_1. For elucidation of the stereochemistry, see *Lit.*[3].

Lit.: [1] J. Chem. Soc. C **1968**, 2560; Sydowia **37**, 284 (1987). [2] Zechmeister **51**, 154–162 (review). [3] Tetrahedron: Asymmetry **1**, 621 (1990); Nat. Prod. Rep. **11**, 82f. (1994).
gen.: Aust. J. Chem. **48**, 1 (1995) (review). – *[CAS 19953-89-0 (F. A)]*

Flavomentins.

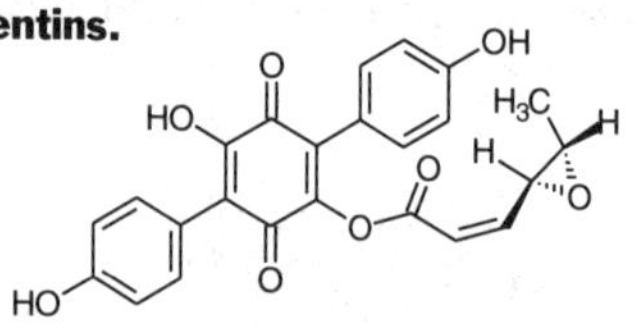

Flavomentin B

Yellow pigments from the paxilli *Paxillus atrotomentosus* (Velvet-footed Pax) and *P. panuoides* (Stalkless Paxillus, Basidiomycetes). F. are mono- and diesters of *atromentin with derivatives of (2*Z*,4*E*)-2,4-hexadienoic acid. The ^{1}H-NMR spectra of *F. B* ($C_{24}H_{18}O_8$, M_R 434.40, orange cryst., mp. 173 °C, dihydrate) and other monoesters of atromentin show equivalence of the aromatic signals down to about –70 °C, this is due to rapid synchronous migrations of the acyl group and the opposing proton to the respective neighboring carbonyl group with tautomerization of the quinone ring. The F. are probably formed in the fungi from *leucomentins and serve as precursors of the violet *spiromentins.

Lit.: Justus Liebigs Ann. Chem. **1989**, 803 ▪ Zechmeister **51**, 21–24 (review). – *[CAS 121254-49-7 (F. B.)]*

Flavones. A class of, generally yellow *plant pigments belonging to the *flavonoids having as a common feature the basic skeleton of flavone, hydroxylated in the 3-position in the case of flavonols. Not only the lipophilic aglycones of yellow heartwood pigments but also many *flower pigments are derived from flavone. Similar to the chemically closely related anthocyanidins, these pigments form oxonium salts on treatment with mineral acids; they are used for mordant dyeing of wool (yellow or brown color). Since F. often occur as copigments with *anthocyanins (intermolecular stacking), frequently yellow and red flowers occur on the same plant or even red or blue and yellow colorations in the same flower. With the exception of flavone itself, all F. are hydroxylated and carry methyl or sugar ether groups, preferentially at the positions 3,5,7,3′, and 4′. For occurrence see also under the individual compounds.

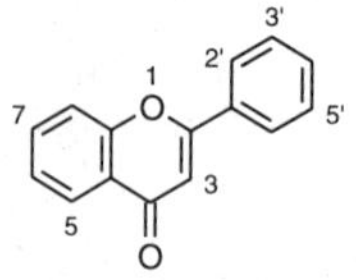

Flavone

Table: Structure of Flavones and Flavonols.

	3	5	7	3′	4′	+OH
Flavonol	OH					
*Chrysin	OH	OH				
*Galangin	OH	OH	OH			
*Apigenin	OH	OH		OH		
*Fisetin	OH		OH	OH	OH	
*Luteolin	OH	OH	OH	OH		
*Kaempferol	OH	OH	OH		OH	
*Quercetin	OH	OH	OH	OH	OH	
*Morin	OH	OH	OH		OH	2′
*Robinetin	OH		OH	OH	OH	5′
*Gossypetin	OH	OH	OH	OH	OH	8
Myricetin	OH	OH	OH	OH	OH	5′

Lit.: Acta Chem. Scand., Ser. B **42**, 303–308 (1988) ▪ Flavonoids: Adv. Res. **1982**, 189–259 ▪ Harborne (1988), p. 233–328 ▪ Nat. Prod. Chem. **2**, 131–254 (1975) ▪ Phytochemistry **28**, 681–694 (1989) ▪ Stud. Org. Chem. **19**, 660–706 (1985) ▪ Ullmann (5.) **B2**, 23–26 ▪ see also flavonoids. – *[HS 2932 90]*

Flavonoids. F. (from Latin *flavus* = yellow), which occur in all higher plants, are important phenylpropane derivatives possessing the C_{15} basic skeleton of flavane. They occur mainly in the vacuoles of plant cells in water-soluble glycosylated form and are often esterified with aliphatic and/or aromatic acids, e. g., with malonic acid or caffeic acid. Non-glycosylated, lipophilic F. occur as non-volatile components in essential oils, accumulate in wood parenchyma (*dye-woods), or are excreted on the epidermis (cuticula) of leaves. The aglycones of F. are classified according to the degree of oxidation of their central pyran ring into the following groups: anthocyanidins (see anthocyanins), *aurones, *catechins, *chalcones, deoxyanthocyanidins, flavanols, flavanones, *flavones, *isoflavones, flavonols, *leucoanthocyanidins. The *biflavonoids, built up of two F. units, are also widely distributed, e. g., the *amentoflavones and the *proanthocyanidins. The members of the classes of F. differ in the number of hydroxy and methoxy substituents. To date more

than 5000 structures are known which vary in the nature, number, and arrangement of their nonacylated and acylated sugar residues.
Use: In the past F.-containing plant extracts were used to dye wool yellow, e.g., with extracts from the wood of dyer's oak (*Quercus velutina*, Fagaceae), the herbage of dyer's weld (*Reseda luteola*, Resedaceae), or the wood of Dyer's mulberry (*Morus tinctoria*, Moraceae) (see dye plants). Today, herbal drugs containing the so-called "bioflavonoids" are used in homeopathic medicine for (a) protection against venous diseases (e.g., the flavonol derivative *rutin or the flavanone derivatives hesperidin and *naringin), (b) improving coronary perfusion as well as peripheral vascular resistance [e.g. F. in hawthorn (*Crataegus*, Rosaceae) drugs], (c) for protection and in some cases also for curing diseases of hepatic parenchyma (e.g., *silybin in the fruit of Our-Lady's thistle, a rare "flavolignan fraction" in which the components are formed by oxidative coupling of the flavanol taxifolin with coniferyl alcohol), (d) diuretic activity, e.g., the F. in birch leaves, or (e) spasmolytic activity, e.g., the F. in chamomile (*Chamomilla*, Asteraceae) drugs.
Biosynthesis: The initial step in the biosynthesis of F., the conversion of phenylalanine to (*E*)-cinnamic acid, is catalyzed by phenylalanine ammonia lyase (PAL). In the following "hydroxycinnamic acid pathway" the four most important hydroxycinnamic acids 4-coumaric, caffeic, ferulic, and sinapic acids are formed by sequential hydroxylation and methylation reactions. 4-Coumaric acid is a central building block in the biosynthesis of F. and condenses as the coenzyme A ester with three molecules of malonylcoenzyme A under catalysis by chalcone synthase via loss of three molecules of CO_2 to furnish the first C_{15}-F. structure, trihydroxychalcone. This is transformed by a chalcone flavanone isomerase (formation of the pyran ring) to naringenin, from which the various classes of F. arise through oxidation reactions on the central pyran ring.
Lit.: Agrawal (ed.), ^{13}C-NMR of Flavonoids, Stud. Org. Chem. **39**, Amsterdam: Elsevier 1989 ▪ Antioxid. Health Dis. **7**, 111–136, 199–219, 221–251 (1998) (antioxidant properties) ▪ Farkas & Kallas, Flavonoids and Bioflavonoids, Amsterdam: Elsevier 1986 ▪ Harborne & Baxter, Handbook of the Natural Flavonoids, vol. 1 and 2, New York: Wiley 1999 ▪ Karrer, No. 1435–1673, 4595–4681 ▪ Phytochemistry **43**, 921–982 (1996) (prenylated F.) ▪ Zechmeister **25**, 150–174; **27**, 158 ff.; **31**, 153–216; **34**, 269–280.

Flavonols see flavones.

Flavopiridol see rohitukine.

Flavor compounds (aroma compounds). Term for volatile compounds in food that are perceived by osmoreceptors (see aroma chemicals) either directly in the nose (smelling, nasal perception) or in the pharyngeal space on eating or drinking (retronasal perception). Together with the generally non-volatile *taste compounds* (sour, sweet, bitter, salty, or spicy tasting compounds). F. c. make a decisive contribution to the taste of a food, the consistency of the food also contributes to the complete sensory impression.
Modern methods of isolation and identification of volatile compounds in foods (review, see *Lit.*[1]) have resulted in an increase in the number of F. c., described in the literature up to 1994 to over 7000[2]. Not all of the F. c. found in a food are formed by genuine biosynthesis[3] (so-called primary F. c.); most are formed after destruction of the cell aggregate, e.g., by enzymatic processes as well as by fermentative or thermal preparation of the food (so-called secondary F. c.); in the latter case especially by the Maillard reaction. As a result of the numerous precursors and routes of formation, F. c. include many classes of compounds, especially alcohols, aldehydes and ketones, esters, lactones, sulfides, and heterocyclic species, e.g., furan compounds, *pyrazines, thiazoles or thiophenes. The most important precursors are *amino acids, *fatty acids, sugars, and isoprenoids.
The aroma of a food often contains well over 100 (up to 800) components, of which only few have sensory significance. The so-called *impact compounds which pronounce the character of some aromas have in general been known for a long time. Only recently have methods been developed to identify the aroma-active key components of complex aromas.
Natural and natural-identical F. c. are used in the commercial production of aromas. The latter are F. c. that have been identified as components of foods and then produced synthetically. For the biotechnical production of natural F. c., see *Lit.*[4]. The world market of aroma and flavor substances in 1994 amounted to 9.7 billion US $, the most important individual compounds being (consumption in t per year): 2-phenylethanol and its esters (7000), *musk aromas (6200), *linalool or its esters (6000), ester of lower *fatty acids (5800) and vanilla (see 3,4-dihydroxybenzaldehydes) (5500)[5]. Using tools of molecular biology, scientists are beginning to figure out how the olfactory sense works and how flavor impression is formed[6].
Lit.: [1]Maarse, p. 1–39. [2]TNO list (6.), Suppl. 1–5. [3]Müller-Lamparsky, p. 101–126; Schreier, Chromatographic Studies of Biogenesis of Plant Volatiles, Heidelberg: Hüthig 1984. [4]Perfum. Flavor. **20** (5), 5–14 (1995). [5]Chem. Ind. (London) **1996**, 170 ff. [6]Cell **87**, 675 (1996); Chem. Eng. News, 23. 12. **1996**, 18 ff.; 25. 10. **1999**, 38 ff.
gen.: ACS Symp. Ser. **596**, 1–279 (1995) (F. c. from fruits); **610** (1996) (flavor technology); **674** (1997) (flavor and lipid chemistry of seafoods); **705** (1998) (flavor analysis); **714** (1998) (wine flavor) ▪ Chem. Rev. **96**, 3201–3240 (1996) (structure and activity) ▪ Fenaroli (2.), **1** u. **2** ▪ Helv. Chim. Acta **75**, 1341–1415 (1992) ▪ Kirk-Othmer (4.) **11**, 1–60 ▪ Morton u. MacLeod (eds.), Food Flavours, Part A, Amsterdam: Elsevier 1982 ▪ Tetrahedron **54**, 7633–7703 (1998) (review) ▪ Ziegler u. Ziegler, Flavourings, Production, Composition, Applications, Regulation, Weinheim: Wiley VCH 1998.

Flavoxanthin [(3*S*,3′*R*,5*R*,6′*R*,8*R*)-5,8-epoxy-5,8-dihydro-*β*,*ε*-carotine-3,3′-diol].

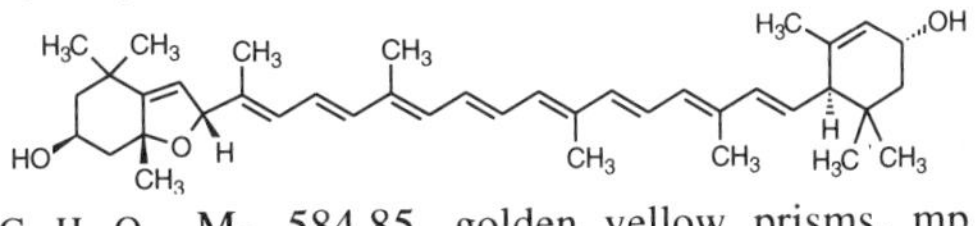

$C_{40}H_{56}O_3$, M_R 584.85, golden yellow prisms, mp. 184 °C, $[\alpha]_D^{20}$ +190° (benzene), uv$_{max}$ 430, 459 nm ($CHCl_3$); soluble in acetone; less soluble in alcohols. F. is a xanthophyll (see carotinoids), occurring as the flower pigment of e.g., buttercup (*Ranunculus acer*, Ranunculaceae), dandelion (*Taraxacum officinale*, Asteraceae), broom species [including common broom

(*Cytisus scoparius*, Fabaceae)], pansies, and pollen of acacia.

Lit.: Beilstein E V **17/5**, 490 ▪ Helv. Chim. Acta **61**, 783 (1978) ▪ Karrer, No. 1847. – *[CAS 512-29-8]*

Flavylium cation see anthocyanins.

Flexilin.

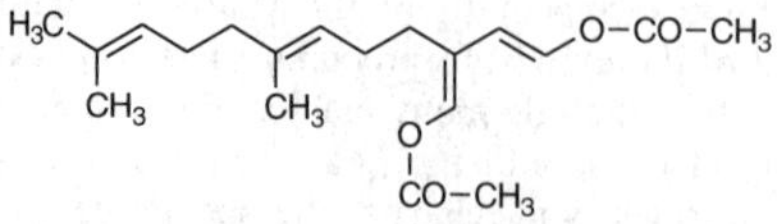

$C_{19}H_{28}O_4$, M_R 320.43, colorless oil, bp. 100 °C (13.3 Pa). An acyclic *sesquiterpene from the green alga *Caulerpa flexilis* (Chlorophyta) endemic to Tasmania. The 1,4-diacetoxy-1,3-butadiene system on which F. is based is found in many other sesqui- and diterpenoids of marine green alga. The compound is easily hydrolyzed to toxic dialdehydes and thus serves as a defence against predators. The hydrolysis products are general cell toxins and inhibit cell division in fertilized sea urchin eggs; effective dose: ED_{100} = 16 µg/mL.

Lit.: Scheuer II **1**, 1–29, 64, 124 ▪ Tetrahedron Lett. **33**, 3063 (1978) ▪ J. Chem. Soc. **1994**, 161 (14-oxo-F.). – *[CAS 69625-33-8]*

Flexirubin.

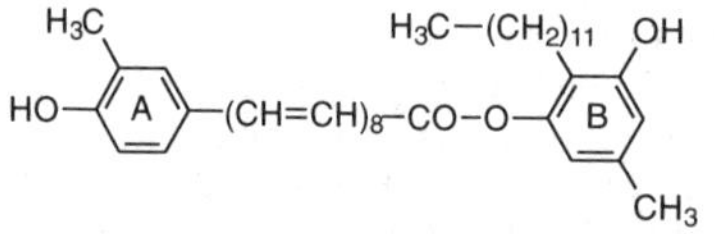

Major pigment of the gliding bacterium *Flexibacter elegans*, $C_{43}H_{54}O_4$, M_R 634.90, violet-red needles, mp. 174–176 °C from acetone, uv_{max} 453 nm (log ε=4.97), shifted to 492 nm on addition of bases. F. is not, as first believed, a carotenoid (which, however, are also formed) but rather a polyenecarboxylate of a dialkylresorcinol. The minor components (about 30) differ in the length of the polyene chain, the alkyl groups on the two benzene rings, or carry chlorine substituents. The biosynthesis proceeds by the polyketide pathway with the A ring originating from tyrosine via *p*-hydroxyphenylacetic acid and the B ring from acetate.

Lit.: Chem. Ber. **109**, 2490 (1976) ▪ Tetrahedron **39**, 175 (1983). – *Synthesis.:* Angew. Chem. Int. Ed. Engl. **16**, 191 (1977) ▪ Zechmeister **52**, 73–109. – *[CAS 54363-90-5]*

Flies. Together with the midges, flies comprise the insect order of two-winged insects (Diptera) with ca. 85 000 species worldwide; their jaws are fitted with sucking and often penetrating apparatus. Many flies are blood-sucking and disease-carrying insects. The ubiquitous house fly (*Musca domestica*) is known to be a carrier of tuberculosis, typhus, poliomyelitis, and other diseases. Tsetse flies (*Glossina* spp.) carry sleeping sickness and nagana (livestock disease). The larvae of many flies live as parasites in the internal organs of animals. Numerous F. are known to be plant pests, e. g., species of the genera *Dacus* and *Bactrocera*. Fruit flies of the genus *Drosophila* are widely used as model organisms for molecular biology research. The *pheromones of many flies have been identified [1]; their chemical structures are manifold and include alkenes such as (*Z*)-9-tricosene (*muscalure), alkadienes such as *(7Z,11Z)-7,11-heptacosadiene* ($C_{27}H_{52}$, M_R 376.71; the sex pheromone of *Drosophila melanogaster* [2]), branched alkanes such as *15,19-* and *17,21-dimethylheptatriacontane* ($C_{39}H_{80}$, M_R 549.06) or *15,19,23-trimethylheptatriacontane* ($C_{40}H_{82}$, M_R 563.09; pheromones of the tsetse flies, *Glossina* spp.) [3] as well as *15,19-dimethyltritriacontane* ($C_{35}H_{72}$, M_R 492.96; sexual stimulant of *Stomoxys calcitrans*) [4], *spiroacetals [1], alkylated pyrazines [1], amides [1]. Furthermore, a series of long-chain aliphatic compounds such as *(S)-2-tridecanol acetate* ($C_{15}H_{30}O_2$, M_R 242.40) and *(S)-2-pentadecanol acetate* [5] ($C_{17}H_{34}O_2$, M_R 270.46), *(Z)-11-eicosen-1-ol acetate* [6] ($C_{22}H_{42}O_2$, M_R 338.57) or *(Z)-10-heptadecen-2-one* [5] ($C_{17}H_{32}O$, M_R 252.44) occurs in *Drosophila*; see also anastrephin, olean, suspensolide.

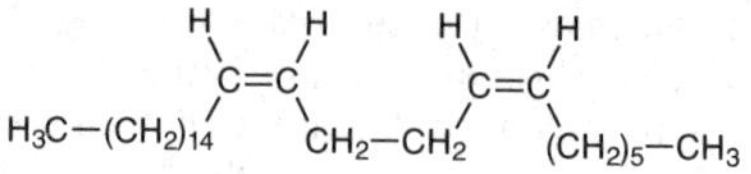

(7*Z*,11*Z*)-7,11-Heptacosadiene (**1**)

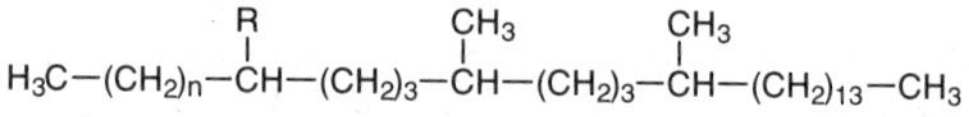

R = H , n = 13 : 15,19-Dimethylheptatriacontane (**2**)
R = CH3 , n = 13 : 15,19,23-Trimethylheptatriacontane (**3**)
R = H , n = 9 : 15,19-Dimethyltritriacontane (**4**)

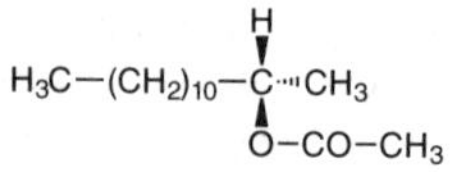

(*S*)-2-Tridecanol acetate (**5**)

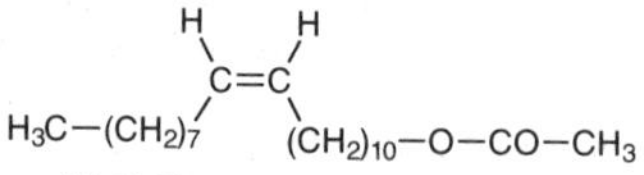

(*Z*)-11-Eicosen-1-ol acetate (**6**)

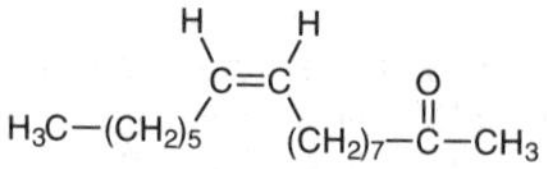

(*Z*)-10-Heptadecen-2-one (**7**)

Lit.: [1] Chem. Rev. **95**, 789–828 (1995). [2] Synthesis **1989**, 936. [3] Science **201**, 750–753 (1978). [4] J. Chem. Ecol. **10**, 771–781 (1984). [5] J. Chem. Ecol. **15**, 399–411, 2577–2588 (1989). [6] J. Chem. Ecol. **15**, 265–273 (1989). – *[CAS 100462-58-6 (1); 56987-91-8 (2); 67979-80-0 (3); 56987-80-5 (4); 120876-13-3 (5); 66731-37-1 (6); 58257-63-9 (7)]*

Flowering hormone (florigen). Plants form substances which stimulate them to form buds and to bloom. The regulatory genes in the meristem are activated by these substances. As plant hormones (phytohormones), these substances can also be transferred to other plants, but little is known about their composition. *Gibberellin A3 can trigger the formation of flowers in many long-day plants also under short-day conditions. Certain gibberellins act exclusively in the induction of blooming. The phytohormone possibly consists of a mixture of growth and inhibiting substances. Then certain concentration ratios between, e. g., *auxins, *cytokinins and gibberellins and the F. h. are significant for the transition from the vegetative to the flowering states.

Lit.: Biol. Plant **27**, 292–302 (1985) ▪ Bopp (ed.), Plant Growth Subst. 1985, p. 303–307, Berlin: Springer 1986 ▪ J. Am. Soc. Hortic. **116**, 450 ff. (1991) ▪ Kamanik, Mok & Zazimalowa (eds.), Physiol. Biochem. Cytokinins Plants Symp. 1990, p. 347–351, 389 ff., Den Haag: SPB Acad. Publ. 1992 ▪ Plant Physiol. Biochem. (Paris) **27**, 443–450 (1989).

Flower pigments. Term for a subgroup of *plant pigments responsible for the manifold colors of flowers. The most important F. p. are the *anthocyanins, *flavonoids, *carotinoids, and *betalains. Yellow colors are caused by flavonoids, carotenoids, and betaxanthins while anthocyans and betacyans (see betalains) are responsible for red, violet, and blue colors. The many color tones result on the one hand from the simultaneous presence of flavonoids and anthocyans (intermolecular copigmentation) and on the other hand from intramolecular copigmentation and the formation of chelates between metal salts and anthocyans.
Lit.: Czygan ▪ Harborne, Introduction to Ecological Biochemistry, New York: Academic Press 1994 ▪ Schweppe ▪ see also natural dyes.

Fluopsins (Thioformins). The F. are isolated from culture broths of *Pseudomonas fluorescens* and show broad antibiotic activity against Gram-negative and -positive bacteria as well as fungi. The brown to black F. are only soluble in highly polar solvents and decompose on heating.

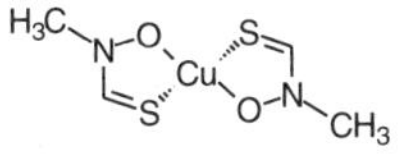

F.C

F. C was identified as a 1 : 2 copper complex of *N*-hydroxy-*N*-methylthioformamide, F. F as a 1 : 3 iron complex. *F. C:* $C_4H_8CuN_2O_2S_2$, M_R 243.79, mp. 182 °C; *F. F*: $C_6H_{12}FeN_3O_3S_3$, M_R 326.21. The free ligands can be obtained by treatment with hydrogen sulfide and used in complexation reactions with other two- and three-valent heavy metal ions.
Lit.: Antimicrob. Agents Chemother. **15**, 384 (1979) ▪ J. Am. Chem. Soc. **100**, 2251 (1978) ▪ J. Antibiot. **23**, 267, 546 (1970). – *[HS 2941 90; CAS 31323-25-8 (F. C); 31323-26-9 (F. F)]*

Fluorescyanin see ichthyopterin.

Fluoro... see fluoroorganic natural products.

Fluoroacetic acid.

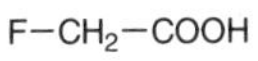

Fluoroacetic acid

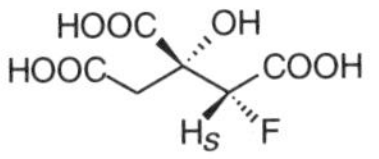

(2*R*,3*R*)-2-Fluorocitric acid

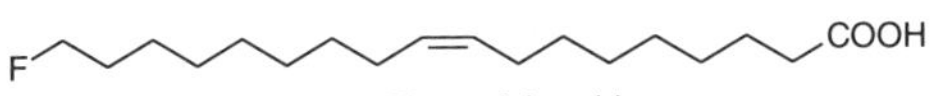

ω-Fluorooleic acid

$C_2H_3FO_2$, M_R 78.04, needles, mp. 35 °C, bp. 165 °C. The highly toxic F. was first discovered in 1943 as the toxic principle of the South African plant *Dichapetalum cymosum* (Dichapetalaceae); which is responsible for the death of a large number of cattle. Since then F. has been detected in many other plants, in particular in African *Dichapetalum* species (*D. braunii* from Tanzania contains up to 8 mg/g dry weight) and in poisonous Australian Fabaceae such as *Acacia georginae*, *Gastrolobium* sp., and *Oxylobium* sp.. It is of interest that kangaroos are less sensitive to poisoning than introduced grazing animals. High concentrations of F. are also found in the Brazilian Rubiaceae *Palicourea marcgravii* (rat weed). The oral LD_{50} in rats is 3–5 mg/kg, the lethal dose for humans is 2–10 mg/kg (Na salt). Contact with F. leads to severe irritation, damage to the eyes, the respiratory organs (pulmonary edema), and the skin (F. is absorbed through the skin). Death occurs through circulatory arrest and respiratory paralysis. The Na salt acts as a rat poison but is not authorized for use in Germany.
In the cellular metabolism F. reacts with citrate synthase to form *(2R,3R)-2-fluorocitrate* ($C_6H_7FO_7$, M_R 210.12), which inhibits transport of citrate through mitochondrial membranes. The "lethal synthesis" of fluorocitrate, however, only contributes to a small extent to the toxic activity. *D. toxicarium* stores F. in its extremely toxic seeds in the form of ω-fluorofatty acids, principally *ω-fluorooleic acid* ($C_{18}H_{33}FO_2$, M_R 300.46). Fluoroacetyl-CoA functions here as a starter unit. The biosynthesis of F. from fluoride ions has not been elucidated.
Lit.: Beilstein E IV **2**, 446 f. ▪ Chem. Br. **1992**, 785 (review) ▪ Hayes et al. (eds.), Handbook of Pesticide Toxicology, p. 1273, 1277, London: Academic Press 1991 (toxicology) ▪ Merck-Index (12.), No. 4205 ▪ Nat. Prod. Rep. **11**, 123 (1994) (review) ▪ Phytochemistry **26**, 2293 (1987) (occurrence) ▪ Sax (8.), FIC 000 (toxicology). – *[CAS 144-49-0 (F.); 357-89-1 (2-fluorocitric acid)]*

Fluorocitric acid see fluoroacetic acid.

Fluoroorganic Natural Products. Natural products containing fluorine are rare – up to 1998 only 16 naturally occurring organofluorine metabolites have been identified (11 of them fluorinated fatty acids) and practically nothing is known about their origins or the enzymatic processes for generating C–F bonds. The most famous and well investigated metabolite is the highly toxic fluoroacetic acid (inhibitor of the citric acid cycle). It was first isolated in 1944 from the South African plant *Dichapetalum cymosum* (called gifblaar = poison leaf), which has long been recognized as a hazard to livestock in this region. Most of the fluorinated fatty acids are found in the plant *Dichapetalum toxicarium*. S-Fluoropolyoxin L and 4-Fluoropyrrolnitrin are produced in bacterial submersion cultures only on feeding with organofluorine precursors (figure see p. 236, table p. 237).
Lit.: [1] Nat. Prod. Rep. **11**, 123–133 (1994); Zechmeister **68**, 108; Sax (8.), FFF 000–FIC 000. [2] J. Org. Chem. **53**, 2991–2999 (1988). [3] J. Chem. Soc., Chem. Commun. **1997**, 1471; J. Chem. Soc., Perkin Trans. 1 **1998**, 759–768. [4] J. Antimicrob. Chemother. **30**, 261–272 (1992). [5] Agric. Biol. Chem. **51**, 1183 (1987). [6] Biochem. Pharmacol. **12**, 627–632 (1963). [7] Nature (London) **201**, 611 (1964). [8] J. Chem. Soc., Perkin Trans. 1 **1975**, 2523. [9] Phytochemistry **44**, 1129 (1997). [10] Tetrahedron Lett. **31**, 7661 (1990).
gen.: Chem. Brit. **28**, 785–788 (1992).

Flurithromycin see fluoroorganic natural products.

$F-CH_2-COOH$

Fluoroacetic acid

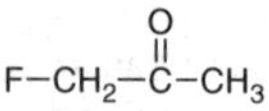

Fluoroacetone

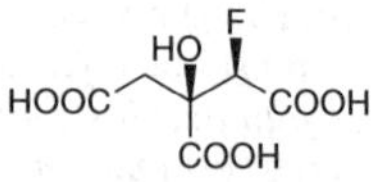

(*R,R*)-Fluorocitric acid

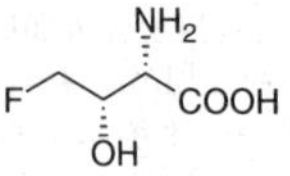

4-Fluoro-L-threonine*

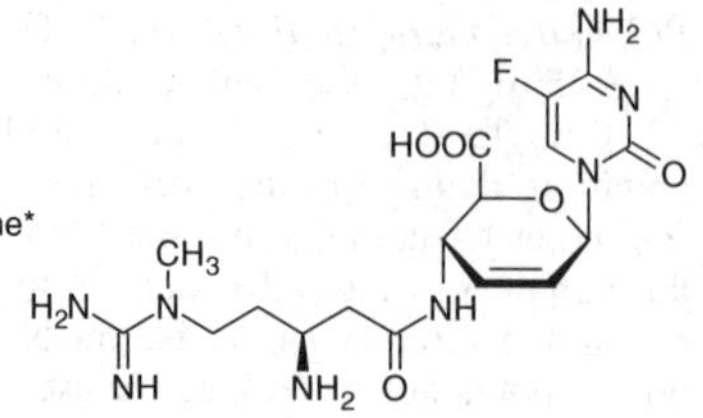

5-Fluoroblasticidin S*

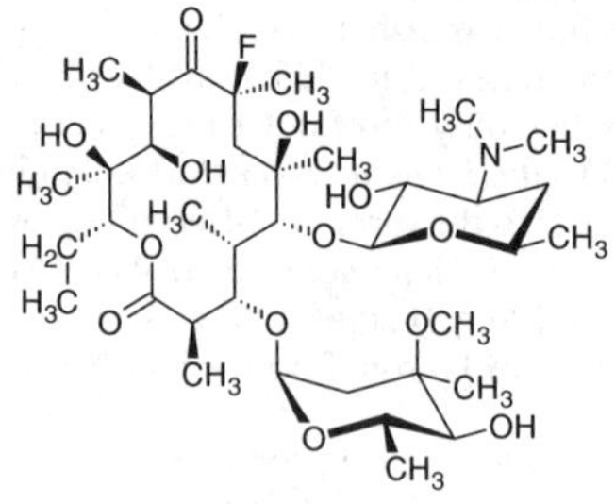

8-Fluoroerythromycin (Flurithromycin)*

$F-(CH_2)_n-COOH$

ω-Fluoroalkanoic acids (n = 3, 9, 13, 15, 17, 19)

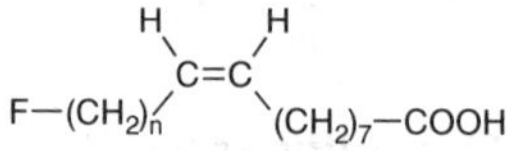

n = 6 : 16-Fluoropalmitoleic acid
n = 8 : 18-Fluorooleic acid

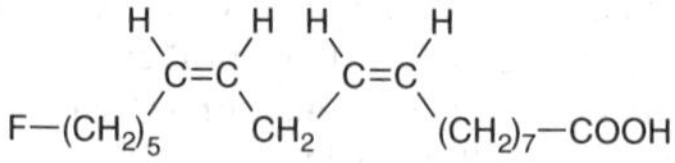

18-Fluorolinoleic acid

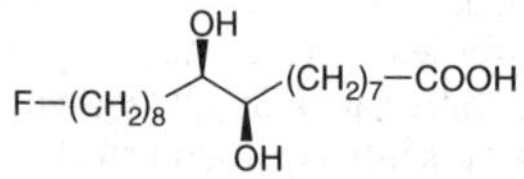

threo-18-Fluoro-9,10-dihydroxy-octadecanoic acid (relative configuration)

* produced by feeding bacterial cultures with *inorganic* fluoride
Figure: Selected structures of fluoroorganic natural products.

Flustramines.

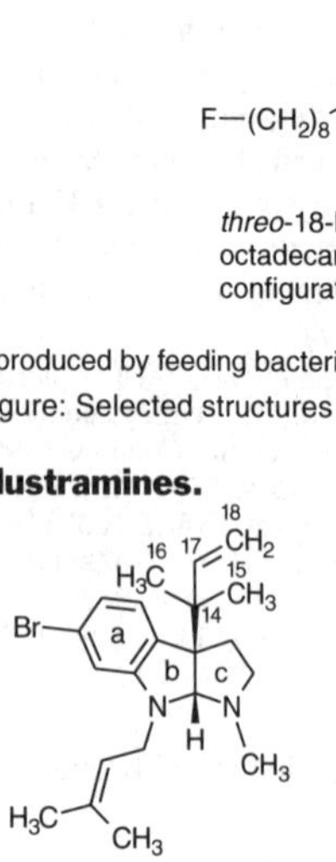

Flustramine A

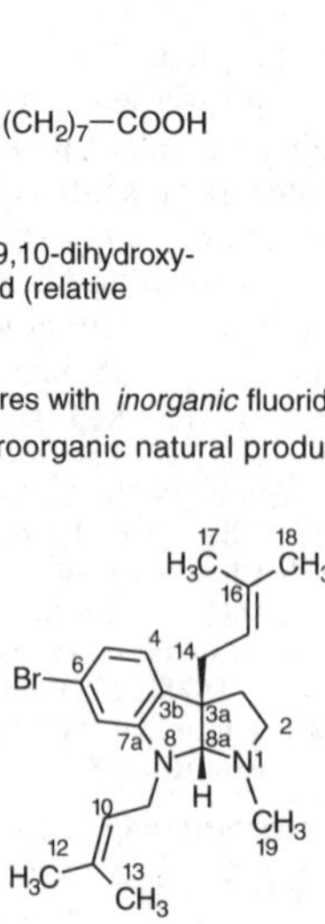

Flustramine B

(relative configuration)

The bryozoon *Flustra foliacea* contains bromoindole alkaloids (*bryozoan alkaloids) with the *physostigmine skeleton; this is also characteristic for a group of alkaloids from the calabar bean (*Physostigma venenosum*): *F. A*, $C_{21}H_{29}BrN_2$, M_R 389.38; *F. B*, $C_{21}H_{29}BrN_2$, M_R 389.38.

Lit.: Chem. Pharm. Bull. **31**, 1806 (1983) (synthesis) ▪ J. Org. Chem. **45**, 1586 (1980); **59**, 5543 (1994) ▪ Tetrahedron Lett. **37**, 7525 (1996) (synthesis F. C). – *[HS 2939 90; CAS 71239-64-0 (F. A); 71239-65-1 (F. B)]*

Fly agaric constituents.

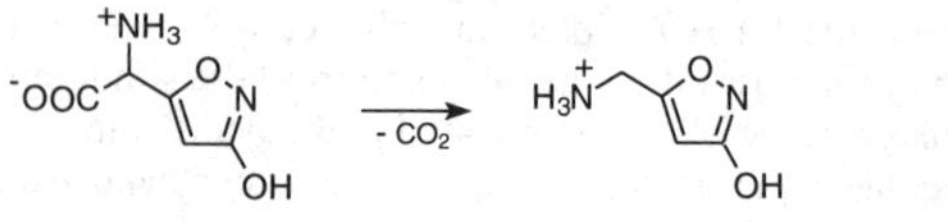

Ibotenic acid (**1**) Muscimol (**2**)

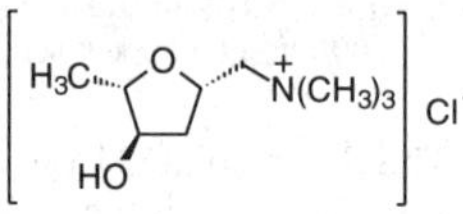

Muscarine (**3**)

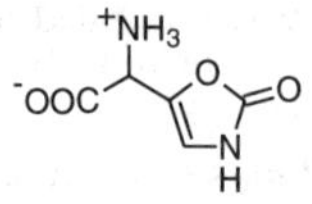

Muscazone (**4**)

On account of its conspicuous appearance (see fly agaric pigments) the fly agaric (*Amanita muscaria*) is one of the most popular toadstools. The name is based on the legend that tissue of the fungi placed in milk is able to attract and kill flies; in fact, however, the flies are merely anesthetized. Fly agaric is also a good-luck charm; this is based on the fact that it has been in use for thousands of years as a narcotic drug. Thus, it is now known, after many years of uncertainty about its plant origin, that the "soma" drink described in the Indian "Rig-Weda" (1500–1000 B. C.) is fly agaric juice. The effects of the psychoactive drug have been described in various ways; it is similar to an alcohol intoxication. Although *muscarine*[1] ($C_9H_{20}ClNO_2$, M_R 209.72, hygroscopic needles, mp. 181.5–182 °C) is highly toxic [LD_{50} (mouse i. v.) 0.23 mg/kg] and thus causes nausea and vomiting in the early stage of a fly agaric intoxication, *ibotenic acid* [2] ($C_5H_6N_2O_4$, M_R 158.11, cryst., mp. 151–152 °C) and, especially its decarboxylation product *muscimol* [$C_4H_6N_2O_2$, M_R 114.11, mp. 155–156 °C (monohydrate)] are responsible for the hallucinogenic effects. It is probable that ibotenic acid and muscimol react with GABA receptors as potential agonists to cause the state of intoxication. Another hallucinogenic amino acid, *muscazone* ($C_5H_6N_2O_4$, M_R 158.11, mp. 190 °C) is probably formed by photochemical rearrangement of ibotenic acid. Muscarine also occurs in various Inocybes and Clitocybes.

Lit.: [1] Acta Crystallogr. Sect. C **46**, 1279 (1990); Beilstein E V **18/10**, 220 f.; Hager (5.) **3**, 849 ff. – *Synthesis:* Can. J. Chem. **70**, 2726 (1992); Carbohydr. Res. **288**, 241 (1996); Collect. Czech. Chem. Commun. **63**, 1522 (1998); Synlett **1994**, 295; Sax (8.), MRW 250; Chem. Pharm. Bull. **43**, 1067 (1995); J. Chem. Soc., Perkin Trans. 1 **1992**, 3023. [2] Zechmeister **27**, 261–321; Eur. J. Med. Chem. **31**, 515–537 (1996); Nature Medicine **49**, 354 (1995); Sax (8.), AKG 250. – *Synthesis:* Chem. Pharm. Bull. **19**, 46 (1971); Synth. Commun. **22**, 1939 (1992); Tetrahedron Lett. **1965**, 2081.

gen.: Bresinsky & Besl, Giftpilze, p. 71, 98–106, Stuttgart: Wissenschaftl. Verlagsges. 1985 ▪ J. Nat. Prod. **44**, 422 (1981)

Table: Fluoroorganic natural products.

	molecular formula	M_R	mp. (bp.) [°C]	producing organism	CAS	*Lit.*
Fluoroacetic acid	$C_2H_3FO_2$	78.04	35.3	e.g. *Dichapetalum cymosum*	144-49-0	1
Fluoroacetone	C_3H_5FO	76.07	75–77	*Acacia georginae*	430-51-3	2
(*R,R*)-Fluorocitric acid	$C_6H_7FO_7$	210.11	amorphous solid	soybean	357-89-1	1
4-Fluoro-L-threonine	$C_4H_8FNO_3$	137.11	182–183	*Streptomyces cattleya*	102130-93-8	3
Flurithromycin	$C_{37}H_{66}FNO_{13}$	751.92	183–184	*Streptomyces erythraeus*	82664-20-8	4
5-Fluoroblasticidin "S"	$C_{17}H_{25}FN_8O_5$	440.43	225–227	*Streptomyces griseochromogenes*	102865-76-9	5
4-Fluorobutanoic acid	$C_4H_7FO_2$	106.10	(78–79, 0.8 kPa)	*Streptomyces cerevisiae*	462-23-7	6
10-Fluorodecanoic acid	$C_{10}H_{19}FO_2$	190.26	49–50	*Dichapetalum toxicarium*	334-59-8	7
14-Fluoromyristic acid	$C_{14}H_{27}FO_2$	246.36	amorphous solid	*Dichapetalum toxicarium*	151277-21-3	7
16-Fluoropalmitoleic acid	$C_{16}H_{31}FO_2$	274.41	74.5–75.5	*Dichapetalum toxicarium*	3109-58-8	7, 8
18-Fluorooleic acid	$C_{18}H_{33}FO_2$	300.45	53–54 (*E*)	*Dichapetalum toxicarium*	1478-37-1	9
18-Fluorostearic acid	$C_{18}H_{35}FO_2$	302.47	amorphous solid	*Dichapetalum toxicarium*	408-37-7	9
18-Fluorolinoleic acid	$C_{18}H_{31}FO_2$	298.44	amorphous solid	*Dichapetalum toxicarium*	188893-23-4	9
20-Fluoroarachidic acid	$C_{20}H_{39}FO_2$	330.53	amorphous solid	*Dichapetalum toxicarium*	188893-24-5	9
threo-18-Fluoro-9,10-dihydroxyoctadecanoic acid	$C_{18}H_{35}FO_4$	334.47	amorphous solid	*Dichapetalum toxicarium*	132891-45-3	10
16-Fluoro-9-hexadecenoic acid	$C_{16}H_{29}FO_2$	272.40	amorphous solid	*Dichapetalum toxicarium*	78417-38-6	9

▪ Naturwissenschaften **69**, 326 (1982) ▪ Schultes & Hofmann, Plants of the Gods, p. 82–85, New York: Alfred van der Marck Editions 1979 ▪ Stud. Org. Chem. **18**, 61 (1984) ▪ Tetrahedron **35**, 2843 (1979). – *[CAS 2552-55-8 (1); 2763-96-4 (2); 2303-35-7 (3); 300-54-9 ((2S,4R,5S)-form of 3); 2255-39-2 (4)]*

Fly agaric pigments.

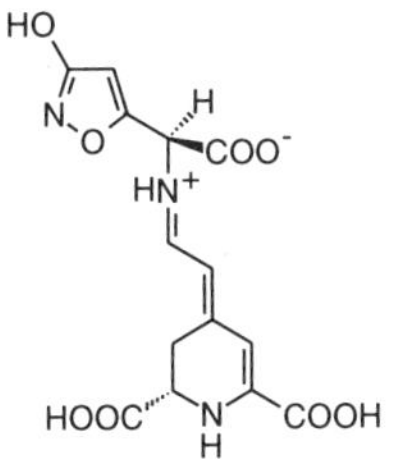

Muscaaurin-I (**1**)

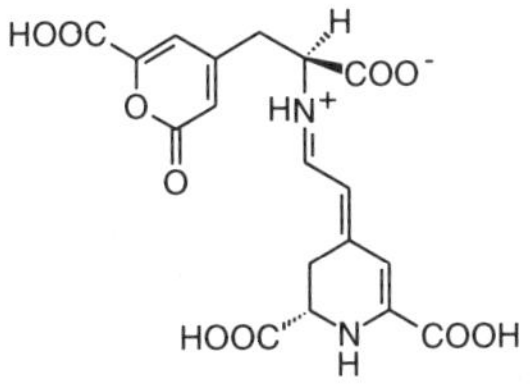

Muscaaurin-II (**2**)

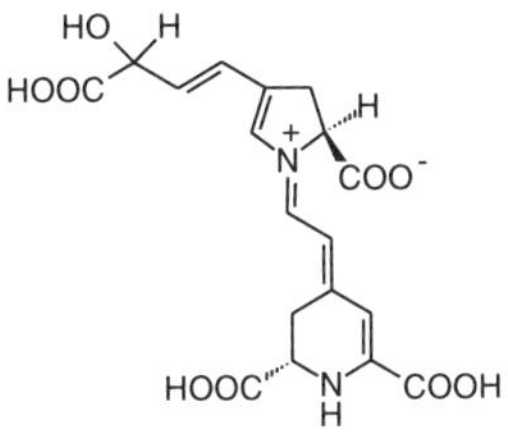

Muscapurpurin (**3**)

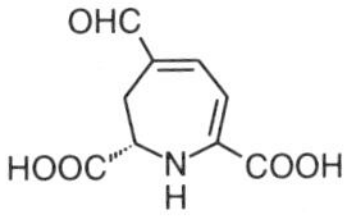

Muscaflavin (**4**)

The cap skin of the fly agaric (*Amanita muscaria*) contains yellow, orange, and purple pigments which together are responsible for the brilliant red color. The orange-colored *muscaaurins* belong to the *betalains and contain *betalamic acid as well as some unusual amino acids such as ibotenic acid (*muscaaurin-I*, $C_{14}H_{13}N_3O_8$, M_R 351.27) or *stizolobic acid (*muscaaurin-II*, $C_{18}H_{16}N_2O_{10}$, M_R 420.33). The chromophore of the purple-colored *muscapurpurin* ($C_{18}H_{18}N_2O_9$, M_R 406.35) is extended by two double bonds. Besides the betalains the yellow dihydroazepine pigment *muscaflavin* {$C_9H_9NO_5$, M_R 211.17, unstable yellow crystals, mp. of the dimethyl ester 98 °C, $[\alpha]_D^{20}$ –270° (CH_3OH)} is also present. The latter, like the isomeric betalamic acid is formed biosynthetically by extradiol[1] cleavage from L-Dopa. Muscaflavin[2] also occurs in *Hygrocybe* species where it exists as Schiff's bases formed by coupling via the aldehyde group to various amino acids (so-called hygroaurins).

Lit.: [1] Luckner (3.), p. 83. [2] Beilstein E V **22/7**, 171; Justus Liebigs Ann. Chem. **1981**, 2164; Helv. Chim. Acta **62**, 1231 (1979); Z. Naturforsch. C **29**, 637 (1974).

gen.: Justus Liebigs Ann. Chem. **1982**, 254 ▪ Naturwissenschaften **69**, 326 (1982) ▪ Tetrahedron **35**, 2843 (1979) ▪ Zechmeister **51**, 75 ff. (reviews). – *[CAS 52012-51-8 (1); 12624-17-8 (2); 12624-19-0 (3); 12624-18-9 (4)]*

FMN. Abbreviation for flavin mononucleotide; see riboflavin.

Foliamenthin.

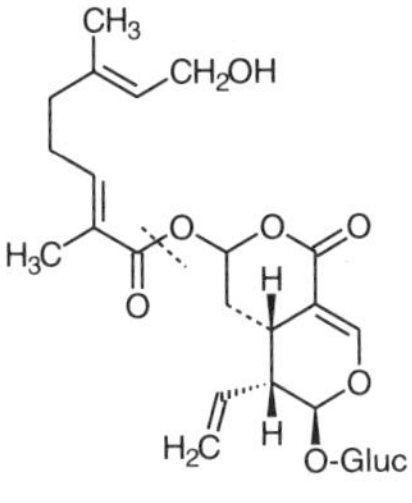

Foliamenthin

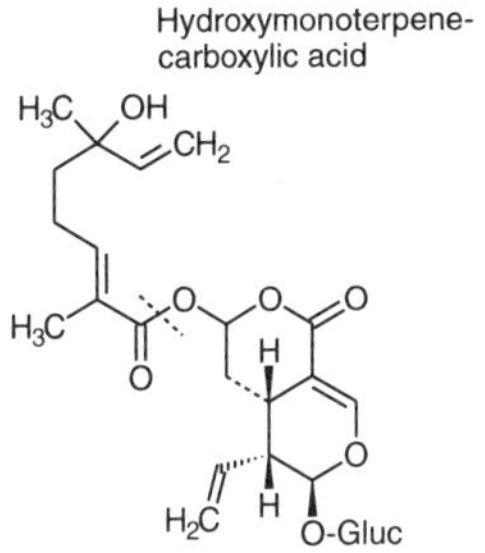

Menthiafolin

$C_{26}H_{36}O_{12}$, M_R 540.56, mp. 194–196 °C, $[\alpha]_D$ –63° (CH_3OH). Isomer: *menthiafolin*, mp. 186 °C, $[\alpha]_D$ –68° (CH_3OH). Both natural products possess the *secoiridoid skeleton of secologanic acid in which the OH group is esterified by a hydroxymonoterpenecarboxylic acid. F. can be formed by an allyl rearrangement from menthiafolin. F. and menthiafolin occur in the leaves of buckbean (bog bean, water trefoil) (*Menyanthes trifoliata*, Menyanthaceae)[1]. The dried leaves (Fol. trifolii fibrini) were frequently used in the past in veterinary medicine as a feeding stimulant.

Lit.: [1] J. Chem. Soc., Chem. Commun. **1968**, 1276, 1277.

gen.: Planta Med. **57**, 181 (1991). – *[CAS 21848-66-8 (F.); 19351-64-5 (menthiafolin)]*

Folic acid (International generic name for *N*-pteroyl-L-glutamic acid, PteGlu, coenzyme F).

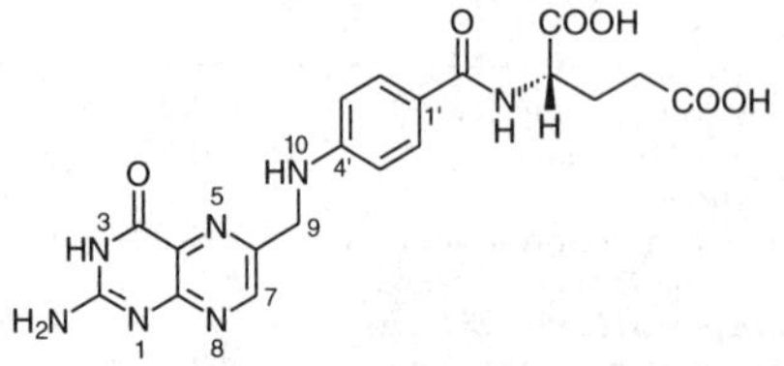

$C_{19}H_{19}N_7O_6$, M_R 441.40, $[\alpha]_D^{25}$ +25° (0.1 N NaOH), $[\alpha]_D^{27}$ +19.9° (0.1 N NaOH). Yellow to orange platelets, mp. >250 °C (charring). F. occurs in liver, kidney, yeasts, fungi, grain crops, and green leaves (e.g., spinach), mainly as a conjugate with poly-γ-L-glutamic acid (*pteroylpolyglutamic acid,* "vitamin B_C"). It is a growth factor for many microorganisms and functions as a vitamin in the human organism. F. was previously known as vitamin B_9 or M. Adults have a daily requirement for about 200 µg of bioavailable folate.

Biochemistry: In organisms F. exists in equilibrium with 7,8-*dihydrofolic acid* (H_2folate; older abbreviation: FH_2) with participation of nicotinamide adenine dinucleotide phosphate (NADP) and the enzyme *dihydrofolate reductase* (EC 1.5.1.3). H_2folate, in turn, is formed in plants and some microorganisms via several intermediates from guanosine 5′-triphosphate and reacts in the presence of the above-mentioned dihydrofolate reductase to form (6*S*)-5,6,7,8-*tetrahydrofolic acid (H_4folate; older abbreviation: FH_4, THF) which is the actual physiologically active form of folic acid. H_4folate is a transport metabolite for 1-C units with the transfer proceeding through 5-methyl-H_4folate, 5,10-methylene-H_4folate ("active formaldehyde"), 5-formyl-H_4folate (*folinic acid), 5-formimidoyl-H_4folate, 10-formyl-H_4folate ("activated formic acid") or 5,10-methylene-H_4folate. The one-carbon building blocks are required, among others, in the biosyntheses of *glycine and *methionine, as well as purine nucleotides and deoxythymidine 5′-monophosphate, i.e., precursors of deoxyribonucleic acid (DNA). F. is synthesized by plants and microorganisms but not by humans and animals. A deficiency of H_4folate leads to anemic symptoms with excessively large, short-lived erythrocytes (megalocytes) as well as impaired formation of platelets and granulocytes. On the other hand, the F. antagonists and dihydrofolate reductase inhibitors aminopterine and methotrexate are employed in chemotherapy of leukemia. The bactericidal activities of sulfonamides are based on their properties as antimetabolites of *4-aminobenzoic acid and the thus resulting inhibition of the biosynthesis of F. *Acetyl-CoA is synthesized in many anaerobic bacteria with the help of methyl-H_4folate [1].

Lit.: [1] Blakely & Benkovic, Folates and Pterins, vol. 1: Chemistry and Biochemistry of Folates, Chichester: Wiley 1984.
gen.: Beilstein E V **26/18**, 237 f. ▪ Hager (5.) **8**, 283 ff. ▪ J. Chem. Soc., Chem. Commun. **1987**, 470 (synthesis) ▪ Methods Enzymol. **66**, 468–483 (1980); **122**, 309 (1986) ▪ Merck-Index (12.), No. 4253 ▪ Nat. Prod. Rep. **3**, 395–419 (1986) (biosynthesis). – *[HS 2936 29; CAS 59-30-3 (F.); 75708-92-8 (dihydrate)]*

Folicanthine see pyrroloindole alkaloids.

Folimycins see concanamycins.

Folinic acid {*N*-[(6*S*)-5-formyl-5,6,7,8-tetrahydropteroyl]-L-glutamic acid, (6*S*)-5-formyltetrahydrofolic acid, leucovorin}.

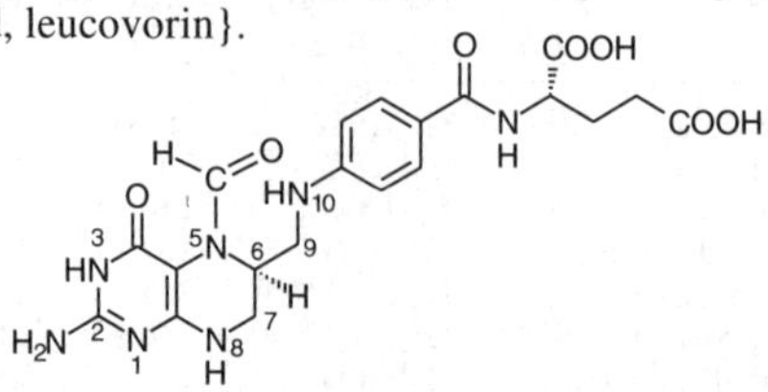

$C_{20}H_{23}N_7O_7$, M_R 473.45, pale yellow crystals +3 H_2O, mp. 248–250 °C (decomp.), $[\alpha]_D^{25}$ +16.8° (aqueous hydrogen carbonate). F. occurs in various microorganisms in which it acts as a growth factor; e.g., for the lactic acid bacterium *Leuconostoc citrovorum*, therefore, it is also known as *citrovorum factor*. F. functions as an active 1-C unit (cf. folic acid) and formyl group transfer agent in the metabolism of folic acid *tetrahydrofolic acid. F. is accessible by synthesis and its calcium salt is used therapeutically as an antidote for folic acid antagonists such as methotrexate.

Lit.: Hager (5.) **8**, 283 f. ▪ J. Chem. Soc., Perkin Trans. 1 **1993**, 871 (synthesis) ▪ Merck-Index (12.), No. 4254 ▪ see folic acid. – *[HS 2936 29; CAS 58-05-9]*

Fomannosin.

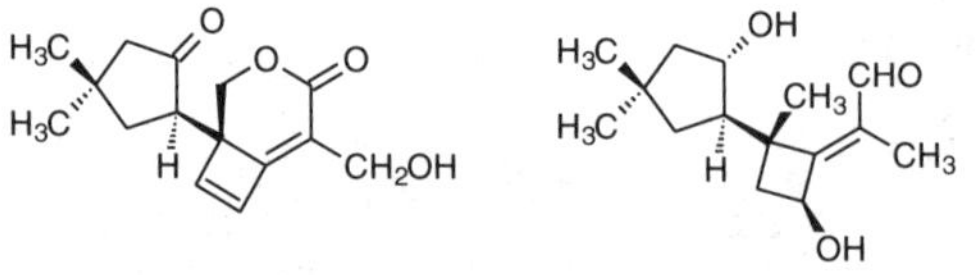

Fomannosin Illudosin

$C_{15}H_{18}O_4$, M_R 262.31, unstable wax, optically active. F. is formed by mycelium cultures of the wood-rotting fungus (*Heterobasidion annosum = Fomes annosus*, Basidiomycetes). The biosynthesis proceeds from a *protoilludane precursor by oxidative ring cleavage in which *illudosin* {$C_{15}H_{24}O_3$, M_R 252.35, $[\alpha]_D$ +52.4° ($CHCl_3$)} has been discussed as an intermediate [1]. F. is toxic to pine seedlings.

Lit.: [1] J. Chem. Soc., Perkin Trans. 1 **1991**, 1787.
gen.: Tetrahedron Lett. **21**, 437 (1980) (biosynthesis). – *Synthesis:* Chem. Lett. **1977**, 1491 ▪ J. Am. Chem. Soc. **104**, 747 (1982) ▪ Tetrahedron Lett. **1974**, 1545. – *Review:* Beilstein E V **18/3**, 293 ▪ Tetrahedron **37**, 2199 (1981) ▪ Turner **2**, 242, 244 f. – *[CAS 18885-59-1 (F.); 137360-24-8 (illudosin)]*

Fomannoxin.

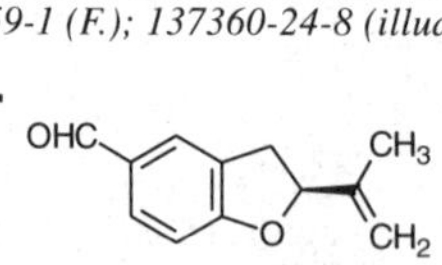

$C_{12}H_{12}O_2$, M_R 188.23, oil, $[\alpha]_D$ +89° ($CHCl_3$). A toxic metabolite from the fungus *Heterobasidion annosum = Fomes annosus* [1]. The *fungus* causes white rot and is a major pest in forestry since it penetrates the roots of living conifers and kills the trees. F. is also isolated from cultures of the basidiomycete *Laurilia taxodi* [2].

Lit.: [1] Tetrahedron Lett. **18**, 651 (1977); Chem. Lett. **1980**, 1581. [2] Gazz. Chim. Ital. **122**, 245 (1992).
gen.: Beilstein E V **17/10**, 208 ▪ Bull. Chem. Soc. Jpn. **55**, 2500 (1982) (synthesis) ▪ Phytochemistry **41**, 111 (1996) (F. acid). – *[CAS 63587-64-4]*

Fomentariol.

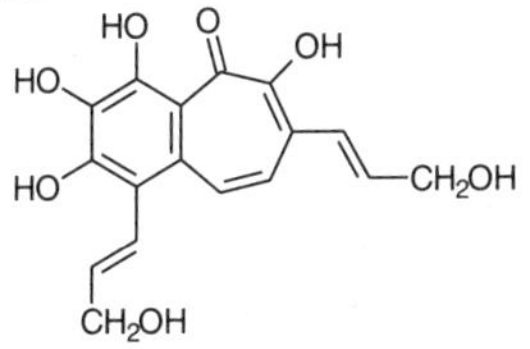

$C_{17}H_{16}O_7$, M_R 332.31, brown crystals, mp. >350 °C. The red-brown purpurogallin derivative F. is localized in the crustaceous layer of the tinder fungus (*Fomes fomentarius*, Aphyllophorales). On treatment with alkali it shows a blood-red color reaction by which the fungus can easily be distinguished from similar polypores. F. can be prepared biomimetically by oxidative dimerization of 2,3,4-trihydroxycinnamyl alcohol with the aqueous fungus extract or potassium iodate. *F. fomentarius* occurring on beech trees was used in the Middle Ages to produce tinder. The fungus was soaked with nitric acid solution, allowed to dry, and then used with flints to light fires.

Lit.: Chem. Ber. **111**, 3939 (1978) ▪ Zechmeister **51**, 87–88 (review). – *[CAS 53948-12-2]*

Formic acid. HCOOH, CH_2O_2, M_R 46.03, clear, volatile colorless liquid with a pungent odor in the anhydrous state, D. 1.22, mp. 8 °C, bp. 101 °C. Miscible with water, ethanol, ethers, and glycerol. The vapors are a powerful irritant for eyes and airways. Contact with dilute solutions of F. also leads to injury of the skin and eyes. The strongly cytotoxic F. was the first natural product to be isolated from insects; it was observed by Fischer as early as 1670 and isolated for the first time by Marggraf in 1749 by distillation from ants. F. a. is used as a defence substance by *ants (up to 60% solution), butterfly larvae (up to 25%), beetles (up to 75%), and other arthropods[1]. Often, the presence of accompanying substances such as hydrocarbons facilitates penetration of the attacker's cuticula. Some ants use F. as a pheromone. F. also occurs in some plants such as stinging nettles and in pine needles.

Biosynthesis: Starting from *glycine or *L-serine and *tetrahydrofolic acid via 10-formyltetrahydrofolic acid, from which F. is liberated in the poison glands[2].

Lit.: [1] Blum, Chemical Defenses of Arthropods, New York: Academic Press 1981. [2] Biochim. Biophys. Acta **201**, 454–455 (1978).

gen.: Beilstein EIV **2**, 3 ▪ J. Toxicol.-Toxin Rev. **11**, 115–164 (1992) – *[HS 2915 11; CAS 64-18-6]*

Formic acid ethyl ester [ethyl formate] see fruit esters.

***N*-Formyl-L-kynurenine** [(*S*)-amino-4-(2-formylaminophenyl)-4-oxobutanoic acid].

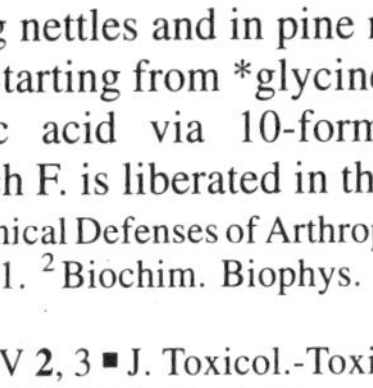

$C_{11}H_{12}N_2O_4$, M_R 236.23; occurs as an intermediate in the oxidative degradation of L-*tryptophan in many organisms and plays a role in the biosynthesis of *nicotinamide and the *ommochromes.

Lit.: Haslam, Shikimic Acid: Metabolism and Metabolites, p. 238–245, Chichester: Wiley 1993 ▪ J. Chromatogr. **661**, 93–104 (1994) (synthesis) ▪ Kerkut & Gilbert (eds.), Comprehensive Insect Physiology, Biochemistry and Pharmacology, vol. 10, p. 367–415, Oxford: Pergamon 1985 ▪ Luckner (3.), p. 342. – *Pharmacology:* Infect. Immun. **62**, 1131–1136 (1994) ▪ Melanoma Res. **1**, 177–185 (1991). – *[CAS 1022-31-7 (F.); 3978-11-8 (L-form); 32999-58-9 ((±)-form)]*

Foromacidin see spiramycin.

Forskolin (boforsin, colforsin).

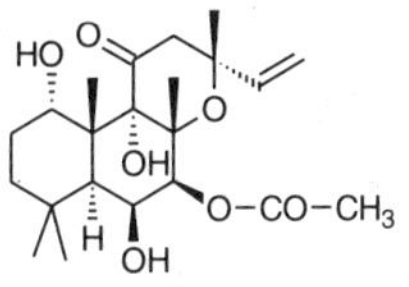

$C_{22}H_{34}O_7$, M_R 410.51, cryst., mp. 230–232 °C, $[\alpha]_D$ −26.2° ($CHCl_3$). Cardioactive diterpenoid (*labdane type) from the Indian medicinal plant *Coleus forskolii* (Lamiaceae). F. has a positive inotropic effect, it lowers blood pressure and has general vasodilatatory activity, especially on intravenous administration. The activity results from a direct stimulation of adenylate cyclase. F. is the archetype for all substances with this mechanism of action. In laboratory experiments it has been used successfully in therapy of congestive cardiomyopathy, glaucoma, or asthma. Several F. derivatives with improved solubility in water have been prepared[1].

Lit.: [1] Chem. Pharm. Bull. **44**, 2274 (1996); Indian J. Chem. B **37**, 979 (1998).

gen.: Adv. Cyclic Res. **17**, 81–89, 101–109 (1984); **20**, 1–38 (1986) ▪ Annu. Rep. Med. Chem. **19**, 293–302 (1984) ▪ Beilstein E V **18/5**, 56 ▪ Bull. Soc. Chim. Fr. **134**, 203–222 (1997) ▪ J. Am. Chem. Soc. **109**, 8115 f. (1987); **110**, 3670 ff. (1988) ▪ J. Chem. Soc., Chem. Commun. **1989**, 512 ff. ▪ J. Med. Chem. **26**, 436–439; 2439–2448 (1983); **31**, 1872–1879 (1988) ▪ J. Org. Chem. **54**, 2505 f. (1989) ▪ Merck-Index (12.), No. 2539 ▪ Phytochemistry **29**, 821–824 (1990) ▪ Phytotherapy Res. **3**, 91 (1989) ▪ Planta Med. **51**, 473–477 (1985); **54**, 200–204 (1988) ▪ Sandler & Smith, Design of Enzyme Inhibitors as Drugs, p. 650 ff., Oxford: Univ. Press 1989 ▪ Tetrahedron **45**, 763 (1989); **48**, 963–1037 (1992) (synthesis, review); **56**, 1081–1095 (2000) ▪ Tetrahedron Lett. **37**, 1015–1024 (1996) (synthesis) ▪ Trends Pharmacol. Sci. **10**, 442–447 (1989) ▪ Ullmann (5.) **A 5**, 285 ▪ Zechmeister **62**, 1–74. – *[CAS 66575-29-9]*

Fosfazinomycin A.

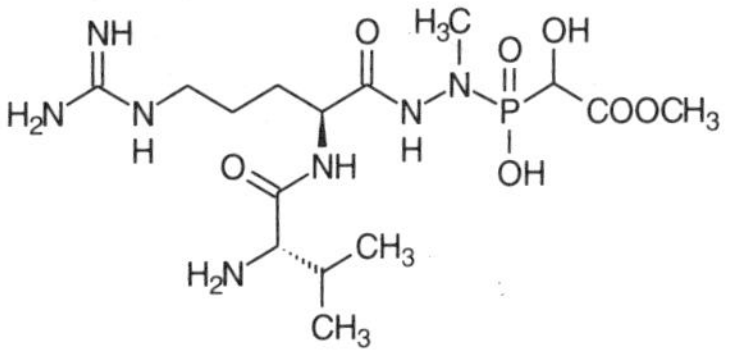

$C_{15}H_{32}N_7O_7P$, M_R 453.44, hygroscopic powder, mp. 157–161 °C (decomp.), $[\alpha]_D$ +14.7° (H_2O). An unusual *N'*-L-arginylphosphonic acid hydrazide from cultures of *Streptomyces lavendofoliae*. It also occurs without the L-valyl residue as *F. B* ($C_{10}H_{23}N_6O_6P$, M_R 354.30). The F. have antifungal activities.

Lit.: Agric. Biol. Chem. **47**, 2061 (1983) ▪ Bull. Korean Chem. Soc. **12**, 127 (1991) (synthesis) ▪ Tetrahedron Lett. **24**, 2283 (1983). – *[HS 2941 90; CAS 87423-10-7 (F. A); 87423-11-8 (F. B)]*

Fosfomycin (phosphomycin).

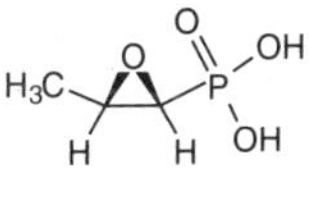

$C_3H_7O_4P$, M_R 138.06, crystals, mp. 94 °C, $[\alpha]_D$ –14° (H_2O); soluble in water, *antibiotic with a wide spectrum of activity produced by *streptomycetes and pseudomonads. For the unusual biosynthesis, see *Lit.*[1]. F. prevents the biosynthesis of bacterial cell walls by inhibiting pyruvate-uridine diphospho-*N*-acetylglucosamine transferase (EC 2.5.1.7). The calcium and ammonium salts are used in therapy. F. achieves high urine levels after oral administration and single doses are used for urinary tract infections.

Lit.: [1] Angew. Chem. Int. Ed. Engl. **33**, 341 (1994); Barton-Nakanishi **1**, 865–880.
gen.: J. Antibiot. **49**, 502 (1995) ▪ J. Am. Chem. Soc. **117**, 2931 (1995) (synthesis) ▪ J. Org. Chem. **54**, 1470 ff. (1989) (asymmetric synthesis) ▪ Justus Liebigs Ann. Chem. **1992**, 1201 ff. (biosynthesis). – *Pharmacology:* Drugs **54**, 637 (1997) (review) ▪ Drugs Exp. Clin. Res **6**, 281–288 (1980) ▪ Med. Actual **23**, 151–158 (1987). – *[HS 2941 90; CAS 23155-02-4]*

Fosfonochlorin (chloroacetylphosphonic acid).

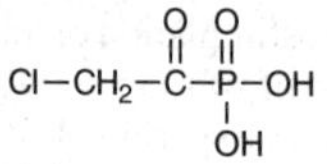

$C_2H_4ClO_4P$, M_R 158.48, white powder. The sodium salt of F. is isolated from cultures of *Talaromyces flavus* and various *Fusarium* species. In some bacteria it effects the formation of spheroblasts; this is attributed to an inhibition of cell wall biosynthesis.

Lit.: J. Antibiot. **42**, 198 (1989). – *[CAS 89699-33-2]*

Fostriecin.

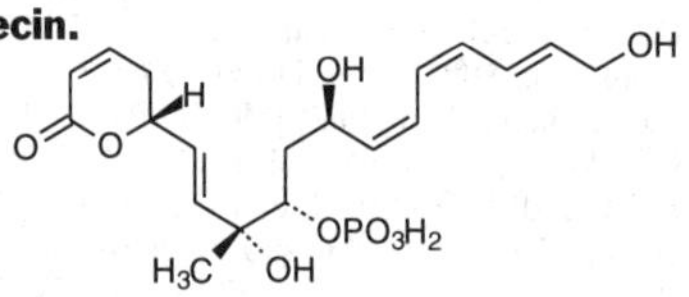

$C_{19}H_{27}O_9P$, M_R 430.39, $[\alpha]_D$ +33° (buffer pH 7), a polyene lactone produced by *Streptomyces pulveraceus* with activity against yeasts and tumor cells (especially leukemia). It is a specific inhibitor of topoisomerase II (EC 5.99.1.2), but does not bind to DNA and does not effect strand cleavage.

Lit.: Biochem. Pharmacol. **37**, 4063 (1988) ▪ Fundam. Appl. Toxicol. **15**, 258 (1990) ▪ J. Antibiot. **36**, 1595, 1601 (1983); **39**, 1465 (1986) ▪ J. Org. Chem. **50**, 462 (1985); **62**, 1748 (1997) (absolute configuration) ▪ Tetrahedron Lett. **29**, 753 (1988). – *[CAS 87810-56-8]*

FR-900411 see FK-409.

FR 900848.

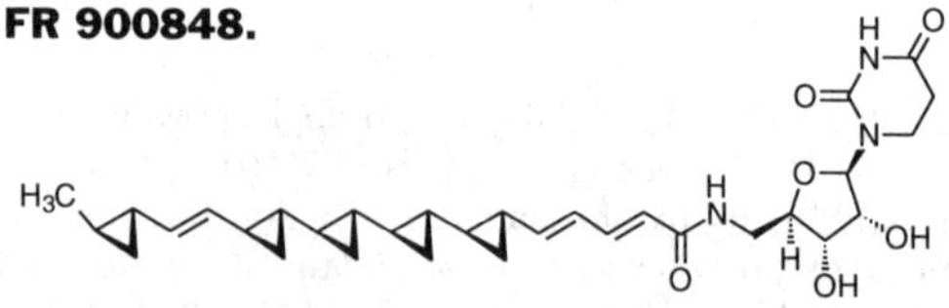

$C_{32}H_{43}N_3O_6$, M_R 565.71, needles, mp. 198–201 °C (decomp.), LD_{50} (mouse p.o.) >1 g/kg; $[\alpha]_D^{20}$ –36.2° (DMSO). An antifungal nucleoside antibiotic from cultures of *Streptoverticillium fervens* with an unusual accumulation of cyclopropane rings[1].

Lit.: [1] J. Antibiot. **43**, 748 (1990).
gen.: J. Org. Chem. **61**, 3280 (1996) (absolute configuration) ▪ Nachr. Chem. Tech. Lab. **43**, 435 f. (1995). – *Synthesis:* J. Am. Chem. Soc. **118**, 6096, 11030 (1996) ▪ J. Chem. Soc., Chem. Commun. **1996**, 325 f.; **1997**, 1693 (absolute configuration) ▪ see also U 106305. – *[CAS 120500-69-8]*

Fragarin (*pelargonidin 3-*O*-galactoside, 3-*β*-D-galactopyranosyl-4′,5,7-trihydroxyflavylium).
$C_{21}H_{21}O_{10}^+$, M_R 433.39, chloride: orange red crystals with 4 mol water. F. belongs to the *anthocyanins and is isolated from leaves of *Fagus sylvatica* var. *purpurea* (Fagaceae). F. also occurs in members of the genera *Cornus* (Cornaceae), *Cotoneaster* (Rosaceae), and *Fragaria* (Rosaceae). Together with the corresponding glucopyranoside (*callistephin), F. occurs in the fruits of strawberry.

Lit.: Beilstein E IV **17**, 3856 ▪ Karrer, No. 1707.

Fragilin see salicyl alcohol.

Fragranol [2-(*trans*-2-isopropenyl-1-methyl-cyclobutyl)ethanol].

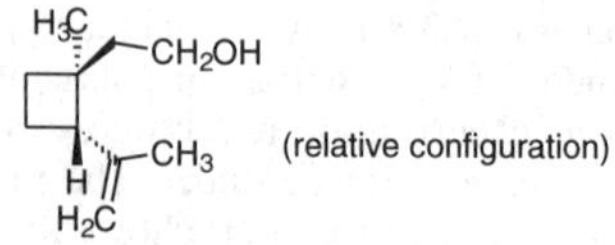

$C_{10}H_{18}O$, M_R 154.25, oil, bp. 120 °C (1.2 kPa). *Monoterpene with cyclobutane structure, C-1 epimer of *grandisol, known as the sexual pheromone of the male boll weevil. It is not known if F. possesses the same biological activity. F. occurs in the roots of the mugwort species *Artemisia fragrans* (Asteraceae), content: 0.0033%[1].

Lit.: [1] Chem. Ber. **106**, 2904 (1973).
gen.: Chem. Pharm. Bull. **41**, 861 (1993) ▪ Justus Liebigs Ann. Chem. **1992**, 257 ▪ Tetrahedron **46**, 3077 (1990); **47**, 8259 (1991); **50**, 3235 (1994). – *[CAS 30346-21-5]*

Frambinone see zingerone.

Framycetin see neomycin.

Frangufoline (sanjoinine A). A *cyclopeptide alkaloid isolated from sanjoin, the seeds of *Zizyphus vulgaris* (Rhamnaceae) with strong sedative activity: $C_{31}H_{42}N_4O_4$, M_R 534.69, needles, mp. 244 °C, $[\alpha]_D^{22}$ –299° ($CHCl_3$). F. also occurs in the bark of other *Zizyphus* spp., in leaves of *Rhamnus frangula*, *Euonymus* spp. and *Melochia* spp., and in roots of *Discaria* spp.

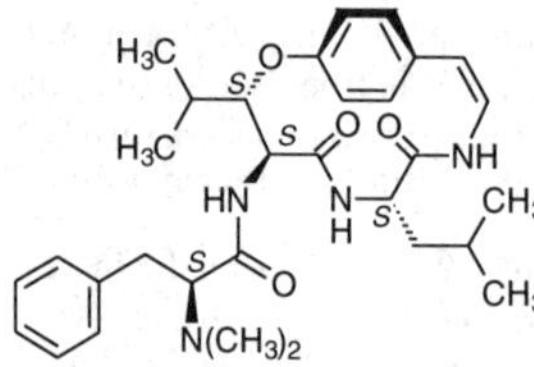

Lit.: Tetrahedron Lett. **39**, 9631 (1998). – *[CAS 19526-09-1]*

Frangulanine see cyclopeptide alkaloids.

Frangulins. Anthraquinone glycosides from root, bark, and seeds of the black alder (*Rhamnus frangula*, Rhamnaceae) and the American buckthorn (*Cascara sagrada*, Rhamnaceae) as well as rhubarb roots. F. are glycosides of *emodin.

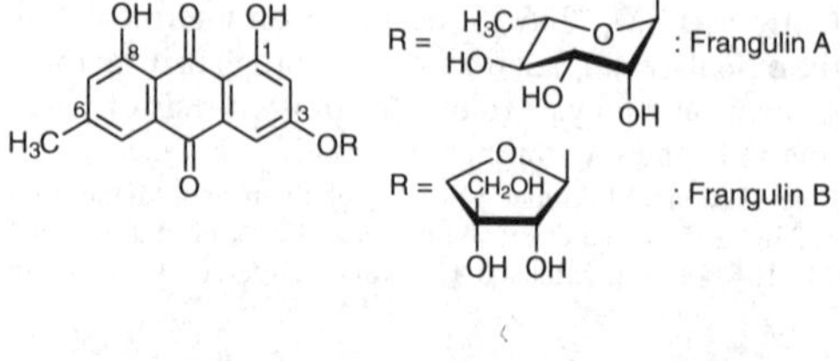

F. A [R=α-L-rhamnopyranosyl, $C_{21}H_{20}O_9$, M_R 416.38, orange crystals, mp. 228 °C, uv_{max} 430 nm (lg ε 4.05)] and *F. B* (R=D-apio-β-D-furanosyl, $C_{20}H_{18}O_9$, M_R 402.36, orange cryst., mp. 196 °C) co-occur. The corresponding 1-*O*-β-D-glucopyranosides of F. A and F. B (*glucofrangulin A*, $C_{27}H_{30}O_{14}$, M_R 578.53 *glucofrangulin B*, $C_{26}H_{28}O_{14}$, M_R 564.50) are also isolated from the bark of black alder and are in part responsible for the laxative effects of extracts of the bark.

Lit.: Beilstein E V **17/6**, 267 ▪ Hager (5.) **6**, 400, 405 ▪ Merck-Index (12.), No. 4288 ▪ Sci. Pharm. **55**, 275 (1987). – *[CAS 521-62-0 (F. A); 14101-04-3 (F. B); 21133-53-9 (glucofrangulin A); 14062-59-0 (glucofrangulin B)]*

Fredericamycin A.

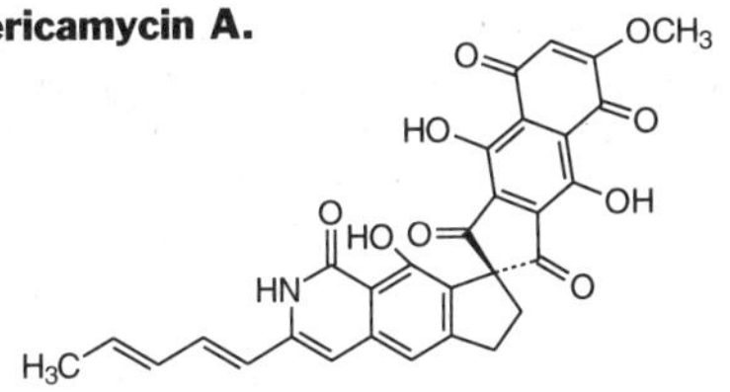

$C_{30}H_{21}NO_9$, M_R 539.50, red powder, mp. >350 °C. A very effective cytotoxic antitumor *antibiotic with additional activity against Gram-positive bacteria and fungi. F. A and the minor components *F. B* and *F. C* are isolated from cultures of *Streptomyces griseus*. The antitumor activity of F. A is assumed to be due to the formation of free radicals which effect DNA strand cleavage. The water-soluble potassium salt of F. A has been developed for cancer therapy. In contrast to other antineoplastic substances F. A does not show any mutagenicity in the Ames test.

Lit.: Biochemistry **24**, 478–486 (1985) (biosynthesis); **25**, 5533–5539 (1986); **28**, 748 (1989) ▪ Heterocycles **28**, 71 ff. (1989) ▪ J. Am. Chem. Soc. **110**, 6471–6480 (1988) ▪ J. Antibiot. **34**, 1389–1407 (1981); **40**, 786–802 (1987) (structure); **41**, 976 (1988) ▪ J. Org. Chem. **53**, 5519–5527 (1988) ▪ Pure Appl. Chem. **60**, 1645–1654 (1988) ▪ Tetrahedron Lett. **28**, 451, 455 (1987); **30**, 827, 2037 (1989). – *Stereoselective synthesis:* Angew. Chem. Int. Ed. Engl. **38**, 683 (1999) ▪ Atta-ur-Rahman **16**, 27–74 ▪ J. Am. Chem. Soc. **116**, 11275–11286 (1994); **117**, 11839 (1995) ▪ J. Heterocycl. Chem. **33**, 1519 (1996). – *[HS 2941 90; CAS 80455-68-1 (F. A); 80450-64-2 (F. B); 80450-65-3 (F. C)]*

Friedelanes (A-friedooleananes).

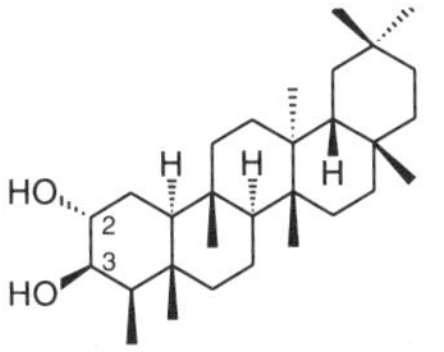

Pachysandiol A

F. is the systematic name for the skeleton of a group of pentacyclic *triterpenes. F. are components of many plants (*Catha*, *Euonymus*, *Castanopsis*, *Prionostemma*, etc.), in which they predominately occur in hydroxylated form, e.g., 2α,3β-friedelanediol {*pachysanediol A*, $C_{30}H_{52}O_2$, M_R 444.74, mp. 291 °C, $[\alpha]_D$ +14° ($CHCl_3$)} from *Pachysandra terminalis* (Buxaceae).

Lit.: Aust. J. Chem. **22**, 597 (1969); **23**, 1651 (1970) ▪ Phytochemistry **19**, 1989 (1980) ▪ Tetrahedron Lett. **1971**, 3807; **29**, 6971 (1988). – *[CAS 559-73-9 (F.); 17946-96-2 (2α,3β-friedelanediol)]*

Friedelin (A-friedooleanan-3-one, 3-friedelanone).

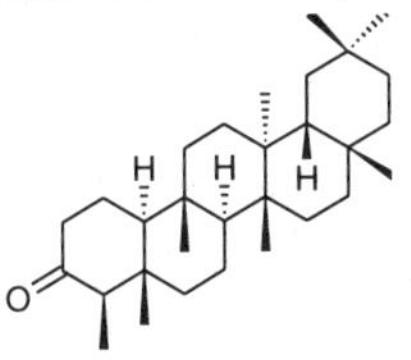

$C_{30}H_{50}O$, M_R 426.73, needles, mp. 267–269 °C, $[\alpha]_D$ –27.8° ($CHCl_3$). A pentacyclic *triterpene widely distributed in many higher plants and lichens; it can be obtained as the main component from the cork of the cork oak (*Quercus suber*, Fagaceae) by extraction with hot ethanol. F. has diuretic activity.

Lit.: ApSimon **6**, 86, 112–132 (synthesis) ▪ Beilstein E IV **7**, 1168 ▪ Phytochemistry **40**, 1227 (1995) (NMR). – *[CAS 559-74-0]*

Frog toxins. Like toads, frogs can also excrete physiologically active secretions from the skin glands. The *toxins of the colored frogs (Dendrobatidae, see also Dendrobates alkaloids) are used as arrow poisons by South American Indians. These are mainly alkaloids, some of which are highly toxic such as, e.g., *batrachotoxins with steroid skeletons, *pumiliotoxins with indolizidine or decahydroquinoline skeletons, and spiropiperidines of the *histrionicotoxin type as well as numerous other compounds, some of which are unique to frogs. Besides the about 300 currently known alkaloids, other substances such as peptides or biogenetic amines have been identified in frog secretions.

Lit.: *Review:* Alkaloids: Chem. Biol. Perspect. **13**, 1–161 (1999) ▪ Manske **43**, 185–288 ▪ Zechmeister **41**, 205–340. – *Synthesis:* Manske **21**, 139 ▪ Pelletier **4**, 1–274 ▪ Zechmeister **41**, 205 ▪ see also amphibian venoms, toad poisons, animal toxins.

Frondosins. Sesquiterpene hydroquinone derivatives from *Euryspongia* spp. (marine sponges) with HIV-inhibitory activity, e.g., *F. A*: $C_{21}H_{28}O_2$, M_R 312.45, powder, mp. 111–113 °C, $[\alpha]_D$ +31.5° (CH_3OH). F. also show IL-8 inhibition [1].

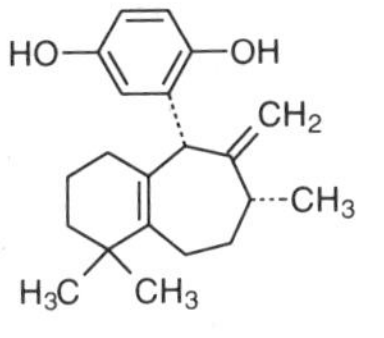

F. A
(relative configuration)

Lit.: [1] Tetrahedron **53**, 5047–5060 (1997).
gen.: Nat. Prod. Lett. **11**, 153 (1998). – *[CAS 189514-05-4 (F. A)]*

Frontalin [(1*S*)-1,5-dimethyl-6,8-dioxabicyclo[3.2.1]-octane].

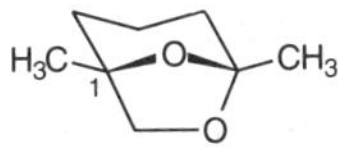

$C_8H_{14}O_2$, M_R 142.20; oil, bp. 99–100 °C (15.99 kPa), $[\alpha]_D^{23}$ –55.5° (diethyl ether). F. is one of the aggregation pheromones of bark beetles of the genus *Dendroctonus*. This attractant is used to promote an effective colonization of a tree first chosen as nesting site by one insect by controlling an optimal ratio of the sexes. Males of *D. brevicomis* have pure (1*S*)-(–)-F., the pheromone of *D. frontalis* (Southern Pine Beetle) is an

85:15 mixture of the (1*S*)- and (1*R*)-enantiomers; see also brevicomin, multistriatin. Many enantioselective syntheses have been described.

Lit.: An. Quim. **91**, 103–112 (1995) (synthesis) ▪ ApSimon **4**, 68, 152; **9**, 381–393 ▪ Beilstein E V **19/1**, 228 ▪ Can. J. Chem. **68**, 782 (1990) ▪ Chem. Pharm. Bull. **46**, 217 (1998) (synthesis) ▪ Eur. J. Org. Chem. **1998**, 233 (synthesis) ▪ J. Chem. Ecol. **14**, 113–122, 1217–1225 (1988); **21**, 1043–1063 (1995) ▪ Justus Liebigs Ann. Chem. **1989**, 9–18; **1992**, 1191 ff. ▪ Nachr. Chem. Tech. Lab. **33**, 392–396 (1985); **37**, 480 (1989) ▪ Synthesis **1997**, 156 [synthesis (*S*)-(–)-F.] ▪ Tetrahedron **44**, 6633–6644 (1988); **45**, 5469 (1989) ▪ Tetrahedron: Asymmetry **9**, 2611 (1998) (synthesis) ▪ Tetrahedron Lett. **31**, 1039 f. (1990); **40**, 1425 (1999). – *[CAS 28401-39-0 (F.); 57917-96-1 ((+)-F.)]*

D-Fructose (D-*arabino*-2-hexulose, levulose, fruit sugar).

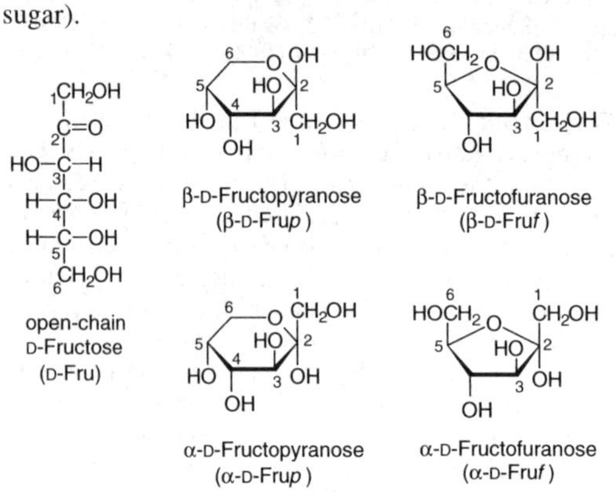

$C_6H_{12}O_6$, MR 180.16, sweet-tasting prisms or needles, D. 1.60, mp. 102–104 °C (decomp.), $[\alpha]_D^{20}$ –133 → –92° (H_2O), D-F. reduces Fehling's solution. D-F. is a ketose (ketohexose, hexulose), existing as *β*-fructopyranose in the crystalline state, and in aqueous solution (deuterium oxide) as an equilibrium of 57% *β*-fructopyranose, 3% *α*-fructopyranose, 31% *β*-fructofuranose, and 9% *α*-fructofuranose (see figure). The mutarotation in aqueous solution is attributed to a conversion of the pyranoside from to the furanoside form; the amount of the open-chain keto form (see figure) in the equilibrium is less than 1%. Since the natural form can be degraded to D-glycerol aldehyde it is assigned to the D-series irrespective of the fact that in aqueous solution it rotates the plane of polarized light to the left (thus the name levulose from Latin: laevus = left; in contrast to dextrose for D-*glucose, from Latin: dexter = right).

Occurrence: D-F. occurs in the free form both in the plant kingdom and in mammalian sperm, it also occurs in bound form as *di-, *oligo-, and *polysaccharides; the most important sources are *sucrose and *inulin. In the human organism D-F. is found only in trace amounts in blood, sperm, and amniotic fluid. D-F. is used as a sugar substitute in pharmacy and dietetics for diabetics.

Lit.: Beilstein E IV **1**, 4401–4411 ▪ Collins-Ferrier, p. 50–57 (synthesis) ▪ Merck-Index (12.), No. 4295–4298. – *[HS 1702 50; CAS 57-48-7]*

Fruit esters. General term for esters of short- to medium-chain fatty acids with alcohols; previously also known as fruit ethers. The name is derived from their abundant occurrence in *fruit flavors and their mostly fruity odors. More than 400 F. e. have been identified as components of aromas and flavors [1]. Ethanol and acetic acid are particularly frequent components, followed by acids and alcohols with 3–6 C-atoms (see table, p. 243). The *biogenesis proceeds through catabolism of *amino and *fatty acids [2]. For production, see *Lit.* [3].

Lit.: [1] TNO list (6.), Suppl. 5, p. 261 ff. [2] Maarse, p. 283 ff. [3] Perfum. Flavor. **20** (5), 5–14 (1995).

gen.: Bauer et al. (2.), p. 14–18 ▪ Ullmann (5.) **A 9**, 567; **A11**, 152.

Fruit flavors. One of the most important features of the various types of fruit (pomaceous, stone, berry fruits as well as wild and tropical fruits) are their typical flavors and aromas, which generally, however, only form during the ripening phase. F. f. are complex mixtures of substances, often with a high content of *fruit esters and lactones (see alkanolides), together with alcohols, aldehydes, ketones, carboxylic acids, heterocyclic compounds, and many other classes of compounds. The concentration of the entire flavor amounts to ca. 10–100 mg/kg fresh weight, usually less than 30 mg/kg. The number of *flavor compounds identified to date in individual F. f. is often over 200, in strawberry flavor, e. g., ca. 360. The individual components are present in ppm down to ppt amounts. Composition and concentration of the F. f. vary widely depending on the variety, climate, position, and degree of ripeness. Changes in flavor may also occur on destruction of the cellular structure when secondary compounds are formed by enzymatic hydrolysis or oxidation processes (see apple flavor). Commercially available F. f. are produced from natural or nature-identical flavor substances and are used to improve the aromatic character of drugs, beverages, and many foods, often in combination with natural fruit extracts or concentrates. Only few of the key components are listed in the following selection of F. f.:

Apple flavor: The sometimes marked differences between aromas of individual varieties of apples are mainly due to quantitative variations in the composition of apple flavor substances [1]. Key components are ethyl (+)-2-methylbutanoate and other esters of 2-methylbutanoic acid, in addition to ethyl and hexyl butanoates, hexyl acetate, (*E*)-2- and (*Z*)-3-hexenyl acetates (see fruit esters) and *β-damascenone. (*E*)-2-*Hexenal, (*E*)-2-*hexen-1-ol, and hexanal (see alkanals) play a special role in A. f. These are trace aroma substances in intact apples. When the fruit cells are destroyed, the concentration of the C_6 units increase strongly due to enzymatic processes. They are the main aroma components of apple juice. Accordingly, the aromas of fresh apples and apple juice differ markedly.

Apricot flavor: The typical aroma is due to the combined effects of numerous components with flowery and fruity characters [2]; these include *linalool, 1-terpinen-4-ol, *α*-terpineol (see *p*-menthenols), *2-phenylethanol, *α*- and *β*-*ionones, *β*-*damascenone, and *(*Z*)-jasmone for the flowery part together with *fruit esters and lactones, e. g., 4-octanolide, 4- and 5-decanolide, 4-dodecanolide (see alkanolides), hexyl acetate and hexyl butanoate for the fruity part, rounded off by *benzaldehyde.

Table: Characteristics and occurrence of important Fruit esters.

name (trivial name)	molecular formula M_R	formula	D. bp. [°C]	HS [CAS]	flash point [°C]	occurrence (selection)
Ethyl formate	$C_3H_6O_2$ 74.08	$HCOOC_2H_5$	0.917 54.3	2915 13 [109-94-4]	–20	apple, peach, rum
Ethyl acetate (Acetic ester)	$C_4H_8O_2$ 88.11	$H_3C–COOC_2H_5$	0.900 77	2915 31 [141-78-6]	–4	ubiquitous in fruit, spirits
Butyl acetate	$C_6H_{12}O_2$ 116.16	$H_3C–COO–(CH_2)_3–CH_3$	0.882 126.5	2915 31 [123-86-4]	22	apple, apricot, pear, thyme
3-Methylbutyl acetate (Isoamyl acetate)	$C_7H_{14}O_2$ 130.19	$H_3C–COO–(CH_2)_2–CH(CH_3)_2$	0.867 142	2915 39 [123-92-2]	36	apple, banana, whiskey, wine
Hexyl acetate	$C_8H_{16}O_2$ 144.21	$H_3C–COO–(CH_2)_5–CH_3$	0.867 171.6	2915 39 [142-92-7]	62	apple, apricot, banana, crisped mint
(*E*)-2-Hexenyl acetate	$C_8H_{14}O_2$ 142.20	$H_3C–COO–CH_2–CH=CH–(CH_2)_2–CH_3$	0.898 165–166	2915 39 [2497-18-9]	62	apple, strawberry, peach, grape
(*Z*)-3-Hexenyl acetate	$C_8H_{14}O_2$ 142.20	$H_3C–COO–(CH_2)_2–CH=CH–C_2H_5$	0.850 156	2915 39 [3681-71-8]	57	apple, guava, peach, strawberry
Ethyl propanoate	$C_5H_{10}O_2$ 102.13	$H_3C–CH_2–COOC_2H_5$	0.892 99	2915 50 [105-37-3]	12	apple, guava, tomato, wine, brandy, rum
Ethyl butanoate	$C_6H_{12}O_2$ 116.16	$H_3C–(CH_2)_2–COOC_2H_5$	0.879 121	2915 60 [105-25-4]	26	apple, banana, pear, strawberry, juice from citrus fruit
Butyl butanoate	$C_8H_{16}O_2$ 144.21	$H_3C–(CH_2)_2–COO–(CH_2)_3–CH_3$	0.871 166	2915 60 [109-21-7]	60	apricot, banana, cherimoya, papaya, citrus fruit oils
3-Methylbutyl butanoate (Isoamyl butanoate)	$C_9H_{18}O_2$ 158.24	$H_3C–(CH_2)_2–COO–(CH_2)_2–CH(CH_3)_2$	0.865 179	2915 60 [106-27-4]	61	banana, cherimoya, mango, cheese
Hexyl butanoate	$C_{10}H_{20}O_2$ 172.27	$H_3C–(CH_2)_2–COO–(CH_2)_5–CH_3$	0.865 208	2915 60 [2639-63-6]	80	apple, apricot, cherimoya, plum
Ethyl 2-methyl-butanoate	$C_7H_{14}O_2$ 130.19	$H_3C–CH_2–CH(CH_3)–COOC_2H_5$	0.869 133	2915 60 [7452-79-1]	31	apple and pear: (S)-(+)-E. [CAS 10307-61-6]; strawberry, grape
Ethyl 3-methyl-butanoate (Ethyl isovalerate)	$C_7H_{14}O_2$ 139.19	$(H_3C)_2CH–CH_2–COOC_2H_5$	0.866 135	2915 60 [108-64-5]	36	apple, strawberry, cheese, grape
3-Methylbutyl 3-methylbutanoate (Isoamyl iso-valerate)	$C_{10}H_{20}O_2$ 172.27	$(H_3C)_2CH–CH_2–COO–(CH_2)_2–CH(CH_3)_2$	0.858 193	2915 60 [659-70-1]	80	banana, peppermint, grape
Ethyl hexanoate	$C_8H_{16}O_2$ 144.21	$H_3C–(CH_2)_4–COOC_2H_5$	0.871 168	2915 90 [123-66-0]	60	pineapple, strawberry, guava, clove, rum, juice from citrus fruit
Ethyl octanoate	$C_{10}H_{20}O_2$ 172.27	$H_3C–(CH_2)_6–COOC_2H_5$	0.869 208	2915 90 [106-32-1]	120	apricot, guava, cheese, plum, brandy
Ethyl (2*E*,4*Z*)-2,4-decadienoate	$C_{12}H_{20}O_2$ 196.29	$H_3C–(CH_2)_4–CH=CH–CH=CH–COOC_2H_5$	70–72 (6 Pa)	2916 19 [3025-30-7]	116	pear brandy, apple

Banana flavor: Of the 350 *flavor compounds identified to date in B. f. 3-methylbutyl acetate as well as 3-methylbutyl butanoate and 3-methylbutyl 3-methylbutanoate (see fruit esters) are of particular sensory relevance. Fatty-fruity and exotic-fruity notes result mainly from aroma substances with the (*Z*)-4-configuration, e. g., (*Z*)-*4-hepten-2-one* ($C_7H_{12}O$, M_R 112.17, CAS [90605-45-1]) and (*Z*)-*4-hepten-2-ol* ($C_7H_{14}O$, M_R 114.19, CAS [34146-55-9]), as well as their acetates and butanoates. *Eugenol, elemicin (see safrole), and *O*-methyleugenol are responsible for the spicy aroma.

Bilberry flavor: The aroma of the European bilberry (*Vaccinium myrtillus*) is mainly due to (*E*)-2-*hexenal, ethyl 2- and 3-methylbutanoates (see fruit esters) as well as *ethyl 3-hydroxy-3-methylbutanoate* ($C_7H_{14}O_3$, M_R 146.19, CAS [18267-36-2]).

Cherry flavor: Seven compounds with very high aroma values have been found in the juice of sour cherries: *benzaldehyde, *linalool, hexanal (see alkanals), (*E*)-2-*hexenal, *phenylacetaldehyde, (*E,Z*)-2,6-*nonadienal, and *eugenol. Benzaldehyde is reported to be the impact compound of cherry flavor.

Grape flavor: The quantitative relationships between free terpenes, terpene alcohols (e. g., *geraniol, *linalool, *hotrienol* ($C_{10}H_{16}O$, M_R 152.24, CAS [29957-43-5]), terpene oxides, and C_{13}-norisoprenoids determine the varying flavors of the different varieties of *Vitis vinifera*; however, the dominating components are the terpenes bound as glycosides or diphosphates. The impact compounds in the flavor of the Concord grape (*Vitis labrusca*) are considered to be *methyl anthranilate, *2-aminoacetophenone* (C_8H_9NO, M_R 135.17, CAS [551-93-9]), and *ethyl 3-mercaptopropanoate*

($C_5H_{10}O_2S$, M_R 134.19, CAS [5466-06-8]), compare *wine flavor.

Guava flavor: G. f., resembling pear and quince flavor, contains β-*caryophyllene, numerous *fruit esters, e. g., ethyl butanoate and ethyl hexanoate, cinnamyl acetate (see cassia oil) and the green notes from hexanal, (*E*)-2-*hexenal, and (*Z*)-3-*hexen-1-ol. Important trace components are C_6–C_{10}-dienals, *Furaneol®, some *pyrazines and thiazoles, as well as *6-mercapto-1-hexanol* ($C_6H_{14}OS$, M_R 134.24, CAS [1633-78-9])[3].

Kiwi flavor: The quantitatively dominating aroma component is (*E*)-2-*hexenal (ca. 80%), together with other 2- and 3-hexenals and 2- and 3-*hexen-1-ols. In addition K. f. is rounded off by terpene hydrocarbons such as α-/β-*pinenes and limonene (see *p*-menthadienes), as well as mesifuran (see Furaneol®), β-*damascenone and ethyl butanoate.

Mango flavor: The flavor depends strongly on the variety and differs widely: e. g., Alphonso mango (fresh, green-fruity), Baladi (melon-like, caramel-like, woody), and Kensington (sweet, fruity, peach-like). Dominating components are terpene hydrocarbons such as 3-*carene, *myrcene, (*Z*)-*ocimene, and limonene (see *p*-menthadienes), also 4- and 5-*alkanolides, *Furaneol®, and β-*damascenone are generally present. Typical components of Baladi mangos are ethyl esters of the even-chain fatty acids C_2–C_{16}; Alphonso mangos contain (*Z*)-3-hexenyl and other esters of alkanoic acids.

Melon flavor: The typical M. f. of the various sorts are due to secondary *flavor compounds from the degradation of linolic and linolenic acids, e. g., hexenals, (*Z*)-6-nonenal (see alkenals), *(Z,Z)-3,6-nonadien-1-ol* ($C_9H_{16}O$, M_R 140.23, CAS [53046-97-2], water melon odor), *(Z,Z)-3,6-nonadienyl acetate* ($C_{11}H_{18}O_2$, M_R 182.26, CAS [130049-88-2]), and (*Z*)-1,5-octadien-3-one (see tea flavor). Sweet melon (muskmelon; *Cucumis melo*) also contains *fruit esters, especially 2-methylbutanoates, *alkanolides, *anisaldehyde, *eugenol, and some (methylthio)carboxylates with sensory relevance[4].

Papaya flavor: The main components are *linalool (68%) and benzyl isothiocyanate (see mustard oils) (13%), besides *fruit esters, e. g., ethyl butanoate, and *alkanolides. The green-metallic note of the fresh fruits is attributed to metabolites of linalool.

Passion fruit: Red P. f. (*Passiflora edulis*): The flowery-fruity flavor is due in particular to (*Z*)-3-hexenyl esters, (*Z*)-3-octenyl acetate, citronellyl and geranyl acetate. Yellow P. f. (*P. edulis flavicarpa*): *2-Methyl-4-propyl-1,3-oxathiane, *3-mercapto-1-hexanol* ($C_6H_{14}OS$, M_R 134.24, CAS [51755-83-0]) and the corresponding esters are mainly responsible for the exotic-fruity flavor. In both sorts about 30 C_{13}-norisoprenoids, including β-*ionone, β-*damascenone and edulane (2,5,5,8a-tetramethyl-1-benzopyran) make major contributions to the aroma.

Peach flavor: The impact compound in P. f. is 4-decanolide (see alkanolides) besides other γ- and δ-lactones (C_5–C_{12}), marmelolactone (see quince flavor), γ-*jasmin(e) lactone, C_{13}-norisoprenoids such as β-*damascenone, β-*ionone, and dihydroionones, as well as *fruit esters, e. g., (*Z*)-3-hexenyl acetate and ethyl octanoate[5].

Pear flavor: Ethyl and methyl (2*E*,4*Z*)-2,4-decadienoates (see fruit esters) are the characteristic *impact compound of "Williams Christ" pears. Other fruit esters in B. are butyl and hexyl acetate, ethyl butanoate and ethyl (+)-2-methylbutanoate

Pineapple flavor: Numerous *fruit esters, including ethyl 2-methylbutanoate and hexanoate, as well as some 4- and 5-*alkanolides are responsible for the general fruity character. Key components include *Furaneol®, mesifuran (see Furaneol®) and some sulfur-substituted esters, especially *methyl 3-(methylthio)propanoate* ($C_5H_{10}O_2S$, M_R 134.19, CAS [13532-18-8]) and *ethyl 3-(methylthio)propanoate* ($C_6H_{12}O_2S$, M_R 148.22, CAS [13327-56-5]). Both substances resemble pineapple at very high dilutions. Remarkable is the occurrence of 1,3,5-undecatriene (see galbanum oil) in fresh pineapple.

Plum flavor: An important factor is the correct quantitative ratio of a few major components such as *linalool, *benzaldehyde, methyl cinnamate (see cinnamic acid), *2-phenylethanol, and 4-decanolide (see alkanolides), in addition to green notes such as (*Z*)-3-*-hexen-1-ols and 2-*hexenals, as well as *fruit esters, e. g., ethyl octanoate.

Quince flavor: Major components of quince essential oil (ca. 25%) are *cis*- and *trans-marmelolactones* ($C_{10}H_{14}O_2$, M_R 166.22, *cis*-: CAS [74183-60-1], *trans*: CAS [74133-35-0]). Other important components are *cis*- and *trans-marmelooxides* ($C_{10}H_{16}O$, M_R 152.24, *cis*-: CAS [89103-56-0], *trans*-: CAS [92343-93-6]), *cis*- and *trans-quince oxepanes* ($C_{12}H_{20}O$, M_R 180.29, *cis*-: CAS [130021-99-7], *trans*-: CAS [131320-18-8]) and *quince oxepin* ($C_{12}H_{18}O$, M_R 178.27, CAS [130021-98-6]). A number of flavor-active C_{13}-norisoprenoids are formed from non-volatile precursors, especially from glycosides, on steam distillation[6].

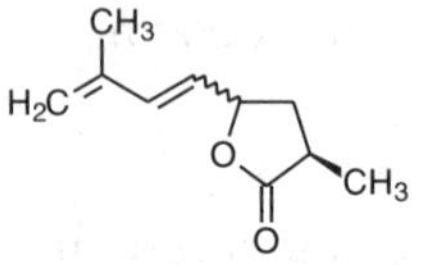

cis / trans-
Marmelolactone

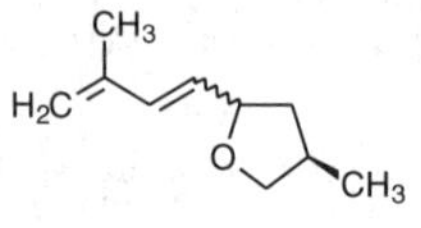

cis / trans-
Marmelooxide

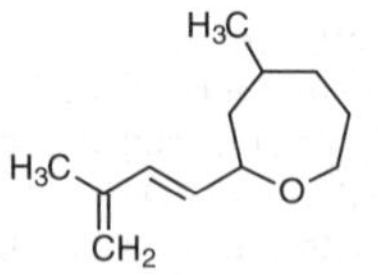

quince oxepane

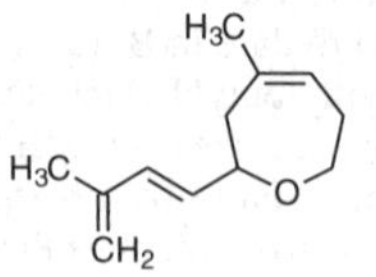

quince oxepin

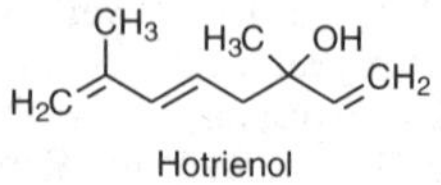

Hotrienol

Raspberry flavor: More than 250 flavor compounds are currently known. The aroma is mainly due to the *impact compound 4-(4-hydroxyphenyl)-2-butanone (so-called raspberry ketone, see zingerone) in combination with α- and β-*ionones. Additional contributions come from β-*damascenone, (*Z*)-3-*hexen-1-ol, and the terpene alcohol *geraniol, *linalool, and α-terpineol (see *p*-menthenols) with their flowery and fruity notes. The flavor of wild raspberries contains 5-decanolide (see alkanolides) and *Furaneol®. α-Ionone exists in the (*R*)-(+)-form.

Red currant flavor: The flavor of red currants is principally due to (*E*)-2-*hexenal, (*E*)-2-*hexen-1-ol, and hexanal (see alkanals). In black currants the *impact compound *4-methoxy-2-methyl-2-butanethiol* ($C_6H_{14}OS$, M_R 134.24, CAS [94087-83-9]) is responsible for the typical "feline" note. Terpene compounds such as 1,8-*cineole, *linalool, α-terpineol (see *p*-menthenols), *citronellol and β-*damascenone, *fruit esters (e.g., ethyl butanoate), the butter note of *2,3-butanedione, *4-methoxyacetophenone* ($C_9H_{10}O_2$, M_R 150.18, CAS [100-06-1]) and 2-isopropyl-3-methoxypyrazine (see pyrazines) complete the aroma. In the aroma industry *buchu leaf oil is used to generate the black currant note.

Strawberry flavor: The "strawberry" aroma is made up of caramel-like [*impact compounds: *Furaneol®, mesifuran (see Furaneol®)], fruity [*fruit esters, 4-decanolide (see alkanolides)], green [(*E*)-2-*hexen-1-ols, (*E*)-2-*hexenals, 2-hexenyl acetate (see fruit esters)], and flowery notes (*linalool, *methyl anthranilate). Particularly important fruit esters are ethyl butanoate and hexanoate as well as the corresponding methyl esters. Free acids such as *2-methylbutanoic acid* ($C_5H_{10}O_2$, M_R 102.13, CAS [116-53-0]) and trace substances, e.g., dimethyl sulfide and vanillin, round off the aroma.

Lit.: [1] Erwerbsobstbau **33**, 36–41 (1991); Phytochemistry **43**, 145 (1996); Synthesis **1995**, 1263 ff. [2] Erwerbsobstbau **33**, 174 (1991). [3] J. Agric. Food Chem. **33**, 138–143 (1985). [4] J. Agric. Food Chem. **40**, 1385–1388 (1992). [5] J. Agric. Food Chem. **39**, 778–781 (1991); Fischer et al., in Schreier & Winterhalter (eds.), Progress in Flavour Precursor Studies, p. 287–298, Carol Stream (USA): Allured 1993. [6] J. Agric. Food Chem. **36**, 560 ff. (1988); Schreier (ed.), Bioflavour '87, p. 255–273, Berlin: de Gruyter 1988.
gen.: ACS Symp. Ser. **596** (1995) ▪ Maarse, p. 283–409 ▪ Morton & MacLeod, Food Flavours, Part C, The Flavour of Fruits, p. 1–326, Amsterdam: Elsevier 1990.

Frullanolide.

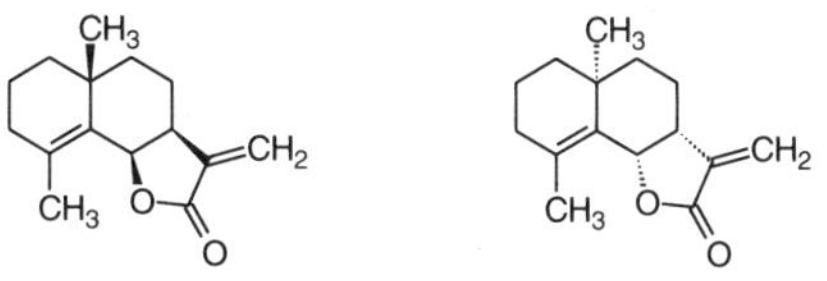

(-)-Frullanolide (+)-Frullanolide

(+)-F.: $C_{15}H_{20}O_2$, M_R 232.32, needles, mp. 75–76 °C, soluble in acetone, ether. *(–)-F.:* mp. 76.5–77 °C, $[\alpha]_D$ –114° ($CHCl_3$). An allergenic sesquiterpene lactone from liverworts of the genus *Frullania* living epiphytically on forest trees; both enantiomers are responsible for the contact dermatitis afflicting forestry workers. The allergenic activity results from the α-methylene-γ-butyrolactone ring.

Lit.: Bull. Chem. Soc. Jpn. **66**, 2298 (1993) ▪ J. Chem. Res. (S) **1990**, 158 ▪ J. Chem. Soc., Perkin Trans. 1 **1990**, 2385 ▪ J. Org. Chem. **55**, 1096 (1990) ▪ Phytochemistry **37**, 1337 (1994) ▪ Tetrahedron **49**, 4761 (1993). – *[CAS 40776-40-7 ((+)-F.); 27579-97-1 ((–)-F.)]*

Fucoidan (fucosidan, fucan sulfates).

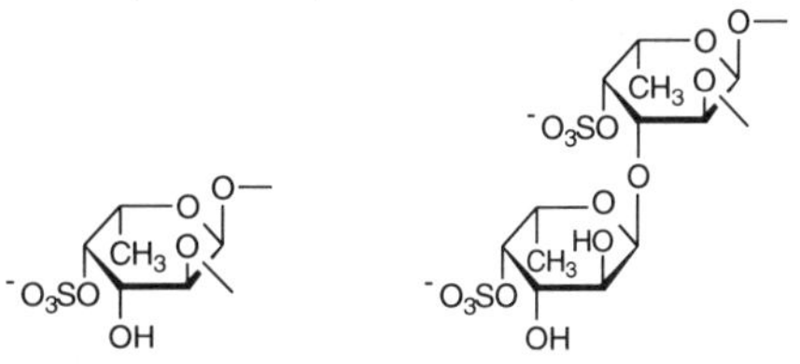

1,2 1,2,3

A *polysaccharide from brown algae (especially *Fucus* species), which contain up to 5–20% dry weight F., M_R ca. 100 000 to 150 000, hygroscopic powder, forms colloids. F. consists mainly of sulfated L-*fucose in 1,2-α-glycosidic linkages; in addition, smaller amounts of galactose, xylose, and uronic acids. Acidic hydrolysis gives L-fucose. The polysaccharides, also known as fucans (other sugars in addition to L-fucose may also be present) are excreted on the surface of the algae at low tide and thus prevent them from drying out; they are thus responsible for the slimy properties of algal masses. F. and related sulfated polysaccharides possess anti-coagulative and fibrinolytic properties [1] and are active against AIDS viruses [2] and cancer cells [3].

Lit.: [1] Kitasato Arch. Exp. Med. **60**, 105–121 (1987). [2] Hydrobiologia **1987**, 151, 497; Jpn. J. Exp. Med. **58**, 145 (1988). [3] Hydrobiologia **1987**, 491; Int. J. Cancer **39**, 82 (1987). – *[CAS 9072-19-9]*

Fucose (6-deoxygalactose, galactomethylose).

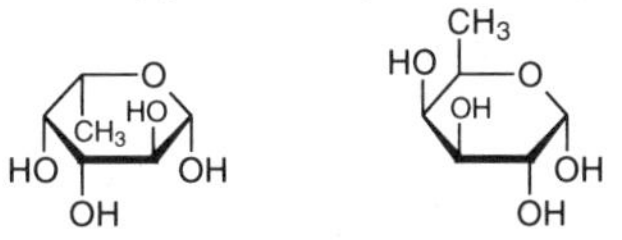

L-Fucose β-pyranose form D-Fucose α-pyranose form

$C_6H_{12}O_5$, M_R 164.16. D-(+)-F. {*rhodeose*, mp. 144 °C, $[\alpha]_D^{20}$ +153→+76° (H_2O)} occurs in various higher plants and tree woods. L-(–)-F. {mp. 145 °C, $[\alpha]_D^{20}$ –153→–76° (H_2O)} is, among others, a component of the polysaccharides from marine brown algae (especially *Fucus* species), it is also found in the glycoproteins of blood group substances, milk oligosaccharides, and mucopolysaccharides.

Lit.: Adv. Carbohydr. Chem. Biochem. **39**, 279–345 (1981) ▪ Beilstein E IV **1**, 4265 f. ▪ Carbohydrate Res. **126**, 165 (1984); **270**, 93 (1995) (synthesis) ▪ J. Carbohydrate Chem. **7**, 277 (1988) ▪ Karrer, No. 592, 593. – *[HS 2940 00; CAS 3615-37-0 (D-(+)-F.); 2438-80-4 (L-(–)-F.)]*

Fucoserratene see algal pheromones.

Fucosterol see sterols.

Fucoxanthin see peridinin.

Fugu. Japanese name for the puffer (globefish). Porcupine fish (burrfish), trunkfish (cowfish), and globefish (Diodontidae, Ostracionidae, and Tetrodontidae) belong to a systematically inhomogeneous group of fish, the meat and especially intestines of which can contain large amounts of the poisonous *tetrodotoxin depending on the time of the year. The skin and muscles of the back do not contain any toxin. The fish are much valued as luxury food in Japan; their preparation is restricted to specially trained personnel.

Fujimycin see FK-506.

Fuligorubin A.

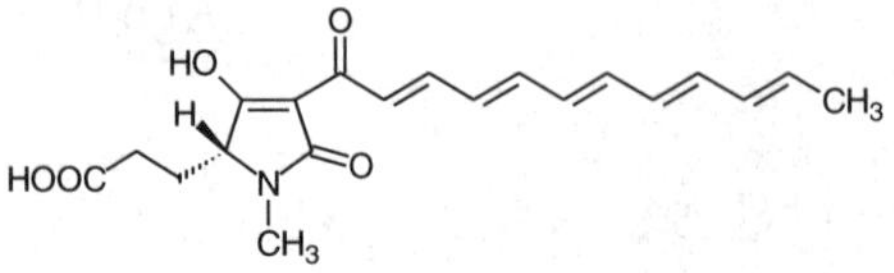

$C_{20}H_{23}NO_5$, M_R 357.41, red crystals, mp. 150 °C (dihydrate), an optically active *tetramic acid derivative from the yellow plasmodia of the slime mold *Fuligo septica* (Myxomycetes). F. exists in *Fuligo* as a stable calcium complex. F. is assumed to possess photoreceptor properties and to play a role in the phototaxis and sporulation of the slime mold.

Lit.: Angew. Chem. Int. Ed. Engl. **26**, 586 (1987) ▪ Tetrahedron **48**, 1145 (1992) (synthesis). – *[CAS 108343-55-1]*

Fulvic acids see humic acids.

Fulvoferruginin.

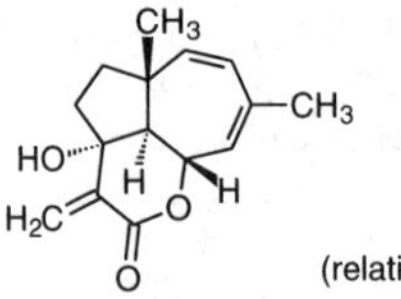

(relative configuration)

$C_{15}H_{18}O_3$, M_R 246.31, crystals, mp. 127 °C, $[\alpha]_D$ −65.4° (CH_3OH). A cytotoxic and antibiotically active sesquiterpenoid of the carotane type from mycelial cultures of the agaric *Marasmius fulvoferrugineus* (Basidiomycetes). At medium concentrations F. is especially active against *Paecilomyces varioti* (MIC 1–5 μg/mL).

Lit.: Z. Naturforsch. C **46**, 989 (1991). – *[CAS 130100-07-1]*

Fulvoic acids see humic acids.

Fumagillin.

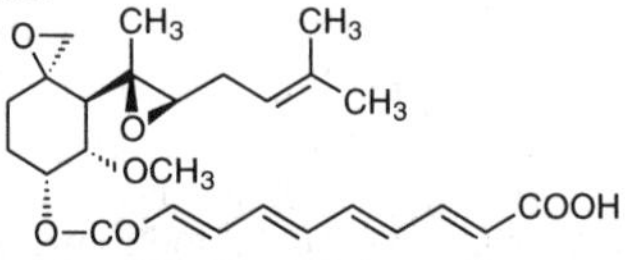

$C_{26}H_{34}O_7$, M_R 458.55, bright yellow needles, mp. 189–194 °C, $[\alpha]_D$ −26.2° (C_2H_5OH). F. is produced by *Aspergillus fumigatus* and *Penicillium nigricans*. It exhibits weak antibiotic but strong amoebicidic activity and is also active against bacteriophages. F. acts against *Nosema apis* infections in bees and inhibits tumor growth (Lewis lung carcinoma, B16 melanoma) in mice. Mechanistically, F. acts on angiogenesis, inhibiting the growth of blood vessels in new tumor tissue [1]. F. is in clinical phase III testing as treatment for microsporidiosis in AIDS patients. The biosynthesis proceeds from mevalonic acid (sesquiterpene) and acetate (tetraene side-chain). Cleavage of the acyl side chain yields *Fumagillol* (Deacyl-F.), found in *Penicillium jensenii* with immunosuppressive activity, $C_{16}H_{26}O_4$, M_R 282.38, mp. 56 °C, $[\alpha]_D^{23}$ −68° (C_2H_5OH). A derivative of F.: TNP-470 is under development as antiangiogenesis drug for cancer therapy (target: it binds to and inhibits protein-methionine aminopeptidase –2) [2].

Lit.: [1] Angew. Chem. Int. Ed. Engl. **38**, 3228 (1999). [2] Bioorg. Med. Chem. Lett. **10**, 39 (2000); Nature Biotechnol. **15**, 542 (1997); Proc. Natl. Acad. Sci. USA **94**, 6099 (1997).

gen.: Beilstein E V **19/3**, 247 ▪ Cole-Cox, p. 810 ▪ J. Chem. Soc. C **1969**, 1473 (biosynthesis) ▪ Nachr. Chem. Tech. Lab. **39**, 6 (1990) ▪ Nature (London) **348**, 555 (1990) ▪ Sax (8.), FOZ 000. – *Synthesis:* Angew. Chem. Int. Ed. Engl. **38**, 971 (1999) (fumagillol) ▪ J. Am. Chem. Soc. **121**, 5589 (1999) ▪ Tetrahedron Lett. **38**, 4437 (1997); **40**, 4797 (1999). – *[HS 2941 90; CAS 23110-15-8 (F.); 108102-51-8 (fumagillol)]*

Fumarprotocetraric acid.

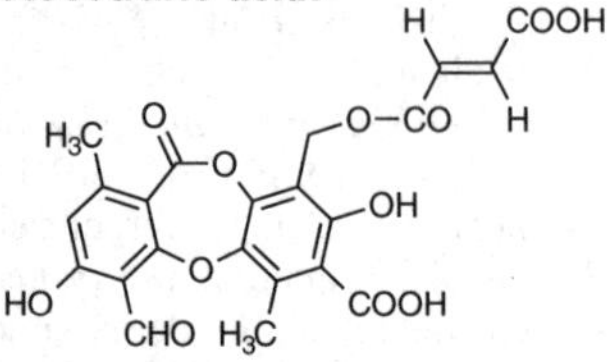

$C_{22}H_{16}O_{12}$, M_R 472.36, needles, decomposing between 250 and 260 °C with evolution of gas and formation of fumaric acid; poorly soluble in most organic solvents, very bitter taste. A *depsidone with antibiotic activity from the lichen *Cetraria islandica* and various *Cladonia* species.

Lit.: Asahina-Shibata, p. 140 ▪ Karrer, No. 1062. – *[CAS 489-50-9]*

Fumigatin (3-hydroxy-2-methoxy-5-methyl-*p*-benzoquinone).

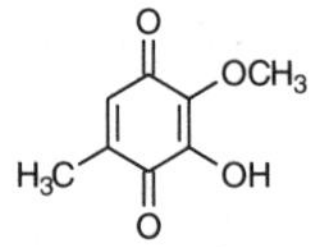

$C_8H_8O_4$, MR 168.15, chestnut-brown needles or plates, mp. 114–116 °C (subl.). A toxic, antibiotically active metabolite from cultures of the mold fungus *Aspergillus fumigatus*.

Lit.: Chem. Pharm. Bull. **36**, 178–189 (1988) ▪ Cole-Cox, p. 773–780 ▪ J. Med. Chem. **17**, 371 (1974) (biosynthesis) ▪ Thomson **3**, 14 f. ▪ Turner **1**, 92–97; **2**, 59, 65. – *[CAS 484-89-9]*

Fumigatin methyl ether see ubiquinones.

Fumiquinazoline see fungi.

Fumitremorgins. A group of neurotoxic 6-methoxyindole alkaloids from *Aspergillus*- and *Penicillium* species. *Lanosulin*, formed by *Penicillium lanosum* is identical with F. B. The F. belong to the *mycotoxins that have tremorgenic effects on animals (=induce quivering). Other compounds with this property are, e. g., *aflatrems, *janthitrems, *penitrems (tremortins), *paspalitrems, *territrems, and *verrucosidin.

Table: Data of Fumitremorgins.

toxin	molecular formula M_R	mp. [°C]	$[\alpha]_D$	CAS
F. A	$C_{32}H_{41}N_3O_7$ 579.69	206–209 (acetone)	+61°	12626-18-5
F. B	$C_{27}H_{33}N_3O_5$ 479.58	211–212 ($CHCl_3$)	+24°	12626-17-4
F. C	$C_{22}H_{25}N_3O_3$ 379.46	125–130		118974-02-0
F. C-Diol	$C_{22}H_{25}N_3O_5$ 411.46		+18.4°	111427-99-7
TR-2	$C_{22}H_{27}N_3O_6$ 429.47	150–152		51177-07-2
Verruculogen	$C_{27}H_{33}N_3O_7$ 511.57	233–235 ($CHCl_3$)	−27.7°	12771-72-1

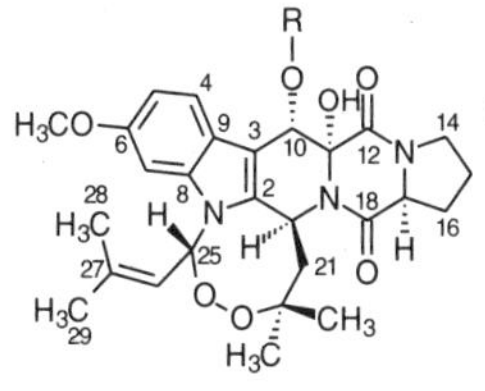

R= H : Verruculogen
R = CH_2—CH=C(CH_3)$_2$: F. A

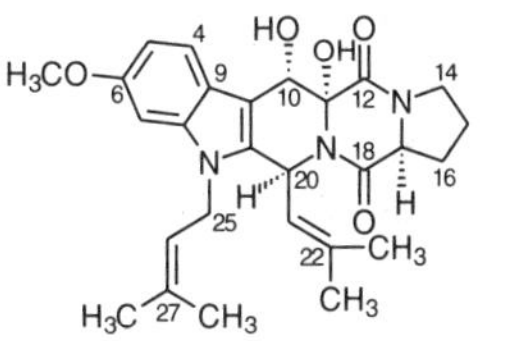

F. B (Lanosulin)

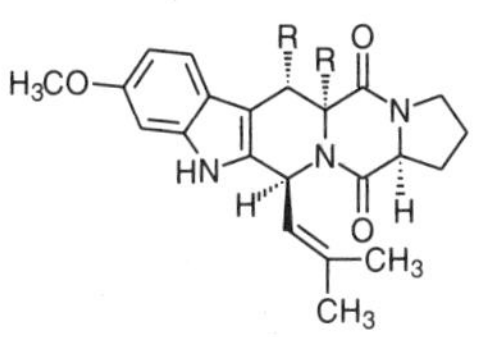

R = H : F. C (SM-Q)
R = OH : F. C-Diol

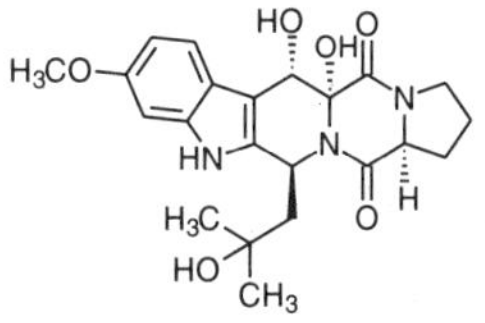

TR-2

With the exception of the territrems and verrucosidin, all these substances are biosynthetically derived from tryptophan. F. A and B. at doses of 1 mg/animal (i. p.) induce tremors in mice, 5 mg/animal lead to 70% mortality. Verruculogen leads to quivering even at 10 μg/animal. Grazing animals are especially affected by these toxins (see darnel, ryegrass). *Penicillium piscarium*, isolated from grass, produces verruculogen and F. B.; *Aspergillus fumigatus* from moldy corn silages produces verruculogen, F. A, F. B, and TR-2.

Lit.: Betina, chap. 16 ▪ Chem. Pharm. Bull. **37**, 23–32 (1989) ▪ Cole-Cox, p. 335–509 ▪ Heterocycles **46**, 673–704 (1997) (synthesis review) ▪ Steyn & Vleggar, Mycotoxins and Phycotoxins, p. 399–408, 501–512, Amsterdam: Elsevier 1986 ▪ Steyn (ed.), The Biosynthesis of Mycotoxins, p. 204–209, New York: Academic Press 1980 ▪ Tetrahedron **44**, 359–377, 1991–2000 (1988); **45**, 1941 (1989) ▪ Tetrahedron Lett. **28**, 1131 (1987); **29**, 1323 (1988) ▪ Turner **2**, 410–413, 535ff. ▪ Zechmeister **48**, 54–60.

Fumonisins. A group of *mycotoxins and phytotoxins formed by the ascomycetes *Gibberella fujikuroi* (asexual form: *Fusarium moniliforme* =*F. verticillioides*), a corn pathogen, and *Alternaria alternata* f. sp. *lycopersici*, a tomato pathogen (AAL).

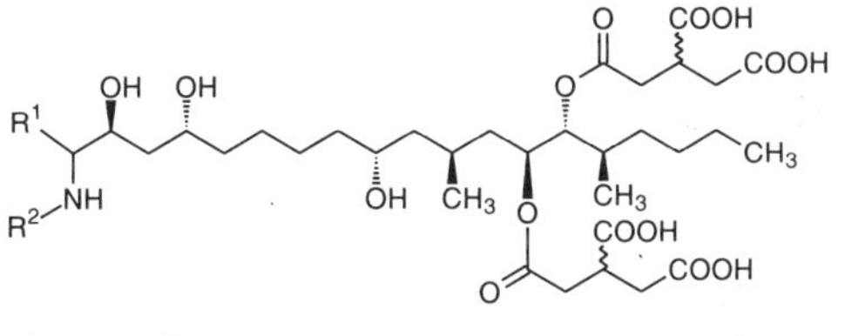

R^1 = CH_3 , R^2 = CO—CH_3 : F. A_1
R^1 = CH_3 , R^2 = H : F. B_1
R^1 = R^2 = H : F. C_1

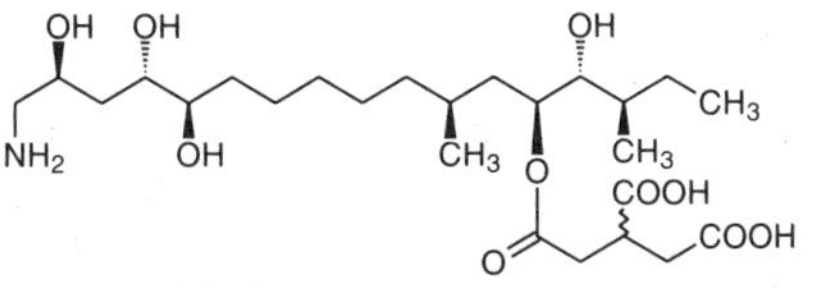

AAL Toxin TA_1

F. have been associated with the high incidence of esophageal cancer in the Transkei, the occurrence of leukoencephalomalacia in horses, the development of pulmonary edema in pigs, as well as with liver and esophageal carcinomas in experimental animals after feeding with contaminated corn. In sensitive tomato plants AAL toxin and F. B_1 elicit the symptoms of "stem cancer" and leaf necrosis. F. B_1 is also a virulence factor in corn. From 1 kg pasture grass infected with *Fusarium verticillioides* 9 mg F. B_1 and 4 mg F. B_1 methyl ester have been isolated; up to 300 mg/kg F. B_1 has been obtained from corn. F. B_1 inhibits the biosynthesis of sphingosine. F. of the A series and *N*-acetyl derivatives of F. B_1 or F. B_2, respectively, are presumed to be artefacts of processing. F. are not destroyed by boiling, baking, and beer brewing.

Table: Data of Fumonisins.

Fumonisin	molecular formula	M_R	CAS
A_1	$C_{36}H_{61}NO_{16}$	763.81	117415-48-2
B_1	$C_{34}H_{59}NO_{15}$	721.84	116355-83-0
C_1	$C_{33}H_{57}NO_{15}$	707.81	152726-69-7
AAL Toxin TA_1	$C_{25}H_{45}NO_{10}$	521.65	79367-52-5

Lit.: Annv. Rev. Phytopathol. **31**, 233–252 (1993) ▪ Appl. Environ. Microbiol. **54**, 1806 (1988); **58**, 3929 (1992); **59**, 2673 (1993); **61**, 79 (1995) ▪ Carcinogenesis **12**, 1247 (1991) ▪ Dev. Plant Pathol. **13**, 81–90 (1998) ▪ J. Agric. Food Chem. **38**, 1900 (1990); **39**, 1958 (1991) ▪ J. Am. Chem. Soc. **116**, 9409 (1994) (absolute configuration) ▪ J. Biol. Chem. **266**, 1486 (1991) ▪ J. Chem. Soc., Chem. Commun. **1988**, 743 ▪ J. Nat. Prod. **56**, 1630 (1993); **59**, 970 (1996) (new F.); **61**, 367 (1998) ▪ J. Org. Chem. **62**, 5666 (1997) (synthesis F. B_2) ▪ Mycopathologia **117**, 11, 23, 47 (1992) ▪ Phytopathology **82**, 353 (1992) ▪ Tetrahedron Lett. **22**, 2723 (1981) (AAL toxin); **39**, 3803 (1998) (synthesis F. B_1); **40**, 4515 (1999) (structure).

Fungal toxins see toadstools.

Fungi (marine). Fungi colonizing marine substrates are known to produce interesting active principles. For example, *phomactin A*[1] ($C_{20}H_{30}O_4$, M_R 334.46, oil, $[\alpha]_D$ +175°) is isolated from culture media of the marine fungus *Phoma* sp. (Deuteromycetes) that lives on the shell of the crab *Chionoecetes opilio*. It inhibits PAF (*platelet activating factor)-induced platelet aggregation (IC_{50} 10 μmol/L) and prevents the bonding of the factor to the receptors (IC_{50} 2.3 μmol/L). *Fellutamides A*[2] ($C_{27}H_{49}N_5O_8$, M_R 571.71, $[\alpha]_D$ −12.7°) and *B* ($C_{27}H_{49}N_5O_7$, M_R 555.72) are found in culture filtrates of *Penicillium fellutanum* living symbiotically in the gastrointestinal tract of the marine fish *Apogon endekataenia*. Fellutamides A and B are cytotoxic towards murine leukemia P388 cells (IC_{50} 0.2 and 0.1 μg/mL), L1210 cells (IC_{50} 0.8 and 0.7 μg/mL), and skin cancer cells (IC_{50} 0.5 and 0.7 μg/mL, *in vitro*), and stimulate the synthesis of NGF (nerve growth factor)[3]. The culture filtrates of *Aspergillus fumigatus*, occurring in the gastrointestinal tract of the marine fish *Pseudolabrus japonicus*, contain the quinazolidine alkaloids *fumiquinazolines A*[4] ($C_{24}H_{23}N_5O_4$, M_R 445.48, mp. 178–182 °C, $[\alpha]_D$ −215.5°), to G[5] which are moderately cytotoxic towards P388 leukemia cells. Further constituents of marine fungi are, e. g., the *o*-quinone *leptosphaerodione*[6] [$C_{21}H_{22}O_5$, M_R 354.40, dark red cryst., $[\alpha]_D^{20}$ −53.3° (CH_3OH)] from *Leptosphaeria oraemaris* or the cytotoxic *octalactin A*[7] [$C_{19}H_{32}O_6$, M_R 356.46, mp. 155–157 °C, $[\alpha]_D$ −14° ($CHCl_3$)] from a

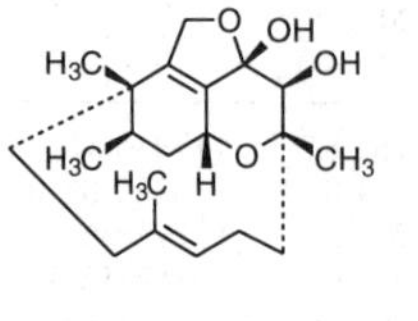

Phomactin A (1)

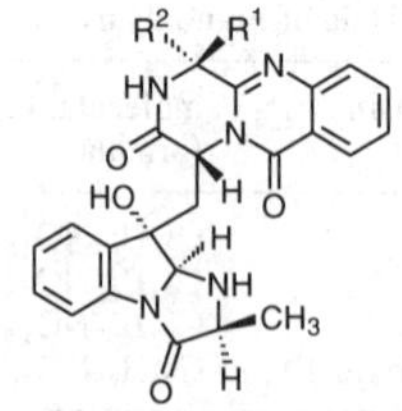

$R^1 = CH_3$, $R^2 = H$: Fumiquinazoline A (2)
$R^1 = H$, $R^2 = CH_3$: Fumiquinazoline B (3)

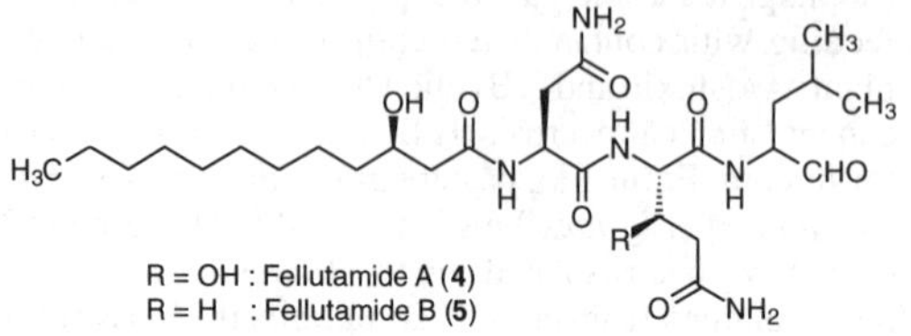

R = OH : Fellutamide A (4)
R = H : Fellutamide B (5)

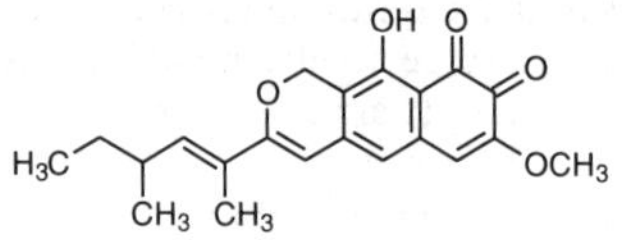

Leptosphaerodione (6)

(relative configuration)

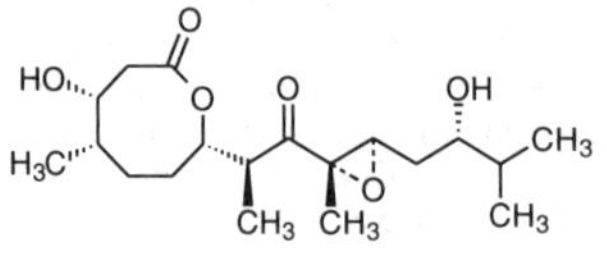

Octalactin A (7)

Streptomyces sp. living on the surface of the sea fan *Pacifigorgia* sp.

Lit.: [1] J. Am. Chem. Soc. **113**, 5463 (1991); Tetrahedron Lett. **37**, 7107 (1996) (P. D). [2] Tetrahedron **47**, 8529 (1991). [3] J. Biosci. Biotech. Biochem. **57**, 195 (1993). [4] Tetrahedron Lett. **33**, 1621 (1992). [5] J. Org. Chem. **63**, 2432 (1998); Synlett. **1997**, 483. [6] Helv. Chim. Acta **74**, 1445 (1991); Tetrahedron Lett. **30**, 3483 (1989). [7] Chem. Lett. **1997**, 117; J. Am. Chem. Soc. **113**, 4682 (1991); **116**, 5511, 8378 (1994); Synlett **1998**, 735. – *[CAS 130595-24-3 (1); 140715-85-1 (2); 140852-71-7 (3); 138682-08-3 (4); 138752-26-8 (5); 138039-33-5 (6); 133473-06-0 (7)]*

Fungicidin see nystatin.

Fungi imperfecti (Deuteromycetes). Anamorphous asco- and basidiomycetes as well as similar fungi which, in contrast to perfect fungi (Fungi perfecti), show exclusively asexual fruiting. This group includes about 30% of all known fungi. In most cases the F. i. form conidia as embryos although sometimes chlamydospores, bulbils, or sclerotia can also serve as propagating units. The conidia stages of F. i. are very similar to those known for the ascomycetes. It is thus assumed that F. i. represent conidia stages of ascomycetes for which the ascus stage (final stage of sexual reproduction, main fruit form with ascospores) has not yet been discovered or has been lost in the course of evolution. However, parasexuality like that occurring in the basidio- and ascomycetes has also been demonstrated for F. i. A complete assignment of F. i. in the system of the asco- and basidiomycetes is not yet possible. Thus the classification and nomenclature of the F. i. are based on the state of the conidia and external characteristics.

Lit.: J. Chem. Technol., Biotechnol. **74**, 479 (1999) (review) ▪ Müller & Löffler, Mykologie, 5. edn., p. 318 ff., Stuttgart: Thieme 1992.

Funtumine see Apocynaceae steroid alkaloids.

Furaneol® [4-hydroxy-2,5-dimethyl-3(2*H*)-furanone].

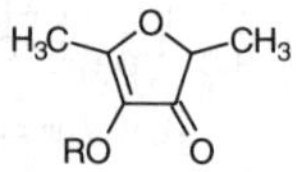

R = H : Furaneol®
R = CH_3 : Mesifuran

$C_6H_8O_3$, M_R 128.13; cryst., mp. 79–80 °C; odor resembling caramel, in dilute solution pineapple- and strawberry-like; olfactory threshold in water 40 ppt [1]. LD_{50} (rat p. o.) 1660 mg/kg. An important *flavor compound, occurring naturally in pineapples, strawberries, raspberries, etc., sometimes as the glucoside. It is also formed thermally from deoxyhexoses [1], e. g., in coffee. For production, see *Lit.* [2–4].

Mesifuran [4-methoxy-2,5-dimethyl-3(2*H*)-furanone], $C_7H_{10}O_3$, M_R 142.15, often occurs together with F. and is responsible for the characteristic aroma of the blackberry species *Rubus arcticus* [5].

Lit.: [1] Fenaroli (2.) **2**, 265. [2] J. Org. Chem. **38**, 123 (1973); DE.P 2812713 (1978), Polak's Frutal Works, Inv.: Cohen; Synthesis **1981**, 709. [3] Helv. Chim. Acta **56**, 1882–1894 (1973). [4] EP-A 398417 (1989), Quest International: Inv. DeCnop, VanDort, DeHey. [5] Maarse, p. 354.

gen.: ApSimon **6**, 170 ▪ J. Org. Chem. **57**, 5023 (1992) (synthesis) ▪ Maarse, p. 111, 336 ff., 340, 364, 636 ▪ TNO list (6.) Suppl. 5, p. 373. – *[HS 2932 19; CAS 3658-77-3 (F.); 4077-47-8 (mesifuran)]*

Furan fatty acids see F-acids.

Furfurylmercaptan see 2-furylmethanethiol.

Furochromone see furocoumarins and khellin.

Furocoumarins.

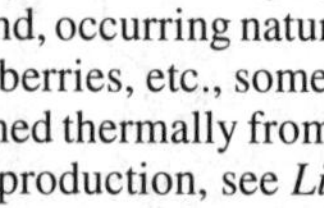

Psoralen

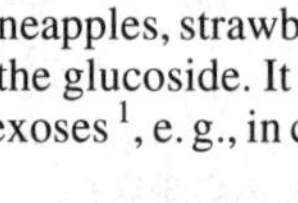

Angelicin

Name for a group of *coumarins occurring in plants with a furan ring condensed at C-6 and C-7 or C-7 and C-8. The parent compounds are *psoralen* (7*H*-furo[3,2-*g*]-1-benzopyran-7-one), $C_{11}H_6O_3$, M_R 186.17, needles, mp. 163–164 °C (169–179 °C), and *angelicin* (isopsoralen, 2*H*-furo[2,3-*h*]-1-benzopyran-2-one), mp. 139 °C. Numerous members of this class of natural compounds contain methoxy substituents (see bergapten and pimpinellin). The F. partial structure is also present in the *aflatoxins. Both types of compounds usually occur in many plants, e. g., in Umbelliferous plants (Apiaceae) such as *Ammi*, *Pimpinella* (pimpernel), *Angelica*, and *Heracleum* as well as Rutaceae (bergamot, lemon, clove), Fabaceae, Moraceae, and in the Asian Fabaceae *Psoralea corylifolia*. In higher plants they serve as important *phytoalexins for defence against predatory insects and fungal attack [1]. Structural relatives of the F. are the furochromones (5*H*-furo-[3,2-*g*]-1-benzopyran-5-ones, e. g., *khellin). ***Toxicology:*** Angelicin: LD_{50} (rat p. o.) 322 mg/kg. F. possess genotoxic, mutagenic, and carcinogenic activities, thus precluding their use in longer duration therapies.

Activity: Many F. induce a brown pigmentation in the skin on exposure to sunlight. They are photosensitizing, phototoxic, and can induce allergic dermatitis (Berloque's dermatitis), possibly with severe symptoms such as blister formation. The effect can also arise when, for example, scented toilet waters whose plant components still contain F. are applied to the skin.
Use: In photochemotherapy for vitiligo, psoriasis, atopic dermatitis, mycosis fungoides. Photoactivated F. bind covalently as haptens to proteins as well as pyrimidines of DNA[2] and thus have antimitotic effects. For synthesis, see *Lit.*[3].
Biosynthesis: See umbelliferone; the furan ring originates from activated isoprene (dimethylallyl pyrophosphate)[4]. Key step in psoralene biosynthesis is the oxidative dealkylation of (+)marmesine, which in one step is converted to psoralene and acetone by a cytochrome P450 enzyme.

Lit.: [1]Science **212**, 927–929 (1981). [2]Eur. J. Med. Chem. Chim. Ther. **17**, 95f. (1982). [3]J. Am. Chem. Soc. **106**, 6735–6740 (1984); **110**, 7419–7434 (1988). [4]Tetrahedron **53**, 17699 (1997).
gen.: Annu. Rev. Biochem. **54**, 1151–1194 (1985) ▪ Fitzpatrick, Psoralens, Proc. Int. Congr. "Psoralens 1988", Paris: Libbey 1989 ▪ Hager (5.) **5**, 665 f. ▪ Luckner (3.), p. 388 ff. ▪ NATO ASI Ser., Ser. H **15**, 269–278 (1988) ▪ Nat. Prod. Rep. **12**, 477–505 (1995) ▪ Photomed. Photobiol. **10**, 1–46 (1988) (review) ▪ Phytochemistry **12**, 1657–1667 (1973) (biosynthesis); **18**, 139 f. (1979) (^{13}C NMR) ▪ Zechmeister **35**, 199–430. – *Toxicology:* J. Photochem. Photobiol. B **8**, 235–254 (1991) ▪ Sax (8.), FQD 000, FQC 000. – *[CAS 66-97-7 (psoralen); 523-50-2 (angelicin)]*

Furostan(e) derivatives. The name furostan is equivalent to 16*β*,22-epoxycholestane.

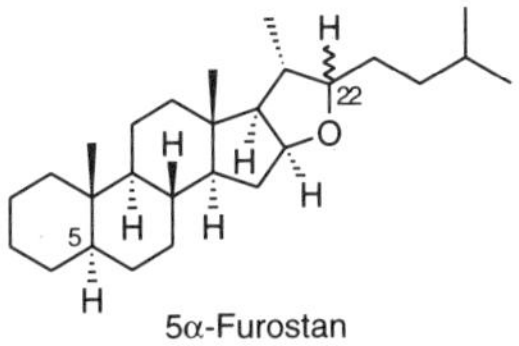

5α-Furostan

F. occur occasionally in nature, for example as 26-*O*-glycosylated furostan-22,26-diol derivatives (see steroid saponins and Solanum steroid alkaloid glycosides).

2-Furylmethanethiol (furfurylmercaptan).

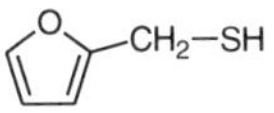

C_5H_6OS, M_R 114.16. Liquid with an unpleasant, sulfur-like odor, coffee-like odor at high dilution (0.1–1 ppb); bp. 160 °C, D. 1.132. F. is long known as one of the important aroma constitutents of *coffee flavor[1], but also contributes to the aroma of *meat flavor[2]. The olfactory threshold is 5 ppt[2]. F. is formed by the Maillard reaction and is thus considered as a secondary *flavor compound. For production, see *Lit.*[3]. Use in meat and coffee flavors.

Lit.: [1]Morton & McLeod (eds.), Food Flavours, Part A, p. 215, Amsterdam: Elsevier 1982. [2]TNO list (6.), Suppl. 2, p. 35, 56, & 106; Lebensm. Unters. Forsch. **190**, 3 (1990). [3]Bauer et al. (2.), p. 124.
gen.: Beilstein EV **17/3**, 351 ▪ Fenaroli (2.) **2**, 209. – *[HS 2932 19; CAS 98-02-2]*

Fusarenone-X see tric(h)othecenes.

Fusaric acid (5-butylpyridine-2-carboxylic acid).

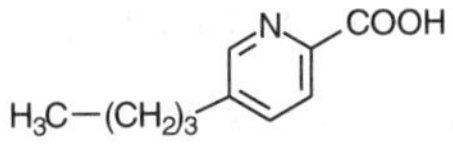

$C_{10}H_{13}NO_2$, M_R 179.22, mp. 108–109 °C, crystals. F. is produced by various *Fusarium* species and is one of the toxins responsible for wilting [see wilting (withering) agents]. F. has herbicidal, insecticidal, and antibacterial activities, it lowers blood pressure, inhibits dopamine *β*-hydroxylase, and has emetic effects in dogs (on oral administration from 10 mg/kg). The LD_{50} for mice is about 2 mg/animal (i. p.).

Lit.: Beilstein EV **22/2**, 384 ▪ Heterocycles **29**, 2249 (1989); **38**, 1595 (1994) ▪ Pharmazie **39**, 155 (1984) ▪ Sax (8.), BSI 000. – *[HS 2933 39; CAS 536-69-6]*

Fusarin C.

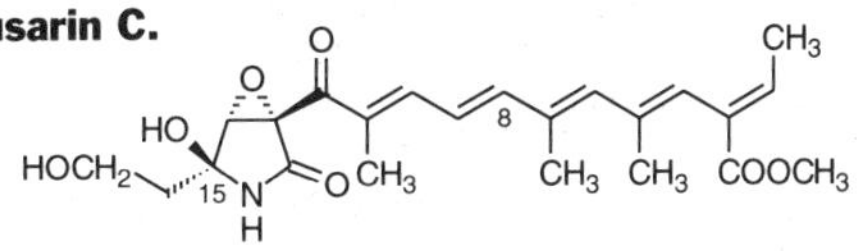

$C_{23}H_{29}NO_7$, M_R 431.49, yellow oil, $[\alpha]_D$ +47.04° (CH_3OH). A mutagenic metabolite from various *Fusarium* species, especially *F. moniliforme* (asexual form of *Gibberella fujikuroi*). This ascomycete is a ubiquitous plant pathogen. F. C was previously held to be responsible for the carcinogenic effects of infected corn but it is now known that this is due to the *fumonisins. F. C is labile under thermal conditions and exposure to UV radiation; it is destroyed on boiling. The mutagenicity depends on the arrangement of the substituents at C_{15} and the epoxide ring (C_{13}–C_{14}). When the ring is shifted to another position the mutagenicity disappears.

Lit.: App. Environ. Microbiol. **55**, 649–655 (1989) ▪ Chelkowski (ed.), Fusarium Mycotoxins, Taxonomy and Pathogenicity, chap 2, Amsterdam: Elsevier 1989 ▪ J. Agric. Food Chem. **32**, 1064 (1984); **34**, 963 (1986) ▪ Phytochemistry **30**, 2259 (1991). – *[CAS 79748-81-5]*

Fusariotoxin T-2 see T-2 toxin.

Fusarium toxins. Toxins formed by phytopathogenic *Fusarium* species that cause damage to plants, e. g., *fusaric acid (*wilting (withering) agents) or toxins of the *tric(h)othecene group.

Lit.: Chelkowski (ed.), Fusarium, Mycotoxins, Taxonomy and Pathogenicity, Amsterdam: Elsevier 1989.

Fusarochromanones.

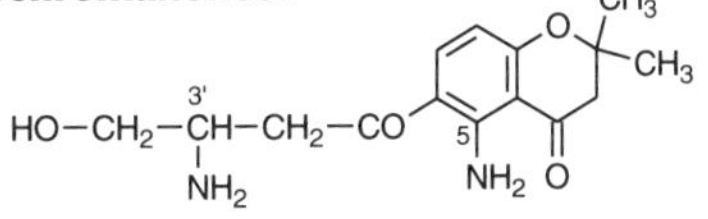

TDP-1

A group of water-soluble *mycotoxins derived from fusarochromanone (TDP-1: $C_{15}H_{20}N_2O_4$, M_R 292.33, mp. 132–134 °C; TDP-2 = 3′-*N*-acetyl-TDP-1). F. are produced by *Fusarium equiseti* sometimes in considerable amounts (up to 1.8 g/kg), especially on cereal and pea plants. F. are supposed to be responsible for tibial dyschondroplasia (TDP), a disease affecting fowl in which bone formation in the legs is impaired. A further symptom is a reduced hatching rate of eggs. The disease occurs worldwide in the fowl breeding industry. Similar bone and joint malformations are also known in pigs, horses, cattle, and dogs. F. have been mentioned in the context of Kashin-Beck disease in children.

Lit.: Appl. Environ. Microbiol. **55**, 794, 3184 (1989); **56**, 2946 (1990) ▪ Can. J. Chem. **64**, 1208 (1986) ▪ J. Agric. Food Chem. **39**, 1757 (1991) ▪ J. Nat. Prod. **54**, 1165 (1991). – *[CAS 104653-89-6 (TDP-1)]*

Fusicoccin H.

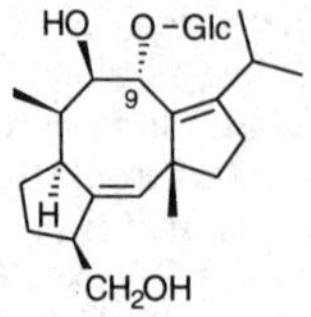

$C_{26}H_{42}O_8$, M_R 482.61, mp. 125 °C, $[\alpha]_D$ +20° (C_2H_5OH), a diterpenoid from *Fusicoccum amygdali.*

Lit.: J. Chem. Soc., Chem. Commun. **1984**, 1695–1696; **1999**, 367 ▪ J. Chem. Soc., Perkin Trans. 1 **1976**, 2221–2228 ▪ Tetrahedron **38**, 3169–3172 (1982). – *[CAS 50906-51-9]*

Fusidic acid [(17*Z*)-16*β*-acetoxy-3*α*,11*α*-dihydroxy-(4*α*,8*α*,9*β*,13*α*,14*β*)-29-nordammara-17(20),24-diene-21-carboxylic acid]. $C_{31}H_{48}O_6$, M_R 516.72, cryst., mp. 192–193 °C, $[\alpha]_D$ −9° ($CHCl_3$), a tetracyclic triterpene acid with a stereoisomeric 29-nordammarane skeleton (fusidane). F. is isolated from culture filtrates of *Fusidium coccineum* and other *Fusidium* species; F. is also produced by dermatophytes. F. has bacteriostatic activity against Gram-positive pathogens and is used in the treatment of multiresistant

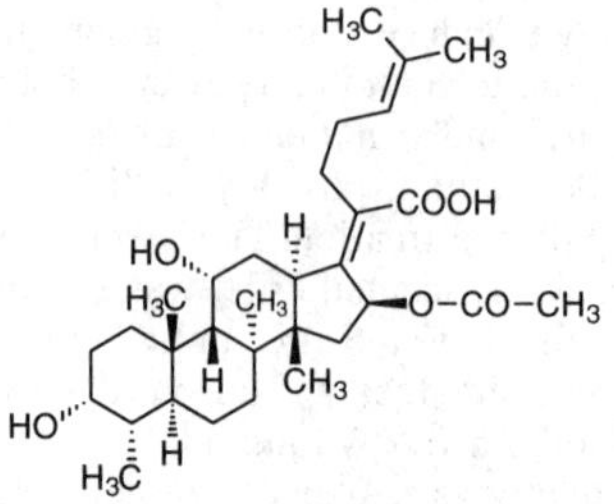

staphylococci infections. An immunosuppressive activity has also been reported.

Lit.: Merck-Index (12.), No. 4340. – *Activity:* Chemotherapy **40**, 201 (1994) ▪ Cytokine **2**, 423 (1990) ▪ J. Antimicrob. Chemother. **20**, 467 (1987); **25** (Suppl. B), 1 (1990) ▪ Lyon Pharm. **45**, 337 (1994). – *[CAS 6990-06-3]*

Fusidienol A.

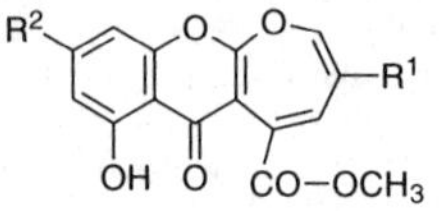

R^1 = H, R^2 = CH_3 (Fusidienol A)
R^1 = CH_2OH, R^2 = H (Fusidienol)

$C_{16}H_{12}O_6$, M_R 300.27, yellow cryst., mp. 168–170 °C, a Ras farnesyl protein transferase inhibitor from fungal cultures of *Phoma* spp. and *Fusidium griseum.*

Lit.: J. Org. Chem. **62**, 7485 (1997) ▪ Tetrahedron Lett. **35**, 4693 (1994). – *[CAS 157173-61-0 (F. A)]*

G

GABA see 4-aminobutanoic acid.

Gabaculine [(*S*)-5-amino-1,3-cyclohexadiene-1-carboxylic acid].

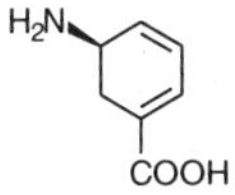

$C_7H_9NO_2$, M_R 139.15, amorphous powder, $[\alpha]_D$ −454° (H_2O). Synthetic racemate, mp. 196–197 °C; antibiotic from *Streptomyces toyocaensis* [1], G. inhibits the enzyme GABA aminotransferase by formation of a Schiff's base and subsequent (irreversible) aromatization with formation of a *m*-anthranilic acid derivative. For racemic synthesis, see *Lit.*[3], asymmetric or regioselective syntheses, see *Lit.*[4–6].

Lit.: [1]Tetrahedron Lett. **17**, 537 (1976). [2]J. Org. Chem. **43**, 1448–1455 (1978). [3]J. Org. Chem. **44**, 3451–3457 (1979). [4]J. Org. Chem. **49**, 2496-2498 (1984). [5]Tetrahedron Lett. **25**, 281–284 (1984). [6]Tetrahedron Lett. **27**, 1411–1414 (1986). – *[CAS 59556-29-5 (S); 59556-18-2 (racemate)]*

Gabosins.

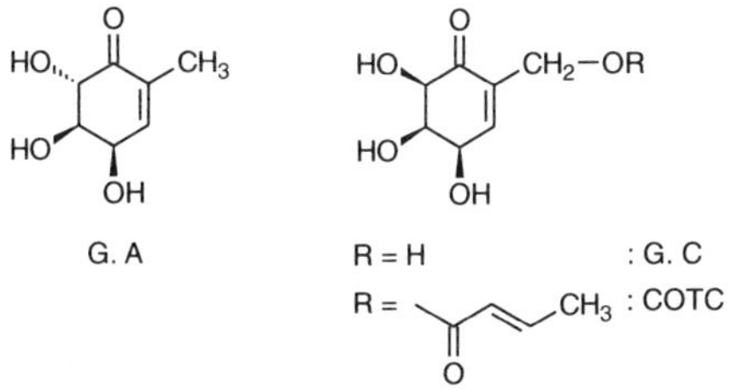

Polar cyclitols (G. A–K) isolated from various *streptomycetes in which the substituents on the carbocyclic C_7 skeleton resemble those of hexoses, while the ring oxygen atom of the pyranose form is replaced by a carbon atom (pseudosugar, carbasugar). The G. differ mainly in the configurations of the hydroxy groups. *G. A:* $C_7H_{10}O_4$, M_R 158.15, mp. 182 °C, $[\alpha]_D$ −132° (CH_3OH); *G. C:* $C_7H_{10}O_5$, M_R 174.15, mp. 113–114 °C. The crotonate of *G. C* (COTC, $C_{11}H_{14}O_6$, M_R 242.23, mp. 181 °C) inhibits glyoxylase I.

Lit.: Eur. J. Org. Chem. **2000**, 149 ▪ J. Antibiot. **28**, 737–748 (1975); **35**, 1222–1229 (1982) ▪ Justus Liebigs Ann. Chem. **1993**, 241–250 ▪ Tetrahedron **46**, 6575 (1990). – *[CAS 127545-53-3 (G. A); 40980-53-8 (G. C); 57449-30-6 (COTC)]*

Galactitol (dulcitol).

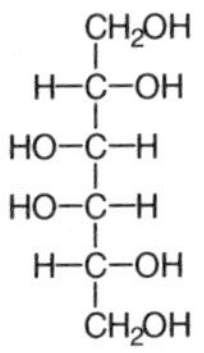

$C_6H_{14}O_6$, M_R 182.17, sweet-tasting rods, mp. 188–190 °C, bp. 275–280 °C (130 Pa), optically inactive. The sugar alcohol G. is present in the urine of patients with galactosemia. G. occurs in plants and algae and is obtained from Madagascan manna (*Melampyrum nemorosum*, Scrophulariaceae); commercial production by catalytic hydrogenation of galactose. G. is used in bacteriology.

Lit.: Beilstein E IV **1**, 2844 ▪ Brimacombe, The Carbohydrates, vol. 1A, p. 479, New York: Academic Press 1972 ▪ Carbohydr. Res. **229**, 17 (1992) (structure) ▪ Phytochemistry **11**, 1705 (1972) (biosynthesis). – *[HS 2905 49; CAS 608-66-2]*

Galactonojirimycin see galactostatin.

D-Galactosamine (chondrosamine, 2-amino-2-deoxy-D-galactose, abbr.: GalN).

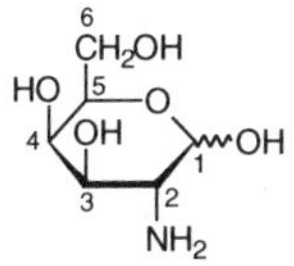

$C_6H_{13}NO_5$, M_R 179.17, hydrochloride: mp. 185 °C (decomp.), $[\alpha]_D^{20}$ +125→+98° (H_2O); GalNAc: mp. 172–173 °C, $[\alpha]_D^{20}$ +115→+86° (H_2O). D-G. is an *amino sugar; the *N*-acetylated derivative (*N-acetylgalactosamine*, abbr.: GalNAc) is a building block of chondroitin sulfates and is a widely distributed component of cartilage (Greek: chondros, hence the name chondrosamine), connective tissue, and tendons and also occurs in glycoproteins and lipids.

Lit.: Beilstein E IV **4**, 2024 ▪ J. Chem. Soc., Perkin Trans. 1 **1972**, 277; **1992**, 235 (synthesis). – *[HS 2922 50; CAS 7535-00-4]*

Galactose (cerebrose; abbreviation: Gal).

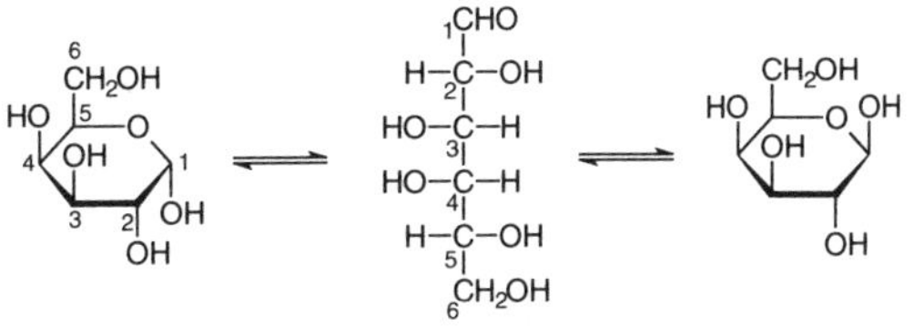

α-D-Galactopyranose (α-D-Gal*p*) | open-chain D-Galactose | β-D-Galactopyranose (β-D-Gal*p*)

$C_6H_{12}O_6$, M_R 180.16; D-Gal: α-anomer: mp. 167 °C, $[\alpha]_D^{20}$ +157→+80° (H_2O); β-anomer: mp. 167 °C, $[\alpha]_D^{20}$ +53→+80° (H_2O); α-D-Gal. H_2O: mp. 118–120 °C. L-Gal: α-anomer: mp. 163–165 °C, $[\alpha]_D^{20}$ −120→−78° (H_2O). An aldohexose, epimeric with *glucose. The mutarotation of G. in aqueous solution is based on the equilibrium shown in the figure. D-G. is a component of various oligo- and polysaccharides, glycoproteins and -lipids as well as other *galactosides* such as, e. g., the disaccharide *lactose, and others such as *raffinose, melibiose, *stachyose, galactans, guar flour, gummi arabicum, and (together with D-galacturonic acid and its methyl ester) *pectins. D-G. is found connected to D-glucosamine in almost all glycoproteins; the important *cerebrosides in the cerebrum and gangliosides contain chemically bound D-galactose. In lactic glands (hence the name from Greek: gala, genitive galaktos = milk) D-G. is formed enzymatically in a reversible reaction in the form of uridine 5′-diphosphate(UDP)-D-galactose from UDP-D-glucose and then linked with a further molecule of D-glucose to give galactose with liberation of UDP. Elevated urinary excretion of D-G. (*galactosuria*) and high levels in blood (*galactosemia*) are signs of liver disease. L-G. also frequently occurs in nature, e. g., in polysaccharides in agar, red algae, or galactans from snails.

Lit.: Carbohydr. Res. **191**, 150 (1989) (synthesis) ▪ Merck-Index (12.), No. 4353 ▪ Science **220**, 949 (1983). – *[HS 2940 00; CAS 59-23-4 (D-G.); 15572-79-9 (L-G.)]*

Galactostatin (*galacto*-nojirimycin, galactonojirimycin, 5-amino-5-deoxy-D-galactose).

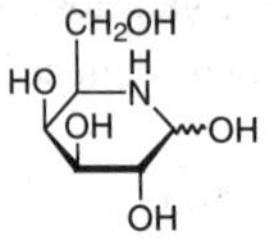

An α- and β-galactosidase inhibitor from cultures of *Streptomyces lydicus*. $C_6H_{13}NO_5$, M_R 179.17, amorphous powder, mp. 94–98 °C, $[\alpha]_D^{23}$ +85.6° (H_2O).

Lit.: Heterocycles **30**, 783 (1990); **41**, 2271 (1995) ▪ J. Antibiot. **40**, 122 (1987) ▪ J. Chem. Soc., Chem. Commun. **1991**, 1576; **1994**, 1247 ▪ J. Org. Chem. **56**, 815 (1991); **60**, 4749 (1995). – *[CAS 107537-94-0]*

Galangin (3,5,7-trihydroxyflavone). Formula see flavones. $C_{15}H_{10}O_5$, M_R 270.24, yellow needles (C_2H_5OH), mp. 214–215 °C. G. occurs in galangal (*Alpinia officinarum*) and poplar buds (*Populus nigra*). G. is used in the photometric determination of uranium (uv_{max} 450 nm), zirconium (uv_{max} 410 nm at pH 4.5), and thorium (uv_{max} 410 nm). G. occurs as the 3-*O*-glucoside (*galanginin*), $C_{21}H_{20}O_{10}$, M_R 432.38, in *Cephalaria procera* and *Datisca cannabina*.

Lit.: Karrer, No. 1490 ▪ Merck-Index (12.), No. 4356. – *[HS 2938 90; CAS 548-83-4 (G.); 68592-14-3 (galanginin)]*

Galanthamine, galanthine see Amaryllidaceae alkaloids.

Galapagin.

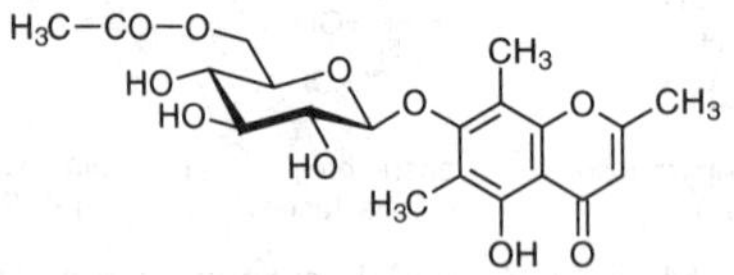

$C_{20}H_{24}O_{10}$, M_R 424.40, yellowish needles, mp. 163–165 °C, which give an ink blue color with $FeCl_3$. A chromone glucoside from the dyer's lichen *Roccella galapagoensis*; furnishes 8-methyleugenitol and glucose on acid hydrolysis.

Lit.: Z. Naturforsch. B **47**, 449–451 (1992). – *[CAS 41666-58-4]*

Galbanum oil or resin. Obtained by steam distillation or solvent extraction (e. g., with hexane) from G. rubber; yield 20% oil or 40–50% resin. The oil is a readily mobile, colorless to pale yellow liquid with a unique green-spicy, leafy, balsamy odor. The resin is a gold-brown, more or less viscous mass with an odor similar to that of the oil but more woody-resinous and balsamy and less terpene-like.

G. rubber is the dried exudate from the roots of *Ferula* species (Apiaceae), e. g., *Ferula gummosa* (syn. *F. galbaniflua*) growing mainly in Iran. In order to obtain the rubber the upper parts of the roots are uncovered and cut during the vegetative pause. The resinous, rubber-like juice is scraped off and collected after some time.

Composition [1]: Main components of the oil are monoterpene hydrocarbons such as β- (>50%) and α-**pinenes* (ca. 10%) and *3-*carene* (ca. 15%). The typical odor is mainly due to the ca. 1% content of (3*E*,5*Z*)-*1,3,5-undecatriene* ($C_{11}H_{18}$, M_R 150.26) together with decisive contributions from other trace components such as *2-isobutyl-3-methoxypyrazine* ($C_9H_{14}N_2O$, M_R 166.22) and *S-sec-butyl 3-methyl-2-butenethioate* ($C_9H_{16}OS$, M_R 172.29).

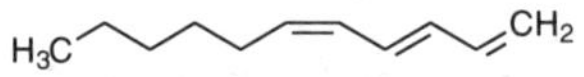

(3*E*,5*Z*)-1,3,5-Undecatriene (**1**)

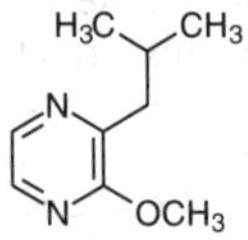

2-Isobutyl-3-methoxypyrazine (**2**)

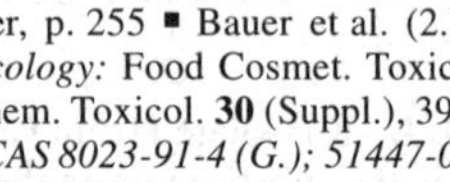

S-sec-Butyl-3-methyl-2-butenethioate (**3**)

Use: Mainly in the production of perfumes to give a fresh-green note in flowery, herby, and coniferous compositions.

Lit.: [1] Perfum. Flavor. **7** (6), 21 (1982); **14** (6), 98 (1989).
gen.: Arctander, p. 255 ▪ Bauer et al. (2.), p. 155 ▪ Ohloff, p. 175. – *Toxicology:* Food Cosmet. Toxicol. **16**, 765 (1978) (oil) ▪ Food Chem. Toxicol. **30** (Suppl.), 39 S (1992) (resin). – *[HS 3301 29; CAS 8023-91-4 (G.); 51447-08-6 (1); 24683-00-9 (2); 34322-09-3 (3)]*

Galbonolides. Macrolide antibiotics produced by *Micromonospora chalcea* and *Streptomyces galbus*. G. A and G. B are structurally related to rustmicin. Both show antifungal activity, e. g., *G. A*: $C_{21}H_{32}O_6$, M_R 380.48, needles, mp. 68 °C, $[\alpha]_D$ –231° (acetone).

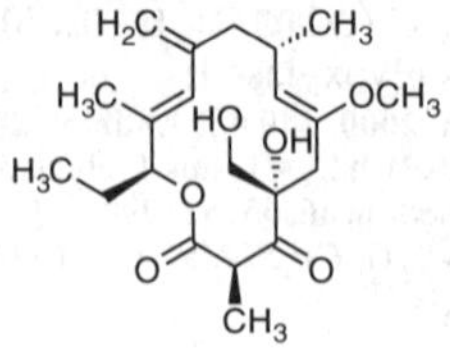

G. A

Lit.: J. Antibiot. **51**, 837 (1998) ▪ J. Org. Chem. **62**, 3236 (1997). – *[CAS 100227-57-4; 100938-28-1 (G. A)]*

Galegine [(3-methyl-2-butenyl)-guanidine].

$$HN{=}\underset{\displaystyle NH_2}{\underset{|}{C}}{-}NH{-}CH_2{-}CH{=}C(CH_3)_2$$

$C_6H_{13}N_3$, M_R 127.19, mp. 60–65 °C (sulfate: 223–225 °C). G. is a guanidine derivative occurring mainly in the seeds (ca. 5%) of goat's rue *Galega officinalis* (Fabaceae). G. has blood-pressure lowering activity but is too toxic for therapeutic use[1].
Lit.: [1] Aust. Vet. J. **70**, 169 (1993).
gen.: Beilstein EIII **4**, 465 ▪ Merck-Index (12.), No. 4358 ▪ Mothes et al., p. 99 ▪ Phytochemistry **21**, 97 (1982). – *[CAS 543-83-9]*

Gallic acid (3,4,5-trihydroxybenzoic acid).

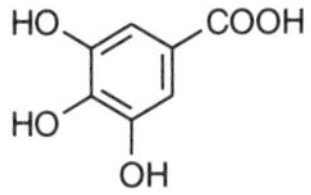

$C_7H_6O_5$, M_R 170.12, needles as hydrate with 1 mol water, mp. 253 °C (decomp.), pK_{a1} 3.13, pK_{a2} 8.84, pK_{a3} 12.4 (20 °C, 0.1 M KNO_3), light-sensitive, poorly soluble in cold, readily soluble in warm water, alcohol, acetone. G. forms a blue-black precipitate with iron salts (ferrogallic ink). Alkali salts of G. are readily oxidized by oxygen with a brown coloration. G. reduces the mutagenic action of nitrosamines[1]. G. was isolated for the first time in 1786 by Scheele from gallnuts and is widely distributed, especially in the form of *gallates* with sugars, in the so-called hydrolysable *tannins[2], in particular in oak bark, divi-divi, pomegranate roots, sumac, and tea. In tannins G. also often exists as a *depside (*m-digallic acid, 3-O-galloylgallic acid*, for formulae, see tannins) which can be obtained in crystalline form after partial hydrolysis or cleavage by the enzyme tannase (from mold fungi).
Use: For production of ferrogallic ink, as antioxidant, in sun blockers and dyes (anthracene brown, gallocyanine, rufigallic acid, etc.), also as a reducing agent and in medicine.
Biosynthesis: There are at least three pathways to G.[2,3]: dehydrogenation of *3-dehydroshikimic acid, side-chain degradation of hydroxylated *hydrocinnamic acid, and hydroxylation of oxidized benzoic acids.
Lit.: [1] Biol. Plant. **28**, 386–390 (1986); Folia Microbiol. **32**, 55–62 (1987). [2] J. Chem. Soc., Perkin Trans. 1 **1982**, 2515–2524; Phytochemistry **21**, 1049–1062 (1982); Rec. Adv. Phytochem. **20**, 163 (1986); Zechmeister **41**, 1–46. [3] Nat. Prod. Rep. **1**, 451–469 (1984).
gen.: Acta Cryst. **49**, 810 (1993) (structure) ▪ Beilstein EIV **10**, 1993 ▪ Hager (5.) **4**, 27 f., 257 f., 727 f. ▪ Karrer, No. 903, 904, 909, 910 ▪ Luckner (3.), p. 172, 397–398 ▪ Martindale (31.), p. 1116 ▪ Phytochemistry **26**, 1147–1152 (1987); **27**, 3004 f. (1988); **29**, 267–270 (1990) ▪ Zechmeister **35**, 73–119. – *[HS 2918 29; CAS 149-91-7]*

Gallidermin see epidermin.

Gallotannic acid see tannins.

Gallotannins see tanning agents and tannins.

Gambierol.

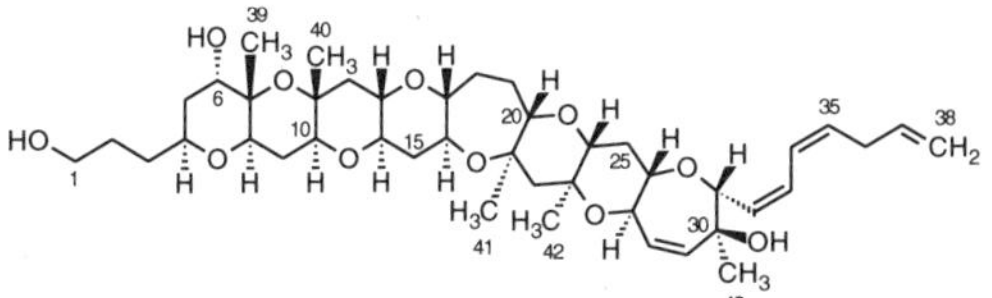

A marine polyether toxin from the dinoflagellate *Gambierdiscus toxicus*, together with *brevetoxins, *ciguatoxins, *maitotoxin and *yessotoxin. G. is responsible for ciguatera poisoning. $C_{43}H_{64}O_{11}$, M_R 756.97, amorphous solid.
Lit.: J. Am. Chem. Soc. **115**, 361 (1993) (isolation) ▪ Tetrahedron Lett. **39**, 6365, 6369, 6373 (1998) (synthesis); **40**, 97 (1999) (absolute configuration). – *[CAS 146763-62-4]*

Gamboge (gamboge resin). The dried, latex-like wound secretion (exudate) from *Garcinia* species, especially *G. hanburyi* or *G. morella* (Hypericaceae [Guttiferae]). G. consists to 15–20% of rubber and to 70–80% of a yellow pigment resin composed mainly of *β-guttilactone* ("β-guttic acid", $C_{29}H_{34}O_6$, M_R 478.59, mp. 80–90 °C) and *α-gambogic acid* ($C_{38}H_{44}O_8$, M_R 628.76).

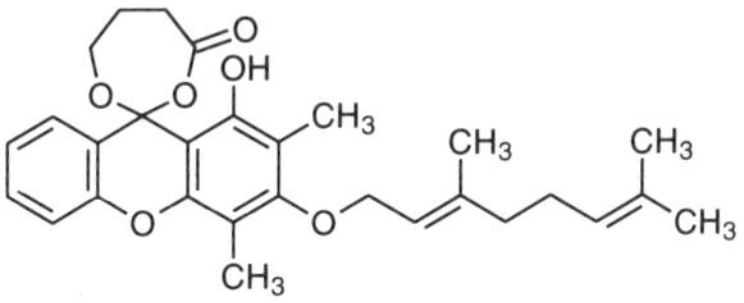

β-Guttilactone

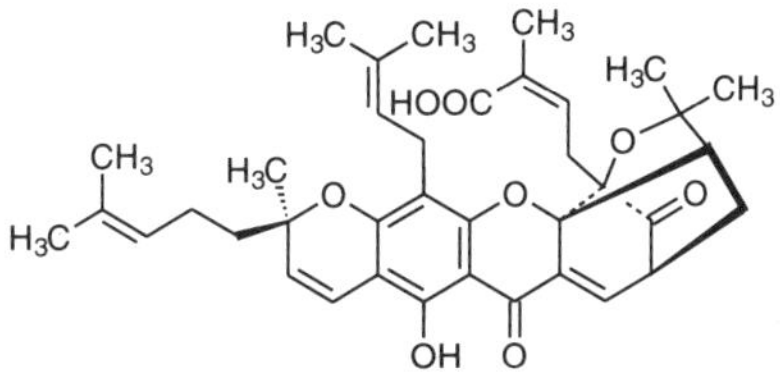

α-Gambogic acid

The pigment resin dissolves in ethanol and diethyl ether to give yellow solutions that turn deep red on addition of alkali. In the past solutions of G. were used in the paint industry as yellow spirit varnish.
Use: A strong laxative used only in veterinary medicine with no major significance today.
Lit.: Magn. Res. Chem. **31**, 340 (1993) ▪ Tetrahedron **21**, 1453 (1965); **23**, 4201 (1967). – *[HS 1301 90; CAS 2752-65-0 (α-gambogic acid)]*

***α*-Gambogic acid** see gamboge.

Ganoderic acids. The polypore *Ganoderma lucidum* („Ling Chih", Aphyllophorales) plays an important role in traditional Chinese and Japanese medicine as a stimulant for vitality and long life. To date more than 80, mostly highly oxidized, triterpenoids of the lanostane type and their degradation products have been isolated from the fruitbodies and mycelium cultures of this fungus, among them 45 G. The compounds have been named by several groups and their nomenclature is rather confusing. Thus, various ganoderic acids, ganoderenic acids, ganodermic acids, ganoderals, ganoderiols, ganolucidic acids, lucidic acids, etc. have been described. The bitter *G. A* {$C_{30}H_{44}O_7$, M_R 516.68, amorphous powder, $[\alpha]_D$ +153.8° ($CHCl_3$); methyl ester: cryst., mp. 196–197 °C} and *G. T* {$C_{36}H_{52}O_8$, M_R 612.80, needles, mp. 209–210 °C, $[\alpha]_D$ +23°} are mentioned as examples. Some of these *Ganoderma* compounds not only inhibit histamine release in rat mast cells but also angiotensin converting enzyme (ACE). Furthermore, cytotoxic activities against hepat-

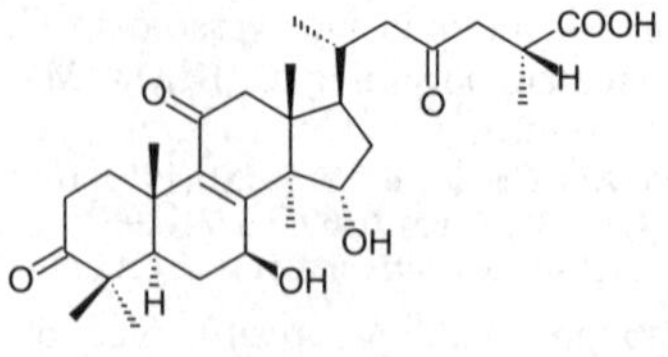

Ganoderic acid A (fruit bodies)

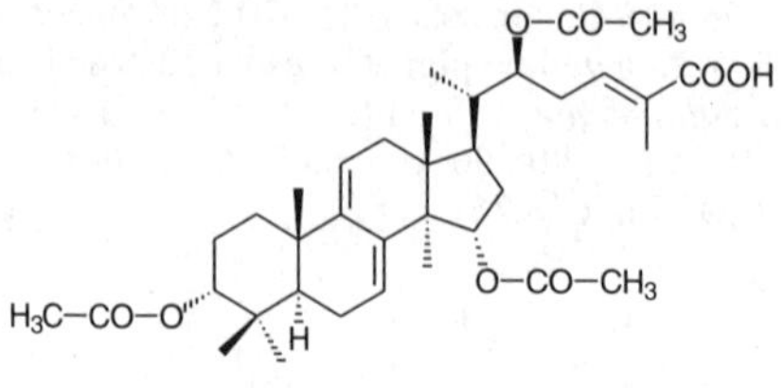

Ganoderic acid T (mycelium)

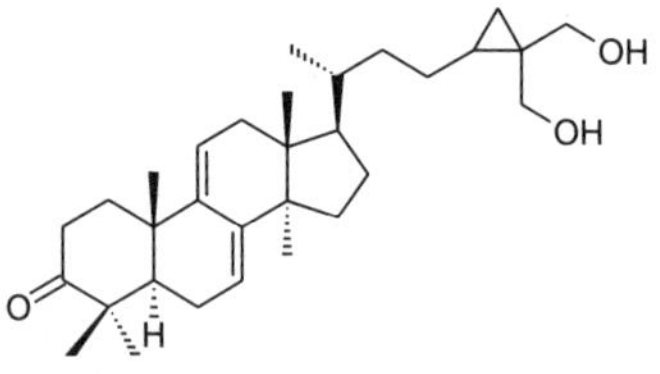

Ganoderiol F

oma cells have been demonstrated *in vitro* [1]. The cyclopropane *ganoderiol F* is active against HIV (inhibits protease) [2].

Lit.: [1] Phytother. Res. **1**, 17 (1987); Planta Med. **55**, 385 (1989). [2] Phytochemistry **49**, 1651 (1998).
gen.: Biological activity: Chem. Pharm. Bull. **33**, 1367 (1985); **34**, 2282 (1986) ▪ J. Nat. Prod. **51**, 918 (1988) ▪ Tetrahedron Lett. **24**, 1081 (1983). – *Biosynthesis & distribution:* J. Chem. Soc., Perkin Trans. 1 **1990**, 2751 ▪ Phytochemistry **29**, 3767 (1990); **32**, 766 (1993); **35**, 1305 (1994). – *Review:* J. Chin. Chem. Soc. (Taipei) **39**, 669 (1992) ▪ Nat. Prod. Rep. **1**, 53 (1984); **3**, 421 (1986); **6**, 476–478 (1989); **11**, 93–95 (1994). – *[CAS 81907-62-2 (G. A); 103992-91-2 (G. T)]*

Garcifurans. Constituents of the roots of *Garcinia kola* (Guttiferae), a traditional herbal medicine from southeast Asia; e.g., *G. A* (garcinol): $C_{16}H_{14}O_5$, M_R 286.28, liquid; *G. B*: $C_{15}H_{12}O_4$, M_R 256.25, oil.

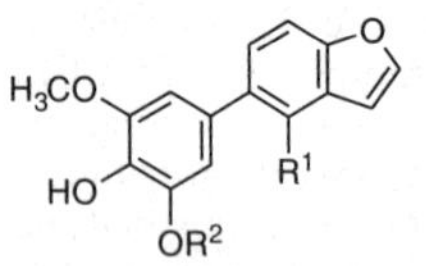

R^1 = OH, R^2 = CH_3 : G. A
R^1 = H, R^2 = H G. B

Lit.: Heterocycles **38**, 1071, 1927 (1994) (G. A) ▪ J. Org. Chem. **62**, 428 (1997) (G. B). – *[CAS 149064-55-1 (G. A); 156162-11-7 (G. B)]*

Garcinol.

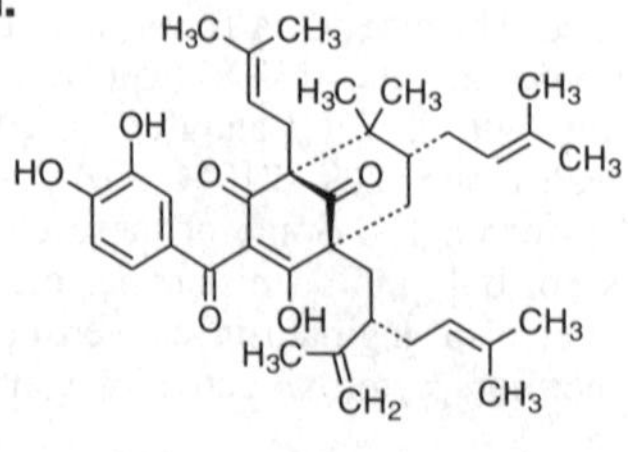

$C_{38}H_{50}O_6$, M_R 602.78, pale yellow needles, mp. 123–124 °C, $[\alpha]_D^{24}$ –138° (CHCl₃). A polyprenylated benzophenone derivative from *Garcinia purpurea* (Guttiferae) with remarkable antibacterial activity against methicillin-resistant *Staphylococcus aureus* (*in vitro* comparable with *vancomycin).

Lit.: Biol. Pharm. Bull. **19**, 311–314 (1996) ▪ Phytochemistry **28**, 1233 (1989).

Gardenin see crocin.

[Common] Garden spider. *Araneus diadematus*, a harmless orbitelous spider, has a white cross formed of guanine crystals on its back. These crystals are enclosed in guanocytes which are visible through the cuticule; see also spider venoms.

Lit.: Foelix, Biology of Spiders, Oxford: Oxford University Press 1997.

Garlic constituents see allium.

Garryine. $C_{22}H_{33}NO_2$, M_R 343.51, mp. 74–82 °C, $[\alpha]_D$ –84° (C_2H_5OH). A diterpene alkaloid from the bark of *Garrya veatchii* (Garryaceae).

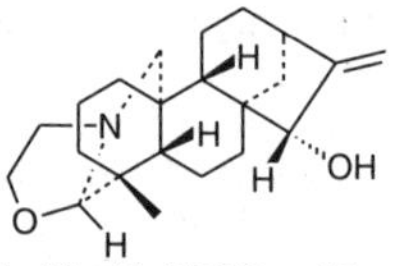

Lit.: J. Nat. Prod. **43**, 41 (1980) ▪ Manske **18**, 99–216. – *[CAS 561-51-3]*

Gaultherin see primeverose.

Geiparvarin.

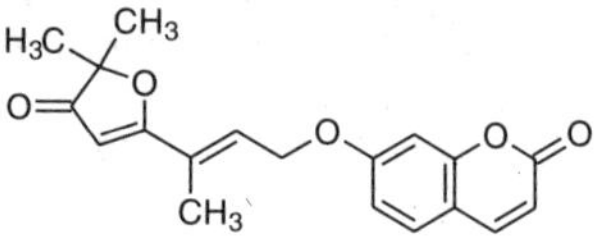

$C_{19}H_{18}O_5$, M_R 326.35, crystals, mp. 157–158 °C (*E*-form). Antitumor principle from the leaves of *Geijera parviflora* (Rutaceae). A dihydro form with a hydrogenated furanone ring, $C_{19}H_{20}O_5$, M_R 328.36, also isolated from *G. parviflora* and possessing cytostatic activity is known as well [1].

Biosynthesis: 7-Hydroxycoumarin (*umbelliferone) is formed on the shikimic acid pathway and *geraniol is the precursor of the side chain.

Lit.: [1] Phytochemistry **11**, 763–767 (1972).
gen.: Beilstein E V **18/1**, 409 ▪ Fitoterapia **68**, 115 (1997) (review) ▪ Helv. Chim. Acta **82**, 191 (1999) (synthesis) ▪ J. Med. Chem. **32**, 284–288 (1989) ▪ J. Org. Chem. **50**, 3997–4005 (1985) ▪ Tetrahedron **44**, 1267–1272 (1988) ▪ Tetrahedron Lett. **32**, 683–686 (1991); **38**, 537 (1997) (synthesis). – *[CAS 36413-91-9 (G.); 74796-44-4 (Z-form); 16850-99-0 ((±)-dihydro-form); 36413-97-5 ((+)-dihydro-form)]*

Geissoschizine.

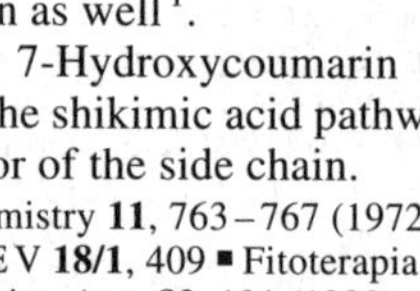

$C_{21}H_{24}N_2O_3$, M_R 352.43, crystals, mp. 194–196 °C, $[\alpha]_D$ +115° (C_2H_5OH), soluble in alcohol, chloroform, ether, practically insoluble in water. An indole or, respectively, *quinolizidine alkaloid from *Rhazya*

stricta, *Tabernaemontana siphilitica* (=*Bonafousia tetrastachya*), and *Rauvolfia volkensii* (Apocynaceae). The methyl ether of G. ($C_{22}H_{26}N_2O_3$, M_R 366.46, prisms, mp. 190–192 °C) is isolated from *Uncaria rhynchophylla*. 19*Z*- and 19*E*-isomers of G. have been synthesized[1]. G. or 4,21-dehydrogeissoschizine, respectively, which represents the precursor of *cathenamine, plays a key role in the biosynthesis of *monoterpenoid indole alkaloids[2].

Lit.: [1] Justus Liebigs Ann. Chem. **1986**, 1262; Org. Lett. **1**, 79 (1999); Tetrahedron **52**, 6803–6810 (1996); Tetrahedron Lett. **37**, 9105 (1996); **38**, 5307 (1997). [2] J. Chem. Soc., Chem. Commun. **1979**, 1016.

gen.: Angew. Chem. Int. Ed. Engl. **18**, 862 (1979) ▪ Beilstein EV **25/8**, 107 ▪ Chem. Ber. **109**, 3825, 3837 (1976) ▪ J. Am. Chem. Soc. **110**, 5925 (1988); **111**, 300 (1989) ▪ J. Org. Chem. **54**, 1166 (1989) ▪ Justus Liebigs Ann. Chem. **1985**, 175. – *[HS 2939 90; CAS 18397-07-4]*

Gelsemine.

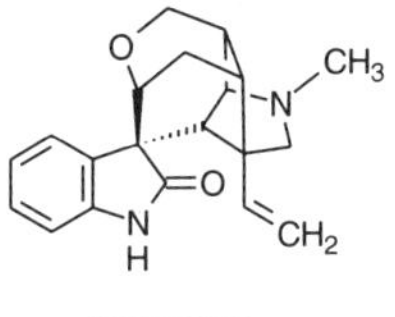

Gelsemine

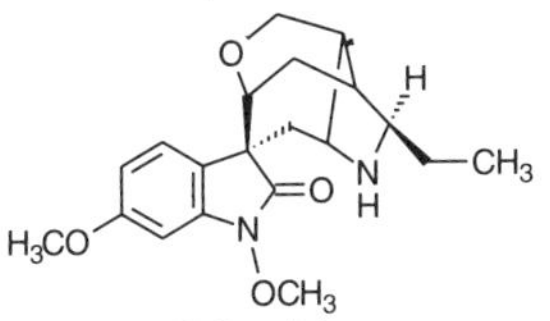

Gelsemicine

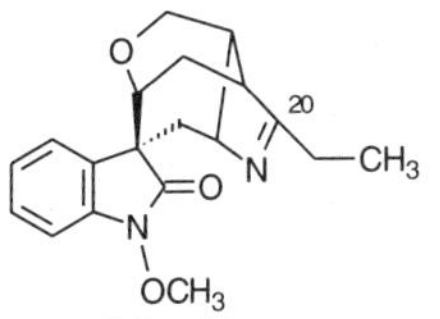

Gelsenicine

$C_{20}H_{22}N_2O_2$, M_R 322.41, crystals, mp. 178 °C, $[\alpha]_D$ +13° ($CHCl_3$), as hydrochloride +5° (H_2O); soluble in organic solvents and dilute acids. A polycyclic *indole alkaloid from the roots of *Gelsemium* species (*G. sempervirens* – syn. wild jasmine), a North American climbing shrub that also occurs in the highlands of Mexico and Guatemala. Other *Gelsemium* species such as *G. elegans* are indigenous to South East Asia; G. is also contained in *Mostuea stimulans* (all Loganiaceae). G. is the toxic major alkaloid and is accompanied by other toxic *Gelsemium* alkaloids such as gelsemicine, gelsenicine, gelsiverine, gelsedine, koumine, koumidine, taberpsychine and the indole-quinolizidine *sempervirine. *Gelsemium* plants are used in China, Vietnam, Borneo as a suicide agent, in Burma as a fish poison, in Chinese traditional medicine as an analgesic, spasmolytic, and for treatment of skin lesions[1]. *G. sempervirens* is responsible for spontaneous abortions and lethal poisonings among grazing animals; the drug is also used as treatment for neuralgias and migraines[2] as well as asthma[3].

Activity: At first a stimulating and then a subduing effect on the CNS, the analgesic/antispastic activity is typical.

Toxicology: The toxic effects depend on the mode of administration, in addition the effect on humans varies widely from person to person, as was shown by a historical epidemic poisoning by G. due to spirits contaminated with *Gelsemium* bark. Symptoms of a G. poisoning are loss of appetite, abdominal pain, internal hemorrhages, muscular weakness, paralysis of the limbs, respiratory distress, and respiratory paralysis; the circulation is impaired (cardiac arrhythmia), blood pressure drops, dilation of pupils, lowered body temperature; death through respiratory paralysis, the subject remains conscious for a long time; the pattern of intoxication is similar to that of botulism (see botulism toxin)[4]. LD_{50} (mouse i. v.) 133 mg/kg, (p. o.) 1.2 g/kg. *Gelsenicine*, $C_{19}H_{22}N_2O_3$, M_R 326.40 (mp. 169–171 °C) is the most toxic alkaloid of *G. elegans* with LD_{50} (mouse i. p.) 0.18 mg/kg[4]. *Gelsemicine* {$C_{20}H_{26}N_2O_4$, M_R 358.44, mp. 161–164 °C, $[\alpha]_D^{26}$ –149° ($CHCl_3$)} MLD (rabbit i. v.) ~0.05 mg/kg, for comparison, gelsemine 180 mg/kg. G. has antagonistic activity against *pilocarpine, *physostigmine (eserine), but is not antagonistic for *atropine, *adrenaline.

Biosynthesis: As *monoterpenoid indole alkaloid from *strictosidine.

Lit.: [1] J. Nat. Prod. **52**, 588–594 (1989). [2] Manske **7**, 93–117. [3] Leyel (ed.), Modern Herbal, vol. I, p. 345 ff., London: Jonathan Cape 1975. [4] Acta Chim. Sinica **40**, 1137–1141 (1982).

gen.: Synthesis: Alvarez & Joule, in Saxton (ed.), The Monoterpenoid Indole Alkaloids, p. 236–259, Suppl. to vol. 25, Heterocyclic Compounds, Chichester: Wiley 1994 ▪ Angew. Chem., Int. Ed. Engl. **38**, 2214, 2934 (1999) (gelsedine) ▪ J. Am. Chem. Soc. **116**, 6943 f. (1994); **118**, 7426 f. (1996); **119**, 6226 (1997) ▪ J. Chem. Soc., Chem. Commun. **1994**, 765 f., 767 f. ▪ Manske **33**, 83–140; **49**, 1–78 ▪ Pure Appl. Chem. **69**, 501 (1997). – *Toxicology, pharmacology:* Hager (5.) **8**, 330 ff. ▪ Sax (8.), No. GCK 000. – *[HS 2939 90; CAS 509-15-9 (G.); 6887-28-1 (gelsemicine); 82354-38-9 (gelsenicine)]*

Genipin. G. is a C_{10}-*iridoid-carboxylic acid methyl ester occurring in nature in the free form (*genipin*) and as its glucoside (*geniposide*). *Geniposidic acid* is the saponification product of geniposide.

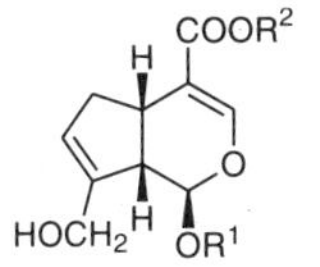

R^1 = H, R^2 = CH_3 : Genipin
R^1 = Gluc, R^2 = CH_3 : Geniposide
R^1 = Gluc, R^2 = H : Geniposidic acid

Table: Genipin and derivatives.

trivial name	Genipin	Geniposide	Geniposidic acid
molecular formula	$C_{11}H_{14}O_5$	$C_{17}H_{24}O_{10}$	$C_{16}H_{22}O_{10}$
M_R	226.23	388.37	374.34
mp. [°C]	120–121	163–164	amorphous
$[\alpha]_D$	+135° (CH_3OH)	+7.5° (C_2H_5OH)	+19.3° (CH_3OH)
CAS	6902-77-8	24512-63-8	27741-01-1

Seed extracts of the Mexican tree *Genipa americana* (Rubiaceae), which is purported to show antibiotic activity[1], contain G.[2] and geniposidic acid[3] together with caffeine[4]. Geniposide is isolated from the fruits of *Gardenia jasminoides* (Rubiaceae)[5]. An inhibitory action of geniposide on benzo[*a*]pyrene/TPA-induced skin tumors (see tigliane) has been demonstrated[6]. G. derivatives show activity against Hepatitis B virus[7].

Lit.: [1]Tetrahedron **20**, 1781 (1964). [2]J. Nat. Prod. **59**, 1169 (1996); Phytochemistry **25**, 2515 (1986); Tetrahedron Lett. **1978**, 3803. [3]Tetrahedron Lett. **1972**, 5125. [4]Hegnauer **VI**, 171. [5]Tetrahedron Lett. **1969**, 2347. [6]Anticancer Res. **15** (2), 411–416 (1995). [7]Pat. WO 9817663 (published 30.4.1998). – *[HS 2938 90]*

Genistein see isoflavones and sparteine.

Gentianin (*p*-coumaric acid ester of *delphinidin 3-*O*-glucoside). $C_{30}H_{27}O_{14}^+$, M_R 611.54, blue-violet powder. G. occurs in the flowers of blue gentian (*Gentiana verna*). The major representative of the *Gentiana* alkaloids, a monoterpenoid iridoid, also has the name gentianine.
Lit.: Karrer, No. 1737.

Gentianose (β-D-fructofuranosyl-6-*O*-β-D-glucopyranosyl-α-D-glucopyranoside).

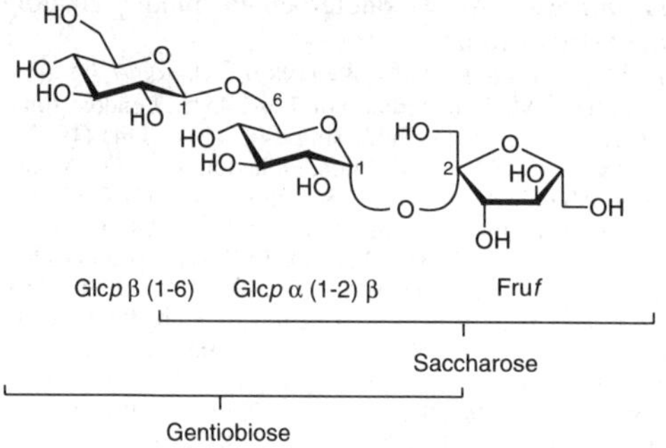

$C_{18}H_{32}O_{16}$, M_R 504.44, mp. 209–211 °C, $[\alpha]_D^{20}$ +33° (H_2O). G. is a non-reducing trisaccharide present in the roots of gentian.
Lit.: Carbohydr. Res. **21**, 451 ff. (1972). – *[CAS 25954-44-3]*

Gentiobiose (6-*O*-β-D-glucopyranosyl-D-glucose). Formula, see gentianose. $C_{12}H_{22}O_{11}$, M_R 342.30, bitter-tasting, very hygroscopic crystals, mp. of α-pyranose form 189–195 °C, mp. of β-pyranose form 190–195 °C, $[\alpha]_D^{20}$ −0.8→+1° (H_2O). G. is a component of *gentianose, α-crocin (see crocetin), and amygdalin (see cyanogenic glycosides), from which it can be liberated by acid hydrolysis or treatment with invertase.
Lit.: Am. Biotechnol. Lab. **5**, 38, 40 ff., 44, 48 ff. (1987) (chromatography) ▪ Beilstein E V **17/7**, 203 ▪ Lett. Bot. **1987**, 359–363 (isolation). – *Synthesis:* Can. J. Chem. **53**, 1970 (1975). – *[HS 2940 00; CAS 554-91-6 (G.); 5995-99-3 (α-pyranose form); 5996-00-9 (β-pyranose form)]*

Gentiopicroside (gentiopicrin).

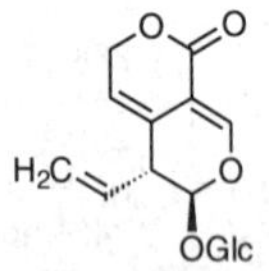

$C_{16}H_{20}O_9$, M_R 356.32, needles, mp. 191 °C (anhydrous), as hydrate 121 °C, soluble in water and alcohols, $[\alpha]_D$ −196° (H_2O). G. is an extremely bitter component of gentian species (Gentianaceae) and centaury (*Centaurium*).
Lit.: Acta Crystallogr., Sect. C **41**, 798 (1985) ▪ Beilstein E V **19/5**, 512 ▪ Karrer, No. 1127 ▪ Phytochemistry **23**, 908 (1984); **45**, 1483 (1997) (NMR) ▪ Zechmeister **17**, 124–182, especially 129–132. – *[CAS 20831-76-9]*

Gentisein (1,3,7-trihydroxyxanthone).

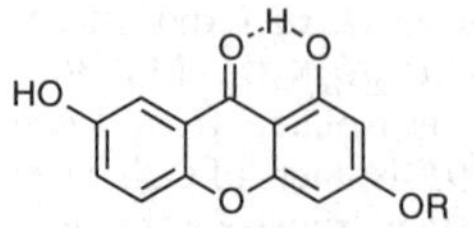

R = H : Gentisein
R = CH_3 : Gentisin

$C_{13}H_8O_5$, M_R 244.20, orange needles, mp. 321–323 °C, occurs both in the free form and in methylated and/or glycosylated forms in yellow gentian (*Gentiana lutea*), *Chironia krebsii*, *Swertia petiolata*, and *Polygala tenuifolia*. The roots of yellow gentian (gentian root) are used as a bitter cordial and sedative. The pharmacologically active component is the 3-*O*-primveroside (=6-*O*-β-D-xylopyranosyl-β-D-glucopyranoside) of 7-*O*-methyl-G. (*gentioside*).
Lit.: Arch. Pharm. (Weinheim, Ger.) **323**, 163–169 (1990) (synthesis) ▪ Luckner (3.), p. 401 ▪ Phytochemistry **44**, 191 (1997) (review) ▪ Synth. Commun. **19**, 1641, 1647 (1989) (synthesis, NMR) ▪ Tetrahedron **35**, 2035–2038 (1979) (^{13}C NMR). – *[HS 2932 90; CAS 529-49-7]*

Gentisic acid (2,5-dihydroxybenzoic acid, 2,5-DHBA).

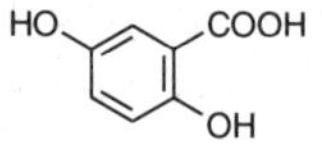

$C_7H_6O_4$, M_R 154.12, needles or prisms, mp. 205 °C (sublimes at 200 °C), soluble in water, insoluble in chloroform. G. is produced by *Penicillium* spp. and occurs in the form of ethers and esters in many higher plants, including gentian (*Gentiana* sp.). G. often occurs as glycosides: *G. 5-O-glucoside*, $C_{13}H_{16}O_9$, M_R 316.26, $[\alpha]_D$ −53.2° (CH_3OH); *trichocarpin* (benzyl ester, 2-*O*-β-D-glucopyranoside), $C_{20}H_{22}O_9$, M_R 406.39, mp. 134–136 °C, $[\alpha]_D^{20}$ −46.3° (CH_3OH); see also primeverin.
Activity: The antipyretic activity of G. is lower than that of *salicylic acid; like the latter it inhibits prostaglandin biosynthesis and was used in the past as an antirheumatic drug.
Biosynthesis: In *Penicillium urticae* from 6-methylsalicylic acid → *m*-cresol → *m*-hydroxybenzyl alcohol → gentisic acid. In plants possibly by hydroxylation of benzoic acid.
Lit.: Appl. Environm. Microbiol. **45**, 1939 (1982) (biosynthesis) ▪ Beilstein E IV **10**, 1441 ▪ Chem. Pharm. Bull. **33**, 527 (1985); **41**, 1491 (1993) (glucosides) ▪ J. Med. Chem. **31**, 1039–1043 (1988) ▪ Luckner (3.), p. 397–398 ▪ Nat. Prod. Rep. **1**, 281–297 (1984) ▪ Negwer (6.), No. 647 ▪ Phytochemistry **26**, 1147–1152 (1987). – *Toxicology:* Sax (8.), GCU 000. – *[HS 2918 29; CAS 490-79-9 (G.); 1820-89-9 (G. 5-O-glucoside); 10590-85-9 (trichocarpin)]*

Gentisin (gentianin, gentiin, 1,7-dihydroxy-3-methoxyxanthone). Formula, see gentisein. $C_{14}H_{10}O_5$, M_R 258.23, yellow needles, mp. 273–275 °C, uv_{max} 410 nm (CH_3OH). Pigment – together with other xanthones – from the root of yellow gentian (*Gentiana lutea*); in combination with alum mordant, G. dyes wool bright yellow.
Lit.: Beilstein E V **18/4**, 497 ▪ Hager (5.) **5**, 243 f. ▪ Karrer, No. 1675, 1675b ▪ Phytochemistry **44**, 191 (1997) (review). – *[CAS 437-50-3.]*

Geogenin see cortisalin.

Geoporphyrins (petroporphyrins). To date about 80 different *porphyrins have been identified in sediments and crude oil.

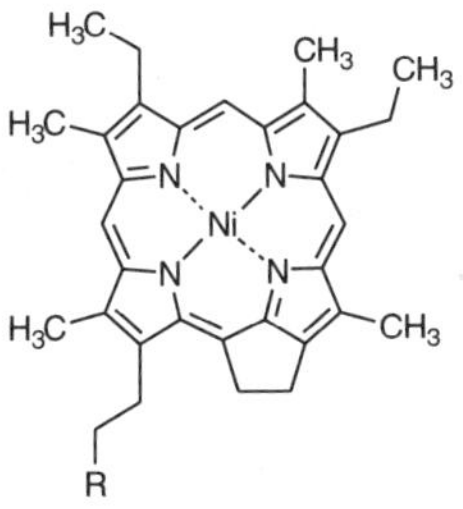

a : R = H
b : R = COOH

The most frequent members of the G. are *nickel deoxophylloerythroetioporphyrin* (formula a) and *nickel deoxophylloerythrin* (formula b); also the corresponding vanadylporphyrins are widely distributed in sediments. In addition, metal-free porphyrins have been detected. The substitution pattern of the fossil porphyrins generally allows conclusions to be drawn about the biological origin of the structures and is thus an important tool for evolution research and the organic geochemistry subsection of geology.

Lit.: ACS Symp. Ser. (Met. Complexes Fossil Fuels) **344**, 173 (1987) ▪ Aust. J. Chem. **48**, 1873–1885 (1995) ▪ Dolphin **III**, 451 f. ▪ Energy and Fuels **4**, 635 (1990) ▪ Smith, p. 172 f., 785.

Geosmin.

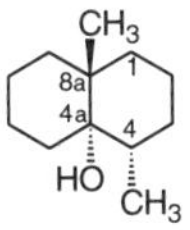

$C_{12}H_{22}O$, M_R 182.31, oil with an earthy odor, bp. 270 °C, $[\alpha]_D^{25}$ −16.5°, decomposes on acid treatment to odorless components. G. is produced by various *Streptomyces* species and myxobacteria and has also been detected in beet juice [1] and beetroots. G. is responsible for the characteristic (earthy-musty) odor (olfactory threshold 0.1 ppb) of freshly plowed soil and can be experienced sensorily at even higher dilutions in soil or water. G. is probably formed biogenetically from a sesquiterpene after prior cleavage of an isopropyl group [2]. For synthesis, see *Lit.* [3].

Lit.: [1] J. Agric. Food Chem. **26**, 1466 (1978). [2] Helv. Chim. Acta **76**, 2547 (1993). [3] Acta Chem. Scand. Sect. B **44**, 1036 (1990); **46**, 103 (1992); Helv. Chim. Acta **76**, 1949 (1993); Tetrahedron **48**, 5497 (1992); Tetrahedron: Asymmetry **7**, 2923 (1996).
gen.: Chem. Ind. (London) **1975**, 973 ▪ Merck-Index (12.), No. 4408. – *[CAS 19700-21-1]*

Gephyrotoxin.

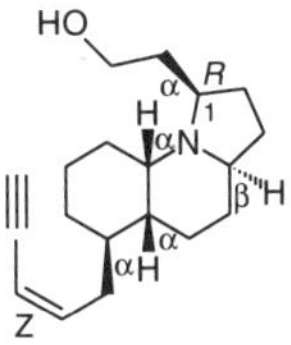

$C_{19}H_{29}NO$, M_R 287.45, mp. 230–232 °C (decomp.), $[\alpha]_D$ −51.5° (C_2H_5OH). G. (287C) and its dienyl isomer dihydro-G. (289B) occur in only a few populations of the South American arrow poison frog *Dendrobates histrionicus*. In comparison with other *Dendrobates alkaloids it has a relatively low toxicity [1]. Like many other lipophilic frog alkaloids, it non-competitively blocks nicotinic receptors [2].

Lit.: [1] Manske **43**, 185–288. [2] Neurochem. Res. **16**, 489 (1991).
gen.: Beilstein E V **21/3**, 202 ▪ Pelletier **4**, 1–274 ▪ Zechmeister **41**, 283–299. – *Synthesis:* Tetrahedron **39**, 49 (1983) ▪ Tetrahedron Lett. **22**, 4197 f. (1981); **30**, 6661 (1989). – *[HS 2939 90; CAS 55893-12-4]*

Geranial see citrals.

Geranic acid (3,7-dimethyl-2,6-octadienoic acid). Formula, see geraniol. $C_{10}H_{16}O_2$, M_R 186.23. An unsaturated, acyclic *monoterpene carboxylic acid occurring in the (2*E*)- (G.) and (2*Z*)-configurations (*nerolic acid*): G., bp. 85–87 °C (3 Pa); nerolic acid, bp. 90 °C (10 Pa). G. and nerolic acid occur in *lemongrass oil (*Cymbopogon flexuosus*, Poaceae) produced in Taiwan.

Lit.: Bee World **65**, 175 (1984) ▪ Beilstein E IV **2**, 1734 ▪ Food Cosmet. Toxicol. **17**, 785 (1979) ▪ J. Chem. Soc., Perkin Trans 1 **1975**, 897 ▪ Karrer, No. 750 ▪ Org. Prep. Proced. Int. **30**, 79 (1998) (synthesis). – *[CAS 4698-08-2 (G.); 4613-38-1 (nerolic acid)]*

Geraniol [(2*E*)-3,7-dimethylocta-2,6-dien-1-ol].

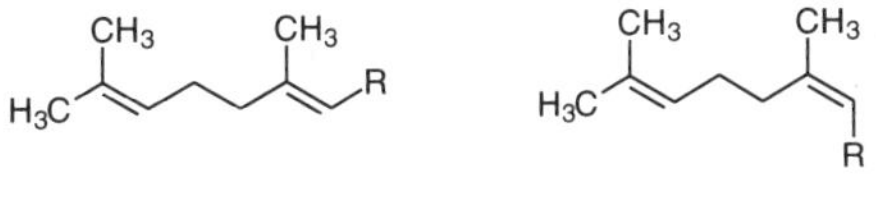

R = CH_2OH : Geraniol	Nerol
R = CHO : Geranial = (2*E*)-*Citral	Neral = (2*Z*)-Citral
R = COOH : *Geranic acid	Nerolic acid

$C_{10}H_{18}O$, M_R 154.24, liquid with a pleasant, flower-like odor. Unsaturated, acyclic *monoterpene alcohols [(2*E*)-form: G.; (2*Z*)-form: *nerol*]. G., bp. 230 °C; nerol, bp. 224–225 °C. G. and nerol undergo slowly changes on exposure to atmospheric oxygen. Geranyl pyrophosphate is an important intermediate in the biosynthesis of *terpenes and *steroids.

Occurrence: G. and nerol occur in nature in the free form and as esters, G. in *palmarosa oil (up to 85%) and *geranium oil (40–50%), nerol in the essential oil of strawflower, *Helichrysum italicum* (= *H. angustifolium*, Asteraceae) 30–50%.

For biosynthesis see *Lit.* [1], synthesis *Lit.* [2]. G. and nerol are used in the perfume, cosmetic, and luxury food industries.

Lit.: [1] Tetrahedron Lett. **38**, 3889 (1997). [2] J. Chem. Soc., Perkin Trans. 1 **1990**, 2715.
gen.: Beilstein E IV **1**, 2277 ▪ Karrer, No. 118, 119 ▪ Merck-Index (12.), No. 4411 ▪ Ullmann (5.) **A 11**, 154. – *[HS 2905 22; CAS 106-24-1 (G.); 106-25-2 (nerol)]*

Geranium oil. Yellowish-green to amber oil with a leafy-green, sweet rose-like odor with pronounced fruity-minty undertones.

Production: By steam distillation from the flowering herb *Pelargonium graveolens*. Main areas of cultivation are China, Egypt, and Réunion. Smaller amounts are produced in India and Israel. In 1990, the yearly production world-wide was ca. 250 t.

Composition [1]***:*** Depending on the geographical location, there are quantitiative differences in the composition. Components: **rose oxide*, **linalool*, *menthone/isomenthone* (see *p*-menthanones), **citronellol*, **geraniol*, *citronellyl formate* ($C_{11}H_{20}O_2$, M_R 184.28), and *geranyl formate* ($C_{11}H_{18}O_2$, M_R 182.26). These components are also mainly responsible for the odor. Criteria for differentiating between the oils from China and Réunion ("bourbon" type) and those of the so-called "African" type are the components *(−)-6,9-guaiadiene* ($C_{15}H_{24}$, M_R 204.36) and *10-epi-γ-eudesmol* ($C_{15}H_{26}O$, M_R 222.37). The former compound is characteristic of the bourbon type and the latter is only found in African oils.

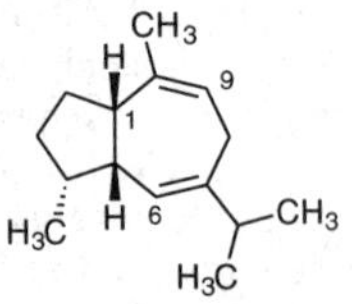

(-)-6,9-Guaiadiene

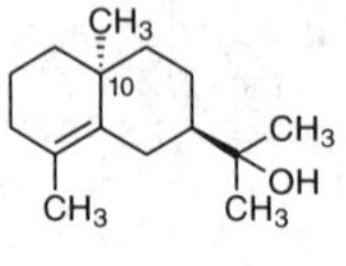

10-*epi*-γ-Eudesmol

Use: For the production of perfumes; G. is a natural odor building block with a wide scope of use.

Lit.: [1] Perfum. Flavor. **13** (5), 65 (1988); **17** (2), 46; (6), 59 (1992); **19** (1), 40 (1994).
gen.: Arctander, p. 262 ▪ Bauer et al. (2.), p. 156 ▪ ISO 4731 (1978) ▪ Ohloff, p. 157. – *Toxicology:* Food Cosmet. Toxicol. **12**, 883 (1974); **13**, 451 (1975). – *Eudesmol:* Tetrahedron **50**, 4755 (1994) (synthesis) ▪ Phytochemistry **47**, 1085 (1998) (spectroscopy). – *[HS 3301 21; CAS 8000-46-2 (G.); 105-85-1 (citronellyl formate); 105-86-2 (geranyl formate); 36577-33-0 (6,9-guaiadiene); 15051-81-7 (10-epi-γ-eudesmol)]*

Geranylgeraniol [(*E,E,E*)-3,7,11,15-tetramethyl-2,6,10,14-hexadecatetraen-1-ol].

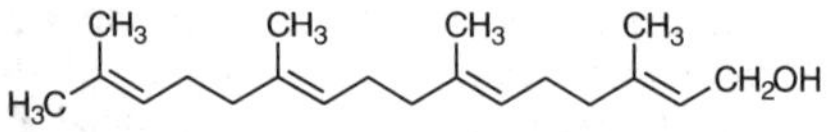

$C_{20}H_{34}O$, M_R 290.49, oil, bp. 152–153 °C (9.3 Pa); D_4^{18} 0.8930. G. is the biogenetic precursor of all *diterpenes and the *carotinoids, it is formed biosynthetically by the addition of two isoprene units to geraniol. G. also occurs as the side chain (sometimes hydrogenated) of *ubiquinones, *plastoquinones, rhodoquinones, in *boviquinone-4, *suillin, and the K vitamins, etc.

Lit.: Beilstein E IV **1**, 2351 ▪ J. Chem. Ecol. **14**, 1153 (1988) ▪ Karrer, No. 5927 ▪ Synthesis **1997**, 202 ▪ Tetrahedron Lett. **36**, 5669 (1995) (synthesis); **39**, 2033 (1998) (biosynthesis). – *[CAS 24034-73-9]*

Germacranes.

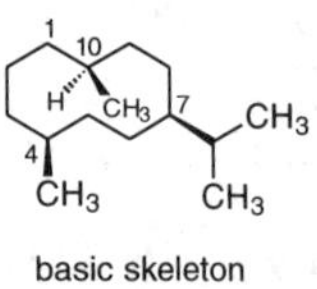

basic skeleton

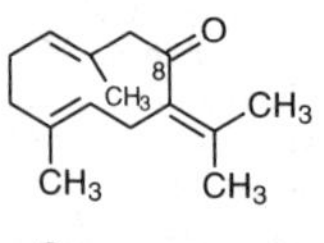

Germacrone

Name for a group of about 100 *sesquiterpenes with the 4,10-dimethyl-7-isopropylcyclodecane skeleton. Under the influence of oxygen or light the cyclodecadiene system of the germacrenes can undergo various Cope rearrangements to furnish other mono-, bi- and tricyclic sesquiterpene skeletons (*elemenes, *eudesmanes or *guaianes, copaenes)[1]. Most G. are lactones (see germacranolides), but also ketones (*germacrone*, $C_{15}H_{22}O$, M_R 218.34, crystal, mp. 56 °C)[2] and hydrocarbons (see germacrenes) have been found in nature. Thus, e. g., *periplanone B is the sex pheromone of the cockroach *Periplaneta americana*. The biosynthesis proceeds through cyclization of farnesyl diphosphate.

Lit.: [1] Angew. Chem. Int. Ed. Engl. **12**, 793–806 (1973); Scheuer I **5**, 179 ff. [2] Tetrahedron Lett. **24**, 3489 (1983); Phytochemistry **26**, 312 (1987).
gen.: Tetrahedron **55**, 2115–2146 (1999) (synthesis, review). – *[CAS 6902-91-6 (germacrone)]*

Germacranolides. Bicyclic *sesquiterpene lactones derived from the *germacranes. The lactone ring is usually in the 7β,8α-position as in pyrethrosin[1]; it can also be in the 7β,6α-position but is only rarely seen in the 4,6-position as in linderalactone.

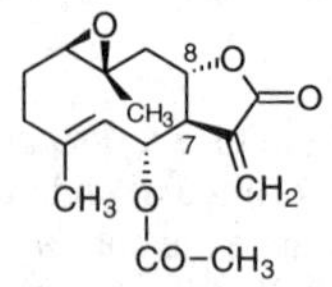

Pyrethrosin

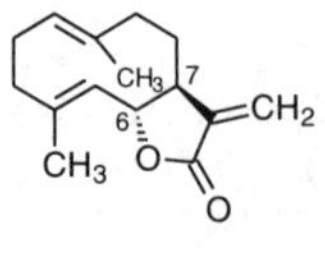

Costunolide

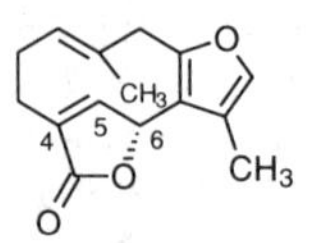

Linderalactone

Table: Data of Germacranolides.

	molecular formula	M_R [°C]	mp.	$[\alpha]_D$	CAS
Pyrethrosin	$C_{17}H_{22}O_5$	306.36	201	−31° (CHCl₃)	28272-18-6
Costunolide	$C_{15}H_{20}O_2$	232.32	106	−128° ($CHCl_3$)	553-21-9
Linderalactone	$C_{15}H_{16}O_3$	244.29	140	−102° (dioxane)	728-61-0

The first germacrane derivative, *pyrethrosin*, was isolated in 1934 from flowers of *Chrysanthemum cinerariaefolium*[2]. Pyrethrosin and *costunolide*[3] are contact allergens and repeated contact can lead to dermatitis. The cause of this is the exocyclic methylene group on the γ-lactone ring which takes part in a *Michael* reaction with nucleophilic groups such as, e. g., the SH groups of proteins and enzymes and alkylates them. The cytotoxic activity also extends to tumor cells[4]. *Linderalactone*[5], a representative of the furanogermacranolides, is isolated from *Lindera strychnifolia*.

Lit.: [1] J. Chem. Soc., Chem. Commun. **1971**, 559; Beilstein E III/IV **19**, 2525. [2] J. Physiol. (London) **8**, 167 (1934). [3] Phytochemistry **39**, 839 (1995); Beilstein E III/IV **17/6**, 5028; J. Nat. Prod. **59**, 1128 (1996). [4] Phytochemistry **15**, 1573 (1976). [5] Acta Crystallogr., Sect. C **49**, 341 (1993); Aust. J. Chem. **1992**, 445; Beilstein E V **19/4**, 416.
gen.: Tetrahedron **49**, 4761 (1993) (synthesis).

Germacrenes (germacratrienes). $C_{15}H_{24}$, M_R 204.36. Isomeric, monocyclic *sesquiterpenes that are difficult to isolate on account of their sensitivity. *G. A* {oil, $[\alpha]_D^{25}$ −3.2° ($CHCl_3$)} is isolated from the sea fan *Eunicea mammosa* and readily undergoes isomerization to β-elemene[1] (see elemenes). It is an alarm pheromone of the aphid *Therioaphis maculata*[2]. The achiral *G. B* is found in the peel of *Citrus junos*[3] and, together with G. A, in larvae of the butterfly *Papilio protenor*[4]. The achiral *G. C* (precursor of δ-elemene) is isolated from

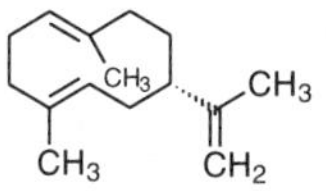

Germacrene A

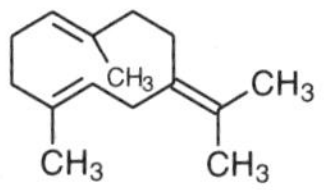

Germacrene B

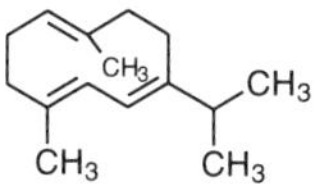

Germacrene C

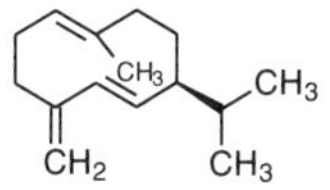

Germacrene D

the dried fruits of *Kadsura japonica* (Schisandraceae)[5]; *G. D* ($[\alpha]_D^{23}$ −240°) occurs in the leaves of the same plant and is also a constituent of the essential oil of *Pseudotsuga japonica*[6].

Lit.: [1]Tetrahedron Lett. **1970**, 497; **1978**, 2903; **26**, 2171 (1985). [2]J. Insect Physiol. **23**, 697 (1977); J. Chem. Ecol. **3**, 349 (1977). [3]Tetrahedron Lett. **1969**, 3097, 1799. [4]J. Insect. Physiol. **26**, 39 (1980). [5]Tetrahedron Lett. **1969**, 1799. [6]J. Chem. Soc., Perkin Trans. 1 **1997**, 2065 (biosynthesis); J. Org. Chem. **49**, 3281 (1984) (synthesis).
gen.: Angew. Chem. Int. Ed. Engl. **37**, 1400–1402 (1998) (biosynthesis); Plant Physiol. **117**, 1381 (1998) (biosynthesis). – *[CAS 28387-44-2 (G. A); 15423-57-1 (G. B); 34323-15-4 (G. C); 23986-74-5 (G. D)]*

Germacrone see germacranes.

Germicidins.

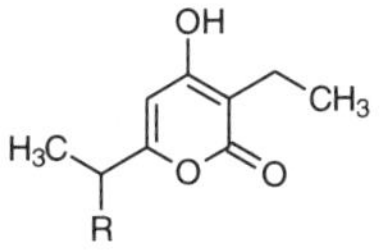

R = CH_3 : G. A
R = C_2H_5 : G. B

Secondary metabolites from *Streptomyces viridochromogenes. G. B* ($C_{11}H_{16}O_3$, M_R 196.11, yellow solid, mp. 80 °C) inhibits the germination of *streptomycetes and plants at low concentrations (40 ng/L). G. B represents the first autoregulative inhibitor of spore germination in streptomycetes. *G. A* ($C_{10}H_{14}O_3$, M_R 182.09, yellow solid, mp. 138 °C) does not possess these activities.

Lit.: J. Antibiot. **46**, 1126–1138 (1993) ▪ Tetrahedron **55**, 4783 (1999) [synthesis (±-G.)]. – *[CAS 151271-57-7 (G. B)]*

Germine, germitrine see Veratrum steroid alkaloids.

Gibberellic acid (gibberellin A_3).

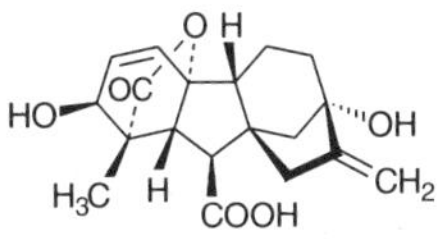

$C_{19}H_{22}O_6$, M_R 346.38, crystals, mp. 233–235 °C, $[\alpha]_D^{19}$ +92° (C_2H_5OH), soluble in methanol, ethanol, acetone, poorly soluble in ether and water. G. is one of the most important and most studied diterpenoids and was originally isolated from the Japanese fungus *Gibberella fujikuroi* that damages rice by causing an excessive growth of young rice shoots. G. and other *gibberellins are responsible for this activity as *plant growth substances (phytohormones). In the meantime G. has been isolated from many other plants in which it acts as a growth regulator. G. is used as an experimental mutagenic agent.

Lit.: Beilstein E V **18/9**, 269 ▪ Phytochemistry **32**, 781 (1993). – *Synthesis:* Lindberg **1**, 22–70 ▪ Merck-Index (12.), No. 4426 ▪ Tetrahedron Lett. **30**, 971–974 (1989) ▪ see also gibberellins. – *[HS 2932 99; CAS 77-06-5]*

Gibberellins.

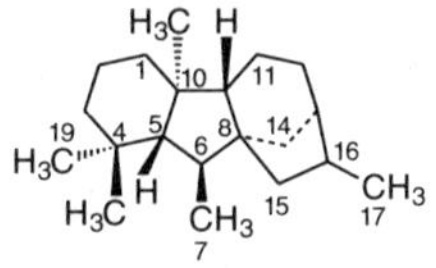

Gibberellane

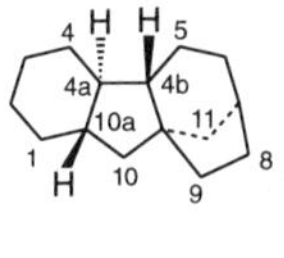

Gibbane

The G. are a group of *plant growth substances (phytohormones). Since 1938 when a G. was isolated for the first time in Japan from the culture filtrate of the Japanese fungus *Gibberella fujikuroi*, over 120 different G. have been identified and labelled with 1,2 or 3-digit numbers, e.g. G. A_1, G. A_{10}, G. A_{108}. Their structures are based on that of *gibberellane* ($C_{20}H_{34}$, 1,1,4α,8,10β-pentamethylgibbane, see formula), a tetracyclic diterpenoid. The biosyntheses of the G. mostly proceed from geranylgeranyl diphosphate via kaurene (see kauranes). Kauranoids occur together with the G. in plants. Kaurene is sequentially oxidized. Ring contraction of ring B gives rise to G. A_{12}-aldehyde, which is then oxidized to the acid; for biosynthesis and metabolism of G., see *Lit.*[1]. The most important G. is *gibberellic acid or G. A_3. G. are found not only in many higher plants but also in ferns, mosses, fungi, algae, and bacteria[2] in concentrations ranging from a few µg to several mg/kg plant material. Similar to the *cytokinins G. influence the growth regulation in plants in many ways.

Activity: The G. influence seed germination[3], longitudinal growth[4] by stimulation of cell growth and division, they can act as flowering hormones, the effects of G. A_4 and G. A_7 appear to be particularly important in this context[5], they also have an effect of the development of fruit[6]. When used in vineyards certain types of grapes become larger. G. inhibit the degradation of *chlorophyll. The effects of G. may be controlled by the use of artificial G. antagonists, e. g., chlorocholine chloride which is known to be an opposing growth regulator for plants[7]. For investigations on G. receptors, see *Lit.*[8], for the commercial use of G., *Lit.*[9]. G. occur not only in the free form but also as glucosides[10]. For analysis of G., see *Lit.*[11]. The effect of herbicides on the formation and metabolism of G. is discussed in *Lit.*[12].

Lit.: [1]Annu. Rev. Plant Physiol. **38**, 419–465 (1987); Sponsel, in Davies (ed.), Plant Hormones, p. 43–75, Dordrecht: Nijhoff 1987. [2]J. Basic Microbiol. **26**, 483–497 (1986). [3]Annu. Rev. Plant Physiol. **36**, 517–568 (1985). [4]Metraux, in Davies (ed.), Plant Hormones and Their Role in Plant Growth and Development, p. 296–317, Dordrecht: Nijhoff 1987. [5]For. Ecol. Manage. **19**, 65–84 (1987); Tetrahedron Lett. **31**, 423 (1990). [6]Sci. Hortic. (Amsterdam) **22**, 1–8 (1984). [7]Naturwiss. Rundsch. **32**, 18–21 (1979). [8]NATO ASI Ser., Ser. H **10**, 199–227, 285–289 (1987). [9]Martin, in Crozier (ed.), Biochem. Physiol. Gibberellins, vol. 2, p. 395–444, New York: Praeger 1983. [10]Schneider, in Crozier (ed.), Biochem. Physiol. Gibberellins, vol. 1, p. 389–456, New York: Praeger 1983. [11]Semin. Ser. – Soc. Exp. Biol. **23**, 1–16 (1984). [12]Wilkinson, in Woodbine (ed.), Antimicrob. Agric., Proc. Int. Symp. Antibiot. Agric.: Benefits Malefits, p. 117–125, London: Butterworth 1984.

gen.: Beilstein EV **18/9**, 67–355 ▪ Turner **1**, 240–248; **2**, 280–288. – *Synthesis:* Aust. J. Chem. **50**, 289 (1997) ▪ Chem. Rev. **92**, 573–612 (1992) ▪ J. Am. Chem. Soc. **119**, 3828 (1997) ▪ J. Chem. Soc., Perkin Trans. 1 **1997**, 751 ▪ Lindberg **1**, 21–70 ▪ Nat. Prod. Rep. **5**, 541–579 (1988); **7**, 41–60 (1990) ▪ Pure Appl. Chem. **70**, 351 (1998) ▪ Synform **2**, 197 (1984) (review) ▪ Synlett **1995**, 105 ff. (G. A_1, A_3) ▪ Tetrahedron Lett. **37**, 719, 723 (1996) ▪ see also cytokinins, plant growth substances, gibberellic acid.

Gigartinine (N^5-guanidinocarbonyl-L-ornithine).

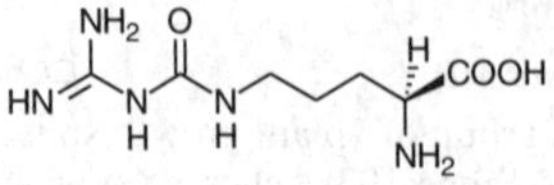

$C_7H_{15}N_5O_3$, M_R 217.23; nitrate, mp. 197 °C, $[\alpha]D$ +7.5° (H_2O). A non-proteinogenic amino acid in *Gymnogongrus flabelliformis* and *Chondrus crispus* (Rhodophyllaceae).

Lit.: Can. J. Biochem. **55**, 27–30 (1977) ▪ Nature (London) **211**, 417 (1966) ▪ Tetrahedron **40**, 235 (1984). – *[CAS 7536-90-5]*

Gilmicolin see mycorrhizin.

Gilvocarcins. *C*-Glycoside antibiotics from cultures of *Streptomyces gilvotanareus* and *S. coerulescens*. G. are active against Gram-positive bacteria and tumor cells, e.g., *G. M* (anandimycin B, toromycin B): $C_{26}H_{26}O_9$, M_R 482.47, yellowish-green crystals, mp. 245–248 °C, $[\alpha]_D^{20}$ –209° (DMSO).

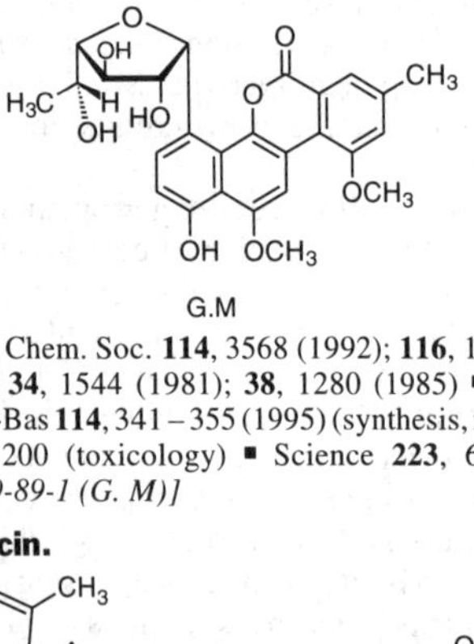

G.M

Lit.: J. Am. Chem. Soc. **114**, 3568 (1992); **116**, 1004 (1994) ▪ J. Antibiot. **34**, 1544 (1981); **38**, 1280 (1985) ▪ Recl. Trav. Chim. Pays-Bas **114**, 341–355 (1995) (synthesis, review) ▪ Sax (8.), GEO 200 (toxicology) ▪ Science **223**, 69 (1984). – *[CAS 77879-89-1 (G. M)]*

Gilvusmycin.

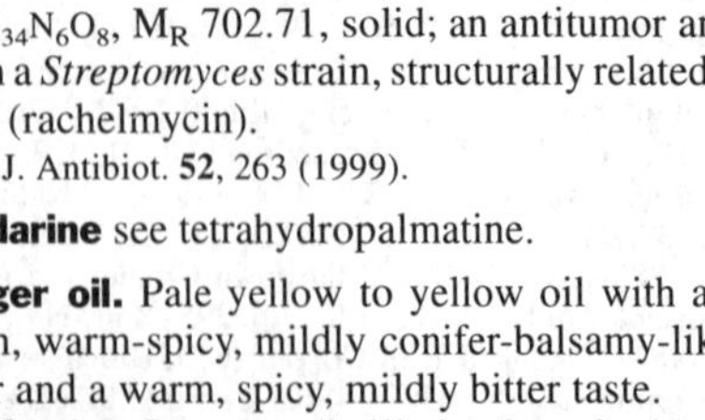

$C_{38}H_{34}N_6O_8$, M_R 702.71, solid; an antitumor antibiotic from a *Streptomyces* strain, structurally related to *CC-1065 (rachelmycin).

Lit.: J. Antibiot. **52**, 263 (1999).

Gindarine see tetrahydropalmatine.

Ginger oil. Pale yellow to yellow oil with a typical fresh, warm-spicy, mildly conifer-balsamy-like citrus odor and a warm, spicy, mildly bitter taste.

Production: By steam distillation from freshly ground, unpeeled, dried ginger roots.

Composition [1]***:*** G. consists mainly of sesquiterpene hydrocarbons: the major component is **zingiberene* (ca. 20–40%); together with *β-*bisabolene, (+)-ar-curcumene* [=(*S*)-1-(1,5-dimethyl-4-hexenyl-4-methylbenzene], *(E,E)-α-*farnesene*, and *β-sesquiphellandrene* ($C_{15}H_{24}$, M_R 204.36), usually in relatively large amounts. The typical citrus odor is due to **citral* (2–15%, depending on origin).

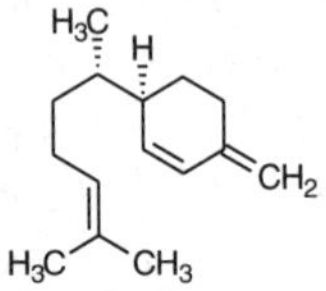

β-Sesquiphellandrene

Use: In small amounts in the perfume industry; mainly to improve the aromatic character of foods, especially refreshing beverages. In large amounts in the food industry as ginger oleoresin, prepared from ginger roots by various extraction procedures. In addition to the volatile components it also contains *gingerol, the hot principle of ginger.

Lit.: [1] Perfum. Flavor. **9** (5), 1 (1984); **13** (4), 69 (1988); **15** (1), 66 (1990); **16** (6), 53 (1991); Dev. Food. Sci. **34**, 579–594 (1994); **20** (2), 54 (1995).

gen.: Arctander, p. 276 ▪ Bauer et al. (2.), p. 157. – *Toxicology:* Food Cosmet. Toxicol. **12**, 901 (1974). – *[HS 3301 29; CAS 8007-08-7 (G.); 20307-83-9 (β-sesquiphellandrene)]*

Gingerol [6-gingerol, 5-hydroxy-1-(4-hydroxy-3-methoxyphenyl)-3-decanone].

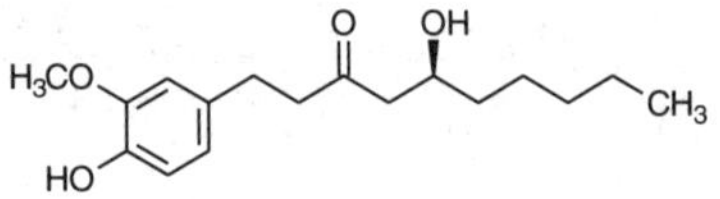

(*S*)-form

$C_{17}H_{26}O_4$, M_R 294.39, (*S*)-form, yellow oil, D_4^{20} 1.0713, $[\alpha]_D$ +26.5° ($CHCl_3$), soluble in organic solvents, the main spice component and peppery constituent of ginger (*Zingiber officinale*). (*R*)-Form, yellow oil, $[\alpha]_D$ –25.1° ($CHCl_3$). The racemate has taste properties similar to those of the (*S*)-enantiomer but lacks the typical ginger odor. *Norgingerol*=*O*-demethylgingerol ($C_{16}H_{24}O_4$, M_R 280.36) acts as an inhibitor of 5-lipoxygenase [1]. G. is used as a flavor substance. For synthesis, see *Lit.* [2].

Biosynthesis: From ferulic acid (see caffeic acid) by chain lengthening with malonyl-CoA and hexanoyl-CoA. *Dehydrogingerdione* is an intermediate.

Lit.: [1] Chem. Pharm. Bull. **38**, 842–844 (1990). [2] Chem. Ber. **112**, 3703–3714 (1979).

gen.: Beilstein EIV **8**, 2768 ▪ Chem. Pharm. Bull. **32**, 1676 (1984) (synthesis) ▪ Enantiomer **3**, 45 (1998) (synthesis) ▪ J. Chem. Soc., Perkin Trans. 1 **1980**, 2637–2644 (biosynthesis) ▪ J. Org. Chem. **58**, 2181 (1993) (synthesis) ▪ Synth. Commun. **29**, 1933 (1999) (synthesis). – *[CAS 39886-76-5 (±); 23513-14-6 (S); 72749-01-0 (R); 122858-51-9 (nor-G.); 61871-71-4 (dehydrogingerdione)]*

Ginkgo extract (EGB). An extract prepared from leaves of the "fossil" tree *Ginkgo biloba* indigenous to China and Japan, containing glucosides of *flavones – in particular quercetin and isoquercetin glycosides, kaempferol 3-rhamnoside, luteolin glycoside, and sitosterol glycoside (see also flavonoids), *bilobalide, and *ginkgolides. G. e. promotes cranial perfusion, increases the blood sugar level, and increases energy metabolism in the brain[1]. In animal experiments G. e. appreciably increased the survival rate of rats suffering from hypoxia. G. e. has a favorable effect on clinical situations such as hypoglucemia and ischemia. This is supported by the vasodilating properties of G. e. In medical practice G. e. is used to treat disorders of cranial and peripheral perfusion. An antitumor effect of G. e. was described by Itokawa et al.[2]. The active principles are alkylated phenols {e.g., *ginkgol*, 3-((Z)-8-pentadecenyl)phenol, $C_{21}H_{34}O$, M_R 302.50} and phenolcarboxylic acids {e.g., *ginkgolic acid, 2-hydroxy-6-((Z)-8-pentadecenyl)benzoic acid*, $C_{22}H_{34}O_3$, M_R 346.51, crystals, mp. 230–233 °C, *ginkgolides, and *bilobalide}. G. e. is commercially available under various names, e.g., as Tebonin® (G. e. sales 1998: 380 Mio. DM) in Germany.

Lit.: [1] Fortschr. Med. **107**, 90 (1989); Life Sci. **39**, 2327–2334 (1986); Münch. Med. Wochenschr. **131**, 10–15 (Suppl.) (1989); Pharmazie **52**, 735 (1997). [2] Chem. Pharm. Bull. **35**, 3016 (1987).
gen.: Anal. Chem. **71**, 2929 (1999) (analytics) ▪ Die Offizin **1989**, 111 ff. ▪ Fitoterapia **69**, 195–244 (1998) (review) ▪ Lancet II **1989**, 1513 ▪ Merck-Index (12.), No. 4432 ▪ Presse Med. **15**, 1455 ff. (1986). – *[HS 1302 19; CAS 501-26-8 (ginkol); 22910-60-7 (ginkgolic acid)]*

Ginkgolides. Hexacyclic *diterpenes that decompose above 300 °C; very resistant to acids but readily attacked by bases.

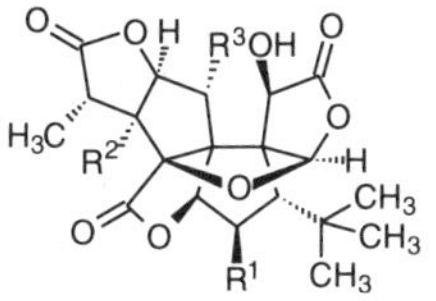

R^1	R^2	R^3	
H	OH	H	G.A
H	OH	OH	G.B
OH	OH	OH	G.C
OH	OH	H	G.J
OH	H	OH	G.M

Table: Data of Ginkgolides.

Ginkgolide	molecular formula	M_R	$[\alpha]_D$ (dioxane)	CAS
A [1,2]	$C_{20}H_{24}O_9$	408.41	–39°	15291-75-5
B [1]	$C_{20}H_{24}O_{10}$	424.40	–63°	15291-77-7
C [1]	$C_{20}H_{24}O_{11}$	440.40	–19°	15291-76-6
J [3]	$C_{20}H_{24}O_{10}$	424.40	– 2.5°	107438-79-9
M [1]	$C_{20}H_{24}O_{10}$	424.40	–39° [–36°] [4]	15291-78-8

Occurrence: In *Ginkgo extract as bitter-tasting components of the leaves and roots of *Ginkgo biloba*; G. J occurs only in the leaves , G. M only in the roots. For synthesis, see *Lit.*[5,6], for biosynthesis see *Lit.*[7,8]. For use, see Ginkgo extract. G. B exhibits the activity of a PAF antagonist[9]. Cerebroprotective effects have been demonstrated for G. A and B[10].

Lit.: [1] Tetrahedron Lett. **1967**, 299–326. [2] Chem. Commun. **1967**, 259. [3] Justus Liebigs Ann. Chem. **1987**, 521. [4] Justus Liebigs Ann. Chem. **1993**, 287, 1023. [5] J. Am. Chem. Soc. **121**, 10249 (1999). [6] Tetrahedron Lett. **19**, 3201, 3205 (1988). [7] J. Am. Chem. Soc. **93**, 3546 (1971); Barton-Nakanishi **2**, 367–400. [8] Diss. ETH Zürich No. 10951 (1994). [9] Drugs of the Future **12**, 7, 643 (1987). [10] Eur. J. Pharm. Sci. **3**, 39 (1995).
gen.: Braquet, Ginkgolides – Chemistry, Pharmacology, Biology and Clinical Perspectives, Barcelona: J. R. Prons Science Publications 1988 ▪ Hager (5.) **5**, 269–295 ▪ J. Nat. Prod. **60**, 735 (1997) (isolation) ▪ Lindberg **3**, 89–119 ▪ Nicolaou, p. 451 ▪ Tetrahedron **52**, 4505 (1996) (crystal structure).

Ginseng. A herbaceous, anemone-like Araliaceae *Panax ginseng*, growing wild and being cultivated in Korea and China. The roots contain steroid derivatives, they are considered in Chinese medicine to be a life-lengthening and aphrodisiac tonic, and are also used to treat stomach ulcers. The constituents isolated to date include sitosterol (see sterols) and more than 10 *ginsenosides*. The ginsenosides are a group of tetracyclic *triterpene glycosides (*sapogenins). These sapogenins belong to the dammarane type; see table. The sugar components are D-glucopyranose, L-arabinopyranose, L-arabinofuranose, D-xylofuranose, and D-xylopyranose as well as L-rhamnopyranose. The ginsenosides are purported to have anti-ischemic, immunostimulating, antithrombotic (*PAF antagonist.), hypoglycemic, and cholesterol-lowering properties. G. extract is supposed to stimulate protein and nucleic acid metabolism. The recently described ginsenoside RF is the first monosubstance to exhibit all the effects of the crude ginseng extract[1].

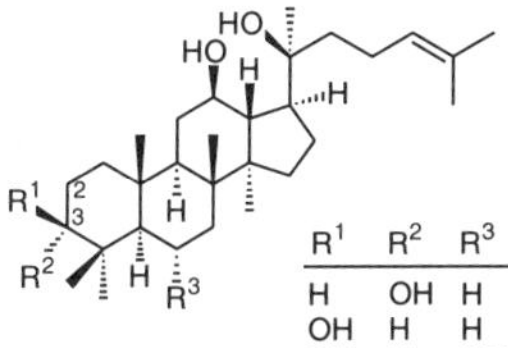

R^1	R^2	R^3	
H	OH	H	Betulafolientriol (3α,12β,20*S*)
OH	H	H	Protopanaxadiol (3β,12β,20*S*)
OH	H	OH	Protopanaxatriol (3β,6α,12β,20*S*)

Table: Ginsenosides.

name	molecular formula	M_R	mp. [°C]	CAS
Betula-folientriol	$C_{30}H_{52}O_3$	460.74	236–238	6892-79-1
Proto-panaxadiol	$C_{30}H_{52}O_3$	460.74	199–200	7755-01-3
Proto-panaxatriol	$C_{30}H_{52}O_4$	476.74	not crystalline	1453-93-6

Use: As a roborant and geriatric remedy.

Lit.: [1] Chem. Ind. (London) **1995**, 723; Chem. Pharm. Bull. **45**, 1039–1062, 1097 (1997); Proc. Natl. Acad. Sci. USA **92**, 8739 (1995).
gen.: Biochim. Biophys. Acta **990**, 315 (1989) ▪ Chemtech. **28**, 26–32 (1998) (review) ▪ Chem. Pharm. Bull. **45**, 1039 (1997) ▪ Planta Med. **59**, 76 (1993); **62**, 179 (1996) ▪ Tetrahedron **55**, 7157 (1999) (synthesis panaxatriol). – *[HS 1211 20]*

Ginsenosides see ginseng.

Giracodazole see girolline.

Girolline [(1*S*,2*S*)-3-amino-1-(2-amino-1*H*-imidazol-4-yl)-2-chloro-1-propanol].

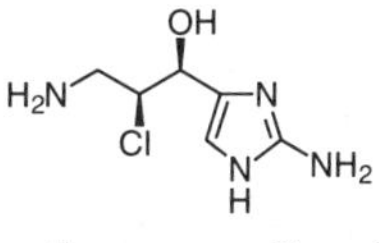

An alkaloid from the sponge *Pseudaxinyssa cantharella*, $C_6H_{11}ClN_4O$, M_R 190.63, as dihydrochloride (*giracodazole*): powder, $[\alpha]_D^{20}$ +8° (CH_3OH). G. inhibits protein biosynthesis and has antineoplastic activity.

Lit.: Biochem. Pharmacol. **43**, 1717 (1992) ▪ Tetrahedron **45**, 6713 (1989); **48**, 4327 (1992) ▪ Tetrahedron Lett. **32**, 1419,

4905 (1991); **39**, 8085 (1998) [synthesis (±)-G.]. – *[CAS 110883-46-0 (G.); 135824-74-7 (giracodazole)]*

Gitoxigenin, gitoxin see Digitalis glycosides.

Glaucine see aporphine alkaloids.

Glidobactins. Cyclic tripeptide antibiotics substituted with various 2*E*,4*E*-diene fatty acids. G. are isolated from cultures of *Polyangium brachysporum*. G. A, B, C exhibit broad *in vitro* activity against fungi and yeasts. A significant activity against leukemia cells has been observed in a mouse model. The components A to H have been distinguished with A to C being the more active G. occurring in larger amounts. They are all readily soluble in alcohols and dimethyl sulfoxide but insoluble in water and *n*-hexane.

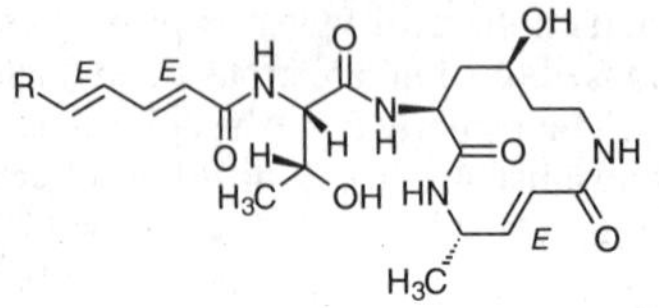

Table: Data of Glidobactins.

Glidobactin	A	B	C
molecular formula	$C_{27}H_{44}N_4O_6$	$C_{29}H_{46}N_4O_6$	$C_{29}H_{48}N_4O_6$
M_R	520.65	546.70	548.71
mp. [°C]	259–261	232–234	273–275
$[\alpha]_D^{24}$ (CH_3OH)	111°	–92°	–104°
LD_{50} (mg/kg)	8.1	–	25
MIC[a] (mg/L)	1.6	1.6	0.8
MIC[b] (mg/L)	1.6	3.1	0.8
CAS	108351-50-4	108351-51-5	108351-52-6

[a] *Candida albicans*; [b] *Aspergillus fumigatus*

Lit.: J. Antibiot. **41**, 1331–1350, 1358–1365, 1812–1822, 1906–1909 (1988) ▪ J. Chem. Soc., Chem. Commun. **1992**, 1687 (G. A). – *[HS 2941 90]*

Gliotoxin (aspergillin).

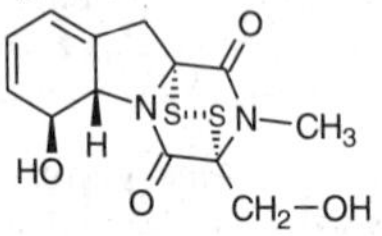

Gliotoxin

$C_{13}H_{14}N_2O_4S_2$, MR 326.39, monoclinic crystals, mp. 221 °C, $[\alpha]_D$ –290° ($CHCl_3$), epidithiodioxopiperazine antibiotic from *Gliocladium fimbriatum* as well as *Aspergillus* and *Penicillium* species. G. exhibits immunomodulatory, antiviral, and antifungal activities. Cultures of *Aspergillus fumigatus* also yield the structurally related *G. E* ($C_{13}H_{14}N_2O_4S_3$, M_R 358.46, mp. 172–173 °C) and *G. G* ($C_{13}H_{14}N_2O_4S_4$, M_R 390.51, powder, mp. 160–165 °C), these compounds possess an S_3- or an S_4-bridge, respectively, in place of the S_2-bridge and show similar activities [1]. On account of its high toxicity G. has not yet been used in therapy; however, it is being tested as a drug to reduce rejection reactions after organ transplantations [2] and it can be used as a disinfectant for seeds.

Lit.: [1] Tetrahedron Lett. **27**, 735 (1986); Aust. J. Chem. **40**, 991 (1987). [2] Drug News (29.7.) **1996**, 3.
gen.: Beilstein E III/IV **27**, 8902 ▪ Cole-Cox, p. 571, 575. ▪ Sax (8.), ARO 250. – *Biosynthesis:* J. Chem. Soc., Perkin Trans. 1 **1975**, 383. – *Synthesis:* Tetrahedron **37**, 2045 (1981) ▪ Tetrahedron Lett. **27**, 5539 (1986). – *[CAS 67-99-2 (G.); 101623-21-6 (G. E); 53348-47-3 (G. G)]*

Globularifolin see monomelittoside.

Globularimin.

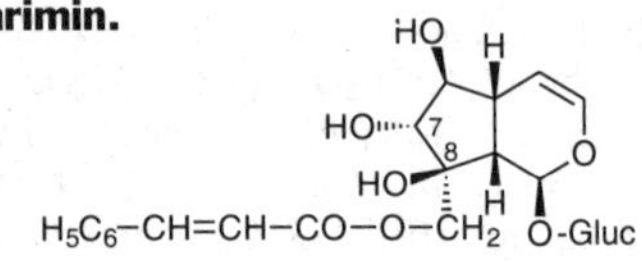

Globularimin

G. is a cinnamic acid ester of an *iridoid glucoside, its 7,8-epimer is known as *globularinin*. G. is formed biosynthetically by reductive cleavage of the epoxide ring of globularin (scutellarioside I, see catalpol). G., $C_{24}H_{30}O_{12}$, M_R 510.50, amorphous, $[\alpha]_D$ –106° (CH_3OH). Globularinin, amorphous, $[\alpha]_D$ –84.5° (CH_3OH). G. and globularinin occur together with globularin in the south European globe daisy (*Globularia alypum*, Globulariaceae) [1]. G., globularinin, and globularin cause colics and vertigo. Accordingly, the use of leaves of *G. alypum* as a substitute for senna leaves ("Séné de Provence"), *Cassia senna*, is not advisable [2].

Lit.: [1] Tetrahedron Lett. **1979**, 3149. [2] Hegnauer **IV**, 208.
gen.: Helv. Chim. Acta **64**, 3 (1981) ▪ J. Nat. Prod. **43**, 649, No. 56, 57 (1980). – *[HS 2938 90; CAS 73366-19-5 (G.); 73343-11-0 (globularinin)]*

Gloeosporone [(4*S*,7*R*,13*R*)-4,7-epoxy-4-hydroxy-5-oxo-13-octadecanolide].

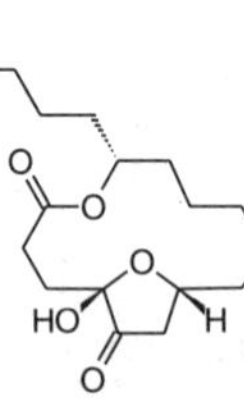

$C_{18}H_{30}O_5$, M_R 326.43, crystals, mp. 108–110 °C. An endogenous germination inhibitor (autoinhibitor) against various lower fungi, isolated from *Colletotrichum gloeosporioides*, and structurally related to colletodiol and grahamimycin. For synthesis, see *Lit.* [1].

Lit.: [1] J. Am. Chem. Soc. **110**, 6210–6218 (1988); **119**, 9130 (1997); Tetrahedron Lett. **27**, 6133 (1986); **28**, 3747 (1987).
gen.: Chem. Pharm. Bull. **40**, 524 (1992) ▪ Helv. Chim. Acta **70**, 281 (1987) ▪ Justus Liebigs Ann. Chem. **1989**, 1233. – *[CAS 88936-02-1]*

Glomerine.

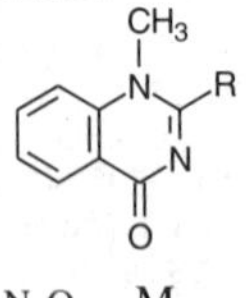

R = CH_3 : Glomerine
R = C_2H_5 : Homoglomerine

$C_{10}H_{10}N_2O$, M_R 174.20, yellow needles, mp. 209–211 °C. A quinazoline alkaloid from the millipede *Glomeris marginata* (Diplopoda), which uses it together with *homoglomerine* ($C_{11}H_{12}N_2O$, M_R 188.23, mp. 149 °C) as a defence secretion [1]. The biosynthesis proceeds via anthranilic acid [2].

Lit.: [1]Z. Naturforsch. B **21**, 552 (1966); Science **154**, 390 (1966). [2]Tetrahedron Lett. **1967**, 1815.
gen.: Zechmeister **46**, 202, 208–209. – *Synthesis:* Heterocycles **9**, 1585–1591 (1978); **14**, 1469 (1980) ▪ Monatsh. Chem. **100**, 948 (1969). – *[HS 2939 90; CAS 7471-65-0 (G.); 10553-05-5 (homo-G.)]*

Glowworms and fire-flies see luciferases, luciferins.

Glucans (D-glucans). General term (alternative name polyglucosans) for the, mostly naturally occurring, linear and branched polymers of *glucose; *Examples:* *cellulose, *amylose, *glycogen[1]; *Laminarin* (a linear glucan with β1,3- and a few β1,6-glycosidic bonds) from algae, *pachyman* from wood-rotting fungi, and yeast glucans with β1,3-bonds; *nigeran*: a polysaccharide from *Aspergillus niger* (α1-3+α1-4 Glcp)$_n$; *lichenin*: a water-soluble β-glucan from Iceland moss with β1-3- and β1-4-bonds (~3:7). Certain G. from the cell walls of fungi are toxic for higher plants and thus act as so-called *elicitors* (triggers) of *phytoalexin synthesis in the plants[2]. The excretion of G. by streptococci in the oral flora, e.g., *S. mutans* is partially responsible for the formation of dental plaque[3]. G. possesses diverse immunomodulatory properties, β-glucan receptors are present on cell surfaces in the human immune system[4].
Lit.: [1]Adv. Carbohydr. Chem. **17**, 371–430 (1962). [2]Science **201**, 364 (1978). [3]Doyle et al. (eds.), Glucosyltransferases, Glucans, Sucrose and Dental Caries, Washington D. C.: IRL Press 1983. [4]Pathol. Immunopathol. Res. **5**, 286–296 (1987); Trends Pharmacol. Sci. **4**, 85–96 (1984); **5**, 78 (1985).
gen.: Food Technol. (Chicago) **40**, 90–98 (1986) (biosynthesis) ▪ Kirk-Othmer (4.) **13**, 61. – *[CAS 9012-72-0]*

D-Glucitol (D-sorbitol).

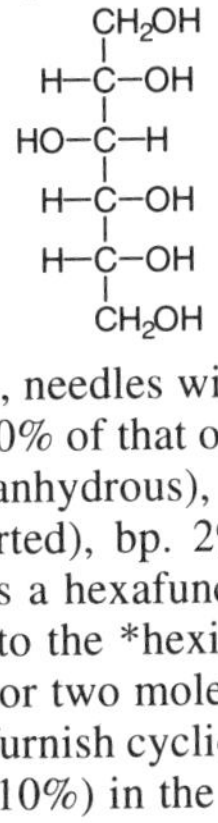

$C_6H_{14}O_6$, M_R 182.17, needles with a sweet taste (relative sweetness ca. 50% of that of *sucrose). D. 1.489, mp. 110°–112 °C (anhydrous), 75 °C (hydrate, 95 °C has also been reported), bp. 295 °C (470 Pa), $[\alpha]_D^{20}$ –1.9° (H_2O). D-G. is a hexafunctional alcohol (sugar alcohol) belonging to the *hexitols, the intramolecular cleavage of one or two molecules of water occurs relatively easily to furnish cyclic ethers. G. is particularly abundant (ca. 10%) in the fruit of mountain ash (rowan, *Sorbus aucuparia*) and thus also has the name sorbitol. Hawthorn (*Crataegus oxyacantha* = *C. laevigata*) also has a rather high content of G. It is also found in smaller amounts in apples, apricots, pears, cherries, medlars, plums, etc. G. is practically not affected by human oral bacteria and thus is not converted to acids which can promote the formation of caries. Since G. is also practically not fermentable, it does not cause heartburn even in people with sensitive stomachs. In larger amounts it has a mild laxative effect.
Use: In the food industry for dietetic foods, e.g., as sugar substitute for diabetic patients, in the cosmetic and pharmaceutical industries, as a medicament, in the production of *ascorbic acid as well as technical applications in the paper, leather, textile industries for the production of glues and gelatine, and as a starting material for the synthesis of polyethers, detergents, and varnishes.
Lit.: Beilstein E IV **1**, 2839 ▪ Dang. Prop. Ind. Mater. Rep. **8**, 73–77 (1988) ▪ Food Sci. Technol. **17**, 165–183 (1986) ▪ Hager (5.) **5**, 751; **6**, 766; **9**, 636 ▪ Martindale (31.), p. 1314, 1369, 1375 ▪ Merck-Index (12.), No. 8873 ▪ Pharm. Ind. **49**, 495–503 (1987) ▪ Ullmann (5.) **A 5**, 90; **A 11**, 571 ff., **A 24**, 222. – *[HS 2905 44; CAS 50-70-4]*

Glucoaubrietin, glucobrassicin, glucocapparin see glucosinolates.

Glucocorticoids see corticosteroids.

Gluconasturtiin see glucosinolates.

Glucopsychosine see sphingosine.

D-Glucosamine (2-amino-2-deoxy-D-glucose, abbreviation: D-GlcN; the pyranose form is shown).

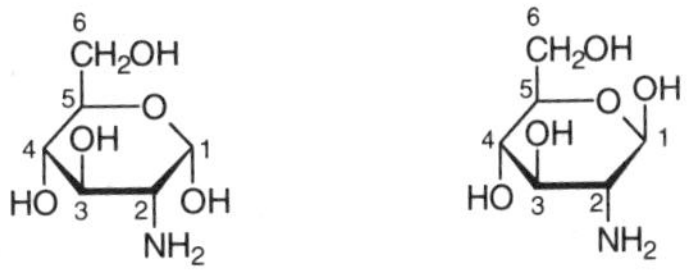

α-form β-form

$C_6H_{13}NO_5$, M_R 179.17. crystals, α-D-GlcN: mp. 88 °C, $[\alpha]_D^{20}$ +100→+47.5° (H_2O); β-D-GlcN: mp. 110–111 °C, $[\alpha]_D^{20}$ +28→+47° (H_2O). With hydrochloric acid D-G. forms well crystallizing, readily soluble salts; like the closely related D-glucose, it reduces Fehling's solution. Similar to other *amino sugars D-G. occurs only in bound form, mostly as *N*-acetyl-D-G. [recommended abbreviation: D-GlcNAc; also abbreviated as NAG; α-D-GlcNAc: mp. 205 °C, $[\alpha]_D^{20}$ +64→+41° (H_2O); β-D-GlcNAc: mp. 182–183.5 °C, $[\alpha]_D^{20}$ –22→+41° (H_2O)], as in, e.g., mucoproteins, glycosaminoglycans (mucopolysaccharides) such as chondroitin sulfates and *hyaluronic acid, in *chitin, as well as in murein, a component of bacterial cell walls. In addition, D-G. is a component of aminoglycoside antibiotics, e.g., gentamycin
Lit.: Beilstein E IV **1**, 2017–2021 ▪ Carbohydr. Res. **162**, 181 (1987) ▪ Merck Index (12.), No. 4466. – *[HS 2932 90; CAS 3416-24-8]*

D-Glucose (dextrose, grape sugar, abbreviation: D-Glc).

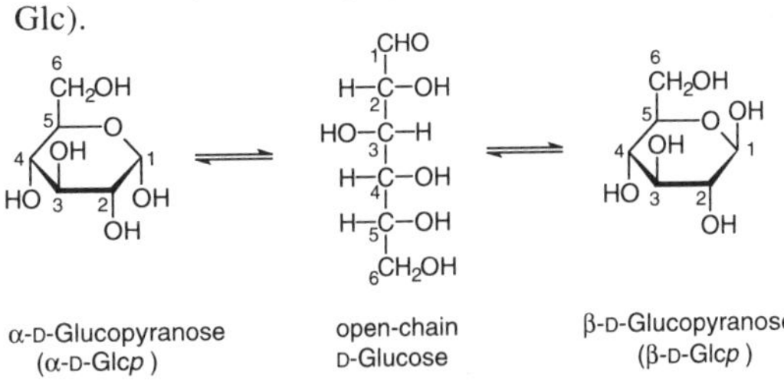

α-D-Glucopyranose (α-D-Glc*p*) open-chain D-Glucose β-D-Glucopyranose (β-D-Glc*p*)

$C_6H_{12}O_6$, M_R 180.16. Sweet-tasting crystals, D. 1.56; α-D-G. monohydrate (dextrose hydrate) has mp. 83–86 °C; anhydrous mp. 146 °C (decomp.), $[\alpha]_D^{20}$ +111→+53° (H_2O); β-D-G. mp. 150 °C, $[\alpha]_D^{20}$ +19→+53° (H_2O). The dextrorotatory aqueous solution of D-G. exhibits mutarotation, indicative of the formation of both anomers (see figure) which differ in the configuration at carbon atom 1. Besides the open-chain and the *pyranoside* 6-membered ring forms (shown in the Haworth projection in the figure), sugars also exist in the *furanoside* 5-membered ring forms, However, this is only of minor significance for G. G. normally exists as the glucopyranose and the aldehyde group of the open-chain form only reacts as such under certain circumstances, e. g., when Fehling's solution is reduced by G.

Occurrence: D-G. is widely distributed in practically all sweet fruits (free and bound with D-*fructose as *sucrose), especially in grapes from which the name grape sugar is derived. It also participates in the formation of *di- and *polysaccharides. In addition, D-G. is involved in the construction of physiologically important glycolipids and *glycoproteins; phosphorylated derivatives of D-G. are important metabolic intermediates – under consideration of its occurrence in polysaccharides (*glucans: especially cellulose and starch) and in *glycosides D-G. is the most widely distributed organic compound on Earth. D-G. is fermented by microorganisms to ethanol, acetic, lactic, and butanoic acids; it takes part in Maillard reactions when heated with amino acids; this is highly significant, e. g., for the formation of bread aroma and other flavors.

Physiology: Human blood usually contains about 0.08–0.11% of dissolved D-G. (blood sugar). The maintenance of constant blood sugar concentrations is achieved by regulatory mechanisms in which insulin and glucagon participate as well as the so-called *glucose tolerance factor* (GTF) which contains chromium. An insulin deficit can lead to diabetes. Furthermore, G. can cross-link proteins to non-degradable products through non-enzymatic reactions (so-called *glycation*, involved in tissue damage, cell aging) [1].

Metabolism: For uptake and transport of G., see *Lit.* [2–6]. D-G. plays a central part in the carbohydrate metabolism. It is degraded to smaller molecules in complicated reaction sequences (glycolysis) with release of energy – one example is pyruvic acid, which can enter the citric acid cycle via *acetyl-CoA – or (pentose phosphate pathway) can be converted to derivatives of other sugars for biosynthetic purposes under the concomitant availability of reduction equivalents. Alternatively D-G. can be stored in the liver and muscles as a reserve substance *glycogen [7] (in plants: *starch). An antimetabolite of D-G. is 5-thio-D-glucose. For detection, see *Lit.* [8].

Use: D-G. is produced by acidic hydrolysis of potato or corn starch and is marketed in the pure crystalline form as dextrose as well as in 5–50% solutions as G. preparations for parenteral nutrition in hospitals. G. is also used in chemical synthesis and for technical purposes. A considerable amount is used in the form of G. syrup for the manufacture of confectionery products, for the latter purpose increasing amounts of so-called *isosyrup* (G. syrup partially isomerized enzymatically to *fructose) are now being used.

Historical: Lowitz reported in 1792 on a sugar from grapes ("grape sugar") that differed from sugar from sugar cane. Kirchhoff obtained this sugar in 1811 by hydrolysis of starch, Braçonnot in 1819 by hydrolysis of cellulose products, and in 1838 the name G. was generally adopted for this carbohydrate.

Lit.: [1] Sci. Am. **256**, No. 5, 82–88 (1987); Trends Biochem. Sci. **11**, 311 ff. (1986). [2] Trends Biochem. Sci. **13**, 226–231 (1988). [3] Annu. Rev. Biochem. **55**, 1059–1089 (1986). [4] Nature (London) **333**, 183 ff. (1988). [5] Nature (London) **340**, 70–74 (1989). [6] Nature (London) **330**, 379 ff. (1987). [7] Trends Biochem. Sci. **11**, 136 ff. (1986); **13**, 329 f. (1988). [8] Hager (5.) **8**, 355–357. – *[HS 1702 30; CAS 492-62-6 (α-D-G.); 492-61-5 (β-D-G.)]*

Glucosidase inhibitors see glycosidase inhibitors.

Glucosinalbin see glucosinolates.

Glucosinolates (isothiocyanate glucosides). Constituents of crucifers (Brassicaceae) such as rape, cabbage sorts, mustard, radish, etc. They have major toxicological relevance in the food industry. The G. are biogenetic precursors of the, also frequently toxic, *mustard oils; for the removal of G. from foods by, e. g., boiling, enzymatic cleavage with myrosinase (EC 3.2.3.1), and extraction with aqueous media, see *Lit.* [1] and *goitrin. The content of G. in new strains of rape, the so-called double-zero types, has been reduced to <1%. G. often function in plants as *phytoalexins [2].

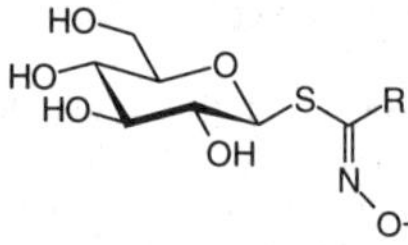

1 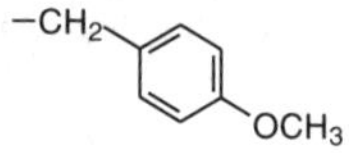

5 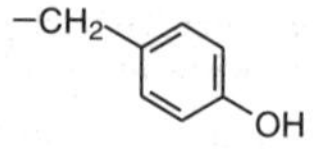

2 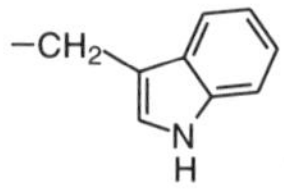

6 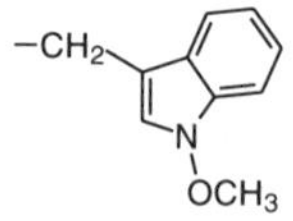

3 $-CH_3$

7 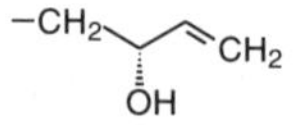

4 $-(CH_2)_2-C_6H_5$

8 $-CH_2-CH=CH_2$

Table: Structure and data of Glucosinolates.

name	R	molecular formula	M_R	CAS
Glucoaubrietin	1	$C_{15}H_{21}NO_{10}S_2$	439.45	499-27-4
Glucobrassicin	2	$C_{16}H_{20}N_2O_9S_2$	448.46	4356-52-9
Glucocapparin	3	$C_8H_{15}NO_9S_2$	333.33	497-77-8
Gluconasturtiin	4	$C_{15}H_{21}NO_9S_2$	423.45	499-30-9
Glucosinalbin	5	$C_{14}H_{19}NO_{10}S_2$	425.43	19253-84-0
Neoglucobrassicin	6	$C_{17}H_{22}N_2O_{10}S_2$	478.49	5187-84-8
Progoitrin	7	$C_{11}H_{19}NO_{10}S_2$	389.33	585-95-5
Sinigrin	8	$C_{10}H_{17}NO_9S_2$	359.37	534-69-0

Progoitrin, ((*R*)-2-hydroxy-3-butenyl glucosinolate), occurs in rape (*Brassica napus*) and other *Brassica* species. The (*S*)-epimer *epiprogoitrin*, potassium salt: amorphous solid, $[\alpha]_D^{25}$ – 14.8° (H_2O), is the main glucosinolate in the seeds of sea kale (*Crambe* spp.), e. g., *C. maritima* (white sea kale) and *C. hispanica* (Spanish sea kale). – *Glucosinalbin* (4-hydroxybenzyl glucosinolate) occurs in *Brassica* seeds. The salt with *sinapine is called *sinalbin*, $C_{30}H_{42}N_2O_{15}S_2$, M_R 734.83, cryst. · 5 H_2O, mp. 83–84 °C, 139 °C (anhydrous), $[\alpha]_D$ –8.23° (H_2O); it occurs in the seeds of white mustard (*Sinapis alba*) and other crucifers. – *Glucoaubrietin* is found in *Aubrieta* species. – *Glucobrassicin* (3-indolylmethyl glucosinolate) and *neoglucobrassicin* (*N*-methoxyglucobrassicin) occur in radish (*Raphanus*) and *Brassica* species (see also brassinin). – *Glucocapparin* (methyl glucosinolate), the potassium salts crystallizes as needles, mp. 207–209 °C, $[\alpha]_D^{22}$ –28° (H_2O); it is widely distributed in the family Capparidaceae, e.g., in the spider flower (*Cleome spinosa*) and in the caper tree (*Capparis spinosa*) indigenous to the Mediterranean region and is the flavor substance of the flower buds (capers). – *Gluconasturtiin* (2-phenylethyl glucosinolate), potassium salt: cryst., mp. 171 °C, $[\alpha]_D^{20}$ –20.7° (H_2O); from watercress (*Nasturtium officinalis*) and *Barbarea vulgaris*. – *Sinigrin* (allyl glucosinolate), occurs in the form of the potassium salt, $C_{10}H_{16}KNO_9S_2$, M_R 397.46, colorless cryst., mp. 125–127 °C, $[\alpha]_D^{28}$ –17° (H_2O), in the seeds of black mustard (*Brassica nigra*) as well as in types of cabbage, hedge garlic (*Alliaria officinalis*), and other crucifers.

Activity: When G. is metabolized not only the corresponding mustard oils but also isothiocyanates are formed, the latter inhibit the accumulation of iodine and thus the formation of thyroid hormones; G. may thus lead to hyperthyroidism and goiter, see also goitrin. Growth disorders and loss of feathers are observed in fowl. For biosynthesis, see *Lit.*[3].

Lit.: [1] Br. J. Nutr. **72**, 455–466 (1994); J. Sci. Food Agric. **62**, 259–265 (1993). [2] ACS Symp. Ser. **380**, 155–181 (1988); Can. J. Plant Sci. **74**, 595–601 (1994); J. Exp. Bot. **44**, 963–970 (1993); New Phytol. **127**, 617–633 (1994) (review). [3] J. Plant Physiol. **102**, 609–613, 1307–1312 (1993); **144**, 17–21 (1994).

gen.: ACS Symp. Ser. **546**, 181–196 (1994) (review) ▪ Biol. Plant. **34**, 451 ff. (1992) (sinigrin) ▪ J. Carbohydr. Chem. **15**, 109 (1996) (synthesis sinigrin) ▪ J. Sci. Food Agric. **65**, 201–207 (1994) ▪ Karrer, No. 2316, 3198 ▪ Nat. Prod. Rep. **10**, 327–348 (1993); **12**, 101–133 (1995) ▪ Tetrahedron Lett. **31**, 1417 (1990); **34**, 7967–7970 (1993). – *Toxicology:* J. Sci. Food Agric. **61**, 245–252 (1993) ▪ Sax (8.), AGH 125 (sinigrin). – *[HS 2918 90 (sinalbin); 2938 90 (sinigrin); CAS 19237-18-4 (epiprogoitrin); 20196-67-2 (sinalbin); 3952-98-5 (sinigrin)]*

Glucovanillin see 3,4-dihydroxybenzaldehydes.

D-Glucuronic acid (abbr.: D-GlcUA).

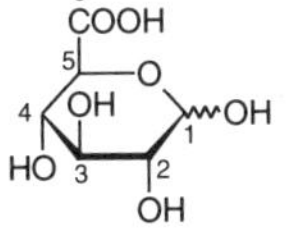

$C_6H_{10}O_7$, M_R 194.14, *β*-D-G.: soluble in water and alcohol, needles, mp. 167 °C, $[\alpha]_D^{20}$ +36.3° (H_2O). The pyranose form is shown in the figure. D-G. belongs to the uronic acids and is formed in the liver from D-glucose – after prior activation as uridine 5′-diphospho(UDP)-D-glucose[1]. The glycosides of D-G. are excreted in amounts of 0.3–0.4 g per day. This is a *detoxification reaction*, with the purpose of making the toxic phenols, frequently formed, especially in the small intestine, more soluble in water for easier excretion. Steroid hormones are removed in the same way; however, carcinogens can also be formed in the organism by glucuronidation (*intoxication*). D-G. is a building block of glycosaminoglycans (mucopolysaccharides) such as *hyaluronic acid, *teichuronic acids, and chondroitin sulfate; in most animals (except primates) it is the starting material for the biosynthesis of L-*ascorbic acid. It is also a building block of arabic acid (gummi arabicum).

Lit.: [1] Collins-Ferrier, p. 46 ff.

gen.: Beilstein E IV **3**, 1996 ff. ▪ Carbohydr. Res. **83**, 135 (1980) ▪ J. Chromatogr. **487**, 125 (1989). – *[HS 2932 90; CAS 6556-12-3]*

D-Glucurono-6,3-lactone (generic names: glucuro(no)lactone, glucurone, glucurono-6,3-lactone, dicurone).

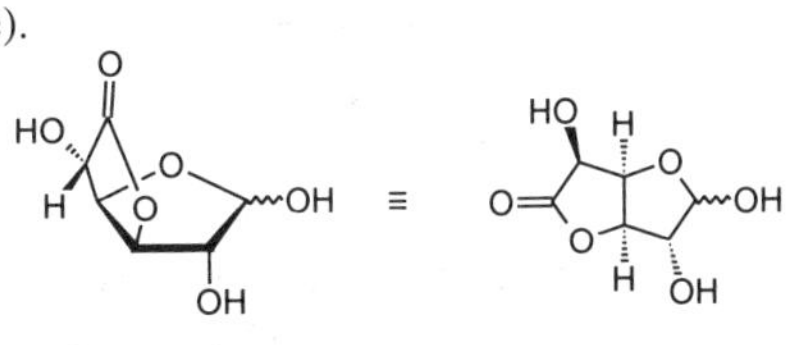

$C_6H_8O_6$, M_R 176.12, cryst., mp. 176–178 °C, $[\alpha]_D^{20}$ +20° (H_2O). G. occurs in many plant slimes, animal fibers, and connective tissue. It is used as a detoxifying therapeutic agent in cases of hepatitis, sciatica, and arthrosis. It reduces the toxicity of sulfonamides.

Lit.: Beilstein E V **18/5**, 34 ▪ Carbohydr. Res. **39**, 53 (1975); **92**, 51 (1981); **101**, 255 (1982) ▪ Justus Liebigs Ann. Chem. **1975**, 752. – *[HS 2932 29; CAS 32449-92-6]*

Glutamic acid (2-aminopentanedioic acid, 2-aminoglutaric acid, abbr.: Glu or E; Glu+Gln: Glx or Z).

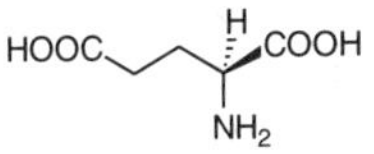

(2*S*)-, L-form

$C_5H_9NO_4$, M_R 147.13, subl. 200 °C, mp. 247–249 °C (decomp.), $[\alpha]_D^{20}$ +17° (H_2O), +46.8° (5 M HCl); pK_a 2.19, 4.25, 9.67, pI 3.22. A widely distributed proteinogenic amino acid. Genetic code: GAA, GAG.

Occurrence: In practically all proteins, especially those of wheat, corn, soybean, egg, and milk (casein). The average content is about 6.2%[1]. It was first isolated from wheat gluten. When heated G. undergoes lactam ring closure to furnish *pyroglutamic acid. Free Glu plays a central role in nitrogen metabolism.

Metabolism: Glu is formed from 2-oxoglutaric acid (an intermediate of the citric acid cycle) and various amino acids under the action of aminotransferases (EC 2.6.1 group) or in mitochondra by reductive amination. Glu is metabolized catabolically via glutamic acid-5-semialdehyde and represents the precursor of the amino acids ornithine and proline. Glu acts as a transport mediator in the transport of aspartic acid through mitochondrial membranes into the cytoplasm. Excess ammonia from G. is bound in L-*glutamine and excreted via the kidneys in the urine.

G. is involved in the biosynthesis of *glutathione and numerous *γ*-glutamyl compounds and acts as an "amino acid carrier" in the degradation steps of the *γ*-glutamyl cycle. In the CNS G. is an inhibitory neuro-

transmitter and precursor of *4-aminobutanoic acid with excitatory activity.
The monosodium salt of G. is produced in large amounts, especially in Japan, and used as a flavor substance for fish products and to intensify the flavors of soup concentrates.

Lit.: [1]Biochem. Biophys. Res. Commun. **78**, 1018–1022 (1977).
gen.: Adv. Enzymol. **62**, 37–92; 315–374 (1989) ▪ Beilstein E IV **4**, 3028–3032 ▪ Nature (London) **321**, 772 f. (1986); **332**, 452 f. (1988) ▪ Trends Biochem. Sci. **9**, 300 (1984) ▪ Ullmann (5.) **A 16**, 711–716; **A11**, 574. – *[HS 2922 42; CAS 56-86-0 (L-form); 6893-26-1 (D-form)]*

Glutamine [glutamic acid-5-amide, abbr.: Glu(NH_2), Gln, or Q].

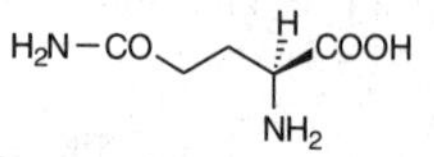

(2*S*)-, L-form

$C_5H_{10}N_2O_3$, M_R 146.15, mp. 185–186 °C, $[\alpha]_D^{19}$ +6.1° (H_2O), pK_a 2.17, 9.13, pI 5.65. Genetic code: CAA, CAG. A widely distributed proteinogenic amino acid, occurring in most proteins with an average content of 3.9% [1]. The free form is found frequently in plants, animals, fungi, and bacteria and plays a central role in nitrogen metabolism. Like asparagine, G. is stored in plant seeds. In the mammalian organism G. is formed from glutamic acid and the ammonia originating from deamination of amino acids (ammonia detoxification).
Metabolism: Gln may be considered as the first product of ammonia assimilation in plants. It is formed from endogenous Glu, NH_3, and ATP by the action of glutamine synthetase (EC 6.3.1.2). The amide group of G. is transaminated by glutamate synthase (EC 6.3.1.2) to furnish 2-oxoglutaric acid (glutamate synthase cycle). Gln acts in several enzymatic reactions as an amino group transfer reagent.

Lit.: [1]Biochem. Biopys. Res. Commun. **78**, 1018–1024 (1977).
gen.: Acta Crystallogr., Sect., C **52**, 1313 (1996) ▪ Beilstein E IV **4**, 3037 ff. ▪ Kvamme, Glutamine und Glutamate in Mammals, Boca Raton: Academic Press 1989 ▪ Ullmann (5.) **A 2**, 58–62, 70, 85. – *[HS 2924 10; CAS 56-85-9]*

γ-Glutamylamino acids. Numerous γ-G. are known in plants and fungi; with a few exceptions such as *glutathione or *lentinic acid, they are mostly dipeptides. Some γ-G. serve as the storage form for ammonia or nitrogen in plant seeds, onion, garlic, etc. γ-G. probably play an important role in the γ-glutamyl cycle for the energy-dependent transport of conjugated amino acids in the cell.
γ-Glutamyltranspeptidases with differing specificities have been characterized. The enzymatic activity was determined spectroscopically for the transfer of glutamic acid from γ-glutamyl-*p*-nitroanilide to other amino acids or amines.

Lit.: Annu. Rev. Biochem. **45**, 559–604 (1976) ▪ Zechmeister **39**, 173.

γ-Glutamylmarasmine.

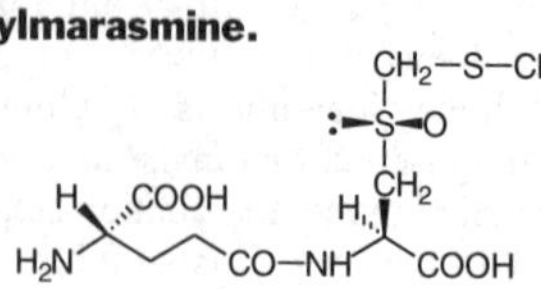

$C_{10}H_{18}N_2O_6S_2$, M_R 326.38, cryst., mp. 162–165 °C, $[\alpha]_D$ 0° (H_2O). A dipeptide from the agarics *Marasmius alliaceus*, *M. scorodonius*, and *M. prasiosmus* (Basidiomycetes) with garlic-like odors.

Lit.: J. Org. Chem. **52**, 1511 (1987) ▪ Phytochemistry **15**, 1717 (1976). – *[CAS 106565-95-1]*

Glutathione [glutathione-SH, GSH, *N*-(*N*-L-γ-glutamyl-L-cysteinyl)glycine].

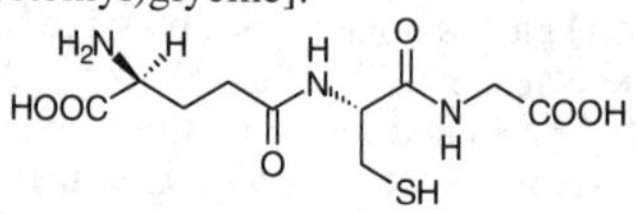

$C_{10}H_{17}N_3O_6S$, M_R 307.32, cryst., mp. 190–192 °C, $[\alpha]_D^{25}$ –18.9° (H_2O); pK_{a1} 2.21, pK_{a2} 3.53, pK_{a3} 8.66, pK_{a4} 9.12; disulfide (oxidized form) $C_{20}H_{32}N_6O_{12}S_2$, M_R 612.63, mp. 123 °C, $[\alpha]_D^{20}$ –108° (H_2O). G. is a tripeptide first isolated in 1921 by Hopkins and Kendall from muscle and yeast cells.
Biosynthesis: GSH is formed in the liver from L-*glutamic acid and L-*cysteine under the action of γ-glutamylcysteine synthetase (EC 6.3.2.2) as well as from *glycine under the action of glutathione synthetase (EC 6.3.2.3). The energy for synthesis is supplied by ATP.
Biological importance: GSH is transformed by reversible oxidation to the disulfide GSSG and thus represents a buffer system for redox states in the cell. Peroxides and radicals are reduced under catalysis by the selenium-containing *glutathione peroxidase* (EC 1.11.1.9). Essential thiol groups of proteins are protected by GSH or reformed by reduction:

Protein-S–S-Protein + 2 GSH ⇌ (Glutathione Transhydrogenases (EC 1.8.4)) 2 Protein-SH + GSSG

GSH is maintained in the reduced state by the NAD(P)H-dependent, flavin-containing *glutathione reductase* [1] (EC 1.6.4.2; 2 identical subunits, M_R depending on the source 100 000–125 000) and serves as prophylaxis against oxidative stress. Inhibitors of glutathione reductase are under consideration as anti-malarial drugs [2]. GSH can also be regenerated by L-*ascorbic acid:

2 GSSG + L-Ascorbic acid ⇌ (Glutathione Dehydrogenase (EC 1.8.5.1)) 2 GSH + Dehydro-L-ascorbic acid

GSH enables the transport of amino acids and low-molecular-weight peptides through cell membranes by transfer of γ-L-glutamyl residues to amino acids and certain dipeptides [catalyzed by *γ-glutamyl transpeptidase* (γ-glutamyl transferase, abbreviation: γ-GT, EC 2.3.2.2)] . The *detoxification* of electrophilic xenobiotic substances by elimination as – mostly water-soluble, non-toxic – *S*-substituted acetylcysteine derivatives (so-called *mercapturic acids*) is initiated by *glutathione S-transferases* [3] (abbreviation: GST, EC 2.5.1.18). For further information on the biochemistry, see *Lit.* [4].

Lit.: [1]Biochemistry **27**, 4465–4474 (1988); Eur. J. Biochem. **178**, 693–703 (1989); FEBS Lett. **250**, 72–74 (1989); Nature (London) **343**, 38–43 (1990). [2]Nachr. Chem. Tech. Lab. **37**, 1026–1033 (1989). [3]CRC Crit. Rev. Biochem. **23**, 283–337 (1988). [4]Free Radical Biol. Med. **12**, 337 (1992).

gen.: Annu. Rev. Biochem. **52**, 711–760 (1983); **58**, 743–764 (1989) ▪ Beilstein EIV **4**, 3165 ff. ▪ Dolphin, Coenzymes and Cofactors: Glutathione. Chemical, Biochemical and Medical Aspects, Parts A and B, New York: Wiley 1989 ▪ Sies & Kletterer, Glutathione Conjugation, Mechanisms and Biological Significance, London: Academic Press 1988. – *Reviews:* Free Radical Biol. Med. **12**, 337 (1992) ▪ J. Biol Chem. **263**, 17205–17208 (1988) ▪ J. Chem. Soc., Perkin Trans. 1 **1990**, 1753 (synthesis) ▪ J. Indian Chem. Soc. **72**, 495 (1995) ▪ Naturwissenschaften **76**, 57–64 (1989) ▪ Prog. Ind. Microbiol. **24**, 259 (1986). – *Toxicology:* Sax (8.), GFW 000. – *[HS 2930 90; CAS 70-18-8]*

Glycans see polysaccharides.

Glyceollins.

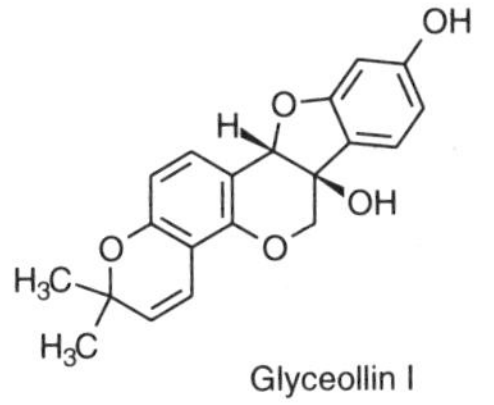

Glyceollin I

6a-Hydroxy-*pterocarpans with substituents derived biogenetically from prenyl residues that occur as *phytoalexins in Fabaceae such as the cultivated soybean (*Glycine max*), its wild form (*G. soja*), *G. canescens*, *Psoralea* sp., and *Psophocarpus tetragonolobus*, e. g., *glyceollin I*, $C_{20}H_{18}O_5$, M_R 338.36.

Lit.: Harborne (1994), p. 166–180 ▪ Nat. Prod. Rep. **12**, 101–133 (1995) ▪ Phytochemistry **19**, 1203–1207 (1980); **20**, 795–798 (1981); **22**, 2729–2733 (1983) (biosynthesis); **30**, 3233–3236 (1991) (isolation) ▪ Plant Physiol. **77**, 591–601 (1985) (occurrence, activity) ▪ Zechmeister **43**, 152–155 (1–266). – *[CAS 57103-57-8 (G. I)]*

Glycerides (acylglycerols). General name for fatty acid esters of glycerol. Depending on the number of fatty acids esterified with one molecule of glycerol the G. are classified as mono-, di-, and *triglycerides.

Glycine (aminoacetic acid, abbr.: Gly or G). H_2N-CH_2-COOH, $C_2H_5NO_2$, M_R 75.07, mp. 233 °C (decomp.), pK_a 2.34, 9.60, pI 5.97; monohydrochloride, hygroscopic, mp. 185 °C. The simplest of the non-essential, proteinogenic amino acids with a sweet taste, G. occurs in most proteins. The average content in over 200 proteins is the second highest after alanine [1]. Genetic code: GGU, GGC, GGA, GGG. Gly was the first amino acid to be isolated from the acid hydrolysate of a protein (gelatin). Proteins of the collagen or elastin types have high concentrations of G., e. g., silk fibroin 43%. Gly also occurs in the free form and is a component of smaller peptides such as *glutathione.

Biosynthesis: From serine with tetrahydrofolate (THF) and serine hydroxymethyltransferase (EC 2.1.2.1) or from CO_2, NH_4^+, THF, and NADH by glycine synthase (EC 2.1.2.10). Gly is converted to serine and further to pyruvic acid by L-serine dehydratase (EC 1.4.1.7). Gly is involved in the biosynthesis of *heme, *purines, *porphyrins, etc. It is a synaptic modulator [2].

Lit.: [1] Biochem. Biophys. Res. Commun. **78**, 1018–1024 (1977). [2] Nature (London) **338**, 337, 422–427 (1989).

gen.: Beilstein EIV **4**, 2349–2354 ▪ Greenstein & Winitz, Chemistry of the Amino Acids, vol. 3, p. 1955–1970, New York: Wiley 1961 ▪ Ullmann (5.) **A 2**, 58–63, 70. – *[HS 2922 49; CAS 56-40-6]*

Glycocholic acids see bile acids.

Glycogen. $(C_6H_{10}O_5)_n$, M_R up to 16000000. G. is an α1-4-*glucan with branching by α1-6-bonds, a *polysaccharide built up exclusively of D-*glucose units which (in contrast to, e. g., *cellulose) are α-glycosidically linked. G., occurring as a reserve substance (animal starch) especially in the liver but also in other animal cells, is a white, tasteless powder, that first swells in water and then forms an opalescent colloidal solution. This solution does not reduce Fehling's solution and gives a brown or violet-brown to violet-red color with iodine (distinction from *starch: blue). The structure of G. is similar to that of the plant glucan *amylopectin but more highly branched. It is one of the largest known molecules and can consist of as many as 100000 D-glucose units.

Formation and degradation of G. are controlled by a hormonal (adrenaline and glucagon) regulatory cascade.

Lit.: Carbohydr. Polym. **16**, 37 (1991) ▪ Carbohydr. Res. **220**, 1 (1991) ▪ Crit. Rev. Biochem. Mol. Biol. **24**, 69–99 (1989) ▪ Nature (London) **336**, 202 f., 215–221 (1988) ▪ Stryer 1995, p. 581–602. – *[HS 3913 90; CAS 9005-79-2]*

Glycopeptides see glycoproteins.

Glycoproteins. Name made up from glyco... + protein for a group of compounds widely distributed in the plant and animal kingdoms containing both *carbohydrates and proteins in the same molecule. The carbohydrate components which, with few exceptions, comprise less than 50% of the total G. are linked to the peptide part by *O*-, *N*-, or ester glycosidic bonds. The mostly branched *oligosaccharides are made up of core oligosaccharides of only few types, the peripheral oligosaccharides (mostly repeating β-D-*galactose- and *N*-acetyl-β-D-glucosamine residues), and the terminal monosaccharide residues, which vary and contribute mainly to the specificity of the G. The terminal residues of the individual G. also vary according to the tissue and developmental state of the cells. Building blocks of the carbohydrates (*monosaccharides) found in G. are hexoses (galactose, mannose, less abundant glucose), *N*-acetyl-D-hexosamine, *N*-acylneuramic acids, L-fucose, etc. The higher molecular weight proteoglycans which are also made up of carbohydrates and proteins albeit in other constitutions and with other biological functions are considered as a separate class of compounds. G. also occur in viral envelopes as well as in bacteria; examples are the cell surface G. and flagellin of the halobacteria (bacteria living in salt water) [1]. In the field of microbial secondary products there are also low-molecular-weight G. (M_R <2000) with biological activities (glycopeptide antibiotics). These G. often contain unusual, non-proteinogenic amino acids (some with D-configurations) and unusual hexoses. As far as the activity is concerned it generally holds that the peptide part is responsible for the affinity to specific target molecules while the sugar building block modulates this activity. Examples that are used in tumor therapy include *bleomycins and *vancomycin. For synthesis, see *Lit.* [2].

Biosynthesis: The G. are produced by *posttranslational modification* in the endoplasmatic reticulum and

in the Golgi apparatus, i. e., by introduction of the sugar residues after biosynthesis of the polypeptide chain by means of specific glycosyltransferases[3], different types of which are formed (expressed) depending on the cell type. In some cases dolichyl phosphate (see dolichols) serves as an intermediate carrier of the oligosaccharides.
Occurrence: The group of G. includes practically all serum proteins, many plasma proteins (e. g., α-fetoprotein), the blood group determinants, many enzymes, receptors and proteohormones, antibodies, mucines (*mucoproteins*, components of body mucus), lectins (phytagglutinins), the *antifreeze proteins of polar fish, P-glycoprotein which is responsible for the multidrug-resistance of cancer cells, and many others. Some G. play a part in the pathogenicity of viruses and other pathogens as membrane or surface proteins. In this case, as for other receptor-specific, cellular interactions, and also protein transport processes (e. g., in the Golgi apparatus, where the newly synthesized G. are sorted according to their sugar chains which are then modified), the carbohydrate components are responsible for the recognition processes at the molecular level (for molecular recognition during egg fertilization, see *Lit.*[4]; for their roles as differentiation and tumor antigens, see *Lit.*[5]; in the pathogenesis of meningitis, see *Lit.*[6]). A review on G. of the cytoplasm and the cell nucleus is given in *Lit.*[7].

Lit.: [1] Annu. Rev. Biochem. **58**, 173–194 (1989). [2] Pure Appl. Chem. **65**, 1223 (1993); Angew. Chem. Int. Ed. Engl. **33**, 336 (1994). [3] J. Biol. Chem. **264**, 17615 ff. (1989). [4] Annu. Rev. Biochem. **57**, 415–442 (1988). [5] Trends Biochem. Sci. **10**, 24–29 (1985). [6] Trends Biochem. Sci. **10**, 129 ff. (1985). [7] Annu. Rev. Biochem. **58**, 841–874 (1989); Trends Biochem. Sci. **13**, 380–384 (1988).
gen.: Adv. Carbohydr. Chem. Biochem. **57**, 157–223 (1980) ▪ Angew. Chem. Int. Ed. Engl. **29**, 823–839 (1990) ▪ Annu. Rev. Biochem. **54**, 631–664 (1985); **56**, 915–944 (1987) ▪ Beeley, Glycoprotein and Proteoglycan Techniques (Laboratory Techniques in Biochemistry and Molecular Biology, vol. 16), Amsterdam: Elsevier 1985 ▪ Collins-Ferrier, p. 498–504 ▪ von Döhren, in Vining & Stuttard (eds.), Genetics and Biochemistry of Antibiotic Production, p. 121–222, Boston: Butterworth-Heinemann 1995 ▪ Ivatt, The Biology of Glycoproteins, New York: Plenum Press 1984 ▪ Kirk-Othmer (4.) **4**, 921 f. ▪ Trends Biochem. Sci. **8**, 287–291 (1983); **9**, 198–202 (1984); **10**, 78–82 (1985); **14**, 272–276 (1989) ▪ Ullmann (5.) **A 13**, 96.

Glycosidase inhibitors. Mostly low-molecular-weight principles that specifically inhibit glycolytic enzymes (e. g., glucosidases, mannosidases). Examples include especially polyhydroxylated mono- and bicyclic alkaloids such as *nojirimycin, *deoxynojirimycin, *mannojirimycin, *deoxymannojirimycin, *castanospermine, and *swainsonine. G. i. inhibit, e. g., glucosidase and mannosidases that are important in the processing ("ripening") of glycoproteins and thus have potential therapeutic uses, e. g., for viral infections such as HIV as well as for metabolic diseases such as diabetes and adiposity.
The α-mannosidase inhibitors *mannostatin A* ($C_6H_{13}NO_3S$, M_R 179.24, mp. 121 °C) and *mannostatin B* ($C_6H_{13}NO_4S$, M_R 195.24), isolated from *Streptoverticillium verticillus* var. *quintum*, represent a new structural type.

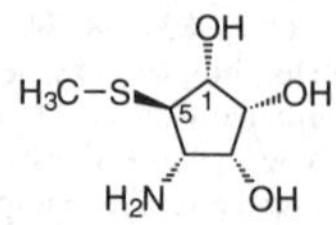
Mannostatin A

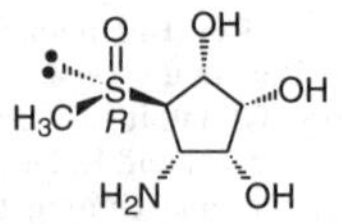
Mannostatin B

A further structural type of G. i. are the pseudooligosaccharides. These include as α-glucosidase inhibitors the *amylostatins* (e. g., *acarbose), *adiposins*, *oligostatins*, and *trestatins*.
These substances, produced by streptomycetes, reduce the postprandial (= after meal) glucose level by delaying the cleavage of complex polysaccharides in the stomach.

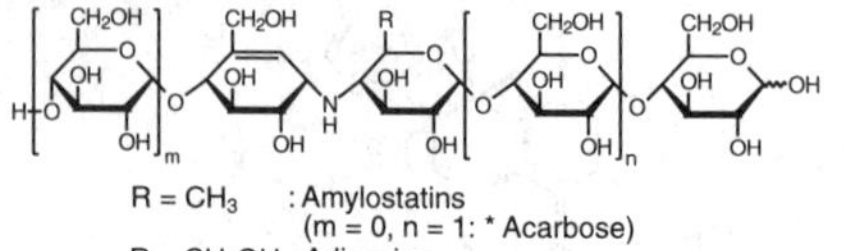
R = CH_3 : Amylostatins
(m = 0, n = 1: * Acarbose)
R = CH_2OH : Adiposins

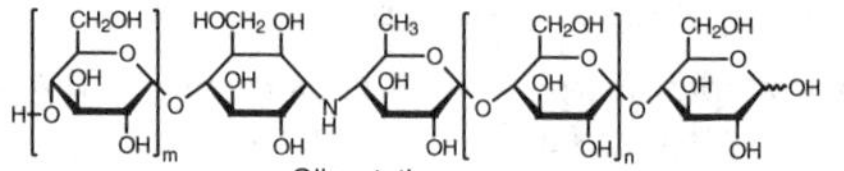
Oligostatins

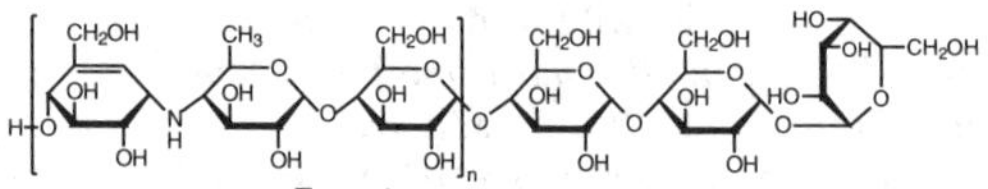
Trestatins

The branched aminocyclitol *valiolamine* (a pseudoamino sugar, $C_7H_{15}NO_5$, M_R 193.20) from *Streptomyces hygroscopicus* subsp. *limoneus* is also a strong α-glucosidase inhibitor.

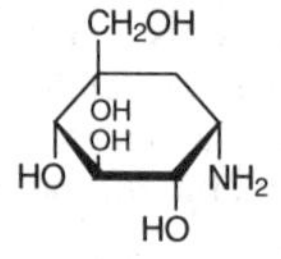
Valiolamine

The α-amylase inhibitor *tendamistat* (Hoe 467A, $C_{345}H_{523}N_{93}O_{116}S_4$, M_R 7957.77) is isolated from *Streptomyces tendae*. This polypeptide has potent antidiabetic activity.

Lit.: Agric. Biol. Chem. **46**, 1941, 2021 (1982) ▪ Barton-Nakanishi **3**, 129–160 (review) ▪ Carbohydr. Res. **204**, 131–139 (1990); **242**, 141–151 (1993) (trestatin) ▪ Chem. Rev. **99**, 779–844 (1999) ▪ Hoppe Seyler's Z. Physiol. Chem. **364**, 1347 (1983) ▪ J. Antibiot. **42**, 883, 1008 (1989) ▪ J. Mol. Biol. **206**, 677 (1989) ▪ J. Org. Chem. **63**, 6072 (1998) (mannostatins, synthesis) ▪ Jpn. Pharmacol. Ther. **22**, 3491 (1994) ▪ J. Takeda Res. Lab. **52**, 1 (1993) ▪ Nachr. Chem. Tech. Lab. **42**, 1119 (1994) ▪ Nat. Prod. Rep. **11**, 135 (1994). – *[CAS 102822-56-0 (mannostatin A); 102822-66-2 (mannostatin B); 83465-22-9 (valiolamine); 86596-25-0 (tendamistat)]*

Glycosides. General term for an extensive group of natural and synthetic compounds built up from *carbohydrates (mono- or oligosaccharides) and aglycones (i. e., non-sugar components). In G. the aglycone is *glycosidically bound* at the anomeric center (C1) to form an intact acetal. As a result of the ring structures of the monosaccharide (pyranose or furanose form), this C center is asymmetric; the two possible diastereomeric G. are distinguished as α- and β-anomeric G. (for definition of the configuration cf. D-glucose). For an ex-

ample of the β-glycosidic linkage, see figure under cellulose; the figure under starch shows an α-glycosidic linkage, see also figure in *Lit.*[1].
The *glucosinolates, usually occurring as precursors of the *mustard oils are examples of the *S*-G. in which bonding is via sulfur. The *amino sugars, i.e., aminoglycosides (e.g., aminoglycoside antibiotics) are of particular biological and pharmaceutical interest. The *N*-G., in which the carbohydrate residue is D-ribose or 2-deoxy-D-ribose and the aglycone is a derivative of a nitrogen base (usually pyrimidine or purine) having a C–N bond to the sugar, are of major biological significance. These G. are well known as *ribonucleosides* and *deoxyribonucleosides*; they are not only building blocks of the nucleic acids but also – in the form of so-called *sugar nucleotides* – cosubstrates of numerous biosyntheses. The number of natural G. increases dramatically if the glycolipids, *glycoproteins, and proteoglycans are included.
The carbohydrate of the *O*-G. is usually a *monosaccharide such as, e.g., glucose (most frequently); in addition carbohydrates are known that occur exclusively in G., for example, digitalose, cymarose, etc. The aglycones are mainly hydroxy compounds (phenols, *steroids). When classifying the G. into *S*-G., *N*-G., and *O*-G. the latter and largest group is subdivided into simple alcohol and phenol G., benzopyran G. and related compounds (e.g., *anthocyanins, *coumarin, and *flavone G.), *lignan G., anthra G. and dianthrone G., *di- and *triterpene G. (e.g., saponins), *steroid saponin (e.g., *Solanum steroid alkaloid glycosides, *bufadienolide G. and *cardenolides-=*Digitalis glycosides), *cyanogenic glycosides, aminoglycosides.
The G. have widely differing biological significance in plants; in some cases they serve as sugar reservoirs, in others they stabilize highly sensitive plant pigments or aroma substances or detoxify hazardous metabolic products (phenols). Many plant seeds and tree barks are protected against rot by antimicrobially active G. In general the G. can be characterized as mostly well crystalline, often bitter-tasting compounds that are readily soluble in alcohol, acetone, and hot water. For structural elucidation, the G. are usually cleaved, and the aglycone part identified, mass spectrometric procedures have facilitated the analyses.[2].
For glycosylation, see *Lit.*[3]. Many G. exhibit pronounced physiological activities. For the stimulating effects on the taste cells of insects, see *Lit.*[1].

Lit.: [1] Naturwiss. Rundsch. **29**, 73–81 (1976). [2] Helv. Chim. Acta **64**, 297–303 (1981). [3] Angew. Chem. Int. Ed. Engl. **13**, 157–170 (1974).
gen.: ACS Symp. Ser. **380**, 130–142, 403–416 (1988); **560** (1994) (entire vol.) ▪ Adv. Carbohydr. Chem. Biochem. **33**, 111–188 (1976); **50**, 21–123 (1994) ▪ Angew. Chem. Int. Ed. Engl. **25**, 212–235 (1986); **34**, 1ff. (1995) ▪ Atta-ur-Rahman **14**, 201–266 (1994) ▪ Bochkov & Zaikov, Chemistry of the O-Glycosidic Bond, Oxford: Pergamon 1979 ▪ Carbohydr. Chem. **16**, 17–50 (1985); **17**, 15–43 (1985) ▪ Chem. Rev. **93**, 1503–1531 (1993); **96**, 443–473, 683–720 (1996) ▪ Contemp. Org. Synth. **3**, 173–200 (1996) ▪ Curr. Opt. Struct. Biol. **2**, 674 (1992); **3**, 694 (1993) ▪ Drug Discovery Today **1**, 331–342 (1996) ▪ Erdmann et al., Cardiac Glycosides 1785–1985, Darmstadt: Steinkopff 1986 ▪ Glycoconjugate J. **13**, 5–17 (1996) ▪ Ikan, Naturally Occurring Glycosides, Chichester: Wiley 1999 ▪ Lockhoff in: Houben-Weyl, Methoden der Organischen Chemie, vol. E 14a/3, p. 621–1080, Stuttgart: Thieme 1992 ▪ Kontakte (Merck Darmstadt) **1992**, 3–10 ▪ Modern Synthetic Methods **1995**, 283–330 ▪ Nachr. Chem. Tech. Lab. **35**, 930–935 (1987); **40**, 828–834 (1992) ▪ Ogura et al., Carbohydrates **1992** (entire vol.) ▪ Pure Appl. Chem. **67**, 1647–1662 (1995) ▪ Rodd's Chem. Carbon Compd. **2**, 509–554 (1994) ▪ Synthesis **1994**, 1–20 ▪ Tetrahedron **52**, 1095–1121 (1996) ▪ Tetrahedron, Org. Chem. Ser. **12**, 252–311 (1995) ▪ Zechmeister **25**, 150–174. – *[HS 2938 10, 2938 90]*

Glycyrrhet(in)ic acid (3β-hydroxy-11-oxoolean-12-ene-30-carboxylic acid).

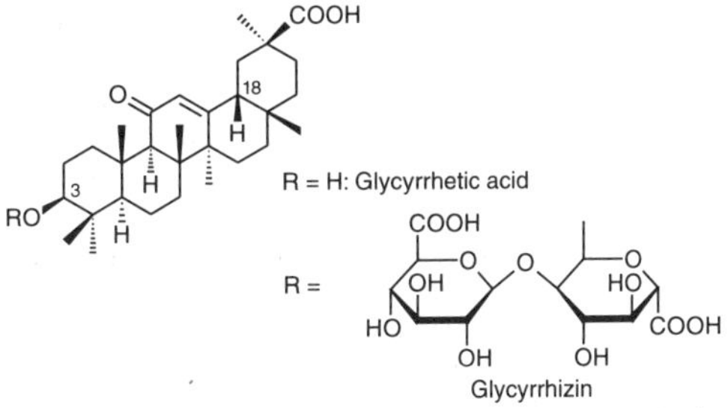

$C_{30}H_{46}O_4$, M_R 470.69. Besides the naturally occurring 18β-form {needles, mp. 300–304 °C, $[\alpha]_D$ +161° ($CHCl_3$)} there is also an 18α-form ("β-G.", platelets, mp. 330–335 °C), soluble in alcohol, chloroform. G., isolated from licorice (liquorice; *Glycyrrhiza glabra*, Fabaceae), is the aglycone of *glycyrrhizin. It is used in medicine as an anti-inflammatory agent, the 3-hemisuccinate is used in ulcer therapy (carbenoxolone). LD_{50} (mouse p.o.) 560 mg/kg. For side effects, see glycyrrhizin.

Lit.: Beilstein EIV **10**, 3775 ▪ Chem. Pharm. Bull. **36**, 3264 (1988); **42**, 1016 (1994) ▪ Karrer, No. 2006 ▪ Seifen, Öle, Fette, Wachse **112**, 990 (1996); **124**, 920ff. (1998) ▪ Zechmeister **73**, 1–155 (review). – *[HS 2918 90; CAS 471-53-4]*

Glycyrrhizin [glycyrrhizic acid, glycyrrhet(in)ic acid glycoside, 2-β-glucuronido-α-glucuronide of glycyrrhet(in)ic acid (formula, see there), liquorice sugar]. A *triterpene, $C_{42}H_{62}O_{16}$, M_R 822.94, colorless, very sweet-tasting crystals, soluble in hot water and alcohol, G. occurs (up to 14%) as potassium and calcium salts in the roots of the liquorice plants (*Glycyrrhiza glabra*, *Gl. glandulifera*, and *Gl. typica*, Fabaceae) cultivated in Europa and the Middle East. G. is a sweetening agent with the 50-fold sweetness of saccharose and a pronounced liquorice taste; it is used in tobacco products and in drugs. G. has anti-inflammatory activity and is used in the form of liquorice juice (Sirupus liquiritiae) as a cough mixture and expectorant. G., present in considerable amounts in liquorice, and glycyrrhet(in)ic acid resemble in their stereostructure steroids such as hydrocortisone and prednisone, this is probably the reason for the toxicity. They cause peudoaldosteronism by blockade of 5β-reductase. The daily dose of G. consumed should not exceed 100 mg[1].

Lit.: [1] Sax (8.), GIE 100, GIG 000.
gen.: Beilstein EIII/IV **18**, 5156 ▪ Fitoterapia **55**, 279 (1984) ▪ Food Chem. Toxicol. **26**, 435 (1988); **31**, 303 (1993) ▪ Steroids **59**, 121 (1994) ▪ Zechmeister **73**, 1–155 (review). – *[HS 2938 90; CAS 1405-86-3]*

Gnidicin. Formula, see daphnetoxin. $C_{36}H_{36}O_{10}$, M_R 628.68, $[\alpha]_D$ +86.5° ($CHCl_3$), a *diterpene of the daphnane type from *Gnidia lamprantha* with antileukemic properties.

Lit.: J. Am. Chem. Soc. **97**, 672 (1975) ▪ Sax (8.), MDJ 250. – *[CAS 55319-39-6]*

Goitrin (5-vinyloxazolidine-2-thione).

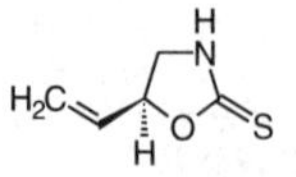

(*S*)-(-)-form

C_5H_7NOS, M_R 129.18, prisms, mp. 50 °C, $[\alpha]_D$ –70.5° (CH_3OH), pK_a 10.5, occurs in the leaves and seeds of numerous crucifers (Brassicaceae), e. g., rape (*Brassica napus*) and cabbage varieties. Epi-G.: mp. 47–48 °C, $[\alpha]_D^{20}$ +75.1° ($CHCl_3$) is isolated from *Isatis indigotica*. G., like the biogenetic precursor progoitrin (see glucosinolates), is an unwanted component of foods (cabbages, rapeseed oil) on account of its toxicity. White cabbage contains ca. 20 mg/kg G. in autumn but only 1 mg/kg in spring.

Activity: G. reduces the activity of 5′-monodeiodinase in the thyroid and liver, resulting in insufficient levels of triiodothyronine (T_3) and thyroxine (T_4) with thyroid overfunction and development of goiter ("cabbage goiter") or enlargement of the liver.

Biosynthesis: G. or epi-G. is formed by enzymatic cleavage (myrosinase, EC 3.2.3.1) from progoitrin or epiprogoitrin, respectively, via an unstable isothiocyanate, see glucosinolates.

Lit.: Beilstein E V **27/10**, 212 f. ▪ Heredity **72**, 594–598 (1994) (metabolism) ▪ J. Heterocycl. Chem. **27**, 811 f. (1990) (synthesis) ▪ Phytochemistry **26**, 669–673 (1987) (biosynthesis) ▪ Sax (8.), VQA 100 (toxicology) ▪ Tetrahedron: Asymmetry **5**, 1157–1160 (1994) (synthesis). – *Pharmacology:* Hager (5.) **4**, 540, 558 ▪ J. Anim. Physiol. Anim. Nutr. **66**, 12–27 (1991) ▪ J. Nutr. **123**, 1554–1561 (1993). – *[HS 2934 90; CAS 500-12-9 (S); 1072-93-1 ((R), epi-G.); 13997-13-2 (±)]*

Gomphidic acid.

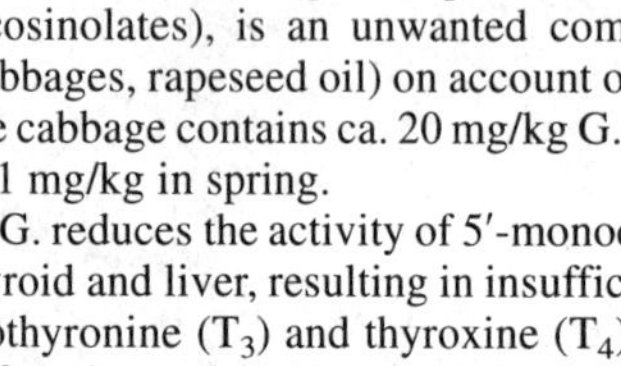

$C_{18}H_{12}O_9$, M_R 372.29, orange red crystals, mp. 300–302 °C (decomp.). A *pulvinic acid derivative from the mushroom *Gomphidius glutinosus* (Basidiomycetes). Together with *xerocomic acid G. is responsible for the yellow color of the stem base.

Lit.: Beilstein E V **18/9**, 411 ▪ Tetrahedron Lett. **27**, 403 (1986) (review) ▪ Zechmeister **51**, 43, 44, 47 (review). – *[CAS 25328-77-2]*

Gomphilactone see benzene-1,2,4-triol.

Gomphrenins. G. I–VIII belongs to the betacyans (see betalains). The name G. encompasses the 6-*O*-glucoside (G. I) and 6′-*O*-acyl derivatives (esterified with hydroxycinnamic acids, e. g., 4-coumaric and ferulic acids) of *betanidin. G. occurs in the flowers of *Gomphrena globosa* (Amaranthaceae). G. I has also been found in other members of the Caryophyllales where it possibly serves as a precursor of the 6-*O*-sophoroside of betanidin (see bougainvilleins).

Lit.: Manske **39**, 16 ▪ Phytochemistry **37**, 761 (1994). – *[CAS 17008-59-2 (G. I)]*

Gonane see steroids.

Gondoic acid see rapeseed oil.

Goniopedaline see oxindole alkaloids.

Goniothalamin see styrylpyrones.

Gonyauline [((1*S*)-*cis*-2-carboxycyclopropyl)dimethylsulfonium betaine].

$C_6H_{10}O_2S$, M_R 146.20, $[\alpha]_D$ +214° (CH_3OH). G. is isolated from the dinoflagellate *Gonyaulax polyedra* (10 mg/1 g wet cells). It shortens the period of the circadian rhythm of bioluminescence in this organism.

Lit.: Experientia **47**, 103 (1991) ▪ J. Chem. Soc., Perkin Trans. 1 **1990**, 3219 ▪ Tetrahedron Lett. **33**, 2821 (1992) (synthesis) ▪ Tetrahedron **53**, 9067 (1997) (biosynthesis). – *[CAS 125559-51-5]*

Gonyautoxins see saxitoxin.

Gorgosterol see sterols.

Gorlic acid see chaulmoogric acid.

Gossypetin (3,3′,4′,5,7,8-hexahydroxyflavone). Formula, see flavones. $C_{15}H_{10}O_8$, M_R 318.24, mp. 300–310 °C (decomp.). G. occurs as the 8-*O*-glucoside [(*gossypin*) $C_{21}H_{20}O_{13}$, M_R 480.38, golden yellow needles, mp. 228–230 °C (decomp.)] in yellow mallows and cotton flowers. It can be obtained by hydrolysis of gossypin.

Lit.: Beilstein E V **18/5**, 657 ▪ Karrer, No. 1558, 1560 ▪ Phytochemistry **27**, 1491 (1988) ▪ see also flavones. – *[CAS 489-35-0 (G.); 652-78-8 (gossypin)]*

Gossypol (thespesin).

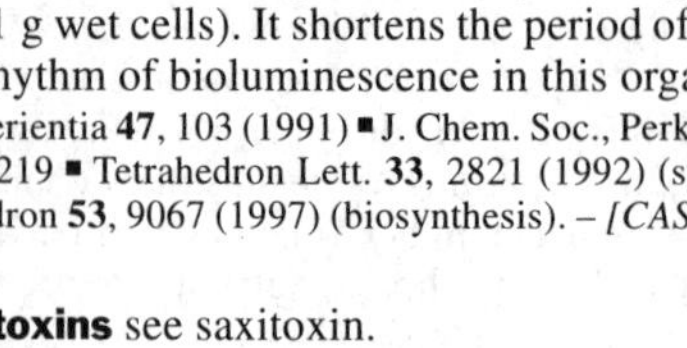

$C_{30}H_{30}O_8$, M_R 518.56, canary yellow crystals, mp. 181–184 °C {(+)-form, $[\alpha]_D$ +445° or 371° ($CHCl_3$)} and 199 or 214 °C (racemate), respectively. An atropisomeric, aromatic dimeric *sesquiterpene, the (+)-form or racemate of which is contained as the main yellow pigment of cotton seeds, occurs not only in cotton seed oil but also in the root bark (ca. 1.2%) of the cotton plant (*Gossypium* spp., Malvaceae). G. is mildly toxic and has been developed in Brazil as a contraceptive ("pill for the man") on account of its inhibitory activity on lactate dehydrogenase and thus also spermatogenesis, in addition activity against Herpes genitalis has been described [1].

Lit.: [1] Am. J. Obstet. Gynecol. **142**, 593 (1982); Drugs **38**, 333 (1989); **48**, 851 (1994); J. Neuro-Oncol. **19**, 25 (1994).
gen.: Beilstein E V **19/3**, 730 ▪ Helv. Chim. Acta **72**, 353 (1989) (synthesis) ▪ J. Chem. Soc. Chem. Commun. **1997**, 1573 [synthesis (*S*)-(+)-G.] ▪ Phytochemistry **33**, 335 (1993) (biosynthesis) ▪ Tetrahedron **54**, 10493–10511 (1998) [synthesis (*S*)-(+)-G.] ▪ Ullmann (5.) **A 10**, 182, 224. – *[CAS 303-45-7 (G.); 20300-26-9 (+); 90141-22-3 (–)]*

Gossypose see raffinose.

Gracilins. Components of the sponge *Spongionella gracilis* (G. A–F), e. g., *G. B*, $C_{22}H_{28}O_8$, M_R 420.46, mp.

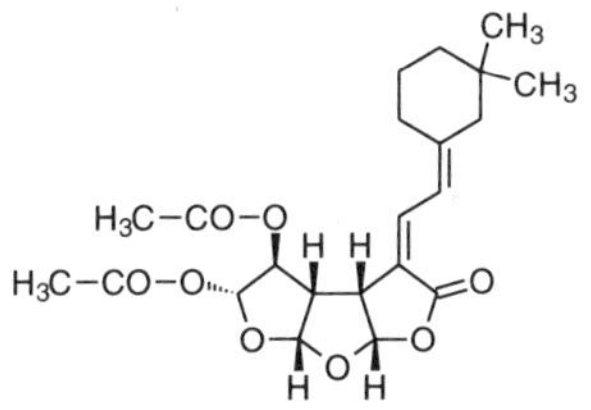

Gracilin B

167–168 °C, $[\alpha]_D$ +191° ($CHCl_3$). A synthesis of G. B has been reported[1].

Lit.: [1] J. Am. Chem. Soc. **117**, 9616 (1995).

gen.: Aust. J. Chem. **43**, 1713 (1990) ▪ J. Nat. Prod. **49**, 823 (1986) ▪ Tetrahedron **42**, 5369 (1986). – *[CAS 96313-95-0 (G. B)]*

Grahamine see monocrotaline.

Gramicidins.

OHC–X–Gly–Ala–D-Leu–Ala–D-Val–Val–D-Val–Trp–D-Leu (X = position 1)

HO–CH_2–CH_2–NH–Trp–D-Leu–Trp–D-Leu–Y (Trp = position 15, Y = position 11)

	X	Y
Valine-Gramicidin A	Val	Trp
Isoleucine-Gramicidin A	Ile	Trp
Valine-Gramicidin B	Val	Phe
Isoleucine-Gramicidin B	Ile	Phe
Valine-Gramicidin C	Val	Tyr
Isoleucine-Gramicidin C	Ile	Tyr

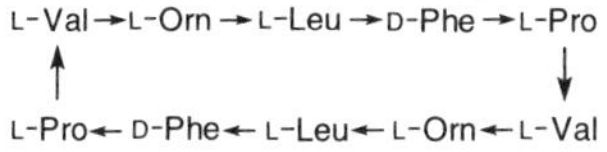

Gramicidin S

A group of peptide*antibiotics from the culture broth of *Bacillus brevis* (some have been synthesized) that are only active against Gram-positive bacteria. The open-chain G. (A, B, C, obtained by fractionation of G. D) consist of 15 amino acids (*pentadecapeptides*) and differ only in the type of amino acid at positions 1 and 11; *Example: G. A* ($C_{99}H_{140}N_{20}O_{17}$, M_R 1882.33) consists of 2 molecules each of L-valine and D-valine, L-alanine, 4 molecules each of D-leucine and L-tryptophan, and one of glycine, cf. *Lit.*[1]. The content of D-amino acids is noteworthy. In contrast, *G. S* (name for *Soviet G.,* since it was first isolated in 1942 in the former USSR) is a cyclopeptide ($C_{60}H_{92}N_{12}O_{10}$, M_R 1141.47) made up of two identical pentapeptides. The components G. A, G. B, G. C as well as G. S_1, G. S_2, and G. S_3 are described in *Lit.*[2]. For the biosynthesis of G. S see *Lit.*[3]. Together with the *tyrocidines G. are constituents of the antibiotic tyrothricin (ratio 80:20), produced on a technical scale by fermentation. The activity of G. is based on the destruction of proton gradients (e.g., of oxidative phosphorylation) whereby a channel made up of 2 molecules of G. transports specific cations (Na^+ and K^+) through the cell membrane (ionophoresis)[4]. According to an X-ray analysis the crystal structure involves a counter-clockwise, antiparallel, double-stranded helical dimer with a total length of 31 Å and an average channel diameter of 4.8 Å. The channels contain three pouch-like expansions: the potential ionic binding sites which migrate through the channel by means of cooperative conformational changes of the participating amino acids[5].

Lit.: [1] Martindale (30.), p. 90. [2] Kirk-Othmer (4.) **3**, 275 ff. [3] Chem. Abstr. **97**, 195394 (1982). [4] Angew. Chem. Int. Ed. Engl. **39**, 900 (2000); Annu. Rev. Physiol. **46**, 531–548 (1984). [5] Science **241**, 182, 188 (1988).

gen.: Beilstein EV **26/19**, 649 ▪ Bull. Chem. Soc. Jpn. **61**, 3925–3929 (1988) ▪ Curr. Top. Membr. Transp. **33**, 15–33, 35–50, 91–111, 113–130 (1988) ▪ Hager (5.) **8**, 382 f. ▪ J. Am. Chem. Soc. **115**, 10492 (1993) ▪ J. Bioenerg. Biomembr. **19**, 655–676 (1987) ▪ Nachr. Chem. Tech. Lab. **36**, 984 (1988) ▪ Q. Rev. Biophys. **20**, 173–200 (1987) ▪ Sax (8.), GJO 000, GJO 025 ▪ Snell-Hilton **5**, 466 f., 540 ff. ▪ Ullmann (5.) **A 2**, 498 ff., 527. – *[HS 2941 90; CAS 11029-61-1 (G. A); 11041-38-6 (G. B); 9062-61-7 (G. C); 113-73-5 (G. S_1); 92278-12-1 (G. S_2); 92278-13-2 (G. S_3)]*

Gramine (donaxine).

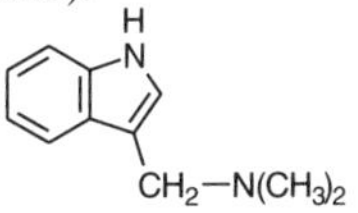

$C_{11}H_{14}N_2$, M_R 174.25, crystals, mp. 139 °C, insoluble in water, soluble in alcohol, ether, chloroform. The name is derived from Gramineae (Poaceae, grasses). G. was the first alkaloid to be isolated from grasses and belongs to the class of indolylalkylamines (for biological activity, see there). G. acts as an antioxidant[1].

Biosynthesis: The first alkaloid whose biosynthesis was investigated by *in vivo* feeding with radioactively labelled compounds[2]. 3-(Aminomethyl)indole and 3-[(methylamino)methyl]indole serve as precursors of G. *Tryptophan is a specific precursor providing the 3-methylindole part as well as the β-nitrogen. A cleavage of the side chain of tryptophan is feasible but has not yet been proven[3]. The main site of synthesis in the Poaceae (e.g., barley) appears to be the basal region of the seed leaves; G. is transported from there to the leaf tips where it accumulates. High concentrations of G. are observed under conditions of thermal stress.

Lit.: [1] Plaste Kautsch. **24**, 401–404 (1977); Tetrahedron Lett. **1971**, 4047. [2] Can. J. Chem. **29**, 1037 (1951). [3] Phytochemistry **14**, 471 (1975).

gen.: Beilstein EIV **22**, 4302; EV **22/10**, 25 f. ▪ Hager (5.) **4**, 440; **9**, 1112 ▪ J. Antibiotics **47**, 724 (1994) ▪ Mothes et al., p. 273 ff. (biosynthesis). – *[HS 2933 90; CAS 87-52-5]*

Graminin see polyfructosans.

Granatan-3-one see pseudopelletierine.

Granaticins.

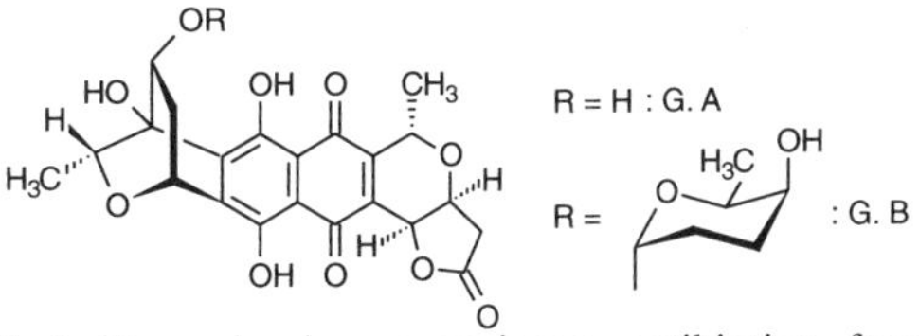

Red *benzoisochromanequinone antibiotics from streptomycetes (e.g., *Streptomyces violaceoruber*), the color changes to blue under alkaline conditions (indicator). *G. A* (litmomycin): $C_{22}H_{20}O_{10}$, M_R 444.39, cryst., mp. 223–225 °C. In the biosynthesis the skeleton originates from an octaketide (*polyketides, *acetogenins), a 2,6-deoxy sugar is condensed to the quinone in a subsequent step. *G. B* ($C_{28}H_{30}O_{12}$, M_R 558.54) is additionally glycosylated with α-L-rhodinose. The G. inhibit the initiation step of RNA biosynthesis and are active against Gram-positive bacteria and tumor cells (*in vitro*).

Lit.: Beilstein E V **19/10**, 689 ▪ J. Antibiot. **41**, 512, 570 (1988) ▪ Justus Liebigs Ann. Chem. **1987**, 751 ▪ Merck-Index (12.), No. 4555 ▪ Tetrahedron **33**, 673 (1977). – *Biosynthesis:* Angew. Chem. Int. Ed. Engl. **28**, 146–174 (1989) ▪ J. Am. Chem. Soc. **101**, 701 (1979). – *Synthesis:* J. Am. Chem. Soc. **109**, 3402 (1987) ▪ J. Chem. Soc., Chem. Commun. **1989**, 354. – *Toxicology:* Sax (8.), GJS 000. – *[HS 2941 90; CAS 19879-06-2 (G. A); 19879-03-9 (G. B)]*

Grandisol.

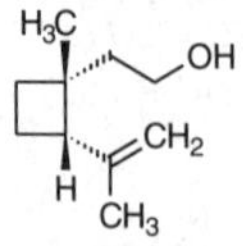

$C_{10}H_{18}O$, M_R 154.25, oil, bp. 50–60 °C (133 Pa), $[\alpha]_D^{21.6}$ 22.1° (hexane). Sexual pheromone of the male boll weevil (*Anthonomus grandis*). G. is also a component of the *pheromone system of bark beetles of the genus *Pityophthorus*. In addition it forms, together with the aldehyde (*grandisal*; $C_{10}H_{16}O$, M_R 152.24), the aggregation pheromone of weevils of the genus *Pissodes*.

Lit.: J. Chem. Ecol. **21**, 1043–1063 (1995) ▪ J. Insect. Physiol. **20**, 1825 (1974) (biosynthesis) ▪ Science **166**, 1010 (1969); **194**, 139–148 (1976) ▪ Ullmann (5.) **A 9**, 432 ▪ Zechmeister **37**, 18–29. – *Synthesis:* ApSimon **4**, 80, 120; **9**, 303–312 ▪ J. Org. Chem. **60**, 7256–7266 (1995) ▪ Org. Lett. **2**, 163 (2000) ▪ Tetrahedron **52**, 1279–1292, 3879–3888 (1996) ▪ Tetrahedron Lett. **35**, 9211 f. (1994); **40**, 7569 (1999). – *[CAS 26532-22-9 (G.); 63623-52-9 (grandisal)]*

Grape flavor see fruit flavors.

Grapefruit aroma see *p*-menth-1-ene-8-thiol and eremophilanes.

Grapefruit bitter principle see naringin.

Grape sugar see D-glucose.

Grayanotoxins. Toxic diterpenoids from leaves of Ericaceae species (Ericaceae diterpenes), e. g. *Rhododendron maximum* or *Leucothoe grayana*. G. also occur in the nectar of rhododendron flowers. Mass poisonings through contaminated honey even occurred in Antiquity. About 25 G. have been isolated. The most important are G. I, II, and III (see table).

G. have neurotoxic activity – similar to *batrachotoxins – through hyperpolarization of sodium ion channels in neurons.

Table: Data of Grayanotoxins.

	G. I	G. II	G. III
synonyms	Andromedo-toxin	Deacetyl-anhydro-andromedo-toxin	Andromedol, Deacetyl-andromedo-toxin
molecular formula	$C_{22}H_{36}O_7$	$C_{20}H_{32}O_5$	$C_{20}H_{34}O_6$
M_R	412.52	352.47	370.49
mp. [°C]	267–270	197–200	218 (decomp.)
optical activity	$[\alpha]_D$ –8.8° (C_2H_5OH)	$[\alpha]_D^{28}$ –41.9°	$[\alpha]_D^{15}$ –12°
LD_{50} (mg/kg)	1.3 (mouse i. p.)	26 (mouse i. p.)	0.8 (mouse i. p.)
CAS	4720-09-6	4678-44-8	4678-45-9

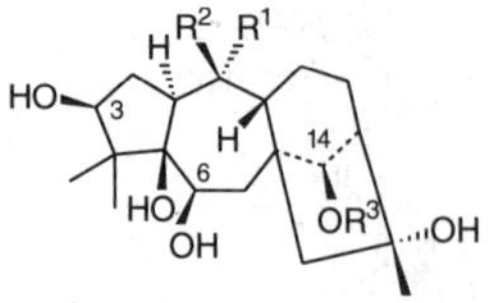

R^1	R^2	R^3	
OH	CH_3	CO–CH_3	G. I
CH_2		H	G. II
OH	CH_3	H	G. III

Lit.: Beilstein E IV **6**, 7930, 7895 ▪ Hager (5.) **3**, 72, 73; **4**, 988; **5**, 608 ▪ J. Nat. Prod. **53**, 131 (1990) ▪ J. Org. Chem. **59**, 5532 (1994) ▪ Karrer, No. 3678, 3680, 3679 ▪ Merck-Index (12.), No. 4566 ▪ Phytochemistry **19**, 2159–2162 (1980) ▪ Sax (8.), AOO 375.

Grevillins. Characteristic pigments of the fungal genus *Suillus* (Basidiomycetes), especially *S. grevillei* (Larch suillus: G. A, B, C), *S. luteus* (Slippery Jack: G. A, B, C), *S. granulatus* (Dotted-Stalk Suillus: G. B, C, D). They are orange to orange red compounds of the 2*H*-pyran-2,5(6*H*)-dione type and differ in the hydroxylation pattern on the phenyl rings.

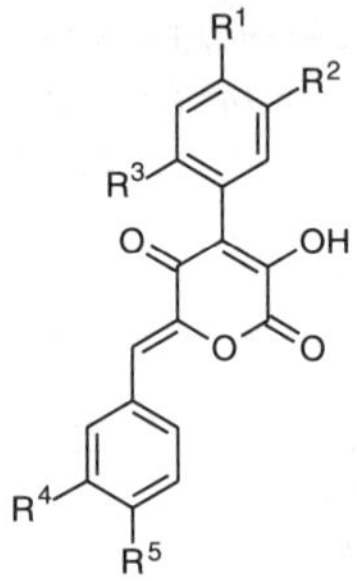

R^1	R^2	R^3	R^4	R^5	
OH	H	H	H	OH	G.A
OH	H	H	OH	OH	G.B
OH	OH	H	OH	OH	G.C
H	OH	OH	OH	OH	G.D

Table: Data of Grevillins.

name	molecular formula	M_R	mp. [°C]	CAS
G. A	$C_{18}H_{12}O_6$	324.29	197–199 (peracetate)	41744-32-5
G. B	$C_{18}H_{12}O_7$	340.29	184–186 (peracetate)	41744-33-6
G. C	$C_{18}H_{12}O_8$	356.29	175–178 (peracetate)	41744-34-7
G. D	$C_{18}H_{12}O_8$	356.29	300 (decomp.)	54707-49-2

The G. are easily recognized by their violet color reaction with concentrated sulfuric acid and are thus distinguished from *pulvinic acids. The G. in fungi are formed by condensation of 2 molecules of 4-hydroxyphenylpyruvic acid and subsequent hydroxylation steps (see also terphenylquinones).

Lit.: Zechmeister **51**, 4–12 (review). – *Synthesis:* Aust. J. Chem. **43**, 1495 (1990) ▪ J. Chem. Soc., Perkin Trans. 1 **1991**, 2363 ▪ Justus Liebigs Ann. Chem. **1986**, 177.

Grifolin.

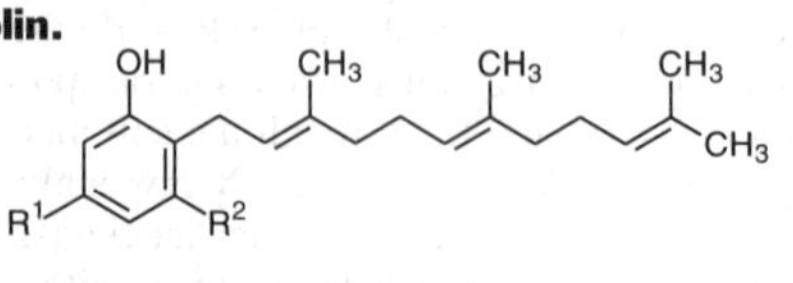

R^1 = CH_3, R^2 = OH : Grifolin
R^1 = OH, R^2 = CH_3 : Neogrifolin

$C_{22}H_{32}O_2$, M_R 328.49, crystals, mp. 43 °C. Farnesylphenol from *Albatrellus confluens* and other *A.* species

(Basidiomycetes). G. exhibits antibiotic and cytostatic activity and is accompanied by the isomeric *neogrifolin* (oil).

Lit.: Beilstein EIV **6**, 6631 ▪ Bull. Chem. Soc. Jpn. **67**, 1178 (1994) ▪ Chem. Pharm. Bull. **36**, 2239 (1988) (synthesis) ▪ Justus Liebigs Ann. Chem. **1981**, 2099 ▪ J. Nat. Prod. **58**, 324 (1995). – *[CAS 6903-07-7 (G.); 23665-96-5 (neo-G.)]*

Grindelic acid see labdanes.

Grisemin see streptothricins.

Griseochelin see zincophorin.

Griseofulvin.

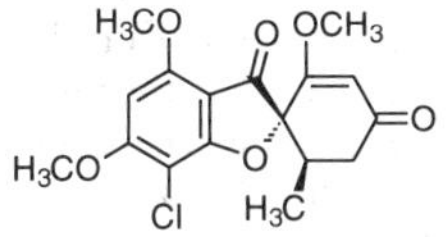

$C_{17}H_{17}ClO_6$, M_R 352.77, colorless crystals, mp. 225–226 °C, $[\alpha]_D$ +370° ($CHCl_3$). A fungistatic antibiotic discovered in 1939[1] in cultures of *Penicillium griseofulvum*; the structure was elucidated in 1952[2]. G. acts in fungi as an inhibitor of mitosis by influencing cell division during formation of the spindle apparatus. It is used as an oral antimycotic agent in human medicine, but is too expensive for wide use in crop protection (production by fermentation of *P. patulum*). In animal experiments there are indications of cancerogenic activity in some species.

Lit.: [1] Biochem. J. **33**, 240 (1939). [2] J. Chem. Soc. **1952**, 3949. *gen.:* Beilstein E V **18/5**, 150 ▪ Berdy, in Rehm & Reed (eds.), Biotechnology, vol. 4, p. 483–486, Weinheim: Verl. Chemie 1986 ▪ Martindale (30.), p. 323 ▪ Mutat. Res. **195**, 91 (1988). – *Synthesis:* Chem. Pharm. Bull **45**, 327, 2011 (1997) ▪ J. Am. Chem. Soc. **101**, 7018 (1979); **113**, 8561 (1991). – *[CAS 126-07-8]*

Griseolic acids.

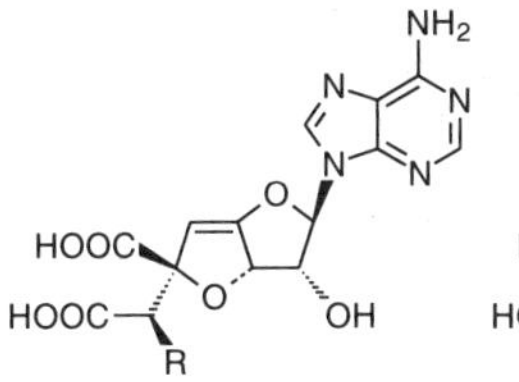

R = OH : G. A
R = H : G. B

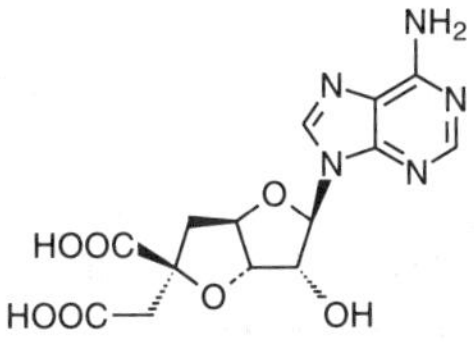

G. C

Table: Data of Griseolic acids.

G.	molecular formula	M_R	$[\alpha]_D$ (DMSO)	CAS
A	$C_{14}H_{13}N_5O_8$	379.29	+ 6.9°	79030-08-3
B	$C_{14}H_{13}N_5O_7$	363.29	–50.7°	98890-01-8
C	$C_{14}H_{15}N_5O_7$	365.30	+13.2°	100242-49-7

Bicyclic nucleoside derivatives formed biosynthetically by linkage of oxaloacetate to adenosine. The three components (G. A to C) are produced by *Streptomyces griseoaurantiacus*. They are competitive inhibitors of cyclic AMP phosphodiesterase (K_i=0.16 mM). G. A stimulates glycogen mobilisation in the liver, in the mouse this leads to a higher blood glucose level.

Lit.: Chem. Pharm. Bull. **35**, 1036 (1987) ▪ J. Am. Chem. Soc. **117**, 7008 ff. (1995) (synthesis G. A) ▪ J. Antibiot. **38**, 823, 830 (1985); **41**, 705, 1711 (1988).

Grossularines.

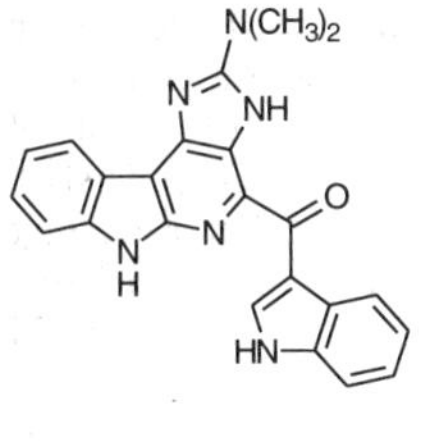

Grossularine 1

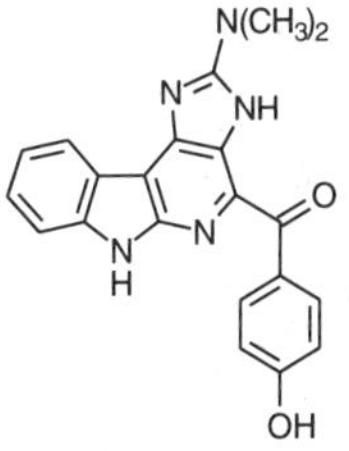

Grossularine 2

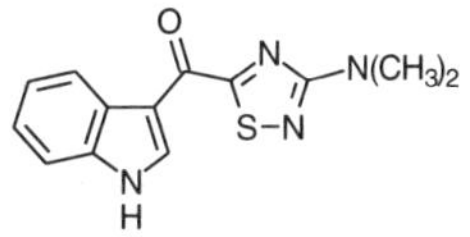

Dendrodoine

G. 1 ($C_{23}H_{18}N_6O$, M_R 394.44, yellow, amorphous, mp. 350 °C) and *G. 2* ($C_{21}H_{17}N_5O_2$, M_R 371.40, mp. 281 °C) are the first α-carbolines isolated from natural sources (tunicate *Dendrodoa grossularia* occurring in the North and Baltic Seas)[1]. G. are cytotoxic[2]. A further constituent is the thiadiazole *dendrodoine*[3] ($C_{13}H_{12}N_4OS$, M_R 272.32, mp. 280–285 °C).

Lit.: [1] J. Org. Chem. **60**, 5899 (1995); Tetrahedron **45**, 3445 (1989); Tetrahedron Lett. **36**, 2615 (1995) (synthesis); Synlett **1995**, 147 (synthesis). [2] Cancer Biochem. Biophys. **9**, 271 (1987). [3] Tetrahedron Lett. **21**, 1457 (1980); Tetrahedron **40**, 681 (1984) (synthesis). – *[CAS 94935-97-4 (G. 1); 102488-58-4 (G. 2); 75351-10-9 (dendrodoine)]*

Growth regulators see plant growth substances.

Growth substances. Collective name for substances that promote growth: the *plant growth substances and regulators (plant hormones,auxins) in plants, some vitamins in microorganisms, hormones in animals – in humans, e.g., somatotropin, *insect hormones in insects – as well as individual vitamins and a large number of other growth factors, some of which have only recently been discovered.

GSH see glutathione.

Guaiacol (2-methoxyphenol, catechol monomethyl ether).

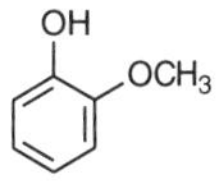

$C_7H_8O_2$, M_R 124.14, mp. 30–32 °C, bp. 205 °C, D_4 1.128, forms prismatic crystals with a strong spicy odor, poorly soluble in water, readily soluble in organic solvents, light-sensitive, weakly toxic [LD_{50} (rat) 0.7 g/kg, (cat) 1.5 g/kg], G. occurs in beech tar oil and was isolated for the first time in 1926 from the distillation products of *guaiac(um).

Use: An important intermediate in organic synthesis, e.g., for preparation of vanillin and G. glycerol ether. G. is used as a flavor substance in foods, as a disinfectant and expectorant for coughs and influenza.

Lit.: Beilstein EIV **6**, 5563 ▪ Food Chem. Toxicol. **20**, 697 (1982) ▪ Hager (5.) **9**, 735 ▪ Karrer, No. 180 ▪ Kirk-Othmer (3.) **9**, 542 ▪ Negwer (6.), p. 1307 ▪ Sax (8.), GKI 000. – *[HS 2909 50; CAS 90-05-1]*

Guaiac(um) (gum guaiac, resin guaiac). The heartwood of *Guaiacum officinale* (pockwood tree, lignum vitae, Zygophyllaceae), brown to brown-green lumps, mp. 85–90 °C, soluble in ethanol, chloroform, insoluble in water; LD_{50} (rat p.o.) >5 g/kg. G. consists to 70% of α- and β-guaiaconic acids (the latter is a mixture of several substances), 11% guaiac resin acid, 10–15% guaiacinic acid, *guaiol* [$C_{15}H_{26}O$, M_R 222.37, mp. 89–90 °C, $[\alpha]_D$ −41° ($CHCl_3$)] and *guaiacol, pyroguaiacin, guaiene, 1% resins, small amounts of essential oil, vanillin, saponins, and pigments (guaiac yellow). G. is used in the food industry as an antioxidant, as a reagent for oxidizing substances (blood pigments in urine and feces, oxidases and peroxidases in pharmacognostic studies). The oxidation of α-guaiaconic acid (= furoguaiacin) to furoguaiacin blue is the basis for the test for occult blood in feces.
Use: In homeopathic medicine for joint rheumatism, dry angina, and pleuritis.
Lit.: Food Cosmet. Toxicol. **2**, 327 (1968) ▪ Hager (5.) **1**, 544; **2**, 699. – *[HS 1301 90; CAS 8016-23-7 (G.); 489-86-1 (guaiol)]*

(−)-6,9-Guaiadiene see geranium oil.

Guaianes.

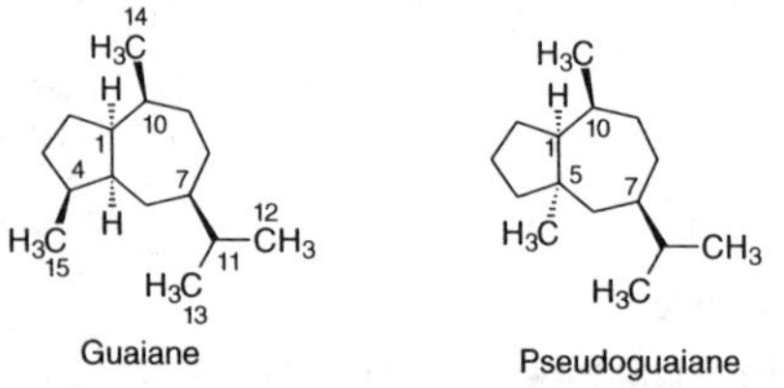

Guaiane Pseudoguaiane

Name for a group of *sesquiterpenes with the 7-isopropyl-1,4-dimethyldecahydroazulene skeleton. Besides the pure hydrocarbons (guaiazulene, see azulenes), ketones, and alcohols, ca. 60 G. lactones (*guaianolides*) are known, some of which are cytotoxic. Most of these are unsaturated 6α,7β-lactones, occurrence of the lactone ring in the 7β,8α- or 7β,8β-positions is much less common. *Absinthin from vermouth is a Diels-Alder product of two guaianolides. In addition, there are ca. 70 *pseudoguaianes* (lactones) that are methyl-substituted at C-5 instead of at C-4.
Lit.: Angew. Chem. Int. Ed. Engl. **12**, 793–806 (1973) ▪ Beilstein E III/IV **5**, 287 ff.; E IV **5**, 1751–1755 ▪ J. Org. Chem. **62**, 7346 (1997) (synthesis) ▪ Synthesis **1972**, 517 ▪ Tetrahedron **44**, 4575 (1988) ▪ Zechmeister **38**, 47–390.

(−)-Guaiaretic acid see lignans.

Guaiazulene see azulenes.

Guaiol see guaiac(um).

Guanine see purines.

Guava flavor see fruit flavors.

Guinesines. Pyrrolidine alkaloids from the bark of *Cassipourea guianensis* (Rhizophoraceae) with insecticidal activity, $C_8H_{15}NOS_2$, M_R 205.35, mixture of diastereomers.

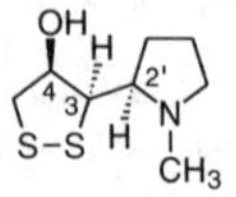

G.A: (*rel*-2'*R*, 3*R*, 4*R*)
(relative configuration)

cis-erythro form

G. A: cis-erythro (*rel*-2'*R*,3*R*,4*R*), oil, $[\alpha]_D^{24}$ +80.5° ($CHCl_3$); *G. B: cis-threo* (*rel*-2'*R*,3*S*,4*S*), needles, mp. 61–62 °C, $[\alpha]_D^{24}$ −36.5° ($CHCl_3$); *G. C: trans-erythro* (*rel*-2'*R*,3*R*,4*S*), needles, mp. 76–77 °C, $[\alpha]_D^{24}$ −5° ($CHCl_3$).
Lit.: J. Heterocycl. Chem. **27**, 1361 (1990) ▪ Tetrahedron Lett. **30**, 3671 (1989). – *[CAS 121702-91-8 (G. A); 121702-92-9 (G. B); 121702-93-0 (G. C)]*

Gutta-percha. The name, derived from Malayan getah pertcha = latex of the percha tree, for a natural rubber (structure, see there) from the gutta-percha trees (*Palaquium gutta* and *P. oblongifolia*, Sapotaceae) with properties similar to those of *balata. In Sumatra, Java, and south east India, the rapidly coagulating latex of incised trees is collected, rapidly kneaded, and marketed as raw G. Pure G. is the *all-trans*-isomer of polyisoprene, related to balata; molecular mass ca. 100 000. In contrast to the *cis*-isomeric *natural rubber, G. is hard and less elastic but not brittle, it softens at 25–30 °C, becomes plastic at 60 °C, and melts at >100 °C with decomposition and formation of a sticky mass. For uses, see *literature*.
Lit.: Hager (5.) **1**, 576 ▪ Kirk-Othmer (3.) **20**, 488 ▪ Ullmann (5.) **A 23**, 225 ff. – *[HS 4001 30; CAS 9000-32-2]*

β-Guttilactone see gamboge.

Guvacine, guvacoline see Areca alkaloids.

Gymnastatins. Cytotoxic spirocyclic compounds produced by a strain of *Gymnascella dankaliensis*, which was isolated from the sponge *Halichondria japonica*, e.g., *G. B*: $C_{24}H_{35}Cl_2NO_5$, M_R 488.44, powder, $[\alpha]_D$ −122° ($CHCl_3$).

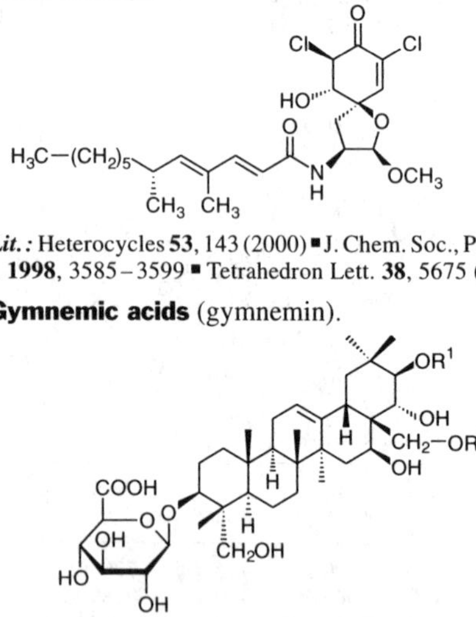

G. B

Lit.: Heterocycles **53**, 143 (2000) ▪ J. Chem. Soc., Perkin Trans. 1 **1998**, 3585–3599 ▪ Tetrahedron Lett. **38**, 5675 (1997).

Gymnemic acids (gymnemin).

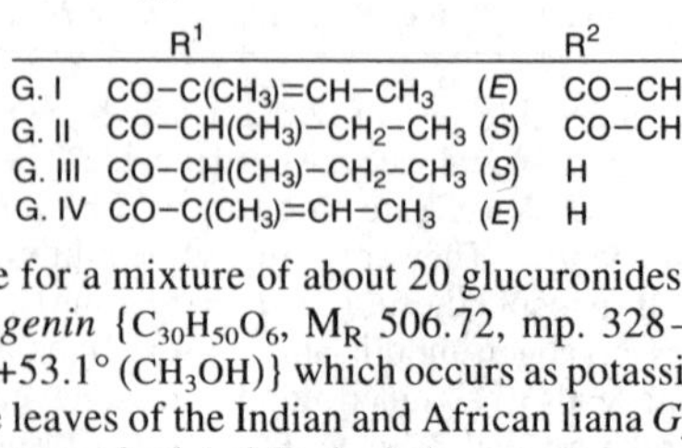

	R^1	R^2
G. I	CO–C(CH₃)=CH–CH₃ (*E*)	CO–CH₃
G. II	CO–CH(CH₃)–CH₂–CH₃ (*S*)	CO–CH₃
G. III	CO–CH(CH₃)–CH₂–CH₃ (*S*)	H
G. IV	CO–C(CH₃)=CH–CH₃ (*E*)	H

Name for a mixture of about 20 glucuronides of *gymnemagenin* {$C_{30}H_{50}O_6$, M_R 506.72, mp. 328–335 °C, $[\alpha]_D$ +53.1° (CH_3OH)} which occurs as potassium salts in the leaves of the Indian and African liana *Gymnema sylvestre* and related Asclepiadaceae. G. is a yellow to brown, amorphous, bitter-tasting powder, soluble in

Table: Data of Gymnemic acids.

G.	molecular formula	M_R	mp. [°C]	$[\alpha]_D$ (CH_3OH)	CAS
G.I	$C_{43}H_{66}O_{14}$	806.99	211–212	+36.7°	122168-40-5
G.II	$C_{43}H_{68}O_{14}$	809.00	212–213	+36.3°	122144-48-3
G.III	$C_{41}H_{66}O_{13}$	766.97	218–219	+ 7.6°	122074-65-1
G.IV	$C_{41}H_{64}O_{13}$	764.95	220–221	+ 8.8°	121903-96-6

ethanol, insoluble in water. G. completely suppresses taste sensitivity for "bitter" or "sweet" for several hours, sour, astringent, or spicy substances, however, can still be tasted. The potassium salt of G. is a red-brown crystalline mass, soluble in water and ethanol. For structure elucidation, see *Lit.*[1]. G. derivatives can inhibit glucose absorption from food (anti-diabetic effect)[2].

Lit.: [1]Chem. Pharm. Bull. **37**, 852–854 (1989); **40**, 1366–1375, 1779ff. (1992); **41**, 1730 (1993); **44**, 469 (1996); **45**, 1671; 2034 (1997); Tetrahedron Lett. **30**, 361, 1103, 1547 (1989). [2]U. S. Pat. 5843909 (1. 12. 1998).

gen.: Atta-ur-Rahman **18**, 649–676 (review) ▪ Nachr. Chem. Tech. Lab. **37**, 580 (1989). – *[CAS 22467-07-8 (gymnemagenin)]*

Gymnocolin.

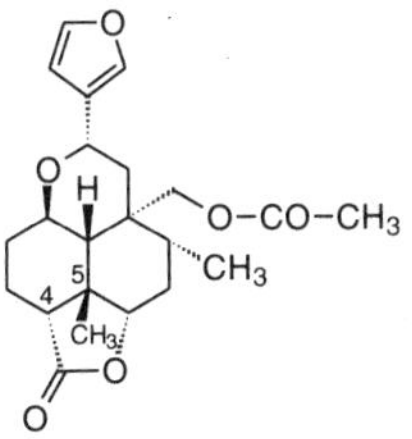

$C_{22}H_{28}O_6$, M_R 388.46, square prisms, mp. 196–197 °C. G., a very bitter-tasting *diterpene of the 4α,5β-*clerodane type from the liverwort *Gymnocolea inflata*, shows a number of biological activities[1,2]. Thus antifeedant effects on larvae of Japanese *Pieris* species, inhibition of root growth (10^{-3}–10^{-4} Mol) in the cress (*Lepidium sativum*), and promotion of germination (10^{-6}–10^{-7} Mol) of wheat seeds have been reported.

Lit.: [1]J. Hattori Bot. Lab. **39**, 215 (1975). [2]Zinsmeister & Mues (eds.), Bryophytes, their Chemistry and Chemotaxonomy (Proc. Phytochem. Soc. Eur. 29), p. 399, 404, Oxford: Clarendon Press 1990.

gen.: Phytochemistry **10**, 3279 (1971) ▪ Tetrahedron Lett. **24**, 115 (1983). – *[CAS 37299-07-3]*

Gymnodimine.

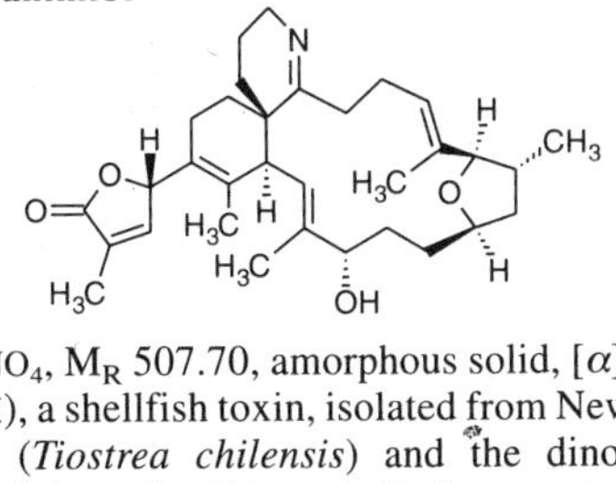

$C_{32}H_{45}NO_4$, M_R 507.70, amorphous solid, $[\alpha]_D^{25}$ –10.4° (CH_3OH), a shellfish toxin, isolated from New Zealand oysters (*Tiostrea chilensis*) and the dinoflagellate *Gymnodinium* cf. *mikimotoi*. G. shows potent ichtyotoxicity and is structurally related to the *pinnatoxins.

Lit.: Tetrahedron Lett. **38**, 4889 (1997). – *[CAS 173792-58-0]*

Gymnodinium breve toxins. Toxins from the unicellular alga *Gymnodinium breve* (*Ptychodiscus brevis*) which belong to the polycyclic polyethers and interact with molecular ion channels. Especially during massive algal blooms ("red tides"), G. lead to poisonings of fish, examples: *brevetoxins A–C.

Lit.: Baden, in Tu (ed.), Marine Toxins and Venoms, p. 259–278, New York: Dekker 1988 ▪ J. Am. Chem. Soc. **108**, 514 (1986); **111**, 4186, 6476 (1989) ▪ Nat. Prod. Rep. **1**, 251, 551 (1984); **3**, 1 (1986); **4**, 539 (1987) ▪ Scheuer I **1**, 1–42 ▪ Scheuer II **1**, 67 ▪ Science **257**, 1476 (1992).

Gymnomitranes (barbatanes).

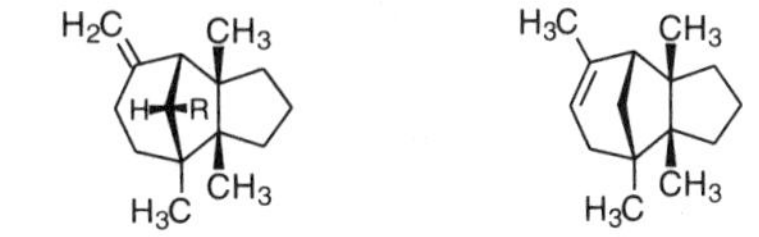

R = H : Gymnomitrene
R = OH : Gymnomitrol

α-Barbatene

Table: Data of Gymnomitranes.

	α-Barbatene	Gymnomitrene	Gymnomitrol
molecular formula	$C_{15}H_{24}$	$C_{15}H_{24}$	$C_{15}H_{24}O$
M_R	204.36	204.36	220.35
mp. [°C]		114–116	
$[\alpha]_D^{20}$	+48–78°	–14–16°	+7°
CAS	53060-59-6	39863-73-5	41410-53-1

A class of *sesquiterpenes occurring in numerous liverwort species (=Hepaticae) but not yet found in other plants. For synthesis, see *Lit.*[1]; biosynthesis *Lit.*[2].

Lit.: [1]Zechmeister **42**, 34; J. Org. Chem. **45**, 4080 (1980); **62**, 1970 (1997) (gymnomitrol). [2]J. Chem. Soc. Chem. Commun. **1998**, 169.

gen.: J. Chem. Res. (S) **1988**, 380 ▪ J. Chem. Soc., Chem. Commun. **1972**, 1320 ▪ Phytochemistry **27**, 2161 (1988).

Gymnoprenols.

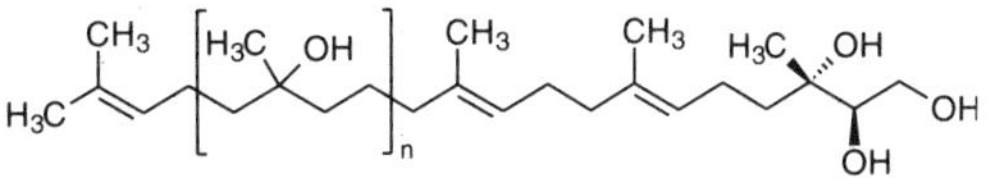

n = 5,6 : Gymnoprenol A_9, A_{10}

G. occur as mixture of homologues and sometimes as 1-*O*-(*S*)-(3-hydroxy-3-methylglutaric acid)-semiesters (gymnopilines, neurotoxic)[1] in fruit bodies of the hallucinogenic agaric *Gymnopilus spectabilis* (big-laughter mushroom, ohwaraitake) (Basidiomycetes). The G. are responsible for the bitter taste of the mushrooms.

Lit.: [1]Phytochemistry **31**, 4355 (1992).

gen.: Phytochemistry **34**, 661 (1993) ▪ Tetrahedron Lett. **24**, 1731, 1735, 1991 (1983); **25**, 1371, 3783, 4023 (1984). – *[CAS 86989-11-9 (G. A_9); 86989-10-8 (G. A_{10})]*

Gyrinidal. $C_{14}H_{18}O_3$, M_R 234.30. Defence substance of water beetles (Carabidae) of the genera *Gyrinus*[1] and *Dineutes*[2,3] produced in the so-called pygidial glands

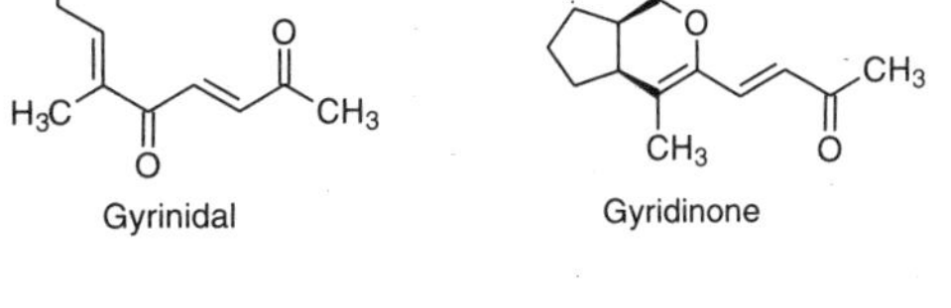

Gyrinidal Gyridinone

of the insects. The (9Z)-isomer (*isogyrinidal*) is also known[3] as well as the cyclic hemiacetal *gyrinidone* [$C_{14}H_{20}O_3$, M_R 236.31, $[\alpha]^{25}_{400}$ −69° (C_2H_5OH)][4].

Lit.: [1] Justus Liebigs Ann. Chem. **1972**, 155–161. [2] Proc. Natl. Acad. Sci. USA **69**, 1208ff. (1972). [3] J. Chem. Ecol. **1**, 59–82 (1975). [4] J. Am. Chem. Soc. **94**, 7589f. (1972). – *[CAS 36518-11-3 (G.); 39013-24-6 (gyrinidone)]*

Gyrocyanin.

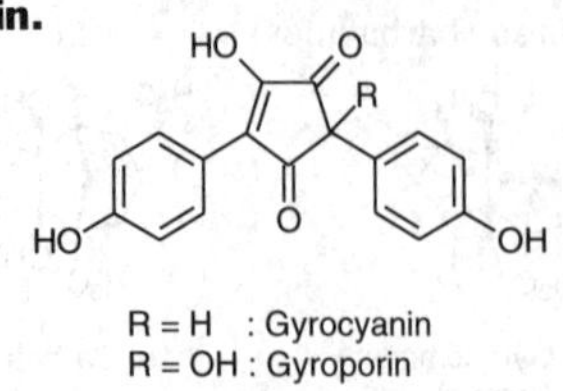

$C_{17}H_{12}O_5$, M_R 296.28, lemon-yellow prisms, mp. ~240 °C. An oxidation-sensitive cyclopentanetrione derivative from the mushrooms *Gyroporus cyanescens* (Basidiomycetes), *Leccinum aurantiacum*, and the gasteromycete *Chamonixia caespitosa*. G. is responsible for the blue color rapidly acquired by these fungi upon injury; see also chamonixin. *Gyroporin* ($C_{17}H_{12}O_6$, M_R 312.28, needles, mp. 209–210 °C, racemate) is formed on oxidation and also occurs in the above-mentioned species as well as in numerous other basidiomycetes; it can be easily recognized on account of the dark blue color reaction with concentrated sulfuric acid.

Lit.: Chem. Ber. **106**, 3223 (1973) ▪ Zechmeister **51**, 63–70 (review) ▪ Z. Naturforsch. C **32**, 46 (1977). – *[CAS 52591-12-5 (G.); 52077-14-2 (gyroporin)]*

Gyromitrin (acetaldehyde formylmethylhydrazone).

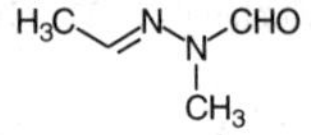

$C_4H_8N_2O$, M_R 100.12, unstable crystals, mp. 19.5 °C. Toxin from the false morel (*Gyromitra esculenta*, Ascomycetes). It occurs in the fresh toadstool as a mixture with higher homologues. G. can be steam distilled which, on the one hand can result in detoxification of the fungus on cooking while, on the other hand, gives rise to toxic vapors. G. is strongly cytotoxic, especially neuro- and hepatotoxic. G. and the parent compound *N*-formyl-*N*-methylhydrazine are strongly carcinogenic (liver cancer).

Lit.: Besl & Bresinsky, Giftpilze, p. 62–70, Stuttgart: Wissenschaftliche Verlagsges. 1985 ▪ Hirono, Naturally Occurring carcinogens of Plant Origin, p. 127–138, Amsterdam: Elsevier 1987 ▪ J. Agric. Food Chem. **31**, 1117 (1983) ▪ Sax (8.), AAH 000 ▪ Tetrahedron Lett. **1967**, 1893. – *[CAS 16568-02-8]*

Gyrophoric acid.

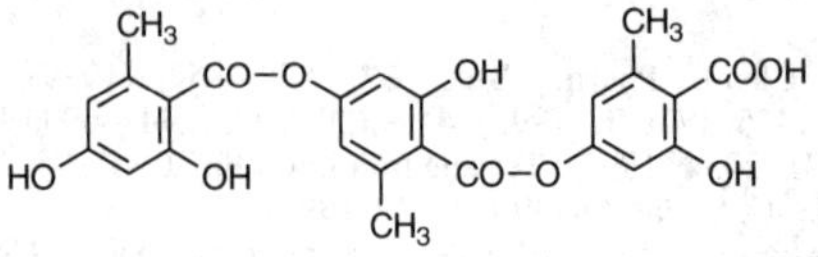

$C_{24}H_{20}O_{10}$, M_R 468.42, needles, mp. 220–225 °C (decomp.); soluble in acetone. A tridepside which, on hydrolysis with concentrated sulfuric acid, gives 3 molecules of *orsellinic acid. It occurs in numerous lichens, especially species from the families Umbilicariaceae, Parmeliaceae, and Roccellaceae.

Lit.: Aust. J. Chem. **48**, 1761 (1995) ▪ Culberson, p. 119–120 ▪ Karrer, No. 1048 ▪ Phytochemistry **42**, 839 (1996). – *[CAS 548-89-0]*

Gyroporin see gyrocyanin.

H

Haedoxans. H. A–H are insecticidal sesquilignans from the Asian herb *Phryma leptostachya* (Verbenaceae). The most potent H. A is able to displace *avermectins from their receptor (LD_{50} 0.25 ng/fly).

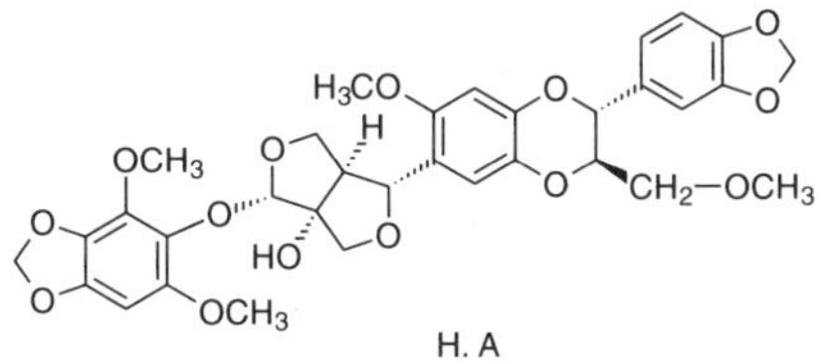

H. A

H. A: $C_{33}H_{34}O_{14}$, M_R 654.61, rods, mp. 158–159 °C, $[\alpha]_D^{18}$ +125° (C_2H_5OH/CH_2Cl_2).
Lit.: Agric. Biol. Chem. **53**, 631–634, 1565–1573 (1989) ▪ Phytochemistry **49**, 613–622 (1998). – *[CAS 123619-26-1 (H. A)]*

Haem... see hem...

Haemanthamine, haemanthidine see Amaryllidaceae alkaloids.

Haematein see haematoxylin.

Haematopodin.

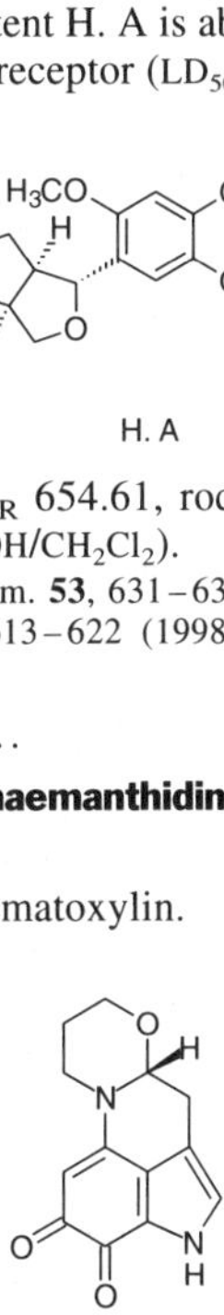

$C_{13}H_{12}N_2O_3$, M_R 244.25, black lustrous crystals (monohydrate), mp. >350 °C, $[\alpha]_D$ +571° (CH_3OH). Pigment from the blood-red latex of the fungus *Mycena haematopus* (Basidiomycetes). It is formed from an unstable precursor during chromatographic work-up.
Lit.: Angew. Chem. Int. Ed. Engl. **32**, 1087 (1993) ▪ Justus Liebigs Ann. Chem. **1996**, 1117–1120 (synthesis). – *[CAS 151964-21-5]*

Haematoxylin (hydroxybrasilin, 3,4′,5′,7,8-pentahydroxy-2′,3-methyleneneoflavan, C. I. natural black 1, C. I. 75290).

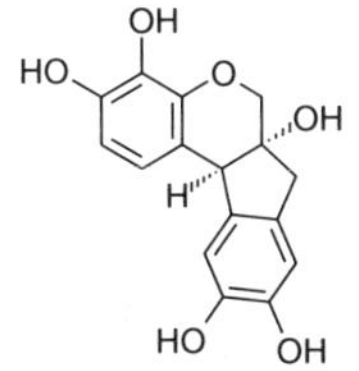

Haematoxylin [(±)-form]

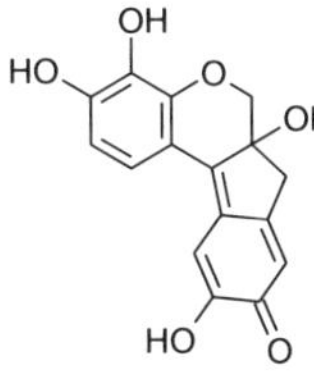

Haematein

$C_{16}H_{14}O_6$, M_R 302.28. Colorless to yellowish crystals, turning reddish in the light; mp. 210–212 °C, soluble in hot water, ethanol, and alkali hydroxide solutions. Under the action of atmospheric oxygen H. is oxidized to the genuine pigment *haematein* [$C_{16}H_{12}O_6$, M_R 300.27, mp. 250 °C (decomp.)]. Haematein forms as brown-red, dark green metallic shining crystals that are poorly soluble in hot water, ethanol, and ether. H. occurs in the heartwood of the logwood or campeachy tree (*Haematoxylum campechianum*, Fabaceae) in glycosidic form.
Use: Used in the past to dye natural fibers, paper, and leather, now only for dyeing historical objects.
Lit.: Beilstein E V **17/8**, 469 ▪ Hager (5.) **1**, 527 ff. ▪ J. Chem. Soc., Perkin Trans. 1 **1995**, 2447 (synthesis). – *[HS 3203 00; CAS 1621-46-1 (H.); 475-25-2 (haematein)]*

Haemocorin aglycone see phenalenones.

Haemofluorone A see phenalenones.

Haemoventosin.

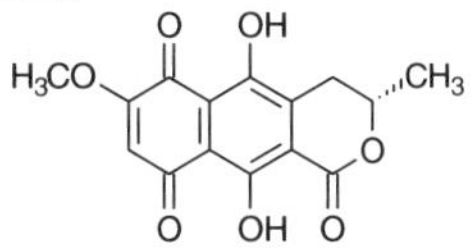

$C_{15}H_{12}O_7$, M_R 304.26, red brown crystals, mp. 202–204 °C. A naphthoquinone from the blood red apothecia of the crustaceous lichen *Ophioparma ventosa*.
Lit.: Acta Chem. Scand. **25**, 483 (1971) ▪ Khim. Prir. Soedin. **1990**, 400 ▪ Z. Naturforsch. B **50**, 1557 (1995). – *[CAS 32638-04-3]*

Halenaquinone.

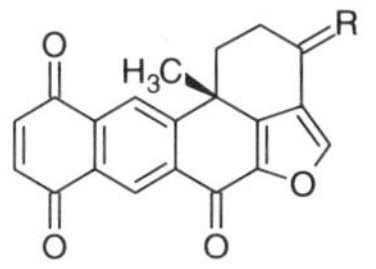

R = O : Halenaquinone
R = H_2 : Xestoquinone

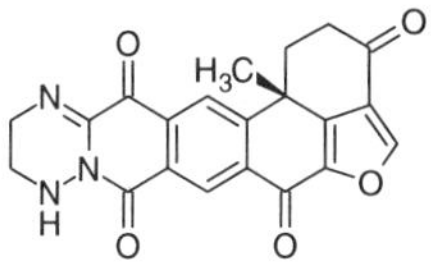

Noelaquinone

$C_{20}H_{12}O_5$, M_R 332.31, yellow solid, mp. >250 °C (decomp.), $[\alpha]_D$ +22.2° (CH_2Cl_2). Meroterpenoid of the sesquiterpene hydroquinone class, biogenetically related to *avarol, *ilimaquinone, and many other compounds from sponges and brown algae [1]. H. was isolated from the pacific sponges *Xestospongia exigua* and *X. sapra* [2], where it is present as the corresponding hydroquinone *halenaquinol* [$C_{20}H_{14}O_5$, M_R 334.33, yellow solid, $[\alpha]_{577}$ +179° (acetone)]. *Xestoquinone* [$C_{20}H_{14}O_4$, M_R 318.33, yellow powder, mp. 212–214 °C, $[\alpha]_D$ +17.2° (CH_2Cl_2)], co-occurs with H. in *Xestospongia* species [3].
H. (and to a lesser degree xestoquinone) are potent irreversible inhibitors of both the oncogenic protein

tyrosine kinase pp60$^{v\text{-}src}$ encoded by the Rous sarcoma virus [and the human epidermal growth factor kinase (EGF)][4]. The weakly antibiotic xestoquinone is a potent cardiotonic agent resulting from its unique positive inotropic effect on cardiac muscle[5] and has been used as a specific biochemical probe for the elucidation of structure and function of the muscle contraction machinery[6]. An unprecedented triazine derivative *noelaquinone*[7] [$C_{21}H_{15}N_3O_5$, M_R 389.37, yellow solid, decomp. >300 °C, $[\alpha]_D$ +53.6° (CH_2Cl_2)] co-occurs with H. in an Indonesian *Xestospongia* species.

Lit.: [1] Tetrahedron **48**, 6667 (1992). [2] J. Am. Chem. Soc. **105**, 6177 (1983); Chem. Pharm. Bull. **33**, 1305 (1985). [3] Chem. Lett. **1985**, 713; J. Org. Chem. **55**, 3158 (1990). [4] J. Org. Chem. **58**, 4871 (1993). [5] J. Pharm. Exp. Ther. **257**, 90 (1991). [6] Biochemistry **34**, 12570 (1995). [7] Heterocycles **49**, 355 (1998).
gen.: Syntheses: (+)-H., J. Am. Chem. Soc. **110**, 8483 (1988) ▪ J. Org. Chem. **61**, 4876 (1996) ▪ Synthesis **1998**, 581 ▪ (+)-X., J. Am. Chem. Soc. **118**, 10766 (1996) ▪ Tetrahedron **54**, 13073 (1998) ▪ (+)-Xestoquinone, J. Org. Chem. **62**, 2330. – *[CAS 86690-14-4 (H.); 97743-96-9 (xestoquinone); 220503-29-7 (noelaquinone)]*

Halichondramide.

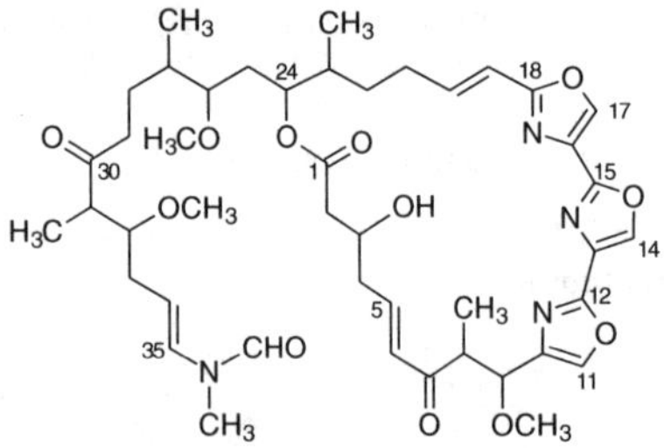

$C_{44}H_{60}N_4O_{12}$, M_R 836.97, mp. 66–68 °C, $[\alpha]_D$ –101° (CH_3OH). A macrolide antibiotic with antifungal activity from *Halichondria* sponges. The structure is similar to those of *ulapualides.

Lit.: J. Nat. Prod. **60**, 150 (1997) (related halishigamides A–D) ▪ J. Org. Chem. **53**, 5014–5020 (1988). – *[CAS 113275-14-2]*

Halichondrin B see halistatins.

Halistatins. Polyether macrolides from Pacific sponges (Porifera) of the genus *Phakellia* with strong antitumor activities[1]. The H. are present to merely 10^{-7}–10^{-8}% in the organisms, e.g., *H. 1*: $C_{60}H_{86}O_{20}$, M_R 1127.33, amorphous, $[\alpha]_D^{25}$ –58.4° (CH_3OH). The ED_{50} for growth inhibition of various human tumor cell lines (brain, lung, ovaries, kidneys) are ca. 10^{-5}–10^{-6} µg/mL.

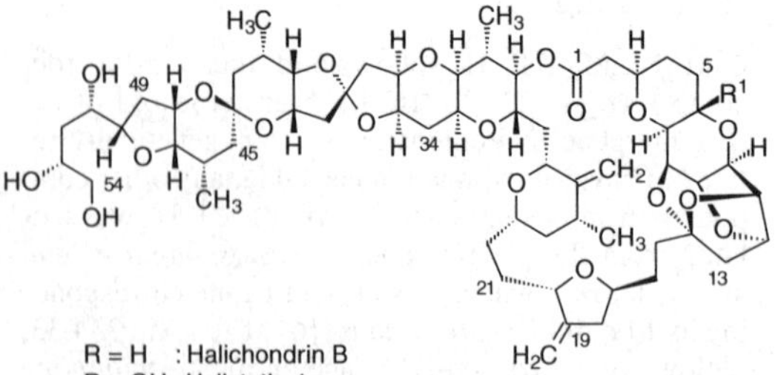

R = H : Halichondrin B
R = OH : Halistatin 1

Phakellia, *Axinella*, and *Halichondria* contain the structurally related halichondrines[2], e.g., *halichondrine B*: $C_{60}H_{86}O_{19}$, M_R 1111.33, crystals, mp. 164–166 °C, $[\alpha]_D^{25}$ –58.9° (CH_3OH), which are also cytotoxic. H. 1 and halichondrin are undergoing preclinical development as anticancer drugs.

Lit.: [1] J. Chem. Soc., Chem. Commun. **1995**, 383; J. Org. Chem. **58**, 2538 (1993). [2] Chem. Ind. (London) **1994**, 226; Gazz. Chim. Ital. **123**, 371 (1993); J. Am. Chem. Soc. **114**, 3162 (1992); Pure Appl. Chem. **58**, 701 (1986).
gen.: Chem. Pharm. Bull. **45**, 1265–1281, 1558–1572 (1997) ▪ Chem. Rev. **95**, 2041 (1995) ▪ Tetrahedron Lett. **38**, 8965 (1997). – *[CAS 147427-91-6 (H. 1); 103614-76-2 (halichondrin B)]*

Hallachrome.

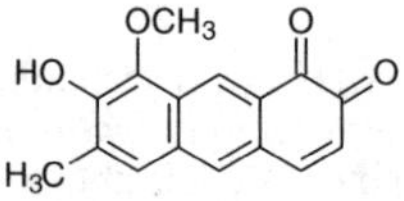

$C_{16}H_{12}O_4$, M_R 268.27, dark red prisms, mp. 224–226 °C (decomp.). The marine polychaete worm *Halla parthenopeia* (Eunicidae) contains the unusual 1,2-anthraquinone H.[1]. It also occurs in traces in *Lumbriconereis impatiens*[2].

Lit.: [1] Experientia **27**, 15 (1971); J. Chem. Soc., Perkin Trans. 1 **1972**, 1614. [2] J. Nat. Prod. **48**, 828 (1985).
gen.: Justus Liebigs Ann. Chem. **1993**, 905 (synthesis). – *[CAS 38393-67-8]*

Hallactone A see nagilactone C.

Halogen compounds (natural). H. c. occur frequently and in numerous structural variations in nature[1,2]. At present about 2500 halogenated natural products are known[1]. The opinion, expressed a few years ago, that chlorine-containing compounds are "unnatural" and that their production should not be allowed is without any justification. Halogenated compounds are produced in particular by marine organisms for which chloride, bromide, and iodide are abundantly available from sea water. *Fluoro-organic natural products, in contrast, are very rare. Chlorine-containing compounds are also frequently found in terrestrial microorganisms and lichens which take up chloride from their surroundings. In addition, halogenated compounds are broadly distributed in widely differing organisms and even occur as hormones in mammals.

CH_3Cl **1** $CHBr_3$ **2**

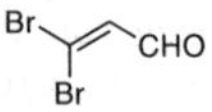

3

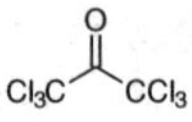

4

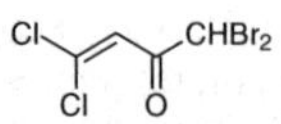

5

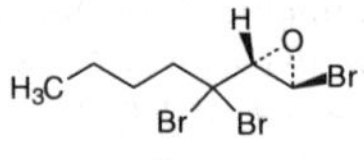

6

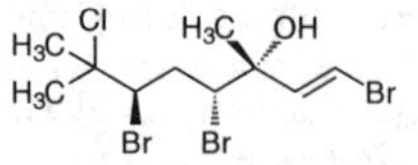

7

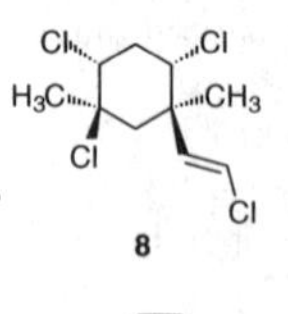

8

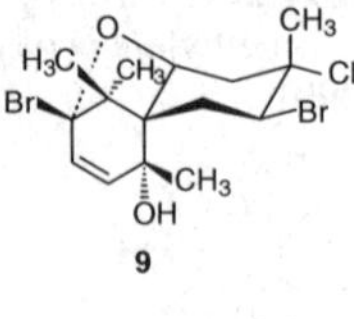

9

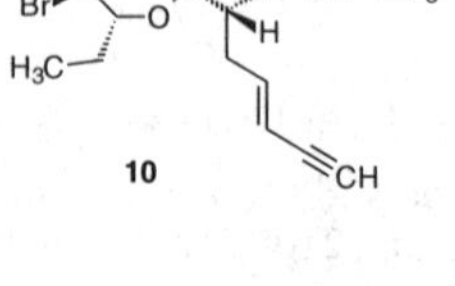

10

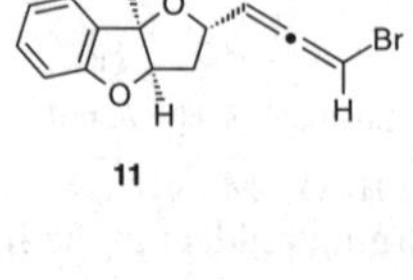

11

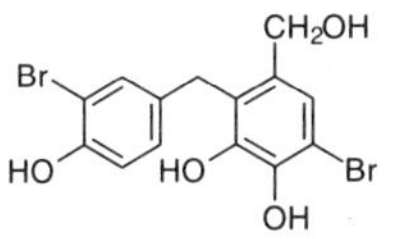

12

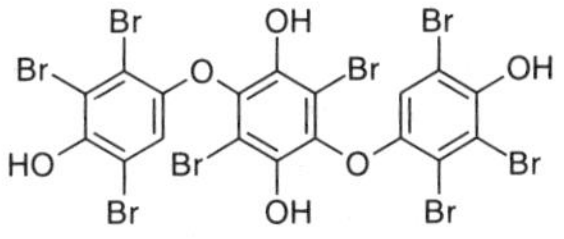

13

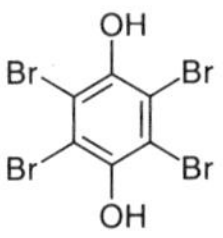

14

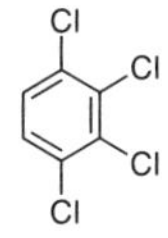

15

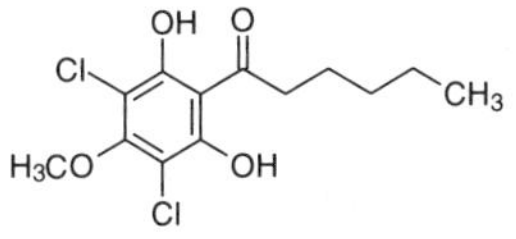

16

Table: Data of Halogen compounds.

no.	molecular formula	M_R	mp. [°C]	$[\alpha]_D$	CAS
3	$C_3H_2Br_2O$	213.86			18328-08-0
4	C_3Cl_6O	264.75			116-16-5
5	$C_4H_2Br_2Cl_2O$	296.77			59228-10-3
6	$C_7H_{11}Br_3O$	350.88		+78.6° ($CHCl_3$)	66002-41-3
7	$C_{10}H_{16}Br_3ClO$	427.40		−64° ($CHCl_3$)	52194-66-8
8	$C_{10}H_{14}Cl_4$	276.03	86−88	+9.4° ($CHCl_3$)	119945-08-3
9	$C_{15}H_{21}Br_2ClO_2$	428.59	148−150.5		33880-90-9
10	$C_{17}H_{23}BrO_3$	355.27	73−74	+70.2° ($CHCl_3$)	3442-58-8
11	$C_{15}H_{15}BrO_2$	307.19		+382°	66389-39-7
12	$C_{14}H_{12}Br_2O_4$	404.05			87402-66-2
13	$C_{18}H_6Br_8O_6$	957.47			74076-55-4
14	$C_6H_2Br_4O_2$	425.70	244		2641-89-6
15	$C_6H_2Cl_4$	215.89			634-66-2
16	$C_{13}H_{16}Cl_2O_4$	307.17			111050-72-1

Chloromethane (**1**) is the most frequent natural H. c., and is responsible for ca. 25% of the chlorine in the atmosphere. It is estimated that natural sources such as wood-destroying fungi, marine algae, phytoplankton, plants, volcanic eruptions, and bush fires produce about 5 million t/a chloromethane while that originating from anthropogenic sources contributes with merely 26000 t/a[3]. Other halogenated alkanes such as bromoform (**2**) are also found in surprising concentrations in sea water, where marine algae are even able to produce trichloroethene and perchloroethene. From the red alga *Asparagopsis taxiformis* ("limu kohu"), which is considered as a delicacy by inhabitants of Hawaii, over 100 different halogenated compounds have been isolated besides **2**, these include all possible halogenated methanes, dibromoacetaldehyde, 3,3-dibromoacrolein (**3**), and halogenated ketones such as hexachloroacetone (**4**) and 4,4-dibromo-1,1-dichlorobut-1-en-3-one (**5**)[1,4]. The red alga *Bonnemaisonia hamifera* contains the α-bromoepoxide (**6**) as the main component[5]. Halogenated cyclopentane and cyclohexane derivatives are produced by various plants (e.g., *cryptosporiopsin, *lachnumone, *mycorrhizin, *mycenone). Polyhalogenated terpenes such as **7** are transferred in the food chain from red algae of the genus *Plocamium* to the lumpfish *Aplysia californica* which uses them as a defence against predators[6]. Polyhalogenated terpenoids often have unusual carbon skeletons such as, e.g., aplysiaterpenoid A (**8**) from *Aplysia kurodai*, an extremely potent antifeedant with insecticidal activity against cockroaches and mosquito larvae exceeding that of lindane[7]. Halogenated sesquiterpenoides and C_{15}-*acetogenins are produced in a remarkable variety by red algae of the genus *Laurencia*[1]. Typical representatives are the *chamigrene derivative pacifenol (**9**)[8], the oxocine laurencin (**10**)[9], and the bromoallene panacene (**11**)[10] which protects the sea hare *Aplysia brasiliana* from sharks (see also aplysi..., okamurallene). Numerous chlorinated *sesquiterpene lactones and *polyynes occur in the Asteraceae where the halogen is usually present as chlorohydrin and α-chloroenone units[1].

A wide variety of halogenated and polyhalogenated phenols and polyphenols is found in marine algae, e.g., avrainvilleol (**12**) which even in minute concentrations deters herbivorous fish from the tropical green alga *Avrainvillea rawsoni*[11]. The acorn worm *Ptychodera flava* (Pentacoela) living on the sea bed contains tetrabromohydroquinone (**14**) and the octabrominated diphenyl ether **13** as main metabolites[12]. Similar polychlorophenyl ethers have been isolated from fruit bodies of fungi (*russuphelins). The oil of the rush *Juncus roemerianus* (Juncaceae) growing in Mississippi salt marshes contains 1.2% 1,2,3,4-tetrachlorobenzene (**15**)[13]. Sea fans living at a depth of 350 m produce chlorinated *azulenes. Marine sponges are able to metabolize tyrosine in many ways via bromination and subsequent transformations (*aerothionin, *discorhabdins, *eudistomins). Cyanobacteria (hapalindoles), bryozoa (*chartelline A, *flustramines), and tunicates use tryptophan in similar diverse ways to build halogenated alkaloids. Chlorine bound to aromatic or heteroaromatic rings is often found in secondary metabolites of streptomycetes and fungi, e.g. in the Armillaria sesquiterpenoids, *calicheamicins, *drosophilin A, *griseofulvin, *maytansinoids, *pyrrolomycins, *pyoluteorin, *rebeccamycin, *tetracyclines, *vancomycin, *virantmycin. Lichen acids such as buellolide, *chloroatranorin, *diploicin, eriodermin, *pannarin, *sordidone, and the *lichen xanthones often contain chlorine. Antimetabolically active chlorinated amino acids are present in the fruit bodies of some agarics (*2-amino-4-pentenoic acid).

Biologically important halogenated compounds include the thyroid hormone thyroxine (see iodoamino acids) and 1-(3,5-dichloro-2,6-dihydroxy-4-methoxyphenyl)-1-hexanone (**16**) which regulates differentiation of the slime mold *Dictyostelium discoideum*[14]. Two chlorinated steroids serve as aggregation pheromones for *cockroaches. For the occurrence of fluorinated natural products, see fluoroacetic acid.

The introduction of halogen atoms to an organic compound proceeds with the aid of bromo- and chloroperoxidases, halogenides, and H_2O_2. Human haloperoxidases (myeloperoxidases), which have been detected, e.g., in blood (neutrophils) and saliva, play an important role in the defence against microorganisms[1,14]. Enormous amounts of chlorinated compounds are produced in nature each year through biohalogenation of *humic acids by microbial haloperoxidases[15].

Lit.: [1]Zechmeister **68**, 1−498. [2]Phytochemistry **23**, 1449 (1984). [3]Nature (London) **315**, 55 (1985); **328**, 811 (1987). [4]Acc. Chem. Res. **10**, 40 (1997). [5]Phytochemistry **19**, 233

(1980). [6]Comp. Biochem. Physiol. **49B**, 25 (1974). [7]Justus Liebigs Ann. Chem. **1988**, 1191; Pestic. Biochem. Physiol. **37**, 275 (1990); see also Phytochemistry **23**, 1449 (1984). [8]J. Am. Chem. Soc. **95**, 972 (1973). [9]J. Chem. Soc. B **1969**, 559. [10]Tetrahedron Lett. **1977**, 3913. [11]Phytochemistry **22**, 743 (1983). [12]Comp. Biochem. Physiol. **65B**, 525 (1980). [13]Phytochemistry **12**, 1399 (1973). [14]Nature (London) **328**, 811 (1987). [15]Risk. Anal. **14**, 143 (1994).
gen.: Review: Acc. Chem. Res. **31**, 141 (1998) ▪ Chem. Soc. Rev. **28**, 335–346 (1999).

Halomon [(3*S*,6*R*)-6-bromo-3-bromomethyl-2,3,7-trichloro-7-methyl-1-octene].

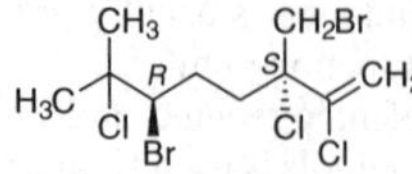

$C_{10}H_{15}Br_2Cl_3$, M_R 401.39, mp. 49–50 °C, $[\alpha]_D$ +206° (CH_2Cl_2). A cytotoxic antitumor compound from the red alga *Portieria hornemannii*. H. shows selective cytotoxicity against different tumor cell lines.
Lit.: Angew. Chem. Int. Ed. Engl. **37**, 2085 (1998) (synthesis) ▪ J. Med. Chem. **35**, 3007 (1992) (isolation). – *[CAS 142439-86-9]*

Haminol A see navenone.

12-*epi*-Hapalindole E-isonitrile see indole alkaloids from cyanobacteria.

Hapalindoles. A group of at present 23 *indole alkaloids from cyanobacteria of the genus *Hapalosiphon*. The compounds are structurally related to *lysergic acid and have antibacterial and antimycotic activities, e.g., *hapalindole A*, one of the main components, $C_{21}H_{23}ClN_2$, M_R 338.88, yellow platelets, mp. 141 °C, 160–167 °C, $[\alpha]_D$ –78° (CH_2Cl_2), soluble in dichloromethane. The biosynthesis of H. A presumably proceeds via a coupling of tryptophan with a monoterpene unit. The ambiguines isonitriles A–F are isolated from *Fischerella ambigua*, *ambiguine isonitrile D*[1]: $C_{26}H_{30}ClN_2O_3$, M_R 453.19, cryst., > 300 °C (decomp.), $[\alpha]_D$ –30.3°. *Hapalosiphon hibernicus* also produces *microcystin LA[2].

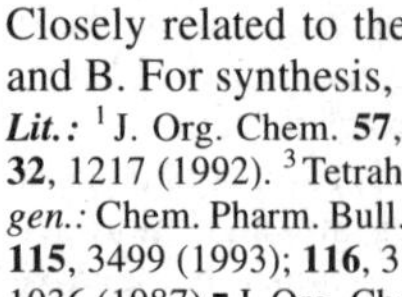

H. A

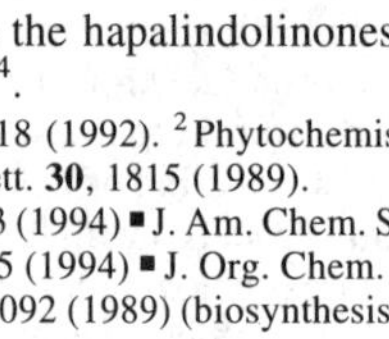

Ambiguine isonitrile D

Closely related to the H. are the hapalindolinones A and B. For synthesis, see *Lit.*[4].
Lit.: [1]J. Org. Chem. **57**, 857, 2018 (1992). [2]Phytochemistry **32**, 1217 (1992). [3]Tetrahedron Lett. **30**, 1815 (1989).
gen.: Chem. Pharm. Bull. **42**, 1393 (1994) ▪ J. Am. Chem. Soc. **115**, 3499 (1993); **116**, 3125, 9935 (1994) ▪ J. Org. Chem. **52**, 1036 (1987) ▪ J. Org. Chem. **54**, 2092 (1989) (biosynthesis). – *[CAS 92219-95-9]*

Hapalosin.

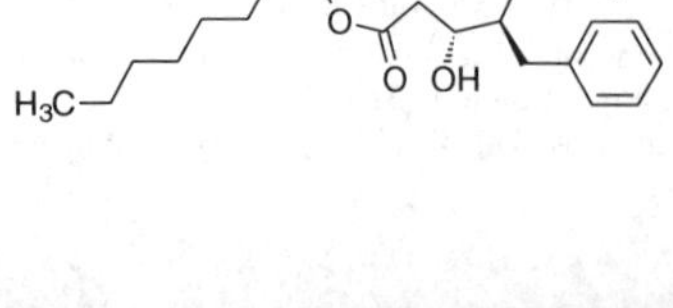

$C_{28}H_{43}NO_6$, M_R 489.65, $[\alpha]_D$ –49.2° (CH_2Cl_2). A cyclic depsipeptide antibiotic from the blue alga (cyanobacterium) *Hapalosiphon welwitschii*. H. is able to counteract the resistance of certain tumor cells to cytostatic agents and thus (experimentally) potentiates the action of anti-cancer drugs.
Lit.: Angew. Chem. Int. Ed. Engl. **35**, 74–76 (1996) ▪ J. Org. Chem. **59**, 7219 (1994). – *Synthesis:* Synlett **1999**, 1118 ▪ Tetrahedron **52**, 14723–14734 (1996) ▪ Tetrahedron Lett. **37**, 6557 (1996).

Harderoporphyrin see porphyrins.

Hardwickiic acid.

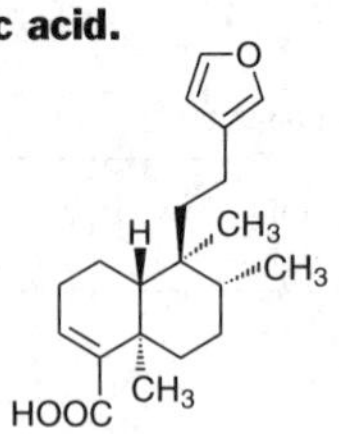

(–)H. a.

$C_{20}H_{28}O_3$, M_R 316.44, mp. 106–107 °C, $[\alpha]_D$ – 114.7° ($CHCl_3$). A *diterpene with the *clerodane structure from *Hardwickia* species (Fabaceae).
Lit.: Magn. Reson. Chem. **36**, 542 (1998) (NMR) ▪ Phytochemistry **30**, 2991 (1991); **38**, 451 (1995). – *[CAS 1782-65-6]*

Harmalin see harmans.

Harmalol see alkaloidal dyes.

Harmans (harman alkaloids, *β*-carbolines).

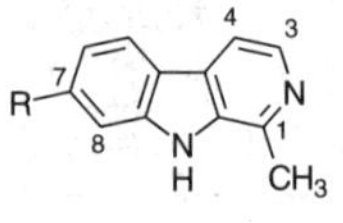

R = H : Harman
R = OH : Harmol
R = OCH_3 : Harmine
R = H, 3-COOH : Harman-3-carboxylic acid
3,4-Dihydroharmine : Harmaline

H. are indole alkaloids from passion flowers (*Passiflora incarnata*, *P. caerulea*), *Peganum harmala*, or from South American *Banisteriopsis* species (Rutaceae). They belong to the natural product group of the pyridoindoles and also occur in mammals (brain, lungs, plasma, urine). *Harman*[1] ($C_{12}H_{10}N_2$, M_R 182.22, mp. 237–238 °C) is the parent compound of the class. The N^2-oxide is harmanine, the N^2-methochloride melinonine F. Harman also occurs in tobacco smoke through pyrolysis of tryptophan and in smoked chicory roots. *Harmine*[2] [$C_{13}H_{12}N_2O$, M_R 212.25, mp. 262–264 °C (decomp.), prisms] is the main alkaloid in *Banisteriopsis* species (up to ~6% dry weight) and the main component of ayahuasca (see below). It is less active than *harmaline*[3] ($C_{13}H_{14}N_2O$, M_R 214.27, mp. 250–251 °C).
Use & activity: H. are used in pharmacological research, they may have significance in the development of alcoholism, and are considered to be metabolites of tryptamine. Plasma levels of harman in heroin-dependent and methadone-substituted patients are elevated by ca. 200%. H. have hallucinogenic and narcotic activities. Extracts of the above-mentioned plants are thus used by traditional healers in central Asia. South American Indians and mestizos use the macerated bark of *B. caapi* as a hallucinogenic drink (ayahuasca = liana of the soul, also Yajé, Caapi)[4], the effects begin with nervousness, sweating, nausea, and then intoxication states (visions in brilliant colors), dream-like

fantasies, sleep, but also aggressive states[5]. High doses of harmine (~400 mg/kg p.o.) result in nausea, vomiting, and finally collapse. The seeds of *P. harmala* are used as a sedative or anthelmintic. H. selectively and competitively inhibits monoaminoxidase (MAO inhibitor), increases the concentration of biogenic amines in the brain[6], and is reversibly bound to benzodiazepine receptors. Harman inhibits DNA replication, is cytotoxic, an inhibitor of plant growth, and potentiates the mutagenicty of aromatic amines but is itself not mutagenic.

Biosynthesis: The N-containing precursor is the amino acid *tryptophan; after decarboxylation to the amine tryptamine the latter cyclizes by condensation with acetate (possibly acetaldehyde), but more probably with pyruvic acid.

Lit.: [1]Helv. Chim. Acta **51**, 203 (1976); Ther. Umschau **50**, 178 (1993); Biochem. Biophys. Res. Commun. **86**, 124 (1979); Bioorg. Chem. **15**, 213 (1987); Tetrahedron **49**, 3325 (1993). [2]Beilstein E V **23/12**, 237; Heterocycles **29**, 521 (1989); Sax (8.), HAI 500. [3]Life Sci. **27**, 893 (1980); J. Chem. Soc., Perkin Trans. 2 **1992**, 1049. [4]J. Ethnopharmacol. **10**, 195 (1984). [5]Schultes & Hofmann, Pflanzen der Götter – Die magischen Kräfte der Rausch- u. Giftgewächse, Bern: Hallwag 1981. [6]J. Ethnopharmacol. **28**, 1 (1990). [7]Mothes et al., p. 277.

gen.: Agric. Biol. Chem. **51**, 921 (1987) ▪ Beilstein E V **23/12**, 237 ff. ▪ Manske **30**, 223–249 ▪ Ullmann (5.) **A 1**, 384 f. – *Pharmacology:* Eur. J. Pharmacol. **70**, 409–416 (1981) ▪ Hager (5.) **4**, 458 f.. – *[HS 2939 90; CAS 486-84-0 (harman); 442-51-3 (harmine); 304-21-2 (harmaline)]*

Harmi . . . see harmans.

Harmonine.

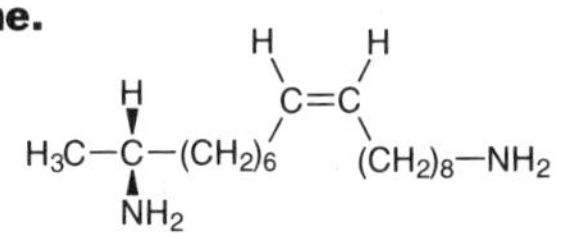

(17*R*, 9*Z*)

$C_{18}H_{38}N_2$, M_R 282.51, yellowish oil, $[\alpha]_D^{23}$ +3.95° (benzene). A defence substance of ladybird beetles (Coccinellidae) of the genus *Harmonia* occurring in the hemolymph of mature insects. On attack by a predator, H. is liberated by so-called reflex bleeding from the knee axilla.

Lit.: Experientia **41**, 519–521 (1985) ▪ Justus Liebigs Ann. Chem. **1991**, 569–574. – *[CAS 133576-26-8]*

Harpagide. H. is an *iridoid glucoside occurring with the tertiary 8-OH group esterified by various carboxylic acids[1].

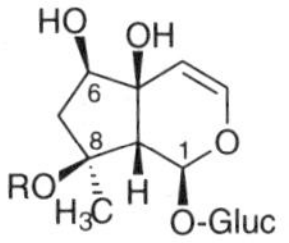

R = H : Harpagide
R = $COCH_3$: Harpagide-8-acetate
R = CO–CH=CH–C_6H_5 (*E*) : Harpagoside
R = 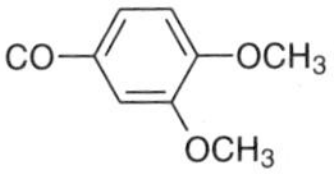 : Tecoside

Occurrence: H. and harpagoside were isolated for the first time, together with procumbide, from the roots of the grapple plant (*Harpagophytum procumbens*, Pedaliaceae)[2]. H. also occurs in the figwort[3] (*Scrophularia nodosa*, Scrophulariaceae) and harpagoside in the dark mullein[4] (*Verbascum nigrum*, Scrophulariaceae). Harpagide 8-acetate has as yet only been found in the roots of bastard balm (*Melittis melissophyllum*, Lamiaceae)[5]. 8-*O*-Acetylharpagide is a nonsteroidal ecdysteroid agonist from *Ajuga reptans* (Lamiaceae)[6]. Tecoside is isolated from *Tecomella undulata* (Bignoniaceae)[7].

Table: Harpagide and derivatives.

trivial name	Harpagide	Harpagide 8-acetate	Harpa-goside	Tecoside
molecular formula	$C_{15}H_{24}O_{10}$	$C_{17}H_{26}O_{11}$	$C_{24}H_{30}O_{11}$	$C_{24}H_{32}O_{13}$
M_R	364.35	406.39	494.50	528.51
mp. [°C]	amorphous	154–156	amorphous	139-142
$[\alpha]_D$	–154° (C_2H_5OH)	–132° (CH_3OH)	–42.6° (CH_3OH)	–159.3° (CH_3OH)
CAS	6926-08-5	6926-14-3	19210-12-9	72933-37-0

Use: Cut pieces of the dried roots of *H. procumbens* are known in Europe as a herbal drug (harpago tea). The tea is a traditional medicine, probably because of the very bitter taste of harpagoside. No pharmacological activities of pure H. and harpagoside have as yet been detected.

Lit.: [1]Justus Liebigs Ann. Chem. **1985**, 1063. [2]Tetrahedron Lett. **1964**, 835; Helv. Chim. Acta **49**, 1552 (1966). [3]Justus Liebigs Ann. Chem. **1978**, 1968. [4]Helv. Chim. Acta **65**, 1678 (1982). [5]Tetrahedron Lett. **1965**, 3439. [6]Insect Biochem. **26**, 519 (1996). [7]J. Chem. Soc., Perkin Trans. 1 **1979**, 2473.

gen.: Chem. Pharm. Bull. **31**, 2296 (1983) ▪ Hager (5.) **4**, 153 ff.; **5**, 385 f.; **6**, 930 ff. ▪ J. Nat. Prod. **43**, 646, No. 23–25, 234 (1980) ▪ Phytochemistry **35**, 621 (1994) (biosynthesis). – *[HS 2938 90]*

Harringtonine see Cephalotaxus alkaloids.

Harveynone.

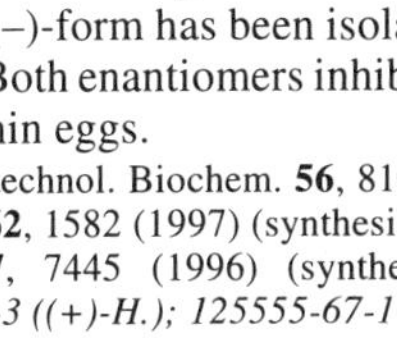

(+)-H.

$C_{11}H_{10}O_3$, M_R 190.20, $[\alpha]_D^{25}$ +147° (CH_3OH), light yellow oil. A phytotoxic fungal metabolite of *Pestalotiopsis theae*. The (–)-form has been isolated from *Curvularia harveyi*. Both enantiomers inhibit spindle formation in sea urchin eggs.

Lit.: Biosci. Biotechnol. Biochem. **56**, 810 (1992) (isolation) ▪ J. Org. Chem. **62**, 1582 (1997) (synthesis of (–)-H.) ▪ Tetrahedron Lett. **37**, 7445 (1996) (synthesis of (±)-H.). – *[CAS 142435-66-3 ((+)-H.); 125555-67-1 ((–)-H.)]*

Hashish. The Arabian word H. (originally weed, grass) is used for the narcotic drug originating from the resin of flower buds of a west Asian hemp variety (*Cannabis sativa* var. *indica*). It is usually smoked on its own or mixed with tobacco or opium, it is less frequently taken orally or drunk in the form of a decoction or tea. The consumption of H. has a long tradition in Middle Eastern and Asian societies and is nowadays abused worldwide. The official international name is *cannabis*. Because of its psychoactive component Δ^9-*tetrahydrocannabinol (THC) cannabis is misused as a narcotic drug. The concentration of THC varies according to the preparation: 1–3% in *marihuana (mainly small pieces of plant material), 3–6% in H. (resin of the female inflorescence), and 30–50% in H. oil (cannabis extract). In addition to tetrahydrocannabinol, cannabis contains 60 other *cannabinoids as well as ca. 360 further components such as *sterols, terpenes, alkaloids, flavonoids, and furan derivatives. On account of its lipophilicity THC is rapidly cleared from the blood, thus the detection of metabolites in urine is used for forensic purposes.

Activity: The effects of H. vary widely from individual to individual and can include euphoria, restlessness, and changed sensory perception. Chronic consumption can lead to depressions, disorientation, and disorders of mental state. For details, see *Lit.*[1]. H. principally damages the central nervous system, lungs, gonads, and the immune system. In pregnant women the fetus is also affected. Following heavy consumption of H. withdrawal symptoms such as nausea, vomiting, sweating, tremor are reported. H. is not classified as an addictive drug but is the first step towards "harder" drugs[2]. H. is mainly cultivated in the following countries: Jamaica, Columbia, Turkey, Marocco, Lebanon, Nigeria, Afghanistan, Pakistan, and Thailand; the world production of cannabis is estimated at 150 000 t; see also cannabinoids and tetrahydrocannabinols.

Lit.: [1]Schweiz. Med. Wochenschr. **119**, 1173–1176 (1989); Dtsch. Ärztebl. **78**, 117–126 (1981). [2]Suchtgefahren **36**, 1–17 (1990); Med. Res. Rev. **3**, 119 (1983); J. Heterocycl. Chem. **21**, 121 (1984).

gen.: Martindale (30.), p. 884 ▪ Nat. Prod. Rep. **16**, 131 (1999) (review) ▪ Redda et al. (eds.), Cocaine, Marijuana, Designer Drugs: Chemistry, Pharmacology, and Behaviour, p. 113, Boca Raton: CRC Press 1989 (pharmacology, toxicology) ▪ Täschner, Haschisch, Stuttgart: Hippokrates 1987 – *[HS 5302 10, 5302 90]*

Hasubanan alkaloids. H. a. are a subgroup of the *morphinan alkaloids occurring only in *Stephania* species (Menispermaceae), e. g. hasubanonine.

Lit.: J. Am. Chem. Soc. **120**, 8259 (1998) [synthesis (+)-cepharamine] ▪ Manske **16**, 393–430; **33**, 307–347 ▪ Mothes et al., p. 223 ff. ▪ Synthesis **1998**, 653 (synthesis) ▪ Ullmann (5.) **A 1**, 378.

Hasubanonine see morphinan alkaloids.

Hazelnut flavor see filbert flavor.

Hb. Abbreviation for *hemoglobin.

HC toxins.

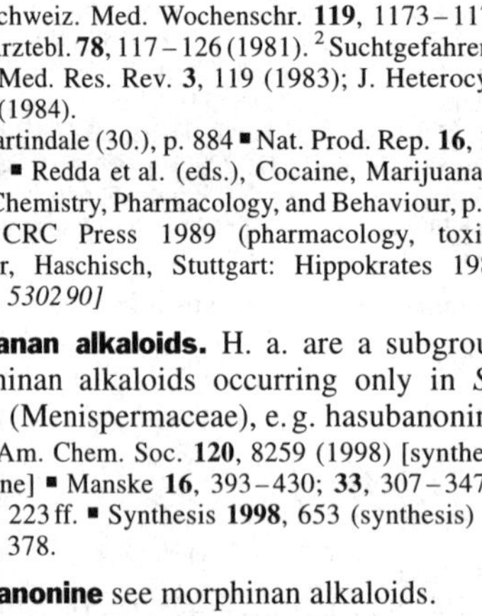

$R^1 = CH_3$, $R^2 = H$: HC toxin I
$R^1 = R^2 = H$: HC toxin II
$R^1 = CH_3$, $R^2 = OH$: HC toxin III

HC toxin I: $C_{21}H_{32}N_4O_6$, M_R 436.51, fine needles, mp. 150 °C; *HC toxin II* (demethyl-HC toxin I): $C_{20}H_{30}N_4O_6$, M_R 422.48; *HC toxin III* (hydroxy-HC toxin I): $C_{21}H_{32}N_4O_7$, M_R 452.51. Cyclic tetrapeptides with an epoxyoctyl side chain from the fungus *Helminthosporium carbonum* (*Bipolaris zeicola*) pathogenic to corn; a host-specific toxin. Host-specific toxins represent virulence factors and are also produced by other phytopathogenic fungi (e. g., *H. victoriae*, *Alternaria* species). Plants with resistance to the fungal infection are not affected.

Lit.: Biochemistry **22**, 3502 (1983) ▪ Biochem. Biophys. Res. Commun. **107**, 785 (1982) ▪ Goodman, Kiraly & Wood, Biochemistry and Physiology of Plant Disease, chap. 9, Columbia: Univ. Missouri Press 1986 ▪ Greenhalgh & Roberts (eds.), Pesticide Science and Biotechnology, p. 89–96, Boston: Blackwell Scientific Publ. 1987 ▪ Heterocycles **24**, 3423 (1986) ▪ J. Nat. Prod. **62**, 143 (1999) (biosynthesis) ▪ Tetrahedron **38**, 45 (1982) ▪ Tetrahedron Lett. **26**, 969 (1985). – *[CAS 83209-65-8 (I); 106973-32-4 (II); 106894-13-7 (III)]*

Hebestatis toxins.

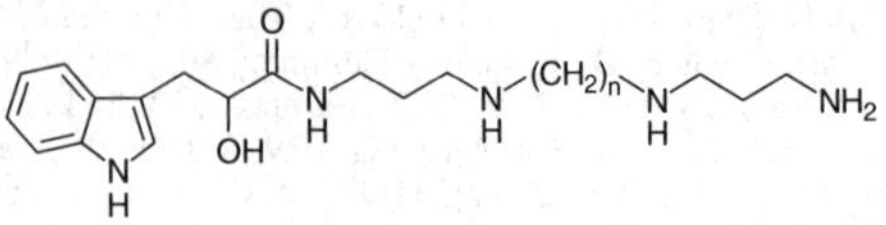

Het 389 : n = 4
Het 403 : n = 5

The two polyamines *Het 389* ($C_{21}H_{35}N_5O_2$, M_R 389.54) and *Het 403* ($C_{22}H_{37}N_5O_2$, M_R 403.57) have been identified in the venom of the trap door spider *Hebestatis theveniti* and a species of tarantula (*Harpactirella* sp.). In contrast to other polyamines from spiders they do not contain amino acids.

Lit.: Toxicon **28**, 541–546 (1990). – *[CAS 128941-10-6 (Het 389); 128941-11-7 (Het 403)]*

Hebevinosides.

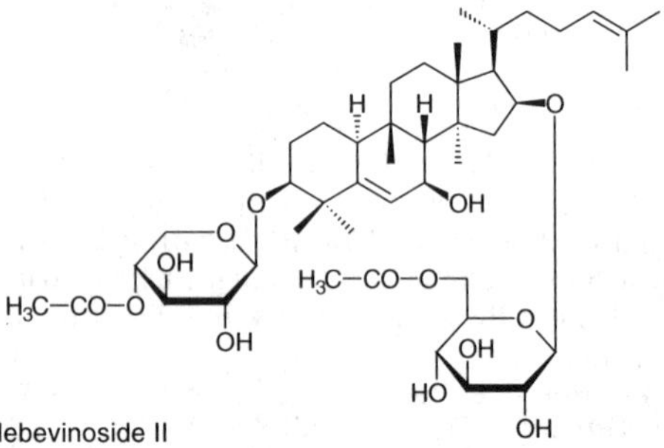

Hebevinoside II

Neurotoxically active *cucurbitacin derivatives from the poisonous Japanese agaric *Hebeloma vinosophyllum* (Basidiomycetes). An example is *hebevinoside II* {$C_{45}H_{72}O_{14}$, M_R 837.06, amorphous, $[\alpha]_D$ +30° (acetone)}, a diglycoside of cucurbita-5,24-diene-3β,7β,16β-triol.

Lit.: Chem. Pharm. Bull. **34**, 88 (1986); **35**, 2254 (1987); **40**, 869 (1992). – *[CAS 89456-98-4 (H. II)]*

Hecogenin see steroid sapogenins.

α-Hederin (sapindoside A, calopanaxsaponin A, helixin). $C_{41}H_{66}O_{12}$, M_R 750.97, needles, mp. 256–259 °C, $[\alpha]_D$ +16.5° (CH_3OH), soluble in alcohol, glacial acetic acid, dilute potassium hydroxide, insoluble in water. A *triterpene from the leaves and fruits of ivy (*Hedera helix*, Araliaceae). The aglycone *hederagenin* [$C_{30}H_{48}O_4$, M_R 472.71, mp. 334 °C, $[\alpha]_D$ +82° (pyridine)] is obtained from α-H. by acidic hydrolysis

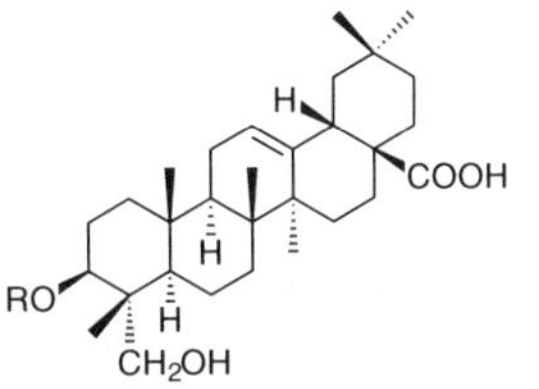

R = H : Hederagenin
R = [α-L-Rhamnopyranosyl-(1→2)-α-L-arabinopyranosyl] : α-Hederin

(e.g., with 5% sulfuric acid). α-H. exhibits strong hemolytic activity.

Lit.: Karrer, No. 1996. – *Isolation:* Chem. Pharm. Bull. **34**, 2209 (1986) ▪ Phytochemistry **15**, 781 (1976); **22**, 1045 (1983) ▪ Planta Med. **53**, 62 (1987). – *[HS 2938 90; CAS 27013-91-8 (α-H.); 465-99-6 (hederagenin)]*

Hegoflavones.

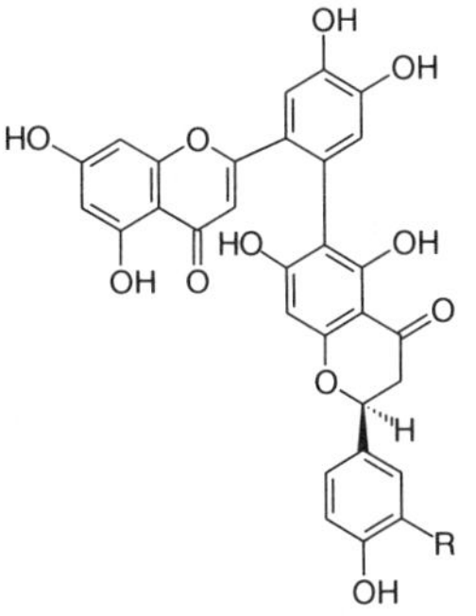

R = H : Hegoflavone A
R = OH : Hegoflavone B

H. A: $C_{30}H_{20}O_{11}$, M_R 556.48, yellow needles, mp. 224–225 °C (decomp.), $[\alpha]_D^{28}$ +12.4°. *H. B:* $C_{30}H_{20}O_{12}$, M_R 572.48, yellow needles, mp. 214–215 °C (decomp.), $[\alpha]_D^{21}$ +7.8°. *Biflavonoids from the tree fern *Alsophila spinulosa* (=*Cyathea spinulosa*) exhibiting a different linkage of the two flavone monomers from that of the other known biflavonoids.

Lit.: Chem. Pharm. Bull. Jpn. **33**, 4182 (1985). – *[CAS 100288-22-0 (H. A); 100288-21-9 (H. B)]*

Helenalin.

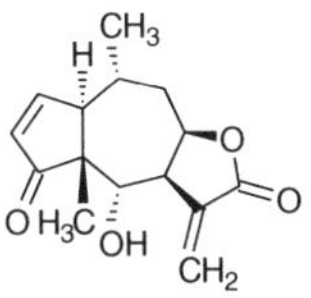

$C_{15}H_{18}O_4$, M_R 262.31, cryst., mp. 168 °C, $[\alpha]_D^{20}$ –102° (acetone), soluble in alcohol, chloroform, poorly soluble in water. A toxic *sesquiterpene lactone that induces sneezing from *Helenium* and *Arnica* species such as, e.g., *Helenium autumnale* and *Arnica montana* [LD_{50} (mouse p.o.) 150 mg/kg]. After intoxication, the symptoms are nasal irritation, vomiting, diarrhea, vertigo, palpitation, respiratory impairment, and finally collapse with a weak, very rapid pulse. In the past deaths have been reported after consumption of *Arnica* preparations. Contact with the skin leads to dermatitis.

Use: H. is used as an antipyretic and antirheumatic agent as well as for venous complaints. *Arnica* tinctures are used externally for sprains, contusions, hematomas, and for wound treatment. *Arnica* extracts are components of hair lotions, shampoos, and cosmetics. *Arnica* preparations are taken internally as heart tonics and analeptics. In the past the drug was misused to induce abortions on account of its uterus-stimulating activity. H. exhibits antitumor properties [1].

Lit.: [1] Cancer Chemother. Pharmacol. **34**, 344 (1994).
gen.: Beilstein E V **18/3**, 295 ▪ Karrer, No. 1928a ▪ Zechmeister **25**, 106 ff. – *Synthesis:* Tetrahedron **52**, 6307 (1996) ▪ Tetrahedron Lett. **30**, 523 (1989). – *Toxicology:* Sax (8.), HAK 300. – *Activity:* Drug Des. Discovery **11**, 23 (1994) ▪ Planta Med. **61**, 199 (1995) (review). – *[CAS 6754-13-8]*

Helenien (*O,O'*-dipalmitoyl-*lutein). $C_{72}H_{116}O_4$, M_R 1045.71, red needles (C_2H_5OH), mp. 92 °C. H., occurring in petals of the yellow marigold (*Tagetes*) and in *Helenium autumnale* (Asteraceae) as well as in orange peel, is authorized for use (E 161b) as a food colorant for butter and bakery products. It has also been isolated from certain cells of the retina where it serves to promote adaptation of the eye to darkness and to restore the normal sensitivity after blinding by bright light. It is also used pharmaceutically for these purposes.

Lit.: Beilstein E IV **6**, 7018 ▪ Hager (5.) **5**, 409 ▪ Int. Rev. Biochem. **14**, 51 (1977) ▪ Karrer, No. 1845 ▪ see also carotinoids. – *[HS 1302 19; CAS 547-17-1]*

Helenine (alantolactone).

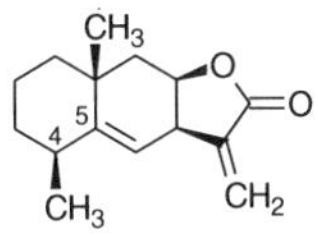

$C_{15}H_{20}O_2$, M_R 232.32, colorless crystals, mp. 82 °C, bp. 275 °C, $[\alpha]_D$ +197° ($CHCl_3$), almost insoluble in water, soluble in alcohol and ether. The name of this tricyclic *sesquiterpene lactone is derived from the Asteraceae *Inula helenium* originally indigenous to Central Asia. H. was isolated for the first time from the essential oil of the roots (*Inula* oil). H. is also an older synonym for *helenalin as well as the name for a metabolite from *Penicillium funiculosum* (ribonucleoprotein). Because of its α-methylene-γ-lactone structure H. is a contact allergen.

Use: H. has antibiotic activity and, like the roots of *I. helenium*, is used as an anthelmintic. In homeopathic medicine *Inula* teas are used for treatment of chronic bronchial catarrh, respiratory distress, and asthma.

Lit.: Beilstein E V **17/10**, 97 ▪ Chem. Nat. Compd. (Engl. Trans.) **26**, 251 (1990) (review) ▪ Hager (5.) **5**, 246 ▪ J. Nat. Prod. **47**, 1013 (1984) ▪ Karrer, No. 1900, 1928a ▪ Phytochemistry **17**, 1165 (1978). – *Synthesis:* J. Am. Chem. Soc. **88**, 3408 (1966). – *[CAS 546-43-0]*

Helicobasidin [2,5-dihydroxy-3-methyl-6-((*S*)-1,2,2-trimethylcyclopentyl)-*p*-benzoquinone].

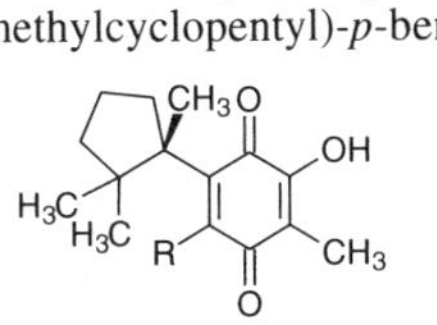

R = OH : Helicobasidin
R = H : Helicobasin

$C_{15}H_{20}O_4$, M_R 264.32, orange red needles, mp. 190–192 °C, uv_{max} 430 nm (C_2H_5OH), optically active pigment from the fungus *Helicobasidium mompa.* H. is formed biosynthetically from farnesyl diphosphate. The 5-deoxy compound, *helicobasin,* $C_{15}H_{20}O_3$, M_R 248.32, yellow needles, mp. 194–195 °C, is isolated from the same fungus.

Lit.: Tetrahedron Lett. **1973**, 4723 (biosynthesis) ▪ Thomson **3**, 30 ▪ Turner **1**, 235; **2**, 255. – *[HS 291469; CAS 13491-25-3 (H.); 13491-69-5 (helicobasin)]*

Helicocerin see cerulenin.

Heliotridine see necine(s).

Heliotropin see piperonal.

Helminthosporal.

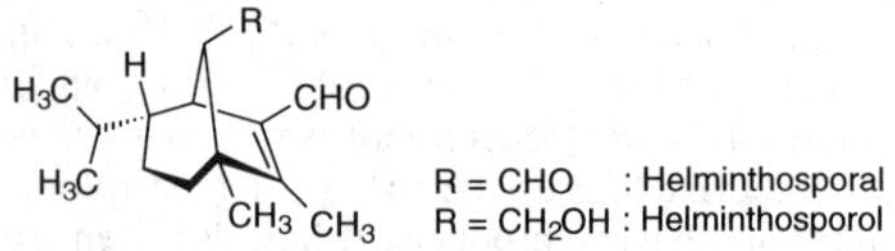

$C_{15}H_{22}O_2$, M_R 234.34, mp. 56–59 °C, $[\alpha]_D$ −49° ($CHCl_3$); a toxic *sesquiterpene dialdehyde from phytopathogenic fungi, mainly *Cochliobolus sativus* and its asexual form, *Helminthosporium sativum* (*Bipolaris sorokiniana*); both are major pests for cereal crops. The reduced form, *helminthosporol,* is a natural plant growth regulator.

Lit.: *Isolation:* Can. J. Chem. **43**, 1357 (1965). – *Synthesis:* J. Am. Chem. Soc. **114**, 644 (1992) ▪ J. Chem. Soc., Chem. Commun. **1986**, 63; **1987**, 1136 ▪ Tetrahedron Lett. **32**, 6741 (1991). – *Activity:* Biochem. Biophys. Res. Commun. **28**, 878 (1967). – *[CAS 723-61-5 (H.); 1619-29-0 (helminthosporol)]*

Hematin see heme.

Hematoporphyrin (haematoporphyrin IX).

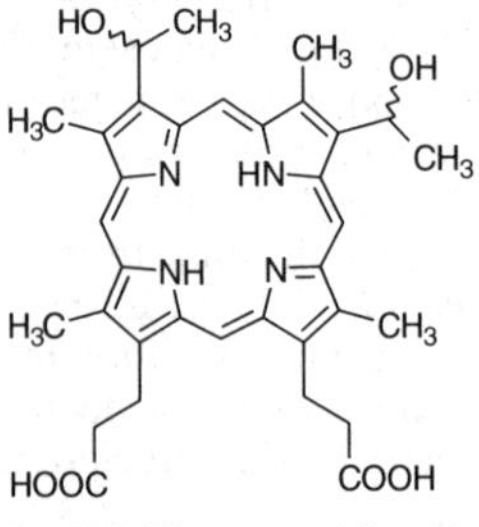

$C_{34}H_{38}N_4O_6$, M_R 598.70; uv_{max} (dimethyl ester in pyridine) 623, 596, 569.2, 532, 499.5, 402 nm. H. belongs to the structural class of the *porphyrins.

Occurrence: H. is formed by acid hydrolysis of *hemoglobin during which the heme is cleaved from the protein. Heme is transformed into H. via loss of the central iron ion and hydration of the vinyl groups. Since the hydration of the vinyl groups is not stereoselective H. occurs as a mixture of 4 stereoisomers. On account of its photophysical properties (fluorescence; sensitizer for singlet oxygen formation) H. is used as a fluorescence indicator; similarly, H. derivatives such as Photofrin® are used as sensitizers in photodynamic tumor therapy [1].

Lit.: [1] Nachr. Chem. Tech. Lab. **33**, 582 (1985).
gen.: Beilstein E V **26/15**, 396 ▪ Dolphin **I**, 297–298; **VI**, 347 ▪ Henderson et al., Photodynamic Therapy, New York: M. Dekker 1992 ▪ Photochem. Photobiol. **39**, 851 (1984); **55**, 145 (1992) (reviews) ▪ Smith, p. 771 f. – *[HS 2933 90; CAS 14459-29-1]*

Heme (from Greek: haima = blood). According to recommendations of IUPAC/IUB H. is an arbitrary iron-porphyrin complex [1]. Here the name H. is used for the iron(II) complex (protoheme) of *protoporphyrin. Other H. are discussed under separate entries.

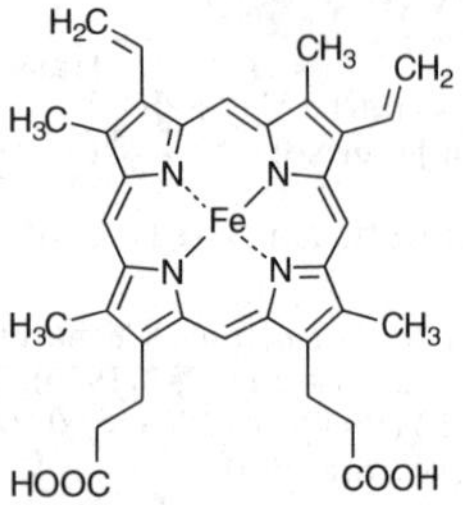

$C_{34}H_{32}FeN_4O_4$, M_R 616.50. Thin brown crystalline needles that are readily oxidized to hematin. H. belongs to the structural class of the *porphyrins.

Occurrence: H. is the prosthetic group of the blood pigment *hemoglobin and related heme proteins (e.g., *cytochromes, myoglobin) and is responsible for the red color. The coordination sites of the iron lying above and below the plane of the porphyrin ring are available for complexation with proteins or other ligands, e.g., oxygen. The iron(II) ion in H. is easily oxidized to furnish hematin [2] or *hemin.

Biosynthesis & synthesis: The biosynthesis from protoporphyrin proceeds through chelatization of an iron(II) ion catalyzed by *ferrochelatase* (protoheme ferrolyase, EC 4.99.1.1). The H. in red blood particles is degraded to *bile pigments after an average life-time of ca. 120 d.

H. was first synthesized in 1928 by H. Fischer who received the Nobel Prize for this work.

Lit.: [1] Pure Appl. Chem. **59**, 820 (1987). [2] Dolphin **I**, 460 f.; **IV**, 742; Hager (5.) **4**, 460; Smith, p. 63.
gen.: Ann. N. Y. Acad. Sci. **504**, 151 (1987) ▪ Barton-Nakanishi **4**, 61–108 (biosynthesis) ▪ Beilstein E V **26/15**, 285 ▪ Dolphin **I**, 45, 65, 317; **II**, 121–126 ▪ J. Chem. Soc., Perkin Trans. 1 **1997**, 2099–2138 (review synthesis, stereochemistry) ▪ Jordan, p. 3 ff.; 211 ff. ▪ Heterocycles **26**, 1947 (1987) (synthesis review) ▪ Lever & Gray, Iron Porphyrins, Weinheim: VCH Verlagsges. 1989 ▪ Nat. Prod. Rep. **6**, 171 (1989) ▪ Smith, p. 123–153. – *[CAS 14875-96-8 (H.); 15489-90-4 (hematin)]*

Heme d.

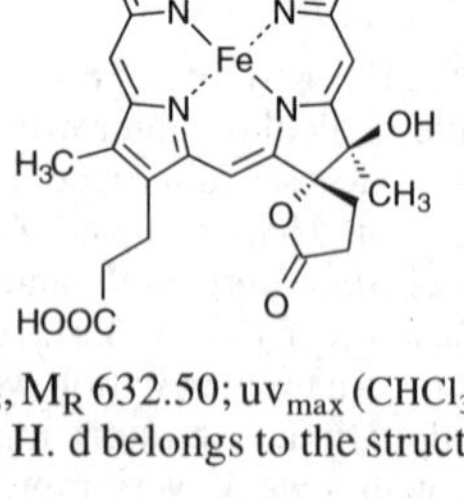

$C_{34}H_{32}FeN_4O_5$, M_R 632.50; uv_{max} ($CHCl_3$) 752, 603, 477, 382, 274 nm. H. d belongs to the structural class of the *chlorins.

Occurrence: H. d is the prosthetic group of *cytochrome d, which occurs in the respiratory chain of *Escherichia coli* and other bacteria. Cytochrome d pos-

sesses a higher affinity for oxygen than other cytochromes. Thus, the bacteria can still maintain their metabolism even at low oxygen concentrations. The relative and absolute configurations of H. d have not been elucidated, hence H. d may be a stereoisomer of the structure shown. For synthesis, see *Lit.*[1].
Lit.: [1] J. Am. Chem. Soc. **110**, 2264 (1988).
gen.: Chem. Rev. **94**, 327 (1994) ▪ Synform **7**, 179 (1989) (review). – *[CAS 60318-31-2]*

Heme d_1.

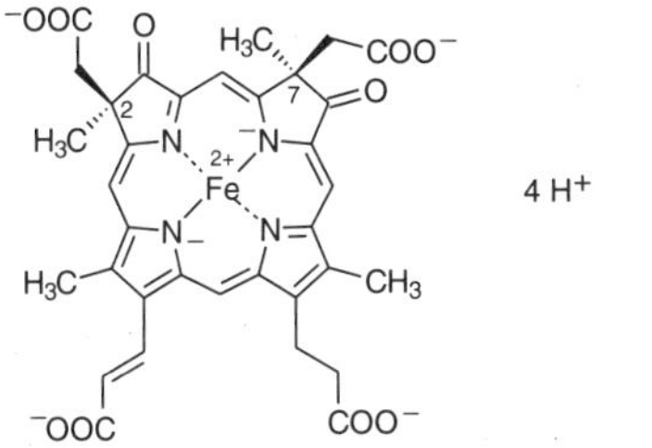

$C_{34}H_{30}FeN_4O_{10}$, M_R 710.48; uv_{max} (tetramethyl ester) 660, 611, 570, 532, 446, 422 nm. H. d_1 belongs to the structural class of the *isobacteriochlorins.
Occurrence: H. d_1 is one of the two prosthetic groups of *cytochrome cd_1. Cytochrome cd_1, and thus also H. d_1, participates in the reduction of nitrite to N_2O in chemoautotrophic bacteria such as *Pseudomonas aeruginosa*, *Paracoccus denitrificans*, and *Thiobacillus denitrificans*. H. d_1 is isolated from such denitrificating bacteria, which play an important role in the global nitrogen cycle.
Biosynthesis & synthesis: The biosynthesis of H. d_1 has not yet been clarified; however, because of the substitution pattern it may be assumed that H. d_1 is formed by the usual tetrapyrrole biosynthesis route and that the methyl groups in the 2- and 7-positions originate from **S*-adenosylmethionine. Total syntheses of *porphyrin d_1 have been carried out in several laboratories and the configuration of H. d_1 has also be determined.
Lit.: Chem. Rev. **94**, 327 (1994) ▪ J. Chem. Soc., Chem. Commun. **1993**, 277. – *Synthesis:* J. Am. Chem. Soc. **109**, 3149 (1987) ▪ J. Chem. Soc., Perkin Trans. 1 **1997**, 2111, 2123 ▪ Tetrahedron Lett. **33**, 765 (1992). – *[CAS 59948-35-5]*

Heme derivatives. Name for derivatives of *heme, i.e., for iron complexes of the *porphyrins and *hydroporphyrins, in the strictest sense of *protoporphyrins. These H. d. are the prosthetic groups of hemoproteins such as the *cytochromes, cytochrome oxidase, peroxidases, catalases, *leghemoglobin, nitrite reductases, sulfite reductase; see also heme d, heme d_1, and siroheme. In spite of their similar names *hemerythrin, *hemocyanin, hemocuprein, and *hemovanadins do not belong to the H. d.
Lit.: see heme and individual entries.

Hemerythrin. The iron-containing blood pigment of various marine worms (*Sipunculus*, *Phascolopsis*) as well as the genus *Lingula* of the brachiopods dating back to the Silurian period (ca. 400 million years ago) and still present in the oceans today. H. is – in contrast to the *heme derivatives – a non-heme iron protein (M_R 108 000) made up of 8 subunits, each with two Fe(II) ions directly bound to the protein. Each subunit (M_R 13 500) consists of 113 amino acid residues and is able to bind 1 molecule O_2. H. itself is colorless whereas the oxidized form, *oxyhemerythrin*, exists as red-violet crystals. H. contains about three times as much iron as *hemoglobin.
Lit.: J. Am. Chem. Soc. **111**, 4688 (1989) ▪ J. Mol. Biol. **204**, 155 (1988).

Hemicelluloses see polyoses.

Hemin (chlorohemin). According to the recommendations of IUPAC/IUB H. is an arbitrary chloroiron(III)-porphyrin complex (see also heme). Here H. is specifically the chloroiron(III) complex of *protoporphyrin (protohemin): $C_{34}H_{32}ClFeN_4O_4$, M_R 651.95; uv_{max} (pyridine) 557, 526, 418.5 nm. Long thin platelets or oblique prisms, first isolated from blood by Teichmann in 1852 and thus also known as *Teichmann's crystals*. H. is the usually isolated, stable form of *heme.
Lit.: see heme. – *[HS 2934 90; CAS 16009-13-5]*

Hemin a see cytohemin.

Hemlock alkaloids. Name for the poisonous *piperidine alkaloids of the spotted hemlock (*Conium maculatum*, Apiaceae). The spotted hemlock, endemic throughout Europe, contains alkaloids in all plant parts but especially in the fruits. The alkaloid concentration in the fruits (especially pericarp) can amount to 3.5%. The main alkaloid is coniine (ca. 90%), accompanied by minor alkaloids such as γ-coniceine, conhydrine, pseudoconhydrine, and *N*-methylconiine (figure).

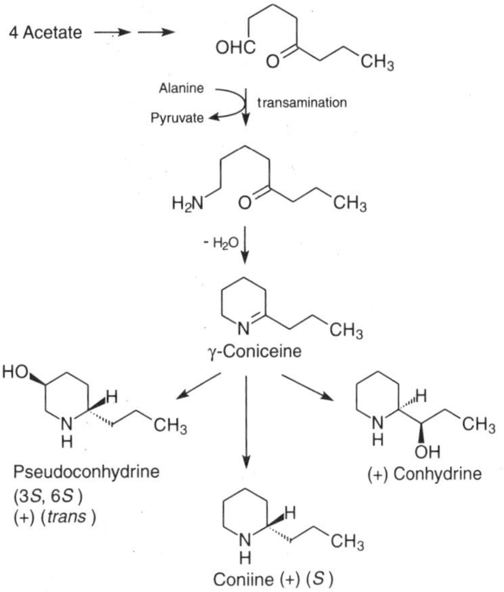

Figure: Biosynthesis of hemlock alkaloids. Introduction of nitrogen occurs by way of an alanine-dependent transaminase.

The H. a. are pseudoalkaloids, their carbon skeletons do not originate from amino acid metabolism but rather from polyketide metabolism (figure). Coniine is a highly toxic alkaloid and is rapidly resorbed by mucous membranes and intact skin. It acts at first as a stimulant for motoric nerve endings followed by *curare type paralysis of striated musculature. Death follows after 0.5–5 h in full consciousness through par-

Table: Data of Hemlock alkaloids.

	γ-Coniceine	(+)-Conhydrine	Pseudoconhydrine	Coniine
molecular formula	$C_8H_{15}N$	$C_8H_{17}NO$	$C_8H_{17}NO$	$C_8H_{17}N$
M_R	125.21	143.23	143.23	127.23
mp.	171–172	121	105–106	–2
bp. [°C]				166–167
optical activity		$[\alpha]_D$ +10° (C_2H_5OH)	$[\alpha]_D^{15}$ +11° (C_2H_5OH)	$[\alpha]_D$ +8.0° ($CHCl_3$)
LD_{50} (mouse p.o.)	12 mg/kg			100 mg/kg
(human)				0.5–1 g/kg
CAS	1604-01-9	495-20-5	140-55-6	458-88-8

alysis of thorax musculature – the toxic symptoms were described by Plato for Socrates' death in 399 BC. Coniine was recognized by Mody et al.[1] as the paralytic principle of the insect-eating American pitcher plant (*Sarracenia flava*). It was discovered in 1826 by Giesecke[2] and synthesized for the first time in 1886 by Ladenburg[3]. For the synthesis of coniine, see *Lit.*[4].

Lit.: [1]Experienta **32**, 829 (1976). [2]Arch. Pharm. (Weinheim, Ger.) **20**, 97 (1827). [3]Ber. Dtsch. Chem. Ges. **19**, 2578 (1886). [4]Chem. Lett. **1994**, 21; Heterocycles **33**, 17 (1992); J. Org. Chem. **58**, 7732 (1993); **62**, 746 (1997); **63**, 6344 (1998); Justus Liebigs Ann. Chem. **1993**, 173; Org. Lett. **2**, 155 (2000); Synthesis **1997**, 1151; Tetrahedron **49**, 1749 (1993); Tetrahedron Lett. **35**, 2223 (1994); **38**, 9073 (1997).

gen.: Beilstein E IV **20/3**, 1970 f.; E V **20/4**, 206; E V **21/1**, 138 ▪ Hager (5.) **3**, 343 ff. ▪ Kirk-Othmer (4.) **1**, 1083 ▪ Manske **1**, 211–218; **11**, 473–477; **26**, 163 f. ▪ Merck-Index (12.), No. 2564, 2566, 2569, 8099 ▪ Pelletier **3**, 49–54 ▪ Sax (8.), PNT 000 ▪ Ullmann (5.) **A 1**, 359. – *Synthesis:* Carbohydr. Chem. **4**, 129 (1985) ▪ Chem. Lett. **1984**, 1101 ▪ J. Org. Chem. **60**, 7084 (1995) ▪ Justus Liebigs Ann. Chem. **1997**, 1267 ▪ Synlett **1997**, 905, 1179 ▪ Tetrahedron: Asymmetry **7**, 3047 (1996) ▪ Tetrahedron Lett. **19**, 2051 (1978); **20**, 771 (1979); **24**, 4577 (1983); **25**, 1555 (1984); **30**, 6395 f. (1989); **37**, 5543 (1996). – *[HS 2939 90]*

Hemochromes (hemochromogens). Complex compounds of *heme with bases such as primary amines, ammonia, pyridine, nicotine, imidazole, etc.; when Fe(III) is the central atom instead of Fe(II), the term is hemichromes or ferrihemochromes.

Hemocyanin. Blood pigment of many invertebrates (such as octopus, snails, crabs, mussels) with the same functions as *hemoglobin in the vertebrates. H. is iron-free and has copper as central atom. With bound oxygen and Cu(II) it has a blue color (oxyhemocyanin), oxygen-free [with Cu(I)], in contrast, it is colorless (deoxyhemocyanin). H. is freely dissolved in blood and not bound to blood particles. It is obtained by saturating blood with ammonium sulfate and subsequent dialysis. The molecular masses of H. are, in lobsters up to 770 000, in marine polyps up to 2 780 000, in snails up to 6 700 000, and thus among the largest naturally occurring molecules. The Cu can be expelled from snail H. by treatment with a cyanide buffer at pH 8.4 in the presence of Ca ions (apohemocyanin) and subsequently be re-introduced, whereupon the binding ability for oxygen is restored.

Lit.: Eckert, Tierphysiologie, 2. edn., p. 536, Stuttgart: Thieme 1993.

Hemoglobin (abbr.: Hb). Hb is the red blood pigment of vertebrates, contained in the erythrocytes (red blood particles) of blood. Hb is a tetrameric iron protein, the monomer of which consists of a globin chain with a molecule of *heme as prosthetic group. As well as in vertebrates, Hb also occurs in other organisms, although not always bound to erythrocytes, e. g., in insects and in the root tubercules of nitrogen-fixing Fabaceae (*leghemoglobin). A very similar molecule – containing chlorocruoroheme (see chlorocruoroporphyrin) – erythrocruorin occurs in the free form in the blood of snails, worms, and some insect larvae as a high-molecular-weight respiratory pigment; in sea cucumbers, polychaete worms, mussels, primitive vertebrates it may be monomeric or cell-bound. Hb serves to take up the inspired oxygen and to transport it to the muscles, where myoglobin serves for further transport, and to other respiratory tissues. Hb plays a key role in blood pressure regulation by the reversible binding of *nitrogen monoxide[1].

Lit.: [1]Nature (London) **380**, 221 (1996).

gen.: Ann. N. Y. Acad. Sci. **504**, 151 (1987) ▪ Chir. Forum Exp. Klin. Forsch. (Suppl. 1) **1996**, 97 (use) ▪ Dolphin **IV**, 324–346; **VII**, 437–441 ▪ Heterocycles **26**, 1947 (1987) (review) ▪ Merck-Index (12.), No. 4682 – *[HS 3002 10]*

Hemovanadins. Green components of certain blood cells of marine *tunicates, consisting of a polypeptide with 24 vanadium ions (V^{3+}) in the molecule. H. are not oxygen transporters and their function is unknown but is purported to be related to the formation of cellulose. 100 000 m^3 of sea water contain merely ca. 1 g of vanadium which is enriched in the tunicates. Vanadium can be extracted technically from some types of crude oil (containing the remains of prehistoric tunicates).

Hemozoin. The malaria pathogen *Plasmodium falciparum* is a parasite of red blood cells where it degrades *hemoglobin. The thus released *heme, which is toxic for the parasite, is detoxified by *P. falciparum* through polymerization to insoluble H., a crystalline compound with a polymeric structure in which the iron atom of one heme molecule is bound to a carboxylate group of the next heme molecule. Chemotherapeutic agents against malaria, such as chloroquine, prevent the polymerization and the free heme kills the parasite.

Lit.: Chem. Eng. News (29.1.) **1996**, 32 ▪ Science **271**, 219 (1996).

Heneicosanedioic acid see Japan wax.

(*Z*)-6-Heneicosen-11-one see butterflies.

Henna. The powdered leaves of the henna plant (Egyptian dyer's shrub, *Lawsonia inermis*, *L. alba*, Lythraceae) growing from East and North Africa to India and being cultivated in the Middle East. The dyeing principle of H. is *lawsone (dyes orange-yellow, content in H. ca. 1%), formed by hydrolysis and autoxidation from the genuine components hennosides A, B, and C[1]. H. is suitable for dyeing hair from orange to fox red. In combination with the powdered leaves of the Persian indigo plant (*Indigofera argentea*), further tones from blond to black are possible. The color is permanent but not harmful and the dyeing process is tedious. A dye mixture consisting of H., gallnut powder, iron filings, iron sulfide, copper and cobalt salts is in use under the name rasti(c)k. The mummy of an Egyptian princess from the 18th dynasty (1400 BC) reveals hair dyed with H. In the Middle East H. is also used to color the skin, especially the palms of the hands and the soles of the feet. H. exhibits weak allergenic potential.

Lit.: [1] J. Nat. Prod. **32**, 518 (1969).
gen.: Br. Med. J. **II**, 944 (1960 ▪ Ullmann (4.) **10**, 735; **11**, 114; **12**, 565. – *[HS 1404 10]*

Hennoxazoles. Bisoxazoles from marine sponges of the genus *Polyfibrospongia* with antiviral activities, e.g., *H. A:* $C_{29}H_{42}N_2O_6$, M_R 514.66, yellow oil, $[\alpha]_D^{25}$ –47° ($CHCl_3$); activity against HSV-I: IC_{50} 0.6 µg/mL.

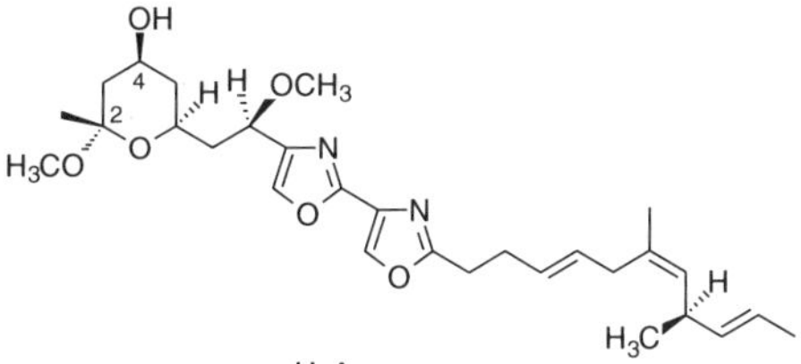

H. A
(absolute configuration)

Lit.: Chimia **50**, 157–167 (1996) (synthesis) ▪ J. Am. Chem. Soc. **113**, 3173 (1991) (isolation); **117**, 558 (1995) (synthesis); **120**, 2204 (1998) (stereochemistry); **121**, 4924 (1999) (synthesis) ▪ Synlett **1995**, 415. – *[CAS 132564-95-5 (H. A)]*

(8*Z*,42*Z*)-24,25-Epoxy-8,42-henpentacontadiene see cockroaches.

Hentriacontanoic acid. $H_3C–(CH_2)_{29}–COOH$, $C_{31}H_{62}O_2$, M_R 466.83, mp. 93–93.2 °C, isolated from the leaf wax of sisal (*Agave sisalana*) and other plants.

Lit.: Gunston, Harwood & Padley, The Lipid Handbook, 2d. edn., London: Chapman & Hall 1994 ▪ Indian J. Chem. Sect. B **26**, 208 (1987) (synthesis). – *[CAS 38232-01-8]*

1-Hentriacontanol. $H_3C–(CH_2)_{29}–CH_2OH$, $C_{31}H_{64}O$, M_R 452.84, mp. 87 °C. H. is a component of various plant waxes. In older literature *1-triacontanol (melissyl or myricyl alcohol) is often wrongly named as H. The same holds for *hentriacontanoic acid.

Lit.: see 1-triacontanol. – *[HS 2905 19; CAS 544-86-5]*

Hepialone [(*R*)-2-ethyl-2,3-dihydro-6-methyl-4*H*-pyran-4-one].

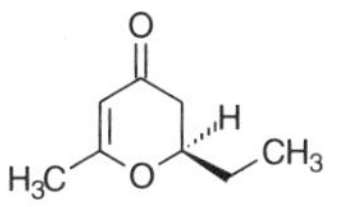

Hepialone

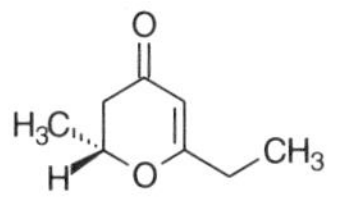

a

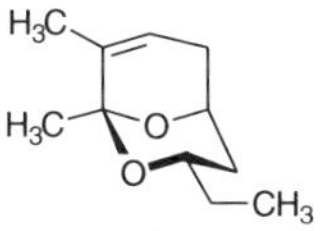

b

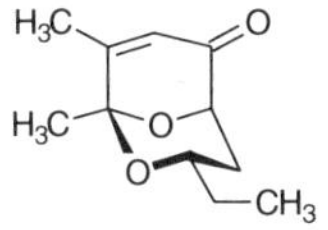

c

$C_8H_{12}O_2$, M_R 140.18, liquid, $[\alpha]_D^{22}$ +236° (C_2H_5OH). Sexual attractant of the male butterfly *Hepialus californicus* [1]. *H. hecta* produces the isomeric (*R*)-*6-ethyl-2,3-dihydro-2-methyl-4H-pyran-4-one* [2] ($C_8H_{12}O_2$, M_R 140.18; **a**). Besides the homologous *2,6-diethyl-2,3-dihydro-4H-pyran-4-one* ($C_9H_{14}O_2$, M_R 154.21) the *pheromone of *H. hecta* also contains the bicyclic acetal *(1R)-3-exo-3-ethyl-1,8-dimethyl-2,9-dioxabicyclo[3.3.1]non-7-ene* ($C_{11}H_{18}O_2$, M_R 182.26; **b**) and the corresponding *6-ketone* ($C_{11}H_{16}O_3$, M_R 196.25; **c**).

Lit.: [1] Tetrahedron Lett. **26**, 563 (1985). [2] J. Chem. Ecol. **16**, 3511–3521 (1990).
gen.: ApSimon **9**, 367–369 ▪ Synth. Commun. **25**, 1531 (1995) (synthesis). – *[CAS 95833-18-4 (H.); 96998-54-8 (a)]*

Hepoxilins (Hx). Acyclic *eicosanoids; the name is derived from the chemical structure (*h*ydroxy-*epox*ide) and biological activity (release of *insulin*). A distinction is drawn between an A- (hydroxy group at C-8) and a B-series (hydroxy group at C-10) as well as various peptide-H., consisting of cystein-*S*-yl derivatives of H. (cf. leukotrienes); the index indicates the number of C=C double bonds in the molecule (see figure). H. are formed in various organs and tissues. Their biosynthesis proceeds from *arachidonic acid which is converted by 12-lipoxygenase to (*S*)-12-hydroperoxy-5*Z*,8*Z*,10*E*,14*Z*-eicosatetraenoic acid (12-HpETE) and then mainly to 11*S*,12*S*-epoxy-8*RS*-hydroxy-5*Z*,9*E*,14*Z*-eicosatrienoic acid (*hepoxilin A_3* or HxA_3, $C_{20}H_{32}O_4$, M_R 336.47, epimers) and (*S*)-12-hydroxy-5*Z*,8*Z*,10*E*,14*Z*-eicosatetraenoic acid (12-HETE) under the action of hepoxilin synthase. 11*S*,12*S*-Epoxy-10*RS*-hydroxy-5*Z*,8*Z*,14*Z*-eicosatrienoic acid (*hepoxilin B_3* or HxB_3, $C_{20}H_{32}O_4$, M_R 336.47, epimers) is principally formed by catalytic conversion of 12-HpETE in the presence of hemin. Hydrolysis of the epoxy ring of H. leads to trioxilins (trihydroxy derivatives), reaction with the SH-group of glutathione furnishes peptide hepoxilins.

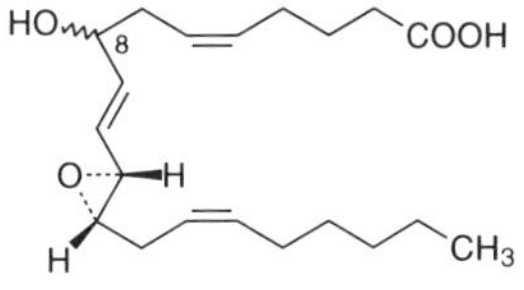

Hepoxilin A_3 (HxA_3)

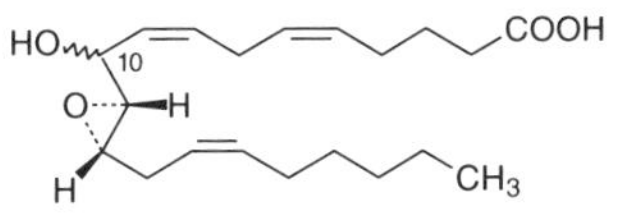

Hepoxilin B_3 (HxB_3)

Little is known about their biological activities, e.g., they participate in the export of potassium and calcium ions from the cell, also in the development of skin inflammations, and they promote the release of insulin from pancreatic islet cells.

Lit.: Agents Actions (Suppl.) **45**, 291 (1995) ▪ Biochim. Biophys. Acta **1215**, 1 (1994) ▪ Gen. Pharmacy **24**, 805 (1993) ▪ Lipids **30**, 107 (1995) ▪ Phytochemistry **36**, 1233 (1994) (isolation). – *Synthesis:* Tetrahedron **49**, 4099 (1993) ▪ Tetrahedron Lett. **30**, 2545 (1989). – *[CAS 94161-11-2 (H. A_3); 94161-10-1 (H. B_3)]*

(7*Z*,11*Z*)-Heptacosa-7,11-diene see flies.

(*Z*)-10-Heptadecen-2-one see flies.

Heptanoic acid see (o)enanthic acid.

2-Heptanone see ants.

(*Z*)-4-Heptenal see alkenals.

(Z)-4-Hepten-2-one, (Z)-4-hepten-2-ol see fruit flavors (banana flavor).

Heptoses see monosaccharides.

2-Heptyl-1-hydroxy-4(1*H*)-quinolinone. The principal member of a family of 2-alkylquinolines (*pseudans*) from *Pseudomonas pyocyanea, P. aeruginosa*, and *P. methanica*, sometimes abbreviated as HQNO (for 2-heptyl-4-quinolinol *N*-oxide)[1]. Besides the 2-nonyl and 2-undecyl homologues, the corresponding 4-quinolone (**2**) and 3-hydroxy-4-quinolone (**3**) have been isolated. The *N*-methylquinolone (**4**)[2] is obtained from *Ruta graveolens*.

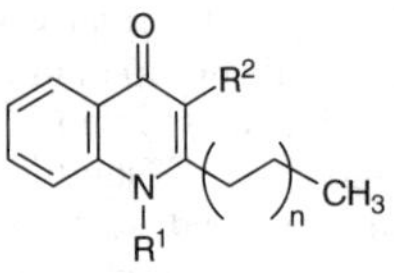

	R^1	R^2	n
1	OH	H	3
2	H	H	3
3	H	OH	3
4	CH_3	H	4

Table: Data of HQNO.

	molecular formula	M_R	mp. [°C]	CAS
1	$C_{16}H_{21}NO_2$	259.3	158–160	341-88-8
2	$C_{16}H_{21}NO$	243.3	146–147	40522-46-1
3	$C_{16}H_{21}NO_2$	259.3	182–185	69808-31-7
4	$C_{19}H_{27}NO$	285.4	71– 75	68353-24

HQNO (**1**) and *N*-deoxy-HQNO (**2**) have weak activity against Gram-positive bacteria, **1** inhibits the mitochondrial respiratory chain in the cytochrome bc_1 complex, 5-lipoxygenase, and is an antagonist to the antibacterial activity of dihydrostreptomycin.

Lit.: [1] J. Antibiot. **31**, 1160 (1986). [2] Phytochemistry **18**, 1768 (1979).

Herbertenes.

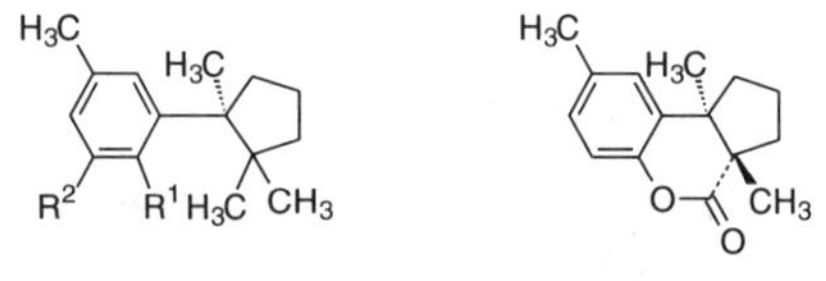

$R^1 = R^2 = H$: Herbertene (**1**)
$R^1 = OH, R^2 = H$: α-Herbertenol (**2**)
$R^1 = R^2 = OH$: Herbertenediol (**3**)

Herbertenolide (**4**)

Table: Data of Herbertenes.

	molecular formula	M_R	mp. [°C]	$[\alpha]_D^{20}$	CAS
1	$C_{15}H_{22}$	202.34	(oil)	–48.3°	80322-46-9
2	$C_{15}H_{22}O$	218.34	143–144 (oil)	–55°	81784-10-3
3	$C_{15}H_{22}O_2$	234.34	90.5–91.5	–46.5°	86996-95-4
4	$C_{15}H_{18}O_2$	230.31	95.5–96.5	–86.4°	86996-94-3

Fungitoxic sesquiterpenes with the herbertane skeleton (for examples, see table) isolated from species of the liverwort genera *Herberta* and *Mastigophora*. For synthesis, see *Lit.*[1].

Lit.: [1] J. Chem. Soc., Perkin Trans. 1 **1999**, 2479; Tetrahedron Lett. **33**, 329 (1992); **35**, 3353 (1994); **40**, 4733 (1999) (herbertenol); J. Org. Chem. **64**, 1741 (1999); J. Am. Chem. Soc. **121**, 2762 (1999).

gen.: Chem. Lett. **4**, 463 (1982); **5**, 1041 (1983) ▪ J. Chem. Soc., Perkin Trans. 1 **1986**, 701; **1988**, 2485; **1991**, 2737.

Herbicidins. Nucleoside antibiotics (H. A–G) with herbicidal activity from cultures of *Streptomyces saganonensis*, e.g., *H. B:* $C_{18}H_{23}N_5O_9$, M_R 453.41, cryst., mp. 155 °C (decomp.), $[\alpha]_D^{20}$ +63° (CH_3OH).

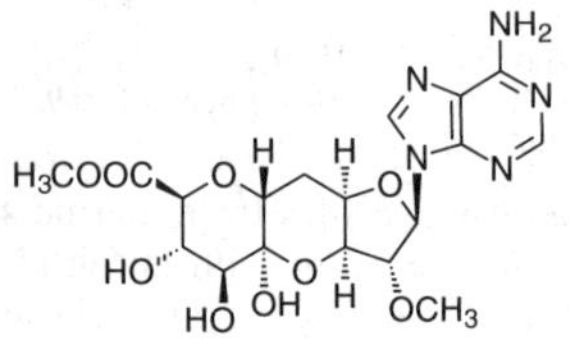

H.B

Lit.: J. Am. Chem. Soc. **114**, 6430 (1993); **121**, 10270 (1999) (synthesis) ▪ J. Antibiot. **29**, 836, 863, 870 (1976); **32**, 857, 862 (1979); **35**, 1711 (1982) (isolation); **36**, 30 (1983) (biosynthesis); **41**, 1711 (1988) (review). – *[CAS 55353-32-7 (H. B)]*

Herbicolins. Cyclic depsiglycopeptides isolated from *Erwinia herbicola*. They show good activities against yeasts and fungi as well as immunosuppressive effects.

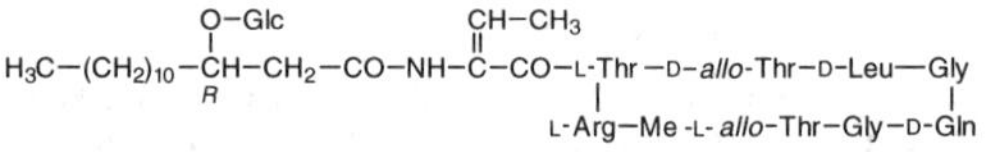

H. A

H. A: $C_{58}H_{101}N_{13}O_{20}$, M_R 1300.52, soluble in water and lower alcohols, insoluble in ethyl acetate or ether. *H. B* is the aglycone of H. A, $C_{52}H_{91}N_{13}O_{15}$, M_R 1138.37.

Lit.: Chem. Pept. Proteins **3**, 307 (1986) ▪ J. Antibiot. **33**, 353–358 (1980). – *[CAS 74188-23-1 (H. A); 74188-24-2 (H. B)]*

Herbimycin (herbimycin A).

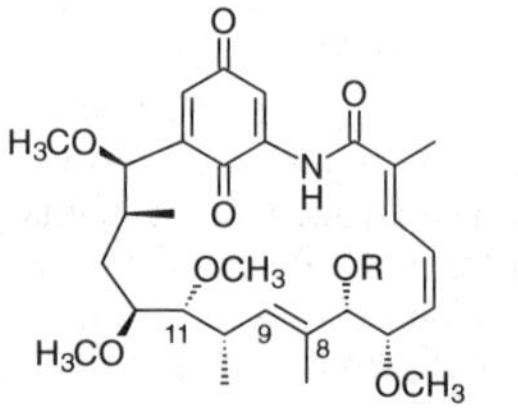

R = CO–NH_2

$C_{30}H_{42}N_2O_9$, M_R 574.67, yellow columns, mp. 230 °C, $[\alpha]_D^{20}$ +137° ($CHCl_3$), LD_{50} (mouse i.p.) 19 mg/kg. An *ansamycin antibiotic with herbicidal effects and activity against the tobacco mosaic virus isolated from cultures of *Streptomyces hygroscopicus*.

Lit.: Bull. Chem. Soc. Jpn. **65**, 2974–2991 (1992) ▪ J. Antibiot. **39**, 1630 (1986). – *[CAS 70563-58-5]*

Herboxidiene.

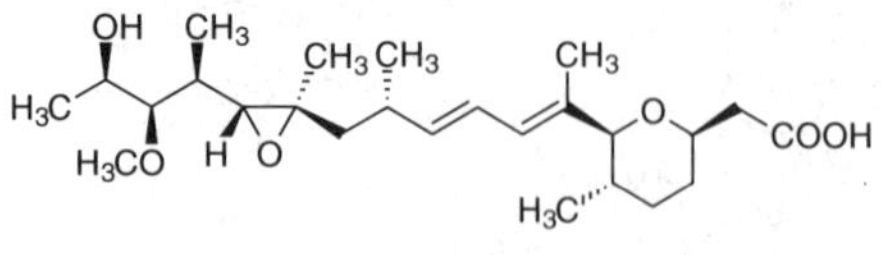

$C_{25}H_{42}O_6$, M_R 438.60, amorphous powder. A selective herbicide from cultures of *Streptomyces chromofuscus*.

Lit.: J. Antibiotics **45**, 914–921 (1992) (isolation) ▪ J. Chem. Soc., Perkin Trans. 1 **1999**, 955 (synthesis) ▪ Synthesis **1996**, 652–666 ▪ Tetrahedron **53**, 2785–2802 (1997) (synthesis). – *[CAS 142861-00-5]*

Hercynin (N^{α},N^{α},N^{α}-trimethylhistidinium betaine).

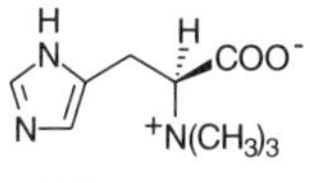

(2*S*)-form

$C_9H_{15}N_3O_2$, M_R 197.23, mp. 237–238 °C (decomp.), $[\alpha]_D^{22}$ +44.5° (5 m HCl). An intermediate product in the biosynthesis of *ergothioneine from *histidine. H. occurs in the milk juice of the rubber tree (*Hevea brasiliensis*), in various fungi, e. g., cépe (*Boletus edulis*), fly agaric (*Amanita muscaria*), field mushroom (*Agaricus campestris*), and also in the king crab (*Limulus polyphemus*).

Lit.: Beilstein E V **25/16**, 408 ▪ Merck-Index (12.), No. 4702. – *[CAS 507-29-9]*

Herical.

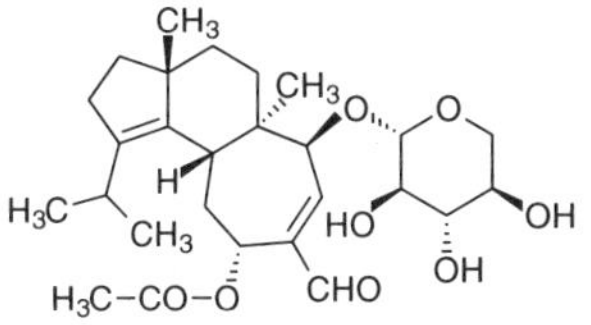

Herical

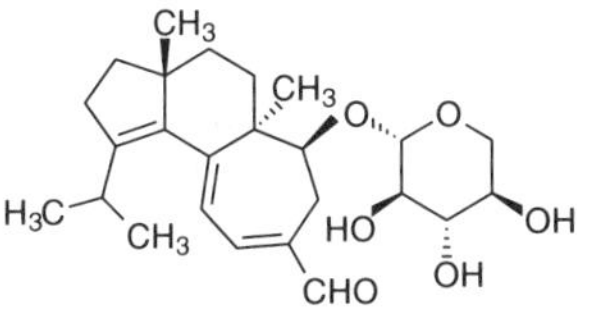

Erinacine A

$C_{27}H_{40}O_8$, M_R 492.61, oil, $[\alpha]_D$ –35.9° ($CHCl_3$). Diterpenoid xyloside of the *cyathane type from cultures of the fungus *Hericium clathroides* (=*H. ramosum*, Basidiomycetes). Through feeding experiments it has been shown that H. is the biosynthetic precursor of *striatal A. Closely related compounds such as *erinacin A* ($C_{25}H_{36}O_6$, M_R 432.56, cryst., mp. 74–76 °C) are isolated from cultures of *H. erinaceum*. The erinacins stimulate the synthesis of nerve growth factor (NGF) [1]. Erinacin E acts as a kappa opioid receptor agonist. It is found in *H. ramosum* cultures [2].

Lit.: [1] J. Am. Chem. Soc. **118**, 7644 (1996); J. Org. Chem. **63**, 306, 4732 (1998). [2] J. Antibiot. (Tokyo) **51**, 983 (1998); Tetrahedron Lett. **37**, 7399 (1996).

gen.: DECHEMA Monogr. **129**, 3–13 (1993) ▪ Tetrahedron Lett. **35**, 1569 (1994). – *[CAS 156101-08-5 (E. A)]*

Hericenones.

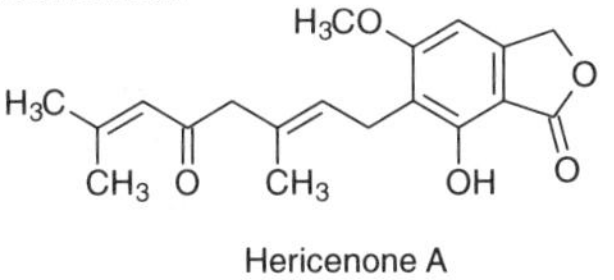

Hericenone A

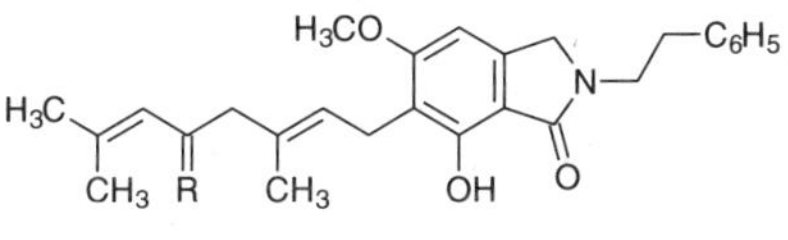

R = H,H : Hericerin
R = O : Hericenone B

Cytotoxic components from fruit bodies of the fungus *Hericium erinaceum* (Basidiomycetes). *H. A*: $C_{19}H_{22}O_5$, M_R 330.38, cryst., mp. 100–102 °C. Other components are the cytotoxic *H. B* ($C_{27}H_{31}NO_4$, M_R 433.55, cryst., mp. 136–138 °C) and *hericerin* [$C_{27}H_{33}NO_3$, M_R 419.56, needles (hemihydrate), mp. 138–140 °C] which inhibits pollen growth. Other H. stimulate the synthesis of nerve growth factor (NGF).

Lit.: Agric. Biol. Chem. **55**, 2673 (1991) ▪ J. Nat. Prod. **57**, 602 (1994) ▪ Phytochemistry **32**, 175 (1993) ▪ Tetrahedron Lett. **31**, 373 (1990); **33**, 4061 (1992) (synthesis). – *[CAS 126654-52-2 (H. A); 126654-53-3 (H. B); 140381-53-9 (hericerin)]*

Hernandulcin.

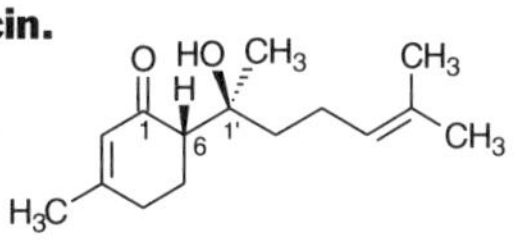

$C_{15}H_{24}O_2$, M_R 236.35, yellowish oil, bp. 130–140 °C (12 Pa), $[\alpha]_D^{20}$ +126° (C_2H_5OH), monocylic *sesquiterpene with the *bisabolene skeleton; a natural sweetener with a taste ca. 1000 times sweeter than that of *sucrose, isolated from *Lippia dulcis* (Verbenaceae) [1]. This plant was known by the Aztecs under the name *Tzonpelic xihuitl* ('sweet plant') and was first described by the Spanish physician Francisco Hernández in 1570 [1,2]. H. does not show any mutagenic activity in the Ames test and is non-toxic to mice up to an oral dose of 2 g/kg body weight [1,3]. Derivatives of H. that are modified at the C-1 carbonyl and C-1′ hydroxy groups no longer have a sweet taste. The sensory properties of these compounds can be explained on the basis of the geometry of these functional groups using the Shallenberger model [4].

Lit.: [1] J. Nat. Prod. **55**, 1136 (1992); Phytochemistry **44**, 1077 (1997); Science **227**, 417 (1985). [2] Chem. Br. **1985**, 330. [3] Chem.-Biol. Interact. **43**, 323 (1983); Pharm. Ind. **48**, 1416 (1986). [4] Nature (London) **221**, 555 (1969); Experientia **44**, 447 (1988). – *Synthesis:* Tetrahedron **42**, 5895 (1986); J. Agric. Food Chem. **35**, 273 (1987). – *[CAS 95602-94-1]*

Herniarin see umbelliferone.

Heroin see morphine and morphinan alkaloids.

Herperal® see distamycins.

Hesperetin [(2*S*)-3′,5,7-trihydroxy-4′-methoxyflavanone].

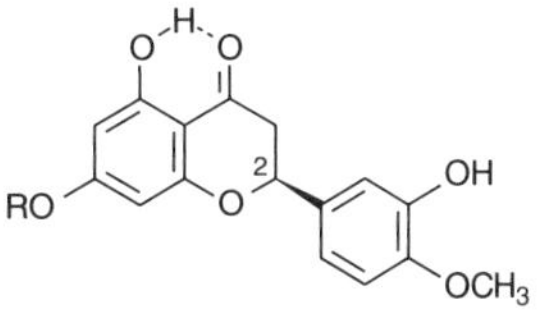

R = H : Hesperetin
R = β-D-Rutinosyl : Hesperidin
R = β-D-Neohesperidosyl : Neohesperidin

$C_{16}H_{14}O_6$, M_R 302.28; platelets, mp. 216–218 °C, $[\alpha]_D$ –37.6° (C_2H_5OH). H. is the aglycone of the flavanone glycoside hesperetin 7-*O*-rutinoside {*hesperidin*, $C_{28}H_{34}O_{15}$, M_R 610.57, hygroscopic needles, mp. 251 °C (decomp.), $[\alpha]_D$ –47.4° (pyridine)} present in the skin of unripe oranges. *Neohesperidin* {$C_{28}H_{34}O_{15}$, M_R 610.57, bitter-tasting crystals, mp. 244 °C, $[\alpha]_D$ –100° (pyridine)} occurs in the sour orange (bigarade, Seville orange, *Citrus aurantium*) and other *Citrus*-

and *Mentha* species. Gluco-H. {7-*O*-β-D-glucopyranosyl-H., $C_{22}H_{24}O_{11}$, M_R 464.43, mp. 206 °C, $[\alpha]_D^{19}$ −53.9° (pyridine)} is the biogenetic precursor of neohesperidin [1].

Activity: H. has anti-inflammatory [2] and antiviral (influenza) activities and is an antioxidant [3]; phosphoric acid esters are used for venous complaints.

Lit.: [1] J. Agric. Food. Chem. **41**, 1916–1924 (1993). [2] J. Pharm. Pharmacol. **46**, 118–122 (1994). [3] Biochem. Pharmacol. **46**, 1257–1271 (1993); J. Am. Chem. Soc. **116**, 4846–4851 (1994).

gen.: Biosci. Biotechnol. Biochem. **58**, 521–525, 1479–1485 (1994) (synthesis) ▪ Harborne (1994), p. 413 ff., 607 ff. ▪ Karrer, No. 1626, 1628, 1629 ▪ Nat. Prod. Rep. **12**, 101–133 (1995). – *Pharmacology:* Mutagenesis **9**, 101–106 (1994) ▪ Planta Med. **60**, 99–100 (1994). – *[CAS 520-33-2 ((S)-form); 41001-90-5 ((±)-form); 520-26-3 (hesperidin); 13241-33-3 (neohesperidin); 2500-68-7 (glucohesperetin)]*

Hesperidin see hesperetin.

Het 389, Het 403 see Hebestatis toxins.

HETE see hydroxyfatty acids.

Heteroauxin see 3-indolylacetic acid.

Heterodendrin see cyanogenic glycosides.

Heteroglycans see polysaccharides.

Heteroyohimbine alkaloids. A group of *indole alkaloids with the *yohimbine structure but with an oxygen function in ring E (see formula under yohimbine); H. occur mainly in the Apocynaceae. Typical examples are *ajmalicine (raubasine), 19-*epi*-ajmalicine, tetrahydroalstonine, rauniticine (all with the 3α-configuration), as well as serpentine and alstonine with an aromatic ring C.

Biosynthesis: Very well known, the first example of a complete elucidation at the enzymatic level for a group of alkaloids. It proceeds from tryptamine and secologanin (see secoiridoids) via *strictosidine, 4,21-dehydrogeissoschizine, and *cathenamine.

Lit.: Brown, in Bartmann & Winterfeldt (eds.), Stereoselective Synthesis of Natural Products, p. 62–70, Amsterdam: Excerpta Medica 1979 ▪ Heterocycles **48**, 1483 (1998) (biomimetic interconversions) ▪ Phillipson & Zenk (eds.), Indole and Biogenetically Related Alkaloids, p. 113–141, 142–184, 185–200, London: Academic Press 1980. – *[HS 2939 90]*

Hetisine (delatine).

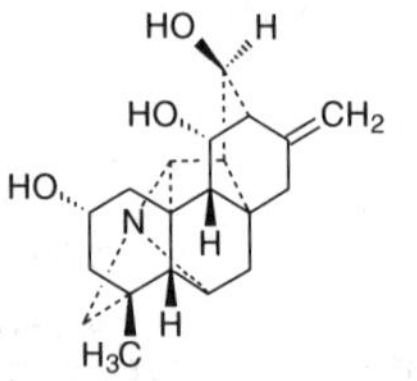

$C_{20}H_{27}NO_3$, M_R 329.44, mp. 261–264 °C, $[\alpha]_D^{27}$ +10.9° ($CHCl_3$). A diterpene alkaloid from the roots of *Aconitum heterophyllum* and *Delphinium tatsienense*, the seeds of *D. elatum*, and the aerial parts of *D. cardinale* (Ranunculaceae).

Lit.: Heterocycles **24**, 1605 (1986); **27**, 185 (1988) ▪ Manske **34**, 95–179. – *[CAS 10089-23-3]*

Hexanal see alkanals and alarm substances.

Hexanol see alarm substances.

Hexenals (2- and 3-hexenals).

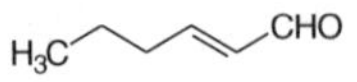

(*E*)-2-H.

(*Z*)-3-H.

$C_6H_{10}O$, M_R 98.14. Like the *hexen-1-ols, the H. are formed by autoxidation or enzymatic degradation principally from *linoleic acid in plant material cut into small pieces [1].

1. (*E*)-*2-H.* (leaf aldehyde): an oil with a hot, herby-green odor, pleasant fruity-green in dilute solution; olfactory threshold in water [2] 17 ppb, bp. 146 °C, LD_{50} (rat p.o.) 0.78 g/kg.

Occurrence: Widely distributed [3] in *essential oils, especially of spice plants, *fruit, *vegetable, and *meat flavor; particularly important in the flavors of apple juice (*impact compound), kiwis, tea, and fresh tomatoes; occurs also as an *alarm or defence substance in various insects, e.g., bugs [4]. For preparation, see *Lit.* [5]; used mainly in fruit and vegetable flavors.

2. (*Z*)-*3-H.*: odor of freshly cut leaves, olfactory threshold [2] 0.25 ppb, bp. 36 °C (2.7 kPa). Occurs frequently together with (*E*)-2-H. in fruit and vegetable flavors; isomerizes readily to (*E*)-2-H. and thus is not used widely.

The rarer isomers (*Z*)-*2-H.* [bp. 48 °C (2.7 kPa), occurrence in, e.g., apples, kiwis, tea, and tomatoes] and (*E*)-*3-H.* [bp. 35 °C (2.7 kPa), occurrence in, e.g., apples, guavas, and kiwis] are described in *Lit.* [6].

Lit.: [1] Phytochemistry **33**, 253–280 (1993). [2] Perfum. Flavor. **16** (1), 16 (1991). [3] TNO list (6.) Suppl. 5, p. 398. [4] Angew. Chem. Int. Ed. Engl. **20**, 164–184 (1981). [5] Ullmann (5.) **A 1**, 335. [6] Z. Naturforsch., C **47**, 183–189 (1992).

gen.: Arctander, No. 1598, 1599 ▪ Bauer et al. (2.), p. 13 ▪ Bedoukian (3.), p. 29–33. – *[HS 2912 19; CAS 6728-26-3 ((E)-2-H.); 6789-80-6 ((Z)-3-H.); 16635-54-4 ((Z)-2-H.); 69112-21-6 ((E)-3-H.)]*

Hexen-1-ols (2- and 3-hexen-1-ols).

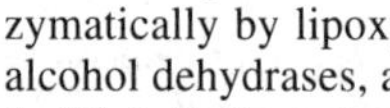

(*Z*)-3-H.

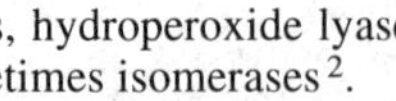

(*E*)-2-H.

$C_6H_{12}O$, M_R 100.16. All isomers of H. occur in nature but only (*Z*)-3-H. [frequently accompanied by (*E*)-2- and (*E*)-3-H.] is practically ubiquitous, especially in fresh leaves and fruits cut into small pieces [1]. They are formed, e.g., from *linoleic or *arachidonic acid enzymatically by lipoxygenases, hydroperoxide lyases, alcohol dehydrases, and sometimes isomerases [2].

1. *(Z)-3-H.* (leaf alcohol): oil with an odor of freshly cut grass or leaves, bp. 157 °C, LD_{50} (rat p.o.) 4.7 g/kg; olfactory threshold (water) 70 ppb.

Occurrence: In the essential oil of mulberry leaves (up to 50%), green tea (up to 30%), and thyme (ca. 0.4%), as well as in many *fruit and *vegetable flavors.

Use: To create green notes in flavors and perfumes, for production of the esters used for similar purposes.

2. *(E)-2-H.:* has a green-fruity, sweet odor although markedly weaker than that of (*Z*)-3-H. [3], bp. 155 °C, LD_{50} (rat p.o.) 3.5 g/kg. Used mainly in fruit flavors. The sensory and physical data as well as the production of the less frequently found isomers (*Z*)-2-H. (occurrence in, e.g., papaya, endive, and kiwi flavors),

(*E*)-3-H., (*E*)-4-H. (detected in banana, passion fruit, and kiwi flavors), and 5-H. (in apple flavor) are described in *Lit.*[3].

Lit.: [1]TNO list (6.) Suppl. 5, p. 400. [2]Müller-Lamparsky, p. 101 ff.; Maarse, p. 182. [3]Z. Naturforsch. C **47**, 183–189 (1992); J. Agric. Food Chem. **19**, 1111 (1971).
gen.: Arctander, No. 1604, 1606, 1607 ▪ Bauer et al. (2.), p. 9 ▪ Perfum. Flavor. **15** (4), 47–52 (1990). – *[HS 2905 29; CAS 928-96-1 ((Z)-3-H.); 928-95-0 ((E)-2-H.); 928-94-9 ((Z)-2-H.); 928-97-2 ((E)-3-H.); 928-92-7 ((E)-4-H.); 928-91-6 ((Z)-4-H.); 821-41-0 (5-H.)]*

Hexitols. General term for hexafunctional alcohols (polyols, so-called sugar alcohols, in the stricter sense *alditols*) of the general formula $HOCH_2$–$(CHOH)_4$–CH_2OH, relatives of the corresponding hexoses and accessible from the latter by reduction. They are sweet-tasting, generally well crystallizing substances that cannot be fermented and do not reduce Fehling's solution. The most important H. are sorbitol (D-*glucitol), *mannitol, and dulcitol (*galactitol), widely distributed in plants, some of them find use as sugar substitutes (diabetics sugar); Further H. are allitol, altritol, and *iditol.

Lit.: Kirk-Othmer (3.) **1**, 754–778. – *[HS 2905 49]*

Hexosans see polyoses.

Hexyl rhizoglyphinate see mites.

Heyneanine, heyneatine see Iboga alkaloids.

Hierridin (2-heneicosyl-4,6-dimethoxyphenol).

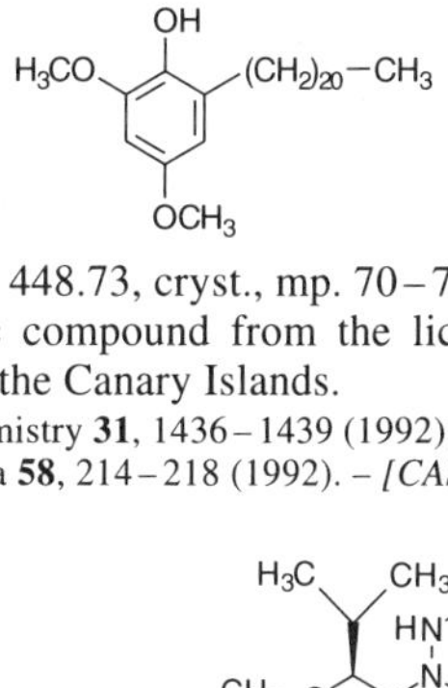

$C_{29}H_{52}O_3$, M_R 448.73, cryst., mp. 70–72 °C. Monocyclic aromatic compound from the lichen *Ramalina hierrensis* of the Canary Islands.

Lit.: Phytochemistry **31**, 1436–1439 (1992); **49**, 2383 (1998) ▪ Planta Medica **58**, 214–218 (1992). – *[CAS 142382-20-5]*

Himastatin.

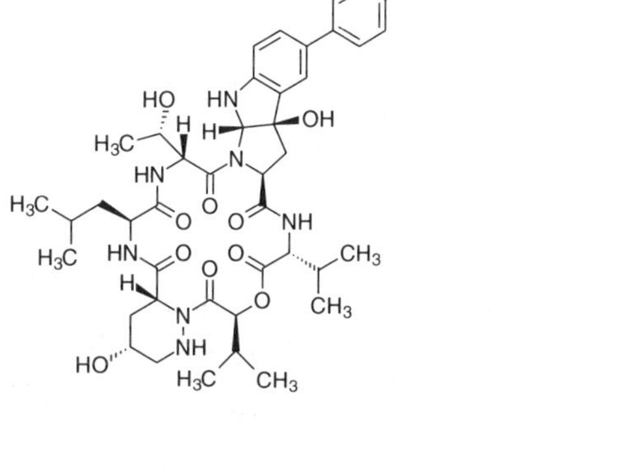

$C_{72}H_{104}N_{14}O_{20}$, M_R 1485.68, cryst., mp. 200 °C (decomp.), $[\alpha]_D^{20}$ –34° (CH_3OH). A dimeric cyclohexadepsipeptide antibiotic produced by *Streptomyces hygroscopicus*. H. shows activity against Gram-positive bacteria and tumors. It is structurally related to certain *omphalotins.

Lit.: Angew. Chem. Int. Ed. Engl. **37**, 2993–2995, 2995–2998 (1998) ▪ J. Antibiot. **43**, 956–960, 961–966 (1990); **49**, 299–311 (1996). – *[CAS 126775-74-4]*

Himbacine.

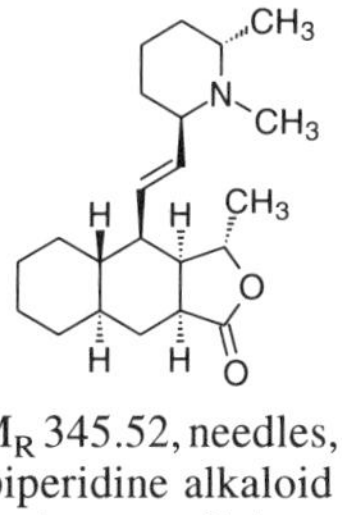

$C_{22}H_{35}NO_2$, M_R 345.52, needles, mp. 132 °C, $[\alpha]_D^{14}$ +63° ($CHCl_3$). A piperidine alkaloid from the bark of *Galbulimima* species, e.g. *G. baccata* (Himantandraceae, New Guinea)[1]. It is a muscarinic M_2-receptor-antagonist and has spasmolytic activity. H. and its synthetic analogues are being investigated as drugs for treatment of Alzheimer's disease.

Lit.: [1]Aust. J. Chem. **16**, 112 (1963); **18**, 569 (1965).
gen.: J. Am. Chem. Soc. **117**, 9369 (1995); **118**, 9812 (1996) ▪ J. Org. Chem. **62**, 5023 (1997); **64**, 1932 (1999) ▪ Synthesis **1998**, 479 [synthesis (+)-H.] ▪ Tetrahedron Lett. **36**, 7515 (1995); **40**, 3399 (1999) (synthesis). – *[CAS 6879-74-9]*

Hindarine see tetrahydropalmatine.

Hinokiflavone.

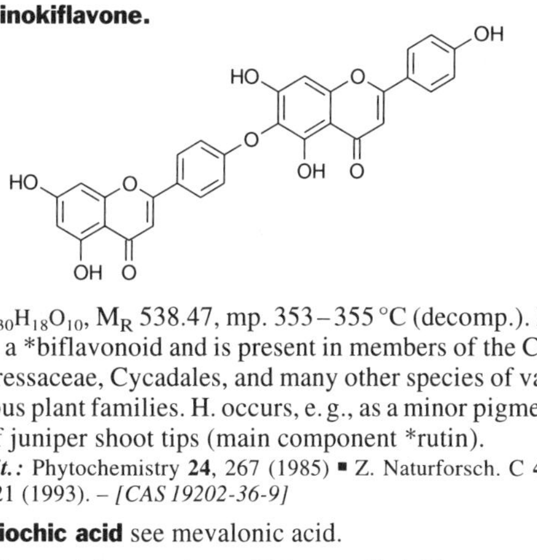

$C_{30}H_{18}O_{10}$, M_R 538.47, mp. 353–355 °C (decomp.). H. is a *biflavonoid and is present in members of the Cupressaceae, Cycadales, and many other species of various plant families. H. occurs, e.g., as a minor pigment of juniper shoot tips (main component *rutin).

Lit.: Phytochemistry **24**, 267 (1985) ▪ Z. Naturforsch. C **48**, 821 (1993). – *[CAS 19202-36-9]*

Hiochic acid see mevalonic acid.

Hippeastrine see Amaryllidaceae alkaloids.

Hippocasine.

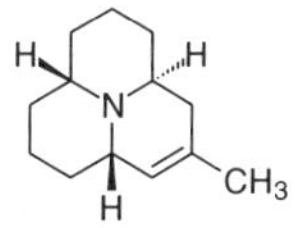

(relative configuration)

An alkaloid from the defense secretion of the West Canadian insect *Hippodamia caseyi*; $C_{13}H_{21}N$, M_R 191.32, oily substance. *Hippocasine N-oxide hydrochloride:* mp. >220 °C (decomp.), $[\alpha]_D^{25}$ +14.8° (CH_3OH). Further components are the structurally related hippodamine and convergine (see coccinelline). The insect *Coccinella transversoguttata* produces the stereoisomeric precoccinelline and *coccinelline.

Lit.: Can. J. Chem. **54**, 1807 (1976) ▪ J. Org. Chem. **49**, 2217 (1984) ▪ Heterocycles **7**, 685, 693 (1977). – *[CAS 60022-25-5]*

Hippuric acid (*N*-benzoylglycine).

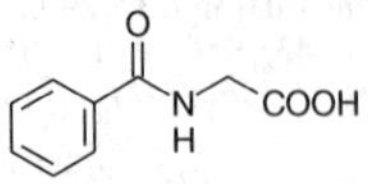

$C_9H_9NO_3$, M_R 179.17, mp. 187–188 °C. H. occurs in the urine of most mammals, especially herbivorous animals. It was first isolated by J. v. Liebig from horse urine (=·Greek: hippou ouron). The biosynthesis proceeds from benzoic acid or related aromatic compounds and glycine under the action of hippuricase (EC 3.5.1.14).

Lit.: Beilstein E IV **9**, 778 ▪ Can. J. Chem. **53**, 3175 (1975) ▪ Hager (5.) **9**, 1147 f. – *[HS 2924 29; CAS 495-69-2]*

Hiptagenic acid, hiptagin see 3-nitropropanoic acid.

Hirsutanes.

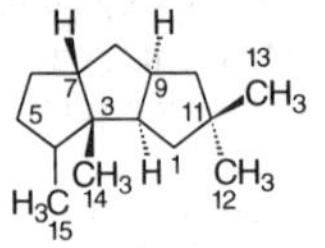

Hirsutane

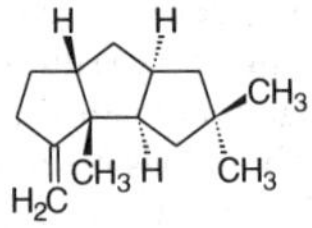

Hirsutene

Sesquiterpenoids of the triquinane type produced by fungi. The key step in their biosynthesis is the cyclization of *humulene to *hirsutene*[1] ($C_{15}H_{24}$, M_R 204.36, oil) which is isolated from cultures of *Coriolus consors* (Basidiomycetes). Hirsutene has been synthesized by more than 20 different routes. Functionalized hirsutanes such as *complicatic acid, *hirsutic acid, hypnophilin, *pleurotellic acid, and the *coriolins isolated from basidiomycete cultures all possess the absolute configuration shown for hirsutene.

Lit.: [1] Tetrahedron Lett. **1976**, 195; **36**, 6851 (1995).
gen.: Synthesis: J. Org. Chem. **55**, 2725 (1990); **57**, 1968 (1992) ▪ Justus Liebigs Ann. Chem. **1993**, 403 ▪ Tetrahedron **46**, 1859 (1990); **48**, 1911, 6897 (1992); **49**, 7931, 11189 (1993); **50**, 415 (1994). – *Review:* Nicolaou, p. 381–420 ▪ Tetrahedron **37**, 2199 (1981) ▪ Turner **2**, 246. – *[CAS 59372-72-4 (hirsutene)]*

Hirsutic acid (hirsutic acid C).

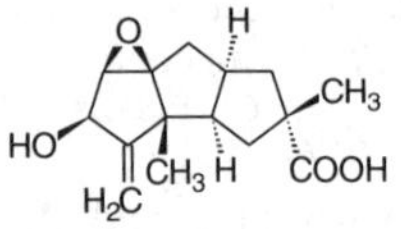

$C_{15}H_{20}O_4$, M_R 264.32, prisms, mp. 182 °C, $[\alpha]_D$ +116° ($CHCl_3$). A sesquiterpenoid of the *hirsutane type from cultures of the fungus *Stereum hirsutum* (basidiomycetes), frequently found on dead hardwood (see also complicatic acid).

Lit.: Beilstein E V **18/7**, 337 ▪ Chem. Pharm. Bull. **38**, 3230 (1990) [(+)-H.] (synthesis) ▪ J. Org. Chem. **50**, 3957 (1985) [(+)-H.]; **51**, 2742 (1986) [*rac.*-H.] ▪ Tetrahedron **37**, 2202 (1981) (review). – *[CAS 3650-17-7]*

Hirsutin(e). A name used for four different compounds: a) 1-isothiocyanato-8-(methylsulfinyl)octane, see mustard oils. – b) A *corynantheine alkaloid from *Mitragyna* and *Uncaria* species, $C_{22}H_{28}N_2O_3$, M_R 368.48, cryst., mp. 101 °C, $[\alpha]_D$ +68.6° ($CHCl_3$). The

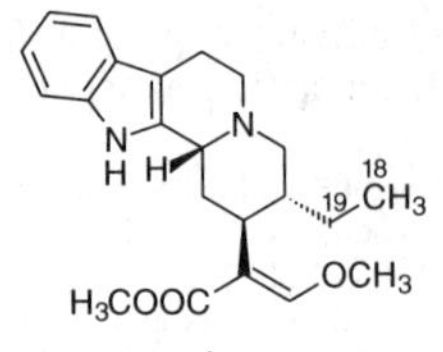

b

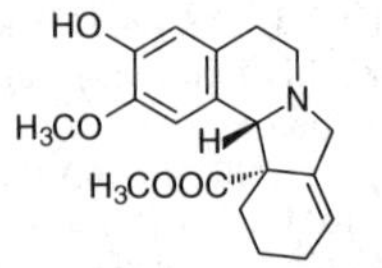

c

N-oxide[1] of H. occurs in *Uncaria tomentosa*. H. with a Δ^{18}-double bond is known as hirsuteine[2]. H. has blood vessel and muscle relaxing properties[3]. H. also shows activity against influenza A viruses. – c) An isoquinoline alkaloid from *Cocculus* species, $C_{19}H_{23}NO_4$, M_R 329.40. – d) 3,5-Bis-*β*-D-glucopyranosyloxy-4′-hydroxy-3′,5′,7-trimethoxyflavylium, $C_{30}H_{37}O_{17}^+$, M_R 669.61

Lit.: [1] Life Sci. **50**, 491 (1992). [2] Heterocycles **26**, 1739–1742 (1987). [3] Jpn. J. Pharmacol. **33**, 463–471 (1983).
gen. (to b): Angew. Chem. Int. Ed. Engl. **38**, 2045–2047 (1999) ▪ Heterocycles **49**, 445 (1998) (synthesis) ▪ J. Chem. Soc., Chem. Commun. **1984**, 847 f. ▪ Tetrahedron Lett. **38**, 1455 (1997). – *(to c):* J. Nat. Prod. **54**, 582 (1991). – *(to d)*: Phytochemistry **20**, 1971 (1981). – *[CAS 7729-23-9 (b); 55176-58-4 (N-oxide); 35467-43-7 (hirsuteine); 135250-40-7 (c); 32221-58-2 (d)]*

Hirudin (lepirudin). Protein from the head and gullet rings of the medicinal leech (*Hirudo medicinalis*) (ca. 3 mg/leech). H is a heparinoid that delays blood coagulation (thrombin inhibitor). The M_R of H. is ca. 9060 (other reports: 10800), it has a high content of acidic amino acid residues. H. is a colorless powder, soluble in water, insoluble in ethanol. The anticoagulative activity of H. is based on the formation of a compound with thrombin, so that the latter cannot exert its catalytic effect[1]. Commercially available for use in the treatment of thrombosis, hematomas, etc. In Europe, recombinant H. (Refludan®) is marketed for heparin-associated thrombocytopenia.

Lit.: [1] Angew. Chem. Int. Ed. Engl. **34**, 867–880 (1995).
gen.: Pharmazie **43**, 737 ff. (1988). – *[CAS 8001-27-2]*

Hispidin see styrylpyrones.

Histamine.

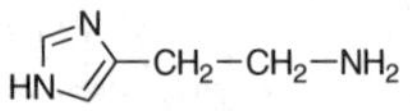

$C_5H_9N_3$, M_R 111.15. Deliquescent crystals, mp. 84 °C, bp. 210 °C (2.4 kPa), readily soluble in water and ether. H. is a biogenic amine, formed by decarboxylative decomposition of proteins. As a tissue hormone it occurs especially in the skin, lungs, and mast cells. Many insect toxins (e. g., *wasp and *bee venom, stinging hairs of caterpillars) contain H., as does the venom of stinging nettle. H. also causes fish poisonings when it is formed by putrefactive bacteria from the muscle protein of dead fish. Although mice tolerate relatively high doses of H., even small amounts are lethal to hamsters. Doses of merely 4 µg in humans result in a lowering of blood pressure, dilation and increased permeability of blood vessels, stimulation of the smooth intestinal muscles, constriction of the bronchi through binding to the so-called H_1 receptors as well as increased secretion of gastric juice and increased heart rate by action on the H_2 receptors. H_3 receptors inhibit H. release from various cell types[1]. Since the capillary blood vessels become more permeable under the effect of H., blood plasma migrates into the surrounding tissue. In mast cells and basophilic leukocytes H. is bound in a loose complex to heparin and other proteins. It can be released from these by scratching or warming areas of the skin; this causes irritation and reddening. H. occurs in larger amounts in the course of allergies or anaphylaxis when it is ejected from mast cells on account of the binding of the allergen to membrane-bound immunoglobulin E. In cases of insufficient endogenous degradation by diaminooxidase (histaminase), the adverse effects of H. – it also acts as a mediator of pain sensation – are reduced by administration of antihistaminic drugs such as mepyramin (H_1 receptor antagonist), ranitidin or cimetidin (H_2 receptor antagonists).

Lit.: [1] Trends Pharmacol. Sci. **10**, 159–162 (1989).
gen.: Barrett, Chemistry and Biochemistry of the Amino Acids, London: Chapman and Hall 1985 ▪ Beilstein E V **25/9**, 521–523 ▪ Luckner (3.), p. 319–323. – *[HS 2933 29; CAS 51-45-6]*

Histidine (2-amino-3-imidazol-4-ylpropanoic acid, abbreviation: His or H).

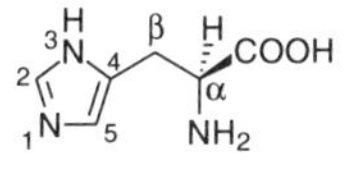

(2*S*)-, L-form

$C_6H_9N_3O_2$, M_R 155.16, mp. 287 °C (decomp.), $[\alpha]_D^{25}$ –38.5° (H_2O), pK_a 1.78, 5.97 (Im), 8.97 (NH_3^+), pI 7.47, sweet taste; monohydrochloride, mp. 251–252 °C (decomp.), $[\alpha]_D^{26}$ +8.0° (3 m HCl). H. is a proteinogenic amino acid, essential for rats but not for humans. The average content of H. in more than 200 different proteins is 2.1% [1]. Genetic code: CAU, CAC.

Biosynthesis: His is formed from ATP and 5-phospho-α-D-ribofuranosyl pyrophosphate (PRPP) by a multistep process. The C_5 chain originates from PRPP, N-1 and C-2 come from N-1 and C-2 of ATP and N-3 is supplied by N-5 of *glutamine. The concomitantly formed degradation product of ATP, 5-amino-1-(5-phospho-β-D-ribofuranosyl)-4-imidazolecarboxamide is used in the synthesis of purine nucleotides.

Catabolism: The biological degradation of His proceeds through urocanic acid [histidine ammonia lyase (EC 4.3.1.3)] and 4-imidazolone-5-propanoic acid [urocanic acid hydratase (EC 4.2.1.49)]. The imidazolone ring is cleaved by imidazolone propionase (EC 3.5.2.7) and furnishes *N*-formiminoglutaminic acid. Finally, Glu is formed by glutamate formiminotransferase (EC 2.1.2.5) with the cofactor tetrahydrofolic acid.

Use: In chemically defined diets and as infusion solution for volume supplementation. H. supports child growth and is used to treat allergies, arteriosclerosis, anemia, and rheumatoid arthritis.

Lit.: [1] Biochem. Biophys. Res. Commun. **78**, 1018–1024 (1977).
gen.: Beilstein E V **25/16**, 363–380 ▪ Merck-Index (12.), No. 4758 ▪ Ullmann (5.) **A 2**, 58, 62, 70, 85. – *[HS 2933 29; CAS 71-00-1 (L-H.)]*

Histrionicotoxins. Alkaloids from arrow poison frogs (Dendrobatidae) of the genera *Dendrobates*, *Epipedobates*, and *Phyllobates* with a 1-azaspiro[5.5]undecane ring system. The 16 presently known H. have unsaturated side chains R^1 and R^2 with *cis*-configurated double bonds but differing lengths. Triple bonds and allene structures may also be present in the side chains. In comparison to other *Dendrobates alkaloids they only exhibit low toxicity. The H. are non-competitive inhibitors of nicotinic receptors. They also attack sodium and potassium ion channels [1]. For the main alkaloids, see table.

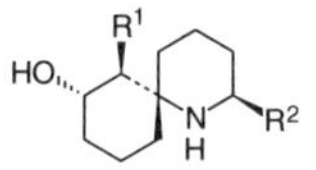

	R^1	R^2
H. (283 A)	CH=CH–C≡CH	CH_2–CH=CH–C≡CH
Allodihydro-H. (285 C)	CH=CH–C≡CH	$(CH_2)_3$–C≡CH
Isodihydro-H. (285 A)	CH=CH–C≡CH	CH_2–CH_2–CH=C=CH_2
Octahydro-H. (291 A)	CH_2–CH_2–CH=CH_2	$(CH_2)_3$–CH=CH_2
235 A	CH=CH_2	CH_2–CH=CH_2
259 A	CH=CH–C≡CH	CH_2–CH=CH_2

Table: Data of Histrionicotoxins.

	molecular formula	M_R	mp. [°C]	$[\alpha]_D^{25}$	CAS
283 A	$C_{19}H_{25}NO$	283.41	79–80	–96.3° (C_2H_5OH, hydrochloride)	34272-51-0
285 C	$C_{19}H_{27}NO$	285.43	247–250	–43.4° ($CHCl_3$)	63983-63-1
285 A	$C_{19}H_{27}NO$	285.43	240–243 (hydrochloride)	–35.3° (C_2H_5OH, hydrochloride)	34272-52-1
291 A	$C_{19}H_{33}NO$	291.48	151–154 (hydrochloride)		55475-50-8
235 A	$C_{15}H_{25}NO$	235.37		–38.6 ($CHCl_3$)	67217-84-9
259 A	$C_{17}H_{25}NO$	259.39			67217-83-8

Lit.: [1] Manske **43**, 185–288.
gen.: Daly et al., in Spande, Alkaloids: Chemical and Biological Perspectives, vol. 4, p. 1, New York: Wiley 1986 ▪ Zechmeister **41**, 205, 247–276. – *Synthesis:* J. Am. Chem. Soc. **111**, 4852–4856 (1989); **112**, 5875 (1990); **121**, 4900 (1999) ▪ J. Chem. Soc., Chem. Commun. **1998**, 2509 (perhydro-H.) ▪ J. Org. Chem. **64**, 5485 (1999) ▪ Tetrahedron Lett. **26**, 5887–5890 (1985) ▪ see also Dendrobates alkaloids. – *[HS 2939 90]*

Hitachimycin (stubomycin).

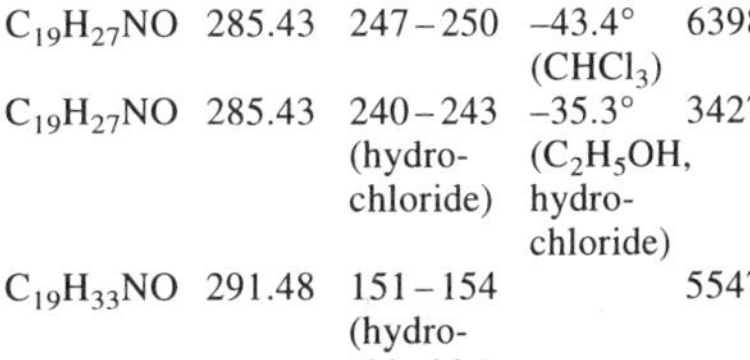

$C_{29}H_{35}NO_5$, M_R 477.60, colorless cryst., mp. 253–256 °C (decomp.), $[\alpha]_D^{20}$ +246° (DMSO), macrocyclic lactam antibiotic produced by *actinomycetes that inhibits the growth of cancer cells and protozoa.

Lit.: J. Am. Chem. Soc. **114**, 8003 (1992) ▪ J. Antibiot. **34**, 259–263 (1981) ▪ Pure Appl. Chem. **61**, 405 ff. (1989) ▪ Tetrahedron Lett. **23**, 4713 ff. (1982). – *Synthesis:* J. Am. Chem. Soc. **114**, 8008–8022 (1992); **115**, 4947 (1993) ▪ J. Org. Chem. **55**, 1133 ff., 1136 ff. (1990). – *[HS 2941 90; CAS 77642-19-4]*

Hodgkinsine see pyrroloindole alkaloids.

Holadienine, Holarrhena alkaloids see Apocynaceae steroid alkaloids.

Holocalin see cyanogenic glycosides.

Holomycin.

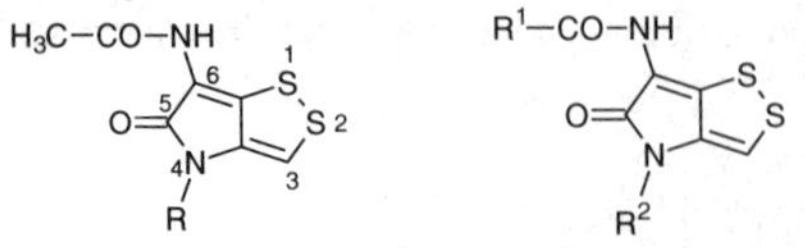

R = H : Holomycin
R = CH_3: Thiolutin

$R^1 = C_2H_5$, $R^2 = CH_3$: Aureothricin
$R^1 = (CH_2)_4\text{–}CH_3$, R^2 = H: Xenorhabdin 1

$C_7H_6N_2O_2S_2$, M_R 214.26, orange-yellow flakes, mp. 268–270 °C (decomp.). Antibiotic from a *Streptomyces griseus* strain, active against Gram-positive and -negative bacteria, fungi, and protozoa [1]. A series of further 1,2-dithiolo[4,3-*b*]pyrrol-5(4*H*)-ones have since been isolated which differ in the acyl group at N(6) and are sometimes methylated at N(4). On account of their biological activities, *thiolutin* {$C_8H_8N_2O_2S_2$, M_R 228.28, yellow needles, mp. 273–276 °C (decomp.)} [2], *aureothricin* {$C_9H_{10}N_2O_2S_2$, M_R 242.31, golden yellow cryst., mp. 260–270 °C (decomp.)} [3], and *xenorhabdin 1* ($C_{11}H_{14}N_2O_2S_2$, M_R 270.37, mp. 192–193 °C) [4] are mentioned as examples. Thiolutin isolated from various strains of *Streptomyces albus* and *S. pimprina* inhibits the growth of microbes in beer. It is phototoxic in high doses and inhibits aggregation of blood platelets. Aureothricin from *Streptomyces* spp. is reported to have antibacterial and antifungal activities, xenorhabdin 1 from *Xenorhabdus* spp. has antimicrobial and insecticidal activities.

Lit.: [1] ACS Symp. Ser. **504**, 384 (1992); J. Antibiot. **22**, 231 ff. (1969); **30**, 334 (1977); Bull. Chem. Soc. Jpn. **47**, 1484 (1974); J. Org. Chem. **42**, 2891 (1977) (synthesis). [2] Chem. Pharm. Bull. **28**, 3157 (1980); Curr. Sci. **53**, 659 (1984). [3] J. Antibiot. **35**, 1367 (1982). [4] J. Nat. Prod. **54**, 774 (1991).
gen.: Beilstein E IV **27**, 6758. – *[HS 2941 90; CAS 488-04-0 (H.); 87-11-6 (thiolutin); 574-95-8 (aureothricin); 92680-94-9 (xenorhabdin I)]*

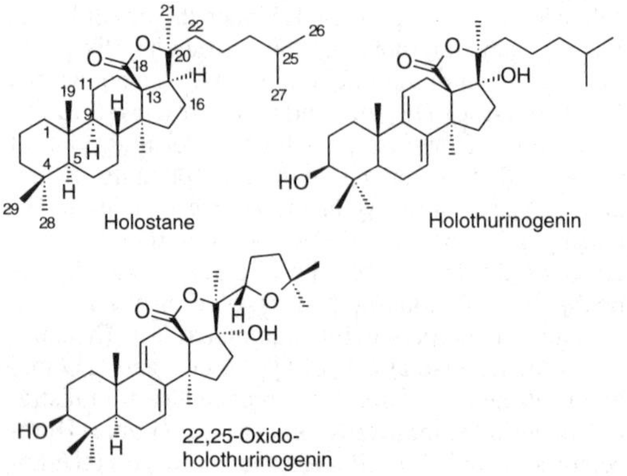

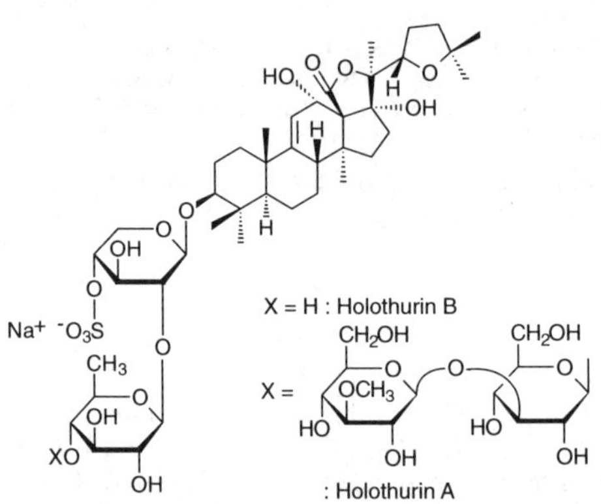

Holothurins (holotoxins). General name for toxins of sea cucumbers (Holothuroidea). They have hemolytic and local anesthetic activity, anesthetic and lethal effects on fish, and affect the nervous systems of small mammals in a manner similar to *cocaine or *physostigmine. In mice they inhibit the growth of sarcomas. They are occasionally used medicinally in homeopathic doses as cardiac tonics. Chemically, H. are steroid glycosides (saponins); the parent structure of the aglycone is holostane [(20*S*)-20-hydroxy-5α-lanostan-18-oic acid lactone]. H. coming into contact with human skin by way of Cuvier's tubes of seacucumbers cause burning pain, contact with the eyes can in severe cases lead to blindness, further symptoms are muscle spasms and paralysis.
Since sea cucumbers are a culinary delicacy ("Trepang" or "béche de mer"), the Cuvier's tubes must be removed completely, otherwise consumption will result in severe digestive complaints through to death by respiratory paralysis [LD_{50} (mouse i.v.) 7.5 mg/kg]. The H. inhibit the growth of fungi (MIC: 2–17 μg/mL solution). Examples for H. are the aglycones *holothurinogenin* ($C_{30}H_{46}O_4$, M_R 470.69, mp. 277 °C) and *22,25-oxidoholothurinogenin* ($C_{30}H_{44}O_5$, M_R 484.68, mp. 315–316 °C) as well as the saponins *H. A* ($C_{54}H_{85}NaO_{27}S$, M_R 1221.30, $[\alpha]_D$ –14.9°, needles) and *H. B* ($C_{41}H_{64}O_{17}S$, M_R 861.01) from *Holuthuria leucospilota*.

Lit.: Beilstein E V **19/6**, 550 ▪ Bull. Soc. Chim. Fr. **2**, 124 (1985) ▪ J. Nat. Prod. **58**, 172 (1995) (biosynthesis) ▪ Scheuer I **5**, 324–378 ▪ Scheuer II **1**, 132 ▪ Toxicon Suppl. **3**, 179 (1983) ▪ Zechmeister **27**, 322–339. – *[CAS 25495-63-0 (holothurinogenin); 6853-99-2 (22,25-oxidoholothurinogenin); 11052-32-7 (H. B)]*

Homaline. H. is a *polyamine alkaloid from the leaves of some *Homalium* species (Flacourtiaceae).

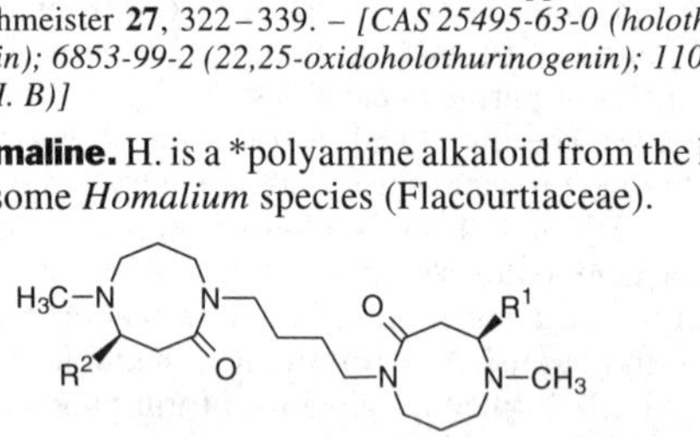

$R^1 = R^2$ = phenyl : Homaline (**1**)
R^1 = heptyl, R^2 = pentyl : Hopromine (**2**)
R^1 = 2-hydroxyheptyl, R^2 = pentyl : Hopromminol (**3**)
R^1 = 2-hydroxyheptyl, R^2 = phenyl : Hopromalinol (**4**)

Table: Data of Homaline and derivatives.

	molecular formula	M_R	mp. [°C]	$[\alpha]_D$	CAS
1	$C_{30}H_{42}N_4O_2$	490.69	134	–34° ($CHCl_3$)	20410-93-9
2	$C_{30}H_{58}N_4O_2$	506.82	–	–10°	49620-03-3
3	$C_{30}H_{58}N_4O_3$	522.82	–	–19°	49620-04-4
4	$C_{31}H_{52}N_4O_3$	528.78	–	–17°	49620-05-5

Structurally related compounds include *hopromine*, *hoprominol*, and *hopromalinol* from *Homalium pronyense*. These alkaloids all contain *spermine as a building block.

Lit.: J. Chem. Soc., Perkin Trans. 1 **1993**, 2047, 2055 (synthesis) ▪ Tetrahedron **29**, 1001 (1973); **39**, 2459 (1983) ▪ Tetrahedron Lett. **27**, 5147 (1986). – *[HS 2939 90]*

Homoaporphine alkaloids see phenethyltetrahydroisoquinoline alkaloids.

L-Homoarginine [N^6-(aminoiminomethyl)-L-lysine, N^6-amidino-L-lysine].

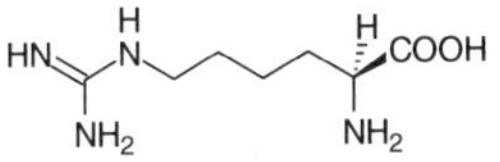

$C_7H_{16}N_4O_2$, M_R 188.23, monohydrochloride, mp. 207–209 °C, $[\alpha]_D^{22}$ +13° (1 m HCl). Non-proteinogenic amino acid in *Lathyrus* spp.[1,2] and *Lotus helleri*[3] which inhibits the growth of *Staphylococcus aureus* and *Candida albicans*. H. is probably a competitive antimetabolite of arginine.

Lit.: [1]Biochem. J. **83**, 225–229 (1962); Nature (London) **193**, 1078–1079 (1962). [2]Biochemistry **2**, 298–300 (1963). [3]Bull. Acad. Pol. Sci., Ser. Sci. Biol. **18**, 603 (1970).
gen.: Beilstein E IV **4**, 2727 ▪ Rosenthal, Plant Nonprotein Amino and Iminoacids, p. 114 f., New York: Academic Press 1982. – *[CAS 156-86-5]*

(22*S*,23*S*)-Homobrassinolide see brassinosteroids.

Homocysteine see cystathionine.

Homoerythrinan alkaloids see Cephalotaxus alkaloids.

Homofuraneol see hydroxyfuranones.

Homogentisic acid [(2,5-dihydroxyphenyl)acetic acid].

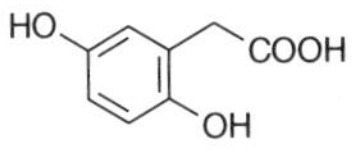

$C_8H_8O_4$, M_R 168.15, cryst. with 1 mol H_2O, mp. 152–154 °C. H., which is easily dehydrated to the corresponding lactone, is an important intermediate in the degradation of the aromatic amino acids L-*phenylalanine and L-*tyrosine. The acids are degraded by homogentisate 1,2-dioxygenase (EC 1.13.11.5) via opening of the aromatic ring and through various other intermediates to fumaric acid and acetoacetic acid; the latter compound can then enter the citric acid cycle. In cases of a congenital deficiency of H. dioxygenase there is a high excretion rate of H. in urine; in this alkaline medium H. is converted in the presence of air to a black-brown quinoid pigment (so-called alkaptonuria). H. occurs in the form of esters or glycosides in the seeds of *Entada phaseoloides* , e.g., as *phaseoloidin* {2-O-*β*-D-glucopyranoside, $C_{14}H_{18}O_9$, M_R 330.29, mp. 207–209 °C, $[\alpha]_D^{25}$ –41.13° (H_2O)} which is effective as a fish poison at a concentration of 125 mg/L. *4,6-Dibromohomogentisic acid amide* ($C_8H_7Br_2NO_3$, M_R 324.96, mp. 170 °C) occurs in the marine sponge *Verongia aurea* and has bacteriostatic activity[1]. For synthesis, see *Lit.*[2].
Biosynthesis: From 4-hydroxy*phenylpyruvic acid. The multi-enzyme complex *p*-hydroxyphenylpyruvic acid dioxygenase catalyzes the decarboxylation as well as the oxidation of the aromatic ring and the rearrangement of the side chain. The enzyme complex contains copper and requires *ascorbic acid as cofactor.

Lit.: [1]Tetrahedron Lett. **1975**, 507–510, 4667 f. [2]Synthesis **1978**, 66 f.
gen.: Beilstein E IV **10**, 1506 ▪ Phytochemistry **27**, 3259–3261 (1988) (phaseoloidin); **30**, 3749–3752 (1991) ▪ Stryer 1995, p. 647 f. – *[HS 2918 29; CAS 451-13-8 (H.); 118555-82-1 (phaseoloidin), 55895-97-1 (dibromohomogentisic acid amide)]*

Homoglycans see polysaccharides.

Homoharringtonine see Cephalotaxus alkaloids.

Homolycorenine see Amaryllidaceae alkaloids.

Homomycin see hygromycins.

Homoserine (2-amino-4-hydroxybutanoic acid).

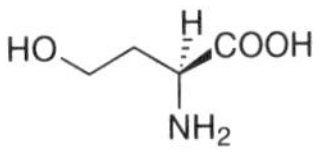

(2*S*)-, L-form

$C_4H_9NO_3$, M_R 119.12, mp. 203 °C (decomp.), $[\alpha]_D^{26}$ –8.8° (H_2O), $[\alpha]_D^{26}$ +18.3° (2 m HCl). The optical activity is lost on standing on account of formation of the levorotatory *γ*-lactone. L-*Homoserine lactone*, monohydrochloride, $[\alpha]_D^{26}$ –27.0° (H_2O). Non-proteinogenic amino acid and intermediate in the biosynthesis of some proteinogenic amino acids.
Occurrence: In young pea shoots (*Pisum sativum*), other green plants, and in methionine-free mutants of *Neurospora*.
Biosynthesis: H. is formed from aspartic acid semialdehyde by homoserine dehydrogenase (EC 1.1.1.3). H., in turn, is converted either to threonine via *O*-phosphohomoserine or to methionine through *O*-succinylhomoserine, *cystathionine, and homocysteine.

Lit.: Beilstein E IV **4**, 3187 f. ▪ Merck-Index (12.), No. 4777. – *[CAS 498-19-1; 672-15-1 (L)]*

Homoterpenes. 1) General name for *terpenes with an additional carbon atom; terpenes with one carbon atom less are known as *norterpenes*.

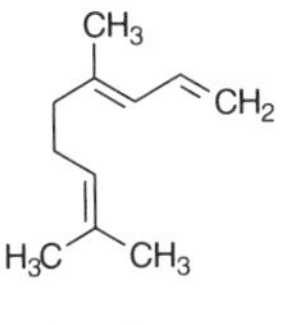

1

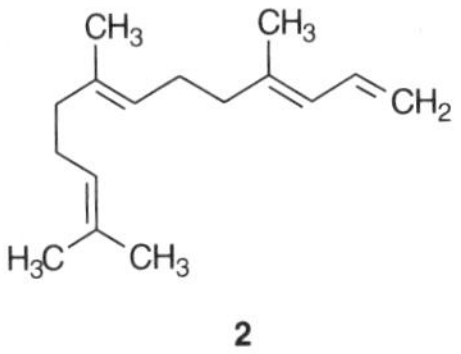

2

2) Name for two acyclic, unsaturated hydrocarbons, the *synomones *(E)-4,8-dimethylnona-1,3,7-triene* (**1**, $C_{11}H_{18}$, M_R 150.26, bp. 190 °C) and *(E,E)-4,8,12-trimethyltrideca-1,3,7,11-tetraene* (**2**, $C_{16}H_{26}$, M_R 218.38) which have been detected in the flower odor of many plants[1]. **1** is also liberated by beans (*Phaseolus lunatus*) when they are attacked by predators (herbivores, e.g., *Tetranychus urticae*). Carnivores (e.g., *Phytoseiulus persimilis*) detect this signal and thus are led to a source of food where they reduce the herbivore population on the plant[2]. For synthesis, see *Lit.*[3].
Biosynthesis: By oxidative bond cleavage of the terpene alcohols nerolidol or geranyllinalool, respectively[4,5].

Lit.: [1]Müller-Lamparsky, p. 213. [2]Neth. J. Zool. **38**, 148 (1988); Naturwissenschaften **79**, 368 (1992). [3]Bull. Chem. Soc. Jpn. **53**, 1698 (1980). [4]Helv. Chim. Acta **72**, 247 (1989); **74**, 1773 (1991). [5]J. Plant Physiol. **143**, 473 (1994).
gen.: Tetrahedron **54**, 14725 (1998) (biosynthesis). – *[CAS 19945-61-0 (1); 101427-55-8 (2)]*

Homothallin I see isocyanides.

Honey flavor. Over 300 (partly species-specific) components of H. from different origins are known. Of general importance are aromatic carboxylic acids such as *cinnamic acid and *phenylacetic acid* ($C_8H_8O_2$, M_R 136.15) and esters, in addition *phenylacetaldehyde, *β-damascenone, and phenolic compounds such as *p-anisaldehyde, *eugenol, and *4-vinylguaiacol. A typical consitituent of lime honey[1] is *linden ether* ($C_{10}H_{14}O$, M_R 150.22), of heather honey *dehydrovomifoliol* ($C_{13}H_{18}O_3$, M_R 222.28).

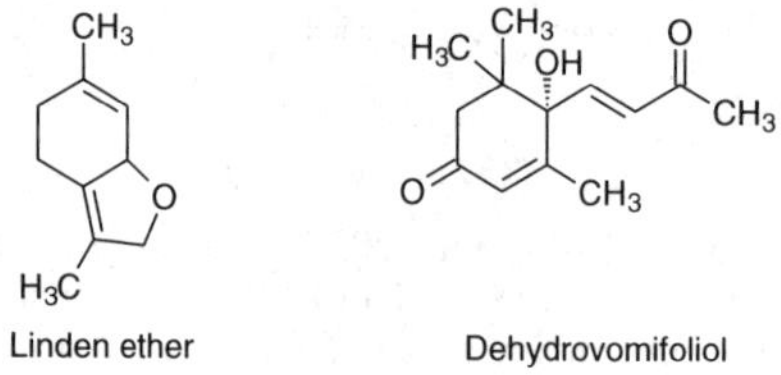

Lit.: [1]Helv. Chim. Acta **73**, 1250–1257 (1990); J. Braz. Chem. Soc. **7**, 237 (1996) (synthesis).
gen.: TNO list (6.), p. 741 ff. – *[CAS 103-82-2 (phenylacetic acid); 125811-37-2 (linden ether); 39763-33-2 (dehydrovomifoliol)]*

Hopanes (*A*′-neogammaceranes).

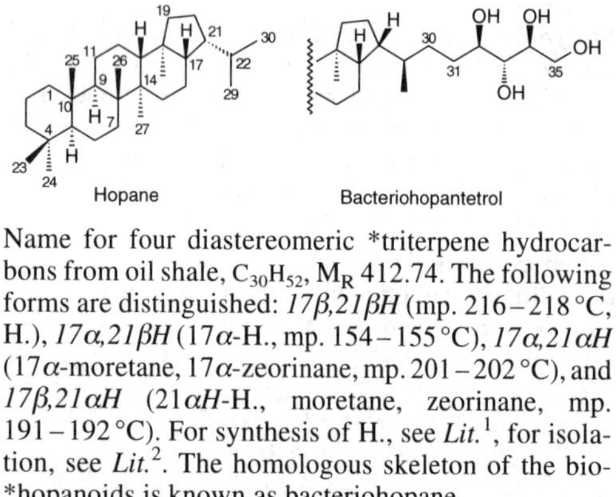

Name for four diastereomeric *triterpene hydrocarbons from oil shale, $C_{30}H_{52}$, M_R 412.74. The following forms are distinguished: *17β,21βH* (mp. 216–218 °C, H.), *17α,21βH* (17α-H., mp. 154–155 °C), *17α,21αH* (17α-moretane, 17α-zeorinane, mp. 201–202 °C), and *17β,21αH* (21αH-H., moretane, zeorinane, mp. 191–192 °C). For synthesis of H., see *Lit.*[1], for isolation, see *Lit.*[2]. The homologous skeleton of the bio-*hopanoids is known as bacteriohopane.

Lit.: [1]J. Chem. Soc. C **1967**, 1622; **1971**, 1885. [2]Geochim. Cosmochim. Acta **41**, 499 (1977); Tetrahedron Lett. **1978**, 1575.
gen.: Org. Geochim. **13**, 665–669 (1988) ▪ Ullmann (5.) **A 23**, 495 ▪ see also hopanoids. – *[CAS 471-62-5 (H.); 13849-96-2 (17α-H.); 33281-23-1 (17α,21αH-H.); 1176-44-9 (21αH-H.)]*

Hopanoids. Derivatives of *hopanes contained in biogenic sediments. They are also known as *geo-H.* The total amount of H. contained especially in younger sediments (oil shale, age up to 500 million years) amounts to ca. 10^{12} t. Thus, besides *cellulose, they belong to the quantitatively most significant organic substances on earth. Even so, they were first discovered only at the end of the 1960's by Ourisson[1]. The geo-H. consist of 31 to 36 carbon atoms. In comparison to hopane, they contain additional *n*-alkyl chains at C-30. Furthermore, *bio-H.* are distinguished which, like geo-H., contain additional alkyl side chains; however, these chains are hydroxylated at C-32–35. Bio-H. can replace *cholesterol in bacterial cell walls. Bio-H. occur in many microorganisms, e. g., *Acetobacter xylinum*, bacilli, and cyanobacteria, but not in *archaebacteria. The wide occurrence of geo-H. is easily explained: dead organic material is degraded to bio-H. by bacteria. When these bacteria also die under the depositing sediment, the hydroxy groups of bio-H are reduced and geo-H. are formed. The most important bio-H., isolated from membranes of *Acetobacter xylinum* and *Rhodopseudomonas acidophila*, is (32*R*, 33*R*,34*S*)-bacteriohopane-32,33,34,35-tetraol[2] ($C_{35}H_{62}O_4$, M_R 546.87, formula, see hopanes). Numerous higher and lower plants form oxidized hopanes of the composition $C_{30}H_{52}O_x$ (x = 1–5) as well as hopenes, which must also be assigned to the H. *Zeorin (6α,22-hopanediol) is a widely distributed component of lichens; for biosynthesis of H. see *Lit.*[3].

Lit.: [1]Pure Appl. Chem. **51**, 709 (1979). [2]J. Chem. Soc., Chem. Commun. **1989**, 1471 (biosynthesis); Nature (London) **400**, 554 (1999). [3]Angew. Chem. Int. Ed. Engl. **37**, 2237–2240 (1998); Tetrahedron Lett. **38**, 3905 (1997).
gen.: Acc. Chem. Res. **25**, 398–408 (1992) ▪ ACS Symp. Ser. **562**, 131–143 (1994) ▪ Annu. Rev. Microbiol. **41**, 301–333 (1987) ▪ ApSimon **2**, 590–594 ▪ Nachr. Chem. Tech. Lab. **34**, 8–14 (1986) ▪ Prog. Drug. Res. **37**, 272–285 (1991) ▪ Ullmann (5.) **A 9**, 606. – *Bacteriohopanetetraol:* Agric. Biol. Chem. **50**, 1345 (1986) ▪ J. Org. Chem. **54**, 2958 (1989). – *[CAS 51024-98-7 (bacteriohopanetetraol)]*

Hopromalinol, hopromine, hoprominol see homaline.

Hordenine {anhaline, eremursine, peyocactine, 4-[2-(dimethylamino)ethyl]-phenol}. For formula, data, and occurrence, see β-phenylethylamine alkaloids. H. is biosynthesized from phenylalanine or tyrosine via tyramine and *N*-methyltyramine. H. is a sympathicomimetic. It has diuretic effects, at higher doses it increases blood pressure, and is generally similar to ephedrine and tyramine. In addition H. is an antifeedant for locusts. H. is used as a cardiac stimulant of low toxicity and as a disinfectant in cases of dysentery.

Lit.: Acta Crystallogr., Sect. C **47**, 1450 (1991) ▪ Beilstein E IV **13**, 1790 ▪ Hager (5.) **5**, 708 f. ▪ J. Nat. Prod. **50**, 422 (1987); **53**, 882 (1990) ▪ Karrer, No. 2471 ▪ see also phenylethylamine alkaloids.

Hormaomycin.

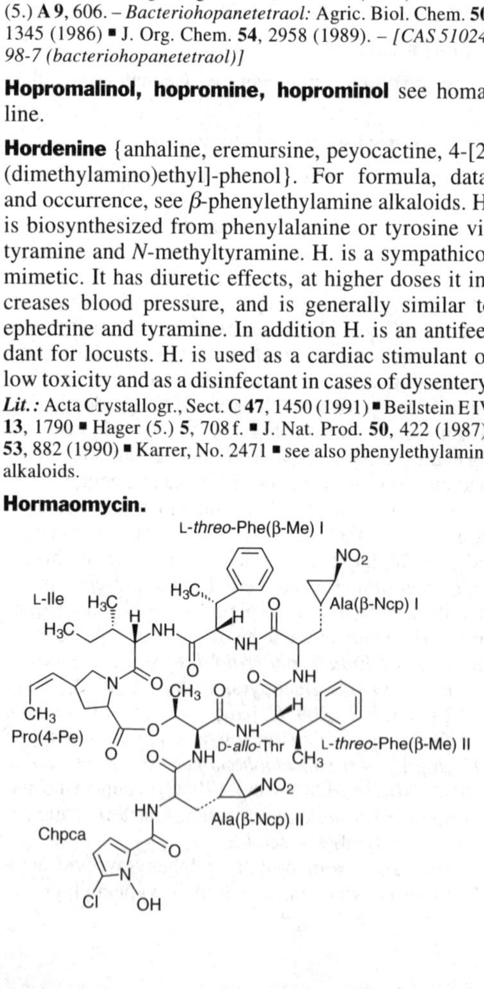

$C_{55}H_{69}ClN_{10}O_{14}$, M_R 1129.68, colorless powder, mp. 166–168 °C, $[\alpha]_D$ +20.8° (CH_3OH), soluble in chloroform, methanol, ethyl acetate, acetone.
Peptide lactone produced by *Streptomyces griseoflavus* with an amide side chain at D-*allo*-threonine. H. contains unusual amino acids, including β-(2-nitrocyclopropyl)alanine, found here for the first time in nature, and occurring twice with different configuration at the α-C atoms. H. is an intercellular signalling substance, it influences the formation of aerial mycelia and production of secondary substances in *S. griseoflavus* and some other *streptomycetes. In this respect H. is comparable to the non-peptidic *A-factor. H. has activity against Gram-positive bacteria, *Arthrobacter* species are extremely sensitive. It is presumed to interact with cytoplasmatic membranes.
Lit.: Angew. Chem. Int. Ed. Engl. **29**, 64 (1990) ▪ DECHEMA Monogr. **129**, 53–72 (1993) ▪ Helv. Chim. Acta **72**, 426–437 (1989) ▪ Tetrahedron. Lett. **34**, 1917 ff. (1993) ▪ Z. Naturforsch. C **45**, 851 (1990). – *[CAS 121548-21-8]*

Hormosirene see algal pheromones.

Hornet venom, poison. H. consists mostly of biogenic amines (*histamine, *serotonin, *acetylcholine), enzymes (phospholipases A and B, hyaluronidase), and hornet kinin. In contrast to popular opinion, hornet stings are less dangerous than bee stings, and deaths are very rare. Besides the amines, the hornet kinins also contribute to the painful effects. – *[HS 0510 00]*

Horse tail (family Equisetaceae). The plant class of the Equisetales belonging to the *pteridophytes; ferns (Filicatae), reached the apex of its development in the Carbonaceous era with tree-like members but is now extinct except for the family Equisetaceae. This comprises the single genus *Equisetum* (horse tail) with about 30 species distributed from the tropics to the polar regions. They are all shrub-like, herbaceous plants and can grow up to between 0.1 and a maximum of 12 m (*E. giganteum*).
Components: The shoots contain large amounts of silicic acid in the cell walls (up to 8%). The fatty acids found in H. t. contain 23–30 C-atoms. Secondary constituents of *Equisetum* species include alkaloids (e. g., *palustrine), numerous *flavones (e. g., 6-chloroapigenin), and flavonols as well as their dihydro derivatives, phenol- and hydroxycinnamic acids (e. g., 2,3-*O*-di-(*E*)-caffeoyl-*meso*-tartaric acid [1] in *E. arvense*), indanone derivatives and novel *styrylpyrones such as, e. g., equisetumpyrone [2]. The secondary product pattern reveals interspecific and intraspecific differences whereby the latter depend on the state of development [3].
Use: The field H. t. (*E. arvense*) is an officinal plant (*Equiseti herba*) and is used as tea or in tea mixtures or phytopharmaceuticals for bacteriuria and inflammations of the renal pelvis. The diuretic activity is attributed to *flavonoids and *saponins. Other activities reported are: hemostatic, stabilizing effects in the early stages of tuberculosis through strengthening of pulmonary tissue, and external use as an astringent (additives in aftershaves and hair waters). The fresh herb was used in the past as an agent for cleaning tin on account of its content of silicic acid. Extracts of S. were formerly also used in dying to achieve green tones; today they are used in biological agriculture as agents to strengthen plants by increasing the resistance to fungal diseases and to promote growth of fruit, vegetable, and ornamental plants.
Lit.: [1] Phytochemistry **30**, 527 (1991). [2] Phytochemistry **32**, 1029 (1993); **39**, 915 (1995). [3] Phytochemistry **38**, 881 (1995). – *gen.:* Hager (5.) **1**, 660 f.; **5**, 65–72 ▪ J. Am. Chem. Soc. **111**, 8223, 8231 (1989). – *[HS 1211 90]*

(–)-Horsfiline see oxindole alkaloids.

Hotrienol see fruit flavors (grape flavor).

HpETE see leukotrienes, hydroxyfatty acids.

HQNO see 2-heptyl-1-hydroxy-4(1*H*)-quinolinone.

HT-2 toxin [15-acetoxy-12,13-epoxy-8α-(3-methylbutyryloxy)trichothec-9-ene-3α,4β-diol].

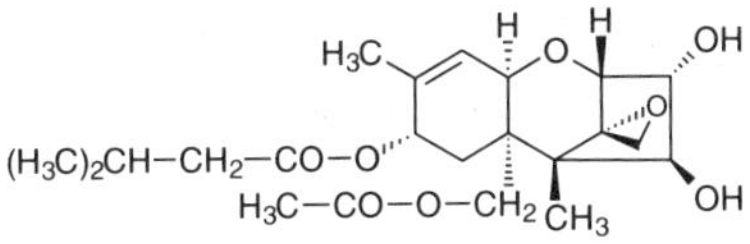

$C_{22}H_{32}O_8$, M_R 424.49, light yellow oil. *Mycotoxin of the *tric(h)othecene group formed by various *Fusarium* species. It is also formed in the metabolism of *T-2 toxin. In comparison to the latter, H. is about 3- to 10-times less toxic. Otherwise it exhibits the typical properties of the trichothecenes. LD_{50} (mouse i. p.): 9 mg/kg.
Lit.: Chelkowski (ed.), Fusarium – Mycotoxins, Taxonomy and Pathogenicity, p. 139–165, Amsterdam: Elsevier 1989 ▪ Cole-Cox, p. 181–184 ▪ J. Agric. Food. Chem. **32**, 1420, 1423 (1984) ▪ J. Nat. Prod. **44**, 324 (1981) ▪ Sax (8.), THI 250 ▪ Tetrahedron Lett. **27**, 4133 (1986). – *[CAS 26934-87-2]*

Human pheromones see pheromones.

Humic acids. Chocolate brown, dust-like powder, poorly soluble in water with a large increase in volume, soluble in aqueous alkalis to give brown solutions, and in conc. nitric acid to give red solutions. H. forms a heteropolycondensate with M_R of 2000–500 000 (mostly 20 000–50 000), mp. >300 °C. They consist of a polycyclic core and loosely bound polysaccharides, proteins, simple phenols, and chelated metal ions which are linked to the core through carboxyl and carbonyl groups [1]. The latter usually have aromatic character. The H. are strongly acidic (hydroxy- and polyhydroxycarboxylic acids) and mostly occur as salts. For analysis, see *Lit.* [2].
Biosynthesis: H. are formed together with the neutral or weakly acidic humines from dead plant material (probably from the lignins) in the course of compost formation in the soil by chemical and biological processes (*humification*). Their composition is not uniform and depends on the type of phenol- and amine-containing precursors. Furthermore, the more strongly acidic *fulvic acids* (fulvinic acids) are derived from H. but have markedly lower molecular masses. H. occur in normal field soil to 1–2%, in chernozem (black soil) to 2–7%, in meadow soil 10%, and in marshy soils to 10–20%. They improve the physical structure of the

soil. H. with low M_R dissolve in surface water and give it a brown color (especially intense in marshy areas). H. form complexes with metals such as copper, zinc, manganese, and especially iron. Certain herbicides (triazines and paraquat) are also complexed by H.
Use: H. additives in animal fodder can reduce the uptake of heavy metals such as cadmium and lead from food since the H. complexes are excreted unchanged. They are also employed in veterinary medicine as antibiotics. H. are applied as components of peat, peat mold, mud baths, and in Cassel brown [a sodium hum(in)ate]; H. qualities with, e. g., M_R 600–1000 are commercially available.

Lit.: [1]Gieseking (ed.), Soil Components, vol. 1, p. 1–211, New York: Springer 1975. [2]Adv. Chem. Ser. **219**, 3–23 (1989).
gen.: ACS Symp. Ser. **225**, 215–229 (1983); **651** (Humic and Fulvic Acids, Isolation, Structure and Environmental Role) (1996) ▪ Angew. Chem. Int. Ed. Engl. **28**, 555–570 (1989) ▪ Int. J. Environ. Anal. Chem. **11**, 105–115 (1982) ▪ Kontakte (Darmstadt) **1989**, Heft 3, 37–41 ▪ Org. Mass. Spectrom. **23**, 622f. (1988) ▪ Tan, Principles of Soil Chemistry, New York: Dekker 1982 ▪ Toxicol. Environ. Chem. **4**, 209–295 (1981); **6**, 127–171, 231–257 (1983) ▪ Ziechmann, Huminstoffe, Weinheim: Verl. Chemie 1980.

Humulene (α-humulene, α-caryophyllene).

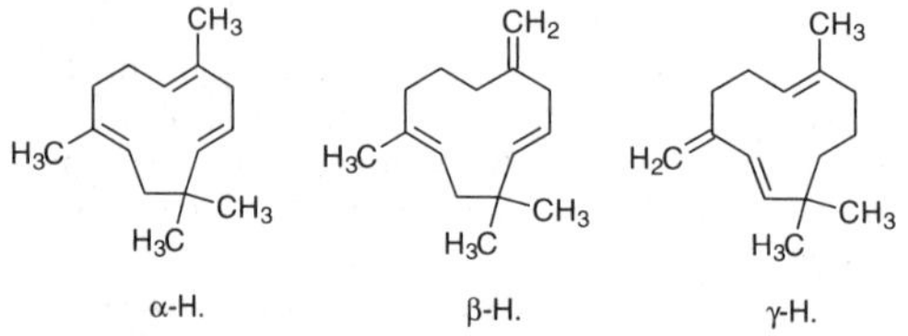

α-H. β-H. γ-H. (Isohumulene)

$C_{15}H_{24}$, M_R 204.36, colorless oil, bp. 123 °C (1.33 kPa). Monocyclic *sesquiterpene from the essential oils of many plants such as, e.g., cloves, hops, and various *Didymocarpus* species. α-H. is often accompanied by the isomeric β-humulene. The completely hydrogenated, monocyclic derivative of H. is *humulane*. H. is the biosynthetic precursor of various sesquiterpenoids, see hirsutanes, protoilludanes, etc. formed by basidiomycetes.
Biosynthesis: By cyclization of farnesyl diphosphate, rearrangement, and subsequent elimination.

Lit.: Beilstein EIV **5**, 1171 ▪ Chem. Ind. (London) **1977**, 30 ▪ Karrer, No. 1929 ▪ Zechmeister **19**, 1–32. – *Biosynthesis:* Angew. Chem. Int. Ed. Engl. **12**, 793–806 (1973) ▪ Arch. Biochem. Biophys. **233**, 838 (1984). – *Synthesis:* ApSimon **2**, 280 ▪ Chem. Lett. **1983**, 281 ▪ Tetrahedron **43**, 5489–5498 (1987) ▪ Tetrahedron Lett. **23**, 2723 (1982); **34**, 3675 (1993). – *[HS 2902 19; CAS 6753-98-6 (α-H.); 116-04-1 (β-H.); 26259-79-0 (γ-H.); 430-19-3 (humulane)]*

Humulone. $C_{21}H_{30}O_5$, M_R 362.47, cryst., mp. 66 °C, $[\alpha]_D^{20}$ −212° (96% alcohol). H. is a *bitter principle from the resin of the fruiting parts of female hop plants (*Humulus lupulus*). H. is structurally closely related to the β-hopic acids (*lupulone, etc.), which are not bitter. On the basis of structural features a distinction is made between the so-called humulone and lupulone groups. The most important members of the humulone group (see formula) are *cohumulone* ($C_{20}H_{28}O_5$, M_R 348.44), *adhumulone* ($C_{21}H_{30}O_5$, M_R 362.47), *prehumulone* ($C_{22}H_{32}O_5$, M_R 376.47), *posthumulone* ($C_{19}H_{26}O_5$, M_R 334.41). H. and the other α-hopic acids

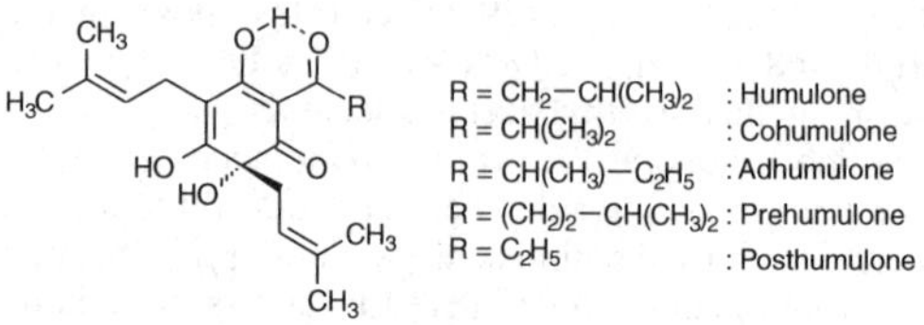

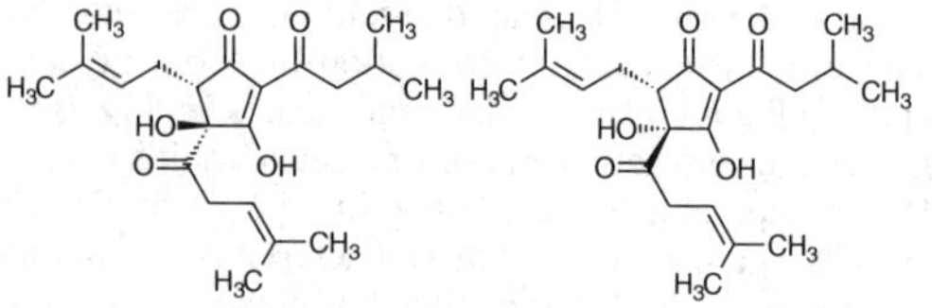

cis-Isohumulone *trans*-Isohumulone

are only present in very small amounts in beer. At the pH value of beer they rearrange to the cyclopentane-1,3-diones, existing in the enol form in beer, e. g., *cis-* and *trans*-isohumulone ($C_{21}H_{30}O_5$, M_R 362.47) as well as other derivatives that are responsible for the bitter taste of beer. H. are active against Gram-positive bacteria [1]. H. are used as bitter substance in brewing. For synthesis, see *Lit.*[2].
Biosynthesis: H. belongs to the group of meroterpenes, i.e., isoprene units are linked with components derived from other key building blocks. The biosynthesis of the ring structure of the hop bitter principles H. and lupulone proceeds through the *polyketide pathway [3]. The isoprene units are formed on the deoxyxylulose pathway.

Lit.: [1]Can. J. Microbiol. **21**, 205 (1975). [2]Bull. Chem. Soc. Jpn. **62**, 3034f. (1989). [3]Phytochemistry **38**, 77 (1995); **44**, 1047 (1997); **49**, 2315 (1998).
gen.: Angew. Chem. Int. Ed. Engl. **33**, 1454 (1994) ▪ Hager (5.) **5**, 450 ▪ Karrer, No. 541 ▪ Luckner (3.), p. 167 ▪ Phytochemistry; **15**, 1689, 1695ff. (1976) (biosynthesis) ▪ Zechmeister **25**, 63–89. – *[HS 2914 50; CAS 23510-81-8 (H.); 26472-41-3 (R-form); 142628-20-4 (co-H.); 142628-21-5 (ad-H.); 59122-94-0 (pre-H.); 1534-03-8 (cis-iso-H..); 467-72-1 (trans-iso-H.)]*

Huntericine see eburnamonine.

Huperzine A (selagine).

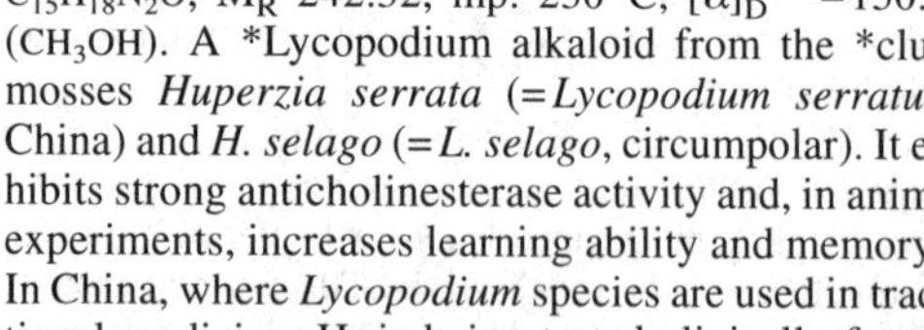

Huperzine A Huperzine B

$C_{15}H_{18}N_2O$, M_R 242.32, mp. 230 °C, $[\alpha]_D^{24.5}$ −150.4° (CH_3OH). A *Lycopodium alkaloid from the *club-mosses *Huperzia serrata* (=*Lycopodium serratum*, China) and *H. selago* (=*L. selago*, circumpolar). It exhibits strong anticholinesterase activity and, in animal experiments, increases learning ability and memory [1]. In China, where *Lycopodium* species are used in traditional medicine, H. is being tested clinically for the treatment of age-related loss of memory and Alzheimer's disease [2]. The toxic effects of H. A and other *Lycopodium alkaloids [e. g., *lycopodine, *huperzine B* ($C_{16}H_{20}N_2O$, M_R 256.35)] can cause the death of grazing animals. Several syntheses of H. A [3,4] and H. B [5] have been reported.

Lit.: [1] Can. J. Chem. **64**, 837 (1986); J. Org. Chem. **56**, 4636 (1991). [2] New Drugs Clin. Remed. **5**, 260 (1986); Chem. Eng. News (20.9.) **1993**, 35f.; (1.6.) **1998**, 45f. [3] Acc. Chem. Res. **32**, 641 (1999); Bioorg. Med. Chem. Lett. **6**, 1927 (1996) (analogues); Angew. Chem., Int. Ed. Engl. **39**, 1775 (2000); Heterocycles **46**, 27 (1997); J. Am. Chem. Soc. **111**, 4116f. (1989); **113**, 4695 (1991); J. Chem. Soc., Chem. Comm. **1993**, 860; J. Org. Chem. **58**, 7660 (1993); Tetrahedron Lett. **30**, 2089f. (1989); **31**, 6159 (1990); Tetrahedron **54**, 5471, 5485 (1998); **55**, 848 (1999) (analogues). [4] Manske **45**, 233–266. [5] J. Org. Chem. **62**, 5978 (1997).
gen.: Pelletier **3**, 185–240. – *[HS 2939 90; CAS 102518-79-6 (H. A); 103548-82-9 (H. B)]*

Huratoxin.

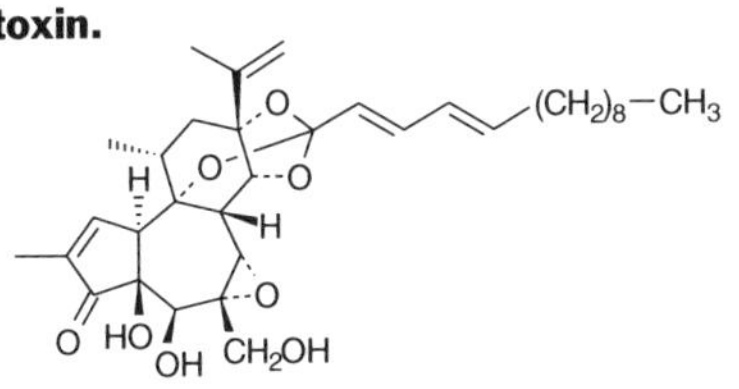

$C_{34}H_{48}O_8$, M_R 584.75, $[\alpha]_D$ +55.1° ($CHCl_3$), a toxic *diterpene with *daphnetoxin structure from *Hura crepitans* (Euphorbiaceae). H. irritates the eyes and skin and can lead to blindness. As a fish toxin H. is ten times more potent than *rotenone. It has insecticidal activity.
Lit.: J. Nat. Prod. **47**, 482 (1984) ▪ Phytochemistry **32**, 141 (1993). – *[CAS 33465-16-6]*

Hx see hepoxilins.

Hyaluronic acid. A glucosaminoglycan occurring in the vitreous body of the eye, the synovial liquid of the joints, and in the skin; it is, together with chondroitin sulfates and dermatan sulfate, a component of all connective tissues (except the cornea). At low concentrations H. forms highly viscous aqueous solutions that serve the joints for lubrication and to absorb shocks. In cases of arthritis H. in the joints is depolymerized by hyaluronidases. H. is a high-molecular-weight compound with an M_R between 50 000 and several millions. The basic building block of H. is an aminodisaccharide composed of D-*glucuronic acid and *N*-acetyl-D-glucosamine linked by a β1-3-glycosidic bond; this unit is joined to the next by a β1-4-glycosidic bond:

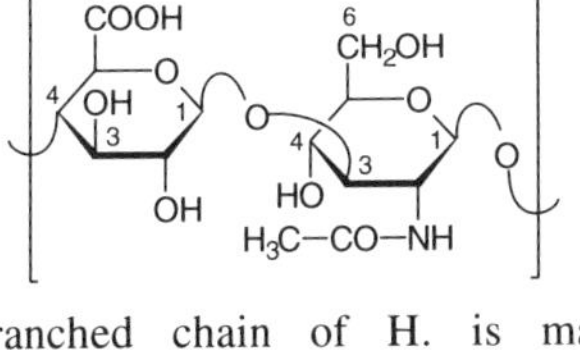

The unbranched chain of H. is made up of 2000–10 000 of these units. H. forms high-molecular aggregates with proteoglycans and so-called linkage proteins. The β-glycosidic bonds of H. are hydrolysed by hyaluronidases [1] and the molecule is thus degraded to smaller units. The commercially available H. – mostly as potassium salt – is isolated from human umbilical cords or cockscombs but is slowly being replaced by preparations produced biotechnologically through bacterial fermentations.
Lit.: [1] Merck-Index (12.), No. 4794.
gen.: Annu. Rev. Biochem. **55**, 539–568 (1986); **60**, 443–476 (1990) ▪ Hager (5.) **8**, 458f. ▪ Merck-Index (12.), No. 4793 ▪ Polysaccharides **1998**, 313–334 (review). – *[HS 3913 90; CAS 9004-61-9]*

Hybrid antibiotics. Name for plant and microbial biologically active compounds from the secondary metabolism formed with the aid of genetically engineered biosynthesis genes. On the basis of the work of Hopwood on the genetics of secondary metabolite formation in *streptomycetes, various strategies have been developed for the use of recombinant DNA techniques ("genetic engineering") for the production of new secondary metabolites:

1) Individual sections are removed or their order changed in the cluster of the biosynthesis genes of a producer A. This changes the sequence of enzymatic reactions without development of new enzymatic activities while still making the production of new products possible.

2) Individual gene sections of a producer A are incorporated in the genome of a producer B, the sections of A then occur additionally in B or replace similar gene sections there. In this way enzyme qualities of different origins are mixed during the biosynthesis and, when successful, this leads to larger changes in the structures of the secondary products.

Using these techniques (combinatorial biosynthesis) with streptomycetes, the *polyketides have now been investigated, including not only the *macrolides (e.g., *erythromycin) but also polycyclic aromatic compounds (e.g., *actinorhodin, *tetracenomycins). The formation of hybrids can alter not only the size of the polyketide skeleton, its stereochemistry or its functionality but also enzyme systems of the later steps of biosynthesis such as, e.g., oxygenases or glycosyltransferases. In practice major difficulties arise because each intermediate in the biosynthetic sequence is a substrate for the following enzyme; thus if a changed substrate is not accepted by the respective enzyme the biosynthesis breaks down.

Secondary products produced by biotransformation, precursor-directed biosynthesis, or mutasynthesis should not be confused with H.-A. However, the separation of the terms often remains unclear.

Lit.: Angew. Chem. Int. Ed. Engl. **34**, 881–885, 1107–1110 (1995) ▪ Annu. Rev. Microbiol. **47**, 875–912 (1993); **49**, 201–238 (1995) ▪ Chem. Biol. **2**, 355–362, 583–589 (1995) ▪ J. Nat. Prod. **57**, 557–573 (1994) ▪ Nat. Prod. Rep. **10**, 265–289 (1993) ▪ Nature (London) **375**, 549–554 (1995) ▪ Nature Biotech. **14**, 335–338 (1996). – *Combinatorial biosynthesis:* Chem. Eng. News (14.9.) **1998**, 29f.; (2.11.) **1998**, 27f. (polyketides); (4.1.) **1999** (sec. metabolites from microorganisms) ▪ Chemtracts. Org. Chem. **11**, 1–15 (1998) (review) ▪ Comb. Chem. Mol. Diversity Drug Discovery **1998**, 401–417 (review) ▪ Modern Drug Discovery **2**, No. 2, 22–29 (1999) ▪ Science **281**, 428 (1998) (beetles).

Hydantocidin. A metabolite [1,2] isolated from culture broths of *Streptomyces hygroscopicus* (SANK 63584) with a potent, non-selective herbicidal activity against annual and perennial monocotyledonous and dicotyledonous weeds.

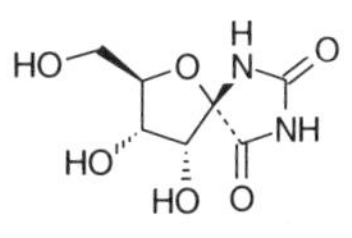

$C_7H_{10}N_2O_6$, M_R 218.17, mp. 187–189 °C, $[\alpha]_D^{25}$ +28.8° (H_2O), needles, soluble in water.

Biological activity: Herbicidal activity, on pretreatment at 2.5 kg/ha with complete activity against *Setaria faberi*, *Echinochloa crus-galli*, *Amaranthus retroflexus*, *Sida spinosa*, and other weeds.

Lit.: [1] J. Antibiot. **44**, 293–300 (1991). [2] J. Chem. Soc., Perkin Trans. 1 **1991**, 1637–1640.
gen.: ACS Symp. Ser. **551**, 74–84 (1994). – *Biosynthesis:* Angew. Chem. Int. Ed. Engl. **38**, 3160 (1999). – *Synthesis:* Agric. Biol. Chem. **55**, 1105–1109 (1991) ▪ Bioorg. Med. Chem. **6**, 911–923 (1998) ▪ Tetrahedron **47**, 2111–2154 (1991); **51**, 6669, 12563 (1995); **52**, 1177–1194 (1996) ▪ Tetrahedron: Asymmetry **6**, 1143–1150 (1995) ▪ Tetrahedron Lett. **34**, 6289, 7391 (1993); **36**, 2145, 2149 (1995). – *[CAS 130607-26-0]*

Hydnocarpic acid see chaulmoogric acid.

Hydnuferrugin, hydnuferruginin.

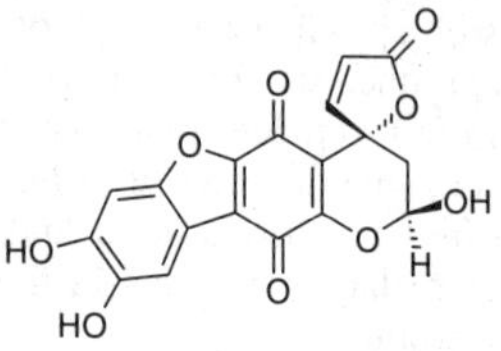

Hydnuferrugin

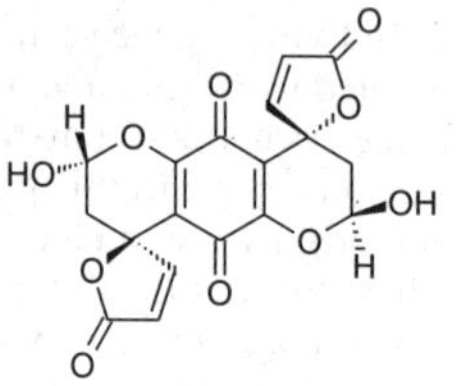

Hydnuferruginin

$C_{18}H_{10}O_9$, M_R 370.27, dark violet microcrystals, decomposition without melting at >200 °C. A violet pigment from fruit bodies of the fungi *Hydnellum ferrugineum* and *H. concrescens* (=*Hydnum zonatum*, Basidiomycetes). H. is accompanied by the yellow *hydnuferruginin* ($C_{18}H_{12}O_{10}$, M_R 388.29, platelets, no mp.). Both pigments are probably formed biosynthetically by oxidative ring cleavage of a *terphenylquinone precursor.

Lit.: Acta. Chem. Scand. Ser. B **35**, 513 (1981) ▪ Zechmeister **51**, 26–30 (review). – *[CAS 53274-37-6 (H.); 80234-50-0 (hydnuferruginin)]*

(–)-*β*-Hydrastine see phthalide-isoquinoline alkaloids.

Hydrazine derivatives (natural). *N*-Methylhydrazine is formed by hydrolysis of the toxin in the false morel (*Gyromitra esculenta*, Ascomycetes) (see gyromitrin). Various phenylhydrazine derivatives occur in mushrooms (*Agaricus* species) [*agaritine, *4-(hydroxymethyl)benzenediazonium ion]. Complex acid hydrazides with activity against Gram-negative bacteria are formed by streptomycetes (negamycins) [1].

Lit.: [1] J. Antibiot. **32**, 531 (1979).

Hydrocinnamaldehyde (3-phenylpropionaldehyde, 3-phenylpropanal). $H_5C_6–CH_2–CH_2–CHO$, $C_9H_{10}O$, M_R 134.18. Prisms, mp. 47 °C, bp. 223 °C, bp. 104 °C (1.7 kPa), soluble in alcohol. A component of Ceylonese *cinnamon leaf (bark) oil used in the perfume industry. For synthesis, see *Lit.* [1].

Lit.: [1] Bull. Chem. Soc. Jpn. **50**, 2148–2152 (1977); Synthesis **1992**, 123.
gen.: Beilstein E IV **7**, 692. – *Toxicology:* Sax (8.), HHP 000. – *[HS 2912 29; CAS 104-53-0]*

Hydrocinnamic acid (3-phenylpropanoic acid). $H_5C_6–CH_2–CH_2–COOH$, $C_9H_{10}O_2$, M_R 150.18, prisms, mp. 47–48 °C, bp. 280 °C, bp. 125 °C (0.8 kPa), soluble in hot water, chloroform, steam-volatile. H. is produced by *Clostridium butyricum* [1] and is a volatile component of the tail gland secretion of red deer of the species *Cervus elaphus* [2]. H. is used as a fixative in the perfume industry. It is technically synthesized by reduction of *cinnamic acid with sodium amalgam.

Lit.: [1] Biochem. Z. **2**, 186–190 (1966). [2] J. Chem. Ecol. **9**, 513–520 (1984).
gen.: Beilstein E IV **9**, 1752 f. ▪ J. Chem. Ecol. **9**, 513 (1984). – *Toxicology:* Sax (8.), HHP 100. – *[HS 2916 39; CAS 501-52-0]*

Hydrocinnamyl alcohol (3-phenyl-1-propanol). $H_5C_6–CH_2–CH_2–CH_2OH$, $C_9H_{12}O$, M_R 136.19. Liquid with a pleasantly sweet, reseda-like odor, bp. 253 °C, bp. 119 °C (1.6 kPa), soluble in water and alcohol. H. esterified with *cinnamic acid occurs in benzoin gum, *peru balsam and *storax. H. is used in many fragrances.

Lit.: Sax (8.), HHP 050, PGA 750. – *[HS 2906 29; CAS 122-97-4]*

Hydrocortisone see corticosteroids.

Hydrolaccol see urushiol(s).

Hydroporphyrins. H. are derived from completely unsaturated *porphyrins by reduction of the double bonds in the *porphine skeleton. The spectrum of naturally occurring H. ranges from the dihydro derivatives *chlorin and *phlorin(s) through the tetrahydro derivatives *bacteriochlorin and *isobacteriochlorins, the hexahydro derivatives *porphyrinogens and *precorrins to the more highly reduced *corphins and *factor F 430.

Lit.: Angew. Chem. Int. Ed. Engl. **27**, 5–39 (1988); **32**, 1223 (1993) ▪ Chem. Rev. **94**, 327 (1994) ▪ Dolphin **II**, 23–68.

Hydroquinone (1,4-dihydroxybenzene).

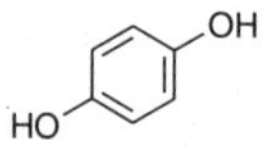

$C_6H_6O_2$, M_R 110.11, needles or prisms, mp. 173–174 °C (subl.), bp. 285–287 °C, pK_{a1} 9.91, pK_{a2} 11.65 (20 °C). H. are easily oxidized by numerous oxidizing agents including atmospheric oxygen to *p*-*benzoquinones (E_0=+0.703). H. occurs in nature as its glucoside (see arbutin) or ethers: *mequinol* (4-methoxyphenol), $C_7H_8O_2$, M_R 124.14, mp. 53 °C, bp. 243 °C, pK_a 10.12; and as *pheromone of the butterfly *Ascia manuste phileta*. It is found in the free form in *Pyrola rotundifolia* (wintergreen). In addition, H. is a component of the defensive secretion of some insects, especially *bombardier beetles. When threatened, these insects mix highly concentrated hydrogen peroxide with H. in an abdominal gland (pygidial gland), as a result of the exothermic reaction a hot mixture of H_2O_2 and benzoquinone is ejected in the direction of the predator [1]. Quinone/H. redox systems have important functions in biochemistry, e.g., in the respiratory chain and photosynthesis (see ubiquinones, plastoquinones).

Use: H. is used as an antioxidant, an inhibitor of polymerization, and mainly in photographic development, *mequinol* can be used to depigment the skin.

Biosynthesis: From *(*E*)-*p*-coumaric acid by oxidative degradation of the side chain.

Toxicology: LD_{50} (rat p. o.) 320 mg/kg. H. irritates the skin, eyes, and airways. Lethal dose 5–12 g (humans); maximum working place concentration 2 mg/m^3, possibly cancerogenic. Its use in cosmetics is allowed with limitations.

Lit.: [1] Angew. Chem. Int. Ed. Engl. **9**, 1–9 (1970).
gen.: Beilstein E IV **6**, 5712 ff. ▪ Can. J. Chem. **67**, 525–534 (1989) (^{13}C NMR, mequinol) ▪ Karrer, No. 203, 205, 208 ▪ Luckner (3.), p. 400 ff. ▪ Negwer (6.), No. 422 ▪ Turner **2**, 10 (mequinol) ▪ Ullmann (5.) **A 8**, 500 f. (mequinol). – *Toxicology:* Sax (8.), EFA 100, HIH 000, MFC 700. – *[HS 2907 22; CAS 123-31-9 (H.); 150-76-5 (mequinol)]*

***p*-Hydroxyacetophenone** (piceol) see picein.

Hydroxyalkanoic acid lactones see alkanolides.

3-Hydroxyanthranilic acid (2-amino-3-hydroxybenzoic acid).

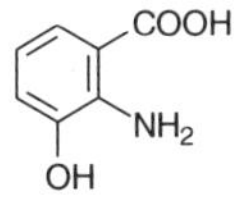

$C_7H_7NO_3$, M_R 153.14, platelets, mp. 164 °C, metabolite of *tryptophan [1] occurring in humans, rats, and fungi (*Claviceps purpurea*). 3-H. is also isolated from cultures of *Klebsiella pneumoniae*, from *Streptomyces* species as an intermediate of *actinomycin biosynthesis [2], and from kohlrabi (*Brassica oleracea*, var. *gongylodes*, Brassicaceae). The methylated derivative *damascenine* (methyl 3-methoxy-2-methylaminobenzoate), $C_{10}H_{13}NO_3$, M_R 195.22, mp. 24–26 °C, bp. 270 °C (decomp.), bp. 147–148 °C (1.3 kPa), is an alkaloid from the seeds of *Nigella damascena* (Ranunculaceae). 3-H. acts as a radical scavenger, is mutagenic and probably carcinogenic. For synthesis, see *Lit.* [4].

Lit.: [1] Biosci. Biotechnol. Biochem. **57**, 858 f. (1993); J. Chromatogr. **661**, 101–104 (1994); J. Nutr. Sci. Vitaminol. **37**, 269–283 (1991). [2] J. Am. Chem. Soc. **111**, 7932–7938 (1989); J. Biol. Chem. **267**, 11745–11752 (1992). [3] Proc. Natl. Acad. Sci. U. S. A. **87**, 2506–2510 (1990). [4] J. Heterocycl. Chem. **29**, 263 f. (1992); Tetrahedron **47**, 5051–5070 (1991).
gen.: Carcinogenesis (London) **12**, 1409–1415 (1991) (pharmacology) ▪ J. Gen. Microbiol. **138**, 85–89 (1992) (biosynthesis) ▪ Luckner (3.), p. 342 ▪ Sax (8.), AKE 750 (toxicology). – *[HS 2922 50; CAS 548-93-6 (3-H.); 483-64-7 (damascenine)]*

4-Hydroxybenzoic acid.

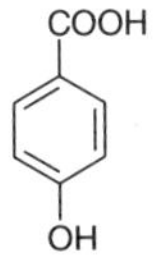

$C_7H_6O_3$, M_R 138.12, cryst., mp. 215 °C. An important biosynthetic precursor of the *ubiquinones and fungal pigments of the *boviquinone type. An elegant synthesis of labelled 4-hydroxy-[1-^{13}C]-benzoic acid for biosynthetic studies has been published [1].

Lit.: [1] Org. Synth. **78** (2000); Synthesis **1998**, 1047.
gen.: Beilstein E IV **10**, 345 ▪ Nature (London) **332**, 354 (1988) ▪ Zechmeister **51**, 105–109 (review). – *[HS 2918 29; CAS 99-96-7]*

Hydroxybrasilin see haematoxylin.

Hydroxychelidonic acid see poppy acid.

25-Hydroxycholecalciferol see calcitriol.

(*Z*)-2-Hydroxycinnamic acid see coumarinic acid.

(*E*)-2-/4-Hydroxycinnamic acid see (*E*)-*p*-coumaric acid and coumarinic acid.

4-Hydroxycinnamyl alcohol (*p*-coumaryl alcohol).

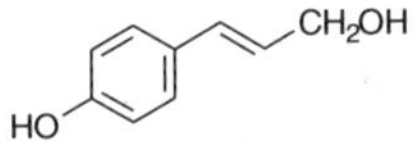

$C_9H_{10}O_2$, M_R 150.18, mp. 124 °C. Widely distributed as the (4-hydroxycinnamyl)-β-D-glucopyranoside *triandrin*, $C_{15}H_{20}O_7$, M_R 312.32, mp. 177–179 °C, $[\alpha]_D^{20}$ –60.5° (H_2O), in *Salix* species (willow and poplar). The *lignin from monocotyledones contains *coniferyl alcohol and *sinapyl alcohol units as well as H. monomers as building blocks.

Lit.: Beilstein E IV **6**, 6332 ▪ Chem. Pharm. Bull. **39**, 803 f. (1991) ▪ Heterocycles **31**, 1409–1412 (1990) ▪ Holzforschung **42**, 221–224 (1988) ▪ Phytochemistry **28**, 2115–2125 (1989) ▪ Tetrahedron Lett. **32**, 2475 f. (1991) (synthesis) ▪ Z. Naturforsch. C **35**, 942–948 (1990). – *[CAS 3690-05-9, 206-40-5 (E); 124076-60-4 (Z); 19764-35-3 (triandrin)]*

4-Hydroxycoumarin (benzotetronic acid).

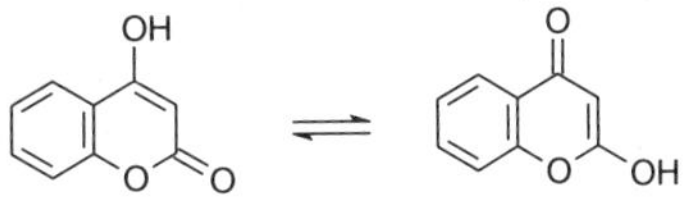

$C_9H_6O_3$, M_R 162.14, needles, mp. 206 °C (232–233 °C), tautomerism in solution (2-hydroxy-4*H*-1-benzopyran-4-one), is formed by *Penicillium jensenii* and occurs in *Magnolia salicifolia*. *Setarin* (4-allyloxycoumarin, $C_{12}H_{10}O_3$, M_R 202.21, cryst., mp. 104–105 °C) is a constituent of *Setaria italica*. 4-H. and the 4-methyl ether ($C_{10}H_8O_3$, M_R 176.17, cryst., mp. 124 °C) have antiviral activity [1]; 4-H., like other coumarins (see dicoumarol), is an anticoagulant [2]. For synthesis, see *Lit.* [3], for biosynthesis, see *Lit.* [4].

Lit.: [1] Acta Virol. **37**, 41–250 (1993). [2] J. Anal. Toxicol. **17**, 56–61 (1993); Lyon Pharm. **40**, 23–33 (1989) (review). [3] Synthesis **1988**, 257; Tetrahedron **45**, 6867–6874 (1989). [4] Xenobiotica **24**, 795–803, 893–907 (1994).
gen.: Chem. Pharm. Bull. **39**, 3265–3271 (1991) (isolation) ▪ J. Chem. Ecol. **18**, 1287–1297 (1992) (pharmacology) ▪ Luckner (3.), p. 390 ▪ Phytochemistry **30**, 3826 f. (1991) (setarin) ▪ Sax (8.), HJY 000 (toxicology). – *[HS 2932 29; CAS 1076-38-6 (4-H.); 31005-07-9 (setarin); 20280-81-3 (4-methyl ether)]*

(*E*)-9-Hydroxy-2-decenoic acid see queen substance.

4-Hydroxy-2,5-dimethyl-3(2*H*)furanone see Furaneol®.

(3*S*,11*S*)-29-Hydroxy-3,11-dimethyl-2-nonacosanone see cockroaches.

***N*′-Hydroxy-*N*′,*N*′-dimethylurea** see connatin.

***N*^{4}-(2-Hydroxyethyl)-L-asparagine** see asparagine.

2-(2-Hydroxyethyl)clavam see clavams.

Hydroxyfatty acids. *Fatty acids containing an OH function. In plants H. a, occur mainly as esters, e.g., in triglycerides (ricinus oil) and waxes (e.g., *carnauba and *beeswax). In plants and animals 2-H. are components of *cerebrosides. On the other hand, esters and glycosides of 3-H. [e.g. *(R)-3-hydroxypalmitic acid* from *Rhodotorula* yeasts, $C_{16}H_{32}O_3$, M_R 272.43, mp. 78–79 °C, $[\alpha]_D$ – 12.9°] are typical components of bac-

teria and yeasts. ω-H. [e.g., *sabinic acid* (12-hydroxylauric acid) from the needles of creeping juniper (*Juniperus sabina*, Cupressaceae), $C_{12}H_{24}O_3$, M_R 216.32, mp. 78–79 °C (84–85 °C)] are also widely distributed in the lipids of plants and microorganisms. Unsaturated H. such as, e. g., *ricinoleic acid occur as esters in the seed oil of ricinus (*Ricinus communis*) as well as in the fungus *Claviceps purpurea*, see ergot. Di-H. [e.g., *(R,R)-9,10-dihydroxyoctadecanoic acid*, $C_{18}H_{36}O_4$, M_R 316.48, mp. 99.5 °C, $[\alpha]_D^{20}$ +23.5°] are isolated from microorganisms such as, e. g., *Claviceps sulcata*. *Aleuritic acid* [9,10,16-trihydroxypalmitic acid, $C_{16}H_{32}O_5$, M_R 304.43, mp. 102 °C] is obtained from *shellac (shellac wax). *Mycolic acids also belong to the H.
In animals, hydroxy derivatives of polyunsaturated fatty acids occur as intermediates in the biosynthesis of *eicosanoids, e. g., hydroxy derivatives of *arachidonic acid and also *oxylipins. The action of 5-, 12-, and 15-lipoxygenases on polyunsaturated fatty acids containing a 1,4-diene unit furnishes the corresponding (5*S*)-, (12*S*)-, and (15*S*)-hydroperoxyfatty acids (HpETE) regio- and stereoselectively as precursors of these H. (see figure at lipoxins), subsequent action of peroxidase gives the hydroxyeicosatetraenoic acids ($C_{20}H_{32}O_3$, M_R 320.47): (5*S*,6*E*,8*Z*,11*Z*,14*Z*)-HETE (**1**), (12*S*,5*Z*,8*Z*,10*E*,14*Z*)-HETE (**2**), and (15*S*,5*Z*,8*Z*,11*Z*,13*E*)-HETE (**3**).

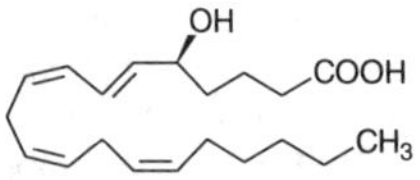

1

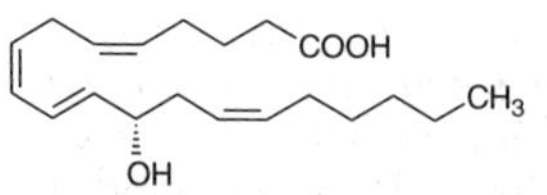

2

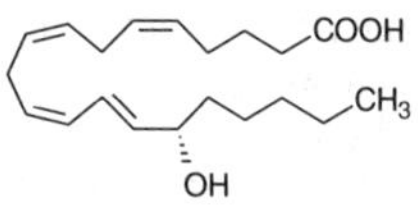

3

In addition to their biological significance as intermediates in eicosanoid biosynthesis, the HETE themselves are physiologically active, e.g., in the regulation of intracellular calcium concentration, cell proliferation, phospholipase activity, and prostaglandin formation. Physiologically active, polyunsaturated dihydroxy-fatty acids are formed by the action of epoxygenases or various lipoxygenases on polyunsaturated fatty acids.

Lit.: Adv. Lipid Res. **21**, 47 (1985) ▪ Annu. Rev. Biochem. **55**, 69 (1986) ▪ Gunston-Harwood-Padley, The Lipid Handbook, 2nd. edn., p. 15 f., 29 f., 55 f., 616, London: Chapman & Hall 1994 ▪ Kolattukudy (ed.), Chemistry and Biochemistry of Natural Waxes, p. 318 ff., 369 ff. Amsterdam: Elsevier 1976 ▪ Pharmacol. Rev. **40**, 229 (1988) ▪ Prog. Lipid Res. **27**, 271 (1988); **33**, 329 (1994) ▪ Pure Appl. Chem. **59**, 269 (1987) ▪ Ratledge-Wilkinson (ed.), Microbial Lipids, vol. 1, p. 448, & vol. 2, p. 22 ff., 672 ff., New York: Academic Press, 1988, 1989 ▪ Tetrahedron **54**, 3929 (1998) (synthesis 9-HETE). – *[HS 2918 19; CAS 20595-04-4 ((R)-3-hydroxypalmitic acid); 505-95-3 (sabinic acid); 10067-09-1 ((R,R)-9,10-dihydroxyoctadecanoic acid); 17941-34-3 (aleuritic acid); 70608-72-9 (1); 54397-83-0 (2); 54845-95-3 (3)]*

Hydroxyfuranones. H. are *flavor compounds of moderate stability formed thermally or enzymatically from sugars with fruity-sweet (**1**) or rusty-spicy (**2**) caramel-like odors that are typical for compounds with the enol-oxo structure (see 2-hydroxy-3-methyl-2-cyclopenten-1-one and maltol). The most important compound of this group is *Furaneol®. *Norfuraneol* ($C_5H_6O_3$, M_R 114.10) occurs in the flavors of soy sauce, bouillon, guavas, and raspberries[1]. *Homofuraneol* [$C_7H_{10}O_3$, M_R 142.15, bp. 63–64 °C (2 Pa), mp. 36–37 °C (A)[2], LD_{50} (mouse) 2.8 g/kg] is found, e. g., in the flavors of cheese, melon, and soy sauce[1]. *Sotolone* ($C_6H_8O_3$, M_R 128.13, mp. 26–29 °C) is important for the flavor of sake and also occurs in the extract from the seeds of fenugreek, *Trigonella foenum-graecum*, coffee, tee, and sherry[1]. *Abhexone* ($C_7H_{10}O_3$, M_R 142.15, mp. 31–35 °C) is an aroma component of coffee[1]. For preparation, see *Lit.*[2,3] (4-H.) and *Lit.*[4] (3-H.); chirospecific analysis of H., see *Lit.*[5]. Olfactory thresholds in water (ppb)[2]: norfuraneol 8300; furaneol 158; homofuraneol 21; sotolone 1–5; abhexone 0.024.

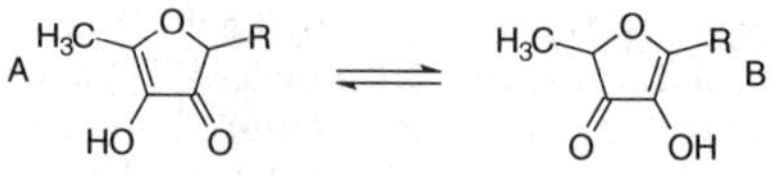

1

4-Hydroxy-3(2*H*)-furanones
R = H : Norfuraneol
R = CH_3 : Furaneol®
R = C_2H_5 : Homofuraneol (Homofuronol)

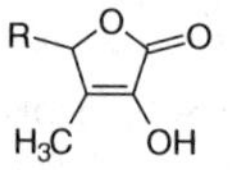

2

3-Hydroxy-2(5*H*)-furanones
R = CH_3 : Sotolone
R = C_2H_5 : Abhexone

Lit.: [1]Chem. Eur. J. **1998**, 311 (sotolone); Chimia **46**, 403 (1992); J. Org. Chem. **57**, 5023 (1992); Tetrahedron Lett. **33**, 5625 (1992); TNO list (6.), Suppl. 5, p. 373. [2]Perfum. Flavor. **17**(4), 15 ff. (1992). [3]Helv. Chim. Acta **56**, 1882 ff. (1973); **70**, 369 (1987). [4]Tetrahedron Lett. **34**, 2753–2756 (1993). [5]J. High Resolut. Chromatogr. **15**, 590–593 (1992); **16**, 101–105 (1993).
gen.: Ohloff, p. 38 f. – *[CAS 19322-27-1 (norfuraneol); 27538-10-9 (homofuraneol (1A)); 27538-09-6 (homofuraneol (1B)); 28664-35-9 (sotolone); 698-10-2 (abhexone)]*

4-Hydroxyglutamic acid (2-amino-4-hydroxypentanedioic acid).

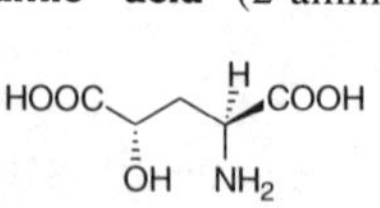

(2*S*, 4*S*)-form

$C_5H_9NO_5$, M_R 163.13, mp. 183–185 °C, $[\alpha]_D^{26}$ –1.38° (H_2O). H. occurs in *Phlox decussata*[1] and *Lunaria vulgaris*. For synthesis, see *Lit.*[2].

Lit.: [1]Acta Chem. Scand. **16**, 513 (1962). [2]Bull. Chem. Soc. Jpn. **51**, 1261 f. (1978).
gen.: Beilstein EIV **4**, 3251 f. ▪ Tetrahedron Lett. **28**, 1277 (1987). – *[CAS 3913-68-6 (2S,4S-form)]*

2-Hydroxy-4-imino-2,5-cyclohexadienone. $C_6H_5NO_2$, M_R 123.11, only stable in highly dilute solutions. 2-H. exists in equilibrium with 4-amino-*o*-ben-

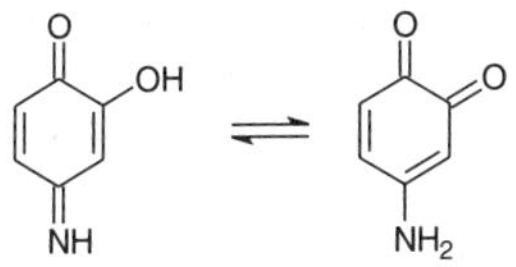

zoquinone. H. is a red pigment from sporulating gill tissue of the cultivated mushroom *Agaricus bisporus* (Basidiomycetes). On the basis of its UV/Vis absorption at 490 nm, 2-H. is also known as "490-quinone". 2-H. is formed in fungal tissue from N^5-(4-hydroxyphenyl)-L-glutamine (see agaritine). It is a strong inhibitor of enzymes containing SH groups and prevents the premature germination of spores.
Lit.: J. Biol. Chem. **255**, 4766 (1980) ▪ J. Org. Chem. **45**, 3540 (1980). – *[CAS 74331-93-4]*

8-Hydroxylabdane-15-carboxylic acid see labdanum absolute/oil/resinoid.

Hydroxylubimin see vetispiranes.

13-Hydroxylupanine (13α-hydroxy-2-sparteinone).

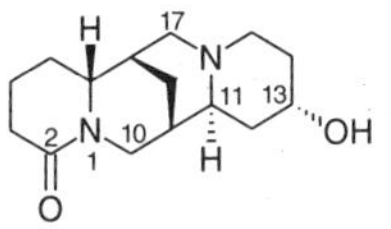

$C_{15}H_{24}N_2O_2$, M_R 264.37, mp. 178–180 °C, $[\alpha]_D$ +51.8° (C_2H_5OH). A *quinolizidine alkaloid of the sparteine type from many genera of the Fabaceae [1]. It is mildly toxic [LD_{50} (rat i.p.) 199 mg/kg] [2]. 13-H. is purported to have antiarrhythmic and hypotensive activities [3]. Not only the 13β- but also the 13α-isomer occur as natural products in plants [4], the latter being more abundant.
Lit.: [1]Waterman **8**, 197–239. [2]J. Appl. Toxicol. **7**, 51 ff. (1987). [3]Pelletier **2**, 105–148. [4]Phytochemistry **31**, 4343 (1992).
gen.: Beilstein E V **25/1**, 223. – *[HS 2939 90; CAS 15358-48-2]*

5-Hydroxylysine (2,6-diamino-5-hydroxyhexanoic acid).

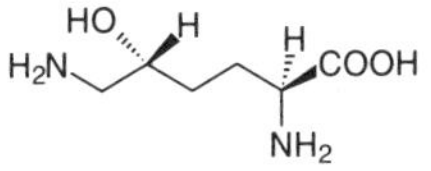

erythro-L-, (*2S,5R*)-form

$C_6H_{14}N_2O_3$, M_R 162.19, *erythro*-L-, $[\alpha]_D^{25}$ +17.8° (6 m HCl); hydrochloride, $[\alpha]_D^{25}$ +14.5° (6 m HCl). H. was first isolated from fish gelatine. It is present in proteins, especially collagen, and is important for stabilization of the fiber-like structure of the protein. Hydroxylation by a hydroxylase happens after formation of the peptide chain.
Lit.: Beilstein E IV **4**, 3239 f. ▪ Greenstein & Winitz, Chemistry of the Amino Acids, vol. 3, p. 1996–2017, New York: J. Wiley 1961 ▪ J. Chromatogr. **378**, 67–76 (1986). – *[CAS 1190-94-9 (erythro-L)]*

2-(4-Hydroxy-3-methoxyphenyl)ethanol see queen substance.

4′-(Hydroxymethyl)azoxybenzene-4-carboxylic acid see azoxybenzene-4,4′-dicarboxylic acid.

4-(Hydroxymethyl)benzenediazonium ion.
$C_7H_7N_2O^+$, M_R 135.15, chloride: $C_7H_7ClN_2O$, M_R 170.60. 4-H. has been detected by azo-coupling in the

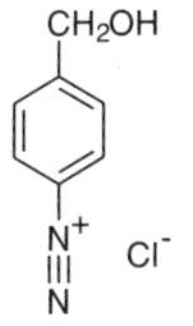

4-(Hydroxymethyl)-benzenediazonium chloride

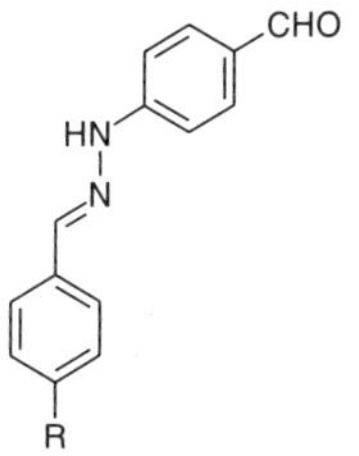

R = H : Agaricus hydrazone A
R = OH : Agaricus hydrazone B

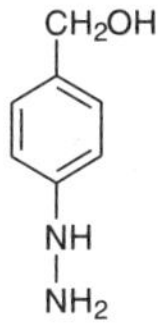

4-Hydrazino-benzyl alcohol

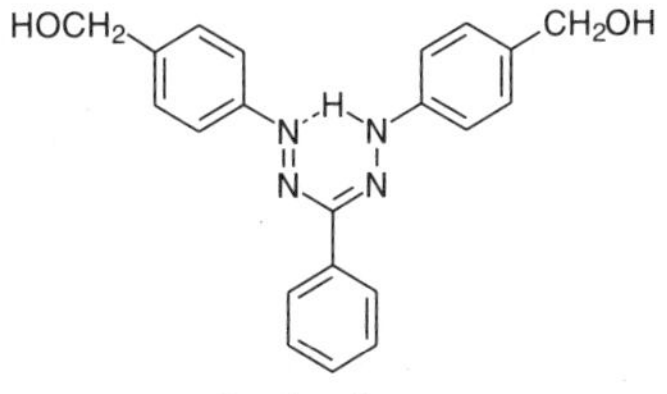

Agaricus formazane

stem base of the cultivated mushroom *Agaricus bisporus* [1]. It is probably formed from *agaritine present in the mushroom by enzymatic cleavage with release of 4-hydrazinobenzyl alcohol [2]. The diazonium ion is also responsible for the formation of *Agaricus formazane* ($C_{21}H_{20}N_4O_2$, M_R 360.42, blood-red solid) which can be isolated on work-up of yellowing *Agaricus* species together with the *hydrazones A* ($C_{14}H_{12}N_2O$, M_R 224.26, wax-like solid) and *B* ($C_{14}H_{12}N_2O_2$, M_R 240.26, lemon-yellow powder, mp. 211 °C). Hydrazones of this type are responsible for the orange-red color reaction (Schäffer's reaction) resulting when aniline and concentrated nitric acid are applied in a cross-like manner to the cap skin of certain mushrooms, the color occurs at the intersection of the two strokes. It occurs in the yellowing horse and the Prince mushrooms but not in some similar inedible species (cf. 4-diazo-2,5-cyclohexadien-1-one, xanthodermine).
Lit.: [1]Biochim. Biophys. Acta **63**, 212 (1962). [2]J. Biol. Chem. **239**, 2274 (1964); Bresinsky & Besl, A colour atlas of poisonous fungi, Boca Raton: Manson publishing 1989.

4-Hydroxy-5-methylcoumarin.

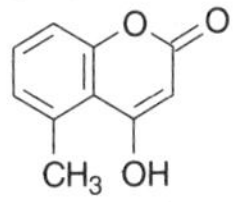

$C_{10}H_8O_3$, M_R 176.17. cryst., mp. 233–234 °C, occurs in free and glycosidic forms in gerbera (*Gerbera anandria*, *Gerbera jamesonii*, Asteraceae); a number of *O*-prenylated derivatives has been isolated from the Bolivian plant *Mutisia orbignyana* (Asteraceae) [1].
Biosynthesis: In contrast to the numerous coumarins produced on the shikimate pathway, 4-hydroxy-5-methylcoumarin is derived from a pentaketide.
Lit.: [1]Phytochemistry **27**, 891–897 (1988).
gen.: Luckner (3.), p. 172 ff. ▪ Nat. Prod. Rep. **12**, 1–32 (1995) ▪ Phytochemistry **30**, 333 ff. (1991) ▪ Tetrahedron **34**, 1221–1224 (1978) (synthesis, biosynthesis). – *[CAS 24631-87-6]*

2-Hydroxy-3-methyl-2-cyclopenten-1-one (Cyclotene®).

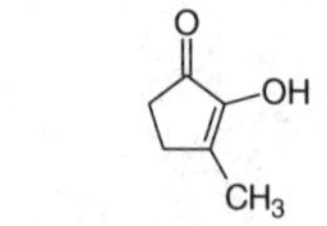

$C_6H_8O_2$, M_R 112.13; cryst. with a spicy, caramel-like odor resembling that of liquorice, mp. (monohydrate) 106 °C; soluble in alcohol (25%), water (2.5%), but hardly soluble in petroleum ether; it is formed from *glucose by the Maillard reaction[1] and is present in the flavor of thermally prepared foods[2] such as maple syrup, bread, coffee, roasted nuts, and roast pork. It is mainly obtained from beech wood tar and used as a *flavor compound, e. g., for the above-mentioned aromas.

Lit.: [1] Hopp & Mori (eds.), Recent Developments in Flavor and Fragrance Chemistry, p. 172, Weinheim: VCH Verlagsges. 1993. [2] TNO list (6.), Suppl. 5, p. 336.
gen.: Arctander, No. 1987 ▪ Bauer et al. (2.), p. 69 ▪ Chimia **46**, 399 (1992) (synthesis) ▪ Synth. Commun. **21**, 1783 (1991) (synthesis). – *[HS 2914 49; CAS 80-71-7]*

5-Hydroxy-4-oxo-L-norvaline [(*S*)-2-amino-5-hydroxy-4-oxopentanoic acid, RI 331, HON].

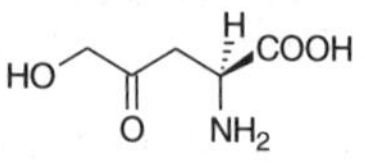

$C_5H_9NO_4$, M_R 147.13, mp. >100 °C (decomp.); *N*-benzoyl derivative, mp. 135–136 °C, $[\alpha]_D^{20}$ – 15°. An antibiotic from *Streptomyces akiyoshiensis*[1,2].
H. has antimycotic activity against pathogenic yeasts through inhibition of protein biosynthesis. Its biosynthesis proceeds from acetyl-CoA and activated aspartic acid[3].

Lit.: [1] Chem. Pharm. Bull. **8**, 1071–1083 (1960). [2] J. Antibiot. **14**, 39–43 (1961). [3] Can. J. Chem. **72**, 1645 (1994); J. Am. Chem. Soc. **110**, 8228 f. (1988).
gen.: Ann. N. Y. Acad. Sci. **544**, 188 f. (1988) ▪ Drugs Exp. Clin. Res. **14**, 467–472 (1988). – *Synthesis:* Synthesis **1992**, 482 ▪ Synlett **1997**, 691 ▪ Tetrahedron Lett. **34**, 5879 (1993). – *[CAS 26911-39-7]*

(*R*)-3-Hydroxypalmitic acid see hydroxyfatty acids.

4-(*p*-Hydroxyphenyl)-2-butanone see insect attractants.

4-Hydroxyphenylpyruvic acid see phenylpyruvic acid.

4-/5-Hydroxypipecolic acid [4-/5-hydroxy-2-piperidinecarboxylic acid].

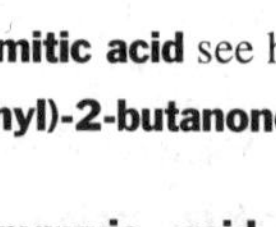

$C_6H_{11}NO_3$, M_R 145.16. *4-H.*: (2*S*,4*S*)-form (*trans*-L-), mp. 294 °C (decomp.), $[\alpha]_D^{20}$ – 13° (H_2O), +2.7° (5 m HCl). (2*S*,4*R*)-form (*cis*-L-), mono- or dihydrate, mp. 265 °C (decomp.), $[\alpha]_D^{23}$ – 17.0° (H_2O), occurs in the wood and leaves of acacias (*Acacia* sp.).
The (2*S*,5*R*)-form (*trans*-L-) of *5-H.* occurs is *Rhapis excelsa* (Palmae) and ripe dates, *Leucaena glauca*, *Gymnocladus dioicus*, mulberry leaves (*Morus alba*), and in the seeds of *Caesalpinia* species. Mp. 210–215 °C (decomp.), $[\alpha]_D^{20}$ – 10.9° (H_2O). The (2*S*,5*S*)-form (*cis*-L-) is a component of *Gymnocladus dioicus*, mulberry leaves, and seeds of *Lathyrus japonicus*. Mp. 230–235 °C (decomp.), $[\alpha]_D^{22}$ – 29.5° (H_2O).

Lit.: Beilstein E V **22/5**, 23 ff. ▪ J. Chem. Soc. Chem. Commun. **1996**, 349 ▪ J. Org. Chem. **61**, 2226 (1996) (synthesis) ▪ Med. Chem. Res. **2**, 491 (1992) (synthesis) ▪ Phytochemistry **16**, 387 f., 1041 f. (1977); **17**, 1127 (1978). – *[CAS 14228-16-1 (4-H. (2S,4S)); 50439-45-7 (5-H. (2S,5R)); 63088-78-8 (5-H. (2S,5S))]*

4-/3-Hydroxyproline (4-/3-hydroxy-2-pyrrolidinecarboxylic acid).

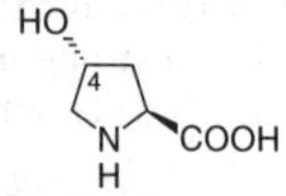

(2*S*, 4*R*)-, *trans*-L-form, 4 Hyp, Hyp

trans-4-Hydroxy-L-proline: $C_5H_9NO_3$, M_R 131.13, mp. 274 °C, $[\alpha]_D^{26}$ – 76.5° (H_2O), pK_a 1.82, 9.65, pI 5.74. Sweetish taste. Allohydroxyproline (*cis*-Form), mp. 238–241 °C, $[\alpha]_D^{18}$ – 58.1° (H_2O).
4-H. was first isolated from gelatin and is a component of plant glycoproteins[1]. The *cis*- and *trans*-forms have both been detected in collagen hydrolysates[2]. The free *cis*-form occurs in *Santalum album*. Hyp is formed posttranslationally by hydroxylation of proline in polypeptide chains under the action of protocollagen proline dioxygenase (EC 1.14.11.2). The major proportion of bovine collagen consists of the amino acid sequence Gly-X-Y. X is mostly Pro and Y Hyp, so that both presumably play a decisive role in the stabilization of the collagen triple helix. Hyp occurs very frequently in the *β*-loops of proteins.

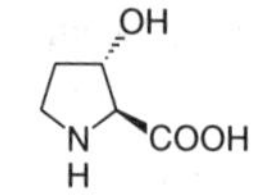

(2*S*,3*S*)-, *trans*-L-form, 3 Hyp

trans-3-Hydroxy-L-proline: $C_5H_9NO_3$, M_R 131.13, mp. 228–235 °C, $[\alpha]_D^{20}$ – 15.3° (H_2O), +17.4° (1 m HCl). Component of collagen. The (2*S*,3*R*)- and (2*S*,3*S*)-forms are components of marine sponges and cyclopeptide antibiotics, e. g., telomycin.

Lit.: [1] Trends Biochem. Sci. **5**, 245–248 (1980). [2] Anal. Biochem. **137**, 151–155 (1984).
gen.: Beilstein E V **22/5**, 4–9 ▪ Merck-Index (12.), No. 4887 ▪ Ullmann (5.) **A 2**, 58, 62. – *[HS 2933 90; CAS 51-35-4 (4 Hyp); 4298-08-2 (3 Hyp)]*

Hydroxystilbenes. A large group of natural products, mostly *phytoalexins; many from conifers, often occurring as glycosides and/or methyl ethers (see table). Important examples:
Pinosylvin (3,5-stilbenediol): (*E*)-form: needles, insoluble in water, soluble in acetone. (*E*)- and (*Z*)-isomers occur together with the corresponding mono- and dimethyl ethers in pines, firs, and conifers. Pinosylvin has antifungal and antibacterial activities and protects the wood against rot, as a disinfectant it is 30-times stronger than phenol and is highly toxic for fish. As a result of condensation with *lignin, pinosylvin prevents sulfite dissolution of wood. *Pinosylvin monomethyl ether* acts as an antifeedant for certain rabbit species[1].

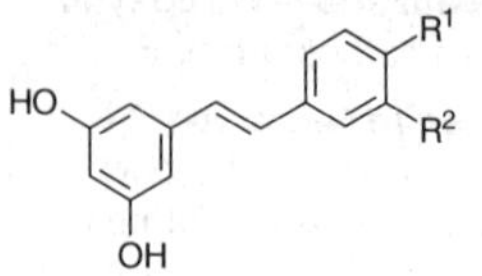

$R^1 = R^2 = H$: Pinosylvin
$R^1 = OH$, $R^2 = H$: Resveratrol
$R^1 = R^2 = OH$: Piceatannol

Table: Data of Hydroxystilbenes.

compound	molecular formula	M_R	mp. (bp.) [°C]	CAS
Pinosylvin	$C_{14}H_{12}O_2$	212.25		102-61-4
(*E*)-form			155.5–156	22139-77-1
(*Z*)-form			(123 [2.7 Pa])	106325-78-4
Pinosylvin-mono-methyl-ether	$C_{15}H_{14}O_2$	226.27	122–123	5150-38-9 (*E*-form)
Resveratrol	$C_{14}H_{12}O_3$	228.25		
(*E*)-form			266–267	501-36-0
(*Z*)-form				61434-67-1
Trihydroxy-diphenyl-ethane	$C_{14}H_{14}O_3$	230.26	156–157	58436-28-5
Piceatannol	$C_{14}H_{12}O_4$	244.25		4339-71-3
(*E*)-form			231–232	10083-24-6
(*Z*)-form				106325-86-4

Resveratrol (3,4′,5-stilbenetriol) is a phytoalexin from the roots of *Veratrum grandiflorum* and the bark of *Pinus sibirica*, it also occurs in *Eucalyptus*, *Polygonum*, *Nothofagus*, and other species. The resveratrol contained in red wine (see wine, phenolic constituents of) is supposed to be beneficial for health. In animal experiments, resveratrol inhibits monoamine oxidase A (MAO-A)[2]. *Dihydroresveratrol* (α,β-dihydro-3,4′,5-stilbenetriol) and its 3-methyl ether are constituents of *Cannabis sativa*; the latter shows estrogenic activity. *Piceatannol* (3,3′,4,5′-stilbenetetrol, astringenin): (*E*)-form: needles, is isolated from spruce trees, it is fungitoxic, inhibits plant growth, and has ichthyotoxic activity.

Biosynthesis: From hydroxycinnamoyl-CoA and malonyl-CoA with stilbene synthase. Tobacco plants into which the gene for stilbene synthase (isolated from the grape, *Vitis vinifera*), has been transferred are resistant to fungal infection with *Botrytis cinerea*[3].

Lit.: [1]Science **222**, 1023–1025 (1983). [2]Arch. Pharmacol. Res. **11**, 230–239 (1988). [3]Nature (London) **361**, 153–156 (1993).
gen.: Curr. Opin. Biotechnol. **5**, 125–130 (1994) ▪ J. Nat. Prod. **47**, 89–92 (1984) (synthesis); **53**, 498 f. (1990) ▪ Luckner (3.), p. 470 ▪ Nat. Prod. Rep. **10**, 233–263 (1993) ▪ Phytochemistry **24**, 321–324 (1985); **32**, 1561–1565 (1993) ▪ Plant Mol. Biol. **15**, 325–335 (1990) (biosynthesis) ▪ Tetrahedron **42**, 2725–2730 (1986) (synthesis). – *Pharmacology:* Arch. Pharmacol. Res. **13**, 132–135 (1990) ▪ Chem. Pharm. Bull. **32**, 213–219 (1984).

Hydroxysuccinic acid see malic acid.

5-Hydroxytryptamine see serotonin.

Hygrine [1-(1-methyl-2-pyrrolidinyl)-2-propanone].

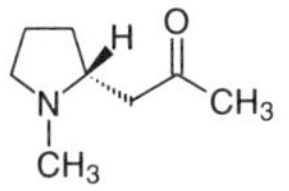

(+)-form, (*R*)-form

$C_8H_{15}NO$, M_R 141.21, liquid, bp. 77 °C (1.5 kPa), *R*-form: $[\alpha]_D^{22}$ +45° (C_2H_5OH), *S*-form: $[\alpha]_D^{16}$ −57.2° (C_2H_5OH). H. is a readily racemizing alkaloid from some *Dendrobium* species (Orchidaceae), bindweeds (Convolvulaceae), *coca leaves, and roots of many plants containing *tropane alkaloids.

Lit.: Beilstein E V **21/6**, 511 ▪ Hager (5.) **4**, 432 ▪ Manske **27**, 270 ff., 294 ff.; **44**, 115–187 (biosynthesis) ▪ Merck-Index (12.), No. 4899. – *Synthesis:* Heterocycles **20**, 671 (1983); **29**, 155 (1989) ▪ Tetrahedron **40**, 2879 (1984). – *[HS 2939 90; CAS 496-49-1 (R); 65941-22-2 (S); 45771-52-6 (racemate)]*

Hygrocybe pigments (hygroaurines). Yellow, orange, and red azomethine pigments from agarics of the genus *Hygrocybe* related to the *betalains. H. are formed biosynthetically by oxidative ring opening and recyclization from *α-3,4-dihydroxyphenylalanine and subsequent condensation with various amino acids. In contrast to the betalains, the H. contain muscaflavine (see fly agaric constituents) as a component of their chromophores in place of betalamic acid (see betalains).

Lit.: Zechmeister **51**, 75–86 (review).

Hygromycins.

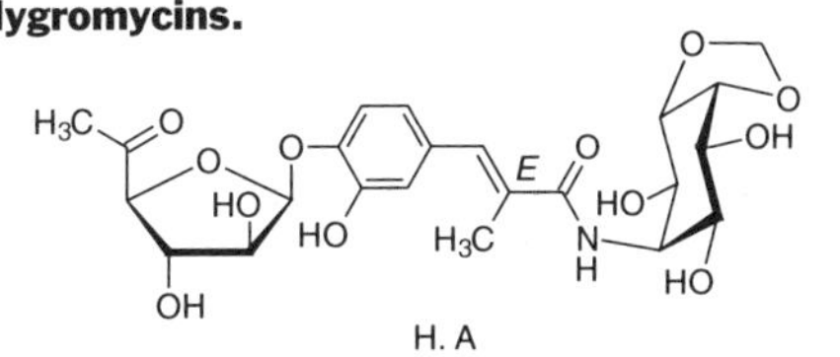

H. A

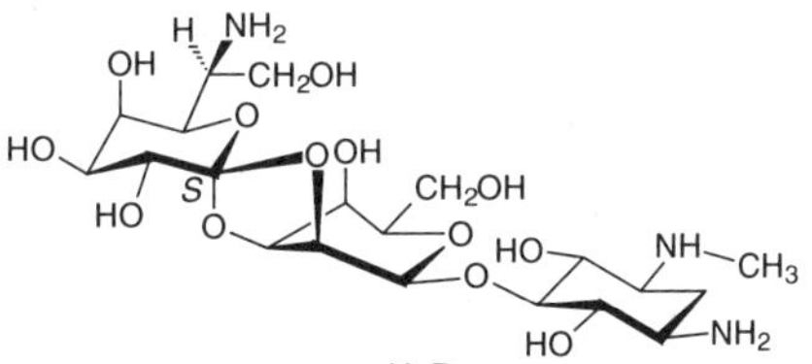

H. B

A group of aminoglycoside antibiotics with inhibitory activity on protein synthesis [1,2]. *H. A* [*homomycin, totomycin*, $C_{23}H_{29}NO_{12}$, M_R 511.48, mp. 105–109 °C, $[\alpha]_D^{25}$ −126° (H_2O); pK_a 8.9; LD_{50} (mouse i.p.) 1067 mg/kg], isolated from culture filtrates of *Streptomyces hygroscopicus*, *S. noboritoensis*, and *Corynebacterium* sp.[3–5], is active against a series of Gram-positive and Gram-negative bacteria including *Mycobacteria*, *Entamoeba*, and *Leptospira*. H. A has anthelmintic activity and has recently attracted interest because of its non-classical effects against adhesin and fimbria formation in animal-pathogenic bacteria. *H. B* (*macromycin*, $C_{20}H_{37}N_3O_{13}$, M_R 525.53, mp. 160–180 °C) is produced by *S. hygroscopicus*[6] and acts as a broad-spectrum *antibiotic; H. B is used in veterinary medicine as an anthelmintic, especially against ascarids.

Lit.: [1]Eur. J. Biochem. **157**, 409 (1980). [2]Biochim. Biophys. Acta **521**, 459 (1978). [3]Antibiot. Chemother. (Washington, D. C.) **3**, 1279 (1953). [4]J. Antibiot. **10A**, 21 (1957). [5]J. Antibiot. **33**, 695 (1980). [6]J. Am. Chem. Soc. **80**, 2714 (1958).
gen.: Agr. Biol. Chem. **42**, 279 (1978) ▪ Helv. Chim. Acta **53**, 2314 (1970) ▪ J. Chem. Soc., Chem. Commun. **1989**, 436 (synthesis) ▪ J. Org. Chem. **56**, 2976 (1991); **59**, 1224 (1994) ▪ Merck-Index (12.), No. 4900, 4901. – *[HS 2941 90; CAS 6379-56-2 (H. A); 31282-04-9 (H. B)]*

Hymenialdisine see oroidin.

Hymenoquinone. $C_{13}H_8O_6$, M_R 260.20, orange-red powder, mp. 228–232 °C (decomp.). The main pigment of the blood-red fruit body of the wood-rotting fungus *Hymenochaete mougeotii* (Aphyllophorales)

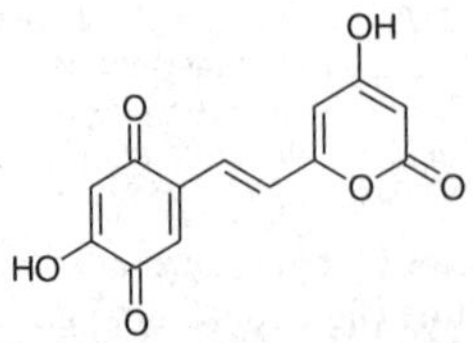

growing on silver fir. In addition to H., the fungus also contains the corresponding leuko compound and hispidin (see styrylpyrones).

Lit.: Chem. Ber. **110**, 1063 (1977). – *[CAS 62750-97-4]*

Hymenoxon(e) (hymenovin).

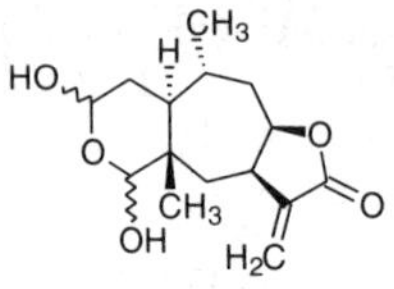

Hymenoxon

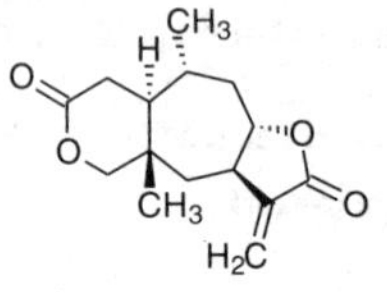

Vermeerin

$C_{15}H_{22}O_5$, M_R 282.34, isolated as an epimeric mixture, mp. 135–142 °C, soluble in organic solvents, poorly soluble in water. A *sesquiterpene lactone from leaves of *Hymenoxys odorata*[1] (Asteraceae) [LD_{50} (mouse p.o.) 150 mg/kg], it is especially toxic for cattle and sheep. In Texas mass poisonings of grazing animals occur each year, resulting in severe economic losses[2]. H. and the structurally similar *vermeerin* ($C_{15}H_{20}O_4$, M_R 264.32) from South African *Geigeria* species influence the microflora of the rumen of ruminants and thus their vital functions[3]. H. shows mutagenic and probably also carcinogenic activities[4].

Lit.: [1]Biochem. Syst. Ecol. **13**, 403 (1985); Phytochemistry **29**, 551 (1990); Res. Commun. Chem. Pathol. Pharmacol. **11**, 647 (1975). [2]Res. Commun. Chem. Pathol. Pharmacol. **28**, 189 (1980). [3]Phytochemistry **15**, 1573 (1976). [4]Food Cosmet. Toxicol. **15**, 225 (1977); Toxicol. Appl. Pharmacol. **45**, 629 (1978). – *[CAS 57377-32-9 (H.); 16983-23-6 (vermeerin)]*

Hyocholic acid, hyodeoxycholic acid see bile acids.

Hyoscyamine [(*S*)-tropic acid 3*endo*-tropyl ester, (–)tropyl tropate].

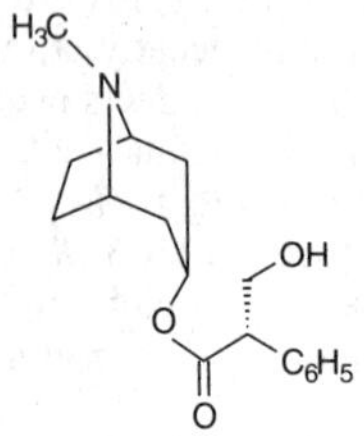

$C_{17}H_{23}NO_3$, M_R 289.37, needles, mp. 106–108 °C, $[\alpha]_D^{15}$ –22° (50% C_2H_5OH/H_2O). A *tropane alkaloid form numerous Solanaceae genera[1] such as, e.g.: *Atropa* (deadly nightshade), *Datura* (thorn apple), *Duboisia*, *Hyoscyamus* (henbane), and *Scopolia* (banewort). H. racemizes slowly in alcoholic solution, rapidly in alkaline and acidic solutions and in molten state to *atropine. It has the same effects on the central nervous system as atropine and a twice as strong toxic effect on the peripheral nervous system [LD_{50} (mouse i.v.) 95 mg/kg]. Symptoms of poisoning include dry mouth and throat. For isolation from cell cultures, see *Lit.*[2]. For biosynthesis, see tropane alkaloids.

Lit.: [1]Manske **44**, 1–114. [2]Acta Pharm. Fenn. **95**, 49–58 (1986); J. Am. Chem. Soc. **111**, 1141 f. (1989).
gen.: Anal. Profiles Drug Subst. **23**, 153 (1994) ▪ Beilstein E V **21/1**, 235 f. ▪ Hager (5.) **3**, 682 f.; **4**, 423–440, 1140; **8**, 511–515 ▪ Merck-Index (12.), No. 4907 ▪ Sax (8.), p. 1966 ▪ Ullmann (5.) **A 1**, 360 f. – *Synthesis:* Chem. Pharm. Bull. **19**, 2603 (1971) ▪ Manske **9**, 269–303. – *[HS 2939 90; CAS 101-31-5]*

Hypacrone see pterosins.

Hyperforin. Extracts of St. John's wort (*Hypericum perforatum*) are licensed in Germany to treat anxiety and sleep and depressive disorders. Clinical trials have demonstrated the activity of the extract. The main antidepressive principle in *H. perforatum* seems to be hyperforin, a meroterpene of the *humulone type: $C_{35}H_{52}O_4$, M_R 536.78, cryst., mp. 79–80 °C, $[\alpha]_D^{18}$ +41° (C_2H_5OH).

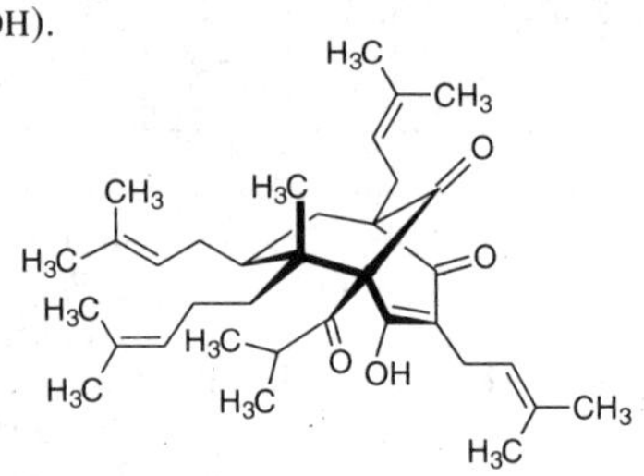

Lit.: Chem. Eng. News **76** (49) (1998, Dec. 7) ▪ Dtsch. Apoth. Ztg. **138**, 4754–4760 (1998) ▪ J. Nat. Prod. **62**, 770 (1999) ▪ Tetrahedron Lett. **23**, 1299 (1982). – *[CAS 11079-53-1]*

Hypericin.

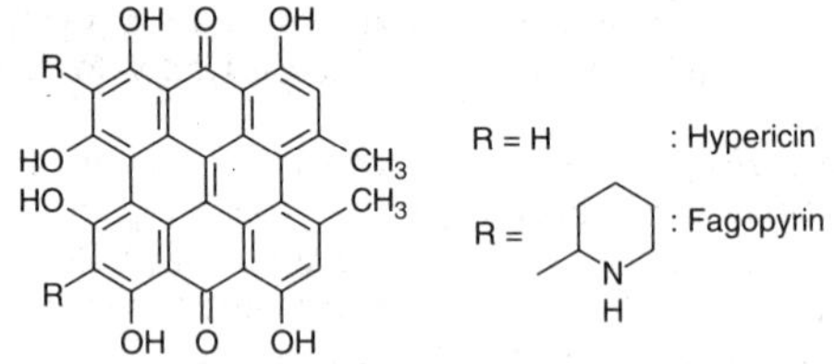

$C_{30}H_{16}O_8$, M_R 504.45, blue-black needles, mp. 320 °C (decomp.), uv_{max} 590 nm (CH_3OH), soluble in organic bases with cherry-red color and red fluorescence, soluble in aqueous alkalis with green color, above pH 11.5 with brick-red fluorescence. H. occurs in mealy bugs (*Nipaecoccus aurilanatus*), Saint-John's-wort (*Hypericum perforatum*, Hypericaceae), and other *Hypericum* species, it sensitizes the skin to UV light and can cause erythemas and necrosis of the skin[1]. H. interacts with digoxin (causing dosage problems). The same activities are shown by the structurally related *fagopyrin* ($C_{40}H_{34}N_2O_8$, M_R 670.72), occurring in the flowers of buckwheat (*Fagopyrum esculentum*, Polygonaceae)[2]. The consumption of Saint-John's-wort by non-pigmented grazing animals (white goats, sheep, or horses) or administration of H. to white mice and rats with subsequent exposure to sunlight leads to severe hemolytic clinical situations (hypericism) on account of the photodynamic effects of hypericin. H. has also been detected in the agaric *Dermocybe austroveneta*[3].

Use: Saint-John's-wort harvested in the flowering season is used for diarrhea diseases on account of its content of tannins. It also possesses anthelmintic activity; the ruby-red Saint-John's-wort oil, prepared from crushed fresh flowers by maceration with olive oil, is claimed to have magic healing power and anti-inflammatory activity[4]. H. exhibits antiretroviral (clinical testing against AIDS)[5] and insecticidal[6] activities. Recently, a meroterpene of the *humulone-type: *Hyperforin has been assumed to be responsible for the proven antidepressive effects of St. John's wort.

Biosynthesis: Emodin-9-anthrone (=10-deoxoemodin) formed by the polyketide pathway (*acetyl-CoA and 7 mol malonyl-CoA) dimerizes to the heptacyclic protohypericin, which is then dehydrogenated and transformed to H.[7,8]. For synthesis, see *Lit.*[9].

Lit.: [1] ACS Symp. Ser. **339**, 265–270 (1987); J. Chem. Ecol. **15**, 875–885 (1989); Photochem. Photobiol. **43**, 677–680 (1986); Photochem. Photobiol. Rev. **5**, 229–255 (1980). [2] Justus Liebigs Ann. Chem. **575**, 53 (1952). [3] J. Nat. Prod. **51** 1255–1260 (1988); Zechmeister **51**, 152. [4] Med. Res. Rev. **15**, 111–119 (1995). [5] Proc. Natl. Acad. Sci. USA **85**, 5230–5234 (1988); **86**, 5963–5967 (1989); SCRIP **1995** (19.4.). [6] J. Chem. Ecol. **15**, 855–862 (1989). [7] Zechmeister **14**, 141–185. [8] Luckner (3.), p. 176. [9] Angew. Chem. Int. Ed. Engl. **12**, 79 (1973); **16**, 46 (1977); Monatsh. Chem. **124**, 339 (1993).

gen.: Dtsch. Apoth. Ztg. **138**, 1783, 4754 (1998); **139**, 1741, 4718 (1999) (reviews) ▪ Hager (5.) **5**, 474–495 ▪ Naturwiss. Rundsch. **32**, 67 (1979) ▪ Roth, Hypericum-Hypericin, Landsberg: ecomed 1990. – *[CAS 548-04-9 (H.); 72393-03-4 (fagopyrin)]*

Hypholomins.

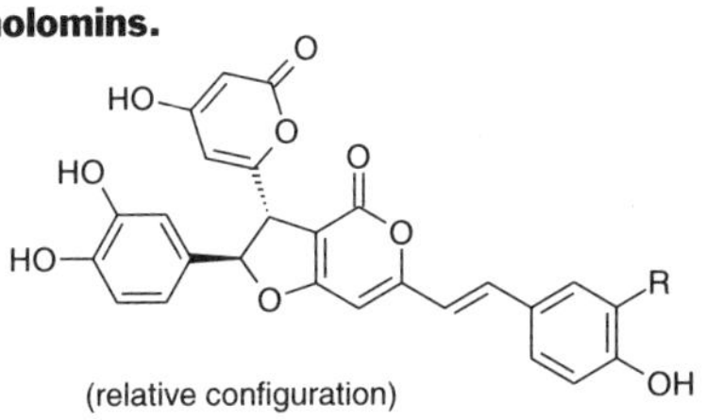

(relative configuration)
R = H : Hypholomin A
R = OH : Hypholomin B

H. A ($C_{26}H_{18}O_9$, M_R 474.42; tetramethyl ether: pale yellow needles, mp. 155–158 °C, optically active) and *H. B* ($C_{26}H_{18}O_{10}$, M_R 490.42; pentamethyl ether: yellow needles) are responsible for the yellow color and green-yellow UV fluorescence of the toadstool *Hypholoma fasciculare* (Agaricales). The H. are derived biosynthetically from the *styrylpyrone hispidin and are, like the closely related *fasciculins, widely distributed in *Hypholoma*- and *Pholiota* species (Basidiomycetes). They also occur in wood-rotting fungi of the genera *Inonotus* and *Phellinus*.

Lit.: Chem. Ber. **110**, 1047 (1977) ▪ Zechmeister **51**, 92–96 (review). – *[CAS 62350-92-9 (H. A); 62350-94-1 (H. B)]*

Hypnogenols.

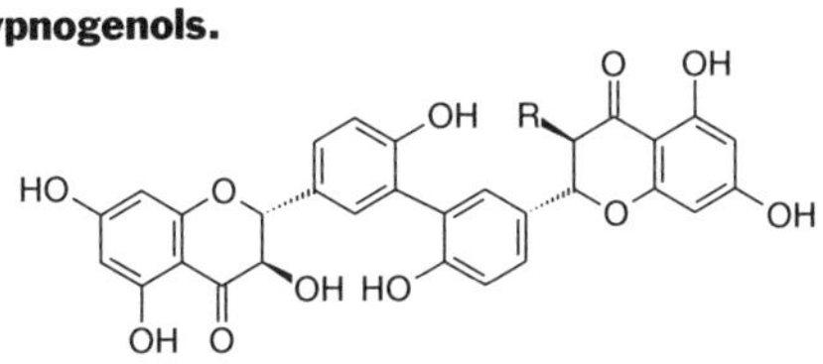

R = OH : Hypnogenol A
R = H : Hypnogenol B

H. A: $C_{30}H_{22}O_{12}$, M_R 574.50, pale yellow, amorphous solid, mp. >250 °C (decomp.), $[\alpha]_D^{20}$ −9.04° (CH_3OH), soluble in methanol. The main flavonoid (bidihydroflavonol) of the moss *Hypnum cupressiforme*. It belongs to a novel *biflavonoid structural type (C–C-linkage of both B rings of the monomeric dihydroflavonol building blocks).

Besides H. A, very small amounts of *H. B* ($C_{30}H_{22}O_{11}$, M_R 558.50) and other biflavonoids with this structure type are also present.

Lit.: Phytochemistry **31**, 3223 (1992); **33**, 795 (1994). – *[CAS 144506-12-7 (H. A); 144506-13-8 (H. B); 156250-61-2 (H. B₁)]*

Hypocrellin (hypocrellin A).

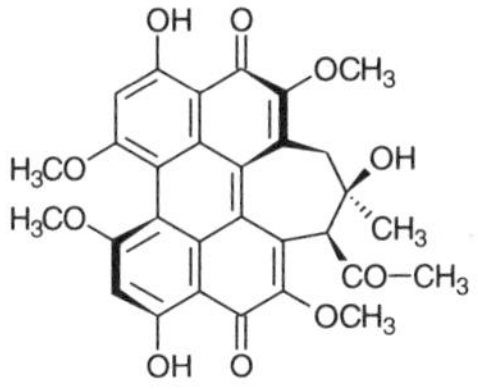

Hypocrellin

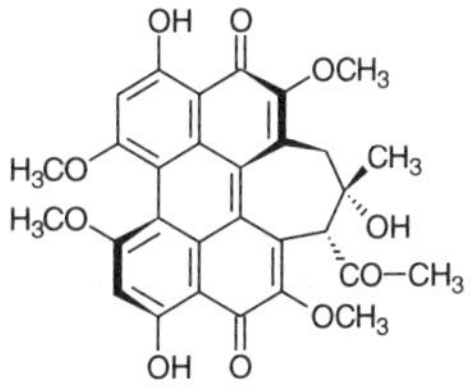

Shiraiachrome A

$C_{30}H_{26}O_{10}$, M_R 546.53, dark red crystals, mp. 245–250 °C (after softening at 214 °C). A perylenequinone from the fruit bodies of the ascomycete *Hypocrella bambusae* growing on bamboo with photodynamic activity against microorganisms. *Shiraiachrome A* is a stereoisomer of H. (deep red cryst., mp. 247–250 °C) and occurs together with the enantiomer of H. (shiraiachrome B) and its dehydrogenation product shiraiachrome C in mycelia of *Shiraia bambusicola*.

Lit.: Can. J. Chem. **75**, 99 (1997) ▪ Chin. Sci. Bull. **38**, 703 (1993) ▪ J. Nat. Prod. **52**, 948 (1989) ▪ Justus Liebigs Ann. Chem. **1981**, 1880 ▪ Planta Med. **57**, 376 (1991). – *[CAS 77029-83-5 (H.); 124709-39-3 (shiraiachrome A)]*

Hypoglycines. *H. A* [3-((*R*)-methylenecyclopropyl)-L-alanine], a non-proteinogenic amino acid with hypoglycemic and teratogenic activity from unripe akee plums (*Blighia sapida*, Sapindaceae).

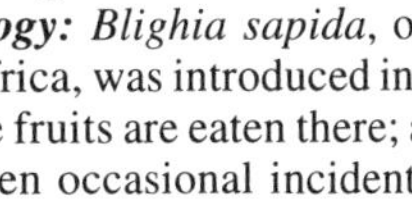

(2*S*,4*R*)-form

$C_7H_{11}NO_2$, M_R 141.17, yellow plates, mp. 280–284 °C, $[\alpha]_D^{26}$ +11° (H_2O), +49.0° (H_2O, pH 2), pK_a 2.40, 9.37. The content of H. A is ca. 0.1%. It occurs naturally as a mixture of diastereomers, with the (2*S*,4*R*)-form predominating[1].

Toxicology: *Blighia sapida*, originally indigenous to West Africa, was introduced in Jamaica 200 years ago. The ripe fruits are eaten there; after consumption there have been occasional incidents of vomiting, spasms, and loss of consciousness. H. can also be a lethal poison[2]. The LD_{50} value is ca. 40 mg/kg (human).

H. B [*N*-γ-glutamyl-3-((*R*)-methylenecyclopropyl)-L-alanine], yellow needles, mp. 194–200 °C, $[\alpha]_D^{32}$ +9.6° (H_2O), uv_{max} 359 nm. H. A and its γ-glutamylpeptide H. B have been found in various species of *Blighia*, *Billia*, and in *Acer pseudoplatanus*.

A lower homologue, α-(methylenecyclopropyl)glycine, is isolated from *Litchi chinensis* (Sapindaceae) and *A. pseudoplatanus*[3]. It also shows hypoglycemic activity.

Although H. A lowers the glucose content in blood it has practically no effect on the aerobic catabolism of acetate and glucose. In contrast, it blocks the formation of CO_2 from fatty acids[4], presumably by inhibiting the enzymes of β-oxidation. H. A has teratogenic activity in rats[5].

Lit.: [1] J. Chem. Soc., Chem. Commun. **1992**, 1249 ff. (synthesis). [2] Trends Pharmacol. Sci. **7**, 186 (1986); Sax (8.), MJP 500. [3] Phytochemistry **9**, 2423 f. (1970); **12**, 1677–1681 (1973). [4] Biochem. Pharmacol. **3**, 305 (1969); Biochem. Biophys. Res. Commun. **34**, 340–347 (1969). [5] Nature (London) **217**, 471 (1968).

gen.: Beilstein E IV **14**, 998 f. ▪ J. Org. Chem. **57**, 2471 (1992); **58**, 5915 (1993) ▪ Tetrahedron **46**, 2231 (1990); **50**, 12015 (1994). – *[CAS 156-56-9 (H. A); 502-37-4 (H. B); 31137-88-9 (α-(methylenecyclopropyl)glycine)]*

Hypothallin [*N*-benzoyl-L-leucine (*N'*-benzoyl-L-phenylalaninol ester)].

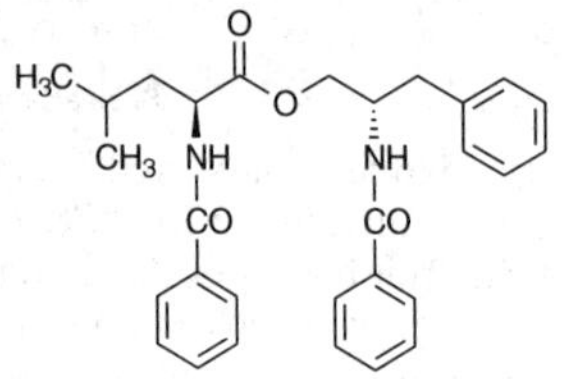

$C_{29}H_{32}N_2O_4$, M_R 472.58, needles, mp. 175–176 °C, $[\alpha]_D^{24}$ –22.5° ($CHCl_3$). An amino acid-aminoalcohol ester (leucine – phenylalaninol) from the crustaceous lichen *Schismatomma hypothallinum*.

Lit.: Z. Naturforsch. C **47**, 785–790 (1992). – *[CAS 146668-71-5]*

Hypoxanthine see purines.

Hypoxylone.

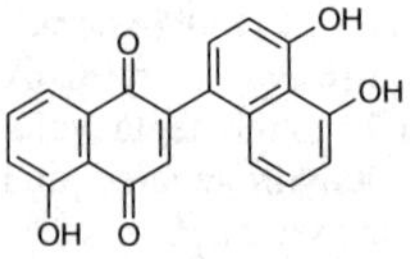

Hypoxylone

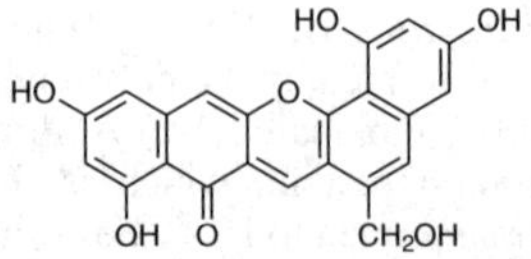

Hypoxyxylerone

$C_{20}H_{12}O_5$, M_R 332.31, violet crystals, mp. 179 °C (decomp.). Pigment from the fruit bodies of the wood-rotting fungus *Hypoxylon sclerophaeum* (Ascomycetes)[1]. H. belongs, like *bulgarein, to the ascomycete pigments that are biogenetically derived from *1,8-naphthalenediol. The green *hypoxyxylerone* ($C_{22}H_{14}O_7$, M_R 390.35, dark green metallic lustrous crystals, mp. >300 °C) is isolated from cultures of *H. fragiforme* and its structure has been confirmed by X-ray crystallography[2].

Lit.: [1] Phytochemistry **22**, 2579 (1983). [2] J. Chem. Soc., Chem. Commun. **1991**, 1009. – *[CAS 89701-95-1 (H.); 137361-21-8 (hypoxyxylerone)]*

I

IAA. Abbreviation for *3-indolylacetic acid.

Iboga alkaloids. A group of *monoterpenoid indole alkaloids from the iboga plant *Tabernanthe iboga*, a 1.5–2 m high shrub growing in the wet tropical regions of western central Africa. The shrub is the only Apocynaceae used by the native inhabitants on account of its hallucinogenic constituents, the I. a. The psychoactive alkaloids are mainly found in the light yellow roots, the bark, and also in the leaves. The bark or roots are grated and eaten or a decoction is made and drunk. The drug *Iboga* promotes strength and endurance and also seems to have aphrodisiac properties. Motoric activities are strongly reduced. Hunters in the bush can remain practically motionless for days with full consciousness under its influence, the sense of time is lost. As "plant of the gods" iboga enables the natives to "contact their ancestors" in trance while the "soul leaves the body". Iboga also has social significance (iboga cult, biwiti cult) to increase the community spirit in a tribe [1].

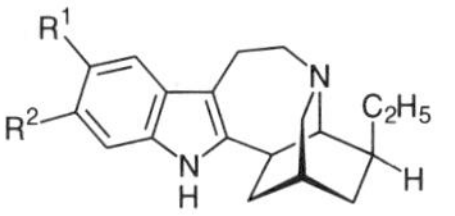

$R^1 = OCH_3$, $R^2 = H$: Ibogaine (**1**)
$R^1 = R^2 = H$: Ibogamine (**2**)

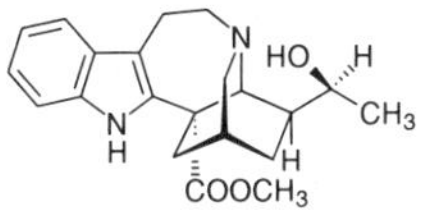

Heyneanine (**3**)

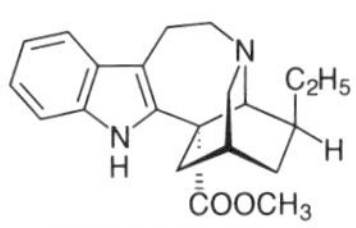

Coronaridine (**4**)

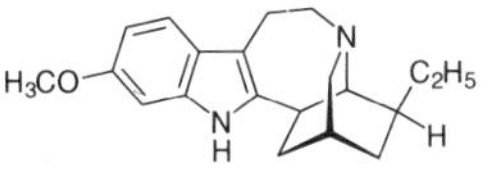

Tabernanthine (**5**)

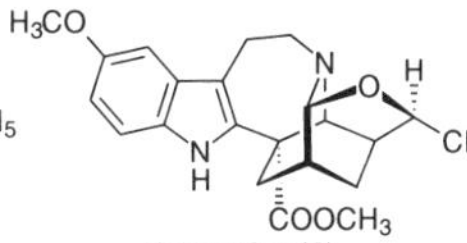

Heyneatine (**6**)

The important alkaloids occurring in the drug have different biological activities. *Ibogaine* {$C_{20}H_{26}N_2O$, M_R 310.44, needles, mp. 152–153 °C, $[\alpha]_D$ –53° (C_2H_5OH)} and *tabernanthine* {$C_{20}H_{26}N_2O$, M_R 310.44, needles, mp. 211–212 °C, $[\alpha]_D$ –40° (acetone)} act similarly to *cocaine and at high doses induce a state of tremor [2–4]. *Ibogamine* {$C_{19}H_{24}N_2$, M_R 280.41, cryst., mp. 162–164 °C, $[\alpha]_D$ –54.7° (C_2H_5OH)} has antibacterial activity and is weakly cytotoxic and lowers blood pressure. In addition, *heyneanine* {$C_{21}H_{26}N_2O_3$, M_R 354.45, cryst., mp. 105–107 °C or 160–162 °C} and the amorphous *heyneatine* {$C_{22}H_{26}N_2O_4$, M_R 382.46} are of interest for their cytotoxic activities [5,6]. *Coronaridine* ($C_{21}H_{26}N_2O_2$, M_R 338.45) is also a typical representative of the I. a., occurrence in *Tabernaemontana*, *Conopharyngia*, *Ervatamia*, *Stemmadenia*, and *Voacanga*. Its activity is similar to that of catharanthine [3,7,8]. A number of I. a. appear to be artifacts of isolation, e. g., the oxopropyl derivatives of heyneanine, coronaridine, voacangine [9]. Pandoline [10] and voacristine [11] also demonstrate the structural diversity of the I. alkaloids. Their biogenetic precursor is *tryptophan. Secologanin is supposed to supply the iridoid part, thus *strictosidine may also be considered as a precursor. For synthesis of ibogamine, see *Lit.* [12–15].

Lit.: [1] Schultes & Hofmann, Pflanzen der Götter, p. 112–115, Bern: Hallwag 1987. [2] Saxton (ed.), The Monoterpenoid Indole Alkaloids, Suppl. Vol. 25, Heterocyclic Compounds, p. 731, Chichester: Wiley 1994. [3] Brain Res. **571**, 242 (1992). [4] Pharmacol. Biochem. Behav. **43**, 1121 (1992). [5] Tetrahedron Lett. **1965**, 3873. [6] Phytochemistry **19**, 1213 (1980). [7] J. Ethnopharmacol. **13**, 165 (1985). [8] J. Nat. Prod. **51**, 528 (1988). [9] Chem. Pharm. Bull. **40**, 2075 (1992). [10] J. Nat. Prod. **52**, 1279 (1989). [11] Phytochemistry **30**, 1740 (1991). [12] J. Am. Chem. Soc. **100**, 3920 (1978). [13] J. Org. Chem. **50**, 1460, 1464 (1985). [14] Chem. Pharm. Bull. **33**, 4202 (1985). [15] J. Org. Chem. **57**, 1752 (1992); Tetrahedron Lett. **37**, 8289 (1996).

gen.: Beilstein E V **23/8**, 375; E V **23/12**, 283 ▪ Chem. Heterocycl. Compd. **25**, 467–537 (1983) ▪ Creasey, in Saxton (ed.), The Monoterpenoid Indole Alkaloids, p. 783, Chichester: Wiley & Sons 1983 ▪ Eur. J. Pharmacol. **140**, 303 (1987) ▪ Hager (5.) **6**, 890 f. ▪ Manske **8**, 203–235; **11**, 79–98; **52**, 197–232 ▪ Saxton (ed.), The Monoterpenoid Indole Alkaloids, Suppl. Vol. 25, Heterocyclic Compounds, p. 487–521, Chichester: Wiley & Sons 1994 ▪ Ullmann (5.) **A 1**, 393. – *Occurrence:* Fitotherapie **60**, 141 (1989) ▪ J. Nat. Prod. **47**, 478 (1984) ▪ Phytochemistry **30**, 3785 (1991) ▪ Planta Med. **54**, 519 (1988). – *[HS 2939 90; CAS 83-74-9 (1); 481-87-8 (2); 4865-78-5 (3); 467-77-6 (4); 83-94-3 (5); 76129-65-2 (6)]*

Ibotenic acid see fly agaric constituents.

Iceland moss. Common name for *Cetraria islandica* (Parmeliaceae), a lichen (not a member of the *mosses) from northern countries and exported from Iceland, Norway, and Sweden. About 60% by weight of the powdered lichen dissolves on boiling in highly dilute sodium hydrogen carbonate solution, on cooling the solution turns to a gel. The extract consists of a polysaccharide of lichenin and isolichenin, a number of bitter-tasting lichen acids (*fumarprotocetraric, protocetraric and cetraric acids) as well as *protolichesterinic acid, which is converted to *lichesterinic acid during work-up, and *usnic acid as antibiotically active *lichen pigment.

Uses: As mucolytic agent for catarrhs and diarrhea, as a bitter tonic, externally for poorly healing wounds, in cosmetic products, as hair-fixing material, additives to biscuit flour, etc..

Lit.: Lichen islandicus (Monographie 43/02.03.84), Köln: Bundesanzeiger-Verlagsges. 1989 ▪ Schönfelder u. Schönfelder, Der Kosmos-Heilpflanzenführer, p. 239f., Stuttgart: Franckh 1988. – *[HS 1211 20]*

Ichthyopterin (fluorescyanin).

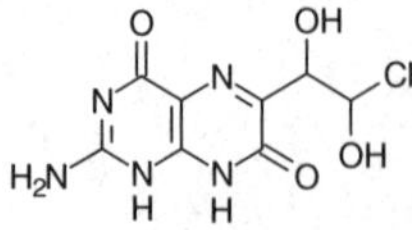

$C_9H_{11}N_5O_4$, M_R 253.22, yellow rosettes, that decompose on warming. The monoacetate ($C_{11}H_{13}N_5O_5$, M_R 295.25) melts at 143–153 °C, the diacetate at 188–196 °C (decomp.). I. is a pigment from the scales of various fish (component of chromoproteins), see also pteridines.

Lit.: Beilstein E III/IV **26**, 4038 ▪ J. Biochem. (Tokyo) **63**, 127 (1968) ▪ Pteridines **2**, 151–156 (1991); **3**, 165f. (1993). – *[CAS 490-58-4 (I.); 18503-57-6 (monoacetate)]*

ICI A 5504 see strobilurins.

Icosa... see eicosa...

Id(a)ein (idein, 3-β-D-galactopyranosyloxy-3′,4′,5,7-tetrahydroxyflavylium). $C_{21}H_{21}O_{11}^+$, M_R 449.39. The *cyanidin glycoside I. belongs to the *anthocyanins and occurs in cranberries (*Vaccinium vitis-idaea*, Ericaceae) and apples (*Malus sylvestris*, Rosaceae) as well as in the leaves of hornbeam (*Carpinus betulus*, Betulaceae).

Lit.: Beilstein E V **17/8**, 475 ▪ Karrer, No. 1714 – *[CAS 27661-36-5 (chloride)]*

Iditol.

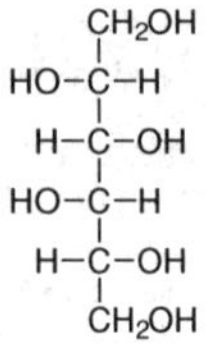

D-form

$C_6H_{14}O_6$, M_R 182.17, D-I.: mp. 73–74 °C, $[\alpha]_D^{20}$ +3.5° (H_2O), a rare sugar alcohol. L-Form: mp. 73 °C, $[\alpha]_D^{20}$ −3.5° (H_2O), occurs together with D-*glucitol in the mountain ash, rowan (*Sorbus aucuparia*).

Lit.: Beilstein E IV **6**, 30 ▪ Carbohydr. Res. **14**, 207 (1970). – *[HS 2905 49; CAS 25878-23-3 (D); 488-45-9 (L)]*

L-Iduronic acid.

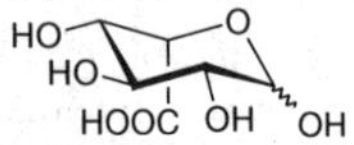

$C_6H_{10}O_7$, M_R 194.14. The figure shows the pyranoid form. Name for a rare uronic acid, mp. 131–132 °C, derived from L-idose and occurring especially in chondroitin sulfate B, heparin, and dermatan sulfate. Biosynthetically I. is formed from the isomeric D-*glucuronic acid from which it differs only in the configuration at C-5. This epimerase-catalyzed reaction takes place in the intact glycosaminoglycan.

Lit.: Beilstein E IV **3**, 2000 ▪ Carbohydr. Res. **77**, 281 (1979) ▪ J. Biol. Chem. **250**, 3419 (1975) (biosynthesis). – *[CAS 2073-35-0]*

Ikarugamycin. $C_{29}H_{38}N_2O_4$, M_R 478.63, cryst., mp. 228–229 (252–255)°C (decomp.); $[\alpha]_D$ +390° (DMF). A tetramic acid derivative from cultures of *Streptomyces phaeochromogenes* with antibacterial and antiprotozoic properties.

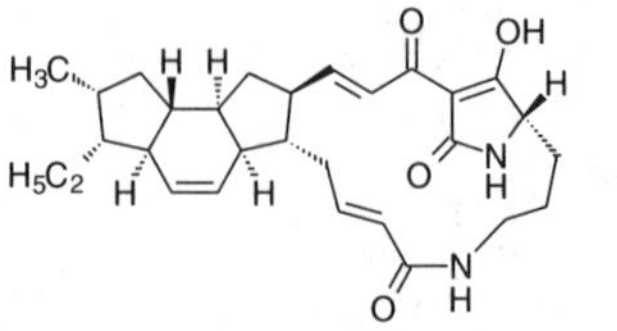

Lit.: Bull. Chem. Soc. Jpn. **50**, 1813 (1977) ▪ Tetrahedron Lett. **1972**, 1181, 1185, 2557. – *Synthesis:* J. Am. Chem. Soc. **111**, 8036f. (1989); **112**, 9285, 9292 (1990); **116**, 2151 (1994) ▪ J. Org. Chem. **49**, 731 (1984). – *[CAS 36531-78-9]*

Ilimaquinone.

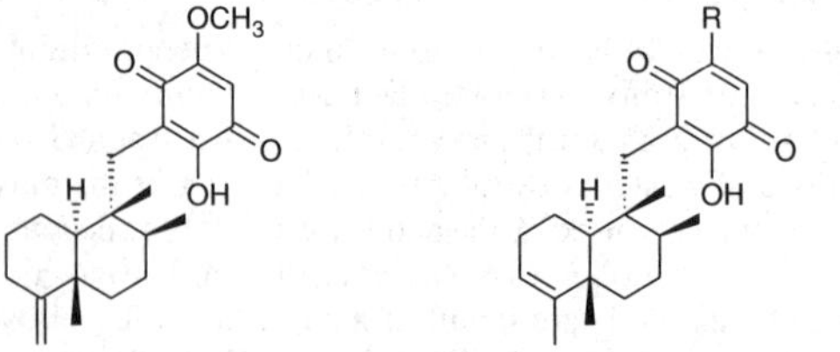

Ilimaquinone

R = OCH_3 : Isospongiaquinone
R = L-NH—CH(CH_2OH)COOH : Nakijiquinone C

$C_{22}H_{30}O_4$, M_R 358.48, orange needles (hexane), mp. 113–114 °C, $[\alpha]_D$ −23.2° ($CHCl_3$).

Meroterpenoid with a rearranged drimane-benzoquinone skeleton from the Hawaiian sponge *Hippiospongia metachromia*[1] and *Smenospongia* spp. The absolute configuration was proven by oxidative degradation[2]. I. exhibits anti-HIV activity[3] and induces the vesiculation of Golgi membranes[4]. Some of these effects can be explained by the interaction of I. with methylation enzymes[5]. The isomeric *isospongiaquinone* [yellow needles (hexane), mp. 135.5–136 °C, $[\alpha]_D$ +64.8° ($CHCl_3$)] occurs in the sponge *Stelospongia conulata*[6]. The red nakijiquinones A–D were obtained from an Okinawan marine sponge[7]. They are derived from I. by substitution of the methoxy group with amino acids, e.g. *nakijiquinone C* [$C_{25}H_{33}NO_6$, M_R 431.53, red amorphous solid, $[\alpha]_D$ +62° (C_2H_5OH)[7]] with L-serine. The nakijiquinones are selective inhibitors of Her-2/Neu kinase and exhibit pronounced cytotoxicity against L-1210 murine leukemia cells and human KB-epidermoid carcinoma cells[7].

Lit.: [1]Tetrahedron **35**, 609 (1979). [2]J. Org. Chem. **52**, 5059 (1987). [3]Antimicrob. Agents Chemother. **34**, 2009 (1990). [4]Cell **73**, 1079 (1993); J. Cell Biol. **129**, 577 (1995). [5]Chemistry Biology **6**, 639 (1999). [6]Aust. J. Chem. **31**, 2685 (1978). [7]Tetrahedron **50**, 8347 (1994); **51**, 10867 (1995).

gen.: Syntheses: J. Org. Chem. **60**, 1114 (1995) (ilimaquinone) ▪ Angew. Chem. Int. Ed. Engl. **38**, 3710 (1999) [(−)-isospongiaquinone, (+)-nakijiquinone C]. – *[CAS 71678-03-0 (I.); 69672-66-8 (isospongiaquinone); 169438-43-1 (nakijiquinone C)]*

Illudalic acid see illudinine.

Illudanes. Sesquiterpenoids with the illudane skeleton occur in basidiomycetes and ferns (*pterosins). They are derived biosynthetically from *humulene. Numerous derivatives of I. have been isolated from cultures of *Omphalotus illudens* (=*Clitocybe illudens*) and *Lampteromyces japonicus*, e.g., the toxic *illudin M* {$C_{15}H_{20}O_3$, M_R 248.32, cryst., mp. 130–131 °C, $[\alpha]_D$ −126° (C_2H_5OH)} and *illudin S* {$C_{15}H_{20}O_4$, M_R 264.32,

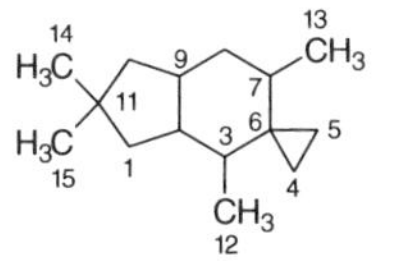

Illudane

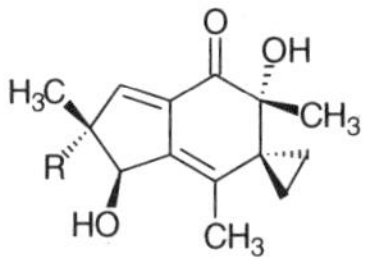

R = CH_3 : Illudin M
R = CH_2OH : Illudin S

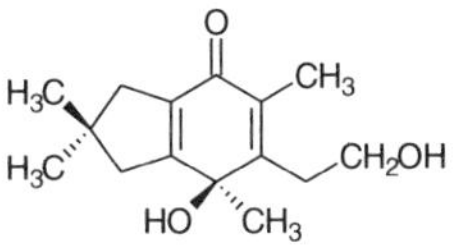

Illudalenol

cryst., mp. 137–138 °C (from C_2H_5OH), $[\alpha]_D$ −165° (C_2H_5OH)}, which show strong cytotoxicity to human adenocarcinoma cells and are active in antitumor tests. In *illudalenol* {$C_{15}H_{22}O_3$, M_R 250.34, oil, $[\alpha]_D$ +6° ($CHCl_3$)} and the isoquinoline derivative *illudinine the cyclopropane ring has been opened (illudalane skeleton).

Lit.: Cancer Res. **47**, 3186 (1987) ▪ Gazz. Chim. Ital., **121**, 345 (1991) ▪ J. Antibiot. **40**, 1643 (1987); **49**, 821 (1996) (I. C_2, C_3) ▪ J. Chem. Soc., Perkin Trans. 1 **1991**, 733 ▪ J. Nat. Prod. **52**, 380 (1989) ▪ Phytochemistry **21**, 942 (1982) ▪ Tetrahedron Lett. **30**, 3537 (1989). – *Biosynthesis:* J. Chem. Res. (S) **1985**, 396 ▪ J. Chem. Soc., Perkin Trans. 1 **1982**, 2445 ▪ Tetrahedron Lett. **26**, 4755 (1985). – *Synthesis:* ApSimon **2**, 534 ▪ J. Am. Chem. Soc. **116**, 2667 (1994) ▪ J. Org. Chem. **62**, 1317 (1997). – *Review:* Tetrahedron **37**, 2199 (1981) ▪ Turner **1**, 228; **2**, 244, 246. – *[CAS 1146-04-9 (illudin M); 1149-99-1 (illudin S); 134857-04-8 (illudalenol)]*

Illudinine.

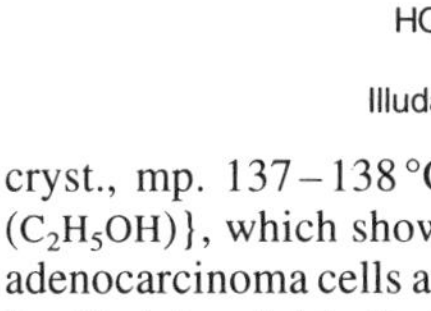

Illudinine

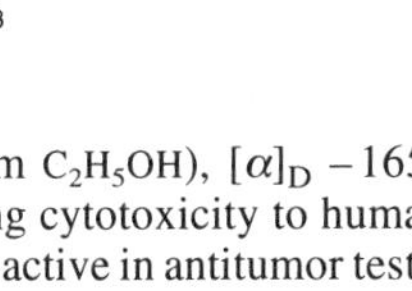

Illudalic acid

$C_{16}H_{17}NO_3$, M_R 271.32, cryst., mp. 228–229 °C (decomp.). An isoquinoline derivative derived biosynthetically from an *illudane precursor and isolated from cultures of the fungus *Omphalotus illudens* (Basidiomycetes). I. is accompanied by *illudalic acid* ($C_{15}H_{16}O_5$, M_R 276.29, cryst., mp. >200 °C, optically inactive) and other illudalane derivatives.

Lit.: J. Org. Chem. **34**, 240 (1969). – *Synthesis:* Indian J. Chem., Sect. B **32**, 1209 (1993) ▪ J. Am. Chem. Soc. **99**, 8007 (1977) ▪ J. Org. Chem. **38**, 4305 (1973). – *[CAS 18500-63-5 (I.); 18508-77-5 (illudalic acid)]*

Illudol see protoilludanes.

Illudosin see fomannosin.

Imbricatine. $C_{24}H_{26}N_4O_7S$, M_R 514.55, a *benzyl(tetrahydro)isoquinoline and *imidazole alkaloid from the starfish *Dermasterias imbricata*. I. is as yet the only alkaloid of this class that does not originate from a plant source. I. induces an escape reaction in the sea anemone *Stomphia coccinea* when it comes into contact with the starfish[1]. I. has cytotoxic activity against leukemia cell lines.

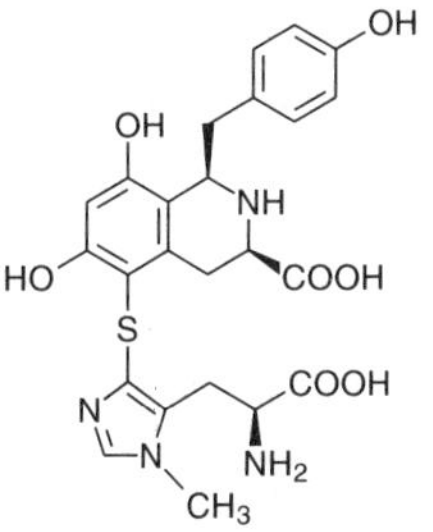

Lit.: [1] Biol. Bull. **176**, 73 (1989).
gen.: J. Am. Chem. Soc. **108**, 8288 (1986) (structure) ▪ Can. J. Chem. **69**, 20 (1991) (synthesis) ▪ Chem. Pharm. Bull. **47**, 83 (1999) (synthesis) ▪ Tetrahedron **55**, 4999 (1999) [synthesis (+)-tri-O-methyl-I.]. – *[CAS 105372-70-1]*

Imidazole alkaloids. A group of natural products of plant and animal (especially marine organisms) origin containing the imidazole ring system.

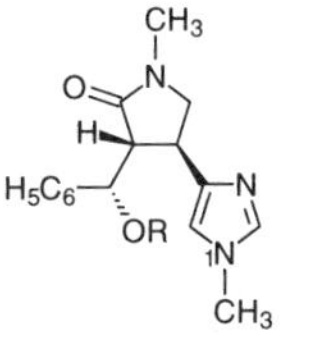

R = H : Cynometrine (**1**)
R = CO–C_6H_5 : Cynodine (**2**)

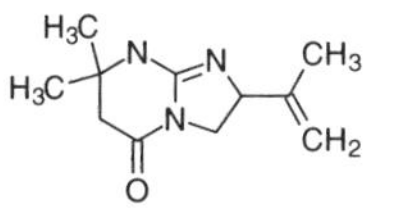

Alchornine (**4**)

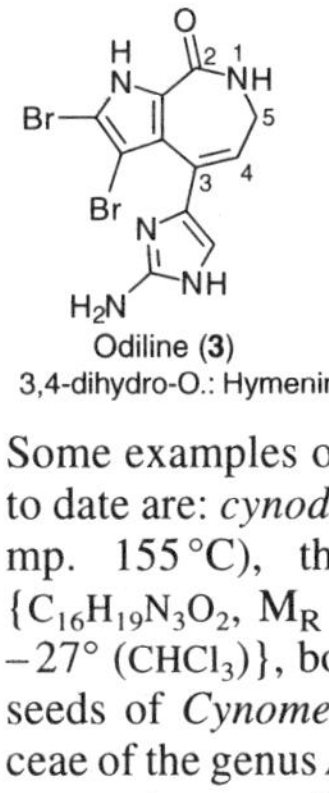

Odiline (**3**)
3,4-dihydro-O.: Hymenin [6]

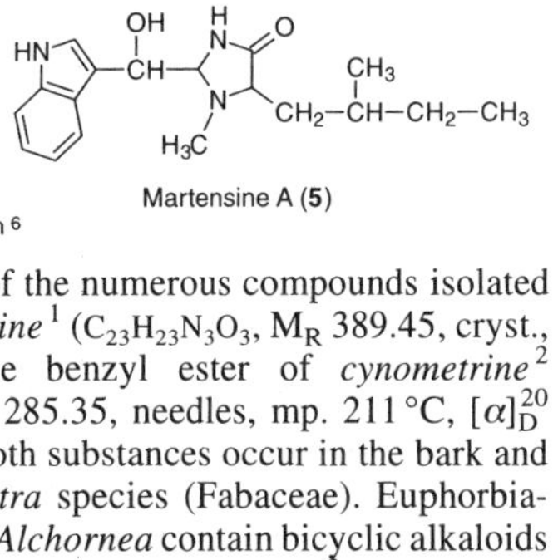

Martensine A (**5**)

Some examples of the numerous compounds isolated to date are: *cynodine*[1] ($C_{23}H_{23}N_3O_3$, M_R 389.45, cryst., mp. 155 °C), the benzyl ester of *cynometrine*[2] {$C_{16}H_{19}N_3O_2$, M_R 285.35, needles, mp. 211 °C, $[\alpha]_D^{20}$ −27° ($CHCl_3$)}, both substances occur in the bark and seeds of *Cynometra* species (Fabaceae). Euphorbiaceae of the genus *Alchornea* contain bicyclic alkaloids possessing a condensed pyrimidine ring[3], such as, e. g., *alchornine* {$C_{11}H_{17}N_3O$, M_R 207.28, mp. 134–135 °C, $[\alpha]_D$ +74° ($CHCl_3$)}. The starfish *Dermasterias imbricata* contains *imbricatine, the red alga *Martensia fragilis martensine A*[4] ($C_{18}H_{25}N_3O_2$, M_R 315.42) possessing a 3-indolyl unit. The marine sponge *Pseudaxinyssa* contains the bromine-substituted *odiline*[5] (stevensine, $C_{11}H_9Br_2N_5O$, M_R 387.03). Further examples: *Oroidin and related heterocycles, the fluorescent parazoanthoxanthins (see zoanthoxanthins), *pilocarpine, the *roquefortines C and D., and the *topsentins. Some marine pyrrole imidazole alkaloids biosynthetically seem to be the product of combinatorial folding.

Lit.: [1] Tetrahedron Lett. **1973**, 1757. [2] Phytochemistry **20**, 2765 (1981); Tetrahedron **38**, 2687 (1982). [3] Aust. J. Chem. **23**, 1679 (1970). [4] Tetrahedron Lett. **24**, 2087 (1983). [5] J. Org. Chem. **50**, 4163 (1985). [6] J. Org. Chem. **62**, 456 (1997).
gen.: Alkaloids (London) **2**, 49–104 (1984) ▪ J. Nat. Prod. **60**, 180 (1997) ▪ J. Org. Chem. **52**, 5638 f. (1987); **56**, 4304 (1991) ▪ Justus Liebigs Ann. Chem. **1997**, 1525 ▪ Manske **22**, 281–333 ▪ Nat. Prod. Rep. **5**, 351–361 (1988) ▪ Tetrahedron Lett. **28**,

3003 (1987); **38**, 8935 (1997) ▪ Ullmann (5.) **A 1**, 400 ▪ Zechmeister **68**, 137 ff. – *[CAS 50656-83-2 (1); 50656-84-3 (2); 99102-22-4 (3); 25819-91-4 (4); 87168-35-2 (5); 105748-62-7 (hymenin)]*

Immunomycin see ascomycin.

Impact compound. The name for *flavor compounds that decisively influence the character of an aroma or flavor, e.g., (–)-*carvone in *spearmint oil, *citral in *lemon oil, *filbertone in *filbert flavor, or vanillin in *vanilla. For further examples, see literature.

Lit.: J. Food Sci. **29**, 158 ff. (1964) ▪ Maarse, p. 237, 292, 456.

Imperatorin [9-(3-methyl-2-butenyloxy)-7*H*-furo-[3,2-*g*][1]benzopyran-7-one].

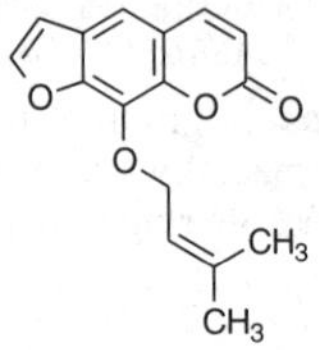

$C_{16}H_{14}O_4$, M_R 270.28, crystals, mp. 102 °C, contents: up to 0.5% in *Angelica archangelica*, masterwort (*Peucedanum ostruthium*), hogweed (*Heracleum* spp.) and other Apiaceae. I. belongs to the group of *furocoumarins and exhibits antimicrobial activity.

Lit.: Beilstein E V **19/6**, 16 ▪ Karrer, No. 1377 ▪ Tetrahedron **40**, 5225 ff. (1984). – *Pharmacology:* Ghana J. Sci. **10**, 82 (1970). – *Toxicology:* Sax (8.), IHR 300. – *[HS 2932 99; CAS 482-44-0]*

Inandenines.

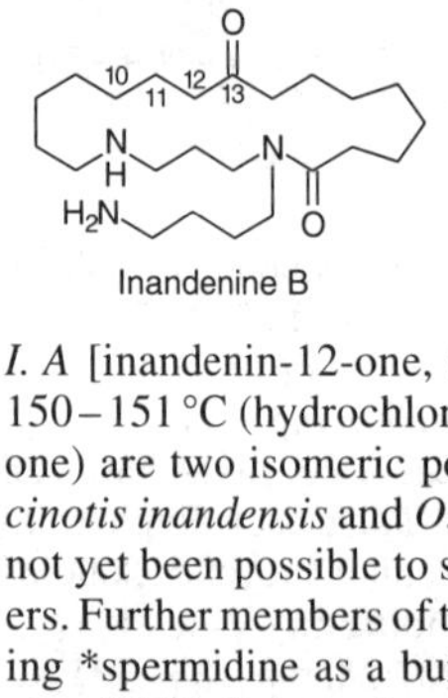

Inandenine B

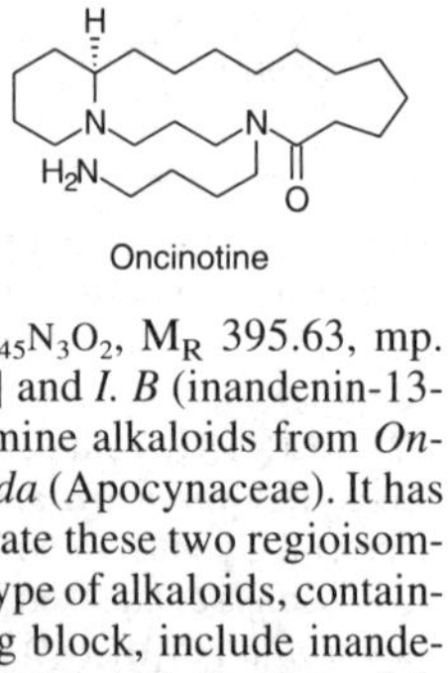

Oncinotine

I. A [inandenin-12-one, $C_{23}H_{45}N_3O_2$, M_R 395.63, mp. 150–151 °C (hydrochloride)] and *I. B* (inandenin-13-one) are two isomeric polyamine alkaloids from *Oncinotis inandensis* and *O. nitida* (Apocynaceae). It has not yet been possible to separate these two regioisomers. Further members of this type of alkaloids, containing *spermidine as a building block, include inandenine-10,11-diol and *oncinotine* {$C_{23}H_{45}N_3O$, M_R 379.63, oil, $[\alpha]_D$ –29° ($CHCl_3$)}.

Lit.: Helv. Chim. Acta **57**, 414, 434 (1974); **59**, 3013, 3026 (1976) ▪ J. Am. Chem. Soc. **104**, 6881 (1982) ▪ J. Org. Chem. **61**, 1023 (1996) (synthesis, oncinotine). – *[CAS 29579-65-5 (I. A); 29579-66-6 (I. B); 21008-79-7 (oncinotine)]*

Incaflavin.

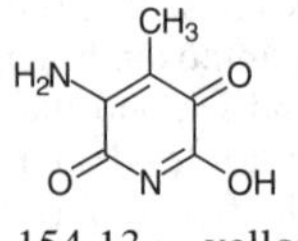

$C_6H_6N_2O_3$, M_R 154.13, yellow crystals, mp. 267–272 °C (decomp.). A yellow azaquinone from the alcoholic extract of the agaric *Entoloma incanum* (Basidiomycetes), in which it occurs together with red, blue, and colorless, strongly fluorescent components. The fungus contains a chromogen that transforms to a blue pigment when the fruit body is injured. On standing in aqueous solution, this pigment is oxidized by atmospheric oxygen to I.

Lit.: Zechmeister **51**, 231–234.

Indanomycin (X-14547 A).

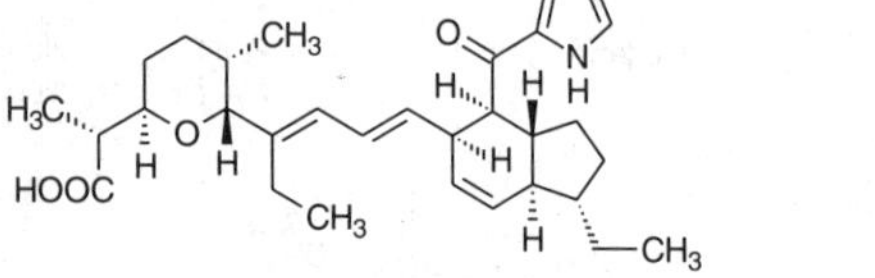

$C_{31}H_{43}NO_4$, M_R 493.69, mp. 138–141 °C, $[\alpha]_D$ –328° ($CHCl_3$), a pyrrole ether antibiotic with ionophoric properties from *Streptomyces antibioticus*. I. inhibits the growth of Gram-positive bacteria and possesses antihypertensive properties, it promotes the growth of ruminants.

Lit.: J. Am. Chem. Soc. **100**, 6784–6786 (1978) ▪ J. Antibiot. **32**, 95–99, 100–107 (1979); **41**, 1170–1177 (1988). – *Synthesis:* J. Org. Chem. **50**, 1440–1456 (1985); **59**, 332 (1994) ▪ Tetrahedron Lett. **26**, 1163–1166 (1985). – *[HS 2941 90; CAS 66513-28-8]*

Indian yellow (C. I. 75320).

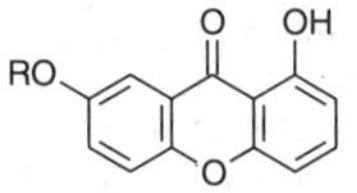

Mg(Ca) salt of *euxanthic acid* (1,7-dihydroxy-9*H*-xanthen-9-one 7-*O*-*β*-D-glucuronide, $C_{19}H_{16}O_{10}$, M_R 404.33), a derivative of *euxanthone* ($C_{13}H_8O_4$, M_R 228.20, yellow needles, mp. 240 °C), obtained in Bengal from the urine of cattle fed almost exclusively with leaves of the mango tree (*Mangifera indica*, Anacardiaceae). Up to about 1920 I. was a frequently used pigment for artists' paints.

Lit.: Bull. Chem. Soc. Jpn. **30**, 629 (1957) ▪ Feller, Artists Pigments, vol. 1, p. 17, Cambridge: University Press 1986 ▪ Synth. Commun. **19**, 1641 (1989). – *[HS 3203 00; CAS 525-14-4 (euxanthic acid); 529-61-3 (euxanthone)]*

Indican (3-indolyl-*β*-D-glucopyranoside).

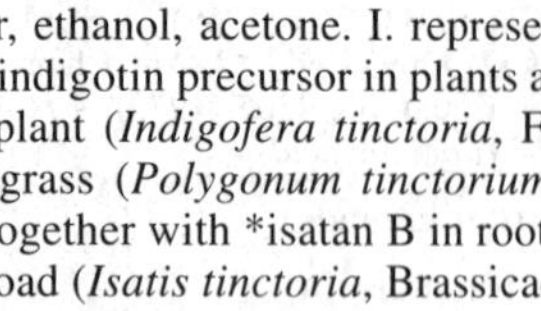

$C_{14}H_{17}NO_6$, M_R 295.29, colorless crystals, mp. 57–58 °C (trihydrate), 178–180 °C (decomp.), soluble in water, ethanol, acetone. I. represents the most important *indigotin precursor in plants and occurs in the indigo plant (*Indigofera tinctoria*, Fabaceae), in dyer's knotgrass (*Polygonum tinctorium*, Polygonaceae), and together with *isatan B in roots and leaves of dyer's woad (*Isatis tinctoria*, Brassicaceae).[1]

Lit.: [1] Biochem. Physiol. Pflanz. **184**, 321 (1989).

gen.: Beilstein E V **21/3**, 6 ▪ Hager (5.) **4**, 467 ▪ Phytochemistry **31**, 2695 (1992) (biosynthesis) ▪ Ullmann (5.) **A 14**, 149 ff. – *[HS 2938 90; CAS 487-60-5]*

Indicaxanthin (condensation product of *betalamic acid and L-proline). $C_{14}H_{16}N_2O_6$, M_R 308.29, orange crystals, mp. 160–162 °C (decomp.). I. is an optically

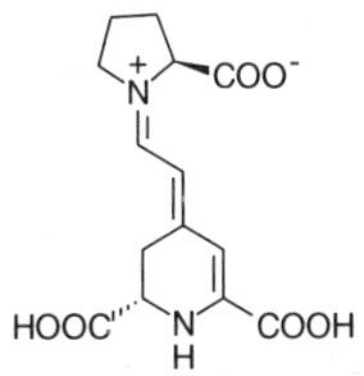

active pigment from the Indian fig (*Opuntia ficus-indica*, Cactaceae) as well as flowers of *Mirabilis jalapa* (Nyctaginaceae) and *Portulaca grandiflora* (Portulacaceae), belonging to the betaxanthine group. I. undergoes a rapid *E/Z*-isomerization in solution, see also betalains.

Lit.: Beilstein E V **22/7**, 172 ▪ Helv. Chim. Acta **67**, 1793 (1984) ▪ Phytochemistry **11**, 2499 (1972). – *[CAS 2181-75-1]*

Indicine.

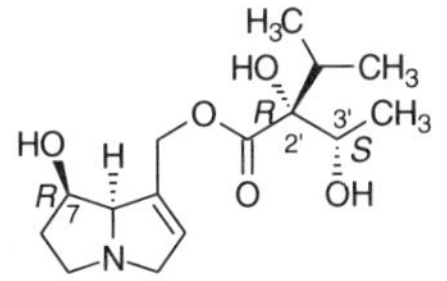

$C_{15}H_{25}NO_5$, M_R 299.37, mp. 97–98 °C, $[\alpha]_D$ +22.3° (C_2H_5OH). A *pyrrolizidine alkaloid from some *Heliotropium* species (Boraginaceae). I. exists in the plant as the *N*-oxide and exhibits a strong antitumor activity, probably as a purine antagonist. It is too toxic for therapeutic use. *Echinatine* is the (7*S*,2′*S*)-isomer (mp. 109–110 °C, $[\alpha]_D$ +12.8°). It occurs in Boraginaceae and Asteraceae species. *Intermedine*, the (2′*S*,3′*R*)-isomer {mp. 140–142 °C, $[\alpha]_D$ +7.8° (C_2H_5OH)} and *lycopsamine*, the (2′*S*)-isomer (mp. 132–134 °C, $[\alpha]_D$ +3.3°) often occur together in Boraginaceae species. *Rinderine*, the (7*S*,2′*S*,3′*R*)-isomer {mp. 100–101 °C, $[\alpha]_D$ +24.6° (C_2H_5OH)} is found in Boraginaceae species of the genus *Rindera* and Asteraceae species of the genus *Eupatorium*. More than 100 further alkaloids derived from these five positionally isomeric structures are known to date.

Lit.: Beilstein E V **21/4**, 402 ▪ Pelletier **9**, 155–233 ▪ Sax (8.), ICD 100. – *Synthesis:* Tetrahedron **49**, 1571 (1993) ▪ Tetrahedron Lett. **27**, 4323 (1986); **30**, 4985 (1989). – *[HS 2939 90; CAS 480-82-0 (I.); 41708-76-3 (N-oxide); 480-83-1 (echinatine); 10285-06-0 (intermedine); 10285-07-1 (lycopsamine); 6029-84-1 (rinderine)]*

Indigoidin.

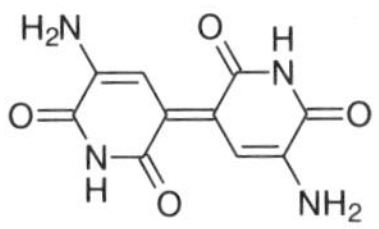

$C_{10}H_8N_4O_4$, M_R 248.20. A blue extracellular pigment dye with the indigoid system from *Corynebacterium insidiosum*, *Pseudomonas indigofera*, *Arthrobacter atrocyaneus*, and *A. polychromogenes*, uv_{max} 605 nm. Its insolubility in many organic solvents is remarkable. I. dissolves in hot sulfuric acid with orange-brown color whereas solutions in concentrated nitric acid are indigo blue. I. is decomposed by sodium hydroxide.

Lit.: Beilstein E V **25/16**, 134 ▪ Chem. Ber. **98**, 2139 (1965). – *[CAS 2435-59-8]*

Indigo red (2,3′-biindolinylidene-2′,3-dione, indirubin, C. I. 75790).

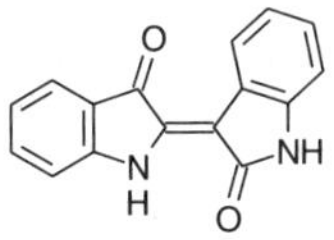

$C_{16}H_{10}N_2O_2$, M_R 262.27, red cryst., mp. >430 °C, uv_{max} 561 nm (xylene), soluble in ethanol and ether, insoluble in water. Red isomer of *indigotin present in small amounts in synthetic indigo and sometimes in considerable amounts in natural indigo (2–4% in Bengal indigo, up to 15% in Java indigo).

Lit.: Beilstein E V **24/8**, 507 ▪ Tetrahedron Lett. **1974**, 609; **1977**, 2625 ▪ Ullmann (4.) **13**, 177. – *[HS 3203 00; CAS 479-41-4]*

Indigotin, indigo [2,2′-biindolinylidene-3,3′-dione, C. I. natural blue 1].

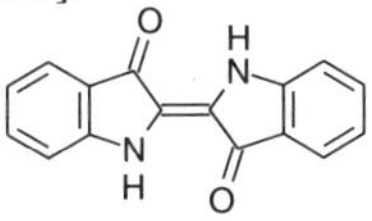

$C_{16}H_{10}N_2O_2$, M_R 262.27, dark blue crystals with a copper-red gloss, mp. 390–392 °C (decomp.), uv_{max} 590 nm (xylenes), insoluble in water, alcohol, ether, and dilute acids, soluble in glacial acetic acid, pyridine, dimethylformamide, and dimethyl sulfoxide.

Occurrence: I. is one of the longest known and most important organic dyes. Indigo-dyed textiles have been found in 4000 years old Egyptian mummies. I. occurs in the leaves and stems of east Indian *Indigofera* species, especially the indigo plant (*Indigofera tinctoria*, Fabaceae), in dyer's knotgrass (*Polygonum tinctorium*, Polygonaceae) and in the roots of the dyer's woad cultivated in Europe in the past (*Isatis tinctoria*, Brassicaceae) in the form of *indican, whereas in the leaves of dyer's woad the precursor of I., *isatan B, dominates.

Synthesis: After A. von Baeyer achieved the first synthesis of I. in 1878 and the exact structure was elucidated in 1883, further syntheses of I. were developed, the best of which was that by Heumann in 1890. Like natural I., synthetic I. contains the isomeric *indigo red, the content of the latter ranges from 2–4% in Bengal I. to 15% in Java I. The biosynthesis proceeds from indole and/or L-tryptophan and has not been completely clarified[1,2].

Use: I. is used to dye both animal as well as plant fibers. It is marketed in the paste form as indigo vat or indigo white, which consist of the sodium salt obtained by reduction of I. with sodium dithionite (indigo vat) or, respectively, the free acid (indigo white, leuko-indigo). When wool or cotton are dipped in a vat containing 0.15–0.2% of dye, the textiles are spontaneously dyed deep blue on hanging in the air, while atmospheric oxygen reoxidizes the leuko-indigo to I.; see also purple. For review of history and technology see *Lit.*[3]. Recently, a bacterium has been found, that reduces indigo[4].

Lit.: [1] Z. Naturforsch. B **23**, 572 (1968). [2] Phytochemistry **29**, 817–819 (1990). [3] Melliand Textilber. **78**, 418–422 (1997). [4] Nature (London) **396**, 225 (1998).

gen.: Beilstein E V **24/8**, 503; **25/9**, 236; **25/16**, 134 ▪ Chem. Unserer Zeit **31**, 121–128 (1997) (industrial synthesis) ▪ Hager (5.) **4**, 463 ff. ▪ Merck-Index (12.), No. 4977 ▪ Ullmann (4.) **13**,

177–182; (5.) **A14**, 149f. ▪ Vogt, Farben u. ihre Geschichte, p. 38–41, Stuttgart: Franckh 1973 ▪ Zollinger, Color Chemistry (2.), Weinheim: VCH Verlagsges. 1991. – *[HS 3204 15; CAS 482-89-3]*

Indolactams. Tumor-promoting compounds, similar to the *teleocidins and lyngbyatoxins (see majusculamides). I. activate protein kinase C (PKC) isoenzymes, e.g., *I. V* from *Streptoverticillium blastmyceticum*: $C_{17}H_{23}N_3O_2$, M_R 301.38, needles, mp. 130–165 °C, $[\alpha]_D^{27}$ –170° (C_2H_5OH).

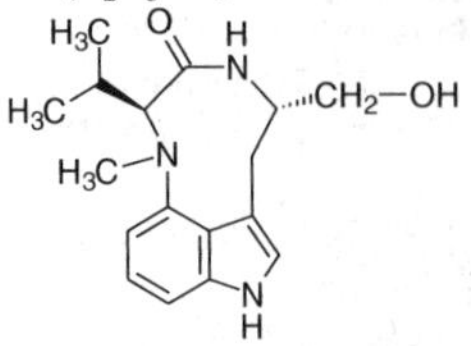

I. V

Lit.: J. Am. Chem. Soc. **118**, 10733–10743 (1996) ▪ J. Med. Chem. **40**, 1316–1326 (1997); **42**, 3436 (1999) (activity) ▪ Tetrahedron Lett. **35**, 8549 (1994). – *[CAS 90365-57-4 (I. V)]*

Indole (1*H*-benzo[*b*]pyrrole).

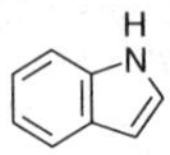

C_8H_7N, M_R 117.15, platelets, unpleasant odor at high concentrations, flower-like odor in high dilution, D. 1.22, mp. 52 °C, bp. 253 °C, bp. 124 °C (0.66 kPa), readily soluble in alcohol, relatively well soluble in water, skin irritant. LD_{50} (rat p.o.) 1 g/kg. I. is steam distillable and can be detected with the so-called *pine splinter reaction*. I. occurs in the free state in jasmine flower oil, in the flowers of the bastard acacia (*Robinia pseudoacacia*), and in cuckoo plant (*Arum maculatum*). The odor of flowering rape fields is due to I., as is also the town gas odor of the toadstool *Tricholoma sulfureum* and other agarics. In cabbage leaves I. is bound to *ascorbic acid. It is also present in the pit coal tar fraction boiling between 240–260 °C (0.2%) and in feces – together with *skatole – as a biological degradation product of *tryptophan contained in most proteins. A partial I. structure biogenetically derived from tryptophan is present in many natural products, see, e.g., indolylalkylamines, indigotin, indole alkaloids, melanins, and serotonin.

Use: I. is used in the perfume industry for the creation of artificial jasmine and neroli oils. It is also a starting material for the synthesis of numerous pharmaceutical agents and pigment dyes.

Lit.: Beilstein E V **20/7**, 5 ▪ J. Am. Chem. Soc. **99**, 3532–3534 (1977) (synthesis) ▪ Keith & Walters, Compendium of Safety Data Sheets for Research and Industrial Chemicals, Part VII, p. 3846f., Weinheim: VCH Verlagsges. 1989 ▪ Ullmann (5.) **A 1**, 384; **A11**, 210 ▪ Zechmeister **17**, 248–297. – *Toxicology:* Sax (8.), ICM000. – *[HS 2933 90; CAS 120-72-9]*

3-Indoleacetic acid see 3-indolylacetic acid.

Indole alkaloids. The I. a. are widely distributed and occur in microorganisms, fungi, animals, and also higher plants. Besides the *isoquinoline alkaloids, the I. a. represent the largest group of alkaloids. In the plant kingdom they occur almost exclusively in Apocynaceae (*Aspidosperma, Catharanthus, Corynanthe, Iboga, Rauvolfia, Vinca*), Loganiaceae (*Strychnos, Gardneria, Gelsemium*), and Rubiaceae (*Uncaria*), all belonging to the order Gentianales, as well as in the Fabaceae (*Physostigma*). Exceptions: indolalkylamines and *ergot alkaloids (from the ascomycete *Claviceps purpurea*, ergot). Parent structures of the I. a. are *indole, indolenine, and indoline, to which further rings may be condensed in the 2,3-positions. The most important groups are: indole-ethylamines, *pyrroloindole alkaloids, *Iboga alkaloids, *ergot alkaloids, β-*carboline derivatives (*harmans, *Vinca and *Rauvolfia alkaloids), carbazole derivatives (Catharanthus, *Aspidosperma, *Strychnos, and Calebasse *curare alkaloids).

Biosynthesis: The indole ring system is supplied by the amino acid *tryptophan, from which the indolalkylamines are directly derived (e.g., *gramine). The diversity of the I. a. arises from additional substituents. In the *harmans this fragment consists of only 2 carbon atoms. However, most I. a. contain iridoid C_{10} fragments, which may sometimes be degraded to 9 C-atoms. On the basis of the additional fragments, the I. a. can be classified into alkaloids with intact (yohimban type) or rearranged (seco-)loganin skeletons (*Aspidosperma or *Iboga alkaloids). The biosynthesis of I. a. without rearranged monoterpene units has been investigated in detail at the enzymatic level. Complete biosynthesis sequences are available for Catharanthus and Rauvolfia alkaloids and have been elucidated on the basis of the isolated enzymes, see monoterpenoid indole alkaloids. In the ergoline alkaloids the basic heterocyclic structure originates from tryptophan and isopentenyl pyrophosphate. The syntheses of I. a. have played a major role for their structure elucidation [2]; the work of Carl Djerassi led to the development of mass spectroscopy as a major method for structural analysis. Many I. a. exhibit pharmacological activities and are used in therapy [3].

Lit.: [1] Heterocycles **25**, 617–640 (1987); Nat. Prod. Chem. **3**, 257–273 (1988). [2] Heterocycles **27**, 1253–1268 (1988); J. Am. Chem. Soc. **110**, 5925 (1988); Manske **31**, 1–28; Nat. Prod. Chem. **3**, 187–213 (1988). [3] Farmaco Ed. Sci. **43**, 1097–1114 (1988); Med. Res. Rev. **8**, 231–308 (1988).

gen.: Alkaloids (London) **13**, 205–276 (1983) ▪ Curr. Org. Chem. **2**, 63–90 (1998) ▪ Hesse, Indolalkaloide, Weinheim: Verl. Chemie 1974 ▪ Luckner (3.), p. 340–359 ▪ Manske **52**, 104–196 ▪ Mothes et al., p. 273–307 ▪ Phillipson et al. ▪ Rodd's Chem. Carbon Compds. (2.) **4B**, 69–164 (1997) ▪ Zechmeister **24**, 13–53; **26**, 1–51; **31**, 469–520; **50**, 27–56.

Indole alkaloids from cyanobacteria. A series of related indole alkaloids bearing isonitrile and isothiocyanate groups has been isolated from terrestrial cyanobacteria (blue-green algae), e.g., hapalindoles [1], Fischer indoles [2], welwitindolinones [3], that are biogenetically derived from *3-((*Z*)-2-isocyanovinyl)indole [4]. Addition of the vinyl group of the latter to a cyclic chloronium ion of a linear monoterpene gives rise to *12-epi-hapalindole-E isonitrile* (a) {$C_{21}H_{23}ClN_2$, M_R 338.88, $[\alpha]_D$ +42.9° (CH_2Cl_2)} which can then be transformed to other indoles and oxindoles [3]. *N-Methylwelwitindolinone-C isothiocyanate* (b) {$C_{22}H_{21}ClN_2O_2S$, M_R 412.93, $[\alpha]_D$ –278° (CH_2Cl_2)} eliminates the multiple-drug resistance of vinblastine-resistant adenocarcinoma cells and has insecticidal activity [3]. *12-epi-Fischer indole-G isonitrile* (c) {$C_{21}H_{23}ClN_2$, M_R 338.88, $[\alpha]_D$ +67° (CH_2Cl_2)} [3], 12-*epi*-hapalindole-E isonitrile,

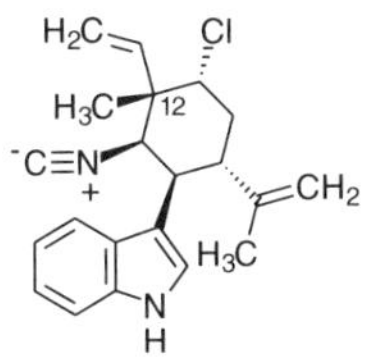

12-*epi*-Hapalindole-E isonitrile (a)

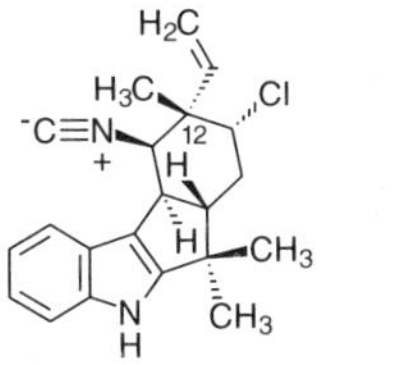

12-*epi*-Fischer indole-G isonitrile (c)

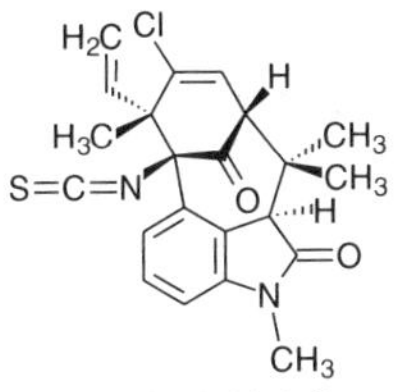

N-Methylwelwitindolinone-C isothiocyanate (b)

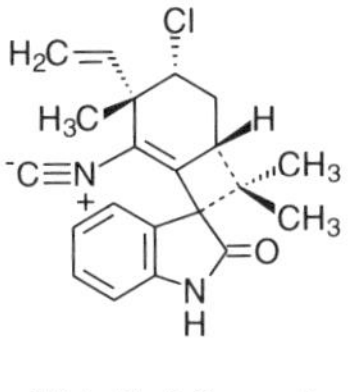

Welwitindolinone-A isonitrile (d)

and *welwitindolinone-A isonitrile* (d) {$C_{21}H_{21}ClN_2O$, M_R 352.86, $[\alpha]_D$ +377° (CH_2Cl_2)}[3] are potent fungicides; see hapalindoles.

Lit.: [1] J. Org. Chem. **52**, 1036, 3704 (1987); J. Ind. Microbiol. **5**, 113 (1990). [2] J. Am. Chem. Soc. **106**, 6456 (1984); J. Org. Chem. **52**, 1036 (1987); Tetrahedron Lett. **33**, 3257 (1992). [3] J. Am. Chem. Soc. **116**, 9935 (1994). [4] J. Antibiot. **19**, 850 (1976). – *[CAS 159249-51-1 (a); 159189-05-6 (b); 159249-50-0 (c); 159934-03-9 (d)]*

Indolizidine alkaloids. A group of *alkaloids from various classes based on the bicyclic ring system of indolizidine. They differ in structure, biogenesis, and occurrence; thus, the indolizidine ring system has very little value as a characterizing structural. The I. a. include the *pumiliotoxins (Dendrobates alkaloids), polyhydroxy alkaloids such as *swainsonine and *castanospermine, *Elaeocarpus alkaloids, *Securinega alkaloids, *Tylophora alkaloids, and the *Ipom(o)ea alkaloids.

Lit.: Manske **28**, 183–308; **31**, 193–315; **44**, 189–257 ▪ Nat. Prod. Rep. **16**, 697 (1999). – *Synthesis:* Heterocycles **25**, 659–700 (1987) ▪ J. Org. Chem. **62**, 8182 (1997) ▪ Stud. Nat. Prod. Chem. **1**, 227–304 (1988).

Indolmycin.

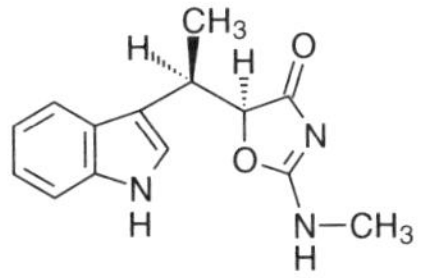

$C_{14}H_{15}N_3O_2$, M_R 257.29, prisms, mp. 209–210 °C, $[\alpha]_D$ −214° (CH_3OH), an antibiotic isolated from *Streptomyces albus* or *S. griseus*. I. is active against Gram-positive bacteria (including mycobacteria), it is an antagonist of L-tryptophan at the stage of aminoacyl-tRNA synthesis.

Lit.: Exp. Opin. Ther. Patents **9**, 1021 (1999) ▪ J. Org. Chem. **35**, 3519 ff. (1970). – *Biosynthesis:* J. Am. Chem. Soc. **99**, 273 f. (1977) ▪ J. Antibiot. **34**, 551–554 (1981). – *Synthesis:* Chem. Pharm. Bull. (Jpn.) **38**, 323–328 (1990) ▪ J. Org. Chem. **51**, 4920 (1986) ▪ Tetrahedron Lett. **37**, 6447 (1996). – *[HS 2941 90; CAS 21200-24-8]*

Indolo[2,3-*a*]carbazoles see Arcyria pigments.

Indolo[2,1-*a*]isoquinoline alkaloids see dibenzopyrrocoline alkaloids.

3*H*-Indol-3-one (dehydroindoxyl).

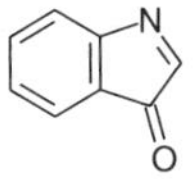

C_8H_5NO, M_R 131.13, dark, orange-red prisms, mp. 110–112 °C (decomp.). The Asian edible fungus *Pleurotus salmoneostramineus* (Basidiomycetes) needs light for its growth. Its attractive pink color is due to a *chromoprotein (uv_{max} 496, 344, 311, 272 and 217 nm) with the previously unknown 3*H*-I. as chromophore. In addition, the pigment contains Zn, Fe, and Cu as well as galactose chains. The chromoprotein on illumination generates oxygen from water; this may be of physiological importance for the fungus (photosynthesis without chlorophyll!). An electron transfer from coordinated water to photochemically excited 3*H*-I. in an Fe(III) complex appears to proceed, through which oxygen is formed from water.

Lit.: J. Am. Chem. Soc. **116**, 8849 (1994). – *[CAS 67285-12-5]*

3-Indolylacetic acid (3-indoleacetic acid, heteroauxin, IAA).

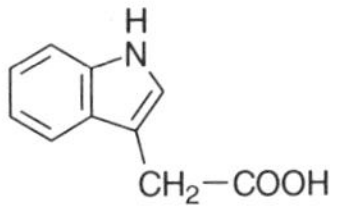

$C_{10}H_9NO_2$, M_R 175.19, mp. 166 °C. I. is a natural growth substance of higher plants to promote longitudinal growth (see auxins, plant growth substances).

Biological activity: Besides promoting longitudinal growth, I. also plays a role in root formation and in the development of plant diseases. Corn roots react to as little as 10^{-12} g 3-indolylacetic acid. When the optimal effective concentration is exceeded an inhibitory reaction occurs. Recently, the so-called "auxin-binding proteins" have been controversially discussed as possible auxin receptors. I. effects a change in the gene expression pattern by an as yet unknown signal cascade.

I. is used as a plant growth regulator. For synthesis, see *Lit.*[1]. The biosynthetic precursor of I. is considered to be L-*tryptophan[2]. Biosynthetic studies have been performed with *Agrobacterium tumefaciens* transformed plant cell cultures[3]. For detection, see *Lit.*[4].

Lit.: [1] J. Heterocycl. Chem. **29**, 953–958 (1992); J. Am. Chem. Soc. **116**, 3127 (1994). [2] Aust. J. Plant Physiol. **20**, 527–539 (1993). [3] Plant Growth Regul. **10**, 313–327 (1991). [4] Pharis, Rood & Stewart (eds.), Plant Growth Subst., p. 441–449, Berlin: Springer 1990.

gen.: Beilstein E V **22/3**, 65 f. ▪ Sax (8.), p. 1982 f. (toxicology) ▪ Ullmann (4.) **13**, 207 f.; (5.) **A20**, 415, 417 ▪ Zechmeister **17**, 248–297. – *Plant physiology:* Annu. Rev. Phytopathol. **31**, 253–273 (1993) ▪ Bot. Mag. **103**, 345–370 (1990) ▪ Curr. Plant Sci. Biotechn. Agric. **13**, 1–12 (1992) ▪ Kung & Wu (eds.), Transgenic Plants, vol. 1, p. 195–223, San Diego: Academic Press 1993 ▪ Plant Growth Regul. **13**, 77–84 (1993) ▪ Proc. Natl. Acad. Sci. USA **90**, 11442 ff. (1993). – *[HS 2933 90; CAS 87-51-4]*

Indolylalkylamines.

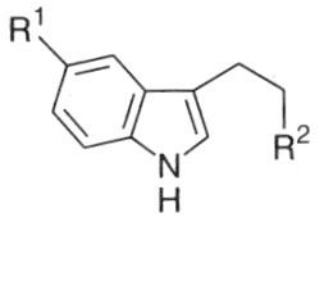

	R^1	R^2
Tryptamine (**1**)	H	NH_2
*Serotonin (**2**)	OH	NH_2
Bufotenine (**3**)	OH	$N(CH_3)_2$
Bufotenidine (**4**)	O^-	$N(CH_3)_3^+$
Bufoviridine (**5**)	OSO_3H	$N(CH_3)_2$

Table: Data of Indolylalkylamines.

	molecular formula	M_R	mp. [°C]	CAS
1	$C_{10}H_{12}N_2$	160.22	284 (hydrochloride)	61-54-1
2	$C_{10}H_{12}N_2O$	176.22	167–168 (hydrochloride)	50-67-9
3	$C_{12}H_{16}N_2O$	204.27	146–147	487-93-4
4	$C_{13}H_{18}N_2O$	218.30	209 (hydroiodide)	487-91-2
5	$C_{12}H_{16}N_2O_4S$	284.34		16369-08-7

I. are a group of natural products with interesting biological properties occurring in fungi, higher plants, and animals; they are closely related to the hallucinogenic components of the Mexican toxic fungus *Psilocybe mexicana* (see psilocybine). I. are also widely distributed in economically important and food plants, especially in fruits such as pineapples, oranges, plums, tomatoes. *Serotonin occurs in the hairs of the stinging nettle, as well as in Central and South American Euphorbiaceae, and in Fabaceae (e. g., the tropical velvet bean *Mucuna pruriens*).

Lit.: Sax (8.), DPG 109. – *Bufotenine:* Br. J. Pharmacol. **69**, 597 (1980) ▪ Martindale (30.), p. 1345.

Indospicine [(*S*)-6-amidino-2-aminohexanoic acid, 6-amidino-L-norleucine].

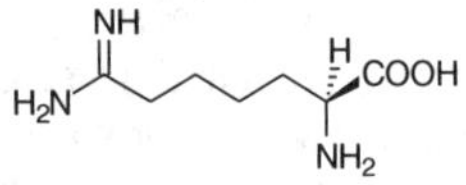

$C_7H_{15}N_3O_2$, M_R 173.22, monohydrate of the monohydrochloride, mp. 131–134 °C, $[\alpha]_D^{22}$ +18° (5 M HCl). A toxic amino acid from the creeping indigo plant (*Indigofera spicata*) and related species (Fabaceae)[1]. It is hepatotoxic and teratogenic, potently inhibits the aminoacylation of arginine[2], and is a competitive inhibitor of arginase[3]. It also prevents the incorporation of thymidine in the DNA in cultured human red blood cells. For synthesis, see *Lit.*[4].

Lit.: [1]Nature (London) **217**, 354 f. (1968); Aust. J. Biol. Sci. **23**, 831–842 (1970); J. Chromatogr. **135**, 377–384 (1977). [2]Biochem. Pharmacol. **19**, 853–857 (1970). [3]Biochem. Pharmacol. **19**, 2391 f. (1970). [4]Bioorg. Med. Chem. Lett. **6**, 111 (1996); J. Labelled Compd. Radiopharm. **24**, 1273 (1987).
gen.: J. Dermatol. **14**, 35–38 (1973) ▪ Rosenthal, Plant Nonprotein Amino and Imino Acids, p. 91–95, New York: Academic Press 1982. – *[CAS 16377-00-7]*

Ineketone see rosanes.

Infochemicals see semiochemicals.

Infractines.

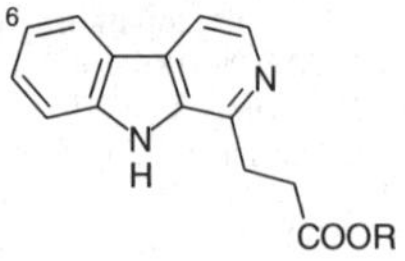

R = H, CH3

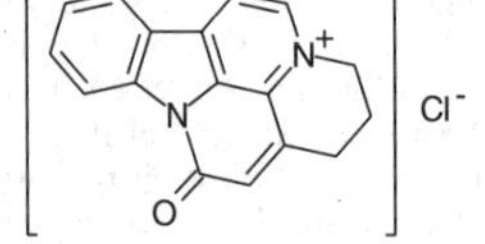

Infractopicrine

A group of carboline alkaloids, e. g., *β-carboline-1-propanoic acid* (a) ($C_{14}H_{12}N_2O_2$, M_R 240.26, needles, mp. 215 °C) from the wood of *Picrasma quassioides* (Simaroubaceae)[1] and its methyl ester *I.* (b) ($C_{15}H_{14}N_2O_2$, M_R 254.29, cryst., mp. 145–146 °C), from the fruit bodies of the agaric *Cortinarius infractus* (Basidiomycetes) have been isolated[2]. Besides I. which shows light blue fluorescence under UV light, *C. infractus* also contains the yellow fluorescent *6-hydroxyinfractine* and the bitter principle *infractopicrine* ($C_{17}H_{13}ClN_2O$, M_R 296.76)[2], which possesses two additional rings.

Lit.: [1]Chem. Pharm. Bull. **32**, 3579 (1984); Phytochemistry **23**, 453 (1984). [2]Tetrahedron Lett. **25**, 2341 (1984).
gen.: Synthesis: Justus Liebigs Ann. Chem. **1993**, 153–159 ▪ Pharmazie **50**, 182 (1995). – *[CAS 89915-39-9 (a); 91147-07-8 (b, cation); 91147-09-0 (infractopicrine)]*

Ingenane see Euphorbiaceae and Thymelaeaceae poisons.

Ingenol (3*β*,4*β*,5*β*,20-tetrahydroxy-1,6-ingenadien-9-one).

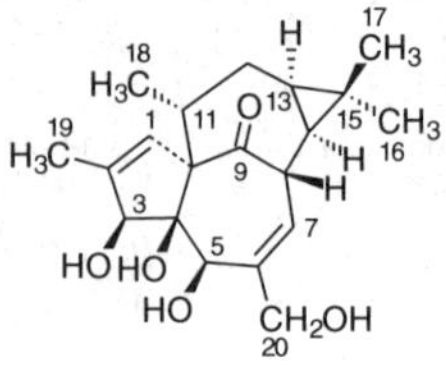

$C_{20}H_{28}O_5$, M_R 348.44, an amorphous, tetracyclic, unsaturated diterpene with one keto and four hydroxy groups, derived from ingenane (see Euphorbiaceae and Thymelaeaceae poisons).

Occurrence: I. occurs as the O-3-, O-5-, or O-20-monoesters and as the 3,20-*O*-diester in many Euphorbiaceae, e. g., *Euphorbia ingens*. The acid components are mainly highly unsaturated carboxylic acids. Esters of 5- and 20-deoxyingenol, 13- and 16-hydroxyingenol, and of 16-hydroxy-20-deoxyingenol have been isolated.

Biological activity: Like phorbol, I. does not exhibit any irritating and cocarcinogenic activities. The esters with the acid component at O-3, on the other hand, are effective cocarcinogens. In the phorbol esters it has been demonstrated that the free allylic hydroxy group at C-20 is important for the biological activity; however, the esters of 16-hydroxy-20-deoxyingenol are also cocarcinogenic. In this case, the quasi-allylic hydroxy group at C-16 must take over the biological function of the 20-hydroxy group.

Lit.: J. Nat. Prod. **49**, 386 (1986) ▪ Z. Naturforsch. B **46**, 1425 (1991) ▪ Zechmeister **44**, 58. – *[CAS 30220-46-3]*

Inositols (cyclohexane-1,2,3,4,5,6-hexaol, abbreviation: Ins).

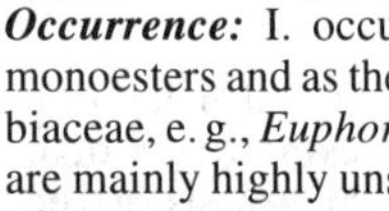

myo-Inositol

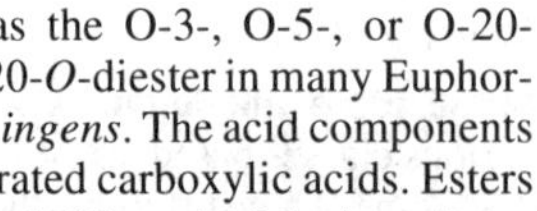

Phytic acid

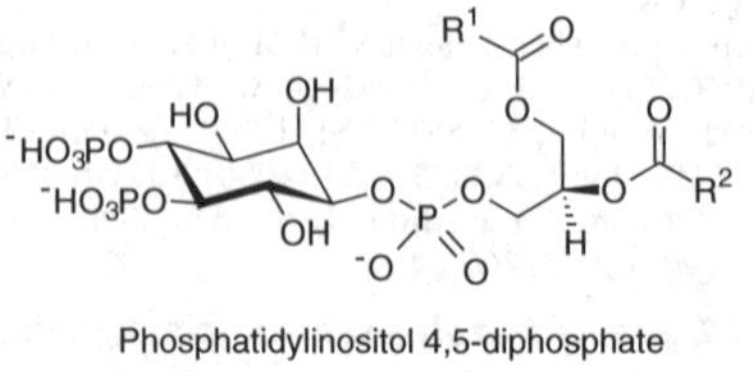

Phosphatidylinositol 4,5-diphosphate
(R1, R2 = saturated and unsaturated long-chain fatty acids)

$C_6H_{12}O_6$, M_R 180.16. A group of hexafunctional cyclic alcohols (cyclitols) comprising 9 stereoisomers. The most important is *myo-inositol* (sweet-tasting cryst., mp. 225–227 °C) which occurs in the free form in muscle tissue (name from Greek: 'mys', genitive 'myos' and 'is', genitive 'inos', both: muscle) as well as in many plants and animal organs. The human body contains ca. 40 g; *myo*-I. is important for many organisms as a growth factor. Human requirements are mainly covered from fruit and cereal products in which *myo*-I., as in other plant organs, is present in the form of *myo*-I. hexakisphosphate (InsP6, *phytic acid*, $C_6H_{18}O_{24}P_6$, M_R 660.03). InsP6 accelerates the release of oxygen from hemoglobin. Other *myo*-I. phosphates, especially *phosphatidylinositol 4,5-diphosphate*, act in animal cells as *second messengers*, phosphatidylinositols are contained as phospholipides in membranes.

Lit.: Annu. Rev. Biochem. **56**, 159–193 (1987); **61**, 225–250 (1992) ▪ Beilstein E IV **6**, 7919 ff., 7927 ▪ Hager (5.) **4**, 8, 397, 545 f.; **5**, 662 ▪ Karrer, No. 285 ▪ Nat. Prod. Rep. **7**, 1–24 (1990). – *[HS 2919 00, 2906 13; CAS 39907-99-8 (myo-I.); 83-86-3 (phytic acid)]*

Insect attractants. I. a. are sex and aggregation hormones produced by the insects themselves and belong to the *pheromones such as *anastrephin, *bombykol, *brevicomin, *chalcogran, *cucujolides, *disparlure, *dominicalure, *epoxy polyenes, *ferrulactone, *frontalin, *grandisol, *hepialone, *ipsdienol, *lardolure, *lineatin, *matsuone, *muscalure, *multistriatin, *olean, *periplanones, *pityol, *quadrilure, *rhynchophorol, *serricornin, *stegobinone, *sulcatol, verbenol (see pinenols), etc. The I. a. excreted as odor substances by females or males of various insect species have attracting effects on sexual partners even at high dilutions and they are able to follow such traces to the origin by means of their highly developed sense of smell (aggregation pheromones effect both sexes). It is not uncommon that, besides their species-specific effects, I. a. also have an attracting effect on other species including even predatory species – One then speaks of *kairomones. Certain orchid species are interesting from a coevolutionary point of view. They use I. a. to induce the insects to copulate on their flowers which are thus pollinated.

Chemically, I. a. are often alcohols or carboxylic acids and esters derived from alkenes or alkadienes; but also terpenoid or heterocyclic compounds may have I. a. activity. In many cases the ratio of *cis/trans*-isomers or, in the case of chiral compounds, the enantiomer ratio often plays a decisive role for the effectiveness of I.a. Many other substances occurring in the environment also have an attracting effect on insects, especially those associated with the search for food; *examples:* the typical odors of certain plants (e.g., arum), flower and fruit odors, emanations from animals and humans [CO_2, lactic acid, *(–)-(*R*)-1-octen-3-ol] or carrion.

The actions of I. a. are used in crop protection: as decoy substances they attract insects either for subsequent eradication (insect traps) or for population control before or after treatment with insecticides or for collection for scientific purposes. In pest control with I. a. both species-specific pheromones and natural or synthetic foreign substances, provided that they possess insect-attracting properties, are used. *o*-Methyleugenol, anisylacetone, *geraniol/*eugenol have proved to be useful in various mixtures as I. a. for certain species. *4-(p-Hydroxyphenyl)-2-butanone* ($C_{10}H_{12}O_2$, M_R 164.20) and its acetate are commercially available as I. a. for the Mediterranean fruit fly *Ceratitis capitata*; see also pheromones, allomones, synomones, kairomones, semiochemicals.

Lit.: Chem. Br. **1990**, 124–127 ▪ Jutsum & Gordon (eds.), Insect Pheromones in Plant Protection, Chichester: Wiley 1989 ▪ Kydonieus & Beroza (eds.), Insect Suppression with Controlled Release Pheromone Systems, vols. I & II, Boca Raton: CRC-Press 1982 ▪ Nachr. Chem. Tech. Lab. **37**, 478–483 (1989) ▪ Naturwissenschaften **69**, 457 (1982) ▪ Planta Med. **55**, 333–338 (1989) ▪ Ridgway, Silverstein & Inscoe (eds.), Behavior-modifying Chemicals for Insect Management, New York: Dekker 1990 ▪ Tetrahedron **45**, 3233–3298 (1989) ▪ Ullmann (5.) **A 14**, 308 f. – *[CAS 5471-51-2 (4-(p-hydroxyphenyl)-2-butanone)]*

Insect defence substances see alarm substances, allelochemicals, semiochemicals, warning substances (odors), defensive secretions.

Insect hormones. Of particular importance are those I. h. that regulate the sloughing (molting) processes (metamorphosis of insects). Sloughing is initiated by expression of a neurosecretion; this so-called *prothoracotropic hormone* from brain ganglions stimulates a thorax gland to express *ecdysone* (see ecdysteroids), the genuine *sloughing hormone* which triggers pupal sloughing (larva → pupa) and the imaginal sloughing (pupa → insect). Since it was first assumed that different I. h. were responsible for these two stages terms such as molting hormone and hatching hormone were coined. However, the two functions are in fact due to the same sloughing hormone (*ecdysone, ecdysterone, inokosterone*, see ecdysteroids). Ecdysterone is identical with the sloughing hormone of crabs (*crustecdysone*). *Juvenile hormone formed in the corpora allata first acts against ecdysone allowing the larva to change its skin (small larva → large larva) but inhibiting metamorphosis. Formation of juvenile hormone decreases in the last stage of the larva development and the now dominating secretion of ecdysone effects metamorphosis (larva → pupa). When insect development is complete juvenile hormone is again formed which stimulates as *gonadotropic* hormone the ripening of eggs; see also juvenile hormones, insect attractants, pheromones. For a review of insect neuropeptides with hormonal activity that represent targets for novel insecticides, see *Lit.*[1].

Lit.: [1] Zechmeister **71**, 1–128.
gen.: Chem. Labor Biotech. **45**, 11–16 (1994) ▪ Gilbert, Tata & Atkinson (eds.), Metamorphosis, p. 59–107, 223–251, San Diego: Academic Press 1996.

Insect venoms. Substances excreted by many species of insects for protection or predatory purposes that are more or less toxic for other organisms. They are produced in various skin or body glands and used when required for attack or defence (see also defensive secretions). The compounds are used in many different ways: some species have penetrating (e.g., mosquitoes, bugs) or biting (ants) mouth parts or modified egg-laying stylets on the abdomens (hornets, wasps, bees, see hornet venom, wasp venom, bee venom). Many species of beetle give off a defensive secretion on contact, sometimes from glands (also explosively, see bombardier beetle) or by the direct transfer of hemolymph loaded with venom. The toxic body hairs of caterpillars often contain *histamine. In many species of termites, soldier classes have developed with defence glands sometimes comprising up to 70% of their body volume. Many different classes of compounds are found as active principles, e.g., polypeptides, amino acids, biogenic amines, alkaloids, steroids, quinones (e.g., *hydroquinone), terpenes, aliphatic compounds, etc. Frequently the substances are not produced by the insects themselves but are taken up from plants (e.g., cardenolides, *pyrrolizidine alkaloids) and are sometimes transformed. In other cases the total biosynthesis proceeds in the insect.

Lit.: Bettini, Arthropod Venoms, Berlin: Springer 1977 ▪ Blum, Chemical Defenses of Arthropods, New York: Academic Press 1981 ▪ Whitman et al., in Evans & Schmidt (eds.), Insect Defenses, p. 289–420, Albany: SUNY Press 1990 ▪ see also defensive secretions.

Integerrimine see senecionine.

Intermedine see indicine.

Inthomycin A see phthoxazolin A.

Intybin see lactucin.

Inulin. A linear polymer of ca. 30 β,2-1-linked fructose units. The chain is probably terminated by *glucose (total content: 2–3%); the M_R is ca. 5000. I. is found alone or together with starch as a reserve carbohydrate in dahlia bulbs, artichokes, topinambour tubers, chicory roots, dandelion roots, in the cells of *Inula* species, and other Asteraceae, but less frequently in related plant families (Campanulaceae, Lobeliaceae).

Lit.: Carbohydr. Res. **26**, 401 (1973); **48**, 1 (1976) ▪ Diabetologia **20**, 268–273 (1981) ▪ Ind. Obst Gemüseverwert. **72**, 467 ff. (1987) ▪ Karrer, No. 679 ▪ Martindale (30.), p. 776 ▪ Proc. Biochem. **15**, 2,4,32 (1980) ▪ Stärke/Starch **38**, 91–94 (1986); **39**, 335–343 (1987) ▪ Ullmann (5.) **A 12**, 471. – *[HS 1108 20; CAS 9005-80-5]*

Inumakilactone A see nagilactone C.

Invictolide [(3*R*)-tetrahydro-3α,5β-dimethyl-6α-((*R*)-1-methylbutyl)-2*H*-pyran-2-one].

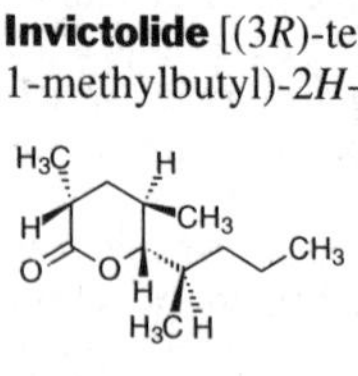

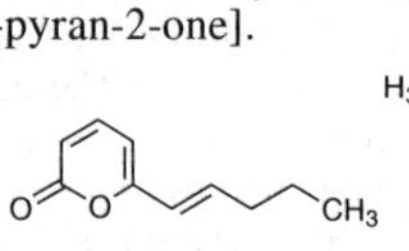

1

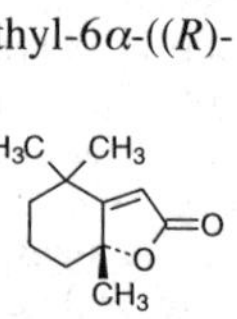

2

$C_{12}H_{22}O_2$, M_R 198.31, mp. 28–28.5 °C, $[\alpha]_D^{25}$ −101° ($CHCl_3$). I., in mixture with *(E)-6-(1-pentenyl)-2H-pyran-2-one* ($C_{10}H_{12}O_2$, M_R 164.20; **1**) and *(R)-dihydroactinidiolide* [$C_{11}H_{16}O_2$, M_R 180.25, $[\alpha]_D^{20}$ −119.9° ($CHCl_3$); **2**], constitutes the recognition signal of the fire ant *Solenopsis invicta*. I. is structurally related with *stegobinone and *serricornin. The biogenesis of I. is presumably based on the coupling of 4 propionate units or 4 methylmalonate units.

Lit.: Naturwissenschaften **82**, 142 (1995). – *Synthesis:* ApSimon **9**, 227 ▪ Heterocycles **26**, 1761 (1987) ▪ J. Org. Chem. **58**, 338 (1993) ▪ Synlett **1993**, 191 ▪ Tetrahedron **52**, 12177 (1996). – *[CAS 103619-04-1 (I.); 38273-56-2 (2)]*

Involutin.

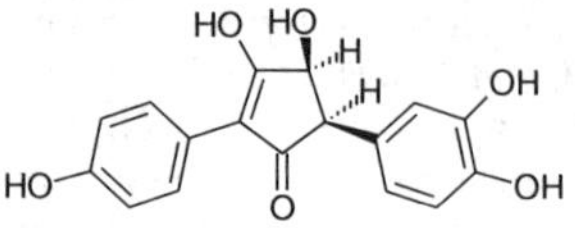

(absolute configuration)

$C_{17}H_{14}O_6$, M_R 314.29, pale yellow needles, mp. 171–174 °C (decomp.), $[\alpha]_D$ −23°. I. occurs in the fruit bodies of *Paxillus involutus* (Basidiomycetes) and is responsible for the brown color that develops when the fungus is bruised; see also chamonixin and gyrocyanin.

Lit.: J. Chem. Soc., Perkin Trans. 1 **1973**, 1529 ▪ Zechmeister **51**, 63–70 (review). – *[CAS 13677-78-6]*

Iodinin (1,6-phenazinediol 5,10-dioxide).

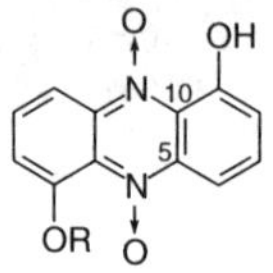

R = H : Iodinin
R = CH_3 : Myxin

$C_{12}H_8N_2O_4$, M_R 244.21, red cryst., mp. 236 °C (decomp.), soluble in DMSO; insoluble in hexane, ethanol. Stable towards acids, unstable towards bases. Antibiotic from *Chromobacterium iodinum* with activity against Gram-positive bacteria, fungi, and yeasts. *Myxin*, the monomethyl ether from *Pseudomonas* sp. [$C_{13}H_{10}N_2O_4$, M_R 258.23 [1], red needles, mp. 120–130 °C (decomp.)] is also active against Gram-negative bacteria. It can explode on drying. The copper complex is used in medicine (*cuprimyxin*).

Biosynthesis: I. is formed from shikimic acid [2]; the 1,6-phenazinediol is formed first and is oxidized in subsequent enzymatic steps to the 5-oxide and then to the 5,10-dioxide. These steps do not occur when the fermentation is performed without a temperature shift (27 °C → 4 °C).

Lit.: [1] Agric. Biol. Chem. **52**, 301–306 (1988). [2] Tetrahedron Lett. **1974**, 4201–4204.

gen.: Beilstein EV **23/13**, 363. – *[HS 2941 90; CAS 68-81-5 (I.); 13925-12-7 (myxin)]*

Iodoamino acids.

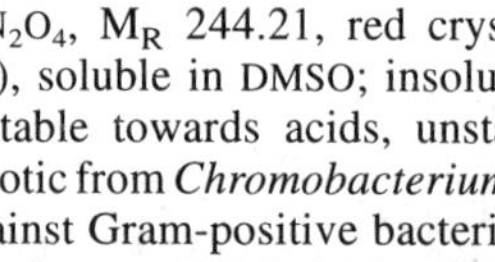

3,5-Diiodo-L-tyrosine

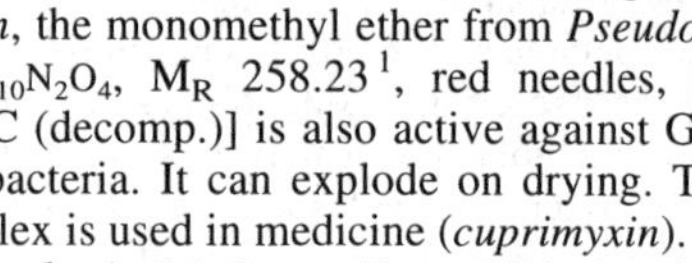

$R^1 = R^2 = H$: 3,5-Diiodo-L-thyronine
$R^1 = I$, $R^2 = H$: 3,3',5-Triiodo-L-thyronine
$R^1 = R^2 = I$: L-Thyroxine

General name of iodine-containing amino acids. The most important members are the *thyroid hormones* 3,3',5-triiodo-L-thyronine (T_3, triiodothyronine) and 3,3',5,5'-tetraiodo-L-thyronine (T_4, thyroxine).

The first I. isolated was *3,5-diiodo-L-tyrosine* [iodogorgic acid, $C_9H_9I_2NO_3$, M_R 432.98, mp. ca. 200 °C (decomp.)] from the alkaline hydrolysate of coral polyps, it is the biosynthetic precursor of T_3 and T_4. A further derivative is *3,5-diiodo-L-thyronine* [$C_{15}H_{13}I_2NO_4$, M_R 525.08, mp. 256–257 °C (decomp.)] with activity similar to that of thyroxine. Pure thyroxine is isolated from the $Ba(OH)_2$ hydrolysate of thyroid gland tissue. Although various synthetic I. are known, *L-thyroxine* ($C_{15}H_{11}I_4NO_4$, M_R 776.87, mp. 235–236 °C) and *triiodo-L-thyronine* ($C_{15}H_{12}I_3NO_4$, M_R 650.98, mp. 236–237 °C) are the most important thyroid hormones. T_3 is present in a much smaller amount as T_4 but has a five-times higher activity. Aromatic substitution by iodine occurs in the polypeptide chain. The major part of the thyroid hormones exists as protein-bound "thyroxine-binding globulin" (TBG) and is proteolysed and released into the blood when required after hormonal stimulation.

Activity: Thyroid hormones influence growth and maturation processes in animals. Synthetically prepared I. are used therapeutically for thyroid diseases.

Lit.: Biochem. Int. **10**, 803 (1985) ▪ Cooke et al., Hormones and their Actions I (New Comprehensive Biochemistry), vol. 18A, p. 61 ff., Amsterdam: Elsevier 1988 ▪ Hager (5.) **7**, 1246; **8**, 729–735; **9**, 907 ff. ▪ Neurol. Neurobiol. **85**, 425 (1984). – *[HS 2922 50, 2937 99; CAS 66-02-4 (3,5-diiodotyrosine); 534-51-0 (3,5-diiodothyronine); 6893-02-3 (triiodothyronine); 51-48-9 ((S)-thyroxine)]*

Ionomycin.

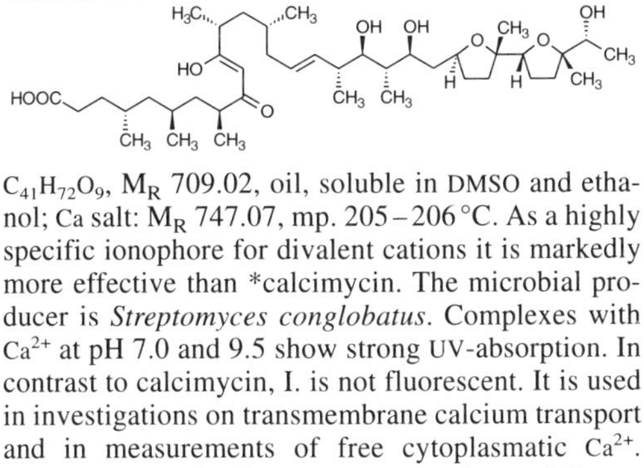

$C_{41}H_{72}O_9$, M_R 709.02, oil, soluble in DMSO and ethanol; Ca salt: M_R 747.07, mp. 205–206 °C. As a highly specific ionophore for divalent cations it is markedly more effective than *calcimycin. The microbial producer is *Streptomyces conglobatus.* Complexes with Ca^{2+} at pH 7.0 and 9.5 show strong UV-absorption. In contrast to calcimycin, I. is not fluorescent. It is used in investigations on transmembrane calcium transport and in measurements of free cytoplasmatic Ca^{2+}. The ion specificities are reported as follows: $Ca^{2+}>Mg^{2+}\gg Sr^{2+}\sim Ba^{2+}$.

Lit.: J. Am. Chem. Soc. **101**, 3344 (1979); **112**, 5276–5313 (1990) ▪ J. Antibiot. **31**, 815 (1978). – *[CAS 56092-81-0 (I.); 56092-82-1 (Ca salt)]*

Ionones.

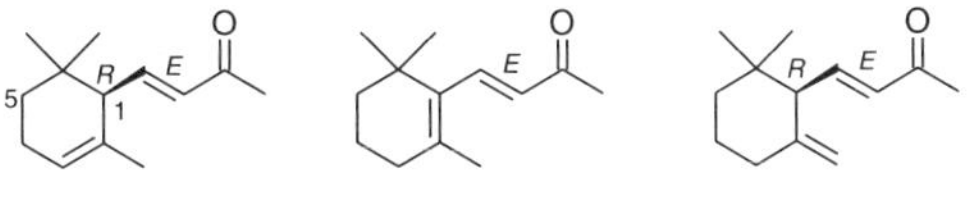

(+)-α-Ionone β-Ionone (Boronione) γ-Ionone

$C_{13}H_{20}O$, M_R 192.30, a group of natural products with a violet-like odor formed by oxidative degradation of tetraterpenoids (carotenes)[1] (see also irones).

I. are widely distributed in vegetables and fruits, especially in berries, tea, and tobacco[2]. I. are the main components (22%) of violet flower oil. Both optical antipodes are found[3]. The dominating form in essential oils of costus roots, *Lawsonia* species, and in flower oil of *Boronia megastigma* is β-ionone. γ-I. is a component of the oil of *Tamarindus indica.* The I. are, together with the *damascones and *damascenones, organic compounds with the strongest odors, β-I. can still be detected in air at 3×10⁹ mol. per mL (0.1 ppb)[4]. There is at present still no convincing explanation for the drastic differences in the olfactory properties of the isomers. For synthesis, see *Lit.*[5–7]. Synthetic I. under various proprietary names are used in the manufacturing of soap, perfumes, and products with a violet-like odor. Furthermore, the readily available I. are used as starting materials, e. g., for the syntheses of vitamin A, damascones[8], and isomethylionones. Among the compounds in this group, perfumists attribute the finest iris/violet odor to α-isomethylionone[9].

Table: Data of Ionones.

	bp. °C (Pa)	$[\alpha]_D^{20}$	CAS
α-I.	80.2 (133)	+347°	24190-29-2
β-I.	110 (718)	–	79-77-6
γ-I.	–	–15.6° (C_2H_5OH)	24190-32-7

Lit.: [1] Zechmeister **35**, 431. [2] Karrer, No. 531, 532; Ohloff, p. 117; Theimer, p. 285; Beilstein E IV **7**, 361; Food Cosmet. Toxicol. **13**, 549 (1975). [3] Agric. Biol. Chem. **51**, 1271 (1987); Helv. Chim. Acta **75**, 1023 (1992). [4] Perfum & Flavor **3**, 11 (1978). [5] Theimer, p. 292 ff. [6] Perfum & Flavor **9**, 19 (1984/85). [7] Theimer, p. 300 ff.; Aust. J. Chem. **48**, 145 (1995); J. Chem. Soc., Perkin Trans. 1 **1998**, 4129 (enantiomers of α-I.); **1999**, 271; Tetrahedron Lett. **34**, 3417 (1993) (γ-I.). [8] Helv. Chim. Acta **56**, 310 (1973). [9] Ohloff, p. 118; Food Cosmet. Toxicol. **14**, 329 (1976).

Ionophores. Name introduced by Pressman in 1967 from ion and ...phore (Greek: phorós = carrying) for mostly macrocyclic compounds with M_R usually < 2000 that reversibly form chelates or complexes with ions and, as a result of their relatively hydrophobic structural features, carry the ions through otherwise ion-impermeable biological membranes. Naturally occurring I. include some macrolides and *peptide antibiotics (e.g., *enniatins, *nonactin, *valinomycin), polyether antibiotics (such as *lasalocids, *monensin, *nigericin, *salinomycin®), and *siderophores. Synthetically prepared I. are even more effective for many purposes, e.g., crown ethers, open-chain polyethers, and derivatives of ethylenediaminetetraacetic acid. While the above-mentioned macrolide antibiotics are able to incorporate metal ions, some other I. (e. g., *gramicidins A[1], *melittin) form oligomeric channels in which the ions can be passed through membranes – these are referred to as *quasi-ionophores.* They differ from the membrane-permeable ion channels consisting of (glyco-)proteins mainly in terms of their lower relative molecular masses (see table, p. 320).

Lit.: [1] Science **241**, 145, 230 (1988).

gen.: Alberts et al., Molecular Biology of the Cell, 3rd edn., p. 511 f., New York: Garland 1994 ▪ Annu. Rev. Biophys. Chem. **19**, 127 (1990) ▪ Arch. Biochem. Biophys. **287**, 18 (1991) ▪ Doblem, Ionophores and Their Structures, New York: Wiley 1981 ▪ Westley, Polyether Antibiotics, vol. I, II, New York: Dekker 1983.

Ipalbidine, ipalbine see ipom(o)ea alkaloids.

Table: Properties of the most important Ionophores.

Ionophore	M_R	solubility	ion selectivity
*Calcimycin (A23187) [*Streptomyces chartreusensis*]	523.63	DMSO, methanol, chloroform, ethyl acetate	high selectivity for divalent cations: $Mn^{2+}>Ca^{2+}>Mg^{2+}\gg Sr^{2+}>Ba^{2+}>Li^+>Na^+>K^+$
Alamethicin [*Trichoderma viride*]	1964.33	methanol	channel-forming peptide ionophore for monovalent cations: $H^+>Cs^+=Rb^+=Na^+=Li^+$
*Enniatin B [*Fusarium orthoceras* var. *enniatum*]	639.83	chloroform, petroleum ether	ionophore antibiotic binding to calmodulin and inhibiting phosphodiesterase in a Ca^{2+}-dependent manner
*Gramicidin A [*Bacillus brevis*]	1883.33	methanol, ethanol, pyridine, acetone	antibiotic peptide ionophore forming channels for H^+ and alkali cations
*Indanomycin (X-14547 A) [*S. antibioticus*]	493.69	ethyl acetate, methanol	ionophore particularly for K^+; but Rb^+ and Cs^{2+} are transported across the membrane as well; binding mono-, di-, and trivalent cations
Ionomycin, acid [*S. conglobatus*]	709.01	DMSO, ethanol	highly specific for divalent cations: $Ca^{2+}>Mg^{2+}\gg Sr^{2+}=Ba^{2+}$
*Monensin, Na salt [*S. cinnamonensis*]	692.86	methanol, ethanol, ethyl acetate, chloroform, DMSO	forms stable complexes with monovalent cations: $Na^+>K^+>Rb^+>Cs^+>Li^+>NH_4^+$
*Nigericin, Na salt [*S. hygroscopicus*]	746.95	chloroform	destroys membrane potential; $K^+>Rb^+>Cs^+>Na^+$
*Palytoxin [*Palythoa caribaeorum*]	2680.17	DMSO, methanol, ethanol, pyridine	polyether toxin acting as ionophore for divalent cations; induces depolarization of excited membranes, inhibits Na^+-K^+-ATPase
*Salinomycin, Na salt [*S. albus*]	772.99	ethanol, DMF	polyether antibiotic exhibiting a rare tricyclic spiroketal ring system; K^+-specific ionophore
*Valinomycin [*S. fulvissimus*]	1111.33	ether, chloroform, acetone	cyclodepsipeptide ionophore: $Rb^+>K^+>Cs^+>Ag^+>NH_4^+>Na^+>Li^+$

Ipecac alkaloids.

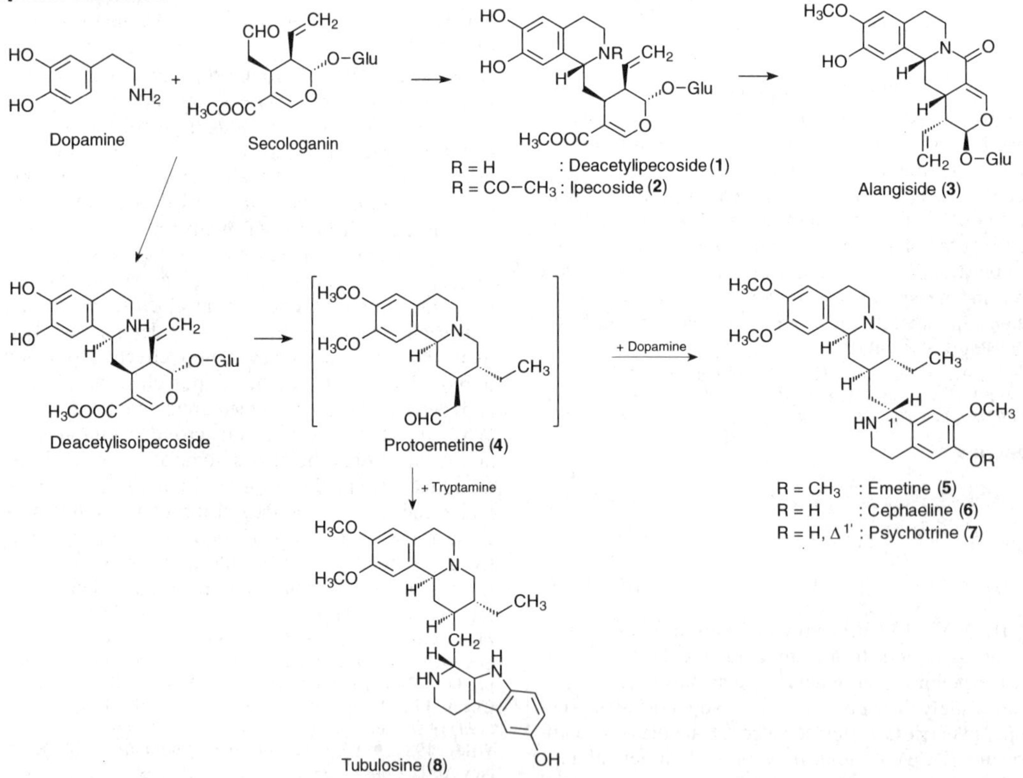

Figure: Biosynthesis of Ipecac alkaloids.

The I. a. occur mainly in the rhizomes and roots of the South American *Cephaelis ipecacuanha* (ipecac, Rubiaceae) or the Central American *Psychotria granadensis*. *Cephaeline* is also found together with alkaloid glucosides in the Indian *Alangium lamarckii* (Alangiaceae). The I. a. are readily soluble in ethanol, acetone, ethyl acetate, ether and chloroform. Parenterally, *emetine* causes damage to the cardiac muscles through to paralysis and leads to hemorrhagic diarrhea. On absorption severe liver damage, paralysis of the spine and brain have been described. It exhibits antiviral, antitumor, and antibacterial activities and inhibits protein biosynthesis by preventing attachment of the aminoacyl t-RNA molecule to the 60 S subunits of eukaryotic ribosomes. It is a highly toxic oral emetic, the lethal dose for humans is ca. 1 g; LD_{50} (mouse i.p.) 12 mg/kg. Cephaeline has similar activities. Tubulosine from *Alangium lamarckii* is toxic and has amoebicidic and antitumor activities.

Table: Data of Ipecac alkaloids.

	molecular formula	M_R	mp. [°C]	$[\alpha]_D$	CAS
2	$C_{27}H_{35}NO_{12}$	565.57	174–175		15401-60-2
3	$C_{25}H_{31}NO_{10}$	505.52	187	−105° (CH_3OH)	34482-51-4
4	$C_{19}H_{27}NO_3$	317.43			549-91-7
5	$C_{29}H_{40}N_2O_4$	480.65	74	−46.55° ($CHCl_3$)	483-18-1
6	$C_{28}H_{38}N_2O_4$	466.62	115–116	−43.4° ($CHCl_3$)	483-17-0
7	$C_{28}H_{36}N_2O_4$	464.60	138	+80.2° (CH_3OH)	7633-29-6
8	$C_{29}H_{37}N_3O_3$	475.63	239–261		2632-29-3

For synthesis and structural elucidation of emetine, see *Lit.*[1–3].

Biosynthesis[4]***:*** The I. a. consist of one or two isoquinoline units and one monoterpenoid unit. Accordingly, geraniol – via loganin and secologanin – and dopamine – after condensation with secologanin and formation of deacetylipecoside and deacetylisoipecoside – have been identified as precursors. Deacetylisoipecoside is the precursor of cephaeline and emetine, probably with protoemetine as an intermediate. The 1*β*-epimer, deacetylipecoside, is the precursor of the alkaloid glucosides ipecoside and alangiside. A similar route has been found for the biosynthesis of tubulosine in young *Alangium lamarckii*. Deacetylisoipecoside is formed from tyrosine via condensation of dopamine with secologanin and then converted with a tryptamine unit to deoxytubulosine and tubulosine.

Lit.: [1]Beilstein EV **23/13**, 611f.; Chem. Pharm. Bull. **34**, 3530–3533 (1986); **36**, 1343–1350 (1988). [2]Heterocycles **24**, 571 (1986); J. Org. Chem. **56**, 6873–6878 (1991). [3]Nature (London) **162**, 524 (1948). [4]Mothes et al., p. 197f.
gen.: Florey **10**, 289–335 ▪ Hager (5.) **4**, 771–788; **8**, 18ff. ▪ Manske **22**, 1–50; **25**, 48–57; **51**, 271–323 ▪ Sax (8.), CCX 125, EAL 500, EAN 000, EAN 500, IGF 000 ▪ Shamma, p. 427–457 ▪ Shamma-Moniot, p. 355–363. – *[HS 2939 90]*

Ipohardine see Ipom(o)ea alkaloids.

Ipomeamarone. Formula see ipomeanin. $C_{15}H_{22}O_3$, M_R 250.34, $[\alpha]_D$ +27° (C_2H_5OH), a *phytoalexin of the class of furan *sesquiterpenes. I. was first isolated from sweet potatoes (*Ipomoea batatas*) afflicted by black rot[1]. The amount of I. reaches ca. 2 weight-% of the plant after a short time, it then acts as a highly active fungicide[2].

Lit.: [1]J. Chem. Soc., Chem. Commun. **1983**, 352. [2]Science **121**, 216 (1953).
gen.: Beilstein EV **19/4** 195 ▪ Cole-Cox, p. 734 ▪ Hager (5.) **5**, 537–542 ▪ Karrer, No. 1792 ▪ Sax (7.), p. 2007. – *Biosynthesis:* J. Chem. Soc., Chem. Commun. **1984**, 272 ▪ Phytochemistry **20**, 647 (1981). – *Synthesis:* Chem. Pharm. Bull. **43**, 890 (1995) ▪ Heterocycles **43**, 1287 (1996) ▪ Tetrahedron Lett. **34**, 509 (1993) ▪ see also myoporone. – *[CAS 494-23-5]*

Ipomeanin(e) [1-(3-furanyl)-1,4-pentanedione].

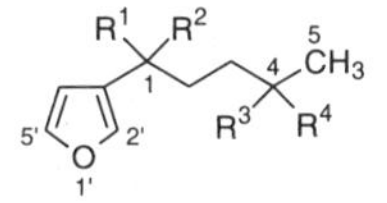

R^1	R^2	R^3	R^4	
O		O		Ipomeanin (**1**)
H	OH	O		1-Ipomeanol (**2**)
O		H	OH	4-Ipomeanol (**3**)
H	OH	H	OH	1,4-Ipomeadiol (**4**)

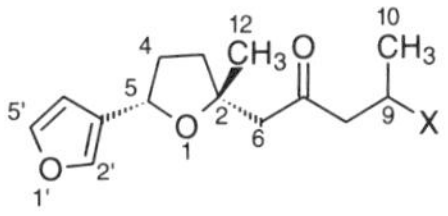

X = CH_3 : Ipomeamarone (**5**)
X = CH_2OH : Ipomeamaronol (**6**)

Table: Data of Ipomeanin derivatives.

	molecular formula	M_R	LD_{50} (mouse p.o.) mg/kg	CAS
1	$C_9H_{10}O_3$	166.18	26	496-06-0
2	$C_9H_{12}O_3$	168.19	79	34435-70-6
3	$C_9H_{12}O_3$	168.19	38	32954-58-8
4	$C_9H_{14}O_3$	170.21	104	53011-73-7
5	$C_{15}H_{22}O_3$	250.34		494-23-5
6	$C_{15}H_{22}O_4$	266.34		26767-96-4

A hepatotoxic and nephrotoxic, cytostatic metabolic product formed in sweet potatoes under conditions of stress, in particular on infection with *Ceratocystis fimbriata* or *Fusarium solani*. The *ipomeanols* are derived from I.: 1-ipomeanol, 4-ipomeanol, 1,4-ipomeadiol, ipomeamaronol, and *ipomeamarone. The latter two compounds are *sesquiterpenes and cause atypical pneumonia in animals.

Lit.: Agric. Biol. Chem. **37**, 2443 (1973) ▪ Beilstein EV **17/11**, 137 ▪ Bull. Chem. Soc. Jpn. **55**, 3225 (1982) ▪ Cole-Cox, p. 717–737 ▪ Sax (8.), IGF 300.

Ipom(o)ea alkaloids. *Indolizidine alkaloids from bindweeds (Convolvulaceae) of the genus *Ipomoea*; the plants grow, among others, in Mexico and are used there as hallucinogenic drugs. Recently, young people in North America have started using the seeds of *I. violacea* as a narcotic drug (morning glory seeds). 100–150 g seeds induce a hallucinogenic effect corresponding to that of ca. 100 µg LSD; the LD_{50} on i.p. administration of aqueous extracts is ca. 200 mg/kg (mouse). *Ipomoea* species cultivated in Central Europe as an ornamental plant contain only small amounts of the corresponding active alkaloids.

The main alkaloids are the glycoside *ipalbine* ($C_{21}H_{29}NO_6$, M_R 391.46, cryst., mp. 118 °C) and *ipo-*

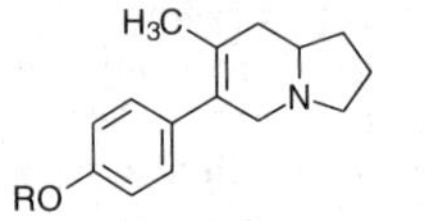

R = H : Ipalbidine
R = Glc : Ipalbine

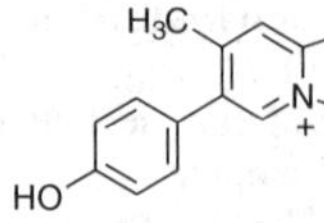

Ipohardine

mine ($C_{30}H_{35}NO_8$, M_R 537.61), the glucose residues at O-6 of which are esterified with 4-hydroxycinnamic acid; the corresponding aglycone is known as *ipalbidine* ($C_{15}H_{19}NO$, M_R 229.32, cryst., mp. 144–146°, racemate). Quaternary ammonium bases such as *ipohardine* ($C_{15}H_{16}NO^+$, M_R 226.30) also occur. Further alkaloids are *lysergic acid derivatives and, especially, ergoline derivatives, e.g., elymoclavine, *agroclavine (see also ergot alkaloids).

Lit.: Hager (5.) **5**, 534–545 ▪ J. Nat. Prod. **50**, 152 (1987); R. D. K. (3.), p. 390 f. ▪ Ullmann (5.) **A 1**, 363. – *Synthesis:* Helv. Chim. Acta **69**, 2048 (1986) ▪ Heterocycles **50**, 31 (1999) [(±)-ipalbidine] ▪ J. Chem. Res. (S) **1979**, 1 ▪ J. Chem. Soc., Perkin Trans. 1 **1985**, 261 ▪ J. Org. Chem. **51**, 3915 (1986); **62**, 438 (1997) (ipalbidine) ▪ Tetrahedron **33**, 1733 (1977) ▪ Tetrahedron Lett. **1981**, 2127. – *[HS 2939 90; CAS 23544-46-9 (ipalbine); 65370-71-0 (ipomine); 26294-41-7 (ipalbidine); 108937-65-1 (ipohardine)]*

Ipomoea resins.

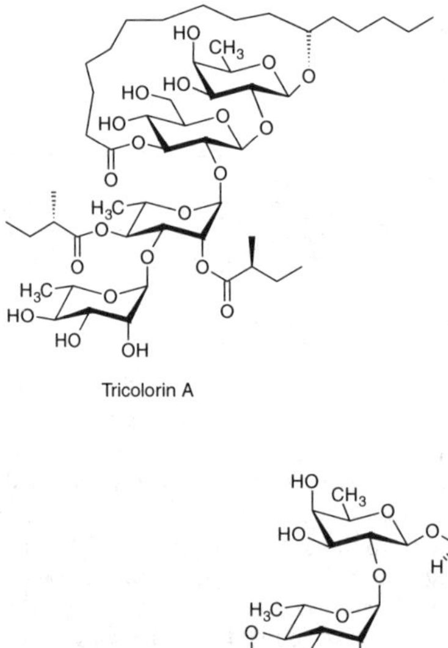

Tricolorin A

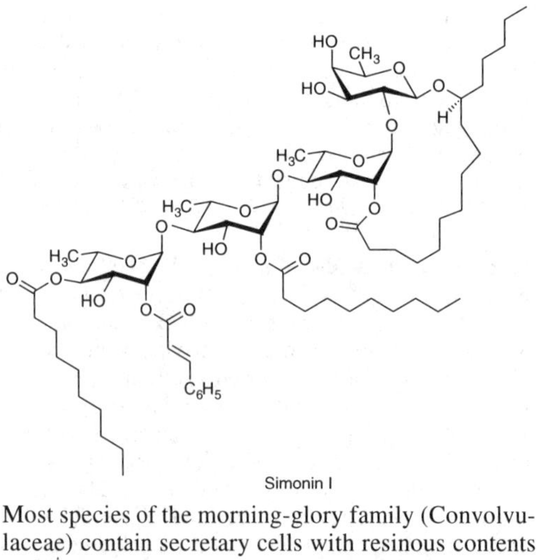

Simonin I

Most species of the morning-glory family (Convolvulaceae) contain secretary cells with resinous contents in foliar tissues, roots and rhizomes. These resin glycosides are responsible for the purgative properties of some medicinal plants of the Convolvulaceae, e. g. *Ipomoea purga, I. orizabensis* and *Convolvulus scammonia*. All resin glycosides contain *jalapinolic acid* [(*S*)-11-hydroxyhexadecanoic acid, mp. 68–69 °C, $[\alpha]_D$ +0.79°] as the aglycone which is usually incorporated in a macrolactone ring spanning two or more units of the saccharide backbones. Resin glycosides are rich in deoxy sugars like D-fucose, L-rhamnose and D-quinovose. From the numerous compounds of this type known [1], *tricolorin A* [$C_{50}H_{86}O_{21}$, M_R 1023.22, needles, mp. 118–120 °C, $[\alpha]_D$ –30.3° (CH_3OH)] from *Ipomoea tricolor* exhibits significant cytotoxic activity against cultured P-388 and human breast cancer cell lines [2]. Together with other tricolorins it is responsible for the allelopathic potential of this plant [3], which is used in traditional agriculture in Mexico as a cover crop for the protection of sugar cane. It is a potent natural uncoupler and inhibitor of the photosystem II acceptor side of spinach chloroplasts [4]. *Simonin I* [$C_{69}H_{112}O_{21}$, M_R 1277.63, powder, mp. 114–116 °C, $[\alpha]_D$ –9.3° (CH_3OH)] and some of its analogues have been isolated from roots of *Ipomoea batatas*, a plant used in Brazilian folk medicine that has been introduced as a health food in Japan [5]. On damage, the plant produces the toxic *ipomeamarone. T. A has been the target of several total syntheses [6], the most efficient using a ring-closing metathesis for formation of the macrolactone ring.

Lit.: [1] *Reviews:* J. Am. Chem. Soc. **121**, 7814 (1999) (and references given therein); Rev. Latinoam. Quim. **25**, 97 (1997). [2] J. Nat. Prod. **56**, 571 (1993); Tetrahedron **52**, 13063 (1996); Tetrahedron **53**, 9007 (1997); Z. Kristallogr. **215**, 114 (2000) (X-ray structure). [3] J. Chem. Ecol. **21**, 289, 1085 (1995). [4] Physiol. Plant. **106**, 246 (1999). [5] Chem. Pharm. Bull. **40**, 3163 (1992). [6] Angew. Chem. Int. Ed. Engl. **34**, 2520 (1995); **36**, 2344 (1997); J. Org. Chem. **61**, 5208 (1996); **62**, 8400, 8406 (1997); **63**, 424 (1998); J. Am. Chem. Soc. **121**, 7814 (1999). – *[CAS 149155-65-7 (tricolorin A); 151310-50-8 (simonin I)]*

Ipsdienol (2-methyl-6-methylene-2,7-octadien-4-ol).

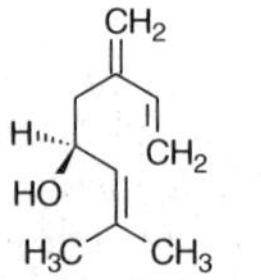

(*S*)-Ipsdienol

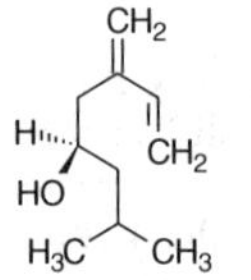

(*R*)-Ipsenol

$C_{10}H_{16}O$, M_R 152.24, oil. The (*R*)-enantiomer {$[\alpha]_D$ –13.6° (CH_3OH)} is a *pheromone of the bark beetle *Ips confusus*, a pest of *Pinus ponderosa*, while the (*S*)-enantiomer is a pheromone of *Ips paraconfusus* [1]. *2,3-Dihydro-I.* [ipsenol, $C_{10}H_{18}O$, M_R 154.25, bp. 86–88 °C (1.99 kPa)] is also a pheromone of bark beetles of the genus *Ips* [2]. Not only the ratio of I. to ipsenol but also the respective enantiomer ratios are species-specific and are used by the beetles to distinguish between species. With regard to the biogenesis not only the oxidation of the monoterpene hydrocarbon *myrcene [3] but also *de novo* synthesis seem to be involved [4]. A survey of the numerous pheromones from various bark beetles is given in *Lit.* [1]. I. is also found in plants [5]; see also pheromones.

Lit.: [1] Phytochemistry **27**, 2759 (1988). [2] Acta Chem. Scand., Ser. B **37**, 1 (1983); **41**, 442 (1987); **43**, 777–782 (1989); Chem. Br. **1990**, 124–127; Helv. Chim, Acta **73**, 353–358 (1990); J. Chem. Soc., Chem. Commun. **1989**, 295; **1995**, 2391; Synthesis **1993**, 1086; Synth. Commun. **27**, 1029 (1997); Tetrahedron Lett. **34**, 3781 (1993); **36**, 3353 (1995). [3] Nature (London) **284**, 485 f. (1980). [4] Insect Biochem. **23**, 655–662 (1993). [5] J. Chem. Ecol. **21**, 1043–1063 (1995).

gen.: ApSimon **4**, 76 ff., 113, 467 ff.; **9**, 294–303 (synthesis) ▪ Tetrahedron: Asymmetry **10**, 4281 (1999) ▪ Tetrahedron **55**, 13205 (1999). – *[CAS 35628-00-3 ((S)-I.); 60894-97-5 ((R)-I.); 35628-05-8 ((S)-ipsenol)]*

Ircinal A see manzamines.

Irenolone see phenalenones.

Iridals.

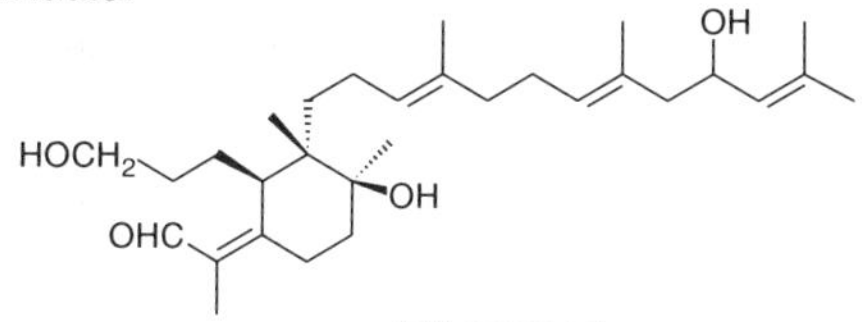

Iridogermanal

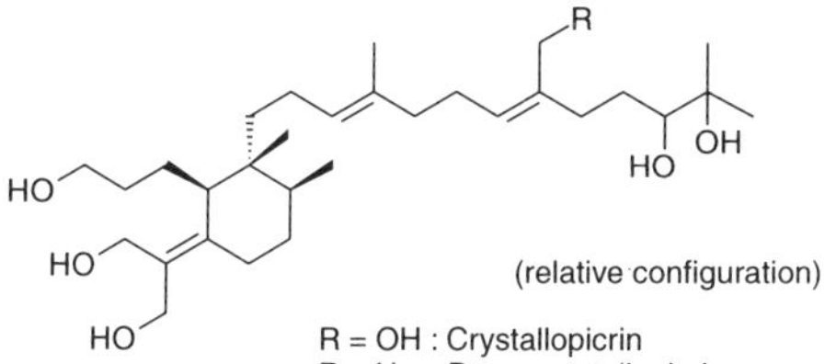

(relative configuration)

R = OH : Crystallopicrin
R = H : Deoxycrystallopicrin

Unusual triterpenes from various *Iris* species. The skeleton, as in the case of, e.g., *iridogermanal* {$C_{30}H_{50}O_4$, M_R 474.73, waxy solid, $[\alpha]_D$ +41° (CH_2Cl_2)}, consists of a highly substituted cyclohexane ring and a homofarnesyl side chain. The hydroxy group at C-3 in I. can be esterified by fatty acids, it is assumed that, like steroids, they are components of the cell membrane. They are biogenetic precursors of the *irones, the fragrance substances of iris oil and are formed from *squalene [1]. Interestingly, the bitter substances *crystallopicrin* {$C_{30}H_{54}O_6$, M_R 510.76, solidified syrup, $[\alpha]_D$ +26° (CH_3OH), extremely bitter} and *deoxycrystallopicrin* {$C_{30}H_{54}O_5$, M_R 494.76, amorphous, $[\alpha]_D$ +28.5° (CH_3OH)} from the toadstool *Cortinarius vibratilis* (Basidiomycetes) and the related agarics *C. crystallinus* and *C. croceocoeruleus* have the same skeleton [2].

Lit.: [1] Helv. Chim. Acta **71**, 1331 (1988); **73**, 433 (1990). [2] Ciba Found. Symp. **154**, 56 (1990); Schering Lectures **24**, 8 (1994).

gen.: J. Nat. Prod. **62**, 89 (1999) ▪ J. Org. Chem. **47**, 2531 (1982) ▪ Justus Liebigs Ann. Chem. **1990**, 563; **1992**, 269 ▪ Phytochemistry **41**, 1281 (1996) ▪ Zechmeister **50**, 1. – *[CAS 81456-98-6 (iridogermanal)]*

Iridin see irigenin.

Iridodial.

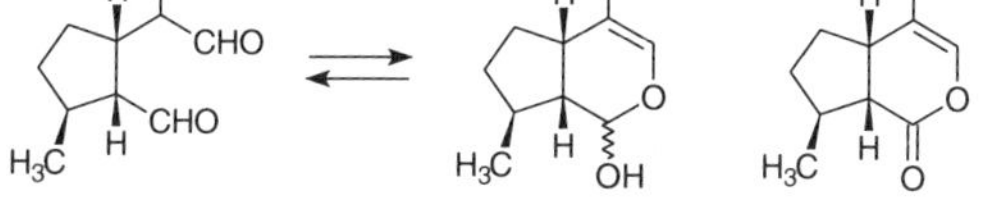

Iridodial Nepetalactone

Basic skeleton of the C_{10}-*iridoids. I. is in equilibrium with the dialdehyde-lactol form. $C_{10}H_{16}O_2$, M_R 168.24, oil, bp. 90–92 °C (100 Pa), $[\alpha]_D$ +67.7° (CH_3OH). Oxidation of the hemiacetal group gives rise to *nepetalactone* {$C_{10}H_{14}O_2$, M_R 166.22, bp. 71–72 °C (5 Pa), $[\alpha]_D$ +11.1° ($CHCl_3$)}.

Occurrence: I. was first isolated from the Australian ant *Iridomyrmex detectus* [1] and the hunting beetle (*Staphylinus olens*, Coleoptera: Staphylinidae) [2]. It was later also found in the essential oil of a not exactly defined Australian species of *Myoporum* [3]. Nepetalactone occurs in the essential oil (51%) of the genuine catmint (*Nepeta cataria*, Lamiaceae) [4] and is an attractant for cats.

Lit.: [1] Aust. J. Chem. **9**, 288 (1956); **13**, 296 (1960); **18**, 1989 (1965). [2] Tetrahedron Lett. **1971**, 4037. [3] Hegnauer **V**, 135. [4] Heterocycles **33**, 117 (1992); Phytochemistry **26**, 1200, 2311 (1987).

gen.: Bioorg. Med. Chem. **4**, 351 (1996) (synthesis, biosynthesis) ▪ J. Nat. Prod. **43**, 649 (1980) ▪ Karrer, No. 1132 ▪ Tetrahedron Lett. **27**, 2896 (1986); **28**, 4431 (1987) (biosynthesis). – *Synthesis:* Agric. Biol. Chem. **52**, 2369 ff. (1988) ▪ ApSimon **9**, 364 (nepetalactone) ▪ Chem. Pharm. Bull. **36**, 172–177 (1988) ▪ J. Chem. Soc., Chem. Commun. **1987**, 1020. – *[CAS 69252-84-2 (I.); 21651-53-6 ((+)-nepetalactone)]*

Iridogermanal see iridals.

Iridoids (older name "pseudoindicans" because they give a blue color on treatment with acids). Cyclopentanoid natural products formed by intramolecular 2,6-cyclization of an acyclic *monoterpene, thus belonging to the class of cyclic monoterpenes. At present ca. 300 I. are known. In the plant organism *geraniol (**1a**) and nerol (**1b**) are important biogenetic intermediates in the formation of cyclic *monoterpenes. For the 2,6-cyclization of **1a** and **1b** C-6 must be transformed to an electrophilic center. Specific oxidation of **1a** with monooxygenases furnishes the dialdehyde **2a** which cyclizes to *iridodial (**3**), that is existing in equilibrium with its lactol (=hemiacetal) form **4**. With glucose **4** preferentially forms an acetal **5**, so that most I. occur in nature as glucosides. The cyclopenta[*c*]pyran ring is numbered as shown in **5**. The *secoiridoids (**6**) are formed from **5**.

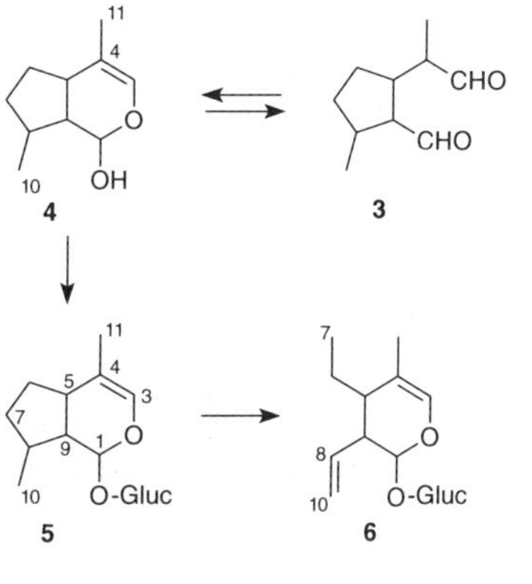

3 is the precursor of numerous I. with a methyl group at C-4[1]. For the biosynthesis of I. oxidized at C-10 or C-11, see *Lit.*[2]. I. glucosides with a C_9-carbon skeleton are formed by oxidative degradation of C-10 or C-11. *Examples:* Deutziol and deutzioside (lacking C-10); *catalpol, *harpagide, *matatabiether, etc. (lacking C-11). A C_8-carbon skeleton is present in unedoside and stilbericoside (lacking C-10 and C-11).
The parent compound of the secoiridoids is secologanin (see secoiridoids), the most important intermediate in the biogenesis of *alkaloids that are not derived from an amino acid. This includes most *indole alkaloids, the *ipecac, the *Cinchona, and the pyrroloquinoline alkaloids as well as simple monoterpene alkaloids[3].
The best known biological property of the I. is their bitter taste. However, bitter principles are mostly not used as pure substances, instead alcoholic extracts are preferred to stimulate appetite (increased secretion of gastric juice). Furthermore, bitter substances are used to modify the taste of pharmaceutical products. Some I. exert various effects on the central nervous system as a consequence of their volatility and lipophilicity, e.g., nepetalactone, *iridodial, *teucrium lactones, and *valepotriates.

Lit.: [1] Planta Med. **33**, 193 (1978). [2] J. Chem. Soc., Chem. Commun. **1970**, 823. [3] Angew. Chem. Int. Ed. Engl. **22**, 828–841 (1983).
gen.: ApSimon **7**, 275–454 (synthesis) ▪ Bobbitt & Segebarth, in Taylor & Battersby (eds), Cyclopentanoid Terpene Derivatives, chap. 1, New York, N. Y.: M. Dekker Inc. 1969 ▪ J. Nat. Prod. **43**, 649 (1980); **54**, 1173–1246 (1991) (review) ▪ Nat. Prod. Rep. **5**, 419–464 (1988) ▪ Phytochemistry **40**, 773–792 (1995) ▪ Tetrahedron **53**, 14507–14545 (1997) (synthesis review) ▪ Zechmeister **50**, 169–237 (biosynthesis).

Iridomyrmecin.

Iridomyrmecin

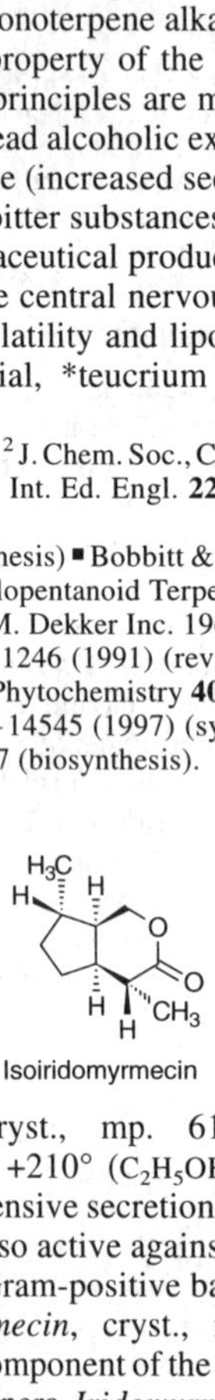
Isoiridomyrmecin

$C_{10}H_{16}O_2$, M_R 168.24, cryst., mp. 61 °C, bp. 104–108 °C (200 Pa), $[\alpha]_D$ +210° (C_2H_5OH). I. is a component of the toxic *defensive secretion of the ant *Iridomyrmex humilis*. It is also active against DDT-resistant insects and inhibits Gram-positive bacteria.
The isomeric *isoiridomyrmecin*, cryst., mp. 57.5–58 °C, $[\alpha]_D^{28}$ −56° (CCl_4), component of the defensive secretions of ants of the genera *Iridomyrmex*, *Dolichoderus*, and *Tapinoma*, is formed in the ants' anal glands; see also iridoids.

Lit.: Beilstein E V **17/9**, 237 ▪ J. Chem. Soc., Chem. Commun. **1994**, 479 ff. ▪ Nat. Prod. Lett. **1**, 217 ff. (1992) ▪ Pure Appl. Chem. **66**, 2071 (1994) ▪ Russ. Chem. Bull. **46**, 1606 (1997) ▪ Synlett **1997**, 657 (synthesis) ▪ Tetrahedron **52**, 3435 (1996) (synthesis) ▪ Tetrahedron Lett. **33**, 1543 ff., 2823 f. (1992). – *[CAS 485-43-8 (I.); 107538-14-7 (iso-I.)]*

Irigenin [5,7-dihydroxy-3-(3-hydroxy-4,5-dimethoxyphenyl)-6-methoxy-4*H*-1-benzopyran-4-one].
$C_{18}H_{16}O_8$, M_R 360.32, pale yellow crystals, mp. 185 °C, occurs, together with the 7-*O*-*β*-glucopyranoside *iridin*, $C_{24}H_{26}O_{13}$, M_R 522.46, colorless needles, mp. 208 °C, in the roots of *Iris* species and in *Belamcanda*

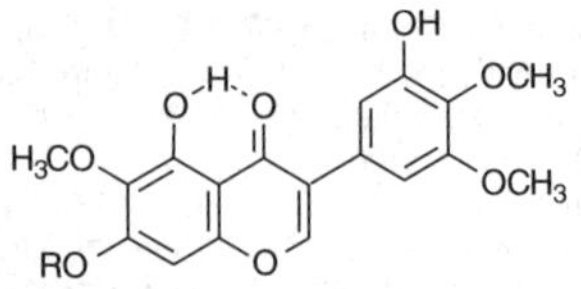
R = H : Irigenin
R = β-D-Glucopyranosyl: Iridin

chinensis. The name iridin is also used for a soft resin from the alcoholic extract of *Iris versicolor* as well as for a protamine from the sperm of rainbow trout.
Activity: I. has antifungal activity. The glycoside is used to bleach hair.

Lit.: Beilstein E V **18/5**, 679 ▪ Bull. Chem. Soc. Jpn. **62**, 2450–2454 (1989) ▪ Karrer, No. 1666 ▪ Nat. Prod. Rep. **12**, 101–133 (1995) ▪ Phytochemistry **36**, 807 ff. (1994) ▪ Planta Med. **56**, 335 (1990). – *Pharmacology:* J. Nat. Prod. **51**, 345–348 (1988) ▪ Prog. Clin. Biol. Res. **280**, 369–374 (1988). – *[HS 2938 90; CAS 548-76-5 (I.); 491-74-7 (iridin)]*

Irinotecan see camptothecin.

Irones (2-methylionones).

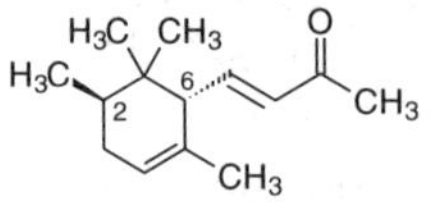
(-)-*trans*-α-Irone (2*R*,6*R*) (**1**)

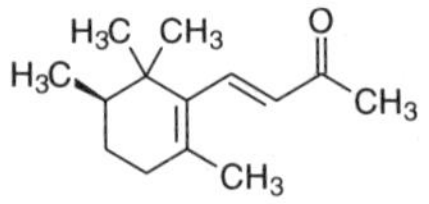
(+)-β-Irone (2*R*) (**2**)

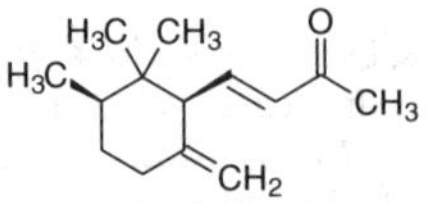
(+)-*cis*-γ-Irone (2*R*,6*S*) (**3**)

$C_{14}H_{22}O$, M_R 206.32, viscous oils, soluble in dichloromethane. Violet-like fragrance substances from *orris (root) oil, also occurring in snowdrops, stock flowers, wallflowers, daphne, iris, and some oak mosses[1]. The biosynthesis[2] proceeds from *iridals. Used as fragrance substances for valuable perfumes. The purest iris note comes from (−)-*trans*-*α*-I. {$[\alpha]_D^{20}$ −410° ($CHCl_3$)}. (+)-*β*-I. {bp. 85–90 °C (13 Pa), $[\alpha]_D^{20}$ +59° (CH_2Cl_2)} has the strongest odor and is the most common isomer. (+)-*cis*-*γ*-I. {bp. 85–88 °C (8 Pa), $[\alpha]_D^{20}$ +2° (CH_2Cl_2)} is reported to have a dry root note with a weak iris-like odor[3].

Lit.: [1] Theimer, Fragrance Chemistry, p. 305–308, New York: Academic Press 1982. [2] Naturwiss. Rundsch. **43**, 395 f. (1990). [3] Ohloff, p. 162 ff.
gen.: Beilstein E IV **7**, 377 ff. ▪ Helv. Chim. Acta **75**, 1023 (1992); **76**, 2070 (1993) (synthesis *α*-I.); **75**, 759 (1992) (*β*-I.) ▪ J. Org. Chem. **61**, 6021 (1996) (synthesis *γ*-I.) ▪ Karrer, No. 534–537 ▪ Ullmann (5.) **A 11**, 173 f., 236 ▪ Zechmeister **8**, 146–206; **50**, 1–26. – *[CAS 35124-14-2 (1); 35124-15-3 (2); 35124-16-4 (3)]*

Isatan B [indoxyl 5-keto-D-gluconate, 3-indolyl (5*R*)-D-*xylo*-5-hexulofuranosonate].

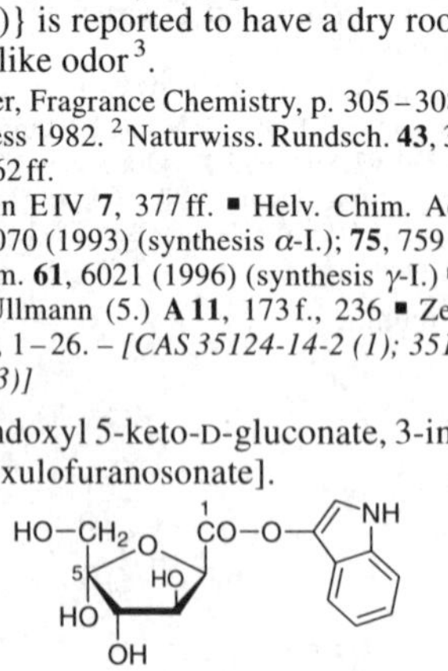

$C_{14}H_{15}NO_7$, M_R 309.28, an unstable *indigotin precursor occurring especially in the leaves of dyer's woad

(*Isatis tinctoria*, Brassicaceae) while the roots contain merely 1–3% I. together with 97–99% *indican [1].
Lit.: [1] Biochem. Physiol. Pflanz. **184**, 321 (1989).
gen.: Phytochemistry **31**, 2695 (1992) (biosynthesis) ■ Z. Naturforsch. B **23**, 572 (1968). – *[CAS 20307-14-6]*

γ-Isatropic acid see truxillic and truxinic acids.

Ishwarane.

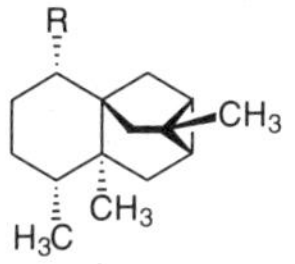

R = H : Ishwarane
R = OH : Ishwarol

$C_{15}H_{24}$, M_R 204.36, oil, bp. 80–82 °C (133 Pa), $[\alpha]_D$ –40° ($CHCl_3$). Example of the small group of tetracyclic *sesquiterpenes, first isolated from the roots of the Indian birthwort (*Aristolochia indica*, Aristolochiaceae). The plant also contains *ishwarol* ($C_{15}H_{24}O$, M_R 220.35).
Lit.: *Isolation:* Tetrahedron **26**, 2371 (1970). – *Synthesis:* ApSimon **5**, 493-498 ■ Can. J. Chem. **58**, 2613 (1980) ■ J. Chem. Soc., Chem. Commun. **1971**, 479; **1977**, 587, 880. – *[CAS 26620-70-2 (I.); 28957-57-5 (ishwarol)]*

Islandicin (1,4,5-trihydroxy-2-methylanthraquinone).

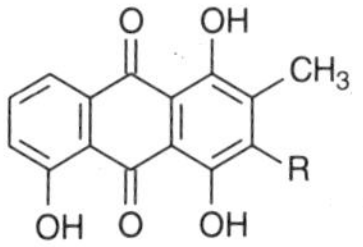

R = H : Islandicin
R = OH : 3-Hydroxyislandicin

$C_{15}H_{10}O_5$, M_R 270.24, dark red platelets, mp. 220–221 °C, uv_{max} 527 nm (ethanol). Octaketide from the mycelium of *Penicillium islandicum* with antibiotic activity. It is also isolated from the fruit of *Cassia occidentalis* (Fabaceae) and from the lichen *Asahinea chrysantha*. 3-Hydroxy-I., *catenarin* ($C_{15}H_{10}O_6$, M_R 286.24, red crystals, mp. 233 °C) occurs in the root bark of *Ventilago calyculata* (Rhamnaceae) and in *Helminthosporium* species as well as imperfect fungi [1].
Lit.: [1] Turner **1**, 158; **2**, 515; Zechmeister **51**, 125.
gen.: Justus Liebigs Ann. Chem. **1981**, 2106–2247 (biosynthesis). – *Isolation:* Beilstein EIV **8**, 3572 ■ Karrer, No. 1278. – *Synthesis:* Can. J. Chem. **62**, 1922 (1984) ■ J. Chem. Soc., Chem. Commun. **1981**, 108; **1987**, 883 ■ J. Org. Chem. **48**, 5373 (1983); **50**, 5433 (1985). – *[CAS 476-56-2 (I.); 8970-80-4 (3-hydroxy-I.)]*

Isoandalusol see labdanes.

Isoavenaciolide see avenaciolide.

Isobacteriochlorins.

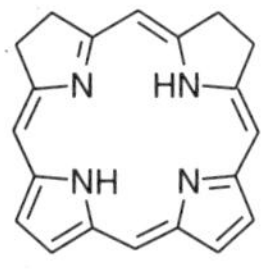

The structural skeleton of I. (tetrahydroporphyrins) is formally derived from *porphine by reduction of two peripheral double bonds in neighboring pyrrole rings of the porphyrin skeleton. I. are constitutional isomers of the *bacteriochlorins, from which their name is derived. Natural members of this structural type are *heme d_1, *siroheme, and sirohydrochlorin.
Lit.: Angew. Chem. Int. Ed. Engl. **32**, 1223 (1993) ■ Chem. Rev. **94**, 327 (1994). – *[CAS 67883-10-7]*

(–)-Isobicyclogermacrenal.

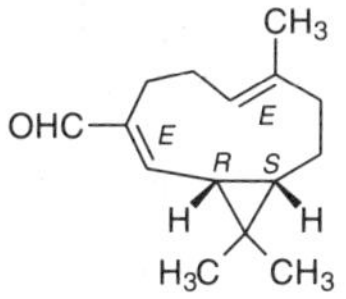

$C_{15}H_{22}O$, M_R 218.34, oil, $[\alpha]_D$ –168°. A sesquiterpene aldehyde isolated from the liverwort *Lepidozia vitrea* that induces growth inhibition in the leaves and roots of rice shoots (IC_{50} 0.2–1 mmol/L).
Lit.: Chem. Lett. **1981**, 1097 ■ J. Chem. Soc., Chem. Commun. **1980**, 1220 ■ J. Chem. Soc., Perkin. Trans. 1 **1984**, 203 ■ Phytochemistry **26**, 1529 (1987). – *Synthesis:* J. Chem. Soc., Chem. Commun. **1987**, 1196. – *[CAS 73256-82-3]*

Isoboldine see aporphine alkaloids.

2-Isobutylthiazole.

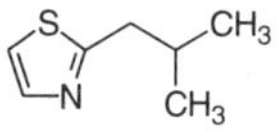

$C_7H_{11}NS$, M_R 141.23, bp. 180 °C; with an odor like tomato leaves, olfactory threshold in water 3 ppb [1]. In contrast to many thermally formed, aroma-active thiazoles [1,2,3] (see also meat, coffee, and cocoa flavor) I. is a typical *flavor compound of fresh tomatoes [4] and is probably formed from *leucine and *cysteine [1]. For synthesis, see *Lit.* [3].
Lit.: [1] Z. Naturforsch. C **43**, 731 (1988). [2] Maarse, p. 157–160. [3] Bedoukian (3.), p. 190–195; Beilstein EV **27/5**, 75 f. [4] Maarse, p. 249; J. Agric. Food Chem. **35**, 540 ff. (1987). – *[CAS 18640-74-9]*

Isochavicine see piperine.

Isochorismic acid [(5*S*)-*trans*(1-carboxyvinyloxy)-6-hydroxy-1,3-cyclohexadiene-1-carboxylic acid].

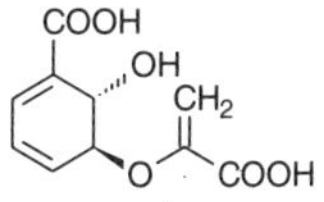

$C_{10}H_{10}O_6$, M_R 226.19, important intermediate in the biosynthesis of *salicylic acid, *vitamin K_1, and *siderophores. The biosynthesis proceeds through rearrangement of *chorismic acid by means of isochorismate synthase (EC 5.4.99.6); see also *Lit.* [1].
Lit.: [1] Can. J. Microbiol. **37**, 276–280 (1991); Tetrahedron **47**, 5979–5990 (1991).
gen.: CRC Crit. Rev. Biochem. Mol. Biol. **25**, 307–384 (1990) (review) ■ Haslam, Shikimic Acid, Metabolism and Metabolites, p. 219–227, New York: Wiley 1993 ■ Luckner (3.), p. 326, 329 f. ■ Nat. Prod. Rep. **11**, 173–203 (1994). – *[CAS 22642-82-6 (trans); 85506-20-3 (±)]*

Isocomene (berkheyaradulene).

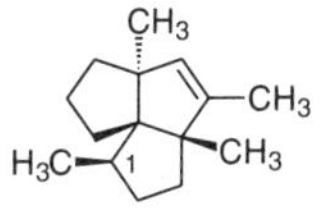

$C_{15}H_{24}$, M_R 204.36, cryst., mp. 60–63 °C, bp. 65–70 °C (46.6 Pa), $[\alpha]_D^{24}$ –85° ($CHCl_3$). A natural triquinane from *Isocoma wrightii*, Asteraceae of South west USA that is toxic for cattle and sheep, and the roots of South African *Berkheya* species [1] and *Senecio isatideus* [2]. On account of its unusual structure composed of three condensed five-membered rings with three quaternary and one tertiary asymmetric centers, I. has been the target of numerous syntheses[3]; see also modhephene.

Lit.: [1] J. Chem. Soc., Chem. Commun. **1977**, 456; J. Nat. Prod. **42**, 96 (1979). [2] Z. Naturforsch. C **37**, 5 (1982). [3] Angew. Chem. Int. Ed. Engl. **30**, 1492 f. (1991); Bull. Chem. Soc. Jpn. **71**, 231 (1998); Helv. Chim. Acta **79**, 1026 (1996); J. Org. Chem. **56**, 5281 (1991); Nicolaou, p. 221–227; Pure Appl. Chem. **68**, 675 (1996); Tetrahedron **51**, 8835–8852 (1995); **53**, 8975 (1997). *gen.:* Atta-ur-Rahman **3**, 3 ▪ J. Org. Chem. **58**, 6171 (1993) (absolute configuration). – *[CAS 65372-78-3]*

α-Isocomene see triquinanes.

Isocorydine see aporphine alkaloids.

Isocorypalmine see protoberberine alkaloids.

Isocyanides (isonitriles, natural).

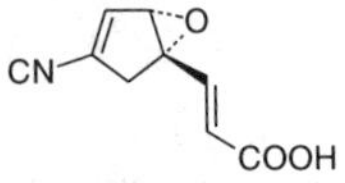

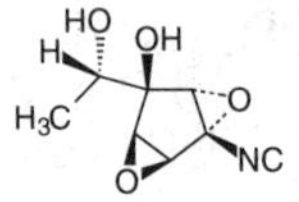

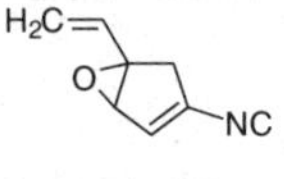

Dermadin (**1**) Trichoviridin (**2**) Homothallin I (**3**)

I. (isonitriles) are produced in nature by fungi, bacteria, and marine organisms. Thus, I. of the *xanthocillin type are found in cultures of various *Penicillium* and *Aspergillus* species while *Trichoderma viride* and related species produce cyclopentyl isocyanides such as *dermadin* {$C_9H_7NO_3$, M_R 177.16, mp. 120–125 °C, $[\alpha]_D^{22}$ +133° ($CHCl_3$)} and *trichoviridin* {isonitrin C, $C_8H_9NO_4$, M_R 183.16, mp. 102–104 °C, $[\alpha]_D^{25}$ –35.8° (CH_3OH)}. These compounds have antibiotic activities against Gram-positive and -negative bacteria as well as against numerous fungi. Interestingly, the volatile *homothallin I* (C_8H_7NO, M_R 133.15) from *T. koningii* is able to induce the production of homothallic oospores of the soil fungus *Phytophthora cinnamomi* that attacks tree roots. Both types of compound are formed biosynthetically from tyrosine, the origin of the isocyanate carbon is still unknown.

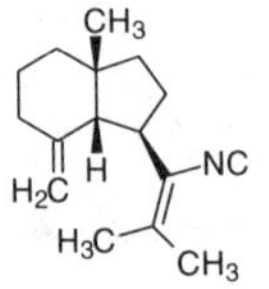

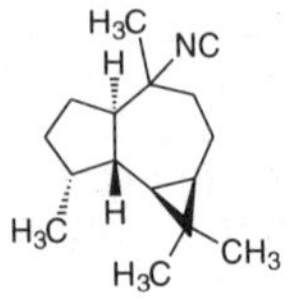

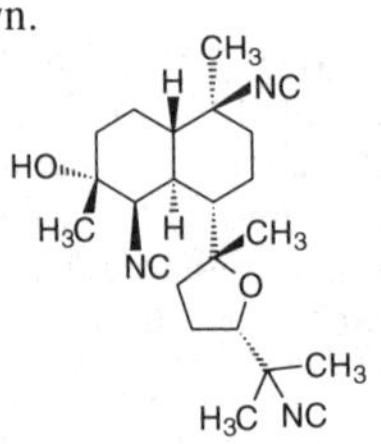

Axisonitrile-IV (**4**) Axisonitrile-II (**5**) Kalihinol-F (**6**)

Sesquiterpenoid I. such as *axisonitrile-IV* {$C_{16}H_{23}N$, M_R 229.36, mp. 56–58 °C, $[\alpha]_D$ +51.4° ($CHCl_3$)} and *axisonitrile-II* {$C_{16}H_{25}N$, M_R 231.38, oil, $[\alpha]_D$ +29° ($CHCl_3$)} occur together with the corresponding isothiocyanates and *N*-formylamino compounds in the marine sponge *Axinella cannabina*. Numerous compounds of these types with different carbon skeletons have been isolated from sponges, e.g., the triisocyanide *kalihinol-F* {$C_{23}H_{33}N_3O_2$, M_R 383.53, mp. 176–178 °C, $[\alpha]_D$ +8° ($CHCl_3$)} from an *Axinella* species, member of the family of kalihinols A–Z[1]. *Diisocyanoadociane from an *Adocia* species possesses a diterpene skeleton, like various other I. from sponges. The often brilliantly green-colored nudibranchia (mollusks) are able to protect themselves from predators by means of toxic I. taken up with the food. Thus, the ichthyotoxic 9-isocyanopupukeanane probably originates from a sponge of the genus *Hymeniacidon* which is eaten by the nudibranchi *Phyllidia varicosa*. I. have also been isolated from terrestrial cyanobacteria (see indole alkaloids from cyanobacteria).

As shown by biosynthetic studies, marine isothiocyanates and *N*-formyl derivatives are formed from I. and the latter is not formed from its accompanying substances. The origin of the isocyano group was demonstrated by the incorporation of [^{14}C]cyanide in diisocyanoadociane after feeding experiments with a sponge of the genus *Amphimedon*[2].

Lit.: [1] J. Am. Chem. Soc. **109**, 6119 (1987); Tetrahedron **52**, 2359 (1996) (isolation). [2] J. Chem. Soc., Chem. Commun. **1986**, 35.

gen.: Acc. Chem. Res. **25**, 433–439 (1992) ▪ J. Am. Chem. Soc. **120**, 13285 (1998) [synthesis of (–)-isonitrin B] ▪ J. Chem. Soc., Chem. Commun. **1996**, 41 f. (synthesis, trichoviridin) ▪ Nat. Prod. Rep. **5**, 229 (1988). – *Synthesis, axisonitrils:* J. Org. Chem. **61**, 473 (1996); Tetrahedron Lett. **36**, 3365 (1995). – *[CAS 83374-67-8 (1); 56283-32-0 (2); 62078-10-8 (4); 55907-33-0 (5); 93426-91-6 (6)]*

9-Isocyanoneopupukeanane see diisocyanoadociane.

3-((*Z*)-2-Isocyanovinyl)indole.

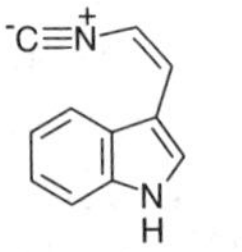

$C_{11}H_8N_2$, M_R 168.20. An isonitrile antibiotic from cultures of a *Pseudomonas* species[1]. The compound is in discussion as a biosynthetic precursor of various complex *indole alkaloids from cyanobacteria[2].

Lit.: [1] J. Antibiot. **29**, 850 (1976); Justus Liebigs Ann. Chem. **1984**, 600. [2] J. Am. Chem. Soc. **116**, 9941 (1994). – *[HS 2939 90; CAS 61168-06-7]*

Isoeugenol [2-methoxy-4-(1-propenyl)phenol].

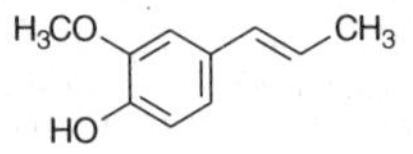

$C_{10}H_{12}O_2$, M_R 164.20, oil, mp. –10 °C, bp. 266 °C, bp. 128 °C (1.0 kPa). I. exists in the (*E*)-form, mp. 33 °C, bp. 140 °C (1.6 kPa) and in the (*Z*)-form, liquid, bp. 133 °C (1.7 kPa). I. has a weak but more pleasant odor than the isomeric *eugenol. I. occurs in various essential oils (*ylang-ylang oil, *nutmeg oil, champaca flower oil), mostly as a mixture of (*E*)- and (*Z*)-forms or with the methyl ether named *methylisoeugenol* [$C_{11}H_{14}O_2$, M_R 178.23, mp. 16 °C, bp. 138–140 °C (1.6 kPa)].

Use: I. as well as the methyl ether are used in perfumery (carnation odor). I. is also employed as a conservation agent and an intermediate in the synthesis of vanillin.

Lit.: Beilstein EIV **6**, 6324 ▪ J. Chem. Soc., Perkin Trans. 1 **1977**, 359–363 (synthesis) ▪ Karrer, No. 187 ▪ Merck-Index (12.), No. 5186 ▪ Sax (8.), IKQ 000. – *[HS 2909 50; CAS 97-54-1, 5932-68-3 (E); 5912-86-7 (Z); 6379-72-2 (methylisoeugenol)]*

Isoflavones. A group of mostly yellow *plant pigments, sometimes also designated as isoflavonoids, belonging to the *flavonoids and derived from isoflavone (3-phenylchromone); however, the I. are of less importance than their isomers, the *flavones. The parent compound of the isoflavones, isoflavone (3-phenylchromone, 3-phenyl-4*H*-1-benzopyran-4-one) occurs in clover species.

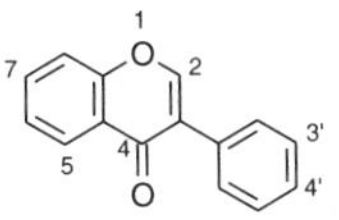

	5	7	3'	4'
1 Isoflavone	H	H	H	H
2 Daidzein	H	OH	H	OH
3 Genistein	OH	OH	H	OH
4 Prunetin	OH	OCH_3	H	OH
5 Biochanin A	OH	OH	H	OCH_3
6 Orobol	OH	OH	OH	OH
7 Santal	OH	OCH_3	OH	OH
8 Pratensein	OH	OH	OH	OCH_3

Table: Data of Isoflavones.

compound	molecular formula	M_R	mp. [°C]	CAS
1	$C_{15}H_{10}O_2$	222.24	150	574-12-9
2	$C_{15}H_{10}O_4$	254.25	323 (decomp.)	486-68-8
3	$C_{15}H_{10}O_5$	270.24	301–302	446-72-0
4	$C_{16}H_{12}O_5$	284.27	240	552-59-0
5	$C_{16}H_{12}O_5$	284.27	212–216	491-80-5
6	$C_{15}H_{10}O_6$	286.24	271	480-23-9
7	$C_{16}H_{12}O_6$	300.27	223	529-60-2
8	$C_{16}H_{12}O_6$	300.27	272–273	2284-31-3

Some of the better known I. are: *daidzein* (4',7-dihydroxy-I.), occurring as the 7-*O*-glucoside daidzin in soy bean flour; *genistein* from soy beans and red clover; *prunetin* (4',5-dihydroxy-7-methoxy-I.) from the bark of plum trees; *biochanin A* (5,7-dihydroxy-4'-methoxy-I.) from chick-peas, red clover, and clover species; *orobol* (3',4',5,7-tetrahydroxy-I.); *santal* (3',4',5-trihydroxy-7-methoxy-I.) from sandalwood, redwood, and other woods; *pratensein* (3',5,7-trihydroxy-4'-methoxy-I.) from fresh red or meadow clover. The I. occurring in some of these clover species and Fabaceae such as alfalfa, etc., have estrogenic effects on grazing animals and can possibly lead to reproduction disorders. The daily uptake of I. with human food exceeds 1 g per day. Very little is known about the effects on human fertility[1]. In addition, derivatives of isoflavanone (see formula, hydrogenated at C-2 and C-3) are known; in the broader sense, the *rotenoids are also related to the I. For metabolism of I. in the living plant, see *Lit.*[2]. The biosyntheses of I. proceeds from the corresponding *chalcone derivatives, for synthesis, see *Lit.*[3]. Some I. have antiarteriosclerotic[4] and antifungal activities, genistein shows weak activity against Gram-positive bacteria.

Lit.: [1]Environ. Health Persp. **103**, 346–351 (1995). [2]Naturwissenschaften **67**, 40f. (1980); Zechmeister **34**, 198–214. [3]Synthesis **1976**, 326; **1978**, 843. [4]Stud. Org. Chem. **23**, 321–324 (1986).

gen.: Chem. Ind. (London) **1995**, 412 (genistein) ▪ J. Chem. Soc., Chem. Commun. **1995**, 1317 f. ▪ Karrer, No. 1644–1673, 4670–4683 ▪ Nat. Prod. Rep. **12**, 321–338 (1995); **15**, 241–260 (1998) ▪ Phytochemistry **46**, 921 (1997) (daidzein) ▪ Zechmeister **28**, 1–73; **40**, 105–152; **43**, 1–266 ▪ Z. Naturforsch. C **45**, 147–153 (1990).

Isojurubidine see Solanum steroid alkaloids.

Isolactaranes.

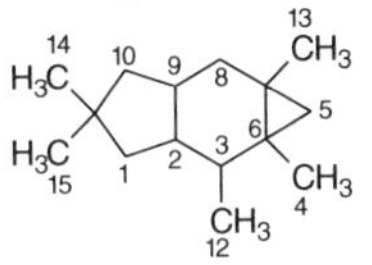

Isolactarane

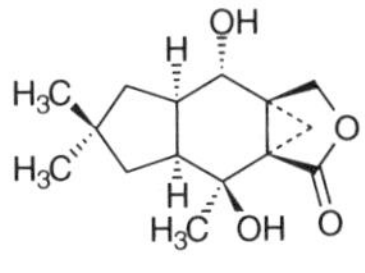

Isolactarorufin

Sesquiterpenoids of the I. type are biosynthetically closely related with the*sterpuranes with which they occur concomitantly in, e. g., cultures of *Merulius tremellosus* (Basidiomycetes) (*merulidial). *Isolactarorufin* {$C_{15}H_{22}O_4$, M_R 266.34, crystals, mp. 81 °C, $[\alpha]_D$ +8.4° (C_2H_5OH)} is isolated from fruit bodies of the agaric *Lactarius rufus*, *sterepolide from cultures of the resupinate polypore *Stereum purpureum.*

Lit.: Tetrahedron **37**, 2199 (1981) (review) ▪ Turner **2**, 252 f. – *[CAS 62024-77-5 (isolactarorufin)]*

Isolation (of natural products). The separation of a pure natural product starting from its biological producer (e. g., plants, fruit bodies of fungi, marine organisms, fermentation cultures). In general a wide selection of chromatographic procedures are employed.

The first step is the *preparation of a crude extract.* This may involve certain preselection steps: – fractions with differing lipophilicity/hydrophilicity (polarity), – adsorption on appropriate synthetic adsorbent resins (e. g., Amberlite XAD-16, Diaion HP-20), – adsorption on active charcoal and stepwise elution (hydrophilic/polar substances), – Soxhlet extraction, – percolation, – continuous liquid-liquid extraction, – extraction with supercritical gases (e. g., CO_2).

The first *gross separation* depends on the following parameters: – molecular mass (gel permeation chromatography, e. g. on Sephadex LH-20), – lipophilicity/hydrophilicity (adsorption chromatography, e. g., on silica gel, aluminum oxide, reversed phases), – charge (ion exchangers), – affinity (affinity chromatography), – distribution coefficients (counter-current distribution chromatography).

Some *special technological methods* used are: – normal pressure liquid chromatography (NPLC), – medium pressure liquid chromatography (MPLC), – high pressure (performance) liquid chromatography (HPLC), – distribution chromatography (e. g., DCCC, RLCC), – preparative thin layer chromatography (PTLC), – thin layer electrophoresis (TLE).

The last stage in an isolation is usually an HPLC step and, if possible, a recrystallization.

After isolation of a natural product its authenticity with regard to the starting organism must be confirmed with the aid of suitable analytical methods (HPLC, MS, etc.).

Lit.: Colegate & Molyneux, Bioactive Natural Products: Detection, Isolation and Structural Determination, Boca Raton: CRC Press 1993 ▪ Egerer, in Präve (ed.), Jahrbuch-Biotechnologie, Chromatographische Methoden in der Aufarbeitung von Naturstoffen, München: Hanser 1986/1987 ▪ Mann et al., Natural Products – Their Chemistry and Biological Significance, Harlow: Addison Wesley Longman 1996 ▪ Nat. Prod. Rep. **8**, 391 (1991) ▪ Natori et al. (ed.), Advances in Natural Products Chemistry, A Halsted Press Book, Tokyo: Kodanska 1981 ▪ Raphael, Natural Products – A Laboratory Guide, San Diego: Academic Press 1991 ▪ Schwedt, Chromatographische Trennmethoden, Stuttgart: Thieme 1986 ▪ Verrall, Discovery and Isolation of Microbial Products, Chichester: Ellis Horwood 1985 ▪ Wagman u. Cooper, Natural Products Isolation, Amsterdam: Elsevier 1989.

L-Isoleucine [(2*S*,3*S*)-2-amino-3-methylpentanoic acid, abbr.: Ile or I].

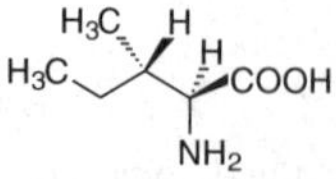

$C_6H_{13}NO_2$, M_R 131.17, mp. 285–286 °C (decomp.), $[\alpha]_D^{25}$ +12.2° (H_2O), +36.7° (1 M HCl), pK_a 2.26, 9.62, pI 5.94. An essential proteinogenic amino acid. Genetic code: AUU, AUC, AUA. Average content in proteins 4.6% [1]. Free I. occurs in plants, e.g., shoots of common vetch (*Vicia sativa*).

Biosynthesis: (1-Hydroxyethyl)-TPP formed from pyruvic acid condenses with 2-oxobutanoic acid to give 2-acetyl-2-hydroxybutanoic acid. The latter is rearranged and reduced by acetolactate mutase (EC 5.4.99.3.) and NAD(P)H-dependent reductase to 2,3-dihydroxy-3-methylpentanoic acid. Dehydration then furnishes 2-oxo-3-methylpentanoic which is finally transaminated to Ile.

Use: In microbiology, in amino acid infusion solutions, and in chemically defined diets.

Lit.: [1] Biochem. Biophys. Res. Commun. **78**, 1018–1024 (1977).

gen.: Beilstein EIV **4**, 2774–2778 ▪ Greenstein & Winitz, Chemistry of the Amino Acids, vol. 3, p. 2043–2074, New York: J. Wiley 1961 ▪ Merck-Index (12.), No. 5196 ▪ Ullmann (5.) **A 2**, 58, 62, 70, 85. – *[HS 2922 49; CAS 73-32-5]*

L-Isoleucine methyl ester, methyl L-isoleucinate see scarabs.

Isoliquiritigenin see chalcones.

Isolubimin see vetispiranes.

Isomaltulose see palatinose.

Isomasticadienonic acid see mastic.

(+)-Isomenthol see 3-*p*-menthanols.

Isomenthones see *p*-menthanones.

Isomycin see anemonin.

Isonitriles see isocyanides.

Isonitrin C see isocyanides.

Isonootkatone see vetivone.

Isopavine alkaloids see pavine and isopavine alkaloids.

Isopentenyl acetate see alarm substances.

Isopilosine see pilocarpine.

Isoplumericin see Plumeria iridoids.

Isoprene rule. In the light of the fact that the then known mono- and sesquiterpenes were built up of two or three isoprene units, respectively, O. Wallach [1] in 1887 proposed his simple "isoprene rule". In 1953 L. Ruzicka [2] formulated his "biogenetic isoprene rule" which proved to be a milestone in the chemistry of the terpenoids. According to this rule, all terpenoids are derived from acyclic parent structures: *monoterpenes are formed from geranyl pyrophosphate (GPP, C_{10}), *sesquiterpenes from farnesyl pyrophosphate (FPP, C_{15}), *diterpenes from geranyl geranyl pyrophosphate (GGPP, C_{20}), and *sesterterpenoids from geranyl farnesyl pyrophosphate (GFPP, C_{25}). In the case of triterpenes, FPP is dimerized through presqualene to *squalene (C_{30}), from which the cyclic triterpenes and steroids are derived. Analogously, the tetraterpenes (*carotinoids, C_{40}) originate from two units of GGP. Ring systems of differing complexity can be built up from simple prenyl pyrophosphates under the action of cyclases in processes in which cationic centers react under stereocontrol with the neighboring double bonds. In these processes 1,2- and 1,3-H-shifts as well as methyl and alkyl group migrations lead to products that are not compatible with the simple isoprene rule. The long sought "active isoprene" proved to be an enzymatic equilibrium between the monoterpenes isopentenyl pyrophosphate (IPP) and dimethylallyl pyrophosphate (DMAPP). Their significance in the biosynthesis of linear prenyl precursors was first recognized by F. Lynen [3]. Besides the classical acetate/(*R*)-mevalonate pathway, IPP and DMAPP can also be formed by a second route (deoxyxylulose pathway) in which the triose phosphate couples with activated acetaldehyde to give 1-deoxy-D-xylulose 5-phosphate that subsequently rearranges to the isoprene precursor. The second route was discovered by M. Rohmer [4] for the microbial *hopanoids, and has since been observed in green algae and plants [5–7].

Lit.: [1] Justus Liebigs Ann. Chem. **239**, 48 (1887). [2] Experientia **9**, 357 (1953); Helv. Chim Acta **38**, 1890 (1995); Proc. Chem. Soc. London **1959**, 341; Pure Appl. Chem. **6**, 493 (1963); Tetrahedron **7**, 82 (1959); Quart. Rev. **21**, 331 (1967). [3] Angew. Chem. **70**, 738 (1958). [4] Pure Appl. Chem. **65**, 1293 (1993); Biochem. J. **295**, 517 (1993); J. Am. Chem. Soc. **118**, 2564 (1996). [5] Proc. Natl. Acad. Sci. USA **93**, 6431 (1996). [6] Z. Naturforsch. C **52**, 15–23 (1997). [7] Chem. Biol. **5**, 221–233 (1998) (review).

gen.: Barton-Nakanishi **2**, 1–400 ▪ Herbert, The Biosythesis of Secondary Metabolites (2.), 63–95, London: Chapman & Hall 1989 ▪ Mann, Secondary Metabolism (2.), p. 95–171, Oxford: Univ. Press 1987 ▪ Merck-Index (12.), No. 6250 (mevalonic acid) ▪ Porter & Spurgeon, Biosynthesis of Isoprenoid Compounds (2 vols.), Chichester: Wiley 1981.

Isoprenoids (isopentenoids). The name for a group of natural products made up of isoprene units (e.g., *sesqui-, *di-, and *triterpenes, *iridoids, *carotinoids, *steroids, *natural rubber, etc.). Many non-isoprenoid compounds, however, do possess isoprenoid side chains, e.g., *tocopherols, *ubiquinones, *chlorophyll, or contain isoprenoid structures incorporated into their skeletons, e.g., *monoterpenoid indole alkaloids, *penitrems, *Cinchona alkaloids.

Lit.: ACS Symp. Ser. **562** (1994) ▪ Barton-Nakanishi **2**, 1–400.

2-Isopropyl-4,5-dimethylthiazole see peanut flavor.

3-Isopropyl-2-methoxypyrido[2,3-*b*]pyrazine see pyridopyrazines.

Isopsoralene see furocoumarins.

(−)-Isopulegol see *p*-menthenols.

(−)-Isopulegone see *p*-menthenones.

Isoquinoline alkaloids. The I. a. constitute a very large group of alkaloids that is divided into various subgroups on the basis of the structural diversity. With few exceptions (bacteria, fungi) I. a. are found solely in plants. The biosynthesis of the isoquinoline ring system proceeds through an enzyme-catalyzed Mannich reaction from *dopamine and (4-hydroxyphenyl)acetaldehyde, both formed in turn from tyrosine. The initial product is (*S*)-norcoclaurine which is *O*- and *N*-methylated and then hydroxylated to furnish (*S*)-reticuline, from which other alkaloids are derived. The structural diversity of the I. a. arises from the possibility for oxidative couplings due to the presence of phenolic hydroxy groups both in the isoquinoline part and in the benzyl residue. The most important subgroups of I. a. are: benzylisoquinoline type, protoberberine type, phthalide isoquinoline type, pavine type, proaporphine type, aporphine type, and the thebaine-morphine type. For more information, see benzyl(tetrahydro)isoquinoline, bisbenzylisoquinoline, protoberberine, phthalide isoquinoline, aporphine, morphinan, pavine and isopavine alkaloids. Other important subgroups are: *cularine, *spirobenzylisoquinoline alkaloids, emetine and related I. a., *rhoeadine, *Erythrina, *cactus, *Corydalis, Colchicum, Ancistrocladus, and *ipecac alkaloids. The I. a. are strongly irritating toxins. After consumption they first cause central excitation, nausea, vomiting, hemorrhagic diarrhea, uresiesthesis, vertigo, somnolence, spasms, respiratory distress, collapse, and then paralysis, especially of the respiratory center. On contact with the skin they cause burning sensation, erythema, blister formation, and possibly necrosis. The *opium alkaloids exhibit strong analgesic activities. Tetrahydro-I. a. can be formed from biogenic amines and acetaldehyde not only in plants but also in mammalian organisms. This is related to the narcotic effects and the addictive potential of ethanol leading to alcoholism.

Lit.: J. Nat. Prod. **48**, 725–738 (1985) ▪ Manske **21**, 255–327; **29**, 141–184; **41**, 1–40 ▪ Mothes et al., p. 188–271 ▪ Nat. Prod. Rep. **1**, 355–370 (1984); **2**, 81–96 (1985); **3**, 153–169 (1986); **4**, 677–702 (1987); **5**, 265–292 (1988); **6**, 405–432 (1989); **8**, 339–366 (1991); **9**, 365–389 (1992); **10**, 449–470 (1993); **11**, 555–576 (1994) ▪ Phillipson et al. ▪ Planta Med. **55**, 163 ff. (1989) ▪ Rodd's Chem. Carbon. Compd. (2.) **4F/G**, 507–587 (1998) ▪ Shamma ▪ Shamma-Moniot ▪ Ullmann (5.) **A 1**, 367–383.

Isoretronecanol see necine(s).

Isorubijervine see Solanum steroid alkaloids.

Isoscopoletin see 6,7-dihydroxycoumarins.

Isosepiapterin.

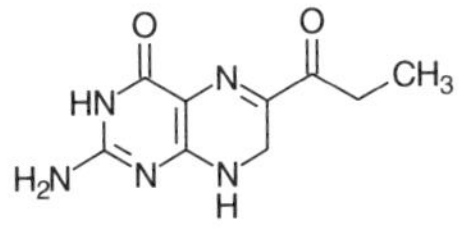

$C_9H_{11}N_5O_2$, M_R 221.22. An eye pigment of the fruit fly *Drosophila melanogaster*, occurs also in the cyanobacterium *Anacystis nidulans*.

Lit.: Helv. Chim. Acta **71**, 531 (1988) ▪ Insect Biochem. **24**, 907 (1994) (biosynthesis). – *[CAS 1797-87-1]*

Isospongiaquinone see ilimaquinone.

Isotetracenones see angucyclins.

Isotetrandrine see bisbenzylisoquinoline alkaloids.

Isothebaine see aporphine alkaloids.

Isotrichodermin see scirpenols.

Isotrilobine see bisbenzylisoquinoline alkaloids.

Isousnic acid.

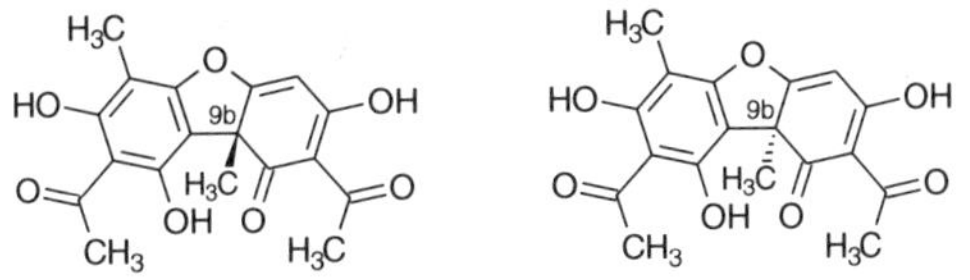

(+)-(9b*R*)- (–)-(9b*S*)-

Isousnic acid

$C_{18}H_{16}O_7$, M_R 344.32, yellow prisms, mp. 150–152 °C, $[\alpha]_D^{21}$ +490° or –490° (dioxan). A dibenzofuran derivative from lichen: the (+)-form occurs in *Cladonia mitis* and *Sphaerophorus ramulifer*, the (–)-form in *Cladonia pleurota*. I. is considerably less common than *usnic acid. The biosynthesis involves phenol oxidation of methylphloracetophenone.

Lit.: Chem. Pharm. Bull. **18**, 374–378 (1970) ▪ Tetrahedron Lett. **22**, 351 f. (1981). – *[CAS 18058-86-1 (+)]*

Isovalencenol see vetiver oil.

Isovelleral.

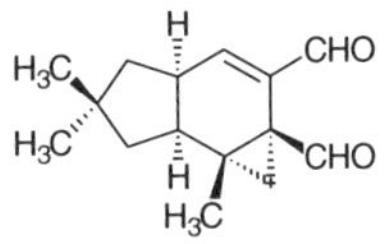

$C_{15}H_{20}O_2$, M_R 232.32, cryst., mp. 105–106 °C, $[\alpha]_D$ +293° ($CHCl_3$). A biologically active *marasmane derivative, together with *velleral, formed enzymatically from *stearoylvelutinal on bruising of the fruit bodies of milky caps (*Lactarius*), hot-tasting russulas (*Russula*), and other basidiomycetes. Its hot, burning taste protects the fruit body from predators. I. has mutagenic and cytotoxic activity and is a strong antibiotic.

Lit.: Tetrahedron Lett. **13**, 1105 (1972); **24**, 4631 (1983) (abs. configuration); **32**, 2541 (1991) (biosynthesis) ▪ Phytochemistry **45**, 1569 (1997) (biosynthesis). – *Synthesis:* J. Chem. Soc., Chem. Commun. **1990**, 865, 1260 ▪ J. Org. Chem. **55**, 3004 (1990); **57**, 5979 (1992). – *[CAS 37841-91-1]*

Isovincoside see strictosidine.

Isowillardiine [*β*-uracil-3-yl-L-alanine].

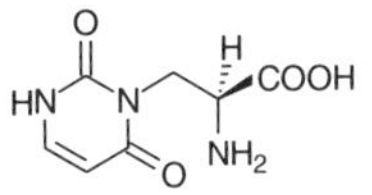

(α*S*)-form

$C_7H_9N_3O_4$, M_R 199.17. Isolated together with *willardiine from growing pea shoots [1]. Uracil has a substrate inhibiting effect on the biosynthesis of the two amino acids [2]. For toxicity, see *Lit.* [2].

Lit.: [1] Acta Chem. Scand. **17**, 641 (1963); Biochem. Biophys. Res. Commun. **32**, 474–479 (1968). [2] Phytochemistry **16**, 223–227 (1977).

gen.: Beilstein EV **24/6**, 87 ▪ Chem. Pharm. Bull. **28**, 2748 (1980) (synthesis). – *[CAS 21381-33-9]*

Isoxanthopterin see pteridines.

Ivermectin (MK-933, mectizan, ivomec). I. is a mixture of two very similar macrolide *antibiotics, that can be prepared semisynthetically from *avermectins.

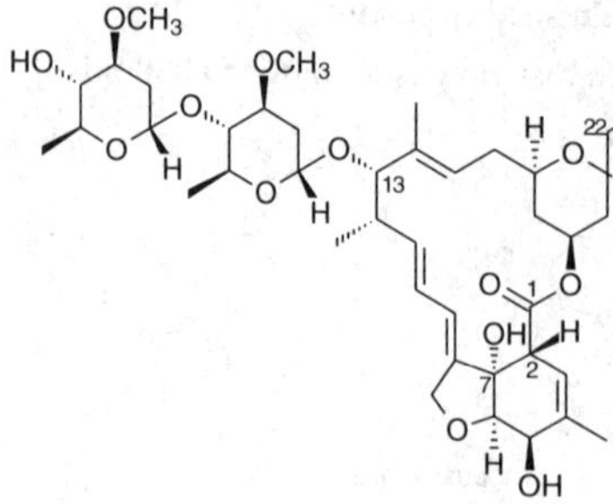

R = C(H)(S)(CH$_3$)–C$_2$H$_5$: Ivermectin B$_{1a}$, 22,23-Dihydro-avermectin B$_{1a}$

R = CH(CH$_3$)$_2$: Ivermectin B$_{1b}$, 22,23-Dihydro-avermectin B$_{1b}$

I. B$_{1a}$ ($C_{48}H_{74}O_{14}$, M_R 891.10) and *I. B$_{1b}$* ($C_{47}H_{72}O_{14}$, M_R 877.08); the mixture containing at least 80% I. B$_{1a}$ is applied. At very low doses I. acts as a selective antibiotic against numerous nematodes and parasitic arthropods (anthelmintic, insecticide, acaricide). I. was originally developed for veterinary medicine but in the meantime it is the drug of choice for treatment of onchocercosis (filariasis, "river blindness") in humans, especially in Africa and Central America. On account of its long biological half-life I. can be administered as a single dose every 6–12 months. I. can be considered as one of the most important breakthroughs in the control of parasitic diseases in the developing countries.

Lit.: Anal. Profiles Drug Subst. **17**, 155–184 (1988) ▪ Campbell, Ivermectin, Abamectin, New York, N. Y.: Springer 1989 ▪ Chem. Eng. News **1988** (14. Nov.), 44 ▪ Chemosphere **18**, 1565–1572 (1989) ▪ Merck-Index (12.), No. 5264 ▪ Sax (8.), p. 2067 (toxicology) ▪ Trop. Med. Parasitol. **38**, 8 (1987) ▪ Ullmann (5.) **A 2**, 197; **A4**, 94f. – *Synthesis:* Heterocycles **27**, 45 (1988) ▪ Lindberg **2**, 221–261 ▪ Tetrahedron Lett. **28**, 5977 (1987). – *Pharmacology:* Annu. Rev. Microbiol. **45**, 445–474 (1991) ▪ Med. Res. Rev. **13**, 61–79 (1993) ▪ Parasitol. Today **9**, 154–159 (1993). – *[HS 2941 90; CAS 70288-86-7 (I.); 71827-03-7 and 70161-11-4 (I. B$_{1a}$); 70209-81-3 (I. B$_{1b}$)]*

J

Jacobine.

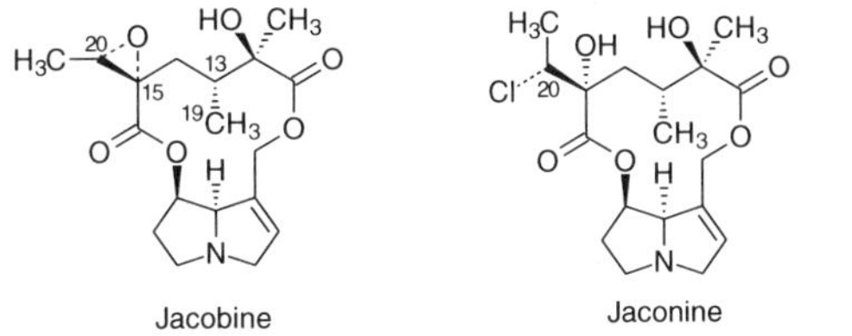

$C_{18}H_{25}NO_6$, M_R 351.399, mp. 228 °C, $[\alpha]_D^{17}$ −40° ($CHCl_3$). J. and further related *pyrrolizidine alkaloids occur in some *Senecio* species (e.g., *S. cineraria* and *S. jacobaea*). J. exists in the plants as the *N*-oxide and is, like all 1,2-unsaturated *pyrrolizidine alkaloids, toxic. Closely related alkaloids are the 13,19-didehydro compound *jacozine* {$C_{18}H_{23}NO_6$, M_R 349.38, mp. 228 °C, $[\alpha]_D$ −140° ($CHCl_3$)}, the 15,20-*R*-diol *jacoline* {$C_{18}H_{27}NO_7$, M_R 369.41, mp. 221.5 °C, $[\alpha]_D$ +48° ($CHCl_3$)} formed by formal addition of water to the epoxy group, and the C-20 chloro-substituted *jaconine* {$C_{18}H_{26}ClNO_6$, M_R 387.86, mp. 147 °C, $[\alpha]_D$ +52.5° (C_2H_5OH)} formed by formal addition of HCl to the epoxy group.

Lit.: Acta Crystallogr. **44**, 1942 (1988) ▪ Beilstein E III/IV **27**, 6661, 6673, 6827 f. ▪ Pelletier **9**, 155–233. – *[HS 2939 90; CAS 6870-67-3 (J.); 5532-23-0 (jacozine); 480-76-2 (jacoline); 480-75-1 (jaconine)]*

Jalapinolic acid see ipomoea resin.

Janthitrems. The J. are octacyclic, tremorigenic *mycotoxins from *Penicillium janthinellum*, a fungus occurring on grasses, e.g., rye. They are responsible for the symptoms of "ryegrass staggers" (see Darnel, ryegrass), a disease affecting grazing animals. The J. are intensely fluorescent solids, their complex structures have not been completely elucidated, and they are related with the *penitrems, lolitrems, and *paspalitrems.

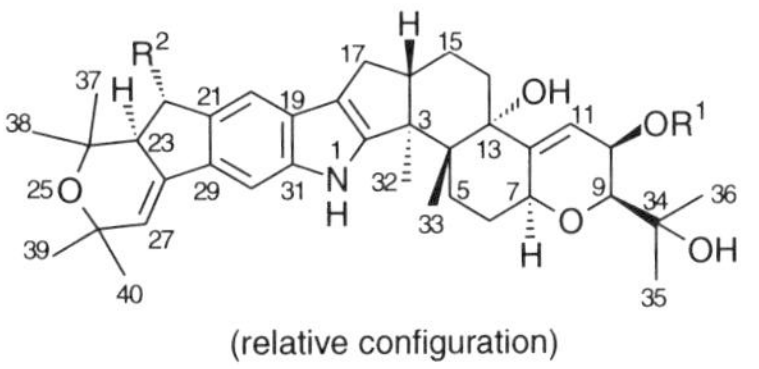

(relative configuration)

R^1 = H, R^2 = OH : J. E
R^1 = CO−CH_3, R^2 = OH : J. F
R^1 = CO−CH_3, R^2 = H : J. G

Table: Data of Janthitrems.

Toxin	molecular formula	M_R	CAS
J. E	$C_{37}H_{49}NO_6$	603.80	90986-50-8
J. F	$C_{39}H_{51}NO_7$	645.84	90986-52-0
J. G	$C_{39}H_{51}NO_6$	629.84	90986-51-9

Lit.: App. Environ. Microbiol. **39**, 272 (1980) ▪ Betina, chap. 16 ▪ Bioact. Mol. **1**, 501–511 (1986) ▪ J. Agric. Food Chem. **40**, 1307 (1992) ▪ J. Chem. Soc., Perkin Trans. 1 **1984**, 697–701 ▪ J. Chromatogr. **248**, 150–154 (1982); **392**, 333–347 (1987); **404**, 195–214 (1987) ▪ J. Food Prot. **44**, 715–722 (1981) ▪ Phytochemistry **32**, 1431 (1993) ▪ Tetrahedron **51**, 3959 (1995) ▪ Zechmeister **48**, 20–25, 38–43.

Japan lacquers (Japan varnishes). Genuine J. l. are obtained from exudates (injury juices) of *Rhus verniciflua* (Anacardiaceae), the Japanese lacquer tree (varnish tree), (Latin: verniciflua = excreting varnish). J. l. are soluble in 60–80% ethanol; they are used traditionally in Japan and Asia as varnish. Since genuine J. l. may contain allergens they are hardly used elsewhere. *Japan wax is produced as a by-product from the seeds of the plant. In the paint industry oil and enamel paints are also generally known as J. l.

Lit.: Hoechst High Chem Magazin **16**, 61 (1994) ▪ Ullmann (4.) **8**, 395; **12**, 527. – *[HS 3205 00]*

Japan wax (Japan tallow, cera japonica). Plant fat isolated from the seeds and seed husks of sumac species (*Rhus*, Anacardiaceae), especially *R. succedanea* and *R. verniciflua*. The latter is the so-called Japanese lacquer tree, compare Japan lacquers. Mp. 48–52 °C; soluble in many organic solvents, insoluble in water. The seeds furnish 7.5–15% J. w., which is bleached with active charcoal. The main constituents of J. w. are glycerol esters (ca. 90%), especially of palmitic acid (76–82%); Thus J. w. are more correctly named as Japan tallow. J. w. also contains 3–7% esters of dicarboxylic acids such as, e.g., *japanic acid* (heneicosanedioic acid, $C_{21}H_{40}O_4$, M_R 356.55, mp. 112–113 °C), *phellogenic acid* [docosanedioic acid, $C_{22}H_{42}O_4$, M_R 370.57, mp. 122–124 °C (126–128 °C)], and *tricosanedioic acid* ($C_{23}H_{44}O_4$, M_R 384.60, mp. 127.5 °C). J. w. is processed for use in candles, furniture polishes, floor waxes, and serves as a substitute for *beeswax.

Lit.: Hamilton (ed.), Waxes: Chemistry, Molecular Biology and Functions, p. 257–310, Dundee: The Oily Press 1995 ▪ Kirk-Othmer (3.) **24**, 470. – *[HS 1515 90; CAS 8001-39-6 (J.); 505-56-6 (phellogic acid)]*

Japonic acid see Japan wax.

Japonilure [(4*R*,5*Z*)-5-tetradecen-4-olide].

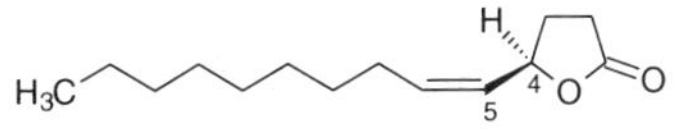

$C_{14}H_{24}O_2$, M_R 224.34, oil, bp. 140–141 °C (93.32 Pa), $[\alpha]_D^{21}$ −70.4° ($CHCl_3$). A sexual pheromone (see pheromones) of the female Japanese beetle *Popillia japonica* (Scarabaeidae)[1]. The attractant effect is completely suppressed by less than 3% of the (*S*)-enantiomer. The pure (*S*)-enantiomer was recently identified as a pheromone of *Anomala osakana* distributed in Japan. J. is used commercially in the USA for the mass capture of beetles. J. or, respectively, its homologue with 2 carbon atoms less also occurs in other Scarabaeidae[2].

Lit.: [1]Science **197**, 789–792 (1977). [2]Naturwissenschaften **80**, 181 ff. (1993).

gen.: Synthesis: ApSimon **4**, 141 ff.; **9**, 238–242 ▪ Tetrahedron **49**, 5961 (1993) ▪ Z. Naturforsch. B **50**, 1537 (1995). – *[CAS 64726-91-6]*

Jasmin(e) absolute. A dark orange to red-brown, somewhat viscous oil with an intense, sweet-flowery, warm, honey-like odor with fruity-spicy undertones.

Production: By extraction of jasmin(e) flowers (*Jasminum grandiflorum*, Oleaceae) with, e.g., hexane, precipitating the wax with alcohol. J. a. is produced, e.g., in southern France, Morocco, Egypt, and India. In China and India an absolute is prepared from another jasmin(e) species, *J. sambac*. The yearly worldwide production amounts to about 10 t.

Composition[1]***:*** J. consists to about 50% of more volatile components of which **benzyl acetate* with ca. 30% is the main component. In addition, many other substances such as, e.g., **linalool* (ca. 5–10%), *p-cresol* (C_7H_8O, M_R 108.14) (ca. 2%), **indole* (ca. 3%), **eugenol* (ca. 3%), **(Z)-jasmone* (ca. 3%), and **methyl jasmonate* contribute to the typical odor.

Use: J. a is one of the most expensive raw materials for perfumes and is now used only in small amounts in luxury perfumes. The odor complex "jasmin(e)" which resembles the natural product still finds very wide use in perfume compositions and is the most widely used type of flower odor.

Lit.: [1]Perfum. Flavor **2** (6), 37 (1977); **13** (2), 69 (1988); **17** (3), 68 (1992); **19** (2), 64 (1994); **20** (4), 39 (1995).

gen.: Arctander, p. 309 ▪ Bauer et al. (2.), p. 159 ▪ Ohloff, p. 149. – *Toxicology:* Food Cosmet. Toxicol. **14**, 331 (1976). – *[HS 3301 22; CAS 8022-96-6; 106-44-5 (p-cresol)]*

Jasmin(e) lactone [(*Z*)-7-decen-5-olide, jasmolactone].

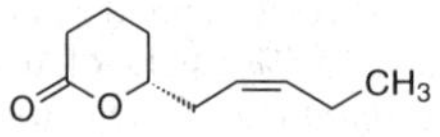

(-)-(5*R*)-δ-J.

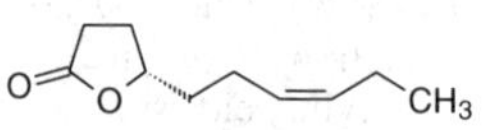

(+)-(4*R*)-γ-J.

$C_{10}H_{16}O_2$, M_R 168.24. A liquid with a fruity-flowery odor, bp. 95–96 °C (40 Pa). (–)-δ-J. ($[\alpha]_D$ –30.4°, neat), found, among others, in *jasmin(e) absolute and *tea flavor, (+)-δ-J. occurs in *tuberose absolute, peach and mango flavor[1]. For general and stereoselective syntheses, see *Lit.*[1,3]. The peach-like smelling γ-J. in the (+)-form contributes to the flavors of peach, mango, and yellow passion fruit[2]. For analysis and preparative chromatographic separation of the enantiomers, see *Lit.*[2]; biosynthesis *Lit.*[4]. The jasmolactones A–D are bicyclic structures[5].

Lit.: [1]Theimer (ed.), Fragrance Chemistry, p. 357, 385 ff., San Diego: Academic Press 1982. [2]Schreier & Winterhalter (eds.), Progress in Flavour Precursor Studies, S. 287 ff., Carol Stream: Allured 1993. [3]Bull. Chem. Soc. Jpn. **65**, 1257 (1992); Helv. Chim. Acta **74**, 787 ff. (1991). [4]Helv. Chim. Acta **79**, 2088 (1996). [5]J. Nat. Prod. **52**, 1060 (1989).

gen.: Beilstein E V **17/9**, 219 ▪ Ohloff, p. 149 ff. – *[HS 2932 29; CAS 25524-95-2 (δ); 66972-29-0 ((±)-δ); 136173-91-6 ((+)-δ); 136173-93-8 ((–)-δ); 63095-33-0 (γ); 155682-86-3 ((+)-γ)]*

Jasmolin, (+)-jasmolone see pyrethrum.

(*Z*)-Jasmone.

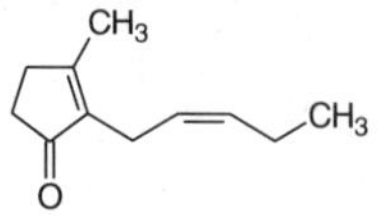

$C_{11}H_{16}O$, M_R 164.25. A pale yellow oil with a jasmin(e) odor resembling that of celery. Bp. 134–135 °C (1.6 kPa), LD_{50} (rat p.o.) 5 g/kg, soluble in alcohol and other organic solvents, practically insoluble in water.

Occurrence: With ca. 3% content in *jasmin(e) absolute J. is one of the decisive compounds for jasmin(e) odor and is also contained, among others, in *orange flower absolute, *peppermint oils, and *tea flavor. For synthesis, see *Lit.*[1,2].

Biosynthesis: Pathway for the inactivation and the disposal of the plant stress hormone jasmonic acid to the gas phase[3].

Lit.: [1]Theimer (ed.), Fragrance Chemistry, p. 357–373, San Diego: Academic Press 1982; Bedoukian (3.), p. 247 ff. [2]Biosci., Biotechnol. Biochem. **58**, 1181 (1994); J. Chem. Res. (S) **1992**, 418; Tetrahedron Lett. **37**, 5735 (1996). [3]Helv. Chim. Acta **80**, 838 (1997).

gen.: Arctander, No. 1790 ▪ Bauer et al. (2.), p. 66 ▪ Ohloff, p. 150. – *[HS 2914 29; CAS 488-10-8]*

Jasmonic acid.

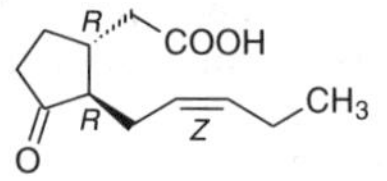

$C_{12}H_{18}O_3$, M_R 210.27, viscous oil, bp. (0.13 Pa) 125 °C, $[\alpha]_D^{20}$ –83.5° ($CHCl_3$); n_D^{19} 1.4885. In higher concentrations a phytotoxic compound, structurally similar to the *prostaglandins of animals, isolated from *Jasminum* spp. Low concentrations of J. stimulate the defensive system of plants (*phytoalexin) by inducing the formation of hydrocarbons that attract parasitic wasps which feed on plant-destroying caterpillars.

Lit.: Annu. Rev. Plant Physiol. **44**, 569–589 (1993) ▪ Barton-Nakanishi **1**, 117–138 (review) ▪ J. Chem. Soc., Perkin Trans. 1 **1997**, 3549–3559 ▪ J. Org. Chem. **62**, 6006 (1997) ▪ Nature (London) **399**, 686 (1999) ▪ Synth. Commun. **22**, 1283–1291 (1992); **27**, 2931 (1997) ▪ Synlett **1997**, 618 ▪ Tetrahedron **53**, 8181 (1997). – *[CAS 6894-38-8]*

Jaspamide (jasplacinolide).

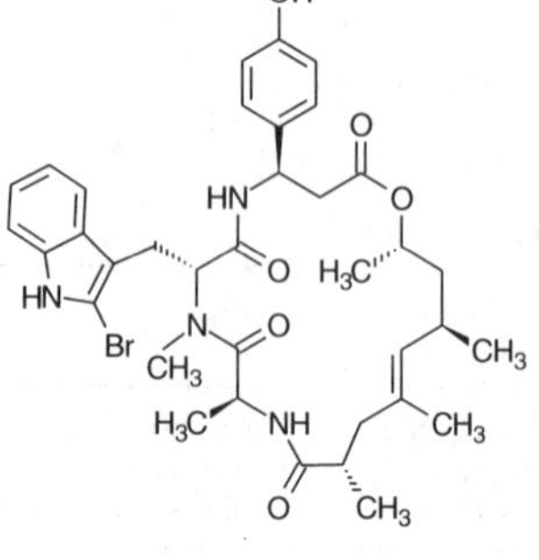

$C_{36}H_{45}BrN_4O_6$, M_R 709.68, oil, cyclodepsipeptide from marine sponges of the genus *Jaspis*[1]. J. has antifungal, insecticidal (*Heliothis virescens*, LC_{50} 4 ppm), and cytotoxic (epithelial laryngeal carcinoma, IC_{50} 0.32 µg/mL) activities. J. is related to the geodiamolides[2].

Lit.: [1] J. Am. Chem. Soc. **108**, 3123 (1986); J. Nat. Prod. **62**, 332 (1999); Pept. Chem. **1987**, 355; Tetrahedron Lett. **27**, 2797 (1986); **34**, 7085 (1993); **35**, 591 (1994). [2] J. Org. Chem. **52**, 3091 (1987); **56**, 5196 (1991); Tetrahedron Lett. **29**, 6465 (1988); **35**, 591 (1994) (synthesis).
gen.: Attaway, Marine Biotechnology, vol. 1, p. 353 f., New York: Plenum Press 1993 ▪ Heterocycles **39**, 603–612 (1994) ▪ Synthesis **1995**, 199–206 (synthesis). – *[CAS 102396-24-7]*

Jatropham [(5*R*)-1,5-dihydro-5-hydroxy-3-methyl-2*H*-pyrrol-2-one].

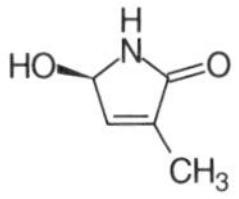

$C_5H_7NO_2$, M_R 113.12, natural *R*-form: crystals or needles, mp. 131–132 °C (119–123 °C), $[\alpha]_D^{25}$ −76.2° (CH_3OH), $[\alpha]_{549}^{25}$ −62° (H_2O), occurs, like the 5-*O*-*β*-glucopyranoside, $C_{11}H_{17}NO_7$, M_R 275.26, needles, mp. 178–183 °C, $[\alpha]_D^{25}$ −15.6° (CH_3OH), in the tubers and flowers of *Lilium martagon* (Turk's-cap lily, Liliaceae) and *L. hansonii*.

Activity: J. possesses antitumor activity against P-388 lymphocytic leukemia.

Lit.: Alkaloids (London) **12**, 36 (1982) ▪ Beilstein E V **21/12**, 64 ▪ Chem. Pap. **41**, 835 ff. (1987); **45**, 709 ff. (1991) (isolation, derivatives) ▪ Chem. Pharm. Bull. **36**, 4841–4848 (1988) ▪ Heterocycles **22**, 1733 ff. (1984) (synthesis) ▪ Phytochemistry **31**, 1084 f.; 2767–2775 (1992) ▪ Tetrahedron Lett. **36**, 4201 (1995) (synthesis). – *[CAS 50656-76-3]*

Jatrophane see diterpenes and jatrophone.

Jatrophatrione.

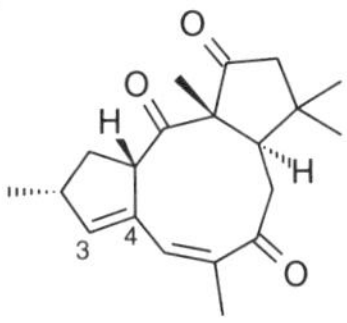

$C_{20}H_{26}O_3$, M_R 314.42, mp. 148–150 °C, $[\alpha]_D^{25}$ −187° ($CHCl_3$). A *diterpene of the cyclojatrophane type from *Jatropha macrorhiza* with antitumor properties.

Lit.: J. Nat. Prod. **51**, 749 (1988) ▪ J. Org. Chem. **41**, 1855 (1976). – *[CAS 58298-76-3]*

Jatrophone.

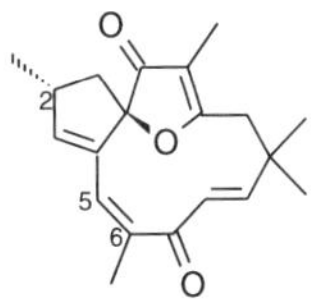

$C_{20}H_{24}O_3$, M_R 312.41, needles, mp. 152–153 °C, $[\alpha]_D^{24}$ +292° (C_2H_5OH). A macrocyclic diterpene of the jatrophane type from *Jatropha gossypiifolia* (Euphorbiaceae). In animal experiments J. is active against various types of cancer. *2β-Hydroxy-J.* ($C_{20}H_{24}O_4$, M_R 328.41) and 2*α*-hydroxy-J. from the roots of the same plant are also active against tumor cells. Other *Jatropha* species are known for their nettle toxins[1].

Lit.: [1] Angew. Chem. Int. Ed. Engl. **20**, 164–184 (1981).
gen.: Beilstein E V **17/11**, 473 ▪ Hager (4.) **5**, 310 ff. ▪ J. Am. Chem. Soc. **111**, 6648–6656 (1989); **114**, 7692 (1992) (synthesis) ▪ Lindberg **1**, 224. – *[CAS 29444-03-9 (J.); 85201-83-8 (2β-hydroxy-J.); 85152-62-1 (2α-hydroxy-J.)]*

Jatrorrhizine see protoberberine alkaloids.

Jervine see Veratrum steroid alkaloids.

Jesaconine see aconine.

Jesuit's bark see Cinchona alkaloids.

Jietacin A.

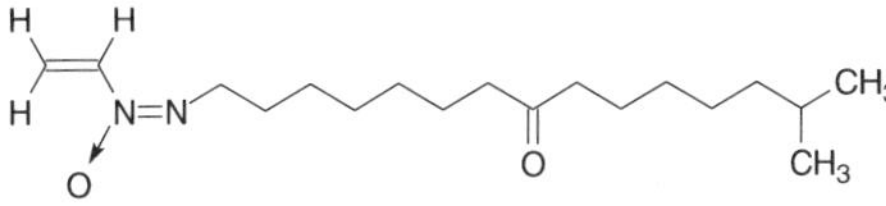

$C_{18}H_{34}N_2O_2$, M_R 310.48. An unsaturated azoxy compound from cultures of a *Streptomycetes* strain. J. A has strong activity against nematodes and a moderate antifungal activity.

Lit.: J. Antibiot. **40**, 623 (1987); **42**, 156 (1989) ▪ Justus Liebigs Ann. Chem. **1989**, 671 (synthesis). – *[HS 2941 90; CAS 109766-61-2]*

Jojoba (jojoba oil). A perennial, dioecious plant (*Simmondsia chinensis*, Buxaceae) indigenous to California and Mexico. Oil content of seeds: 45–50%, protein up to 30%. *J. oil*: mp. 6.8–7 °C, bp. 389 °C. J. oil is an exception among the plant oils since it does not contain triglycerides but rather a liquid wax ester. The main components of J. oil are docosenyl eicosenoate (37%), eicosenyl eicosenoate (24%), and eicosenyl docosenoate (11%), thus mainly wax esters with 40 and 42 C atoms. The wax esters of J. oils are very stable to oxidation (see linoleic acid) since they consist to >95% of mono-unsaturated fatty alcohols and acids of the *ω*9-series.

Use: Especially for cosmetics (as a substitute for cetaceum and sperm whale oil); J. oil is partially hydrogenated and isomerized to give semi-solid, creamy products. Hydrogenated J. oils serve as a carrier for pharmaceuticals (e.g., penicillin) and as a lubricant (partly as sulfur derivatives, e.g., high-pressure gear oil, cutting and drilling oils), softener, transformer oil, leather care products, and furniture polishes.

Lit.: Chem. Ind. (London) **1993**, 94 f. ▪ Ullmann (5.) **A 4**, 102; **A10**, 236 ▪ Wisniak, The Chemistry and Technology of Jojoba Oil, Champaign: Am. Oil. Chem. Soc. 1987. – *[HS 1515 60; CAS 61789-91-1]*

Jolkinol C see lathyranes.

Joro spider toxin see JSTX.

Josamycin see leucomycin.

Joubertiamine see mesembrine alkaloids.

JSTX (joro spider toxin). A group of (acyl-)*polyamines from the tropical orb weaving spider *Nephila clavata* (joro spider, Araneidae), in which *clavamine and the NPTX-polyamines (nephilatoxins) also occur (*spider venoms). The latter have a 3-indolylacetyl or 3-(6-hydroxyindolyl)acetyl group in place of the (2,4-dihydroxyphenyl)acetyl group in JSTX. JSTX[1] and NPTX[2] were the first compounds of this type to be iden-

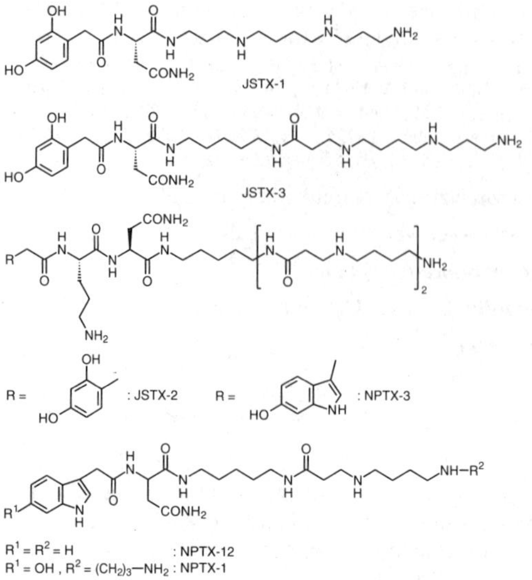

Table: Data of JSTX and NPTX.

JSTX	molecular formula	M_R	CAS
JSTX-1	$C_{22}H_{38}N_6O_5$	466.58	133698-34-7
JSTX-2	$C_{36}H_{64}N_{10}O_8$	764.97	133658-52-3
JSTX-3	$C_{27}H_{47}N_7O_6$	565.71	112163-33-4
NPTX-1	$C_{29}H_{48}N_8O_5$	588.75	119686-54-9
NPTX-3	$C_{38}H_{65}N_{11}O_7$	788.00	119613-49-9
NPTX-12	$C_{26}H_{41}N_7O_4$	515.65	119613-55-7

tified. Like all neurotoxic polyamines they block the calcium-dependent ion channels of glutamate receptors.
Lit.: [1] Proc. Jpn. Acad. Ser. B **62**, 359–362 (1986). [2] Tetrahedron Lett. **38**, 8297 (1997); **39**, 6479 (1998); Heterocycles **47**, 171 (1998).
gen.: Biomed. Res. **9**, 421 (1988) ▪ J. Chem. Ecol. **19**, 2411–2451 (1993) ▪ Pept. Chem. **1989**, 833 ▪ Pharmacol. Ther. **52**, 245–268 (1991) ▪ Tetrahedron **46**, 3813–3818 (1990). – *Synthesis:* Tetrahedron **47**, 3305–3312 (1991); **51**, 10687–10698 (1995) ▪ Tetrahedron Lett. **28**, 3509–3514 (1987); **30**, 2337 (1989); **33**, 2833–2840 (1992); **36**, 5231 (1995).

Juglomycins.

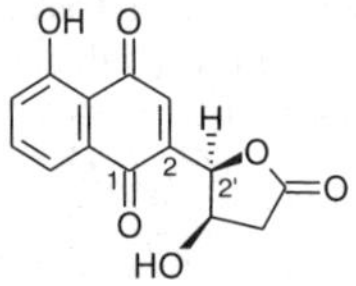

J. A (2'*R*)
J. B (2'*S*)

J. A: $C_{14}H_{10}O_6$, M_R 274.23, yellow-brown needles, mp. 172 °C (decomp.), $[\alpha]_D$ −51.9° (DMSO), −71.4° (CH_3OH). *J. B:* 2′-epimer of J. A, mp. 202 °C (decomp.), $[\alpha]_D$ +227° (DMSO). Naphthoquinone antibiotics (J. A–Z) from *Streptomyces* (*Streptomyces diastatochromogenes* var *juglonensis*) cultures with antibacterial and cytostatic activities. The chiral side chains on the *juglone chromophore can occur in the form of open chains or ring-closed as γ-lactones.
Lit.: J. Antibiot. **24**, 197, 222 (1971); **47**, 1116 (1994) ▪ Z. Naturforsch. B **44**, 345, 353 (1989). – *Synthesis:* J. Org. Chem. **61**, 2986f. (1996) (J. A) ▪ Synth. Commun. **26**, 4501, 4507 (1996) ▪ Tetrahedron Lett. **32**, 6417 (1992). – *[HS 2941 90; CAS 38637-88-6 (J. A); 38637-89-7 (J. B)]*

Juglone (5-hydroxy-1,4-naphthoquinone, C. I. natural brown 7, C. I. 75500).

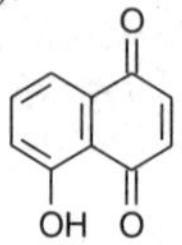

$C_{10}H_6O_3$, M_R 174.16, yellow needles, mp. 164–165 °C (subl.), uv_{max} 422 nm (ethanol), poorly soluble in hot water, readily soluble in chloroform, alcohol, ether, and aqueous alkali (to give a red solution); J. can be steam distilled. It is formed from the 4-*O*-β-D-glucoside of 1,4,5-naphthalenetriol (hydrojuglone; see plant germination inhibitors) contained in the leaves and unripe nutshells of *Juglans* species (walnuts, butternuts). J. is also found in the leaves and nuts of *Carya illinoensis* (Juglandaceae) and in *Penicillium diversum*. J. has allelopathic, antimicrobial, and hemostatic properties.
Biosynthesis: J. is formed from isochorismic acid and 2-oxoglutarate in the 2-succinylbenzoate pathway [1–3].
Lit.: [1] Luckner (3.), p. 330. [2] Z. Naturforsch. B **23**, 259 (1968). [3] Prog. Bot. **54**, 218 (1993).
gen.: J. Med. Chem. **21**, 26 (1978) ▪ Justus Liebigs Ann. Chem. **1981**, 2285 ▪ Phytother. Res. **4**, 11 (1990) ▪ Synthesis **1985**, 781; **1995**, 780 ▪ Synth. Commun. **20**, 2907 (1990) ▪ Turner **1**, 133. – *[HS 2914 69; CAS 481-39-0]*

Junipene see longifolene.

Juniper berry oil. Colorless to light yellow oil with a green-herby, somewhat earthy balsamy odor of conifers and a bitter aromatic taste.
Production: By steam distillation from ripe, dried juniper berries (*Juniperus communis*). *Origin:* mainly South Eastern Europe.
Composition [1]***:*** The main components are monoterpene hydrocarbons such as α-*pinenes* (25–40%), *sabinene* (see thujenes) (4–9%), **myrcene* (8–12%). Important contributors to the organoleptic impression are *terpinen-4-ol* (see *p*-menthenols) (2–10%) and *bornyl acetate* ($C_{12}H_{20}O_2$, M_R 196.29) (0.3–0.8%).
Use: In small amounts in the perfume industry, mainly for masculine compositions; to improve the aromatic character of alcoholic beverages, in medicine in antirheumatic ointments, urologic and balneotherapeutic agents.
Lit.: [1] Perfum. Flavor. **9** (3), 38 (1984); **12** (6), 59 (1987); **15** (4), 67 (1990).
gen.: Arctander, p. 316 ▪ Bauer et al. (2.), p. 159 ▪ ISO 8897 (1991). – *Toxicology:* Food Cosmet. Toxicol. **14**, 333 (1976). – *[HS 3301 29; CAS 8012-91-7 (J. b. o.); 76-49-2 (bornyl acetate)]*

Jurubidine see Solanum steroid alkaloids.

Jurubine see Solanum steroid alkaloid glycosides.

Juvabione (methyl todomatuate).

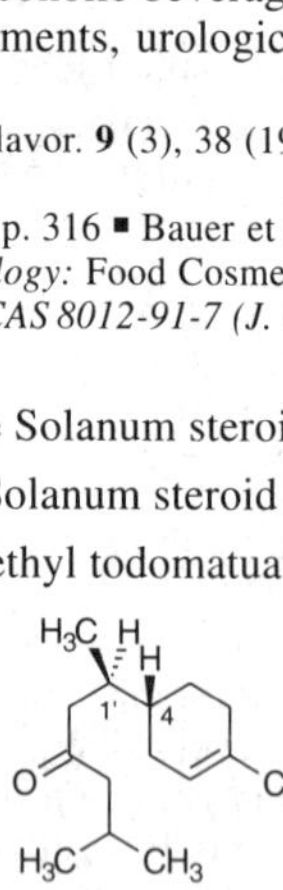

$C_{16}H_{26}O_3$, M_R 266.38, oil, bp. 209–212 °C (1.99 kPa), $[\alpha]_D^{50}$ +93° (acetone), a *sesquiterpene from the wood of the balsam fir *Abies balsamea* with *juvenile hormone activity. J. prevents larvae from entering into metamorphosis and forming the mature adult form. Instead a further larval stage occurs in which the organisms develop abnormally and die. J. acts selectively on insects of the Pyrrhocoridae family only; it is known under the name *paper factor* (in paper from the wood of the balsam fir). J. was the first compound from plants for which juvenile hormone activity was detected. For synthesis, see *Lit.*[1,2].

Lit.: [1] ApSimon **2**, 253–262; Biosci. Biotechnol. Biochem. **59**, 90 ff. (1995); J. Chem. Soc., Perkin Trans. 1 **2000**, 97. [2] J. Chem. Soc., Chem. Commun. **1995**, 2403; J. Am. Chem. Soc. **92**, 366 (1970); Tetrahedron Lett. **33**, 589 (1992); **40**, 4207 (1999).
gen.: Beilstein E III **10**, 2950 ▪ Science **154**, 1020 (1966) ▪ Ullmann (5.) **A 4**, 89. – *[CAS 17904-27-7]*

Juvenile hormones.

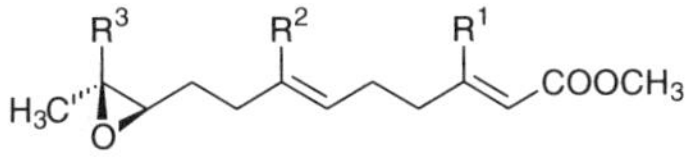

$R^1 = R^2 = R^3 = C_2H_5$: (+)-JH 0, C_{19}-JH
$R^1 = CH_3$, $R^2 = R^3 = C_2H_5$: (+)-JH I, C_{18}-JH
$R^1 = R^2 = CH_3$, $R^3 = C_2H_5$: (+)-JH II, C_{17}-JH
$R^1 = R^2 = R^3 = CH_3$: (+)-JH III, C_{16}-JH

Molting hormones secreted by endocrine glands (corpora allata) of insects for the purpose of triggering larval molting; for details on the specific action, see under insect hormones. *JH I* [C_{18}-JH, methyl (2*E*,6*E*,10*R*,11*S*)-10,11-epoxy-7-ethyl-3,11-dimethyl-2,6-tridecatrienoate, $C_{18}H_{30}O_3$, M_R 294.43, $[\alpha]_D^{23}$ +14.9° ($CHCl_3$)] was isolated from the silkworm; the J. h. of grain beetles, tobacco moths, etc. is methyl (*E,E*)-epoxyfarnesenoate 10,11-(*R*)-oxide (JH III, $C_{16}H_{26}O_3$, M_R 266.38, $[\alpha]_D^{24}$ +6.71° (CH_3OH), the latter compound inhibits biological transmethylation. The (natural) (+)-isomers are ca. 10^4-fold more active than their enantiomers. In the meantime a series of further *farnesene derivatives are known, some of which have similar structures and J. h. activity (so-called *juvenoids*), e. g., derivatives of 4-(1,5-dimethylhexyl)-benzoic acid and *juvabione. A series of substances is characterized by a specific inhibitory activity on the biosynthesis of J. h. in the corpora allata. These substances, known as *anti*-J. h., are produced by plants for self-protection. They serve as leads for new plant protection agents. They can exert their activity in two ways: a deficiency of J. h. results in premature metamorphosis of not adequately developed individuals, while an excess of J. h. induces an additional molting to a giant imago that is not capable of reproduction or life; for details see insect hormones. The juvenoid *Pro-Drone* disrupts differentiation in the fire ant (*Solenopsis invicta*). In 1975 the first substance with a J. h.-like activity, the sesquiterpene methoprene, was authorized for use as a larvicide in the USA.

Lit.: Menn & Beroza, Insect Juvenile Hormones, New York: Academic Press 1972 ▪ Merck-Index (12.), No. 5287 ▪ Tetrahedron Lett. **34**, 8369 (1993). – *Synthesis:* Adv. Asymmetric Synth. **1**, 211–269 (1995) ▪ Helv. Chim Acta **77**, 502–508, 561–568 (1994) ▪ Justus Liebigs Ann. Chem. **1990**, 369 ▪ Zool. Stud. **33**, 237–245 (1994). – *Biosynthesis:* J. Org. Chem. **62**, 3529 (1997) ▪ Tetrahedron **52**, 6869 (1996). – *[CAS 13804-51-8 ((+)-JH I); 22963-93-5 (JH III)]*

K

K252a see staurosporine.

Kabiramide see ulapualides.

Kadsurenone (denudatin B).

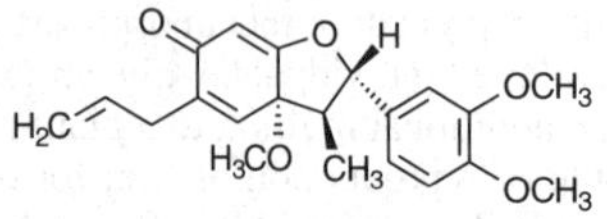

$C_{21}H_{24}O_5$, M_R 356.42, cryst., mp. 62–65 °C, $[\alpha]_D^{22}$ +3.2°; a *lignan from the Chinese pepper species *Piper futokadsura*, with activity as an antagonist of *platelet activating factor, it is undergoing clinical testing as therapy for inflammatory, asthmatic, and cardiovascular diseases.

Lit.: J. Chem. Res. (S) **1998**, 168 ▪ J. Lipid Mediators **1**, 125–137 (1989) ▪ Phytochemistry **24**, 2079 (1985) ▪ Proc. Natl. Acad. Sci. (USA) **82**, 672–678 (1985) ▪ Tetrahedron Lett. **27**, 309 (1986); **29**, 6689 (1988); **40**, 5443 (1999) (synthesis). – *[CAS 95851-37-9]*

Kaempferol (3,4′,5,7-tetrahydroxyflavone). Formula see under flavones. $C_{15}H_{10}O_6$, M_R 286.24, yellow needles (aqueous C_2H_5OH), mp. 276–278 °C, uv_{max} 365 nm. K. is a flavonol which, besides *quercetin, in the form of various glycosides represents the most frequently occurring *flavonoid in plants. 0.3% aqueous ethanol solutions are used for the photometric determination of Ga, In, and Sn(IV) (uv_{max} 430 nm, log ε 4.6).

Lit.: Karrer, No. 1497–1509 ▪ Merck-Index (12.), No. 5288. – *[HS 2932 90; CAS 520-18-3]*

Kahweofuran see coffee flavor.

Kahweol see coffee flavor.

Kainic acid see domoic acid.

Kairomones. Name, derived from the Greek kairos (= favorable condition, advantage + hormone), for signalling substances (*semiochemicals) acting between individuals of different species with attractant effects, in contrast to within-species *pheromones (see also allelochemicals, allomones, insect attractants), providing an advantage for the recipient. *Example:* Predators of bark beetles locate their prey on the basis of the attractants produced by the latter to attract members of their own species.

Lit.: Agosta, Bombardier Beetles and Fever Trees, Chemical Warfare in Animals and Plants, New York: Addison-Wesley 1995.

Kalafungin (kalamycin). $C_{16}H_{12}O_6$, M_R 300.27, orange cryst., mp. 163–166 °C, 1*R*,3*R*,4*R*-form: $[\alpha]_D$ +159° ($CHCl_3$). *Benzoisochromanequinone antibiotic from *Streptomyces tanashiensis*. K. inhibits the growth of

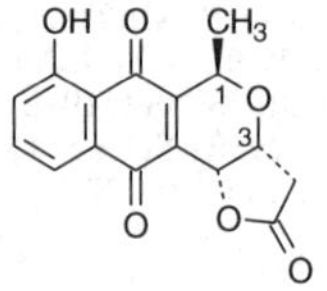

Gram-positive bacteria and fungi. The antipode (1*S*,3*S*,4*S*: *nanaomycin D) is produced by another *Streptomyces* strain. The biosynthesis of K. proceeds through the polyketide pathway.

Lit.: Justus Liebigs Ann. Chem. **1974**, 1100–1125. – *Biosynthesis:* J. Antibiot. **33**, 711–716 (1980); **43**, 391 (1990); **44**, 995–1005 (1991). – *Synthesis:* Bull. Chem. Soc. Jpn. **58**, 1699 (1985) ▪ J. Antibiot. **38**, 680–682 (1985) ▪ J. Org. Chem. **59**, 7572 (1994); **60**, 1154 (1995) ▪ Synth. Commun. **17**, 1021 (1987). – *[HS 2941 90; CAS 11048-15-0]*

Kalihinol-F see isocyanides.

Kamala. An orange-red to brown-red powder composed of the external glands, fruit hairs, and leaves of *Mallotus philippinensis* (Euphorbiaceae), a 4–10 m high tree indigenous to India, Burma, and Sri Lanka. The main colored components of this pigment are the red-yellow chalcone dye *rottlerin besides other phloroglucinol dyes. K. was used in England and India to dye silk and wool (orange) and, on account of its mild laxative activity, as an anthelmintic in veterinary medicine.

Lit.: Hager (5.) **1**, 768. – *[HS 3203 00]*

Kamassine see quebrachamine.

Kanamycins.

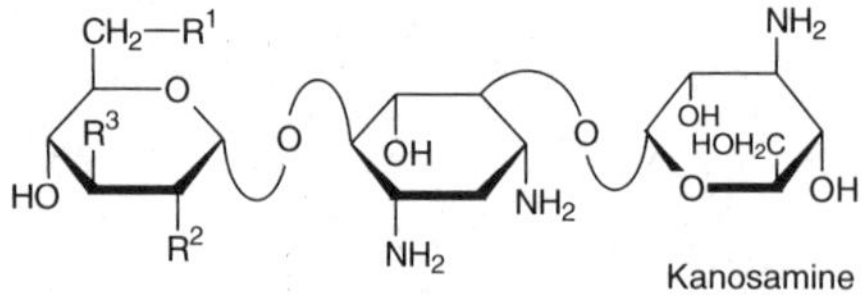

R^1	R^2	R^3	
NH_2	OH	OH	K. A
NH_2	NH_2	OH	K. B
OH	NH_2	OH	K. C
NH_2	NH_2	H	Tobramycin

International generic name for aminoglycoside antibiotics from cultures of *Streptomyces kanamyceticus*. K. were discovered in 1957 by Umezawa [1] and separated into the components *K. A* (main component), *K. B*, and *K. C* [2]. The major constituent, also designated as K., consists of deoxystreptamine, 6-amino-6-deoxy-D-glucose, and 3-amino-3-deoxy-D-glucose (kanosamine) [3]. K. inhibit protein synthesis by interaction with the *30S* subunit of ribosomes [4] and have a broad activity against numerous Gram-positive and Gram-nega-

Table: Data of Kanamycins.

compound	molecular formula	M_R	mp. [°C]	$[\alpha]_D$ (H_2O)	CAS
K. A	$C_{18}H_{36}N_4O_{11}$	484.50	>250 (sulfate)	+149°	59-01-8
K. B	$C_{18}H_{37}N_5O_{10}$	483.52	178–182	+130°	4696-76-8
K. C	$C_{18}H_{36}N_4O_{11}$	484.50	>270	+145°	2280-32-2
Tobramycin	$C_{18}H_{37}N_5O_9$	464.52	–	+131°	32986-56-4

tive pathogens. They are used in human medicine against staphylococci and mycobacteriea as well as in treatment for *Salmonella* and *Shigella* infections. Furthermore, the K. have antiviral, membranotropic, ototoxic, and nephrotoxic properties, LD_{50} (mouse i. v.) 225 mg/kg (K. C). The deactivation of K. by resistant pathogens proceeds through phosphorylation, adenylation, and acetylation.

Lit.: [1] J. Antibiot. **A10**, 181 (1957). [2] J. Antibiot. **A10**, 228 (1957). [3] J. Am. Chem. Soc. **85**, 1547 (1968). [4] J. Antibiot. **21**, 162 (1968).

gen.: Bull. Chem. Soc. Jpn. **56**, 1149 (1983) ▪ J. Chromatogr. **766**, 133 (1997) (analysis) ▪ Martindale (30.), p. 177 ▪ Med. Clin. North Am. **72**, 581 (1988) ▪ Sax (8.), KAL 000, KAM 000, KAV 000. – *[HS 2941 90]*

Karakin see 3-nitropropanoic acid.

Karamu see morindone.

Karnamicins.

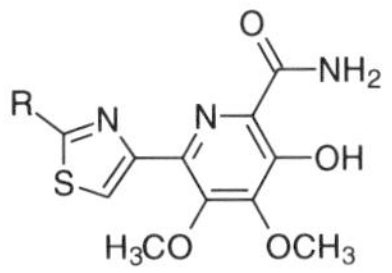

side chain R:

pentyl:
1,4-dihydroxy : K. A_1
4-oxo : K. B_1
4-hydroxy : K. B_2
3-hydroxy : K. C_1
1-hydroxy : K. C_5

4-methylpentyl:
1,4-dihydroxy : K. A_2
4-hydroxy : K. B_3
5-hydroxy : K. C_2

3-methylpentyl:
4-hydroxy : K. C_3
4-oxo : K. C_4
3-hydroxy : K. D_3

hexyl:
5-oxo : K. D_2
4-oxo : K. D_4

5-methylhexyl:
1,5-dihydroxy : K. A_3
5-hydroxy : K. D_1

The karnamicins A_1 to A_3, B_1 to B_3, C_1 to C_5 and D_1 to D_4 are antifungal antibiotics from *Saccharothrix aerocolonigenes*; they vary in the structure of the side chain, e.g., *K. A_1*: $C_{16}H_{21}N_3O_6S$, M_R 383.42, powder, mp. 56–59 °C, $[\alpha]_D^{25}$ −11° ($CHCl_3$), *K. B_1*: $C_{16}H_{19}N_3O_5S$, M_R 365.41, needles, mp. 151–153 °C.

Lit.: J. Antibiot. **42**, 852–868 (1989) (isolation) ▪ Tetrahedron Lett. **38**, 4811 (1997) (synthesis). – *[CAS 122535-48-2 (K. A_1); 122535-51-7 (K. B_1)]*

Kasugamycin. $C_{14}H_{25}N_3O_9$, M_R 379.37. K. is an aminoglycoside *antibiotic isolated from cultures of *Streptomyces kasugaensis*. K. inhibits ribosomal protein synthesis (specifically the *50S* subunit) [1]. In addi-

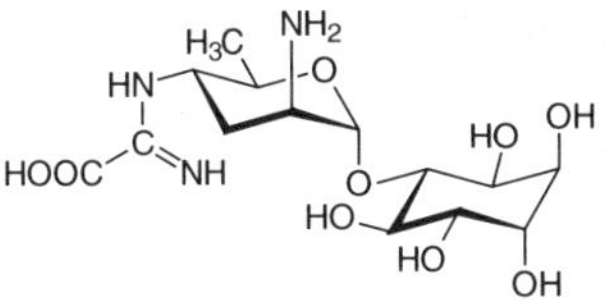

tion to bactericidal properties K. also shows antifungal activity and is thus used in Japan to combat rice mildew (*Pyricularia oryzae*) [2].

Lit.: [1] FEBS Lett. **117**, 119 (1984). [2] Agrochemicals Handbook (3.) A 501, London: R. Soc. Chem. 1992.

gen.: Agrochemicals Handbook (3.), A 501 (1992) ▪ Pesticide Manual, A World Compendium (9.), No. 7610, Farnham: The British Crop Protection Council 1991 ▪ Tetrahedron Lett. **1971**, 1239; **1979**, 2517. – *Biosynthesis:* J. Antibiot. **21**, 50 (1968). – *[HS 2941 90; CAS 6980-18-3]*

Kat (khat, Abyssinian tea). The name K. is used for the consumable tips of twigs and leaves of the K. shrub (*Catha edulis*, Celastraceae) cultivated in Yemen, Ethiopia, Somalia, Kenya, and Madagascar. The appearance of the K. shrub is similar to that of the tea plant (*Thea* sp.). K. is chewed and has an aromatic, bitter taste with a mild anesthetic effect. The plant contains various substances with narcotic activities. The main active principle of the drug is (−)-cathinone (see phenylethylamine alkaloids), a substance related structurally to amphetamine with similar pharmacological properties. The effects of K. chewing by humans have been described as follows: tiredness disappears, a mild stimulation induces a tendency to speak freely so that the subject feels confident, especially in company. Excessive doses cause elevated blood pressure, headache, hyperthermia, palpitations, and sleeplessness (insomnia). When the drug is dried, endogenous plant enzymes reduce most of the cathinone to norpseudoephedrine, whereupon the cerebral stimulating effect is reduced to one fifth.

Lit.: Hager (5.) **3**, 886; **4**, 730 ff. ▪ Manske **39**, 139–164 ▪ Naturwiss. Rundsch. **34**, 19–21 (1981); **39**, 293–297 (1986) ▪ Pure Appl. Chem. **66**, 2183–2188 (1994).

Kauranes. Tetracyclic *diterpenes with the kaurane skeleton as common feature. The hydrocarbons *kaurane* {$C_{20}H_{34}$, M_R 274.49, mp. 85–87 °C, $[\alpha]_D^{25}$ −22.9° ($CHCl_3$)} and *kaurene* {$C_{20}H_{32}$, M_R 272.47, mp. 50 °C, $[\alpha]_D$ −80° ($CHCl_3$)} occur, among others, in the fungus *Gibberella fujikuroi* and in the kauri pine. Kaurenes are important intermediates in the biosynthesis of *gibberellins. Kaurene is formed in the kaurene synthetase reaction, in which geranylgeranyl diphosphate is cyclized to a bicyclic precursor under proton initiation; a complex series of electrocyclic additions, rearrangements of the carbon skeleton, and a subsequent proton abstraction then lead to kaurene.

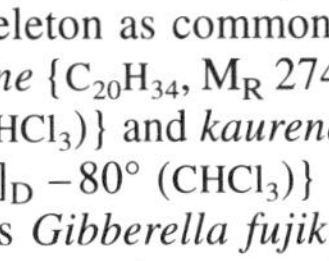

Kaurane

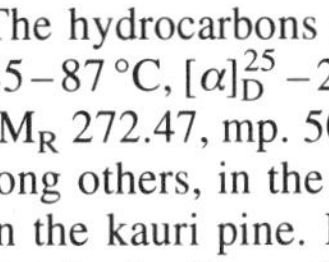

Kaurene

Lit.: Aust. J. Chem. **40**, 469 (1987) (synthesis, kaurene) ▪ Fitoterapia **68**, 303 (1997) (biological activity, review) ▪ Phytochemistry **29**, 660 (1990) (kaurene). – *[CAS 469-84-1 (16β-K.); 6714-12-1 (16α-K.); 562-28-7 ((−)-16-kaurene)]*

Kava-kava, kawain see styrylpyrones.

KDO (ketodeoxyoctulosonic acid, 3-deoxy-D-*manno*-2-octulosonic acid).

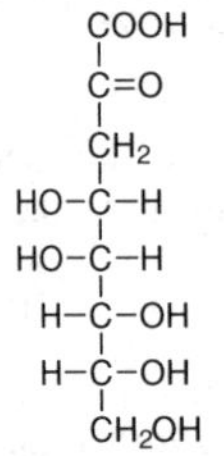

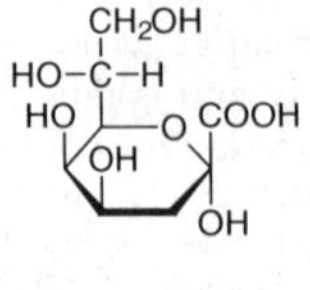

α-pyranose form

$C_8H_{14}O_8$, M_R 238.19, exists not only in open-chain and α- and β-pyranoid but also in furanoid forms. Mp. of the ammonium salt: 125–126 °C. KDO is a sugar acid from Gram-negative bacteria and effects the linkage of carbohydrate and lipid subunits in the biosynthesis of lipopolysaccharides. KDO has also been found in the cell walls of the green alga *Tetraselmis striata* (Prasinophyceae)[1]. Derivatives of KDO are undergoing testing in antibacterial chemotherapy.

Lit.: [1] Eur. J. Biochem. **182**, 153–160 (1989).
gen.: Adv. Carbohydr. Chem. Biochem. **50**, 211–276 (1994). – *Analysis:* Methods Enzymol. Tl. B **41**, 32 ff. (1975). – *Biosynthesis:* J. Biol. Chem. **264**, 6956–6966 (1989). – *Pharmacology:* J. Immunol. **142**, 185–194 (1989). – *Synthesis:* Helv. Chim. Acta **72**, 213–223 (1989) ▪ Tetrahedron **53**, 3325–3346 (1997) ▪ Tetrahedron Lett. **38**, 6415 (1997). – *[CAS 10149-14-1]*

Kedarcidin. A further *enediyne antibiotic with antitumor activity besides the *calicheamicins, *esperamycins, *dynemicins, and *neocarzinostatins.

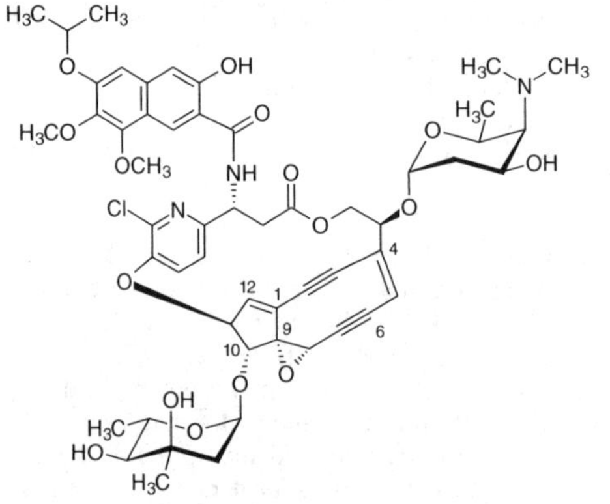

$C_{53}H_{60}ClN_3O_{16}$, M_R 1030.53, brown, amorphous solid. K. shows potent activity in a mouse leukemia model (P 388) with an ED_{50} of 0.2 mg/kg. K. has the same core structure as *neocarzinostatin.

Lit.: J. Am. Chem. Soc. **114**, 7946 (1992); **119**, 12012 (1997) (revised structure) ▪ Nachr. Chem. Tech. Lab. **41**, 175 (1993) ▪ Tetrahedron Lett. **39**, 9633 (1998); **40**, 8281 (1999) (synthesis). – *[CAS 128512-39-0]*

Kempene see termites.

Keracyanin [sambucin, 3-*O*-rhamnoglucoside of *cyanidin, 3-(6-*O*-α-L-rhamnopyranosyl-β-D-glucopyranosyloxy)-3′,4′,5,7-tetrahydroxyflavylium]. $C_{27}H_{31}O_{15}^+$, M_R 595.53, uv_{max} 532 (ethanolic HCl); as chloride: $C_{27}H_{31}ClO_{15}$, M_R 630.99, red needles, mp. 175 °C (co-crystallized with 3.5 molecules of water). Pigment from sweet cherries (*Prunus avium*), elder (*Sambucus*), plums (*Prunus domestica*), tulips (*Tulipa*), and blackthorn (*Prunus spinosa*). K. is purported to have, among others, a regenerative effect on visual purple (rhodopsin).

Lit.: Beilstein EV **17/8**, 476 ▪ Drugs Today **15**, 395 (1979) ▪ Karrer, No. 1717 ▪ Z. Naturforsch. C **33**, 475 (1978). – *[HS 2938 10; CAS 18719-76-1 (chloride)]*

Keramamides see theonellamides.

Kermesic acid (3,5,6,8-tetrahydroxy-1-methylanthraquinone-2-carboxylic acid, C. I. 75460). Formula see carminic acid; $C_{16}H_{10}O_8$, M_R 330.25, red needles, mp. 320 °C (decomp.), uv_{max} 538 nm (C_2H_5OH). The main component of *kermes* (false cochineal), a pigment from dried female scale insects (*Kermes vermilio*) which are parasites of the so-called scarlet oak (*Quercus coccifera*, Fagaceae) growing in the Middle East and Mediterranean region. K. is also obtained from the Polish cochineal (*Porphyrophora polonica*). Kermes was used in the Middle East as much as 3000 years ago as a scarlet dye since it colors wool orange in acidic solution, blue-red when mixed with alum mordant, and scarlet red when mixed with tin mordant.

Lit.: Aust. J. Chem. **34**, 2401 (1981) ▪ J. Chem. Soc., Perkin Trans. 1 **1997**, 3637 (synthesis) ▪ J. Org. Chem. **52**, 5469 (1987) ▪ Merck-Index (12.), No. 5304. – *[HS 3203 00; CAS 18499-92-8]*

Kestoses.

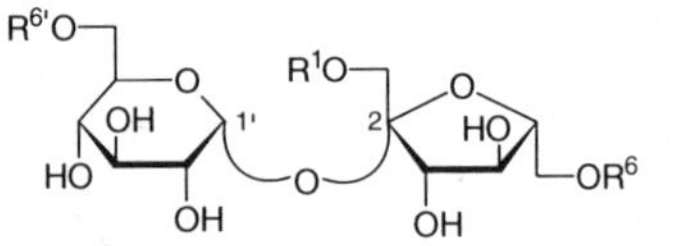

	R^1	R^6	$R^{6'}$
1-Kestose	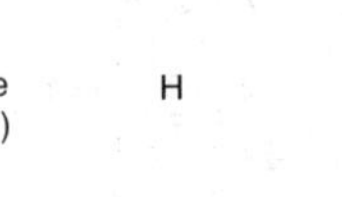	H	H
6-Kestose	H	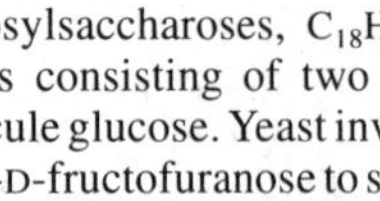	H
Neokestose (6′-Kestose)	H	H	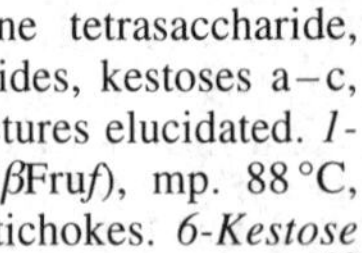

Isomeric fructofuranosylsaccharoses, $C_{18}H_{32}O_{16}$, M_R 504.44. Trisaccharides consisting of two molecules fructose and one molecule glucose. Yeast invertase catalyses the transfer of β-D-fructofuranose to saccharose. To date, four disaccharides, one tetrasaccharide, and the three isomeric trisaccharides, kestoses a–c, have been isolated and their structures elucidated. *1-Kestose* (Glc*p*α1 ↔ 2βFru*f*1 ← 2βFru*f*), mp. 88 °C, $[\alpha]_D^{20}$ +29.2° (H_2O) occurs in artichokes. *6-Kestose* (Glc*p*α1 ↔ 2βFru*f*6 ← 2βFru*f*), mp. 145 °C, $[\alpha]_D^{20}$ +27.3° (H_2O). *Neokestose* (Fru*f*β2 → 6Glc*p*α1 ↔ 2-βFru*f*), amorphous, $[\alpha]_D^{20}$ +22.2° (H_2O).

Lit.: Beilstein EV **17/8**, 416 ▪ Carbohydr. Res. **217**, 43 (1991) (NMR) ▪ Karrer, No. 3357, 3358. – *[CAS 470-69-9 (1-K.); 562-68-5 (6-K.); 3688-75-3 (Neokestose)]*

Ketodeoxyoctonate see KDO.

Khat see kat.

Khellactone see visnadin.

Khellin (4,9-dimethoxy-7-methyl-5*H*-furo[3,2-*g*][1]-benzopyran-5-one).

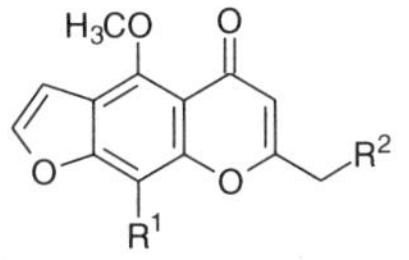

$R^1 = OCH_3$, $R^2 = H$: Khellin
$R^1 = H$, $R^2 =$ Gluc : Khellinin
$R^1 = H$, $R^2 = OH$: Khellol
$R^1 = R^2 = H$: Visnagin

$C_{14}H_{12}O_5$, M_R 260.25. Bitter tasting needles, mp. 154–155 °C, bp. 180–200 °C (6 Pa), LD_{50} (rat p.o.) 80 mg/kg. K. is a furochromone (benzopyranone; see furocoumarins) from the fruits of *Ammi visnaga* (bishop's-weed, Arabian: khella), an Apiaceae from the Mediterranean region. The fruits also contain the glucoside *khellinin* (khellol glucoside, khelloside), $C_{19}H_{20}O_{10}$, M_R 408.36, cryst., mp. 179 °C, cryst. with 2 molecules H_2O, mp. 142 °C, $[\alpha]_D^{17} -1.69°$. The aglycone of khellinin, *khellol* ($C_{13}H_{10}O_5$, M_R 246.22, mp. 179 °C) is a constituent of *Cimicifuga simplex. Visnagin* ($C_{13}H_{10}O_4$, M_R 230.22, mp. 144 °C) can be isolated from the seeds of *Ammi visnaga.*

Activity: K. and khellinin have vasodilatatory and spasmolytic activities. K. is used in cases of angina pectoris, bronchial asthma, and skin diseases, e.g., vitiligo; khellinin for severe attacks of angina pectoris [1]. K. has cholesterol-lowering and phototoxic properties [2] (see furocoumarins).

Lit.: [1] Farmaco Ed. Sci. **43**, 333–346 (1988); J. Invest. Dermatol. **90**, 720 (1988). [2] Arzneim. Forsch. **35**, 1257 (1985).
gen.: Acta Crystallogr., Sect. C **53**, 1949 (1997) ▪ Arch. Pharm. (Weinheim, Ger.) **320**, 823–829 (1987) ▪ Beilstein EV **19/6**, 320 ▪ Florey **9**, 371–396 ▪ Hager (5.) **8**, 677 ff. ▪ Heterocycles **27**, 1159 ff (1988) ▪ J. Am. Chem. Soc. **107**, 5823 f. (1985) ▪ J. Org. Chem. **54**, 3625–3634 (1989) ▪ Negwer (6.), No. 3052 ▪ Tetrahedron Lett. **26**, 1385–1388 (1985) (synthesis). – *Pharmacology:* Martindale (30.), p. 1024 ▪ Planta Med. **54**, 131–135 (1988); **60**, 101–105 (1994). – *Toxicology:* Sax (8.), AHK 750. – *[HS 2932 90, 2938 90; CAS 82-02-0 (K.); 17226-75-4 (khellinin); 478-79-5 (khellol); 82-57-5 (visnagin)]*

Khusimol see vetiver oil.

Kifunensine.

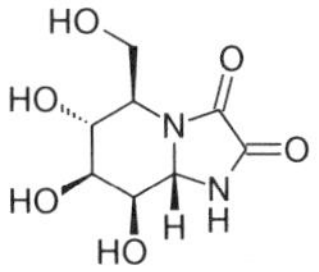

$C_8H_{12}N_2O_6$, M_R 232.20. Immunomodulator with α-mannosidase inhibiting activity from cultures of *Kitasatosporia kifunense*. The absolute configuration of K. corresponds to that of *deoxymannojirimycin.

Lit.: Chem. Pharm. Bull. **39**, 1378, 1392, 1397 (1991). – *[CAS 109944-15-2]*

Kinamycins. A family of antibiotics from streptomycetes with activity against Gram-positive bacteria, however, their use is limited by their high toxicity. At first, an *N*-cyanobenzo[*b*]carbazole structure was

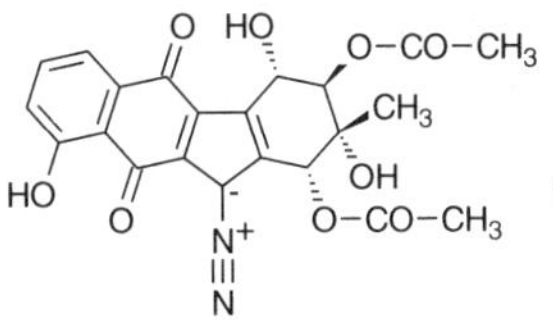

Kinamycin D

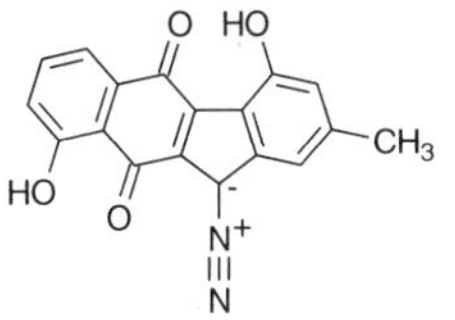

Prekinamycin

assumed [1] for the compounds such as, e.g., *K. D* {$C_{22}H_{18}N_2O_9$, M_R 454.39, orange cryst., mp. 170–175 °C (decomp.), $[\alpha]_D$ −37° ($CHCl_3$)} isolated from *Streptomyces murayamaensis* and a *Saccharothrix* species and appeared to be confirmed by X-ray crystallography [2]. However, later biosynthetic investigations revealed that K. are actually 5-diazobenzo[*b*]fluorenes [3]. Biosynthetic precursor of K. is the purple-colored *prekinamycin* [$C_{18}H_{10}N_2O_4$, M_R 318.29, mp. >300 °C (decomp.)] [3,4].

Lit.: [1] J. Antibiot. **23**, 315 (1970); **24**, 353 (1971). [2] Isr. J. Chem. **10**, 173 (1972). [3] J. Am. Chem. Soc. **116**, 2207, 2209 (1994) (bibliography); J. Org. Chem. **61**, 5720 (1996). [4] J. Antibiot. **42**, 179, 189 (1989); Chem. Rev. **97**, 2499 (1997); J. Org. Chem. **61**, 5722 (1996) (synthesis). – *[HS 2941 90; CAS 35303-14-1 (K. D); 120796-24-9 (prekinamycin)]*

Kinetin [*N*-(2-furanylmethyl)-7*H*-purin-6-amine; N^6-furfuryladenine].

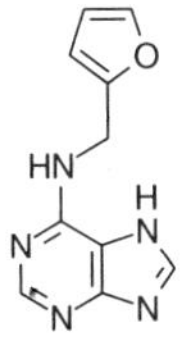

$C_{10}H_9N_5O$, M_R 215.21, mp. 267 °C. K. belongs to the *cytokinin group and is a derivative of adenine; it occurs in various higher plants and yeasts and acts as a cell division factor. Like the closely related *zeatin, K. is still active at million-fold dilutions and is thus considered to be a plant hormone. K. is isolated from autoclaved, aqueous slurries of deoxyribonucleic acids and is sometimes considered to be an artefact [1]. K. is used as a plant growth regulator and also for inducing callus formation in plant cell cultures, etc. [1].

Lit.: [1] Plant Cell, Tissue Organ Cult. **36**, 73–79 (1994).
gen.: Beilstein EV **26/16**, 248 ▪ Merck-Index (12.), No. 532 f. ▪ Ullmann (5.) **A 20**, 421. – *[HS 2934 90; CAS 525-79-1]*

King Kong peptide (conotoxin KK-0). A peptide (from the cone snail *Conus textile*) composed of 27 amino acids. It induces muscular spasms in snails. It has a remarkable effect on crabs: small crabs do not act in the usual submissive way towards larger members of the species but rather commandingly erect their head parts and curl their tails up like a scorpion. The toxin was given its name on the basis of this unexpected activity. In spite of large structural similarities with other *Conus* toxins (e.g., the same arrangement of cysteine building blocks) it has a completely different effect the reason for which is still unknown. Sequence: H-Trp-Cys-Lys-Gln-Ser-Gly-Glu-Met-Cys-Asn-Leu-Leu-Asp-Gln-Asn-Cys-Cys-Asp-Gly-Tyr-Cys-Ile-Val-Leu-Val-Cys-Thr-OH.

Lit.: Biochemistry **28**, 358 (1989) ▪ Brain Res. **640**, 48 (1994) ▪ Chem. Rev. **93**, 1923 (1993) ▪ Eur. J. Biochem. **202**, 589 (1991) ▪ Naturwiss. Rundsch. **43**, 29 (1989). – *[CAS 128194-63-8]*

Kirromycin (mocimycin, delvomycin).

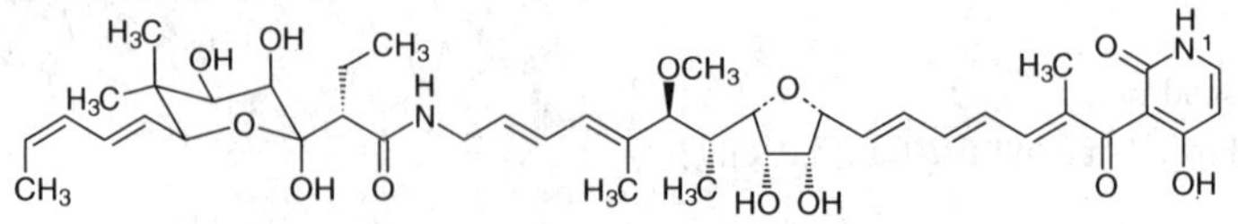

$C_{43}H_{60}N_2O_{12}$, M_R 796.95, yellow powder, mp. 141 °C, $[\alpha]_D$ −60° (CH_3OH), a narrow-spectrum *antibiotic from *streptomycetes (e. g., *Streptomyces collinus*) that only has growth inhibitory activity against some Gram-positive and Gram-negative bacteria, especially clostridia. K. is an inhibitor of protein biosynthesis through specific interactions with elongation factor Tu (EF-Tu). K. is the parent compound of a family of antibiotics known as the *elfamycins*. These include, e. g., the N^1-methyl derivative goldinomycin (see efrotomycin). K. and goldinomycin are used in the fowl breeding industry because they effect a better utilization of food and have low toxicity (nutritional effect).

Lit.: J. Antibiot. **46**, 1175 (1993); **48**, 1312 (1995) ▪ Magn. Res. Chem. **27**, 748 (1989) ▪ Parmegani & Sander, in Sammes (ed.), Topics of Antibiotic Chemistry, p. 159–222, Chichester: Ellis Horwood 1980. – *[CAS 50935-71-2]*

Kitamycin see leucomycin.

Kitol see retinol.

Kiwi flavor see fruit flavors.

Knipholone. Biphenylic constituent of the rizomes and bulbs of *Kniphofia foliosa* and *Bulbine* spp. (Asphodelaceae), respectively. *Bulbine* species are used in traditional medicine for the treatment of various ailments that probably arise from bacterial and fungal infections. The milk decoction of the roots of *B. capitata* are used for the treatment of body rash and sexually transmitted diseases in Botswana.

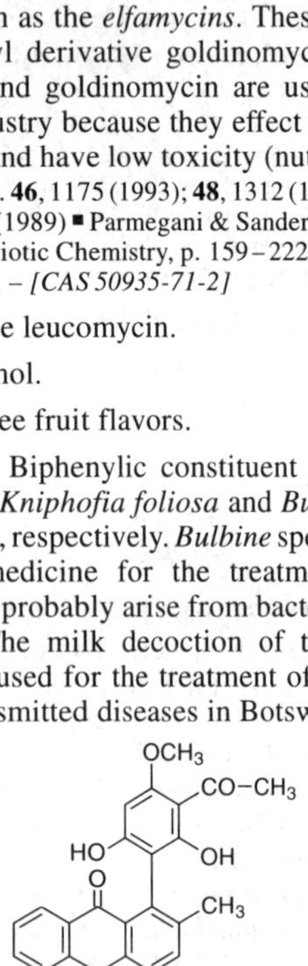

K.: $C_{24}H_{18}O_8$, M_R 434.39, red needles, mp. 237 °C, $[\alpha]_D^{22}$ +80° (CH_3OH).

Lit.: Phytochemistry **35**, 685 ff. (1994); **37**, 525 (1994); **46**, 1063 (1997). – *[CAS 94450-08-5]*

Kojic acid [5-hydroxy-2-hydroxymethyl-4*H*-4-pyrone].

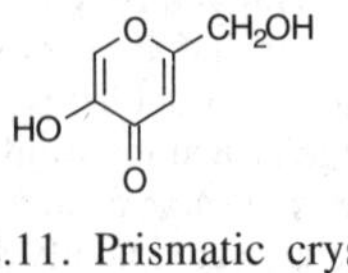

$C_6H_6O_4$, M_R 142.11. Prismatic crystal needles, mp. 153–154 °C, soluble in water; K. is formed by *Aspergillus* species (e. g., *A. albus*, *A. flavus*, *A. oryzae*, *A. fumigatus*), *Penicillium daleae* and bacteria of the genus *Acetobacter* in carbohydrate rich nutrient media, it can also be synthesized from glucose and other hexoses. K. has antibiotic and insecticidal activities and forms chelates. It inhibits polyphenol oxidases (tyrosinase, laccase) and thus the formation of Dopa melanin. The name has a Japanese origin: koji = cultures of *A. oryzae* on rice.

Lit.: Agric. Biol. Chem. **43**, 1337 (1979) ▪ Appl. Environ. Microbiol. **32**, 298 f. (1976) ▪ Beilstein E V **18/2**, 516 ▪ Biosci. Biotech. Biochem. **56**, 987 (1992) ▪ J. Sci. Ind. Res. **41**, 185–194 (1982) ▪ Rehm, Industrielle Mikrobiologie, p. 271 f., Berlin: Springer 1980 ▪ Sax (8.), HLH 500 ▪ Turner **1**, 28. – *[HS 2941 90; CAS 501-30-4]*

Kopsine.

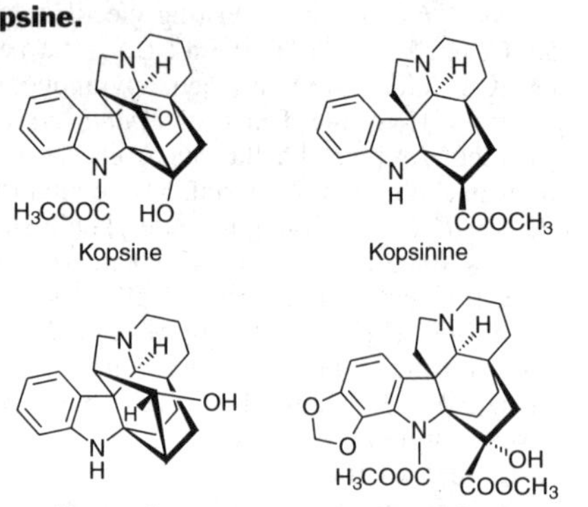

Kopsine, Kopsinine, Kopsanol, Kopsamine

$C_{22}H_{24}N_2O_4$, M_R 380.44, cryst. from alcohol, mp. 217–218 °C, $[\alpha]_D^{27}$ −14.3° ($CHCl_3$), readily soluble in chloroform, poorly soluble in ethanol. K. is a *monoterpenoid indole alkaloid with a heptacyclic skeleton belonging to the group of kopsan alkaloids (over 50 members known to date). It is isolated from the leaves of the dogbane (Apocynaceae) *Kopsia fruticosa*, which contains numerous structurally related alkaloids. *K. fructicosa* is a 1.5–4 m high bush indigenous to Indochine and cultivated in Java as an ornamental plant. On account of its structural similarities with comparable hexacyclic alkaloids of *aspidofractinine* (e. g., with kopsinine), K. is included in the aspidofractinine group, a subgroup of the *Aspidosperma alkaloids. Similar alkaloids are *kopsinine* ($C_{21}H_{26}N_2O_2$, M_R 338.45), *kopsanol* ($C_{20}H_{24}N_2O$, M_R 308.42), and *kopsamine* ($C_{24}H_{28}N_2O_7$, M_R 456.50), an alkaloid of the kopsaline group with a methylenedioxy group, a rather rare feature for the *indole alkaloids [1]. For biosynthesis of K., see *Lit.* [2–4].

Lit.: [1] Planta Med. **48**, 280 (1983). [2] J. Am. Chem. Soc. **105**, 2086 (1983). [3] J. Am. Chem. Soc. **106**, 2105 (1984). [4] J. Am. Chem. Soc. **111**, 6707 (1989).

gen.: Beilstein E V **25/2**, 292; **25/9**, 172; **25/7**, 134 ▪ Hager (5.) **6**, 1125 ▪ Helv. Chim. Acta **65**, 2548 (1982) ▪ Lindberg **1**, 110 ▪ Manske **11**, 244–259 ▪ Saxton (ed.), The Monoterpenoid Indole Alkaloids, Chichester: Wiley 1983; Suppl Vol. 25, Heterocyclic Compounds, p. 357–436, Chichester: Wiley 1994 ▪ Ullmann (5.) **A 1**, 392 f. – *[HS 2939 90; CAS 559-48-8 (K.); 559-51-3 (kopsinine); 6582-68-9 (kopsanol); 1358-62-9 (kopsamine)]*

Korupensamine A see naphthyl-(tetrahydro)isoquinoline alkaloids.

Kosins.

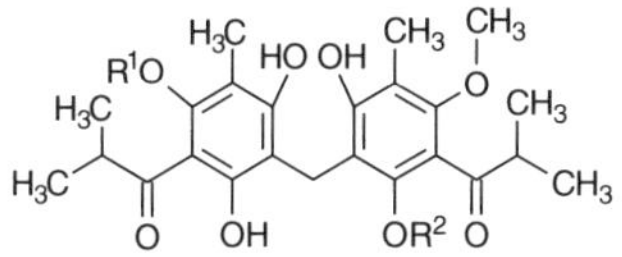

$R^1 = CH_3$, $R^2 = H$: α-Kosin
$R^1 = H$, $R^2 = CH_3$: β-Kosin

Flower pigments of the East African koso tree *Hagenia abyssinica* (Rosaceae), occurring in the plant as a mixture of *α-K.* ($C_{25}H_{32}O_8$, M_R 460.52, yellow needles, mp. 160 °C), and *β-K.* ($C_{25}H_{32}O_8$, M_R 460.52, mp. 120 °C). K. are the decomposition products of the so-called koso toxin and prototoxin. These compounds are muscle toxins for lower animals. Koso also has a muscle-paralysing effect in warm-blooded animals and causes death by respiratory paralysis. In humans, local irritations of the gastrointestinal tract, visual disorders, and severe states of collapse have been observed. K. slows cardiac activity. It is used as an anthelminthic.

Lit.: Merck-Index (12.), No. 5333 ▪ Planta Med. **34**, 153 (1978) ▪ Z. Naturforsch. B **40**, 669 (1985). – *[CAS 568-50-3 (α-K.); 1400-15-3 (β-K.)]*

Kreysigin(in)e, kreysiginone see phenethyltetrahydroisoquinoline alkaloids.

Kryptosterin see lanosterol.

Kuanoniamines. Pyrido[2,3,4-*kl*]acridinone and -acridine alkaloids from the mollusc *Chelynotus semperi* which feeds on tunicates.

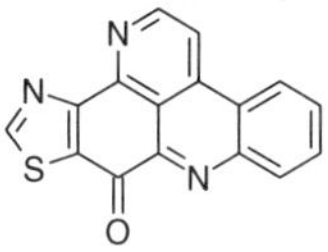

K. A

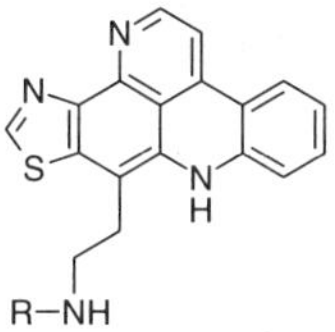

R = $CO-CH_2-CH(CH_3)_2$: K. B
R = $CO-CH_2-CH_3$: K. C
R = $CO-CH_3$: K. D

K. are cytotoxic; yellow needles, e.g., *K. A:* $C_{16}H_7N_3OS$, M_R 289.31, mp. 255–258 °C; *K. B:* $C_{23}H_{22}N_4OS$, M_R 402.51, mp. 300 °C; see also Lissoclinum alkaloids.

Lit.: J. Am. Chem. Soc. **117**, 12460–12469 (1995) (synthesis) ▪ J. Chem. Soc., Perkin Trans. 1 **1999**, 437 (synthesis K. A) ▪ J. Nat. Prod. **61**, 301 (1998) (isolation) ▪ J. Org. Chem. **55**, 4427 (1990); **57**, 1523 (1992) (isolation). – *[CAS 133401-10-2 (K. A); 133401-11-3 (K. B)]*

Kurchi alkaloids see Apocynaceae steroid alkaloids.

L-Kynurenine [(*S*)-2-amino-4-(2-aminophenyl)-4-oxobutanoic acid].

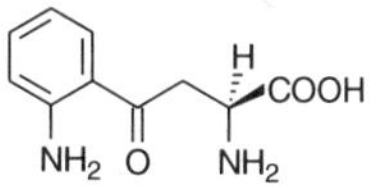

$C_{10}H_{12}N_2O_3$, M_R 208.22, mp. 194 °C (decomp.), $[\alpha]_D^{27}$ –30.5° (H_2O), pK_a 2.38, 9.39, pI 5.89. K. occurs in the urine of various animals as an intermediate product in the metabolism of *tryptophan (the name is derived from Greek: kynos ouron = dog's urine and Latin: renes = kidney).

Metabolism: *N*-Formylkynurenine formed from Trp by tryptophan 2,3-dioxygenase (EC 1.13.11.11) is degraded by arylformamidase (EC 3.5.1.9.) to K. Further degradation of K. proceeds through 3-hydroxykynurenine and by kynureninase (EC 3.7.1.3) to 3-hydroxyanthranilic acid. The benzene ring of the latter is oxidatively cleaved by 3-hydroxyanthranilate 3,4-dioxygenase (EC 1.13.11.6), followed by decarboxylation, reduction, and deamination to give 2-oxoadipic acid by way of which *lysine is also degraded to *acetyl-CoA. 3-Hydroxykynurenine is also an intermediate in the formation of the *ommochromes in crabs and insects.

Lit.: Beilstein E IV **14**, 2562 f. ▪ Gazz. Chim. Ital. **97**, 3 (1967) (synthesis) ▪ Naturwissenschaften **75**, 141 f. (1988). – *[HS 2922 50; CAS 2922-83-0]*

L

L-783,281. $C_{32}H_{30}N_2O_4$, M_R 506.60, amorphous solid.

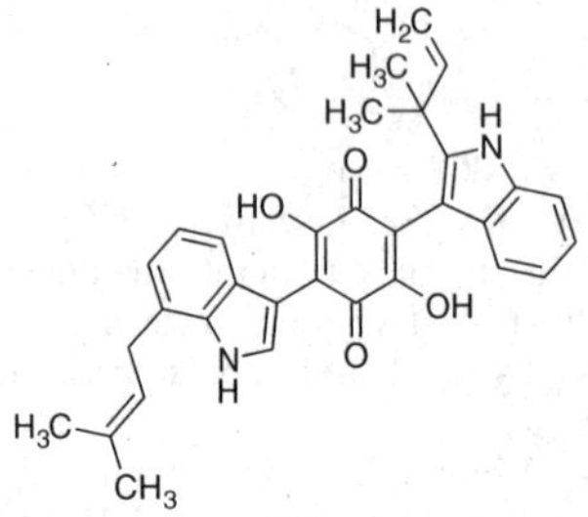

In a screen for small molecules that activate the human insulin receptor tyrosine kinase, a nonpeptidyl fungal metabolite from cultures of *Pseudomassaria* sp. (L-783,281) was identified that acted as an insulin mimetic in several biochemical and cellular assays. The compound was selective for insulin receptor versus insulin-like growth factor I (IGFI) receptor and other receptor tyrosine kinases (insulinreceptor agonist). Oral administration of L-783,281 to two mouse models of diabetes resulted in significant lowering in blood glucose levels. L. and analogues of L. are intensively investigated as new lead structures for oral diabetes therapy.

Lit.: Nature Med. **5**, 614 (1999) ▪ Science **284**, 886, 974 (1999). – *[CAS 78860-34-1]*

Labdane-8,15-diol see labdanum absolute/oil/resinoid.

Labdanes. A large group of diterpenes with the skeleton (**1**). Both enantiomers occur in nature. L. were first isolated from conifers [1]. They also occur in tobacco [2], Asteraceae [3], Lamiaceae [4], Cistaceae [5], Apocynaceae [6], etc. Some of the compounds contain further rings. The L. are mostly derived from the grindelane diterpenes; examples: grindelic acid [7] (**2**), sclareol (**3**) [8], larixol (**4**) [9], agathadiol (**5**) [10], and isoandalusol (**6**) [11].

Table: Data of Labdanes.

no.	molecular formula	M_R	mp. [°C]	$[\alpha]_D$	CAS
2	$C_{20}H_{32}O_3$	320.47	100–101	–102° ($CHCl_3$)	1438-57-9
3	$C_{20}H_{36}O_2$	308.50	105–106	–6.3° (C_2H_5OH)	515-03-7
4	$C_{20}H_{34}O_2$	306.49	103–104	+57° ($CHCl_3$)	1438-66-0
5	$C_{20}H_{34}O_2$	306.49	107–108	+31° ($CHCl_3$)	1857-24-5
6	$C_{20}H_{34}O_3$	322.49	102–105	–34° (CH_3OH)	77887-59-3

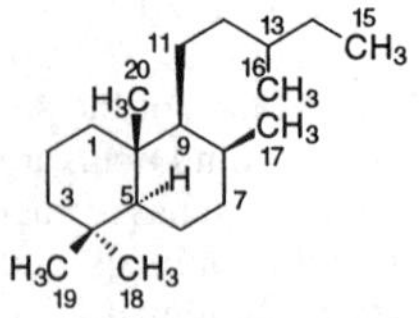

1

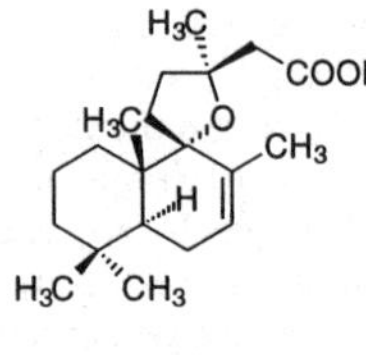

2

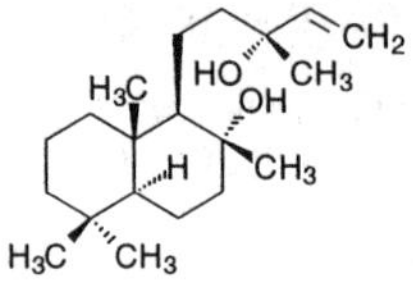

3

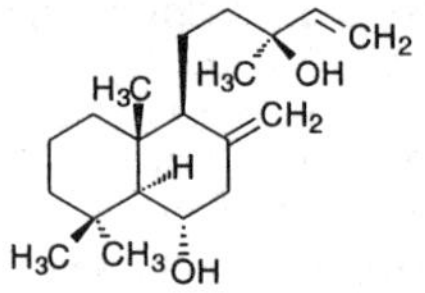

4

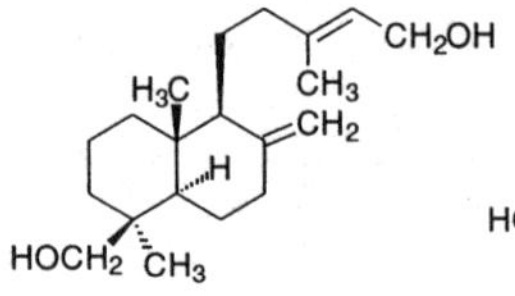

5

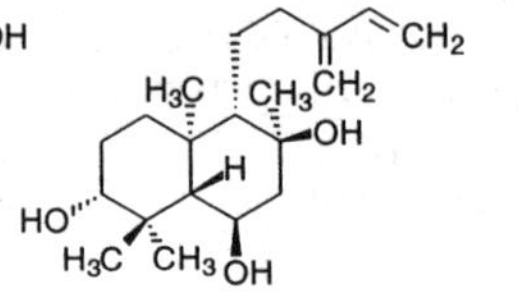

6

Some L. have been reported to have cytotoxic, fungicidal, and antiviral activities. Sclareol, for example, is used in the perfume industry for the synthetic production of ambergris fragrances such as *ambrox.

Lit.: [1] Indian J. Chem., Sect. B **26**, 453 (1987). [2] Acta Chem. Scand., Sect. B **42**, 708 (1988); Phytochemistry **21**, 395 (1982). [3] Phytochemistry **18**, 115, 1533 (1979); **19**, 111, 977, 2475 (1980); **20**, 275, 1613, 2383 (1981); **21**, 173, 1103 (1982); **22**, 1294 (1983); **27**, 2953, 2994 (1988); **28**, 1463 (1989). [4] S. Afr. J. Chem. **41**, 124 (1988). [5] Phytochemistry **28**, 557 (1989). [6] Phytochemistry **27**, 2255 (1988). [7] J. Org. Chem. **61**, 5352 (1996); Tetrahedron Lett. **36**, 6005 (1995). [8] Phytochemistry **29**, 2145 (1990); Tetrahedron **49**, 10405 (1993). [9] Tetrahedron Lett. **1979**, 3263. [10] Phytochemistry **37**, 1109 (1994). [11] Phytochemistry **19**, 2405 (1980).

Labdanum (absolute/oil/resinoid). Absolute and resinoid are dark brown, highly viscous masses. L. oil is a golden-brown to brown liquid. The odor is warm, spicy, ambergris-like erogenic and somewhat balsamy dry, wood-like.

Production: Starting material for all these products is L. resin, sometimes falsely called L. rubber. L. resin is obtained from the Cistus bushes (*Cistus ladanifer*) characteristic for the Mediterranean maqui: the twigs are boiled with water, the resin separates from the water on cooling but still consists to about 1/3 of water; yield: ca. 5–10%. *Origin:* mainly Spain. Extraction of

L. resin with an appropriate solvent (e.g., toluene) leads to the so-called L. resinoid (sometimes called L rubber resinoid or L. rubber concrete), yield: ca. 50%. Precipitation of wax-containing components with alcohol leads to the so-called L. absolute, yield: 80–90%; this product is also prepared by direct extraction of the resin with alcohol. So-called "colorless", "light", or "clear" L. absolutes can be obtained from the normally dark colored L. absolute by dissolving it in hexane. The essential L. oil is obtained from the resin by steam distillation, yield: 0.1–0.2%; this product should not be confused with *Cistus oil*, obtained by steam distillation of young leaves and twigs of Cistus bushes; the latter is of less significance for perfumery.

Composition [1]**:** The resinous components of L. absolute and resinoid are mainly diterpenes with the *labdane skeleton, e.g., *8-hydroxylabdane-15-carboxylic acid* ($C_{20}H_{36}O_3$, M_R 324.50) and *labdane-8,15-diol* ($C_{20}H_{38}O_2$, M_R 310.52)[2], the oxidative degradation products of which contribute decisively to the odor of L. products[3]. L. oil has no main constituents but consists rather of numerous oxygen-containing monoterpenes, sesquiterpene hydrocarbons, and alcohols. In contrast cistus oil contains mainly monoterpene hydrocarbons, e.g., ca. 50% α-*pinene.

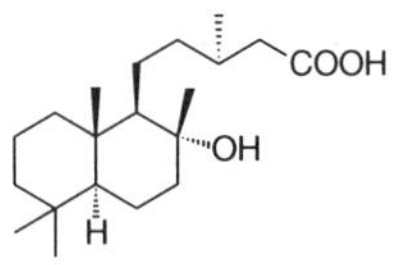

8-Hydroxylabdane-15-carboxylic acid

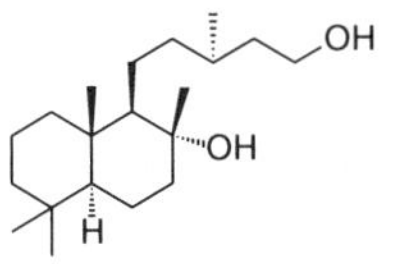

Labdane-8,15-diol

Use: L. products have many uses in the perfume industry, the absolute and the resinoid are excellent fixatives. L. oil is a valuable raw material and is used in small amounts only.

Lit.: [1] Perfum. Flavor. **9** (1), 49 (1984); **11** (4), 73 (1986); **15** (2), 78 (1990). [2] Bull. Chem. Soc. Jpn. **61**, 4023 (1988); Tetrahedron Lett. **26**, 5717 (1985). [3] Ohloff, p. 177.
gen.: Arctander, p. 326 ▪ Bauer et al. (2.), p. 159 ▪ Nat. Prod. Lett. **6**, 285 (1995). – *Toxicology:* Food Cosmet. Toxicol. **14**, 335 (1976). – *[HS 1301 90; CAS 8016-26-0 (L.); 469-11-4 (8-hydroxylabdane-15-carboxylic acid); 10267-22-8 (labdane-8,15-diol)]*

(+)-Laburnine see necine(s).

Laccaic acids (C. I. natural red 25). A group of red anthraquinone pigments excreted from glands of scale insect species (e.g., *Laccifer lacca, Tachardia lacca*) living in South East Asia on *Butea monosperma* (Fabaceae) and *Zizyphus mauritiana* (Rhamnaceae) as host plants and used as pigments for cosmetics and foods.

Table: Data of Laccaic acids.

compound	molecular formula	M_R	CAS
L. A	$C_{26}H_{19}NO_{12}$	537.44	15979-35-8
L. B	$C_{24}H_{16}O_{12}$	496.38	17249-00-2
L. C	$C_{25}H_{17}NO_{13}$	539.41	23241-56-7
L. E	$C_{24}H_{17}NO_{11}$	495.40	–
L. D	$C_{16}H_{10}O_7$	314.25	18499-84-8

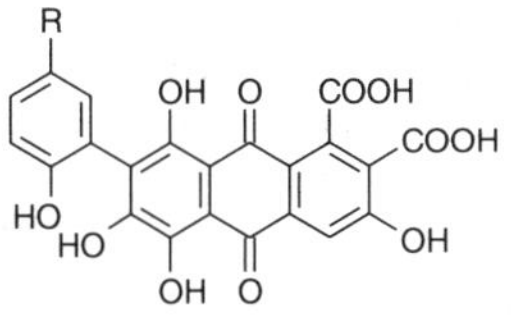

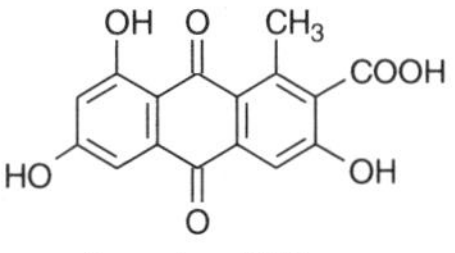

Laccaic acid D

R = $(CH_2)_2$–NH–CO–CH_3 : L. A
R = CH_2–CH_2–OH : L. B
R = CH_2–CH(NH_2)–COOH : L. C
R = CH_2–CH_2–NH_2 : L. E

L. D (xanthokermesic acid) [yellow needles, mp. >300 °C (decomp.), uv_{max} 345 nm] exhibits large structural differences to the L. A–C and also occurs in *Rheum-* and *Cassia* species [1].

Lit.: [1] Aust. J. Chem. **34**, 2401 (1981); J. Chem. Soc., Chem. Commun. **1978**, 688; Chem. Pharm. Bull. **22**, 1159 (1974); Phytochemistry **17**, 895 (1978); Tetrahedron Lett. **28**, 1199 (1987).
gen.: Aust. J. Chem. **31**, 2651 (1978); **35**, 1469 (1982) ▪ Cell. Biol. Toxicol. **1**, 111–125 (1984) ▪ Curr. Sci. **57**, 30 (1988) ▪ Merck-Index (12.), No. 5342 ▪ Thomson **3**, 440.

Laccarin.

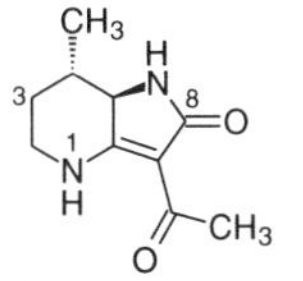

relative configuration

$C_{10}H_{14}N_2O_2$, M_R 194.22, powder, $[\alpha]_D^{31}$ +188° ($CHCl_3$). Alkaloid from the mushroom *Laccaria vinaceoavellanea* (Basidiomycetes). L. is an inhibitor of cyclic AMP phosphodiesterase.

Lit.: Heterocycles **43**, 685 (1996). – *[CAS 175669-28-0]*

Lacceric acid see shellac.

Lachnumone.

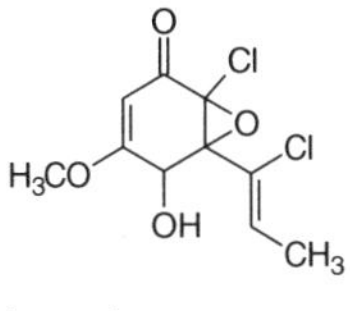

In submersion culture the ascomycete *Lachnum papyraceum* produces not only *mycorrhizin but also varous biogenetically related chlorine compounds. The α-chloroepoxide L. [$C_{10}H_{10}Cl_2O_4$, M_R 265.09, cryst., mp. 70 °C (decomp.)] exhibits nematicide, antimicrobial, and cytotoxic activities.

Lit.: J. Antibiot. **46**, 961, 968 (1993); **48**, 149, 154, 158 (1995). – *[CAS 150671-02-6]*

Lactacystin.

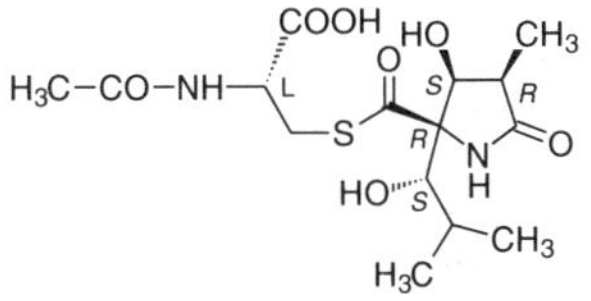

$C_{15}H_{24}N_2O_7S$, M_R 376.43, mp. 236–238 °C (decomp.), $[\alpha]_D^{23}$ +72° (CH_3OH). A neurotrophic (stimulating the growth of nerve cells) γ-lactam thioester from streptomycete cultures. The biosynthesis proceeds from isobutyrate/L-valine, L-leucine, and L-cysteine [1]. L. is a selective irreversible inhibitor of 20S proteasome, a large protein that effects degradation of certain proteins. L. serves as a tool to understand processes like proteolysis, apoptosis, cystic fibrosis[2].

Lit.: [1]Pure Appl. Chem. **66**, 2411 (1994); J. Antibiot. **48**, 1015–1020 (1995). [2]Cell **73**, 1251 (1993).
gen.: Synthesis: Angew. Chem. Int. Ed. **37**, 1676 (1998); **38**, 1093 (1999) ▪ Chem. Commun. **1998**, 1929 ▪ Chem. Pharm. Bull. **47**, 1–10 (1999) (review) ▪ J. Am. Chem. Soc. **118**, 3584–3590 (1996); **120**, 2330 (1998) ▪ Synlett **1998**, 1045 ▪ Tetrahedron **53**, 16287 (1997) ▪ Tetrahedron Lett. **39**, 7475 (1998). – *[CAS 133343-34-7]*

β-Lactam antibiotics. A general name for *antibiotics with a four-membered β-lactam ring in their skeleton, their bactericidal activities are based on the inhibition of bacterial cell wall synthesis[1]. The natural β-L. a. can be classified according to structures into 5 classes (see figure).

basic structure		Antibiotics
Penam	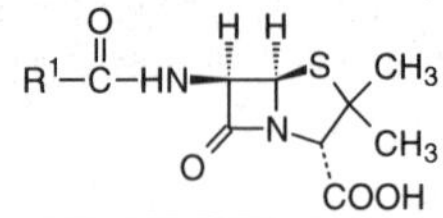	*Penicillins
Ceph-3-em	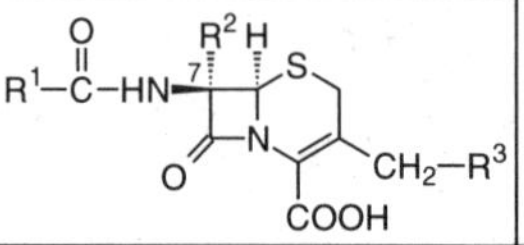	*Cephalosporins 7-Methoxy-cephalosporins ($R^2 = OCH_3$)
Clavam	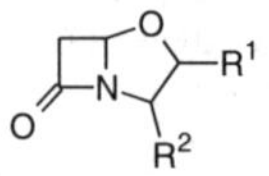	*Clavulanic acid
Carbapenem	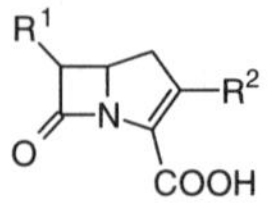	*Thienamycins Olivanic acids Epithienamycins PS compounds Asparenomycins Carpetimycins
Monolactam	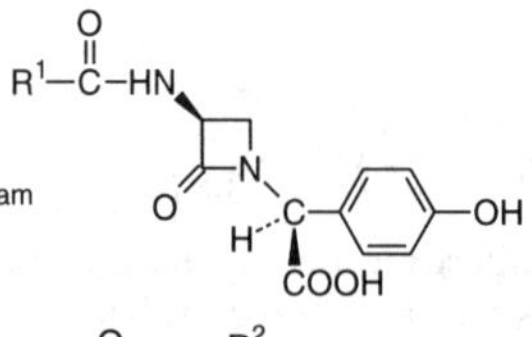	*Nocardicins
	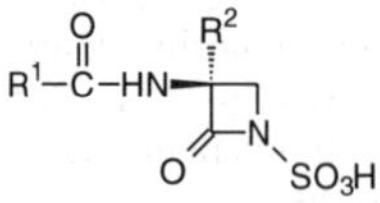	*Monobactams

Figure: The basic structures of naturally occurring β-Lactam antibiotics.

The *penicillins[2], discovered in 1929 by Fleming together with the *cephalosporins[3], found in 1953 are named as the classic β-L. a. With the exception of the *cephamycins[4] (7-methoxycephalosporins) discovered in 1971, they are formed by fungi. Thousands of semisynthetic derivatives of the classical β-L. a. have been examined in order to improve the pharmacokinetic properties and to increase the resistance against the action of β-lactamases[5] (bacterial enzymes that cleave the lactam ring of β-L. a. and thus inactivate them); some of which have become the most efficient therapeutic agents available today for use against infectious diseases (especially infections by Gram-positive bacteria). Improvements in screening methods in the past 15 years have resulted in the isolation of *clavulanic acid[6] from *actinomycetes with good β-lactamase inhibitory activity and the *carbapenem compounds (olivanic acid[7], *thienamycins[8], epithienamycins, PS compounds[9], asparenomycins[10], puracidomycins[11], and carpetimycins[12]) having not only broad antibacterial activities but also β-lactamase inhibiting properties.
The monocyclic lactams include the *nocardicins[13] formed by actinomycetes and *monobactams[14] formed mostly by bacteria and possessing antibiotic activities with differing sensitivities to β-lactamases. For biosynthesis of clavulanic acid see *Lit.*[15].

Lit.: [1]Annu. Rev. Microbiol. **33**, 113 (1979). [2]Br. J. Exp. Pathol. **10**, 226 (1929). [3]Biochem. J. **62**, 651 (1955). [4]J. Am. Chem. Soc. **93**, 2308 (1971). [5]Hamilton-Miller & Smith, Beta-Lactamases, London: Academic Press 1979. [6]J. Chem. Soc., Chem. Commun. **1976**, 266f. [7]J. Antibiot. **29**, 668 (1976). [8]J. Am. Chem. Soc. **100**, 6491 (1978). [9]J. Antibiot. **32**, 262 (1979). [10]J. Antibiot. **35**, 1237 (1982). [11]J. Antibiot. **35**, 536 (1982). [12]J. Antibiot. **33**, 1388 (1980). [13]J. Antibiot. **29**, 492 (1976). [14]J. Antimicrob. Chemother. **8**, Suppl. E 1–16 (1981). [15]Chem. Commun. **1997**, 1025.
gen.: O'Sullivan & Sykes, in Rehm & Reed (eds.), Biotechnology, vol. 4, p. 247–281, Weinheim: Verl. Chemie 1986 ▪ Page (ed.), The Chemistry of β-Lactams, London: Blackie 1992 ▪ Swarz, in Moo-Young (ed.), Comprehensive Biotechnology, vol. 3, p. 7–47, Oxford: Pergamon Press. – *[HS 2941 90]*

Lactaranes.

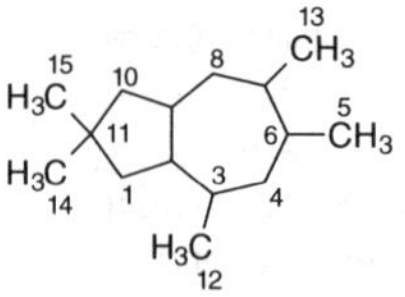

Lactarane

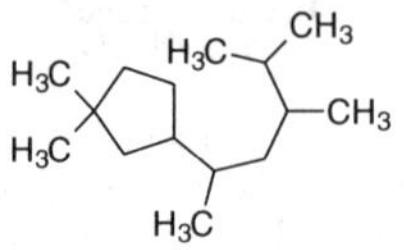

Secolactarane

*Sesquiterpenes of the L. type with a wide structural diversity are found in the fruit bodies of the milkcaps (*Lactarius*), russulas (*Russula*), and related basidiomycetes (blennins, furanether, furoscrobiculins, furosardonins, lactarorufins, lactaroscrobiculides, piperalol, etc.). Some of these compounds are artefacts formed by enzymatic and oxidative changes from sesquiterpene esters of the type *stearoylvelutinal during work-up of the toadstools. The burning, hot-tasting L. dialdehydes *velleral, *piperdial, and chrysorrhedial protect the fungi from feeding animals and microbes. Enzymatic cleavage of the seven-membered ring leads to the formation of secolactaranes such as lactaral and the hot-tasting *lactardial.

Lit.: Pol. J. Chem. **66**, 791 (1992) ▪ Tetrahedron **37**, 2199 (1981) (review); **40**, 2757 (1984); **42**, 4277 (1986) (abs. configuration); **50**, 1211 (1994) ▪ Tetrahedron Lett. **33**, 6863 (1992) ▪ Turner **2**, 247–252 (review) ▪ Zechmeister **77**, 69–180.

Lactardial. $C_{15}H_{22}O_3$, M_R 250.34, oil, $[\alpha]_D$ +23.8° (ether). A pungent-tasting secolactarane (see lactaranes from injured fruit bodies of *Lactarius vellereus*, *L. pergamenus*, *L. piperatus*, and other lacteous fungi

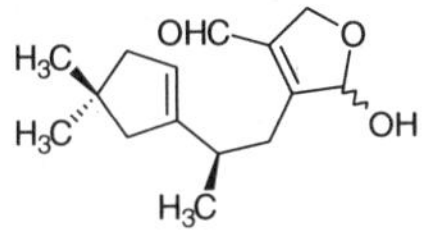

Lactardial

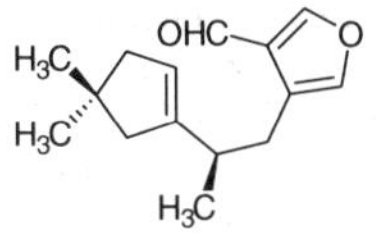

Lactaral

(Basidiomycetes). L. is formed, like *lactaral* {$C_{15}H_2O_2$, M_R 232.32, oil, $[\alpha]_D$ −7.6° ($CHCl_3$)} and other secolactaranes, in the fungi by enzymatic and chemical transformations from marasmane and lactarane precursors.

Lit.: Acta Chem. Scand., Ser. B **28**, 265 (1974) (synthesis) ▪ Tetrahedron **30**, 1431 (1974); **42**, 4277 (1986) (abs. configuration) ▪ Tetrahedron Lett. **23**, 5509 (1982) (synthesis); **26**, 3163 (1985). – *[CAS 108944-81-6 (L.), 54462-53-2 (lactaral)]*

Lactarius pigments.

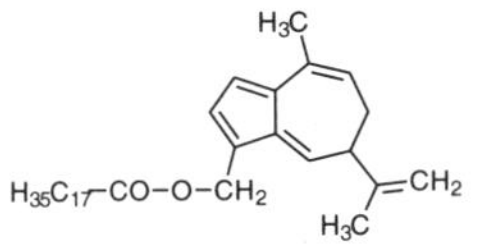

15-Stearoyloxyguaia-1,3,5,9,11-pentaene (e.g. *Lactarius deliciosus*)

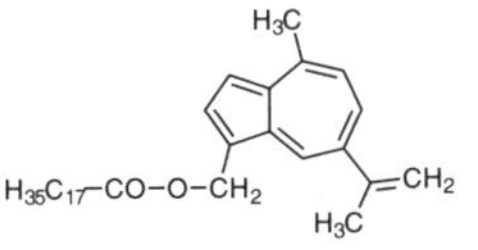

15-Stearoyloxy-11,12-didehydroguaiazulene (*Lactarius indigo*)

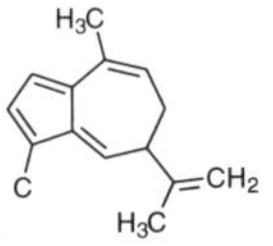

Delicial

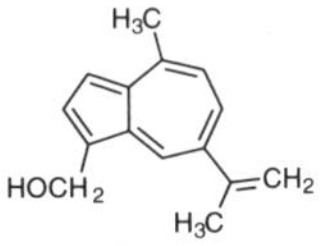

Deterrol

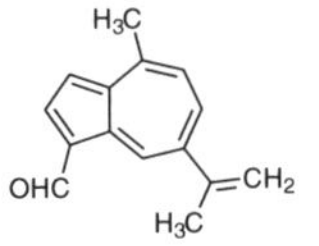

Lactaroviolin

Various species of the agaric genus *Lactarius* (milk-caps) contain colored milk juices or milk juices that become colored on contact with air. Most popular is the saffron milk cap (*L. deliciosus*) which, like related species of the section Dapetes (*L. sanguifluus*, *L. deterrimus*), contains a carrot-red milk juice that turns green within a few minutes when the fruit body is injured. The chemical mechanism of this discoloration is based on the enzymatic conversion of the yellow *15-stearoyloxyguaia-1,3,5,9,11-pentaene* ($C_{33}H_{52}O_2$, M_R 480.77, orange, unstable resin) to compounds of the *azulene type such as *delicial* ($C_{15}H_{16}O$, M_R 212.29, yellow unstable oil), *deterrol* ($C_{15}H_{16}O$, M_R 212.29, dark blue needles, mp. 100–101 °C), and *lactaroviolin* ($C_{15}H_{14}O$, M_R 210.28, violet, glossy prisms, mp. 58 °C). 15-Stearoyloxy-11,12-didehydroguaiazulene is responsible for the brilliant blue color of the milk juice of *Lactarius indigo*. The yellow color acquired by the white milk juice from injured fruit bodies of *L. chrysorrheus* and *L. scrobiculatus* arises from the enzymatic transformation of fatty acid esters of velutinal (see stearoylvelutinal) to *chrysorrhealactone, whereas in *L. fuliginosus* and *L. picinus* oxidation products of a prenylphenol [*4-methoxy-2-(3-methyl-2-butenyl)-phenyl stearate] are responsible for the pink coloration of the milk juice.

Lit.: Collect. Czech. Chem. Commun. **35**, 1296 (1970) ▪ Experientia **36**, 54 (1980) ▪ Phytochemistry **27**, 97 (1988); **28**, 2501 (1989) ▪ Zechmeister **51**, 185–187 (review). – *[CAS 113393-58-1 (delicial); 113393-59-2 (deterrol); 85-33-6 (lactaroviolin)]*

Lactaroscrobiculide A see chrysorrhealactone.

Lactaroviolin see Lactarius pigments.

Lactic acid (2-hydroxypropanoic acid, E 270).

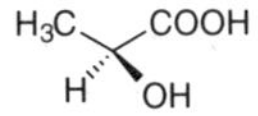

(*S*)-, L-form

$C_3H_6O_3$, M_R 90.08. The racemate and the enantiomers occur in nature. (*R*)-Form, mp. 53 °C, $[\alpha]_D^{20}$ −2.6° (H_2O), pK_a 3.83. Prismatic platelets, soluble in water, ethanol, and ether, insoluble in chloroform. (*R*)-(−)-L. is formed in the fermentation of glucose by *Lactobacillus leichmannii* and *L. delbrueckii*; with 1- and 2-valent metal ions it forms dextrorotatory salts and laevorotatory salts with 3-valent metal ions, e. g., zinc D-(+)-lactate · $2H_2O$, $[\alpha]_D^{14}$ +8.18° (H_2O). (*S*)-form (sarcolactic acid, paralactic acid), mp. 53 °C, $[\alpha]_D^{15}$ +3.82° (H_2O), pK_a 3.79 (25 °C), is highly hygroscopic, occurs in blood, muscles serum, bile, kidneys, and other organs. The content of L. increases after strenuous muscle activity (lactate acidosis). The racemate, oil, mp. 17 °C, bp. 122 °C (1.86 kPa), steam distillable, is widely distributed in nature, e. g., in sour milk products, in molasses as a result of partial fermentation of the sugars from apples and other fruits. For biochemistry and preparation, see *Lit.*[1].

Use: Various uses include application as caustic, dermatologic, antiseptic and as digestive. Used in the resolution of alcohols. Aq. soln. used as complexing agent in ion-exchange separations of alkaline earth metals. Used in foodstuffs and pharmaceuticals (preservative, flavour, acidulant), plasticisers, mordants. In brewing, manuf. of cheese and confectionery. Depilatory for hides.

Toxicology: LD_{50} (racemate; rat p.o.) 3750 mg/kg.

Lit.: [1] Fieser, Fieser, Reagents for Organic Synthesis, vol. 12, p. 226; 15, p. 181, New York: Wiley 1985, 1990.

gen.: Beilstein EIV **3**, 633–637 ▪ Florey **22**, 263 (1993) (review) ▪ Holten, Lactic Acid, Weinheim: Verl. Chemie 1971 ▪ Inorg. Chem. **16**, 1220 (1977) (synthesis) ▪ Karrer, No. 774 ▪ Kirk-Othmer (3.) **13**, 80 ▪ Stryer 1995, p. 497, 577. – *Toxicology:* Sax (8.), LAG 000, LAQ 000. – *[HS 2918 11; CAS 50-21-5; 10326-41-7 (R); 79-33-4 (S)]*

Lactivicin.

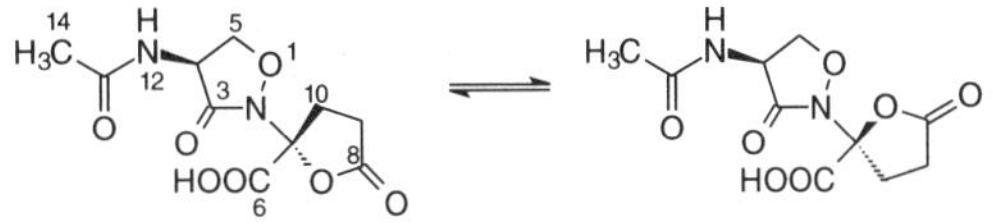

$C_{10}H_{12}N_2O_7$, M_R 272.21; $[\alpha]_D$ −24.1° (H_2O); epimerizes in aqueous solution; a cyclic peptide antibiotic from *Empedobacter lactamgenus* and *Lysobacter albus*. At a first glance L. seems to have no structural similarity with the *penicillins or *cephalosporins, but it does have a similar sensitivity to lactamases and binds to penicillin-binding proteins[1]. The partial structures of the acetamido group, the 3-isoxazolidinone ring and the carboxy group including the absolute configuration at C-4 correlate with the active center of, e. g., penicillin. L. is converted to an open-chain compound by cephalosporinase. It is active against Gram-positive and -negative bacteria and may be a new lead structure for non-β-lactam antibiotics[2]. For synthesis, see *Lit.*[3].

Lit.: [1]Nature (London) **325**, 179 (1987). [2]Tetrahedron **44**, 3221–3240 (1988). [3]J. Chem. Soc., Chem. Commun. **1987**, 62.

gen.: J. Antibiot. **42**, 84–93 (1989) ▪ Nachr. Chem. Tech. Lab. **37**, 130 (1989) ▪ Spec. Publ.-R. Soc. Chem. **70**, 119 (1989) (review) ▪ Tetrahedron **44**, 6589–6606 (1988) ▪ Tetrahedron Lett. **27**, 6229–6232 (1986). – *[HS 2941 90; CAS 107167-31-7]*

Lactobacillic acid [10-((1*R*)-*cis*-2-hexylcyclopropyl)-decanoic acid, (11*R*,12*S*)-11,12-methylenoctadecanoic acid].

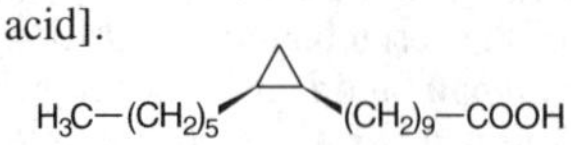

$C_{19}H_{36}O_2$, M_R 296.49, mp. 33.6–35 °C. A lipid component in *Lactobacillus arabinosus*, *L. casei*, *Agrobacterium tumefaciens*, and *Streptococcus* spp. L. occurs as phospholipide ester in the cell membrane. The biosynthesis of the cyclopropane ring proceeds through alkylation of the 11,12-double bond of an unsaturated fatty acid phospholipid by *S-adenosylmethionine.

Lit.: Beilstein EIV **9**, 106 ▪ Can. J. Chem. **59**, 828 (1981) (biosynthesis) ▪ Tetrahedron Lett. **40**, 6689 (1999) (synthesis). – *[CAS 19625-10-6]*

γ- and δ-Lactones see alkanolides.

Lactose (milk sugar, lactobiose, 4-*O*-β-D-galactopyranosyl-D-glucose).

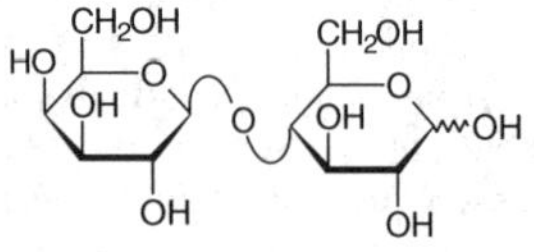

$C_{12}H_{22}O_{11}$, M_R 342.30. As monohydrate, weakly sweet tasting monoclinic crystals, β-form: mp. 252 °C, $[\alpha]_D^{20}$ +34.9→+55.4° (H_2O); α-form monohydrate: mp. 202 °C, $[\alpha]_D^{20}$ +85→+55.4° (H_2O). L. reduces Fehling's solution. L. is a *disaccharide made up of D-*galactose and D-*glucose occurring in milk (human milk ca. 6.7%, cow milk ca. 4.5%). L. is formed in the mammary glands of mammals from uridinediphosphate-D-galactose (UDP-D-galactose) and D-glucose under catalysis by *lactose synthase* (EC 2.4.1.22). α-L. and β-L. are formed in the ratio 2:3; at temperatures below 93 °C the α-form crystallizes from aqueous solutions while the β-form crystallizes at higher temperature. L. occurs in various milk products; it is fermented during the production of cheese, sour milk, etc.; most bacteria utilizing L. have a special transport system for its uptake in the cell membrane[1]. L. has approximately the same nutritional value as *sucrose and can be hydrolytically cleaved by mineral acids and by *β-galactosidase*, previously known as lactase. This can be used as a detection reaction. β-Galactosidase is formed abundantly by babies and young children; in adults it is often insufficiently available (Europeans) or practically completely absent (Asians, Africans), so that uncontrolled consumption of milk can lead to digestive problems (cramps, diarrhea). L. is obtained technically from whey; it is used mainly for the production of dietetic foods, in pharmaceutical agents as a binder, filler, or adsorbent. In the laboratory it can serve as a starting material in the synthesis of heterocyclic compounds and industrially it is used for the production of emulsifiers and detergents.

Lit.: [1]Annu. Rev. Biophys. Biophys. Chem. **15**, 279–319 (1986); Trends Biochem. Sci. **8**, 404–408 (1983).

gen.: Adv. Carbohydr. Chem. **16**, 159–206 (1961) ▪ Anal. Profiles Drug Subst. **20**, 369 (1991) (review) ▪ Beilstein EV **17/7**, 196 ▪ Hager (5.) **8**, 688 ▪ Merck-Index (12.), No. 5356. – *[HS 1702 10; CAS 63-42-3]*

Lactucin.

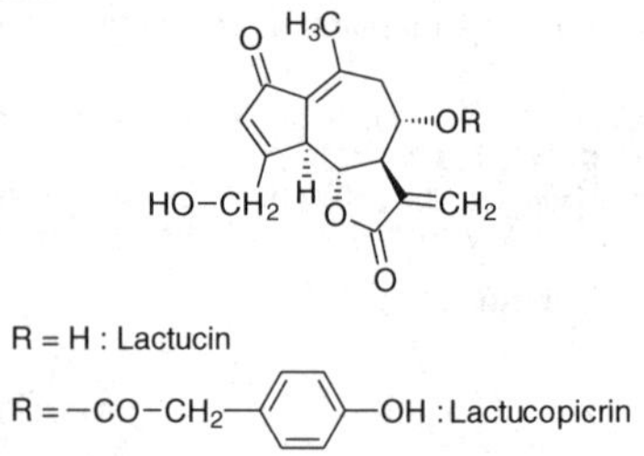

$C_{15}H_{16}O_5$, M_R 276.29, cryst., mp. 224–228 °C, $[\alpha]_D^{17}$ +77.9° (pyridine), soluble in water, ethanol, and ethyl acetate. A toxic *sesquiterpene lactone from *Lactuca* species (Asteraceae), it is found especially in prickly lettuce (*Lactuca virosa*) together with *lactucopicrin* {lactupicrin, intybin, $C_{23}H_{22}O_7$, M_R 410.42, cryst., mp. 148–151 °C, $[\alpha]_D^{17.5}$ +67.3° (pyridine)} in relevant amounts (ca. 3.5% in latex), it is also present in low concentrations in *L. sativa* (lettuce), chicory, endives, succhory. The sesquiterpenes are responsible for the bitter taste of these vegetables. In the past *L. virosa* was cultivated as a medicinal herb to harvest its latex. The dried latex under the name Lactucarium was used as a cough and sleeping agent; L. and lactucopicrin are responsible for the sedative properties of this drug. Poisonings have occurred through overdoses of Lactucarium, consumption of the leaves of *L. virosa* as a salad or of *Lactuca* juice.

Lit.: Beilstein EV **18/4**, 421 ▪ Hager (5.) **3**, 723; **4**, 866 ff. ▪ Merck-Index (12.), No. 5359 ▪ Phytochemistry **20**, 2371 (1981); **21**, 1163 (1982); **40**, 1659 (1995) (biosynthesis); **45**, 365 (1997) (isolation) ▪ Planta Med. **57**, 190 (1991) ▪ Pol. J. Chem. **63**, 297 (1989). – *Toxicology:* Pharmazie **3**, 469 (1948) ▪ Liebenow & Liebenow, Giftpflanzen, Jena: G. Fischer 1981. – *[CAS 1891-29-8]*

Lactucopicrin see lactucin.

(–)-Laevojunenol see eudesmanes.

Lagopodins.

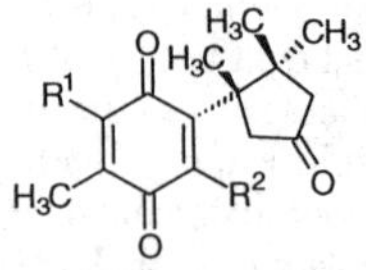

$R^1 = R^2 = H$: Lagopodin A
$R^1 = H, R^2 = OH$: Lagopodin B
$R^1 = R^2 = OH$: Hydroxylagopodin B

Sesquiterpene quinones of the cuparane type from cultures of the inky caps *Coprinus lagopus*, *C. cinereus*, and *C. macrorhizus* (Basidiomycetes). The wild forms of these mushrooms do not form lagopodins. *L. A* {$C_{15}H_{18}O_3$, M_R 246.31, yellow cryst., $[\alpha]_D$ –10° ($CHCl_3$)}, *L. B* ($C_{15}H_{18}O_4$, M_R 262.31), and *hydroxy-L. B* ($C_{15}H_{18}O_5$, M_R 278.30, orange red cryst., mp. 184–186 °C) have been described. Both L. A and L. B are unstable in aqueous solution and exhibit antibiotic activity against Gram-positive bacteria.

Lit.: Acta Chem. Scand., Ser. B **28**, 492 (1974) (synthesis) ▪ Phytochemistry **14**, 1433 (1975); **15**, 2004 (1976). – *Reviews:* Thomson **3**, 30 ▪ Zechmeister **51**, 187 ff. – *[CAS 62185-66-4 (L. A); 62512-03-2 (L. B); 56973-45-6 (hydroxy-L. B)]*

Laherradurin see annonaceous acetogenins.

Lambdamycin see chartreusin.

Lamellarins.

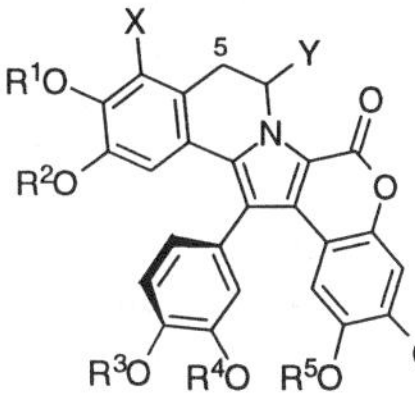

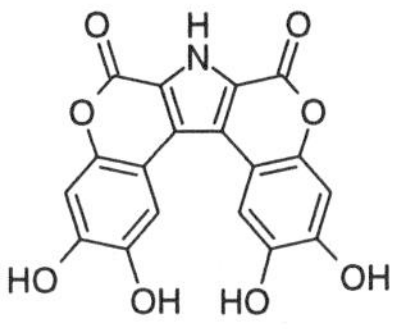

Ningalin A

	R^1	R^2	R^3	R^4	R^5	R^6	X	Y	Δ-5
L. A	CH_3	CH_3	H	CH_3	CH_3	H	OCH_3	OH	-
L. B	CH_3	CH_3	H	CH_3	CH_3	H	OCH_3	H	+
L. C	CH_3	CH_3	H	CH_3	CH_3	H	OCH_3	H	-
L. D	H	CH_3	H	CH_3	CH_3	H	H	H	+
L. G	H	CH_3	CH_3	H	H	CH_3	H	H	-
L. H	H	H	H	H	H	H	H	H	+
L. I	CH_3	CH_3	CH_3	CH_3	CH_3	H	OCH_3	H	-
L. K	CH_3	CH_3	H	CH_3	CH_3	H	OH	H	-
L. L	H	CH_3	CH_3	H	CH_3	H	H	H	-
L. M	CH_3	CH_3	H	CH_3	CH_3	H	OH	H	+

Table: Data of Lamellarins and Ningalin A.

compound	molecular formula	M_R	mp. [°C]	CAS
L. A	$C_{30}H_{27}NO_{10}$	561.54	168–172 pale-yellow prisms	97614-62-5
L. B	$C_{30}H_{25}NO_9$	543.53	258–259 pale-yellow needles	97614-63-6
L. C	$C_{30}H_{27}NO_9$	545.55	225–230 needles	97614-64-7
L. D	$C_{28}H_{21}NO_8$	499.49	pale-yellow powder	97614-65-8
L. G	$C_{28}H_{23}NO_8$	501.49	263–265 prisms	115982-21-3
L. I	$C_{31}H_{29}NO_9$	559.57	218–220 white prisms	149355-75-9
L. K	$C_{29}H_{25}NO_9$	531.52	196–198 amorphous powder	149378-56-3
L. L	$C_{28}H_{23}NO_8$	501.49	285–287 amorphous powder	149378-57-4
L. M	$C_{29}H_{23}NO_9$	529.50	amorphous powder	149378-58-5
Ningalin A	$C_{18}H_9NO_8$	367.27	amorphous yellow	188111-67-3

The L. are a class of marine alkaloids with the 5-oxa-6b-aza-dibenzo[*a,i*]-9-fluoren-6-one skeleton. Since the first isolation of lamellarins A–D in 1985 from a prosobranch mollusc of the genus *Lamellaria* [1], more than 35 representatives of this class of alkaloids have been isolated from ascidians [2,3]. The individual L. differ mainly in the arrangement of the hydroxy and methoxy substituents at the aromatic rings and can in addition carry *O*-sulfate groups [4] or a double bond or OH-substituent in the piperidine ring. The L. P, Q and R from the Australian marine sponge *Dendrilla cactos* possess simpler structures [5] and are structurally related to the lukianols [6]. The central 3,4-diarylpyrrole core unit of the L. is also present in other marine alkaloids like the ningalins from an Australian ascidian of the genus *Didemnum* [7], *purpurone or the *storniamides. Some L. inhibit the growth of several tumor cell lines and especially L. K reverts the P-glycoprotein mediated multidrug resistance (MDR) of tumor cells at very low concentrations [8]. L. I, K and L showed significant cytotoxicity against P388 and A549 cell lines in culture ($IC_{50} \approx 0.25$ µg/mL) [3]. Consequently, these natural products possess a considerable potential for the development of anti-tumor drugs as well as of non-toxic modulators of the MDR phenotype [8].

Biosynthetically the L. appear to be formed by oxidative coupling of two molecules of arylpyruvic acid, followed by condensation of the resulting 1,4-diketone with a 2-arylphenylethylamine to yield a 3,4-diarylpyrrole-2,5-dicarboxylic acid, which finally undergoes two consecutive oxidative cyclisations. This sequence was followed in biomimetic syntheses of L. G trimethyl ether and L. L [9]. L. sulfatus inhibit HIV-1-integrase of AIDS viruses [10].

Lit.: [1] J. Am. Chem. Soc. **107**, 5492 (1985). [2] J. Org. Chem. **53**, 4570 (1988); Tetrahedron Lett. **37**, 363 (1996); J. Magn. Reson. Ser. A **118**, 282 (1996); Aust. J. Chem. **49**, 711 (1996). [3] Aust. J. Chem. **46**, 489 (1993). [4] Tetrahedron **53**, 3457 (1997); J. Nat. Prod. **62**, 419 (1999). [5] Aust. J. Chem. **47**, 1919 (1994); **48**, 1491 (1995); J. Nat. Prod. **62**, 419 (1999). [6] Helv. Chim. Acta **75**, 1721 (1992). [7] J. Org. Chem. **62**, 3254 (1997). [8] Br. J. Cancer **74**, 677 (1996). [9] Angew. Chem. Int. Ed. Engl. **36**, 155 (1997) (L. G trimethyl ether); Chem. Eur. J. **2000**, 1147 (L. L); Tetrahedron **53**, 5951 (1997) (L. D, L. H); Chem. Commun. **1997**, 207, 2259 (L. K); J. Am. Chem. Soc. **121**, 54 (1999) (L. O, Ningalin A). [10] J. Med. Chem. **42**, 1901 (1999).

Lamiide. An *iridoid glucoside with a C_{10} aglycone occurring bound to various aromatic carboxylic acids.

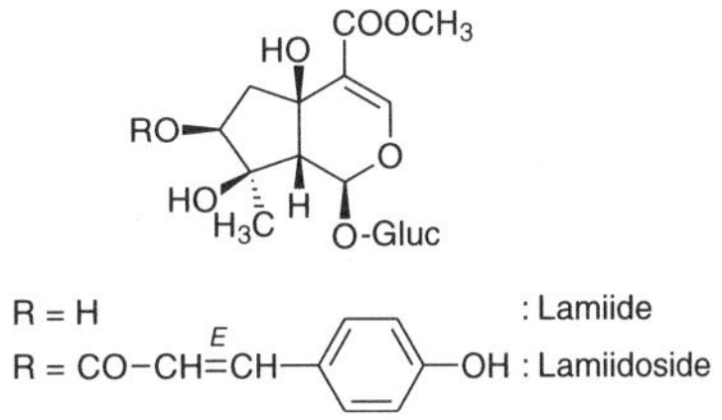

Table: Lamiide and Lamiidoside.

trivial name*	molecular formula	M_R	$[\alpha]_D$ (CH_3OH)	CAS
Lamiide	$C_{17}H_{26}O_{12}$	422.39	–127°	27856-54-8
Lamiidoside	$C_{26}H_{32}O_{14}$	568.53	–80°	64597-22-4

* amorphous solids.

L. and its derivatives occur in Lamiaceae and Verbenaceae which therefore are related in a biochemical sense [1]. The name L. is derived from its occurrence in the nettle *Lamium amplexicaule*, Lamiaceae [2]. L. and lamiidoside are isolated from the extract of the Jerusalem sage *Phlomis fruticosa* (Lamiaceae) [3].

Lit.: [1] Hegnauer **IV**, 346. [2] Gazz. Chim. Ital. **99**, 1150 (1969). [3] Gazz. Chim. Ital. **107**, 67 (1977).
gen.: J. Nat. Prod. **43**, 649, No. 99–103 (1980) ▪ Phytochemistry **31**, 3839 (1992) (biosynthesis). – *[HS 2938 90]*

Laminarin see glucans.

Lamoxirene see algal pheromones.

Lampteroflavin.

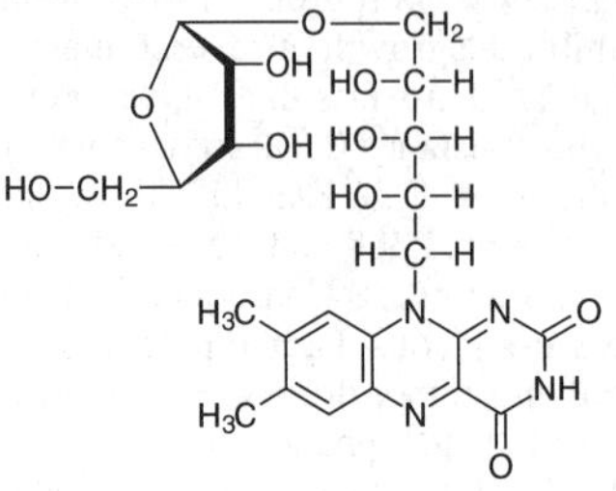

$C_{22}H_{28}N_4O_{10}$, M_R 508.49. The Japanese "moon night mushroom" *Lampteromyces japonicus* (Tsukiyo-take, Basidiomycetes) shows a green glowing on the underside of its cap [uv_{max} (emission) 524 nm]. L. is responsible for this bioluminescence, and is isolated from the fruit bodies of the freshly collected toadstool. The strongly fluorescent compound is very unstable and decomposes rapidly in aqueous solution under formation of *riboflavin(e).

Lit.: Bioorg. Chem. **17**, 474 (1989) ▪ Tetrahedron **46**, 1367 (1990); **47**, 6215 (1991) (synthesis). – *[CAS 114590-52-2]*

Lanatosides see Digitalis glycosides.

Lanosterol (lanosterin, kryptosterin, lanosta-8,24-dien-3*β*-ol).

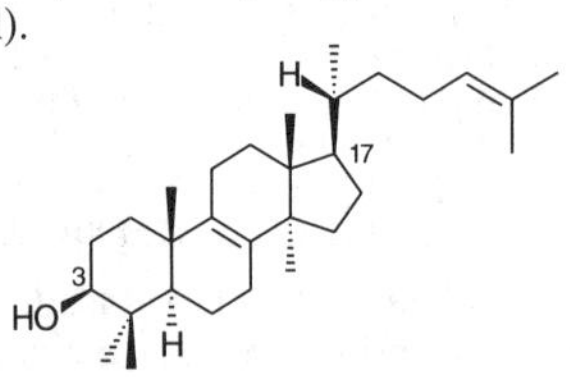

$C_{30}H_{50}O$, M_R 426.73, mp. 140–141 °C, $[\alpha]_D$ +58° ($CHCl_3$). An alcohol from the group of tetracyclic *triterpenes. L. can be isolated in larger amounts from the wool fat of sheep (lanolin). It has also been detected in yeasts, other fungi, as well as in Euphorbiaceae and some other plants, it is present in animal and human organisms as well. L. is the first isolable cyclization product of (3*S*)-squalene 2,3-epoxide in the course of steroid biosynthesis in animal and fungal organisms (see steroids). In contrast, steroid biosynthesis in photosynthetically active plants and algae proceeds through *cycloartenol which is structurally isomeric with L.

Lit.: Beilstein EIV **6**, 4188 ▪ J. Am. Chem. Soc. **117**, 11819 (1995) (biosynthesis) ▪ Merck-Index (12.), No. 5372 ▪ Luckner (3.), p. 204–215 ▪ Zeelen, p. 43–71. – *[HS 2906 13; CAS 79-63-0]*

Lanosulin see fumitremorgins.

Lanthionine [*S*-(2-amino-2-carboxyethyl)-cysteine, 3,3′-thiodialanine].

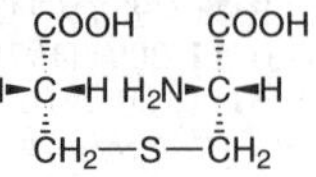

(*R,R*)-, L-form

$C_6H_{12}N_2O_4S$, M_R 208.23, (*R,R*)-, L-form, mp. 295–296 °C (decomp.), $[\alpha]_D^{22}$ +8.6° (2.4 m NaOH); (*S,S*)-, D-form, mp. 293–295 °C (decomp.), $[\alpha]_D^{21}$ −8.0° (2.4 m NaOH); (*RS,RS*)-form, (±)-, mp. 286–292 °C (decomp.); (*RS,SR*)-form, *meso*-, mp. 304 °C (decomp.). A secondary hydrolysis product of some proteins. It was first isolated from the hydrolysate of wool. L. is an oxidation product of *cysteine. Cys in hair proteins, in feathers, lactalbumin, or insulin as well as free Cys are converted to lanthionine.

Lit.: Acta Crystallogr., Sect. C **46**, 627 (1990) ▪ Beilstein EIV **4**, 3152 ▪ Davies, Amino Acids and Peptides, p. 212, London: Chapman & Hall 1985 ▪ J. Am. Chem. Soc. **96**, 1925 (1974). – *[HS 2930 90; CAS 922-55-4 (R,R)-form]*

Lantibiotics. L. are antibacterial peptides from Gram-positive bacteria which are characterized by a high content of the non-proteinogenous thioether aminoacids *lanthionine and 3-methyllanthionine. Many compounds of this group cause short-lived tiny pores in the bacterial cytoplasmic membrane (e.g., *epidermin, *nisin) by inhibiting the peptidoglycan synthesis on the level of the translocase II reaction, the step prior to transglycosylation. Mersacidin acts differently.

Lit.: Barton-Nakanishi **4**, 275–304 (biosynthesis) ▪ Jack, Bierbaum & Sahl (eds.), Lantibiotics and Related Peptides, Berlin: Springer 1998.

Lapachol [2-hydroxy-3-(3-methyl-2-butenyl)-1,4-naphthoquinone, C. I. 75490].

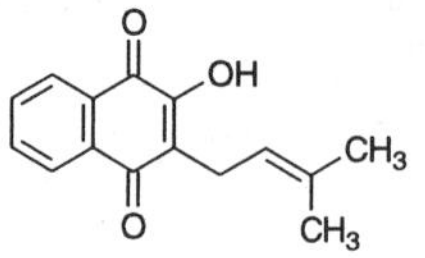

Lapachol

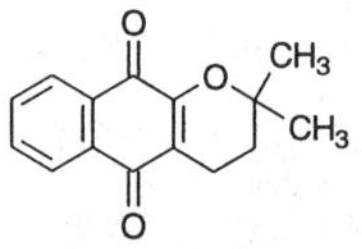

α-Lapachone

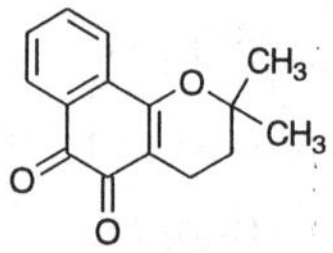

β-Lapachone

$C_{15}H_{14}O_3$, M_R 242.27, yellow prisms, mp. 140 °C, uv_{max} 331 nm (C_2H_5OH), soluble in ethanol, chloroform, benzene, acetic acid, poorly soluble in ether and hot water. L. and its derivatives are widely distributed prenylated naphthoquinones from the heartwood of *Paratecoma peroba* (Bignoniaceae) and teak wood (*Tectona grandis*, Verbenaceae). Like the structurally related *vitamin K L. inhibits blood coagulation [1], inhibits tumors [2], and has abortive, teratogenic [3] and antiviral [4] activities. Important naturally occurring derivatives of L. are *lomatiol, *α-lapachone* {$C_{15}H_{14}O_3$, M_R 242.27, yellow needles, mp. 119 °C, uv_{max} 375 nm (C_2H_5OH)}, and *β-lapachone* {$C_{15}H_{14}O_3$, M_R 242.27, orange needles, mp. 155–156 °C, uv_{max} 431 nm (C_2H_5OH)}. It is used to dye wool and hair [5].

Lit.: [1] Arch. Biochem. Biophys. **234**, 405–412 (1984). [2] J. Environ. Sci. Health A **19**, 533–577 (1984); Cancer Chemother. Rep. Part 5, **4**, 11 (1974). [3] Rev. Port. Farm. **38**, 21 (1988). [4] Rev. Microbiol. **14**, 21–26 (1983). [5] Aerosol Cosmet. **8**, 4, 6–13 (1986); J. Heterocycl. Chem. **29**, 1457 (1992).
gen.: Beilstein EIV **8**, 2487 ▪ Karrer, No. 1218, 1222 ▪ Merck-Index (12.), No. 5376 ▪ Mutat. Res. **255**, 155 (1991) ▪ Rowe, p. 849 ▪ S. Afr. J. Chem. **43**, 96 (1990) ▪ Sax (8.), HLY 500 ▪ Tetrahedron Lett. **39**, 8221 (1998) (synthesis) ▪ Thomson **3**, 139-142. – *[CAS 84-79-7 (L.); 4707-33-9 (α-lapachone); 4707-32-8 (β-lapachone)]*

Lardolure [(1*R*,3*R*,5*R*,7*R*)-1,3,5,7-tetramethyldecyl formate].

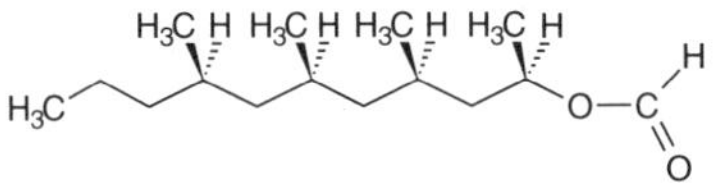

$C_{15}H_{30}O_2$, M_R 242.40, $[\alpha]_D^{23}$ −3.4° (hexane). An aggregation pheromone of the mite *Lardoglyphus konoi*, that causes serious trouble in the storage of dried meat and flour. The biosynthesis presumably involves linkage of propionate units.
Lit.: Agric. Biol. Chem. **46**, 2283 (1982). – *Synthesis:* ApSimon **9**, 116 f. ▪ Justus Liebigs Ann. Chem. **1995**, 2001 ▪ Tetrahedron **42**, 5539, 5545 (1986). – *[CAS 83540-84-5]*

Larixol see labdanes.

Lasalocids.

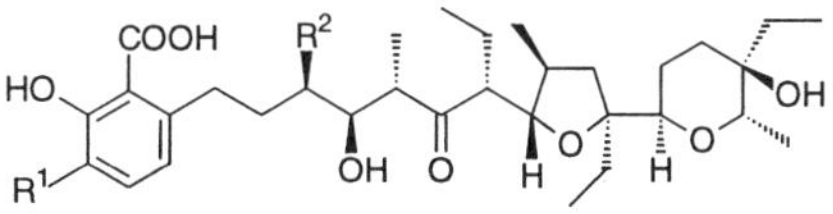

Table: Structures and data of Lasalocids.

L.	R^1	R^2	molecular formula	M_R	CAS
A	CH_3	CH_3	$C_{34}H_{54}O_8$	590.80	2599-31-9
B	C_2H_5	CH_3	$C_{35}H_{56}O_8$	604.82	55051-86-0
C	CH_3	C_2H_5	$C_{35}H_{56}O_8$	604.82	55051-84-8

A mixture of ionophoric *polyether antibiotics produced by *Streptomyces lasaliensis*, from which the components A to E have been separated. L. exert antibacterial and antiviral (HIV) activities, LD_{50} (mouse p.o.) 146 mg/kg. *L. A* {mp. 110–114 °C, $[\alpha]_D$ −7.5° (CH_3OH)}, which preferentially forms complexes with divalent cations, is formed biosynthetically by the *polyketide pathway from five acetate units, four propionate units, and three butanoate units, the benzene ring arises through cyclization. L. A (Bovatec®) in the form of its sodium salt (Avatec®) is used in fowl breeding as a coccidostatic.
Lit.: J. Am. Chem. Soc. **112**, 3659 (1990) ▪ Westley (ed.), Polyether Antibiotics, vols. 1 & 2, New York: Dekker 1983. – *Biosynthesis:* Birch & Robinson, in Vining & Stuttard (eds.), Genetics and Biochemistry of Antibiotic Production, p. 443–476, Boston: Butterworth-Heinemann 1995 ▪ Zechmeister **58**, 1–82 ▪ J. Antibiot. **39**, 1270 (1986). – *Synthesis:* J. Org. Chem. **53**, 1046 (1988) ▪ Tetrahedron **49**, 5979, 5997 (1993). – *Activity:* Antimicrob. Agents Chemother. **36**, 492 (1992).

Lascivol. $C_{17}H_{28}N_2O_7$, M_R 372.42, cryst., mp. 138 °C (monohydrate), $[\alpha]_D$ −89.7° (CH_3OH). A bitter princi-

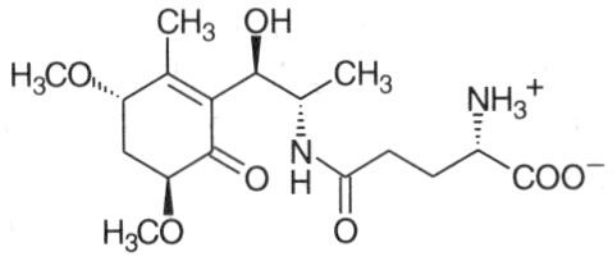

Lascivol

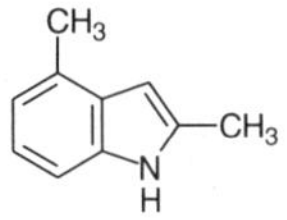

2,4-Dimethylindole

ple from the toadstool *Tricholoma lascivum* (Basidiomycetes) [1]. The cyclohexenone derivative L. makes up ca. 2% of the dried fungus. It reacts with methanolic hydrochloric acid to give 5-methoxy-2,4-dimethylindole. Indoles with a similar substitution pattern, e. g., *2,4-dimethylindole* ($C_{10}H_{11}N$, M_R 145.20) are isolated from *Tricholoma sciodes* and *T. virgatum*, this is indicative of a biosynthetic relationship to L. (see also peronatins) [2].
Lit.: [1] Justus Liebigs Ann. Chem. **1990**, 1115. [2] Tetrahedron **50**, 3571 (1994). – *[CAS 129421-88-1]*

Lasubines see Lythraceae alkaloids.

Lathyranes.

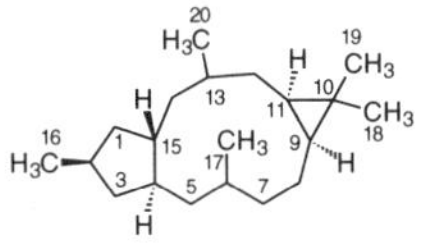

1

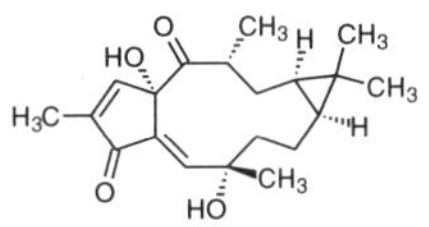

2

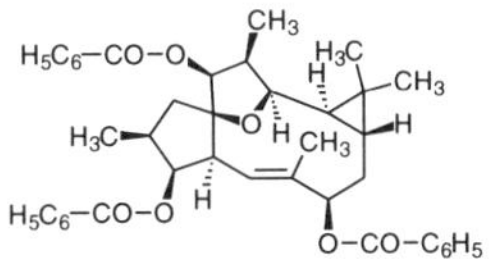

3

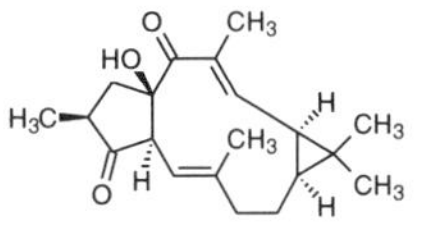

4

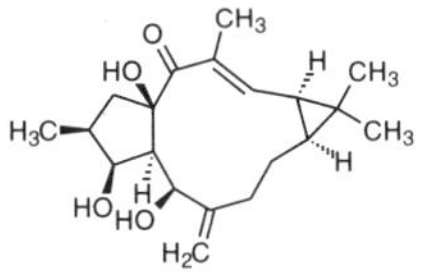

5

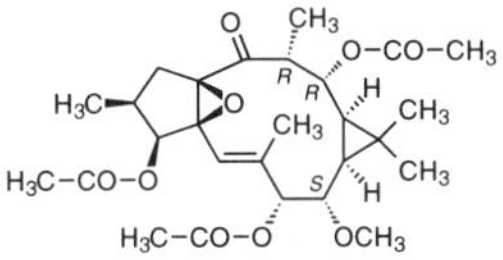

6

Table: Data of Lathyranes.

no.	molecular formula	M_R	mp. [°C]	$[\alpha]_D$	CAS
2	$C_{20}H_{28}O_4$	332.44	cryst. 168–170	−38.5° (CH_2Cl_2)	106644-31-9
3	$C_{41}H_{44}O_7$	648.80	oil	−19.7° ($CHCl_3$)	95852-24-7
4	$C_{20}H_{28}O_3$	316.44	oil	+5° ($CHCl_3$)	62820-13-7
5	$C_{20}H_{30}O_4$	334.46	cryst. 168–169		34420-19-4
6	$C_{27}H_{38}O_9$	506.59	needles 148–150	+16.1° ($CHCl_3$)	77573-15-0

A group of diterpenes with the skeleton (**1**). L. occur mainly in various Euphorbiaceae, examples, e. g., euphohelionone (**3**), jolkinol C (**4**), lathyrol (**5**), and tirucalicin (**6**).

Curculathyrane A (**2**) can be obtained from *Jatropha curcus*. Lathyrane is formally a 6,10-cyclocasbane. The diacetate-benzoate and the triacetate of lathyrol are isolated from the seed oil and latex of *Euphorbia lathyris* in which the mixed esters of 7-hydroxylathyrol and 6,17-epoxylathyrol also occur. Various esters of L. are skin irritants and carcinogenic [1].

Lit.: [1] Chem. Labor Betr. **42**, 121 (1991).
gen.: Experientia **27**, 1393 (1971) ▪ Heterocycles **27**, 2851 (1988) ▪ Phytochemistry **28**, 3421 (1989); **29**, 2025 (1990); **31**, 3479 (1992) ▪ Planta Med. **60**, 588 (1994) ▪ Tetrahedron Lett. **27**, 5675 (1986).

Lathyrine [(*S*)-α,2-diamino-4-pyrimidinepropanoic acid].

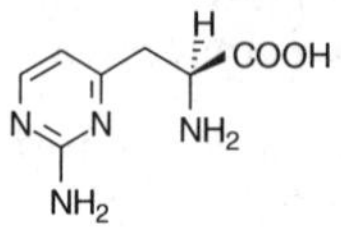

$C_7H_{10}N_4O_2$, M_R 182.18, mp. 215 °C (decomp.), $[\alpha]_D^{21}$ −55.9° (H_2O), pK_a 2.4, 4.1, 9. A non-proteinogenic amino acid in *Lathyrus* spp., e. g. *L. tingitanus* [1] (content: 2.11% of dry weight of seeds).
Biosynthesis: Hydroxyhomoarginine [2,3] and uracil [4] are possible precursors of lathyrine.

Lit.: [1] Nature (London) **194**, 91 f. (1962). [2] Biochem. J. **94**, 35P (1965). [3] Phytochemistry **12**, 119–124 (1973). [4] Biochem. J. **164**, 589–594 (1977).
gen.: Chem. Commun. **1997**, 1757 (synthesis) ▪ Rosenthal, Plant Nonprotein Amino and Iminoacids, p. 117 ff., New York: Academic Press 1982. – *[CAS 13089-99-1]*

Lathyrism see 3-cyano-L-alanine.

Lathyrol see lathyranes.

Laudanidine, laudanosine see benzyl(tetrahydro)isoquinoline alkaloids.

Laurel leaf oil (sweet or Mediterranean bay oil). Pale yellow to yellow oil with a fresh, sweet-spicy, slightly camphor-like odor and a fresh delicate-spicy, sweet taste, sometimes slightly bitter.
Production: By steam distillation of leaves of the laurel tree *Laurus nobilis* growing mostly in the Mediterranean region.
Composition [1]***:*** The main components are *1,8-*cineole* (ca. 50%) and *α-terpinyl acetate* (ca. 10%; see oil of cardamom). Furthermore, **eugenol* (ca. 1%), *methyleugenol* (see eugenol) (2–4%), and **linalool* (4–6%) are important for the organoleptic impression.
Use: In the perfume industry for spicy fragrances (masculine notes); to improve the aromatic character of foods such as sauces, pickles, meat and fish products.

Lit.: [1] Perfum. Flavor. **11** (3), 52 (1986); **12** (4), 71 (1987); **15** (3), 67 (1990); **18** (3), 65 (1993); **20** (1), 51 (1995).
gen.: Arctander, p. 337 ▪ Bauer et al. (2.), p. 160. – *Toxicology:* Food Cosmet. Toxicol. **14**, 337 (1976). – *[HS 3301 29; CAS 8002-41-3]*

Laurene. Sesquiterpene from the red alga *Laurencia glandulifera*. L. co-occurs with many oxidized metabolites which have characteristic odors.

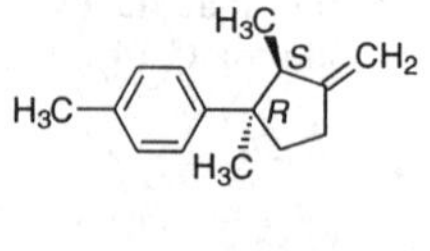

$C_{15}H_{20}$, M_R 200.32, oil, bp.$_{21}$ 131–133 °C, $[\alpha]_D^{23}$ +48.7° (C_2H_5OH).

Lit.: *Synthesis:* J. Chem. Soc., Perkin Trans. 1 **1997**, 3127 f. ▪ J. Org. Chem. **60**, 6511, 7791 (1995). – *[CAS 18452-41-0]*

Laurifin(in)e see morphinan alkaloids.

Lavandin [oil]. A light yellow, mobile oil with a fresh, flowery-herby odor but more earthy and camphor-like than *lavender oil.
Production: By steam distillation from the flowering herbage of Lavandin (*Lavandula angustifolia* × *Lavandula latifolia*), a sterile cross breed between genuine lavender (see lavender oil) and spike-lavender [see spike-lavender oil]. Lavandin is mainly cultivated in southern France, departments Alpes de Haute Provence, Drôme, and Vaucluse. The L. plant is more resistant than lavender and thus cultivation is not limited to high plateaus as is the case for genuine lavender but rather L. is distributed widely even into lower parts of the Rhône valley. Since L. also gives an appreciably higher yield of oil, the production of L. oil now exceeds by far that of lavender oil although the latter does have a higher value for perfumery. The world-wide production probably amounts to ca. 1000 t L. oil with ca. 75% originating from the "grosso" variety. In addition to the essential L. oil, appreciable amounts of extracts (L. concrete, L. absolute) are also produced. The odor of these dark green pasty to viscous products is very natural, sweeter, and more persistent than that of L. oil.
Composition [1]***:*** A typical quantitative composition of the oil of the "grosso" variety is as follows: *1,8-*cineole*: 4–7%, **camphor*: 6–8%, (−)-**linalool*: 25–35%, (−)-*linalyl acetate* (see linalool): 28–38%, (+)-*terpinen-4-ol* (see *p*-menthenols): 2–4%, **borneol*: 1.5–3%; (−)-*lavandulol* ($C_{10}H_{18}O$, M_R 154.25): 0.3–0.5%, (−)-*lavandulyl acetate* ($C_{12}H_{20}O_2$, M_R 196.29): 1.5–3%.

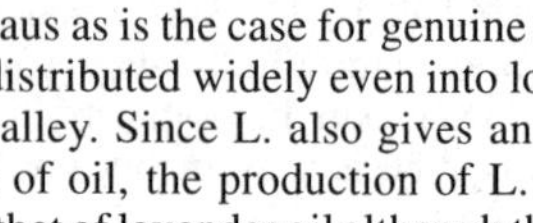

R = H : (-)-Lavandulol
R = CO–CH3: (-)-Lavandulyl acetate

Use: L. oil is a natural raw material for perfumes with a wide scope of applications, it is used especially for perfuming less expensive products for a broad range of purposes. L. concrete is mainly used to perfume soaps, while L. absolute is used in Eaux de Cologne.

Lit.: [1] Perfum. Flavor. **20** (3), 23 (1995).
gen.: Arctander, p. 338, 341, 343 ▪ Bauer et al. (2.), p. 161 ▪ ISO 3054 ("abrialis") (1987); 8902 ("grosso") 1987 ▪ J. Org. Chem. **62**, 734 (1997) (synthesis lavandulol) ▪ Ohloff, p. 142. – *Toxicology:* L. oil: Food Cosmet. Toxicol. **14**, 447 (1976); L. absolute: Food Chem. Toxicol. **30** (Suppl.), 65 (1992). – *[HS 3301 29; CAS 8022-15-9 (L.); 498-16-8 (lavandulol); 20777-39-3 (lavandulyl acetate)]*

(−)-Lavandulol, (−)-lavandulyl acetate see lavandin [oil].

Lavendamycin. $C_{22}H_{14}N_4O_4$, M_R 398.38, dark red crystals, mp. >300 °C. A quinone *antibiotic from *Strep-*

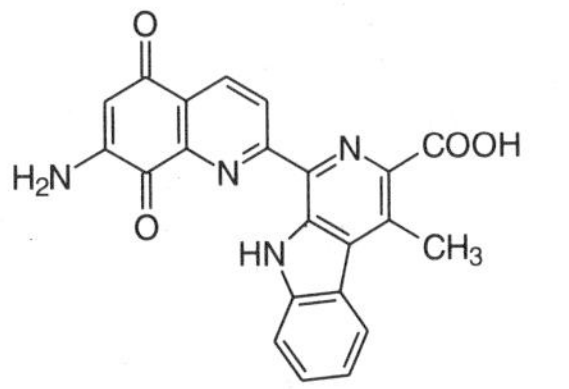

tomyces lavendulae with antineoplastic and antifungal activities. L. is similar to *streptonigrin (biosynthetic precursor of streptonigrin) both in activity and structure.

Lit.: Foye (ed.), Cancer Chemotherapeutic Agents, p. 645–651, Washington: ACS 1995 ▪ Heterocycles **21**, 91–106 (1984); **24**, 1067–1073 (1986) ▪ J. Antibiot. **35**, 259–265 (1982) ▪ J. Org. Chem. **58**, 7089 ff. (1993); **61**, 6552 (1996) ▪ Lindberg **2**, 1–56 ▪ Tetrahedron Lett. **35**, 1453 (1994). – *[HS 2941 90; CAS 81645-09-2]*

Lavender oil. Pale yellow to light yellowish-brown oil with a typical fresh, flowery-herby, sweet, balsamy, mildly woody odor.

Production: By steam distillation of flowering lavender herbage, *Lavandula angustifolia*. About half of the world production, estimated to be ca. 200–250 t, comes from Bulgaria, ca. 50 t come from the classical cultivation areas in southern France where lavender grows at an altitude of between 600 and 1500 m in the departments Alpes de Haute Provence, Drôme, and Vaucluse. 75% of the French oil originates from so-called "population lavender", grown from seeds of wild-growing plants, the remaining 25% mainly come from the cloned variety "maillette" which provides a higher yield of oil with a less favorable odor quality. Smaller cultivation areas are scattered all over the world, e. g., in regions of the former USSR, in Tasmania, etc.

Composition [1]***:*** Oil from French population lavender: *cis-*ocimene*: 5–9%; *trans-ocimene*: 3–5%; 1,8-*cineole*: <1%; **camphor*: <0.4%; (–)-**linalool*: 27–35%; (–)-*linalyl acetate* (see linalool): 30–40%; (+)-*terpinen-4-ol* (see *p*-menthenols): 3–4%; (–)-*lavandulyl acetate* (see lavandin [oil]): 3–4%. Bulgarian oil contains mostly 45–50% (–)-*linalyl acetate* [2].

Use: L. has many uses in the perfume industry, e. g., in lavender waters, in masculine perfumes or perfumes with masculine notes, etc.; in medicine in carminatives and balneotherapeutics.

Lit.: [1] Perfum. Flavor. **20** (3), 23 (1995). [2] Parfum. Cosmet. Arom. **110**, 58 (1993).

gen.: Arctander, p. 347 ▪ Bauer et al. (2.), p. 160 ▪ ISO 3515 (1987) ▪ Ohloff, p. 142. – *Toxicology:* Food Cosmet. Toxicol. **14**, 4512 (1976). – *[HS 3301 23; CAS 8000-28-0]*

Lawsone (2-hydroxy-1,4-naphthoquinone).

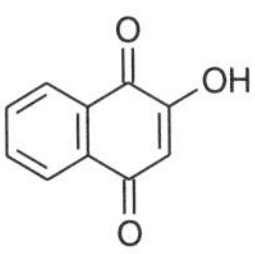

$C_{10}H_6O_3$, M_R 174.16. Yellow prisms, mp. 195–196 °C (decomp.), pK_{a1} –5.6, pK_{a2} 4.00 (25 °C). L. occurs in various *Lawsonia* and *Impatiens* species and is the dyeing constituent of *henna. *Lawsone methyl ether* (2-methoxy-1,4-naphthoquinone, $C_{11}H_8O_3$, M_R 188.18, light yellow needles, mp. 183 °C) is isolated from the leaves of *Impatiens glandulifera* and other *Impatiens* species as well as from Caribbean soils that are rich in montmorillonite clays. L. is formed from *isochorismic acid and 2-oxoglutaric acid via 2-succinylbenzoic acid and subsequent ring closure. L. is used in hair dyes and sun blockers; it exhibits fungicidal and bactericidal activities.

Lit.: Karrer, No. 1207 ▪ Luckner (3.), p. 330 ▪ Martindale (31.), p. 1719 ▪ Pediatrics **7**, 707 (1996) (toxicology) ▪ Pharmacology **51**, 356 (1995) ▪ Phytochemistry **14**, 801 (1975) ▪ Thomson **2**, 202; **3**, 137. – *[HS 2914 69; CAS 83-72-7 (L.); 2348-82-5 (L. methyl ether)]*

Leaf beetles (Chrysomelidae). Species-rich family of, in part brightly-colored beetles with close ecological relationships to the plants on which the adult insects and larvae feed. Some species are major pests (e. g., Colorado beetle). L. b. possess a wide spectrum of *defensive secretions [1]. The females of various economically significant *Diabrotica* species produce chiral, methyl-branched ketones and esters as sexual pheromones. *(R)-10-Methyl-2-tridecanone* ($C_{14}H_{28}O$, M_R 212.38) is the *pheromone of *D. undecimpunctata* [2], while *(6R,12R)-6,12-dimethyl-2-pentadecanone* ($C_{17}H_{34}O$, M_R 254.46) is that of *D. balteata* [3]. Different diastereomers of *1,7-dimethylnonyl propanoate* ($C_{14}H_{28}O_2$, M_R 228.38) are the main components of the pheromones of *D. virgifera*, *D. longicornis*, *D. porracea*, and *D. barberi* [4].

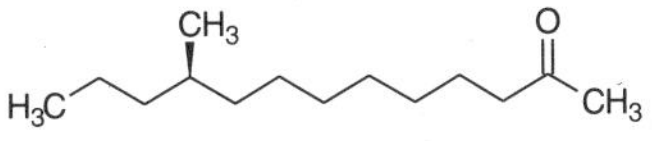

(10*R*)-10-Methyl-2-tridecanone (**1**)

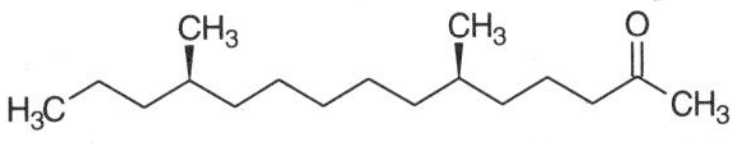

(6*R*,12*R*)-6,12-Dimethyl-2-pentadecanone (**2**)

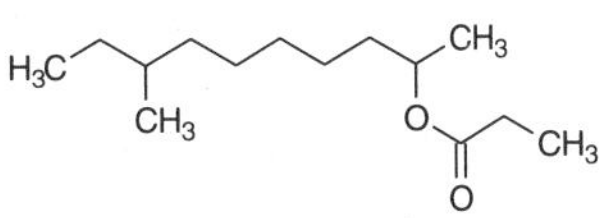

1,7-Dimethylnonylpropanoate (**3**)

Lit.: [1] Barbosa & Letourneau (eds.), Novel Aspects of Insect-Plant Interactions, p. 235–272, New York: Wiley 1988. [2] J. Chem. Ecol. **9**, 1363–1375 (1983). [3] J. Chem. Ecol. **13**, 1601–1616 (1987). [4] J. Chem. Ecol. **10**, 1123–1131 (1984); **11**, 21–26, 1371–1387 (1985). – *[CAS 82621-53-2 (1); 114907-67-4 (2); 83375-82-0 (3)]*

Leaf-Movement Factors (LMF). A term introduced by Schildknecht [1] for compounds in plants that cause a response to an external stimulus, in particular the folding of leaves. Relevant stimuli include warmth (so called thermonasty), touch (thigmonasty), impact (seismonasty), wounding (traumonasty), day and night rhythms (nyctinasty) or chemicals (chemonasty). Most plants of the Fabaceae family close their leaves in the evening, as if to sleep, and open them in the morning. This circadian rhythmic movement (nyctinasty) relies on the interplay of a leaf-closing and a leaf-opening substance and is controlled by the biological clock [2].

Leaf-opening substances Leaf-closing substances

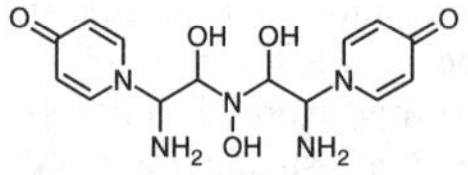

Mimopudine

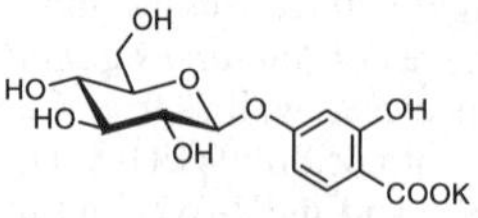

Potassium 5-O-β-D-glucopyranosylgentisate

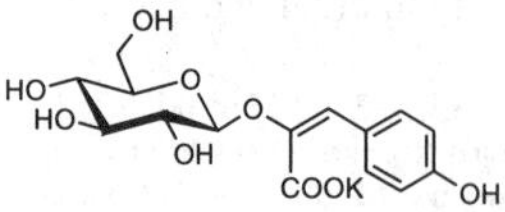

Potassium lespedezate

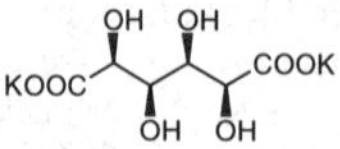

Potassium-D-idarate

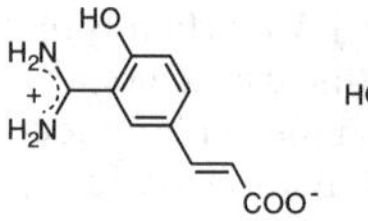

Phyllurine

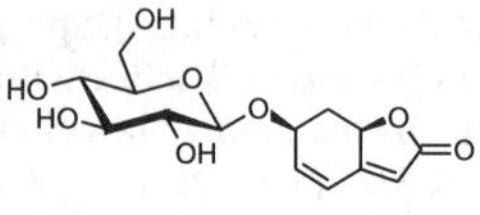

Phyllanthurinolactone

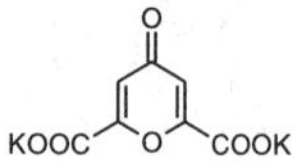

Potassium chelidonate

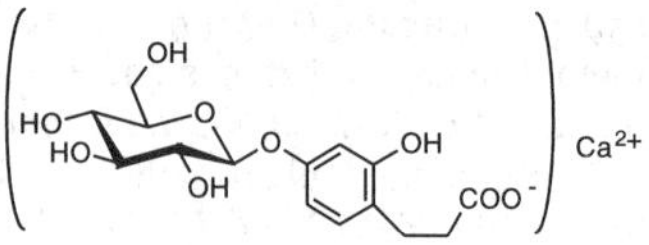

Calcium 4-O-β-D-glucopyranosyl- cis-p-coumarate

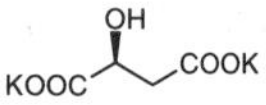

Potassium L-malate

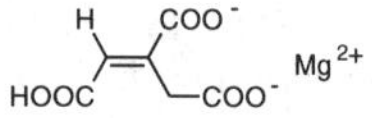

Magnesium *trans*-aconitate

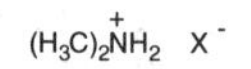

Dimethylammonium salt

The leaf-opening and leave-closing substances differ from species to species. This is contrary to the former opinion that the leaf-movements of nyctinastic plants depend on the concentration of one single leaf-closing factor, *turgorin*, which exists in all nyctinastic plants [1]. In the case of the humble plant, *Mimosa pudica*, *potassium 5-O-β-D-glucopyranosylgentisate* ($C_{13}H_{15}KO_9$, M_R 354.36, yellow sirup) has been identified as the slow leaf-closing substance (active at 5×10^{-5} mol) [3] and *mimopudine* ($C_{14}H_{20}N_5O_5$, M_R 338.34, yellow powder, unstable in aqueous solution) [4] as the leaf-opening substance (0.5 mg from 12.2 kg of leaves). The glucoside is not responsible for the rapid thigmonastic movement of *M. pudica* leaves which is induced by a mixture of *potassium L-malate, magnesium potassium trans-aconitate* and *dimethylammonium salts* (10^{-8} ~ 10^{-9} mol) [5]. Potassium 5-O-β-D-glucopyranosylgentisate is hydrolysed by L-glucosidase to yield the biologically inactive aglucone in the morning. The activity of the enzyme is controlled by the biological clock of the plant. Similarly, in *Lespedeza cuneata* (Fabaceae) the nyctinastic movement depends on *potassium D-idarate* (leaf-closing) and *potassium lespedezate* (leaf-opening) (8×10^{-7} mol) [2,6]. The movement is controlled by enzymatic cleavage of potassium lespedezate in the evening, which affords the inactive 4-hydroxyphenylpyruvate. *Cassia mimosoides* uses *potassium chelidonate* and *calcium 4-O-β-D-glucopyranosyl-(Z)-p-coumarate* as a pair of regulators [7], and in the leaves of *Albizia julibrissin* and *Aeschynomene indica* (Fabaceae) *cis-p-coumaroyl* **agmatine* and **trigonellin*, respectively, have been identified as leaf-closing substances. Potassium lespedezate and *cis-p*-coumaroylagmatine are not active in *Mimosa pudica*. In *Phyllurus urinaria* (Euphorbiaceae) *phyllurine* [7] controls the leaf-opening (2.5×10^{-5} mol) and *phyllanthurinolactone* ($C_{14}H_{18}O_8$, M_R 314.29, yellow powder, $[\alpha]_D$ −6° (H_2O)) [8] the leaf closing process. The lactone glycoside is hydrolysed by a β-glucosidase whose activity is regulated by a biological clock. The LMF act on the H^+, Ca^{2+}, K^+ ion flows and thus effect changes in the membrane potentials of the pulvinus cells. The resultant change in turgor is the actual reason for the sudden folding of the leaves.

Lit.: [1] Angew. Chem. Int. Ed. Engl. **22**, 695 (1981); **24**, 689 (1983). [2] Tetrahedron **54**, 12173 (1998). [3] Tetrahedron Lett. **40**, 2981 (1999). [4] Tetrahedron Lett. **40**, 353 (1999). [5] Tetrahedron **55**, 10937 (1999). [6] Tetrahedron Lett. **38**, 2497 (1997); **40**, 3757 (1999); Tetrahedron **46**, 383 (1990); Chem. Lett. **1998**, 179. [7] Tetrahedron Lett. **39**, 9731 (1998); Tetrahedron **55**, 5781 (1999). [8] Tetrahedron Lett. **36**, 6267 (1995).

gen.: Synthesis: Eur. J. Org. Chem. **1998**, 57 (phyllanthurinolactone) ▪ Tetrahedron Lett. **30**, 6389 (1989) (potassium lespedezate). – *Review:* Angew. Chem. Int. Ed. Engl. **39**, 1400–1414 (2000) ▪ Phytochemistry **53**, 39 (2000). – *[CAS 220345-84-6 (phyllurine); 168180-12-9 (phyllanthurinolactone); 221056-52-6 (mimopudine); 123955-02-2 (potassium lespedezate)]*

Leaianafulvene.

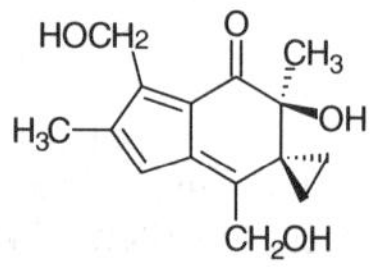

$C_{15}H_{18}O_4$, M_R 262.31, orange crystals, mp. 142 °C, $[\alpha]_D$ +306° (CH_3OH). An orange-yellow pigment with a fulvene chromophore from cultures of the mushroom *Mycena leaiana* (Basidiomycetes). L. has a modified *illudane skeleton and exhibits weak antibacterial and stronger cytotoxic activity.

Lit.: Phytochemistry **29**, 3932 (1990). – *[CAS 132998-73-3]*

Lecanoric acid.

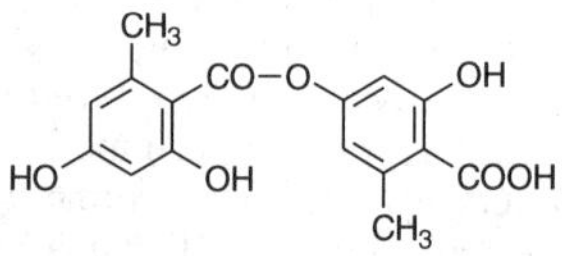

$C_{16}H_{14}O_7$, M_R 318.28, needles, mp. 184 °C (decomp.). A *depside from numerous lichens, especially Parmeliaceae and Roccellaceae. L. was synthesized for the first time by E. Fischer in 1913. On acidic hydrolysis L. furnishes *orsellinic acid, while alkaline hydrolysis gives *orcinol.

Lit.: Aust. J. Chem. **41**, 1789 (1988) (synthesis) ▪ Beilstein EIV **10**, 1527 ▪ Culberson, p. 121–123 ▪ Karrer, No. 1024. – *[CAS 480-56-8]*

Lectins (from latin: legere = to choose). Term for certain proteins or glycoproteins which specifically recognize and bind to (poly)saccharides which may even be bound to lipids or carbohydrates. Important examples are *abrin, *concanavalin A, *ricin and phasin. L. are abundant in most organisms, very high contents are found in beans and slugs. The M_R of L. ranges from

8500 to about 300 000, the smallest L. yet is isolated from stinging nettles, it binds *chitin and has fungistatic activity. In the human organism, L. act as hemagglutinins (causing aggregation of erythrocytes), therefore many L. are toxic. Mistletoe l. has potential in cancer chemotherapy.

Lit.: Franz (ed.), Advances in Lectin Research (5 vols.), Berlin: Springer 1988, 1989, 1990, 1991, Wiesbaden: Ullstein Medical 1992 ▪ Sharon and Lis, Lectins, London: Chapman & Hall 1989 ▪ Van Damme et al., Handbook of Plant Lectins: Properties and Biomedical Applications, New York: Wiley 1998.

Ledol (ledum camphor).

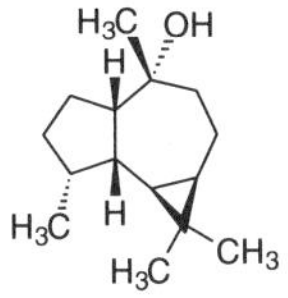

$C_{15}H_{26}O$, M_R 222.37, cryst., mp. 105 °C, bp. 282 °C, $[\alpha]_D^{20}$ +28° ($CHCl_3$), soluble in organic solvents, insoluble in water. A strongly irritating *sesquiterpene from the leaves of *Ledum palustre* (marsh tea, Ericaceae). On oral consumption L. causes vomiting, sweating, increased pulse, pain in muscles and joints as well as cramps and states of excitation. Historically, the leaves of *L. palustre* were added to beer to increase the exciting effect.

Use: In homeopathic medicine for joint rheumatism, ischias, contusions, and insect stings. In therapy *Ledum* products are used for whooping cough, bronchial catarrh, eczemas, and rheumatic diseases.

Lit.: Beilstein E IV **6**, 426 ▪ Karrer, No. 1908 ▪ Merck-Index (12.), No. 5454 ▪ Phytochemistry **42**, 677 (1996); **44**, 1393 (1997) (NMR) ▪ Tetrahedron Lett. **37**, 949 (1996) (synthesis). – *[CAS 577-27-5]*

Leek flavor see vegetable flavors.

Leghemoglobin (legoglobin). A red substance isolated from root tubercules of legumes (Leguminosae) resembling the myoglobin of vertebrates. By reversible binding of oxygen L. ensures the high oxygen requirements in nitrogen fixation by root nodule bacteria. The apoprotein is formed by the plant cells and the *heme by the bacteria.

Lit.: Endeavour **31**, 139–142 (1972) ▪ Merck-Index (12.), No. 5456 ▪ see also hemoglobin.

Leinamycin.

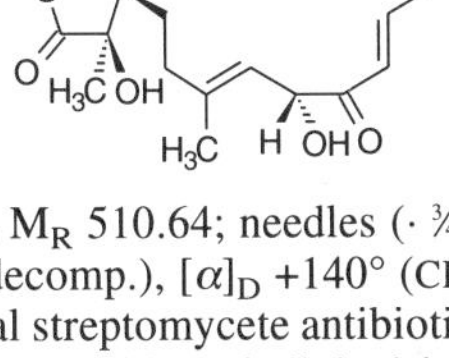

$C_{22}H_{26}N_2O_6S_3$, M_R 510.64; needles (· ¾ ethyl acetate): mp. 155 °C (decomp.), $[\alpha]_D$ +140° (CH_3OH). A structurally unusual streptomycete antibiotic with a 1,2-dithiolane-3-one 1-oxide unit linked in a spirane-like manner to an 18-membered, highly unsaturated lactam ring. L. exhibits strong activity against various experimental tumors. Two possible mechanisms of thiol-activated DNA cleavage (oxidative and alkylative) seem to be involved [1].

Lit.: [1] Chem. Eng. News (8. 12.) **1997**, 23; J. Am. Chem. Soc. **118**, 6802 (1996); **119**, 11691 (1997).

gen.: Biochemistry **29**, 5676 (1990) (activity) ▪ J. Antibiot. **42**, 333, 1768 (1989). – *Synthesis:* J. Am. Chem. Soc. **115**, 8451 (1993) ▪ J. Heterocycl. Chem. **29**, 607 (1992) ▪ J. Med. Chem. **42**, 1330 (1999) ▪ Synlett **1993**, 215. – *[HS 2941 90; CAS 120500-15-4]*

Lelobanidine see Lobelia alkaloids.

Lemmasterone see ecdysteroids.

Lemongrass oil. Light yellow to orange-yellow oil with a fresh, strong odor of lemons but somewhat hot-bitter, herby, leaf-like.

Production: By steam distillation of the tropical grasses *Cymbopogon flexuosus* (so-called East Indian L.) or *C. citratus* (so-called West Indian L.). The main areas of cultivation for the East Indian variety are India and for the West Indian variety Central and South America.

Composition [1]*:* The main component is **citral* (30–35% neral; 40–50% geranial).

Use: Mainly to perfume household products (soaps); in the past also for the isolation of citral. The significance of L. has declined in the past decades as a result of competition not only from synthetically produced citral but also from *Litsea cubeba* oil as another natural source of citral. The annual production world-wide is now probably only ca. 1000 t.

Lit.: [1] Perfum. Flavor. **19** (2), 29 (1994).

gen.: Arctander, p. 352 ▪ Bauer et al. (2.), p. 152 ▪ ISO 3217 (1974), 4718 (1981). – *Toxicology:* Food Cosmet. Toxicol. **14**, 455, 457 (1976). – *[HS 3301 29; CAS 8007-02-1]*

Lemon oil. Light, mobile, pale yellow to greenish-yellow oil with a fresh, sparkling odor of grated lemon peel and fresh, bitter-tart taste.

Production: By mechanical processing ("pressing", see citrus oils) from the peel of citrus fruit (*Citrus limon*). Main producing countries are Italy, USA, Argentina, Israel, and Spain. The annual world-wide production amounts to 2000–3000 t.

Composition [1]*:* Main components are (+)-*limonene* (see *p*-menthadienes) (ca. 65%), *β-*pinenes* (ca. 10%), and *γ-terpinene* (see *p*-menthadienes) (ca. 10%). The major odor and taste component is **citral* (ca. 3–5%), others are geranial and neral.

Use: C. has wide uses in the perfume industry, especially in compositions with particularly refreshing tones such as Eau de Toilette and Eaux de Cologne. Because of a low content of *furocoumarins, however, C. can be used in perfume oils with restrictions only [2]. The major amount of C. produced is used for aromatization of foodstuffs such as refreshing beverages, sweets, confectionery products, etc. In general concentrated C., which contains a small amount of terpene hydrocarbons, is used(*citrus oils). C. is also used medicinally in, e. g., carminatives and rhinological preparations.

Lit.: [1] Perfum. Flavor. **16** (2), 17 (1991); **17** (1), 45 (1992); **19** (3), 64; (6), 29 (1994); Flavour Fragr. J. **9**, 105 (1994); **10**, 33 (1995). [2] J. Chromatogr. **A 672**, 177 (1994).

gen.: Arctander, p. 354 ▪ Bauer et al. (2.), p. 147 ▪ ISO 855 (1981) ▪ Ohloff, p. 133. – *Toxicology:* Food Cosmet. Toxicol. **12**, 725 (1974). – *[HS 3301 13; CAS 8008-56-8]*

Lenthionine (1,2,3,5,6-pentathiepane).

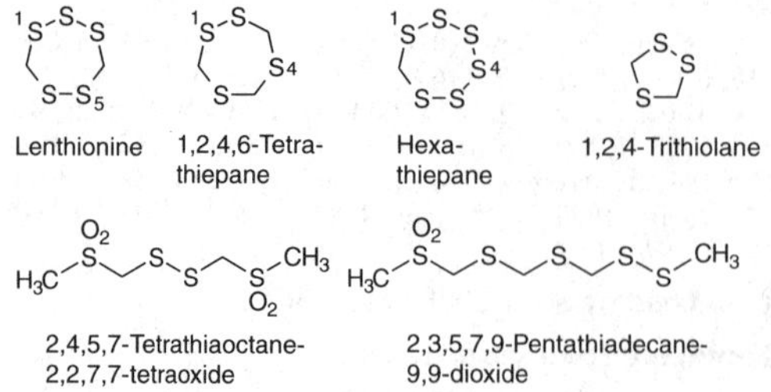

$C_2H_4S_5$, M_R 188.35, cryst., mp. 60–61 °C. L. is a component of the East Asian edible mushroom shiitake (*Lentinus edodes*, Basidiomycetes) and, together with the polythiepanes and 1,2,4-trithiolane shown in the formula scheme, is responsible for the characteristic odor of the mushroom [1]. Precursors of this flavor substance are 2,4,5,7-tetrathiaoctane 2,2,7,7-tetraoxide and 2,3,5,7,9-pentathiadecane 9,9-dioxide (SE-3), the structures of which were confirmed by synthesis [2] (see also dysoxysulfone). The cyclic flavor substances can be prepared simply by reaction of dichloromethane with $Na_2S_{2.5}$ (from Na_2S and sulfur) [1]. Shiitake mushrooms – cultivated on dead wood – are becoming increasingly popular in Europe and North America on account of their excellent flavor. L. is also found as a volatile component of cooked mutton [3] and in the alga *Chondria californica* [4].

Lit.: [1] Tetrahedron Lett. **1966**, 573; **22**, 1939 (1981); Chem. Pharm. Bull. **15**, 988 (1967); J. Food Sci. **32**, 559 (1967); Zechmeister **36**, 251. [2] J. Org. Chem. **59**, 2273 (1994). [3] J. Agric. Food Chem. **27**, 355 (1979). [4] J. Org. Chem. **41**, 2465 (1976). *gen.:* Beilstein E V **19/12**, 251 ▪ J. Chem. Soc., Perkin Trans. 1 **1990**, 509 (synthesis). – *[CAS 292-46-6]*

Lentinellic acid.

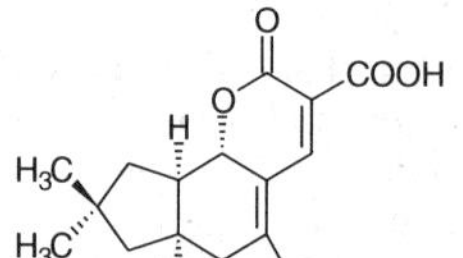

$C_{18}H_{20}O_5$, M_R 316.35, yellow cryst., mp. 174–176 °C, $[\alpha]_D$ −187.1° ($CHCl_3$). An antibacterially and cytotoxically active sesquiterpenoid of the *protoilludane type from mycelia cultures of *Lentinellus omphalodes* and *L. ursinus* (Basidiomycetes). The biosynthesis probably involves condensation of malonic acid with a protoilludane aldehyde (*Armillaria sesquiterpenoids).

Lit.: Z. Naturforsch. C **43**, 177 (1988). – *[CAS 115219-90-4]*

Lentinic acid [(*R*)-2-(L-γ-glutamylamino)-4,6,8,10,10-pentaoxo-4,6,8,10-tetrathiaundecanoic acid].

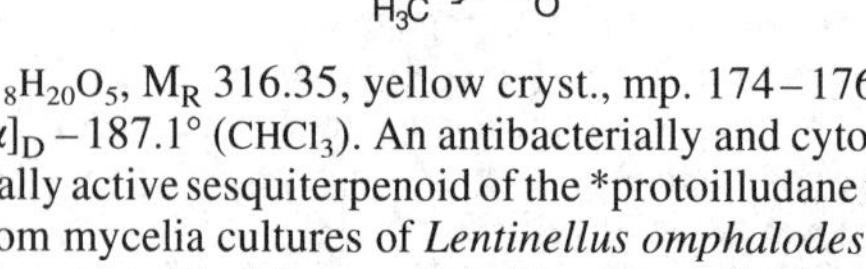

$C_{12}H_{22}N_2O_{10}S_4$, M_R 482.55, mp. 186 °C (decomp.), $[\alpha]_D^{22}$ +27.0° (0.1 m $NaHCO_3$). L. occurs in the East Asian edible mushroom shiitake, *Lentinus edodes* as a flavor precursor. L. was first isolated as a precursor of the flavor compound *lenthionine (1,2,3,5,6-pentathiepane [1]) and other cyclic polythiepanes from *L. edodes* [2]. L. also occurs in *Micromphale perforans* [3] and *Collybia hariolorum* [3].

Metabolism: Deglutamyllentinic acid formed with the aid of γ-glutamyltranspeptidase is further degraded by C-S lyase to ammonia, pyruvic acid, and a labile intermediate. This leads through several intermediates to various flavor substances including lenthionine [4,5].

An epimer, *epilentinic acid*, is isolated from other aromatic fungi such as *Micromphale foetidum*, *M. cauvetii*, and *Collybia impudica*. Mp. 221 °C, $[\alpha]_D^{22}$ +58.8° (0.1 m $NaHCO_3$).

Lit.: [1] Chem. Pharm. Bull. **15**, 756–760, 988–993 (1967). [2] Agric. Biol. Chem. **35**, 2059–2069 (1971). [3] Tetrahedron Lett. **1976**, 3129–3132. [4] Agric. Biol. Chem. **35**, 2070–2080 (1971). [5] Phytochemistry **19**, 553–557 (1980). *gen.:* Turner **2**, 469, 486, 488. – *[CAS 12705-98-5]*

Leontidine see camoensine.

Lepicidins see spinosyns.

Lepiochlorin [5-chloromethyl-5-hydroxy-3-methyl-2(5*H*)-furanone].

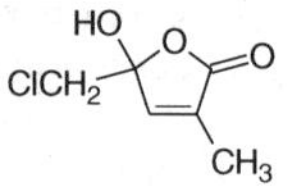

$C_6H_7ClO_3$, M_R 162.57, cryst., mp. 68–70 °C. A metabolite from cultures of a *Lepiota* species (Basidiomycetes) cultivated by leaf-cutting ants. L. inhibits the growth of *Staphylococcus aureus*.

Lit.: Phytochemistry **18**, 326 (1979). – *Synthesis:* Bull. Chem. Soc. Jpn. **56**, 2183 (1983) ▪ Tetrahedron Lett. **21**, 2771 f. (1980); **24**, 3959 (1983). – *[CAS 71339-41-8]*

Lepistine.

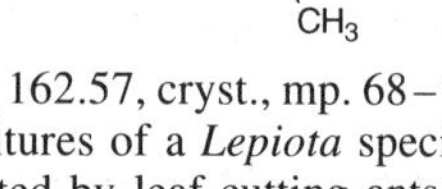

$C_{10}H_{16}N_2O_2$, M_R 196.25, hygroscopic liquid, bp. 140–150 °C, mp. of the hydrochloride 242 °C. An unusual alkaloid from the Asian agaric *Clitocybe fasciculata* (=*Lepista caespitosa*, Basidiomycetes).

Lit.: Tetrahedron Lett. **1975**, 269. – *[CAS 55623-00-2]*

Lepistirone.

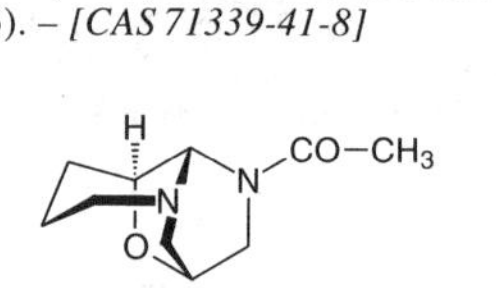

$C_{15}H_{22}O_2$, M_R 234.34, oil, $[\alpha]_D$ +129.5° (CH_3OH). A sesquiterpenoid from cultures of the agaric *Lepista irina* (Basidiomycetes).

Lit.: Z. Naturforsch. C **46**, 169 (1991). – *[CAS 134984-20-6]*

Lepranthin.

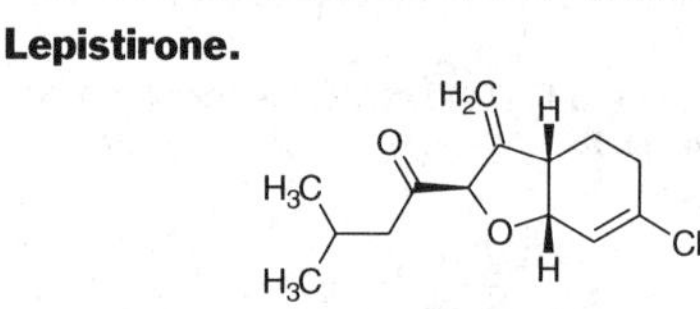

Ac = CO−CH₃

$C_{32}H_{52}O_{14}$, M_R 660.76, platelets, mp. 185 °C, $[\alpha]_D$ +70° ($CHCl_3$). The first macrocyclic bislactone isolated in 1904 by Zopf from a lichen (*Arthonia impolita*); its structure, however, was not confirmed by X-ray crystallography until 1995.

Lit.: Justus Liebigs Ann. Chem. **336**, 46 (1904) ▪ Z. Naturforsch. B **50**, 1111–1114 (1995). – *[HS 2932 29; CAS 166334-55-0]*

Lepraric acid.

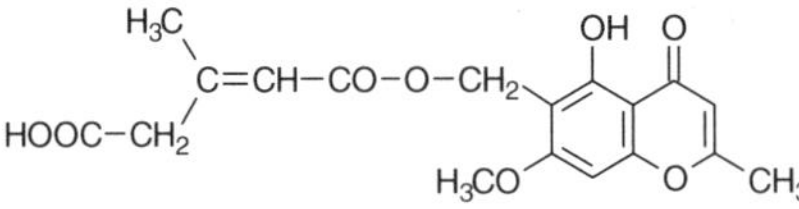

$C_{18}H_{18}O_8$, M_R 362.34, prisms, mp. 158–160 °C, L. shows the following color reactions: KOH yellow, KOH+NaOCl green, and $FeCl_3$ red, turning green. A chromone from the lichens *Lecanactis latebrarum* and *Roccella fuciformis*.

Lit.: J. Chem. Soc. C **1969**, 704–707. – *[CAS 22399-41-3]*

Leprocybe pigments (fluorescence substances).

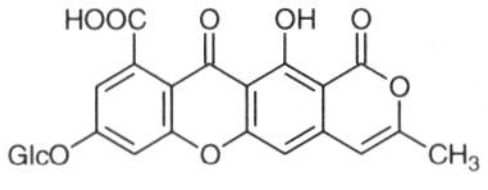

Leprocybin

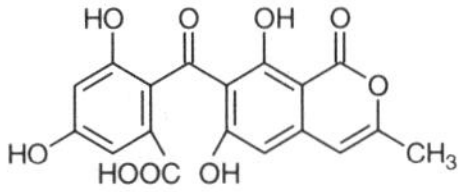

Leprophenone

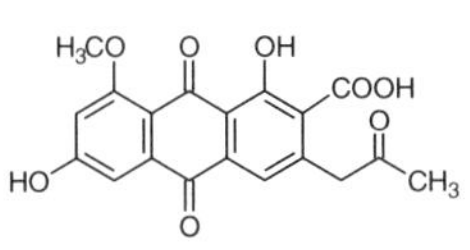

Leprolutein

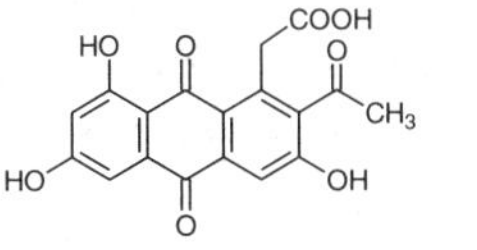

Leprovenetin

Xanthone pigments from the agaric genus *Cortinarius*, subgenus *Leprocybe*, showing yellow-green fluorescence under UV light. The main fluorescent compound of the genus is the nonaketide *leprocybin* {$C_{24}H_{20}O_{13}$, M_R 516.42, yellowish-green powder, mp. >240 °C (decomp.), $[\alpha]_D$ –60° (H_2O)}, a *β*-D-glucoside with the aglycone leprocyboside. The biosynthesis of leprocybin presumably proceeds through ring closure from the diaryl ketone *leprophenone* ($C_{18}H_{12}O_9$, M_R 372.29, colorless cryst., mp. 198 °C). The toadstools also contain small amounts of nonaketide anthraquinone pigments (see dermochrysone) such as *leprolutein* ($C_{19}H_{14}O_8$, M_R 370.32, red cryst., mp. >350 °C), which is closely related to the biosynthetic precursor of leprophenone, and the unusually substituted *leprovenetin* [$C_{18}H_{12}O_8$, M_R 356.29, yellow cryst., mp. 270 °C (decomp.)].

Lit.: Justus Liebigs Ann. Chem. **1982**, 1280 ▪ Zechmeister **51**, 174–179. – *[CAS 82850-45-1 (leprocybin)]*

Leptine I, leptinines see Solanum steroid alkaloid glycosides.

Leptinidine see Solanum steroid alkaloids.

Leptosidin, leptosin see maritimetin.

Leptosphaerodione see fungi.

Leucenine, leucenol see mimosine.

L-Leucine (2-amino-4-methylpentanoic acid, abbreviation: Leu or L).

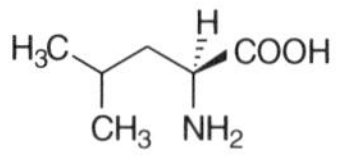

(2*S*)-, L-form

$C_6H_{13}NO_2$, M_R 131.17, mp. 293–295 °C (decomp.), $[\alpha]_D^{25}$ –10.8° (H_2O), $[\alpha]_D^{15}$ +17.3° (20% HCl), pK_a 2.36, 9.60, pI 5.98. A widely distributed amino acid and an essential amino acid for humans and animals. Genetic code: UUA, UUG, CUU, CUC, CUA, CUG. Average content in proteins 7.5% [1]. L-Leu is isolated from gluten, casein, keratin, and hemoglobin. The D-form occurs in some polypeptide antibiotics.

Biosynthesis: Leu is formed from pyruvic acid: →2-acetolactic acid [acetolactate synthase (EC 4.1.3.18.)+(1-hydroxyethyl)-TPP] → 2,3-dihydroxyisovaleric acid [reductase+NAD(P)H] → 2-oxoisovaleric acid [dihydroxyacid dehydratase] → 2-isopropylmalate [2-isopropylmalate synthase + acetyl-CoA (EC 4.1.3.12)] → 3-isopropylmalate [isopropylmalate dehydratase (EC 4.2.1.33): –H_2O+H_2O] → 2-oxoisocaproate [3-isopropylmalate dehydrogenase (EC 1.1.1.85) + NAD^+] → L. [leucine aminotransferase (EC 2.6.1.6)].

Lit.: [1] Biochem. Biophys. Res. Commun. **78**, 1018–1024 (1977).

gen.: Beilstein E IV **4**, 2738 ▪ Greenstein & Winitz, Chemistry of the Amino Acids, vol. 3, p. 2075–2096, New York: J. Wiley 1961 ▪ Merck-Index (12.), No. 5475. – *[HS 2922 49; CAS 7005-03-0]*

Leucoanthocyanidins.

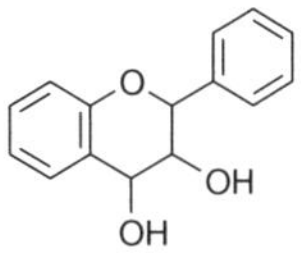

Leucoanthocyanidin
(basic structure)

L. are flavan-3,4-diols belonging to the *flavonoids. They are transformed to colored anthocyanidins by the action of acid and are thus also known as *proanthocyanidins. They occur especially in wood, bark, and fruit skins.

Leucomelone see cycloleucomelone.

Leucomentins.

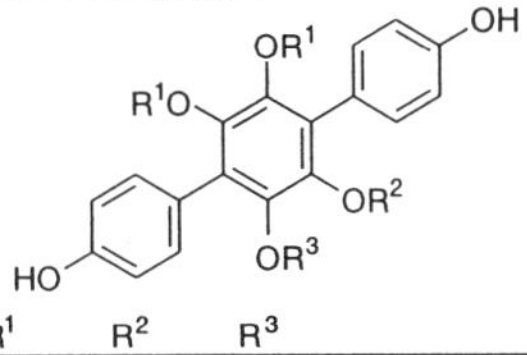

R^1	R^2	R^3	
EPH	H	H	Leucomentin-2
EPH	EPH	H	Leucomentin-3
EPH	EPH	EPH	Leucomentin-4
H	H	H	Leuco-atromentin

EPH =

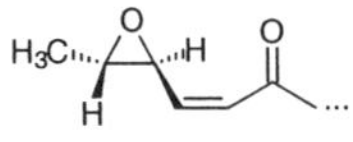

Coloress storage forms of the long-known *terphenylquinone *atromentin from the toadstool *Paxillus atrotomentosus* (Basidiomycetes) [1]. L. are esters of leucoatromentins with (2*Z*, 4*S*,5*S*)-4,5-epoxy-2-hexenoic acid that are cleaved under the action of alkali and oxidized to atromentin. According to the number of ester groups a distinction is drawn between *L.-2* ($C_{30}H_{26}O_{10}$, M_R 546.53), *L.-3* (main component, $C_{36}H_{32}O_{12}$, M_R 656.64), and *L.-4* [$C_{42}H_{38}O_{14}$, M_R 766.76, colorless cryst., mp. 180°C, $[\alpha]_D$ +104.4° (CH_3OH)]. Acidic hydrolysis converts the 4,5-epoxy-2-hexenoic acid residues of L. to (+)-*osmundalactone, which has proven antifeedant activity on caterpillars of the butterfly *Eurema hecabe mandarina*[2]. It is possible that L. and their degradation products are responsible for the reduced insect attacks on *Paxillus atrotomentosus*. The L. are accompanied by yellow *flavomentins and violet *spiromentins. Methylether acetates of leuco-atromentin are the butlerins[3].

Lit.: [1] Justus Liebigs Ann. Chem. **1989**, 797. [2] Appl. Entomol. Zool. **18**, 129 (1983); Chem. Pharm. Bull. **32**, 2815 (1984). [3] Aust. J. Chem. **49**, 1247 (1996). – *[CAS 121254-42-0 (L.-2); 121254-43-1 (L.-3); 121254-44-2 (L.-4)]*

Leucomycin.

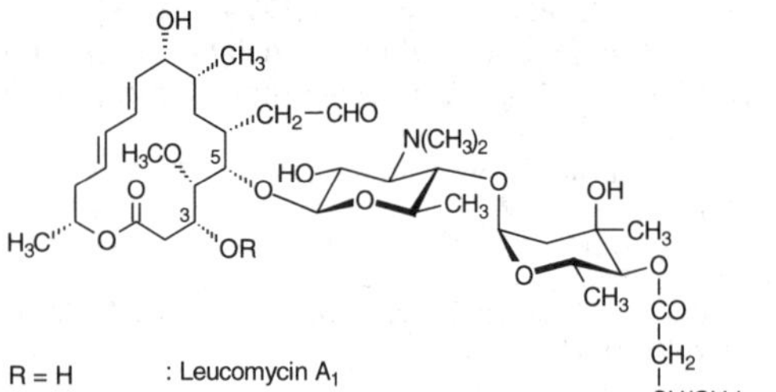

R = H : Leucomycin A_1
R = CO—CH_3 : Josamycin (Leucomycin A_3)

Table: Data of Leucomycin components.

	molecular formula	M_R	CAS
L. V	$C_{35}H_{59}NO_{13}$	701.85	22875-15-6
L. A_1	$C_{40}H_{67}NO_{14}$	785.97	16846-34-7
L. A_3	$C_{42}H_{69}NO_{15}$	828.02	16846-24-5
L. A_5	$C_{39}H_{65}NO_{14}$	771.94	18361-45-0
L. B	5 components of unknown structure		

A complex, 16-membered, optically active *macrolide antibiotic from *streptomycetes (e.g., *Streptomyces kitasatoensis*) discovered in 1953. The parent compound L.V exhibits only weak biological activities and does not have acyl groups at 3-OH and at the end of the disaccharide unit consisting of β-D-mycaminose and α-L-mycarose. The L. components differ in the acyl groups at the mentioned positions, L. A_1 (see formula) has an isovaleryl group at the sugar, L. A_3 {international generic name *josamycin*, mp. 120–121 °C, $[\alpha]_D$ –58° (CH_3OH)} has an additional acetyl group at 3-OH. Only the esters significantly inhibit the growth of Gram-positive and some Gram-negative bacteria. Because of its excellent tolerability, josamycin finds pediatric use. Many semisynthetic esters have been prepared. In Japan a mixture of L. (international generic name: *Kitamycin*) is used clinically. The aglycone (*leuconolide*) differs from other classical macrolides by having an aldehyde side chain (derived biosynthetically from butyrate) and the fact that in the *polyketide biosynthesis of C-3/C-4 a C_2-unit originates from glycolate and not from acetate.

Lit.: Drug Action Drug Resist. Bact. **1**, 267 (1971) ▪ J. Antibiot. A **21**, 199–203 (1968); **40**, 1851 (1987) ▪ J. Chem. Soc., Perkin Trans. 2 **1999**, 529 ▪ Martindale (30.), p. 178 ▪ Omura (ed.), Macrolide Antibiotics, New York: Academic Press 1984 (review) ▪ Pharmazie **39**, 414 (1984) ▪ Tetrahedron **28**, 2839–2848 (1972); **41**, 3569 (1985). – *[HS 2941 90]*

Leuconolide see leucomycin.

Leucophleol see pimaranes.

Leucopterin see pteridines.

Leucovorin see folinic acid.

Leuhistin [(2*R*,3*S*)-3-amino-2-hydroxy-2-(1*H*-imidazol-4-ylmethyl)-5-methylhexanoic acid].

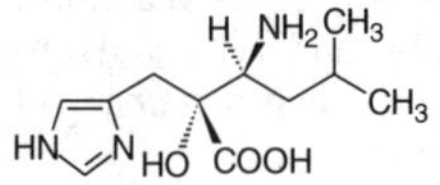

$C_{11}H_{19}N_3O_3$, M_R 241.29, needles, mp. 180–183 °C, $[\alpha]_D^{25}$ –51.4° (CH_3OH), soluble in water, methanol. L. is produced by *Bacillus laterosporus* and inhibits aminopeptidase M (IC_{50} 0.2 µg/mL). L. is formed biosynthetically by coupling of L-leucine to L-histidine under elimination of the carboxy group of L-leucine and the amino group of L-histidine[1].

Lit.: [1] J. Antibiot. **44**, 573–581, 683 f. (1991). – *[CAS 129085-76-3]*

Leukaemomycin see daunorubicin.

Leukotoxin B see vernolic acid.

Leukotrienes (LT). The name L. is used for various biologically highly active compounds derived biosynthetically from *arachidonic acid and other multiply unsaturated C_{20} fatty acids (see eicosanoids). The name L. is derived from their origin in *leukocytes* and the three conjugated C=C double bonds (*triene* system) in the molecule. 5-Lipoxygenase[1] is a key enzyme in the biosynthesis of L. Its action on arachidonic acid [(*all-Z*)-5,8,11,14-eicosatetraenoic acid] leads first to (5*S*)-hydroperoxy-6*E*,8*Z*,11*Z*,14*Z*-eicosatetraenoic acid (5-HpETE, $C_{20}H_{32}O_4$, M_R 336.47) and then via the intermediate 5,6-epoxide [(5*S*,6*S*)-epoxy-7*E*,9*E*,11*Z*,14*Z*-eicosatetraenoic acid, LTA_4, $C_{20}H_{30}O_3$, M_R 318.46] to the 5,12-dihydroxy derivative [(5*S*,12*R*)-dihydroxy-6*Z*,8*E*,10*E*,14*Z*-eicosatetraenoic acid, LTB_4, $C_{20}H_{32}O_4$, M_R 336.47] or by transfer of a glutathione residue to (5*S*,6*R*)-6-*S*-glutathionyl-5-hydroxy-7*E*,9*E*,11*Z*,14*Z*-eicosatetraenoic acid (LTC_4, $C_{30}H_{47}N_3O_9S$, M_R 625.78) (see figure). LTC_4 is then converted to LTD_4 {*N*-[*S*-[(*R*)-1-((*S*)-4-carboxy-1-hydroxybutyl)-2*E*,4*E*,6*Z*,9*Z*-pentadecatetraenyl]-L-cysteinyl]glycine, $C_{25}H_{40}N_2O_6S$, M_R 496.66} and LTE_4 {(5*S*,6*R*)-6-((*R*)-2-amino-2-carboxyethylthio)-5-hydroxy-7*E*,9*E*,11*Z*,14*Z*-eicosatetraenoic acid, $C_{23}H_{37}NO_5S$, M_R 439.61} containing cysteine residues, as well as further eicosanoids of the L. group (the index 4 shows the number of double bonds in the molecule). LTC_4, LTD_4, and LTE_4 are also known as peptide-L. or cysteinyl-L.

LTB_4 effects the aggregation of leukocytes and their adhesion to the vessel walls, it also promotes the re-

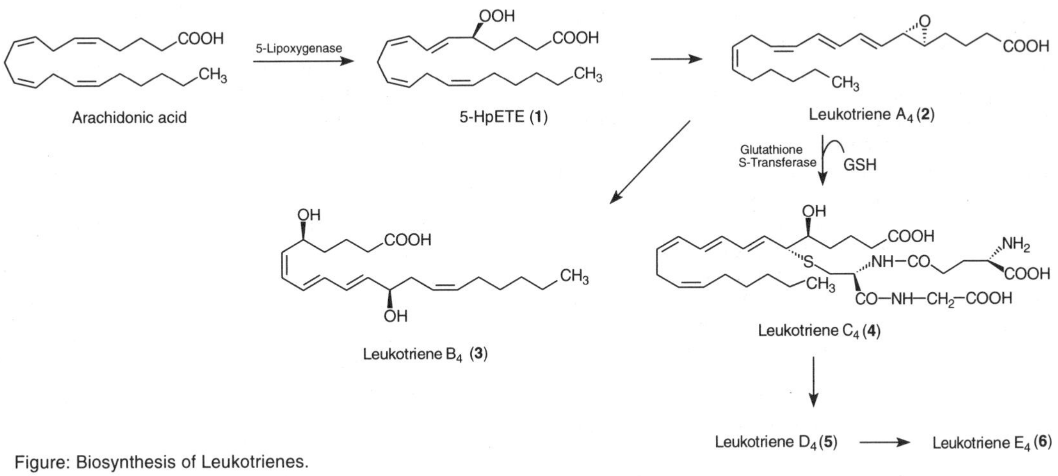

Figure: Biosynthesis of Leukotrienes.

lease of oxidizing enzymes and the formation of peroxide radicals which lead to inflammation and tissue damage. This is the reason for the key role played by L. in the development of inflammations. New therapeutic concepts are thus aimed at an inhibition of the biosynthesis of L. using 5-lipoxygenase inhibitors and L. antagonists for possible use in the therapy of chronic diseases such as asthma, rheumatism, psoriasis, and arthritis. The so-called *SRS-A* ("slow reacting substance of anaphylaxis"), first identified in human lungs in 1938, is identical with the cysteinyl-L. LTC_4–LTE_4. It can trigger allergic reactions and lead to asthma by restricting the airways.

Lit.: [1] Annu. Rev. Biochem. **63**, 383 (1994); Biochim. Biophys. Acta **1128**, 117 (1992).

gen.: Annu. Rev. Biochem. **55**, 69 (1986) ▪ Gunston, Harwood & Padley, The Lipid Handbook, 2. edn., p. 605 ff., London: Chapman & Hall 1994 ▪ J. Org. Chem. **58**, 3516 (1993) (synthesis LTB_4) ▪ Pediatric Res. **37**, 1 (1995) ▪ Rokach, Leukotrienes and Lipoxygenases, Amsterdam: Elsevier 1989 ▪ Tetrahedron **52**, 6635 (1996) (synthesis 5-HpETE). – *Journal:* Adv. Prostaglandin, Thromboxane, Leukotriene Research. – *[CAS 71774-08-8 (1); 72059-45-1 (2); 73151-67-4 (3); 72025-60-6 (4); 73836-78-9 (5); 75715-89-8 (6)]*

Leurocristine see vincristine.

Levarterenol see noradrenaline.

Levomenol see (–)-α-bisabolol.

Levorin A see candicidin.

Lichenin see glucans.

Lichen pigments. In the case of L. p. a distinction must be made between the colored compounds of the lichen themselves and the pigments arising from chemical transformations of colorless *lichen substances. The first group includes *usnic and *isousnic acid (yellow), *pulvinic acid derivatives (yellow and orange), xanthones (yellow), *naphthoquinones (red, violet), *anthraquinones (red), phenanthrenequinones (violet), *polyporic acid (red), *thelephoric acid (violet), and *carotinoids (yellow, red). The second group of pigments are complex mixtures of polymeric components with 7-hydroxy-3-phenoxazone chromophores. These phenoxazones are formed from *depsides (lecanoric acid, erythrin, gyrophoric acid) by hydrolysis and subsequent reaction with ammonia. The dyeing of textiles with L. p. is no longer of practical relevance. It is possible that the pigments deposited in the cortex of the thallus serve to protect the symbiontic fungus from excessive exposure to light.

Lit.: Culberson.

Lichen substances. Organic compounds occurring in lichens that are in part characteristic for this group of symbiontic organisms (algae and fungus). L. s. (about 650) can be classified into four large groups on the basis of their biogenesis: 1) acetyl-CoA derivatives; – 2) terpenoids derived from mevalonic acid; – 3) compounds whose biosynthesis proceeds through shikimic acid and amino acids, and – 4) sugar alcohols, oligo- and polysaccharides.

The 1st group includes *paraconic acid derivatives, macrocyclic compounds, monocyclic aromatic compounds, *depsides, *depsidones, *depsones, dibenzofurans, chromones, xanthones, *naphthoquinones, *anthraquinones, diphenyls, and diphenyl ethers. The 2nd groups comprises diterpenoids, sesterterpenoids, triterpenoids, *steroids, and *carotinoids, while the 3rd group contains *pulvinic acid derivatives, amino acid aminoalcohol esters, and cyclic peptides. According to recent results, most L. s. are produced by the fungal partner. Some lichens accumulate surprisingly large amounts of L. s.; for example, the crustaceous lichen *Pertusaria aleianta* from the Cape Verde Islands contains up to 50% dry weight of a mixture of chloroxanthones. The L. s. are deposited both in the cortex and in the medullary and are of major significance for the taxonomy of lichens, since many taxa cannot be distinguished by morphology but do reveal chemical differences. The presence of specific lichen substances can be demonstrated with simple reagents; the cortex and medullary are separately spotted with solutions of specific chemicals and any color reactions occurring are recorded:

a) Potassium hydroxide (10% solution, abbreviated as K in the lichen literature); causes deep blue or deep violet colorations in the presence of naphtho- and anthraquinones;

b) Sodium hypochlorite (5% solution, abbreviated as C or Cl in lichen literature); causes a blood red color

in the presence of depsides having two free hydroxy groups in the *meta* positions (the color often fades rapidly);
c) Potassium hydroxide solution followed by sodium hypochlorite (abbreviation K+C or Cl); potassium hydroxide cleaves ester bonds of depsides and depsidones and the resultant *m*-dihydroxy compounds react with NaOCl to give a red color. Lichens containing usnic and isousnic acids show a deep yellow color;
d) *p*-Phenylenediamine (2% solution in ethanol, short-lived, must always be freshly prepared; abbreviated as PD); reacts with lichen substances containing an aldehyde group (e. g., *atranorin, *psoromic acid, or *stictic acid) with formation of yellow or orange-red Schiff's bases. A special procedure for identifying lichen substances is to introduce pieces of the lichen directly into the ion source of a mass spectrometer[1].

Lit.: [1] Phytochemistry **29**, 2277–2283 (1990); Symbiosis **11**, 193–206 (1991).
gen.: Bryologist **73**, 177–377 (1970) ▪ Culberson ▪ Culberson et al., Second Supplement to Chemical and Botanical Guide to Lichen Products, St. Louis: The Am. Bryological and Lichenological Soc. 1977 ▪ Phytochemistry **29**, 2277–2283 (1990) ▪ Symbiosis **11**, 193–206, 225–248 (1991) ▪ Ullmann (4.) **11**, 99–130 ▪ Zechmeister **29**, 209–306; **45**, 103–234.

Lichen xanthones. Compounds found in numerous lichens that are mostly derived from *norlichexanthone* (occurring, e. g., in *Lecanora straminea*). The occurrence of mono-, di-, tri-, and tetrachloroxanthones such as, e. g., *thuringione* (from *Lecidella carpathica*) or *thiophanic acid* (from *Lecanora rupicola* and numerous other lichens) is remarkable.

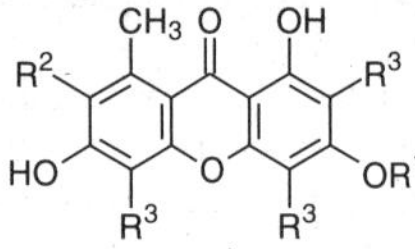

$R^1 = R^2 = R^3 = H$: Norlichexanthone (**1**)
$R^1 = CH_3$, $R^2 = H$, $R^3 = Cl$: Thuringione (**2**)
$R^1 = H$, $R^2 = R^3 = Cl$: Thiophanic acid (**3**)

Table: Characterization of selected Lichen xanthones.

no.	molecular formula	M_R	mp. [°C]	CAS
1	$C_{14}H_{10}O_5$	258.23 yellow needles	274–275	20716-98-7
2	$C_{15}H_9Cl_3O_5$	375.59 yellow needles	278–279	22105-34-6
3	$C_{14}H_6Cl_4O_5$	396.01 yellow needles	242–243	7584-33-0

The biosynthesis of lichen xanthones proceeds through the corresponding polyketides and their cyclization.
Lit.: Tetrahedron **34**, 2491–2502 (1978) ▪ Tetrahedron Lett. **1966**, 3547 ff. ▪ Zechmeister **45**, 103–234.

Lichesterinic acid.

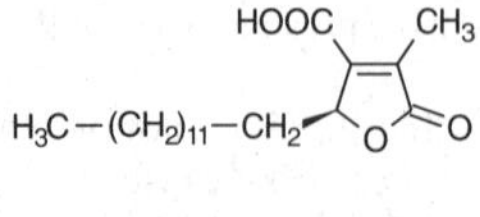

$C_{19}H_{32}O_4$, M_R 324.46, mother-of-pearl-like platelets, mp. 123–124 °C, $[\alpha]_D^{15}$ −32.6° ($CHCl_3$). A *paraconic acid derivative from lichens, e. g., *Cetraria islandica*, usually accompanied by the isomeric *protolichesterinic acid; has antibiotic and growth regulating (in plants) activities.
Lit.: Culberson, p. 102–103 ▪ Karrer, No. 1094 ▪ Z. Naturforsch. B **47**, 842–854 (1992). – *[CAS 22800-25-5]*

Lichesterol (lichesterin, ergosta-5,8,22*E*-trien-3*β*-ol).

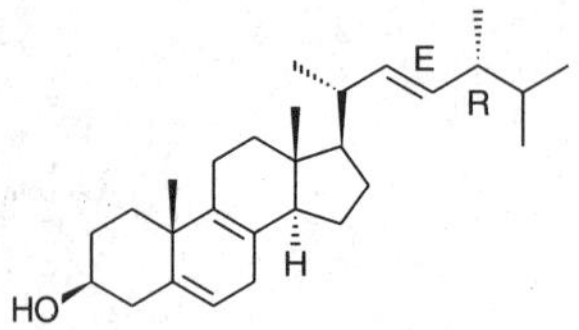

$C_{28}H_{44}O$, M_R 396.66, platelets, mp. 114–115 °C, $[\alpha]_D$ −27° ($CHCl_3$). A phytosterin occurring in lichens, e. g., *Umbilicaria cylindrica* and *Xanthoria parietina*.
Lit.: J. Chem. Soc., Perkin Trans. 1 **1981**, 2125 f. – *[CAS 50657-31-3]*

Licochalcones. Chalcone metabolites from roots of *Glycyrrhiza glabra* and *G. inflata*, e.g., *L. A:*

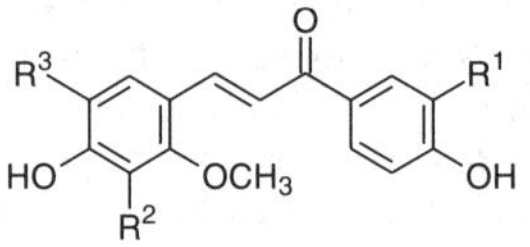

$R^1 = H$, $R^2 = H$, $R^3 = C(CH_3)_2{-}CH{=}CH_2$: L. A
$R^1 = H$, $R^2 = OH$, $R^3 = H$: L. B
$R^1 = H$, $R^2 = CH_2{-}CH{=}C(CH_3)_2$, $R^3 = H$: L. C
$R^1 = CH_2{-}CH{=}C(CH_3)_2$, $R^2 = OH$, $R^3 = H$: L. D

$C_{21}H_{22}O_4$, M_R 338.40, yellow needles, mp. 101–102 °C. L. show anti-protozoal (antileishmanial, antimalarial) activity. They are in preclinical development against malaria tropica.
Lit.: Antimicrob. Agents Chemother. **38**, 1339, 1470 (1994) ▪ Bioorg. Med. Chem. Lett. **5**, 449 (1995) (synthesis). – *[CAS 58749-22-7 (L. A)]*

Licorice, liquorice see glycyrrhizin.

Lignans. Low-molecular-weight natural products occurring in plants and some higher fungi formed biosynthetically by oxidative coupling of aryl-C_3 units(*coniferyl alcohol, *4-hydroxycinnamyl alcohol, *sinapyl alcohol). They are classified according to structural features.
Guaiaretic acid is isolated from the *guaiac(um) of *Guaiacum officinale*; it serves as reference compound for determination of the absolute configurations of the L. (see also nordihydroguaiaretic acid). Cubebin is present in the unripe fruits of *Piper cubeba*. Pinoresinol occurs in *Picea*, *Pinus*, and *Abies* spp. Many L. act as antagonists of *platelet activating factor. Steganacin, from *Steganotaenia araliacea* (Apiaceae), has antileukemic activity. In the past *neo-L.* (example: eusiderin A), compounds not coupled via the *β,β'*-positions of the side chains, were also distinguished. Trimers are known as *sesqui-L.* (example: americanin B),

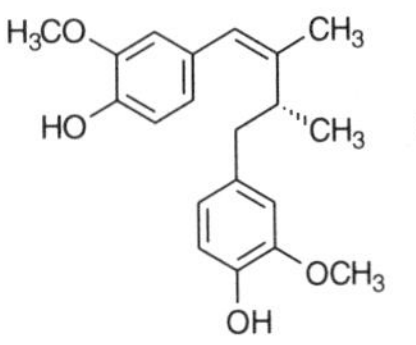

(–)-Guaiaretic acid (**1**)

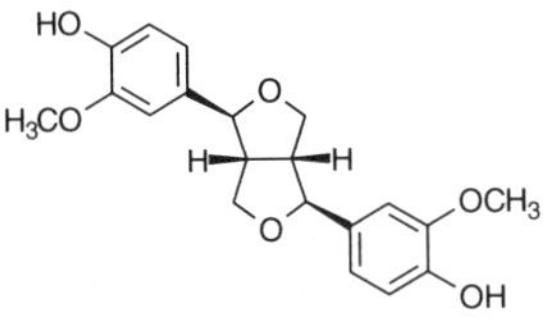

(+)-Pinoresinol (**2**)

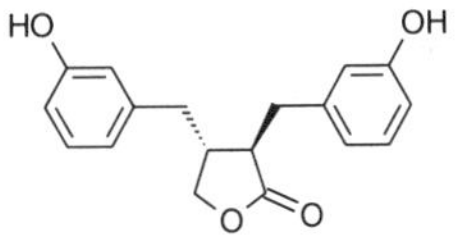

Enterolactone (**3**)

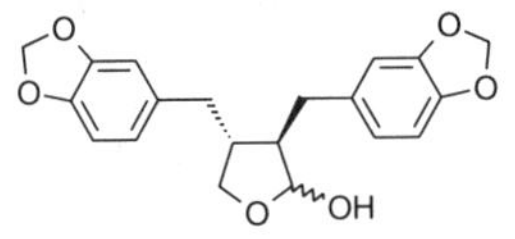

(–)-Cubebin (**4**)

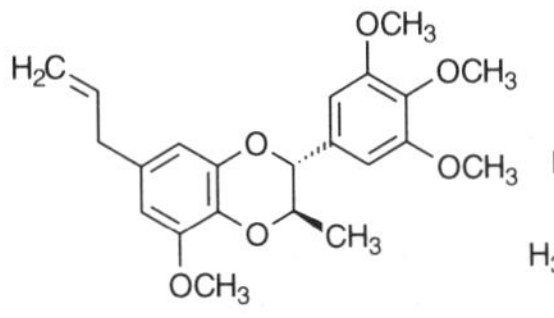

(–)-Eusiderin A (**5**)

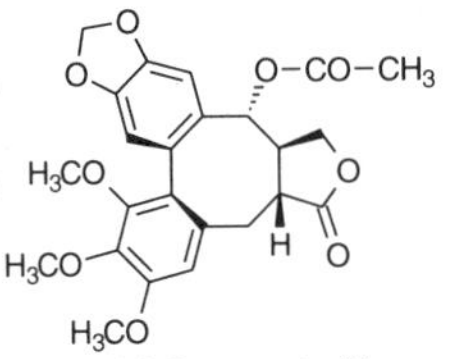

(–)-Steganacin (**6**)

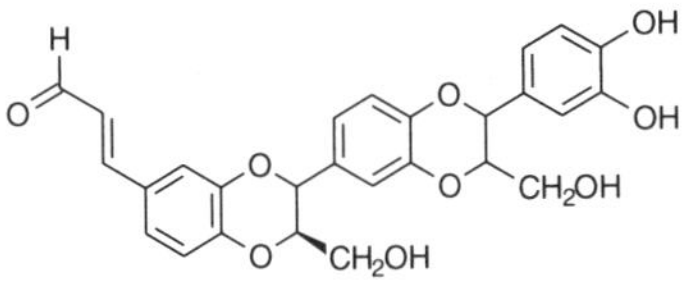

Americanin B (**7**)

tetramers as *di-L.* Some nitrogen-containing benzodioxane dimers and trimers of *N*-acetyldopamine (see *N*-acylcatecholamines) are contained in the cuticules of insects. L. occur in mammals including humans [1]. They are formed in the digestive tract by microbial transformation from lignin taken up as food. Enterolactone belonging to this group is reported to have both carcinogenic as well as cytostatic properties. Although it was earlier assumed that L. are intermediates in the biosynthesis of *lignin, this is contradicted by the fact that they are optically active while lignin is not. Coupling of phenylpropane units with *flavonoids gives rise to *flavono-L.* (see silybin).

Lit.: [1]Nature (London) **287**, 738–742 (1980); Tetrahedron Lett. **22**, 349–350 (1981).

gen.: Barton-Nakanishi **1**, 639–712 (biosynthesis, function) ▪ Chem. N. Z. **52**, 109–111, 115 (1988) ▪ Chem. Pharm. Bull. **40**, 252 (1992) (americanol) ▪ Exp. Opin. Ther. Patents **6**, 547–554 (1996) (activity) ▪ Fitotherapie **60**, 3–35 (1989) ▪ Front. Gastrointest. Res. **14**, 165–176 (1988) ▪ Holzforschung **42**, 375–384 (1988) ▪ J. Nat. Prod. **53**, 396–406 (1990); **61**, 1447 (1998) (antiviral activity) ▪ Justus Liebigs Ann. Chem. **1989**, 1147–1151 ▪ Luckner (3.), p. 395–397 ▪ Nat. Prod. Rep. **12**, 183–205 (1995); **16**, 75–96 (1999) (isolation) ▪ Pharmacol. Ther. **59**, 163 (1993) (review steganacin) ▪ Phytochemistry **23**, 1207–1220 (1984) ▪ Phytother. Res. **1**, 97–106 (1987) ▪ Planta Med. **55**, 531–535 (1989) ▪ Zechmeister **35**, 1–72.

Lignin. A high-molecular-weight, aromatic substance filling the spaces between cell membranes in lignifying plants to form the wood (lignification). In this way a mixed unit of pressure-resistant L. and stretch-resistant *cellulose is formed. L. is also bound to other polysaccharides (polyoses). The content of L. in dried plant material amounts to 27–33% in conifers and 22% in deciduous tress.

Structure: L. consists of phenylpropane units. Depending on the type of wood, the phenyl ring carries one to two methoxy groups and the propane unit hydroxy substituents. Conifers contain exclusively the guaiacyl type (see guaiacol) while deciduous tree wood also contains the syringyl (see sinapyl alcohol) and coumaryl types. The various linkage possibilities also give rise to *lignan and *coumarin structures, cyclic ethers, lactones, etc.

Isolation: A small part of L. is soluble in water; depending on the extraction medium, the insoluble part is differentiated as milled wood (MW) L. (extraction medium: acetone) and dioxane lignin. It has a molecular mass of ca. 5000–10000, pinewood L.: $C_9H_7O_2 \cdot H_2O \cdot (OCH_3)_{0.94}$, M_R ca. 10000, D. 1.3–1.4. The methoxy content of L. from different plant types varies. L. has a cream color and becomes thermoplastic after treatment with hot water.

Biosynthesis: L. is formed from *coniferyl alcohol (or syringaaldehyde) under the action of laccase (phenol dehydrase). L. is degraded in the soil by the lignases of bacteria and fungi to *humic acids.

Lit.: ACS Symp. Ser. **697** (Lignin and Lignan Biosynthesis 1998) ▪ Angew. Chem. Int. Ed. Engl. **38**, 1283 (1999) (biosynthesis) ▪ Barton-Nakanishi **3**, 617–746 ▪ Int. Symp. Wood Pulping Chem. (8th) **1995**, 2, 29–34 ▪ Kirk-Othmer (4.) **15**, 268–289 ▪ Ullmann (5.) **A 15**, 305–315 ▪ Z. Chem. **30**, 233–239 (1990) (biosynthesis). – *[CAS 8068-00-6, 9005-53-2]*

Ligustrazine (tetramethylpyrazine).

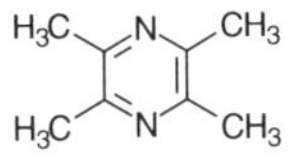

Table: Data of Lignans.

no.		structural type	molecular formula	M_R	mp. [°C]	$[\alpha]_D$	CAS
1	(–)-Guajaretic acid	dibenzylbutane	$C_{20}H_{24}O_4$	328.41	99–101	–94° (C_2H_5OH)	500-40-3
2	(+)-Pinoresinol	furofuran	$C_{20}H_{22}O_6$	358.39	122	+84.4° (acetone)	487-36-5
3	Enterolactone	dibenzylbutyrolactone	$C_{18}H_{18}O_4$	298.34			78473-71-9
4	(–)-Cubebin	dibenzylbutyrolactol	$C_{20}H_{20}O_6$	356.38	131–132	–17.1° (acetone); –45.7° ($CHCl_3$)	18423-69-3
5	(–)-Eusiderin A	benzodioxane	$C_{22}H_{26}O_6$	386.44	94	–25.4°	59332-00-2
6	(–)-Steganacin	dibenzocyclooctane	$C_{24}H_{24}O_9$	456.45		–114° ($CHCl_3$)	41451-68-7
7	Americanin B	sesquilignan	$C_{27}H_{24}O_9$	492.48			

$C_8H_{12}N_2$, M_R 136.20. An antihypertensive component of the Chinese medicinal plant *Ligusticum chuanxiong* (Apiaceae). L. inhibits platelet aggregation by expelling Ca^{2+} from the platelet membranes. This is also the reason for its antithrombotic activity.

Lit.: Acta Pharm. Sin. **20**, 334, 689 (1985) ▪ Chem. Pharm. Bull. **40**, 954 (1992) ▪ Clin. Sci. **75** (1988); Suppl. 1955 ▪ Med. Chem. Res. **2**, 434–442 (1992) ▪ Tang & Eisenbrand (eds.), Chinese Drugs of Plant Origins, p. 609f., Berlin: Springer 1992. – *[CAS 1124-11-4]*

Ligustroside.

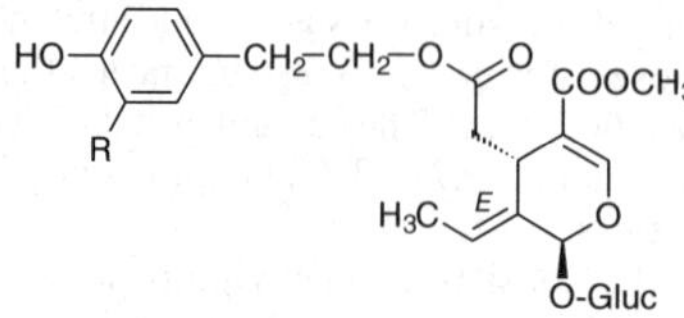

R = H : Ligustroside
R = OH : Oleuropein

An ester of secoxyloganin (see secoiridoids, with an (*E*)-ethylidene group instead of a vinyl group) with 2-(4-hydroxyphenyl)- or 2-(3,4-dihydroxyphenyl)ethanol. *L.*, $C_{25}H_{32}O_{12}$, M_R 524.52, amorphous, $[\alpha]_D -180°$ (C_2H_5OH); *oleuropein*, $C_{25}H_{32}O_{13}$, M_R 540.52, crystals, mp. 87–89 °C, $[\alpha]_D -168°$ (CH_3OH). L. is isolated from the ethanol extract of the North American white ash (*Fraxinus americana*, Oleaceae)[1] and the Japanese privet (*Ligustrum obtusifolium*, Oleaceae)[2]. Oleuropein occurs in the fruit of the olive tree (*Olea europaea*, Oleaceae)[3]. The blood-pressure lowering activity of the olive leaf extract is due to its content of oleuropein[3].

Lit.: [1] J. Am. Chem. Soc. **98**, 3007 (1976). [2] Chem. Lett. **1972**, 141. [3] Gazz. Chim. Ital. **90**, 1449 (1960).
gen.: Chem. Pharm. Bull. **45**, 367 (1997) ▪ Hager (5.) **5**, 188, 937, 945 ▪ J. Nat. Prod. **55**, 760 (1992) ▪ Phytochemistry **31**, 4197 (1992); **34**, 1291 (1993); **40**, 785 (1995). – *[CAS 35897-92-8 (L.); 32619-42-4 (oleuropein)]*

Lime oil. L. of varying qualities is obtained from the lime species *Citrus aurantiifolia*, cultivated in Mexico, Peru, and the West Indies ("Mexican lime, Key lime"), and *Citrus latifolia*, cultivated in Florida and Brazil ("Persian lime, Tahiti lime"). The yearly production of L. world-wide amounts to between 500 and 1000 t.

1. ***Distilled lime oil:*** Colorless to light yellow, readily mobile oil with a unique, fresh-sparkling, turpentine-like odor and a fresh turpentine taste of lime juice.
Production: By steam distillation of the oil/juice emulsion obtained by comminutation of whole, unripe (green) fruits.
Composition [1]***:*** The main components are *limonene* (50–60%) and *γ-terpinene* (see *p*-menthadienes) (10–20%). In contrast to pressed L. the distilled oil contains many components that are formed during the production process by action of the acidic juice on the oil liberated from of the fruit skin. Thus, under the conditions of distillation components such as *p*-cymol (see cymenes), 1,4- and 1,8-*cineole, *fenchol, *borneol, α- and β-terpineol (see *p*-menthenols), etc. are formed. All these compounds are either not present or are present in much lower concentrations in pressed oil. Accordingly, the amounts of acid-labile components such as, e. g., the *pinenes and *citral are reduced. The content of the latter in distilled L. is merely ca. 0.1–0.2% (see below).
Use: Mainly to improve the aromatic character of foods and, especially, refreshing beverages. Distilled oil is an essential component of the cola aroma.

2. ***Pressed lime oil:*** Yellow-green to dark green oil with a fresh odor and taste of lemon peel.
Production: By mechanical processes from the skins of unripe lime fruits, e. g., by comminutation of the whole fruit with addition of water and centrifugation of the obtained oil/water emulsion or puncturing or grating the skin (*citrus oils).
Composition [1]***:*** Pressed L. has a very similar composition to that of *lemon oil. Main components are *limonene* (ca. 50%), *β-*pinenes* (ca. 10–20%), and *γ-terpinene* (ca. 10%). The component responsible for the typical odor and taste is **citral* (3–5%).
Use: In the perfume industry, e. g., for fresh notes in Eaux de Cologne. However, on account of the content of *furocoumarins (see also citrus oils) pressed L. is used in limited amounts only. Pressed L. is used to improve the aromatic character of foods similar to lemon oil and is usually employed in combination with lemon oil.

Lit.: [1] Perfum. Flavor. **16** (2), 17, 60 (1991); Flavour Fragr. J. **10**, 33 (1995).
gen.: Arctander, p. 372, 374 ▪ Bauer et al. (2.), p. 148 ▪ ISO 3519 (1976), 3809 (1987) ▪ Ohloff, p. 138. – *Toxicology:* Food Cosmet. Toxicol. **12**, 729, 731 (1974) ▪ Food Chem. Toxicol. **31**, 331 (1993). – *[HS 3301 14; CAS 8008-26-2 (1.); 90063-52-8 (2.)]*

Limonene see *p*-menthadienes.

Limonin

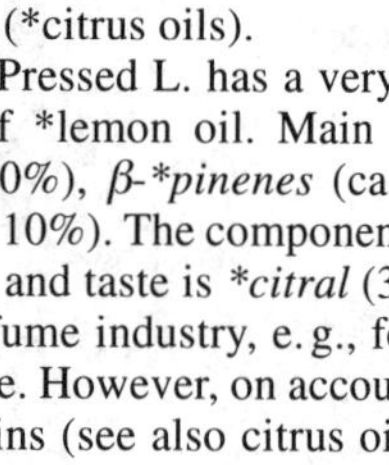

(evodin, limonic acid di-δ-lactone). A nortriterpene, $C_{26}H_{30}O_8$, M_R 470.52, mp. 298 °C, $[\alpha]_D -125°$ (acetone), soluble in alcohol, glacial acetic acid, poorly soluble in water and ether. The dilactone (bitter principle of lemon and orange seeds) occurs in *Citrus* species and other Rutaceae; it is formed by oxidative degradation from tetracyclic triterpene precursors. The olfactory threshold for the bitter taste of L. amounts to 0.75 μMol/L. The separation of L. in the production of orange juice from certain fruit types is very difficult[1]. L. has insecticidal activity.

Lit.: [1] Food Biotechnol. **1**, 249–261 (1987).
gen.: Acta Crystallogr., Sect. C **46**, 425 (1990) ▪ Beilstein E V **19/12**, 475 ▪ Karrer, No. 3903 ▪ Phytochemistry **24**, 2911 (1985) (biosynthesis). – *[CAS 1180-71-8]*

Limonoids. A group of oxidized *triterpenes of the plant families Meliaceae, Rutaceae, and Cneoraceae, closely related to the *quassin(oid)s. Like the latter, the L. have antifebrile properties and show activity in the treatment of malaria. In traditional medicine the leaves of the corresponding plants are used to prepare teas. The L. *azadirachtin has strong antifeedant activity on insects. Certain L. have ecdysone-like activity (see ecdysteroids) on insect larvae [1].

Lit.: [1] ACS Symp. Ser. **296**, 206–219 (1986); **330**, 396–415 (1987).
gen.: Can. J. Chem. **67**, 257–260 (1989) ▪ Chem. Br. **1990**, 31 ▪ J. Agric. Food. Chem. **38**, 1400–1403 (1990) ▪ Phytother. Res. **4**, 29–35 (1990) ▪ Zechmeister **45**, 1–102.

Linalool (β-linalool, 3,7-dimethyl-1,6-octadien-3-ol).

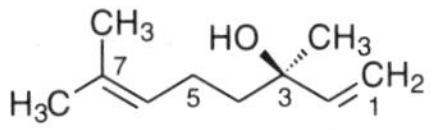

(3*R*)-(-)-form

$C_{10}H_{18}O$, M_R 154.25, liquid with an odor like lily of the valley. An unsaturated acyclic *monoterpene alcohol, both enantiomers of which occur in nature: (3*R*)-(–)- and (3*S*)-(+)-L., bp. 198–199 °C (760 mbar), $[\alpha]_D$ ±20.6° (neat).
Occurrence: (–)-L. in Brazilian rosewood oil (Brazilian "linaloe oil", *Aniba rosaeodora*, Lauraceae) 80–85%; (+)-L. (coriandrol) in *coriander oil (60–80%) and in Mexican "linaloe oil" (*Bursera delpechiana*=*B. penicillata*, Burseraceae) 60–65%. L. also occurs bound to various carboxylic acids, e.g., *linalyl acetate* [1] (3-acetoxy-3,7-dimethylocta-1,6-diene, $C_{12}H_{20}O_2$, M_R 196.29). The linalyl esters are synthesized from L. since they have various applications in perfumery. For synthesis of L., see *Lit.* [2]. L. and L. esters are used in the perfume industry and for the synthesis of *tocopherols.

Lit.: [1] Beilstein EIV **2**, 204 f. [2] J. Chem. Soc., Perkin Trans. 1 **1990**, 2715; J. Org. Chem. **51**, 2599 (1986).
gen.: Karrer, No. 120 ▪ Merck-Index (12.), No. 5520 ▪ Tetrahedron Lett. **40**, 3803 (1999) (biosynthesis of linalyl acetate) ▪ Ullmann (5.) **A 11**, 156. – *[HS 2905 22; CAS 126-91-0 ((–)-L.); 126-90-9 ((+)-L.); 115-95-7 (linalyl acetate)]*

Linalyl acetate see linalool.

Linamarin see cyanogenic glycosides.

Linatine (1-L-γ-glutamylamino-D-proline).

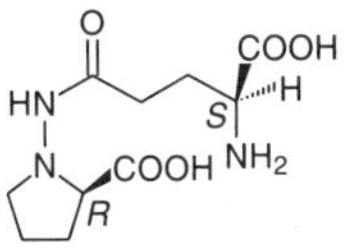

$C_{10}H_{17}N_3O_5$, M_R 259.26, amorphous powder, $[\alpha]_D^{25}$ +46.4° (H_2O). A γ-glutamyl compound from linseed (*Linum usitatissimum*) [1]. The toxic hydrazine derivative acts as a pyridoxine antagonist [2].

Lit.: [1] Biochemistry **6**, 170–177 (1967); Tetrahedron Lett. **1974**, 1799 (synthesis). [2] Methods Enzymol. **62**, 483 (1979).
gen.: Beilstein EV **22/1**, 210 ▪ Zechmeister **39**, 230. – *[CAS 10139-06-7]*

Lincomycins.

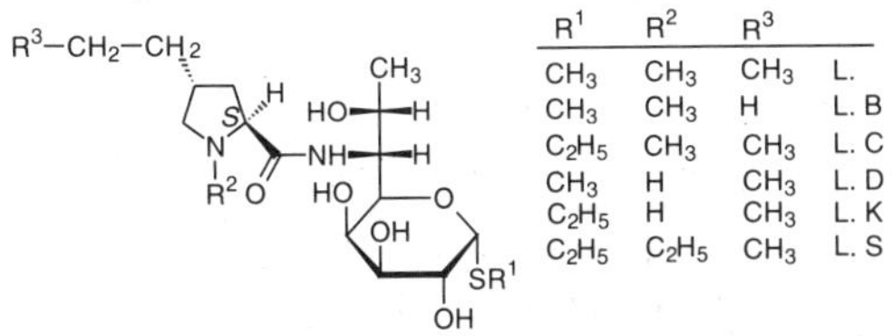

Table: Data of Lincomycins.

compound	molecular formula	M_R	CAS
L.	$C_{18}H_{34}N_2O_6S$	406.54	154-21-2
L. B	$C_{17}H_{32}N_2O_6S$	392.52	2520-24-3
L. C	$C_{19}H_{36}N_2O_6S$	420.57	14042-43-4
L. D	$C_{17}H_{32}N_2O_6S$	392.52	2256-16-8
L. K	$C_{18}H_{34}N_2O_6S$	406.54	
L. S	$C_{20}H_{38}N_2O_6S$	434.60	21085-65-4

A family of *streptomycetes antibiotics from cultures of *Streptomyces lincolnensis*. *Lincomycin* is used clinically as a broad-spectrum *antibiotic. It does not show any cross-resistance with other antibiotics but isolated cases of liver and kidney damage have been reported [1]. For synthesis, see *Lit.* [2]. The other members of the L. family also show biological activity against Gram-positive bacteria.

Lit.: [1] J. Antimicrob. Chemother., (Suppl. A) **7**, 11 (1981); Kirk-Othmer (4.) **3**, 159. [2] J. Org. Chem. **55**, 1632 (1990).
gen.: Anal. Profiles Drug Subst. **23**, 269 (1994). – *[HS 2941 90]*

Linden ether see honey flavor.

Linderalactone see germacranolides.

Lineatin.

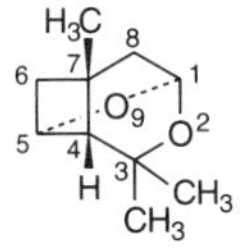

(+)-Lineatin

$C_{10}H_{16}O_2$, M_R 168.24, bp. 68 °C (6.7 hPa). An aggregation *attractant produced by female bark beetles (ambrosia beetles) of the genus *Trypodendron* [1]. L. in *T. lineatum* has the (1*R*,4*S*,5*R*,7*R*)-configuration, $[\alpha]_D^{22}$ +36° (pentane) [2]. L. has the same carbon skeleton as *grandisol.

Lit.: [1] J. Chem. Ecol. **3**, 549–561 (1977). [2] Naturwissenschaften **69**, 602 f. (1982).
gen.: ACS Symp. Ser. **190**, 87–106 (1982) ▪ Acta Chem. Scand. **47**, 1232 (1993) ▪ ApSimon **4**, 85 ff., 488 ff.; **9**, 436–444 (synthesis) ▪ Beilstein E V **19/1**, 327 ▪ Synform **5**, 125–144 (1987) ▪ Tetrahedron **50**, 3235 (1994) (synthesis). – *[CAS 71899-16-6 (L.); 65035-34-9 ((+)-L.)]*

Linocinnamarin see (*E*)-*p*-c(o)umaric acid.

Linoleic acid [(*Z*,*Z*)-9,12-octadecadienoic acid].

$C_{18}H_{32}O_2$, M_R 280.45, mp. –5.0 to –5.2 °C, bp. 229–230 °C (1.6 kPa); soluble in organic solvents. L. is an essential *fatty acid occurring as glycerol esters in practically all natural *fats and oils (see table); the recommended daily uptake for an adult [1]: ca. 10 g. On account of its vitamin characteristics L. together with some other multiply unsaturated fatty acids used to be called vitamin F. L. was first isolated from linseed oil (*lini oleum*) (*Linum usitatissimum*). Lecithins and other phospholipids are often especially rich in linoleic acid. In animal organisms L. is converted by the introduction of two further double bonds and chain extension by 2 C atoms to *arachidonic acid. L. is particularly sensitive to oxidation; hydroperoxides and their subsequent products are formed on exposure to air causing the oil to turn rancid. The rates of oxidation of various fatty acids in relation to the oxidation of stearic acid (= 1) are as follows: oleic acid, 10; linoleic acid, 1200; and linolenic acid 2500.

Table: Contents of Linoleic acid in fats and oils.

fat/oil	L. (weight %)	fat/oil	L. (weight %)
olive oil	3–20	sunflower oil	20–75
rapeseed oil	18–30	thistle oil (safflower oil)	55–81
grapeseed oil	58–78	lard	3–16
maize germ oil	34–62	beef tallow	0.5–5

Use: In dietetic products, production of soaps, in oleochemistry as starting material for the production of dimeric acids.

Lit.: [1] Fat Sci. Technol. **96**, 34 (1994).
gen.: Merck-Index (12.), No. 5529 ▪ Ullmann (5.) **A 10**, 248 ▪ see linolenic acid. – *[HS 2916 15; CAS 60-33-3]*

Linolenic acid [α-linolenic acid, (*all-Z*)-9,12,15-octadecatrienoic acid].

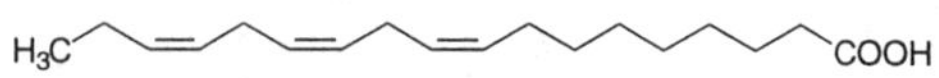

$C_{18}H_{30}O_2$, M_R 278.44, mp. –11.3 °C, bp. 230–232 °C (2.3 kPa); colorless liquid, soluble in organic solvents, insoluble in water. L. is a triply unsaturated essential *fatty acid, occurring as glycerol esters in many plant oils (see table). Larger amounts (up to ca. 60%) are present in the seed oil (*lini oleum*) of flax (linseed oil, *Linum usitatissimum*).

Table: Contents of Linolenic acid in fats and oils.

fat/oil	L. (weight %)	fat/oil	L. (weight %)
soybean oil	4–11	linseed oil	35–56
rapeseed oil	5–16	hempseed oil	28
sunflower oil	<1	lard	<1.5
olive oil	<1.5	horsefat	ca. 30

L. is a biochemical precursor for various *eicosanoids. On account of its three double bonds L. is extremely sensitive to oxidation (see linoleic acid). Thus, plant oils containing L. and other polyene fatty acids are suitable for use as so-called drying oils in the production of paints and varnishes as well as in linoleum. The positional isomer *γ-L.* [(*all-Z*)-6,9,12-octadecatrienoic acid] occurs in a few other plant oils as glycerol esters, e.g., in the seed oil of evening primrose (*Oenothera biennis*, ca. 10%), black currants (*Ribes nigrum*, ca. 15%), borrage (*Borago officinalis*, 20–25%) and other Boraginaceae as well as in alga-like fungi (phycomycetes). It shows activity against neurodermatitis. γ-L. is a biochemical precursor of *arachidonic acid. α-L. and γ-L. are used in the manufacture of dietetic products.

Lit.: Ching Kuang Chow, Fatty Acids in Food and their Health Implications, p. 429ff., New York: Dekker 1992 ▪ Prog. Lipid. Res. **20**, 581, 609 (1981); **25**, 177 (1986) ▪ Synthesis **1995**, 271 ▪ Ullmann (5.) **A 10**, 248. – *[HS 2916 15; CAS 463-40-1 (L.); 506-26-3 (γ-L.)]*

Linustatin see cyanogenic glycosides.

Lipid A.

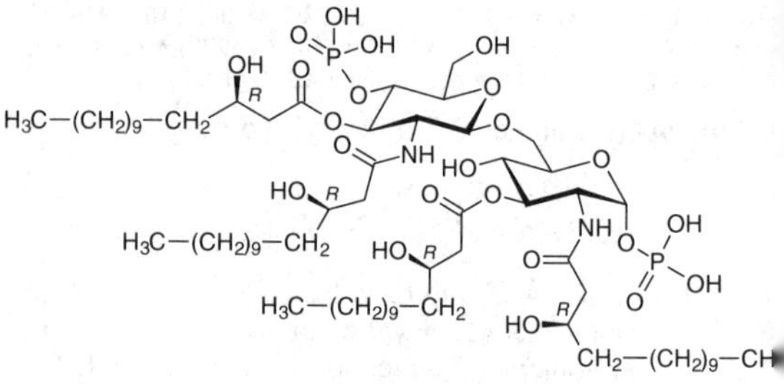

L.A of a *Salmonella* mutant

$C_{68}H_{130}N_2O_{23}P_2$, M_R 1405.73, amorphous solid. Component of glycophospholipids of Gram-negative bacteria. L. A is important for the toxicity, pyrogenicity, and adjuvant activities of the bacteria. Variants of the formula shown from *Escherichia* and *Proteus* are acylated at the OH groups of the tetradecanoyl units.

Lit.: Chem. Pharm. Bull. **35**, 4436, 4517 (1987); **45**, 312 (1997) ▪ J. Am. Chem. Soc. **116**, 3637 (1994) ▪ J. Endotoxin Res. **5**, 46 (1999) (synthesis) ▪ J. Org. Chem. **62**, 3654 (1997) (synthesis) ▪ Microbiology (Washington, D. C.) **1986**, 5 (review). – *[CAS 95991-05-2]*

Lipid X.

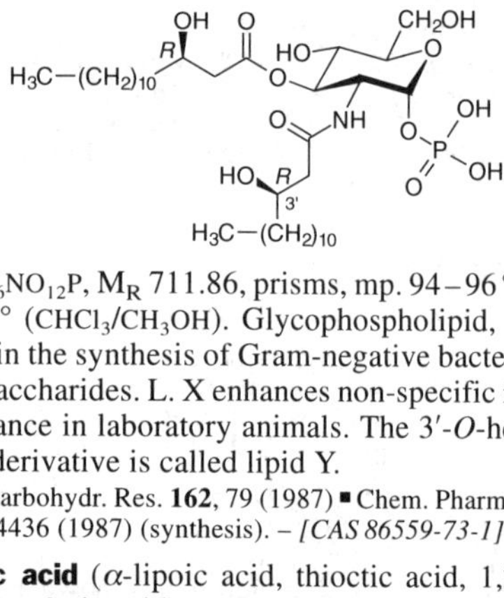

$C_{34}H_{66}NO_{12}P$, M_R 711.86, prisms, mp. 94–96 °C, $[\alpha]_D^{22}$ +14.6° ($CHCl_3/CH_3OH$). Glycophospholipid, intermediate in the synthesis of Gram-negative bacterial lipopolysaccharides. L. X enhances non-specific infection resistance in laboratory animals. The 3′-*O*-hexadecanoyl derivative is called lipid Y.

Lit.: Carbohydr. Res. **162**, 79 (1987) ▪ Chem. Pharm. Bull. **35**, 1383, 4436 (1987) (synthesis). – *[CAS 86559-73-1]*

Lipoic acid (α-lipoic acid, thioctic acid, 1,2-dithiolane-3-valeric acid).

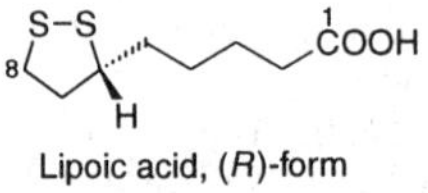

Lipoic acid, (*R*)-form

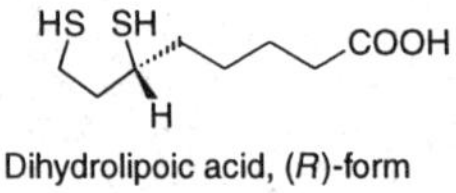

Dihydrolipoic acid, (*R*)-form

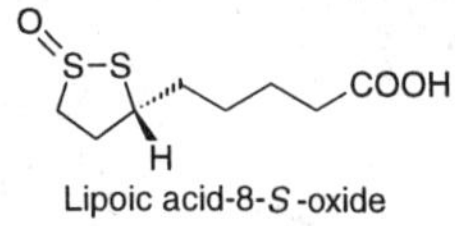

Lipoic acid-8-*S*-oxide

$C_8H_{14}O_2S_2$, M_R 206.32, insoluble in water, soluble in methanol and chloroform. (*R*)-form, cryst., mp. 46–48 °C, $[\alpha]_D^{23}$ +104° (benzene), pK_a 5.4. (*R*)-L. is widely distributed in nature and is a growth factor for microorganisms (protogen A). It is sometimes incorrectly designated as a vitamin of the B group. In metabolic processes (*R*)-L. acts as a *coenzyme in pyruvate dehydrogenase- (EC 1.2.4.1) and in the 2-oxoglutarate dehydrogenase complex (EC 1.2.4.2) where it is bound to the ε-amino group of lysine as an amide. Enzymatic reduction leads to (*R*)-*dihydrolipoic acid* [(*R*)-6,8-dimercaptooctanoic acid], $C_8H_{16}O_2S_2$, M_R 208.33, which is oxidized under the influence of dihydroliponamide dehydrogenase (EC 1.8.1.4) to (*R*)-L. The redox couple L./dihydro-L. is involved in metabolic and other electron transfer reactions. (*S*)-form, cryst., mp. 45–48 °C, $[\alpha]_D^{23}$ –113° (benzene); racemate, pale yellow needles, mp. 60 °C, bp. 160–165 °C, bp. 85–90 °C (3.3 kPa), pK_a 4.7. The 8-*S-oxide of* L., $C_8H_{14}O_3S_2$, M_R 222.32, is known as β-L. or protogen B. β-L. is formed by bacteria (e.g., *Escherichia* sp.).

Use: For treatment of heavy metal poisonings. The racemate of L. is used in the treatment of liver diseases, including liver necrosis resulting from consumption of the deadly amantia (*Amanita phalloides*). L. is applied for therapy of diabetic neuropathy. For asymmetric synthesis of (*R*)-α-L., see *Lit.*[1].

Lit.: [1] Eur. J. Org. Chem. **1998**, 1949; J. Chem. Soc., Chem. Commun. **1995**, 1563; Synth. Commun. **25**, 1531 (1995); Tetrahedron Lett. **26**, 2535–2538 (1985).

gen.: Beilstein EV **19/7**, 237 f. ▪ J. Am. Chem. Soc. **100**, 5243–5244 (1978) (biosynthesis) ▪ J. Chem. Soc., Perkin Trans. 1 **1990**, 1615–1618, 1897–1900 (synthesis) ▪ Methods Enzymol. **62**, 129–145 (1979) ▪ Negwer (6.), No. 996. – *Pharmacology:* Borbe & Ulrich, Thioctsäure: Neue biochemische, pharmakologische u. klinische Erkenntnisse zur Thioctsäure, Frankfurt: pmi 1989. – *Toxicology:* Sax (8.), DXN 800. – *[HS 2934 90; CAS 62-46-4 (±); 1200-22-2 ((R)-L.); 119365-69-4 ((R)-dihydro-L.); 1077-27-6 ((S)-dihydro-L.); 1077-28-7 (racemate); 6992-30-9 (8-S-oxide)]*

Lipoxins (LX). Acyclic *eicosanoids with conjugated tetraene or pentaene structures and three hydroxy groups. The first identified compounds were designated as L. A and L. B with the index 4 or 5 to indicate the number of their C=C-double bonds. Biosynthetically L. are formed from *arachidonic acid under the action of various oxidizing enzymes, especially 5- and 15-lipoxygenases. (5*S*,6*R*,15*S*)-Trihydroxy-7*E*,9*E*,11*Z*,13*E*-eicosatetraenoic acid (*lipoxin A* or LXA_4, $C_{20}H_{32}O_5$, M_R 352.47) and (5*S*,14*R*,15*S*)-trihydroxy-6*E*,8*Z*,10*E*,12*E*-eicosatetraenoic acid (*lipoxin B* or LXB_4, $C_{20}H_{32}O_5$, M_R 352.47) are formed through hydroperoxy- and dihydroperoxy-fatty acid intermediates as well as (5*S*,6*S*)-epoxy-LXA_4. Another biosynthetic route to L. proceeds through leukotriene A_4 (see leukotrienes). Numerous L. of the A- and B-series are formed in small amounts in leukocytes, granulocytes, macrophages, blood platelets, and other types of cell. Similar to the leukotrienes, L. are also involved in the development of inflammations and allergic reactions. Other biological activities include arterial dilatation, activation of phosphokinase C, inactivation of killer cells, acceleration of degranulation, stimulation of thromboxane biosynthesis, suppression of leukotriene B_4-induced inflammation, and bronchial constriction.

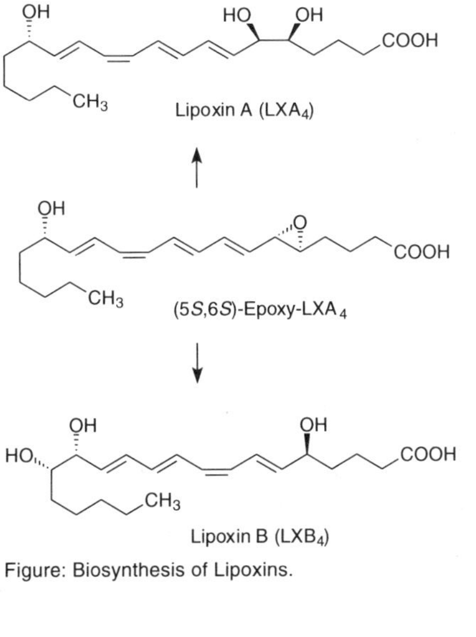

Figure: Biosynthesis of Lipoxins.

Lit.: Angew. Chem. Int. Ed. Engl. **30**, 1100–1116 (1991) ▪ Actual. Chim. Ther. **18**, 77–97 (1991) ▪ Adv. Prostagl. Thrombox. Leukotr. Res. **16**, 99 (1986); **19**, 1, 122 (1989) ▪ J. Bioenerg. Biomembr. **23**, 105 (1991) ▪ J. Biol. Chem. **261**, 16340 (1986) ▪ J. Clin. Invest. **85**, 772 (1990) ▪ J. Immunol. **138**, 266 (1987) ▪ Proc. Natl. Acad. Sci. USA **83**, 1983 (1986) ▪ Science **237**, 1171 (1987) ▪ Synlett **1993**, 217 ▪ Tetrahedron: Asymmetry **8**, 2949 (1997) (synthesis, LXB_4) ▪ Tetrahedron **48**, 2441–2452 (1992) (synthesis) ▪ Tetrahedron Lett. **39**, 101 (1998); **41**, 823 (2000) (synthesis, LXA_4, LXB_4) ▪ Wong et al., Lipoxins: Biosynthesis, Chemistry and Biological Activities, New York: Plenum 1988. – *[CAS 89663-86-5 (LXA₄); 98049-69-5 (LXB₄)]*

Lipstatin.

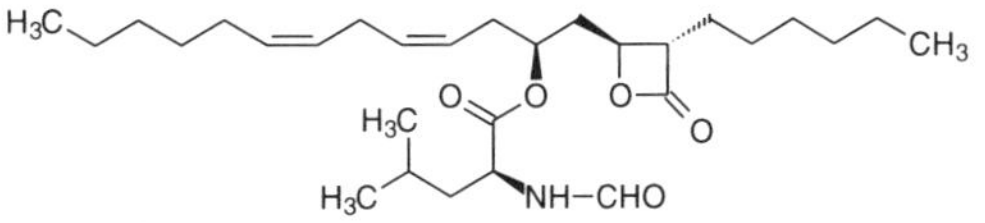

$C_{29}H_{49}NO_5$, M_R 491.71, light yellow oil, $[\alpha]_D$ –19° ($CHCl_3$), a lipophilic ester with a central β-lactone ring and *N*-formyl-L-leucine as side chain produced by *Streptomyces toxytricini*. L. is a specific inhibitor of triacylglycerol lipases, especially pancreas lipase and thus reduces the digestion of nutritional fats without affecting the absorption of free fatty acids. *Tetrahydrolipstatin* {THL, orlistat, Xenical®, $C_{29}H_{53}NO_5$, M_R 495.74, mp. 43 °C, $[\alpha]_D$ –33° ($CHCl_3$)}, which can be synthesized easily, has recently (1998/1999) been marketed very successfully (sales 1999: above 300 mio. $). The enzyme inhibition by L. or THL is due to the fact that one serine residue of the active center is esterified by the reactive β-lactone.

Lit.: J. Antibiot. **40**, 1081, 1086 (1987). – *Synthesis:* Chem. Commun. **1999**, 1743 ▪ J. Chem. Soc., Perkin Trans. 1 **1993**, 1549; **1998**, 2679 ▪ J. Org. Chem. **56**, 4714 (1991); **60**, 7334 (1995) ▪ Org. Lett. **1**, 753 (1999) ▪ Synthesis **1995**, 729–744 ▪ Tetrahedron Lett. **40**, 393 (1999). – *Activity:* Biochem. J. **256**, 357 (1988) ▪ Helv. Chim. Acta **75**, 1593–1603 (1992) ▪ Int. J. Obes. **11**, 35 (1987). – *[CAS 96829-59-3 (L.); 96829-58-2 (THL)]*

Lissoclinotoxin A see varacin.

Lissoclinum alkaloids. Pyrido[2,3,4-*kl*]acridinone (e.g., diplamine, cf. *kuanoniamines) and *harman alkaloids from tropical tunicates of the genus *Lissoclinum*.

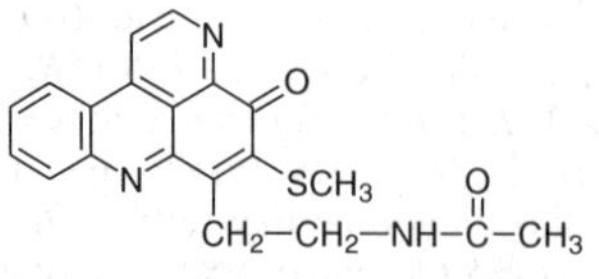

Diplamine

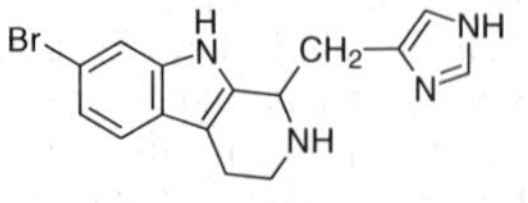

Lissoclines C

Diplamine[1]: $C_{20}H_{17}N_3O_2S$, M_R 363.43, orange solid, mp. 202–204 °C, cytotoxic.
Lissocline C[2]: $C_{15}H_{15}BrN_4$, M_R 331.21, amorphous solid.
Lit.: [1]J. Am. Chem. Soc. **117**, 12460–12469 (1995) ▪ Tetrahedron Lett. **36**, 4709 (1995). [2]J. Org. Chem. **59**, 6600 (1994). *gen.:* Alkaloids: Chem. Biol. Perspect. **12**, 187–228 (1998) (review) ▪ Synthesis **1999**, 1520 (synthesis). – *[CAS 123794-30-9 (diplamine); 158761-14-9 (lissocline C)]*

Lithocholic acid see bile acids.

Litlure A see prodlure.

Litmomycin see granaticins.

Litmus. A blue pigment that can be obtained from various lichens, e. g., *Variolaria, Roccella* and *Lecanora.* In the past it was used particularly in Holland for bluing linen and for coloring luxury foods, cosmetics, and sugar paper. It is not suitable as a textile dye on account of its color change in acid and alkaline media. Today L. is used exclusively as an acid-base indicator (red at pH 4.5, blue at pH 8.3). The main component of L. is a polymer made up of 7-hydroxy-3-phenoxazinone chromophores, which explains its relationship with *orcein.
Historical: L. was first discovered as a chemical reagent by the physician and alchemist Arnaldus de Villanova in 1300 AD. The name has Dutch origin (Lakmoes, the meaning remains unclear).
Lit.: Ullmann (4.) **11**, 116; (5.) **A14**, 130. – *[HS 3203 00; CAS 573-56-8]*

Lituarines.

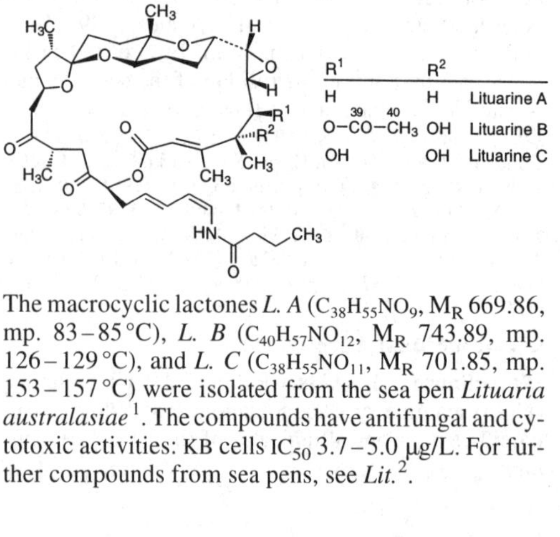

The macrocyclic lactones *L. A* ($C_{38}H_{55}NO_9$, M_R 669.86, mp. 83–85 °C), *L. B* ($C_{40}H_{57}NO_{12}$, M_R 743.89, mp. 126–129 °C), and *L. C* ($C_{38}H_{55}NO_{11}$, M_R 701.85, mp. 153–157 °C) were isolated from the sea pen *Lituaria australasiae*[1]. The compounds have antifungal and cytotoxic activities: KB cells IC_{50} 3.7–5.0 μg/L. For further compounds from sea pens, see *Lit.*[2].

Lit.: [1]J. Org. Chem. **57**, 5857 (1992). [2]J. Nat. Prod. **47**, 155 (1984); Natural Prod. Rep. **2**, 578 (1984). – *[CAS 143621-75-4 (L. A); 143621-76-5 (L. B); 143621-77-6 (L. C)]*

LMF. Abbreviation for *leaf-movement factors.

Lobaric acid.

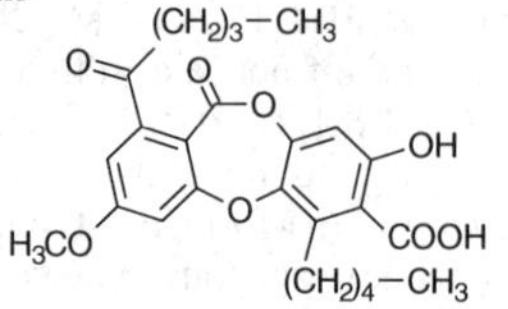

$C_{25}H_{28}O_8$, M_R 456.49, needles, mp. 196–197 °C. One of the few *depsidones with an acyl group on the benzene ring. It occurs, e. g., in the crustaceous lichen *Lecanora badia.*
Lit.: Asahina-Shibata, p. 120 ▪ Aust. J. Chem. **50**, 763 (1997) (NMR) ▪ Culberson, p. 137 f. ▪ Karrer, No. 1056. – *[CAS 522-53-2]*

Lobelia alkaloids. Name for *piperidine alkaloids from *Lobelia inflata* (Lobeliaceae) and some related species.

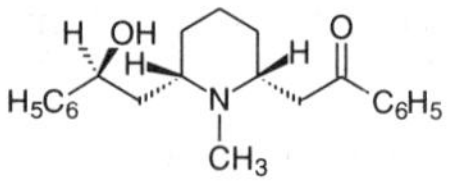

(-)-Lobeline (**1**)
racemate: Lobelidine (**2**)

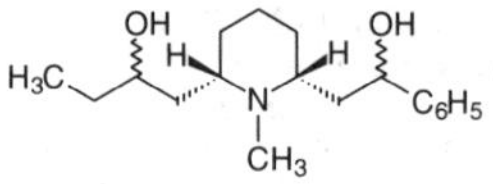

Lelobanidine (**3**)

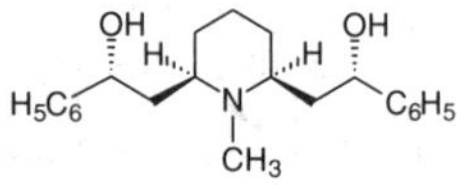

Lobelanidine (**4**)

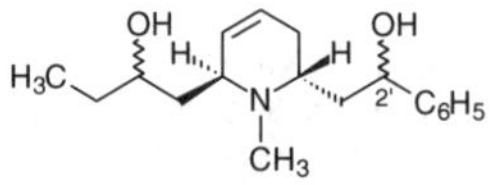

Lobinanidine (**5**)
2'-ketone: Isolobinine (**6**)

Table: Data of Lobelia alkaloids.

no.	molecular formula	M_R	mp. [°C]	optical activity	CAS
1	$C_{22}H_{27}NO_2$	337.46	130–131	$[\alpha]_D^{15}$ –42.8° (C_2H_5OH)	90-69-7
2	$C_{22}H_{27}NO_2$	337.46	110		134-65-6
3	$C_{18}H_{29}NO_2$	291.43			492-48-8
4	$C_{22}H_{29}NO_2$	339.48			552-72-7 (*meso*-form)
5	$C_{18}H_{27}NO_2$	289.42	95	$[\alpha]_D$ –120°	530-11-0
6	$C_{18}H_{25}NO_2$	287.40	78		530-12-1

L. inflata (Indian tobacco) is indigenous to the Eastern and Central states of USA and Canada. The plant contains ca. 0.3% of alkaloids. The L. a. are 2,6-disubstituted piperidine derivatives. Among the ca. 20 L. a., *lobeline* is the major alkaloid. When administered parenterally (3–10 mg) lobeline stimulates respiration and was used in the past as a respiratory analeptic agent for asthma, collapse, and narcosis incidents. When taken orally it is rapidly degraded and thus not effective. Since lobeline potentiates the action of nicotine and induces nausea and revulsion it has been developed clinically in depot form in antismoking preparations[1]. – The piperidine ring is biogenetically derived from lysine and the substituents from phenylalanine.

Lit.: [1] Drug News (25.3) **1996**, 6.
gen.: Anal. Profiles Drug Subst. **19**, 261 (1990) ▪ Beilstein E V **21/12**, 627 f. ▪ Hager (5.) **3**, 744 f. ▪ J. Chem. Soc., Perkin Trans. 1 **1975**, 415 (biosynthesis) ▪ J. Med. Chem. **42**, 3726 (1999) (synthesis lobeline) ▪ Manske **26**, 112–113 ▪ Merck-Index (12.), No. 5577–5580 ▪ Pelletier **3**, 59–63 ▪ Sax (8.), p. 2133 ▪ Sci. Pharm. **48**, 365–368 (1980).

Lochneram, lochnerine see sarpagine.

Locustol.

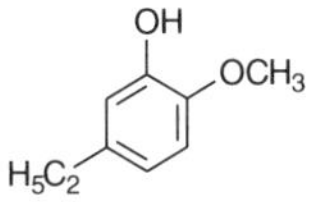

$C_9H_{12}O_2$, M_R 152.19, a *pheromone isolated from the feces of the desert locust *Schistocerca gregaria* that influences aggregation (formation of swarms) and is supposed to have other physiological activities [1]. However, in recent studies, the substance could not be detected, at least in *S. gregaria*, instead a mixture of various phenols and *phenylacetonitrile* (C_8H_7NO, M_R 133.15) was reported [2]. For general information on the pheromones of grasshoppers, see *Lit.* [3].

Lit.: [1] J. Insect Physiol. **19**, 1547–1554 (1973). [2] J. Chem. Ecol. **20**, 2077–2087 (1994). [3] Chapman & Joern (eds.), The Biology of Grasshoppers, p. 337–353 and 357–391, New York: Wiley 1990. – *[CAS 2785-88-8]*

Loganin. An important *iridoid glucoside from which secologanin (see secoiridoids), a key compound in the biosynthesis of *indole alkaloids, is formed [1]: After oxidation of the methyl group to an alcohol and its phosphorylation, ring opening occurs to give secologanin, which then reacts with tryptamine to form the indole alkaloids. L. also occurs as the free acid (*loganic acid*) and as the 7-oxo derivative (*ketologanin*).

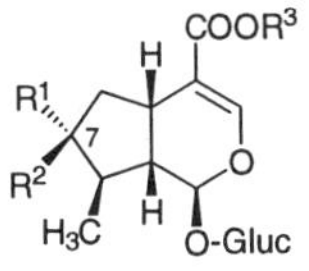

$R^1 = H, R^2 = OH, R^3 = CH_3$: Loganin
$R^1 = H, R^2 = OH, R^3 = H$: Loganic acid
$R^1, R^2 = O, R^3 = CH_3$: Ketologanin

Table: Loganin and derivatives.

trivial name	Loganin	Loganic acid	Keto-, Oxo-loganin (Dehydro-loganin)
molecular formula	$C_{17}H_{26}O_{10}$	$C_{16}H_{24}O_{10}$	$C_{17}H_{24}O_{10}$
M_R	390.39	376.36	388.37
mp. [°C]	222–223	amorphous	194–195
$[\alpha]_D$	-82.8° (H_2O)	–	-110° (H_2O)
CAS	18524-94-2	22255-40-9	152-91-0

L. was first isolated in 1884 from an extract of the poison nut tree (*Strychnos nux-vomica*, Loganiaceae). The poison nut tree is indigenous to tropical India and Sri Lanka. L. also occurs in European bog bean (water trefoil) (*Menyanthes trifoliata*, Menyanthaceae) and in those plants that contain indole alkaloids. Loganic acid and ketologanin have been detected in an American *Swertia* species (*Swertia caroliniensis* = *Frasera caroliniensis*, Gentianaceae). For synthesis, see *Lit.* [2].

Lit.: [1] Chem. Commun. **1968**, 136, 138; Phytochemistry **34**, 1291 (1993). [2] J. Org. Chem. **58**, 4756 (1993); **64**, 659 (1999); Tetrahedron **42**, 4035 (1986); **48**, 9495 (1992).
gen.: ApSimon **2**, 81; **4**, 494–507 ▪ Beilstein E V **18/7**, 478 ▪ J. Nat. Prod. **43**, 649, No. 81, 82, 85 (1980) ▪ Karrer, No. 974b ▪ Planta Med. **1986**, 327 ff. – *[HS 2938 90]*

Lolines. The name L. is used to define a small group of alkaloids that may be considered as *N*-alkyl and *N*-acyl derivatives of 1-aminopyrrolizidine. On account of the common ring system the L. are often included in the *pyrrolizidine alkaloids which, however, are rather esters of 1-hydroxymethylpyrrolizidine.

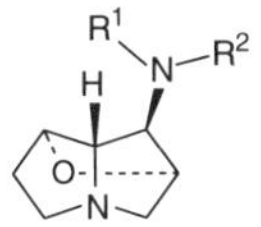

The occurrence of L. is limited to some sweet grasses (Poaceae, genera *Lolium*, *Festuca*) and the genus *Adenocarpus* (Fabaceae, tribe Genisteae). In *Adenocarpus* the L. (especially decorticasine) occur preferentially in the flowers and seeds [1]. The L. of grasses have attracted attention since their occurrence coincides with an infection by fungal endophytes belonging to the ascomycetes (*Acremonium*, family Clavicipitaceae) [2]. They cannot be detected either in uninfected grass or in the isolated cultivated fungus. In addition to the L., endophyte-infected grasses contain two further classes of alkaloids: ergopeptides (see ergot alkaloids) the formation of which is due to the fungus and the diazaphenanthrene alkaloids (e.g., *perloline) formed by the plants. Endophyte-infected grasses have a higher resistance to herbivorous insects, this is assumed to be due to the L. [3] The poisonings observed in cattle and other grazing animals after consumption of endophyte-infected meadow grass is probably not caused by the L. but rather by the ergopeptides [4] (see table, p. 366).

Lit.: [1] Biochem. Syst. Ecol. **20**, 69–73 (1992). [2] Crop. Sci. **22**, 941–943 (1983); **23**, 1136–1140 (1983); J. Chem. Ecol. **16**, 3301–3316 (1990). [3] Biotechnology **1**, 189–191 (1983); J. Agric. Chem. **37**, 354–357 (1989). [4] Cheeke (ed.), Toxicants of Plant Origin, vol. 1, p. 281–289, Boca Raton: CRC Press 1989.
gen.: Merck-Index (12.), No. 5591 ▪ Pelletier **9**, 155–233. – *[HS 2939 90]*

Lomatiol.

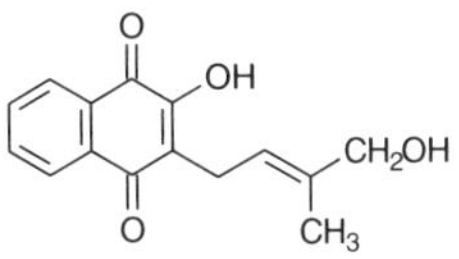

$C_{15}H_{14}O_4$, M_R 258.27, yellow needles, mp. 128 °C, uv_{max} 333 nm (C_2H_5OH), occurs in the seeds of the Australian plant *Lomatia ilicifolia* and other *Lomatia* species (Proteaceae) as well as in the heartwood of *Paratecoma peroba* (Bignoniaceae).

Lit.: Karrer, No. 1222 ▪ J. Nat. Prod. **44**, 562 (1981) ▪ Tetrahedron Lett. **24**, 1905 (1983). – *[CAS 523-34-2]*

Table: Lolines known from the plant kingdom.

name	R^1	R^2	molecular formula	M_R	mp. [bp.] [°C]	$[\alpha]_D$	CAS
Norloline	H	H	$C_7H_{12}N_2O$	140.19	[94–95 (0.67 kPa)]	+15.1°	4839-19-4
Loline	H	CH_3	$C_8H_{14}N_2O$	154.21	[103 (0.67 kPa)]	+6.2° (2 HCl/ H_2O)	25161-91-5
N-Methylloline	CH_3	CH_3	$C_9H_{16}N_2O$	168.24	[90–91 (0.27 kPa)]	+9.31° (CH_3OH)	22143-50-6
N-Formylnorloline	H	CHO	$C_8H_{12}N_2O_2$	168.20	oil	+31.3° (acetone)	
N-Acetylnorloline	H	$COCH_3$	$C_9H_{14}N_2O_2$	182.22	oil	+49.8° ($CHCl_3$)	38964-35-1
N-Formylloline	CH_3	CHO	$C_9H_{14}N_2O_2$	182.22	93–94	+47.9° ($CHCl_3$)	38964-33-9
N-Acetylloline (Lolinine)	CH_3	$COCH_3$	$C_{10}H_{16}N_2O_2$	196.25	73	+36.9	4914-36-7
Decorticasine	H	$CO–C_2H_5$	$C_{10}H_{16}N_2O_2$	196.25			1380-03-6
N-Butyrylnorloline	H	$CO–CH_2–C_2H_5$	$C_{11}H_{18}N_2O_2$	210.28	206 (picrate, decomp.)		
N-Isobutyrylnorloline	H	$CO–CH(CH_3)_2$	$C_{11}H_{18}N_2O_2$	210.28	127		
N-Isovalerylnorloline	H	$CO–CH_2–CH(CH_3)_2$	$C_{12}H_{20}N_2O_2$	224.30	241 (picrate, decomp.)		

Longhorn beetles (Cerambycidae). Slender beetles with long, bent antennae. The larvae, mostly living in wood, can be a major pest. The long-horned beetle *Hylotrupes bajulus* is a dangerous pest for wood in buildings. Chiral α-hydroxyketones with 6 or 8 carbon atoms have been identified as *pheromones in some B.[1]. The sex pheromone of the female sugarcane borer *Migdolus fryanus* is *(S)-2-methyl-N-(2-methylbutyl)-butyramide* ($C_{10}H_{21}NO$, M_R 171.28)[2]. The musk beetle *Aromia moschata* contains considerable amounts of the strong smelling *rose oxide as well as *iridodial[3] which is possibly used as a *defensive secretion against enemies. The ten-membered, methyl-substituted lactone *phoracantholide* ($C_{10}H_{18}O_2$, M_R 170.25), which is isolated from the metasternal glands of the eucalyptus borer *Phoracantha synonyma* and considered to be one of its defence substances, possesses an unbranched carbon skeleton with a distribution pattern of oxygen substituents similar to that of the *queen substance of the honeybee[4]. The natural product has the (*R*)-configuration[5].

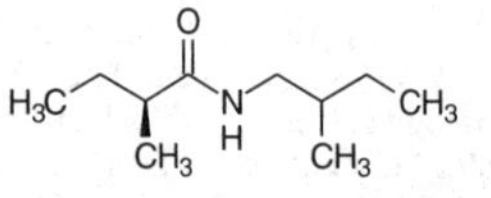

(*S*)-2-Methyl-*N*-(2-methylbutyl)-butyramide

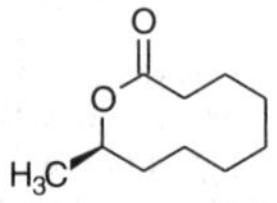

Phoracantholide

Lit.: [1] Appl. Entomol. Zool. **21**, 21–33 (1986); Justus Liebigs Ann. Chem. **1994**, 1211–1218; Proc. Natl. Acad. Sci. USA **92**, 1038–1042 (1995). [2] Experientia **50**, 853–856 (1994). [3] Tetrahedron Lett. **1973**, 465–468. [4] Aust. J. Chem. **29**, 1365–1374 (1976); Justus Liebigs Ann. Chem. **1995**, 1127 f. [5] Agric. Biol. Chem. **47**, 389–393 (1983). – *[CAS 61448-27-9 (phoracantholide)]]*

Longifolene (junipene, kuromatsuene).

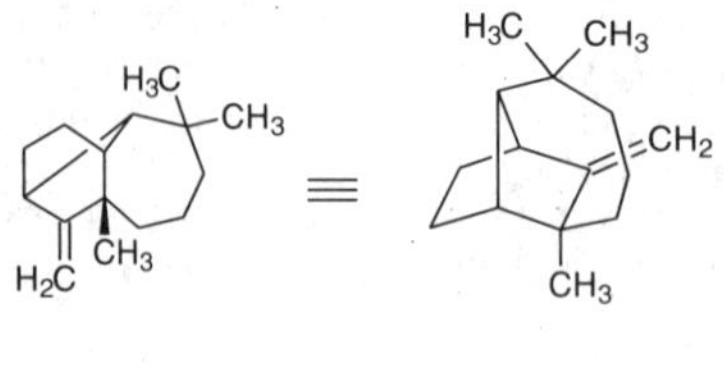

$C_{15}H_{24}$, M_R 204.36, oil, bp. 126 °C (1.99 kPa), $[\alpha]_{578}$ +47° (acetone), D_4^{18} 0.9319, insoluble in water. L. is a *sesquiterpene widely distributed in essential oils. The content of L. in Indian turpentine oil amounts to up to 20%. L. was first isolated in 1920 from *Pinus ponderosa*.

Use: L. is used as an additive to solvents and is the starting material for the production of fragrance substances. It can also be used as a borane derivative (*dilongifolylborane*) for asymmetric hydrogenations[1]. For synthesis, see *Lit.*[2].

Lit.: [1] J. Org. Chem. **46**, 2988 (1981). [2] ApSimon **2**, 517; **5**, 446; Can. J. Chem. **70**, 1375 (1992); J. Org. Chem. **50**, 915 (1985); **58**, 2186 (1993); J. Am. Chem. Soc. **112**, 4609 (1990). *gen.:* Beilstein E IV **5**, 1192 ▪ Karrer, No. 1930 a ▪ Ullmann (4.) **22**, 547 ▪ Zechmeister **40**, 49–104. – *Biosynthesis:* Pure Appl. Chem. **41**, 219 (1975). – *Toxicology:* Food Chem. Toxicol. **30**, 67s (1992). – *[CAS 475-20-7 ((+)-L.); 16846-09-6 ((–)-L.); 38142-68-6 (dilongifolylborane)]*

Longithorones.

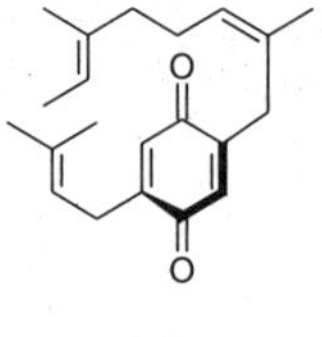

L. B

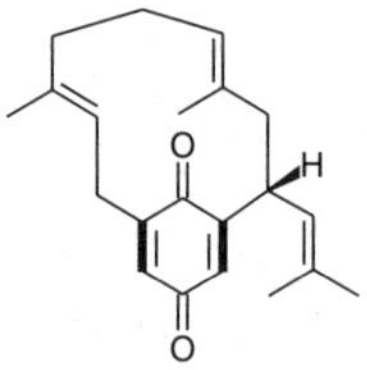

L. C

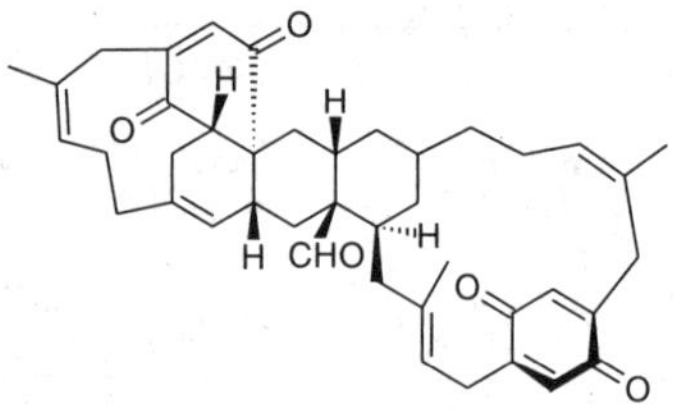

L. A

The tunicate *Aplidium longithorax* produces several meroterpenoid ansaquinones, e. g., *L. B* [$C_{21}H_{26}O_2$, M_R 310.44, mp. 80–82 °C, $[\alpha]_D$ –92.4° (CH_2Cl_2)] and *L. C* [$C_{21}H_{26}O_2$, M_R 310.44, mp. 127–129 °C, $[\alpha]_D$ –305.3°

(CH_2Cl_2)]. The compounds are biogenetically derived from farnesylhydroquinone and are accompanied by dimers like *L. A* [$C_{42}H_{46}O_5$, M_R 630.82, mp. 195–196 °C, $[\alpha]_D$ −87.5° ($CHCl_3$)] which may originate from modified monomers by enzymatic Diels-Alder reactions (compare endiandric acids). The L. are optically active due to atropisomerism caused by restricted rotation around the C–C-bonds connecting the benzoquinone ring.

Lit.: J. Am. Chem. Soc. **116**, 12125 (1994) ▪ J. Org. Chem. **62**, 3810 (1997). – *Synthesis* (of L. B): Tetrahedron Lett. **40**, 1941 (1999). – *[CAS 190837-20-8 (L. B); 159736-39-7 (L. A); 190913-23-6 (L. C)]*

Loniceroside see secoiridoids.

Lophophora alkaloids. Alkaloids from the Mexican *Lophophora* cacti (*peyote) cultivated as ornamental cacti in Europe, see cactus alkaloids. The plants contain ca. 5% of alkaloids, mainly *phenylethylamine alkaloids and simple isoquinoline derivatives. Their best known property is the generation of colored hallucinations and other sensory changes. The most toxic alkaloid of this group is lophophorine (see Anhalonium alkaloids).

Lit.: J. Am. Chem. Soc. **93**, 6248 (1971). – *[HS 2939 90]*

Lophophorine see Anhalonium alkaloids.

Lophotoxin.

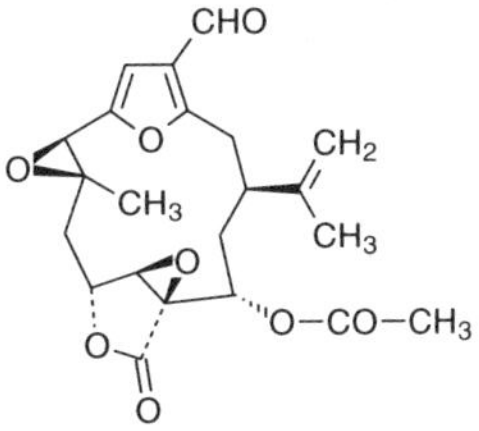

$C_{22}H_{24}O_8$, M_R 416.43, needles, mp. 164–166 °C. Furanocembranolide from seafans and flagellated corals of the genus *Lophogorgia*. L. is a diterpene with the 14-membered carbocylic ring of the *cembranoids. L. is a potent neuromuscular toxin that causes an irreversible postsynaptic blockade of acetylcholine receptors. In this respect it resembles α-*bungarotoxin, a cobra venom, LD_{50} (mouse s.c.) 8 mg/kg.

Lit.: Tetrahedron **41**, 981 (1985) (review) ▪ Toxicon **19**, 825 (1981). – *Pharmacology:* Brain Res. **359**, 233 (1985) ▪ J. Biol. Chem. **263**, 18568 (1988) ▪ J. Exp. Biol. **137**, 603 (1988) ▪ Neuroscience **20**, 875 (1987). – *synthesis:* Chem. Lett. **1987**, 1491 ▪ J. Org. Chem. **51**, 765 (1986) ▪ Tetrahedron Lett. **28**, 2525 (1987). – *[CAS 78697-56-0]*

Lotaustralin see cyanogenic glycosides.

Lovastatin see mevinolin.

LT. Abbreviation for *leukotrienes.

Lubimin see vetispiranes.

Lucidine B see Lycopodium alkaloids.

Luciferases (EC 1.13.12.5–1.13.12.8, 1.14.14.3, 1.14.99.21). General term for structurally different enzymes belonging to the oxidoreductases from various organisms with the common property of inducing *bioluminescence by oxidation of *luciferins or other compounds. In marine organisms these reactions occur in so-called scintillones. The system in the firefly *Photinus pyralis* has been investigated in depth; here, an L. with M_R 62 000 (polypeptide chain, 550 amino acid residues, EC 1.13.12.7) catalyzes the formation of the oxidized Photinus luciferin with evolution of light and conversion of ATP to AMP.

The enzyme has been cloned, the photoyield of the reaction amounts to 88%, and it can be used as a sensitive method for the detection of oxygen and ATP (down to 10^{-11} mol%) as well as other biochemical substrates. The sensitive detection of the enzyme itself also favors its use as a marker enzyme for gene expression.

Lit.: Anal. Biochem. **175**, 5, 14 (1988).

Luciferins.

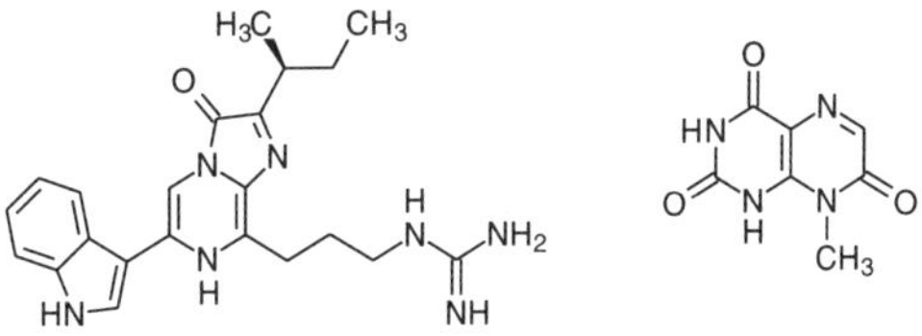

Cypridina L. Luciopterin

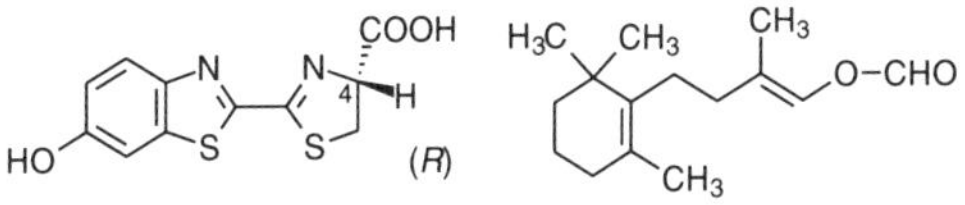

Photinus L. Latia L.

Name for natural compounds that exhibit *bioluminescence under the action of appropriate enzymes (*luciferases). The L. occurring in different organisms (bacteria, crabs, mussels, jellyfish, deep-sea fish, worms, beetles, etc.) have completely different structures. *Cypridina* species (ostracods), e. g., *Cypridina hilgendorfii*, living on the Pacific coasts contain a pyrazine derivative [1] (*Cypridina L.*): $C_{22}H_{27}N_7O$, M_R 405.50 as L. Cypridina L. exists, depending on the pH value, in three tautomeric forms. The kinetics of Cypridina luminescence are of first order, this represents the simplest bioluminescence system:

$$\text{L.} + O_2 \xrightarrow{\text{Luciferase}} \text{product(s)} + h\nu.$$

Cypridina L. is also able to exhibit chemiluminescence, however, with a much lower photoyield.

The lamp jellyfish *Aequorea aequorea* exhibits a green bioluminescence under the action of Ca^{2+}, this is due to *aequorin. The marine species *Renilla reniformis* also contains a L. which requires oxygen for the emission of light. This L. is bound to a protein, to a certain degree in a storage form. Similar photoprotein systems have also been found in other anthozoa. The Renilla L. is *coelenterazine, which is also present in the deep-sea decapods *Sergia lucens* and *Oplophorus gracilorostris*. Lantern fish contain either coelenterazine or Cypridina luciferin. The L. of the marine fireflies have not yet been identified. The surface luminescence of the oceans is mostly due to dinoflagellates (e.g., *Pyrocystis lunula*), the L. of which are related to bile pigments [2]. A thiazole derivative {(*R*)-4,5-dihydro-2-(6-hydroxy-2-benzothiazolyl)-4-thiazolecarboxylic acid (*Photinus L.*), $C_{11}H_8N_2O_3S_2$, M_R 280.32, pale yellow crystals, mp. 190 °C (decomp.)} has been isolated from the American firefly *Photinus pyralis*; this compound also occurs in the European glowworms (*Lampyris noctiluca, Phausis splendidula*)[3]. The Japanese glowworm *Luciola cruciata* contains a pteridine derivative (*luciopterin*) as its L. ($C_7H_6N_4O_3$, M_R 194.15)[4]. The water snail *Latia neritoides* contains a sesquiterpene (*Latia L.*) ($C_{15}H_{24}O_2$, M_R 236.35) that also acts as an L.[5]. Synthetic analogues of L. have been described[6].

Lit.: [1] Tetrahedron Lett. **1972**, 2747. [2] J. Am. Chem. Soc. **110**, 2683 (1988); **111**, 7607 (1989); Nachr. Chem. Tech. Lab. **38**, 159 (1990); J. Biolumin. Chemilumin. **4**, 12 (1989); Methods Enzymol. **133**, 307 (1986). [3] J. Chem. Soc., Chem. Commun. **1976**, 32, 153. [4] Tetrahedron Lett. **1968**, 2847. [5] Biochemistry **7**, 1734 (1968); J. Chem. Soc., Chem. Commun. **1978**, 297; Tetrahedron Lett. **1979**, 707. [6] Bull. Chem. Soc. Jpn. **65**, 2604 (1992).

gen.: Adv. Oxygenated Processes **1**, 123–178 (1988) ▪ Chem. Br. **1994**, 300 ff. ▪ J. Appl. Biochem. **5**, 197 (1983) ▪ J. Biolumin. Chemilumin. **4**, 289 (1989) ▪ J. Mol. Catal. **47**, 315 (1988) ▪ Methods Enzymol. **133**, 51 (1986) ▪ Scheuer I **3**, 180 ▪ see also coelenterazine. – *[CAS 7273-34-9 (Cypridina L.); 20240-21-5 (Photinus L.); 19845-00-2 (luciopterin); 21730-91-6 (Latia L.)]*

Luciopterin see luciferins.

Lumazine see pteridines.

β-Lumicolchicine see phenethyltetrahydroisoquinoline alkaloids.

Lunarine.

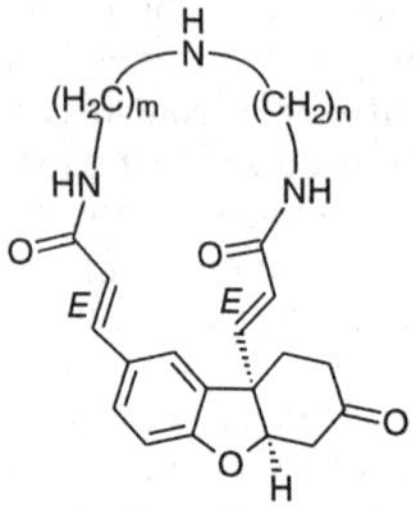

m = 3, n = 4 : Lunarine
m = 4, n = 3 : Lunaridine

$C_{25}H_{31}N_3O_4$, M_R 437.54, mp. 239–240 °C (232–235 °C), $[\alpha]_D^{20}$ +291° ($CHCl_3$). L. and its isomer *lunaridine* are polyamine alkaloids from the seeds of the penny flower [*Lunaria annua* (=*L. biennis*), *L. rediviva*; Brassicaceae]. They contain *spermidine as building block, which is released on hydrolysis.

Lit.: Beilstein E IV **27/13**, 9415. – *Synthesis:* Heterocycles **17**, 537 (1982) ▪ Pure Appl. Chem. **53**, 1141 (1981) ▪ Tetrahedron Lett. **38**, 2443 (1997). – *[CAS 24185-51-1 (L.); 34340-56-2 (lunaridine)]*

Lunularic acid

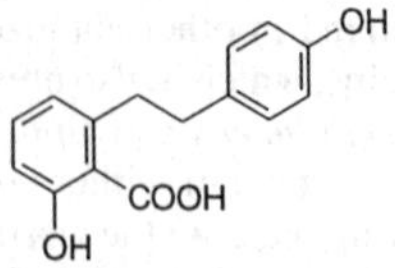

[6-(4-hydroxyphenethyl)salicylic acid]. $C_{15}H_{14}O_4$, M_R 258.27, mp. 201–202 °C, cryst., first isolated from the liverwort *Lunularia cruciata*[1]. L. is widely distributed in liverworts (see mosses), often occurs in algae but is only rarely found in mosses. In lower plants L. exerts a development regulating[2] function similar to that of *abscisic acid in higher plants. For synthesis, see *Lit.*[3]. Some derivatives of L. (esters, amides) have considerable molluscicidal activities.

Lit.: [1] Nature (London) **223**, 1176 (1969). [2] J. Exp. Bot. **21**, 138 (1970). [3] Justus Liebigs Ann. Chem. **1996**, 2107; Synthesis **1988**, 88.

gen.: Zinsmeister & Mues (eds.), Bryophytes, their Chemistry and Chemical Taxonomy (Proc. Phytochem. Soc. Eur. 29), p. 171, Oxford: Clarendon Press 1990. – *[HS 2902 90; CAS 23255-59-6]*

Lupan(e) see triterpenes.

Lupanidine see matrine.

Lupanine (2-oxosparteine).

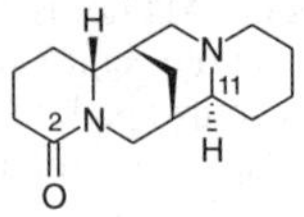

(+)-form

$C_{15}H_{24}N_2O$, M_R 248.37, (+)-form: mp. 40 °C, $[\alpha]_D$ +61.4° (acetone); (–)-form: viscous oil, $[\alpha]_D$ –61° (acetone); racemate: mp. 98–99 °C. A *quinolizidine alkaloid of the sparteine type that is the most widely distributed and often the major alkaloid in many genera of the Fabaceae (e.g., *Cytisus, Genista, Lupinus*). It also occurs in some species of the genus *Leontice* (Berberidaceae)[1]. In plants L. occurs not only in the (+)-form but also as the (–)-form and as the racemate. It is accompanied by numerous minor alkaloids (see also anagyrine, angustifoline, baptifoline, sparteine, 13-hydroxylupanine). L. is weakly toxic [LD_{50} (rat i.p.) 177 mg/kg, (p.o.) 1644 mg/kg][2] and is considered to be, besides other quinolizidine alkaloids, the main cause of poisonings among grazing animals. It is known to have anti-arrhythmic, hypotensive, and hypoglycemic activities although they are weaker than those of *sparteine[3,4].

Lit.: [1] Waterman **8**, 197–239. [2] J. Appl. Toxicol. **7**, 51–53 (1987). [3] Arzneim.-Forsch./Drug Res. **24**, 7753–7759 (1974). [4] Pelletier **2**, 105–148.

gen.: Beilstein E V **24/2**, 295 ff. ▪ J. Chem. Soc., Chem. Commun. **1984**, 1477 (biosynthesis) ▪ Manske **47**, 1–115 ▪ Phytochemistry **29**, 1297 (1990) ▪ Sax (8.), p. 2138. – *[HS 2939 90; CAS 550-90-3 ((+)-form); 486-88-4 ((–)-form); 4356-43-8 (racemate)]*

Lupeol [lup-20(29)-en-3β-ol].

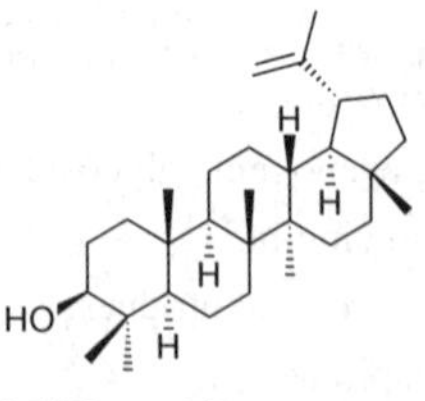

$C_{30}H_{50}O$, M_R 426.73, needles, mp. 215–216 °C, $[\alpha]_D$ +26.4° ($CHCl_3$), soluble in ether, warm ethanol, insoluble in water. L. is a widely distributed *triterpene in plants. L. occurs in the bark of many plants (Apocynaceae, Fabaceae), it was first isolated from the husk of lupin seeds (*Lupinus luteus*).

Lit.: Beilstein E IV **6**, 4200 ▪ Karrer, No. 2023 ▪ Phytochemistry **40**, 1227 (1995); **41**, 1437 (1996); **47**, 311 (1998). – *[CAS 545-47-1]*

Lupine alkaloids see quinolizidine alkaloids.

Lupinic acid.

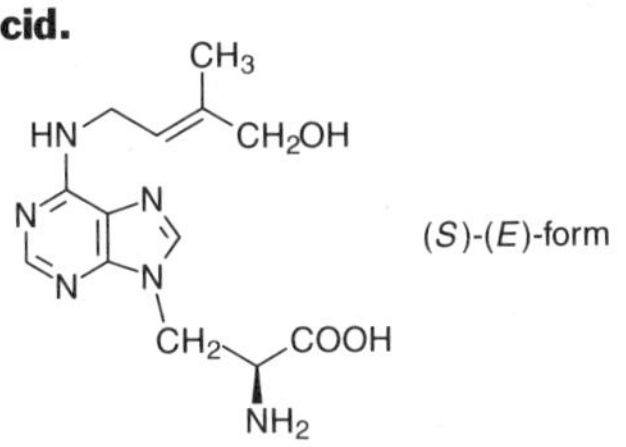

$C_{13}H_{18}N_6O_3$, M_R 306.30, cryst., mp. 216–217 °C, $[\alpha]_D^{20}$ –25° (H_2O). A purine derivative from shoots of *Lupinus angustifolius* (a lupine species with small leaves), it is formed from *zeatin.

Lit.: Aust. J. Chem. **31**, 1291, 1301 (1978) ▪ Chem. Pharm. Bull. **25**, 520 (1977) ▪ J. Chem. Soc., Chem. Commun. **1975**, 809 ▪ Tetrahedron **46**, 913–920 (1990). – *[CAS 58137-33-0]*

Lupinine.

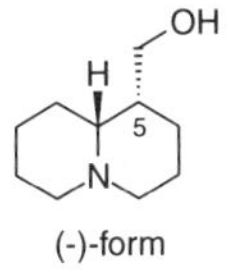

$C_{10}H_{19}NO$, M_R 169.27, mp. 68.5–69.2 °C, bp. 269–270 °C, $[\alpha]_D^{28}$ –21° (C_2H_5OH); racemate, mp. 63–64 °C (59 °C). L. and the 5-epimer *epi-lupinine* {mp. 77–78 °C, $[\alpha]_D$ +32° (C_2H_5OH)} are *quinolizidine alkaloids of which only few members contain the bicyclic lupinine type structure. They occur in several Fabaceae genera [1]. L. is also a component of the Chenopodiaceae *Anabasis aphylla* [1]. In addition, esters of *epi*-lupinine with acetic, coumaric, and ferulic acid are known as natural products. Several syntheses, some enantioselective, of these simple quinolizidine alkaloids have been reported [2].

Lit.: [1] Waterman **8**, 197–239. [2] J. Chem. Soc., Perkin Trans. 1 **1994**, 2903; J. Am. Chem. Soc. **110**, 289 (1988); Tetrahedron **45**, 6161 (1989); Heterocycles **30**, 885 (1990); J. Org. Chem. **55**, 1148 (1990); Tetrahedron: Asymmetry **1**, 147 (1990); Tetrahedron **54**, 10349–10362 (1998); Tetrahedron Lett. **34**, 215 (1993).
gen.: Beilstein E V **21/1**, 338 ▪ Manske **47**, 1–115 ▪ Phytochemistry **50**, 189 (1999). – *Biosynthesis:* Can. J. Chem. **63**, 2707 (1985). – *[HS 2939 90; CAS 486-70-4 ((–)-form); 10248-30-3 (racemate); 486-71-5 (epi-lupinine)]*

Lupulone.

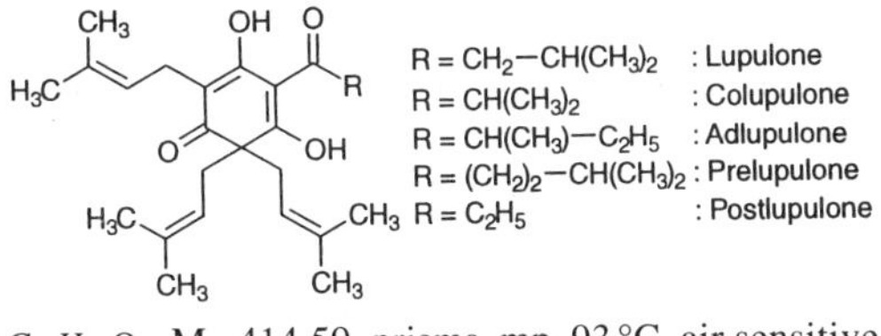

$C_{26}H_{38}O_4$, M_R 414.59, prisms, mp. 93 °C, air sensitive, soluble in methanol, ethanol, petroleum ether, the sodium salt of L. is soluble in water. L., which does not have a bitter taste, and the other *β*-lupulinic acids from hops are not soluble at the usual pH value of the mash in beer brewing and do not isomerize during the boiling process. They are thus removed with the plant remains. On longer storage of the hops L. and its derivatives (see formula) are oxidized to the so-called *β*-soft resins, the latter are soluble in the mash and generate the pleasant bitter taste [1]. L. is active against Gram-positive bacteria. For biosynthesis, see humulone.

Lit.: [1] Eur. Brew. Conv. Proc. Congr. **15**, 141–152 (1975).
gen.: Angew. Chem. Int. Ed. Engl. **33**, 1454 (1994) ▪ Arch. Mikrobiol. **94**, 159–171 (1973) ▪ Beilstein E IV **7**, 2854 ▪ Int. Rev. Biochem. **14**, 51 (1977) ▪ Luckner (3.), p. 167 ▪ Monatsschr. Brauwiss. **39**, 259–262 (1986); **41**, 252–255 (1988) ▪ Pollock (ed.), Brewing Science, p. 279–323, London: Academic Press 1981. – *[CAS 468-28-0]*

Lutein [(3*R*,3′*R*,6′*R*)-*β*,*ε*-carotene-3,3′-diol].

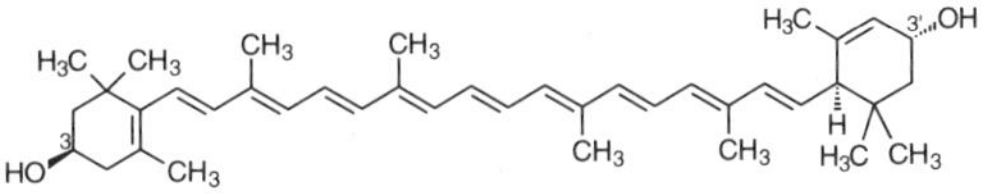

$C_{40}H_{56}O_2$, M_R 568.88. Yellow or garnet red prisms or plates. Mp. 96 °C, insoluble in water, soluble in fat solvents and fats, uv_{max} 428.456, 487 nm ($CHCl_3$).

Occurrence: In alfalfa grass, stinging nettle leaves, algae, lucern, palm oil, egg yolk, etc.; besides *carotene and *chlorophyll, L. is the most widely distributed leaf pigment in green plants and becomes apparent especially in autumn. Apart from the green plant parts, L. occurs in numerous yellow flower petals and pollens, in fruits, and in many lower plants. The yellow pigment of egg yolk contains L. and ca. 30% of the isomeric *zeaxanthin. In spite of its close relationship to *β*-carotene L. does not exhibit any provitamin A activity, see retinol. In the form of its dipalmitate *helenien, however, L. is also found in the retina of the eye where it promotes light-dark adaptation. For the qualitative analysis of L., e. g., to demonstrate the presence of chicken egg in pastry products, the material is extracted with ether. The yellow color caused by L. disappears on addition of aqueous HNO_2 solution (Weyl's reaction). L. is authorized for use as a pigment in foods (C. I. 75136, E 161 b).

Historical: Berzelius (1837) named the pigment xanthophyll (Greek: xanthós = yellow, phýllon = leaf; leaf yellow). This was later proven to be identical with L. (Latin: luteus = yellow) isolated from egg yolk by Willstätter and Escher in 1912, the structure of the latter was elucidated in 1930 by Karrer. Today the name xanthophylls is used generally as a group name for oxygenated carotinoids (hydroxy compounds, epoxides, some oxo compounds).

Lit.: Acta Chem. Scand. **50**, 637 (1996) (synthesis) ▪ Ann. N. Y. Acad. Sci. **691**, 246 (1993) (pharmacology) ▪ Beilstein E IV **6**, 7017 ▪ Karrer, No. 1844 ▪ Straub, Key to Carotinoids (2.), p. 133, 232, 302 Basel: Birkhäuser 1987. – *[HS 3203 00; CAS 127-40-2]*

Luteolin (3′,4′,5,7-tetrahydroxyflavone; C. I. 75590, natural yellow 2). Formula see flavones. $C_{15}H_{10}O_6$, M_R 286.23, fine yellow crystals with silk-like luster, mp. 329 °C (decomp.), poorly soluble in water, soluble in alkalis to give deep yellow solutions. The bitter-tasting L. has anticonvulsant activity. L. occurs in leaves, flowers, and stems of dyer's weld (*Reseda luteola*), in the flowers of the yellow foxglove (*Digitalis lutea*; thus also the name *digitoflavone* for L.), and in other flowers. In the past L. was once the most important yellow dye for wool and silk but has now been completely replaced by synthetic dyes. L. shows antiviral and antitumor activities.

Lit.: Beilstein E V **18/5**, 296 ▪ Cancer Lett. **87**, 107 (1994) (pharmacology) ▪ Hager (5.) **6**, 933 f. ▪ Justus Liebigs Ann. Chem. **1995**, 1711 (synthesis) ▪ Karrer, No. 1470 ▪ Kirk-Othmer (3.) **8**, 358f. ▪ Planta Med. **61**, 570 (1995) (NMR) ▪ see also flavonoids. – *[CAS 491-70-3]*

Luteolinidin see anthocyanins.

Luteoskyrin (*ent*-8,8′-dihydroxyrugulosin).

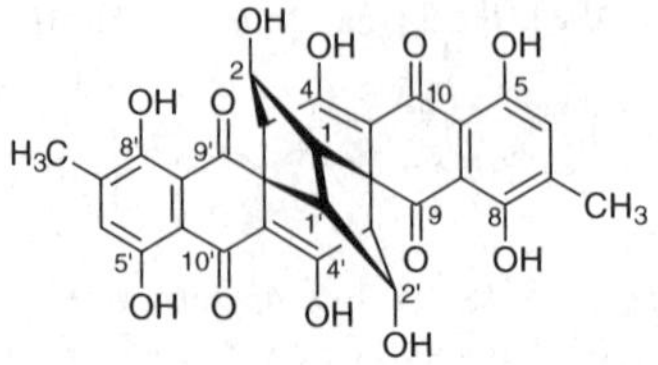

$C_{30}H_{22}O_{12}$, M_R 574.50, mp. 278 °C, $[\alpha]_D$ −830°; yellow, hepatotoxic *mycotoxin from *Penicillium islandicum*. L. belongs to the bianthraquinones such as, e.g., *rugulosin and *skyrin. The biosynthesis proceeds on the *polyketide pathway. On being fed daily with 1 mg toxin per 20 g animal, mice develop liver tumors within 6 months. The toxin accumulates in hepatic mitochondria. In the presence of Mg ions L. binds to double- and single-stranded DNA and inhibits the DNA-dependent RNA polymerases. L. has antimicrobial and cytotoxic properties and is active against protozoa. Chromosome breaks are induced in cell cultures.

Lit.: Betina, chap. 15 ▪ Cole-Cox, p. 702–707 ▪ CRC Crit. Rev. Toxicol. **14**, 99 (1985) ▪ IARC Monogr. **40**, 99, Suppl. 7, 65, 71 (1986) ▪ Sax (8.), LIV 000. – *[CAS 21884-44-6]*

LX. Abbreviation for *lipoxins.

Lyaloside, lyalosidic acid see Ophiorrhiza alkaloids.

Lyclavatol see α-onocerin.

Lycoctonine see browniine.

Lycodine. Formula see Lycopodium alkaloids.
$C_{16}H_{22}N_2$, M_R 242.36, mp. 118 °C, $[\alpha]_D$ −10° (C_2H_5OH). A *Lycopodium alkaloid of the $C_{16}N_2$ type from several clubmoss species (Lycopodiaceae).

Lit.: Beilstein E V **23/7**, 448 ▪ J. Am. Chem. Soc. **104**, 1054 (1982) ▪ Justus Liebigs Ann. Chem. **1983**, 220–225 ▪ Merck-Index (12.), No. 5647. – *[HS 2939 90; CAS 20316-18-1]*

Lycomarasmine.

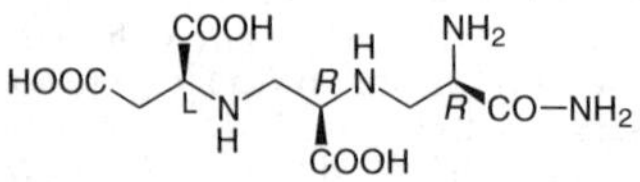

$C_9H_{15}N_3O_7$, M_R 277.23. Fine needles or microscopic powder, mp. 227–229 °C (decomp.). A plant toxin formed by various *Fusarium* species, e.g., *F. oxysporum* and *F. lycopersici* and made up from *aspartic acid, glycine, and pyruvate. L. acts as a *wilting (withering) agent and is considered to be the cause of the withering of tomato plants. The free tetracarboxylic acid of L. (aspergillomarasmine B) is a natural degradation product, also possessing phytotoxic activity. It is also formed by other fungi such as *Aspergillus flavus*, *Colletotrichum gloeosporioides*, *Paecilomyces* sp., and *Pyrenophora teres*. In addition, aspergillomarasmines A and B are potent inhibitors of endothelium converting enzyme (ECE).

Lit.: Biosci. Biotech. Biochem. **57**, 1944 (1993) ▪ Can. J. Chem. **51**, 3943 (1973) ▪ Experientia **32**, 608 (1976) ▪ Turner **1**, 334; **2**, 423 f. – *[CAS 7611-43-0 (L.); 3262-58-6 (L. acid)]*

Lycopene (ψ,ψ-carotene, E 160 d, C. I. 75125).

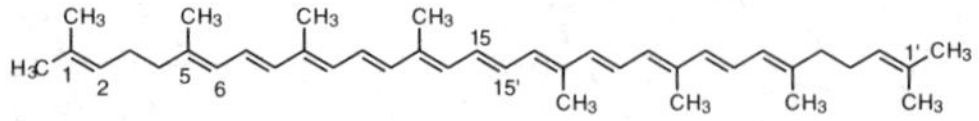

$C_{40}H_{56}$, M_R 536.88, long, dark red needles (petroleum ether), mp. 175 °C, soluble in chloroform. L. is a *carotinoid from the tomato (*Lycopersicon esculentum*, 1 kg tomatoes contains ca. 20 mg L.), rose hips, and other fruits, where it occurs together with its isomers α-, β-, and γ-*carotenes as well as the 1,2-epoxide and the 5,6-epoxide. L. is also present in chanterelles (*Cantharellus cibarius*), butter, serum, and liver. L. is authorized for use as a cosmetic and food pigment (E 160 d). With concentrated sulfuric acid, antimony trichloride, and trichloroacetic acid L. gives a dark blue color. The 7,7′,8,8′,11,11′,12,12′,15,15′-decahydro derivative is called *lycopersin* ($C_{40}H_{66}$, M_R 546.96).

Lit.: Beilstein E IV **1**, 1165 ▪ Helv. Chim. Acta **67**, 964 (1984); **75**, 1848 (1992) ▪ Karrer, No. 1818 ▪ Tetrahedron Lett. **28**, 5751 (1987) ▪ Turner **1**, 268 ▪ see also carotinoids. – *[HS 2901 29; CAS 502-65-8 (L.); 51599-09-8 (1,2-epoxide); 51599-10-1 (5,6-epoxide); 502-62-5 (lycopersin)]*

Lycoperdic acid [3-((*S*)-2-carboxy-5-oxo-tetrahydro-2-furyl)-L-alanine].

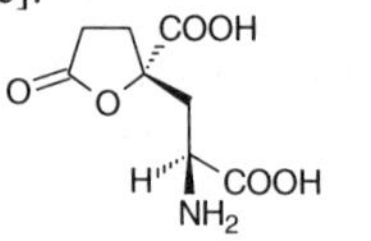

$C_8H_{11}NO_6$, M_R 217.18, $[\alpha]_D^{20}$ +14.9° (H_2O), +36.5° (1 m HCl). A non-proteinogenic amino acid from fruit bodies of *Lycoperdon perlatum* (Gasteromycetes, puffballs)[1].

Lit.: [1] Phytochemistry **18**, 482–484 (1979).
gen.: Chem. Pharm. Bull. **43**, 1617 (1995) (synthesis) ▪ Tetrahedron Lett. **33**, 8103 (1992). – *[CAS 69086-72-2]*

Lycopersin see lycopene.

Lycopodine. Formula see Lycopodium alkaloids.
$C_{16}H_{25}NO$, M_R 247.38, mp. 116 °C, $[\alpha]_D^{26}$ −24.5° (C_2H_5OH). A *Lycopodium alkaloid of the $C_{16}N$ type from many club moss species (Lycopodiaceae). It is the most widely distributed Lycopodium alkaloid and usually occurs as the major alkaloid. It was first isolated over 100 years ago from a clubmoss species[1]. L. is toxic and exerts a *curare-like paralysing activity[2]. Several syntheses have been reported[3].

Lit.: [1] Justus Liebigs Ann. Chem. **208**, 363 (1881). [2] Manske **26**, 241 ff. [3] J. Am. Chem. Soc. **107**, 4341 f. (1985); **120**, 5128 (1998); J. Chem. Soc., Chem. Commun. **1984**, 714 f.; J. Org. Chem. **62**, 78 (1997); Heterocycles **25**, 377 (1987).
gen.: Beilstein E V **21/7**, 482 f. ▪ Merck-Index (12.), No. 5652. – *[HS 2939 90; CAS 466-61-5]*

Lycopodium alkaloids. L. a. is a general name for alkaloids whose occurrence is limited to *pteridophytes; ferns (Filicatae) from the phylogenetically very old family of the *clubmosses. About 500 species are distributed world-wide and are today usually divided into the 4 genera (*Lycopodium*, *Diphasiastrum*, *Lycopodiella* and *Huperzia*). L. a. have been found in all of the about 50 species examined so far. The about 90 known L. a. are formally classified according to the number of C and N atoms in their skeletons into three structural types (figure). The $C_{16}N$ type represents the largest group of ca. 60 alkaloids which can be arranged according to the type of ring formation of the lycopodine or fawcettimine groups. Typical examples include the ubiquitous *lycopodine as well as *annotinine and *fawcettimine* ($C_{16}H_{25}NO_2$, M_R 263.38). The $C_{16}N_2$ type is represented by the lycodine group (ca. 20 structures); e. g., *lycodine and *cernuine. The complex $C_{27}N_3$ type encompasses only a few compounds like *lucidine B* ($C_{30}H_{49}N_3O$, M_R 467.74) shown in the figure. In some species the L. a. are accompanied by small amounts of *nicotine.

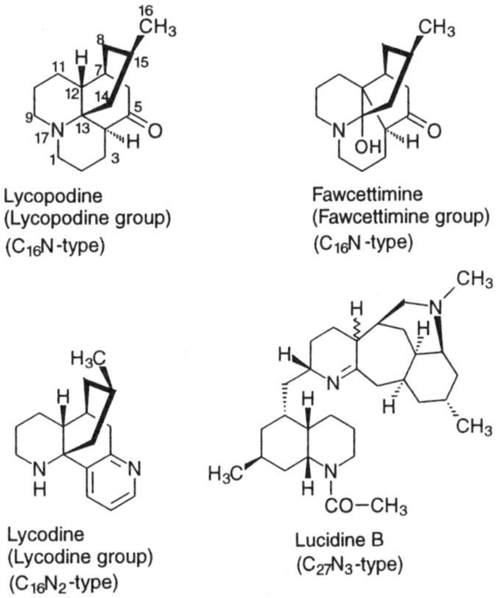

Figure: Structural types of the Lycopodium alkaloids.

Biosynthesis: The biosynthesis of the L. a. has not been completely elucidated. Since *Lycopodium* species cannot or can only be cultivated with difficulty, tracer experiments must be performed at the natural locations. It is certain that the biogenetic building blocks for the complex ring systems are lysine and acetate.

Pharmacological activity: The L. a. are moderately toxic. Some L. a. are purported to be potent inhibitors of acetylcholine esterase. *Huperzine A has attracted attention on account of its stimulating effect on learning and memory behavior in laboratory animals. In Chinese traditional medicine various *Lycopodium* species are used to treat skin diseases and as a tonic.

Synthesis: Numerous syntheses have been described (see huperzine A, lycopodine, lycodine).

Lit.: Manske **26**, 241–298; **45**, 233–266. – *Biosynthesis:* Herbert, The Biosynthesis of Secondary Metabolites (2.), p. 127, London: Chapman and Hall 1989 ▪ J. Am. Chem. Soc. **115**, 3020 (1993) ▪ Mothes et al., p 140. – *Synthesis:* Alkaloids: Chem. Biol. Perspectives **3**, 185–240 (1985) ▪ Eur. J. Org. Chem. **1999**, 1925 ▪ J. Am. Chem. Soc. **115**, 2992 (1993) ▪ Nat. Prod. Rep. **8**, 455 (1992) ▪ Pelletier **3**, 185–240 ▪ Synthesis **1998**, 1803 ▪ Tetrahedron **54**, 7865–7882 (1998). – *Occurrence:* Hegnauer **I**, 224; **VII**, 406 ▪ Phytochemistry **19**, 803 (1980). – *[HS 2939 90; CAS 15228-74-7 (fawcettimine); 71384-23-1 (lucidine B)]*

Lycopsamine see indicine and pyrrolizidine alkaloids.

Lycorenine, lycorine see Amaryllidaceae alkaloids.

Lycoricidinol see narciclasine.

Lymphostin.

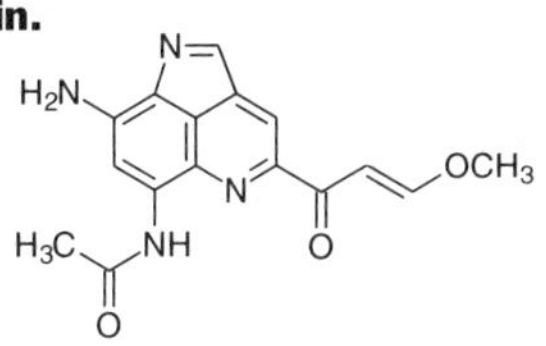

$C_{16}H_{14}N_4O_3$, M_R 310.31, orange powder, mp. 275–277 °C, immunosuppressant antibiotic from cultures of *Streptomyces* sp. KY 11783.

Lit.: J. Antibiotics **50**, 537–542, 543 ff. (1997). – *[CAS 191474-39-2]*

Lyngbyatoxins see majusculamides.

Lyophyllin.

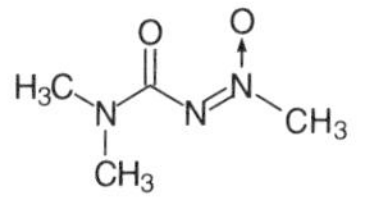

$C_4H_9N_3O_2$, M_R 131.13, cryst., mp. 27–28 °C. (*Z*)-Azoxy compound from the mushroom *Lyophyllum connatum* (Basidiomycetes). L. is formed biosynthetically from *N*′-hydroxy-*N,N*-dimethylurea (see connatin) and *N*-methylhydroxylamine. The former is also present in the fungus.

Lit.: Angew. Chem. Int. Ed. Engl. **23**, 72 (1984) (isolation) ▪ Tetrahedron Lett. **38**, 8013 (1997) (biosynthesis). – *[CAS 88245-13-0]*

Lysergic acid.

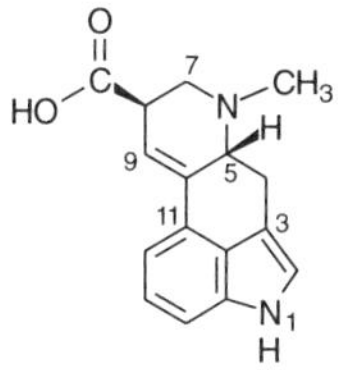

$C_{16}H_{16}N_2O_2$, M_R 268.32, platelets, mp. 238–240 °C, $[\alpha]_D$ +40° (pyridine). *Isolysergic acid*: 8α-COOH-epimer, mp. 218 °C, $[\alpha]_D$ +281° (pyridine). Both acids in the form of various amides are present in most of the *ergot alkaloids. In addition L. and Iso-L. are also present in *ipomoea resin and other twining plants (Convolvulaceae). For pharmacological activity of L. derivatives, see ergot alkaloids. Synthetic lysergic acid diethylamide is known as the psychotropic drug LSD.

Lit.: Beilstein E V **25/5**, 125 ff. ▪ Chem. Pharm. Bull. **34**, 442 (1986); **35**, 4793 (1987) (synthesis) ▪ Hager (5.) **3**, 750; **4**, 911 ff.; **8**, 61 ▪ Hofmann & Schultes, Pflanzen der Götter, p. 105–162, Bern: Hallwag 1980 ▪ Pape & Rehm (eds.), Biotechnology, vol. 4, chap. 18, Weinheim: Verl. Chemie 1986 ▪ Sax (8.), DJO 000, LJF 000-LJM 000 ▪ Tetrahedron Lett. **29**, 3117 (1988) ▪ Zechmeister **9**, 114–175; **20**, 17–21. – *[HS 2939 60; CAS 82-58-6 (L.); 478-95-5 (iso-L.)]*

L-Lysine [(*S*)-2,6-diaminohexanoic acid, abbreviation: Lys or K].

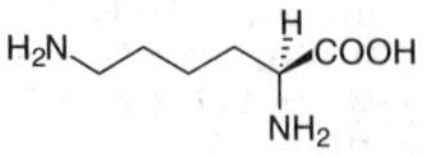

$C_6H_{14}N_2O_2$, M_R 146.19, mp. 224–225 °C (decomp.), $[\alpha]_D^{20}$ +14.6° (H_2O), $[\alpha]_D^{23}$ +25.9° (6 m HCl), pK_a 2.20, 8.90, 10.28, pI 9.59. dihydrochloride, mp. 193 °C, $[\alpha]_D^{20}$ +15.3° (H_2O). Ubiquitous amino acid essential for humans and many animals. Genetic code: AAA, AAG. Average content in proteins 7.0%[1], lower in plant proteins, including grain proteins.

Biosynthesis & catabolism: Lys is formed in plants and bacteria from *meso*-2,6-*diaminopimelic acid by diaminopimelate decarboxylase (EC 4.1.1.20). The catabolism proceeds through eleven enzymatic steps to acetoacetic acid (acetyl-CoA). L. is a precursor of the *cadaverines. Because of its two amino groups it has a cross-linking function in polypeptides such as collagen and elastin, see also 5-hydroxylysine. L. is used as a fodder additive.

Lit.: [1] Biochem. Biophys. Res. Commun. **78**, 1018 (1977).
gen.: Beilstein E III/IV **4**, 2717 f. ▪ Bull. Chem. Soc. Jpn. **64**, 613 (1991) ▪ Greenstein & Winitz, Chemistry of the Amino Acids, vol. 3, p. 2097–2124, New York: J. Wiley 1961 ▪ Ullmann (5.) **A 2**, 58, 63, 86; **A14**, 449; **A22**, 292. – *[HS 2922 41; CAS 56-87-1]*

Lythraceae alkaloids. Characteristic alkaloids from a series of genera (e. g., European loosestrife species, *Lythrum*) of the Lythraceae family. The more than 50 L. a. can be divided into five structural classes (see figure). Biogenetic building blocks of the L. a. are *cadaverine, which supplies the piperidine ring as a decarboxylation product of lysine and one (**1**) to two (**2**–**5**) phenylpropane units (C_6–C_3 derivatives), which provide the C_6–C_4-element occurring in all structural types via chain lengthening.

Little is known about the pharmacological activities of the L. a. Some alkaloids have demonstrated anti-inflammatory, blood pressure lowering, diuretic, and fungicidal activities in animal experiments. *Heimia salicifolia* has been used for centuries by the natives of Mexico under the name "Sinicuiche" for numerous medicinal purposes and is used to prepare a mildly intoxicating beverage. It is not known for certain whether the L. a. are responsible for the intoxicating effects.

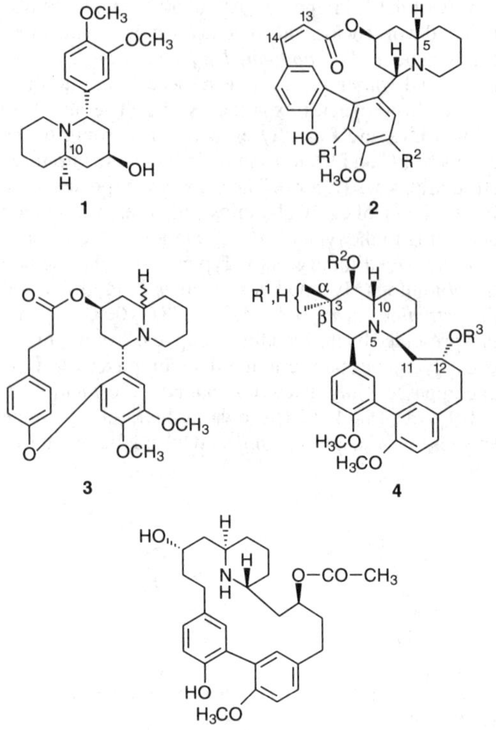

Figure: Structural types of the Lythraceae alkaloids; **1**: phenylquinolizidine type (10αH=lasubine I, 10βH=lasubine II); **2**: quinolizidine-biphenyl lactone type (10αH=vertine, 10βH= lythrine); **3**: quinolizidine-diphenyl ether lactone type (10αH =vertaline, 10βH=decaline); **4**: quinolizidine-metacyclophane type (lythrancines); **5**: piperidine-metacyclophane type (lythranine).

Table: Data and structures of Lythracae alkaloids.

	R^1	R^2	R^3	molecular formula	M_R	mp. [°C]	$[\alpha]_D$ ($CHCl_3$)	CAS
Lythrancine I (**4**)	α OH	H	H	$C_{27}H_{35}NO_5$	453.58	–	+65°	32209-71-5
Lythrancine II (**4**)	α OH	H	CO–CH_3	$C_{29}H_{37}NO_6$	495.61	274–275	+125°	32209-72-6
Lythrancine III (**4**)	α O–CO–CH_3	H	CO–CH_3	$C_{31}H_{39}NO_7$	537.65	134–136	+38°	32209-73-7
Lythrancine IV (**4**)	α O–CO–CH_3	CO–CH_3	CO–CH_3	$C_{33}H_{41}NO_8$	579.69	237–238	+27°	32209-74-8
Lythrancine V (**4**)	β O–CO–CH_3	CO–CH_3	CO–CH_3	$C_{33}H_{41}NO_8$	579.69	133–134	+91°	40179-98-4
Lythrancine VI (**4**)	β O–CO–CH_3	CO–CH_3	H	$C_{31}H_{39}NO_7$	537.65	oil	+25.5°	40179-99-5
Lythrancine VII (**4**)	β O–CO–CH_3	H	CO–CH_3	$C_{31}H_{39}NO_7$	537.65	oil	+101.5°	40180-00-5
Lythrine (**2**)	H	OCH_3	–	$C_{26}H_{29}NO_5$	435.52	243–245	+32.5°	5286-10-2
(–)-Decinine (13,14-Dihydro) (**2**)	H	OCH_3	–	$C_{26}H_{31}NO_5$	437.54	222–224	–142°	10183-64-9
Dehydro-decodine (**2**)	OH	H	–	$C_{25}H_{27}NO_5$	421.49	181–183		35323-19-4
Decodine (13,14-Dihydro) (**2**)	OH	H	–	$C_{25}H_{29}NO_5$	423.51	193–197	–97°	26996-01-0
Lasubine I (**1**)	–	–	–	$C_{17}H_{25}NO_3$	291.39	120.5–122	–8.8° (CH_3OH)	68622-77-5
Lasubine II (**1**)	–	–	–	$C_{17}H_{25}NO_3$	291.39	oil	–34.7° (CH_3OH)	68681-73-2

Some data for several L. a. are given in the table. The lythrancines I–VII[1] belong to structural type **4**, lythrine[2], (–)-decinine, and decodine to structural type **2**. For the synthesis of lasubines (structural type **1**), see *Lit.*[3].

Lit.: [1] J. Chem. Soc., Perkin Trans. 1 **1972**, 2141; **1973**, 297, 301, 306. [2] Manske **35**, 155–176. [3] Heterocycles **48**, 507 (1998); J. Chem. Soc., Chem. Commun. **1983**, 1143; J. Org. Chem. **60**, 717 (1995); Chem. Lett. **1985**; 1117; Tetrahedron: Asymmetry **9**, 4361 (1998); Tetrahedron Lett. **34**, 2729 (1993). *gen.:* Manske **18**, 263–322; **28**, 183–308; **35**, 215–257. – *Biosynthesis:* Mothes et al., p. 155. – *Synthesis:* Nat. Prod. Rep. **11**, 649 (1994). – *Occurrence:* Hegnauer, **IV**, 441; **VIII**, 691. – *[HS 2939 90]*

Lythrancines, lythrine see Lythraceae alkaloids.

Lyxoflavin see riboflavins.

M

Maackiains see pterocarpans.

Macarpine see benzo[*c*]phenanthridine alkaloids.

Macbecins. Ansamycin-type antibiotics from *Nocardia* sp., e.g., *M. I*: $C_{30}H_{42}N_2O_8$, M_R 558.67, yellow cryst., mp. 187–188 °C, decomp., $[\alpha]_D^{25}$ +350° (CH_3OH).

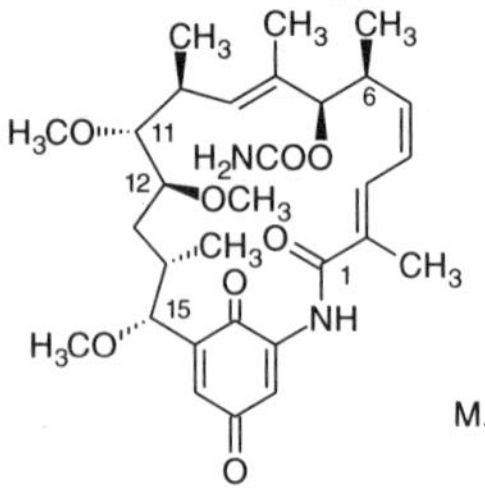

M. I

M. show antibacterial, antifungal and antineoplastic activities.

Lit.: J. Am. Chem. Soc. **120**, 4113 (1998) ▪ J. Org. Chem. **57**, 1067, 1070 (1992); **58**, 471 (1993) ▪ Tetrahedron **52**, 3229 (1996). – *[CAS 73341-72-7 (M. I)]*

Macelignan see nordihydroguaiaretic acid.

Mace oil see nutmeg oil.

Maclurin (2,3′,4,4′,6-pentahydroxybenzophenone).

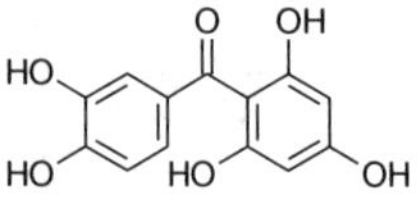

$C_{13}H_{10}O_6$, M_R 262.22, yellow needles, mp. 222–222.5 °C, soluble in hot water, ethanol, and alkalis. M. is the main pigment of the yellow heartwood of the dyer's mulberry tree (*Chlorophora tinctoria*, Moraceae).

Lit.: Karrer, No. 476 ▪ Phytochemistry **14**, 1674, 2517 (1975); **36**, 501 (1994) ▪ Rowe, p. 848. – *[CAS 519-34-6]*

Macrolactins.

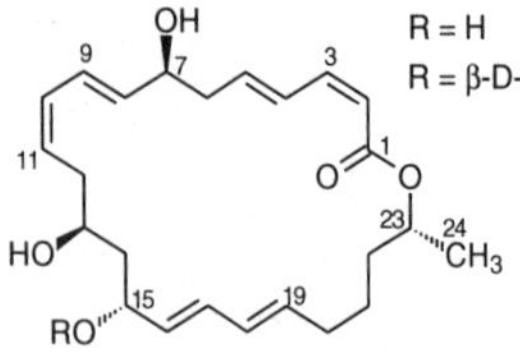

R = H : Macrolactin A
R = β-D-Glucopyranosyl : Macrolactin C

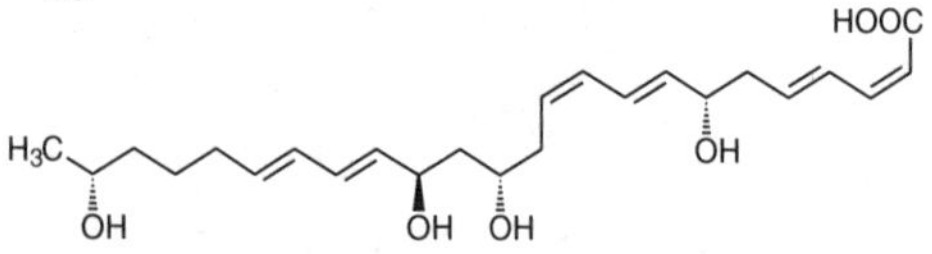

Macrolactinic acid

A group of antiviral and cytotoxic macrolides from an undefined deep-sea bacterium, isolated from sediments. In culture this bacterium produces a series of 24-membered macrocyclic lactones and open-chain compounds of acetogenic origin. The parent compound *M. A* {$C_{24}H_{34}O_5$, M_R 402.53, cryst., mp. 75–78 °C, $[\alpha]_D$ −9.6° (CH_3OH)} is active against various human pathogenic viruses such as *Herpes simplex* type I and II and protects lymphoblasts from infection by HIV viruses. Other compounds are *M. C:* $C_{30}H_{44}O_{10}$, M_R 564.67, $[\alpha]_D$ −21° (CH_3OH) and *macrolactinic acid:* $C_{24}H_{36}O_6$, M_R 420.55, $[\alpha]_D$ −13.9° (CH_3OH).

Lit.: Angew. Chem. Int. Ed. Engl. **37**, 1261 (1998) (synthesis) ▪ J. Am. Chem. Soc. **111**, 7519 (1989); **114**, 671 (1992); **118**, 13095 (1996) (synthesis, M. A); **120**, 3935–3948 (M. A, M. E, macrolactinic acid) ▪ Nachr. Chem. Tech. Lab. **38**, 159 (1990) ▪ Synth. Commun. **26**, 559 (1996) ▪ Tetrahedron Lett. **37**, 3501, 8949 (1996) (synthesis). – *[CAS 122540-27-6 (M. A); 122540-29-8 (M. C); 122540-33-4 (macrolactinic acid)]*

Macrolides (macro... = large; ...olide = lactone). The term coined by R. B. Woodward for natural compounds containing a lactone ring with more than 10 ring atoms. Many M. contain one or more O-glycosidically bound rare sugars (deoxyhexoses, aminodeoxyhexoses). The aglycone, the macrolactone ring, is formed on the *polyketide biosynthetic pathway. More than 500 compounds of this type are known, most are biologically active, and the spectrum of activity is very broad. Because of their structural diversity, the M. are separated into subgroups: The *classical M.* produced by *streptomycetes contain 12-, 14-, or 16-membered lactone rings and 1–3 sugar building blocks. Examples include *carbomycin, *erythromycin, *leucomycin, *oleandomycin which are used clinically. They have low toxicity and are principally active against Gram-positive bacteria, some also against spirochetes, Rikkettsia and/or viruses. They inhibit protein biosynthesis by binding to the 50S subunit of prokaryotic 70S ribosomes. On the other hand there are the *non-classical M.*, formed by streptomycetes and fungi with comparable ring sizes but without sugar building blocks, however, these are still biologically active (e.g., albocyclin, *soraphen(s), *brefeldin A). The *polyene M.* formed by streptomycetes contain 20–38-membered lactone rings, exhibiting a varying number of conjugated double bonds (4–7) and a poly-(1-hydroxy-1,2-ethanediyl) chain. They are often glycosylated by an amino sugar (e.g., D-mycosamine). As amphiphilic molecules, the polyene M. exhibit interactions with the cell membranes of eucaryotic cells, show significant (nephro)toxicity, and are all antifungally active; some of them are used clinically for treatment of filamentous fungal and yeast infections (e.g., *amphotericin B, *nystatin, *filipin). The *spiro-M.* group includes *avermectins and *milbemycins that are of economic importance as agents against endo- and ectoparasites in animals and as insecticides in crop protection. The *pleco-M.* contain a folded side chain as well as the lactone ring and are highly active inhibitors of ATPase (e.g., *concanamycin A, *bafilomycin A_1). In addition, there is an ever-increasing number of unusual macrolactones with up to 60-membered rings (quinolidomycin A_1) originating from various organisms including marine species. In the stricter sense, the *macrodiolides* (e.g., *elaiophylin, grahamimycin, *antimycins, *boromycin) and *macrotetrolides* (e.g., *nonactin), built up of repeating subunits, should be distinguished from the M. Natural compounds containing additional lactam bonds do not belong to the M., e.g., the immunomodulators *rapamycin and *FK-506 (tsukubaenolide).

Lit.: Angew. Chem. Int. Ed. Engl. **30**, 1302 (1991) ▪ Annu. Rev. Microbiol. **49**, 201–238 (1995) ▪ Chem. Biol. **2**, 355–362 (1995) ▪ Justus Liebigs Ann. Chem. **1994**, 305–312 ▪ Lukacs, Recent Progress in the Chemical Synthesis of Antibiotics and Related Microbial Products, p. 1–120, Berlin: Springer 1993 ▪ Med. Res. Rev. **19**, 543–558 (1999) (genetic engineering) ▪ Omura (ed.), Macrolide Antibiotics, New York: Academic Press 1984 ▪ Pape & Rehm (eds.), Biotechnology, vol. 4, p. 359–391, 431–463, Weinheim: VCH Verlagsges. 1986.

Macromerine see *β*-phenylethylamine alkaloids.

Macronine see Amaryllidaceae alkaloids.

Macrozamin see cycasin.

Mactraxanthin [(3*S*,3′*S*,5*R*,5′*R*,6*R*,6′*R*)-5,5′,6,6′-tetrahydro-*β*,*β*-carotene-3,3′,5,5′,6,6′-hexaol].

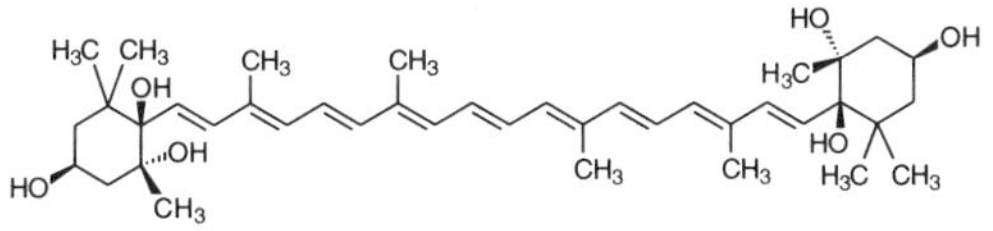

$C_{40}H_{60}O_6$, M_R 636.91, dark orange needles, mp. 232–233 °C. M. is a *carotinoid from the edible Japanese mussel *Mactra chinensis* living in the tidal zone.

Lit.: Helv. Chim. Acta **67**, 2043 (1984) (synthesis) ▪ Nachr. Chem. Tech. Lab. **31**, 165 (1983) ▪ Tetrahedron Lett. **24**, 911 (1983). – *[CAS 86105-69-3]*

Maculotoxin see tetrodotoxin.

Madder (dyer's madder, *Rubia tinctorum*, Rubiaceae). M. was cultivated in the Mediterranean region in antiquity and is also widely distributed in East Asia and North America, it is a 50–80 cm high, perennial *dye plant with fleshy-yellow rhizomes that turn red on drying. The main colored components of M. are *ruberythric acid, the primveroside of rubiadin (1,3-dihydroxy-2-methyl-anthraquinone), and purpurin-3-carboxylic acid (pseudopurpurin, 1,3,4-trihydroxy-anthraquinone-2-carboxylic acid) as well as *alizarin(e), rubiadin, *purpurin, and munjistin (1,3-dihydroxyanthraquinone-2-carboxylic acid). M. was used in antiquity by the advanced civilizations (Egyptians, Greeks, Romans, Indians, Persians, and Turks) to dye wool, silk, and in particular for dying carpets[1]. For centuries, M. – as well as *indigotin – was the most important dye. At the end of the 16th century M. was introduced to Germany from Italy and later cultivated, e.g., in France and Holland in large amounts for dyeing purposes. Apart from M., other *Rubia* and *Galium* species as well as *Asperula tinctoria* (Rubiaceae) belong to the dyer's plants containing alizarine and/or pseudopurpurin or purpurin as the main pigments.

Lit.: [1] Chem. Ind. (London) **1994**, 28.
gen.: Dragoco-Rep. **25**, 21–27 (1978). – *[HS 1404 10]*

Mad honey. Common name for *Rhododendron* honey containing tetracyclic diterpene toxins (*grayanotoxins) that can lead to *atropine-like poisonings.

Madindolines. Mixture of indolines from a *Streptomyces* sp., $C_{22}H_{27}NO_4$, M_R 369.46, amorphous solid.

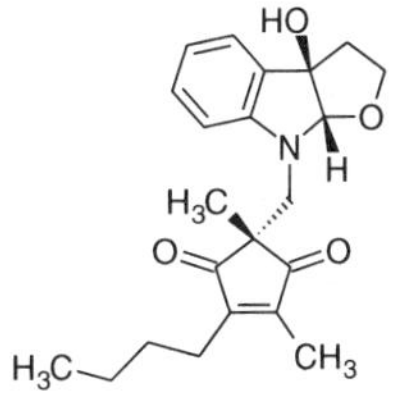

(+)-M. A (absolute configuration)

M. are specific inhibitors of IL-6 activity and show antitumor potential against myelomas and sarkomas.

Lit.: Immunity **5**, 449 (1996) ▪ J. Am. Chem. Soc. **122**, 2122 (2000) (stereochemistry, synthesis) ▪ J. Antibiot. **49**, 1091 (1996); **50**, 1069 (1997). – *[CAS 184877-65-4]*

Maduramicin *α*.

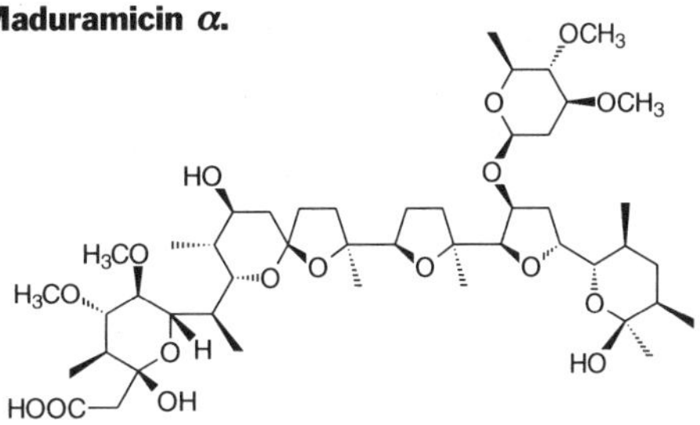

$C_{47}H_{80}O_{17}$, M_R 917.14, Na salt: mp. 193–194 °C, $[\alpha]_D^{25}$ +40.6° ($CHCl_3$). Ionophoric (Na^+), glycosylated *polyether antibiotic from *Nocardia* strains and *Actinomadura yumaense* with activity against nematodes, coccides, and malaria pathogens. M. *α* is used in animal breeding as a growth promoter and coccidiostatic agent. M. *α* is formed biosynthetically by the *polyketide pathway, with epoxidation of the intermediate double bonds and subsequent cyclization. M. *α* should not be confused with *maduramycin*[1] [$C_{28}H_{24}O_{10}$, M_R 520.49], a quinone antibiotic isolated from *Actinomadura rubra*.

Lit.: [1] Z. Allg. Mikrobiol. **18**, 389, 603 (1978).
gen.: J. Antibiot. **36**, 343 (1983); **40**, 94 (1987). – *Biosynthesis:* Birch & Robinson, in Vining & Stuttard (eds.), Genetics and Biochemistry of Antibiotic Production, p. 443–476, Boston: Butterworth-Heinemann 1995 ▪ J. Chem. Soc., Perkin Trans. 1 **1988**, 3195–3207. – *Review:* Zechmeister **58**, 1–82. – *Activity:* Antimicrob. Agents Chemother. **39**, 854 (1995); Vet. Res. Commun. **16**, 45 (1992). – *[CAS 79356-08-4 (M. α); 61991-54-6 (maduramycin)]*

Magainins.

H-Gly-Ile-Gly-Lys-Phe-Leu-His-Ser-Ala-Gly(10)-Lys-Phe-Gly-
Lys-Ala-Phe-Val-Gly-Glu-Ile-Met-Lys(22)-Ser-OH

M. I

M. II = Lys in position 10, Asn in position 22

Name for peptides (M_R ≤2500) derived from the Hebrew word for protection. The first chemical defence system against infections discovered in vertebrates, besides the immune system. The skin of the South African frog *Xenopus laevis* contains at least ten structurally similar, antimicrobially active peptides that are membrane active (disruption of bacterial cell walls). Analogues of M. are undergoing clinical testing against bacterial skin infections. Magainin Pharmaceuticals Locilex® cream (Cytolex®) (pexiganan acetate) for the treatment of infections in diabetic food ulcers is in the registration process. Locilex is a 22 amino acid analogue of magainin II.

Lit.: Annu. Rev. Biochem. **59**, 395–414 (1990) ▪ Biochem. Pharmacol. **39**, 625 (1990) ▪ Biol. Pharm. Bull. **20**, 267 (1997) ▪ Chem. Lett. **1989**, 749 (synthesis) ▪ J. Biol. Chem. **266**, 19851 (1991) ▪ Nature (London) **328**, 478 (1987); **329**, 494 (1987). – *[CAS 113041-69-3; 108433-99-4 (M. I); 108433-95-0 (M. II)]*

Magnamycin see carbomycin.

Magnesidin.

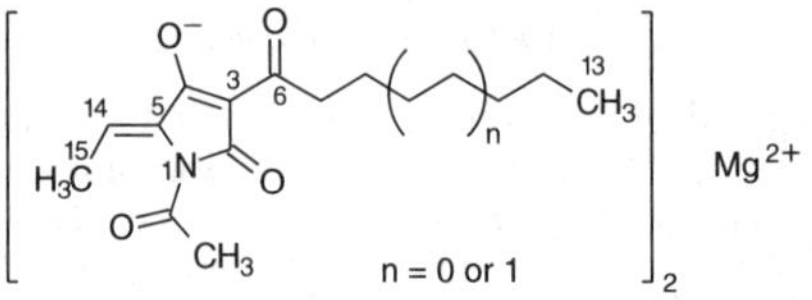

A (1:1) mixture of the colorless Mg^{2+} complex with 3-hexanoyl- and 3-octanoyltetramic acid derivatives as chelators produced by the marine bacterium *Pseudomonas magnesiorubra*. M. belongs to the large group of *tetramic acid antibiotics and is active against Gram-positive bacteria. Homogeneous M. A (n = 1), $C_{32}H_{44}MgN_2O_8$, M_R 609.01, and its magnesium-free form have been isolated from *Vibrio gazogenes*.

Lit.: J. Antibiot. **26**, 797 (1973); **47**, 257–261 (1994) ▪ Tetrahedron Lett. **1974**, 983. – *[HS 2941 90; CAS 52081-52-4]*

Magnesium *trans*-aconitate see leaf-movement factors.

Magnoflorine see aporphine alkaloids.

Magnolioside see 6,7-dihydroxycoumarins.

Maitansinoids see maytansinoids.

Maitotoxin (MTX).

$C_{164}H_{256}Na_2O_{68}S_2$, M_R 3425.90 (Na salt). M.[1] is the strongest known non-proteinogenic poison, LD_{50} (mouse i.p.) 50 ng/kg. 25 mg M. are isolated from 5000 L culture filtrate of the dinoflagellate *Gambierdiscus toxicus* (strain GII-1). M. activates phospholipases A2 and C. The ion channel stimulated by M. is selective and allows passage of more Ca^{2+} ions than Na^+ ions; ratio 50 : 1[2]. Related compounds giving rise to ciguatera type symptoms (see ciguatera toxin) from *G. toxicus* are gambierol[3] and the gambieric acids[4], with antifungal activities 2000-fold more potent than those of *amphotericin B.

Lit.: [1] Angew. Chem. Int. Ed. Engl. **35**, 1672, 1675 (1996) (absolute configuration); J. Am. Chem. Soc. **114**, 6594 (1992); **115**, 2060 (1993); Tetrahedron Lett. **36**, 9007, 9011 (1995). [2] Biochem. Pharmacol. **39**, 1633 (1990); Mol. Pharmacol. **41**, 487 (1992). [3] J. Am. Chem. Soc. **115**, 361 (1993). [4] J. Am. Chem. Soc. **114**, 1102 (1992); J. Org. Chem. **57**, 5448 (1992).
gen.: Chem. Rev. **93**, 1897–1909 (1993) ▪ Gazz. Chim. Ital. **123**, 367 (1993) ▪ J. Am. Chem. Soc. **118**, 7946 (1996); **119**, 7928 (1997) (absolute configuration) ▪ Nachr. Chem. Tech. Lab. **41**, 173 (1993); **44**, 696 (1996). – *[CAS 59392-53-9 (M.); 131594-69-9 (Na salt)]*

Majusculamides. The diastereomeric *M. A* [$C_{28}H_{45}N_3O_5$, M_R 503.68, mp. 96 °C, $[\alpha]_D^{26}$ +19.3° (C_2H_5OH)] and *M. B* [mp. 102 °C, $[\alpha]_D^{26}$ +14.6° (C_2H_5OH)] as well as *M. C* [(lyngbyastatin, $C_{50}H_{80}N_8O_{12}$, M_R 985.23, $[\alpha]_D$ −96° (CH_2Cl_2) cytotoxic and with activity against phytopathogenic fungi] are lipopeptides or, respectively, cyclodepsipeptides from the marine cyanobacterium *Lyngbya majuscula*, see also microcolin A. *Malyngamide A*[2] {oil, $C_{29}H_{45}ClN_2O_6$, M_R 553.14, $[\alpha]_D$ −6.5°} has a characteristic polyketide chain with an *O*-methyl-*tetramic acid residue. Lyngbyatoxins are the toxins from *L. majuscula*, e.g., *lyngbyatoxin A*[3] ($C_{27}H_{39}N_3O_2$, M_R 437.62) and are closely related to *teleocidin B[4] from *Streptomyces* sp. The activities as tumor promotors are comparable with those of phorbol esters (12-*O*-tetradecanoylphorbol 13-acetate, TPA type, antagonist: *sarcophytol A). *L. majuscula* also contains the antibacterially active *malyngolide*[5], $C_{16}H_{30}O_3$, M_R 270.41, cryst., mp. 36 °C, $[\alpha]_D^{20}$ −13° ($CHCl_3$). *Debromoaplysiatoxin*[6] ($C_{32}H_{48}O_{10}$, M_R 592.73, mp. 105.2–107 °C, $[\alpha]_D$ +60.6°, needles) was first isolated from the slug

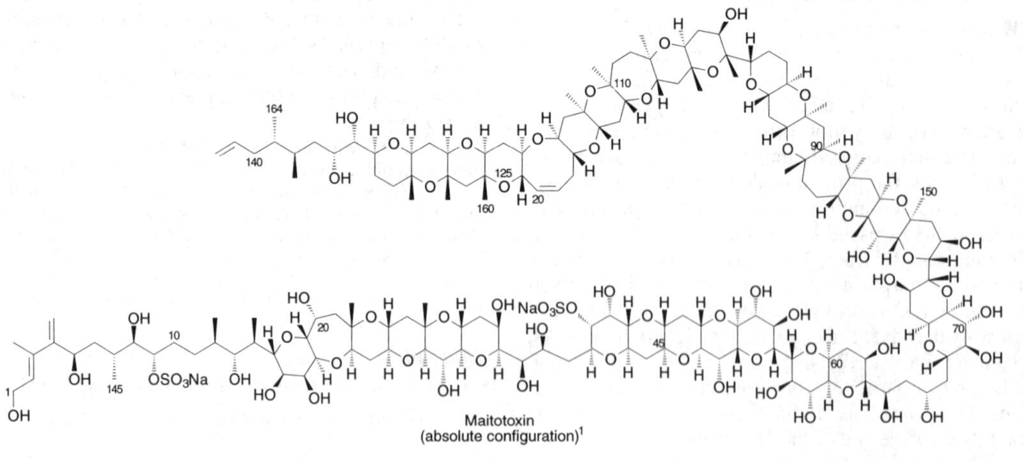

Maitotoxin
(absolute configuration)[1]

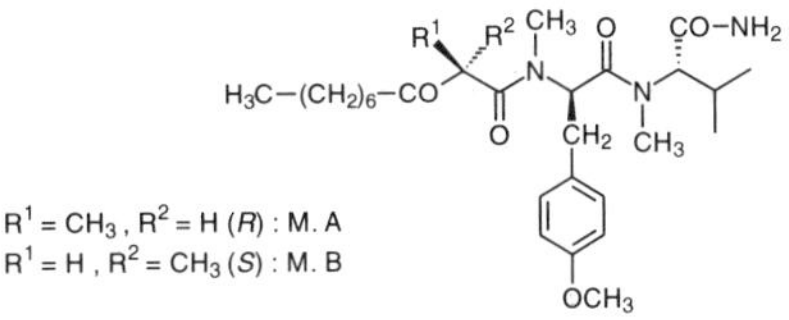

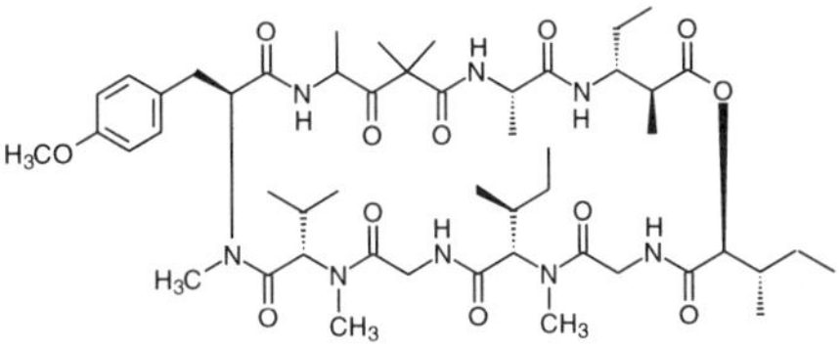
M. C

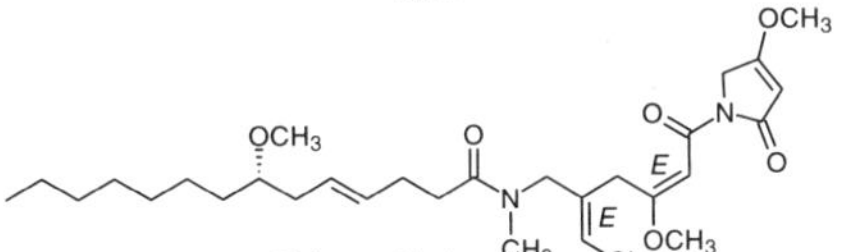
Malyngamide A

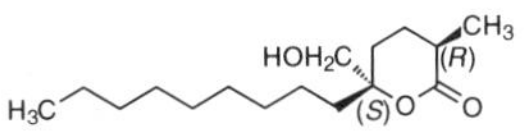
Malyngolide

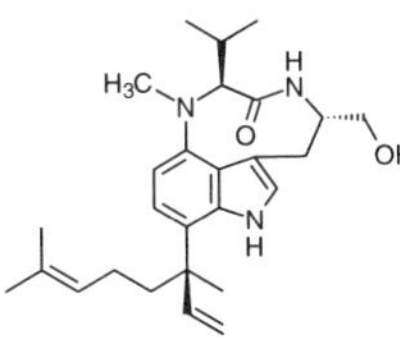
Lyngbyatoxin A

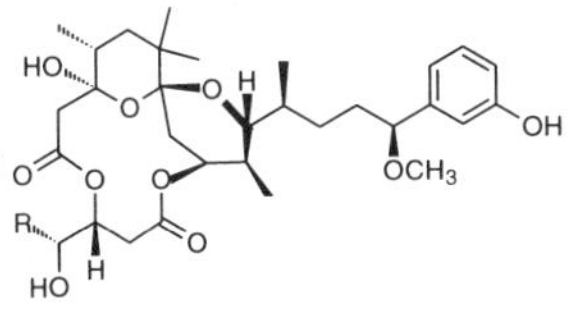
R = CH3 : Debromoaplysiatoxin
R = H : Oscillatoxin A

Aplysia (see aplysi...), but could then be traced back to *L. majuscula.* It causes strong inflammations and is responsible for severe contact dermatitis in swimmers who come into contact with this alga ("swimmer's itch"); tumor promoter (TPA type). The closely related *oscillatoxin A*[7] [$C_{31}H_{46}O_{10}$, M_R 578.70, $[\alpha]_D$ +67° (C_2H_5OH)] occurs in the marine form of the cyanobacterium *Oscillatoria* sp. which also lives in fresh water.

Lit.: [1] J. Nat. Prod. **56**, 545 (1993); **61**, 1221 (1998) (M. C); J. Org. Chem. **49**, 236 (1984). [2] J. Am. Chem. Soc. **101**, 240 (1979); Tetrahedron Lett. **36**, 7837 (1995). [3] Tetrahedron **46**, 2773 (1990) (biosynthesis); **47**, 8535, 8545, 8559 (1991) (synthesis). [4] Tetrahedron Lett. **27**, 5219 (1986). [5] Heterocycles **48**, 1213 (1998); Tetrahedron **54**, 13087 (1998); Justus Liebigs Ann. Chem. **1996**, 147 (synthesis); J. Chem. Soc., Chem. Commun. **1997**, 1755; J. Chem. Soc., Perkin Trans. 1 **1998**, 293; Tetrahedron **52**, 5805–5818 (1996); **54**, 10329–10338 (1998) (synthesis); Tetrahedron Lett. **38**, 6023 (1997). [6] J. Org. Chem. **49**, 2484 (1984); Cancer Res. **43**, 1529 (1983); **50**, 7085 (1990); Gazz. Chim. Ital. **123**, 309 (1993). Tetrahedron **49**, 10531 (1993). [7] J. Org. Chem. **43**, 2301 (1978); **49**, 2484 (1984); Scheuer I **1**, 83; **2**, 374; Scheuer II **1**, 154, **3**, 7; Chem. Rev. **93**, 1685 (1993). – *[CAS 52423-28-6 (debromoaplysiatoxin); 70497-14-2 (lyngbyatoxin A); 62758-06-9 (M. A); 62840-08-8 (M. B); 83712-17-8 (M. C); 69778-82-1 (malyngamide A); 66671-95-2 (oscillatoxin A)]*

Makisterones see ecdysteroids.

Malformins. Mixtures of cyclic pentapeptides with phytotoxic activity formed by fungi of the *Aspergillus niger* group. The main component is M. A_1; M. C, M. A_3, and M. B_{1b} are identical.

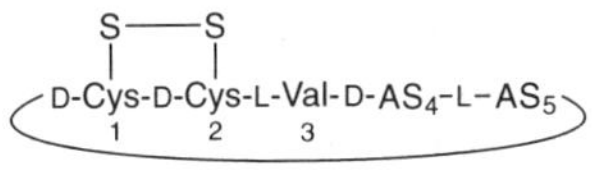

Table: Data of Malformins.

	D-AS4	L-AS5	molecular formula	MR	CAS
M. A_1	Ile	Leu	$C_{23}H_{39}N_5O_5S_2$	529.71	3022-92-2
M. A_2	Val	Leu	$C_{22}H_{37}N_5O_5S_2$	515.69	83680-20-0
M. A_3 = M. C	Leu	Leu	$C_{23}H_{39}N_5O_5S_2$	529.71	59926-78-2

M. cause malformations of bean plants and root curling in corn.

Lit.: Biosci. Biotech. Biochem. **57**, 240, 787 (1993) ▪ Cole-Cox, p. 670–682 ▪ Tetrahedron Lett. **31**, 4337 (1990); **32**, 6715 (1991) ▪ Turner **1**, 339; **2**, 431, 436.

Malic acid (hydroxysuccinic acid).

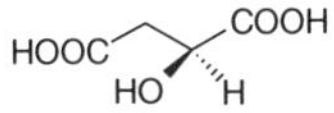

$C_4H_6O_5$, M_R 134.09. The racemate melts at 131–132 °C. The rarely occurring (*R*)-(+) form, $[\alpha]_D$ +5.2° (acetone), forms crystals of mp. 101 °C that are soluble in water, alcohol, and acetone. It is isolated from *Hibiscus sabdariffa*. The (*S*)-(–) form, $[\alpha]_D^{21}$ –2.3° (H_2O), pK_{a1} 3.46, pK_{a2} 5.10, forms deliquescent crystals, mp. 100 °C (decomp.), which are readily soluble in water and alcohol but poorly soluble in ether. LD_{50} 1600 mg/kg (rat p.o.). M. irritates the skin and eyes. (*S*)-(–)-M. was first isolated from apple juice by Scheele in 1785 and its constitution elucidated by Liebig in 1832. It is also found in the free state in barberries, quinces, gooseberries, grapes, and rowanberries. In the human organism it is an intermediate in the citric acid cycle and in gluconeogenesis.

Use: The racemate is used to impregnate packing materials for cheese and other foodstuffs (against mold fungi) and, in place of citric acid, as souring agent for bread and confectionery as well as beverages. The enantiomers find use in racemate separation.

Lit.: Acta Chem. Scand. **26**, 2349–2359 (1972) (abs. configuration) ▪ Acta Crystallogr., Sect. C **45**, 1406–1408 (1989) (crystal structure) ▪ Beilstein E IV **3**, 1123–1125 ▪ Bergmeyer, Methods of Enzymatic Analysis (3.), vol. 7, p. 39–59, Weinheim: Verl. Chemie 1985 (Detection) ▪ Luckner (3.), p. 126–128, 134–135. – *Toxicology:* Sax (8.), MAN 000. – *[HS 2918 19; CAS 636-61-3 (R)-(+); 97-67-6 (S)-(–)]*

Maltol (3-hydroxy-2-methyl-4*H*-pyran-4-one).

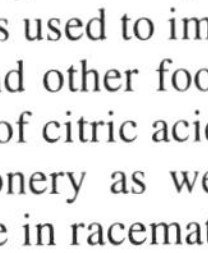

$C_6H_6O_3$, M_R 126.11; crystals with a sweet-fruity caramel odor, mp. 164 °C (subl. at 93 °C); poorly soluble in water and alcohol. LD_{50} (rat p.o.) 1.4 g/kg. M. is formed from *lactose or maltose by the Maillard reaction[1], it was recognized as an odor substance in malt in 1894[2] and occurs, e.g., in *bread, *coffee, *cocoa, *meat and sake aroma[3], as well as in larch bark and pine needles. M. can be obtained from the latter by extraction[4], but is mainly obtained from birch tar oil, molasses, or is prepared synthetically[5]. M. is used in flavors for caramel notes and to intensify tastes, e. g., in *fruit flavors (in which it is active even below the olfactory threshold of ca. 5 ppm[6]) or in lemonades as a partial substitute for sugar[7]. The world market in 1997 amounted to ca. 150 t.

Lit.: [1] Hopp & Mori (eds.), Recent Developments in Flavor and Fragrance Chemistry, p. 172, Weinheim: VCH Verlagsges. 1993. [2] Ber. Dtsch. Chem. Ges. **27**, 806 (1894). [3] TNO list (6.), Suppl. 5, p. 501. [4] US. P. 5 221 756 (1993), Florasynth Inc. [5] J. Org. Chem. **45**, 1109 ff. (1980); Org. Prep. Proced. Int. **24**, 95 (1992) (synthesis). [6] Arctander, No. 1831. [7] Ullmann (5.) **A 11**, 205.
gen.: J. Nat. Prod. **60**, 472 (1997) (isolation) ▪ Bauer et al. (2.), p. 115 f. ▪ Beilstein E V **18/1**, 114 ▪ Ohloff, p. 38 f. – *[HS 2932 90; CAS 118-71-8]*

Malvalic acid (2-octyl-1-cyclopropene-1-heptanoic acid).

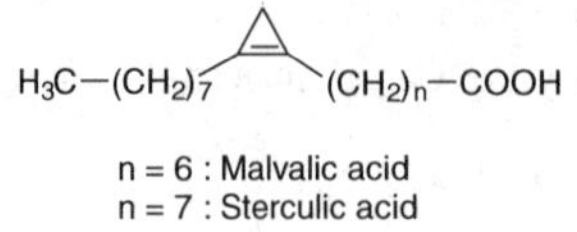

$C_{18}H_{32}O_2$, M_R 280.45, mp. 10.3–10.5 °C. M. and *sterculic acid* [$C_{19}H_{34}O_2$, M_R 294.48, mp. 18.2–18.3 °C] are components of the triglycerides in the seeds of *cotton (*Gossypium hirsutum*) and other Malvaceae. M. and sterculic acid are toxic and carcinogenic fatty acids which, however, are converted to non-toxic products during the hardening (hydrogenation) of cotton seed oil.

Lit.: Bull. Chem. Soc. Jpn. **64**, 3084 (1991) ▪ Gunston, Harwood & Padley, The Lipid Handbook, 2nd. edn., p. 13 f., 51 f., 64, London: Chapman & Hall 1994 ▪ Toxicol. Appl. Pharmacol. **84**, 3 (1986). – *[CAS 503-05-9 (M.); 738-87-4 (sterculic acid)]*

Malvidin (3′,5′-di-*O*-methyldelphinidin, 3,4′,5,7-tetrahydroxy-3′,5′-dimethoxyflavylium). Formula see anthocyanins. $C_{17}H_{15}O_7^+$, M_R 331.30; the chloride ($C_{17}H_{15}ClO_7$, M_R 366.75, mp. >300 °C) forms dark brown prisms or needles; see also anthocyanins.

Lit.: Karrer, No. 1746 ▪ Tetrahedron **39**, 3005 (1983). – *[CAS 643-84-5 (chloride)]*

Malvin (primulidin, *malvidin 3,5-di-*O*-glucoside). $C_{29}H_{35}O_{17}^+$, M_R 655.59, mp. of the chloride 165 °C (decomp.), reddish brown prisms and needles with green luster. Widespread occurrence as flower pigments.

Lit.: Karrer, No. 1749 ff. ▪ Phytochemistry **33**, 1227 (1993). – *[CAS 16727-30-3 (chloride)]*

Malyngamide A, malyngolide see majusculamides.

Mammalian alkaloids. The few *alkaloids detected in mammals, including humans, usually originate biogenetically from dopamine and tryptamine; e. g., tetrahydroisoquinolines such as salsolinol and 3-carboxysalsolinol (the occurrence of *morphine as an endogenous opiate has been proposed), *harmans and other β-carbolines. Isatin has been detected in human urine.

Lit.: Manske **43**, 119–185.

Mammantaquinone see avarol.

Mancinellin. $C_{36}H_{52}O_8$, M_R 612.80, ester of 6,7-epoxy-4,5,9,13,20-pentahydroxy-1-tiglien-3-one, a diterpene of the *tigliane type. M. can be isolated from the *Hippomane mancinella* (Euphorbiaceae) growing in

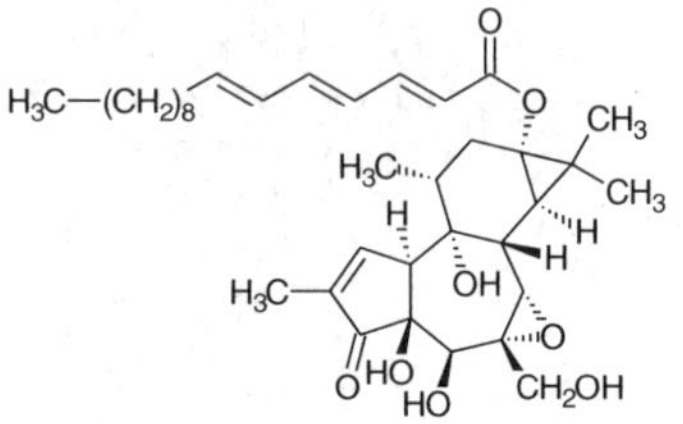

Central America and the West Indies. M. is carcinogenic.

Lit.: Aust. J. Chem. **32**, 2459 (1979) ▪ Beilstein E V **18/5**, 421 ▪ Planta Med. **39**, 3 (1983) ▪ Tetrahedron Lett. **1975**, 1587. – *[CAS 57672-76-1]*

Mandaratoxin. A *wasp venom from *Vespa mandarina* structurally related to *mastoparan.

Lit.: Biochemistry **21**, 1693–1697 (1984) ▪ Shinkei Kenkyu No Shimpo **35**, 584–590 (1991). – *[CAS 82545-77-5]*

Mandarin oil. Depending on the degree of ripeness of the fruits, a greenish to red-orange oil with a bluish fluorescence. A sparkling, fresh, slightly sweet, typical mandarin odor, with a characteristic sweet-sour taste.
Production: By cold pressing (see citrus oils) of the peel of the mandarin *Citrus reticulata. Origin:* Mainly Italy, besides Spain, Argentina, and Brazil.
Composition[1]: Main components are *limonene* (ca. 70%) and *γ-terpinene* (see *p*-menthadienes) (ca. 20%). Trace components such as *decanal* (see alkanals) (0.08%), *α-*sinensal* (0.3%), *thymol* (see cymenols) (0.06%), *2-dodecenal* (see alkenals) (0.02%) as well as *methyl N-methylanthranilate* (ca. 0.5%), which causes the blue fluorescence, are important for the organoleptic impression.
Use: In the perfume industry, principally in fresh Eaux de Toilette; to improve the aromatic character of foods such as refreshing beverages, liquors, confectionery, etc., usually in combination with orange flavor.

Lit.: [1] Perfum. Flavor. **16** (1), 53 (1991); **17** (4), 53 (1992); **19** (6), 29 (1994); Flavour Fragr. J. **9**, 105 (1994); **10**, 33 (1995).
gen.: Arctander, p. 394 ▪ Bauer et al. (2.), p. 148 ▪ ISO 3528 (1977) ▪ Ohloff, p. 133. – *Toxicology:* Food Chem. Toxicol. **30** (Suppl.), 69S (1992). – *[HS 3301 19; CAS 8008-31-9]*

Mango aroma see fruit flavors.

Manicone [(4*E*,6*S*)-4,6-dimethyl-4-octen-3-one].

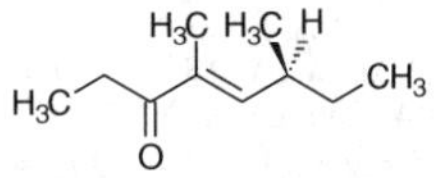

$C_{10}H_{18}O$, M_R 154.25, $[\alpha]_D^{20}$ +38.8° (CH_2Cl_2). An alarm *pheromone of ants of the genus *Manica*. The corresponding methyl ketone *normanicone* ($C_9H_{16}O$, M_R 140.23) and the 4,6-dimethyl-4-nonene *homomanicone* ($C_{11}H_{20}O$, M_R 168.28) have also been identified, see also ants.

Lit.: ApSimon **9**, 179 ff. (synthesis) ▪ Justus Liebigs Ann. Chem. **1988**, 55–60 ▪ Synthesis **1991**, 560 ▪ Tetrahedron **52**, 12815 (1996) (synthesis (*S*)-(+)-M.). – *[CAS 59686-13-4]*

Maniladiol see oleanane.

Maniwamycins.

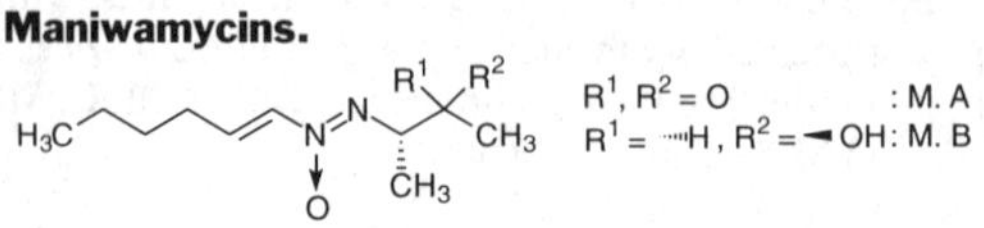

The *M. A* {$C_{10}H_{18}N_2O_2$, M_R 198.27, oil, $[\alpha]_D$ −144° ($CHCl_3$)} and *M. B* {$C_{10}H_{20}N_2O_2$, M_R 200.28, oil, $[\alpha]_D$ +108° ($CHCl_3$)} are antifungal azoxy compounds from cultures of *Streptomyces prasinopilosus*.
Lit.: J. Antibiot. **42**, 1535, 1541 (1989) ▪ Tetrahedron Lett. **32**, 1067 (1991); **34**, 6095 (1993) (synthesis). – *[HS 2941 90; CAS 122566-70-5 (M. A); 122547-71-1 (M. B)]*

Mannans. Group name for glycans made up of *mannose units. In higher terrestrial plants such as, e.g., the ivory nut – *Phytelephas macrocarpa* (Arecaceae), linear β1-4-M. occurs as a practically pure, amorphous solid, $[\alpha]_D^{20}$ −42.5° (NaOH). Historically, M. were important for button production. In contrast, yeast mannans are highly branched and contain α-glycosidic bonds, e.g., baker's yeast contains α1-6-, α1-2-, and α1-3-bonds in a ratio of ca. 2:3:1. Mannopyranose building blocks are present in all M. *Glucomannan*, possibly with *O*-acetyl and galactosyl side chains (*galactoglucomannan*), frequently occurs in yeasts, yeast extract, and orchid tubers. *Galactomannans* from the fruits of the carob tree and from guar flour are used as thickeners in the food industry and as additives for tablet production in the pharmaceutical industry. They consist of β1-4-M. chains carrying short, α1-6-linked galactopyranoses. Hydrolysis of the M. furnishes mannose.
Lit.: Adv. Carbohydr. Chem. Biochem. **31**, 241–312 (1975); **33**, 398 (1976); **35**, 341–376 (1978); **41**, 68–104 (1983) ▪ Pharm. Ind. **42**, 1292 ff. (1980); **43**, 570 f., 672 ff., 1238 ff. (1981). – *[CAS 37251-47-1 (ivory nut M.); 11078-31-2 (gluco-M.); 11078-30-1 (galacto-M.)]*

Mannitol (D-mannitol).

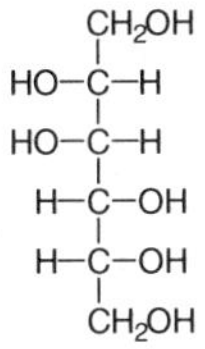

$C_6H_{14}O_6$, M_R 182.17, sweet tasting crystals, mp. 166–168 °C, bp. 290–295 °C (0.47 kPa), $[\alpha]_D^{20}$ −0.5° (H_2O), $[\alpha]_D$ +28.61° (12.8% borax solution). The sugar alcohol M. belongs to the *hexitols. It occurs in numerous plants and is the main component (40–60%) of manna. In summer, brown algae (e.g., *Laminaria cloustoni*) consist of up to 40% dry weight of mannitol. Marine kelps form M. as an early product of photosynthesis. M. is stored in algae, fungi, and lichens as a reserve substance [1]. Coccidia (parasitic microorganisms) possess an M. cycle for their energy supply [2].
Use: As additive for soldering, for electrolyte capacitors, for the production of synthetic resins, as filler in the pharmaceutical industry [2], in nutrient broths for bacteria, as reagent for boric acid titrations, for the synthesis of the vasodilatory mannitol hexanitrate, as starting reagent for asymmetric synthesis [2], as sweetener [3], laxative, for prophylaxis of anuria, in the food industry for maintaining the powder form of hygroscopic substances (M. is not hygroscopic), as lubricant, and stabilizer.
Lit.: [1] Physiol. Veg. **23**, 95–106 (1985). [2] Parasitol. Today **5**, 205–208 (1989). [3] Dev. Sweeteners **2**, 1–25 (1983); Food Chem. **16**, 231–241 (1985); Food Sci. Technol. **17**, 165–183 (1986).
gen.: Hager (5.) **8**, 812 ff. ▪ Ullmann (5.) **A 5**, 90. – *Synthesis:* Collins-Ferrier, p. 53 f. ▪ J. Chem. Soc., Chem. Commun. **1980**, 930 ▪ Starch (Stärke) **37**, 136–141 (1985). – *[HS 1702 90, 2905 43; CAS 69-65-8]*

Mannojirimycin (5-amino-5-deoxy-D-mannose).

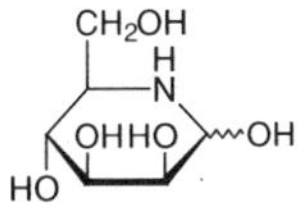

$C_6H_{13}NO_5$, M_R 179.17. A mannosidase inhibitor isolated, together with *deoxymannojirimycin, from cultures of *Streptomyces lavendulae*. Both compounds exhibit antiviral activity.
Lit.: J. Org. Chem. **58**, 985 (1993) ▪ Tetrahedron **49**, 2939 (1993) (synthesis). – *[CAS 62362-63-4]*

Mannosamine (2-amino-2-deoxy-D-mannose).

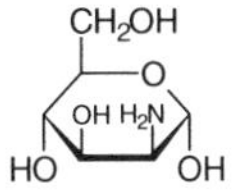

α-pyranose form

$C_6H_{13}NO_5$, M_R 179.17; D-M. · HCl: mp. 178–180 °C, $[\alpha]_D^{20}$ −3° (H_2O). M. occurs in animal mucolipids and proteins. *N*-Acetylmannosamine (ManNAc) is obtained on preparative scale by the enzymatic isomerization of *N*-acetylglucosamine (GlcNAc) with GlcNAc 2-epimerase (EC 5.1.3.8).
Lit.: Angew. Chem. Int. Ed. Engl. **30**, 827 f. (1991) ▪ Carbohydr. Res. **175**, 311 (1988). – *[HS 2940 00; CAS 14307-02-9 (D-M.); 2636-92-2 (racemic M.)]*

Mannose (D-mannose, seminose, carubinose, abbreviation: Man).

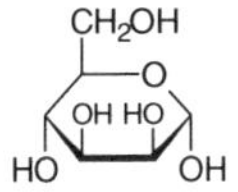

α-form

$C_6H_{12}O_6$, M_R 180.16, sweet tasting powder. M. exists in the α-pyranose {mp. 133 °C, $[\alpha]_D^{20}$ +30→+15° (H_2O)} and the β-pyranose forms {mp. 132 °C, $[\alpha]_D^{20}$ −16→+15° (H_2O)}, it reduces Fehling's solution. M. is fermented by baker's yeast. M. belongs to the hexoses and is the 2-epimer of D-*glucose. M. only occasionally occurs in the free form (e.g., in orange peel), but frequently in the glycosidically bound form and is widely distributed in complex carbohydrates, the so-called *mannans (e.g., in the ivory nut, carob tree, lucern seeds, guar flour, orchid tubers, and sea kelp). M. is toxic to bees [1].
Lit.: [1] Science **131**, 297f. (1960).
gen.: Beilstein E IV **1**, 4328 ▪ Bull. Chem. Soc. Jpn. **66**, 2268 (1993) (synthesis) ▪ Collins-Ferrier, p. 53 f. (synthesis) ▪ J. Biol. Chem. **250**, 8069 (1975) ▪ Karrer, No. 612 ▪ Merck-Index (12.), No. 5791 ▪ Science **220**, 949 (1983). – *[HS 2940 00; CAS 31103-86-3 (M.); 3458-28-4 (D-M.); 7296-15-3 (α-pyranose form); 7322-31-8 (β-pyranose form)]*

Mannosidase inhibitors see glycosidase inhibitors.

Mannostatin A and B see glycosidase inhibitors.

Manoalide. $C_{25}H_{36}O_5$, M_R 416.56, amorphous solid, as 25-acetate: mp. 117–119 °C. A sesterterpenoid marine natural product from the sponge *Luffariella variabilis*, the name is derived from the Manoa valley, Oahu, Ha-

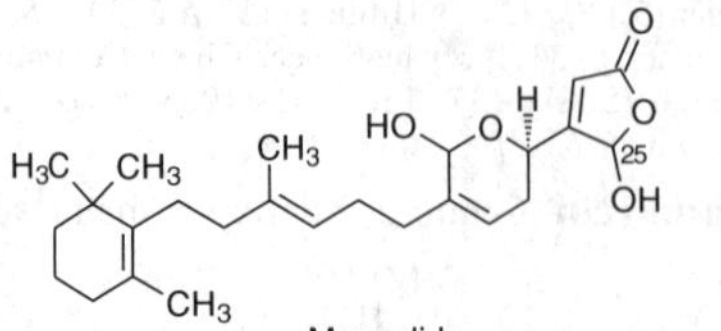

Manoalide

waii. M. has anti-inflammatory, analgesic, and immunosuppressive activities. M. irreversibly inhibits the enzyme phospholipase A2 and the liberation of calcium ions in various types of cells[1]. It is currently undergoing clinical testing for treatment of skin inflammations[2]. Dehydrosecomanoalide, an artefact formed during the isolation of M., has potent immunosuppressive activity. The hydrophilic core of M. can be considered bioisosteric with the tetracyclic steroidal skeleton[3].

Lit.: [1]Biochem. Pharmacol. **37**, 2899, 3639 (1988); Biochim. Biophys. Acta **917**, 258 (1987); J. Am. Chem. Soc. **114**, 5093 (1992); J. Biol. Chem. **264**, 8520 (1989); Mol. Pharmacol. **36**, 782 (1989). [2]Drugs of the Future **15**, 460 ff. (1990); J. Med. Chem. **41**, 3232 (1998). [3]Synthesis **1998**, 1367.

gen.: Chem. Commun. **1997**, 1139 (synthesis) ▪ Chem. Pharm. Bull. **42**, 265 (1994) (absolute configuration) ▪ Curr. Med. Chem. **6**, 415–431 (1999) (review) ▪ J. Nat. Prod. **51**, 326 (1988) ▪ Synthesis **1996**, 171–177 ▪ Tetrahedron **41**, 981 (1985); **50**, 8793 (1994) ▪ Tetrahedron: Asymmetry **10**, 4481 (1999) ▪ Tetrahedron Lett. **29**, 1173, 2401 (1988) (synthesis). – *[CAS 75088-80-1]*

Manumycins.

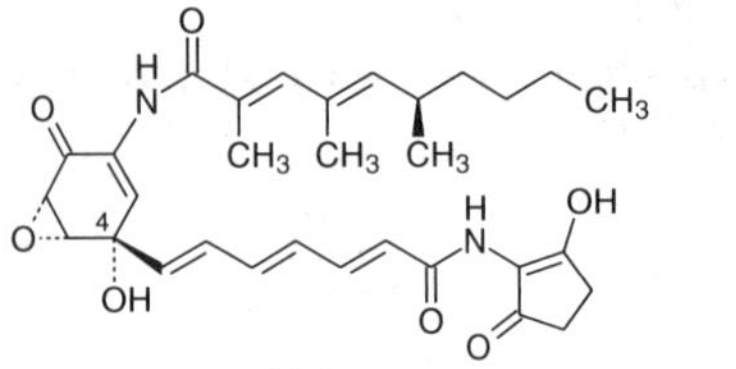

M. A

Antibiotics (components A–G) isolated from *streptomycetes (e.g., *Streptomyces parvulus*). *M. A:* $C_{31}H_{38}N_2O_7$, M_R 550.65, pale yellow cryst., mp. 139–141 °C (decomp.), $[\alpha]_D$ –185° ($CHCl_3$). M. A is active against Gram-positive bacteria and also has antifungal, insecticidal, and cytotoxic properties (L1210: IC_{50} 0.93 µg/mL). M. A inhibits polymorphonuclear leukocyte elastase and Ras farnesyl transferase. These enzymes are involved in metastasis or, respectively, uninhibited growth of tumor cells, their selective inhibition would provide a tumor therapy in which DNA is not the site of attack. M. differ in their side chains (number of C atoms and double bonds, branching) bound as amides to the six-membered ring as well as in the configuration at C-4. The biosynthesis of M. is complicated: the six-membered ring originates from a C_7N-starter unit (from succinate and glycerol), two polyketide synthases are involved, oxygenases supply the two O atoms on the six-membered ring, and amidases effect the linkage of the side chain with the C_5N part of the skeleton. In precursor-directed biosyntheses the C_7N starter unit can be replaced by amino- and hydroxybenzoic acids to produce M. with an aromatic central unit. Antibiotics (e.g., asukamycin, colabomycins, nisamycin) related to the M. are included in the M. group. The absolute configuration of M. A has been corrected in 1998[1], for total synthesis see *Lit.*[2]; M. B *Lit.*[3].

Lit.: [1]J. Org. Chem. **63**, 3526 (1998). [2]Tetrahedron **55**, 3707 (1999). [3]Chem. Commun. **1999**, 421.

gen.: J. Antibiot. **40**, 1530–1554 (1987); **47**, 324–333 (1994); **49**, 1212 (1996) ▪ J. Org. Chem. **58**, 6583–6587 (1993). – *Biosynthesis:* Appl. Microbiol. Biotech. **41**, 309–312 (1994) ▪ J. Am. Chem. Soc. **112**, 3979–3987 (1990); **119**, 4301 (1997) ▪ J. Chem. Soc., Perkin Trans. 1 **1989**, 851–855; **1988**, 2123–2127. – *Activity:* Methods Enzymol. **250**, 43–51 (1995) ▪ Proc. Natl. Acad. Sci. USA **90**, 2281–2285 (1993). – *Review:* Nat. Prod. Rep. **15**, 221 (1998). – *[HS 2941 90; CAS 52665-74-4 (M. A)]*

Manzamines.

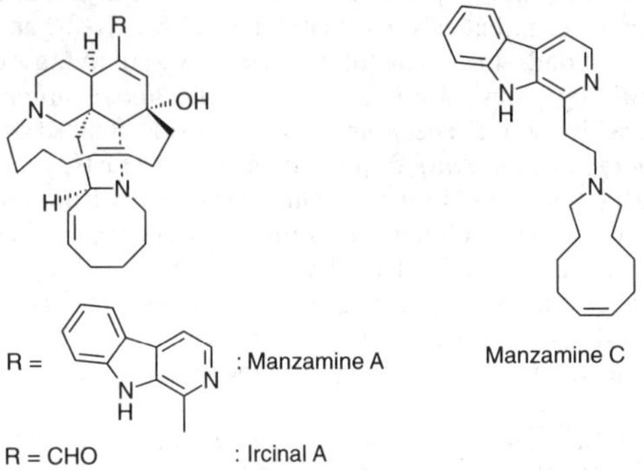

A group of structurally related alkaloids from marine sponges, their biosynthesis can be envisaged as proceeding from ammonia, acrolein units, as well as (*Z*)-5-decenedialdehyde[1]. Key step of the proposed biosynthesis would be an intramolecular endo-Diels-Alder cycloaddition of a bisdihydropyridine[3]. The simple *β*-carboline alkaloid *M. C* ($C_{23}H_{29}N_3$, M_R 347.50, platelets, mp. 77–82 °C) contains an unusual azacycloundecene ring[2]. It occurs in *Haliclona* sp. together with the octacyclic *M. A* {$C_{36}H_{44}N_4O$, M_R 548.77; hydrochloride: mp. >240 °C (decomp.), $[\alpha]_D$ +50° ($CHCl_3$)}[2]. M. A has been isolated from sponges of the genus *Pellia*. It is prepared synthetically by Pictet-Spengler condensation of *ircinal A* {$C_{26}H_{38}N_2O_2$, M_R 410.50, mp. 70 °C, $[\alpha]_D$ +48° ($CHCl_3$)} with tryptamine and subsequent dehydrogenation with dichlorodicyanobenzoquinone (DDQ)[3]. Ircinal A was discovered in a sponge of the genus *Ircinia* near Okinawa. The M. and ircinal A exhibit antileukemic and antibacterial activities.

Lit.: [1]Tetrahedron Lett. **33**, 2059 (1992). [2]Tetrahedron Lett. **28**, 5493 (1987). [3]J. Org. Chem. **57**, 2480 (1992); Angew. Chem. Int. Ed. Engl. **37**, 2661 (1998); Heterocycles **46**, 765–794 (1997).

gen.: Structure: Heterocycles **31**, 999 (1990); **50**, 485 (1999) ▪ J. Am. Chem. Soc. **108**, 6404 (1986) ▪ Tetrahedron Lett. **28**, 621 (1987); **29**, 3083 (1988). – *Synthesis:* Bioorg. Med. Chem. Lett. **6**, 2565 (1996) ▪ Chem. Commun. **1999**, 1757 (review) ▪ J. Am. Chem. Soc. **120**, 6425 (1998); **121**, 866 (1999) ▪ Pure Appl. Chem. **66**, 2131 (1994) ▪ Tetrahedron Lett. **35**, 691, 3191, 6005 (1994). – *Review:* Heterocycles **46**, 765 (1998) ▪ Tetrahedron **54**, 6201–6258 (1998). – *[CAS 112693-24-0 (M. C); 104196-63-1 (M. A); 139975-55-6 (ircinal A)]*

Maracenins. The M.: *Maracin A*, $C_{19}H_{24}O_3$, M_R 300.40, colourless oil and *Maracen A*, $C_{19}H_{25}ClO_3$, M_R 336.86, colourless oil can be isolated from cultures of

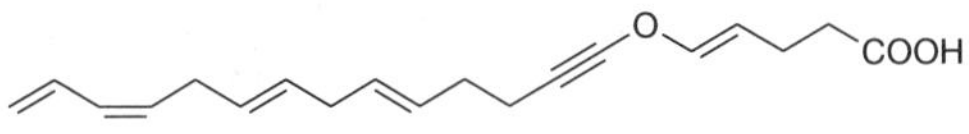

Maracin A

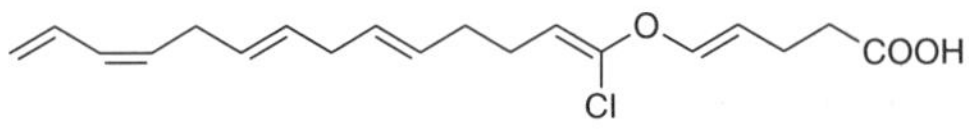

Maracen A

a myxobacterium (*Sorangium cellulosum*) strain. They show interesting activity against *Mycobacterium tuberculosis*.

Lit.: Angew. Chem. Int. Ed. **37**, 1253 (1998).

Marasmal see marasmenes.

Marasmanes.

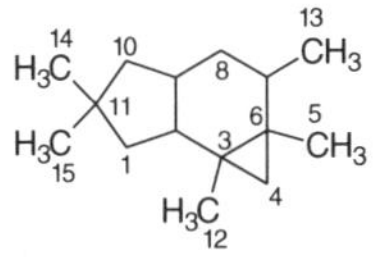

Sesquiterpenoids derived biosynthetically from *humulene and isolated from the fruit bodies and cultures of basidiomycetes, the best known examples are *isovelleral, *marasmic acid, and *stearoylvelutinal. *Marasmenes are sesquiterpenes with the *drimane skeleton.

Lit.: Tetrahedron **37**, 2199 (1981) (review) ▪ Turner **2**, 246.

Marasmenes.

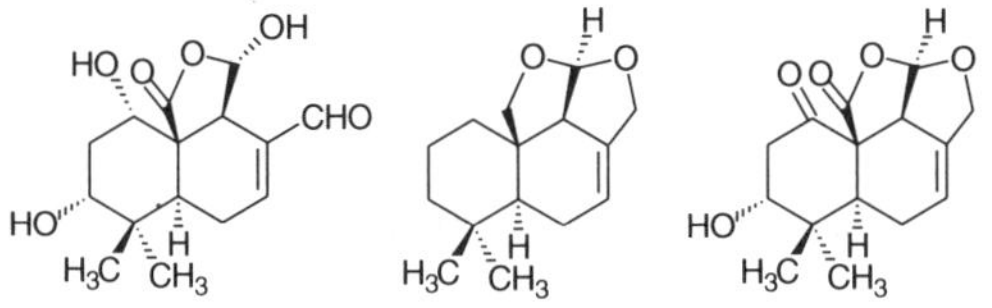

Marasmal Marasmene Marasmone

Numerous sesquiterpenoids, some with highly oxidized *drimane skeletons, isolated from cultures of *Marasmius oreades* (Basidiomycetes). Examples include: *marasmal* {$C_{15}H_{20}O_6$, M_R 296.32, cryst., mp. 195–200 °C, $[\alpha]_D$ −73.1° (CH_3OH)}, *marasmene* {$C_{15}H_{22}O_2$, M_R 234.34, needles, mp. 70–74 °C, $[\alpha]_D$ +57° (CH_3OH)}, and *marasmone* {$C_{15}H_{18}O_5$, M_R 278.30, foam, $[\alpha]_D$ +74.6° (CH_3OH)}. The choice of name of these compounds is unfortunate because of confusion with sesquiterpenoids of the *marasmane type. The *mniopetals are structural relatives.

Lit.: Can. J. Chem. **67**, 773, 1371 (1989) ▪ J. Antibiotics (Tokyo) **47**, 733, 1017 (1994). – *[CAS 124869-10-9 (marasmal); 124869-12-1 (marasmene); 122458-04-2 (marasmone)]*

Marasmic acid.

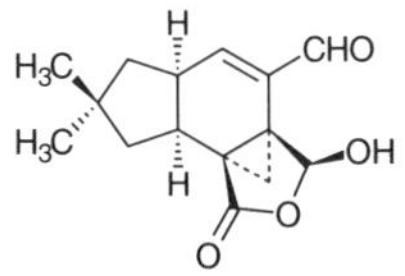

$C_{15}H_{18}O_4$, M_R 262.31, cryst., mp. 173–174 °C, $[\alpha]_D$ +182° ($CHCl_3$). A potent antibacterial and mutagenic sesquiterpenoid of the *marasmane type from cultures of *Marasmius conigenus*, *Flagelloscypha pilatii*, *Lachnella villosa*, *Peniophora laeta*, and other basidiomycetes. It has been the target of many total syntheses.

Lit.: J. Antibiot. **36**, 155 (1983); **41**, 1752 (1988) ▪ Mutat. Res. **188**, 169 (1987) ▪ Tetrahedron **37**, 2199 (1981) (review) ▪ Z. Naturforsch. C **44**, 1 (1989). – *Synthesis:* J. Am. Chem. Soc. **102**, 7146 (1980); **104**, 3216 (1982); **112**, 775 (1990) ▪ Tetrahedron **36**, 3367 (1980) ▪ Tetrahedron Lett. **1970**, 349. – *[CAS 2212-99-9]*

Marasmin(e)s see wilting (withering) agents.

Marasmone see marasmenes.

Marcfortines.

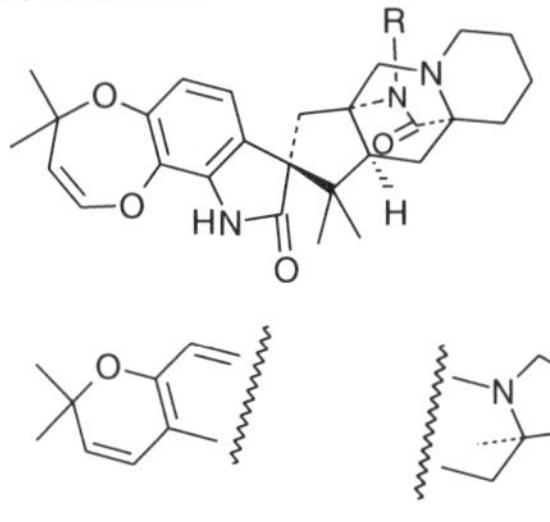

R = CH_3 : M. A
R = H : M. B

R = H : M. C R = CH_3 : Paraherquamide

Table: Data of Marcfortines.

	molecular formula	M_R	mp. [°C]	CAS
M. A	$C_{28}H_{35}N_3O_4$	477.60	242–244 cryst.	75731-43-0
M. B	$C_{27}H_{33}N_3O_4$	463.58	178–180	75789-29-6
M. C	$C_{27}H_{33}N_3O_3$	447.58	264–267	75789-30-9

Indole alkaloids formed, together with *roquefortine C, clavine alkaloids, and *PR toxin by *Penicillium roqueforti*, used in the production of Roquefort cheese. The main component is M. A, which is accompanied by M. B and M. C. The strains used for cheese production were investigated in detail some years ago and found to be free of toxins. The M. are derived biosynthetically from 2 units of dimethylallyl pyrophosphate, tryptophan, and *pipecolic acid. *Paraherquamide*[1] formed by *Penicillium paraherquei* has a very similar structure (paraherquein, $C_{28}H_{35}N_3O_5$, M_R 493.60, prisms, mp. 224–247 °C). It differs only in the second amino acid unit and contains 2-hydroxy-2-methylproline as biosynthetic building block in place of pipecolic acid. The M. are of interest for veterinary medicine on account of their antiparasitic activities, particularly against helminths and arthropods.

Lit.: [1] Angew. Chem. Int. Ed. Engl. **38**, 786 (1999) (biosynthesis); J. Am. Chem. Soc. **115**, 9323 (1993); **118**, 557–579 (1996).

gen.: J. Antibiot. **49**, 1006 (1996) (biosynthesis) ▪ J. Am. Chem. Soc. **121**, 1763 (1999) (biosynthesis) ▪ J. Chem. Soc., Chem. Commun. **1980**, 601 ▪ Mycologia **81**, 837 (1989) ▪ Tetrahedron Lett. **22**, 1977 (1981); **35**, 1135 (1994). – *[CAS 77392-58-6 (paraherquamide)]*

Marchantins. The M. are the largest group (ca. 20) of macrocyclic bisbibenzyls in which the monomeric units are linked by diphenyl ether bridges and are hydroxylated at various positions. They are polyphenols, with the exception of *M. N* ($C_{29}H_{24}O_6$, M_R 468.51)

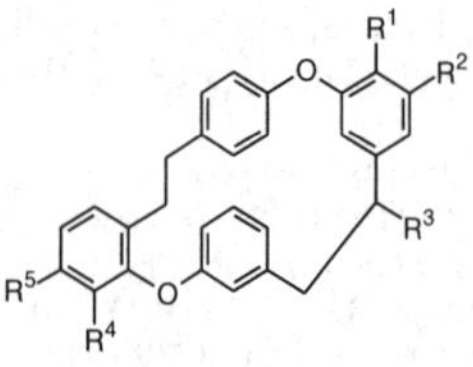

R¹	R²	R³	R⁴	R⁵	
OH	OH	H	OH	H	: M.A
OH	OH	H	OH	OH	: M.B
OH	H	H	OH	H	: M.C
OH	OH	OH	OH	H	: M.D
OH	OH	OCH_3	OH	H	: M.E

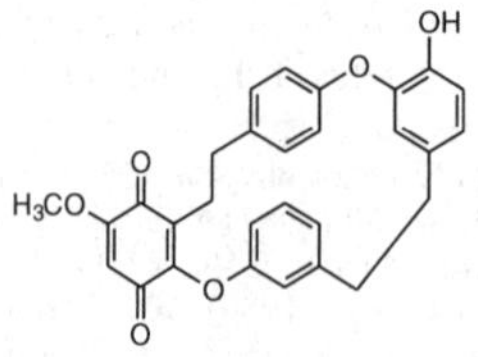

Marchantin N

which contains a 1,4-benzoquinone structure. They are found in *Marchantia* and other liverwort species but not in true mosses and higher plants. Many of the compounds are conspicuous for their biological activity. *M. A* ($C_{28}H_{24}O_5$, M_R 440.50, oil) and *M. B* ($C_{28}H_{24}O_6$, M_R 456.49) are cytotoxic. M. A, *M. D* ($C_{28}H_{24}O_6$, M_R 456.49), and *M. E* ($C_{29}H_{26}O_6$, M_R 470.52, crystalline) are potent inhibitors of 5-lipoxygenase and calmodulin. M. A also exhibits antifungal and antibacterial activities. M. A has been suggested as a vasodilator on account of its perfusion-promoting effects. M. A can be isolated as the main component of *Marchantia paleacea*: 6.7 kg fresh liverwort furnish 80 g M. A. For synthesis, see *Lit.*[1–3].

Lit.: [1] J. Org. Chem. **53**, 72 (1988). [2] Justus Liebigs Ann. Chem. **1989**, 1141. [3] J. Chem. Res. **1991**, 137.

gen.: Chem. Pharm. Bull. **42**, 52 (1994) (structure) ▪ Experientia **34**, 971 (1978) ▪ Hager (5.) **5**, 774 ff. ▪ J. Hattori Bot. Lab. **53**, 283 (1982); **56**, 215 (1984) ▪ Phytochemistry **26**, 1811 (1987); **29**, 1577 (1990); **36**, 73 (1994); **39**, 91 (1995) ▪ Zechmeister **65** ▪ Zinsmeister & Mues (eds.), Bryophytes, their Chemistry and Chemotaxonomy (Proc. Phytochem. Soc. Eur. 29), p. 369, Oxford: Clarendon Press 1990. – *[CAS 88418-46-6 (M. A); 88418-48-8 (M. B); 98093-92-6 (M. D); 98093-91-5 (M. E); 165337-68-8 (M. N)]*

Marcomycin see hygromycins.

Margatoxin. A toxin from the venom secretion of the scorpion *Centruroides margaritatus*. M. is a small protein containing 39 amino acids, its tertiary structure is known[1]. M. is a highly specific inhibitor of the voltage dependent potassium channel of human T lymphocytes. It inhibits lymphocyte activation and formation of interleukin 2. M. is undergoing preclinical testing as a new immunosuppressive agent. A synthesis has been described[2], see also charybdotoxin and scorpion venoms.

Lit.: [1] Biochemistry **33**, 15061–15070 (1994). [2] Biochem. Biophys. Res. Commun. **198**, 619 (1994).

gen.: Exp. Opin. Ther. Patents **5**, 827 f. (1995) ▪ J. Biol. Chem. **268**, 18866–18874 (1993) ▪ J. Exp. Med. **177**, 637–645 (1993) ▪ Perspect. Drug Disc. Design **2**, 233–248 (1994). – *[CAS 145808-47-5]*

Marginalin [5-hydroxy-3-(4-hydroxybenzylidene)-2(3*H*)-benzofuranone].

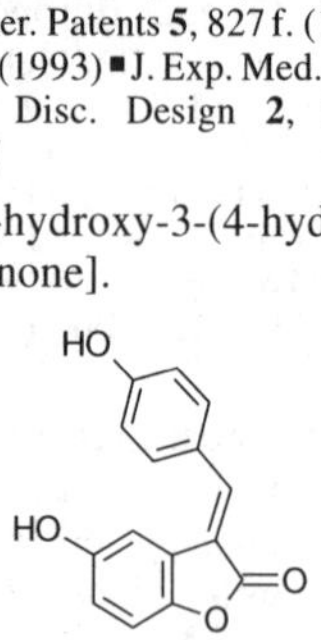

$C_{15}H_{10}O_4$, M_R 254.24, yellow cryst., mp. 245.5 – 247 °C, occurs in the defence secretion of the water beetle *Dytiscus marginalis*.

Lit.: Beilstein E V **18/4**, 102 ▪ Justus Liebigs Ann. Chem. **1970**, 116–125; **1987**, 545 f. (synthesis); **1991**, 393 f. – *[CAS 27439-06-1 (E); 107680-50-2 (Z)]*

Margosa oil see neem oil.

Marihuana (marijuana, "Mary Jane"). Name for a narcotic drug, generally mixed with tobacco and smoked, but also eaten or drunk as a tea. It is obtained by grinding the flowers and bracts of Indian hemp (*Cannabis sativa*) which contains the same aroma substances and constituents as *hashish. The content of Δ^9-*tetrahydrocannabinol, mainly responsible for the hallucinogenic properties, amounts to 0.5–4% (*Lit.*[1]); for clinical and psychosocial effects, see hashish and tetrahydrocannabinols. The name M. is synonymous for hemp cultivated in Central and South America and could be derived from the Indian: malihua or Portuguese: Maranguano = the intoxified or from Maria Johanna since it was originally but wrongly assumed that only the female plants had psychotropic activity.

Lit.: [1] Harvey, Marihuana 84, p. 37, Oxford: IRL Press 1985.

gen.: J. Forensic Sci. **33**, 1385 (1988) ▪ Nature (London) **346**, 561 (1990) ▪ Redda, Walker & Barnett, Cocaine, Marijuana, Designer Drugs: Chemistry, Pharmacology and Behavior, Boca Raton: CRC Press 1989.

Marindinin see styrylpyrones.

Marine fungi. Ascomycetes and deuteromycetes are involved in the degradation of algin- and cellulose-containing substrates, cause formation of galls when living as parasites on brown algae, or living symbiotically in the receptacles of algae, the rhizomes or leaves of sea grasses, or in the gastrointestinal tract of fish; marine basidiomycetes are also found as smuts in sea grasses or as saprophytes on wood; e. g., phomactin A (see fungi).

Lit.: Pharmazie **50**, 583–588 (1995) ▪ Riedl, Fauna u. Flora des Mittelmeeres, Hamburg: Parey 1983.

Marine natural products. The history of natural products from marine organisms ranges from the *purple of the ancient Phoenicians who had a flourishing industry around 1600 BC to the most complicated low-molecular-weight structures known to date: the polyether toxins *maitotoxin and *palytoxin.

Although the successes of natural product chemistry after World War II are impressively reflected in particular in the development of the *antibiotics, medical-pharmaceutical research has been directed at marine organisms as a source of interesting secondary metabolites only since about 1970. The structural diversity of these compounds is equal to that of the metabolites of the terrestrial *streptomycetes. Up to 1999, about 16000 m. n. p. have been described. 120 of them made it into drug development, which is a high success rate. The rapidly increasing number of publications and patents on the subject, especially since 1980, illustrates the ever-increasing importance of this field in medical research.

The *brevetoxins, *conotoxins, *okadaic acid, *palytoxin, *tetrodotoxin, *saxitoxin, etc. are important tools of biochemical research (on ion channels), this is also the case for the bioluminescent substances *aequorin and the *luciferins. Numerous heterocyclic compounds such as the *cephalostatins, *didemnins, *dolastatins are interesting antitumor principles, an insecticide (Cartap®) for crop protection has been developed from *nereistoxin. *Agar and *alginic acid products are used as nutrient broths for bacteria and as food additives. Numerous *algal pheromones are known and have provided new information on the communication between members of the same species. Coinciding with the composition of sea water, numerous natural *halogen compounds, often with very unusual structures, have been found in marine organisms. Metal ions are also often enriched to a very high (million-fold) degree such as, e. g., nickel in *tunichlorin, vanadium in the *tunichromes of tunicates, or copper in the *hemocyanin of octopus species, snails, and mussels. Marine natural products from *sponges (Porifera) show an enormous diversity.

Lit.: ACS Symp. Ser. **262** (seafood toxins) (1984); **418** (marine toxins) (1990) ▪ Angew. Chem. Int. Ed. Engl. **32**, 1 ff. (1993); **110**, 2280–2297 (1998) (sponges) ▪ Barton-Nakanishi **8**, 415–650 (review) ▪ Chem. Ind. (London) **1996**, 54–58 ▪ Chem. Rev. **92**, 613–631 (1992); **93**, 1671–1944 (1993) ▪ Exp. Opin. Ther. Pat. **9**, 1207–1222 (1999) (review, pharmacolog. activity) ▪ Jefford et al., Pharmaceuticals and the Sea, Lancaster: Technomic Publ. Co. Inc. 1988 ▪ Nachr. Chem. Tech. Lab. **46**, 1175 (1998) ▪ Nat. Prod. **1987**, 539–576; **1990**, 269–309; **1995**, 223–269 ▪ Nat. Prod. Rep. **13**, 75–125 (1996); **16**, 155–198 (1999) (review) ▪ Sarma, Daum & Müller, Secondary Metabolites from Marine Sponges, Berlin: Ullstein Mosby 1993 ▪ Scheuer I ▪ Scheuer II ▪ Scrip Magazine, June 1998, p. 9 f. ▪ Tetrahedron **51**, 4571–4618 (1995) ▪ Tu, Marine Toxins and Venoms, New York: Dekker 1988 ▪ Winterfeldt, Marine Natural Products, ESRF Lectures **34**, Berlin: Hellmich Pre Press 1999 ▪ Zaborsky & Attaway, Marine Biotechnology, New York: Plenum Press (since 1993) ▪ Zechmeister **49**, 151–363.

Marine phosphorescence see bioluminescence and luciferins.

Marinobufagin.

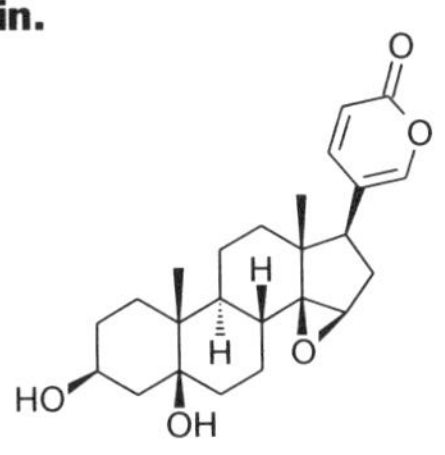

$C_{24}H_{32}O_5$, M_R 400.52, prisms, mp. 224–225 °C, $[\alpha]_D^{25}$ +10° ($CHCl_3$), LD_{50} (mouse i.v.) 0.152 mg/kg. A *bufadienolide from the secretion of the Brazilian toad *Bufo marinus*, which also contains the corresponding 5-deoxy compound resibufogenin. In contrast to other bufogenins, these two epoxy compounds have only weak pharmacological activities; see also bufadienolides and toad poisons.

Lit.: *Synthesis:* Chem. Pharm. Bull. **28**, 1559–1562 (1980) ▪ J. Org. Chem. **39**, 3003–3006 (1974). – *[CAS 470-42-8]*

Maritidine see Amaryllidaceae alkaloids.

Maritimetin (3′,4′,6,7-tetrahydroxyaurone). A *flower pigment from the tickseed *Coreopsis tinctoria*, *Zinnia angustifolia* (= *Z. linearis*) and other Asteraceae (see table).

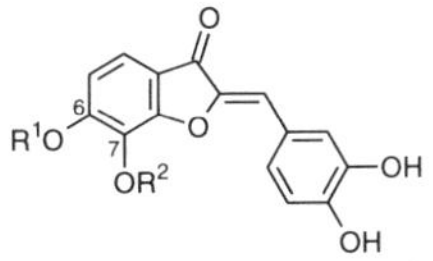

	R^1	R^2
Maritimetin (1)	H	H
Maritimein (2)	β-D-glucopyranosyl	H
Leptosidin (3)	H	CH_3
Leptosin (4)	β-D-glucopyranosyl	CH_3

Table: Data of Maritimetin and derivatives.

	molecular formula	M_R	mp. [°C]	$[\alpha]_D^{20}$	CAS
1	$C_{15}H_{10}O_6$	286.24	292 (decomp.) orange needles		576-02-3
2	$C_{21}H_{20}O_{11}$	448.38	208–214 256–272 yellow cryst.	−98° (DMSO) −82.5° (CH_3OH)	490-54-0
3	$C_{16}H_{12}O_6$	300.27	254 (decomp.) orange-yellow needles		486-24-8
4	$C_{22}H_{22}O_{11}$	462.41	229–231 (decomp.) yellow needles		486-23-7

M. is an inhibitor of thyroxine 5′-deiodinase[1]; maritimein inhibits fatty acid peroxidase from horseradish[2]. M. is used as a flavor substance.

Biosynthesis: M. belongs to the group of the *aurones, which are derived from *chalcones, intermediates of *flavonoid biosynthesis.

Lit.: [1] Stud. Org. Chem. **23**, 411–421 (1986). [2] J. Pharm. Belg. **44**, 325 ff. (1989); Planta Med. **60**, 288 (1994).
gen.: Ann. Pharm. Fr. **47**, 95 ff. (1989) ▪ Chem. Pharm. Bull. **39**, 709 ff. (1991) ▪ Harborne (1994), p. 399 ff. ▪ J. Nat. Prod. **46**, 190 ff. (1983) ▪ Luckner (3.), p. 413 ▪ Phytochemistry **22**, 2741 f. (1983). – *Pharmacology:* Plant. Med. Phytother. **21**, 131–137 (1987).

Sweet marjoram oil. Yellow to yellowish green oil with a unique spicy-aromatic, earthy-camphor like odor and a typical warm-spicy, aromatic, somewhat bitter taste.

Production: By steam distillation of the flowering marjoram shrub *Origanum majorana* cultivated mainly in the Mediterranean region.

Composition[1]***:*** Main components are the *sabinenes* (see thujenes) (ca. 5%), α- (5–10%) and *γ-terpinenes* (see *p*-menthadienes) (10–15%), *cis-sabinene hydrate* (see thujanoles) (10–15%), and *terpinen-4-ol* (see *p*-menthenols) (20–30%); the latter two compounds are mainly responsible for the organoleptic properties of sweet marjoram oil. The (genuine) oil should not be confused with the so-called Spanish oil obtained from *Thymus mastichina* which contains between 40 and 70% 1,8-**cineole*.

Use: In the production of perfumes as herby-spicy component, especially for masculine tones, to improve the aromatic character of foods such as sausage products, meat products, sauces, and salads.

Lit.: [1] Perfum. Flavor. **12** (4), 73 (1987); **14** (1), 32 (1989); **19** (4), 39 (1994).
gen.: Arctander, p. 400 ▪ Bauer et al. (2.), p. 163. – *Toxicology:* Food Cosmet. Toxicol. **14**, 469 (1976). – *[HS 3301 29; CAS 8015-01-8]*

Marmelo lactone, marmelo oxide see fruit flavors (quince flavor).

Marmesin.

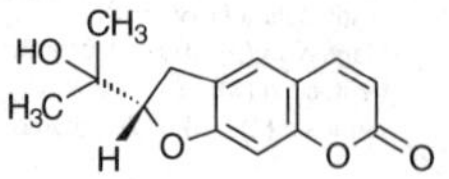

(*S*)-(+)-form

$C_{14}H_{14}O_4$, M_R 246.26, platelets, mp. 189.5 °C (the racemate melts at 166 °C), $[\alpha]_D^{34}$ +26.8° ($CHCl_3$). A *furocoumarin from bael bark (*Aegle marmelos*), *Ammi majus*, *Poncirus trifoliata*, and other plants. The *O*-β-D-glucopyranoside (*ammajin*, marmesinin), $C_{20}H_{24}O_9$, M_R 408.41, platelets, mp. 259–260 °C, $[\alpha]_D$ –60° (aqueous C_2H_5OH) is also a component of *Ammi majus*. The (*R*)-form (*nodakenetin*, prangeferol), mp. 186.5 °C, $[\alpha]_D$ –25.2° ($CHCl_3$) occurs in *Angelica* species, the roots of *Prangos ferulacea*, and the leaves of *Scaevola frutescens*. The *O*-β-D-glucopyranoside (*nodakenin*), prisms or platelets, mp. 215–219 °C (259–260 °C, decomp.), $[\alpha]_D^{30}$ +56.6° (H_2O), is a component of the roots of *Peucedanum decursivum* (Chinese herbal drug "Zi-Hua Qian-Hu") .

Activity: Nodakenetin glycosides inhibit *in vitro* the ADP-dependent aggregation of blood platelets [1].

Biosynthesis: M. and nodakenetin are intermediates in the biosynthesis of *furocoumarins.

Lit.: [1] Planta Med. **50**, 488–492 (1984); **52**, 135 f. (1986). *gen.:* Bull. Singapore Natl. Inst. Chem. **17**, 21–29 (1989) ▪ Cryst. Res. Technol. **23**, 1465–1470 (1988) (crystal structure) ▪ Eur. J. Biochem. **171**, 369–375 (1988) (biosynthesis) ▪ Fitoterapia **61**, 88 (1990) ▪ Nat. Prod. Rep. **12**, 477–505 (1995) ▪ Phytochemistry **25**, 505 ff. (1986); **29**, 1137–1142 (1990) (biosynthesis). – *[CAS 13849-08-6 (S); 495-32-9 (R); 13710-70-8 (racemate); 495-30-7 (ammajin); 495-31-8 (nodakenin)]*

Marmin see umbelliferone.

Marrubiin.

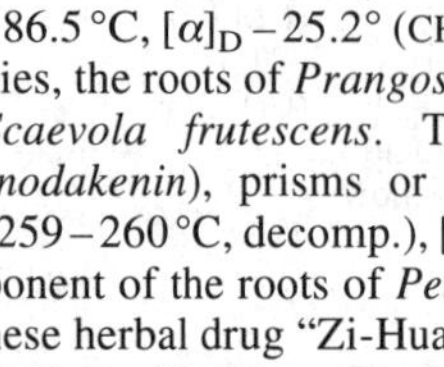

Marrubiin

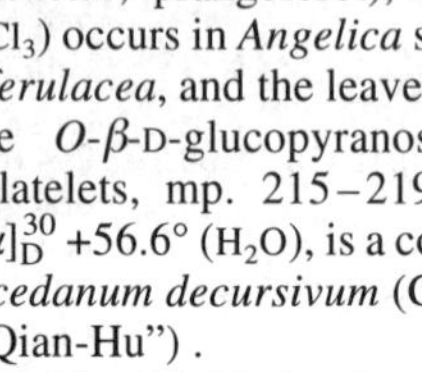

Premarrubiin

$C_{20}H_{28}O_4$, M_R 332.44, crystalline, mp. 160 °C, $[\alpha]_D$ +33.3° ($CHCl_3$). A diterpene of the*labdane type isolated from *Marrubium vulgare* and *Leonotis leonurus* (Lamiaceae). M. is formed in the plant from *premarrubiin* [$C_{20}H_{28}O_4$, M_R 332.44, oil; $[\alpha]_D$ +29° ($CHCl_3$)] during the isolation process.

Lit.: J. Chem. Soc. C **1968**, 807; **1969**, 2015 ▪ J. Chromatogr. **605**, 124 (1989) ▪ Plant Physiol. Biochem. **32**, 785 (1994) ▪ Tetrahedron Lett. **1965**, 4337. – *Synthesis:* Tetrahedron **28**, 611 (1972). – *[CAS 465-92-9 (M.); 72059-02-0 (premarrubiin)]*

Masoprocol see nordihydroguaiaretic acid.

Mast cell degranulating peptide.

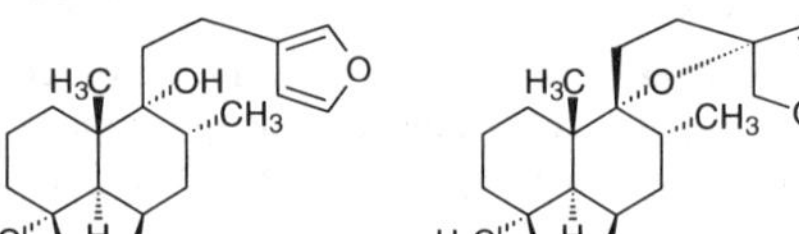

A peptide of *bee venom (content up to 2%). It consists of 22 amino acids and has two disulfide bridges ($C_{110}H_{192}N_{40}O_{24}S_4$, M_R 2587.24, amorphous). LD_{50} (mouse i. v.) 40 mg/kg. M. liberates *histamine from mast cells and is thus mainly responsible for the pain caused by a bee sting. In addition, it selectively blocks K^+ ion channels.

Lit.: Int. J. Pept. Prot. Res. **33**, 86 (1989) (synthesis) ▪ J. Pharm. Pharmacol. **42**, 457–461 (1990) ▪ Rev. Physiol. Biochem. Pharmacol. **115**, 93 (1990). – *[CAS 32908-73-9; 83856-13-7]*

Mastic (pistachio resin).

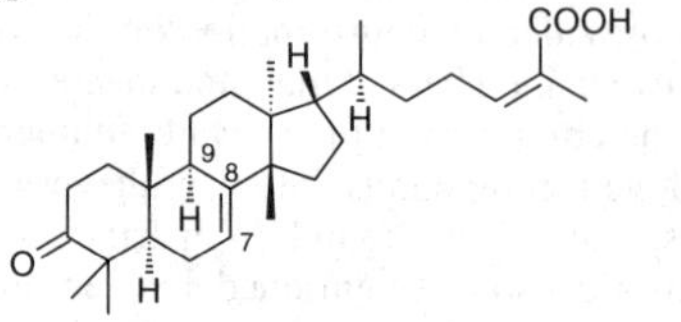

Masticadienonic acid
8,9-C=C : Isomasticadienonic acid

Resin-like exudate (wound secretion) from pistachio trees (*Pistacia lentiscus*, Anacardiaceae) indigenous in the Mediterranean region; soluble in diethyl ether, partly soluble in ethanol and turpentine oil; mp. 105–120 °C. The resin fraction contains triterpene acids such as *masticadienonic acid* ($C_{30}H_{46}O_3$, M_R 454.69) and *isomasticadienonic acid* ($C_{30}H_{46}O_3$, M_R 454.69) as well as the triterpene alcohol *tirucallol* (see triterpenes). M. is used to glaze foods, in paints, glues, plastics, and dental cement. In Greece it is used to flavor Retsina wine.

Lit.: Hager (5.) **1**, 656; **6**, 627 f. ▪ Kirk-Othmer (3.) **17**, 386 f.; **20**, 198, 205 ▪ Phytochemistry **30**, 3709 (1991); **37**, 1409 (1994) ▪ Riv. Ital. Essenze, Profumi, … **56**, 245–252 (1974) ▪ Ullmann (5.) **A 23**, 78. – *[HS 1301 90; CAS 61789-92-2 (M.); 514-49-8 (masticadienonic acid); 5956-26-3 (isomasticadienonic acid)]*

Masticadienonic acid see mastic.

Mastigophorenes. The M. are atropisomeric dimeric sesquiterpenoid biaryls from the liverwort *Mastigophora diclados*, co-occurring with their monomeric phenolic precursors, related with *herbertene derivatives. *M. A*: $C_{30}H_{42}O_4$, M_R 466.65, crystalline, mp. 258–261 °C, $[\alpha]_D^{19}$ –65.3° ($CHCl_3$). M. B is the atropisomer, mp. 210 °C, $[\alpha]_D^{19}$ –39.1° ($CHCl_3$).

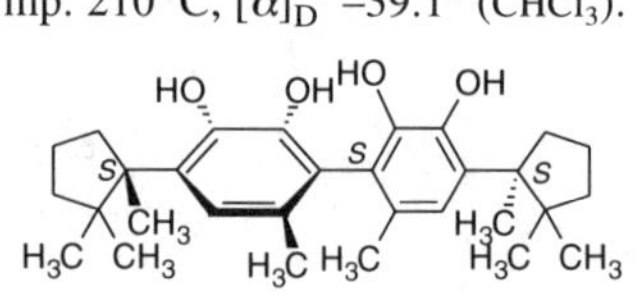

M. A

Lit.: *Isolation:* J. Chem. Soc., Perkin Trans. 1 **1991**, 2737. – *Synthesis:* J. Am. Chem. Soc. **121**, 2762–2769 (1999) ▪ Tetrahedron **54**, 1425–1438 (1998). – *[CAS 136088-03-4 (M. A); 136087-30-4 (M. B)]*

Mastoparans. M. I: $C_{70}H_{131}N_{19}O_{15}$, M_R 1478.93, colorless powder, a peptide from the *wasp venom of *Vespula lewisii*; M. X: $C_{73}H_{126}N_{20}O_{15}S$, M_R 1556.0 from *Vespa xanthoptera* with the structural formulae:

M. I: H–Ile–Asn–Leu–Lys–Ala–Leu–Ala–Ala–Leu–Ala–Lys–Lys–Ile–Leu–NH_2.
M. X: H–Ile–Asn–Trp–Lys–Gly–Ile–Ala–Ala–Met–Ala–Lys–Lys–Leu–Leu–NH_2.

M. and related peptides promote, similar to the *mast cell degranulating peptide of bee venom, the release of *histamine.

Lit.: Biopolymers **25** (Suppl.), S115–S121 (1986) ▪ Chem. Lett. **1999**, 193 (synthesis) ▪ Trends Pharmacol. Sci. **11**, 358–362 (1990). – *[CAS 72093-21-1 (M. I); 72093-22-2 (M. X)]*

Matatabiether.

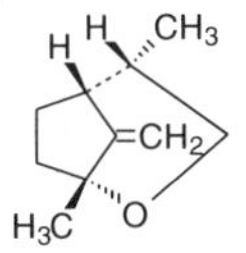

(-)-form

$C_{10}H_{16}O$, M_R 152.24, bp. 67 °C (1.6 kPa), $[\alpha]_D$ –150° (CCl_4). A C_{10}-*iridoid with an unusual structure. (–)-M. is isolated from the leaves and galls of the East Asian plant *Actinidia polygama* (Actinidiaceae). The name of the compound is derived from the Japanese name for the plant "Matatabi". The plant has an attracting effect on cats (Felidae) and lacewing flies (Chrysopidae).

Lit.: Heterocycles **30**, 341 (1990) ▪ J. Nat. Prod. **43**, 649, No. 178 (1980) ▪ Tetrahedron **36**, 3115 (1980). – *[CAS 21700-60-7]*

Matricarin see proazulenes.

Matricin (prochamazulene).

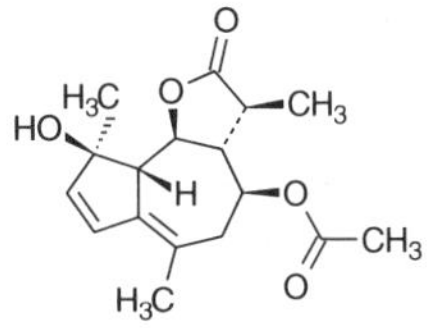

$C_{17}H_{22}O_5$, M_R 306.36, cryst., mp. 158–160 °C. Flowers of chamomile (*Matricaria chamomilla*, Asteraceae) contain ca. 0.15% of *proazulenes, especially M., which can be obtained by chloroform extraction. When extracts containing M. are heated, elimination of water and acetic acid occurs to furnish chamazulene-carboxylic acid which, as an α-aryl-carboxylic acid, easily undergoes decarboxylation to chamazulene (see azulenes).

Lit.: Karrer, No. 3643 ▪ Phytochemistry **21**, 2555 (1982); **28**, 3526 (1989); **29**, 3575 (1990); **31**, 4361 (1992) ▪ Z. Naturforsch. B **37**, 508–511 (1982). – *[CAS 29041-35-8]*

Matrine (matridin-15-one, lupanidine).

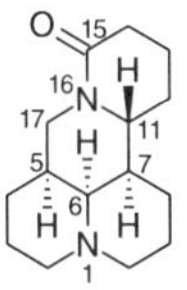

(+)-form

$C_{15}H_{24}N_2O$, M_R 248.37, mp. 77 °C, $[\alpha]_D^{15}$ +40.9° (H_2O). M. is the parent compound of the matrine type *quinolizidine alkaloids. It is the main alkaloid of *Sophora flavescens* and occurs also in three further Fabaceae genera (*Euchresta*, *Goebelia*, *Vexibia*)[1]. M. exhibits antiulcer, antineoplastic, and antibacterial activity. The skeleton of M. contains four chiral C atoms, resulting in eight different diastereomers. Besides M., four are known as natural alkaloids: *sophoridine* (5β-M.), *isomatrine* (5β,6β,7β-M.), *allomatrine* (6β-M.), and *darvasamine* (5β,11α-M.), that are used in Chinese traditional medicine as Kuh Seng drug or, respectively, in Japanese traditional medicine as Shinkyo-gan drug. In addition, *matrine N^1-oxide* and several hydroxymatrines are also known as natural products[1].

Lit.: [1]Waterman **8**, 197–239.

gen.: Beilstein E V **24/2**, 301 ff. ▪ Merck-Index (12.), No. 5799. – *Biosynthesis:* Can. J. Chem. **59**, 106 (1981) – *Synthesis:* Angew. Chem. Int. Ed. Engl. **37**, 1128 (1998) ▪ Chem. Pharm. Bull. **34**, 2018 (1986) ▪ J. Chem. Soc., Chem. Commun. **1986**, 905. – *[HS 2939 90; CAS 519-02-08 (M.); 16837-52-8 (M. N-oxide); 641-39-4 (allomatrine); 36284-98-7 (darvasamine); 17801-36-4 (isomatrine); 6882-68-4 (sophoridine)]*

Matsuone [(2*E*,4*E*)-4,6,10,12-tetramethyl-2,4-tridecadien-7-one].

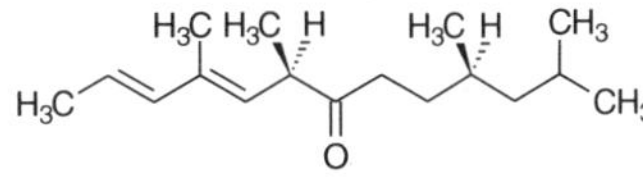

$C_{17}H_{30}O$, M_R 250.42, liquid, $[\alpha]_D^{18}$ –318° ($CHCl_3$). A sexual pheromone of female scales of the genus *Matsucoccus*, in particular *M. resinosae*, *M. matsumurae*, and *M. thunbergianae*[1]. The natural attractant has the (6*R*,10*R*)-configuration[2]. The same partial structure as the unsaturated chain, namely (2*E*,4*E*)-4,6-dimethyl-2,4-dien-7-one, is also present in the *pheromones of *M. feytaudi*[3] and *M. josephi*[4].

Lit.: [1]J. Chem. Ecol. **15**, 1645–1659 (1989). [2]Justus Liebigs Ann. Chem. **1993**, 993–1001; Tetrahedron Lett. **34**, 5931 (1993). [3]Tetrahedron Lett. **31**, 6633–6636 (1990). [4]J. Chem. Ecol. **21**, 331–341 (1995).

gen.: ApSimon **9**, 184 (synthesis). – *[CAS 121981-51-9]*

Matsutake alcohol see (–)-(*R*)-1-octen-3-ol.

Maytansinoids (maitansinoids). A group of macrolide antibiotics with antileukemic, antimitotic, and cytotoxic properties from the Ethiopian shrub *Maytenus ovatus*, other Celastraceae, and *Trewia nudiflora* (Euphorbiaceae) as well as the Japanese mosses *Isothecium subdiversiforme* and *Thamnobryum sandei* (Neckeraceae).

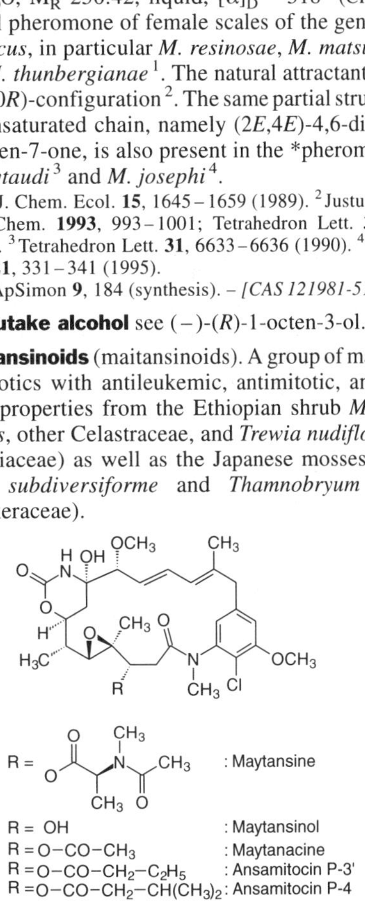

M. are derived from maytansine, the first known ansamycin of plant origin; several derivatives (see table) have been isolated from Celastraceae. The closely related *ansamitocins* have been found not only in *Nocardia* sp.[1] but also in Oregon mosses and their associated actinomycetes[2], thus, the M. isolated from the plants may be of microbial origin (see table, p. 386).

All the listed compounds have pronounced pharmacological properties[3].

Table: Data of Maytansinoids.

	molecular formula	M_R	mp. [°C]	optical activity	CAS
Maytansine	$C_{34}H_{46}ClN_3O_{10}$	692.21	171–172 cryst.	$[\alpha]_D^{26}$ −145° ($CHCl_3$)	35846-53-8
Maytansinol (Ansamitocin P-0)	$C_{28}H_{37}ClN_2O_8$	565.06	173–174	$[\alpha]_D^{23}$ −309° ($CHCl_3$)	57103-68-1
Maytanacine (Ansamitocin P-1)	$C_{30}H_{39}ClN_2O_9$	607.10	234–237 ($CHCl_3$)	$[\alpha]_D$ −121°	57103-69-2
Ansamitocin P-3′	$C_{32}H_{43}ClN_2O_9$	635.15	182–185 (decomp.)	$[\alpha]_D$ −134° ($CHCl_3$)	66547-09-9
Ansamitocin P-4	$C_{33}H_{45}ClN_2O_9$	649.18	177–180 (decomp.)	$[\alpha]_D$ −142° ($CHCl_3$)	66547-10-2

Lit.: [1] Nature (London) **270**, 721 (1977); Tetrahedron **35**, 1079 (1979). [2] Experientia **46**, 117 (1981). [3] Pharmacol. Ther. **55**, 31 ff. (1992) (review).
gen.: Agric. Biol. Chem. **48**, 1721–1729 (1984) (ansamitocins) ▪ Arch. Pharm. (Weinheim, Ger.) **326**, 853–856 (1993) ▪ J. Am. Chem. Soc. **101**, 4732 (1979); **102**, 6597, 6613 (1980) (synthesis) ▪ J. Antibiot. **35**, 1415 ff. (1982) (biosynthesis) ▪ J. Nat. Prod. **51**, 845–850 (1988) ▪ J. Org. Chem. **61**, 7133 (1996) ▪ Manske **23**, 71–204; **25**, 142–155. – *Toxicology:* Quant. Struct. Act. Rel. **10**, 306–332 (1991); Sax (8.), MBU 820, APE 529.

Maytine, maytoline see agarofurans.

MCD peptide see mast cell degranulating peptide.

Meat flavor. Although raw meat has only a weak aroma and a somewhat salty, bitter (*lactic acid), and sweet taste, on heating (boiling roasting, grilling) numerous volatile *flavor compounds (more than 1000 have been identified) are formed, among others, by Maillard reactions, lipid oxidation, and thiamine degradation. Sulfur compounds such as *2-methyl-3-furanthiol, bis-(2-methyl-3-furyl) disulfide (see 2-methyl-3-furanthiol), *methional, *2-furylmethanethiol, *2-acetylthiazole* (1) (C_5H_5NOS, M_R 127.16), and *2-acetyl-2-thiazoline* (2) (C_5H_7NOS, M_R 129.18) contribute significantly to the aroma, in addition to degradation products of fat such as *alkanals, *alkenals, and alkadienals [especially (*E,E*)-*2,4-decadienal] as well as *2-acetyl-1-pyrroline, *furaneol® and *pyrazines (especially on roasting and grilling, e.g., 2-ethyl-3,5-dimethylpyrazine and 2,3-diethyl-5-methylpyrazine). The differences in the various meat flavors are mainly due to a shift in the quantitative composition of ingredients. A typical component of pork flavor is *5-ethyl-2,4-dimethylthiazole* (3) ($C_7H_{11}NS$, M_R 141.23) with an earthy, rusty odor; *12-methyltridecanal* (4) ($C_{14}H_{28}O$, M_R 212.38) with a suety-bouillon-like aroma formed from plasmalogens is found in boiled beef flavor.

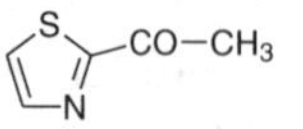

2-Acetylthiazole

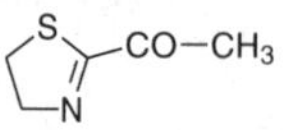

2-Acetyl-2-thiazoline

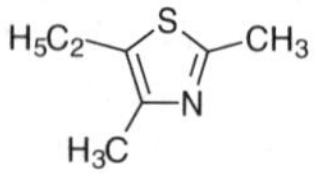

5-Ethyl-2,4-dimethylthiazole

Lit.: Maarse, p. 107–177 ▪ TNO list (6.), p. 421–474. – *[CAS 24295-03-2 (1); 29926-41-8 (2); 38205-61-7 (3); 75853-49-5 (4)]*

Mecambrine see proaporphine alkaloids.

Mecocyanin [*cyanidin 3-*O*-sophoroside, 3-(2-*O*-*β*-D-glucopyranosyl-*β*-D-glucopyranosyloxy)-3′,4′,5,7-tetrahydroxyflavylium]. $C_{27}H_{31}O_{16}^+$, M_R 611.53, dark red crystals with a green luster. M. occurs in the fruits of sour cherry (*Prunus cerasus*, Rosaceae) and in the flowers of the corn poppy (*Papaver rhoeas*, Papaveraceae).
Lit.: Beilstein E III/IV **17**, 3900. – *[CAS 4453-78-5 (ion)]*

Meconic acid see poppy acid.

Mectizan see ivermectin.

Medicagenic acid (2*β*,3*β*-dihydroxyolean-12-ene-23,28-dioic acid).

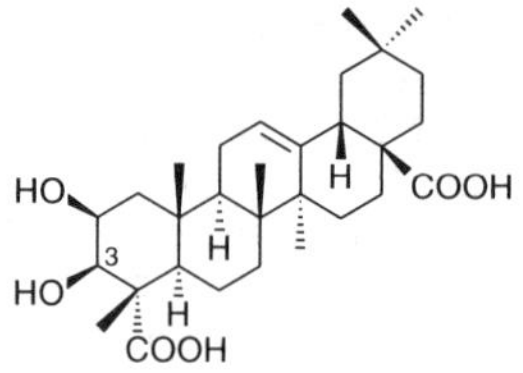

(2β,3β)-form

$C_{30}H_{46}O_6$, M_R 502.69, orthorhombic cryst., mp. 352–353 °C, $[\alpha]_D$ +106° (C_2H_5OH). The 2*α*,3*β*-form is called *barringtogenic acid* and is isolated from *Barringtonia* species. M. is the aglycone of numerous *saponins from the roots of lucerne (alfalfa, *Medicago sativa*) and other Fabaceae (e.g., *Dolichos kilimandscharicus*), for example, the 3-*O*-*β*-D-glucopyranoside {$C_{36}H_{56}O_{11}$, M_R 664.83, mp. 253–255 °C, $[\alpha]_D^{23}$ +71.4° (C_2H_5OH)}. M., glycosides derived from M., and synthetic derivatives of M. show, in comparison with other saponin extracts, good activity against animal and plant fungal diseases; glycosides of M. also have hemolytic activity and delay the development of wheat shoots.
Lit.: Acta Crystallogr., Sect. C **45**, 341 (1989) ▪ Beilstein E IV **10**, 2321 ▪ Carbohydr. Res. **193**, 115–123 (1989) ▪ J. Agric. Food Chem. **34**, 960 (1986) ▪ Phytochemistry **28**, 1379 f. (1989) ▪ Ullmann (5.) **A 23**, 493. – *[CAS 599-07-5 (M.); 49792-23-6 (3-O-β-D-glucopyranoside); 471-58-9 (barringtogenic acid)]*

Megovalicins see myxovirescins.

Melampyroside see aucubin.

Melanins. Brown or black pigments, that occur in humans and vertebrates in melanocytes (cells under the epidermis) and are principally responsible for the coloration (pigmentation) of skin, eyes, hair, and feathers. Although humans of different skin colors have the same number of melanocytes, their distribution and concentration differ. M. are complex aggregations of quinoid substances with the empirical formula $C_8H_3NO_2$, they are derived from indole and are formed from aromatic chromogens under the influence of enzymes (e.g., tyrosinase occurring in melanosomes). Important intermediates in the formation of M. are the 1,2-quinones dopachrome (2-carboxy-5,6-indolinedione), indolequinone (5,6-indoledione), and adrenoquinone (3-hydroxy-5,6-indolinedione). In addition, hydroxylation, oxidation, and cyclization reactions are involved in their formation. Polymerized M., which contains ca. 10–15% protein (melanoprotein), is structurally related with *humic acids and *lignin. In the skin it acts to trap radicals generated on exposure to sunlight. The enzymes synthesizing M. are activated (browning) only on exposure to sunlight (also alpha or X-rays). M. is also present in the exoskeleton of arthropods and as coloring component of octopus ink. The name is derived from the Greek: mélan = black.

Lit.: Angew. Chem. Int. Ed. Engl. **28**, 555–570 (1989) ▪ Med. Res. Rev. **8**, 525–556 (1988) ▪ Shinkei Kenkyu No Shimpo **35**, 584–590 (1991) ▪ Zechmeister **64**, 93–148.

Melanthioidine see phenethyltetrahydroisoquinoline alkaloids.

Melatonin (*N*-acetyl-5-methoxytryptamine, MLT).

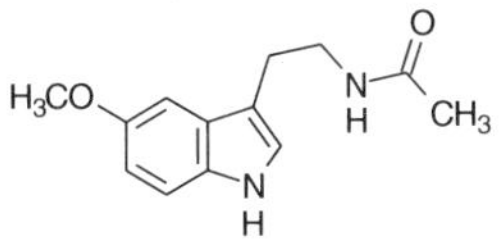

$C_{13}H_{16}N_2O_2$, M_R 232.28, pale yellow platelets, mp. 116–118 °C. M. belongs to the tissue hormones that are synthesized in the epiphysis (pineal body) of vertebrates from *serotonin. *N*-Acetyltransferase (EC 2.3.1.5) regulates the biosynthesis of M., it is active in the dark. Methylation occurs under the action of *N*-acetylserotonin *O*-methyltransferase (EC 2.1.1.4).

Activity: One of the major functions of M. is to transduce environmental information (i.e., photoperiod) into neuroendocrine signals. M. is assumed to control the circadian rhythm, i.e., the "internal biological clock". In addition, M. inhibits the secretion of ACTH and gonadotropic hormones and thus also acts on the gonads in dependence on the length of the days and accordingly the season of the year. Besides the above, M. is associated with a number of other activities, it is thus sometimes referred to as "the wonder drug of the 1990's". Among others, the consumption of M. is supposed to compensate for "jet lag" [1] in air travel. In amphibians and fish M. acts as an opponent to melanotropin and leads to a lightening of the skin by way of a contraction of the pigment cells (melanophores). M. receptors have been characterized in the brain and retina [2]. For putative synergistic activities with 1α,25-dihydroxycholecalciferol, see *Lit.*[3]. M. is supposed to possess immunoenhancing and antineoplastic activities [4].

Lit.: [1] Chem. Ind. (London) **1996**, 637–640; Chem. Eng. News (23.12.) **1996**, 27. [2] FASEB J. **2**, 2765–2773 (1988). [3] Naturwissenschaften **75**, 247–251 (1988). [4] Drugs of Today **33**, 25–39 (1997).

gen.: Pharmacology: Comp. Biochem. Physiol. **94A**, 467 ff. (1989) ▪ Endocr. Rev. **12**, 151 (1991); **18**, 91 (1992) ▪ J. Am. Chem. Soc. **120**, 6195 (1998) ▪ Martindale (30.), p. 1385 ▪ Science **214**, 821 ff. (1981); **242**, 78 (1988). – *Review:* CRC Crit. Rev. Clin. Lab. Sci. **25**, 231–253 (1987) ▪ Experientia **46**, 120–128 (1990); **49**, 642 (1993) ▪ Life Sci. **28**, 1975–1986 (1981) ▪ Physiol. Res. (Prag) **40**, 11–24 (1991). – *Synthesis of M. and analogues:* J. Med. Chem. **40**, 1990 (1997) ▪ Synth. Commun. **28**, 3681 (1998). – *[HS 2933 90; CAS 73-31-4]*

Melbex see mycophenolic acid.

Melianol.

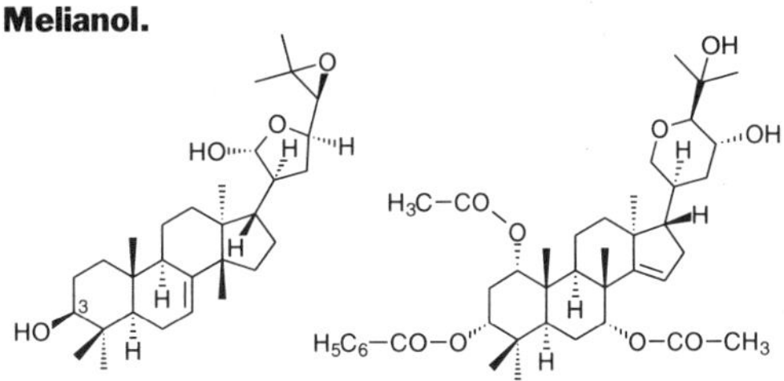

Melianol Melianin A

$C_{30}H_{48}O_4$, M_R 472.71, mp. 194–195 °C, $[\alpha]_D$ −38° ($CHCl_3$). A *triterpene from the fruit of the Persian lilac (*Melia azedarach*), where it occurs together with *melianone* (3-ketone, $C_{30}H_{46}O_4$, M_R 470.69), *melianin A* ($C_{41}H_{58}O_9$, M_R 694.91) and the epoxide ring-opened 24,25-diol *meliantriol* ($C_{30}H_{50}O_5$, M_R 490.72), which has an antifeedant effect on grasshoppers.

Lit.: Beilstein EV **18/3**, 386 ▪ Chem. Pharm. Bull. **34**, 100 (1986) ▪ J. Chem. Soc., Chem. Commun. **1975**, 517. – *[CAS 16838-01-0 (M.); 6553-27-1 (melianone); 57589-59-0 (melianin A); 25278-95-9 (meliantriol)]*

Meliatoxins. Toxic components of the *limonoid group (triterpenes) from Meliaceae species, e.g., the Persian lilac (*Melia azedarach*) and *Trichilia roka* (see figure). Structural relatives of the M. are the trichilins, isolated from *M. azedarach* and from *T. roka*[1]. M. A_1 is also known as trichilin D; except for the fact that the

Table: Data of Meliatoxins.

compound	molecular formula	M_R	mp. [°C]	CAS
M. A_1	$C_{35}H_{46}O_{12}$	658.74	148–154 (decomp.)	87725-70-0
M. A_2	$C_{34}H_{44}O_{12}$	644.72	powder 155–160 (decomp.)	87617-82-1
M. B_1	$C_{35}H_{46}O_{12}$	658.74	powder 140–150 (decomp.)	87617-81-0
M. B_2	$C_{34}H_{44}O_{12}$	644.72	powder 155–162 (decomp.)	87617-80-9
Trichilin A	$C_{35}H_{46}O_{13}$	674.74	cryst. 191–192 (decomp.)	77182-69-5
Trichilin B	$C_{35}H_{46}O_{13}$	674.74	oil	77210-33-4
Trichilin C	$C_{35}H_{46}O_{13}$	674.74	oil	77182-68-4
Azadirachtanin	$C_{32}H_{40}O_{11}$	600.66	cryst. 225–226	98798-58-4
Sendanin	$C_{32}H_{40}O_{12}$	616.66	cryst. 254–255	62078-28-8
Amoorastatin	$C_{28}H_{36}O_9$	516.59	cryst. 205	68985-96-6

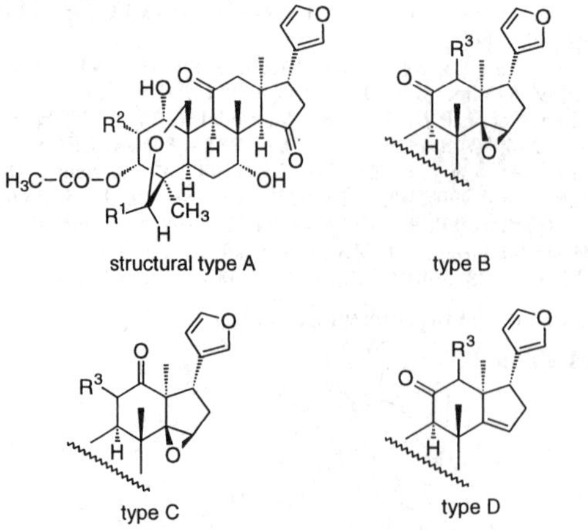

compound	structural type	R^1	R^2	R^3
M. A_1 (Trichilin D)	B	O–CO–CH(CH_3)–C_2H_5	O–CO–CH_3	H
M. A_2	B	O–CO–CH$(CH_3)_2$	O–CO–CH_3	H
M. B_1	A	O–CO–CH(CH_3)–C_2H_5	O–CO–CH_3	-
M. B_2	A	O–CO–CH$(CH_3)_2$	O–CO–CH_3	-
Trichilin A	B	O–CO–CH(CH_3)–C_2H_5	O–CO–CH_3	β-OH
Trichilin B	B	O–CO–CH(CH_3)–C_2H_5	O–CO–CH_3	α-OH
Trichilin C	C	O–CO–CH(CH_3)–C_2H_5	O–CO–CH_3	β-OH
Azadirachtanin	D	O–CO–CH_3	H	α-O–CO–CH_3
Sendanin	B	O–CO–CH_3	H	α-O–CO–CH_3
Amoorastatin	B	OH	H	H

Figure: Structure of Meliatoxins.

configurations of the cyclic acetal and the 2-methylbutyrate residues (R^1) have not been checked for identity, the structures are identical. *Azadirachtanin* {$[\alpha]_D$ +8.7° ($CHCl_3$)} and *sendanin* {$[\alpha]_D^{16}$ +4.3° ($CHCl_3$)} occur in Indian lilac (*Antelaea azadirachta*, syn. *Melia azadirachta*, *Azadirachta indica*, neem tree); the cytotoxic component *amoorastatin* occurs in *Aphanamixis grandiflora*.

Activity: M., trichilins, and azadirachtanin have strong antifeedant effects on insects, they cause growth impairments in, e.g., caterpillars of the forest pest *Spodoptera*; in mammals M. cause severe poisoning symptoms of vomiting, cardiac weakness, nervous disorders, or death.

Lit.: [1] Bioorg. Med. Chem. **4**, 1355 (1996); J. Nat. Prod. **61**, 179 (1998); Heterocycles **36**, 725 (1993); **38**, 2407 (1994); Phytochemistry **41**, 117 (1996).
gen.: Bull. Chem. Soc. Jpn. **67**, 2468 (1994) (M. A_2) ▪ Phytochemistry **22**, 531 (1983) (M. B) ▪ Zechmeister **45**, 1–102 ▪ see also azadirachtin.

Melilotic acid (2-hydroxyhydrocinnamic acid, hydrocoumaric acid).

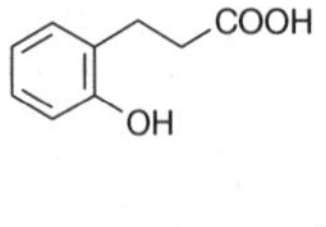

$C_9H_{10}O_3$, M_R 166.18, cryst., monohydrate, mp. 82–83 °C, pK_{a1} 4.75 (25 °C), occurs in the orchid *Angraecum fragans*, white sweet clover *Melilotus albus*, *Dipteryx odorata* and other plants in the free form or as the 2-*O*-*β*-D-glucoside, $C_{15}H_{20}O_8$, M_R 328.32. The lactone *melilotin* (3,4-dihydrocoumarin), $C_9H_8O_2$, M_R 148.16, cryst., mp. 25 °C, bp. 145 °C (2.3 kPa), LD_{50} (rat p. o.) 1460 mg/kg, isolated from *Melilotus*- and *Artemisia* species as well as the urine of red deer (*Cervus elaphus*)[1], is used in the perfume industry for sweet-woody fragrances. M. and the glucoside show antiulcer activities[2].

Lit.: [1] Comp. Physiol. **97A**, 427–431 (1990). [2] Planta Med. **55**, 245–248 (1989).
gen.: Acta Crystallogr., Sect. C **48**, 1076 (1992) ▪ Aust. J. Chem. **42**, 1235–1248 (1989) ▪ Flavour Fragr. J. **6**, 63–68 (1991) ▪ J. Org. Chem. **55**, 5867–5877 (1990) ▪ Luckner (3.), p. 386, 390 ▪ Synthesis **1991**, 739 f. ▪ Xenobiotica **21**, 499–514 (1991). – *Pharmacology:* Contact Derm. **15**, 289–294 (1986) ▪ Mutagenesis **5**, 3–14, 425–432 (1990); 6, 423 ff. (1991) ▪ Toxicol. Appl. Pharmacol. **97**, 311–323 (1989). – *Toxicology:* Sax (8.), HHR 500. – *[CAS 495-78-3 (M.); 24696-05-7 (2-O-β-D-glucoside); 119-84-6 (melilotin)]*

Melilotoside see coumarinic acid.

Melissic acid (triacontanoic acid).
$H_3C–(CH_2)_{28}–COOH$, $C_{30}H_{60}O_2$, M_R 452.81, mp. 93.5–94 °C. The saturated M. a. is a component of bee, wool, and *carnauba wax as well as other plant waxes. In the past hentriacontanoic acid was also named as M. as a result of analytical inaccuracies. The same was true for 1-hentriacontanol.

Lit.: Beilstein E IV **2**, 1321 ▪ J. Org. Chem. USSR (Engl. Trans.) **26**, 1402 (1990) (synthesis) ▪ Karrer, No. 713 ▪ Prog. Lipid Res. **28**, 147 (1989) ▪ Ullmann (5.) **A 10**, 248. – *[HS 2915 90; CAS 506-50-3]*

Melissyl alcohol see 1-triacontanol.

Melitriose see raffinose.

Melittin.

H–Gly–Ile–Gly–Ala–Val–Leu–Lys–Val–Leu–Thr–Thr–Gly–Leu–Pro–Ala–Leu–Ile–Ser–Trp–Ile–Lys–Arg–Lys–Arg–Gln–Gln–NH_2

$C_{131}H_{229}N_{39}O_{31}$, M_R 2846.50, yellowish, water-soluble powder, LD_{50} (mouse i. v.) 3.5 mg/kg. The strongly basic M. (isoelectric point at pH = 10), composed of 26 amino acids, is the main component of *bee venom (50% of the dry weight). The molecule is ambiphilic since a non-polar amino acid occurs at the N-terminal end while a basic polar amino acid is at the C-terminal end. Thus, M. has a high membrane activity and, as ionophore, promotes the formation of ion channels in the membrane.

Lit.: Biochemistry **26**, 6627 (1987) (synthesis) ▪ Biochim. Biophys. Acta **1031**, 143–161 (1990) ▪ J. Am. Chem. Soc. **118**, 8989 (1996) (synthesis) ▪ Merck-Index (12.), No. 5867. – *[CAS 20449-79-0]*

Melittoside see monomelittoside.

Melledonal A see Armillaria sesquiterpenoids.

Mellein (ochracin, 8-hydroxy-3-methyl-1-isochromanone).

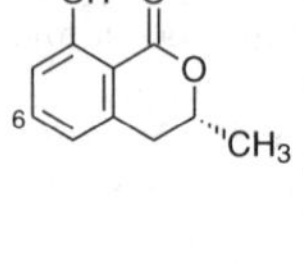

(*R*)-form

$C_{10}H_{10}O_3$, M_R 178.19, cryst., mp. 58 °C. M. occurs in nature both in the (*R*)- and in the (*S*)-form in many mold fungi (e.g., *Aspergillus ochraceus*, *A. melleus*), phytopathogenic fungi (*Septoria nodorum*), and ascomycetes (*Pezicula livida*, *Hypoxylon* species). M. has weak antibacterial and antifungal activities. The wax moth *Aphomia sociella* uses M., arising from *A. ochraceus* associated with the insect, as a pheromone precursor. 6-Methoxymellein occurs in carrots as a phytoalexin where its synthesis is induced by fungal attack or stress. The defence secretions of ants and termites also contain mellein, e.g., M. acts as a trail pheromone component of the ant *Lasius fuliginosus* [1].

Lit.: [1] J. Chem. Ecol. **23**, 779–792 (1997).
gen.: ACS Symp. Ser. **330**, 295 (1987) ▪ Agric. Biol. Chem. **50**, 997 (1986) ▪ Beilstein EV **18/1**, 274 ▪ Gazz. Chim. Ital. **121**, 455 (1991) ▪ Heterocycles **33**, 357 (1992) (synthesis) ▪ Tetrahedron **41**, 5295 (1985) ▪ Tetrahedron: Asymmetry **8**, 2153 (1997) ▪ Zechmeister **49**, 1–50. – *[CAS 17397-85-2 (M.); 480-33-1 (R-form); 62623-81-1 (S); 1200-93-7 (racemate)]*

Mellitoxin see coriamyrtin.

Melon flavor see fruit flavors.

Meloxin see bergapten.

Menaquinone, menadione see vitamin K_1.

***p*-Menthadienes.** Doubly unsaturated, monocyclic *monoterpenes derived from *p*-menthane. $C_{10}H_{16}$, M_R 136.23, liquids, sensitive to light, air, and heat.
Occurrence: *p*-M. are widely distributed in nature. The table (see below) lists those plants from which larger amounts of *p*-M. can be isolated from the essential oils. *β*-Terpinene and (−)-*β*-phellandrene occur in very small amounts only.
Use: *α*-Terpinene as an inhibitor of self-polymerization of tetrafluoroethylene. Terpinols as additives in shoe cleaning products and furniture polishes. Phellandrene and limonene as perfume in cheap soaps. Di-

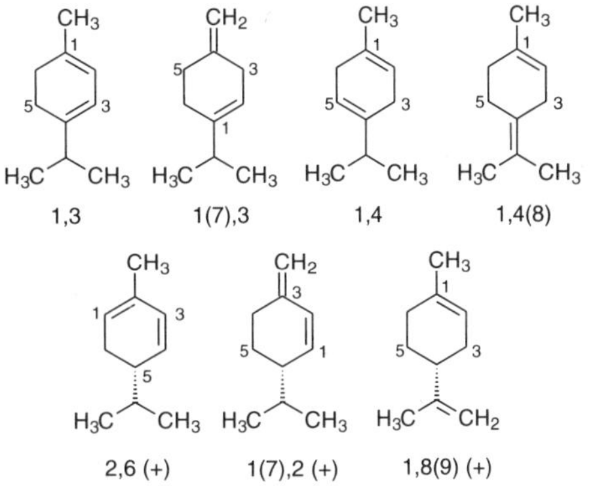

The numbering shown in the formulae is the Chemical Abstracts nomenclature, the *p*-menthadiene nomenclature is given below the formulae.

pentene as solvent and thinner in the paint and varnish industry.

Lit.: Hegnauer **IV**, 351 ▪ Karrer, No. 46–55. – *Limonenes:* IARC Monogr. **56**, 135 (1993) ▪ J. Org. Chem. **58**, 3998 (1993) (biosynthesis) ▪ Nat. Prod. Rep. **6**, 291 (1989). – *β-Phellandrene*: J. Chem. Ecol. **16**, 2519 (1990). – *[HS 2902 19]*

***p*-Menthane** [1-methyl-4-(1-methylethyl)-cyclohexane].

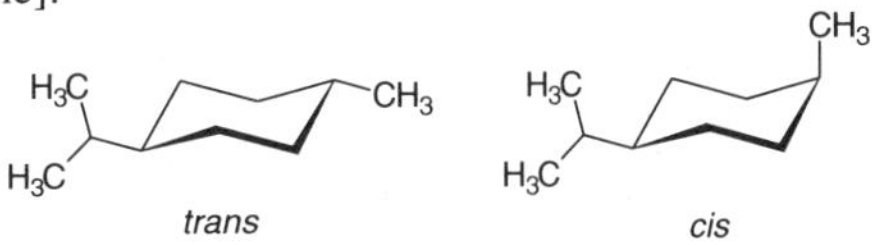

$C_{10}H_{20}$, M_R 140.27, liquid with a fennel-like odor. A monocyclic *monoterpene, bp. *trans*-form 170.6 °C, *cis*-form 172 °C. *p*-M. occurs in the essential oils of the Asian plants *Gardenia taitensis* (Rutaceae), *Syzygium cuminii* (Myrtaceae), and *Hibiscus syriacus* (Malvaceae), detected by GC/MS only.

Table: Physical data and occurrence of *p*-Menthadienes.

	trivial name	bp. [°C]	$[\alpha]_D$	CAS	occurrence	yield [%]
1,3-M.	*α*-Terpinene	173–175	–	99-86-5	*Litsea ceylanica* (Lauraceae)	20
1(7),3-M.	*β*-Terpinene	75 (22 mbar)	–	99-84-3	pittosporum oil (*Pittosporum tenuifolium*, Pittosporaceae)	
1,4-M.	*γ*-Terpinene	183	–	99-85-4	ajowan seed oil (*Carum copticum*, Apiaceae)	20–30
1.4(8)-M.	Terpinolene	100 (18 mbar)	–	586-62-9	common parsnip oil (*Pastinaca sativa*, Apiaceae)	40–70
2,6-M.	(+)-(*S*)-*α*-Phellandrene (−)-(*R*)-*α*-Phellandrene	173–175	±177.4°	2243-33-6 4221-98-1	*Ridolfia segetum* *Zanthoxylum alatum* (Rutaceae)	85 50
1(7),2-M.	(+)-(*S*)-*β*-Phellandrene (−)-(*R*)-*β*-Phellandrene	172–174	±74.4°	6153-16-8 6153-17-9	water fennel oil (= *Oenanthe aquatica* = *Phellandrium aquaticum*) (Apiaceae)	80
1.8(9)-M.	(+)-(*R*)-Limonene (−)-(*S*)-Limonene	175–176	±123.3°	5989-27-5 5989-54-8	neroli oil (*Citrus nobilis* var. *deliciosa* Rutaceae) turpentine oil (*Pinus serotina*, Pinaceae)	90 80
	(±)-Limonene (= Dipentene)	177–178	–	7705-14-8	*Zanthoxylum piperitum* (Rutaceae)	54

Lit.: Beilstein E IV **5**, 151 ▪ J. Chem. Soc. (C) **1966**, 946 (synthesis) ▪ Karrer, No. 3519, 3520. – *[HS 2902 19; CAS 99-82-1 (p-M.); 1678-82-6 (trans-p-M.); 6069-98-3 (cis-p-M.)]*

***p*-Menthane-1,8-diol** see terpin.

3-*p*-Menthanols [5-methyl-2-(1-methylethyl)cyclohexanols].

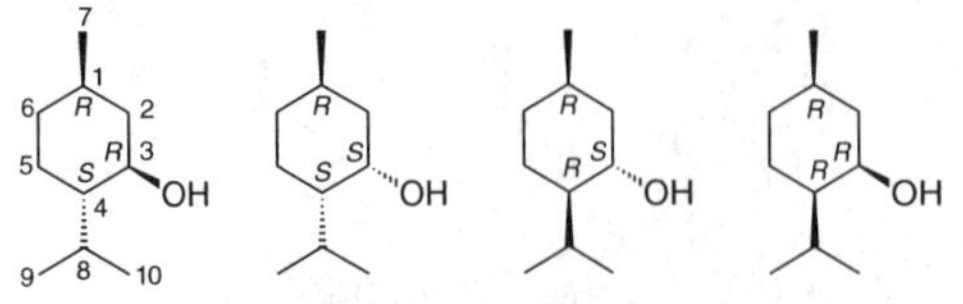

(-)-Menthol (1) (+)-Neomenthol (2) (+)-Isomenthol (3) (+)-Neoisomenthol (4)

Table: *p*-Menthanols.

	mp. [°]	bp. [°]	$[\alpha]_D$	CAS
1	43	216.5	–50° (C_2H_5OH)	2216-51-5
2	–	212	+19.6° (undiluted)	2216-52-6
3	82	218	+28.5° (ether)	23283-97-8
4	–	214	+2.2° (C_2H_5OH)	20752-34-5

$C_{10}H_{20}O$, M_R 156.27. Monocyclic *monoterpenes with three centers of chirality. Only four of the eight possible optically active forms occur in nature.
Occurrence: (–)-Menthol (85%) and (+)-neomenthol (3%) in the essential oil of field mint (*Mentha arvensis*, Lamiaceae). (+)-Isomenthol and (+)-neoisomenthol in *peppermint oils (*Mentha piperita*, Lamiaceae).
Use: (–)-Menthol exhibits particular physiological properties not shown by (+)-M. and the other diastereoisomers. Thus, only enantiomerically pure (–)-M. should be used: in alcoholic solution (–)-M. has antiseptic activity, it increases blood pressure, has spasmolytic effects in the stomach, intestines, and bile ducts, and relieves from itching. On application to the skin (–)-M. generates a pleasant cooling feeling in cases of migraine and other pain symptoms as a result of its local anesthetic and stimulating effects on cold-sensitive nerves. It is thus used medicinally as a component of creams and ointments. It is also used in oral hygiene on account of its refreshing effects. Large amounts are used in the liquor, confectionery, and tobacco industries. (–)-M. is applied in organic chemistry for enantioselective syntheses.
Lit.: Beilstein E IV **6**, 150 ▪ Karrer, No. 292, 293 ▪ Martindale (31.), p. 1724 ▪ Nicolaou, p. 343–380 (synthesis, review of menthol) ▪ Tetrahedron **50**, 3639 (1994) (manufacture) ▪ Top. Catal. **4**, 271 (1998) (synthesis menthol) ▪ Ullmann (5.) **A 11**, 167. – *[HS 2906 11]*

***p*-Menthanones** [5-methyl-2-(1-methylethyl)-cyclohexanones].

(-)-Menthone (+)-Isomenthone

Table: *p*-Menthanones.

trivial name	bp. [°C]	$[\alpha]_D$	CAS
(–)-Menthone (+)-Menthone	209–210	±29.9°	14073-97-3 3391-87-5
(+)-Isomenthone (–)-Isomenthone	212	±93.8°	1196-31-2 491-07-6

$C_{10}H_{18}O$, M_R 154.25, liquid. Monocyclic *monoterpene ketones, the four optically active forms occur in nature. (–)-M. (ca. 35%) occurs in the essential oil of field mint (*Mentha arvensis*, Lamiaceae), (+)-M. in *Micromeria biflora* (Lamiaceae) ca. 56%, (+)-iso-M. in *Micromeria abyssinica* ca. 42%, and (–)-iso-M. in geranium oil (*Pelargonium* spp., Geraniaceae). *p*-M. is used in perfumes as a mixture with other fragrance substances.
Lit.: Agric. Biol. Chem. **53**, 2517 (1989) (synthesis, menthone) ▪ Beilstein E IV **7**, 87 ▪ Chem. Commun. **1998**, 221 (biosynthesis menthone, isomenthone) ▪ Food Cosmet. Toxikol. **14**, 315 (1976) (isomenthone) ▪ Karrer, No. 545, 546 ▪ Tetrahedron Lett. **38**, 3889 (1997) (biosynthesis).

***p*-Mentha-1,3,8-triene** see parsley leaf/seed oil.

***p*-Menth-1-ene-8-thiol.**

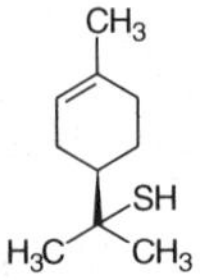

(+)-(*R*)-form

$C_{10}H_{18}S$, M_R 170.31, oil, bp. 40 °C (0.13 Pa). The *impact compound in the flavor of grapefruit juice with an extremely low olfactory threshold of 0.02 ppt for the (+)-(*R*)- and 0.08 ppt for the (–)-(*S*)-forms.
Lit.: Helv. Chim. Acta **65**, 1785–1794 (1982) ▪ Nachr. Chem. Tech. Lab. **30**, 920 (1982) ▪ Ohloff, p. 135. – *[CAS 71159-90-5]*

***p*-Menthenols.** Unsaturated monocyclic *monoterpene alcohols with the *p*-menthane structure. $C_{10}H_{18}O$, M_R 154.25, colorless liquids.
p-M. can be isolated from the essential oils of the plants listed in the table. Acetylated *p*-M. are used in the per-

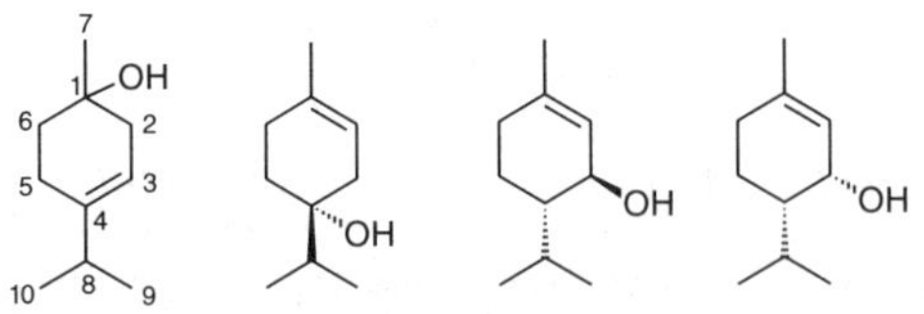

1 (-)-**2** **3 a** **3 b**

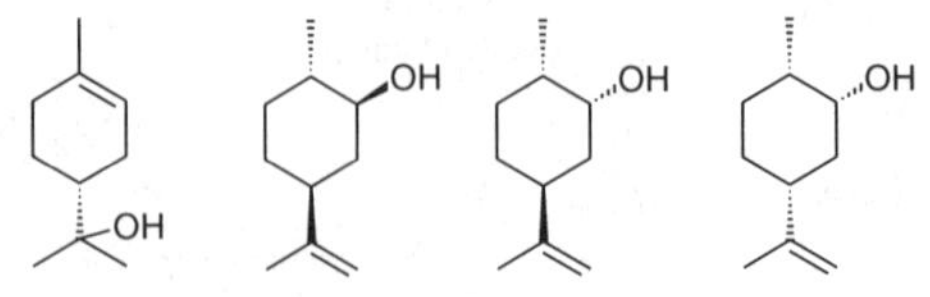

(+)-(R)-**4** (+)-**5 a** (-)-**5 b** (-)-**5 c**

Table: Physical data and occurrence of *p*-Menthenols.

	trivial name	bp. [°C]	$[\alpha]_D$	CAS	occurrence
1	Terpinen-1-ol	208–210	–	586-82-3	origano (*Origanum vulgare*, Lamiaceae)
2	(–)-Terpinen-4-ol	70–71 (300 Pa)	±48.3°	20126-76-5	*Melaleuca linariifolia* (Myrtaceae)
	(+)-Terpinen-4-ol			2438-10-0	*Eucalyptus australiana*
3a	(–)-*trans*-Piperitol	57 (15 Pa)	–28°	25437-28-9	*Andropogon* sp. (unknown species) (Poaceae)
b	(+)-*cis*-Piperitol	55 (10 Pa)	±30–40°	65733-27-9	*Eucalyptus radiata* (Myrtaceae)
c	(–)-*cis*-Piperitol			34350-53-3	
4	(+)-α-Terpineol	104 (1.5 kPa)	±98.4°	7785-53-7	*pine oil
	(–)-α-Terpineol			10482-56-1	(*Pinus palustris*, Pinaceae)
5a	(+)-Dihydrocarveol	106–107 (1.5 kPa)	+33.5°	22567-21-1	caraway
b	(–)-Neodihydrocarveol	214–215	–32.5°	22567-22-2	(*Carum carvi*, Apiaceae)
c	(–)-Neoisodihydrocarveol	104 (900 Pa)	–28.0°	53796-80-8	
6a	(–)-Isopulegol	94 (1.4 kPa)	–23.5°	89-79-2	round-leaved mint
b	(+)-Neoisopulegol	68 (300 Pa)	+1.2°	21290-09-5	(*Mentha rotundifolia*, Lamiaceae)
7	*trans*-Shisool	225	–	22451-48-5	2.2–8.2% in *Perilla acuta* (Lamiaceae) including a very small amount of *cis*-Shisool
	cis-Shisool	?	–	22521-57-9	

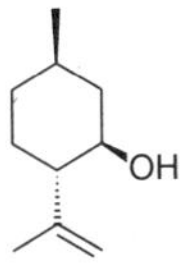
(–)-**6 a**

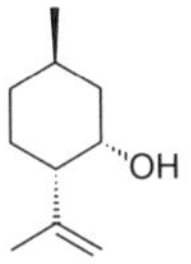
(+)-**6 b**

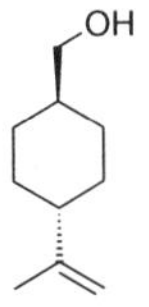
trans-**7**

fume industry since they have varying scents, e.g., α-terpineol has a pleasant, lilac-like odor.

Lit.: Aust. J. Chem. **46**, 1869–1879 (1993) ▪ Helv. Chim. Acta **57**, 2062 (1974) (synthesis, terpinenol) ▪ J. Indian Chem. Soc. **48**, 993 (1971) (synthesis, terpineol) ▪ Karrer, No. 296–298, 303–306 ▪ Phytochemistry **47**, 1117 (1998) (NMR) ▪ Tetrahedron **32**, 1437 (1976) (synthesis, piperitol) ▪ Tetrahedron Lett. **39**, 1997 (1998) (synthesis isopulegol, neoisopulegol) ▪ Ullmann (5.) **A 11**, 170.

***p*-Menthenones.** Unsaturated, monocyclic *monoterpene ketones with the *p*-menthane structure. $C_{10}H_{16}O$, M_R 152.24, colorless liquids. For occurrence, see table. *p*-M. are used in the synthesis of *3-*p*-menthanols.

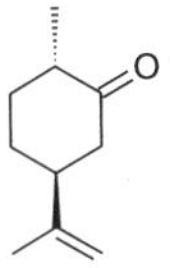
(–)-**1**

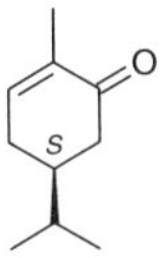
(+)-**2**

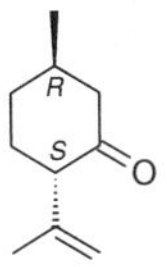
(–)-**3**

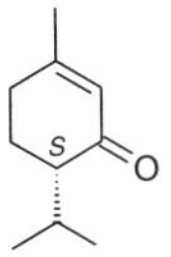
(+)-**4**

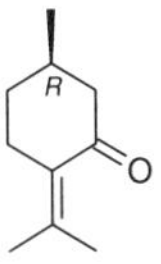
(+)-**5**

Lit.: Karrer, No. 548–552. – *Isopulegone:* J. Indian Chem. Soc. **53**, 50 (1976) (synthesis). – *Pulegone:* J. Org. Chem. **41**, 380 (1976) (synthesis) ▪ Tetrahedron Lett. **38**, 3889 (1997) (biosynthesis). – *[HS 2906 14, 2906 19]*

Menthiafolin see foliamenthin.

(+)-Menthofuran see peppermint oils.

Table: Physical data and occurrence of *p*-Menthenones.

	trivial name	bp. [°C]	$[\alpha]_D$	CAS	occurrence	yield [%]
1a	(–)-*trans*-Dihydrocarvone	220	–16.8°	619-02-3	caraway	
b	(–)-*cis*-Dihydrocarvone	221	–23.7°	53796-79-5	(*Carum carvi*, Apiaceae)	<1
2	(+)-Carvotanacetone	227 (70.7 kPa)	±59.6°	499-71-8	*Blumea malcolmii* (Asteraceae)	82
	(–)-Carvotanacetone			33375-08-5	*Eucalyptus deglupta* (Myrtaceae)	1
3	(–)-Isopulegone	102 (1.7 kPa)	–13.5°	17882-43-8	*Hedeoma pulegioides* (Lamiaceae)	10–18
4	(+)-Piperitone	233	±67.8°	6091-50-5	*Cymbopogon sennaarensis* (Poaceae)	45
	(–)-Piperitone			4573-50-6	*Orthodon perforatum* (Lamiaceae)	55
5	(+)-Pulegone	104–108 (2.4 kPa)	±22.5°	89-82-7	*Mentha rotundifolia* (Lamiaceae)	95
	(–)-Pulegone			3391-90-0	*Agastache formosana* (Lamiaceae)	80

(–)-Menthol see 3-*p*-menthanols.

(–)-Menthone see *p*-menthanones.

(–)-Menthyl acetate see peppermint oils.

Mentzeloside see deutzioside.

Mequinol see hydroquinone.

3(6)-Mercapto-1-hexanol see fruit flavors (passion fruit, guava flavors).

(+)-*trans*-8-Mercapto-*p*-menthan-3-one see buchu leaf oil.

3-Mercapto-3-methylbutyl formate see coffee flavor.

Meridine.

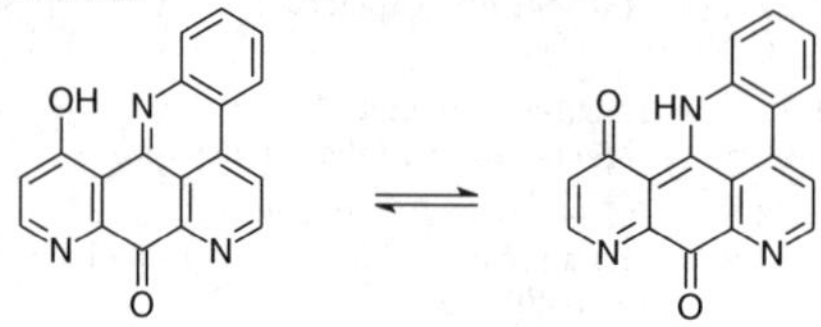

$C_{18}H_9N_3O_2$, M_R 299.29, an alkaloid (both tautomers isolable) from the ascidian *Amphicarpa meridiana*, yellow solid, mp. >250 °C, with antifungal activity.

Lit.: Chem. Pharm. Bull. **42**, 1363 (1994); Tetrahedron **53**, 1743 (1997); **54**, 8421 (1998) (synthesis). – *[CAS 129722-90-3]*

Merulidial.

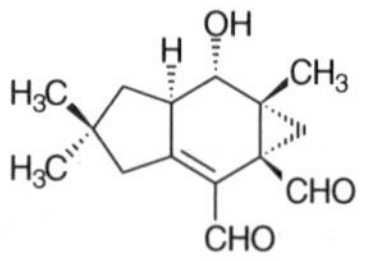

$C_{15}H_{20}O_3$, M_R 248.32, cryst., mp. 99 °C, $[\alpha]_D$ –154° (CH_3OH). A sesquiterpenoid of the *isolactarane type from cultures of *Merulius tremellosus* (Basidiomycetes). M. is active against bacteria, fungi, and yeasts. It also has cytotoxic activity against ascites cells of the Ehrlich carcinoma and shows strong mutagenicity in the Ames test.

Lit.: J. Antibiot. **31**, 737 (1978); **42**, 738 (1989) (activity) ▪ Tetrahedron **42**, 3579 (1986); **46**, 2389 (1990) ▪ Tetrahedron Lett. **33**, 4541 (1992) (synthesis). – *[CAS 68053-32-7]*

Mescaline (3,4,5-trimethoxyphenylethylamine). Formula and data, see *β*-phenylethylamine alkaloids. M. is soluble in water, alcohol, and chloroform. With acids, M. forms water-soluble crystalline salts. The orange color with formaldehyde in sulfuric acid solution is used as a quick test. M. is the most important *Lophophora alkaloid from the Mexican cactus *Lophophora williamsii* (*peyote). Its pharmacological properties are similar to those of anhalonidine, lophophorine, pellotine (see Anhalonium alkaloids), etc.: M. has a paralysing effect on the CNS, in larger doses (>400 mg) it acts hypotensively, causes bradycardia, respiratory depression, and vasodilatation. In high concentrations M. induces a persisting paralysis. The best known activities of M. (doses of 100 – 200 mg), however, are the generation of visual, colored hallucinations, changes in the senses, in thinking, judgement, schizophrenia, loss of sense of time and space, etc. In this respect M. is comparable to other hallucinogenic drugs such as LSD and *psilocybine. Similar to LSD, M. also occasionally stimulates the vomiting center of the CNS. It causes dilatation of the pupils and recontraction does not occur even in bright light, thus Mexican Indians consume *peyote at night only. On acute oral consumption M. is weakly toxic [LD_{50} (mouse p.o.) 800 mg/kg].

Historical: M. is one of the oldest known hallucinogenic drugs; it was used in pre-Columbian times by tribes in Central America. The name is derived from the Mexican: mexcalli (intoxicating drink from plants).

Lit.: Beilstein E IV **13**, 2919 ▪ Hager (5.) **3**, 775 ff.; **5**, 708 f. ▪ Martindale (30.), p. 1388 ▪ Merck-Index (12.), No. 5965 ▪ Sax (8.), MDI 500, MDI 750 ▪ Ullmann (5.) **A 1**, 357. – *[HS 2939 90; CAS 54-04-6]*

Mesembrine alkaloids.

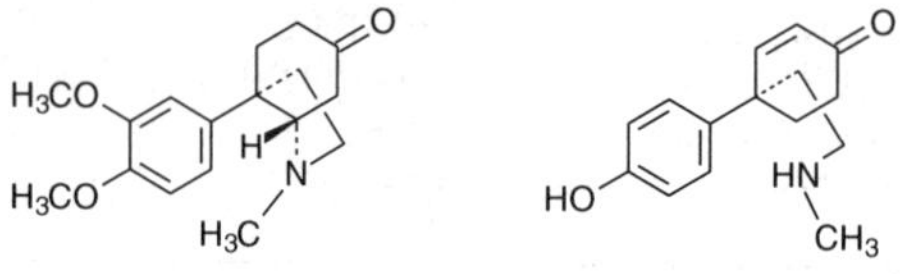

A group of *indole alkaloids from *Mesembryanthemum* species. (–)-*Mesembrine*, [$C_{17}H_{23}NO_3$, M_R 289.37, oil, bp. 186 – 190 °C (40 Pa), $[\alpha]_D^{20}$ –55.4° (CH_3OH)] has a cocaine-like activity. The plants are used in South West Africa to prepare the drug Channa. *Joubertiamine* ($C_{15}H_{19}NO_2$, M_R 245.32) occurs together with mesembrine. The biosynthesis of M. a. is similar to that of Amaryllidaceae alkaloids of the crinine group and proceeds from *phenylalanine *via* cinnamic acid, *p*-coumaric acid, and 3-(4-hydroxyphenyl)propanoic acid. The latter is then coupled with *N*-methyltyramine.

Lit.: Beilstein E V **21/13**, 169 f. ▪ Manske **9**, 467 – 481. – *Biosynthesis:* J. Chem. Soc., Chem. Commun. **1977**, 60 ▪ Mothes et al., p. 246 ▪ Phytochemistry **17**, 719 (1978). – *Synthesis:* Chem. Lett. **1989**, 1963 ff. ▪ Heterocycles **42**, 135 (1996) (mesembrine) ▪ J. Org. Chem. **60**, 6785 (1995); **62**, 1675, 3263 (1997) ▪ Tetrahedron: Asymmetry **4**, 1409 (1993) ▪ Tetrahedron **49**, 8503 (1993) (joubertiamine) ▪ Tetrahedron Lett. **26**, 4083 (1985); **33**, 6023, 6999 (1992); **35**, 6499 (1994); **38**, 1893 (1997); **39**, 7747, 8979 (1998). – *[HS 2939 90; CAS 24880-43-1 (mesembrine); 28379-30-8 (joubertiamine)]*

Mesifuran see Furaneol®.

Mesobilirubin (mesobilirubin IX*α*).

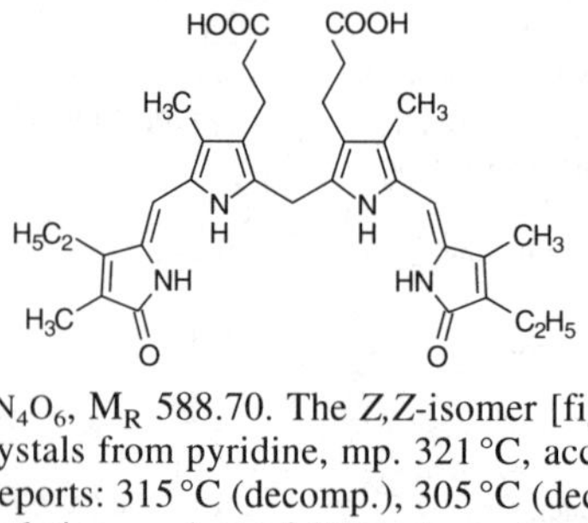

$C_{33}H_{40}N_4O_6$, M_R 588.70. The *Z,Z*-isomer [figure; yellow crystals from pyridine, mp. 321 °C, according to other reports: 315 °C (decomp.), 305 °C (decomp.)] is a degradation product of *bile pigments which, like *bilirubin, *urobilin, and *stercobilin belongs to the linear *tetrapyrroles. It differs from bilirubin by having two ethyl groups in place of two vinyl groups.

Lit.: Beilstein E V **26/15**, 519 ▪ Falk, The Chemistry of Linear Oligopyrroles and Bile Pigments, p. 13, 35, 75, Wien: Springer 1989. – *[CAS 16568-56-2]*

Mesoporphyrin see porphyrins.

Metanephrine see β-phenylethylamine alkaloids.

Methanofuran.

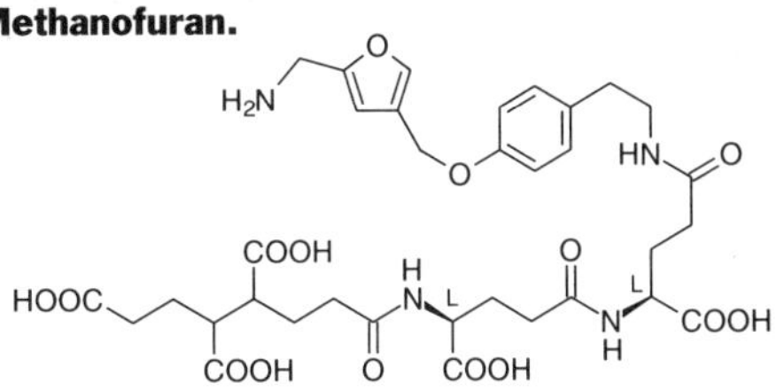

$C_{34}H_{44}N_4O_{15}$, M_R 748.74. M. is the carbon dioxide reducing factor of the archaebacterium *Methanobacterium thermoautotrophicum*.

Lit.: J. Am. Chem. Soc. **115**, 6646 (1993) (synthesis) ▪ J. Biol. Chem. **258**, 7536 (1983) (isolation); **267**, 17574 (1992) (biosynthesis). – *[CAS 89873-36-9]*

***cis*-3,4-Methano-L-proline** see L-proline.

Methional [3-(methylthio)propanal].
$H_3CS-(CH_2)_2-CHO$, C_4H_8OS, M_R 104.17, bp. 165 °C; LD_{50} (rat p. o.) 0.75 g/kg. A Strecker aldehyde [1] formed thermally or enzymatically from methionine with an odor resembling, when diluted, bouillon and boiled potatoes; olfactory threshold 0.2 ppb. M. occurs in many foods prepared thermally or by fermentation and is important for, e. g., the flavor of bread, meat, potatoes (boiled, chips), cheese, and tomato (paste) [1]. It is used as a flavor substance especially for meat and cheese flavors.

Lit.: [1] TNO list (6.) Suppl. 5, p. 481.
gen.: Arctander, No. 2244 ▪ Beilstein E IV **1**, 3974. – *[HS 2930 90; CAS 3268-49-3]*

L-Methionine [(*S*)-2-amino-4-(methylthio)butanoic acid, abbr.: Met, M].

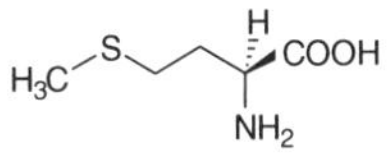

$C_5H_{11}NO_2S$, M_R 149.21, mp. 280–282 °C (decomp. in sealed capillary), $[\alpha]_D^{25}$ –6.87° (H_2O), +22.5° (1 m HCl), pK_a 2.28, 9.21, pI 5.74. Essential proteinogenic amino acid. Genetic code: AUG (important for the initiation of protein biosynthesis on ribosomes). Average content in many proteins 1.7% [1]. Although L-Met has no taste, the D-form is sweet. Met is isolated, e.g. from casein hydrolysate.

Biosynthesis: Like Thr and Lys M. belongs to the Asp group formed from aspartic acid semialdehyde. Degradation of Met gives first *S*-adenosylmethionine (SAM) and then homocysteine via *S*-adenosylhomocysteine; the homocysteine condenses with serine to give *cystathionine, the latter finally decomposes to 2-oxobutanoate and cysteine. SAM is a methyl group-transfer reagent in many enzymatic processes. Met is the starting material for ethylene, a plant hormone (via *1-amino-cyclopropanecarboxylic acid). M. is used as an additive in food and fodder.

Lit.: [1] Biochem. Biophys. Res. Commun. **78**, 1018 (1977).
gen.: Annu. Rev. Biochem. **52**, 187–222 (1983) ▪ Annu. Rev. Nutr. **6**, 179 (1986) ▪ Beilstein E IV **4**, 3189 ▪ Chem. Ind. (Düsseldorf) **35**, 651 f. (1983) (synthesis) ▪ CRC Crit. Rev. Biochem. **23**, Suppl. 1, S1–S42 (1988) (biosynthesis) ▪ Lewis, Food Additives Handbook, MDT 750, New York: Van Nostrand Reinhold 1989 ▪ Trends Biotechnol. **8**, 156–160 (1990) (gene technology). – *[HS 2930 40; CAS 63-68-3]*

Methoxatin see pyrroloquinoline quinone.

4-Methoxyacetophenone see fruit flavors (red currant flavor).

4-Methoxybenzaldehyde see *p*-anisaldehyde.

Methoxybrassenin, methoxybrassinins see brassinin.

7α-Methoxycephalosporins see cephamycins.

2-Methoxycinnamaldehyde see cassia oil.

11-Methoxydiaboline see Wieland-Gumlich aldehyde.

4-Methoxy-2-methyl-2-butanethiol see fruit flavors (red currant flavor).

4-Methoxy-2-(3-methyl-2-butenyl)-phenyl stearate.

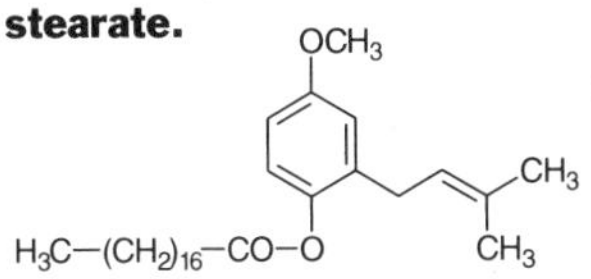

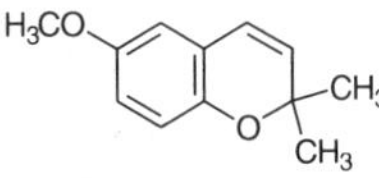

6-Methoxy-2,2-dimethyl-2*H*-chromene

$C_{30}H_{50}O_3$, M_R 458.73, mp. 100–101 °C. A component of the latex from the toadstools *Lactarius fuliginosus* and *L. picinus* (Basidiomycetes). When the fruit bodies are injured the mild-tasting ester is enzymatically saponified to 4-methoxy-2-(3-methyl-2-butenyl) phenol which is responsible for the bitter taste of the fungi. The phenol is unstable and after some time is oxidized to 6-methoxy-2,2-dimethyl-2*H*-chromene, 2-(1-hydroxy-1-methylethyl)-5-methoxybenzofuran, dimeric compounds, and red pigments. M. is thus probably responsible for the pink discoloration of the milk juices on bruising these toadstools.

Lit.: Tetrahedron **35**, 7331 (1992). – *[CAS 144705-42-0]*

Methyl anthranilate (methyl 2-aminobenzoate).

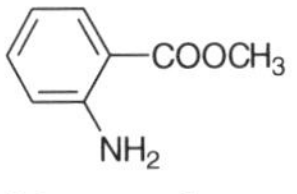

$C_8H_9NO_2$, M_R 151.16. crystals or pale yellow, blue fluorescent liquid with an odor like orange flowers; soluble in ethanol, poorly soluble in water; bp. 237 °C, mp. 24.5 °C, LD_{50} (rat p. o.) 2.9 g/kg.

Occurrence: In many flowers (oils), e. g., wallflower, *jasmin(e) absolute, *orange oils, *tuberose absolute, and *ylang-ylang oil as well as the juice of the Concord grape (*Vitis labrusca*). M. is also responsible for the characteristic odor of the fruit bodies of *Cortinarius odoratus* and *Hygrophoropsis olida*.

Use: In flower perfumes and to prepare Schiff's bases [1], that are also used as fragrance substances.

Lit.: [1] Bauer et al. (2.), p. 96.
gen.: Arctander, No. 1910 ▪ Bedoukian (3.), p. 48 ▪ Beilstein E IV **14**, 1008 ▪ Fenaroli (2.) **2**, 346 ▪ TNO list (6.), Suppl. 5, p. 277. – *[HS 2922 49; CAS 134-20-3]*

(Methyl-*ONN*-azoxy)-methanol see cycasin.

Methyl benzoate see ylang-ylang oil.

Methyl-1,4-benzoquinone (methyl-*p*-benzoquinone, toluquinone).

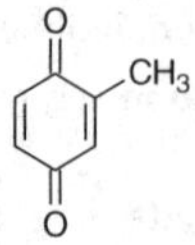

$C_7H_6O_2$, M_R 122.12, yellow cryst., mp. 69 °C. A widely distributed component of the *defensive secretions of insects and millipedes. In some species of beetles the repellent activity is enhanced by solvents (hydrocarbons, lactones, carboxylic esters) which facilitate penetration of M. through the predator's cuticle[1]. For biosynthesis, see ethyl-1,4-benzoquinone.

Lit.: [1] Entomol. Gener. **15**, 275–292 (1991); J. Nat. Prod. **56**, 1700 (1993).

gen.: Biochem. Syst. Ecol. **21**, 143–162 (1993) ▪ Blum, Chemical Defenses of Arthropods, p. 181–195, New York: Academic Press 1981 ▪ J. Org. Chem. **52**, 5053 (1987) (synthesis). – *[HS 2914 69; CAS 553-97-9]*

2-,3-Methylbutanal see alkanals.

Methylbutanoates see fruit esters.

2-Methylbutanoic acid see fruit flavors (strawberry flavor).

3-Methylbutyl acetate see bees.

***S*-Methyl-L-cysteine (*S*-oxide).**

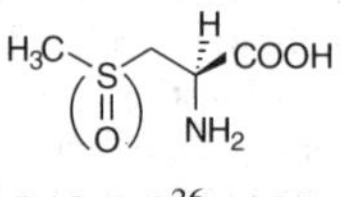

$C_4H_9NO_2S$, M_R 135.18, $[\alpha]_D^{26}$ –32° (H_2O), *N*-formyl derivative, mp. 118–119 °C, $[\alpha]_D^{25}$ –13.5° (H_2O).
Non-proteinogenic amino acids in plants and the simplest of the natural, *S*-substituted Cys derivatives. Isolated from runner beans (*Phaseolus vulgaris*) and other plants[1].
S-Methyl-L-cysteine S-oxide ($C_4H_9NO_3S$, M_R 151.18) occurs in white beet (*Brassica napus*) and in cabbage, mp. 164 °C (decomp.), $[\alpha]_D$ +125° (H_2O), odor of cabbage.

Lit.: [1] Nature (London) **178**, 593 (1956).

gen.: Beilstein E IV **4**, 3145 ▪ J. Chromatogr. **625**, 183–190 (1990) ▪ Phytochem. Anal. **5**, 4–9 (1994). – *[HS 2930 90; CAS 7728-98-5 (M.); 32726-14-0 ((S)-S-oxide)]*

Methyl (2*E*,4*Z*)-2,4-decadienoate see chalcogran.

4-Methyleneglutamic acid (2-amino-4-methylenepentanedioic acid).

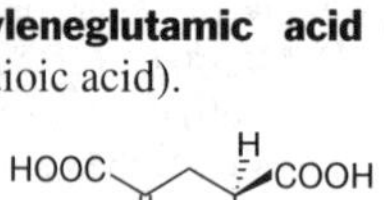

(2*S*)-, L-form

$C_6H_9NO_4$, M_R 159.14, mp. 195–197 °C, $[\alpha]_D$ +14.0° (11% HCl).
Occurrence: Non-proteinogenic amino acid, widely distributed in plants, e.g., in peanut shoots (*Arachis hypogaea*) together with its 4-amide, in shoots of *Amorpha fruticosa*, tulips (*Tulipa gesneriana*), *Phyllitis scolopendrium* (Pteridophytae), *Tetrapleura tetraptera* (Mimosaceae), in the tubers of *Lilium candidum*, and in fruit bodies of *Mycena pura* (Tricholomataceae)[1].
Synthesis: From substituted aziridines and Wittig reagents[2]. For stereospecific synthesis, see *Lit.*[3].
4-M. 5-amide (4-methyleneglutamine), $C_6H_{10}N_2O_3$, M_R 158.16, amorphous, $[\alpha]_D^{18}$ 0° (H_2O), +21° (6 m HCl).
(E)-4-Ethylidene-L-glutamic acid, (2*S*,4*E*)-form, L-*cis*-, $C_7H_{11}NO_4$, M_R 173.16, monohydrate, mp. 198–201 °C, $[\alpha]_D^{25}$ +24.3° (H_2O). Occurs in tulips, *Guilandina crista*, *T. tetraptera*, and *M. pura*[1,3].
(E)-4-Propylidene-L-glutamic acid, $C_8H_{13}NO_4$, M_R 173.17, mp. 162–163 °C (decomp.), $[\alpha]_D^{23}$ +35° (H_2O). Non-proteinogenic amino acid from fruit bodies of *M. pura*[1,3].

Lit.: [1] Phytochemistry **14**, 1434 ff. (1975). [2] J. Chem. Soc., Chem. Commun. **1987**, 153 ff. [3] Tetrahedron Lett. **34**, 4667–4670 (1993). – *[HS 2922 42; CAS 16804-57-2 (M.); 16804-56-1 (4-ethylideneglutamic acid); 56973-40-1 (4-propylideneglutamic acid)]*

4-Methyl(ene)proline see L-proline.

Methylenomycins.

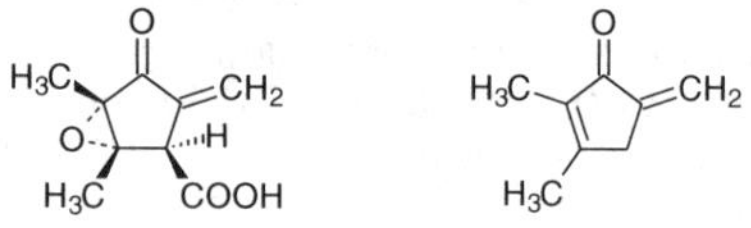

M. A M. B

Cyclopentenoid antibiotics from *Streptomyces violaceoruber* that inhibit the growth of Gram-positive and Gram-negative bacteria as well as KB-tumor cells. *M. A*: $C_9H_{10}O_4$, M_R 182.18, cryst., mp. 115 °C, $[\alpha]_D$ –42.3° ($CHCl_3$), is additionally active against *Proteus vulgaris* and lung tumors in the mouse. *M. B*: $C_8H_{10}O$, M_R 122.17, oil, decomposes. M. A is also formed by *Streptomyces coelicolor* A3(2) (*actinorhodin producer), the model strain for streptomycete genetics. A plasmid-coded biosynthesis has been demonstrated for M. A.

Lit.: J. Antibiot. **38**, 1061–1067 (1985). – *Biosynthesis:* Hopwood, Chater & Bibb, in: Vining & Stuttard (eds.), Genetics and Biochemistry of Antibiotic Production, p. 65–102, Boston: Butterworth-Heinemann 1995 ▪ J. Gen. Microbiol. **98**, 239–252 (1979). – *Synthesis:* J. Org. Chem. **54**, 46–51 (1989); **56**, 1217–1223 (1991); **59**, 5292, 5305 (1994) ▪ Mathew, in: Lukacs (ed.), Recent Progress in the Chemical Synthesis of Antibiotics and Related Microbial Products, vol. 2, p. 435–474, Heidelberg: Springer 1993 ▪ Synlett **1993**, 237 ▪ Tetrahedron Lett. **33**, 2265–2268 (1992). – *[HS 2941 90; CAS 52775-76-5 (M. A); 52775-77-6 (M. B)]*

2-Methyl-3-furanthiol.

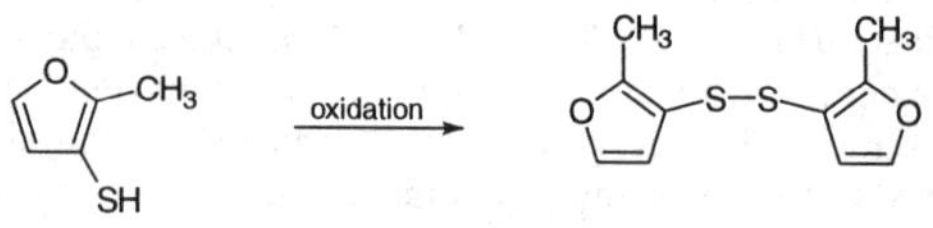

2-Methyl-3-furanthiol Bis(2-methyl-3-furyl)disulfide

C_5H_6OS, M_R 114.16. Liquid with an extremely strong, thiamine-like odor; bouillon-like under high dilution. Bp. (5.5 kPa) 55–56 °C, extremely low olfactory threshold in water: 0.4 ppt; LD_{50} (mouse p.o.) 100 mg/kg. M. is a secondary *flavor compound, which is formed during boiling or roasting of meat by hydrolysis of *thiamin(e) or in a Maillard reaction from, e.g., *cysteine, presumably via 5-hydroxy-3-mercapto-2-pentanone[1]. For synthesis, see *Lit.*[2]. Used as a meat flavor, as is *bis(2-methyl-3-furyl) disulfide* [$C_{10}H_{10}O_2S_2$, M_R 226.31], olfactory threshold in water: 0.02 ppt. The disulfide is an accompanying substance of M. from which it is easily formed by oxidation.

Lit.: [1] Schreier & Winterhalter (eds.), Progress in Flavor Precursor Research, p. 361ff., Carol Stream IL, USA: Allured 1993. [2] Helv. Chim. Acta **76**, 2528–2537 (1993); US. P. 3922288 (1975), International Flavors & Fragrances, Inv.: Evers, Heinsohn, Mayers.
gen.: Beilstein EV **17/3**, 331 ▪ Fenaroli (2.) **2**, 365. – *[CAS 28588-74-1 (thiol); 28588-75-2 (disulfide)]*

4-Methyl-3-heptanol.

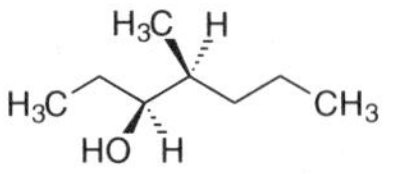

(3*S*,4*S*)-form

$C_8H_{18}O$, M_R 130.23, bp. 160–161 °C. The (3*S*,4*S*)-isomer, $[\alpha]_D^{23}$ −23.3° (hexane), an important component of the aggregation pheromone (see pheromones) of bark beetles of the genus *Scolytus*[1], e.g. the smaller European bark beetle. *S. multistriatus* is a carrier of the fungus *Ceratocystis ulmi* causing death of elm trees. It also occurs in ants. The (3*R*,4*S*)-stereoisomer is a trail pheromone of the ant *Leptogenys diminuta*[2]; see also multistriatin.

Lit.: [1] J. Chem. Ecol. **8**, 477–492 (1982). [2] Naturwissenschaften **75**, 315f. (1988).
gen.: ApSimon **9**, 101–108 ▪ J. Org. Chem. **57**, 3867 (1992) ▪ Synlett. **1996**, 511 (synthesis) ▪ Beilstein EIV **1**, 1789. – *[CAS 68509-48-8 (3S,4S); 63782-91-2 (3S,4R)]*

4-Methyl-3-heptanone.

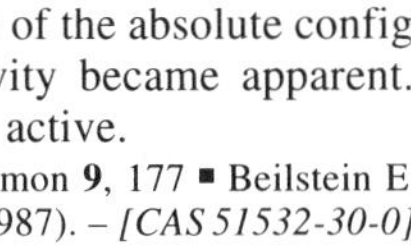

(*S*)-form

$C_8H_{16}O$, M_R 128.21, bp. 63–67 °C, (5.33 kPa) $[\alpha]_D^{20}$ +21.7° [(4*S*)-enantiomer, hexane]. A widely distributed alarm *pheromone in leaf-cutting ants, e.g. *Atta texana*. M. was the first pheromone in which the significance of the absolute configuration for the biological activity became apparent. Only the (4*S*)-enantiomer is active.

Lit.: ApSimon **9**, 177 ▪ Beilstein EIV **1**, 3347 ▪ Org. Synth. **65**, 183 (1987). – *[CAS 51532-30-0]*

6-Methyl-5-hepten-2-one see ants.

Methyl 4-hydroxybenzoate see queen substance.

2-Methylionones see irones.

Methyl jasmonate.

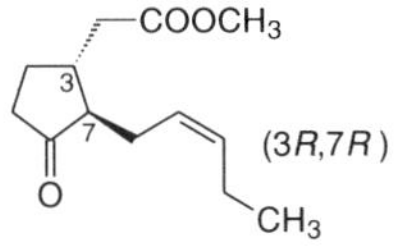

(3*R*,7*R*)

(-)-M. [(-)-*trans*-(*Z*)-M.]
1

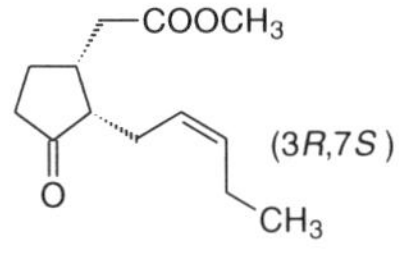

(3*R*,7*S*)

(+)-epi-M. [(+)-*cis*-(*Z*)-M.]
2

$C_{13}H_{20}O_3$, M_R 224.30. Oil with an odor of jasmin(e), bp. 110 °C (0.27 kPa), soluble in organic solvents including alcohol; $[\alpha]_D$ (CH_3OH): (**1**) −76.5°, (**2**) +57.3°; LD_{50} (rat p.o.) 5 g/kg. Compound (**1**) discovered in 1962[1] in *jasmin(e) absolute and synthesized as the (±)-compound has the (3*R*,7*R*) absolute configuration[2]. (**1**) is the thermodynamically preferred epimer and is formed during the processing by epimerization of (**2**) present in the jasmin(e) flowers [according to headspace analysis[3] 97% (**2**) and 3% (**1**)], which is actually responsible for the odor[4].

Occurrence: Apart for jasmin(e) absolute [ca. 1% (**1**) and 0.1% (**2**)], M. occurs in *tea flavor and in various plants where it plays an important role as signal substance[5,6].

Activity: Via the gas phase [10 nL (±)-(1)/L] it is a potent inducer of proteinase-inhibitor proteins in some species of Solanaceae and Fabaceae[7,8]. After uptake in the cells it is presumably rapidly hydrolysed to *jasmonic acid* [$C_{12}H_{18}O_3$, M_R 210.27, viscous oil, bp. 125 °C (0.13 Pa)], a ubiquitous, intracellular, plant signal substance which stimulates the synthesis of special proteins and induces the plant's defensive reaction via a gene activation[9]. *De novo* synthesis of (3*R*,7*S*)-(+)-jasmonic acid and an increase in the intracellular concentration of (3*R*,7*S*)-(+)-jasmonic acid from 1–10 ng/g (dry weight) by one to three orders of magnitude occur within a few minutes, e.g., after contact with microbial cell walls[10], wounding[11], and in shoot spiralization[12]. Exogenous application of (methyl) jasmonate induces, e.g., the synthesis of plant secondary substances, shoot spiralization, new formation of roots, formation of tubers in potatoes, ethylene synthesis, senescence, and inhibits seed germination, callus growth, formation of flowers, and longitudinal growth of roots and shoots[13].

Biosynthesis: From α-linolenic acid by the so-called jasmonate cascade. For synthesis, see *Lit.*[16].

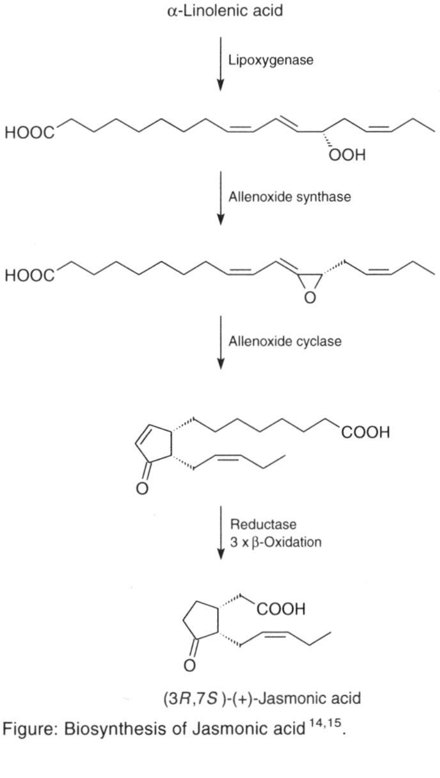

Figure: Biosynthesis of Jasmonic acid[14,15].

Use: As a result of its high price, preferentially in fine perfumes and aromas. The similarly smelling but cheaper *methyl dihydrojasmonate*[17] ($C_{13}H_{22}O_3$, M_R 226.32), occurring, among others, in *tea flavor, is used in larger amounts.

Lit.: [1] Helv. Chim. Acta **45**, 675, 685, 692 (1962). [2] Tetrahedron **21**, 1501 (1965). [3] Teranishi et al. (eds.), Bioactive Volatile Compounds from Plants, p. 188, Washington: ACS 1993. [4] J. Agric. Food Chem. **33**, 425 (1985). [5] Ohloff, p. 151; Phytochemistry **31**, 1111 (1992). [6] Theimer (ed.), Fragrance Chemistry, p. 349–384, San Diego: Academic Press 1982; Müller-Lamparsky, p. 596. [7] Proc. Natl. Acad. Sci. USA **87**, 7713 (1990). [8] Plant Mol. Biology **26**, 1423 (1994). [9] Trends in Cell Biology **2**, 236 (1992). [10] Proc. Natl. Acad. Sci. USA **89**, 2389 (1992). [11] Planta **191**, 86 (1993). [12] Phytochemistry **32**, 591 (1993). [13] Annu. Rev. Plant Physiol. Plant Mol. Biol. **44**, 569 (1993); Nat. Prod. Rep. **15**, 533 (1998) (review, plant defence). [14] Plant Physiol. **75**, 458 (1984). [15] Biochim. Biophys. Acta **1165**, 1 (1992). [16] Justus Liebigs Ann. Chem. **1990**, 751; Synth. Commun. **22**, 1283 (1992); Synlett **1995** (5) 463; **1997**, 618. [17] Bauer et al. (2.), p. 71. – *[HS 2918 90; CAS 1211-29-6 (1); 95722-42-2 (2); 20073-13-6 (±)-(1); 53369-26-9 (±)-(2); 6894-38-8 ((1R)-trans-jasmonic acid); 62653-85-4 ((1R)-cis-jasmonic acid); 24851-98-7 (methyl dihydrojasmonate)]*

S-Methyl-L-methioninesulfonium chloride (cabagin, vitamin U, *S*-methyl-L-methioninium chloride).

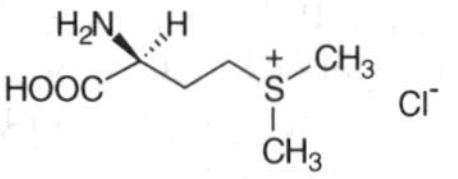

$C_6H_{14}ClNO_2S$, M_R 199.70. Hygroscopic, fine cryst., mp. 139 °C (decomp.), readily soluble in water, pH-value of a 10% solution: ca. 4.5; occurs in cabbage, lettuce, celery, tomatoes, bananas, carrots, asparagus, and yeast.

Activity: M. has a protective effect on the liver and the mucous membranes of the gastrointestinal tract, wherefore it is also known as ulcer protecting factor (vitamin U).

Use: The iodide (as racemate), $C_6H_{14}INO_2S$, M_R 291.15, mp. 156–157 °C, is used for therapy of rheumatic diseases.

Lit.: Arzneim.-Forsch. **28**, 1711 (1978); **29**, 1517 (1979) ▪ Die Nahrung **21**, 32 ff. (1977) ▪ Karrer, No. 2419 ▪ Tetrahedron Lett. **28**, 3605–3608 (1987) (synthesis) ▪ Ullmann (5.) **A 2**, 60, 87. – *[HS 2930 90; CAS 3493-12-7; 1115-84-0 (S); 3493-11-6 (iodide)]*

Methyl N-methylanthranilate.

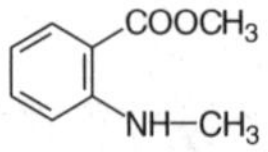

$C_9H_{11}NO_2$, M_R 165.19. Pale yellow, fluorescent oil with a mandarin-like odor, bp. 256 °C, mp. 18.5–19.5 °C.

Occurrence: As the main component in petitgrain oil from mandarin leaves, in smaller amounts also in *mandarin oil and other *citrus oils, as well as in carambola, the fruit of *Averrhoa carambola* (Oxalidaceae).

M. is used in artificial mandarin aroma and in the perfume industry.

Lit.: Bauer et al. (2.), p. 95 f. ▪ Bedoukian (3.), p. 48–54 ▪ Beilstein E IV **14**, 1016 ▪ TNO list (6.), Suppl. 5, p. 277. – *[CAS 85-91-6]*

(S)-2-Methyl-N-(2-methylbutyl)-butyramide see longhorn beetles.

7-Methyl-3-methylene-7-octenyl propanoate see scales.

S-Methyl 3-(methylthio)thiopropanoate see vegetable flavors (asparagus).

4-(Methylnitrosamino)-benzaldehyde.

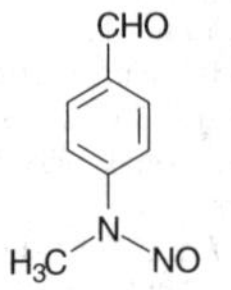

$C_8H_8N_2O_2$, M_R 164.16, air- and light-sensitive, straw-yellow cryst., mp. 81 °C. A nitrosamine from cultures of the fungus *Clitocybe suaveolens* (Basidiomycetes).

Lit.: Hoppe-Seyler's Z. Physiol. Chem. **326**, 13 (1961).

3-Methyl-2,4-nonanedione see tea flavor.

Methyl β-orcinate, orcinol monomethyl ether see oak moss absolute.

Methyl oxepine-2-carboxylate.

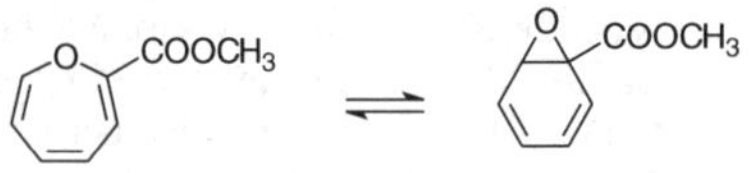

$C_8H_8O_3$, M_R 152.15, liquid. A metabolite from cultures of the fungus *Phellinus tremulae* (Basidiomycetes). According to its ^1H NMR spectrum the compound exists as an equilibrium mixture with its valence tautomer methyl 1,6-epoxy-2,4-cyclohexadiene-1-carboxylic acid (methyl benzoate 1,2-oxide).

Lit.: Tetrahedron Lett. **34**, 1589 (1993). – *[CAS 67490-10-2]*

5-Methyl-2-phenyl-2-hexenal see cocoa flavor.

S-Methyl 2-propenethioate see vegetable flavors (asparagus).

2-Methyl-4-propyl-1,3-oxathiane.

(-)-(2*R*,4*S*)-form (-)-(2*S*,4*S*)-form

$C_8H_{16}OS$, M_R 160.27; discovered and synthesized in 1976 as the typical *flavor compound of the yellow passion fruit (*Passiflora edulis*) (see fruit flavors)[1]. Racemic *cis-/trans*-mixture 1:10, oil with an exotic fruity odor, bp. 85–86 °C (1.6 kPa). Synthesis and sensory evaluation of the four stereoisomers[2] demonstrated that the (+)-(2*S*,4*R*)-form was the compound with the typical passion fruit character, olfactory threshold in water 2 ppb, $[\alpha]_D^{20}$ +56.1° (CCl_4). According to chiral analysis[3], however, the passion fruit contains the (4*S*)-epimer pair with an untypical, sulfurous, herby-green or, respectively flowery odor: (–)-(2*R*,4*S*), $[\alpha]_D^{20}$ –56.1°, and (–)-(2*S*,4*S*), $[\alpha]_D^{20}$ –117.6°. For biosynthesis, see *Lit.*[4].

Lit.: [1] Beilstein E V **19/1**, 117; Helv. Chim. Acta **59**, 1613 ff. (1976); **67**, 947 (1984). [2] Tetrahedron Lett. **25**, 507 (1984); Helv. Chim. Acta **67**, 947 ff. (1984); Justus Liebigs Ann. Chem. **1985**, 1185 ff.; **1987**, 451 [3] J. Agric. Food Chem. **43**, 2438 ff. (1995). [4] Parliment & Croteau (eds.), Biogeneration of Aromas, p. 124 f., ACS Symp. Ser. No. 137, Washington: ACS 1986. – *[CAS 67715-80-4 (racemic cis/trans-mixture); 90243-*

47-3 ((+)-(2S,4R)-M.); 90243-46-2 ((−)-(2R,4S)-M.); 90243-45-1 ((−)-(2S,4S)-M.)]

Methyl pulvinate see vulp(in)ic acid.

6-Methylpurine.

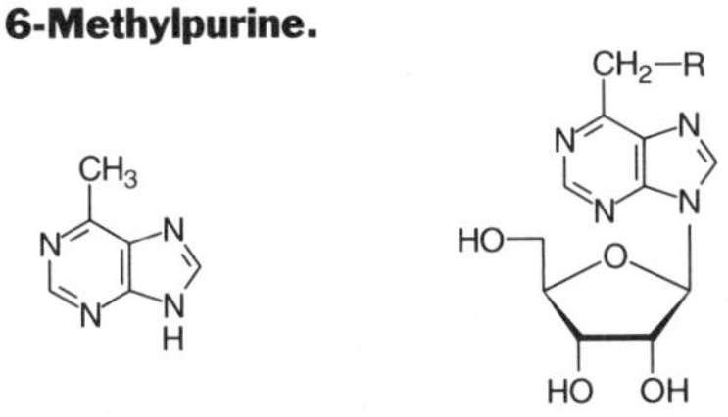

6-Methylpurine R = H : 6-Methyl-9-β-D-ribofuranosylpurine
R = OH : 6-Hydroxymethyl-9-β-D-ribofuranosylpurine

In culture, the fungus *Collybia maculata* (Basidiomycetes) produces the antiviral purine derivatives *6-M.* ($C_6H_6N_4$, M_R 134.14, cryst., mp. 235 °C), *6-methyl-9-β-D-ribofuranosylpurine* ($C_{11}H_{14}N_4O_4$, M_R 266.26, cryst., mp. 210 °C), and *6-hydroxymethyl-9-β-D-ribofuranosylpurine* ($C_{11}H_{14}N_4O_5$, M_R 282.26, cryst., mp. 152–156 °C). The ribosides are effective inhibitors of adenosine deaminase.

Lit.: Z. Naturforsch. C **42**, 420 (1987). – *[CAS 2004-03-7 (6-M.); 14675-48-0 (6-methyl-9-β-D-ribofuranosylpurine)]*

4-Methylpyrrole-2-carboxylic acid see pheromones.

Methyl(*Z*)-5-tetradecenoate see scarabs.

Methyltheobromine see caffein(e).

4-Methylthio-1,2-dithiolane see charamin.

3-(Methylthio)propanoates see fruit flavors (pineapple flavor).

5-Methylthio-1,2,3-trithiane see charamin.

12-Methyltridecanal see meat flavor.

(*R*)-10-Methyl-2-tridecanone see leaf beetles.

Methyl 2,6,10-trimethyltridecanoate see bugs.

***N*-Methyltyramine** see β-phenylethylamine alkaloids.

4-Methyl-5-vinylthiazole see cocoa flavor.

(*R*)-(+)-Methysticin see styrylpyrones.

Mevalonic acid [(MVA, (*R*)-3,5-dihydroxy-3-methylpentanoic acid, hiochic acid].

$C_6H_{12}O_4$, M_R 148.16, oil, natural (–)-(*R*)-form, readily soluble in water and polar organic solvents. M. exists in equilibrium with its δ-lactone, *mevalonolactone.

Biosynthesis: Starts with ester condensation of two mol of acetyl-CoA and subsequent aldol addition of another mol of acetyl-CoA to give hydroxymethylglutaryl-CoA (HMG-CoA). In a two-step reduction, HMG-CoA yields M. All steps proceed stereospecifically. Phosphorylation of MVA and decarboxylation finally gives isopentenyl pyrophosphate, the so-called "active isoprene", compare the formula scheme. For further biosyntheses of *terpenes, see also isoprene rule.

Pharmacology: Hypercholesterolemia can be reduced by HMG-CoA reductase inhibitors, e.g., *compactin, lovastatin (see mevinolin), pravastatin, simvastatin, atorvastatin, and cerivastatin. Of lesser importance are inhibitors of terpenoid biosynthesis for crop protection purposes, e.g., the longitudinal growth of plants can be reduced by influencing *gibberellin biosynthesis (plant growth inhibitors).

History: While conceiving his isoprene rule in 1953, Ružička postulated M. as the precursor of the C_5 unit in the biosynthesis of terpenes. In the same year, Tavormina demonstrated the complete incorporation of M. into *cholesterol (loss of carbon dioxide, rat liver experiments). The first synthesis and determination of the absolute configuration were achieved by Tschesche et al. in 1960. Other renowned scientists connected to M. and terpenoid biosynthesis are Bloch, Lynen, Cornforth and others.

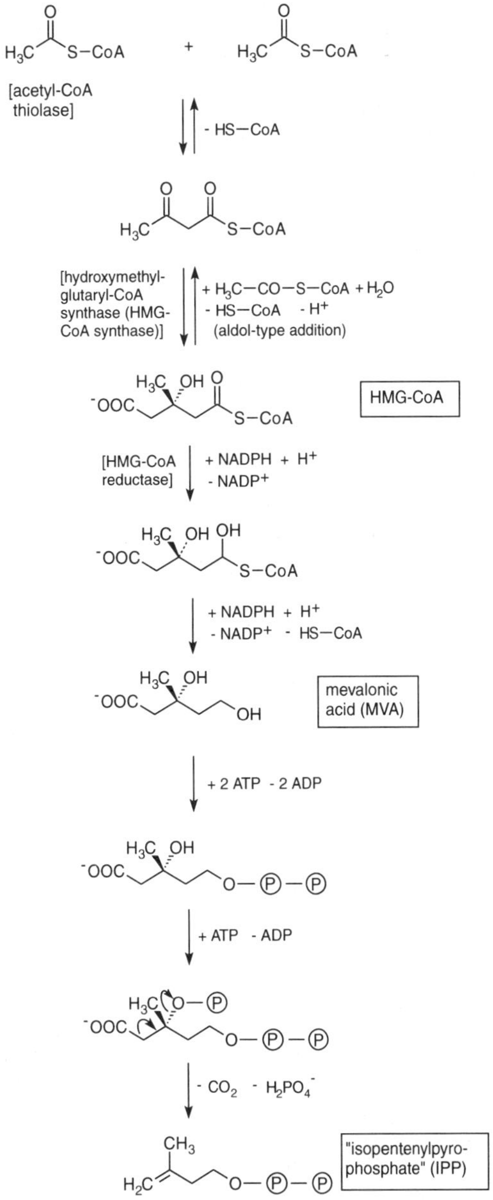

Scheme: Biosynthesis of M. and monoterpenoid precursors.

Lit.: Beilstein E IV **3**, 1068 ▪ Karrer, No. 1130 ▪ Merck Index (12.), No. 6250 ▪ Nature (London) **343**, 425–430 (1990) ▪ Plant

Physiol. Biochem. (Paris) **25**, 163–178 (1987) ▪ Sabine, 3-Hydroxy-3-methylglutaryl-Coenzyme A Reductase, Boca Raton: CRC Press 1984. – *Synthesis:* J. Org. Chem. **61**, 3923 (1996) ▪ Tetrahedron **48**, 9427 (1992) ▪ Tetrahedron: Asymmetry **8**, 2519 (1997). – *[CAS 150-97-0 (racemate); 17817-88-8 (R-form)]*

(R)-Mevalonolactone [mevalolactone, (*R*)-tetrahydro-4-hydroxy-4-methyl-2*H*-pyran-2-one].

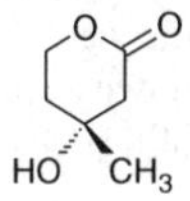

$C_6H_{10}O_3$, M_R 130.14, cryst., mp. 28 °C, bp. 110 °C (10 Pa), $[\alpha]_D^{20}$ –23° (ethanol), readily soluble in water and organic solvents. M. is the δ-lactone of the biosynthetically very important *mevalonic acid and occurs glycosidically bound in many plant species.

Lit.: Beilstein EV **18/1**, 18 f. – *Synthesis:* J. Am. Chem. Soc. **120**, 5427 (1998) ▪ J. Org. Chem. **60**, 6148 (1995); **61**, 3923 (1996) ▪ Synlett **1994**, 754. – *[CAS 19115-49-2]*

Mevastatin see compactin.

Mevinolin (Lovastatin).

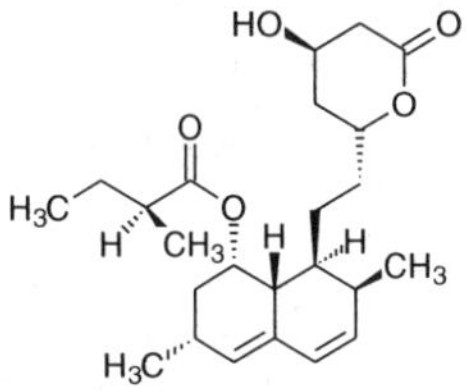

$C_{24}H_{36}O_5$, M_R 404.55, cryst., mp. 174 °C, $[\alpha]_D^{25}$ +323° (CH_3CN), a *polyketide. M. is a potent inhibitor (K_i= 1 nM) of HMG-CoA-reductase, the key enzyme in the biosynthesis of higher terpenes and steroids such as, e. g., *cholesterol. It is produced by *Aspergillus terreus* and various *Monascus* species. Thus, e. g., the plasma cholesterol concentration (a major risk factor for the occurrence of arteriosclerosis) decreases by ca. 50% in patients under medication with M. In the terpene metabolism HMG-CoA-reductase reduces 3-hydroxy-3-methylglutaryl-CoA to mevalonate. M. mimics the substrate and thus leads to inhibition of the enzyme. M. is commercially available under the tradename Mevacor®. M. was the lead structure for numerous synthetic HMG-CoA-reductase inhibitors that are now available or are being developed (Atorvastatin, Cerivastatin, Fluvastatin, Pravastatin, Simvastatin). In these derivatives the hexahydronaphthalene structure is replaced by heterocylic ring systems, see also compactin.

Lit.: Atta-ur-Rahman **11**, 335–377 (1992) (synthesis) ▪ Biotechnol. Bioeng. **56**, 671 (1997) (biotechnology) ▪ Chem. Ind. (London) **1996**, 85–89 (activity) ▪ Clin. Ther. **16**, 2 (1994) (review) ▪ Indian J. Chem., Sect. **B35**, 1125–1143 (1996) (synthesis) ▪ J. Chem. Soc., Perkin Trans. 1 **1996**, 2357 ▪ J. Org. Chem. **57**, 5596 (1992) (synthesis); **61**, 2613–2623 (1996) (biosynthesis) ▪ Lukacs, Recent Progress in the Chemical Synthesis of Antibiotics and Related Microbial Products, p. 829–938, Berlin: Springer 1993 (synthesis, review) ▪ Merck-Index (12.), No. 5616. – *[CAS 75330-75-5]*

Mezerein. $C_{38}H_{38}O_{10}$, M_R 654.71, cryst., mp. 265–269 °C (decomp.), $[\alpha]_D^{25}$ +117.5° ($CHCl_3$). An ester of 12β-hydroxy-*daphnetoxin, diterpene of the daphnane type. M. is isolated from the seeds of com-

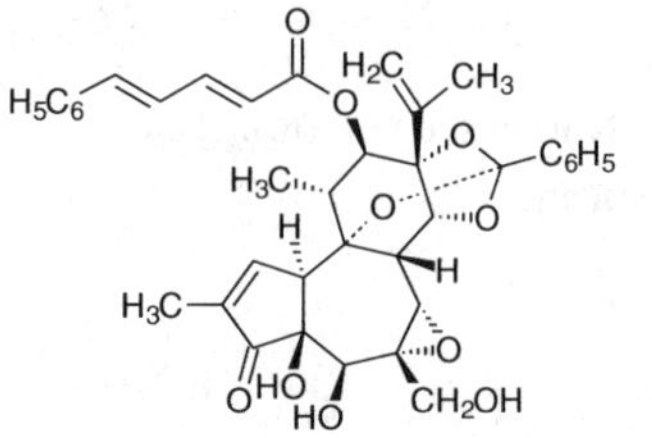

mon mezereon (*Daphne mezereum*). It is toxic and causes skin damage such as erythema, blisters, finally ulcerous decomposition of the skin. In addition, it shows cocarcinogenic activity.

Lit.: Angew. Chem. Int. Ed. Engl. **20**, 164 (1981) ▪ Beilstein E V **19/12**, 120 ▪ Hager (5.) **3**, 829 f. ▪ Nature (London) **257**, 824 (1975) ▪ Phytochemistry **29**, 3633 (1990) ▪ Sax (8.), MDJ 250 ▪ Science **187**, 652 (1975). – *[CAS 34807-41-5]*

Michellamines. Atropisomeric dimeric *naphthyl-(tetrahydro)isoquinoline alkaloids from the tropical lianas *Ancistrocladus korupensis* and *A. abbreviatus.* $C_{46}H_{48}N_2O_8$, M_R 756.89.

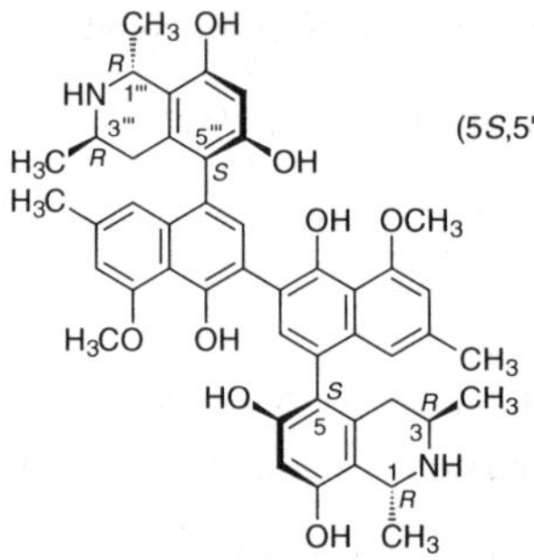

(5*S*,5‴*S*)-Atropisomer : M.A

Table: Data of Michellamines.

Atropisomers	$[\alpha]_D^{20}$ (CH_3OH)	CAS
M. A (5*S*,5‴*S*)	–10.5°	137793-81-8
M. B (5*R*,5‴*S*)	–14.8°	137893-48-2
M. C (5*R*,5‴*R*)	–16.8°	143168-23-4

The M. A–F are active against HIV viruses of types 1 and 2.

Lit.: Angew. Chem. Int. Ed. Engl. **32**, 1190 (1993) ▪ J. Chromatogr. **810**, 231 (1998) (LC/MS) ▪ J. Chem. Soc., Chem. Commun. **1996**, 923 f. (asymmetric synthesis, M. B) ▪ J. Heterocycl. Chem. **33**, 1371 (1996) ▪ J. Med. Chem. **34**, 3402 (1991); **37**, 1740 (1994) ▪ J. Nat. Prod. **60**, 677 (1997) (isolation M. D–F) ▪ J. Org. Chem. **63**, 1090 (1998); **64**, 7184 (1999) ▪ Justus Liebigs Ann. Chem. **1996**, 2045 (synthesis) ▪ Synlett **1997**, 965 (synthesis) ▪ Tetrahedron **50**, 7807, 9643 (1994).

Miconidin see primin.

Micrococcin P₁. $C_{48}H_{49}N_{13}O_9S_6$, M_R 1144.36, mp. 252 °C (decomp.), $[\alpha]_D^{21}$ +116° (90% C_2H_5OH). A peptide antibiotic isolated from the culture broth of *Bacillus pumilus* as main component of the micrococcin P complex. For synthesis see *Lit.*[1]. Closely related with M. is the antibiotic nosiheptide[2].

Lit.: [1] Heterocycles **48**, 1319 (1998). [2] Bull. Chem. Soc. Jpn. **71**, 1391 (1998); **72**, 1561, 2483 (1999); Org. Lett. **1**, 1843 (1999). *gen.:* Antibiotics **3**, 480 (1975) ▪ J. Chem. Soc., Chem. Commun. **1977**, 706; **1978**, 256 ▪ Tetrahedron Lett. **32**, 4263 (1991). – *[HS 2941 90; CAS 67401-56-3]*

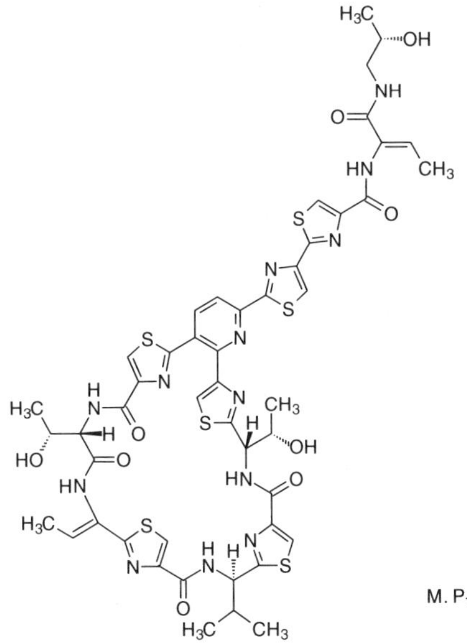

M. P_1

Microcolin A.

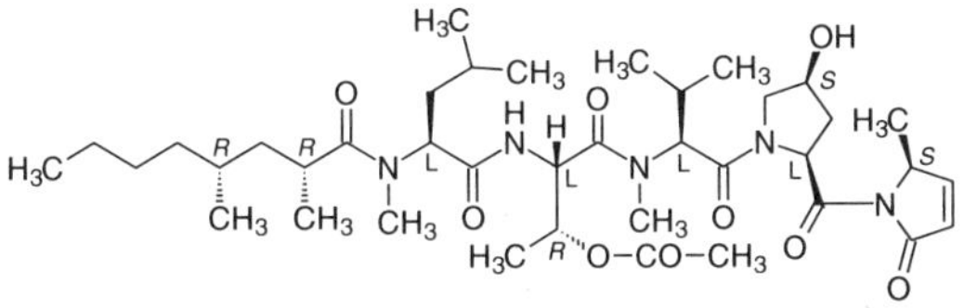

$C_{39}H_{65}N_5O_9$, M_R 747.96, $[\alpha]_D^{25}$ –145° (C_2H_5OH). Highly active immunosuppressive peptide from the Venezuelan cyanobacterium *Lyngbya majuscula*, structurally related with the cytotoxic *majusculamides. M. B is the deoxy compound (Pro for Hypro).

Lit.: J. Med. Chem. **37**, 3181 (1994) (synthesis) ▪ J. Nat. Prod. **55**, 613 (1992) (isolation) ▪ J. Org. Chem. **61**, 3534–3541 (1996); **62**, 5542 (1997) (synthesis M. B). – *[CAS 141205-31-4 (M. A); 141205-32-5 (M. B)]*

Microcystins (cyanoginosins).

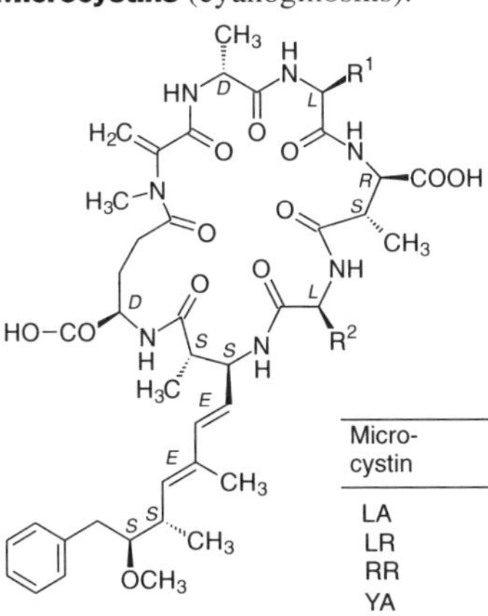

Microcystin	amino acid with R^1	R^2
LA	L-Leu	L-Ala
LR	L-Leu	L-Arg
RR	L-Arg	L-Arg
YA	L-Tyr	L-Ala
YM	L-Tyr	L-Met
YR	L-Tyr	L-Arg

Group of ca. 20 heptapeptides formed by *Microcystis aeruginosa*, a cyanobacterium living in fresh and stagnant waters and *Anabaena* sp. M. often act as hepatotoxins and are thus responsible for the death of livestock. They inhibit protein phosphatases 1 and 2A. A particular building block, 3-amino-9-methoxy-2,6,8-trimethyl-10-phenyl-4,6-decadienoic acid (*Adda*), is probably responsible for the toxic activity. There are several M. The main toxin is M. RR [$C_{49}H_{75}N_{13}O_{12}$, M_R 1038.21, $[\alpha]_D$ –100° (CH_3OH)], LD_{50} (mouse i.p.) 0.5 mg/kg.

Lit.: Bioorg. Med. Chem. **7**, 543–564 (1999) ▪ Bioorg. Med. Chem. Lett. **6**, 2113 (1996) ▪ Chem. Eng. News (26.10) **1992**, 26–32 ▪ J. Am. Chem. Soc. **110**, 8557 (1988); **118**, 11759 (1996) ▪ J. Chromatogr. **749**, 271 (1996); **799**, 155 (1998) (analysis) ▪ J. Nat. Prod. **61**, 851 (1998) ▪ J. Org. Chem. **55**, 6135 (1990); **57**, 866 (1992) ▪ Stud. Nat. Prod. Chem. **20**, 887–920 (1998) (review) ▪ Tetrahedron Lett. **29**, 11 (1988); **30**, 4245 (1989) ▪ Tetrahedron **54**, 637 (1998). – *[CAS 111755-37-4 (M. RR)]*

Microlin see mycorrhizin.

Milbemycins.

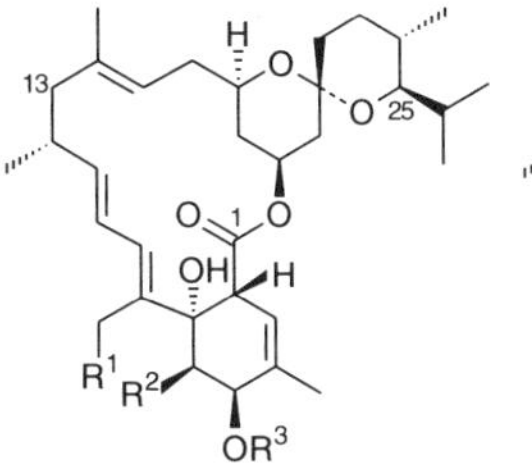

M. D : R^1, R^2 = —O—, R^3 = H
M. E : R^1 = OH, R^2 = H , R^3 = CH_3

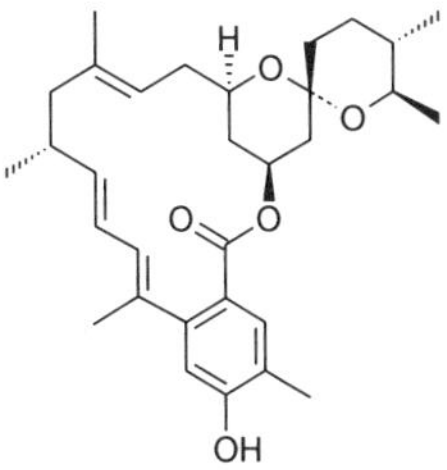

M. β_3

A group of more than 20 naturally occurring macrolide *antibiotics from *Streptomyces hygroscopicus* var. *aureolacrimosus* with antiparasitic, insecticidal, and acaricidal activities. Common structural features are the differently substituted, 16-membered lactone ring and a spiroketal unit. In terms of structure and activity, the M. are closely related with the *avermectins. Examples of this class of compounds are *M. D* ($C_{33}H_{48}O_7$, M_R 556.74, needles, mp. 186–188 °C), *M. E* ($C_{34}H_{52}O_7$, M_R 572.78, amorphous) and *M. β_3* ($C_{31}H_{42}O_5$, M_R 494.67)[1].

Lit.: [1] Synform **4**, 1–25, 26–33 (1986); J. Org. Chem. **51**, 4840 (1986).

gen.: Nat. Prod. Rep. **3**, 87–121 (1986) ▪ Sax (8.), p. 2424 ▪ Stud. Nat. Prod. Chem. **1**, 435–496 (1988). – *Biosynthesis/biotransformation:* J. Antibiot. **48**, 831–837 (1995); **52**, 109 (1999). – *Isolation:* J. Antibiot. (Tokio) **33**, 1120 (1980); **36**, 438, 502, 509, 980, 991 (1983); **49**, 272 (1996). – *Synthesis:* ACS Symp. Ser. **355**, 251–259 (1987); **443** (Synth. Chem. Agrochem. II), 436–447 (1991); **504** (Synth. Chem. Agrochem. III), 226–238 (1992) ▪ J. Am. Chem. Soc. **118**, 7513–7528 (1996) ((+)-M.) ▪ J. Antibiot. **47**, 233–242 (1994) ▪ J. Chem. Soc., Chem. Commun. **1989**, 1250; **1995**, 2519 ▪ J. Chem. Soc., Perkin Trans. 1 **1997**, 371, 381, 391 (synthesis M. E) ▪ Nachr. Chem. Tech. Lab. **34**, 444 (1986) ▪ Reissig, Milbemycin β_3, in Org. Synth. Highlights, Weinheim: VCH Verlagsges. 1991 ▪ Spec. Publ. – Royal Soc. Chem. Ser. **79**, 69–89, 99–124 (1990) ▪ Synform **7**, 49–96 (1989); **8**, 276–284 (1990) ▪ Tetrahedron **45**, 7161–7194 (1989). – *[HS 2941 90; CAS 77855-81-3 (M. D); 83204-48-2 (M. E); 56198-39-1 (M. β_3)]*

Mildiomycin. $C_{19}H_{30}N_8O_9$, M_R 514.49, hygroscopic, cryst., mp. >300 °C (monohydrate, decomp.), $[\alpha]_D^{20}$ +100°. A *nucleoside antibiotic from *Streptoverticillium rimofaciens*[1] of low toxicity with activity against

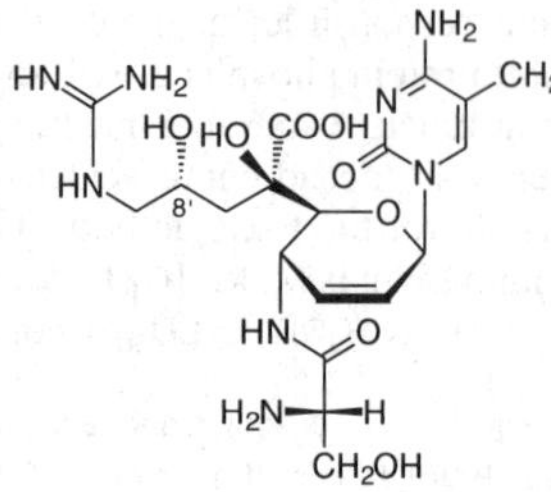

mildew and other pathogenic fungi, Gram-positive and Gram-negative bacteria as well as yeasts.

Lit.: [1] Agric. Biol. Chem. **48**, 881 (1984); J. Antibiot. **38**, 415 (1985); J. Ferment. Technol. **63**, 17 (1985); Sax (8.), MQU 000. – *[CAS 67527-71-3]*

Milk sugar see lactose.

Mimopudine see leaf-movement factors.

Mimosamycin.

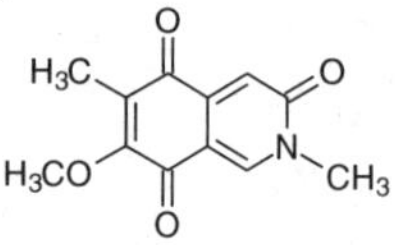

$C_{12}H_{11}NO_4$, M_R 233.22. Antibiotic from sponges (*Reniera, Petrosia*) and *Streptomyces lavendulae*. Yellow prisms, mp. 227–231 °C, $[\alpha]_D^{24}$ –1.8° ($CHCl_3$).

Lit.: J. Chem. Res. (S) **1994**, 282 ▪ J. Nat. Prod. **59**, 973 (1996) ▪ J. Org. Chem. **53**, 2847 (1988). – *[CAS 59493-94-6]*

Mimosine [3-(3-hydroxy-4-oxo-1(4*H*)-pyridyl)-L-alanine, leucenol, leucenine].

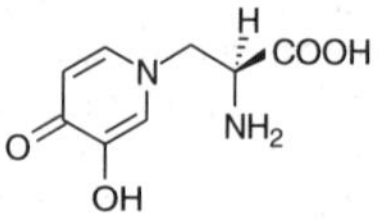

$C_8H_{10}N_2O_4$, M_R 198.18, mp. 225 °C, $[\alpha]_D^{22}$ –20° (H_2O). A non-proteinogenic amino acid in seeds and leaves of *Leucaena* spp. and *Mimosa* spp. (Mimosaceae). M. is identical with leucenol. Leaves of *L. leucocephala* (*L. glauca*) contain 2–5% M. (dry weight). Seeds of *L. leucocephala* and *M. pudica* also contain the 3′-*O*-*β*-D-glucosyl derivative (mimoside).

Biosynthesis: The pyridone ring originates from Lys and the Ala side chain from Ser. Hydroxylation appears to occur after formation of the pyridone ring.

Activity & toxicity: M. is a powerful inhibitor of cystathionine synthase and cystathionase [1]. The growth of shoots of *Phaseolus aureus* is inhibited by 3,4-dihydroxypyridine and mimosine with equal activity. The inhibitory activity is neutralised by addition of pyridoxal phosphate. M. causes loss of hair in sheep, horses, and pigs; goats and cattle, in contrast, are not affected in this way. M. effects growing hair, impairs the estrogen cycle in rats [2], and DNA synthesis [3].

M. is used as a depilatory agent in sheep.

Lit.: [1] J. Agric. Food Chem. **17**, 492–496 (1969). [2] Biochem. Pharmacol. **14**, 1167 ff. (1965). [3] Aust. J. Biol. Sci. **29**, 189–196 (1976).

gen.: Beilstein E V **21/12**, 107 ▪ Chem. Pharm. Bull. **34**, 1473 (1986) (synthesis) ▪ Karrer, No. 2411 ▪ Hager (4.) **5**, 492 f., 867 f. ▪ Rosenthal, Plant Nonprotein Amino and Iminoacids, p. 78–87, New York: Academic Press 1982 ▪ Zechmeister **28**, 137. – *[CAS 500-44-7]*

Mineralocorticoids see aldosterone or, respectively, corticosteroids.

Mintlactone.

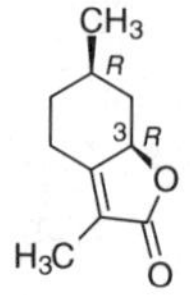

$C_{10}H_{14}O_2$, M_R 166.22, oil, $[\alpha]_D^{20}$ –51.8° (C_2H_5OH). A constituent and aroma component of *peppermint oils from *Mentha piperita*.

Lit.: *Synthesis:* Indian J. Chem., Sect. B **36**, 226 (1997) ▪ Synth. Commun. **27**, 3471 (1997) ▪ Tetrahedron **48**, 9789 (1992); **49**, 6429, 6433 (1993); **51**, 5831 (1995) (synthesis) ▪ Tetrahedron Lett. **31**, 6789 (1990); **32**, 5191 (1991); **33**, 4589, 4605 (1992). – *Biosynthesis:* Phytochemistry **30**, 485 (1991). – *[CAS 38049-04-6]*

Miotoxins see baccharinoids.

Mirabazole B see tantazoles.

Miraxanthins. A group of betaxanthins (see betalains) occurring in the flowers of marvel-of-Peru (*Mirabilis jalapa*, Nyctaginaceae). *M. I* ($C_{14}H_{18}N_2O_7S$, M_R 358.37) is a condensation product of *betalamic acid with methionine sulfoxide, *M. II* ($C_{13}H_{14}N_2O_8$, M_R 326.26) with aspartate, *M. III* ($C_{17}H_{18}N_2O_5$, M_R 330.34) with tyramine.

Lit.: Manske **39**, 21 ff. ▪ Phytochemistry **4**, 817 (1965). – *[CAS 5296-79-7 (M. I); 5375-63-3 (M. II); 5589-85-5 (M. III)]*

Miriamide see oviposition deterring pheromones.

Miriquidic acid.

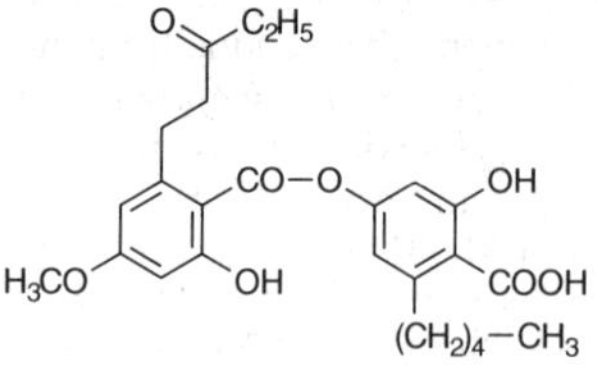

$C_{25}H_{30}O_8$, M_R 458.51, needles, mp. 140–141 °C. A *depside with an unusual 3-oxopentyl side chain from the crustaceous lichen *Miriquidica leucophaea*. M. is named after the site where the lichen was found in Germany (Erzgebirge = ore mountains, the old Germanic name being Miriquidi). On the basis of the occurrence of M. the genus *Miriquidica* with 14 species was separated from the Lecanoraceae – an example of chemotaxonomy.

Lit.: Aust. J. Chem. **39**, 791 (1986) (synthesis) ▪ Mitt. Bot. Staatssamml. München **23**, 377–392 (1987) ▪ Z. Naturforsch., B **26**, 1357–1364 (1971). – *[CAS 35578-70-2]*

Miroestrol.

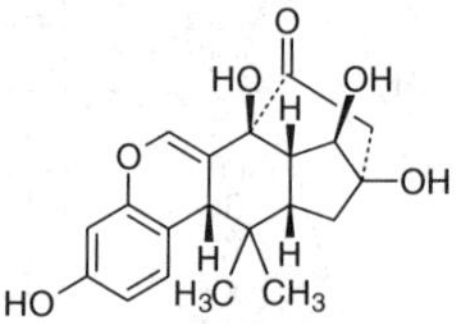

$C_{20}H_{22}O_6$, M_R 358.39, rectangular plates, mp. 268–270 °C, $[\alpha]_D^{17}$ +301° (C_2H_5OH). A phenol with estrogen-like activity from the Siamese plant *Pueraria*

mirifica ("kwaokeur") used in traditional medicine [1]. For synthesis, see *Lit.* [2].

Lit.: [1] J. Chem. Soc. **1960**, 3685, 3696. [2] J. Am. Chem. Soc. **115**, 9327 f. (1993).

gen.: Sax (5.), p. 836. – *[CAS 2618-41-9]*

Miserotoxin see 3-nitropropanoic acid.

Mites (Acaridae). An extremely diversiform group of arachnids (ca. 30 000 species), some of which are microscopically small, living as parasites on animals and plants; of economic significance as pests on plant material and animal products. The M. bore into the skin where they, or their excretion products, cause scabies in humans and mange (scabies) in animals; they can also induce allergies and asthma. The *pheromones of some M. species are known. These are mostly unbranched alkenes such as *(Z)-6-pentadecene* ($C_{15}H_{30}$, M_R 250.29), derivatives of salicylaldehyde such as *hexyl rhizoglyphinate* ($C_{14}H_{18}O_4$, M_R 250.29), or oxygenated monoterpenes such as *neral* and *geranial* (see citrals) or, respectively, *neryl* and *geranyl formates*, *neral (2S,3S)-2,3-epoxide* {(2*S*,3*S*)-2,3-epoxy-3,7-dimethyl-6-octenal, $C_{10}H_{16}O_2$, M_R 168.24}, *α-acaridial* {(*E*)-(4-methyl-3-pentenyl)-2-butenedial, $C_{10}H_{14}O_2$, M_R 166.22}, *robinal* {3-oxo-*p*-mentha-1,4(8)-dien-7-al, $C_{10}H_{12}O_2$, M_R 164.20}, and furanoterpenoids such as *perillene* {3-(4-methyl-3-pentenyl)-furan, $C_{10}H_{14}O$, M_R 150.22} or *rosefuran* {3-methyl-2-(3-methyl-2-butenyl)-furan, $C_{10}H_{14}O$, M_R 150.2}. Structures with 1,3-dimethyl-branching as in *lardolure are rare. Pheromones have also been described for the dangerous bee parasite *Varroa jacobsoni* (Varroa mite).

(*Z*)-6-Pentadecene (**1**)

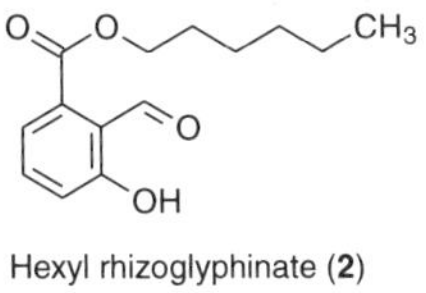

Hexyl rhizoglyphinate (**2**)

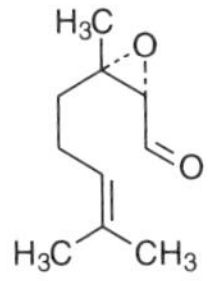

Neral-(2S,3S)-epoxide (**3**)

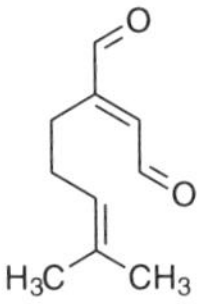

α-Acaridial (**4**)

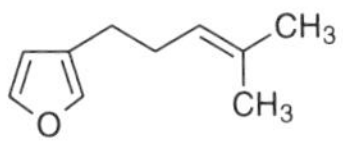

Perillene (**5**)

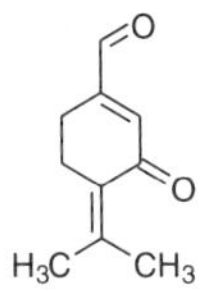

Robinal (**7**)

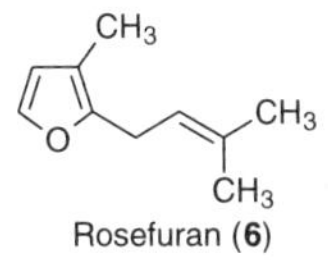

Rosefuran (**6**)

Lit.: Harris, Natural Toxins, p. 56–71, Oxford: Oxford Science Publ. 1986 ▪ Tu, Handbook of Natural Toxins, vol. 2, p. 371–396, New York: Marcel Dekker 1984. – *Pheromones:* Biosci., Biotechnol., Biochem. **56**, 1510 (1992) (robinal) ▪ Chem. Lett. **1984**, 1261 (perillene) ▪ Dusbábek & Bukva, Modern Acarology, vol. 1, p. 43–52, 53–58, The Hague: SPB Academic Publishing 1991 ▪ J. Chem. Soc., Chem. Commun. **1997**, 1083 (rosefuran) ▪ J. Org. Chem. **58**, 3602 (1993); **59**, 1078 (1994) ▪ Tetrahedron **51**, 7721 (1995); **53**, 3497 (1997) (rosefuran). – *Varroa mites:* J. Chem. Ecol. **20**, 2437–2454 (1994) ▪ Science **245**, 638 f. (1989) ▪ see also ticks. – *[CAS 74392-29-3 (1); 131524-42-0 (2); 61063-18-1 (3); 121325-68-6 (4); 539-52-6 (5); 15186-51-3 (6); 131190-65-3 (7)]*

Mithramycin see aureolic acid.

Mitomycins.

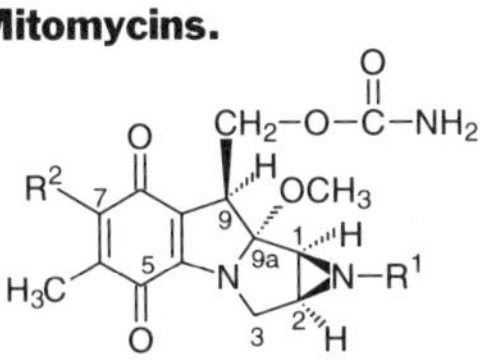

R^1	R^2	
H	OCH_3	M. A
H	NH_2	M. C
CH_3	NH_2	Porfiromycin

Table: Data of Mitomycins.

compound	molecular formula	M_R	mp. [°C]	$[\alpha]_D$	CAS
M. A	$C_{16}H_{19}N_3O_6$	349.34	160	–143° (CH_3OH)	4055-39-4
M. C	$C_{15}H_{18}N_4O_5$	334.33	>360	+34.7° ($CHCl_3$)	50-07-7
Porfiromycin	$C_{16}H_{20}N_4O_5$	348.36	201	+275° (CH_3OH)	801-52-5

General name for antitumor antibiotics with *benzoquinone and aziridine structural units from cultures of streptomycetes (e.g., *Streptomyces lavendulae*, *S. verticillatus*, *S. caespitosus*). There are more than 15 natural variants containing a pyrrolo[1,2-*a*]indole system and differing in R^1 and R^2, with OH in place of OCH_3 at C-9a, and different stereochemistries at C-9. The violet to blue-violet, optically active M. crystallize readily and are soluble in water, methanol, and acetone. A cluster of 47 bacterial genes that control biosynthesis of M. C has been characterised [1]. The biosynthetic building block of the benzoquinone (C_7N) unit is 3-amino-5-hydroxybenzoic acid, several total syntheses have been reported.

The antibacterial, antiviral, and cytotoxic properties of the M. result from an inhibition of DNA replication through alkylation and cross-linking of the DNA double helix. The aziridine ring is activated for nucleophilic attack at C-1 (e.g., of guanine) by elimination of methanol (C-9/C-9a). The formation of free radicals has been discussed in the context of the mechanism of action, leading to DNA strand cleavage. The enzyme mitomycin C resistance protein A (MCRA) protects M. producing organisms against their own antibiotic [2]. M. C is used clinically in combination with other chemotherapeutics for the treatment of lymphatic leukemia and pancreatic carcinoma but has considerable side effects. Numerous derivatives have been produced to improve the properties and some are undergoing clinical trials.

Lit.: [1] Chem. Biol. **6**, 251 (1999). [2] J. Am. Chem. Soc. **119**, 2584 (1997).

gen.: Isolation & structure: Acta Crystallogr., Sect. C **52**, 1866, 2272 (1996) ▪ J. Am. Chem. Soc. **105**, 7199 (1983); **108**, 4648 (1986); **109**, 7224 (1987) ▪ J. Antibiot. **41**, 869 (1985). – *Reviews:* Carter & Crooke, Mitomycin C, Orlando: Academic Press 1979 ▪ Chem. Biol. **2**, 575–579 (1995) ▪ Heterocycles **24**, 2104 (1986) ▪ Lukacs & Ohno (eds.), Recent Progress in the Chemical Synthesis of Antibiotics, p. 415–445, Berlin: Springer 1993 ▪ Remers & Iyengar, Antitumor Antibiotics, in Foye (ed.), Cancer Chemotherapeutic Agents, p. 584–592, Washington: Am. Chem. Soc. 1995. – *Synthesis:* Angew.

Chem. Int. Ed. Engl. **31**, 915 ff. (1992) ▪ J. Am. Chem. Soc. **111**, 8303 (1989) ▪ Nachr. Chem. Tech. Lab. **40**, 235 ff. (1992) ▪ Stud. Nat. Prod. Chem. **13A**, 433–471 (1993) ▪ Synlett **1992**, 778–790 ▪ Tetrahedron Lett. **37**, 6049 (1996) (M.K). – *Toxicology:* Sax (8.), AHK 500. – *[HS 2941 90]*

Mitoquinones see ubiquinones.

Mitorubrin. An azaphilone pigment, the (*R*)-(–)-form of which has been isolated from *Penicillium rubrum* and *P. funiculosum* where it occurs together with *mitorubrinol*, *mitorubrinic acid*, and *mitorubrinic acid B*.

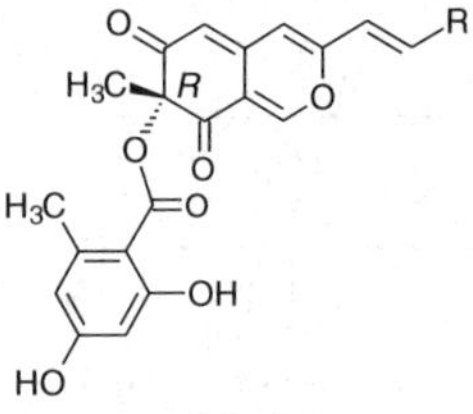

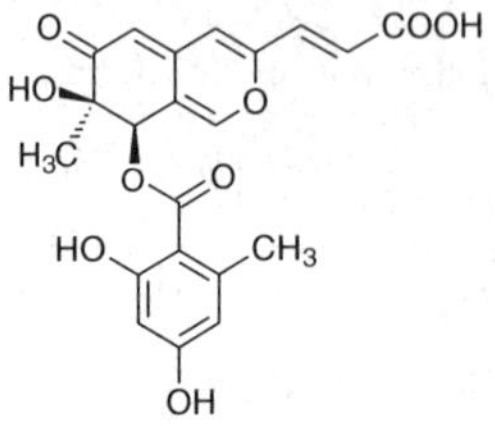

R = CH_3 : Mitorubrin (**1**)
R = CH_2–OH : Mitorubrinol (**2**)
R = COOH : Mitorubrinic acid (**3**)

Mitorubrinic acid B (**4**)

Table: Data of Mitorubrin and derivatives.

	molecular formula	M_R	mp. [°C]	$[\alpha]_D$	CAS
1	$C_{21}H_{18}O_7$	382.37	218 (*R*-form) 220–225 (racemate)	–405° (dioxane)	3403-71-2
2	$C_{21}H_{18}O_8$	398.37	230 (decomp.)		3215-47-2
3	$C_{21}H_{16}O_9$	412.35	222–225 (*R*-form) 225–227 (*S*-form)	–450° +500°	58958-07-9
4	$C_{21}H_{18}O_9$	414.37	133–136	+241°	98770-62-8

P. wortmanni contains the acetylated wortmin in addition to M. The (*S*)-forms of M., mitorubrinol, and mitorubrinic acid occur in the fruit bodies of *Hypoxylon fragiforme*. Azaphilones (affinity for nitrogen) react with ammonia and amines to form red or dark red compounds. Mitorubrinic acids and lunatic acids formed by *Cochliobolus lunatus*, belong to the morphogenous substances that induce the formation of thick-walled mycelia. All compounds exhibit weak antifungal activities.

Lit.: Agric. Biol. Chem. **49**, 2517 (1985) ▪ Beilstein E V **18/3**, 440 ▪ Collect. Czech. Chem. Commun. **57**, 408 (1992) (isolation) ▪ J. Chem. Soc., Perkin Trans. 1 **1976**, 1366 ▪ Phytochemistry **39**, 719 (1995) ▪ Turner **1**, 137 ff.; **2**, 112 ff. ▪ Zechmeister **51**, 120.

Mitragyna alkaloids. Alkaloids (*monoterpenoid indole alkaloids) from the genus *Mitragyna* (Rubiaceae), containing more than 10 species of trees found in the tropical and subtropical regions of Africa and Asia.

Use: The leaves, the bark and roots are used in traditional medicine: in West Africa for the treatment of leprosy wounds, blood poisonings, colics, and as an emetic and diuretic [1]; also as yellow dye (bark, *M. africana*). In India the dried leaves are smoked (like opium) and are supposed to be a substitute for opium

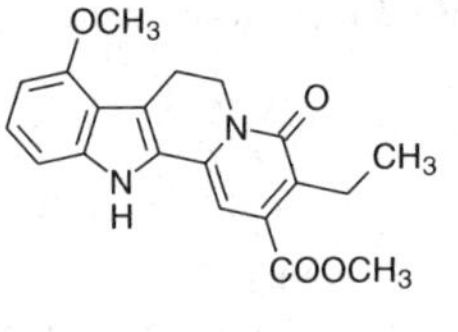

Mitragynaline

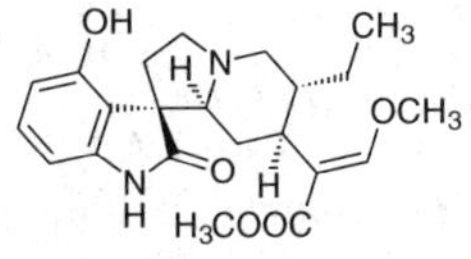

Mitragynol

or, respectively, to have an opium-like activity [2]; in Malaysia the leaves are used to promote wound healing. It is noteworthy that *Mitragyna* plants are used by all these cultures for attacks of fever. To date, more than 50 structurally different indole alkaloids, including N-oxides [3] and oxindoles, e.g., Mitragynaline and Mitragynol have been isolated from *Mitragyna* species.

Lit.: [1] Watt & Breyer-Brandwijk, The Medicinal and Poisonous Plants of Southern and Eastern Africa, Edinburgh: Livingstone 1962. [2] Manske **8**, 59–91; Phytochemistry **30**, 347 (1991). [3] Pelletier **1**, 348.

gen.: Synthesis: J. Org. Chem. **64**, 1772 (1999). – *[HS 2939 90]*

Mniopetals.

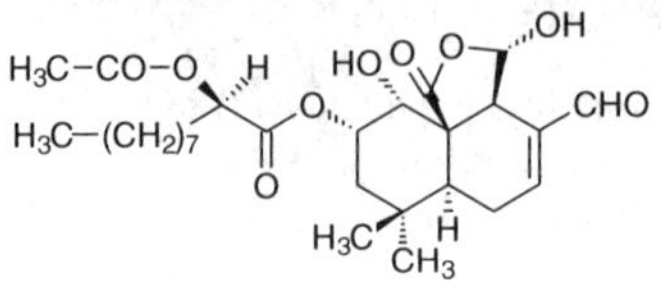

Mniopetal A

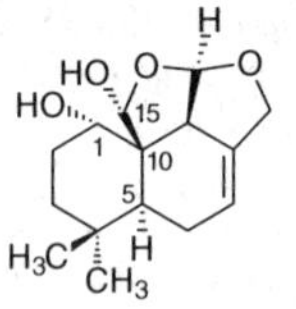

1α,15-Dihydroxymarasmene

Varous inhibitors of RNA-dependent DNA-polymerases (reverse transcriptases) have been isolated from culture filtrates of a Canadian *Mniopetalum* species (Basidiomycetes), e.g., *M. A* {$C_{27}H_{40}O_9$, M_R 508.61, oil, $[\alpha]_D$ –63° ($CHCl_3$)} (see also marasmenes). The absolute configuration of the *drimane derivative was defined on the basis of 1α,15-dihydroxymarasmene {$C_{15}H_{22}O_4$, M_R 266.34, cryst., mp. 150–154 °C, $[\alpha]_D$ +92° (CH_3OH)}.

Lit.: J. Antibiot. **47**, 733, 1017 (1994) ▪ Tetrahedron: Asymmetry **5**, 1229 (1994) ▪ Tetrahedron Lett. **40**, 7835 (1999) (synthesis). – *[CAS 158760-98-6 (M. A); 124869-04-1 (1α,15-dihydroxymarasmene)]*

Mocimycin see kirromycin.

Modhephene.

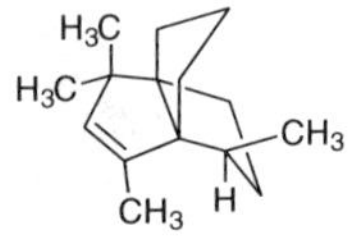

$C_{15}H_{24}$, M_R 204.36, oil, bp. 65–70 °C (33.3 Pa), $[\alpha]_D$ –15.8° ($CHCl_3$). A natural triquinane hydrocarbon with a [3.3.3]propellane structure from *Isocoma wrightii* (Asteraceae), an indigenous species of south west USA toxic to cattle and sheep, from South African *Berkheya* species, and the roots of *Senecio isatideus* [1]. Derivatives of M. show antiproliferative activity [2].

Lit.: [1] Z. Naturforsch. C **37**, 5 (1982). [2] Chem. Pharm. Bull. **36**, 542 (1988).

gen.: Isolation: J. Org. Chem. **42**, 96 (1979) ▪ Phytochemistry **24**, 505 (1985); **25**, 1133 (1986). – *Synthesis:* Angew. Chem. Int. Ed. Engl. **30**, 1492 f. (1991) ▪ Bull. Chem. Soc. Jpn. **71**, 231 (1998) ▪ Bull. Korean Chem. Soc. **20**, 269 (1999) ▪ Chem. Commun. **1997**, 2381 ▪ J. Am Chem. Soc. **104**, 5805 (1982);

112, 5601 (1990) ▪ Tetrahedron **45**, 4945 (1989); **49**, 755–770 (1993); **51**, 8835–8852 (1995) ▪ Tetrahedron Lett. **37**, 7295 (1996). – *[CAS 68269-87-4]*

Moenomycin (bambermycin). A glycophospholipid *antibiotic complex. The main component M. A is also the main component of the antibiotic Flavomycin® used in veterinary medicine and in animal breeding to improve production. The antibiotic activity of M. A is due to its interaction with penicillin-binding protein 1 b (PBP 1 b). The transglycosylase activity of this bi-functional enzyme and thus the formation of the high-molecular weight peptidoglycan of the bacterial cell wall from a disaccharide precursor are inhibited[1]. M. is obtained by fermentation from cultures of *Streptomyces bambergiensis*. The main components are M. A, C_1, C_3, and C_4. For biosynthesis see *Lit.*[2].

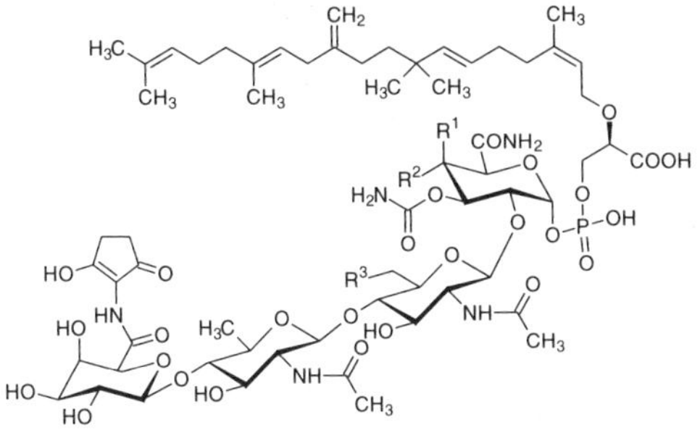

$R^1 = CH_3$, $R^2 = OH$, $R^3 = OGlc$: M.A
$R^1 = OH$, $R^2 = R^3 = H$: M.C_1
$R^1 = CH_3$, $R^2 = OH$, $R^3 = H$: M.C_3
$R^1 = CH_3$, $R^2 = R^3 = OH$: M.C_4

Table: Daten of the Moenomycin complex.

M. molecular formula	M_R	mp. [°C]	$[\alpha]_D^{20}$ (H_2O)	CAS
A $C_{69}H_{108}N_5O_{34}P$	1582.60	184–185	+4° (Na salt)	76095-39-1
C_1 $C_{62}H_{96}N_5O_{28}P$	1390.43	178–179	+4°	–
C_3 $C_{63}H_{98}N_5O_{28}P$	1404.46	amorphous	–	123589-03-7
C_4 $C_{63}H_{98}N_5O_{29}P$	1420.46	amorphous	–	149633-74-9

Lit.: [1]Tetrahedron **48**, 8401–8418 (1992); **49**, 1635–1648, 3091–3100, 7667–7678 (1993); **51**, 1931 (1995). [2]Tetrahedron Lett. **39**, 1, 13 (1998).
gen.: Angew. Chem. Int. Ed. Engl. **20**, 121 (1981); **38**, 3703 (1999) (mechanism of action) ▪ Magn. Reson. Chem. **36**, 615 (1998) (^{1}H NMR) ▪ Sax (8.), MRA 250 ▪ Tetrahedron **39**, 1583, 2219 (1983); **46**, 1557 (1990).

Momilactones A and B see pimaranes.

Mompain [(2,5,7,8-tetrahydroxy-1,4-naphthoquinone)].

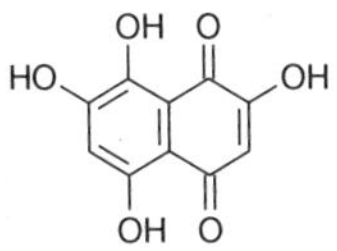

$C_{10}H_6O_6$, M_R 222.15, red platelets, mp. >300 °C (decomp.). A metabolite from *Helicobasidium mompa*.
Lit.: Biochemistry **14**, 3138 (1975) (biosynthesis) ▪ Chem. Pharm. Bull. **11**, 1343 (1963); **13**, 633 (1965) (isolation) ▪ Karrer, No. 6943 ▪ Russ. Chem. Bull. **48**, 1010 (1999) (synthesis) ▪ Tetrahedron **24**, 2969 (1968) (synthesis) ▪ Tetrahedron Lett. **38**, 4219 (1997) (synthesis, M. trimethyl ether). – *[CAS 2473-16-7]*

Monactin see nonactin.

Monatin [(2*S*,4*S*)-4-hydroxy-4-(indol-3-ylmethyl)-L-glutamic acid].

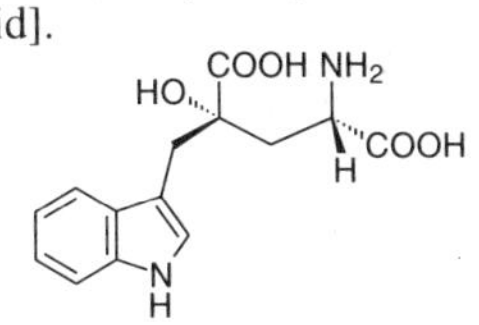

$C_{14}H_{16}N_2O_5$, M_R 292.27, mp. 216–220 °C, $[\alpha]_D^{20}$ –7.6° (1 N HCl). A natural sweetener with a 1300-fold higher sweetness than saccharose, isolated from the South African (Transvaal) shrub *Sclerochiton ilicifolius* (Acanthaceae, bark and roots); see also glycyrrhizin, stevioside, thaumatin, osladin.
Lit.: J. Chem. Soc., Perkin Trans. 1 **1992**, 3095–3099 (isolation) ▪ Synth. Commun. **24**, 3197 (1994) (synthesis) ▪ US P. 5 994 559 (30. 11. 1999). – *[CAS 136440-37-4 (rac.); 146142-94-1 (2S,4S)]*

Monellin. The sweet-tasting principle of the West African fruit *Dioscoreophyllum cumminsii* (Menispermaceae). 1 kg of these berries furnish ca. 15 g M. M., first discovered in 1967, consists of two polypeptide chains that are not covalently joined to each other; both are essential for the sweet taste. They consist of 44 and 50 amino acids, respectively[1], and together have an M_R of 11 500. M. is soluble in water and has a 2000–2500-fold higher sweetness than *sucrose, it is not registered for use as a sweetener in Germany. The sweet taste of M. is still detectable at a dilution of 10^{-8} Mol/L[2]. The name M. is derived from the laboratory where it was characterized for the first time (Monell Chemical Senses Center, Univ. of Pennsylvania, U.S.A.)[3]. M. was the first protein with a sweet taste for humans to be discovered. The stability is, however, not sufficient for practical uses in food technology since M. hydrolyses rapidly and changes in the conformation result in loss of the sweet taste. The stability can be improved by covalent linkages between the two peptide chains, but then the sweetness decreases markedly. In spite of the widely differing amino acid sequences, antibodies against M. are also active against *thaumatin and vice versa[4] (immunological cross-reaction). For synthesis see *Lit.*[4].
Lit.: [1]Z. Physiol. Chem. **357**, 585 (1976). [2]Nabors (ed.), Alternative Sweeteners, p. 310, New York: Dekker 1986. [3]J. Biol. Chem. **248**, 534 (1973); Science **181**, 32 (1973). [4]Agric. Biol. Chem. **54**, 1521, 2219 (1990); Biosci., Biotechnol., Biochem. **62**, 2043 (1998) (solid phase).
gen.: Dobbing, Sweetness, Berlin: Springer 1987 ▪ Hudson (ed.), Developments in Food Proteins (4.), p. 219–245, London: Elsevier 1986 ▪ Merck-Index (12.), No. 6328 ▪ Nutrition **12**, 206 (1988) ▪ Trends Biochem. **13**, 13 (1988) ▪ Ullmann (4.) **22**, 364. – *[CAS 121337-41-5]*

Monensin.

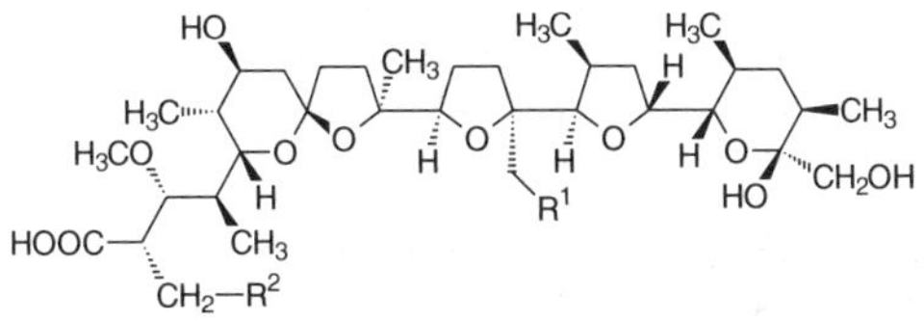

Table: Structure and data of Monensines.

M.	R^1	R^2	molecular formula	M_R	CAS
A	CH_3	H	$C_{36}H_{62}O_{11}$	670.90	17090-79-8
B	H	H	$C_{35}H_{60}O_{11}$	656.85	30485-16-6
C	CH_3	CH_3	$C_{37}H_{64}O_{11}$	684.91	31980-87-7

Ionophoric *polyether antibiotic produced as a mixture by *Streptomyces cinnamonensis* with the main component being *M. A* {cryst., mp. 103–105 °C, $[\alpha]_D^{20}$ +47.7° (CH_3OH)}. M. A specifically binds Na^+ ions (crystalline complex) and is used in studies on cell membranes. The selectivity of M. B between Na^+ and K^+ is appreciably smaller. M. is active against protozoa (coccidia, *Plasmodium falciparum*), Gram-positive bacteria, mycobacteria, fungi, and viruses (HIV); it is used in fowl breeding as a fodder additive for protection against coccidiosis and to promote growth. M. irritates the skin and eyes and is cardiotoxic, LD_{50} (rat p. o.) 100 mg/kg. M. A is formed biosynthetically on the *polyketide pathway from five units of acetate, seven of propionate, and one of butyrate; the rings are formed in the initially linear chain by inclusion of atmospheric oxygen.

Lit.: Antimicrob. Agents Chemother. **36**, 492 ff. (1992); **40**, 602–608 (1996) (activity) ▪ Kirk-Othmer (4.) **3**, 310 ▪ Sax (8.), MRE 225, 230 ▪ Trends Biochem. Sci. **9**, 313 (1984). – *Biosynthesis:* Angew. Chem. Int. Ed. Engl. **34**, 298 ff. (1995) ▪ Birch & Robinson, in Vining & Stuttard (eds.), Genetics and Biochemistry of Antibiotic Production, p. 443–476, Boston: Butterworth-Heinemann 1995 ▪ J. Antibiot. **48**, 1280–1287 (1995) ▪ Zechmeister **58**, 1–82. – *Synthesis:* J. Am. Chem. Soc. **115**, 7152, 7166 (1993) ▪ Nicolaou, p. 185–209, 227–248.

Moniliformin (hydroxycyclobutenedione potassium salt, potassium semisquarate).

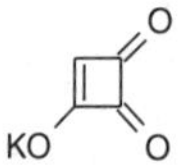

C_4HKO_3, M_R 136.15, brilliant yellow cryst.; a *mycotoxin from *Fusarium moniliforme* (*Gibberella fujikuroi*), *F. fusarioides*, *F. sporotrichioides* and others (mold fungi on fruits and corn). LD_{50} (chick, duckling p. o.) 4 mg/kg; 40 mg/kg are lethal within 45 min. M. is phytotoxic and has growth regulating activity in various plant systems. Pyruvate and α-ketoglutarate oxidases in mitochondria are inhibited at low concentrations (IC_{50} <5 mg/L). M. is found world-wide in infected corn and other grain crops.

Lit.: Angew. Chem. Int. Ed. Engl. **23**, 996 (1984) (biosynthesis) ▪ Appl. Environ. Microbiol. **53**, 196 (1987) ▪ Ber. Landwirtsch. **63**, 257 (1985) ▪ Cole-Cox, p. 893–897 ▪ J. Org. Chem. **55**, 1177 (1990); **59**, 1149 (1994) ▪ Sax (8.), MRE 250 ▪ Synthesis **1990**, 237, 583; **1995**, 571 ▪ Zentralbl. Mikrobiol. **144**, 3–12 (1989). – *[CAS 52591-22-7; 31876-38-7 (free acid)]*

Monoamines. General name for naturally occurring alkyl- and aralkylamines with one amino group. Aliphatic M. such as, e. g., isobutyl-, isoamyl-, or hexylamine occur as fragrance substances in flowers of various plants pollinated by flies [including hawthorn (*Crataegus*), elder (*Sambucus nigra*), arum (*Arum maculatum*)] as well as in some fungi and many marine red algae. The aliphatic M. are formed biogenetically by decarboxylation of the corresponding amino acids (e. g., in red algae) or by transamination of aliphatic aldehydes (demonstrated in flowering plants). *Histamine, the decarboxylation product of histidine, is found in many plants, e. g., together with *serotonin in the stinging hairs of stinging nettle (*Urtica* sp.). In animal organisms, both M. act as important mediators or, respectively, neurotransmitters. Tryptamine, the decarboxylation product of *tryptophan, is the biogenetic precursor of many *indole alkaloids and building block of the *ergot alkaloids. Phenylalkylamines (see β-phenylethylamine alkaloids) such as phenylethylamine (from phenylalanine), tyramine (from *L-tyrosine), dopamine (3,4-dihydroxyphenylethylamine) are widely distributed in the plant kingdom. Tyramine and dopamine are biogenetic precursors of *isoquinoline alkaloids, *betalains, and *melanins.

Lit.: Bell & Charlwood (eds.), Encyclopedia of Plant Physiology, New Ser. vol. 8, p. 433, Heidelberg: Springer 1980 ▪ Conn (ed.), The Biochemistry of Plants, vol. 7, p. 249, New York: Academic Press 1981 ▪ Mothes et al., p. 99.

Monobactams. A group of monocyclic *β-lactam antibiotics, characterized by a core skeleton of a four-membered ring without any other condensed ring units; see also sulfazecin.

Lit.: J. Antibiot. **34**, 621–627 (1981) ▪ Nature (London) **291**, 489 ff. (1981) ▪ see also nocardicins.

Monocrotaline.

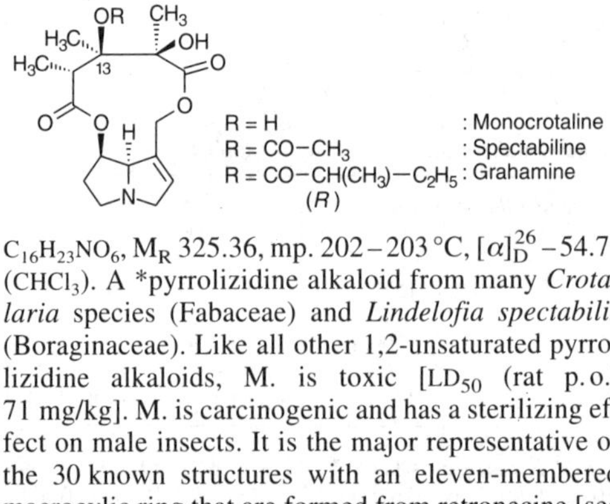

$C_{16}H_{23}NO_6$, M_R 325.36, mp. 202–203 °C, $[\alpha]_D^{26}$ –54.7° ($CHCl_3$). A *pyrrolizidine alkaloid from many *Crotalaria* species (Fabaceae) and *Lindelofia spectabilis* (Boraginaceae). Like all other 1,2-unsaturated pyrrolizidine alkaloids, M. is toxic [LD_{50} (rat p.o.) 71 mg/kg]. M. is carcinogenic and has a sterilizing effect on male insects. It is the major representative of the 30 known structures with an eleven-membered macrocylic ring that are formed from retronecine [see necine(s)] and differently substituted pentanedioic acids. These alkaloids are principally isolated from *Crotalaria* species; e. g., *spectabiline*[1] {$C_{18}H_{25}NO_7$, M_R 367.40, mp. 185.5–186 °C, $[\alpha]_D^{20}$ +121° ($CHCl_3$)} and *grahamine*[2] {$C_{21}H_{31}NO_7$, M_R 409.48, mp. 163 °C, $[\alpha]_D^{20}$ +100° (C_2H_5OH)}.

Lit.: [1] Aust. J. Chem. **10**, 474 (1957); **11**, 97 (1958). [2] Aust. J. Chem. **22**, 1773–1777 (1969).

gen.: Beilstein E III/IV **27/9**, 6660 ▪ Bull. Chem. Soc. Jpn. **52**, 3329–3336 (1979) (synthesis) ▪ J. Chem. Soc., Perkin Trans. 1 **1974**, 2082–2086 (biosynthesis) ▪ Merck-Index (12.), No. 6333 ▪ Pelletier **9**, 155–233 ▪ Sax (8.), MRH 000 ▪ Tetrahedron **48**, 10531 (1992) (synthesis). – *[HS 2939 90; CAS 315-22-0 (M.); 520-55-8 (spectabiline); 24583-56-0 (grahamine)]*

Monomargine.

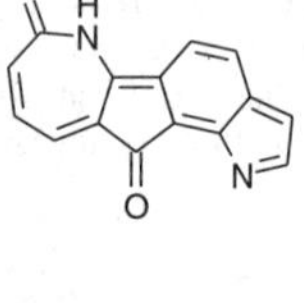

$C_{15}H_8N_2O_2$, M_R 248.24, orange solid. M. is an unusual pigment from the bark of *Monocarpia marginalis* (Annonaceae). *In vitro* M. is cytotoxic to various cell lines.

Lit.: Tetrahedron Lett. **34**, 1795 (1993). – *[CAS 148717-63-9]*

Monomelittoside. M. is an *iridoid glucoside occurring in nature either with an additional glucose unit at C-5 (*melittoside*) or as the benzoyl ester at C-10 (*globularifolin*).

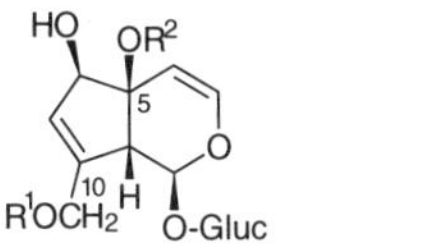

$R^1 = R^2 = H$: Monomelittoside
$R^1 = H$, $R^2 = Gluc$: Melittoside
$R^1 = Benzoyl$, $R^2 = H$: Globularifolin

Table: Monomelittoside and derivatives.

trivial name	Mono-melittoside	Melittoside	Globulari-folin
molecular formula	$C_{15}H_{22}O_{10}$	$C_{21}H_{32}O_{15}$	$C_{22}H_{26}O_{11}$
M_R	362.33	524.48	466.44
mp. [°C]	amorphous	167–168	amorphous
$[\alpha]_D$	–180° (H_2O)	–29° (H_2O)	–122° (CH_3OH)
CAS	20633-72-1	19467-03-9	73248-91-6

M. and melittoside are isolated in small amounts together with *harpagide and harpagide 8-acetate from extracts of the bastard balm (*Melittis melissophyllum*, Lamiaceae)[1]. M. and globularifolin occur in *Globularia cordifolia* (Globulariaceae)[2].

Lit.: [1]Gazz. Chim. Ital. **97**, 1209 (1967). [2]Helv. Chim. Acta **63**, 117 (1980).

gen.: J. Nat. Prod. **43**, 649, No. 52, 54, 222 (1980). – *Acetyl derivatives of M.:* J. Nat. Prod. **54**, 626 (1991); Phytochemistry **29**, 3938 (1990). – *[CAS 20633-72-1]*

Monomorines.

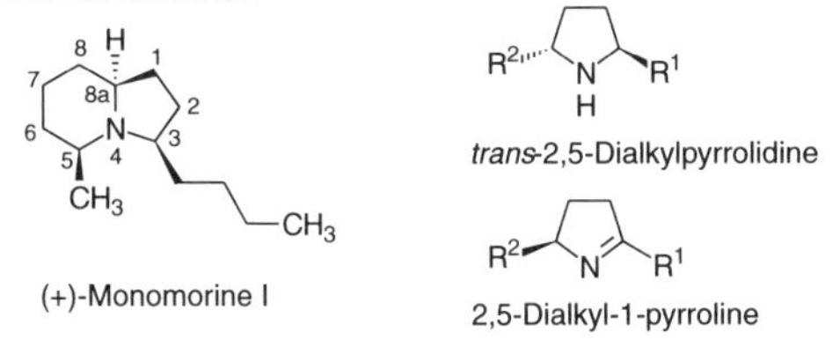

(+)-Monomorine I

trans-2,5-Dialkylpyrrolidine

2,5-Dialkyl-1-pyrroline

Term originally for (3*R*,5*S*,8a*S*)-3-butyl-5-methyloctahydroindolizine (M. I)[1]. *M. I* ($C_{13}H_{25}N$, M_R 195.35, oil) occurs in the *defensive secretion of the Pharaoh ant *Monomorium pharaonis*. Also the general name for a group of *trans*-configurated 2,5-dialkylpyrrolidine and 2,5-dialkyl-1-pyrrolines from ants of the genera *Monomorium* and *Solenopsis* (fire ants). The alkyl-substituents are unbranched, saturated, or bear a single double bond at the end of the chain, and have a chain length of 4 to 7 carbon atoms. 2,6-Dialkylpiperidines are known as *solenopsins. M. are also used by the ants to paralyse their prey. For synthesis of (+)-M. I, see *Lit.*[2].

Lit.: [1]J. Org. Chem. **50**, 670–673 (1985); **52**, 2094–2096 (1987); **54**, 4088–4097 (1989). [2]ApSimon **9**, 478; Eur. J. Org. Chem. **1999**, 3277; Heterocycles **36**, 2777 (1993); J. Am. Chem. Soc. **117**, 5399 f. (1995); J. Chem. Soc. Perkin Trans. 1 **1994**, 351 f.; **1997**, 113, 1315; Tetrahedron Lett. **31**, 5065 (1990); **34**, 3985 (1993); **37**, 111 (1996).

gen.: Acta Chem. Scand. **51**, 1024 (1997) (rac. synthesis) ▪ Atta-ur-Rahman **6**, 421–466 ▪ J. Chem. Ecol. **10**, 1233–1249 (1984); **14**, 2197–2212 (1988) ▪ J. Nat. Prod. **52**, 779 (1989); **53**, 375, 429 (1990). – *[CAS 53447-44-2 ((+)-M. I)]*

Monorden see radicicol.

Monosaccharides. M. are linear polyhydroxyaldehydes (aldoses) or polyhydroxyketones (ketoses). Most important among M. are the pentoses ($C_5H_{10}O_5$) and hexoses ($C_6H_{12}O_6$). Important aldopentoses include, e. g., D-*ribose, D-*xylose, and L-*arabinose. Important aldohexoses include D-*glucose, D-*mannose, and D-*galactose; the major ketohexoses are D-*fructose and *sorbose. The 6-*deoxy sugars L-*fucose and L-*L-rhamnose are also widely distributed hexoses. M. with more carbon atoms (*heptoses*: 7 carbon atoms, octoses, etc.) or less carbon atoms (*trioses:* 3 carbon atoms) do not occur in the free form in organisms but do play a role in carbohydrate metabolism as phosphate esters; *tetroses* (4 carbon atoms): erythrose, threose are relatively rare.

Stereochemistry: The D-configuration is by far the more frequent in the naturally occurring M. The large number of asymmetric C-atoms makes the M. an ideal source of synthetic building blocks for asymmetric synthesis. In the schematic representation of the relative configurations of tri-, tetr-, pent-, and hexoses where the horizontal lines represent OH-groups while A and B designate organic residues, only members of the D-series are shown; *example:* D-*galacto*-hexose = D-galactose. Syntheses of all L-hexoses are given in *Lit.*[1].

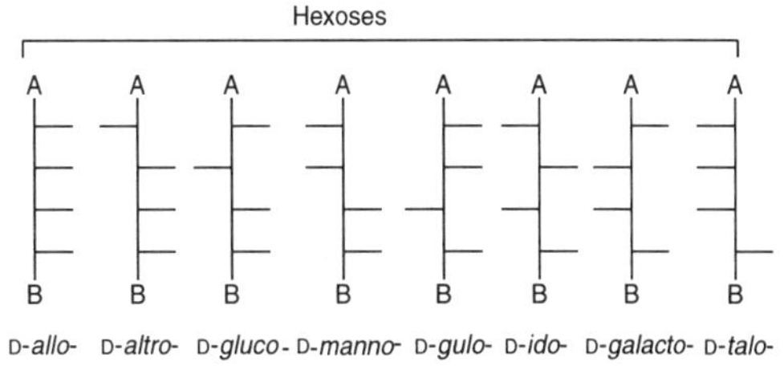

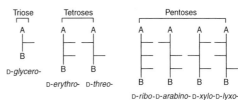

Figure: Relative configurations of Monosaccharides.

Whenever possible, M. form intramolecular hemiacetals giving the ring structures of the pyran- (*pyranoses*) and furan-types (*furanoses*):

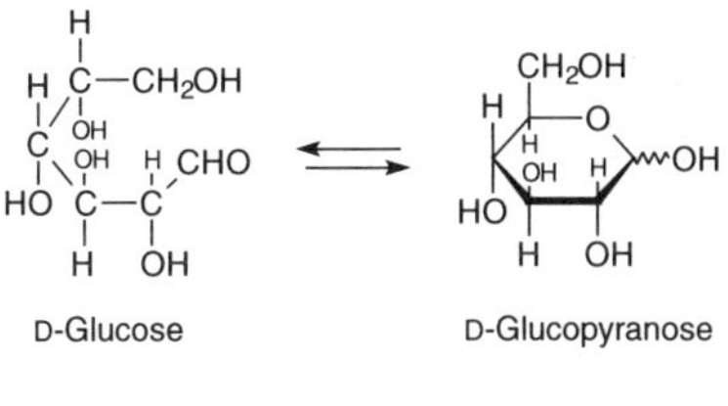

Smaller rings are unstable, larger rings are only stable in aqueous solution. A further asymmetric C-atom (the so-called anomeric C-atom) is formed by the cyclization, its configuration is designated by the prefixes α and β[2]. The formation of the hemiacetals is a dynamic process depending on various factors such as temperature, solvent, pH value, etc. Usually mixtures of both anomeric forms occur, in some cases also mixtures of the furanose and pyranose forms.
The epimerization ($\alpha \rightleftarrows \beta$) proceeding through ring opening is also the reason for the occurrence of mutarotation; ^{13}C-NMR spectroscopy is useful for determination of the proportion of the carbonyl form in aqueous sugar solutions[3]. For investigations on conformations and configurations of M. as well as the nomenclature of the M. conformations, see *Lit.*[4].
The carbon chain in most known M. is not branched; however, many pharmaceutically relevant natural products, especially antibiotics, contain not only amounts of branched sugars (*examples:* apiose, hamamelose, streptose), but also of 6-deoxy-, amino-, and hydroxyaminosugars, see calicheamicins, enediyne antibiotics, esperamycins. D-*Glucosamine is widely distributed in nature as a component of, e.g., *chitin, mucopolysaccharides, and blood group substances. For a review article on rare sugars, see *Lit.*[5]. For nomenclature according to the IUPAC rules, see *Lit.*[6,7].
The M. show the following common reactions: 1) copper, silver, and bismuth salts are reduced in solution. – 2) M. react with bases to form alcoholates, with concentrated alkali resinification occurs to furnish yellow and dark-colored products. – 3) No reaction occurs with hydrogen sulfite or fuchsin/sulfurous acid because the aldehyde group is masked. – 4) Reduction leads to multifunctional alcohols, e.g., *hexitols. – 5) Oxidation gives rise to mono- and dicarboxylic acids (sugar acids). – 6) The alcohol groups can be esterified. – 7) Addition of hydrogen cyanide occurs. – 8) Osazones are formed with phenylhydrazine.
Suitable analytical separation procedures include thin layer chromatography[8] and HPLC[9], after derivatization furanoids and pyranoids as well as enantiomers can be separated by gas chromatography[10]. For identification with boranes, see *Lit.*[11].

Lit.: [1]Science **220**, 949 ff. (1983). [2]Pure Appl. Chem. **54**, 1517–1526 (1982). [3]Collins-Ferrier. [4]Annu. Rev. Biochem. **41**, 953–996 (1972); Pure Appl. Chem. **53**, 1901–1905 (1981); **55**, 1269–1272 (1983). [5]Angew. Chem. Int. Ed. Engl. **8**, 401–409 (1969); **10**, 236–248 (1971); **11**, 159–173 (1972). [6]J. Biol. Chem. **247**, 613–635 (1972). [7]Pure Appl. Chem. **54**, 207–215 (1982). [8]J. Chromatogr. **127**, 133–162 (1976); **180**, 1–16 (1979). [9]Adv. Carbohydr. Chem. Biochem. **46**, 17–72 (1988); Carbohydr. Res. **183**, 11–17 (1988); LABO **12**, 732 ff. (1981). [10]Angew. Chem. Int. Ed. Engl. **20**, 693 f. (1981). [11]Pure Appl. Chem. **49**, 765–789 (1977).

gen.: Adv. Carbohydr. Chem. **40**, 1–131 (1982) ▪ Adv. Carbohydr. Chem. Biochem. **41**, 27–66 (1983) ▪ Carbohydr. Chem. **19**, 1–303 (1987); **20**, 1–274 (1988) ▪ Collins-Ferrier ▪ El Khadem, Carbohydrate Chemistry: Monosaccharides and their Oligomers, San Diego: Academic Press 1988 ▪ Györgydeak, Pelyvás (eds.), Monosaccharide Sugars, San Diego: Academic Press 1998 ▪ Nachr. Chem. Tech. Lab. **30**, 934–938 (1982) ▪ Shallenberger, Advanced Sugar Chemistry, Westport: Avi 1983 ▪ Ullmann (5.) **A 5**, 80–82. – *Synthesis:* Angew. Chem. Int. Ed. Engl. **26**, 15–23 (1987) ▪ Tetrahedron **46**, 245–264 (1990).

Monoterpenes. M. may be considered formally as dimerization products of isoprene (2-methyl-1,3-butadiene, C_5H_8) and have a C_{10} skeleton. For biogenesis, see terpenes. The class of M. includes not only acyclic but also mono- and bicyclic hydrocarbons and their oxidation products (alcohols, aldehydes, ketones, and acids). Tri- and tetracyclic M. are very rare. The simplest acyclic M. are *geraniol and nerol (see geraniol), from which mono- and bicyclic M. are formed by intramolecular cyclization. The most important carbon skeletons of the monocyclic M. are illustrated in formulae **1–6**.

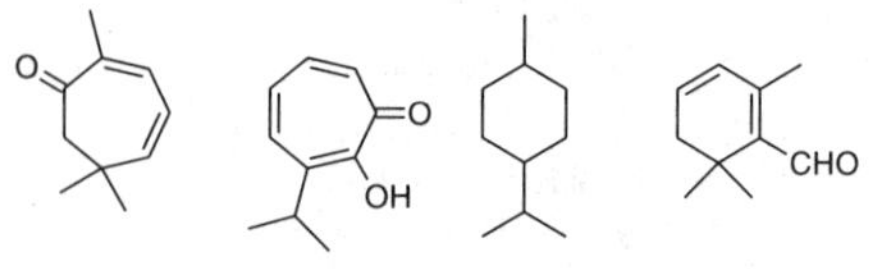

*Eucarvone **1** · *α-Thujaplicin **2** · *p-Menthane **3** · *Safranal **4**

Iridodial **5** · *Fragranol **6**

The most important carbon skeletons of the bicyclic M. are illustrated by the examples of formulae **7–12**, the hydrocarbons of which are known by the trivial names given under each formula.

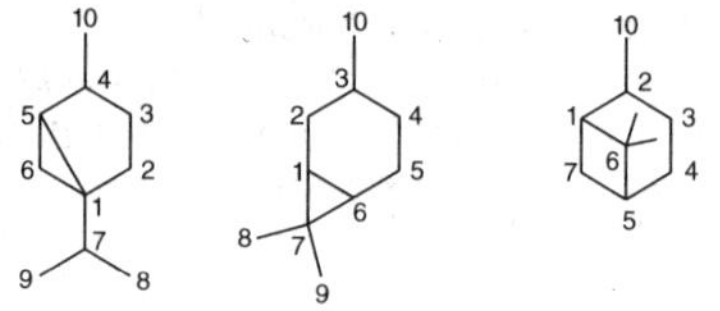

*Thujane **7** · Carane **8** · Pinane **9**

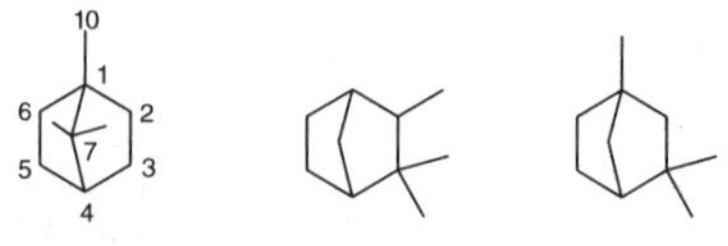

Bornane **10** · Isocamphane **11** · Fenchane **12**

Occurrence: M. are widely distributed in the plant kingdom. They are obtained by steam distillation or by extraction from crushed plant parts (flowers, leaves, fruits, bark, roots) as essential (=volatile) oils. Since the M. are volatile the amounts of M. emitted by plants in tropical climate zones are considerable. In animal organisms, M. are present only in small amounts as pheromones or defence substances.
Use: Technically important M. are *geraniol, α-*pinenes, limonene, cymenoles, 3-*carenes, and *camphor. M. alcohols usually occur in esterified form, these esters are used as fragrances, flavor, and spice substances in the perfume, cosmetic, and food industries. Some enantiomerically pure M. are used in stereospecific syntheses of certain natural products.

Lit.: Connolly & Hill, Dictionary of Terpenoids (3 vols.), London: Chapman & Hall 1992 ▪ Ermann, Chemistry of the Monoterpenes (2 vols.), New York: Dekker 1985 ▪ Nat. Prod. Rep.

9, 531–556 (1992); **13**, 195–225 (1996) ▪ Ullmann (5.) **A 11**, 154ff.

Monoterpenoid indole alkaloids. A large group of plant alkaloids encompassing ca. 2000 different structures derived almost exclusively from the decarboxylation products of *tryptophan, tryptamine, and the monoterpene glucoside secologanin (see secoiridoids). The majority of these alkaloids are found in only three plant families: the Rubiaceae, the Apocynaceae, and the Loganiaceae. Many of the alkaloids or, respectively, their carbon skeleton types can be used as taxonomic markers[1]. Many of the alkaloids are formed by skeletal rearrangements: In the past, only three larger subgroups were distinguished for simplicity: the Corynanthe/Strychnos type, the Iboga type, and the Aspidosperma type. The compounds can also be conveniently classified according to their occurrence, e.g., *Alstonia, *Amsonia, *Aspidosperma, *Catharanthus, *Cinchona, *Rauvolfia, *Strychnos, *Vinca alkaloids.
Many of the compounds are of great pharmacological or toxicological importance, see also ajmalicine, ajmaline, brucine, camptothecin, quinine, curare, reserpine, strychnine, toxiferines, vinblastine, vincamine, vincristine, and yohimbine.
Biosynthesis: The biosynthesis of these alkaloids begins with the key reaction, the condensation of tryptamine with secologanin to give *strictosidine catalyzed by the enzyme *strictosidine synthase. The biosynthesis was elucidated mainly by the use of plant cell suspension cultures. Cell cultures of *Catharanthus* and *Rauvolfia* species have been the subject of intense phytochemical studies.
The biosynthesis of ajmaline is illustrated as an example for the biosynthesis of a monoterpenoid indole alkaloid (see figure).

Lit.: [1] Pelletier **1**, 211–376.
gen.: Manske **47**, 115–172; **50**, 415–452; **52**, 261–376 ▪ Nat. Prod. Rep. **14**, 559–590 (1997) ▪ Szántay, in Saxton (ed.), Monoterpenoid Indole Alkaloids, Supp. vol. 25, Heterocyclic Compounds, p. 478–480, 733–736, Chichester: Wiley 1994 ▪ Zechmeister **55**, 89. – *[HS 2939 90; CAS 1617-90]*

Monotropein. M. is a C_{10}-*iridoid glucoside occurring as the free C-11 carboxylic acid (M.) and as the methyl ester (*monotropein methyl ester*). The compound esterified at O-10 of M. with *p*-coumaric acid is *vaccinoside* (formula see p. 408).
M. and its derivatives occur in all Ericaceae where the bitter taste of the iridoid glucoside protects the plants from predators. M. was first isolated from genuine pinesap (*Monotropa hypopitys*, Monotropaceae)[2]. M. and its methyl ester occur in the Japanese plant *Tripetaleia paniculata* (=*Elliottia paniculata*, Ericaceae)[3]. Vaccinoside derives its name from its isolation from fruits of the Japanese *Vaccinium bracteatum* (Erica-

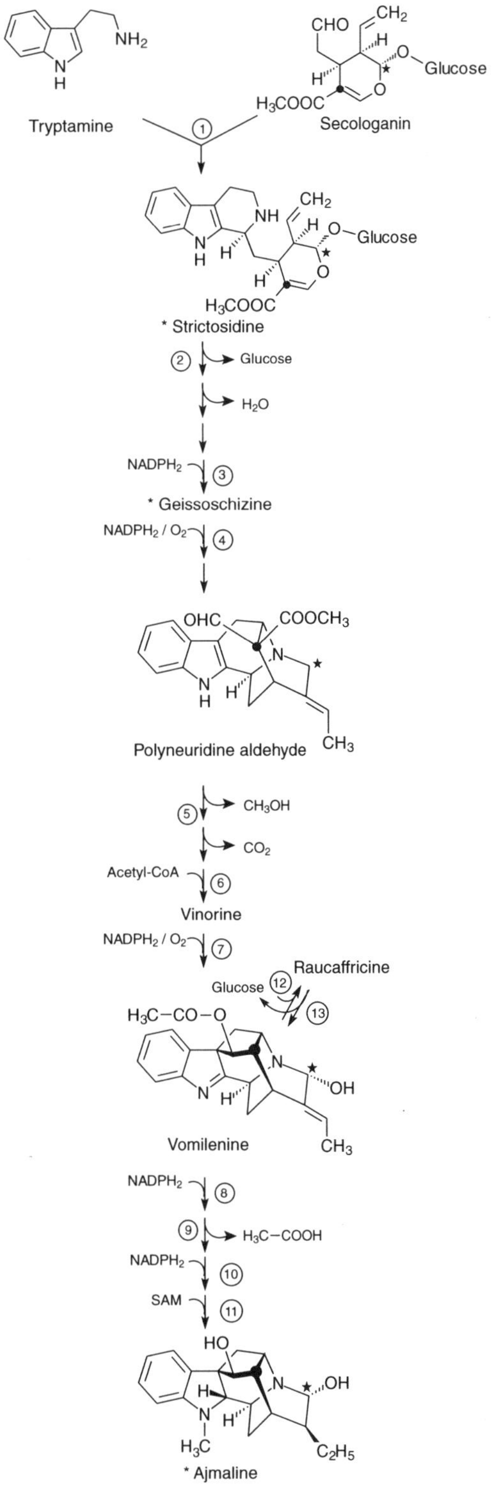

Figure: Simplified scheme for the biosynthesis of the antiarrhythmic agent ajmaline. Only the key steps are shown in the scheme. The cofactors given on the left side participate in individual enzymatic reactions. The following enzymes were identified in the 15-step sequence leading to the final product ajmaline: 1: strictosidine synthase; 2: strictosidine glucosidase; 3: dehydrogeissoschizine reductase; 4: sarpagine bridging enzyme; 5: polyneuridine aldehyde esterase; 6: vinorine synthase; 7: vinorine hydroxylase; 8: vomilenine reductase; 9: methylajmalan esterase; 10: 19,20-reductase; 11: norajmaline *N*-methyltransferase; 12: raucaffricine synthase; 13: raucaffricine glucosidase ▶

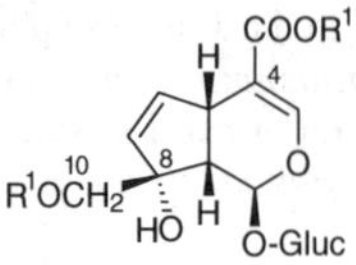

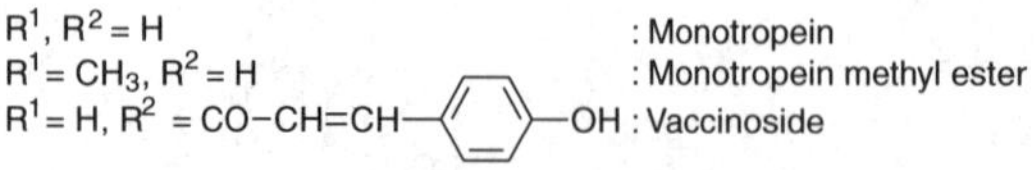

Table: Monotropein and derivatives.

trivial name	Monotropein[1]	Monotropein methyl ester	Vaccinoside
molecular formula	$C_{16}H_{22}O_{11}$	$C_{17}H_{24}O_{11}$	$C_{25}H_{28}O_{13}$
M_R	390.34	404.37	536.49
mp. [°C]	161–163	amorphous	150–153
$[\alpha]_D$	–130.7° (H_2O)		–72.2° (H_2O)
CAS	5945-50-6	54712-59-3	36138-58-6

ceae)[4]. It has also been detected in the wild rosemary (*Andromeda polifolia*, Ericaceae)[5].

Lit.: [1]Tetrahedron Lett. **1967**, 2367. [2]C. R. Acad. Sci. **176**, 1742 (1923). [3]Yakukagu Zasshi **94**, 1634 (1974) [C. A. **82**, 108822b (1975)]. [4]Chem. Pharm. Bull. **19**, 1979 (1971). [5]Arch. Pharm. (Weinheim, Ger.) **313**, 702 (1980).

gen.: Fitoterapia **62**, 176 (1991) (NMR) ▪ Hager (5.) **4**, 326 ff.; **5**, 220 f., 223 f. ▪ J. Nat. Prod. **43**, 649, No. 125–127 (1980). – *[HS 2938 90]*

Montagnetol see erythrin.

Montanol see zoapatanol.

Moracin(s). A group of *phytoalexins from the mulberry tree (*Morus alba*). They are formed after infection of the tree with the fungus *Fusarium solani*. All 26 of the currently known M. (M. A to Z) have the 2-phenylbenzofuran skeleton bearing hydroxy, methoxy, and prenyl substituents.

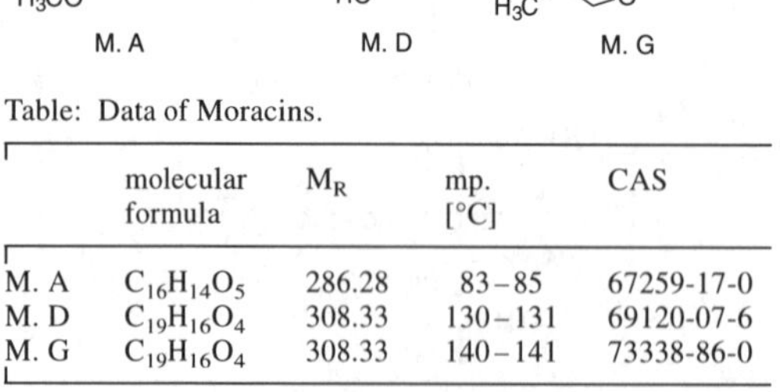

Table: Data of Moracins.

	molecular formula	M_R	mp. [°C]	CAS
M. A	$C_{16}H_{14}O_5$	286.28	83–85	67259-17-0
M. D	$C_{19}H_{16}O_4$	308.33	130–131	69120-07-6
M. G	$C_{19}H_{16}O_4$	308.33	140–141	73338-86-0

Lit.: Chem. Lett. **1978**, 1239 f. ▪ Heterocycles **29**, 807 (1989) ▪ J. Chem. Res. (S) **1985**, 34 ▪ J. Chem. Soc., Perkin Trans. 1 **1998**, 3453 (synthesis M. C) ▪ Justus Liebigs Ann. Chem. **1984**, 734–741 (synthesis) ▪ Nat. Prod. Rep. **12**, 101–133 (1995); **11**, 205–218 (1994) ▪ Zechmeister **53**, 109–114.

Moretane see hopanes.

Morin (2′,3,4′,5,7-pentahydroxyflavone). Formula see flavones. $C_{15}H_{10}O_7$, M_R 302.24, light yellow needles, mp. 303–304 °C. M. occurs in various woods, e. g. in *Morus alba* (mulberry tree, Moraceae). M. was first isolated in 1830 and is used as a reagent in spot reactions to detect Al, Be, Zn, Ga, In, and Sc salts, in photometric analysis[1] and as a textile dye. It shows anti-HIV activity[2].

Lit.: [1]Onishi (ed.), Photometric Determination of Traces of Metals, Part II b, p. 765, 4. edition, New York: Wiley 1989. [2]Eur. J. Biochem. **190**, 469 (1990).

gen.: Chem. Ind. (London) **1984**, 881 ▪ Cody, Plant Flavonoids in Biology and Medicine II, New York: A. R. Liss 1988 ▪ Eur. J. Biochem. **190**, 469 (1990) ▪ J. Mol. Struct. **317**, 89 (1994) ▪ Karrer, No. 1572 ▪ see also flavones. – *[HS 2932 90; CAS 480-16-0]*

Morindone (1,2,5-trihydroxy-6-methylanthraquinone, C. I. 75430).

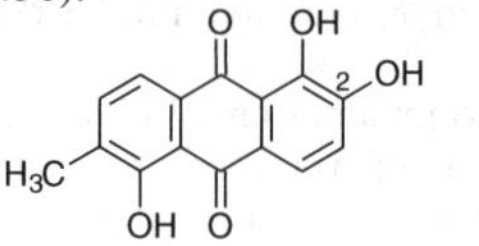

$C_{15}H_{10}O_5$, M_R 270.24, orange-red needles, mp. 285 °C, uv$_{max}$ 432 nm (lg ε 3.78/CH_3OH). The anthraquinone pigment M. occurs in the free form and as the 2-*O*-β-primeveroside (morindin) or 2-*O*-β-rutinoside in the heartwood, roots, and leaves of *Morinda* species (Rubiaceae). In the past, roots of *Morinda* species were used, especially in India, as a red dye for cotton. The Maoris used the "karamu", the M.-containing barks of *Coprosma grandiflora*, *C. areolata*, and *C. lucida* (Rubiaceae) to dye flax fibers. The biosynthesis proceeds from isochorismate, 2-oxoglutarate, and mevalonic acid lactone (see anthranoids).

Lit.: Aust. J. Chem. **30**, 1553 (1977) ▪ Karrer, No. 1260 ▪ Phytochemistry **19**, 119 (1980); **23**, 2307 (1984) ▪ Zechmeister **49**, 120 ▪ Tetrahedron **40**, 3455 (1984) ▪ Z. Naturforsch. B **36**, 1180 (1981). – *[CAS 478-29-5]*

Morphinan alkaloids. The best known class of *isoquinoline alkaloids containing the morphinan system as skeleton. The most important M.a. is *morphine which is obtained together with a series of other benzylisoquinoline alkaloids including *codeine and thebaine from opium or *Papaver somniferum* (Papaveraceae). For other occurrences in plants, see the table. The M. a. are soluble in ethanol, acetone, ethyl acetate, poorly soluble in ether and chloroform. Heroin is a synthetic M. alkaloid. Neopine, the 8,14-double bond isomer of codeine, is a minor alkaloid in *opium and has similar pharmacological properties to codeine. Protostephanine is an antihypertensive agent. Salutaridine is an intermediate in the biosynthesis of morphine and shows antitumor activity. Sebiferine exhibits a morphine-like central antinociceptive activity. Thebaine is a highly toxic opium alkaloid. It has a stronger narcotic but a weaker analgesic effect than morphine, chronic use leads to addiction.

It is interesting to note that the ability to biosynthesize morphine appears not to be limited to plants of the Papaveraceae; traces of morphine (1–10 μg/g) can also be detected, e. g., in hay and lettuce[1]. In addition,

Table: Data and occurrence of Morphinan alkaloids.

name	molecular formula	M_R	mp. [°C]	$[\alpha]_D$	occurrence	CAS
Amurine	$C_{19}H_{19}NO_4$	325.36	213–215	+10° ($CHCl_3$)	*Papaver nudicaule*	4984-99-0
Codeine	$C_{18}H_{21}NO_3$	299.37	155	–137.8° (C_2H_5OH)	*P. somniferum*	76-57-3
Codeinone	$C_{18}H_{19}NO_3$	297.35	181.5–182.5	–205° (C_2H_5OH)	*P. somniferum*	467-13-0
Flavinantine	$C_{19}H_{21}NO_4$	327.38	130–132	–14.5° (C_2H_5OH)	*Croton flavens*	19777-82-3
Hasubanonine	$C_{21}H_{27}NO_5$	373.45	116–117	–214° (CH_3OH)	*Stephania japonica*	1805-85-2
Heroin	$C_{21}H_{23}NO_5$	369.42	173		*P. somniferum*	561-27-3
Laurifine	$C_{19}H_{23}NO_3$	313.40			*Cocculus laurifolius*	56261-28-0
Laurifinine	$C_{19}H_{23}NO_3$	313.40	179–181		*C. laurifolius*	56261-29-1
Morphine	$C_{17}H_{19}NO_3$	285.34	254–256	–130.9° (CH_3OH)	*P. somniferum*	57-27-2
Neopine	$C_{18}H_{21}NO_3$	299.37	127.5–127.8	–28.1° ($CHCl_3$)	*P. somniferum*, *P. bracteatum*	467-14-1
Oripavine	$C_{18}H_{19}NO_3$	297.35	201–202	–216.9° ($CHCl_3$)	*P. somniferum*, *P. orientale*, *P. bracteatum*	467-04-9
Protostephanine	$C_{21}H_{27}NO_4$	357.45	73–74		*Stephania japonica*	549-28-0
Salutaridine	$C_{19}H_{21}NO_4$	327.38	197–198	+111° (C_2H_5OH)	*Croton-*, *Papaver-* and *Glaucium* species	1936-18-1
Sebiferine	$C_{20}H_{23}NO_4$	341.41	112–113		*Litsea sebifera*, *Alseodaphne perakensis*, *C. laurifolius*, *Rhizocarya racemifera*	23979-25-1
Sinoacutine	$C_{19}H_{21}NO_4$	327.38	198	–112° (C_2H_5OH)	*Sinomenium acutum*, *Corydalis-*, *Glaucium-* and *Nandina* species	4090-18-0
Sinomenine	$C_{19}H_{23}NO_4$	329.40	182	–71° (C_2H_5OH)	*S. acutum*	115-53-7
Thebaine	$C_{19}H_{21}NO_3$	311.38	192.5	–221° (CH_3OH)	*Papaver* species, esp. *P. bracteatum*	115-37-7

codeine and morphine have also been found in mammals, including humans [2,3]. There is evidence that morphine is indeed synthesized in mammals – i.e. is not of dietetic/plant origin – and that its formation follows the plants biosynthetic route: intermediates of the plant biosynthesis of morphine such as reticuline, salutaridine, and thebaine can be metabolized *in vitro* and presumably also *in vivo* by rats to morphine or morphine precursors [4,5,6]. In addition, a highly specific and stereoselective key enzyme of plant morphine biosynthesis, salutaridine synthase, has been unequivocally identified in various tissues of many mammals [7].

Biosynthesis: The biosynthesis in poppy proceeds from the central intermediate of the benzylisoquinolines, namely (*S*)-norcoclaurine, via (*S*)-reticuline [8]. (*S*)-Reticuline can isomerize through 1,2-dehydroreticuline as intermediate to (*R*)-reticuline [9]. (*R*)-Reticuline undergoes an oxidative phenol coupling with formation of salutaridine [10] which, after reduction to salutaridinol [11] forms thebaine by ring closure [12]. This can be converted to morphine by one of two biosynthetic routes [13,14]. With the exception of two steps, the conversion of salutaridinol 7-*O*-acetate to thebaine [12] and of neopinone to codeinone [15], all steps are catalyzed by enzymes.

Routes similar to those for morphine, codeine, and thebaine are also being considered for the biosyntheses of *sinoacutine* and *sinomenine*. Sinoacutine is the enantiomer of salutaridine; the biosynthesis proceeds accordingly from (*S*)-reticuline. While morphine and morphine derivatives as well as sinoacutine are formed by an *ortho-para*-coupling of (*R*)- or (*S*)-reticuline, the

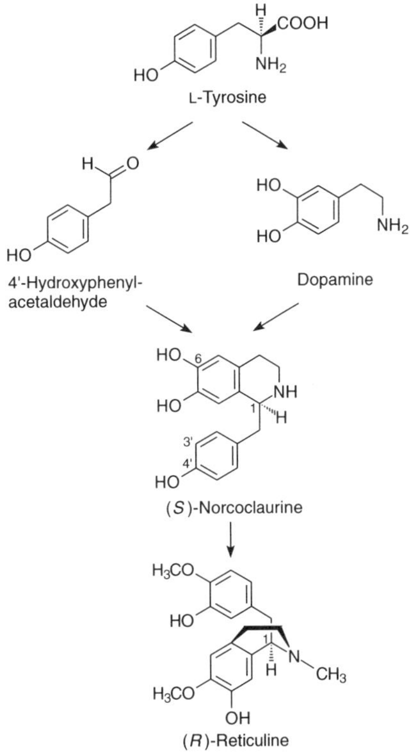

Figure 1: Biosynthesis of (*R*)-Reticuline from L-tyrosine in *Papaver somniferum*.

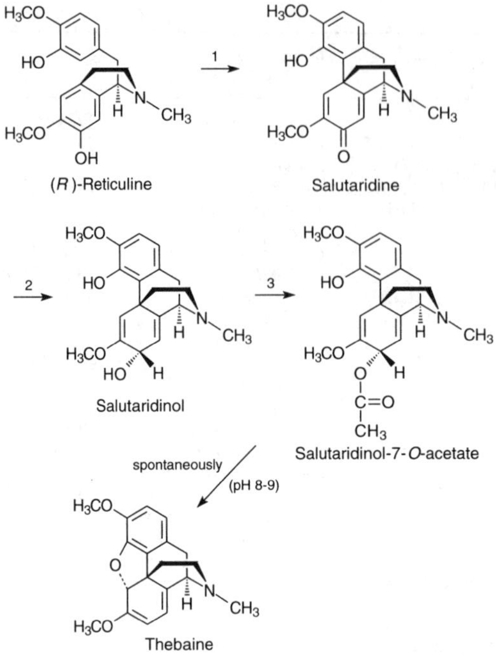

Figure 2: Established enzymatic formation of Thebaine from (*R*)-Reticuline in the poppy (*Papaver somniferum*). The participating enzymes are: 1=salutaridine synthase; 2=NADPH-dependent salutaridine oxidoreductase; 3=salutaridinol 7-*O*-acetyltransferase.

corresponding *para-para*-coupling leads to *flavinantine*, *sebiferine*, and *amurine* [16–19].

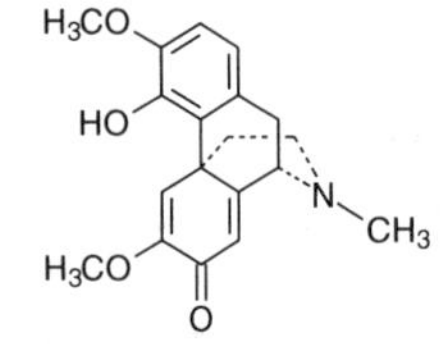

Sinoacutine

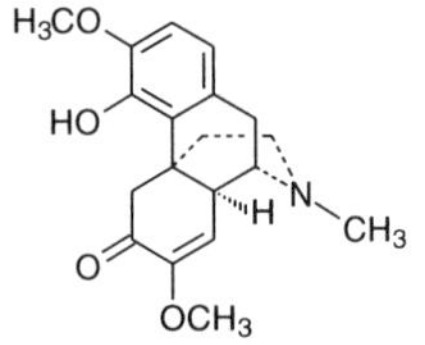

Sinomenine

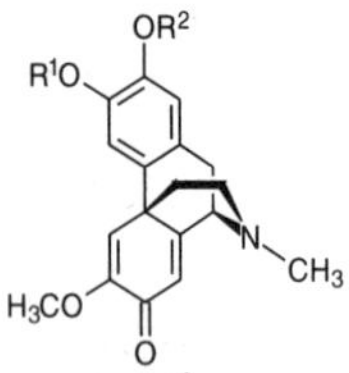

$R^1 = H$, $R^2 = CH_3$: Flavinantine
$R^1 = R^2 = CH_3$: Sebiferine
$R^1 + R^2 = CH_2$: Amurine

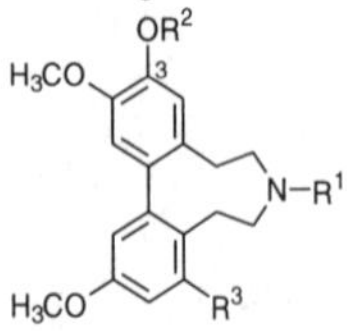

$R^1 = R^2 = CH_3$, $R^3 = OCH_3$: Protostephanine
$R^1 = R^3 = H$, $R^2 = CH_3$: Laurifine
$R^1 = CH_3$, $R^2 = R^3 = H$: Laurifinine

Protostephanine and other dibenz[*d,f*]azonine alkaloids as well as the *hasubanan alkaloids (e.g., hasubanonine) must also be included in the M. a. since they are biosynthetically derived from morphinans.

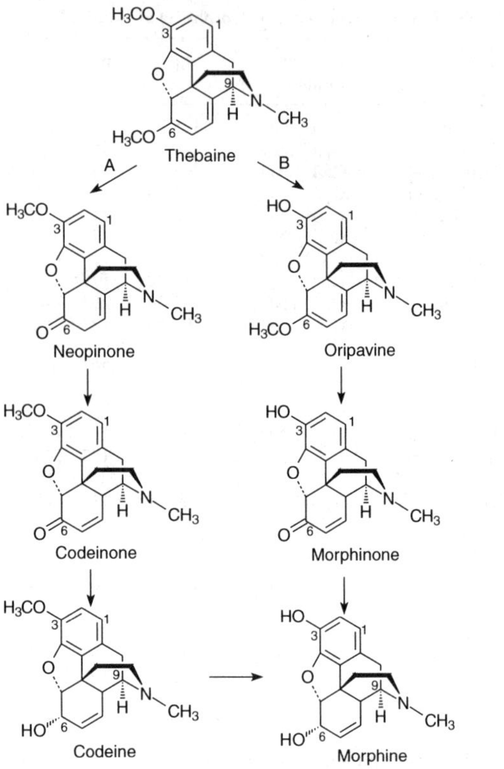

Figure 3: Two established biosynthetic routes, starting from thebaine, to morphine in *Papaver somniferum*: A) the route via neopinone and codeinone; B) the route via oripavine and morphinone.

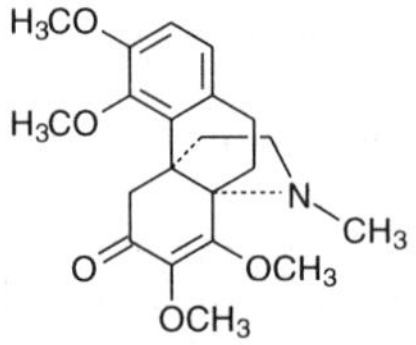

Hasubanonine

Lit.: [1] Science **213**, 1010 (1981). [2] Proc. Natl. Acad. Sci. USA **82**, 5203 (1985). [3] Life Sci. **40**, 301 (1987). [4] Nature (London) **330**, 674 (1987). [5] Proc. Natl. Acad. Sci. USA **83**, 4566 (1986). [6] Proc. Natl. Acad. Sci. USA **85**, 1267 (1988). [7] Tetrahedron Lett. **32**, 3675 (1991). [8] Justus Liebigs Ann. Chem. 555 (1990); Tetrahedron Lett. **31**, 7591 (1990). [9] Phytochemistry **31**, 813 (1992). [10] Phytochemistry **32**, 79 (1993). [11] Tetrahedron Lett. **33**, 2443 (1992). [12] Tetrahedron Lett. **35**, 3897 (1994). [13] J. Am. Chem. Soc. **94**, 1276 (1972). [14] Planta Medica **50**, 343 (1984). [15] Drug Metab. Dispos. **19**, 895 (1991); Tetrahedron Lett. **34**, 5703 (1993). [16] Mothes et al., p. 215–220. [17] Prog. Bot. **51**, 115–133 (1989). [18] Justus Liebigs Ann. Chem. **1990**, 555–562. [19] J. Nat. Prod. **56**, 973–975 (1993).

gen.: Beilstein E III/IV **27**, 2186–2263; E V **27/9**, 310–368 ■ Florey **10**, 93–138, 357–403; **17**, 259–366 ■ Hager (5.) **2**, 256–301; **3**, 662–664, 843–847, 911; **7** 1067–1073; **8**, 1040–1049 ■ Manske **16**, 393–430; **17**, 385–544; **33**, 307–347; **35**, 177–214; **45**, 128–232 ■ Opium Poppy: Botany, Chemistry and Pharmacology, Pharmaceutical Products Press 1995 ■ Phillipson et al., p. 38–46, 171–203, 229–239 ■ Sax (8.), CNF 500, CNF 750, CNG 500, CNG 675, CNG 750, DBH 400, DNO 700, ENK 000, MRO 500, MRO 750, MRP 000, MRP 100, MPP 250, SDY 600, TDX 835, TEN 000, TEN 100. – *[HS 2939 10]*

Morphine. Formula and data, see morphinan alkaloids. M. is the most important morphine (opium) alkaloid. M. is poorly soluble in boiling water and chloroform, soluble in alcohol, aqueous calcium and magnesium hydroxides and phenol. M. is obtained by extraction from the poppy *Papaver somniferum*. M. can be detected by several color reactions (e. g. violet color with Fe^{3+} in alkaline solution), particularly sensitive procedures are gas chromatography, mass spectroscopy, and radio-immuno assays.
Use & activity: M. is used as a powerful analgesic for severe cases of pain. The effective analgesic dose for an adult is 10 mg. In most humans M. generates a positive mood (euphoria) as a result of its effect on the limbal system. This can also be the triggering factor for the abuse of M. A sedative-hypnotic effect of M. is observed in most patients at the normal dose. At higher doses a narcosis-like state develops. The respiration depressing effect of M. can lead to respiratory arrest at doses of >50 mg. Rodents are considerably less sensitive: LD_{50} (mouse p.o.) 524 mg/kg, (mouse i.p.) 140 mg/kg. M. passes through the placental barrier and thus must not be used as a pain-killer in obstetrics. M. has an antitussive effect through dampening of reflexive stimulation of the coughing center. This property is more pronounced in the M. derivative *codeine. M. also subdues the vomiting center. In addition to central activities, M. inhibits gastric, bladder, and intestinal emptying. In cases of acute M. poisoning and threatening respiratory arrest morphine analogues in which the methyl group at the N atom is replaced by an allyl group (e. g., naloxone) may be given as an antidote. The exceptionally specific activities of M. and its analogues have been explained by the detection of opiate receptors. Heroin, which is degraded in the body to M., is a strongly addictive drug. It influences the central nervous system: states of anxiety and pain are blocked, sensory perception is reduced. It is originally planned that heroin should replace M. as a pain-killer since it was then thought to be a substitute without any addictive potential. It was even used in cough mixtures and prescribed as therapy for tuberculosis.
Historical: M. was first isolated in the pure form from opium in 1806 by Sertürner who introduced the name morphium (from Morpheus, the Greek god of sleep). The structure was elucidated 120 years later by Robinson and Schöpf. The first synthesis of the M. skeleton was realized by Grewe, the first total synthesis of M. was achieved by Gates in 1956; see also opium. Three families of brain peptides that bind to opiate receptors have been identified since 1975 – endorphins, enkephalins and dymorphins. Peptide ligands that have a high affinity and high specificity for the opiate receptor that binds M. have been isolated from bovine brain[2].

Lit.: [1] Geschichte der Pharmazie **47**, 55–60 (1995); **48**, 18 ff. (1996). [2] Nature (London) **386**, 499 (1997).
gen.: Angew. Chem. Int. Ed. Engl. **35**, 2830 (1996) (synthesis) ▪ Atta-ur-Rahman **18**, 43–154 (1996) (synthesis, review) ▪ Beilstein E III/IV **27**, 2223–2227; E V **27/9**, 316 f. ▪ Chirality **11**, 475 (1999) (synthesis) ▪ Hager (5.) **3**, 662 ff.; 911, 843 ff.; **7**, 1068 ff.; 1301; **8**, 1040 ff. ▪ J. Am. Chem. Soc. **114**, 9688 (1992); **115**, 11028 (1993) ▪ J. Org. Chem. **62**, 5250 (1997) (synthesis) ▪ Nachr. Chem. Tech. Lab. **41**, 1120–1128 (1993) (synthesis, review) ▪ Sax (8.), MRO 500 ▪ Synthesis **1998**, 653 ▪ Tetrahedron Lett. **35**, 3453 (1994) (synthesis) ▪ see also morphinan alkaloids. – *[HS 2939 10]*

Morphium see morphine.

Mosquito/gnat/midge toxin/venom. The insects penetrate the skin with their piercing proboscis and release anticoagulant enzymes. The saliva contains *histamine, *putrescine, *spermine, and *spermidine, which can lead to the unpleasant itching. Equally unpleasant are stings of the smaller blackflies (buffalo gnats; ca. 4 mm, *Simulium* sp.) or even smaller species (0.5–3 mm, *Ceratopogon* sp.). In tropical regions the insects are carriers of many infectious diseases.

Lit.: Arch. Insect Biochem. Physiol. **27**, 27–38 (1984) ▪ Biochem. Biophys. Res. Commun. **193**, 699–693 (1993) ▪ Eisei Dobutsu **36**, 315–326 (1985) ▪ Infect. Immun. **63**, 182–190 (1995).

Mosses (Bryophyta, bryophytes). M. constitute the second largest group of terrestrial plants (max. 15000 species). Together with the *pteridophytes; ferns (Filicatae) they are also the oldest group of terrestrial plants, fossils have been found dating back to the Devonian period (ca. 300 million years ago). With regard to the morphological organization they are intermediates between the more primitive Thallophyta (=storage plants) and the more developed Cormophyta (= shoot plants). Their size varies between ca. 0.5 to somewhat more than 700 mm, epiphytically growing M. and water mosses can be some meters long. M. are ubiquitous, the largest species diversity is found in tropical mountain forests. The M. are divided into the classes Anthocerotae, Hepaticae (*liverworts*), and Musci. In contrast to the other classes, the Hepaticae possess oil bodies as typical organelles[1,2]. These contain very complex essential oils with high proportions of terpenes consisting mainly of mono-, sesqui-, and diterpenes. *Calypogeia azurea* is colored dark blue to violet on account of its azulene derivatives. M. are relatively rarely eaten by animals or attacked by microbial pests. They are the only group of plants free from viruses.
Metabolites: Some moss species contain considerable amounts of *arachidonic acid. M. form a surprising diversity of previously unknown secondary metabolites. Phenolic compounds and terpenes dominate although clear differences between the individual classes of M. are apparent. Anthocerotae are characterized by *lignans, liverworts mainly by terpenes – mostly sesquiterpenes (often enantiomers of those occurring in higher plants) – and bibenzyl or bisbibenzyl derivatives, Musci by the occurrence of *biflavonoids, *sphagnorubins (peat mosses) and coumarin derivatives (Polytrichaceae). Alkaloids are very rare. Recently sterile cultures (*in vitro* cultures[3]) have been introduced to obtain the secondary metabolites from M. The production of substances corresponds qualitatively to that of M. in natural locations.
Use: In Chinese traditional medicine the diuretic effects of some Musci species are reported. The ash of the haircap moss (*Polytrichum commune*) is purported to promote hair growth. North American Indians use pastes and ointments prepared from Musci for wound

treatment. Peat moss was used in the first World War to produce wound dressings[4]. It is now also used in sanitary towels. Because M., especially Musci species, are able to take up large amounts of pollutants without any visible signs of damage, the are used as markers of pollution (biomonitors) for inorganic (e. g., Pb, Cd, Zn) and organic (e. g., *o,p'*-DDT, fluoranthenes, 3,4-benzopyrenes) contaminants[5].

Lit.: [1] Hoppe-Seyler's Z. Physiol. Chem. **45**, 299 (1905). [2] Proc. K. Ned. Akad. Wet. Ser. C **80**, 378 (1977). [3] BIOForum **18**, 494–499 (1995). [4] Schultze-Motel (ed.), Advances in Bryology, vol. 2, p. 135, Vaduz: Cramer 1984. [5] Zinsmeister & Mues (ed.), Bryophytes, their Chemistry and Chemotaxonomy (Proc. Phytochem. Soc. Eur. 29), p. 319, Oxford: Clarendon Press 1990.
gen.: Angew. Chem. Int. Ed. Engl. **30**, 130–147 (1991) ▪ Schuster (ed.), New Manual of Bryology, vol. 2, p. 1071, Nichinan: Miyazaki Hattori Bot. Lab. 1984 ▪ Zechmeister **65**. – *[HS 1211 90]*

Motuporin see theonellamides.

MTX. Abbreviation for *maitotoxin.

Mucidin see strobilurins.

Mucocin.

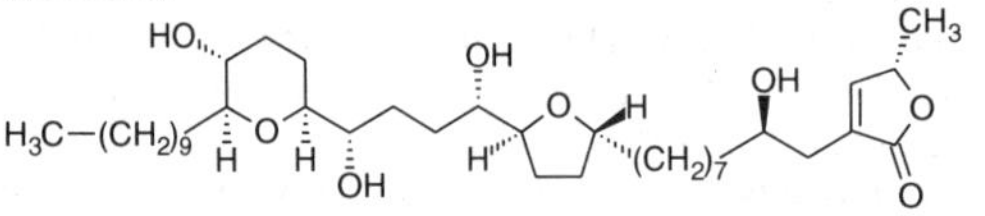

$C_{37}H_{66}O_8$, M_R 638.92, waxy solid, mp. 57–58 °C, $[\alpha]_D^{23}$ −10.8° (CH_2Cl_2). M., a constituent of the leaves of *Rollinia mucosa* is structurally related to the *annonaceous acetogenins, a group of acetogenic natural products. It shows highly specific activity against lung carcinoma and pancreas carcinoma cell lines (up to 10,000 fold higher than *adriamycin).

Lit.: Angew. Chem. Int. Ed. Engl. **38**, 1263 (1999) (synthesis) ▪ J. Am. Chem. Soc. **117**, 10409 (1995) (isolation) ▪ Tetrahedron Lett. **40**, 723, 727 (1999) (synthesis). – *[CAS 170591-47-6]*

Mugic acid see nicotianamine.

Mulberrofurans.

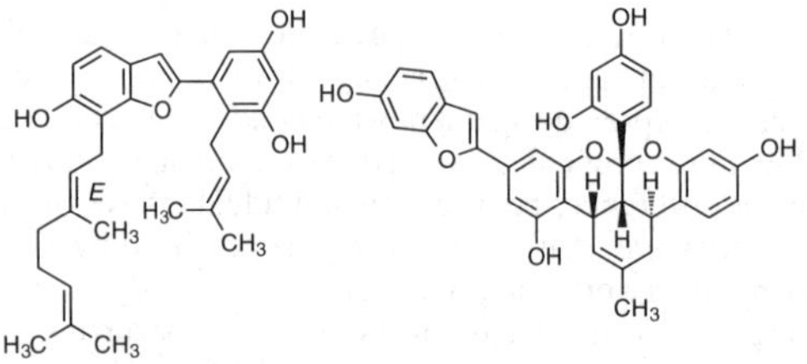

M.D M.G

A group of ca. 18 amorphous, sometimes colored *lignans and *lignin partial structures (red, violet, blue), occurring mainly in the bark of mulberry tree species (*Morus alba, M. australis, M. lhou, M. bombycis*, Moraceae). A common feature of all M. is a 6-hydroxybenzofuran system, sometimes with a geranyl side chain in the 7 position. C-2 bears a complicated ring system which originates from intermediate products of the biosynthesis of lignin and is formed by condensation of aryl-C_3 building blocks (e. g., *coniferyl alcohol and 4-hydroxycinnamyl alcohol), aryl-C_1 building blocks, and phenols; as *examples* for M., see the formulae of *M. D* ($C_{29}H_{34}O_4$, M_R 446.59, mp. 116–120 °C) and *M. G* ($C_{34}H_{26}O_8$, M_R 562.58). M. show activity against Gram (+) bacteria.

Lit.: Chem. Pharm. Bull **33**, 4175, 4288, 5294 (1985); **41**, 1238 (1993) ▪ Heterocycles **19**, 1855 (1982); **24**, 1381, 1807 (1986); **26**, 759 (1987); **29**, 2035 (1989) ▪ Nat. Prod. Rep. **11**, 205–218 (1994) ▪ Planta Med. **46**, 28 (1982) ▪ Tang u. Eisenbrand, Chinese Drugs of Plant Origin, p. 669–696, Berlin: Springer 1992. – *[CAS 83474-71-9 (M. D); 87085-00-5 (M. G)]*

Multhiomycin see nosiheptide.

Multifidene see algal pheromones.

Multifidin.

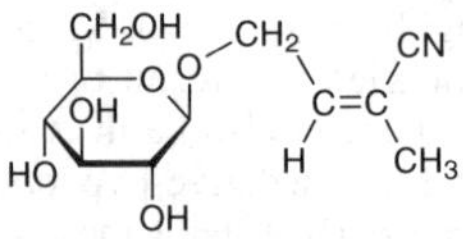

$C_{11}H_{17}NO_6$, M_R 259.25, amorphous solid. A *glycoside from the juice of the Euphorbiacee *Jatropha multifida*. Although it contains a cyano group in the aglycone, in contrast to the *cyanogenic glycosides, cyanide is not formed on hydrolysis.

Lit.: Phytochemistry **40**, 597 f. (1995).

Multifloramine see phenethyltetrahydroisoquinoline alkaloids.

Multiflorine (4-oxo-2,3-didehydrosparteine).

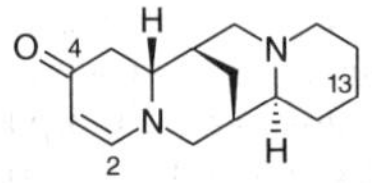

$C_{15}H_{22}N_2O$, M_R 246.35, mp. 108–109 °C, $[\alpha]_D^{22}$ −317° (CH_3OH). A *quinolizidine alkaloid of the sparteine type from some *Lupinus* species[1] and *Cadia ellisiana* (Fabaceae). Analogous to *lupanine a *13α-hydroxy derivative* ($C_{15}H_{22}N_2O_2$, M_R 262.35) of M. is also known. It is found in *Lupinus albus* and *L. cosentinii*. A depressive activity of M. on the central nervous system has been described[2].

Lit.: [1] Waterman **8**, 197–239. [2] Pelletier **2**, 105–148.
gen.: Beilstein E V **24/2**, 561 f. ▪ J. Agric. Food Chem. **31**, 934 (1983) ▪ Manske **47**, 1–115 ▪ Phytochemistry **46**, 365 (1997). – *[HS 2939 90; CAS 529-80-6 (M.); 71657-64-2 (13α-hydroxy-M.)]*

Multistriatin.

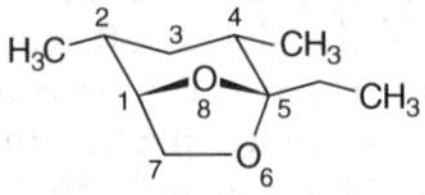

$C_{10}H_{18}O_2$, M_R 170.25, oil. α-M. The (1*S*)-2*endo*,4*endo*-form [bp. 90 °C (933 Pa), $[\alpha]_D^{25}$ −47° (hexane)] is a component of the aggregation pheromone of the elm bark beetle (*Scolytus multistriatus*) one of the carriers of the disease causing death of elm trees. The non-stereospecific synthesis of M. gives rise to further stereoisomers: β-M. [(1*S*)-2*exo*,4*exo*-form], γ-M. [(1*S*)-2*endo*,4*exo*-form], and δ-M. [(1*S*)-2*exo*-form] which is also a component of the aggregation pheromone of *S. multistriatus*. For synthesis, see *Lit.*[1].

Lit.: [1] J. Am. Chem. Soc. **117**, 3653 (1995).
gen.: ApSimon **4**, 74 f., 154–163; **9**, 428–436 (synthesis) ▪ Beilstein E V **19/1**, 252 ▪ Experientia **33**, 845 ff. (1977) ▪ Isr. J. Chem. **36**, 185 (1996) ▪ J. Chem. Ecol. **10**, 373–385 (1984);

12, 583–608 (1986) ▪ J. Chem. Soc., Chem. Commun. **1996**, 1477 (synthesis) ▪ J. Org. Chem. **60**, 5127–5134 (1995) ▪ Justus Liebigs Ann. Chem. **1995**, 1011 ff. [(–)β-M.] ▪ Tetrahedron Lett. **36**, 2595 (1995). – *[CAS 59014-03-8 (α-M.); 59014-05-0 (β-M.); 54832-21-2 ((±)-γ-M.); 54832-22-3 ((±)-δ-M.)]*

Munitagine see pavine and isopavine alkaloids.

Mupirocin see pseudomonic acids.

Muramic acid [2-amino-3-*O*-((*R*)-1-carboxyethyl)-2-deoxy-D-glucose, abbreviation: Mur].

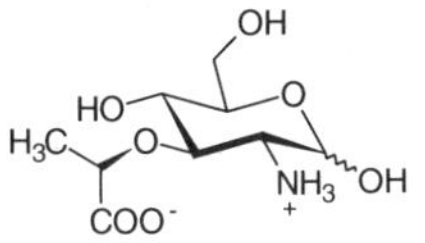

$C_9H_{17}NO_7$, M_R 251.23, mp. 150–151 °C (decomp.), $[\alpha]_D^{20}$ +144° (H_2O), a *glucosamine with an ether-bonded D-lactic acid unit at position 3. *N*-Acetylmuramic acid is a building block of the cell walls of Gram-positive bacteria [1]. As in *chitin the sugars have β1-4-interglycosidic bonds and several amino acids have peptide bonds to the lactic acid side chain, thus allowing cross-linking. For synthesis, see *Lit.* [2].

Lit.: [1] Carbohydr. Res. **191**, 144 (1989). [2] Carbohydr. Res. **79**, C 20 (1980); J. Org. Chem. **30**, 448 (1965).

gen.: Ann. N. Y. Acad. Sci. **235**, 29 (1974) ▪ Arch. Biochem. Biophys. **110**, 341 (1965) (biosynthesis) ▪ Beilstein EIV **4**, 2029 f. – *[CAS 1114-41-6 (M.); 10597-89-4 (N-acetyl-M.)]*

Muramine see protopine alkaloids.

Murrayin see 6,7-dihydroxycoumarins.

Muscaaurins, muscaflavin see fly agaric pigments.

Muscalure [(*Z*)-9-tricosene].

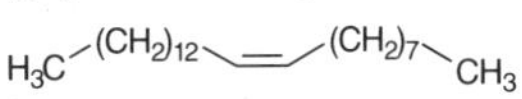

$C_{23}H_{46}$, M_R 322.62, oil, bp. 157–158 °C (13.33 Pa). M. is the sexual pheromone of the female house fly *Musca domestica*. M. has also been detected in various other insects [1]. The strongest *pheromone activity has been observed at doses of 10–20 µg/fly on 2–3 day-old male *flies [2]. M. is formed together with other C_{23}-alkenes in 2 day-old flies, the content increases with increasing age. It is transferred to the male during copulation and oxidized to various metabolites [3]. M. can be used in lures to combat house flies. In admixture with the isomeric (*Z*)-7-tricosene M. is also a component of the female sexual pheromone of the beetle *Aleochara curtula*. Freshly hatched males produce this mixture as a camouflage in order to protect themselves from aggression from older males.

Lit.: [1] Justus Liebigs Ann. Chem. **1989**, 1123 ff. [2] J. Chem. Ecol. **15**, 1475–1490 (1989). [3] J. Insect Physiol. **35**, 775–780 (1989).

gen.: Arch. Insect Biochem. Physiol. **12**, 173–186 (1989) (biosynthesis). – *Synthesis:* ApSimon **4**, 20 ff.; **9**, 25–28 ▪ Indian J. Chem., Sect. B **34**, 718 (1995); **35**, 724 (1996) ▪ J. Chem. Ecol. **13**, 1993–2008 (1987) ▪ Org. Prep. Proced. Int. **26**, 680 (1994) ▪ Tetrahedron **44**, 7423–7254 (1988). – *[HS 2901 29; CAS 27519-02-4]*

Muscapurpurin see fly agaric pigments.

Muscarine, muscazone see fly agaric constituents.

Muscimol see fly agaric constituents.

Muscone (3-methylcyclopentadecanone).

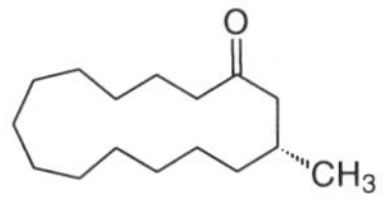

(-)-(*R*)-M.

$C_{16}H_{30}O$, M_R 238.41. oil, bp. 328 °C (128 °C (0.16 kPa)), $[\alpha]_D^{22}$ –13.9° (neat), LD_{50} (rats p.o.) >5 g/kg. (–)-M. is the odor component of natural *musk and was isolated therefrom by Walbaum in 1906 [1]. For structure elucidation and synthesis, see *Lit.* [2–5]. Used for synthetic musk and in fine perfumes.

Lit.: [1] J. Prakt. Chem. **73**, 488 (1906). [2] Chem. Pharm. Bull. **46**, 1484 (1998); Chem. Lett. **1997**, 1291. [3] Justus Liebigs Ann. Chem. **512**, 164 (1934). [4] J. Chem. Soc., Perkin Trans. 1 **1998**, 2253. [5] Theimer (ed.), Fragrance Chemistry, p. 444–469, San Diego: Academic Press 1982; Ohloff, p. 200–205; Helv. Chim. Acta **73**, 896–901 (1990); J. Am. Oil Chem. Soc. **69**, 837 (1992); J. Am. Chem. Soc. **115**, 1593 (1993); Justus Liebigs Ann. Chem. **1995**, 1381; Rev. Heteroaromat. Chem. **7**, 149–170 (1992).

gen.: Beilstein EIV **7**, 118 ▪ Ullmann (5.) **A 11**, 178. – *[HS 2914 29; CAS 541-91-3 (M.); 10403-00-6 ((–)-M.)]*

Mushroom aroma constituents. The typical fungus odor is due to *(–)-(*R*)-1-octen-3-ol. In addition, many species have individual odors for which often simple aromatic compounds are responsible. Examples are *benzaldehyde (*Agaricus augustus*), **p*-anisaldehyde (*Russula laurocerasi*), 2-aminobenzaldehyde (*Hebeloma sacchariolens*), methyl 4-methoxybenzoate (*Cortinarius odorifer*), *methyl anthranilate (*Cortinarius odoratus, Hygrophoropsis olida*), methyl cinnamate (*Inocybe corydalina, I. pyriodara*), 1,3-dimethoxybenzene (*Entoloma icterinum*), *coumarin, herniarin (see umbelliferone), and limettin (*Hydnellum suaveolens*). The town gas odor of *Tricholoma sulfureum* is caused by *indole and *skatole, the phenolic odor of injured fruit bodies of *Agaricus xanthoderma* is caused by phenol. In the stinkhorn (*Phallus impudicus*) a mixture of hydrogen sulfide, methanethiol, and *phenylacetaldehyde serves to attract flies which then distribute the sticky spore mass. Perigórd truffles (*Tuber nigrum*) are able to attract pigs to the underground locations with the help of androst-16-en-3α-ol, the pigs then dig out the truffles, eat them, and spread the spores in their feces [1]. Androst-16-en-3α-ol is the sexual pheromone of the hog.

Lit.: [1] Experientia **37**, 1178 (1981).

Musk. Black-brown, grainy mass with ammonia-animalic odor. The actual erogenic-animalic, dry woody M. odor develops on preparation of a tincture in ca. 70–80%, slightly alkaline ethanol.

Production/extraction: M. is the secretion from an abdominal gland of the musk deer (*Moschus moschiferus*), a small, hornless deer species living in high plateaus of East Asia (Himalayas, Siberia). The male animals use the secretion to mark their territory and to attract females. In the past the animals were killed and the sac containing the secretion cut out. These sacs were dried and marketed. The black-brown mass contained in the sacs was dissolved in alcohol ("M. tincture") for use in the production of perfumes. About 40 M. sacs are necessary to obtain 1 kg of grainy M. As a consequence of intensive hunting for decades the musk deer is in danger of extinction. The hunting of musk deers is thus forbidden by the Washington species protection agreement. The export of M. is now strictly controlled. Samples appearing in the East Asian markets achieve prices of between US $ 30 000 and 50 000/kg.

Composition [1]***:*** The major odor substance is (–)-**muscone*.

Use: In the past M. was used as an erogenic component in expensive, luxury perfumes. Genuine M. is no longer used in Europe and the USA. It has been completely substituted by synthetic M.

Lit.: [1] Perfum. Flavor. **1** (5), 12 (1976); Theimer, Fragrance Chemistry, p. 436, New York: Academic Press 1982.
gen.: Arctander, p. 422 ▪ Ohloff, p. 195, 200, 208 ▪ Parfum. Cosmet. Arom. **88**, 63 (1988); **90**, 81 (1990) ▪ Synthesis **1999**, 1707 (synthesis) ▪ Theimer, Fragrance Chemistry, p. 433, 495, 509, New York: Academic Press 1982. – *Toxicology:* Food Chem. Toxicol. **21**, 865 (1983). – *[HS 051000; CAS 68991-41-3]*

Mustard oil glucosides see glucosinolates.

Mustard oils. Name for organic isothiocyanates, R–N=C=S, occurring in the essential oils of plants, mainly Brassicaceae, as the components responsible for sharp odors and pungent tastes. They exist in the plant in glycosidically bound form (see glucosinolates) and are released from the latter by thioglycosidases (myrosinases) and subsequent rearrangements.

M. often serve the plants as defensive substances [1], they have microbicidal and fungistatic activities. They are relevant for food toxicology because they often have irritant and/or toxic properties and, like the glucosinolates, can induced the development of goiter. Some S. such as sulforaphane and sulforaphene from broccoli are purported to provide protection from cancer [2]. Sulforaphene has antibiotic activity.

Toxicology: Allyl isothiocyanate is a strong irritant of mucous membranes, has mutagenic activity, and is toxic, LD_{50} (rat p. o.) 108.5 mg/kg. For synthesis, see *Lit.* [3–6].

Table: Data of Mustard oils.

name	R	molecular formula	M_R	bp. [°C] [mp.]	optical activity	CAS	occurrence
Allyl isothiocyanate	$CH_2–CH=CH_2$	C_4H_5NS	99.16	151 (oil, pungent smell) 44 (1.6 kPa) [–80]		57-06-7	species of cabbage (*Brassica* spp.), horseradish
Butyl isothiocyanate	$(CH_2)_3–CH_3$	C_5H_9NS	115.19	167 58–59 (1.2 kPa)		592-82-5	species of cabbage
3-Butenyl isothiocyanate	$(CH_2)_2–CH=CH_2$	C_5H_7NS	113.18	78.5 (3.5 kPa) 57,5–58.5 (1.5 kPa)		3386-97-8	seeds of rape (*Brassica napus*)
Erucin	$(CH_2)_4–S–CH_3$	$C_6H_{11}NS_2$	161.28	136 (1.6 kPa)		4430-36-8	seeds of *Eruca sativa*
Erysolin	$(CH_2)_4–SO_2–CH_3$	$C_6H_{11}NO_2S_2$	193.29	[60–60.5]		504-84-7	*Erysimum perovskianum*
Hirsutin	$(CH_2)_8–SO–CH_3$	$C_{10}H_{19}NOS_2$	233.39	122 (40 Pa)	$[\alpha]_D^{23}$ –47° (aqueous C_2H_5OH)	31456-68-5	roots of *Rorippa sylvestris*, seeds of *Arabis hirsuta* and *Sibara virginica*
Isopropyl isothiocyanate	$CH(CH_3)_2$	C_4H_7NS	101.17 (67 Pa)	55–58		2253-73-8	horseradish (*Armoracia rusticana*, *Armoracia lapathifolia*), *Lunaria* sp., *Sisymbrium* sp. and *Patranjiva* sp.
β-Phenylethyl isothiocyanate	$CH_2–CH_2–C_6H_5$	C_9H_9NS	163.24	143–145 (1.6 kPa) 106 (0.3 kPa)		2257-09-2	Brassicaceae
Sulforaphane	$(CH_2)_4–SO–CH_3$	$C_6H_{11}NOS_2$	177.28	130–135 (4 Pa)	$[\alpha]_D$ –79.3° ($CHCl_3$)	4478-93-7	leaves of *Lepidum draba* and *Brassica*, *Eruca* and *Iberis* species
Sulforaphene (Raphanin)	$(CH_2)_2–CH=CH–SO–CH_3$	$C_6H_9NOS_2$	175.26	oil	$[\alpha]_D$ –108° ($CHCl_3$)	2404-46-8	seeds of radish (*Raphanus sativus* var. *alba*), radish and stock *Matthiola bicornis*

Lit.: [1] ACS Symp. Ser. **380**, 155–181 (1988); Ann. Appl. Biol. **123**, 155–164 (1993); Plant Soil **129**, 277–281 (1990). [2] Chem. Eur. J. **1997**, 713; Proc. Natl. Acad. Sci. U.S.A. **89**, 2399–2403 (1992); **91**, 3147 (1994); J. Med. Chem. **37**, 170 (1994). [3] Org. Prep. Proced. Int. **26**, 555 ff. (1994). [4] Tetrahedron Lett. **26**, 1661–1664 (1985). [5] Tetrahedron Lett. **32**, 3503–3506 (1991). [6] Can. J. Chem. **64**, 940 ff. (1986).
gen.: Agric. Biol. Chem. **53**, 3361 f. (1989); **54**, 1587 (1990) ▪ Karrer, No. 2296–2329, 5385–5398 ▪ Nat. Prod. Rep. **10**, 327–348 (1993) ▪ Pesticide Outlook, April 1997, p. 28–32 (insect resistance) ▪ Sax (8.), AGJ 250, ISP 000 (toxicology). – *[HS 3301 29]*

Mutilin see pleuromutilin.

MVA. Abbreviation for *mevalonic acid.

Mycalamides.

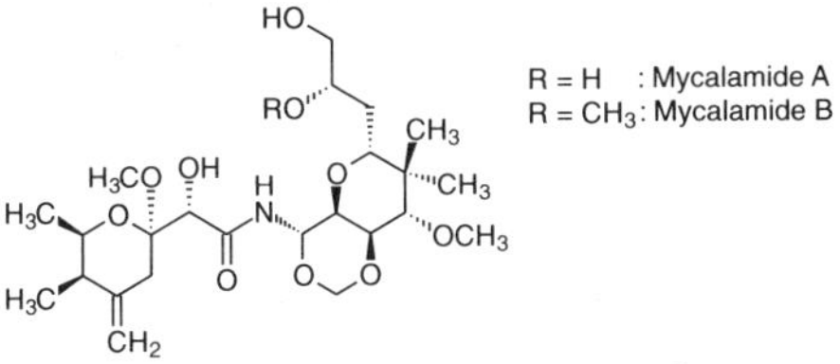

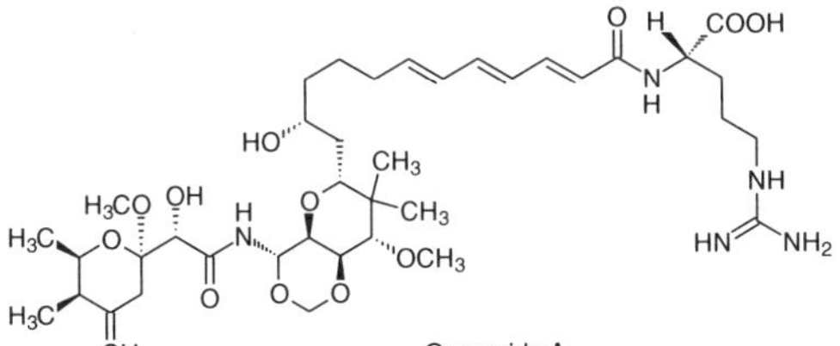

Onnamide A

The New Zealand marine sponge genus *Mycale* contains heterocyclic compounds with inhibitory properties against human and murine tumor cell lines and antiviral properties. These are *M. A* ($C_{24}H_{41}NO_{10}$, M_R 503.59, oil, $[\alpha]_D$ +110°) and *M. B* ($C_{25}H_{43}NO_{10}$, M_R 517.28, oil, $[\alpha]_D$ +110°) [1]. They inhibit protein biosynthesis. The M. are structurally related with *onnamide A* [2] (from sponges of the genus *Theonella*, $C_{39}H_{63}N_5O_{12}$, M_R 793.96, $[\alpha]_D$ +99.1°, amorphous) and *pederin. Further effective anti-tumor compounds from marine organisms include, e.g., *didemnin B, *bryostatins, *jaspamide and arabinosides from the Mediterranean coral *Eunicella cavolini*. The arabinosides exhibit antiviral properties: *ara-A* (vidarabin, 9-*β*-D-arabinofuranosyladenine, $C_{10}H_{13}N_5O_4$, M_R 267.24) has been in use as a drug against *Herpes* encephalitis since the end of the 1970's [3].

Lit.: [1] Bioorg. Med. Chem. Lett. **7**, 2081 (1997); J. Chem. Soc., Perkin Trans. 1 **1997**, 1647; Heterocycles **42**, 159 (1996); J. Am. Chem. Soc. **110**, 4850 (1988); J. Org. Chem. **55**, 223, 4242 (1990); **61**, 1797 (1996); Pol. J. Chem. **68**, 2249 (1994) (synthesis); Synlett **1998**, 869. [2] J. Am. Chem. Soc. **110**, 4850 (1988); J. Nat. Prod. **56**, 976 (1993); Tetrahedron **48**, 8369 (1992). [3] Scheuer II **1**, 98; **13**, 149; Cancer Res. **49**, 2935 (1989); Florey **15**, 647; J. Antibiot. **41**, 1711 (1988) (pharmacology). – *[CAS 115185-92-7 (M. A); 124512-46-5 (M. B); 115204-07-4 (onnamide A); 5536-17-4 (ara-A)]*

Mycarose (2,6-dideoxy-3-*C*-methyl-L-*ribo*-hexose).

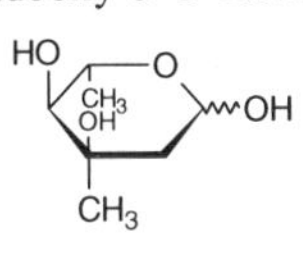

$C_7H_{14}O_4$, M_R 162.18, mp. 128–129 °C, $[\alpha]_D^{20}$ −31° (H_2O), M. · H_2O: mp. 88–89 °C, $[\alpha]_D^{20}$ −25° (H_2O). M. belongs to the group of C-alkyl branched *deoxy sugars and is present in macrolide antibiotics such as, e.g., *carbomycin.

Lit.: Adv. Carbohydr. Chem. Biochem. **42**, 69–134 (1984) ▪ Carbohydr. Res. **136**, 187 (1985); **222**, 173 (1991) (synthesis) ▪ Collins-Ferrier, p. 281, 302. – *[CAS 6032-92-4]*

Mycenone.

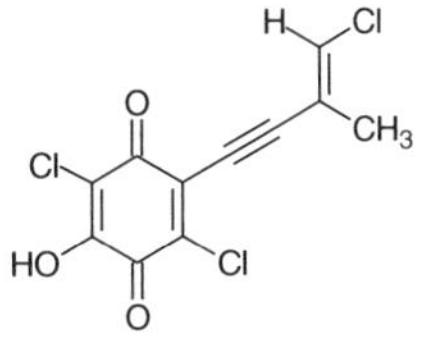

$C_{11}H_5Cl_3O_3$, M_R 291.52, red cryst., mp. 118 °C. A chlorinated benzoquinone alkaloid from cultures of a *Mycena* species (Basidiomycetes). M. inhibits isocitrate lyase and is structurally related with *siccayne.

Lit.: J. Antibiot. **43**, 1240 (1990). – *[CAS 131651-40-6]*

Mycobacillin.

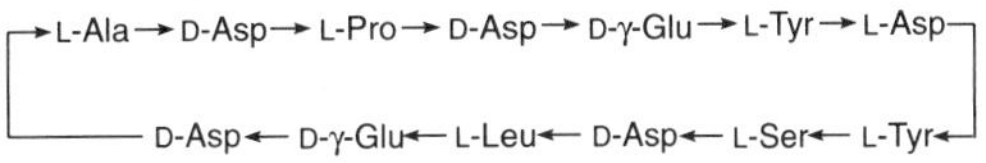

$C_{65}H_{85}N_{13}O_{30}$, M_R 1528.46, mp. 235–240 °C. Polypeptide antibiotic from *Bacillus subtilis* made up from 13 amino acids. M. is active against phytotoxic fungi and yeasts.

Lit.: Antibiotics **2**, 271, 445 (1967) (review) ▪ Biochem. J. **121**, 839 (1971); **230**, 785 (1985) ▪ J. Antibiot. **26**, 257–260 (1973); **30**, 420–422 (1977) ▪ Merck-Index (12.), No. 6406. – *[HS 2941 90; CAS 18524-67-9]*

Mycobactins.

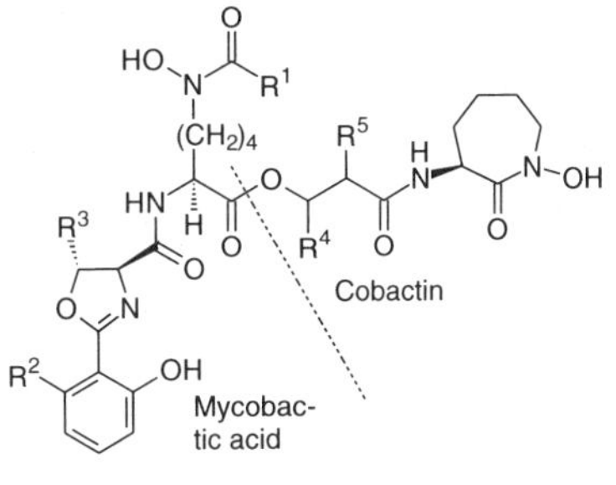

R^1, R^4 = CH_3, C_2H_5 or long chain
R^2, R^3, R^5 = H or CH_3

Growth factors (in concentrations of ca. 30 ng/mL) for *Mycobacterium paratuberculosis* (*M. johnei*), the microorganism responsible for Johne's disease in livestock. M. are iron chelators (*siderophores). 9 M. (M. A, F, H, M, N, P, R, S, and T) have been isolated from various non-pathogenic mycobacteria. They are used as taxonomic markers for mycobacteria. All M. form extremely stable hexacoordinate complexes in which the iron(III) is bound to two hydroxamic acid and one 2-(2-hydroxyphenyl)-2-oxazoline units. M. are stable towards heat and acids but unstable towards bases. On saponification the M. are cleaved to mycobactic acid and cobactin. M. show an apple green fluorescence under UV light. They are very soluble in chloroform

and less soluble in alcohols. M. are made up of six components arising from different biosynthetic pathways. Most important is *M. S*: $C_{44}H_{69}N_5O_{10}$, M_R 828.06.
Lit.: Bacteriol. Rev. **34**, 99–125 (1970) ▪ Biochem. J. **108**, 593 (1968); **111**, 785 (1969); **115**, 1031–1045 (1969) ▪ J. Am. Chem. Soc. **105**, 240 (1983); **119**, 3462 (1997) (synthesis M. S) ▪ Merck-Index (12.), No. 6405. – *[CAS 1400-46-0 (M. complex); 26769-11-9 (M. S)]*

Mycolic acids. Group name for complex, long-chain *hydroxy fatty acid compounds that are typical constituents of some bacterial species, e. g., mycobacteria, corynebacteria, and nocardia. M. are hydrophobic components of the bacteria's envelope in which they are esterified with arabinogalactans (see polysaccharides). The M. consist of 22–74 C-atoms and contain hydroxy as well as other functional groups, e. g., double bonds, keto, ester, epoxy, and methoxy groups, cyclopropane rings, and methyl branching groups. The M. are divided into the following groups on the basis of their functional groups: α- and α'-M., methoxy- and (ω-1)-methoxy-M., epoxy-, keto-, and wax ester mycolic acids.
Lit.: Minnikin, in: Weber & Mukherjee (eds.), Analysis of Lipids, p. 339–348, Boca Raton: CRC Press 1993 ▪ Minnikin, in: Ratledge & Stanford (eds.), The Biology of the Mycobacteria, p. 95–184, London: Academic Press 1982.

Mycolutein see aureothin.

Mycomycin see polyynes.

Mycophenolic acid (melbex, NSC 129185).

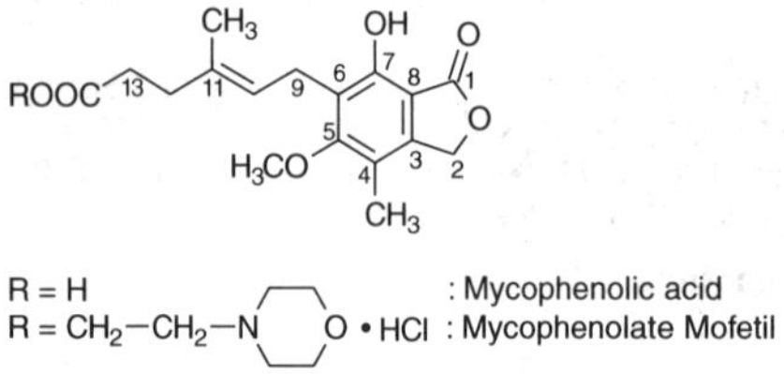

$C_{17}H_{20}O_6$, M_R 320.34, needles, mp. 141 °C. First isolated in 1896 by B. Gosio from *Penicillium brevi-compactum*, M. occurs in many other strains of fungi. It has antibacterial, antiviral, antifungal, and immunosuppressive properties and is active against tumor cells showing low toxicity. Side effects of high doses are liver necrosis, heart failure, hemorrhagic diarrhea. M. is used to treat psoriasis. The derivative of M., *mycophenolate mofetil* ($C_{23}H_{31}NO_7$, M_R 433.50, mp. 93 °C) is marketed since 1995 as an immunosuppressive agent (Cellcept®) to combat rejections after kidney transplantations [1]. The biosynthesis proceeds via a methylated tetraketide with a farnesyl side chain that is later oxidatively cleaved.
Mode of action: M. inhibits the biosynthesis of guanosine nucleotides at the stage of inosine monophosphate dehydrogenase.
Lit.: [1] Drugs **51**, 278 (1996); Drugs of Today **33**, 224 f. (1997); Agents Actions Suppl. **44**, 165 (1993).
gen.: Can. J. Chem. **75**, 641 (1997) (synthesis) ▪ Cole-Cox, p. 866 ▪ Hager (5.) **6**, 59–63 ▪ J. Am. Chem. Soc. **108**, 806–810 (1986) ▪ J. Biol. Chem. **261**, 8363 (1986) (enzyme inhibition) ▪ J. Chem. Soc., Perkin Trans. 1 **1982**, 365 (biosynthesis) ▪ J. Med. Chem. **33**, 833 (1990); **39**, 46, 1236 (1996) ▪ J. Org. Chem. **60**, 4542; 5717 (1995) ▪ Mycologia **81**, 837 (1989) ▪ Sax (8.), MRX 000 ▪ Tetrahedron **49**, 4789–4798 (1993); **53**, 3383, 3395 (1997) ▪ Tetrahedron Lett. **39**, 2881 (1998) (synthesis). – *[HS 2932 29; CAS 24280-93-1 (M.); 128794-94-5 (Cellcept®)]*

Mycorrhizin (mycorrhizin A).

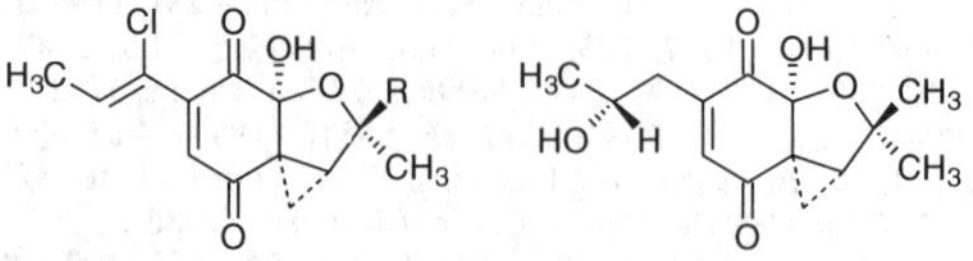

R = CH_3 : Mycorrhizin
R = CH_2OH : Microlin

Gilmicolin

$C_{14}H_{15}ClO_4$, M_R 282.72, pale yellow cryst., mp. 163–165 °C, $[\alpha]_D$ +33.3° (C_2H_5OH). Metabolite from cultures of the hyphomycete *Gilmaniella humicola* living in mycorrhiza with the chlorophyll-less pinesap (*Monotropa hypopitys*). M. inhibits the growth of the fungus *Heterobasidion annosum* (=*Fomes annosus*) which damages tree roots. Close relatives are the antifungal metabolites *gilmicolin* {$C_{14}H_{18}O_5$, M_R 266.29, mp. 120 °C, $[\alpha]_D$ –48° (CH_3OH)} and *microlin* {$C_{14}H_{15}ClO_5$, M_R 298.72, cryst., mp. 113–114 °C, $[\alpha]_D$ +135.6° (CH_3OH)}, produced by the same fungus. Microlin and gilmicolin exist partly as tautomers with 2 hemi-acetal rings.
Lit.: Acta Crystallogr., Sect. B **33**, 870 (1977) ▪ Helv. Chim. Acta **59**, 1809, 1821 (1976); **61**, 2002 (1978); **62**, 1129 (1979) ▪ J. Am. Chem. Soc. **104**, 2659 (1982) ▪ J. Antibiotics **48**, 149, 154, 158 (1995) (isolation) ▪ Tetrahedron **33**, 875 (1977). – *Synthesis:* Aust. J. Chem. **38**, 1339 (1985) ▪ J. Org. Chem. **50**, 1342, 1343 (1985) ▪ Tetrahedron Lett. **28**, 3659, 3663 (1987). – *[CAS 64356-85-0 (M.); 70901-36-9 (gilmicolin); 60958-71-6 (microlin)]*

Mycosporins.

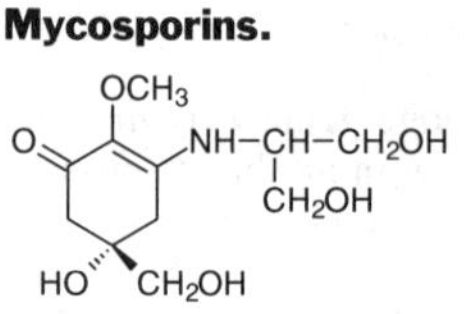

Mycosporin 1

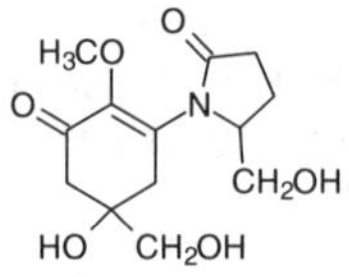

Mycosporin 2

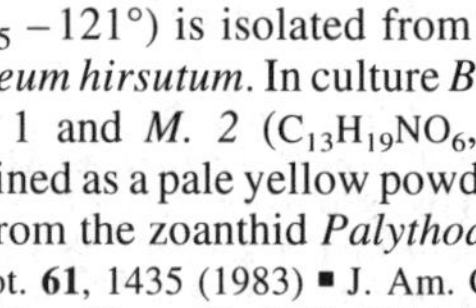

Mycosporin-gly

Strongly fluorescent cyclohexenones from various fungi and marine organisms. *M. 1* ($C_{11}H_{19}NO_6$, M_R 261.27, $[\alpha]_{365}$ –121°) is isolated from *Nectria galligena* and *Stereum hirsutum*. In culture *Botrytis cinerea* produces M. 1 and *M. 2* ($C_{13}H_{19}NO_6$, M_R 285.30). *M. gly* is obtained as a pale yellow powder ($C_{10}H_{15}NO_6$, M_R 245.23) from the zoanthid *Palythoa tuberculosa*.
Lit.: Can. J. Bot. **61**, 1435 (1983) ▪ J. Am. Chem. Soc. **111**, 8970 (1989) ▪ J. Org. Chem. **60**, 3600–3611 (1995) (synthesis) ▪ Nat. Prod. Rep. **15**, 159 (1998). – *[CAS 59719-29-8 (M. 1); 63720-26-3 (M. 2); 65318-21-0 (M.-gly)]*

Mycosterols see sterols.

Mycotoxins. Low-molecular weight fungal metabolites which occur in food and fodder and cause the symptoms of a mycotoxicosis (e. g., Saint Anthony's fire, ergotism, Kashin-Beck disease) in humans and animals. If animals destined for slaughter consume fodder containing M., the toxins can also occur in eggs, meat, or milk, what is known as "carry-over". Some of the M. are also active against other microorganisms so that a separation from the *antibiotics is not always clear. The most strongly effected foodstuffs are grain and grain products, nuts, peanut butter, pressed cakes from oil plants (usually only <10% of the toxins go over to the oil), rice, corn, malt, fruit juices (especially apple juice), milk and cheese. Producers of M. include *Aspergillus, Penicillium, Fusarium, Claviceps* (*ergot alkaloids), and *Rhizopus* species. M. may be formed in the field or on storage (also in a refrigerator). Well known M. are the *aflatoxins (the most potent natural carcinogens), *byssochlamic acid, *citrinin, *citreoviridins, *fumonisins, *patulin, *ochratoxins, *sterigmatocystin, *ergot alkaloids, *ergochromes, *cytochalasins, *penicillic acid, *zearalenone, *penitrems, *tric(h)othecenes, etc.

The most secure protection against M. poisoning is the avoidance of moldy food. Crop protection and preserving agents prevent attack by molds. There are legally defined highest permitted amounts for some M., e. g. the highest permissible aflatoxin content in food is defined by legislation in 60 countries. The limiting values in the various laws range from 0.05 to 5 μg/kg. M. of the trichothecene type have attracted attention as biological weapons [1]. For analytical determination of M., see *Lit.*[2].

Lit.: [1]Nachr. Chem. Tech. **32**, 598 (1984); Nature (London) **309**, 207 (1984). [2]J. Chromatogr. **815** (1) (1998) (whole issue); Nachr. Chem. Tech. Lab. **47**, 553 (1999).

gen.: Chem. Ind. (London) **1995**, 260–264 ▪ Food Rev. Int. **6**, 115 (1990) ▪ Fördergemeinschaft Integrierter Pflanzenbau (eds.), Natürliche Gifte in Getreide, Bonn: Rheinischer Landwirtschafts-Verl. 1990 ▪ Forum Mikrobiol. **138** (1981) ▪ Rev. Environ. Contam. Toxicol. **127**, 69–94 (1992) ▪ Samson & van Reenen-Hoekstra, Introduction to Food-borne Fungi, Baarn: CBS 1988.

Mycotrienins see ansatrienins.

Myoporone.

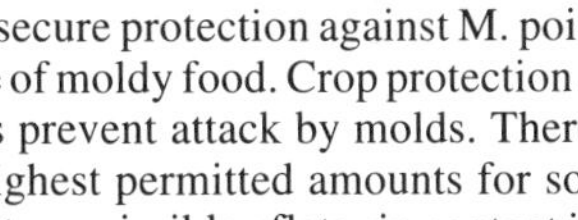

$C_{15}H_{22}O_3$, M_R 250.34, cryst. (CH_3OH), mp. 15.5–16.5 °C, $[\alpha]_D$ −7° (CH_3OH). Occurs together with the (9*R*)-alcohol dihydro-M., $C_{15}H_{24}O_3$, M_R 252.35, oil, $[\alpha]_D^{20}$ −3.6° (CH_3OH), as stress metabolite in the sweet potato (*Ipomoea batatas*) and in the Australian bush *Myoporum deserti*. M. is hepatotoxic, see also ipomeanine and ipomeamarone.

Lit.: J. Nat. Prod. **60**, 493 (1997) (synthesis 6-myoporol) ▪ Phytochemistry **18**, 873 (1979) (isolation) ▪ Synthesis **1994**, 1327 (synthesis) ▪ Tetrahedron **49**, 6515 (1993) (abs. configuration). – *[CAS 19479-15-3 (M.); 72145-16-5 (dihydro-M.)]*

Myosmine [3-(3,4-dihydro-2*H*-pyrrol-5-yl)-pyridine].

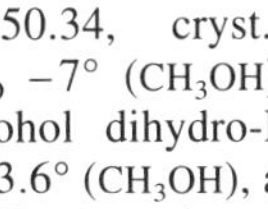

$C_9H_{10}N_2$, M_R 149.19, mp. 40.5–42 °C. A toxic minor alkaloid from the group of the *tobacco alkaloids. It occurs in *Nicotiana* species. M. is formed biosynthetically by dehydrogenation of nornicotine.

Lit.: Beilstein E V **23/6**, 466f. ▪ Manske **26**, 133–135 ▪ Pelletier **1**, 85–152; **3**, 23 ▪ Synthesis **1977**, 242 – *[HS 2939 90; CAS 532-12-7]*

Myrcene (*β*-myrcene, 7-methyl-3-methylene-1,6-octadiene).

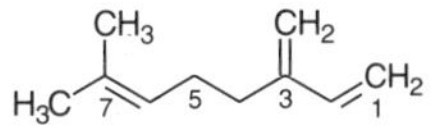

$C_{10}H_{16}$, M_R 136.24, bp. 93 °C (7 kPa), isomeric with *ocimene. M. occurs in the essential oil of the sitka spruce (*Picea sitchensis*, Pinaceae), content 95% [1] and in many other plant oils. The red alga *Chondrococcus* (=*Desmia*) *hornemanni* (Rhodophytae) contains larger amounts of mono-, di-, and trihalogenated M. (e. g., 2-chloromyrcene)[2]. M. is used in the synthesis of other starting materials for perfumes (e. g., *geraniol, *citronellol) and as an auxiliary substance in plastics. *α*-M. is the 7,8-double bond isomer.

Lit.: [1]Phytochemistry **13**, 2167 (1974). [2]Chem. Lett. **1974**, 1333.

gen.: Beilstein E IV **1**, 1108 ▪ Karrer, No. 32 ▪ Merck-Index (12.), No. 6413 ▪ Sax (8.), MRZ 150. – *[HS 2901 29; CAS 123-35-3]*

Myricetin [3-(3,3′,4′,5,5′,7-hexahydroflavone). Formula, see under flavones. $C_{15}H_{10}O_8$, M_R 318.24, yellow needles, mp. 357–360 °C. M. is a flavonol and is widely distributed as glycosides in plants, e. g., in seeds, flowers, and shoots. M. 3-*O*-rhamnoside (*myricitrin*) occurs in the bark of *Myrica esculenta* (Myricaceae) and in the leaves of *M. gale*. A 0.1% solution (acetone) is used for the fluorimetric determination of scandium.

Lit.: Karrer, No. 1575 ▪ Merck Index (12.), No. 6414 ▪ see also flavones. – *[HS 2932 90; CAS 529-144-2]*

Myriocin (thermozymocidin).

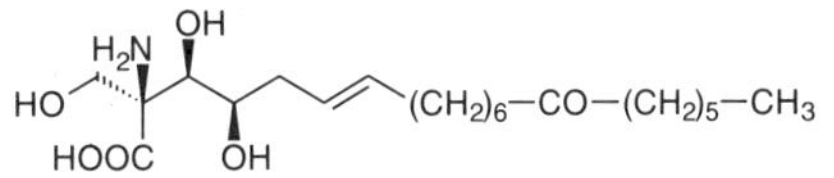

$C_{21}H_{39}NO_6$, M_R 401.54, mp. 183–184 °C, $[\alpha]_D^{24}$ +10.3° (CH_3OH). Antifungal antibiotic with immunosuppressive activity from *Myriococcus albomyces, Melanconis flavovirens*, and *Isaria sinclairii*.

Lit.: Chem. Pharm. Bull. **42**, 994 (1992) ▪ Gazz. Chim. Ital. **122**, 51 (1982) (biosynthesis) ▪ J. Antibiot. **25**, 109 (1972); **47**, 208, 216 (1994) (isolation) ▪ Tetrahedron **47**, 6079 (1991) ▪ Tetrahedron Lett. **38**, 7887 (1997) (synthesis). – *[CAS 35891-70-4]*

Myristic acid (tetradecanoic acid). $H_3C–(CH_2)_{12}–COOH$, $C_{14}H_{28}O_2$, M_R 228.38, mp. 54.4 °C, bp. 142 °C (133 Pa); soluble in organic solvents, insoluble in water. The seed fat of nutmeg (*Myristica fragrans*) contains 70–80% M. as the glycerol ester. The seed fats of nutmeg and Lauraceae plants as well as the kernel fat of coconut and oil palms are rich in M. *triglycerides. These also occur in smaller amounts in many plant and animal *fats and oils, e. g., up to 12% in milk fat. M. and palmitic acid are made responsible for increased cholesterol levels in blood (hypercholesterolemia)[1]. M. is used in the manufacture of soaps, cosmetics, and lubricants.

Lit.: [1]Am. J. Clin. Nutr. **56**, 895 (1992); **59**, 841 (1994); J. Nutr. **125**, 466 (1995).

gen.: Beilstein E IV **2**, 1126 ▪ Handbook of Pharmaceutical Excipients (2.), Editor Wade et al., p. 243, Am. Pharm. Association/Pharmaceutical Press 1994 ▪ Merck-Index (12.), No. 6416 ▪ Ullmann (5.) **A 10**, 247. – *[HS 1519 19, 2915 90; CAS 544-63-8]*

Myristicin see safrole.

Myrmicacin.

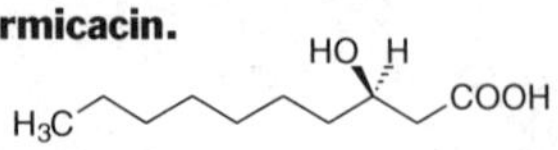

(–)-(*R*)-form

$C_{10}H_{20}O_3$, M_R 188.27, cryst., mp. 48–49 °C, soluble in chloroform, $[\alpha]_D^{20}$ –20.8° ($CHCl_3$). M. is secreted by the South American leaf-cutting ant (*Atta sexdens*, Myrmicinae). It serves to protect the fungi gardens cultivated by the ants from attack by other microorganisms. The germination of storage grain collected by harvesting ants is prevented by M.[1]. M. and closely related 3-hydroxycarboxylic acids also arising from acetate metabolism have been detected in solitary bees, honeybees, and bumblebees.

Lit.: [1] Angew. Chem. Int. Ed. Engl. **10**, 124 (1971).
gen.: Agric Biol. Chem. **42**, 879 (1978) (synthesis) ▪ Tetrahedron Lett. **26**, 4665 (1985) (synthesis) ▪ Z. Naturforsch. C **51**, 409–422 (1996). – *[CAS 33044-91-6]*

Myrmicarins. Pyrrolo[2,1,5-*cd*]indolizine alkaloids from the poison gland secretion of the African ant *Myrmicaria opaciventris*; e.g., the monomeric M. 213 B (**1**); heptacyclic M. 430 A (**2**), and the most complex M. 663 (**3**).

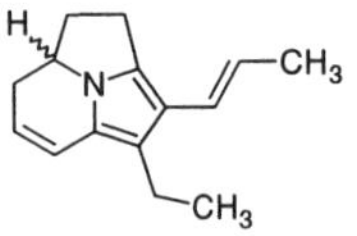

M. 213 B
1

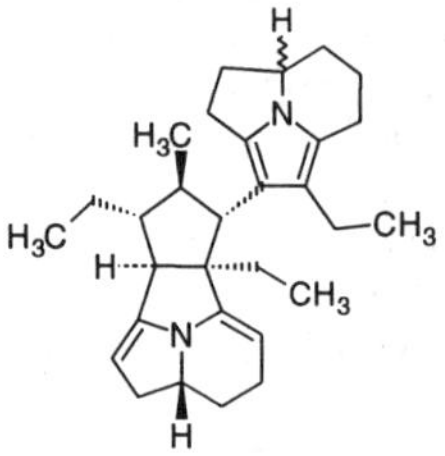

M. 430 A
2

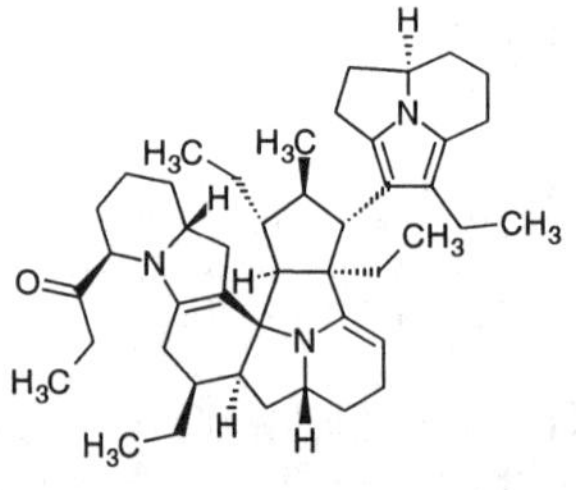

M. 663
3

Figure: Structures of M., relative configuration shown.

Table: Data of Myrmicarins (selection).

M.	molecular formula	M_R	CAS
213 B	$C_{15}H_{19}N$	213.32	183254-01-5
430 A	$C_{30}H_{42}N_2$	430.68	183122-49-8
663	$C_{45}H_{65}N_3O$	663.99	183127-15-3

The toxins serve as chemical weapons to fight off attacking termites. Compounds (**2**) and (**3**) are short-lived. They decompose readily in air. The composition of the secretion varies between *Myrmicaria* species.

Lit.: Angew. Chem. Int. Ed. Engl. **36**, 77 (1997) ▪ J. Chem. Soc., Chem. Commun. **1996**, 2139 ▪ Tetrahedron **52**, 13539–13546 (1996); **54**, 5259 (1998).

Myrtenals see pinenones.

Myrtenols see pinenols.

Mytilotoxin see saxitoxin.

Myxalamides.

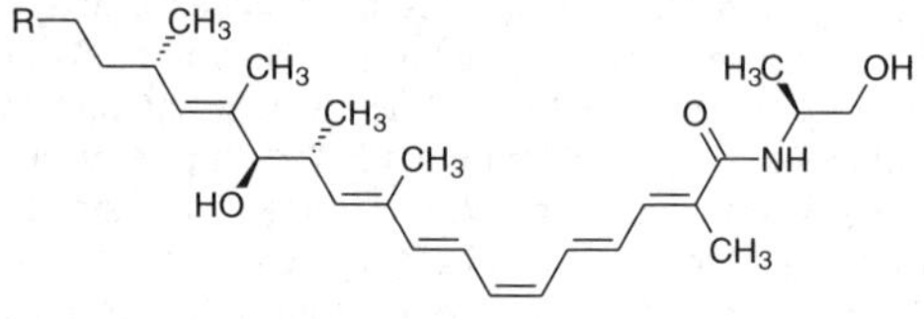

R = H : Myxalamide A
R = C_6H_5 : Phenalamide A_1 (Stipiamide)

Table: Data of Myxalamide A (**1**) and Phenalamide A_1 (**2**).

	molecular formula	M_R	λ_{max} (lg ε)	$[\alpha]_D^{23}$ (CH_3OH)
1	$C_{26}H_{41}NO_3$	415.62	357 nm (4.59)	–71.6°
2	$C_{32}H_{45}NO_3$	491.71	356 nm (4.77)	–115.1°

A family of polyenecarboxylic acid amides with L-alaninol isolated from myxobacteria. Variation of the acyl group gives the homologues M. A–D from *Myxococcus canthus* and the phenalamides A–C (= stipiamides) from *M. stipitatus*. M. and the phenalamides are unstable yellow oils that isomerize to the all-*trans*-isomers with light and slowly decompose under hydroxylation in the presence of atmospheric oxygen. All M. and phenalamides are active against fungi, yeasts, and some Gram-positive bacteria by inhibition of NADH oxidation in complex I of the respiratory chain. Cellular respiration and growth of mammalian cells are also effectively inhibited, thus explaining the described cytostatic and anti-HIV-acivities of phenalamides. The acute toxicity (LD_{50}) of M. B (isopropyl in place of *sec*-butyl) in mice is between 1 and 3 mg/kg (s. c.) or 3 and 10 mg/kg (p. o.). The biogenesis of M. and phenalamides proceeds from L-alanine, acetate, and propionate, the respective starter units of the polyketide synthesis originate from the corresponding aliphatic amino acids for the M. and from phenylalanine for the phenalamides.

Lit.: J. Antibiot. **36**, 1150 (1983); **44**, 553 (1991) ▪ J. Chem. Soc., Perkin Trans. 1 **1991**, 1901, 1907 ▪ Justus Liebigs Ann. Chem. **1983**, 1081; **1984**, 78; **1992**, 659 ▪ J. Org. Chem. **64**, 23 (1999) (synthesis M. A). – *[CAS 86934-09-0 (M. A); 135383-02-7 (phenalamide A_1)]*

Myxin see iodinin.

Myxochelin A see azotochelin.

Myxomycete pigments. In the life cycles of genuine slime molds (Myxomycetes, Myxomycotina) amoeboid mobile plasma masses without cell walls (plasmodia) develop into fruit bodies (sporangia) the spores of which send out myxamoeba or myxoflagellates which finally fuse to give new plasmodia. Not only the plasmodia but also the sporangia are often brightly colored. To date, polyenes such as *fuligorubin A, *physarochrome A, or *ceratiopyrones have been isolated as the yellow plasmodia pigments. The bright colors of the sporangia of *Metatrichia* species are due to naphthoquinones such as *trichione and *vesparione, while those of *Arcyria* species contain *arcyria pigments of the bisindolylmaleimide or indolocarbazole type.

Lit.: Pure Appl. Chem. **61**, 281 (1989) ▪ Zechmeister **51**, 210–215 (reviews).

Myxopyronins.

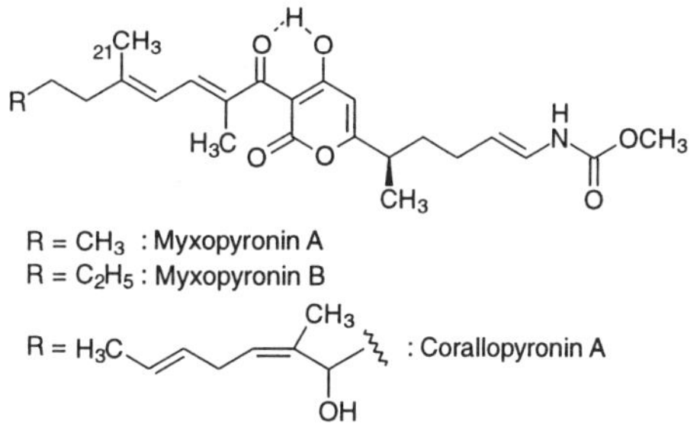

Table: Data of the Myxopyronins.

	molecular formula, M_R	$[\alpha]_D^{20}$ λ_{max} (lg ε)	CAS
M. A	$C_{23}H_{31}NO_6$ 417.50	–73.5° 298 nm (4.31)	88192-98-7
M. B	$C_{24}H_{33}NO_6$ 431.53	–74.9° 297 nm (4.31)	88192-99-8
Corallo-pyronin A	$C_{30}H_{41}NO_7$ 527.66	–95.8° 296 nm (4.54)	93195-32-5

A family of pyrone antibiotics from *Myxococcus fulvus* and *Corallococcus coralloides* (Myxobacteria). A common feature of the compounds occurring as colorless oils is a 4-hydroxy-2-pyrone with an *N*-methoxycarbonyl-enamine side chain in an ε-position. The biosynthesis proceeds through a head-to-head condensation of two polyketides, one of which starts with glycine. Methyl branchings arise (with the exception of C-21 which comes from the C-2 of acetate) from methionine. Both M. and the corallopyronins are active against Gram-positive bacteria in the concentration range of 1 ng/L. The mechanism of action has been identified as inhibition of RNS polymerase from *Escherichia coli*. Fungi and yeasts as well as the polymerase II from wheat kernels are not sensitive. A total synthesis of (±)-M. A and some derivatives starting from dehydracetic acid has been reported.

Lit.: BIOForum **1996** (6), 262 ▪ J. Antibiot. **36**, 1651 (1983); **38**, 145 (1985) ▪ J. Org. Chem. **63**, 2401 (1998) (synthesis M. A, B) ▪ Justus Liebigs Ann. Chem. **1983**, 1656; **1984**, 1088; **1985**, 822 ▪ WO P. 9934793 (15.7.1999); US. p. 6,022,983 (8.2.2000) (synthesis). – *[HS 2941 90]*

Myxothiazol.

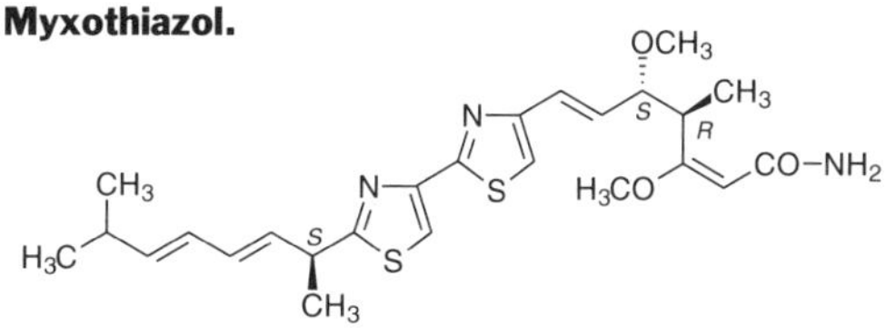

Antifungal antibiotic from *Myxococcus fulvus* and other Myxobacteria, $C_{25}H_{33}N_3O_3S_2$, M_R 487.7, mp. 79 °C, $[\alpha]_D^{25}$ +43.4° (CH_3OH). M. exhibits strong antifungal (MIC 0.01–3 µg/mL), cytotoxic (0.5 ng/mL), and insecticidal activities, the reason being inhibition of the respiratory chain at the cytochrome b_{c1} complex. Like the *strobilurins and *oudemansins, M. has a β-methoxyacrylate pharmacophore, albeit not in the α-position as in the former, but in the β-position to which the rest of the molecule is linked. In addition, M. is highly toxic, LD_{50} (mouse p.o.) 2 mg/kg.

Lit.: Biochim. Biophys. Acta **636**, 282 (1981) ▪ J. Antibiot. **1980**, 1474, 1480 ▪ Justus Liebigs Ann. Chem. **1986**, 93 (biosynthesis) ▪ Tetrahedron Lett. **22**, 3829 (1981); **34**, 5151 (1993) (synthesis). – *[HS 2941 90; CAS 76706-55-3]*

Myxovirescins (megovalicins).

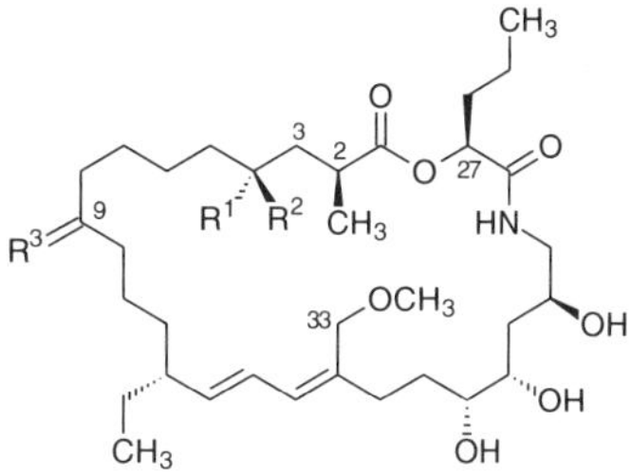

	R^1	R^2	R^3
M.A1 (Megovalicin C)	CH_3	H	O
M.A2	H	CH_3	O
M.B (Megovalicin B)	CH_3	H	O 2,3-didehydro
M.C1 (Megovalicin H)	CH_3	H	H,H

Table: Data of Myxovirescins.

	molecular formula, M_R	mp. [°C]	$[\alpha]_D^{20}$ λ_{max} (lg ε)	CAS
M. A_1	$C_{35}H_{61}NO_8$ 623.87	44	+27.3° 238 nm (4.38)	85279-97-6
M. A_2	$C_{35}H_{61}NO_8$ 623.87	110	+28.5° 238 nm (4.39)	85279-98-7
M. B	$C_{35}H_{59}NO_8$ 621.86	109–111	+32.9° 232 nm (4.46)	89759-26-2
M. C_1	$C_{35}H_{63}NO_7$ 609.89	–	+26.0° 238 nm (4.30)	115932-37-1

Highly diverse family of ca. 35 macrolide antibiotics from *Myxococcus virescens*, *M. xanthus*, and *M. flavescens*. The M. are active *in vitro* against Gram-positive and Gram-negative bacteria, especially enterobacteria. Natural and semisynthetic structural variants show that the activity depends strongly on the length and configurations of the side chain at position 27. No antibacterial activity or acute toxicity has been observed *in vivo*. The biosynthesis of M. proceeds through the polyketide pathway with acetate and glycine as starter units; the methyl groups are derived from methionine, the ethyl side chain and C-33 from the C-2 of acetate. For synthesis, see *Lit.*[1].

Lit.: [1] Helv. Chim. Acta **74**, 2112 (1991).

gen.: J. Antibiot. **35**, 788, 1454 (1982); **41**, 433, 439 (1988) ▪ J. Org. Chem. **55**, 3457 (1990) ▪ Justus Liebigs Ann. Chem. **1985**, 1629; **1989**, 345; **1990**, 69; **1994**, 701, 719, 731 ▪ Tetrahedron Lett. **35**, 5113 (1994) (synthesis M. A_1). – *[HS 2941 90]*

N

NAD, NADH, NADP, NADPH. Abbreviations for nicotinamide adenine dinucleotide [phosphate] in the oxidized and reduced forms, see nicotinamide coenzymes.

Naematolin.

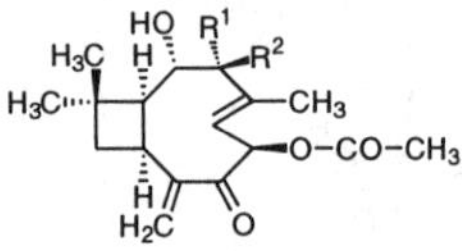

R^1 = OH, R^2 = H : Naematolin
R^1, R^2 = O : Naematolone

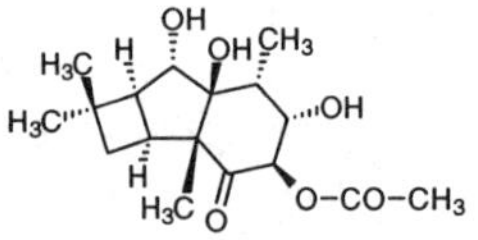

Naematolin C

$C_{17}H_{24}O_5$, M_R 308.37, cryst., mp. 145 °C, $[\alpha]_D$ −370° ($CHCl_3$). A sesquiterpenoid with the rare *cis*-caryophyllane structure from cultures of various fungal species (genus *Hypholoma* = *Naematoloma*, Basidiomycetes). The initially assumed *trans*-ring linkage [1] was later corrected by X-ray crystallography [2]. N. is a bitter substance with cytotoxic, antiviral, and vasodilatory activities. It occurs together with *naematolone* {$C_{17}H_{22}O_5$, M_R 306.36, oil, $[\alpha]_D$ +116° ($CHCl_3$)} which has a somewhat stronger cytotoxic activity. *N. C* {$C_{17}H_{26}O_6$, M_R 326.39, needles, $[\alpha]_D$ − 19° (dioxan)} is isolated from cultures of *Hypholoma fasciculare* [3].

Lit.: [1] Justus Liebigs Ann. Chem. **1984**, 1332. [2] Bull. Chem. Soc. Jpn. **59**, 1921 (1986). [3] J. Chem. Soc., Chem. Commun. **1990**, 725.

gen.: Chem. Lett. **1986**, 653. – *[CAS 11054-16-3 (N.); 92121-62-5 (naematolone); 130009-40-4 (N. C)]*

Naematolone see naematolin.

Nagilactone C.

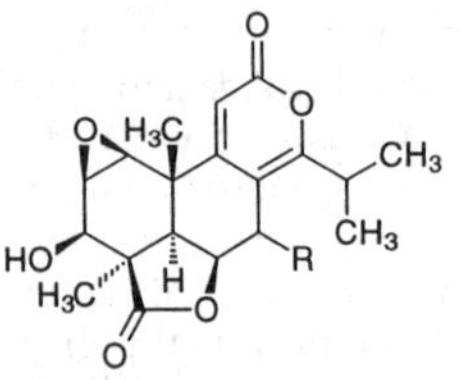

R = H : Hallactone A
R = β-OH : Nagilactone C

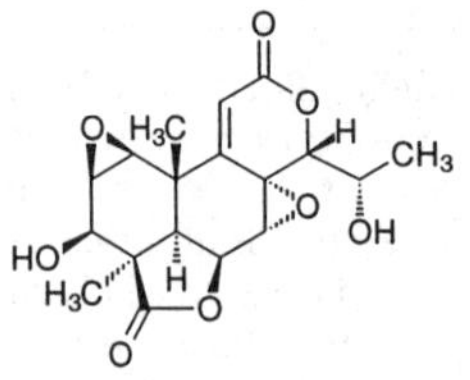

Inumakilactone A

$C_{19}H_{22}O_7$, M_R 362.38, crystalline, mp. 290 °C (decomp.), $[\alpha]_D$ +111°. A norditerpene of the *nagilactone group, occurring in *Podocarpus nagi* [1], *P. nubigenus* [2], and *P. purdieanus* [3] and others. N. C is a plant growth regulator [4]. In addition it has antifeedant effects on herbivorous mammals [5]. It shows potent cytotoxicity [6]. To date nagilactones A – G, I and J have been described [7]. The related *hallactone A* [8] {$C_{19}H_{22}O_6$, M_R 346.38, mp. 266 – 268 °C (decomp.)} from *Podocarpus hallii* is toxic to insects. *Inumakilactone A* [9] {$C_{18}H_{20}O_8$, M_R 364.35, mp. 251 – 253 °C (decomp.)} is the bitter principle from the wood and seeds of *Podocarpus macrophyllus*.

Lit.: [1] Tetrahedron Lett., **1968**, 2071; **1969**, 2951. [2] Phytochemistry **12**, 883 (1973). [3] Phytochemistry **13**, 1991 (1973). [4] Phytochemistry **30**, 455, 1967 (1991). [5] Biosci. Biotechnol. Biochem. **56**, 1302 (1992). [6] Phytochemistry **30**, 3476 (1991). [7] Beilstein E V **19/5**, 164; **19/6**, 536; **19/7**, 25; **19/10**, 354, 531, 538, 610. [8] J. Chem. Soc. **1973**, 166. [9] Tetrahedron Lett. **1972**, 3385.

gen.: Aust. J. Chem. **28**, 745 (1975) – *Synthesis:* J. Org. Chem. **47**, 18 (1982) ▪ Tetrahedron **55**, 7289 (1999). – *[CAS 24338-53-2 (N. C); 41787-72-8 (hallactone A); 19885-83-7 (inumakilactone A)]*

Nagilactones.

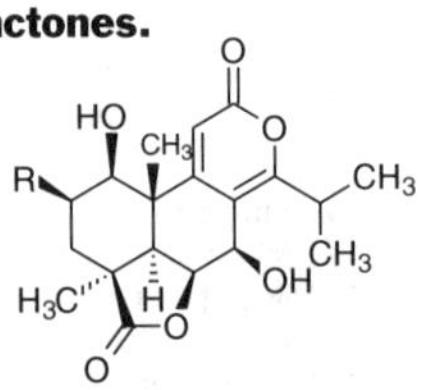

R = H : N. A
R = OH : N. B

The genus *Podocarpus* is a rich source of diterpenoids with various biological activities, e.g. antifeedant, antimicrobial and antitumor properties. To date, more than 30 diterpene dilactones, most of them crystalline compounds, have been reported from *Podocarpus nagi*, e.g. *N. A*: $C_{19}H_{24}O_6$, M_R 348.39 and *N. B*: $C_{19}H_{24}O_7$, M_R 364.39. They occur in glycosidically bound form as the nagilactosides.

Lit.: Aust. J. Chem. **50**, 841 – 848 (1997) (synthesis) ▪ Phytochemistry **34**, 1107 (1993); **39**, 1143 (1995) (isolation) ▪ Tetrahedron **55**, 7289 (1999) (synthesis). – *[CAS 19891-50-0 (N. A); 19891-51-1 (N. B)]*

Nagstatin.

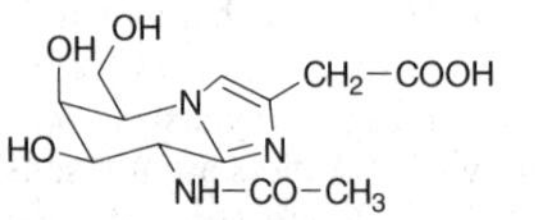

$C_{12}H_{17}N_3O_6$, M_R 299.28, powder, mp. 190 – 195 °C, $[\alpha]_D$ +46.2° (H_2O). An *N*-acetylgalactosamine analogue with a condensed ring from cultures of *Streptomyces amakusaensis*. N. is an inhibitor of *N*-acetyl-*β*-D-glucosaminidase (IC_{50} 1.2 μg/L).

Lit.: Bull. Chem. Soc. Jpn. **70**, 427 (1997) (synthesis) ▪ J. Antibiot. **45**, 1404, 1557 (1992); **48**, 286 (1995) ▪ Tetrahedron Lett. **36**, 6721 (1995) (synthesis). – *[CAS 126844-81-3]*

Nakijiquinones see ilimaquinone.

Namenamicin. $C_{43}H_{62}N_2O_{14}S_5$, M_R 991.29, amorphous solid. An antitumor agent, related to the *calicheamicins (*enediyne antibiotics) from the marine ascidian

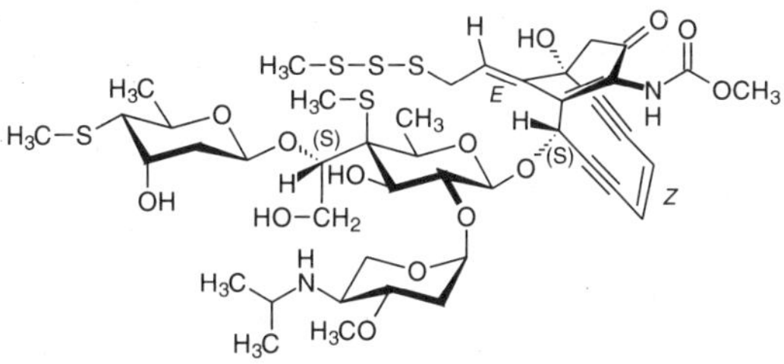

Polysyncraton lithostrotum. 1 mg was obtained from 1 kg of tissue. N. exhibits strong *in vitro* cytotoxicity with a mean IC_{50} of 3.5 ng/mL and *in vivo* antitumor activity in mice (P388 leukemia model) with ED_{40} 3 µg/kg.

Lit.: J. Am. Chem. Soc. **118**, 10898 (1996) (isolation) ▪ J. Chem. Soc., Perkin Trans. 1 **1999**, 545–558 (synthesis). – *[CAS 184349-17-5]*

Nanaomycins.

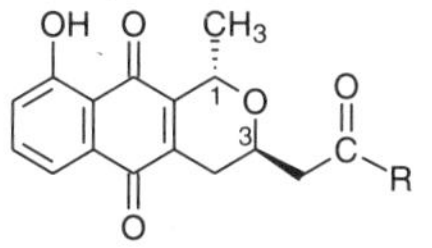

R = OH : N. A
R = NH_2 : N. C

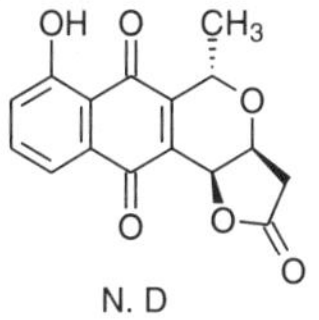

N. D

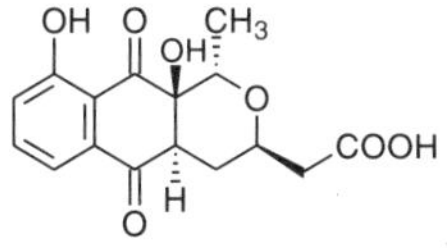

N. B

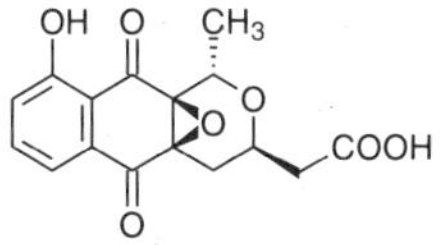

N. E

Table: Data of Nanaomycins.

N.	molecular formula	M_R	mp. [°C]	$[\alpha]_D$ (CH_3OH)	CAS
A	$C_{16}H_{14}O_6$	302.28	178–180	–27.5°	52934-83-5
B	$C_{16}H_{16}O_7$	320.30	84–86	–74.5°	52934-85-7
C	$C_{16}H_{15}NO_5$	301.30	222–224	–2°	58286-55-8
D	$C_{16}H_{12}O_6$	300.27	170–173	–278°	60325-08-8
E	$C_{16}H_{14}O_7$	318.28	172–174	+89°	72660-52-7

A family of *benzoisochromanequinone antibiotics from *Streptomyces rosa notoensis* var. *notoensis*. The N. are orange to yellow in color. N. D is the antipode of *kalafungin. The biosynthesis proceeds through the polyketide pathway; the genetics have been investigated. N. D is formed first and is transformed to other N. in subsequent biosynthetic processes. The N. inhibit the growth of Gram-positive bacteria, fungi, mycoplasms, and yeasts. N. A is registered for use in Japan against bovine dermatophytosis.

Lit.: Gene **142**, 31–39 (1994) ▪ J. Antibiot. **27**, 363–365 (1974); **28**, 860, 868, 925 (1975); **33**, 711–716 (1980); **36**, 1268–1274 (1983); **42**, 1186–1188 (1989) ▪ Sax (8.), RMK 200 (toxicology). – *Synthesis:* J. Org. Chem. **48**, 2630–2632 (1983); **52**, 1273–1280 (1987) ▪ Tetrahedron Lett. **31**, 3913–3916 (1990). – *[HS 2941 90]*

Nannochelin A see aerobactin.

Napelline. $C_{22}H_{33}NO_3$, M_R 359.51, mp. 117–118 °C, $[\alpha]_D^{21}$ –13° (CH_3OH). A diterpene alkaloid from *Aconitum napellus* and other *Aconitum* species (Ranunculaceae).

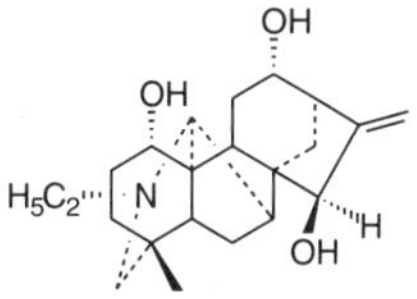

Lit.: Can. J. Chem. **52**, 2355 (1974) (synthesis) ▪ Manske **34**, 95–179. – *[HS 2939 90; CAS 5008-52-6]*

1,8-Naphthalenediol.

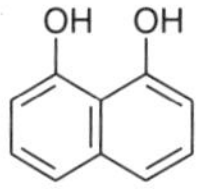

$C_{10}H_8O_2$, M_R 160.17, colorless platelets, mp. 140 °C. A pentaketide from which typical metabolites of ascomycetes are formed by oxidative coupling and other transformations. This family includes, e. g., *bulgarein, *4,9-dihydroxy-3,10-perylenedione, and *hypoxylone. Oxidative polymerization of 1,8-N. leads to brown to black fungal melanins, e. g., in the fruit bodies of the tree fungus *Daldinia concentrica.* Simple derivatives of 1,8-N., namely the monomethyl ether ($C_{11}H_{10}O_2$, M_R 174.20, cryst., mp. 55–56 °C) and the dimethyl ether ($C_{12}H_{12}O_2$, M_R 188.23, platelets, mp. 158–161 °C) were isolated from cultures of *D. concentrica.*

Lit.: Beilstein EIV **6**, 6560 ▪ J. Org. Chem. **57**, 271 (1986) ▪ Zechmeister **51**, 117–120 (review). – *[HS 2907 29; CAS 569-42-6]*

Naphthomycins.

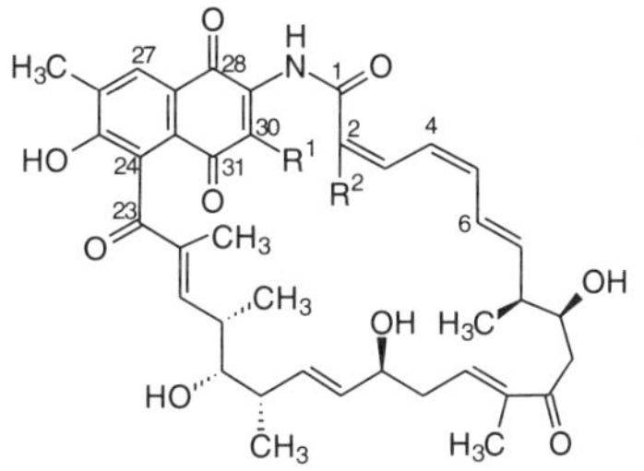

N. A series

N.	R^1	R^2
A	Cl	CH_3
D	OH	CH_3
E	H	CH_3
F	S–CH_2–CH(NH–CO–CH_3)–$COOCH_3$ (*R*)	CH_3
G	S–CH_2–CH(NH–CO–CH_3)–COOH (*R*)	CH_3
H	Cl	H

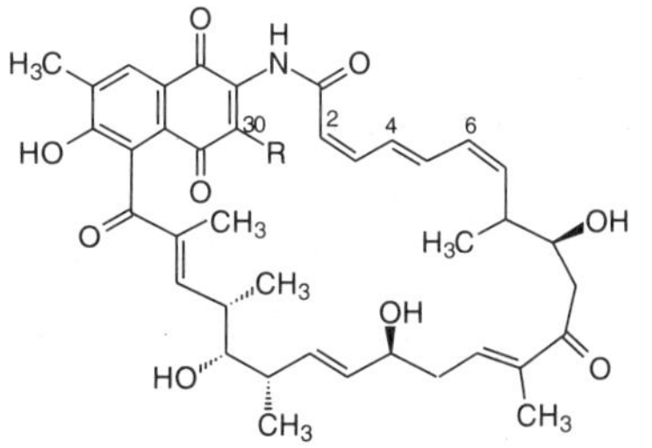

N.	R
B	Cl
C	H

N. B series

Table: Data of Naphthomycins.

	molecular formula	M_R	mp. [°C]	$[\alpha]_D^{20}$	CAS
N. A	$C_{40}H_{46}ClNO_9$	720.26	142	+432°	55557-40-9
N. B	$C_{39}H_{44}ClNO_9$	706.23	156–165 (decomp.)	+412°	86825-88-9
N. C	$C_{39}H_{45}NO_9$	671.79	>318 (decomp.)	+117.7°	86825-87-8
N. D	$C_{40}H_{47}NO_{10}$	701.81			105225-04-5
N. E	$C_{40}H_{47}NO_9$	685.81			105225-03-4
N. F	$C_{46}H_{56}N_2O_{12}S$	861.02			105225-01-2
N. G	$C_{45}H_{54}N_2O_{12}S$	847.00	188		105225-02-3
N. H	$C_{39}H_{44}ClNO_9$	706.23	162–169 (decomp.)	+218°	98525-20-3

A group of *ansamycin antibiotics from various *streptomycete strains. They include *N. A* (yellow, wax-like needles), *N. B* (yellow, wax-like needles), *N. C* (green-yellow powder), *N. D* (orange powder), *N. E*, *N. F* (amorphous), *N. G* (orange prisms), and *N. H* (*naphthoquinomycin C*, light yellow crystals). N. A, B, C, F have antibacterial and antifungal activities. The N. are subdivided structurally on the basis of the configuration of the triene chain into an N. A (2*Z*,4*Z*,6*E*)- and an N. B (2*Z*,4*E*,6*Z*)-series; for N. A, see *Lit.*[1], for N. B and N. C, see *Lit.*[2], for N. D, N. E, N. F, N. G, see *Lit.*[3], for N. H, see *Lit.*[4].

Lit.: [1] J. Antibiot. **39**, 157, 1357 (1986). [2] J. Antibiot. **36**, 484 (1983). [3] Helv. Chim. Acta **69**, 1356 (1986). [4] J. Antibiot. **38**, 948 (1985).
gen.: Can. J. Chem. **72**, 182 (1994) (biosynthesis).

Naphthoquinomycin C see naphthomycins.

Naphthoquinones. Derivatives of hydroxy-1,4-naphthoquinones, often modified by prenylation that are widely distributed in nature as the yellow to red *natural dyes (pigments) of plants, fungi, mold fungi, and bacteria. Important plant representatives of this group are *shikonin, *alkannin, *lawsone, *lapachol, *juglone, *lomatiol, *plumbagin. In the animal kingdom N. pigments are found, e.g., in the form of *spinochromes and *echinochrome in sea urchins. *Vitamin K_1 and related compounds are also derived from N.
Biosynthesis: N. are formed not only from polyketides (e.g., *mompain, *plumbagin), but also from isochorismic acid and 2-oxoglutaric acid (e.g., *juglone, *lawsone), as well as from 4-hydroxybenzoic acid and various polyprenyl chains (e.g., *vitamin K_1 and homologues).

Lit.: Luckner (3.), p. 330, 402, 708 ▪ Phytochemistry **47**, 935–959 (1998) (N. from fungi) ▪ Thomson **3**, 137–253 ▪ Turner **1**, 130ff.

Naphthyl(tetrahydro)isoquinoline alkaloids. A group of biaryl alkaloids (ca. 70 reported up to 1998) from lianas of the genera *Ancistrocladus* (Ancistrocladaceae) and *Triphyophyllum* (Dioncophyllaceae) growing in the tropical rain forests of Africa and Asia; see also michellamines. The N. a. have diverse biological activities, e.g., antiviral and spasmolytic activities. Some N. a. are effective against the malaria pathogen *Plasmodium falciparum*[1].
The N. a. are formed biosynthetically as polyketides from acetate building blocks. For synthesis, see *Lit.*[2].

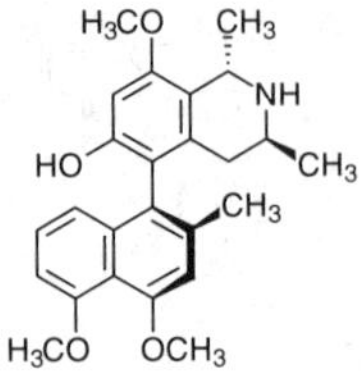

Ancistrocladine (**1**)

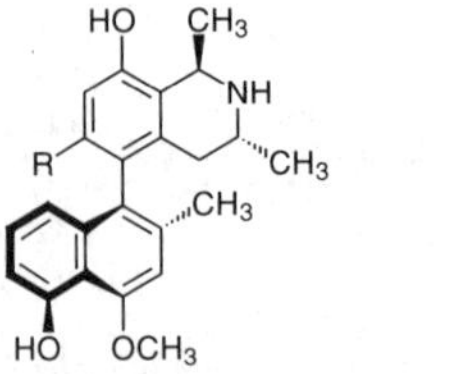

R = H : Dioncophylline C (**4**)
R = OH : Korupensamine A (**5**)

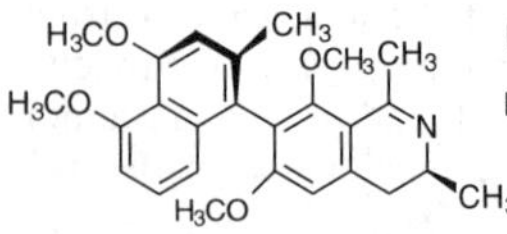

Ancistrocladisine (**2**)

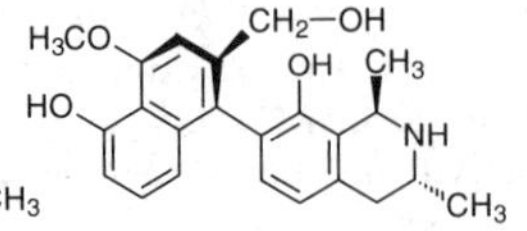

Dioncopeltatine A (**6**)
(Dioncopeltine A)

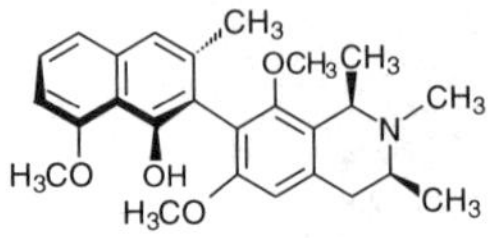

Ancistrotectorine (**3**)

Table: Data of Naphthyl(tetrahydro)isoquinoline alkaloids.

no.	molecular formula	M_R	mp. [°C]	$[\alpha]_D^{20}$	CAS
1	$C_{25}H_{29}NO_4$	407.51	265–267	–20.5° ($CHCl_3$)	32221-59-3
2	$C_{26}H_{29}NO_4$	419.52	178–180	–16.1° ($HCl/CHCl_3$)	41787-65-9
3	$C_{26}H_{31}NO_4$	421.54	134–140	0° ($CHCl_3$)	98985-59-2
4	$C_{23}H_{25}NO_3$	363.46	246	+19.2° ($CHCl_3$)	146471-75-2
5	$C_{23}H_{25}NO_4$	379.46	amorphous	–75.5° (CH_3OH)	158182-18-4
6	$C_{23}H_{25}NO_4$	379.46	241	–96° (pyridine)	60158-81-8

Lit.: [1] Antimicrob. Agents Chemother. **16**, 710 (1979); J. Ethnopharmacol. **46**, 115 (1995). [2] Angew. Chem. Int. Ed. Engl. **25**, 913 (1986); **28**, 1672 f. (1989); Atta-ur-Rahman **20**, 407–456 (1998); J. Org. Chem. **64**, 7184–7201 (1999); Justus Liebigs Ann. Chem. **1985**, 2105–2134; Heterocycles **28**, 137 (1983); **39**, 503 (1994); J. Chem. Soc., Perkin Trans. 1 **1991**, 2773–2781; Synlett **1998**, 1294; Tetrahedron Lett. **37**, 3097, 3099 (1996); Z. Naturforsch. B **51**, 144 (1996).
gen.: Beilstein E V **21/6**, 211, 219 ▪ Bull. Soc. Chim. Belg. **105**, 601–613 (1996) (review) ▪ Manske **29**, 141–184; **46**, 127–271. – *Isolation:* J. Nat. Prod. **59**, 854 (1996) ▪ J. Org. Chem. **59**, 6349 (1994) ▪ Nat. Prod. Lett. **6**, 315 (1995) ▪ Phytochemistry **30**, 1691 (1991); **31**, 3297, 4019 (1992); **33**, 1511 (1993); **35**, 259, 1461 (1994); **36**, 1057 (1994); **39**, 701 (1995); **45**, 1287 (1997); **47**, 37 (1998) ▪ Tetrahedron **52**, 13419 (1996); **55**, 423, 1731 (1999) ▪ Tetrahedron Lett. **36**, 4753 (1995). – *Synthesis:* J. Org. Chem. **61**, 7101 (1996) ▪ Stud. Nat. Prod. Chem. **20**, 407–455 (1998) ▪ Tetrahedron **54**, 497–512 (1998) (synthesis dioncophylline C). – *Biosynthesis:* Angew. Chem. Int. Ed. Engl. **39**, 1464 (2000).

Naphthyridinomycin A.

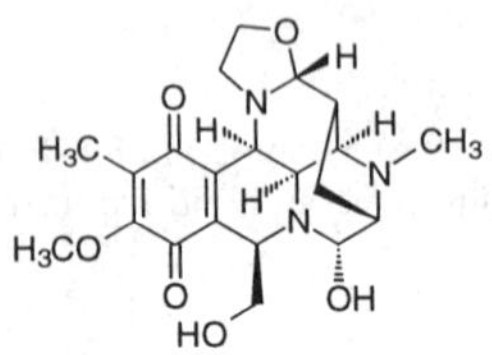

(relative configuration)

$C_{21}H_{27}N_3O_6$, M_R 417.46, ruby red cryst., mp. 108–110 °C, $[\alpha]_D$ +69.4° ($CHCl_3$), An alkaloid-like, hexacyclic *benzoquinone derivative from cultures of *Streptomyces lusitanus*. Biosynthetic precursors include tyrosine and glycine, the methyl groups originate from methionine. N. A has broad activity against Gram-positive and Gram-negative bacteria and is cytotoxic through inhibition of DNA synthesis. N. A is closely related to the bioxalomycins and cyanocyclines.

Lit.: Adv. Heterocycl. Nat. Prod. Synth. **2**, 189–249 (1992) (synthesis, review) ▪ J. Am. Chem. Soc. **104**, 4969 ff. (1982) (biosynthesis) ▪ J. Antibiot. **35**, 642 ff. (1982); **47**, 1417 (1994) ▪ J. Org. Chem. **59**, 4045 ff. (1994) (abs. configuration). – *[CAS 54913-26-7]*

Narasin [narasin A, (4*S*)-4-methylsalinomycin].

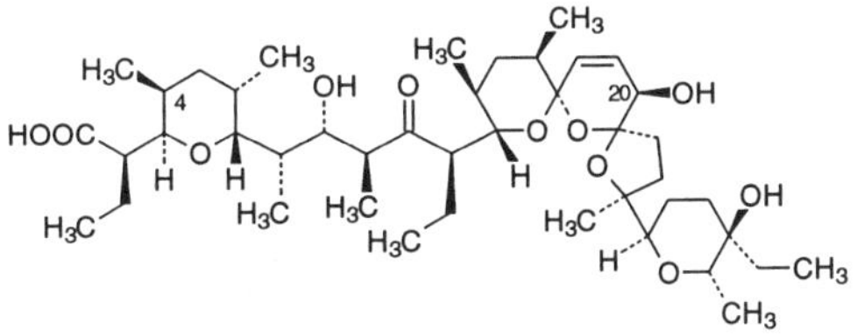

$C_{43}H_{72}O_{11}$, M_R 765.04, cryst., double mp. 98–100 °C and 198–200 °C, $[\alpha]_D^{25}$ –54° (CH_3OH). A polyether antibiotic from *Streptomyces* species with particularly useful activity against coccidial infections in poultry. Some derivatives of M. are used as fodder additives.

Lit.: ApSimon **8**, 645 ▪ Beilstein E V **19/12**, 553 ▪ J. Antibiot. **30**, 530 (1977); **31**, 1 (1978) ▪ Kirk-Othmer (4.) **3**, 315 ▪ Merck-Index (12.), No. 6506 ▪ Shepherd (ed.), Food Contaminants: Sources and Surveillance, p. 143, London: R. Soc. Chem. 1991 ▪ Tetrahedron **47**, 10109 (1991). – *[HS 2941 90; CAS 55134-13-9]*

Narciclasine (lycoricidinol).

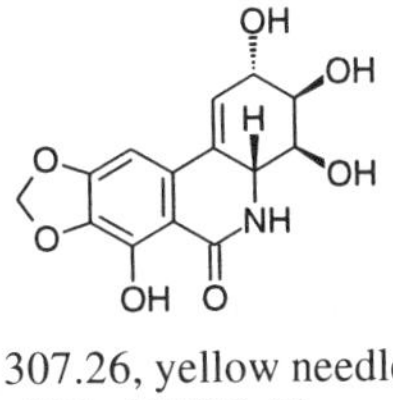

$C_{14}H_{13}NO_7$, M_R 307.26, yellow needles with green fluorescence, mp. 232–234 °C (decomp.), $[\alpha]_D^{20}$ +145° (C_2H_5OH), LD_{50} (mouse s. c.) 5 mg/kg. An *Amaryllidaceae alkaloid from *Narcissus* species, e. g., *N. incomparabilis*. N. has several biological properties. It exhibits strong antineoplastic and antiviral activities and, like for example *thujaplicins and *pimpinellin inhibits chlorophyll biosynthesis.

Lit.: Chem. Ber. **108**, 445 (1975) ▪ J. Am. Chem. Soc. **119**, 12655 (1997); **121**, 5176 (1999) (synthesis) ▪ J. Chem. Soc., Chem. Commun. **1972**, 239 (biosynthesis) ▪ Manske **25**, 205 ▪ Sax (8.), No. LJC 000 ▪ Tetrahedron Lett. **38**, 7955 (1997) (synthesis); **40**, 3077 (1999) [synthesis (+)-N.]. – *[CAS 29477-83-6]*

Narcissidine see Amaryllidaceae alkaloids.

Narcotine see phthalide isoquinoline alkaloids.

Naringin [naringenin-7-*O*-(2-*O*-α-L-rhamnosyl-β-D-glucoside]. $C_{27}H_{32}O_{14}$, M_R 580.54, bitter tasting cryst., mp. 82 °C (octahydrate) or, respectively, 172 °C (dihydrate), $[\alpha]_D^{10}$ –82° (C_2H_5OH), soluble in alcohol and hot water; N. is a glycoside from the flowers, fruits (espe-

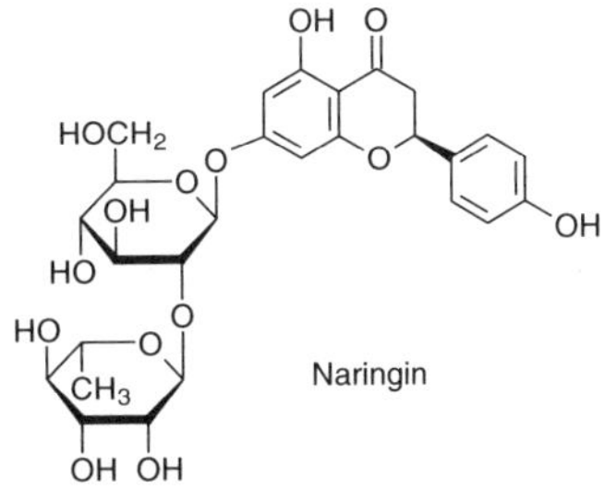

Naringin

cially unripe), and bark of the grapefruit tree (*Citrus paradisi*, Rutaceae) and other citrus plants. N. is the bitter principle of grapefruit juice. It can still be tasted at a dilution of 1:10 000 in water. The aglycone of N., the flavanone *naringenin* {floribundigenin, naringetol, salipurol, $C_{15}H_{12}O_5$, M_R 272.26, mp. 250–251 °C, $[\alpha]_D^{27}$ –22.5° (CH_3OH), (*S*)-Form)}, acts as an antagonist of *gibberellins in resting peach flowers. The bitterness can be removed from grapefruit juice by means of the enzyme naringinase. Hydrogenation of N. leads to a dihydrochalcone, which is ca. 300-fold sweeter than saccharose.

Lit.: Beilstein E V **18/4**, 528 ▪ Hager (5.) **4**, 84 f.; **5**, 962 ▪ Karrer, No. 1605 ▪ Kirk-Othmer (3.) **9**, 194 f.; **22**, 458–464 ▪ Merck-Index (12.), No. 6511, 6512 ▪ Ullmann (4.) **10**, 515; (5.) **A11**, 507, 578. – *Isolation:* J. Nat. Prod. **45**, 635 (1982) ▪ Phytochemistry **21**, 1464 (1982); **22**, 625 (1983); **23**, 1338 (1984); **34**, 843 (1993). – *[HS 2938 90; CAS 10236-47-2 (N.); 480-41-1 (naringenin)]*

Narwedine see Amaryllidaceae alkaloids.

Nasutins see ellagic acid.

Natamycin.

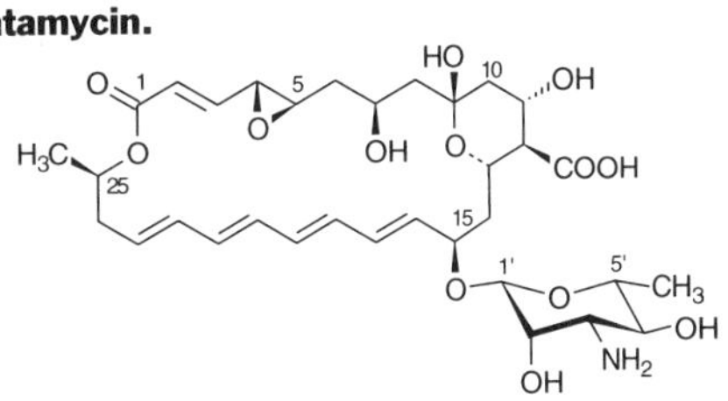

International free name for *pimaricin*, a polyene *macrolide of the tetraene type from cultures of *Streptomyces natalensis*, the aglycone is formed biosynthetically through the *polyketide pathway from one propanoate unit and 12 acetate units and is *O*-glycosidically linked to the amino sugar D-mycosamine. $C_{33}H_{47}NO_{13}$, M_R 665.74, mp. 200 °C (decomp.). Light-sensitive crystals, very poorly soluble except in *N*-methyl-2-pyrrolidone and dimethylformamide. N. has a broad spectrum of activity against fungal diseases (skin infections with *Candida*), it is registered in some countries (e. g., USA) for use as a fungicide for treating the surfaces of hard and sliced cheese and as conservation agent for sausages. N. also has antiviral and immunostimulating activities.

Lit.: Beilstein E V **19/8**, 321 ▪ Biochim. Biophys. Acta **864**, 257 (1986) ▪ J. Am. Chem. Soc. **112**, 4060 f., 7079 ff. (1990) (abs. configuration) ▪ J. Chem. Soc., Chem. Commun. **1986**, 1241–1244 ▪ Med. Actual. **14**, 254 (1978) ▪ Omura (ed.), Macrolide Antibiotics, p. 351–404, New York: Academic Press 1984 ▪ Pesticide Manual **1991**, No. 9760. – *[HS 2941 90; CAS 7681-93-8]*

Natural dyes (pigments). Collective name for the ubiquitous, naturally occurring, organic pigments of plant, animal, and microbial origin. The color effects are due either to their chemical structures (pigment colors) or to the physical properties of the surfaces (structure colors). N. d. can be classified according to various criteria: 1) their occurrence (*plant pigments, animal pigments, insect pigments, algal pigments, fungal pigments, *lichen pigments, bacterial pigments); – 2) their chemical structure (*carotinoids, *benzoquinones, *naphthoquinones, *anthranoids, indigo dyes, *flavonoids, *anthocyanins, *betalains, *neoflavonoids, xanthones, *alkaloidal dyes, benzophenone dyes, *pteridines, *pyrrole pigments, *ommochromes, *ergochromes); – 3) their functions (attractant, repellent, or camouflage pigments).
The following functions have been discussed for the pigments of higher plants: 1) absorption, transport, and conversion of light energy to effect the endogenous processes of photosynthesis, – 2) communication between plants and animals (pigments in fruits and flowers), – 3) absorption of damaging UV light. The physiological significance of many N. d. as well as other secondary natural products is still unknown. N. d. have been used for thousands of years to dye various materials, e. g., textiles and wool for carpets. N. d. with the longest history include *alizarin(e), *indigotin, Tyrian purple, as well as *saffron, kermes, *cochineal and flavonoid- or neoflavonoid-containing *dyewoods. The use of N. d. has declined greatly with the development of synthetic dyes which usually have better stabilities and other advantages over the natural dyes. The cell culture technique is employed to produce N. d. for use as pigments in cosmetics (*shikonin) or foods.

Lit.: Atta-ur-Rahman **20**, 719–788 (1998) (review) ▪ Chem. Eng. News (11.8.) **1997**, 36–40 (review) ▪ Counsell, Natural Colours for Food and other Uses, Barking: Appl. Sci. Publ. 1981 ▪ Goodwin II ▪ Harborne (1975, 1982, 1988, 1993) ▪ Isler ▪ Roth, Kormann & Schweppe, Färbepflanzen Pflanzenfarben, Landsberg/Lech: ecomed 1992 ▪ Rowe, p. 843 ff. ▪ Vevers, The Colours of Animals, London: Arnold 1982 ▪ SÖFW J. **125**, 1–11 (1999) (legal colours for food dying) ▪ Zechmeister **51**, 75–285 ▪ Zollinger, Color Chemistry, Weinheim: VCH Verlagsges. 1987.

Natural organoarsenic compounds see arsenic in natural products.

Natural products. General term for substances from animals, plants, or microorganisms, generally arising from their secondary metabolism but also in the wider sense from the primary metabolism. The primary substances are ubiquitous in nature, the occurrence of secondary substances is mostly limited to specific organisms. The classification into primary and secondary natural products was introduced by the physiologist A. Kessel.
Although the term N. p. is now used almost exclusively for specific organic compounds of natural origin, in the past substances of mineral origin were also included. Natural products can be classified, for example, according to a) their *chemical structure:* e. g., alkaloids, amines, amino acids, aromatic compounds, fatty acids, flavonoids, hydrocarbons, lipids, nucleotides, mono-, oligo-, and polysaccharides, phenylpropane derivatives, steroids, terpenoids, etc.; – b) their *biogenetic origin:* e. g., acetogenins, eicosanoids, isoprenoids, polyketides, shikimic acid derivatives; – c) their *function or activity:* e. g., antibiotics, antibodies, dyes, enzymes, toxins, hormones, defence, signal, and growth substances, odor and storage compounds, vitamins.
N. p. are characterized by an enormous structural diversity. As a result of recent improvements in instrumental analysis and separation techniques as well as knowledge of biogenetic relationships, the structural elucidation of a complicated N. p. can often be accomplished within a few days or weeks whereas in the past years or decades would have been required. In particular, microorganisms (bacteria and fungi) as well as marine organisms (invertebrates) have recently proved to be especially fruitful sources of new parent structures.
N. p. possess a wide range of biological activities. The significance and function of most primary substances is now understood, but even today only very little is known about the function or ecological relevance of most secondary substances. Many N. p. were and are used in traditional medicine (ethnopharmacology). New drugs [1] and plant protection agents [2] have been developed from N. p. by chemical or enzymatic modification of the original structure (N. p. as lead structures; more than 40% of all drugs have been developed from N. p. as derivatives or analogues). N. are used in research as indispensable tools for the elucidation of biological mechanisms, e. g., control of the cell cycle [3]. Classical biotechnological and new genetic engineering methods are becoming more important for the production of complicated natural products or derivatives, the total syntheses of which are too expensive or not possible; *examples:* fermentation of antimicrobially active principles; isolation of cardiac glycosides from *Digitalis* species and quinine alkaloids from *Cinchona* (bark); valuable plant components from plant cell cultures, human peptide hormones and proteins such as insulin, growth hormone, or factor VIII from genetically engineered microorganisms and mammalian cell cultures. Combinatorial biosynthesis is used to investigate the possibilities of obtaining new natural products for biological testing (see antibiotics).
The chemistry of natural products encompasses their isolation, structure elucidation, partial and total synthesis, elucidation of their biogenesis, and the biomimetic synthesis of N. p. Major breakthroughs in analysis were, e. g., the structural clarifications of morphine, lignin, insulin, estrones, and cholesterol as well as the elucidation of the biosyntheses of terpenoids, morphine, penicillin, chlorophyll, and vitamin B_{12}. Major advances in synthetic chemistry were, e. g., the total syntheses of camphor, hemin, quinine, saccharose, tropine, strychnine, chlorophyll, vitamin B_{12}, erythromycin, taxol and palytoxin. Numerous N. p. of the so-called "chiral pool" are used as starting materials for the synthesis of optically active compounds or serve (in the form of their derivatives) as catalysts for enantioselective syntheses.

The development of chemotaxonomy on the basis of the occurrence of secondary products has made a major contribution to the clarification of the relationships among animals, plants, and microorganisms.
In some fields of application, e.g., the food or cosmetic industries, distinctions are made between natural additives and "nature-identical" compounds, when the latter are produced by total synthesis.
The traditional demarcation of the chemistry of natural products from, e.g., the rest of organic chemistry, physics, biosciences and, in particular, biotechnology, molecular biology, and research on active principles is no longer possible. Thus the entire field of the extensively widened topic of natural products chemistry is now often referred to as "bioorganic chemistry". The nomenclature principles of natural products – use of semisystematic names – are discussed in Section F of IUPAC rules, The Compendium of IUBMB, Biochemical Nomenclature, London: Portland Press 1992, and in the Chemical Abstracts Index Guide.

Lit.: [1] ACS Symp. Ser. **534**, (1993); Ciba Found. Symp. **185**: Ethnobotany and the Search of New Drugs, Chichester: Wiley 1994; Grabley & Thiericke, Drug Discovery from Nature, Berlin: Springer 1999; Harvey, Drugs from Natural Products, New York: Ellis Horwood 1993. [2] ACS Symp. Ser. **551** (1994). [3] Chem. & Biol. **3**, 623–639 (1996).
gen.: ACS Symp. Ser. **691** (Phytomedicines of Europe) (1998); **723** (Biopolymers) (1999) ▪ Angew. Chem. Int. Ed. Engl. **38**, 1903 (1999) (solid phase synthesis of N. p.); **39**, 44–122 (2000) (monograph on total synthesis); 1538–1559 (2000) (synthetic problems) ▪ Atta-ur-Rahman **20** (1998) (review) ▪ Barton-Nakanishi **9** ▪ Cannell, Methods in Biotechnology, vol. 4, Natural Products Isolation, Totowa: Humana Press 1998 ▪ Colegate & Molyneux (eds.), Bioactive Natural Products, Boca Raton: CRC Press 1993 ▪ Dewick, Medicinal Natural Products (A Biosynthetic Approach), Chichester: Wiley 1997 ▪ Mann et al., Natural Products: Their Chemistry and Biological Significance, Essex: Longman Scientific & Technical 1994 ▪ Med. Res. Rev. **18**, 383–402 (1998) ▪ Milne, Yan & Zhou, Traditional Chinese Medicines, Aldershot: Asgate 1999 ▪ Phytochemistry **40**, 1585–1612 (1995) ▪ Rimpler, Biogene Arzneistoffe (2.), Stuttgart: Dtsch. Apoth. Verlag 1999 ▪ Spec. Publ. Royal Soc. Chem. **240** (1990) ▪ Thomson (ed.), The Chemistry of Natural Products, London: Blackie Academic & Professional 1993 ▪ Torssell, Nat. Prod. Chem. (2.), Stockholm: Swedish Pharmaceutical Press 1997 ▪ Verrall, Downstream Processing of Natural Products, Chichester: Wiley 1996 (isolation techniques). – *History:* Barton-Nakanishi **1** – Further Lit. see list of frequently cited texts at the beginning of this book.

Natural resins. Collective name for resins of plant or animal origin – the sole example of the latter that has attained technical importance is *shellac (shellac wax). The plant natural resins are based on secretions (*exudates*) of particular plants, mostly trees, from natural or artificial (cuts in the bark) injuries. These exudates normally occur in the form of sticky masses that harden in the air as a result of evaporation of volatile components and/or polymerization and oxidation reactions.
The N. r. include benzoin gum, Canada balsam, China or *Japan lacquers, *dammar resin, *amber, *labdanum, *mastic, incense, sandarac, and *storax.
The N. r. range from liquid to solid, amorphous, and non-crystalline products with average molar masses of less than 2000. Their states of aggregation allow a further classification into hard and soft resins and rubbers, or mucous resins.
N. r. can be classified on the basis of their solubility properties into oil-soluble or alcohol-soluble N. r.
Most N. r. are yellow to brown in color. Their densities are in the range 0.9–1.3, their hardness between that of plaster and rock salt. Hard N. r. melt at a temperature between 40 °C (*Asa foetida*) and 360 °C (amber). N. r. are mostly odor- and tasteless; some contain *essential oils.
At present mainly terpenes and aromatic compounds have been isolated and identified as components of resins with *diterpenes (less frequently *sesquiterpenes and *triterpenes) and phenylpropane derivatives usually dominating.
Use: N. r. were used in antiquity to embalm dead bodies. Today the resins and their derivatives are used, for example, in the production of oil resin paints, linoleum, colored printing inks and resin soaps, as auxiliaries in paper processing, as starting materials for sealants, leather and shoe cleaning agents, glues, or coating masses. Some N. r. are indispensable in the preparation of fragrance substances and *flavor compounds in the form of oleoresins (see also essential oils). The importance of N. r. has decreased greatly with the increasing production of synthetic resins.

Lit.: Kirk-Othmer (3.) **20**, 197–230; (4.) **12**, 842–862 ▪ Ullmann (5.) **A 23**, 73 f., 107 f. – *[HS 1301 10, 1301 20, 1301 90]*

Natural rubber. The white milk juice (latex) from rubber or para rubber trees (*Hevea brasiliensis*, Euphorbiaceae), cultivated in large plantations in practically all tropical regions of Africa, Asia, and South America. Among the other rubber producing plants and organisms, only the gutta-percha tree (*gutta-percha), the guayule bush (*Parthenium argentatum*), *Mimusops balata* (syn. *Manilkara bidentata*) (*balata), and kok-saghyz (*Taraxacum kok-saghyz*, Asteraceae) are used to a minor extent to produce N. r. An average-size rubber tree (height ca. 15–20 m) provides about 7 g latex per day.
Latex is an emulsion of 0.5–1 μm droplets of N. r. in water, with proteins serving as a protective colloid. 100 g latex contain about 30–35 g N. r., proteins, sterols, fats, carbohydrates (together 4.5–5 g) and 0.5 g mineral components, the rest is water.
N. r. is a polyisoprene with

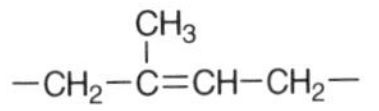

as basic unit existing in the *cis*-1,4- (a; *all-Z-*, hevea-NR) or *trans*-1,4-configuration (b; *all-E*-NR, balata and gutta-percha):

a) *cis-1,4 = all-Z*

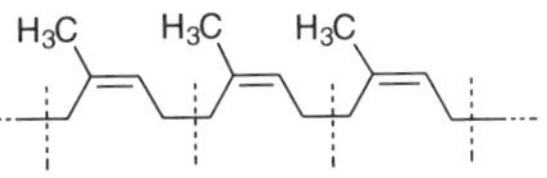

Hevea rubber (NR)

b) trans-*1,4 = all-E*

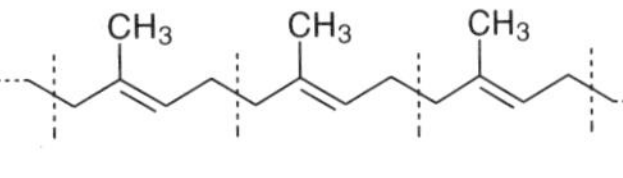

Balata Guttapercha

For the biosynthesis of N. r. by polymerization of monomeric isopentenyl diphosphate ("active isoprene") which proceeds as head-to-tail process both via farnesyl diphosphate and via other oligomeric precursors, see *Lit.*[1]. *Hevea* NR, balata, and gutta-percha additionally differ in their degrees of polymerization, amounting to ca. 8000–30 000 for N. r., and to ca. 1500 for the other rubbers. Crude N. r. undergoes disadvantageous changes on longer storage in light and air resulting from cross-linking and oxidation reactions. N. r. is converted to a valuable technical product by vulcanization[2].
World-wide production and consumption: in 1992 ca. 5 400 000 t N. r. were produced world-wide, 3 975 000 t by the major producing countries Thailand, Indonesia, and Malaysia, exact data are given in *Lit.*[3]. Information on the production and use of N. r. (as well as synthetic rubbers) is provided by the International Institute of Synthetic Rubber Producers (IISRP) and published regularly in specialist journals.
Use: N. r. are used in unmodified form as derivatives and as vulcanizate; in small amounts also for the production of adhesive tapes, rubber solutions, plasticine, etc. By far the major portion is used in vulcanized form for processing to soft or hard rubber. Main field of use of the vulcanizate are the production of vehicle tyres, rubber springs and rubber buffers, manufacture of thin-walled soft articles of high strength (balloons, surgical gloves, rubber sanitary articles), conveyor bands, belts, technical articles such as piping, seals, membranes, engine mounts as well as consumer articles like rubber boots, shoe soles and heels, gloves, sponges, rubber bands, etc.; see also rubber.

Lit.: [1]Luckner (3.), p. 235; J. Plant Physiol. **136**, 257–263 (1990). [2]Ullmann (4.) **13**, 583–594. [3]Gummi, Asbest+ Kunstst. **46**, 10–17 (1993).
gen.: Morton, Rubber Technology, 3. edn., New York: Van Nostrand Reinhold 1987 ▪ Roberts, Natural Rubber Science and Technology, Oxford: Oxford University Press 1988 ▪ Tanaka, Rubber and Related Polyprenols, in Charlwood & Banthorpe (eds.), Methods in Plant Biochemistry, vol. 7, London: Academic Press 1991 ▪ Ullmann (5.) **A23**, 225–237.

Navenone.

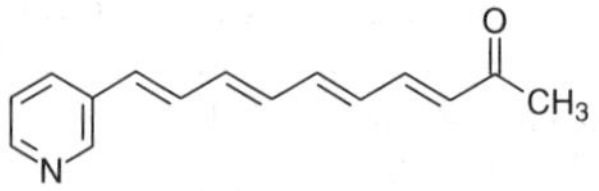

Navenone A

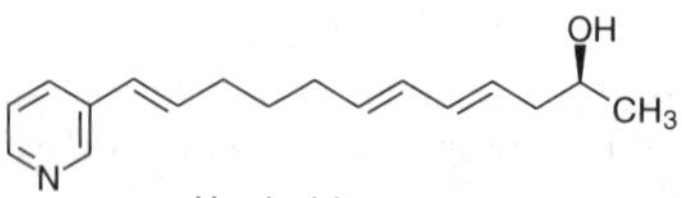

Haminol A

Opistobranchia (molluscs) usually live in the tidal regions of the oceans and find prey and partners by chemoreception of trail hormones (see pheromones) composed of mucopolysaccharides. In case of danger the snails emit an alarm pheromone that induces following animals to seek a different route. The Californian *Navanax inermis* (Cephalospidea) secretes the brilliant yellow N.[1] (*N. A.*, $C_{15}H_{15}NO$, M_R 225.29, mp. 144–145 °C); the Mediterranean *Haminoea navicula* emits, among others, *haminol A*[2] (oil, $C_{17}H_{23}NO$, M_R 257.37, $[\alpha]_D$ +5°).

Lit.: [1]Agric. Biol. Chem. **43**, 117 (1979); J. Am. Chem. Soc. **99**, 2367 (1977); Pure Appl. Chem. **51**, 1865 (1979); Nat. Prod. Lett. **4**, 203 (1994) (synthesis). [2]Experientia **47**, 61 (1991); Tetrahedron: Asymmetry **9**, 3065 (1998) (synthesis). – *[CAS 62695-67-4 (N.); 133412-13-2 (haminol A)]*

Neamine see neomycin.

Nebularine (9-*β*-D-ribofuranosyl-9*H*-purine).

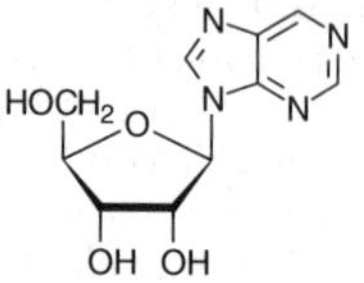

$C_{10}H_{12}N_4O_4$, M_R 252.23, cryst., mp. 182–183 °C, $[\alpha]_D^{35}$ −46.8° (H_2O). A *nucleoside antibiotic from the fungus *Clitocybe nebularis*[1], an agaric frequently found in Central European woods in autumn. N. has also been found in a *Streptomyces* species[2]. N. has antimitotic and tuberculostatic activities, LD_{50} (rat/guinea-pig s.c.) 220/15 mg/kg.

Lit.: [1]Tetrahedron **44**, 7001 (1988). [2]J. Antibiot. Ser. A **13**, 270 (1960).
gen.: Beilstein E V **26/11**, 198 ▪ J. Antibiot. **41**, 1711 (1988) ▪ Merck-Index (12.), No. 6521 ▪ Sax (8.), RJF 000 ▪ Ullmann (5.) **A 2**, 492. – *Synthesis:* Synthesis **1984**, 401; **1988**, 848 ▪ Tetrahedron **44**, 7001 (1988). – *[HS 2941 90; CAS 550-33-4]*

Necatorone (5,10-dihydroxy-6*H*-pyrido[4,3,2-*kl*]acridin-6-one).

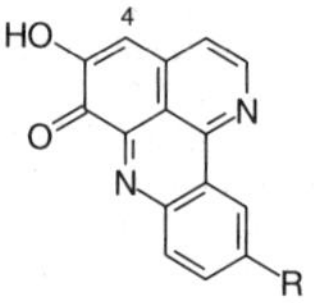

R = OH : Necatorone
R = H : 10-Deoxynecatorone

$C_{15}H_8N_2O_3$, M_R 264.24, red needles, mp. >360 °C. N. and *4,4′-binecatorone* ($C_{30}H_{14}N_4O_6$, M_R 526.46, dark brown, poorly soluble cryst., mp. >360 °C) are responsible for the dark, olive brown color of the milkcap *Lactarius necator* (Basidiomycetes) and account for the change in color of its fruit bodies to violet in the presence of ammonia[1]. The North American *L. atroviridis* also contains *10,10′-dideoxy-4,4′-binecatorone* ($C_{30}H_{14}N_4O_4$, M_R 494.47, black-green cryst., mp. >360 °C) which is responsible for the bottle-green color of the toadstool[2]. As shown by feeding experiments, N. is formed biosynthetically from *L-tyrosine and *anthranilic acid. N. is mutagenic in the Ames *Salmonella* test.

Lit.: [1]Tetrahedron Lett. **25**, 3575 (1984); **26**, 5975 (1985) (synthesis); Zechmeister **51**, 227–229 (review). [2]Phytochemistry **28**, 3519 (1989). – *[CAS 92631-70-4 (N.); 126624-08-6 (4,4′-binecatorone); 126647-32-3 (10,10′-dideoxy-4,4′-binecatorone)]*

Necine(s). The name necine is used for the bicyclic amino alcohol 1-pyrrolizidinemethanol. It also occurs, under the name necine base, as the alcohol part, generally esterified, of *pyrrolizidine alkaloids (more than 250 natural products described to date). The N. are saturated or have 1,2-unsaturation and often carry an additional hydroxy group in the 7-position, less frequently in the 2- or 6-position.

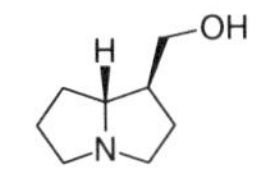

(+)-Laburnine (**1**)
(+)-Trachel-anthamidine (**2**)

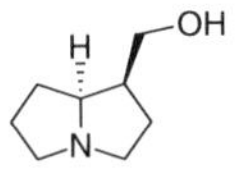

(-)-Isoretro-necanol (**3**)

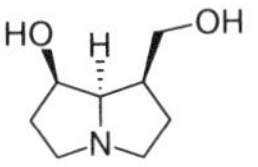

(-)-Platynecine (**4**)

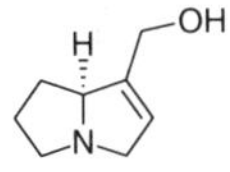

(-)-Supinidine (**5**)

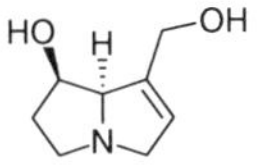

(+)-Retronecine (**6**)

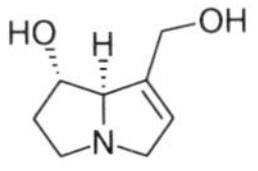

(+)-Heliotridine (**7**)

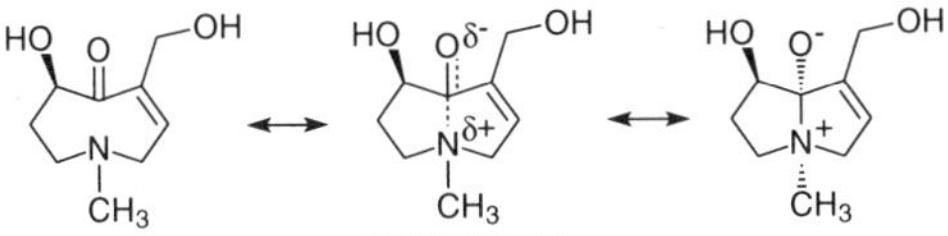

Otonecine (**8**)

The figure shows some N. bases. The most frequently occurring representatives are: *laburnine*[1] {$C_8H_{15}NO$, M_R 141.21, bp. 134 °C (1.6 kPa), $[\alpha]_D^{20}$ +17.01° (C_2H_5OH)}, (+)-isomer of the two enantiomeric trachelanthamidines. Laburnine is a pyrrolizidine alkaloid from golden chain (*Laburnum anagyroides*). Together with the second pair of enantiomers *isoretronecanol* these four N. occur mostly as the base part of more than 40 ester alkaloids isolated mainly from orchids and Boraginaceae. Of the corresponding 1,2-dehydro compounds,(−)-*supinidine* ($C_8H_{13}NO$, M_R 139.20) is the mostly frequently occurring enantiomer in nature. About 10 ester alkaloids of this N. are known. For synthesis of the mentioned compounds, see *Lit.*[2,3].
Retronecine[4] {$C_8H_{13}NO_2$, M_R 155.20, mp. 121–122 °C, $[\alpha]_D^{26}$ +50.2° (C_2H_5OH)} is the N. base of most of the known toxic pyrrolizidine alkaloids (see callimorphine, indicine, jacobine, monocrotaline, retrorsine, senecionine) and also occurs in the free form. For synthesis, see *Lit.*[5,6].
Otonecine[7] {$C_9H_{15}NO_3$, M_R 185.22, oil; as hydrochloride: mp. 146–148 °C, $[\alpha]_D$ −18.5° (C_2H_5OH)} is the N. base of about 30 known pyrrolizidine alkaloids, most of which are of the *senecionine structural type (see also senkirkine) and a few of which are of the *monocrotaline structural type. For synthesis, see *Lit.*[8].
Platynecine[9] {$C_8H_{15}NO_2$, M_R 157.21, mp. 148–148.5 °C, $[\alpha]_D$ −56.8° ($CHCl_3$)} is the N. base of almost 20 pyrrolizidine alkaloids from plants of the family Asteraceae (see platyphylline). For synthesis, see *Lit.*[5].
Heliotridine[10] {$C_8H_{13}NO_2$, M_R 155.20, mp. 116–118 °C, $[\alpha]_D^{20}$ +31°} is the N. base of more than 20 mono- and diester derivatives of the pyrrolizidine alkaloids, e.g., echinatine (see indicine), heliosupine, heliotrine. For synthesis, see *Lit.*[6].
The N. bases are mostly esterified by aliphatic mono- or dicarboxylic acids, the so-called necine acids. In the subfamily (tribe) Senecioeae of the Asteraceae the alkaloids occur almost exclusively as 12-membered macrocyclic diesters of variously substituted hexanedioic acids or as mono- and diesters of C_5 acids (angelic, tiglic, and senecioic acids) with retronecine. In the subfamily Eupatorieae of the Asteraceae and the Boraginaceae, besides some mono- and diesters of the C_5 acids, esters of viridifloric and trachelanthic acids or their derivatives with retronecine and heliotridine dominate. Some syntheses of these acid components have been reported[11]. In the Fabaceae genus *Crotalaria* 11-membered macrocyclic diesters of unusually substituted pentanedioic acid with retronecine are frequent.

Lit.: [1] Beilstein E V **21/1**, 288 ff. [2] J. Am. Chem. Soc. **110**, 289 (1988); Tetrahedron Lett. **29**, 6133 (1988); J. Org. Chem. **55**, 1148 (1990). [3] J. Chem. Soc., Perkin Trans. 1 **1981**, 909. [4] Beilstein E V **21/4**, 398; J. Org. Chem. **50**, 2170 (1985). [5] Synlett **1994**, 277 f.; J. Org. Chem. **62**, 435 (1997). [6] J. Chem. Soc., Chem. Commun. **1988**, 685; J. Chem. Soc., Perkin Trans. 1 **1990**, 571–577. [7] Beilstein E III/IV **21**, 2504; J. Am. Chem. Soc. **120**, 3613 (1998). [8] Tetrahedron Lett. **24**, 5731 f. (1983). [9] Beilstein E V **21/4**, 389; Tetrahedron Lett. **38**, 603 (1997); Bull. Chem. Soc. Jpn. **70**, 2541 (1997). [10] Beilstein E V **21/4**, 398; Heterocycles **27**, 253 (1988). [11] Chem. Ber. **116**, 3413 (1983).
gen.: Hager (5.) **3**, 1079 f. ▪ J. Chem. Soc., Perkin Trans. 1 **1997**, 2089 ▪ Pelletier **9**, 155–233. – *[HS 2939 90; CAS 3348-73-0 (1); 526-63-6 (3); 520-62-7 (4); 551-59-7 (5); 480-85-3 (6); 520-63-8 (7); 6887-34-9 (8)]*

Necrodols.

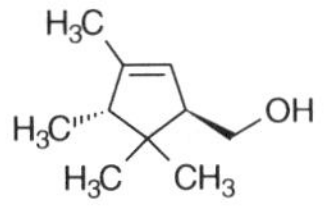

(-)-α

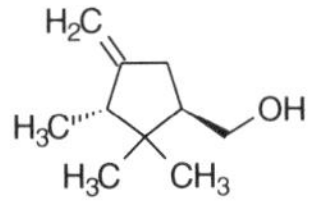

(-)-β

α-N., $C_{10}H_{18}O$, M_R 154.25, $[\alpha]_D$ −129.7° ($CHCl_3$) and *β-N.*, $C_{10}H_{18}O$, M_R 154.25, $[\alpha]_D$ −18.1° ($CHCl_3$). The defence secretion ejected from an abdominal gland of the tropical dung beetle (*Necrodes surinamensis*, Silphidae) consists of a mixture of aliphatic acids and monoterpene alcohols. The cyclopentanoid monoterpenes *α*- and *β*-N. are principally responsible for the repellent action of the secretion.

Lit.: J. Org. Chem. **55**, 4047–4062 (1990) ▪ Phytochemistry **36**, 43 (1994) ▪ Tetrahedron Lett. **40**, 4401 (1999). – *[CAS 104104-38-3 (α-N.); 104086-70-6 (β-N.)]*

Neem oil (margosa oil). Yellow plant oil from the crushed seeds of the *neem tree (*Azadirachta indica*) originally indigenous to India but now also cultivated in Africa, Australia, and Central and South America. The composition of the oil depends strongly on its origin. The most important components are various stereoisomers and derivatives of *azadirachtin (50–4000 ppm). It also contains numerous *limonoids as well as various disulfides, the latter being responsible for the garlic-like odor of the oil.
Use: N. can be used in plant protection as a natural antifeedant and insecticide. These uses are due to azadirachtin. The oil is superior in action to the pure active principle; this is due to the stabilizing effect of the oil and to its content of other active substances. Aqueous seed extracts or emulsions of the oil have traditional uses in India as insecticides and represent an important alternative to the use of synthetic neurotoxins (for mechanism of action, see azadirachtin(s)). In India the yearly production of the oil amounts to ca. 80 000 t/a N.; however, the larger part is used for the production of soaps.

Lit.: ACS Symp. Ser. **296**, 220–235 (1986); **387**, 70 f., 112–126 (1989) ▪ Entomol. Exp. Appl. **72**, 77–84 (1994) ▪ J.

Agric. Food Chem. **43**, 507–512 (1995) ▪ Pesticide Outlook, Oct. 1997, p. 32 ▪ Sax (8.), NBS 300. – *[CAS 8002-65-1]*

Neem tree. Originally indigenous to India, the neem tree *Antelaea* or *Melia azadirachta* (*Azadirachta indica*, Meliaceae) is widely distributed in tropical regions; the seed oil (*neem oil) with its unpleasant, garlic-like odor contains *azadirachtin which has an antifeedant and repellent effect on insects. Terpenoids with similar activities are found in other Meliaceae (*Lit.*[1]). Various bitter principles (*limonoids) are isolated from the bark, e.g., *nimbin*[2] {$C_{30}H_{36}O_9$, M_R 540.61, bitter tasting cryst., mp. 205 °C, $[\alpha]_D$ +167.5° ($CHCl_3$)} and *nimbiol* [$C_{18}H_{24}O_2$, M_R 272.36, mp. 248–252 °C]. Bark extracts are used in agents for skin and oral hygiene.

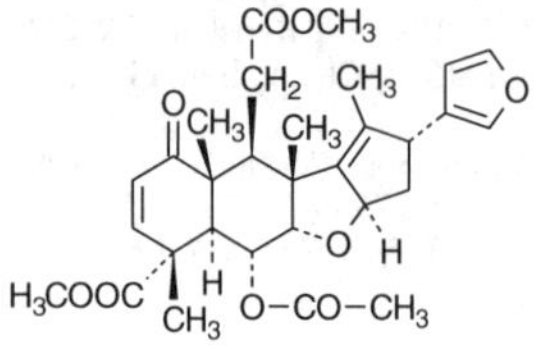
Nimbin

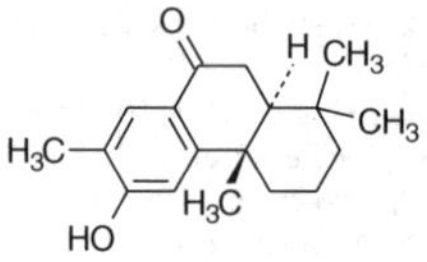
Nimbiol

Lit.: [1] Angew. Chem. Int. Ed. Engl. **17**, 452 (1978). [2] Indian Drugs **25**, 526f. (1988); Indian J. Chem., Sect. B **24**, 1105 (1985); Tetrahedron **24**, 1517 (1968); **46**, 775–782 (1990). *gen.:* J. Chem. Soc., Perkin Trans. 1 **1998**, 1423 [synthesis (+)-nimbiol] ▪ Karrer, No. 3683 ▪ Schmutterer, The Neem Tree, Weinheim: VCH Verlagsges. 1995 ▪ see also azadirachtin. – *[CAS 5945-86-8 (nimbin)]*

Nematophin.

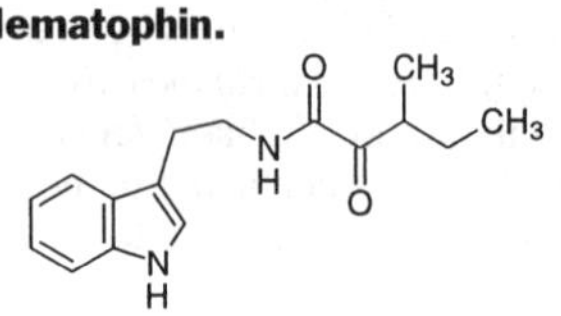
(+)-form, easily racemized

$C_{16}H_{20}N_2O_2$, M_R 272.34. An antibiotic indole derivative from culture broths of *Xenorhabdus nematophilus* (Enterobacteriaceae), a bacterial symbiont of the entomopathogenic nematode *Steinernema carpocapsae*. It shows activity against different strains of *Staphylococcus aureus*.

Lit.: Bioorg. Med. Chem. Lett. **7**, 1349–1352 (1997). – *[CAS 183314-23-0]*

Nemertelline see quaterpyridine.

Neocarzilins.

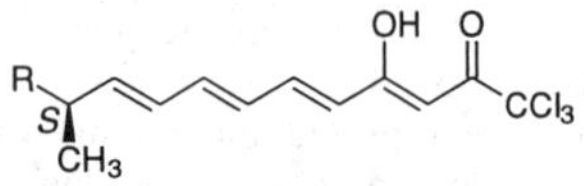
R = C_2H_5 : N. A
R = CH_3 : N. B

Chlorinated polyenones from *Streptomyces carzinostaticus*. *N. A* [$C_{14}H_{17}Cl_3O_2$, M_R 322.03, yellow oil, $[\alpha]_D^{29}$ +45.7° ($CHCl_3$)] exhibits high activity against leukemia cells [IC_{50} 0.06 µg/mL, LD_{50} (mouse) 88.1 mg/kg] and antibacterial activity against Gram-positive bacteria. No such activities have been reported for *N. B* ($C_{13}H_{15}Cl_3O_2$, M_R 308.02, yellow oil).

Lit.: Tetrahedron Lett. **33**, 7547 ff. (1992) (isolation); 7551 f. (synthesis). – *[CAS 124958-29-8 (N. A); 125002-00-8 (N. B)]*

Neocarzinostatins. Antitumor antibiotics produced by *Streptomyces carzinostaticus* [*N. A* (zinostatin): M_R ca. 11 000] composed of the globular, inactive protein consisting of 113 amino acids and a labile enediyne chromophore which is stabilized by the protein and has enhanced activity. The protein and chromophore can be separated and reconstituted to the active complex.

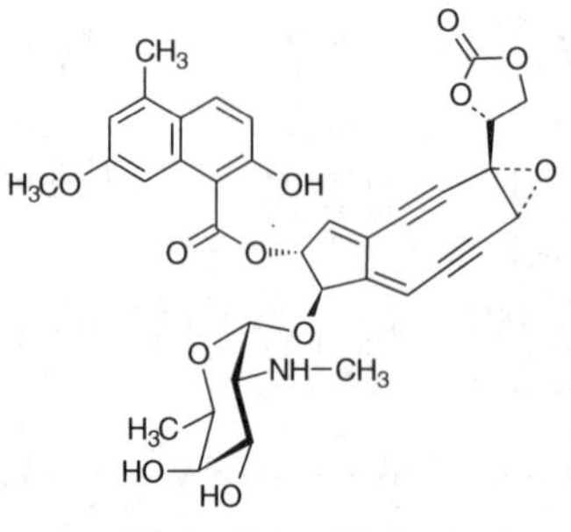
Chromophore of N. A

N. A chromophore: $C_{35}H_{33}NO_{12}$, M_R 659.65, $[\alpha]_D$ –171° (CH_3OH). The C_{14}-enediyne skeleton is formed biosynthetically from acetate (presumably via oleate), the naphthoic acid from acetate, it also contains *N*-methyl-D-fucosamine. The N. are suitable agents in the therapy for leukemia, stomach, bladder, and pancreatic cancers and are currently undergoing clinical testing. N. A. (zinostatin) has been available as a cancer drug since 1996, first in Japan. The N. exert their activities on DNA, on cycloaromatization the enediyne chromophore forms radicals that effect strand cleavage, preferably at the deoxyribose of thymine. The N. are rather toxic [LD_{50} mouse (i. v.) 0.96 mg/kg, (i. p.) 1.7 mg/kg, (s. c.) 7.2 mg/kg, (p. o.) 1.05 g/kg].

Lit.: J. Am. Chem. Soc. **110**, 7212 (1988) ▪ J. Antibiot. **39**, 1615 (1986). – *Biosynthesis:* J. Am. Chem. Soc. **111**, 3295 (1989). – *Review:* Angew. Chem. Int. Ed. Engl. **30**, 1387 (1991) ▪ Borders & Doyle (eds.), Enediyne Antibiotics as Antitumor Agents, New York: Dekker 1995 ▪ Edo, Koide: Neocarcinostatin **1997**, p. 23–45 (activity) (CAS **129**, 216814). – *Synthesis:* J. Am. Chem. Soc. **112**, 5369 (1990); **113**, 694 (1991); **118**, 10006 (1996); **120**, 5319 (1998) ▪ Lukacs, Recent Progress in the Chemical Synthesis of Antibiotics and Related Microbial Products, p. 293–330, Berlin: Springer 1993 ▪ Science **262**, 1042 (1993) (structure). – *Activity:* Cancer Res. **50**, 3897 (1990) ▪ Foye (eds.), Cancer Chemotherapeutic Agents, p. 617–625, Washington: ACS 1995 ▪ Krohn, Kirst & Maag, Antibiotics and Antiviral Compounds, p. 289–301, Weinheim: VCH Verlagsges. 1993 ▪ Nachr. Chem. Tech. Lab. **38**, 160 (1990) ▪ Proc. Natl. Acad. Sci. USA **90**, 5881 (1993). – *[HS 2941 90; CAS 9014-02-2 (N. A); 79633-18-4 (N. A chromophore)]*

Neoflavonoids. The name N. is used for a group of C_{15} compounds that are structurally and biogenetically related to the *flavonoids. Their biosynthesis is not known, it is possible that arylcoumarins act as central intermediates. The basic skeleton of the N. is that of neoflavan.

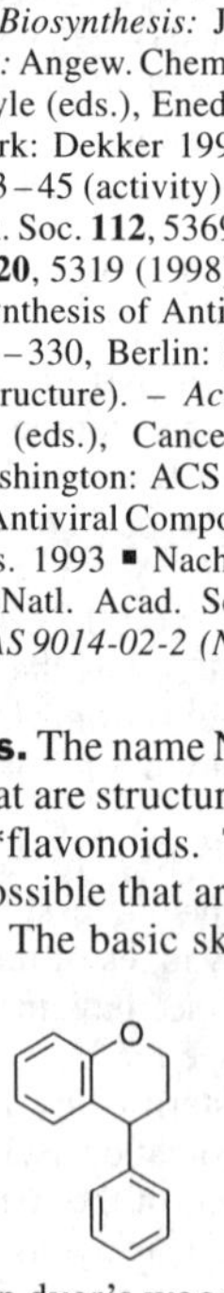
Neoflavan

The N. occur in dyer's woods, examples are *brazilin, the main pigment in red woods and the blue wood pigment *haematoxylin. They are colorless precursors of the neoflavonoid pigments. Brazilin is oxidized to brazilein (see brazilin) and hematoxylin to hematein (see haematoxylin), mordant dyes important in the past for coloring textiles.

Lit.: Harborne (1988, 1994) ▪ Nat. Prod. Rep. **12**, 321–338 (1995).

***A'*-Neogammaceranes** see hopanes.

Neoglucobrassicin see glucosinolates.

Neohesperidin see hesperetin.

(+)-Neoisomenthol see 3-*p*-menthanols.

(+)-Neoisopulegol see *p*-menthenols.

Neolaquinone see halenaquinone.

(+)-Neomenthol see 3-*p*-menthanols.

Neomycin.

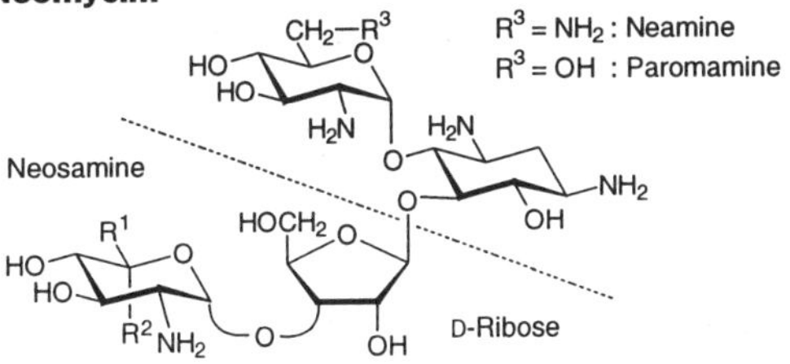

$R^1 = H$, $R^2 = CH_2NH_2$, $R^3 = NH_2$: Neomycin B
$R^1 = CH_2NH_2$, $R^2 = H$, $R^3 = NH_2$: Neomycin C
$R^1 = H$, $R^2 = CH_2NH_2$, $R^3 = OH$: Paromomycin I
$R^1 = CH_2NH_2$, $R^2 = H$, $R^3 = OH$: Paromomycin II

International free name for an aminoglycoside *antibiotic isolated from cultures of *Streptomyces fradiae*. The substance, first isolated by Waksman in 1949 is made up of three components: N. A, B (major components), and C. N. A is a degradation product of the two stereoisomers N. B and N. C.

Table: Data of Neomycins.

N.	molecular formula	M_R	$[\alpha]_D^{25}$	CAS
N. A (Neamine)	$C_{12}H_{26}N_4O_6$	322.36	+112.8° (H_2O)	3947-65-7
N. B (Framycetin, Streptothricin B_2, Dekamycin III)	$C_{23}H_{46}N_6O_{13}$	614.67	+83° (0.2 N H_2SO_4)	119-04-0
N. C	$C_{23}H_{46}N_6O_{13}$	614.67	+121° (0.2 N H_2SO_4)	66-86-4
Paromomycin I (Paromomycin)	$C_{23}H_{45}N_5O_{14}$	615.65	+65° (H_2O)	7542-37-2

N. is active against Gram-positive and Gram-negative bacteria through inhibition of protein synthesis by preventing the function of the 30S subunits. N.-resistant strains of bacteria produce aminoglycoside-modifying enzymes that lead to its inactivation (by phosphorylation, adenylation, acetylation). The biosynthesis of N. B and N. C starts from the separate formation of the individual subunits neosamine B or C, *neamine*, D-*ribose, and 2-deoxystreptamine which are subsequently linked by glycosidic bonds. The biogeneses of neosamine and 2-deoxystreptamine from D-*glucose proceed by two different routes.

Use: Not only the mixture (as neomycin sulfate) but also the sulfate of N. B (framycetin sulfate) are used therapeutically. Since N. and the closely related *paromomycin* (with the same activity properties as N.)[1] have proved to be toxic on parenteral administration, only topical (external) use against bacterial infections of the skin, eyes, ears, etc. is possible (also in veterinary medicine). Since N. is not absorbed, it can be used for intestinal disinfection prior to operations; $LD_{50\,(N.\,B)}$ (mouse s. c.) 220 mg/kg, (mouse p. o.) 1250 mg/kg.

Lit.: [1] Beilstein EV **18/11**, 72f.; J. Antibiot. **39**, 1598 (1986).
gen.: Beilstein EV **18/10**, 509 ▪ J. Chem. Soc., Chem. Commun. **1983**, 20 (biosynthesis) ▪ J. Antibiot. **45**, 984 (1992) (biosynthesis) ▪ Kirk-Othmer (4.) **2**, 904 ▪ Kleemann-Engel, p. 1450 ▪ Sax (8.), NCE 500. – *[HS 2941 90; CAS 1404-04-2]*

Neopine see morphinan alkaloids.

Neopterin.

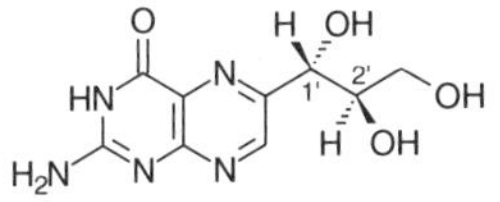

(1'*S*, 2'*R*-form)

$C_9H_{11}N_5O_4$, M_R 253.22, pale yellow crystals. N. belongs to the *pteridines and occurs in four stereoisomeric forms. The name N. was originally used for the (1'*S*,2'*R*)-form, the D-*erythro*-form, $[\alpha]_D$ +55° (0.1 M HCl) which, like the enantiomeric (1'*R*,2'*S*)-form, the L-*erythro*-form, occurs in human and primate urine. The other two stereoisomers also occur as natural products. N. is a precursor in the biosynthesis of *biopterin. Like all other pteridines, it is formed biosynthetically from guanosine triphosphate. The urine of patients with malignant tumors and virus infections, including AIDS, contains elevated concentrations of N.[1], presumably produced by human macrophages stimulated by γ-interferon. Very high concentrations of N. are often indicative of a poor prognosis for the clinical situation, they are also observed in cases of cerebral infections and multiple sclerosis. The N. values are highest in the acute phases of autoimmune diseases[2]. N. has, in general, major significance in clinical chemical analysis.

Lit.: [1] Hoppe Seyler's Z. Physiol. Chem. **373**, 1061 (1992). [2] Immunology Today **9**, 150–155 (1988).
gen.: Adv. Clin. Chem. **27**, 81–141 (1989) ▪ Beilstein EV **26/18**, 427f. – *[CAS 670-65-5 (N.); 2009-64-5 (1'S,2'R); 2277-43-2 (1'R,2'S)]*

Neopyrrolomycin.

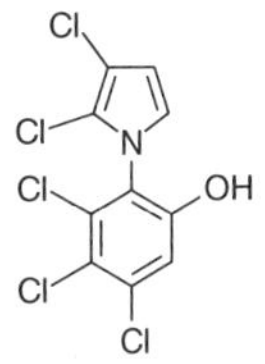

$C_{10}H_4Cl_5NO$, M_R 331.41, pale yellow syrup, $[\alpha]_D$ +40° ($CDCl_3$); K salt: dihydrate, needles, mp. 70–72 °C, $[\alpha]_D$ –30° (CH_3OH). A halogenated pyrrole derivative from cultures of a *Streptomyces* species occurring in optically active form as a result of hindered rotation about the *N*-aryl bond. It is active against Gram-positive and Gram-negative bacteria as well as some fungi.

Lit.: Bull. Chem. Soc. Jpn. **67**, 1449 (1994) (synthesis) ▪ J. Antibiot. **43**, 1192f. (1990) ▪ Tetrahedron Lett. **34**, 8443 (1993). – *[HS 2941 90; CAS 131956-34-8]*

Neosolaniol (solaniol).

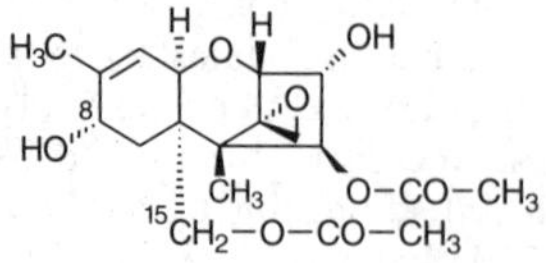

$C_{19}H_{26}O_8$, M_R 382.41, mp. 171–172 °C, a *mycotoxin of the *tric(h)othecene type formed by various *Fusarium* species (including *F. culmorum, F. solani, F. poae*). In ducklings N. causes vomiting (0.2 mg/kg s.c.), is a skin irritant, LD_{50} (mouse i.p.) 14.5 mg/kg, in chick embryos 5 µg/egg. N. inhibits protein biosynthesis in reticulocytes above 0.25 mg/L. In addition to N., *F. sporotrichioides* forms a series of derivatives, e.g., N. 8-acetate, propanoate, isobutyrate, butanoate, and hexanoate.

Lit.: Betina, chap 10 ▪ Chelkowski (ed.), Fusarium – Mycotoxins, Taxonomy and Pathogenicity, Amsterdam: Elsevier 1989 ▪ Cole-Cox, p. 175–177 ▪ J. Nat. Prod. **50**, 953 (1987); **54**, 1303 (1991) ▪ Sax (8.), NCK 000. – *[CAS 36519-25-2]*

Neosurugatoxin.

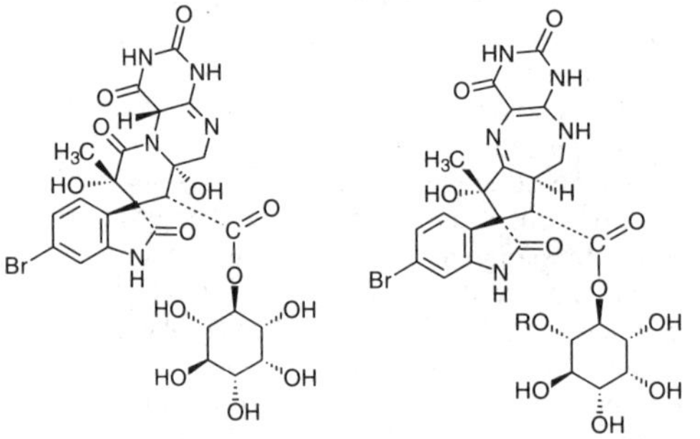

Surugatoxin

R = β-D-Xylopyranosyl : Neosurugatoxin
R = H : Prosurugatoxin

$C_{30}H_{34}BrN_5O_{15}$, M_R 784.53. N. and *prosurugatoxin* ($C_{25}H_{26}BrN_5O_{11}$, M_R 652.41) are water-insoluble, toxic alkaloids from the Japanese ivory snail *Babylonia japonica*. The actual producers of the toxins are coryneform bacteria[1] living in the digestive organs of the snails. Poisonings have occurred after consumption of snails from particular regions, the symptoms are unquenchable thirst, numbness in the lips, reduction of the field of vision, speech impairments, and dysuria. N. and prosurugatoxin are about 5000-fold more active as ganglion-blockers than the drugs mecamylamine or hexamethonium. Since they specifically inhibit nicotine receptors, they are excellent tools for investigations on the nervous system and brain. N. and prosurugatoxin rearrange to the less toxic *surugatoxin* ($C_{25}H_{26}BrN_5O_{13}$, M_R 684.41, prisms, mp. >300 °C). The cyclopentanimine structure of N. is oxidatively opened in surugatoxin and recyclized to the piperidone.

Lit.: [1] Chem. Pharm. Bull. **33**, 3059 (1985); Chem. Rev. **93**, 1685 (1993).
gen.: Pharmacology: Chem. Pharm. Bull. **30**, 3255 (1982) ▪ J. Neurochem. **42**, 1491 (1984) ▪ Methods Neurosci. **8**, 311 (1992) ▪ Tetrahedron Lett. **27**, 5225 (1986). – *Synthesis:* Chem. Pharm. Bull. **37**, 791 (1989) ▪ Tetrahedron **50**, 2729, 2753 (1994) ▪ Tetrahedron Lett. **27**, 5225 (1986); **29**, 1547 (1988). – *[CAS 80680-43-9 (N.); 40957-92-4 (surugatoxin); 99102-40-6 (prosurugatoxin)]*

(1*S*,3*R*,4*R*,5*R*)(–)-Neothujol see thujanols.

Neotigogenin see steroid sapogenins.

Neoxanthin [(3*S*,3′*S*,5*R*,5′*R*,6*R*,6′*S*,9′*Z*)-6,7-didehydro-5′,6′-epoxy-5,5′,6,6′-tetrahydro-*β*,*β*-carotene-3,3′,5-triol].

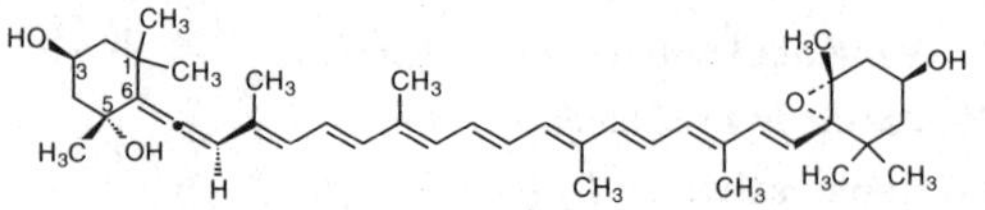

$C_{40}H_{56}O_4$, M_R 600.88, mp. 200 °C. N. belongs to the xanthophylls (see carotinoids) and occurs in *Capsicum annuum* (Solanaceae), *Medicago sativa* (Fabaceae), *Acer* species (Aceraceae), and *Trollius europaeus* (Ranunculaceae). N. is thought to have two key roles: as part of light-harvesting complexes (photosynthetic proteins of higher plants) and as a precursor to the plant growth hormone *abscisic acid. In the parasitic plant *Cuscuta reflexa*, N. is replaced by lutein 5,6-epoxide[1].

Lit.: [1] Proc. Natl. Acad. Sci. USA **96**, 1135 (1999).
gen.: Britton, Liaaen-Jensen & Pfander, Carotenoids, vols. 1 & 2, Basel: Birkhäuser 1995 ▪ Helv. Chim. Acta **75**, 773 (1992); **77**, 909 (1994) (synthesis) ▪ Phytochemistry **21**, 2859 (1982) (biosynthesis). – *[CAS 30743-41-0]*

Nepetalactone see iridodial.

Nephilatoxins see JSTX.

Neplanocins.

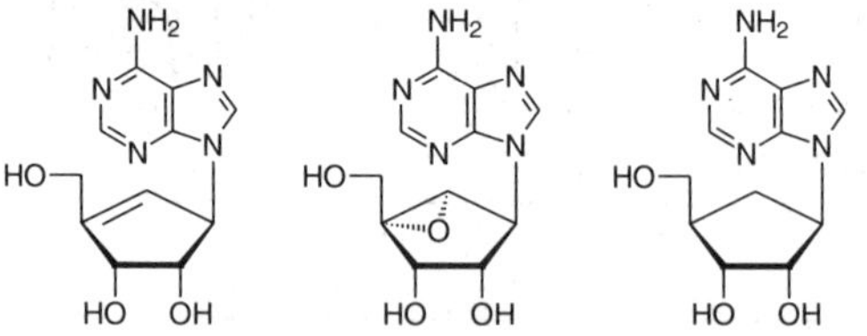

N.A N.C Aristeromycin

Carbocyclic *nucleoside antibiotics from *Ampullariella regularis* cultures exhibiting antifungal and antiviral properties[1]. Thus, *N. C* {$C_{11}H_{13}N_5O_4$, M_R 279.26, mp. 226 °C (decomp.), $[\alpha]_D$ –43.6°} has been developed as a drug against HIV and Epstein-Barr virus. *N. A* ($C_{11}H_{13}N_5O_3$, M_R 263.26, mp. 220–222 °C, $[\alpha]_D$ –157°) additionally shows strong antitumor activity. It exists in the cell in the non-phosphorylated form. The intact molecule inhibits **S*-adenosylmethionine hydrolase (EC. 3.3.1.2) and accordingly, through the accumulation of *S*-adenosylhomocysteine, inhibits cellular and viral methyltransferases in the synthesis of macromolecules[2]. For synthesis, see *Lit.*[3]. The good activity of N results from the double bond or, respectively, the epoxide unit, as shown by a comparison with *aristeromycin*[4] ($C_{11}H_{15}N_5O_3$, M_R 265.27) which possesses a saturated ring system. For biosynthesis, see *Lit.*[5].

Lit.: [1] Nucleic Acids Symp. Ser. **8**, 565 (1980); J. Antibiot. **34**, 359 (1981). [2] J. Antibiot. **41**, 1711 (1988); J. Med. Chem. **39**, 3847 (1996); Nucleosides, Nucleotides **15**, 1157 (1996). [3] J. Am. Chem. Soc. **105**, 4059 (1983); Tetrahedron Lett. **28**, 4131 (1988); **36**, 1537 (1995); **38**, 1707 (1997); **39**, 4677 (1998); J.

Chem. Soc., Perkin Trans. 1 **1988**, 3133; J. Org. Chem. **55**, 4712 (1990); **57**, 2071 (1992); Angew. Chem. Int. Ed. Engl. **29**, 99 (1990); Tetrahedron **48**, 571 (1992); Synlett **1994**, 491. [4]Bioorg. Med. Chem. Lett. **7**, 247 (1997); J. Org. Chem. **62**, 3976 (1997); **63**, 5447 (1998); J. Chem. Soc., Chem. Commun. **1999**, 807; Ullmann (5.) **A 2**, 493. [5]J. Am. Chem. Soc. **117**, 5391 (1995). – *[CAS 121548-44-5 (N.); 72877-50-0 (N. A); 72877-48-6 (N. C); 19186-33-5 (aristeromycin)]*

Neral see citrals.

Neral (2*S*,3*S*)-epoxide see mites.

Nereistoxin.

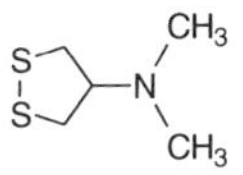

$C_5H_{11}NS_2$, M_R 149.27, odorless oil, bp. 212–213 °C. The 1,2-dithiolane N. occurs in the approximately 40 cm long marine annelid *Lumbriconereis heteropoda* used by anglers as bait. It has long been known that insects such as ants or flies die after contact with the worm. N. was identified as the toxic principle. Injections of the neurotoxin cause myosis, an elevated tone in smooth muscles, lacrimation, and tremor, LD_{50} (mouse p. o.) 38 mg/kg, (rabbit s. c.) 180 mg/kg. As a result of its interesting activity, N. has been a lead structure for the synthesis of insecticides (e. g., Cartap, Padane®, which gets metabolized to N., or Bensultap, Bancol®).

Lit.: Beilstein EV **19/8**, 347 ▪ Chem. Pharm. Bull. **13**, 253 (1965) (synthesis) ▪ J. Pharm. Technol. **3**, 136 (1987) ▪ Tu, Marine Toxins and Venoms, p. 366, New York: Dekker 1988. – *[CAS 1631-58-9]*

Neristatin 1 see bryostatins.

Nerol see geraniol.

Nerolic acid see geranic acid.

Nerolidol, neroli oil see orange flower absolute (oil).

Nervonic acid see erucic acid.

Neuraminic acid (5-amino-3,5-dideoxy-D-*glycero*-D-*galacto*-2-nonulosonic acids, abbreviation: Neu).

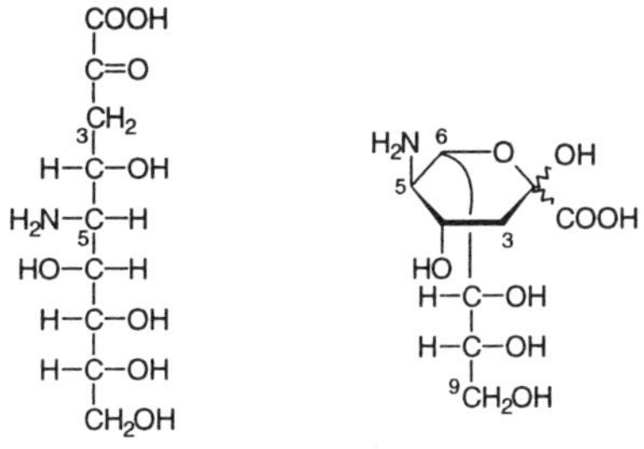

openchain form pyranoside form

$C_9H_{17}NO_8$, M_R 267.25. An amino sugar acid isolated from gangliosides and named by Klenk. N. is unstable and cyclizes spontaneously to furnish a pyrroline derivative. However, it is widely distributed in *N*- and *O*-acylated forms. *N-Acetylneuraminic acid* (Neu5Ac, Neu*N*Ac, NANA): $C_{11}H_{19}NO_9$, M_R 309.27, mp. 185–187 °C, $[\alpha]_D$ −33° (H_2O).

Collectively, these derivatives are known as acylneuraminic acids (*sialic acids*); they are, mostly glycosidically bound – the bond can be cleaved by neuraminidase – components of many bacterial and animal gly-

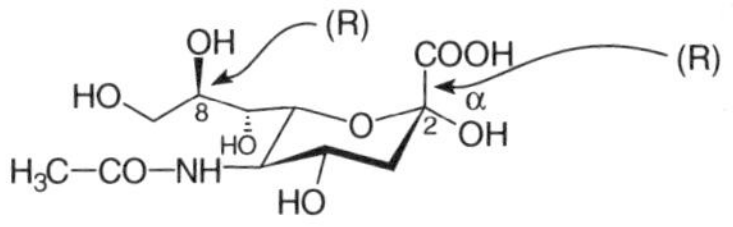

Neu5Ac

colipids, *glycoproteins, and glycosaminoglycans, where they prevail at the non-reducing end.

Isolation: From eggs, milk, swallow's nests, submandibular mucins of various mammalian and bacterial N. polymers (colomic acids). Their biochemical formation involves enzymatic aldol addition of *N*-acetylmannosamine with pyruvic acid and transfer to the growing carbohydrate chains of the glycoproteins and glycolipids by sialyltransferases. For synthesis, see *Lit.*[1,2].

Lit.: [1] Justus Liebigs Ann. Chem. **616**, 221–225 (1958). [2] Angew. Chem. Int. Ed. Engl. **30**, 827–8 (1991).

gen.: Adv. Carbohydr. Chem. Biochem. **40**, 131–234 (1981) ▪ Ann. N. Y. Acad. Sci. **1995**, 300–305 ▪ Chem. Commun. **2000**, 227 (synthesis NANA) ▪ Curr. Med. Chem. **4**, 185–210 (1997) (review) ▪ Schauer, Sialic Acids: Chem., Metab., Function, p. 346 ff. Berlin: Springer 1982. – *[CAS 114-04-5 (N.); 131-48-6 (N-acetyl-N.)]*

Neurolathyrism see 3-cyano-L-alanine.

Neurosporene (7,8-dihydrolycopin; 7,8-dihydro-ψ,ψ-carotene).

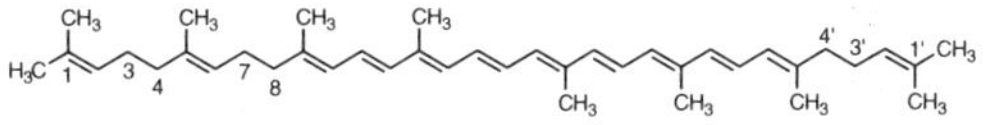

$C_{40}H_{58}$, M_R 538.90, orange-yellow cryst., mp. 124 °C. N. belongs to the *carotenes and occurs not only in plants, e. g,. in *Ananas sativus* (syn. *A. comosus*, Bromeliaceae), but also in microorganisms, e. g., in *Neurospora crassa* and *Rhodotorula rubra*.

Lit.: Britton, Liaaen-Jensen & Pfander, Carotenoids, vols. 1 and 2, Basel: Birkhäuser 1995 ▪ Karrer, No. 18176 ▪ Phytochemistry **23**, 1707 (1984). – *[CAS 502-64-7]*

Neurotoxins. General name for compounds that damage nerves. Natural N. include reptilian and *amphibian venoms, such as *snake venoms (*crotoxin, *cobratoxins), *frog toxins (*batrachotoxins, *pumiliotoxins), *spider venoms (*argiopinins, *argiotoxins, *JSTX, latrotoxin), *bee venom, *wasp venom (*apamin, *philanthotoxin), *scorpion venoms (*margatoxin), fungal toxins (muscarine, *ergot alkaloids, pentitrems), bacterial toxins (tetanus toxin, *botulism toxin), plant toxins, and toxins from marine organisms (*anatoxins A, *saxitoxin, *tetrodotoxin).

Lit.: Chubb & Geffen, Neurotoxins: Fundamental and Clinical Advances, Adelaide: Union Press 1979 ▪ Oliver, Neurotoxins in Neurochemistry, Chichester: Ellis Horwood 1988 ▪ Tu, Handbook of Natural Toxins, New York: M. Dekker 1982–1991.

Niazinins. Thiocarbamates from the South and South-East Asian plant *Moringa oleifera* ("sahjna"), Morin-

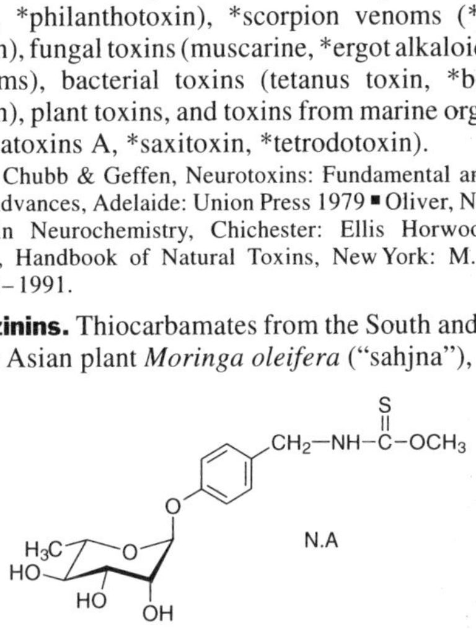

N.A

gaceae, which finds wide use as a vegetable. In addition to flavonoids, steroids, and other components, the plants contain the N., hypotensive mustard oil glycosides (*glucosinolates), e.g., *N. A:* $C_{15}H_{21}NO_6S$, M_R 343.38, oil.

Lit.: J. Chem. Soc., Perkin Trans. 1 **1992**, 3237. – *[CAS 148152-07-2]*

Nicandrenone see withanolides.

Nicotiana alkaloids see tobacco alkaloids.

Nicotianamine {(*S*)-1-[(*S*)-3-((*S*)-3-amino-3-carboxypropylamino)-3-carboxypropyl]-2-azetidinecarboxylic acid}.

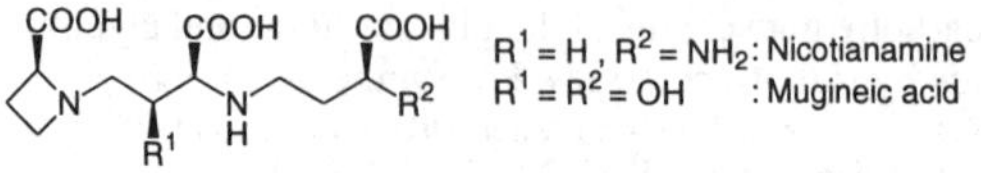

$C_{12}H_{21}N_3O_6$, M_R 303.32, monohydrate, mp. >240 °C, $[\alpha]_D^{23}$ –60.5° (H_2O), pK_a 6.97, 9.13, 9.75 (0.1 m $KClO_4$, 25 °C). A component of tobacco leaves (*Nicotiana tabacum*)[1], seeds of *Fagus sylvatica* and related species[2]. N. regulates the biosynthesis of chlorophyll and is possibly involved in iron transport in plants[3], like *mugineic acid* ($C_{12}H_{20}N_2O_8$, M_R 320.30)[4] which is excreted from oat and rice roots, N. binds iron ions, and transports them into the plant.

Lit.: [1]Tetrahedron Lett. **1971**, 2017–2019. [2]Phytochemistry **13**, 2791–2798 (1974). [3]Phytochemistry **19**, 2295–2297 (1980). [4]Proc. Jpn. Acad. Ser. B. **54**, 469–473 (1978).
gen.: Beilstein E V **22/1**, 21 ▪ Experientia **40**, 794 (1984) (review) ▪ Hager (5.) **5**, 723 ▪ Heterocycles **44**, 519–530 (1997) (synthesis) ▪ Nachr. Chem. Tech. Lab. **37**, 1577 ff. (1990) (review) ▪ Tetrahedron Lett. **39**, 89 (1998) (synthesis) ▪ Z. Chem. **28**, 336 (1988) (synthesis). – *[CAS 34441-14-0 (N.); 69199-37-7 (mugineic acid)]*

Nicotianine [1-((*S*)-3-amino-3-carboxypropyl)-3-carboxypyridinium betaine].

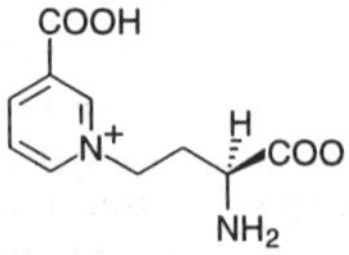

$C_{10}H_{12}N_2O_4$, M_R 224.22, mp. 241–243 °C (decomp.), $[\alpha]_D^{24}$ +24° (H_2O). A non-proteinogenic amino acid from tobacco leaves[1]; it also occurs in the fruit body of the Japanese edible fungus shii-take (*Lentinus edodes*)[2].

Lit.: [1]Phytochemistry **7**, 1861–1866 (1968). [2]Agric. Biol. Chem. **41**, 213–214 (1977). – *[CAS 19701-79-2]*

Nicotimine see anabasine.

Nicotinamide (3-pyridinecarbamide, PP factor, anti-pellagra factor, vitamin B_3).

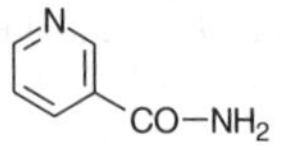

$C_6H_6N_2O$, M_R 122.13, mp. 129–130 °C, bp. 150–160 °C (0.07 Pa), needles. N. occurs in plants, yeasts, and fungi and is ubiquitous as a component of nicotinamide adenine dinucleotide (see nicotinamide coenzymes).

Use: As additive in food and fodder to combat N.-deficiency diseases such as dermatitis and pellagra. Also as reference substance in elemental analysis. *N-Hydroxynicotinamide* (*nicoxamate*, $C_6H_6N_2O_2$, M_R 138.13, prisms, mp. 168 °C, pK_a 8.3) and *N-(hydroxymethyl)nicotinamide* ($C_7H_8N_2O_2$, M_R 152.15, mp. 141 °C) have antimicrobial and anti-inflammatory activities and are used in liver therapy.

Biosynthesis: In animals: *tryptophan → *kynurenine → 3-hydroxyanthranilic acid → quinolinic acid (pyridine-2,3-dicarboxylic acid) → nicotinic acid → nicotinamide. In bacteria and higher plants from L-*aspartic acid and a C_3-unit, probably D-glyceraldehyde 3-phosphate. Also by cleavage of NAD.

Lit.: Beilstein E V **22/2**, 80 ff. ▪ Luckner (3.), p. 295–296 ▪ Negwer (6.), No. 410 ▪ Ullmann (4.) **23**, 706–712; (5.) **A12**, 5146 (*N*-hydroxymethyl-N.). – *[HS 2936 29; CAS 98-92-0 (N.); 5657-61-4 (N-hydroxy-N.); 3569-99-1 (N-hydroxymethyl-N.)]*

Nicotinamide coenzymes. Collective name for the *coenzymes of numerous hydrogen-transferring enzymes occurring in all animal and plant cells. They are composed of adenine, *nicotinamide, D-*ribose, and phosphate. The hydrogen transfer proceeds according to the reaction shown in the figure, always highly stereoselectively, i.e., either the pro-*R* or the pro-*S* hydrogen atom is transferred (H_{Re} or H_{Si}).

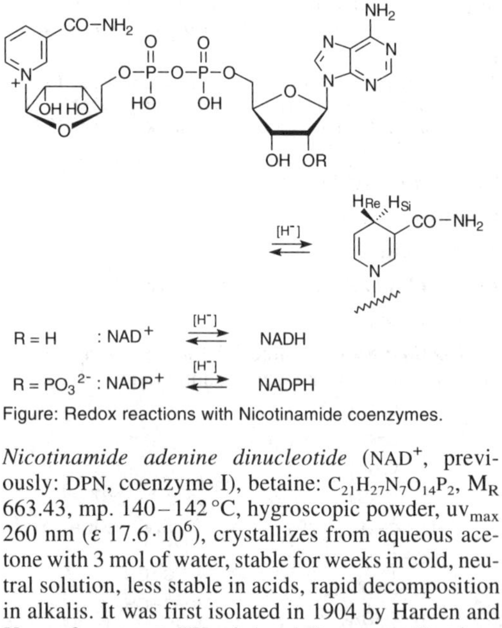

Figure: Redox reactions with Nicotinamide coenzymes.

Nicotinamide adenine dinucleotide (NAD^+, previously: DPN, coenzyme I), betaine: $C_{21}H_{27}N_7O_{14}P_2$, M_R 663.43, mp. 140–142 °C, hygroscopic powder, uv_{max} 260 nm (ε 17.6 · 10^6), crystallizes from aqueous acetone with 3 mol of water, stable for weeks in cold, neutral solution, less stable in acids, rapid decomposition in alkalis. It was first isolated in 1904 by Harden and Young from yeast ("Harden and Young ester"), where its content is ca. 25 mg/kg. – *Dihydronicotinamide adenine dinucleotide* (NADH, DPNH), $C_{21}H_{29}N_7O_{14}P_2$, M_R 665.45, readily soluble in water, yellowish powder, uv_{max} 340 nm (ε 6.2 · 10^6). NADH is easily autoxidized. – *Nicotinamide adenine dinucleotide phosphate* ($NADP^+$, previously: TPN, coenzyme II), betaine: $C_{21}H_{28}N_7O_{17}P_3$, M_R 743.41, pK_{a1} 3.9, pK_{a2} 6.1, gray-white powder, uv_{max} 260 nm (ε 18 · 10^6); widely distributed, especially in liver and red blood particles. First isolated in 1931 by Warburg. Properties similar to those of NAD. – *Dihydronicotinamide adenine dinucleotide phosphate* (NADPH, TPNH), $C_{21}H_{30}N_7O_{17}P_3$, M_R 745.43, uv_{max} 340 nm (ε 6.2 · 10^6).

Biosynthesis: Phosphoribose pyrophosphate + nicotinic acid → nicotinic acid ribonucleotide. Coupling with ATP under cleavage of diphosphate furnishes deamido-NAD^+. The amide nitrogen originates from glutamine.

Lit.: ACS Symp. Ser. **185**, 205 (1982) (enzymatic synthesis, review) ▪ Annu. Rev. Biochem. **47**, 881–931 (1978) ▪ J. Am. Chem. Soc. **106**, 234–239 (1984) (synthesis) ▪ J. Chem. Soc., Chem. Commun. **1999**, 729 ▪ Methods Enzymol. **66**, 11 (1980) ▪ Stryer 1995. – *[CAS 53-84-9 (NAD+); 58-68-4 (NADH); 53-59-8 (NADP+); 53-57-6 (NADPH)]*

Nicotine [3-((*S*)-1-methyl-2-pyrrolidinyl)pyridine].

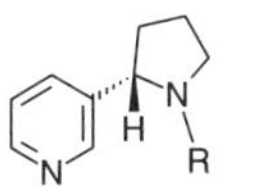

(–)-(*S*)-form
R = CH_3 : Nicotine
R = H : Nornicotine

$C_{10}H_{14}N_2$, M_R 162.23, hygroscopic oil, bp. 246–247 °C, steam distillable. $[\alpha]_D$ –166 → –169°. The *tobacco alkaloid has a pyridine-like odor and a burning, scratching taste; it forms salts with many acids and also double salts with metal salts (e.g., zinc chloride double salt monohydrate). N. occurs in widely differing concentrations in tobacco plants (*Nicotiana* sp.), as well as in other Solanaceae species and numerous plants of other families [including *Acacia* species (Fabaceae), *Sedum* species (Crassulaceae), *Equisetum* species (Equisetaceae), *Lycopodium* species (Lycopodiaceae)] as a trace component. Virginia tobacco contains ca. 0.05% N., the strong "Burley" variety 3–4%, the parent tobacco plant of the Russian "machorka" up to 7.5%. N. is enzymatically degraded to *nornicotine in some tobacco species. N. is a potent human poison when swallowed, on contact with the skin, or on subcutaneous [LD_{50} (rat s.c.) 50 mg/kg], parenteral, and intravenous [LD_{50} (mouse i.v.) 300 µg/kg] administration, it was found to be teratogenic in animal experiments[1]. The lethal dose on oral administration for adults is estimated to be 40–60 mg, the maximum allowed working place concentration is 0.5 mg/m^3 or 0.07 ppm.

Activities[2]***:*** *On the central nervous system:* At first central stimulation followed by rapid paralysis of the medulla oblongata and m. spinalis, death results through respiratory paralysis. This occurs very rapidly at sufficiently high doses. *Peripheral activities:* In small doses like acetylcholine, in higher doses as a ganglion blocker, hypersecretion of the body glands, weakening of cardiac activity, constriction of the coronary vessels, stimulation of peristalsis, increase in blood pressure, uterine contractions (especially during pregnancy). The stimulating and concomitant calming effects of N. are probably the main reasons for tobacco smoking[3]. N. shows therapeutic potential for therapy of brain disorders[4]. N. also has strong insecticidal and anthelmintic activities and was thus used in the early days of plant protection in the 19th century as a pesticide. N. shows activity in treatment of inflammatory bowel diseases (Crohns disease, ulcerative colitis)[5].

Historical: The name is due to the French diplomat Jean Nicot (1530–1600) who introduced tobacco to Europe from America as a purported medicinal plant. For biosynthesis, see tobacco alkaloids.

Lit.: [1] Sax (8.), p. 2500 f. [2] Ann. N. Y. Acad. Sci. **562**, 211–240 (1989). [3] Br. Assoc. Psychopharmacol. Monogr. **1988**, 27–49; Chem. Ind. (London) **1994**, 221–224. [4] Chem. Eng. News (27. 3.) **2000**, 23–26. [5] Pat. WO 9728801 (7. 2. 1997).
gen.: Beilstein E III/IV **23**, 999 ff.; E V **23/6**, 64–68 ▪ Gorrod (ed.), Nicotine and Related Alkaloids, London: Chapman & Hall 1993 ▪ Hager (5.) **3**, 870 ff. ▪ Manske **12**, 522 f.; **26**, 121–151 ▪ Ullmann (5.) **A 1**, 360; **A 14**, 271 ▪ Zechmeister **28**, 109–161; **34**, 44–55. – *Pharmacology:* Martin et al., Advances in Behavioral Biology, vol. 31, Tobacco Smoking and Nicotine, A Neurobiological Approach, New York: Plenum Press 1987 ▪ Med. Chem. Res. **2**, 509, 522 (1993) ▪ Prog. Brain Res. (Nicotine Receptors CNS) **79** (1989) (entire vol.) ▪ Prog. Drug Res. **33**, 9–41 (1989) ▪ Rand & Thurau, The Pharmacology of Nicotine, Oxford: IRL Press 1988. – *[HS 2939 70; CAS 54-11-5 (N.); 25162-00-9 (R-form); 22083-74-5 (racemate)]*

Nicotyrine [3-(1-methyl-2-pyrrolyl)-pyridine, *β*-nicotyrine].

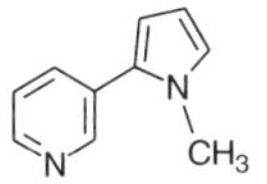

$C_{10}H_{10}N_2$, M_R 158.20, easily decomposing oil, bp. 281 °C, 150–151 °C (2.1 kPa). N. is a *tobacco alkaloid but differs from the most important member of this class, *nicotine, in that the pyrrolidine ring is aromatized to a pyrrole ring. It is probably first formed in tobacco leaves after harvesting.

Lit.: Beilstein E V **23/7**, 227 ▪ Hager (5.) **3**, 871 ▪ J. Org. Chem. **54**, 2476 (1989); **56**, 1822 (1991) (synthesis) ▪ Manske **26**, 140 ▪ Pelletier **1**, 85–152; **3**, 19–34 ▪ Sax (8.), p. 2503 ▪ Ullmann (5.) **A 1**, 360. – *[HS 2933 39; CAS 487-19-4]*

Nidulal.

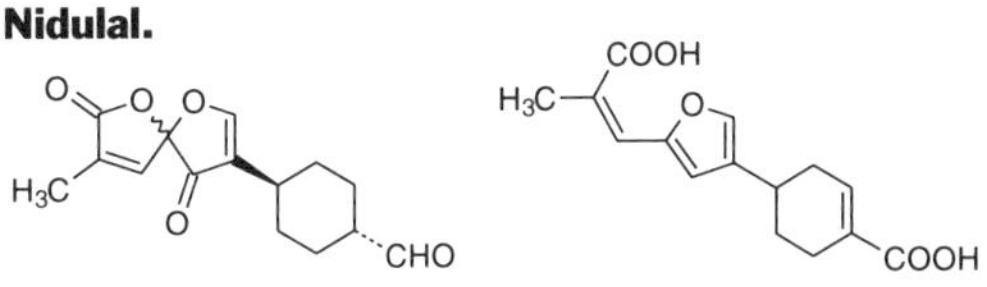

Nidulal (probably racemic)

Niduloic acid [(–)-form]

$C_{15}H_{16}O_5$, M_R 276.28, oil, $[\alpha]_D^{22}$ 0° ($CHCl_3$). N. is an inducer of differentiation of human HL-60 promyelocytic leukemia cells. It was isolated from fermentations of the basidiomycete *Nidula candida*, together with *niduloic acid*: $C_{15}H_{16}O_5$, M_R 276.28, yellow oil, $[\alpha]_D^{22}$ –53° ($CHCl_3$). The name niduloic acid is also used for (+)-3-hydroxy-5-(4-hydroxyphenyl)pentanoic acid.

Lit.: J. Antibiot. **49**, 1189–1195 (1996) (isolation). – *[CAS 185853-14-9 (N.); 185853-15-0 (niduloic acid)]*

Niduloic acid see nidulal.

Nigakilactone D see quassin(oids).

Nigellidine.

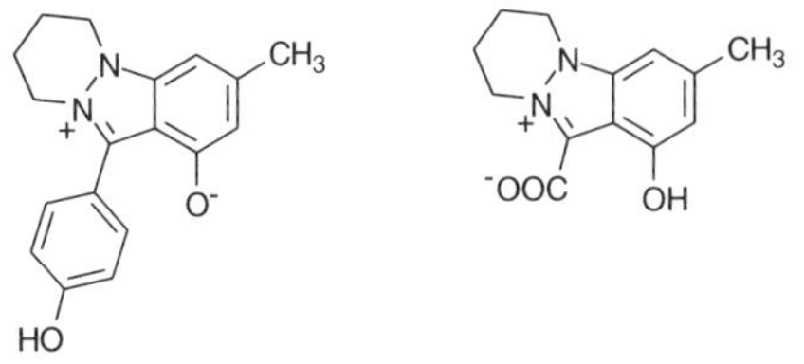

Nigellidine Nigellicine

$C_{18}H_{18}N_2O_2$, M_R 294.35. An unusual indazole alkaloid from the seeds of the genuine nutmeg (*Nigella sativa*, Ranunculaceae)[1] used in the past as a spice and herbal remedy. Oxidation of the 4-hydroxyphenyl ring to a carboxy group gives *nigellicine* ($C_{13}H_{14}N_2O_3$, M_R 246.27, yellow cryst.)[2].

Lit.: [1]Tetrahedron Lett. **36**, 1993 (1995). [2]Tetrahedron Lett. **26**, 2759 (1985). – *[HS 2939 90; CAS 120993-86-4 (N.); 98063-20-8 (nigellicine)]*

Nigellinic acid see abscisic acid.

Nigericin (polyetherin A, azalomycin M, helixin C).

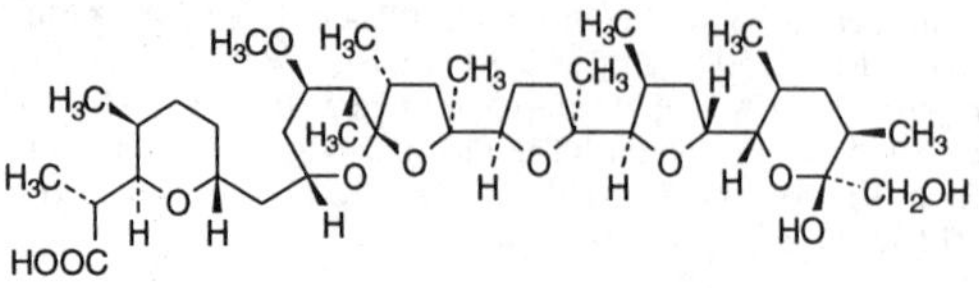

$C_{40}H_{68}O_{11}$, M_R 724.97, mp. 183.5–185 °C. A *polyether antibiotic from cultures of *Streptomyces hygroscopicus* and other streptomycetes. N. acts as an ionophore for monovalent cations in mitochondria, is active against Gram-positive bacteria and is used as a coccidiostatic agent

Lit.: Beilstein E V **19/12**, 552 ▪ J. Antibiot. **48**, 1011 ff. (1995) (biosynthesis) ▪ J. Org. Chem. **54**, 98–108 (1989) (synthesis). – *[HS 2941 90; CAS 28380-24-7]*

Nikkomycins.

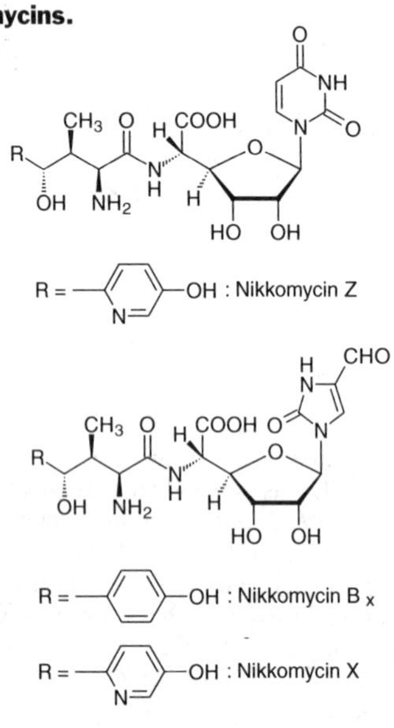

Table: Data of Nikkomycins.

	molecular formula	M_R	mp. [°C]	$[\alpha]_D^{20}$ (H_2O)	CAS
N. Z	$C_{20}H_{25}N_5O_{10}$	495.45	194–197	+49.5°	59456-70-1
N. B_x	$C_{21}H_{26}N_4O_{10}$	494.46		+56.7°	75410-71-8
N. X	$C_{20}H_{25}N_5O_{10}$	495.45	204 (decomp.)	+24°	72864-26-7

A group of aminoacylamino-*nucleoside antibiotics (more than 25 components) from cultures of *Streptomyces tendae*, see also polyoxins. The N. exhibit acaricidal, insecticidal, and antifungal activities. The activity is based on an inhibition of *chitin biosynthesis as a result of the structural analogy to UDP-*N*-acetyl-D-glucosamine. Synergistic effects against pathogenic yeasts and primary pathogenic fungi have been observed with antimycotically active azoles[1]. N. Z is in phase I of clinical development[2].

Lit.: [1]Antimicrob. Agents Chemother. **34**, 587–593 (1990); Clin. Dermatol. **7**, 325 (1993); J. Infect. Dis. **157**, 212 ff. (1988). [2]Curr. Opin. Antiinfect. Invest. Drugs **1**, 346 (1999).
gen.: Agric. Biol. Chem. **44**, 1709 (1980) ▪ Biopolymers **29**, 1297 (1990) (configuration) ▪ Chem. Commun. **1997**, 2085 (analogues) ▪ J. Antibiot. **42**, 711–717, 913–918 (1989); **43**, 43–48 (1990); **52**, 582 (1999) (isolation) ▪ Pflanzenschutz-Nachr. Bayer **38**, 305–348 (1985) ▪ Ullmann (5.) **A 2**, 492. – *Synthesis:* Curr. Pharm. Des. **5**, 73–99 (1999) ▪ J. Org. Chem. **61**, 1192 f. (1996) ▪ Justus Liebigs Ann. Chem. **1985**, 2165–2177; 2465 ff.; **1987**, 803–807 ▪ Tetrahedron Lett. **34**, 3267 (1993). – *Biotechnological synthesis:* J. Antibiot. (Tokyo) **52**, 102 (1999). – *[CAS 86003-55-6 (group)]*

Nimbin, nimbiol see neem tree.

Ningalin C, D see purpurone.

Ningalins see lamellarins.

Nisin.

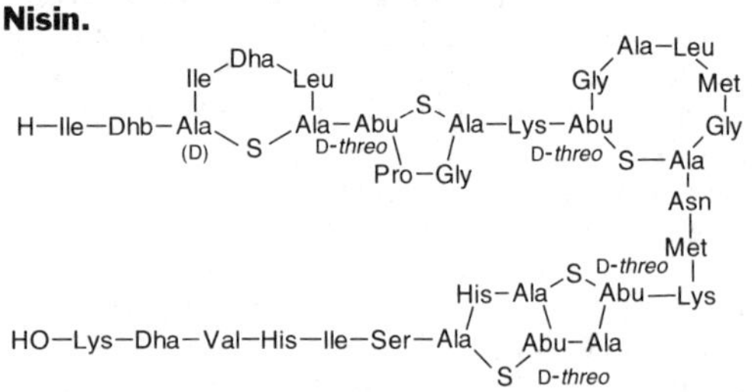

Abu : 2-Aminobutyric acid
Dha : Dehydroalanine (2-Aminoacrylic acid)
Dhb : Dehydrobutyrin (2-Amino-2-butenoic acid)

$C_{143}H_{230}N_{42}O_{37}S_7$, M_R 3354.08; a high-molecular weight peptide antibiotic, resistant to heat in acidic media; its 34 amino acids are linked to form cyclic structural units by 5 disulfide bonds.

N. is formed in lactic bacteria such as *Streptococcus lactis* and *S. cremoris*. In contrast to smaller peptide antibiotics, the biosynthesis occurs at the ribosomes. N. is active against numerous Gram-positive bacteria, especially *Bacillus* and *Clostridium*; it reduces the heat resistance of bacterial spores by attacking the cytoplasmic membrane directly after germination of the spores.

Use: In some countries (e.g., USA) N. is used for food conservation, especially in cheese production but also for fruit and vegetable products. Human gastrointestinal proteases hydrolyse N. completely, the Gram-negative strains of intestinal flora are not damaged.

Lit.: Adv. Appl Microbiol. **27**, 85 (1981) ▪ Eur. J. Biochem. **201**, 581 (1991) ▪ Experientia **44**, 266 (1988) ▪ J. Biol. Chem. **263**, 16260 (1988) ▪ J. Food Sci. Technol. **20**, 231 (1983) ▪ Kirk-Othmer (4.) **3**, 270, 285, 295 ▪ Tetrahedron Lett. **29**, 795 (1988). – *Review:* Angew. Chem. Int. Ed. Engl. **30**, 1051 (1991) ▪ Food Sci. **10**, 327 (1983). – *Synthesis:* Bull. Chem. Soc. Jpn. **65**, 2227–2240 (1992). – *[HS 2941 90; CAS 1414-45-5]*

Nitidine see benzo[*c*]phenanthridine alkaloids.

Nitidone.

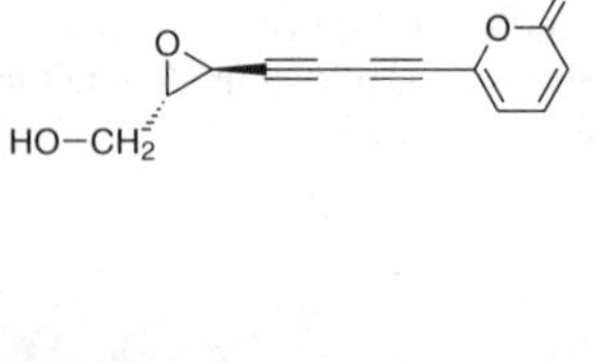

(relative configuration)

$C_{12}H_8O_4$, M_R 216.19, cryst., mp. 115–117 °C, $[\alpha]_D$ –34° ($CHCl_3/CH_3OH$). A pyranone derivative from cultures of the basidiomycete *Junghuhnia nitida*. N. exhibits antibiotic and cytotoxic activities and induces morphological and physiological differentiation of tumor cells at nanomolar concentrations.
Lit.: Z. Naturforsch. **C 53**, 89–92 (1998) (isolation).

Nitracidomycins see enteromycin.

Nitramarine [1-(2-quinolinyl)-*β*-carboline].

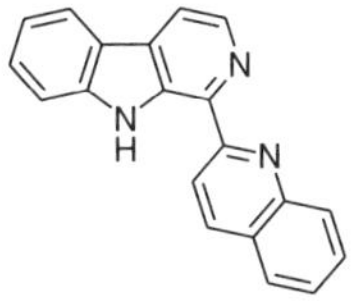

$C_{20}H_{13}N_3$, M_R 295.34, needles, mp. 180–182 °C. A *quinoline alkaloid and *β-carboline alkaloid from *Nitraria komarovii*. In animal experiments N. promotes the onset of sleep.
Lit.: Chem. Pharm. Bull. **35**, 2261 (1987) ▪ J. Indian Chem. Soc. **74**, 891 (1997) ▪ Justus Liebigs Ann. Chem. **1993**, 837 ▪ Tetrahedron **49**, 3325 (1993); **50**, 12329 (1994) ▪ Tetrahedron Lett. **35**, 8851 (1994). – *[CAS 95360-17-1]*

Nitramine [2-azaspiro(5.5)undecan-7-ol].

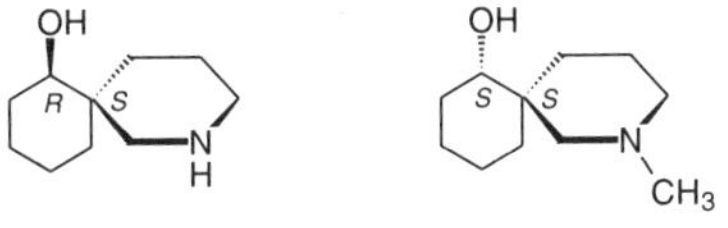

(+)-Nitramine (-)-Sibirine

$C_{10}H_{19}NO$, M_R 169.27; (+)-form: a *piperidine alkaloid from *Nitraria schoberi* (Zygophyllaceae), oil, $[\alpha]_D^{20}$ +16.5° ($CHCl_3$); (–)-*N*-methyl-N.: *sibirine*[1] ($C_{11}H_{21}NO$, M_R 183.29), from *Nitraria sibirica*, oil, $[\alpha]_D^{20}$ –22.5° ($CHCl_3$).
Lit.: [1] Angew. Chem. Int. Ed. Engl. **37**, 104 (1998) (synthesis); J. Chem. Soc., Chem. Commun. **1991**, 1409; Synlett **1994**, 285; **1996**, 925.
gen.: Heterocycles **46**, 413 (1997) (synthesis nitramine) ▪ J. Indian Chem. Soc. **74**, 891 (1997) ▪ J. Org. Chem. **60**, 3916 (1995). – *[CAS 49620-06-6 (N.); 83023-77-2 (sibirine)]*

3-(Nitramino)-L-alanine.

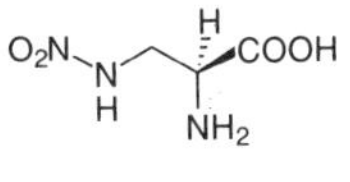

$C_3H_7N_3O_4$, M_R 149.11, mp. 196–198 °C (decomp.), $[\alpha]_D^{20}$ –36° (H_2O), –10° (3 m HCl). A non-proteinogenic amino acid in *Agaricus silvaticus* and *A. subrutilescens*. Both species also contain the decarboxylation product *N*-nitroethylenediamine.
Lit.: J. Chem. Soc. Perkin Trans. 1 **1984**, 2809 (biosynthesis) ▪ Phytochemistry **14**, 2291 f. (1975) ▪ Trans. Mycol. Soc. Jpn. **22**, 213–217 (1981). – *[CAS 58130-89-5]*

1-Nitroaknadinine. $C_{20}H_{24}N_2O_7$, M_R 404.42, yellow needles, mp. 206 °C (decomp.), $[\alpha]_D$ –186.7° (C_2H_5OH). An unusual example of a nitroalkaloid occurring in *Stephania sutchuenensis* (Menispermaceae) together with the not nitrated *aknadinine* {$C_{20}H_{25}NO_5$, M_R 359.42, yellow amorphous powder, mp. 206 °C

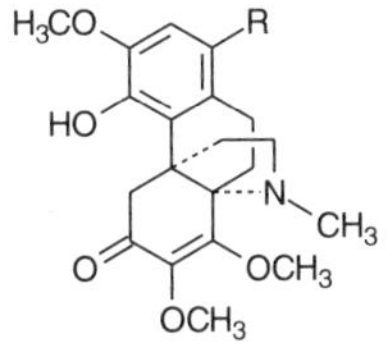

R = NO_2 : 1-Nitroaknadinine
R = H : Aknadinine

(decomp.), $[\alpha]_D$ –233.9° (C_2H_5OH)}. This is indicative of a biological nitration process in the biosynthesis which can be mimicked *in vitro* with concentrated HNO_3.
Lit.: Phytochemistry **35**, 263 (1994). – *[HS 2939 90; CAS 154739-06-7 (1-N.); 24148-86-5 (aknadinine)]*

Nitro compounds (natural). N. c. are widely distributed in nature. The best known natural nitro compound is probably the antibiotic *chloramphenicol. Chlorine-containing aromatic nitro compounds are also produced by basidiomycetes (*drosophilin A). Further aromatic nitro compounds are the *aristolochic acids from plants, *aureothin from streptomycetes, and 2-nitrophenol which *ticks produce as a pheromone. The polycyclic nitropolyzonamine (*polyzonimine) is a component of the defence secretion of a millipede. *(E)-1-Nitro-1-pentadecene*, $H_3C–(CH_2)_{12}–CH=CH–NO_2$ ($C_{15}H_{29}NO_2$, M_R 255.40, oil) is used for defence by the termite *Prorhintermes flavus*[1]. *3-Nitropropanoic acid occurs in free and glucose-bound forms in toxins of Leguminosae (Fabaceae) and as a metabolite of *Aspergillus*- and *Penicillium* species. Nitramines have been isolated from mushrooms (*Agaricus* sp.) [*3-(nitramino)-L-alanine]. In most cases the nitro group is formed by oxidation of primary amines, hydroxylamines, or *nitroso compounds (*chloramphenicol, *3-nitropropanoic acid, *protorubradirin), but a direct substitution on an aromatic ring also seems to occur (*1-nitroaknadinine, *pyrrolomycins).
Lit.: [1] Tetrahedron Lett. **1974**, 1463; **23**, 3047 (1982); J. Chem. Ecol. **16**, 685 (1990). – *[CAS 53520-53-9 ((E)-1-nitro-1-pentadecene)]*

Nitrogen monoxide (nitric oxide, endothelium-derived relaxing factor, EDRF). NO, M_R 30.006, mp. –163.6 °C, bp. –151.7 °C, half-life 5–10 s. By means of investigations on blood vessels in 1980 R. F. Furchgott demonstrated that a very small molecule diffused into neighboring muscle layers and triggered a relaxation reaction when acetylcholine receptors of endothelial cells were activated. The insufficiently characterized substance – named as endothelium-derived relaxing factor (EDRF) – was later identified as NO. NO and L-*citrulline are formed in the enzymatic reaction of L-*arginine with NO synthase (NOS, EC 1.14.13.39). In the first step of the reaction, a two-electron oxidation, N^6-hydroxy-L-arginine is formed as an enzyme-bound intermediate. In the subsequent three-electron oxidation the latter is converted to NO and L-citrulline. NO synthases belong to a family of enzymes that are related to the cytochrome P_{450}-flavoheme reductases. Three NOS-isoforms are distinguished: neuronal (nNOS = NOS I, M_R 160000, e.g., in neurons), inducible (iNOS = NOS II, M_R 130000, e.g., in macrophages), endothelial (eNOS = NOS III, M_R 133000,

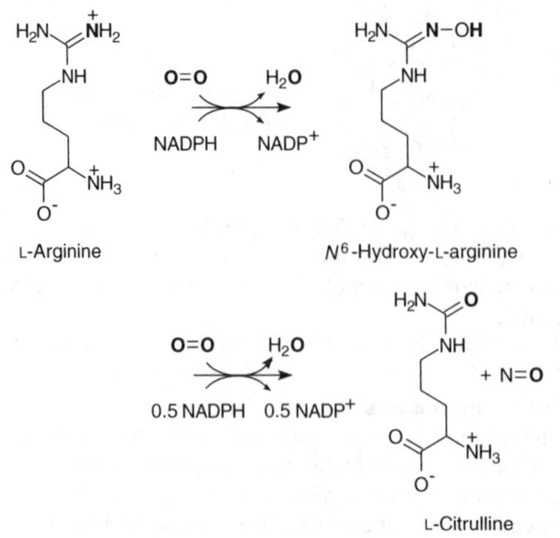

e. g., in endothelial cells). Five prosthetic groups are necessary for the enzymatic reaction: flavine adenine dinucleotide (FAD, see riboflavin), flavine mononucleotide (FMN, see riboflavin), *heme, *tetrahydrobiopterin (H_4B), and calmodulin (CaM); Ca^{2+} is essential for NOS I and NOS III.
The physiological functions of NO range from regulation of blood pressure, inhibition of blood platelet aggregation, wound-healing, neuronal signal transmission (brain and peripheral autonomous nerves) to inactivation of bacteria, parasites, and tumor cells. Toxic concentrations of NO are held to contribute to a number of diseases, e. g., septic shock, stroke, arthritis, migraine, chronic inflammations.
Therapeutic interventions are currently possible principally by means of substrate inhibitors, e.g., N^6-methyl-L-arginine, L-6-thiocitrulline derivatives. As an activating agent NO binds to nitrogen of the heme group of the cytoplasmatic guanylate cyclase, which forms the intracellular signal substance (second messenger) cGMP from GTP. Peroxynitrite has been identified as a further highly reactive subsequent product in the reaction of NO with the superoxide anion: $NO^{\bullet} + O^{\bullet -}_2 \rightarrow ONOO^-$. The substitution therapy with NO-donating compounds plays a central role in pharmakotherapy of heart disease. Glycerol trinitrate and other organic nitrates, which release NO in the body, are indispensible standard therapeutics for long-term treatment of angina pectoris (myocardial ischemia) and heart insufficiency, but also for life-safing therapy of acute infarction. NO also acts as peripheral neurotransmitter, e. g. in male erection. This activity is increased by drugs against male impotence. NO-donating compounds may also serve for therapy of urinary incontinence.
Toxic effects of NO find their expression in infections and inflammations (septic shock, rheumatoid arthritis, e. g.) by induction of NO-synthases. The aim of drug therapy therefore is to develop selective inhibitors for NO-synthase (NOS), induced by cytokines [1].
The elucidation of the physiological functions of NO has been honored with the Nobel price for physiology and medicine in 1998.

Lit.: [1] Angew. Chem., Int. Ed. **38**, 1714–1731, 1856–1868, 1870–1880, 1882–1892 (1999); Exp. Opin. Ther. Patents **9**, 537, 549 (1999); **10**, 559–574 (2000).
gen.: Annu. Rev. Biochem. **63**, 175 (1994) ▪ Chem. Eng. News (6.5.) **1996**, 38–42 ▪ Feelish & Stammler (eds.), Methods in Nitric Oxide Research, Chichester: Wiley 1996 ▪ Ignarro & Murad (eds.), Nitric Oxide – Biochemistry, Molecular Biology and Therapeutic Implications, San Diego: Academic Press 1995 ▪ J. Med. Chem. **38**, 4343–4362 (1995) ▪ Moncada et al., The Biology of Nitric Oxide, vol. 1–5, London: Portland Press 1996 ▪ Nachr. Chem. Tech. Lab. **41**, 1130 (1993) ▪ Pharm. Unserer Zeit **28**, 197–207 (1999) ▪ Sax (8.), NEG 100 ▪ Zechmeister **76**, 1–186 (review). – *[HS 2811 29]*

10-Nitrophenanthrene-1-carboxylic acid see aristolochic acids.

2-Nitrophenol see ticks.

1-Nitro-2-phenylethane see vegetable flavors (tomato).

Nitropolyzonamine see polyzonimine.

3-Nitropropanoic acid.

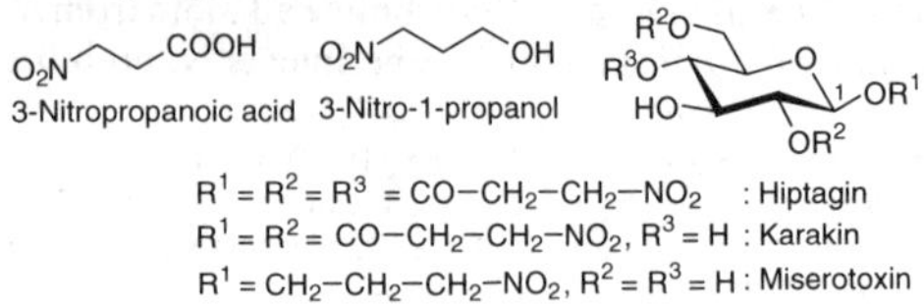

$C_3H_5NO_4$, M_R 119.08, long needles, mp. 65–66 °C. 3-N. is a toxic metabolite from Leguminosae (Fabaceae) [1] and is also produced by fungi of the genera *Penicillium* and *Aspergillus* [2]. 3-N. occurs in plants both in the free form and as esters with glucose. Glucose esters of 3-N. are also present in the toxic seeds of the New Zealand karaka tree (*Corynocarpus laevigatus*, Corynocarpaceae) [3] which are eaten by the Maoris after detoxification through storage or heating in water. The most important toxic constituent is *karakin* {$C_{15}H_{21}N_3O_{15}$, M_R 483.34, platelets, mp. 124–125 °C, $[\alpha]_D$ +4.5°} [4], known since 1871 and also present in many species of the Fabaceae genera *Indigofera*, *Astragalus*, *Lotus*, etc. The 1-epimer occurs as *coronillin* (cryst., mp. 112.5–114 °C) [5] in *Coronilla varia*. A close relative is *hiptagin* {endecaphyllin X, $C_{18}H_{24}N_4O_{18}$, M_R 584.40, fine needles, mp. 110 °C, $[\alpha]_D$ +3.5° (acetone)} [6] from the bark of *Hiptage madablota* (Malpighiaceae) and various Fabaceae.
Glucose esters of 3-N. (*endecaphyllins*) are responsible for the toxicity of *Indigofera endecaphylla* for grazing animals [1]. 3-N. functions as suicide substrate of succinate dehydrogenase where the formed 3-nitroacrylic acid adds to the active thiol group of the enzyme [7]. The compound is formed biosynthetically in *Penicillium atrovenetum* from L-aspartic acid which is oxidized to the α-nitrosuccinate and then decarboxylated to furnish the *aci*-nitro form of 3-N. [8]. Malonate and *N*-hydroxymalonamate appear to be intermediates in *Indigofera spicata*.
Miserotoxin {$C_9H_{17}NO_8$, M_R 267.24, cream-colored powder, $[\alpha]_D$ −22° (H_2O)} [10], the β-D-glucoside of *3-nitro-1-propanol* ($C_3H_7NO_3$, M_R 105.09, thick oil with a weakly astringent odor) from *Astragalus* species (tragacanth, Fabaceae) is considered to be responsible for animal poisonings by release of nitrite. The LD_{50} (p.o.) for rats amounts to >2500 mg/kg, the aglycone also present in the plants is considerably more toxic (LD_{50} 77 mg/kg).

Lit.: [1] Phytochemistry **31**, 2392 (1992). [2] Turner **1**, 303. [3] Phytochemistry **35**, 901 (1994). [4] J. Sci. Food Agric. **2**, 54 (1951); J. Nat. Prod. **33**, 491 (1968); Phytochemistry **18**, 111 (1979). [5] Phytochemistry **16**, 375 (1977). [6] J. Pharm. Sci. **54**, 1136 (1965); Phytochemistry **14**, 1421, 2306 (1975); **25**, 409 (1986). [7] Toxicol. Lett. **27**, 83 (1985). [8] J. Chem. Soc., Perkin Trans. 1 **1992**, 2495. [9] Biochemistry **8**, 182 (1969). [10] Phytochemistry **11**, 1117 (1972); **14**, 2306 (1975); **16**, 1438 (1977) (isolation, activity); Synthesis **1986**, 535 (synthesis); Toxicol. Lett. **19**, 171 (1983); Sax (8.), NIY 500 (toxicology).
gen.: Beilstein EIV **2**, 771 f. (N.); E V **17/7**, 33 (miserotoxin); E V **17/7**, 325 (karakin); E IV **17**, 3288 (hiptagin). – *[CAS 504-88-1 (3-N.); 1400-11-9 (karakin); 63368-43-4 (coronillin); 19896-10-7 (hiptagin); 24502-76-9 (miserotoxin); 25182-84-7 (3-nitro-1-propanol)]*

Nitroso compounds (natural). The *C*-nitroso sugar in *protorubradirin is easily converted to the corresponding nitro compound even during work-up. Naturally occurring nitrosamines have been isolated from cultures of a basidiomycete, namely **N*-(methylnitrosamino)benzaldehyde and the strongly antineoplastic *streptozocin produced by *Streptomyces achromogenes*. On treatment with sodium hydroxide solution both compounds furnish diazomethane. *N*-Nitrosohydroxylamines such as *alanosine, *dopastin and the 5-lipoxygenase inhibitor *nitrosoxacin have been isolated from cultures of streptomycetes.
Lit.: Antibiotics (N. Y.) **4**, 325 (1981) (review) ■ Helv. Chim. Acta **52**, 2555 (1969) ■ J. Org. Chem. **35**, 245 (1970) (synthesis).

Nitrosoxacin. [*N*-(14-methylpentadecyl)-*N*-nitrosohydroxylamine].

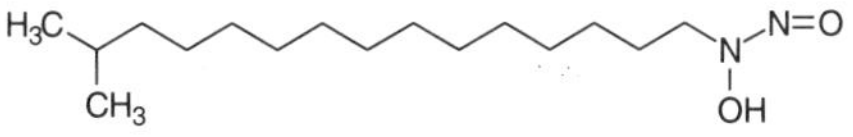

$C_{16}H_{34}N_2O_2$, M_R 286.46, amorphous powder, mp. 44–44.5 °C. An *N*-nitrosohydroxylamine from streptomycetes with inhibitory activity against 5-lipoxygenase (IC_{50} 1.7 μM). In *nitrosoxazin C* ($C_{14}H_{30}N_2O_2$, M_R 258.40) the chain is shortened by two CH_2 groups while N. B contains an unbranched hexadecyl chain.
Lit.: J. Antibiot. **46**, 193 (1993). – *[CAS 147317-96-2 (N. A); 147317-97-3 (N. B); 147317-95-1 (N. C)]*

Nivalenol.

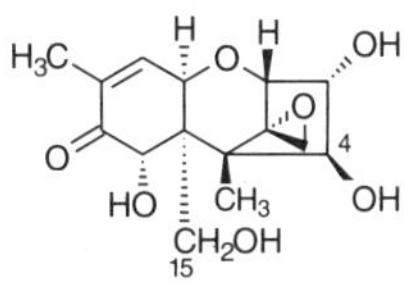

$C_{15}H_{20}O_7$, M_R 312.32, cryst., mp. 223–335 °C, a *mycotoxin of the *tric(h)othecene type, produced by *Fusarium nivale*, *F. sporotrichioides*, and *Gibberella zeae* growing on cereal plants (oats, wheat, barley). N. has strong hemorrhagic and skin irritating activities, blisters form and tissue necrosis sets in; consumption leads to dazed states, vomiting, and possibly lethal internal hemorrhages. LD_{50} (mouse i. p.) 4.1 mg/kg, (chick embryo) 4 μg/egg; in ducks 1 mg/kg leads to vomiting. Like all trichothecenes N. inhibits protein biosynthesis. The 4,15-diacetate of N., known as *saubinine* I ($C_{19}H_{24}O_9$, M_R 396.39) with antineoplastic and virostatic activities has been isolated from *Gibberella saubinetii* (anamorph: *Fusarium sulphureum*) and *F. nivale*. Nivalenol 4-acetate (fusarenone-X, see tric(h)othecenes) is also formed by *F. nivale* and other *Fusarium* species.
Lit.: Beilstein E V **19/7**, 14 ■ Chelkowski (ed.), Fusarium – Mycotoxins, Taxonomy and Pathogenicity, Amsterdam: Elsevier 1989 ■ Cole-Cox, p. 206 ■ Phytochemistry **14**, 2469 (1975) ■ Sax (8.), NMV 000. – *[CAS 23282-20-4 (N.); 14287-82-2 (saubinine I)]*

NO see nitrogen monoxide.

Nocardicins.

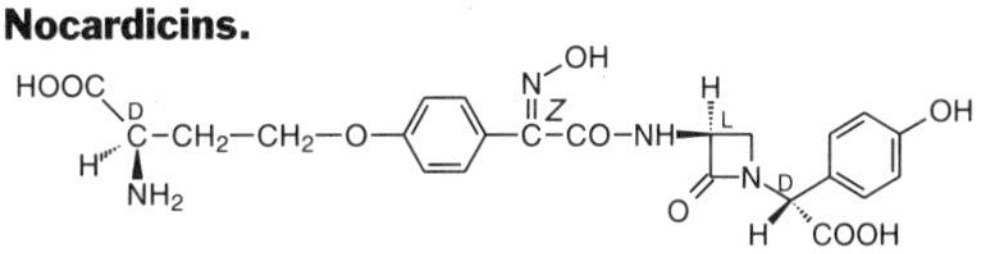

N. A

Trivial name for a group of *β-lactam antibiotics (N. A–G), isolated from culture media of *Nocardia uniformis* which inhibit construction of the bacterial cell wall. The most important member is *N. A:* $C_{23}H_{24}N_4O_9$, M_R 500.46, $[\alpha]_D^{25}$ −135° (Na salt, H_2O). N. A is built up from the precursors D-*p*-hydroxyphenylglycine, L-serine, and D-homoserine, with the former originating from L-tyrosine.
Lit.: Angew. Chem. Int. Ed. Engl. **21**, 810–819 (1982) ■ Chem. Biol. β-Lactam Antibiot. **2**, 165–226 (1982) ■ J. Am. Chem. Soc. **112**, 760 (1990) (synthesis) ■ J. Antibiot. **38**, 133–138, 139–144 (1985) ■ Sax (8.), APT 750. – *[HS 2941 90; CAS 39391-39-4 (N. A)]*

Nodakenetin, nodakenin see marmesin.

Nogalamycin.

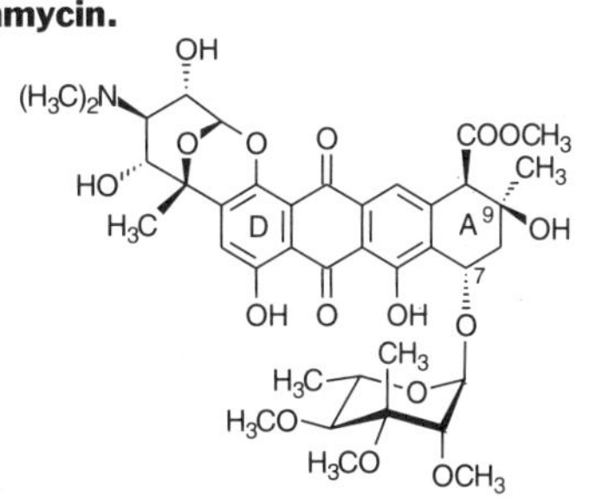

$C_{39}H_{49}NO_{16}$, M_R 787.81, yellow-red powder, mp. 195–196 °C, $[\alpha]_D$ +479° ($CHCl_3$), soluble in acids (red) and bases (violet). Antitumor antibiotic related to the *anthracyclines isolated from *Streptomyces nogalater*, it is also active against Gram-positive bacteria. Acid hydrolysis gives rise to the sugar L-nogalose and the aglycone *nogalarol*. N. differs from the anthracyclines in that an amino sugar (3,6-dideoxy-3-dimethylamino-α-L-glucopyranose) is linked to ring D by *O*-glycosidic and *C-C* bonds and that the stereochemistry of ring A is 7*S*,9*S*,10*R*. N. binds to DNA and is an inhibitor of helicase. A derivative, 7-*O*-methylnogarol is undergoing clinical testing. Other antitumor antibiotics of the N. type are viriplanins, arugomycin, decilorubicin, avidinorubicin.
Lit.: Biochemistry **25**, 4349 (1986); **30**, 1364–1372 (1991) ■ J. Am. Chem. Soc. **99**, 542–549 (1977); **105**, 1328–1332 (1983) (structure) ■ Pharmacol. Ther. **51**, 239 (1991) (review) ■ Priebe (ed.), Anthracycline Antibiotics, ACS Symp. Ser. 574, Washington: ACS 1995 ■ Sax (8.), NMV 500 ■ Tetrahedron **44**, 5695, 5713, 5727, 5745 (1988) (synthesis). – *[HS 2941 30; CAS 1404-15-5]*

Nojirimycin (5-amino-5-deoxy-D-glucose).

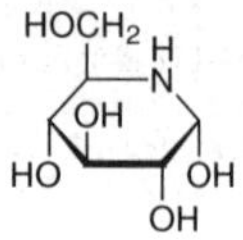

α-Piperidinose form

$C_6H_{13}NO_5$, M_R 179.17, cryst., mp. 126–130 °C, $[\alpha]_D^{20}$ +63° (H_2O; final value). The amino sugar N. is present in the culture broths of various *Streptomyces* species (e. g., *Streptomyces lavendulae*). It exhibits very good antibacterial activity against Gram-positive bacteria. N. is a competitive inhibitor for specific glycosidases and simulates carbohydrate substrates (α- and β-glucosidases). It is accessible by synthesis [1].

Lit.: [1] Chem. Pap. **51**, 147 (1997); C. R. Acad. Sci. Ser. B **311**, 521 ff. (1990); J. Chem. Soc., Chem. Commun. **1989**, 1230 f.; Tetrahedron **49**, 2939–2956 (1993); Tetrahedron Lett. **30**, 755–758 (1989).
gen.: ApSimon **6**, 206–211 ▪ Beilstein E V **21/6**, 223 ▪ Nat. Prod. Rep. **11**, 135 (1994). – *[HS 2941 90; CAS 15218-38-9]*

***galacto*-Nojirimycin** see galactostatin.

Nonactin.

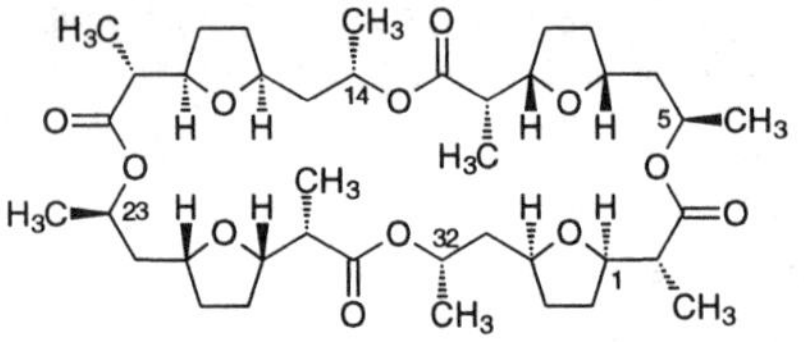

Nonactin

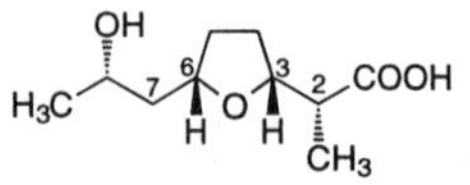

Nonactic acid (racemate)

$C_{40}H_{64}O_{12}$, M_R 736.93, colorless needles, mp. 147–148 °C, an ionophoric macrotetrolide antibiotic produced by *streptomycetes, it is made up of four molecules of *nonactic acid* ($C_{10}H_{18}O_4$, M_R 202.25) arranged in such a way that that the (+)- and the (−)-forms occupy opposing positions, N. is thus a *meso*-form and so optically inactive. The carbon skeleton of nonactic acid is formed biosynthetically from 2 acetate (C-7 to C-9), 1 propionate, and 1 succinate (C-3 to C-6) units. Decarboxylation occurs at C-7 so that this carbon atom originates from the methyl group of an acetate unit. When the starter unit of nonactic acid is not acetate but propionate the methyl groups at C-5, -14, -23, and -32 are replaced by ethyl groups. This can take place in succession, accordingly the four homologues are known: *monactin* (5-ethyl, $C_{41}H_{66}O_{12}$, M_R 750.97), *dinactin* (5,23-diethyl, $C_{42}H_{68}O_{12}$, M_R 764.99), *trinactin* (5,14,23-triethyl, $C_{43}H_{70}O_{12}$, M_R 779.02), and *tetranactin* (5,14,23,32-tetraethyl, $C_{44}H_{72}O_{12}$, M_R 793.05). N. forms complexes with monovalent cations (K^+/Na^+-selectivity: 210) and is used in investigations on ion transport through the cell membrane. N. inhibits the growth of Gram-positive bacteria, it also exhibits insecticidal, coccidiostatic, and immunosuppressive activities; several syntheses have been reported.

Lit.: Zechmeister **26**, 161–189. – *Biosynthesis:* Bioorg. Med. Chem. Lett. **7**, 2187 (1997) ▪ J. Chem. Soc., Chem. Commun. **1985**, 408, 1568 ▪ Zechmeister **58**, 1–82. – *Synthesis:* Atta-ur-Rahman **18**, 229–268 ▪ J. Am. Chem. Soc. **106**, 5304 (1984) ▪ J. Chem. Soc., Perkin Trans. 1 **1998**, 2733–2748 (nonactin) ▪ Synlett **1997**, 159 (nonactic acid) ▪ Tetrahedron **52**, 571–588 (1996) ▪ Tetrahedron Lett. **36**, 3361 (1995). – *[CAS 6833-84-7]*

Nonadienals.

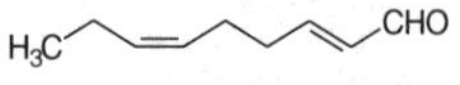

(*E*,*Z*)-2,6-N.

(*Z*,*Z*)-3,6-N.

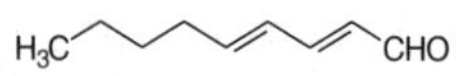

(*E*,*E*)-2,4-N.

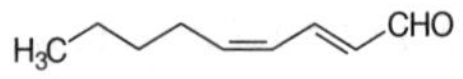

(*E*,*Z*)-2,4-N.

$C_9H_{14}O$, M_R 138.21. Strongly smelling degradation products of *linolenic and *arachidonic acid [1], olfactory threshold in water [2] ca. 0.01–0.09 ppb.

1. *(E,Z)-2,6-N.* (violet leaf aldehyde): bp. 88 °C (1.3 kPa), LD_{50} (rat p. o.) >5 g/kg; fatty-green odor of violet leaves in which it was first detected in 1925 [3], (*E,Z*)-structure assigned in 1944 [4]. It also occurs in *vegetable flavors (*impact compound in cucumber), *fruit flavors (guava, melon, mango), *meat, and seafood flavors [5]. For synthesis, see *Lit.* [6].
2. *(Z,Z)-3,6-N.*: odor of cucumbers and melons in which it is also present [5].
3. *(E,E)-2,4-N.*: bp. 90 °C (0.8 kPa), fatty-oily odor, is also found in guava, melon, and chicken meat flavors [5].
4. *(E,Z)-2,4-N.*: occurs, among others, in pea, endive, and *tea flavor. For syntheses of 2–4, see *Lit.* [7–9].

Use: (*E,Z*)-2,6-N. in small amounts in fragrance and perfume oils; (*E,E*)-2,4-N. especially in meat flavors.

Lit.: [1] Maarse, p. 179 ff. [2] Perfum. Flavor. **16** (1), 18 (1991). [3] J. Prakt. Chem. [2] **124**, 55 ff. (1930). [4] Helv. Chim. Acta **27**, 1561 (1944). [5] TNO list (6.), Suppl. 5, p. 432. [6] Bedoukian (3.), p. 26 ff. [7] Agric. Biol. Chem. **39**, 1617–1621 (1975). [8] J. Org. Chem. **23**, 1580 ff. (1958). [9] Helv. Chim. Acta. **57**, 1309 (1974). – *[HS 2912 19; CAS 557-48-2 (1); 21944-83-2 (2); 5910-87-2 (3); 21661-99-4 (4)]*

(*Z,Z*)-3,6-Nonadien-1-ol, (*Z,Z*)-3,6-nonadienyl acetate see fruit flavors (melon flavor).

Nonanal see alkanals.

4-Nonanolide see alkanolides.

Nonenals see alkenals.

(*E*)-2-Nonen-1-ol see scarabs.

(+)-Nootkatone see eremophilanes.

Nopaline [*N*-((*S*)-1-carboxy-4-guanidinobutyl)-D-glutamic acid].

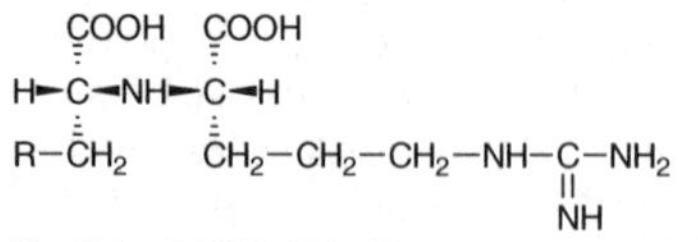

R = CH_2–COOH : Nopaline
R = H : Octopine

$C_{11}H_{20}N_4O_6$, M_R 304.30, mp. 183 °C (hemihydrate), $[\alpha]_D^{23}$ +16.3° (H_2O). Component of plant root neck galls [1].

Biosynthesis: The biosynthesis from L-Arg and 2-oxoglutaric acid occurs in plant tumors after infection with *Agrobacterium tumefaciens* [2,3], see also octopine. For synthesis, see *Lit.* [4–6].

Lit.: [1]C. R. Acad. Sci., Ser. D **268**, 852 (1969). [2]Mol. Gen. Genet. **145**, 177–181 (1976). [3]J. Bacteriol. **129**, 101–107 (1977). [4]Biochem. Biophys. Res. Commun. **75**, 1066–1070 (1977). [5]Org. Prep. Proc. Int. **9**, 99 (1977). [6]Phytochemistry **21**, 225 ff. (1982).
gen.: J. Gen. Microbiol. **136**, 97–103 (1990) ▪ Mol. Gen. Genet. **241**, 65–72 (1993). – *[CAS 22350-70-5]*

Noradrenaline [4-((*R*)-2-amino-1-hydroxyethyl)-brenzcatechol, international free name: norepinephrine, previously: arterenol].

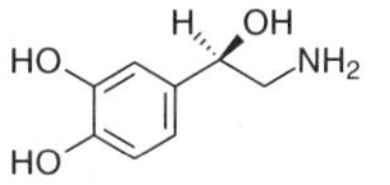

(–)-(*R*)-Form

$C_8H_{11}NO_3$, M_R 169.18, (–)-(*R*)-form (*levarterenol*), mp. 216–218 °C, pK_{a1} 8.6, pK_{a2} 9.8, pK_{a3} 12.0 (25 °C). LD_{50} (mouse p.o.) 20 mg/kg. hydrochloride, $C_8H_{12}ClNO_3$, M_R 205.64, cryst., mp. 145–146 °C, $[\alpha]_D^{25}$ –40° (H_2O). (*R*)-N. occurs in many higher plants, e.g., in Fabaceae, passion flowers, roses, Rutaceae, or buttercups (see also phenylethylamine alkaloids). In the human organism both (*R*)-N. and (*R*)-*adrenaline are formed as hormones in the adrenal medulla as well as in the brain and the region of the sympathetic ganglions. N. is a neurotransmitter of the adrenergic system. The (+)-(*S*)-form, mp. 215 °C, has no physiological activity.
Biochemistry: The biosynthesis proceeds from *L-tyrosine via L-Dopa and *dopamine to (*R*)-N. and (*R*)-adrenaline. Hydroxylation occurs stereospecifically at the dopamine stage by means of dopamine *β*-monooxygenase (EC 1.14.17.1). (*R*)-N. is inactivated through methylation of the phenolic hydroxy groups by catechol *O*-methyltransferase (EC 2.1.1.6) and monoaminoxidase (EC 1.4.3.4); the product excreted in urine is 3-methoxy-4-hydroxymandelic acid. In the metabolism the sympathicomimetic (*R*)-N. stimulates glycolysis in liver, adipose tissue, and skeletal muscle. The physiological, blood pressure elevating activity involves constriction of peripheral vessels (as vasoconstrictor, see also (*R*)-adrenaline), it relaxes smooth musculature and stimulates cardiac activity. The vasoconstricting effect is based on an opening of calcium ion channels [1].
Use: (*R*)-N. as well as the racemate, mp. 191 °C (decomp.) and the hydrochloride are used as adrenergic, bronchodilatory, sympathicomimetic agents and as a vasopressor. For detection, see *Lit.*[2].
Lit.: [1]Nature (London) **336**, 382–385 (1988). [2]Bergmeyer, Methods of Enzymatic Analysis (3.), vol. 8, p. 543–555, Weinheim: Verl. Chemie 1985.
gen.: Fillenz, Noradrenergic Neurons, Cambridge: Cambridge University Press 1990 ▪ Florey **11**, 555 f. (review) ▪ Luckner (3.), p. 366 ▪ Phytochemistry **16**, 9–18 (1977) (review). – *Pharmacology:* Heal & Marsden, The Pharmacology of Nordrenaline in the Central Nervous System, Oxford: Oxford University Press 1990. – *Toxicology:* Sax (8.), NNO500. – *[HS 2937 99; CAS 51-41-2 (R); 149-95-1 (S); 138-65-8 (racemate); 329-56-6 (hydrochloride)]*

(+)-Norambreinolide see clary sage oil.

Norbaeocystine see psilocybine.

Norbixin (6,6′-diapo-6,6′-carotenedioic acid, E 160 b, C. I. 75120).

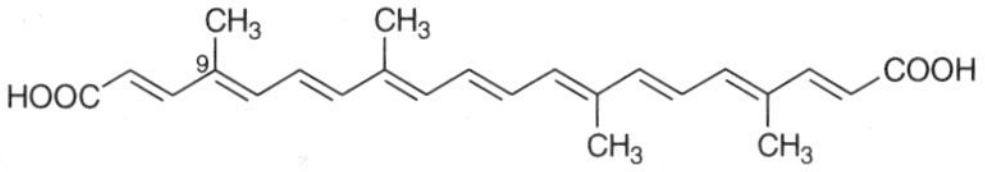

(9*E*)-form

$C_{24}H_{28}O_4$, M_R 380.48, mp. of (9*E*)-form (*β*-N.) >300 °C (decomp.), mp. of (9*Z*)-form (*α*-N.) 256 °C (decomp.), uv_{max} 482 nm. N. is used together with *bixin or alone to dye cheese, fruit juices, soups, bakery products, etc.
Lit.: Britton, Liaaen-Jensen & Pfander, Carotenoids, vols. 1 and 2, Basel: Birkhäuser 1995 ▪ J. Chem. Soc. (C) **1970**, 235 (synthesis) ▪ Phytochemistry **46**, 1379 (1997) (isolation). – *[CAS 542-40-5 (β-N.); 626-76-6 (α-N.)]*

Nordihydroguaiaretic acid (NDGA).

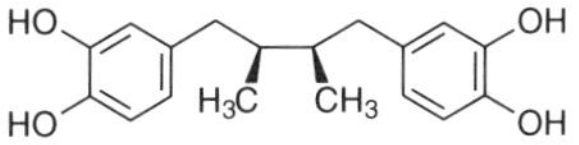

$C_{18}H_{22}O_4$, M_R 302.37, cryst., mp. 184–185 °C, soluble in alcohol, ether, acetone, and in aqueous alkalis (red solution), insoluble in petroleum ether. The *lignan N. occurs in *guaiac(um) and in the leaves of evergreen shrubs (*Larrea* species, e.g., *Larrea tridentata*). In addition to free N., the Australian flora in particular contains methylenedioxy derivatives and methyl ethers. The 3′-*O*-methyl-3,4-*O*-methylene derivative of (8*R*,8′*S*)-N., known as *macelignan*, $C_{20}H_{24}O_4$, M_R 328.41, cryst., mp. 70–72 °C, $[\alpha]_D^{20}$ +5.28° ($CHCl_3$), from *Myristica fragrans* and the root bark of *Kadsura longipedunculata* is active against P388 leukemia and hepatitis B [1]. The racemate of N., *masoprocol*, is available in the USA under the name Actinex® for the treatment of actinic keratoses (skin diseases)[2].
Use: As an antioxidant for fats. After addition of 0.01% N. lard can be stored for 19 months at 20 °C without becoming rancid.
Lit.: [1]Phytochemistry **26**, 1542 f. (1987). [2]Cancer Res. **50**, 2068 (1990); J. Am. Acad. Dermatol. **24**, 738 (1991); **31**, 295 (1994).
gen.: Aust. J. Chem. **28**, 81–90 (1975) ▪ Beilstein E IV **6**, 7771 ▪ C. A. **128**, 22676p (synthesis) ▪ Chem. Pharm. Bull. **36**, 648 (1988) ▪ J. Chem. Soc., Perkin Trans. 1 **1978**, 1147–1150 (synthesis) ▪ J. Nat. Prod. **53**, 212 (1990) ▪ J. Pharm. Sci. **63**, 1905–1907 (1974) ▪ Karrer, No. 1167 ▪ Phytochemistry **23**, 2647–2652 (1984); **26**, 1513–1515 (1987) ▪ Tetrahedron **51**, 12203 (1995). – *Toxicology:* Sax (8.), NBR 000. – *[HS 2907 29; CAS 500-38-9 (N.); 107534-93-0 (macelignan)]*

Norephedrine see phenylethylamine alkaloids.

Norepinephrine see noradrenaline.

Norfuraneol see hydroxyfuranones.

Norharman see *β*-carboline.

Norhyoscine see scopolamine.

Norlichexanthone see lichen xanthones.

Nornicotine [3-(2-pyrrolidinyl)-pyridine]. Structural formula, see nicotine. $C_9H_{12}N_2$, M_R 148.21, hygroscopic oil with an odor of amines, that rapidly decomposes in the air, bp. 270 °C [131 °C (1.5 kPa)]; (*S*)-form: $[\alpha]_D^{22}$ –88.8°; (*R*)-form: $[\alpha]_D^{20}$ +86.3°. A *tobacco alkaloid and, in contrast to *nicotine, hardly volatile in steam. N. occurs in tobacco plants (*Nicotiana* sp.) and in tobacco, presumably formed by demethylation of nicotine. The (*R*)-form is also known in *Duboisia hopwoodii*. N., like *nicotine, is highly toxic

[LD_{50} (rabbit i. v.) 3 mg/kg], and has insecticidal activity.

Lit.: Beilstein E V **23/6**, 62 ▪ Hager (5.) **3**, 870 ▪ Manske **26**, 121–140 ▪ Merck-Index (12.), No. 6807 ▪ Pelletier **3**, 21 f. ▪ Sax (8.), p. 2606 ▪ Ullmann (5.) **A 14**, 271. – *[HS 2939 90; CAS 494-97-3 ((S)-N.); 7076-23-5 ((R)-N.)]*

Nornuciferine see aporphine alkaloids.

Norphytane see pristane.

Norsteroids see androgens, estrogens, progestogens, and steroids.

Norstictic acid.

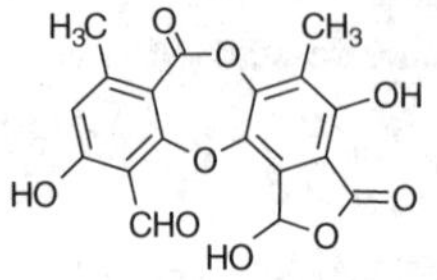

$C_{18}H_{12}O_9$, M_R 372.29, needles, mp. 286–287 °C (decomp.), poorly soluble in most organic solvents. When N. is treated with concentrated KOH solution the colorless acid first dissolves to a yellow solution which changes to red after a few seconds and precipitates the deep-red potassium complex in the form of microscopic, fine needles. N. occurs in many lichen, e. g., *Parmelia acetabulum*, *Graphis scripta*, and *Aspicilia radiosa*.

Lit.: Asahina-Shibata, p. 137 ▪ Karrer, No. 1063. – *[CAS 571-67-5]*

Nortrachelogenin see wikstromol.

Noscapine see phthalide isoquinoline alkaloids.

Nosiheptide (multhiomycin).

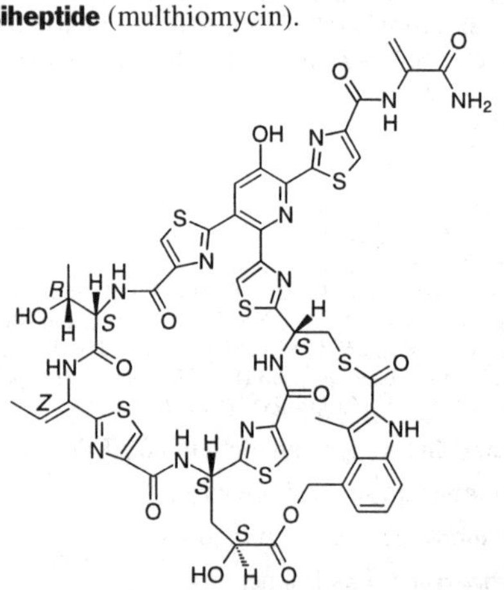

$C_{51}H_{43}N_{13}O_{12}S_6$, M_R 1222.38, yellow powder, mp. 310–312 °C, a broad spectrum *antibiotic produced by *streptomycetes (e.g., *Streptomyces actuosus*), used as a fodder additive for livestock to promote growth. N. is related to *thiostrepton. Like the latter, it inhibits protein biosynthesis through binding to the *23S* ribosomal RNA and the protein L-11, in this way the GTP hydrolysis activity of the *50S* subunit is inhibited. The producer protects itself by methylation of *23S* rRNA (host site resistance).

Lit.: J. Heterocyclic Chem. **32**, 1309 (1995) (synthesis). – *Biosynthesis:* J. Am. Chem. Soc. **115**, 7557 (1993) ▪ Strohl & Floss, in Vining & Stuttard (eds.), Genetics and Biochemistry of Antibiotic Production, p. 223–238, Boston: Butterworth-Heinemann 1995. – *Review:* Experientia **36**, 414 (1980) ▪ J. Antibiot. **42**, 1643 (1989) ▪ Kirk-Othmer (4.) **3**, 282. – *[CAS 56377-79-8]*

Nostocine A.

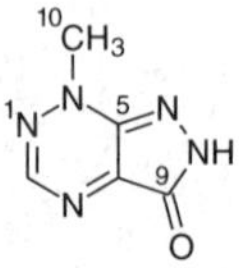

$C_5H_5N_5O$, M_R 151.13, black needles, mp. 171–172 °C or violet powder. Black-violet pigment from the freshwater cyanobacterium *Nostoc spongiaeforme*. It exhibits broad growth inhibiting action.

Lit.: Heterocycles **43**, 1513 (1996). – *[CAS 180128-24-9]*

Nostoclides.

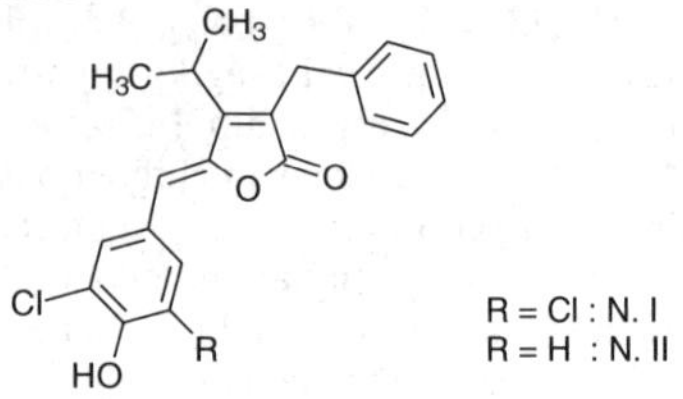

N. I: $C_{21}H_{18}Cl_2O_3$, M_R 389.28, prisms, mp. 186–187 °C. *N. II*: $C_{21}H_{19}ClO_3$, M_R 353.83, yellow prisms, mp. 132–133 °C. N. I and II are the first extracellular metabolites from the symbiotic Nostoc-cyanobacterium (blue alga) system of the lichen *Peltigera canina*.

Lit.: Tetrahedron Lett. **34**, 761–764 (1993). – *[CAS 147714-57-6 (N. I); 147714-58-7 (N. II)]*

Nostocyclamide.

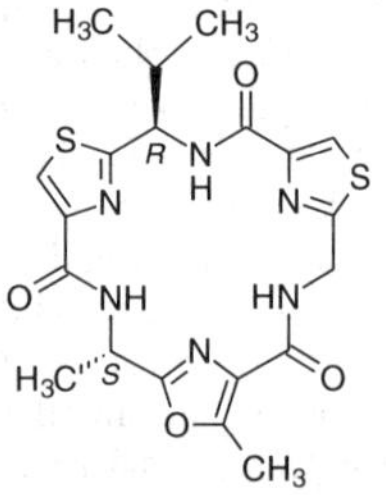

The cyclic amide N. has been isolated from a nitrogen-fixing marine cyanobacterium (*Nostoc* sp.): $C_{20}H_{22}N_6O_4S_2$, M_R 474.56, mp. 257–260 °C, $[\alpha]_D^{19}$ +51° ($CHCl_3$), crystals.

Lit.: J. Chem. Soc., Perkin Trans. 1 **1998**, 601–607 (synthesis) ▪ J. Org. Chem. **60**, 7891 (1995) (isolation) ▪ Synlett **1996**, 1171. – *[CAS 170129-43-8]*

Nostocyclin.

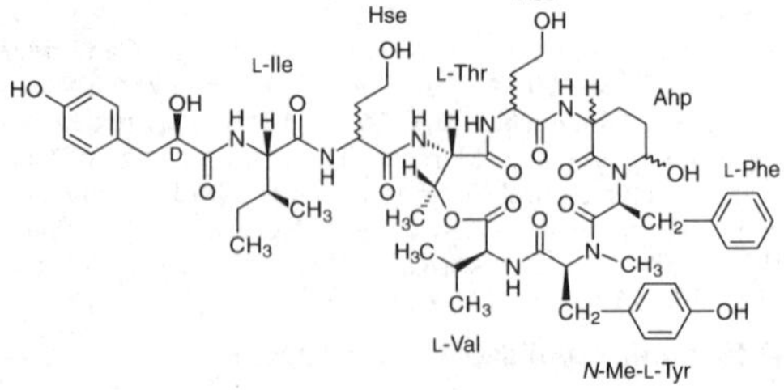

A cyclic depsipeptide from a hepatotoxic strain of the marine cyanobacterium *Nostoc* sp. N. inhibits protein phosphatase-1 at high concentrations *in vitro*:

$C_{56}H_{76}N_8O_{16}$, M_R 1117.25, amorphous solid (see also microcystins).
Lit.: Tetrahedron Lett. **37**, 6725 (1996) (isolation). – *[CAS 181622-50-4]*

Nostocyclophanes.

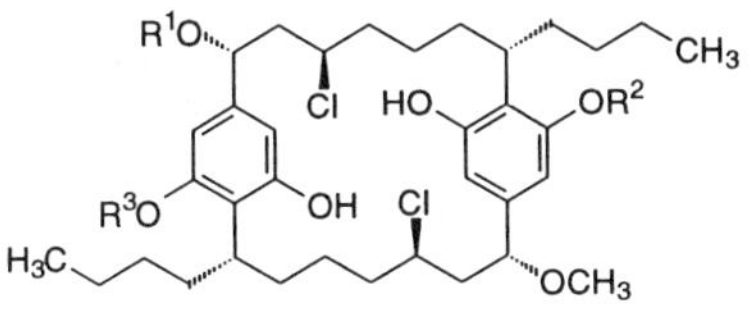

N.A : $R^1 = CH_3$, $R^2 = R^3$ = βDGlc*p*
N.B : $R^1 = CH_3$, R^2 = H, R^3 = βDGlc*p*
N.C : $R^1 = R^2 = R^3$ = H
N.D : $R^1 = CH_3$, $R^2 = R^3$ = H

Table: Data of Nostocyclophanes.

N.	molecular formula	M_R	consistency	$[\alpha]_D^{25}$ (CH₃OH)	CAS
A	$C_{48}H_{74}Cl_2O_{16}$	978.01	oil	−12.0°	134237-84-6
B	$C_{42}H_{64}Cl_2O_{11}$	815.87	oil	−3.7°	134237-85-7
C	$C_{35}H_{52}Cl_2O_6$	639.70	oil	−5.53°	134208-58-5
D	$C_{36}H_{54}Cl_2O_6$	653.73	mp. 242–243 °C	+10.8°	126693-93-4

N. are the first, naturally occurring paracyclophanes (see table). They are isolated from the cyanobacterium *Nostoc linckia*. Four [7.7]paracyclophanes have been identified. They all have cytotoxic activity (IC_{50} 1–2 μg/mL) and antitumor properties. For isolation, see *Lit.*[1], for structure elucidation, see *Lit.*[2], for biosynthesis, see *Lit.*[3].
Lit.: [1]J. Am. Chem. Soc. **112**, 4061–4063 (1990). [2]J. Org. Chem. **56**, 4360–4364 (1991). [3]Tetrahedron **49**, 7615 (1993).

Nostodione A.

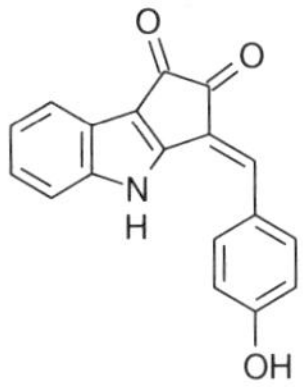

$C_{18}H_{11}NO_3$, M_R 289.29, orange-colored powder, mp. 320 °C (decomp.). An unusual indole derivative from the terrestrial blue alga *Nostoc commune*. N. inhibits the formation of spindles during mitosis in sea urchin eggs.
Lit.: Z. Naturforsch., C **49**, 464 (1994). – *[CAS 158182-28-6]*

Notholaenic acid.

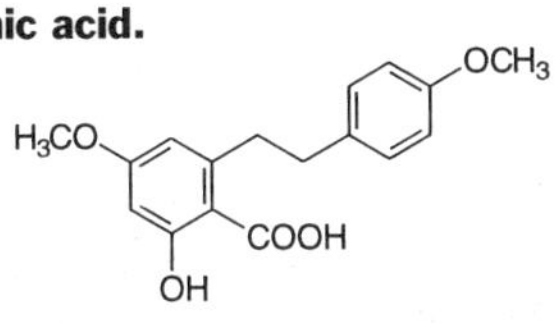

$C_{17}H_{18}O_5$, M_R 302.33, needles, mp. 148–150 °C, soluble in polar solvents. The dihydrostilbene N. is the main component of the whitish, flour-like covering on the leaf frond of two fern species of the genus *Notholaena*. It influences photosynthetic activity of chloroplasts (decoupling of electron transport of phosphorylation) [1] and also exhibits weak antimicrobial activity [2]. For synthesis, see *Lit.*[3].
Lit.: [1]Phytochemistry **19**, 2059 (1980). [2]J. Pharm. Sci. **70**, 951 (1981). [3]J. Nat. Prod. **48**, 293 (1985).
gen.: Phytochemistry **18**, 1243 (1979); **33**, 611 (1993). – *[CAS 72578-97-3]*

Novobiocin.

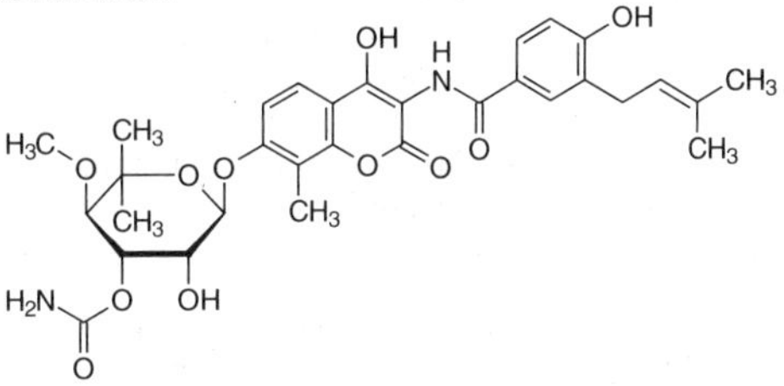

$C_{31}H_{36}N_2O_{11}$, M_R 612.65, mp. 152–156 °C. An antibiotically active *coumarin glycoside from culture filtrates of *streptomycetes (e.g., *Streptomyces spheroides*). Sensitive to light, soluble in alcohols, pyridine, acetone, ethyl acetate, water (pH>7.5), insoluble in acids, chloroform, ether. N. is active against Gram-negative bacteria and staphylococci, as well as some Gram-positive pathogens. N. is a leukotriene B_4 antagonist. N. acts as a specific inhibitor of DNA gyrase during DNA replication but does not show any cross-resistance with other antibiotics; LD_{50} (mouse, p.o.) 1500 mg/kg. On account of adverse reactions such as hypertensive or hemopoietic effects it is rarely used in human medicine and then in combination with *tetracyclines, rifampicin or fusidic acid. However, it is used in veterinary medicine.
Lit.: Beilstein EV **18/12**, 105 ▪ J. Antibiot. **45**, 1958 ff. (1992) ▪ J. Heterocycl. Chem. **36**, 365 (1999) (spectra) ▪ J. Pharm. Pharmacol. **24**, 972 (1972) ▪ Martindale (30.), p. 188 ▪ Murray, The Natural Coumarins, New York: Wiley 1982 ▪ Sax (8.) NOB 000. – *Biosynthesis:* J. Antibiot. **51**, 676 (1998) ▪ Tetrahedron Lett. **39**, 2717 (1998). – *[HS 2941 90; CAS 303-81-1]*

NPTX see JSTX.

NSC 129185 see mycophenolic acid.

NT-1(2) see tric(h)othecenes.

Nuciferal.

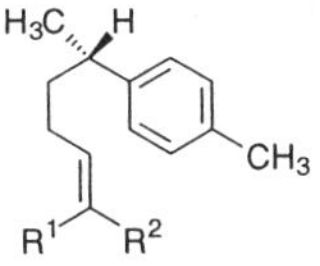

$R^1 = CH_3$, R^2 = CHO : Nuciferal
$R^1 = CH_2OH$, $R^2 = CH_3$: Nuciferol

$C_{15}H_{20}O$, M_R 216.32, oil, bp. 108 °C (4 Pa), $[\alpha]_D$ +37.6° ($CHCl_3$), a *sesquiterpene aldehyde with the *bisabolene skeleton, isolated from *Torreya nucifera*. The corresponding alcohol, *nuciferol* ($C_{15}H_{22}O$, M_R 218.34) is also present in the plant.
Lit.: *Synthesis:* ApSimon **5**, 40 ▪ Chem. Pharm. Bull. **35**, 913 (1987) ▪ J. Chem. Soc., Perkin Trans. 1 **1988**, 3163 ▪ Tetrahedron Lett. **23**, 5567 (1982). – *[CAS 25532-74-5 (N.); 39599-18-3 (nuciferol)]*

Nuciferine see aporphine alkaloids.

Nucleocidin (4′-*C*-fluoroadenosine 5′-sulfamate). $C_{10}H_{13}FN_6O_6S$, M_R 364.31, mp. >190 °C (monohydrate, decomp.), $[\alpha]_D$ −33.3° (C_2H_5OH/HCl). A fluorine-con-

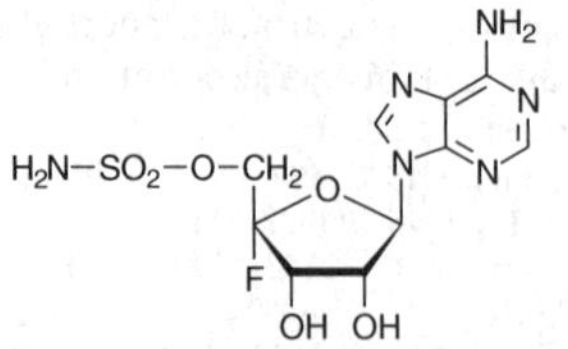

taining *nucleoside antibiotic from cultures of *Streptomyces calvus*. N. shows antitrypanosomal activity, inhibits protein biosynthesis, and is a potent cell toxin. The original cultures of *S. calvus* have since lost their ability to biosynthesize N., so that the compound must now be prepared synthetically[1].

Lit.: [1]J. Chem. Soc., Perkin Trans. 1 **1993**, 1795 (synthesis, isolation).

gen.: Beilstein E V **26/16**, 509 ▪ J. Am. Chem. Soc. **91**, 1535 (1969) (structure); **93**, 4323 (1971); **98**, 3346 (1976). – *[HS 2941 90; CAS 24751-69-7]*

Nucleoside antibiotics. A large group of microbial secondary metabolites produced mainly by *actinomycetes and derived structurally from nucleosides. The N. a. group includes individual nucleobases (8-azaguanine, *bacimethrin), simple, modified nucleosides, and N. a. that are linked to additional structural elements. Classical members of the simple, modified nucleosides contain, e. g., unusual sugars (angustmycin, *oxetanocin, *neplanocins, Ara-A), altered nucleobases (*tubercidin, coformycin), or *C*-glycosidic bonds (*showdomycin). The N. a. bearing additional substituents can be further divided into, among others, the acylnucleosides (*blasticidin S, *puromycin), the peptidylnucleosides (*polyoxins, *nikkomycins), the glycosylnucleosides (amicetin, *tunicamycins), and the nucleotide analogues (*thuringiensin). N. a. act as antimetabolites in numerous metabolic processes and thus exhibit an unusually broad spectrum of activities. These include inhibitory action against cancer cells. Bacteria, fungi, trypanosomes, viruses, and insects as well as mutagenic, immunostimulating and -suppressing properties. Particular interest has been directed to the antiviral (oxetanocin) and cancer inhibiting (*neplanocins) N. a., although the compounds must often first be activated by phosphorylation within the cell. N. a. are used, among others, in agriculture as fungicides (blasticidin S) or insecticides (nikkomycins, thuringiensin). Biosynthetic investigations have been concentrated mainly on the *C*-glycosidic N. a. and the peptidylnucleosides.

Lit.: Isono, in Vining & Stuttard (eds.), Genetics and Biochemistry of Antibiotic Production, p. 597–617, Boston: Butterworth-Heinemann 1995 ▪ J. Antibiot. **38**, 144 (1985); **41**, 1711–1739 (1988) ▪ Remers & Iyengar, in Cancer Chemotherapeutic Agents, p. 631–636, Washington: ACS 1995 ▪ Suhadolnik, in Nucleosides as Biological Probes, New York: Wiley 1979.

Nuphar alkaloids. Alkaloids occurring in the genus *Nuphar* (water lilies) of the family Nymphaeaceae. About 30 compounds are known. In biogenetic sense the N. a. are not genuine alkaloids but rather pseudoalkaloids since their carbon skeleton does not originate from the amino acid metabolism but is derived entirely from the terpene metabolism. They contain the carbon skeleton of the *sesquiterpenes (figure) and are thus

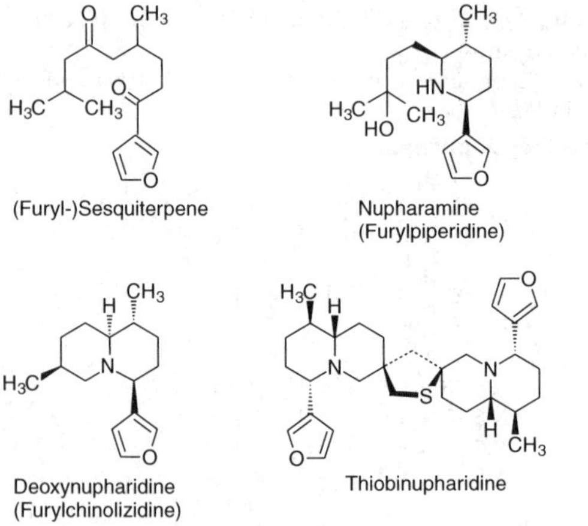

Figure: Structural types of the Nuphar alkaloids and their biogenetic relationship to the sesquiterpenes.

often called sesquiterpene alkaloids. The N. a. can be assigned to three structural classes (figure): the furylpiperidines, e. g.: *nupharamine* {$C_{15}H_{25}NO_2$, M_R 251.37, bp. 130 – 134 °C (0.133 kPa), $[\alpha]_D$ −35.4° (C_2H_5OH)}, the furylquinolizidines, e. g., *deoxynupharidine* {$C_{15}H_{23}NO$, M_R 233.35, mp. 21–22 °C, $[\alpha]_D$ −112.5°}, and the dimeric, sulfur-containing furylquinolizidines, e. g., *thiobinupharidine* {$C_{30}H_{42}N_2O_2S$, M_R 494.74, mp. 129–130 °C, $[\alpha]_D$ +7.8° (CH_3OH)}. Since the quinolizidine ring system is typical for some of the N. a. they are often placed in a relationship to the *quinolizidine alkaloids of the Fabaceae and the *Lycopodium alkaloids, However, there are no biogenetic connections between these groups. Castoreum contains the simple N. a., "castoramine". It is assumed that this is a metabolic product of N. a., taken up by Canadian beavers with their food.

Little is known about the biological activities of the N. a. Alkaloid extracts of various *Nuphar* species have antibiotic activity and inhibit the growth of pathogenic – also for humans – bacteria and fungi (e. g., *Candida albicans*).

Lit.: J. Chem. Soc., Chem. Commun. **1994**, 499 ▪ Manske **28**, 183–308; **35**, 215–257. – *Biosynthesis:* Mothes & Schütte (eds.), Biosynthese der Alkaloide, p. 606, Berlin: VEB Deutscher Verlag der Wissenschaft 1969 ▪ Mothes et al., p. 360. – *Synthesis:* J. Org. Chem. **64**, 3736 (1999) ▪ Nat. Prod. Rep. **11**, 28, 649 (1994) ▪ Tetrahedron Lett. **32**, 4325 (1991). – *Occurrence:* Hegnauer, **V**, 209; **IX**, 135. – *[HS 2939 90; CAS 17812-38-3 (nupharamine); 1143-54-0 (deoxynupharidine); 30343-72-7 (thiobinupharidine)]*

Nupharamine see Nuphar alkaloids.

Nutmeg oil (mace oil). Colorless to light yellow oil with a fresh herby, aromatic, slightly woody odor of nutmeg.

Production: By steam distillation of the seeds (nutmeg) and seed husks (mace) of *Myristica fragrans*.

Composition[1]***:*** The oils contain as main components monoterpene hydrocarbons such as *α-*, *β-*pinenes* and *sabinene* (see thujenes), ca. 20–30% of each. The components responsible for the odor and taste include small amounts of *terpinenol-4* (see *p*-menthenols), **safrole*, **eugenol*, and **isoeugenol*. A characteristic component is *myristicin* (see safrole), present to about 5 – 10% in Indonesian oil which contains more higher boiling constituents.

Use: In small amounts in the perfume industry (masculine notes) and to improve the aromatic character of foods. It is also used medicinally as a component of expectorants for inhalation, antirheumatic ointments, in carminatives, and rhinological medications.

Lit.: [1] Perfum. Flavor. **10** (4), 47 (1985); **15** (6), 62 (1990); **17** (5), 146 (1992).
gen.: Arctander, p. 442 ▪ Bauer et al. (2.), p. 166 ▪ ISO 3215 (1974), 4734 (1981). – *Toxicology:* Food Cosmet. Toxicol. **14**, 631 (1976); **17**, 851 (1979). – *[HS 3301 29; CAS 8007-12-3]*

Nystatin (fungicidin).

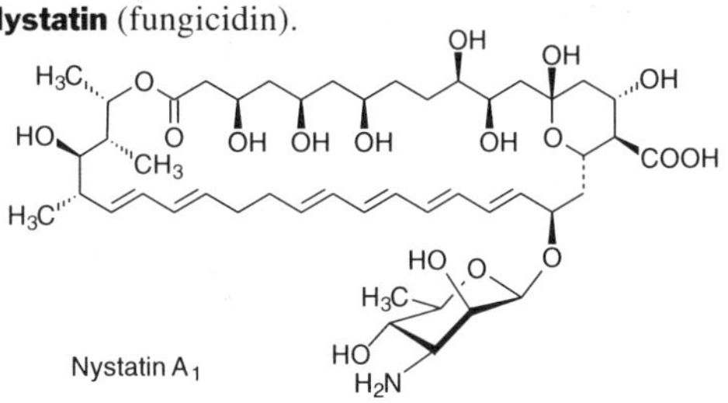

Nystatin A₁

A polyene *macrolide complex with D-mycosamine as amino sugar and *N. A*$_1$ {$C_{47}H_{75}NO_{17}$, M_R 926.11, yellow powder, $[\alpha]_D$ +21° (pyridine), partial decomp. >160 °C, mp. >250 °C} as main component. Further components from *Streptomyces noursei* are N. A_2 and A_3 (polyfungin A_2 and A_3), as well as polyfungin B. N. is practically insoluble in water and non-polar organic solvents, poorly soluble in lower alcohols, and readily soluble in dimethylformamide and dimethyl sulfoxide. The aglycone is formed biosynthetically from 3 propionate and 16 acetate units. N. is used as an antimycotic agent against yeast and other fungal infections. It attacks the membrane steroids of sensitive eukaryotes and increases permeability by forming ion channels. The higher affinity of N. for fungal ergosterol in comparison to the cholesterol of invertebrates is the reason for its low toxicity. N. is used in medicine, including pediatric medicine, to treat *Candida* infections and for treatment of leishmaniasis.

Lit.: Beilstein E V **18/10**, 524 ▪ Can. J. Chem. **63**, 77 – 85 (1985) ▪ Florey **6**, 341 ▪ Holz, Antibiotics 5, vol. 2, p. 313 – 340, New York: Springer 1979 ▪ J. Antibiot. **41**, 1289 ff. (1988) ▪ Kirk-Othmer (4.) **3**, 475 ▪ Tetrahedron Lett. **30**, 4517 – 4520, 4521 – 4524 (1989) ▪ Sax (8.), NOH 500. – *[HS 2941 90; CAS 1400-61-9; 34786-70-4 (N. A₁); 37371-05-4 (polyfungin B)]*

O

Oak moss absolute. Dark green to dark brown viscous mass. The odor is earthy-mossy, sweet, herby and of longer persistence.
Production: From the lichen *Evernia prunastri* (Usnaeaceae) growing on oak trees by extraction (e.g., with toluene) which first furnishes wax-containing oak moss concretes. Precipitation of the waxes with alcohol gives E.; oak moss is collected in southern France (Auvergne), Morocco (Atlas), and Macedonia.
Composition*[1]*: Main components of the volatile part are phenol derivatives such as *orcinol monomethyl ether* ($C_8H_{10}O_2$, M_R 138.17), *ethyl everninate* ($C_{11}H_{14}O_4$, M_R 210.23), and *methyl β-orcinol carboxylate* ($C_{10}H_{12}O_4$, M_R 196.20), which are partly formed under the processing conditions from the *depsides present in the lichen. The latter compound is carrier of the typical oak moss odor; it is also produced synthetically in large quantity and used as a fragrance substance.

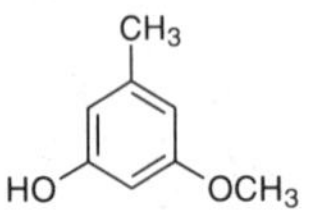

Orcinol monomethyl ether (**1**)

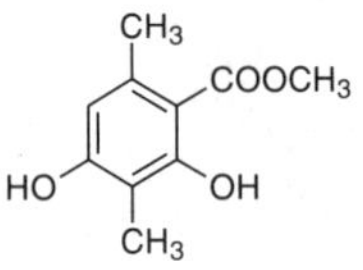

Methyl β-orcinolcarboxylate (**2**)

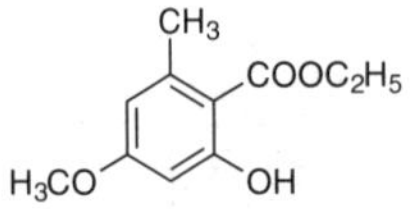

Ethyl everninate (**3**)

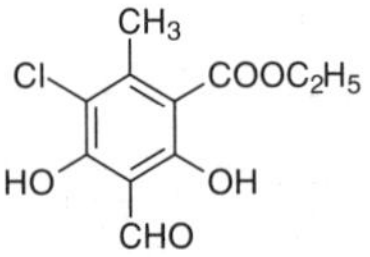

Ethyl chloroh(a)ematommate (**4**)

Use: For perfume production; E. is an essential component of Chypre and Fougere nuances and widely used on account of its strong fixating action. Because of a possible sensitization effect, the use of E. in perfumes is restricted to certain concentration limits. Triggering substance is ethyl chloroh(a)ematommate ($C_{11}H_{11}ClO_5$, M_R 258.66)[3], present in small amounts but which can be removed from E. by various processes[4,5]. Related to oak moss is the tree moss *Evernia furfuracea* growing on conifers, the extracts of which have a similar composition and find similar uses.

Lit.: [1] Perfum. Flavor. **1** (5), 12 (1976); Parfum. Cosmet. Arom. **16**, 37 (1977); J. Chromatogr. **466**, 301 (1989). [2] Z. Naturforsch., B **44**, 1283 (1989). [3] Proc. of the 10th Int. Congress of Essential Oils, p. 685, Amsterdam: Elsevier 1988. [4] E. P. 468189 (1990), Givaudan-Roure. [5] PCT WO 93/23509 (1993), Givaudan-Roure.
gen.: Arctander, p. 446, 627 ▪ Bauer et al. (2.), p. 167 ▪ Moxham, in Brunke (ed.), Progress in Essential Oil Research, p. 491, Berlin: de Gruyter 1986 ▪ Ohloff, p. 185. – *Toxicology:* Food Cosmet. Toxicol. **13**, 891 (1975). – *[HS 330129; CAS 9000-50-4 (resin); 68917-10-2 (absolue); 3209-13-0 (1); 4707-47-5 (2); 6110-36-7 (3); 57857-81-6 (4)]*

Oasomycins.

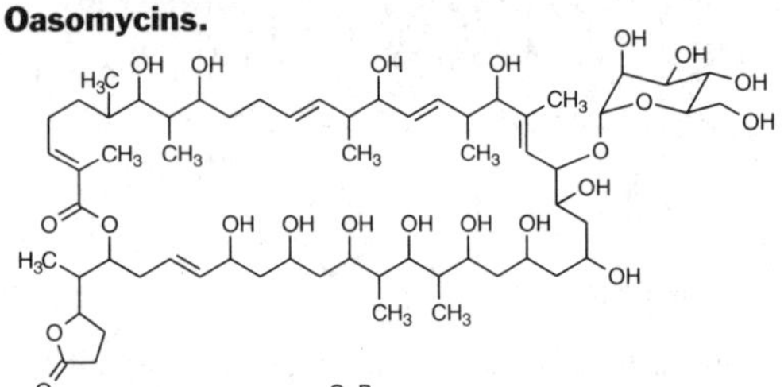

O. B

A group of macrolactones produced by *actinomycetes (e.g., *Streptoverticillium baldaccii*); the main components are O. A and B. *O. A:* $C_{55}H_{94}O_{17}$, M_R 1027.34, colorless powder, mp. 146 °C, $[\alpha]_D$ −13.1° (CH_3OH). *O. B:* $C_{61}H_{104}O_{22}$, M_R 1189.48, colorless powder, mp. 157 °C, $[\alpha]_D$ +24.6° (CH_3OH), soluble in water, methanol, and DMSO. The α-D-mannopyranosyl group is absent in O. A. O. C to F are carboxylic acids formed by cleavage of the γ-lactone ring. In O. C and D the 42-membered macrolactone ring is expanded to a 44-membered ring by transesterification. The O. are *polyketides, the starter unit originates from ornithine. *Desertomycin A is a biosynthetic precursor of the O. (group name: marginolactones). In cell tests O. A inhibits the biosynthesis of cholesterol and *in vivo* has lipid-regulating properties. The O. have weak activities against protozoa, but not against fungi and yeasts, in contrast to the related desertomycins.

Lit.: J. Chem. Soc., Perkin Trans. 1 **1993**, 2525–2531 ▪ J. Org. Chem. **59**, 6986–6993 (1994) ▪ Justus Liebigs Ann. Chem. **1993**, 573–579. – *[HS 294190; CAS 143436-50-4 (O. A); 143452-11-3 (O. B)]*

Obafluorin [*N*-(2,3-dihydroxybenzoyl-4-(4-nitrophenyl)-L-threonine lactone].

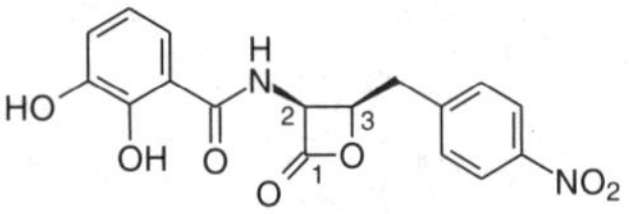

$C_{17}H_{14}N_2O_7$, M_R 358.31. A β-lactone *antibiotic from *Pseudomonas fluorescens*, crystallizes from acetonitrile/water, mp. 109–113 °C, $[\alpha]_D^{25}$ +116° (CH_3CN). O. exhibits a broad but weak antibacterial activity, presumably on account of the rapid hydrolysis of the lactone ring (half-life: a few minutes at basic pH or in methanol)[1]. For synthesis, see *Lit.*[2].
Biosynthesis*[3]*: According to labelling experiments with ^{13}C-enriched precursors, the 4-nitrobenzyl unit

including C-3 originates from 4-aminophenylalanine (see chloramphenicol) which, in turn, like the 2,3-dihydroxybenzoyl unit, is derived from *shikimic acid. Glycine provides C-1 and C-2 by way of glyoxylate. Further β-lactone antibiotics related to O. are SQ 26 517 (*N*-acetyl-L-threonine lactone) and SQ 27 012 (*N*-phenylacetyl-L-serine lactone)[4]; see also papulinone.

Lit.: [1]J. Antibiot. **37**, 802f. (1984); J. Org. Chem. **50**, 5491–5495 (1985). [2]J. Org. Chem. **57**, 10f. (1992); **59**, 3642 (1994). [3]J. Chem. Soc. Perkin Trans. 1 **1992**, 103–113. [4]J. Antibiot. **35**, 814–821, 900f. (1982). – *[HS 294190; CAS 92121-68-1 (O.); 83112-05-4 (SQ 26517); 83112-06-5 (SQ 27012)]*

Obtusallene I.

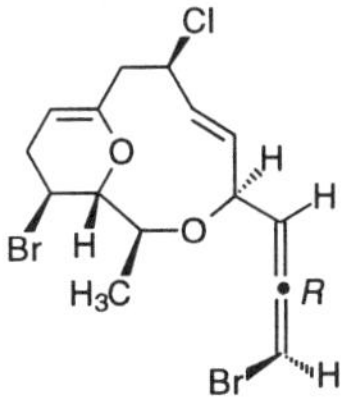

(relative configuration)

$C_{15}H_{17}Br_2ClO_2$, M_R 424.56, colorless cryst., mp. 165–167 °C, $[\alpha]_D$ −257.6° ($CHCl_3$). The allene O. I was first isolated from the alga *Laurencia obtusa* but also occurs in other *Laurencia* species. It is accompanied by numerous other halogenated C_{12}–C_{20} fatty acid derivatives with antifungal activities. O. exists as the 3,13-dioxabicyclo[7.3.1]trideca-5,9-diene as a result of the formation of ether bridging bonds.

Lit.: Acta Crystallogr., Sect. B **38**, 1368 (1982) ▪ Chem., A Eur. J. **3**, 1223 (1997) (conformation) ▪ Scheuer I **5**, 247 ▪ Tetrahedron **47**, 2273 (1991) ▪ Tetrahedron Lett. **23**, 579 (1982) ▪ see also okamurallene. – *[CAS 81920-18-5]*

(+)-Occidentalol see eudesmanes.

Ochotensimine, ochotensine see spirobenzylisoquinoline alkaloids.

Ochracin see mellein.

Ochratoxins. *Mycotoxins from *Aspergillus ochraceus*, *A. melleus*, *A. glaucus*, and other *Aspergillus* and *Penicillium* species. Most of these fungi also produce other mycotoxins, e. g., *penicillic acid or *citrinin. In lodged grain, the most important producer is *P. viridicatum*. Fungi of the *A. glaucus* group are xerophilic and belong to the "pacemaker flora" which facilitates the colonization of stored grain by other fungi.

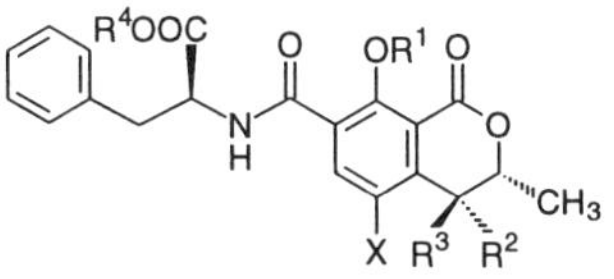

	X	R[1]	R[2]	R[3]	R[4]
Ochratoxin A (**1**)	Cl	H	H	H	H
Ochratoxin B (**2**)	H	H	H	H	H
Ochratoxin C (**3**)	Cl	H	H	H	C_2H_5
O-Methylochratoxin C (**4**)	Cl	CH_3	H	H	C_2H_5
(4*R*)-4-Hydroxyochratoxin A (**5**)	Cl	H	OH	H	H
(4*S*)-4-Hydroxyochratoxin A (**6**)	Cl	H	H	OH	H

O. A shows high acute toxicity [LD_{50} (rat p.o.) 22 mg/kg, (dog p.o) 1 mg/kg, (pig p.o.) 0.62 mg/kg],

Table: Data of Ochratoxins.

	molecular formula	M_R	mp. [°C]	$[\alpha]_D$	CAS
1	$C_{20}H_{18}ClNO_6$	403.82	169–173	−118° ($CHCl_3$)	303-47-9
2	$C_{20}H_{19}NO_6$	369.38	221	−35° (C_2H_5OH)	4825-86-9
3	$C_{22}H_{22}ClNO_6$	431.87		−100° (C_2H_5OH)	4865-85-4
4	$C_{23}H_{24}ClNO_6$	445.89			
5	$C_{20}H_{18}ClNO_7$	419.82			
6	$C_{20}H_{18}ClNO_7$	419.82	216–218		35299-87-7

on chronic consumption it has kidney damaging and tumor promoting effects in rats and mice. The acute toxicity can be neutralized by phenylalanine. O. A inhibits the loading of tRNS with Phe by Phe-tRNS synthase and is considered to be the inducer of "endemic Balkan nephropathy", a kidney disease affecting people in Bulgaria, Yugoslavia, and Romania. A high incidence of cancers of the pelvis region, ureter, and bladder is associated with the disease. O. are formed biogenetically from phenylalanine and a dihydroisocoumarin part (pentaketide). Some countries have legally defined the highest tolerable amounts, e. g., Denmark, Romania, Brazil. O. A has a high "carry-over" effect. When O. A.-containing fodder is given to pigs or chickens, the substance accumulates especially in the kidneys and also in blood and muscles. Beef does not contain O. A. because it is metabolized in the rumen.

Lit.: Beilstein E V **18/9**, 55 ff. ▪ Betina, chap. 8 ▪ Cole-Cox, p. 128–151 ▪ DFG (ed.), Ochratoxin A – Vorkommen u. toxikolog. Bewertung, Weinheim: VCH Verlagsges. 1990 ▪ J. Med. Chem. **42**, 3075 (1999) (activity) ▪ Nat. Tox. Program Report No. 358, Toxicology and Carcinogenesis Studies of Ochratoxin A in F344/N Rats, US Department of Health and Human Services, NIH Publ. 89-2813 (1989) ▪ Sax (8.), CHP 250 ▪ Stormer, Ochratoxin A – a Mycotoxin of Concern, in: Bhatnagar, Lillehoj & Arora (eds.), Handbook of Applied Mycology, vol. 5, chap. 16, New York: Dekker 1992.

Ocimene [3,7-dimethyl-1,3,6(7)-octatriene]. An unsaturated, acyclic *monoterpene occurring mostly as a mixture of isomers. $C_{10}H_{16}$, M_R 136.24, colorless oil, bp. ca. 60 °C.

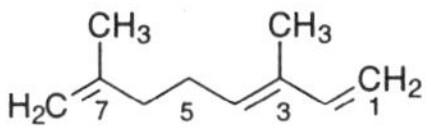

(*E*)-α-Ocimene

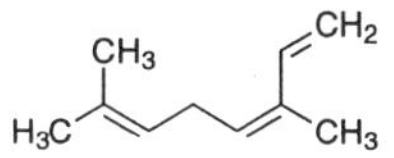

(*Z*)-β-Ocimene

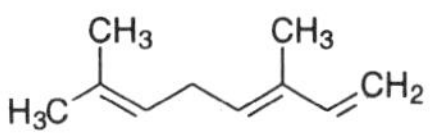

(*E*)-β-Ocimene

Occurrence: α-O. in genuine *lavender oil 3%[1]. β-O. in *sweet basil oil (*Ocimum basilicum*) and the essential oil of *Tagetes minuta* (Asteraceae) 41%[2].

Lit.: [1]Helv. Chim. Acta **43**, 1619 (1960). [2]Phytochemistry **10**, 1359 (1971).

gen.: Beilstein E IV **1**, 1108 ▪ Karrer, No. 30 ▪ Merck-Index (12.), No. 6837 ▪ Ullmann (5.) **A 11**, 154. – *[CAS 6874-10-8 ((E)-α-O.); 3779-61-1 ((E)-β-O.); 3338-55-4 ((Z)-β-O.)]*

Ocimenone [(*E*)- and (*Z*)-2,6-dimethyl-2,5,7-octatrien-4-one]

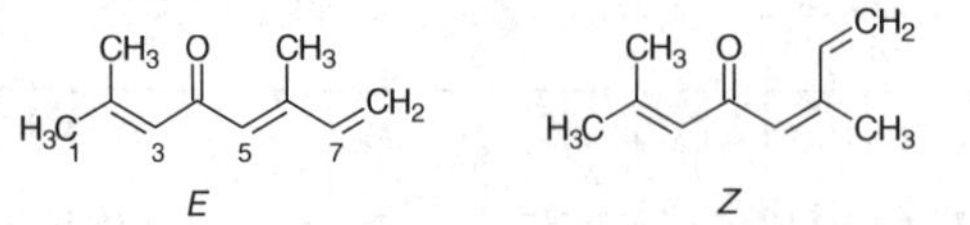

An unsaturated acyclic *monoterpene ketone occurring in the *E*- and *Z*-configurations. $C_{10}H_{14}O$, M_R 150.22, colorless oil, bp. 72–74 °C (2 kPa, mixture of isomers).
Occurrence: Mixtures of (*E*)- and (*Z*)-O. in the essential oil of *Tagetes minuta* (Asteraceae)[1]. The 2,3-dihydro derivatives of O., (*E*)- and (*Z*)-*tagetones* ($C_{10}H_{16}O$, M_R 152.24) are, together with O. and *ocimene, the characteristic natural components of the essential oils of *Tagetes* species[2], (*Z*)-O. is present in the essential oil of *Lippia asperifolia* (Verbenaceae; up to ca. 5%)[3]. For synthesis, see *Lit.*[4].

Lit.: [1]Phytochemistry **10**, 1359 (1971). [2]Hegnauer **III**, 457. [3]Helv. Chim. Acta **31**, 29 (1948). [4]Tetrahedron Lett. **1976**, 1625.
gen.: Karrer, No. 438. – *[CAS 33746-72-4 ((E)-O.); 33746-71-3 ((Z)-O.); 6752-80-3 ((E)-tagetone); 3589-18-9 ((Z)-tagetone)]*

(*Z*)-6-Octadecenoic acid see petroselinic acid.

(*Z*)-1,5-Octadien-3-one see tea flavor.

Octalactin A see fungi.

Octanal see alkanals.

4-Octanolide see alkanolides.

(*E*)-2-Octenal see alkenals.

(−)-(*R*)-1-Octen-3-ol (Matsutake alcohol). $C_8H_{16}O$, M_R 128.21, bp. 175 °C, $[\alpha]_D^{20}$ −20.2°; LD_{50} (rat p.o.) 340 mg/kg. A typical *mushrooms aroma constituent (olfactory threshold 1 ppb[1]) with an earthy-fungal odor; *impact compound in mushroom and Camembert flavor (see cheese flavor), but also occurring in many *essential oils and flavors[2], e.g., in *lavender, *peppermint, *rosemary, and *thyme oil, as well as seafood flavors. Only the (*R*)-enantiomer occurring in 90–97% enantiomeric excess possesses the pure fungus odor, the (*S*)-compound has a more vegetable-like odor.
Biosynthesis: (−)-(*R*)-1-O. is formed in the cultivated mushroom (*Agaricus bisporus*) by oxidation of *linoleic acid. In the first step a lipoxygenase-catalyzed oxidation gives (8*E*,10*S*,12*Z*)-10-hydroperoxy-8,12-octadienoic acid which is cleaved by hydroperoxide lyase into (−)-(*R*)-1-O. and (*E*)-10-oxo-8-decenoic acid:

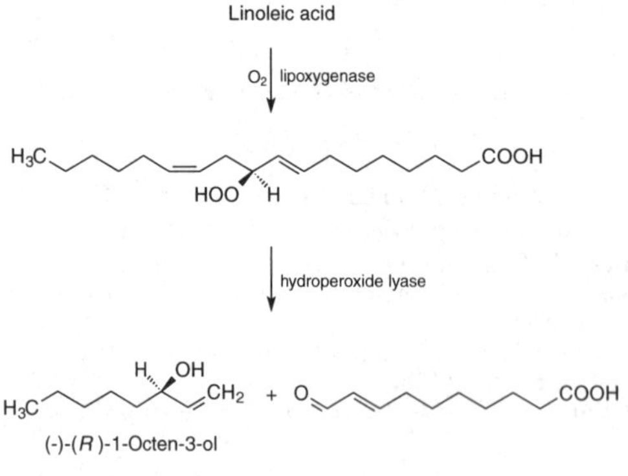

For activity as an insect attractant for tsetse flies, mosquitoes, and corn weevils, see *Lit.*[3].

Lit.: [1]Perfum. Flavor. **16** (1), 1–19 (1991). [2]TNO list (6.), Suppl. 5, p. 452. [3]Bull. Entomol. Res. **75**, 209 ff. (1985); J. Am. Mosqu. Control Assoc. **5** (3.), 311 (1989); J. Chem. Ecol. **17**, (3), 581–597 (1991).
gen.: Synthesis: Chem. Pharm. Bull. **40**, 3073 (1992) ▪ J. Org. Chem. **58**, 718 (1993). – *Isolation:* Aust. J. Chem. **35**, 373 (1982); Phytochemistry **32**, 1235 (1993). – *Use:* SOFW J. **120**, 577 (1994). – *[CAS 3687-48-7 (R)]*

1-Octen-3-one. $H_3C–(CH_2)_4–CO–CH=CH_2$, $C_8H_{14}O$, M_R 126.20. Liquid with a metallic-fungus-like odor, bp. 63–65 °C (1.8 kPa); olfactory threshold 0.1 ppb[1]. O. is a degradation product of *linoleic acid and occurs as an *mushrooms aroma constituents and in *lavender oil, as well as in *bread, *cheese, seafood, *tea flavor, etc.[2].

Lit.: [1]Ohloff, p. 62. [2]TNO list (6.), Suppl. 5, p. 453.
gen.: J. Org. Chem. **56**, 5924 (1991) (synthesis) ▪ SOFW J. **120**, 577 (1994). – *[CAS 4312-99-6]*

Octicidin see phaseolotoxin.

Octopine [N^2-((*R*)-1-carboxyethyl)-arginine]. Formula, see nopaline. $C_9H_{18}N_4O_4$, M_R 246.27, mp. 281–282 °C, $[\alpha]_D^{17}$ +20.94° (H_2O). O. was first found in the muscles of mollusks, e.g., in scallops (*Patinopecten yessoensis*) and octopodes (*Eledone moschata*). It also occurs together with *nopaline in root neck galls of plants.
Biosynthesis: In plants O. is formed by condensation of Arg and pyruvate after infection with *Agrobacterium tumefaciens*. O. and nopaline are designated as opines. Upon infection, the gene for opine formation by Ti-plasmid (tumor-inducing) is integrated in the plant's genome. The opines serve *A. tumefaciens* as a source of energy.

Lit.: Beilstein E IV **4**, 2653 ▪ Bull. Chem. Soc. Jpn. **55**, 261 (1982) (synthesis) ▪ Pühler (ed.), Mol. Genet. Bact.-Plant Interact. (symposium), p. 303–321, Berlin: Springer 1983. – *[CAS 34522-32-2]*

Octosyl acids.

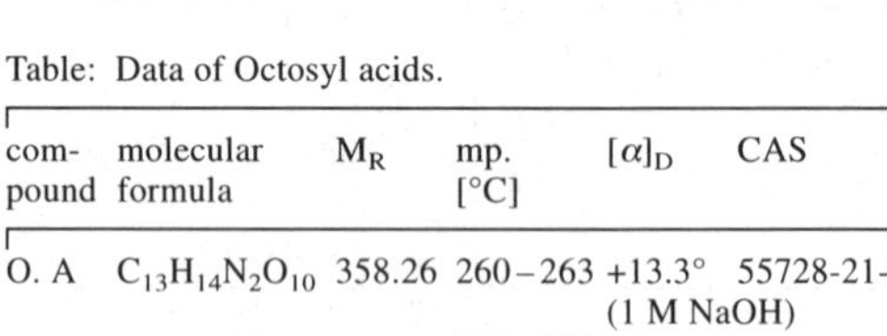

R = COOH : O. A
R = CH_2OH : O. B
O. C

Table: Data of Octosyl acids.

compound	molecular formula	M_R	mp. [°C]	$[\alpha]_D$	CAS
O. A	$C_{13}H_{14}N_2O_{10}$	358.26	260–263 290–295 (monohydrate, decomp.)	+13.3° (1 M NaOH)	55728-21-7
O. B	$C_{13}H_{16}N_2O_9$	344.28	200 (decomp.)		55728-22-8
O. C	$C_{13}H_{12}N_2O_{10}$	356.25	192–198 (monohydrate)		55728-23-9

*Nucleoside antibiotics with a bicyclic carbohydrate unit from *Streptomyces cacaoi* var. *asoensis*. Distinctions are made between *O. A* (pK_{a1} 3.0; pK_{a2} 4.3; pK_{a3} 9.4), *O. B*, and *O. C* (pK_{a1} 3.1; pK_{a2} 4.5; pK_{a3} 9.9). O. represent a large group of *polyoxin-like antibiotics with antifungal activities.

Lit.: J. Am. Chem. Soc. **97**, 943 (1975) ▪ Tetrahedron Lett. **1979**, 3441. – *Review:* J. Antibiot. **41**, 1711 (1988). – *Synthesis:* ACS Symp. Ser. **386**, 64 (1989) ▪ J. Am. Chem. Soc. **110**, 7434 (1988).

Odiline see imidazole alkaloids.

(O)enanthic acid (heptanoic acid). $H_3C–(CH_2)_5–COOH$, $C_7H_{14}O_2$, M_R 130.19, mp. –7.5 °C, bp. 223 °C; oil with a rancid odor, it is toxic and shows bactericidal activity, it is soluble in organic solvents. Esters of O. obtained from fusel oils (Greek: oinanthe = grapevine) are used as components of perfumes.

Lit.: Beilstein EIV **2**, 958 ▪ Karrer, No. 693 ▪ Merck-Index (12.), No. 4695. – *[HS 2915 90; CAS 111-14-8]*

Oenin (*malvidin 3-*O*-glucoside, 3-*β*-D-glucopyranosyloxy-4′,5,7-trihydroxy-3′,5′-dimethoxyflavylium). $C_{23}H_{25}O_{12}^+$, M_R 493.44, chloride: violet crystals with yellow-metallic lustre in transmitted light and bronze-color in reflected light. O. belongs to the *anthocyanins and occurs as the main pigment of blue grapes (*Vitis vinifera*, Vitaceae, Greek: oinos = wine).

Lit.: Karrer, No. 1747 ▪ Tetrahedron **39**, 3005 (1983) ▪ Tetrahedron Lett. **22**, 3621 (1981). – *[CAS 7228-78-6]*

Oestr... see estr...

Oil bodies see mosses.

Oil of cardamom (cardamom oil). Colorless to light yellow oil with a penetrating, fresh-aromatic, slightly camphor-like spicy odor and persisting, hot-bitter, aromatic taste.

Production: By steam distillation of cardamom (cardamum, *Elettaria cardamomum*, *E. major*, and *Amomum compactum*).

Composition[1]**:** Main components are *1,8-*cineole* and *α-terpinyl acetate* ($C_{12}H_{20}O_2$, M_R 196.29) (ca. 20–50% each). The exact composition depends on the location (India, Sri Lanka, Central America).

Use: For improving the aromatic flavors of foods and alcoholic beverages; in small amounts in the perfume industry.

Lit.: [1] Perfum. Flavor. **11** (1), 30 (1986); **14** (6), 87 (1989); **16** (1), 51 (1991); **17** (6), 51 (1992).

gen.: Arctander, p. 126 ▪ Bauer et al. (2.), p. 141 ▪ ISO 4733 (1981). – *Toxicology:* Food Cosmet. Toxicol. **12**, 837 (1974). – *[HS 3301 29; CAS 8000-66-6 (C.); 80-26-2 (α-terpinyl acetate)]*

Oil of Chenopodium (wormseed oil). Essential oil obtained from *Chenopodium ambrosioides* var. *anthelminticum* (goosefoot), previously used in humans and animals against worm parasites (eelworms and hookworms). O.C. is no longer used in human medicine on account of its side effects. O.C. contains mainly *ascaridole (60–75%), ascaridole glycol, *p*-cymol (20–30%, see cymenes), *α*-terpinene, limonene.

Lit.: Hager (4.) **3**, 842–851. – *[HS 3301 29]*

Oil of mace see nutmeg oil.

Oils see fats and oils.

Okadaic acid (phytoxin II).

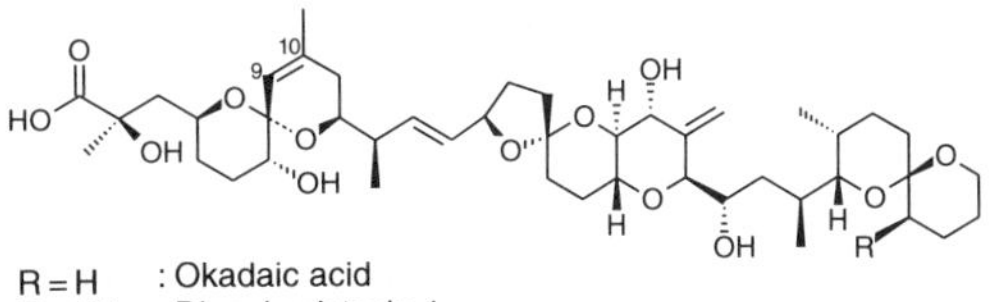

R = H : Okadaic acid
R = CH_3 : Dinophysistoxin-1

$C_{44}H_{68}O_{13}$, M_R 805.02, cryst., mp. 164–166 °C, $[\alpha]_D^{25}$ +28° ($CHCl_3$). The polyether fatty acid O. was first isolated from the marine sponges *Halichondria okadai* and *H. melanodocia*[1], later also from various dinoflagellates *Prorocentrum lima*, *P. concavum*[2] and *Dinophysis* sp.[3]. As active filterers, sponges and mussels accumulate O. through uptake of algae. O. is a highly potent toxin [LD_{50} (mouse i.p.) 192 µg/kg] and is considered to be one of the substances responsible for diarrhetic shellfish poisoning (DSP). Consumption of mussels containing O. leads to severe diarrhea and numerous deaths have been reported. Other polyether toxins, besides O., responsible for this poisoning are, e.g., *dinophysistoxin 1*[3] [$C_{45}H_{70}O_{13}$, M_R 819.04, mp. 128–130 °C; $[\alpha]_D^{20}$ +15.5° ($CHCl_3$)] from toxic mussels and culture filtrates of *P. lima* and *acanthifolicin*[1] (= O. 9*S*,10*R*-episulfide, $C_{44}H_{68}O_{13}S$, M_R 837.08, mp. 167–169 °C, $[\alpha]_D$ +25.3°, amorphous) from the sponge *Pandoras acanthifolium*.

Activity: O. functions as a tumor promoter; in contrast to the phorbol esters which activate protein kinase C, O. inhibits protein phosphatases 1 (PP1) and 2A (PP2A), resulting in a 2.5- to 3-fold increase in the degree of phosphorylation of proteins in hepatic and adipose cells which, in turn, causes persisting contractions of the smooth musculature.

The consequence of this is an increased secretion of glucose from hepatic cells and an inhibition of fatty acid biosynthesis. The high degree of phosphorylation (see cholera toxin) leads to an elevated secretion of gastric juices in the intestines which cannot be resorbed, and thus to dehydratation of the body. It is interesting to note that the toxin *calyculin A[4] from the sponge *Discodermia calyx* has a similar activity to O. although its structure is different. Calyculin has a 10- to 100-fold stronger inhibitory effect on PP1 than O.

Lit.: [1] J. Am. Chem. Soc. **103**, 2467, 2469 (1981); Scheuer II **1**, 141; Tetrahedron **41**, 1019 (1985); Trends Biochem. Sci. **15**, 98–102 (1990). [2] Toxicon **28**, 371 (1990). [3] Agric. Biol. Chem. **50**, 2853 (1986); Chem. Ind. (London) **1995**, 214; J. Nat. Prod. **54**, 1487 (1991) (also biosynthesis); J. Chem. Soc., Chem. Commun. **1992**, 39; **1995**, 597; Tetrahedron Lett. **35**, 1441 (1994). [4] J. Am. Chem. Soc. **108**, 2780 (1986).

gen.: *Pharmacology:* Adv. Protein Phosphatases **5**, 219, 579 (1989) ▪ FEBS Lett. **217**, 81 (1988) ▪ Nature (London) **337**, 78 (1989). – *Structure:* Tetrahedron **51**, 12229–12238 (1995). – *Synthesis:* Angew. Chem. Int. Ed. Engl. **38**, 2258 (1999) ▪ Chem. Rev. **93**, 1897 (1993) ▪ J. Am. Chem. Soc. **119**, 8381 (1997); **120**, 2523, 2534 (1998) ▪ J. Chem. Soc., Perkin Trans. 1 **1998**, 3907 ▪ Tetrahedron **43**, 4737, 4749, 4759, 4767 (1987). – *[CAS 77739-71-0 (acanthifolicin); 81720-10-7 (dinophysistoxin 1); 78111-17-8 (okadaic acid)]*

Okamurallene.

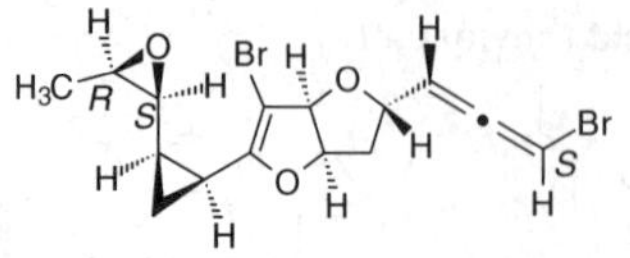

$C_{15}H_{16}Br_2O_3$, M_R 404.10, oil, $[\alpha]_D$ +160° ($CHCl_3$). An acetogenin from the brown algae *Laurencia okamurai* and *L. intricata* with an unusual structure.

Lit.: Chem. Lett. **1991**, 33 ▪ Phytochemistry **28**, 2145 (1989) ▪ Scheuer I **5**, 235 ▪ see also obtusallene I. – *[CAS 80539-33-9]*

Oleamide [(*Z*)-9-octadecenoic acid amide].

$C_{18}H_{35}NO$, M_R 281.48, mp. 75–76 °C. O. was first isolated from cerebrospinal fluid of cats that had been deprived of sleep for a longer time. The compound has also been detected in the cerebrospinal fluids of humans and rats. Injections of O. in animals artificially induce sleep (*"sleep lipids"*). O. is converted in the brain to oleic acid by enzymatic hydrolysis and thus deactivated. Another physiologically active fatty acid amide occurring in cerebrospinal fluid is *erucic acid amide* [(*Z*)-13-docosenoic acid amide, $C_{22}H_{43}NO$, M_R 337.59, mp. 94 °C].

Lit.: Angew. Chem. Int. Ed. Engl. **34**, 2362 (1995) ▪ Chem. Ind. (London) **1995**, 445 ▪ J. Am. Chem. Soc. **118**, 5938 (1996) ▪ Science **268**, 1506 (1995). – *[HS 2924 10; CAS 301-02-0 (O.); 112-84-5 (erucic acid amide)]*

Olean (1,7-dioxaspiro[5.5]undecane).

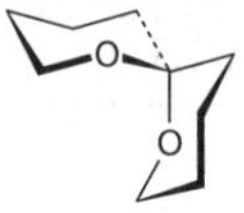

(*R*)

$C_9H_{16}O_2$, M_R 156.22, bp. 77–78 °C (1.73 kPa); liquid with a flower-like odor. O. is a chiral sex hormone secreted by female olive flies *Dacus oleae* [1]. The natural product is racemic [2]; the (*R*)-enantiomer, $[\alpha]_D^{21}$ –121.6° (pentane), attracts males of the species, the (*S*)-enantiomer, $[\alpha]_D^{23}$ +109.3° (pentane), has an attractive effect on the females. The *spiroacetal O. also occurs in other fruit flies of the genus *Dacus* [3].

Lit.: [1] J. Chem. Soc., Chem. Commun. **1980**, 52 ff. [2] J. Chem. Ecol. **12**, 1559–1568 (1986). [3] Chem. Rev. **95**, 789–827 (1995).

gen.: Synthesis: ApSimon **9**, 462–470 ▪ Tetrahedron Lett. **33**, 1799 (1992). – *Biosynthesis:* Chem. Commun. **1998**, 863. – *[CAS 180-84-7 (O.); 90839-16-0 ((R)-O.); 90839-15-9 ((S)-O.)]*

Oleanane.

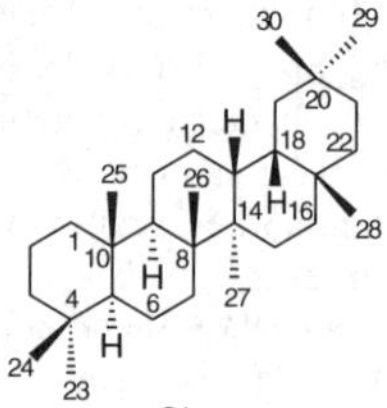

Oleanane

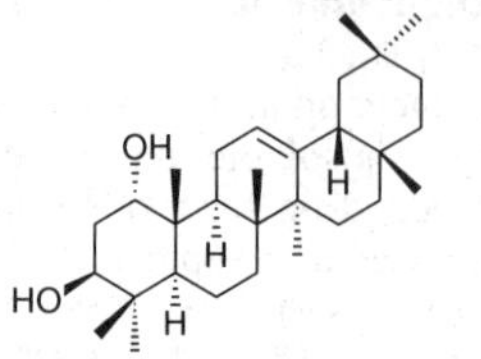

Castanopsol

$C_{30}H_{52}$, M_R 412.74, 18α-O.: mp. 210 °C, $[\alpha]_D$ +40.3° ($CHCl_3$), a pentacyclic *triterpene hydrocarbon occurring in crude oil. O. represents the saturated skeleton of β-*amyrin. Numerous related triterpenes are derived from O.: e. g., the hydrocarbons *11,13(18)-oleanadiene*, ($C_{30}H_{48}O$, M_R 424.71, mp. 226–227 °C), *12-oleanene* ($C_{30}H_{50}$, M_R 410.73, mp. 162–164 °C), and *18-oleanene* (mp. 174–175 °C) from polypody species (*Polypodium*), as well as various polyols (see table) that are widely distributed as glycosides (*saponins).

Lit.: Anal. Chem. **47**, 1617 (1975) ▪ Nature (London) **228**, 355 (1970) ▪ Tetrahedron **32**, 3051 (1976). – *[CAS 30759-92-3 (18α-O); 471-67-0 (18β-O); 54411-26-6 (11,13-oleanadiene); 471-68-1 (12-oleanene); 432-11-1 (18-oleanene)]*

Oleandomycin. $C_{35}H_{61}NO_{12}$, M_R 687.87, mp. 110 °C, $[\alpha]_D$ –65° (CH_3OH). International free name for a 14-membered *macrolide antibiotic with a broad spectrum of antimicrobial activity isolated from cultures of *Streptomyces antibioticus*. O. is readily soluble in methanol, ethanol, butanol, acetone, less soluble in wa-

Table: Polyols of the Oleanane type.

compound	trivial name	molecular formula	M_R	CAS	occurrence
12-Oleanene-1α,3β-diol	Castanopsol	$C_{30}H_{50}O_2$	442.73	66088-16-2	*Castanopsis indica*
12-Oleanene-3β,6β-diol	Daturadiol	$C_{30}H_{50}O_2$	442.73	41498-79-7	*Datura innoxia*
12-Oleanene-3β,16β-diol	Maniladiol	$C_{30}H_{50}O_2$	442.73	595-17-5	Manila elemi
12-Oleanene-3β,29-diol	Paniculatadiol	$C_{30}H_{50}O_2$	442.73	58167-03-6	*Celastrus paniculatus*
12-Oleanene-3β,16α,28-triol	Primulagenin A	$C_{30}H_{50}O_3$	458.73	465-95-2	*Primula officinalis*
12-Oleanene-3β,19α,28-triol	19α-Hydroxy-erythrodiol	$C_{30}H_{50}O_3$	458.73	65063-30-1	*Gardenia gummifera*
12-Oleanene-2β,3β,23,28-tetraol	Castanogenol	$C_{30}H_{50}O_4$	474.72	26553-62-8	*Castanospermum australe*
12-Oleanene-3β,15,22α,28-tetraol	Acergenin	$C_{30}H_{50}O_4$	474.72	52591-07-8	*Acer caesium*
12-Oleanene-3β,15α,16α,22α,28-pentaol	Barrigenol A_1	$C_{30}H_{50}O_5$	490.72	15448-03-0	*Barringtonia asiatica* and others
12-Oleanene-3β,16α,22α,23,28-pentaol	Camelliagenin C	$C_{30}H_{50}O_5$	490.72	14440-27-8	*Camellia japonica*
12-Oleanene-3β,16β,21β,22α,23,28-hexaol	Gymnemagenin	$C_{30}H_{50}O_6$	506.72	22467-07-8	*Gymnema sylvestre* cf. Gymnemic acid
12-Oleanene-3β,16α,21β,22α,24,28-hexaol	Protoaescigenin	$C_{30}H_{50}O_6$	506.72	20853-07-0	*Aesculus hippocastanum* cf. escin

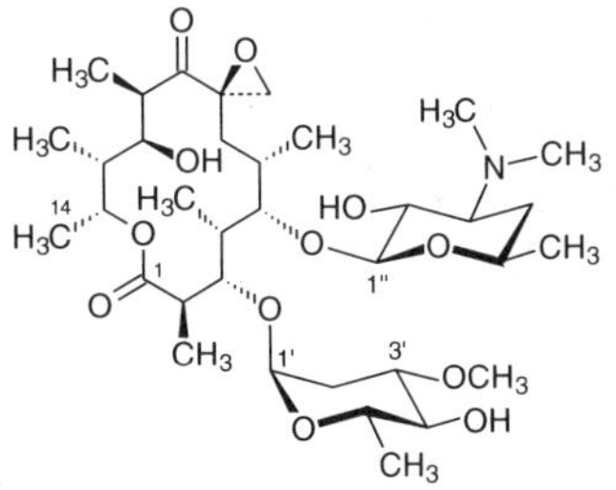

ter, and insoluble in hexane. The spectrum of activity and uses of O. and its triacetyl derivative (semisynthetic, international free name troleandomycin) in human medicine and as a fodder additive are very similar to those of *erythromycin although the activity of O. is weaker. Further derivatives are diproleandomycin (semisynthetic) and O. B and O. Y. (= *O*-demethyl-O.) isolated from *S. antibioticus*. Experiments on mutasynthesis with *S. antibioticus* and the mutasynthon erythronolide A oxime gave rise to a novel O. 9-oxime derivative [1].

Lit.: [1] Sebek, Antibiotics, in Rehm & Reed (eds.), Biotechnology, vol. 6a, p. 239–276, Weinheim: VCH Verlagsges. 1984; J. Antibiot. **36**, 1439–1450 (1983).
gen.: Clin. Pharmacokinet. **16**, 193 (1989) ▪ J. Am. Chem. Soc. **116**, 3623 (1994); **118**, 11323 (1996) (synthesis) ▪ Kirk-Othmer (4.) **3**, 177 ▪ Tetrahedron **41**, 3569–3624 (1985) ▪ Tetrahedron Lett. **31**, 709–712 (1990). – *[HS 2941 90; CAS 3922-90-5 (O.); 2751-09-9 (troleandomycin)]*

L-Oleandrose (2,6-dideoxy-3-*O*-methyl-L-*arabino*-hexose).

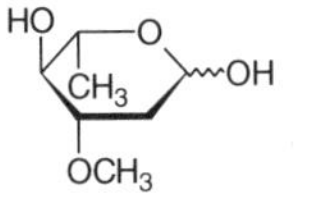

$C_7H_{14}O_4$, M_R 162.19, cryst., mp. 62–63 °C, $[\alpha]_D^{20}$ +13° (H_2O). The deoxy sugar of the toxic glycoside oleandrin (oleandrigenin α-L-oleandroside, from oleander leaves) and of macrolide antibiotics such as the *avermectins and *oleandomycin.

Lit.: Beilstein EIV **1**, 4193 ▪ Carbohydr. Res. **189**, 113 (1989) ▪ Nachr. Chem. Tech. Lab. **32**, 798–802 (1984) ▪ Synth. Commun. **22**, 2459 (1992) ▪ Tetrahedron **38**, 3067 (1982) ▪ Ullmann (5.) **A 5**, 275. – *[HS 2940 00; CAS 87037-59-0]*

Oleanolic acid (3β-hydroxy-12-oleanene-28-carboxylic acid).

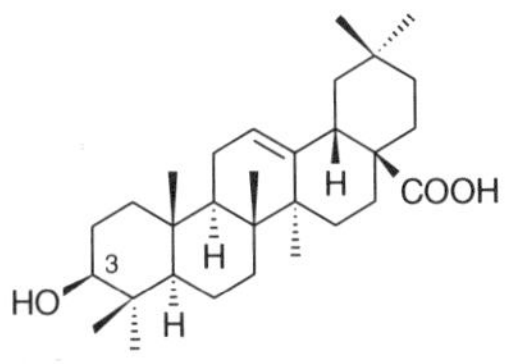

$C_{30}H_{48}O_3$, M_R 456.71, mp. 306–308 °C, $[\alpha]_D$ +79.5° ($CHCl_3$), insoluble in water, soluble in ether. The *triterpene O. occurs in the free and acetylated forms as well as in many glycosides of various plants such as mistletoe, clove, sugar beet, olive leaves, beeches, etc. O. has the *oleanane skeleton.

Lit.: Acta Crystallogr., Sect. C **53**, 349 (1997) ▪ Beilstein EIV **10**, 1164 ▪ J. Am. Chem. Soc. **115**, 8873 (1993) ▪ Karrer, No. 1982 ▪ Merck-Index (12.), No. 6964 ▪ Phytochemistry **31**, 1321 (1992) ▪ Simonsen et al., The Terpenes, vol. 5, p. 221–285, Cambridge: University Press 1957. – *[HS 2938 90; CAS 508-02-1]*

Oleuropein see ligustroside.

Oligomycins.

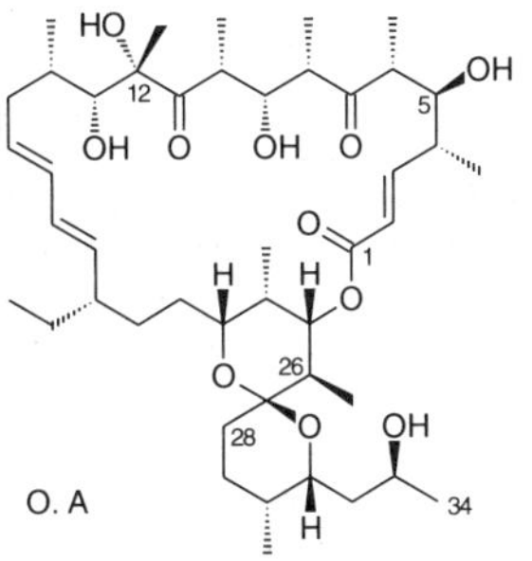

O. A

The O. are a group of 26-membered *macrolide antibiotics in which the side chain is folded to give a spiroketal unit. The components A to F are known and differ slightly at C-12, C-26, C-28, and C-34: *O. A* {$C_{45}H_{74}O_{11}$, M_R 791.07, cryst., mp. 140–141 °C, $[\alpha]_D$ –54.5° (dioxane)}, *O. B* (28-oxo-O. A, $C_{45}H_{72}O_{12}$, M_R 803.06), *O. C* (12-deoxy-O. A, $C_{45}H_{74}O_{10}$, M_R 775.08), *O. D* (*rutamycin A*, 26-demethyl-O. A, $C_{44}H_{72}O_{11}$, M_R 777.05), *O. E* (26-hydroxy-28-O. A, $C_{45}H_{72}O_{13}$, M_R 821.06), and *O. F* (34-methyl-O. A, $C_{46}H_{76}O_{11}$, M_R 805.10). The O. are *polyketides formed by *streptomycetes (e. g., *Streptomyces diastatochromogenes*). They have cytotoxic and antifungal activities and are also inhibitors of mitochondrial ATPases of the F-type. The main activity of O. E is against Gram-positive bacteria while O. F exhibits pronounced immunosuppressive activity.

Lit.: Helv. Chim. Acta **67**, 1208–1216 (1984) ▪ J. Antibiot. **40**, 1053–1057 (1987); **46**, 1334–1341 (1993) ▪ J. Chem. Soc., Chem. Commun. **1978**, 318 f. ▪ J. Org. Chem. **51**, 4264–4271 (1986); **55**, 6260 (1990); **63**, 4572 (1998) (synthesis) ▪ Magn. Reson. Chem. **23**, 676–683 (1985) ▪ Methods Enzymol. **55**, 472 (1979); **156**, 1–25 (1988). – *[CAS 579-13-5 (O. A); 11050-94-5 (O. B); 11052-72-5 (O. C); 1404-59-7 (O. D)]*

Oligosaccharides. General name for molecules formed by condensation of 2 to ca. 10 *monosaccharide units. O. can have linear, branched, or cyclic structures. In contrast to the *polysaccharides, the properties of the O. still correspond well with those of the monosaccharides. The compounds are classified as *disaccharides, trisaccharides (name of O. with 3 monosaccharide units, occurring rarely in nature in the free form, e. g., *gentianose, *kestoses, *raffinose), tetrasaccharides (name for O. with 4 monosaccharide units, e. g., *acarbose, sialyl-Lewis X, stachyose, secalose, lychnose), etc. The oligomerization of sugars theoretically leads to a very large number of possible stereoisomers, but only a few of them occur in nature. For nomenclature of O., see *Lit.* [1].

Occurrence: O. occur in the free form mainly in the plant kingdom. They consist mainly of hexoses. Animal sources are, in particular, milk and honey. The most widely distributed, non-reducing O. are *sucrose, raffinose, and *trehalose. Some O. exhibit antibiotic activity, the most important group of these antibiotics are the aminoglycosides (see also kanamycins, neomycin, streptomycin) containing *amino sugars. O. often occur as components of various glycosides in the animal and plant kingdoms, e. g., *Digitalis glycosides and *strophanthins, glycolipids, *glycoproteins, glycohemoglobins. The O. moieties in these substances are essential for cell to cell recognition and interactions. The O. units function as receptors for proteins, hormones, and viruses and play a decisive role in immune reactions.

Biosynthesis & synthesis: As a consequence of the structural diversity of the O. (linkage sites, branching sites, α-β-coupling of the glycoside units) "zip" processes like those for protein biosynthesis are not possible for O. In general, the formation of a saccharide bond requires a specific enzyme, a glycosyl transferase. For synthesis, see *Lit.*[2].

Lit.: [1] Pure Appl. Chem. **54**, 1517–1522 (1982); **55**, 605–622 (1983); **56**, 1031–1048 (1984). [2] ACS Symp. Ser. **560**, 2–18 (1994); Adv. Carbohydr. Chem. Biochem. **42**, 193–226 (1984); Angew. Chem. Int. Ed. Engl. **21**, 155–173 (1982); **25**, 212–235 (1986); **26**, 555 ff. (1987); **34**, 1432 (1995); **35**, 1380–1419 (1996); Chem. Eng. News (23. 9.) **1996**, 62–66; (22. 3.) **1999**, 30 f.; Chem. Rev. **92**, 1167–1195 (1992); Contemp. Org. Synth. **3**, 173–200 (1996); Curr. Opin. Biotechnol. **10**, 616–624 (1999) (biocatalytic); Drug Discovery Today **1**, 331–342 (1996); Fuhrhop & Penzlin, Organic Synthesis: Concepts, Methods, Starting Materials, 2nd edn., p. 269 ff., Weinheim: VCH Verlagsges. 1994; Front. Nat. Prod. Res. **1**: Methods in Carbohydr. Synthesis, Harwood Publ. 1996; Modern Synthetic Methods **1995**, 283–330; Pozsgay in: Ogura, Carbohydrates (1992), p. 188–227; Nachr. Chem. Tech. Lab. **32**, 6–16 (1984); Prep. Carbohydr. Chem. **1997**, 283–544; Carbohydr. Chem. **30**, 62–89 (1998); Science **260**, 1307 (1993) (solid phase synthesis); Tetrahedron **52**, 1095–1121 (1996); **55**, 1807–1850 (solid phase synthesis).
gen.: Adv. Carbohydr. Chem. Biochem. **50**, 21–24 (1994) ▪ Angew. Chem. Int. Ed. Engl. **25**, 212–234 (1986) ▪ Chem. Soc. Rev. **13**, 15–36 (1984); **18**, 347–374 (1989) ▪ Collins-Ferrier, p. 415–430 (synthesis) ▪ El Khadem, Monosaccharides and Their Oligomers, New York: Academic Press 1988 ▪ Ginsburg & Robbins, Biology of Complex Carbohydrates, vol. 2, New York: Wiley 1984 ▪ J. Am. Chem. Soc. **121**, 734 (1999) (one pot synthesis) ▪ Progr. Nucl. Magn. Res. Spec. **27**, 445–474 (1995) (NMR) ▪ Shallenberger, Advanced Sugar Chemistry, Chichester: Horwood & Westport, AVI 1982 ▪ Ullmann (5.) **A 5**, 83.

Oligosaccharines. Name for *oligosaccharides that regulate the growth, development, and defensive responses of plants. Plants biosynthesize and accumulate *phytoalexins after microbial infection or treatment with elicitors (compounds inducing a defensive response). Partial hydrolysis of the mycelia wall of *Phytophthora megasperma*, a pathogenic fungus for soybean, gives rise to a mixture of ca. 300 structurally different oligoglucosides of which only one is highly active while the others have no or only low elicitor activity. The active compound is the heptaglucoside:

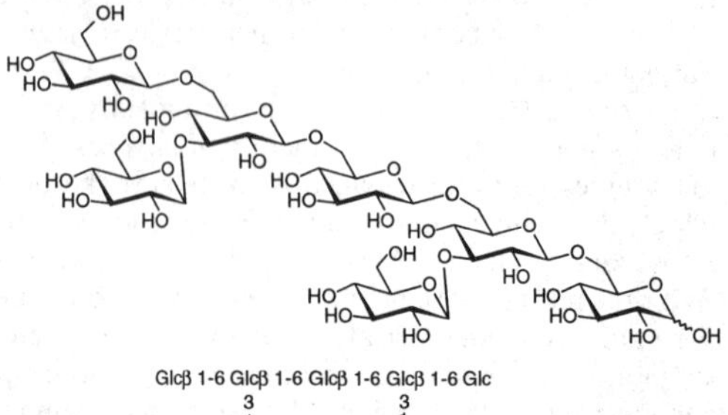

Glcβ 1-6 Glcβ 1-6 Glcβ 1-6 Glcβ 1-6 Glc
3 3
1 1
Glcβ Glcβ

For isolation and structure elucidation, see *Lit.*[1], for synthesis, see *Lit.*[2].

Lit.: [1] J. Biol. Chem. **259**, 11312–11320, 11321–11336 (1984). [2] Angew. Chem. Int. Ed. Engl. **22**, 793 f. (1983); J. Biol. Chem. **259**, 11337 ff. (1984).
gen.: Acc. Chem. Res. **25**, 77–83 (1992).

Oligostatins see glycosidase inhibitors.

Olivil.

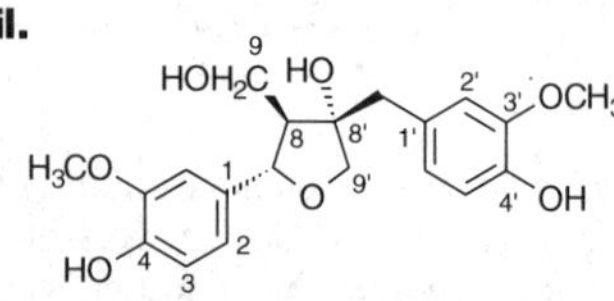

(–)-Olivil

$C_{20}H_{24}O_7$, M_R 376.41, cryst. The monohydrate has the mp. 127 °C, anhydrous mp. 142–143 °C, $[\alpha]_D^{12}$ –127°, soluble in hot water, alcohol, acetic acid, and oils. The lignan O. occurs in the roots of the olive tree *Olea europaea*; the resinous exudate of this plant contains 50% O. The substance has also been isolated from the roots of *Vladimiria sonliei*. O. is also present in glycosidically bound form in *Stauntonia chinensis* as 9-*O*-[6-*O*-D-apio-β-D-furanosyl-β-D-glucopyranoside] (*yemuoside* YM_2) $C_{31}H_{42}O_{16}$, M_R 670.66, powder, mp. 124 °C, $[\alpha]_D^{23}$ –48.7° (CH_3OH).

Lit.: Beilstein E V **17/8**, 509 ▪ J. Nat. Prod. **52**, 342 (1989) (yemuoside YM_2) ▪ Planta Med. **56**, 475 (1990) ▪ Tetrahedron Lett. **1979**, 3773 ff. – *[CAS 2955-23-9 (O.); 122127-68-8 (yemuoside YM_2)]*

Olivomycins.

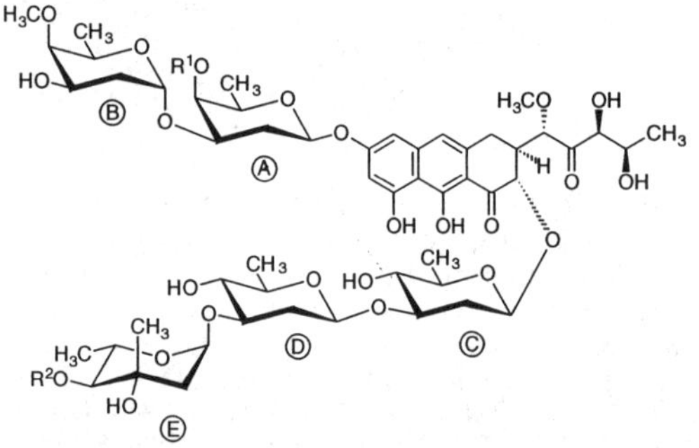

R^1 = CO–CH_3, R^2 = CO–$CH(CH_3)_2$: O. A
R^1 = CO–CH_3, R^2 = CO–CH_3 : O. B
R^1 = H, R^2 = CO–$CH(CH_3)_2$: O. C

O. A: $C_{58}H_{84}O_{26}$, M_R 1197.27, yellow crystals, mp. 160–165 °C, $[\alpha]_D^{25}$ –35.5° (C_2H_5OH), LD_{50} (mouse, i.p.) 6.6 mg/kg. Members of the *aureolic acid family. O. are isolated from cultures of *Streptomyces olivoreticuli*.

O. A, O. B and O. C are highly toxic antineoplastic antibiotics.

Lit.: Chem. Eng. News (15. 3.) **1999**, 11 ▪ J. Am. Chem. Soc. **111**, 2984–2995 (1989); **121**, 1990, 4092 (1999) ▪ Sax (8.), OIS 000, OIU 000. – *[CAS 6988-58-5 (O. A); 6992-69-4 (O. B); 6988-59-6 (O. C)]*

Olivoretins see teleocidins.

Ommochromes. A group of acidic, yellow and red phenoxazine pigments occurring in the eyes, wings, hatching secretions, and skin of arthropods, especially

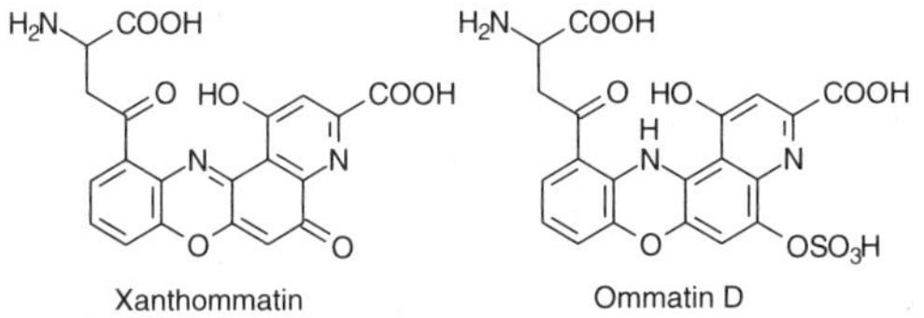

Xanthommatin Ommatin D

crabs and insects, as well as other invertebrates. The O. are often associated with proteins (*chromoproteins) and are very poorly soluble in water and neutral organic solvents. In spite of their relatively wide distribution they were not discovered until 1957 (by Butenandt). The O. are a heterogeneous class of substances which can be divided into two groups: the yellow and red, low-molecular-weight *ommatins* [1] that are labile in alkaline solution and the red-violet, sulfur-containing *ommins* [2] that usually occur as mixtures of five to six substances. The most thoroughly characterized compounds are the ommatins *xanthommatin* [3] ($C_{20}H_{13}N_3O_8$, M_R 423.34, in the hatching secretion of the painted lady, *Vanessa urticae*) and *ommatin D* ($C_{20}H_{15}N_3O_{11}S$, M_R 505.41). The ommins mainly occur as pigments of vision in insects and crustaceans and as skin pigments in arthropods and cephalopods where they are responsible for their rapid changes in color (mimicry), see also the pigments of vision of the vertebrates: melanins, opsin (see pigments of vision), rhodopsin. The O. are derived biogenetically from tryptophan [4].

Lit.: [1] Scheuer I **3**, 148–151. [2] Scheuer I **3**, 151. [3] Insect Biochem. **20**, 785–792 (1990); **21**, 785–794 (1991); **22**, 561–589 (1992); J. Biol. Chem. **266**, 21392–21398 (1991). [4] Adv. Insect. Physiol. **10**, 117–246 (1974).
gen.: Biol. Crustacea **9**, 301–394 (1985) ▪ J. Heterocycl. Chem. **25**, 1243 ff. (1988) ▪ Naturwissenschaften **54**, 259–267 (1967) ▪ Prax. Naturwiss. Biol. **44**, 1–7 (1995). – *[CAS 521-58-4 (xanthommatin); 28991-26-6 (ommatin D)]*

Omphalone.

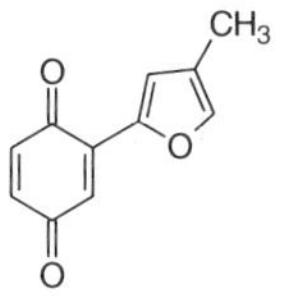

$C_{11}H_8O_3$, M_R 188.18, dark red cryst., mp. 98–100 °C. An antibiotically active benzoquinone from mycelia cultures of the agaric *Lentinellus omphalodes* (Basidiomycetes). O. is strongly cytotoxic and inhibits the germination of lettuce seeds (*Lactuca sativa*) at a concentration of 100 µg/mL.

Lit.: Z. Naturforsch., C **46**, 989 (1991). – *[HS 2941 90; CAS 107320-38-7]*

Omphalotins. Potent nematicidal cyclopeptides (O. A–D) from submerged cultures of *Omphalotus olearius* (Basidiomycetes), e.g., *O. B*: $C_{74}H_{123}N_{13}O_{18}$, M_R 1482.86, crystalline, mp. 180–183 °C, $[\alpha]_D^{22}$ –246° (CH_3OH) (see formula on the right).

Lit.: Nat. Prod. Lett. **10**, 25, 33 (1997) ▪ Tetrahedron **54**, 5345–5352 (1998). – *[CAS 208394-46-1 (O. B)]*

Oncinotine see inandenine.

Onion constituents see Allium.

Onnamide A see mycalamides.

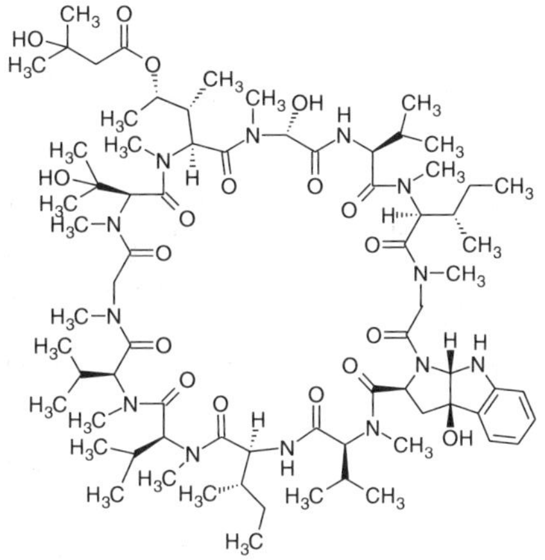

Omphalotin B

α-Onocerin.

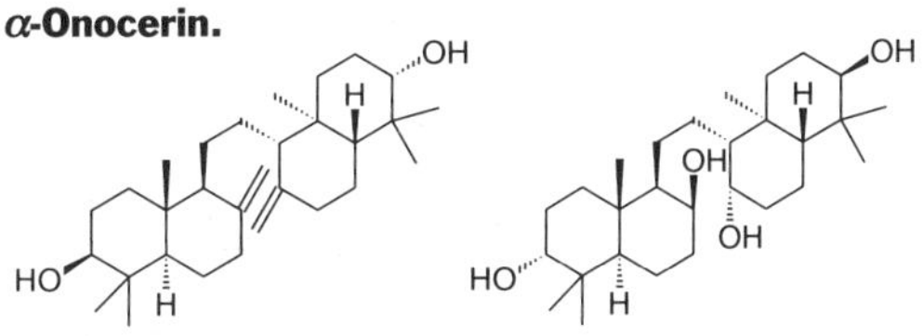

α-Onocerin Lyclavatol

$C_{30}H_{50}O_2$, M_R 442.73, mp. 206.5–208 °C, $[\alpha]_D^{15}$ +17.0°. The main *triterpene of several species of the genus *Lycopodium* (= clubmoss) which, like *lyclavatol* ($C_{28}H_{50}O_4$, M_R 450.70, previously known as clavatol), has an onocerane skeleton. α-O. was first isolated from the Fabaceae *Ononis spinosa* (thorny rest-harrow) [1].

Lit.: [1] J. Prakt. Chem. **65**, 419 (1855).
gen.: Arch. Pharm. (Weinheim, Ger.) **311**, 318 (1978) ▪ Chem. Nat. Compd. (Engl. Trans.) **28**, 568 (1992) (structure) ▪ Chem. Pharm. Bull. **23**, 1784 (1975); **29**, 3424 (1981). – *[CAS 511-01-3 (α-O.); 33044-79-0 (lyclavatol)]*

Oogoniol see antheridiol.

Ooporphyrin see protoporphyrins.

Oosporein (iso-oosporein, chaetomidin).

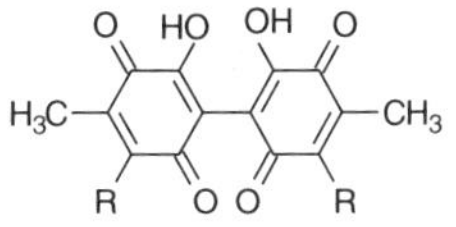

R = H : Phoenicin
R = OH : Oosporein

$C_{14}H_{10}O_8$, M_R 306.23, bronze-colored cryst., mp. 290–295 °C (as tetraacetate yellow needles, mp. 190 °C), a toxic benzoquinone pigment from fungi (*Oospora colorans*, *Chaetomium aureum*, *C. trilaterale*, *Beauveria*, *Cordyceps*, *Penicillium*, *Verticillium*, and *Acremonium* species). Among the basidiomycetes O. has been found in *Phlebia* sp. The corresponding dideoxy derivative *phoenicin* ($C_{14}H_{10}O_6$, M_R 274.22) has been found in *Penicillium* species. On account of its wide occurrence in fungi that infect food crops, O. can be found as a contaminant in animal fodder. It does not have chronic toxicity but shows acute toxic effects [LD_{50} (chick p. o.) 6.12 mg/kg]. O. inhibits malate synthase, a key enzyme in the glyoxylate cycle.

Lit.: Agric. Biol. Chem. **48**, 1065 (1984) ▪ Behrens & Driesel (eds.), Dechema-Biotechnology Conferences, vol. 5B, p. 1073–1076, Weinheim: VCH Verlagsges. 1992 ▪ Beilstein E IV **8**, 3742 ▪ Cole-Cox, p. 830–833 ▪ Poult. Sci. **63**, 251–259 (1984) ▪ Turner **1**, 105; **2**, 71. – *[CAS 475-54-7 (O.); 128-68-7 (phoenicin)]*

Ophiobolins. Phytotoxic *sesterterpenoids derived from the unusual tricyclic skeleton of ophiobolane ($C_{25}H_{46}$, M_R 346.62). The O. A–M are produced by plant-pathogenic fungi such as *Ophiobolus, Helminthosporium, Cephalosporium,* and *Drechslera* species as well as by *Aspergillus ustus* (O. H and O. K). O. M is a potent nematocidal agent.

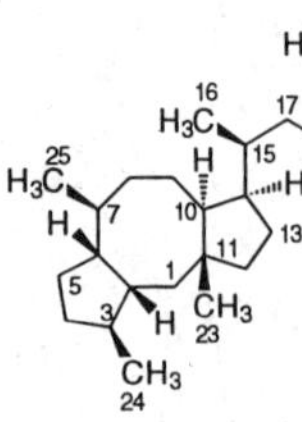

Ophiobolane

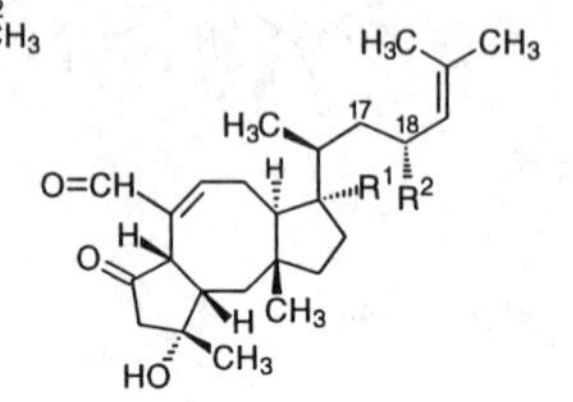

$R^1 = R^2 =$ —O— : O. A
$R^1 = R^2 = H$: O. C
$R^2 = R^2 = H$; (Z)-17,18-C=C : O. K
$R^2 = H$; 13,14-Dehydro : O. M

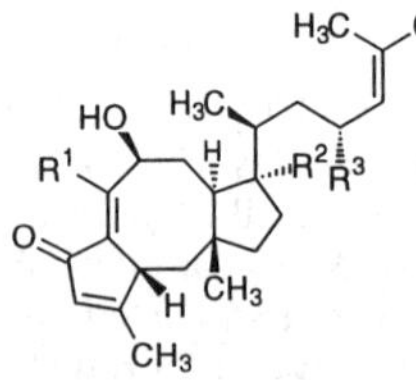

$R^1 = COOH$, $R^2 = R^3 = H$: O. D
$R^1 = CH_2OH$, $R^2 = R^3 =$ —O— : O. J

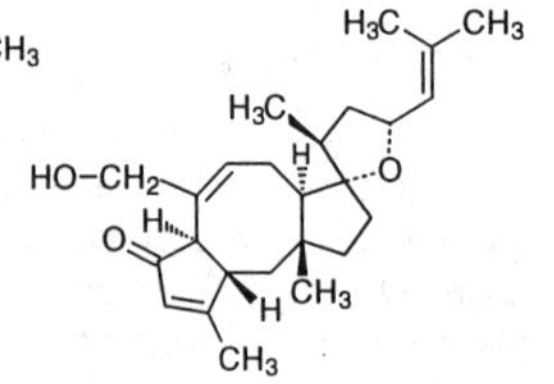

O. I

Table: Data of Ophiobolins.

Ophiobolins	molecular formula	M_R	mp. [°C]	CAS
O. A (Cochliobolin A)	$C_{25}H_{36}O_4$	400.56	182	4611-05-6
O. C (Zizanin)	$C_{25}H_{38}O_3$	386.57	121	19022-51-6
O. D (Cephalonic acid)	$C_{25}H_{36}O_4$	400.56	139	24436-08-6
O. I	$C_{25}H_{36}O_3$	384.56	oil	110013-85-9
O. J	$C_{25}H_{36}O_4$	400.56	oil	114058-47-8
O. K	$C_{25}H_{36}O_3$	384.56	80–82	138057-90-6
O. L	$C_{25}H_{36}O_4$	400.56	oil	
O. M	$C_{25}H_{36}O_3$	384.56	oil	177988-78-2

Lit.: Agric. Biol. Chem. **48**, 803 (1984) ▪ Beilstein E V **18/3**, 513 ▪ J. Agric. Food Chem. **32**, 778 (1984) ▪ J. Am. Chem. Soc. **111**, 2737 (1989) ▪ Phytochemistry **27**, 1653 (1988) ▪ Proc. Natl. Acad. Sci. USA **84**, 3091 (1987) ▪ Sax (8.), CNF 109 ▪ Tetrahedron **47**, 6931 (1991). – *Structure:* Bioorg. Med. Chem. **4**, 531 (1996) (O. M) ▪ J. Nat. Prod. **58**, 74 (1995) (O. L) ▪ J. Org. Chem. **53**, 2170 (1988).

Ophiocarpine see protoberberine alkaloids.

Ophiocordin see balanol.

Ophiorrhiza alkaloids. A group of gluco-indole alkaloids from various *Ophiorrhiza* species (Rubiaceae) such as *O. pumila, O. japonica,* or *O. kuroiwai.* The genus *Ophiorrhiza* is indigenous to tropical Asia, more than 150 different species are known to botanists. They are usually inconspicuous, shrub-like plants. The alkaloids occurring in these plants include: chaboside (the first naturally occurring gluco derivative of *camptothecin to be described), pumiloside, deoxypumiloside, *ophiorine A* ($C_{26}H_{28}N_2O_9$, M_R 512.52, mp. 217–219 °C, $[\alpha]_D$ +51.0°) and the epimeric *ophiorine B* (mp. 188–191 °C, $[\alpha]_D$ +18.2°), *lyaloside* {$C_{27}H_{30}N_2O_9$, M_R 526.54, mp. 168–169 °C, $[\alpha]_D^{20}$ –202° (CH_3OH)}, *lyalosidic acid* {$C_{26}H_{28}N_2O_9$, M_R 512.52, $[\alpha]_D$ –151.7° (CH_3OH)}.

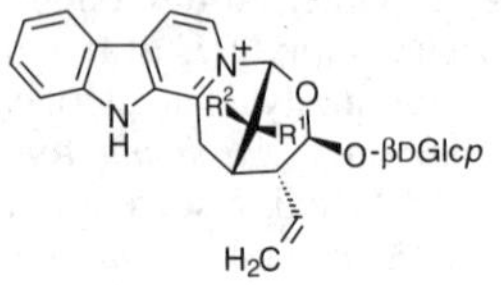

$R^1 = H$, $R^2 = COO^-$: Ophiorine A **1**
$R^1 = COO^-$, $R^2 = H$: Ophiorine B **2**

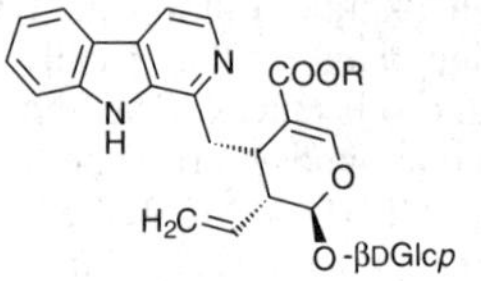

R = H : Lyalosidic acid **3**
R = CH_3 : Lyaloside **4**

The O. a. (especially chaboside) are of interest on account of their possible role as biosynthetic precursors of *camptothecin.

Lit.: Chem. Pharm. Bull. **34**, 3064 (1986); **40**, 2588 (1992) ▪ Phytochemistry **37**, 1471 (1994) ▪ Tetrahedron Lett. **26**, 5299 (1985); **30**, 4991 (1989); **36**, 5169 (1990). – *[HS 2939 90; CAS 99615-91-5 (ophiorine A); 99631-26-2 (ophiorine B); 56021-85-3 (lyaloside); 104821-27-4 (lyalosidic acid)]*

Ophiotoxins see snake venoms.

Opines see octopine.

Opium. O. is the air-dried milk juice obtained from cut, unripe fruits of *Papaver somniferum* with a *morphine content of at least 9.5%. Crude O. obtained from the poppy is marketed in the form of roundish brown lumps or as a dark brown powder and is characterized by an unusual odor and a very bitter, rather sharp taste. Chemically O. is a complex mixture of ca. 30 different alkaloids comprising 20-30 weight% with *morphine as the major alkaloid (ca. 12%). The composition varies widely depending on the origin, age, and pretreatment. *Detection:* A red chelate of meconic acid (3-hydroxy-4-oxo-4*H*-pyran-2,6-dicarboxylic acid) is formed with iron(III) chloride solution. Specific detection reactions are usually performed to demonstrate the presence of morphine; this is also the cause of the red color with formaldehyde/sulfuric acid (Marquis test). O. for smoking (chandu, chandoo) contains 6–8% morphine. Preparations made from raw O. for medical purposes include: stabilized O. (O. powder stabilized and adjusted to 10% morphine with lactose), O. extract (extracted with water, 19.6–20.4% morphine), and O. tincture or *laudanum* (name introduced by Paracelsus for the macerate with aqueous alcohol, 0.95–1.05% morphine). "*Dover's powder*" (named after the English physician and pirate Thomas Dover, 1660–1742) used in the past as an expectorant, consists of a mixture of O. and *Ipecacuanha* powders. O. is the starting material for the production of heroin. O. has a long history of use in medicine as a potent narcotic and anal-

gesic (morphine activity) and as an effective remedy for diarrheas, it also subdues coughing (narcotine activity). At present the world-wide annual production is ca. 12000 t, mainly from India, Turkey, and Yugoslavia. In drug abuse O. is injected, smoked, eaten, or drops of O. tincture are drunk. On account of its addictive properties the use O. is subject to strict regulations in most countries.
Historical: The use of poppy and O. has a long cultural history. Capsules and seeds of poppy have been found in over four thousand-year-old pile buildings on some Swiss lakes, they were used as food and possibly also for narcotic effects. Poppy is mentioned as a drug in six thousand-year-old Sumerian ideograms (plant of pleasure). Poppy was probably first cultivated in Egypt and spread from there through the Middle East, Mediterranean countries, Persia, India, and China. Poppy was described by Hesiod in the 8th century BC; he named the Greek town of Sikyon as Mekone, the poppy town (from which the old name of meconim for opium is derived). The Greeks gave poppy juice its now usual name (opos = juice). In the 6th and 7th centuries Arabs introduced opium (under the name afyum) to Persia, India, and China. Opium smoking was elevated to an art in Persia. In China the drug achieved its greatest significance as a narcotic for the general population (smoking). However, at first it was not smoked but eaten. The change over to smoking is possibly related to the prohibition of tobacco decreed by the Chinese Emperor Tsung Cheng in 1644. The consumption of O. was forbidden in China in 1792. The consumption of opium became popular in Europe in the 19th century. Morphine was isolated from opium for the first time by the German apothecary Sertürner (1806). The name morphine was introduced by Gay-Lussac, its constitution was established by L. Knorr, R. Robinson, and C. Schöpf.
Lit.: Chem. Ind. (London) **1988**, 146–153 ▪ Hager (5.) **1**, 640 ff.; **2**, 1020; **3**, 912 f. ▪ Sax (8.), OJG 000 ▪ Seefelder (ed.), Opium: Eine Kulturgeschichte (3rd edn.), Landsberg: Ecomed 1996. – *[HS 2939 10]*

Opium alkaloids. General name for the alkaloids occurring in the various poppy species and in *opium. These are the main alkaloid *morphine and the *morphinan alkaloids derived therefrom as well as alkaloids of the papaverine type. Both groups of alkaloids are *isoquinoline alkaloids. As a consequence of their strong analgesic activities, the morphine alkaloids were used as lead structures for synthetic morphinans.
Lit.: Chem. Unserer Zeit **34**, 99–112 (2000) (review) ▪ Geschichte der Pharmazie **47**, 55–60 (1995) ▪ Phillipson et al., p. 38–46 ▪ see opium. – *[HS 2939 10]*

Opsin see pigments of vision.

Orange flower absolute (oil) (neroli oil). Both products are obtained from freshly harvested flowers of the bitter orange tree (Seville orange), *Citrus aurantium* ssp. *aurantium* [see also orange oils]. *Origin:* Southern France, Spain, Morocco, Tunisia.
1. ***Orange flower absolute:*** Red-brown to dark brown liquid with an intense, warm flowery odor with bitter, herby-spicy tones and a bitter aromatic taste.
Production: Extraction with a suitable solvent (usually hexane) furnishes the concrete (yield 0.2%), extraction of which with ethanol gives the absolute (total yield ca. 0.1%).
Composition [1]***:*** The main components and contributors to the odor are **linalool* (ca. 50%), *linalyl acetate* (ca. 20%, see linalool), *nerolidol* (4–8%) ($C_{15}H_{26}O$, M_R 222.37, *S*-form: $[\alpha]_D$ +10.7° ($CHCl_3$)), **farnesol* (7–12%), **indole* (2–5%), and **methyl anthranilate* (2–5%).

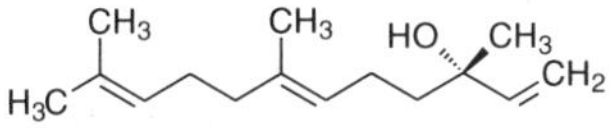

Nerolidol

Use: Because of the laborious production O. absolute is one of the most precious raw materials for perfume production. It is thus used only in minute amounts in expensive Eaux de Toilette and flower-like perfumes.
2. ***Neroli oil:*** Yellowish liquid with a weak blue fluorescence, sweetish, terpene-spicy, mildly bitter flowery odor and bitter aromatic taste.
Production: By steam distillation.
Composition [1]***:*** In comparison to the absolute, the distilled oil contains more monoterpene hydrocarbons such as *β-*pinenes* (7–15%), *limonene* (11–13%, see *p*-menthadienes), and others. The main component is **linalool* (30–40%). The remaining constituents are qualitatively similar to those of the absolute.
Use: An essential component of the classical Eaux de Toilette (Eau de Cologne), on account of its price, however, it is used in very small amounts only. The distillation water ("O. water") occurring as a by-product of the distillation is also much used for aromatization of products. Extraction with hexane yields the so-called orange flower water absolute (absolue des fleurs d'oranger de l'eau), which is mainly used to improve the aromatic character of foods, e. g., liquors.
Lit.: [1] Perfum. Flavor. **16** (6), 1 (1991); Chromatographia **39**, 529 (1994).
gen.: Arctander, p. 436, 474, 477, 478 ▪ Bauer et al. (2.), p. 166 ▪ ISO 3517 (1975). – *Toxicology:* Food Cosmet. Toxicol. **14**, 813 (1976); **20**, 785 (1982). – *Nerolidol:* Justus Liebigs Ann. Chem. **1992**, 1325 (structure) ▪ Phytochemistry **39**, 785 (1995) (biosynthesis). – *[HS 3301 19; CAS 8030-28-2 (1.); 8016-38-4 (2.); 142-50-7 (nerolidol)]*

Orange oils. Two different types of O. are used in the perfume and fragrance industries.
1. ***Sweet orange oil:*** Yellow to reddish yellow oil with a bright, fresh-fruity, sweet odor of freshly grated orange peel and a similar taste.
Production: By mechanical processes ("pressing", *citrus oils) from the peel of the sweet orange, *Citrus sinensis*; *origin:* Brazil, USA (California, Florida), Israel, Italy, Spain. O. is usually obtained as a by-product in the production of orange juice. With an annual world-wide production of ca. 20000 t, O. is quantitatively the second most important essential oil after turpentine oil.
Composition [1]***:*** The main component is (+)-*limonene* (see *p*-menthadienes) (usually >90%). The organoleptic properties are mainly due to components such as *octanal, decanal* (see alkanals), **citral* (all <0.5%), *α*- and *β-*sinensals* as well as *nootkatone* (see eremophilanes) (all <0.05%).
Use: For the production of perfume, for fresh citrus odors in many applications; to improve the aromatic character of refreshing beverages and bakery products;

for the latter purpose concentrated oils (citrus oils) with a low content of terpene hydrocarbons are normally used; and for the production of pure (+)-limonene.
2. ***Bitter orange oil:*** Yellow to yellow-brown oil with an odor less aldehyde-like but more herby, more flowery-fresh than the sweet oil and a marked orange-like but dry bitter taste.
Production: By mechanical processes ("pressing", citrus oils) from the peel of the bitter orange, *Citrus aurantium* ssp. *aurantium*. *Origin:* e. g., Italy, Spain. The amounts produced are low in comparison to the sweet oil.
Composition[2]***:*** The composition is very similar to that of the sweet oil, the main component is also (+)-*limonene* (>90%).
Use: In the perfume industry for, e. g., Eaux de Cologne and fresh Eaux de Toilette. Because of the low content of furocoumarins which can induce phototoxic reactions, the bitter oil is used in limited amounts in perfume oils only. To improve the aromatic character of liquors as well as confectionery and bakery products.

Lit.: [1]Perfum. Flavor. **16** (2), 17 (1991); **17** (5), 133 (1992); **19** (4), 35; (6), 29 (1994); Flavour Fragr. J. **9**, 105 (1994), **10**, 33 (1995). [2]Perfum. Flavor. **16** (6), 1; (2) 17 (1991); **19** (5), 83; (6), 29 (1994); Flavour Fragr. J. **10**, 33 (1995).
gen.: Arctander, p. 480, 485 ▪ Bauer et al. (2.), p. 149 ▪ ISO 3140 (1990), 9844 (1991) ▪ Ohloff, p. 132 ▪ Perfum. Flavor. **16** (6), 1 (1991). – *Toxicology:* Food Cosmet. Toxicol. **12**, 733, 735 (1974). – *[HS 3301 12; CAS 8008-57-9 (1.); 68916-04-1 (2.)]*

Orbiculamide A see theonellamides.

Orcein.

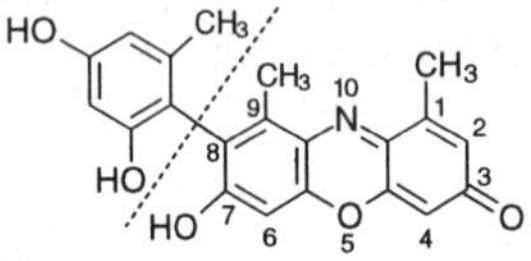

Orcein component

The principle responsible for the dying properties of various lichen pigments used in the 19th century under names such as Orseille, Archil, Cudbear, Persio, Pourpre Française, Tournesol, etc. and obtained from the practically colorless *Roccella*, *Lecanora*, and *Variola* lichens by treatment with urine or ammonia and air; in the Middle Ages they were used besides alizarin and indigo to dye silk and wool. However, the dyes were not permanently stable to light and are no longer in use. O. obtained from Orseille or *orcinol is a brown-red microcrystalline powder. It is practically insoluble in water, benzene, chloroform, or ether, soluble in alcohol, acetone, or acetic acid to give red solutions, and soluble in dilute alkalis to give blue-violet solutions. At least 14 components can be isolated by chromatography; they are all derived from phenoxazone (*3H-phenoxazin-3-one*, see figure) by substitution in the 8-position by an orcinol group, possibly in the 2-position by another such group, in the 7-position by OH or NH_2, and in the 3-position by NH (in place of O). The main components are red-violet in neutral solution, red in acidic solution, and deep violet in alkaline solution, thus they exhibit indicator properties similar to those of the related pigments of *litmus. O. is no longer permitted for use as a colorant for foods.
Use: Mainly in microscopy to stain elastic fibers, acidic mucines, cartilage, etc. The names orcein, orcinol, and Orseille are purported to originate via Orcela from the name of the Florentine merchant family Rocela who possessed a monopoly for lichen dyes in the 14th century.

Lit.: Beilstein EIII/IV **27**, 5546 ▪ Endeavour **33**, 149–155 (1974) ▪ Merck-Index (12.), No. 6994 ▪ Ullmann (5.) **A 3**, 229; **A24**, 570, 573 ▪ Zechmeister **45**, 103–234. – *[CAS 1400-62-0]*

Orcinol (5-methylresorcinol).

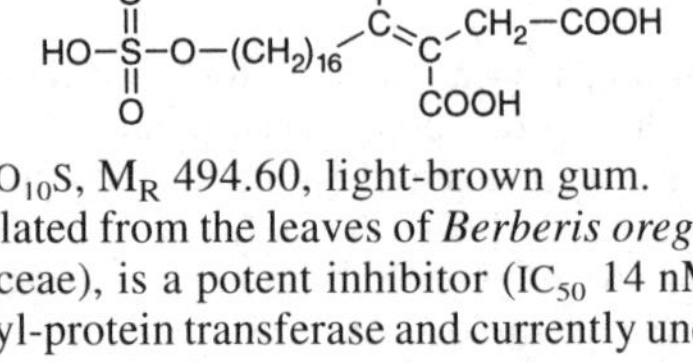

$C_7H_8O_2$, M_R 124.14, colorless crystals that turn red on exposure to air, with a sweet but unpleasant taste, mp. 58 °C (hydrate), 107 °C (anhydrous), bp. 290 °C [147 °C (0.66 kPa)], soluble in water, alcohol, ether. The tetraketide O. occurs naturally in lichens (e. g., *Evernia prunastri*, oak moss, and *Pseudevernia furfuracea*), lower fungi (*Aspergillus fumigatus*), and as a component of lichen dyes (*litmus, Orseille, see orcein). In addition to O., the lichens also contain *β-O.* (2,5-dimethylresorcinol, $C_8H_{10}O_2$, M_R 138.17) as well as its monomethyl ether and methyl *β*-orcinate (see oak moss absolute); the latter can be prepared synthetically from *β*-O. (Kolbe-Schmitt synthesis) and used as a substitute for oak moss extract in the perfume industry. O. itself is used as a reagent for pentoses, lignin, saccharose, arabinose, diastase.

Lit.: Beilstein EIV **6**, 5982 ▪ J. Chem. Soc., Perkin Trans. 1 **1988**, 755 ▪ Karrer, No. 198 ▪ Turner **1**, 92; **2**, 59, 196 ▪ Ullmann (5.) **A 11**, 202, 235. – *[HS 2907 29; CAS 504-15-4 (O.); 488-87-9 (β-O.)]*

Oreganic acid [(*E*)-19-sulfooxy-2-nonadecene-1,2,3-tricarboxylic acid].

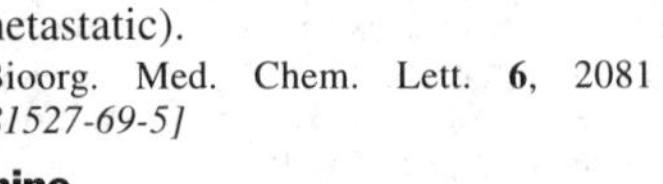

$C_{22}H_{38}O_{10}S$, M_R 494.60, light-brown gum.
O., isolated from the leaves of *Berberis oregana* (Berberidaceae), is a potent inhibitor (IC_{50} 14 nM) of Ras farnesyl-protein transferase and currently under investigation as a lead structure for antitumor compounds (anti-metastatic).

Lit.: Bioorg. Med. Chem. Lett. **6**, 2081 (1996). – *[CAS 181527-69-5]*

Orellanine.

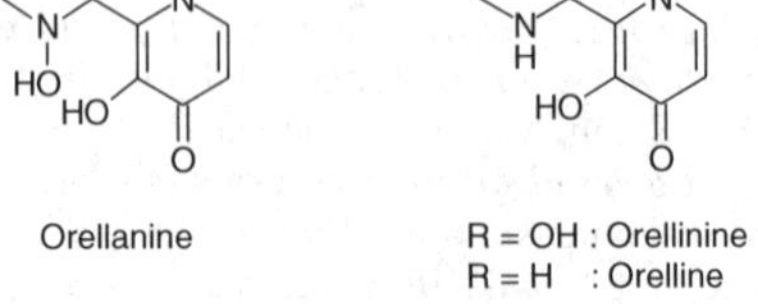

$C_{10}H_8N_2O_6$, M_R 252.18, colorless crystals with blue fluorescence, decomp. >150 °C, explosive at 267 °C with evolution of O_2. A lethal fungal toxin from *Cortinarius orellanus*, *C. speciosissimus*, and the North American *C. raineriensis* (Basidiomycetes). *C. orellanus* has been the cause of numerous, sometimes fatal mushroom poisonings. The lethal dose is contained in 50–100 g of fresh fungi. O. is a potent neurotoxin [LD_{50} (cat, guinea-pig, mouse p.o.) 3–5 mg/kg] with a long latency period of 2–17 d depending on the severity of the poisoning [1]. The content of O. in the fungus amounts to ca. 3% (referred to the dry weight) [2]. Photochemical or thermal reactions of O. lead to the alkaloids *orellinine* ($C_{10}H_8N_2O_5$, M_R 236.18) and *orelline* ($C_{10}H_8N_2O_4$, M_R 220.18) that also occur in the fungus [3]. The fictive cyclopeptide ("cortinarine") which was assumed to be responsible for the toxic properties of *Cortinarius* fungi does not exist.

Lit.: [1] Besl & Bresinsky, Colour Atlas of Poisonous Fungi, Boca Raton: CRC Press 1989. [2] Mycopathologia **74**, 65 (1981); **108**, 155 (1989); Arch. Toxicol. **53**, 87 (1983); **62**, 81 (1988). [3] Experientia **41**, 769 (1985); **43**, 462 (1987); Tetrahedron **49**, 8373 (1993).
gen.: Bull. Soc. Chim. Belg. **95**, 1009 (1986) ▪ Experientia **43**, 462 (1988). – *Synthesis:* Helv. Chim. Acta **71**, 957 (1988) ▪ Justus Liebigs Ann. Chem. **1987**, 857 ▪ Tetrahedron **42**, 1475 (1986); **49**, 8373 (1993) ▪ Tetrahedron Lett. **26**, 4903 (1985). – *[CAS 37338-80-0 (O.); 98726-96-6 (orellinine); 72016-31-0 (orelline)]*

Oreodine see rhoeadine alkaloids.

Orientaline see benzyl(tetrahydro)isoquinoline alkaloids.

Orientalinone see proaporphine alkaloids.

Origanum (oregano) oil. Yellowish red to dark brown oil with a phenolic-spicy, tart-herby, leathery odor and taste.
Prod.: By steam distillation of *Origanum* and related Lamiaceae, mainly *Origanum vulgare* ssp. *hirtum* (so-called Greek origanum), and *Coridothymus capitatus* (syn. *Thymbra capitata*, so-called Spanish origanum) growing principally in the Mediterranean region.
Composition [1]***:*** The main component is *carvacrol* (50–70%), generally accompanied by *thymol* (see cymenols) (0–20%).
Use: In the perfume industry for tart-herby compositions, in condiments to improve the aromatic character of foods.

Lit.: [1] Perfum. Flavor. **13** (4), 75 (1988); **14** (1), 39 (1989); **18** (1), 53 (1993).
gen.: Arctander, p. 492 ▪ Bauer et al. (2.), p. 168 ▪ Dev. Food Sci. **34**, 439–456 (1994). – *Toxicology:* Food Cosmet. Toxicol. **12**, 945 (1974). – *[HS 3301 29; CAS 8007-11-2]*

Oripavine see morphinan alkaloids.

Orleane see bixin.

Orlistat see lipstatin.

Ormosanine (piptamine).

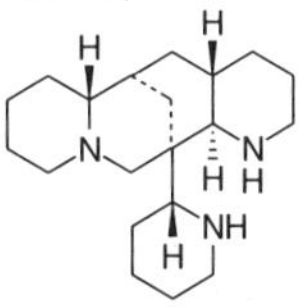

(–)-form

$C_{20}H_{35}N_3$, M_R 317.52, mp. 178–179 °C, $[\alpha]_D$ –9°; racemate, mp. 183–184 °C. O. is the parent *quinolizidine alkaloid of the ormosanine type and occurs together with several isomers and other structures derived from the same skeleton in some South American Fabaceae genera (e.g., *Ormosia, Podopetalum*) [1]. The 6-epimer is called *piptanthine* {(–)-form, mp. 142–146 °C (136–137 °C), $[\alpha]_D$ –24°}.

Lit.: [1] Waterman **8**, 197–239.
gen.: Beilstein E V **26/1**, 383 ▪ Can. J. Chem. **59**, 34 (1981) (abs. configuration, isolation). – *[HS 2939 90; CAS 5001-21-8 ((–)-form); 34429-31-7 (racemate); 7344-67-4 (piptanthine)]*

L-Ornithine [(*S*)-2,5-diaminopentanoic acid, abbreviation Orn].

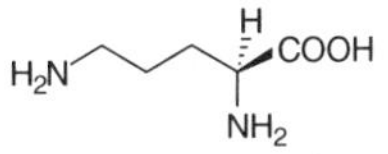

$C_5H_{12}N_2O_2$, M_R 132.16, mp. 140 °C, $[\alpha]_D^{25}$ +11.5° (H_2O); monohydrochloride, mp. 233 °C (decomp.), $[\alpha]_D^{23}$ +11.0°; dihydrochloride, mp. 230–232 °C (decomp.), $[\alpha]_D^{25}$ +16.7° (H_2O), pK_a 1.94, 8.65, 10.76, pI 9.70. O. is not an essential amino acid for mammals but is an intermediate in the O. cycle. Orn was first isolated in 1877 from L-ornithuric acid (N^2,N^5-dibenzoyl-L-ornithine) from the excrement of chickens that had been fed with benzoic acid. O. is found in the urine of patients with Fanconi's syndrome and in bacterial cell walls. L-Orn is a constituent of cyclopeptide antibiotics such as *gramicidin S and *tyrocidins. Gramicidin J also contains the D-form of ornithine.
Metabolism: O. is formed during detoxification of ammonia in the body by the so-called O. cycle (urea cycle): 1) 2 ATP + HCO_3^- + NH_3 → carbamoyl phosphate + 2 ADP + P_i [carbamoyl phosphate synthetase I (CPS I, EC 6.3.4.16)]. – 2) Carbamoyl phosphate + Orn → citrulline + phosphate (ornithine carbamoyltransferase, EC 2.1.3.3). – 3) Cit + Asp + ATP → argininosuccinate + AMP + PP_i (argininosuccinate synthetase, EC 6.3.4.5). – 4) argininosuccinate → fumarate + Arg (argininosuccinate lyase, EC 4.3.2.1). – 5) Arg + H_2O → urea + Orn (arginase, EC 3.5.3.1). The final product Orn again serves for binding of carbamoyl phosphate. The enzymatic reactions 1 and 2 proceed in mitochondria while reactions 3 to 5 take place in cytosol. O. is formed from L-glutamic acid 5-semialdehyde by δ-transamination and is thus an intermediate of L-arginine biosynthesis. Putrefactive bacteria form *putrescine (1,4-butanediamine) from Orn under the action of ornithine decarboxylase (PLP-enzyme, EC 4.1.1.17).

Lit.: Beilstein E IV **4**, 2644 f. ▪ Bull. Chem. Soc. Jpn. **64**, 613 (1991) (synthesis) ▪ Phytochemistry **27**, 1571–1574 (1988) ▪ Stryer 1995, p. 634 ff., 723 ▪ Synthesis **1990**, 223 (synthesis) ▪ Ullmann (5.) **A 2**, 58, 60, 63, 86. – *[HS 2922 49; CAS 70-26-8]*

Orobol see isoflavones.

Oroidin. Sponges of the genus *Agelas* produce some interesting pyrrole-imidazole compounds derived from the various cyclization possibilities of the parent alkaloid O.: *O.* [1] [$C_{11}H_{11}Br_2N_5O$, M_R 389.05, mp. 262–265 °C (*N*-acetyl compound)], cytotoxic to KB cells, EC_{50} 0.3 mg/L, antimicrobial; *(–)-dibromophakellin* [2] from *Phakellia flabellata* [$C_{11}H_{11}Br_2N_5O$, M_R 389.05, mp. 237–245 °C (decomp.)], weak antibacterial activity; *hymenialdisine* [3] from *Hymeniacidon aldis, Axinella verrucosa*, and *Acanthella aurantiaca* [$C_{11}H_{10}BrN_5O_2$, M_R 324.14, yellow needles, mp.

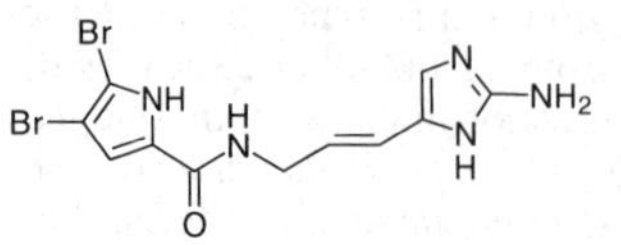

Oroidin (**1**)

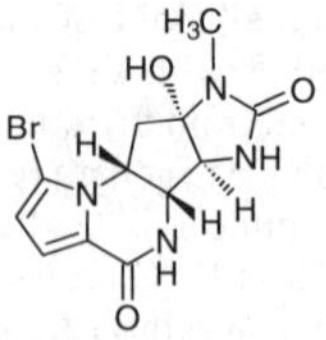

Agelastatin A (**2**)

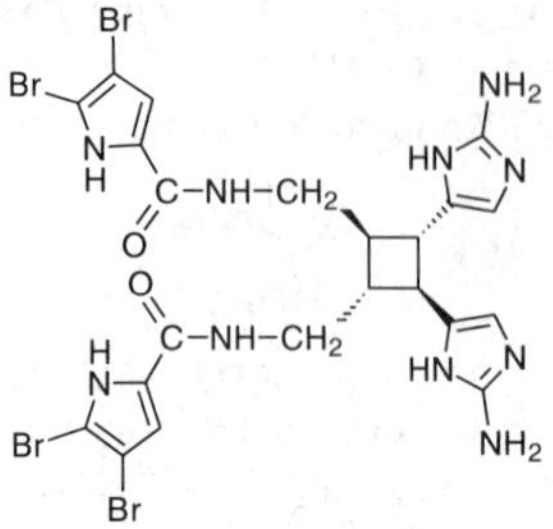

Dibromosceptrin (**3**)

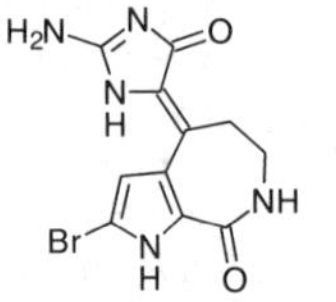

Hymenialdisine (**4**)

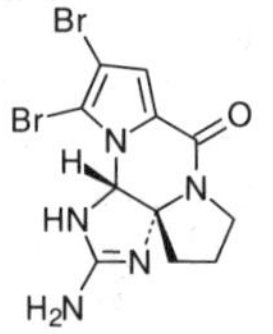

(–)-Dibromophakellin (**5**)

160–164 °C (decomp.)], moderately cytotoxic *in vitro* to KB cells; *agelastatin A*[4] from *Agelas dendromorpha* ($C_{12}H_{13}BrN_4O_3$, M_R 341.16), cytotoxic to KB cells, EC_{50} 0.1–0.5 mg/L; *dibromosceptrin*[5] from *Agelas conifera* ($C_{22}H_{22}Br_4N_{10}O_2$, M_R 778.10), actomyosin ATPase activator.

Lit.: [1] J. Chem. Soc. Chem. Commun. **1973**, 78; Zechmeister **33**, 1; Bull. Soc. Chim. Fr. **1986**, 813; Tetrahedron **53**, 7605 (1997); J. Org. Chem. **63**, 1248 (1998). [2] J. Org. Chem. **42**, 4118 (1977); Scheuer II **2**, 24; J. Am. Chem. Soc. **104**, 1776 (1982); Tetrahedron **41**, 6019 (1985). [3] J. Org. Chem. **62**, 456 (1997); Nat. Prod. Lett. **9**, 57 (1996); J. Pharmacol. Exp. Ther. **283**, 955 (1997); Tetrahedron Lett. **36**, 413 (1995). [4] J. Am. Chem. Soc. **121**, 9574 (1999); Helv. Chim. Acta **77**, 1895 (1994); **79**, 727 (1996); J. Nat. Prod. **61**, 158 (1998). [5] J. Org. Chem. **56**, 2965, 6728 (1991); J. Am. Chem. Soc. **103**, 6772 (1981); Tetrahedron **46**, 5579 (1990). – *[CAS 34649-22-4 (1); 152406-28-5 (2); 117417-71-7 (3); 82005-12-7 (4); 31954-96-8 (5)]*

Orotic acid (uracil-6-carboxylic acid, 1,2,3,6-tetrahydro-2,6-dioxo-4-pyrimidinecarboxylic acid).

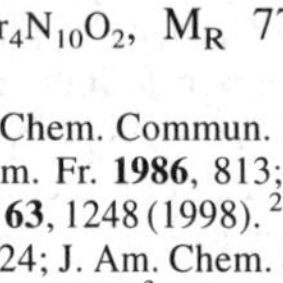

$C_5H_4N_2O_4$, M_R 156.10. cryst., mp. 345–346 °C, monohydrate: mp. 322–325 °C, pK_{a1} 1.8, pK_{a2} 9.55 (25 °C). O. occurs in cow milk [56 mg/L (winter) to 66 mg/L (summer)]; human milk does not contain orotic acid. O. functions as a vitamin for some organisms and has bacteriostatic and cytostatic activities. The free nucleoside 3-*β*-ribofuranosylorotic acid (*orotidine*, 6-carboxyuridine), $C_{10}H_{12}N_2O_8$, M_R 288.21, needles, mp. >200 °C (brown coloration at 200 °C) can be isolated from some fungi (e. g., *Neurospora crassa*) and plants (*Phaseolus vulgaris, Xanthium pennsylvanicum*). LD_{50} (mouse p. o.) 2 g/kg, (mouse i. v.) 770 mg/kg.

Biochemistry: O. is a biosynthetic precursor of *pyrimidine nucleotides. The biosynthesis proceeds from L-*aspartic acid and carbamoyl phosphate with subsequent dehydrogenation. Reaction with 5-phospho-*α*-D-ribofuranose 1-diphosphate and loss of pyrophosphate furnishes OMP and, after decarboxylation, uridine 5′-monophosphate. This *de novo* synthesis of pyrimidine bases is an important biochemical process in the biosynthesis of nucleic acids (see *Lit.*[1]). For synthesis, see *Lit.*[2].

Use: O. as well as its derivatives and salts are used in medicine as uricosuric agents, for electrolyte substitution, in cardio- and hepatoprotective agents, in dietetics and geriatrics as well as food additives in veterinary medicine, e. g., in combination with lysine and methionine. O. inhibits dihydroorotase and dihydroorotate dehydrogenase of the malaria pathogen *Plasmodium berghei* and, in combination with other principles, may be a potential antimalaria agent[3].

Lit.: [1] Biochem. Med. Metab. Biol. **49**, 338–350 (1993); Int. J. Parasitol **22**, 427–433 (1992); J. Appl. Bacteriol. **77**, 1–8 (1994). [2] Biosci. Biotechnol. Biochem. **57**, 1014 f. (1993). [3] Biochem. Pharmacol. **43**, 1295–1301 (1992).
gen.: Ann. Acad. Sci. Fenn., Ser. A2 **217**, 39 (1988) (review) ▪ Beilstein E V **25/8**, 224–228, 238 f. ▪ J. Chem. Soc., Perkin Trans. 2 **1994**, 319 ff. (synthesis) ▪ Karrer, No. 2550 ▪ Luckner (3.), p. 441 ▪ Pharmazie **40**, 377–383 (1985) (review) ▪ Sax (8.), No. OJV 500 (toxicology) ▪ Stryer 1995, p. 746 ▪ Ullmann (5.) **A 12**, 156 (synthesis). – *Pharmacology:* Int. J. Pharm. **61**, 43–49 (1990) ▪ Williams (ed.), Orotic Acid in Cardiology, p. 68–74, Stuttgart: Thieme 1992. – *[HS 2933 59; CAS 65-86-1 (O.); 314-50-1 (orotidine)]*

Orris (root) oil. Yellow to yellow-brown waxy material, also known as iris butter or (incorrectly) iris concrete, with a mild fatty, sweet, fruity-flowery, warm-woody odor of violets; melting range 38–50 °C. The main component of iris butter, *myristic acid, can be removed by precipitation with alcohol or extraction with alkali reagents, resulting in a 10–20% yield of a liquid product, the so-called "ten-fold" orris oil or iris absolute.

Production: By steam distillation after two to three years storage of the dried, ground roots of the iris species *Iris pallida* and *Iris germanica* cultivated mainly in Italy and Morocco. Since this mode of production is tedious and gives a correspondingly expensive product, the cheaper iris resin is finding increasing use; this is produced by various solvent extraction procedures.

Composition[1]***:*** Iris butter consists to 80–90% of alkanoic acids including the main component **myristic acid*. The liquid orris oil contains **irones* (ca. 20–30% *cis-α*-irone and ca. 30–40% *cis-γ*-irone) as the main components, which are also mainly responsible for the odor and taste. The oil from *I. pallida* is of more sensory value and contains predominately the dextrorotatory enantiomers while that from *I. germanica* contains the laevorotatory enantiomers[2]. The irones are not present in fresh iris roots, they are first formed during storage by oxidative degradation of *iridals[3]. In order to avoid this laborious process microbiological techniques have recently been introduced to convert the iridals to irones[4].

Use: Orris oil is one of the most valuable natural products. It is thus used only in minute doses, e.g., in high quality perfumes or to improve the aromatic character of luxury foods such as liquors, confectionery, and bakery products.

Lit.: [1] Riv. Ital. Essenze Profumi Piante Off. **63**, 141 (1981); Helv. Chim. Acta **72**, 1400 (1989). [2] J. Essent. Oil Res. **5**, 265 (1993). [3] Zechmeister **50**, 1. [4] Phytochemistry **34**, 1313 (1993).
gen.: Arctander, p. 494 ▪ Bauer et al. (2.), p. 168 ▪ Ohloff, p. 162 ▪ Parfüm. Kosmet. **71**, 411, 532 (1990). – *Toxicology:* Food Cosmet. Toxicol. **13**, 895 (1975). – *[HS 3301 29; CAS 8002-73-1 (oil); 8023-76-5 (resin)]*

Orsellinic acid (2,4-dihydroxy-6-methylbenzoic acid).

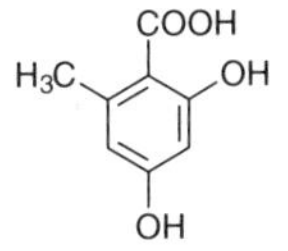

$C_8H_8O_4$, M_R 168.15, needles, mp. 176 °C (anhydrous), 186–189 °C (monohydrate), soluble in water and alcohol. The phenolcarboxylic acid O. represents an important intermediate in the biosynthesis of aromatic natural products from *polyketides.

Lit.: J. Chem. Soc., Chem. Commun. **1992**, 646 (biosynthesis) ▪ J. Chem. Soc., Perkin Trans. 1 **1988**, 755 (synthesis) ▪ J. Gen. Microbiol. **139**, 153 (1993) (isolation) ▪ Karrer, No. 916. – *[CAS 480-64-8]*

Orthosomycins see avilamycin.

Oryzarol see eugenol.

Oscillariolide. $C_{41}H_{69}BrO_{11}$, M_R 817.90. O. is a macrolide from cultures of an *Oscillatoria* sp. (cyanobacterium). At 0.5 µg/mL, O. inhibits the cell division in fertilized starfish eggs.

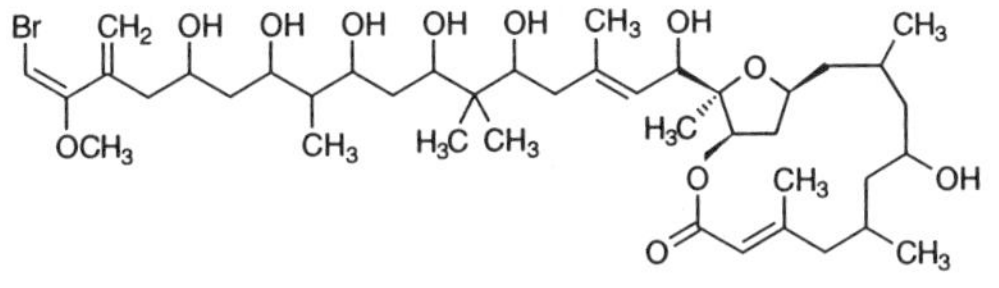

Lit.: Tetrahedron Lett. **32**, 2391–2394 (1991). – *[CAS 135097-77-7]*

Oscillatoxin see majusculamides.

Osladin.

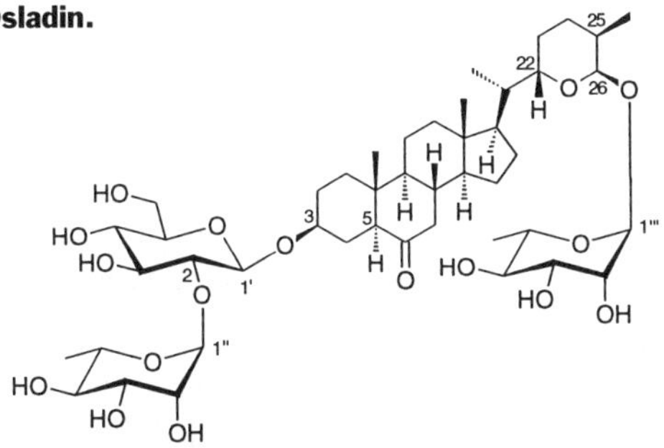

$C_{45}H_{74}O_{17}$, M_R 887.07, cryst. (CH_3OH), mp. 198–199 °C. Very sweet (500-fold sweeter than saccharose) tasting *saponin from the fern *Polypodium vulgare* (Polypodiaceae). The total synthesis has been reported [1].

Lit.: [1] Synlett **1993**, 54 ff.; **1995**, 785–793.
gen.: Adv. Exp. Med. Biol. **1996**, 405 ▪ Stud. Plant Sci. **6**, 360–372 (1999) (review) ▪ Tetrahedron Lett. **33**, 4009 (1992) ▪ see also glycyrrhizin, stevioside. – *[CAS 33650-66-7]*

Osmundalactone.

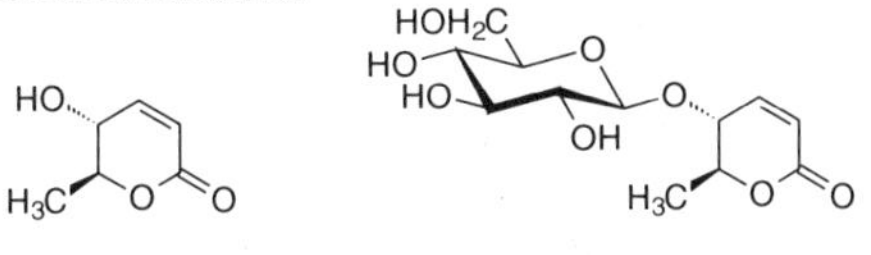

Osmundalactone Osmundalin

$C_6H_8O_3$, M_R 128.13, platelets, mp. 82 °C, $[\alpha]_D^{22}$ –71° (H_2O). O. occurs as the monoglucoside *osmundalin* ($C_{12}H_{18}O_8$, M_R 290.27, oil) in the leaves (more than 1.5% of the dry weight) of the fern *Osmunda japonica* (Osmundaceae) [1]. The enantiomeric (+)-O. is also formed on hydrolysis of *leucomentins and acts as an antifeedant for larvae of the butterfly *Eurema hecarbe mandarina* [2].

Lit.: [1] Phytochemistry **40**, 1251 (1995); Tetrahedron **30**, 2307 (1974). [2] Appl. Entomol. Zool. **18**, 129 ff. (1983).
gen.: Synthesis: ApSimon **6**, 166 ▪ Beilstein E V **18/1**, 59 ▪ Heterocycles **51**, 1503 (1999) ▪ J. Chem. Res. (S) **1999**, 488 ▪ Justus Liebigs Ann. Chem. **1989**, 797–801, 803–810. – *[HS 2932 29; CAS 5426-92-5 ((+)-O.); 54826-92-5 (O.); 54835-71-1 (osmundalin)]*

Ostreocin D see palytoxin.

Ostreogrycins s. streptogramins.

Otonecine see necine(s).

Ouabagenin, ouabain see strophanthins.

Oudemansins.

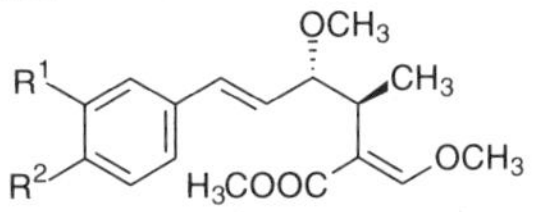

$R^1 = R^2 = H$: Oudemansin A
$R^1 = OCH_3$, $R^2 = Cl$: Oudemansin B
$R^1 = H$, $R^2 = OCH_3$: Oudemansin X

The strongly antifungal O. are isolated from cultures of several basidiomycetes, e.g., *O. A* {$C_{17}H_{22}O_4$, M_R 290.36, cryst., mp. 44 °C, $[\alpha]_D$ –17° (C_2H_5OH)} from *Oudemansiella mucida*, *O. B* {$C_{18}H_{23}ClO_5$, M_R 354.83, yellow oil, $[\alpha]_D$ –8.3° ($CHCl_3$)} from *Xerula longipes* and *X. melanotricha*, and *O. X* {$C_{18}H_{24}O_5$, M_R 320.39, oil, $[\alpha]_D$ –20.4° (C_2H_5OH)} from *Oudemansiella radicata*. Like the structurally closely related *strobilurins and *myxothiazol the O. have an (*E*)-*β*-methoxyacrylate structure (Moa) and specifically inhibit the respiration of fungi (IC_{50} 0.1 µg/mL). Interaction with the bc_1 complex of the respiratory chain leads to an interruption of electron transport. The O. are strongly cytotoxic but exhibit no antibacterial activity. They are especially active against phytopathogenic fungi and thus O. and the strobilurins were used as lead structures for novel fungicides for crop protection [1].

Lit.: [1] Nachr. Chem. Tech. Lab. **38**, 1236–1240 (1990); Steglich & Anke in: Grabley & Thiericke: Drug Discovery from Nature, p. 320–334, Berlin: Springer 1999.
gen.: J. Antibiot. **32**, 1112 (1979); **36**, 661 (1983). – *Synthesis:* Chem. Lett. **1992**, 687 ▪ Chem. Pharm. Bull. **32**, 1242 (1984); **43**, 1111, 1162 (1995) ▪ J. Am. Chem. Soc. **107**, 1429 (1985) ▪ J. Antibiot. **43**, 1010 (1990) ▪ J. Chem. Soc., Chem. Commun. **1992**, 1218 ▪ J. Chem. Soc., Perkin Trans. 1 **1993**, 1957; **1995**, 2159 ▪ J. Org. Chem. **52**, 4303 (1987) ▪ Synlett **1995**, 869 ▪ Tetrahedron: Asymmetry **4**, 757 (1993) ▪ Tetrahedron Lett. **24**, 2009 (1983); **27**, 2443, 5397 (1986). – *Isolation:* J. Antibiot. **52**, 332 (1999). – *[CAS 73341-71-6 (O. A); 87081-56-9 (O. B); 130640-32-3 (O. X)]*

Oudenone.

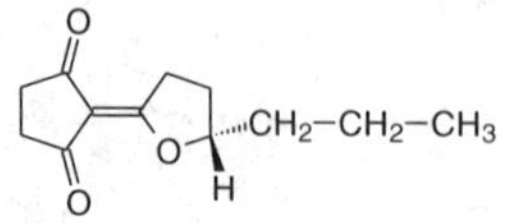

$C_{12}H_{16}O_3$, M_R 208.26, needles, mp. 77–78 °C, $[\alpha]_D$ –10.8° (C_2H_5OH). An inhibitor of tyrosine hydroxylase from the fungus *Oudemansiella radicata* (Basidiomycetes). O. exhibits hypertensive activity *in vivo*.

Lit.: J. Antibiot. **35**, 458 (1982). – *Synth.:* Bull. Chem. Soc. Jpn. **60**, 1161 (1987) ▪ Chem. Pharm. Bull. **25**, 2775 (1977) ▪ Heterocycles **6**, 261 (1977) ▪ J. Org. Chem. **60**, 6922–6929 (1995) (biosynthesis). – *[CAS 31323-50-9]*

Ovalicin.

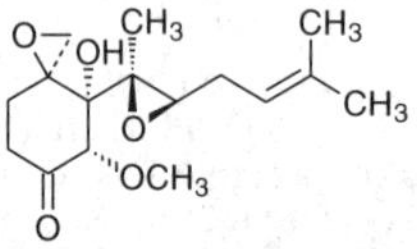

$C_{16}H_{24}O_5$, M_R 296.36, cryst., $[\alpha]_D^{20}$ –117° ($CHCl_3$), mp. 94–95 °C, soluble in chloroform, poorly soluble in water. O. was isolated for the first time in 1968 from culture filtrates of *Pseudeurotium ovalis* [1]. O. is identical to graphinone [2] and has interesting properties as an immunosuppressive agent [3].

Biosynthesis: Farnesyl diphosphate is converted to *trans*-*β-bergamotene by cyclization of an intermediate bisabolyl cation. Oxidative ring opening followed by three specific oxidation reactions leads to O.[4].

Lit.: [1] Helv. Chim. Acta **51**, 1395, 1402 (1968). [2] Planta Med. **34**, 231 (1978). [3] Nature (London) **222**, 773 (1969); J. Am. Chem. Soc. **98**, 1184 (1976); **99**, 6132 (1977); Tetrahedron Lett. **1974**, 4417. [4] Tetrahedron Lett. **28**, 6545 (1987).

gen.: Beilstein E V **19/6**, 186. – *Synthesis:* J. Am. Chem. Soc. **107**, 256 (1985); **116**, 12109 (1994) ▪ J. Chem. Soc., Perkin Trans. 1 **1995**, 1551. – *[CAS 19683-98-8]*

Oviposition deterring pheromones (host marking pheromones).

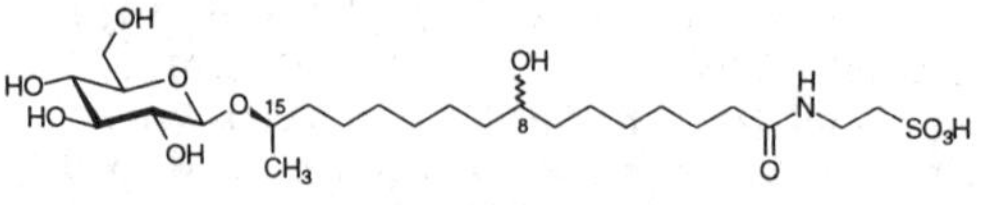

O. d. p. of cherry maggot

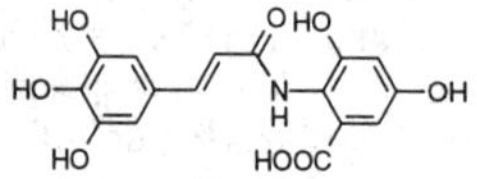

Miriamide

Pheromones with which, especially, insects mark the position of egg deposits in order, e. g., to avoid double occupancy of fruits. A presumably widely distributed phenomenon which, however, with regard to identification of relevant compounds has not been investigated with success. The compound used by the cherry maggot *Rhagoletis cerasi* is *2-[15-(β-D-glucopyranosyloxy)-8-hydroxyhexadecanoylamino]ethanesulfonic acid* (**1**) ($C_{24}H_{47}NO_{11}S$, M_R 557.70) [1]. The natural product is apparently a diastereomeric mixture with (8*RS*,15*R*) configuration [2]. Simple fatty acid esters have been reported for the corn borer *butterfly *Ostrinia nubilalis* [3]. The O.d.p of the cabbage butterfly (*Pieris brassicae*) is *miriamide* (**2**)[4] ($C_{16}H_{13}NO_8$, M_R 347.28).

Lit.: [1] Experientia **43**, 157–160 (1987); Helv. Chim. Acta **72**, 165–171 (1989). [2] Justus Liebigs Ann. Chem. **1993**, 291–295. [3] Naturwissenschaften **78**, 132–133 (1991). [4] J. Nat. Prod. **57**, 90 (1994).

gen.: Chem. Ind. (London) **1994**, 370 ff. – *[CAS 107959-84-2 (1); 153698-89-6 (2)]*

Oviposition pheromone see oviposition deterring pheromones.

Ovothiols. A group of mercaptohistidines from sea urchin and starfish eggs.

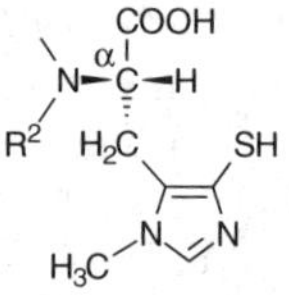

Table: Data of Ovothiols.

	R^1	R^2	molecular formula	M_R	CAS
O. A	H	H	$C_7H_{11}N_3O_2S$	201.24	108418-13-9
O. B	CH_3	H	$C_8H_{13}N_3O_2S$	215.27	108418-14-0
O. C	CH_3	CH_3	$C_9H_{15}N_3O_2S$	229.30	105496-34-2

The O. function in fertilized eggs as a non-enzymatic redox system. High concentrations of peroxide radicals, which are hazardous for the embryo, are generated to form a varnish-like protective layer of dityrosyl residues by radical polymerization. With a pK_a of 2.3 the O. are stronger nucleophiles than glutathione (GSH), they thus facilitate reduction of the peroxide radicals in the vicinity of the embryo.

Lit.: Biochemistry **26**, 4028 (1987) ▪ BIOFactors **1**, 85 (1988) (review) ▪ J. Org. Chem. **52**, 4421 (1987); **54**, 4570 (1989).

Ovulation inhibitors see estrogens and progestogens.

***N*²-, *N*³-Oxalyl-2,3-diaminopropanoic acid.**

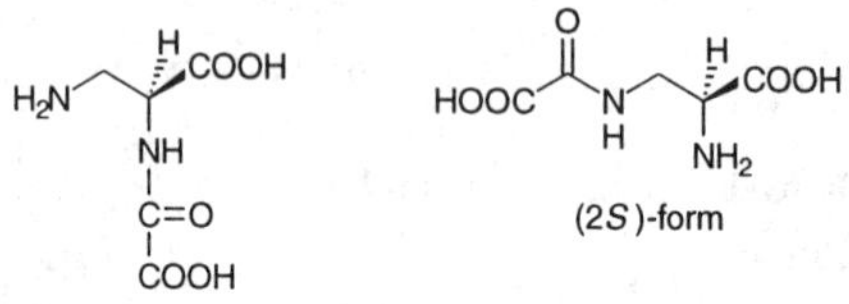

N^2-O. N^3-O.

$C_5H_8N_2O_5$, M_R 176.13. Potent neurotoxin in the seeds of *Lathyrus sativus* as well as other *L.*, *Crotalaria*, and *Acacia* species [1,2] (Fabaceae). The biosynthesis proceeds through L-2,3-diaminopropanoic acid which presumably originates from Asn and is acylated by oxalyl-CoA [3].

Lit.: [1] Nature (London) **218**, 197 (1968). [2] Phytochemistry **16**, 477 ff. (1977). [3] Phytochemistry **9**, 1603–1610 (1970).

gen.: Rosenthal, Plant Nonprotein Amino and Iminoacids, p. 69–74, New York: Academic Press 1982.

Oxapenems see clavulanic acid.

Oxetanocin {oxetanocin A, 9-[(2*R*,3*R*,4*S*)-3,4-bis-(hydroxymethyl)-2-oxetanyl]adenine}. $C_{10}H_{13}N_5O_3$, M_R 251.24, needles (×H_2O), mp. 197 °C, $[\alpha]_D^{20}$ –44.3° (pyridine), a *nucleoside antibiotic from culture broths of *Bacillus megaterium* exhibiting antibacterial and potent antiviral activities. In the name *O. A* the A is in-

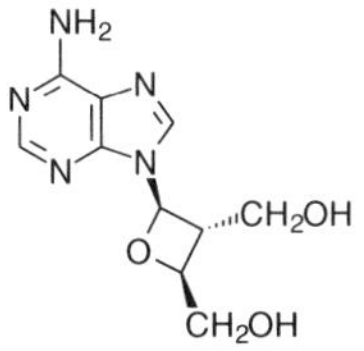

ɿnocin A

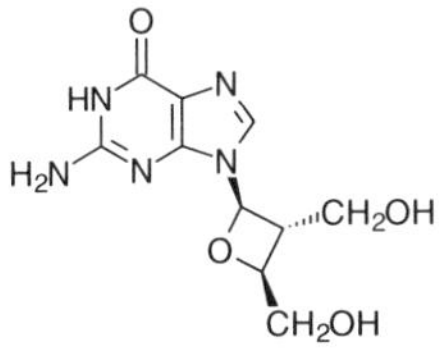

Oxetanocin G

dicative of adenine for distinction from other, synthetic derivatives such as *O. G* (guanine) or *O. H* (hypoxanthine). O. and its synthetic derivatives (other natural bases, cyclobutane ring) inhibit the reverse transcriptase of retroviruses and have thus been tested against AIDS, cytomegalovirus (CMV), hepatitis B virus, and herpes simplex virus (HSV) (*O. G:* $C_{10}H_{13}N_5O_4$, M_R 267.24). The remarkable anti-HIV activity is synergistically enhanced by *mycophenolic acid which is not active itself.

Lit.: Antimicrob. Agents Chemother. **32**, 1053 (1988); **33**, 773 ff.; **34**, 287–294 (1990) ▪ J. Antibiot. **42**, 644 ff., 1308–1311, 1854–1859 (1989). – *Synthesis:* Atta-ur-Rahman **10**, 585–627 (1992) ▪ Chem. Rev. **92**, 1745 (1992) ▪ J. Am. Chem. Soc. **110**, 7217 f. (1988) ▪ J. Chem. Soc., Chem. Commun. **1989** 1919 ff. ▪ Tetrahedron: Asymmetry **1**, 527–530 (1990) ▪ Tetrahedron Lett. **28**, 3967–3970 (1987); **29**, 4739–4742, 4743–4746 (1988); **30**, 2269 f. (1989); **31**, 5445, 6931 (1990); **32**, 3531 (1991). – *[HS 2941 90; CAS 103913-16-2 (O. A); 113269-46-8 (O. G)]*

2-Oxetanones see ebelactones, lipstatin, obafluorin, panclicins, papulinone.

Oxetin [(2*R*)-*cis*-3-amino-2-oxetanecarboxylic acid].

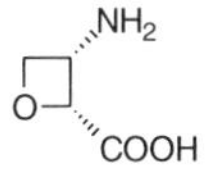

$C_4H_7NO_3$, M_R 117.10, mp. 185–190 °C (decomp.), $[\alpha]_D^{25}$ +56.4° (H_2O). A non-proteinogenic amino acid with antibacterial and herbicidal activities produced by a *Streptomyces* species. O. is a non-competitive inhibitor of glutamine synthetase [1]. Although only the natural (2*R*)-*cis*-form is active against bacteria, the other isomers also function as inhibitors of glutamine synthetase [2].

Lit.: [1] J. Antibiot. **37**, 1324–1332 (1984). [2] Chem. Pharm. Bull. **34**, 3102–3110 (1986). – *Synthesis:* Justus Liebigs Ann. Chem. **1997**, 2265. – *[CAS 94818-85-6]*

Oxindole alkaloids. A widely distributed group of *monoterpenoid indole alkaloids and other alkaloid groups, all of which have *oxindole* (1,3-dihydro-2*H*-indol-2-one) as core structure.

While the monoterpenoid O. a. occur in plants of the families Rubiaceae, Naucleaceae and Loganiaceae as well as Apocynaceae, compounds of the other O. a. groups also occur in members of the families Aristolochiaceae (e.g., *goniopedaline*, $C_{17}H_{13}NO_4$, M_R 295.29, mp. 218 °C [1,2]), Annonaceae (e.g. enterocarpams II [3]), or also Myristicaceae (e.g. (–)-*horsfiline*, $C_{13}H_{16}N_2O_2$, M_R 232.28, $[\alpha]_D^{20}$ –7.2° (CH_3OH) [4]). The *phytoalexins *spirobrassinin* [5] (see also brassinin) [$C_{11}H_{10}N_2O_2S_2$, M_R 250.35, mp. 158–159 °C, $[\alpha]_D$ –69.5° ($CHCl_3$)] or *dioxybrassinin* [6] [$C_{11}H_{12}N_2OS_2$, M_R 268.36, $[\alpha]_D$ –7.6° (CH_3OH)] formed by *Brassica* species when the plants are infected with *Pseudomonas*

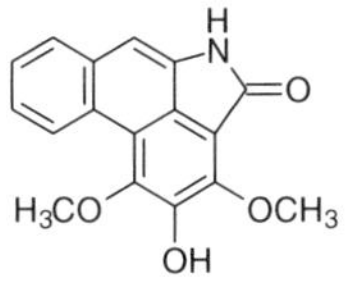

Goniopedaline (**1**)

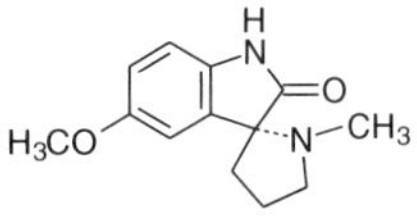

(–)-Horsfiline (**2**)

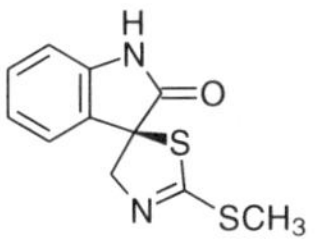

Spirobrassinin (**3**)

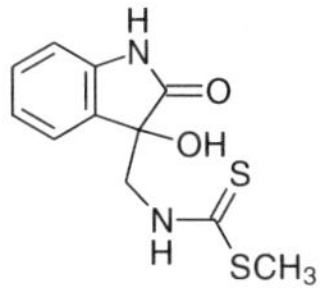

Dioxybrassinin (**4**)

also belong in this group of alkaloids. O. a. have also been found in cyanobacteria (e.g,. hapalindolinone A [7]) or in marine organisms. Typical examples of O. a. of monoterpenoid origin are those from *Gelsemium*, *Pseudocinchona*, *Mitragyna*, and *Uncaria* species; e.g. *gelsemine, humantenirine [9], mitraphylline, the uncarines A–F, or corynoxeine.

Lit.: [1] J. Nat. Prod. **50**, 843 (1987). [2] Phytochemistry **27**, 903 (1988). [3] Phytochemistry **25**, 965 (1986). [4] Heterocycles **38**, 725 (1994); J. Org. Chem. **64**, 1699 (1999); Tetrahedron: Asymmetry **5**, 1979 (1994); **7**, 1 (1996). [5] Chem. Lett. **1987**, 1631. [6] Phytochemistry **30**, 2915 (1991). [7] J. Org. Chem. **52**, 3704 (1987). [8] J. Org. Chem. **53**, 3276 (1988); **56**, 3403 (1991); **57**, 3636 (1992); Tetrahedron Lett. **34**, 7917 (1993); **35**, 8851 (1994). [9] J. Nat. Prod. **52**, 588 (1989). – *[HS 2939 90; CAS 112501-43-6 (1); 136247-72-8 (2); 113866-40-3 (3); 133761-64-5 (4)]*

(*E*)-9-Oxo-2-decenoic acid see queen substance.

3-Oxohexanal, (*E*)-4-oxo-2-hexenal see bugs.

Oxyacanthine see bisbenzylisoquinoline alkaloids.

Oxyblepharismin C see blepharismins.

Oxylipins. Collective name for oxygen derivatives of fatty acids, especially hydroperoxy and *hydroxy fatty acids, formed by the action of mono- or dioxygenases on polyene fatty acids in plants and animals. The O. include not only the *eicosanoids (with 20 C atoms) but also similar compounds with more or less carbon atoms (e.g. "octadecanoids" with 18 C atoms).

Lit.: Arch. Biochem. Biophys. **290**, 436 (1991) ▪ Lipids **28**, 783 (1993).

Oxypterine [(+)-4-chloro-1-chloromethyl-5-methyleneazocane].

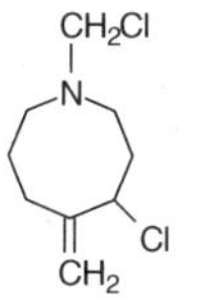

$C_9H_{15}Cl_2N$, M_R 208.13, cryst., mp. 127–129 °C, $[\alpha]_D$ +19.6° ($CHCl_3$). An unusual, chlorinated azacyclooctane alkaloid from the Fabaceae genus *Lotononis*, subsection *Rostrata*. O. is possibly biogenetically related to the *pyrrolizidine alkaloids occurring in the same genus.

Lit.: Phytochemistry **31**, 1029 (1992). – *[HS 2939 90; CAS 143114-89-0]*

Oxytetracycline see tetracyclines.

P

PABA see 4-aminobenzoic acid.

Pachycarpine see sparteine.

Pachymic acid see tumulosic acid.

Pachysandiol A see friedelanes.

Paederoside (paederosidic acid) see asperuloside.

PAF. Abbreviation for *platelet activating factor.

Pahutoxin [*O*-(*S*)-3-acetoxyhexadecanoylcholine chloride].

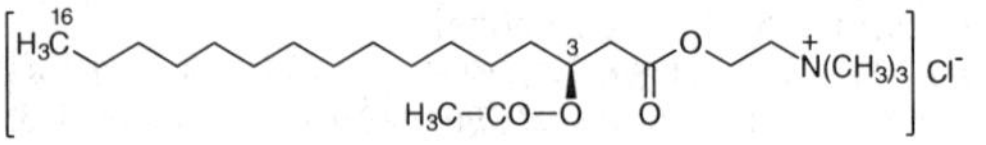

$C_{23}H_{46}ClNO_4$, M_R 436.08, needles, mp. 74–75 °C. A fish toxin obtained from the boxfish (*Ostracion lentiginosus*, known in Hawaii as pahu). On being touched and under other stimulations the fish secretes a slime containing P. through its skin. The secretion is not only extremely toxic to smaller fish but also to the trunkfish itself.

Lit.: Science **155**, 52 (1967) ▪ Toxicon **26**, 651 (1988). – *Synthesis:* Agric. Biol. Chem. **53**, 37 (1989) ▪ Scheuer II **2**, 119 ▪ Zechmeister **27**, 325. – *[CAS 27742-14-9]*

Palatinose (isomaltulose, 6-*O*-α-D-glucopyranosyl-D-fructose).

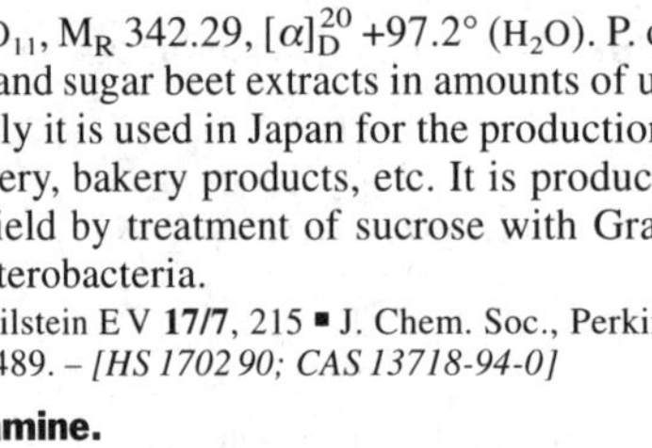

$C_{12}H_{22}O_{11}$, M_R 342.29, $[\alpha]_D^{20}$ +97.2° (H_2O). P. occurs in honey and sugar beet extracts in amounts of up to 1%; currently it is used in Japan for the production of confectionery, bakery products, etc. It is produced in ca. 90% yield by treatment of sucrose with Gram-negative enterobacteria.

Lit.: Beilstein E V **17/7**, 215 ▪ J. Chem. Soc., Perkin Trans. 2 **1990**, 1489. – *[HS 1702 90; CAS 13718-94-0]*

Palauamine.

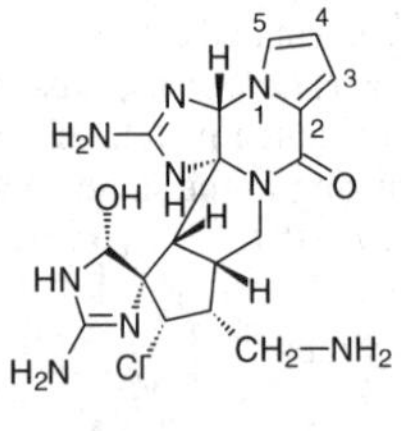

(relative configuration)

$C_{17}H_{22}ClN_9O_2$, M_R 419.87, amorphous, $[\alpha]_D^{24}$ –45.2° (CH_3OH). Alkaloid from the sponge *Stylotella aurantium* with cytotoxic and immunosuppressive activities. The less active 4-bromo and 4,5-dibromo derivatives have also been isolated.

Lit.: J. Am. Chem. Soc. **115**, 3376 (1993) ▪ J. Org. Chem. **63**, 3281 (1998). – *[CAS 148717-58-2]*

Palmarosa oil. A pale yellow oil with a sweet-flowery, rose-like odor with herby and bread-like undertones.

Production: By steam distillation from the tropical grass *Cymbopogon martinii* var. *motia*. The major producing country is India, smaller amounts of the grass are cultivated in Guatemala, Brazil, and Madagascar.

Composition [1]***:*** The main component is *geraniol* (75–85%). Oil of the Guatemala quality differs and contains ca. 60% *geraniol* and 15% *nerol* (see geraniol), the latter is present in traces only in other oils. The second most abundant component is *geranyl acetate* [$C_{12}H_{20}O_2$, M_R 196.28, bp. 128–129 °C (2.1 kPa)] (ca. 10%).

Use: In the perfume industry for flowery, rose-like notes for all applications. Smaller amounts of P. are used to produce geraniol with a high quality odor for fine perfumes.

Lit.: [1] Perfum. Flavor. **19** (2), 29 (1994).
gen.: Arctander, p. 501 ▪ Bauer et al. (2.), p. 152 ▪ ISO 4727 (1988). – *Toxicology:* Food Cosmet. Toxicol. **12**, 947 (1974). – *[HS 3301 29; CAS 8014-19-5 (P.); 105-86-3 (geranyl acetate)]*

Palmatine see protoberberine alkaloids.

Palm kernel oil (palm seed oil or fat). An oil obtained from the seeds (kernels) of the palm oil tree (*Elaeis guineensis*) and related palm species, the fruit pulp of the palm oil tree furnishes *palm oil. The worldwide production of P. in 1993 amounted to 1.8 million tons. Mp. 23–30 °C. On account of its high contents of esterified, saturated *fatty acids of medium chain length, e. g. *lauric acid* (ca. 50%) and **myristic acid* (ca. 15%), P. resembles coconut fat (so-called "lauric" fats and oils that are rich in lauric acid and other medium-chain length fatty acids); other components: caproic acid (hexanoic acid) (5%), caprylic acid (octanoic acid) (3%), palmitic acid (6–9%), stearic acid (2–3%), oils (10–18%), and *linoleic acid (1–3%). For the composition of the seed oils of other palm species, see *Lit.* [1].

Use: The refined oil is used in the production of edible fats, margarines, and special margarine products [MCT margarine: see triglycerides] for consumption, in particular, by people with metabolic diseases, as well as cocoa butter substitute fats. It is also used for cooking and as a frying fat. Fractions of the oil, e.g., palm kernel stearin (mp. 30–32 °C) are used alone or as mixtures with coconut stearin in glazing fats or as fillers in confectionery and bakery products. The hydrogenated oil is used as a fat for deep-frying. In the oleochemical industry the oil is used in the production of detergents.

Lit.: [1] Chem. Phys. Lipids **4**, 96 (1970).
gen.: Applewhite (ed.), Proceedings of the World Conference on Lauric Oils: Sources, Processing, and Applications, Champaign: AOCS Press 1994 ▪ Inform **5**, 145 (1994) ▪ Ullmann (5.) **A 10**, 176, 221. – *[HS 151321, 151329; CAS 8023-79-8]*

Palm oil (palm fat). An oil obtained from the fruit pulp tissue (endosperm) of the palm oil tree (*Elaeis guineensis*); *palm kernel oil is the corresponding oil from the seeds. Mp. 30–37 °C; color: orange-yellow to red-brown ("red palm oil") because of the high content of *carotinoids (approx. 500 mg/kg). The fatty acid compositions of the two oils are very different: P. is rich in *palmitic acid* (ca. 42%) and *oleic acid* (ca. 41%), while palm kernel oil is rich in lauric acid (ca. 50%). Main areas of cultivation: Malaysia, Indonesia, and West Africa. World-wide production (1992): 12000000 tons; thus palm oil is second only to soya oil as the economically most important plant oil.
Use: For its various uses the oil is usually fractionated into low-melting palmolein (ca. 70% of P., rich in oleic acid used especially for deep-frying and baking fats; shortenings [1]) and into palmstearin with mp. 41–50 °C (ca. 30% of P., rich in palmitic acid, used especially in the manufacture of margarines).

Lit.: [1] Fat Sci. Technol. **90**, 375 (1989).
gen.: Chem. Ind. (London) **1989**, 244 ▪ Gunston, Palm Oil, Chichester: Wiley 1987 ▪ Ullmann (5.) **A 10**, 176, 218. – *[HS 151110, 151190; CAS 8002-75-3]*

Palominol see dolabellanes.

Palustric acid ($\Delta^{8,13}$-abietic acid).
$C_{20}H_{30}O_2$, M_R 302.46, mp. 162–167 °C, cryst., $[\alpha]_D$ +71.6° (C_2H_5OH). A diterpene of the *abietane group.

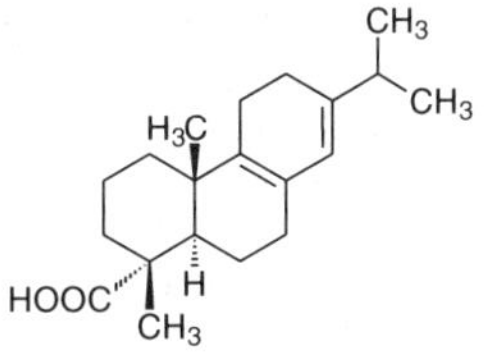

It is obtained mainly from pine balsam, balsam, and root resins where its contents range between 10 and 20%.

Lit.: Arzneim.-Forsch. **44**, 17 (1994) ▪ Beilstein E IV **9**, 2174 ▪ Hager (5.) **6**, 168 f., 179 f. ▪ Ullmann (5.) **A 23**, 82 ff. – *Synthesis:* Can. J. Chem. **59**, 2224 (1972). – *[CAS 1945-53-5]*

Palustrine.

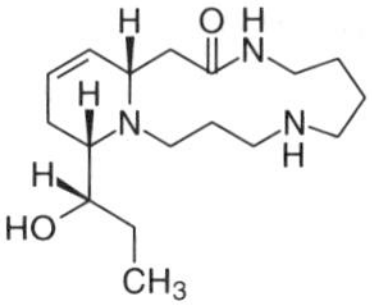

$C_{17}H_{31}N_3O_2$, M_R 309.45, mp. 188–190 °C (decomp.), $[\alpha]_D^{18}$ +15.8° (H_2O), readily soluble in H_2O, practically insoluble in petroleum ether. The major macrocyclic alkaloid [1] from *Equisetum palustre* (marsh horsetail). It is usually accompanied by its *N*-formyl derivative, *palustridine* ($C_{18}H_{31}N_3O_3$, M_R 337.46, mp. 204 °C, $[\alpha]_D^{19}$ +50.2° (H_2O) and other minor alkaloids. Recent investigations have shown that a number of other *Equisetum* species (*E. fluviatile*, *E. hyemale*, *E. giganteum*, etc.) also contain P. and accompanying alkaloids. The contents range between 0.3 and 1.3 g/kg dry weight. The toxic P. can cause respiratory paralysis [LD_{50} (mouse s.c.) 1 mg/kg] [2]. Because of its toxicity *E. palustre* is a much feared weed in grazing pastures. The structure was corrected in 1984.

Lit.: [1] Helv. Chim. Acta **61**, 921 (1978). [2] Hegnauer **I**, 246.
gen.: Chem. Pharm. Bull. **32**, 3789 (1984) ▪ J. Org. Chem. **62**, 776 (1997) (synthesis) ▪ Tetrahedron Lett. **25**, 2391 (1984). – *[HS 293990; CAS 22324-44-3 (P.); 22324-43-2 (palustridine)]*

Palytoxin (PTX).

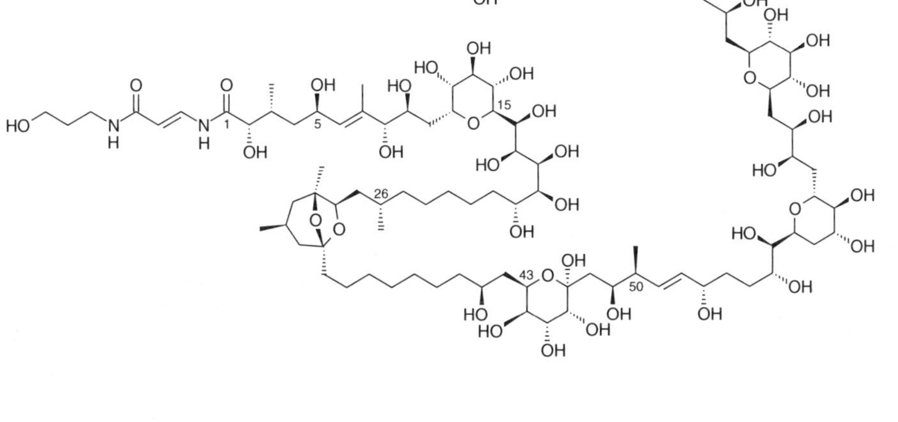

Palytoxin

$C_{129}H_{223}N_3O_{54}$, M_R 2680.18, hygroscopic solid, decomp. >300 °C. P. is isolated from the Pacific crustaceous anemone *Palythoa toxica* (order Zoanthidae)[1] and presumably originates from a bacterial symbiont of *Palythoa*; it is one of the most toxic natural products known to date: LD_{50} (mouse i. v.) 0.15 µg/kg or (common crabs) 0.06 µg/kg; (rabbit) 0.025 µg/kg; and is about 25 times more toxic than *tetrodotoxin. *Palythoa* extracts containing P. were used by the native population of Hawaii to poison the tips of spears used for hunting and for warfare ("limu-make-o-Hana", lethal sea plant from Hana). Four similarly toxic minor components have been isolated from *P. toxica*, *P. tuberculosa*, and *P. mammilosa*: homopalytoxin, bishomopalytoxin, neopalytoxin, deoxypalytoxin. P. has also been detected in the red alga *Chondria armata*, crabs of the genera *Demania* and *Lophozozymus*[2], and in fish; it is supposed to be responsible for numerous deaths after consumption of the animals. The dinoflagellate *Ostreopsis siamensis* contains 3,26-bis-demethyldeoxypalytoxin (ostreocin D)[3].

Activity: P. and the minor components form pores in the cell membrane (cytolysins). Nanomolar concentrations of P. increase the membrane permeability of vascular muscle cells for Na^+ and K^+ ions. At higher concentrations the membranes also become permeable for molecules such as, e. g., ATP. The increased permeability for Na^+ ions leads to depolarization and contraction of every susceptible organ. P. has the highest coronary contracting activity of all compounds known to date. P. is a model substance for investigations on membrane processes including the genesis of angina pectoris. P. was tested for cancer therapy and as a local anesthetic but cannot be used on account of its extreme toxicity.

Lit.: [1]J. Am. Chem. Soc. **103**, 2491 (1981); **104**, 3776, 7362–7372 (1982); Tetrahedron **41**, 1007 (1985). [2]Toxicon **26**, 105 (1988). [3]J. Am. Chem. Soc. **117**, 5389 (1995). – *Pharmacology:* Toxicon **26**, 105 (1988); **27**, 1171–1187 (1989); Pure Appl. Chem. **54**, 1963 (1982).
gen.: ACS Symp. Ser. **418**, 202–255 (1990) ▪ Adv. Cancer Res. **49**, 223–264 (1987) ▪ Chem. Rev. **93**, 1897 (1993) ▪ Nachr. Chem. Tech. Lab. **32**, 115 (1984) ▪ Naturwiss. Rundsch. **36**, 406 (1983) ▪ Pure Appl. Chem. **61**, 313 (1989) ▪ Sax (8.), PAE 875, PAF 000 ▪ Scheuer II **1**, 109; II **2**, 25 ▪ Zechmeister **48**, 81–202. – *Synthesis:* J. Am. Chem. Soc. **111**, 7525 (1989) ▪ Nicolaou, p. 711–730. – *[CAS 77734-91-9]*

Pamamycins.

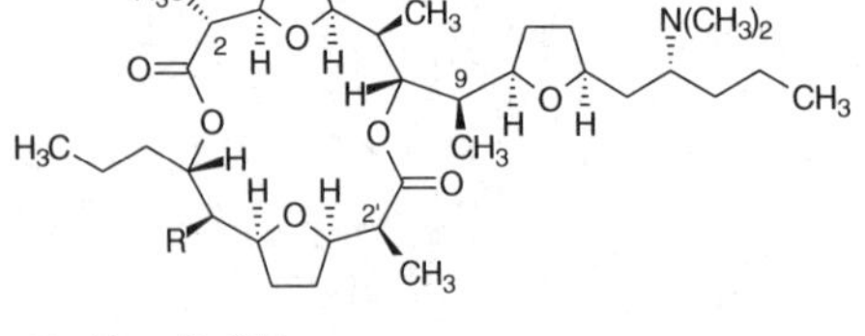

R = H : P.- 607
R = CH_3 : P.- 621

Endogenous autoregulators (*differentiation factors) of *Streptomyces alboniger* and *S. aurantiacus*. The P. stimulate the formation of aerial mycelia, inhibit the formation of substrate mycelia at higher concentrations, and also have antibacterial and antifungal activities. In two-phase systems, they mediate anion transport into the lipophilic phase under neutral and acidic conditions. P. occur as mixtures of eight homologues (M_R 593–691) with two major components: *P.-607:* $C_{35}H_{61}NO_7$, M_R 607.87, oil, $[\alpha]_D$ –22.8° (CH_3OH); *P.-621:* $C_{36}H_{63}NO_7$, M_R 621.90. The P. are macrodiolides made up of differing polyketide building blocks. The methyl groups at C-2, C-9, and C-2′ can be replaced by ethyl groups.

Lit.: Annu. Rev. Microbiol. **46**, 377–398 (1992) (review) ▪ J. Antibiot. **32**, 673 (1979); **41**, 1196 (1988) ▪ Nat. Prod. Lett. **3**, 265 (1994) ▪ Tetrahedron Lett. **32**, 3087 (1987). – *[CAS 100905-89-3 (P.-607); 71892-94-9 (P. complex); 149146-57-6 (P.-621)]*

Panamine.

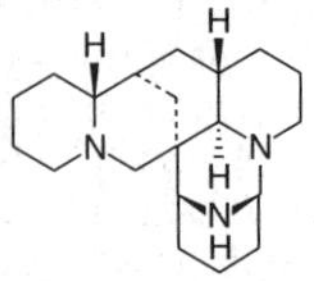

(–)-form

$C_{20}H_{33}N_3$, M_R 315.50, oil, $[\alpha]_D$ –11°. A *quinolizidine alkaloid of the ormosanine type. It occurs together with several isomers and other structures derived from the same skeleton in some South American Fabaceae species of the genus *Ormosia*[1] (see also ormosanine).

Lit.: [1]Waterman **8**, 197–239.
gen.: Beilstein E V **26/2**, 85 ▪ Can. J. Chem. **54**, 97 (1976). – *[HS 2939 90; CAS 2448-27-3]*

Panaxosides. Triterpene glycosides from the *ginseng root (*Panax ginseng*). The P. are also known as ginsenosides.

Panaxynol see falcarinol.

Panaxytriol (falcarintriol) see falcarinol.

Panclicins.

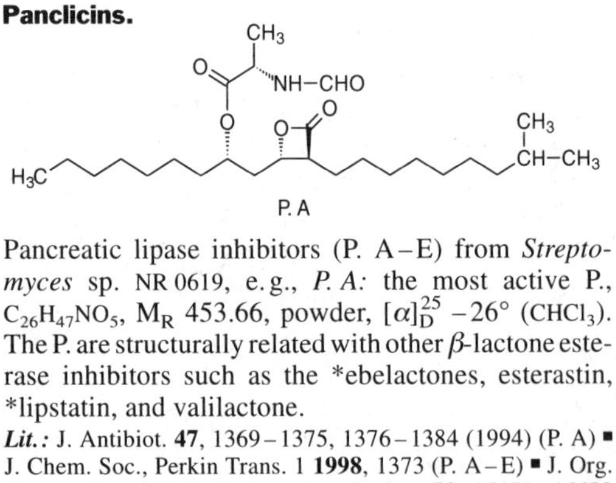

P. A

Pancreatic lipase inhibitors (P. A–E) from *Streptomyces* sp. NR 0619, e.g., *P. A:* the most active P., $C_{26}H_{47}NO_5$, M_R 453.66, powder, $[\alpha]_D^{25}$ –26° ($CHCl_3$). The P. are structurally related with other β-lactone esterase inhibitors such as the *ebelactones, esterastin, *lipstatin, and valilactone.

Lit.: J. Antibiot. **47**, 1369–1375, 1376–1384 (1994) (P. A) ▪ J. Chem. Soc., Perkin Trans. 1 **1998**, 1373 (P. A–E) ▪ J. Org. Chem. **62**, 4 (1997) (P. D) ▪ Tetrahedron **53**, 16471–16488 (1997) (synthesis P. D).

Pancratistatin.

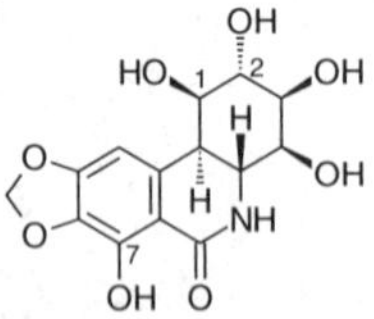

Alkaloid from the roots and buds of *Pancratium littorale* and the buds of *Haemanthus kalbreyeri* (Amaryllidaceae). $C_{14}H_{15}NO_8$, M_R 325.27, mp. 322–324 °C, $[\alpha]_D^{34}$ +48° (DMSO). P. exhibits highly promising antineoplastic activities (leukemia, ovarial carcinoma). In addition *H. kalbreyeri* also contains derivatives of P.: 7-deoxy-P. and O²-β-D-glucopyranosyl-P. (pancratiside).

Lit.: Anticancer Drug Design **10**, 243–250 (1995) ▪ J. Am. Chem. Soc. **111**, 4829 (1989) ▪ J. Nat. Prod. **49**, 995 (1986); **58**, 37 (1995) ▪ Phytochemistry **28**, 611 (1989). – *Synthesis(+)P.:* Chemtracts: Org. Chem. **9**, 101–109 (1996) ▪ Heterocycles **43**, 1385 (1996) ▪ J. Am. Chem. Soc. **117**, 3643, 7289, 10143 (1995); **118**, 10752–10765 (1996); **120**, 5341 (1998) ▪ J. Heterocycl. Chem. **35**, 343 (1998) (analogues) ▪ J. Org. Chem. **63**, 9164 (1998); **64**, 4465 (1999) ▪ Pat. WO 9716,453 (analogues) ▪ Tetrahedron **53**, 11153–11170 (1997); **54**, 15509–15524 (1998). – *[HS 2939 90; CAS 96203-70-2 (P.); 83603-30-9 (7-deoxy-P.); 121849-70-5 (O²-β-D-glucopyranosyl-P.)]*

Panepoxydone.

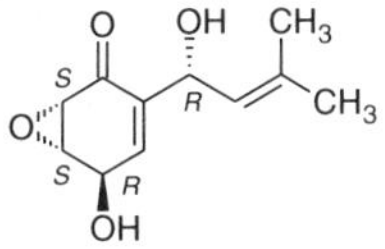

$C_{11}H_{14}O_4$, M_R 210.23, oil, $[\alpha]_D^{20}$ −61° (CH_2Cl_2), a constituent of *Lentinus crinitus* (Tricholomataceae) and *Panus* spp. P. inhibits NF-κB dependent signal transduction.

Lit.: Biochem. Biophys. Res. Commun. **226**, 214–221 (1996) ▪ Helv. Chim. Acta **53**, 1577–1597 (1970). – *[CAS 31298-54-1]*

Paniculatadiol see oleanane.

Pannaric acid.

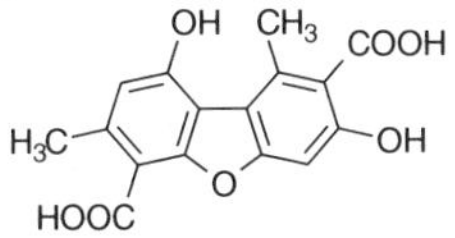

$C_{16}H_{12}O_7$, M_R 316.27, needles, mp. 243–245 °C. A dibenzofuran derivative from lichens, e.g., *Leproloma membranaceum.*

Lit.: Aust. J. Chem. **45**, 785 (1992); **46**, 407 (1993) ▪ Culberson, p. 168. – *[CAS 479-46-9]*

Pannarin.

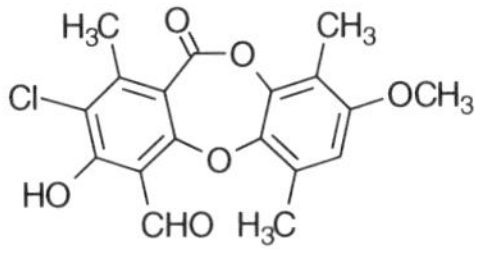

$C_{18}H_{15}ClO_6$, M_R 362.77, needles, mp. 216–217 °C. A chlorine-containing *depsidone from lichens, e.g., *Lecanora subaurea.*

Lit.: Aust. J. Chem. **31**, 1383 (1978); **39**, 719 (1986) ▪ Karrer, No. 1066 ▪ Rev. Latinoam. Quim. **30**, 79 (1988) (review). – *[CAS 55609-84-2]*

Pantetheine, pantoic acid see pantothenic acid.

(−)(*R*)-Pantolactone see sherry flavor.

Pantothenic acid [*N*-((*R*)-2,4-dihydroxy-3,3-dimethylbutyl)-β-alanine, D-pantoyl-β-alanine, vitamin B_3, vitamin B_5].

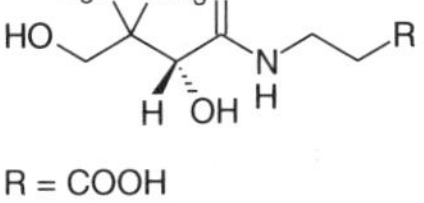

R = COOH : Pantothenic acid
R = CO−NH−CH₂−CH₂−SH : Pantetheine

$C_9H_{17}NO_5$, M_R 219.24, (*R*)-form, unstable, viscous, very hygroscopic oil, $[\alpha]_D^{25}$ +37.5°. P. is distributed ubiquitously in human, animal, and plant tissues. The average concentration amounts to 1–10 mg/kg; liver and yeast cells are particularly rich in P. as are also the adrenals, kidneys, and brain. The highest concentration of 110–320 mg/kg is found in *royal jelly. The sodium salt (*Panthoject®*): $C_9H_{16}NNaO_5$, M_R 241.22, very hygroscopic crystals, is stable in sealed ampoules only, $[\alpha]_D^{25}$ +27.1°; the calcium salt (*Pantholin®*, calpanate, BAN, USAN), $C_{18}H_{32}CaN_2O_{10}$, M_R 476.54, needles with a sweet taste and bitter aftertaste, decomp. 195 °C, $[\alpha]_D^{25}$ +28.2° (H_2O), is less hygroscopic than P., not sensitive to light, and is stable in the air and in solutions at pH 5–7. The 4′-*O*-β-D-glucopyranoside, $C_{15}H_{27}NO_{10}$, M_R 381.38, occurs in tomato juice and acts as a growth factor in microorganisms. Synthetic (*S*)-P., viscous oil, $[\alpha]_D^{21}$ −26.7° (H_2O) has no biological activity.

Biological function: P. is a biosynthetic precursor of coenzyme A. P. is present in coenzyme A bound to cysteamine as *pantetheine*, $C_{11}H_{22}N_2O_4S$, M_R 278.37, $[\alpha]_D^{20}$ +12.9° (H_2O). P. is considered to belong to the vitamin B complex; deficiency symptoms in chickens are loss of feathers.

Use: In hair cosmetics, vitamin preparations, and in medications for wound-healing and skin disorders.

Biosynthesis: 2-Oxo-3-methylbutanoic acid → 4-hydroxy-3,3-dimethyl-2-oxobutanoic acid; ketopantoaldolase (EC 4.1.2.12) and 5,10-methylenetetrahydrofolic acid (see folic acid) participate in this reaction. Hydrogenation to *D-pantoic acid* [(*R*)-2,4-dihydroxy-3,3-dimethylbutanoic acid], $C_6H_{12}O_4$, M_R 148.16, is catalyzed by 2-dehydropantoate reductase (EC 1.1.1.169) with nicotinamide adenine dinucleotide phosphate (see nicotinamide coenzymes) as coenzyme. In the presence of pantothenate synthetase (EC 6.3.2.1) and adenosine 5′-triphosphate (ATP) D-pantoic acid condenses with β-alanine to furnish P.

Historical: P. was first recognized as a growth factor in 1933 by R. J. Williams and – since it is found in practically all plant and animal tissues – named as P. (Greek: pantothen = from every side).

Lit.: Beilstein E IV **4**, 2569 ff. ▪ Biotechnol. Lett. **12**, 357 (1990) ▪ Food Sci. Technol. **13**, 437 (1984) (review) ▪ Hager (5.) **9**, 14 ff. ▪ J. Org. Chem. **51**, 1955 (1986) (synthesis) ▪ Karrer, No. 2613 ▪ Methods Enzymol. **62**, 227–236 (1976) ▪ Negwer (6.), No. 1356. – *Toxicology:* Sax (8.), CAU 750. – *[HS 2936 24; CAS 79-83-4; 37138-77-5 (S); 137-08-6 (Ca salt); 29493-59-3 (glucopyranoside); 496-65-1 (pantetheine); 1112-33-0 (pantoic acid)]*

Papaverine see benzyl(tetrahydro)isoquinoline alkaloids.

Papaya flavor see fruit flavors.

Paper factor see juvabione.

Papiliochrome II.

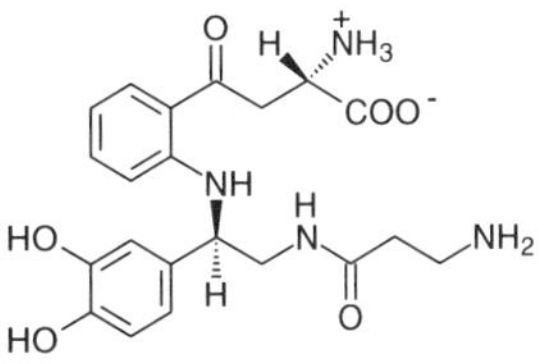

$C_{21}H_{26}N_4O_6$, M_R 430.46, yellow powder. P. II is a pigment in the wings of the swallowtail butterfly *Papilio xuthus* and other *Papilio* species, it is composed of L-*kynurenine, *noradrenaline, and *β-alanine. The yellow color is due to a charge-transfer interaction between the two aromatic rings.

Lit.: Amino Acids **2**, 181 (1992) (synthesis) ■ FEBS Lett. **279**, 145 (1991) (biosynthesis) ■ Insect Biochem. **19**, 673–678 (1989) ■ Progress Tryptophan Serotonin Research, p. 743–746, Berlin: De Gruyter 1984 ■ Z. Naturforsch. C **33**, 498–503 (1978). – *[CAS 60650-78-4 (P. IIa); 60650-79-5 (P. IIb); 68335-21-7 (stereoisomers)]*

Paprika flavor see vegetable flavors.

Papuamine.

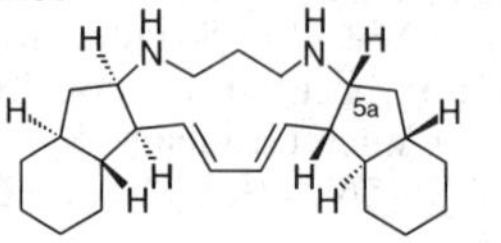

(-)-P.

$C_{25}H_{40}N_2$, M_R 368.61, mp. 168–169 °C, $[\alpha]_D$ –150° (CH_3OH). An alkaloid from the marine sponge *Haliclona* sp. with antifungal activity. For synthesis, see *Lit.*[1].

Lit.: [1] J. Chem. Soc., Chem. Commun. **1994**, 1881; J. Am. Chem. Soc. **116**, 9789 (1994); **117**, 10905-10913 (1995); **119**, 22 (1997); J. Org. Chem. **61**, 685–699, 700–709 (1996); Org. Prep. Proced. Int. **30**, 3 (1998).
gen.: J. Am. Chem. Soc. **110**, 965 (1988) (isolation) ■ New Sci. (18.6.) **1994**, 16. – *[CAS 112455-84-2]*

Papulacandins.

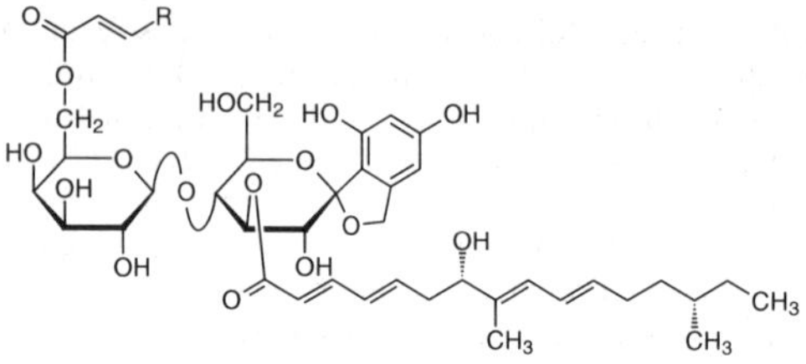

R = –CH=CH–(CH₂)₄–CH₃ : P. A

R = –CH=CH–CH=CH–CH(OH)–C₂H₅ (*Z,E*) : P. B
(*E,E*) : P. C

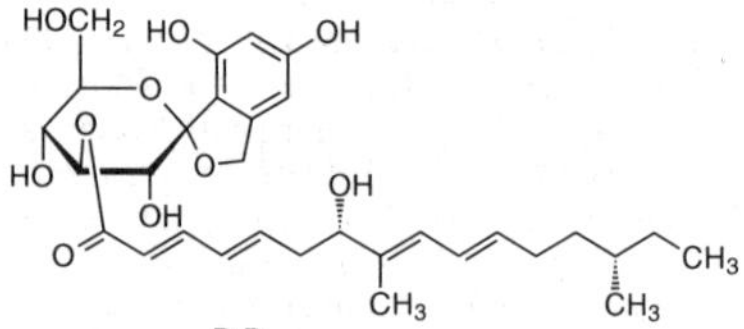

P. D

Tricyclic spiroketal *glycosides from cultures of *Papularia sphaerosperma*. The P. have antibiotic activities. Similar to the lipopeptide antibiotic *echinocandin B, they inhibit the biosynthesis of β-*glucans in yeasts.

Table: Data of Papulacandins.

P.	molecular formula	M_R	mp. [°C]	$[\alpha]_D^{22}$ (CH_3OH)	CAS
A	$C_{47}H_{66}O_{16}$	887.03	171–173	+30°	61036-46-2
B	$C_{47}H_{64}O_{17}$	901.01	193–197	+50°	61032-80-2
C	$C_{47}H_{64}O_{17}$	901.01	140–150	+33°	61036-48-4
D	$C_{31}H_{42}O_{10}$	574.67	127–130	+7°	61036-49-5

Lit.: Biochim. Biophys. Acta **884**, 550–558 (1986) ■ Carbohydr. Res. **223**, 157 (1992) ■ J. Am. Chem. Soc. **112**, 2746ff. (1990) ■ Tetrahedron **44**, 7163–7169 (1988). – *Synthesis: P. D:* J. Chem. Soc., Chem. Commun. **1995**, 1147f. ■ J. Org. Chem. **61**, 1082–1100 (1996).

Papulinone [methyl 2-(1-hydroxy-2-phenylethyl)-4-oxooxetane-2-carboxylate].

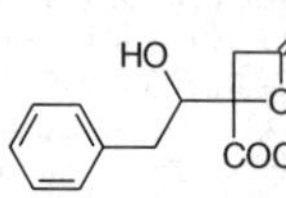

$C_{13}H_{14}O_5$, M_R 250.25, $[\alpha]_D^{25}$ +6.7° ($CHCl_3$). A β-lactone from *Pseudomonas syringae* pv. *papulans*. P. is weakly cytotoxic (2–6 mg/mL show chlorotic activity while 1 mg/mL does not) and is possibly responsible for various diseases of apple and pear trees ("apple blister spots"). In addition to P. the biogenetically related methyl esters of 3-phenyllactic acid bearing hydroxy and nitro substituents on the phenyl ring are also formed. Further α-oxetanones are, e. g., *ebelactones, esterastin, *obafluorin, *lipstatin, tetrahydrolipstatin, and valilactone.

Lit.: Phytochemistry **29**, 1491–1497 (1990) ■ Synthesis **1995**, 729 (review). – *[CAS 128232-41-7]*

Papyracillic Acid.

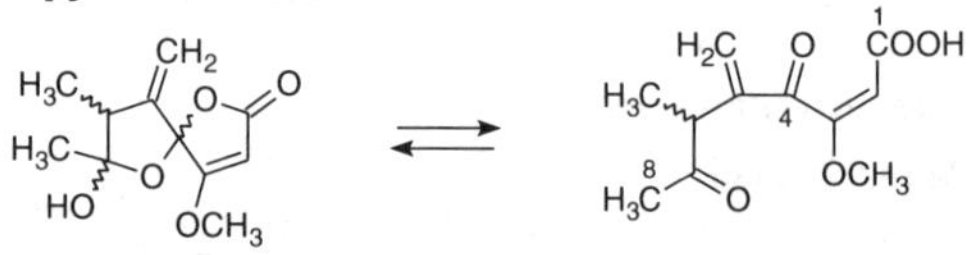

bicyclic form acyclic form

$C_{11}H_{14}O_5$, M_R 226.23, cryst., mp. 97–99 °C, racemic. An antibiotic isolated from cultures of *Lachnum papyraceum* (Ascomycetes), structurally related to other 5-membered cyclic lactones such as *patulin and *penicillic acid. P. acts as a fibrinogen antagonist.

Lit.: Tetrahedron **52**, 10249–10254 (1996); **53**, 6209–6214 (1997). – *[CAS 179308-49-7 (bicyclic form); 190075-60-6 (acyclic form)]*

Parabactin see agrobactin.

Paraconic acids.

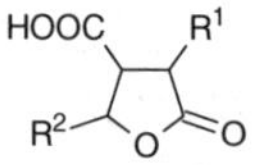

I

5-Oxo-3-furancarboxylic acids of the general formula I. Numerous derivatives of P. occur in lichens, e. g.: *allo*-pertusaric acid, *allo*-*protolichester(in)ic acid, constipatic acid, dihydropertusaric acid, isomuronic acid, *lichesterinic acid, murolic acid (from the widely distributed and common wall lichen *Lecanora muralis*), muronic acid, neodihydromurolic acid, nephromopsic acid, nephrosteranic acid, nephrosterinic acid, neuropogolic acid, praesorediosic acid, protoconstipatic acid, *protolichester(in)ic acid (from the shrub lichen *Cetraria islandica*, *Iceland moss), and *roccellaric acid.

Lit.: Symbiosis **11**, 245–248 (1991) ■ Zechmeister **29**, 209–306; **45**, 103–234.

Paracyclophanes see nostocyclophanes.

Paraherquein, paraherquamide see marcfortines.

Paralytic shellfish poison(ing) see PSP.

Paravallarine see Apocynaceae steroid alkaloids.

Parietin see physcion.

Parillin see steroid saponins.

Paromomycin see neomycin.

Parsley camphor see apiole.

Parsley leaf/seed oil. Two types of essential oil obtained from the parsley plant *Petroselinum crispum* are used in the perfume and food industries.
A) ***Parsley leaf oil:*** Yellow to yellowish-green oil with a strong odor and taste of fresh parsley: fresh, herby-leafy but also warm and spicy.
Production: By steam distillation of the aerial parts of the parsley plant, including the unripe seeds.
Composition[1]***:*** A major part of the oil consists of monoterpene hydrocarbons such as *α*- and *β*-**pinenes*, **myrcene*, *limonene*, *β-phellandrene* (see *p*-menthadienes). The occurrence of *p-mentha-1,3,8-triene* (usually between 20 and 30%) ($C_{10}H_{14}$, M_R 134.22) is characteristic and mainly responsible for the organoleptic impression. According to recent investigations a series of other trace components with widely varying structures also contributes to the typical parsley odor[2]. The leaf oil also contains lower concentrations of those compounds that constitute the main components of parsley seed oil (see below).
Use: To improve the aromatic character of foods such as sauces, soups, fish and meat preparations, pickles, etc.
B) ***Parsley seed oil:*** Yellow to yellowish-brown, viscous oil with a warm spicy, woody, sweet herby odor and an aromatic warm spicy, bitter taste.
Production: By steam distillation of the ripe fruits (seeds) of the parsley plant.
Composition[1]***:*** Parsley seed oil contains less monoterpene hydrocarbons than the leaf oil with only *α*- and *β-pinenes* being present in appreciable amounts. Characteristic main components are **apiole*, *myristicin*, *elemicin* (see safrole), and *1-allyl-2,3,4,5-tetramethoxybenzene* ($C_{13}H_{18}O_4$, M_R 238.28).

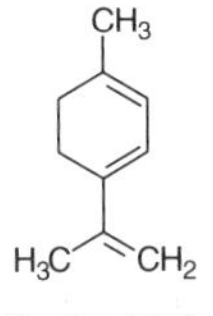

p-Mentha-1,3,8-triene

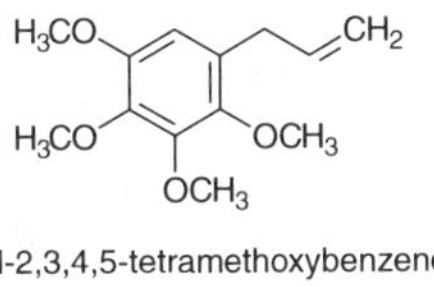

1-Allyl-2,3,4,5-tetramethoxybenzene

Use: In the perfume industry in small amounts for herby-spicy compositions.

Lit.: [1] Perfum. Flavor. **14** (5), 54 (1989); **16** (5), 81, 82 (1991); **19** (2), 72 (1994); Dev. Food Sci. **34**, 457–467 (1994). [2] Lebensm. Wiss. Technol. **25**, 55 (1992).
gen.: Arctander, p. 505 ▪ Bauer et al. (2.), p. 169 ▪ ISO 3527 (1975). – *Toxicology:* Leaf oil: Food Chem. Toxicol. **21**, 871 (1983); Seed oil: Food Cosmet. Toxicol. **13**, 897 (1975). – *[HS 3301 29; CAS 8000-68-8 (P.); 18368-95-1 (p-mentha-1,3,8-triene); 15361-99-6 (1-allyl-2,3,4,5-tetramethoxybenzene)]*

Parthenolide see tansy.

Parviflorin. a) A phenanthro[4,5-*bcd*]pyran from *Vanda parviflora* (Orchidaceae). P. is related to *tesselatin* from *V. tessalata* (syn. *V. roxburghii*) and *aeridin* from *Aerides crispum*.

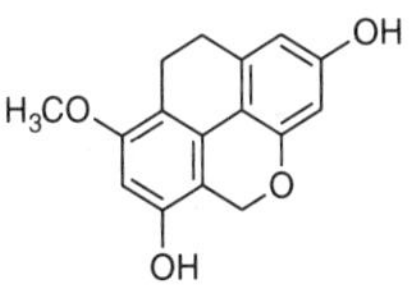

Parviflorin (from *Vanda parviflora*)

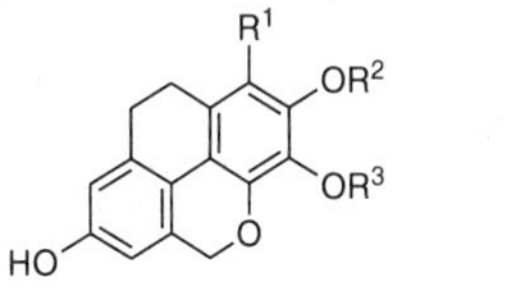

R^1	R^2	R^3	
OCH_3	H	CH_3	Aeridin
H	CH_3	H	Tesselatin

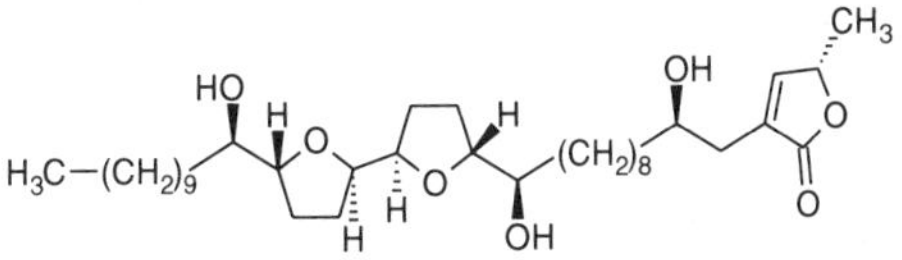

Parviflorin (from *Asimina parviflora*)

Table: Data of Parviflorin and related phenanthro[4,5-*bcd*]pyrans.

name	molecular formula	M_R	mp. [°C]	CAS
Aeridin	$C_{17}H_{16}O_5$	300.31	56	210046-68-7
Parviflorin	$C_{16}H_{14}O_4$	270.28	142	210047-46-4
Tesselatin	$C_{16}H_{14}O_4$	270.28	125	210046-67-6

b) The name parviflorin is also used for an annonin (see annonaceous acetogenins) type natural product from Annonaceae.

Lit.: Angew. Chem., Int. Ed. Engl. **36**, 2632–2635 (1997) (b) ▪ Phytochemistry **48**, 181, 183, 185 (1998) (a).

Paspalitrems.

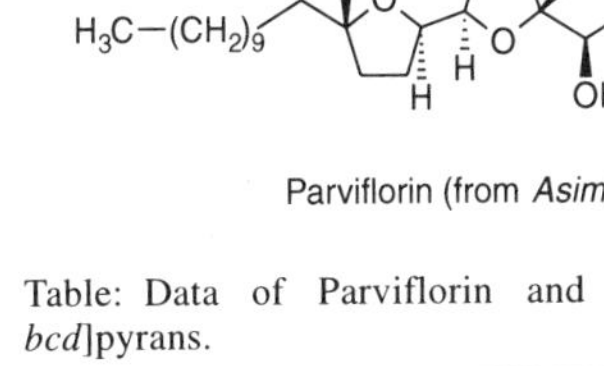

$R^1 = R^2 = H$: Paspalicin
$R^1 = H$, $R^2 = OH$: Paspalinin
$R^1 = CH_2{-}CH{=}C(CH_3)_2$, $R^2 = OH$: Paspalitrem A

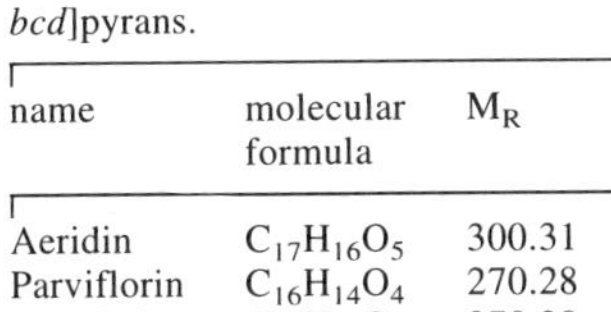

Paspalin

A group of tremorigenic *mycotoxins from *Claviceps paspali*, *Aspergillus flavus*, and *A. nomius*. The compounds are formed in submersion cultures but also occur in the *sclerotia of the fungi where they are assumed to have an antifeedant effect on insects. In addition, these neurotoxins are also found in pasture grasses (*Paspalum dilatatum*) infected with *Claviceps paspali*, where they cause "paspalum staggers" (see Darnel, rye-grass) in grazing animals. The compounds include P. A, B, and C, paspalin, paspalinin, as well as paspalicin, e.g., *P. A*, $C_{32}H_{39}NO_4$, M_R 501.67, amorphous; *paspalin*, $C_{28}H_{39}NO_2$, M_R 421.62, mp. 264 °C; *paspalicin*, $C_{27}H_{31}NO_3$, M_R 417.55, mp. 230–240 °C. *A. nomius* forms 4-hydroxypaspalinin and 14-(*N,N*-dimethyl-L-valyloxy)paspalinin.

Lit.: Betina, p 366–371 ▪ Cole-Cox, p. 390–409 ▪ J. Agric. Food Chem. **29**, 293 (1981); **32**, 1069 (1984) ▪ Tetrahedron Lett. **21**, 231, 235 (1980); **34**, 2569 (1993) ▪ Zechmeister **48**, 32–45 ▪ see also janthitrems and penitrems. – *[CAS 63722-90-7 (P. A); 11024-56-9 (paspalin); 11024-55-8 (paspalicin)]*

Passion fruit see fruit flavors.

Patchoulene, patchoulenol, patchoulenone, patchouli alcohol see patchouli oil.

Patchouli oil. Reddish-brown to brown, viscous oil with a typical woody, camphor-like earthy, balsamy sweet odor.

Production: By steam distillation of the dried (fermented) leaves of the patchouli shrub, *Pogostemon cablin*, syn. *P. patchouli* (Lamiaceae), growing mainly in Indonesia but also in China, Brazil, and Russia. Solvent extraction of patchouli leaves furnishes the valuable patchouli resin. The annual production worldwide probably amounts to 800 t.

Composition[1]***:*** Main component and responsible for the woody odor is (–)-*patchouli alcohol*[2] (ca. 40–50%) {$C_{15}H_{26}O$, M_R 222.37, colorless crystals with a pleasant odor, mp. 55–56 °C, $[\alpha]_D^{27}$ –129° ($CHCl_3$), soluble in organic solvents}. The unnatural (+)-enantiomer has a much weaker, uncharacteristic odor not like patchouli. The synthetic racemate has a mixed odor in which the sensory impression of (–)-P. is stronger. Dehydrogenation furnishes the pharmacologically important guaiazulene (see azulenes). The earthy notes arise from degradation products of patchouli alcohol that are present in small amounts, e.g., *patchoulenol*[3] {$C_{15}H_{24}O$, M_R 220.35, colorless cryst., mp. 74 °C, $[\alpha]_D^{30}$ –54.23° ($CHCl_3$)}, *patchoulenone*[4] ($C_{15}H_{22}O$, M_R 218.34), *norpatchoulenol* ($C_{14}H_{22}O$, M_R 206.33), and *nortetrapatchoulol* ($C_{14}H_{22}O$, M_R 206.33). The remaining components of P. o. are sesquiterpene hydrocarbons, which do not contribute essentially to the odor, such as e.g., **seychellene* (~6%), *α-patchoulene*[5] (~3%) [$C_{15}H_{24}$, M_R 204.36, bp. 53.5–54.5 °C (18.62 Pa)], and *β-patchoulene* (~4%) {$C_{15}H_{24}$, M_R 204.36, bp. 66.8 °C (79.8 Pa), $[\alpha]_D^{30}$ –42.6° ($CHCl_3$)}, *α-guaiene* (~10%) ($C_{15}H_{24}$, M_R 204.36), and *α-bulnesene* (~12%) ($C_{15}H_{24}$, M_R 204.36). The sensory contribution of the minor sesquiterpenoid alkaloids patchoulipyridine and guaiapyridine to the essential oil is considered to be unpleasant[6].

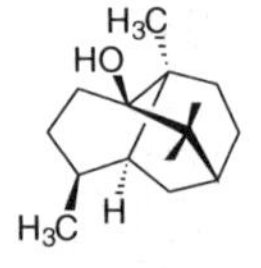
Patchouli alcohol (**1**)

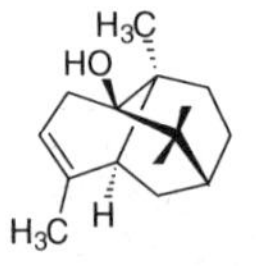
Patchoulenol (**2**)

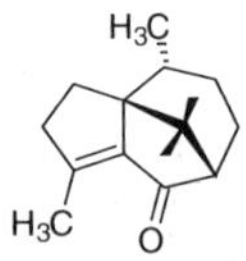
Patchoulenone (**3**)

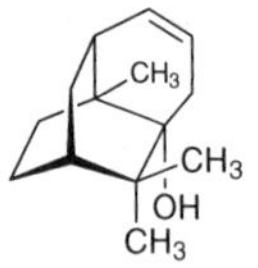
Norpatchoulenol (**4**)

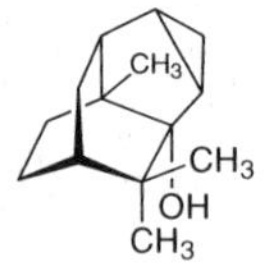
Nortetrapatchoulol (**5**)

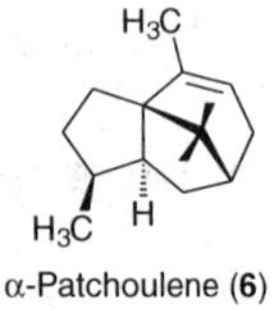
α-Patchoulene (**6**)

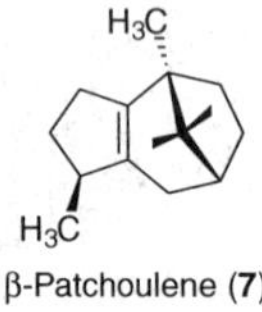
β-Patchoulene (**7**)

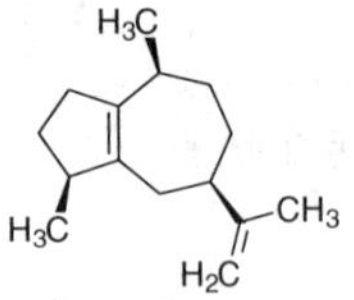
α-Guaiene (**8**)

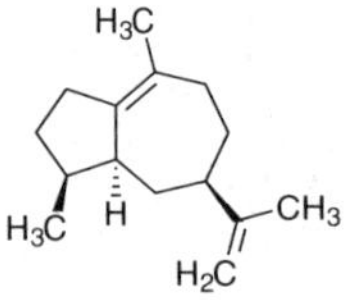
α-Bulnesene (**9**)

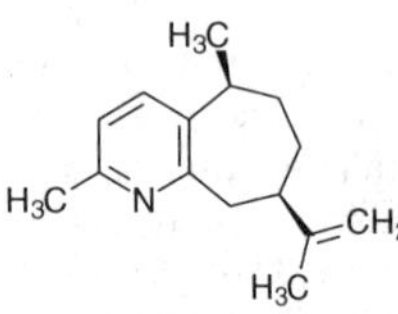
Guaiapyridine (**10**)

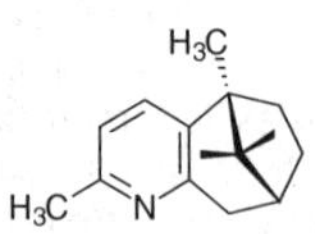
Patchoulipyridine (**11**)

Use: P. is an essential starting material for perfumes with masculine and oriental odors.

Lit.: [1] Perfum. Flavor. **1** (2), 20 (1976); **6** (4), 73 (1981); **15** (2), 76 (1990); **20**, (3), 67 (1995); Ohloff, p. 147. [2] Helv. Chim. Acta **73**, 1730 (1990) (synthesis); J. Chem. Soc., Perkin Trans. 1 **1997**, 1385; Phytochemistry **27**, 2105 (1988) (biosynthesis); Tetrahedron Lett. **36**, 7607 (1995) (synthesis). [3] Chem. Pharm. Bull. **16**, 43 (1968) (synthesis); Phytochemistry **40**, 125 (1995). [4] Chem. Commun. **1998**, 1851; J. Nat. Prod. Rep. **11**, 548 (1994); Magn. Reson. Chem. **30**, 295 (1992); Phytochemistry **38**, 1037 (1995). [5] J. Am. Chem. Soc. **83**, 927 (1961); **84**, 1307 (1962) (synthesis); Arch. Biochem. Biophys. **256**, 56 (1987); Phytochemistry **26**, 2705 (1987) (biosynthesis) [6] J. Am. Chem. Soc. **88**, 3109 (1966).

gen.: Arctander, p. 508 ▪ Bauer et al. (2.), p. 169 ▪ ISO 3757 (1978). – *Toxicology:* Food Cosmet. Toxicol. **20**, 791 (1982). – *[HS 3301 29; CAS 8014-09-3 (P.); 5986-55-0 (1); 17806-54-1 (2); 5986-54-9 (3); 41429-52-1 (4); 62731-84-4 (5); 560-32-7 (6); 514-51-2 (7); 3691-12-1 (8); 3691-11-0 (9)]*

Patellamides. Numerous cyclopeptides with anti-tumor properties have been isolated from the tunicate *Lissoclinum patella* and contain as characteristic structural elements thiazole and oxazoline-amino acids: *patellamides* A–D[1] (see table); according to an X-ray crystallographic analysis P. D exists in a highly folded conformation in which the two thiazole rings are almost parallel. *Ulicyclamide*[2], $C_{33}H_{39}N_7O_5S_2$, M_R 677.84, oil; *ulithiacyclamides*[3]: two structural variants are known, namely ulithiacyclamide (ulithiacyclamide A, $C_{32}H_{42}N_8O_6S_4$, M_R 762.98, oil) and an analogue with a benzyl group in place of the isobutyl group (ulithiacyclamide B, $C_{35}H_{40}N_8O_6S_4$, M_R 796.99). P. A and ulithiacyclamide are acute inhibitors of leukemia T-cells, ED_{50} 0.028 mg/L and 0.01 mg/L. Further cyclopeptides are the *patellins*[4] (patellin 2: $C_{37}H_{60}N_6O_7S$, M_R 732.98, cryst., $[\alpha]_D$ –110°) with *O*-(1,1-dimethylallyl)-L-threonine residues.

Lit.: [1] J. Org. Chem. **48**, 2302 (1983); Chem. Pharm. Bull. **41**, 1686 (1993); J. Nat. Prod. **55**, 376 (1992); **58**, 594–597 (1995).

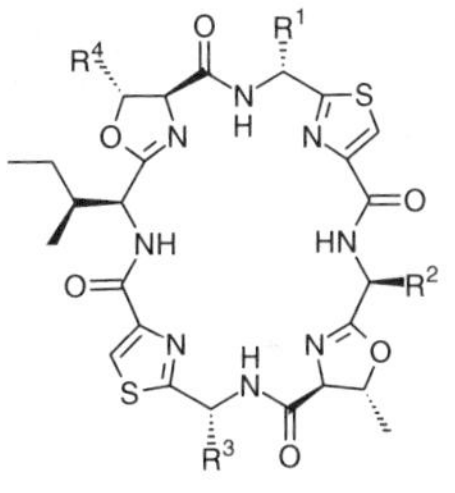

Patellamides

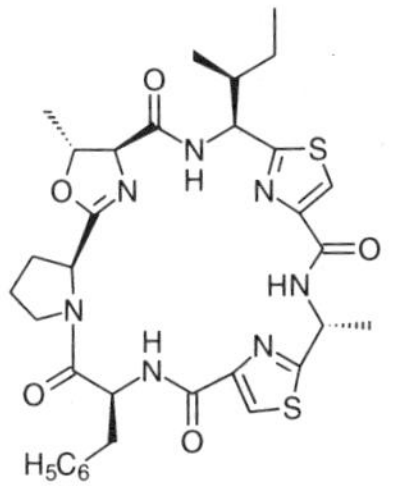

Ulicyclamide

R^1	R^2	R^3	R^4	
$CH(CH_3)_2$	$(S)CH(CH_3)-C_2H_5$	$CH(CH_3)_2$	H	P. A
$CH_2-C_6H_5$	$CH_2-CH(CH_3)_2$	CH_3	CH_3	P. B
$CH_2-C_6H_5$	$(S)CH(CH_3)_2$	CH_3	CH_3	P. C
$CH_2-C_6H_5$	$CH(CH_3)-C_2H_5$	CH_3	CH_3	P. D

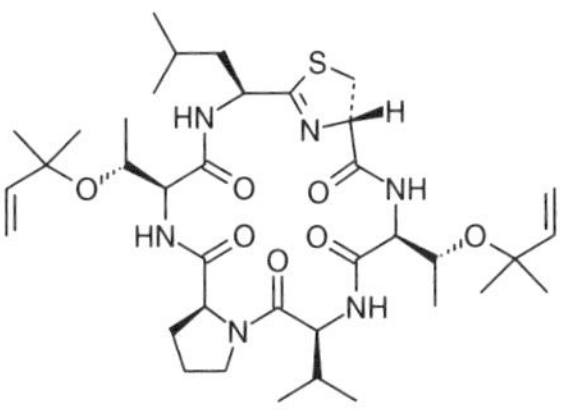

Patellin 2

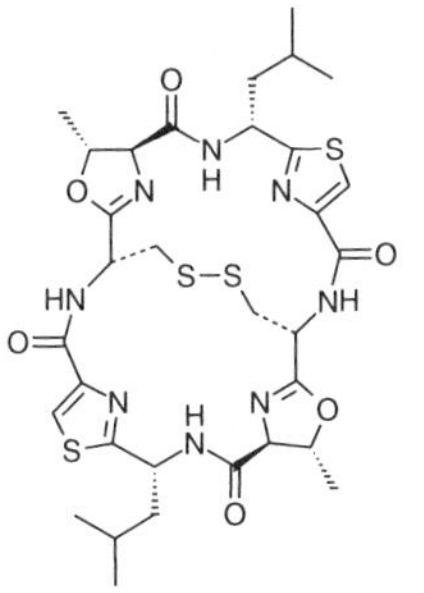

Ulithiacyclamide A

Table: Data of Patellamides.

	molecular formula	M_R	mp. [°C]	CAS
P. A	$C_{35}H_{50}N_8O_6S_2$	742.95	228–229	81120-73-2
P. B	$C_{38}H_{48}N_8O_6S_2$	776.98	–	81098-23-9
P. C	$C_{37}H_{46}N_8O_6S_2$	762.95	–	81120-74-3
P. D	$C_{38}H_{48}N_8O_6S_2$	776.98	144–145	120853-15-8

– *Synthesis:* Tetrahedron Lett. **26**, 5155, 5159, 6501 (1985) (corrected structure); **27**, 163, 179 (1986); Scheuer II **1**, 150; **3**, 29. [2] J. Org. Chem. **48**, 4445 (1983); Angew. Chem. Int. Ed. Engl. **24**, 569 (1985) (synthesis). [3] J. Nat. Prod. **52**, 732 (1989); **61**, 1524, 1547 (1998); Nachr. Chem. Tech. Lab. **37**, 1040 (1989). [4] Aust. J. Chem. **49**, 659 (1996); J. Am. Chem. Soc. **112**, 8080 (1990); Chem. Rev. **93**, 1771 (1993). – *[CAS 129216-76-8 (patellin 2); 74839-81-9 (ulicyclamide); 74847-09-9 (ulithiacyclamide A); 122759-67-5 (ulithiacyclamide B)]*

Patulin {4-hydroxy-4*H*-furo[3,2-*c*]pyran-2(6*H*)-one, claviformin, expansin}.

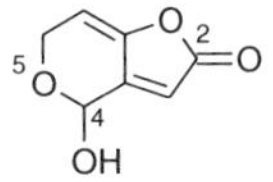

$C_7H_6O_4$, M_R 154.12, mp. 111 °C, compact prisms or thick plates. As a metabolic product (*mycotoxins) of many different mold fungi such as *Penicillium patulum*, *P. urticae*, *P. chrysogenum*, *P. roqueforti*, *Aspergillus clavatus*, *A. terreus*, and *Byssochlamys nivea* (*Paecilomyces niveus*), that are found in grain, bakery, and sausage (salami, uncooked sausages) products as well as fruits, P. is responsible, among others, for food poisonings. In mice and rats P. exhibits carcinogenic and teratogenic activities, it is also active against bacteria and fungi so that it may be considered as an antibiotic, it is even used in nasal sprays for colds and influenza complaints! P. is found particularly often in apple, grape, and pear juices. Peaches, cherries, apricots, tomatoes, and paprika can also be afflicted. Moldy apples or apple juice can contain up to 1 g P. per kg. *Penicillium expansum* is considered to be the pathogen responsible for brown mold in apples and pears. P. is destroyed during fermentation or upon addition of SO_2; wine and cider are thus free of patulin. According to WHO recommendations, food containing more than 50 µg/kg P. should be withdrawn from the market. P. has a destructive effect, especially on DNA (strand cleavage), in addition, it reacts with nucleophiles such as SH groups and thus inhibits many enzymes. The biosynthesis of the tetraketide proceeds through 6-methylsalicylic acid. P. is often formed together with *citrinin.

Lit.: Beilstein EV **18/3**, 5 f. ▪ Betina, p. 246–253 ▪ Chem. Pharm. Bull. **34**, 3534 (1986); **42**, 2167 (1994) ▪ J. Environ. Pathol., Toxicol. Oncol. **10**, 254–259 (1990) ▪ Merck-Index (12.), No. 7185 ▪ Rodricks, Hesseltine, & Mehlman (eds.), Mycotoxins in Human and Animal Health, p. 609–624, Park Forest South: Pathotox Publ. 1977 ▪ Sax (8.), CMV 000 ▪ Steyn & Vleggaar (eds.), Mycotoxins and Phycotoxins, p. 293–304, Amsterdam: Elsevier 1986 ▪ Tetrahedron Lett. **36**, 7175 (1995) (synthesis). – *[HS 2941 90; CAS 149-29-1]*

Pavine and isopavine alkaloids. The P. a. are derivatives of *benzyl(tetrahydro)isoquinoline alkaloids that occur in various species of, in particular, the Papaveraceae, Berberidaceae, Ranunculaceae, Lauraceae, and Menispermaceae (see table). Argemonine exhibits its curare-like properties.

Table: Data of Pavine and isopavine alkaloids.

name, molecular formula, M_R	mp. [°C] $[\alpha]_D$ (solvent)	occurrence	CAS
Amurensine $C_{19}H_{19}NO_4$ 325.36	221–223 −178° (CH_3OH)	*Papaver* species	10481-92-2
Argemonine (*N*-Methyl-P). $C_{21}H_{25}NO_4$ 355.43	155.5–156.5 −187.93° ($CHCl_3$)	*A. gracilenta*, *B. buxifolia*, *T. revolutum*, *T. strictum*	6901-16-2
Dinorargemonine $C_{19}H_{21}NO_4$ 327.38	254–255 −244° ($CHCl_3$)	*A. hispida*, *A. munita*, *C. longifolia*, *Eschsch.*, *T. dasycarpum*	6808-63-5
Eschscholtzine $C_{19}H_{17}NO_4$ 323.35	128 −220.2° (C_2H_5OH)	*C. chinensis*, *Eschsch.*	4040-75-9
Munitagine $C_{19}H_{21}NO_4$ 327.38	167–169 −239° ($CHCl_3$)	*A. gracilenta*, *A. munita*	7691-07-8
Pavine $C_{20}H_{23}NO_4$ 341.41	224	*Papaver* species	529-79-3

A. = *Argemone*; *B.* = *Berberis*; *C.* = *Cryptocarya*; *Eschsch.* = *Eschscholtzia californica, douglasii, glauca*; *T.* = *Thalictrum*

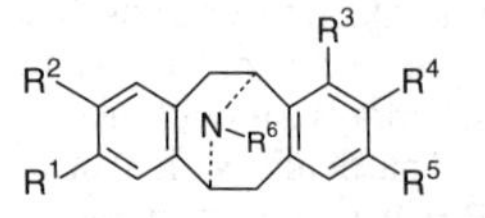

	R^1	R^2	R^3	R^4	R^5	R^6
Argemonine	OCH_3	OCH_3	H	OCH_3	OCH_3	CH_3
Dinorargemonine	OH	OCH_3	H	OCH_3	OH	CH_3
Eschscholtzine	O–CH_2–O		H	O–CH_2–O		CH_3
Munitagine	OH	OCH_3	OH	OCH_3	H	CH_3
Pavine	OCH_3	OCH_3	H	OCH_3	OCH_3	H

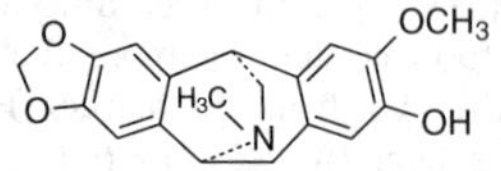

Amurensine

The biosynthesis proceeds via cyclization of reticuline.

Lit.: Hager (5.) **5**, 110–115 ▪ J. Chem. Soc., Chem. Commun. **1982**, 1113 (argemonine) ▪ J. Nat. Prod. **49**, 922 (1986) (eschscholtzine) ▪ Manske **17**, 433–439; **31**, 317–389 ▪ Shamma, p. 97–114 ▪ Shamma-Moniot, p. 61–70 ▪ Ullmann (5.) **A 1**, 370. – *[HS 2939 29]*

PB toxin (PB-1, diphenyl *N*-cyclooctylphosphoramidate).

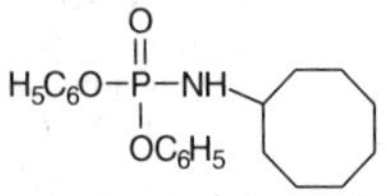

$C_{20}H_{26}NO_3P$, M_R 359.41. Toxin from the dinoflagellate *Ptychodiscus brevis*, a constituent of the so-called *red tide, a massive reproduction of dinoflagellates that sometimes leads to the extensive death of fish and shellfish. The toxicity of PB t. for the common guppy, *Lebestes reticulatus,* amounts to 1 ppm (LD_{100}). PB t. acts as an inhibitor of acetylcholinesterase and belongs to the same structural type as many commercial insecticides or neurotoxins.

Lit.: Nachr. Chem. Tech. Lab. **31**, 165 (1983) ▪ Science **257**, 1476 (1992) ▪ Tetrahedron Lett. **24**, 855 (1983). – *[CAS 86126-37-6]*

Pea aroma see vegetable flavors.

Peach flavor see fruit flavors.

Peanut flavor. The aroma of roasted peanuts is mainly due to thermally formed *flavor compounds. Major constituents include *pyrazines, e.g., methyl-, 2,5-dimethyl-, trimethyl-, and 2-ethyl-3,5(6)-dimethylpyrazine, as well as pyrroles, thiophenes, oxazoles, and thiazoles, e.g., *2-isopropyl-4,5-dimethylthiazole* ($C_8H_{13}NS$, M_R 155.26), also contributing to the typical P. f. Also important are alkyl-substituted *dihydro-4*H*-1,3,5-dithiazines [1].

Lit.: [1] Food Rev. Int. **8**, 391–442 (1992).
gen.: Maarse, p. 680–683. – *[CAS 53498-30-9 (2-isopropyl-4,5-dimethylthiazole)]*

Pear aroma see fruit flavors.

Pear esters see fruit esters.

Pecilocin see variotin.

Pectenotoxins (PTX). Cyclic polyether macrolides from the scallop *Patinopecten yessoensis*. Up to now 8 different P. (PTX1 to PTX7 and PTX10) have been isolated, most of them have been characterized. One example is *PTX2* {$C_{47}H_{70}O_{14}$, M_R 859.05, amorphous, $[\alpha]_D^{20}$ +16° (CH_3OH)}. PTX are involved in diarrhetic shellfish poisoning, see also yessotoxin.

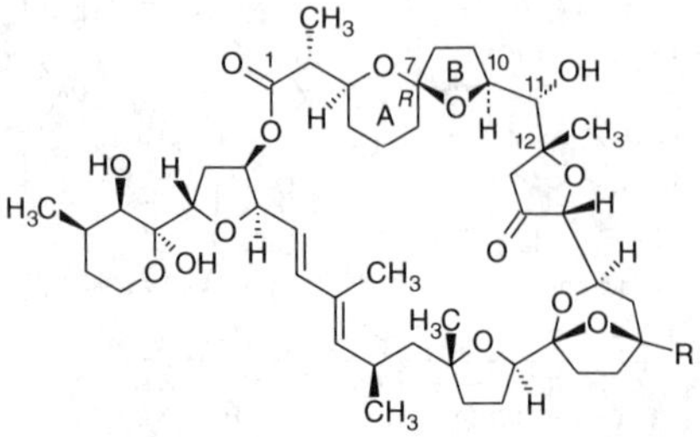

R = CH_2OH : PTX 1
R = CH_3 : PTX 2
R = CHO : PTX 3
R = CH_2OH, (7*S*)-epimer) : PTX 4
unidentified : PTX 5
R = COOH : PTX 6
R = COOH, (7*S*)-epimer : PTX 7

expanded ring B:

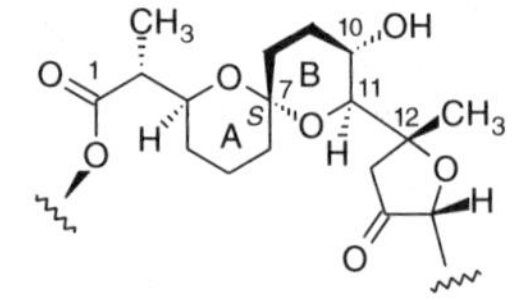

R = CH_2OH : PTX 8
R = COOH : PTX 9
unidentified : PTX 10

Lit.: Chem. Lett. **1998**, 653 ▪ J. Nat. Prod. **58**, 1722–1726 (1995) ▪ J. Org. Chem. **63**, 2475–2480 (1998). – *[CAS 97564-91-5 (PTX2)]*

Pectins (Greek.: pēktikós = coagulating). High-molecular weight glycosidic plant constituents (M_R 10 000–500 000) that are very widely distributed in fruits, roots, and leaves. The P. consist principally of a chain of 1,4-α-glycosidically linked galacturonic acid units; the ester groups are esterified to between 20 and 80% by methanol; a distinction is made between *highly esterified* (>50%) and *low esterified* P. (<50%). Hydrolysis of the ester groups yields the *pectic acids* {M_R 35 000–100 000, $[\alpha]_D^{20}$ +280° (dil. NaOH)} and their salts, e.g., calcium pectate. The salts formed from native P. by reaction at the free carboxy groups are called *pectinates*.

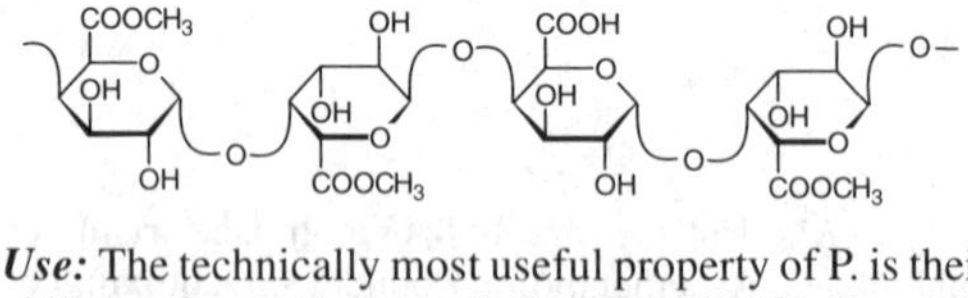

Use: The technically most useful property of P. is their ability to form gels (so-called *hydrocolloids*). In acidic aqueous solution (pH 3–3.5) the highly esterified P. on addition of ca. 60–65% sugar (and if necessary acids) form clear, solid gels for use especially in food technology.

Lit.: ACS Symp. Ser. **310** (1986) ▪ Aspinall, The Polysaccharides vol. 2, chap 3, p. 97–193, New York: Academic Press 1983 ▪ Carbohydr. Polym. **12**, 79–99 (1990) ▪ Kirk-Othmer (4.) **11**, 812 f., 828, 1091 ▪ Merck-Index (12.), No. 7194 ▪ Ullmann (4.) **10**, 514 f.; **19**, 237–244; (5.) **A5**, 88; **A9**, 394; **A11**, 570 f. – *[HS 1302 20; CAS 9000-69-5]*

Pederin.

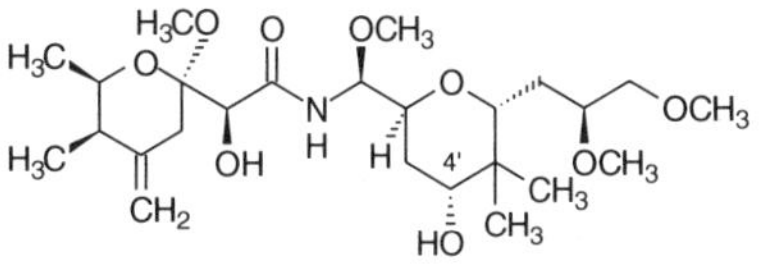

$C_{25}H_{45}NO_9$, M_R 503.63, mp. 113 °C, LD_{50} (mouse i. v.) 2 µg/kg. Highly toxic amide from the hemolymph of the blister beetle *Paederus fuscipes* and related species. Contact with P. leads to severe skin diseases. P. and related compounds such as pederone (4′-Oxo-P.) inhibit the biosynthesis of proteins and mitosis and also have cytostatic activity. Similar compounds have been isolated from marine sponges (*mycalamides, onnamides) [1].

Lit.: [1] Tetrahedron **44**, 7063–7080 (1988); **46**, 1757–1782 (1990).

gen.: Beilstein E V **18/7**, 471 ▪ Biochem. Syst. Ecol. **21**, 143–162 (1993) ▪ J. Med. Ent. **24**, 155–191 (1987). – *Synthesis:* Lindberg **3**, 199–236 ▪ Synthesis **1991**, 1191–1200 ▪ Synlett **1998**, 1432 ▪ Tetrahedron **44**, 7063 (1988). – *[CAS 27973-72-4]*

Peduncularine see Aristotelia alkaloids.

Peganine (vasicine, (*R*)-1,2,3,9-tetrahydropyrrolo-[2,1-*b*]quinazolin-3-ol).

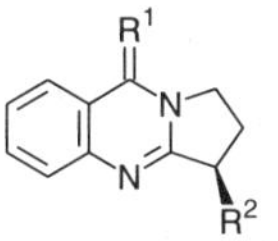

$R^1 = H_2$, $R^2 = OH$: Peganine = Vasicine
$R^1 = O$, $R^2 = OH$: Vasicinone
$R^1 = O$, $R^2 = H$: Deoxyvasicinone

$C_{11}H_{12}N_2O$, M_R 188.23, (–)-(*R*)-form: mp. 211–212 °C, $[\alpha]_D^{14}$ –254° ($CHCl_3$). P. is isolated from *Peganum harmala* (Zygophyllaceae), *Linaria* species, and goat's rue (*Galega officinalis*, Fabaceae) [1]. Structurally related or accompanying alkaloids are *vasicinone* ($C_{11}H_{10}N_2O_2$, M_R 202.21) and *deoxyvasicinone* ($C_{11}H_{10}N_2O$, M_R 186.21). P. undergoes racemization relatively easily in acidic solution and on heating. The racemate also occurs in nature (in *Anisotes sessiliflorus*). P. belongs to the rather small group of the quinazoline alkaloids. This group contains about 70 members and is further divided into the simple quinazolinones, pyrroloquinazolines, pyrido[2,1-*b*]quinazolines, and indoloquinazolines.

Activity: P. acts as a bronchodilator and stimulates respiration, it was a lead structure for the development of secretolytically active substances of the ambroxol and bromohexine types. This confirms the Indian traditional medicinal use of corresponding phytopreparations as expectorants and secretolytic agents.

Biosynth.: Anthranilic acid is most certainly one of the starting materials for the biosynthesis of P. and structurally related alkaloids but the nature of the other reaction partner is still unknown.

Lit.: [1] Mothes et al., p. 328 f.

gen.: Beilstein E V **23/12**, 40 ▪ Manske **29**, 99–140 ▪ Tetrahedron Lett. **32**, 7131 (1991) ▪ Tetrahedron: Asymmetry **7**, 25 (1996) ▪ Zechmeister **46**, 158–229. – *[HS 2939 90; CAS 6159-55-3 (–); 6159-56-4 (±); 486-64-6 (vasicinone)]*

Pelargonidin (3,4′,5,7-tetrahydroxyflavylium). Formula, see anthocyanins, $C_{15}H_{11}O_5^+$, M_R 271.25, as chloride ($C_{15}H_{11}ClO_5$, M_R 306.70) red-brown prisms, mp. >350 °C (decomp.), uv_{max} 530 nm.

Lit.: Karrer, No. 1705–1711 ▪ Tetrahedron **39**, 3005 (1983). – *[CAS 134-04-3]*

Pelargonin (monardin, salvinin, *pelargonidin 3,5-di-*O*-glucoside). $C_{27}H_{31}O_{15}^+$, M_R 595.53, red needles with a green luster (CH_3OH+HCl), mp. >350 °C. An *anthocyanin from the flowers of *Pelargonium zonale* (Geraniaceae), orange-colored dahlias, cornflowers, sage, etc.; dark violet powder, poorly soluble in water, soluble in alcohol and alkali solutions. The name is derived from the botanic name of the genus *Pelargonium* (Greek.: pelargos = stork). P. is usually isolated as the chloride [fine scarlet red needles, mp. 180–184 °C (decomp.)], $[\alpha]_D^{20}$ –291° (CH_3OH/HCl), uv_{max} 269, 505 nm. It is authorized for use as a colorant for foods (E 163) when isolated from natural sources.

Lit.: Beilstein E V **17/6**, 552 ▪ Karrer, No. 1709 ▪ Merck-Index (12.), No. 7198 f. – *[CAS 17334-58-6]*

Pelletierine [1-((*R*)-2-piperidinyl)-2-propanone].

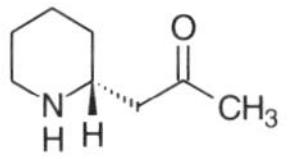

(*R*)-form

$C_8H_{15}NO$, M_R 141.21, $[\alpha]_D^{23}$ –18.1° (C_2H_5OH), oil, readily racemizes, bp. 195 °C, 102–107 °C (1.46 kPa). P. is a toxic *piperidine alkaloid [LD_{50} (rabbit i. v.) 40 mg/kg] of the pomegranate tree (*Punica granatum*, Punicaceae). The occurrence of the optically active base has not been confirmed with certainty because of the ready racemization. P. is highly toxic for tapeworms and is used as an anthelmintic. The first written testimony of this is an ancient Egyptian papyrus (ca. 1550 BC) describing the use of pomegranate roots. P. is an oxo derivative of coniine (see hemlock alkaloids) and a precursor in the biosynthesis of the more complex *Lycopodium alkaloids.

Lit.: Beilstein E V **21/6**, 522 f. ▪ Heterocycles **29**, 155 (1989) ▪ J. Org. Chem. **44**, 573–578 (1979) ▪ Kirk-Othmer (4.) **1**, 1052 ▪ Merck-Index (12.), No. 7200 ▪ Pelletier **3**, 54 ff. ▪ Sax (8.), PAO 500 ▪ Tetrahedron: Asymmetry **7**, 3047 (1996) (synthesis (–)-P.). – *[HS 2939 90; CAS 2858-66-4 (R-form); 539-00-4 (racemate)]*

Pellotine see Anhalonium alkaloids.

Peltatin.

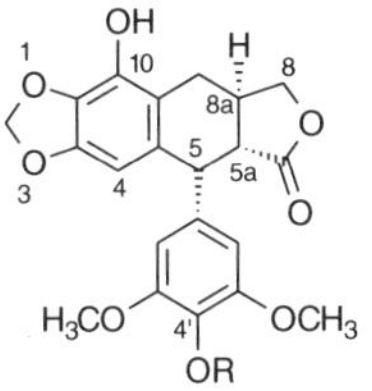

R = H : α-Peltatin
R = CH_3 : β-Peltatin

A *lignan of the aryltetralin type; it occurs both in the free form and as glycosides in species of the Lamiaceae and Berberidaceae (e. g., *Podophyllum peltatum*). *α-P.* (a), $C_{21}H_{20}O_8$, M_R 400.39; prisms, mp. 242–246 °C (decomp., sinters at 236 °C), $[\alpha]_D^{20}$

–124.8° ($CHCl_3$). *α-P. 10-O-β-D-glucoside* (b), $C_{27}H_{30}O_{13}$, M_R 562.53, needles, mp. 168–171 °C (decomp.), $[\alpha]_D^{20}$ –124.8° ($CHCl_3$). *β-P.* (c), $C_{22}H_{22}O_8$, M_R 414.41, prisms, mp. 231–238 °C (238–241 °C) (decomp.), $[\alpha]_D^{20}$ –122.0° ($CHCl_3$). *β-P. 10-O-β-D-glucoside:* (d), $C_{28}H_{32}O_{13}$, M_R 576.55, white amorphous powder, mp. 156–159 °C (decomp.), $[\alpha]_D^{20}$ –122.7° (CH_3OH). The 10-*O*-methyl ether of *β*-P. (*β-P. A methyl ether*) (e), $C_{23}H_{24}O_8$, M_R 428.44, prisms, mp. 124–126 °C, $[\alpha]_D^{20}$ –116.0° ($CHCl_3$), isolated from *Bursera fagaroides*, exhibits antitumor activity. The 5a-epimer (*β-P. B methyl ether*) (f), cryst., mp. 184 °C, $[\alpha]_D^{24}$ +3.3 ($CHCl_3$), occurs in *Juniperus sabina*. *α*-P. is a skin irritant and, like *β*-P., inhibits mitosis. For biosynthesis, see lignans.

Lit.: Can. J. Chem. **70**, 1082 (1992) ▪ Chem. Pharm. Bull. **39**, 192 ff. (1991) (synthesis) ▪ Luckner (3.), p. 395 ff. ▪ Phytochemistry **25**, 2089–2092 (1986) (biosynthesis); **29**, 1335–1338 (^{1}H and ^{13}C NMR); 3839–3844 (1990) (^{1}H NMR, glucoside). – *[CAS 568-53-6 (a); 11024-58-1 (b); 518-29-6 (c); 11024-59-2 (d); 23978-65-6 (e); 38943-35-0 (f)]*

Pemptoporphyrin see porphyrins.

Penaresidins. Azetidine alkaloids with actomyosin ATPase activating activity from the marine sponge *Penares* sp. (from Okinawa): $C_{19}H_{39}NO_3$, M_R 329.52 (P. A and P. B).

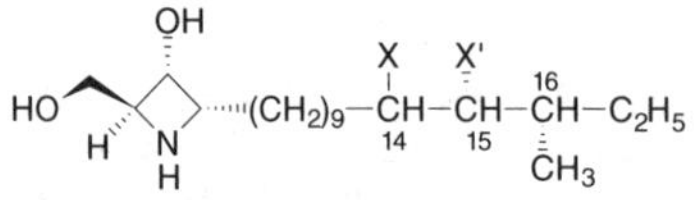

P. A : X = H, X' = OH
P. B : X = OH, X' = H

Lit.: Heterocycles **47**, 337 (1998) ▪ J. Chem. Soc., Perkin Trans. 1 **1991**, 1135 (isolation); **1997**, 97–111 (synthesis) ▪ Tetrahedron Lett. **36**, 7689 (1995); **38**, 3283, 3813 (1997); **40**, 337 (1999) (synthesis); **37**, 6775 (1996) (abs. configuration). – *[CAS 135574-62-8 (P. A); 135574-63-9 (P. B)]*

Penicillic acid.

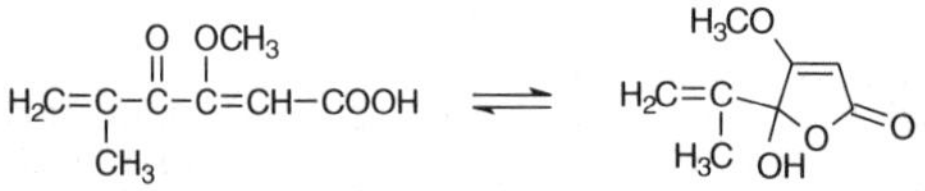

$C_8H_{10}O_4$, M_R 170.17, cryst., mp. 87 °C (anhydrous). A *mycotoxin formed by many fungi of the genera *Penicillium*, *Paecilomyces*, and *Aspergillus*. P. was first isolated in 1913 from cultures of *Penicillium puberulum*. P. is not a component of penicillin! P. has antibacterial, antifungal, cytotoxic, antiviral, mutagenic, and – in animal experiments (rats and mice) – cancerogenic activities. The LD_{50} for mice on subcutaneous administration is 2.2 mg/animal. P. reacts rapidly with SH groups, e. g., with glutathione or cysteine with loss of the biological activities (detoxification). Thus, no penicillic acid is found in protein-rich foods. The biosynthesis, like that of *patulin, proceeds through 6-methylsalicylic acid (tetraketide). P. often occurs together with patulin. P. occurs in moldy grains and grain products.

Lit.: Annu. Rev. Microbiol. **34**, 235 (1980) ▪ Beilstein E III **3**, 1467 ▪ Biosci. Biotechnol. Biochem. **60**, 1375 (1996) ▪ Cole-Cox, p. 520–525 ▪ Milchwissenschaft **33**, 16 (1978) ▪ Mycologia **81**, 837 (1989) ▪ Sax (8.), PAP 750 ▪ Tetrahedron Lett. **1978**, 3987 (synthesis). – *[HS 2918 90; CAS 90-65-3]*

Penicillide see purpactins.

Penicillins. Collective name for a group of bactericidic *antibiotics from culture broths of various genera of mold fungi, especially *Penicillium notatum* and *P. chrysogenum*. The basic skeleton of P., penam, consists of the (5*R*)-4-thia-1-azabicyclo[3.2.0]heptan-7-one system, a bicyclic system with a *β*-lactam and a condensed thiazolidine ring (see figure). The skeleton is very similar to that of the *cephalosporins, which are classified together with the P. in the group of **β*-lactam antibiotics. Naturally occurring P. are active against numerous Gram-positive bacteria, are unstable to acids, and are inactivated by penicillinases through cleavage of the *β*-lactam ring.

Biosynthesis: The biosynthesis proceeds by a non-ribosomal process via a dipeptide of D-*α*-aminoadipic acid (D-*α*-AAA) and L-cysteine, followed by linkage to L-valine with epimerization of L-valine to give the Arnstein tripeptide, *δ*-(D-*α*-aminoadipyl)-L-cysteinyl-D-valine, in which D-*α*-AAA forms the side chain which is exchanged in the further course of the synthesis of P., in contrast to that of the cephalosporins. The numerous natural and synthetic P. differ in the substituents on the side chain and the carboxy group; for nomenclature, see the table.

Production: Depending on the preparation, a distinction is in general made between the *natural P.*, formed by fermentation without additives for the side chain, e. g. P. G or P. F. as originally discovered by Fleming, and the *biosynthetic P.*, by which a specific P. is formed through additional feeding with the side chain. More than 100 biosynthetic P. have been prepared in this way, although only P. G obtained by addition of phenylacetic acid (ca. 80% of total P.), and P. V (phenoxyacetic acid) are of economic importance. For the synthesis of semisynthetic P., mainly P. G is cleaved to *6-aminopenicillanic acid (6-APA) chemically or by immobilized penicillin acylase[1] (penicillin amidase) which, in turn, is chemically reacylated. The semisynthetic P. have found wide use in therapy on account of their improved properties (stability towards acids, resistance to plasmid-coded and chromosomal-coded *β*-lactamases, expanded antimicrobial spectrum)[2]. In the modern production of P. yields of ca. 50 g/L are obtained by use of high-yielding strains[3] and optimized fermentation conditions in submersion procedures in large-scale fermentors of up to 300 m^3 capacity. For comparison, at the start of the industrial preparation of P. (1941) yields of 6–12 mg/L were obtained in resting surface cultures. A completely synthetic production of P. on a technical scale is not economic. A survey of the production techniques for P. and the related *carbapenems (CH_2 group in place of the *S* atom; *example*: *thienamycin) is given in *Lit.*[4,5,6]. P. were the first antibiotics to be prepared industrially and are still among the most frequently used antibiotics, still ahead of quinolones and *macrolides.

Activity: The mechanism of action of P. is mainly based on a blockade of cell wall synthesis or the participating enzymes in growing bacteria. It was found that P. bind to transpeptidases and carboxypeptidases so that, as a result of competitive inhibition, the cross-linking

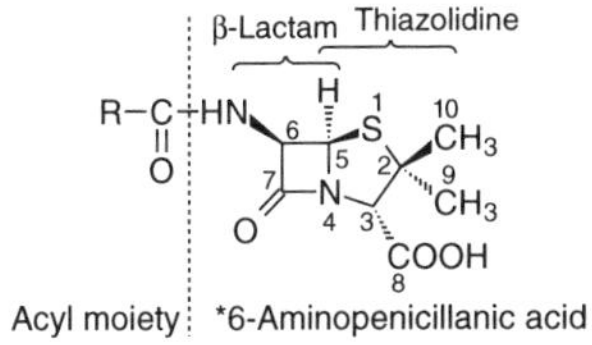

Table: Systematics of the Penicillins.

	R	short name, generic name	short name USA	short name Engl.
natural	$H_3C-CH_2-CH=CH-CH_2-$	2-Pentenyl-P.	I	F
	$H_5C_6-CH_2-$	Benzyl-P.	II	G
	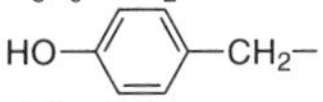	4-Hydroxy-benzyl-P.	III	X
	$H_3C-(CH_2)_6-$	Heptyl-P.	IV	K
	R $HOOC-CH(NH_2)-(CH_2)_3-$	Synnematin B		N
biosynthetic	$H_5C_6-O-CH_2-$	Phenoxymethyl-P.		V
	$H_2C=CH-CH_2-S-CH_2-$	Allylthiomethyl-P.	O	AT
	$H_3C-(CH_2)_3-S-CH_2-$	Butylthiomethyl-P.		BT
	$H_3C-C(Cl)=CH-CH_2-S-CH_2-$	(3-Chloro-2-butenyl-thiomethyl)-P.		S
semisynthetic	R $H_5C_6-CH(NH_2)-$	Ampicillin		
	$H_5C_6-CH(COOH)-$	Carbenicillin		
	$H_5C_6-O-CH(CH_3)-$	Phenethicillin		
	$H_5C_6-O-CH(C_2H_5)-$	Propicillin		
	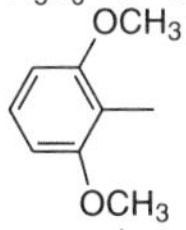	Meticillin		
	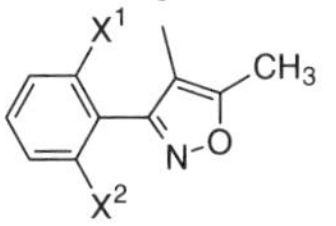	Oxacillin ($X^1 = X^2 = H$) Cloxacillin ($X^1 = H, X^2 = Cl$) Dicloxacillin ($X^1 = X^2 = Cl$) Flucloxacillin ($X^1 = Cl, X^2 = F$)		
		Ciclacillin		
	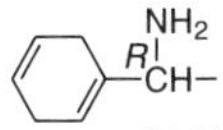	Epicillin		
	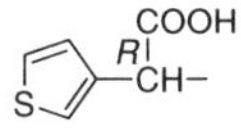	Ticarcillin		

of murein (the supporting substance for bacterial cell walls) is prevented. Many microorganisms, however, are resistant to P.[7,8,9], this is due to the presence of P.-cleaving enzymes – especially the *penicillinases*. Only the more recent semisynthetic P. with spatially demanding side chains such as oxacillin, cyclacillin, etc. are not yet affected by these enzymes.

Use: P. are marketed solely in the form of their more stable K or Na salts or in depot form and are used against numerous infectious diseases or their pathogens, especially against various Gram(+) cocci and spirochetes, e.g., pulmonary inflammations and meningitis. On account of their low toxicity P. can be given in high doses, only a very small portion of the patients (0.5 – 2%) suffer from allergic reactions.

Lit.: [1]Moss, in Wisemann (ed.), Topics in Enzyme and Fermentation Technology, vol. 1, p. 110 – 131, Chichester: Ellis Horwood 1977. [2]Queener & Schwartz, in Rose (ed.), Economic Microbiology, vol. 3, p. 35 – 122, London: Academic Press 1979. [3]Ball, in Hütter et al. (eds.), Antibiotics and other Secondary Metabolites, p. 165 – 176, FEMS Symp. No. 5, London: Academic Press 1978. [4]Nachr. Chem. Tech. Lab. **27**, 127 (1979). [5]Angew. Chem. Int. Ed. Engl. **24**, 180 – 202 (1985); Nicolaou, p. 41 – 54. [6]Barton-Nakanishi **4**, 239 – 274; Page (ed.), The Chemistry of β-Lactams, London: Blackie 1992. [7]Science **264**, 375 – 383 (1994). [8]Science **264**, 382 – 388 (1994). [9]Science **264**, 388 – 393 (1994).
gen.: Biotechnol. Bioeng. **21**, 261 (1979); **22**, 289 (1980) ▪ Enzyme Microbiol. Technol. **2**, 313 (1980) ▪ Eur. J. Appl. Microbiol. Biotechnol. **12**, 205 (1981) ▪ J. Antibiot. **48**, 1195 – 1212 (1995) ▪ Nachr. Chem. Tech. Lab. **38/5**, 616 ff. (1990) ▪ Reuben & Wittkoff, Pharmaceutical Chemicals in Perspective, New York: Wiley 1989 ▪ Swartz, in Moo-Young (ed.), Comprehensive Biotechnology, vol. 3, p. 7 – 48, Oxford: Pergamon Press 1985 ▪ see also β-lactam antibiotics. – *[HS 2941 10; CAS 1404-05-9]*

Penitrems (tremortins). A group of tremorigenic nona-, deca-, and undecacyclic *mycotoxins from *Penicillium crustosum*, *P. glandicola*, and other *Penicillium* species. Many of the originally named producers such as *P. palitans* are now included in *P. crustosum*. The P. have neurotoxic activities. In grazing animals, the P. cause shivering, staggering movements, and convulsions with a lethal result in severe cases. In the central nervous system of vertebrates P. effect an increased spontaneous release of amino acids with neurotransmitter functions such as *glutamic, *aspartic, and *4-aminobutanoic acid. On the other hand, the formation of glycine is inhibited. These effects can be counteracted by administration of mephenesin. The insecticidal activity of P., on the other hand, is increased in the presence of mephenesin. The LD_{50} of P. A for mice is 20 μg/animal. P. can occur in meat and sausage products as well as fresh cheese on infection with the respective producers. *Pennigritrem* is a structurally related toxin from *Penicillium nigricans*. The biosynthesis proceeds from tryptophan and a triterpene residue. The following P. have been described:

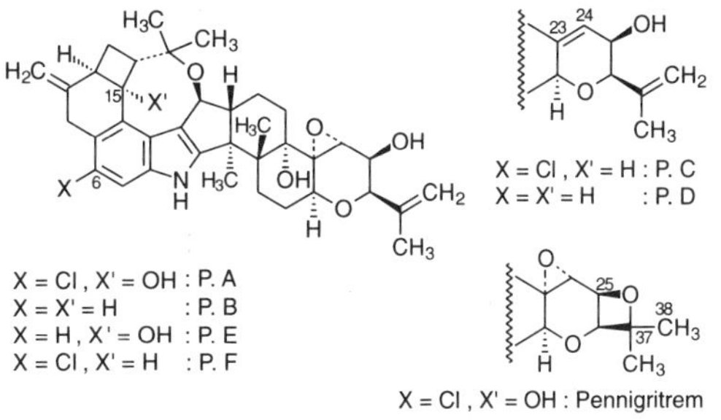

Table: Data of Penitrems and derivatives.

Penitrem	molecular formula	M_R	mp. [°C]	CAS
A	$C_{37}H_{44}ClNO_6$	634.21	237 – 239	12627-35-9
B	$C_{37}H_{45}NO_5$	583.77	185 – 195	11076-67-8
C	$C_{37}H_{44}ClNO_4$	602.21	amorphous	37318-84-6
D	$C_{37}H_{45}NO_4$	567.77	>300	78213-64-6
E	$C_{37}H_{45}NO_6$	599.77	amorphous	78213-66-8
F	$C_{37}H_{44}ClNO_5$	618.21	amorphous	78213-65-7
Penni-gritrem	$C_{37}H_{44}ClNO_6$	634.21	amorphous	139682-30-7

Lit.: Betina, chap. 16 ▪ Chem. Express **8**, 177 (1993) ▪ Cole-Cox, p. 382 – 385 ▪ J. Antibiot. **41**, 1868 (1988) ▪ J. Chem. Soc.,

Perkin Trans. 1 **1989**, 1539; **1992**, 23 ▪ J. Nat. Prod. **54**, 207 (1991) ▪ Mycologia **81**, 837 (1989) (chemotaxonomy) ▪ Sax (8.), PAR 250 ▪ Steyn & Vleggaar (eds.), Mycotoxins and Phycotoxins, p. 501–511, Amsterdam: Elsevier 1986 ▪ Zechmeister **48**, 1–80.

15-Pentadecanolide (Exaltolide®, 15-hydroxypentadecanoic acid lactone).

15-P.

$C_{15}H_{28}O_2$, M_R 240.39. cryst. with a fine, musk-like odor, mp. 37–38 °C, bp. 176 °C (2 kPa), LD_{50} (rat p.o.) >5 g/kg. The structure of 15-P. occurring in *angelica root/seed oil (ca. 1%) was elucidated in 1927 by Kerschbaum[1]. For synthesis, see *Lit.*[2,3].

Use: To an increasing extent as a nature-identical musk fragrance substance in perfume oils and, above all, in fine perfumes.

Lit.: [1]Ber. Dtsch. Chem. Ges. **60**, 902 (1927). [2]Helv. Chim. Acta **11**, 1159 (1928). [3]Bauer et al. (2.), p. 119 f.; Ohloff, p. 199–207; DE. P. 41 15 182 (1991), Haarmann & Reimer; Tetrahedron Lett. **34**, 6107 (1993).

gen.: Bedoukian (3.), p. 301–322 ▪ Beilstein E V **17/9**, 106 ▪ Synth. Commun. **25**, 3457 (1995) ▪ Theimer (ed.), Fragrance Chemistry, p. 433–494, San Diego: Academic Press 1982 ▪ Ullmann (5.) **A 11**, 207 f. – *[HS 2932 29; CAS 106-02-5]*

(*Z*)-6-Pentadecene see mites.

Pentadin. Sweet-tasting protein (M_R 12 000) from the plant *Pentadiplandra brazzeana* (Pentadiplandraceae) having a 500-fold higher sweetness than *sucrose. In contrast to other, protein-based sweeteners, P. is stable to heat.

Lit.: Chem. Senses **14**, 75–79 (1989) ▪ Phys. Unserer Zeit **21**, 101–108 (1990).

Pentalenene.

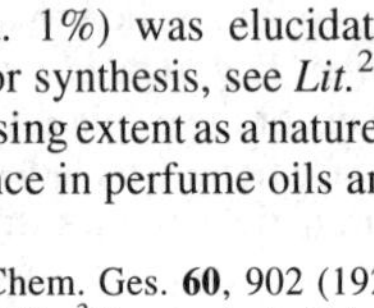

$C_{15}H_{24}$, M_R 204.36, oil, a *sesquiterpene hydrocarbon of the triquinane type from cultures of *Streptomyces griseochromogenes*[1] and *S. UC 5319*[2]. It is the first isolable biosynthetic intermediate[3] in the pathway from farnesyl pyrophosphate to the antibiotically active *pentalenolactones[2].

Lit.: [1]J. Antibiot. **33**, 92 (1980). [2]J. Org. Chem. **57**, 844 (1992). [3]J. Am. Chem. Soc. **110**, 5922 (1988); **112**, 4513 (1990).

gen.: Biosynthesis: Bioorg. Chem. **12**, 312 (1984) ▪ Can. J. Chem. **72**, 118 (1994). – *Synthesis:* J. Am. Chem. Soc. **105**, 7358 (1983) ▪ Tetrahedron **43**, 5637, 5685 (1987); **53**, 2111 (1997) ▪ Tetrahedron Lett. **33**, 3879 (1992); **41**, 403 (2000). – *[CAS 73306-73-7]*

Pentalenic acid see triquinanes.

Pentalenolactones. A group of *sesquiterpene antibiotics, P. A–P, from *Streptomyces chromofuscus*, *S. arenae*, *S. omiyaensis*, *S. UC 5319*, and others[1].

The P. have antineoplastic and antiviral activities[2]; they also irreversibly inhibit the glyceraldehyde 3-phosphate dehydrogenase of glycolysis by alkylating – through opening of the epoxide ring – the four cys-

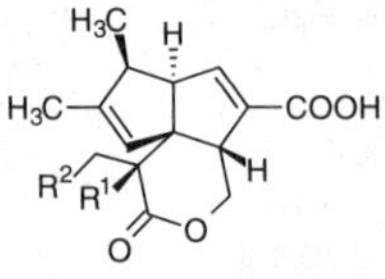

R^1, R^2 = —O— : Pentalenolactone (Arenemycin E, Antibiotic PA 132)

$R^1 = R^2$ = OH : P. O (Arenemycin D)

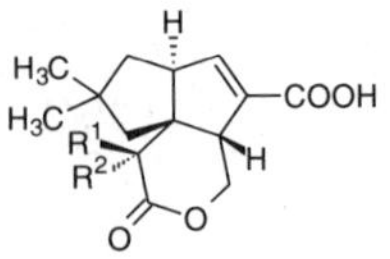

$R^1 = CH_3, R^2 = H$: P. D

$R^1, R^2 = CH_2$: P. E

R^1, R^2 = —O–CH_2—: P. F

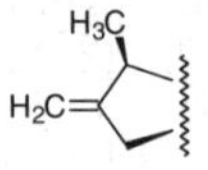

R^1, R^2 = —O— : P. B

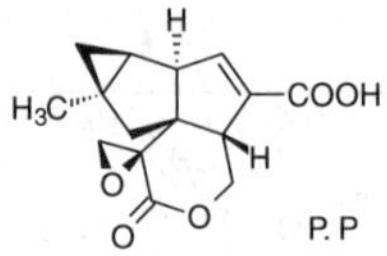

P. P

Table: Data of Pentalenolactones.

	molecular formula	M_R	mp. [°C]	$[\alpha]_D^{23}$	CAS
P. (Arenemycin E)	$C_{15}H_{16}O_5$	276.29	61–62	–172° (CH_3OH)	31501-48-1
P. B	$C_{15}H_{16}O_5$	276.29			138433-48-4
P. D	$C_{15}H_{20}O_4$	264.32			138458-84-1
P. E	$C_{15}H_{18}O_4$	262.31		–70.6° ($CHCl_3$)	72715-03-8
P. F	$C_{15}H_{18}O_5$	278.30	128–130 (methyl ester)		85416-36-0
P. O (Arenemycin D)	$C_{15}H_{18}O_6$	294.30	203–205	–160.1° ($CHCl_3$)*	93361-64-9
P. P	$C_{15}H_{16}O_5$	276.29			93361-68-3

* $[\alpha]_D^{25}$

teine residues at the active center of the tetrameric enzyme[3]. Numerous syntheses of the tricyclic system have been reported[4]. The biosynthesis[5] proceeds through *pentalenene as the first isolable intermediate.

Lit.: [1]J. Antibiot. **36**, 226 (1983); **37**, 816, 1076 (1984); **41**, 130 (1988); Tetrahedron Lett. **1978**, 923, 4411; J. Org. Chem. **57**, 844 ff. (1992). [2]J. Antibiot. **38**, 111 (1985); Tetrahedron Lett. **1970**, 4901 (mechanism of action). [3]Nature (London) **282**, 535 (1979); Mol. Biochem. Parasitol. **19**, 223 (1986); J. Bacteriol. **153**, 930 (1983); Arch. Biochem. Biophys. **270**, 50, (1989) (synthesis). [4]Eur. J. Org. Chem. **1998**, 257, 275; J. Am. Chem. Soc. **108**, 8015 (1986); **114**, 7387 (1992); J. Org. Chem. **53**, 227 (1988); Tetrahedron **44**, 2835 (1988). [5]J. Am. Chem. Soc. **112**, 4513 (1990); Bioorg. Chem. **12**, 312 (1984); J. Antibiot. (Tokyo) **39**, 266 (1986); Synform **2**, 136 (1989).

2,3-Pentanedione see butter flavor.

1-Penten-3-one see vegetable flavors (tomato).

Pentosans see polyoses.

Pentostatin.

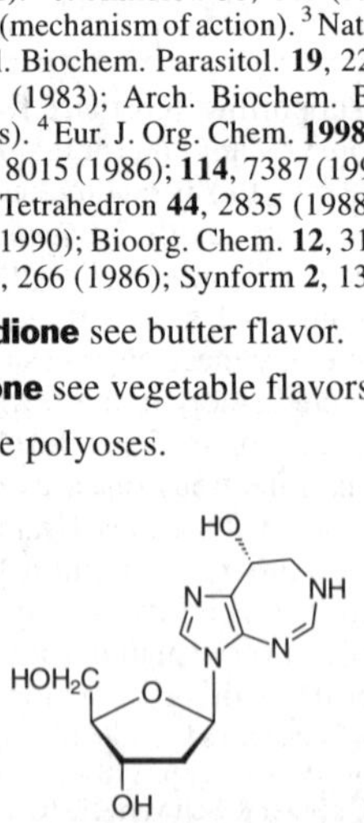

$C_{11}H_{16}N_4O_4$, M_R 268.27, mp. 220–225 °C, $[\alpha]_D^{20}$ +76.4° (H_2O). An *antibiotic from cultures of *Streptomyces*

antibioticus and *Aspergillus nidulans* with antineoplastic activity. P. is an adenosine deaminase inhibitor and is used in the treatment of hair cell leukemia, chronic lymphatic leukemia, and mycosis fungoides, however, in therapeutic doses it exhibits adverse reactions on the CNS, kidneys, and liver and has mutagenic properties.

Lit.: Biochemistry **27**, 5790 (1988) ▪ Cancer Treat. Rep. **17**, 213 (1990) ▪ J. Nat. Cancer Inst. **80**, 765 (1988) ▪ J. Org. Chem. **47**, 3457 (1982) ▪ Pharmacol. Rev. **44**, 459 (1992) ▪ Sax (8.), PBT 100. – *[CAS 53910-25-1]*

2-Pentylpyridine see sesame seeds (roasted) flavor.

Peonidin (cyanidin 3′-methyl ether, 3,4′,5,7-tetrahydroxy-3′-methoxyflavylium). Formula see anthocyanins. $C_{16}H_{13}O_6^+$, M_R 301.28, as chloride ($C_{16}H_{13}ClO_6$, M_R 336.73) reddish-brown needles; see peonin and anthocyanins.

Lit.: Hager (5.) **5**, 297; **6**, 1–6 ▪ Karrer, No. 1723–1726 ▪ Tetrahedron **39**, 3005 (1983). – *[CAS 134-01-0 (chloride)]*

Peonin (*peonidin 3,5-di-*O*-glucoside). $C_{28}H_{33}O_{16}^+$, M_R 625.56, reddish-violet cryst., mp. 165–167 °C (decomp.). P. is the typical *anthocyanin component of the violet-red flowers of the peony (*Paeonia*, Paeoniaceae), but also occurs in blue grapes (*Vitis vinifera*, Vitaceae), sweet cherries (*Prunus avium*, Rosaceae), and many other plants.

Lit.: Hager (5.) **6**, 1 ▪ Karrer, No. 1725. – *[CAS 132-37-6 (chloride)]*

Pep 5 see epidermin.

Pepper (*Piper nigrum*, Piperaceae). The P. plant has been used as a source of genuine pepper since antiquity. It is a climbing shrub indigenous to southern India and is now cultivated in all tropical regions of the world. The berry fruits have a diameter of approx. 3–5 mm. Three commercial forms are classified according to the type of pretreatment: (1) black pepper consists of the fully grown seeds harvested before ripening and then dried; (2) white pepper is produced from the ripe, red fruits by removing the outer parts of the fruit husk; (3) green pepper encompasses the harvested unripe fruits immersed in brine or vinegar for marketing. Pepper contains *essential oils (black pepper: 1.5–3.5%; white pepper: 1.0–2.4%), that are responsible for the aromatic taste, and hot-tasting amides (5–10%), mainly *piperine[1]. The hot taste results from a stimulation of thermoreceptors whereupon saliva and gastric juice secretion are stimulated by a reflex mechanism. Pepper is an important condiment for meat, sausage, and fish products as well as cheese. Pepper extract is used as a topical analgesic (see resiniferatoxin).

Lit.: [1] Can. J. Chem. **74**, 419–432 (1996); Phytochemistry **46**, 597–673 (1997) (review).
gen.: Hager (5.) **3**, 387; **6**, 627.

Peppermint oils. The following p. are distinguished on the basis of the mint plant involved:

1) ***Genuine peppermint oil*** (oil of *Mentha piperita*): Light yellow to light greenish yellow oil with a fresh, minty, herby sweet, balsamy odor and a sweet, fresh minty, cooling taste.

Production: By steam distillation of the flowering peppermint plant, *Mentha* ×*piperita*, mainly of the variety "Black Mitcham". Main producers are the USA, where in 1997: 4500 tons of oil were produced in the states of Washington, Oregon, and Idaho, and India (750 tons).

Composition [1]***:*** Main components and decisive for the organoleptic character are (–)-*menthol* (see 3-*p*-menthanols) (25–45%), (–)-*menthone* (see *p*-menthanones) (20–30%), (–)-*menthyl acetate* (ca. 5%; $C_{12}H_{22}O_2$, M_R 198.31), and (+)-*menthofuran* (2–10%; $C_{10}H_{14}O$, M_R 150.22). The exact composition depends on the site of cultivation.

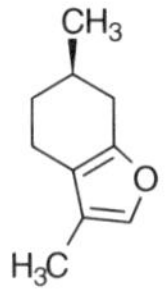

(+)-Menthofuran

Use: To improve the aromatic character of confectionery products, chewing gum, oral hygiene products (toothpaste, mouth waters), and tobacco. Medicinal uses in carminatives, rhinological, urological, mouth and throat therapeutic agents, in antirheumatic ointments and bronchologicals.

2. ***Mentha arvensis oil:*** Colorless to light yellow oil with a sharp, fresh, minty odor and a cooling, minty, somewhat bitter-sharp taste which is markedly less sweet and full than that of peppermint oil.

Production: By steam distillation of the so-called Japanese mint, *Mentha arvensis* var. *piperascens*, a cultivated form of field mint. Main producers are China and India, where about 8000 t *Mentha arvensis* oil were produced in 1992, this corresponds to 3000 t dementholized oil and 5000 t menthol.

Composition [2]***:*** The crude oil contains between 70 and 80% (–)-*menthol* and is solid at temperatures below 20 °C. It is not marketed as such but is used rather for the production of natural (–)-menthol by crystallization. The remaining, dementholized oil still contains approx. 35–40% (–)-*menthol* together with ca. 20% (–)-*menthone* and 10% (+)-*isomenthone* (see *p*-menthanones).

Use: Similar to p. as a cheaper but poorer quality alternative.

Lit.: [1] Perfum. Flavor. **13** (5), 66 (1988); **14** (6), 21 (1989) (distinction of Mentha arvensis oils); **18** (4), 60 (1993). [2] Perfum. Flavor. **8** (2), 61 (1983); **14** (1), 29 (1989); **19** (6), 58 (1994).
gen.: Arctander, p. 412, 513 ▪ Bauer et al. (2.), p. 163 f. ▪ Dtsch. Apoth. Ztg. **138**, 3793–3799 (1998) ▪ Heterocycles **33**, 73 (1992); **36**, 345 (1993) (menthofuran) ▪ ISO 856 (1997) ▪ Ph. Eur. **1997** ▪ USP XXII and 23/NF 18, BP 93. – *Toxicology:* Food Cosmet. Toxicol. **13**, 771 (1975). – *[HS 3301 29; CAS 8006-90-4 (1.); 68917-18-0 (2.); 2623-23-6 ((–)-menthyl acetate); 17957-94-7 ((+)-menthofuran)]*

Pepstatin A.

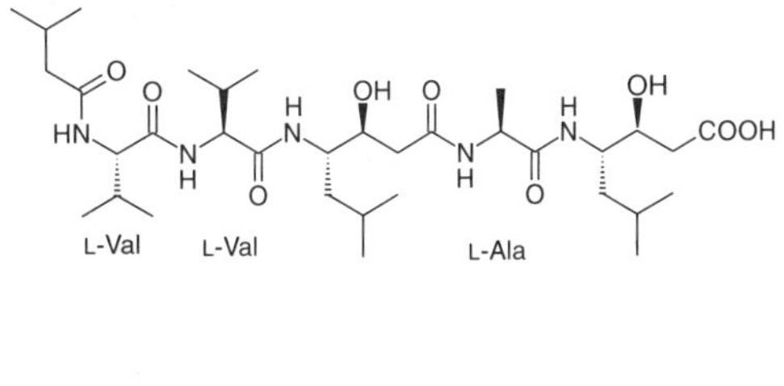

$C_{34}H_{63}N_5O_9$, M_R 685.90, *N*-acylated peptide isolated from culture broths of various streptomycetes (*Streptomyces testaceus*, *S. argenteolus*); needles, mp. 228–229 °C (decomp.). The substance acts as an inhibitor of aspartate proteinases (carboxy proteinases), e. g.,pepsin, cathepsin D (a lysosomal protease), and is also active against renin, formed in excessive amounts in some forms of kidney disease. There are also the pepstatins B and C, containing an *N*-hexanoyl or *N*-(4-methylpentanoyl) group, respectively, in place of the *N*-(3-methylbutanoyl) group.

Lit.: Merck-Index (12.), No. 7290. Methods Enzymol. **45**, 689–693 (1976). – *[CAS 26305-03-3]*

Pepticinnamin E.

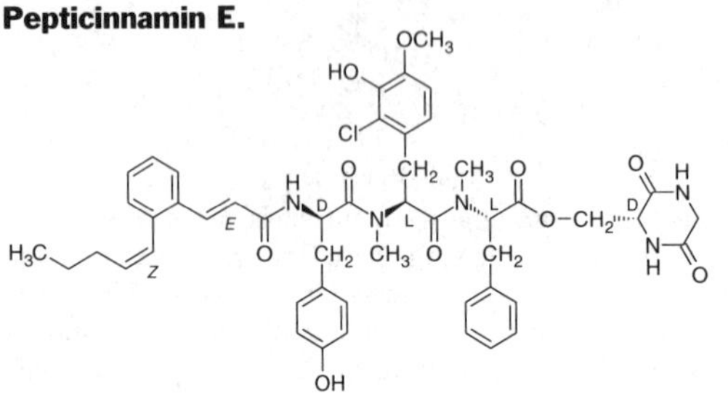

$C_{49}H_{54}ClN_5O_{10}$, M_R 908.43, powder, mp. 143–146 °C, $[\alpha]_D^{26}$ –207° (CH_3OH), a farnesyl transferase inhibitor from *Streptomyces* sp.

Lit.: Angew. Chem., Int. Ed. Engl. **37**, 1236–1239 (1998) ▪ Chem.-Eur. J. **5**, 227–236 (1999) ▪ J. Antibiot. **46**, 222–228, 229–234 (1993). – *[CAS 147317-36-0]*

Peptide alkaloids. *Alkaloids containing amino acids linked by peptide bonds; examples include the *ergot alkaloids and the *cyclopeptide alkaloids (Rhamnaceae alkaloids) as well as the *amanitins and *phallotoxins of the death cup. In spite of their structural similarities, the *peptide antibiotics and *β-lactam antibiotics are not included in the peptide alkaloids.

Lit.: Alkaloids: Chem. Biol. Perspect. **12**, 187–228 (1998) ▪ Manske **26**, 299–326 ▪ Zechmeister **28**, 162–203.

Peptide antibiotics. Antibiotics with linear or cyclic oligopeptide structures containing not only L-*amino acids but often also unusual, non-proteinogenic amino acids. *Homopeptides* are built up exclusively of amino acids (e. g., *gramicidins), the so-called *heteropeptides* also contain other structural elements, e. g., hydroxamic acids. Macrocyclic P.a. containing one or more ester bonds in the ring are known as peptide lactones (e. g., *hormaomycin) or depsipeptides (e. g., *valinomycin).

The P. a. exhibit a broad spectrum of biological activities. These include antibacterial, antifungal, antiviral, insecticidal, and herbicidal activities as well as antitumor and immunosuppressive properties. P. a. are also used as preservation agents in food technology.

P. a. can be classified according to their biosynthesis routes: 1) P. a., formed ribosomally from the 20 proteinogenic amino acids, these include the P. a. from *Escherichia coli* and *Bacillus subtilis*. – 2.) Sequential linkage of amino acids by soluble enzymes which activate and couple the amino acids. This mechanism is poorly specific and is only used to produce di- and tripeptides. – 3) Non-ribosomal synthesis at a multienzyme complex via a "protein thiotemplate" mechanism. Non-proteinogenic amino acids can also be incorporated in the peptide by this route. The amino acids are activated as aminoacyladenylates and transferred to the enzyme by thioester bonds. This sequence depends on the structure of the multienzyme complex. The resulting P. a. consist of up to ca. 30 amino acids.

Lit.: Kleinkauf & von Döhren, Biochemistry of Peptide Antibiotics, Berlin: De Gruyter 1990 ▪ Pure Appl. Chem. **54**, 2409–2424 (1982) ▪ Rimpler et al., Biogene Arzneistoffe, p. 596 ff., Stuttgart: Thieme 1990 ▪ Vining & Stuttard (eds.), Genetics and Biochemistry of Antibiotic Production, p. 121–171, Boston: Butterworth-Heinemann 1995.

Peramine.

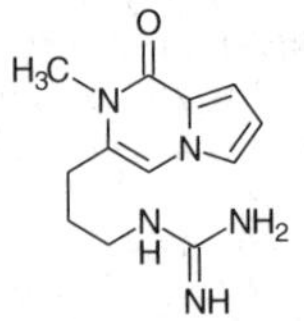

$C_{12}H_{17}N_5O$, M_R 247.30, mp. 242–243 °C. A *mycotoxin alkaloid from the Australian rye grass *Lolium perenne*, formed after infection with the fungus *Acremonium loliae*. P. has a potent antifeedant effect on some insect species. It contains the unusual pyrrolo[1,2-*a*]pyrazin-1(2*H*)-one ring system.

Lit.: J. Chem. Ecol. **12**, 647–658 (1986) ▪ J. Chem. Soc., Chem. Commun. **1986**, 935; **1988**, 978 ▪ J. Chem. Soc., Perkin Trans. 1 **1990**, 311 ▪ J. Nat. Prod. **52**, 193 (1989) ▪ J. Org. Chem. **53**, 4650 (1988). – *[HS 2939 90; CAS 102482-94-0]*

Perezone [2-((*R*)-1,5-dimethyl-4-hexenyl)-3-hydroxy-5-methyl-*p*-benzoquinone].

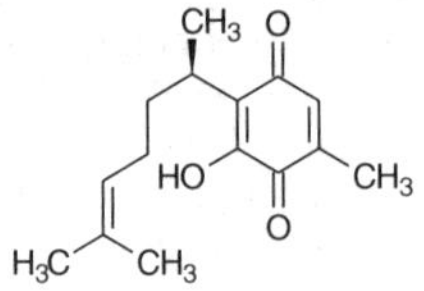

$C_{15}H_{20}O_3$, M_R 248.32, golden yellow platelets, mp. 104–106 °C, $[\alpha]_D^{20}$ – 17° (ether). A *sesquiterpene of the *bisabolene type with laxative activity from *Perezia* (Asteraceae) and *Jungia* species.

Lit.: Karrer, No. 1198. – *Synthesis:* Chem Pharm. Bull. **43**, 553 (1995) ▪ Heterocycles **37**, 181 (1994) ▪ J. Org. Chem. **55**, 1177 (1990). – *[CAS 3600-95-1]*

Peridinin (sulcatoxanthin).

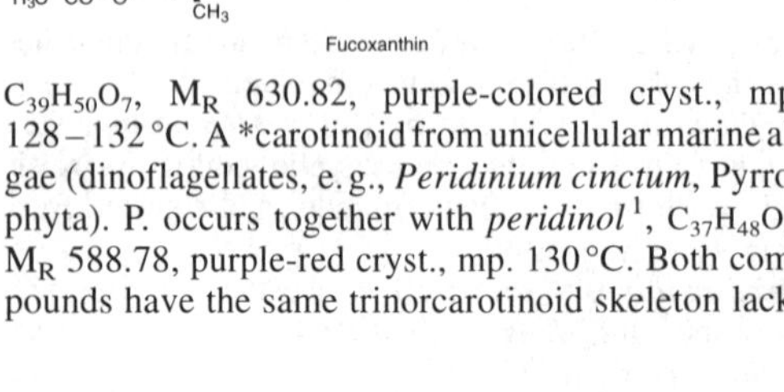

$C_{39}H_{50}O_7$, M_R 630.82, purple-colored cryst., mp. 128–132 °C. A *carotinoid from unicellular marine algae (dinoflagellates, e. g., *Peridinium cinctum*, Pyrrophyta). P. occurs together with *peridinol*[1], $C_{37}H_{48}O_6$, M_R 588.78, purple-red cryst., mp. 130 °C. Both compounds have the same trinorcarotinoid skeleton lacking 3 carbon atoms in the middle of the chain. *Fucoxanthin*[2], $C_{42}H_{58}O_6$, M_R 658.92, deep red prisms or needles, mp. 168 °C, is also found in many algae, marine sponges, snails, and bird feathers. Fucoxanthin and P. are *carotinoids with an unusual allene structure. In the biosynthesis of P. a C_3 fragment is removed from the middle of the chain of the C_{40} carotenoid precursor *zeaxanthin by an as yet unknown mechanism.

Lit.: [1] Acta Chem. Scand., Ser. B **30**, 157 (1976); Beilstein E V **19/6**, 696; Phytochemistry **21**, 2859 (1982). [2] Beilstein E V **18/4**, 673; J. Chem. Soc., Perkin Trans. 1 **1995**, 1895–1904.
gen.: Goodwin, The Biochemistry of the Carotenoids, 2. edn., London: Chapman & Hall 1980 ▪ Scheuer I **2**, 2 ▪ Thomson, The Chemistry of Natural Products, p. 187, London: Chapman & Hall 1985. – *Synthesis:* J. Chem. Soc., Perkin Trans. 1 **1993**, 1599–1610. – *[CAS 33281-81-1 (P.); 54369-14-1 (peridinol); 3351-86-8 (fucoxanthin)]*

Perilla alcohol [4-(1-methylethenyl)-1-cyclohexene-1-methanol]. Formula, see perillaldehyde. $C_{10}H_{16}O$, M_R 152.24, oil. A monocyclic *monoterpene alcohol occurring in both enantiomeric forms: (*R*)(+)- and (*S*)(−)-P., bp. 119–121 °C (1.1 kPa), $[\alpha]_D$ ±68°. (+)-P. occurs in "delft grass oil"[1] (*Cymbopogon polyneuros*, Poaceae) a grass indigenous to the Nilgiri mountains (Southern India) and the mountains of Sri Lanka; (−)-P. occurs in false lavender oil (*Lavandula hybrida*, Lamiaceae)[2]. For synthesis, see *Lit.*[3]. P. induces apoptosis. It is in clinical development against breast and prostate cancer.

Lit.: [1] Hegnauer **II**, 188. [2] Helv. Chim. Acta **28**, 1220 (1945). [3] J. Am. Chem. Soc. **91**, 6473 f. (1969).
gen.: Karrer, No. 314. – *[CAS 57717-97-2 ((+)-P.); 18457-55-1 ((−)-P.)]*

Perilla ketone see egomaketone.

Perillaldehyde [4-(1-methylethenyl)-1-cyclohexene-1-carbaldehyde].

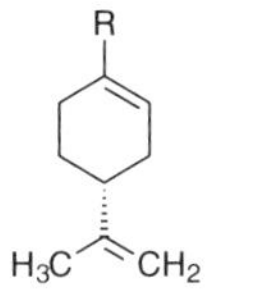

R = CHO : Perillaldehyde [(*R*)-form]
R = CH_2OH : Perilla alcohol [(*R*)-form]

$C_{10}H_{14}O$, M_R 150.22, oil. Monocyclic *monoterpene aldehyde occurring in both enantiomeric forms: (*R*)-(+)- and (*S*)-(−)-P., bp. 104 °C (900 Pa), $[\alpha]_D$ ± 150.7°. LD_{50} (rat p.o.) 2500 mg/kg. (−)-P. constitutes ca. 50% of "Perilla oil" from *Perilla arguta* and *P. frutescens* var. *crispa* (Lamiaceae), (+)-P. constitutes ca. 60% of the Philippine "Sulpicia oil" from *Sulpicia orsuami* and also occurs in *Siler trilobum*, *Sium latifolium*, *Citrus reticulata*, and other plants. The (*E*)-oxime of (*S*)-(−)-P. is used in Japan as a sweetener "Perilla sugar" and has a 2000-fold higher sweetness than saccharose.

Lit.: Beilstein E IV **7**, 316 ▪ Karrer, No. 416 ▪ Merck-Index (12.), No. 7308 ▪ Science **197**, 573 (1977) ▪ Synth. Commun. **18**, 1905 (1988) ▪ Ullmann (5.) **A 10**, 228. – *[HS 2912 29; CAS 5503-12-8 ((+)-P.); 18031-40-8 ((−)-P.)]*

Perillene see mites.

Periplanones. The P. A–D are germacranoid *sesquiterpene *pheromones of females of the American cockroach *Periplaneta americana*[1].

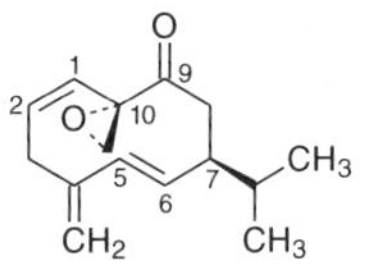

Periplanone A

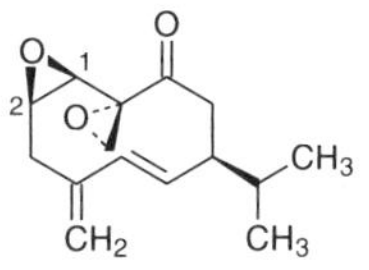

Periplanone B

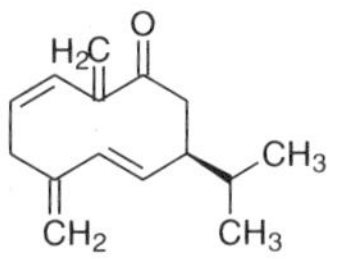

Periplanone C
(= P. D_1)

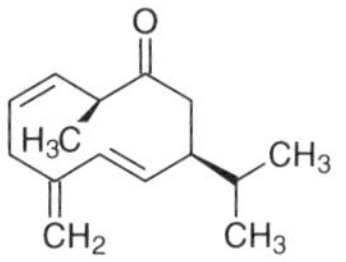

Periplanone D
(= P. D_2)

Table: Data of Periplanones.

compound	molecular formula	M_R	mp. [°C]	CAS
P. A[a)]	$C_{15}H_{20}O_2$	232.32	47–50	112608-82-9
P. B[b)]	$C_{15}H_{20}O_3$	248.32	57–57.5	61228-92-0
P. C	$C_{15}H_{20}O$	216.15		123163-72-4
P. D	$C_{15}H_{22}O$	218.34	55–57	123062-72-6

a) $[\alpha]_D^{25}$ −290° (hexane)
b) $[\alpha]_D^{22}$ −553° (hexane)

P. A and B induce running and mating behavior in male cockroaches. P. C (previously D_1) and D (previously D_2) are about 100 times less active[2]. The P. are found mostly in the insects excrements from which they can be isolated (0.2 mg of P. B are obtained from 75 000 female insects)[1]. The P. only act over short distances, possibly only on contact. The biologically most active P. A and P. B occur in *P. americana* in a ratio of ca. 10:1. The unnatural epoxy epimer of P. A is about 10 000-fold less active. The structures were elucidated by synthesis[3] and comparison with the electroantennograms (EAG) of natural samples[2]. The conformation of the flexible, 10-membered germacranoids has been studied in detail by NMR spectroscopy and X-ray diffraction[4]. The P. are added to insecticide preparations for the protection of stored products[5]. P. J is the sex pheromone of the Japanese cockroach *Periplaneta japonica*[6].

Lit.: [1] Agric. Biol. Chem. **54**, 575 (1990); Tetrahedron Lett. **28**, 1791 (1987); **30**, 2367 (1989); Chem. Lett. **1988**, 517; J. Chem. Ecol. **8**, 439 (1982). [2] Tetrahedron Lett. **27**, 1315 (1986); **30**, 2367 (1989); **31**, 1747–1750 (1990); Anim. Behav. **33**, 591 (1985); Comp. Biochem. Physiol. **74a**, 909 (1983). [3] J. Am. Chem. Soc. **101**, 2493, 2495 (1979); Synform **6**, 150 (1988); Synlett **1999**, 744; Tetrahedron Lett. **29**, 1971 (1988); **31**, 1747 (1990); **33**, 369 (1992); **35**, 4505 (1994); Tetrahedron **46**, 5101, 8083 (1990); **53**, 7209 (1997). [4] J. Chem. Soc., Perkin Trans. 1 **1990**, 1769. [5] Comp. Pharmacol Toxicol. **92C**, 193 (1989); Environ. Entomol. **13**, 448 (1984); Pest Control **54**, 40 (1986). [6] Heterocycles **47**, 139 (1998).
gen.: Synthesis: ApSimon **9**, 336-344 ▪ Nat. Prod. Lett. **1**, 129 (1992) ▪ Nicolaou, p. 211–220, 333–343 ▪ Synlett **1995**, 255.

Periwinkle. 1. *Tropical periwinkle*: small, 40–80 cm high, subshrub with glossy leaves and violet, red, or white flowers (Apocynaceae). The tropical P. (also *Catharanthus roseus*, previously named as *Vinca rosea*) was originally indigenous to Madagascar but is

now distributed throughout the tropical regions and is even available as an ornamental indoor plant. P. is cultivated as an ornamental and medicinal plant in many tropical and subtropical regions. The flowers and roots contain over 100 different *indole alkaloids (*monoterpenoid indole alkaloids, *Catharanthus alkaloids, also incorrectly known as *Vinca alkaloids). The total alkaloid content amounts to <1% with the major alkaloid being *vindoline. The therapeutically used dimeric bases, the minor alkaloids *vinblastine and *vincristine, are isolated from the plant on an industrial scale. Both alkaloids are very important for cancer therapy, they cause severe damage to cells (cytotoxic) and, like colchicine, inhibit cell division.
2. *Small periwinkle* (*Vinca minor*): genus *Vinca* (Apocynaceae), is indigenous to central and eastern Europe. Of the over 25 *Vinca* species, only two (*V. major* and *V. herbacea*) are found in Europe. The small P. is a persevering, small – ca. 20 cm high – plant with leathery, dark green leaves and mostly blue flowers. It is often planted as a ground cover (parks, graveyards) on account of its robust nature. To date over 150 indole alkaloids (*monoterpenoid indole alkaloids, *Vinca alkaloids) have been isolated from *Vinca* species and assigned to many structural groups including the *ajmalicine, *ajmaline, *Aspidosperma alkaloids, eburnamine (see eburnamonine), *oxindole alkaloids, picraline, *quebrachamine, and *strychnine types.
V. minor contains more than 40 alkaloids [1,2]. The alkaloid content of central European *Vinca* species reaches ca. 1% with *vincamine as the main component [3]. The German BGA (federal drug agency) revoked its authorization for the use of all P.-containing preparations with immediate effect on 20.7.1987. Reason: animal experiments had revealed a blood-damaging action in the sense of an immunosuppressive activity which was attributed not to the main alkaloid but rather to the accompanying components. This restriction did not apply to pure *Vinca* alkaloid preparations or to homeopathic products. In the past the drug *Vincae minoris herba* was used to treat headaches and migraine and was also widely used as an antihypertensive agent.
Lit.: [1] Planta Med. **55**, 188 (1989). [2] Khim. Prit. Soedin. **3**, 434 (1989). [3] Pharmazie **24**, 273 (1969).
gen.: Hager (5.) **3**, 259; **6**, 1123–1134; **6c**, 446–456. – *[HS 1211 90]*

Perloline (as hydroxide).

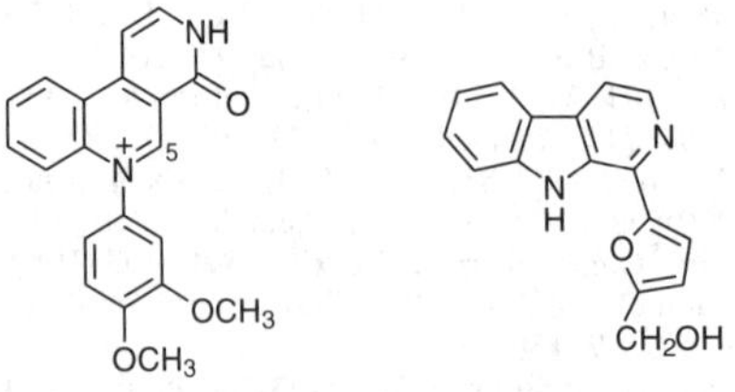

Perloline Perlolyrine

$C_{20}H_{17}N_2O_3^+$, M_R 333.37, "hydroxide": $C_{20}H_{18}N_2O_4$, M_R 350.37, cryst., mp. 154–173 °C, 181 °C, 248–258 °C (decomp.). A *quinoline alkaloid with an ionic structure from *Lolium* species (sweet grasses, Poaceae) cultivated as grazing grasses in the Mediterranean region, Australia, and South Africa (*L. rigidum*). P. occurs together with the *harman derivative *perlolyrine* ($C_{16}H_{12}N_2O_2$, M_R 264.28, yellow cryst., mp. 183 °C). *Lolium* grasses often cause poisonings in grazing animals but these must be attributed to corynetoxins (see tunicamycins) as the plants are frequently infected with corynebacteria.
Lit.: Aust. J. Chem. **40**, 631 (1987) ▪ Beilstein E V **25/2**, 140 ▪ Tetrahedron Lett. **27**, 3399 (1986) ▪ see also peramine. – *[HS 2939 90; CAS 7344-94-7 (P. cation); 29700-20-7 (perlolyrine)]*

Peronatins.

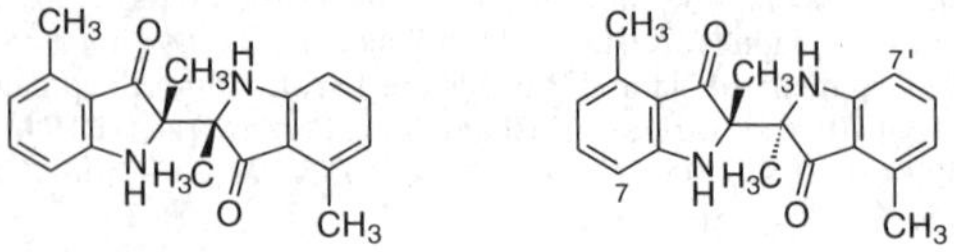

Peronatin A (racemate) Peronatin B

2,2′-Biindoline-3,3′-diones from injured fruit bodies of the fungus *Collybia peronata*, Agaricales. *P. A* ($C_{20}H_{20}N_2O_2$, M_R 320.39, yellow oil) occurs as the racemate but rearranges on standing in chloroform solution to the more stable *meso*-form *P. B* (yellow cryst., mp. 213 °C). *Tricholoma scalpturatum* contains 7,7′-dimethoxyperonatin B and the corresponding 7-hydroxy-7′-methoxy derivative. The P. are probably formed by dimerization of the corresponding indolin-3-ones which are structurally related with *lascivol. It is not yet known if precursors of P. are responsible for the burning taste of *C. peronata*.
Lit.: J. Nat. Prod. **57**, 852 (1994) (isolation) ▪ J. Org. Chem. **62**, 4756 (1997) (synthesis). – *[CAS 157536-38-4 (P. A); 157536-39-5 (P. B)]*

Perrottetianal A see sacculatanes.

Perrottetins.

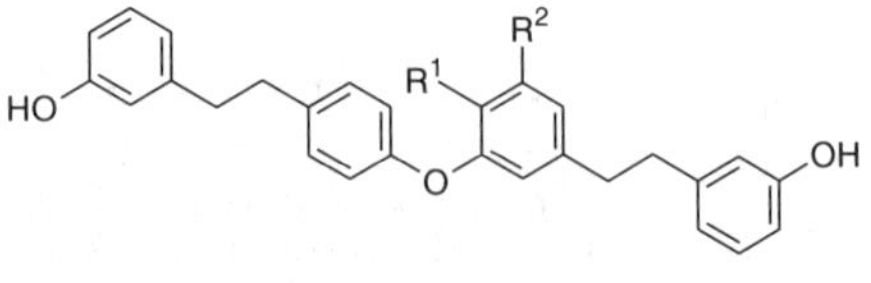

Table: Data of Perrottetins.

	R^1	R^2	molecular formula	M_R	CAS
P. E	OH	H	$C_{28}H_{26}O_4$	426.51	89911-97-7
P. F	OH	OH	$C_{28}H_{26}O_5$	442.51	89911-98-8
P. G	OCH_3	OH	$C_{29}H_{28}O_5$	456.54	104290-38-2

P. E–G are open-chain bisbibenzyls with linear linkages of the two bibenzyl units through an ether bridge. They are considered to be biogenetic precursors of cyclic bisbibenzyls (see riccardins and marchantins). In contrast, the P. A–D [1] are prenylated bibenzyls. The occurrence of P. is limited to liverworts. P. E., isolated from *Radula perrottetii* (which also contains F and G) and other liverworts, is cytotoxic to KB cells [2]. For synthesis, see *Lit.* [3].
Lit.: [1] Phytochemistry **21**, 2481 (1982). [2] J. Hattori Bot. Lab. **56**, 215 (1984). [3] Tetrahedron Lett. **26**, 6097 (1985); Justus Liebigs Ann. Chem. **1989**, 401; Angew. Chem. Int. Ed. Engl. **30**, 130–147 (1991).

gen.: Zinsmeister & Mues (eds.), Bryophytes, their Chemistry and Chemotaxonomy (Proc. Phytochem. Soc. Eur. 29), p. 186, Oxford: Clarendon Press 1990.

Pertussis toxin (whooping cough toxin). *Whooping cough* (pertussis) is a bacterial infection caused by *Bordetella pertussis* with characteristic symptoms of coughing attacks which can be life-threatening in infants and possibly accompanied by complications. P. t. secreted by the pathogenic bacteria is responsible for these symptoms. P. t. is an AB toxin consisting of a functional and a binding unit. The complete toxin molecule (holotoxin) consists of six polypeptide subunits (S1, S2, S3, two S4, and S5). The enzymatically active subunit is identical with S1 (A-subunit); the other subunits form the binding component, the B oligomer. The A-subunit is composed of two functional regions: the first 187 amino acids that determine the enzymatic activity and the remaining amino acids (188–234) that are apparently important for the structure of the toxin. The B oligomer is made up of two dimers (S2–S4 and S3–S4) linked through one S5 subunit. The B oligomer contains the binding sites for the receptors with which the toxin binds to the surface of eukaryotic cells. The receptor is probably a glycoprotein. For the mechanism of action of P. t. as inhibitor of a series of signal transmitting systems and a large number of cells types, see *Lit.*[1].

Lit.: [1] Microbiol. Sci. **5**, 285 ff. (1988); Naturwiss. Rundsch. **42**, 327 (1989).

gen.: Biochem. J. **255**, 1–13 (1988) ■ Birnbaumer et al. (eds.), G Proteins, p. 267–294, San Diego: Academic Press 1990 ■ Med. Immunol. **13**, 431–436 (1987) ■ Pharmacol. Ther. **19**, 1–53 (1983) ■ Sekura et al. (eds.), Proc. Pertussis Toxin Conf. (1984), Orlando: Academic Press 1985 ■ TIPS **4**, 289 f. (1983); **5**, 277 ff. (1984); **7**, 429 f. (1986) ■ see also cholera toxin, diphtheria toxin. – *[CAS 82248-93-9]*

Peru balsam, Peru oil. P. balsam is a dark red, viscous but still fluid mass. It is an oil-containing resin secreted when the bark of the indigenous Central American tree species *Myroxylon pereirae* (Fabaceae) is injured. P. oil is a yellow to light brown, somewhat viscous oil from which the contained cinnamic acid partially crystallizes and has an adhering, sweet, balsamy odor and a warm bitter taste.

Production: P. oil is produced by high-vacuum distillation from the balsam.

Composition[1]***:*** The main components of the oil are *benzyl benzoate* (see benzoic acid) and *benzyl cinnamate* (H_5C_6–CH=CH–COO–CH_2–C_6H_5, $C_{16}H_{14}O_2$, M_R 238.29). Minor components include *nerolidol*, **cinnamic acid*, **benzoic acid*, and *vanillin* (see 3,4-dihydroxybenzaldehydes).

Use: P. oil is used in the production of perfumes, mainly as a fixative, e.g., in flowery, oriental compositions. P. b. is a strong contact allergen and is thus no longer used, the oil is free of sensitizers and thus can be used with some restrictions. P. oil is also used to improve the aromatic character of foods.

Lit.: [1] Fenaroli (2.) **1**, 435.

gen.: Arctander, p. 521, 523 ■ Bauer et al. (2.), p. 171. – *Toxicology:* Food Cosmet. Toxicol. **12**, 951, 953 (1974). – *[HS 1301 10, 3301 29; CAS 8007-00-9 (P.); 103-41-3 (benzyl cinnamate)]*

Pestalotin.

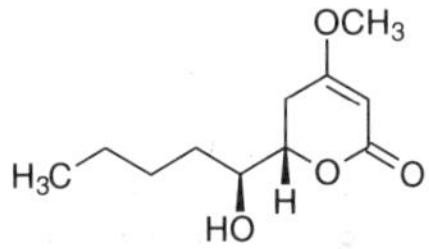

$C_{11}H_{18}O_4$, M_R 214.26, mp. 88–89 °C, $[\alpha]_D^{20}$ −86° (CH_3OH). A dihydro-α-pyrone from the phytopathogenic fungus *Pestalotia cryptomeriaecola* with synergistic activity to the *gibberellins. P. stimulates plant growth (see plant growth substances).

Lit.: Chem. Pharm. Bull. **39**, 1866 (1991) ■ J. Chem. Soc., Perkin Trans. 1 **1991**, 2627; **1992**, 693–700 (synthesis) ■ Tetrahedron: Asymmetry **8**, 3393 (1997) (synthesis). – *[CAS 34565-32-7]*

Petasalbin see eremophilanes.

Petasin.

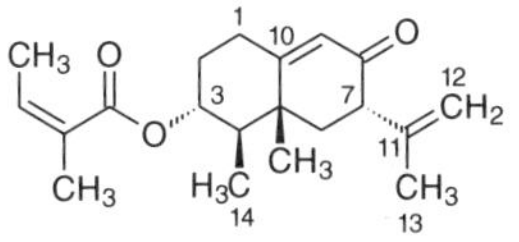

$C_{20}H_{28}O_3$, M_R 316.44, mp. 65–68 °C, $[\alpha]_D$ +49° ($CHCl_3$). Extracts of the Indian butterbur (*Petasites hybridus*) have been used for centuries as a cure for inflammations and spasms. Therapeutic applications were described in the first century by Dioskurides, the physician of Claudius and Nero, in his "De materia medica". The spasmolytically active components of the plants have been identified as the *eremophilane derivative P. and its 7,11-double bond isomer *isopetasin*. Both compounds inhibit the biosynthesis of *leukotrienes.

Lit.: Phytochemistry **11**, 3235 (1972); **22**, 1619 (1983) ■ Synthesis **1997**, 107 (synthesis (+) P.) ■ Tetrahedron Lett. **36**, 3325 (1995). – *[CAS 26577-85-5 (P.); 469-26-1 (iso-P.)]*

Petroporphyrins see geoporphyrins.

Petroselinic acid (*Z*)-6-octadecenoic acid]. H_3C–$(CH_2)_{10}$–CH=CH–$(CH_2)_4$–COOH, $C_{18}H_{34}O_2$, M_R 282.47, mp. 32–33 °C, bp. 215–217 °C (0.2–0.3 kPa); soluble in organic solvents. P. was first found as the glycerol ester in the seed oil of parsley (*Petroselinum crispum*); it is a characteristic component of the seed oils of umbelliferous and ivy plants. The fatty acids of these seed oils contain up to 87% P. (e.g. fat coriander oil, 75–80%). P. is also used as a starting material in oleochemistry. For nutritional-physiological aspects, see *Lit.*[1].

Lit.: [1] J. Nutrition **125**, 1563 (1995).

gen.: Acta Crystallogr., Sect. C **48**, 1054, 1057 (1992) ■ Fat Sci. Technol. **92**, 463 (1990) ■ J. Am. Oil Chem. Soc. **59**, 29 (1982) ■ Merck-Index (12.), No. 7330 ■ Murphy (eds.), Designer Oil Crops,. Weinheim: VCH Verlagsges. 1994 ■ Progr. Lipids Res. **33**, 155 (1994) (biosynthesis). – *[HS 2916 19; CAS 593-39-5]*

Petrosin.

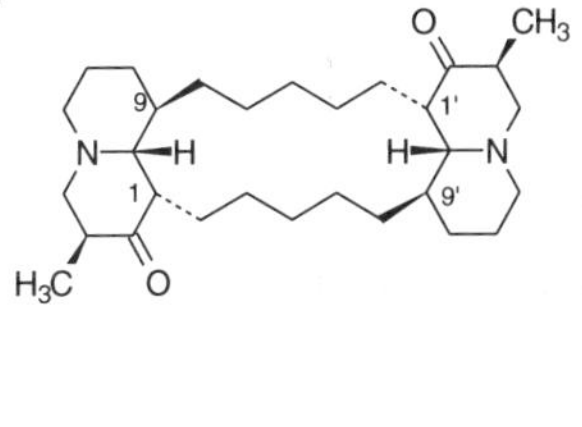

$C_{30}H_{50}N_2O_2$, M_R 470.74, mp. 215–216 °C. A *quinolizidine alkaloid from the sponge *Petrosia seriata* that is toxic to fish. P. occurs in 2 epimeric forms [1,9-diepimer (P. B) and the optically inactive *meso*-form existing as the 1,3,9,10-tetraepimer (P. A)].

Lit.: Bull. Soc. Chim. Belg. **93**, 941 (1984); **97**, 519 (1988) ▪ J. Am. Chem. Soc. **116**, 8853 (1994) (synthesis) ▪ J. Org. Chem. **63**, 5001, 5013–5030 (1998) (synthesis) ▪ Tetrahedron Lett. **23**, 4277 (1982); **30**, 4149 (1989). – *[CAS 84679-41-4 (P.); 95189-04-1 (P. B); 95189-03-0 (P. A)]*

Petunia steroids. The leaves of *Petunia* hybrids (Solanaceae) contain a tetrahydroxylated ergosta-1,4-dien-3-one existing as the orthoester of 2-(methylthiocarbonyl)acetic acid and named as *petuniasterone A* {$C_{32}H_{46}O_6S$, M_R 558.77, mp. 130–135 °C, $[\alpha]_D$ +52.1° ($CHCl_3$)}[1]. This compound is responsible for the potent growth retarding effects of the plant on the insect *Heliothis zea.*

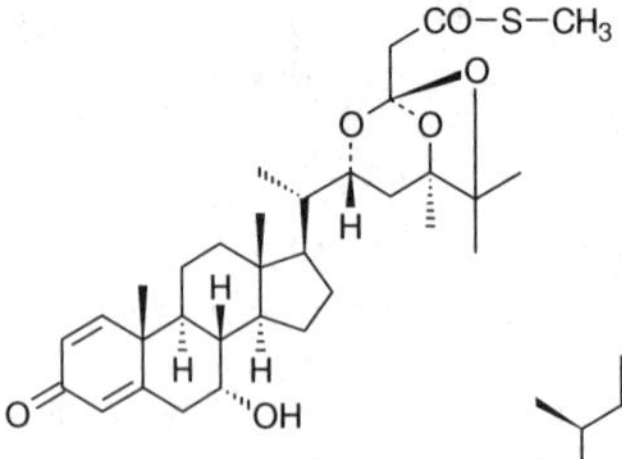

Petuniasterone A

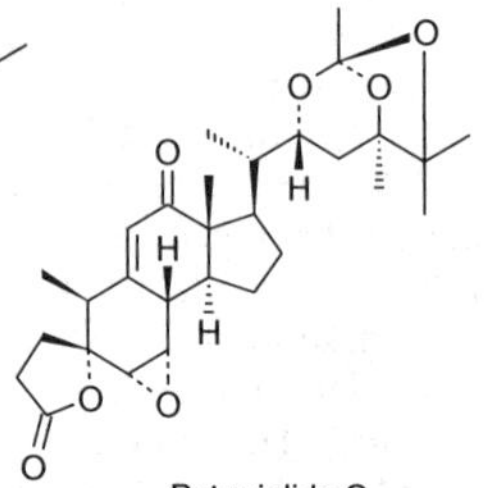

Petuniolide C

A compound with an even stronger inhibitory effect on insect growth, *petuniolide C* {$C_{29}H_{40}O_7$, M_R 500.63, mp. 235–238 °C, $[\alpha]_D$ +7° ($CHCl_3$)}, an *A*-nor-1,10-secosteroid lactone, has been found in *P. parodii* and *P. integrifolia*[2]. To date, about 25 P. s. with varying structures and the 7 petuniolides A–G have been isolated.

Lit.: [1] J. Chem. Soc., Perkin Trans. 1 **1988**, 711; Phytochemistry **27**, 3597 (1988); **29**, 2853 (1990). [2] J. Chem. Soc., Perkin Trans. 1 **1990**, 525.

gen.: Zeelen, p. 265–281. – *[CAS 114175-99-4 (petuniasterone A); 128255-52-7 (petuniolide C)]*

Petunidin (delphinidin 3′-methyl ether, 3,3′4′,5,7-pentahydroxy-5′-methoxyflavylium). Formula see anthocyanins, $C_{16}H_{13}O_7^+$, M_R 317.28, as chloride ($C_{16}H_{13}ClO_7$, M_R 352.73) gray-brown platelets or prisms (aqueous HCl). P. is isolated from *Petunia* hybrids; see anthocyanins.

Lit.: Karrer, No. 1742 ▪ Merck-Index (12.), No. 7331 ▪ Phytochemistry **38**, 1293 (1995) ▪ Tetrahedron **39**, 3005 (1983). – *[CAS 1429-30-7 (chloride)]*

Petunin (*petunidin 3,5-di-*O*-glucoside). $C_{28}H_{33}O_{17}^+$, M_R 641.56, violet platelets with a copper like gloss (aqueous HCl), mp. 178 °C, uv_{max} 540 nm (methanolic HCl). The *anthocyanin P. occurs in the grapevine (*Vitis rotundifolia*, Vitaceae), in the flowers of *Petunia* hybrids (Solanaceae), in blue grapes (*Vitis vinifera*, Vitaceae), in bilberries (*Vaccinium myrtillus*, Ericaceae), and other plants.

Lit.: Karrer, No. 1743. – *[CAS 25846-73-5]*

Petuniolide C see Petunia steroids.

Peyote (payote, peyotl, from Aztec.: nahuatl peyotl = caterpillar). A small woolly-hairy hedgehog cactus (*Lophophora williamsii*, Cactaceae) widely distributed in northern central Mexico. The most important constituents responsible for the hallucinogenic activity of P. include *Anhalonium or *cactus alkaloids such as *mescaline and the *isoquinoline alkaloids lophophorine and pellotine (see Anhalonium alkaloids). Slices of the plant ("mescal buttons") are eaten or extracts therefrom are drunk to achieve the narcotic state. The "Christian Native Church" uses P. for ritual purposes.

Lit.: Hager (5.) **3**, 775 ff. ▪ J. Nat. Prod. **36**, 1, 9, 36, 46 (1973) ▪ Naturwiss. Rundsch. **23**, 5–18 (1970) ▪ Rimpler et al., Biogene Arzneistoffe, p. 378, Stuttgart: Thieme 1990.

PG, PGI. Abbreviations for *prostaglandins and *prostacyclins.

PHA, PHB, PHF see poly(*β*-hydroxyalkanoates).

Phalaenopsine see pyrrolizidine alkaloids.

Phallotoxins. Together with the amatoxins (see amanitins) the P. are the toxic principles of the green and white death cups (*Amanita phalloides* and *A. verna*). The P. are responsible for the chronic components of death cup poisonings. They are amorphous, water-soluble solids. Seven naturally occurring P. are currently known.

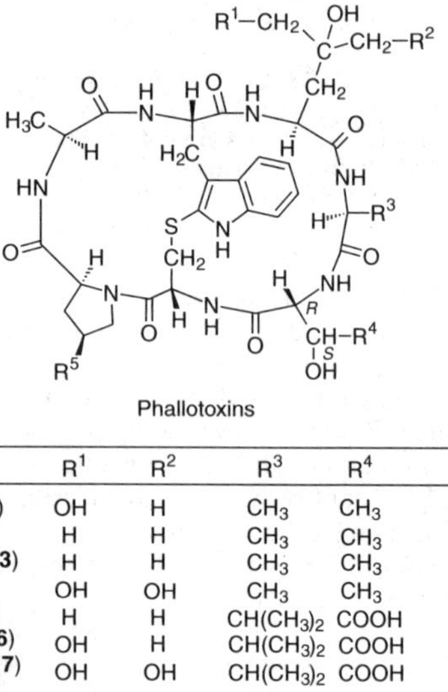

Phallotoxins

	R^1	R^2	R^3	R^4	R^5
Phalloidin (1)	OH	H	CH_3	CH_3	OH
Phalloin (2)	H	H	CH_3	CH_3	OH
Prophalloin (3)	H	H	CH_3	CH_3	H
Phallisin (4)	OH	OH	CH_3	CH_3	OH
Phallacin (5)	H	H	$CH(CH_3)_2$	COOH	OH
Phallacidin (6)	OH	H	$CH(CH_3)_2$	COOH	OH
Phallasacin (7)	OH	OH	$CH(CH_3)_2$	COOH	OH

Table: Data of Phallotoxins.

P.	molecular formula	M_R	LD_{50} (mouse i.p.) [mg/kg]	CAS
1	$C_{35}H_{48}N_8O_{11}S$	788.87	2.0	17466-45-4
2	$C_{35}H_{48}N_8O_{10}S$	772.87	1.5	28227-92-1
3	$C_{35}H_{48}N_8O_9S$	756.87	>100	67739-84-8
4	$C_{35}H_{48}N_8O_{12}S$	804.87	2.5	19774-69-7
5	$C_{37}H_{50}N_8O_{12}S$	830.91	1.5	53568-40-4
6	$C_{37}H_{50}N_8O_{13}S$	846.91	1.5	26645-35-2
7	$C_{37}H_{50}N_8O_{14}S$	862.91	4.5	58286-46-7

Phalloidin reversibly binds to cardiac actin and – in small doses – might be beneficial for treatment of heart muscle weakness.

Lit.: Bresinsky & Besl, Colour Atlas of Poisonous Fungi, Boca Raton: CRC Press 1989 ▪ Crit. Rev. Biochem. **5**, 185–260 (1978) ▪ J. Am. Chem. Soc. **112**, 3719 (1990) (conformation) ▪ J. Chromatogr. **462**, 442 (1989) (analytics) ▪ Justus Liebigs Ann. Chem. **1991**, 179 (conformation) ▪ Manske **40**, 189 (review) ▪ Naturwissenschaften **74**, 367 (1987) (review) ▪ Sax (8.), PCU 350 ▪ Wieland, Peptides of Poisonous Amanita Mushrooms, Berlin: Springer 1986 ▪ Zechmeister **25**, 214–250 ▪ see also amanitins.

Phaseolin (phaseollin).

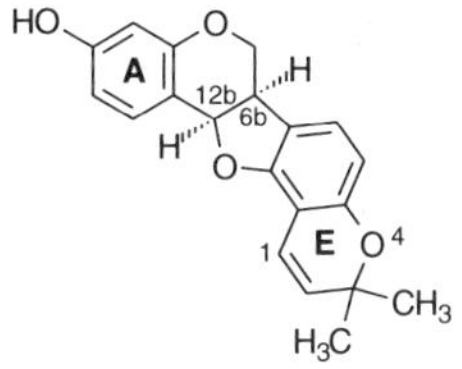

$C_{20}H_{18}O_4$, M_R 322.36, mp. 117–118 °C, $[\alpha]_D$ −145° (C_2H_5OH), pK_a 9.13. A *phytoalexin isolated from beans (*Phaseolus vulgaris*), *Erythrina abyssinica*, and *Vigna unguiculata*. P. is active against bacteria, yeasts, and fungi and is biogenetically related to the *pterocarpans derived from isoflavonoids. Ring E contains an isoprene unit.

Lit.: Beilstein EV **19/9**, 412 ▪ Hager (4.) **6a**, 561 ▪ Harbone (1994), p. 166–180 ▪ J. Chem. Soc., Perkin Trans. 1 **1987**, 431 (synthesis) ▪ Luckner (3.), p. 413–414 ▪ Phytochemistry **21**, 1599–1603 (1982) (biosynthesis). – *[CAS 13401-40-6]*

Phaseoloidin see homogentisic acid.

Phaseolotoxin {N^5-[amino(sulfoamino)phosphoryl]-L-ornithyl-L-alanyl-L-homoarginine}.

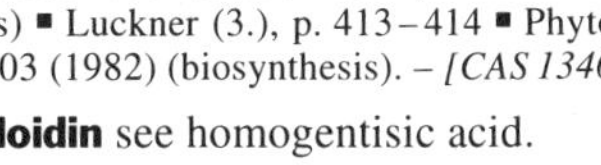

R = L-Ala-L-Homoarg-OH : Phaseolotoxin
R = OH : Octicidin

A phytotoxin from cultures of *Pseudomonas syringae* pv. *phaseolicola*; $C_{15}H_{34}N_9O_8PS$, M_R 531.52. In infected bean plants P. leads to chlorosis and accumulation of *L-ornithine. The reason for this is the irreversible inhibition of ornithine carbamoyltransferase (OCT), which converts ornithine to *citrulline. P. is enzymatically degraded by the plant to *octicidin* ($C_5H_{15}N_4O_6PS$, M_R 290.24), which has a ca. 20-fold stronger inhibitory effect on OCT and thus accelerates the damage.

Lit.: Tetrahedron Lett. **25**, 3931 (1984). – *[CAS 62249-77-8 (P.); 93289-64-6 (octicidin)]*

Phaseolunatin. Synonym for linamarin see cyanogenic glycosides.

Phellandrenes see *p*-menthadienes.

Phellodonic acid.

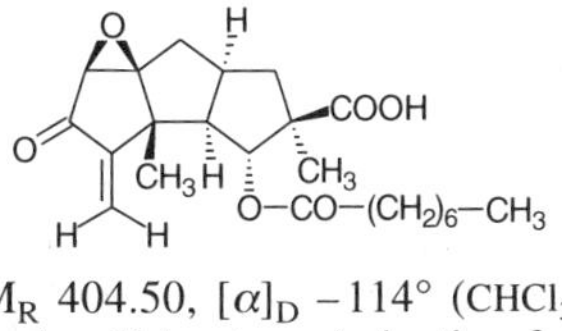

$C_{23}H_{32}O_6$, M_R 404.50, $[\alpha]_D$ −114° ($CHCl_3$). An antibiotically active *hirsutane derivative from mycelia cultures of the fungus *Phellodon melaleucus* (Basidiomycetes). The cytotoxicity as well as the strong antimicrobial and algicidal activities of P. result from the addition of nucleophiles to the activated methylene group.

Lit.: Z. Naturforsch. C **48**, 545 (1993). – *[CAS 152613-17-7]*

Phellogenic acid see Japan wax.

Phenalamides see myxalamides.

Phenalenones. A group of pigments occurring in plants of the family Haemodoraceae (Haemodoreae and Conostyleae) principally living in the southern hemisphere and, restricted to 4 genera of the Hyphomycetes (Fungi imperfecti) and the genus *Roesleria* within the class Discomycetes (Ascomycotina). The 4-phenylphenalenone *irenolone* as well as *4'-hydroxyanigorufone* and some related 9-phenylphenalenones are *phytoalexins of the banana plant (*Musa paradisiaca*) and can be isolated from leaves and banana skins of plants infected with the fungi [1]. The basic skeleton of fungal P. (e.g., *atrovenetin, herqueinone, herqueichrysin) is formed biosynthetically from acetate units, while mevalonate is the precursor of the annelated dihydrofuran ring. The plant P., in contrast to the fungal P., are derived structurally from 9-phenyl-1*H*-phenalen-1-one.

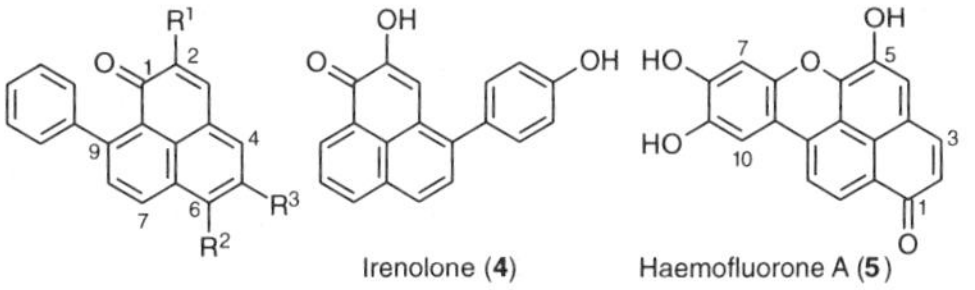

$R^1 = R^2 = R^3 = H$: 9-Phenyl-1*H*-phenalen-1-one
$R^1 = OH$, $R^2 = R^3 = H$: Anigorufone (**1**)
$R^1 = R^3 = OH$, $R^2 = OCH_3$: Anigozanthin (**2**)
$R^1 = R^2 = OH$, $R^3 = OCH_3$: Haemocorin aglycone (**3**)

Table: Data of Phenalenones.

	molecular formula	M_R	mp. [°C] (color)	λ_{max} [nm]	CAS
1	$C_{19}H_{12}O_2$	272.30	123–125 (orange)	424 (C_2H_5OH)	56252-32-5
2	$C_{20}H_{14}O_4$	318.33	149–151 (carmine red)	–	56252-33-6
3	$C_{20}H_{14}O_4$	318.33	277–278 (purple red)	505 (dioxane)	22138-90-5
4	$C_{19}H_{12}O_3$	288.30	285 (red needles)	390 (CH_3OH)	149184-19-0
5	$C_{19}H_{10}O_5$	318.29	>370 (black)	578 (CH_3OH)	72458-11-8

Haemocorin was isolated from *Haemodorum corymbosum* in 1955 as the first P. from a higher plant.

P. are toxic towards microorganisms [2] and animals. Haemocorin has tumor-inhibiting [3] and antibacterial activities [4].

Biosynthesis: Two phenylpropane units are linked through C-2 of acetate to furnish a diarylheptanoid which is incorporated entirely in plant P. through intramolecular Diels-Alder cyclization [5].

Lit.: [1] J. Org. Chem. **58**, 4306 (1993); Phytochemistry **41**, 753 (1996); **45**, 47 (1997). [2] Biochim. Biophys. Acta **286**, 88 (1972). [3] Arzneim. Forsch. **12**, 1143 (1962). [4] Current Sci. **37**, 288 (1968). [5] J. Chem. Soc., Chem. Commun. **1995**, 525.

gen.: Aust. J. Chem. **32**, 1841 (1979); **34**, 587 (1981) ▪ Biosci. Biotechnol. Biochem. **62**, 95 (1998) ▪ J. Org. Chem. **58**, 4306

(1993) ▪ Tetrahedron **34**, 3005 (1978) ▪ Turner **2**, 134–137 ▪ Zechmeister **40**, 153–190.

Phenanthroquinolizidine alkaloids see Tylophora alkaloids.

Phenazinomycin.

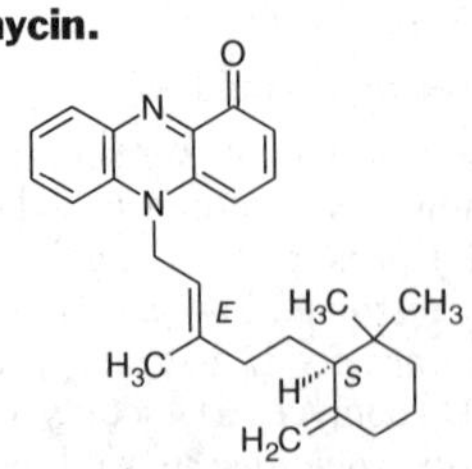

$C_{27}H_{32}N_2O$, M_R 400.56, dark blue needles, mp. 113–118 °C, $[\alpha]_D^{25}$ −49° (CH_3OH). The first example of a phenazinone sesquiterpene alkaloid, isolated from *Streptomyces* sp.

Lit.: J. Antibiot. **42**, 1037–1042 (1989) (isolation) ▪ Tetrahedron Lett. **38**, 4993–4996 (1997) (synthesis). – *[CAS 122898-63-9]*

Phenethyltetrahydroisoquinoline alkaloids. While the *benzyl(tetrahydro)isoquinoline alkaloids are built up form a C_6–C_2 unit and dopamine, the phenethyltetrahydroisoquinolines are formed from dopamine and a C_6–C_3 unit. Similar to the benzylisoquinolines, the compounds can undergo a series of secondary reactions to furnish the corresponding homo compounds. They can be classified into 7 different groups:

1) simple P. a., e. g. autumnaline
2) Homo-morphinandienones, e. g., kreysiginine and androcymbine
3) Bisphenethyltetrahydroisoquinolines, e. g., melanthioidine
4) Homoproaporphines, e. g., kreysiginone
5) Homoaporphines [1], e. g., floramultine, multifloramine, kreysigine
6) Homoerythrina alkaloids see Cephalotaxus alkaloids
7) Dibenz[*d,f*]azecines (see Cephalotaxus alkaloids) have been discussed as intermediates (formula II in the biosynthesis of cephalotaxin) of homoerythrinan-Cephalotaxus alkaloids.

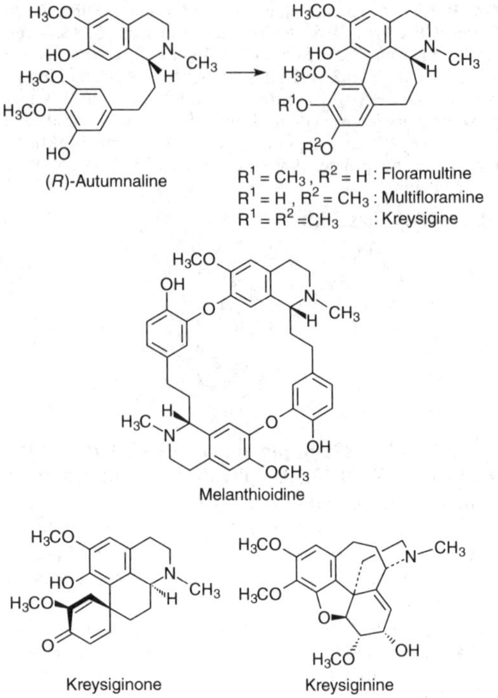

Figure 1: Various Phenethylisoquinoline derivatives and the biosynthesis of homoaporphines.

Their occurrence is limited to the Liliaceae, e. g., *Colchicum*, *Kreysigia*, and *Androcymbium* species (see table). They are readily soluble in water, chloroform, and ethanol. Colchicine is one of the few natural tropolone derivatives, see, e. g., fomentariol. It occurs together with other, structurally related alkaloids (colchamine, colchicoside, colchiceine, colchiciline, colchifoline) in meadow saffron (autumnal crocus, naked lady, *Colchicum autumnale*). Colchicine is a highly potent mi-

Table: Data of Phenethyltetrahydroisoquinoline alkaloids.

name	molecular formula	M_R	mp. [°C]	$[\alpha]_D$	occurrence	CAS
Androcymbine	$C_{21}H_{25}NO_5$	371.43	199–201	−260° ($CHCl_3$)	*Androcymbium melanthioides*	2115-98-2
Autumnaline	$C_{21}H_{27}NO_5$	373.45	166–168		*Colchicum cornigerum*, *C. visianii*, *C. latifolium*, *C. ritchii*	23068-65-7
Colchicine	$C_{22}H_{25}NO_6$	399.44	155–157	−121° ($CHCl_3$)	*C. autumnale* and other *Colchicum* species, *Merendera* species, *Gloriosa superba*	64-86-8
Demecolcine	$C_{21}H_{25}NO_5$	371.43	186	−129° ($CHCl_3$)	*C. speciosum*, *Merendera jolantae*, *M. persica*, *A. melanthioides*	477-30-5
Floramultine	$C_{21}H_{25}NO_5$	371.43	215–216	−111.5° ($CHCl_3$)	*Kreysigia multiflora*	21305-36-2
Kreysigine	$C_{22}H_{27}NO_5$	385.46	112–113	−65° ($CHCl_3$)	*C. cornigerum*	23117-57-9
Kreysiginine	$C_{21}H_{27}NO_5$	373.45	149	+89° (C_2H_5OH)	*K. multiflora*	19741-86-7
Kreysiginone	$C_{20}H_{23}NO_4$	341.41	193–194		*K. multiflora*	17441-87-1
β-Lumicolchicine	$C_{22}H_{25}NO_6$	399.44	183		*Colchicum* species	6901-13-9
Melanthioidine	$C_{38}H_{42}N_2O_6$	622.76	142–144	−63° ($CHCl_3$)	*A. melanthioides*	4085-28-3
Multifloramine	$C_{21}H_{25}NO_5$	371.43	215–216	−115.5° (CH_3OH)	*K. multiflora*, *C. szovitsii*	22324-15-8

tosis toxin [LD_{50} adults 20 mg, children 5 mg (corresponds to one seed)] and causes vomiting, nausea, paralysis of the central nervous system, and respiratory arrest some hours after consumption.
A colchicine poisoning is similar to a cholera-like diarrhea with vomiting. A special characteristic is the long incubation time. The intestinal effects are followed by an ascending paralysis. Colchicine interferes with the aggregation of tubulin and microtubilin which is important for mitosis. It can cause an elevated chromosome number (polyploidy) and giant growth in plants. It is used in the treatment of liver cirrhosis. Low doses have analgesic and anti-inflammatory effects in acute episodes of gout. Demecolcine is an antineoplastic and antimitotic agent. It inhibits cell division in the meta phase by destroying the microtubuli. Lumicolchicine is a photoisomer of colchicine. For synthesis, see *Lit.*[2].
Biosynthesis[3]***:*** Simple P. a. such as autumnaline are formed starting from tyrosine by combination of dopamine and a C_6–C_3 unit from phenylalanine via cinnamic acid. For the biosynthesis of colchicine, (*S*)-autumnaline in conformation I undergoes a phenol oxidative *para,para*-coupling to give the homomorphinedienone androcymbine (figure 2). Hydroxylation to II is followed by ring expansion and ring opening to *N*-formyldemecolcine, hydrolysis to demecolcine, and acetylation to colchicine.

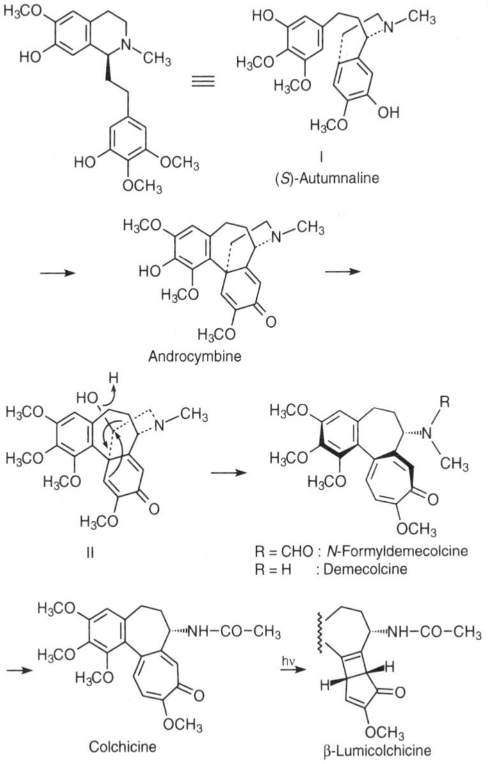

Figure 2: Biosynthesis of Colchicine.

The homoaporphines, e.g., floramultine, multifloramine, and kreysigine as well as the homoproaporphines are formed, similar to the bis-phenethylisoquinolines, by phenol oxidation from phenethyltetrahydroisoquinolines.

Lit.: [1] J. Nat. Prod. **52**, 909–922, 1055 (1989). [2] Aust. J. Chem. **47**, 957 (1994); J. Chem. Soc., Chem. Commun. **1992**, 974; J. Chem. Soc. Perkin Trans. 1 **1992**, 1415–1426. [3] Mothes et al., p. 234–238; J. Chem. Soc., Perkin Trans 1 **1998**, 2979, 2989, 2995, 3003; Tetrahedron Lett. **38**, 7357 (1997).
gen.: Beilstein EIV **14**, 946 (colchicine) ▪ Chem. Listy **86**, 445–460 (1992) (colchicine) ▪ Florey **10**, 139–182 ▪ Hager (5.) **3**, 338–339; **4**, 946–955; **7**, 336–339 ▪ Handb. Exp. Pharmacol. **72**, 569–612 (1984) ▪ Helv. Chim. Acta **82**, 323 (1999 (abs. configuration colchicine) ▪ J. Chem. Soc., Chem. Commun. **1992**, 974 ▪ J. Emergency Med. **12**, 171 (1994) (toxicology) ▪ J. Org. Chem. **63**, 2804 (1998) (synthesis colchicine) ▪ Manske **23**, 1–70; **36**, 172–223; **41**, 125–176; **53**, 288–362 (review colchicine) ▪ Pharmacother. **11**, 196 (1991); **12**, 171 (1994) (pharmacology) ▪ Sax (8.), CNG 830, CNX 800, MIW 500 ▪ Shamma, p. 458–482 ▪ Shamma-Moniot, p. 365–380. – *[HS 2939 90]*

Phenoxan.

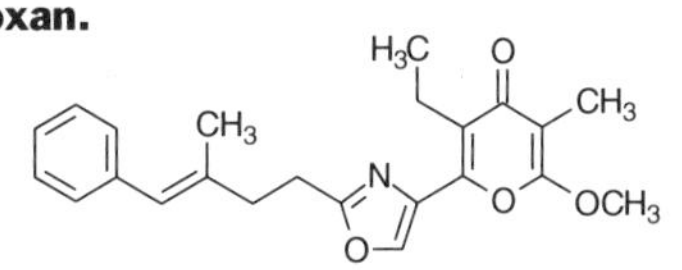

Antifungal and cytotoxic α-methoxy-γ-pyrone derivative from *Polyangium* sp. (Myxobacteria), $C_{23}H_{25}NO_4$, M_R 379.46. P. crystallizes from butyl methyl ether as colorless crystals with mp. 92–93 °C, uv_{max} 244 nm, 252 nm. The pyrone part of P. has a close relationship to *aureothin and spectinabilin from streptomycetes. In addition to the antifungal activity, e. g., against *Ustilago maydis* (MHK 19 µg/L) a cytotoxic activity has also been observed which mimics an antiviral activity in cellular assays. The reason for this is an inhibition of NADH ubiquinone oxidoreductase in complex I of the respiratory chain.

Lit.: Antivir. Chem. Chemother. **3**, 189 (1992) ▪ Eur. J. Biochem. **219**, 691 (1994) (activity) ▪ J. Antibiot. **45**, 1549 (1992) ▪ Justus Liebigs Ann. Chem. **1991**, 707. – *Synthesis:* J. Heterocycl. Chem. **34**, 1061 (1997) (analogues) ▪ J. Org. Chem. **61**, 4853 (1996) ▪ Tetrahedron Lett. **37**, 2997 (1996). – *[CAS 134332-63-1]*

Phenoxazin-3-one see cinnabarin.

Phenoxazone pigments. A group of naturally occurring pigments formally derived from 3*H*-phenoxazin-3-one (phenoxazone) by substitution and/or annelation of further rings, e. g., *actinomycins, *cinnabarin, *litmus, and *orcein pigments, *ommochromes.
Lit.: Ullmann (5.) **A 3**, 224, 228.

Phenylacetaldehyde. H_5C_6–CH_2–CHO, C_8H_8O, M_R 120.15. Oil with a green-flowery, hyacinth-like odor, easily undergoes polymerization or oxidation to phenylacetic acid; bp. 195 °C, LD_{50} (rat p.o.) 3.89 g/kg. P. occurs in small amounts in many *essential oils and flavors[1] including *tea flavor.
Use: In the perfume industry[2], in flavors, for the production of drugs, insecticides, and disinfectants.
Lit.: [1] TNO list (6.), Suppl. 5, p. 259. [2] Food Cosmet. Toxicol. **14**, 197 (1976); **17**, 377 (1979).
gen.: Bauer et al. (2.), p. 78 ▪ Bedoukian (3.), p. 360-369 ▪ Beilstein EIV **7**, 664 ▪ Ullmann (5.) **A 1**, 341. – *[HS 2912 29; CAS 122-78-1]*

Phenylacetic acid see honey flavor and ants.

Phenylacetonitrile see locustol.

L-Phenylalanine [(*S*)-2-amino-3-phenylpropanoic acid, abbr. Phe or F].

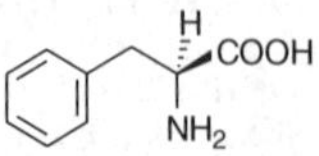

$C_9H_{11}NO_2$, M_R 165.19, monohydrate, mp. 283–284 °C (decomp.), sublimes in vacuum, $[\alpha]_D^{25}$ −34.5° (H_2O), −4.5° (5 m HCl), pK_a 1.83, 9.13, pI 5.48. An essential proteinogenic amino acid. Genetic code: UUU, UUC. Average content in proteins 3.5%[1]. P. is produced industrially from ovalbumin, lactalbumin, zein, and fibrin. The free form occurs in most plants and fungi.
Biosynthesis: In plants and microorganisms the aromatic amino acids Tyr, Phe, and Trp are formed via *shikimic acid. Phe is hydroxylated by phenylalanine 4-monooxygenase (EC 1.14.16.1) to tyrosine. Tyrosine is degraded to fumarate and acetoacetate. A genetic deficit of phenylalanine 4-monooxygenase causes phenylketonuria and pathologically reduced intelligence. The reason for this is the transamination of Phe to phenylpyruvate that is excreted. Furthermore, a genetic deficit of homogentisate 1,2-dioxygenase causes alkaptonuria which manifests as inflammations of the joints. The urine acquires a black color through oxidation of *homogentisic acid.
Under the action of phenylalanine ammonia lyase (EC 4.3.1.5) Phe is transformed to *trans*-*cinnamic acid. This plays an important role in plants as a precursor of numerous phenolic secondary products such as *flavonoids, *lignin, etc. D-Phe is a component of *gramicidin S and *tyrocidines. L-Phe tastes weakly bitter, D-Phe tastes sweet.
Lit.: [1]Biochem. Biophys. Res. Commun. **78**, 1018–1024 (1977).
gen.: Beilstein E IV **14**, 1552–1555 ▪ Gazz. Chim. Ital. **119**, 215 (1989) ▪ J. Org. Chem. **45**, 2010 (1980) ▪ Naturwissenschaften **70**, 115–118 (1983) ▪ Ullmann (5.) **A 2**, 58, 63, 72, 86; **A9**, 423–429. – *[HS 2922 49; CAS 63-91-2]*

2-Phenylbenzofuran see moracins.

2-Phenylethanethiol see sesame seeds (roasted) flavor.

2-Phenylethanol (phenethyl alcohol).
H_5C_6–CH_2–CH_2–OH, $C_8H_{10}O$, M_R 122.17. Liquid with a rose-like odor, bp. 219.8 °C, flash point 101 °C; soluble in most organic solvents and to 1.7% in water; LD_{50} (rat p.o.) 2.23 g/kg, olfactory threshold in water ca. 1 ppm[1]; weak bactericidal activity[2].
Occurrence: The main component of *rose absolute and rose water; occurs in small amounts in many flavors and *essential oils, e. g., *beer, *cocoa, and *tea flavor, as well as *geranium, neroli [see orange flower absolute (oil)], and *ylang-ylang oil. Frequently, P. occurs in glycosidic form[3].
Use: As an *aroma chemical in perfumes with rose and flower notes as well as a *flavor compounds.
Lit.: [1]Perfum. Flavor. **16** (1), 19 (1991). [2]Ullmann (5.) **A 8**, 555. [3]Biosci., Biotechnol., Biochem. **59**, 738 (1995); J. Agric. Food Chem. **42**, 1732 (1994); Nat. Prod. Lett. **10**, 39 (1997).
gen.: Bauer et al. (2.), p. 74 f. ▪ Bedoukian (3.), p. 370–382 ▪ Beilstein E IV **6**, 3067 ff. ▪ Fenaroli (2.) **2**, 461. – *[HS 2906 29; CAS 60-12-8]*

β-Phenylethylamine. Formula, data, and occurrence, see β-phenylethylamine alkaloids. Liquid with a fish-like odor, pK_a 9.96 (H_2O, 19 °C), it absorbs carbon dioxide from the air. β-P. is widely distributed in plants as a precursor of *benzyl(tetrahydro)isoquinoline alkaloids[1]. It occurs in bitter almond oil. It can also be detected as a biogenic amine in humans, especially in the brain and urine. β-P. has been associated with the development of feelings of passion and happiness. It has a blood-pressure lowering activity. *Tyramine* (*p*-hydroxy-β-phenylethylamine) (formula, data, and occurrence, see β-phenylethylamine alkaloids) is also widely distributed as a biogenic amine in plants. Tyramine is excreted in the urine of patients with Parkinson's disease. In insects it occurs in the free form as a neurotransmitter in the nervous system, and as a precursor of *N-acylcatecholamine in the form of the *O*-β-D-glucoside and the *N*-acetyl derivative as well as the peptide *N*-β-alanyltyramine in the hemolymph.
Biosynthesis: From L-*phenylalanine or L-*tyrosine by decarboxylation.
Lit.: [1]Nat. Prod. Rep. **11**, 555–576 (1994).
gen.: Adv. Insect Physiol. **15**, 317–473 (1980) ▪ Chem. Pharm. Bull. **29**, 3387 (1981) ▪ J. Med. Chem. **36**, 1711 (1993) ▪ J. Org. Chem. **56**, 457–459 (1991) (synthesis, IR, ^1H NMR) ▪ Mosnaim & Wolf, Noncatecholic Phenylethylamines (2 vols), New York: Dekker 1978, 1980 ▪ Science **215**, 1127 ff. (1982). – *[HS 2921 49; CAS 64-04-0]*

β-Phenylethylamine alkaloids. P. a. are found in many plant families; the Cactaceae, Fabaceae, Ephedraceae, Ranunculaceae, Celastraceae, Taxaceae, Malvaceae, Papaveraceae, Poaceae, and Amaryllidaceae (see table) are particularly rich in P. a. The P. a. are physiologically active compounds and thus of major importance. Epinephrine (*adrenaline) is well known as an adrenal hormone. The best known naturally occurring P. a. are the transmitters of the sympathetic nervous system, namely *adrenaline, *noradrenaline, and *dopamine. *β-Phenylethylamine has been connected

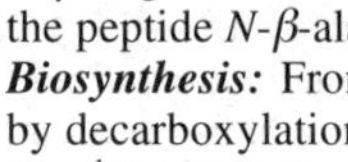

	R^1	R^2	R^3
β-Phenylethylamine	H	H	H
Hordenine	CH_3	CH_3	OH
N-Methyltyramine	H	CH_3	OH
Tyramine	H	H	OH

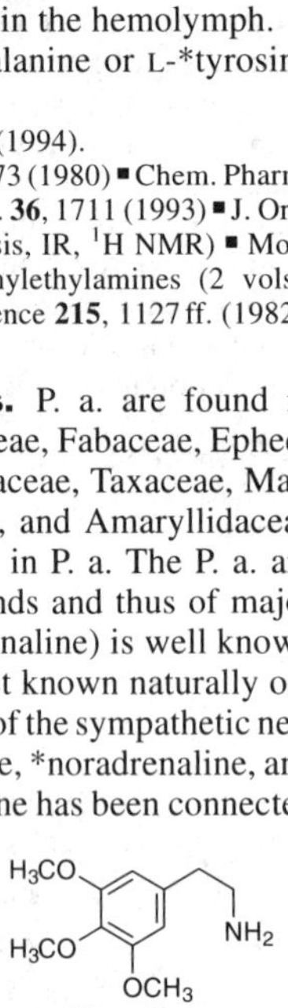

Mescaline

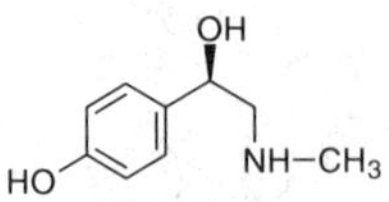

Synephrine

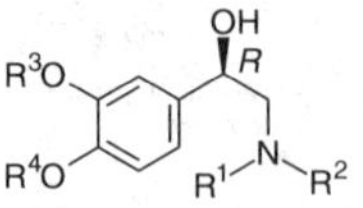

	R^1	R^2	R^3	R^4
Epinephrine (Adrenaline)	H	CH_3	H	H
Macromerine	CH_3	CH_3	CH_3	CH_3
Metanephrine	H	CH_3	CH_3	H
Normacromerine	H	CH_3	CH_3	CH_3
Norepinephrine (Noradrenaline)	H	H	H	H

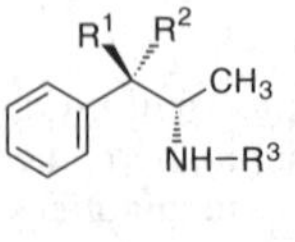

	R^1	R^2	R^3
Cathinone	=O		H
Ephedrine	OH	H	CH_3
Pseudoephedrine	H	OH	CH_3
Norephedrine	OH	H	H

Table: Data and occurrence of Phenylethylamine alkaloids.

name	molecular formula	M_R	mp. [°C]	$[\alpha]_D$	occurrence	CAS
Cathinone	$C_9H_{11}NO$	149.19		–26.5° (CH_2Cl_2)	*Catha edulis*, *Ephedra gerardiana*	71031-15-7
*Ephedrine	$C_{10}H_{15}NO$	165.24	40/225	–6.3° (C_2H_5OH)	*Ephedra* species, *Aconitum napellus*, *Catha edulis*, *Taxus baccata*, *Sida cordifolia*, *Roemeria refracta*	299-42-3
Epinephrine (*Adrenaline)	$C_9H_{13}NO_3$	183.21	216	–53° (1 M HCl)	*Coryphantha macromeris*	51-43-4
Hordenine	$C_{10}H_{15}NO$	165.24	118		*Anhalonium fissuratum*, *Hordeum vulgare*	539-15-1
Macromerine	$C_{12}H_{19}NO_3$	225.29	66–67.5	–147° ($CHCl_3$)	*Coryphantha macromeris*, *Cactus* species	19751-75-8
*Mescaline	$C_{11}H_{17}NO_3$	211.26	35–36		*Lophophora williamsii*, *Opuntia cyclindrica*, *Trichocereus* species	54-04-6
Metanephrine (Methyladrenaline)	$C_{10}H_{15}NO_3$	197.23	158–159		*Coryphantha macromeris*	5001-33-2
N-Methyltyramine	$C_9H_{13}NO$	151.21	130–131		*Hordeum vulgare*, *Panicum miliaceum*, *Lophophora williamsii*, *Coryphantha* and *Trichocereus* species	370-98-9
Norephedrine	$C_9H_{13}NO$	151.21	51	–15° (C_2H_5OH)	*Ephedra vulgaris*, *Catha edulis*	492-41-1
Norepinephrine (*Noradrenaline)	$C_8H_{11}NO_3$	169.18	216.5–218	–37.3° (C_2H_5OH)		51-41-2
Normacromerine	$C_{11}H_{17}NO_3$	211.26	101–103	–38.4° ($CHCl_3$)	*Coryphantha macromeris*, *C. calipensis*, *C. greenwoodii*, *Desmodium tiliaefolium*, *Dolichothele longimamma*	41787-64-8
*β-Phenylethylamine	$C_8H_{11}N$	121.18	197–198		*Acacia*, *Crataegus*, Fabaceae, Rosaceae, Cactaceae species	64-04-0
Pseudoephedrine	$C_{10}H_{15}NO$	165.24	117–118	+52° (C_2H_5OH)	*Ephedra vulgaris*, *Roemeria refracta*, *Sida cordifolia*	90-82-4
Synephrine	$C_9H_{13}NO_2$	167.21	162–164	–55.6° (1 M HCl)	*Citrus* and *Coryphantha* species, *Haloxylon salicornicum*	614-35-7
Tyramine	$C_8H_{11}NO$	137.18	164–164.5 (161°)		*Magnolia* and *Desmodium* species, *Pisum sativum*, *Hordeum vulgare*, *Lophophora williamsii*	51-67-2

with the etiology of depression and migraine. Many of the plants containing *β*-P. a. are used as food so that their consumption may induce physiologically significant effects such as migraine. The same holds for tyramine. Synephrine is a sympathicomimetic agent. The *β*-oxidized macromerine is purported to have hallucinogenic properties. Pseudoephedrine and norephedrine are sympathicomimetic substances with effects similar to those of *ephedrine. *Mescaline is one of the oldest known hallucinogenic substances. Cathinone is responsible for the stimulating properties of *kat. The LD_{50} for ephedrine are 400 mg/kg (mouse p.o.), 74 mg/kg (mouse i. v.) and 350 mg/kg (mouse i. p.); the LD_{50} for epinephrine (adrenaline) is approx. 50 mg/kg (mouse p.o.).

Biosynthesis[1]**:** Proceeds from tyrosine via tyramine or *dopamine to *hordenine or *mescaline, respectively, and macromerine. For the formation of ephedrine a C_6–C_1 unit reacts with an as yet unknown C_2 residue.

Lit.: [1] Mothes et al., p. 191–194.

gen.: Florey **7**, 193–229; **8**, 489–507; **15**, 233–281 ▪ Hager (5.) **1**, 740; **3**, 521 ff., 775 ff., 957; **5**, 46–57; **8**, 39–42, 45–48, 1195–1201 ▪ Heal & Marsden, The Pharmacology of Noradrenaline in the Central Nervous System, Oxford: University Press 1990 ▪ Manske **35**, 77–154 ▪ Sax (8.), AES 000 – AES 625, EAW 000 – EAY 500, MDI 500 – MDJ 000, NNM 000 – NNP 050 ▪ Ullmann (5.) **A 1**, 317. – *[HS 293940, 293990]*

β-Phenylethyl isothiocyanate see mustard oils.

3-Phenyl-1-propanol see hydrocinnamyl alcohol.

3-Phenylpropyl cinnamate see storax (oil).

Phenylpyruvic acid (2-oxo-3-phenylpropanoic acid).

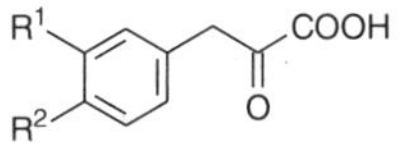

$R^1 = R^2 = H$: Phenylpyruvic acid
$R^1 = H$, $R^2 = OH$: 4-Hydroxyphenylpyruvic acid
$R^1 = R^2 = OH$: 3,4-Dihydroxyphenylpyruvic acid

$C_9H_8O_3$, M_R 164.16. Platelets, mp. 158 °C (decomp.), P. occurs in plants and microorganisms. It is excreted

in larger amounts by human patients with the hereditary disease phenylketonuria.
4-*Hydroxy-P.* [3-(4-hydroxyphenyl)-2-oxopropanoic acid, $C_9H_8O_4$, M_R 180.16, mp. 220 °C (sodium salt; decomp.)] is isolated from bacterial cultures of *Aerobacter aerogenes* and *Escherichia coli* and is also widely distributed in the plant kingdom. The oxime occurs in the marine sponge *Hymeniacidon sanguinea* [1]. 4-Hydroxy-P. and the related *3,4-dihydroxy-P.*, $C_9H_8O_5$, M_R 196.16, are central intermediates in the biosyntheses of numerous aromatic natural products, especially *alkaloids and constituents of fungi [2].
Biosynthesis: From *chorismic acid by enzyme-catalyzed rearrangement to *prephenic acid with subsequent dehydratation and decarboxylation by bifunctional chorismate mutase (EC 5.4.99.5 and 4.2.1.51). Oxidative decarboxylation of *chorismic acid leads to 4-hydroxyphenylpyruvic acid. Enzymatic transamination of L-*L-tyrosine or L-*DOPA furnishes 4-hydroxy-P. or 3,4-dihydroxyphenylpyruvic acid, respectively.

Lit.: [1] Experientia **31**, 756f. (1975). [2] Zechmeister **51**, 1–317.
gen.: Beilstein E IV **10**, 2760f. ▪ J. Chem. Soc., Perkin Trans. 1 **1975**, 2340–2348 (synthesis of 4-hydroxy-P.); **1984**, 1531–1537 (synthesis of P., IR, MS) ▪ Luckner (3.), p. 363 ff. ▪ Nat. Prod. Rep. **12**, 101–133 (1995). – *[CAS 156-06-9 (P.); 156-39-8 (4-hydroxy-P.); 4228-66-4 (3,4-dihydroxy-P.)]*

1-Phenyltetrahydroisoquinoline alkaloids. Rare structures that only occur in orchids (Orchidaceae), e.g., *cryptostyline I* {$C_{19}H_{21}NO_4$, M_R 327.4, mp. 101–102 °C, $[\alpha]_D$ +56° ($CHCl_3$)} from *Cryptostylis erythroglossa.*

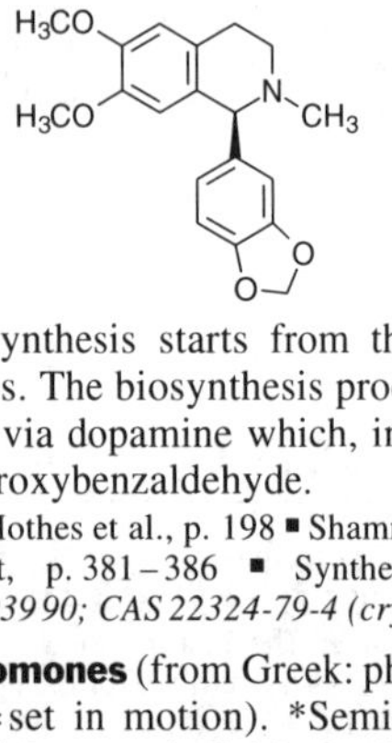

(+)-Cryptostyline I

The synthesis starts from the corresponding benzamides. The biosynthesis proceeds from tyrosine and Dopa via dopamine which, in turn, reacts with *3,4-dihydroxybenzaldehyde.

Lit.: Mothes et al., p. 198 ▪ Shamma, p. 490–494 ▪ Shamma-Moniot, p. 381–386 ▪ Synthesis **1982**, 486 (C. I). – *[HS 2939 90; CAS 22324-79-4 (cryptostyline I)]*

Pheromones (from Greek: pherein = to carry and hormon = set in motion). *Semiochemicals that are released by one individual of a species and cause a change in behavior or a physiological reaction in another member of the same species: In contrast to the *allomones P. have species-specific activities. P. occur in the communication systems of practically all organisms from unicellular species to mammals. Up to 1995 ca. 800 P. structures were known. Bitches on heat are supposed to give off *methyl p-hydroxybenzoate* as a sexual pheromone. The same substance also acts as an *attractant for certain swallowtail butterflies. The sexual pheromone of the hog is *5α-androst-16-en-3-one*, that of the elephant 7-dodecenyl acetate (see tetradecenyl acetate). In insects, in particular, P. play a central role and have thus been most intensively investigated. Extent and significance of pheromonal communication (especially that of the above mentioned steroid derivative which has also been detected in increased concentrations in axillary perspiration of human males) in humans is still a subject of controversy. The most effective human P. are probably *androsta-4,16-dien-3-one* and *1,3,5(10),16-estratetraene* [1]. P. are often mixtures of several substances the individual components of which have only relatively small effects (bouquet effect, synergism). In the case of chiral P. the enantiomeric ratio can have a decisive significance for the biological activity. The chemical structures of the P. are highly diverse and can originate from completely different biogenetic pathways. The following classes are known: *sexual pheromones* [2], that are secreted to attract a mating partner (especially well investigated for *butterflies among which P. are mostly produced by the females) and *aggregation pheromones* which are produced by only one gender of a species but attract both sexes, e.g., for the joint colonization of a habitat or to form groups (e.g., among bark beetles for the massive invasion of a host tree). *Trail pheromones* are used especially by *ants and *termites to mark the route between a source of food and the nest; they mostly consist of several components which may have widely differing chemical structures. Well investigated examples are *faranal and the methyl ester of *4-methylpyrrole-2-carboxylic acid* ($C_6H_9NO_2$, M_R 127.14) derived from leucine used by various leaf-cutting ants of the genus *Atta* [3] and *2-ethyl-3,6-dimethylpyrazine* ($C_8H_{12}N_2$, M_R 136.20) of *Atta sexdens rubropilosa* and some *Myrmica* species [4]. In the case of *Solenopsis* species (fire ants) sesquiterpenes and homo-sesquiterpenes play an important role. *Mellein and other dihydrocoumarins are components of the trail pheromone of *Formica*, *Lasius*, and *Campanotus* species [5]. Other examples of trail pheromones are the traces made by certain marine snails (*Navax* species), the compounds secreted by mammals to mark trails or territories, and the substances used by wild bees and bumblebees to mark areas as orientational aids (*terrestrol). P. may also be secreted as a warning signal (*alarm pheromones* of *ants and *bees) or as *decoy substances* in order to keep members of the same species away and thus prevent the overpopulation of an area. Synthetic P. of insect pests have been used with success in integrated plant protection schemes. P. may serve as a bait to attract the insects or in higher concentrations to disorientate the insects (for details, see insect attractants) [6]. For neuronal activity of P. and P. binding proteins see *Lit.* [7].

Lit.: [1] Dragoco Report **46**, no. 1, 4–28 (1999); Nature (London) **392**, 177 (1998); EP (application) 562843. [2] Arn, Toth & Priesner, List of Sex Pheromones and Sex Attractants (2.), Montfavet: Int. Org. for Biological Control. 1992. [3] J. Insect. Physiol. **18**, 809 (1972). [4] Naturwissenschaften **68**, 374 (1981). [5] Angew. Chem. Int. Ed. Engl. **31**, 795 (1992); Z. Naturforsch. C **49**, 865 (1994). [6] Chirality Agrochem. **1998**, 199–257. [7] Biochem. Biophys. Res. Commun. **250**, 217 (1998).
gen.: Agosta, Bombardier Beetles and Fever Trees, Chemical Warfare in Animals and Plants, New York: Addison Wesley 1995 ▪ Agosta, Chemical Communication, the Language of Pheromones, Oxford: Freeman 1992 ▪ Agosta, Dialog der Düfte, Heidelberg: Spektrum, Akadem. Verl. 1994 ▪ Albone, Mammalian Semiochemistry, The Investigation of Chemical Signals between Mammals, New York: Wiley 1984 ▪ ApSimon **4**, 1–183; **9** (synthesis) ▪ Asymm. Synth. **1**, 211–269 (1995)

▪ Barton-Nakanishi **8**, 197–262, 415–650 ▪ Bell & Cardé (eds.), Chemical Ecology of Insects, London: Chapman & Hall 1984 ▪ Brown & McDonald (eds.), Social Odours in Mammals (2 vols.), Oxford: Univ. Press 1984 ▪ Chem. Commun. **1997**, 1153 ▪ Chem. Ind. (London) **1994**, 370–373 ▪ Collect. Czech. Chem. Commun. **63**, 899–954 (1998) (synthesis) ▪ Eur. J. Org. Chem. **1998**, 13 ▪ Hummel & Miller (eds.), Techniques in Pheromone Research, Berlin: Springer 1984 ▪ Insect. Biochem. **29**, 481–514 (1999) (review) ▪ Lewis (ed.), Insect Communication, London: Academic Press 1984 ▪ Mayer & McLaughlin, CRC-Handbook of Insect Pheromones and Sex Attractants, Boca Raton: CRC-Press 1991 ▪ Prestwich & Blomquist (eds.), Pheromone Biochemistry, Orlando: Academic Press 1987 ▪ Vandenberg (ed.), Pheromones and Reproduction in Mammals, New York: Academic Press 1983 ▪ Zechmeister **37**, 1–190.

Phidolopin.

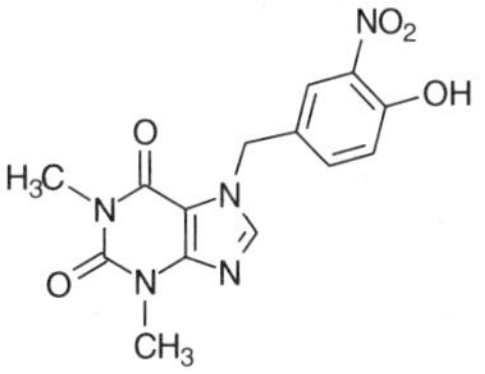

An unusual nitrophenol ($C_{14}H_{13}N_5O_5$, M_R 331.29, mp. 226.227 °C) isolated from the bryozoa *Phidolopora pacifica*. P. has antifungal and antibacterial activities and is also active against algae (e. g. the diatom *Cylindrotheca fusiformis*).

Lit.: Comp. Biochem. Physiol. **84B**, 43 (1986) ▪ Indian J. Chem. Sect. B **35**, 437 (1996) (synthesis) ▪ J. Org. Chem. **49**, 3869 f. (1984) ▪ Tetrahedron Lett. **26**, 2355 (1985) (synthesis). – *[CAS 92014-27-2]*

δ-Philanthotoxin (PTX-4.3.3).

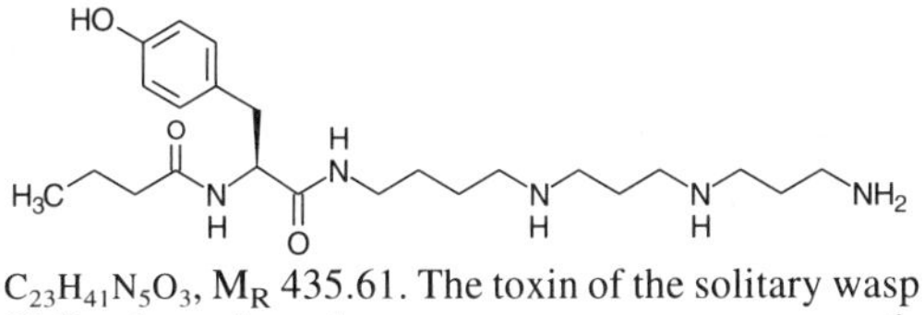

$C_{23}H_{41}N_5O_3$, M_R 435.61. The toxin of the solitary wasp *Philanthus triangulum* acts as a potent, non-competitive glutamate receptor antagonist (neurotoxic effect on prey insects, e. g., grasshoppers). The main component is the polyamine δ-P., which closely resembles the polyamines occurring in *spider venoms.

Lit.: J. Med. Chem. **39**, 515–521 (1996) ▪ Proc. Natl. Acad. Sci. USA **85**, 4910–4913 (1988) ▪ Pure Appl. Chem. **62**, 1223–1230 (1990) ▪ Tetrahedron **46**, 3267–3286 (1990); **49**, 5777–5790 (1993) ▪ Tetrahedron Lett. **36**, 9401 (1995). – *[CAS 115976-91-5]*

Phlebiakauranol.

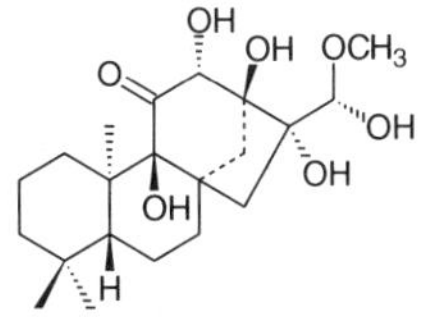

$C_{21}H_{34}O_7$, M_R 398.50, cryst., optically active. An antibacterial diterpenoid with kaurane skeleton from cultures of the basidiomycete *Phlebia strigoso-zonata*.

Lit.: J. Chem. Soc., Chem. Commun. **1975**, 406. – *[CAS 57743-94-9]*

Phlebiarubrone. $C_{19}H_{12}O_4$, M_R 304.30, red needles, mp. 248–250 °C. A *terphenylquinone from cultures

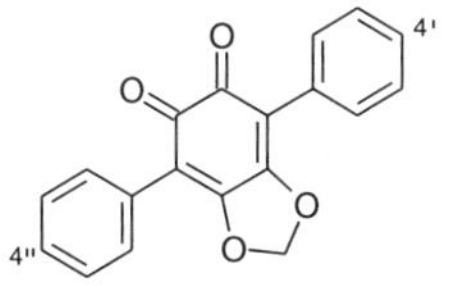

of *Phlebia strigoso-zonata* and *Punctularia atropurpurascens* (Basidiomycetes). P. is derived biosynthetically from *polyporic acid. The violet 4′-hydroxy-, 4′,4″-dihydroxy-, and 3′,4′,4″-trihydroxy derivatives of P. isolated from cultures of *P. atropurpurascens* are formed from the parent compound by sequential hydroxylation. The hydroxy derivatives are cytotoxic and show weak antibiotic activities.

Lit.: DECHEMA-Monogr. **129**, 9 (1993) ▪ Tetrahedron Lett. **1974**, 1645 ▪ Z. Naturforsch. C **39**, 695 (1984) ▪ Zechmeister **51**, 18 f. (review). – *[CAS 7204-23-1]*

Phlegmacins.

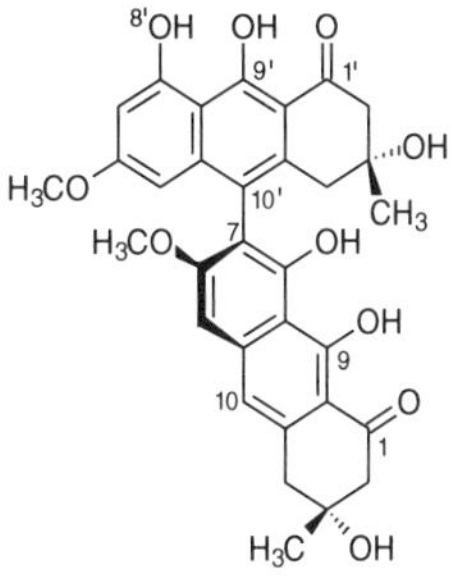

Phlegmacin A_1

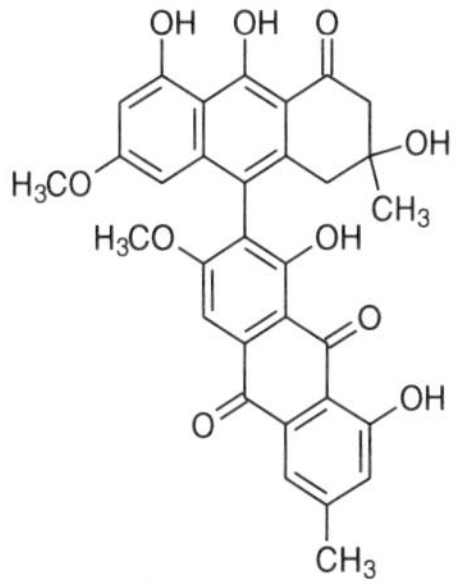

Anhydrophlegmacin-9,10-quinone

Dimeric *preanthraquinone pigments from plants (*Cassia torosa*, Fabaceae)[1] and agarics of the genus *Cortinarius*, subgenus *Phlegmacium* (Basidiomycetes)[2,3]. The two halves of the molecule (*torosachrysone) are linked by a 7,10′-bond (in contrast to the 7,7′-linked *flavomannins). On account of the two chiral centers and the chiral axis, 8 stereoisomers are possible ($C_{32}H_{30}O_{10}$, M_R 574.58, yellow powder); among these the stereoisomers with the same absolute configurations at the diaryl bond can be correlated on the basis of their CD spectra (A-series: negative Cotton effect to longer wavelengths and positive Cotton effect to shorter wavelengths, B-series: reversed)[3]. Pure P. B_1 occurs in *Cortinarius aureoturbinatus*, a mixture of the atropisomers A_1 and B_1 in *C. odorifer*. *Cassia torosa* contains the atropisomers P. A_2 and B_2, which are enantiomeric with the fungus pigments B_1 and A_1. *O*-Methylation of P. in the 8′-position, dehydratation with formation of an anthrone, oxidation to the 9,10-anthraquinone (e.g., *anhydrophlegmacin-9,10-quinone*, $C_{32}H_{26}O_{10}$, M_R 570.55, orange crystals, optically active) and dimerization at the 10-position leads to further compounds of this type.

Lit.: [1] Z. Naturforsch. B **27**, 1286 (1972); C **28**, 354 (1973); Sydowia **37**, 284 (1987). [2] Phytochemistry **16**, 999 (1977). [3] J. Chem. Soc., Perkin Trans. 1 **1999**, 119; Zechmeister **51**, 150–167 (reviews). – *[CAS 40501-61-9 (P.); 64233-77-8 (P. A_1); 64233-78-9 (P. B_1); 64233-73-4 (P. A_2); 64233-72-3 (P. B_2)]*

Phloeodictines. The sponge *Phloeodictyon* sp. contains ca. 15 cyclic aminoketals with a 2,3,4,6-tetrahy-

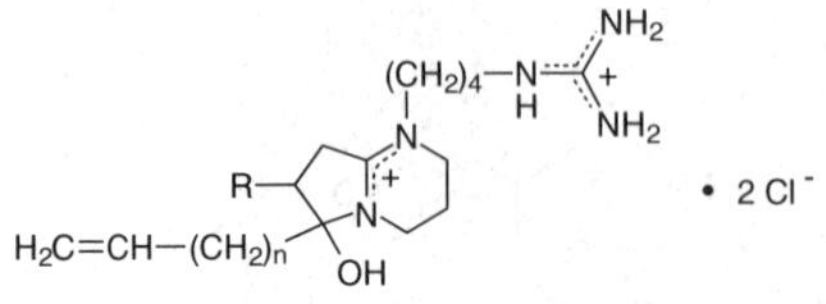

P. A : n = 12, R = H

P. B : n = 10, R = 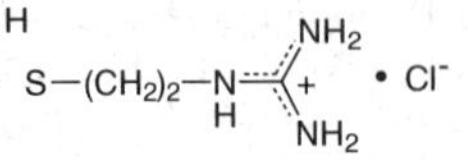

dro-*6H*-pyrrolo[1,2-*a*]pyrimidinium ring system. The compounds are cytotoxic to KB nasopharyngeal tumor cells (IC_{50} 1.5 μg/mL or 11.2 μg/mL). *P. A*, amorphous, $C_{26}H_{49}Cl_2N_5O$, M_R 518.61; *P. B*, $C_{27}H_{53}Cl_3N_8OS$, M_R 644.19.

Lit.: J. Org. Chem. **57**, 3832 (1992) ▪ Tetrahedron **50**, 3415 (1994). – *[CAS 142260-80-8 (P. A dichloride); 142260-81-9 (P. B trichloride)]*

Phloretin [dihydronaringenin, phloretol, 3-(4-hydroxyphenyl)-1-(2,4,6-trihydroxyphenyl)-1-propanone].

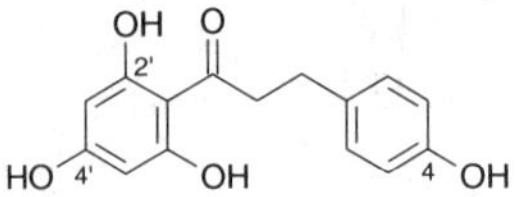

A dihydrochalcone from apples (*Malus* spp.), Japanese lavender heather (*Pieris japonica*), $C_{15}H_{14}O_5$, M_R 274.26, needles, mp. 262 °C, and other plants; the 2′-*O*-*β*-glucopyranoside is called *phloridzin*, $C_{21}H_{24}O_{10}$, M_R 436.42, sweet tasting needles, as dihydrate, mp. 110 °C, $[\alpha]_D^{25}$ −52° (96% alcohol). In apples the compounds, like *chlorogenic acid, *quercetin, *catechin, play a role in ripening processes and browning [1]. They inhibit the membrane transport of glucose [2], normalize an elevated glucose level in blood [3], and support the renal excretion of glucose. They may thus be useful in the treatment of diabetes, certain tumors [4], and malaria [5].

Biosynthesis: Reduction of the chalcone glycoside.

Lit.: [1] J. Agric. Food Chem. **38**, 945 ff. (1990); **39**, 1050 (1991). [2] J. Plant Growth Regul. **10**, 33 (1991); Plant Physiol. **85**, 711 (1987). [3] J. Clin. Invest. **87**, 561 (1991). [4] Anticancer Res. **13**, 2287–2299 (1993); Mutat. Res. **287**, 261–274 (1993). [5] Mol. Pharmacol. **45**, 446 (1994).

gen.: Harborne (1994), p. 401–406 ▪ Luckner (3.), p. 413, 470 ▪ Nat. Prod. Rep. **12**, 101–133 (1995) ▪ Synth. Commun. **19**, 119–123 (1989). – *[CAS 60-82-2 (P.); 60-81-1 (phloridzin)]*

Phloridzin see phloretin.

Phlorin(s).

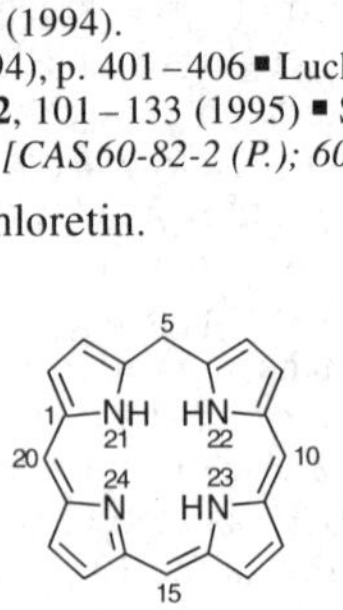

Phlorin

P. is the 5,22-dihydro-*porphyrin skeleton isomeric to *chlorin in which the conjugation is interrupted; it plays a role in the biosynthesis of porphyrin and the total synthesis of *chlorophyll. Close relatives of P. are the porphomethenes in which the conjugation is also interrupted and which represent to the *bacteriochlorins and *isobacteriochlorins isomeric tetrahydroporphyrins. The term P. is also used for the glucopyranoside of 1,3,5-benzenetriol.

Lit.: Dolphin **II**, 1–90; **V**, 36 ff., 140 ▪ Smith, p. 614 f., 704. – *[CAS 26660-92-4]*

Phloroglucinol (1,3,5-trihydroxybenzene).

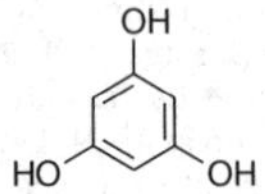

$C_6H_6O_3$, M_R 126.11; rhombic crystals (with 2 mol of water), mp. (anhydrous) 218–221 °C, dihydrate: 117 °C, soluble in ether, ethanol, pyridine, poorly soluble in water. In the light P. changes its color to brown, then black, it reduces Fehling's solution when warmed, and irritates the skin and mucous membranes. P., which is isomeric with pyrogallol, can be obtained by degradation of plant polyphenols (*tanning agents, *flavones, and *anthocyanins). P. occurs in the free form in *Eucalyptus* and *Acacia* species (*E. kino*, *A. arabica*). P. has also been isolated from streptomycetes cultures.

Use: For the detection of *lignin, pentoses, free HCl in gastric juices, nitriles, methylenedioxy groups, metals and to differentiate between aromatic and aliphatic aldehydes, intermediate in the synthesis of black pigments, drugs, liquid crystals, and plastics, additive in galvanizing baths, in diazo printing, in microscopic studies to decalcify bones, and as an optical brightener.

Lit.: Beilstein E IV **6**, 7361 ▪ Helv. Chim. Acta **68**, 1251–1275 (1985) ▪ Kirk-Othmer (3.) **18**, 684–691 ▪ Merck-Index (12.), No. 7482 ▪ Ullmann (5.) **A 19**, 347 f. – *[HS 2907 29; CAS 108-73-6]*

Phoenicin see oosporein.

Phomactin A see fungi.

Phomins see cytochalasins.

Phomopsins.

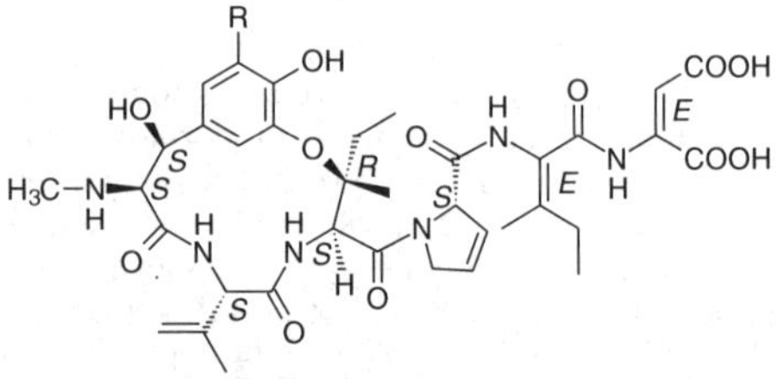

R = Cl : Phomopsin A
R = H : Phomopsin B

P. A: $C_{36}H_{45}ClN_6O_{12}$, M_R 789.24, mp. (decomp.) 205 °C; *P. B:* $C_{36}H_{46}N_6O_{12}$, M_R 754.79. P. A and B are formed by *Phomopsis leptostromiformis*, an endophytic fungus living on lupine species both in the plant and in culture. P. are responsible for the poisoning of grazing animals ("lupinosis"). The animals lose their appetite and develop jaundice with enlarged, fatty livers. Long-duration administration of lower doses, on the other hand, leads to liver cirrhosis. At the molecular biology level an interaction with tubulin or, respectively, a depolymerization of microtubuli is observed.

Lit.: J. Antibiotics **50**, 890 (1997) ▪ J. Appl. Toxicol. **9**, 83 (1989) ▪ J. Chem. Soc., Chem. Commun. **1983**, 1259; **1986**, 1219 ▪ Sax (8.), PGR 775 ▪ Steyn & Vleggaar (eds.), Mycotoxins and Phycotoxins, p. 169–184, Amsterdam: Elsevier

1986 ▪ Tetrahedron **45**, 2351 (1989). – *[CAS 64925-80-0 (P. A); 64925-81-1 (P. B)]*

Phoracantholide see longhorn beetles.

Phorbol see tiglianes.

Phorboxazoles. *P. A* {($C_{53}H_{71}BrN_2O_{13}$, M_R 1024.04, pale yellow solid, $[\alpha]_D^{44}$ +44.8° (CH_3OH)} and its (13*S*)-epimer *P. B* are a new class of biologically highly active macrolides from marine sponges (*Phorbas*). Together with the *spongistatins (altohyrtins) and the *bryostatins they are amongst the most cytotoxic natural products. P. inhibit the growth of tumor cells at sub-nanomolar concentrations (mean GI_{50} 1.58×10^{-9} mol/L). Unlike antimitotic products such as the *taxols and *epothilones the P. arrest the cell cycle during the S phase.

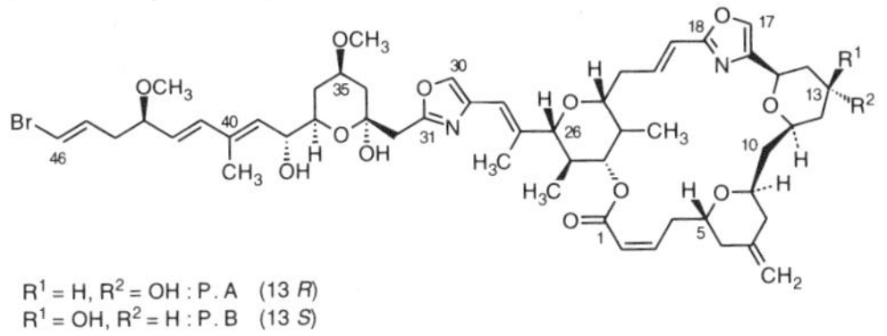

Lit.: J. Am. Chem. Soc. **117**, 8126–8131 (1995). – *Synthesis:* J. Am. Chem. Soc. **120**, 5597 f. (1998) ▪ Synthesis **1999**, 797 ▪ Tetrahedron **55**, 1905–1914 (1999) ▪ Tetrahedron Lett. **37**, 7879 (1996); **40**, 2291, 4527–4530 (1999). – *[CAS 165883-76-1 (P. A); 165689-31-6 (P. B)]*

Phosphatidylinositol 4,5-diphosphate see inositols.

Phosphinothricin see bialaphos.

Phosphoenolpyruvic acid (2-phosponooxyacrylic acid, abbr.: PEP).

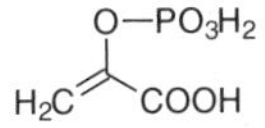

$C_3H_5O_6P$, M_R 168.04, exists in physiological solution as the dianion, it is a central metabolic product of glycolysis, gluconeogenesis, and the biosynthesis of *shikimic acid. Free PEP is not stable; stable derivatives include the crystalline metal salts, barium silver PEP trihydrate, monosodium PEP hydrate, calcium bis-PEP dihydrate.

Lit.: Arch. Microbiol. **162**, 187–192 (1994) ▪ Beilstein EIV **3**, 977 f. ▪ Chemtracts Org. Chem. **1**, 482 ff. (1988) (review) ▪ FEMS Microbiol. Rev. **63**, 193–200 (1989) (review) ▪ J. Am. Chem. Soc. **111**, 8920 f., 9233 f. (1989) (synthesis) ▪ J. Biochem. **25**, 1–5 (1993) ▪ Justus Liebigs Ann. Chem. **1992**, 1201 ff. ▪ Tetrahedron: Asymmetry **3**, 673 ff. (1992) ▪ Voet & Voet, Biochemie, p. 404–410, 561–568, Weinheim: VCH Verlagsges. 1992. – *[CAS 138-08-9]*

Phosphomycin see fosfomycin.

Phosphonothrixin.

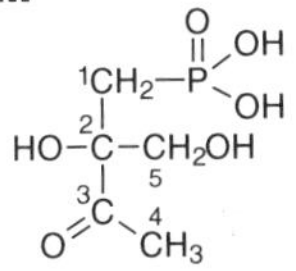

$C_5H_{11}O_6P$, M_R 198.11, syrup, $[\alpha]_D^{22}$ –4° (H_2O) as the sodium salt. Natural P. has the (*S*) configuration. A herbicidal antibiotic from cultures of *Saccharothrix* sp. P. is formed biosynthetically as an isoprenoid. A further herbicidal phosphonic acid is phosphinothricin (see bialaphos and ciliatine).

Lit.: J. Antibiot. **48**, 1124, 1130, 1134 (1995) (isolation) ▪ Phosphorus, Sulfur, Silicon Relat. Elements **1999**, 144, 613 (activity synthesis) ▪ Tetrahedron Lett. **38**, 437 (1997); **39**, 6621 (1998) (synthesis). – *[CAS 161803-17-4]*

Phosphoramidon {*N*-[*N*-[(6-deoxy-α-L-mannopyranosyloxy)hydroxyphosphoryl]-L-leucyl]-L-tryptophan}.

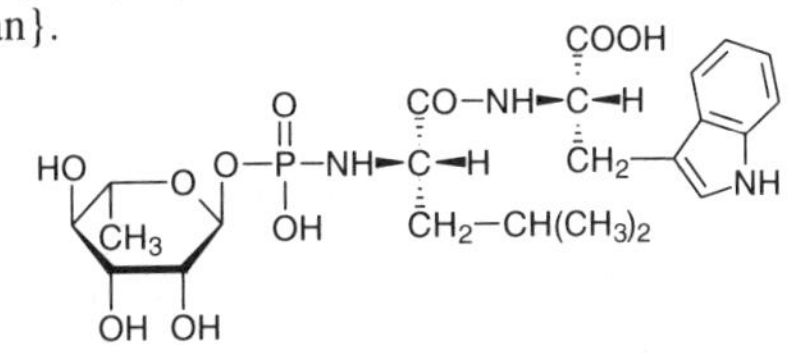

$C_{23}H_{34}N_3O_{10}P$, M_R 543.51, mp. of Na salt: 173–178 °C (decomp.). An antibiotic from cultures of *Streptomyces tanashiensis* and other actinomycetes. P. is an inhibitor of thermolysin, encephalinase, and endothelin converting enzyme. It inhibits the spontaneous metastasis of certain tumor cells. P. is related to talopeptin, a metalloproteinase inhibitor. For total synthesis, see *Lit.*[1].

Lit.: [1] Tetrahedron Lett. **36**, 1435 (1995).
gen.: Anticancer Res. **4**, 221 (1984). – *[HS 2941 90; CAS 36357-77-4]*

Phosphorus-carbon compounds (natural products with a C–P bond). Many natural products contain phosphorus-oxygen-bonds, espec. metabolites of the primary metabolism. A few only contain the C–P bond, many of those show antibiotic activities. The first natural product of the C–P-type was aminoethylphosphonic acid (AEP), discovered in 1959. Others, covered in this book are *bialaphos, *fosfazinomycin, *fosfomycin, *fosfonochlorin, phosphinothricin, *phosphonothrixin and the *rhizocticins. For a complete overview, see *Lit.*[1].

Lit.: [1] Tetrahedron **55**, 12237–12273 (1999).
gen.: Nat. Prod. Rep. **16**, 589 (1999).

Photodermatitis see furocoumarins.

Photoprotoporphyrin see chlorocruoroporphyrin.

Phrymarolins. Sesquilignans from the Asian herb *Phryma leptostachya* with insecticidal activity. They are structurally related to the *haedoxans.

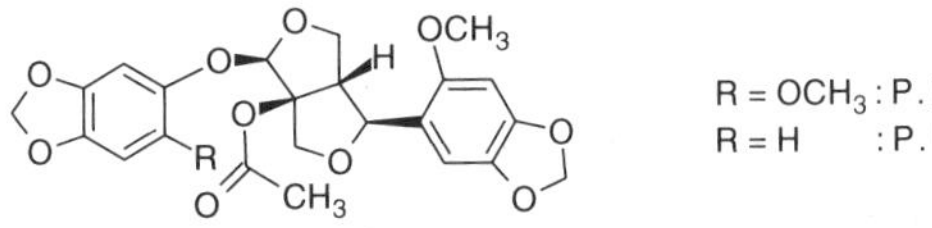

P. I: $C_{24}H_{24}O_{11}$, M_R 488.44, cryst., mp. 155–157 °C, $[\alpha]_D$ +131.3° (dioxane); *P. II:* $C_{23}H_{22}O_{10}$, M_R 458.42, cryst., mp. 161–162 °C, $[\alpha]_D$ +117.6° (dioxane).

Lit.: Bull. Chem. Soc. Jpn. **61**, 4361 (1988) ▪ Chem. Lett. **1989**, 313. – *[CAS 38303-95-6 (P. I); 23720-86-7 (P. II)]*

Phthalide isoquinoline alkaloids. The P. a. contain a tetracyclic system with a γ-lactone ring. They are formed principally by plants of the Berberidaceae, Fumariaceae, Papaveraceae, and Ranunculaceae families. Thus, for example, *(–)-β-hydrastine* {$C_{21}H_{21}NO_6$, M_R 383.40, mp. 133–135 °C, $[\alpha]_D$ –68° ($CHCl_3$)} occurs in *Hydrastis canadensis* and *Berberis laurina*, while *narcotine* {$C_{22}H_{23}NO_7$, M_R 413.43, mp. 176 °C, $[\alpha]_D$ –200° ($CHCl_3$)} is mainly obtained from *Corydalis* species as well as *Papaver somniferum* and other

Papaver species. The P. a. are soluble in hot acetone and readily soluble in chloroform.

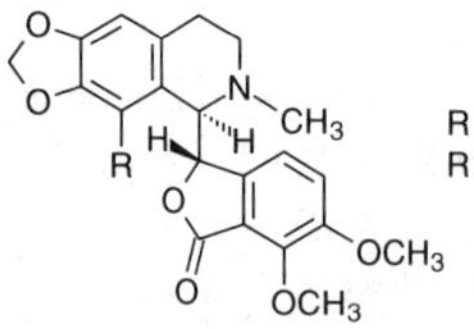

R = H : (–)-β-Hydrastine
R = OCH_3 : Narcotine (Noscapine)

Hydrastine effects a moderate slowing down of cardiac activity as a result of its sympathicolytic properties. In ophthalmology it is used to dilate the pupils and as a local anesthetic. On account of its vasoconstricting effects it was used in medicine as a hemostatic agent. At higher doses it causes *strychnine-like toxic symptoms [LD_{50} (rat p.o.) 1 g/kg, (rat i.p.) 104 mg/kg]. The activity of narcotine, a major alkaloid of *opium, resembles that of thebaine, but with the reflex-stimulating properties being more pronounced than the narcotic activity.

Biosynthesis[1]**:** P. a. are formed from *benzyl(tetrahydro)isoquinoline alkaloids such as reticuline via the *protoberberine alkaloid (*S*)-scoulerine, probably by way of ophiocarpine.

Lit.: [1]Mothes et al., p. 230 f.

gen.: Acta Crystallogr., Sect. C **49**, 504 (1993) ▪ Florey **11**, 407–461 ▪ Hager (5.) **8**, 1214 ff. ▪ J. Org. Chem. **48**, 1621 (1983) ▪ Manske **17**, 385–544; **24**, 253–286 ▪ Mutagenesis **7**, 205 (1992); **8**, 311 (1993) (narcotine) ▪ Sax (8.), HGQ 500, HGR 000, HGR 500, NBP 275, NOA 000, NOA 500 ▪ Shamma, p. 360–379 ▪ Shamma-Moniot, p. 307–323 ▪ Tetrahedron Lett. **30**, 869 (1989) (hydrastine). – *[HS 2939 90; CAS 118-08-1 (hydrastine); 128-62-1 (narcotine)]*

Phthiocol (2-Hydroxy-3-methyl-1,4-naphthoquinone).

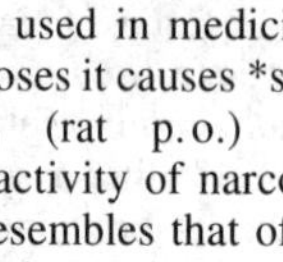

$C_{11}H_8O_3$, M_R 188.18, orange needles, mp. 179–180 °C, poorly soluble in water, soluble in organic solvents. An antibiotic from various bacteria (e. g., *Mycobacterium tuberculosis*), which is presumably derived from menaquinone (see vitamin K_1); it has also been described as a metabolite of higher plants (e. g., *Asplenium indicum*).

Lit.: Analysis **10**, 390 (1982) ▪ Curr. Sci. **45**, 44 (1976) ▪ J. Nat. Prod. **47**, 901 (1984). – *Synthesis:* Beilstein E IV **8**, 2375 ▪ Synth. Commun. **25**, 1669 (1995). – *[HS 2941 90; CAS 483-55-6]*

Phthoxazolin A (inthomycin A).

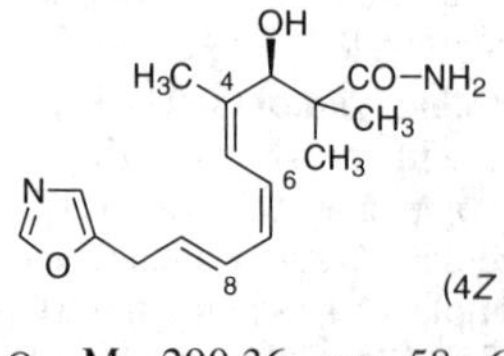

(4*Z*,6*Z*,8*E*)

$C_{16}H_{22}N_2O_3$, M_R 290.36, mp. 58–62 °C, $[\alpha]_D^{20}$ +82° (acetone). A specific inhibitor of the biosynthesis of *cellulose from *Streptomyces* sp. OM-5714. P. is a substructure of oxazolomycin.

Lit.: J. Antibiot. **43**, 1034 (1990); **46**, 1208, 1214 (1993); **48**, 714 (1995) ▪ Justus Liebigs Ann. Chem. **1991**, 367 ▪ Org. Lett. **1**, 1137 (1999) (synthesis). – *[CAS 130288-22-1]*

Phycarone see elsinochromes.

Phycobilins.

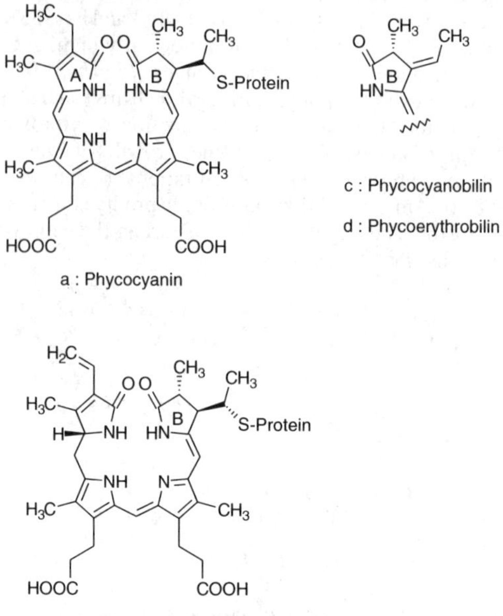

Collective name for the chromophores in *antenna pigments* (accessory pigments, light-harvesting pigments) of *photosystems* II (pigment system II, photosynthesis) of blue and red algae. The P. are open chain *tetrapyrroles and are structurally related to *bile pigments; according to the substitution pattern, a distinction is made between *phycocyanobilin* (phycobiliverdin) and *phycoerythrobilin* (phycobiliviolin). In the phycobiliproteins the P. are linked through thioether bridges to the proteins. These chromoproteins – like phytochrome belong to the biliproteins – are classified according to their colors as the blue *phycocyanin* and the red *phycoerythrin*. They are found closely packed in the phycobilisomes, ca. 40 nm large particles on the thylakoids (membrane system of photosynthesis) in the algae. For the mechanism of energy transfer, see *Lit.*[1], for X-ray structure of a hexameric phycocyanin, see *Lit.*[2], for biosynthesis, see *Lit.*[3], absolute configuration, see *Lit.*[4].

Lit.: [1]Eur. J. Biochem. **187**, 283–305 (1990). [2]J. Mol. Biol. **188**, 651–676 (1986). [3]Chem. Rev. **93**, 785–802 (1993). [4]Enantiomer **1**, 109 (1996).

gen.: Angew. Chem. Int. Ed. Engl. **28**, 848–869, 829–847 (1989) ▪ Beilstein E V **26/15**, 521 ▪ Falk, The Chemistry of Linear Oligopyrroles and Bile Pigments, p. 33, 517, Wien: Springer 1989 ▪ van den Hoek, Algen, Stuttgart: Thieme 1993.

Phyllanthocin. $C_{25}H_{30}O_7$, M_R 442.51, mp. 126–127 °C, $[\alpha]_D^{24}$ +25.2° ($CHCl_3$) P. is the methyl ester of phyllanthocic acid and is obtained together with phyllanthose by methanolysis of the antileukemic sesquiterpene glycoside phyllanthoside. *Phyllanthoside*, $C_{40}H_{52}O_{17}$, M_R 804.84, amorphous solid, mp. 125 °C, $[\alpha]_D^{24}$ +19.2° ($CHCl_3$) is isolated from *Phyllanthus*

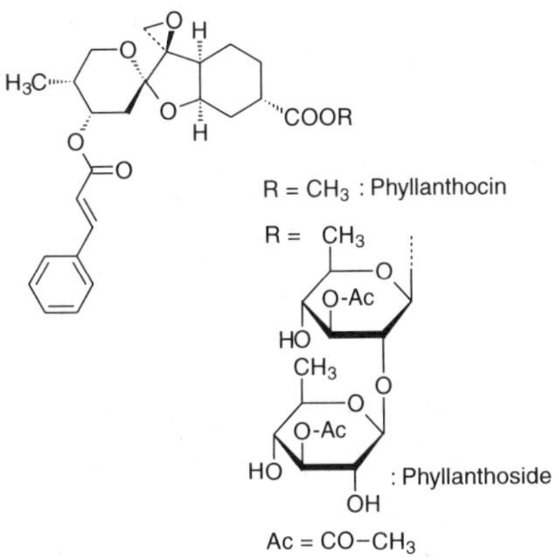

brasiliensis (Euphorbiaceae). Phyllantoside has antineoplastic activity.

Lit.: *Isolation & structure:* Can. J. Chem. **60**, 544, 939 (1982) ▪ J. Am. Chem. Soc. **98**, 7092 (1976); **99**, 3199 (1977) ▪ J. Org. Chem. **49**, 4258 (1984); **50**, 5060 (1985). – *Synthesis:* Angew. Chem. Int. Ed. Engl. **29**, 520 (1990) ▪ ApSimon **8**, 601 – 612 ▪ J. Am. Chem. Soc. **109**, 1269 (1987); **113**, 2071, 2092, 2112 (1991) ▪ J. Org. Chem. **50**, 3420 (1985); **55**, 2138 (1990); **62**, 7806 (1997) ▪ Tetrahedron Lett. **32**, 1613 (1991). – *[CAS 62948-37-2 (P.); 63166-73-4 (phyllanthoside)]*

Phyllanthurinolactone see leaf-movement factors.

Phylloquinone see vitamin K_1.

Phyllurine see leaf-movement factors.

Physal(a)emin.

H–(5–Oxo–Pro)–Ala–Asp–Pro–Asn–Lys–Phe–Tyr–Gly–Leu–Met–NH_2

$C_{58}H_{84}N_{14}O_{16}S$, M_R 1265.45. An undecapeptide with kinin-like activity (tachykinin) isolated from the skin of the frog *Physalaemus fuscumaculatus*. It is related to eledoisin and *substance P and has strong blood-pressure lowering, vasodilatory, and salivation-stimulating activities.

Lit.: Eur. J. Pharmacol. **86**, 59 – 64 (1982); **97**, 171 – 189 (1984) ▪ Merck-Index (12.), No. 7537 ▪ Peptides (N. Y.) **8**, 169 (1987); **9**, 619 (1988) (synthesis, pharmacology) ▪ Science **219**, 79 (1983). – *[CAS 2507-24-6]*

Physalien see zeaxanthin.

Physarochrome (physarochrome A).

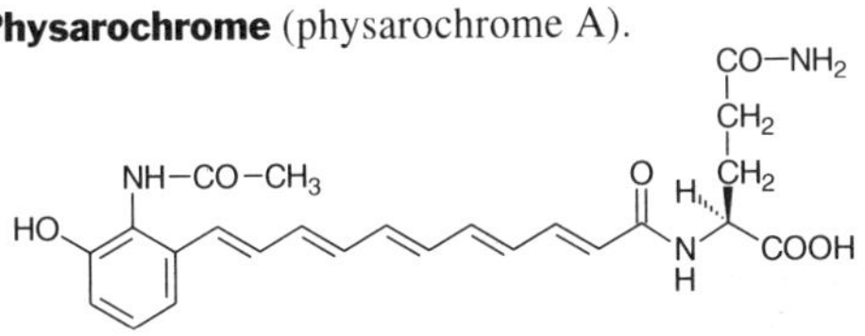

$C_{24}H_{27}N_3O_6$, M_R 453.49, yellow powder, $[\alpha]_D$ +7.2° (CH_3OH). A pentaene pigment from plasmodia of the slime mold *Physarum polycephalum*. P. and related compounds appear to play a role as photoreceptors in the life cycle of this myxomycete (see also fuligorubin A).

Lit.: Tetrahedron Lett. **28**, 3667 (1987); see also myxomycete pigments. – *[CAS 114670-90-5]*

Physcion (parietin, 1,8-dihydroxy-3-methoxy-6-methylanthraquinone, emodin 3-*O*-methyl ether).

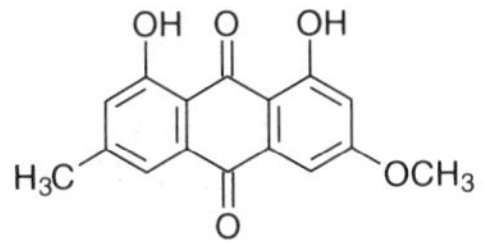

$C_{16}H_{12}O_5$, M_R 284.27, orange needles, mp. 209 – 210 °C, uv_{max} 431 nm (C_2H_5OH). An octaketide that occurs in senna leaves (*Cassia senna*, Fabaceae), rhubarb roots, in chrysarobinum (see chrysarobin), lichens (*Parmelia* species), and fungi (*Aspergillus* and *Penicillium* species), as well as in glycosidically bound form in *Cassia* species. It has a weak spasmolytic activity.

Lit.: Acta Crystallogr., Sect. C **47**, 1879 (1991) (structure) ▪ Fitoterapia **57**, 271 (1986) ▪ Heterocycles **43**, 969 (1996) ▪ Karrer, No. 1274 ▪ Phytochemistry **25**, 1115 (1986) ▪ Planta Med. **45**, 48 (1981); **46**, 159 (1982) ▪ Synthesis **1994**, 255 (synthesis) ▪ Turner **1**, 365; **2**, 144. – *[CAS 521-61-9]*

Physodic acid.

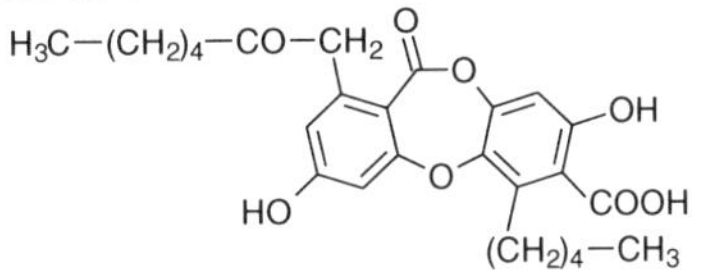

$C_{26}H_{30}O_8$, M_R 470.52, needles, mp. 205 °C (decomp.). A *depsidone with antibiotic properties from the foliose lichen *Hypogymnia physodes*.

Lit.: Aust. J. Chem. **49**, 1175 (1996) ▪ Culberson, p. 139 f. ▪ Karrer, No. 1057 ▪ Environ. Mutagen. **6**, 757 (1984); **18**, 35 (1991) (activity). – *[CAS 84-24-2]*

Physostigmine (eserine).

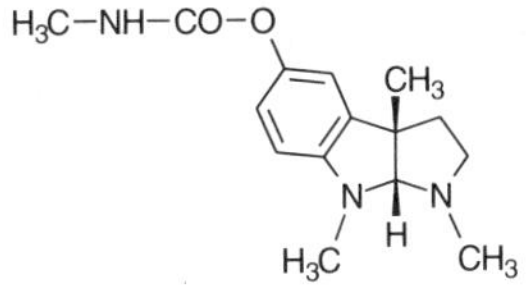

$C_{15}H_{21}N_3O_2$, M_R 275.35, orthorhombic crystals or platelets occurring in two forms: stable, mp. 105 – 106 °C, $[\alpha]_D$ −76° ($CHCl_3$); unstable, mp. 86 – 87 °C, $[\alpha]_D$ − 120° (benzene), readily soluble in alcohol, chloroform. P. is unstable under the action of air, light, and heat and in the presence of metals, and changes its color to red in the crystalline or dissolved state. P. is a tricyclic indoline alkaloid and the main alkaloid of the calabar bean (*Physostigma venenosum*, Fabaceae), content ca. 0.1% (total alkaloid content: ca. 0.5%), further minor alkaloids are physovenine and eseramine. *Physostigma* grows in the coastal regions of the Gulf of Guinea, from Angola to the Canary Islands. On account of its toxicity the calabar bean is used in West Africa as an ordeal poison (to obtain a judgement of the gods)[1]. P. itself is also used in suicide attempts. The alkaloid also occurs in streptomycetes.

Toxicology: Lethal dose on oral consumption 6 – 10 mg for humans, the toxicity is even higher on parenteral uptake. On incorrect use of eyedrops P. can reach the pharyngeal spaces and then causes nausea, vomiting, diarrhea. Severe poisonings lead to convulsions, respiratory paralysis, and death. Low doses of *atropine sulfate or *scopolamine can be used as an antidote.

Use: P. is isolated from the beans on an industrial scale. Its activity is similar to that of muscarine (see fly agaric constituents) and *pilocarpine as an indirect parasympathomimetic through inhibition of acetylcholine esterase; it is used in ophthalmology as a miotic agent to contract the pupils after they have been dilated by administration of *atropine, as an antiglaucoma drug, and to reduce intraocular pressure. P. can be used experimentally for impairments of brain function and symptoms of dementia. Derivatives of P. are under study for treatment of Alzheimer's disease. It is an important antidote for atropine and diphenhydramine poisonings. P. was a model alkaloid for the development of completely synthetic derivatives such as neostigmine, pyridostigmine, distigmine.

Lit.: [1] Z. Phytother. **11**, 7 (1989).
gen.: Beilstein E V **23/11**, 401 f. ▪ Clin. Neuropharmacol. **1976**, 63–79 ▪ Drug Rev. Eval. **15**, 1–9 (1989) ▪ Hager (5.) **1**, 730 f.; **9**, 193–200 ▪ Manske **10**, 383; **13**, 213 ▪ Negwer (7.), p. 4767 ▪ Stöckel, Das zentralanticholinergische Syndrom: Physostigmin in der Anästhesiologie u. Intensivmedizin, Stuttgart: Thieme 1982. – *Synthesis:* Aust. J. Chem. **49**, 171 (1996) ▪ Chem. Lett. **57** (1991) ▪ Chem. Pharm. Bull. **44**, 715 (1996) ▪ Heterocycles **31**, 411 (1990); **39**, 519 (1994) ▪ J. Am. Chem. Soc. **114**, 556 (1992); **120**, 6500 (1998) ▪ J. Org. Chem. **56**, 872, 5982 (1991); **58**, 6949 (1993) ▪ Tetrahedron **52**, 5339 (1996) ▪ Tetrahedron: Asymmetry **5**, 111, 363 (1994) ▪ Tetrahedron Lett. **35**, 6087 (1994). – *Toxicology:* Sax (8.), PIA 500, PIB 000. – *[HS 2939 90; CAS 57-47-6]*

Phytal see phytol.

Phytic acid see inositols.

Phytoalexins. A term introduced by K. O. Müller and Börger in 1940, derived from Greek: phyton = plant and alexein = to ward off, for those compounds that are formed by plants as a response to an infection (postinfectional) and for defence against the attacking organism – usually fungi. The accumulation of such P. is believed to be responsible, among others, for the resistance of certain plants towards pathogenic fungi. Thus, P. may be compared with the interferons of mammals: both are host specific, are formed in response to an external stimulus, but have rather unspecific activities. For the role of P. within the plant's "survival strategy", see *Lit.* [1–3]. The P. belong to widely differing classes of compounds (examples in parentheses): terpenoids and sesquiterpenoids (*gossypol, *rishitin from potatoes, *capsidiol from paprika fruits), isoflavonoids (*pisatin, *phaseolin, sativan, *glyceollins from soybeans), furanoterpenoids (*ipomeamarone), polyynes (safinol from safflower, wyerone and wyeronic acid from broad beans), dihydrophenanthrenes (orchinol), *furocoumarins (xanthotoxin see bergapten), and other groups [4], see also volicitin. Some of these compounds that are important for the plants' immune systems are also effective as preinfectional defensive substances (see phytoncides; both groups were sometimes combined in the past as phytoantibiotics), so that an assignment is not also unambiguous. Triggers for the formation of P. are not only fungal infections but also the action of bacteria, viruses, nematodes, as well as injuries, cold, UV radiation, heavy metals, etc. In isolated cases, the synthesis of P. is initiated by specific polysaccharides of the fungal cell wall. Compounds of this type, for which Keen introduced the name *elicitors* (from Latin: elicere = to attract) in 1972 include glucans, glycoproteins, polypeptides, and enzymes [5]. It is assumed that elicitors act via enzyme induction. On the other hand, the fungi have their own mechanisms for overcoming plant resistance, e. g., by blocking the biogenesis of P. [6] or by transformation of the P. into inactive products. The genetic principles for the manipulation of antimicrobially active plant secondary compounds are now available in part [8]. The expression of exogenous P. in cultured plants has been realized [9].

Lit.: [1] Angew. Chem. Int. Ed. Engl. **20**, 164 (1981). [2] ACS Symp. Ser. **439** (Microbes and Microbial Products as Herbicides), 114–129 (1990); **449** (Naturally Occurring Pest Bioregulators), 198–207 (1991). [3] Ciba Found. Symp. **154** (Bioact. Comp. Plants), 126–139, 140–156, 213–228 (1990). [4] Angew. Chem. Int. Ed. Engl. **17**, 635 (1978); Zechmeister **34**, 187–248. [5] Naturwissenschaften **68**, 447–457 (1981); Pure Appl. Chem. **53**, 79–88 (1981); ACS Symp. Ser. **439**, 114–129 (1990). [6] Naturwissenschaften **64**, 643 f. (1997); **67**, 310 f. (1980). [7] Weltring, Pytoalexins in the Relation between Plants and their Fungal Pathogens, in Stahl & Tudzynski (eds.), Molecular Biology of Filamentous Fungi, Proc. EMBO-Workshop, Meeting Date 1991, p. 111–124, Weinheim: VCH Verlagsges. 1992. [8] Dixon, Prospects for the Genetic Manipulation of Antimicrobial Plant Secondary Products, in BCPC Monogr. **55**, 113–118 (1993). [9] Curr. Opin. Biotechnol. **5**, 125–130 (1994).
gen.: ACS Symp. Ser. **62** (1977); **268** (1985); **296** (1986); **380** (1988); **439** (1990); **449** (1991); **658** (1997) ▪ Angew. Chem. Int. Ed. Engl. **37**, 1213 (1998) ▪ NATO ASI Ser. H **4** (Microbe-Plant Interact.) (1986) ▪ Photochemistry **53**, 161–176 (2000) (Cruciferae) ▪ Rice, Allelopathy, New York: Academic Press 1984 ▪ Sharma & Salunkhe, Mycotoxins Phytoalexins, Boca Raton, Fla.: CRC 1991. – *Reviews:* Annu. Rev. Phytopathol. **27**, 143–164 (1989) ▪ Ciba Found. Symp. **133**, 92–108 (1987); **137** 178–198 (1988).

Phytobacteriomycin D see streptothricins.

Phytochrombilin see phytochrome.

Phytochrome, phytochromobilin.

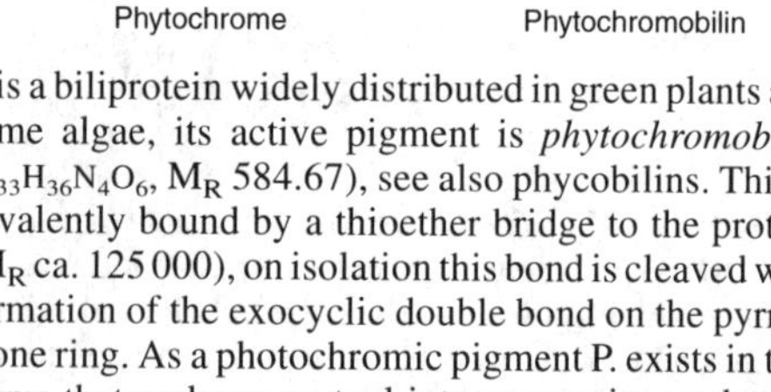

Phytochrome Phytochromobilin

P. is a biliprotein widely distributed in green plants and some algae, its active pigment is *phytochromobilin* ($C_{33}H_{36}N_4O_6$, M_R 584.67), see also phycobilins. This is covalently bound by a thioether bridge to the protein (M_R ca. 125 000), on isolation this bond is cleaved with formation of the exocyclic double bond on the pyrrolidone ring. As a photochromic pigment P. exists in two forms that undergo mutual interconversion under the action of light and can be distinguished on the basis of their absorption maxima as P_{660} or P_r (r = red) and P_{730} or P_{fr} (fr = far red, dark red). In the dark a plant embryo forms only P_r, which is stable in the dark and physiologically inactive. On exposure to bright red light (sunlight) P_r transforms to the more energy-rich, physiologically active P_{fr} which absorbs in the dark red and in the dark or on irradiation with light of 730 nm wavelength reverts to the inactive form P_r.

$$P_r \underset{\lambda = 730\ nm}{\overset{\lambda = 660\ nm}{\rightleftharpoons}} P_{fr} \longrightarrow \text{physiological effect}$$

Since daylight contains both wavelengths, during the day P_r and P_{fr} are present in green plants in the ratio 1:2. P_{fr} serves as light receiver (photoreceptor) in *photomorphogenesis*, the development of plants under the action of light. It is thus analogous as a photosensor pigment to *rhodopsin (processes of vision in vertebrates). As a morphogenetically active *effector* molecule, probably acting as a light-activated transcription factor, P_{fr} during germination induces not only the area growth of embryonic leaves and enhances cellular respiration, the synthesis of proteins and ribonucleic acids, and inhibits hypocotyl growth but also has a regulatory effect on the formation of chloroplasts and participates as a regulator in the electron transport chain of photosynthesis. The P. system also has a controlling function in the *photoperiodicity* ("internal clock"). In some plants the light red/dark red system of the P. is supplemented by a *blue light system*, which possibly contains an isoalloxazine (flavin) as chromophore [1]. Experimental studies on the action of P. and other plant physiological processes are performed in so-called *phytotrons* in which the various metabolic and climatic conditions can be simulated.

Lit.: [1] Senger, Blue Light Effects in Biological Systems, Berlin: Springer 1984.
gen.: Angew. Chem. Int. Ed. Engl. **30**, 1216 (1991) ▪ Annu. Rev. Plant Physiol. **44**, 617 (1993) ▪ Falk, The Chemistry of Linear Oligopyrroles and Bile Pigments, p. 38–41, 55 f., Wien: Springer 1989 ▪ J. Biol. Chem. **267**, 14790 (1992) (isolation) ▪ J. Org. Chem. **62**, 2894, 2907 (1997) (synthesis) ▪ Naturwissenschaften **75**, 132–139 (1988) ▪ Pigment of the Imagination. A History of Phytochrome Research, San Diego: Academic Press 1992 ▪ Trends Biochem. Sci. **9**, 450–453 (1984) ▪ Trends Genet. **4**, 37–42 (1988). – *[CAS 78249-71-5 (phytochrombilin)]*

Phytoecdysones, phytoecdysteroids see ecdysteroids.

Phytoene (7,7′,8,8′,11,11′,12,12′-octahydro-ψ,ψ-carotene).

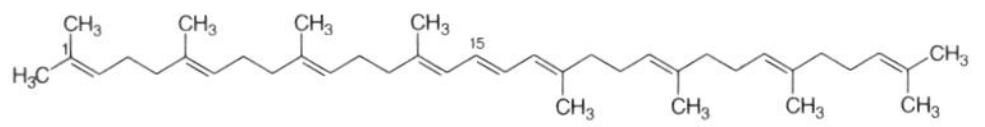

$C_{40}H_{64}$, M_R 544.95, a viscous, strongly fluorescing oil. P. is a partially saturated, colorless *tetraterpene from *Mycobacterium phlei* and other bacteria and plants. It represents a biosynthetic precursor of the *carotinoids. The (15*Z*)-isomer has been isolated from tomatoes.

Lit.: Beilstein EIV **1**, 1155 ▪ J. Chem. Soc., Perkin Trans. 1 **1993**, 1869 ▪ Phytochemistry **23**, 1707 (1984). – *[CAS 540-04-5; 13920-14-4 (15Z)]*

Phytofluene [(15*Z*)-7,7′,8,8′,11,12-hexahydro-ψ,ψ-carotene].

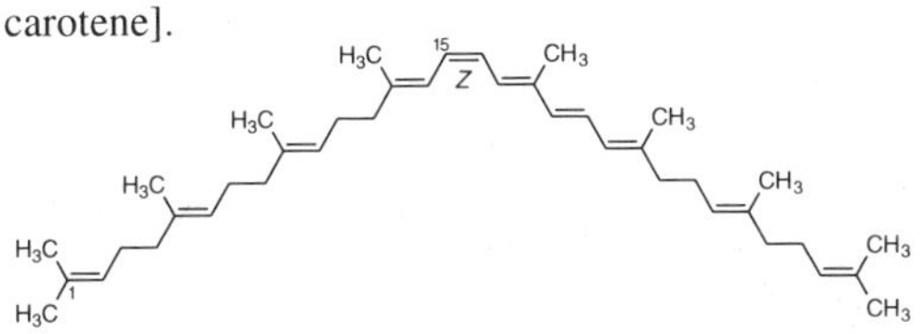

$C_{40}H_{62}$, M_R 542.93, light yellow, viscous oil with an intense green fluorescence in UV light; λ_{max} 332, 348, 367 nm; the *carotinoid P. is widely distributed in plants.

Lit.: Karrer, No. 1817 ▪ Phytochemistry **23**, 1707 (1984). – *[CAS 27664-65-9]*

Phytohormones see plant growth substances.

Phytol (2-phyten-1-ol)

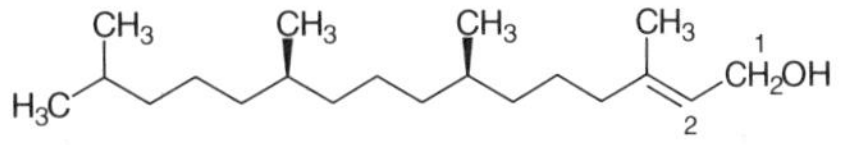

$C_{20}H_{40}O$, M_R 296.54, colorless oil, bp. 202–204 °C (1.3 kPa), 140–141 °C (4 Pa), $[\alpha]_D^{18}$ +0.2°. A *diterpene of the phytane type. P. occurs as an ester in *chlorophyll and in pheophytin and is thus present in all green plants. The biological degradation of the P. residue proceeds through phytanic acid and pristanic acid. The phytyl group also constitutes the lipophilic part in vitamin E (*tocopherols) and *vitamin K_1; accordingly P. is used in the synthesis of the latter. The isomeric *cis*-phytol with the (2*Z*,7*R*,11*R*)-configuration and the 1-aldehyde of phytol: *phytal* ($C_{20}H_{38}O$, M_R 294.52) are also found as natural products.

Lit.: Beilstein EIV **1**, 2208 ▪ J. Chem. Res. (S) **1990**, 154 ▪ Merck-Index (12.), No. 7545 ▪ Methods Geochem. Geophys. **24**, 1 (1986) ▪ Phytochemistry **21**, 1361 (1982); **25**, 509 (1986). – *Synthesis:* Helv. Chim. Acta **65**, 684 (1982) ▪ Synlett **1990**, 279 ▪ Tetrahedron **43**, 4481 (1987). – *Biosynthesis:* Chem. Commun. **1995**, 2529 ▪ J. Chem. Soc., Perkin Trans. 1 **1997**, 261. – *[CAS 150-86-7 (P.); 5492-30-8 (2Z,7R, 11R); 13754-69-3 (phytal)]*

Phytolacca red (pokeweed red, pokeberry red). Betacyans (see betalains) from the berry fruits of the pokeweeds (*Phytolacca decandra, Ph. americana*, Phytolaccaceae). The main component is betanin (see betanidin). In the past P. was used in France and Portugal to color red wine and sugar products, it is also suitable for coloring collagen which it stains red and the other tissue elements yellow. The Indians in Virginia and New England used pokeweed concentrates to dye baskets, furs, and leather. The roots of *Ph. americana* contain a mitogenic lectin, other species (*Ph. dodecandra*) molluscicidic saponins, which makes them useful in the control of the snail vectors of schistosomiasis (bilharziosis).

Lit.: Helv. Chim. Acta **67**, 1310–1315 (1984) ▪ Roth & Daunderer, Giftliste, Landsberg: ecomed since 1981 ▪ Sax (8.), p. 2829. – *[HS 3203 00]*

Phytomenadione see vitamin K_1.

Phytoncides. A term introduced by Tokin (1956), in analogy to names such as herbicides, fungicides, bactericides, etc. although somewhat misleading, for the solid, liquid, and gaseous defensive substances with antibiotic activities which plants use to protect themselves from predators, pests, and parasites. These substances, formed preinfectionally in active or activatable form by plants and given off continuously or as required into the environment, are named, sometimes per se and sometimes together with the *phytoalexins, as phytoantibiotics. Secondary plant substances with P. activity include phenols, unsaturated lactones, *steroid saponins, *glucosinolates, *cyanogenic glycosides and volatile terpenoids, most of which exist (except in essential oils) as glycosides; they are liberated enzymatically by the plant after injury and thus prevent an infection by the pathogens. The antibiotic activities of numerous medicinal plants have been used empirically for thousands of years. The wound gases that escape from damaged membranes – especially ethylene, C_6- and C_9-aldehydes and -alcohols (the most frequent is *trans*-2-*hexenal) – can also be included among the P. Schildknecht has proposed the name *phytones* for substances of this type possessing not only microbicidal but in part also growth factor properties. In the broader sense one could also include the allelopathic defence substances of higher plants among the P. These allelopathic agents with which plants mutually defend their living space are generally *plant growth substance antagonists, and have also been held responsible for certain forms of soil exhaustion. Some of these inhibitors also act as *plant germination inhibitors and thus enable isolated individuals within the "chemical ecology of plants" to secure their position against other plants by chemical means. Such compounds are sometimes also referred to as phytoaggressins of the plants' or "chemical weapons".

Lit.: see phytoalexins.

Phytoregulators see plant growth substances and plant germination inhibitors.

Phytosterols see sterols.

Phytoxin II see okadaic acid.

Phytuberin.

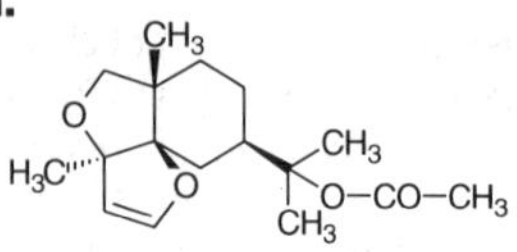

A sesquiterpene *phytoalexin, $C_{17}H_{26}O_4$, M_R 294.39, oil, $[\alpha]_D^{24}$ −35.9° ($CHCl_3$) formed, together with *rishitin and many other components (including *vetispiranes), by potatoes (*Solanum tuberosum*) after infection by the putrefactive bacterium *Erwinia carotovora*. Deacetylation furnishes *phytuberol*, $C_{15}H_{24}O_3$, M_R 252.35, oil. P. can also be isolated from the culture medium of potato cell cultures after elicitation with *Phytophthora infestans*. Some sorts of tobacco (*Nicotiana tabacum*) produce P. after infection with various *Pseudomonas* spp. or the tobacco mosaic virus; pretreatment with a cellulase shows the same effect.

Lit.: Isolation & structure: J. Chem. Soc., Perkin Trans. 1 **1977**, 53 ▪ Phytochemistry **27**, 133, 365 (1988). – *Synthesis:* Synlett **1999**, 1395. – *[CAS 37209-50-0 (P.); 56857-64-8 (phytuberol)]*

Piceatannol see hydroxystilbenes.

Picein.

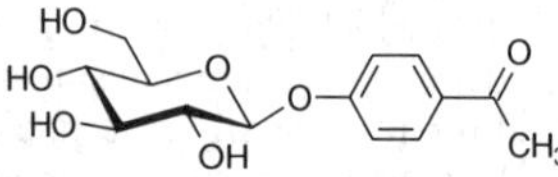

$C_{14}H_{18}O_7$, M_R 298.29, bitter tasting needles, as monohydrate mp. 195–196 °C, $[\alpha]_D^{23}$ −88° (H_2O). P. occurs in softwoods, willow bark, and mistletoe as well as the bearberry (*Arctostaphylos uva-ursi*). Treatment of P. with dilute acids or emulsin affords D-glucose and *p*-hydroxyacetophenone (*piceol*), $C_8H_8O_2$, M_R 136.15, mp. 109 °C, bp. 148 °C (4 hPa), pK_a 8.05 (25 °C); the latter has also been isolated in the free form from, among others, *Picea glehnii*.

Lit.: Phytochemistry **28**, 2115–2125, 2071–2078 (isolation), 3009 ff. (1989) (biosynthesis) ▪ Planta Med. **53**, 307 f. (1987) ▪ Tetrahedron Lett. **1974**, 3029 ff. (synthesis) ▪ Z. Anal. Chem. **327**, 535–538 (1987). – *[CAS 530-14-3 (P.); 99-93-4 (piceol)]*

Picrocrocin see safranal.

Picrolichenic acid.

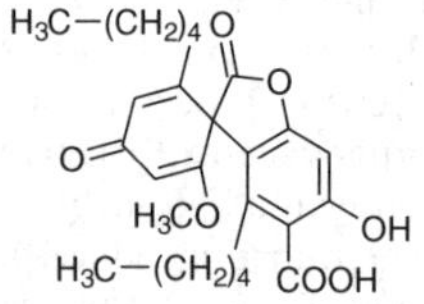

$C_{25}H_{30}O_7$, M_R 442.51, prisms, mp. 184–187 °C (decomp.) with an extremely bitter taste. P. a. belongs to the small group of *depsones; it occurs in some lichens of the genus *Pertusaria* (e. g., *Pertusaria amara*, on the bark of old deciduous trees). In the past it was used successfully in doses of 0.5–1 g for the treatment of intermittent fever.

Lit.: Aust. J. Chem. **44**, 1487 (1991); **47**, 1345 (1994) ▪ Culberson, p. 140 f. ▪ Zopf, Die Flechtenstoffe in chemischer, botanischer, pharmakologischer und technischer Beziehung, p. 384, Jena: G. Fischer 1907. – *[CAS 466-34-2]*

Picromycin.

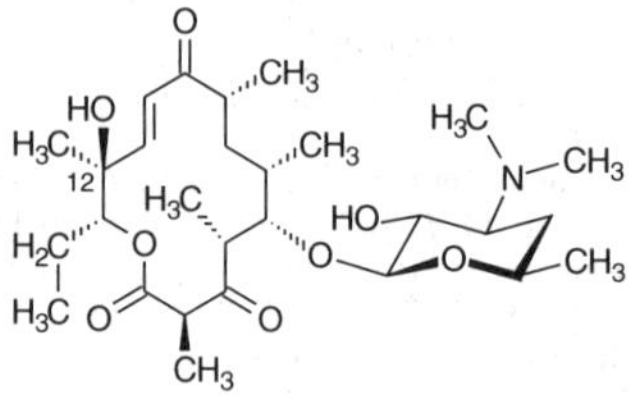

$C_{28}H_{47}NO_8$, M_R 525.68, mp. 169.5 °C, $[\alpha]_D$ −33.5° ($CHCl_3$), crystals with a very bitter taste, readily soluble in acetone, benzene, chloroform. P. was the first *macrolide antibiotic to be isolated from *streptomycetes (e. g., *Streptomyces felleus*), by Brockmann in 1950. The biosynthesis of the 14-membered ring aglycone proceeds from 1 acetate and 6 propionate units, the *β*-O-glycosidically bound sugar is D-desosamine. P. can be prepared by biotransformation (hydroxylation) from narbomycin with the help of streptomycetes. P. is active against Gram-positive bacteria including mycobacteria, but is too toxic for medical use.

Lit.: Heterocycles **46**, 105 (1997) (synthesis picronolide) ▪ J. Am. Chem. Soc. **108**, 4645 ff. (1986). – *Biosynthesis:* Chem. Pharm. Bull. **25**, 2385 (1977) ▪ J. Antibiot. **29**, 316 f. (1976). – *Toxicology:* Sax (7.), p. 2231. – *[HS 2941 90; CAS 19721-56-3]*

Picroroccellin.

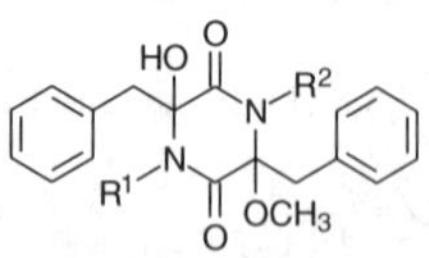

R^1, R^2 = H, CH_3

$C_{20}H_{22}N_2O_4$, M_R 354.41, rectangular prisms, mp. 190–220 °C, $[\alpha]_D$ +12.5°. A piperazinedione derivative with a bitter taste from the dyer's lichen *Roccella fuciformis*.

Lit.: Aust. J. Chem. **38**, 1785 (1985) ▪ Karrer, No. 2535 ▪ Tetrahedron Lett. **24**, 1445–1448 (1983). – *[CAS 87291-18-7]*

Picrosalvin see carnosol.

Picroside I see catalpol.

Picrotoxin (cocculin). P. is long known as the active component of the dried fruits (fishberries, Indian berries, Fructus cocculi) of the false myrtle *Anamirta cocculus* (*Menispermum cocculus*) indigenous to Indonesia[1]. The bitter principle P., $C_{30}H_{34}O_{13}$, M_R 602.59, mp. 203–204 °C, $[\alpha]_D^{12}$ –30° (H_2O), isolated from the plant is an equimolar mixture of the two sesquiterpene dilactones *picrotoxinin*, $C_{15}H_{16}O_6$, M_R 292.29, mp. 210 °C, $[\alpha]_D$ –5.85° (H_2O) and the non-toxic *picrotin*, $C_{15}H_{18}O_7$, M_R 310.30, mp. 255° $[\alpha]_D^{16}$ –70° (C_2H_5OH).

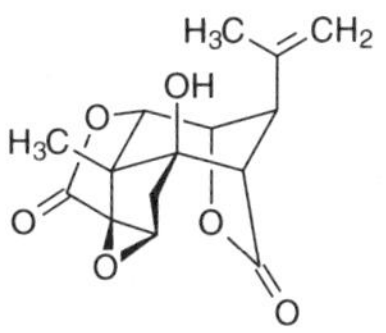

Picrotoxinin

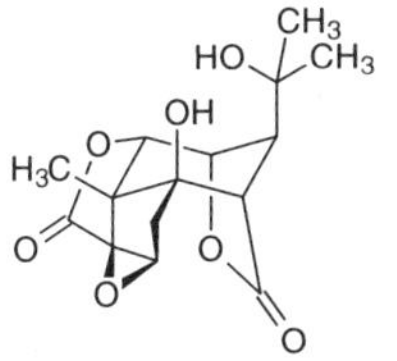

Picrotin

Picrotoxinin, soluble in boiling water and alcohol, is highly toxic to fish. It stimulates the central nervous system and acts as a potent GABA antagonist [LD_{50} (mouse i. p.) 3 mg/kg]. The lethal dose for humans is about 20 mg picrotoxinin or 2–3 g of fishberries. In low doses P. has a similar analeptic activity as *strychnine, in higher doses it induces convulsions. It is used as an antidote for barbiturate poisonings. On account of its very effective inhibition of presynaptic inhibitory mechanisms, P. is used in neurochemical research.

Lit.: [1] Aust. J. Chem. **36**, 2111, 2219 (1983); J. Am. Chem. Soc. **101**, 5841 (1979); **111**, 3728 (1989); ACS Symp. Ser. **356**, 44–70 (1987); Karrer, No. 3629 ff.
gen.: Beilstein E V **19/10**, 530, 605. – *Activity:* Adv. Biochem. Psychopharmacol. **41**, 67 (1986) ▪ Brain Res. Bull. **5**, 919 (1980) ▪ Hager (5.) **3**, 319; **4**, 268 f.; **9**, 201 ▪ J. Physiol. (London) **447**, 191 (1992) ▪ Med. Biol. **64**, 301 (1986) ▪ Merck-Index (12.), No. 7569–7571 ▪ Neurotoxicology **6**, 139 (1985) ▪ Phytochemistry **50**, 1365 (1999) ▪ Sax (8.), PIE 500, PIE 510 ▪ Toxicol. Lett. **42**, 117 (1988). – *Synthesis:* J. Am. Chem. Soc. **111**, 3728 (1989) ▪ Tetrahedron Lett. **21**, 1823 (1980). – *[CAS 21416-53-5 (picrotin); 124-87-8 (picrotoxin); 17617-45-7 (picrotoxinin)]*

Pigments of vision. In the process of vision in higher organisms, the conversion of light stimuli to electrical signals (transduction) is mediated by the pigments of vision contained in cone cells and rod cells. On absorption of light the pigments of vision undergo a change of configuration. The most widely distributed compound is *rhodopsin (visual purple), a *chromoprotein, composed of a pigment component, the chromophore *retinal and the glycoprotein *opsin*, a polypeptide chain with 2 oligosaccharide side chains. In most animals the chromophore of the pigment of vision is retinal (retinal 1, vitamin A_1-aldehyde) or retinal 2 (3-dehydroretinal, vitamin A_2-aldehyde). The rod cells of the human eye, responsible for night vision, contain a rhodopsin with an absorption maximum at 500 nm. They transmit light stimuli of different colors only as different gray tones. Three different types of rod cells exist with pigments of vision with maxima at 445, 535, and 570 nm, respectively (iodopsin), the combined action of these types is responsible for color vision. Some freshwater fish and amphibians have pigments of vision containing retinal 2 which, in combination with rod cell opsin gives a substance sensitive to purple-red light (porphyropsin) and in combination with cone cell opsin a substance sensitive to blue light, *cyanopsin with an absorption maximum at 620 nm. Lower organisms also possess light-sensitive cells in which purely phototactic reactions are induced. In a series of phototactically active algae, the pigments of vision occur in the "eyespots". For the physicochemical foundation of the process of vision, see rhodopsin and *Lit.*[1].

Lit.: [1] Chem. Ind. (London) **1995**, 735–739.
gen.: J. Am. Chem. Soc. **119**, 3619 (1997) ▪ Proc. Natl. Acad. Sci. USA **94**, 7937 (1997).

Pilocarpine [(3*S*)-*cis*-3-ethyl-4-(1-methyl-5-imidazolylmethyl)-dihydro-2(3*H*)-furanone].

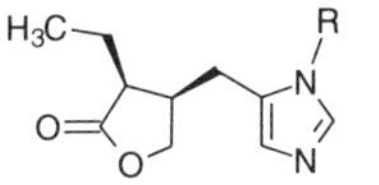

R = CH_3 : Pilocarpine
R = H : Pilocarpidine

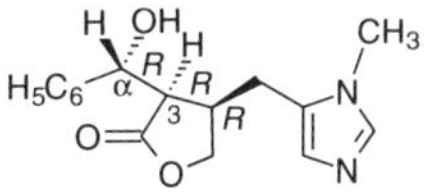

Pilosine (Carpidine)
(3*S*)-Epimer: Isopilosine (Carpiline)
(3*S*,α*S*)-Epimer: Epiisopilosine

$C_{11}H_{16}N_2O_2$, M_R 208.26, oily liquid or crystals, mp. 34 °C (as hydrochloride: 193–205 °C, nitrate: 174 °C), bp. 260 °C (0.7 kPa), $[\alpha]_D^{18}$ +106° (H_2O), soluble in water, ethanol, chloroform; forms well crystallizing salts with acids. P. is the main alkaloid of the South American jaborandi tree (*Pilocarpus jaborandi*). It is isolated from the leaves. P. is also easily accessible by synthesis and is used as its salts in medicine as a cholinergic parasympathicomimetic, miotic (for glaucoma) agent and as an atropine antagonist; it is also used in veterinary medicine for colics and constipation. Detection by means of Helch's reaction.
Pilocarpus species contain further *imidazole alkaloids (see table).

Table: Data of Pilocarpus alkaloids.

Alkaloid	molecular formula	M_R	mp. [°C]	CAS
Pilocarpidine	$C_{10}H_{14}N_2O_2$	194.23	oil	127-67-3
Pilosine	$C_{16}H_{18}N_2O_3$	286.33	101–104 (· 2 H_2O) 171–172 (anhydrous)	13640-28-3
Isopilosine	$C_{16}H_{18}N_2O_3$	286.33	182–183	491-88-3
Epiisopilosine	$C_{16}H_{18}N_2O_3$	286.33	179–180 (182–184)	38993-92-9

Pilosine occurs in *Pilocarpus microphyllus* and *P. jaborandi*. The natural product consists of a 1:1 mixture of the epimers pilosine {*carpidine*, $[\alpha]_D^{20}$ +83.9° (C_2H_5OH)} and isopilosine {*carpiline*, $[\alpha]_D^{20}$ +37.6° (C_2H_5OH)}. Epiisopilosine {$[\alpha]_D^{20}$ –44° (C_2H_5OH)} occurs in *P. jaborandi*. The effects of pilosine in mammals are similar but not so pronounced as those of pilocarpine.

Lit.: Beilstein E III/IV **27**, 7463 ▪ Florey **12**, 385–432 ▪ Hager (5.) **3**, 972 ff.; **6**, 127–135; **9**, 204–210 ▪ Martindale (29.),

p. 1335 f. ▪ Progr. Drug. Res. **25**, 421–460 (1981). – *Biosynthesis:* Planta Med. **28**, 1 (1975) ▪ Ullmann (5.) **A 18**, 140–145. – *Synthesis:* J. Org. Chem. **58**, 62 ff., 1159 (1993) ▪ Tetrahedron Lett. **30**, 2145 ff. (1989); **33**, 2447 (1992); **38**, 5213 (1997). – *Toxicology:* Sax (8.), PIF 000–500. – *Activity:* Annu. Rev. Med. **35**, 185–205 (1984) ▪ Drugs **49**, 143 (1995) (review) ▪ J. Ocul. Pharmacol. **5**, 119–125 (1989) ▪ Kleemann-Engel, p. 1518 f. ▪ Synapse (N. Y.) **3**, 154–171 (1989). – *[HS 2939 90; CAS 92-13-7 (P.); 54-71-7 (hydrochloride)]*

Pilocarpus alkaloids see pilocarpine.

Pimaranes.

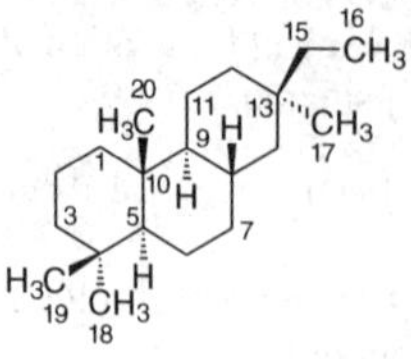

1

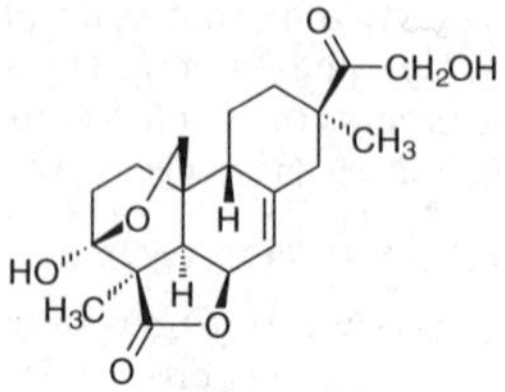

2 Annonalide

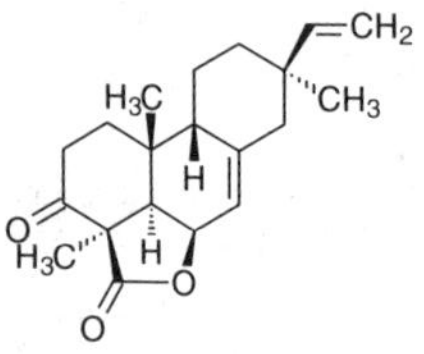

3 Momilactone A

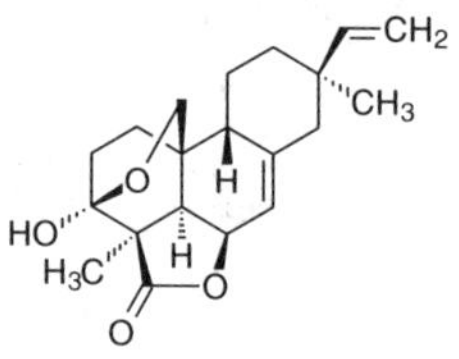

4 Momilactone B

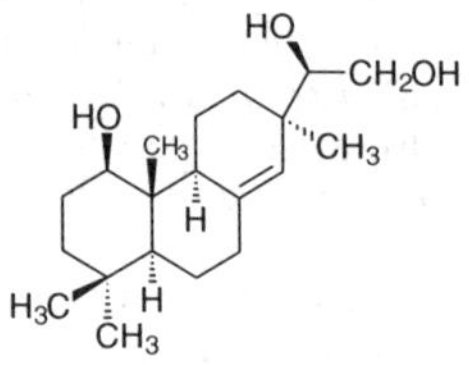

5 Leucophleol

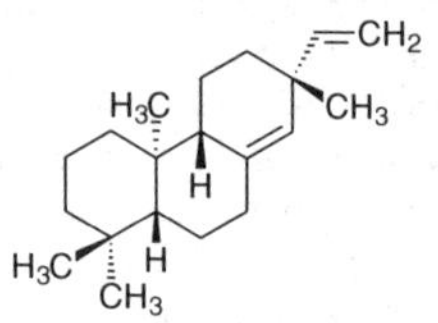

6 *ent*-8(14),15-Pimaradiene

Table: Data of different Pimaranes.

no.	molecular formula	M_R	mp. [°C]	$[\alpha]_D$	CAS
2	$C_{20}H_{26}O_6$	362.42	cryst. 261–263	–174° (pyridine)	52691-15-3
3	$C_{20}H_{26}O_3$	314.42	cryst. 235–236	–277° ($CHCl_3$)	51415-07-7
4	$C_{20}H_{26}O_4$	330.42	cryst. 242	–185° ($CHCl_3$)	51415-08-8
5	$C_{20}H_{34}O_3$	322.49	cryst. 176–178	+6.5° (C_2H_5OH)	77063-89-9
6	$C_{20}H_{32}$	272.47	oil	–115° ($CHCl_3$)	19882-10-1

A group of *diterpenes with the pimarane skeleton (**1**). In P., the methyl group at C-10 is *cis* to the ethyl group at C-13, whereas in the isopimaranes it is in the *trans*-position. Both enantiomers of not only the P. but also of the isopimaranes occur in nature. P. are formed by a further cyclization of the *labdane skeleton. Examples are annonalide, momilactones A and B (both growth inhibitors for plants[1]), leucophleol and *ent*-8(14),15-pimaradiene. In the past the P. were also known as sandaracopimaranes.

Lit.: [1] Phytochemistry **16**, 45 (1977); **30**, 1579 (1991). *gen.:* Arch. Biochem. Biophys. **293**, 320 (1992) (biosynthesis) ▪ Chem. Ber. **111**, 843 (1978) ▪ J. Chem. Soc., Perkin Trans. 1 **1973**, 2551; **1977**, 250 ▪ Phytochemistry **19**, 1979 (1980); **22**, 1645 (1983) ▪ Tetrahedron Lett. **1977**, 1085.

Pimaricin see natamycin.

Pimpifolidine see Solanum steroid alkaloids.

Pimpinellin (5,6-dimethoxy-2*H*-furo[2,3-*h*]-1-benzopyran-2-one).

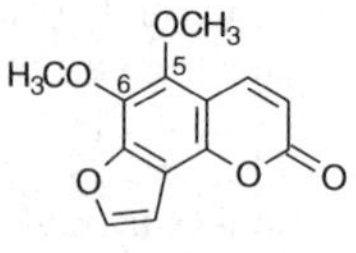

Pimpinellin

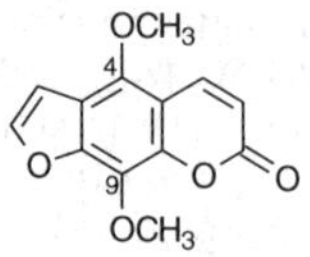

Isopimpinellin

$C_{13}H_{10}O_5$, M_R 246.22, needles, mp. 117–119 °C. P. is a *furocoumarin from Apiaceae and occurs, e. g., in the roots of the greater and lesser saxifrages *Pimpinella major* and *P. saxifraga*) and in numerous *Heracleum* species. P. is a *phytoalexin in celery and parsley. It is often accompanied by *isopimpinellin* (4,9-dimethoxy-7*H*-furo[3,2-*g*][1]benzopyran-7-one, 4,9-dimethoxypsoralen), golden yellow needles, mp. 147–148 °C.

Lit.: Beilstein E V **19/6**, 319 ▪ J. Chem. Ecol. **17**, 1859–1870 (1991) (MS) ▪ J. Nat. Prod. **50**, 997–998 (1987) ▪ J. Org. Chem. **53**, 4166–4171 (1988) (synthesis) ▪ Luckner (3.), p. 390 ▪ Nat. Prod. Rep. **6**, 591–624 (1989) ▪ Phytochemistry **22**, 2207–2209 (1983) (biosynthesis). – *[CAS 131-12-4 (P.); 482-27-9 (iso-P.)]*

Pineapple flavor see fruit flavors.

Pine needle oils see fir and pine needle oils.

Pinenes (α-P. = 2,6,6-trimethylbicyclo[3.1.1]hept-2-ene; β-P. = 6,6-dimethyl-2-methylenebicyclo[3.1.1]-heptane).

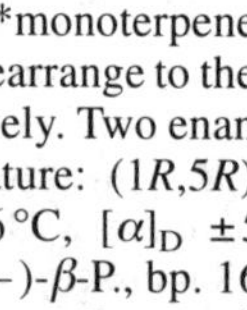

(+)-α-Pinene

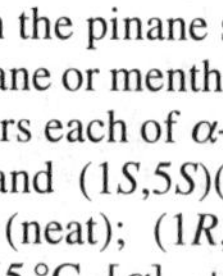

(+)-β-Pinene

$C_{10}H_{16}$, M_R 136.24, easily inflammable liquids with a turpentine-like odor (flash point 33 °C). Unsaturated, bicyclic *monoterpenes with the pinane structure that readily rearrange to the bornane or menthane skeleton, respectively. Two enantiomers each of α- and β-P. exist in nature: (1*R*,5*R*)(+)- and (1*S*,5*S*)(–)-α-P., bp. 155–156 °C, $[\alpha]_D$ ±52.4° (neat); (1*R*,5*R*)(+)- and (1*S*,5*S*)(–)-β-P., bp. 164–165 °C, $[\alpha]_D$ ±22.7° (neat).

Occurrence: P. occur in the essential oil from needles and cones of most conifers. On the industrial scale P. are obtained from *turpentine oil (essential oil of *Pinus* species, Pinaceae).

Use: For the synthesis of *camphor and insecticides, as solvents for waxes, and as softeners for plastics. Oxidative cleavage of the double bond in α-P. leads to optically active cyclobutane derivatives that can be used as starting materials in the synthesis of cyclobutanoid natural products.

Lit.: J. Chem. Ecol. **18**, 1693 (1992) ▪ J. Org. Chem. **54**, 1764 (1989) ▪ Karrer, No. 62, 63 ▪ Merck-Index (12.), No. 7599, 7600. – *[HS 2902 19; CAS 7785-70-8 ((+)-α-P.); 7785-26-4 ((–)-α-P.); 19902-08-0 ((+)-β-P.); 18172-67-3 ((–)-β-P.)]*

Table: Physical data and occurrence of Pinenols.

trivial name	bp. [°C]	$[\alpha]_D$	CAS	occurrence
(−)-*trans*-Pinocarveol	209–210	−73.4° (undiluted)	547-61-5	*Eucalyptus globulus* (Myrtaceae)
(−)-*trans*-Chrysanthenol	–	−55° ($CHCl_3$)	38043-83-3	*Chrysanthemum* spp. (Asteraceae)
(+)-*trans*-Verbenol	88 (900 Pa)	+141° ($CHCl_3$)	22339-08-8	frankincence, gum olibanum (*Boswellia sacra*, syn. *B. carterii*, Burseraceae)
(+)-*cis*-Verbenol	mp. 69	+7.6° (acetone)	13040-03-4	
(+)-Myrtenol			6712-78-3	sweet orange (*Citrus sinensis*, Rutaceae)
	221–222	±49.7° (undiluted)		
(−)-Myrtenol			19894-97-4	*Cyperus articulatus* (Cyperaceae)

Pinenols.

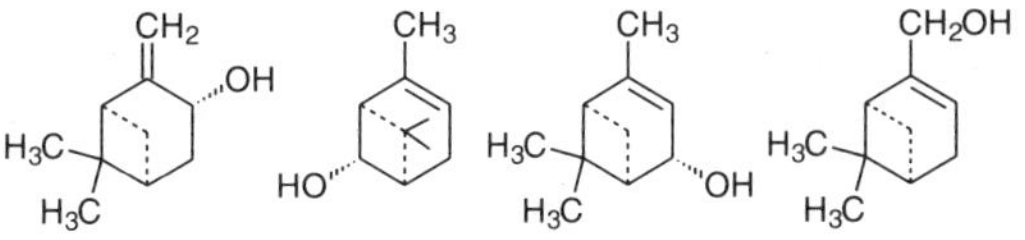

(−)-*trans*-Pinocarveol (−)-*trans*-Chrysanthenol (+)-*trans*-Verbenol (+)-Myrtenol

$C_{10}H_{16}O$, M_R 152.23. Unsaturated, bicyclic *monoterpene alcohols with the pinane structure. Pinocarveol and chrysanthenol occur in nature in the above shown configurations while all four possible stereoisomers of verbenol and the two possible enantiomers of myrtenol are known. The table lists plants from which P. can be isolated from the essential oils. P. also occur as the acetates in nature. (+)-*trans*- and (+)-*cis*-verbenol are components of the aggregation pheromone of the conifer bark beetle *Ips typographus* and the American *Ips confusus,* a pest of the yellow pine (*Pinus ponderosa*, Pinaceae). After a tree has been colonized by a sufficient number of individuals further beetles are diverted to other trees by means of verbenone (see pinenones). The other P. do not exhibit any particular biological activities.

Lit.: Food Chem. Toxicol. **30**, 935 (1992) (review, myrtenol) ▪ Karrer, No. 316, 318, 319 ▪ Synth. Commun. **24**, 2923 (1994) (synthesis, myrtenol). – *[HS 2906 19]*

Pinenones.

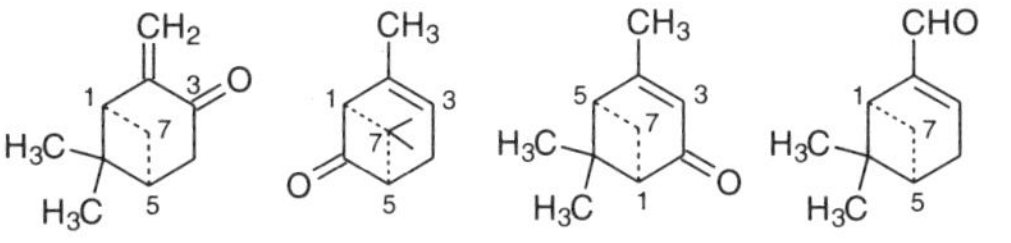

(−)-Pinocarvone (+)-Chrysanthenone (+)-Verbenone (+)-Myrtenal

Table: Physical data and occurrence of Pinenones.

trivial name	bp. [°C]	$[\alpha]_D$ (undiluted)	CAS	occurrence	yield [%]
(−)-Pinocarvone	222–223	−68.2°	34413-88-2	wormwood (*Chenopodium ambrosioides*, Chenopodiaceae)	57
(+)-Verbenone	227–228	+249°	18309-32-5	verbena from Spain (*Verbena triphylla*, Verbenaceae)	1
(+)-Chrysanthenone			38301-80-3	'chrysanthemum' (*Dendranthema indicum*, Asteraceae)	40
	88–89 (1.2 kPa)	±108°			
(−)-Chrysanthenone			58437-73-3	*Chrysanthemum sinense* (Asteraceae)	33
(+)-Myrtenal			23727-16-4	*Hernandia peltata* (Hernandiaceae)	
	89–92 (1.1 kPa)	±16.9°			
(−)-Myrtenal			18486-69-6	*Cyperus articulatus* (Cyperaceae)	0.67

$C_{10}H_{14}O$, M_R 150.22. Unsaturated bicyclic *monoterpene ketones with the pinane structure. Pinocarvone and verbenone occur in nature only in the shown configurations while both enantiomers of chrysanthenone and myrtenal have been isolated. The table lists the plants from which P. can be isolated from the essential oils. For the biological properties of (+)-verbenone, see pinenols.

Lit.: Beilstein EIV **7**, 327f. ▪ Karrer, No. 566, 567, 3540 ▪ Merck-Index (12.), No. 2311, 10094 ▪ Synth. Commun. **24**, 2923 (1994). – *[HS 2914 29]*

Pine oil. Colorless to yellowish oil with a tart-fresh, earthy, somewhat bitter odor.

Production: By distillation of crude wood *turpentine oil. The major portion of the commercially marketed P. o. today is prepared synthetically by hydrogenation of turpentine oil (α-*pinenes).

Composition: The main component is *α-terpineol* (see *p*-menthenols), together with smaller amounts of mainly monoterpene alcohols such as **borneol*, **fenchol*, *β-terpineol*, *terpinenol*.

Use: For lower-priced perfumes and, especially for technical purposes; also for the production of pure α-terpineol.

Lit.: Arctander, p. 537 ▪ Ullmann (4.) **22**, 561. – *Toxicology:* Food Chem. Toxicol. **21**, 875 (1983). – *[HS 3805 20; CAS 8002-09-3]*

Pinidine see pinus alkaloids.

Pinnatins A – E. A group of highly functionalized diterpenoid polycyclic α,γ-disubstituted α,β-unsaturated γ-lactones, building a unique bicyclo[11.1.0]tetradecane skeleton, the gersolane ring system. P. A – E were isolated from the Caribbean gorgonian *Pseudopterogorgia bipinnata. P. A* ($C_{20}H_{24}O_4$, M_R 328.40, cryst., mp. 189–211 °C, $[\alpha]_D^{24}$ +225.4°) and *P. B* show significant antitumor activity. The term pinnatin is also used for a furochromone [1].

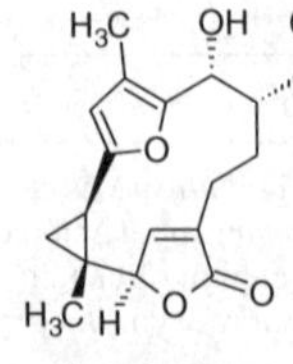

Lit.: [1]Phytochemistry **22**, 800 (1983).
gen.: J. Org. Chem. **63**, 4425–4432 (1998). – *[CAS 209969-38-0 (P. A); 1232-43-5 (pinnatin)]*

Pinnatoxins. Mussel toxins from the genus *Pinna* (*Pinna attenuata, P. pectinata*). These mussels are precious food in the coastal regions of Japan and China and contain varying amounts of P. Poisonings are common but only rarely lethal. The P. are spirocyclic acetals, e.g., *P. A:* $C_{41}H_{61}NO_9$, M_R 711.94, $[\alpha]_D$ +2.5° (CH_3OH), LD_{99} (mouse i.p.) 180 µg/kg; isolated from *Pinna muricata*; see also PSP.

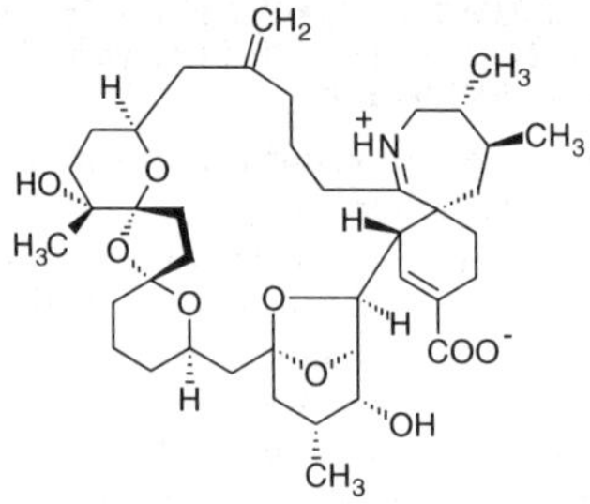

P. A

Lit.: J. Am. Chem. Soc. **117**, 1155 (1995); **120**, 7647 (1998) (synthesis (–)-P. A) ▪ Synlett **1999**, 692, 695 ▪ Tetrahedron Lett. **37**, 4023, 4027 (1996). – *[CAS 160759-36-4 (P. A); 167228-67-3 (P. D)]*

(–)-*trans*-Pinocarveol see pinenols.

(–)-Pinocarvone see pinenones.

Pinopalustrin see wikstromol.

(+)-Pinoresinol see lignans.

Pinosylvin see hydroxystilbenes.

Pinus alkaloids. For a long time, *pinidine* [1] {(2*R*)-*cis*-2-methyl-6-((*E*)-1-propenyl)piperidine, $C_9H_{17}N$, M_R 139.24, $[\alpha]_D^{25}$ –10.5°} isolated from the American pine *Pinus sabiniana* was the only alkaloid known from the Pinaceae.

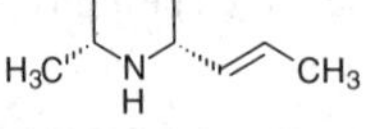

Pinidine

Recently, some further, structurally related derivatives were found and it was demonstrated that 2,6-disubstituted piperidines occur in all investigated species of *Pinus* (pine) and *Picea* (spruce) [2]. The P. a. are found in all parts of the plants. First observations suggest that the potentially toxic alkaloids (structural similarities with coniine) serve as a defence against herbivores [3].
Lit.: [1]J. Am. Chem. Soc. **78**, 4467 (1956). [2]J. Org. Chem. **58**, 4813 (1993); Phytochemistry **35**, 951 (1994); **40**, 401 (1995). [3]J. Nat. Prod. **54**, 905 (1991).
gen.: Beilstein EV **20/4**, 395 ▪ J. Org. Chem. **40**, 2151 (1975) (synthesis) ▪ Tetrahedron Lett. **1975**, 3779 (biosynthesis). – *[HS 2939 90; CAS 501-02-0]*

L-Pipecolic acid [(*S*)-2-piperidinecarboxylic acid, (*S*)-hexahydropicolinic acid, L-homoproline].

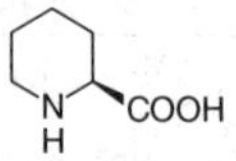

$C_6H_{11}NO_2$, M_R 129.16, mp. 259–260 °C (decomp.), $[\alpha]_D^{20}$ –30.6° (H_2O); monohydrochloride, mp. 256–258 °C (decomp.), $[\alpha]_D^{23}$ –10.3° (H_2O). A non-proteinogenic amino acid in plants. Very widely distributed in plants, e. g., in apples, fruits of the date palm, hops, climbing beans (*Phaseolus vulgaris*), white clover (*Trifolium repens*).
Biosynthesis: From lysine via 2-aminoadipic acid hemialdehyde and 2,3,4,5-tetrahydropyridine-2-carboxylic acid [1] or via 3,4,5,6-tetrahydropyridine-2-carboxylic acid [2].
L-P. is a component of some macrolide antibiotics such as, e. g., *rapamycin. For synthesis, see *Lit.* [3].
Lit.: [1]J. Exp. Bot. **11**, 302–315 (1960). [2]Phytochemistry **9**, 2329–2334 (1970). [3]Bull. Chem. Soc. Jpn. **64**, 620 (1991); J. Org. Chem. **58**, 2904 (1993); **59**, 2075, 3769 (1994).
gen.: Beilstein EV **22/1**, 220f. ▪ Karrer, No. 2398. – *[HS 2933 39; CAS 3105-95-1]*

Piperdial.

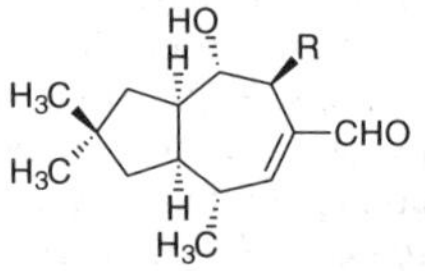

R = CHO : Piperdial
R = CH_2OH : Piperalol

$C_{15}H_{22}O_3$, M_R 250.34, oil, $[\alpha]_D$ +77° (ether). An antibacterially active defensive substance of the *lactarane type with a burning hot taste from the latex containing agarics *Lactarius piperatus, L. torminosus* and *Russula queletii* (Basidiomycetes). P. is formed together with *velleral and lactardial by enzymatic transformation of *stearoylvelutinal when the fruit body is injured. P. is converted within a few minutes by the fungus into the mild *piperalol* {$C_{15}H_{24}O_3$, M_R 252.35, oil, $[\alpha]_D$ +57° (ether)}.
Lit.: Phytochemistry **27**, 3315 (1988) ▪ Tetrahedron Lett. **26**, 3163 (1985). – *[CAS 100288-36-6 (P.); 100288-35-5 (piperalol)]*

Piperidine alkaloids. Collective name for widely differing alkaloids containing a piperidine ring. As a con-

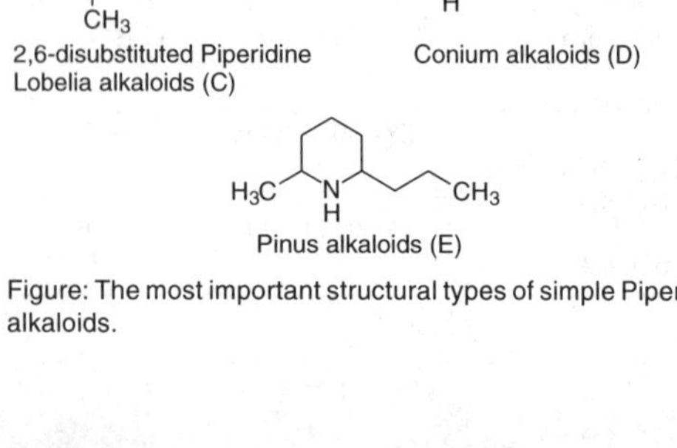

Figure: The most important structural types of simple Piperidine alkaloids.

sequence of their heterogenicity, most P. a. are classified according to their occurrence and biogenetic origins and treated as separate groups. The most important groups (see figure) are the hot principles of pepper (see piperine), 2-monosubstituted piperidines of the *Sedum alkaloid or *pelletierine types, the 2,6-disubstituted piperidines of the *Lobelia alkaloid type; the *hemlock alkaloids, and the *Pinus alkaloids, the importance of which has only recently been recognized.

Lit.: Heterocycles **52**, 313 (2000) ▪ Mothes et al., p. 128 ▪ Nat. Prod. Rep. **9**, 581 (1992) ▪ Tetrahedron: Asymmetry **8**, 477 (1997); **9**, 1597, 2201 (1998) (review, anatabine, cassaine). – *[HS 2939 90]*

Piperine (1-piperoylpiperidine).

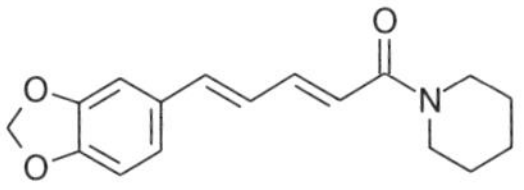

$C_{17}H_{19}NO_3$, M_R 285.34; yellowish or colorless, monoclinic prisms, mp. 128–129 °C. P. is present to 5–9% in *pepper (*Piper nigrum*) as the most important pungent principle [LD_{50} (rat p. o.) 514 mg/kg] and in many other *Piper* species (Piperaceae). Besides P. over 50 other, analogous amides, mostly with piperidine, pyrrolidine, and isobutylamine as base component are known in *Piper* species. The hot pepper taste is also the reason for its use in stomachics and in alcoholic beverages (brandy). Like many *Piper* alkaloids, P. exhibits antimicrobial properties. Under the action of light P. isomerizes to a mixture of stereoisomers, consisting of 18% P., 27% of the *Z,Z*-isomer (*chavicine*, oil), 22% of the *Z,E*-isomer (*isopiperine*, mp. 89 °C), and 33% of the *E,Z*-isomer (*isochavicine*, mp. 110 °C).

Lit.: Beilstein E V **20/3**, 469 f. ▪ Hager (5.) **6**, 199 f. ▪ J. Agric. Food Chem. **29**, 942 (1981) ▪ Karrer, No. 1021, 1022 ▪ Merck-Index (12.), No. 7625 ▪ Pelletier **3**, 41–46 ▪ Phytochemistry **45**, 1617 (1997) (pipericine); **46**, 597–673 (1997) (review) ▪ Sax (8.), PIV 600. – *Synthesis:* Justus Liebigs Ann. Chem. **1981**, 1725 ▪ Synlett **1994**, 607. – *[HS 2939 90; CAS 94-62-2 (P.); 495-91-0 (chavicine); 30511-76-3 (isopiperine); 30511-77-4 (isochavicine)]*

Piperitol see *p*-menthenols.

Piperitone see *p*-menthenones.

Piperonal (piperonylaldehyde, heliotropin, 3,4-methylenedioxybenzaldehyde, 1,3-benzodioxole-5-carbaldehyde). $C_8H_6O_3$, M_R 150.13, mp. 37 °C, bp. 263 °C; soluble in organic solvents, poorly soluble in water. Aromatic aldehyde occurring in flower oils. P. is used as a fragrance and flavor substance.

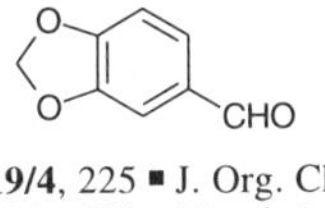

Lit.: Beilstein E V **19/4**, 225 ▪ J. Org. Chem. **49**, 1830 (1984) (synthesis) ▪ Karrer, No. 401 ▪ Merck-Index (12.), No. 7628 ▪ Mölleken & Bauer, in: Croteau, Fragrance and Flavor Substances, p. 75–81, Pattensen: D & PS 1980 ▪ Ullmann (5.) **A 11**, 200. – *[HS 2932 90; CAS 120-57-0]*

Piptamine, piptanthine see ormosanine.

Piptoporic acid.

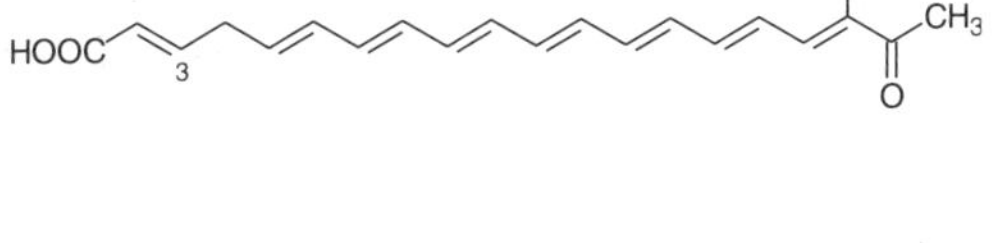

$C_{21}H_{24}O_3$, M_R 324.42, yellow oil. P. occurs in the bright orange brackets of the Australian tree fungus *Piptoporus australiensis* (Aphyllophorales). The polyenecarboxylic acid is accompanied by its methyl ester and the (3*R*)-3-acetoxy-2,3-dihydro derivative.

Lit.: J. Chem. Soc., Perkin Trans. 1 **1982**, 1449. – *[CAS 83016-06-2]*

Pironetin.

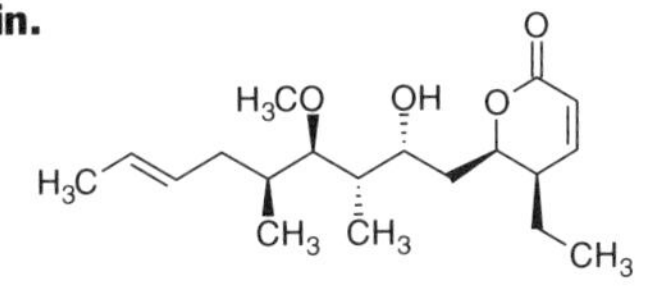

$C_{19}H_{32}O_4$, M_R 324.5, needles, mp. 78–79 °C, $[\alpha]_D^{20}$ –136.6° ($CHCl_3$). A plant growth regulator from *Streptomyces prunicolor* with antitumor and immunosuppressant activity.

Lit.: Heterocycles **45**, 7 (1997) (synthesis) ▪ J. Antibiot. **47**, 697, 703 (1994) (isolation) ▪ J. Chem. Soc., Chem. Commun. **1997**, 1043 ▪ J. Org. Chem. **60**, 7567 (1995) ▪ Tetrahedron Lett. **37**, 6615 (1996). – *[CAS 151519-02-7]*

Pisatin (6a-hydroxy-3-methoxy-8,9-methylenedioxypterocarpan).

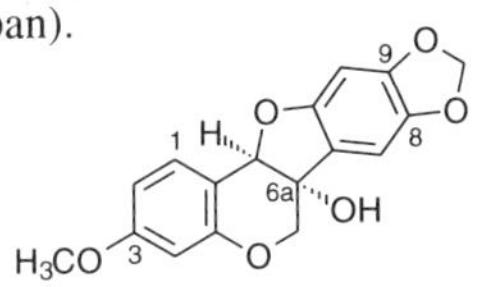

(+)-form

$C_{17}H_{14}O_6$, M_R 314.29, cryst., mp. 61 °C, $[\alpha]_D^{20}$ +280° (C_2H_5OH). The 6a-hydroxy-*pterocarpan (+)-(6a*R*,11a*R*)-P. is isolated from garden peas (*Pisum sativum*) infected with *Sclerotinia* spores and is a *phytoalexin. P. has antimycotic activity. The biosynthesis proceeds from maackiain, see pterocarpans. In pea tissue cultures (–)-(6a*R*,11a*R*)-maackiain is converted to (–)-(6a*S*,11a*S*)-P.

Lit.: Arch. Biochem. Biophys. **290**, 468–473 (1991) (biosynthesis) ▪ Can. J. Bot. **67**, 1698 ff. (1989) ▪ Harborne (1994), p. 166–183, 526–531 ▪ Justus Liebigs Ann. Chem. **1989**, 35–39 (synthesis) ▪ Luckner (3.), p. 414 ▪ Phytochemistry **21**, 2235–2242 (1982); **22**, 1591–1595 (1983) (biosynthesis) ▪ Tetrahedron Lett. **39**, 2739 (1998) (synthesis) ▪ Zechmeister **43**, 121–151. – *[CAS 469-01-2 ((+)-(6aR)-cis); 20186-22-5 ((–)-(6aS)-cis)]*

Pistillarin.

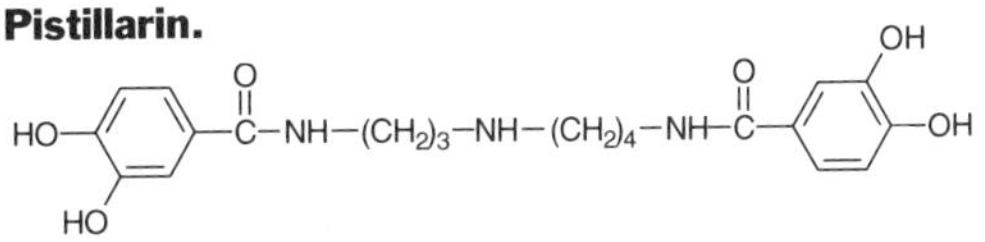

$C_{21}H_{27}N_3O_6$, M_R 417.46, amorphous solid. The fruit bodies of the giant club (*Clavariadelphus pistillaris*, Basidiomycetes) contains the bitter principle P., which is also responsible for the dark green color reaction of the toadstool with iron(III) chloride solution. P. is a spermidine derivative and is also found in bitter species of coral fungi (*Ramaria*, Basidiomycetes).

Lit.: Z. Naturforsch. C **39**, 10 (1984). – *[CAS 89647-69-8]*

Pityol.

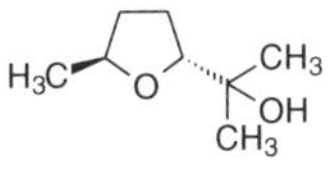

(+)-form

$C_8H_{16}O_2$, M_R 144.21, oil, bp. 89–91 °C (6.4 kPa), $[\alpha]_D^{24}$ +18° ($CHCl_3$). An aggregation pheromone of bark beetles of the genera *Pityogenes*[1] and *Conophthorus*[2]. P. is structurally related with *sulcatol, from which it can be formed by oxidation and ring closure.

Lit.: [1]Naturwissenschaften **74**, 343–345 (1987). [2]J. Chem Ecol. **21**, 143–167, 169–185 (1995).
gen.: ApSimon **9**, 370–371 (synthesis). – *[CAS 16194-32-4]*

Placodiolic acid.

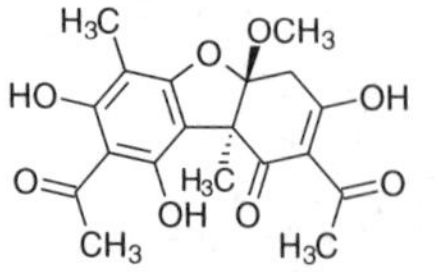

(relative configuration)

$C_{19}H_{20}O_8$, M_R 376.36, pale yellowish platelets, mp. 156–158 °C, $[\alpha]_D^{21}$ –231°, that change to yellow with KOH. An iso-*usnic acid derivative from one chemovariety of the lichen *Rhizoplaca chrysoleuca*.

Lit.: Phytochemistry **23**, 702 (1984) ▪ Tetrahedron Lett. **1981**, 351–352. – *[CAS 38965-63-8]*

Plagiochilins.

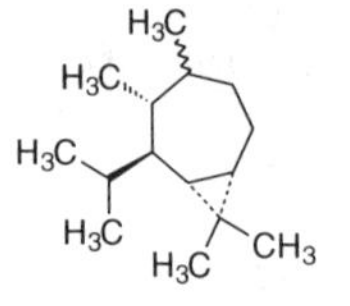

2,3-Secoaromadendrane skeleton

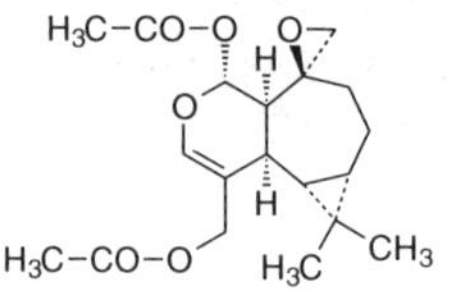

Plagiochilin A

*Sesquiterpenes (P. A–Q) of the 2,3-secoalloaromadendrane type, widely distributed in the liverwort genus *Plagiochila* (not to be confused with the macrocyclic bisbibenzyls, the plagiochins, also occurring in liverworts). Some show a surprising similarity to the monoterpenoid *valepotriates, e. g., valtrate, isolated from *Baldrian* species. The sesquiterpene hemiacetal *P. A*[1] ($C_{19}H_{26}O_6$, M_R 350.41, liquid, opt. activity 0) exhibits a series of biological activities. Thus, it was found to be an anti-feedant for the cotton plant pest known as the "African army worm" (= butterfly species) with a 10- to 100-fold higher activity than substances with similar properties from higher plants ($1-10$ ng · cm^{-2} in the "choice test"). The astringent sharp taste of numerous *Plagiochila* species is also due to this compound. Furthermore, it shows a similar toxicity to fish as sacculatal (see sacculatanes) and antitumor activity in the mouse P-338 lymphocytic leukemia and KB cell tests.

Lit.: [1]Chem. Pharm. Bull **42**, 1542 (1994).
gen.: Phytochemistry **19**, 2623 (1980); **20**, 1065 (1981); **44**, 89 (1997) ▪ Zechmeister **42**, 16. – *[CAS 67779-73-1 (P. A)]*

Plagiogyrins.

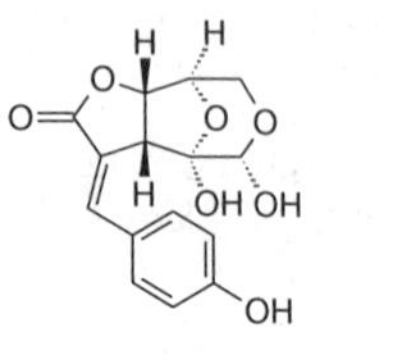

Plagiogyrin A

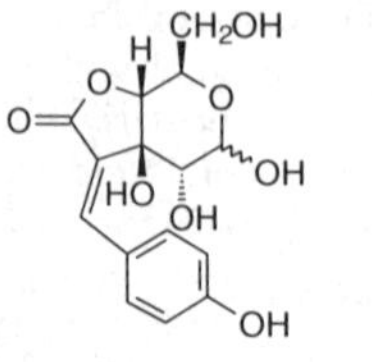

Plagiogyrin B

P. A: $C_{15}H_{14}O_7$, M_R 306.27, needles, mp. 223–225 °C, $[\alpha]_D^{25}$ +438°. *P. B:* $C_{15}H_{16}O_8$, M_R 324.29, oil, $[\alpha]_D^{22}$ +94.1°. Novel hydroxycinnamic acid derivatives isolated from two species of the fern genus *Plagiogyria*. Their biogenetic precursor is probably 4-*O*-*p*-coumaryl-D-glucose.

Lit.: Chem. Pharm. Bull. **32**, 1808, 1815 (1984). – *[CAS 91486-94-1 (P. A); 91486-95-2 (P. B)]*

Plakinic acids.

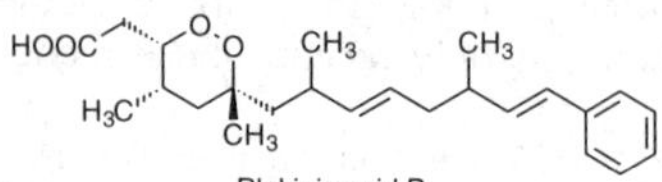

Plakinic acid B

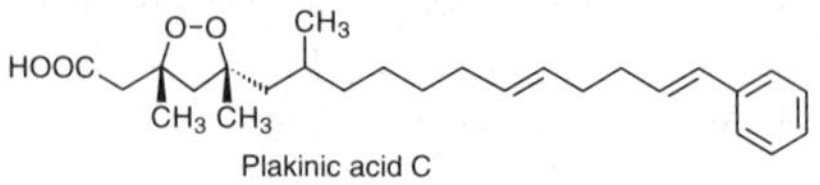

Plakinic acid C

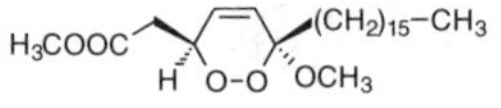

Chondrillin

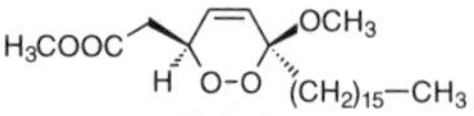

Plakorin

Marine sponges of the genus *Plakortis* produce cyclic peroxides of polyketide origin. Examples are *P. B* {$C_{24}H_{34}O_4$, M_R 386.53; methyl ester: $[\alpha]_D$ –186° ($CHCl_3$)} with antifungal activity and *P. C* {$C_{26}H_{38}O_4$, M_R 414.59; methyl ester: $[\alpha]_D$ +31.7° ($CHCl_3$)} with cytotoxic activity[1]. The stereoisomeric methoxyperoxides *chondrillin* {$C_{24}H_{44}O_5$, M_R 412.61, wax-like solid, mp. 30 °C, $[\alpha]_D$ +144° ($CHCl_3$)}[2] and *plakorin* {solid, mp. 42.5–43.5 °C, $[\alpha]_D$ +30.5° ($CHCl_3$)}[3] were isolated from *P. lita*. Plakorin is a potent activator of the Ca^{2+}-ATPase of the sarcoplasmic reticulum and may be of interest in studies on Ca^{2+} transport. In addition, it inhibits *in vitro* various cancer cell lines (IC_{50} ca. 1 μg/mL), while chondrillin has a markedly lower activity.

Lit.: [1]J. Am. Chem. Soc. **105**, 7735 (1983); J. Org. Chem. **56**, 6722 (1991). [2]J. Org. Chem. **64**, 1789 (1999); J. Nat. Prod. **53**, 926 (1990); J. Am. Chem. Soc. **114**, 1790 (1992); **119**, 3824 (1997) (synthesis); Tetrahedron Lett. **1976**, 2637; Tetrahedron **43**, 263 (1987). [3]Experientia. **45**, 898 (1989); J. Am. Chem. Soc. **119**, 3824 (1997) (abs. configuration); Nat. Prod. Rep. **9**, 289 (1992) (review). – *[CAS 87803-06-3 (P. B, methyl ester); 136707-69-2 (P. C, methyl ester); 61656-50-6 (chondrillin); 124264-01-3 (plakorin)]*

Plakorin see plakinic acids.

Plantain (Indian) see petasin.

Plant germination inhibitors. Plant substances that delay or prevent the germination of the diaspores of mostly alien species (see allelopathy). They thus act as natural herbicides. They are washed out from leaves and needles by rain water and thus reach the soil. Examples are *juglone that is released in its hydroquinone form in the soil by oxidation of the 4-*β*-D-glucoside[1]; *camphor as well as 1,8-*cineole that are excreted by *Salvia leucophylla* and *Artemisia californica* indigenous to the dry regions of California (so-called chaparral vegetation) and when in the soil kill other plants. Increased fungal growth following heavy rain can lead to a temporary degradation of these terpenes. The result is a diverse vegetation. The Australian eucalyptus behaves in a similar manner. Its transfer to other regions thus often leads to the extinction of the local vegetation. A number of other classes of compounds such as (e. g., L-tryptophan[2]), flavonoids[3], and volatile, short-chain (C_6–C_9) aliphatic compounds, alcohols, aldehydes [including (*E*)-2-*hexenals, nonanal] and ketones[4] are also plant germination inhibitors and some are of ecological significance. A further important role of plant germination inhibitors is to delay plant growth until favorable conditions for vegetation occur. Examples include *abscisic acid, *coumarin, *trans*-ferulic acid (see caffeic acid), *salicylic acid, and scopoletin. Vanillic acid and *p*-coumaric acid are endogenous germination inhibitors of soy bean seeds[5]. Germination may be delayed in several ways. Plant germination inhibitors can be eliminated by enzymatic degradation, decomposition under the action of air and light, or by washing out. In moderate climatic zones, a cold period is necessary for the degradation of the inhibitors. *Gloeosporone isolated from the spores of *Colletotrichum gloeosporioides* is an endogenous germination inhibitor (autoregulator) of this fungus[6].

Lit.: [1] Thomson, Naturally Occurring Quinones III, p. 154, London: Chapman & Hall 1987. [2] J. Chem. Ecol. **20**, 309–314 (1994). [3] J. Basic Microbiol. **32**, 241–248 (1992). [4] J. Chem. Ecol. **14**, 1633–1638 (1988); **16**, 645–666 (1990); **17**, 2193–2212 (1991). [5] J. Plant Physiol. **136**, 120f. (1990). [6] Chem. Pharm. Bull. **40**, 524–527 (1992).

Plant growth substances (growth regulators). General name for a chemically heterogeneous group of organic compounds with the common characteristic that they both stimulate general and specific plant growth processes as well as differentiation and development processes and also regulate and promote them by acting as catalysts or effectors[1]. While these growth substances in lower plants and bacteria are often essential milieu factors with vitamin character, higher plants have the ability to synthesize them. They are active even at very low concentrations and the site of formation is usually spatially separated from the site of action. They are thus also known as *plant hormones* (phytohormones) or as plant growth regulators (phytoregulators, growth regulators). Some of the compounds have particular uses in gardening (e.g. promotion of flowering), in agriculture to strengthen stems, crop protection to accelerate fruit ripening, and in storage to inhibit germination of potatoes.

Four main classes of natural and synthetic plant growth substances are distinguished: 1) The *3-indolylacetic acid derivatives (IAA) and other C-3 substituted indoles, previously known as auxins, induce a pronounced longitudinal growth, stimulate the formation of roots and flowers, the activity of the cambium, and influence fruit ripening and numerous other developmental processes in the plant. In the literature many compounds with different constitutions are also designated as auxins if they have activities similar to those of IAA. These include, e. g., 2-naphthyloxyacetic acid, benzoic acid, phenoxycarboxylic acid derivatives.

2) The *gibberellins do not all have the same biological activities[2].

3) The *cytokinins (previously: phytokinins) principally stimulate cell division and, together with *auxins and *gibberellins regulate the developmental and differentiation processes of fruit formation and ripening, bud formation, etc.[3].

4) *Abscisic acid acts as an inhibitor, is an antagonist of the other growth substances, and regulates the coloration, withering, and loss of leaves (senescence), fall of fruit, and winter rest.

5) Ethylene (C_2H_4, M_R 28.05, gas, mp. −169 °C, bp. −104 °C) is often considered as a plant growth substance because it acts like a fruit-ripening hormone. The biosynthesis of ethylene in the plant from *1-aminocyclopropanecarboxylic acid is stimulated by auxins; abscisic acid and cytokinins can – depending on the type of plant – have stimulating or inhibiting effects, and tissue injuries in plants lead to the formation of the so-called "wound ethylene". For information on the numerous roles of ethylene, see *Lit.*[4–6]. Other groups of substances with plant hormone character are the *leaf-movement factors[7], *brassinosteroids[8], jasmonates (see methyl jasmonate)[9], endogenous polyamines[10], and *salicylic acid[11].

In contrast to animal hormones, plant hormones have only low organ and effect specificities. Progress in the biology of plants has been decisively influenced by plant molecular biology[12]. Some genes and cDNAs regulated by plant hormones have been isolated and characterized[13]. However, a detailed understanding of the plant hormone receptors, in analogy to eukaryotic receptors, is still lacking. Even so, some plant hormone binding proteins and their genes have been isolated and characterized[13]; for detection, see *Lit.*[14]. The primary activities of some further physiologically active substances from plants such as the morphactins, blastokolins, *wilting (withering) agents, chlorocholine chloride, etc. have inhibitory rather than growth promoting character. Since the growth of plants results from the interactions between growth factors and inhibitors, both inhibitors and growth factors can be classified as plant hormones or growth regulators. In the living environment (biocenosis), the plants apparently also make use of substances that would be designated here as allelopathic agents, see phytoncides. For the influence of the reversible light-red/dark-red system on plant growth, see phytochrome.

Lit.: [1] Kirk-Othmer (3.) **18**, 16f. [2] Curr. Plant Sci. Biotechnol. Agric. **13**, 13–27 (1992); Proc. Plant Growth Regul. Soc. Am. (17.), **1990**, 94–105. [3] Naturwiss. Rundsch. **29**, 257–262 (1976). [4] Bull. Soc. Chim. Fr. **133**, 441 (1996); Trends Genet. **9**, 356 (1993) (biosynthesis). [5] Naturwissenschaften **71**, 210f. (1984). [6] Naturwiss. Rundsch. **37**, 135–140 (1984). [7] Angew. Chem. Int. Ed. Engl. **22**, 695–710 (1983). [8] ACS Symp. Ser. **474** (brassinosteroids) (1991); Phytochemistry **44**, 609 (1997); Tetrahedron **52**, 1997 (1996); Tetrahedron Lett. **38**, 2837 (1997). [9] Bot. Acta **104**, 446–454 (1991); Curr. Plant Sci. Biotechnol. Agric. **13** (Prog. Plant Growth Regul.), 276–285 (1992); Plant Physiol. **99**, 804 (1992). [10] Slocum, Flores & Hector (eds.), Biochem. Physiol. Polyamines Plants, p. 187–199, Boca Raton: CRC 1991; Curr. Plant Sci. Biotechnol. Agric. **13** (Prog. Plant Growth Regul.), 264–275 (1992). [11] Plant Physiol. **99**, 799 (1992). [12] Annu. Rev. Plant Physiol., Plant Mol. Biol. **42**, 529–551 (1991); Physiol. Plant. **90**, 230–237 (1994). [13] J. Plant Growth Regul. **12**, 197–205, 171–178 (1993). [14] Annu. Rev. Plant Physiol., Plant Mol. Biol. **44**, 107–129 (1993).

gen.: Chem. Soc. Rev. **6**, 261–276 (1977) ▪ J. Plant Growth Regul. **12**, 227–235 (1993) ▪ Naturwiss. Rundsch. **26**, 284-293 (1973); **32**, 152f. (1979); **33**, 480f. (1980) ▪ Plant Growth Regulators, Amsterdam: Hewin Internat. 1985 ▪ Zechmeister **34**, 187–284 ▪ Z. Chem. **20**, 397–406 (1980). – *[CAS 74-85-1 (ethylene).]*

Plant hormones (phytohormones) see plant growth substances.

Plant pigments. Collective name or pigments occurring in higher plants belonging to the *natural dyes that can be classified according to occurrence (*flower pigments, leaf pigments, fruit pigments, pigments of heartwood or roots) or to chemical constitution (*anthranoids, *carotinoids, quinones, *flavonoids, *anthocyanins, *betalains, etc.). In the past natural pigments and, especially plant pigments, were the main sources for textile dying. The longest known and most widely distributed pigments for commercial use include *alizarin(e), *indigotin, *saffron, *henna, pigments of the *dye-woods, etc. which today, except for those registered as cosmetic and/or food colorants, have mostly been replaced by synthetic pigments with better properties.

Lit.: Britton et al., The Biochemistry of Natural Pigments, Cambridge: Univ. Press 1983 ▪ see also natural dyes.

Plant waxes see 1-triacontanol and waxes.

Plasmalogens see ether lipids (alkoxylipids).

Plastoquinones (abbreviation: PQ).

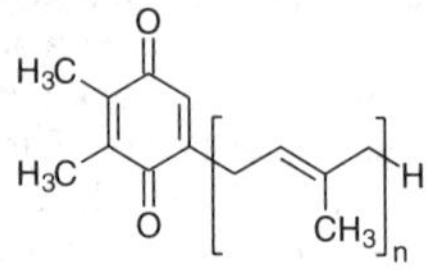

n = 9 : P. 9, PQ-9

Collective name for polyprenylated 2,3-dimethylbenzoquinone derivatives, [e.g., PQ-9 (n=9), $C_{53}H_{80}O_2$, M_R 749.22, bright yellow platelets, mp. 48–49 °C, uv_{max} 314 nm (petroleum ether)], isolated from chloroplasts. The P. play a role as redox substrates in photosynthesis for cyclic and non-cyclic electron transport where they are converted reversibly into the corresponding hydroquinones (plastoquinols). The biosynthesis of P. proceeds from *homogentisic acid, a product of L-tyrosine degradation, through prenylation and methylation (methyl group from L-methionine) to the plastoquinols which are dehydrated to the P.[1].

Lit.: [1] Luckner (3.), p. 415; Mann, Secondary Metabolism, 2nd edn., p. 266, Oxford: Clarendon Press 1987.
gen.: Biochim. Biophys. Acta **301**, 35 (1973) ▪ Eur. J. Biochem. **53**, 15–18 (1975) ▪ Lawlor, Photosynthese, Stuttgart: Thieme 1990. – *[CAS 4299-57-4 (PQ-9)]*

Platanic acid see betulinic acid.

Platelet activating factor (PAF, PAF acether).

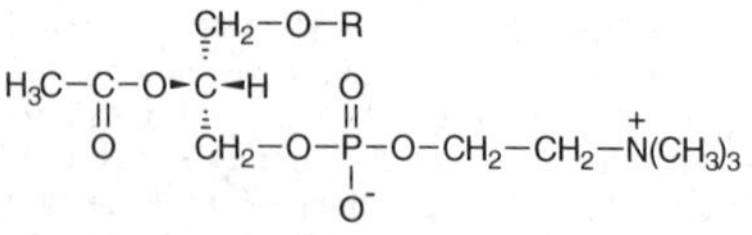

R = Alkyl

(Blood) *p*latelet *a*ctivating *f*actor; a physiologically active *ether lipid (alkoxylipid) that was first isolated from blood leukocytes. PAF is formed in many animal tissues as the result of an external stimulus and has the character of a hormone (so-called mediator). The chemical structure of PAF was elucidated in 1979 as a mixture of 1-*O*-alkyl-2-*O*-acetyl-*sn*-glycero-3-phosphocholines. PAF effects the aggregation of blood platelets (thrombocytes). Numerous physiological reactions, e.g., allergic inflammations and allergic asthma, are induced when PAF is liberated from macrophages after an antigen stimulus and initiates the release of granular components (e.g., *histamine) from white blood cells. PAF also participates in the development of other, especially allergic, clinical situations such as anaphylactic and endotoxin shocks. Because of these harmful activities antagonists have been developed to suppress the biological effects of PAF. These antagonists[1] were prepared synthetically or isolated from plants, e.g., the *ginkgolides from the maidenhair tree (*Ginkgo biloba*). For the mechanism of action of PAF, see *Lit.*[2]. PAF is formed biosynthetically by transfer of an acetyl group to *lyso-PAF* (1-*O*-alkyl-*sn*-glycero-3-phosphocholines). PAF is inactivated by cleavage of the acetyl groups, i.e., conversion to lyso-PAF[3].

Lit.: [1] Annu. Rev. Pharmacol. Toxicol. **27**, 237 (1987); Braquet (ed.), The Ginkgolides: Chemistry, Biology and Clinical Sciences, Barcelona: Prous Science 1987. [2] Pharmacol. Rev. **39**, 97 (1987). [3] Annu. Rev. Biochem. **55**, 483 (1986).
gen.: Benveniste et al., in Mangold & Paltauf (eds.), Ether Lipids: Biochemical and Biomedical Aspects, New York: Academic Press 1983 ▪ Drugs **42**, 9, 176 (1991) ▪ Progr. Lipid Res. **26**, 1 (1987) (review) ▪ Snyder (ed.), Platelet-Activating Factor and Related Lipid Mediators, New York: Plenum 1987 ▪ Synth. Commun. **21**, 1763 (1991) ▪ Winslow & Lee (eds.), New Horizons in Platelet Activating Factor Research, New York: Wiley 1987. – *[CAS 65154-06-5]*

Platynecine see necine(s).

Platyphylline.

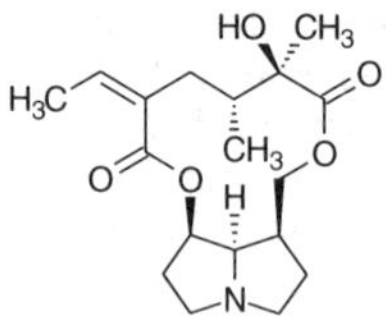

$C_{18}H_{27}NO_5$, M_R 337.42, mp. 129 °C, $[\alpha]_D$ –59° (C_2H_5OH). P. is a *pyrrolizidine alkaloid from various genera of the Asteraceae, tribe Senecioneae (*Adenostyles*, *Senecio*, *Petasites*). It exists in the plants as the *N*-oxide. P. does not exhibit hepatotoxicity, is relatively non-toxic, and has atropine-like activities. It was used in the former USSR for the treatment of stomach diseases.

Lit.: Beilstein E III/IV **27**, 6633 f. ▪ Hager (5.) **6**, 93, 674 ▪ Merck-Index (12.), No. 7691 ▪ Pelletier **9**, 155–233. – *[HS 2939 90; CAS 480-78-4]*

Plaunotol.

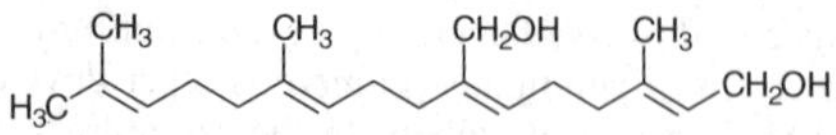

$C_{20}H_{34}O_2$, M_R 306.49, pale yellow, bitter oil, soluble in acetone. A *diterpene of the geranylgeraniol type. It is isolated from the leaves of the Euphorbiaceae *Croton sublyratus* (Plau-noi) and *C. columnaris*. Plau-noi is used in Thai traditional medicine. P. has anti-ulcerative activity with low acute toxicity; LD_{50} (mouse, rat p.o.) >8 g/kg. In Japan, P. is used as peptic ulcer therapeutic (trade name Kelnac®).

Lit.: Chem. Pharm. Bull. **26**, 3117 (1978) ▪ Kleemann-Engel, p. 1555 ▪ Merck-Index (12.), No. 7692. – *Synthesis:* Bioorg. Med. Chem. Lett. **9**, 1347 (1999) ▪ Bull. Chem. Soc. Jpn. **63**,

1328, 1625 (1990) ▪ Chem. Lett. **1988**, 1433 ▪ Tetrahedron Lett. **35**, 1581 (1994). – *[CAS 64218-02-6]*

Pleco-macrolides see macrolides.

Plectaniaxanthin see aleuriaxanthin.

Pleocidin see streptothricins.

Pleuromutilin.

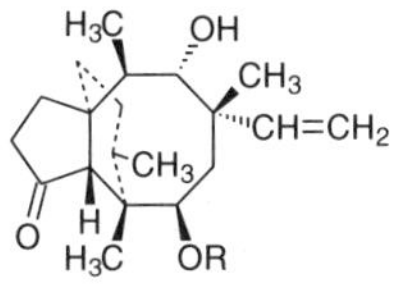

R = CO–CH$_2$OH : Pleuromutilin
R = H : Mutilin

$C_{22}H_{34}O_5$, M_R 378.51, cryst., mp. 170–171 °C, $[\alpha]_D$ +20° (C_2H_5OH/H_2O, 1:1). A diterpene antibiotic from cultures of the basidiomycete *Pleurotus mutilus* and related species. P. was the subject of intense synthetic research. P. and semi-synthetic derivatives inhibit protein biosynthesis. Some derivatives possess broad antibacterial activity against both Gram-positive and Gram-negative bacteria, as well as against *Chlamydia* and *Mycoplasma*. *Thiamulin*, a derivative of P. is applied in animal health. The related *mutilin* [1] ($C_{20}H_{32}O_3$, M_R 320.47, mp. 192 °C) has no antibiotic activity.

Lit.: [1] J. Antibiot. **29**, 125, 132, 915 (1976); Tetrahedron **36**, 1807 (1980).
gen.: Antibiotics (N. Y.) **5**, 344 (1979) (review) ▪ Beilstein E IV **8**, 1920 ▪ J. Org. Chem. **53**, 1441, 1450, 1461 (1988). – *Synthesis:* J. Am. Chem. Soc. **104**, 1767 (1982); **111**, 8284 (1989). – *[CAS 125-65-5 (P.); 6040-37-5 (mutilin)]*

Pleurotellic acid.

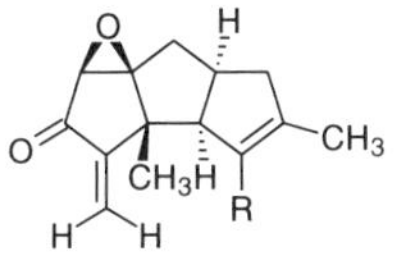

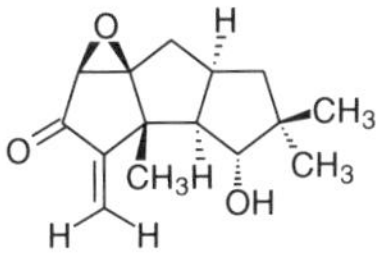

R = COOH : Pleurotellic acid
R = CH$_2$OH : Pleurotellol

Hypnophilin

$C_{15}H_{16}O_4$, M_R 260.29, cryst., mp. 148 °C, $[\alpha]_D$ –70.8° ($CHCl_3$). P. and *pleurotellol* {$C_{15}H_{18}O_3$, M_R 246.31, oil, $[\alpha]_D$ –86.5° ($CHCl_3$)} together with *hypnophilin* {$C_{15}H_{20}O_3$, M_R 248.32, oil, $[\alpha]_D^{20}$ –82.9° ($CHCl_3$)} are produced in cultures of the agaric *Pleurotellus hypnophilus* (Basidiomycetes). P. and pleurotellol possess a modified *hirsutane skeleton and inhibit, like hypnophilin, the growth of bacteria and fungi.

Lit.: Arch. Mikrobiol. **130**, 223 (1981) ▪ J. Am. Chem. Soc. **110**, 5064 (1988) ▪ J. Org. Chem. **52**, 4647 (1987) (synthesis of hypnophilin) ▪ Justus Liebigs Ann. Chem. **1991**, 893 ▪ Tetrahedron **42**, 3587 (1986). – *[CAS 80677-94-7 (P.); 80677-95-8 (pleurotellol); 80677-96-9 (hypnophilin)]*

Pleurotin.

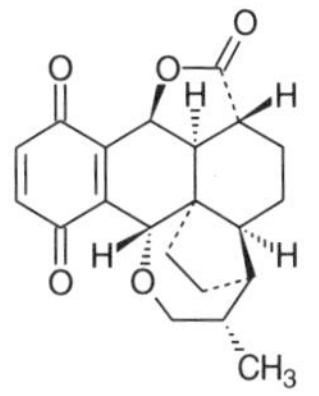

$C_{21}H_{22}O_5$, M_R 354.40, orange needles, mp. 200–215 °C (decomp.), $[\alpha]_D$ –20° ($CHCl_3$). A benzoquinone derivative from cultures of *Pleurotus griseus*, *Hohenbuehelia geogenia*, and *Nematoctonus robustus* (Basidiomycetes). P. and some related metabolites exhibit antibacterial, antifungal, and cytotoxic activities.

Lit.: Beilstein E V **19/5**, 385 ▪ J. Am. Chem. Soc. **110**, 1634 (1988); **111**, 7507–7519 (1989) (synthesis) ▪ Nat. Prod. Lett. **4**, 209 (1994) ▪ Pure Appl. Chem. **17**, 331 (1978). – *[HS 2941 90; CAS 1404-23-5]*

Plicadin see coumestrol.

Plicamycin see aureolic acid.

Plipastatins.

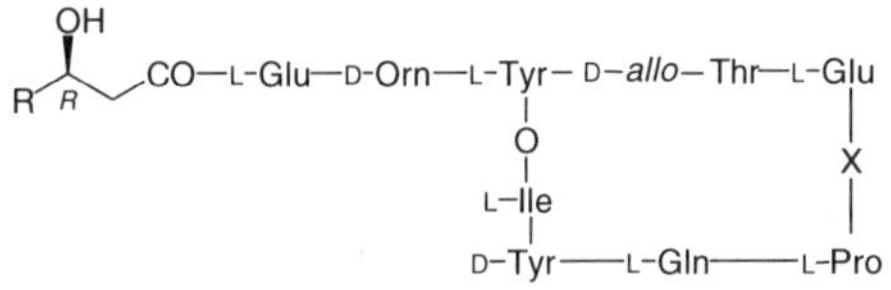

R = –(CH$_2$)$_{12}$–CH$_3$, X = D-Ala : P. A$_1$
R = –(CH$_2$)$_{10}$–CH(CH$_3$)–CH$_2$–CH$_3$ (*S*), X = D-Ala : P. A$_2$
R = –(CH$_2$)$_{12}$–CH$_3$, X = D-Val : P. B$_1$
R = –(CH$_2$)$_{10}$–CH(CH$_3$)–CH$_2$–CH$_3$ (*S*), X = D-Val : P. B$_2$

Table: Data of Plipastatins.

P.	molecular formula	M_R	mp. (K^+ salts) [°C]	CAS
A_1	$C_{72}H_{110}N_{12}O_{20}$	1463.73	193–195	103651-09-8
A_2	$C_{73}H_{112}N_{12}O_{20}$	1477.76	188–190	103955-72-2
B_1	$C_{74}H_{114}N_{12}O_{20}$	1491.79	186–187	103955-73-3
B_2	$C_{75}H_{116}N_{12}O_{20}$	1505.81	184–186	103955-74-4

A lipopeptide antibiotic complex from cultures of *Bacillus cereus*. As phospholipase A$_2$, C, and D inhibitors the P. have antiallergic and bronchodilatory activities. They are yellow solids and have been characterized as the potassium salts, see formula and table.

Lit.: J. Antibiot. (Tokyo) **39**, 737, 745, 755, 860 (1986). – *[HS 2941 90]*

PLMF. Abbreviation for periodic *leaf-movement factors.

Plumbagin (2-methyl-5-hydroxy-1,4-naphthoquinone).

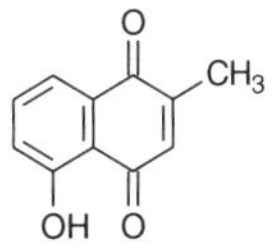

$C_{11}H_8O_3$, M_R 188.18, orange-yellow needles, mp. 78–79 °C, uv$_{max}$ 418 nm (C_2H_5OH). The hexaketide P. occurs in the herbage of the sundew (*Drosera rotundifolia*, Droseraceae) as well as in the roots and leaves of leadwort (*Plumbago*, Plumbaginaceae) and *Diospyros* species (Ebenaceae). P. exhibits bactericidal, fungicidal, and tuberculostatic activities.

Lit.: Chem. Pharm. Bull. **34**, 2810 (1986) ▪ Fitoterapia **1990**, 387 ▪ Indian J. Chem. B **36**, 1044 (1997) (synthesis) ▪ Karrer, No. 1213 ▪ Merck-Index (12.), No. 7697 ▪ Tetrahedron Lett. **39**, 8445 (1998) (biosynthesis). – *[HS 2914 69; CAS 481-42-5]*

Plumeria iridoids. The *Plumeria* plants belong to the family Apocynaceae and grow only in tropical and subtropical regions. They contain *iridoids which are structurally very similar to *allamandin and allamandicin occurring in *Allamanda* species of the Apocynaceae family.

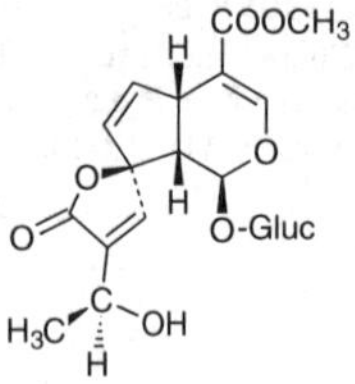

Plumieride (**1**)

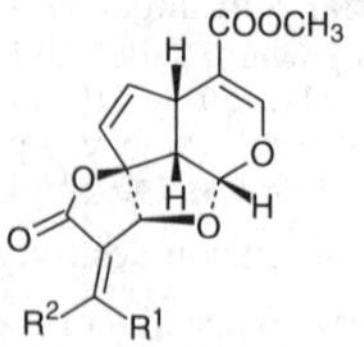

$R^1 = CH_3$, $R^2 = H$: Plumericin (**2**)
$R^1 = H$, $R^2 = CH_3$: Isoplumericin (**3**)

Occurrence: Plumieride is an iridoid glucoside and occurs in the bark of *Plumeria acutifolia* (content: ca. 4%), while the two non-glucosidic iridoids occur in *P. rubra* var. *alba* (contents: 0.05 to 0.18%).

Table: Data of Plumeria iridoids.

no.	molecular formula	M_R	mp. [°C]	$[\alpha]_D$	CAS
1	$C_{21}H_{26}O_{12}$	470.43	224–225	–80° (CH_3OH), –114° (H_2O)	511-89-7
2	$C_{15}H_{14}O_6$	290.27	211.5–212.5	+197.5° ($CHCl_3$)	77-16-7
3	$C_{15}H_{14}O_6$	290.27	200.5–201.5	+216.4° ($CHCl_3$)	31298-76-7

Biosynthesis: Plumieride is a condensation product of an iridoid glucoside with acetoacetic acid.
For synthesis, see *Lit.*[1]. Plumericin has *in vitro* activity against fungi, Gram-positive and Gram-negative bacteria as well as *Mycobacterium tuberculosis* 607.

Lit.: [1] J. Am. Chem. Soc. **108**, 4974 (1986).
gen.: Fitoterapia **68**, 478 (1997) (plumericinic acid) ▪ J. Nat. Prod. **43**, 649 ff. (No. 110, 192–195) (1980) ▪ J. Chem. Soc., Perkin Trans. 1 **1988**, 1119. – *[HS 2938 90]*

Plum flavor see fruit flavors.

Pneumocandins. A group of antimycotically active lipopeptide antibiotics from cultures of *Cryptosporiopsis*, *Pezicula*, *Glarea*, and *Zalerion* sp., e.g., *P.* A_0: $C_{51}H_{82}N_8O_{17}$, M_R 1079.25, powder, mp. 206–214 °C (decomp.). P. are active against the pneumonia pathogen *Pneumocystis carinii*.

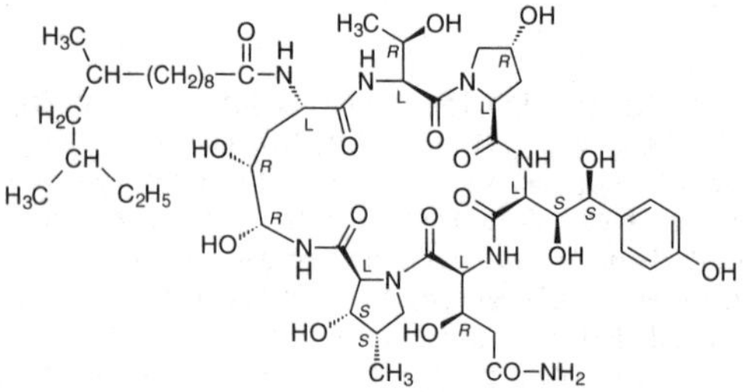

Pneumocandin A_0

The P. are structurally related to *echinocandin B and, like the latter, represent important lead structures for the discovery of new, systemic antimycotic agents. They act as inhibitors of fungal 1,3-*β*-D-glucan synthase (EC 2.4.1.34) and thus inhibit construction of the fungal cell wall.

Lit.: Clin. Dermatol. **7**, 375, 421 (1993) (review) ▪ J. Am. Chem. Soc. **113**, 3542 (1991) (biosynthesis) ▪ J. Antibiot. **42**, 163, 168, 174 (1989); **45**, 1853–1891, 1953 (1992); **47**, 755 (1994) (isolation) ▪ J. Chromatogr. **831**, 217 (1999) (purification). – *[CAS 120692-19-5 (P. A_0)]*

Podocarpic acid.

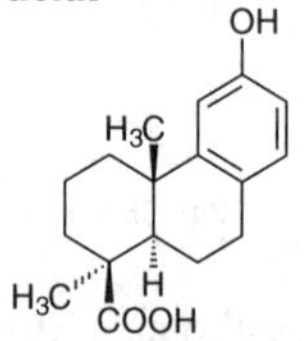

(+)-form

$C_{17}H_{22}O_3$, M_R 274.36, needle-like cryst., mp. 193–194 °C, $[\alpha]_D$ +164° (C_2H_5OH). The compound is isolated mainly from numerous members of the Podocarpaceae (e.g., *Podocarpus cupressina*, *P. totara*, *P. dacrydioides*, *P. hallii*). P. is a major component of *Podocarpus* resin and thus belongs to the group of resin acids.

Lit.: Beilstein E IV **10**, 1140 ▪ Phytochemistry **22**, 1163 (1983). – *Synthesis:* Aust. J. Chem. **46**, 1447–1471 (1993) ▪ Bioorg. Med. Chem. Lett. **9**, 469 (1999) (analogues) ▪ J. Org. Chem. **50**, 3659 (1985); **55**, 3952 (1990) ▪ J. Chem. Soc., Perkin Trans. 1 **1992**, 1505 ▪ Tetrahedron: Asymmetry **8**, 2793 (1997). – *[CAS 5947-49-9]*

Podophyllotoxin (podophyllic acid lactone).

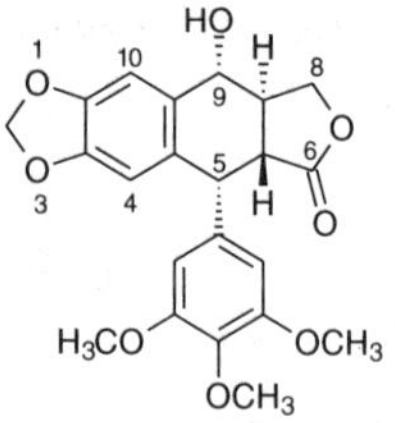

$C_{22}H_{22}O_8$, M_R 414.41, cryst., mp. 183–184 °C, 114–118 °C (monohydrate), $[\alpha]_D^{20}$ –132.7° ($CHCl_3$). A *lignan of the aryltetralin type from rhizomes of *Podophyllum hexandrum* (syn. *P. emodi*, Berberidaceae) and other *Podophyllum* roots (up to 0.2–4% of the dry weight and also present in savine oil). P. exists in the plant as a glycoside, e.g., the 9-*O*-*β*-D-glucoside, $C_{28}H_{32}O_{13}$, M_R 576.55, amorphous solid, mp. 149–152 °C, $[\alpha]_D^{20}$ –65° (H_2O). P. is the main component (up to 20%) of *podophyllin* obtained from the ethanol extract of the roots of *Podophyllum peltatum*, a popular ornamental plant from North America (mandrake, mayapple). Podophyllin precipitates on addition of water to the extract and is obtained as yellow, resinous, bitter-tasting lumps or as a powder. Podophyllin contains other lignans besides P., e.g. *α*- and *β*-*peltatin, *9-deoxypodophyllotoxin* (anthricin), $C_{22}H_{22}O_7$, M_R 398.41, cryst., mp. 170–171 °C, $[\alpha]_D^{19}$ –119° ($CHCl_3$), and *dehydropodophyllotoxin* (5,5*α*,8a,9-tetradehydropodophyllotoxin), $C_{22}H_{18}O_8$, M_R 410.38, needles, mp. 272–278 °C.

Activity: P. is a strong irritant of the skin and mucous membranes, has laxative activity, and is a mitotic toxin since it inhibits topoisomerase II. Like 9-deoxypodophyllotoxin, it is cytotoxic and possibly cancerogenic. P. shows anti-HIV-activity.

Toxicology: LD_{50} (rat i.v.) 8.7 mg/kg, (rat i.p.) 15 mg/kg. P. is also a strong toxin on oral administration: LD_{50} (rat p.o.) 100 mg/kg. In humans P. poisonings (consumption of ca. 2 g) result in severe gastroenteritis, vertigo, and shock states.
Use: P. is used experimentally as an antitumor principle. Like podophyllin, it is suitable for the topical treatment of genital warts [1]. Semisynthetic glycosides of P., namely etoposide and teniposide ($C_{32}H_{32}O_{13}S$, M_R 656.66) are used clinically as cytostatic agents [2].
Lit.: [1] Pat. Appl. EP 351974 (Schering, 24.1.1990); Lancet **1987**, 513; **1989**, 831. [2] Drug Discovery Today **1**, 343–351 (1996); Kleemann-Engel, p. 1830.
gen.: Adv. Enzyme Regul. **27**, 223–256 (1987) ▪ Bioorg. Med. Chem. Lett. **7**, 2897 (1997) ▪ Curr. Med. Chem. **5**, 205–252 (1998) (review activity) ▪ Hager (5.) **3**, 938 ff.; **9**, 277 ff. ▪ J. Chem. Soc., Chem. Commun. **1993**, 1200 ▪ J. Chem. Soc., Perkin Trans. 1 **1996**, 151–155 ▪ J. Org. Chem. **65**, 847 (2000) (synthesis) ▪ Luckner (3.), p. 396, 478 ▪ Nat. Prod. Rep. **12**, 101–133, 183–205 (1995) ▪ Phytochemistry **25**, 2093–2102 (1986) (biosynthesis) ▪ Prog. Drug Res. **33**, 169–266 (1989) (review) ▪ Synthesis **1992**, 719–730 (Synth.) ▪ Tetrahedron **47**, 4675–4682 (1991); **50**, 10829 (1994) ▪ Tetrahedron Lett. **37**, 4791 (1996) (synthesis). – *Toxicology:* Sax (8.), PJJ 225. – *Pharmacology:* Cancer Res. **4**, 626 (1984) ▪ J. Med. Chem. **32**, 604 (1989) ▪ J. Nat. Prod. **52**, 606–613 (1989); **58**, 870 (1995). – *[CAS 518-28-5 (P.); 16481-54-2 (9-O-β-D-glucoside); 19186-35-7 (9-deoxy-P.); 110087-42-8 (dehydro-P.)]*

Podoscyphic acid [(*E*)-4,5-dioxo-2-hexadecenoic acid].

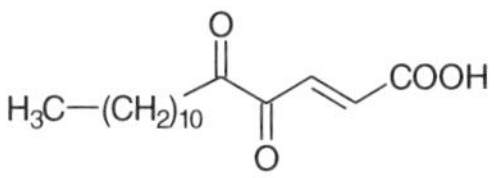

$C_{16}H_{26}O_4$, M_R 282.38, yellowish solid, mp. 105 °C. An unsaturated dioxocarboxylic acid from cultures of a Tasmanian *Podoscypha* species (Basidiomycetes). P. inhibits reverse transcriptases of viruses of the avian myeloblastosis and Moloney murine leukemia types.
Lit.: Z. Naturforsch. C **46**, 442 (1991). – *[CAS 135863-69-3]*

Poecillanosine [(*S*)-1-(hydroxynitrosoamino)-2-heptadecanol-*O*-acetate].

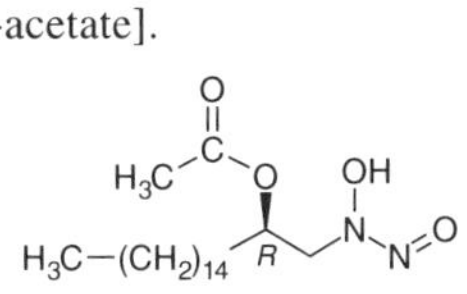

$C_{19}H_{38}N_2O_4$, M_R 358.52, solid, $[\alpha]_D^{25}$ −20.2° (CH_3OH), isolated from the marine sponge *Poecillastra tenuilaminaris*. P. shows antileukemic properties. Other similar natural *N*-hydroxy-*N*-nitrosoamines are the *nitrosoxacins.
Lit.: Tetrahedron Lett. **38**, 8349 (1997). – *[CAS 200564-72-3]*

Poison ivy, poison oak see urushiol(s).

Pokeweed (*Phytolacca americana*, Phytolaccaceae). The fruits of this 1–3 m high plant, indigenous to North America and cultivated as an ornamental plant in North Africa and Europe, contain the betacyan dye betanin. Indians in Virginia and New England prepared a dye concentrate from P. and used it to dye basketwork, furs, and leather. In addition P. juice was misused in Portugal in the past to adulterate red wine.
Lit.: Helv. Chim. Acta **67**, 1310 (1984).

Polyacetylenes see polyynes.

Polyamines. Collective name for naturally occurring aliphatic amines containing more than one amino group, such as, e.g., *putrescine, *cadaverine, *agmatine, *spermidine, *spermine. P. occur in all organisms. In spite of their ubiquitous occurrence their function is only poorly understood. They presumably participate in numerous biological processes in the cell (cell proliferation, DNA stacking, membrane stabilization). In bacteria the abundant P. are often replaced by unusual derivatives; in such cases the P. can be used as taxonomic markers. The name P. alkaloids is applied to those alkaloids that contain P. as building blocks [1]. A typical P. alkaloid of the putrescine type is *aerothionin isolated from marine sponges or penaramide A. P. alkaloids of the spermidine type include the bacterial *siderophores such as the *agrobactins, the oncinotines (see inandenine) found in flowering plants, and *lunarines as well as the *palustrines isolated from horsetails. A P. alkaloid of the spermine type is *homaline. Structurally related to the P. alkaloids are the P. toxins occurring in *spider, *scorpion, and *wasp venom [2]. Spermine was first described in 1678 as the phosphate (crystals, from sperm) [3].
Lit.: [1] Chem. Unserer Zeit **32**, 206–218 (1998); Manske **22**, 85–188. [2] Manske **45**, 1–125. [3] R. Soc. London Philosoph. Trans. **12**, 1048 (1678).
gen.: Gangas, Polyamines in Biomedical Research, New York: Wiley 1980 ▪ Morris & Morton, Polyamines in Biology and Medicine, New York: Dekker 1981. – *Bacteria:* Crit. Rev. Microbiol. **18**, 261–283 (1992). – *Plants:* Annu. Rev. Plant Physiol. **40**, 235–269 (1989) ▪ Galston & Smith, Polyamines in Plants, Dordrecht: Nijhoff 1985 ▪ Plant Physiol. **95**, 406–410 (1990). – *Animal organisms:* J. Org. Chem. **63**, 4838 (1998) ▪ McCann, Pegg & Sjoerdsma (eds.), Inhibition of Polyamine Metabolism, Orlando: Academic Press 1987.

Polyene macrolides see macrolides.

Polyether antibiotics. P. a. comprise a large group of structurally related natural products produced principally by *actinomycetes through the *polyketide pathway. They consist of a linear chain of C atoms as skeleton which is cyclized to form tetrahydrofuran and/or tetrahydropyran rings at certain positions, this can also result in spiroketal structures. There is always a free or esterified carboxy group at one end of the chain. Some P. a. are additionally *O*-glycosylated or contain aromatic or heterocyclic building blocks (e.g., *calcimycin). It has been suggested that the P. a. should be classified according to structural features, e.g., P. a. without a spiroketal unit (*lasalocids), those with a spiroketal unit (*monensin, *nigericin), those with a dispiroketal unit (*salinomycin), further subdivisions can be made according to the presence of sugar residues. The prominent property of the P. a. is their ability to form mostly well crystallizing complexes with mono- or divalent cations. Accordingly the P. a. are also designated as *ionophores. The P. a. can also be classified under consideration of their ion selectivities.
Activity: Numerous biological activities of the P. a. have been described, although these properties often vary greatly between the individual compounds. They have been reported to have activity against bacteria (Gram-positive and Gram-negative), fungi, yeasts, viruses (including HIV), protozoa (coccidi, *Plasmodium falciparum*), insects, worms, and cancer cells.

Use: Medical use is as yet not possible on account of the usually high toxicity of the P. a. On the other hand they have already found wide application in animal breeding (e.g., of poultry). In these cases the P. a. are active against coccidiosis and at the same time are growth promoters. In addition the P. a. are important tools in biochemical investigation of ion transport through the cell membrane and serve as lead structures for the development of synthetic crown ethers. On account of the significant marketing potential of the P. a. this class of compounds has been the subject of intensive research: more than 100 natural products have been isolated, numerous syntheses have been developed, and biosynthetic investigations have considerably added to our knowledge of polyketide metabolism.

Lit.: Antimicrob. Agents Chemother. **36**, 492 (1992); **40**, 602 (1996) (activity) ▪ Westley (ed.), Polyether Antibiotics, vols. 1 & 2, New York: Dekker 1982/83. – *Biosynthesis:* Angew. Chem. Int. Ed. Engl. **34**, 297 (1995) ▪ Birch & Robinson, in Vining & Stuttard (eds.), Genetics and Biochemistry of Antibiotic Production, p. 443–476, Boston: Butterworth-Heinemann 1995 ▪ Zechmeister **58**, 1–81. – *Synthesis:* J. Org. Chem. **59**, 2848–2876 (1994) ▪ Lukacs & Ohno (eds.), Recent Progress in Chemical Synthesis of Antibiotics, vol. 1, p. 447–466, Berlin: Springer 1990 ▪ Nicolaou, p. 185 ff., 227 ff.

Polyetherin A see nigericin.

Polyfibrospongols. Meroterpenoid vanillic acid derivatives from the Taiwanese marine sponge *Polyfibrospongia australis* with antitumor and antibacterial properties. Sesquiterpene quinones could also be useful in the treatment of AIDS, see avarol from *Dysidea avara*. *P. A:* $C_{24}H_{34}O_4$, M_R 386.5, amorphous solid, $[\alpha]_D^{25}$ +5.3° ($CHCl_3$) and *P. B:* $C_{24}H_{34}O_5$, M_R 402.5, amorphous solid, $[\alpha]_D^{25}$ +5.3° ($CHCl_3$).

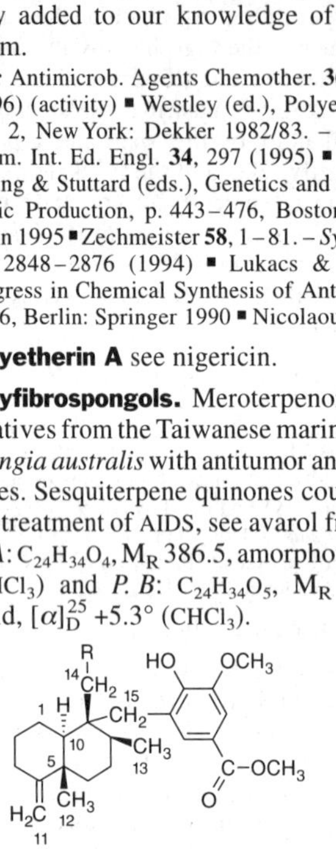

Biosynthesis: P. are sesquiterpene-shikimate metabolites.

Lit.: J. Nat. Prod. **60**, 93 (1997). – *[CAS 185618-01-3 (P. A); 185618-02-4 (P. B)]*

Polyfructosans. Oligosaccharides with ca. 10–40 *fructose units. The P. are *storage carbohydrates* in the underground parts of Asteraceae, grasses, and cereals. Rye contains *graminin* (trifructosan) made up of 10 fructose units, see also inulin.

Lit.: Adv. Carbohydr. Chem. **2**, 253–277 (1946).

Polygodial (tadeonal, drim-7-ene-11,12-dial).

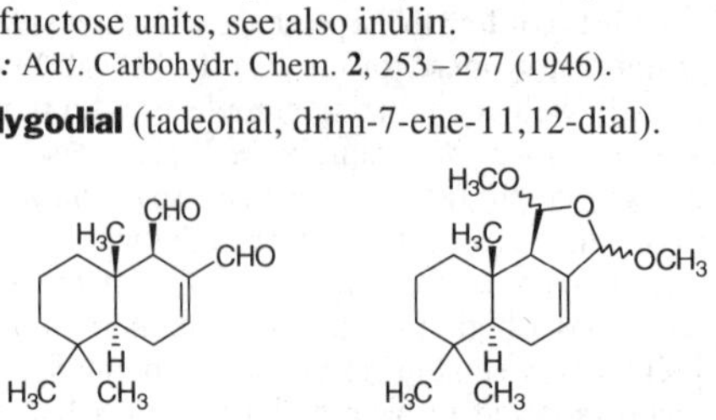

Polygodial Polygodial acetal

The hot tasting P., $C_{15}H_{22}O_2$, M_R 234.34, colorless cryst., mp. 50 °C, $[\alpha]_D^{25}$ –131° ($CHCl_3$), is produced together with *warburganal and some other sesquiterpenes of the *drimane type as the strong insect antifeedant components by Polygonaceae, e.g., *Polygonum hydropiper*, water pepper (leaves and seeds)[1]. P. has antimicrobial activity, inhibits plant growth, is cytotoxic, and is a fish toxin[2]. P. also occurs in East African *Warburgia* species and tree barks[3]. P. together with warburganal and *P. acetal* ($C_{17}H_{28}O_3$, M_R 280.41) is used by marine slugs for chemical defence.

Lit.: [1] Phytochemistry **21**, 2895 (1982); **24**, 1521 (1985); **28**, 1631 (1989); J. Chem. Soc. Chem. Commun. **1976**, 1013; **1981**, 507; Science **219**, 1237 (1983); Agric. Biol. Chem. **52**, 1409 (1988); Tetrahedron Lett. **1971**, 1961; **22**, 2995 (1981). [2] Chem. Ind. (London) **1985**, 735; Tetrahedron Lett. **23**, 3295 (1982). [3] Planta Med. **44**, 134 (1982). [4] J. Org. Chem. **48**, 1866 (1983). *gen.:* Merck-Index (12.), No. 7732 ▪ Tetrahedron **51**, 1845–1860, 7435 (1995); **53**, 6313 (1997) (synthesis). – *[CAS 6754-20-7]*

Polyhydroxy alkaloids. Collective name for mono- and bicyclic *alkaloids that, on account of their good solubility in water, are not susceptible to the usual alkaloid extraction processes and are thus often overlooked. The number of newly discovered P. a. is increasing continuously. A common property of the P. a. is their pronounced activity as glycosidase inhibitors.

Structure & occurrence: The ca. 30 known P. a., occurring frequently in the Fabaceae family, can be classified in five structural types on the basis of their polyhydroxylated ring systems (figure).

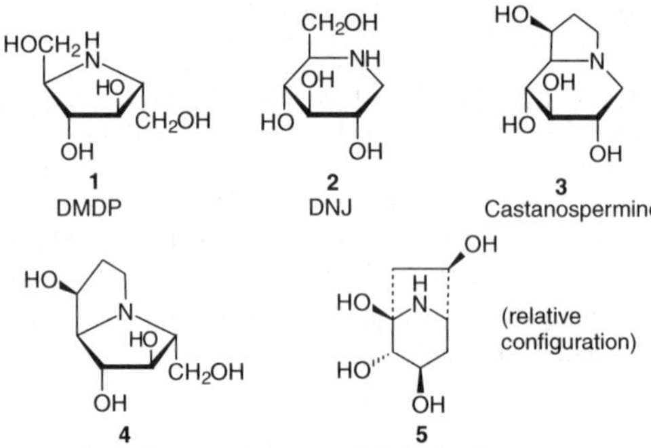

1 DMDP; 2 DNJ; 3 Castanospermine; 4 Australine; 5 Calystegine B_1 (relative configuration)

Figure: Polyhydroxy alkaloids.

1) Pyrrolidines such as, e.g., *(2*R*)-3β,4α-dihydroxy-2α,5β-pyrrolidinedimethanol (DMDP); – 2) Piperidines such as *deoxynojirimycin (DNJ), *deoxymannojirimycin, and *galactostatin first detected in microorganisms and later also found in the mulberry tree; – 3) Indolizidines such as *swainsonine and *castanospermine; – 4) Pyrrolizidines, such as australine, occurring together with castanospermine in the Australian Fabacee *Castanospermum australe*, and *alexine detected in the Brazilian Fabacee *Alexa leiopetala*; – 5) Tropane derivatives (*calystegines) such as calystegine B_1. Calystegines were first found in *Calystegia* (Convolvulaceae), and later in other families, especially the Solanaceae.

Activity: The potential of P. a. as highly effective glycosidase inhibitors becomes apparent when they are drawn as azapyranose or azafuranose derivatives of known monosaccharides (figure). As aza analogues (sugar mimics), the P. a. are not only potent defensive substances for their plant producers but are also of major pharmaceutical-medical interest for the development and production of selective glycosidase inhibitors.

Lit.: New Sci. **15**, 40–41 (1985); **16**, 45–48 (1989) ▪ Pelletier **5**, 1–54 ▪ Phytochem. Analysis **4**, 193–204 (1993) ▪ Rec. Adv. Phytochem. **23**, 395–427 (1989) ▪ Rosenthal & Berenbaum (eds.), Herbivores: their Interactions with Secondary Plant Metabolites, vol. 1, p. 79–121, San Diego: Acad. Press 1991.

Poly(β-hydroxyalkanoates) [poly(β-hydroxyalkanoic acids) PHA, PHAA]. Numerous bacteria form poly(β-hydroxyalkanoates) (PHA) as reserve or storage substances that can serve as sources of carbon and energy when required. Monomeric components of the PHA include β-hydroxybutyric acid, β-hydroxyvaleric acid, and *lactic acid. The linear polymer poly(β-hydroxybutyric acid) (PHB, PHBA) is accumulated intracellularly in the form of grains or granules and is one of the more important PHAA. As the polyester it is soluble in chloroform, insoluble in ether and consists on average of 60 (max. 25000) hydroxybutyric acid units.

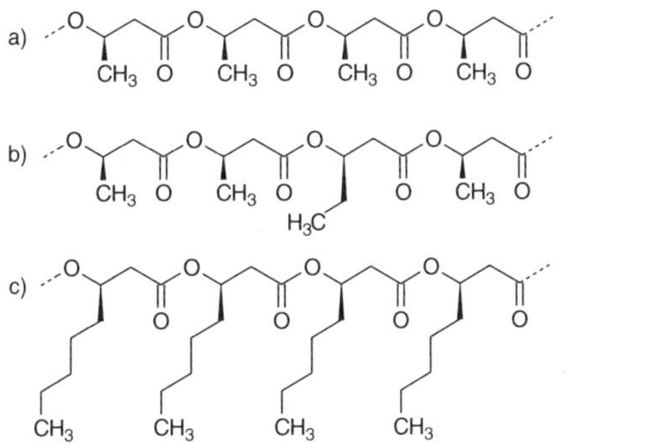

Figure: Poly(β-hydroxyalkanoates): a) poly(β-hydroxybutyric acid), b) copolyester of β-hydroxybutyric acid and β-hydroxyvaleric acid, c) homopolyester of β-hydroxyoctanoic acid.

PHB is only formed in bacteria (e.g., *Bacillus megaterium, Alcaligenes eutrophus*) [1].

Biosynthesis: The biosynthesis proceeds under nitrogen-deficient conditions from acetyl-coenzyme A via β-hydroxybutyryl-CoA. Under specific growth conditions PHB can constitute up to 80% of the dry weight of *A. eutrophus*.

Use: PHB and other PHA are gaining economic significance as completely biologically degradable polymers with thermoplastic properties. The plastic obtained by thermal processing of PHB at ca. 170 °C is brittle. By use of modified growth techniques a copolymer in which 5–20% of the hydroxybutyric acid molecules have been replaced by polyhydroxyvaleric acid (PHV) is obtained. The resultant PHB/PHV ("PHB-V") is less brittle, more flexible, and has a lower mp.(ca. 130 °C) than PHB, its physical properties are similar to those of polypropylene and it can be used to manufacture plastic foils and bags. The polymer commercially available under the name "Biopol" is completely degraded within 4–6 weeks under waste-water processing conditions, however, it is still very expensive in comparison to the plastics obtained from crude oil. Special uses, e.g., in medicine are feasible. Further copolymerizates can be produced by fermentation with other carbon sources. A long-term objective is the incorporation of the bacterial gene for PHB/PHA production in *E. coli* bacteria or crop plants (transgenic plants), in order to cultivate these biopolymers in grain crops, potatoes, etc. (regenerating raw materials).

Lit.: [1] Labor 2000 **1988**, 148–155.

gen.: Angew. Chem. Int. Ed. Engl. **32**, 477 (1993) ▪ Forum Mikrobiol. **1989**, 190–198 ▪ Helv. Chim. Acta **65**, 495–503 (1982) ▪ Nachr. Chem. Tech. Lab. **39**, 1112–1124 (1991) ▪ Naturwiss. Rundsch. **43**, 173 f. (1990) ▪ Science **245**, 1187 (1989) ▪ Ullmann (5.) **A 13**, 515.

Poly(β-hydroxybutyric acid) see poly(β-hydroxyalkanoates).

Polyketides. Collective name for natural products produced biosynthetically by way of poly(β-oxocarboxylic acids). The name was derived in 1907 by Collie on the basis of the hypothesis that natural products may be formed by multiplication of ketene ($H_2C=C=O$) units. The P. chains are constructed on multienzyme complexes (*polyketide synthases*) from acetyl- and malonyl-CoA (*acetogenins) or also by use of propionyl- and/or butyryl-CoA. Depending on the number of building blocks the natural products are classified as triketides (n = 3), tetraketides (n = 4), etc.

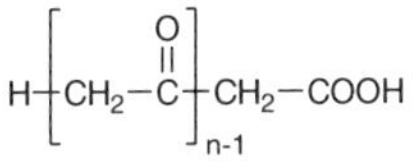

Poly(β-oxocarboxylic acid) from acetate only

The poly(β-oxocarboxylic acids) with n = 3 or more are not stable in the free form, they are partially reduced or cyclized by ester or aldol condensation while still on the enzyme. By variations of the starter units, the sequence of the building blocks acetate/propionate/butyrate, the chain length (up to n > 20), the possible changes in the reduction steps and the chain folding prior to condensation more than 10^8 different P. are possible. In addition, there are also possibilities to alter the basic skeletons released from the P. synthases in the further stages of the biosynthesis, e.g. by oxidation, methylation, rearrangement, and/or glycosylation. P. are typical secondary metabolites of microorganisms and plants. The genetics of P. synthases have been investigated intensively. Novel P. have been obtained by manipulation/new combinations of the biosynthesis genes (combinatory biosynthesis, see hybrid antibiotics) and are undergoing broad testing as potential active principles [1]. *Examples*: *actinorhodin, *avermectin, *macrolides, *anthracyclines, *mycotoxins, *flavonoids.

Lit.: [1] Chem. Biol. **2**, 355–362 (1995); **3**, 827, 833 (1996).

gen.: Angew. Chem. Int. Ed. Engl. **34**, 881–885 (1995) ▪ Annu. Rev. Microbiol. **47**, 875–912 (1993) ▪ Barton-Nakanishi **1** (review) ▪ Chem. Rev. **97**, 2463–2706 (1997) (biosynthesis, review) ▪ Chem. Eng. News (2.11.) **1998**, 27 f. (combinatorial biosynthesis) ▪ Nature (London) **375**, 549–554 (1995) ▪ O'Hagan, The Polyketide Metabolites, Chichester: Ellis Horwood 1991 ▪ Vining & Stuttard (eds.), Genetics and Biochemistry of Antibiotic Production, Boston: Butterworth-Heinemann 1995.

Polymycins see streptothricins.

Polymyxins. Collective name for a complex of *peptide antibiotics, the first member of which was discovered in 1947 in *Bacillus polymyxa* (soil bacteria). Among the P. which are solely active against Gram-negative bacteria (pneumococci, *Pseudomonas*, whooping cough pathogen), only P. B, a mixture of *P. B_1* ($C_{56}H_{98}N_{16}O_{13}$, M_R 1198.45) and *P. B_2* ($C_{55}H_{96}N_{16}O_{13}$, M_R 1184.42, $[\alpha]_D^{22}$ −112.4°), is used in medicine. Other compounds of the P. complex described are P. D_1^1, P. D_2^1, P. F^2, P. M^3, P. S_1^4, and P. T_1^4.

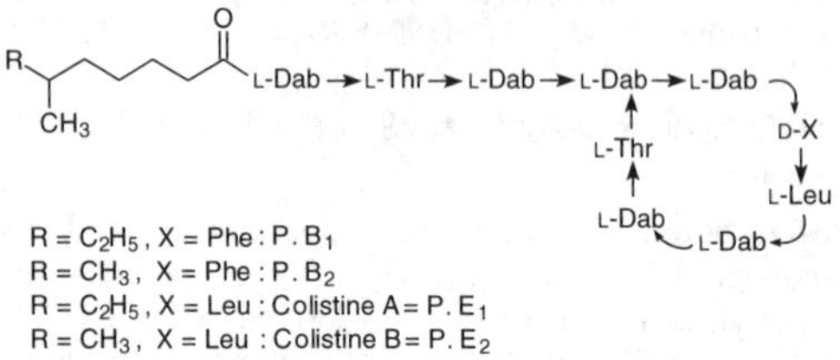

The P. B are cyclopeptides containing L-2,4-diaminobutanoic acid residues (Dab) and a D-amino acid; colistins A and B, which are also known as P. E_1 and E_2, have the same basic structure. P. exert bactericidal effects on growing and resting cells by impairing the permeability of the bacterial membrane. On account of toxic side affects (nephro- and neurotoxicity) the therapeutic use of P. is limited; it is mostly administered locally or orally but rarely parenterally[5].

Lit.: [1] Experientia **22**, 354 (1966). [2] J. Antibiot. **30**, 767 (1977). [3] Antibiotiki **16**, 250 (1971). [4] J. Antibiot. **30**, 427, 1029, 1039 (1977). [5] Hager (5.) **1**, 749 f.; **7**, 1091; **9**, 286 ff.; Kleemann-Engel, p. 1556.

gen.: Antimicrob. Agents Chemother. **39**, 529 (1995) ▪ Arteriosclerosis Thromb. **12**, 503 (1992) ▪ Beilstein E III/IV **26**, 4245 f. ▪ Encycl. Chem. Technol. Wiley **3**, 267 (1992) ▪ J. Antibiot. **37**, 1605 (1984) ▪ Kirk-Othmer (4.) **2**, 895; **3**, 267–272 (review) ▪ see also peptide antibiotics. – *[CAS 1406-11-7 (P. complex); 1404-24-6 (P. A); 1404-26-8 (P. B); 4135-11-9 (P. B_1); 34503-87-2 (P. B_2)]*

Polyoses. A group of *polysaccharides of varying compositions occurring in the plant fibers and cell walls of grasses, grain crops, and other higher plants and, in particular, in wood cell walls where they are accompanied by cellulose and lignin. In the past they were known as *hemicelluloses*, a name still frequently found in the literature. The P. (*wood P.*), like *cellulose, are made up of glycosidically linked sugar units but the chain molecules are more or less branched and the degree of polymerization is lower than that of cellulose; between 50 and 250. Common monomeric components are hexoses (e.g., *galactose, *glucose, *mannose) and pentoses (e.g., arabinose, *D-xylose). The P. are accordingly classified as *hexosans* and *pentosans*. The most important P. are the *xylans (wood gum), arabans, *mannans, and galactans.

Lit.: Fengel & Wegener, Wood, Berlin: De Gruyter 1983 ▪ Kirk-Othmer (4.) **13**, 54–60 ▪ Zechmeister **37**, 221–228. – *[CAS 9034-32-6]*

Polyoxins. Peptide nucleosides from cultures of *Streptomyces* species, including *S. cacaoi*. P. are soluble in water and, although of minor economic significance, are used as fungicides (against *Pellicularia sasakii*) in rice cultivation in Japan. They act as inhibitors of *chitin biosynthesis in fungi. Examples are *P. A*

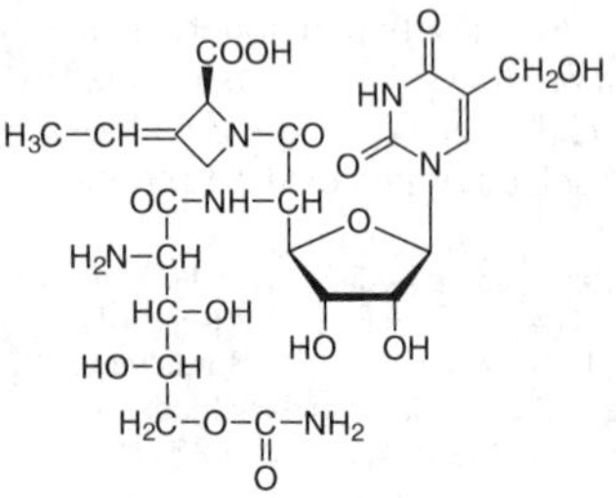

P. A

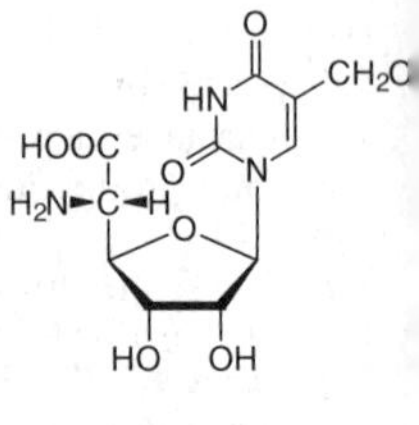

P. C

($C_{23}H_{32}N_6O_{14}$, M_R 616.54, amorphous) and *P. C* ($C_{11}H_{15}N_3O_8$, M_R 317.26, mp. 260–267 °C).

Lit.: Agric. Biol. Chem. **36**, 1229 (1972); **45**, 1901 (1981) ▪ Chem. Pharm. Bull. **25**, 1740 (1977) ▪ Curr. Pharm. Des. **5**, 73–99 (1999) (review) ▪ Exp. Opin. Ther. Patents **4**, 699 f. (1994) (activity) ▪ Heterocycles **47**, 157 (1998) (synthesis (–)-P. J, L) ▪ J. Am. Chem. Soc. **100**, 3937 (1978) ▪ J. Antibiot. **41**, 1711 (1988) ▪ J. Chem. Soc., Chem. Commun. **1995**, 2127 f. ▪ J. Med. Chem. **34**, 174 (1991) ▪ J. Org. Chem. **55**, 3853 (1990); **56**, 4875 (1991); **62**, 5497 (1997) (synthesis P. J); **64**, 2789 (1999) (synthesis (+)-P.) ▪ Synthesis **1999**, 1678 (synthesis P. J, L) ▪ Tetrahedron **46**, 7019 (1990) ▪ Tetrahedron Lett. **34**, 4153 (1993); **37**, 163 (1996) ▪ see also blasticidins and nikkomycins. – *[CAS 19396-03-3 (P. A); 21027-33-8 (P. C)]*

Polypeptide antibiotics. Collective name for antibiotically active, mostly cyclic peptides such as *nisin or *subtilin. About 40–50 P. a. with M_R from 3000 to 7000 are known from *streptomycetes or bacteria; they are mostly made up of proteinogenic L-*amino acids, less frequently from D-amino acids or other non-proteinogenic amino acids. P. a. composed of proteinogenic amino acids are formed by ribosomal peptide synthesis. The P. a. exhibit antibacterial activity with differing mechanisms of action (including membrane damage). They are used as local antibiotics when the pathogens develop resistance to other less toxic agents.

Lit.: Angew. Chem. Int. Ed. Engl. **30**, 1051 (1991).

Polyporenic acids.

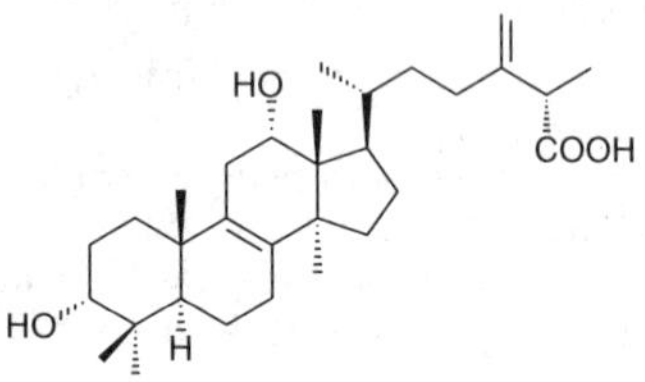

Polyporenic acid A

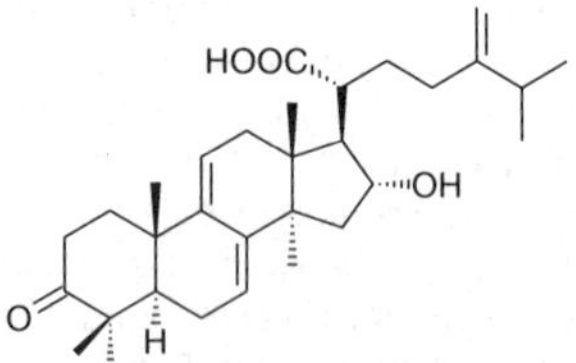

Polyporenic acid C

Triterpene acids with the eburicane skeleton (see eburicoic acid) from tree fungi (Aphyllophorales). *P. A* {$C_{31}H_{50}O_4$, M_R 486.74, needles, mp. 199–200 °C, $[\alpha]_D$ +69° (pyridine)} occurs, sometimes esterified with fatty acids, in the birch polypore (*Piptoporus betulinus*), *P. C* {$C_{31}H_{46}O_4$, M_R 482.70, cryst., mp. 273–276 °C, $[\alpha]_D$ +6° (pyridine)} is widely distributed in polypores.

Lit.: Phytochemistry **20**, 1746 (1981) ▪ Tetrahedron Lett. **25**, 3489 (1984) (structure) ▪ Turner **1**, 260 ff.; **2**, 324 (distribution). – *[CAS 516-25-6 (P. A); 465-18-9 (P. C)]*

Polyporic acid (2,5-dihydroxy-3,6-diphenyl-1,4-benzoquinone, 3′,6′-dihydroxy-*p*-terphenyl-2′,5′-quinone).

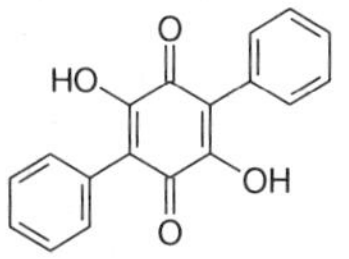

$C_{18}H_{12}O_4$, M_R 292.29, bronze-colored plates, mp. 310–312 °C. P. forms red solvates from pyridine and yellow solvates from dioxan; it has the parent skeleton of the *terphenylquinones. P. is found in some tree fungi of the order Aphyllophorales and in the frondous lichen *Sticta coronata*. The cinnamon-colored fruit bodies of *Hapalopilus rutilans* consist of up to 43.5% dry weight of polyporic acid. P. is formed biosynthetically by condensation of two molecules of phenylpyruvic acid and exhibits weak antileukemic activity. The dimethylether is called betulinan A [1].

Lit.: [1] J. Nat. Prod. **59**, 1090 (1996).
gen.: Beilstein E IV **8**, 3298 ▪ Culberson, p. 208 f. (reviews) ▪ Thomson **2**, 153 f.; **3**, 94 ff. ▪ Zechmeister **51**, 12 ff. – *[CAS 548-59-4]*

Polyquinanes. Polycyclic compounds with a skeleton of only 5-membered carbocyclic rings. Among natural products there are some tri- and tetraquinanes derived from the isoprenoid pathway. Different condensation types exist. *Triquinanes are often condensed in a linear way, such as *capnellene, *complicatic acid, the *coriolins, the *hirsutanes, and *phellodonic acid. Different triquinanes are *isocomene, *modhephene and *pentalenene. The only known natural tetraquinanes are the *crinipellins, isolated from mushrooms.

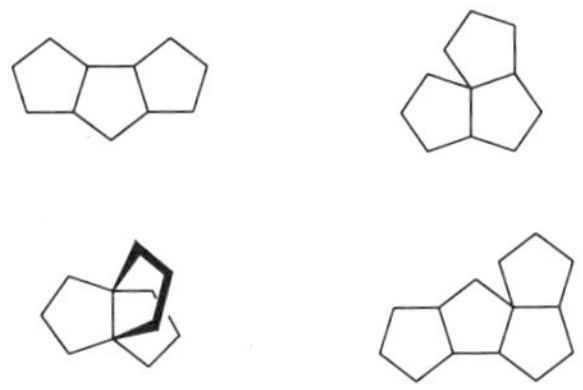

Polyquinane skeletons

Lit.: Chem. Rev. **97**, 671–720 (1997).

Polysaccharides. Name for macromolecular carbohydrates.
Classification: Members of the P. include *starch, *glycogen, and *cellulose, the polycondensation products of D-glucose, *inulin as polycondensate of D-fructose, *chitin, *alginic acid, etc. In contrast to these homoglycans which occur principally in plant gums, body mucus, and connective tissue, *hetero-P.* or *heteroglycans* are made up of *different* monomeric units, e.g,. *pectins, *mannans, galactans, *xylans, and other *polyoses, as well as chondroitin sulfates, heparin, *hyaluronic acid, and other glycosaminoglycans. In some P. the monomeric units carry carboxy and/or sulfonyl substituents, thus providing these P. with polyelectrolyte properties. P.-containing conjugates include some glycoproteins, the proteoglycans, and the peptidoglycans as well as some of the glycolipids and the lipopolysaccharides. For uses of P., see under the individual entries.

Lit.: ACS Symp.. Ser. **737** (Polysaccharide Applications 1999) ▪ ApSimon **8**, 245–310 (synthesis) ▪ Aspinall, The Polysaccharides, vols. 1, 2, & 3, Orlando: Academic Press 1982, 1983, 1985 ▪ Atta-ur Rahman **19E**, 689–746 (1997) ▪ Creszenzi et al., Biomedical and Biotechnological Advances in Industrial Polysaccharides, New York: Gordon and Breach 1989 ▪ Food Sci. Technol. (N. Y.) **87**, 15–36 (1998) ▪ IUPAC/IUBMB, Biochemical Nomenclature, 2nd ed., p. 127–179, London: Portland Press 1992 ▪ Polysaccharides **1998**, 57–100 (supramolecular structures) ▪ Ullmann **A25**, 1–55.

Polystictin see cinnabarin.

Polyterpenes (polyterpenoids). Natural products made up of n C_{10} units (=2 n isoprene building blocks) with n≥4, the biogenesis of which generally obeys the *isoprene rule. The most important P. are the *tetraterpenes (n=4) including *carotinoids, ficaprenols, *natural rubber, *balata, and *gutta-percha. The name is also used for hydrocarbon resins (terpene resins) prepared synthetically by polymerization of monoterpenes.

Lit.: Nat. Prod. Rep. **7**, 223–249 (1991).

Polythiepanes see lenthionine.

Polyynes (polyacetylenes). Compounds with very diverse structures containing several C/C triple bonds are produced mainly by fungi (basidiomycete cultures) and plants of the families Asteraceae, Apiaceae, and Araliaceae. In addition to conjugated triple bonds the P. often also contain C/C double bonds, allene units, thiophene and furan rings. On account of the close biosynthetic relationships between these compounds, the term P. is used as a collective name even when only one C/C triple bond is present in the molecule. As result of the work of Bohlmann, E. R. H. Jones, Sörensen, and others more than one thousand natural P. are now known. The antibiotically active *mycomycin* ($C_{13}H_{10}O_2$, M_R 198.22, mp. 75 °C) from basidiomycete cultures, *dehydromatricaria ester* ($C_{11}H_8O_2$, M_R 172.18, mp. 105–106 °C) from Asteraceae, and the *thiarubrins may be mentioned as typical examples (see also terthienyls).
P. are formed biosynthetically from unsaturated fatty acids such as oleic and linoleic acids by dehydrogenation with subsequent shortening of the carbon chain by β-oxidation and possibly decarboxylation or ω-oxidation. An important intermediate in this process is *crepenynic acid* ($C_{18}H_{30}O_2$, M_R 278.43). The enzymes have been identified, that catalyze the formation of acetylenic and epoxy fatty acids in plants of the genus *Crepis*. Most P. are highly toxic (e. g. the hemlock toxin *cicu(to)toxin and falcarinone) and thus may act as defensive substances. The antifungal principles *safynol* [$C_{13}H_{12}O_2$, M_R 200.24, mp. 122.5–123 °C, $[\alpha]_D^{20}$ −17° (CH_3OH)] and *wyerone* ($C_{15}H_{14}O_4$, M_R 258.27, mp. 63.5–64 °C), are formed by plants of the Asteraceae and Fabaceae, respectively, on invasion by phytogenic fungi (*phytoalexins). 1-Tridecene-3,5,7,9,11-pentayne and the (*E/Z*)-isomeric trideca-1,11-diene-3,5,7,9-tetraynes, on the other hand, stimulate the germination of rust spores. A P. with antifeedant activity towards insects is the enediyne *tonghaosu*

($C_{13}H_{12}O_2$, M_R 200.24, oil) isolated from the Chinese vegetable plant Tonghao (*Chrysanthemum segetum*). The high toxicity and low stability of the P. prevent their use as antibiotics.

$HC{\equiv}C{-}C{\equiv}C{-}CH{=}C{=}CH{-}CH\overset{Z}{=}CH{-}CH\overset{E}{=}CH{-}CH_2{-}COOH$

Mycomycin (**1**)

$H_3C{-}C{\equiv}C{-}C{\equiv}C{-}C{\equiv}C{-}CH\overset{E}{=}CH{-}COOCH_3$

Dehydromatricaria ester (**2**)

$H_3C{-}(CH_2)_4{-}C{\equiv}C{-}CH_2{-}CH\overset{Z}{=}CH{-}(CH_2)_7{-}COOH$

Crepenynic acid (**3**)

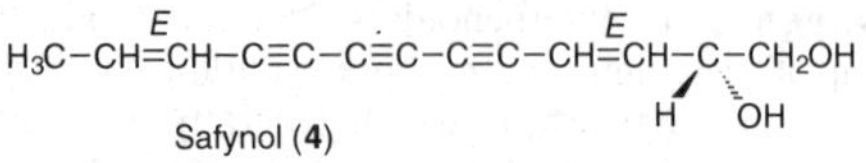

Safynol (**4**)

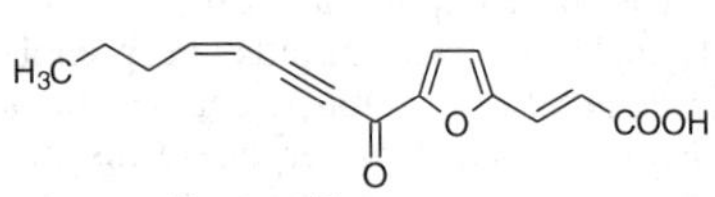

Wyerone (**5**)

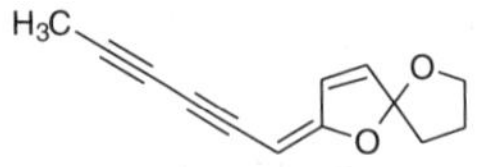

Tonghaosu (**6**)

Lit.: Bohlmann, Burkhardt & Zdero, Naturally Occurring Acetylenes, London: Academic Press 1973 ▪ Jones & Thaller, in Larkin & Lechevalier (eds.), Handbook of Microbiology, vol. 3, p. 63, Cleveland: CRC Press 1974 ▪ Science **280**, 915 (1998) ▪ Tetrahedron **54**, 12523 (1998) (synthesis tonghaosu) ▪ Tetrahedron Lett. **37**, 893 (1993) (synthesis of tonghaosu) ▪ Turner **1**, 66ff.; **2**, 49 ff. – *[CAS 544-51-4 (1); 692-94-4 (2); 2277-31-8 (3); 27978-14-9 (4); 20079-30-5 (5)]*

Polyzonimine.

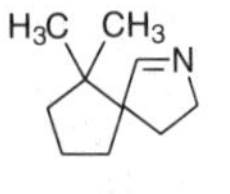

Polyzonimine [(+)-form]

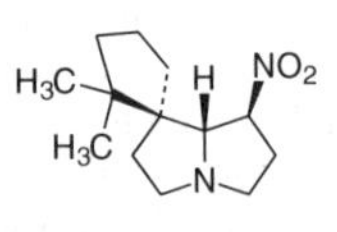

Nitropolyzonamine

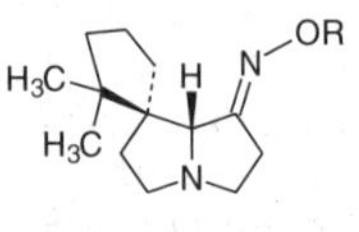

R = H : Alkaloid 222
R = CH_3 : Alkaloid 236

$C_{10}H_{17}N$, M_R 151.25, $[\alpha]_D^{25}$ +3.26° ($CHCl_3$). A terpenoid, spirocyclic imine from the *defensive secretions of the millipede *Polyzonium rosalbum* with an earthy, camphor-like odor. A further component is *nitropolyzonamine* ($C_{13}H_{22}N_2O_2$, M_R 238.33, mp. 65.5–66.5 °C). The imine has also been detected in frogs that feed on soil organisms [1]. P. serves as a defensive substance against ants and even at low concentrations induces an intense scratching behavior. Close relatives are the *pyrrolizidone oximes* 222 ($C_{13}H_{22}N_2O$, M_R 222.33, oil) and 236 ($C_{14}H_{24}N_2O$, M_R 236.36, oil), two minor alkaloids from skin extracts of the Panamanian poisonous frog *Dendrobates pumilio* [2].

Lit.: [1] J. Chem. Ecol. **20**, 943–955 (1994). [2] Tetrahedron **48**, 4247 (1992); **50**, 6129 (1994) (synthesis).
gen.: J. Chem. Soc. Chem. Commun. **1984**, 214–215 ▪ Rodd's Chem. Carbon Compds. (2.) **4B**, 1–19 (pyrrolidine alkaloids) ▪ Science **188**, 734–736 (1975). – *[CAS 55811-47-7 (P.); 57308-84-6 (nitropolyzonamine)]*

Pomacerone. $C_{30}H_{42}O_3$, M_R 450.66, amorphous solid, mp. 216–218 °C, $[\alpha]_D$ +81.5° ($CHCl_3$). A lanostane derivative from the plum agaric (*Phellinus pomaceus*, Basidiomycetes).

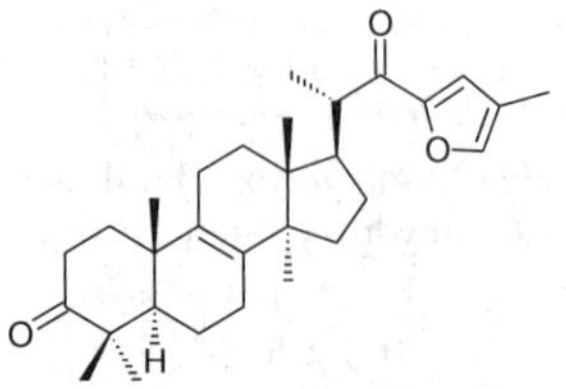

Lit.: Heterocycles **31**, 841 (1990). – *[CAS 129683-98-3]*

Ponasterone A, ponasteroside A see ecdysteroids.

Poppy see opium.

Poppy acid (meconic acid, hydroxychelidonic acid).

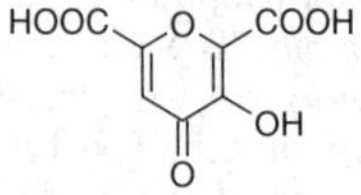

$C_7H_4O_7$, M_R 200.10, cryst. Crystals of p. a., a component of the latex of *opium poppy, have been known since the 19th century to take up a wide range of molecules during growth from solution. Recently, their ability to accommodate structurally dissimilar dyes and to orient the dye molecules similarly within crystal layers has been demonstrated [1]. This ability is useful for studies examining the anisotropic (orientation-dependent) properties of molecules.

Lit.: [1] J. Am. Chem. Soc. **121**, 7020 (1999). – *[CAS 497-59-6]*

Populin see salicyl alcohol.

Porfiromycin see mitomycins.

Poriferastane, poriferasterol see sterols.

Porphine (porphyrin). 21*H*,23*H*-Porphine, $C_{20}H_{14}N_4$, M_R 310.36, formula, see porphyrins. Dark red cryst., mp. >360 °C (decomp.), soluble in pyridine. P. forms salts of, e.g., iron, magnesium, and copper. It is the parent compound of all *porphyrins, the respiratory pigments of animals and plants.

Lit.: Beilstein E V **26/12**, 168–172 ▪ Dolphin **I**, 224–280, 550–584 ▪ J. Porphyrins Phthalocyanines **1**, 305 (1997) (synthesis) ▪ Merck-Index (12.), No. 7758 ▪ Smith, p. 14, 576, 774–778. – *[HS 2939 90; CAS 101-60-0]*

Porphobilinogen (5-aminomethyl-4-carboxymethyl-3-pyrrolepropanoic acid).

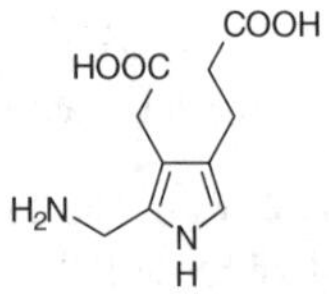

$C_{10}H_{14}N_2O_4$, M_R 226.23, as monohydrate pink cryst., mp. 172–175 °C, poorly soluble in water. P. is a key building block in the biosynthesis of all *porphyrins, *hydroporphyrins, and *corrins. It is formed biogenetically from *5-amino-4-oxovaleric acid and tetramerization proceeds with the aid of the enzyme porphobilinogen deaminase and a cosynthetase to furnish uroporphyrinogen III (see uroporphyrins). The acid-catalyzed, non-enzymatic tetramerization of P. leads to a mixture of the four constitutionally isomeric uroporphyrinogens I–IV. P. is found in the urine of patients with lead poisoning and the metabolic disease porphyria.

Lit.: Angew. Chem. Int. Ed. Engl. **34**, 383 (1995) (review) ■ Beilstein E V **22/14**, 210 ■ Jordan, p. 36–57 ■ Nat. Prod. Rep. **4**, 77–87 (1987) ■ Tetrahedron **53**, 7731–7752 (1997) (synthesis) ■ Tetrahedron Lett. **36**, 9121 (1995) ■ Trends Org. Chem. **6**, 125–144 (1997) (synthesis). – *[CAS 487-90-1]*

Porphyrin see porphine.

Porphyrin biosynthesis. The biosyntheses of all porphinoid natural products are parallel up to the stage of uroporphyrinogen III (see figure).

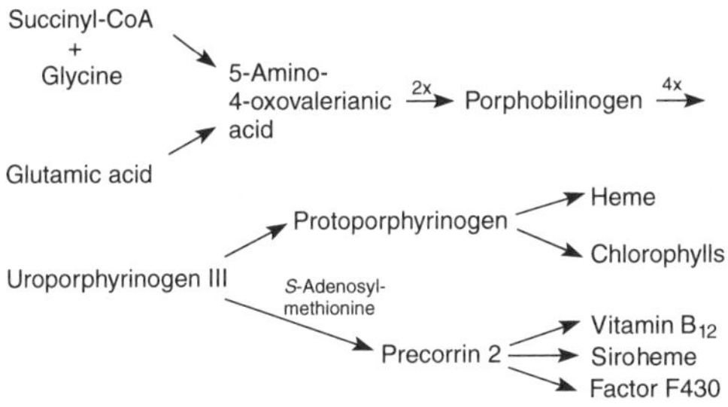

Figure: Biosynthesis of Porphyrins.

Uroporphyrinogen is converted to protoporphyrinogen by degradation and modification of the side chains; incorporation of metals and aromatization then furnishes *heme, alternatively, further modifications of the side chain afford the *chlorophylls. Methylation of the periphery of uroporphyrinogen III with **S*-adenosylmethionine leads to *precorrin 2, from which *vitamin B_{12}, *siroheme, and *factor F 430 are formed by incorporation of metals together with chemical modification of the periphery and the basic skeleton. The courses of some steps in the biosynthesis of vitamin B_{12} and the final steps in the biosynthesis of chlorophyll remain to be clarified. Similarly, the mechanism of degradation of the porphinoid macrocycles, with the exception of the catabolism of heme, is still unknown.

Lit.: Angew. Chem. Int. Ed. Engl. **27**, 5–39 (1988); **32**, 1223 (1993) ■ Dolphin **VI**, 125–178 ■ Dolphin B_{12} **I**, 145–167 ■ Jordan, p. 2, 161–172 ■ J. Chem. Soc., Perkin Trans. 1 **1996**, 2079–2102; **1998**, 1493, 1519, 1531 ■ Science **264**, 1551 (1994) ■ Smith, p. 32–48.

Porphyrinogens.

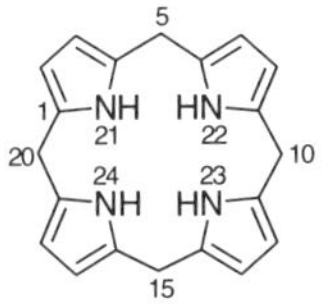

P. are 5,10,15,20,22,24-hexahydro-*porphyrins with four isolated pyrrole rings linked through methylene bridges. P. are formed chemically by condensation of pyrroles with formaldehyde under strictly anaerobic conditions or by reduction of porphyrins. P. are extremely readily oxidized to porphyrins and thus occur only as transient intermediates in *porphyrin biosynthesis. The biologically most important P. is uroporphyrinogen III[1].

Lit.: [1] Angew. Chem. Int. Ed. Engl. **27**, 5–39 (1988).
gen.: Angew. Chem. Int. Ed. Engl. **32**, 1040 (1993) (biosynthesis); **34**, 383 (1995) (review) ■ Dolphin **I**, 91–94 ■ Dolphin B_{12} **I**, 118, 139 ■ Jordan, p. 19f., 44, 50f. ■ Smith, p. 74, 792–795.

Porphyrins. Collective name derived from Greek: porphyra = purple, purple snail, for the widely distributed purple (red) natural pigments formally prepared from the parent skeleton of **porphine* by substitution of the macrocylic tetrapyrrole framework with methyl, vinyl, acetic acid, propanoic acid units, or other substituents.

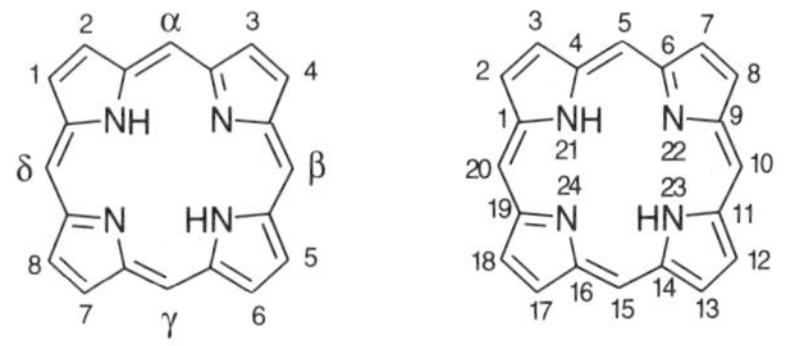

The figure shows the numbering and designation of the methine groups by Greek letters introduced by H. Fischer and used in older literature together with the numbering system recommended in the IUPAC/IUB rules for the semisystematic nomenclature of *tetrapyrroles (including *bile pigments)[1], which is used consequently in a few allowed trivial names. The assessment of the numerous possible constitutional isomers of P.[2] with identical substituents is facilitated for the naturally occurring P. by the fact that all porphinoid compounds occurring in nature are derived biosynthetically from uroporphyrinogen III and thus the substitution patterns can be referred back to *uroporphyrin III. For P. with only two different types of substituents of which each pyrrole ring carries one (*coproporphyrins, etioporphyrins, and uroporphyrins), 4 constitutional isomers are possible and are distinguished by the Roman numerals I–IV; type III is predominant in nature.
If two identical side chains of the types I–IV are changed, 15 isomers are possible. The natural *protoporphyrin was previously called *protoporphyrin IX.* Trivial names are also used for the partially hydrogenated P. derivatives; *examples: *chlorins, *bacteriochlorins, *isobacteriochlorins, *corphins, pyrrocorphin, *phlorins,* and **porphyrinogens.* The derived natural or synthetic compounds are strongly colored, fluorescing substances and, in general, are good sensitizers for the formation of singlet oxygen which is responsible not only for the photodynamic effects and skin lesions in patients with porphyria but also for the efficacy in the photochemotherapy for tumors[3]. The P. readily form stable metal complexes, the so-called *metalloporphyrins.* These are of essential importance in biochemistry (see, e. g., bacteriochlorophylls, chlorophyll, heme, heme d, heme d_1, factor F 430, siroheme, vitamin B_{12}).
The ring system of the P. is characterized by the stability resulting from the 18π-aromaticity of the chromophore and is easily formed from pyrroles. It is assumed that P. were already formed during the prebiological phase of evolution[4]. The stability of the ring system is reflected by the fact that it has been detected in interstellar space and also occurs as the *geoporphyrins in sediments and crude oil. Naturally occurring, metal-free P. are rare. They occur in the form of porphyrinogens in *porphyrin biosynthesis and can be isolated in the oxidized form (see uroporphyrin III, coproporphyrin III, protoporphyrins). Other naturally occurring P.

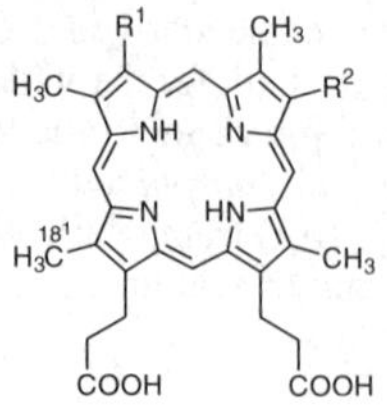

Porphyrin	R^1	R^2
Deuteroporphyrin	H	H
Mesoporphyrin	C_2H_5	C_2H_5
Protoporphyrin	$CH{=}CH_2$	$CH{=}CH_2$
Pemptoporphyrin	H	$CH{=}CH_2$
Harderoporphyrin	$CH{=}CH_2$	$CH_2{-}CH_2{-}COOH$
Porphyrin S-411	$CH{=}CH{-}COOH$	$CH_2{-}CH_2{-}COOH$
Porphyrin a (*Cytoporphyrin)	OH, CH–CH_2–Farnesyl	$CH{=}CH_2$ 18^1-oxo
Porphyrin c	S–CH_2–CH(NH_2)–COOH, CH–CH_3	S–CH_2–CH(NH_2)–COOH, CH–CH_3

are usually the products of an impaired P. metabolism, *corallistin A is an exception. Non-natural derivatives or degradation products of natural P. are *chlorocruoroporphyrin, cytoporphyrin, proto-P., etio-P., meso-P., *hematoporphyrin, phyto-P., phyllo-P., etc. For photosynthesis see *Lit.*[5]. The deep sea dragon fish *Malocosteus niger* has retinal pigments other than *carotinoids, derived from *chlorophyll, that absorb the far-red wavelengths and then pass the light signal onto the visual pigments. Recently, a photostable pigment complex has been isolated from the fish's retina, with an absorption peak at 667 nm, that acts as a photosensitizer, transferring energy from the triplet excited state to the ground state of the visual pigments[6].

Lit.: [1]Eur. J. Biochem. **178**, 277–328 (1988). [2]Ann. N. Y. Acad. Sci. **244**, 327 (1975). [3]Nachr. Chem. Tech. Lab. **33**, 582 (1985). [4]Angew. Chem. Int. Ed. Engl. **27**, 5–39 (1988). [5]Chem. Unserer Zeit **33**, 72–83 (1999). [6]Nature (London) **393**, 423 (1998).

gen.: Chem. Eng. News (26.6.) **1995**, 30 f. ▪ Cubbeddu & Andreoni, Porphyrins in Tumor Phototherapy, New York: Plenum 1984 ▪ Doiron & Gomer, Porphyrin Localisation and Treatment of Tumors, New York: Liss 1985 ▪ Dolphin **I**, 37, 308 f.; **III**, 34–37, 403; **VI**, 125–178 ▪ Hayata & Dougherty, Lasers and Hematoporphyrin Deriv. in Cancer, Tokio: Igaku-Shoin 1983 ▪ J. Chem. Soc., Perkin Trans. 1 **1999**, 2691 (biosynthesis) ▪ Journal of Porphyrins and Phthalocyanines, New York: Wiley (from 1997) ▪ Jordan, p. 1–96 ▪ Karrer, No. 2500–2508, 4362–4368 ▪ Kessel, Methods in Porphyrin Photosensitation, New York: Plenum 1986 ▪ Lavelle, The Chemistry & Biochemistry of N-Substituted Porphyrins, Weinheim: VCH Verlagsges. 1987 ▪ Lever & Gray, Iron Porphyrins, Weinheim: VCH Verlagsges. 1989 ▪ Milgrom, the Colors of Life: Chemistry of the Porphyrins, Oxford: Univ. Press 1997 ▪ Progr. Heterocycl. Chem. **10**, 1–24 (1998) ▪ Pure Appl. Chem. **65**, 1113–1122 (1993) (biosynthesis) ▪ Smith, p. 32–48, 97 ff. ▪ Thomson (2.), p. 329–381 ▪ see also tetrapyrroles.

Porphyropsin. *Pigment of vision (photoreceptor protein) in the retina of fresh water fish and some amphibians. The absorption maximum of P. is 520–530 nm. P. consist of the chromophore 3,4-didehydro-11-*cis*-*retinal and the protein scotopsin. The biological activity is similar to that of *rhodopsin.

Lit.: Merck-Index (12.), No. 7761 ▪ Science **162**, 230–239 (1968) ▪ Sebrell et al., The Vitamins, New York: Academic Press 1967. – *[CAS 9005-58-9]*

Portulacaxanthins (portulaxanthins). The P. belong to the betaxanthines (see betalains). They are immonium conjugates of *betalamic acid with hydroxyproline[1] (P. I), tyrosine[2] (P. II), or glycine[2] (P. III). They occur in the flowers of *Portulaca grandiflora* (Portulacaceae).

Lit.: [1]Rend. Accad. Sci. Fis. Mat. Naples **32**, 55 (1965). [2]Phytochemistry **30**, 1897 (1991).

Portulal.

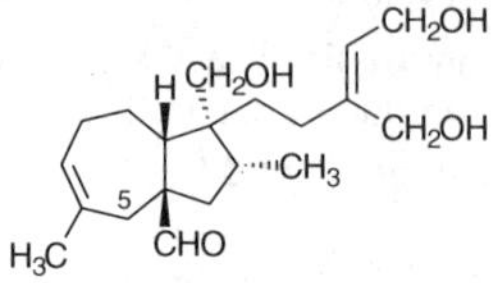

$C_{20}H_{32}O_4$, M_R 336.47, colorless cryst., mp. 113 °C, $[\alpha]_D^{22}$ −50° (C_2H_5OH). The bicyclic diterpene P., like *5α-hydroxyportulal* ($C_{20}H_{32}O_5$, M_R 352.47, colorless oil, $[\alpha]_D^{12.5}$ −79.1°), occurs in *Portulaca grandiflora.* P. is a plant growth regulator[1].

Lit.: [1]Tetrahedron Lett. **1969**, 359.

gen.: Chem. Lett. **1986**, 1585 ▪ Chem. Pharm. Bull. **33**, 2171 (1985). – *Synthesis:* J. Org. Chem **46**, 1828 (1981) ▪ Tetrahedron **36**, 3377 (1980). – *[CAS 22571-65-9 (P.); 98263-95-7 (5α-hydroxy-P.)]*

Posthumolone see humulone.

Potassium chelidonate see leaf-movement factors.

Potassium 5-*O*-*β*-D-glucopyranosylgentisate see leaf-movement factors.

Potassium D-idarate see leaf-movement factors.

Potassium lespedezate see leaf-movement factors.

Potassium L-malate see leaf-movement factors.

Potato flavor see vegetable flavors.

PQ. Abbreviation for *plastoquinones.

PQQ. Abbreviation for *pyrroloquinoline quinone.

Pradimicins. A group of 20 currently known anthracyclinone antibiotics with antimycotic and antiviral activity isolated from various *Actinomadura* strains (Actinomycetes); e.g., *P. A* (*N*-methylbenanomycin B): $C_{40}H_{44}N_2O_{18}$, M_R 840.79, red cryst., mp. 193–195 °C, $[\alpha]_D^{26}$ +685° (0.1 m HCl), solubility in H_2O 17 mg/L, LD_{50} (mouse i. v.) 120 mg/kg. The P. are related to the benanomycins, and are in some cases identical. Clinical development of a P. derivative as a systemic antimycotic was started in 1996 but has now been abandoned because of efficacy and tolerance problems.

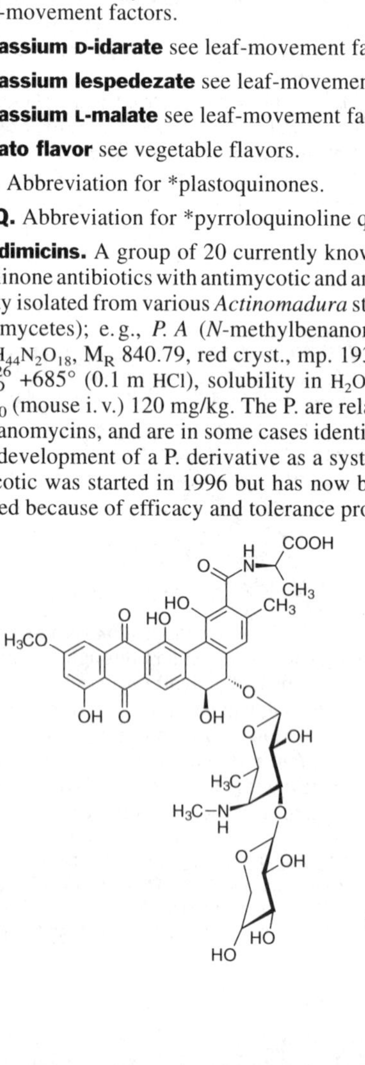

P. A

Lit.: J. Antibiot. **46**, 387–411, 412–432, 589–605, 631, 1447–1457 (1993). – *Synthesis:* Angew. Chem. Int. Ed. Engl. **38**, 1229 (1999). – *[CAS 117704-65-1 (P. A)]*

Prangeferol see marmesin.

Prasterone see androgens.

Pratensein see isoflavones.

Pravastatin (eptastatin, 3β-hydroxycompactinic acid).

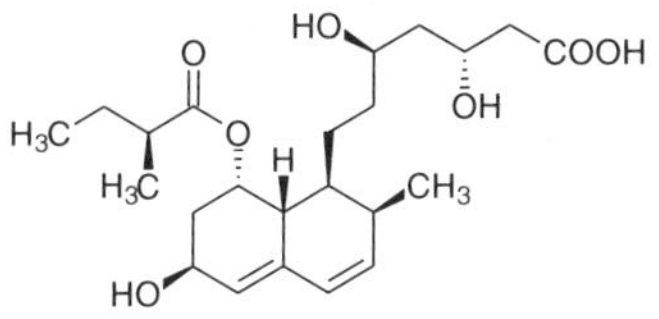

$C_{23}H_{36}O_7$, M_R 424.53, amorphous, colorless solid, soluble in methanol, acetone, as sodium salt (mevalotin, eptastatin sodium), moderately soluble in water, LD_{50} (rat p. o.) >12 g/kg. P. is isolated from the microorganisms *Syncephalastrum nigricans*, *Absidia coerulea*, and *Nocardia autotrophica* and was introduced to the pharmaceutical market in 1991 as the third 3-hydroxy-3-methylglutaryl-coenzyme A (HMG-CoA) reductase inhibitor after *mevinolin and simvastatin. In contrast to the other two hexahydronaphthalenes, P. is not employed as a δ-lactone prodrug, but in the active form as the 3,5-dihydroxycarboxylic acid. P. thus acts as a competitive inhibitor of HMG-CoA reductase (EC 1.1.1.34), the key enzyme in the biosynthesis of cholesterol. As a consequence of this inhibition the intracellular cholesterol concentration (especially in the liver) is lowered. This increases the number of LDL (low density lipoprotein) receptors on the surface of liver cells, which reduces the plasma concentration of cholesterol. P. is active on oral administration. Target indications are hypercholesterolemia and arteriosclerosis. New findings suggest, that P. and other statins may reduce inflammatory processes that are believed to contribute to heart disease. The basic framework has been the subject of much synthetic work.

Lit.: Arzneimitteltherapie **8**, 199 (1990) ▪ Atherosclerosis **61**, 125 (1986) ▪ Clin. Pharm. **11**, 677 (1992) ▪ Drugs **42**, 65 (1991); **53**, 299 (1997) ▪ Kleemann-Engel, p. 1566 ▪ Lancet **1986**, No. 2, 740 ▪ Merck-Index (12.), No. 7894 ▪ Pharm. Ztg. **137**, 38 (1992) ▪ Wood (ed.), Lipid Management: Pravastatin, London: Royal Soc. Med. 1990 ▪ Yuki Gosei Kagaku Kyokaishi **55**, 334 (1997) (fermentation). – *Synthesis:* Chem. Pharm. Bull. **34**, 1459 (1986) ▪ J. Am. Chem. Soc. **108**, 4586 (1986) ▪ J. Med. Chem. **30**, 1858 (1987) ▪ Tetrahedron **42**, 4909 (1986) ▪ Tetrahedron Lett. **31**, 2235 (1990). – *[CAS 81093-37-0 (P.); 81131-70-6 (P. Na salt)]*

Preanthraquinones. Name for biosynthetic precursors of polyketide anthraquinone dyes, in which the aromatic system is not completely formed. The dihydroanthracenone *atrochrysone formed by cyclization with decarboxylation of a linear octaketide precursor can be further transformed in several ways (*austrocortilutein, emodin anthrone, *emodin, *torosachrysone). Dimeric P. are widely distributed in certain Basidiomycetes genera (*Cortinarius*, *Dermocybe*, *Leucopaxillus*, *Tricholoma*), as well as in plants (*Cassia*). Depending on the linkage of the two halves of P. several basic types can be distinguished: *atrovirin (5,5′-linkage), cortinarins (8-*O*,10′-linkage), *flavomannins (7,7′-linkage), *phlegmacins (7,10′-linkage), pseudophlegmacins (5,10′-linkage), and tricolorins (10,10′-linkage).

Lit.: Zechmeister **51**, 143 ff.

Precalciol, precholecalciferol see calcitriol.

Precocenes.

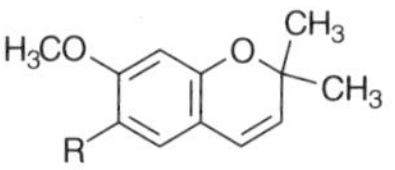

R = H : Precocene I
R = OCH₃ : Precocene II
R = OC₂H₅ : Precocene III (synthetic)

Chromenes from *Ageratum* species of the tropical rain forests of Central and South America: *P. I* (7-methoxy-2,2-dimethyl-2*H*-1-benzopyran), $C_{12}H_{14}O_2$, M_R 190.24, oil, bp. 68 °C (130 Pa); *P. II* (ageratochromene, 6,7-dimethoxy-2,2-dimethyl-2*H*-1-benzopyran), $C_{13}H_{16}O_3$, M_R 220.27, cryst., mp. 47 °C.

Toxicology: P. are assumed to be carcinogenic. P. II: dnd (DNA damage) 25 μM (rat liver).

Biological activity: P. are toxic to the corpora allata of insects and directly inhibit thereby the biosynthesis of *juvenile hormones, and thus the larvae prematurely develop into pupa (Latin praecox = premature). In adults, egg ripening and thus the ability to reproduce is principally affected. The synthetic, so-called *P. III* (proallatocidin, 6-ethoxy-7-methoxy-2,2-dimethyl-2*H*-1-benzopyran), $C_{14}H_{18}O_3$, M_R 234.30, oil, bp. 109 °C (13 Pa), has a 10-fold stronger activity than P. II[1].

Use: P. and especially the synthetic analogues with higher activities may find use as insecticides.

Lit.: [1] Can. J. Zool. **66**, 2295–2300 (1988); Gen. Comp. Endocrinol. **74**, 18–44 (1989); Org. Prep. Proced. Int. **14**, 337–342 (1982).

gen.: Can. J. Chem. **72**, 1866 (1994) ▪ Chem. Ind. (London) **1989**, 725 (synthesis) ▪ Coats (ed.), Insecticide Mode of Action, p. 403–427, New York: Academic Press 1982 ▪ Experientia **44**, 790 ff. (1988) ▪ Heterocycles **27**, 2459–2465 (1988) ▪ J. Org. Chem. **53**, 2364 f. (1988) (synthesis) ▪ Luckner (3.), p. 480 ▪ Nat. Prod. Rep. **10**, 327–348 (1993) ▪ Pestic. Sci. **19**, 185–196 (1987) (metabolism) ▪ Phytochemistry **29**, 2135 (1990) (biosynthesis). – *Toxicology:* Sax (8.), AEX 850. – *[CAS 17598-02-6 (P. I); 644-06-4 (P. II); 65383-74-6 (P. III)]*

Precorrins.

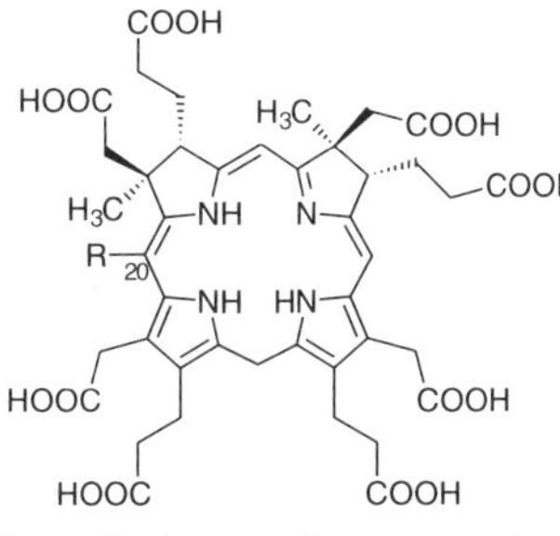

R = H : Precorrin 2
R = CH₃ : Precorrin 3A

P. are the intermediates occurring after uroporphyrinogen in the biosynthetic pathway to *vitamin B_{12}. The most important of the as yet discovered P. is P. 2 which can be transformed into other porphinoid natural products such as *siroheme and *factor F 430. P. 2 is derived from uroporphyrinogen III via P. 1, a tetrahydro form of *factor I, by methylation with *S*-adenosylmethionine (SAM). Methylation to P. 3A and further reaction steps finally lead to vitamin B_{12}, with the carbon atom 20 of P. 3A together with the methyl group being cleaved as acetic acid, as has been demonstrated by isotope labelling experiments.

Lit.: Angew. Chem. Int. Ed. Engl. **27**, 5–39 (1988); **32**, 1223 (1993); **34**, 383 (1995) ▪ J. Am. Chem. Soc. **115**, 10380 (1993) ▪ J. Chem. Soc., Chem. Commun. **1990**, 593 ▪ Jordan, p. 111, 118–120 ▪ Science **264**, 1551 (1994). – *[CAS 82542-92-5 (P. 2); 114019-24-8 (P. 3A)]*

Pregnane derivatives. *Steroids possessing the C_{21}-carbon skeleton of 5α-pregnane.

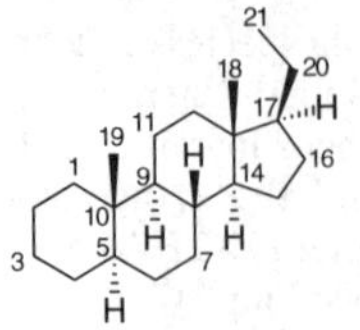

5α-Pregnane

P. d. are widely distributed in nature and often carry important regulatory functions. Thus, *steroid hormones, e. g., progesterone (see progestogens) and the *corticosteroids (see also aldosterone) belong to the P. d. Plant P. d. include, e. g., the *Apocynaceae steroid alkaloids. In general, the biosynthesis of P. d. proceeds through oxidative degradation of the side chain of *cholesterol. Large amounts of P. d. are required as starting materials for the production of the commercially used, hormonally active steroids. These are obtained primarily by Marker-Rohrmann degradation of *steroid sapogenins and chemical side chain degradation of the *sterol stigmasterol (see steroids).

Lit.: Ullmann (5.) **A 13**, 89–163 ▪ Zechmeister **71**, 169–326 (P. glycosides).

Preisocalamenediol see calamus oil.

Prekinamycin see kinamycin.

Prephenic acid (*cis*-1-carboxy-4-hydroxy-α-oxo-2,5-cyclohexadiene-1-propanoic acid).

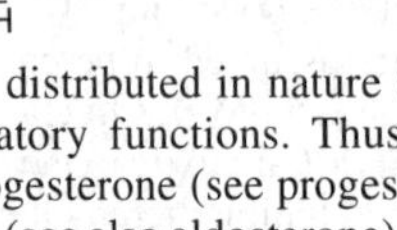

$C_{10}H_{10}O_6$, M_R 226.19. P. is unstable in the free form and undergoes decarboxylation with elimination of water to give *phenylpyruvic acid. It is an intermediate in the biosynthesis of many aromatic natural products by the shikimate pathway (see chorismic acid, shikimic acid). The biosynthesis of P. proceeds from chorismic acid by enzyme-catalyzed [3,3]sigmatropic rearrangement (Claisen rearrangement) under the action of chorismate mutase (EC 5.4.99.5).

Lit.: Beilstein E IV **10**, 4024 ▪ Karrer, No. 4059 ▪ Luckner (3.), p. 363 f. ▪ Nat. Prod. Rep. **12**, 101–133 (1995) ▪ Naturwissenschaften **70**, 115–118 (1983) ▪ Pure Appl. Chem. **56**, 1005–1010 (1984). – *[CAS 87664-40-2]*

L-Pretyrosine [L-arogenic acid, (*S*)-*cis*-α-amino-1-carboxy-4-hydroxy-2,5-cyclohexadiene-1-propanoic acid].

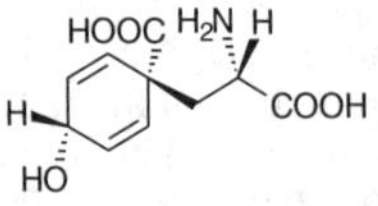

$C_{10}H_{13}NO_5$, M_R 227.22, $[\alpha]_D$ +6.6° (0.1 N NaOH). P. is an intermediate in the biosyntheses of L-*phenylalanine and L-*tyrosine in prokaryotes. P. has been isolated from *Neurospora crassa* and *Pseudomonas aureofaciens* as the disodium, or diammonium salt, respectively. Free P. is unstable and is rapidly converted to phenylalanine in acids. The biosynthesis proceeds via *chorismic acid → *prephenic acid → pretyrosine.

Lit.: J. Am. Chem. Soc. **102**, 4499–4504 (1980); **103**, 1602 f. (1981) ▪ Nat. Prod. Rep. **12**, 101–133 (1995). – *[CAS 53078-86-7]*

Preussine [(2*S*,3*S*,5*R*)-2-Benzyl-1-methyl-5-nonyl-3-pyrrolidinol].

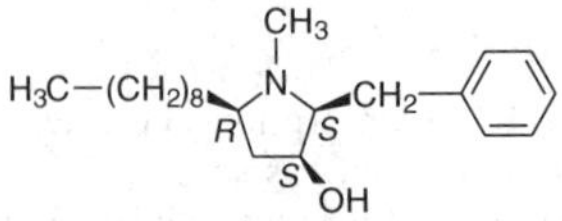

$C_{21}H_{35}NO$, M_R 317.51, yellow oil, $[\alpha]_D^{25}$ + 22.0° ($CHCl_3$). Fungal amino alcohol from a *Preussia* sp., produced also by *Aspergillus ochraceus*. The substance class has antifungal activity and especially *anisomycin shows strong antitumor activity.

Several enantioselective syntheses of (+)-P. have been published [1].

Lit.: [1] Angew. Chem. Int. Ed. Engl. **37**, 3400 ff. (1998); Chin. J. Chem. **16**, 458–467 (1998); Heterocycles **46**, 335–348 (1997); J. Chem. Soc., Chem. Commun. **1997**, 1601 f.; J. Chem. Soc., Perkin Trans. 1 **1999**, 265; J. Org. Chem. **59**, 4721 (1994); **63**, 4660 (1998); Synthesis **1997**, 1296; Tetrahedron **54**, 15673 (1998); Tetrahedron Lett. **39**, 131 (1998). – *[CAS 125356-66-3]*

Preussomerins. Novel class of fungal metabolites (P. A – P. I), isolated from the coprophilous (dung-colonizing) fungus *Preussia isomera* and the endophytic fungus *Harmonema dematioides*, e. g., *P. A* {$C_{20}H_{14}O_7$, M_R 366.33, cryst., mp. 235–240 °C, $[\alpha]_D^{20}$ –212° (CH_3OH)}. All P. share an unusually stable bis-acetal ring system. They are natural antifungal agents and are thought to play an important biological role as part of the survival mechanism of the fungi. They have also been identified as novel inhibitors of Ras farnesyl transferase, an enzyme associated with the regulation of tumor growth.

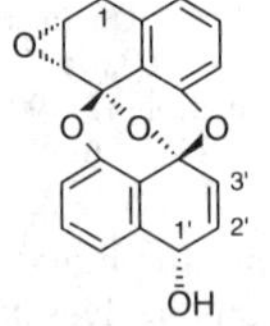

P. A

Biosynthetically, the P. are related to a growing class of natural products that includes the deoxypreussomerins, palmarumycins, diepoxins, *spiroxins, which all have a 1,8-naphthalenediol-derived spiroacetal as a key structural element. These compounds show a wide range of biological activity as antifungal, antibacterial, antitumor and herbicidal agents.

Lit.: J. Org. Chem. **56**, 4355 (1991); **59**, 6296 (1994) ▪ Org. Lett. **1**, 3 ff. (1999). – *[CAS 129240-52-4 (P. A)]*

Previtamin D see calcitriol.

Prianosins see discorhabdins.

Primetin (5,8-dihydroxyflavone).

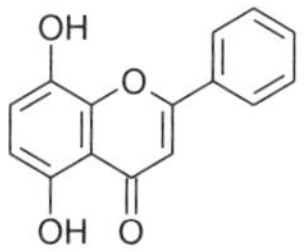

$C_{15}H_{10}O_4$, M_R 254.24, yellow cryst. (C_2H_5OH), mp. 230–232 °C. The *flavone P. is isolated from the leaves of *Primula* species (Primulaceae) and flowers of *Dionysia* species (Primulaceae).

Lit.: Hager (5.) **6**, 271 ▪ Karrer, No. 1437. – *[CAS 548-58-3]*

Primeverin (primeveroside).

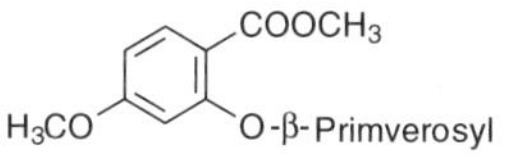

$C_{20}H_{28}O_{13}$, M_R 476.43, mp. 203–204 °C, $[\alpha]_D^{20}$ –71.5° (H_2O). P. is the 2-*O*-glycoside of methyl 2-hydroxy-4-methoxybenzoate with *primeverose. P. occurs together with *primulaverin* (2-*O*-glycoside of methyl 2-hydroxy-5-methoxybenzoate with primeverose, $C_{20}H_{28}O_{13}$, M_R 476.43) in the roots of *Primula officinalis* and other primrose species.

Lit.: Beilstein E III/IV **17**, 2447 ▪ J. Med. Chem. **28**, 717 (1985) ▪ Karrer, No. 894 ▪ Phytochemistry **27**, 1835 (1988). – *[CAS 154-60-9]*

Primeverose (6-*O*-*β*-D-xylopyranosyl-D-glucose).

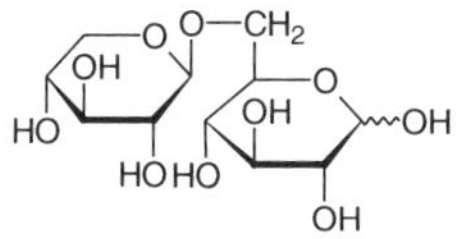

$C_{11}H_{20}O_{10}$, M_R 312.27, cryst., mp. 194–210 °C (decomp.), $[\alpha]_D^{20}$ +23.8 → –3.4° (H_2O), a sweet tasting *disaccharide of xylose and glucose, it reduces Fehling's solution. It can be obtained from a series of glycosides by enzymatic hydrolysis; e.g., from *primeverin, *ruberythric acid, and from *gaultherin = monotropitoside* (glycoside with methyl salicylate in wintergreen oil).

Lit.: Beilstein E III/IV **17**, 2447 ▪ Karrer, No. 633 ▪ Merck-Index (12.), No. 7926 – *[CAS 26531-85-1]*

Primin (2-methoxy-6-pentyl-*p*-benzoquinone).

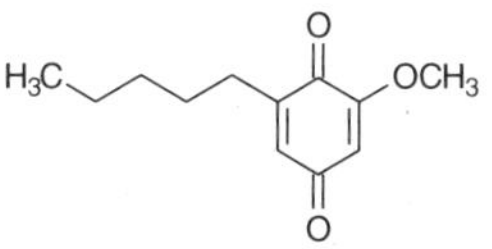

$C_{12}H_{16}O_3$, M_R 208.26, golden-yellow cryst., mp. 67 °C. P. is found in the glandular hairs of leaves of *Primula elatior* and *P. obconica* (Primulaceae) as well as in the root bark of *Miconia* species (Melastomataceae). It is the allergen of primroses and causes severe skin irritations. The corresponding 2-methoxy-6-pentylhydroquinone, *miconidin* ($C_{12}H_{18}O_3$, M_R 210.27) is isolated from *Miconia* species and possesses antimicrobial activity as well as antifeedent effects on insects.

Lit.: Acta Chem. Scand. **27**, 3211 (1973) (synthesis) ▪ Experientia **40**, 1010 (1984) ▪ Hager (5.) **6**, 269–290 ▪ Org. Prep. Proced. Int. **23**, 639 (1991) ▪ Phytochemistry **41**, 451 (1996) (biosynthesis). – *[HS 291469; CAS 15121-94-5 (P.); 34272-58-7 (miconidin)]*

Primulagenin A see oleanane.

Primulaverin see primeverin.

Primulidin see malvin.

Pristane (norphytane, *meso*-2,6,10,14-tetramethylpentadecane).

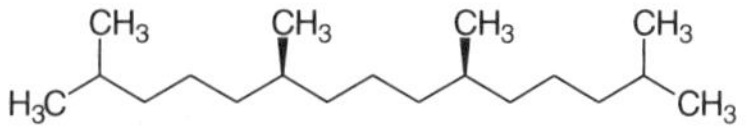

$C_{19}H_{40}$, M_R 268.53, mobile liquid, mp. –100 °C, bp. 296 °C, 158 °C (1.3 kPa), 68 °C (0.13 Pa). P., a linear norditerpene, is very widely distributed in nature. It occurs in crude oil, shark liver, herring oil, as well as in gray *ambergris, in plankton, anise fruits, etc. The relative proportions of P. to its (*R,R*)- and (*S,S*)-diastereoisomers or to other hydrocarbons such as, e. g., phytane can be used to identify the origin of crude oils [1]. Ruminants degrade phytol via phytanoic acid to *pristanic acid* [(6*R*,10*R*)-2,6,10,14-tetramethylpentadecanoic acid, $C_{19}H_{38}O_2$, M_R 298.51, oil] [2], since only in this form is it amenable to the normal fatty acid degradation.

Use: In lubricants, transformer oils, anticorrosion agents.

Lit.: [1] Mar. Pollut. Bull **12**, 78 (1981). [2] Agric. Biol. Chem. **38**, 1859 (1974); Biochim. Biophys. Acta **360**, 166 (1974).

gen.: Beilstein E IV **1**, 562 ▪ Can. J. Chem. **47**, 4359 (1969) ▪ J. Chem. Soc. C **1966**, 2144. – *[CAS 1921-70-6 (P.); 1189-37-3 (pristanic acid)]*

Pristinamycin see virginiamycin.

Proamanullin see amanitins.

Proanthocyanidins (procyanidins). P. are polyflavan-3-ols that can be converted by the action of acid to the colored anthocyanidins (see anthocyanins). They include not only the monomeric flavan-3,4-diols (*leucoanthocyanidins) but also the so-called condensed tanning substances (condensed P., see also tannins). Many of these *tanning agents are derived from *catechin, from its *cis*-isomer epicatechin, and from the leucoanthocyanidins and have dimeric or trimeric structures that often occur in the bark and wood of trees. P. occurring in nature have a molecular mass of up to about 7000 (brown color). On account of their bitter (sometimes adstringent) taste they are of importance for the sensory properties of alcoholic beverages and fruit juices. P. have blood-pressure lowering (ACE inhibition), antiatherosclerotic, and radical scavenging properties. P. from cranberries show activity in urinary tract infections by inhibiting adhesion of *Escherichia coli* bacteria [1].

Lit.: [1] New Engl. J. Med. **339**, 1085 (1998).

gen.: Chem. Eng. News (12.4.) **1999**, 47–50 ▪ Phytochemistry **30**, 2041 (1991); **37**, 533 (1994).

Proaporphine alkaloids. P. a. are precursors of some aporphines and possess a characteristic, spiro-annelated five-membered ring. They can be converted to *aporphine alkaloids by a dienone-phenol rearrangement of, after reduction to a dienol, by a dienol-benzene rearrangement. P. a. occur primarily in the plant families Euphorbiaceae, Lauraceae, Menispermaceae, Monimiaceae, Nymphaeaceae, and Papaveraceae.

Table: Data and occurrence of Proaporphine alkaloids.

name	molecular formula	M_R	mp. [°C]	$[\alpha]_D$	occurrence	CAS
Crotonosine	$C_{17}H_{17}NO_3$	283.33	>300	+180° (CH_3OH)	*Croton linearis*	2241-43-2
Crotsparine	$C_{17}H_{17}NO_3$	283.33	193–195	–30° ($CHCl_3$)	*C. sparsiflorus*	18373-79-0
Crotsparinine	$C_{17}H_{19}NO_3$	285.33	184–185	+215° ($CHCl_3$)	*C. sparsiflorus*	23179-61-5
Mecambrine	$C_{18}H_{17}NO_3$	295.34	178.5–179.5	–116° ($CHCl_3$)	*Meconopsis cambrica, Papaver* species	1093-07-8
N-Methylcrotsparine	$C_{18}H_{19}NO_3$	297.35	223–225	–113° ($CHCl_3$)	*C. sparsiflorus*	
Orientalinone	$C_{19}H_{21}NO_4$	327.38	183–184	–62.6° ($CHCl_3$)	*Papaver orientale*	2689-17-0
Pronuciferine	$C_{19}H_{21}NO_3$	311.38	127–129	+99° ($CHCl_3$)	*Nelumbo nucifera, Stephania glabra, P. caucasicum, M. cambrica, Isolona pilosa*	2128-60-1

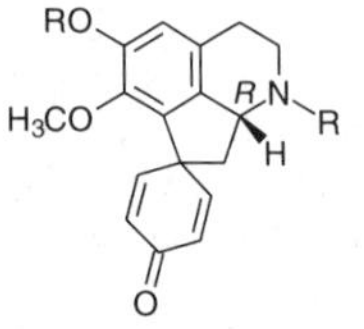
R = H : Crotonosine
R = CH_3 : Pronuciferine

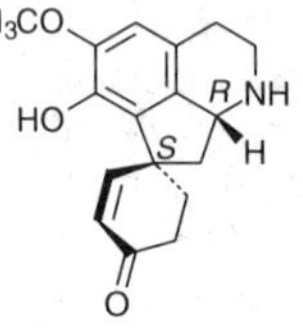
Crotsparinine

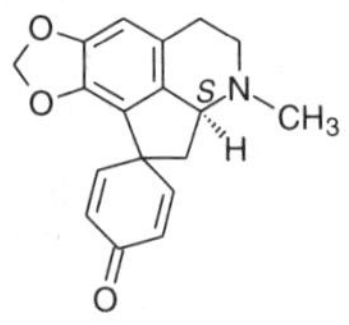
Mecambrine

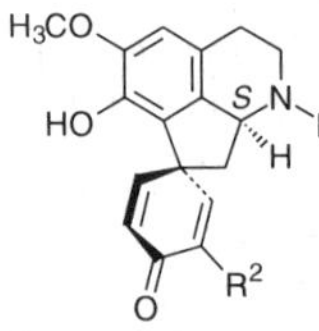
R^1 = H, R^2 = H : Crotsparine
R^1 = CH_3, R^2 = H : *N*-Methylcrotsparine
R^1 = CH_3, R^2 = OCH_3 : Orientalinone

Mecambrine exhibits antihistaminic and hypertensive properties, is toxic with an LD_{50} (mouse p.o.) of 4.1 mg/kg. Pronuciferine shows potent local anesthetic activity. Orientalinone is an intermediate in the biosynthesis of isothebaine.

Biosynthesis [1]**:** By oxidative phenol coupling from benzyltetrahydroisoquinolines.

Lit.: [1] Mothes et al., p. 209–214.

gen.: Beilstein E V **21/13**, 309 (crotonosine & crotsparine); 537 (orientalinone) ▪ Manske **12**, 333–454; **17**, 385–544 ▪ Shamma, p. 178–193 ▪ Shamma-Moniot, p. 117–121 ▪ Tetrahedron **83**, 1765 (1987) ▪ Ullmann (5.) **A 1**, 372. – *[HS 2939 90]*

Proazulenes. General name for *sesquiterpenes that can be converted to *azulene derivatives under conditions ranging from mere steam distillation to drastic thermal dehydration; *example:* sesquiterpene lactones from essential oils such as *chamomile oils, *armoise (artemisia) oil, absinthe oil, or guaiac wood oil. P. are found as colorless biogenetic precursors of the blue guaiazulene and chamazulene (see azulenes) in chamomile, e. g., *matricin and *matricarin* ($C_{17}H_{20}O_5$, M_R 304.34, mp. 193–195 °C) and in the absinthe herb, e. g., *artabsin* ($C_{15}H_{20}O_3$, M_R 248.32, mp. 133–135 °C) and its Diels-Alder dimer *absinthin ($C_{30}H_{40}O_6$, M_R 496.64, mp. 182–183 °C).

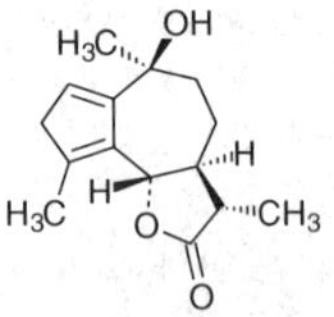
Artabsin

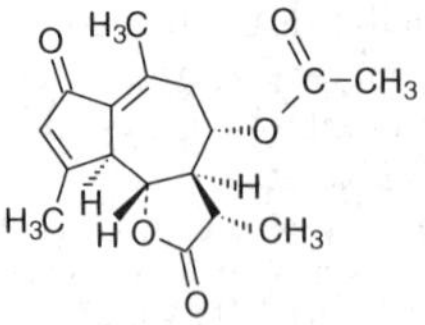
Matricarin

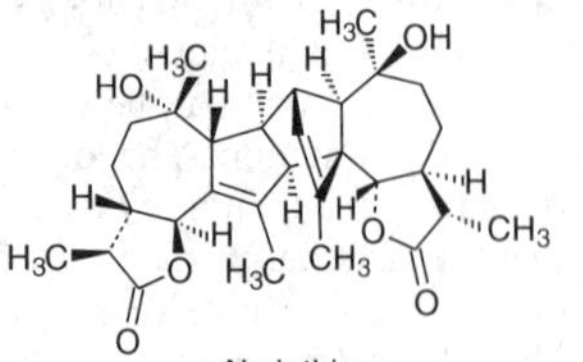
Absinthin

Lit.: see absinthin. – *[CAS 5989-43-5 (matricarin); 24399-20-0 (artabsin); 1362-42-1 (absinthin)]*

Procalciferol, procholecalciferol see calcitriol.

Prochamazulene see matricin.

Procumbide see antirrhinoside.

Procyanidins see proanthocyanidins.

Prodigiosin.

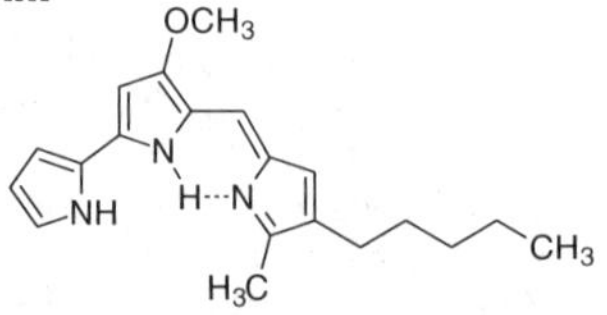

$C_{20}H_{25}N_3O$, M_R 323.44, dark red, pyramidal cryst., mp. 151–152 °C (sinters at 70–80 °C), LD_{50} (mouse i. p.) 18 mg/kg. An antibiotically active *pyrrole pigment from various bacteria (*Serratia marcescens, Beneckea gazogenes, Cladospora* spp.). Besides P. several other analogues have been isolated, including metacycloprodigiosin with a cyclophane structure. In the Middle Ages such bacteria colonizing consecrated wafers were believed to be drops of blood ("bleeding Host"), which lead to persecution of the Jewish population. P. has antifungal and antileukemic activities and is effective against the malaria pathogen *Plasmodium falciparum*. However, it is too toxic for clinical use. For synthesis of analogues, see *Lit.* [1].

Lit.: [1] Synth. Commun. **29**, 35 (1999).

gen.: Beilstein E V **26/3**, 352 ff. ▪ J. Bacteriol. **129**, 124 (1977) (biosynthesis) ▪ Merck-Index (12.), No. 7948. – *Synthesis:* ApSimon **1**, 227–232 ▪ J. Chem. Soc., Chem. Commun. **1990**, 734 ▪ J. Org. Chem. **53**, 1405 (1988) ▪ Tetrahedron Lett. **28**, 2499 (1987); **30**, 1725 (1989). – *[CAS 82-89-3]*

Prodlure [litlure A, (9*Z*,11*E*)-9,11-tetradecadienyl acetate].

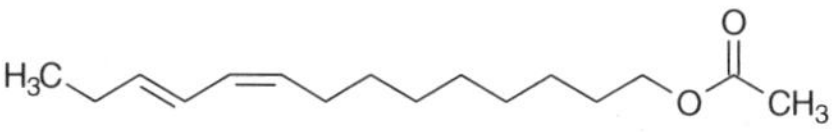

$C_{16}H_{28}O_2$, M_R 252.40, liquid, bp. 147–148 °C (26.66 Pa). The main component of the sexual *pheromone of the cotton pest *Spodoptera littoralis*.

Lit.: ApSimon **4**, 37–38; **9**, 56–59 (synthesis) ▪ Merck-Index (12.), No. 7950. – *[CAS 50767-79-8]*

Progesterone, progestens see progestogens.

Progestogens (progestins, progestagens). The P. are *steroid hormones, formed by human females together with *estrogens in the corpus luteum (yellow body) and in the placenta. The most important endogenous P. is *progesterone* (pregnancy hormone, yellow body hormone). It has the chemical structure of pregn-4-ene-3,20-dione {$C_{21}H_{30}O_2$, M_R 314.47, mp. 121–122 °C and 127–131 °C (dimorphous), $[\alpha]_D$ +192° (dioxane)} and, like the *corticosteroids, belongs to the *pregnane derivatives.

Progesterone is formed biosynthetically from 3*β*-hydroxypregn-5-en-20-one which, in turn, is built up from *cholesterol by enzymatic oxidative side-chain degradation. It is metabolized in the liver to (20*S*)-5*β*-pregnane-3*α*,20-diol, which is then excreted via the kidneys as the glucuronide. At the same time progesterone is an intermediate in the biosynthesis of *androgens and *corticosteroids. It has also been detected in small amounts in plants, namely, in the leaves of *Holarrhena floribunda* and in apple seeds (*Pyrus malus*, Rosaceae).

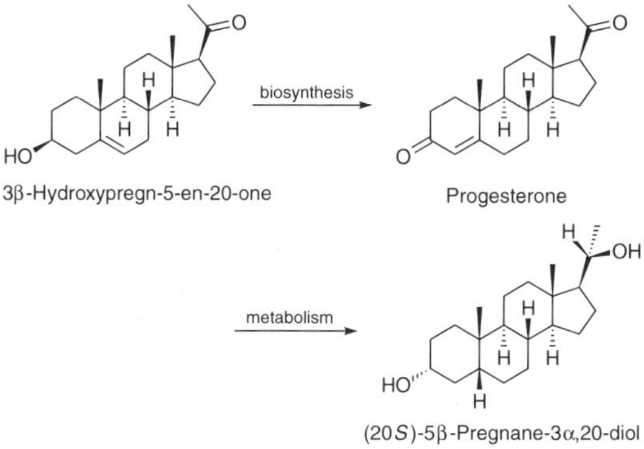

Figure 1: Structure, biosynthesis, and metabolism of Progesterone.

Activity: As an antagonist of estrogens, progesterone induces the nidation and further development of a fertilized egg in the uterine mucous membrane; during pregnancy it prevents the ripening of further follicles and stimulates the development of the functional sections of the lactic glands. A deficiency of progesterone leads to spontaneous abortion.

Use: Progesterone is used therapeutically to prevent premature births and for menstrual disorders. Synthetic P. in combination with estrogens are used as *ovulation inhibitors* (oral contraceptives) to prevent pregnancy and for climactic disorders. Examples of synthetic P. (all 19-*norsteroids*) are *norethisterone* (norethindrone), *levonorgestrel* (racemate = norgestrel). The progesterone antagonist *mifepristone* (RU 486) is authorized in some countries for termination of pregnancies up to the 7th week (induced abortion).

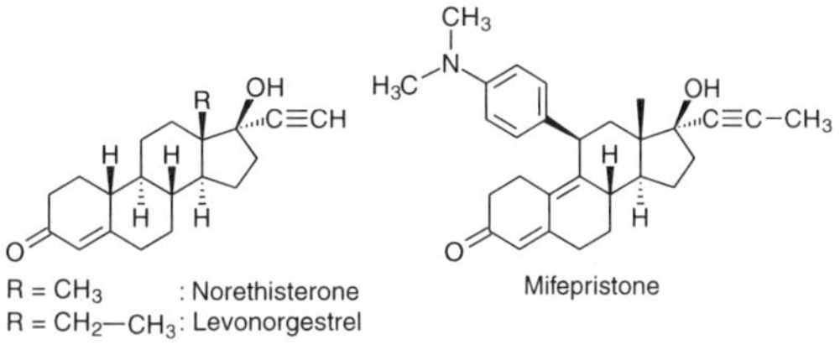

Figure 2: Synthetic Progestogens and Mifepristone.

Lit.: Bohl & Duax (eds.), Molecular Structure and Biological Activity of Steroids, Boca Raton: CRC Press 1992 ▪ Fieser & Fieser, Steroide, p. 588–628, Weinheim: Verl. Chemie 1961 ▪ J. Nat. Prod. **53**, 152 (1990) (synthetic progesterone) ▪ Ullmann (5.) **A 13**, 126–134 ▪ Zeelen, p. 97–123. – *[HS 2937 92; CAS 57-83-0 (progesterone); 68-22-4 (norethisterone); 797-63-7 (levonorgestrel); 6533-00-2 (norgestrel); 84371-65-3 (mifepristone)]*

Progoitrin see glucosinolates.

L-Proline [(*S*)-2-pyrrolidinecarboxylic acid; abbreviation: Pro or P].

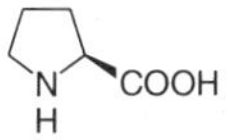

$C_5H_9NO_2$, M_R 115.13, hygroscopic cryst., mp. 220–222 °C (decomp.), $[\alpha]_D^{25}$ −86.2° (H_2O), −60.4° (5 m HCl). Monohydrochloride, mp. 115 °C, pK_a 1.99, 10.60, pI 6.30. A proteinogenic, non-essential amino acid (imino acid). Genetic code: CCU, CCC, CCA, CCG. Average content in proteins 4.6%[1]. Some proteins contain P. in very high concentrations, together with hydroxyproline, e. g., casein (6.7%), prolamines (ca. 20%) and collagen or gelatine (15–30%).

Biosynthesis: Pro is formed from glutamic acid with pyrroline-5-carboxylic acid reductase (EC 1.5.1.2). For synthesis, see *Lit.*[2].

Importance: Pro is able to participate in *α*-helices to a much lower extent than other amino acids. It is thus a "helix breaker" of particular importance for the structure of proteins. Rotation about the amino peptide bond of proline is especially hindered; the correct folding of Pro-containing proteins thus appears to be catalyzed by peptidylprolyl *cis-trans*-isomerases (rotamases, EC 5.2.1.8; see FK-506). On account of its presence in gluten, Pro is involved in the Maillard reaction resulting in the typical bread flavor. The inner salt (betaine) of 1,1-dimethyl-P. is a widely distributed pyrrolidine alkaloid in plants, e.g., in woundwort (*Stachys* spp., *stachydrin*).

4-Methylproline (**1**): A non-proteinogenic amino acid in apples and pears[3]. $C_6H_{11}NO_2$, M_R 129.16, mp. 239–240 °C (decomp.), $[\alpha]_D^{18}$ −56.6° (H_2O), $[\alpha]_D^{20}$ −23.9° (3 m HCl). It also occurs in Japanese medlar fruits (*Eriobotrya japonica*) together with the *trans*-D-form.

4-Methylene-DL-proline (**2**): The free racemate occurs in medlar fruits. $C_6H_9NO_2$, M_R 127.14, mp. 243–245 °C (decomp.). For synthesis, see *Lit.*[4].

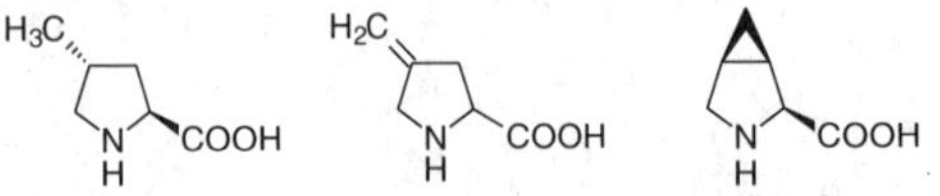

1 (2*S*,4*R*)-, *trans*-L-form 2 racemate 3 (2*S*,3*R*,4*S*)-, *cis*-L-form

cis-3,4-Methano-L-proline (*cis*-3,4-methylene-L-proline) (**3**): Occurs in the seeds of *Aesculus parviflora* (Hippocastanaceae) and *Blighia sapida* (Sapindaceae)[5]. $C_6H_9NO_2$, M_R 127.14, mp. 243–245 °C, $[\alpha]_D^{20}$ −94° (H_2O).

Lit.: [1]Biochem. Biophys. Res. Commun. **78**, 1018–1024 (1977). [2]Tetrahedron **36**, 2961–2965 (1980). [3]Phytochemistry **11**, 751–756 (1972). [4]J. Org. Chem. **57**, 2060 (1992); Tetrahedron: Asymmetry **8**, 1855 (1997). [5]Tetrahedron **46**, 2231 (1990); **53**, 14773 (1997).
gen.: Beilstein E V **22/1**, 30–34 ▪ Bull. Chem. Soc. Jpn. **64**, 620 (1991) ▪ Chem. Lett. **1989**, 1413 ▪ J. Nat. Prod. **59**, 1205 (1996) (review P. analogues) ▪ Karrer, No. 2401 ▪ Nature (London) **329**, 268 ff. (1987) ▪ Ullmann (5.) **A 2**, 58, 63, 72, 86. – *[HS 2933 90; CAS 147-85-3 (P.); 23009-50-9 (1); 2370-38-9 (2); 22255-16-9 (3)]*

Pronuciferine see proaporphine alkaloids.

1,2,3-Propanetricarboxylic acid (tricarballylic acid).

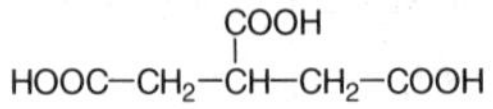

$C_6H_8O_6$, M_R 176.13, prisms, mp. 166 °C, sublimation. P. is formed in sugar beets and sugar maple by reduction of *citric acid, on the technical scale by reduction of *aconitic acid. P. esters with fatty alcohols belong to the so-called inverse fats.

Lit.: Acta Crystallogr., Sect. C **44**, 758 (1988) (structure) ▪ Beilstein E IV **2**, 2366 ▪ Karrer, No. 875. – *[HS 2917 19; CAS 99-14-9]*

Propargylglycine see (*S*)-2-amino-4-pentynoic acid.

Prophalloin see phallotoxins.

Propioxatins. *Peptide antibiotics from cultures of *Kitasatosporia setae* with inhibitory effects on the enzyme enkephalinase B that hydrolyses the Gly-Gly bond of *enkephalins. P. thus reinforces the activity of enkephalins, that are of significance as endogenous analgesic principles (reduction of pain).

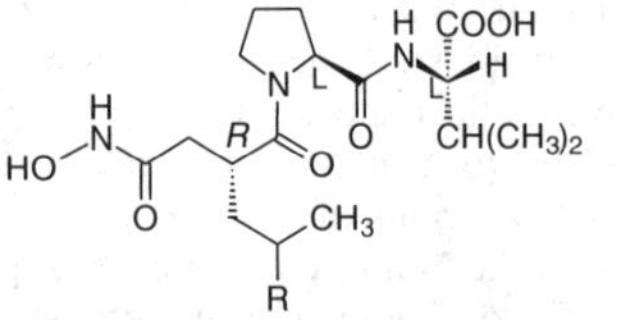

R = H : P. A
R = CH_3 : P. B

Table: Data of Propioxatins.

P.	molecular formula	M_R	mp. [°C]	$[\alpha]_D^{25}$ (H_2O)	CAS
A	$C_{17}H_{29}N_3O_6$	371.43	106–110	-70.9°	102962-94-7
B	$C_{18}H_{31}N_3O_6$	385.46	84–90	-51.3°	102962-95-8

Lit.: Biochim. Biophys. Acta **925**, 27–35 (1987). – *Activity:* J. Antibiot. **39**, 1368–1385 (1986) ▪ J. Biochem. (Tokyo) **104**, 706–711 (1988). – *[HS 2941 90]*

Propolis (hive dross, bee glue). Name derived from Greek: pro = for and polis = state). A dark yellow to light brown, resin-like mass that softens when handled with a spicy-balsamy odor, mp. 50–70 °C, collected principally from buds of poplar, beech, and other trees and used by *bees as a wall covering and to strengthen honeycomb (bee glue). P. contains ca. 10–20% wax, larger amounts of resin, essential oils, and, above all, *flavonoids, especially pinocembrin (5,7-dihydroxyflavanone). These are assumed to be responsible for the general antimicrobial properties of P., while the virostatic activity is due to caffeic acid derivatives[1] (especially the 2-phenylethyl ester). Recently an increasing number of contact allergies due to P. has been reported, caused in particular, by the component 1,1-dimethylallyl caffeate[3].

Use: In popular medicine as an ointment for rheumatism, gout, and for lip and skin hygiene[2], as well as for fumigating purposes.

Lit.: [1]Naturwissenschaften **72**, 659 ff. (1985). [2]Chem. Pharm. Bull. **46**, 1477 (1998). [3]Med. Mo. Pharm. **15**, 343 f. (1992).
gen.: Chem. Pharm. Bull. **47**, 1388, 1521 (1999) ▪ Justus Liebigs Ann. Chem. **1991**, 93–97 ▪ Merck-Index (12.), No. 8021 ▪ Parfums Cosmet. Arômes **82**, 73–77 (1988). – *Egyptian P.:* Z. Naturforsch. C **53**, 201 (1998). – *Brazilian P.:* Z. Naturforsch. C **52**, 676, 828 (1997). – *[CAS 85665-41-4]*

(*E*)-4-Propylidene-L-glutamic acid see 4-methyleneglutamic acid.

Proscillaridin see bufadienolides.

Prostacyclins (prostaglandins I, abbreviation PGI).

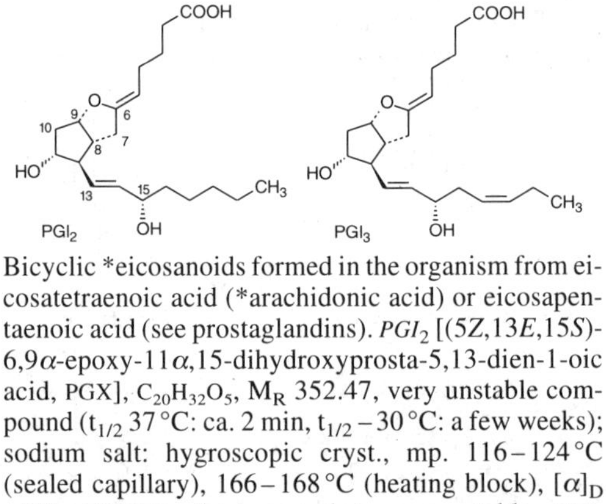

PGI_2 PGI_3

Bicyclic *eicosanoids formed in the organism from eicosatetraenoic acid (*arachidonic acid) or eicosapentaenoic acid (see prostaglandins). *PGI_2* [(5*Z*,13*E*,15*S*)-6,9α-epoxy-11α,15-dihydroxyprosta-5,13-dien-1-oic acid, PGX], $C_{20}H_{32}O_5$, M_R 352.47, very unstable compound ($t_{1/2}$ 37 °C: ca. 2 min, $t_{1/2}$ −30 °C: a few weeks); sodium salt: hygroscopic cryst., mp. 116–124 °C (sealed capillary), 166–168 °C (heating block), $[\alpha]_D$ +88° ($CHCl_3$), the somewhat more stable *PGI_3* [(5*Z*,9*S*,11*R*,13*E*,15*S*, 17*Z*)-6,9-epoxy-11,15-dihydroxyprosta-5,13,17-trien-1-oic acid], $C_{20}H_{30}O_5$, M_R 350.46; sodium salt, powder. PGI are easily hydrolysed non-enzymatically to 6-oxo-PGF.

Activity: PGI_2 has an even stronger vasodilatory activity than prostaglandin E_2. PGI inhibit the aggregation of thrombocytes and are used clinically, in the form of more stable synthetic analogues, in therapy for occlusive arterial diseases (antagonist of *thromboxanes). The inhibition of platelet aggregation is due to the activation of adenylate cyclase and the resultant accumulation of cyclic AMP in the thrombocytes. PGI_2 has cytoprotective activity, e. g., for the stomach wall in gastritis or tissue in organ transplantations.

Biosynthesis: PGI are formed from prostaglandins H under the action of prostacyclin synthetase.

Lit.: Beilstein E V **18/7**, 485 ▪ Justus Liebigs Ann. Chem. **1990**, 1087–1091 (synthesis) ▪ Nachr. Chem. Techn. Lab. **37**, 584–590 (1989) ▪ Thromb. Res. **1987**, 581 ▪ see also prostaglandins. – *[HS 2918 19; CAS 35121-78-9 (PGI_2); 61849-14-7 (PGI_2-Na salt); 68794-57-0 (PGI_3); 68324-96-9 (PGI_3-Na salt)]*

Prostaglandins (abbreviation PG). Name for a group of biologically highly active derivatives of eicosapolyenoic acid ($C_{20}H_{40-2n}O_2$) and thus also included among the so-called *eicosanoids. PG are formed in practically all mammal and human tissues. Some PG [PGA_2, 15-epi-PGA_2, (5*E*)-PGA_2, and 15-epi-PGE_2] also occur in surprisingly high concentrations in the coral *Plexaura homomalla* [1]. PG have also been detected in insects [2].

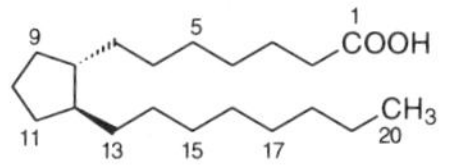

basic skeleton: Prostan(o)ic acid

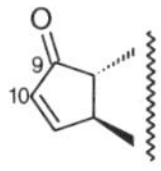
PGA

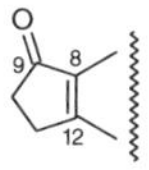
PGB

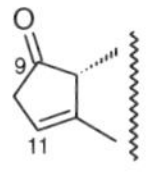
PGC

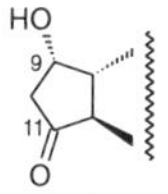
PGD

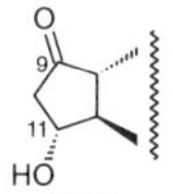
PGE

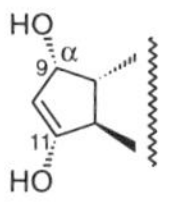
PGFα

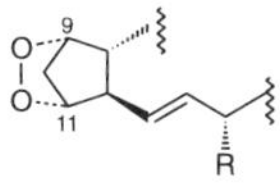
R = OOH : PGG
R = OH : PGH
= PGR

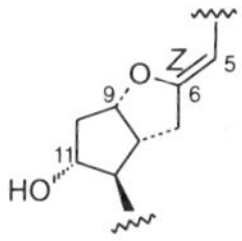
PGI (*Prostacyclins, PGX)

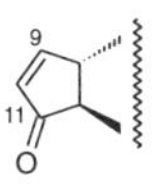
PGJ

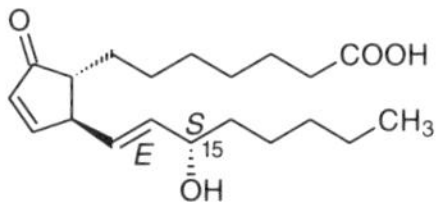
PGA_1

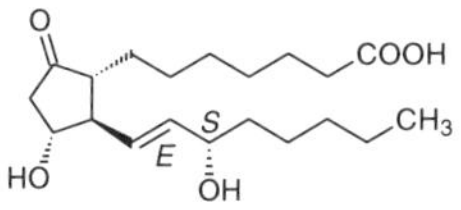
PGE_1 (Alprostadil)

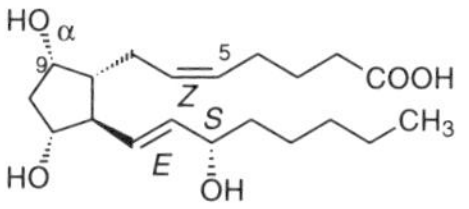
$PGF_{2\alpha}$ (Dinoprost)
9-Epimer : $PGF_{2\beta}$

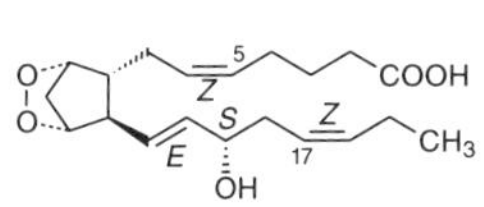
PGH_3 (PGR_3)

Table: Data of Prostaglandins.

PG	molecular formula	M_R	mp. [°C]	$[\alpha]_D$	CAS
PGA_1	$C_{20}H_{32}O_4$	336.47			14152-28-4
PGB_1	$C_{20}H_{32}O_4$	336.47			13345-51-2
PGD_1	$C_{20}H_{34}O_5$	354.49			17968-82-0
PGE_1	$C_{20}H_{34}O_5$	354.49	114–116.5	$[\alpha]_D^{24}$ −53.2° (THF*)	745-65-3
$PGF_{1\alpha}$	$C_{20}H_{36}O_5$	356.60	102–103		745-62-0
PGH_3	$C_{20}H_{30}O_5$	350.46			60114-66-1
PGA_2	$C_{20}H_{30}O_4$	334.45	oil	$[\alpha]_D^{30}$ +140° ($CHCl_3$)	13345-50-1
PGB_2	$C_{20}H_{30}O_4$	334.46			13367-85-6
PGD_3	$C_{20}H_{30}O_5$	350.46	56–57	$[\alpha]_D^{25}$ +9.22° (THF*)	71902-47-1
$PGF_{2\alpha}$	$C_{20}H_{34}O_5$	354.49	25–35	$[\alpha]_D^{25}$ +23.5° (THF*)	551-11-1
$PGF_{2\beta}$	$C_{20}H_{34}O_5$	354.49	94–95	$[\alpha]_{365}^{25}$ +12.7° (CH_3OH)	4510-16-1
PGC_3	$C_{20}H_{28}O_4$	332.44			52590-97-3

* THF = tetrahydrofuran

The PG are named with the letters A–J depending on the type of substitution at the cyclopentane ring. The index number indicates the number of double bonds in the side chain: the index number 1 characterizes derivatives of (*all-Z*)-8,11,14-eicosatrienoic acid, the number 2 derivatives of (*all-Z*)-5,8,11,14-eicosatetraenoic acid (*arachidonic acid), and the number 3 derivatives of (*all-Z*)-5,8,11,14,17-eicosapentaenoic acid (timnodonic acid; often found in marine vertebrates). The orientation of the OH group at C-9 is designated by the index α or β.

PGA_1 occurs in seminal fluid, inhibits platelet aggregation and contraction of smooth musculature; PGB_1 also occurs in seminal fluid and inhibits *in vitro* uterine contractions; PGD_1 has been found in the seminal vesicles of sheep; $PGF_{1\alpha}$, occurs together with 8-epi-$PGF_{1\alpha}$ in the seminal fluid of sheep and stimulates duodenal peristalsis in rabbits; PGH_3 is formed *in vivo* from (*all-Z*)-5,8,11,14,17-eicosapentaenoic acid present in the plasma membrane or taken up with vegetable food; PGA_2 occurs in the coral *Plexaura homomalla* (content 1–3%) together with the stereoisomeric 15-epi- and (5*E*)-PGA_2; PGB_2 isolated from human seminal plasma inhibits *in vitro* uterine contractions; PGD_3 arises as a metabolite of PGH_3; it inhibits blood platelet aggregation; $PGF_{2\alpha}$ is widely distributed, it induces uterine contractions (abortive activity) and stimulates smooth musculature. The 9-epimer ($PGF_{2\beta}$) is a potent bronchodilator; PGC_3 is an intermediate in the conversion of PGA_3 to PGB_3.

Biosynthesis: The endoperoxides PGG_2 and PGH_2 are formed in the so-called *arachidonic acid cascade from eicosatetraenoic acid (arachidonic acid) liberated from membrane phospholipids by phospholipase A_2 (EC 3.1.1.4) (see figure under thromboxanes) under the action of cyclooxygenase. All other PG are then formed by further enzymatic reactions. In contrast to other hormones, PG are not stored in the cell but are always newly synthesized when required. The close relatives of the PG, the *thromboxanes, *prostacyclins, *leukotrienes, and *lipoxins are also generated from eicosapolyenoic acids in the lipoxygenase pathway. For synthesis, see *Lit.* [3].

Historical: At the beginning of the 1930's U. S. von Euler and M. W. Goldblatt independently detected a substance with activity in smooth musculature in seminal fluid, the structure of this compound was elucidated by Bergström in 1958–1964. Further key data: 1968 total synthesis by Corey, beginning of the 1970's elucidation of the mechanism of action of NSAID (nonsteroidal anti-inflammatory drugs), 1975 thromboxane, 1976 prostacyclin, 1979 leukotrienes, 1982 Nobel Prize for Medicine awarded to Vane, Samuelsson, and Bergström [4].

Lit.: [1] Tetrahedron Lett. **1969**, 5185 ff.; J. Am. Chem. Soc. **94**, 2124 (1972). [2] Insect Biochem. Mol. Biol. **24**, 481–492 (1994). [3] Angew. Chem. Int. Ed. Engl. **30**, 455–465 (1991); Corey & Cheng, The Logic of Chemical Synthesis, p. 250–309, New York: Wiley 1989; Synlett **1997**, 734 ($F_{2\alpha}$). [4] Angew. Chem. Int. Ed. Engl. **22**, 741–752, 805–815, 858–866 (1983).
gen.: Bailey (ed.), Prostaglandins, Leukotrienes and Lipoxins: Biochemistry, Mechanism and Clinical Applications, New York: Plenum 1985 ▪ Barton-Nakanishi **5**, 225–262 (bio-

synthesis) ▪ Luckner (3.), p. 163f., 457, 459, 464, 487 ▪ Nat. Prod. Rep. **10**, 593–624 (1993) ▪ Sax (8.), MCA 025 (PGA_2), POC 250 (PGA_1), POC 350 (PGE_1), POC 400 ($PGF_{1\alpha}$), POC 500 ($PGF_{2\alpha}$) (toxicology) ▪ Stryer 1990, p. 1012, 1027f., 1063 ▪ Willis (eds.), CRC Handbook of Eicosanoids, Prostaglandins and Related Lipids, Boca Raton: CRC 1987. – *Journal:* Advances in Prostaglandin, Thromboxane and Leukotriene Research, New York: Raven Press (since 1976). – *[HS 2918 19]*

Prostratine see tigliane.

Protease inhibitors. Low-molecular weight substances produced principally by microorganisms that mostly inhibit specific proteases (see figure below). P. i. are of therapeutic significance for, among others, the following clinical situations: microbial infections, viral infections, parasitic diseases (e.g., malaria), inflammations, asthma, metastases and tumor growth, neurodegenerative diseases (e.g. Alzheimer's); as well as for coagulation processes, immunological interventions, and apoptosis.

Lit.: [1] Aoyagi, in Bushell & Gräfe (eds.), Bioactive Metabolites from Microorganisms, Amsterdam: Elsevier 1989.
gen.: Aoyagi, in Kleinkauf & von Doehren (eds.), Biochemistry of Peptide Antibiotics, p. 311, Berlin: De Gruyter 1990 ▪ Biochimie **73**, 121 (1991) ▪ Biol. Chem. Hoppe-Seyler **377**, 217 (1996) ▪ Int. J. Oncol. **2**, 861 (1993) ▪ Kenezana et al., in Troll & Kennedy (eds.), Protease Inhibitors as Cancer Chemopreventive Agents, p. 131 New York: Plenum 1993.

Prothoracotropic hormone see insect hormones.

Protoaescigenin see oleanane.

Protoberberine alkaloids. Most P. a. occur as tetrahydroprotoberberines or as protoberberinium salts. The P. a. are among the most widely distributed group of the *isoquinoline alkaloids, they occur in at least 9 plant families, especially in Annonaceae, Berberidaceae, Lauraceae, Menispermaceae, Papaveraceae, and Rutaceae, as well as in Fumariaceae (see table).

The P. a. are readily soluble in ethanol, chloroform, and ether. Berberine is a toxic yellow pigment occurring in *Berberis* and *Mahonia* species. Silk, cotton, and leather can be dyed with berberine. It has also been the subject of intensive pharmaceutical research [1,2,3] and exhibits many interesting properties including stimulation of respiration, hypotensive and convulsive activities. It is an inhibitor of choline esterase, tyrosine decarboxylase, and tryptophanase as well as being an antianemic agent. It exhibits antibacterial, antifungal, cytotoxic, and antineoplastic activities. It is used clinically in the treatment of gastrointestinal diseases, cholera, and infantile diarrhea. It has also found use as an antimalarial, antipyretic, bitter stomachic, and carminative agent. LD_{50} (mouse i.p.) 24.3 mg/kg. Columbamine is a butyrylcholine esterase inhibitor. Corydaline exhibits analgesic and antirheumatic activities. Palmatine effects uterine contractions and has bactericidic activity. Tetrahydropalmatine is an analgesic, sedative, and hypnotic agent as well as exhibiting a potent, papaverine-like activity.

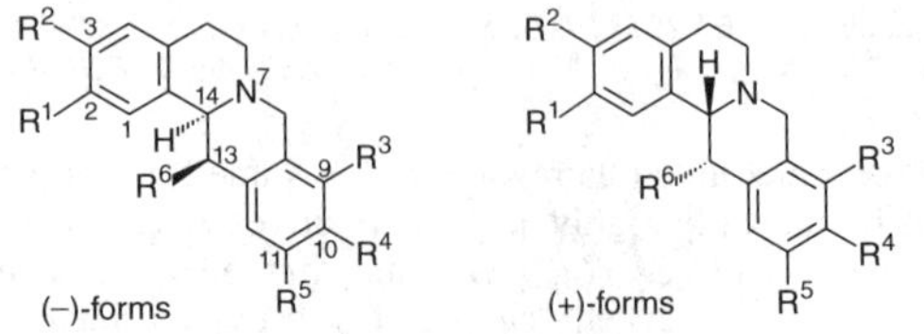

	R^1	R^2	R^3	R^4	R^5	R^6
Canadine	O–CH_2–O		OCH_3	OCH_3	H	H
Cheilanthifoline	OH	OCH_3	O–CH_2–O		H	H
Coreximine	OH	OCH_3	H	OCH_3	OH	H
Corydaline	OCH_3	OCH_3	OCH_3	OCH_3	H	CH_3
Isocorypalmine	OH	OCH_3	OCH_3	OCH_3	H	H
Ophiocarpine	O–CH_2–O		OCH_3	OCH_3	H	OH
Scoulerine	OH	OCH_3	OH	OCH_3	H	H
Stylopine	O–CH_2–O		O–CH_2–O		H	H
Tetrahydropalmatine	OCH_3	OCH_3	OCH_3	OCH_3	H	H

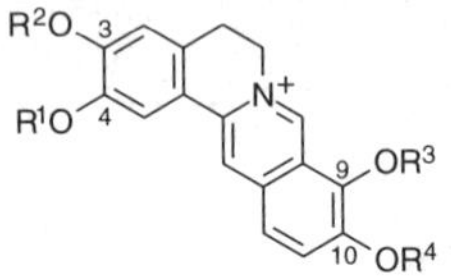

	R^1	R^2	R^3	R^4
Berberine	CH_2		CH_3	CH_3
Columbamine	H	CH_3	CH_3	CH_3
Coptisine	CH_2		CH_2	
Jatrorrhizine	CH_3	H	CH_3	CH_3
Palmatine	CH_3	CH_3	CH_3	CH_3

Biosynthesis: The P. a. are formed from benzylisoquinolines, with the so-called berberine bridge being constructed by oxidative cyclization with participation of the *N*-methyl group [4,5].

Lit.: [1] Eur. J. Pharmacol. **204**, 35–40 (1991). [2] Pharmacology **43**, 329 (1991). [3] J. Pharmacol. **108**, 534–537 (1993); J. Cardiovasc. Pharmacol. **23**, 716 (1994). [4] Mothes et al., p. 226–230. [5] Phytochemistry **36**, 327–332 (1994).

		Peptidase / Proteinase	Inhibitors
Protease	Endo-	Serine- and Cysteine-	Leupeptin, Antipain, *Chymostatin, Elastatinal, Ac-Leu-Argal, Z-Diprotin
		Aspartate-	*Pepstatin A, Pepstanone, Hydroxpepstatin
		Metallo-	*Phosphoramidon, *Steffimycins
	Exo-	Peptidase	
		Amino-	Amastatin, Actinonin, Arphamenine A and B, *Bestatin, *Ebelactone A and B, Formestin A and B, Probestin, Prostatin, *Leuhistin
		Dipeptidylamino-	Ac-Leu-Argal, Antipain, Leupeptin, Diprotin A and B
		Carboxy-	(*S*)-α-Benzyl-malic acid, Histargin
		Dipeptidylcarboxy-	EDDS, Foroxymithine, Histargin

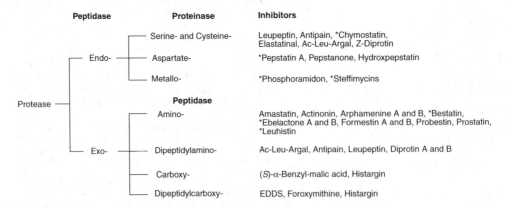

Figure: Proteases and some typical inhibitors (*Lit.*[1]).

Table: Data and occurrence of Protoberberine alkaloids.

name	molecular formula	M_R	mp. [°C]	$[\alpha]_D$	occurrence	CAS
Berberine	$C_{20}H_{18}NO_4^+$	336.37	145		*Berberis* and *Mahonia* species	2086-83-1
(*S*)-Canadine	$C_{20}H_{21}NO_4$	339.39	134	−299° ($CHCl_3$)	*Hydrastis canadensis*, *Corydalis* species	5096-57-1
Cheilanthifoline	$C_{19}H_{19}NO_4$	325.36	184	−311° (CH_3OH)	*Corydalis cheilanthifolia*, *C. scouleri*, *C. sibirica* and *Argemone* species	483-44-3
Columbamine	$C_{20}H_{20}NO_4^+$	338.38			*Jateorhiza palmata*, *Berberis* species	3621-36-1
Coptisine	$C_{19}H_{14}NO_4^+$	320.32	216–218		*Coptis japonica*	3486-66-6
Corexímine	$C_{19}H_{21}NO_4$	327.38	254–256	−420° (pyridine)	*Dicentra eximia*, *Papaver somniferum*, *Corydalis* species	483-45-4
(+)-Corydaline	$C_{22}H_{27}NO_4$	369.46	135	+295° (C_2H_5OH)	*Corydalis tuberosa* and *Corydalis* species	518-69-4
(−)-(*S*)-Isocorypalmine	$C_{20}H_{23}NO_4$	341.41	239–241	−282° (CH_3OH)	*Glaucium fimbrilligerum*, *Hydrastis canadensis*, *Corydalis* and *Thalictrum* species	483-34-1
Jatrorrhizine (Jateorhizine)	$C_{20}H_{20}NO_4^+$	338.38			*Jateorhiza palmata*, *Berberis* and *Mahonia* species	3621-38-3
Ophiocarpine	$C_{20}H_{21}NO_5$	355.39	188	−283° ($CHCl_3$)	*Corydalis* species	478-13-7
Palmatine	$C_{21}H_{22}NO_4^+$	352.41			*Jateorhiza palmata*, *Berberis* and *Mahonia* species	3486-67-7
(*R*)-Scoulerine	$C_{19}H_{21}NO_4$	327.38	195	+284° (CH_3OH)	*Corydalis tuberosa*, *C. cava*	6507-34-2
Stylopine	$C_{19}H_{17}NO_4$	323.35	204	−310° ($CHCl_3$)	*Corydalis* species	84-39-9
(*R*)-Tetrahydropalmatine	$C_{21}H_{25}NO_4$	355.43	143	+291° (C_2H_5OH)	*Pachypodanthium confine*, *Corydalis* species	3520-14-7

gen.: Beilstein EIII/IV **27**, 6539 (berberine); EV **21/6**, 174 (corydaline) ▪ Hager (5.) **3**, 519; **4**, 480–497, 835–848, 1013–1027, 1055–1158; **5**, 110–115, 745–750 ▪ Heterocycles **37**, 897 (1994) (berberine) ▪ J. Chem. Soc., Perkin Trans. 1 **1997**, 2291 (synthesis tetrahydropalmatine) ▪ Manske **28**, 95–181; **33**, 141–230; **49**, 273 ▪ Pelletier **6**, 297–337 ▪ Prog. Bot. **51**, 113–133 (1989) ▪ Sax (8.), BFN 500–BNF 750 ▪ Shamma, p. 269–314 ▪ Shamma-Moniot, p. 209–259 ▪ Ullmann (5.) **A 1**, 373. – *[HS 2939 90]*

Protocatechualdehyde see 3,4-dihydroxybenzaldehydes.

Protocatechuic acid see dihydroxybenzoic acids.

Protocevine see Veratrum steroid alkaloids.

Protoemetine see ipecac alkaloids.

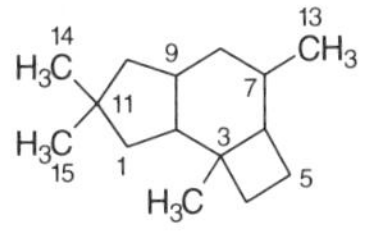

Protoilludane

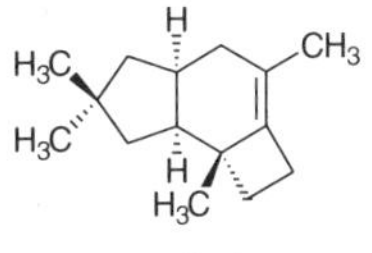

6-Protoilludene

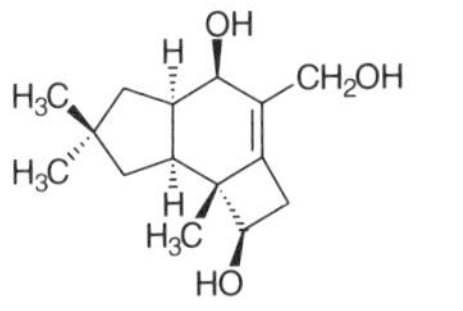

Illudol

Protoilludanes. Sesquiterpenoids with the P. skeleton are produced in mycelia cultures of various basidiomycetes. They are formed biosynthetically by acid-catalyzed cyclization of α-*humulene. *6-Protoilludene* {$C_{15}H_{24}$, M_R 204.36, $[\alpha]_D$ −52° (CH_3OH)} is formed first and also occurs in cultures of *Fomitopsis insularis* and *Omphalotus olearius*. An example of a P. modified by oxidation is *illudol* {$C_{15}H_{24}O_3$, M_R 252.35, cryst., mp. 130–132 °C, $[\alpha]_D$ −116° (C_2H_5OH)} from cultures of *Clitocybe illudens*. Many other P. are produced by *Armillaria mellea* (honey mushroom or agaric). Ring cleavage leads to secoprotoilludanes such as *fomannosin and fomajorin D. Some P. exhibit biological activities, see Armillaria sesquiterpenoids, lentinellic acid, sulcatins B.

Lit.: Tetrahedron **37**, 2199 (1981) (review) ▪ Turner **1**, 226–228; **2**, 244–246. – *Synthesis:* Chem. Pharm. Bull. **33**, 440 (1985) ▪ J. Am. Chem. Soc. **113**, 381 (1991); **119**, 3427 (1997) ▪ Tetrahedron Lett. **27**, 5471 (1986). – *[CAS 64242-89-3 (6-protoilludene); 16981-75-2 (illudol)]*

Protolichester(in)ic acid.

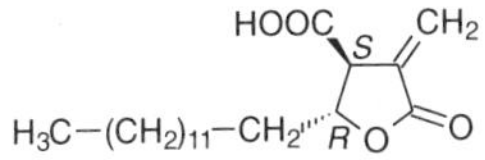

(+)-form

$C_{19}H_{32}O_4$, M_R 324.46, platelets with a mother-of-pearl like glance, mp. 107 °C, $[\alpha]_D^{20}$ +12°. A *paraconic acid derivative from the fruticose lichen *Cetraria islandica.* It undergoes ready isomerization (e. g., on melting) to (+)-*lichester(in)ic acid. The (−)-form occurs in *C. ericetorum*. Diazomethane adds to the *exo*-methylene bond of P. to give a mixture of the two stereoisomeric pyrazolines. P. has antibiotic and growth regulating activities (in plants).

Lit.: Asahina-Shibata, p. 17–23 ▪ J. Org. Chem. **58**, 7537 (1993) ▪ Karrer, No. 1095 ▪ Z. Naturforsch. B **47**, 842–854 (1992). – *[CAS 1448-96-0]*

Protopanaxadiol, protopanaxatriol see ginseng.

Protopine alkaloids. The structures of the P. a. are characterized by an *N*-methyltetrahydroprotoberberine structure in which the quinolizine system has been opened to give a 10-membered azecine ring; they also usually possess a 14-oxo group and can be converted in a transannular reaction to *protoberberine alkaloids. They are widely distributed in the plant families Berberidaceae, Fumariaceae, Papaveraceae, Ranunculaceae, and Rutaceae (see table).

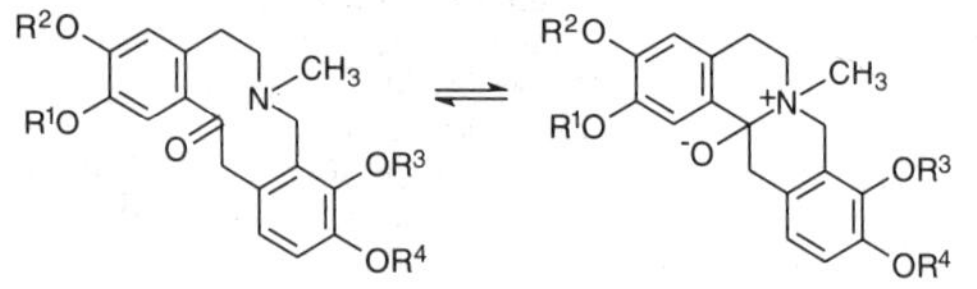

	R¹	R²	R³	R⁴
Allocryptopine	CH_2		CH_3	CH_3
Cryptopine	CH_3	CH_3	CH_2	
Muramine	CH_3	CH_3	CH_3	CH_3
Protopine	CH_2		CH_2	

P. a. are soluble in chloroform, ethanol, ether, and insoluble in water. Allocryptopine is a hypotensive agent, a respiratory stimulant, and a muscle relaxant. Cryptopine is a highly toxic minor alkaloid of opium. Protopine exhibits weak spasmolytic and bactericidal activities.

Table: Data and occurrence of Protopine alkaloids.

name	molecular formula	M_R	mp. [°C]	occurrence	CAS
Allocryptopine	$C_{21}H_{23}NO_5$	369.42	160–161 (α-form) and 169–171 (β-form)	*Dactylicapnos macrocapnos, Argemone bocconia, Eschscholtzia, Glaucium, Macleaya, Meconopsis, Papaver, Sanguinaria, Stylomecon, Chelidonium, Thalictrum* species	485-91-6 (α-form) and 24240-04-8 (β-form)
Cryptopine	$C_{21}H_{23}NO_5$	369.42	221–223	*Corydalis, Dicentra, Argemone, Meconopsis, Papaver, Stylomecon, Thalictrum* species	482-74-6
Muramine	$C_{22}H_{27}NO_5$	385.46	176–177	*Papaver nudicaule* and *Papaver* species, *Glaucium vitellinum, Argemone munita, A. squarrosa*	2292-20-8
Protopine	$C_{20}H_{19}NO_5$	353.37	207	Papaveraceae, Fumariaceae, Berberidaceae, Sapindaceae	130-86-9

The biosynthesis[1] proceeds from reticuline via protoberberine derivatives, e. g., stylopine methochloride.

Lit.: [1]Mothes et al., p. 226–230; Prog. Bot. **51**, 113–133 (1980).

gen.: Hager (5.) **4**, 480–497, 771–788, 1013–1027; **5**, 110–115 ▪ J. Heterocycl. Chem. **27**, 623 (1990) (protopine) ▪ Manske **10**, 467; **12**, 333; **15**, 207; **17**, 385; **18**, 217; **34**, 181–209 ▪ Sax (8.), FOW 000 ▪ Shamma, p. 345–358 ▪ Shamma-Moniot, p. 299–306 ▪ Ullmann (5.) **A1**, 375. – *[HS 2939 90]*

Protoporphyrins. Metal-free biosynthetic precursors of the *porphyrins. In the strictest sense only P. IX, the parent porphyrin of the red blood pigment *heme, should be considered as a P. [formula, see heme, with 2 H in place of Fe(II)]; $C_{34}H_{34}N_4O_4$, M_R 562.67. Brown cryst., uv_{max} (dimethyl ester in $CHCl_3$) 407, 505, 541, 575, 603, 630 nm. In the biosynthesis P. IX is formed from its reduced analogue proto-*porphyrinogen by enzymatic oxidation and then transformed to heme or *chlorophyll by incorporation of iron or magnesium atoms. On account of its occurrence in bird egg shells the pigment was also known as *ooporphyrin* in the past.

Lit.: Aust. J. Chem. **33**, 557 (1980) ▪ Dolphin **I**, 290–299 ▪ Jordan, p. 155, 167 ▪ Smith, p. 104, 107, 771. – *[HS 2933 90; CAS 533-12-8 (P. IX)]*

Protorubradirin.

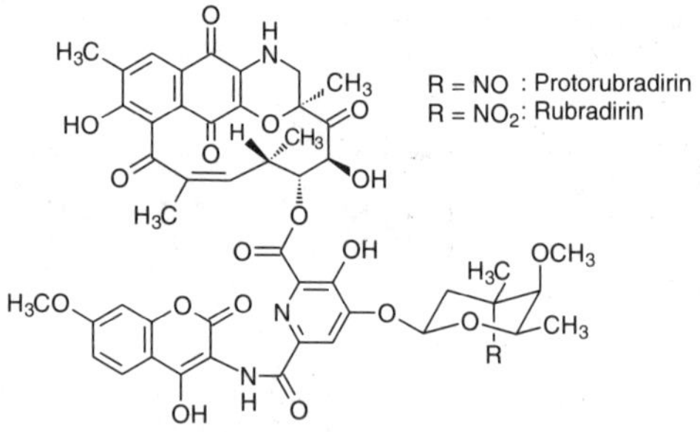

$C_{48}H_{46}N_4O_{19}$, M_R 982.91, amorphous red solid, $[\alpha]_D$ +921° (acetone); a *C*-nitroso sugar-containing antibiotic from cultures of *Streptomyces achromogenes* var. *rubradiris* [1]. On standing in the light P. is oxidatively transformed to the nitro compound *rubradirin* [2] [$C_{48}H_{46}N_4O_{20}$, M_R 998.91, amorphous solid, $[\alpha]_D$ +571° (acetone)], which was originally isolated from the culture media. It is possible that other *C*-nitro sugars of antibiotics are formed from the corresponding nitroso compounds. Like rubradirin, P. exhibits *in vitro* and *in vivo* high activity against Gram-positive bacteria and inhibits protein biosynthesis.

Lit.: [1] J. Antibiot. **45**, 1313–1324 (1992). [2] Antimicrob. Agents Chemother. **91**, 97 (1964); J. Antibiot. **32**, 771 (1979); J. Am. Chem. Soc. **104**, 5173 (1982) [abs. configuration of the sugar not correct]. – *[HS 2941 90; CAS 122525-61-5 (P.); 11031-38-2 (rubradirin)]*

Protosinomenine see benzyl(tetrahydro)isoquinoline alkaloids.

Protostemonine see Stemona alkaloids.

Protostephanine see morphinan alkaloids.

Provitamin D see calcitriol.

Provitamins. Name for precursors of vitamins, which are activated in the organism or *in vitro*. *Examples:* *β-carotene is the P. for *retinal, ergosterol (see sterols) is that for vitamin D_2 (see calcitriol).

PR proteins (pathogenesis related proteins, PRP). PRP include elicitors (see phytoalexins) such as chitosan as well as PRP induced in plants by pathogenic viruses and fungi such as chitinases glucanases, proteins related to thaumatin (TL proteins) with α-amylase and proteinase activities, biosynthesis enzymes for aromatic compounds, hydroxyproline-rich and glycine-rich proteins as well as numerous proteins with unknown functions. For example, PR-1 from tobacco inhibits the growth of *Phytophthora infestans* (potato disease). Research on the phytopathological relationships involves biochemical and genetic methods.

Lit.: Annu. Rev. Phytopathology **28**, 113–136 (1990) ▪ Bol & Linthorst, in Meyer, DeBonde, & Zipf (eds.), Antiviral Proteins in Higher Plants, p. 147–160, Boca Raton: CRC 1995 ▪ Crit. Rev. Plant Sci. **10**, 123–150 (1991) ▪ Plant, Cell Environ. **18**, 865–874 (1995) ▪ Plant Physiol. **108**, 17–27 (1995) ▪ Proc. Natl. Acad. Sci. USA **92**, 4134-4137, 7137 ff. (1995) ▪ Recherche **25**, 334 f. (1994) ▪ Results Probl. Cell Differ. **20** (Plant Promotors and Transcription Factors), 163–179 (1994) ▪ Thaumatin **1994**, 193–199.

PR toxin (PRT).

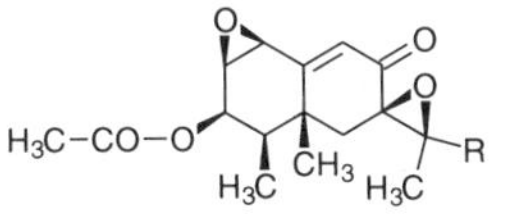

R = CHO : PR-Toxin
R = CH_3 : Eremofortin A
R = CH_2OH : Eremofortin C (12-Hydroxyeremofortin A)
R = $CONH_2$: Eremofortin E

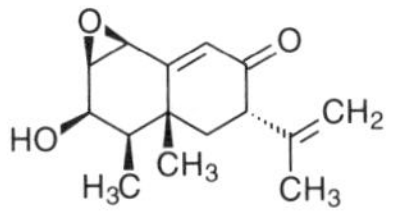

Eremofortin B

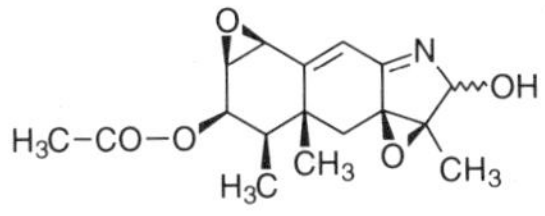

PR imine

A *mycotoxin formed together with structurally related but non-toxic eremofortines and PR imine by *Penicillium roqueforti*. The latter is also formed spontaneously from PRT in aqueous methanolic ammonia solution. All compounds are *sesquiterpenes with the *eremophilane framework. PRT inhibits RNA polymerases and protein biosynthesis, is mutagenic, and probably carcinogenic [1]. It has not been found in blue cheese since it probably reacts with amino acids. LD_{50} (mouse i. p.) 1–5.8 mg/kg, on oral administration the value is between 58 and 100 mg/kg.

Table: Data of PR toxin and Eremofortins.

compound	molecular formula	M_R	mp. [°C]	$[\alpha]_D$ ($CHCl_3$)	CAS
PR toxin	$C_{17}H_{20}O_6$	320.34	155–157	+290°	56299-00-4
PR imine	$C_{17}H_{21}NO_5$	319.36			56349-25-8
Eremofortin A	$C_{17}H_{22}O_5$	306.36	159–161	+205°	62445-06-1
Eremofortin B	$C_{15}H_{20}O_3$	248.32	121–123	+115°	60048-73-9
Eremofortin C	$C_{17}H_{22}O_6$	322.36	122–126		62375-74-0
Eremofortin E	$C_{17}H_{21}NO_6$	335.36	240 (decomp.)	+227°	77732-43-5

Lit.: [1] Cancer Res. **36**, 445 (1976).

gen.: Beilstein E V **19/6**, 233 ▪ Betina, p. 407-411 ▪ Cole-Cox, p. 870–875 ▪ Sax (8.), No. POF 800 ▪ Tetrahedron Lett. **23**, 5359 (1982).

Prulaurasin, prunasin see cyanogenic glycosides.

Prunetin see isoflavones.

Psammaplysin A see aerothionin.

Pseudanes see 2-heptyl-1-hydroxy-4(1*H*)-quinolinone.

Pseudobactin.

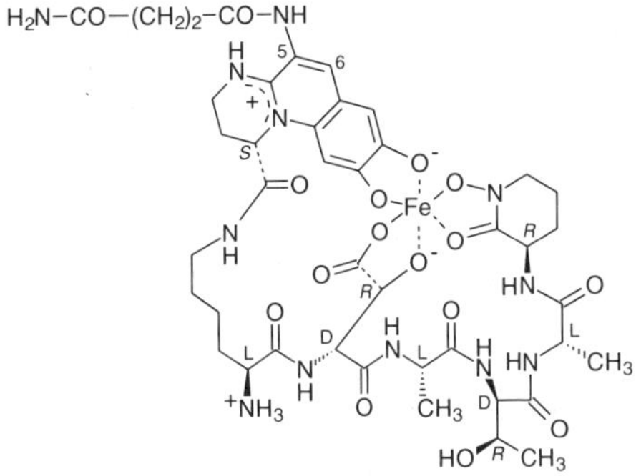

$C_{42}H_{57}FeN_{12}O_{16}$, M_R 1041.83, a water-soluble and fluorescing *siderophore of the hydroxamate/catechol type; P. is produced by cultures of *Pseudomonas fluorescens* under iron deficiency conditions. P. is a plant growth substance but also has antifungal activity against *Fusarium* sp.

Lit.: Biochemistry **20**, 6446, 6457 (1981) ▪ Can. J. Chem. **71**, 1401 (1993) ▪ J. Org. Chem. **55**, 4246 (1990); **59**, 7643 (1994) (synthesis). – *[CAS 76975-04-7]*

Pseudoconhydrine see Hemlock alkaloids.

Pseudocyphellarins. *P. A:* $C_{21}H_{22}O_8$, M_R 402.40, prisms, mp. 173–175 °C. A *depside from the foliose lichen *Pseudocyphellaria vaccina*, occurring together with *P. B* ($C_{21}H_{24}O_8$, M_R 404.42).

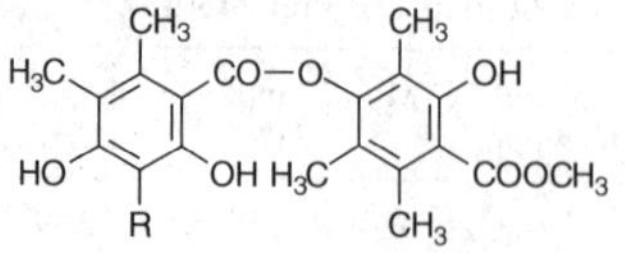

R = CHO : Pseudocyphellarin A
R = CH_2OH : Pseudocyphellarin B

Lit.: Phytochemistry **23**, 431 ff. (1984). – *Synthesis:* Aust. J. Chem. **37**, 2153–2157 (1984); **40**, 2023 (1987) ▪ Helv. Chim. Acta **68**, 1940–1951 (1985). – *[CAS 90685-95-3 (P. A); 90685-96-4 (P. B)]*

Pseudoephedrine see *β*-phenylethylamine alkaloids.

Pseudojervine see Veratrum steroid alkaloids.

Pseudolaric acids. Diterpenoids from the bark of *Pseudolarix kaempferi*, also known as *Tujinpi* in Chinese folk medicine, traditionally used against fungal skin infections. Besides the *in vivo* antifungal activity P. show *in vitro* cytotoxicity against several human tumor cell lines, the most effective is *P. B* ($C_{23}H_{28}O_8$, M_R 432.46, cryst., mp. 165–167 °C).

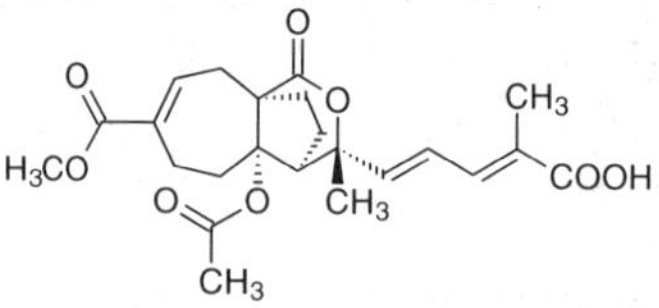

Pseudolaric acid B

Lit.: J. Nat. Prod. **58**, 57–67 (1995) ▪ Planta Med. **47**, 35 (1983) ▪ Tang & Eisenbrand, Chinese Drugs of Plant Origin, p. 793 ff., Berlin: Springer 1992. – *Synthesis:* Tetrahedron Lett. **39**, 9229 (1998). – *[CAS 82508-31-4 (P. B)]*

Pseudomonic acids. Trivial name for the antibiotic mixture isolated from cultures of *Pseudomonas fluorescens*, four components have been identified to date: P. A [1–3], P. B [4], P. C [5,6], and P. D [7]. The P. are polyketides containing a dihydroxylated tetrahydropyran ring with functionalized side chains in the 2- and 5-positions. For synthesis, see *Lit.* [8].

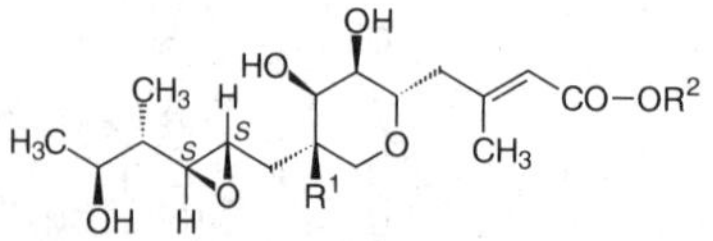

R^1 = H, R^2 = $(CH_2)_8$–COOH : P. A
R^1 = OH, R^2 = $(CH_2)_8$–COOH : P. B
R^1 = H, R^2 = $(CH_2)_4$–CH$\overset{E}{=}$CH–CH_2–CH_2–COOH : P. D

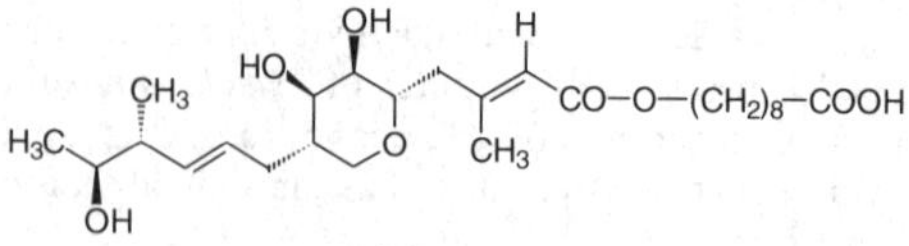

P. C

The main component P. A (international free name *mupirocin*, proprietary name as calcium salt Bactroban®) is primarily active against Gram-positive pathogens and mycoplasma, but poorly active against Gram-negative bacteria (except *Haemophilus* and *Pasteurella*) and corynebacteria. P. A is used for treatment of skin infections and some infections of the nasal mucous membranes (also resistant staphylococci: MRSA). It acts as an inhibitor of bacterial protein synthesis by

Table: Data of Pseudomonic acids.

Pseudomonic acids	molecular formula	M_R	mp. [°C]	CAS
P. A	$C_{26}H_{44}O_9$	500.63	77–78	12650-69-0
P. B (P. I)	$C_{26}H_{44}O_{10}$	516.63		40980-51-6
P. C	$C_{26}H_{44}O_8$	484.63		71980-98-8
P. D	$C_{26}H_{42}O_9$	498.61		85248-93-7

specifically binding to isoleucyl-tRNA synthase. In human serum it is rapidly degraded to the inactive monic acid (for which herbicidal activity has been reported) by ester hydrolysis. P. have recently been investigated as lead structures for herbicides. For analogues, see *Lit.* [9].

Lit.: [1] Nature (London) **234**, 416 (1971). [2] J. Chem. Soc., Perkin Trans. 1 **1978**, 561. [3] Antimicrob. Agents Chemother. **27**, 495 (1985). [4] J. Chem. Soc., Perkin Trans. 1 **1977**, 318. [5] J. Chem. Soc., Perkin Trans. 1 **1982**, 2827. [6] Tetrahedron Lett. **21**, 881 (1980). [7] J. Chem. Soc., Perkin Trans. 1 **1983**, 2655. [8] J. Med. Chem. **39**, 446–457, 3596–3600 (1996); **40**, 2563 (1997); Tetrahedron Lett. **24**, 3661 (1983); **36**, 7631 (1995). [9] Bioorg. Med. Chem. Lett. **7**, 2805 (1997); **9**, 1847 (1999).
gen.: Beilstein E V **19/7**, 644, 663 ▪ Chem. Rev. **95**, 1843–1857 (1995) ▪ Clin. Pharmacol. **7**, 761 (1987) ▪ Exp. Opin. Ther. Patents **6**, 971 f. (1996); **9**, 1021–1028 (1999) ▪ J. Antibiot. **41**, 609 (1988) ▪ Synform **4**, 93 (1986). – *[HS 2941 90]*

Pseudopelletierine (granatan-3-one, 9-methyl-9-azabicyclo[3.3.1]nonan-3-one).

$C_9H_{15}NO$, M_R 153.22, prismatic plates, mp. 54 °C (62–64 °C), bp. 246 °C, a volatile alkaloid from the root bark of the pomegranate tree (*Punica granatum*, Punicaceae).

Lit.: Beilstein E V **21/7**, 59 ▪ Pelletier **3**, 76 f. – *Synthesis:* Tetrahedron **38**, 1959 (1982). – *[HS 2939 90; CAS 552-70-5]*

Pseudopterosins. Diterpene glycosides (P. A–L) from the sea whip *Pseudopterogorgia elisabethae* living in the tropical Atlantic. P. occur as aglycones not only in the form shown but also as the 7*β*-epimers and the enantiomers. The sugar residue at *O*-8 or *O*-9 can be D-xylo-, L-fuco-, and D-arabinopyranosyl which may be acetylated in the 2-, 3-, or 4-position.

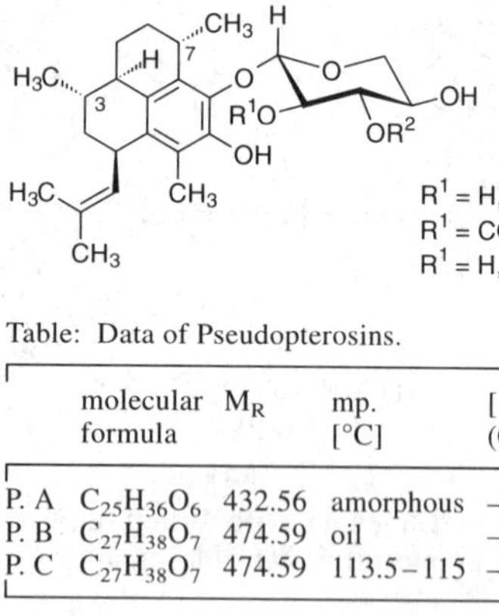

R^1 = H, R^2 = H : P. A
R^1 = CO–CH_3, R^2 = H : P. B
R^1 = H, R^2 = CO–CH_3 : P. C

Table: Data of Pseudopterosins.

	molecular formula	M_R	mp. [°C]	$[\alpha]_D^{20}$ ($CHCl_3$)	CAS
P. A	$C_{25}H_{36}O_6$	432.56	amorphous	−85°	104855-20-1
P. B	$C_{27}H_{38}O_7$	474.59	oil	−55.2°	104855-21-2
P. C	$C_{27}H_{38}O_7$	474.59	113.5–115	−77°	104881-78-9

P. show strong anti-inflammatory activity. P. C is used cosmetically to inhibit aging processes of the skin.

Lit.: Chem. Eng. News (20.11.) **1995**, 42 ▪ J. Chem. Soc., Chem. Commun. **1996**, 1743 ▪ J. Chem. Soc., Perkin Trans. 1 **1998**, 2475 ▪ J. Am. Chem. Soc. **120**, 12777 (1998) (synthesis) ▪ J. Org. Chem. **55**, 4916 (1990) ▪ Synlett **1997**, 1303 ▪ Tetrahedron Lett. **36**, 9129 (1995) (synthesis).

Pseurotin A.

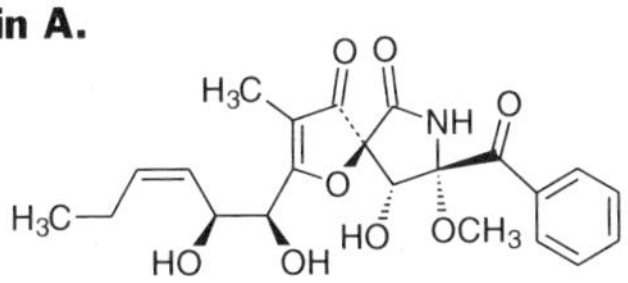

$C_{22}H_{25}NO_8$, M_R 431.44, colorless cryst., mp. 162–163.5 °C, $[\alpha]_D$ −5° (CH_3OH), isolated from cultures of *Pseudeurotium ovalis* and *Aspergillus fumigatus*. P. A contains a substituted 1-oxa-7-azaspiro[4.4]nonane skeleton. No biological activities have been reported. Several natural derivatives are known, they differ from P. A in the aliphatic side chain or do not possess an *O*-methyl group.

Lit.: Biosci. Biotechnol. Biochem. **57**, 961 (1993) ▪ Helv. Chim. Acta **64**, 304–315 (1981); **78**, 1278–1290 (1995) (synthesis) ▪ J. Antibiot. **44**, 382–390 (1991). – *[CAS 58523-30-1]*

Psilocybin {3-[2-(dimethylamino)-ethyl]-4-indolyl dihydrogen phosphate}. P. is an *indole alkaloid from the Mexican narcotic fungus Teonanácatl (*Psilocybe mexicana*); the fungus is indigenous to Mexico and can also be cultivated. P. is an hallucinogenic substance. Chemically it is a tryptamine derivative related to *serotonin. Oral consumption leads to colored visions, a feeling of expanded consciousness, as well as to split personality and a high sensitivity to light. P. and the co-occurring, unesterified *psilocin* [an isomer of bufotenine (see indolylalkylamines)] have ca. 1% of the effects of lysergic acid diethylamide, LSD (see ergot alkaloids). Their toxicities (LD_{50} mouse i. v.: 280 mg/kg, rabbit 12.5 mg/kg) are lower than that of *mescaline.

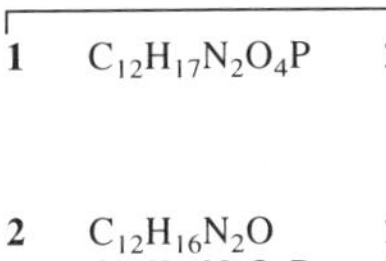

	R^1	R^2	R^3
Psilocybin (**1**)	PO_3H_2	CH_3	CH_3
Psilocin (**2**)	H	CH_3	CH_3
Baeocystin (**3**)	PO_3H_2	CH_3	H
Norbaeocystin (**4**)	PO_3H_2	H	H

Table: Data of Psilocybin and derivatives.

no.	molecular formula	M_R	mp. [°C]	CAS
1	$C_{12}H_{17}N_2O_4P$	284.25	185–195 (anhydrous) 220–228 (hydrate)	520-52-5
2	$C_{12}H_{16}N_2O$	204.27	173–176	520-53-6
3	$C_{11}H_{15}N_2O_4P$	270.22	254–258	21420-58-6
4	$C_{10}H_{13}N_2O_4P$	256.20	188–192 (decomp.)	21420-59-7

Psilocin is the biosynthetic precursor of psilocybin. It is formed in the body as a result of metabolic dephosphorylation from P. and represents the principle with effects on the CNS. In addition to P. and psilocin *Psilocybe* species also contain small amounts of the psilocin derivatives *baeocystin* and *norbaeocystin*. The use of P. is restricted by legislation in many countries.

Lit.: Acta Pharm. Fenn. **96**, 133–145 (1987) ▪ Arch. Pharm. (Weinheim, Ger.) **321**, 487 (1988) (review) ▪ Hager (5.) **6**, 287–293; **9**, 443 f. ▪ Heterocycles **51**, 1131 (1999) (psilocin) ▪ J. Ethnopharmacol. **10**, 249–254 (1984) ▪ J. Heterocycl. Chem. **18**, 175 (1981) (synthesis) ▪ Planta Med. **1986**, 83 ff. ▪ Sax (8.), HKE 000, PHU 500 ▪ Schweiz. Laboratoriumsztg. **50**, 10–14 (1993) ▪ Synthesis **1999**, 935. – *[HS 2939 90]*

Psilotin.

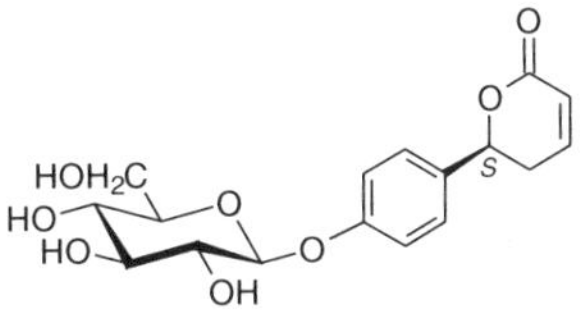

$C_{17}H_{20}O_8$, M_R 352.34, crystalline needles, mp. 130–131 °C, $[\alpha]_D^{25}$ −145.3° (H_2O). A phenyl-α-pyrone derivative from *Psilotum nudum* and *Tmesipteris vieillardii* (Psilotacae) that is cleaved by emulsin to give *psilotinin* ($C_{11}H_{10}O_3$, M_R 190.20) and glucose. P. exhibits germination and plant growth inhibiting activities in plants (turnips, onions, lettuce) [1]. The biosynthesis proceeds from phenylalanine via hydroxycinnamic acid and acetylhydroxycinnamic acid [2].

Lit.: [1] Phytochemistry **15**, 566 (1976). [2] Tetrahedron Lett. **23**, 2365 (1982).

gen.: Justus Liebigs Ann. Chem. **1981**, 2384. – *[CAS 4624-52-6 (P.); 4624-53-7 (psilotinin)]*

Psoralene see furocoumarins.

Psoromic acid.

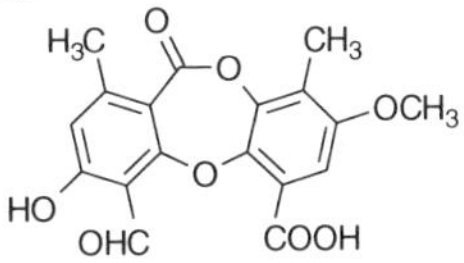

$C_{18}H_{14}O_8$, M_R 358.30, needles, mp. 265 °C (decomp.). A *depsidone occurring in numerous lichens such as, e.g., *Pentagenella fragillima* and *Squamarina cartilaginea*.

Lit.: Aust. J. Chem. **29**, 1059–1067 (1976) ▪ J. Chem. Soc., Perkin Trans. 1 **1979**, 2593–2598 (synthesis) ▪ Karrer, No. 1055. – *[CAS 7299-11-8]*

PSP. Abbreviation for *p*aralytic *s*hellfish *p*oison(ing), paralytic poisoning. Unicellular marine algae, dinoflagellates, and cyanobacteria form highly toxic compounds, such as *saxitoxin, neosaxitoxin, gonyautoxins, *brevetoxins, *anatoxins, that are accumulated in plankton filterers such as mussels and can cause severe poisonings (PSP) in humans when the latter are consumed. Ca. 8% of all cases of poisonings are fatal, the main cause being saxitoxin. The symptoms usually occur within the first 30 min.: Tingling and burning sensation in the tongue, in the face, progressing from the neck and body to the fingers and toes, increasing numbness, balance impairments, general dazed states, weakness, head and muscle pain, thirst and visual disorders. Gastrointestinal symptoms are very rarely seen. Death follows within 12 h through respiratory arrest. The lethal dose for humans is ca. 0.5–1.5 mg. When smaller amounts are consumed there are usually no lasting damages. The various toxins show LD_{50} values in the mouse (i. p.) in the range of 10–500 µg/kg. They act as neuromuscular depolarizers and irreversible choline esterase inhibitors. The toxins are probably not produced by the dinoflagellates themselves but rather by bacteria living in symbiosis with the algae.

Lit.: ACS Symp. Ser. **262**, 9–24, 99–125, 151–216 (1984); **418**, 66–77, 87–106 (1990) ▪ Bioact. Mol. **10**, 425 (1989) ▪ Chem. Rev. **93**, 1897 (1993) ▪ Naturwiss. Rundsch. **42**, 113 (1989) ▪ Pure Appl. Chem. **61**, 7–18 (1989) ▪ Zechmeister **45**, 235–264.

Psychotridine see pyrroloindole alkaloids.

Psychotrine see ipecac alkaloids.

Psylloborine A see coccinelline.

Ptaquiloside see pterosins.

Pteridines (Greek: pteryx = feather, wing).

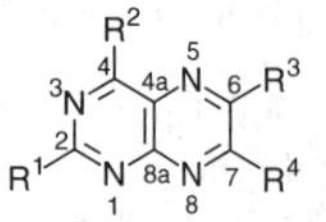

	R^1	R^2	R^3	R^4
Pteridine	H	H	H	H
Pterin	NH_2	OH	H	H
Xanthopterin	NH_2	OH	OH	H
Isoxanthopterin	NH_2	OH	H	OH
Leucopterin	NH_2	OH	OH	OH

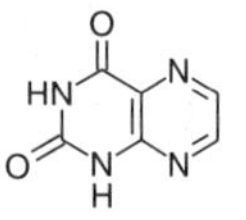

Lumazine

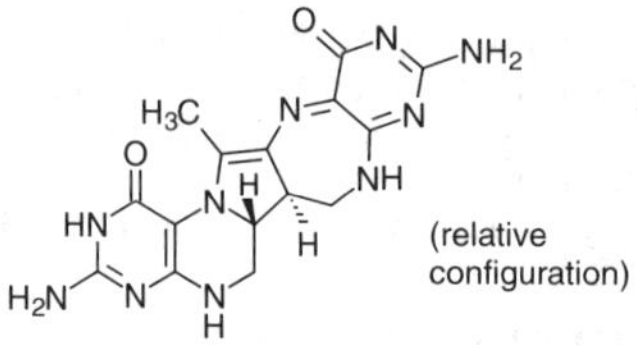

Drosopterin

Table: Data of Pteridines (Biopterin and Neopterin see entries).

name	molecular formula	M_R	mp. (bp.) [°C]	CAS
Pteridine	$C_6H_4N_4$	132.12	140; yellow cryst. (125–130/27 kPa)	91-18-9
Pterin	$C_6H_5N_5O$	163.14	>360; yellow cryst.	2236-60-4
Xantho-pterin	$C_6H_5N_5O_2$	179.14	360 (decomp.); monohydrate: orange-yellow cryst.	119-44-8
Isoxantho-pterin	$C_6H_5N_5O_2$	179.14	310	529-69-1
Leuco-pterin	$C_6H_5N_5O_3$	195.14	fine colorless cryst.	492-11-5

A group of widely distributed natural products occurring mainly as dyes (pigments) especially in the eyes and wings of insects, in particular butterflies, and in the eyes and skin of fish, amphibians, and reptiles. In contrast, the parent compound of the group, *pteridine* (pyrazino[2,3-*d*]pyrimidine), is synthetic. Many properties of the P. (including poor solubility) are similar to those of the structurally closely related *purines from which they can sometimes be obtained by ring expansion. They are formed biogenetically via guanosine phosphates. Derivatives of *pterin* which, like most naturally occurring P. can be formulated in tautomeric forms, include not only *biopterin and *neopterin from the *royal jelly of *bees, lepidopterin, *leucopterin* [1] (blue fluorescence in alkaline solution, occurs in butterfly wings), *xanthopterin* [2] (yellow pigment in butterfly wings, wasps, and other insects as well as in the urine of mammals), *isoxanthopterin* [2] (in the skin of the fire salamander) and other pigments but also a series of growth factors the most important of which is *folic acid (pteroylglutamic acid) which exhibits vitamin character in the human organism.

The group of P. was also known as the *pterins* in the past whereas today this name is restricted to pterin and its derivatives. Other natural P. are derivatives of the so-called *lumazine* [2,4(1*H*,3*H*)-pteridinedione, $C_6H_4N_4O_2$, M_R 164.12, yellow-orange, mp. 349 °C], the benzo derivative of which constitutes the core of riboflavin. Lumazines are also found in higher plants [3] (see russupteridines). A dimeric pentacyclic derivative of P. with a diazepine ring is the eye pigment of the fly *Drosophila melanogaster* (*drosopterin*, $C_{15}H_{16}N_{10}O_2$, M_R 368.36, red cryst., mp. >350 °C). Tetrahydropterins act as coenzymes in enzymatic hydroxylation reactions. Thus, e. g., *tetrahydrobiopterin* (BH_4) is necessary for the function of phenylalanine, tyrosine, and tryptophan hydroxylases. A deficiency of this cofactor leads to phenylketonuria and neurological impairments since the neurotransmitters dopamine and serotonin then cannot be synthesized. *Tetrahydrofolic acid plays an important role in the organism as a C_1-transfer agent. A sulfur-containing P. has been recognized as cofactor of oxidoreductases, its oxidative degradation leads to *urothione*, a P. with a condensed thiophene ring.

Historical: The first P. to be completely characterized were the substances contained in butterfly wings, namely *leucopterin* (cabbage white) and *xanthopterin* (brimstone) by H. Wieland and Schöpf (1924) and Purrmann (1940), respectively: Wieland needed 200 000 cabbage white butterflies to obtain 40 g of leucopterin.

Lit.: [1] J. Chem. Ecol. **13**, 1843–1847 (1987); J. Heterocycl. Chem. **29**, 583 (1992); Zechmeister **4**, 64. [2] Int. J. Biochem. **25**, 1873 (1993); Naturwissenschaften **74**, 563–572 (1987). [3] Helv. Chim. Acta **67**, 550–569 (1984); Zechmeister **51**, 210–215.
gen.: Blair, Chemistry and Biology of Pteridines – Pteridines and Folic Acid Derivatives, Berlin: De Gruyter 1983 ▪ Blakeley et al., Folates and Pterins (3 vols.), New York: Wiley 1984–1986 ▪ Cooper & Whitehead, Chemistry and Biology of Pteridines, Berlin: De Gruyter 1986 ▪ Lovenberg & Levine, Unconjugated Pterins in Neurobiology (Top. Neurochem. Neuropharm. 1), London: Taylor & Francis 1987 ▪ Pfleiderer, Chemistry and Biology of Pteridines, Berlin: De Gruyter 1975. – *Journal:* Pteridines, Berlin: De Gruyter (since 1989). – *[CAS 487-21-8 (lumazine); 33466-46-5 (drosopterin)]*

Pteridophyta see pteridophytes; ferns (Filicatae).

Pteridophytes; ferns (Filicatae) (Pteridophyta). A division of the plant kingdom designated as ferns or vascular spore plants with over 12 000 recent species. They are assigned in the classes Psilophytatae or Psilophytopsida (psilopsids), Psilotatae or Psilotopsida (whisk ferns), Lycopodiatae or Lycopodiopsida (*club-mosses), Equisetatae or Equisetopsida (*horse tails) and Filicatae or Filicopsida (ferns in the stricter sense). The latter comprise the largest group with about 10 000 species. The P. are considered to be a parallel branch to the *mosses that originated from a common group of ancestors. They reached their highest state of development in the Paleozoic era (pteridophytic time!) and represented the dominating group of plants in the Devonian and Carboniferous periods (paleophytic, 400–280 million years ago). In the upper Carboniferous period, tree-like representatives of the P. together with seed ferns and the very original naked-seed plants (Gymnosperms) formed thick forests. These were the origins of enormous hard or pit coal deposits. P. are found in all climatic zones of the Earth and prefer wet locations. The largest species variety and differentiation are found in tropical floral zones.

Components [1,2]**:** The phytochemistry of the individual classes varies enormously [for components of the *clubmosses and *horse tails, see the appropriate entries]. The true ferns (Filicatae) contain saccharose as the main carbohydrate. Large amounts of C_{20}-polyenoic acids, see also *arachidonic acid, and fatty acids with 21–30 C atoms are found among the lipids. Ferns produce a great variety of secondary products such as procyanidins, condensed tannins, flavan derivatives, benzoic and cinnamic acid derivatives, *acylphloroglucinols, *cyanogenic glycosides, *pterosins, *di- and *triterpenes as well as phytoecdysone (see ecdysteroids). Alkaloids have not yet been detected.

Use: The male fern (*Dryopteris filix-mas*) is used in traditional medicine as a remedy for tapeworm. Other species of ferns are used in the perfume industry. Some are also used as foods [3].

Lit.: [1] Prog. Phytochem. **1**, 589 (1968). [2] Jermy et al. (eds.), The Phylogeny and Classification of Ferns, p. 111, London: Academic Press 1973. [3] Jones, Encyclopaedia of Ferns, p. 12, Portland, Oregon: Timber Press 1987.
gen.: Fukarek (ed.), Enzyklopädie Urania-Pflanzenreich, vol. 2 (Moose, Farne, Nacktsamer), p. 78, Leipzig: Urania-Verlagsges. 1992 ▪ Kramer, Schneller & Wollenweber, Farne u. Farnverwandte, p. 159, Stuttgart: Thieme 1995 ▪ Zechmeister **54**, 1. – *[HS 1211 90]*

Pterocarpans.

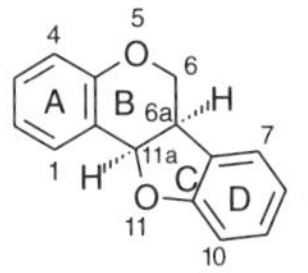

(–)-Pterocarpan

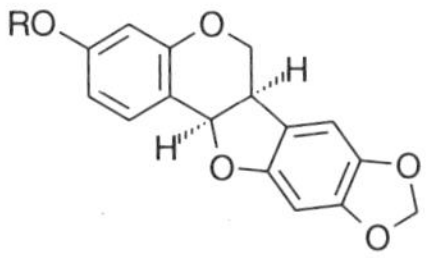

R = H : (–)-Maackiain
R = CH_3: (–)-Pterocarpin

A group of natural products occurring characteristically in Fabaceae with the *cis*-6a,11a-dihydro-6*H*-benzofuro[3,2-*c*][1]benzopyran ring system (pterocarpan, 2′,4-epoxyisoflavan). Important members of the P. include *maackiain* and *pterocarpin* which have been recognized as constitutive components or also as *phytoalexins. Both enantiomers and the racemic forms occur in nature.

(–)-*Maackiain*: $C_{16}H_{12}O_5$, M_R 284.27, mp. 179–181 °C, $[\alpha]_D^{22}$ –260° (acetone), isolated among others from the heartwoods of *Maackia amurensis* and *Sophora* species. (+)-*Maackiain* occurs in *Dalbergia oliveri*, *Derris elliptica*, and other plants. (±)-*Maackiain*, mp. 196 °C, was detected in *Sophora japonica* and *Dalbergia spruceana*. Maackiain has antifungal activity. (–)-*Pterocarpin* [1]: $C_{17}H_{14}O_5$, M_R 298.30, mp. 168–169 °C, $[\alpha]_D$ –214.5° ($CHCl_3$), insoluble in water, soluble in chloroform, occurs in red sandalwood (*Pterocarpus santalinus*), in *Swartzia madagascariensis*, and in *Flemingia chappar*. The *cabenegrins acting as phytoalexins are used in Brazil as antidotes for snake and spider bites.

Biosynthesis: P. are derived biogenetically from *isoflavones.

Lit.: [1] J. Chem. Soc., Perkin Trans. 1 **1998**, 2533.
gen.: Beilstein E V **19/2**, 117 ▪ Harborne (1994), p. 166–180 ▪ J. Chem. Soc., Perkin Trans. 1 **1984**, 2831 (biosynthesis); **1989**, 1219 (synthesis) ▪ J. Nat. Prod. **58**, 1966 (1995) (Maackiain) ▪ J. Org. Chem. **55**, 1248 (1990) (synthesis) ▪ Luckner (3.), p. 413 f. ▪ Merck-Index (12.), No. 8117. – *[CAS 61080-21-5 (P.); 2035-50-9 ((–)-P.); 2035-15-6 ((–)-maackiain); 23513-53-3 ((+)-maackiain); 19908-48-6 ((±)-maackiain); 524-97-0 ((–)-pterocarpin)]*

Pterosins.

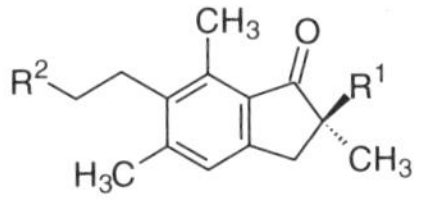

R^1 = H, R^2 = OH : (2*R*)-P. B
R^1 = H, R^2 = Cl : (2*R*)-P. F
R^1 = CH_3, R^2 = Cl : P. H
R^1 = CH_3, R^2 = OH : P. Z

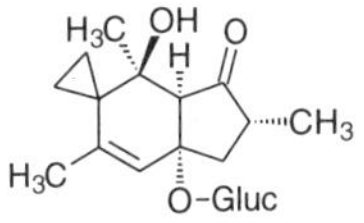

Ptaquiloside

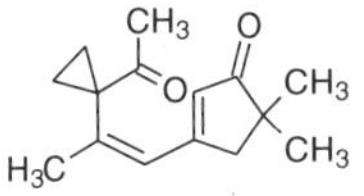

Hypacrone

Table: Data of Pterosins.

	molecular formula	M_R	mp. [°C]	$[\alpha]_D^{20}$	CAS
(2*R*)-P. B	$C_{14}H_{18}O_2$	218.30	109–110	–31.9°	34175-96-7
(2*R*)-P. F	$C_{14}H_{17}ClO$	236.74	66–67	–14.6°	34175-98-9
P. H	$C_{15}H_{19}ClO$	250.77	87.5–88		39004-41-6
P. Z	$C_{15}H_{20}O_2$	232.32	86–88		34169-69-2

1-Indanone derivatives belonging to the illudalane group of *sesquiterpenes, their glycosides (mainly glucosides but also arabinosides) are known as *pterosides*. Distinguished are P. of the B- (C_{14}) and Z-types (C_{15}). More than 100 of these compounds have been isolated from ferns, outside of this plant group they seem to occur only sporadically. Fungi contain biogenetically related sesquiterpenes [1,2]. *Ptaquiloside* [3] ($C_{20}H_{30}O_8$, M_R 398.45, amorphous powder, $[\alpha]_D^{22}$ –188°) from *bracken can be cleaved to P. B and glucose, whereupon its toxicity is lost. The hot principle *hypacrone* [4] ($C_{15}H_{20}O_2$, M_R 232.32, oil, mp. 48–52 °C) of the fern *Hypolepis punctata* furnishes P. H and Z on acid treatment. P. F is an important defensive substance against insects for the bracken fern [5].

Lit.: [1] Tetrahedron **37**, 2199 (1981). [2] Phytochemistry **25**, 1979 (1986). [3] Angew. Chem. Int. Ed. Engl. **37**, 1818 (1998); Phytochemistry **40**, 53 (1995); **44**, 901 (1997); J. Am. Chem. Soc. **115**, 3056 (1993); Tetrahedron Lett. **33**, 6647 (1992). [4] Chem. Lett. **1973**, 63. [5] Syst. Ecol. **7**, 95 (1979).
gen.: Eur. J. Med. Chem. **34**, 953–966 (1999) (synthesis, activity) ▪ J. Chem. Res. (S) **1999**, 406 (synthesis P.), 1561 (synthesis P. F) ▪ Phytochemistry **40**, 53 (1995) ▪ Zechmeister **54**, 51. – *[CAS 87625-62-5 (ptaquiloside); 41059-92-1 (hypacrone)]*

Pterulinic acid and pterulones. P. a. and P. are halogenated oxepin antifungal compounds isolated from fermentation broths of a *Pterula* species, strain 82168 (Basidiomycetes), with little or no cytotoxicity. P. a. and P. act as eukaryotic respiration inhibitors. The same *Pterula* strain produces the *strobilurins and *oudemansins.

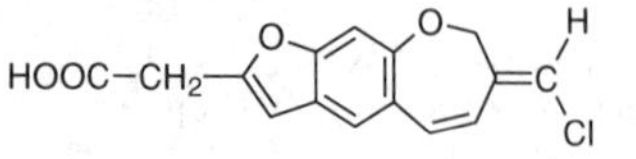

Pterulinic acid (**1**)

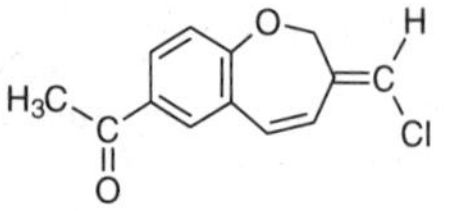

Pterulone (**2**)

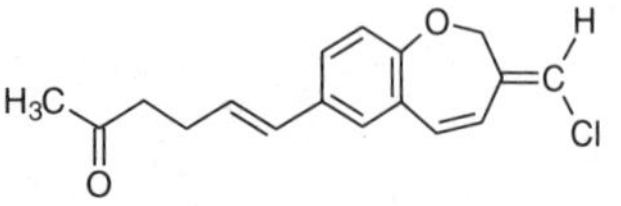

Pterulone B (**3**)

Table: Data of Pterulinic acid and pterulones.

	molecular formula	M_R	mp. [°C]	CAS
1	$C_{15}H_{11}ClO_4$	290.70	yellowish crystals, mp. 152–154°	190143-30-7
2	$C_{13}H_{11}ClO_2$	234.68	yellowish crystals, mp. 122–124°	190143-31-8
3	$C_{17}H_{17}ClO_2$	288.77	yellow oil	209545-10-8

Lit.: J. Antibiot. **50**, 325, 330 (1997) ▪ Z. Naturforsch. C **53**, 318–324 (1998) (P. B).

Ptilocaulin see ptilomycalins.

Ptilomycalins.

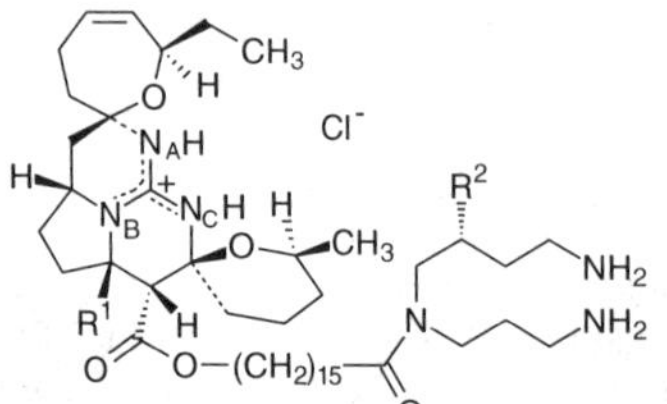

$R^1 = R^2 = H$: P. A
$R^1 = H$, $R^2 = OH$: Crambescidin 800
$R^1 = R^2 = OH$: Crambescidin 816

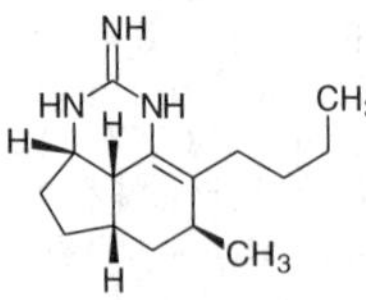

Ptilocaulin

Polycyclic guanidine derivatives isolated from the sponges *Ptilocaulis spiculifer* (Caribbean) and *Hemimycale* sp. (Red Sea). *P. A*[1], $C_{45}H_{80}N_6O_5$, M_R 785.17 (hydrochloride of bistrifluoracetyl derivative: oil, $[\alpha]_D$ –15.8°), is cytotoxic against P388 cells (IC_{50} 0.1 µg/mL), antimicrobial and antifungal against *Candida albicans* (MIC 0.8 µg/mL), and antiviral (HSV) at a concentration of 0.2 µg/mL. *Ptilocaulin*[2], $C_{15}H_{25}N_3$, M_R 247.38, (nitrate: mp. 183–185 °C) also has significant antileukemic and antimicrobial activities. The related *crambescidins*[3] [e.g., crambescidin 816, $C_{45}H_{80}N_6O_7$, M_R 817.17, oil, $[\alpha]_D$ –20.14° (CH_3OH)] from the Mediterranean orange-red sponge *Crambe crambe* have antiviral and cytotoxic activities[4].

Lit.: [1] J. Am. Chem. Soc. **111**, 8926 (1989); **117**, 2657 (1995); Tetrahedron **52**, 2629–2646 (1996). [2] Angew. Chem. Int. Ed. Engl. **35**, 2146–2148 (1996); Chem. Eur. J. **4**, 57–66 (1998); Isr. J. Chem. **31**, 239 (1991); J. Am. Chem. Soc. **103**, 5604 (1981); **106**, 721, 1143 (1984); J. Nat. Prod. **60**, 704 (1997). [3] J. Org. Chem. **56**, 5712 (1991); J. Am. Chem. Soc. **121**, 6944 (1999). [4] U. S. P. 5,756,734
gen.: Synthesis: J. Chem. Soc., Chem. Commun. **1986**, 539 ▪ Nachr. Chem. Tech. Lab. **43**, 1302–1309 (1995) (review, ptilomycalin A) ▪ Tetrahedron **41**, 3463 (1985) ▪ Tetrahedron Lett. **31**, 4759 (1990). – *[CAS 78777-02-3 (ptilocaulin); 124512-47-6 (P. A); 135257-45-3 (crambescidin 816)]*

Ptomaines. Name, derived from modern Greek: ptoma = cadaver, for the so-called *cadaveric poisons* formed from putrefying proteins. The enzymatic decarboxylation products of the amino acids *lysine and *L-ornithine (*cadaverine and *putrescine), previously known as P., are, however, relatively non-toxic biogenic amines. Today the name P. is used for the toxic metabolic products of putrefactive bacteria that colonize rotting, protein-containing foodstuffs such as meat, fish, etc. These P. have widely differing chemical structures: their activities are purported to be similar to those of the plant toxins such as *strychnine, *atropine, etc..

Lit.: Daldrup, Die Aminosäuren des Leichengehirnes, Stuttgart: Enke 1984 ▪ J. Forensic Sci. **22**, 558–572 (1977) ▪ Kauert, Katecholamine in der Agonie, Stuttgart: Enke 1985.

PTX. Abbr. for 1) *pumiliotoxins, 2) pectenotoxins (see also yessotoxin), 3) *palytoxin.

PTX-4.3.3 see δ-philanthotoxin.

Pulegones see *p*-menthenones.

Pulvinic acid.

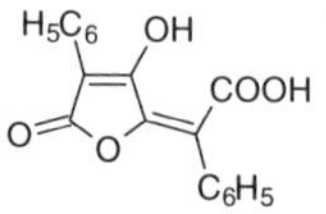

$C_{18}H_{12}O_5$, M_R 308.29, orange prisms, mp. 216–217 °C. A pigment of numerous lichens. The biosynthesis proceeds from *polyporic acid. The methyl ester of P., *vulp(in)ic acid, is a typical lichen acid. Hydroxylated P. derivatives are important mushroom pigments, e. g., *gomphidic acid, *variegatic acid, and *xerocomic acid; the latter two compounds are responsible for the typical blue color reaction of many *Boletus* and *Xerocomus* species on bruising the fruit bodies.

Lit.: Beilstein E V **18/8**, 611 ▪ Chem. Rev. **76**, 625 (1976). – *[CAS 26548-70-9]*

Pumiliotoxins (PTX). Alkaloids isolated from the arrow poison frog *Dendrobates pumilio* with indolizidine (P. A and B) or decahydroquinoline skeletons (P. C). The P. were later also detected in other frogs and toads[1].

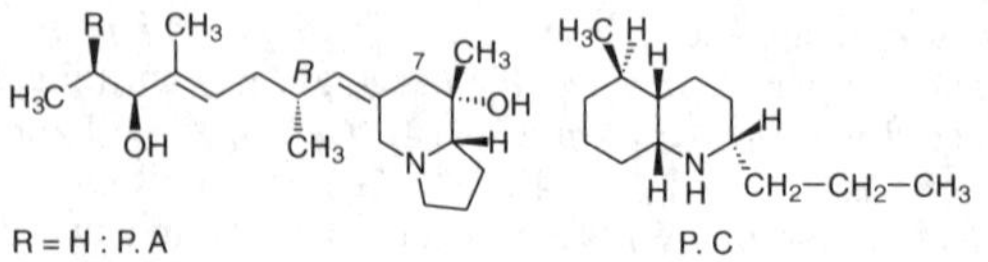

R = H : P. A
R = OH : P. B

P. C

Table: Data of Pumiliotoxins.

	P. A (307 A)	P. B (323 A)	P. C (195 A)
molecular formula	$C_{19}H_{33}NO_2$	$C_{19}H_{33}NO_3$	$C_{13}H_{25}N$
M_R	307.48	323.48	195.35
mp. [°C]	oil	oil	230–240 (decomp.)
$[\alpha]_D^{25}$	+22.7° (C_2H_5OH)	+20.5° (C_2H_5OH)	–13.1° (C_2H_5OH, as hydrochloride)
LD_{50}	2.5 mg/kg	1.5 mg/kg	1.2 mg/kg
CAS	67054-00-6	67016-65-3	27766-71-8

Besides the main alkaloids, about 20 minor P. A with different side chains, 15 allopumiliotoxins (with an OH-group at C-7 of the indolizine system), and 10 homopumiliotoxins (with a quinolizidine system instead of indolizidine as the skeleton) as well as about 25 decahydroquinolines have been found; however, the exact structures of some of them are still unknown. The minor alkaloids differ from the P. in the side chains which may be hydrogenated, oxidized, methylated, or shortened[1]. P. A and B exhibit cardiotonic and myotonic activities and are thus of pharmaceutical interest. Furthermore they have an effect on sodium ion channels. Subcutaneous injections of 50 µg P. A or 20 µg P. B in mice are lethal. The decahydroquinolines on the other hand are less toxic (injection of 400 µg P. C causes death of mice) and influence nicotine receptors as well as sodium and potassium channels.

Lit.: [1] J. Am. Chem. Soc. **115**, 6652 (1993); J. Org. Chem. **60**, 279 (1995); Manske **43**, 185–288.

gen.: Aldrichim. Acta **28**, 107–120 (1995) ▪ Beilstein E V **20/4**, 457, 459 ▪ Pelletier **4**, 1–247 ▪ Proc. Natl. Acad. Sci. USA **85**, 1272–1276 (1988) ▪ Zechmeister **41**, 235–245, 300–310. – *Synthesis:* Chem. Rev. **96**, 505–522 (1996) (P. A) ▪ Eur. J. Org. Chem. **1998**, 2641 (P. C) ▪ J. Am. Chem. Soc. **111**, 3363–3368, 4988 ff. (1989); **113**, 2652–2656 (1991); **118**, 9062, 9073 (1996) (P. A, B); **121**, 9873 (1999) (P. A) ▪ J. Chem. Soc., Perkin Trans. 1 **1996**, 1113–1124 ▪ J. Org. Chem. **61**, 8687 (1996); **63**, 6566 (1998) ((–)-P. C) ▪ Tetrahedron **43**, 643 (1987) ▪ Tetrahedron: Asymmetry **7**, 1595 (1996) ((–)-P. C) ▪ Tetrahedron Lett. **37**, 4401 (1996); **38**, 989, 1505 (1997) ((+)-P. C). – *[HS 2939 90]*

Punctaporonins (punctatins).

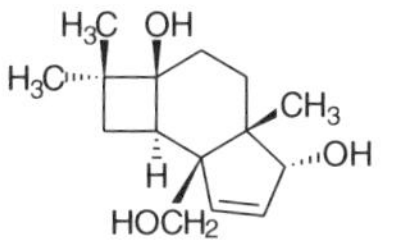

Punctaporonin A

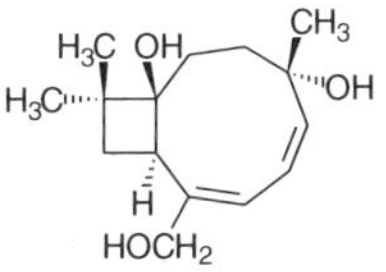

Punctaporonin B

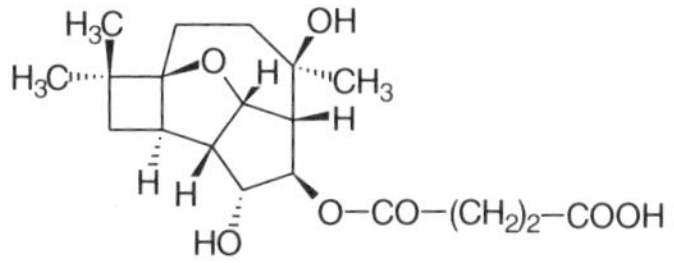

Punctaporonin C

Sesquiterpenoids from cultures the fungus *Poronia punctata* (Ascomycetes) growing on horse dung. *P. A* {$C_{15}H_{24}O_3$, M_R 252.35, cryst., mp. 187–192 °C, $[\alpha]_D$ –26° (CH_3OH)} has an unusual ring system derived biosynthetically from *humulene. Other P. such as *P. B* {$C_{15}H_{24}O_3$, M_R 252.35, cryst., mp. 187.5–188.5 °C, $[\alpha]_D$ –221° (CH_3OH)} are of the caryophyllane type. P. B also occurs in cultures of the ascomycete *Hypoxylon terricola*. *P. C* {$C_{19}H_{28}O_7$, M_R 368.43, cryst., mp. 158–161 °C, $[\alpha]_D$ –14° (CH_3OH)} is oxidatively highly modified and esterified with succinic acid. Most of the P. have antibiotic activities.

Lit.: J. Chem. Soc., Chem. Commun. **1984**, 405 ▪ J. Antibiot. **39**, 167 (1986) ▪ J. Chem. Soc., Perkin Trans. 1, **1988**, 823; **1989**, 1939. – *Synthesis:* J. Am. Chem. Soc. **108**, 3841 (1986) (abs. configuration); **109**, 3017 (1987); **110**, 6265 (1988). – *[CAS 91161-74-9 (P. A); 93697-36-0 (P. B); 93672-19-6 (P. C)]*

Pungent (hot) flavors. Collective name for substances with widely differing chemical structures but with the common property of having a pungent (hot) flavor. The P. f. exert a stimulus on the thermo- and/or pain receptors present on the skin or mucous membranes, leading either directly or indirectly to increased secretion of saliva and gastric juices, increased peristalsis, sensations of warmth, and hyperemia in the skin. Since these effects are not only desirable for the taste of food but are also of therapeutic utility, P. f. or plants containing such compounds find use not only as foods but also as stimulants for therapy; *examples:* for rheumatism and lumbago, as stomachics and carminatives, and in stimulation therapy. Typical P. f. or plants containing them described under separate entries are: *capsaicin from paprika, *piperine from pepper, *gingerol and *zingerone from ginger, myristicin (see safrole) from nutmeg, allicin from garlic (see allium), sedamine from stonecrop (*Sedum alkaloids), sacculatal (see sacculatanes) from liverworts, *chalciporone from the fungus *Suillus piperatus*, *velleral from milkcap (*Lactarius*) species, isothiocyanates (*mustard oils) from mustard, horseradish, cress, and other Brassicaceae.

Punica alkaloids see pelletierine, pseudopelletierine.

Pupating hormone see insect hormones and ecdysteroids.

(+)-2-Pupukeanone. $C_{15}H_{24}O$, M_R 220.35. A fish toxic metabolite from the nudibranch *Phyllidia varicosa* with the isotwistane skeleton. The sponges *Ciocalypta* spp. contain isocyano and *Phycopsis terpnis* thiocyanato derivatives of pupukeanone.

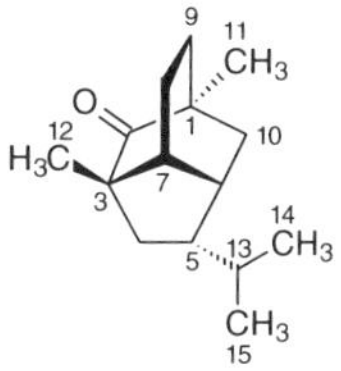

(+)-2-P.

Lit.: *Synthesis:* Indian J. Chem., Sect. B **38**, 766 (1999) ▪ J. Chem. Soc., Perkin Trans. 1 **1997**, 3293 f., 3295, 3387–3392, 3393–3399; **1998**, 2137 ▪ Tetrahedron Lett. **38**, 2003, 2185 (1997). – *[CAS 200720-94-1]*

Purimycin see puromycin.

Purines (purine bases). Name for a group of important compounds that are widely distributed in nature and participate in human, animal, plant, and microbial metabolic processes. They are derived from the parent compound *purine* ($C_5H_4N_4$, M_R 120.11, mp. 217 °C) by substitution with OH, NH_2, SH groups in the in 2-, 6-, and 8-positions (figure 1) and/or with CH_3 in the 1-, 3-, 7-positions (figure 2).

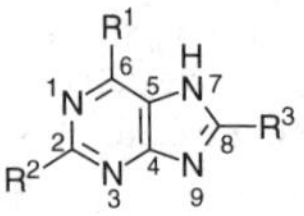

$R^1 = R^2 = R^3 = H$: Purine
$R^1 = NH_2$, $R^2 = R^3 = H$: Adenine
$R^1 = OH$, $R^2 = NH_2$, $R^3 = H$: Guanine
$R^1 = R^2 = R^3 = OH$: *Uric acid
$R^1 = OH$, $R^2 = R^3 = H$: Hypoxanthine
$R^1 = SH$, $R^2 = R^3 = H$: 6-Purinthiol
$R^1 = SH$, $R^2 = NH_2$, $R^3 = H$: 6-Thioguanine
$R^1 = R^2 = OH$, $R^3 = H$: *Xanthine

Figure 1: Purines.

In the natural nucleosides and their derivatives the sugar residue is bound to the nitrogen atom 9.

Biosynthesis & degradation: The biosynthesis of P. proceeds at the stage of the nucleotides from glycine and carbon dioxide as well as from small molecular fragments of L-glutamine, L-aspartic acid, and 10-formyltetrahydrofolic acid. Chemically the P. are closely related to the *pteridines which can be formed biochemically from the P. The degradation of P. proceeds via *xanthine and *uric acid to *allantoin and in further steps to urea and glyoxylic acid. Humans who lack the enzyme uricase must excrete uric acid as the final product (*uricotelics*). Disorders of the P. metabolism can manifest as gout , see also uric acid.

Detection, occurrence: Enzymatic methods, the murexide reaction, and UV spectroscopy are suitable for the detection and analysis of P. Some of the most important P are *adenine* and *guanine* (figure 1), which – together with the *pyrimidines uracil, thymine, and cytosine – are components of nucleic acids, in addition *hypoxanthine, xanthine*, and *uric acid* as metabolic products of humans and animals and the plant P., often designated as *purine alkaloids*, such as *caffeine, *theobromine, and *theophylline (figure 2), that occur in coffee, cocoa, and tea, respectively. *Plant growth substances also belonging to the P. are *zeatin and *kinetin (*cytokinins*). Foodstuffs rich in P. include offal, especially thymus, as well as fish and green peas.

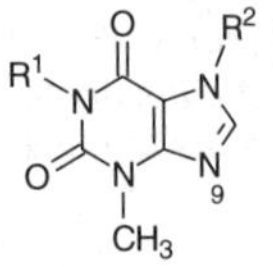

$R^1 = R^2 = CH_3$: Caffeine
$R^1 = H$, $R^2 = CH_3$: Theobromine
$R^1 = CH_3$, $R^2 = H$: Theophylline

Figure 2: Purine alkaloids.

Use: A series of P. derivatives and analogues, the *P.-nucleoside antibiotics*, is used pharmaceutically as antimetabolites. On the other hand *N*-hydroxy-P. and P. *N*-oxides may be carcinogenic. For the role of P. in nociception, see *Lit.*[1].

Lit.: [1] Neuroscience **32**, 557–569 (1989).
gen.: Adv. Exp. Med. Biol. **253**, entire vol. (1989) ▪ Ann. N. Y. Acad. Sci. **451**, entire vol. (1985) ▪ Rodd's Chem. Carbon. Compd. (2.) **4**, K/L, 127–179 (1999) ▪ Wolfram, Genetic and Therapeutic Aspects of Lipid and Purine Metabolism, Berlin: Springer 1989. – *[CAS 120-73-0 (purine)]*

Puromycin [3′-deoxy-6-*N*-dimethyl-3′-(*O*-methyl-L-tyrosinamido)adenosine]. $C_{22}H_{29}N_7O_5$, M_R 471.52, cryst., mp. 175.5–177.0 °C, $[\alpha]_D^{20}$ –11° (C_2H_5OH), generic name proposed by WHO for the *nucleoside antibiotic from *Streptomyces alboniger*[1] previously called

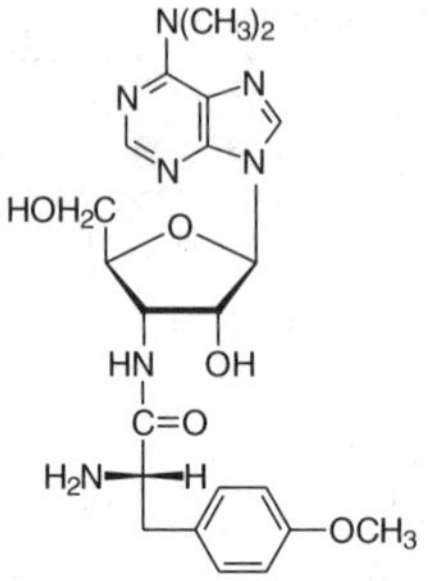

purimycin. P. forms water-soluble dihydrochloride and monosulfate salts. As a structural analogue of the CCA end of aminoacyl-tRNA P. inhibits protein biosynthesis in pro- and eukaryotes through release of the growing peptide chain from ribosomes. It is also active against some tumors as well as amoebae, trypanosomes, and worms but is too toxic for therapeutic use in humans[2].

Lit.: [1] Antibiot. Chemother. **2**, 409 (1952). [2] Antibiotics **1**, 259 (1967); Martindale (29.), p. 143 (1989). – *[HS 2941 90; CAS 53-79-2 (P.); 58-58-2 (dihydrochloride)]*

Purpactins. Metabolites of *Penicillium purpurogenum*, potent inhibitors of acyl-CoA : cholesterol *O*-acyltransferase (ACAT). The related penicillide from *P. purpurogenum* and *P. vermiculatum* lacks the acetyl group at O-1′.

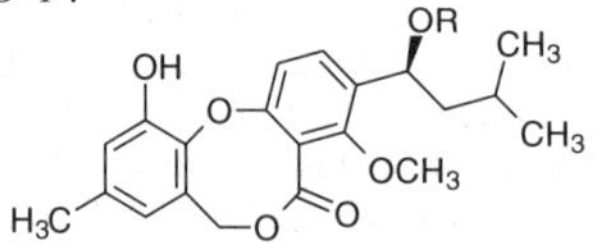

R = CO–CH_3 : P. A
R = H : Penicillide

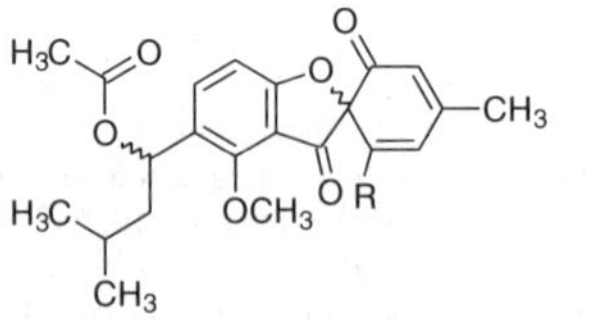

R = CH_2OH : P. B
R = CHO : P. C

Lit.: J. Antibiot. **44**, 136, 144, 152 (1991); **47**, 16 (1994).

Table: Data of Purpactins.

name	molecular formula	M_R	$[\alpha]_D^{18}$ ($CHCl_3$)	mp.	CAS
P.A	$C_{23}H_{26}O_7$	414.45	–57.6°	powder	133806-59-4
P.B	$C_{23}H_{26}O_7$	414.45	–355.6°	yellow oil	133806-60-7
P.C	$C_{23}H_{24}O_7$	412.43	+476.2°	yellow powder	133806-61-8
Peni-cillide	$C_{21}H_{24}O_6$	372.41	+4.9° (C_2H_5OH, 24 °C)	powder	55303-92-9

Purple, Tyrian purple. P. or Tyrian P. is a mixture of indigoid dyes consisting mainly of *6,6′-dibromoindigo* ($C_{16}H_8Br_2N_2O_2$, M_R 420.06, C. I. 75800, C. I. natural violet 1) and smaller amounts of *6,6′-dibromoindirubin* as well as *indigotin and *indigo red.

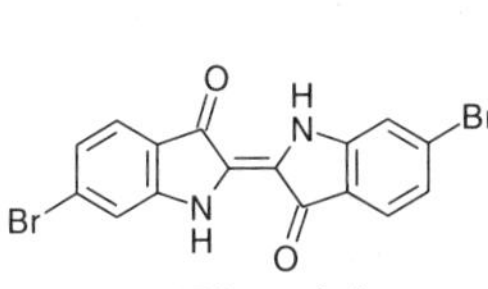
6,6'-Dibromoindigo

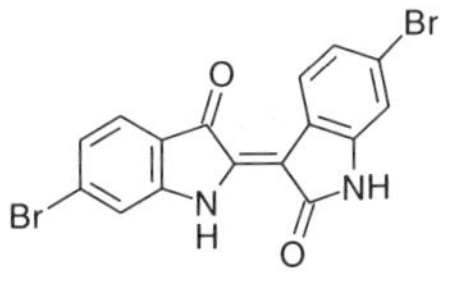
6,6'-Dibromoindirubin

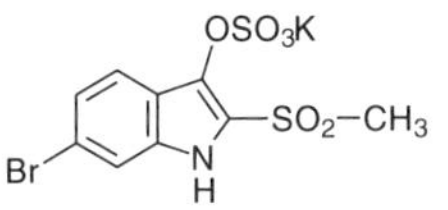
2-Methylsulfonyl-6-bromo-indoxyl sulfate potassium salt

This dye has been appreciated since antiquity (first mentioned in 1439 BC in Phoenicia). The main producing organisms are the purple snails (*Hexaplex trunculus, Murex brandaris, Thais naemastoma, Ocenabra erinacea*) indigenous to the Mediterranean region. The secretions of the hypobronchial glands of the snails contain substituted 6-bromoindoxyl sulfates (colorless) from which the P. pigments are formed under the action of arylsulfatases as well as oxygen and light. They were used to dye wool. The dying procedure was described by Plinius the older. P. was very valuable since ca. 12 000 snails were needed to obtain ca. 1.5 g P. In antiquity P. was considered to be a personification of power. It is reported from the era of Caesar Diokletian (300 AD) that purple-dyed wool was 20 times more valuable than the starting wool. The dye *Tekhelet* mentioned in the 4th book of Moses originated from the slime of a *Murex* snail species. The content of only 50% dibromoindigo explains the blue color [1].

Lit.: [1] Naturwiss. Rundsch. **38**, 386 (1985).
gen.: Chem. Ber. **122**, 1199 (1989) ▪ Experientia **50**, 802 (1994) ▪ Hager (5.) **3**, 207. – *[HS 3203 00; CAS 19201-53-7 (6,6'-dibromoindigo)]*

Purpurea glycosides see Digitalis glycosides.

Purpurin (1,2,4-trihydroxyanthraquinone, C. I. 75410).

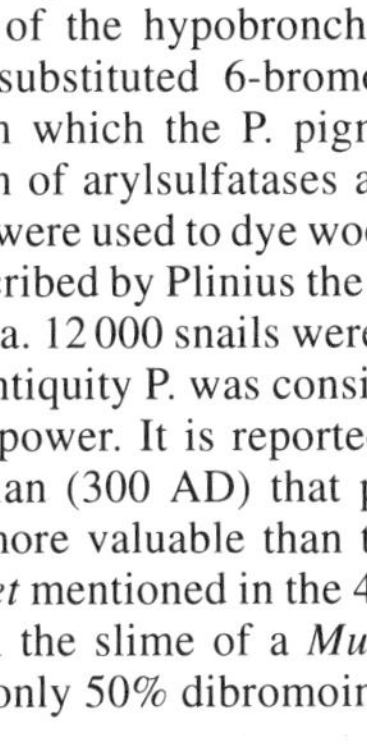

$C_{14}H_8O_5$, M_R 256.21, purple red needles, mp. 262–263 °C, uv_{max} 518 nm (CH_3OH), readily soluble in hot ethanol and acetic acid. P. occurs together with *alizarin in the free form and as the 1-*O*-*β*-primveroside (galiosin) of its biosynthetic precursor purpurin-3-carboxylic acid (1,3,4-trihydroxyanthraquinone-2-carboxylic acid, pseudopurpurin) in the roots of *madder (dyers madder, *Rubia tinctorum*) and various *Galium* species. P. forms red-violet chrome lacquers and dyes alum stained fibers scarlet red; P. is used for nucleus staining in histology. The biosynthesis proceeds from isochorismic acid, 2-oxoglutarate, and mevalonic acid lactone (see anthranoids) [1].

Lit.: [1] Zechmeister **49**, 79–149.
gen.: Beilstein E IV **8**, 3568 ▪ J. Microsc. (Oxford) **166**, 199 (1992) ▪ Karrer, No. 1247 ▪ Kirk-Othmer (3.) **8**, 272 ▪ Merck-Index (12.), No. 8132 ▪ Phytochemistry **42**, 177 (1996) ▪ Sigma-Aldrich Library of Stains, Dyes and Indicators, 592 ▪ Synthesis **1991**, 438. – *[HS 2914 69; CAS 81-54-9]*

Purpurone.

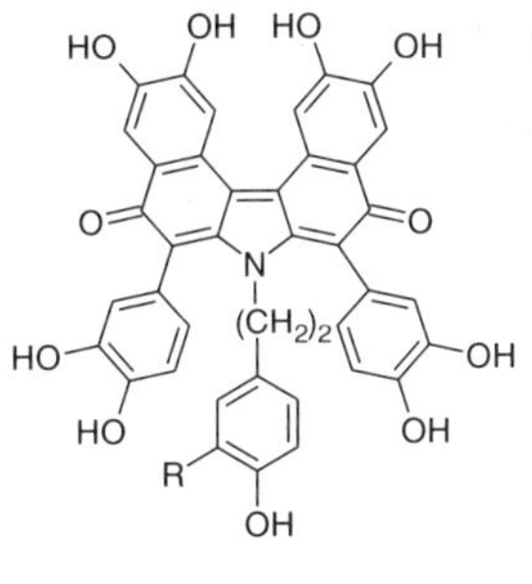
R = H : Purpurone
R = OH : Ningalin D

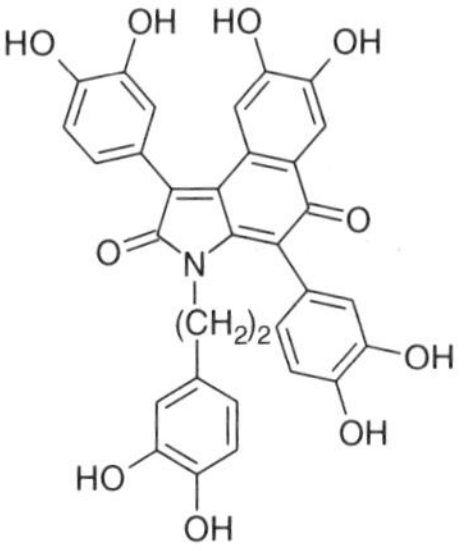
Ningalin C

$C_{40}H_{27}NO_{11}$, M_R 697.65, dark-red glass. Highly symmetrical alkaloid from a marine sponge of the genus *Iotrochota* with inhibitory activity on ATP citrate lyase (EC 4.1.3.8) [1]. Its hydroxy derivative *ningalin D* [$C_{40}H_{27}NO_{12}$, M_R 713.65, amorphous dark-red solid] and the structurally related pyrrolone derivative *ningalin C* [$C_{32}H_{23}NO_{10}$, M_R 581.53, amorphous red solid] occur in an ascidian of the genus *Didemnum* from Western Australia [2]. The compounds are considered to participate in the metal chelation phenomena characteristic of these marine invertebrates.

Lit.: [1] J. Org. Chem. **58**, 2544 (1993). [2] J. Org. Chem. **62**, 3254 (1997). – *[CAS 147362-37-6 (P.); 188111-70-8 (ningalin D); 188111-69-5 (ningalin C)]*

Putrescine (1,4-butanediamine). $H_2N\text{–}(CH_2)_4\text{–}NH_2$, $C_4H_{12}N_2$, M_R 88.15, mp. 27–28 °C, bp. 158–159 °C, soluble in water and all common organic solvents. P. is a biogenic *polyamine formed by decarboxylation of *L-ornithine or degradation of *arginine via *agmatine. P. is the biosynthetic precursor of *spermidine, *spermine and other *polyamines. In plants, P. is the biogenetic precursor of some important alkaloids including the *tropane alkaloids, *tobacco alkaloids, and *pyrrolizidine alkaloids.

Lit.: Beilstein E IV **4**, 1283 ▪ Biochem. J. **202**, 153 (1982) ▪ Merck-Index (12.), No. 8134 ▪ Tetrahedron Lett. **40**, 6233 (1999) (synthesis) ▪ Ullmann (5.) **A 2**, 27. – *[HS 2921 29; CAS 110-60-1; 306-60-5]*

Puwainaphycins. Chlorine-containing cyclic decapeptides from cultures of the cyanobacterium *Anabaena Bq-16-1*. *P. C:* $C_{56}H_{95}ClN_{12}O_{16}$, M_R 1227.9, $[\alpha]_D$ –39° (aq. CH_3CN). P. contain (*E*)-2-amino-2-butenoic acid (*α*-dehydrobutyrine, Dhb), 3-*O*-methyl-L-threonine, N^{α}-methyl-L-asparagine, and (2*R*,3*R*,4*S*)-3-amino-14-chloro-2-hydroxy-4-methylhexadecanoic acid as unusual amino acid building units. P. C (X = Thr)

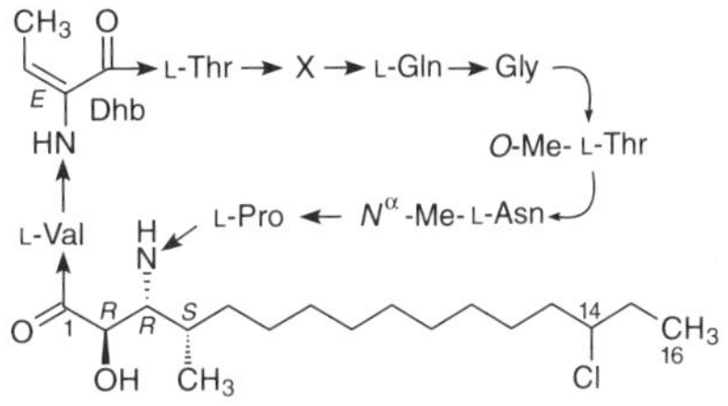

X = L-Thr : P.C
X = L-Val : P.D
Dhb = α-Dehydrobutyrine

induces a strong positive inotropic effect in isolated atria of the mouse heart. P. D (X=Val) on the other hand is almost inactive.

Lit.: J. Am. Chem. Soc. **111**, 6128 (1989) ▪ Tetrahedron **48**, 3727 (1992). – *[CAS 121596-40-5 (P. C)]*

Pycnosanguin see cinnabarin.

Pyocyanine (1-Hydroxy-5-methylphenazinium betaine).

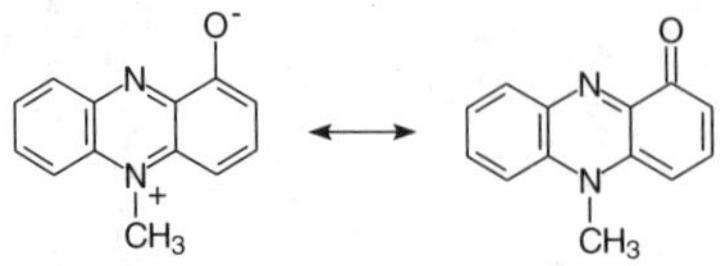

An antibiotically active phenazine pigment from cultures of *Pseudomonas aeruginosa* (=*P. pyocyanea*= *Bacillus pyocyaneus*) and *Streptomyces cyanoflavus*. $C_{13}H_{10}N_2O$, M_R 210.24, dark blue needles (hydrate), mp. 133 °C, soluble in chloroform, hot water, decomposes on standing in the light to give 1-phenazinol. P. has antibiotic activity against Gram-positive and Gram-negative bacteria, fungi, and protozoa; it also improves cell respiration. Since it also has strong toxic activity towards animal cells it cannot be used therapeutically.

Lit.: Antimicrob. Agents Chemother. **20**, 814 (1981) ▪ Beilstein E V **24/4**, 100 ▪ Merck-Index (12.), No. 8137. – *Synthesis:* Anal. Biochem. **95**, 19 (1979) ▪ Tetrahedron Lett. **1970**, 4101. – *[CAS 85-66-5]*

Pyoluteorin [(4,5-dichloro-1*H*-pyrrol-2-yl)-(2,6-dihydroxyphenyl)methanone].

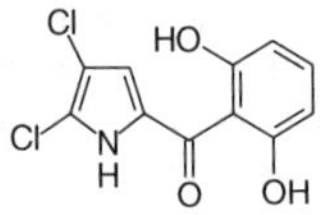

$C_{11}H_7Cl_2NO_3$, M_R 272.09; orange needles, mp. 174–175 °C (decomp.). A pyrrole antibiotic from *Pseudomonas fluorescens* and the *n*-alkane-utilizing *P. aeruginosa*. P. possesses antibacterial, antifungal (e. g., against *Pythium ultimum* on cotton), and herbicidal activities. The biosynthesis proceeds from proline through the polyketide pathway.

Lit.: Agric. Biol. Chem. **42**, 2031 (1978) ▪ Chem. Br. **1992**, 42 ▪ Pestic Sci. **27**, 117 (1989) ▪ Tetrahedron Lett. **1970**, 4879 (synthesis) ▪ Z. Naturforsch. C **41**, 532 (1986) (biosynthesis) ▪ see also pyrrolomycins and pyrrolnitrin. – *[CAS 25683-07-2]*

Pyoverdins.

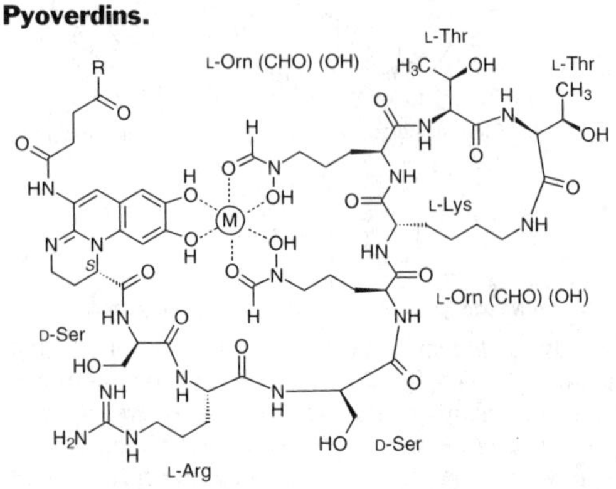

R = COOH : P. C
R = OH : P. D
R = NH_2 : P. E

(M) = metal-binding site

Oligopeptide antibiotics and *siderophores from *Pseudomonas fluorescens* and *P. aeruginosa*. The pyrimido[1,2-*a*]quinoline chromophore is responsible for the yellow-green color and fluorescence of P.

The three major representatives are *P. C* ($C_{56}H_{83}N_{17}O_{23}$, M_R 1362.38), *P. D* ($C_{55}H_{83}N_{17}O_{22}$, M_R 1334.36), and *P. E* ($C_{55}H_{84}N_{18}O_{21}$, M_R 1333.38). The 2,3-diamino-6,7-quinolinediol chromophore corresponds to that of *pseudobactin. For isolation, see *Lit.*[1]; for biosynthesis, see *Lit.*[2].

Lit.: [1] FEMS Microbiol. Rev. **104**, 209–228 (1993); J. Chromatogr. **787**, 195 (1997); Justus Liebigs Ann. Chem. **1989** 375–384; J. Prakt. Chem. **335**, 157–168 (1993); Tetrahedron **43**, 2261–2272 (1987); Z. Naturforsch. C **41**, 497–506 (1986). [2] Z. Naturforsch. C **50**, 622–629 (1995). – *[HS 2941 90; CAS 104022-78-8 (P. C); 104022-79-9 (P. D); 88966-86-3 (P. E); 114616-35-2 (P. I)]*

Pyrazines. Some P. of the types **1–3** are highly active[1] and widely distributed *flavor compounds[2], formed either thermally (especially **1** and **2**) or biosynthetically (**3**)[3]. *Alkylpyrazines* (**1**) (see table) have characteristic roast, nut, and potato odors and occur principally in the flavors of coffee, roasted nuts, meat, potatoes, and sea food. *2-Acetylpyrazine* (**2**) has a bread- and popcorn-like odor and is important for

Table: Characteristics of selected Pyrazines.

Pyrazine	molecular formula	M_R	bp. [°C]	LD_{50} (rat p. o.) [mg/kg]	odor threshold [ppb]	CAS
2-Methylpyrazine	$C_5H_6N_2$	94.12	135.7	1800	60–490	109-08-0
3-Isopentyl-2,5-dimethyl-pyrazine	$C_{11}H_{18}N_2$	178.28	110–120 (1.7 kPa)			18433-98-2
2,5-Dimethylpyrazine	$C_6H_8N_2$	108.14	155	1020	8–38	123-32-0
2-Ethyl-3.5(6)-dimethyl-pyrazine	$C_8H_{12}N_2$	136.20	181	456	4	13925-07-0, 13360-65-1
2-Acetylpyrazine	$C_6H_6N_2O$	122.13	mp. 76–78	>3000	62	22047-25-2
2-Isopropyl-3-methoxypyrazine	$C_8H_{12}N_2O$	152.20	–	–	0.002	93905-03-4
2-Isobutyl-3-methoxypyrazine	$C_9H_{14}N_2O$	166.22	214–215	2000 (mouse)	0.002	24683-00-9
2-sec-Butyl-3-methoxypyrazine	$C_9H_{14}N_2O$	166.22	218–219	1000 (mouse)	0.001	24168-70-5

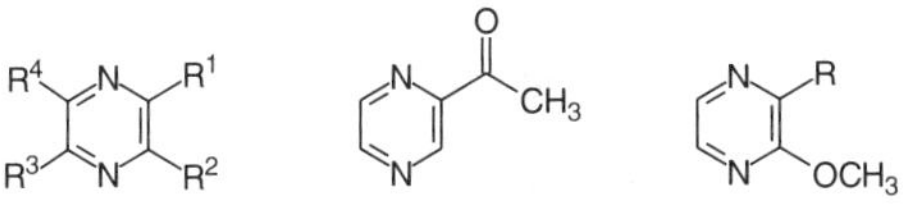

Alkylpyrazines 2-Acetylpyrazine 2-Alkyl-3-methoxypyrazines

1 2 3

*bread flavor, *coffee flavor, and roasted nut flavors. *2-Alkyl-3-methoxypyrazines* (**3**) (see table) have green pea-, pepper-like, or earthy odors and occur in *galbanum oil and in the flavors of green peas and paprikas as well as in wines (Cabernet Sauvignon) and raw potatoes (R = isopropyl). Alkylpyrazines, e. g., *3-isopentyl-2,5-dimethyl-P.*, are also found as *alarm substances in some species of ants[4]. Amino derivatives[4] have a role in bioluminescence (see, e. g., coelenterazine). For production of the P. used in flavors, see *Lit.*; data of important individual compounds are given in the table. *Cephalostatins and *ritterazines are steroids dimerized through a pyrazine structure and have antitumor activity.

Lit.: [1]Food Rev. Int. **1**, 129 ff. (1985); **8**, 479–558 (1992). [2]TNO list (6.), Suppl. 5, p. 493 ff. [3]Food Rev. Int. **4**, 375 ff. (1988). [4]Heterocycles **14**, 477–504 (1980).
gen.: Barlin, The Pyrazines, New York: Wiley 1982 ▪ Bedoukian (3.), p. 201–206 ▪ Beilstein E V **23/5**, 386, 403, 440; E V **23/11**, 182, 188. – *[HS 2933 90]*

Pyrethrosin see germacranolides.

Pyrethrum. An insecticide from the dried flower heads of various *Chrysanthemum* species obtained by pulverization or extraction; the main active principle is made up of the in total 6 optically active esters of (+)-*trans-chrysanthemic acid*[1] {$C_{10}H_{16}O_2$, M_R 168.24, mp. 17–21 °C, $[\alpha]_D^{20}$ +14.2° (C_2H_5OH)} and (+)-*trans-pyrethric acid* {$C_{11}H_{16}O_4$, M_R 212.25, mp. 84 °C, $[\alpha]_D^{18}$ +103.9° (CCl_4)} with the hydroxyketones (+)-*pyrethrolone* {pyrethrolone B_2, $C_{11}H_{14}O_2$, M_R 178.23, oil, bp. (120 Pa) 140–142 °C, $[\alpha]_D^{20}$ +17.8° (ether)}, (+)-*cinerolone* {$C_{10}H_{14}O_2$, M_R 166.22, oil, bp. (133 Pa) 180–184 °C, $[\alpha]_D^{25}$ +9.9° (C_2H_5OH)}, and (+)-*jasmolone* [$C_{11}H_{16}O_2$, M_R 180.25, oil, bp. (6.6 Pa) 140 °C] (see figure).

All six compounds are colorless liquids, soluble in alcohol, petroleum ether, and tetrachloromethane, they are easily oxidized and rapidly lose their activity in light and air and are also sensitive towards moisture and alkali. They are contact poisons that rapidly enter the nervous system and induce the characteristic symptoms in insects (high excitation, followed by impairments of coordination, paralysis, and finally death). The first effect starts rapidly, i. e., the insect is unable to move within a few minutes. This "knock-down" effect, especially in flies, is only achieved by a few insecticides. However, the dose required for this effect is usually not sufficient to cause death since the active principles of P. undergo rapid detoxification by enzymatic oxidation in the insects, thus some of the afflicted insects can recover. This can be prevented by the addition of synergists or active principles of the phosphoric acid ester and carbamate type. P., especially under the recommended conditions of use, is relatively harmless for humans and warm-blooded animals (birds, mammals). The synthetic pyrethroids, devel-

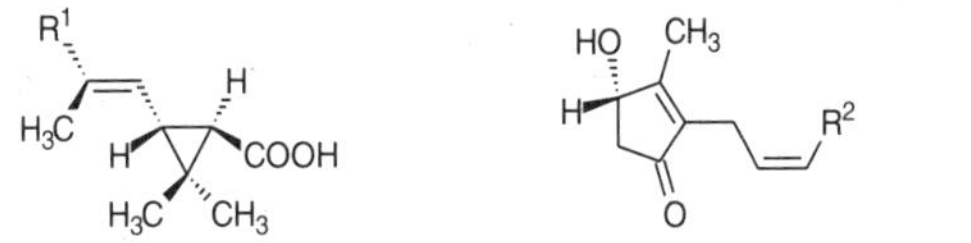

$R^1 = CH_3$: (+)-*trans*-Chrysanthemic acid
$R^1 = COOCH_3$: (+)-*trans*-Pyrethric acid

$R^2 = CH{=}CH_2$: (+)-Pyrethrolone
$R^2 = CH_3$: (+)-Cinerolone
$R^2 = C_2H_5$: (+)-Jasmolone

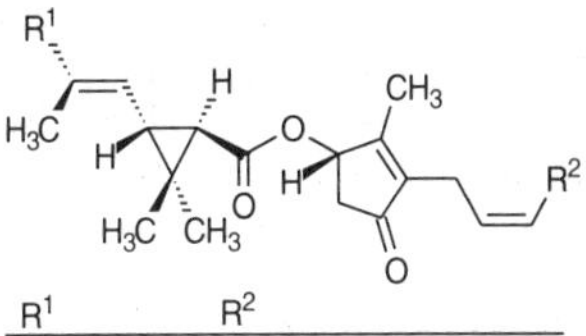

R^1	R^2	
CH_3	$CH{=}CH_2$	Pyrethrin I
$COOCH_3$	$CH{=}CH_2$	Pyrethrin II
CH_3	CH_3	Cinerin I
$COOCH_3$	CH_3	Cinerin II
CH_3	C_2H_5	Jasmolin I
$COOCH_3$	C_2H_5	Jasmolin II

Figure: The insecticidal components of Pyrethrum and their building blocks.

oped in the past 20 years, are even safer than the natural product.

Use: Main exporters of P. are Kenya, Tanzania, Ecuador, Columbia, New Guinea, and Japan. In 1993 Kenya and Tanzania produced ca. 25 000 t, more than 90% of the world harvest. The commercially marketed, pulverized flowers are classified as "dalmatine" and "persian" insect powders. The dalmatine product is obtained from *Tanacetum cinerariifolium* (syn. *Chrysanthemum cinerariaefolium, Pyrethrum cinerariifolium*), the persian from *Tanacetum coccineum* (syn. *Chrysanthemum coccineum, Pyrethrum roseum, P. carneum*). Extraction is performed with solvent mixtures such as methanol/kerosene, petroleum ether/acetonitrile, or petroleum ether/nitromethane. P. has a long history of use as a natural insecticide in Asia and today is still used in numerous agents, especially against insects causing hygienic problems or attacking stored food. The low stability and high production costs prevent an economically reasonable use in agriculture. Thus, selective, highly active compounds with an analogous mechanism of action, the so-called *pyrethroids*, have been developed for broader uses in the fields of veterinary medicine and plant protection.

Mechanism of action: Pyrethrins and pyrethroids increase the sodium influx into nerve cells by extending the opening periods of certain sodium channels. This leads to an overexcitation especially in invertebrates. Resistance phenomena are based, among others, on a change of the site of action of pyrethrins and pyrethroids.

Lit.: [1]J. Org. Chem. **62**, 1886 (1997) (synthesis).
gen.: ACS Symp. Ser. **591** (Molecular Action of Insecticides on Ion Channels), 26–43 (1995) ▪ Handbook of Pesticide Toxicology (Hayes, ed.), p. 585, New York: Academic Press 1991 ▪ Rev. Pestic. Toxicol. **2**, 185–230 (1993) ▪ Sax (8.), p. 2944 f. ▪ Soderlund, Mode of Action of Pyrethrins and Pyrethroids, in Casida & Quistad (eds.), Pyrethrum Flowers, p. 217–233, New York: Oxford Univ. Press 1995. –

[HS 1211 90, 1302 14; CAS 4638-92-0 ((+)-trans-chrysanthemic acid); 497-96-1 (pyrethric acid); 487-67-2 (pyrethrolone B_2); 22054-39-3 ((+)-jasmolone)]

Pyridazomycin.

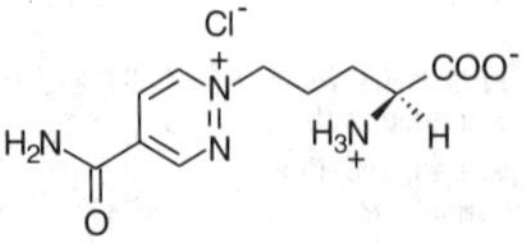

$C_{10}H_{15}ClN_4O_3$, M_R 274.71; monohydrate: hygroscopic, amorphous powder, mp. >119 °C (decomp.), $[\alpha]_D$ +94.6° (aq. CH_3OH). A structurally unusual, antifungally active pyridazinium derivative isolated as the hydrochloride from cultures of *Streptomyces violaceusniger* ssp. *griseofuscus*.

Lit.: J. Antibiot. **41**, 595 (1988). – *[HS 2941 90; CAS 115920-43-9]*

Pyridine alkaloids see tobacco alkaloids.

Pyridopyrazines. Bicyclic alkaloids from the springtail *Tetrodontophora bielanensis* (Collembola). The hemolymph and defensive secretion of this soil insect contains as main components *2,3-dimethoxypyrido[2,3-b]pyrazine* (formula I, $C_9H_9N_3O_2$, M_R 191.19, mp. 125 °C) and *3-isopropyl-2-methoxypyrido[2,3-b]pyrazine* (formula II, $C_{11}H_{13}N_3O$, M_R 203.24, oil). Both compounds have strong repellent effects on predatory beetles.

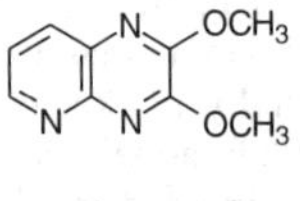

I

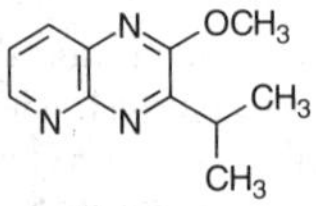

II

Lit.: Collect. Czech. Chem. Commun. **51**, 948–955 (1986) ▪ J. Chem. Ecol. **22**, 1051–1074 (1996). – *[CAS 103518-15-6 (I); 103518-16-7 (II)]*

Pyridoxal (5′-phosphate), pyridoxamine (5′-phosphate) see vitamin B_6.

Pyridoxatin.

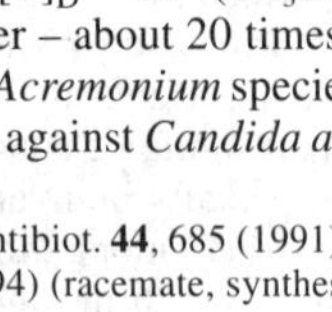

(relative configuration)

$C_{15}H_{21}NO_3$, M_R 263.34, needles, mp. 195–197 °C (decomp.), $[\alpha]_D^{23}$ −23° (CH_3OH). P. is a powerful radical scavenger – about 20 times stronger than vitamin E – from an *Acremonium* species (Hyphomycetes) without – except against *Candida albicans* – any antibiotic activity.

Lit.: J. Antibiot. **44**, 685 (1991) (structure) ▪ J. Org. Chem. **59**, 8065 (1994) (racemate, synthesis). – *[CAS 135529-30-5]*

Pyridoxic acids.

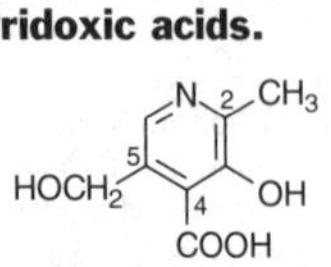

4-Pyridoxic acid

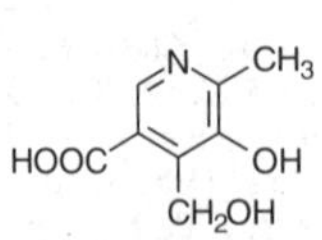

5-Pyridoxic acid

$C_8H_9NO_4$, M_R 183.16. P. represent the final products of *vitamin B_6 metabolism. *4-P.*: cryst., mp. 247 °C. *5-P.*: pale yellow cryst., mp. 280–283 °C (decomp.), can be isolated from *Pseudomonas* cultures and human urine.

Lit.: Beilstein E V **22/5**, 379 f. – *[CAS 82-82-6 (4-P.); 524-07-2 (5-P.)]*

Pyridoxines. An obsolete collective name for pyridine derivatives with *vitamin B_6 activity. According to IUPAC rule M-7 the name P. should only be used for 4,5-bis(hydroxymethyl)-2-methyl-3-pyridinol and its esters and ethers. The name vitamin B_6 is recommended as collective name for P., pyridoxal, pyridoxamine derivatives, etc..

Pyrimidines. The P. include *cytosine, uracil, thymine* (see table), and *orotic acid as well as a few rare compounds (5-methylcytosine, 5-hydroxymethylcytosine, 4-thiouracil).

Table: Pyrimidines and their properties.

	Cytosine	Uracil	Thymine
	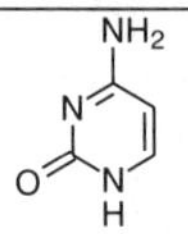	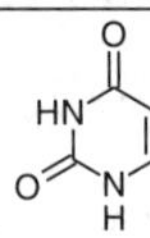	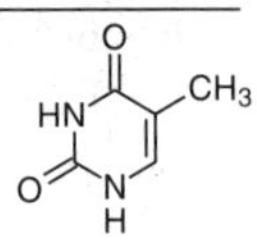
molecular formula	$C_4H_5N_3O$	$C_4H_4N_2O_2$	$C_5H_6N_2O_2$
M_R	111.10	112.09	126.11
mp. [°C]	320–325 (decomp.) platelets	338 needles	335–337 cryst.
occurrence	components of nucleic acids (DNA, RNA)	components of ribonucleic acids (RNA)	components of nucleic acids (DNA)
use	1β-D-Arabinofuranosyl-C. against leukemia	herbicides; 2-thiouracils as thyreostatics	
CAS	71-30-7	66-22-8	65-71-4

The P. nucleus is very important in biology: it forms the basis for many compounds and classes of compounds such as *flavins, *pteridines, and *purines. As components of nucleic acids, nucleotides, and nucleosides P. are essential for all forms of life.

Biosynthesis & degradation: The biosynthesis of P. starts from carbamoyl phosphate and L-aspartate via orotate to orotidine 5′-monophosphate (OMP). OMP is converted to uridine and cytidine 5′-triphosphate (UTP, CTP), by enzymatic reduction to 2′-deoxyuridine 5′-monophosphate (dUMP) and then by methylation to 2′-deoxythymidine 5′-monophosphate (dTMP). P. are degraded to urea and ammonia in most organisms. But they can also be used for the biosynthesis of β-alanine and thus the construction of coenzyme A. The main route for the degradation of *cytosine* proceeds under catalysis by cytosine deaminase (EC 3.5.4.1) to *uracil*, which is then reduced by dihydrouracil dehydrogenase (EC 1.3.1.1) to dihydrouracil. After hydrolytic ring opening the thus formed β-ureidopropionic acid is cleaved to carbon dioxide, ammonia, and β-alanine. The catalyst for this process is β-ureidopropionase (EC 3.5.1.6).

Inhibitors of pyrimidine biosynthesis are of interest as chemotherapeutic agents[1]. For the metabolism of P. analogues, see *Lit.*[2].

Use: Pharmacologically important P. compounds[3] include the derivatives of barbituric acid, vitamins (thiamine, riboflavin), some diuretics, nucleoside antibiotics, and antimetabolites (antipyrimidines) that are used in cancer therapy (e.g., 5-fluorouracil and thiouracil).

Lit.: [1] Med. Res. Rev. **10**, 505–548 (1990). [2] Pharmacol. Therapeut. **48**, 189–222 (1990). [3] Negwer (6.), p. 1735.
gen.: Adv. Exp. Med. Biol. **253** (1989) ▪ Beilstein E V **23/5**, 334–342 ▪ Stryer 1995, p. 745–748. – *[HS 2933 59]*

Pyrindamycins see duocarmycins.

L-Pyroglutamic acid (5-oxo-L-proline, L-glutamic acid lactam, abbreviations: <Glu, Glp, or Pyr).

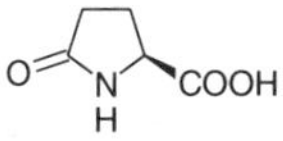

$C_5H_7NO_3$, M_R 129.12, mp. 162–163 °C, $[\alpha]_D^{20}$ –11.9° (H_2O). An internal amide of *glutamic acid. It occurs in fruits, grasses, and molasses; it is the *N*-terminal component of a series of peptide hormones such as gonadoliberin, thyroliberin, neurotensin, and gastrin and also occurs in bombesin and caerulein.

Use: In cosmetics for moisturizing the skin, for racemate separation of amines, and as starting material for asymmetric synthesis. The 3,3,5-trimethylcyclohexyl ester, Crilvastatin® has antihyperlipidemic activity (HMG-CoA reductase inhibitor)[1].

Lit.: [1] Br. J. Pharmacol. **114**, 624 (1995).
gen.: Beilstein E V **22/6**, 7 f. ▪ J. Chem. Soc., Perkin Trans. 1 **1990**, 3363 ff.; **1996**, 507 ▪ J. Heterocycl. Chem. **32**, 1489 (1995) ▪ J. Org. Chem. **59**, 1719 (1994) ▪ Tetrahedron Lett. **32**, 1379 f. (1991). – *[HS 2933 79; CAS 98-79-3]*

Pyrones (pyranones).

α-P., 2-P. γ-P., 4-P.

Ketones derived from the pyrans, $C_5H_4O_2$, M_R 96.09. 2*H*-pyran-2-one (α-P., *cumalin*): colorless liquid with an odor of hay and woodruff, mp. 5 °C, bp. 206 °C. 4*H*-Pyran-4-one (γ-P.): hygroscopic cryst., mp. 32 °C, bp. 217 °C. Substituted P. such as *maltol are formed, e.g., on heating carbohydrates, coffee, cocoa, coffee substitutes; they are also involved in the odor and taste of bread and bakery products; some of them act as taste intensifiers. Other derivatives of 2- or 4-P. such as kawain (see styrylpyrones), chelidonic acid, meconic acid, and *kojic acid, etc. are found as metabolic products in plants and microorganisms; in particular, the condensed ring systems of the chromones and coumarins are present in numerous natural products; *examples:* *flavonoids, *furocoumarins, *aflatoxins, and other *mycotoxins.

Lit.: Angew. Chem. Int. Ed. Engl. **101**, 215 ff. (1989) ▪ Beilstein E V **17/9**, 288, 290 ▪ Karrer, No. 1316–1688, 7208–7298, 7340–7523 ▪ Tetrahedron **33**, 3183–3192 (1978) ▪ Zechmeister **55**, 1–35. – *[CAS 504-31-4 (α-P.); 108-97-4 (γ-P.)]*

Pyrrolams see pyrrolizidine alkaloids.

Pyrrole pigments. A group of biochemically important substances which appear to be colored on account of the presence of several conjugated *pyrrole* rings in a linear or cyclic arrangement. The P. p. include the *porphines and *porphyrins with 16-membered ring systems stabilized by mesomerism as their chromophores in which 4 pyrrole rings are linked by 4 methine bridges. These compounds may contain central metal ions or be free of metal atoms: *heme in *hemoglobin, *protoporphyrins, *hematoporphyrin, mesoporphyrin, etioporphyrins, *uroporphyrins, *coproporphyrins, the photosynthesis pigment *chlorophyll, the *chlorins, *bacteriochlorophylls as well as the corrinoids consisting of 4 partially hydrated pyrrole rings in a cyclic arrangement with 3 methine bridges, e.g., *vitamin B_{12} (*cyanocobalamin*). The *bile pigments are green, yellow, or reddish brown, iron-free linear tetrapyrrole pigments. Their formation in the organisms can be attributed to the oxidative degradation of *hemin; the most important representative is the orange-red *bilirubin. Other linear P. are the green *biliverdin and the bacterial pigment *prodigiosin composed of three rings.

Lit.: Rodd's Chem. Carbon Compds. (2.) **4B**, 277–357 (1997). – *[HS 2933 90, 2936 26, 3203 00]*

Pyrrolizidine alkaloids. Collective name for alkaloids containing the pyrrolizidine ring system (figure).

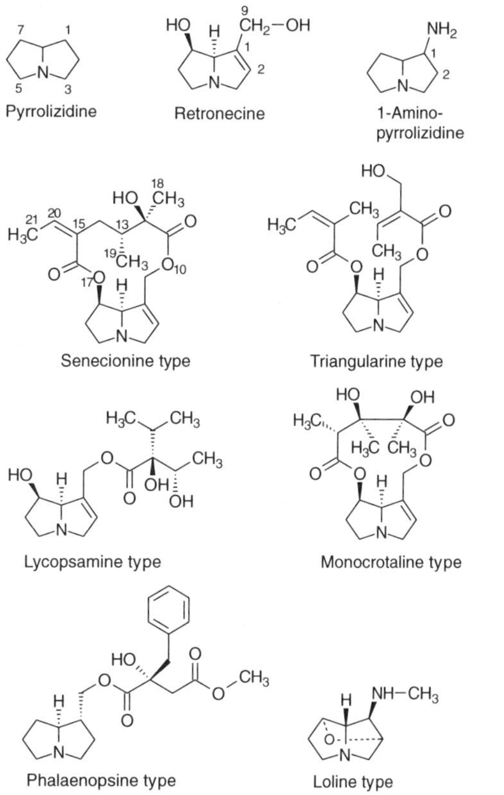

Figure: Pyrrolizidine alkaloids, skeletons, and structural types.

Structure & occurrence: The P. a., widely distributed in the plant kingdom, are almost all ester alkaloids built up from the amino alcohol 1-hydroxymethylpyrrolizidine [see necine(s)] and usually aliphatic mono- or dicarboxylic acids (necinic acids). Most of the ca. 370

Table: Distribution of the Pyrrolizidine alkaloid types within the taxa of Pyrrolizidine alkaloid containing species (see *Lit.* Pelletier).

Taxa	number of P. a. containing species	A* (104)**	B (56)	C (109)	D (33)	E (21)	M (32)	L (11)
Apocynaceae	8	–	–	5	–	1	3	–
Asteraceae								
Eupatorieae	23	1	2	23	–	–	–	–
Senecioneae	231	204	31	–	1	–	7	–
Boraginaceae	145	2	16	128	5	3	8	–
Fabaceae								
Crotalaria	81	19	4	–	58	–	15	–
other genera	37	22	–	–	–	–	8	8
Orchidaceae	35	–	–	–	–	33	2	–
Poaceae	4	1	–	–	–	–	–	3

* A = Senecionine type; B = Triangularine type; C = Lycopsamine type; D = Monocrotaline type; E = Phalaenopsine type; M = other P. a.; L = Loline type
** number of chemical structures within brackets

P. a. currently known from plants can be assigned to one of 6 structural types (figure), that are characteristic for certain plant families (table). The two largest groups are the macrocyclic P. a. of the *senecionine type and esters of the lycopsamine type (see also indicine). P. a. of the triangularine type are closely related to the senecionine type. They consist of open chain diesters in which the macrocyclic ring is not yet closed. In the genus *Crotalaria* of the Fabaceae macrocyclic P. a. of the *monocrotaline type dominate. The most prominent necine base of the mentioned groups is the 1,2-unsaturated retronecine [see necine(s)]; it is absent in P. a. of the phalaenopsine type from Orchidaceae. The *lolines are included in the P. a. for historical reasons only, they differ from all other P. a. in the necine base which is a 1-aminopyrrolizidine group.
Simple lactams with the pyrrolizidine skeleton (*pyrrolams*) are isolated from *Streptomyces olivaceus*[1].
Biosynthesis & physiology: The necine bases are formed from arginine. The necinic acids are mostly derived from the amino acids isoleucine, leucine, and valine. The P. a are formed in the roots of *Senecio* species and are stored intracellularly in the vacuoles of the inflorescence, epidermal tissues, etc. The corresponding *N*-oxides are the transport and storage forms and are easily reduced to the tertiary amines on processing. The genuine tertiary amines are only found in the dried seeds.
Chemical ecology: The amines are components of the plants' chemical defence systems. They have antifeedant or deterrent effects on many animals. Adapted insects of various taxa (Lepidoptera, grasshoppers, beetles, and greenflies) are able to store and utilize the plant P. a. They use them as protective substances against insect-eating predators and display their inedibility through conspicuous warning colors (aposematis)[2]. Some Lepidoptera also use plant P. a. as precursors for their pheromones[3].
Toxicity: All P. a. with a 1,2-unsaturated necine and with at least the primary hydroxy group at C-9 being esterified with a branched aliphatic C_5- to C_7-carboxylic acid are potentially toxic. This accounts for more than 70% of all naturally occurring P. a. Orally consumed P. a. are converted to the actually toxic pyrrole derivatives in the liver by the mixed functional *cytochrome P_{450} dependent oxygenases. The reactive pyrrole esters alkylate nucleophilic groups (NH_2, SH, OH groups) of proteins and nucleic acids. They thus have hepatotoxic, carcinogenic, and mutagenic activities. The highest toxicities are shown by the macrocylic diesters, they are bifunctional alkylating agents and can thus cross-link DNA. The non-cyclic diesters and monoesters follow in order of decreasing toxicity. Chronic poisonings with mutagenic, carcinogenic, and teratogenic results[4] through P. a. have been reported for grazing animals (especially cattle) and domestic animals. Liver diseases with a lethal course are also known among humans[5]. The German health authorities (BGA) have classified all medicinal plants that contain toxic P. a. as well as preparations made therefrom as a potential hazard to health. Under restrictions in the scope of application, a tolerable daily dose of 1 µg P. a. (oral uptake) or of 100 µg (local administration) is allowed. The duration of treatment with P. a. containing products should be limited to 6 weeks[6]. For synthesis, see *Lit.*[7].

Lit.: [1] Justus Liebigs Ann. Chem. **1988**, 387; **1990**, 525; Synlett **1997**, 1179; Tetrahedron: Asymmetry **8**, 515 (1997) **10**, 3827 (1999); Tetrahedron **52**, 869 (1996). [2] Eur. J. Org. Chem. **1998**, 13; Naturwissenschaften **73**, 17–26 (1986); Rosenthal & Berenbaum (eds.), Herbivores: Their Interactions with Secondary Plant Metabolites, vol. 1, p. 79–121, San Diego: Academic Press 1991. [3] J. Chem. Ecol. **16**, 165–185 (1990); Naturwissenschaften **79**, 241–288 (1992). [4] Cheeke (eds.), Toxicants of Plant Origin, vol. 1, p. 23–40, Boca Raton: CRC Press 1989. [5] Dtsch. Apoth. Ztg. **137**, 4066 (1997). [6] Pharmazie **50**, 83–98 (1995). [7] Synlett **1**, 433–440 (1990); J. Am. Chem. Soc. **120**, 7359 (1998); Nat. Prod. Rep. **10**, 487–496 (1993).
gen.: Hager (5.) **3**, 1079 f.; **6**, 664 f. ▪ J. Chem. Soc., Perkin Trans. 1 **1997**, 677 (biosynthesis) ▪ Manske **26**, 327–384; **31**, 193–315 ▪ Mattocks, Chemistry and Toxicology of Pyrrolizidine Alkaloids, London: Academic Press 1986 ▪ Nat. Prod. Rep. **14**, 653 (1997); **15**, 363 (1998); **16**, 499 (1999) (review) ▪ Pelletier **9**, 155–233 ▪ Rizk (ed.), Naturally Occurring Pyrrolizidine Alkaloids, Boca Raton: CRC Press 1990 ▪ Rodd's Chem. Carbon Compds. (2.) **4B**, 21–67 (1997) (review) ▪ Ullmann (5.) **A 1**, 362 f. ▪ Zechmeister **41**, 115–204. – *[HS 293990]*

Pyrrolizidone oxime alkaloids see polyzonimine.

Pyrrolnitrin (rugosin).

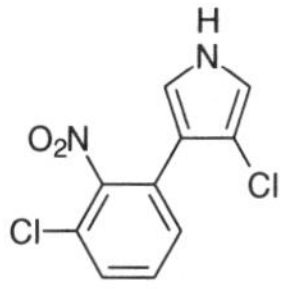

A pyrrole antibiotic formed by various *Pseudomonas* species and *Myxococcus fulvus* (Myxobacteria). $C_{10}H_6Cl_2N_2O_2$, M_R 257.08, pale yellow cryst., mp. 125 °C, that slowly decompose on exposure to light and atmospheric oxygen. In spite of its strong antifungal activity, e. g., towards *Mucor hiemalis* and *Rhizopus stolonifer* (MIC 0.1–0.2 µg/mL), it has surprisingly low toxicity: LD_{50} (mouse p. o.) 1 g/kg. The biosynthesis proceeds from tryptophan by rearrangement and decarboxylation. The 3-cyano-6,7-dichloro and 3-cyano-6,7-difluoromethylenedioxy analogues (Fenpiclonil® and Fludioxomil®) have higher antifungal activities and are more stable to light than P.; they are used in plant protection mainly for prophylactic treatment of seeds (seed dressing).

Lit.: Baker, Fenyes & Steffens (eds.), Synthesis and Chemistry of Agrochemicals, p. 395–404, Washington: Am. Chem. Soc. 1992 ▪ J. Antibiot. **34**, 555 (1981); **35**, 1101 (1982) ▪ Kleemann-Engel, p. 1641. – *[HS 2941 90; CAS 1018-71-9]*

Pyrroloindole alkaloids. Some *indole alkaloids with the pyrrolo[2,3-*b*]indole skeleton are subdivided into the *eserine type* (example: *physostigmine = eserine from *Dioclea* and *Physostigma* species) and the *echitamine-erinine type* (e.g., echitamine from *Alstonia*-, corymine and erinine from *Hunteria* species). Some di-, tri-, and tetrameric indole alkaloids from *Calycanthus* species, e.g., *chimonanthine* ($C_{22}H_{26}N_4$, M_R 346.48) and *folicanthine* ($C_{24}H_{30}N_4$, M_R 374.53) are also included in the P. a.[1].

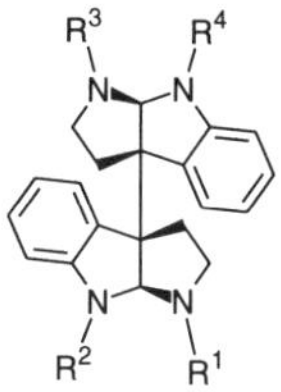

$R^1 = R^3 = CH_3$, $R^2 = R^4 = H$: Chimonanthine
$R^1 = R^2 = R^3 = R^4 = CH_3$: Folicanthine

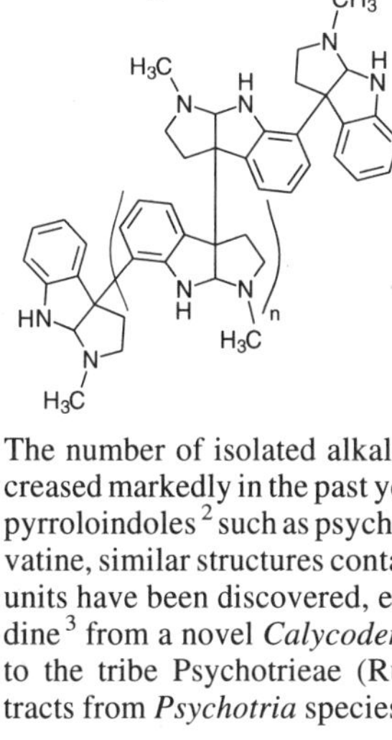

n = 0 : Hodgkinsine
n = 1 : Quadrigemine A
n = 2 : Psychotridine
n = 3 : Vatine
n = 4 : Vatamine
n = 5 : Vatamidine

The number of isolated alkaloids of this type has increased markedly in the past years. Besides pentameric pyrroloindoles[2] such as psychotridine and the hexamer vatine, similar structures containing up to 8 tryptamine units have been discovered, e. g., the octamer vatamidine[3] from a novel *Calycodendron* species belonging to the tribe Psychotrieae (Rubiaceae). Alkaloid extracts from *Psychotria* species exhibit appreciable cytotoxicity towards human leukemia and rat hepatoma cells. "Bioactivity-guided" fractionation of *Psychotria* extracts led to the isolation of tetrameric pyrrolidinoindolines with interesting biological activities[4], which act on the liberation of hypophyseal hormones and their specific binding to somatostatin receptors.

Lit.: [1]Beilstein EIII/IV **26**, 1865; EV **26/12**, 40ff.; Phytochemistry **31**, 317 (1992). [2]Manske **34**, 1. [3]Planta Med. **56**, 212 (1990); **61**, 313 (1995). [4]J. Nat. Prod. **55**, 923 (1992).
gen.: ApSimon **3**, 307–315 ▪ J. Am. Chem. Soc. **116**, 9480 (1994); **118**, 8166 (1996) (synthesis). – *[HS 2939 90; CAS 5545-89-1 (chimonanthine); 6879-55-6 (folicanthine)]*

Pyrrolomycins.

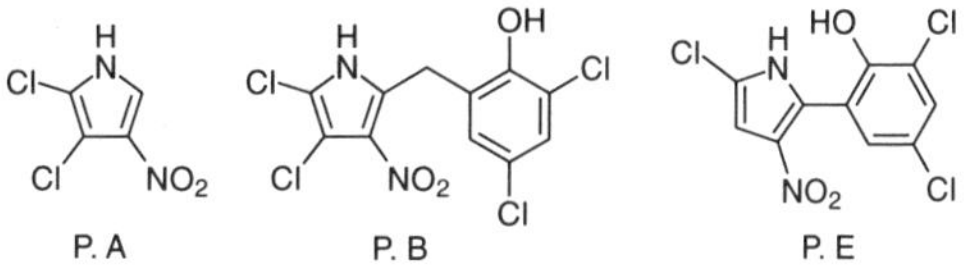

P. A P. B P. E

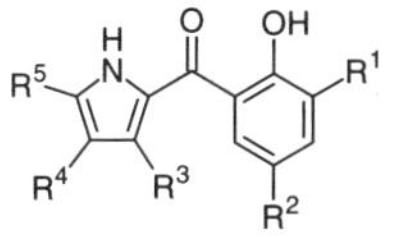

R^1	R^2	R^3	R^4	R^5	
Cl	Cl	H	Cl	Cl	P. C
Cl	Cl	Cl	Cl	Cl	P. D
H	Br	Br	Br	Br	P. F1
H	Br	Br	Cl	Br	P. F2a
H	Br	Br	Br	Cl	P. F2b
H	Br	Br	Cl	Cl	P. F3

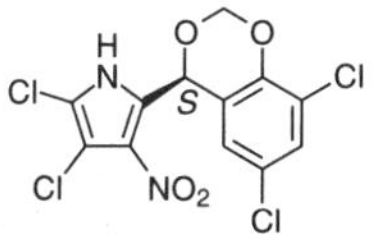

Dioxapyrrolomycin (Pyrroxamycin)

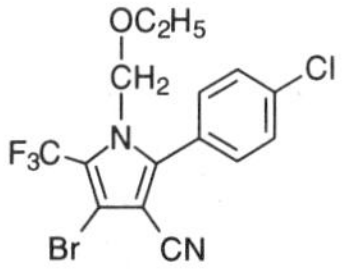

AC 303 630 (synthet. insectizide)

Table: Data of Pyrrolomycins.

Pyrrolomycin	molecular formula	M_R	mp. [°C]	CAS
A	$C_4H_2Cl_2N_2O_2$	180.98	211–213	79763-01-2
B	$C_{11}H_6Cl_4N_2O_3$	355.99	222–225	79763-00-1
C	$C_{11}H_5Cl_4NO_2$	324.98	220–221	81910-06-7
D	$C_{11}H_4Cl_5NO_2$	359.42	195–198	81910-07-8
E	$C_{10}H_5Cl_3N_2O_3$	307.52	>250	87376-16-7
F1	$C_{11}H_5Br_4NO_2$	502.78	188–190	88217-19-0
F2a	$C_{11}H_5Br_3ClNO_2$	458.33	182	88477-78-5
F2b	$C_{11}H_5Br_3ClNO_2$	458.33	206–208	88477-79-6
F3	$C_{11}H_5Br_2Cl_2NO_2$	413.88	192–193	88228-92-6

A group of secondary metabolites from culture broths of, e.g., *Streptomyces* strain SF-2080 with antibiotic activity towards Gram-positive and some Gram-negative bacteria as well as fungi. Cultures of *Streptomyces fumanus* produce *dioxapyrrolomycin* (pyrroxamycin) {$C_{12}H_6Cl_4N_2O_4$, M_R 384.00, pale yellow cryst., mp. 216–220 °C (decomp.), $[\alpha]_D$ −37.2° (acetone)}[1]. Doubly labelled nitrate is incorporated in dioxapyrrolomycin without any changes in the isotopic pattern, this was the first demonstration of a biochemical nitration of a precursor such as pyrrolomycin C[2]. The relatively weak antibiotic has a relatively high oral toxicity for mice (LD_{50} = 13 mg/kg) and served as a lead structure for the completely synthetic insecticide AC 303630, in which toxicity was reduced by a factor of 5 and the insecticidal activity was increased by an order of magnitude. Dioxapyrrolomycin also has anthelmintic activity.

Lit.: [1] J. Antibiot. **36**, 1263, 1483 (1983); **40**, 233, 899, 961 (1987). [2] J. Chem. Soc., Chem. Commun. **1989**, 1271 (biosynthesis).
gen.: ACS Symp. Ser. **524** (Pest Control With Enhanced Environmental Safety), 219–232 (1993). – *Isolation, structure elucidation:* J. Antibiot. **34**, 1363 ff., 1366 ff. (1981); **36**, 1263–1267, 1431–1438, 1483–1489 (1983); **40**, 233 ff. (1987); **45**, 977–983 (1992). – *Pharmacology:* Biochem. Biophys. Res. Commun. **105**, 82–88 (1982) ▪ J. Antibiot. **37**, 1253 ff. (1984); **44**, 533–540 (1991). – *Synthesis:* J. Antibiot. **34**, 1569–1576 (1981) ▪ Justus Liebigs Ann. Chem. **1993**, 847–852 ▪ Tetrahedron Lett. **34**, 8443 (1993). – *[HS 2941 90; CAS 105888-54-8 (dioxapyrrolomycin)]*

Pyrroloquinoline quinone (methoxatin, PQQ).

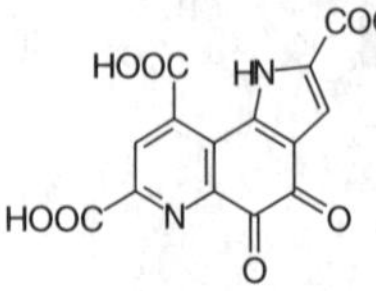

$C_{14}H_6N_2O_8$, M_R 330.21, dark-red solid. Decomposes in aqueous solution; for stabilization by formation of ruthenium complexes, see *Lit.*[1]. Together with the NAD and flavin systems, P. is a coenzyme of the hydrogen and electron transfer processes of oxidoreductases, e.g., bacterial methylamine dehydrogenase (EC 1.4.99.3)[2] or methanol dehydrogenase. On reduction P. is converted to the 4,5-diol. As an *o*-quinone it reacts with amines and other nucleophiles; it probably occurs *in vivo* only in the protein bound form (quinoproteins). A deficiency of P. leads to an inadequate function of lysyl hydrolase and thus influences collagen formation and wound healing. A high concentration of PQQ is found in the adrenals. The relatively late discovery of the cofactor was due to analytical difficulties. Even today the presence of P. in some enzymes is not completely clarified, and in some cases P. was erroneously identified as a cofactor[3]. P. is formed by microorganisms, plants, and some animals. In addition to its coenzyme function, P. also exhibits further biological properties, e.g., as growth factor and as inhibitor of reverse transcriptase and aldose reductase; synthetic analogues are of interest because of these activities[4].

Lit.: [1] Angew. Chem. Int. Ed. Engl. **29**, 78 f. (1990). [2] Acta Cryst. Sect. B-Struct. Sci. **46**, 806–823 (1990); EMBO J. **8**, 2171–2178 (1989); J. Chem. Soc., Perkin Trans. 2 **1992**, 1245. [3] Curr. Biol. **1**, 135 f. (1991); FEBS Lett. **278**, 120 ff. (1991); Nature (London) **350**, 22 f. (1991). [4] Synthesis **1996**, 377–382.
gen.: Annu. Rev. Biochem. **58**, 403–426 (1989) ▪ Chem. Commun. **1995**, 2077 ▪ FEMS Microbiol. Rev. **87**, 221–225 (1990) ▪ Helv. Chim. Acta **76**, 988, 1667 (1993); **79**, 658 (1996) ▪ J. Am. Chem. Soc. **111**, 6822–6828 (1989) ▪ Manske **49**, 79–219 (review) ▪ Methods Enzymol. **280**, 89–98 (1997) (fermentation) ▪ Synthesis **1996**, 377 ▪ Trends Biochem. Sci. **14**, 343–346 (1989); **15**, 96 f., 299 (1990); **16**, 12 (1991). – *[CAS 72909-34-3]*

Pyrylium salts. In contrast to the non-aromatic pyrans, their oxidized products the P. s. possess an aromatic electron sextet. P. s. belong to the oxonium salts but are, in contrast to most other oxonium salts of ethers, stable, especially when the 2,4,6-positions are substituted. The widely distributed natural products, the anthocyanidins (see anthocyanins), see also flavonoids, are isolated as 2-phenyl-1-benzopyrylium chlorides.

Lit.: Adv. Heterocycl. Chem. **50**, 158–254 (1990) ▪ J. Org. Chem. **56**, 126 ff. (1991) ▪ Mitra et al., New Trends in Heterocyclic Chemistry, p. 79–111, Amsterdam: Elsevier 1979.

Pyxinol.

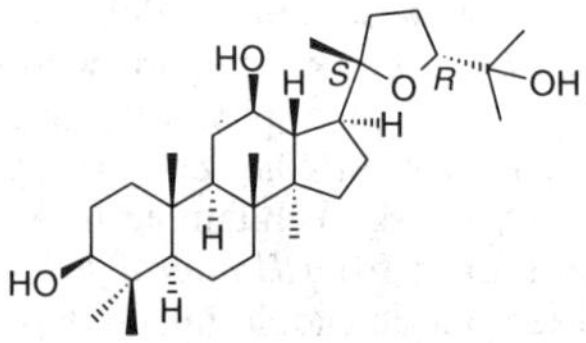

$C_{30}H_{52}O_4$, M_R 476.74, needles, mp. 225–226 °C, $[\alpha]_D$ +62.8° ($CHCl_3$). A dammarane triterpenetriol from the lichen *Pyxine endochrysina.*

Lit.: Phytochemistry **31**, 923 (1992) ▪ Tetrahedron Lett. **1969**, 4245–4248 ▪ Z. Naturforsch. C **51**, 750 (1996). – *[CAS 25330-18-1]*

Q

Qinghaosu (arteannuin, artemisinin, qinhao su, qing hau sau, qinghao, Qhs).

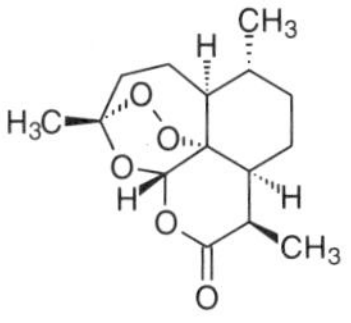

Qinghaosu

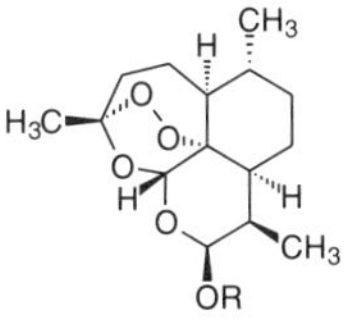

R = H : Dihydroqinghaosu, DHQS
R = CH_3 : Artemether, Paluther®
R = C_2H_5 : Arteether
R = CO–$(CH_2)_2$–COOH : Artesunate®

$C_{15}H_{22}O_5$, M_R 282.34, needles, mp. 156–157 °C, $[\alpha]_D^{17}$ +66.3° ($CHCl_3$), soluble in chloroform, petroleum ether. Qhs is a component of *Artemisia annua* [1]. Parts of this shrub are used in Chinese traditional medicine to reduce fever. Qinhao was mentioned 2000 years ago in the "52 prescriptions" found in the grave of the *Mawangdui Han* dynasty. In the middle of the 4th century its efficacy for malaria treatment was mentioned in *Zhouhou Bei Ji Fang* (Handbook of emergency medicine). In 1972 Qhs was identified as the active principle. Its structure was confirmed by X-ray diffraction and synthesis [2]. *A. annua* contains ca. 0.01–0.6% dry weight of Qhs, that is best extracted with petroleum ether. Qhs is active against multiresistant strains of *Plasmodium falciparum* (pathogen of the most dangerous form of malaria, *malaria tropica*) [3]. Every year ca. 300–500 000 000 cases of malaria are reported worldwide [4]; about 2–3 000 000 people die of *M. tropica*, in part because of the numerous resistances of the pathogen against the older antimalarial drugs (Resochin®, Fansidar®, etc.) [5]. Even nanomolar concentrations of Qhs are active *in vitro* against *P. falciparum*. Q. damages the parasite's membranes (presumably through formation of singlet oxygen) and attacks its nucleic acid metabolism. Recent investigations have shown that Qhs and its derivatives have gametocidic activity, effect the sexually differentiated stage of the developmental cycle of *P. falciparum* and thus prevent transmission. The acute toxicity of Qhs is very low [LD_{50} (mouse p.o) >4000 mg/kg] [6]. The main suppliers of Qhs are China and Vietnam where *A. annua* is cultivated in plantations.

More stable than Qhs are the compounds obtained by reduction of the lactone to *dihydroqinghaosu* [1], DHQHS, $C_{15}H_{24}O_5$, M_R 284.35, cryst., mp. 152–154 °C, $[\alpha]_D^{22}$ +129° ($CHCl_3$), and subsequent derivatization. The methyl ether *artemether*, $C_{16}H_{26}O_5$, M_R 298.38, cryst., mp. 86–87 °C, $[\alpha]_D^{25}$ +163° ($CHCl_3$) has been marketed since 1992 by Rhône Poulenc Rorer under the proprietary name Paluther®, especially in Asia and Africa. The monosuccinate ester is available as Artesunate®. The modifications result in a suppression of hydrolysis of the lactone ring in Qhs so that doses of Artemether® or Artesunate® of ca. 400–600 mg every third day are usually sufficient to cure the patients [3,4] (Qhs: 0.5–3 g daily). The ethyl ether of dihydro-Q., *arteether* [$C_{17}H_{28}O_5$, M_R 312.41, cryst., mp. 80–82 °C] is widely used clinically in China [7]. Qhs and its derivatives are recommended by WHO for treatment of chloroquine- and mefloquine-resistant malaria. For synthesis, see *Lit.* [8,9].

Lit.: [1] Helv. Chim. Acta **67**, 1515 (1984); Planta Med. **1985**, 441; J. Nat. Prod. **51**, 1273 (1988); **52**, 337 (1989). [2] J. Am. Chem. Soc. **105**, 624 (1983). [3] Science **228**, 1049 (1985); Lancet **341**, 603 (1993). [4] Parasitol. Today **12**, 79 (1996). [5] Parasitology Today **7**, 297 (1991). [6] J. Med. Chem. **31**, 645 (1988). [7] Tetrahedron **46**, 1871–1884 (1990); Parasitol. Today **12**, 79 ff. (1996). [8] Acc. Chem. Res. **30**, 73 (1997); J. Am. Chem. Soc. **114**, 974–979 (1992); J. Chem. Soc., Chem. Commun. **1990**, 726; J. Nat. Prod. **61**, 633 (1998); J. Prakt. Chem. **339**, 642 (1997) (derivatives); Synth. Commun. **26**, 321–329 (1996); Tetrahedron **54**, 319–336 (1998); Tetrahedron Lett. **31**, 755 (1990); **34**, 4435 (1993). [9] J. Chem. Soc., Chem. Commun. **1988**, 372; J. Med. Chem. **32**, 1249 (1989); **38**, 607–616 (1995); **39**, 1885–1897 (1996); Tetrahedron **45**, 7287 (1989). *gen.:* Atta-ur-Rahman **3**, 495–527 ▪ Chem. Soc. Rev. **1992**, 85–90 ▪ Heterocycles **32**, 1593–1638 (1991); **42**, 493 (1996) ▪ Med. Res. Rev. **7**, 29–52 (1987) ▪ Tang & Eisenbrand, Chinese Drugs of Plant Origin, p. 159–178, Berlin: Springer 1992 ▪ Zechmeister **72**, 121–214 (review). – *Biosynthesis:* Phytochemistry **25**, 2777 (1986); **26**, 1927 (1987). – *Activity, mechanism:* J. Am. Chem. Soc. **118**, 3537 (1996) ▪ Tetrahedron Lett. **37**, 253, 257 (1996). – *[CAS 63968-64-9 (Q.); 75887-54-6 (arteether); 71963-77-4 (artemether); 88495-63-0 (artesunate)]*

Quadrigemine A see pyrroloindole alkaloids.

Quadrilure [(3*R*,6*E*)-3-acetoxy-7-methyl-6-nonene].

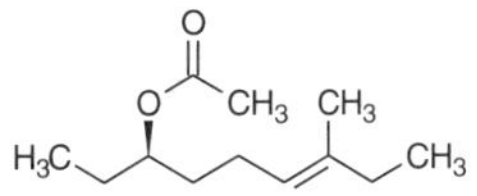

$C_{12}H_{22}O_2$, M_R 198.31, oil, bp. 63–64 °C (1.73 kPa); soluble in chloroform, pentane; $[\alpha]_D^{20}$ +9.59° ($CHCl_3$). An aggregation pheromone of the male corn weevil *Cathartus quadricollis*, a major pest of stored grain crops throughout the world.

Lit.: ApSimon **9**, 285 f. (synthesis) ▪ Biosci., Biotechnol. Biochem. **56**, 999 (1992) ▪ J. Chem. Ecol. **14**, 2169–2184 (1988) ▪ Justus Liebigs Ann. Chem. **1990**, 159 ff. ▪ Tetrahedron: Asymmetry **6**, 463 (1995) ▪ see also pheromones. – *[CAS 100429-35-4]*

Quadrone. $C_{15}H_{20}O_3$, M_R 248.32, cryst., mp. 185–186 °C, $[\alpha]_D$ –50° (C_2H_5OH), a *sesquiterpene produced by *Aspergillus terreus* with anti-tumor activity.

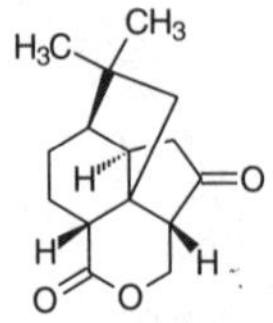

Lit.: Tetrahedron Lett. **19**, 499 (1978) (isolation); **25**, 1119 (1984) (biosynthesis). – *Synthesis:* Can. J. Chem. **66**, 528 ff. (1988) ▪ J. Am. Chem. Soc. **105**, 563 (1983); **113**, 3533–3542 (1991) ▪ J. Chem. Soc., Chem. Commun. **1989**, 1090 ▪ J. Org. Chem. **50**, 4418 (1985); **52**, 1483 (1987) ▪ Tetrahedron Lett. **31**, 485 ff. (1990). – *[CAS 6650-08-1]*

Quassin(oids). A group of structurally complex *triterpenes with various tetra- and pentacyclic C_{18}-, C_{19}-, C_{20}-, and C_{25}-skeletons. The C_{20}-picrasane system is the most widely distributed. Q. occur principally in the wood of the tree *Quassia amara* indigenous to Brazil and Surinam as well as the Caribbean *Picrasma excelsa* (Simaroubaceae). They have antifeedant effects on insects and taste very bitter. The most important representative of the Q. is *quassin* (2,12-dimethoxypicrasa-2,12-diene-1,11,16-trione, nigakilactone D) $C_{22}H_{28}O_6$, M_R 388.46, mp. 221–222 °C, $[\alpha]_D$ +34.5° ($CHCl_3$) – a partially hydrogenated phenanthrene derivative.

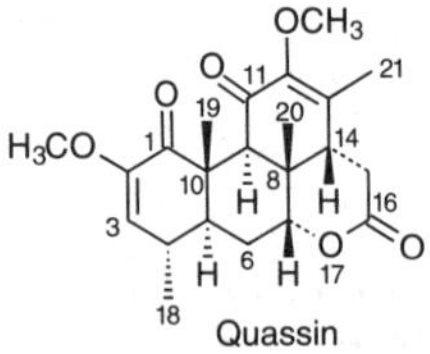

Quassin

Q. is used in the food industry as a *bitter principle, according to legal regulations brandies may contain up to 50 mg/L Q. as bitter component. Q. has a bitter taste even at a dilution of 1:60 000. In mammals Q. can effect a decrease in heart rate and, at higher concentrations, muscular convulsions and paralyses. The commercially available "quassin" is a mixture of quassin, neoquassin, isoquassin, and 18-hydroxyquassin. Q. can be used as a substitute for emetine hydrochloride (see ipecac alkaloids). Some pentacyclic Q. have antiviral, antiparasitic, insecticidal, antifeedant, amoebicidic, and anti-inflammatory activities.

Lit.: Aust. J. Chem. **44**, 927–938 (1991) ▪ Beilstein E V **18/5**, 175 ▪ Karrer, No. 3688, 3711 ▪ Merck-Index (12.), No. 8209. – *Reviews:* Annu. Proc. Phytochem. Soc. Eur. **22**, 247–266 (1983) ▪ Can. Chem. News **47**, 16 f. (1995) ▪ Org. Prep. Proc. Int. **21**, 521–618 (1989) ▪ Zechmeister **30**, 101–150; **47**, 221–264. – *Synthesis:* Atta-ur-Rahman **11**, 71–111 (1992) ▪ J. Am. Chem. Soc. **111**, 6287 (1989); **112**, 9436 (1990) ▪ J. Org. Chem. **63**, 2056, 9576 (1998) ((+)-quassin) ▪ Tetrahedron **46**, 2187 (1990). – *[HS 2932 29; CAS 76-78-8 (quassin)]*

Quaterpyridine (nemertelline).

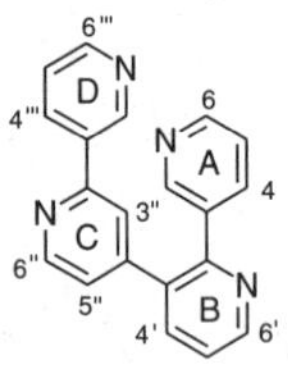

$C_{20}H_{14}N_4$, M_R 310.36, oil. Toxin from ribbon worms of the species *Amphiporus angulatus* with an armed proboscis.

Lit.: Tetrahedron **51**, 11401 (1995). – *[CAS 59697-14-2]*

Quebrachamine (kamassine).

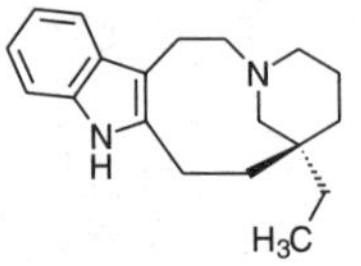

(+)-(*S*)-form

$C_{19}H_{26}N_2$, M_R 282.43, mp. 147–149 °C; soluble in acetone, chloroform, dilute acids. A *monoterpenoid indole alkaloid of the *Aspidosperma* type, Q. occurs in both enantiomeric forms: the (+)-form, $[\alpha]_D$ +98° ($CHCl_3$), in leaves and root bark of *Pleiocarpa* species as well as the bark of *Stemmadenia* species; the (–)-form, $[\alpha]_D$ –100° ($CHCl_3$), in *Aspidosperma quebracho-blanco* and other *Aspidosperma* species as well as *Gonioma*, *Hunteria*, and *Rhazya* species.

Lit.: Beilstein E V **23/8**, 189 f. ▪ Hager (5.) **4**, 402 ff. ▪ Heterocycles **16**, 247 (1981); **29**, 243 (1989) ▪ J. Org. Chem. **55**, 517 (1990); **63**, 6007 (1998) (synthesis (+)-Q.) ▪ Manske **27**, 131–407 ▪ Merck-Index (12.), No. 8212. – *[HS 2939 90; CAS 4850-21-9]*

Quebracho (cortex). The bark of the quebracho tree, *Aspidosperma quebracho-blanco* (Apocynaceae), a large tree (up to 20 m high), indigenous to the west of South America. The bark contains ca. 1% *monoterpenoid indole alkaloids such as *yohimbine, aspidospermine, *quebrachamine. The **Aspidosperma* alkaloids are structurally related to the *Catharanthus* alkaloid *vindoline. Quebracho bark contains over 25 different alkaloids of widely differing types[1].

Use: Q. extracts are used in combination with other extracts in some antitussive and antiasthmatic preparations. The bark is used in traditional South American medicine for treatment of liver disorders, febrile colds, and for anti-allergic purposes; it is recommended as a fever-lowering agent for malaria[2] as well as an additive to phytopharmaceuticals with expectorant properties[3]. Bark extracts are used industrially in South America to modify the tastes of alcoholic and non-alcoholic beverages, milk products, and for other food technological purposes. Cell suspension cultures started from shoots of the plants produce only few alkaloids, including the new alkaloid aspidochibine, with a novel skeletal type, 3-oxo-14,15-dehydrorhazinilam, 11-hydroxytubotaiwine; as well as a new dioxopiperazine[4,5].

Lit.: [1] J. Pharm. Sci. **62**, 218 (1973); Hager (5.) **4**, 401–406. [2] Ref. Farm. (Buenos Aires) **73**, 82 (1932). [3] Rimpler, Biogene Arzneistoffe, Stuttgart: Thieme 1990. [4] Pelletier **9**, 235–245. [5] Heterocycles **38**, 2411 (1994). – *[HS 1404 10, 3201 10]*

Queen bee substance see royal jelly.

Queen substance. A secretion from the mandibular glands of queen bees having various functions: Q. s. inhibits the development of ovaries in worker bees, inhibits the development of queen cells, and attracts male bees to inseminate young queen bees. It consists of five different compounds[1]: *2-(4-hydroxy-3-methoxyphenyl)ethanol* (A); *methyl 4-hydroxybenzoate* (B);

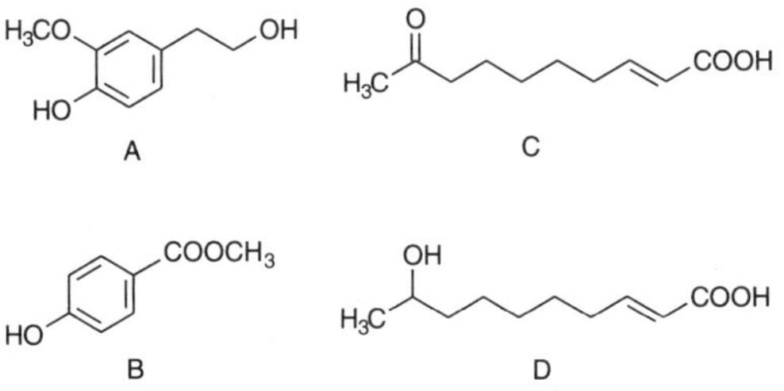

Table: Data of compounds of Queen substance.

compound	molecular formula	M_R	CAS
A	$C_9H_{12}O_3$	168.19	2380-78-1
B	$C_8H_8O_3$	152.15	99-76-3
C	$C_{10}H_{16}O_3$	184.24	334-20-3
D	$C_{10}H_{18}O_3$	186.25	1422-28-2

(E)-9-oxo-2-decenoic acid (C), and the two enantiomers of *(E)-9-hydroxy-2-decenoic acid* (D) in a ratio of 1:10:100:25:10; for data, see the table. In *Lit.*[2] only *9-oxo-2-decenoic acid* is designated as Q. s.

Lit.: [1]Nature (London) **332**, 354 ff. (1989). [2]Merck-Index (12.), No. 8214.
gen.: Apidologie **22**, 205–212 (1991); **24**, 539–548 (1993) ▪ Synth. Commun. **19**, 857–864 (1989); **21**, 1527 ff. (1991) (synthesis) ▪ Tetrahedron Lett. **32**, 7253 (1991) (synthesis).

Quercetin (3,3′,4′,5,7-pentahydroxyflavone; C. I. 75670, natural yellow 10, 13, natural red 1). Formula, see flavones, $C_{15}H_{10}O_7$, M_R 302.24; dihydrate: yellow needles, mp. 316 °C (decomp.), hydrate water is lost at 95–97 °C, LD_{50} (mouse p. o.) 160 mg/kg; poorly soluble in water, readily soluble in boiling ethanol, acetic acid, and dilute sodium hydroxide solution.
Occurrence: Very widely distributed in glycosidic form in the trunk bark of the American dyer's oak, other oak species, and Douglas fir. The pulverized, dirty yellow bark of dyer's oak is called quercitron; in addition to free Q., this contains larger amounts of *quercitrin* (Q. 3-*O*-rhamnoside, $C_{21}H_{20}O_{11}$, M_R 448.38), which forms light yellow needles (mp. 176 °C, soluble in hot water, alcohols, diuretic activity). Furthermore, Q. glycosides are particularly frequent in the bark and skins of many fruits and vegetables as well as their leaves, in yellow flowers (wallflowers, pansies). They are found predominantly in the leaves of *Adonis vernalis*, horse chestnuts, onions, chamomille, hawthorn, hops, apple trees, pansies, oaks, roses, Ericaceae, and many others. Buckthorn berries and other yellow berries contain, among others, rhamnetin (Q. 7-methyl ether) and rhamnazin (Q. 3′,7-dimethyl ether); these compounds were used in China to color the mordant-dyed silk for Mandarins' clothes. *Isoquercitrin* (Q. 3-*O*-*β*-D-glucofuranoside, $C_{21}H_{20}O_{12}$, M_R 464.38) yellow needles, mp. 225–227 °C, occurs in the flowers of cotton, horse chestnut, nasturtium, arnica, etc. *Quercimeritrin* (Q. 7-*O*-glucoside), mp. 247–249 °C, can be detected in cotton flowers and chrysanthemum leaves while *spiraeoside* (Q. 4′-*O*-glucoside), mp. 209–211 °C, occurs in the leaves of spiraea. Q. is also the aglycone of *rutin (Q. 3-*O*-rutinoside), *hyperin* (Q. 3-*O*-galactoside from Saint-John's-wort) and other *flavonoids.

Use: In the past for dying and printing of mordant-treated wool and cotton, as an antioxidant and protective agent against UV radiation, as reagent for the detection of hafnium, zinc, and tin. The activity attributed to Q. and its derivatives hesperidin (see hesperetin) and *rutin (bioflavonoids) against capillary fragility (vein remedy) is doubtful, and the authorization was revoked in 1970. Furthermore, Q. is considered to be an enzyme inhibitor, e. g., of phosphodiesterases and has mutagenic effects on microorganisms and insects; see also flavonoids.

Lit.: Beilstein E V **18/5**, 494 ▪ Hager (5.) **3**, 1024; **4**, 30 f., 59 f.,183 f., 462 f., 618 f., 727 f., 798 f., 1198 f. ▪ Karrer, No. 1522–1552, 4642–4656, 7450–7474 ▪ Merck-Index (12.), No. 5240, 8216–8219 ▪ Sax (8.), QCA 000, QCA 175 ▪ see also flavonoids. – *[HS 2932 90; CAS 117-39-5 (Q.); 21637-25-2 (isoquercitrin); 491-50-9 (quercimeritrin); 20229-56-5 (spiraeoside); 482-36-0 (hyperin)]*

Questiomycin B see 2-aminophenol.

Quillaja saponin.

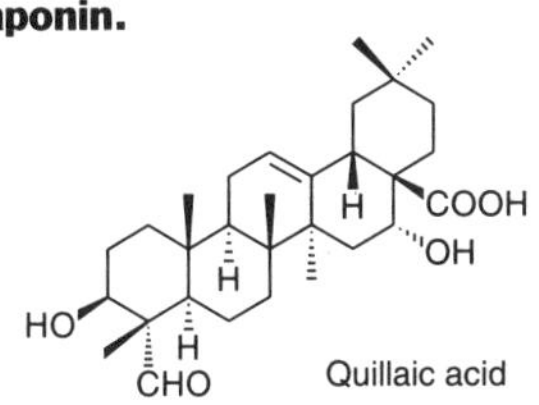

Quillaic acid

A hygroscopic powder that forms foam in water, mp. 292–294 °C (decomp.), it induces a strong sneezing response but is not authorized for use as snuff. The triterpene Q. is present in the bark of the Rosaceae *Quillaja saponaria* (Panama bark, Quillaja bark) to 9–10%; on hydrolysis it furnishes galacturonic acid, glucuronic acid, galactose, and the sapogenin quillaic acid {3*β*,16*α*-dihydroxy-23-oxoolean-12-en-28-oic acid, $C_{30}H_{46}O_5$, M_R 486.69, mp. 294 °C, soluble in alcohol, ether, acetone, glacial acetic acid, $[\alpha]_D$ +56.1° (pyridine)}. *Quillaja bark* is used in shampoos and in the film industry. Q. is used as an immune stimulant and immune adjuvant[1].

Lit.: [1]Chem. Eng. News (11.9.) **1995**, 28–35 (review); PTA heute **13**, 47 ff. (1999).
gen.: Merck-Index (12.), No. 8221 ▪ Phytochemistry **40**, 509 (1995). – *[CAS 631-01-6 (quillaic acid)]*

Quillworts (Merlin's grass) see clubmosses.

Quince flavor, quince oxepane, quince oxepine see fruit flavors (quince flavor).

Quinghaosu see qinghaosu.

Quinic acid (1*α*,3*α*,4*α*,5*β*-tetrahydroxycyclohexanecarboxylic acid).

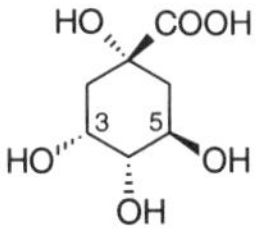

(–)-D-form

$C_7H_{12}O_6$, M_R 192.17. Monoclinic prisms. The (1*R*)-(–)-form shows $[\alpha]_D^{26}$ –42.1° (H_2O), mp. 162–163 °C, pK_a 3.58 (25 °C). At higher temperatures Q. forms a lactone. It has been isolated from numerous plants and also from the bark of *Cinchona* trees, in which it occurs together with gallic acid esters as salts of *Cin-

chona alkaloids, it also exists in coffee beans (O-5 esterified with *caffeic acid, see chlorogenic acid), tobacco leaves, sugar beets, hay, gooseberries, blackberries, and leaves of blueberries and cranberries. Dehydration leads to *shikimic acid.
Activity: For Q. esters from artichokes with cholagogic activity, see cynarin(e). Q. is used in treatment of gout. Q. is formed by reduction of 3-dehydroshikimic acid in the shikimate pathway.

Lit.: Acta Crystallogr., Sect. C **44**, 1287 ff. (1988) (crystal structure) ▪ Beilstein E IV **10**, 2257 ▪ Haslam, Shikimic Acid, Chichester: Wiley 1993 ▪ Helv. Chim. Acta **74**, 807–818 (1991) (synthesis) ▪ J. Chem. Soc., Chem. Commun. **1998**, 2033 (synthesis) ▪ Karrer, No. 989 ▪ Luckner (3.), p. 323 ff. ▪ Nat. Prod. Rep. **12**, 101–133 (1995) ▪ Tetrahedron: Asymmetry **8**, 3515–3545 (1997) ▪ Zechmeister **35**, 80–85. – *[HS 2918 19; CAS 36413-60-2 (±); 77-95-2 (–)]*

Quinidine. Formula see quinine. $C_{20}H_{24}N_2O_2$, M_R 324.42, $[\alpha]_D^{15}$ +230° ($CHCl_3$); a typical *Cinchona alkaloid, the dextrorotatory diastereomer of *quinine from cinchona bark. Crystal plates, poorly soluble in water, better soluble in alcohol, ether, chloroform, mp. 175 °C, obtained from the mother liquor from the isolation of quinine and purified by formation of salts and precipitation. It is also obtained in yields of over 40% semisynthetically from quinine.
Activity [1]***:*** Q. is in use since 1918 as cardiac agent. It blocks Na channels (inhibition of Na influx) and shows antiarrhythmic action for atrial fibrillation (antiarrythmic drug) with delayed transmission of stimuli.
Use [1]***:*** As sulfate for extrasystoles and atrial flutter, also for acute treatment. Not only Q. but also quinine are used in investigations of the mechanisms of "multiple drug resistance" of cancer cells against a series of chemotherapeutic agents [2]. For biosynthesis, see Cinchona alkaloids.

Lit.: [1] Hager (5.) **4**, 876 ff.; **7**, 829–835. [2] J. Clin. Oncol. **10**, 1730 (1992); Manske **34**, 331–398.
gen.: Beilstein E V **23/13**, 395 ▪ Kleemann-Engel, p. 1650 ff. ▪ Sax (8.), QFS 000–QHA 000 ▪ Ullmann (5.) **A 1**, 395; **A 2**, 445. – *[HS 2939 29; CAS 56-54-2]*

Quinine [(8*S*,9*R*)-6′-methoxycinchonan-9-ol].

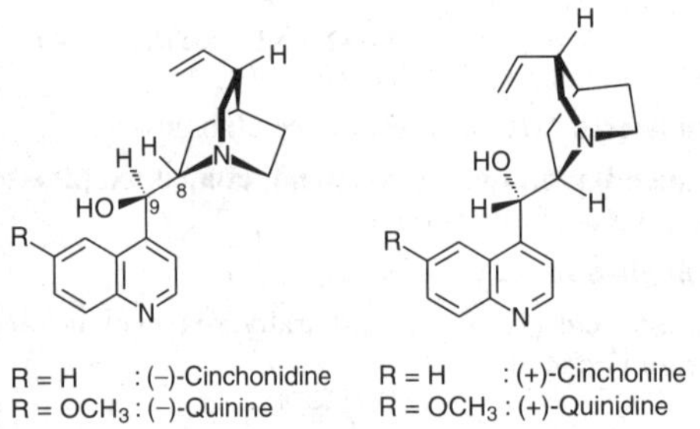

$C_{20}H_{24}N_2O_2$, M_R 324.42; cryst. bitter substance, mp. 177 °C (as trihydrate mp. 57 °C), $[\alpha]_D$ –117° ($CHCl_3$), Q. is a *quinoline alkaloid, poorly soluble in water, readily soluble in organic solvents (alcohol, chloroform). It dissolves in oxygen-containing acids with a blue fluorescence like quinidine. Like other *Cinchona alkaloids Q. forms soluble salts in plants, especially with *quinic acid and tannins, e.g., as quinine citrate. The sulfate and hydrochloride of Q. are used in therapy. Q. is also used in racemate separations. *Quinidine has the opposite configurations at C 8 and C 9. Q. occurs together with more than 30 other Cinchona alkaloids in the drug Cinchonae cortex from which Q. is isolated industrially and purified as its sulfate.
Activity [2]***:*** C. is a protoplasm toxin that acts on all cells at appropriate concentrations since it inhibits numerous enzyme systems. It forms complexes with DNA (like other Cinchona alkaloids) and thus impairs DNA/RNA synthesis. Single-celled organisms (e.g. slipper animalcules/*Paramecium* of a hay infusion) are killed within a few hours by a Q. solution in a dilution of 1 : 20 000. Leukocytes are immobilized at a dilution of 1 : 10 000. Q. destroys the schizonts of the malaria pathogen *Plasmodium falciparum*, thus interrupting the development in human blood (antimalarial drug). Q. also has antipyretic (antifebrile) and analgesic activities for symptomatic treatment of colds as well as local anesthetic and muscle relaxant actions. Q. stimulates smooth musculature, especially of the uterus, and thus induces labor. It is thus used clinically to induce labor, previously it was used to induce abortion. Q. inhibits formation of tumor necrosis factor α [3]. Q. hydrochloride has a bitter value of 200 000 and is prescribed in the German and Swiss pharmacopoeias as standard substance for determining bitter values. Appetite-stimulating beverages (e.g. tonic water) often contain Q., almost half of the commercially obtained Q. is used in the beverage industry. Q. is a stimulant for horses and has been used for doping in horse racing.
Pharmacokinetics: After oral administration Q. is rapidly and almost completely resorbed, ca. 70% are bound to plasma protein. Q. is extensively metabolized, only ca. 5% of a dose can be detected unchanged in urine. The plasma half-life is between 4 to 5 hours, maximum 12 hours. At therapeutic doses neurotoxic side effects (auditory and visual impairments, vertigo) as well as skin reddening and states of confusion (= *quininism, cinchonism*) may occur. Allergic reactions are frequent.
Toxicity: In severe cases of poisoning arrhythmias, respiratory depression, and renal failure are observed. The lethal dose for humans amounts to 5–10 g Q. (therapy: induction of vomiting, administration of charcoal, gastrolavage). Q. has been replaced as antimalarial by chloroquine, plasmochin, resochin, halofantrin. It has recently been reintroduced for treatment of falciparum malaria, especially with chloroquine-resistant strains of *Plasmodium falciparum*. In spite of its long use no malaria pathogens resistant to Q. have appeared as yet.
Historical [4]***:*** The antipyretic effect of Cinchona bark was exploited by the Indians of south America, infusions of the drug were introduced in Europe at the beginning of the 17th century. First isolation of Q. is attributed to Pelletier and Caventou (1820) or to Sertürner (1811). Pictet determined the constitution of Q. in 1912 and Woodward described the first total synthesis in 1944. Today Q. and its quasi-enantiomer quinidine play an important role in the preparation of catalysts for enantioselective aminohydroxylation according to Sharpless and other stereoselective reactions.

Lit.: [1] Florey **12**, 547. [2] Hager (5.) **4**, 876 ff.; **7**, 833–841. [3] Am. J. Respir. Cell. Mol. Biol. **10**, 514 (1994). [4] Pharm. Unserer Zeit **27**, 257–271 (1998); **28**, 11–20, 74–86 (1999).
gen.: Beilstein E V **23/13**, 395–399 ▪ Handb. Exp. Pharmacol. **68**, 61–81, 267–324 (1984) ▪ Manske **3**, 1–63; **14**, 181–223; **34**, 331–398 ▪ Sax (8.), QHJ 000–QMA 000 (toxicology) – *Synthesis:* Chem. Pharm. Bull. **30**, 1925 (1982); **31**, 1551 (1983) ▪ J. Am. Chem. Soc. **100**, 5786 (1978) ▪ Kleemann-Engel, p. 1650 ff. – *[HS 2939 21; CAS 130-95-0 (Q.); 804-63-7 (Q. sulfate)]*

Quinocarcin.

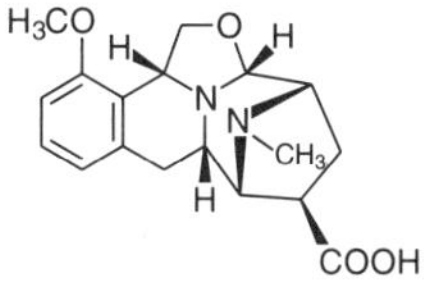

$C_{18}H_{22}N_2O_4$, M_R 330.38, cryst., decomp. above 170 °C, $[\alpha]_D$ −32° (H_2O). An *isoquinoline alkaloid with an anellated 3,8-diazabicyclo[3.2.1]octane system produced by *Streptomyces melanovinaceus*. Q. is active against Gram-positive bacteria, various tumor cell lines and is mutagenic. The alkylating activity is attributed to the iminium ion, formed by opening of the oxazolidine ring, which attacks the 2-amino groups of guanosine residues in DNA. The formation of radicals which lead to oxidative DNA strand cleavage has also been discussed.

Lit.: J. Antibiot. **36**, 463–470 (1983); **37**, 1268–1272 (1984); **44**, 1367 (1991). – *Synthesis:* J. Am. Chem. Soc. **114**, 2767 (1992); **115**, 10742 (1993) ▪ Pure Appl. Chem. **68**, 609 ff. (1996) ▪ Stud. Nat. Prod. Chem. **19E**, 289–350 (1997) ▪ Tetrahedron **50**, 6193–6258 (1994). – *Mechanism of action:* Cancer Res. **55**, 862 (1995) ▪ J. Am. Chem. Soc. **114**, 733–740 (1992) ▪ Tetrahedron **47**, 2629–2642 (1991). – *[CAS 84573-33-1]*

Quinoline alkaloids. General term for ca. 200 alkaloids with the basic skeleton of quinoline that may be altered by hydrogenation or annelation of further rings. These include not only the therapeutically important *Cinchona alkaloids *quinine and *quinidine but also *cinchonine and *cinchonidine, some furoquinoline alkaloids and acridine alkaloids as well as, in the broadest sense, also the ergot alkaloids *strychnine and *brucine containing a hydrogenated quinoline system. Q. a. occur in microorganisms, plants, and animals, in plants mainly in the Rutaceae and Rubiaceae. Q. a. also occur in beetles [1] and sponges [2] and exhibit, among others, bitter, antiseptic, convulsive, and antineoplastic properties. Quinoline is also a partial structure in redox factor PQQ (= *pyrroloquinoline quinone) in the quinoenzymes, see *Lit.*[3]. The 1-(2-quinolinyl)-β-carboline *nitramarine also belongs to the class of Q. alkaloids.
Biosynthesis: In plants the quinoline system can be formed by several different biogenetic routes (biochemical convergence). In the Cinchona alkaloids *tryptophan acts as the precursor while simple quinolines, furoquinolines, and acridines originate from *anthranilic acid. The second biosynthetic partner is often a hemiterpene (for furoquinolines) or an iridoid *monoterpene (secologanin, see secoiridoids) for Cinchona alkaloids.

Lit.: [1] Alkaloids **1**, 33 (1983). [2] J. Am. Chem. Soc. **110**, 4856 (1988). [3] Annu. Rev. Biochem. **63**, 299–344 (1994).
gen.: Alkaloids (London) **7**, 81–91 (1977); **8**, 77–86 (1978); **10**, 74–83, 141–204 (1981); **11**, 71–77 (1981); **12**, 84–93 (1982); **13**, 99–121 (1983) ▪ Manske **32**, 341 ▪ Rodd's Chem. Carbon Compd. (2.) **4** F/G, 423–482 (1999) (review). – *[HS 2939 21–2939 29]*

Quinolizidine alkaloids. General term for alkaloids containing at least one quinolizidine ring system (perhydroquinolizine) (figure).
Structure & occurrence: Q. a. occur mainly in the family of papilionaceous plants (Fabaceae). The Q. a. are so typical for the world-wide distributed genus *Lupinus* (Genisteae) that they are often called *lupine alkaloids*. Most of the over 200 known Q. a. can be classified into 6 structural types (figure, I–VI): the bicyclic *lupinine type (I, with 34 known structures); the *camoensine type (II, 6 structures); the *sparteine type (III, 66 structures), the α-pyridone type (IV, 25 structures), the *matrine type (V, 31 structures), and the *ormosanine type (VI, 19 structures). The various structural types are not uniformly distributed among members of the Fabaceae containing Q. a.[1]. Thus, *sparteine and its many derivatives (type III), which also include *lupanine, *13-hydroxylupanine and its esters as well as partial degradation products such as the tricyclic *angustifoline, as well as α-pyridone alkaloids (type IV), such as *anagyrine and *cytisine are found in most genera of the Q. a-containing tribes. Also occurring in practically all tribes but less widely distributed are *lupinine and its derivatives (type I). The occurrence of matrines (type V) and camoensines (type II) is restricted to the tribe Sophoreae, the ormosanines (type VI) are characteristic, above all, for the genus *Ormosia*. In addition, some dimeric Q. a. have been described. Structurally related Q. a. but differently built and occurring in other taxa are discussed as separate classes, see Lycopodium alkaloids, Lythraceae alkaloids, and Nuphar alkaloids.

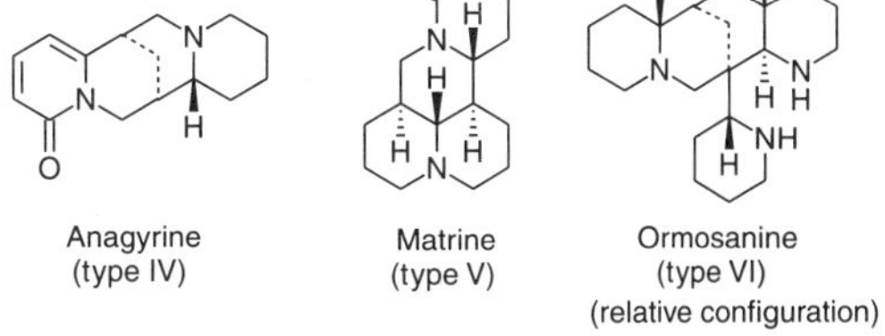

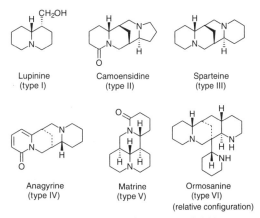

Figure: Structural types of the Quinolizidine alkaloids.

Biosynthesis & physiology: The Q. a. are built up of two to four *cadaverine units formed from the amino acid L-lysine by decarboxylation. The biosynthesis of Q. a of types I and III has been well investigated[2,3]. In lupines the biosynthesis of Q. a. occurs in the chloroplasts of the leaves[3]. The Q. a. are transported in the phloem to their storage sites: flowers, seeds, and epidermal tissue[4] and accumulate intracellularly in the vacuole[5].

Chemical ecology[6]***:*** The Q. a. constitute a part of the chemical defence of plants. They have a bitter taste for humans, act as antifeedants, and are in part toxic for herbivores. Some Q. a. inhibit the growth of plant pathogenic viruses as well as many bacteria and fungi.

Toxicity & pharmacology: Q. a. exhibit numerous actions on warm-blooded organisms. Poisonings of grazing animals (sheep and cattle) by Q. a.-containing plants have been known for a long time. The α-pyridone alkaloids such as *cytisine and *anagyrine are considerably more toxic than the ring-saturated Q. a. of the sparteine type. Toxic symptoms are excitedness, coordination impairments, respiratory distress, cramps, and finally coma leading to death by respiratory paralysis. Anagyrine and other α-pyridone alkaloids as well as *ammodendrine which often occurs together with Q. a. are teratogenic and cause the so-called "crooked calf disease" in cattle, which is discernible as a deformation of the spine and extremities in new born animals[7]. Q. a. poisonings in humans are rare except for laburnum (golden-chain). The only therapeutically used Q. a. is *sparteine.

Lit.: [1] Harborne, Boulter & Turner (eds.), Chemotaxonomy of the Leguminosae, p. 73–178, London: Academic Press 1971. [2] J. Chem. Soc., Perkin Trans. 1 **1986**, 1133; Can. J. Chem. **63**, 2707 (1985). [3] Plant Physiol. **70**, 74 (1982). [4] Planta **161**, 519 (1984). [5] Z. Naturforsch., C **41**, 375 (1986); J. Plant Physiol. **129**, 229 (1987). [6] ACS Symp. Ser. **330**, 524 (1987); Rosenthal & Berenbaum (eds.), Herbivores: their Interactions with Secondary Plant Metabolites, vol. 1, p. 79–121, San Diego: Academic Press 1991; Theor. Appl. Genet. **75**, 225 (1988). [7] Teratology **40**, 423 (1989).

gen.: Alkaloids (New York) **28**, 183–308 (1986) ▪ Bernays (ed.), Focus on Plant Insect Interactions IV, p. 131–166, Boca Raton: CRC Press 1992 ▪ Magn. Res. Chem. **36**, 779–796 (1998) (^{13}C-NMR, review) ▪ Nat. Prod. Rep. **14**, 605–618, 619–636 (1997); **16**, 675, 697 (1999) ▪ Pelletier **2**, 105–148 ▪ Waterman **8**, 197–239. – *Synthesis:* Chem. Pharm. Bull. **34**, 2018 (1986) ▪ Manske **47**, 2–115 (review of lupinane alkaloids) ▪ Nat. Prod. Rep. **11**, 639 (1994) ▪ Tetrahedron Lett. **34**, 215 (1993) (lupinine). – *Isolation:* Phytochemistry **50**, 189 (1999). – *[HS 2939 90]*

Quinones see anthraquinones, benzoquinones, and naphthoquinones.

Quisqualic acid [(*S*)-α-amino-3,5-dioxo-1,2,4-oxadiazolidine-2-propanoic acid].

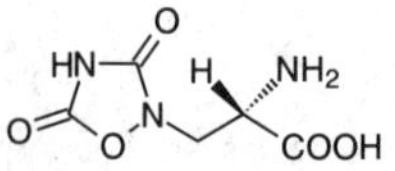

$C_5H_7N_3O_5$, M_R 189.13, mp. 190–191 °C (decomp.), $[\alpha]_D^{20}$ +17.0° (6 m HCl). A neurotoxic, non-proteinogenic amino acid from the seeds of *Quisqualis* spp. such as *Q. fructus*[1], *Q. chinensis*, *Q. indica*. Q. is used in traditional Chinese medicine for its anthelmintic activity. Q. acts in the brain as a so-called excitatory amino acid[2,3] and agonist of the neurotransmitter L-glutamate. The biosynthesis proceeds from 1,2,4-oxadiazolidine-3,5-dione and *O*-acetylserine. For synthesis, see *Lit.*[3]. For mechanisms of action, see *Lit.*[4,5].

Lit.: [1] Yakugaku Zasshi **95**, 176, 326 (1975). [2] Neuropharmacology **13**, 665–672 (1974). [3] Chem. Pharm. Bull. **34**, 1473 (1986); J. Chem. Soc., Chem. Commun. **1984**, 1156 f.; **1985**, 256 f.; Tetrahedron: Asymmetry **4**, 2041 (1993); Tetrahedron Lett. **37**, 5225 (1996). [4] Chem. Pharm. Bull. **22**, 473 ff. (1974). [5] Life Sci. **37**, 1373–1379 (1985); J. Pharmacol. Exp. Ther. **233**, 254–263 (1985). – *[CAS 52809-07-1]*

R

Racemic acid see tartaric acid.

Racemomycins see streptothricins.

Rachelmycin see CC-1065.

Radicicol (monorden).

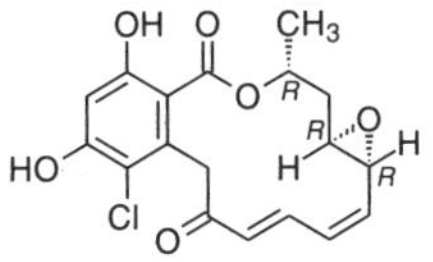

$C_{18}H_{17}ClO_6$, M_R 364.78, prisms, mp. 195 °C, $[\alpha]_D^{20}$ +216° ($CHCl_3$). An antineoplastic and fungicidal macrolide antibiotic from various lower fungi, e. g., *Nectria radicicola*. R. inhibits protein kinases and angiogenesis. Derivatives of R. (dimethyl ether and diacetyl-R.) have similar activities.

Lit.: J. Biol. Chem. **270**, 5418 (1995) ▪ Tetrahedron Lett. **33**, 773, 777 (1992). – *[CAS 12772-57-5]*

Radiosumin.

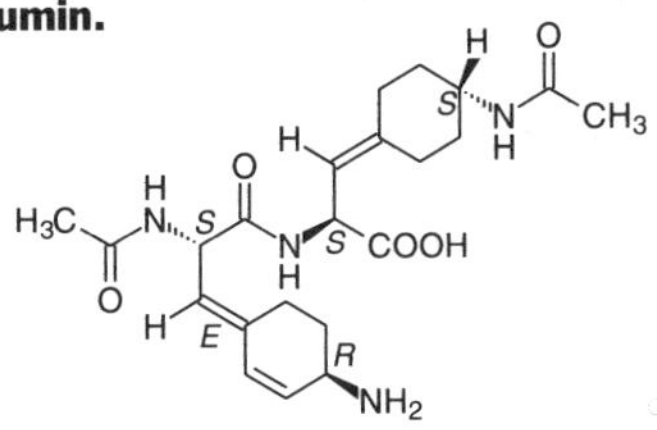

$C_{22}H_{32}N_4O_5$, M_R 432.51, $[\alpha]_D^{20}$ +96° (H_2O), a dipeptide from cultures of the fresh water microalga *Plectonema radiosum*. R. strongly inhibits trypsin.

Lit.: J. Org. Chem. **61**, 8648 (1996) ▪ Tetrahedron Lett. **38**, 2883 (1997). – *[CAS 157744-22-4]*

Raffinose [α-D-galactopyranosyl-(1-6)-α-D-glucopyranosyl-(1-2)-β-D-fructofuranoside, melitriose, gossypose, melitose].

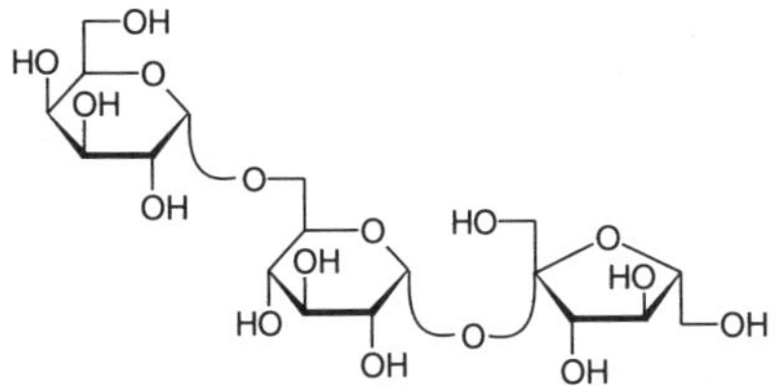

$C_{18}H_{32}O_{16}$, M_R 504.46, prisms. R. pentahydrate: mp. 80 °C, $[\alpha]_D^{20}$ +104° (H_2O); R. (anhydrous): mp. 118–119 °C (decomp.), $[\alpha]_D^{20}$ +123° (H_2O). R. does not reduce Fehling's solution, is stable to alkali, and does not form an osazone. R. is a trisaccharide made up of 1 mol. each of *galactose, *glucose, and *fructose; it is cleaved by dilute acids and invertase to fructose and melibiose, by emulsin and α-galactosidases to saccharose and galactose, and by strong acids to all three monosaccharides. R. is probably the second most abundant oligosaccharide, after saccharose, in the plant kingdom and occurs in sugar beet molasses (<2%), cotton seeds, and *Eucalyptus* manna, but not in sugar cane.

Lit.: Adv. Carbohydr. Chem. Biochem. **33**, 235–294 (1976) ▪ Beilstein E V **17/8**, 403 ▪ Dev. Food Carbohydr. **2**, 145–185 (1980) ▪ Karrer, No. 671 ▪ Merck-Index (12.), No. 8279 ▪ Ullmann (5.) **A 5**, 83, 86; **A14**, 34. – *[HS 1702 90, 2940 00; CAS 512-69-6]*

Rangiformic acid [(2*S*,3*S*)-1,2,3-heptadecanetrioic acid 1-methyl ester].

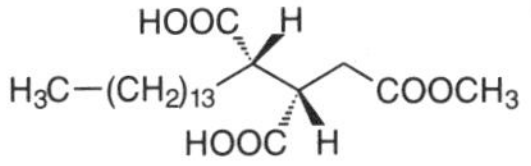

$C_{21}H_{38}O_6$, M_R 386.53, needles, mp. 104–105 °C, $[\alpha]_D^{20}$ +22.7° (C_2H_5OH). An aliphatic carboxylic acid from the fruticose lichen *Cladonia rangiformis*; the isomeric iso-R., platelets, mp. 83–85 °C, $[\alpha]_D^{24}$ +5.2°; occurs in the crustaceous lichen *Lecanora stenotropa*.

Lit.: Culberson, p. 107–108 ▪ Karrer, No. 880 ▪ Phytochemistry **21**, 2407 f. (1982); **47**, 1649 (1998). – *[CAS 29227-63-2 (R.); 85769-30-8 (iso-R.)]*

Rapamycin (Sirolimus®, Rapamune).

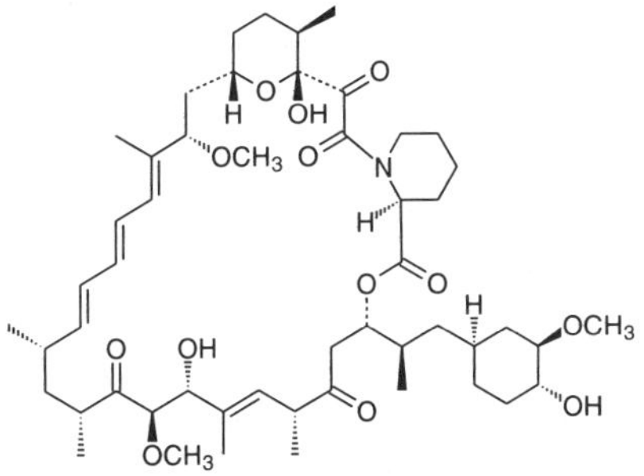

$C_{51}H_{79}NO_{13}$, M_R 914.19, cryst., mp. 183–185 °C, $[\alpha]_D$ –58.2° (CH_3OH), a 31-membered peptide lactone in which a long-chain carboxylic acid is cyclized with L-*pipecolic acid as a bridge and produced by *Streptomyces hygroscopicus*. R. has antifungal, antineoplastic, and immunosuppressive activities, it is structurally related to *FK-506 and exhibits a similar mechanism of action in the development of the immune response. R. was first marketed in 1999 in combination with *cyclosporin and tacrolimus for use in transplantation medicine. The first total synthesis of R. was realized in 1993.

Lit.: *Isolation:* J. Antibiot. **36**, 351 (1983). – *Biosynthesis:* Chem. Rev. **97**, 2611–2629 (1997). – *Structure:* Helv. Chim. Acta **76**, 117–130 (1993) ▪ J. Antibiot. **44**, 688 (1991). – *Synthesis:* Chem. Eng. News (24.5.) **1993**, 28 f. ▪ Chem. Commun. **1997**, 1919 ▪ Chem. Eur. J. **1**, 318–333 (1995) ▪ Contemp. Org. Synth. **3**, 345–371 (1996) ▪ Curr. Org. Chem. **2**, 281–328 (1998) ▪ J. Am. Chem. Soc. **115**, 4419, 7906, 9345 (1993); **117**, 5407 (1995); **119**, 947, 962 (1997) ▪ Nicolaou, p. 563–631. – *Activity:* Angew. Chem. Int. Ed. Engl. **31**, 384 (1992); **33**, 1437–1452 (1994) ▪ Annu. Rep. Med. Chem. **26**, 211 (1991) ▪ J. Antibiot. **51**, 487 (1998) ▪ Nature Medicine **1**, 32 (1995) ▪ Science **251**, 283 (1991); **273**, 239 (1996). – *[CAS 53123-88-9]*

Rapeseed oil (turnip oil). The oil obtained from the seeds of rape (*Brassica napus*) or turnips [*B. rapa* (*B. campestris*) subsp. *oleifera*, Cruciferae]. Rapeseed contains ca. 40–50% oil and ca. 30% protein. Worldwide production (1992): 9400 00 t; in Germany 1 060 000 t; main producing countries: China, India, Germany and other EU countries, Canada. Mp. 0–2 °C. *Old* rape species contain 35–64% *erucic acid, 5–10% gondoic acid [(*Z*)-11-eicosenoic acid, $C_{20}H_{38}O_2$, M_R 310.52, mp. 24–25 °C], and nervonic acid together with 13–38% oleic acid, 10–22% linoleic acid, and 2–10% linolenic acid. Because of the hazardous activities of erucic, gondoic, and nervonic acids on cardiac musculature R. from *old* species (high erucic rapeseed, HEAR) may only be used as a technical oil[1]. *New* rape species usually contain <2% long-chain monoene fatty acids (≥20 C atoms) and furnish oils of high nutritional and physiological value[2]. Recently, the cultivation of new species has resulted in the so-called double zero or "00" species (low-erucic acid rapeseed, LEAR; "Canola") containing only traces of erucic acid and the also toxic *glucosinolates together with large amounts of oleic acid (50–65%), linoleic acid (15–30%), and linolenic acid (5–13%).

Use: Especially in foods (cooking oil, salad oil); partially hardened as margarine, frying and baking fat; smaller amounts for technical purposes (chain-saw oils, drilling oils, etc.) and raw material for oleochemistry; fatty acid methyl esters from R. (so-called rapeseed oil methyl ester, RME) as "biodiesel" fuel.

Lit.: [1] Erucasäure-VO dated 24.5.1977 (BGBl. I, p. 782) vers. dated 26.10.1982 (BGBl. I, p. 1446). [2] Raps **11**, 137 (1993). *gen.:* Kramer, Sauer, & Pigden, High and Low Erucic Rapeseed Oils, New York: Academic Press 1983. – *[HS 1514 10, 1514 90; CAS 8002-13-9 (R.); 5561-99-9 (gondoic acid)]*

Raphanin see mustard oils.

Raspberry flavor see fruit flavors.

Raspberry ketone see zingerone.

Raubasine see ajmalicine.

Raucaffricine (vomilenine 21-*O*-*β*-D-glucopyranoside).

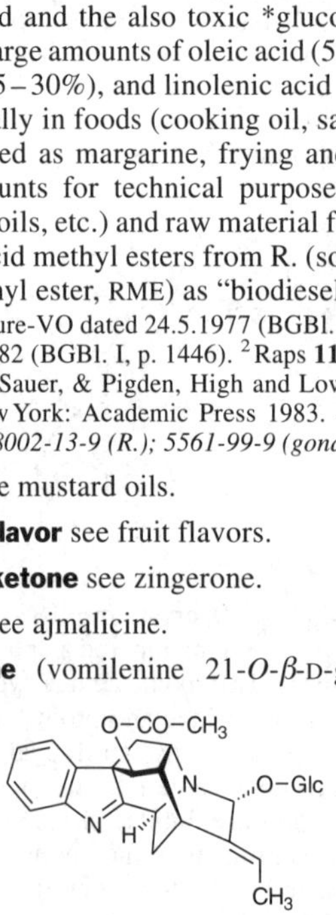

$C_{27}H_{32}N_2O_8$, M_R 512.56, readily soluble in water. An indole alkaloid, $[\alpha]_D$ +14.5° (C_2H_5OH), one of the rare glucoalkaloids, occurring exclusively in a few *Rauvolfia* species, e. g., *R. caffra*, *R. serpentina*. It is the main alkaloid in cell suspension cultures of *Rauvolfia* species, it is easily cleaved into its aglycone, vomilenine, by endogenous glucosidases when the plant material is abraded and thus is only rarely found in dried plant material.

Biosynthesis: R. is derived from vomilenine, an intermediate in the biosynthetic pathway to the antiarrhythmic *ajmaline, by glucosylation. The corresponding glucosyl transferase is membrane-bound, dependent on uridine diphosphate glucose, and has a high substrate specificity. A soluble, substrate-specific glucosidase effects the reversal of this reaction; see also Rauvolfia alkaloids.

Lit.: Helv. Chim. Acta **67**, 2078 (1984); **69**, 538 (1986); **74**, 1707 (1991) ▪ Phytochemistry **28**, 491 (1989). – *[HS 2939 90; CAS 31282-07-2]*

Rauvolfia alkaloids (Rauwolfia alkaloids). Collective name for more than 100 different penta- and hexacyclic *monoterpenoid indole alkaloids isolated from the dried roots and root bark of several one-year-old *Rauvolfia* plants (Apocynaceae) of commercial interest. Of the ca. 40 *Rauvolfia* species only the Indian *Rauvolfia serpentina*, the African *R. vomitoria*, and the Central American *R. tetraphylla* are economically important. While the Indian species is a small, shrub-like plant growing up to a height of ca. 80 cm, the *R. vomitoria* plants are small trees and are easily cultivated in plantations in Africa. The most important R. a. are *ajmaline, *ajmalicine, *reserpine, serpentine, *yohimbine, deserpidine (see reserpine), rescinnamine (Rauwopur®, Moderil®), corynanthine (see yohimbine), which all have therapeutic uses, as well as *sarpagine, reserpic acid, reserpiline, *raucaffricine as one of the few glucoalkaloids in *Rauvolfia*, raufloridine, sandwicoline, sandwicolidine, and vomilenine.

Pharmacology: All R. a. have a general peripheral vasodilatory activity which leads to a reduction of blood pressure. The pure alkaloids as well as *Rauvolfia* extracts are thus of interest as antihypertonic agents (see reserpine and ajmalicine). The R. a. are classified into groups according to their pharmacological activities: alkaloids of the reserpine group have sedative as well as hypotonic activities. They attack the membrane of catecholamine-storing vesicles at adrenergic and dopaminergic nerve endings, where they inhibit the uptake of neurotransmitters into the storage vesicles. This results in a decrease in the concentration of noradrenaline in peripheral and central neurons which, in turn, leads to a long-lasting reduction of the blood pressure. On account of the blood pressure lowering properties of the second group, the serpentine alkaloids, serpentine is often combined with reserpine. The ajmaline group forms a further class of R. a. on account of their *quinidine-like activities. In general *Rauvolfia* total extracts are used in the therapy of high blood pressure and as sedatives; combination preparations with, e. g., *Crataegus* extracts, *Convallaria* glycosides, rutin, or sedatives, e. g., barbituric acid preparations, are also used. The antiarrhythmic activity of the R. a. ajmaline is also exploited therapeutically.

Biosynthesis: The biosynthesis of the R. a. is well understood, especially with regard to the formation of the ajmaline and sarpagine types because it was possible to establish well-growing, alkaloid-producing cell suspension cultures. More than 10 soluble, membrane-bound enzymes participate in the conversion of tryptamine (see indolylalkylamines) and secologanin (see secoiridoids) first to the sarpagan type of R. a. and then to alkaloids of the ajmaline group. In this process, and presumably for all *monoterpenoid indole alkaloids, the enzyme strictosidine synthase and its product – the glucoalkaloid *strictosidine – play a key role.

Historical: *R. serpentina*, the most important of the *Rauvolfia* species, was used as a medicinal plant for therapy in traditional medicine in pre-Vedantic times (ca. 3000 years ago): as treatment for snake bites, febrile colds, and general feelings of illness. In the subsequent Ayurvedic era the drug, then known as "sarpagandha root" was employed systematically; *R. serpentina* was one of the typical drugs in Ayurvedic hospitals long before it was introduced from Arabian countries to Europe by the German botanist and physician Leonhard Rauwolf after whom it was named.

Lit.: Hager (5.) **3**, 31 f.; **6**, 361–380; **9**, 495–502 ▪ Handb. Exp. Pharmacol. **19**, 529–592 (1966); **39**, 77–159 (1977); **55**, 43–58 (1980) ▪ J. Sci. Ind. Res. **27**, 58 (1968) ▪ Kleemann-Engel, p. 1667 ▪ Manske **7**, 62; **8**, 287, 785; **11**, 41; **47**, 115–172 (biosynthesis) ▪ Zechmeister **10**, 390; **13**, 346; **43**, 267–346. – *[HS 2939 90]*

Rauwolfine see ajmaline.

Rebeccamycin.

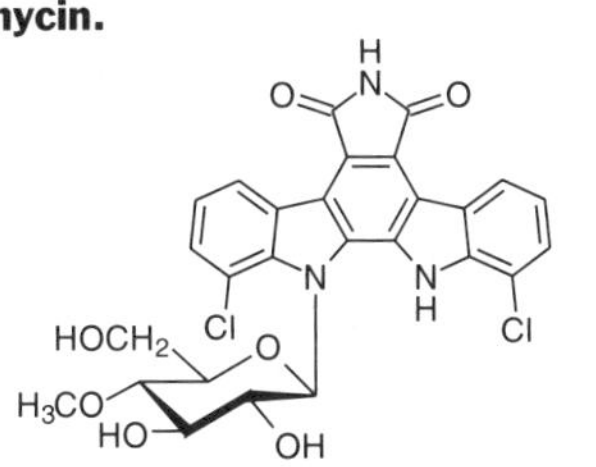

$C_{27}H_{21}Cl_2N_3O_7$, M_R 570.38, mp. 327–330 °C, $[\alpha]_D^{21}$ +137.7° (tetrahydrofuran); isolated from the Ascomycete *Nocardia aerocolonigenes*. R. is composed of a symmetrical indolocarbazole chromophore, to which a 4-*O*-methylglucose unit is linked via a *β*-glycosidic bond. It has potent antitumor activity and acts by causing DNA strand breakage in human pulmonary adenocarcinoma cells. Water-soluble derivatives are undergoing clinical testing. R. is structurally related to *staurosporine and arcyriaflavins (see Arcyria pigments).

Lit.: J. Antibiot. **47**, 792–798 (1994) ▪ J. Med. Chem. **41**, 1631 (1998); **42**, 584, 1816 (1999) (analogs) ▪ J. Org. Chem. **58**, 343–349 (1993); **64**, 2465 (1999) (synthesis) ▪ Tetrahedron Lett. **26**, 4015–4018 (1985). – *[CAS 93908-02-2]*

Red currant flavor see fruit flavors.

Red tide. Name for red colored algal blooms, sometimes used generally for all types of algal blooms. Red-colored algal blooms are caused principally by dinoflagellates (= Pyrrhophyceae), especially those of the genera *Alexandrium*[1] (formerly *Gonyaulax*), *Gymnodinium*[2] (= *Ptychodiscus*), *Ceratium*[3], *Prorocentrum*[1,2,4], and *Noctiluca*[5], less frequently by other algae[4]. The red color of the dinoflagellates in the autotrophic forms is due to *β-carotene as well as *peridinin and other xanthophylls, while in the heterotrophic dinoflagellates it is due to *carotinoids taken up with food[6,7]. R. t., first described in The Bible (Exodus 7, 19–21), have been observed in all oceans[7] and are often toxic for humans and marine organisms[7]. Besides the relatively non-toxic R. t. there are also those that contain *saxitoxin or other not exactly characterized toxins.

Lit.: [1] Marine Biol. **103**, 365 (1989). [2] Taylor & Seliger (eds.), Toxic Dinoflagellate Blooms, p. 139, 215, New York: Elsevier 1979. [3] Helgol. Wiss. Meeresunters. **36**, 393–426 (1983). [4] Bull. Inst. R. Sci. Nat. Belg. Biol. **48** (13), 1 (1972). [5] Meeresforsch. **32**, 77–91 (1988). [6] Br. Phycol. J. **25**, 157–168 (1990). [7] Neth. J. Sea Res. **25**, 101–122 (1990).

gen.: Nagashima et al., Red Tides: Proc. 1st Int. Symp. 1987, New York: Elsevier 1989 ▪ Science **257**, 1476 (1992).

Relomycin see tylosin.

Renghol see urushiol(s).

Renieramycin A see ecteinascidins.

Renierol.

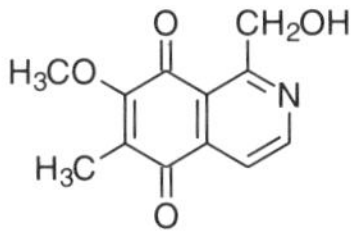

Renierol

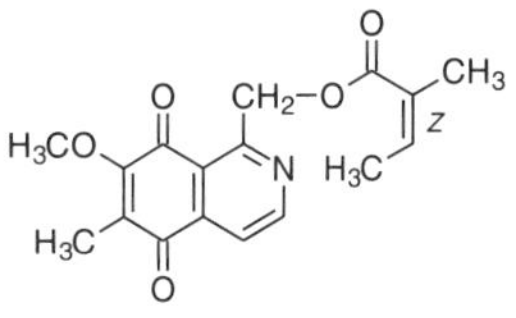

Renierone

$C_{12}H_{11}NO_4$, M_R 233.22, red solid. R. and its angeloyl ester *renierone* occur in Pacific sponges (*Xestospongia, Reniera*) and have broad antibiotic activities.

Lit.: Chem. Pharm. Bull **37**, 1493 (1989); **47**, 1805 (1999) ▪ J. Nat. Prod. **50**, 754 (1987). – *[CAS 77640-19-8 (R.); 73777-65-8 (renierone)]*

Repandiol.

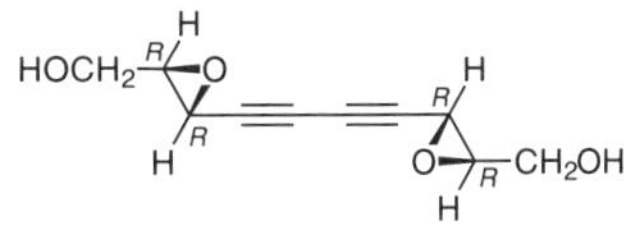

$C_{10}H_{10}O_4$, M_R 194.19, platelets, mp. 168–169 °C, $[\alpha]_D$ +40° (CH_3OH). A cytotoxic component of the fungus *Hydnum repandum* (Basidiomycetes). R. has inhibitory activity against various tumor cell lines, e.g., colon adenocarcinoma cells, this is explained in terms of a double alkylation by the activated diepoxide.

Lit.: Chem. Pharm. Bull. **40**, 3181 (1992). – *[CAS 147921-90-2]*

Repellents. Term derived from Latin: repellere = to drive back, to refute, for compounds that have a defensive or repelling effect on other creatures, pests or predators. Many of the mostly synthetically prepared agents have such an unpleasant smell and taste that animals keep away from, e.g., food sources or certain places.

For many lower and higher plants and animals the development of specific R. is part of the survival strategy. Thus, for example, mollusks produce furanosesquiterpenes as antifeedants against coral fish while a particular fish living in the Pacific Ocean secretes steroid aminoglycosides as protection against sharks (*squalamines). Some mustelids: martens, and in particular *skunks use their anal gland secretions not only to mark their territories but also to drive off other animals. Some animals take up toxic substances with their food (pharmacophagy) and convert them to R. Plants also have a defensive strategy, not only against other plants [allelopathics (see allelopathy), phytoaggressins] but also against feeding animals involving secondary plant substances such as *phytoncides, toxic *alkaloids, *glucosinolates, imitations of *insect hormones (*precocenes), and *pheromones as defensive compounds, see also allomones and defensive secretions.

Lit.: Agosta, Bombardier Beetles and Fever Trees, New York: Addison-Wesley 1996 ▪ Harborne, Ökologische Biochemie, chap. 7 & 8, Heidelberg: Spektrum Akadem. Verl. 1995 ▪ Jacobs & Renner, Biologie u. Ökologie der Insekten, Stuttgart: Fischer 1988 ▪ J. Chem. Ecol. **10**, 1151 (1984) ▪ Schwenke, Die Forstschädlinge Europas, vol. 5, Hamburg: Parey 1986.

Reserpine.

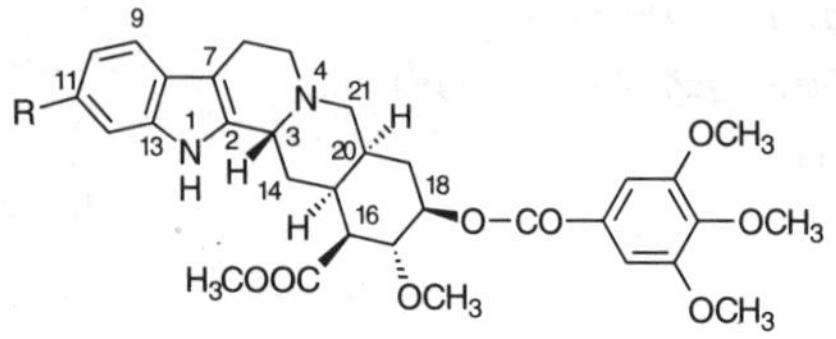

R = H : Deserpidine
R = OCH_3 : Reserpine

International free name for methyl 18-*O*-(3,4,5-trimethoxybenzoyl)-reserpate, $C_{33}H_{40}N_2O_9$, M_R 608.69; prismatic, cryst., decomp. at 265 °C, soluble in chloroform, poorly soluble in water, LD_{50} (mouse i. v.) 20 mg/kg; (dog i. v.) 0.5 mg/kg; (rat p. o.) 420 mg/kg. R. is a *Rauvolfia alkaloid (*indole alkaloids, *monoterpenoid indole alkaloids) from the roots of the Indian *Rauvolfia serpentina*, the Mexican *R. heterophylla*, and the Australian *Alstonia constricta*.

R. has sedative and blood pressure-lowering properties and is thus used in *Rauvolfia* preparations for hypertension; it also has favorable properties for treatment of hyperthyroidism. Not only the *β*-carboline group but also the trimethoxybenzoate group are essential for the biological activity.

The application of R. leads to a reduction in the levels of noradrenaline, dopamine, and serotonin in the synapses [1]. R. binds irreversibly to the transport system of noradrenaline (NA) so that the re-uptake of NA in the vesicles of the synapses is inhibited. Thus, transmission of stimuli is reduced, the blood pressure decreases with a concomitant central sedation. R. thus acts as a neuroleptic agent. The side effects of an overdose may include diarrhea, elevated secretion of saliva and gastric juices, depression, and Parkinson's disease [2]. R. is suspected of being carcinogenic [3,4] since in female patients an increased incidence of breast cancer seems to occur. The roots of *R. serpentina* have been used in India for centuries as a sedative; they also contain *yohimbine. The first total synthesis was achieved by Woodward [5], for stereospecific synthesis, see *Lit.* [6].

Deserpidine {11-demethoxyreserpine, canescine, recanescine, $C_{32}H_{38}N_2O_8$, M_R 578.64, cryst., mp. 228–230 °C, $[\alpha]_D$ –163° (pyridine)} is also a *Rauvolfia* alkaloid; it exists in polymorphic forms and also lowers the blood pressure.

Lit.: [1] Kleemann-Engel, p. 1668. [2] Prakt. Arzt **16**, 109–115 (1979). [3] Fourth Annual Report Carcinogens (NTP 85-002), 117 (1985); J. Am. Coll. Toxicol. **7**, 95–109 (1988). [4] Saxton (ed.), The Monoterpenoid Indole Alkaloids, Suppl. vol. 25, Heterocyclic Compounds, p. 724–725, Chichester: Wiley 1994. [5] J. Am. Chem. Soc. **78**, 2023 (1956). [6] Chem. Commun. **1996**, 1537; J. Org. Chem. **62**, 465 (1997); Pure Appl. Chem. **61**, 439–442 (1989).

gen.: Beilstein E V **25/7**, 42 f. (deserpidine), 104 f. (R.) ▪ Hager (5.) **6**, 361–380 (R.); **7**, 1202 (deserpidine); **9**, 499 f. (R.) ▪ Martindale (29.), p. 498 ▪ Merck-Index (12.), No. 8314 ▪ Saxton (ed.), The Monoterpenoid Indole Alkaloids, Suppl. vol. 25, Heterocyclic Compounds, p. 161, Chichester: Wiley 1994 ▪ Synform **3**, 204–224 (1985) ▪ Toxicol. Pathol. **8**, 1–21 (1980) ▪ Ullmann (5.) **A 1**, 388, 389; **A4**, 244 f. – *[HS 2939 90; CAS 50-55-5 (R.); 131-01-1 (deserpidine)]*

Resiniferatoxin (RTX).

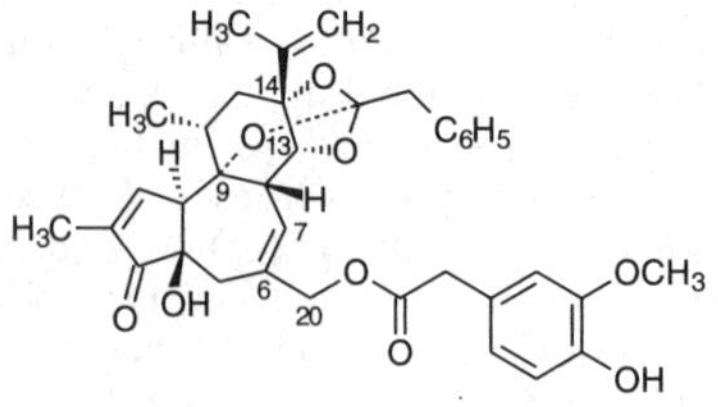

$C_{37}H_{40}O_9$, M_R 628.72, glass-like resin. A diterpene ester of *resiniferonol. R. is a component of many Euphorbiaceae, e. g. *Euphorbia resinifera* and Thymelaeaceae. It is an extremely powerful skin irritant. In contrast to the structurally related phorbol esters, R. is practically inactive in standard tests for tumor promoters [1]. With regard to its pharmacological activity R. is an extremely potent analogue of *capsaicin [2], the hot principle of red pepper (active in picomolar concentrations). The homovanillinic acid unit in the C-20 position is essential for the pain-relieving activity of the so-called vanilloids like *capsaicin, *scutigeral, *zingerone or R. [3], while the tumor promoting properties of phorbol esters are associated with a free OH group at the C-20 position. The RTX containing latex is described in medical literature dating back to the first century as a nose- and skin-irritating agent as well as a treatment for chronic pain. The powdered latex was used by the high society for jokes, e. g. inducing general sneezing of people, gathered for a ball. In pain therapy, unlike treatment with capsaicin, patients experience only itching and mild discomfort. R. is in clinical development for the treatment of neuropathic pain and against bladder incontinence [3,4].

Lit.: [1] J. Cancer Res. Clin. Oncol. **108**, 98 (1984). [2] Mol. Brain Res. **35**, 173 (1996); Neuroscience **30**, 515 (1989). [3] J. Nat. Prod. **45**, 347 (1982); J. Am. Chem. Soc. **119**, 7897, 12976 (1999); Lancet **350**, 640 (1997); Life Sci. **60**, 681 (1997); Z. Naturforsch. B **48**, 364 (1993). [4] Drug News (10.5.) **1999**, 3.

gen.: Chem. Eng. News (26.10.) **1992**, 26–32; (4.3.) **1996**, 30 f. ▪ J. Chem. Soc., Chem. Commun. **1991**, 215 ▪ J. Pharm. Pharmacol. **32**, 373 (1980); **44**, 361 (1992) ▪ Phytother. Res. **3**, 253 (1989). – *Synthesis:* J. Am. Chem. Soc. **119**, 12976 (1997). – *[CAS 57444-62-9]*

Resiniferonol.

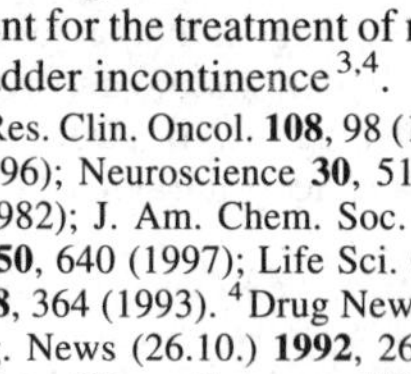

$C_{20}H_{28}O_6$, M_R 364.44. A diterpene of the daphnane type. Various esters and orthoesters of R. are found in many Euphorbiaceae and Thymelaeaceae, e.g., *resiniferatoxin, proresiniferatoxin, and tinyatoxin (demethoxyresiniferatoxin). The above-mentioned compounds are strong skin irritants.

Lit.: J. Nat. Prod. **45**, 347 (1982) ▪ Phytochemistry **15**, 333, 1778 (1976) ▪ Tetrahedron Lett. **1975**, 1595 ▪ Z. Naturforsch., B **48**, 364 (1993). – *[CAS 57444-60-7]*

Resistance inductors. R. i. are compounds which initiate the biosynthesis of *phytoalexins or, respectively, of *PR proteins. Besides certain synthetic compounds [1–3], some low-molecular weight natural compounds such as *salicylic acid [4,5], fatty acids [6], *methyl jasmonate [7,8] as well as the peptide systemin [9] act as resistance inductors in plants and initiate the plant's defensive mechanism against phytopathogens.

Lit.: [1] Métraux et al., Induced Resistance in Cucumber in Response to 2,6-Dichloroisonicotinic Acid and Pathogens, in Hennecke & Verma (eds.), Adv. Mol. Genet. Plant-Microb. Inter. vol. 1, p. 432–439, Dordrecht: Kluwer 1991. [2] J. Pesticide Sci. **15**, 199–203 (1990). [3] Plant Physiol. **106**, 1269–1277 (1994). [4] Science **250**, 1004 ff. (1990). [5] Plant Physiol. **99**, 799–803 (1992). [6] Science **212**, 67 ff. (1981). [7] Trends Cell Biol. **2**, 236–241 (1992). [8] Plant Physiol. **44**, 569–589 (1993). [9] Science **255**, 1570–1573 (1992).

gen.: Biotechnology **10**, 1436–1445 (1992) ▪ Cell **70**, 879–886 (1992).

Resistoflavine see resistomycin.

Resistomycin.

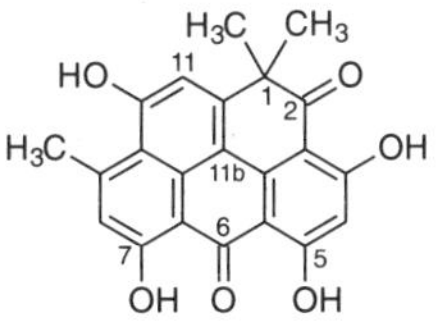

Resistomycin

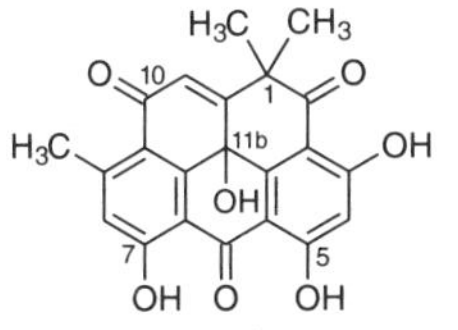

Resistoflavine

$C_{22}H_{16}O_6$, M_R 376.37, yellow needles (dioxan), mp. 315 °C (decomp.), a naphthanthrone derivative from *streptomycetes (e.g., *Streptomyces resistomycificus*). R. is active against Gram-positive bacteria and mycobacteria [LD_{50} (mouse p.o.) 2000 mg/kg], it is an inhibitor of RNA and DNA polymerases as well as HIV-1 protease. A natural derivative of R. is *resistoflavine* {$C_{22}H_{16}O_7$, M_R 392.36, mp. 238–240 °C, $[\alpha]_D$ −96° (pyridine)} possessing a hydroxy group at C-11b and an oxo group at C-10.

Lit.: J. Am. Chem. Soc. **107**, 3879–3884 (1985) (synthesis) ▪ J. Antibiot. **38**, 113 ff. (1985); **47**, 136–142 (1994) ▪ Justus Liebigs Ann. Chem. **1983**, 835. – *[CAS 20004-62-0 (R.); 29706-96-5 (resistoflavine)]*

Resorcylic acids see dihydroxybenzoic acids.

Resveratrol see hydroxystilbenes and wine, phenolic constituents of.

Retamine.

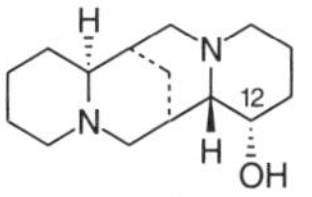

$C_{15}H_{26}N_2O$, M_R 250.38, mp. 166 °C, $[\alpha]_D^{20}$ +46.2° (C_2H_5OH). A *quinolizidine alkaloid of the (+)-*sparteine type. It has been found in four *Genista* species (Fabaceae) [1]. (+)-R. has been reported to exhibit uterotonic, hypotensive, and diuretic activities [2]. Several other hydroxysparteines are known from various genera of the Fabaceae [1].

Lit.: [1] Waterman **8**, 197–239. [2] Manske **9**, 204.

gen.: Beilstein E V **23/11**, 201 f. – *[HS 2939 90; CAS 2122-29-4]*

Reticuline see benzyl(tetrahydro)isoquinoline alkaloids and streptomycin.

Retigeranic acid. $C_{25}H_{38}O_2$, M_R 370.58. *R. A:* cryst., mp. 220–222 °C, $[\alpha]_D^{23}$ −86.5° ($CHCl_3$). *R. B:* cryst., mp. 188–189 °C, $[\alpha]_D^{23}$ −30.4° ($CHCl_3$). A structurally unusual triquinane sesterterpene from the foliose lichen *Lobaria retigera*.

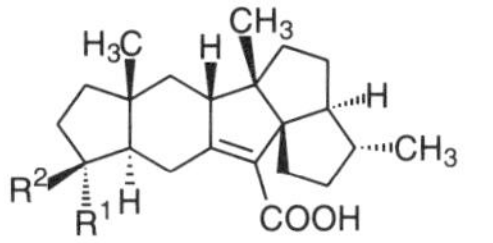

$R^1 = CH(CH_3)_2$, $R^2 = H$: R. A
$R^1 = H$, $R^2 = CH(CH_3)_2$: R. B

Lit.: Chem. Pharm. Bull. **39**, 3051–3054 (1991) (X-ray crystal structure) ▪ J. Am. Chem. Soc. **110**, 5806 (1988); **111**, 6691 (1989) ▪ Lindberg **2**, 91–122 ▪ Tetrahedron Lett. **31**, 2517 (1990) (synthesis). – *[CAS 40184-98-3 (R. A); 108814-51-3 (R. B)]*

Retigeric acid A.

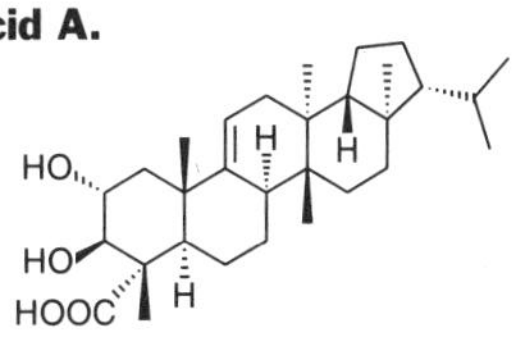

$C_{30}H_{48}O_4$, M_R 472.71, cryst., mp. 296–299 °C, $[\alpha]_D$ +26.5° (pyridine). A fern-9(11)-ene triterpene from the lichen *Lobaria retigera*; not to be confused with *retigeranic acid A.

Lit.: Magn. Reson. Chem. **25**, 503 (1987) ▪ Phytochemistry **11**, 2039–2045 (1972). – *[CAS 35591-41-4]*

Retinal (vitamin A_1 aldehyde).

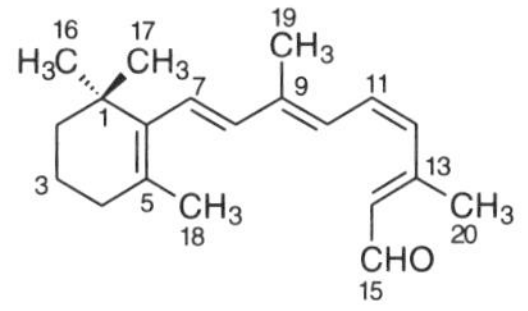

11-*cis*-Retinal

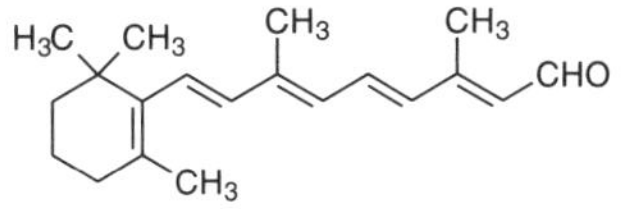

all-trans-Retinal, Retinal

$C_{20}H_{28}O$, M_R 284.44, orange-red cryst., mp. (*all-trans*-R.): 61–64 °C, (11-*cis*-R.): 63.5–64.5 °C. A *diterpene of the cyclophytane type. Soluble in all common organic solvents. All 16 possible stereoisomers, of which the *all-trans*-form is the most stable, are known. In the form of a Schiff base bound to opsins, R. constitutes the pigments of vision *rhodopsin and iodopsin as well as bacteriorhodopsin that has other func-

tions. 11-*cis*-R. occurs only in the eyes of organisms that can see, all phylogenetic tribes that can see – vertebrates, arthropods, and mollusks – use 11-*cis*-R. as a chromophore of vision.
The process of vision is initiated in the retina by a photochemical isomerization: in rhodopsin the Schiff base of 11-*cis*-retinal rearranges to its *all-trans*-antipode, after which *all-trans*-R. and opsin are liberated via several intermediates. For maintenance of vision the 11-*cis*-R. must subsequently be reconstituted, this happens in the vision cycle: *all-trans*-R. is first reduced to *all-trans*-retinol (vitamin A) and then esterified. This *all-trans*-retinyl ester is probably transformed directly to 11-*cis*-retinol with the energy required for the isomerization being obtained by hydrolysis of the ester bond. Oxidation finally leads back to 11-*cis*-R. which reacts again with opsin to afford rhodopsin [1]. R. is formed by oxidative cleavage of carotene. In the older literature R. is often referred to as retinene. Senile blindness has been associated with the formation of a pyridinium bisretinoid ("AZ-E", orange color) [2].

Lit.: [1] Angew. Chem. Int. Ed. Engl. **29**, 461–480 (1990); **35**, 367–387 (1996). [2] J. Am. Chem. Soc. **118**, 1559f. (1996).
gen.: Angew. Chem. Int. Ed. Engl. **23**, 81 (1984) ▪ Chem. Ind. (London) **1995**, 735 ▪ Merck-Index (12.), No. 8331 ▪ Pure Appl. Chem. **63**, 161 (1991) ▪ Sporn et al., The Retinoids, Vol. 1–2, New York: Academic Press 1984. – *Synthesis:* Angew. Chem. Int. Ed. Engl. **37**, 320 (1998) ▪ Tetrahedron **50**, 3389 (1994) ▪ Tetrahedron Lett. **28**, 65 (1987); **29**, 209 (1988); **32**, 4499 (1991); **34**, 319 (1993); **35**, 1209 (1994). – *Series:* Progress in Retinal Research, Oxford: Pergamon Press (9 vols. up to 1990). – *[CAS 116-31-4]*

Retinoic acid see retinol.

Retinol [axerophthol, vitamin A_1, (*all-E*)-3,7-dimethyl-9-(2,6,6-trimethyl-1-cyclohexenyl)-2,4,6,8-nonatetraen-1-ol].

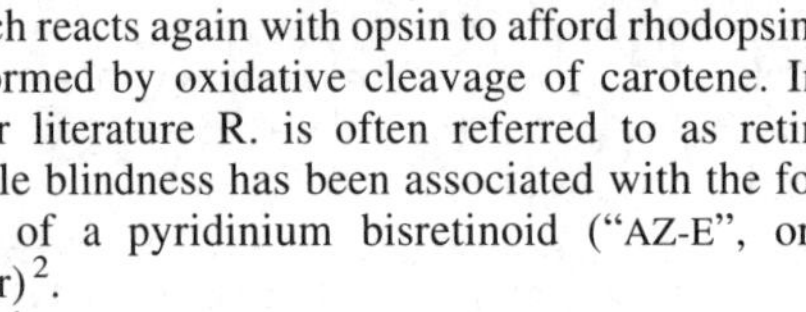

all-E-Retinol, Retinol

$C_{20}H_{30}O$, M_R 286.46, yellow prisms, mp. 63–64 °C, bp. 120–125 °C (0.6 Pa), exhibits a characteristic green fluorescence in solution. The action of UV light converts R. to products without any vitamin A activity. R. is very sensitive to oxygen and heavy metal ions (Cu, Co); singlet oxygen can completely destroy R. by formation of peroxides. R. forms esters with acids, the esters are more stable towards autoxidation.
Occurrence: Of the possible stereoisomers of R. only few occur in nature (e.g., 11-*cis*- and 13-*cis*-R.). R. is very rare in the plant kingdom but its provitamin *β-carotene is widely distributed. The body's requirements of R. are covered by fatty acid esters in milk, butter, egg yolk, and especially fish oil (codliver oil). The latter also contains 3,4-didehydroretinol (vitamin A_2), the aldehyde of which plays a role in the process of vision (see also retinal). Whale liver oil contains a provitamin A (*kitol*, $C_{40}H_{60}O_2$, M_R 572.92, mp. 88–90 °C, a Diels-Alder dimer of R.). For the physiological activities of R., see *Lit.* [1].

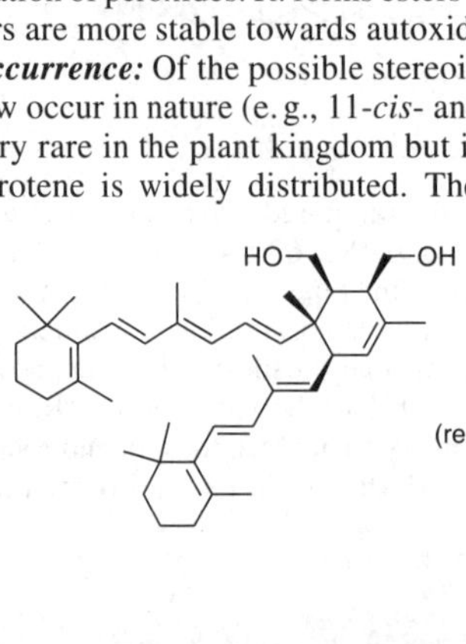

Kitol
(relative configuration)

Detection: By fluorescence measurements, mass spectrometry, or HPLC.
Toxicology: Vitamin A hypervitaminosis can lead to clinical symptoms such as headache, nausea, in chronic cases: disturbances of sleep, loss of appetite, loss of hair, bone swellings in the limbs; however, all symptoms disappear when the consumption of vitamin A_1 is reduced. Acute poisonings have been observed in, e.g., polar explorers after consumption of the extremely vitamin A_1-rich polar bear liver. The metabolite of R., vitamin A acid (*retinoic acid*, tretinoin, $C_{20}H_{28}O_3$, M_R 316.44, mp. 180–182 °C) can cause malformations and thus pregnant women should not consume more than 3.3 mg of vitamin A_1 per day [2].
Synthesis: R. was obtained in the past from fish oils, which have contents of up to 17%, while it is now mostly produced synthetically [3,4].
Retinoids: Retinoids are compounds derived from R. or its synthetic analogues but which do not exhibit any vitamin A activity. Some of them have inhibitory effects on cell growth and differentiation and are used in therapy for acne (e.g., motretinide) and psoriasis (etretinate) [5].
Historical: The structure of R. was elucidated in 1931 by Karrer (Nobel Prize 1937), and its role in the process of vision recognized by Wald in 1935. Isler and coworkers realized the first synthesis of pure, crystalline R. in 1947 while that of 3,4-dehydroretinol was achieved by Jones in 1952.

Lit.: [1] Hager (5.) **9**, 506–510. [2] J. Org. Chem. **61**, 3542 (1996); Teratology **35**, 268–275 (1987); U. S. Pat. 5,808,120. [3] Angew. Chem. Int. Ed. Engl. **16**, 423–429 (1977); **36**, 779 (1997); Pure Appl. Chem. **43**, 527–552 (1975). [4] Justus Liebigs Ann. Chem. **1995**, 717; Kirk-Othmer (3.) **24**, 140–158; Tetrahedron Lett. **28**, 65 (1987); **32**, 4115, 4117 (1991). [5] Chem. Eng. News (25. 5.) **1998**, 40 f.; Sporn & Roberts, Retinoids (2.), New York: Raven Press 1994.
gen.: Angew. Chem. Int. Ed. Engl. **29**, 461–480 (1990) ▪ Bauernfeind, Vitamin A Deficiency and its Control, New York: Academic Press 1986 ▪ Beilstein E IV **6**, 4133 ▪ Kleemann-Engel, p. 1669 ▪ Merck-Index (12.), No. 10150 (vitamin A_1), 10151 (vit. A_2), 8333 (retinolic acid) ▪ Nat. Prod. Rep. **8**, 223–249 (1991) ▪ Pure Appl. Chem. **51**, 581–591 (1979) ▪ Sax (8.), VSK 600, VSK 900, VSP 000 (toxicology). – *[HS 2936 21; CAS 68-26-8 (vitamin A_1); 79-80-1 (vitamin A_2); 4626-00-0 (kitol)]*

Retipolide A. $C_{26}H_{20}O_9$, M_R 476.44, colorless powder, mp. >360 °C (decomp. above 200 °C), $[\alpha]_D$ +160° ($CHCl_3$). A bitter principle from the North American mushroom *Boletus retipes* (Basidiomycetes). The anion of the tautomeric hydroxyaldehyde form is responsible for the yellow color of the fruit body. R. A is probably formed biosynthetically from the oxepinone *R. C* {$C_{26}H_{20}O_9$, M_R 476.44, yellowish cryst., mp. 100–105 °C, $[\alpha]_D$ +112° ($CHCl_3$)}; R. C arises, in turn, by oxidative ring expansion from *R. E* {$C_{26}H_{20}O_8$, M_R

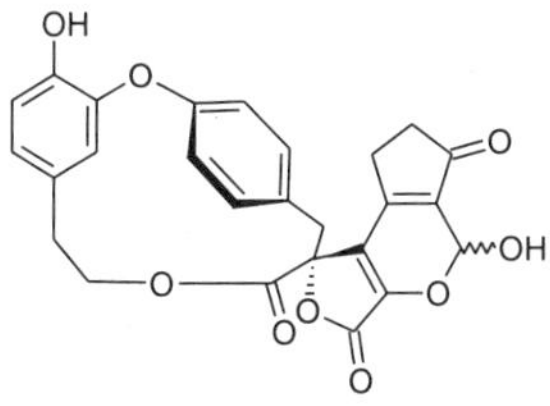

Retipolide A

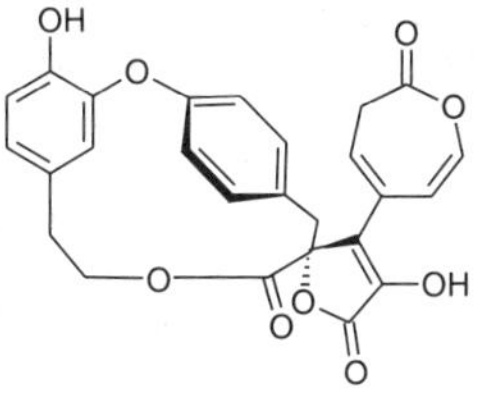

Retipolide C

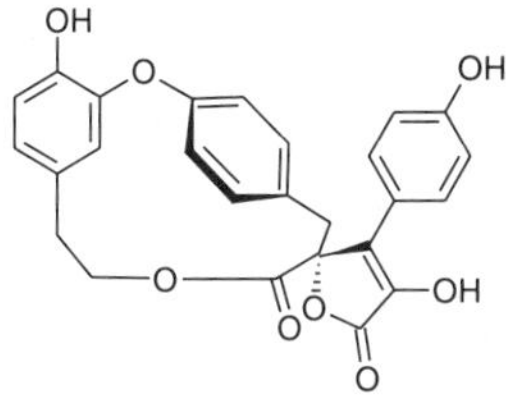

Retipolide E

460.44, colorless cryst., mp. 195–197 °C, $[\alpha]_D$ +83° (CH_3CN)}.
Lit.: Schering Lecture **24**, 9–12 (1994).

Retronecine see necine(s).

Retrorsine.

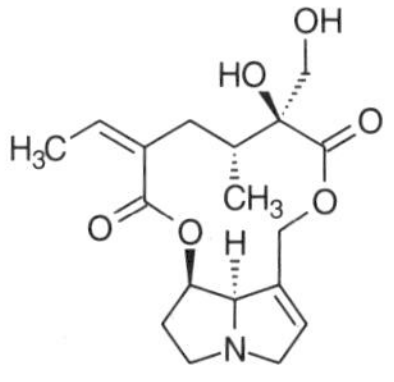

$C_{18}H_{25}NO_6$, M_R 351.40, mp. 217 °C (207–208 °C), $[\alpha]_D^{20}$ –62.4° ($CHCl_3$), $[\alpha]_D$ –17.6° (C_2H_5OH). R. is a *pyrrolizidine alkaloid from Asteraceae (*Senecio*, *Erechtites*) and some *Crotalaria* species (Fabaceae). It exists in the plants as the *N*-oxide and is highly toxic, causing liver and kidney cancer (see pyrrolizidine alkaloids).
Lit.: Aust. J. Chem. **36**, 1203 (1983) ▪ Beilstein E III/IV **27/9**, 6663 f. ▪ Merck-Index (12.), No. 8335 ▪ Pelletier **9**, 155–233. – *[HS 2939 90; CAS 480-54-6]*

Reveromycins. $C_{36}H_{52}O_{11}$, M_R 660.79, powder, mp. 95 °C (R. A), 78–79 °C (R. B), $[\alpha]_D^{20}$ –115° (R. A), –66° (R. B) (CH_3OH). Eukaryotic cell growth inhibitors from a *Streptomyces* species.

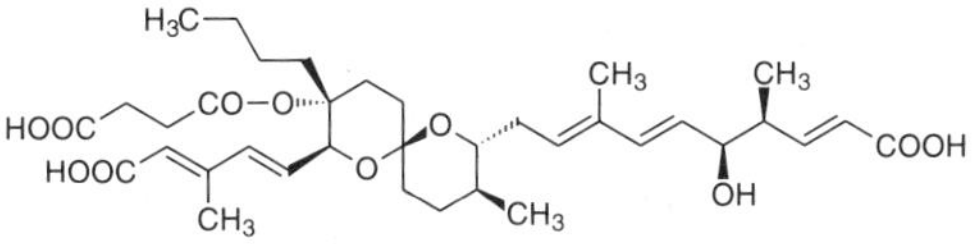

R. A

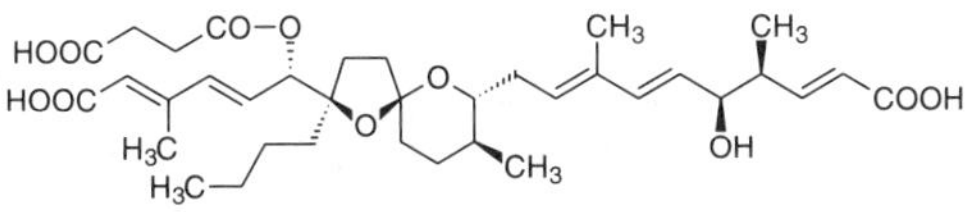

R. B

Lit.: J. Am. Chem. Soc. **121**, 456 (1999) ▪ J. Antibiot. **45**, 1409, 1414, 1420 (1992) ▪ J. Chem. Soc., Chem. Commun. **1994**, 1877 (absolute configuration) ▪ Org. Lett. **1**, 941 (1999); **2**, 191 (2000) (synthesis). – *[CAS 134615-37-5 (R. A); 144860-68-4 (R. B)]*

Rhamnaceae alkaloids see cyclopeptide alkaloids.

L-Rhamnose (6-deoxymannose).

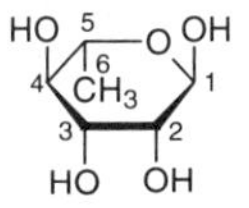

α-L-R.

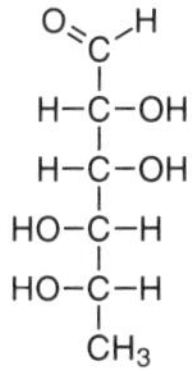

L-R.

$C_6H_{12}O_5$, M_R 164.16. D-R.: (monohydrate) mp. 90–91 °C, $[\alpha]_D^{20}$ –8° (H_2O); is rare and occurs in certain capsule polysaccharides of Gram-negative bacteria. L-R.: initially sweet and then bitter tasting crystals. The formula on the left shows the pyranoid form of α-R. which usually occurs as the monohydrate, mp. 92–94 °C, sublimation at 105 °C (200 Pa), $[\alpha]_D^{20}$ –9→+8° (H_2O). The β-form, mp. 122–126 °C, $[\alpha]_D^{20}$ +38→+9° (H_2O) is hygroscopic and transforms to the α-form (mutarotation) under the action of atmospheric moisture. L-R. is the best known 6-deoxyaldohexose; it occurs in the free form in *poison ivy*, bound as the *rhamnoside* in *rutinose (disaccharide) and in many natural glycosides, e.g., in xanthorhamnin, quercitrin (see quercetin), *strophanthins, *naringin, hesperidin (see hesperetin), various *anthocyanins, etc. L-R. is also present in arabinic acid, the main component of gum arabic, and as rhamnogalacturonan in the polysaccharides of plant cell walls. The name R. is derived from the botanical name of the *buckthorn* species (Rhamnaceae) which contain large amounts of L-R. glycosidically bound to quercetin and its methyl ether.
Use: As nutrient medium for enzymatic and biochemical investigations, and as a synthetic building block ("chiral pool").
Lit.: Adv. Carbohydr. Chem. Biochem. **42**, 15 (1984) ▪ Beilstein E IV **1**, 4260 ▪ Collins-Ferrier, p. 47, 206 (biosynthesis, occurrence) ▪ Karrer, No. 589 ▪ Merck-Index (12.), No. 8338 ▪ Synthesis **1979**, 951. – *[HS 1702 90, 2940 00; CAS 3615-41-6]*

Rhapsamine.

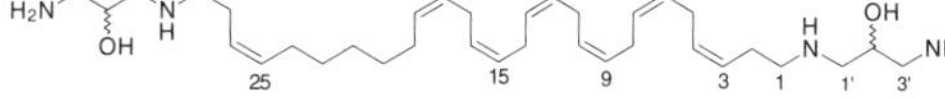

$C_{34}H_{60}N_4O_2$, M_R 556.87, amorphous solid $[\alpha]_D$ 0°, presumably racemic. A cytotoxic alkaloid from the Micronesian sponge *Leucetta leptorhapsis*, also known as rubber sponge.
Lit.: Tetrahedron Lett. **38**, 7507 (1997). – *[CAS 198826-15-2]*

Rhein (4,5-dihydroxyanthraquinone-2-carboxylic acid).

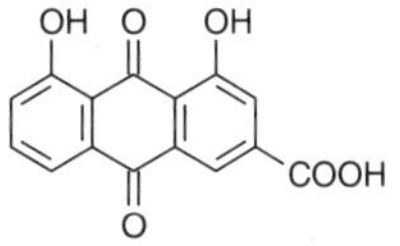

$C_{15}H_8O_6$, M_R 284.22, orange needles, mp. 321 °C, uv_{max} 431 nm (CH_3OH). The octaketide R. occurs in the free form, as β-D-glucosides, and in dimeric from (rheidins and sennidins, see sennosides) in the roots of *rhubarb, sour dock and in senna leaves. In addition to its laxative effects, the anti-inflammatory activity, especially of rhein 1,8-diacetate, is applied in the treatment of arthritis. The biosynthesis follows the polyketide pathway (octaketide, see anthranoids).

Lit.: J. Pharm. Pharmacol. **35**, 262 (1983) ▪ Karrer, No. 1297 ▪ Tetrahedron Lett. **35**, 289 (1994) (synthesis). – *[CAS 478-43-3]*

Rheosmin see zingerone.

Rhinanthin see aucubin.

Rhizobitoxin [(*S*)-(*E*)-2-amino-4-((*R*)-2-amino-3-hydroxypropoxy)-3-butenoic acid].

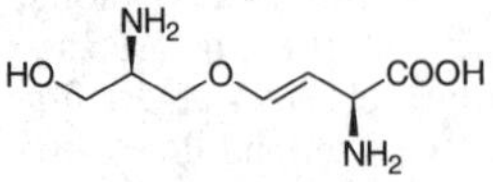

$C_7H_{14}N_2O_4$, M_R 190.20, hygroscopic cryst., mp. 175–179 °C, $[\alpha]_D^{25}$ +78.9° (H_2O). R. is responsible for chlorosis in plants that have been attacked by *Rhizobium japonicum* and *Pseudomonas andropogonis*. R. irreversibly inhibits *β*-cystathionase in bacteria and plants and thus prevents the conversion of *methionine to ethylene in the plants. The biosynthesis proceeds from *aspartic acid via homoserine and 4-hydroxythreonine to rhizobitoxin.

Lit.: Phytochemistry **25**, 2711–2715 (1986); **30**, 1809–1814 (1991) ▪ Tetrahedron **31**, 2629–2636 (1975). – *[CAS 37658-95-0]*

Rhizocticins.

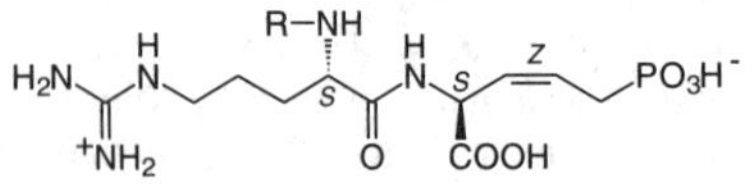

R = H	: R. A
R = L-Valyl	: R. B
R = L-Isoleucyl	: R. C
R = L-Leucyl	: R. D

A phosphorus-containing antibiotic complex from *Bacillus subtilis*. Currently, four R. (A–D) are known with antifungal activities. The R. are hydrophilic and amphoteric. The main components are *R. A* ($C_{11}H_{22}N_5O_6P$, M_R 351.30) and *R. B* ($C_{16}H_{31}N_6O_7P$, M_R 450.43). ○

Lit.: Justus Liebigs Ann. Chem. **1988**, 655–661. – *[CAS 114301-25-6 (R. A); 114301-26-7 (R. B); 114301-27-8 (R. C); 114301-28-9 (R. D)]*

Rhizoferrin see siderophores.

Rhizopogone see tridentoquinone.

Rhizoxin.

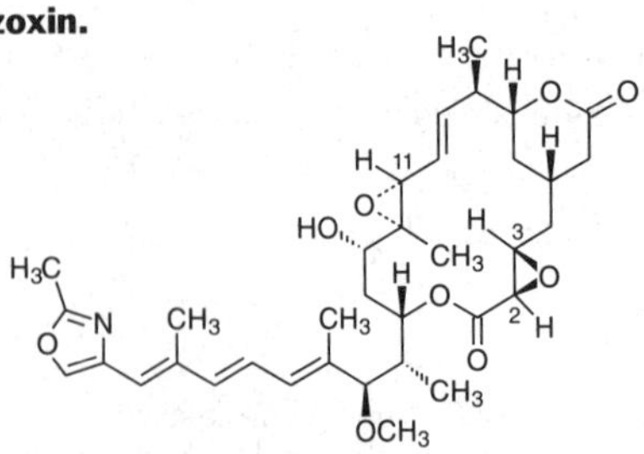

$C_{35}H_{47}NO_9$, M_R 625.76, pale yellow cryst., mp. 131–135 °C, $[\alpha]_D$ +155° (CH_3OH), soluble in methanol. Rhizoxin is an antitumor *macrolide from cultures of *Rhizopus chinensis*, a pathogenic fungus that causes root blight. R. is the main component of the WF1360 complex exhibiting activity against leukemia and melanoma in animal models. R. also has antifungal and phytotoxic activities on rice shoots. The mechanism of action of R. is similar to that of maytansine (see maytansinoids) and the *dolastatins. R. inhibits mitosis, binds to *β*-tubulin at the maytansine/vincristine binding site, and inhibits the aggregation of tubulin to the microtubuli (antimitotic activity); see sorangicins. Limited clinical studies in certain types of cancer are in progress.

Lit.: Biochim. Biophys. Acta **926**, 215–223 (1987) ▪ Cancer Res. **46**, 381 (1986); **50**, 4277–4280 (1990) ▪ Int. J. Pharm. **43**, 191–199 (1988) ▪ J. Antibiot. **37**, 354–362 (1984); **39**, 424–429, 762–772 (1986); **40**, 66–72 (1987) ▪ J. Chem. Soc., Chem. Commun. **1986**, 1701 ff. ▪ J. Org. Chem. **57**, 2235–2244 (1992) ▪ Mem. Sch. Sci. Eng., Waseda Univ. **59**, 167–195 (1995) (synthesis) ▪ Mol. Gen. Genet. **222**, 53–59, 169–175 (1990) ▪ Tetrahedron Lett. **34**, 1035 (1993); **38**, 6825, 7333 (1997); **40**, 4145 (1999) (synthesis) ▪ Yakugaku Zasshi **117**, 486–508 (1997) (synthesis). – *[CAS 90996-54-6]*

Rhodeose see fucose.

Rhodocladonic acid.

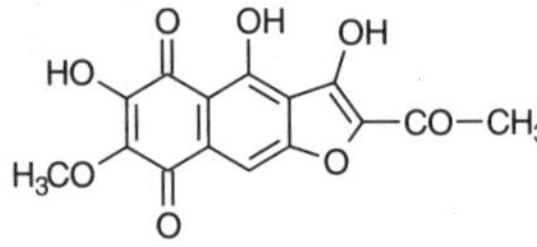

$C_{15}H_{10}O_8$, M_R 318.24, red needles that decompose above 360 °C. A *naphthoquinone derivative from the red apothecia of several *Cladonia* species (Cocciferae group, lichen).

Lit.: Can. J. Chem. **61**, 2055–2058 (1983) (X-ray crystallographic analysis of the triacetate) ▪ Culberson, p. 187–188 ▪ Karrer, No. 1306. – *[CAS 17636-15-6]*

Rhodomycins.

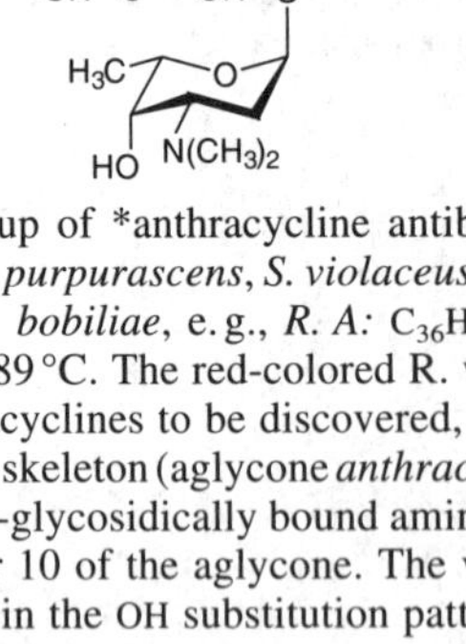

Rhodomycin A

A group of *anthracycline antibiotics from *Streptomyces purpurascens*, *S. violaceus* subsp. *iremyceticus*, and *S. bobiliae*, e.g., *R. A:* $C_{36}H_{48}N_2O_{12}$, M_R 700.78, mp. 189 °C. The red-colored R. were the first natural anthracyclines to be discovered, they contain a tetracyclic skeleton (aglycone *anthracyclinone*) and at least one *O*-glycosidically bound amino sugar at position 7 and/or 10 of the aglycone. The various rhodomycins differ in the OH substitution patterns and the number of sugars.

Activity: The R. have tumorstatic activities; as intercalators, they inhibit DNA and RNA polymerase and topoisomerase II and thus DNA and RNA biosynthesis. As a result of their binding to DNA the R. are toxic.

Biosynthetically, R. represent *polyketides (propionyl-CoA as starter unit and nine elongation steps with malonyl-CoA). The sugar units are connected to the aglycone at a later stage of the biosynthesis.

Lit.: J. Carbohydr. Chem. **9**, 873 (1990) (synthesis) ▪ Präve et al., Handbuch der Biotechnologie (4.), p. 685, München: Oldenbourg Verl. 1994 ▪ Zechmeister **21**, 121 (review). – *[HS 2941 90; CAS 23666-50-4 (R. A)]*

Rhodopsin (visual purple, from Greek: rhódeos = rose-red and ópsis = sight). A labile, deep red *chromoprotein (M_R ca. 28 600) in the layered, slice-like membrane structure of the outer segments of animal retinal rod cells that serve as photoreceptors for the process of vision. The *pigment of vision R. (absorption maximum at 500 nm) consists chemically of the chromophore *retinal as prosthetic group and opsin as protein component. The 5-fold conjugated aldehyde retinal, a derivative of vitamin A (*retinol) switches during the process of vision (see pigments of vision) between the 11-*cis*- and the *all-trans*-forms (formula, see retinal).

The 11-*cis*-form is bound as a Schiff base to an ε-amino group of an L-lysine residue of opsin such that it is on the inside of a cylinder formed by 7 α-helices, spanning the membrane. In the photo-excited state R. interacts with transducin (a G-protein); this interaction serves for transmission of the light signal. After phosphorylation by a rhodopsin kinase R. binds the inhibitor protein arrestin (retinal S-antigen) and loses the ability to activate transducin – a mechanism for adaptation to brightness. A point mutation of R. is held to be responsible for a form of retinitis pigmentosa (a degeneration of the retina) [1]. For the occurrence of R. in a photoactive unicellular organism, see *Lit.*[2]. The pigments of vision *cyanopsin, iodopsin, and *porphyropsin are similar to R. in structure and function. Other related retinal proteins are the light-driven ion pumps bacteriorhodopsin and halorhodopsin. The 7-helix motif is a common feature of many other membrane receptors, e. g., the muscarinic acetylcholine receptor (see acetylcholine), the adrenergic alpha- and beta-receptors, and the dopaminergic D_2-receptor.

Lit.: [1] Chem. Unserer Zeit **33**, 140–151 (1999); Nachr. Chem. Techn. Lab. **38**, 352 f. (1990); Nature (London) **343**, 316 f., 364–366 (1990). [2] Biochemistry (USA) **28**, 819–824 (1989).
gen.: Alberts et al., Molekularbiologie der Zelle, 2nd edn., p. 1330 f., Weinheim: VCH Verlagsges. 1990 ▪ Biochim. Biophys. Acta **1016**, 293–327 (1990) ▪ Chem. Eng. News (13.1.) **1992**, 26 f. ▪ Eur. J. Biochem. **179**, 255–266 (1989) ▪ J. Am. Chem. Soc. **116**, 10165 (1994) ▪ J. Biol. Chem. **266**, 10711–10714 (1991) ▪ Stryer 1995, p. 333–337 ▪ Trends Biochem. Sci. **13**, 388–393 (1988). – *[CAS 9009-81-8]*

Rhodoxanthin (4′,5′-didehydro-4,5′-retro-β,β-carotene-3,3′-dione).

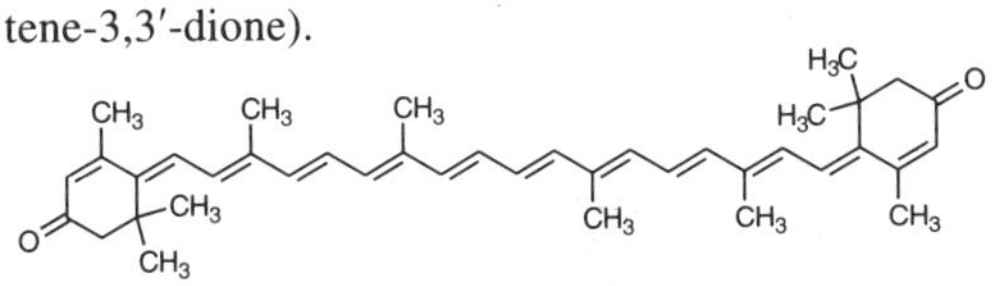

$C_{40}H_{50}O_2$, M_R 562.84, rosette-shaped, blue-black cryst., mp. 219 °C (vacuum), insoluble in petroleum ether, poorly soluble in ethanol, soluble in chloroform. R. is one of the few *carotinoids with a retro configuration. R. (a xanthophyll) is used as a pigment in cosmetics and is also registered as a colorant for foods (E 161 f.); it is widely distributed in low concentrations in nature: e. g., in the seed pods (arils) of yew (*Taxus baccata*, Taxaceae) and in the scarlet-red feathers of *Phoenicurus nigricollis* and the Australian dove *Megaloprepia magnifica*.

Lit.: Beilstein EIII **7**, 4387 ▪ Helv. Chim. Acta **65**, 944–967 (1982) ▪ Karrer, No. 1855 ▪ Merck-Index (12.), No. 8362 ▪ Zechmeister **18**, 177–222. – *[CAS 116-30-3]*

Rhoeadine alkaloids. R. a. contain a cyclic acetal or hemiacetal system. They are easily detected by the characteristic red-brown-green coloration with concentrated sulfuric acid. R. a. occur mainly in the genus *Papaver* (Papaveraceae) as well as in two other genera of this family, namely in *Bocconia* and *Meconopsis*. Thus, the R. a. include alpinigenine in *Papaver alpinum*, *P. bracteatum*, *P. fugax*, and *P. orientale*, oreodine in *P. fugax*, *P. oreophilum*, *P. tauricola*, and *P. triniaefolium*, rhoeadine in various *Papaver* species such as *P. rhoeas* and *P. albiflorum* and also *Bocconia frutescens* and *Meconopsis betonicifolia*. Rhoeagenine is found in many *Papaver* species.

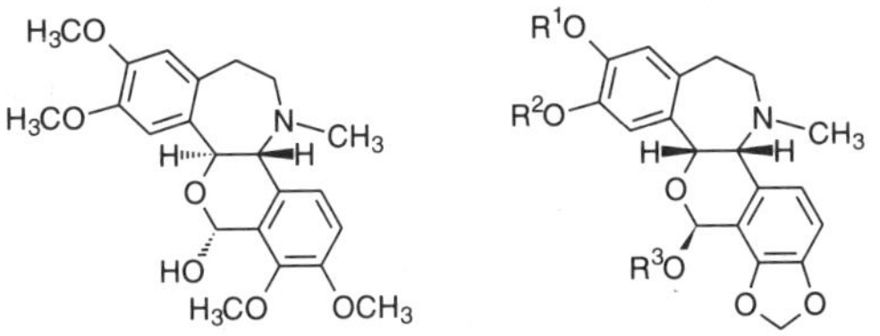

Alpinigenine (**1**)

	R^1	R^2	R^3
Oreodine (**2**)	CH_3	CH_3	CH_3
Rhoeadine (**3**)	$-CH_2-$		CH_3
Rhoeagenine (**4**)	$-CH_2-$		H

Table: Data of Rhoeadine alkaloids.

no.	molecular formula	M_R	mp. [°C]	$[\alpha]_D$	CAS
1	$C_{22}H_{27}NO_6$	401.46	193–195	+286° (CH_3OH)	14028-91-2
2	$C_{22}H_{25}NO_6$	399.44	184–186	+224° ($CHCl_3$)	6516-48-9
3	$C_{21}H_{21}NO_6$	383.40	253–254	+228.5° ($CHCl_3$)	2718-25-4
4	$C_{20}H_{19}NO_6$	369.37	236–238	+134° (pyridine)	5574-77-6

R. a. are insoluble in water, chloroform, ether, poorly soluble in ethyl acetate and methylene chloride. Rhoeadine is weakly toxic [LD_{50} (rats i.p.) 530 mg/kg]. It effects dilatation of the pupils. *Papaver rhoeas* extracts are used as sedatives and expectorants, however, these uses are not attributed to the rhoeadine content.

Biosynthesis[1]***:*** The R. a. are formed from *protoberberine alkaloids.

Lit.: [1] Mothes et al., p. 232 ff.
gen.: Manske **28**, 1–93 ▪ Planta Med. **49**, 43 (1983) ▪ Shamma, p. 400–417 ▪ Shamma-Moniot, p. 337–353 ▪ Ullmann (5.) **A 1**, 376. – *[HS 2939 90]*

Rhubarb. Medicinal rhubarb (*Rheum palmatum*, Polygonaceae), indigenous to North China but also cultivated in Central Europe and North America, is grown as a food crop. Its leaf stems are consumed and its root stock contains 3–5% dry weight of *anthranoids (*emodin, *chrysophanol, *aloe-emodin, *rhein, *physcion, in free and glycosylated forms). These components of R. roots have laxative activities and can be used to dye wool to give yellow and orange colors. Dyer's plants containing principally emodins and chrysophanol include sorrel (*Rumex* sp., Polygonaceae) and *alder buckthorn (*Rhamnus* sp., Rhamnaceae) species.

Lit.: Hager (5.) **1**, 596 ff.; **6**, 432 ff.

Rhynchophorol.

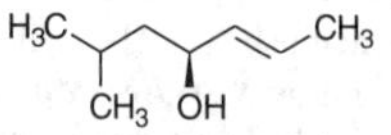

$C_8H_{16}O$, M_R 128.21, oil, bp. 36–37 °C (400 Pa), (*S*)-enantiomer: $[\alpha]_D^{18}$ −11.0° ($CHCl_3$). Aggregation pheromone produced by males of the African palm beetle *Rhynchophorus palmarum* that causes serious economic damage. First tests of R. as a pesticide in integrated plant protection were very promising; see also weevils, pheromones.

Lit.: J. Chem. Ecol. **17**, 1221–1230, 2127–2141 (1991) ▪ Justus Liebigs Ann. Chem. **1992**, 1195 ▪ Naturwissenschaften **79**, 134 f. (1992). – *[CAS 153665-39-5]*

Riboflavin [1-deoxy-1-(3,4-dihydro-7,8-dimethyl-2,4-dioxobenzo[*g*]pteridine-10(2*H*)-yl)-D-ribitol, lactoflavin, vitamin B_2].

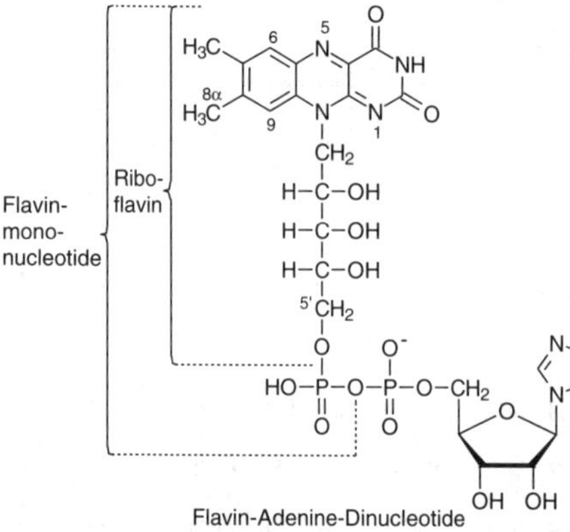

Flavin-Adenine-Dinucleotide

$C_{17}H_{20}N_4O_6$, M_R 376.37, bitter tasting, yellow-orange needles, mp. 275–282 °C (decomp.), $[\alpha]_D$ −9.8° (H_2O), $[\alpha]_D^{25}$ +59° (37% HCl), $[\alpha]_D$ −125° (20 N NaOH), shows green fluorescence in solution. LD_{50} (rats i. p.) 560 mg/kg. R. occurs in the free form only in the retina, in whey, and in urine; it is ubiquitous in tissue and cells in the form of the *coenzymes *flavin mononucleotide* (R. 5′-phosphate, FMN, $C_{17}H_{21}N_4O_9P$, M_R 456.35) and *flavin adenine dinucleotide* (R. 5′-adenosine diphosphate, FAD, $C_{27}H_{33}N_9O_{15}P_2$, M_R 785.56). The heterocyclic ring system is designated as isoalloxazine. R. is a coenzyme of many oxidoreductases. In some enzymes (flavoproteins) it is linked at the 8α-C atom to an L-cysteine or L-histidine residue.

Lyxoflavin (a): the 4′-epimer of R. with arabit-5-yl (= lyxit-1-yl) as carbohydrate residue, $C_{17}H_{20}N_4O_6$, M_R 376.37, the D-form, mp. 276 °C, $[\alpha]_D^{20}$ +46° (0.05 N NaOH), occurs in fish flour and liver products; the L-form (b); mp. 283–284 °C (decomp.), $[\alpha]_D^{20}$ −49° (0.05 N NaOH), is present in human cardiac tissue.

R. 5′-malonic acid ester (c), $C_{20}H_{22}N_4O_9$, M_R 462.42, is the pigment of the blue light receptor and is responsible for phototropism in plants.

R. 5′-carboxylic acid (d) (schizoflavin 1, vitamin B_2 acid), $C_{17}H_{18}N_4O_7$, M_R 390.35, yellow needles, mp. 250 °C, from *Schizophyllum commune* and other basidiomycetes, is active against algae (*Chartonella antigua*).

R. 5′-α-D-ribofuranoside, see lampteroflavin [1].

Use: R. is used as a food additive for prophylaxis against deficiency states and as a colorant (E 101), e. g., in mayonnaise, ice cream, puddings, etc.; L-lyxoflavin is added to animal fodder as growth promotor. For synthesis, see *Lit.*[2].

Biosynthesis: The biosynthesis of R. from purine precursors (especially guanosine) occurs only in plants and microorganisms, humans and animals are not able to synthesize R.[3].

Historical: R. was first isolated in 1932 by R. Kuhn & György and recognized as a vitamin; in 1934 Kuhn and Karrer simultaneously achieved its structure elucidation and synthesis.

Lit.: [1]Tetrahedron **47**, 6215–6222 (1991); Tetrahedron Lett. **31**, 717 (1990) (synthesis). [2]Microbial Process for Riboflavin Production in Microbial Technology, 2nd edn., vol. I, p. 521 f., New York: Academic Press 1979; Rehm, Industrielle Mikrobiologie, 2nd edn., p. 487 f., Berlin: Springer 1980. [3]Tetrahedron **39**, 3471 (1983).
gen.: Adv. Enzymol. **53**, 345–481 (1982) (review) ▪ Beilstein E V **26/14**, 334–356 ▪ Bull. Chem. Soc. Jpn. **64**, 1093 ff. (1991) ▪ Hager (5.) **9**, 510 ff. ▪ Isler et al., Vitamin II, p. 50–159, Stuttgart: Thieme 1988 ▪ Kleemann-Engel, p. 1674 ▪ Nat. Prod. Rep. **3**, 395–419 (1986) ▪ Physiol. Rev. **69**, 1170–1198 (1989) ▪ Prog. Clin. Biol. Res. **321**, 233–248 (1990) – *Toxicology:* Sax (8.), RIK 000. – *[HS 2936 23; CAS 83-88-5 (R.); 482-12-2 (a); 13123-37-0 (b); 88623-79-4 (c); 59224-03-2 (d)]*

D-Ribose (abbr. Rib).

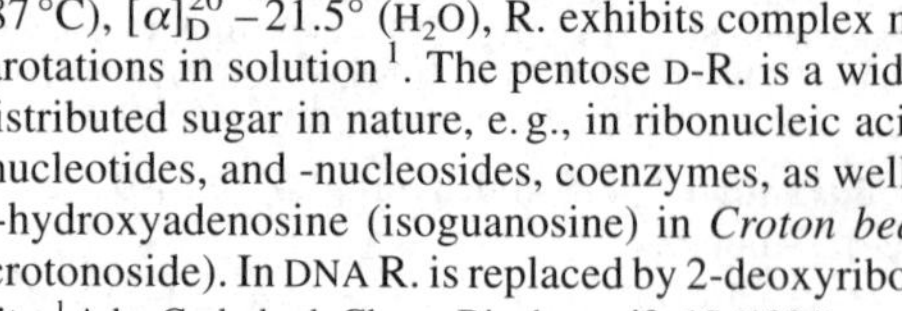

β-D-Ribofuranose

$C_5H_{10}O_5$, M_R 150.13; hygroscopic cryst., mp. 95 °C (87 °C), $[\alpha]_D^{20}$ −21.5° (H_2O), R. exhibits complex mutarotations in solution [1]. The pentose D-R. is a widely distributed sugar in nature, e. g., in ribonucleic acids, -nucleotides, and -nucleosides, coenzymes, as well as 2-hydroxyadenosine (isoguanosine) in *Croton beans* (crotonoside). In DNA R. is replaced by 2-deoxyribose.

Lit.: [1]Adv. Carbohyd. Chem. Biochem. **42**, 15 (1984).
gen.: Althaus et al., ADP-Ribosylation of Proteins, Berlin: Springer 1985 ▪ Beilstein E IV **1**, 4211 ff.; **31**, 21 ▪ Collins-Ferrier, p. 53 (synthesis) ▪ Karrer, No. 580 ▪ Merck-Index (12.), No. 8371 ▪ Miwa et al., ADP-Ribosylation, DNA Repair and Cancer, Utrecht: VNU Sci. Press 1983 ▪ Origins Life Evol. Biosphere **18**, 71–85 (1988) ▪ Tetrahedron Lett. **38**, 4199 (1997) (L-R.). – *[HS 1702 90, 2940 00; CAS 50-69-1 (D-R.); 36468-53-8 (β-D-ribofuranose)]*

Riccardins.

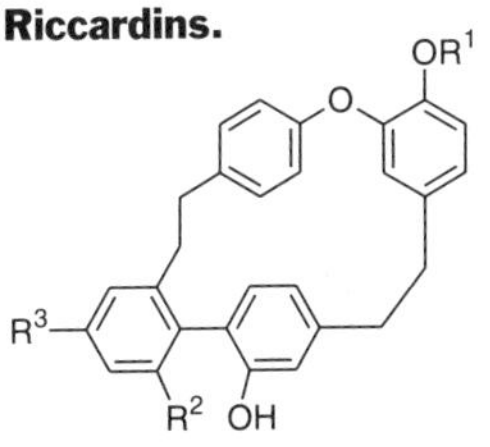

R[1]	R[2]	R[3]	
H	H	OCH_3	R. A
H	H	OH	R. C

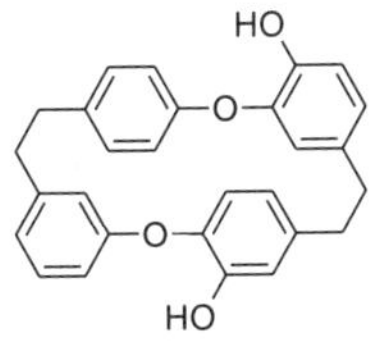

Riccardin B

Table: Data of Riccardins.

	molecular formula	M_R	consis-tency	CAS
R. A	$C_{29}H_{26}O_4$	438.52	resin	85318-25-8
R. B	$C_{28}H_{24}O_4$	424.50	oil	85318-27-0
R. C	$C_{28}H_{24}O_4$	424.50	oil	84575-08-6

Macrocyclic bisbibenzyls isolated from various liverworts of the genera *Riccardia*, *Reboulia*, *Marchantia*, etc.[1–3]. R. A and B are cytotoxic (ED_{50} for KB cells = 10 μg/mL). R. A additionally inhibits 5-lipoxygenase and calmodulin activity.

Lit.: [1] J. Hattori Bot. Lab. **57**, 383 (1984). [2] Phytochemistry **27**, 2603 (1988). [3] J. Hattori Bot. Lab. **74**, 121 (1993).
gen.: Chem. Lett. **1985**, 1587 ▪ J. Chem. Soc., Perkin Trans. 1 **1990**, 315 ▪ J. Org. Chem. **48**, 2164 (1983).

Ricciocarpins.

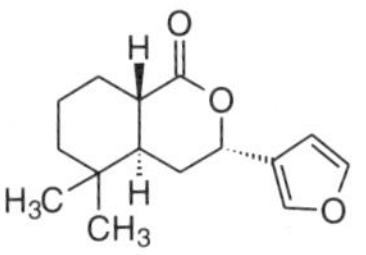

Ricciocarpin A

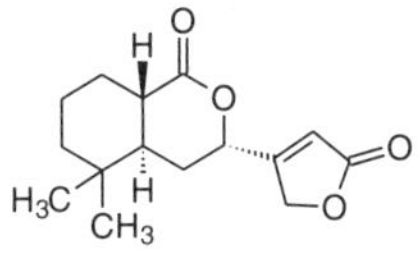

Ricciocarpin B

(relative configuration)

R. A {$C_{15}H_{20}O_3$, M_R 248.32, cryst., mp. 110–111 °C, $[\alpha]_D^{20}$ +17.8° (CH_2Cl_2), soluble in lipophilic solvents} and *R. B* {$C_{15}H_{20}O_4$, M_R 264.32, mp. 160–161 °C, $[\alpha]_D^{20}$ +6.3° (CH_2Cl_2), solubility similar to that of R. A} are sesquiterpene lactones from sterile cultures of the water liverwort *Ricciocarpos natans*. R. A possesses a δ-lactone and a β-substituted furan ring, R. B a δ-lactone and a 2-buten-4-olide ring, a combination of rings not previously observed in nature. Both R. have appreciable molluscicidic activities (LC_{100} R. A 11 ppm, R. B 43 ppm) against the water snail *Biomphalaria glabrata*, a vector snail of *Schistosoma mansoni*, a pathogen of bilharziosis. R. belong to the natural products with the highest known molluscicidic activities. For synthesis, see *Lit.*[1].

Lit.: [1] Angew. Chem. Int. Ed. Engl. **30**, 130–147 (1991); Heterocycles **47**, 277 (1998); J. Chem. Soc., Perkin Trans. 1 **1993**, 2251.
gen.: Phytochemistry **29**, 2565, 2685 (1990). – *[CAS 127350-71-4 (R. A); 127350-72-5 (R. B)]*

Riccionidins. *R. A:* $C_{15}H_9O_6^+$, M_R 285.23, red pigment, soluble in polar solvents (H_2O, CH_3OH, DMSO) on addition of strong acids. Major anthocyanidin (see anthocyanins) of the cell wall (membranochrome) of

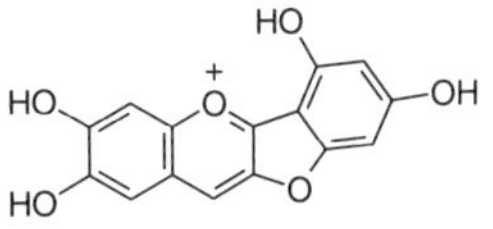

Riccionidin A

some liverwort species (*Ricciocarpos natans*, *Riccia duplex*, *Marchantia polymorpha*, *Scapania undulata*), possessing an unusual, additional ether bridge between the B- and C-rings of the flavan skeleton. Compounds of this type were previously only known from synthetic sources[1]. In aqueous solution the color of R. A is, like those of normal anthocyanidins, strongly dependent on the pH value. R. A is always accompanied by its dimer, *R. B* ($C_{30}H_{17}O_{12}^+$, M_R 569.46).

Lit.: [1] Phys. Chem. **93**, 805 (1989).
gen.: Phytochemistry **35**, 233 (1994). – *[CAS 155518-34-6 (R. A); 155739-85-8 (R. B)]*

Rice flavor. *2-Acetyl-1-pyrroline has been recognized as the impact component in boiled rice, it is more abundant in the brown Asian varieties (e.g., basmati, hieri; ca. 600 ppb in comparison to 6 ppb in white rice). Other *flavor compounds relevant for the sensory impression are (*E*,*E*)-*2,4-decadienal, the *alkanals C_6–C_{10}, (*E*)-2-nonenal (see alkenals) and *4-vinylguaiacol.

Lit.: Food Rev. Intern. **1**, 497–520 (1985–86) ▪ J. Agric. Food Chem. **36**, 1006–1009 (1988) ▪ Maarse, p. 79–89.

Ricin. An extremely toxic protein from the seeds of the castor-oil plant or castor beans (*Ricinus communis*) and other *Ricinus* species. R. consists of peptide chains linked by disulfide bridges, the chains are glycosylated to the extent of 4–8% of their M_R and exhibit sequence homologies with abrins. R. was originally considered to be a uniform substance but is now known to contain two lectins, RCL I and RCL II (hemagglutinins, tetramers of two subunits each, M_R 33000 and 30000, respectively) and two toxins, ricin D and RCL IV (dimer of an A-chain, M_R 30000, and a B-chain, M_R 33000). The R. lectin RCL II (B-chain) binds as a *haptomer* specifically to the terminal galactose residues of receptors on the cell surface and thus activates the *R. toxin*, which then enters the cell by means of endocytosis. The A-chain of the toxin blocks protein synthesis inside the cell by deactivating the 60 S-subunits of the ribosomes (*effectomer*). Conjugates of R. with cell-binding antigens or antibodies (so-called *immunotoxins*) are cytotoxic and may thus find use in cancer therapy. Synergistic effects with *daunorubicin, cisplatin, and *vincristine have been observed in leukemia[1]. Antibody-ricin (toxin, A-chain) conjugates are being tested in AIDS therapy[2]. On the other hand R. was patented in 1962 as a respiratory poison for chemical warfare. The agglutination of erythrocytes by R. is not blood group-specific; the lethal dose is only 1 μg/kg body weight in mice, but even lower doses can suppress tumor growth. R. is resistant to digestive enzymes but is cleaved by strong acids and thus loses its activity. On account of their contents of R. the residues remaining after the pressing of *castor oil cannot be used as animal fodder.

Lit.: [1] Cancer Res. **42**, 2152 (1982). [2] Science **242**, 1166 (1988).
gen.: Carbohydr. Res. **282**, 285 (1996) (review) ▪ Gabins &

Nagel, Lectins, p. 97–112, Berlin: Springer 1988 ▪ Med. Klin. (München) **85**, 555–560 (1990) ▪ Merck-Index (12.), No. 8376 ▪ Methods Enzymol. **112**, 207–225 (1985). – *[CAS 9009-86-3]*

Ricinine.

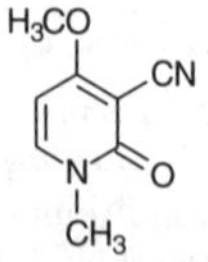

$C_8H_8N_2O_2$, M_R 164.16, mp. 201 °C. A pyridine alkaloid from *Ricinus communis* (Euphorbiaceae). Its content in this plant is 1% and together with the extremely toxic protein *ricin R. is responsible for the toxicity of *Ricinus* seeds. The biosynthesis of R. proceeds through nicotinamide as an intermediate.

Lit.: Acta Crystallogr., Sect. C **45**, 957 (1989) (structure) ▪ Beilstein E V **22/7**, 318 ▪ Manske **26**, 149 ▪ Pelletier **3**, 12–13 ▪ Phytochemistry **17**, 1903 (1978) (biosynthesis). – *[HS 2939 90; CAS 524-40-3]*

Ricinoleic acid [ricinolic acid, (*R*)-12-hydroxy-(*Z*)-9-octadecenoic acid].

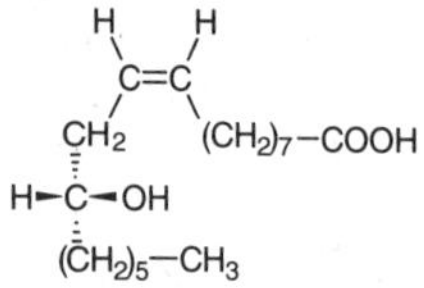

$C_{18}H_{34}O_3$, M_R 298.47, mp. 5.5 °C, bp. 245 °C (1 kPa), $[\alpha]_D^{22}$ +6.67°. R. a. exists in *castor oil as the glycerol ester (80–85%). R. a. is mainly used for technical purposes, e. g., for the production of lubricants, detergents, and auxiliary substances for the textile industry. In oleochemistry R. a. is converted to 10-undecenoic acid and sebacinic acid which serve as starting materials for plastics.

Isoricinolic acid [(*S*)-9-hydroxy-(*Z*)-12-octadecenoic acid] occurs as the glycerol ester in the seed oil of *Wrightia tinctoria* and other species of the Apocynaceae as well as in microorganisms; see also hydroxy fatty acids.

Lit.: Gunston, Harwood, & Padley, The Lipid Handbook, 2nd edn., p. 15f., 55f., 616, London: Chapman & Hall 1994 ▪ J. Am. Chem. Soc. **105**, 7130 (1983) (synthesis) ▪ Merck-Index (12.), No. 8378 ▪ Sax (8.), RJP 000. – *[HS 2918 19; CAS 141-22-0 (R.); 26806-10-0 (isoricinolic acid)]*

Ricinus oil see castor oil.

Rickamicin see sisomicin.

Rifampicin see rifamycins.

Rifamycins. A family of antibiotics discovered in 1959, belonging to the *ansamycins. The R. are produced by rare *actinomycetes, the original strain is now considered to belong to the species *Amycolatopsis mediterranei* (formerly *Streptomyces* or *Nocardia mediterranei*). Under certain culture conditions it produces mainly *R. B* {$C_{39}H_{49}NO_{14}$, M_R 755.82, pale yellow cryst., decomp. above 160 °C, $[\alpha]_D$ –48.7° (CH_3OH)}, mutants of the strain produce mainly *R. SV* [rifamycin, $C_{37}H_{47}NO_{12}$, M_R 697.78, decomp. above 140 °C, $[\alpha]_D$ –4° (CH_3OH)] which has the highest biological activity among the about 20 naturally occurring R.

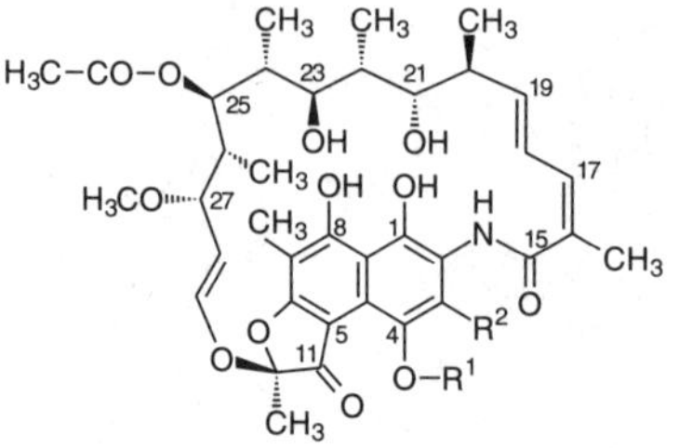

R¹	R²	
H	H	R. SV
CH_2–COOH	H	R. B
H	CH=N–N(piperazine)N–CH_3	Rifampicin

The biosynthesis of R. is stimulated by *B-factor and starts from 3-amino-5-hydroxybenzoic acid (C_7N-unit), from which the skeleton of the naphthohydroquinone bridged by a macrolactam is formed via the polyketide pathway. Insertion of oxygen into the ansa-bridge and further functionalization occur later in the biosynthetic sequence. Syntheses of R. have been reported.

R. SV is mainly active against Gram-positive bacteria and mycobacteria by inhibition of RNA synthesis, in this process it binds to the *β*-subunit of the DNA-dependent RNA-polymerase. The corresponding enzyme in mammals is not inhibited, resistance is easily generated in sensitive bacteria by inducing changes to the enzyme. Numerous semisynthetic derivatives have been prepared, including *rifampicin* (rifampin, R. AMP, $C_{43}H_{58}N_4O_{12}$, M_R 822.95) which possesses better pharmacological properties and a broader spectrum of action. It is used in the treatment of tuberculosis and leprosy, among others.

Lit.: Angew. Chem. Int. Ed. Engl. **24**, 1009 (1985) ▪ Ghisalba, in Vandamme (ed.), Biotechnology of Industrial Antibiotics, p. 282–327, New York: Dekker 1984 ▪ J. Antibiot. **38**, 920 (1985) ▪ Kleemann-Engel, p. 1677 ▪ Lancini, in Pape & Rehm (eds.), Biotechnology, vol. 4, p. 431–463, Weinheim: Verl. Chemie 1986 ▪ Topics Curr. Chem. **72**, 21–49 (1977) ▪ Zechmeister **33**, 231–307. – *Biosynthesis:* Chiao et al., in Vining & Stuttard (eds.), Genetics and Biochemistry of Antibiotic Production, p. 477–498, Boston: Butterworth-Heinemann 1995. – *Synthesis:* J. Am. Chem. Soc. **106**, 5528 (1982) ▪ Pure Appl. Chem. **53**, 1163 (1981) ▪ Tetrahedron **40**, 1289 (1984); **46**, 4629 (1990). – *[CAS 6998-60-3 (R. SV); 13929-35-6 (R. B); 13292-46-1 (rifampicin)]*

Rimuene see rosanes.

Rinderine see indicine.

Rippertenol.

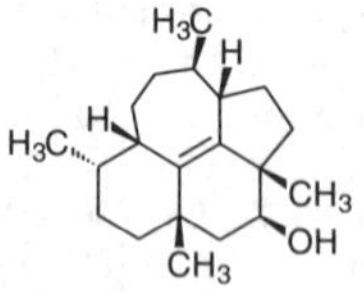

$C_{20}H_{32}O$, M_R 288.47, oil. A component of the defensive secretion of soldier insects of the termite subfamily *Nasutitermitinae* with a unique diterpene structure.

Lit.: Atta-ur-Rahman **1986**, 318–329 ▪ J. Am. Chem. Soc. **102**, 6825 (1980). – *[CAS 76649-58-6]*

Rishitin.

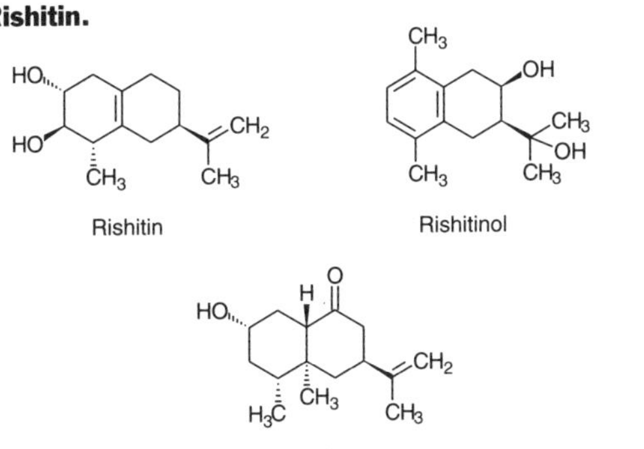

Norsesquiterpene, $C_{14}H_{22}O_2$, M_R 222.33, cryst., mp. 65–67 °C, $[\alpha]_D^{20}$ −8.7° ($CHCl_3$). R.[1] is produced by potatoes (*Solanum tuberosum*), tobacco (*Nicotiana tabacum*), and tomatoes (*Lycopersicon esculentum*) after attack by microorganisms [e.g., *Phytophthora infestans* (fungus) or *Erwinia atroseptica* (bacterium)] together with *rishitinol*[2] {$C_{15}H_{22}O_2$, M_R 234.34, mp. 127–129 °C, $[\alpha]_D$ +47° ($CHCl_3$)}, *rishitinone*[3] ($C_{15}H_{24}O_2$, M_R 236.35, mp. 72–75 °C, $[\alpha]_D$ +10.1°), *phytuberin and some *vetispiranes as *phytoalexins. R. inhibits oxygen uptake in *E. atroseptica* by disturbing the plasmalemma functions[4]. R. can also be isolated from the broths of cell cultures of *S. tuberosum* after pretreatment with spores of *P. infestans*.

Lit.: [1]Agric. Biol. Chem. **49**, 2537 (1985). [2]Arch. Biochem. Biophys. **134**, 34 (1969); Bull. Chem. Soc. Jpn. **45**, 2871 (1972). [3]Chem. Lett. **1980**, 1455; Justus Liebigs Ann. Chem. **1983**, 2042; Can. J. Chem. **61**, 1766 (1983); Phytochemistry **24**, 1219 (1985) (biosynthesis). [4]Physiol. Plant Pathol. **6**, 177 (1975); Hock & Elstner, Schadwirkungen auf Pflanzen, 2nd edn., p. 255 ff., BI Wissenschaftsverl. 1988.
gen.: Agric. Biol. Chem. **49**, 2537 (1985) ▪ Tetrahedron **49**, 4761 (1993) (synthesis) ▪ Tetrahedron Lett. **38**, 1889 (1997) (synthesis (–)-R.) ▪ see also phytoalexins, phaseolin and pisatin. – *[CAS 18178-54-6 (R.); 31316-42-4 (rishitinol); 74299-59-5 (rishitinone)]*

Ritterazines. Group of more than 25 disteroidal alkaloids with a central pyrazine structure from the *tunicate *Ritterella tokioka* with potent cytotoxic and antitumor activities.

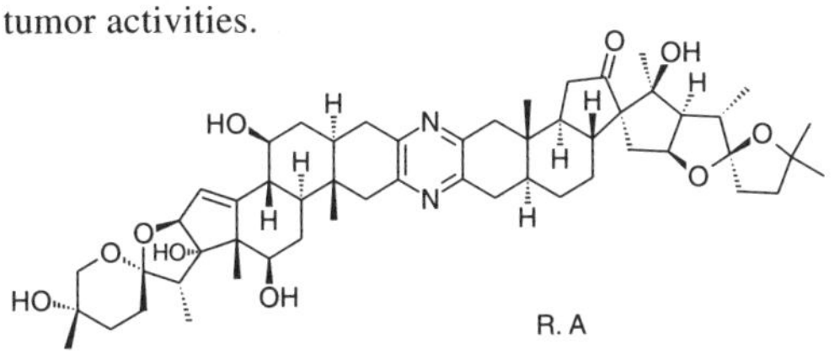

The R., e.g., *R. A:* $C_{54}H_{76}N_2O_{10}$, M_R 913.20, glass, $[\alpha]_D$ +112° (CH_3OH), are closely related to the *cephalostatins.

Lit.: J. Am. Chem. Soc. **117**, 10187 (1995); **121**, 2071–2084 (1999) (synthesis) ▪ J. Org. Chem. **59**, 6164 (1994); **62**, 4484 (1997) (isolation) ▪ Tetrahedron **51**, 6707 (1995) (isolation). – *[CAS 160391-62-8 (R. A)]*

Robinal see mites.

Robinetin (3,3′,4′,5′,7-pentahydroxyflavone). Formula see flavones. $C_{15}H_{10}O_7$, M_R 302.24, yellow-green needles (aqueous acetic acid), mp. 325–330 °C (decomp.). The flavonol (see flavones) R. occurs in *Robinia pseudoacacia* (Fabaceae), *Gleditsia monosperma* (Fabaceae), and other plants. A 1 mM solution of R. in ethanol is used for the photometric detection of Zr (uv_{max} 415 nm).

Lit.: Karrer, No. 1573 ▪ Sax (8.), RLP 000. – *[CAS 490-31-3]*

Robinin [*kaempferol 3-*O*-(6-*O*-rhamnosylgalactoside (=robinobioside))-7-*O*-rhamnoside].

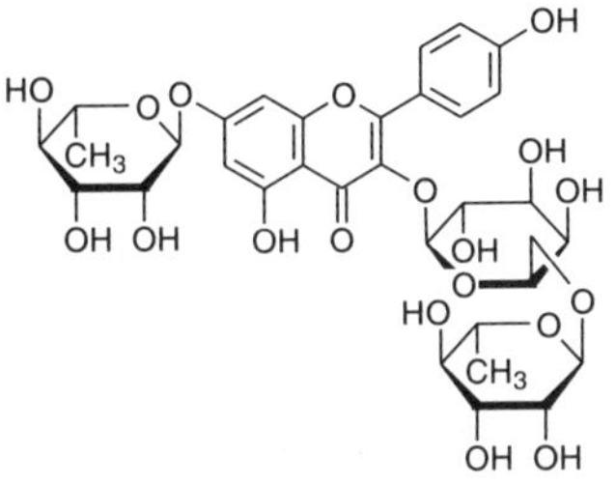

$C_{33}H_{40}O_{19}$, M_R 740.67, light yellow cryst. (C_2H_5OH), mp. 250–254 °C, soluble in hot water; hydrate: light yellow needles, mp. 195–199 °C. R. is a *flavone belonging to the flavonol glycosides occurring in *Robinia* species, e.g., *Robinia pseudoacacia* (Fabaceae) and in *Pueraria* and *Vigna* species (Fabaceae). R. has been reported to have antibacterial activity.

Lit.: Hager (5.) **6**, 1129 ▪ Karrer, No. 1510 ▪ Merck-Index (12.), No. 8404 ▪ Phytochemistry **15**, 215 (1976); **16**, 1811 (1977). – *[CAS 301-19-9]*

Robtein see chalcones.

Robustaflavone.

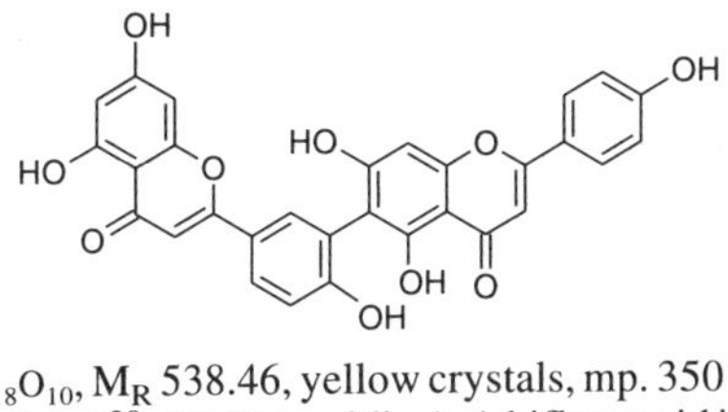

$C_{30}H_{18}O_{10}$, M_R 538.46, yellow crystals, mp. 350 °C (decomp.), $[\alpha]_D^{20}$ −37.5° (pyridine). A biflavonoid isolated from the seed kernel extract of *Rhus succedanea* and the leaf extract of *Agathis robusta*, inhibits hepatitis B viruses, probably via inhibition of DNA polymerase.

Lit.: Bioorg. Med. Chem. Lett. **7**, 2325 (1997) ▪ J. Org. Chem. **63**, 9300 (1998) ▪ Phytochemistry **41**, 1621 (1996); **42**, 1199 (1996). – *[CAS 49620-13-5]*

Rocaglamide.

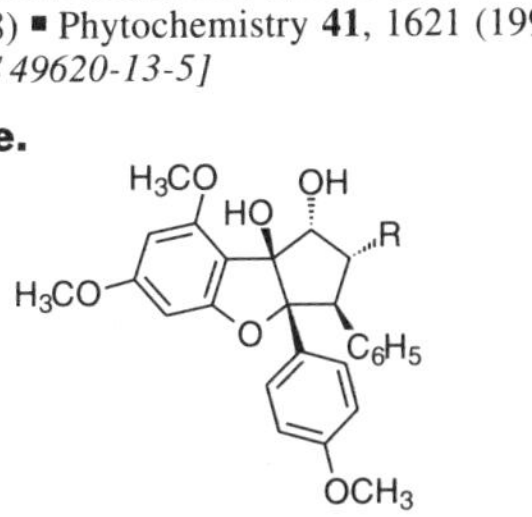

R = COOH : Rocaglatic acid
R = CO–N(CH$_3$)$_2$: Rocaglamide
R = H : Rocaglaol

$C_{29}H_{31}NO_7$, M_R 505.57, cryst., mp. 129–130 °C, $[\alpha]_D^{25}$ −96° ($CHCl_3$). A potent insecticidal and antileukaemic component from the roots and stems of *Aglaia elliptifolia* and *A. odorata* (Meliaceae). The de-carboxamido derivative, *rocaglaol* {$C_{26}H_{26}O_6$, M_R 434.49, cryst., mp. 76–77 °C, $[\alpha]_D^{20}$ −125° ($CHCl_3$)} also shows (weaker) insecticidal activity.

Lit.: J. Am. Chem. Soc. **112**, 9022 (1990) ▪ J. Chem. Soc., Chem. Commun. **1994**, 773 (isolation) ▪ J. Chem. Soc., Perkin Trans. 1 **1992**, 2657 (synthesis) ▪ Phytochemistry **32**, 67, 307 (1993) (isolation); **44**, 1455 (1997); **45**, 1579 (1997); **47**, 1163 (1998) ▪ Tetrahedron **53**, 17625 (1997) (isolation of congeners) ▪ Tetrahedron Lett. **40**, 4279 (1999) (synthesis of analogs). – *[CAS 84573-16-0 (R.); 147059-46-9 (rocaglaol); 136850-46-9 (rocaglatic acid)]*

Roccanin.

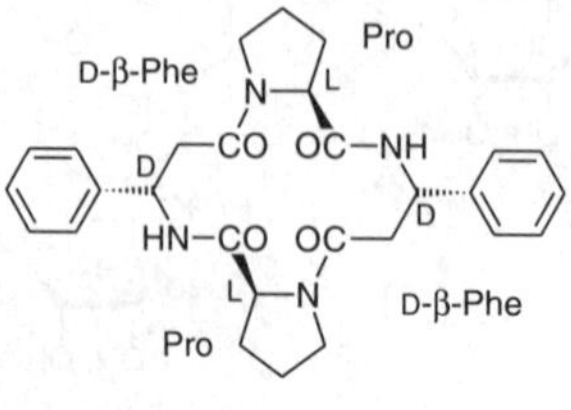

$C_{28}H_{32}N_4O_4$, M_R 488.59, cryst., mp. 320 °C, $[\alpha]_D^{25}$ −92°. A cyclic peptide from the dyer's lichens *Roccella canariensis* and *R. vicentina*.

Lit.: Tetrahedron **28**, 4625–4634 (1972). – *[CAS 30892-06-9]*

Roccellaric acid.

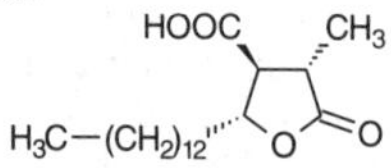

$C_{19}H_{34}O_4$, M_R 326.48, prisms, mp. 110–111 °C, $[\alpha]_D^{20}$ +35° ($CHCl_3$). A paraconic acid derivative from the lichen *Roccellaria mollis* (Roccellaceae).

Lit.: J. Org. Chem. **60**, 5628 (1995) ▪ Synlett **1996**, 343 (synthesis) ▪ Tetrahedron: Asymmetry **4**, 457–471 (1993) ▪ Tetrahedron Lett. **41**, 561 (2000). – *[CAS 19464-85-8]*

Roccellic acid.

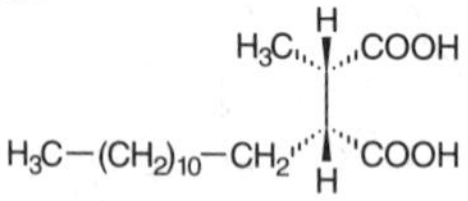

$C_{17}H_{32}O_4$, M_R 300.44, prisms, mp. 129–130 °C, $[\alpha]_D^{26}$ +18° (C_2H_5OH). Aliphatic dicarboxylic acid from various Roccellaceae such as, e.g., *Roccella fucoides*, *R. hypomecha*, *R. montagnei*, and other lichens.

Lit.: Culberson, p. 108 f. ▪ Karrer, No. 843 ▪ Z. Naturforsch., B **49**, 561–568 (1994). – *[CAS 29838-46-8]*

Rodaplutin see blasticidins.

Roemerine see aporphine alkaloids.

Rohitukin.

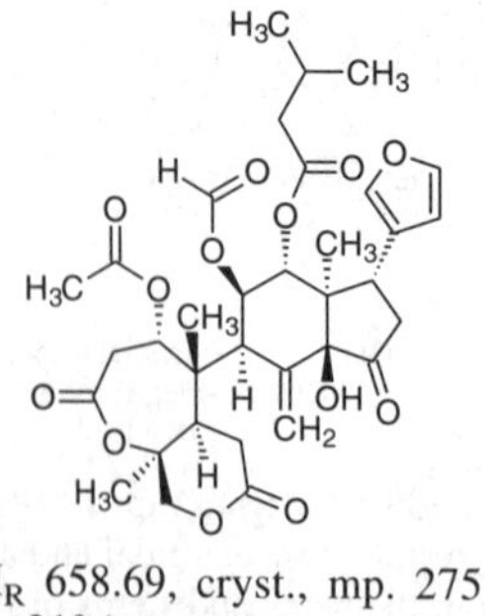

$C_{34}H_{42}O_{13}$, M_R 658.69, cryst., mp. 275–280 °C (decomp.), $[\alpha]_D$ −31° (CH_3OH). A constituent of the seeds of *Aphanamixis polystacha*.

Lit.: Phytochemistry **29**, 215–228 (1990). – *[CAS 62653-92-3]*

Rohitukine.

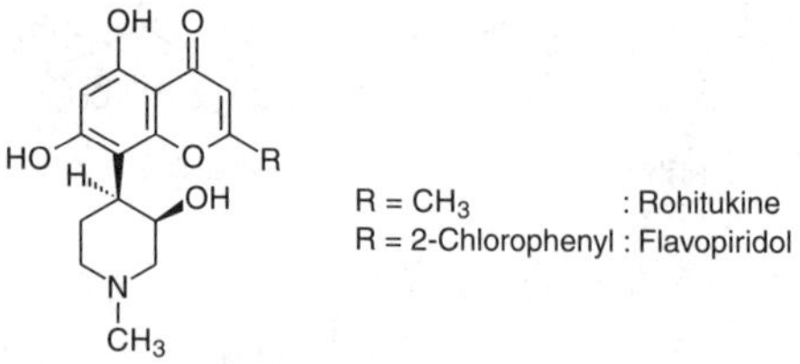

$C_{16}H_{19}NO_5$, M_R 305.33, cryst., mp. 218–219 °C, 227–232 °C, $[\alpha]_D^{20}$ +44.3° (CH_3OH). An alkaloid from leaves, stems, and bark of *Amoora rohituka* and *Dysoxylum binectariferum* (Meliaceae) with anti-inflammatory and immunomodulatory activity, lead structure for cancer therapeutics: e.g. the developmental product flavopiridol (tyrosine kinase antagonist).

Lit.: ACS Symp. Ser. **534**, 331–340 (1993). – *[CAS 71294-60-5]*

Rolliniastatins see Annonaceous acetogenins.

Roquefortines.

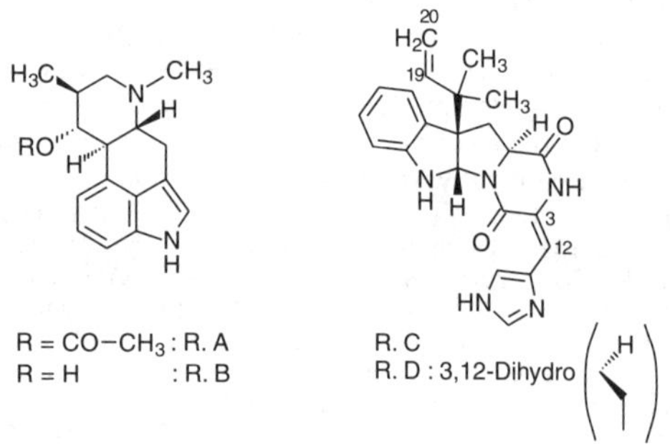

Table: Data of Roquefortines.

Roquefortine	molecular formula	M_R	mp. [°C]	CAS
A	$C_{18}H_{22}N_2O_2$	298.38	182 (decomp.)	58800-19-4
B	$C_{16}H_{20}N_2O$	256.35	222–225	58800-20-7
C	$C_{22}H_{23}N_5O_2$	389.46	195–200	58735-64-1
D	$C_{22}H_{25}N_5O_2$	391.47	153–154	58735-66-3

Neurotoxic *mycotoxins of the *indole alkaloid type from many mold fungi of the genus *Penicillium*, e.g., *P. roqueforti*, *P. cyclopium*, *P. commune*, *P. crustosum*, *P. chrysogenum*, *P. griseofulvum*. The R. occur in blue cheese and cotton seeds and have also been found in beer. The main alkaloid is R. C. R. can lead to abortions in animal experiments. Today, toxin-free strains are used for the production of cheese. R. C and R. D are formed from tryptophan, histidine (diketopiperazine part) and mevalonic acid. R. A and R. B are alkaloids of the clavine type (see ergot alkaloids). The LD_{50} for mice (i. p.) is 15 mg R. C/kg.

Lit.: Agric. Biol. Chem. **43**, 2035 (1979); **44**, 1929 (1980) ▪ Bot. J. Linnean Soc. **99**, 81–95 (1989) (occurrence, chemotaxonomy) ▪ Dev. Food. Sci. **8**, 463 (1984) ▪ Experientia **37**, 472 (1981) ▪ Hager (5.) **6**, 60–65 ▪ J. Agric. Food Chem. **31**, 655 (1983) ▪ J. Chem. Soc., Chem. Commun. **1979**, 225; **1982**, 652 (biosynthesis) ▪ Mycologia **81**, 837–861 (1989) ▪ Tetrahedron Lett. **39**, 8401 (1998) (synthesis R. D).

Roridins.

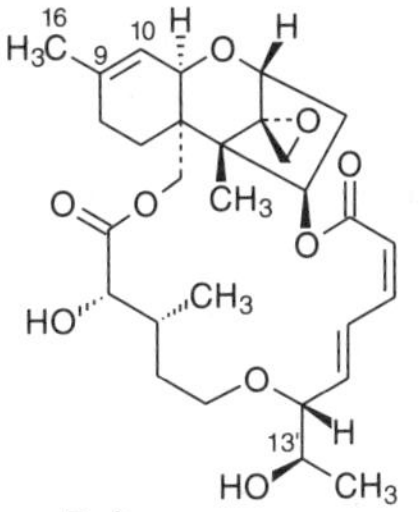

R. A
13'-Epimer: Iso-R. A

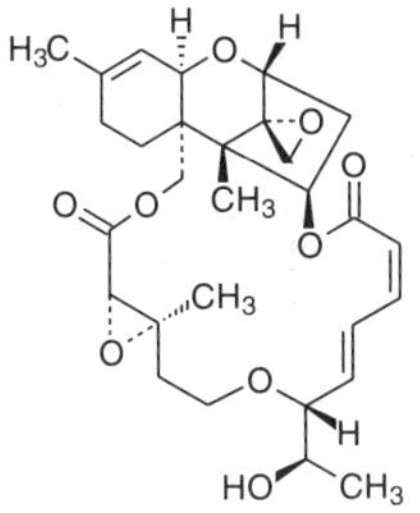

R. D

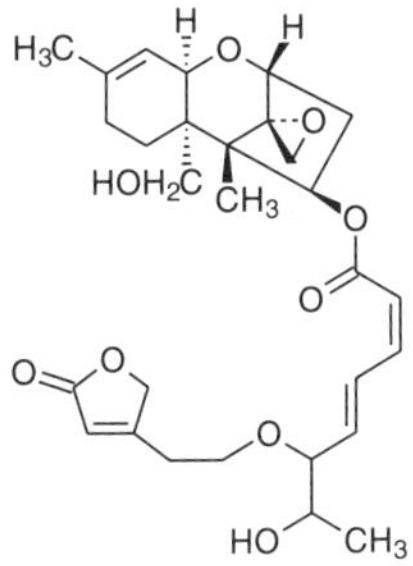

R. L2

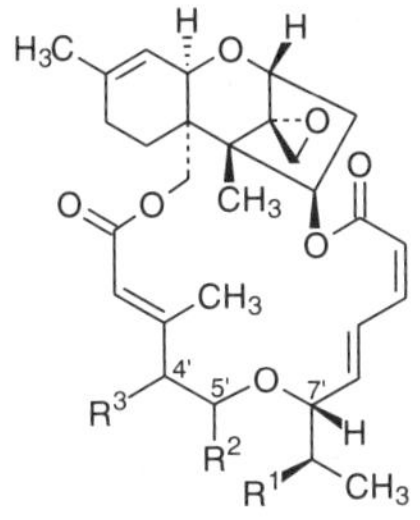

R^1 = OH , R^2 = R^3 = H : R. E
R^1, R^2 = —O—, R^3 = H : R. H
R^1, R^2 = —O—, R^3 = OH : R. J

Table: Data of Roridins.

Roridin	molecular formula	M_R	mp. [°C]	CAS
R. A	$C_{29}H_{40}O_9$	532.63	198–204	14729-29-4
Iso-R. A	$C_{29}H_{40}O_9$	532.63	183–185	84773-08-0
R. D	$C_{29}H_{38}O_9$	530.62	232–235	14682-29-2
R. E	$C_{29}H_{38}O_8$	514.62	183–184	16891-85-3
R. H	$C_{29}H_{36}O_8$	512.60	>325	29953-50-2
R. J	$C_{29}H_{36}O_9$	528.60	281–285	74072-83-6
R. L2	$C_{29}H_{38}O_9$	530.62	93–98	85124-22-7

A group of antifungal, phytotoxic, and cytostatic but highly toxic *mycotoxins of the *tric(h)othecene group in which the trichothecene skeleton is bridged by a macrocyclic ring originating from an acetogenin (exception: R. L2). R. are macrocyclic ether diesters, whereas the triesters belong to the *verrucarins. R. are produced by fungi of the genera *Myrothecium*, *Stachybotrys*, *Dendrodochium*, *Cryptomela*, and *Cylindrocarpon*. R. are precursors of the *baccharinoids. R. were originally isolated as antitumor compounds but proved to be too toxic for clinical use. The LD_{50} (i. v.) for mice is ≈ 1 mg/kg. R. C is identical to trichodermol (see scirpenols), R. E to *satratoxin D, and R. H to verrucarin H.

Lit.: Betina, chap 10 ▪ Cole-Cox, p. 230 ▪ J. Med. Chem. **27**, 239 (1984) ▪ J. Nat. Prod. **45**, 440 (1982); **59**, 254 (1996) ▪ Turner **2**, 228–238 ▪ Zechmeister **31**, 63–117; **47**, 153–219.

Rosanes. A class of diterpenes with the skeleton **1**, members of which have been isolated from various fungi and higher plants. The name is derived from the fungus *Trichothecium roseum*, foods infected with these fungi can cause poisonings.

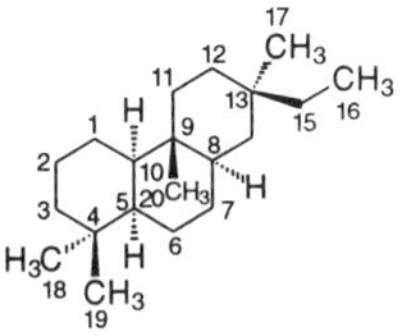

Rosane (**1**)

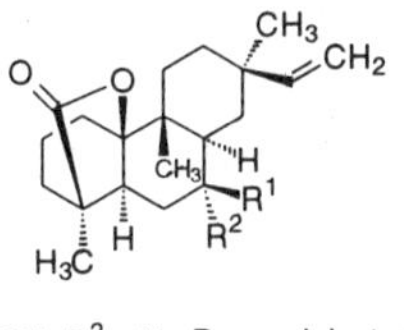

R^1 = OH, R^2 = H : Rosenololactone (**2**)
R^1, R^2 = O : Rosenonolactone (**3**)

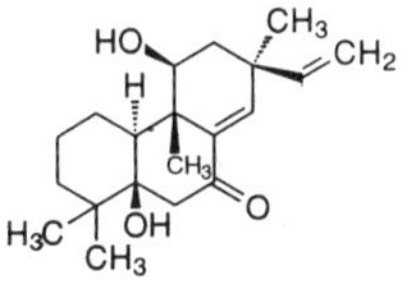

Ineketone (**4**)

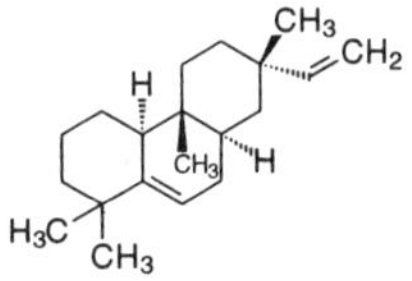

Rimuene (**5**)

Table: Data of Rosanes.

no.	molecular formula	M_R	appearance (mp. [°C])	$[\alpha]_D$ ($CHCl_3$)	CAS
2	$C_{20}H_{30}O_3$	318.46	cryst. (186)	+6.3°	508-69-0
3	$C_{20}H_{28}O_3$	316.44	cryst. (214)	–107.5°	508-71-4
4	$C_{20}H_{30}O_3$	318.46	cryst. (206–209)		62574-18-9
5	$C_{20}H_{32}$	272.47	cryst. (55)	+44.7°	1686-67-5

The responsible compounds include *rosenololactone* **2** and *rosenonolactone* **3**[1]. *Ineketone* **4** also has interesting properties, in rice it leads to a reduced ability to germinate[2]. The biosynthesis of R. proceeds through cyclization of a *labdane species to a *pimarane and subsequent migration of a methyl group.

Lit.: [1]Tetrahedron **24**, 3321 (1968); J. Am. Chem. Soc. **99**, 8327 (1977); Phytochemistry **17**, 572, 1119 (1978). [2]Phytochemistry **16**, 45 (1977).

gen.: Justus Liebigs Ann. Chem. **1984**, 162 ▪ Phytochemistry **23**, 192 (1984); **26**, 2543 (1987); **36**, 77 (1994) ▪ Tetrahedron Lett. **24**, 5043 (1983).

Rosefuran see mites.

Rosemary oil. Colorless to pale yellow, mobile, clear oil with a fresh, herby-camphor-like, bitter-sweet odor.
Production: By steam distillation from the flowering rosemary shrub (*Rosmarinus officinalis*). ***Origin:*** France, Spain, North Africa (Marocco, Tunisia). Annual production world-wide: ca. 250 t.
Composition[1]***:*** The quantitative composition depends on the origin. Spanish oils contain as main constituents *α-*pinenes*, *1,8-*cineole*, and **camphor* (ca. 20% of each), in North African oils the contents are 10, 40, and 10%, respectively.
Use: An important and versatile starting material in the perfume industry, especially for fresh, herby compositions; larger amounts of R. are used for perfuming technical products (soaps, shampoos, bath oils, sauna oils, etc.). In the food industry R. is used to improve the aromatic character of meat products, sauces, etc., and in sausage mixtures. Medicinal uses include rhin-

ologic agents, inhalation media, antirheumatic ointments, and balneotherapeutics.

Lit.: [1]Perfum. Flavor. **14** (2), 49 (1989); **16** (2), 59 (1991); **17** (6), 57 (1992); **20** (1), 47 (1995).

gen.: Arctander, p. 557 ▪ Bauer et al. (2.), p. 173 ▪ ISO 1342 (1988). – *Toxicology:* Food Cosmet. Toxicol. **12**, 977 (1974). – *[HS 3301 29; CAS 8000-25-7]*

Rosenololactone, rosenonolactone see rosanes.

Rose oil/absolute. Two different products obtained from rose flowers are used in the perfume and fragrance industries.

1) ***Rose oil:*** Light yellow, viscous oil with a warm, flowery-rose-like odor with spicy and honey-like nuances and a bitter, only in high dilution pleasant, taste.

Production: By steam distillation from flower petals of the rose species *Rosa x damascena* (Bulgaria, Turkey) and *Rosa centifolia* (Morocco).

Composition [1]***:*** The main components are (–)-**citronellol* (ca. 40%); **geraniol* (ca. 15%), and *nerol* (see geraniol) (ca. 7%). In contrast to rose absolute (see below), R. oil contains very little *2-phenylethanol (ca. 2%). A large number of compounds contributes to the typical odor of R. oil, some of which are only present in traces, e. g., **rose oxide* and *β-*damascenone*.

2) ***Rose absolute:*** Red-brown liquid with an adherent, sweet-balsamy rose-like odor.

Production: By solvent extraction of flowers of the rose species *Rosa centifolia* (southern France: "rose de mai"; Morocco).

Composition [1]***:*** The main component of the volatile fraction is *2-phenylethanol* (ca. 70%).

Use: Because of the laborious production and poor yields, R. oil and R. absolute belong to the most expensive starting materials for perfumes. The annual production world-wide is probably about 10 t. Thus, the natural rose products are used in very small amounts only, e. g., in expensive perfume oils or to improve the aromatic character of confectionery and bakery products.

Lit.: [1]Dtsch. Apoth. Ztg. **139**, 1378 (1999); Perfum. Flavor. **1** (1), 5; (6), 34 (1976); **3** (2), 47 (1978); **4** (2), 56 (1979); **16** (3), 43, 64 (1991); **17** (1), 55 (1992).

gen.: Arctander, p. 551, 561 ▪ Bauer et al. (2.), p. 173 ▪ ISO 9842 (1991) ▪ Ohloff, p. 152. – *Toxicology:* Food Cosmet. Toxicol. **12**, 979, 981 (1974); **13**, 911, 913 (1975). – *[HS 3301 29; CAS 8007-01-0]*

Roseophilin.

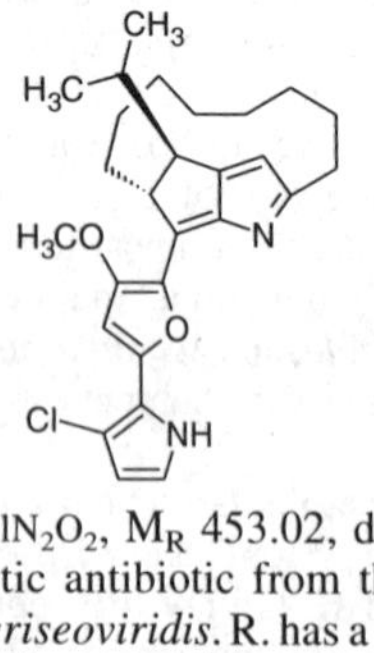

(abs. configuration)

$C_{27}H_{33}ClN_2O_2$, M_R 453.02, dark red powder. An antineoplastic antibiotic from the actinomycete *Streptomyces griseoviridis*. R. has a novel structure composed of pyrrole and furan rings with a branched, long-chain alkane bridge. Another *pyrrole pigment with three pyrrole units but without the furan system and the chlorine atom is *prodigiosin.

Lit.: Chem. Commun. **1999**, 1455 (synthesis) ▪ J. Am. Chem. Soc. **119**, 2944 (1997); **120**, 2817 (1998) ▪ J. Org. Chem. **64**, 2361 (1999) ▪ Tetrahedron Lett. **33**, 2701 (1992) (isolation). – *[CAS 142386-38-7]*

Roseotoxins. Tremorigenic *mycotoxins from *Trichothecium roseum* (on moldy peanuts). *R. B* ($C_{30}H_{49}N_5O_7$, M_R 591.75, mp. 200–202 °C) is a cyclic hexadepsipeptide of the *destruxin group ([1]-*trans*-3-methyl-destruxin A). *R. S* ($C_{23}H_{39}N_3O_8$, M_R 485.58) is a cyclic pentadepsipeptide. R. B has strong cytotoxic and insecticidal activities. The LD_{50} for chicks (p.o.) is 12.5 mg/kg.

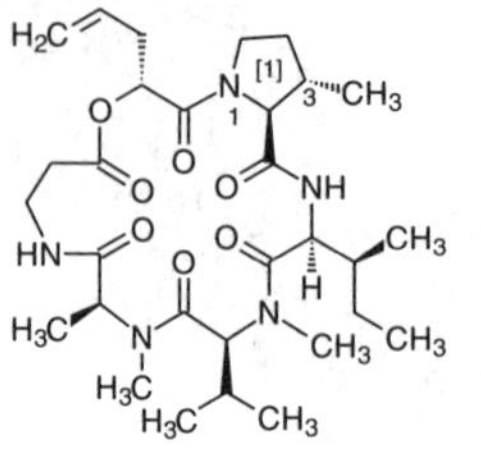

R. B

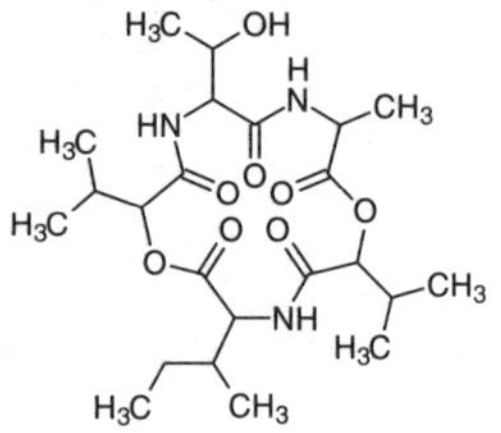

R. S

Lit.: Fresenius Z. Anal. Chem. **330**, 152 (1988) ▪ J. Am. Chem. Soc. **106**, 2388–2400 (1984) ▪ J. Antibiot. **41**, 1868–1872 (1988) ▪ Wein-Wiss. **42**, 111–119 (1987). – *[CAS 55466-29-0 (R. B); 109267-08-5 (R. S)]*

Rose oxide [4-methyl-2-(2-methyl-1-propenyl)tetrahydropyran].

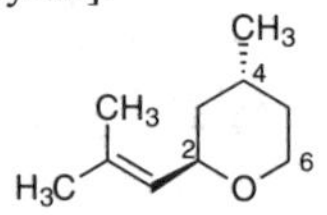

(2*R*,4*R*) (–)

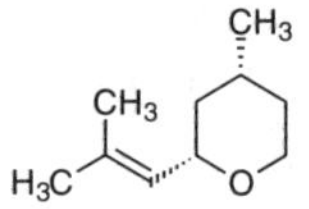

(2*S*,4*R*) (–)

$C_{10}H_{18}O$, M_R 154.25. A *monoterpene with the tetrahydropyran structure, occurring in nature in two C-2-epimeric configurations: (2*R*,4*R*)(–)-R. o., bp. 70–71 °C, $[\alpha]_D$ –41.5° (undiluted). (2*S*,4*R*)(–)-R. o., bp. 75–76 °C (12 hPa), $[\alpha]_D$ –18.0° (undiluted).

Occurrence: In *rose oil/absolute [1] and *geranium oil [2]. In Bulgaria bastard species with an intense odor, known as *Rosa damascena* (Rosaceae), are cultivated for rose oil production [3]. For synthesis, see *Lit.*[4]. R. o. is employed as a specific fragrance component in synthetic rose oil (0.12%) and in fine wood perfume oils.

Lit.: [1]Helv. Chim. Acta **42**, 1830 (1959). [2]Bull. Soc. Chim. Fr. **1961**, 645. [3]Hegnauer **VI**, 111. [4]Helv. Chim. Acta **70**, 1879 (1987); Pat. DE 19645922 (14. 5. 1998); Tetrahedron **48**, 7363 (1992) (synthesis).

gen.: Beilstein EV **17/1**, 256 ▪ Enantiomer **1**, 167 (1996) (biosynthesis). – *[CAS 16409-43-1 (R.); 5258-10-6 ((2R,4R)(–)-R.); 3033-23-6 ((2S,4R)(–)-R.)]*

Rose pigments. The broad color spectrum of rose flowers is due to *carotinoids (yellow color) and *anthocyanins (red color). The carotinoids exhibit a large structural diversity while that of the anthocyanins is small. The rose fruits owe their color to cartinoids, e. g. *rubixanthin.

Lit.: Angew. Chem. Int. Ed. Engl. **30**, 654–672 (1991).

Rosmarinic acid. $C_{18}H_{16}O_8$, M_R 360.32, cryst. (dihydrate), mp. 204 °C (decomp.), $[\alpha]_D^{20}$ +145°, the compound was first isolated from the leaves of *Rosmarinus officinalis* and later found in other species of the

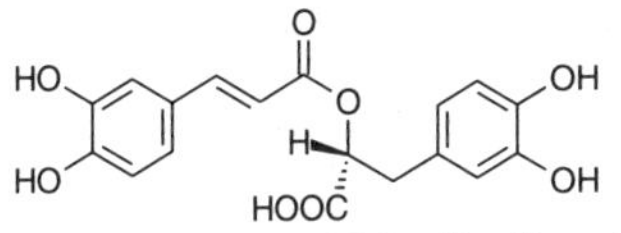

Lamiaceae. R. a. is also widely distributed in other plant families and often occurs together with *chlorogenic acid. R. a. has anti-inflammatory activity and is used in the treatment of rheumatic complaints. For synthesis, see *Lit.*[1].

Lit.: [1]Can. J. Chem. **75**, 1783 (1997); Recl. Trav. Chim. Pay-Bas **110**, 199–205 (1991); Synthesis **1996**, 755.
gen.: Luckner (3.), p. 388 ▪ Planta **137**, 287 (1977); **147**, 163 (1979) (biosynthesis) ▪ Zechmeister **35**, 86 f. – *Pharmacology:* Antiviral Chem. Chemother. **4**, 235 (1993) ▪ Drugs of the Future **19**, 756 (1985) ▪ SÖFW **124**, 635 (1998). – *[CAS 20283-92-5]*

Rotenoids.

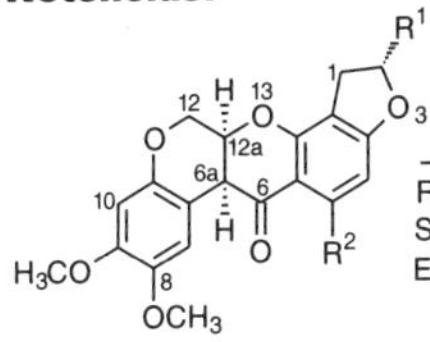

	R^1	R^2
Rotenone	$C(CH_3)=CH_2$	H
Sumatrol	$C(CH_3)=CH_2$	OH
Elliptone	H	H 1,2-didehydro

A group of compounds occurring in the roots and seeds of various *Derris*, *Lonchocarpus*, *Pachyrrhizus*, and *Tephrosia* (Fabaceae) species indigenous to or cultivated in tropical regions. These plants have been used in the Tropics for centuries for vermin control, preparation of arrow poisons, and to daze fish (so that they can be caught by hand). The main components of the dried roots responsible for these effects are shown in the figure: **rotenone*, *sumatrol* ($C_{23}H_{22}O_7$, M_R 410.42, mp. 196 °C), and *elliptone* ($C_{20}H_{16}O_6$, M_R 352.34, mp. 160 °C), other active principles are *deguelin*, *tephrosin*, and *toxicarol* (see tephrosin).
The R. are generally toxic. They are soluble in organic solvents. Chemically the C_{20} or C_{23} compounds are chromanofurochromones (see furocoumarins) or, respectively dichromanopyrones which can be considered as being derived from the *isoflavones; for biosynthesis, see *Lit.*[1]. The activity as an insecticide is based on an impairment of electron transport in the mitochondria caused by an inhibition of the citric acid cycle.

Lit.: [1]Nat. Prod. Rep. **1**, 3–19 (1984); Phytochemistry **49**, 1479–1507 (1998).
gen.: Beilstein E V **19/10**, 577, 642 ▪ J. Chem. Soc., Chem. Commun. **1988**, 1160 f. ▪ Karrer, No. 1422–1434 ▪ Kirk-Othmer (4.) **14**, 531 f. ▪ Sax (8.), p. 3000 ▪ Synthesis **1982**, 337–388 ▪ Tetrahedron **45**, 5895–5906 (1989) ▪ Ullmann (5.) **A 14**, 240 f. ▪ Zechmeister **43**, 98–120 ▪ see also rotenone.

Rotenone. $C_{23}H_{22}O_6$, M_R 394.42 (formula see rotenoids). Orthorhombic plates, mp. 165–166 °C (dimorphous, also 185–186 °C), $[\alpha]_D^{23}$ –177° ($CHCl_3$), readily soluble in alcohol, very readily soluble in chloroform. R., an insecticide and a fish toxin, is a *furocoumarin derivative from the derris root (*Derris elliptica*, Fabaceae) and represents the most important *rotenoid. R. is used as an insecticide with contact, respiratory, and feedant toxicity; in combination with *pyrethrum, piperonyl butoxide, and lindane in plant protection (as a synergist) against sucking and biting insects, caterpillars, worms, flea beetles; in household insecticides; in stores against flies and all types of pests, as an agent for washing animals to protect against warble flies (ectoparasiticide). Under the action of light and air, originally colorless solutions of R. change to yellow, orange, deep red, and finally dark brown as a result of autoxidation. The decomposition proceeds, among others, by hydroxylation at C-6a, dehydrogenation at C-6a/C-12a and oxygenation at C-12 to afford dehydrorotenone and rotenonone.
Activity: R. attacks the respiratory chain, blocks the oxidation of $NADH_2$ coupled with the reduction of *cytochrome b, and interrupts mitochondrial electron transport. R. is not toxic to bees, the LD_{50} (rat p. o.) is 133 mg/kg or (mouse i. p.) 2.8 mg/kg. The maximum allowed working place concentration is 5 mg/m^3. Use of R. is restricted by legislation, the estimated LD_{50} (human p. o.) is 0.3–0.5 g/kg (higher toxicity on inhalation).

Lit.: Beilstein E V **19/10**, 581 ▪ Hayes et al. (eds.), Handbook of Pesticide Toxicology, p. 599 f., New York: Academic Press 1991 ▪ J. Chem. Soc., Perkin Trans. 1 **1992**, 839, 851 (biosynthesis) ▪ J. Nat. Prod. **52**, 1363–1366 (1989) ▪ Kirk-Othmer (4.) **14**, 531 f. ▪ Merck-Index (12.), No. 8427 ▪ Sax (8.), RNZ 000. – *[HS 2932 99; CAS 83-79-4]*

Rottlerin (C. I. 75310).

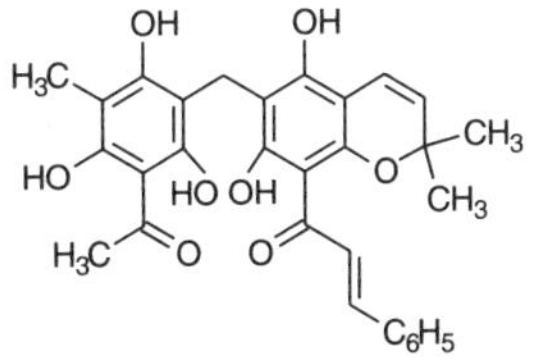

$C_{30}H_{28}O_8$, M_R 516.55, pale yellow to salmon pink cryst., mp. 212 °C. The most important and most toxic component of the Indian pigment drug *kamala.

Lit.: Beilstein E V **18/5**, 695 ▪ Karrer, No. 1696 ▪ Merck-Index (12.), No. 8429 ▪ Sax (8.), DMU 600 ▪ Synthesis **1994**, 255 (synthesis). – *[CAS 82-08-6]*

Rotundial.

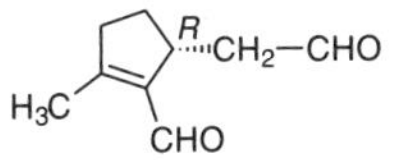

$C_9H_{12}O_2$, M_R 152.19, oil, $[\alpha]_D^{25}$ +39.3° ($CHCl_3$). A dialdehyde from the leaves of *Vitex rotundifolia*, Verbenaceae, with insecticidal activity. The leaves are traditionally used in Japan as a mosquito repellant. R. has the same activity against *Aedes aegypti* as DEET (*N,N*-diethyl-3-methylbenzamide).

Lit.: Biosci. Biotechnol. Biochem. **59**, 1979 (1995) ▪ Eur. J. Org. Chem. **1998**, 229 (synthesis). – *[CAS 64274-27-7]*

Royal jelly. Nutritional paste prepared by nurse bees from head gland secretions and honey stomach contents for rearing queen bee larvae. R. j. contains ca. 24% water, 31% protein, 15% carbohydrates, 15% ether-soluble components as well as 2% ash, and further trace elements such as iron, manganese, nickel, cobalt, etc., and various vitamins. Although synthetically prepared R. j. keeps the larvae alive it does not result in the development of a queen bee. R. j. differs from the breeding nutrition of worker bees by having higher concentrations of *neopterin, *biopterin, and *pantothenic acid[1]. It is doubtful that R. j. has any pharmacological benefit for humans. Even so it is commercially available in various formulations as a remedy for complaints of old age. It should not be confused with the so-called *queen substance.

Lit.: [1] Naturwiss. Rundsch. **26**, 95–102 (1973).
gen.: Am. Bee J. **116**, 560, 565 (1976); **123**, 39–44 (1983) ▪ Bull. Soc. Pharm. Lille **38**, 181–203 (1982) ▪ Donadieu, La Gelée Royale, Paris: Maloine 1975 ▪ Helv. Chim. Acta **64**, 1407–1412 (1981) ▪ Merck-Index (12.), No. 8434. – *[HS 0410 00]*

RTX see resiniferatoxin.

Rubber [Indian rubber, caoutchouc, from (South American) Indianic language: caa = tears and ochu = tree or cahuchu = crying tree]. R. is the name (according to DIN 53 501, 11/1980) for non-cross-linked but cross-linkable (vulcanizable) polymers with rubber-elastic properties at room temperature. At higher temperatures or under the action of deforming forces R. show viscous flow. Thus, under suitable conditions, R. can be processed into specific shapes. They are the starting materials for the manufacture of elastomers and other rubber products.
The cross-linking of R. requires the presence of functional groups, e. g., unsaturated carbon/carbon bonds, hydroxy or isocyanate groups. By means of these groups, the R. molecules can be coupled with each other intermolecularly (cross-linking) on reaction with multifunctional reagents in a process known as vulcanization.
R. are systematically classified into *natural rubber and synthetic rubber. In 1994, the world market for R. amounted to a production volume of ca. 14 500 000 t, of which ca. 9 000 000 t were synthetic R. and 5 400 000 t natural R. from the para rubber tree (*Hevea brasiliensis*)[1].

Lit.: [1] Gummi Fasern Kunstst. **46**, No. 1, 10–17 (1993); Chem. Eng. News (14.8) **1995**, 11–16.
gen.: Hofmann: Rubber Technology Handbook, New York: Hanser 1989 ▪ Kirk-Othmer (4.) **21**, 562–591 ▪ Morton, Rubber Technology, New York: Van Nostrand Reinhold 1987 ▪ Rubber Red Book, New York: Palmeton (annually) ▪ Ullmann (5.) **A 23**, 225–472. – *[HS 4001.., 4002..]*

Ruberythric acid (alizarin 2-*O*-*β*-primveroside).

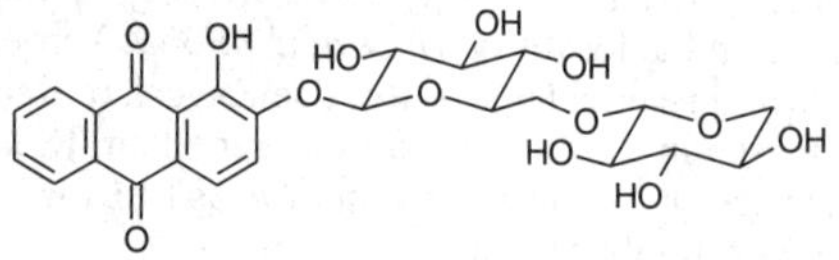

$C_{25}H_{26}O_{13}$, M_R 534.47, yellow prisms or needles, mp. 258–261 °C. R. a., isolated from the roots of *madder and from *Oldenlandia umbellata* (Rubiaceae); represents a storage form of *alizarin.

Lit.: Karrer, No. 1239 ▪ Merck-Index (12.), No. 8437 ▪ Synth. Commun. **26**, 4507 (1996). – *[CAS 152-84-1]*

Rubidomycin see daunorubicin.

Rubijervine see Solanum steroid alkaloids.

Rubixanthin [(3*R*)-*β*,*ψ*-caroten-3-ol, (3*R*)-3-hydroxy-*γ*-carotene, E 161 d].

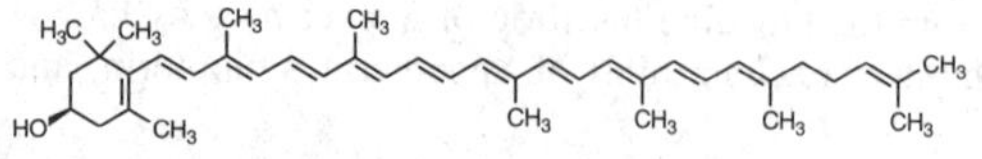

$C_{40}H_{56}O$, M_R 552.88, dark red cryst. with a metallic glance (C_6H_6/CH_3OH), mp. 160 °C, uv_{max} 509, 474, 439 nm. The xanthophyll (see carotinoids) R. occurs not only in plants, e. g., in ripe fruits of *Rosa rubiginosa*, *R. chamaemorus* (Rosaceae), and *Butea capitata* (Fabaceae) but also in microorganisms, e. g., in *Staphylococcus aureus*.

Lit.: Helv. Chim. Acta **66**, 494 (1983) (synthesis) ▪ Karrer, No. 1834 ▪ Merck-Index (12.), No. 8445. – *[CAS 3763-55-1]*

Rubomycin C see daunorubicin.

Rubratoxin B.

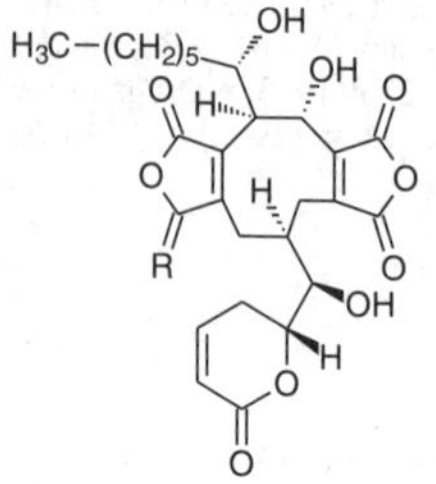

(relative configuration)

The mold fungi (*Penicillium rubrum*, *P. purpurogenum*) form the polycyclic *mycotoxin R. B {R = O, $C_{26}H_{30}O_{11}$, M_R 518.52, mp. 168–170 °C, $[\alpha]_D^{20}$ +67° (acetone)} that occurs as the main component together with *R. A* (R = H, OH, $C_{26}H_{32}O_{11}$, M_R 520.53) in moldy peanuts, grain, sunflower seeds, etc. R. B in moldy fodder consumed by animals can lead to hemorrhages as well as impairments of cell development in the liver, spleen, and kidney. LD_{50} (mouse p. o.) 120 mg/kg, (mouse i. p.) 0.3 and 6 mg/kg. In addition teratogenic effects in mice have been observed. R. B also has herbicidal activity.

Lit.: Arch. Microbiol. **118**, 7 (1978) ▪ Beilstein E V **19/10**, 693 ▪ Betina, chap. 14 ▪ Exp. Mol. Pathol. **50**, 193 (1989) ▪ Food Chem. Toxicol. **26**, 459 (1988) ▪ Sax (8.), RQP 000. – *[CAS 22467-31-8 (R. A); 21794-01-4 (R. B)]*

Rubrifacine.

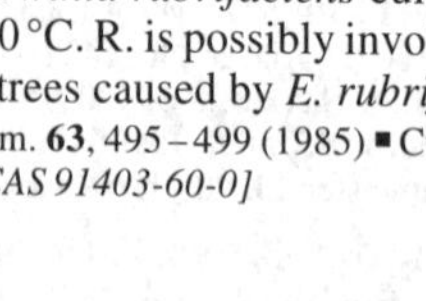

$C_{13}H_{10}N_2O_9$, M_R 338.23, forms a tetrasodium salt; uv_{max} 490 nm (pH 2), 522 nm (pH 9). The red pigment is formed in *Erwinia rubrifaciens* cultures at temperatures below 30 °C. R. is possibly involved in the damage to walnut trees caused by *E. rubrifaciens*.

Lit.: Can. J. Chem. **63**, 495–499 (1985) ▪ Curr. Microbiol. **10**, 169 (1984). – *[CAS 91403-60-0]*

Rubroflavin.

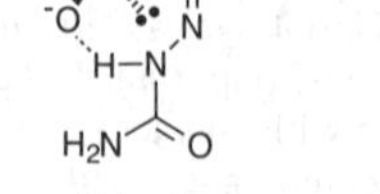

Rubroflavin

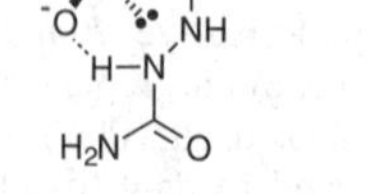

Leucorubroflavin

$C_9H_{11}N_3O_3S_2$, M_R 273.32, orange-red cryst., mp. 184–185 °C, $[\alpha]_D$ −2180° (CH_3OH). The *p*-quinone monosemicarbazone with a chiral sulfoxide group is responsible for the orange color of injured or dried fruit bodies of the North American puffball *Calvatia rubroflava* (Gasteromycetes). Similar compounds occur in *C. craniformis*[1]. On heating R. gives off nitrogen and isocyanic acid in an interesting fragmentation reaction to furnish 3,5-bis(methylthio)phenol. In the intact fungus R. exists as the colorless 4-hydrazinophenol derivative *leucorubroflavin*. Structurally related compounds are *calvatic acid and *xanthodermin.

Lit.: [1] Phytochemistry **45**, 997–1001 (1997).
gen.: Zechmeister **51**, 243–248.

Rubromycins.

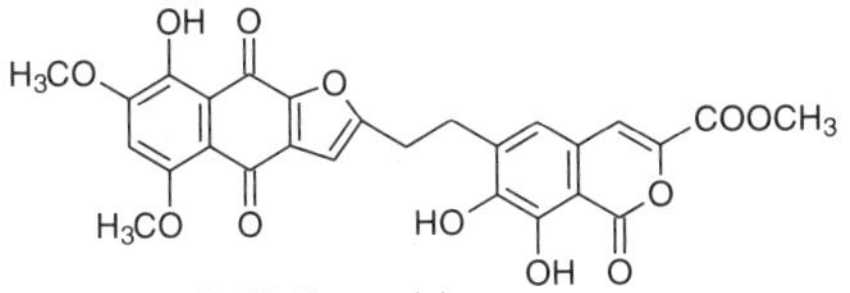

α-R. (Collinomycin)

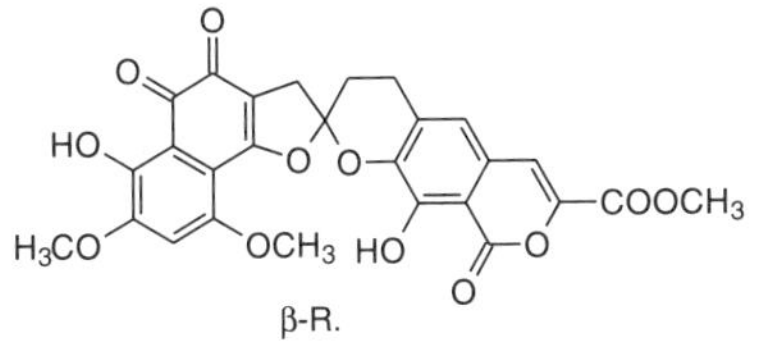

β-R.

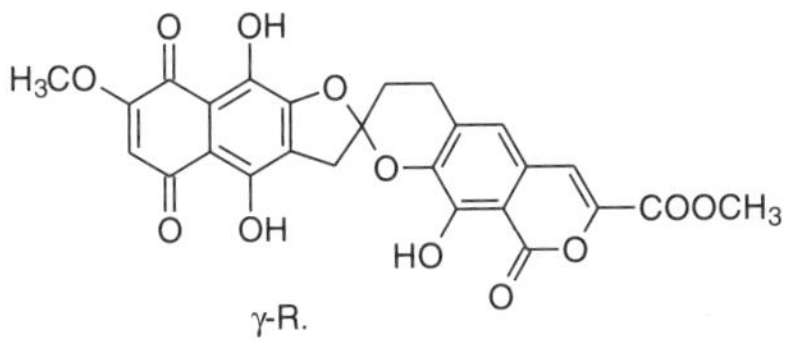

γ-R.

Red, crystalline quinone *antibiotics from cultures of *Streptomyces collinus* and *S. antibioticus* that can be separated into α-R. (*collinomycin*, $C_{27}H_{20}O_{12}$, M_R 536.45, mp. 278–281 °C), β-R. (*rubromycin*, $C_{27}H_{20}O_{12}$, M_R 536.45, mp. 225–227 °C), and γ-R. ($C_{26}H_{18}O_{12}$, M_R 522.42, mp. 235 °C). In addition to activity against cocci, the R. exhibit a selective inhibition of the reverse transcriptase of HIV-1[1].

Lit.: [1] Mol. Pharmacol. **38**, 20–25 (1990).
gen.: Beilstein EV **19/8**, 331; **19/11**, 137, 138 ▪ Helv. Chim. Acta **63**, 1400 (1980). – *[HS 2941 90; CAS 27267-69-2 (α-R.); 27267-70-5 (β-R.); 27267-71-6 (γ-R.)]*

Rubrosterone see androgens.

Rufoolivacins.

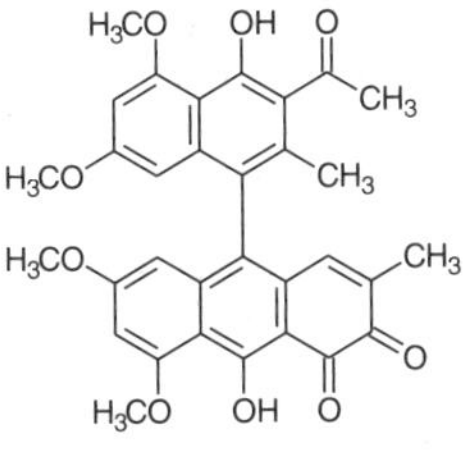

Rufoolivacin A

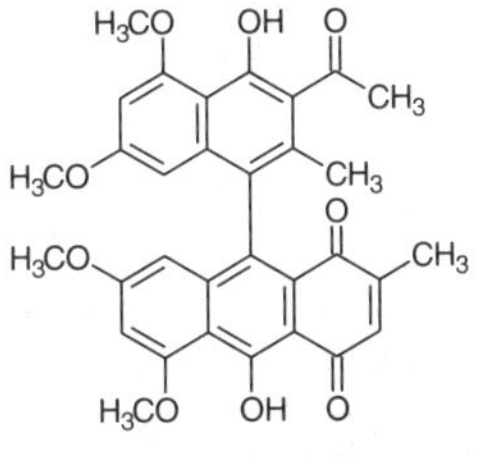

Rufoolivacin B

The red-violet toadstool *Cortinarius rufoolivaceus* (Basidiomycetes) contains violet dimers of *torachrysone 8-*O*-methyl ether and a 1,2- or 1,4-anthracenequinone, known as the R. They are formed from yellow-green precursors by atmospheric oxidation; the precursors contain an anthracene or, respectively, an anthrone unit in place of the anthracenequinone. The main pigments are *R. A* {$C_{32}H_{28}O_9$, M_R 556.57, dark red amorphous powder, mp. 303–307 °C, $[\alpha]_D$ +320° (CH_3OH)} and *R. B* {$C_{32}H_{28}O_9$, M_R 556.57, red amorphous powder, mp. 260 °C, $[\alpha]_D$ +240° (CH_3OH)}.

Lit.: Zechmeister **51**, 168 ff.

Rugosin see pyrrolnitrin.

Rugulosin.

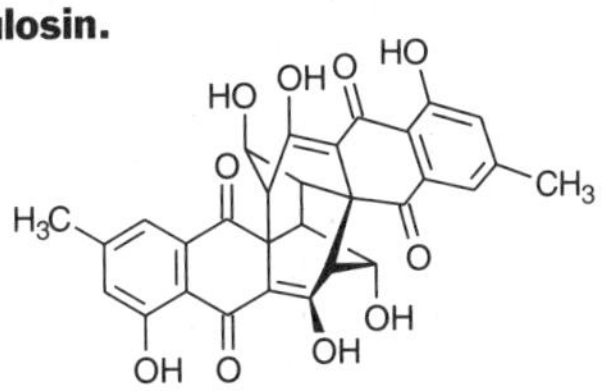

(+)-form

$C_{30}H_{22}O_{10}$, M_R 542.50, mp. 290 °C (decomp.), $[\alpha]_D$ +466° (dioxan). A modified and hepatotoxic *mycotoxin formed by various mold fungi, especially *Penicillium islandicum* [(−)-form] and *P. rugulosum* [(+)-form]. *P. islandicum* also contains 20 bianthraquinones and a series of simple *anthraquinones (e.g., *chrysophanol, *emodin, *islandicin, *skyrin, *luteoskyrin). R. and luteoskyrin are responsible for the symptoms of "yellow rice disease", a mycotoxicosis frequently occurring in Asia. R. is also produced by *Endothia parasitica* (the cause of death of chestnut trees), *Myrothecium verrucaria*, and lichens (*lichen pigments). LD_{50} for (+)-R. (rat i.p.) 44 mg/kg; not only (+)- but also (−)-R. have antibacterial and cytotoxic activities. (+)-R. induces liver cirrhosis and tumors in mice.

Lit.: Betina, chap 14 ▪ Cole-Cox, p. 696–702 ▪ Sax (8.), RRA 000 ▪ Tetrahedron Lett. **1978**, 3375 (biosynthesis). – *[CAS 23537-16-8 ((+)-R.); 21884-45-7 ((−)-R.)]*

Rum flavor. The intense rum flavor develops only on longer storage in barrels through oxidation, formation of esters, and extraction of substances from the oak wood. Esters dominate quantitatively (up to over 2000 ppm) and qualitatively among the ca. 500 known volatile components. The main component is ethyl acetate (see fruit esters) together with the ethyl esters of (C_1–C_{18}) fatty acids (especially even-numbered) and the acetates (C_3–C_5) of fusel alcohols. However, terpenes, sesquiterpenes, terpene and fusel alcohols, acetals and C_{13}-norisoprenoids (e.g., β-*ionones, *β-damascenone), phenols [especially *guaiacol, 4-ethylguaiacol (see coffee flavor), *eugenol] and phenol aldehydes {e.g., vanillin (see 3,4-dihydroxybenzaldehydes) and *syringa aldehyde* (4-hydroxy-3,5-dimethoxybenzaldehyd, $C_9H_{10}O_4$, M_R 182.18}, as well as the *whisk(e)y lactones make decisive contributions to the typical rum flavor.

Lit.: Maarse, p. 547–580 ▪ Morton-McLeod, Food Flavours Part B, p. 267–282, Amsterdam: Elsevier 1986 ▪ TNO list (6.), p. 527–548. – *[CAS 134-96-3 (syringa aldehyde)]*

Russulaflavidin.

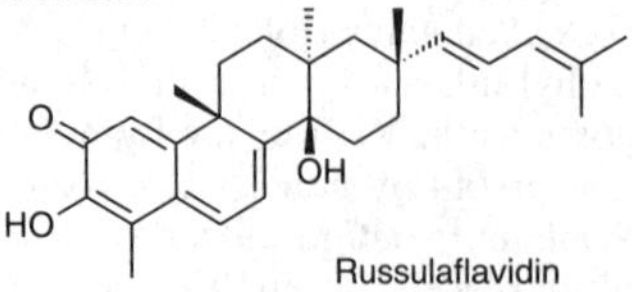

Russulaflavidin

$C_{28}H_{36}O_3$, M_R 420.59, yellow powder, mp. 78–80 °C, $[\alpha]_D$ –130° ($CHCl_3$). An orange-yellow pigment from the lipophilic outer layer of the agaric *Russula flavida* in which it is accompanied by a dihydro derivative. The pigments have the same quinonemethide system (*p*-quinonemethane chromophore) as pentacyclic triterpenoids of the pristimerin type (celastroids).

Lit.: Tetrahedron **51**, 2553 (1995). – *[CAS 162616-75-3]*

Russula pigments. The yellow, red, and violet colors of russula mushrooms (*Russula*, Basidiomycetes) are caused by polar compounds, some of which exhibit intense fluorescence under UV light and are associated with the *riboflavin- and *pteridine metabolisms (*russupteridines). *Russula flavida* is an exception, its fruit bodies have an orange-yellow layer consisting of lipophilic triterpenoids of the *russulaflavidin type.

Lit.: Z. Pilzkd. **39**, 45 (1973) (review) ▪ Zechmeister **51**, 210–215.

Russuphelins.

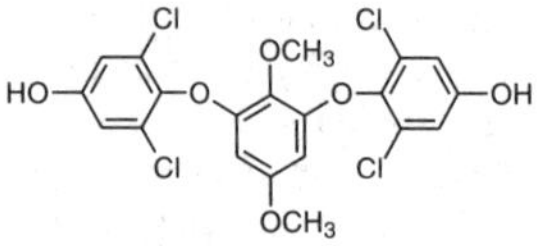

Russuphelin A

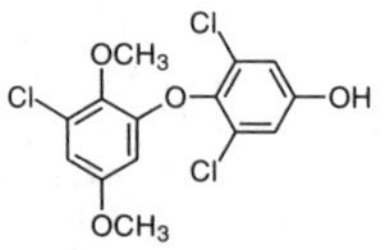

Russuphelin D

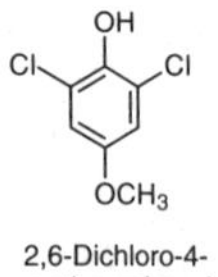

2,6-Dichloro-4-methoxyphenol

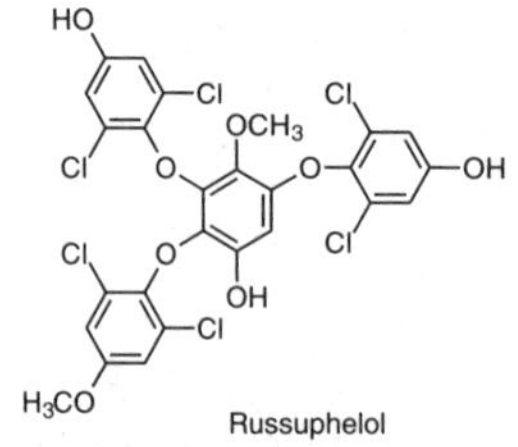

Russuphelol

Cytotoxic, polychlorinated phenol ethers from the toxic Japanese agaric *Russula subnigricans* (Basidiomycetes). 1 kg of fresh toadstools furnishes 240 mg *R. A* ($C_{20}H_{14}Cl_4O_6$, M_R 492.14, needles, mp. 293–294 °C) as well as 20 mg *R. D* ($C_{14}H_{11}Cl_3O_4$, M_R 349.60, needles, mp. 136–138 °C), and 125 mg 2,6-dichloro-4-methoxyphenol ($C_7H_6Cl_2O_2$, M_R 193.03). In addition smaller amounts of compounds with different methyl substitution or fewer *O*-methyl groups as well as a tetramer (russuphelol) have been isolated[1]. The R. exhibit *in vitro* cytotoxic activity against P388 leukemia cells (IC_{50}: 1–15 µg/mL). For total synthesis, see *Lit.*[2].

Lit.: [1]Tetrahedron Lett. **36**, 5223 (1995). [2]Synth. Commun. **27**, 107 (1997).

gen.: Chem. Pharm. Bull. **40**, 3185 (1992); **41**, 1726 (1993). – *[CAS 141794-49-2 (R. A); 155519-91-8 (R. D); 2423-72-5 (2,6-dichloro-4-methoxyphenol)]*

Russupteridines. The brightly colored cap skins of many *Russula* species (Basidiomycetes) contain, in addition to *riboflavin, water-soluble 2,4-pteridinedione derivatives (lumazines) which exhibit a strong fluorescence under UV light and were named as the R. by Eugster. Examples of R. widely distributed in *Russula* species are the violet fluorescent 6-methyl-8-D-ribityl-2,4,7-(1*H*,3*H*,8*H*)-pteridinetrione ($C_{12}H_{16}N_4O_7$, M_R 328.28, pale brownish solid) and *R. yellow I* ($C_{12}H_{16}N_6O_7$, M_R 356.30, yellow-brown needles, mp. >300 °C) which exhibits an intense green-yellow fluorescence in 0.1 N HCl. The red pigments appear to be dimeric pteridines.

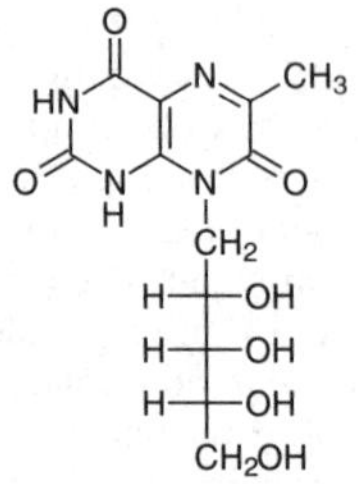

6-Methyl-8-D-ribityl-2,4,7-(1*H*, 3*H*, 8*H*)-pteridinetrione

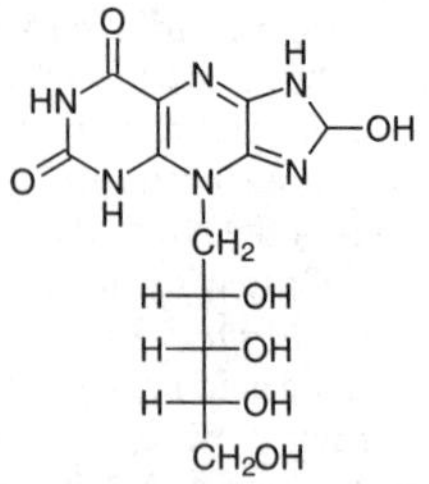

Russupteridine yellow I

Lit.: Helv. Chim. Acta **67**, 550, 570 (1984) ▪ Z. Pilzkd. **39**, 45 (1973) ▪ Zechmeister **51**, 210–215 (review). – *[CAS 70858-26-3 (R. yellow I)]*

Rutacridone see acridone.

Rutamycin A see oligomycins.

Rutin (3,3′,4′,5,7-pentahydroxyflavone 3-*O*-rutinoside, quercetin 3-*O*-rutinoside; C. I. 75730, natural yellow 10).

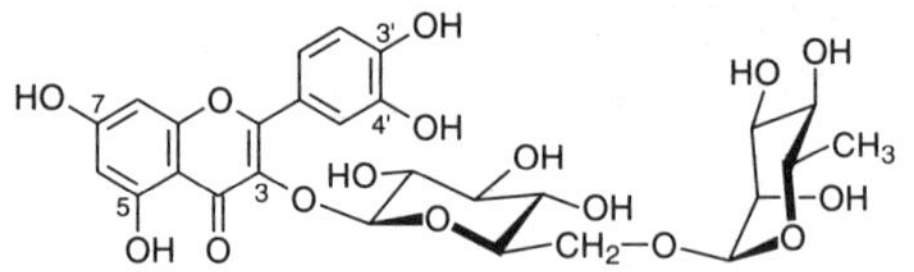

$C_{27}H_{30}O_{16}$, M_R 610.53. Pale yellow to greenish needles (trihydrate); anhydrous R. turns brown at 125 °C and decomposes at 214–215 °C. Anhydrous R. has the properties of a weak acid; it is poorly soluble in boiling water and soluble in alkaline solutions. The flavonol glycoside R. is a glycoside of *quercetin with *rutinose and occurs in many plant species – often accompanying vitamin C, e. g., in *Citrus* species, in yellow pansies, forsythia, acacia species, various *Solanum* and *Nicotiana* species, etc. R. was first isolated in 1842 by the German pharmacist Weiss from rue (*Ruta graveolens*, Rutaceae). It can also be obtained from the leaves of buckwheat, the East Asian dyer's drug *Sophora japonica*, (Fabaceae) which contains 13–27% R. Similarly to hesperidin, R. is used pharmacologically, mostly in the form of the acidic sodium salt, for treatment of capillary bleeding. Synthetic derivatives of R. are often used in the treatment of venous diseases and perfusion disorders[1]; see also flavonoids.

Lit.: [1]Phytother. Res. **5**, 19–23 (1991); Pulvertaft et al., Hydroxyethylrutosides in Vascular Disease, London: Academic Press 1981; Voelter & Jung, O-(*β*-Hydroxyethyl)-rutoside (2 vols.), Berlin: Springer 1978, 1983.

gen.: Beilstein E V **18/5**, 519 ▪ Eur. J. Biochem. **190**, 469 (1990) (activity) ▪ Florey **12**, 623–681 ▪ Hager (5.) **4**, 85 f., 727 f., 1047 f.; **5**, 137 ff., 184 f. ▪ Karrer, No. 1536 ▪ Merck-Index (12.), No. 8456 ▪ Sax (8.), p. 3006. – *[HS 2938 10; CAS 153-18-4]*

Rutinose [6-*O*-(α-L-rhamnopyranosyl)-D-glucose].

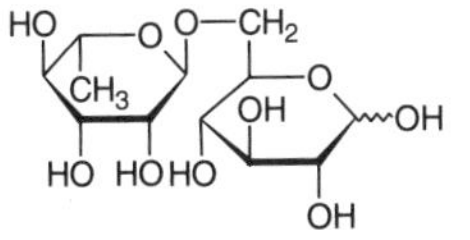

$C_{12}H_{22}O_{10}$, M_R 326.30. Highly hygroscopic powder, mp. 189–192 °C (decomp.), $[\alpha]_D^{10}$ +3.2→ –0.8° (H_2O). Hydrolysis of the *disaccharide with dilute HCl furnishes 1 mol. L-*L-rhamnose and 1 mol. D-*glucose. R. occurs in *rutin and hesperidin (see hesperetin) in glycosidically bound form (as *rutinoside*); other anthocyanidin rutinosides are responsible for the black color of black currants.

Lit.: Beilstein EV **17/6**, 269 ▪ Carbohyd. Res. **5**, 241 (1967); **10**, 105 (1969) ▪ Karrer, No. 639 ▪ Merck-Index (12.), No. 8457. – *[HS 2940 00; CAS 90-74-4]*

Ryanodine [ryanodol 3-(1*H*-pyrrole-2-carboxylate)].

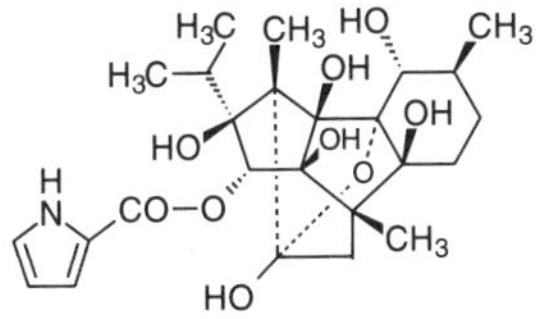

$C_{25}H_{35}NO_9$, M_R 493.55. cryst., mp. 219–220 °C (decomp.), soluble in water, chloroform, insoluble in petroleum ether, $[\alpha]_D^{25}$ +26° (CH_3OH). R. is a pyrrolecarboxylate of *ryanodol*, a pentacyclic, heptahydroxylated diterpene, and the insecticidal main component of the so-called *Ryania powder* (ryanex, ryanicide) from the dried and ground stems and roots of the indigenous South American plant *Ryania pyrifera* (syn. *R. speciosa*, Flacourtiaceae). R. acts both as a contact poison and as a feeding poison towards corn borers, peach moths, apple moths, and some cotton pests. R. is more stable towards light and air than other natural insecticides such as *nicotine, *rotenone, *pyrethrum; LD_{50} (rat p. o.) 750 mg/kg. R. decouples the ATP-ADP-actomyosin cycle in the contraction of longitudinally striped muscles. This is achieved by binding to the so-called R. receptors and blockade of the release of Ca^{2+} from the sarcoplasmatic reticulum. R. is the subject of biochemical investigations for the development of insecticidal crop protection agents with new mechanisms of action [1]. The total synthesis of (+)-ryanodol by Deslongchamps (1979) was a milestone in natural product synthesis [2].

Lit.: [1] J. Med. Chem. **39**, 2331, 2339 (1996). [2] Can. J. Chem. **57**, 3348 (1979).

gen.: Beilstein E III/IV **22**, 226 ▪ Biochemistry **33**, 6074–6085 (1994) ▪ Can. J. Chem. **68**, 115–192 (1990); **74**, 2424 (1999) (new R.) ▪ J. Bioenerg. Biomembr. **21**, 227–246 (1989) ▪ Karrer, No. 3705a ▪ Kirk-Othmer (4.) **14**, 533 ▪ Merck-Index (12.), No. 8459 ▪ Pestic. Biochem. Physiol. **48**, 145–152 (1994) ▪ Pesticide Manual, A World Compendium (9.), No. 10620, Farnham: The British Crop Protection Council 1991 ▪ Sax (8.), No. RSZ 000 ▪ Spec. Publ. R. Soc. Chem. **79** (Rec. Adv. Chem. Insect Control 2), 278–296 (1990) ▪ Trends Neurosci. **11**, 453–457 (1988) ▪ Ullmann (5.) **A 14**, 272 f. – *[CAS 15662-33-6]*

S

Sabadilla (cevadilla) seeds. Brown-black seeds, longish, lance-shaped with irregular edges, length 5–9 mm, thickness up to 2 mm, of *Sabadilla officinalis* (*Schoenocaulon officinale*, Liliaceae), a bulbous plant indigenous to Mexico, Venezuela, Guatemala, and Colombia. Extracts and powdered plant have insecticidal properties. S. s. contain 2–4% of an alkaloid mixture known as veratrine, composed of cevadine and veratridine as well as veratric acid, steroid alkaloids belonging to the *Veratrum steroid alkaloids. Veratridine and veratrine influence the Na^+ channels of muscle fiber membranes and lead to depolarization. An acetic acid extract, known as sabadilla vinegar, has a long history as a remedy for head lice. S. s. extracts have insecticidal activity against house flies, animal ticks, and some plant pests; they act both as contact and ingestive poison. The isolated components have strong irritating effects on the respiratory organs. Use of S. s. as an additive in snuff is forbidden in some countries.
Lit.: J. Agric. Food Chem. **44**, 149–152 (1996) ▪ Kirk-Othmer (4.) **14**, 532f. ▪ Sax (8.), p. 3487 ▪ Toxicon **31**, 1085–1095 (1993). – *[HS 1211 90]*

Sabinane see thujane.

Sabinene see thujenes.

Sabinene hydrate see thujanols.

Sabinic acid see hydroxy fatty acids.

Saccharates. Trivial name for salts of D-glucaric acid, formerly also for the basic compounds that precipitate when *sucrose solutions are treated with calcium or magnesium hydroxides.

Saccharides see carbohydrates, monosaccharides, disaccharides, oligosaccharides, and polysaccharides.

Saccharopine [*N*-((*S*)-5-amino-5-carboxypentyl)-L-glutamic acid].

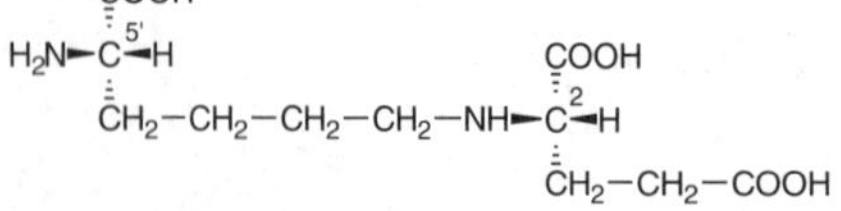

$C_{11}H_{20}N_2O_6$, M_R 276.29, mp. 240–248 °C (decomp.), $[\alpha]_D^{23}$ +33.6° (0.5 m HCl), +8.1° (0.5 m NaOH), pK_a 2.6, 4.1, 9.2, 10.3. A widely distributed non-proteinogenic amino acid in *Saccharomyces cerevisiae*, *Candida utilis*, *Lentinus edodes*, *Agaricus bisporus*, tobacco leaves (*Nicotiana tabacum*), and other plants.
Metabolism: S. is a direct precursor of lysine in the biosynthesis and the first product of lysine catabolism. The responsible enzyme is saccharopine dehydrogenase (EC 1.5.1.8). S. is degraded to 2-aminoadipic acid 6-hemialdehyde by another saccharopine dehydrogenase (EC 1.5.1.10).
Lit.: Agric. Biol. Chem. **42**, 1941 (1978); **43**, 1995 (1979) ▪ Justus Liebigs Ann. Chem. **1986**, 1030 (synthesis) ▪ Phytochemistry **17**, 991 f. (1978). – *[CAS 997-68-2]*

Saccharose. Synonym for *sucrose.

Sacculatanes.

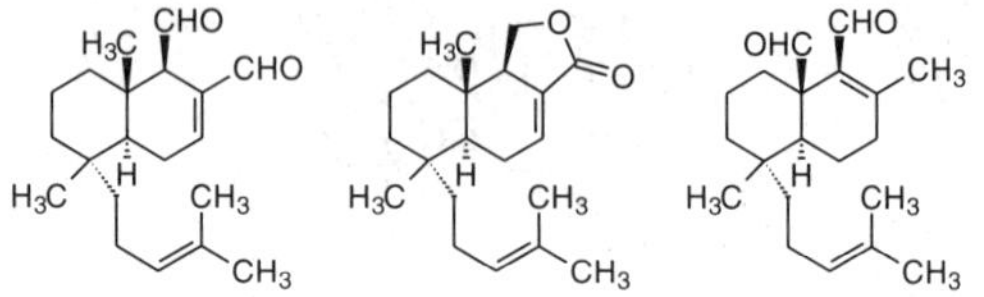

Sacculatal (**1**) Sacculatanolide (**2**) Perrottetianal A (**3**)

*Diterpenes such as *sacculatal*, *sacculatanolide*, and *perrottetianal A* (all: $C_{20}H_{30}O_2$, M_R 302.46) with the sacculatane skeleton have as yet only been found in liverworts. Sacculatal, mp. 65–66 °C, $[\alpha]_D^{20}$ −31.4° ($CHCl_3$) [1,2], is partly responsible for the biting hot taste of the thalli of *Pellia endiviifolia* and *Trichocoleopsis sacculata*. It can cause allergic contact dermatitis. It is also a strong fish poison [3]: LD_{100} 0.4 ppm (carp-like fish *Orizia latipes*) and exerts antifeeding activity on larvae of Japanese *Pieris* species (= white butterflies). It is classified as a tumor-promotor on account of its ornithine decarboxylase inducing activity [4]. Perrottetianal A is the bitter principle of *Porella perrottetiana* and inhibits the germination of rice.
Lit.: [1] Phytochemistry **17**, 153 (1977). [2] Tetrahedron Lett. **17**, 1407 (1977). [3] Phytochemistry **24**, 261 (1985). [4] Rev. Latinoam. Quim. **14**, 109 (1984).
gen.: Bull. Chem. Soc. Jpn. **62**, 624 (1989) (synthesis) ▪ J. Chem. Ecol. **15**, 2607 (1989) (activity) ▪ Zechmeister **42**, 120. – *[CAS 64242-90-6 (1); 64243-43-2 (2); 73483-87-1 (3)]*

Safflower see dye plants.

Safflower oil (thistle oil, dyer's thistle oil). Oil obtained from seeds of safflower (dyer's thistle, *Carthamus tinctorius*, Asteraceae). S. is of economic interest and has some dietetic importance on account of its high content of the essential *linoleic acid (70–80%) in triglyceride form; it also contains in particular palmitic acid (ca. 5%) and oleic acid (10–20%). New strains of safflower have high contents of oleic acid (>60%). – *[HS 1512 11, 1512 19; CAS 8001-23-8]*

Saffron (from Arabic zafaran = to be yellow). The dried, aromatic-smelling flowerheads of the saffron plant (*Crocus sativus*, Iridaceae), indigenous to southern Europe, which contain the yellow pigment crocin (see crocetin) and the bitter principle picrocrocin (saffron bitter, see safranal). Characteristic flavor compounds of S. are *safranal, isophorone and 2,6,6-trimethyl-1,4-cyclohexadiene-1-carboxaldehyde. S. is used as a spice for cooking, in the bakery and essence industries, and as a yellow colorant for liquors, inks, perfumes, hair waters, etc. 100 000 – 200 000 flowers are necessary to obtain 1 kg S.; the flowerheads must be removed manually. This accounts for the high price and the frequent falsifications. Reproduction proceeds only vegetatively through daughter corms. Saffron is one of the oldest textile dyes and is mentioned on an Accadian inscription dated at around 2000 BC.

Lit.: Hager (5.) **1**, 572–582 ▪ J. Agric. Food Chem. **45**, 459 (1997). – *[HS 0910 20]*

Saframycins. An antibiotic complex (to date 22 known components) from cultures of *Streptomyces lavendulae*. The S. contain tetrahydroisoquinoline units dimerized via a piperazine structure. They have antineoplastic activity and are active against Gram-positive bacteria, e. g., *S. B:* $C_{28}H_{31}N_3O_8$, M_R 537.57, orange prisms, mp. 108–109 °C, $[\alpha]_D^{20}$ –54.4° (CH_3OH).

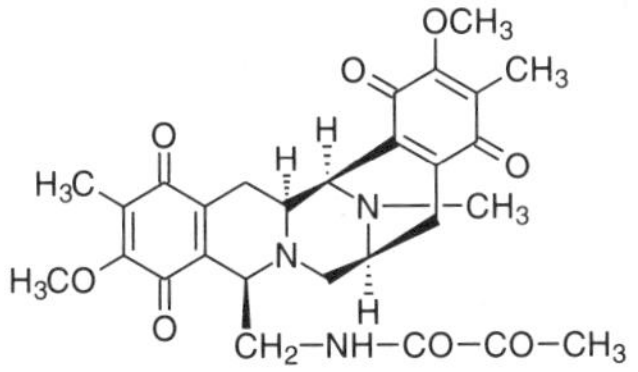

S. B

Lit.: Chem. Pharm. Bull. **43**, 777 (1995) ▪ J. Am. Chem. Soc. **121**, 10828 (1999) (synthesis) ▪ J. Antibiot. **47**, 5643 (1997) ▪ Org. Lett. **1**, 75 (1999) (stereochemistry) ▪ Tetrahedron **46**, 7711 (1990); **47**, 5643 (1991); **51**, 8213–8246 (1995). – *[CAS 66082-28-8 (S. B)]*

Safranal (2,6,6-trimethylcyclohexa-1,3-diene-1-carbaldehyde).

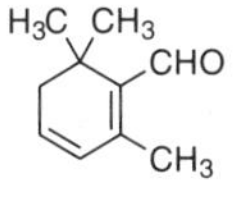

Safranal

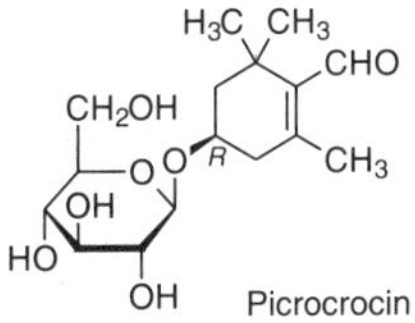

Picrocrocin

$C_{10}H_{14}O$, M_R 150.22, bp. 72 °C (4 hPa). A monocyclic *monoterpene aldehyde. S. is the odor principle of *saffron and is formed from the bitter saffron glucoside *picrocrocin* [$C_{16}H_{26}O_7$, M_R 330.38, mp. 156 °C, $[\alpha]_D^{20}$ –58° (H_2O)]. S. acts as an androtermone, i. e., the substance determining the male gender of the green alga *Chlamydomonas eugametus*. S. occurs in saffron oil (*Crocus sativus*, Iridaceae) [1] and is prepared from, e. g., *citral [2].

Lit.: [1] Phytochemistry **10**, 2755 (1971). [2] Tetrahedron Lett. **1974**, 3175.

gen.: Karrer, No. 415 ▪ Merck-Index (12.), No. 8467 ▪ Tetrahedron **54**, 2753 (1998). – *[CAS 116-26-7 (S.); 138-55-6 (picrocrocin)]*

Safrole [5-(2-propenyl)-1,3-benzodioxole, 4-allyl-1,2-methylenedioxybenzene].

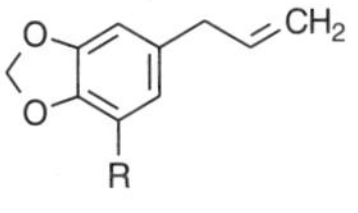

R = H : Safrole
R = OCH_3 : Myristicin

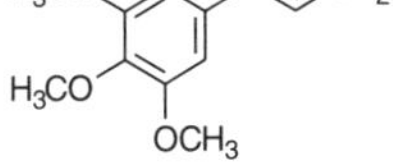

Elemicin

$C_{10}H_{10}O_2$, M_R 162.19, colorless to pale yellow liquid with a saffron-like odor, bp. 232–234 °C, prisms, mp. 11 °C. S. is the main component of sassafras oil (up to 90%) and also occurs in larger amounts in camphor oil. It is present together with smaller amounts of myristicin in many other essential oils, e. g., star anise, bay, fennel oils.

Myristicin [4-methoxy-6-(2-propenyl)-1,3-benzodioxole], $C_{11}H_{12}O_3$, M_R 192.21, oil, bp. 149–149.5 °C (2 kPa), 95–97 °C (27 Pa), soluble in ether and benzene, occurs in *parsley leaf/seed oil, in carrots, in anise oil, and in nutmeg, where it is accompanied by S. and *elemicin* [1,2,3-trimethoxy-5-(2-propenyl)benzene], $C_{12}H_{16}O_3$, M_R 208.26, oil, bp. 144–147 °C (13 Pa). Myristicin is responsible for the typical nutmeg odor.

Toxicology: S.: LD_{50} (rat, mouse p. o.) 1950–2350 mg/kg; myristicin: LD_{50} (rat p. o.) 4260 mg/kg.

Activity: S. has antiseptic and insecticidal activities and is cancerogenic in animal experiments [1], thus it may not be used for the production of food flavors, it can be used in amounts of up to 100 mg/kg in perfumes and cosmetics, or up to 50 mg/kg in dental and oral hygiene preparations. The hallucinogenic activity of nutmeg is attributed to myristicin [2]. Like other methylenedioxybenzene derivatives (*apiol, *dill apiol) it has insecticidal activity and is a synergist with other insecticides.

Use: In the perfume industry and for the denaturization of fats in the soap industry; for the production of *piperonal, isosafrole, piperonyl butoxide, and herniarin (see umbelliferone), as well as *prostaglandins.

Lit.: [1] Food Chem. Toxicol. **28**, 537 (1990); Food Cosmet. Toxicol. **19**, 657–666 (1981); Mutat. Res. **241**, 37–48 (1990). [2] Angew. Chem., Int. Ed. Engl. **10**, 370 (1971).

gen.: ACS Symp. Ser. **330**, 295 (1987) ▪ J. Org. Chem. **47**, 1983 f. (1982) ▪ Luckner (3.), p. 387 ▪ Phytochemistry **31**, 4263 (1992) (elemicin). – *Toxicology:* Food Chem. Toxicol. **28**, 537 (1990) ▪ Nature (London) **337**, 285 (1989) ▪ Sax (8.), MSA 500, SAD 000 ▪ Bioactive Mol. **2**, 139–159 (1987) (review). – *[HS 2932 90; CAS 94-59-7 (S.); 607-91-0 (myristicin); 487-11-6 (elemicin)]*

Safynol see polyynes.

Sage oils. Products named as sage oils are produced from two different sage species.

1. ***Dalmatian (medicinal) sage oil:*** Colorless to greenish-yellow oil with a characteristic strong, fresh, herby-camphor-like, warm spicy odor and a hot, bitter taste.

Production: By steam distillation from the herbage of common sage, *Salvia officinalis*.

Composition [1]***:*** Main components, principally responsible for the odor are *1,8-*cineole* (6–13%), **camphor* (3–9%) as well as *α-* and *β-thujone* (see thujan-3-ones) (8–43%) or 3–9% (cedar leaf oil).

Use: In the perfume industry for fresh, herby-spicy compositions, especially for masculine notes, to improve the aromatic character of liquors and bitters; however, the amounts used are limited by the toxicity of thujone. According to German laws, bitter liquors may only contain 35 ppm of thujone and other beverages and foods only 0.5 ppm. Medicinal uses: in mouth and throat preparations.

2. ***Spanish sage oil:*** Light, yellowish oil with a fresh, camphor-like odor resembling *Eucalyptus*.
Production: By steam distillation from the herbage of the Spanish sage, *Salvia lavandulifolia*.
Composition[2]***:*** Main components, principally responsible for the odor are *1,8-cineole* (ca. 20%) and *camphor* (ca. 30%). In contrast to the Dalmatian oil, the Spanish oil does not contain thujone.
Use: In the perfume industry, mainly for perfuming technical products with fresh notes, in bath oils and pharmaceutical products for rhinological use. In comparison with the Dalmatian oil, the Spanish oil is of lesser economic significance.

Lit.: [1]Perfum. Flavor. **13** (3), 53 (1988); **14** (6), 90 (1989); **16** (4), 51 (1991); **19** (6), 61 (1994). [2]Perfum. Flavor. **13** (1), 48 (1988); **15** (1), 62 (1990); **19** (2), 70 (1994).
gen.: Arctander, p. 570, 572 ▪ Bauer et al. (2), p. 174 f. ▪ ISO 3526 (1991), 9909. – *Toxicology:* Food Cosmet. Toxicol. **12**, 987 (1974); **14**, 857 (1976). – *[HS 3301 29; CAS 8022-56-8]*

Sakyomycins.

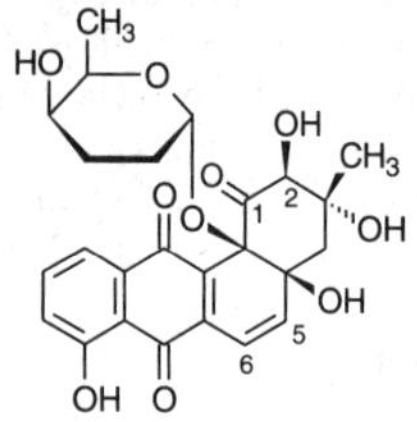

S. A

Table: Data of Sakyomycins.

S.	molecular formula	M_R	mp. [°C]	$[\alpha]_D^{20}$	CAS
A	$C_{25}H_{26}O_{10}$	486.48	205–207	–99.4° (C_2H_5OH)	86413-75-4
B	$C_{19}H_{16}O_8$	372.33		+31.6° (dioxan)	86470-27-1
C	$C_{25}H_{26}O_9$	470.48	143–145	–82.7° (CH_3OH)	86413-76-5
D	$C_{19}H_{18}O_8$	374.35	161–163	–140° (CH_3OH)	86413-77-6
E	$C_{26}H_{28}O_{10}S$	532.56			127395-67-9

Quinone antibiotics of the *angucyclin/angucyclinone type from cultures of *Nocardia* and *Streptomyces* sp. Acid hydrolysis of S. A furnishes S. B as aglycone and D-rhodinose. S. C is 2-deoxy-S. A, S. D is the 5,6-dihydro derivative of the aglycone S. B, S. E is the 5-methylthio derivative of S. A. All S. are yellow-red, optically active compounds with antibacterial activity, S. A is also active against AIDS viruses and S. E has cytostatic properties.

Lit.: J. Antibiot. **37**, 693 (1984) ▪ J. Chem. Soc., Chem. Commun. **1983**, 174 f. – *[HS 2941 90]*

Salamander steroid alkaloids. Toxic compounds from skin gland secretions of salamanders, e. g. the fire salamander (*Salamandra maculosa*) and the alpine salamander (*S. atra*) that are produced as defensive substances against bacteria and fungi. About 10 of these toxins have been reported to date, they are all derived from 3-aza-*A*-homo-5β-androstane. An important example is *samandarine*, which possesses an unusual oxazolidine partial structure. Another major alkaloid is 16-*O-acetylsamandarine* together with the corresponding 16-oxo compound *samandarone* (16-dehydrosamandarine). *Samanine* lacks the 1α,4α-epoxy bridge and *cycloneosamandione* possesses a cyclic 19-carbinolamine function.

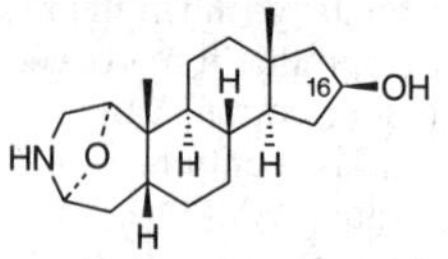

Samandarine

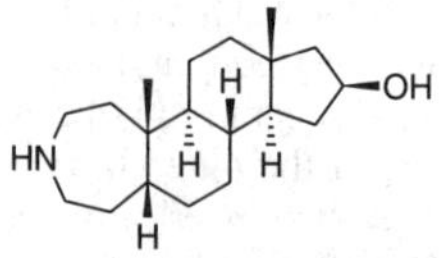

Samanine

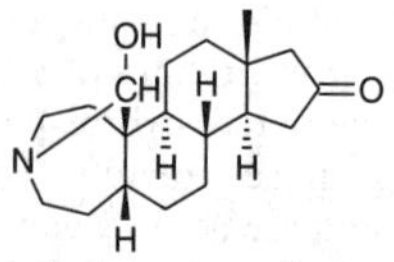

Cycloneosamandione

Table: Data of Salamander steroid alkaloids.

Alkaloid CAS	molecular formula	M_R	mp. [°C]	$[\alpha]_D$
Samandarine 467-51-6	$C_{19}H_{31}NO_2$	305.46	187–188	+43.7° (acetone)
16-*O*-Acetyl-samandarine 1857-07-4	$C_{21}H_{33}NO_3$	347.50	159	
Samandarone 467-52-7	$C_{19}H_{29}NO_2$	303.44	190	–116° (acetone)
Samanine 22614-24-0	$C_{19}H_{33}NO$	291.48	197	
Cycloneo-samandione 3148-28-5	$C_{19}H_{29}NO_2$	303.44	118–119	–207.6° (acetone)

The biosynthetic precursor of these compounds is *cholesterol; the *N*-atom of the alkaloids originates from *glutamine[1]. Samandarine causes respiratory impairment, hypertension, cardiac arrhythmia, and hemolysis of red blood particles. It is an effective local anesthetic but is too toxic for medical use (LD_{50} 1 mg/kg, rabbit).

Lit.: [1]Mothes et al., p. 363–384.
gen.: Manske **9**, 427–440; **43**, 185–288 ▪ Zechmeister **41**, 206–340.

Salazinic acid.

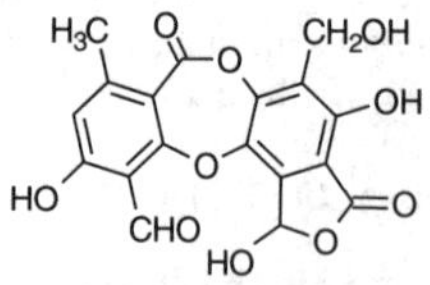

$C_{18}H_{12}O_{10}$, M_R 388.29, needles, mp. 260–280 °C (decomp.). *Depsidone with a 3-hydroxyphthalide group in the alcohol part of the molecule. It occurs in numerous lichens, especially of the Parmeliaceae family.

Lit.: Culberson, p. 161 ff. ▪ Karrer, No. 1065. – *[CAS 521-39-1]*

Salbomycin see elaiophylin.

Salbostatin.

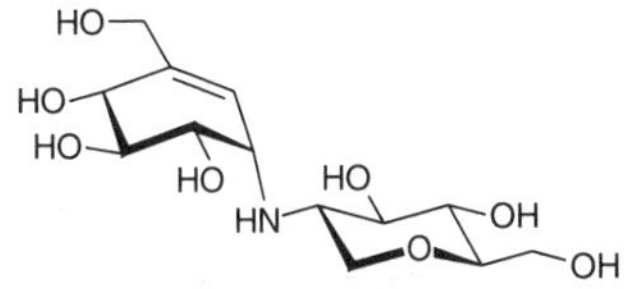

$C_{13}H_{23}NO_8$, M_R 321.33, amorphous, $[\alpha]_D^{20}$ +115° (H_2O), a basic, non-reducing pseudodisaccharide produced by the bacterium *Streptomyces albus*. S. inhibits the enzymes trehalase, saccharase, and maltase; as an inhibitor of glycosidase it is also of interest in therapy of diabetes mellitus (see acarbose). S. is being tested as a plant protection agent.

Lit.: Angew. Chem., Int. Ed. Engl. **33**, 1844 (1994) ▪ Bioorg. Med. Chem. Lett. **5**, 487 ff. (1995) ▪ Chem. Eur. J. **1**, 634 ff. (1995) (synthesis). – *[CAS 128826-89-1]*

Salicin see salicyl alcohol.

Salicortin.

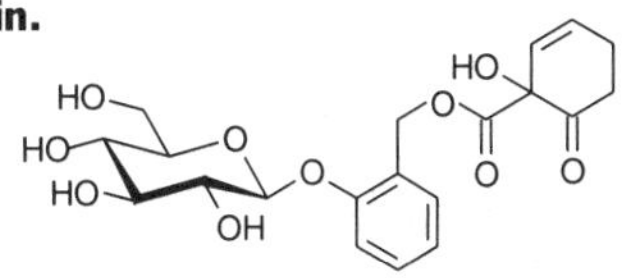

$C_{20}H_{24}O_{10}$, M_R 424.40, mp. 135–137 °C, occurs in the bark of willows (*Salix* spp.) and poplars (*Populus* spp.). It protects the plants from attack by insects (antifeedant)[1,2]. S. is an inhibitor of the β-glucosidase from sweet almonds[3] and kills larvae of the swallowtail *Papilio glaucus*[4].

Lit.: [1]ACS Symp. Ser. **380**, 130–142 (1988); Ecology **69**, 814–822 (1988); Phytochemistry **31**, 2180 f. (1992); Pestic. Biochem. Physiol. **35**, 185–191 (1989). [2]J. Chem. Ecol. **16**, 1941–1959 (1990). [3]Tetrahedron Lett. **31**, 4537 f. (1990). [4]Insect Biochem. **18**, 789 ff. (1988).

gen.: J. Nat. Prod. **52**, 207 ff. (1989); **57**, 808 (1994) ▪ Phytochemistry **28**, 2115–2125 (1989) ▪ Tetrahedron Lett. **31**, 4527–4538 (1990). – *[CAS 29836-41-7]*

Salicyl alcohol (saligenin, 2-hydroxybenzyl alcohol). $C_7H_8O_2$, M_R 124.14, needles (from water) or plates (from ether) with a burning taste, mp. 86 °C, subl. 100 °C; occurs in nature in the form of glycosides, see table.

Aqueous solutions of the glycoside salicin have a bitter taste. Salicin is enzymatically cleaved by emulsin and other glucosidases into glucose and S. S. gives a blue color with iron(III) chloride.

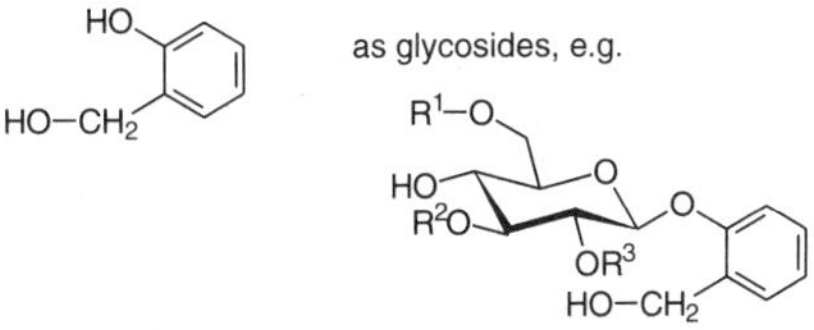

$R^1 = R^2 = R^3 = H$: Salicin
$R^1 = CO{-}CH_3$, $R^2 = R^3 = H$: Fragilin
$R^1 = CO{-}C_6H_5$, $R^2 = R^3 = H$: Populin
$R^1 = R^2 = H$, $R^3 = CO{-}C_6H_5$: Tremuloidin
$R^1 = H$, $R^2 = CO{-}C_6H_5$, $R^3 = H$: Chaenomeloidin

Activity: The antipyretic effects of willow and poplar barks were known in antiquity; these activities are attributable to the mentioned constituents. S. is used in the treatment of acute joint rheumatism and gout as well as in the synthesis of anion exchange resins.

Lit.: Beilstein E IV **6**, 5896 ▪ Biosci. Biotechnol. Biochem. **57**, 1185 ff. (1993) (enzymatic synthesis) ▪ Hager (5.) **9**, 552 f. ▪ Karrer, No. 256, 257, 6669 ▪ Luckner (3.), p. 400 ▪ Phytochemistry **31**, 2909 f. (1992); **33**, 161–164 (1993) ▪ Science **229**, 649 ff. (1985). – *[HS 2907 30; CAS 90-01-7]*

Salicylaldehyde see cassia oil.

Salicylic acid (2-hydroxybenzoic acid).

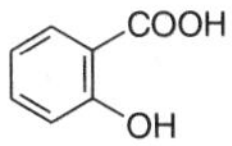

$C_7H_6O_3$, M_R 138.12, odorless crystals with a scratchy, sweet-sour taste and unpleasant aftertaste, mp. 157–159 °C, bp. 211 °C (2.7 kPa), subl. at 76 °C, can be steam distilled, pK_{a1} 2.98, pK_{a2} 13.6 (25 °C), discoloration on exposure to light, undergoes decarboxylation on rapid heating. Free S. occurs in all parts of meadow-sweet (*Filipendula ulmaria*, syn. *Spiraea ulmaria*), in senna leaves, and flowers of the genuine chamomile (*Chamomilla recutita*, syn. *Matricaria chamomilla*); Esters and glycosides of S. are widely distributed in essential oils and tree barks (e. g., gaul-

Table: Glycosides of Salicyl alcohol.

compound	molecular formula	M_R	mp. (bp.) [°C]	optical activity	occurrence	CAS
Salicin (Saligenin-2-*O*-β-glucosid, Salicosid)	$C_{13}H_{18}O_7$	286.28	needles 205–207 (240, decomp.)	$[\alpha]_D^{20}$ –62.56° (H_2O)	*Salix* spp., *Populus* spp.	138-52-3
Fragilin (6′-*O*-Acetyl-salicin)	$C_{15}H_{20}O_8$	328.32	needles 177–179	$[\alpha]_D^{15}$ –38.7° (H_2O)	*Salix fragilis*	19764-02-4
Populin (6′-*O*-Benzoyl-salicin)	$C_{20}H_{22}O_8$	390.39	180	$[\alpha]_D$ –2° (pyridine)	*Salix* spp., *Populus* spp.	99-17-2
Tremuloidin (2′-*O*-Benzoyl-salicin)	$C_{20}H_{22}O_8$	390.39	needles 207–208	$[\alpha]_D^{25}$ +17.1° (pyridine)	*Populus tremuloides*, *Salix* spp.	529-66-8
Chaenomeloidin (3′-*O*-Benzoyl-salicin)	$C_{20}H_{22}O_8$	390.39			*Salix chaenomeloides*	138101-84-

theria oil, wintergreen oil). S. has been detected in 20 of 27 economically significant plants. It often serves as a defensive substance and hormone in plants [1]. For the occurrence of S. in foods, see *Lit.*[2], for synthesis, see *Lit.*[3].
Activity: In plants S. influences the growth of flowers, buds, and roots (phytohormone activity) and, when added to the vase water, delays the withering of cut flowers. S. has antifungal and antibacterial activities and thus, e.g., prevents alcoholic fermentation of sugar, stops milk turning sour, and prevents the formation of acetic acid in alcoholic beverages, in the past it was used to conserve foods. As a keratinolytic agent, S. slowly and painlessly dissolves callous skin; it inhibits the biosynthesis of *prostaglandins and thus also acts as an analgesic, anti-inflammatory, antipyretic agent, and as an antirheumatic drug.
Toxicology: LD_{50} (mouse p.o.) 480 mg/kg, (mouse i.v.) 184 mg/kg.
Use: As intermediate in the industrial production of acetyl-S. (Aspirin) and in the dye industry and a conservation agent for inks, glues, and tanning agents, in the cosmetic industry as an additive to sun blockers and desodorants. Simple esters are used as components of perfumes, and for the photometric determination of iron, zirconium, scandium, ammonia, and fluoride. Some esters of S., similar to S. itself, are widely distributed in plants [4].

Lit.: [1] Plant Physiol. **99**, 799–803 (1992); **104**, 1109–1112 (1994); Trends Cell Biol. **4**, 334–338 (1994). [2] Ernähr. Umsch. **34**, 287–296 (1987); **37**, 108–112 (1990). [3] J. Mol. Catal. **73**, 237–248 (1992); J. Organomet. Chem. **470**, 257–261 (1994). [4] TNO list (6.), Suppl. 5, p. 291.
gen.: Anal. Profiles Drug Subst. Excipients **23**, 421–470 (1994) (review) ▪ Beilstein E IV **10**, 125 ▪ Chem. Unserer Zeit **33**, 213–220 (1999) (review) ▪ Hager (5.) **5**, 184 f.; **9**, 555 f. ▪ Karrer, No. 885 ▪ Kleemann-Engel, p. 1704 ▪ Sax (8.), SAI 000 (toxicology) ▪ Vane & Bottling (eds.), Aspirin and Other Salicylates, London: Chapman & Hall 1992 (pharmacology). – *[HS 2918 21; CAS 69-72-7]*

Saligenin see salicyl alcohol.

Salinomycin.

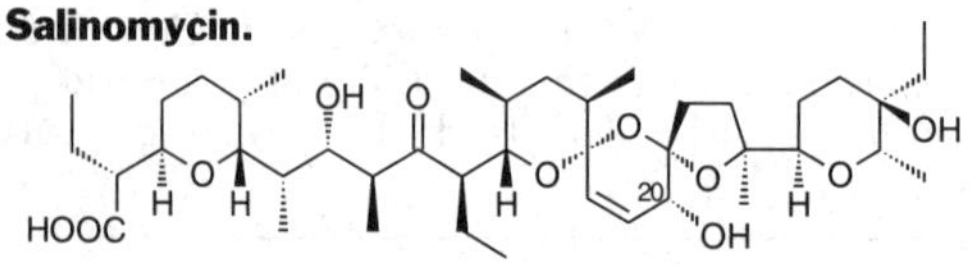

$C_{42}H_{70}O_{11}$, M_R 751.01, cryst., mp. 112.5–113.5 °C, $[\alpha]_D^{25}$ –63° (C_2H_5OH), an ionophoric *polyether antibiotic with a dispiroketal structure produced by *Streptomyces albus*; it preferentially forms complexes with monovalent cations (Na^+). S. is formed biosynthetically on the *polyketide pathway, cyclization in the linear carboxylic acid proceeds on the diene/diepoxide pathway. The 20-deoxy derivative is also formed. S. is used against coccidiosis in poultry breeding and also has antiviral activity (HIV). LD_{50} (mouse p.o.) 50 mg/kg.

Lit.: Antimicrob. Agents Chemother. **36**, 492 (1992) (activity) ▪ J. Am. Chem. Soc. **115**, 8414 (1993) (structure) ▪ Westley (ed.), Polyether Antibiotics, vols. 1 & 2, New York: Dekker 1983. – *Biosynthesis:* Birch & Robinson, in Vining & Stuttard (eds.), Genetics and Biochemistry of Antibiotic Production, p. 443–476, Boston: Butterworth-Heinemann 1995 ▪ Zechmeister **58**, 1–82. – *Synthesis:* Chem. Pharm. Bull. **37**, 1698–1726 (1989) ▪ J. Chem. Soc., Perkin Trans. 1 **1998**, 9–39 ▪ Spec. Publ. Soc. Chem. **198**, 42–60 (1997) ▪ Synlett **1994**, 415–419. – *[CAS 53003-10-4]*

Salsolidine see Anhalonium alkaloids.

Salutaridine see morphinan alkaloids.

SAM. Abbreviation for **S*-adenosylmethionine.

Samandarine, samandarone, samanine see salamander steroid alkaloids.

Sambucin see keracyanin.

Sambunigrin see cyanogenic glycosides.

Sambutoxin.

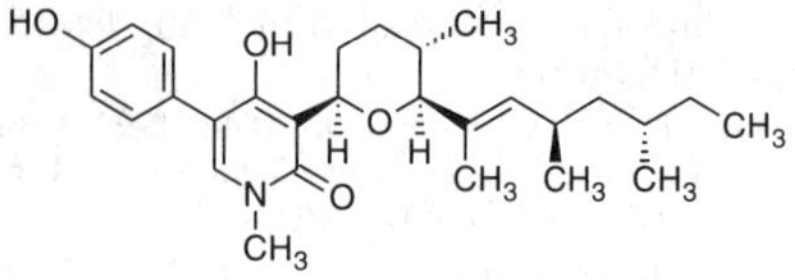

(relative configuration)

Mycotoxin from cultures of different *Fusarium* species, e.g., *F. sambucinum* and *F. oxysporum*: $C_{28}H_{39}NO_4$, M_R 453.62, prisms, mp. 197 °C, $[\alpha]_D^{25}$ –200° (CH_3OH). S. inhibits electron transport (complex III) in isolated mitochondria and is structurally related to funiculosin. It shows antifungal activity. S. is of mixed biosynthetic origin (polyketide chain and aromatic amino acid), see tenellin.

Lit.: J. Org. Chem. **61**, 8083–8088 (1996) ▪ Tetrahedron Lett. **36**, 1047 (1995). – *[CAS 160047-56-3]*

Sampangine.

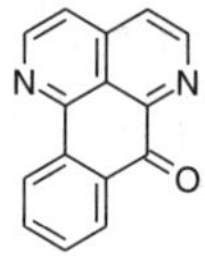

$C_{15}H_8N_2O$, M_R 232.24, yellow needles, mp. 210 °C (decomp.). A naphthonaphthyridine alkaloid from the trunk bark of *Cananga odorata* (ylang-ylang tree, Annonaceae). Related *Eupomatia* and *Cleistopholis* species contain methoxysampangine which, like S., exhibits activity against pathogenic fungi and mycobacteria.

Lit.: Antimicrob. Agents Chemother. **34**, 529 (1990) ▪ Heterocycles **34**, 1089 (1992) ▪ J. Heterocycl. Chem. **34**, 1233 (1997) ▪ Justus Liebigs Ann. Chem. **1989**, 87 ▪ Tetrahedron **53**, 6001 (1997). – *[CAS 116664-93-8]*

Sandalwood oil. Colorless to yellowish, slightly viscous oil with a typical woody-sweet, animalic-balsamy, very adherent odor.
Production: By steam distillation of the wood of the sandalwood tree *Santalum album* (Santalaceae) growing mainly in Southern India. The annual production ranges from 100 to 200 t. Since the sandalwood trees require several decades of growth before oil and wood can be obtained in economically useful amounts and since the distillation is laborious, the amounts of oil commercially available are limited and accordingly very expensive.
Composition [1]***:*** The main components are α- (ca. 50%) and β-**santalols* (ca. 20%). The latter component is responsible for the typical sandalwood odor.
Use: S. is a classical component of perfumes with wide uses. On account of its high price it is only used in very

expensive perfumes. The current requirements for sandalwood odors are also covered by the cheaper synthetic compounds[2].

Lit.: [1] Perfum. Flavor. **1** (1), 5; (5), 14 (1976); **6** (5), 32 (1981); **16** (6), 50 (1991). [2] Helv. Chim. Acta **81**, 1349 (1998); Müller-Lamparsky, p. 298.
gen.: Arctander, p. 574 ▪ Bauer et al. (2.), p. 175 ▪ ISO 3518 (1979) ▪ Ohloff, p. 172. – *Toxicology:* Food Cosmet. Toxicol. **12**, 989 (1974). – *[HS 3301 29; CAS 8006-87-9]*

Sandarac. Light yellow conifer resin consisting of a mixture of terpenoid resin acids, especially pimaric (80%), callitrolic (10%), and sandaricinic acids (10%). S. is obtained from the bark of the North African sandarac cypress [*Tetraclinis articulata* (*Callitris quadrivalvis*)]. Mp. 135 – 140 °C, soluble in many organic solvents and turpentine oil. S. is used in the manufacture of special paints, polishes, putty, dental cement, and as a substitute for *shellac (shellac wax).

Lit.: Merck-Index (12.), No. 8503. – *[HS 1301 90]*

Sandaracopimaranes see pimaranes.

Sanggenon D.

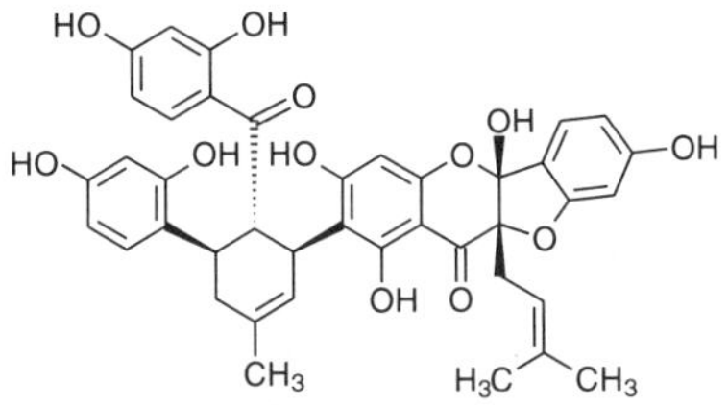

$C_{40}H_{36}O_{12}$, M_R 708.72, amorphous. $[\alpha]_D^{26}$ –145° (CH_3OH). S. D belongs to a group of 19 (S. A – S) currently known flavanone derivatives, the sanggenons, containing as further structural units isoprene residues and usually a *chalcone moiety. They are components of the root bark of the mulberry tree (*Morus* spp.), used in China as the drug "Sang-Bai-Pi". S. D has blood pressure-lowering activity.

Lit.: Heterocycles **23**, 2315 (1988); **45**, 867 (1997) (abs. configuration) ▪ Nat. Prod. Rep. **11**, 205 – 218 (1994) ▪ Planta Med. **47**, 30 (1983). – *[CAS 81422-93-7]*

Sanglifehrins.

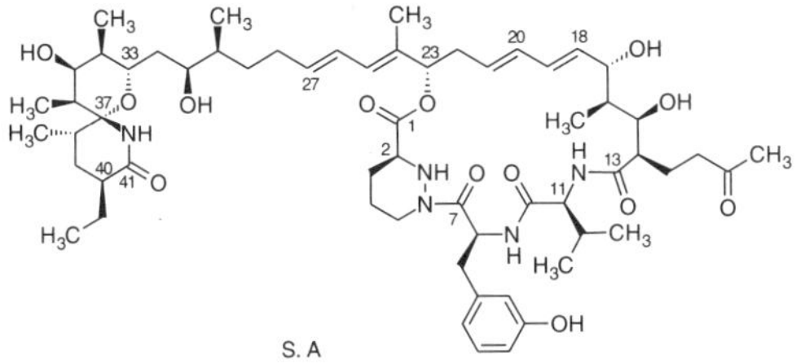

S. A

Group of more than 20 macrolides, isolated from microbial broth extracts of *Streptomyces flaveolus*, binding to the *cyclosporin-binding protein cyclophilin. The affinity of S. A, the most abundant component {$C_{60}H_{91}N_5O_{13}$, M_R 1090.39, white, amorphous powder, $[\alpha]_D^{20}$ –67.3° (CH_3OH)} for cyclophilin (IC_{50} = 2 – 4 nM) is about 20-fold higher than that of *cyclosporin A. It displays potent immunosuppressive activity in an *in vitro* immune response assay. It does not affect T-cell receptor-mediated cytokine production, indicating a mode of action different from that of cyclosporin A. S. are investigated as antiinflammatory and immunosuppressive agents.

Lit.: *Isolation:* J. Antibiot. **52**, 466, 474 (1999) ▪ Pat. WO 9807743 (21.8.1997). – *Synthesis:* Angew. Chem. Int. Ed. Engl. **38**, 2443, 2447 (1999) ▪ Chem. Commun. **1999**, 809 ▪ Tetrahedron Lett. **40**, 2109 (1999).

Sanguinarine see benzo[*c*]phenanthridine alkaloids.

Sanjoinine A see frangufoline.

Santal see isoflavones.

Santalenes.

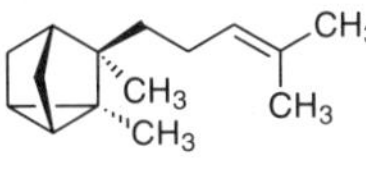

(-)-α-Santalene

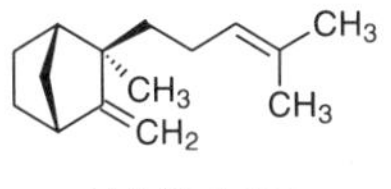

(-)-β-Santalene

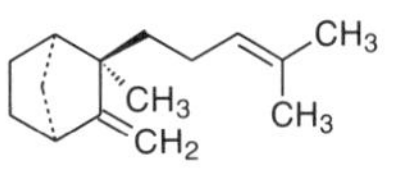

(+)-*epi*-β-Santalene

$C_{15}H_{24}$, M_R 204.36. The weakly smelling (–)-*α*-S. {oil, bp. 252 °C (100.1 kPa), $[\alpha]_D$ –15°}, (–)-*β*-S. with a cedar wood-like odor, {oil, bp. 125 – 127 °C (1.2 kPa), $[\alpha]_D^{20}$ – 107.8° ($CHCl_3$)} and the trace components (+)-*epi*-*β*-S. {oil, bp. 94 – 115 °C (3 – 7 hPa), $[\alpha]_D^{27}$ +26.4° ($CHCl_3$)} are constituents of the East Indian *sandalwood oil; see also santalols.

Lit.: Karrer, No. 1910 ▪ Müller-Lamparsky, p. 557 ff. ▪ Ohloff, p. 172 ff. – *Synthesis:* Chem. Lett. **1995**, 95 ▪ J. Org. Chem. **43**, 2282 (1978) (*α*-S.); **56**, 1434 (1991) (*β*-S.) ▪ Justus Liebigs Ann. Chem. **1994**, 601 ▪ Synth. Commun. **22**, 3159 ff. (1992) ▪ Tetrahedron: Asymmetry **1**, 537 (1990). – *[HS 2902 19; CAS 512-61-8 (α-S.); 511-59-1 (β-S.); 25532-78-9 (epi-β-S.)]*

Santalins.

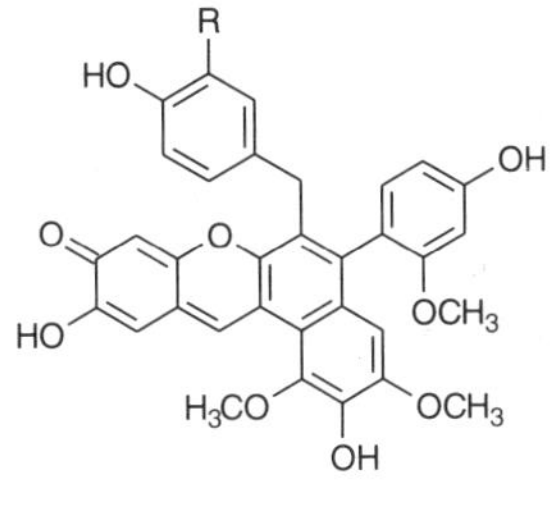

R = OH : Santalin A
R = OCH_3 : Santalin B

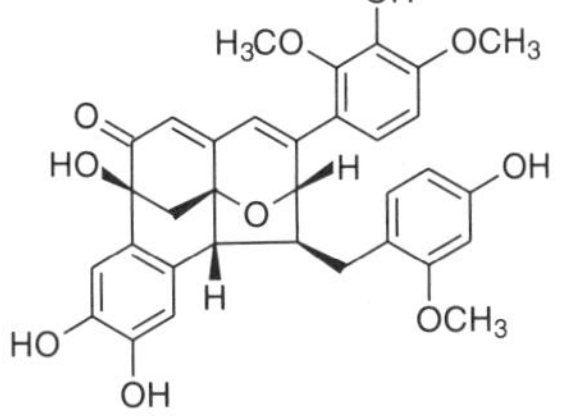

Santalin Y

Xanthene pigment mixtures from the heart wood of the red sandalwood (*Pterocarpus santalinus*, Fabaceae) and other *Pterocarpus* species, used in the past to dye wool and leather. *S. A*, $C_{33}H_{26}O_{10}$, M_R 582.56, red needles, mp. 302 – 303 °C (decomp.), uv_{max} 505 nm; *S. B*, $C_{34}H_{28}O_{10}$, M_R 596.59, red needles, mp. 292 – 294 °C, uv_{max} 505 nm; *S. C*, $C_{34}H_{28}O_{10}$, M_R 596.59, orange needles, mp. >300 °C, is a tetramethyl ether isomer of S.

B with an as yet unknown structure. S. Y has a tetracyclic core structure ($C_{33}H_{30}O_{10}$, M_R 586.60).

Lit.: J. Chem. Soc., Perkin Trans. 1 **1975**, 186; **1977**, 2118 ▪ Tetrahedron Lett. **36**, 5599 (1995). – *[CAS 38185-48-7 (S. A); 51033-46-6 (S. B); 167425-77-6 (S. Y)]*

Santalols.

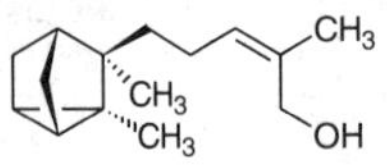

(+)-(*Z*)-α-Santalol

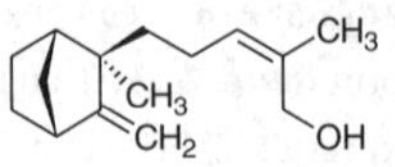

(-)-(*Z*)-β-Santalol

The weaker smelling (+)-(*Z*)-*α*-S., $C_{15}H_{24}O$, M_R 220.35, oil, bp. 166–167 °C (1.86 kPa), $[\alpha]_D^{23}$ +17.5° ($CHCl_3$) and (–)-(*Z*)-*β*-S., oil, bp. 177–178 °C (2.26 kPa), $[\alpha]_D^{21}$ –113° ($CHCl_3$) constitute ca. 70–90% of the expensive, East Indian *sandalwood oil (*Santalum album*) and are responsible for the heavy wood-like character[1]. (–)-*β*-S. is purported to have a urine-like animal tone which, however, cannot be detected by some people; ca. 40% are not able to appreciate this odor. The two alcohols are accompanied by the strongly-smelling aldehydes, the santalals (as *E/Z*-mixture)[2]. The sandalwood note is also expressed by some analogues of S., some terpenylcyclohexanols[3], substituted campholenes[4], ionols[5] as well as some *trans*-decalols. The molecular requirements for the generation of sandalwood notes (sandalwood rule) have been well defined by numerous structure-odor investigations, including computer modeling[6]. Photooxidation of *santalenes furnishes photosantalols with a pronounced sandalwood odor.

Lit.: [1]Helv. Chim. Acta **59**, 737 (1976); Karrer, No. 1911; Merck-Index (12.), No. 8505 f.; Ohloff, p. 172 ff.; Müller-Lamparsky, p. 298 ff., 557 ff. [2]Tetrahedron Lett. **1968**, 1533; **1980**, 2405. [3]Helv. Chim. Acta **47**, 1766 (1964). [4]Müller-Lamparsky, p. 298 ff. [5]Helv. Chim. Acta **68**, 1961 (1985). [6]Flav. Frag. J. **3**, 173 (1988).

gen.: Synthesis: Arch. Pharm. (Weinheim, Ger.) **330**, 112 (1997) ▪ J. Org. Chem. **48**, 1988 (1983) ▪ Justus Liebigs Ann. Chem. **1982**, 1105; **1994**, 601 (*β*-S.) ▪ Tetrahedron: Asymmetry **1**, 537 (1990) (*α*-S.). – *[CAS 115-71-9 (α-S.); 77-42-9 (β-S.)]*

Santiaguine see truxillic and truxinic acids.

Santonin [(11*S*)-3-oxo-1,4-eudesmadien-12,6*α*-olide].

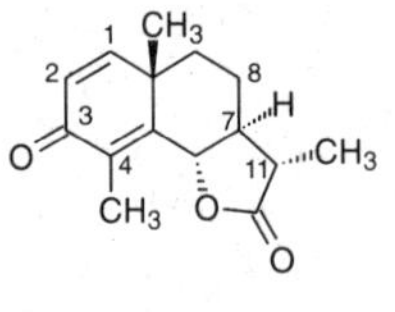

(α-)Santonin

$C_{15}H_{18}O_3$, M_R 246.31, colorless crystals, turning yellow in the light, poisonous, with a bitter taste, mp. 174–176 °C, $[\alpha]_D^{18}$ –173° (C_2H_5OH). S. and its 11-epimer *β*-S., mp. 216–218 °C, $[\alpha]_D^{14}$ –137.2° ($CHCl_3$)] are widely distributed constituents of plants, especially *Artemisia* species[1] (Asteraceae). S. occurs, e. g., together with *artemisin in Levant wormseed (*Artemisia cina*) indigenous to Turkestan and in the sea wormwood (*Seriphidium maritimum* = *Artemisia maritima*). The herbage of the latter contains ca. 1.5% santonin. In the past S. was used in the treatment of nervous diseases and of ascaridosis. Poisonings, sometimes lethal, were reported after medicinal use. A typical symptom is xanthopsia, in which white spots appear violet and finally yellow. The lethal dose for children is 60–300 mg.

Lit.: [1]J. Chem. Soc. B **1969**, 6041; Karrer, No. 1903, 1904, 1905; Zechmeister **38**, 134 ff.; Pure Appl. Chem. **21**, 167 (1970).

gen.: Beilstein E V **17/11**, 314; **18/3**, 304 ▪ Merck-Index (12.), No. 8509 ▪ Sax (8.), SAU 500. – *Synthesis:* ApSimon **2**, 315–324 ▪ Can. J. Chem. **62**, 2813 (1984) ▪ Chem. Pharm. Bull. **42**, 1160 (1994). – *[HS 2932 29; CAS 481-06-1 (S.); 481-07-2 (11-β-S.)]*

Sapindoside A see *α*-hederin.

Sapintoxin A see tiglianes.

Sapogenins. Group name for the aglycones of the *steroid saponins or *triterpene saponins; see also steroid sapogenins.

Saponaceolides.

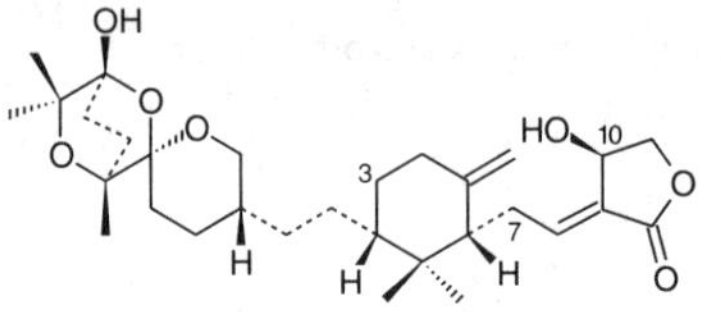

S. A

S. A: $C_{30}H_{46}O_7$, M_R 518.69, needles, mp. 145–146 °C, $[\alpha]_D$ +78.1° ($CHCl_3$). Cytotoxic triterpenoids from the fruit bodies of the toadstool *Tricholoma saponaceum* (Basidiomycetes). The fungus also contains the 10-deoxy derivative *S. B* {$C_{30}H_{46}O_6$, M_R 502.69, cryst., mp. 134–136 °C, $[\alpha]_D$ +17.9° (CH_2Cl_2)}, as well as the (7*S*)-7- and 3*β*-hydroxy derivative S. C and D., respectively. The unusual S. system is probably formed biogenetically by attack of a farnesyl cation on a molecule of farnesyl diphosphate.

S. A and B exhibit *in vitro* activity against human colon adenocarcinomas.

Lit.: Angew. Chem., Int. Ed. **38**, 3664 (1999) (synthesis) ▪ Tetrahedron **44**, 235 (1988); **47**, 7109 (1991). – *[CAS 118584-14-8 (S. A); 123746-67-8 (S. B)]*

Saponins. Name for a group of mostly plant *glycosides that form colloidal, soap-like solutions in water. The S. are classified according to the nature of their aglycones, the sapogenins, into *steroid saponins and *triterpene saponins. The few animal S. include the *echinoderm saponins, e. g., the *holothurins of the sea cucumber.

Lit.: Hostettmann & Marston, Saponins, Cambridge: Univ. Press 1995 ▪ Zechmeister **30**, 461–606; **46**, 1–76.

Sapropterin.

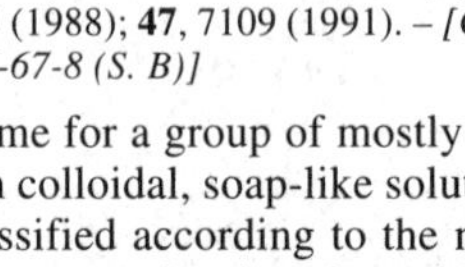

$C_9H_{15}N_5O_3$, M_R 241.25, mp. 245–246 °C, needles (hydrochloride), $[\alpha]_D^{25}$ –6.8° (0.1 m HCl). A cofactor of phenylalanine, tyrosine, and tryptophan hydroxylases (EC 1.14.16.1, 1.14.16.2 u. 1.14.16.4), it is essential for the biosynthesis of catecholamine neurotransmitters. S. is commercially available in Japan under the name Biopten® for the treatment of hyperphenylalaninemia.

Lit.: Ann. N. Y. Acad. Sci. **521**, 129 (1988) (metabolism) ▪ Heterocycles **23**, 3115 (1985) (synthesis) ▪ J. Nutr. Biochem. **2**, 411 (biosynthesis) ▪ Proc. Soc. Exp. Biol. Med. **203**, 1 (1993) (review). – *[CAS 62989-33-7 (S.); 69056-38-8 (hydrochloride)]*

Saquayamycins.

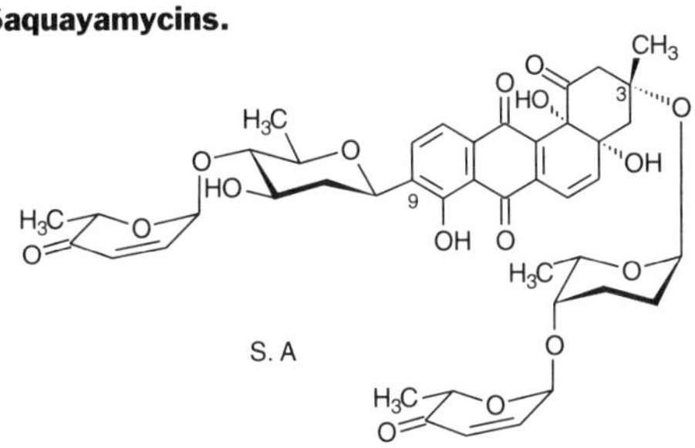

S. A

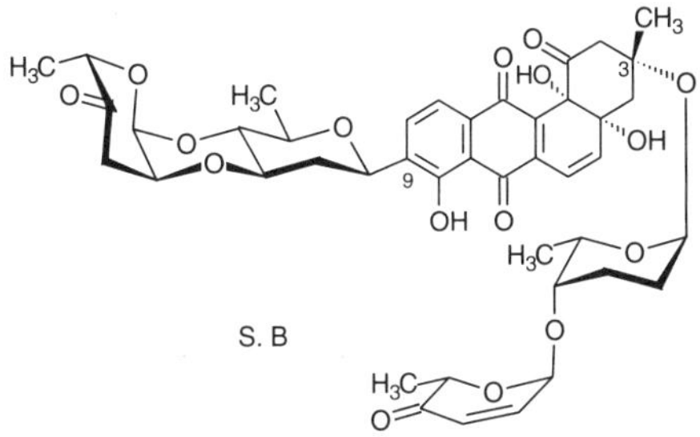

S. B

Table: Data of Saquayamycins.

S.	molecular formula	M_R	mp. [°C]	$[\alpha]_D^{24}$ ($CHCl_3$)	CAS
A	$C_{43}H_{48}O_{16}$	820.84	149–152	+77°	99260-65-8
A_1	$C_{31}H_{32}O_{12}$	596.59			99260-66-9
B	$C_{43}H_{48}O_{16}$	820.84	164–166	+96°	99260-67-0
B_1	$C_{31}H_{32}O_{12}$	596.59			99260-68-1
C	$C_{43}H_{52}O_{16}$	824.88	142–142.5	–53°	99260-70-5
C_1	$C_{31}H_{34}O_{12}$	598.60			99260-69-2
D	$C_{43}H_{50}O_{16}$	822.86	152–155	+10°	99260-71-6
E	$C_{43}H_{50}O_{16}$	822.86	150–153	–11.5°	
F	$C_{43}H_{50}O_{16}$	822.86	148–152	–120°	99260-71-6

Quinonoid antibiotics of the *angucyclin type that furnish the aglycone *aquayamycin on acid hydrolysis. S. A contains 2 mol. L-aculose and 1 mol. L-rhodinose as O-glycosidically bound sugar building blocks while the D-olivose is C-glycosidically bound. In the left-hand part of S. B the OH group of the C-glycosidically bound D-olivose has undergone addition to the double bond of the L-aculose to form a tricyclic system. In S. C the double bonds of the aculose residues of S. A are hydrogenated, in S. D the double bond of the aculose of S. B is hydrogenated. In S. A_1, B_1, and C_1 the disaccharide on O-3 is lacking. The S. are produced by *streptomycetes (e.g., *Streptomyces nodosus*); they are orange, microcrystalline compounds (soluble in $CHCl_3$) and inhibit the growth of Gram-positive bacteria and tumor cells (L1210, HT29, A549).

Lit.: J. Antibiot. **38**, 1171–1181 (1985); **41**, 1913 (1988); **43**, 830–837 (1990); **49**, 487 (1996) (S. E, F).

Saragozic acids see zaragozic acids.

Sarains.

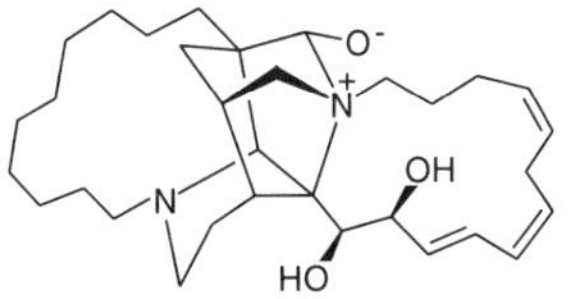

S. A (absolute configuration)

Unusual *quinolizidine and *piperidine alkaloids have been isolated from the Mediterranean sponge *Reniera sarai* (Haplosclerida, Dermospongiae); these compounds are probably related biosynthetically to *petrosin from *Petrosia ficiformis*. *S. A*, $C_{32}H_{50}N_2O_3$, M_R 510.38, amorphous powder, $[\alpha]_D$ +66.0° (acetate in $CHCl_3$).

Lit.: J. Nat. Prod. **53**, 1519 (1990) ▪ Org. Prep. Proced. Int. **30**, 3 (1998) (review) ▪ Pure Appl. Chem. **61**, 535 (1989) ▪ Tetrahedron **52**, 8341 (1996) (abs. configuration). – *[CAS 123117-93-1 (S. A)]*

α-Sarcin. A fungal *toxin from *Aspergillus giganteus* with ribonuclease activity; as a ribosome-inactivating protein it inhibits eukaryotic protein biosynthesis by cleaving the phosphodiester bond at the 3′-side of the guanosine residue 4325 of ribosomal 28-S-rRNA. It is being used in the development of immunotoxins [1].

Lit.: [1] Eur. J. Biochem. **196**, 203–209 (1991).

gen.: J. Biol. Chem. **265**, 2216–2222 (1990) ▪ Trends Biochem. Sci. **9**, 14–17 (1984). – *[HS 3002 90; CAS 86243-64-3]*

Sarcodictyins. Like eleutherobin (see eleuthosides), the S., isolated from the soft corals *Eleutherobia aurea* and *Sarcodictyon roseum* have become important synthetic targets due to their novel molecular architecture and potential for medicinal applications. They are cytotoxic and stabilize microtubuli (like *taxol), resulting in tumor cell death; e.g. *S. A*: $C_{28}H_{36}N_2O_6$, M_R 496.60, powder, mp. 219–222 °C, $[\alpha]_D^{20}$ –15.2° (C_2H_5OH); *S. C*: $C_{28}H_{36}N_2O_7$, M_R 512.60, powder, mp. 225–227 °C, $[\alpha]_D^{20}$ –16.5° (C_2H_5OH).

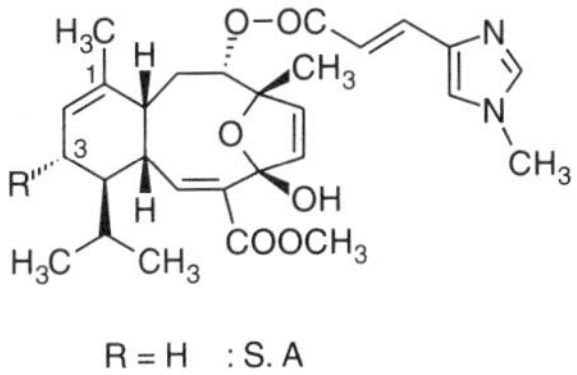

R = H : S. A
R = OH : S. C

Combinatorial S. libraries have been synthesized[1].

Lit.: [1] J. Am. Chem. Soc. **120**, 10814–10826 (1998); Chemtracts **12**, 1006–1012 (1999).

gen.: Angew. Chem. Int. Ed. Engl. **37**, 1418 (1998) (synthesis) ▪ Chem. Pharm. Bull. **47**, 1199 (1999) (review) ▪ J. Am. Chem. Soc. **119**, 11353 (1997); **120**, 8661 (1998) (synthesis) ▪ Helv. Chim. Acta **71**, 964 (1988) (isolation) ▪ J. Nat. Prod. **59**, 873 (1996) (isolation) ▪ Nachr. Chem. Tech. Lab. **47**, 1228 (1999) (review). – *[CAS 113540-81-1 (S. A); 122169-34-4 (S. C)]*

Sarcodonin A.

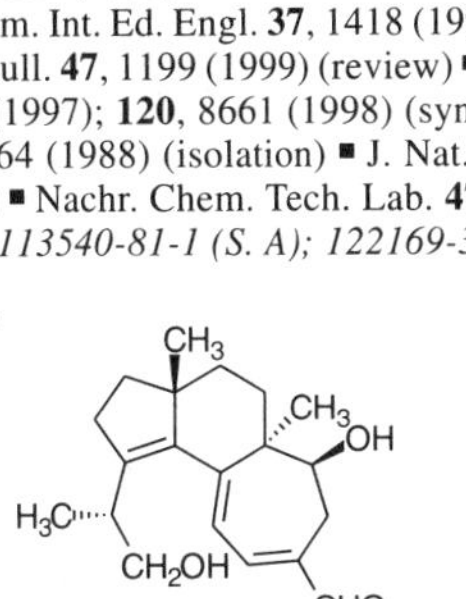

$C_{20}H_{28}O_3$, M_R 316.44, yellow syrup, $[\alpha]_D$ +91.7° ($CHCl_3$). A diterpenoid with the *cyathane skeleton from fruit bodies of the toadstool *Sarcodon scabrosus* (Basidiomycetes). For related diterpenoids, see *Lit.*[1].

Lit.: [1]Biosci., Biotechnol., Biochem. **62**, 2450 (1998).
gen.: Agric. Biol. Chem. **53**, 3373 (1989). – *[CAS 125882-71-5]*

Sarcodontic acid.

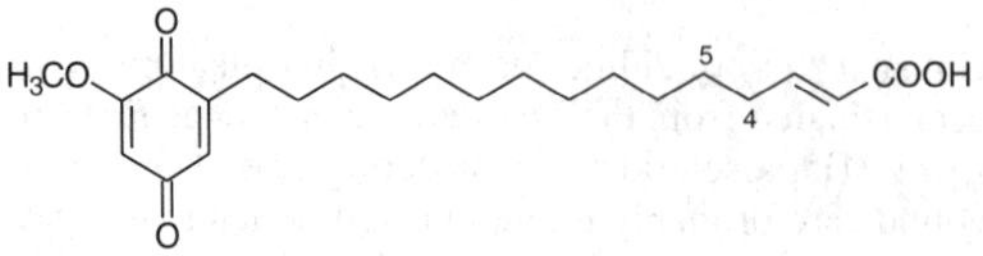

$C_{22}H_{32}O_5$, M_R 376.49, yellow cryst., mp. 122–124 °C. From fruit bodies of the parasitic fungus *Sarcodontia setosa* (Aphyllophorales) growing on old apple trees. S. also occurs in the fungus as the dihydro derivative and a derivative with an additional double bond in the 4,5 position.

Lit.: Collect. Czech. Chem. Commun. **49**, 1622 (1984) ▪ Naturwissenschaften **52**, 591 (1965). – *[CAS 2901-96-4]*

Sarcophytol A.

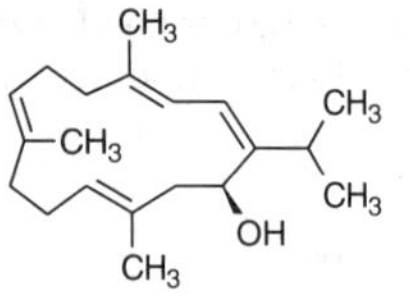

$C_{20}H_{32}O$, M_R 288.47, oil, $[\alpha]_D$ +141°, a cembranoid from the soft coral *Sarcophyton glaucum*[1]; it is an antagonist of the tumor-promotors lyngbyatoxin and debromoaplysiatoxin (see majusculamides) etc., and also inhibits the metastazation of tumors of the colon, lungs, liver, and pancreas[2]. The calcium antagonist *16-deoxysarcophin*[3] ($C_{20}H_{30}O_2$, M_R 302.46, mp. 77–78 °C, $[\alpha]_D$ +157°) is isolated from *Sarcophyton* sp. An exact analysis revealed that the actual producer is the dinoflagellate *Symbiodinium* sp. living in symbiosis with *Sarcophyton*. For further terpenes from coelenterates (*Alcyonaria*), see *Lit.*[4]. For synthesis, see *Lit.*[5].

Lit.: [1]Chem. Pharm. Bull. **27**, 2382 (1979). [2]J. Cancer Res. Clin. Oncol. **115**, 25 (1989); Kuroda, Shankel, & Waters (eds.), Antimutagenesis and Anticarcinogenesis Mechanisms II, Basic Life Science, vol. 52, p. 205, New York: Plenum Press 1990. [3]Experientia **39**, 67 (1983). [4]Scheuer I **2**, 247. [5]J. Org. Chem. **59**, 2700 (1994); Tetrahedron Lett. **34**, 8453 (1993); **40**, 965 (1999). – *[CAS 72629-69-7 (S. A); 74841-41-1 (deoxysarcophin)]*

Sarcosine [*N*-methylglycine, (methylamino)acetic acid, abbr.: Sar]. $H_3C–NH–CH_2–COOH$, $C_3H_7NO_2$, M_R 89.09, mp. 212–213 °C (decomp.). Hydrochloride, mp. 171 °C (decomp.), pK_a 2.23, 10.01, pI 6.12; sweet taste. S. occurs, among others, in sea stars, sea urchins, *Chondria* sp. (red algae), acid hydrolysates of peanut protein, lobsters, crabs, and is a component of many antibiotics, e. g., *actinomycins D and *cyclosporins.
Metabolism: Degradation product of creatine and choline. S. is converted to Gly by S. dehydrogenase (EC 1.5.99.1) or S. oxidase (EC 1.5.3.1). Formaldehyde formed in the process is transferred to (6*S*)-5,6,7,8-tetrahydrofolic acid. For synthesis, see *Lit.*[1].
Use: In the form of *sarcosinates* and the *N*-acyl derivatives as wetting and dispersion agents, corrosion inhibitors in oils, in pharmaceutical and cosmetic products, for production of toothpaste, and for stabilizing diazo compounds in the dye industry.

Lit.: [1]J. Heterocycl. Chem. **20**, 49 (1983).
gen.: Beilstein EIV **4**, 2363 f. – *[HS 2922 49; CAS 107-97-1]*

Sarkomycin A.

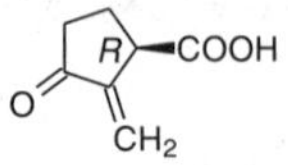

$C_7H_8O_3$, M_R 140.14, liquid, $[\alpha]_D^{25}$ –32.5° (CH_3OH), soluble in water, alcohols, ethyl acetate. A carboxylic acid from *Streptomyces erythrochromogenes* with potent antitumor and teratogenic activities as well as weak antibacterial properties.

Lit.: J. Org. Chem. **47**, 3306, 3333 (1982); **48**, 3581 (1983); **50**, 2965 (1985) ▪ Sax (8.), SJS 500 ▪ Synthesis **1997**, 356 (synthesis review) ▪ Tetrahedron **45**, 7023 (1989). – *[HS 2941 90; CAS 489-21-4]*

Sarpagine (raupine, sarpagan-10,17-diol).

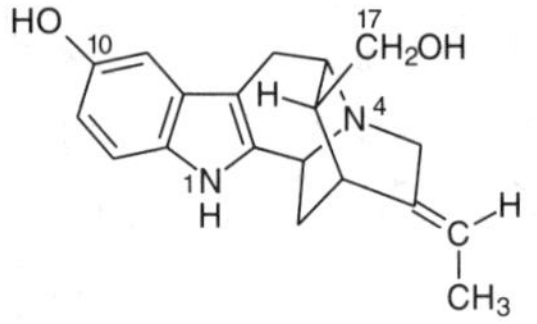

$C_{19}H_{22}N_2O_2$, M_R 310.40, needles, mp. 320 °C (decomp.), $[\alpha]_D$ +54° (pyridine). This *Rauvolfia alkaloid is a typical example of the group of sarpagan alkaloids (ca. 50 naturally occurring members). The sarpagan type is the biogenetic precursor of the ajmalan group including *ajmaline as a therapeutically important alkaloid. Important derivatives of S. are: *N^1-Methylsarpagine* ($C_{20}H_{24}N_2O_2$, M_R 324.42), N^4, O^{17}-dimethylsarpaginium (*lochneram*[1], cation, $C_{21}H_{27}N_2O_2^+$, M_R 339.46), and O^{10}-methyl-S. (*lochnerine*[2], $C_{20}H_{24}N_2O_2$, M_R 324.42, mp. 202.5–203.5 °C, potent blood pressure-lowering activity). For a review of the sarpagan alkaloids, see *Lit.*[3].

Lit.: [1]Helv. Chim. Acta **40**, 705 (1957). [2]Manske **47**, 115–172. [3]Manske **52**, 104–196.
gen.: Zechmeister **43**, 267–346; see also Rauvolfia alkaloids. – *[HS 2939 90; CAS 482-68-8 (S.); 17801-05-7 (N^1-methyl-S.); 6901-26-4 (lochneram)]*

Sarsaparilloside see steroid saponins.

Sarsasapogenin see steroid sapogenins.

Sativene.

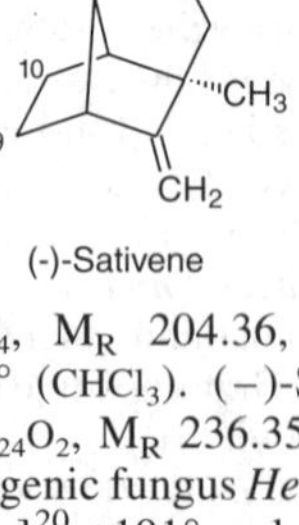

(-)-Sativene

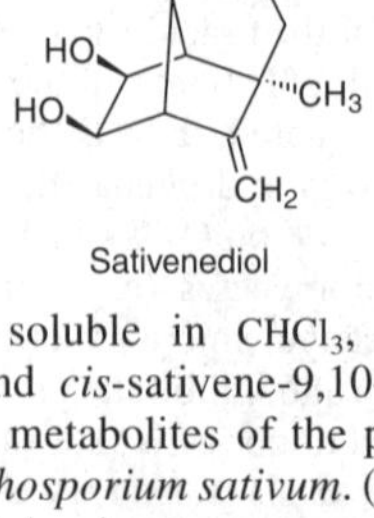

Sativenediol

$C_{15}H_{24}$, M_R 204.36, oil, soluble in $CHCl_3$, $[\alpha]_D^{20}$ –186° ($CHCl_3$). (–)-S.[1] and *cis*-sativene-9,10-diol[2] ($C_{15}H_{24}O_2$, M_R 236.35) are metabolites of the phytopathogenic fungus *Helminthosporium sativum*. (+)-S., oil, $[\alpha]_D^{20}$ +191° and the C-4 epimer *copacamphene*, $[\alpha]_D$ +28° ($CHCl_3$) are accessible from copaborneol by partial synthesis. (+)-S. is present in turpentine oil from various *Abies* (fir) species. Sativenediol is a promotor of plant growth.

Lit.: [1] J. Am. Chem. Soc. **97**, 2542 (1975). [2] Helv. Chim. Acta **71**, 788 (1988); J. Am. Chem. Soc. **114**, 644 (1992).
gen.: Synthesis: Chem. Lett. **1994**, 1415 ▪ Tetrahedron **42**, 3277 (1986) ▪ Tetrahedron Lett. **29**, 5973 (1988). – *Biosynthesis:* Pure Appl. Chem. **41**, 219 (1975). – *[CAS 6813-05-4 (S.); 55556-01-9 (cis-sativene-9,10-diol); 3650-28-0 ((+)-S.); 16641-59-1 (copacamphene)]*

Satratoxins. A group of macrocyclic *tric(h)othecenes formed by *Stachybotrys atra* (newer name *S. chartarum*) and responsible for animal poisonings. S. are structurally closely related to the *roridins and *verrucarins that are also produced by *S. chartarum*, a fungus growing especially on cellulose-containing substrates.

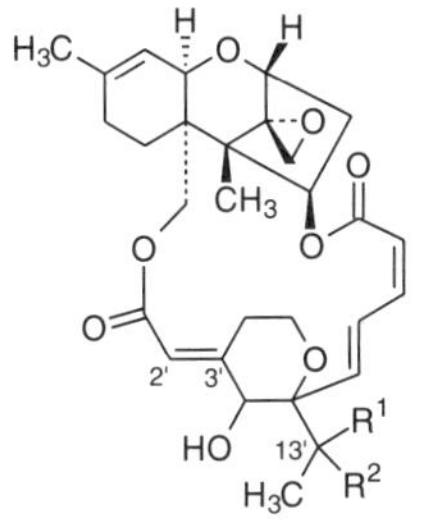

	R^1	R^2	
S. F :	=O		2',3'-epoxide
S. G :	H	OH	2',3'-epoxide
S. H :	H	OH	

Table: Data of Satratoxins.

S.	molecular formula	M_R	mp. [°C]	CAS
S. F	$C_{29}H_{34}O_{10}$	542.58	140–143	73513-01-6
S. G	$C_{29}H_{36}O_{10}$	544.60	167–170	53126-63-9
S. H	$C_{29}H_{36}O_{9}$	528.60	162–166	53126-64-0

The so-called stachybotryotoxicosis occurs mainly in Ukraine, Hungary, and other East European countries. The symptoms in horses, pigs, cattle, and poultry are skin necrosis, inner hemorrhages, states of shock, irritations of stomach mucous membranes, reduced leukocyte count, nervousness. In severe cases the animals die within 1–3 d. Farm-workers who handle infected hay and straw may also show signs of poisoning such as skin and mucous irritations, cough, nosebleeds, fever, etc. Like all trichothecenes the S. inhibit eukaryotic protein biosynthesis.

Lit.: Appl. Environ. Microbiol. **51**, 915 (1986) ▪ Betina, chap. 14 ▪ J. Am. Chem. Soc. **106**, 260 (1984) ▪ J. Org. Chem. **45**, 2522 (1980) ▪ Rodricks, Hesseltine, & Mehlman (eds.), Mycotoxins in Human and Animal Health, p. 277–284, Park Forest South: Pathotox Publishers 1977.

Saubinin I see nivalenol.

Saudin.

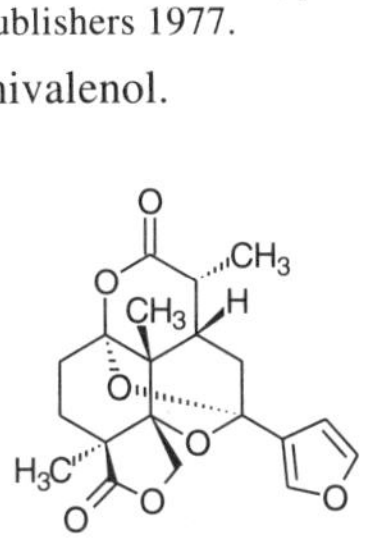

$C_{20}H_{22}O_7$, M_R 374.39, cryst., $[\alpha]_D$ –12° ($CHCl_3$), mp. 202–203 °C. S. was isolated in 0.06% yield from *Cluytia richardiana* (Euphorbiaceae) indigenous to Saudi-Arabia and showed hypoglycemic activity in animal experiments. The saudin skeleton is formed biosynthetically by cleavage of a labdane precursor between C_6 and C_7.

Lit.: Int. J. Crude Drug Res. **26**, 81 (1988) ▪ J. Am. Chem. Soc. **121**, 7425 (1999) (synthesis) ▪ J. Org. Chem. **50**, 916 (1985) ▪ Tetrahedron Lett. **29**, 3627 (1988). – *[CAS 94978-16-2]*

Saussureamines.

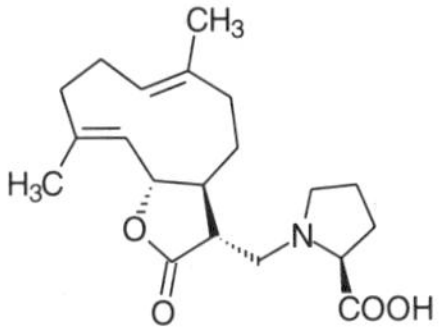

Saussureamine A

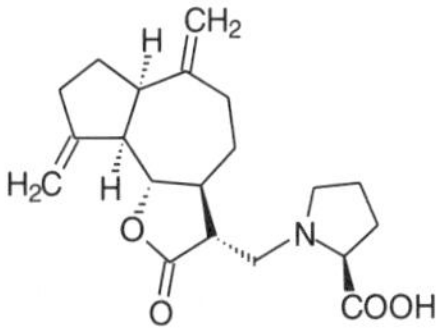

Saussureamine B
(S. C : L-Asn instead of L-Pro)

The S. A–E are components of the dried, Chinese *Saussurea* root (*Saussurea costus*, syn. *S. lappa*) which is used in Chinese and Japanese traditional medicine for stomach complaints. *S. A*, $C_{20}H_{29}NO_4$, M_R 347.45, mp. 115–117 °C, $[\alpha]_D$ +36.7° (CH_3OH), corresponds to the Michael adduct from costunolide and proline; S. B and C represent the amino acid conjugates of proline and asparagine, respectively, with the corresponding α,β-unsaturated dehydrocostus lactone. In animal experiments, S. A exhibits a pronounced protective effect against stress-induced stomach ulcers.

Lit.: Chem. Pharm. Bull. **33**, 1285 (1985); **40**, 2239 (1992); **41**, 214 (1993). – *[CAS 148245-82-3 (S. A); 126209-82-3 (S. B); 148245-83-4 (S. C)]*

Sawwort see dye plants.

Saxitoxin (mytilotoxin, gonyaulax toxin, PSP, STX).

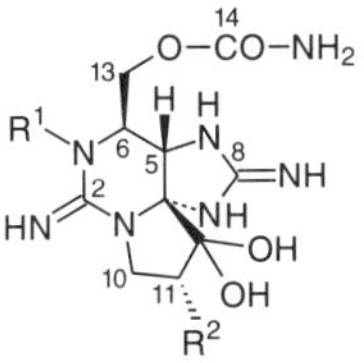

R^1 = H , R^2 = H : Saxitoxin
R^1 = OH , R^2 = H : Neosaxitoxin
R^1 = H , R^2 = SO_3H : Gonyautoxin II

Perhydropurine derivative, $C_{10}H_{17}N_7O_4$, M_R 299.29, the dihydrochloride is crystalline, $[\alpha]_D$ +130°, relatively stable in acidic solution, rapid decomposition in alkaline solution. It is a very potent neurotoxin (LD_{50} mouse i. p. 10, p. o. 263, and i. v. 3.4 µg/kg) from the dinoflagellates (single-celled algae) *Alexandrium cantenella* (formerly *Gonyaulax*) and *A. tamarense* and has also been found in the fresh water alga *Aphanizomenon flos-aquae* [1]. The actual producer of the toxin is the marine bacterium *Moraxella* sp. which has been isolated from *A. tamarense* [2]. More than 12 compounds have so far been isolated and can be classified into two groups: saxitoxins (STX) and neosaxitoxins (neoSTX). The compounds of the two groups are further distinguished by the presence of 11-*O*-sulfates (e.g., *gonyautoxin II*, GTX2, $C_{10}H_{17}N_7O_8S$, M_R 395.35), *N*-sulfates and carbamoyl or hydroxy groups at C-13. S. are formed in mussels (see also shellfish poisons) from neoSTX by reduction of the *N*-hydroxy group. S. can also reach and affect humans via the food chain: it can be enriched in edible mussels, e. g., in North American *Saxidomus* species and European oysters (*Ostrea edulis*) as well as edible mussels (*Mytilus edulis, M. californianus*). For the symptoms of poisoning, see PSP. Like *tetrodotoxin S. also blocks Na^+ ion channels. It is formed biosynthetically from arginine, methionine, and acetate[3].

Lit.: [1] Agric. Biol. Chem. **52**, 1075 (1988); J. Am. Chem. Soc. **97**, 1238, 6008 (1975); Scheuer I **1**, 1; **4**, 10, 78; Zechmeister **45**, 235; Chem. Rev. **93**, 1685 (1993). [2] Mar. Ecol. Prog. Ser. **61**, 203 (1990). [3] J. Am. Chem. Soc. **106**, 6433 (1984); J. Chem. Soc., Chem. Commun. **1989**, 1421.

gen.: Beilstein E III/IV **26**, 4028. – *Pharmacology:* Ann. N. Y. Acad. Sci. **479**, 52, 68, 96, 385, 402 (1986) ▪ Naturwiss. Rundsch. **42**, 113 (1989) ▪ Naturwissenschaften **73**, 459–470 (1986) ▪ Sax (8.), SBA 500, SBA 600 ▪ Toxicon **16**, 595 (1978) ▪ Toxicon Suppl. **3**, 211 (1983). – *Synthesis:* J. Am. Chem. Soc. **106**, 5594 (1984) ▪ Pure Appl. Chem. **58**, 257 (1986). – *Review:* ACS Symp. Ser. **418**, Marine Toxins (1990). – *[CAS 35523-89-8 (S.); 60508-89-6 gonyautoxin II); 64296-20-4 (neo-S.)]*

Scales (scale insects, coccids).

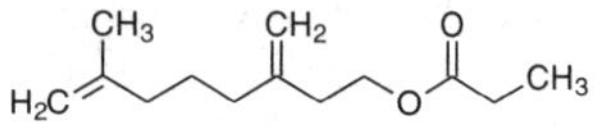

7-Methyl-3-methylene-7-octenyl propanoate (**1**)

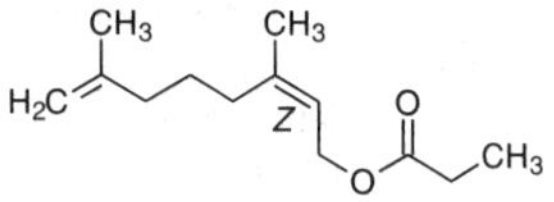

(*Z*)-3,7-Dimethyl-2,7-octadienyl propanoate (**2**)

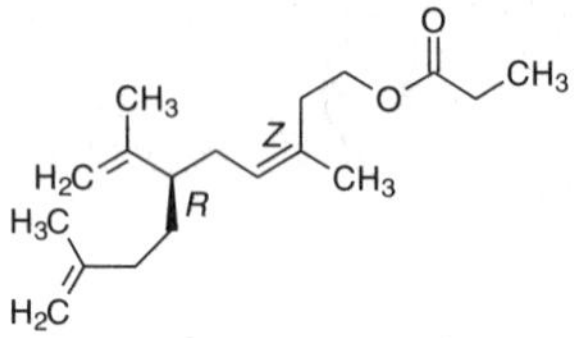

(3*Z*,6*R*)-6-Isopropenyl-3,9-dimethyl-3,9-decadienyl propanoate (**3**)

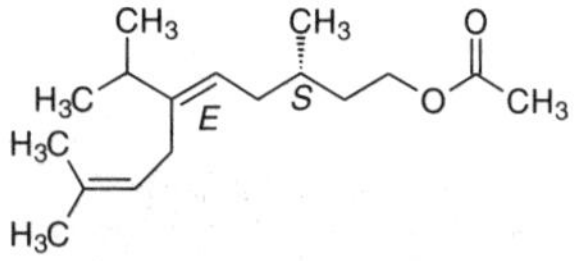

(3*S*,5*E*)-6-Isopropyl-3,9-dimethyl-5,8-decadienyl acetate (**4**)

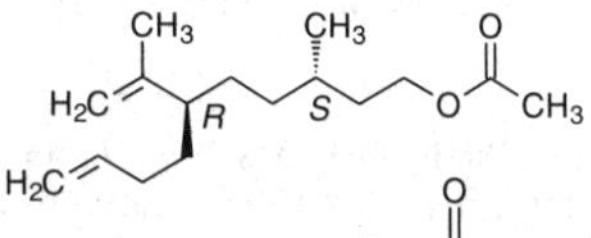

(3*S*,6*R*)-6-Isopropenyl-3-methyl-9-decenyl acetate (**5**)

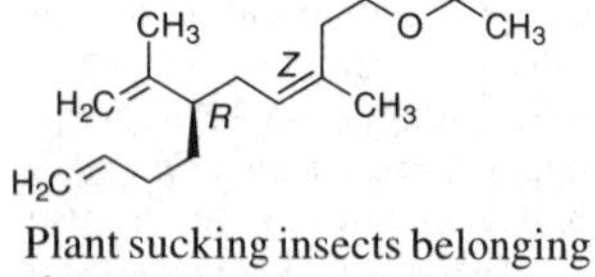

(3Z,6*R*)-6-Isopropenyl-3-methyl-3,9-decadienyl acetate (**6**)

Plant sucking insects belonging to the Homoptera with short proboscis and long piercing stylets. Some species such as the nopal scales and ilex scales produce valuable red pigments (*cochineal, kermes), others are important producers of certain *waxes and *shellac (shellac wax) (wax S., lac S.). Many S. are economically important plant pests. The *pheromones of the females of some species have been identified. These are esters of open-chain monoterpene alcohols such as *7-methyl-3-methylene-7-octenyl propanoate* ($C_{13}H_{22}O_2$, M_R 210.32) and *(Z)-3,7-dimethyl-2,7-octadienyl propanoate* ($C_{13}H_{22}O_2$, M_R 210.32), the sexual hormone of the San José S. (*Quadraspidiotus perniciosus*)[1]. Esters of branched, open-chain sesquiterpene alcohols such as *(3Z,6R)-6-isopropenyl-3,9-dimethyl-3,9-decadienyl propanoate* ($C_{18}H_{30}O_2$, M_R 278.44) and *(3S,5E)-6-isopropyl-3,9-dimethyl-5,8-decadienyl acetate* ($C_{17}H_{30}O_2$, M_R 266.42) are sexual pheromones of the white peach S. *Pseudaulacaspis pentagona*[2] and the yellow S. *Aonidiella citrina*[3]. Pheromone components of the "California red scale" *Aonidiella aurantii* also have similar structures, namely *(3S,6R)-6-isopropenyl-3-methyl-9-decenyl acetate* ($C_{16}H_{28}O_2$, M_R 252.4) and *(3Z,6R)-6-isopropenyl-3-methyl-3,9-decadienyl acetate* ($C_{16}H_{26}O_2$, M_R 250.4)[4]; see also matsuone.

Lit.: [1] J. Chem. Ecol. **5**, 891–900 (1979). [2] J. Chem. Ecol. **5**, 941–953 (1979). [3] J. Chem. Ecol. **5**, 773–779 (1979). [4] J. Chem. Ecol. **7**, 695–706 (1981).

gen.: Synthesis: ApSimon **4**, 93 ff.; **9**, 291 ff., 320–330. – *[CAS 73214-62-7 (2); 22810-05-5 (3); 71524-58-8 (4); 64309-03-1 (5); 64309-04-2 (6)]*

Scarabs (Lamellicornia). Beetles such as rhinoceros beetle, stag beetle, may beetle, or dung beetle (tumblebug, Scarabaeus). Some of the larvae living underground are serious plant pests. The chemical structures of the female sex *pheromones are extremely diverse. Besides lactones like *japonilure one also finds *(E)-2-nonen-1-ol* ($C_9H_{18}O$, M_R 142.24)[1], *(E)/(Z)*-mixtures of *7-tetradecen-2-one* ($C_{14}H_{26}O$, M_R 210.36)[2], and *methyl (Z)-5-tetradecenoate* ($C_{15}H_{28}O_2$, M_R 240.39)[3] as well as *L-isoleucine methyl ester* ($C_7H_{15}NO_2$, M_R 145.20)[4].

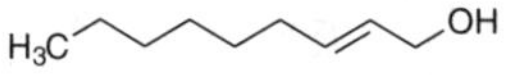

(*E*)-2-Nonen-1-ol (**1**)

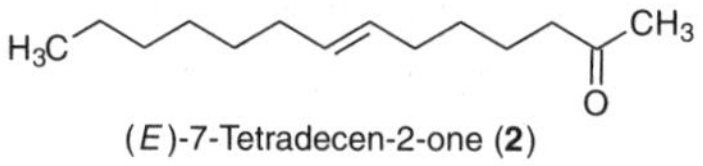

(*E*)-7-Tetradecen-2-one (**2**)

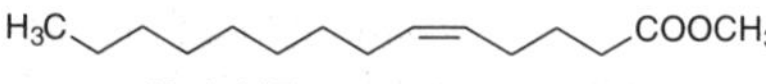

Methyl (*Z*)-5- tetradecenoate (**3**)

Lit.: [1] Naturwissenschaften **79**, 518 f. (1992); J. Chem. Ecol. **20**, 2481–2487 (1994). [2] Naturwissenschaften **80**, 86 f. (1993). [3] Appl. Entomol. Zool. **20**, 359–366 (1985); J. Chem. Ecol. **20**, 2415–2428 (1994). [4] Naturwissenschaften **79**, 184 f. (1992). – *[CAS 31502-14-4 (1); 156733-41-4 (2); 2577-46-0 (isoleucine methyl ester)]*

Scarlet lice see scales.

Sceptrin.

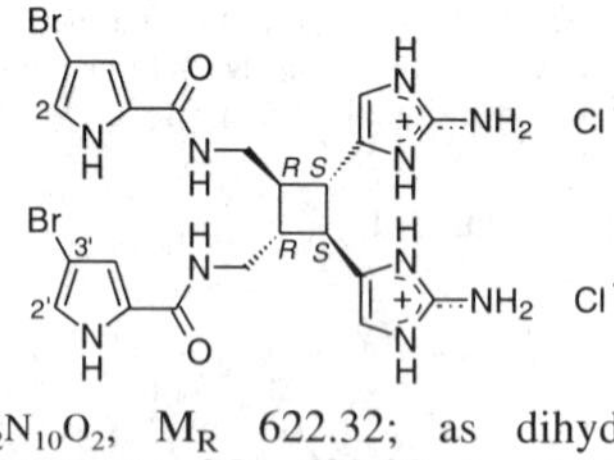

$C_{22}H_{26}Br_2N_{10}O_2$, M_R 622.32; as dihydrochloride: $C_{22}H_{26}Br_2Cl_2N_{10}O_2$, M_R 693.23, cryst. (H_2O), mp. 215–225 °C (decomp.), $[\alpha]_D$ −7.4° (CH_3OH). A cyclobutane derivative from the Pacific sponge *Agelas sceptrum* with antibiotic activity.

Lit.: J. Nat. Prod. **55**, 1366 (1992) ▪ J. Org. Chem. **56**, 2965, 6728 (1991). – *[CAS 79703-25-6]*

Schelhammericine, schelhammerine see Cephalotaxus alkaloids.

Schizokinen see aerobactin.

Schizostatin.

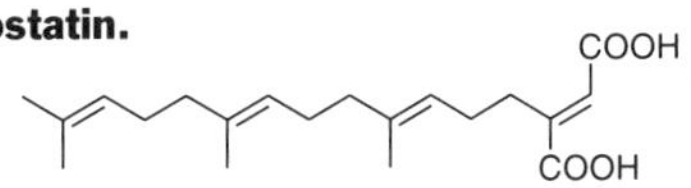

$C_{20}H_{30}O_4$, M_R 334.46, cryst., mp. 119–122 °C. A diterpenoid dicarboxylic acid from the basidiomycete *Schizophyllum commune* with an inhibitory effect on squalene synthase (IC_{50} 0.84 μM), a key enzyme of cholesterol biosynthesis.

Lit.: J. Antibiot. **49**, 617, 624 (1996) (synthesis) ▪ Tetrahedron Lett. **36**, 6301 (1995). – *[CAS 163564-55-4]*

Scillarenin see bufadienolides.

Scirpenols.

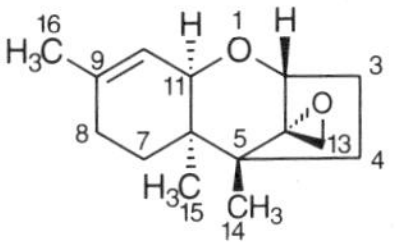

12,13-Epoxy-9-trichothecene (Scirpene)

A group of *mycotoxins of the *tric(h)othecene type with the skeleton of 12,13-epoxy-9-trichothecene (scirpene) which is known as a metabolite of *Trichothecium roseum*. The following positions can be hydroxylated: C-3, C-4, C-7, C-8, C-15. The OH groups are often esterified (see table). They are intermediates in the biosyntheses of the macrocyclic trichothecenes (OH groups at C-4 and C-15) and the *trichoverroids. The most important compound is *diacetylscirpenetriol* (DAS, anguidin), produced by many species of the genus *Fusarium* and the main fruiting forms of the genus *Gibberella*. Within the trichothecene class the compounds belong to the group A (compounds without an oxygen function at C-8) and group B (compounds with an oxygen function at C-8). Members of these trichothecene groups are also produced by species of the genera *Trichoderma* (including the main fruiting form *Hypocrea*) and *Myrothecium* (see also neosolaniol).

Lit.: Agric. Biol. Chem. **50**, 2667 (1986) ▪ Appl. Environ. Microbiol. **47**, 130 (1984); **50**, 1225 (1985) ▪ Betina, chap. 14 ▪ Cole-Cox, p. 152–224 ▪ J. Agric. Food Chem. **32**, 1261 (1984); **34**, 98 (1986); **35**, 125, 137 (1987) ▪ J. Nat. Prod. **58**, 1817 (1995) ▪ J. Org. Chem. **47**, 1117 (1982); **52**, 4405 (1987); **54**, 4663 (1989); **55**, 4403 (1990) ▪ Org. Lett. **1**, 517 (1999) (synthesis scirpene) ▪ Sax (7.), p. 2633 ▪ Tetrahedron Lett. **28**, 2661 (1987). – *Synthesis verrucarol:* J. Org. Chem. **63**, 2679 (1998) ▪ Synthesis **1998**, 619 ▪ Tetrahedron Lett. **38**, 8297 (1997).

Sclareol see labdanes.

Sclerotia (singular: sclerotium). S. are the survival forms of fungi which can start growing after several years of dormancy in the soil or in plant material. According to size a distinction is made between microsclerotia (e. g., of *Penicillium sclerotigenum*, *P. sclerotiorum*, *Aspergillus sclerotiorum*, and *Verticillium* species) that are smaller than 1 mm and S. like those of *Claviceps purpurea* (*ergot) which can be up to 4 cm long. Producers of *mycotoxins often also form S., e. g., *Aspergillus flavus*, *A. niger*, *A. ochraceus* as well as phytopathogenic fungi such as, e. g., *Botrytis*, *Rhizoctonia*, *Sclerotium*, and *Sclerotinia* species. S. are frequently protected through toxic metabolites (*ergot alkaloids, *aflatoxins, *aflatrem) from being eaten by insects while dark pigments (*ergochromes) protect them from UV radiation. S. consist of closely packed, thick-walled, hypha cells. They are mostly resistant to heat, desiccation, and injury and contain reserve substances (fats).

Lit.: Can. J. Bot. **60**, 525 (1982) ▪ Can. J. Microbiol. **29**, (1983) ▪ Mycologia **71**, 415 (1979) ▪ Trans. Br. Mycol. Soc. **83**, 299 (1984).

Table: Structure and data of Scirpenols.

trivial name	systematical name	molecular formula	M_R	mp. [°C]	CAS
3-Scirpenol, Isotrichodermol	12,13-Epoxy-9-trichothecen-3α-ol	$C_{15}H_{22}O_3$	250.34		104155-10-4
Isotrichodermin	3α-Acetoxy-12,13-epoxy-9-trichothecene	$C_{17}H_{24}O_4$	292.38		91423-90-4
Trichodermol	12,13-Epoxy-9-trichothecen-4β-ol	$C_{15}H_{22}O_3$	250.34	116–119	2198-93-8
Trichodermin	4β-Acetoxy-12,13-epoxy-9-trichothecene	$C_{17}H_{24}O_4$	292.38	45–46	4682-50-2
3,15-Scirpendiol	12,13-Epoxy-9-trichothecene-3α,15-diol	$C_{15}H_{22}O_4$	266.34		38818-51-8
15-Deacetylcalonectrin	3α-Acetoxy-12,13-epoxy-9-trichothecen-15-ol	$C_{17}H_{24}O_5$	308.37	184–186	38818-66-5
Calonectrin	3α,15-Diacetoxy-12,13-epoxy-9-trichothecene	$C_{19}H_{26}O_6$	350.41	83–85	38818-51-8
7-Hydroxytrichodermol	12,13-Epoxy-9-trichothecene-4β,7-diol	$C_{15}H_{22}O_4$	266.34	214–215	99624-08-5
Verrucarol	12,13-Epoxy-9-trichothecene-4β,15-diol	$C_{15}H_{22}O_4$	266.34	155–158	2198-92-7
Diacetylverrucarol	4β,15-Diacetoxy-12,13-epoxy-9-trichothecene	$C_{19}H_{26}O_6$	350.41	148–150	2198-94-9
Scirpentriol	12,13-Epoxy-9-trichothecene-3α,4β,15-triol	$C_{15}H_{22}O_5$	282.34	189–191	2270-41-9
Monoacetylscirpentriol	15-Acetoxy-12,13-epoxy-9-trichothecene-3α,4β-diol	$C_{17}H_{24}O_6$	324.37	172–173	2623-22-5
Diacetylscirpentriol (Anguidin)	4β,15-Diacetoxy-12,13-epoxy-9-trichothecen-3α-ol	$C_{19}H_{26}O_7$	366.41	161–162	2270-40-8
Diacetylscirpentetrol	4β,15-Diacetoxy-12,13-epoxy-9-trichothecene-3α,7-diol	$C_{19}H_{26}O_8$	382.41	201–203	59121-84-5

Scoparone see 6,7-dihydroxycoumarins.

Scopolamine (hyoscine).

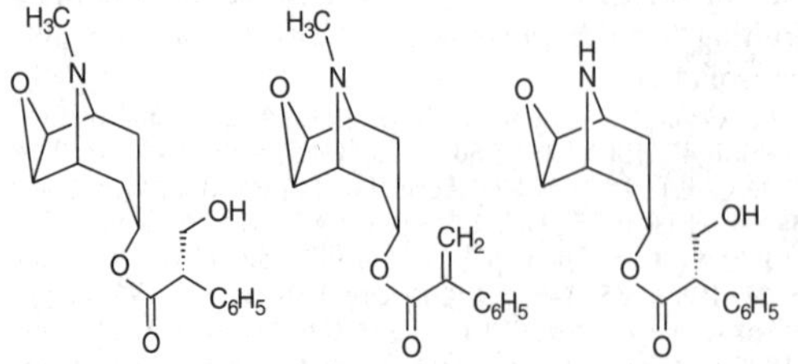

Scopolamine Aposcopolamine Norhyoscine

$C_{17}H_{21}NO_4$, M_R 303.36, viscous oil, the monohydrate is crystalline, mp. 59 °C, $[\alpha]_D -20°$ (H_2O). S. racemizes in alkaline solution to *atroscine* (oil, the monohydrate is crystalline, mp. 56–57 °C, dihydrate mp. 37–39 °C).

S. occurs as the (–)-form (hyoscine) in several Solanaceae genera[1] [e. g.: *Atropa belladonna* (deadly nightshade), *Datura* species (thorn apple), *Duboisia* species, *Hyoscyamus* species (henbane), *Mandragora officinalis* (mandrake), and *Scopolia* species (banewort)]. Racemic S. (atroscine) is a natural product of the Solanaceae *Datura innoxia*, *Hyoscyamus niger*, and *Scopolia carniolica*. S. is easily hydrolyzed in acidic and alkaline solution to tropic acid and scopine. S. differs from *atropine in its pharmacological properties in that it has a subduing effect rather than a stimulating effect on the central nervous system. For this reason it is used in transdermal systems (scopolamine plaster) to suppress the vomiting stimulus (e. g., travel sickness). The peripheral effects of S. and *atropine are qualitatively equal but differ quantitatively. Thus, the mydriatic and secretion inhibiting activities of S. are stronger while the spasmolytic and heart rate-increasing properties are weaker. In the Middle Ages, weak beers were occasionally "strengthened" by addition of S.-containing henbane seeds. Derivatives of S. are a) *aposcopolamine* (apohyoscine, $C_{17}H_{19}NO_3$, M_R 285.34, mp. 97 °C), isolated from some Solanaceae genera (*Datura, Hyoscyamus*, and *Anthocercis*) and, analogous to the formation of *apoatropine is formed under acidic conditions from S., b) S. *N*-oxide[2] [*genoscopolamine*, $C_{17}H_{21}NO_5$, M_R 319.36, mp. 80 °C, hydrochloride 125–130 °C (decomp.)], also occurring in Solanaceae, and c) *norhyoscine*[3], ($C_{16}H_{19}NO_4$, M_R 289.33) from *Datura suaveolens*. The content of S. in certain medicinal plants can be increased with concomitant reduction of the hyoscyamine by means of genetic engineering[4].

Lit.: [1]Manske **44**, 1–114. [2]Merck-Index (12.), No. 8551. [3]Phytochemistry **11**, 3293 (1972). [4]Proc. Natl. Acad. Sci. USA **89**, 11799 (1992).

gen.: Beilstein EIII/IV **27**, 1790 ff.; E V **27/8**, 53 ▪ Hager (5.) **3**, 1073 ff.; **4**, 424 ff., 1138 f.; **5**, 461, 765; **9**, 581 f. ▪ Manske **6**, 145–177 (synthesis) ▪ Merck-Index (12.), No. 8550 ▪ Sax (8.), p. 3022. – *Biosynthesis:* Manske **44**, 115–187 ▪ Planta Med. **56**, 339–352 (1990). – *Pharmacology:* Drugs **29**, 189–207 (1985) ▪ Florey **19**, 477 (1990) ▪ Kleemann-Engel, p. 1711 ▪ Martindale (30.), p. 425 ▪ Nat. Prod. Rep. **8**, 608 (1990); **10**, 203 f. (1991); **11**, 447 f. (1994). – *[HS 2939 90; CAS 51-34-3 (S.); 138-12-5 ((±)-S.); 97-75-6 (b); 535-26-2 (a); 4684-28-0 (c)]*

Scopoletin see 6,7-dihydroxycoumarins.

Scopolin. Trivial name for 7-(*β*-D-glucopyranosyloxy)-6-methoxycoumarin (see 6,7-dihydroxycoumarins).

Scopoline. Trivial name for 3*α*,7*α*-epoxy-6*β*-tropanol (see tropane alkaloids).

Scorpion venoms. Scorpions that may be dangerous to humans occur in Latin America and Africa (especially in Mexico, Brazil, Tunisia, Algeria, and Morocco). The venoms are polypeptides composed of ca. 60 amino acids that principally exert neurotoxic activities, attacking mainly sodium channels of excitable membranes (*charybdotoxin: 37 amino acids). This leads to the liberation of neurotransmitters and gives rise to diverse symptoms: The pain occurring after a sting is caused by release of *serotonin and subsides after some time; whereas S. venoms can cause permanent damage such as speech and visual impairments.

Lit.: Polis, The Biology of Scorpions, p. 415–444, Stanford: Stanford University Press 1990 ▪ Proc. Natl. Acad. Sci. USA **92**, 6404 (1995) ▪ Tu, Handbook of Natural Toxins, vol. 2, p. 530–790, New York: Dekker 1984.

Scoulerine see Protoberberine alkaloids.

Screening. Name for the search procedure applied on mixtures of substances (e. g., crude extracts from biological sources) and substance libraries with various detection systems for finding particular properties (chemical, spectroscopic, biochemical, pharmacological). In the so-called *chemical S.* new substances are sought after TLC separation by means of UV absorptions or specific color reactions with spray reagents. In chemical S. the search for potential biological applications is started after isolation and structure elucidation by means of subsequent *in vitro* test systems. Chemical S. has been applied with great success in the field of *alkaloids. In *target-oriented* S. the objective is to find new lead substances and active principles with defined biological/pharmacological properties. Screenings for the purpose of utilizing the molecular diversity of secondary metabolites in biological sources play a central part in pharmaceutical research. As a result of advances in the elucidation of pathobiochemical processes with the aid of applied molecular biology, cell biology, and biochemistry it is now possible to design and develop mechanism-oriented targets for screening systems. The objective of such a pharmacologically oriented S. is to find the causal relationship between the molecular target and the clinical situation. In addition to the classical test systems for antibacterial, antifungal, antiviral, and carcinostatic principles, cloned proteins (receptors, ligands, enzymes) obtained with the help of molecular methods and cells modified by genetic engineering are now used as pathophysiologically relevant *in vitro* S. models. Measurable parameters include enzyme activities, receptor/ligand, protein/protein, and protein/nucleic acid interactions as well as cellular systems (gene regulation, reporter genes, cytotoxicity, etc.). Detection systems employed include photometry, fluorescence, chemiluminescence, and radioactivity. S. systems with intact organisms are generally used in the search for

new active principles in plant protection (fungicides, herbicides, insecticides, nematicides).
A high degree of automation is necessary for successful S. (miniaturization, robotics, data processing). With the so-called *high-throughput screening* (HTS) processing of several thousand samples per day is standard; *ultra-high throughput screening* (UHTS) means processing of more than 100 000 samples per day.
Natural product S. encompasses the entire interdisciplinary process of searching for new, biologically active natural products.

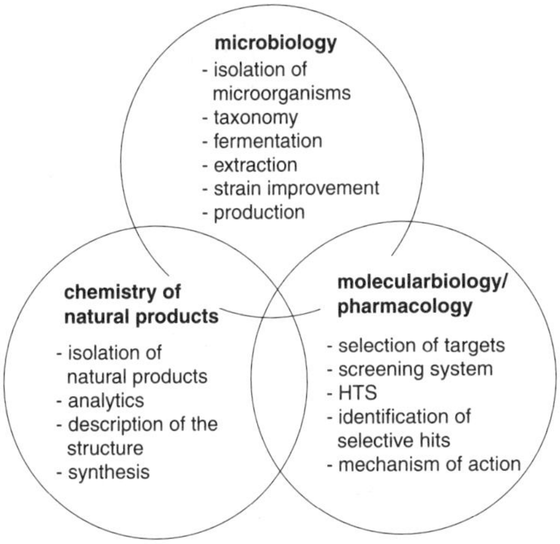

Figure: Integrated scheme for natural product screening for the example of pharmacologically active microbial secondary metabolites.

Lit.: Böger & Sandmann (eds.), Target Sites of Herbicide Action, Boca Raton: CRC-Press 1989 ▪ Colegate & Molyneux (eds.), Bioactive Natural Products: Detection, Isolation, Structural Determination, Boca Raton: CRC-Press 1993 ▪ Duke (ed.), Weed Physiology, vol. II: Herbicide Physiology, Boca Raton: CRC-Press 1985 ▪ Drug & Market Development **6**, 148 (1995) ▪ J. Antibiot. **46**, 535 (1993) ▪ J. Biotechnol. **37**, 89 (1994) ▪ Köller (ed.), Target Sites of Fungicide Action, Boca Raton: CRC-Press 1992 ▪ Microbiol. Rev. **50**, 259 (1986) ▪ Nakagawa, in Omura (ed.), The Search for Bioactive Compounds from Microorganisms, Chemical Screening, p. 263, Berlin: Springer 1992.

Scutigeral.

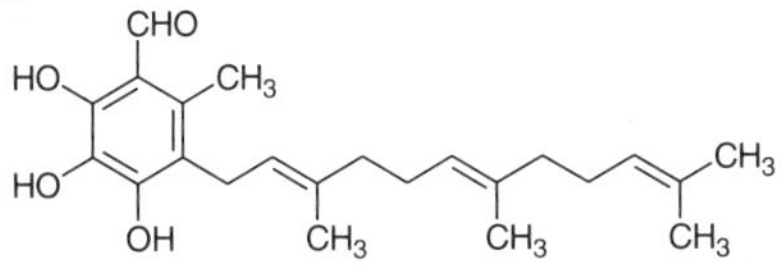

$C_{23}H_{32}O_4$, M_R 372.50, light yellow , mp. 81 °C. Farnesylphenol from fruit bodies of *Albatrellus ovinus* (Basidiomycetes) and *A. subrubescens*. S. is responsible for the yellow color acquired by the mushroom after pressure injury or spotting with alkaline solutions. The non-pungent S. stimulates rat dorsal root ganglion neurons via interaction at vanilloid receptors (see capsaicin)[1].

Lit.: [1] Br. J. Pharmacol. **126**, 1351 (1999); Mol. Pharmacol. **56**, 581 (1999).
gen.: Chem. Ber. **110**, 3770 (1977). – *[CAS 65195-50-8]*

Scymnol (bile alcohol) see bile acids.

Sebiferine see morphinan alkaloids.

Secale alkaloids see ergot alkaloids.

Secalonic acids see ergochromes.

Secoiridoids.

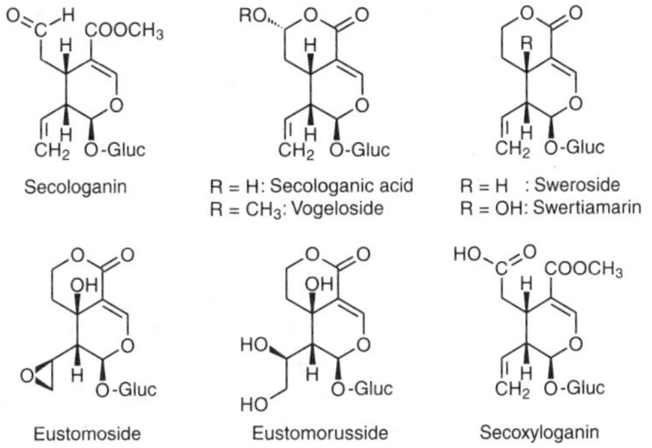

The parent compound of the S. is *secologanin*, a key compound in the biosyntheses of most *indole alkaloids, the *Cinchona, *ipecac, and pyrroloquinoline alkaloids as well as simple monoterpene alkaloids. Over 1000 indole alkaloids are formed from secologanin *in vivo* – comprising almost a quarter of this large group of natural products.
Occurrence: Secologanin is assumed to be present in all plants that form indole alkaloids. However, its occurrence depends on the season of the year. Secologanin was first isolated in 1970 from leaves of the Japanese *Lonicera morrowii* (Caprifoliaceae) harvested in spring, thus, the compound was first called "loniceroside". Secologanin was, however, synthesized earlier in 1969 from menthiafolin (see foliamenthin) by hydrolysis and subsequent esterification with diazomethane and recognized as a key compound in biosynthesis.

Lit.: Beilstein E V **18/9**, 31 ▪ J. Nat. Prod. **43**, 649 (1980) ▪ Justus Liebigs Ann. Chem. **1990**, 253 (synthesis of secologanin)

Table: Data of Secoiridoids.

trivial name	molecular formula	M_R	mp. [°C]	$[\alpha]_D$	CAS
Secologanin (Loniceroside)	$C_{17}H_{24}O_{10}$	388.37	amorphous	–105° (CH_3OH)	19351-63-4
Secologanic acid	$C_{16}H_{22}O_{10}$	374.34	amorphous	–	60077-46-5
Vogeloside	$C_{17}H_{24}O_{10}$	388.37	amorphous	–	60077-47-6
Sweroside	$C_{16}H_{22}O_9$	358.35	amorphous	–236° (H_2O)	14215-86-2
Swertiamarin (Swertiamaroside)	$C_{16}H_{22}O_{10}$	374.34	113–114	–127° (C_2H_5OH)	17388-39-5
Eustomoside	$C_{16}H_{22}O_{11}$	390.34	amorphous	–123° (CH_3OH)	74213-75-5
Eustomorusside	$C_{16}H_{24}O_{12}$	408.36	amorphous	–73° (CH_3OH)	74213-77-7
Secoxyloganin	$C_{17}H_{24}O_{11}$	404.37			58822-47-2

■ Nat. Prod. Rep. **8**, 69 (1991) ■ Phytochemistry **28**, 2971 (1989) (biosynthesis of secologanin); **37**, 1649 (1994) (sweroside) ■ Tetrahedron **44**, 7145 (1988) ■ Tetrahedron Lett. **29**, 3511 (1988).

Secolactarane see lactaranes.

Secologanin, secologanic acid, secoxyloganin see secoiridoids.

Secondary metabolism. Collective name for all side branches of the general metabolism (primary metabolism). The intermediate and final products of secondary metabolism are known as *secondary metabolites* (*natural products). In contrast to the primary metabolism, which follows a similar course in all organisms and is characteristic for terrestrial life (carbohydrates, amino acids, proteins, lipids), the course of the secondary metabolism in various organisms is very individual and leads to compounds that are characteristic for particular species (and can thus serve for the chemotaxonomic definition of the species). The secondary metabolites are of lesser significance for the survival of an individual than the essential, primary metabolites but they are of greater significance for the maintenance of the species (e. g. *pheromones, *phytoalexins, storage substances). They are generally formed only after the active growth phases and in much lower concentrations than the primary metabolites. Generally, only little is known about the functions of the secondary metabolites. The almost infinite variety of products presents an enormous challenge to organic chemistry and has stimulated significant advances in analytic and synthetic techniques. Secondary metabolites and their synthetic analogues are widely used (plant protection, medicine). The most important synthetic routes to secondary metabolites include the biosyntheses of acetogenins (*fatty acids and *polyketides), isoprenoids, *shikimic acid derivatives, and the metabolites of amino acids (*alkaloids).

Lit.: Davies, Secondary Metabolites: Their Function and Evolution, Ciba Found. Symp. **171**, New York: Wiley 1992 ■ Herbert, The Biosynthesis of Secondary Metabolites (2.), London: Chapman & Hall 1989 ■ Luckner (3.) ■ Mann (2.), Secondary Metabolism, Oxford: University Press 1987.

14,15-Seco-tigliane see Euphorbiaceae and Thymelaeaceae poisons.

Securamines.

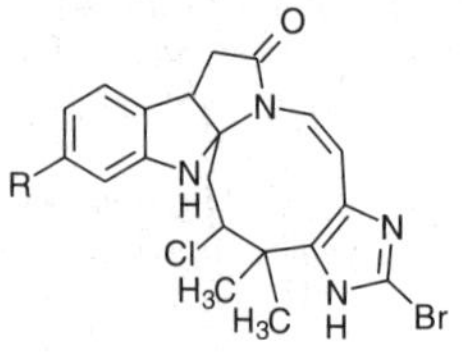

R = H : S. A
R = Br : S. B

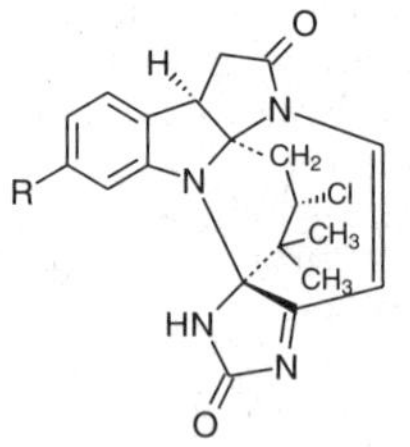

R = Br : S. C
R = H : S. D

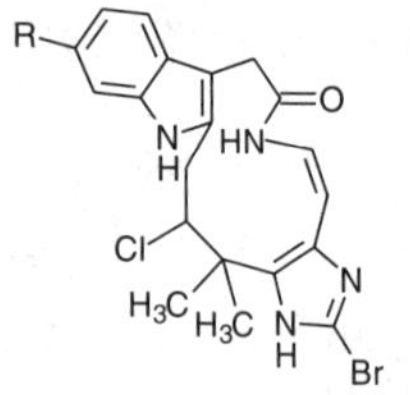

R = H : Securine A
R = Br : Securine B

The bryozoan *Securiflustra securifrons* (Flustridae) produces the halogenated indole-imidazole alkaloids S. A–G. S. A and B exist in equilibrium with the macrocyclic alkaloids securines A and B. *S. A:* $C_{20}H_{20}BrClN_4O$, M_R 447.76, cryst., mp. >200 °C, $[\alpha]_D$ −87.5° ($CHCl_3$); *S. B:* $C_{20}H_{19}Br_2ClN_4O$, M_R 526.66; *S. C:* $C_{20}H_{18}BrClN_4O_2$, M_R 461.74, yellow solid, $[\alpha]_D^{20}$ −434° ($CHCl_3$); *S. D:* $C_{20}H_{19}ClN_4O_2$, M_R 382.85, green solid, $[\alpha]_D^{20}$ −320° ($CHCl_3$); see also chartelline A.

Lit.: J. Nat. Prod. **60**, 175 (1997) (S. E–G) ■ J. Org. Chem. **61**, 887 ff. (1996) (S. A–D). – *[CAS 173220-55-8 (S. A); 173220-56-9 (S. B)]*

Securinega alkaloids. A group of *indolizidine alkaloids from the leaves and roots of *Securinega suffruticosa* and other *Securinega* species (Euphorbiaceae) growing in the Chinese-Russian Ussuri region. Important members are *securinine* and *allosecurinine* (2-*epi*-securinine), which occur together in *S. suffruticosa* and their optical antipodes *virosecurinine* and *viroallosecurinine* from *S. virosa*, *securinol A* (14,15-dihydro-15-hydroxysecurinan-11-one).

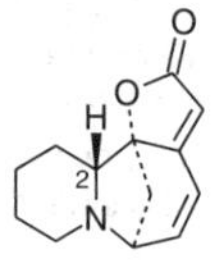

(−): Securinine (**1**)
(+): Virosecurinine (**2**)

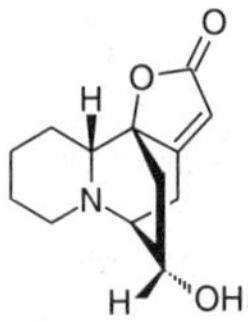

Securinol A (**3**)

Table: Data of some Securinega alkaloids.

	molecular formula	M_R	mp. [°C]	$[\alpha]_D$ ($CHCl_3$)	CAS
1	$C_{13}H_{15}NO_2$	217.27	142–143	−1106°	5610-40-2
2	$C_{13}H_{15}NO_2$	217.27	141–142	+1148°	6704-68-3
3	$C_{13}H_{17}NO_3$	235.28	135–136	+58.2°	5008-48-0

Securinine is a GABA antagonist (see 4-aminobutanoic acid). It has a stimulating effect on the central nervous system with strychnine-like activity but lower toxicity.

Lit.: Acta Crystallogr., Sect. C **51**, 127 (1995) ■ Angew. Chem., Int. Ed. **39**, 237 (2000) (synthesis) ■ Brain Res. **330**, 135–140 (1985) ■ J. Chem. Soc., Chem. Commun. **1975**, 144 (biosynthesis) ■ J. Chem. Soc. Perkin Trans. 1 **1991**, 1863 ff. (revision of structure) ■ J. Heterocycl. Chem. **33**, 1437 (1996) ■ J. Indian Chem. Soc., Sect. B **29**, 801 ff. (1990) ■ J. Nat. Prod. **47**, 677–681 (1984); **49**, 614–620 (1986) ■ Tetrahedron **43**, 2915–2924 (1987). – *[HS 2939 90]*

(−)-Sedamine see Sedum alkaloids.

Sedanenolide see celery seed oil.

Sedoheptulose (D-*altro*-2-heptulose).

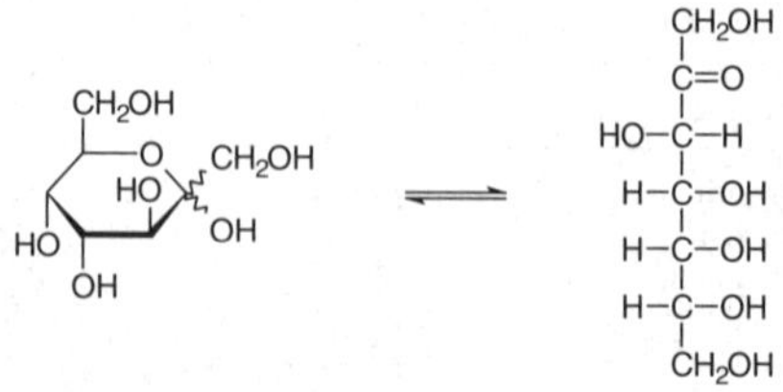

$C_7H_{14}O_7$, M_R 210.18, heptoketose, mp. 100–102 °C, $[\alpha]_D$ +2.5° (H_2O). S. occurs in many plants, especially *Sedum spectabile* (Crassulaceae). S. plays an important role as intermediate product (S. 7-phosphate) in the Calvin cycle of photosynthesis.

Lit.: Adv. Carbohydr. Chem. Biochem. **42**, 15 (1984) ▪ Beilstein EIII/IV **1**, 4446 ▪ Carbohydr. Res. **320**, 31–36 (1999) (synthesis) ▪ Chem. Pharm. Bull. **20**, 2723–2742 (1972) ▪ Karrer, No. 623. – *[CAS 3019-74-7]*

Sedum alkaloids. C-2 monosubstituted *piperidine alkaloids from *Sedum* species (Crassulaceae). *Sedamine* is responsible for the hot taste of the wall pepper (*Sedum acre*).

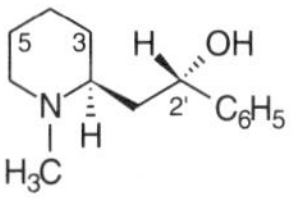

(–)-Sedamine

$C_{14}H_{21}NO$, M_R 219.33, mp. 75–76 °C (89 °C), $[\alpha]_D$ –88.4° (C_2H_5OH). Compounds related to sedamine include 5-hydroxysedamine[1], the 2′-epimer allosedamine[2], its *N*-demethyl derivative norallosedamine, and the corresponding two hydroxy derivatives 3-hydroxyallo-sedamine[1] and 3-hydroxynorallosedamine[1].
The S. a. are formed biogenetically from lysine via enzyme-bound intermediates[3]. The side chain originates from phenylalanine. Sedamine is accompanied by structurally related *piperidine alkaloids such as (±)-*pelletierine and 2,6-disubstituted piperidines of the *Lobelia alkaloid type (including sedinine, sedacrine)[4]. The acute toxicity of the S. a. is low, unambiguous pharmacological activities have not yet been confirmed[4].

Lit.: [1]Tetrahedron **43**, 935-945 (1987) (synthesis). [2]Helv. Chim. Acta **77**, 554 (1994); Heterocycles **32**, 889 (1991); Synth. Commun. **21**, 2213 (1991). [3]J. Am. Chem. Soc. **95**, 4715 (1973); Phytochemistry **18**, 771 (1979). [4]Hager (5.) **6**, 650–656.
gen.: Heterocycles **47**, 263 (1998) (synthesis (–)-sedamine) ▪ J. Org. Chem. **55**, 1086 (1990); **58**, 5035 (1993) (synthesis) ▪ Mothes et al., p. 135 ▪ Pelletier **3**, 56–59. – *[HS 2939 90; CAS 497-88-1 (sedamine); 497-89-2 (allosedamine)]*

Selachyl alcohol see batyl alcohol.

Selagin(e). Synonym for *huperzine A and *selgin.

Selaginella (Selaginellales). Members of the order Lycopodiales (*clubmosses) with the sole family Selaginellaceae to which about 700 mostly tropical species of the genus *Selaginella* belong. S. have partly creeping, upright, or climbing, fork-like branched shoots with small to scale-like leaflets. For metabolites, see clubmosses.

Lit.: Fukarek (ed.), Enzyklopädie Urania-Pflanzenreich, vol. 2 (Moose, Farne, Nacktsamer), p. 131, Leipzig: Urania-Verlagsges. 1992 ▪ Zechmeister **65**. – *[HS 1211 90]*

Selectomycin see spiramycin.

L-Selenocysteine (3-selenyl-L-alanine, (*R*)-2-amino-3-selenylpropanoic acid, abbr.: Sec).

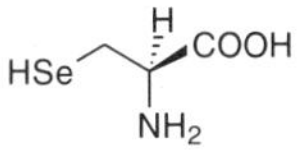

$C_3H_7NO_2Se$, M_R 168.05, an unstable *cysteine analogue in which the sulfur atom has been replaced by selenium. Component of some proteins.
S. is present in the active center of mammalian glutathione peroxidase (EC 1.11.1.9), two formate dehydrogenases from bacteria (EC 1.2.1.2), and a hydrogenase from *Desulfovibrio baculatus*. A hepatic and renal iodothyronine deiodase also contains S.[1]. This can – like other amino acids – be incorporated into proteins during translation with the help of an L-selenocysteinyl transfer ribonucleic acid that recognizes the *opal*-termination codon (UGA) and is formed *in vivo* from an L-seryl transfer ribonucleic acid[2]. The so-called SELB protein (M_R 68 000) is required as a specific translation factor and exhibits sequence homologies to G proteins.

Lit.: [1]Nature (London) **349**, 483–440 (1991). [2]Nature (London) **342**, 453–456 (1989).
gen.: Annu. Rev. Biochem. **59**, 111–127 (1990) ▪ J. Chem. Soc., Perkin Trans. 1 **1997**, 2443 (synthesis) ▪ Mol. Microbiol. **5**, 515–520 (1991) ▪ Nachr. Chem. Tech. Lab. **39**, 966–971 (1991) ▪ Nucleic Acids Res. **19**, 939–943 (1991) ▪ Tetrahedron **42**, 4983 (1986) (synthesis) ▪ Trends Biochem. Sci. **13**, 419–421 (1988). – *[CAS 10236-58-5]*

Selgin.

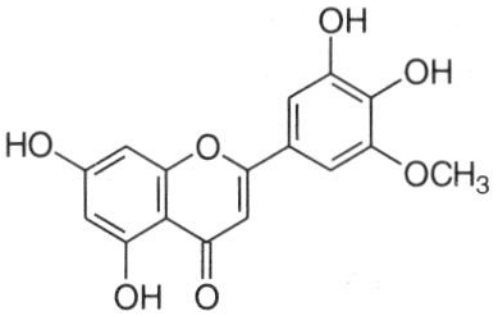

$C_{16}H_{12}O_7$, M_R 316.27, mp. 320–325 °C, yellow cryst., soluble in 80% acetone, DMF, DMSO, practically insoluble in water. 5,7,3′,4′-Tetrahydroxy-5′-methoxyflavone (tricetin 3′-methyl ether), first isolated from a *clubmoss species[1] and designated as *selagin*, a name already assigned to a clubmoss alkaloid, (later renamed as S.[2]). S. is also found in liverworts, other members of the Lycopodiatae, and higher plants[3]. S. exists in the plants as the *O*- or *C*-glycosides.

Lit.: [1]Phytochemistry **15**, 840 (1976). [2]Syst. Ecol. **6**, 95 (1978). [3]Pharmazie **36**, 577 (1981); J. Nat. Prod. **47**, 384 (1984). – *[CAS 60303-28-8]*

Selinanes see eudesmanes.

Selinenes (eudesmenes).

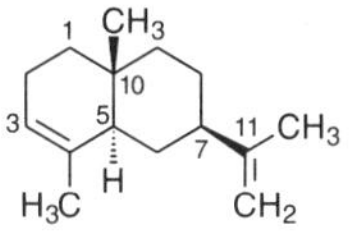

(-)-α-Selinene

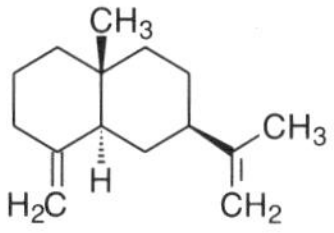

(+)-β-Selinene

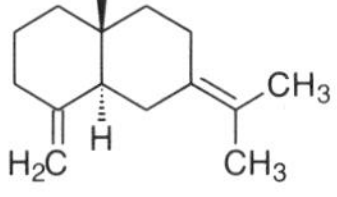

γ-Selinene

α-S. (3,11-eudesmadiene), $C_{15}H_{24}$, M_R 204.36, oil, bp. 268–272 °C, $[\alpha]_D^{20}$ –14.5° ($CHCl_3$), is a component of the essential oil of Indian hemp (*Cannabis sativa*) and hops (*Humulus lupulus*)[1]. (+)-*β*-S. [4(14),11-eudesmadiene], oil, bp. 121–122 °C (798 Pa), $[\alpha]_D$ +61°, occurs in *celery seed oil and the essential oils of cypress grasses, (–)-*β*-S.[1] on the other hand is present in the essential oils of *Libanotis transcaucasica* and *Seseli indicum*. *γ*-S.: oil, $[\alpha]_D$ +2.5° ($CHCl_3$) is a constituent of essential oil of hops. Many other isomers of the eudesmadiene group occur in essential oils. A 4,7-eudesmadiene has been isolated from the red alga *Laurencia nidifica*[2].

Lit.: [1] Tetrahedron Lett. **23**, 3047 (1982). [2] J. Org. Chem. **43**, 1613 (1978).
gen.: Karrer, No. 1890 ▪ Phytochemistry **40**, 1633 (1995). – *Synthesis:* Chem. Pharm. Bull. **44**, 1603 (1996) (γ-S.) ▪ J. Nat. Prod. **57**, 1189 (1994) ▪ J. Org. Chem. **48**, 4380 (1983); **53**, 4124 (1988) ▪ Justus Liebigs Ann. Chem. **1992**, 139. – *[CAS 473-13-2 (α-S.); 17066-67-0 ((+)-β-S.)]*

Semiochemicals (from Greek semion = signal). Collective name for signal substances that carry intraspecific and interspecific messages between organisms. *Intraspecific S.* are classified as *pheromones. These are subdivided into substances that influence the behavior of the receiver (e.g., sexual attractants) and those that change the physiological state of the receiver [primers, e.g., the *queen substance secreted by the queen bee suppresses ovarian development in female worker bees]. *Interspecific S.* are classified as *allelochemicals. These are subdivided into *kairomones, that are useful for the receiver, *allomones, that are useful for the sender, and *synomones, that are advantageous for both sender and receiver. S. are also sometimes known as *infochemicals.*
Lit.: ACS Symp. Ser. **658**, Phytochemicals for Pest Control (1997) ▪ Agosta, Bombardier Beetles and Fever Trees, New York: Addison-Wesley 1995 ▪ Chem. Ind. (London) **1994**, 370 ff. ▪ Eur. J. Org. Chem. **1998**, 1479 (review) ▪ Nordlund, Jones, & Lewis, Semiochemicals, their Role in Pest Control, New York: Wiley 1981 ▪ Rice, Pest Control with Nature's Chemicals: Allelochemicals and Pheromones in Gardening and Agriculture, Norman: Univ. Oklahoma Press 1983.

Sempervirine (2,3,4,13-tetrahydro-1*H*-benz[*g*]indolo[2,3-*a*]quinolizin-6-ium zwitterion).

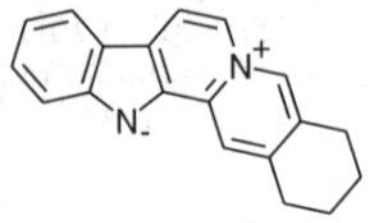

$C_{19}H_{16}N_2$, M_R 272.35, mp. 258–260 °C, an indoloquinolizidine alkaloid from the *Gelsemium* species *G. sempervirens* and *G. elegans* (Loganiaceae). For synthesis, see *Lit.*[1]. S. has antitumor activity[2] and shows a tendency for DNA intercalation (cytotoxicity).
Lit.: [1] Monatsh. Chem. **127**, 1259 (1996); Synth. Commun. **26**, 4409 (1996); Tetrahedron **44**, 3195 (1988). [2] Oncology **43**, 198 (1986). – *[HS 2939 90; CAS 6882-99-1 (cation)]*

Sendanin see meliatoxins.

Senecio alkaloids see pyrrolizidine alkaloids.

Senecionine. Formula see pyrrolizidine alkaloids. $C_{18}H_{25}NO_5$, M_R 335.40, mp. 232–233 °C, $[\alpha]_D$ –56° ($CHCl_3$). S. is a pyrrolizidine alkaloid from many Asteraceae genera of the tribe Senecioeae (*Senecio, Emilia, Erechtites, Petasites*). S. exists in the plants as the *N*-oxide. Antitumor activity has been demonstrated but its high toxicity (see pyrrolizidine alkaloids) prevents medical use. It is hepatotoxic and causes pathological changes in the lungs. S. is the main representative of more than 80 known structures derived therefrom containing a 12-membered macrocyclic diester system formed from retronecine, otonecine (see senkirkine), and platynecine (see platyphylline) as necine bases [see necine(s)] and a variously substituted hexanedioic acid. The (15*E*)-isomer is *integerrimine* {mp. 172 °C, $[\alpha]_D$ +4° (CH_3OH)}. For synthesis of integerrimine, see *Lit.*[1], for synthesis of S., see *Lit.*[2]. 13,19-Didehydrosenecionine is called *seneciphylline* {$C_{18}H_{23}NO_5$, M_R 333.38, mp. 217–218 °C (decomp.), $[\alpha]_D$ –139° ($CHCl_3$)}.
Lit.: [1] J. Org. Chem. **51**, 5492 (1986); **57**, 2270 (1992); Tetrahedron **48**, 393 (1992). [2] Bull. Chem. Soc. Jpn. **67**, 1990 (1994).
gen.: Beilstein E III/IV **27**, 6636 ff. ▪ Hager (5.) **3**, 1079 f.; **6**, 92 f., 662–680 ▪ Merck-Index (12.), No. 8595 ▪ Pelletier **9**, 155–233 ▪ Sax (8.), ARS 500, IDG 000. – *[HS 2939 90; CAS 130-01-8 (S.); 480-79-5 (integerrimine); 480-81-9 (seneciphylline)]*

Seneciphylline see senecionine.

Senkirkine.

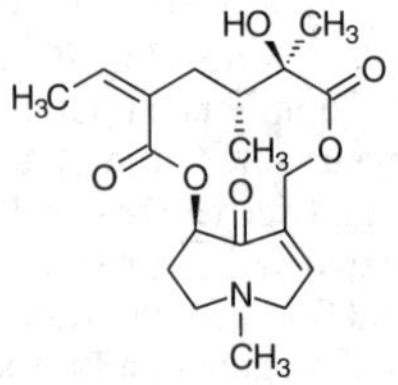

$C_{19}H_{27}NO_6$, M_R 365.43, mp. 196.5–197.5 °C, $[\alpha]_D^{25}$ –16° (CH_3OH). S. is a *pyrrolizidine alkaloid from various Asteraceae genera (*Senecio, Nardosmia, Farfugium*), and *Crotalaria* (Fabaceae). It is hepatotoxic, presumably carcinogenic, and genotoxic. S. is an analogue of *senecionine with otonecine [see necine(s)] as the alcohol component. More than 20 alkaloids derived from S. containing otonecine are known.
Lit.: Beilstein E III/IV **27**, 6663 ▪ Hager (5.) **3**, 1080; **6**, 83, 93, 1019 ▪ Pelletier **9**, 155–233 ▪ Phytochemistry **25**, 1151 (1986); **30**, 1734 (1991) ▪ Sax (8.), DMX 200. – *[HS 2939 90; CAS 2318-18-5]*

Sennosides. A group of yellow, homo- and heterodimeric anthrone glycosides, the aglycones (*sennidins*) of which are derived from the monomers *rhein and *aloe-emodin.

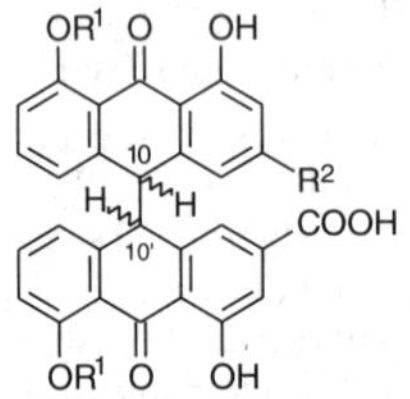

Sennoside A, A' u. B
R^1 = β-D-Glc, R^2 = COOH
Sennoside C u. D
R^1 = β-D-Glc, R^2 = CH_2OH

S. A–G occur in the leaves and fruits of *Cassia* species (*C. senna, C. angustifolia*, Fabaceae) and in rhubarb roots, sometimes as oxalyl derivatives (sennosides E, F). S. are mostly formed from the corresponding anthrone glycosides when the senna leaves are dried. Drugs containing S. are used in therapy for acute constipation[1] (see table, p. 581).
Lit.: [1] Hager (5.) **9**, 597–605.
gen.: Chem. Pharm. Bull. **30**, 1550 (1982) ▪ Merck-Index (12.), No. 8600 ▪ Pharmacology, Suppl. **36**, 1 (1988); **47**, 1 (1993) ▪ Zechmeister **7**, 258.

Senoxepin.

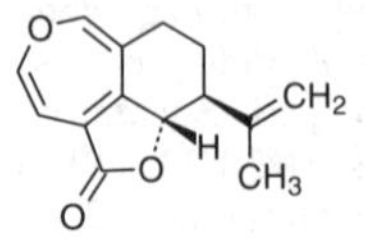

Table: Data of Sennosides.

Sennoside	molecular formula	M_R	stereo-chemistry 10/10′	mp. [°C] (decomp.)	$[\alpha]_D$	CAS
S. A	$C_{42}H_{38}O_{20}$	862.75	*R/R*	220–243	–164° (60% aqueous acetone)	81-27-6
S. A′	$C_{42}H_{38}O_{20}$	862.75	*S/S*	162–176	–110° (70% aqueous dioxan)	66575-30-2
S. B	$C_{42}H_{38}O_{20}$	862.75	*R/S*	209–212	–100° (70% aqueous acetone)	128-57-4
S. C	$C_{42}H_{40}O_{19}$	848.77	*R/R*	204–206	–123° (70% aqueous acetone)	37271-16-2
S. D	$C_{42}H_{40}O_{19}$	848.77	*R/S*	210–220	+3.1° (H_2O)	37271-17-3

$C_{14}H_{14}O_3$, M_R 230.26, yellow oil, $[\alpha]_D$ – 126.5° ($CHCl_3$). An oxepine of sesquiterpenoid origin from *Senecio* sp. (Asteraceae).
Lit.: Phytochemistry **16**, 965 (1977). – *[CAS 64197-44-0]*

Sepedonin.

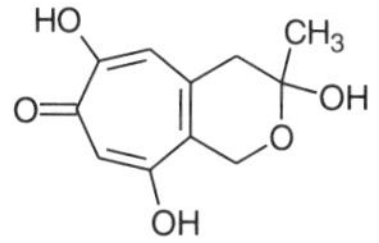

$C_{11}H_{12}O_5$, M_R 224.21, pale yellow prisms, mp. 190–205 °C (decomp.). A tropolone derivative from cultures of the fungus *Sepedonium chrysospermum* (Ascomycetes).
Lit.: Can. J. Chem. **48**, 2702 (1970); **50**, 821 (1972) ▪ J. Chem. Soc., Chem. Commun. **1971**, 325. – *[CAS 3621-27-0]*

Sepiomelanin (C. I. natural brown 9). Gray-black pigment from the dried accessory gland of the hindgut of cephalopods of the genus *Sepia*, (e.g., the Mediterranean common octopus, *Sepia officinalis*) obtained by dissolution in alkali and precipitation with hydrochloric acid. S. is probably a copolymer of indole-5,6-quinone and 5,6-dioxo-5,6-dihydroindole-2-carboxylic acid; it is used, among others, in aquarelle paints.
Lit.: Nicolaus, Melanins, p. 68–91, Paris: Hermann 1968 ▪ Ullmann (4.) **11**, 116. – *[HS 3203 00]*

(+)-Septicine see Tylophora alkaloids.

L-Serine [(*S*)-2-amino-3-hydroxypropanoic acid, abbreviation: Ser or S].

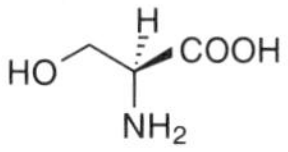

$C_3H_7NO_3$, M_R 105.09, mp. 228 °C (decomp.), sublimes at 150 °C (<0.01 Pa), $[\alpha]_D^{20}$ –6.83° (H_2O), +14.45° (5.6 m HCl), pK_a 2.21, 9.15, pI 5.68. A sweet tasting, non-essential amino acid occurring in practically all proteins. Genetic code: UCU, UCC, UCA, UCG. Average content in proteins 7.1%[1]. Ser can be isolated in high yield from the acid hydrolysate of silk fibroin. In glycoproteins *oligosaccharides are bound to Ser-, Thr, or in the case of collagen to 5-hydroxylysine residues. Phosphoserine is a component of some proteins. Besides proteins, Ser is a component of phospholipids. D-Ser is part of some antibiotics such as echinomycin (quinomycin A) and *polymyxin D.
Ser plays an important role in the stabilization of the three-dimensional structures of proteins by forming hydrogen bonds. Ser is present in the active centers of serine proteases (enzyme group EC 3.4.21, e.g., trypsin, chymotrypsin, elastase, thrombin, plasmin, etc.) and serine carboxypeptidase II (EC 3.4.16.1).
Metabolism: Ser is formed from 3-phosphoglycerate by 3-phosphoglycerate dehydrogenase (EC 1.1.1.95) and NAD^+, phosphoserine aminotransferase (EC 2.6.1.52), and phosphoserine phosphatase (EC 3.1.3.3). In the metabolic degradation, Ser transfers its hydroxymethyl group by way of serine hydroxymethyltransferase (EC 2.1.2.1) to tetrahydrofolic acid, and forms glycine; it thus participates in a one-carbon-atom-cycle[2]. On the other hand the action of serine dehydratase (EC 4.2.1.13) on Ser gives pyruvate. The mitochondrial photorespiration in green plants produces 2 molecules of glycine from Ser and carbon dioxide.
Production: The commercial yearly requirements of ca. 50 t Ser are either extracted from protein hydrolysates or produced by fermentation from glycine, racemic Ser is also synthesized from glycol aldehyde. L-Ser is used in the manufacturing of amino acid infusion solutions, DL-Ser for the production of L-tryptophan and cycloserine.
Non-proteinogenic D-Serine, prisms, mp. 228 °C (decomp.), $[\alpha]_D^{20}$ +6.9° (H_2O) is synthesized in mammals' brains and acts as a neurotransmitter there. In concert with glutamate, it activates the *N*-methyl-D-aspartate (NMDA) receptor. Activation of this receptor has been linked to memory and learning, among other brain processes, as well as to nerve-cell death in stroke[3].
Lit.: [1]Biochem. Biophys. Res. Commun. **78**, 1018–1024 (1977). [2]Chem. Rev. **90**, 1275–1290 (1990). [3]Chem. Eng. News (15.11.) **1999**, 9; Proc. Natl. Acad. Sci. USA **96**, 13409 (1999).
gen.: Beilstein E IV **4**, 3118 ff. ▪ Lewis, Food Additives Handbook No. SCA 350, SCA 355, New York: Van Nostrand 1989 ▪ Stryer 1995. – *[HS 2922 50; CAS 56-45-1 (S.); 312-84-5 (D-S.)]*

Serotonin [5-hydroxytryptamine, usual abbr.: 5-HT]

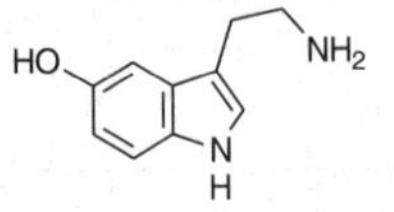

$C_{10}H_{12}N_2O$, M_R 176.22, pK_{a1} 9.98, pK_{a2} 11.26 (20 °C), stable as hydrochloride, $C_{10}H_{13}ClN_2O$, M_R 212.68, hygroscopic, light-sensitive cryst., mp. 167–168 °C, aqueous solutions are stable at pH 2–6.4. S. belongs to the biogenic amines.

Occurrence: S. is widely distributed in the animal and plant kingdoms, especially in pineapple, bananas, stinging nettle toxin, and *coelenterates as well as in the skin glands of amphibians. In humans S. occurs in the gastrointestinal mucous membranes, in the brain, and in thrombocytes. It is stored in the latter and liberated during blood coagulation. In rodents and cattle it is liberated by mast cells in cases of allergic reactions. S. is formed in larger amounts by tumors of enterochromaffinic cells (carcinoid). S. functions as a neurotransmitter in the central and peripheral nervous systems as well as a tissue hormone. For further details of its biological activities, see *Lit.*[1].

Biosynthesis: S. is formed from L-*tryptophan by hydroxylation (tryptophan 5-monooxygenase, EC 1.14.16.4) and decarboxylation (aromatic L-amino acid decarboxylase, EC 4.1.1.28, see dopamine).

Use: The main problem with S. concerning its therapeutic use is tachyphylaxia (rapid reduction of action). *Rauvolfia alkaloids (e.g., *reserpine) potentiate the liberation of S., MAO inhibitors prevent its degradation; these facts may be utilized in therapy for depression.

Lit.: [1]Merck-Index (12.), No. 8607.
gen.: Annu. Rev. Neurosci. **14**, 335–360 (1991) ▪ Beilstein E V **22/12**, 16 ff. ▪ Hager (5.) **9**, 603 ff. ▪ Heterocycles **46**, 91 (1997) (synthesis) ▪ J. Org. Chem. **42**, 3769 ff. (1977) ▪ Karrer, No. 4382 ▪ Luckner (3.), p. 457 ff. ▪ Phytochemistry **16**, 171–175 (1977). – *[HS 2933 90; CAS 50-67-9]*

Serpentene.

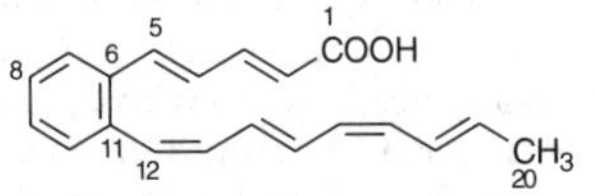

$C_{20}H_{20}O_2$, M_R 292.38, yellow amorphous powder, mp. 118 °C. An unstable polyenecarboxylic acid with an *ortho*-substituted benzene ring and possessing (*E*)- and (*Z*)-double bonds in the side chains, produced by *Streptomyces* sp.

Lit.: Justus Liebigs Ann. Chem. **1993**, 433 ff. – *[CAS 149598-73-2]*

Serratenediol see triterpenes.

Serricornin [(4*S*,6*S*,7*S*)-7-hydroxy-4,6-dimethyl-3-nonanone]. $C_{11}H_{22}O_2$, M_R 186.29, liquid; acetate: $[\alpha]_D^{21.5}$ +18.2° (hexane). The sexual pheromone of females of the tobacco beetle *Lasioderma serricorne* which causes serious economic damage. The open chain form exists in equilibrium with its cyclic semiacetal[1].

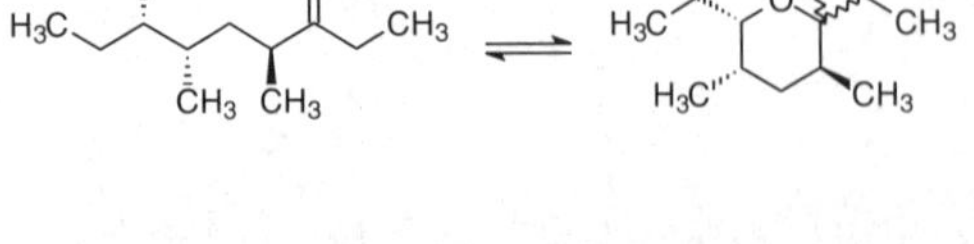

The unnatural (4*S*,6*S*,7*R*)-diastereomer has potent repellent activity. Traps filled with S. are used to monitor the population development of the beetles in tobacco storage.

Lit.: [1]Tetrahedron **41**, 3423 ff. (1985); Synth. Commun. **18**, 981–993 (1988); **24**, 233 (1994).
gen.: Synthesis: ApSimon **9**, 193–200 ▪ J. Braz. Chem. Soc. **9**, 571 (1998) ▪ J. Org. Chem. **63**, 4466 (1998) ▪ Tetrahedron **47**, 2991–2998 (1991); **50**, 2703 (1994) ▪ Tetrahedron: Asymmetry **4**, 1573 (1993). – *[CAS 72598-35-7]*

Serricorole [(2*S*,3*R*,1′*S*,2′*S*)-2-ethyl-2,3-dihydro-6-(2-hydroxy-1-methylbutyl)-3,5-dimethyl-4*H*-pyran-4-one].

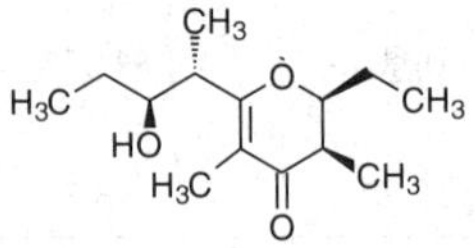

$C_{14}H_{24}O_3$, M_R 240.3, $[\alpha]_D^{24}$ –113° ($CHCl_3$). A pyran derivative formed biogenetically from propionate units, it is structurally related to the pheromones *invictolide and stegobiol (see stegobinone), it forms, together with *serricornin and *serricorone* ($C_{14}H_{22}O_3$, M_R 238.3), in which the hydroxy function of the side chain of S. is oxidized to an oxo group, part of the pheromone complex of the female tobacco beetle *Lasioderma serricorne*.

Lit.: Agric Biol. Chem. **51**, 2925–2931 (1987) ▪ ApSimon **9**, 380 ▪ Helv. Chim. Acta **76**, 1275–1291 (1993) ▪ Synthesis **1998**, 386 (synthesis). – *[CAS 86797-90-2 (S.); 86797-89-9 (serricorone)]*

Sesame oil. Oil obtained from the seeds of sesame, *Sesamum indicum* (*S. orientale*, Pedaliaceae). Sesame is cultivated as an oil plant mainly in India and China. The triglycerides of S. contain mainly oleic acid (35–46%) and linoleic acid (35–48%), together with 7–9% palmitic acid and 4–6% stearic acid. In India and East Asia S. is used as an edible oil, in Europe it is only of minor significance. The characteristic components of S. are the phenolic compounds sesamol, sesamolin, sesaminol, etc. (see sesamol).

Lit.: Bockisch, Nahrungsfette u. -öle, Stuttgart: Ulmer 1993 ▪ Ullmann (5.) **A 10**, 176, 227 ▪ see also fats and oils. – *[HS 1515 90; CAS 8008-74-0]*

Sesame seed (roasted) flavor. According to aroma extract dilution analysis, *2-furylmethanethiol, *guaiacol, *2-phenylethanethiol* ($C_8H_{10}S$, M_R 138.23), and *Furaneol® have the highest dilution factors in the flavor of roasted sesame seeds[1], besides *4-vinylguaiacol, *2-pentylpyridine* ($C_{10}H_{15}N$, M_R 149.24), *2-acetyl-1-pyrroline, 2-acetylpyrazine, and 2-ethyl-3,5-dimethylpyrazine (see pyrazines).

Lit.: [1]Schieberle, in Schreier & Winterhalter (eds.), Progress in Flavour Precursor Studies, p. 343–360, Carol Stream (USA): Allured 1993.
gen.: TNO list (6.), p. 831 ff. – *[CAS 4410-99-5 (2-phenylethanethiol); 2294-76-0 (2-pentylpyridine)]*

Sesamin(ol) see sesamol.

Sesamol (3,4-methylenedioxyphenol, 1,3-benzodioxol-5-ole). The fat-soluble furolignans *sesamin* {$C_{20}H_{18}O_6$, M_R 354.36, mp. 123–124 °C, $[\alpha]_D^{20}$ +78.4° ($CHCl_3$)} and *sesamolin* {$C_{20}H_{18}O_7$, M_R 370.36, mp. 94 °C, $[\alpha]_D$ +220° ($CHCl_3$)} are the main components

of the non-saponifiable portion of *sesame oil. Phenolic derivatives of the two compounds *S.* ($C_7H_6O_3$, M_R 138.12, mp. 65–66 °C) and *sesaminol* {$C_{20}H_{18}O_7$, M_R 370.36, mp. 146–148 °C, $[\alpha]_D$ +42.7° ($CHCl_3$)}, also occur in trace amounts in native sesame, but are principally produced during the bleaching of sesame oil on acidic fuller's earth in the process of fat refining.
S. is also cleaved on heating sesamolin. S. and sesaminol have antioxidant properties as a result of their phenolic character; they thus contribute to the stability of sesame oil towards oxidation. – Color test: S. forms a red pigment with furfural and hydrochloric acid in ethanol (Baudouin's reaction), which can be used to identify sesame oil in fats and oils [1].

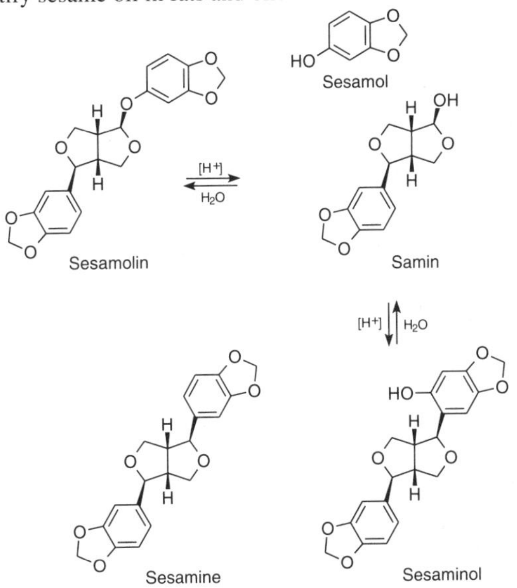

Lit.: [1] Pardun, Analyse der Nahrungsfette, p. 30, Berlin: Parey 1976.
gen.: J. Am. Oil Chem. Soc. **67**, 508 (1990); **71**, 141 (1994) ▪ J. Chem. Soc., Perkin Trans. 1 **1992**, 185 ▪ Phytochemistry **35**, 773 (1994) ▪ Synlett **1992**, 479. – *[CAS 533-31-3 (S.); 607-80-7 (sesamine); 526-07-8 (sesamolin); 74061-79-3 (sesaminol)]*

Sesbanimides.

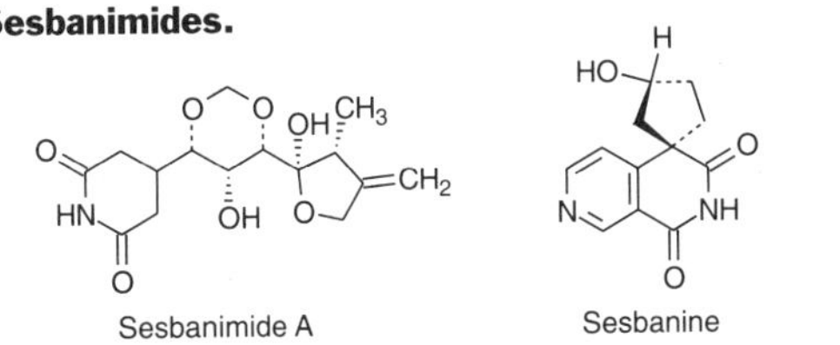

Toxic alkaloids from seeds of *Sesbania* species (Fabaceae) indigenous to Texas and South Africa. Besides other isomeric structures, *S. A* {$C_{15}H_{21}NO_7$, M_R 327.33, cryst., mp. 158–159 °C, $[\alpha]_D^{23}$ −5.6° (CH_3OH)} is the main alkaloid and also has the strongest biological activities. S. are cytotoxic and have general antineoplastic activities. *Sesbanine* [1] {$C_{12}H_{12}N_2O_3$, M_R 232.24, mp. 240–243 °C, $[\alpha]_D^{23}$ +14.6° (CH_3OH)} obtained from *Sesbania drummondii* has similar activities.
Lit.: [1] Tetrahedron **44**, 4351 (1988).
gen.: Atta-ur-Rahman **1**, 305–322 ▪ Bull. Chem. Soc. Jpn. **61**, 2123–2131 (1988) ▪ J. Antibiot. **51**, 64 (1998) (isolation) ▪ J. Chem. Soc., Chem. Commun. **1992**, 368 ▪ J. Org. Chem. **59**, 3055 (1994) ▪ Recl. Trav. Chim. Pays-Bas **110**, 414 (1991) ▪ Tetrahedron **44**, 4721–4736 (1988) ▪ Tetrahedron: Asymmetry **5**, 247 (1994). – *[HS 2939 90; CAS 85719-78-4 (S. A); 70521-94-7 (sesbanine)]*

β-Sesquiphellandrene see ginger oil.

Sesquiterpene alkaloids see Nuphar alkaloids.

Sesquiterpenes (sesquiterpenoids). A structurally highly diverse class of terpenoids with 15 carbon atoms skeleton derived biosynthetically from farnesyl pyrophosphate (FPP) (*farnesol, *isoprene rule, *terpenes). More than 70 different ring systems are formed by enzyme-catalyzed cyclization of the linear parent structure; these cyclic structures can be further modified by 1,2- and 1,3-hydride shifts, renewed cyclizations, hydroxylations, and other subsequent reactions. S. are widely distributed in plants, fungi, and animals but are less common in bacteria. Specific biosynthetic routes are often characteristic for certain organisms. Thus, basidiomycetes preferentially use *humulene as the basis for the syntheses of *protoilludanes, *illudanes, *lactaranes, *hirsutanes, and related S. skeletons. Individual S. systems are also known for liverworts and marine organisms. In addition, liverworts often contain the optical antipodes of S. known from plants.
S. play important roles as *active principles*. Plants, fungi, insects, and marine organisms use S. as *defensive substances* (*polygodial, *stearoylvelutinal, *isovelleral) while plants form them as *stress metabolites* (*phytoalexins, *capsidiol, *ipomeamarone, *phytuberin). Sesquiterpenoid *bitter principles* such as *santonin also protect the plants from herbivores. Cytotoxic methylene lactones with *antitumor activities* occur in the Asteraceae (e.g., *vernolepin). Examples of *toxins* are *coriamyrtin, *picrotoxin, and the alkaloid dendrobine (see Dendrobium alkaloids). *Strigol is a *germination stimulant* for *Striga* species living as parasites on cotton plants; *abscisic acid is widely distributed in plants as a *plant hormone*. In insects and mold fungi S. serve as *sexual pheromones* (*periplanones) and *hormones* [*juvenile hormones, *juvabione]. Various S. from fungal cultures are *antibiotics* (*fumagillin, *ovalicin) or play a role as *mycotoxins* (*tric(h)othecenes). Many S. are responsible for the characteristic odors of *essential oils* and thus find important uses in the perfume industry, e. g., *caryophyllene, davanone (see davana oil), patchouli alcohol (see patchouli oil), *santalenes and *sinensals. Many biologically active, halogenated S. have been isolated from marine organisms (*halogen compounds, natural).
Lit.: *Biosynthesis:* Chem. Rev. **90**, 1089 (1990) ▪ Nat. Prod. Rep. **1**, 443 (1984); **2**, 513 (1985); **4**, 157 (1987); **5**, 247 (1988); **7**, 25, 387 (1990); **8**, 441 (1991); **12**, 507 (1995). – *Phytoalexins:* Nat. Prod. Rep. **8**, 367 (1991). – *Review:* Atta-ur-Rahman **17**, 153–206 ▪ Nat. Prod. Rep. **1**, 105 (1984); **2**, 147 (1985); **3**, 273 (1986); **4**, 473 (1987); **5**, 497 (1988); **7**, 61, 515 (1990); **8**, 69 (1991); **9**, 217, 557 (1992); **10**, 311 (1993); **11**, 451 (1994); **12**, 303 (1995); **13**, 307 (1996); **15**, 73–92 (1998); **16**, 21–38 (1999) ▪ Turner **1**, 219; **2**, 228 ▪ Zechmeister **64**, 1–210.

Sesterterpenoids. A small group of *terpenes formed biosynthetically by head-to-tail addition of 5 isoprene units to geranyl farnesyl pyrophosphate (see also isoprene rule). Examples of the S. include the luffariellins, *manoalide, *ophiobolins, *retigeranic acid, and *variabilin.

Lit.: Bioorg. Med. Chem. Lett. **8**, 2093 (1998) ▪ Chem. Lett. **1999**, 61 ▪ Chem. Pharm. Bull. **44**, 690, 695 (1996) ▪ J. Nat. Prod. **59**, 869 (1996) ▪ Nat. Prod. Rep. **9**, 481 (1992); **13**, 529 (1996) ▪ Zechmeister **33**, 28; **48**, 203–269.

Setamycin see bafilomycins.

Setarin see 4-hydroxycoumarin.

Seudenol.

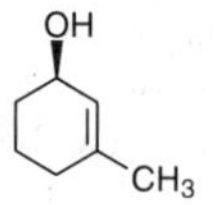

(+)-(*R*)-form

$C_7H_{12}O$, M_R 112.17, liquid, bp. 83–84 °C (2.73 kPa); (*R*)-enantiomer: $[\alpha]_D^{20}$ +96° ($CHCl_3$). Aggregation pheromone of the American bark beetle *Dendroctonus pseudotsugae* (Douglas fir beetle). The natural product exists as an enantiomeric mixture containing 55–65% (+)-(*R*)-S.

Lit.: ApSimon **9**, 315f. (synthesis) ▪ J. Chem. Ecol. **18**, 1201–1208 (1992) ▪ Justus Liebigs Ann. Chem. **1988**, 903 (synthesis). – *[CAS 21378-21-2 (S.); 20053-41-2 ((+)-S.)]*

Sexual hormones see androgens, antheridiol, estrogens, and progestogens.

Sexual pheromones see pheromones.

Seychellene.

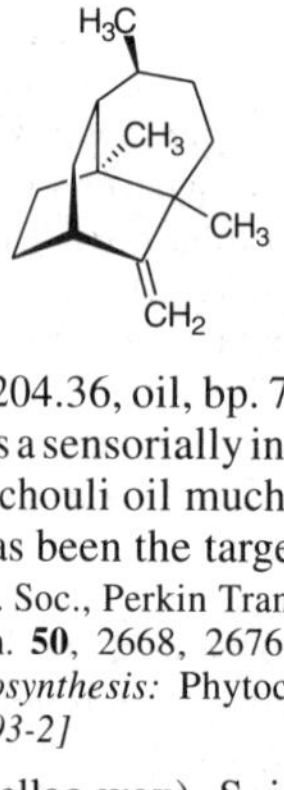

(-)-Seychellene

$C_{15}H_{24}$, M_R 204.36, oil, bp. 70 °C (13.3 Pa), $[\alpha]_D^{20}$ –72° ($CHCl_3$), S. is a sensorially insignificant component (ca. 8%) of *patchouli oil much valued in the perfume industry. S. has been the target of many syntheses.

Lit.: J. Chem. Soc., Perkin Trans. 1 **1990**, 2109; **1993**, 2813 ▪ J. Org. Chem. **50**, 2668, 2676 (1985) ▪ Recherches **19**, 85 (1974). – *Biosynthesis:* Phytochemistry **27**, 2105 (1988). – *[CAS 20085-93-2]*

Shellac (shellac wax). S. is a hard, tough resin (mp. 65–85 °C) obtained from secretions of the female lac insect (*Kerria lacca*, Coccideae). Chemically S. consists of esterified carboxylic acids, especially *hydroxy fatty acids (e.g., aleuritic acid), as well as the terpenoid *shellolic acid* {$C_{15}H_{20}O_6$, M_R 296.32, mp. 204–207 °C, $[\alpha]_D$ +26° ($CHCl_3$)}. S. is used in coatings for foods, cosmetics, and pharmaceuticals.

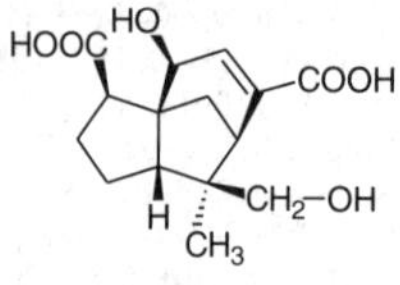

Shellolic acid

Shellac wax (mp. 72–82 °C) is a component (ca. 6%) of S., from which it is obtained by extraction with petroleum ether. It contains 80–82% esters of mostly saturated fatty acids and alcohols, e.g., ceryl lignocerate and ceryl cerotate (=hexacosyltetra- and hexacosanoates), as well as 10–14% free fatty acids, e.g., *cerotic acid, *melissic acid (triacontanoic acid), and *lacceric acid* [dotriacontanoic acid, $C_{32}H_{64}O_2$, M_R 480.86, mp. 96.1–96.3 °C], and hydrocarbons. It is used as a hard wax, e.g., in cosmetic creams and shoe polish.

Lit.: Hamilton (ed.), Waxes: Chemistry, Molecular Biology and Functions, Dundee: The Oily Press 1995 ▪ Seifen Öle Fette Wachse **116**, 221 (1990); **120**, 3 (1994). – *[HS 1301 10; CAS 9000-59-3 (S.); 3625-52-3 (lacceric acid); 4448-95-7 (shellolic acid)]*

Shellfish poisons. Substances of widely differing types from marine plankton that are generally accumulated in mussels, thus the name *"algal toxins"* being more correct. Furthermore, *PSP toxins (paralytic shellfish poisoning) have also been detected in crabs. The M. include, e.g., *saxitoxin, neosaxitoxin, gonyautoxin, *domoic acid, surugatoxin. The PSP toxins have a relatively high thermal stability so that they are destroyed to a small extent only by cooking[1]. In contrast to the PSP toxins, DSP toxins (diarrhetic shellfish poisoning) mostly leads to rather harmless gastrointestinal complaints for which usually *okadaic acid as well as dinophysistoxins 1 and 3 are responsible. For structure-activity relationships of DSP toxins, see *Lit.*[2]. ***Analysis:*** For the determination of DSP toxins (especially okadaic acid), see *Lit.*[3,4], of the PSP toxin domoic acid, see *Lit.*[5,7]. For the analysis of other PSP toxins, see *Lit.*[6–9]. Saxitoxin can be detected by HPLC with post-column derivatization and fluorescence detection.

Lit.: [1]J. Food Protect. **48**, 659 (1985). [2]Carcinogenesis **11**, 1837 (1990). [3]Agric. Biol. Chem. **51**, 877 (1987). [4]Chem. Ind. (London) **1995**, 214; Tetrahedron **51**, 12229 (1995). [5]J. Chromatogr. **462**, 349 (1989). [6]J. Assoc. Off. Anal. Chem. **69**, 547 (1986). [7]J. Assoc. Off. Anal. Chem. **72**, 674 (1989). [8]J. Assoc. Off. Anal. Chem. **68**, 13 (1985). [9]Agric. Biol. Chem. **48**, 2783 (1984).

gen.: ACS Symp. Ser. **418** (1989) ▪ Anal. Chem. **61**, 1053 (1989) ▪ Science **257**, 1476f. (1992).

Shellolic acid see shellac.

Shermilamines.

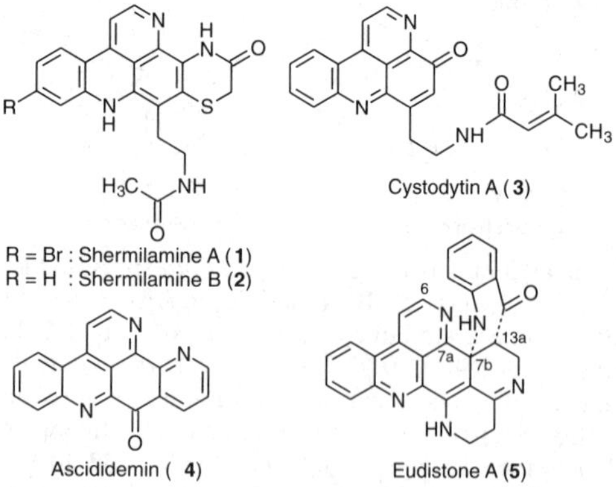

Certain sponges and tunicates contain pyrido[2,3,4-*kl*]acridines of pharmacological interest, e.g., *S. B*[1] from the tunicates *Trididemnum* sp. and *Cystodytes dellechiajei* ($C_{21}H_{18}N_4O_2S$, M_R 390.47, orange prisms, mp. 254 °C); *cystodytin A*[2] ($C_{22}H_{19}N_3O_2$, M_R 357.41); *ascididemin*[3] from *Didemnum* sp. ($C_{18}H_9N_3O$, M_R 283.29) cytotoxic to L1210 cells *in vitro* (IC_{50} 0.4 µg/mL); the octacyclic *eudistone A*[4] from *Eudistoma* sp. ($C_{27}H_{19}N_5O$, M_R 429.48). Ascididemin effects the release of calcium in the sarcoplasmatic reticulum 7-fold more effectively than caffeine.

Lit.: [1] J. Org. Chem. **54**, 4231 (1989); **63**, 4601 (1998). – *Biosynthesis:* Tetrahedron **49**, 6223 (1993). – *Synthesis:* Tetrahedron Lett. **36**, 4709 (1995). [2] J. Org. Chem. **53**, 1800 (1988). – *Synthesis:* J. Am. Chem. Soc. **113**, 8016 (1991). [3] Tetrahedron **56**, 497 (2000). – *Synthesis:* Synthesis **1994**, 239 ff.; Synth. Commun. **27**, 2587 (1997); Tetrahedron **48**, 3589–3602 (1992). [4] J. Org. Chem. **56**, 5369 (1991). – *Review:* Chem. Rev. **93**, 1825 (1993). – *[CAS 116302-28-4 (1); 122271-41-4 (2); 113321-71-4 (3); 114622-04-7 (4); 135340-00-0 (5)]*

Sherry flavor. The barrel maturation of sherry proceeds in the presence of air with the development of the typical oxidative flavor due to, among others, the formation of acetaldehyde (see alkanals), acetals, esters, *acetoin, hydroxycarboxylic acids, and lactones with consumption of alcohol and lower fatty acids. 4-Ethylphenol, 4-ethylguaiacol, *4-vinylguaiacol as well as substances extracted from the wood like vanillin (see 3,4-dihydroxybenzaldehydes), syringa aldehyde (see rum flavor), *piperonal, and *whisk(e)y lactone also add to the flavor. *(–)-(R)-Pantolactone* {$C_6H_{10}O_3$, M_R 130.14, $[\alpha]_D^{25}$ –50.7° (H_2O), cryst., mp. 92 °C}, *sherry lactone* ($C_6H_{10}O_3$, M_R 130.14, $[\alpha]_D$ –120°), and *solerone* {$C_6H_8O_3$, M_R 128.13, $[\alpha]_D^{25}$ +13.4° (CH_3OH)} contribute wine-like notes, sotolone (see hydroxyfuranones) adds a caramel-walnut-like nuance.

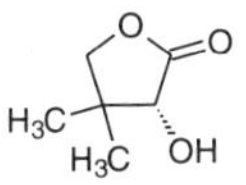

(-)-(*R*)-Pantolactone

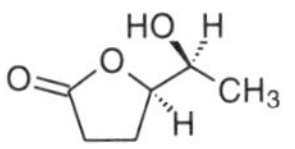

Sherry lactone (Solerol)

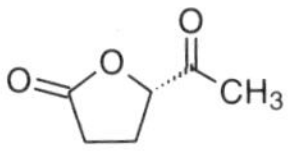

Solerone

Lit.: *Pantolactone:* Org. Prep. Proc. Int. **27**, 107 (1995) ▪ Tetrahedron Lett. **33**, 5323 (1992); **35**, 5735 (1994). – *Sherry lactone:* Phytochemistry **40**, 1251 (1995).
gen.: TNO list (6.), p. 584–592. – *[CAS 599-04-2 (pantolactone); 54656-51-8 (sherry lactone); 29393-32-6 (solerone)]*

Shiga toxin (*Shigella* toxin). A toxic protein from *Shigella dysenteriae*, the pathogen of bacillary dysentery (shigellosis). S. t. inhibits protein biosynthesis (translation) in afflicted cells by inactivation of ribosomes. It appears that certain strains of the normally harmless intestinal bacterium *Escherichia coli* can form shiga-like toxins[1] after attack by phages.

Lit.: [1] Proc. Natl. Acad. Sci. USA **85**, 2568–2572 (1988).
gen.: J. Biol. Chem. **256**, 8739–8744 (1981) ▪ Microbiol. Rev. **51**, 206–220 (1987) ▪ Microb. Pathogen. **8**, 235–242 (1990) ▪ Nature (London) **403**, 669 (2000) (mechanism of action).

Shikalkin see shikonin.

Shikimic acid [(3*R*,4*S*,5*R*)-3,4,5-trihydroxy-1-cyclohexene-1-carboxylic acid].

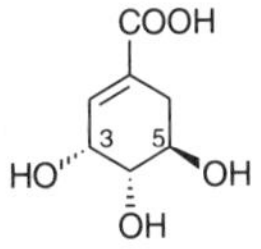

$C_7H_{10}O_5$, M_R 174.15. needles, D. 1.6, mp. 178–180 °C, $[\alpha]_D^{22}$ – 157° (H_2O), pK_a 4.15 (14.1 °C), soluble in water. S. is a widely distributed component of plants and occurs especially in fruits of the star anise (*Illicium anisatum*, syn. *I. religiosum*, Illiciaceae; Japanese: shikimi-no-ki). S. is a key intermediate of the so-called *shikimic acid pathway* which includes the biosynthesis of the aromatic amino acids phenylalanine, tyrosine, and tryptophan. These, in turn, are precursors of numerous alkaloids, *flavonoids, and *lignans, as well as *4-amino- and *4-hydroxybenzoic acid, *gallic acid, *tetrahydrofolic acid, *ubiquinones, *vitamin K_1, and nicotinic acid. The synthetic racemate melts at 191–192 °C.

Biosynthesis [1]***:*** *3-Deoxy-D-*arabino*-2-heptulosonic acid 7-phosphate → 3-dehydroquinic acid (see quinic acid) → *3-dehydroshikimic acid → shikimic acid.

Lit.: [1] Zechmeister **69**, 157–260.
gen.: Acta Crystallogr., Sect. C **44**, 1204–1207 (1988) ▪ Beilstein EIV **10**, 1946 ▪ Chem. Commun. **1998**, 2033 (synthesis) ▪ Chem. Lett. **1996**, 987 (synthesis) ▪ Haslam, Shikimic Acid Metabolism and Metabolites, New York: Wiley 1993 ▪ J. Chem. Soc., Perkin Trans. 1 **1997**, 1805 ▪ Karrer, No. 988 ▪ Luckner (3.), p. 323 ff. ▪ Merck-Index (12.), No. 8625 ▪ Nat. Prod. Rep. **12**, 101–133 (1995) (biosynthesis, metabolites of S.) ▪ Synthesis **1993**, 179–193 (synthesis) ▪ Tetrahedron **54**, 4697 (1998); **55**, 8443 (1999) (synthesis) ▪ Tetrahedron: Asymmetry **8**, 1623 (1997) (synthesis (+)-S.) ▪ Tetrahedron Lett. **38**, 57 (1997) (synthesis). – *[HS 2918 19; CAS 138-59-0]*

Shikonin [(*R*)-5,8-dihydroxy-2-(1-hydroxy-4-methyl-3-pentenyl)-1,4-naphthoquinone, C. I. 75535]. Formula, see alkannin. $C_{16}H_{16}O_5$, M_R 288.30, $[\alpha]_D^{20}$ +135° (benzene), red-brown needles, mp. 148 °C, uv_{max} 555 nm (C_2H_5OH). Pigment from the roots of the Chinese and Japanese murasaki (*Lithospermum erythrorhizon*), gromwell (*L. officinale*, Boraginaceae) and *Arnebia nobilis* (Boraginaceae) with antimycotic and antitumor activities. S. occurs in the plants primarily as the 1′-*O*-acetyl derivative (Tokyo violet). The (*S*)-form of S. is *alkannin. The racemate is called *shikalkin* (mp. 148 °C). The pigment S. is produced commercially in plant cell cultures[1]. S. is used to dye textiles and in lipsticks. Even in antique times, root extracts of *Alkanna tinctoria* and *L. erythrorhizon* have been used as natural purple dyes and for wound healing, as described by Dioscorides[2].

Biosynthesis: From 4-hydroxybenzoic acid and geranyl diphosphate.

Lit.: [1] Fujita, in Bajaj (ed.), Biotechnology in Agriculture and Forestry, vol. 4, Medicinal and Aromatic Plants, p. 225–236, Berlin: Springer 1988. [2] Gunther, in The Greek Herbal of Dioscorides, vol. IV, p. 421, New York: Hafner 1968.
gen.: Angew. Chem., Int. Ed. **37**, 839 (1998) (synthesis); **38**, 271–300 (1999) (review) ▪ Beilstein E III **8**, 4088 ▪ Bull. Chem. Soc. Jpn. **60**, 205 (1987); **68**, 2917 (1995) ▪ Chem. Pharm. Bull. **29**, 116 (1981) (pharmacology) ▪ J. Org. Chem. **52**, 1437 (1987) ▪ Justus Liebigs Ann. Chem. **1991**, 1157 (chemical synthesis) ▪ Karrer, No. 1221 ▪ Tetrahedron Lett. **38**, 7263 (1997) (chemical synthesis); **39**, 2721 (1998) (biosynthesis). – *[HS 3203 00; CAS 517-89-5 (S.); 11031-58-6 (shikalkin)]*

Shiraiachrome see hypocrellin.

Shisool see *p*-menthenols.

Showdomycin.

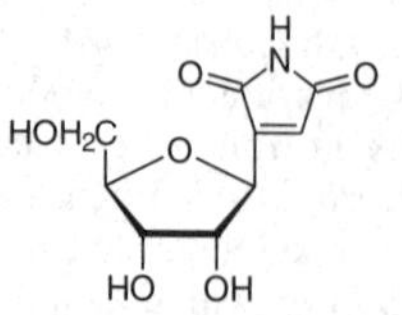

$C_9H_{11}NO_6$, M_R 229.19, platelets, mp. 160–161 °C, $[\alpha]_D^{20}$ +49.9° (H_2O), soluble in water, ethanol, acetone, unstable in neutral or alkaline solution. S. is a *nucleoside antibiotic from *Streptomyces showdoensis* with antitumor properties.

Lit.: J. Am. Chem. Soc. **112**, 881 (1990) ▪ J. Antibiot. **41**, 1711 (1988) ▪ Tetrahedron **44**, 3715 (1988) ▪ Tetrahedron Lett. **29**, 351 (1988). – *Synthesis:* J. Org. Chem. **64**, 5427 (1999) ▪ Tetrahedron Lett. **36**, 4089 (1995). – *[CAS 16755-07-0]*

Shyobunone see calamus oil.

Sialic acids see neuraminic acid.

Sialyl-Lewis x Tetrasaccharide (S Le^x).

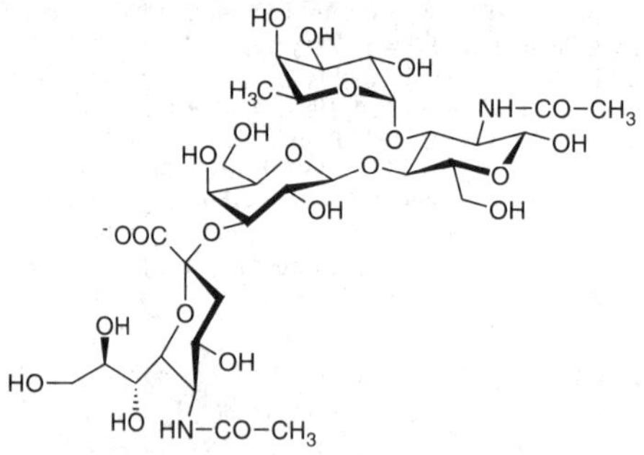

$C_{31}H_{52}N_2O_{23}$, M_R 820.75, solid, $[\alpha]_D^{25}$ +5.8° (CH_3OH). S. is present at the terminus of glycolipids attached to the surface of white blood cells (neutrophils or leukocytes) and to the receptor ELAM-1, a glycoprotein of the selectin family. S. functions as a ligand for cell surface receptors of selectins and plays an important role in inflammatory response by mediating the binding of the leukocytes to the injured tissue. Inflammatory responses begin when SLe^x groups on leukocytes bind to selectin receptors on the endothelial cells that line blood capillaries. SLe^x has been investigated to treat inflammations, but it can only be applied in case of acute symptoms, because it is unstable in blood and inactive on oral application, stimulating synthetic efforts to prepare SLe^x mimics.[1] The mucins in the sera of patients with certain cancers exhibit high levels of SLe^x, therefore it is more important for diagnostic purposes.

Lit.: [1] Angew. Chem. Int. Ed. Engl. **35**, 88 (1996); J. Am. Chem. Soc. **118**, 6826 (1996); **119**, 8152 (1997); J. Carbohydr. Chem. **16**, 11–24 (1997); Tetrahedron Lett. **37**, 9037 (1996); **38**, 4059 (1997); **39**, 3273 (1998); Pat. WO 9808854 (26. 8. 1997).
gen.: Chem. Br. **1994**, 117 (review) ▪ J. Am. Chem. Soc. **114**, 9283 (1992) (synthesis); 5452 (1992) (structure); **116**, 1135 (1994) (synthesis) ▪ J. Carbohydr. Chem. **16**, 983 (1997) (synthesis) ▪ Justus Liebigs Ann. Chem. **1997**, 1303 (synthesis). – *[CAS 98603-84-0 (oxoform)]*

Siastatin B.

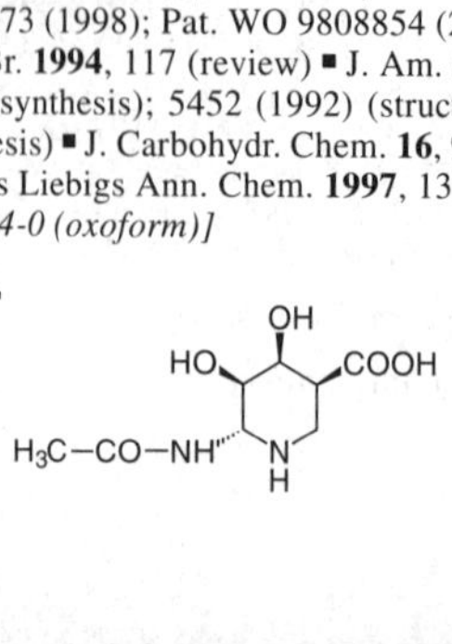

$C_8H_{14}N_2O_5$, M_R 218.21, needles, mp. 137 °C (decomp.), $[\alpha]_D^{25}$ +57.2° (H_2O). A neuraminidase inhibitor from cultures of *Streptomyces verticillus*, var. *quintum*. S. is undergoing testing as therapy for virus infections and tumors caused by viruses.

Lit.: Atta-ur-Rahman **16**, 75–121 (1995) ▪ Bull. Chem. Soc. Jpn. **65**, 978–986 (1992). – *[CAS 54795-58-3]*

(–)-Sibirine see nitramine.

Sibiromycin see anthramycins.

Sibyllimycin.

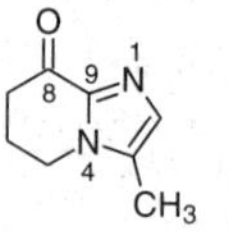

$C_8H_{10}N_2O$, M_R 150.18, mp. 122–123 °C. An indolizidine alkaloid from a *Thermoactinomyces* species (Bacillaceae) found in a hot spring of Lake Tanganyika.

Lit.: Angew. Chem., Int. Ed. Engl. **35**, 545 ff. (1996).

Siccanin.

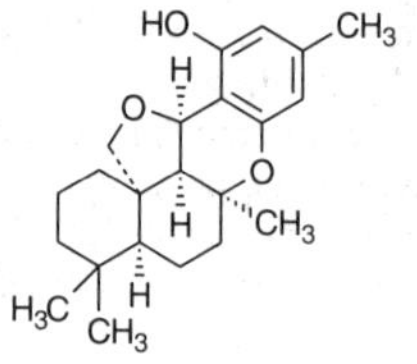

$C_{22}H_{30}O_3$, M_R 342.48, pale yellow cryst., mp. 139–140 °C, $[\alpha]_D$ –136° ($CHCl_3$), soluble in organic solvents. A metabolite from cultures of the phytopathogenic fungus *Helminthosporium siccans*. The structure of S. consists of a *drimane sesquiterpene linked to orcinol. S. has broad antifungal activity, especially against *Trichophyton* species that cause skin infections.

Lit.: J. Am. Chem. Soc. **103**, 2434 (1981). – *Synthesis:* Bull. Chem. Soc. Jpn. **61**, 1991–1998 (1988) ▪ J. Am. Chem. Soc. **118**, 5146 (1996) ▪ Synthesis **1990**, 935. – *[HS 2941 90; CAS 22733-60-4]*

Siccayne.

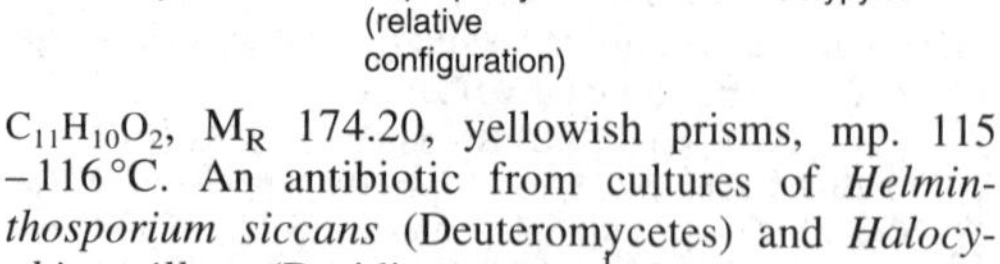

Siccayne — Asperpentyne (relative configuration) — Eutypyne

$C_{11}H_{10}O_2$, M_R 174.20, yellowish prisms, mp. 115–116 °C. An antibiotic from cultures of *Helminthosporium siccans* (Deuteromycetes) and *Halocyphina villosa* (Basidiomycetes)[1]. S. is cytotoxic and inhibits the growth of a series of plants and Gram-positive bacteria. Structurally related active compounds are *asperpentyne* {$C_{11}H_{12}O_3$, M_R 192.21, oil, $[\alpha]_D$ –20° (acetone)}[2] and the phytotoxin *eutypyne* (dying-arm disease of grapevines, $C_{12}H_{10}O_2$, M_R 186.21)[3], see also mycenone.

Lit.: [1] Helv. Chim. Acta **76**, 425 (1993); J. Antibiot. **34**, 298 (1981); Synthesis **1990**, 935 (synthesis). [2] Phytochemistry **27**,

3853 (1988). [3] Phytochemistry **30**, 471 (1991). – *[HS 2941 90; CAS 22944-03-2 (S.); 119483-44-2 (asperpentyne); 121007-17-8 (eutypyne)]*

Sideramines. Obsolete name for microbial iron(III) complexes of the trihydroxamate type with growth promoting properties (e.g., *ferrioxamines, *ferrichromes). The name *siderophores is now established for the low-molecular weight compounds participating in iron transport.

Siderochromes (siderophores). Obsolete collective name for colored iron(III) complexes of the trihydroxamate type isolated from microorganisms. Those S. exhibiting antibiotic activity were known as sideromycins, those exhibiting growth factor properties as *sideramines. This classification is no longer used on account of the structural diversity and broad range of activities of the compounds that participate in iron uptake, transport, and storage in fungi and bacteria. The established group name for these compounds is now *siderophores.

Siderophores.

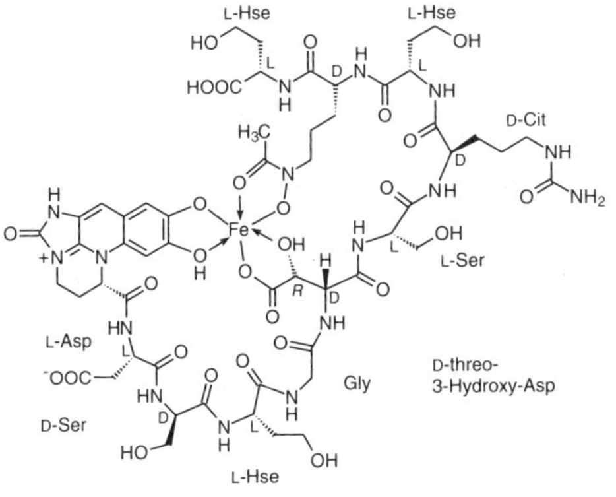

Azotobactin D [Fe(III)-complex]

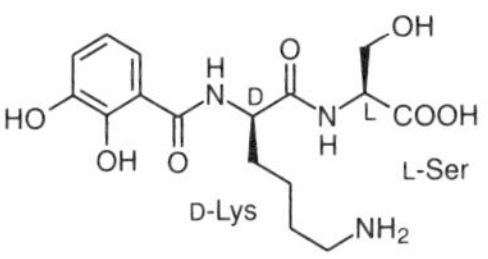

Chrysobactin

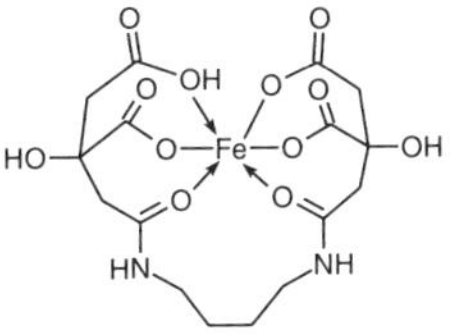

Rhizoferrin [Fe(III)-complex]

Table: Data of some Siderophores.

Siderophore	molecular formula, M_R	producing organism	CAS
Azotobactin D[a]	$C_{55}H_{75}FeN_{16}O_{28}$ 1464.14	*Azotobacter vinelandii*	114844-84-7
Chrysobactin	$C_{16}H_{23}N_3O_7$ 369.37	*Erwinia chrysanthemi*	120124-51-8
Ferrirubin[a,b]	$C_{41}H_{64}FeN_9O_{17}$ 1010.86	*Aspergillus* sp.	23425-25-4
Rhizoferrin	$C_{16}H_{24}N_2O_{12}$ 436.37	*Rhizopus microsporus*	138846-62-5

[a] isolated as Fe(III) complex [b] Structure see ferrichromes

Low-molecular weight, Fe(III)-complexing substances with high affinity for iron that serve for the transport and storage of iron in microorganisms. The formation of S. occurs under iron-deficient conditions in aerobic and facultatively anaerobic microorganisms. The biosynthesis of S. and the membrane-bound iron transport system is regulated by the iron concentration of the medium. Typical partial structures of S. are phenolates, hydroxamates, oxazolines, and α-hydroxycarboxylates. The affinity for S. is strongly reduced on reduction of Fe(III) to Fe(II). For the Fe(III) hydroxamates this mechanism, after enzymatic reduction, is probably the basis for the iron exchange reactions within the cell or, respectively, on the cell surface. The first S., *mycobactin, was characterized in 1949. Depending on the structure and function a distinction is still made between *siderochromes, *sideramines, and sideromycins but they are all known collectively under the name S. Typical S. are shown in the table. For further S., see aerobactin, agrobactin, azotochelin, ferrichromes, ferrimycins, ferrioxamines, ferrithiocin, pseudobactin, and pyoverdins.

Pharmacological fields of application of the S. are treatment of thalassemia and malaria.

Lit.: Biochem. Pharmacol. **43**, 1275 (1992) ▪ FEMS Microbiol. Rev. **104**, 209–228 (1993) ▪ Parasitol. Today **11**, 74 (1995) ▪ Swinburne, Iron, Siderophores and Plant Diseases, New York: Plenum 1986.

Silphinene see triquinanes.

Silybin (silibinin, silymarin I).

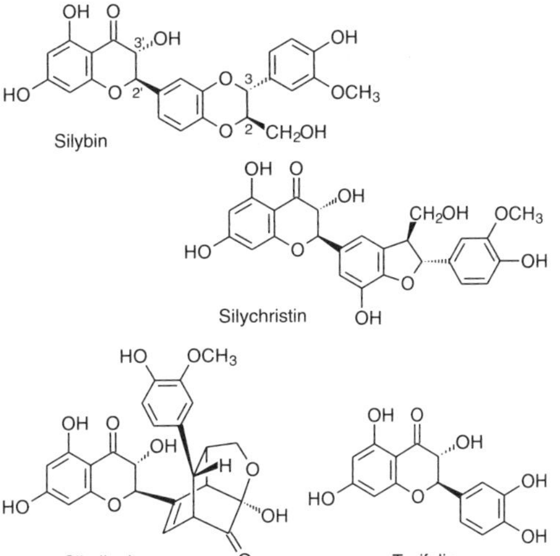

Silybin Silychristin Silydianin Taxifolin

$C_{25}H_{22}O_{10}$, M_R 482.44, cryst., mp. 167 °C (monohydrate), decomp. at 180 °C, $[\alpha]_D^{22}$ +11° (acetone), soluble in acetone, ethyl acetate, methanol, poorly soluble in chloroform. A flavanolignan (see lignans) from the fruits of the milk thistle (*Silybum marianum*, Asteraceae) with hepatoprotective activity. Further components of the milk thistle are: *silychristin* (silymarin II), $C_{25}H_{22}O_{10}$, M_R 482.44, cryst. · H_2O, mp. 174–176 °C, $[\alpha]_D^{23}$ +81.4° (pyridine) and *silydianin*, $C_{25}H_{22}O_{10}$, M_R 482.44, mp. 191 °C, $[\alpha]_D^{24}$ +175° (acetone). These compounds contain the partial structures of the pentahydroxyflavanone *taxifolin*, $C_{15}H_{12}O_7$, M_R 304.26 (see flavonoids), and *coniferyl alcohol.

Use: For therapy of liver diseases, as an antidote for death cap poisonings (see amanitins), and for therapy of brain edemas. Silychristin inhibits peroxidase (from horseradish) and lipoxygenase [1] and, like silydianin, acts as a regulator of plant growth. The compounds act as radical scavengers, thus stabilizing and protecting hepatic cells and preventing the oxidation of membrane lipids.

Lit.: [1] Experientia **35**, 1548–1552 (1979).
gen.: Beilstein E V **19/10**, 690 ▪ Chem. Ber. **108**, 790–802, 1482–1501 (1975) ▪ Chem. Pharm. Bull. **33**, 1419–1423 (1985) (synthesis) ▪ Fitoterapia **6**, 3 (1995) (review) ▪ Hager (5.) **9**, 615 ff.; 620 ff. ▪ Hepatology **6**, 362 (1980) ▪ J. Chem. Soc., Chem. Commun. **1988**, 749 ff. (synthesis) ▪ Martindale (29.), p. 1613 ▪ Nat. Prod. Rep. **12**, 183–205 (1995) ▪ Negwer (6.), No. 7244 ▪ Stud. Org. Chem. **11**, 461–474 (1982) ▪ Zhongguo Yaowu Huaxue Zazhi **7**, 107 (1997) (synthesis) ▪ Z. Naturforsch. C **29**, 82 f. (1974) (biosynthesis). – *[HS 2932 90; CAS 22888-70-6 (S.); 33889-69-9 (silychristin); 29782-68-1 (silydianin); 480-18-2 (taxifolin)]*

Silychristin, silydianin see silybin.

Simonin I see Ipomoea resins.

Simonyellin.

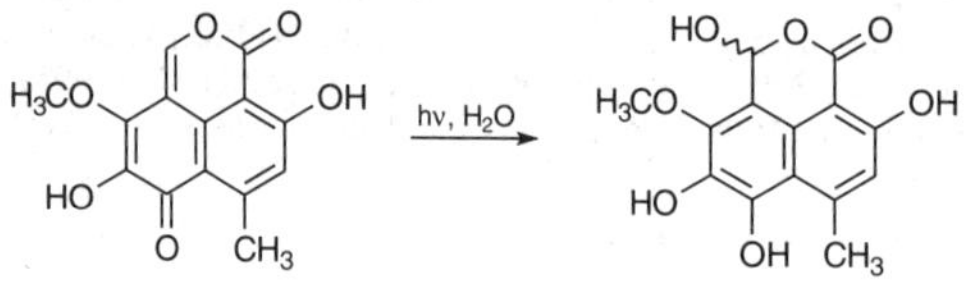

S. S. hydrate

$C_{14}H_{10}O_6$, M_R 274.23, yellow needles with mp. 274–275 °C; in solutions containing water S. is converted to the colorless hydrate under the action of light. S. was the first naphthopyran to be isolated from a lichen (*Simonyella variegata* endemic to Semha Island in the Gulf of Aden).

Lit.: Austral. J. Chem. **48**, 2035 (1995) ▪ J. Hattori Bot. Lab. **32**, 35 (1969).

Sinalbin see glucosinolates.

Sinalexin(e).

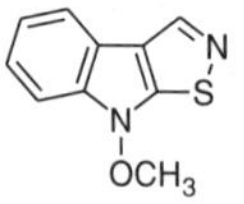

$C_{10}H_8N_2OS$, M_R 204.25, solid. *Phytoalexin from *Sinapis alba* (Cruciferae), formed under biotic or abiotic elicitation.

Lit.: Phytochemistry **46**, 833 (1997). – *[CAS 200192-82-1]*

Sinapaldehyde see whisk(e)y flavor.

Sinapine.

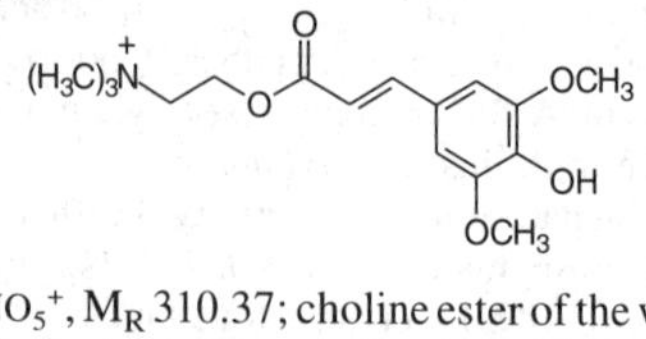

$C_{16}H_{24}NO_5^+$, M_R 310.37; choline ester of the widely distributed plant constituent *sinapic acid*, $C_{11}H_{12}O_5$, M_R 224.21, prisms or needles, mp. 192 °C. S. is unstable as the free base but stable as the iodide, $C_{16}H_{24}INO_5$, M_R 437.27, mp. 185–186 °C. and sulfate, $C_{16}H_{25}NO_9S$, M_R 407.44, mp. 126–127 °C (dihydrate), 186–188 °C (anhydrous). S. occurs in white and black mustard (*Brassica nigra*), as the thiocyanate in crushed rape, and as iodide in the Chinese drug "Ting Li" (*Draba nemorosa*); see also glucosinolates (there sinalbin). At pH 6.2–8.4 a change from colorless to yellow occurs.

Lit.: Dev. Food Proteins **2**, 109–132 (1983) (review) ▪ Karrer, No. 962, 2468 ▪ Phytochemistry **24**, 407–410 (1985) ▪ Zechmeister **35**, 89 ff. – *[CAS 18696-26-9 (S.); 5655-06-1 (S. iodide); 6509-38-2 (S. sulfate); 530-59-6 (sinapic acid)]*

Sinapyl alcohol [4-(3-hydroxy-1-propenyl)-2,6-dimethoxyphenol, syringenin].

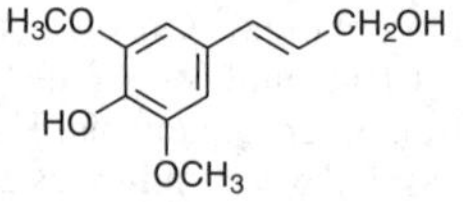

(*E*)-form

$C_{11}H_{14}O_4$, M_R 210.23, needles, mp. 63–65 °C, the principal building block of *lignin from angiosperms. the phenolic β-D-glucopyranoside *syringin*, $C_{17}H_{24}O_9$, M_R 372.37, mp. 191–192 °C, $[\alpha]_D$ −18° (H_2O), is widely distributed in, e. g., *Syringa vulgaris*, *Ligustrum* spp., *Jasminum* spp., *Phillyrea latifolia*, *Forsythia suspensa*, *Fraxinus* spp. The (*Z*)-isomer occurs in the bark of *Fagus grandifolia*; it is not incorporated into the lignin of this plant.

Lit.: Holzforschung **40**, 273–280 (1986) ▪ J. Agric. Food Chem. **40**, 1108 (1992) ▪ Luckner (3.), p. 392 ▪ Nat. Prod. Rep. **12**, 101–133 (1995) ▪ Synthesis **1994**, 369 (synthesis). – *[CAS 20675-96-1 (E); 104330-63-4 (Z); 118-34-3 (syringin)]*

Sinefungin.

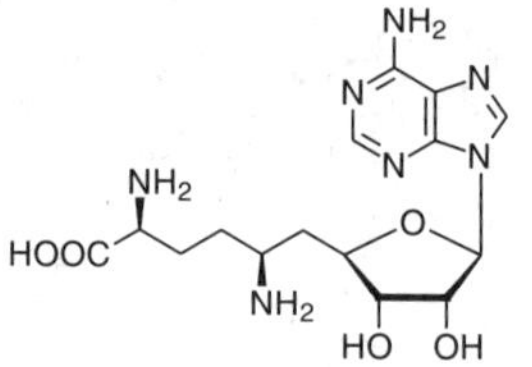

$C_{15}H_{23}N_7O_5$, M_R 381.39, amorphous, LD_{50} (mouse p. o.) 1000 mg/kg. An antiviral and antifungal *nucleoside antibiotic from cultures of *Streptomyces griseolus* and *S. incarnatus*. The development of S. as a fungicide, parasiticide, and human antimycotic was abandoned on account of its nephrotoxic side effects (see also nikkomycins, polyoxins).

Lit.: J. Antibiot. **41**, 1711 (1988) (review) ▪ Kirk-Othmer (4.) **3**, 594. – *Biosynthesis:* J. Am. Chem. Soc. **111**, 8981 (1989) ▪ Tetrahedron **47**, 6069 (1991). – *Synthesis:* J. Chem. Soc., Perkin Trans. 1 **1991**, 981; **1999**, 3597 ▪ J. Org. Chem. **55**, 948 (1990); **61**, 6175 (1996); **62**, 2299 (1997). – *Activity:* J. Med. Chem. **35**, 63 (1992) ▪ Nucleosides, Nucleotides **15**, 1121–1135 (1996). – *[CAS 58944-73-3]*

Sinensals.

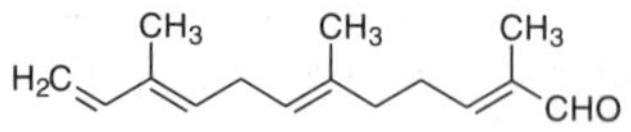

α-Sinensal

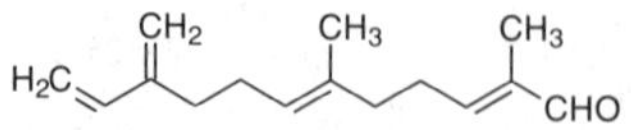

trans-β-Sinensal

α-S., $C_{15}H_{22}O$, M_R 218.34, and the isomeric *trans-β-S.*, bp. 92–95 °C (6.7 Pa), oil, are the odor-determining flavor components (ca. 0.1%) of cold-pressed, orange-peel oil (*Citrus sinensis*) [1]. Mandarin oil (*Citrus reticulata*) has the highest content of α-S. (0.2%). α-S. possesses a pronounced orange character with a very

low olfactory threshold of 0.5 ppb. The odor of *β*-S. is accompanied by metallic, fish-like undertones that are considered to be unpleasant at higher concentrations. Both S. have a tendency to polymerize when exposed to air and light.
Lit.: [1] Ohloff, p. 132.
gen.: Synthesis: Bull. Chem. Soc. Jpn. **61**, 4051 (1988) ▪ J. Org. Chem. **48**, 5183 (1983) ▪ Karrer, No. 5766 ▪ Tetrahedron Lett. **25**, 3213 (1984). – *[CAS 17909-77-2 (α-S.); 3779-62-2 (trans-β-S.); 17909-87-4 (cis-β-S.)]*

Sinigrin see glucosinolates.

Sinoacutine, sinomenine see morphinan alkaloids.

Sinularene see africanol.

Siphonarin A see denticulatins.

Siphulin.

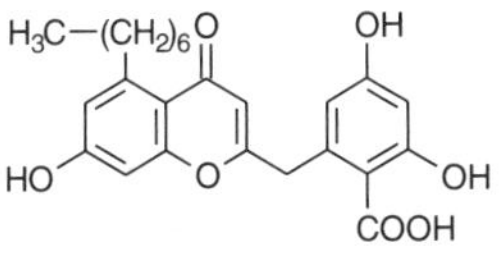

$C_{24}H_{26}O_7$, M_R 426.47, cryst., double mp. 185 °C (decomp.) and 228–229 °C. A chromone from the shrub-like lichen *Siphula ceratites*. A rare example of a polyketide with more than 20 C atoms, probably formed by condensation of an octanoyl starter with 8 malonate units.
Lit.: Acta Chem. Scand. **19**, 1677–1693 (1965) ▪ Acta Chem. Scand., Ser. B **39**, 65 ff. (1985). – *[CAS 4724-05-4]*

Sirenin.

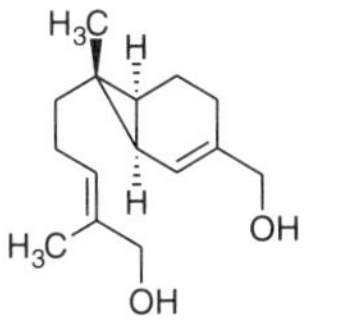

$C_{15}H_{24}O_2$, M_R 236.35, oil, $[\alpha]_D^{23}$ –45° ($CHCl_3$). Highly active pheromone (sperm attractant) of the water fungus *Allomyces* sp., it has been a synthetic target for several scientific groups.
Lit.: J. Org. Chem. **53**, 4877 (1988) ▪ Tetrahedron Lett. **36**, 8745 (1995). – *[CAS 19888-27-8]*

Siroheme.

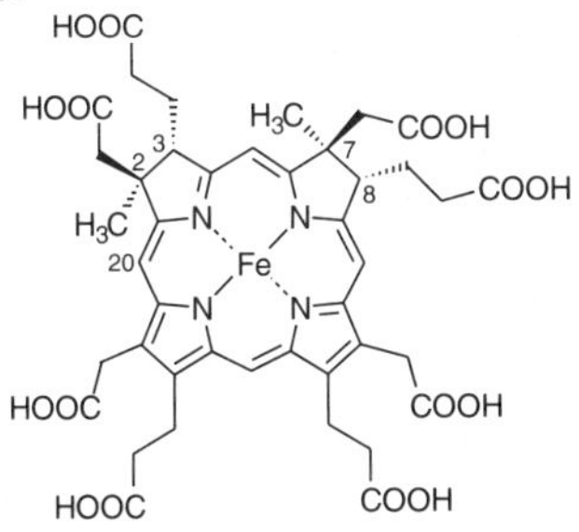

$C_{42}H_{44}FeN_4O_{16}$, M_R 916.68. S. belongs to the structural class of the *isobacteriochlorins.
Occurrence: S. is the prosthetic group of sulfite and nitrite reductases of numerous bacteria and plants. S. plays an important part in the global nitrogen and sulfur cycles. The reduced, metal-free chromophore [*sirohydrochlorin*, $C_{42}H_{46}N_4O_{16}$, M_R 862.84, uv_{max} (octamethyl ester, $CHCl_3$) 362, 377, 482, 514, 546, 589, 634 nm] is, as *precorrin, an essential link in the biosynthetic chain leading to *vitamin B_{12}.
The biosynthesis proceeds from *precorrin 2, the two methyl groups are introduced through methylation by **S*-adenosylmethionine (SAM). For synthesis, see *Lit.*[1].
Lit.: [1] J. Chem. Soc., Chem. Commun. **1985**, 1061; J. Chem. Soc., Perkin Trans. 1 **1987**, 1679.
gen.: Angew. Chem., Int. Ed. **32**, 1223 (1993) ▪ Chem. Rev. **94**, 327 (1994) ▪ Chimia **41**, 277 (1987) ▪ Jordan, p. 102–112 ▪ Science **264**, 1551 (1994). – *[CAS 52553-42-1 (S.); 65207-12-7 (sirohydrochlorin)]*

Sirolimus® see rapamycin.

Sisomicin (sisomycin, rickamicin).

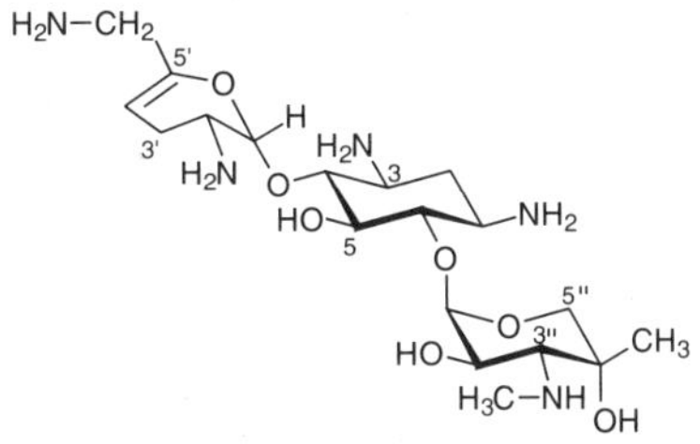

$C_{19}H_{37}N_5O_7$, M_R 447.53, cryst., mp. 198–201 °C, $[\alpha]_D^{26}$ +189° (H_2O). An aminoglycoside *antibiotic from cultures of *Micromonospora inyoensis* with a broad range of activities (especially against pseudomonads), also against mycoplasma. Therapeutic use is accompanied by renal and otological side effects (damage to kidneys and the sense of hearing); LD_{50} (mouse s. c.) 288 mg/kg. S.-resistant pathogens are able to phosphorylate S. The multistep biogenesis of S. is closely related to those of other pseudotrisaccharides (e.g., *streptomycin, gentamycin), with 2-deoxystreptamine as a common precursor.
Lit.: Drugs **27**, 548–578 (1984) ▪ J. Int. Med. Res. **9**, 168–176 (1981) ▪ Kleemann-Engel, p. 1729 ▪ Martindale (30.), p. 201. – *Biosynthesis:* J. Antibiot. **28**, 573 (1975) ▪ Sax (8.), APY 500, GAA 100, SDY 750. – *[HS 2941 90; CAS 32385-11-8]*

Sitophilate [(2*S*,3*R*)-(1-ethylpropyl) 3-hydroxy-2-methylpentanoate].

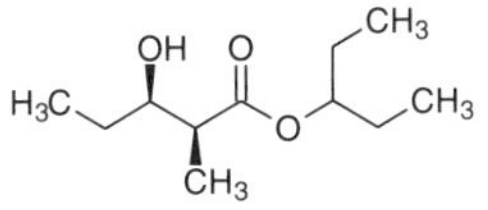

$C_{11}H_{22}O_3$, M_R 202.29, $[\alpha]_D^{24}$ +3.0° ($CHCl_3$), liquid. Aggregation pheromone of the male grain weevil *Sitophilus granarius*; formed from two propionate units; see also sitophilure.
Lit.: ApSimon **9**, 203 ▪ Synth. Commun. **28**, 2399 (1998) ▪ Tetrahedron **45**, 623 (1989); **47**, 7237–7244 (1991) ▪ Tetrahedron: Asymmetry **7**, 3479 (1996) (synthesis (–)-S.). – *[CAS 114607-20-4]*

Sitophilure [(4*R*,5*S*)-5-hydroxy-4-methyl-3-heptanone].

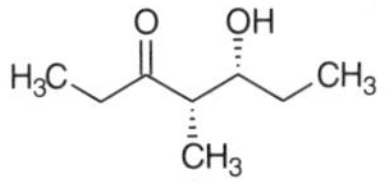

$C_8H_{16}O_2$, M_R 144.2, bp. 80–82 °C (0.8 kPa), $[\alpha]_D^{20}$ –26.7° (diethyl ether), liquid. Aggregation pheromone of the male rice weevil (*Sitophilus oryzae*) and corn weevil (*S. zeamais*); structurally related to *sitophilate.

Lit.: Angew. Chem., Int. Ed. Engl. **27**, 581 (1988) ▪ ApSimon **9**, 189 ff. ▪ Justus Liebigs Ann. Chem. **1988**, 899 ▪ Tetrahedron: Asymmetry **7**, 3479 (1996) (synthesis) ▪ Tetrahedron Lett. **33**, 5567 ff. (1992). – *[CAS 71699-35-9]*

Sitosterin, sitosterol see sterols.

Skatole (3-methyl-1*H*-indole).

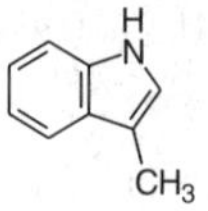

C_9H_9N, M_R 131.18. Flakes, turning brown on exposure to air, mp. 96 °C, bp. 266 °C, soluble in organic solvents; LD_{50} (rat p. o.) 1.75 g/kg; feces-like odor, in high dilution warm animalic odor; olfactory threshold in water 10 ppb. S. is formed from *tryptophan during putrefaction and occurs, among others, in *civet, feces, and pitcoal tar. Used in synthetic civet; in traces also in flavors, e. g., *cheese flavor.

Lit.: Annu. Rev. Pharmacol. Toxicol. **34**, 91 (1994) ▪ Arctander, No. 2846 ▪ Bedoukian (3.), p. 219–224 ▪ J. Org. Chem. **53**, 1323 (1988); **55**, 580 (1990) ▪ Ullmann (5.) **A 14**, 167. – *[HS 2933 90; CAS 83-34-1]*

Skimmin see umbelliferone.

Skunk (*Mephitis*). When threatened, skunks spray an evil-smelling secretion from their anal glands at their enemies. This consists mainly of (*E*)-2-butene-1-thiol, 3-methyl-1-butanethiol, and other thiols as well as their acetates. These acetates are apparently responsible for the persisting effect of the secretion since they are cleaved by water to the thiols.
Thus, dogs that had been sprayed by a skunk weeks ago can suddenly reacquire the unpleasant skunk odor after being out in the rain.

Lit.: Chem. Eng. News **1993**, No. 42, 90 ▪ J. Chem. Ecol. **17**, 1415–1419 (1991); **19**, 837–841 (1993).

Skyrin.

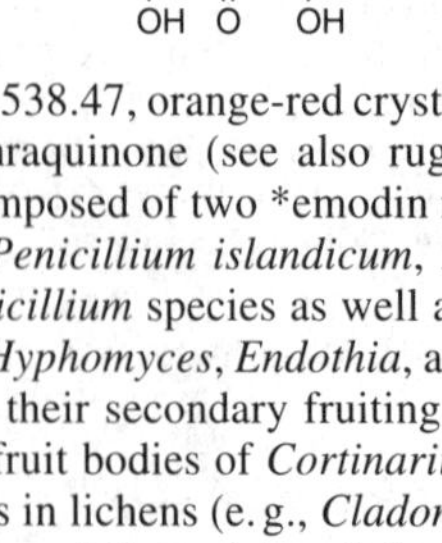

$C_{30}H_{18}O_{10}$, M_R 538.47, orange-red cryst., mp. >380 °C. Dimeric *anthraquinone (see also rugulosin and luteoskyrin), composed of two *emodin molecules, and produced by *Penicillium islandicum*, *P. wortmannii*, and other *Penicillium* species as well as ascomycetes of the genera *Hyphomyces*, *Endothia*, and *Cryphonectria* including their secondary fruiting forms. S. also occurs in the fruit bodies of *Cortinarius* and *Dermocybe* as well as in lichens (e. g., *Cladonia*, *Trypetheliopsis*, *Physcia*, and *Pyxine*) (see lichen pigments). S. and its analogues are formed biosynthetically from the corresponding monomers or their anthrones. The monomers, in turn, are produced on the polyketide pathway (octaketides). In *P. islandicum* cultures emodin anthrone is incorporated more easily into S. than emodin. Many monomeric and dimeric anthraquinones, including S., are decouplers of oxidative phosphorylation in rat liver mitochondria.

Lit.: Beilstein E IV **8**, 3767 ▪ Betina, chap. 14 ▪ Chem.-Biol. Interactions **62**, 179 (1987) ▪ Tetrahedron Lett. **1973**, 2125 (biosynthesis) ▪ Thomson **3**, 425 ▪ Turner **1**, 164 ff.; **2**, 152. – *[CAS 602-06-2]*

Skytanthus alkaloids. Alkaloids derived biogenetically from the terpene metabolism (iridoid monoterpenes).

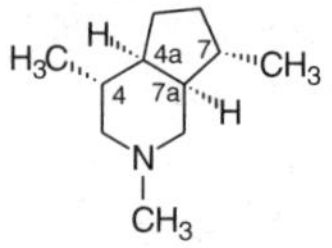

α-Skytanthin

$C_{11}H_{21}N$, M_R 167.29; e. g., *α-skytanthin:* liquid, $[\alpha]_D$ +79°; *β-skytanthin* (4*S*,7a*S*-form): bp. 88 °C (1.33 kPa), $[\alpha]_D$ +16.4°; *γ-skytanthin* (7a*S*-form): synthetic, liquid, $[\alpha]_D$ +59°; *δ-skytanthin* (4*S*-form): $[\alpha]_D$ +9°. S. a. such as skytanthin are isolated from *Skytanthus acutus* (Apocynaceae) and *Tecoma stans* (Bignoniaceae).

Lit.: Mothes et al., p. 357 f. ▪ Phytochemistry **12**, 201 (1973) (biosynthesis). – *Synthesis:* Tetrahedron Lett. **27**, 1141 (1986); **37**, 2463 (1996). – *[HS 2939 90; CAS 2065-32-9 (α-S.); 24282-31-3 (β-S.); 6545-32-0 (γ-S.); 2883-89-8 (δ-S.)]*

Slaframine [(1*S*,6*S*,8a*S*)-6-aminooctahydro-1-indolizinol acetate].

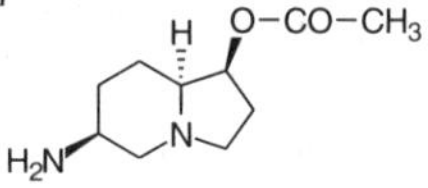

$C_{10}H_{18}N_2O_2$, M_R 198.27, mp. (of *N*-acetyl derivative): 140–142 °C, $[\alpha]_D$ –15.9° (C_2H_5OH). A toxic *indolizidine alkaloid from the phytopathogenic fungus *Rhizoctonia leguminicola*. S. is a potent parasympathicomimetic. In grazing animals S. induces an excessive production of digestive enzymes and saliva: "slobbers or drooling disease". S. is metabolically activated in the liver ("toxification").

Lit.: Beilstein E V **22/11**, 384 ▪ Chin. Chem. Lett. **9**, 999 (1998) (synthesis) ▪ J. Chem. Soc., Perkin Trans. 1 **1997**, 2179 (synthesis) ▪ J. Org. Chem. **49**, 3681 (1984); **56**, 1976 (1991); **57**, 4802 (1992); **64**, 6434, 6518 (1999) ▪ Lukacs, Recent Progress in the Chemical Synthesis of Antibiotics and Related Microbial Products, p. 751–828, Berlin: Springer 1993 ▪ Manske **28**, 269 ▪ Synlett **1999**, 49 (synthesis) ▪ Tetrahedron Lett. **34**, 6607 (1993); **36**, 6957 (1995). – *[HS 2939 90; CAS 20084-93-9 (S.); 30591-15-2 ((+)-S.)]*

Sleeping lipids see oleamide.

Sloughing hormone see insect hormones.

Sloughing or molting hormone see ecdysteroids and insect hormones.

Slow reacting substance of anaphylaxis see leukotrienes.

Smilagenin see steroid sapogenins.

Snake venoms (ophiotoxins, from Greek: ophis = snake). Of the about 2700 snake species ca. 20% are poisonous snakes in the stricter sense. Except for the polar ice caps they occur in all parts of the world and belong to the following snake families: Viperidae (vipers or adders) including the European adder (*Vipera berus*) with the subfamilies Crotalinae (pit viper), to which the American rattlesnakes and the South American fer-de-lance belong (some authors consider these to be a separate family Crotalidae); Atractaspididae (mountain snakes); Elapidae (poisonous adders) with the subfamily Elapinae, to which the cobras, mambas, kraits, and coral snakes belong, and the subfamilies Hydrophiinae (rudder-tailed sea snakes) and Laticaudinae (flat-tailed sea snakes); Colubridae (adders), of which only a few tree snakes (poisonous colubrid snakes) are poisonous.

The venom is usually produced in a venom gland and mostly injected into the prey by biting through grooved or hollow teeth (often no venom is injected in defensive biting!). S. v. belong to the biologically most active natural products. They are complex mixtures of proteins and polypeptides (ca. 90% of the dried crude venom). In addition they contain trace elements (Zn, Ca, Al), nucleotides, amino acids, free and phosphorylated sugars as well as lipids. The composition of the venom shows some common features within the various families and species but is highly species-specific. Sometimes even individuals of a species or certain species populations have a differing venomous secretion. Freshly hatched snakes already possess a complete store of venom. The functions of S. v. include catching prey, digestion (digestion starts even before the prey reaches the stomach), and defence. In small mammals, the preferred prey of snakes, S. v. lead rapidly to paralysis of the musculature or to cardiovascular failure. In the fresh state, S. v. is an aqueous, viscous, colorless to yellowish, clear or milky-turbid liquid with a water content of 50–90% (pH 5.5–7.0). The proteins of S. v. can be classified into two groups (furthermore, there are also proteins with as yet unknown activities). Toxic polypeptides without enzymatic properties often exert specific effects on cell membranes or nerve structures. Some are enzyme inhibitors while others possess synergistic properties. The second group includes enzymes, most of which are hydrolases and amino acid oxidases. Some also have toxic properties such as phospholipase A_2. The symptoms of a poisoning by S. v. result from the combination of toxic and enzymatic reactions.

Neurotoxins are a very important constituent of the toxins of S. v. since they lead to paralysis of the prey. They attack the peripheral nervous system. For example, α-*bungarotoxin binds irreversibly to acetylcholine receptors of the motoric endplates. The often irreversible binding of snake toxins to receptors means that they are also useful auxiliary substances in neurophysiology. In addition there are the cardiotoxins which damage the musculature (especially in Elapidae). The enzyme composition of S. v. can be highly diverse. They are however, exclusively degrading enzymes. Hyaluronidases destroy the connective substance of the cell, hyaluronic acid, and thus facilitate penetration of the tissue. The ester cleaving enzymes such as acetylcholine esterase (only in Elapidae) and phospholipase A_2 (hemolysin, myolysin, and neurotoxin) are also important. Digestion-promoting peptidases and proteases are found in high concentrations, especially in viper venoms. For example, thrombin-like coagulation enzymes (coagulin, anticoagulin) are found. Some of these enzymes are used in the treatment of thrombosis. Hemorrhagins, some of which are also highly specific proteases, destroy the basal membrane and are thus responsible for the typical hemorrhages and tissue destruction following snake bites. Kininogenases lead to the liberation of bradykinin, which then effects a rapid lowering of blood pressure. Finally, yellow snake venoms contain the strongest known L-amino acid oxidase (yellow color due to flavine nucleotides of the enzyme), which has no toxic activity. All protein components act as antigens. The venom of the copperhead snake has shown activity against breast cancer in mice.

The amount of venom injected during a snake bite varies widely depending on the physiological condition of the individual snake. Doses of ca. 10 mg (cottonmouth), 0.55 mg (adder), 0.4 mg (spectacled cobra) down to 0.01 mg (some sea snakes) per kg BW can be lethal. The effects of a snake bite can be roughly classified as damage to nerves and destruction of tissue. Bites by Elapidae and Hydrophiidae show hardly any local symptoms but instead have potent neurotoxic effects: anxiety, respiratory distress, paralysis of transverse striped muscles (unsteady gait, uncoordinated speech, lid ptosis, respiratory paralysis, coma, death through asphyxiation). No sequelae remain after healing. In contrast, bites by Crotalidae (except rattlesnakes) and Viperidae are characterized by pronounced local symptoms: persistent, burning pain; progressive edema in the region of the bite and of the mucous membranes as well as abdominal hemorrhage lead to hypotension, tachycardia and finally cardiovascular failure. Late sequelae from tissue necrosis through to loss of a limb are relatively common. Rattlesnake bites are characterized more by toxic symptoms than local symptoms. Recommended first aid measures for snake bites are rest and possibly bandaging the effected limb; immediate medical care is essential. The death rate of snake bites is, depending on the species, between 2 and 20% (for European species ca. 0%); see also bungarotoxin, crotoxin, crotamine, cobratoxins.

Lit.: Chem. Ind. (London) **1995**, 914 ff. ▪ Harding & Welch, Venomous Snakes of the World, Oxford: Pergamon 1980 ▪ J. Toxicol., Toxin Rev. **9**, 225 (1990) ▪ Lee, Snake Venoms, New York: Springer 1979 ▪ Nachr. Chem. Tech. Lab. **47**, 1120 (1999) (review) ▪ Naturwiss. Rundsch. **44**, 1 (1991) ▪ Pharmacol. Ther. **29**, 353–405 (1985); **30**, 91–113 (1985); **31**, 1–55 (1986); **34**, 403–451 (1987); **36**, 1–40 (1988) ▪ Shier & Mebs, Handbook of Toxicology, New York: Dekker 1990 ▪ Stocker, Medical Use of Snake Venom Proteins, Boca Raton: CRC Press 1990 ▪ Tu, Handbook of Natural Toxins, Vol. **5**, New York: Dekker 1991 ▪ Tu, Rattlesnake Venoms, New York: Dekker 1982.

Sodagnitins. Chromogenic (color change with aq. KOH: colorless → reddish violet) triterpenoids (with modified malabaricane skeleton) from the cap skins of the toadstools *Cortinarius sodagnitus*, *C. fulvoincarnatus* and *C. arcuatorum*, e.g., *S. C*: $C_{36}H_{50}O_{10}$, M_R 642.78, colorless powder, mp. 116–119 °C (dec.), $[\alpha]_D^{21}$ −15° (CH_3OH).

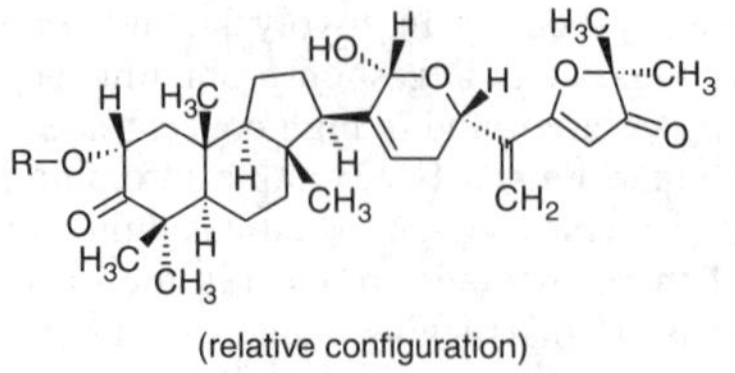

R = H : S. A

R = CO–CH$_2$–C(CH$_3$)(OH)–CH$_2$–COOH : S. C

S. A and C are active against *Bacillus subtilis, B. brevis* and *Nematospora coryli* and cytotoxic against L 1210 tumor cells.

Lit.: Eur. J. Org. Chem. **1999**, 255–260. – *[CAS 220579-22-6 (S. A); 220579-26-0 (S. C)]*

Sola... see Solanum steroid alkaloids and Solanum steroid alkaloid glycosides.

Solaniol see neosolaniol.

Solanone see tobacco flavor.

Solanum steroid alkaloid glycosides. *Solanum steroid alkaloids occur in plants mainly as glycosides which represent a subgroup of the *steroid saponins. Their usually branched chain carbohydrate components mostly consist of 3–4 *monosaccharides (tri- and tetrasaccharides, "tri- and tetraoses").

Tetraosides: Two important tetraoses are β-lycotetraose (from 2 mol D-*glucose and one mol each of D-*galactose and D-*xylose) and β-commertetraose, the analogous tetrasaccharide of *commersonine*, in which the terminal D-xylose of β-lycotetraose has been replaced by D-glucose.

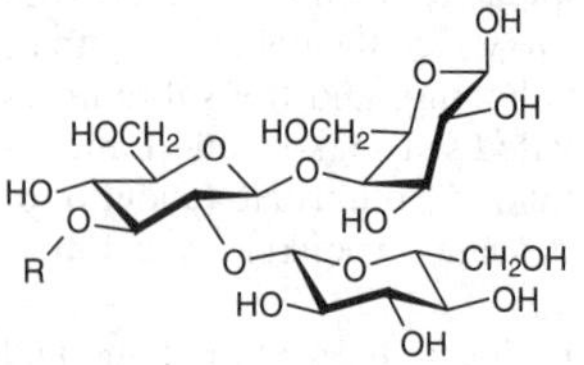

Figure 1: Solanum steroid alkaloid glycoside tetraoses.

Triosides: The trioses β-solatriose and β-chacotriose are the glycoside components of numerous Solanum steroid alkaloids and are thus widely distributed in the Solanaceae genus *Solanum*. They are composed of 1 mol each of D-*galactose, D-*glucose, and L-*rhamnose (β-solatriose) or 1 mol of D-glucose and 2 mol of L-rhamnose (β-chacotriose). The *steroid saponin dioscin is also a β-chacotrioside.

Table: Data and occurrence of Solanum steroid alkaloid glycosides.

glycoside	molecular formula	M$_R$	mp. [°C]	[α]$_D$	occurrence	CAS
Tetraosides						
Tomatine (3-*O*-β-Lycotetraosyltomatidine)	$C_{50}H_{83}NO_{21}$	1034.20	263–267 (decomp.)	–19° (pyridine)	(a)	17406-45-0
Demissine (3-*O*-β-Lycotetraosyldemissidine)	$C_{50}H_{83}NO_{20}$	1018.20	305–308 (decomp.)	–20° (pyridine)	(b)	6077-69-6
Commersonine (3-*O*-β-Commertetraosyldemissidine)	$C_{51}H_{85}NO_{21}$	1048.23	230–232 (decomp.)	–17° (pyridine)	(c)	60776-42-3
Triosides						
Solanine (3-*O*-β-Solatriosylsolanidine)	$C_{45}H_{73}NO_{15}$	868.07	286 (decomp.)	–59° (pyridine)	(d)	20562-02-1
Chaconine (3-*O*-β-Chacotriosylsolanidine)	$C_{45}H_{73}NO_{14}$	852.07	243 (decomp.)	–85° (pyridine)	(d)	20562-03-2
Leptine I (3-*O*-β-Chacotriosyl-23-*O*-acetylleptinidine)	$C_{47}H_{75}NO_{16}$	910.11	230 (vacuum)	–85° (pyridine)	(e)	101030-83-5
Leptinine I (3-*O*-β-Chacotriosylleptinidine)	$C_{45}H_{73}NO_{15}$	868.07	>230 (decomp.)	–90° (pyridine)	(e)	101009-59-0
Leptinine II (3-*O*-β-Solatriosylleptinidine)	$C_{45}H_{73}NO_{16}$	884.07	ca. 255 (decomp.)	–62° (pyridine)	(e)	100994-57-8
Solasonine (3-*O*-β-Solatriosylsolasodine)	$C_{45}H_{73}NO_{16}$	884.07	300–301 (decomp.)	–74° (CH_3OH)	(f)	19121-58-5
Solamargine (3-*O*-β-Chacotriosylsolasodine)	$C_{45}H_{73}NO_{15}$	868.07	301 (decomp.)	–105° (CH_3OH)	(f)	20311-51-7
α-Solamarine (3-*O*-β-Solatriosyltomatidenol)	$C_{45}H_{73}NO_{16}$	884.07	278–281 (decomp.)	–45° (pyridine)	(g)	20318-30-3
β-Solamarine (3-*O*-β-Chacotriosyltomatidenol)	$C_{45}H_{73}NO_{15}$	868.07	275–277 (decomp.)	–85.6° (pyridine)	(g)	3671-38-3
Furostanolglycosides						
Jurubine	$C_{33}H_{57}NO_8$	595.82	212–214	–30.9° (pyridine)	(h)	14256-61-2

(a) *Lycopersicon esculentum, L. pimpinellifolium, Solanum demissum* and numerous other *L.* and *S.* species; (b) *S. demissum, L. pimpinellifolium mut.* and some other *S.* species; (c) *S. commersonii* and *S. chacoense*; (d) *S. tuberosum* and numerous other *S.* species; (e) *S. chacoense*; (f) *S. sodomeum* and numerous other *S.* species; (g) *S. dulcamara* and *S. tuberosum* var. Kennebec; (h) *S. paniculatum* and *S. torvum* (roots).

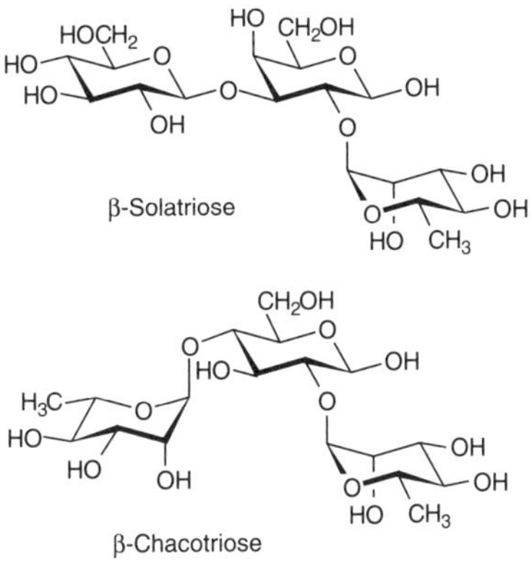

β-Solatriose

β-Chacotriose

Figure 2: β-Solatriose and β-chacotriose.

Furostanol glycosides: Jurubine [(25*S*)-3β-amino-26-(β-D-glucopyranosyloxy)-5α-furostan-22α-ol] is the only compound of this group for which the structure has been elucidated. It is so far the first and only furostanol glycoside to be isolated (see steroid saponins). Enzymatic or acidic hydrolysis of the glucoside furnishes jurubidine (see Solanum steroid alkaloids) after spontaneous and stereospecific cyclization.

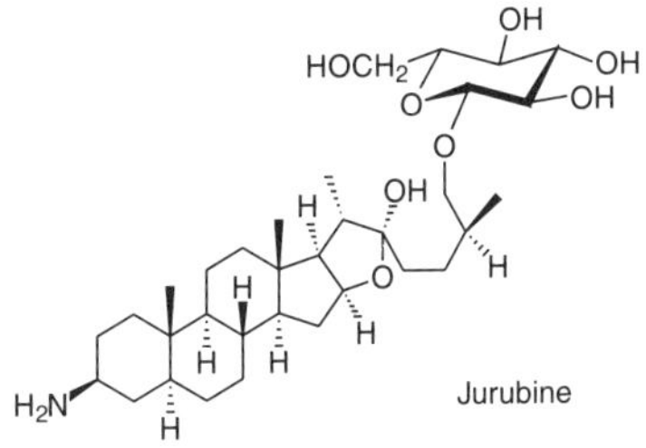

Jurubine

The distribution of the alkaloid glycosides in plants, especially potato plants, has been examined in detail. The highest contents of alkaloids are found in shoots that have been exposed to light (1.9–17.7 g/kg fresh weight) and flowers (2.1–5.0 g/kg fresh weight). In potato tubers for human consumption the alkaloid content should not exceed 60 mg/kg fresh weight [1].

Biological activity: The acute toxicity of some of these alkaloids, especially those of the potato, has been determined, e. g., after oral uptake of *α-solanine*: LD_{50} (rat) 590 mg/kg and of *tomatine* LD_{50} (rat) 900 mg/kg. On parenteral administration to rats the LD_{50} values are 84.0 mg/kg for α-solanine and 67.0 mg/kg for α-chaconine [1,2]. Most of these compounds have antimicrobial and hemolytic activities. The *cholesterol-binding capacity of tomatine (poorly soluble complex) is comparable to that of digitonin (see steroid saponins) so that this property can be used for the isolation and quantitative analysis of cholesterol, other 3β-hydroxysterols, or tomatine. A choline esterase inhibitory activity has been found for *solanine* and *demissine* while solanine also exhibits cytostatic activity towards *ascites* tumor cells in mice. *β-Solamarine* has potent *in vitro* activity against sarcoma 180 in mice [3]. The finding that potatoes infected by the fungus *Phytophthora infestans* (blight and potato disease) apparently have teratogenic effects aroused considerable interest. This activity has been associated with qualitative or quantitative changes in the steroid alkaloid spectrum resulting from the infection [4]. The Solanum steroid alkaloids solanidine, demissidine, and tomatidine, however, proved to be inactive. On the other hand solasodine exhibits weak teratogenicity. The leptines, demissine, and tomatine have a strong antifeedant effect on the Colorado (potato) beetle *Leptinotarsa decemlineata*. This explains why wild potatoes and tomatoes containing these alkaloids are not attacked by the beetle [2].

Biosynthesis: For biosynthesis of the aglycones, see Solanum steroid alkaloids. Their glycosides are formed by stepwise enzymatic glycosylation [5].

Use: Solasonine and solamargine are used as starting materials for the industrial production of hormonally active steroids [6].

Lit.: [1] Risk (ed.), Poisonous Plant Contamination of Edible Plants, p. 117–156, Boca Raton: CRC Press 1990. [2] D'Arcy (ed.), Solanaceae, Biology and Systematics, p. 201–222, New York: Columbia Press 1986; Manske **43**, 1–118. [3] Manske **25**, 1–355. [4] Recent Adv. Phytochem. **9**, 139 (1975); J. Agric. Food Chem. **26**, 566 (1978); Pelletier **4**, 389–425. [5] Mothes et al., p. 363–384. [6] Adv. Agron. **30**, 207–245 (1978).
gen.: Manske **3**, 247–312; **4**, 389; **7**, 343–361; **10**, 1–192; **19**, 81–192 ▪ Tetrahedron **54**, 10771 (1998) ▪ Ullmann (5.) **A 23**, 490–498. – *[HS 2938 90]*

Solanum steroid alkaloids. A subgroup of *steroid alkaloids. The chemical structures of all members contain the intact, unchanged C_{27} carbon skeleton of cholestane, but with differing heterocyclic ring systems. In principle, 5 structural types are distinguished: *solanidanes* (e. g., solanidine), *spirosolanes* (e. g., solasodine), *22,26-epiminocholestanes* (e. g., solacongestidine), *22,26-epimino-16,23-epoxycholestanes* (e. g., solanocapsine), and *3-aminospirostanes* (e. g., jurubidine). To date almost 100 such compounds have been isolated and structurally characterized from more than 350 plant species, mostly of the family Solanaceae (especially *Solanum* and *Lycopersicon* species), but occasionally also from the Liliaceae (see table). The compounds rarely occur as the free alkaloids but are mostly present in the form of glycosides (see Solanum steroid alkaloid glycosides). Thus, they are mostly aglycones (genins) obtained by acid or enzymatic hydrolysis of the glycosides. The alkaloid spectrum of the roots is frequently different from that of the above-ground parts and generally shows larger diversity.

Solanidanes: The solanidane alkaloids are tertiary bases with a heterocyclic indolizidine ring system. In most cases they differ by the presence or absence of a Δ^5-double bond or, respectively, the number and positions of the hydroxy groups.

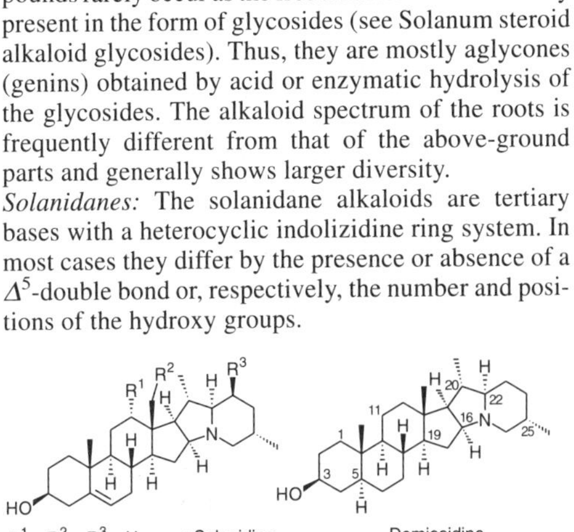

$R^1 = R^2 = R^3 = H$: Solanidine
$R^1 = R^2 = H$, $R^3 = OH$: Leptinidine
$R^1 = OH$, $R^2 = R^3 = H$: Rubijervine
$R^1 = R^3 = H$, $R^2 = OH$: Isorubijervine

Demissidine (5α-Solanidan-3β-ol)

Figure 1: Solanidane alkaloids.

Table: Data and occurrence of Solanum steroid alkaloids.

steroid alkaloid	molecular formula	M_R	mp. [°C]	$[\alpha]_D$	occurrence	CAS
Solanidanes						
Solanidine	$C_{27}H_{43}NO$	397.64	219	–27° ($CHCl_3$)	(a)	80-78-4
Demissidine	$C_{27}H_{45}NO$	399.66	219–220	+28° (CH_3OH)	(b)	474-08-8
Leptinidine	$C_{27}H_{43}NO_2$	413.64	247–248	–24° ($CHCl_3$)	(c)	24884-17-1
Rubijervine	$C_{27}H_{43}NO_2$	413.64	240–242	+19° (C_2H_5OH)	(d)	79-58-3
Isorubijervine	$C_{27}H_{43}NO_2$	413.64	241–244	+9.2° (C_2H_5OH)	(d)	468-45-1
Spirosolanes						
Solasodine	$C_{27}H_{43}NO_2$	413.64	200–201	–108° ($CHCl_3$)	(e)	126-17-0
N-Hydroxysolasodine	$C_{27}H_{43}NO_3$	429.64	206–209	–119.5° ($CHCl_3$)	(f)	142182-57-8
Soladulcidine	$C_{27}H_{45}NO_2$	415.66	209–211	–50° ($CHCl_3$)	(g)	511-98-8
Tomatidenol	$C_{27}H_{43}NO_2$	413.64	235–238	–39° ($CHCl_3$)	(h)	546-40-7
Tomatidine	$C_{27}H_{45}NO_2$	415.66	210–212	+5° ($CHCl_3$)	(i)	77-59-8
Soladunalinidine	$C_{27}H_{46}N_2O$	414.68	145–153	+1.3° ($CHCl_3$)	(k)	66934-59-6
22,26-Epiminocholestanes						
Solacongestidine	$C_{27}H_{45}NO$	399.66	169–174	+35.6° ($CHCl_3$)	(l)	984-82-7
Solafloridine	$C_{27}H_{45}NO_2$	415.66	172–175	+123° ($CHCl_3$)	(m)	2385-18-4
Verazine	$C_{27}H_{43}NO$	397.64	176–178	–91° ($CHCl_3$)	(d)	14320-81-1
Etioline	$C_{27}H_{43}NO_2$	413.64	153–155	–92° (C_2H_5OH)	(n)	29271-49-6
22,26-Epimino-16,23-epoxycholestanes						
Solanocapsine	$C_{27}H_{46}N_2O_2$	430.67	213–215	+24.9° (CH_3OH)	(o)	639-86-1
3-Deamino-3*β*-hydroxysolanocapsine	$C_{27}H_{45}NO_3$	431.66	204	+20.1° ($CHCl_3$)	(p)	35865-62-4
Pimpifolidine (Solanocardinol)	$C_{27}H_{45}NO_3$	431.66	200–203	–1.9° (pyridine)	(q)	138665-46-0
22-Isopimpifolidine	$C_{27}H_{45}NO_3$	431.66	200–204	–13.6° (pyridine)	(q)	152322-51-5
3-Aminospirostanes						
Jurubidine	$C_{27}H_{45}NO_2$	415.66	182–186	–78.7° ($CHCl_3$)	(r)	6084-44-2
Isojurubidine	$C_{27}H_{45}NO_2$	415.66	185–187	–63° ($CHCl_3$)	(s)	32562-75-7

(a) *Solanum tuberosum* and numerous other *S.* species, *Cestrum elegans* (*C. purpureum*, Solanaceae), *Veratrum album*, *Fritillaria camtschatcensis*, *Rhinopetalum bucharicum* and *R. stenantherum* (Liliaceae); (b) *S. demissum* and other *S.* species, *Lycopersicon pimpinellifolium*; (c) *S. chacoense*; (d) *V. album* and other *V.* species; (e) *S. nigrum*, *S. dulcamara*, *S. melongena*, *S. laciniatum*, *S. aviculare*, *S. marginatum*, *S. khasianum* and numerous other *S.* species, *C. elegans* and *Cyphomandra betacea* (Solanaceae) as well as *V. album* (Liliaceae); (f) *S. robustum* (roots); (g) *S. dulcamara* and other *S.* species, *L. pimpinellifolium*; (h) *S. tuberosum*, *S. dulcamara* and numerous other *S.* species, *L. pimpinellifolium* and *F. camtschatcaensis*; (i) *L. esculentum*, *S. demissum* and numerous other *L.* and *S.* species; (k) *S. dunalianum*; (l) *S. congestiflorum*; (m) *S. congestiflorum*, *S. umbellatum* and *S. verbascifolium*; (n) *V. grandiflorum* and other *V.* species, *S. havanense*, *S. spirale* and other *S.* species; (o) *S. pseudocapsicum* and other *S.* species; (p) *S. aculeatum* (roots); (q) *L. pimpinellifolium* (roots); (r) *S. paniculatum* and *S. torvum* (roots); (s) *S. paniculatum* (roots).

Spirosolanes: The spirosolane alkaloids can be divided into two series that differ in their stereochemistry at C-22 and C-25 [(22*R*,25*R*)-series as in *solasodine* and the (22*S*,25*S*)-series as in *tomatidenol*]. For 3*β*-aminospirosolanes and *N-hydroxysolasodine*, see *Lit.*[1]; formula, see figure 2.

22,26-Epiminocholestanes (16,28-secosolanidanes): The alkaloids of this group can be assigned to two main classes which again differ in their stereochemistry at C-25 (solacongestidine: 25*R*, verazine: 25*S*). The epimeric 16*β*-hydroxy compounds cyclize spontaneously and stereospecifically to give the corresponding spirosolanes, this is not possible for the 16*α*-compounds for steric reasons. Of the large number of reported compounds four are shown below as examples.

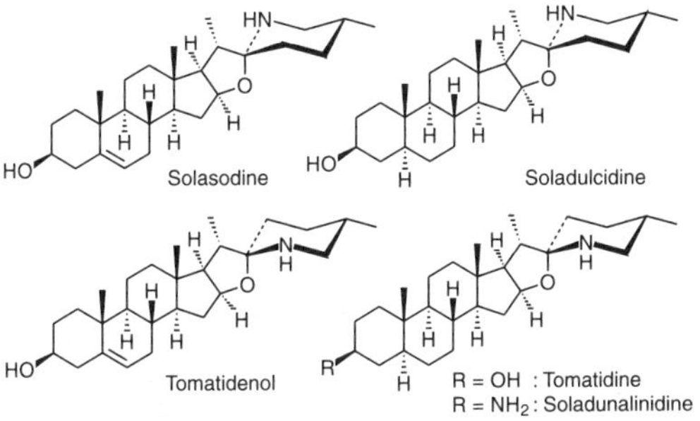

Figure 2: Spirosolane alkaloids.

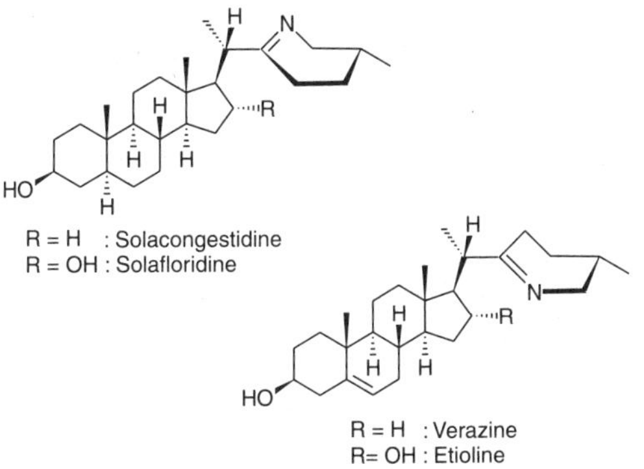

Figure 3: 22,26-Epiminocholestane alkaloids.

22,26-Epimino-16,23-epoxycholestanes (16,23-epoxy-16,28-secosolanidanes): Alkaloids with this skeleton have as yet only been isolated from a few *Solanum* and *Lycopersicon* species. One member known since 1929 is *solanocapsine*. It occurs only as the free alkaloid. In *3-deamino-3β-hydroxysolanocapsine* the 3β-amino group has been replaced by a 3β-hydroxy group. Two 16,23,25-isomers of this compound are *pimpifolidine* (solanocardinol) and *22-isopimpifolidine* (isosolanocardinol, with additional *22β-H*)[2].

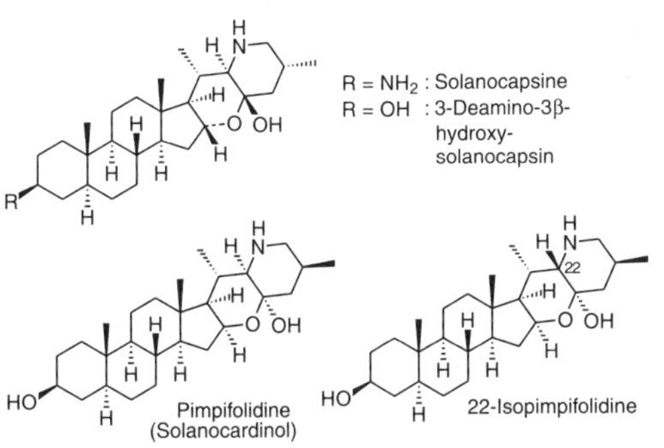

Figure 4: 22,26-Epimino-16,23-epoxycholestane alkaloids.

3-Aminospirostanes: Jurubidine is identical with 3-deoxyneotigogenin-3β-amine (see steroid sapogenins). It is formed by hydrolysis of jurubine [(25*S*)-3β-amino-26-(β-D-glucopyranosyloxy)-5α-furostan-22α-ol] (see Solanum steroid alkaloid glycosides) isolated from the roots of *Solanum paniculatum* and *S. torvum*. The corresponding (25*R*)-stereoisomer of jurubidine (*isojurubidine*) and the 6α- or 9-hydroxylated compounds are found in the same plants.

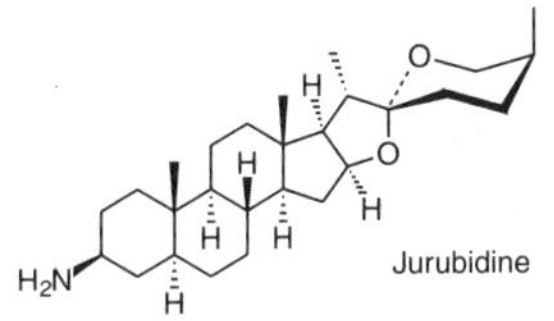

Biosynthesis: The biosynthesis proceeds analogously to that of the steroid sapogenins and is often coupled with the latter via *cycloartenol, *cholesterol (see steroids), and (25*R*)- or (25*S*)-26-aminocholest-5-en-3β-ol (steroid alkaloids) or (25*R*)- or (25*S*)-cholest-5-ene-3β,26-diol (steroid sapogenins). For details, see *Lit.*[3]. Steroid sapogenins with the same configuration at C-25 generally occur in the plants together with Solanum steroid alkaloids. For synthesis, see *Lit.*[4].

Use: Solasodine and tomatidenol are used in the commercial synthesis of hormonally active steroids. Solasodine is obtained on an industrial scale from *Solanum laciniatum* (e. g., New Zealand, Australia, Mexico, the former USSR), *S. marginatum* (Ecuador), and *S. khasianum* (India)[5]. For toxicity and pharmacological activity of some compounds, see *Lit.*[6] and Solanum steroid alkaloid glycosides.

Lit.: [1] Phytochemistry **31**, 1837 ff. (1992). [2] Phytochemistry **35**, 813 ff. (1994). [3] Mothes et al., p. 363–384. [4] Manske **10**, 1–192; **19**, 81–192. [5] Adv. Agron. **30**, 207–245 (1978). [6] D'Arcy (ed.), Solanaceae, Biology and Systematics, p. 201–222, New York: Columbia Univ. Press 1986.
gen.: Alkaloids: Chem. Biol. Perspect. **12**, 103–185 (1998) ▪ Chem. Pharm. Bull. **38**, 827 ff. (1990) ▪ Hager (5.) **3**, 1091–1095; **5**, 725 f.; **6**, 734–750 ▪ Manske **3**, 247–312; **7**, 343–361; **10**, 1–192; **19**, 81–192 ▪ Modern Drug Discovery Jan./Feb. **1999**, 9 f. ▪ Sax (8.), SKQ 500–SKS 100 ▪ Ullmann (5.) **A 1**, 399 f. – *[HS 2939 90]*

Solavetivone see vetispiranes.

Solenopsins.

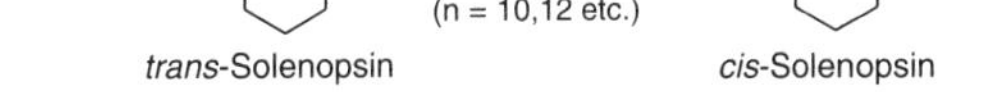

Components of the venom of fire ants of the genus *Solenopsis*, e. g., *S. invicta* which consists of a complex mixture of 2-methyl-6-alkylpiperidines. In contrast to the *monomorines both the *cis*- and the *trans*-isomers occur in nature. The latter have the (2*R*,6*R*)-configuration while the *cis*-alkaloids exist in the (2*R*,6*S*)-configuration.

Lit.: Naturwissenschaften **83**, 222 (1996) ▪ Org. Prep. Proced. Int. **28**, 501–543 (1996) (review) ▪ Recl. Trav. Chim. Pays-Bas **114**, 153 ff. (1995) ▪ Tetrahedron **50**, 8465–8478 (1994) ▪ Tetrahedron: Asymmetry **9**, 2419 (1998) (synthesis 2*R*,6*R*-S. A = *trans*-S.) ▪ Tetrahedron Lett. **34**, 2911 (1993). – *[CAS 28720-60-7 (trans-S.); 35285-24-6 (cis-S.)]*

Solerol, solerone see sherry flavor.

Solorinic acid.

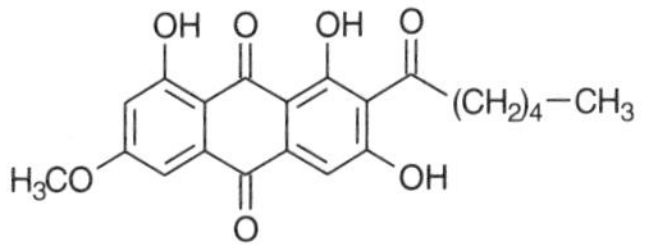

$C_{21}H_{20}O_7$, M_R 384.39, orange-red cryst., mp. 201 °C. An *anthraquinone from the foliose lichen *Solorina crocea*, the undersides of the thalli are colored orange. S. is used as a reference substance for HPLC analysis of lichen substances.

Lit.: Culberson, p. 189 ▪ J. Agric. Food Chem. **23**, 1132 (1975) ▪ Karrer, No. 1307. – *[HS 2914 61; CAS 10210-21-6]*

Sorangicins.

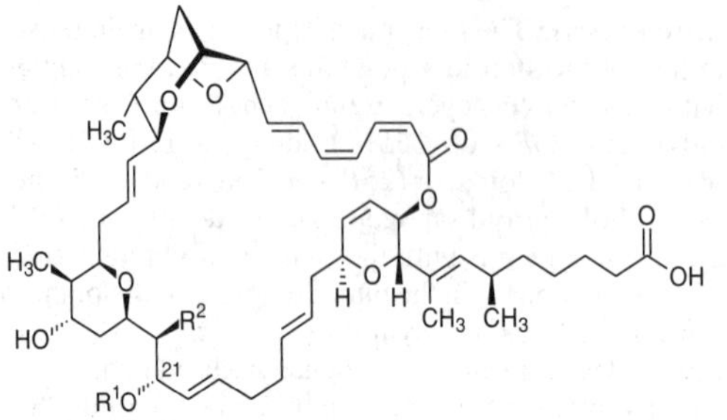

R^1 = H, R^2 = OH : Sorangicin A (**1**)
R^1 = R^2 = H : Sorangicin B (**2**)
R^1 = β-D-Glucopyranosyl, R^2 = OH : Sorangioside A (**3**)
R^1 = β-D-Glucopyranosyl, R^2 = H : Sorangioside B (**4**)

Table: Data of Sorangicins.

no.	molecular formula	M_R	mp. [°C]	$[\alpha]_D^{22}$ (CH_3OH)	CAS
1	$C_{47}H_{66}O_{11}$	807.03	105–107	+60.9°	100415-25-6
2	$C_{47}H_{66}O_{10}$	791.03	–	+49.1°	107745-54-0
3	$C_{53}H_{76}O_{16}$	969.18	–	+42.0°	111727-66-3
4	$C_{53}H_{76}O_{15}$	953.18	–	+5.1°	111749-38-3

A separate group of >20 *macrolide antibiotics from *Sorangium cellulosum* (Myxobacteria), the main components of which are S. A and S. B as well as their 21-*O*-glucosides. A conspicuous feature is the bicyclic ether system, otherwise only known from *palytoxin. The biosynthesis proceeds via a linear polyketide made up of 20 acetate residues and a malonyl-CoA starter unit (C-1 to C-3) as well as four methyl groups from methionine. S. A and S. B have bactericidal activity against Gram-positive and Gram-negative pathogens with minimal inhibitory concentrations in the range 4–100 ng/mL and 3–30 µg/mL, respectively. Good *in vivo* activity and no acute toxicity up to 0.3 g/kg in mice have been observed. The 21-*O*-glucosides (*sorangiosides*) are inactive; some of the many semisynthetic derivatives have higher activities than the parent compound. S. A efficiently inhibits the initiation step of bacterial RNA polymerases, while polymerase II from wheat shoots is insensitive. Surprisingly, not only the activity and inhibitory spectra are similar to those of rifamycin but also cross resistance occurs.

Lit.: J. Antibiot. **40**, 7 (1987) ▪ Justus Liebigs Ann. Chem. **1989**, 111, 213, 309; **1990**, 975; **1993**, 293 ▪ Tetrahedron Lett. **26**, 6031 (1985).

Soraphens. A family of 18-membered, antifungal *macrolides from *Sorangium cellulosum* (Myxobacteria). The main and concomitantly the most active component *S. $A_{1\alpha}$* is obtained from the cell extract by crystallization ($C_{29}H_{44}O_8$, M_R 520.66, $[\alpha]_D^{22}$ –124° (CH_3OH), mp. 101 °C). The S. exhibit very good activities against a broad spectrum of plant pathogenic fungi in field testing. This activity is based on an inhibition of acetyl-CoA carboxylase (ACC) and thus lipid synthesis in the fungi; IC_{50} 3.3 ng S. $A_{1\alpha}$ for 6.5 pmol/mL ACC from *Ustilago maydis*; ACC from rat liver is also sensitive but plant ACCs are not. The biosynthesis of the S. proceeds on the polyketide pathway starting from a benzoyl starter unit. A biogenesis gene cluster consisting of ca. 40 kb has been identified and in part sequenced. S. derivatives have been tested as fungicides for plant protection [1]. For toxicological reasons, the development was terminated.

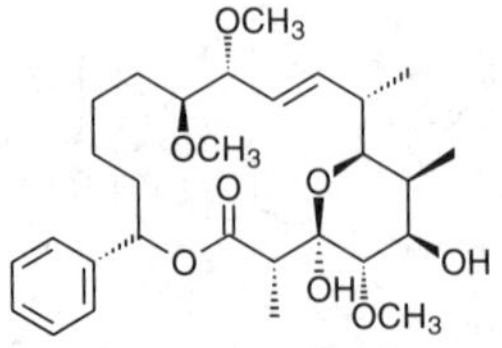

Soraphen $A_{1\alpha}$

Lit.: [1] Europ. Pat. Appl. 711783, 711784 (publ. 15.5.1996).
gen.: Angew. Chem., Int. Ed. **33**, 2466 (1994) ▪ Curr. Genet. **25**, 95 (1994) ▪ J. Antibiot. **47**, 23–31 (1994) ▪ J. Bacteriol. **177**, 3673 (1995) ▪ Justus Liebigs Ann. Chem. **1993**, 1017; **1994**, 759; **1995**, 803, 867 ▪ Synthesis **1999**, 188 (synthesis S. $A_{1\alpha}$). – *[CAS 122547-72-2 (S. $A_{1\alpha}$)]*

Sorbic acid [(*E,E*)-2,4-hexadienoic acid].

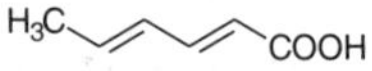

$C_6H_8O_2$, M_R 112.13, mp. 134.5 °C, colorless needles, bp. 228 °C (decomp.); soluble in organic solvents and hot water. S. occurs in the seeds of the rowanberry (mountain ash, *Sorbus aucuparia*, Rosaceae) in the form of a lactone precursor, the so-called *parasorbic acid* [(6*S*)-5,6-dihydro-6-methyl-2*H*-pyran-2-one] and is used for preserving foods.

Lit.: J. Org. Chem. **59**, 4001 (1994) ▪ Org. Prep. Proced. Int. **23**, 441 (1991) ▪ Ullmann (5.) **A18**, 148; **A24**, 507. – *[HS 2916 19; CAS 110-44-1]*

Sorbin see sorbose.

D-Sorbitol see D-glucitol.

Sorbose (*xylo*-2-hexulose, sorbin, sorbinose).

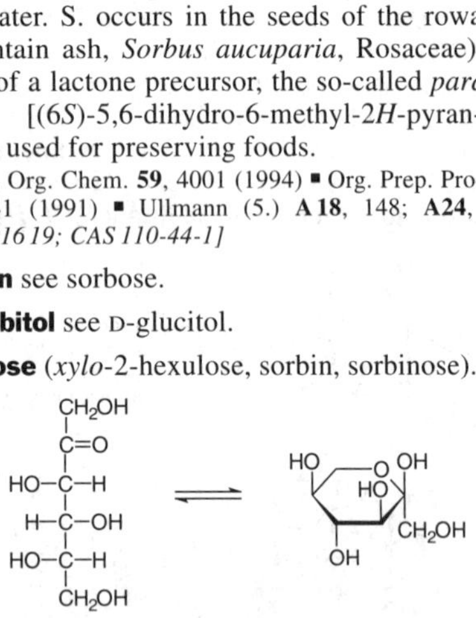

L-form α-L-pyranose form

$C_6H_{12}O_6$, M_R 180.16, L-S.: orthorhombic cryst., mp. 165 °C, $[\alpha]_D^{20}$ –43.2° (H_2O); produced by the bacterium *Acetobacter xylinum* from D-*glucitol. S. exists in the α-L- and β-L-pyranose as well as the corresponding furanose forms. A 4 molar aqueous solution at 31 °C contains 93% α-pyranose, 2% β-pyranose, 4% α-furanose, 1% β-furanose, and 0.3% open chain sorbose. S. reduces Fehling's solution and cannot be fermented. D-S.: mp. 165 °C, $[\alpha]_D^{20}$ +42.9° (H_2O).

Use: As intermediate in the synthesis of *ascorbic acid.

Lit.: Carbohydr. Res. **216**, 141 (1991); **281**, 183 (1996) ▪ Beilstein EIV **1**, 4411 ▪ Karrer, No. 619 ▪ Ullmann (5.) **A 5**, 81. – *[HS 1702 90, 2940 00; CAS 3615-56-3 (D-S.); 87-79-6 (L-S.)]*

Sordarin.

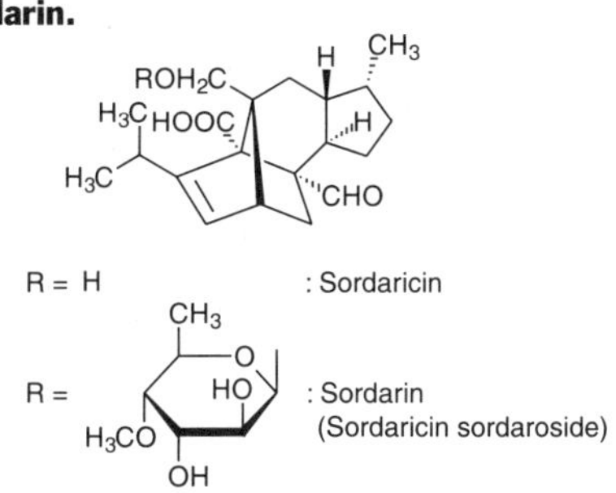

$C_{27}H_{40}O_8$, M_R 492.6, oil, $[\alpha]_D$ –45.2° (CH_3OH), mp. (K salt) 253–255 °C. A diterpene glycoside (aglycone: *sordaricin*, $C_{20}H_{28}O_4$, M_R 332.44), from the ascomycete *Sordaria araneosa*. S. has antifungal and antimycotic activity and exhibits antitumor properties *in vitro*. LD_{50} (mouse) >250 mg/kg. Analogs of S. have been synthesized[1].

Lit.: [1] Bioorg. Med. Chem. Lett. **8**, 2269 (1998).
gen.: Antimicrob. Agents Chemother. **42**, 2274 (1998); **43**, 1716 (1999) ▪ Curr. Opin. Antiinfect. Invest. Drugs **1**, 297 (1999) ▪ J. Org. Chem. **56**, 3595, 5718 (1991). – *[CAS 11076-17-8 (S.); 51493-69-7 (sordaricin)]*

Sordidin.

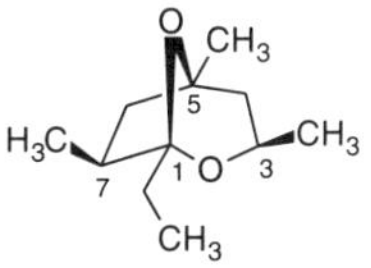

$C_{11}H_{20}O_2$, M_R 184.28, main component of a mixture of four stereoisomers, $[\alpha]_D^{21}$ +26° (ether). Aggregation pheromone, produced by males of the banana weevil *Cosmopolites sordidus*.

Lit.: Bioorg. Med. Chem. **4**, 413 (1996) ▪ Chem. Commun. **1997**, 1153–1158 ▪ Tetrahedron Lett. **36**, 1043 (1995); **37**, 3741 (1996); **38**, 3475 (1997). – *[CAS 162334-33-0 ((±)-3,7-epi-S.); 162428-75-3 ((±)-3-epi-S.); 162428-76-4 ((±)-7-epi-S.); 162490-88-2 (S.)]*

Sordidone.

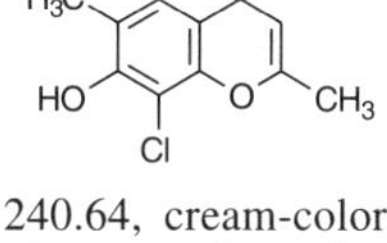

$C_{11}H_9ClO_4$, M_R 240.64, cream-colored needles, mp. 265–266 °C. A chromone from the lichen *Lecanora rupicola*, occurring especially in the apothecia.

Lit.: Culberson, p. 177 ▪ J. Chem. Soc. (C) **1971**, 1324 ff. ▪ Z. Naturforsch. B **24**, 750–756 (1969). – *[CAS 18780-80-8]*

Sotolone see hydroxyfuranones.

Sparsomycin (sparsogenin).

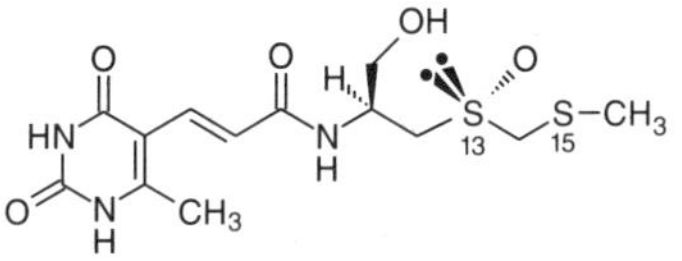

natural 13-*R*(+)-form

$C_{13}H_{19}N_3O_5S_2$, M_R 361.44, mp. 208–209 °C (decomp.), $[\alpha]_D$ +69° (H_2O), LD_{50} (mouse) 2.4 mg/kg (i.p.); 22 mg/kg (p.o.), a *pyrimidine propenamide derivative from *streptomycetes (e.g., *Streptomyces sparsogenes*). S. is active against Gram-negative bacteria, fungi, and viruses. It is an inhibitor of protein biosynthesis and acts by binding to the peptidyl transferase center of the ribosomal *50S*-subunit; however, there is no difference between pro- and eukaryotes. S. is being tested as an additive in tumor therapy since it acts synergistically with cisplatin so that the dosage and hence the side effects of the latter may be reduced. The two epimeric S^{15}-oxides: *sparoxomycins A1* and *A2* ($C_{13}H_{19}N_3O_6S_2$, M_R 377.44) have been isolated as inhibitors of the cell cycle progression. S. is not to be confused with the nucleoside antibiotic *tubercidin that is referred to as *S. A* in the older literature.

Lit.: Anticancer Res. **9**, 923–927 (1989) ▪ Biochem. Biophys. Res. Commun. **166**, 673 (1990) ▪ Bioorg. Med. Chem. Lett. **8**, 3331 (1998) (synthesis) ▪ J. Am. Chem. Soc. **114**, 5946 (1992) ▪ J. Antibiotics (Tokyo) **49**, 65, 1096 (1996) (structure sparoxomycins) ▪ J. Med. Chem. **32**, 2002 (1989) ▪ J. Org. Chem. **50**, 1264 (1985) ▪ Prog. Med. Chem. **23**, 219–268 (1986) (review) ▪ Sax (8.), SKX 000 ▪ Tetrahedron Lett. **35**, 7497 (1994) (biosynthesis). – *[CAS 1404-64-4 (S.); 174390-01-3 (sparoxomycin A1); 174390-02-4 (sparoxomycin A2)]*

(–)-Spartalupine see sparteine.

Sparteine (7aα,14aβ-dodecahydro-7α,14α-methano-2*H*,6*H*-dipyrido[1,2-*a*:1′,2′-*e*][1,5]diazocine).

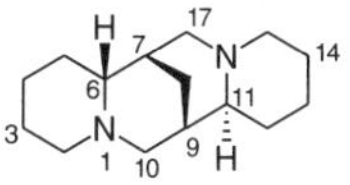

(–)-form

$C_{15}H_{26}N_2$, M_R 234.38, bp. 181 °C (2.4 kPa), $[\alpha]_D$ –17.0° (C_2H_5OH). S. is a viscous, bitter-tasting oil with an aniline-like odor. It is a tetracyclic *quinolizidine alkaloid, and parent compound of the sparteine type, the largest subgroup of this alkaloid family; it occurs in many genera of the Fabaceae (e.g. *Lupinus*, *Cytisus*), the Berberidaceae genus *Leontice*, and the common broom (*Cytisus scoparius*, syn. *Sarothamnus scoparius*). In lower doses S. has stimulating effects on the smooth musculature (e.g. uterus, stomach), in larger doses it causes paralysis (similar to coniine) [LD_{50} (mouse p.o.) 220 mg/kg]. It promotes coronary perfusion by vasodilatation and inhibits the transmission of stimuli; thus, it can be used to treat arrhythmia. The monosulfate is used in therapy for heart diseases, venous complaints, in obstetrics to induce and strengthen labor activity, and in anesthesiology. S. is also used as a chiral catalyst. The enantiomeric (+)-S. (*pachycarpine*) {bp. 173–174 °C (1.1 kPa), $[\alpha]_D$ +17.1° (C_2H_5OH)} also occurs in various Fabaceae and is highly toxic. The racemate of S. also occurs in nature {bp. 119–121 °C (0.133 kPa)}. Epimers of S. are 11β-S. {α-isosparteine, *genisteine*, mp. 96–117 °C (hydrate), $[\alpha]_D$ –56° (CH_3OH)} and 6α-S. {β-isosparteine, (–)-*spartalupine*, mp. 32–32.5 °C, $[\alpha]_D$ –15.4° (CH_3OH)}.

Lit.: Beilstein E V **23/5**, 497 ff. ▪ Hager (5.) **3**, 1096 f. ▪ Manske **47**, 1–115 ▪ Merck-Index (12.), No. 8887, 4395 (genisteine). – *Biosynthesis:* J. Chem. Soc., Chem. Commun. **1984**, 1477. – *Synthesis:* Chem. Pharm. Bull. **35**, 4990 (1987) ▪ J. Chem. Soc., Perkin Trans. 1 **1990**, 2593. – *[HS 2939 90; CAS 90-39-1 (S.);*

4985-24-4 ((±)-S.); 492-08-0 ((+)-S.); 446-95-7 (11β-S.); 24915-04-6 (6α-S.); 492-06-8 ((+)-6α-S.)]

Spatol (15,16;17,18-diepoxy-13-spaten-5-ol).

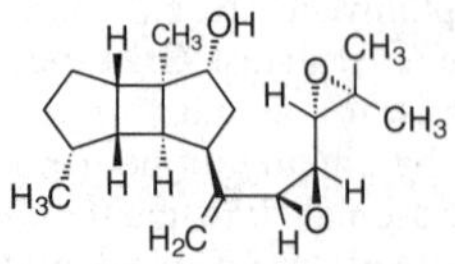

$C_{20}H_{30}O_3$, M_R 318.46, mp. 100–102 °C, $[\alpha]_D$ +45.6° ($CHCl_3$). A diterpenoid from the marine alga *Spatoglossum schmittii*. S. has antineoplastic activity, especially against human melanoma cells.

Lit.: J. Am. Chem. Soc. **113**, 3096 (1991) ▪ Lindberg **3**, 381–416 ▪ Tetrahedron **50**, 12843 (1994) ▪ Tetrahedron Lett. **35**, 517 (1994). – *[CAS 76520-52-0]*

Spearmint oil. Besides *peppermint oils, S. constitutes another important essential oil isolated from *Mentha* species. It has a typical warm, fresh, green-herby, caramel-mint-like odor and a warm, spicy-herby, somewhat bitter astringent taste.

Production: By steam distillation from the flowering shrub of the crisped or curled mint species *Mentha spicata* ("native spearmint") and *Mentha cardiaca* ("Scotch spearmint"). The main areas of cultivation are the American states of Idaho, Oregon, and Washington as well as Indiana, Michigan, and Wisconsin ("Middle West"); oil production in 1992 was ca. 3200 t. It is also cultivated in smaller amounts in China and India; production in 1992 ca. 500 t.

Composition [1]***:*** The main component, typical for the organoleptic impression, is (–)-**carvone* (USA 55–66%; China/India ca. 80%). The further components depend on the type: a Midwestern Scotch oil contains, e.g., *limonene* (see *p*-menthadienes) (15%), *1,8-*cineole* (1.5%), *menthone* (see *p*-menthanones) (1%), *1,6-dihydrocarvone* (see menthenones) (2%), etc.

Use: Mainly to improve the aromatic nature of confectionery products, chewing gum, and toothpaste.

Lit.: [1] Perfum. Flavor. **1** (6), 34 (1976); **5** (4), 6 (1980); **18** (2), 46; (3), 61 (1993).
gen.: Arctander, p. 588 ▪ Bauer et al. (2.), p. 164 ▪ ISO 3033 (1988). – *Toxicology:* Food Cosmet. Toxicol. **16**, 871 (1976). – *[HS 3301 29; CAS 8008-79-5]*

Specionin.

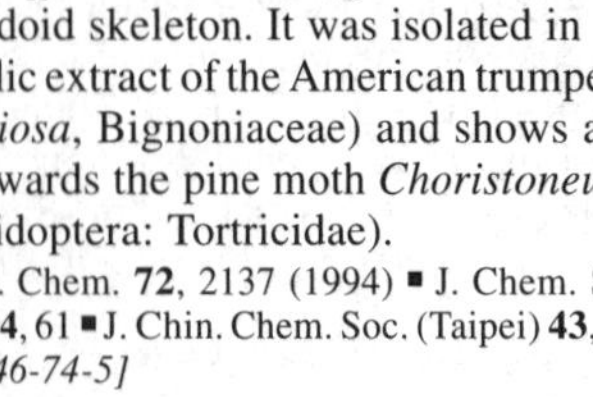

$C_{20}H_{26}O_8$, M_R 394.42, oil, $[\alpha]_D$ –27.2° ($CHCl_3$). S. has the C_9-*iridoid skeleton. It was isolated in 1983 from the alcoholic extract of the American trumpet tree (*Catalpa speciosa*, Bignoniaceae) and shows antifeedant activity towards the pine moth *Choristoneura fumiferana* (Lepidoptera: Tortricidae).

Lit.: Can. J. Chem. **72**, 2137 (1994) ▪ J. Chem. Soc., Perkin Trans. 1 **1994**, 61 ▪ J. Chin. Chem. Soc. (Taipei) **43**, 459 (1996). – *[CAS 87946-74-5]*

Spectabiline see monocrotaline.

Spectinomycin (actinospectacin). $C_{14}H_{24}N_2O_7$, M_R 332.35, mp. 184–194 °C; pK_{a1} 6.95, pK_{a2} 8.7; $[\alpha]_D^{25}$

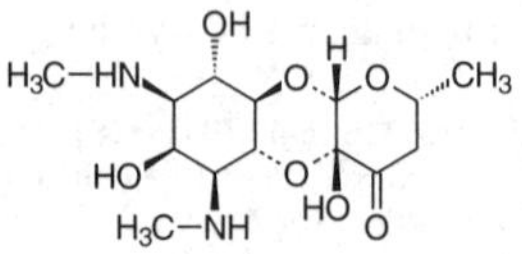

–20°; LD_{50} (rat p.o.) >5000 mg/kg. An aminoglycoside-like antibiotic containing an aminocyclitol residue but lacking an *amino sugar produced by *Streptomyces spectabilis* and *S. flavopersicus*. Similar to the aminoglycosides, the biological activity of S. is based on an inhibition of ribosomal protein synthesis. S. is active against various Gram-positive and Gram-negative bacteria and is used in human therapy mainly for treatment of infections caused by *penicillin-resistant *Neisseria gonorrhoeae* strains. Biogenetically, S. is formed from D-glucose via actinamine [1,3-dideoxy-1,3-bis(methylamino)-*myo*-inositol], which is glycosidically linked to a ketohexose to furnish an intermediate from which S. and related compounds are formed by subsequent dehydration and substitution.

Lit.: Beilstein E V **18/2**, 489 ▪ Curr. Microbiol. **21**, 261 (1990) ▪ Kleemann-Engel, p. 1738 ▪ Kirk-Othmer (4.) **2**, 907 (review) ▪ Med. Clin. North Am. **66**, 169 (1982). – *[HS 2941 90; CAS 1695-77-8]*

Sperabillins

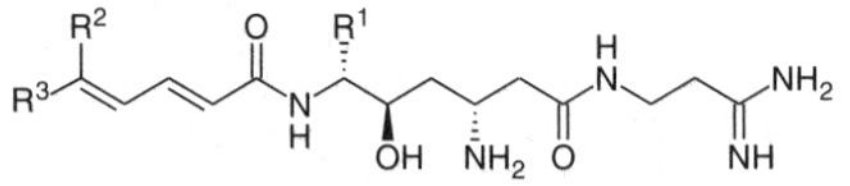

$R^1 = R^3 = H$, $R^2 = CH_3$: S. A (TAN 749A)
$R^1 = R^2 = CH_3$, $R^3 = H$: S. B (TAN 749B)
$R^1 = R^2 = H$, $R^3 = CH_3$: S. C (TAN 749C)
$R^1 = R^3 = Ch_3$, $R^2 = H$: S. D (TAN 749D)

Unconventional depeptide (pseudopeptide) antibiotics[1] with a broad spectrum of antibacterial activity from cultures of *Pseudomonas* spp., e.g., *S. C* (TAN 749C): $C_{15}H_{27}N_5O_3$, M_R 325.41, powder, $[\alpha]_D^{24}$ –11° (H_2O).

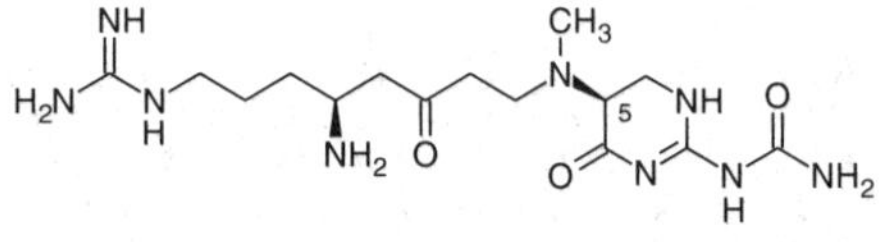

TAN 1057A; (5*R*)-epimer: TAN 1057B

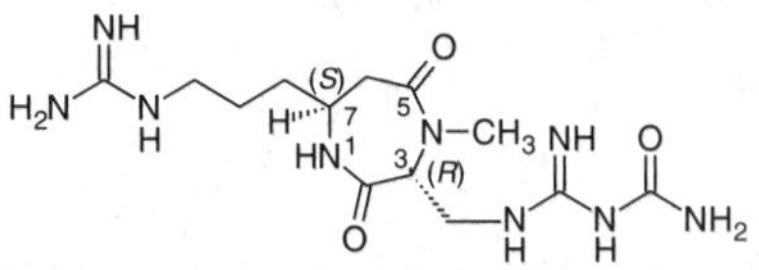

TAN 1057C; (3*S*)-epimer: TAN 1057D

TAN 1057, a different class of pseudopeptides[2], was isolated from *Flexibacter* spp., e.g. *S. C* TAN 1057A $C_{13}H_{25}N_9O_3$, M_R 355.40, powder, $[\alpha]_D^{22}$ –39.1° (dihydrochloride; H_2O), and its (5*R*)-epimer, TAN 1057B, $[\alpha]_D^{24}$ +72.6° (H_2O).

Lit.: [1] Bull. Chem. Soc. Jpn. **66**, 863–869 (1993); J. Antibiot. **45**, 10–19 (1992). [2] Eur. J. Org. Chem. **1998**, 777–783; J. Antibiot. **46**, 606–613 (1993); **51**, 189–201 (1998); Tetrahedron **49**, 13–28 (1993). – *[CAS 111337-84-9 (S. C); 128126-44-3 (TAN 1057A); 128126-45-4 (TAN 1057B)]*

Spergualin {(*S*)-*N*-[4-(3-aminopropylamino)butyl]-2-((*S*)-7-guanidino-3-hydroxyheptanoyl)-2-hydroxyacetamide}.

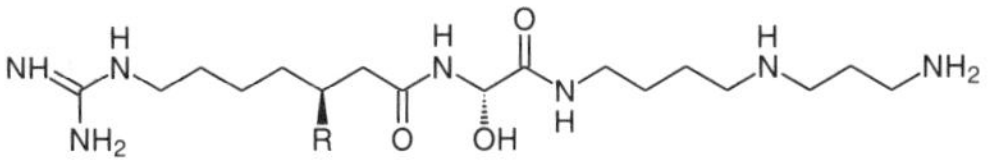

R = OH : Spergualin
R = H : 15-Deoxyspergualin

$C_{17}H_{37}N_7O_4$, M_R 403.52, $[\alpha]_D^{24}$ –11° (H_2O). An antibiotically active compound from *Bacillus laterosporus* exhibiting potent antitumor and also immunosuppressive activity. It is the starting material for the therapeutically used compound *15-deoxyspergualin.

Lit.: Ann. N. Y. Acad. Sci. **696**, 270 (1993) ▪ Biomed. Pharmacother. **41**, 227 (1987) ▪ J. Antibiot. (Tokyo) **38**, 283, 886 (1985); **39**, 1461 (1986) ▪ Transplant. Proc. **26**, 3224 (1994). – *[CAS 80902-43-8]*

Spermidine [*N*-(3-aminopropyl)-1,4-butanediamine]. $H_2N–(CH_2)_3–NH–(CH_2)_4–NH_2$, $C_7H_{19}N_3$, M_R 145.25, oil, bp. 128–130 °C; as trihydrochloride: cryst., mp. 256–258 °C, soluble in water and ethanol. S. is a biogenic *polyamine, formed biogenetically from *putrescine. It is the precursor of *spermine. S. was first found in human sperm but is widely distributed in nature. The S. in sperm is bound to the phosphate groups of nucleic acids. It is essential for cell growth. S. is also a component of the alkaloids *lunarine, *inandenine, and related structures, the spermidine alkaloids [1].

Lit.: [1] Bull. Chem. Soc. Jpn. **71**, 1221 (1998); Chirality **9**, 523 (1997); Tetrahedron Lett. **39**, 257 (1998).
gen.: Beilstein EIV **4**, 1300 ▪ Differentiation **19**, 1–20 (1981) ▪ Dowling, Polyamines in the Gastrointestinal Tract, Dordrecht: Kluwer 1992 ▪ Helv. Chim. Acta **69**, 1012–1016 (1986); **71**, 1708 (1988) ▪ Karrer, No. 3168 f. ▪ Merck-Index (12.), No. 8893 ▪ Morris & Marton, Polyamines in Biology and Medicine, New York: Dekker 1981 ▪ Recent Adv. Phytochem. **23**, 329 (1989). – *[HS 2921 29; CAS 124-20-9]*

Spermine [*N,N'*-bis-(3-aminopropyl)-1,4-butanediamine]. $H_2N–(CH_2)_3–NH–(CH_2)_4–NH–(CH_2)_3–NH_2$, $C_{10}H_{26}N_4$, M_R 202.34, hygroscopic cryst., mp. 55–60 °C, bp. 141–142 °C (66.5 Pa), readily soluble in water and ethanol. S. is a biogenic *polyamine that, like *spermidine, occurs bound to phosphate in seminal fluid and together with the latter is responsible for its typical odor. Furthermore, it is known to be a component of ribosomes and DNA. S. was the first natural product to be characterized by physicochemical methods (crystals of S. phosphate from sperm) [1]. The structure was not elucidated until 1927, 250 years after its discovery.

Lit.: [1] R. Soc. London Philosoph. Trans. **12**, 1048 (1678).
gen.: Beilstein EIV **4**, 1301 ▪ Merck-Index (12.), No. 8894 ▪ see also spermidine. – *[HS 2921 29; CAS 71-44-3]*

Sphagnorubins.

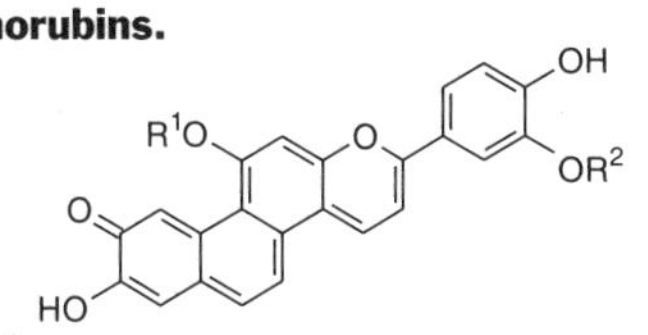

$R^1 = R^2 = H$: Sphagnorubin A
$R^1 = CH_3$, $R^2 = H$: Sphagnorubin B
$R^1 = R^2 = CH_3$: Sphagnorubin C

S. are red cell wall pigments (= membranochromes) of various *Sphagnum* species (peat mosses). They are de-

Table: Data of Sphagnorubins.

	molecular formula	M_R	mp. [°C]	CAS
S. A	$C_{23}H_{14}O_6$	386.36	>350 cryst.	55483-18-6
S. B	$C_{24}H_{16}O_6$	400.39	>350 cryst.	84018-30-4
S. C	$C_{25}H_{18}O_6$	414.41	>350 cryst.	84018-31-5

oxyanthocyanidins with an as yet unique 2-phenylphenanthro[2,1-*b*]pyran skeleton; they are not polar and practically insoluble in weakly polar solvents; their solubility in DMSO amounts to 0.1%. The solutions are sensitive to light and are only moderately stable at pH 1–2. The observed change in color of peat mosses from green to red to red violet at low nocturnal temperatures is due to the formation of sphagnorubins. This can also be achieved experimentally by coldness treatment [1].

Lit.: [1] Biochem. Physiol. Pflanz. **176**, 728 (1981).
gen.: Chem. Ber. **108**, 1116 (1975) ▪ Justus Liebigs Ann. Chem. **1984**, 1024.

Sphagnum acid.

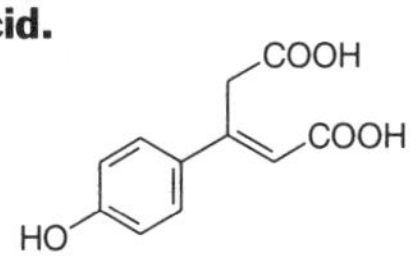

$C_{11}H_{10}O_5$, M_R 222.20, column-shaped cryst., mp. 178–180 °C, very soluble in polar solvents. A cell wall phenol occurring in many *Sphagnum* species (peat mosses). For synthesis, see *Lit.* [1]. The biosynthesis most probably proceeds via the shikimic acid pathway [2].

Lit.: [1] J. Prakt. Chem. **338**, 706 (1996). [2] Proc. Phytochem. Soc. Eur. **29**, 253 (1990).
gen.: Phytochemistry **24**, 745 (1985); **38**, 35 (1995). – *[CAS 57100-28-4]*

Sphenolobanes. *Diterpenes of a new structural type isolated from species of the liverwort genus *Anastrophyllum*. To date 13 different S. are known.

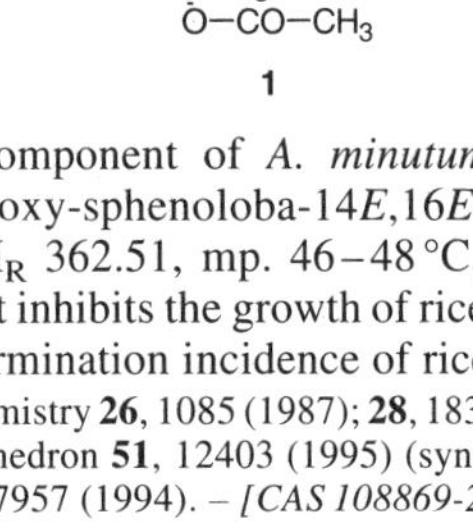

1

The main component of *A. minutum* is 5α-acetoxy-3α,4α-epoxy-sphenoloba-14*E*,16*E*-dien-18-ol (**1**) ($C_{22}H_{34}O_4$, M_R 362.51, mp. 46–48 °C, $[\alpha]_D^{23}$ +22.3°, yellow oil), it inhibits the growth of rice shoots and reduces the germination incidence of rice seeds.

Lit.: Phytochemistry **26**, 1085 (1987); **28**, 183 (1989); **29**, 3243 (1990) ▪ Tetrahedron **51**, 12403 (1995) (synthesis) ▪ Tetrahedron Lett. **35**, 7957 (1994). – *[CAS 108869-25-1 (1)]*

Sphingofungins.

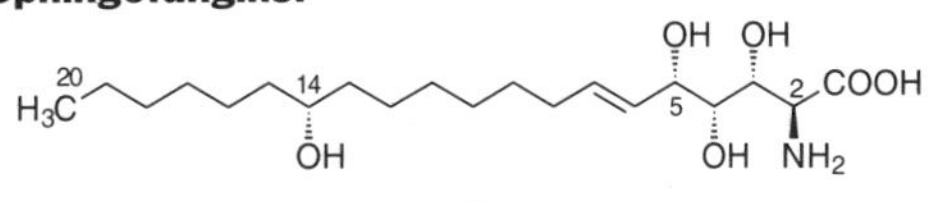

S. B

Antifungal aminopolyhydroxy fatty acids from cultures of *Aspergillus fumigatus* and *Paecilomyces variotii*. S. inhibit serine palmitoyl transferase (EC 2.3.1.50), e.g., *S. B:* $C_{20}H_{39}NO_6$, M_R 389.53, cream-colored solid. A S. derivative has been in preclinical development as an antimycotic agent.

Lit.: Eur. J. Org. Chem. **1999**, 1795 (synthesis) ▪ J. Am. Chem. Soc. **120**, 908, 6818 (1998) (synthesis) ▪ J. Antibiot. **45**, 861, 1692 (1992) ▪ Nat. Prod. Lett. **6**, 295 (1995) ▪ Synlett **1996**, 672; **1997**, 301 (synthesis) ▪ Tetrahedron Lett. **33**, 297 (1992). – *[CAS 121025-45-4 (S. B)]*

Sphingolipids see sphingosine.

Sphingosine [(2*S*,3*R*,4*E*)-2-amino-4-octadecene-1,3-diol].

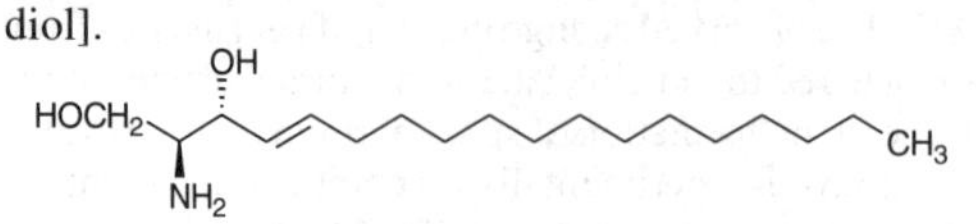

$C_{18}H_{37}NO_2$, M_R 299.50, cryst., mp. 79–81 °C (80–84 °C), (racemate, mp. 65–68 °C), does not occur in nature in the free form, but is an important component of *cerebrosides, gangliosides, and sphingomyelins (sphingolipids)[1]. Fatty acid amides, including *N*-octadecanoyl-S. (a) {$C_{36}H_{71}NO_3$, M_R 565.96, mp. 97–98 °C (91–93 °C), $[\alpha]_D^{25}$ –3.1° ($CHCl_3$)}, *N*-eicosanoyl-S. (b) ($C_{38}H_{75}NO_3$, M_R 594.02), and *N*-13-(*Z*)-docosenoyl-S. (c) ($C_{40}H_{77}NO_3$, M_R 620.06) occur in the blood and brain lipids. 1-*O*-Glucosyl-S. [*glucopsychosine* (d), $C_{24}H_{47}NO_7$, M_R 461.64] occurs as a metabolite of S. in cases of Gaucher's disease. The saturated L-(–)-*threo*-2-amino-1,3-octadecanediol (configuration 2*S*,3*S*) (e) [$C_{18}H_{39}NO_2$, M_R 301.51, mp. 108 °C, $[\alpha]_D^{28}$ –14.1° ($CHCl_3$)] is distributed in cerebrosides and in plant tissue, also D-*ribo*-2-amino-1,3,4-octadecanetriol (configuration 2*S*,3*S*,4*R*) (f) [*phytosphingosine*, 4-hydroxysphinganine, $C_{18}H_{39}NO_3$, M_R 317.51]. For synthesis, see *Lit.*[2].

Biosynthesis: Condensation of L-serine with palmitoyl-CoA and decarboxylation → 2-amino-3-oxo-4-octadecenol → sphinganine → sphingosine.

Lit.: [1] *Review:* Angew. Chem., Int. Ed. **38**, 1532–1568 (1999). [2] Angew. Chem., Int. Ed. **25**, 725 (1986); Carbohydr. Res. **174**, 169–179 (1988); Chem. Eur. J. **1**, 382 (1995); J. Chem. Soc., Chem. Commun. **1991**, 820 f.; J. Org. Chem. **53**, 4395–4398 (1988); **59**, 7944 ff. (1994); Justus Liebigs Ann. Chem. **1991**, 253–257; **1995**, 755–764; **1996**, 2079; Kontakte (Merck Darmstadt) **1992**, 11–28; Spec. Publ. R. Soc. Chem. **180**, 93–118 (1996) (synthesis review); Synthesis **1995**, 868; **1998**, 1075–1091 (review); Synlett **1990**, 665 f.; Tetrahedron **47**, 2835–2842 (1991); Tetrahedron: Asymmetry **8**, 3237 (1997); Tetrahedron Lett. **27**, 481–484 (1986); **29**, 3037–3040 (1988); **35**, 9573 ff. (1994); **39**, 3953 (1998).

gen.: Beilstein EIV **4**, 1894 ▪ J. Lipid Res. **35**, 2305–2311 (1994) ▪ Luckner (3.), p. 151 f. ▪ Stryer 1995, p. 267, 689 f. – *[HS 2922 19; CAS 123-78-4 (S.); 2733-29-1 ((+)-S.); 2304-81-6 (a); 7344-02-7 (b); 54135-66-9 (c); 15639-50-6 (e); 554-62-1 (f)]*

Spiders. Family name for the Araneae, with more than 30000 species, belonging to the arachnids (Arachnida), not to the insects. These arthropods, to which the scorpions (see scorpion venoms), *mites (with ticks and red spider mites), and also daddy longlegs belong, have 4 pairs of legs (insects only 3), 2 mandibles (chelicera), and 2 maxillary palpi (pedipalpi) as well as 1–4 pairs of separate eyes. A tube-like poison gland opens into the chelicera, in the males the reproductive organ is situated in the terminal unit of the pedipalpi. The respiratory organs (trachea) are found on the underside of the hindbody. The spinning nipples with the spinning glands are situated at the end of the hindbody. These can produce up to 10 different types of threads. All S. use spinning threads either to construct webs, as safety lines, or to construct cocoons for their eggs. Spinning thread consists of scleroproteins[1] (spider silk), together with a series of biogenic amines[2], amino acids, pyrrolidone[3], aliphatic methyl ethers[4], and inorganic salts such as potassium hydrogen phosphate. Both chemical and physical factors, especially the water content of the web, are important for the tensile strength (comparable to that of the plastic Kevlar®) and elasticity of some spider's webs. The adhesive properties of the trapping threads are due to the presence of glycoproteins[5]. Many species use vibrational stimuli to attract a partner[6], other also employ chemical signals. Thus, (*R*)-3-hydroxybutanoic acid has been identified as a pheromone of baldachin spiders (*Linyphia*)[7]; other species use butterfly pheromones to attract prey (bola spiders)[8]. Apart from their use of spinning threads, the predatory lifestyle is characteristic for spiders. They trap insects or even small animals, paralyse them with their toxins (see spider venoms), and then suck them out (extraintestinal digestion). In contrast to other arachnids (mites, ticks), most S. are harmless or even useful (reduction of pests) for humans; even so they are mostly persecuted irrationally by men.

Lit.: [1] Angew. Chem., Int. Ed. **36**, 314–327 (1997); Chem. Ind. (London) **1995**, 1009–1012; Science **271**, 84 (1996). [2] Nature (London) **345**, 526 (1990); J. Exp. Zool. **259**, 154–165 (1984). [3] Naturwissenschaften **59**, 98 f. (1972). [4] Tetrahedron **49**, 6805–6820 (1993). [5] Mat. Res. Soc. Symp. Proc. **292**, 9–23 (1994). [6] Witt & Rovner, Spider Communication, Princeton: University Press 1982. [7] Science **260**, 1635 ff. (1993). [8] Science **236**, 964–967 (1987).

gen.: Foelix, Die Biologie der Spinnen, Stuttgart: Thieme 1992 ▪ Nentwig, Ecophysiology of Spiders, Berlin: Springer 1987.

Spider venoms. Predatory spiders bite their prey with their poison gland-containing chelicera. The venoms contain peptides and proteins, which may also be toxic to man, as well as low-molecular weight (acyl)polyamines with unusual structures: one to three amino or aromatic carboxylic acids are linked by peptide bonds to a polyamine chain (see also JSTX, argiopinins, argiotoxins, clavamin). These act as neurotoxins, as do some peptides, e.g. robustoxin, consisting of 423 amino acids, from the Australian funnel-web spider *Atrax robustus* or latrotoxin (M_R 125000) from the black widow (*Latrodectus* sp.)[1] or SNX from *Segestria* species (SNX-482 from the African tarantula *Hysterocrates gigas* is being studied as a selective inhibitor of R-type calcium channels). In addition, necrosis results from the necrotoxins of *Loxosceles* species (sphingomyelinase D) living in Africa and America (brown recluse spider). Finally, the venoms may also contain low-molecular weight neurotransmitters such as serotonin, histamine, 5-hydroxytryptamine, etc. The venom of the tunnel-web spider contains D-serine[2]. Relatively few species are harmful to man since the chelicera usually cannot penetrate human skin. Bites with severe consequences are rather caused by small inconspicuous spiders like the black widow. Other dangerous species are *Atrax*, *Loxosceles*, *Phoneutria* (banana spider), and *Harpactirella*. Bites often cause severe pain but are seldom fatal, even with the most poisonous species, when treated correctly. For Joro spider toxins, see *Lit.*[3].

Lit.: [1]Nat. Struct. Biol. **7**, 48 (2000). [2]Science **266**, 1066 (1994). [3]Heterocycles **47**, 171 (1998); Tetrahedron Lett. **39**, 6479 (1998); Yakugaku Zasshi **117**, 700, 715 (1997).
gen.: Angew. Chem., Int. Ed. **36**, 314–327 (1997) (review) ▪ Biochemistry **34**, 8341 (1995) (SNX) ▪ Chem. Lett. **1993**, 929 (nephilatoxin) ▪ Chem. Pharm. Bull. **44**, 972 (1996) (joramine, spidamine) ▪ Curr. Biol. **3**, 481 ff. (1993) ▪ Helv. Chim. Acta **75**, 1909–1924 (1992) ▪ J. Chem. Ecol. **19**, 2411–2451(1993) ▪ J. Toxicol., Toxin Rev. **10**, 131–167 (1991) ▪ Nentwig, Ecophysiology of Spiders, Berlin: Springer 1987 ▪ Nutting, Mammalian Diseases and Arachnides, Boca Raton: CRC Press 1984 ▪ Pharmacol. Ther. **42**, 115 (1989) ▪ Tetrahedron **49**, 11155–11168 (1993); **51**, 10687–10698 (1995) ▪ Tetrahedron Lett. **36**, 9393 (argiotoxin); 9397 (1995) (JSTX-3) ▪ Tu, Handbook of Natural Toxins, vol. 2, New York: Dekker 1984.

Spike-lavender oil. Colorless to yellowish-green oil with a fresh, herby-camphor, lavender-like odor.
Production: By steam distillation of the flowering herbage of spike lavender, *Lavandula latifolia* mainly growing wild on the Iberian peninsula. Main producing country is Spain, 150–200 t/a.
Composition*[1]*: Main components, principally responsible for the odor, are **camphor* (ca. 15%), *1,8-*cineole* (ca. 25%), and **linalool* (ca. 40%).
Use: In the perfume industry, for soap perfumes, and herby-fresh compositions (e. g., masculine notes).
Lit.: [1]Perfum. Flavor. **20** (3), 23 (1995).
gen.: Arctander, p. 590 ▪ Bauer et al. (2.), p. 161 ▪ ISO 4719 (1983). – *Toxicology:* Food Cosmet. Toxicol. **14**, 453 (1976). – *[HS 3301 23; CAS 8016-78-2]*

Spinochromes. Derivatives of hydroxylated 1,4-*naphthoquinones, occurring together with *echinochromes as orange-red pigments in sea urchins. S. are probably the biosynthetic precursors of the echinochromes.
Lit.: Scheuer I **5**, 82 f.

Spinosyns (Lepicidins). Insecticidal tetracyclic macrolide antibiotics complex, produced in cultures of the actinomycete *Saccharopolyspora spinosa*. S. bind to the nicotinic acetylcholine receptor.
S. A: $C_{41}H_{65}NO_{10}$, M_R 731.97, $[\alpha]_D^{20}$ −135.3° (C_2H_5OH).
S. D: $C_{42}H_{67}NO_{10}$, M_R 745.99, $[\alpha]_D$ −156.7° (C_2H_5OH).

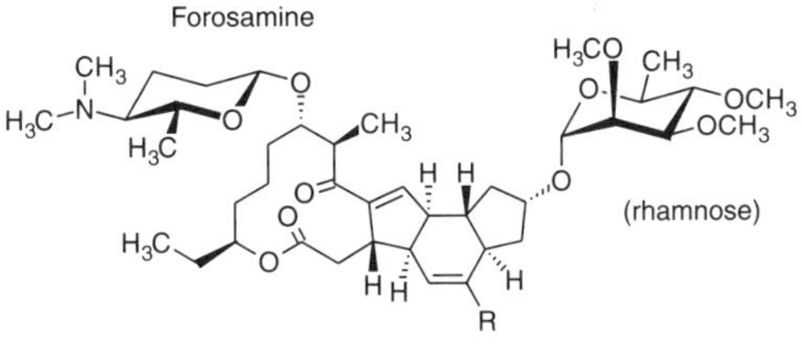

R = H : Spinosyn A
R = CH_3 : Spinosyn D
R = H (85%), CH_3 (15%) : Spinosad (Tracer®)

S. are especially active against moths and butterflies (lepidopterans). Spinosad (Tracer®) is a mixture of S. A and S. D, marketed as a cotton insecticide.

Lit.: ACS Symp. Ser. **504**, 214–225 (1992) ▪ J. Am. Chem. Soc. **114**, 2260 (1992); **115**, 4497–4513 (1993); **120**, 2543–2552, 2553–2562 (1998) (synthesis) ▪ Tetrahedron Lett. **32**, 4839 (1991) (isolation). – *[CAS 131929-60-7 (S. A); 131929-63-0 (S. D)]*

Spinulosin (2,5-dihydroxy-3-methoxy-6-methyl-*p*-benzoquinone).

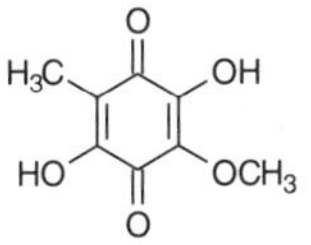

$C_8H_8O_5$, M_R 184.15, purple-red cryst., mp. 203 °C, poorly soluble in water. A tetraketide from *Penicillium spinulosum* and *P. cinerascens* as well as *Aspergillus fumigatus* with antibiotic properties.
Lit.: Chem. Ber. **105**, 614 (1972) (synthesis) ▪ Karrer, No. 1191 ▪ Merck-Index (12.), No. 8902. – *[CAS 85-23-4]*

Spiraeoside see quercetin.

Spiramycin (foromacidin, selectomycin, sequamycin, rovamycin).

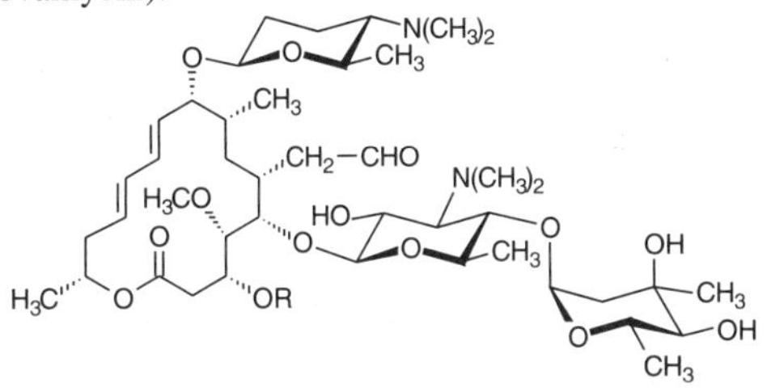

Table: Components of Spiramycin.

	S. I (S. A)	S. II (S. B)	S. III (S. C)
R	H	CO–CH_3	CO–CH_2–CH_3
molecular formula	$C_{43}H_{74}N_2O_{14}$	$C_{45}H_{76}N_2O_{15}$	$C_{46}H_{78}N_2O_{15}$
M_R	843.07	885.10	899.13
mp. [°C]	134–137	130–133	128–131
CAS	24916-50-5	24916-51-6	24916-52-7

International free name for a *macrolide antibiotic produced by *Streptomyces ambofaciens*, it differs from *leucomycin by the 9-*O*-*β*-glycosidically bound sugar forosamine. The other two sugars are D-mycaminose and L-mycarose. S. is a mixture of three to nine components. S. is a colorless, amorphous base, poorly soluble in water, soluble in most organic solvents. It has strong bacteriostatic activity against Gram-positive bacteria and Rickettsia by interference with the ribosomal protein biosynthesis. S. is used in veterinary medicine against gonorrhea, toxoplasmosis, and others.
Lit.: J. Antibiot. **26**, 55 (1974); **49**, 398 (1996) ▪ J. Antimicrob. Chemother. **22**, Suppl. B, 1–213 (1988) ▪ Kleemann-Engel, p. 1739 ▪ Omura (ed.), Macrolide Antibiotics, New York: Academic Press 1984. – *Biosynthesis:* Chem. Pharm. Bull. **27**, 176–182 (1979) ▪ Path. Biol. **37**, 553 (1989). – *[HS 2941 90; CAS 8025-81-8]*

Spiroacetals (spiroketals). A widely distributed class of natural products with characteristic structural features: intramolecular ring closure of a dihydroxy ketone furnishes a bicyclic compound in which one carbon atom (that of the original carbonyl group) is shared by both rings. This so-called spiro center is bound to one carbon atom and one oxygen atom in both rings.

In the S. the ring planes are perpendicular to each other. Many S. have been identified as algal toxins, as antibiotics, and as natural insecticides. S. with unbranched chains of 9 carbon atoms, such as *chalcogran, *conophthorin, and *olean are *pheromones of various insects.

Lit.: ApSimon **8**, 533–690 ▪ Chem. Rev. **89**, 1617–1632 (1989).

Spirobenzylisoquinoline alkaloids.

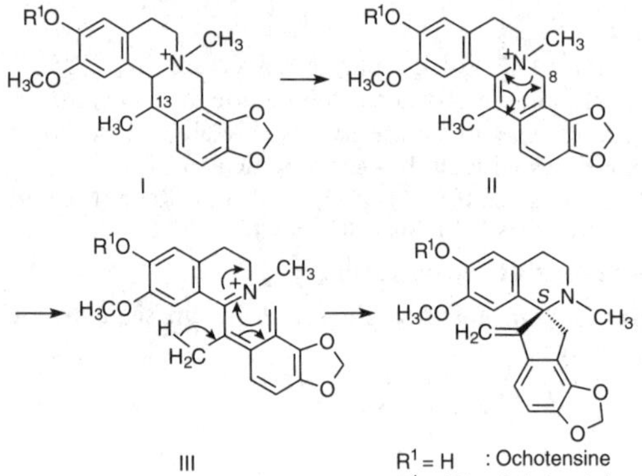

R^1 = H : Ochotensine
R^1 = CH_3 : Ochotensimine

Figure: Biosynthesis of spirobenzylisoquinolines.

The S. a. occur principally in the Fumariaceae family, especially in the genera *Fumaria* and *Corydalis*. Thus, *ochotensine* {$C_{21}H_{21}NO_4$, M_R 351.40, mp. 252 °C, $[\alpha]_D$ +51.7° ($CHCl_3$)} and *ochotensimine* {$C_{22}H_{23}NO_4$, M_R 365.43, $[\alpha]_D$ +46.3° (CH_3OH)}, which has an exocyclic methylene group at C-13, were found in *Corydalis ochotensis* and other *C.* species.

Biosynthesis: *N*-Methyltetrahydroprotoberberinium salts such as I are the precursors of S. a. Dehydrogenation of the tetrahydroberberinium ions to ions II and subsequent electrocyclic cleavage of the N-C-8 bond furnish intermediates III which can cyclize to give the spirobenzylisoquinolines, ochotensimine or ochotensine in the example shown above [1].

Lit.: [1] Mothes et al., p. 226–230.

gen.: Hager (5.) **4**, 1013–1027 ▪ Manske **17**, 385–519; **38**, 157–219 ▪ Shamma, p. 381–398 ▪ Shamma-Moniot, p. 325–335 ▪ Tetrahedron Lett. **38**, 4195 (1997) (synthesis) ▪ Ullmann (5.) **A1**, 376. – *[HS 2939 90; CAS 4829-36-1 (ochotensimine); 4959-88-0 (ochotensine)]*

Spirobrassinin see oxindole alkaloids and brassinin.

Spirographis heme (porphyrin) see chlorocruoroporphyrin.

Spiroketals see spiroacetals.

Spiromacrolides see macrolides.

Spiromentins.

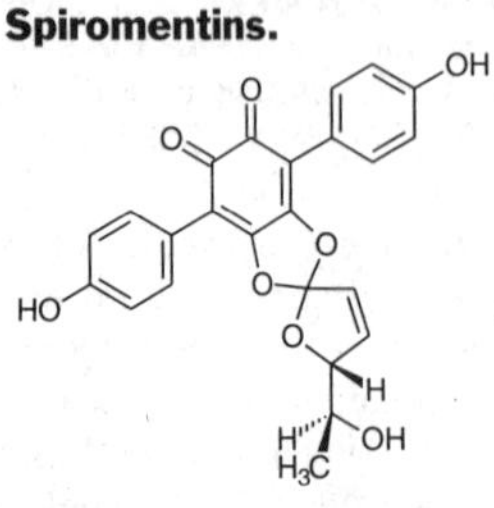

Spiromentin B

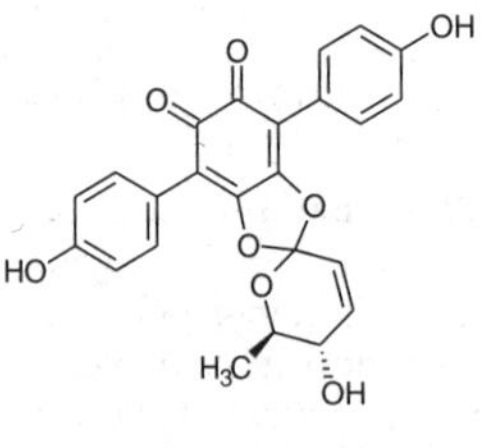

Spiromentin C

Violet pigments (S. A–J) from the toadstool *Paxillus atrotomentosus* (Basidiomycetes) and cultures of *P. panuoides*, e.g., *S. B* ($C_{24}H_{18}O_8$, M_R 434.40) and the isomeric *S. C*. In a biomimetic reaction, the 4,4′-dimethyl ether of (±)-*flavomentin B was cyclized by the action of trifluoroacetic acid to furnish the corresponding derivatives of racemic spiromentins B and C.

Lit.: Justus Liebigs Ann. Chem. **1989**, 803 ▪ Phytochemistry **40**, 1251 (1995) (isolation) ▪ Zechmeister **51**, 21–24 (review). – *[CAS 121254-56-6 (S. B); 121254-57-7 (S. C)]*

Spironolactone see aldosterone

Spirosolanes see Solanum steroid alkaloids.

Spirostans, spirostanols see steroid sapogenins.

Spirotryprostatins. Inhibitors of the cell cycle with potential anticancer activity from fermentation broths of an *Aspergillus fumigatus* strain. The spiroindolinone *S. A* {$C_{22}H_{25}N_3O_4$, M_R 395.44, pale yellow, amorphous powder, $[\alpha]_D^{26}$ –34° ($CHCl_3$)} is biosynthetically derived from tryptophan and inhibits the transitions from the G2- and M-phases to their subsequent phases of the cell cycle.

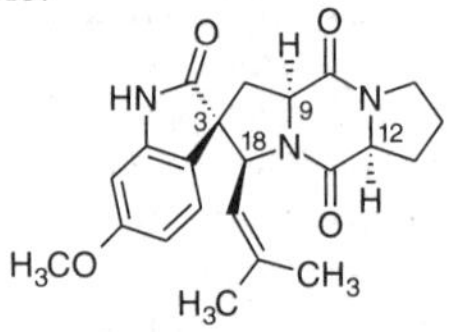

Spirotryprostatin A

Lit.: Angew. Chem. Int. Ed. Engl. **37**, 1138 (1998) (synthesis S. A); **39**, 2175–2178 (2000) (total synthesis S. B) ▪ J. Antibiot. **49**, 832 (1996) ▪ J. Am. Chem. Soc. **121**, 2147 (1999) ▪ Tetrahedron **52**, 12651 (1996) (isolation). – *[CAS 182234-25-9 (S. A)]*

Spirovetivanes see vetispiranes.

Spiroxins. DNA-cleaving antitumor antibiotics from a marine fungus with unusual spirocyclic structure. The S. are 1,1′-binaphthylpiroketal derivatives which, in addition to cytotoxicity, exhibit antibiotic activity and are active in a mouse xenograft model against human ovarian carcinoma. *S. A* [$C_{20}H_9ClO_8$, M_R 412.73, $[\alpha]_D^{25}$ –644° (CH_3OH)] is the major component, produced in culture. Structurally related are the *preussomerins.

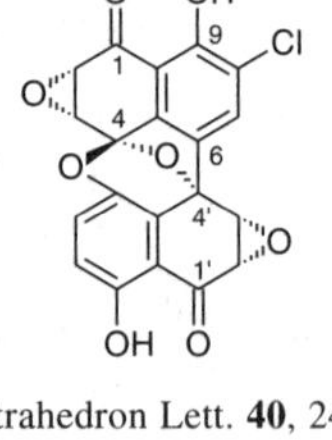

S. A
(relative configuration)

Lit.: Tetrahedron Lett. **40**, 2489 (1999). – *[CAS 225233-83-0 (S. A)]*

Sponges (Porifera). Phylogenetically the simplest phylum of organisms; widely distributed in fresh and sea water; ca. 3000 species have been described to date. The marine S. are characterized by great diversity of secondary metabolites; often halogenated. Examples of natural products described under separate entries and exhibiting in some cases interesting pharmacological properties are: *aaptamine, *aplysi…, *avarol, *bengamides, bengazoles (see bengamides), *calyculins, *calysterol, *dendrolasin, *diisocyanoadociane, *discodermin A, *discorhabdins, *dysidazirine, *dysidiolide, *girolline, *halistatins, *hennoxazoles, *jaspamide, konbamide, latrunculins, *manoalide, *manzamines, *mimosamycin, *mycalamides, nakijiquinones (see ilimaquinone), *okadaic acid, onnamide A (see mycalamides), *palauamine, *papuamine, *penaresidins, *ptilomycalin, *purpurone, *renierol, *swinholides, *theonellamides, *topsentins, xestospongin.

Lit.: Angew. Chem., Int. Ed. **37**, 2162 (1998) (review) ▪ Sarma, Daum, & Müller, Secondary Metabolites from Marine Sponges, Berlin: Ullstein Mosby 1993.

Spongianes.

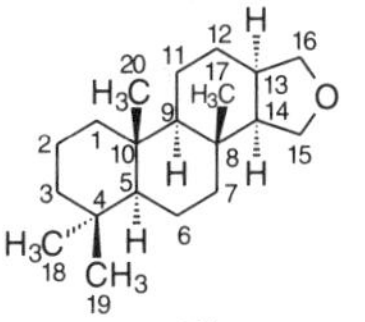

(**1**)

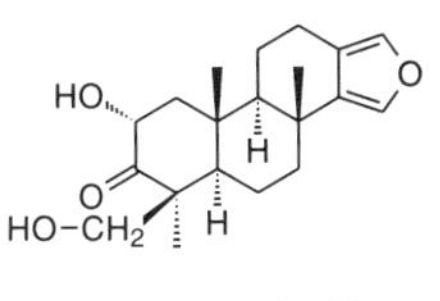

Isospongiadiol (**2**)

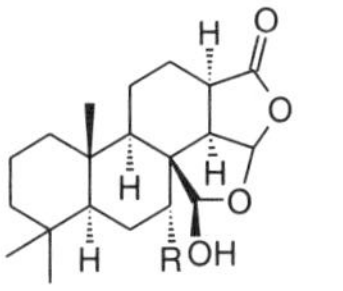

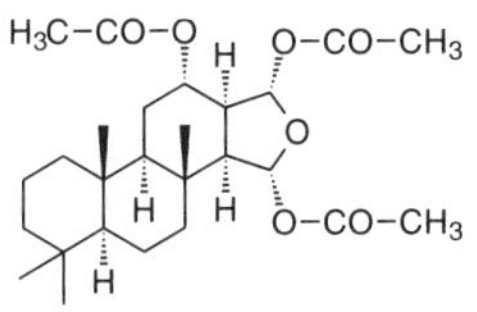

R = O–CO–CH₃: Aplyroseol 2 (**3**)
R = H : Dendrillol 1 (**4**)

Aplysillin (**5**)

Table: Data of Spongianes.

no.	molecular formula	M_R	mp. [°C]	$[\alpha]_D$	CAS
2	$C_{20}H_{28}O_4$	332.44	181–183	–50° (CH_2Cl_2)	111139-69-6
3	$C_{22}H_{32}O_6$	392.49	114–117	–35° ($CHCl_3$)	96999-35-8
4	$C_{20}H_{30}O_4$	334.46	229–231	–31.8° (CH_2Cl_2)	106009-81-8
5	$C_{26}H_{40}O_7$	464.60	169–171	+13° ($CHCl_3$)	71393-11-8

A class of *diterpenes with the spongiane skeleton **1**. S. are isolated from various sponges such as, e.g., *Spongia*, *Aplysilla*, *Dendrilla*, *Hyatella*, and *Igernella* and marine slugs (*Ceratosoma*, *Archidoris*) that feed on the sponges. Isospongiadiol (**2**) has antileukemic and antiviral activity against Herpes simplex viruses.

Lit.: Aust. J. Chem. **39**, 1629, 1643 (1986) ▪ Chem. Lett. **1987**, 1687 ▪ J. Chem. Soc., Perkin Trans. 1 **1998**, 863 (synthesis (+)-spongian-16-one) ▪ J. Nat. Prod. **51**, 293 (1988); **53**, 102 (1990) ▪ J. Org. Chem. **50**, 2862 (1985); **52**, 3766 (1987) ▪ Magn. Reson. Chem. **36**, 325 (1998) (stereochemistry).

Spongistatins (spongipyrans, cinachyrolides, altohyrtins). Class of macrocyclic lactone antibiotics (S. 1–9) from marine sponges *Spongia* sp., *Hyrtios altum*, *Cinachyra* sp., *Spirastrella spinispirulifera*. The S. show antibacterial, antifungal and very potent antineoplastic activities; e.g., *S. 1* (altohyrtin A): $C_{63}H_{95}ClO_{21}$, M_R 1223.88, amorphous powder, mp. 161–162 °C (Na-salt), $[\alpha]_D^{22}$ +26.2° (CH_3OH, Na-salt), IC_{50} (KB and L 1210 tumor cells) 1×10^{-11} g/mL.

S. 2 (altohyrtin C): $C_{63}H_{96}O_{21}$, M_R 1189.44, amorph. solid, mp. 140–141 °C, $[\alpha]_D$ +24.5° (CH_3OH), GI_{50} (60 different tumor cell lines) 8.5×10^{-10} M, IC_{50} (KB cells) 4×10^{-10} g/mL.

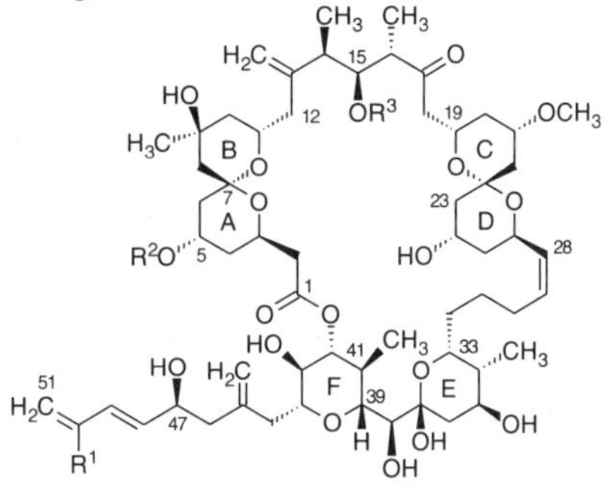

R^1	R^2	R^3	
Cl	CO–CH_3	CO–CH_3	: Altohyrtin A, Spongistatin 1
Br	CO–CH_3	CO–CH_3	: Altohyrtin B
H	CO–CH_3	CO–CH_3	: Altohyrtin C, Spongistatin 2
Cl	H	CO–CH_3	: 5-Deacetylaltohyrtin A, Spongistatin 3
Cl	CO–CH_3	H	: Cinachyrolide A, Spongistatin 4
H	CO–CH_3	H	: Spongistatin 6

The cytotoxic activity is therefore almost as potent as that of the bryostatins.

Lit.: Angew. Chem. Int. Ed. Engl. **36**, 2738, 2741, 2744 (1997); **37**, 187 ff., 190–196 (1998) (synthesis); **37**, 2629 (1998) (review) ▪ Chem. Commun. **1993**, 1166, 1805; **1994**, 1605 (isolation) ▪ Chem. Pharm. Bull. **41**, 989 (1993); **44**, 2142 (1996) (isolation) ▪ J. Org. Chem. **58**, 1302 ((1993); **64**, 8267 (1999) ▪ Nachr. Chem. Tech. Lab. **46**, 170 (1998) (review) ▪ Nat. Prod. Lett. **3**, 239 (1993) (isolation) ▪ Tetrahedron Lett. **39**, 8545 (1998) (synthesis) ▪ Tetrahedron **55**, 8671–8726 (1999) (synthesis S 2). – *[CAS 149715-96-8 (S. 1); 150624-44-5 (S. 2)]*

Sporidesmins.

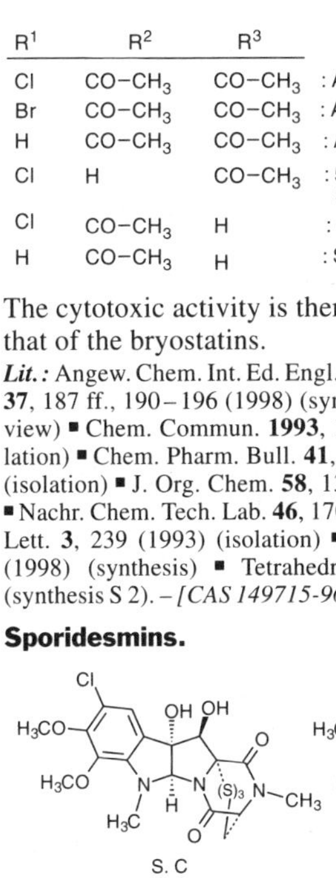

S. C

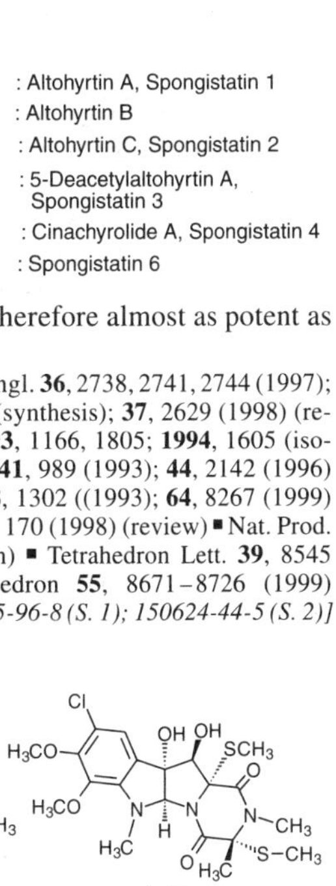

S. D

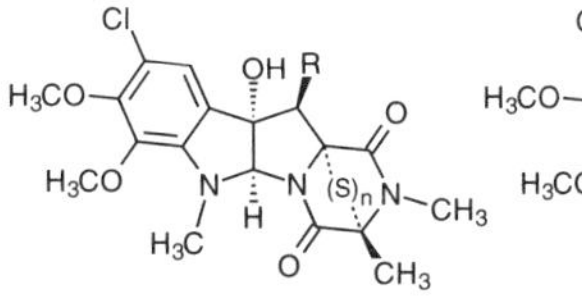

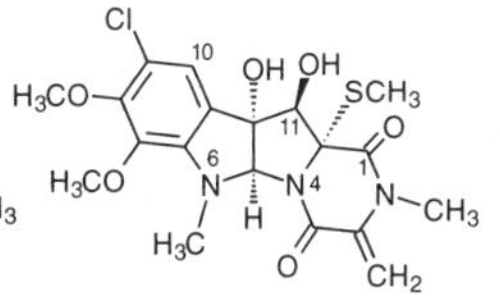

S. A : n = 2, R = OH   S. E : n = 3, R = OH
S. B : n = 2, R = H   S. G : n = 4, R = OH

S. F

Table: Data of Sporidesmins.

Sporidesmin	molecular formula	M_R	mp. [°C]	CAS
A	$C_{18}H_{20}ClN_3O_6S_2$	473.95	179	1456-55-9
B	$C_{18}H_{20}ClN_3O_5S_2$	457.95	183	3351-96-0
C	$C_{18}H_{20}ClN_3O_6S_3$	506.01		7051-84-5
D	$C_{20}H_{26}ClN_3O_6S_2$	504.02	105–107	18180-71-7
E	$C_{18}H_{20}ClN_3O_6S_3$	506.01	180–187	22327-77-1
F	$C_{19}H_{22}ClN_3O_6S$	455.91	65–67	23044-40-8
G	$C_{18}H_{20}ClN_3O_6S_4$	538.07	148–153	32608-68-7

A group of *mycotoxins containing an epipolythiodioxopiperazine part (see gliotoxin) and a pyrrolo[2,3-b]indole part. The S. A–J are formed by *Pithomyces chartarum*, a deuteromycete growing on dead leaves and pasture grasses. S. are responsible for the mycotoxicosis ("facial eczema") occurring in New Zealand sheep that are particularly sensitive to S. The symptoms are edematous swelling of mucus membranes, sensitivity to light, and liver damage. On oral uptake of 1 mg S. A/kg 80% of the animals die within 24 d. S. D and F do not contain an S–S bridge and are almost non-toxic. S. C also occurs as the diacetate. The other S. are highly cytotoxic, concentrations from 0.1 ng/mL have effects on HeLa cells. The biosynthesis proceeds from tryptophan and alanine, the methyl groups originate from methionine.

Lit.: Betina, chap. 17 ▪ Cole-Cox, p. 584–598 ▪ J. Agric. Food Chem. **40**, 701, 2458 (1992) ▪ J. Chem. Soc., Chem. Commun. **1995**, 889 ▪ Manske **26**, 70–85 ▪ Turner **1**, 327–332; **2**, 417-421.

Sporopollenins. *Polyterpenes occurring in pollen, possibly formed from *carotinoids and their esters by oxidative polymerization. Recently the incorporation of L-phenylalanine was demonstrated[1]. S. are extremely resistant to chemical influences (even strong acids) and can only be degraded by oxidation; they are responsible for the high weathering resistance of plant pollen – (even very dry pollen still shows metabolic activity) – and their survival in fossil sediments. Pollen analysis of fossil materials provides information about the vegetative conditions of previous eras (*palynology*=the study of pollen and spores, from Greek: palynein=to spread flour).

Lit.: [1] Bot. Acta **105**, 407 (1992).
gen.: Brooks (eds.), Sporopollenin, Proc. Symp., p. 351–407, London: Academic Press 1971 ▪ Int. Rev. Cytol. **140**, 35–72 (1992) ▪ Nature (London) **220**, 605 (1968) ▪ Wiad. Bot. **24** (4), 249–267 (1980). – *[CAS 12712-72-0]*

Sporotrichiol see tric(h)othecenes.

Spruceanol (5β,10α-cleistanthatetra-8,11,13,15-ene-3α,12-diol).

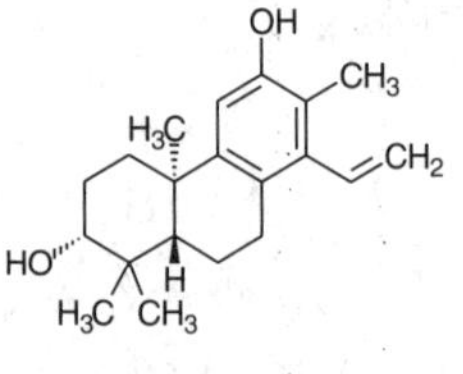

$C_{20}H_{28}O_2$, M_R 300.44, pale-yellow gum, $[\alpha]_D^{25}$ –3.0° ($CHCl_3$). A *diterpene of the cleistanthane group, isolated from the roots of *Cunuria spruceana* (Euphorbiaceae). S. exhibits antitumor properties. The cleistanthanes are formed biosynthetically from *pimaranes or isopimaranes by migration of an ethyl group.

Lit.: J. Nat. Prod. **35**, 658 (1972); **42**, 658 (1979); **44**, 658 (1981). – *[CAS 72963-56-5]*

Squalamine. The broad spectrum antibiotic S. ($C_{34}H_{65}N_3O_5S$, M_R 627.97) from the dogfish shark *Squalus acanthias* contains a steroid skeleton linked to *spermidine. The compound is active against *Candida albicans*, Gram-negative and Gram-positive bacteria, etc. and induces osmotic lysis in protozoa. S. is being studied synthetically as a new antibacterial lead

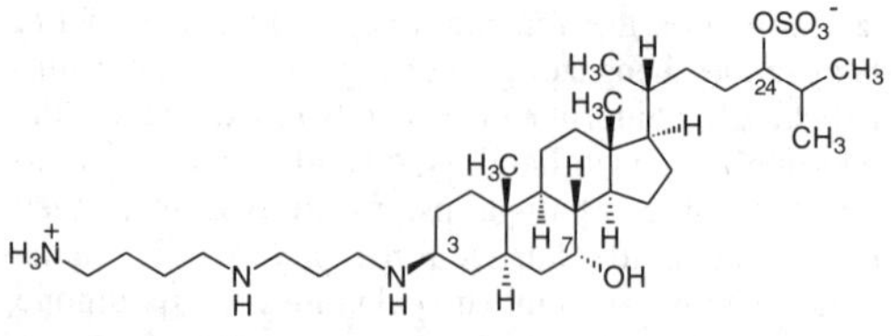

Squalamine

structure[1]. S. is in phase I clinical development as an angiogenesis inhibitor in cancer. Two derivatives are undergoing clinical development as drugs for treatment of brain tumors and AIDS. So far, more than 20 natural S. derivatives have been isolated.

Lit.: [1] Steroids **61**, 565–571 (1996).
gen.: Chem. Ind. (London) **1993**, 136 ▪ J. Am. Chem. Soc. **118**, 8975 (1996) ▪ Proc. Natl. Acad. Sci. USA **90**, 1354-1358 (1993) ▪ Science **259**, 1125 (1993). – *Synthesis:* J. Org. Chem. **60**, 5121–5126 (1995); **63**, 8599 (1998) ▪ Tetrahedron Lett. **35**, 8103 (1994); **36**, 5139 ff. (1995). – *[HS 2941 90; CAS 148717-90-2]*

Squalene [(*all-E*)-2,6,10,15,19,23-hexamethyl-2.6, 10,14,18,22-tetracosahexaene]. The most important aliphatic, acyclic *triterpene, $C_{30}H_{50}$, M_R 410.73, mp. –4.8 to –5.2 °C, bp. 284–285 °C, formula, see steroids. S. was first isolated from fish liver oils and later detected in plant oils and human fat. It is composed of 6 isoprene units and is formed from activated acetate (*acetyl-CoA) via mevalonic acid. It is an intermediate in the biosynthesis of all cyclic triterpenoids and thus also of the *steroids. Its enzymatic cyclization to *lanosterol or *cycloartenol requires molecular oxygen and proceeds through (3*S*)-squalene 2,3-epoxide.

Lit.: Annu. Rev. Biochem. **14**, 555–585 (1982) ▪ Chem. Soc. Rev. **20**, 129–147 (1991) ▪ Karrer, No. 34 ▪ Nat. Prod. Rep. **2**, 525–560 (1985) ▪ Phytochemistry **27**, 628 (1988) (biosynthesis) ▪ Stryer 1995, 692–695. – *[HS 2901 29; CAS 111-02-4]*

(3*S*)-Squalene 2,3-epoxide see steroids.

Squalestatins see zaragozic acids.

Squamocin see Annonaceous acetogenins.

SRS-A. Abbreviation for slow reacting substance of anaphylaxis, see leukotrienes.

Stachyose (Gal*p*α1-6Gal*p*α1-6Glc*p*α1-2βFru*f*). $C_{24}H_{42}O_{21}$, M_R 666.58. A tetrasaccharide, weakly sweet tasting crystalline plates, mp. 167–170 °C, $[\alpha]_D^{20}$ +148° (H_2O), does not reduce Fehling's solution and

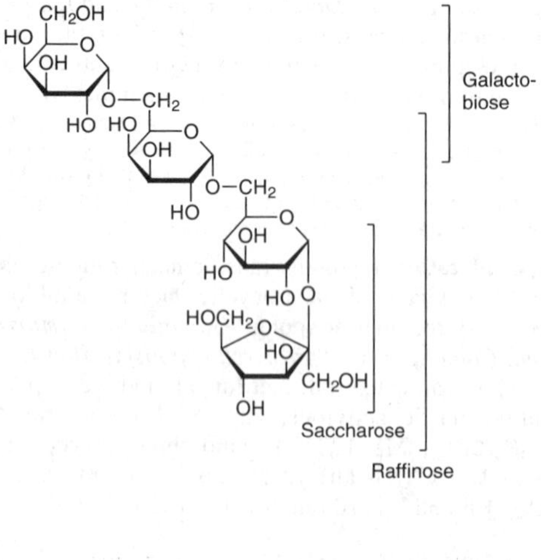

furnishes 2 mol D-*galactose, 1 mol D-*glucose, and 1 mol D-*fructose on hydrolysis. S. is abundant in the root tubers of *Stachys affinis* (*S. tuberifera*, woundwort, Japanese artichoke) and other Lamiaceae, as well as in Fabaceae, e. g., soya beans, etc.
Lit.: Acta Crystallogr., Sect. C **43**, 806 (1987) ▪ Beilstein E V **17/8**, 405 ▪ Carbohydr. Res. **210**, 89 (1991) ▪ Karrer, No. 672. – *[HS 2940 00; CAS 470-55-3]*

Stallimycin see distamycins.

Stanolone see androgens.

St. Anthony's fire see ergot alkaloids.

Staphylococcin see epidermin.

Staphylomycin see virginiamycin.

Star anise oil. Colorless to yellowish oil; freezing point 15 – 18 °C. Intense sweetish, typically somewhat herby odor and taste.
Production: By steam distillation of the fresh fruits of the star anise, *Illicium verum* (Illiciaceae).
Composition[1]***:*** The main component, also responsible for the typical organoleptic impression, is *trans-*anethole* (up to 95%).
Use: In small amounts in the perfume industry for widely differing compositions, to improve the aromatic character of oral hygiene products and foods such as confectionery, alcoholic beverages, etc., and for the production of natural, pure anethole by rectification.
Lit.: [1] Perfum. Flavor. **2** (3), 56 (1977); **9** (2), 26 (1984); **10** (6), 38 (1985); **11** (6), 43 (1986); **17** (2), 49 (1992).
gen.: Arctander, p. 594 ▪ Bauer et al. (2.), p. 176. – *Toxicology:* Food Cosmet. Toxicol. **13**, 715 (1975). – *[HS 3301 29; CAS 68952-43-2]*

Starch. S. is composed of two components with different M_R: ca. 20 – 30% straight-chain amylose (screw-like twisted chain with 300 – 1200 glucose molecules, M_R ca. 50 000 – 150 000) and 70 – 80% branched-chain *amylopectin [branching after ca. 25 glucose building blocks by 1 – 6 bonds to give a branch-like structure (see figure), 1500 – 12 000 molecules of glucose, M_R ca. 300 000 – 2 000 000], together with small amounts of lipids, phosphoric acids, and cations. The so-called *wax starch* (from special varieties of corn, rice, or millet) consists almost exclusively of amylopectin, while the S. from other sorts of corn (amylocorn) and from pea consists mainly of amylose. According to Staudinger the starch molecules are longitudinally stretched and are shaped between the fiber-like cellulose molecules and the spherically-shaped glycogen molecules.

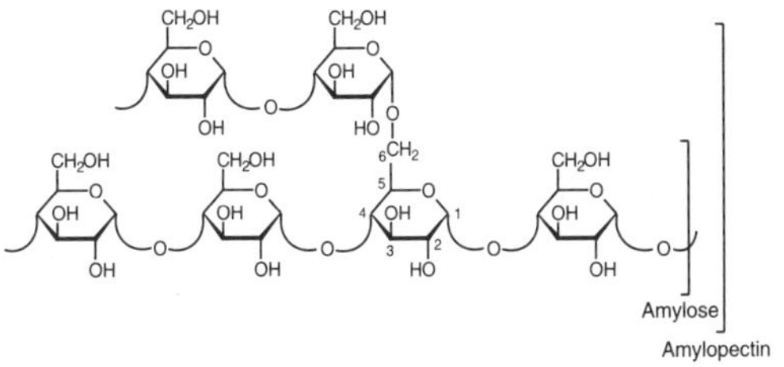

Figure: Structure of the amylopectin part of starch.

S. is a *storage carbohydrate* that many plants accumulate, often in appreciable amounts in the form of starch grains of 5 – 200 µm size. The size and shape of the starch grains is characteristic of their origin, and can be determined by microscopy. S. is formed in the green parts of plants (chloroplasts) by enzymatic synthesis from D-glucose formed in photosynthesis.
For the production, use, and significance of S. for nutritional purposes and as a regenerable raw material, see *Lit.*[1], for structure of S. (review), see *Lit.*[2].
Lit.: [1] Whistler et al., Starch, New York: Academic Press 1984. [2] Starch/Stärke **43**, No. 10, 375 – 384 (1991).
gen.: Aspinall, The Polysaccharides, vol. 3, p. 210 – 282, New York: Academic Press 1985 ▪ Dtsch. Apoth. Ztg. **137**, 770 – 778 (1997) ▪ Galliard, Starch: Properties and Potential, New York: Wiley 1987 ▪ Hager (5.) **9**, 654 ▪ Kirk-Othmer (4.) **22**, 699 – 719 ▪ Tegge, Stärke u. Stärkederivate, Hamburg: Behr 1984 ▪ Ullmann (5.) **A 4**, 102; **A25**, 1 – 58 (1994) ▪ Würzburg, Properties and Uses for Modified Starches, Boca Raton: CRC Press 1986. – *[HS 1108 1; CAS 9005-25-8]*

Statine [(3*S*,4*S*)-4-Amino-3-hydroxy-6-methylheptanoic acid, AHMHA].

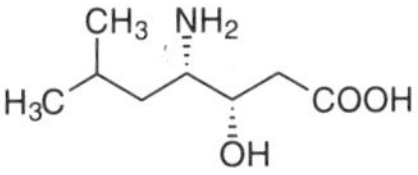

$C_8H_{17}NO_3$, M_R 175.23, mp. 201 – 203 °C (decomp.), $[\alpha]_D^{15}$ –20° (H_2O). An amino acid in *pepstatin A, a pentapeptide mixture and inhibitor of the protease pepsin produced by various *Streptomyces* species. S. and its derivatives are of major importance in the synthesis of pseudopeptidic protease inhibitors as therapeutics for hypertension (as renin inhibitors) and AIDS (as HIV protease inhibitors).
Lit.: Angew. Chem., Int. Ed. Engl. **34**, 1219 (1995) ▪ J. Org. Chem. **55**, 3947 – 3950 (1990); **58**, 1997 (1993); **61**, 5210 (1996); **62**, 2292 (1997) ▪ Merck-Index (12.), No. 8956 ▪ Tetrahedron **53**, 5593 (1997) ▪ Tetrahedron: Asymmetry **10**, 4633 (1999) ▪ Tetrahedron Lett. **31**, 7329, 7359 (1990); **32**, 401 (1991). – *[HS 2922 49; CAS 49642-07-1]*

Staurosporine.

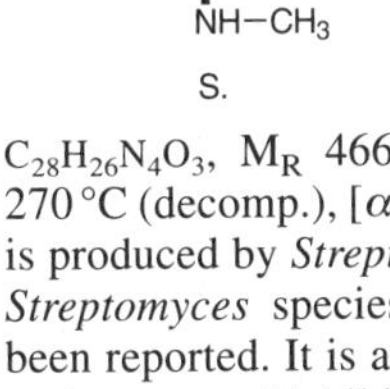

S.

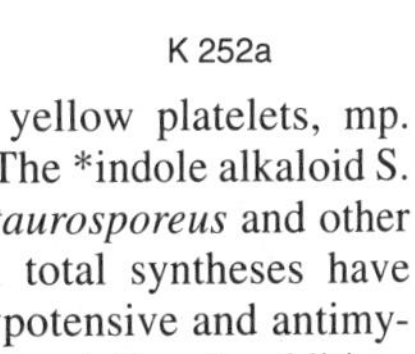

K 252a

$C_{28}H_{26}N_4O_3$, M_R 466.54, pale yellow platelets, mp. 270 °C (decomp.), $[\alpha]_D^{20}$ +35°. The *indole alkaloid S. is produced by *Streptomyces staurosporeus* and other *Streptomyces* species. Several total syntheses have been reported. It is a potent hypotensive and antimycotic agent and inhibits blood coagulation. In addition, S. is a strong but unspecific inhibitor of protein kinases (PKC) and, as a result of this property, has antitumor activity, inhibits contraction of smooth musculature, blocks the cell cycle, and possesses neurothrophic activity. Numerous derivatives of S. have been tested for cancer therapy. Another indolocarbazole alkaloid with similar biological activity is the furanosylated congener K 252a {$C_{27}H_{21}N_3O_5$, M_R 467.48, pale-yellow cryst., mp. 262–273 °C, $[\alpha]_D^{20}$ −23° ($CHCl_3$); +52° (CH_3OH)}[1].

Lit.: [1]J. Antibiot. **39**, 1066, 1072 (1986); J. Am. Chem. Soc. **119**, 9641 (1997); **121**, 6501 (1999); Synthesis **1999**, 1529.
gen.: FEBS Lett. **293**, 169 ff. (1991) ▪ Heterocycles **21**, 309 (1984) ▪ Int. J. Cancer **43**, 851–856 (1989) ▪ J. Am. Chem. Soc. **117**, 552 f. (1995); **118**, 2825–2842, 10656 ff. (1996); **119**, 9652 (1997) (synthesis) ▪ J. Antibiot. **45**, 195, 1428 (1992); **48**, 535–548 (1995) (isolation); **49**, 380–385, 519–526 (1996) ▪ J. Biol. Chem. **263**, 6215–6219 (1988); **266**, 15771–15781 (1991) ▪ J. Nat. Prod. **51**, 884–899 (1988); **59**, 828 (1996); **60**, 230, 788 (1997) (biosynthesis) ▪ J. Org. Chem. **52**, 1177–1189 (1987); **57**, 6327 (1992) ▪ J. Pharmacol. Exp. Therap. **255**, 1218 ff. (1990) ▪ Mol. Pharmacol. **37**, 482–488 (1990) ▪ Synthesis **1995**, 1511–1516; **1999**, 275 (synthesis) ▪ Tetrahedron **47**, 3565 (1991) ▪ Tetrahedron Lett. **35**, 1251 (1994) (abs. configuration). – *[CAS 62996-74-1]*

15-Stearoyloxyguaia-1,3,5,9,11-pentaene see Lactarius pigments.

***O*-Stearoylvelutinal.**

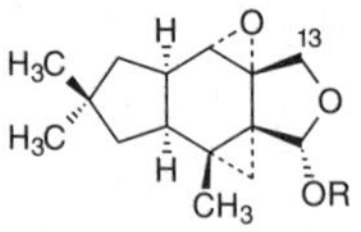

R = −CO−$(CH_2)_{16}$−CH_3 : *O*-Stearoylvelutinal
R = −CO−$(CH_2)_4$−CO−$(CH_2)_{11}$−CH_3 : *O*-(6-Oxostearoyl)velutinal
R = H : Velutinal

$C_{33}H_{56}O_4$, M_R 516.81, amorphous, $[\alpha]_D$ +55° (hexane). Various milkcaps (*Lactarius*, Basidiomycetes) and russulas (*Russula*) with burning tastes possess a chemical defence system against microbes and other food-seeking species[1] based on the enzymatic transformation of fatty acid esters of velutinal[2]. When the fruit body is injured, the biologically inactive *O*-S. contained in the milk juice or tissue of most species (e. g., *L. vellereus*) is converted within seconds to hot-tasting, unsaturated dialdehydes such as *isovelleral, *velleral, *piperdial, and *epi*-piperdial[1]. In addition, seco-*lactaranes such as lactaral and furanoid sesquiterpenes may also be formed. These reaction cascades are initiated by opening of the epoxide ring and hydrolysis of the ester bond[3]. The antimicrobially active dialdehydes are then transformed within minutes to the biologically harmless 13-hydroxyaldehydes (isovellerol, vellerol, piperalol, *epi*-piperalol) via reduction of the 13-aldehyde groups[1]. Remarkably, *L. chrysorrheus* and *L. scrobiculatus* possess a different enzyme system by which *O-(6-oxostearoyl) velutinal* {$C_{33}H_{54}O_5$, M_R 530.76, oil, $[\alpha]_D$ +54.8°} or, respectively, *O*-S. are converted to chrysorrhedial, *chrysorrhealactone, and other compounds[4]. The high reactivity of the velutinal esters explains the isolation of many of the sesquiterpenoid artifacts reported in the literature as being isolated from *Lactarius* and *Russula* species[5]. Depending on the storage, extraction, and work-up conditions of the fungi many different lactaranes and secolactaranes can be obtained, most of which contain lactone and furan rings. For the isolation of native *O*-S. the fresh toadstools must be extracted with hexane at low temperature[1,2]; e. g. in inert atmosphere or addition of dry ice. *O*-S. and its transformation products also occur in the genera *Artomyces* (*Clavicorona*), *Auriscalpium*, *Bondarzewia*, and *Lentinellus*, it is thus a chemotaxonomic marker for the assignment of these basidiomycetes to the order Russulales.

Lit.: [1]J. Chem. Ecol. **10**, 1439 (1983); J. Nat. Prod. **48**, 279 (1985); Tetrahedron Lett. **26**, 3163 (1985). [2]C. R. Acad. Sci. Ser. C. **294**, 1067 (1982); Tetrahedron Lett. **23**, 1907, 4623 (1982); **24**, 1415, 4631 (1983); J. Org. Chem. **50**, 950 (1985); Acta Chem. Scand., Ser. B **42**, 43 (1988). [3]Tetrahedron Lett. **32**, 2541 (1991). [4]Tetrahedron **49**, 1489 (1993). [5]Tetrahedron **37**, 2199 (1981).
gen.: J. Org. Chem. **57**, 5979–5989 (1992) (synthesis) ▪ Turner **2**, 247–252 (review). – *[CAS 86562-14-3 (O-S.); 86533-37-1 ((6-oxostearoyl)velutinal); 83481-29-2 (velutinal)]*

Steffimycin (steffisburgensimycin, NSC 93419, U 20661).

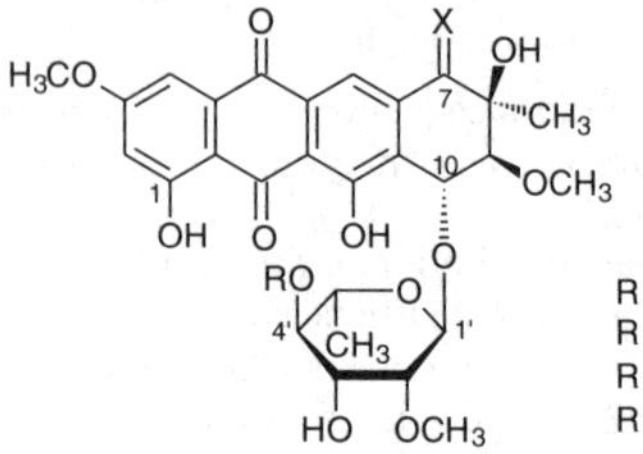

R = H, X = O : S. A
R = CH_3, X = O : S. B
R = CH_3, X = H_2 : S. C
R = H, X = H_2 : S. D

S. (S. A.) $C_{28}H_{30}O_{13}$, M_R 574.54, mp. 257–265 °C, $[\alpha]_D^{25}$ +85° (CH_3OH), orange-yellow crystals. An anthracycline antibiotic from the actinomycetes *Streptomyces steffisburgensis*, *S. elgreteus*, and *Actinoplanes utahensis* exhibiting antibacterial and carcinostatic activities.
S. B (4′-*O*-methylsteffimycin, U 40615): $C_{29}H_{32}O_{13}$, M_R 588.56, mp. 240–246 °C, $[\alpha]_D^{25}$ +94° ($CHCl_3$), orange crystals; from *S. elgreteus*.
S. C (7-deoxo-4′-*O*-methylsteffimycin): $C_{29}H_{34}O_{12}$, M_R 574.58, orange cryst.; from *S. elgreteus*. Active against *Streptococcus pneumoniae*.
S. D (7-deoxosteffimycin): $C_{28}H_{32}O_{12}$, M_R 560.55, orange cryst.; from *S.* sp.; a collagenase inhibitor.

Lit.: J. Antibiot. **38**, 849 (1985); **43**, 1489 (1990); **50**, 496 (1997) ▪ J. Org. Chem. **43**, 3457 (1978); **49**, 3766 (1984). – *[HS 2941 90; CAS 11033-34-4 (S. A); 54526-94-2 (S. B); 98813-22-0 (S. C); 132354-06-4 (S. D)]*

(−)-Steganacin see lignans.

Stegobinone [(2*S*,3*R*,1′*R*)-2,3-dihydro-2,3,5-trimethyl-6-(1-methyl-2-oxobutyl)-4*H*-pyran-4-one].

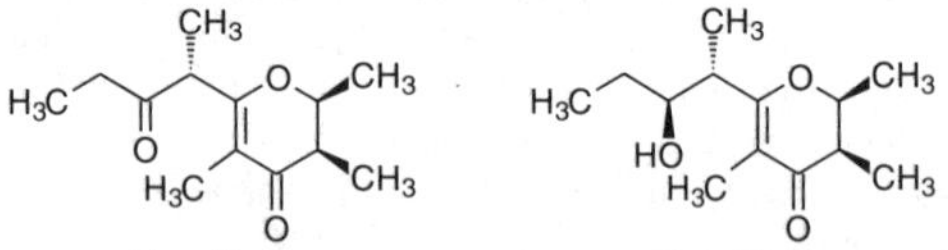

Stegobinone Stegobiol

$C_{13}H_{20}O_3$, M_R 224.30, $[\alpha]_D^{23}$ −282° ($CHCl_3$), liquid. The sexual pheromone of the female drugstore beetle *Stegobium paniceum*[1] and the deathwatch beetle *Anobium punctatum*[2]. The side-chain reduced product (*stegobiol*, $C_{13}H_{22}O_3$, M_R 226.3) has the (2*S*,3*R*,1′*S*,2′*S*)-configuration[3]. Stegobinone and stegobiol are structurally related in the same way as serricorone and *serricorole. For synthesis, see *Lit.*[4].

Lit.: [1]Tetrahedron **34**, 1769–1774 (1978). [2]J. Chem. Ecol. **13**, 1695–1706 (1987). [3]J. Chem. Ecol. **13**, 1871–1879 (1987). [4]Eur. J. Org. Chem. **6**, 1135 (1998); J. Am. Chem. Soc. **118**, 4560 (1996); J. Org. Chem. **58**, 6545 (1993); Synthesis **1998**, 386 ((−)-stegobiol).

gen.: Synthesis: ApSimon **9**, 374–379. – *[CAS 69769-68-2 (S.); 106022-40-6 (stegobiol)]*

Stemofoline see Stemona alkaloids.

Stemona alkaloids. Root extracts of the medicinal plants *Stemona japonica* and *S. tuberosa* (Stemonaceae) have been used in Chinese and Japanese folk medicine as insecticides, anthelmintics, and as drugs for the treatment of tuberculosis, bronchitis and pertussis. Responsible for the activities is a group of more than 20 polycyclic alkaloids with diverse structures, e.g. *protostemonine*: $C_{23}H_{31}NO_6$, M_R 417.50, mp. 172–173 °C, $[\alpha]_D^{20}$ +148° (CH_3OH) and *stemofoline*.

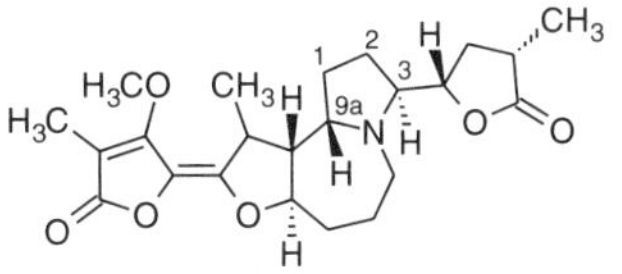

Protostemonine

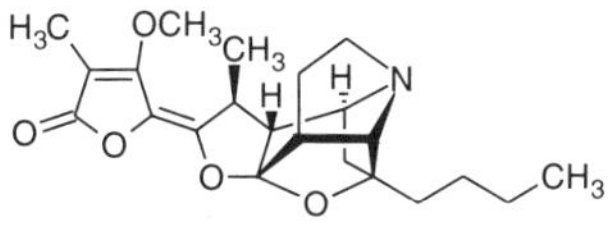

Stemofoline

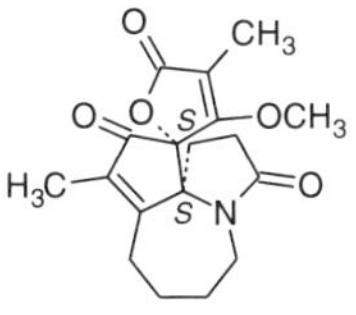

Stemonamide

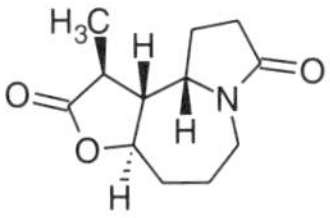

Stemoamide

Other major alkaloids are *stemonamide*[1]: $C_{18}H_{21}NO_5$, M_R 331.37, mp. 183–184 °C, $[\alpha]_D$ –120° (C_2H_5OH), an amide, derived from stemonamine by oxidation of the pyrrolidine ring; and *stemoamide*[2], $C_{12}H_{17}NO_3$, M_R 223.27, mp. 190–191 °C, $[\alpha]_D^{22}$ –28° (CH_3OH, natural), $[\alpha]_D^{26}$ –181° (CH_3OH, synthetic).

Lit.: [1] J. Nat. Prod. **57**, 665 (1994). [2] *Synthesis:* Bull. Chem. Soc. Jpn. **69**, 2063 (1996); J. Am. Chem. Soc. **119**, 3409 (1997); J. Org. Chem. **61**, 8356 (1996); Heterocycles **46**, 287 (1997). *Isolation:* J. Nat. Prod. **55**, 571 (1992).
gen.: Chem. Pharm. Bull. **21**, 451 (1973) ▪ J. Am. Chem. Soc. **117**, 11106 (1995); **121**, 7431 (1999) (synthesis) ▪ J. Chinese Pharm. Sci. **8**, 1 (1999) ▪ Phytochemistry **37**, 1205 (1994). – *[CAS 27495-40-5 (protostemonine); 29881-57-0 (stemofoline); 156953-97-8 (stemonamide)]*

Stentorin.

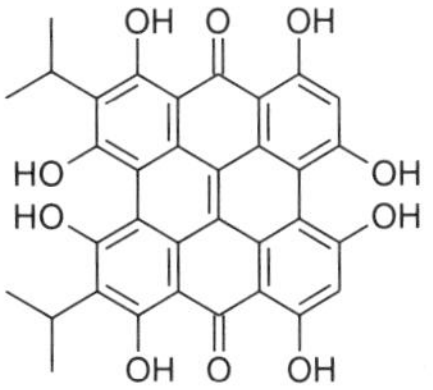

$C_{34}H_{24}O_{10}$, M_R 592.56, dark solid, mp. >350 °C. Pigment related to *hypericin from the unicellular ciliate *Stentor coeruleus*[1]. The compound acts as a photoreceptor and induces photophobic and negative phototactic responses[2]. S. was already described by Lankester in 1873, however, it took more than 100 years until its structure was finally established by synthesis[3]. For similar photoreceptor molecules from ciliates see blepharismins.

Lit.: [1] J. Am. Chem. Soc. **115**, 2526 (1993); Monatsh. Chem. **125**, 955 (1994) (tautomerism and stereochemistry). [2] Photochem. Photobiol. **32**, 781 (1980); Arch. Microbiol. **126**, 181 (1980). [3] Monatsh. Chem. **126**, 1311 (1995); Aust. J. Chem. **50**, 409 (1997).
gen.: Thomson **4**, 569–571. – *Review:* Angew. Chem. Int. Ed. Engl. **38**, 3117 (1999).

Stenusine [1-ethyl-3-(2-methylbutyl)piperidine].

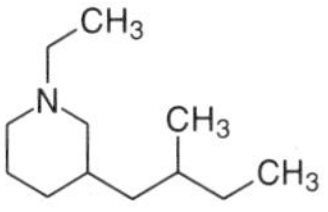

$C_{12}H_{25}N$, M_R 183.34, $[\alpha]_D^{20}$ +5.8° (C_2H_5OH). Together with 6-methyl-5-hepten-2-one and 1,8-*cineole, a component of the pygidial gland secretion of the water beetle *Stenus comma*. On water S. develops extremely high surface tension and thus enables the beetle to move quickly. The dextrorotatory natural product is a mixture of all four stereoisomers.

Lit.: Angew. Chem. Int. Ed. Engl. **39**, 1493 (2000) (biosynthesis) ▪ Beilstein EV **20/4**, 259 ▪ J. Org. Chem. **58**, 4881 (1993) ▪ Tetrahedron Lett. **35**, 2529 (1994). – *[CAS 54985-88-5]*

Stercobilin.

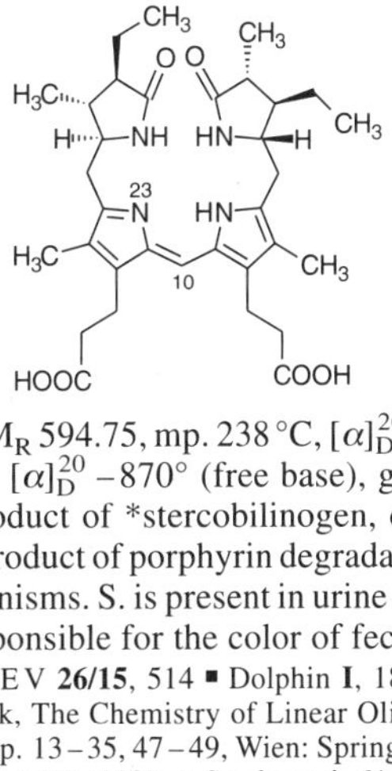

$C_{33}H_{46}N_4O_6$, M_R 594.75, mp. 238 °C, $[\alpha]_D^{20}$ –4000° (hydrochloride), $[\alpha]_D^{20}$ –870° (free base), golden yellow oxidation product of *stercobilinogen, considered to be the final product of porphyrin degradation in warm-blooded organisms. S. is present in urine and feces and is in part responsible for the color of feces.

Lit.: Beilstein EV **26/15**, 514 ▪ Dolphin **I**, 18–20; **VI**, 524, 550–554 ▪ Falk, The Chemistry of Linear Oligopyrroles and Bile Pigments, p. 13–35, 47–49, Wien: Springer 1989 ▪ Helv. Chim. Acta **70**, 2098 (1987) ▪ Synform **4**, 61, 67 (1986) (review). – *[CAS 34217-90-8]*

Stercobilinogen. Formula, see stercobilin (saturated at C-10- and N-23). $C_{33}H_{48}N_4O_6$, M_R 596.77. S. is formed in the intestines by bacterial degradation of the *bile pigment *bilirubin via *urobilinogen and thus represents the actual final product of porphyrin degradation in warm-blooded organisms. It is excreted in urine and feces where it is easily oxidized to *stercobilin.

Lit.: see stercobilin. – *[CAS 17095-63-5]*

Sterculic acid see malvalic acid.

Sterigmatin see austocystins.

Sterigmatocystin.

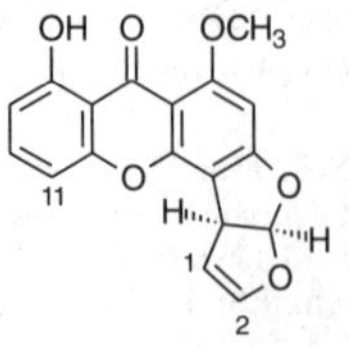

$C_{18}H_{12}O_6$, M_R 324.29, pale yellow needles, mp. 246 °C (decomp.). A *mycotoxin from many mold fungi, e. g., *Aspergillus versicolor*, *A. nidulans*, and *A. flavus* as well as ascomycetes of the genera *Chaetomium*, *Eurotium*, *Emericella*, and *Cochliobolus*. S. occurs in moldy grain and sausage products, it has also been found in moldy hard cheese. S. and its methyl ether (see below) are biosynthetic precursors of *aflatoxin B_1; while the dihydro derivatives are precursors of aflatoxin B_2. Like aflatoxin B_1, S. exerts, after metabolic activation (formation of the 1,2-epoxide that binds covalently to DNS), mutagenic, carcinogenic, and hepatotoxic activities [LD_{50} (rat p. o.) 120 mg/kg, (rat i. p.) 60 mg/kg]. The methyl ether (a, $C_{19}H_{14}O_6$, M_R 338.32), the 1,2-dihydro derivative (b, $C_{18}H_{14}O_6$, M_R 326.31), its methyl ether (c, $C_{19}H_{16}O_6$, M_R 340.33), and 10,11-dimethoxysterigmatocystin (d, $C_{20}H_{16}O_8$, M_R 384.34) have also been isolated from *Aspergillus* species.

Lit.: Agric. Biol. Chem. **55**, 1907 (1991) ▪ Appl. Environ. Microbiol. **58**, 3527 (1992); **59**, 3564 (1993) ▪ Beilstein E V **19/10**, 575 ▪ Cole-Cox, p. 67–94 ▪ Microbiol. Rev. **52**, 274 (1988) (biosynthesis) ▪ Mori, Genotoxicity of Naturally Occurring Metabolites: Structural Analogs of Aflatoxins and Related Metabolites, in Bhatnagar, Lillehoj, & Arora (eds.), Handbook of Applied Mycology, chap. 9, New York: Marcel Dekker 1992 ▪ Mycopathologia **125**, 173 (1994) (synthesis) ▪ Sax (8.), SLP 000. – *[CAS 10048-13-2 (S.); 17878-69-2 (a); 6795-16-0 (b); 21793-91-9 (c); 65176-75-2 (d)]*

Sterins see sterols.

Steroid alkaloids. According to Hegnauer these are, from a biogenetic point of view, not genuine *alkaloids in the stricter sense but rather so-called "pseudoalkaloids" or "alcaloida imperfecta", i. e., simple nitrogen-containing derivatives of widespread nitrogen-free constituents, see, e. g., Solanum, Veratrum, Apocynaceae, and Buxus steroid alkaloids, and in the animal kingdom the salamander steroid alkaloids and the batrachotoxins.

Lit.: Chem. Nat. Compd. **35**, 107–149 (1999) ▪ Hegnauer **3**, 18 ▪ Manske **50**, 61–108; **52**, 233–260 ▪ Nat. Prod. Rep. **16**, 619–635 (1999).

Steroid hormones. Naturally occurring *steroids that exert an endogenous regulatory function in the organism that produces them biosynthetically. They include in humans the male sexual hormones, the *androgens, female sexual hormones, the *estrogens and *progestogens, the *corticosteroids including *aldosterone formed in the adrenal gland, and the *vitamin D metabolite *calcitriol, in addition the *ecdysteroids that act as insect hormones, phytohormones of the *brassinosteroid type, and the sexual hormones of the aquatic fungus *Achlya* (see antheridiol). The term hormonally active steroids also encompasses non-natural, synthetically prepared compounds that are often used in medicine.

Steroid odorants. A few *steroids possess characteristic odors. This holds especially for *5α-androst-16-en-3-one* {$C_{19}H_{28}O$, M_R 272.43, mp. 140–141 °C, $[\alpha]_D$ +38° ($CHCl_3$)} with an intense urine-like odor. It can accumulate up to a concentration of 0.5 ppm in the fat tissue of adult, not castrated male pigs and is responsible for the occasionally found, bad odor of cooked pork, the unpleasant, so-called boar smell. This steroid odorant has been identified as a sexual pheromone for female pigs and is used as an aerosol to facilitate artificial insemination in pig breeding.

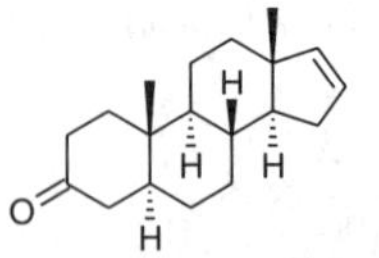

5α-Androst-16-en-3-one

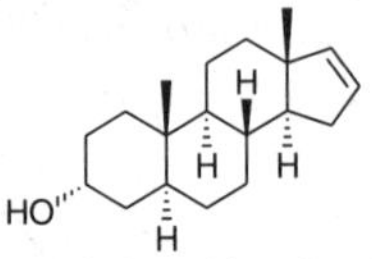

5α-Androst-16-en-3α-ol

An analogous Δ^{16}-unsaturated S. o., namely *5α-androst-16-en-3α-ol* {$C_{19}H_{30}O$, M_R 274.45, mp. 143.5–144 °C, $[\alpha]_D$ +15° ($CHCl_3$)} with a musk-like odor has been detected in relatively high amounts in humans although its physiological function (sexual attractant?) has not been fully clarified. Thus, the average urinary excretion of 5α-androst-16-en-3α-ol in the form of a glycoside amounts to 1.2 mg/d for men and 0.4 mg/d for women. The odor and activity of this compound is strongly coupled to its constitution and stereochemistry. 5α-Androst-16-en-3α-ol is also produced by truffles which thus attracts pigs. Since the spores pass intact through the pig gastrointestinal tract, the distribution of this underground fungus is secured. The biosynthesis proceeds from 3β-hydroxypregn-5-en-20-one (see progestogens).

Lit.: Beilstein E IV **7**, 1132 (androst-16-en-3-one) ▪ Experientia **37**, 1178 (1981) ▪ Helv. Chim. Acta **66**, 192 (1983) ▪ Tetrahedron **40**, 3153 (1984) ▪ Zeelen, p. 197–199. – *[CAS 18339-66-7 (5α-androst-16-en-3-one); 1153-51-1 (5α-androst-16-en-3α-ol)]*

Steroids. The name is derived from *cholesterol (Greek: *cholé* = bile, *stereós* = solid). A large number of crystalline secondary alcohols of this type with C_{27}- to C_{31}-skeletons has been isolated from animals and plants and designated as *sterols. This name gave the basis for the term "steroids" for all compounds possessing the characteristic, tetracyclic carbon skeleton of perhydrogenated cyclopenta[*a*]phenanthrene. Currently over 200000 natural and synthetic S. are purported to be known, including their glycosides and other derivatives (up to 1961 ca. 21000 S. had been described in detail[1]; the *Dictionary of Steroids*[2] published in 1991 covers over 10000 naturally occurring and important synthetic S.). This extensive class of triterpenoid compounds includes many physiologically important groups of natural products. Examples are the sterols and *bile acids, especially the *steroid hormones, such as the male and female sexual hormones (*androgens or, respectively, *estrogens and *progestogens), the *corticosteroids produced in the adrenal gland, the *vitamin D metabolite *calcitriol, the *ecdysteroids that act as insect hormones, and plant hormones of *brassinosteroid type, as well as the many synthetic S. of outstanding medical significance that were derived by use of the natural S. as lead structures. Large groups of S. are found among the secondary metabolites of plants, for example, the *Digitalis glycosides, *strophanthins, *bufadienolides, *withanolides, *steroid sapogenins, *Solanum steroid alkaloids, *Veratrum steroid alkaloids, *Apocynaceae steroid alkaloids, and *Buxus steroid alkaloids, which are of wide interest as components of important medicinal, toxic, or crop plants and play a significant role in the traditional medicine of all cultures. Finally, the S. occurring in lower animals, e.g., the *salamander steroid alkaloids and the highly toxic *batrachotoxins should be mentioned.

Structure & nomenclature: All S. are based on the tetracyclic carbon skeleton of gonane (=perhydrocyclopenta[*a*]phenanthrene) which possesses 6 centers of chirality (asymmetric centers). Thus there are $2^6=64$ theoretically possible stereoisomers but only few of them are realized among the S.

The designation of the rings as A–D and the numbering of the carbon atoms in the S. can be seen from the figure. Accordingly, the three methyl groups, already present in the triterpenoid biosynthetic precursors of the S., at positions C-4 and C-14 are given the numbers 28–30 and the C atom(s) of an additional 24-methyl or 24-ethyl group are numbered as 24^1 and 24^2.

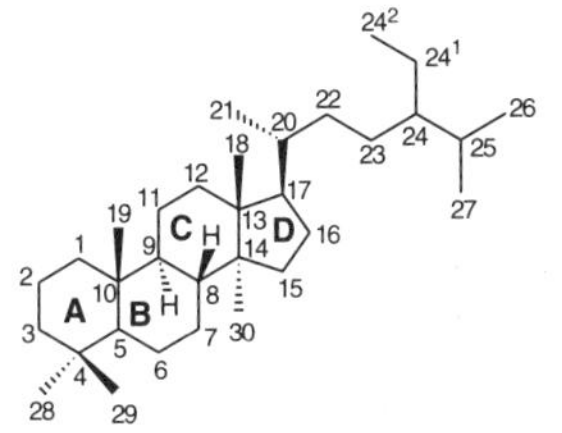

According to the IUPAC-IUB nomenclature rules 3S-1 to 3S-10 from 1989[3] the parent hydrocarbons shown in the table have been defined:

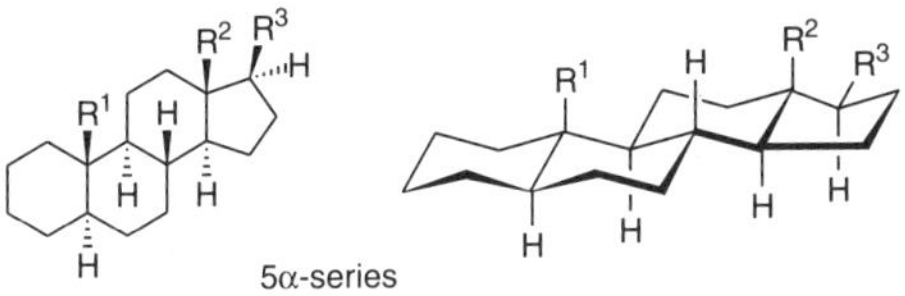

5α-series

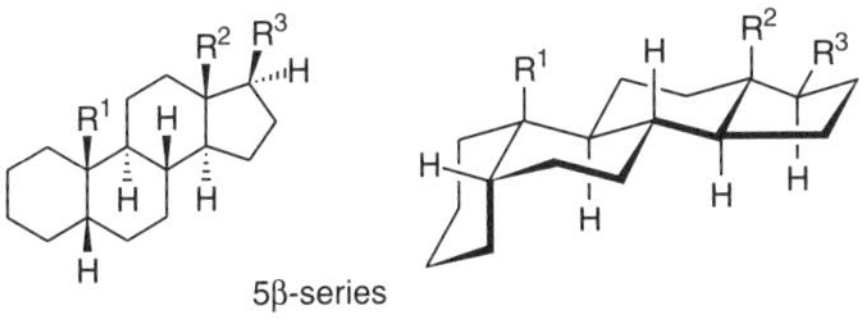

5β-series

parent skeleton		R^1	R^2	R^3
5α/β-Gonane	(C_{17})	H	H	H
5α/β-Estrane	(C_{18})	H	CH_3	H
5α/β-Androstane	(C_{19})	CH_3	CH_3	H
5α/β-Pregnane	(C_{21})	CH_3	CH_3	CH_2-CH_3
5α/β-Cholane	(C_{24})	CH_3	CH_3	$CH(CH_3)-(CH_2)_2-CH_3$
5α/β-Cholestane	(C_{27})	CH_3	CH_3	$CH(CH_3)-(CH_2)_3-CH(CH_3)_2$

Table: Parent hydrocarbons of the Steroids.

The following obsolete names for some of these parent compounds (new names in brackets) are no longer valid and should not be used: sterane (5α-gonane), oestrane (estrane), testane (5α- or 5β-androstane), oetiocholane (5β-androstane), allopregnane (5α-pregnane), allocholane (5α-cholane), and coprostane (5β-cholestane). Numerous naturally occurring S. contain additional heterocyclic rings linked or fused to ring D. Parent names have also been assigned for these constitutionally and sterically fixed species, such as cardanolide (see cardenolides), bufanolide (see bufadienolides), spirostan (see steroid sapogenins), furostan [see furostan(e) derivatives] as well as for the *steroid alkaloids e.g. spirosolane, solanidane (see Solanum steroid alkaloids) and conane (see Apocynaceae steroid alkaloids). In the 5α-series the ring linkages are *trans, anti, trans, anti, trans* and the rings (A–C) exist in the thermodynamically favored chair conformations. In contrast, compounds of the 5β-series have a *cis*-linkage of rings A and B. An α-bond (drawn as a dashed line) is directed below the plane of the ring, a β-bond (shown graphically as a thick solid line) is directed above this plane. α- or β-substituents at a C-atom of the S. ring skeleton can be characterized by differing conformations. The OH-group in, e.g., 3β-hydroxy-5α-androstane has an equatorial, the corresponding 3α-OH-group an axial conformation. The configurations of centers of chirality of the steroid side chains are designated by *R* or *S* according to the sequence rules[4]; the natural 20*R*-configuration is assumed for cholane and cholestane. Deviations from the stereochemistries shown in the above formulae must always be indicated, e.g., by the prefix *ent*. Elimination of methyl groups or ring contractions are defined by the prefix nor (*example:* 19-nor-S.), Ring expansions are indicated by the prefix homo (*example: D*-homo-S.), specially coupled ring rearrangements are designated by the prefix abeo [*example:* 14(13→12)-*abeo*-5α-cholestane in *Veratrum steroid alkaloids] and ring cleavages by the prefix seco (*example:* 9,10-secocholestane in *calcitriol, *miroestrol). In general, the designation of double bonds (abbreviation Δ), functional groups, etc. is carried out according to the general IUPAC nomenclature rules[5].

Biosynthesis: The biosynthesis proceeds by the general pathway to triterpenoids from *acetyl-CoA via mevalonate to farnesyl diphosphate (C_{15}). The tail-to-tail condensation of two molecules of the latter furnishes *squalene[6].

Squalene is converted by squalene epoxidase to (3*S*)-all-*trans*-squalene 2,3-epoxide. The following steps in the biosynthesis to cholesterol are strictly stereospecific and proceed through *lanosterol in animals or *cycloartenol[7] in green plants (see figure, p. 610).

For the biosynthesis of ergosterol and phytosterols (see sterols), see *Lit.*[7]. Cholesterol is the biosynthetic precursor of all human and animal, as well as most plant steroids. The ability of living organisms to cyclize squalene and to synthesize steroids presumably developed very early in the course of evolution. Thus, sterols are purported to occur in some early bacteria. However, in bacteria the pentacyclic *hopanoids seem to fulfill the function of sterols in membranes. Hopanoids are also found in plants as well as in geological sediments and crude oils[8].

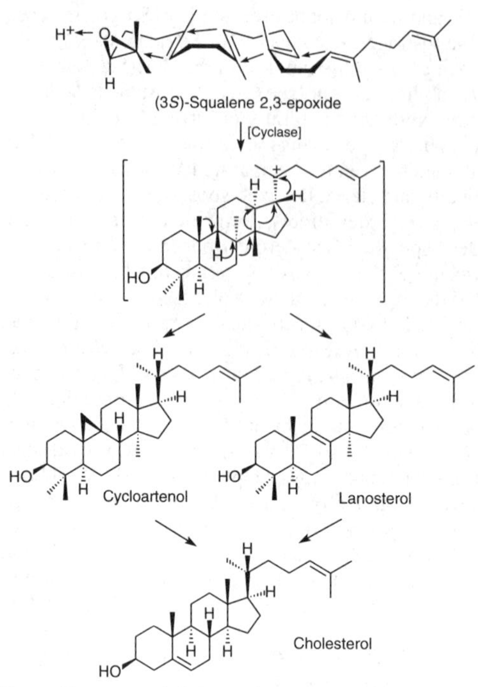

Figure: Precursors of cholesterol.

Microbial transformation and degradation: A complete catabolism of the S. ring system does not seem to occur in animals and higher plants. In humans and vertebrates, S. are generally metabolized in the liver and excreted via the kidneys (in the form of conjugates) and stomach (e. g., bile acids). The degradation is actually a microbial process. Bacteria of the autochthonous soil microflora, such as *Arthrobacter simplex*, the Proactinomycete genus *Nocardia*, various *Bacillus*, *Corynebacterium*, *Mycobacterium*, and *Streptomyces* species, etc. are able, e. g., to oxidatively degrade cholesterol via C_{27}-, C_{24}-, and C_{22}-compounds to androsta-1,4-diene-3,17-dione (C_{19}) and finally to low-molecular weight compounds.

Besides cholesterol, other S. such as pregnane derivatives, bile acids, and phytosterols are also degraded, principally to androsta-1,4-diene-3,17-dione [9]. These microbial processes have been developed in the last two decades to an industrial process for the degradation of cholesterol and sitosterol (about 100000 t/a of each) to androsta-1,4-diene-3,17-dione (ADD), androst-4-ene-3,17-dione (AD), and 9α-hydroxy-AD for use in the partial synthesis of hormonally more active S. [9].

Simple microbial transformations of S., such as hydrogenation, dehydrogenation, hydroxylation, epoxidation, ester cleavage, etc. have been in use for more than 40 years for the production of corticosteroids on an industrial scale [10].

The first total syntheses of non-aromatic S., achieved almost simultaneously in 1951 by Woodward's group ("Harvard synthesis" of cortisone) and Cornforth and Robinson ("Oxford synthesis" of 3-epiandrosterone) were milestones in the history of chemistry. In spite of new strategies to improve the yields, none of the total syntheses of corticoids has yet been able to replace the more economic partial synthesis [11].

However, the total synthesis of 19-nor-S., especially those with an unnatural, angular ethyl group at C-13 is a useful industrial process; the synthesis of the important *progestogen (–)-norgestrel (levonorgestrel) may be mentioned as an example (Schering AG, Berlin, 1967) [12].

Industrial production: There is a large demand for the pharmacologically active, medically used S. such as hormonally active drugs, ovulation inhibitors, spironolactone, anti-inflammatory agents, calciferols, and cardiac glycosides. Although the latter are mainly obtained by cultivation and extraction from *Digitalis* species, the preparation and isolation of pure steroid hormones from animal tissue for medical use and as starting materials for synthetic derivatives is still too expensive, too laborious, and quantitatively inadequate. The method of choice for these S. is thus still partial synthesis from low-priced S. raw materials available in sufficient amounts. Up to the 1970's cortisone was synthesized from progesterone [13]. Prerequisites for this synthesis were firstly the degradation of the steroid sapogenin diosgenin to 3β-acetoxypregna-5,16-diene-20-one developed by Marker u. Rohrmann, the latter is readily converted to progesterone, secondly the discovery of a microbial 11α-hydroxylation of progesterone, and thirdly the fact that a plant with an unusually high diosgenin content (on average 6%), the roots of the Mexican *Dioscorea composita* (barbasco, yam root) was found.

At the end of the 1970's the microbial processes for side chain degradation of cholesterol and sitosterol to androsta-1,4-diene-3,17-dione, androst-4-ene-3,17-dione, and its 9α-hydroxy derivative had been developed to such a state of industrial efficiency that these new raw material sources now serve for the economic production of most of the medically important S. The only exceptions are some corticosteroids, especially those with substituents at C-16, which are still prepared from steroid sapogenins, and a number of 19-nor-S. for which high-yield total syntheses are available. For details on the analysis of S., see *Lit.* [14].

Lit.: [1] Jacques, Kagan, & Ourisson, Tables of Constants and Numerical Data, vol. 14, Selected Constants: Optical Rotatory Power, vol. Ia, Steroids, Oxford: Pergamon Press 1965. [2] Hill et al. (eds.), Dictionary of Steroids, Chemical Data, Structures and Bibliographies, London: Chapman & Hall 1991. [3] IUPAC-IUB Nomenclature of Steroids (Recommendations 1989): Pure Appl. Chem. **61**, 1783–1822 (1989). [4] Angew. Chem., Int. Ed. Engl. **5**, 385–415 (1966). [5] IUPAC Nomenclature of Organic Chemistry, Sections A–F and H (1979 Edition), Oxford: Pergamon Press 1979. [6] Annu. Rev. Biochem. **51**, 555–585 (1982); Porter & Spurgeon (eds.), Biosynthesis of Isoprenoid Compounds, vol. 1, New York: Wiley 1984; Stryer 1995, p. 691–707. [7] Chem. Soc. Rev. **1**, 259–291 (1972); Annu. Rev. Biochem. **43**, 967–990 (1974); **51**, 555–585 (1982); Chem. Rev. **93**, 2189 (1993); Nes & McKean, Biochemistry of Steroids and Other Isopentenoids, Baltimore: Univ. Park Press 1977; Porter & Spurgeon (eds.), Biosynthesis of Isoprenoid Compounds, New York: Wiley 1984. [8] Pure Appl. Chem. **51**, 709–729 (1979); **65**, 1293–1298 (1993); Nachr. Chem. Tech. Lab. **34**, 8–14 (1986). [9] Adv. Appl. Microbiol. **22**, 28–58 (1977); Arima & Sih, Prix Roussel 1980, Romainville: Centre de Recherches Roussel-Uclaf 1980. [10] Cearney & Herzog, Microbial Transformation of Steroids, New York: Academic Press 1967. [11] Blickenstaff, Ghosh, & Word, Total Synthesis of

Steroids, New York: Academic Press 1974; ApSimon **2**, 558–725; **6**, 1–49, 85–139. [12]Justus Liebigs Ann. Chem. **701**, 199–205 (1967); **702**, 141–148 (1967). [13]J. Am. Chem. Soc. **74**, 3711 (1952); **75**, 1286 (1953). [14]Görög, Quantitative Analysis of Steroids, Amsterdam: Elsevier 1983; Görög, Advances in Steroid Analysis (2 vols.), Amsterdam: Elsevier 1982, 1985; Makin et al., Steroids, Weinheim: Wiley VCH 1998 (MS, GC); Nes & Parish (eds.), Analysis of Sterols and Other Biologically Significant Steroids, San Diego: Academic Press 1989; Touchstone, CRC Handbook of Chromatography: Steroids, Boca Raton: CRC Press 1986.
gen.: Bohl & Duax, Molecular Structure and Biological Activity of Steroids, Boca Raton: CRC Press 1992 ▪ Danielsson & Sjövall (eds.), Sterols and Bile Acids, Amsterdam: Elsevier 1985 ▪ Fieser & Fieser, Steroide, Weinheim: Verl. Chemie 1961 ▪ Kirk-Othmer (4.) **13**, 433 f. ▪ Thomson **2**, 140–182 ▪ Ullmann (5.) **A 13**, 101–163 ▪ Zeelen.

Steroid sapogenins (spirostanols). Aglycones (genins) of the *steroid saponins widely distributed in higher plants, especially monocotyledons. All S. have the spirostane skeleton with 5α- or 5β-stereochemistry or a Δ^5-double bond. The methyl group at C-25 can be in an axial [normal or (25*S*)-series, e.g., sarsasapogenin] or equatorial [iso or (25*R*)-series, e.g., diosgenin] position. More than 200 S. s. have been isolated to date and their structures been elucidated, only the most important are mentioned here as examples: diosgenin, the aglycone of many steroid saponins, e.g. dioscin, occurring, among others, in the roots of several *Dioscorea* species and also isolated together with the structurally related *Solanum steroid alkaloids. The corresponding S. s. of the (25*S*)-series, yamogenin, is found, among others, in the potato plant (*Solanum tuberosum*), while tigogenin and neotigogenin occur as aglycones in *Digitalis lanata* and tomato (*Lycopersicon esculentum*), respectively. Smilagenin (from *Smilax ornata*) and sarsasapogenin (from sarsaparilla roots, *Smilax regelii*) are S. s. with the 5β-stereochemistry. Digitogenin occurs in *Digitalis* species, hecogenin in the East African sisal plant (*Agave sisalana*).
Some S. s., especially diosgenin and hecogenin, are used in large amounts as starting materials for the industrial production of *steroid hormones since they can be degraded in good yields by the Marker and Rohrmann process into suitable *pregnane derivatives. On account of its 12-oxo group, hecogenin is particularly suitable for the introduction of an 11-oxygen function, which is present in important *corticosteroids. It can also be used in the synthesis of *cephalostatin analogues [1].

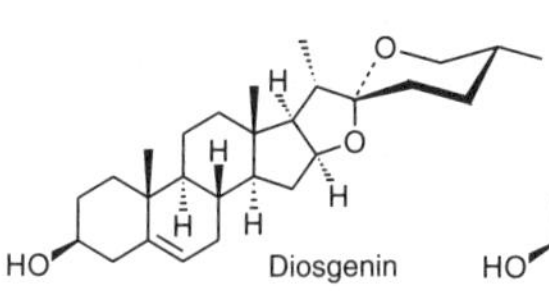
Diosgenin

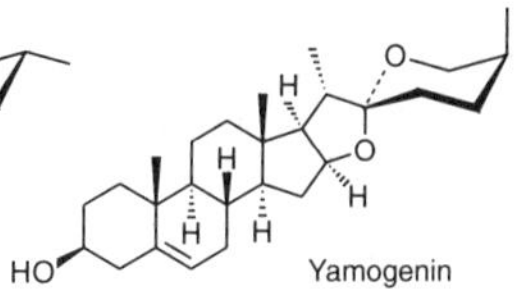
Yamogenin

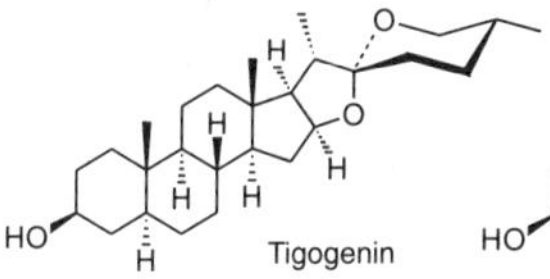
Tigogenin

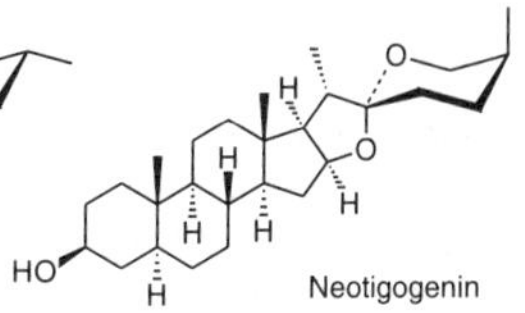
Neotigogenin

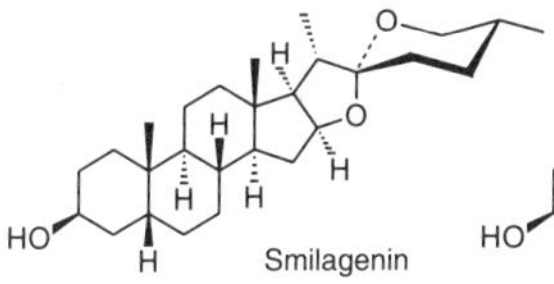
Smilagenin

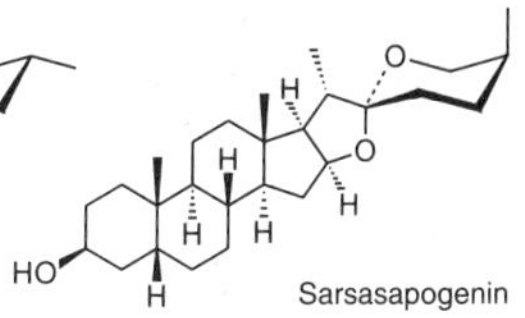
Sarsasapogenin

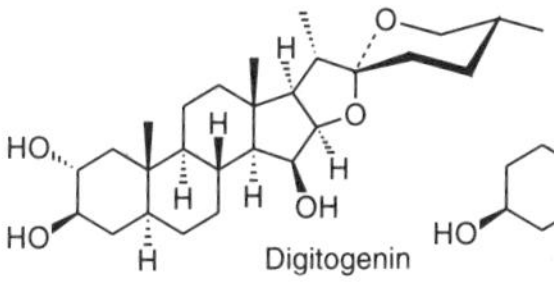
Digitogenin

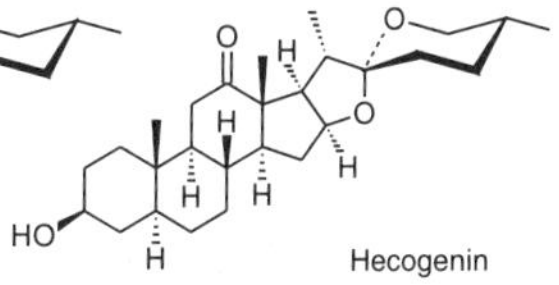
Hecogenin

Lit.: [1] Angew. Chem., Int. Ed. **35**, 1572 (1996); J. Prakt. Chem. **338**, 695–701 (1996).
gen.: Fitoterapia **58**, 67–107 (1987) ▪ Hostettmann & Marston, Chemistry and Pharmacology of Natural Products – Saponins, Cambridge: Cambridge Univ. Press 1995 ▪ Phytochemistry **24**, 2479–2496 (1985) ▪ Ullmann (5.) **A 23**, 488 ff. ▪ Zechmeister **30**, 461–606 ▪ Zeelen, p. 315–350. – *[HS 2932 90, 2938 90]*

Steroid saponins. In the stricter sense the S. s. include the glycosides of spirostanols (see steroid sapogenins), widely distributed in the plant kingdom, and compounds derived therefrom. The basic (*N*-containing) S.-containing spirosolanols or other *steroid alkaloids as aglycone (genin) are discussed elsewhere (see, e.g., Solanum steroid alkaloid glycosides). The carbohydrate components of S. are generally made up of 2–4 monosaccharides, almost exclusively D-*glucose, D-*galactose, D-*xylose, or L-*rhamnose; however, L-*arabinose and rare sugars such as D-chinovose (6-deoxy-D-glucose) also occur occasionally. They often possess branched *oligosaccharide components. When these are only linked to one hydroxy group of the aglycone (usually that at C-3) they are known as *monodesmosidic* S. s., otherwise as *bisdesmosidic*

Table: Data of Steroid sapogenins.

trivial name	systematic name	molecular formula	M_R	mp. [°C]	$[\alpha]_D$ ($CHCl_3$)	CAS
Diosgenin	(25*R*)-Spirost-5-en-3β-ol	$C_{27}H_{42}O_3$	414.63	204–207	–129°	512-04-9
Yamogenin	(25*S*)-Spirost-5-en-3β-ol	$C_{27}H_{42}O_3$	414.63	201	–123°	512-06-1
Tigogenin	(25*R*)-5α-Spirostan-3β-ol	$C_{27}H_{44}O_3$	416.64	205–206	–67.2°	77-60-1
Neotigogenin	(25*S*)-5α-Spirostan-3β-ol	$C_{27}H_{44}O_3$	416.64	202–203	–64.9°	470-01-9
Smilagenin	(25*R*)-5β-Spirostan-3β-ol	$C_{27}H_{44}O_3$	416.64	185	–69°	126-18-1
Sarsasapogenin	(25*S*)-5β-Spirostan-3β-ol	$C_{27}H_{44}O_3$	416.64	200–201.5	–75°	126-19-2
Digitogenin	(25*R*)-5α-Spirostan-2α,3β-15β-triol	$C_{27}H_{44}O_5$	448.64	296 (280–283)	–81°	511-34-2
Hecogenin	(25*R*)-3β-Hydroxy-5α-spirostan-12-one	$C_{27}H_{42}O_4$	430.63	245, 253, 268 (3 cryst. forms)	–10° (dioxan)	467-55-0

steroid saponins. To date, over 150 S. s. have been isolated from plants and their structures been elucidated. In the following only a few important or characteristic examples are discussed. *Dioscin*, isolated principally from *Dioscorea* species, is the branched chain *β*-chacotrioside of *disogenin*, containing 1 molecule of D-glucose and 2 molecules of L-rhamnose. *Digitonin*, also isolated from *Digitalis* species is another branched chain pentaoside of digitogenin containing two molecules each of D-glucose and D-galactose and 1 molecule of D-xylose. *Sarsaparilloside* from *Smilax aristolochiaefolia* (*Radix sarsaparilla*) is a bisdesmosidic steroid saponin composed of 3 molecules of D-glucose and 1 molecule of L-rhamnose, its C-26 hydroxy group is *O*-glycosidically linked to a further molecule of D-glucose. After hydrolysis of the 26-*O*-glucose unit, the resultant furostanetriol glycoside cyclizes spontaneously and stereospecifically with formation of the sarsasapogenin 3-*O*-tetraoside *parillin*. Parillin has also been isolated from *S. aristolochiaefolia*. Bisdesmosidic furostanol glycosides of this type are also known as *prosaponins* of the genuine spirostanol saponins. The Solanum steroid alkaloid glycoside jurubine is a monodesmosidic furostanol glycoside.

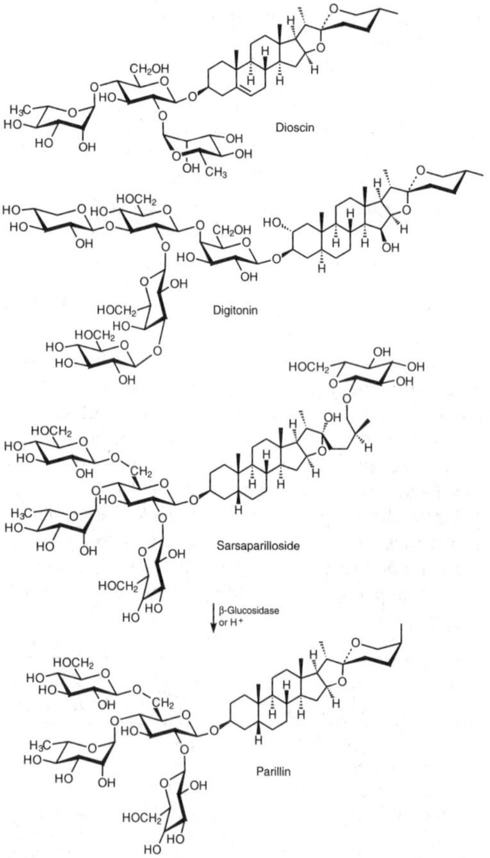

More generally, steroid glycosides without the spirostan or furostan structure can be included in the S. s. Such compounds are widely distributed as secondary metabolites, especially in the plant kingdom, but also in marine animal organisms. These are often glycosylated intermediates of sterol or spirostanol biosynthesis (see steroids).

Table: Data of Steroid saponins.

steroid saponin CAS	molecular formula	M_R	mp. [°C]	$[\alpha]_D$
Dioscin 19057-60-4	$C_{45}H_{72}O_{16}$	869.06	275–277 (decomp.)	–115° (C_2H_5OH)
Digitonin 11024-24-1	$C_{56}H_{92}O_{29}$	1229.33	235–240 (decomp.)	–54.3° (CH_3OH)
Sarsaparilloside 24333-07-1	$C_{57}H_{96}O_{28}$	1229.37	amorphous	–44° (C_2H_5OH)
Parillin 19057-61-5	$C_{51}H_{84}O_{22}$	1049.21	220–223	–64° (C_2H_5OH)

Monodesmosidic S. s. have a high surface tension activity, form strongly foaming, soap-like solutions in water, and on direct introduction into the blood circulation cause hemolysis of red blood particles even at high dilution; they thus act as potent plasma and fish toxins. Digitonin forms poorly soluble 1:1 complexes with *cholesterol and other 3*β*-hydroxy-*sterols, this property is used for the isolation and analysis of digitonin.

Lit.: Hostettmann & Marston, Chemistry and Pharmacology of Natural Products – Saponins, Cambridge: Univ. Press 1995 ▪ J. Prakt. Chem. **338**, 695 (1996) ▪ Phytochemistry **21**, 959–978 (1982); **24**, 2479–2496 (1985) ▪ Ullmann (5.) **A 23**, 485–498 ▪ Zechmeister **30**, 461–606; **46**, 1–76. – *[HS 2938 90]*

Sterols (Sterins).

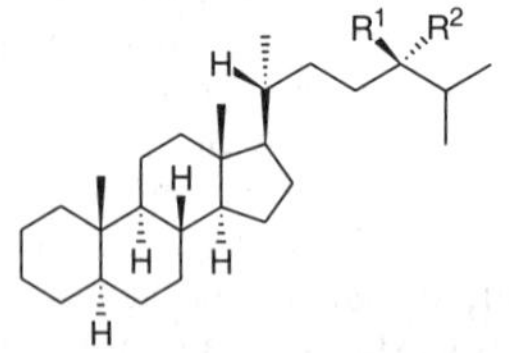

$R^1, R^2 = H$: 5α-Cholestane (20*R*)
$R^1 = H, R^2 = CH_3$: 5α-Ergostane (20*R*, 24*S*)
$R^1 = CH_3, R^2 = H$: 5α-Campestane (20*R*, 24*R*)
$R^1 = H, R^2 = C_2H_5$: 5α-Poriferastane (20*R*, 24*S*)
$R^1 = C_2H_5, R^2 = H$: 5α-Stigmastane (20*R*, 24*R*)

A group of naturally occurring *steroids derived from *cholesterol that possess a 3*β*-hydroxy group and a 17*β*-aliphatic side chain normally with 8–10 carbon atoms. They are in general derived from the parent steroid hydrocarbons 5*α*-cholestane, 5*α*-ergostane, 5*α*-campestane, 5*α*-poriferastane, and 5*α*-stigmastane (figure). S. are widely distributed in the animal and plant kingdoms as cell components either in the free form or as esters or glycosides. Depending on their occurrence, distinguished are *zoosterols* of the animal kingdom, *phytosterols* of the plant kingdom, *mycosterols* in fungi, and *marine S.* in marine fauna and flora. S. have also been detected occasionally in bacteria. Arthropods, e. g., insects and crustaceans require S. as an essential component of their diet since these organisms are not able to biosynthesize steroids.

The most important zoosterol is cholesterol. 7,8-Didehydrocholesterol is formed in the liver and converted photolytically in the skin to cholecalciferol (see vitamin D and calcitriol). Intestinal bacteria convert cholesterol to coprostanol. *Lanosterol, a biosynthetic precursor of cholesterol, is also included in the zoosterols, as are the *ecdysteroids isolated from invertebrates. Ergosterol, the main sterol of yeasts, occurs in most fungi, lichens, and algae. Ergosterol is a provitamin D (see vitamin D and calcitriol). Typical, widely distributed phytosterols are sitosterol and stigmasterol (the non-saponifiable part of soy bean oil contains 12–25% stigmasterol). Campesterol and cholesterol, albeit in smaller amounts, are also generally detectable in plants. A large number of further minor sterols is found both in higher and lower plants and fungi, e.g., brassica-, porifera-, and fucosterols. A wide spectrum of these and other S., some of which have side chains with a complicated structure such as, e.g. gorgosterol[1], see also Echinoderm saponins, are found in lower marine organisms (sponges, starfish, mussels, corals, etc.). *Cycloartenol and the numerous ecdysteroids occurring in plants are phytosterols from a structural point of view.

S. and their fatty acid esters are essential components of the cell membrane. For the biosynthesis of S., see *Lit.*[2] and steroids. S., especially cholesterol, are biosynthetic precursors of all other naturally occurring steroids. Most S. can be detected by numerous color reactions (e.g., Burchardt-Liebermann reaction), the 3β-hydroxysterols are precipitated from ethanol solutions by digitonin (see steroid saponins) or tomatine (see Solanum steroid alkaloid glycosides). Some S. (especially cholesterol, β-sitosterol, stigmasterol, and ergosterol) are used as industrial starting materials for the production of pharmaceuticals, in particular, steroid hormones (*corticosteroids, *progestogens, *androgens, *estrogens, *calcitriol) (see steroids).

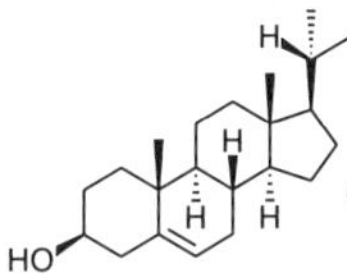
Cholesterol

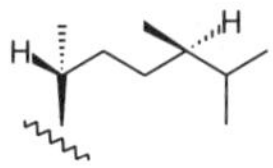
Campesterol

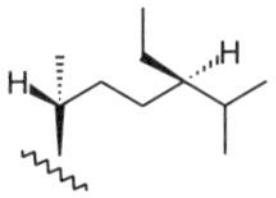
β-Sitosterol

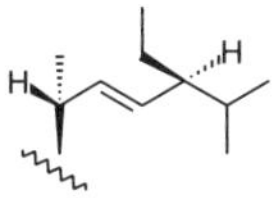
Stigmasterol

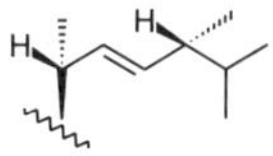
Brassicasterol

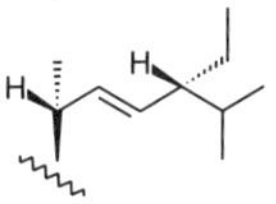
Poriferasterol

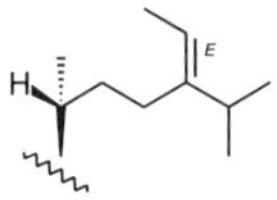
Fucosterol

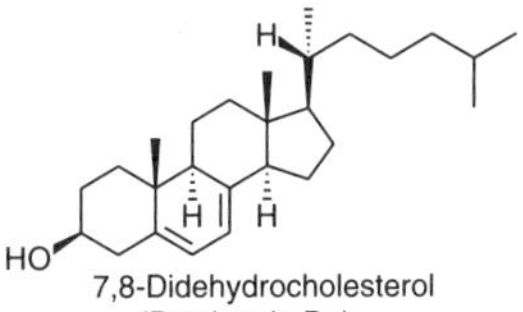
7,8-Didehydrocholesterol (Provitamin D2)

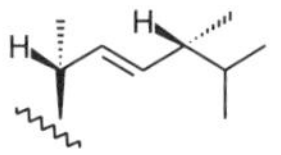
Ergosterol (Provitamin D3)

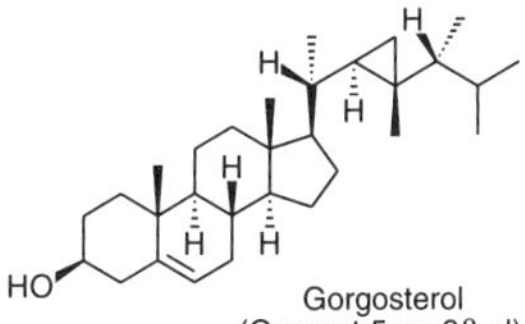
Gorgosterol (Gorgost-5-en-3β-ol)

Lit.: [1]Danielsson & Sjövall (eds.), Sterols and Bile Acids, p. 199–230, Amsterdam: Elsevier 1985; Pure Appl. Chem. **53**, 873–890 (1981). [2]Nes & McKean, Biochemistry of Steroids and Other Isopentenoids, Baltimore: Univ. Park Press 1977; Porter & Spurgeon (eds.), Biosynthesis of Isoprenoid Compounds, New York: Wiley 1984; Zeelen, p. 43–71.

gen.: Annu. Rev. Plant Physiol. **26** 209–236 (1975) ▪ Luckner (3.), p. 205–212 ▪ Nes & Parish (eds.), Analysis of Sterols and Other Biologically Important Steroids, San Diego: Academic Press 1989 ▪ Patterson & Nes (eds.), Physiology and Biochemistry of Sterols, Champaign, Ill.: Am. Oil Chem. Soc. 1991 ▪ Pridham (ed.), Terpenoids in Plants, p. 159–190, London: Academic Press 1967.

Table: Data of Sterols.

	molecular formula	M_R	mp. [°C]	$[\alpha]_D$ ($CHCl_3$)	CAS
7,8-Didehydrocholesterol (Cholesta-5,7-dien-3β-ol)	$C_{27}H_{44}O$	384.65	150–151	−113.6°	434-16-2
Koprostanol (Koprosterol, 5β-Cholestan-3β-ol)	$C_{27}H_{48}O$	388.68	96–100	+26°	360-68-9
Ergosterol [Ergosterin, Provitamin D_2, (22*E*)-Ergosta-5,7,22-trien-3β-ol]	$C_{28}H_{44}O$	396.66	168	−133°	57-87-4
Campesterol (Campesterin)	$C_{28}H_{48}O$	400.69	157–158	−33°	474-62-4
β-Sitosterol (Sitosterin, Stigmast-5-en-3β-ol)	$C_{29}H_{50}O$	414.72	137–138	−35°	83-46-5
Stigmasterol [Stigmasterin, (22*E*)-Stigmasta-5,22-dien-3β-ol]	$C_{29}H_{48}O$	412.70	170	−57°	83-48-7
Brassicasterol [Brassicasterin, (22*E*)-Ergosta-5,22-dien-3β-ol]	$C_{28}H_{46}O$	398.67	148	−64°	474-67-9
Poriferasterol [(22*E*)-Poriferasta-5,22-dien-3β-ol]	$C_{29}H_{48}O$	412.70	156	−46°	481-16-3
Fucosterol [(24*E*)-Stigmasta-5,24(28)-dien-3β-ol]	$C_{29}H_{48}O$	412.70	124	−38.4°	17605-67-3
Gorgosterol [Gorgosterin, (22*R*,23*R*,24*R*)-23,24-Dimethyl-22,23-methylen-cholest-5-en-3β-ol]	$C_{30}H_{50}O$	426.73	186.5–188	−45°	29782-65-8

Sterpuranes.

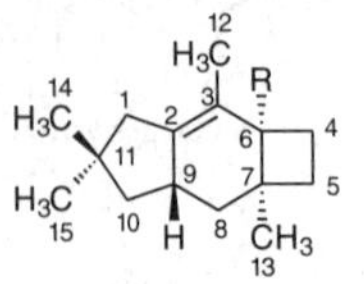

R = H : 2-Sterpurene (**1**)
R = OH : 2-Sterpuren-6-ol (**2**)

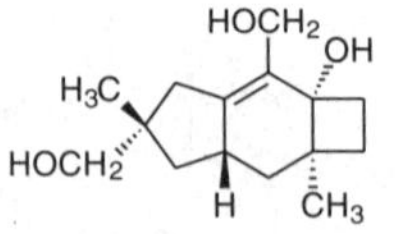

2-Sterpurene-6,12,15-triol (**3**)

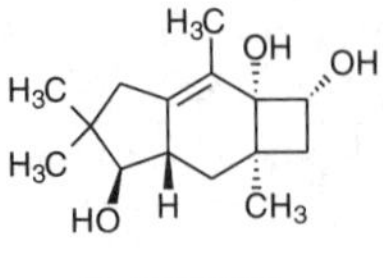

Tremetriol (**4**)

The phytopathogenic wood-rotting fungus *Stereum purpureum* (Basidiomycetes) in culture produces a series of sesquiterpenoids with the sterpurane skeleton. Besides *2-sterpurene* {$C_{15}H_{24}$, M_R 204.36, oil, $[\alpha]_D$ +65.3° ($CHCl_3$)} numerous hydroxylated derivatives such as *2-sterpurene-6,12,15-triol* {$C_{15}H_{24}O_3$, M_R 252.35, cryst., mp. 146–148 °C, $[\alpha]_D$ +64° (CH_3OH)} are also present[1]. The isomeric *tremetriol* {cryst., mp. 160–164 °C, $[\alpha]_D$ +43° (ether)} is a metabolite from *Merulius tremellosus*,[2] which also produces, among others, the isolactarane derivative *merulidial. The acetate of *2-sterpuren-6-ol* {$C_{17}H_{26}O_2$, M_R 262.39, oil, $[\alpha]_D$ +128.6° ($CHCl_3$)} is isolated from the soft coral *Alyconium acaule*[3]. Cleavage of the cyclobutane ring leads to secosterpuranes.

Lit.: [1]Can. J. Chem. **59**, 2536 (1981); **62**, 531 (1984); **75**, 742 (1997); Tetrahedron Lett. **29**, 1985, 4337 (1988); J. Org. Chem. **57**, 2313 (1992). [2]Tetrahedron **46**, 2389 (1990). [3]Tetrahedron **45**, 6479 (1989).

gen.: Synthesis: J. Am. Chem. Soc. **110**, 4062, 5818 (1988) ▪ J. Org. Chem. **55**, 5820 (1990); **57**, 2313 (1992). – *[CAS 79579-56-9 (1); 126201-36-3 (2); 79579-55-8 (3); 129058-90-8 (4)]*

Steviol (13-hydroxykaur-16-ene-19-oic acid).

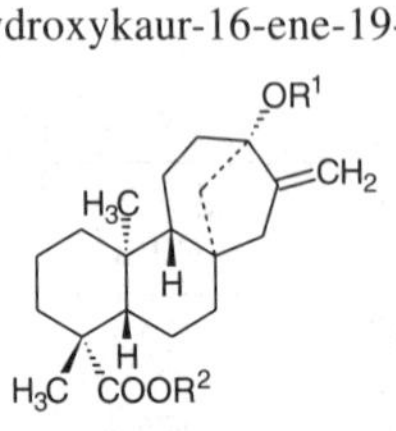

R^1 = H, R^2 = H : Steviol
R^1 = Glc(β1→2)Glcβ, R^2 = β-D-Glc*p* : Stevioside

$C_{20}H_{30}O_3$, M_R 318.46. cryst., mp. 215 °C, $[\alpha]_D$ –65° ($CHCl_3$). A *diterpene of the kaurene type that occurs as the glucoside *stevioside in the plant *Stevia rebaudiana* (Asteraceae). In contrast to the very sweet stevioside, S. is tasteless. It can be obtained from the glucoside by enzymatic hydrolysis only, since it undergoes easy rearrangement to isosteviol (beyerane skeleton) under acidic conditions. There are indications that S. is metabolized in animals to the mutagenic 15-oxosteviol, thus stevioside has not yet been registered for use as a sweetener in Northern America and European countries. Like the structurally similar *gibberellins, S. has a weak growth-promoting effect on plants and is converted to 13-hydroxygibberellins[1] by a mutation of the fungus *Gibberella fujikuroi*. Other biological properties of S. include inhibition of oxidative phosphorylation in rat mitochondria[2] and repellent activity against the greenfly *Schizapis graminum*[3].

Lit.: [1]J. Chem. Soc., Perkin Trans. 1 **1976**, 173, 174. [2]Biochim. Biophys. Acta **118**, 465 (1966). [3]J. Nat. Prod. **50**, 434 (1987).

gen.: Beilstein E V **17/8**, 119 ▪ Hager (5.) **6**, 788 ff. ▪ Nat. Prod. Rep. **10**, 301 (1993). – *Synthesis:* Agric. Biol. Chem. **35**, 918 (1971); **45**, 2115 (1981) ▪ Tetrahedron **28**, 3217 (1970); **32**, 363, 366 (1976); **33**, 373 (1977). – *[CAS 471-80-7]*

Stevioside [β-D-glucopyranosyl 13-(2-*O*-β-D-glucopyranosyl-β-D-glucopyranosyloxy)-kaur-16-ene-19-oate]. Formula, see steviol. $C_{38}H_{60}O_{18}$, M_R 804.88, cryst. (prisms), mp. 238–239 °C, $[\alpha]_D$ –39.3° (H_2O), soluble in water (ca. 0.8 g/L). Diterpene glucoside of the kaurene type obtained from leaves and stems of *Stevia rebaudiana* (Asteraceae), "yerba dulce", indigenous to Paraguay and Brazil. 1 kg dried leaves contain ca. 60 g S., which is cleaved by diastase into 1 molecule of the aglycone *steviol and 3 molecules of glucose.

Use: S. is a natural sweetener with a 100–300-fold higher sweetness than *sucrose; the natives of Paraguay used it to sweeten maté tea. The aglycone steviol is probably mutagenic, however, all investigations to date have been performed with a substance of insufficient purity; thus S. is not authorized for use as a sugar substitute for diabetics in Europe and USA. On the other hand it is used in Brazil, Israel, Korea, China, and Japan, e.g., in 1990 11 producers in Japan marketed ca. 200 t of slightly bitter S. under the name "Steebia" for use in foods, especially Tsukudani (pickled fish) and soy sauce. The annual production in Korea in 1990 amounted to ca. 300 t. These two countries are the market leaders. Further names for S. or sweetening extracts from *Stevia* are Steviosin, Stevix, and Marumillon 50. At present, there are over 200 patent applications concerning the purification and use of stevioside. S. is acutely non-toxic, LD_{50} (rat, mouse p.o.) >8.5 g/kg (other sources: >15 g/kg).

Lit.: Agric. Biol. Chem. **54**, 3137 (1990) ▪ Beilstein E V **17/8**, 119 ▪ Chem. Pharm. Bull. **25**, 3437 (1977); **39**, 12 (1991) ▪ Hager (5.) **6**, 788 ff. ▪ J. Chem. Soc., Perkin Trans. 1 **1990**, 2661 ▪ J. Med. Chem. **24**, 1269 (1981) ▪ Med. Res. Rev. **9**, 91 (1989). – *Review:* Nat. Prod. Rep. **10**, 301 (1993). – *Synthesis:* Tetrahedron **33**, 373 (1977); **36**, 2641 (1980). – *[CAS 57817-89-7]*

Stictane-3β,22α-diol (retigeradiol).

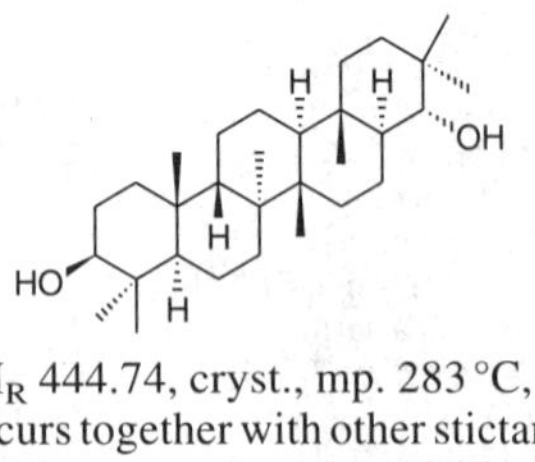

$C_{30}H_{52}O_2$, M_R 444.74, cryst., mp. 283 °C, $[\alpha]_D^{20}$ +12.3° ($CHCl_3$). Occurs together with other stictane triterpenes in lichens of the genus *Pseudocyphellaria* (New Zealand, Chile). Stictane triterpenes have as yet only been found in lichens.

Lit.: Aust. J. Chem. **42**, 243 (1989) ▪ J. Chem. Soc., Perkin Trans. 1 **1973**, 1437–1446; **1978**, 1560. – *[CAS 21688-72-2]*

Stictic acid.

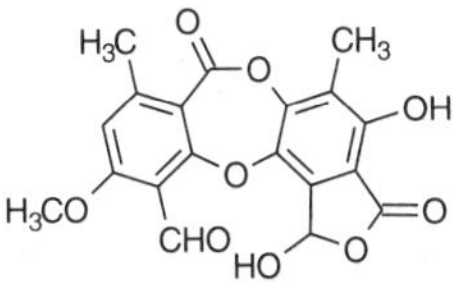

$C_{19}H_{14}O_9$, M_R 386.31, needles, mp. 270–272 °C (decomp.). A *depsidone with a cyclic hemiacylal group in the alcohol part of the molecule; occurs in numerous lichens, especially of the family Parmeliaceae.

Lit.: Culberson, p. 161 ff. ▪ Huneck & Yoshimura, Identification of Lichen Substances, Heidelberg: Springer 1996 ▪ Phytochemistry **25**, 550 (1986). – *[CAS 549-06-4]*

Sticticin.

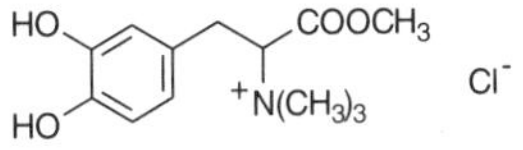

$C_{13}H_{20}ClNO_4$, M_R 289.76, hygroscopic crystals. A quaternary ammonium chloride from lichens (*Lobaria laetevirens* and *Sticta* species), it decomposes under alkaline conditions to *caffeic acid and trimethylamine; thus lichen containing S. develop an unpleasant odor of trimethylamine when spotted with KOH.

Lit.: Phytochemistry **19**, 1967–1969 (1980). – *[CAS 77035-53-1]*

Stigmastane, stigmasterin, stigmasterol see sterols.

Stigmatellin.

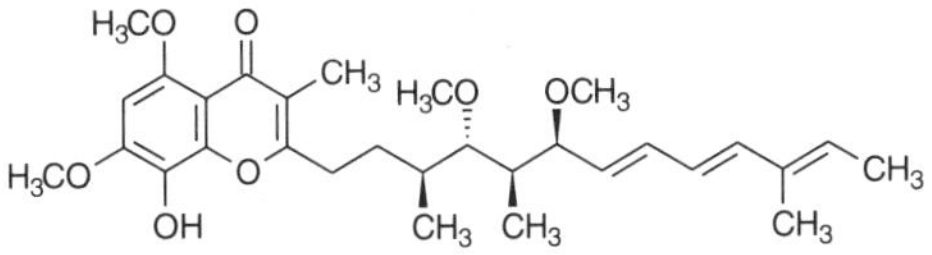

A chromone antibiotic from *Stigmatella aurantiaca* (Myxobacteria) with a unique substitution pattern in the chromone part, previously unknown in flavones, $C_{30}H_{42}O_7$, M_R 514.6. S. forms colorless crystals with mp. 128–130 °C, readily soluble in chloroform and methanol, very poorly soluble in water, $[\alpha]_D^{20}$ +38.5° (CH_3OH). S. is active against fungi and yeasts at concentrations of 0.1–3 µg/mL and is highly toxic in mice, LD_{50} 2 mg/kg. The activity is based on an inhibition of electron flow in the cytochrome bc_1 segment (complex III) of the respiratory chain, similar to antimycin and *myxothiazol. S. inhibits photosynthesis very efficiently, both in photosystem II (IC_{50} 53 nM) and also in the cytochrome b_6/f complex (IC_{50} 59 nM).

Lit.: Biochem. Biophys. Acta **765**, 227 (1984); **807**, 216 (1985) ▪ J. Antibiot. **37**, 454 (1984) ▪ J. Chem. Soc., Chem. Commun. **1993**, 424 (synthesis) ▪ Justus Liebigs Ann. Chem. **1984**, 1883. – *[HS 2941 90; CAS 91682-96-1]*

Stilbericoside see unedoside.

Stipiamides see myxalamides.

Stipitatic acid (6-hydroxytropolone-4-carboxylic acid). $C_8H_6O_5$, M_R 182.13, yellow plates, mp. 302–304 °C (decomp.). A metabolite of *Penicillium stipitatum*.

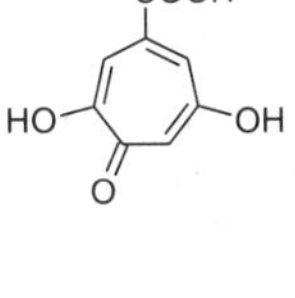

Lit.: Aust. J. Chem. **47**, 203 (1994) (synthesis) ▪ Bioorg. Chem. **23**, 131 (1995) (biosynthesis) ▪ J. Chem. Soc., Perkin Trans. 1 **1993**, 1913 (synthesis). – *[CAS 4440-39-5]*

Stizolobic acid.

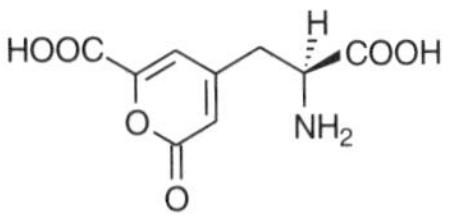

Stizolobic acid

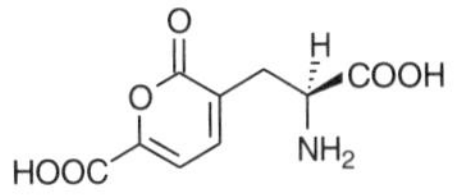

Stizolobinic acid

$C_9H_9NO_6$, M_R 227.17, mp. 231–233 °C, $[\alpha]_D$ +16° (1 m HCl); the isomeric *stizolobinic acid*, mp. >300 °C, $[\alpha]_D^{25}$ +19.5° (3 m HCl). A non-proteinogenic amino acid in *Stizolobium hassjoo* (Fabaceae)[1,2], also present in the seeds of *Mucuna deeringiana* (=*Stizolobium*) and in the toadstools *Amanita pantherina*[3], *A. muscaria* (fly agaric), and *A. gemmata*[4].

Biosynth.: By extradiol cleavage of DOPA and ring closure by S. synthase (EC 1.13.11.30)[5]. For incorporation experiments with ^{14}C- or ^{14}C- and ^3H-doubly labelled precursors, see *Lit.*[6,7].

Lit.: [1] Tetrahedron Lett. **1964**, 3431–3436. [2] Nature (London) **183**, 1116–1117 (1959). [3] Phytochemistry **13**, 1179–1181 (1974). [4] J. Nat. Prod. **39**, 150 (1976). [5] Eur. J. Biochem. **68**, 237–243 (1976); **82**, 385–392 (1978). [6] Z. Naturforsch. C **30**, 659–662 (1975); **31**, 15 (1976); **33**, 793 (1978). [7] Phytochemistry **15**, 489–491 (1976).

gen.: Beilstein E V **18/12**, 414 f. ▪ Tetrahedron Lett. **38**, 2771 (1997) (biomimetic synthesis). – *[CAS 17574-71-9 (S.); 17390-09-9 (stizolobinic acid)]*

Storax, styrax oil (resin). S. balsam is an exudate from the wood and bark of the tree species *Liquidambar styraciflua* growing in Central America ("Storax Honduras"), and *L. orientalis* growing in Turkey and the Middle East ("Turkish storax"). S. balsam should not be confused with the resin secreted from the *Styrax* plants (benzoe, styrax benzoin) used as starting material for the production of benzoeresin. S. balsam itself is not used as a raw material for perfumes but rather the extracts and oils obtained from it. S. resin is obtained by solvent extraction (hexane, ethanol, etc.); yield ca. 50%. S. oil is produced by steam distillation or high vacuum distillations; yields of up to ca. 5% by steam distillation. The resulting products are light yellowish to yellow, viscous liquids with an adherent, sweet, herby-balsamy, hyacinth-like odor. The physical data of the products depend strongly on the chosen enrichment process.

Composition [1]***:*** S. balsam consists of up to ca. 50% triterpene acids, and more than 45% of *cinnamyl cinnamate* (see cinnamic acid) and *3-phenylpropyl cinnamate* (a) ($C_{18}H_{18}O_2$, M_R 266.34), together with small amounts of more volatile compounds such as **cinnamyl alcohol* and **hydrocinnamyl alcohol*. The typical odor notes of S. are determined mainly by the trace content of *styrene* (b) (C_8H_8, M_R 104.15).

Use: In the perfume industry; styrax products are mainly employed for their fixative properties. They are components of many classical flowery, fantasy perfumes. S. balsam itself has allergenic activity due to the content of triterpene acids and is thus not used. The production of extracts can be carried out such that the acids are separated, e. g., by treatment with alkali, to furnish a product free of sensitizers, similar to the distilled oil.

Lit.: [1] J. Essent. Oil Res. **6**, 73 (1994).
gen.: Arctander, p. 597, 600, 601 ▪ Bauer et al. (2.), p. 177. – *[HS 1301 90; CAS 8024-01-9 (S.); 122-68-9 (a); 100-42-5 (b)]*

Storniamides.

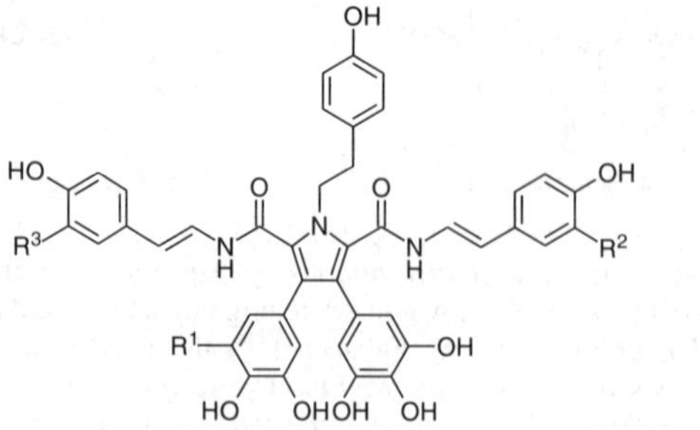

R^1 = OH, R^2 = R^3 = H : S. A
R^1 = R^3 = OH, R^2 = H : S. B
R^1 = H, R^2 = R^3 = OH : S. C
R^1 = R^2 = R^3 = OH : S. D

The S. are a group of marine pyrrole alkaloids biogenetically related to the *lamellarins and *purpurone. They have been isolated from a Patagonian sponge of the genus *Cliona*. *S. A* [$C_{42}H_{35}N_3O_{11}$, M_R 757.752, yellow oil] and the other storniamides exhibit antibiotic activity against several Gram-positive bacteria.
Lit.: Tetrahedron **52**, 2727 (1996). – *Synthesis* (of C. A nonamethyl ether): Tetrahedron Lett. **39**, 9165 (1998) ▪ J. Am. Chem. Soc. **121**, 54 (1999). – *[CAS 174232-38-3 (S. A)]*

Strawberry aroma see fruit flavors.

Streptavidin. A tetrameric protein (M_R 60000) from *Streptomyces avidinii* which, in contrast to *avidin (see there), has no glycoprotein part. It has an appreciably lower isoelectric point (IP = 5.0) than avidin. Substitution of avidin by S. can in some cases markedly reduce non-specific conditions.
Lit.: Angew. Chem. Int. Ed. Engl. **29**, 1269–1285 (1990) ▪ Biochem. J. **269**, 527 (1990) ▪ Biotechniques **1**, 39 (1983); **3**, 494 (1985). – *[CAS 9013-20-1]*

Streptazolin(e).

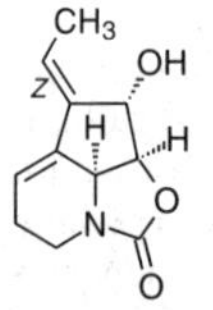

$C_{11}H_{13}NO_3$, M_R 207.23, amorphous, $[\alpha]_D^{25}$ +22° ($CHCl_3$). A tricyclic oxazolidinone from cultures of *Streptomyces viridochromogenes* with a strong tendency for polymerization.
Lit.: J. Am. Chem. Soc. **118**, 1054 (1996) (synthesis) ▪ J. Org. Chem. **58**, 3486 (1993) (biosynthesis). – *[CAS 80152-07-4]*

Streptidine see streptomycin.

Streptindole [2,2-di(3-indolyl)ethyl acetate].

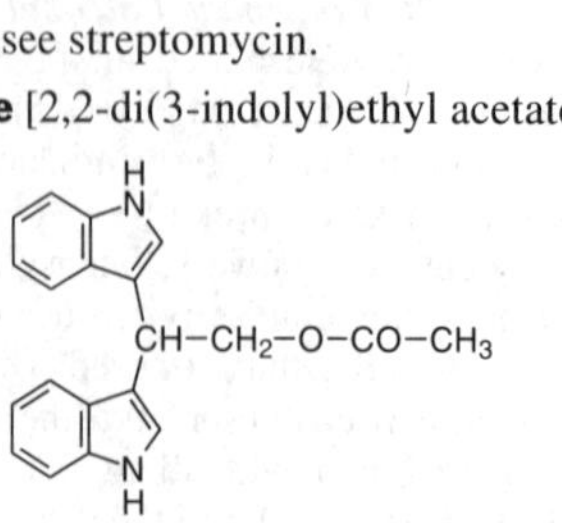

$C_{20}H_{18}N_2O_2$, M_R 318.38. A metabolite from intestinal bacteria (*Streptococcus faecium* IB 37), with genotoxic activity. S. destroys DNA without being mutagenic. A relation with large bowel cancer has not been confirmed. For synthesis, see *Lit.*[1].
Lit.: [1] Synthesis **1984**, 872; Tetrahedron **44**, 5897–5904 (1988). *gen.:* J. Nat. Prod. **57**, 1587 (1994). – *[CAS 88321-08-8]*

Streptobiosamine see streptomycin.

Streptogramins (ostreogrycins). Depsipeptide antibiotics from cultures of *Streptomyces pristinaespiralis, S. virginiae* and *S. olivaceus*; e.g., *S. A*: $C_{28}H_{35}N_3O_7$, M_R 525.60; mp. 200 °C (decomp.), $[\alpha]_D^{23}$ –209° (C_2H_5OH). S. are active against gram-positive bacteria. A mixture of S.: *virginiamycin is marketed as an antibacterial food additive.

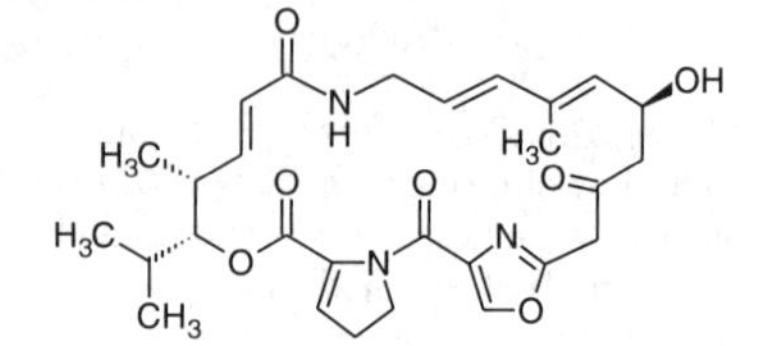

S. A

The first streptogramin antibiotic that has been marketed (in 1999) under the trademark Synercid® is a mixture of the S. quinupristin and dalfopristin, which act synergistically against various methicillin- and *vancomycin-resistant Gram-positive bacteria.[1]
Lit.: [1] R & D Focus **8**, No. 35, 1 (13.9.1999).
gen.: Angew. Chem. Int. Ed. Engl. **39**, 1664 (2000) (griseoviridin) ▪ Antimicrob. Agents Chemother. **37**, 789 (1993) ▪ Curr. Pharm. Des. **4**, 155–180 (1998) ▪ Drugs (Suppl. 1) **51**, 13 (1996) ▪ Drugs. Pharm. Sci. **22**, 695 (1984) ▪ Environ. Sci. Res. **44**, 105 (1992) (biosynthesis) ▪ J. Am. Chem. Soc. **105**, 5106 (1983) (biosynthesis). – *[CAS 21411-53-0 (S. A)]*

Streptolin B see streptothricins.

Streptomycetes. One of the eight main families of the order Actinomycetales. S. are spore-forming, Gram-positive bacteria with a pronounced tendency to form aerial and substrate mycelia. They have been isolated from the soil, from aquatic systems and/or from insects. The S. are producers of important antibiotics (e.g., *erythromycin, *streptomycin). As yet no other group of microorganisms has shown such a diversity of *secondary metabolism.
Lit.: see actinomycetes.

Streptomycin.

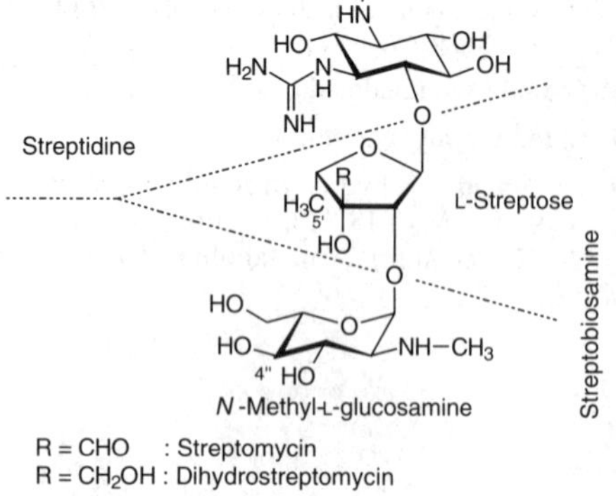

R = CHO : Streptomycin
R = CH_2OH : Dihydrostreptomycin

$C_{21}H_{39}N_7O_{12}$, M_R 581.58. International free name for the aminoglycoside *antibiotic discovered in 1943 by Schatz, Bugie, and Waksman (Nobel Prize for Physiology and Medicine, 1952) from cultures of *Streptomyces griseus*, also known as *S. A.* It is a pseudotrisaccharide consisting of the diguanidinocyclitol *streptidine, N-methyl-L-glucosamine,* and *streptose* (5-deoxy-3-*C*-formyl-L-lyxose). The free compound is a strong base, unstable, $[\alpha]_D$ −86.7° (H_2O), readily soluble in water, and poorly soluble in organic solvents. The most applied salt is the sulfate, a white powder, odorless with a bitter taste, soluble in water, insoluble in ether and acetone. In some cultures of *Streptomyces griseus* S. A occurs together with *S. B* (mannosido-S., $C_{27}H_{49}N_7O_{17}$, M_R 743.72). The antibiotic effect of S. B is 4- to 8-fold lower than that of S. A. S. A is active against Gram-negative pathogens and, especially, against tuberculosis bacteria whereas its effects on Gram-positive bacteria are weaker than those of *penicillins. Other *streptomycetes cultures (e. g., *Streptomyces griseocarneus*) furnish hydroxy-S. (*S. C, reticulin*) which differs from S. A by a hydroxy group at C-5′ in the streptose unit.

Activity: The bactericidal activity of S. A is based on binding to ribosomal protein S12 of the *30S*-subunit of the ribosomes which causes reading errors and inhibition of the elongation step in protein synthesis. S. A damages the bacterial cell membrane. S. A acts as an antimetabolite towards *inositol. On account of its relatively good stability, S. A can also be administered orally. On long-term treatment – similar to other aminoglycoside antibiotics – hypersensitivity reactions, hearing impairments, and kidney damage can occur [further side effects: neurotoxicity, teratogenicity; LD_{50} (rat p. o.) 9000 mg/kg]. Three mechanisms of resistance are known among the S. A-resistant pathogens: Target modification by changes at the *30S*-subunit of ribosomes, changes of the *16S*-rRNA at the *30S*-subunit, modification of the active principle by phosphorylation. The development of S. A-resistant bacteria occurs relatively easily, thus *dihydrostreptomycin*, optionally in combination with 4-aminosalicylic acid (PAS), is often employed. In the biosynthesis of S. A the first steps involve the syntheses of the substructures streptidine, streptose, and *N*-methyl-L-glucosamine in three separate sequences from the common starting material glucose. Subsequently assembly of the activated subunits occurs (phospho- or nucleoside diphospho-derivatives). An intermediate from streptidine phosphate and dTDP-dihydrostreptose is assumed to be formed which then adds *N*-methyl-L-glucosamine to furnish dihydrostreptomycin 6-phosphate. S. A is produced by fermentation processes using high-performance strains (>10 g/L).

Lit.: Anal. Profiles Drug Subst. **16**, 507 (1987) ▪ Beilstein E V **18/11**, 82 ▪ Can. J. Biochem. Cell. Biol. **62**, 231 (1984) ▪ Hahn, Mechanisms of Antibacterial Agents, Antibiotics, vol. 5, 272–303, Berlin: Springer 1979 ▪ Kleemann-Engel, p. 1753 ▪ Science **107**, 233 (1948) ▪ Zechmeister **66**, 128–134, 143. – *Biosynthesis:* J. Am. Chem. Soc. **97**, 4782 (1975). – *Synthesis:* J. Antibiot. **27**, 997 (1974). – *[HS 2941 20; CAS 57-92-1 (S.); 128-45-0 (S. B)]*

Streptonigrin.

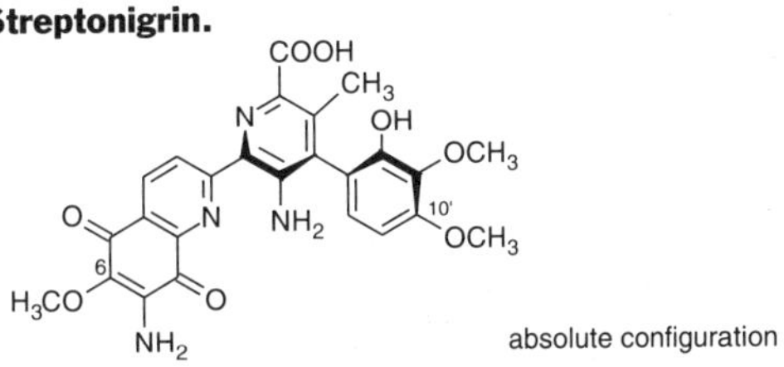

$C_{25}H_{22}N_4O_8$, M_R 506.47, an aminoquinoline quinone *antibiotic isolated for the first time in 1960 from *Streptomyces flocculus* and other *streptomycetes; it is structurally related to *lavendamycin, forms coffee-brown to black platelets (mp. 275 °C) or brown needles (mp. 262–263 °C). The antibiotic activity of S. is based on an impairment of the matrix function of DNA by covalent bonding in the ratio antibiotic : DNA base pairs of 1 : 2000 and single strand cleavages. S. also exhibits antitumor activity and immunosuppressive properties, it is used as an inhibitor of reverse transcriptase of retroviruses; LD_{50} (dog p. o.) 0.5 mg/kg, (mouse p. o.) 2.3 mg/kg.

Lit.: Foye, Cancer Chemotherapeutic Agents, p. 645–651, Washington: ACS 1995 ▪ J. Chem. Soc., Perkin Trans. 1 **1990**, 2611 f. ▪ J. Heterocycl. Chem. **27**, 1437 ff. (1990) ▪ J. Med. Chem. **34**, 1871–1879 (1991) ▪ Lindberg **1**, 325–346; **2**, 1–56 ▪ Sax (8.), SMA 000 ▪ Tetrahedron **53**, 15101 (1997) (abs. configuration) ▪ Ullmann (5.) **A 1**, 385 ▪ Zechmeister **41**, 77–111. – *Biosynthesis:* J. Am. Chem. Soc. **104**, 343, 536 (1982). – *Synthesis:* J. Am. Chem. Soc. **107**, 5745 (1985). – *[CAS 3930-19-6]*

Streptopyrrole. Antimicrobial metabolite from *Streptomyces armeniacus*, $C_{14}H_{12}ClNO_4$, M_R 293.70, needles, mp. 181–183 °C (decomp.).

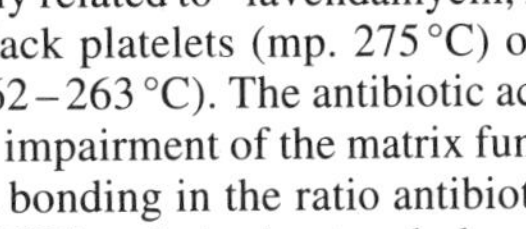

Lit.: Acta Chem. Scand. **52**, 1040 (1998) ▪ J. Antibiot. **53**, 1–11 (2000) (activity). – *[CAS 192819-12-8]*

Streptose see streptomycin.

Streptothricin B₂ see neomycin.

Streptothricins. A mixture of *antibiotics of the nucleoside type first isolated in 1942 by Waksman and Woodruff from *streptomycetes [1]. S. were first considered to be a uniform substance and given the name S. F. The mixture was later found to consist of several, frequently occurring compounds that differ in the number of *β*-lysine units in the side chain.

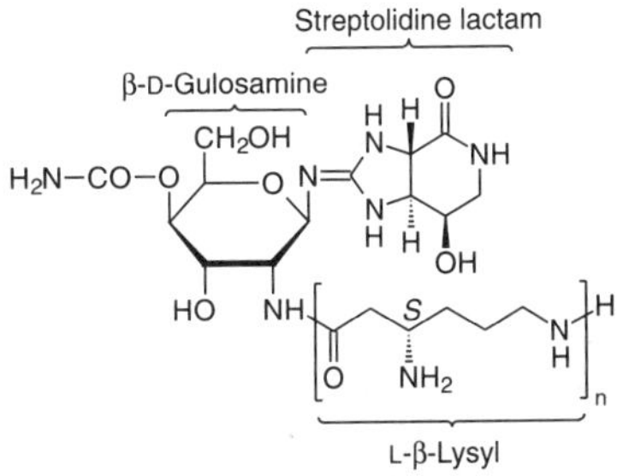

Table: Data of Streptothricins.

S.	synonym	n	molecular formula	M_R	$[\alpha]_D$ (H_2O)	CAS
S. A	Aerosporin, Polymycin A	6	$C_{49}H_{94}N_{18}O_{13}$	1143.40	−9.2°	3484-67-1
S. B	Racemomycin E, Polymycin B	5	$C_{43}H_{82}N_{16}O_{12}$	1015.22	−10°	3484-68-2
S. C	Racemomycin D	4	$C_{37}H_{70}N_{14}O_{11}$	887.05	−17°	3776-36-1
S. D	Racemomycin B, Phytobacteriomycin D, Boseiomycin III	3	$C_{31}H_{58}N_{12}O_{10}$	758.87	−58.6°	3776-37-2
S. E	Racemomycin C, Yazumycin C, Boseiomycin II	2	$C_{25}H_{46}N_{10}O_9$	630.70	−41°	3776-38-3
S. F	Racemomycin A, Yazumycin A, Boseiomycin I, Akimycin, Streptolin B, Pleocidin, Grisemin	1	$C_{19}H_{34}N_8O_8$	502.53	−53°	3808-42-2
S. X		7	$C_{55}H_{106}N_{20}O_{14}$	1271.57		24543-64-4

In addition to a broad spectrum of activities against Gram-positive and Gram-negative bacteria, the S. also exhibit antifungal, antiviral, and ergotropic properties. S. are acutely toxic compounds, LD_{50} (mouse i. v.) 8 mg/kg. S.-resistant pathogens are known. In these cases the active principle is modified by *N*-acetylation at the terminal *β*-lysyl residue. In the biosynthesis of S., the mechanism of non-ribosomal peptide synthesis is coupled with other biogenetic principles: formation of the modified sugar D-gulosamine and the unusual amino acid streptolidine (from L-arginine), the lactam of which is *N*-glycosidically bound. The *β*-lysine units are formed by isomerization of *α*-lysine.

Lit.: [1] Proc. Soc. Exp. Biol. Med. **49**, 207 (1942).
gen.: Drugs Pharm. Sci **22**, 367–384 (1984) ▪ J. Antibiot. **38**, 987 (1985); **40**, 1140 (1987) ▪ Merck-Index (12.), No. 8988 ▪ Zechmeister **66**, 123–130, 134, 137. – *Biosynthesis:* J. Am. Chem. Soc. **103**, 6752 (1981) ▪ J. Chem. Soc., Perkin Trans. 1 **1988**, 2283 ▪ Nat. Prod. Rep. **7**, 105–129 (1982) ▪ Nature (London) **224**, 595 (1969). – *Isolation:* Arch. Pharmacol. Res. **19**, 153–159 (1996). – *Synthesis:* Tetrahedron Lett. **23**, 2961 (1982). – *Activity:* Antibiotics (N. Y.) **6**, 238–247 (1983). – *[HS 2941 90; CAS 54003-27-9]*

Streptovaricin.

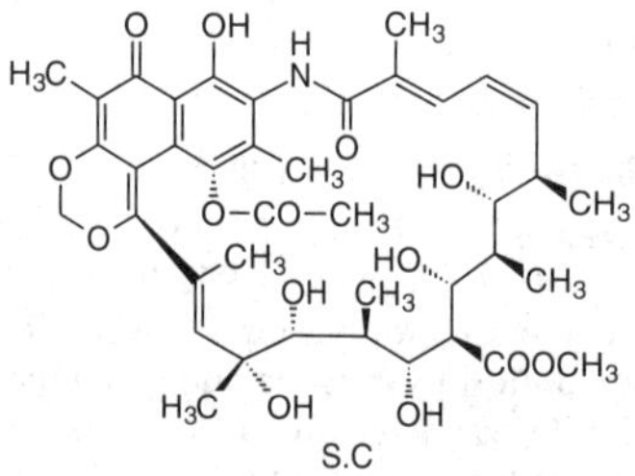

S.C

An *ansamycin antibiotic produced by *Streptomyces spectabilis* with antibacterial (against tuberculosis pathogens), antiviral, and antitumor activities. S. occurs as a multi-component mixture of up to ten individual compounds (S. A to G, J, K, and U) with *S. C* (methyl streptovaricate, $C_{40}H_{51}NO_{14}$, M_R 769.84, amorphous, mp. 189–191 °C, $[\alpha]_D$ +602°) as the main component. The S. are inhibitors of bacterial RNA-polymerase as well as the reverse transcriptase of oncogenic viruses. The biosynthesis of the aromatic core branches off from the *shikimic acid pathway while the alkyl chain is formed on the *polyketide pathway.

Lit.: Zechmeister **33**, 231–307 (Review). – *Biosynthesis:* Chiao et al., in Vining & Stuttard (eds.), Genetics and Biochemistry of Antibiotic Production, p. 477–498, Boston: Butterworth-Heinemann 1995 ▪ J. Antibiot. **44**, 218 (1991). – *Synthesis:* Isobe, in Recent Progress in the Chemical Synthesis of Antibiotics, p. 104–134, Berlin: Springer 1990 ▪ J. Org. Chem. **54**, 5548 (1989) ▪ Tetrahedron Lett. **23**, 4199 (1982); **24**, 269 (1983). – *Activity:* Biochemistry **13**, 861 (1974) ▪ Pharmacol. Ther., Part A **1**, 289 (1977). – *[HS 2941 90; CAS 1404-74-6 (S. complex); 23344-17-4 (S. C)]*

Streptovirudins see tunicamycins.

Streptozocin [streptozotocin, 2-deoxy-2-(3-methyl-3-nitrosoureido)-D-glucose].

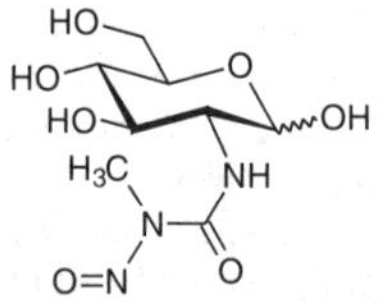

$C_8H_{15}N_3O_7$, M_R 265.22, mp. 115 °C. An aminoglycoside *antibiotic with antifungal and cytostatic activity isolated for the first time in 1960 from *Streptomyces achromogenes*. S. was used for some time as an antitumor agent, but it is carcinogenic and mutagenic and causes diabetes [1,2]. The biogenesis of S. proceeds from D-*glucosamine, as well as *citrulline and *methionine as donors of the carbamoyl and methyl groups, respectively.

Lit.: [1] Trends Pharmacol. Sci. **3**, 376 (1982). [2] Diabetes Metab. **15**, 61 (1989).
gen.: Agarwal, Streptozocin, Fundamentals and Therapy, Amsterdam: Elsevier 1981 ▪ Foye, Cancer Chemotherapeutic Agents, Washington: ACS 1985 ▪ J. Org. Chem. **44**, 9 (1979) ▪ Kleemann-Engel, p. 1755. – *Biosynthesis:* J. Antibiot. **32**, 379 (1979). – *Synthesis:* J. Org. Chem. **35**, 245 (1970). – *Activity:* Cancer Chemother. Rep. **55**, 303 (1971). – *[CAS 18883-66-4]*

Striatals. An unusual group of diterpenoids from cultures of the fungus *Cyathus striatus* (birdsnest fungus, Basidiomycetes) and related species. On standing in methanol the S. furnish cyclic hemiacetals, the *striatins*. *Striatin* A ($C_{28}H_{40}O_8$, M_R 504.62, cryst., mp. 144–145 °C), *Striatin B* ($C_{28}H_{40}O_9$, M_R 520.62, cryst., mp. 143–144 °C), and *Striatin C* ($C_{26}H_{38}O_8$, M_R 478.58, cryst., mp. 144–145 °C) are mentioned as examples. The S. are formed biosynthetically from the *cyathane xyloside *herical by intramolecular condensation of the sugar unit with the cyathane system. S. and the striatins possess antibacterial, strong antifungal, and cytotoxic properties. In Ehrlich ascites tumor cells the incorporation of thymidine, uridine, and leucine in DNA, RNA, and proteins is completely

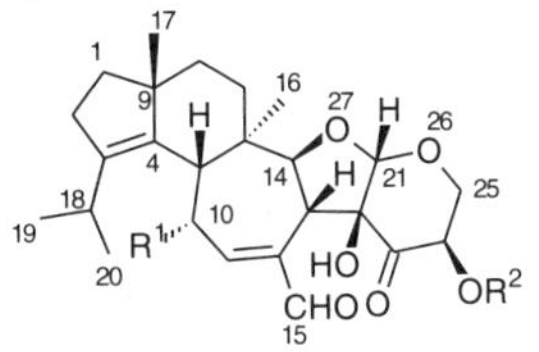

R^1 = H, R^2 = CO−CH_3 : Striatal A (**1**)
R^1 = OH, R^2 = CO−CH_3: Striatal B (**2**)
R^1 = OH, R^2 = H : Striatal C (**3**)

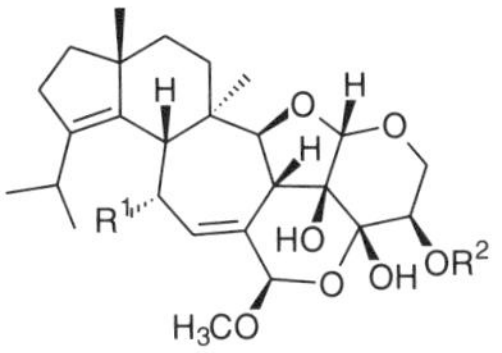

R^1 = H, R^2 = CO−CH_3 : Striatin A (**4**)
R^1 = OH, R^2 = CO−CH_3: Striatin B (**5**)
R^1 = OH, R^2 = H : Striatin C (**6**)

inhibited at concentrations of 2 µg/mL. Positive results were obtained in the treatment of mice with P388 lymphocytic leukemia.

Lit.: Pure Appl. Chem. **53**, 1233 (1981) ▪ Tetrahedron **38**, 1409 (1982). – *Review:* DECHEMA-Monogr. **129**, 3–13 (1993) ▪ Forum Mikrobiol. **11**, 21 (1988). – *[CAS 69075-63-4 (1); 68900-64-1 (2); 62744-72-3 (4); 62744-73-4 (5); 62744-74-5 (6)]*

Strictosidine (isovincoside).

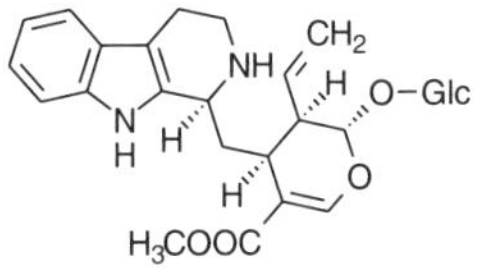

$C_{27}H_{34}N_2O_9$, M_R 530.57, a relatively unstable glucoalkaloid originally isolated from *Rhazya stricta* (Apocynaceae)[1], its occurrence in *Catharanthus roseus* was demonstrated by isotope dilution analysis. The N^6-methyl derivative occurs in *Strychnos gossweileri*. S. is also found in cell suspension cultures of *Rauvolfia serpentina*; it is formed by stereospecific condensation of the monoterpene secologanin (see secoiridoids) with the amine tryptamine under the action of the key enzyme strictosidine synthase[2,3]. S. is probably the universal biosynthetic precursor of all ca. 2000 *monoterpenoid indole alkaloids[2]. In S. the glucoside part functions as a protecting group and, after cleavage by two specific glucosidases[5], liberates a highly reactive aglycone which then enters the biosynthetic processes leading to monoterpenoid indole alkaloids. This precursor role of S. has been demonstrated *in vivo* by feeding experiments for the most important indole alkaloid types[2,6] as well as by cell-free enzymatic biosynthesis for the *heteroyohimbine alkaloids and the *Rauvolfia alkaloids. S. can be synthesized biomimetically by condensation of tryptamine and secologanin together with its epimer *vincoside, it can also be isolated in pure form from cell cultures or be prepared with the help of S. synthase. The gene coding for S. synthase obtained from *Rauvolfia* cell cultures has been cloned and the enzyme has been expressed in *E. coli* as well as in insect cell cultures[3,4].

Lit.: [1] J. Chem. Soc., Chem. Commun. **1968**, 912. [2] J. Chem. Soc., Chem. Commun. **1977**, 646. [3] FEBS Lett. **79**, 233 (1977); **97**, 195 (1979); **237**, 40 (1988). [4] Manske **47**, 115–172; Phytochemistry **35**, 353 (1994). [5] FEBS Lett. **110**, 187 (1980). [6] Tetrahedron Lett. **1978**, 1593.

gen.: J. Nat. Prod. **60**, 69 (1997) (stereochemistry) ▪ Mann, Secondary Metabolism (2.), p. 288–294, Oxford: Clarendon Press 1987 ▪ Methods Enzymol. **136**, 342–350 (1987) ▪ Phytochemistry **18**, 965 (1979); **32**, 493–506 (1993); **44**, 1387 (1997) ▪ Planta Med. **55**, 525 (1989) ▪ Tetrahedron Lett. **1978**, 1593. – *[HS 2939 90; CAS 20824-29-7]*

Strigol.

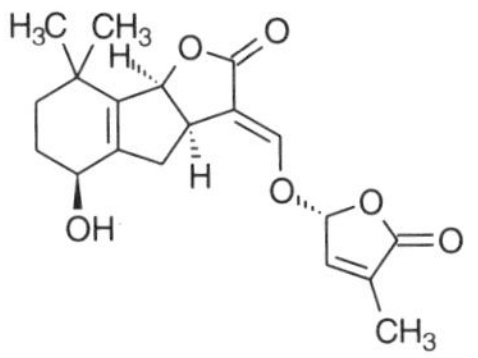

$C_{19}H_{22}O_6$, M_R 346.38, mp. 200–202 °C (decomp.). S. was first isolated from the root extract of the cotton plant (*Gossypium hirsutum*) and later from corn and sorghum[1].

Biological activity: S. has potent germination promoting activity on the root parasite *Striga lutea* (witchweed, Scrophulariaceae)[2]. S. has a germination promoting activity on 50% of the seeds even at a concentration of 10^{-16} Mol/L. *Striga* species are semiparasites that insert absorbing haustoria into the roots of grasses (cereals) and thus constitute an economic problem in agriculture. The seeds of *Striga* can survive in the soil for many years (at least 10–15) and then germinate when roots of host plants appear in the vicinity[3]. For the synthesis of S. and its analogues, see *Lit.*[4–5], for detection, see *Lit.*[6]. For mechanism of action, see *Lit.*[7].

Lit.: [1] J. Agric. Food Chem. **41**, 1486–1491 (1993). [2] Weed Sci. **29**, 101 (1981); **30**, 561 (1982). [3] Annu. Rev. Phytopathol. **18**, 463–489 (1980). [4] J. Org. Chem. **50**, 628, 3779 ff. (1985); J. Am. Chem. Soc. **52**, 1984 (1987); Tetrahedron Lett. **28**, 3091, 3095 (1987); Tetrahedron **47**, 1411–1416 (1991); **50**, 6839 (1994); **54**, 2753 (1998); Eur. J. Org. Chem. **1999**, 2211. [5] J. Chem. Soc., Perkin Trans. 1 **1997**, 767; J. Agric. Food Chem. **40**, 697 (1992); Recl. Trav. Chim. Pays-Bas **111**, 155 (1992); J. Agric. Food Chem. **40**, 1222–1229 (1992); Tetrahedron **49**, 351–360 (1993); **52**, 14827 (1996); **54**, 3439–3456 (1998). [6] Plant Growth Regul. **11**, 91–98 (1992). [7] J. Chem. Soc., Perkin Trans. 1 **1997**, 759.

gen.: ACS Symp. Ser. **355**, 409–432, 445–461 (1987); **443**, 278–287 (1991) ▪ Beilstein E V **18/3**, 104. – *[CAS 11017-56-4]*

Strobilurins. Together with the *oudemansins the S. constitute the important group of antifungal *β*-methoxyacrylate antibiotics[1]. They are produced by numerous basidiomycetes and the ascomycete *Bolinea lutea* in mycelial cultures. At low concentrations (≅ 10 µg/mL) the S. specifically inhibit the respiration of filamentous fungi and yeasts[2] by interrupting the electron transport from ubihydroquinone to cytochrome c in the bc_1-complex of the respiratory chain[3]. This results from a specific binding of the strobilurins to the oxidation center of ubihydroquinone.

S. A from cultures of the pine-cone fungus *Strobilurus tenacellus* is identical to *"mucidin"* originally errone-

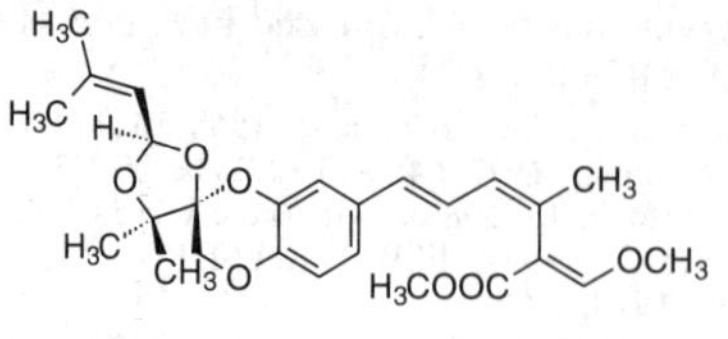

Strobilurin E (relative configuration)

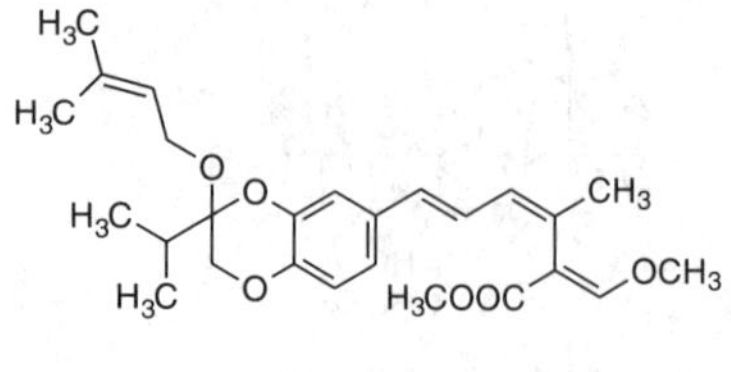

Strobilurin M

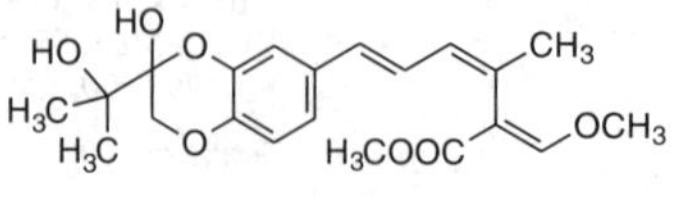

Strobilurin N

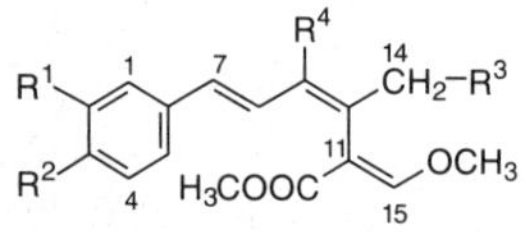

	R^1	R^2	R^3	R^4
Strobilurin A	H	H	H	H
Strobilurin B	OCH_3	Cl	H	H
Strobilurin C	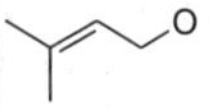	H	H	H
Strobilurin F1	OH	H	H	H
Strobilurin F2	H	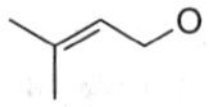	H	H
Strobilurin H	OCH_3	H	H	H
Strobilurin X	H	OCH_3	H	H
Hydroxy-strobilurin A	H	H	OH	H
9-Methoxy-strobilurin A	H	H	H	OCH_3

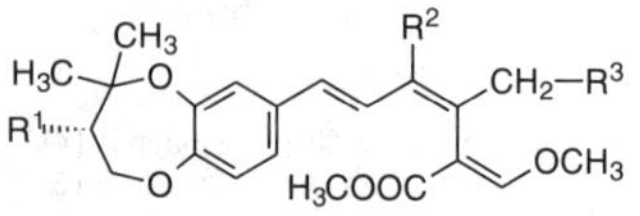

	R^1	R^2	R^3
Strobilurin D (= Strobilurin G)	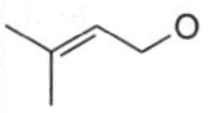	H	H
Strobilurin I	H	H	H
Strobilurin K		H	H
Hydroxy-strobilurin D	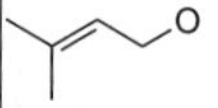	H	OH
9-Methoxy-strobilurin K		OCH_3	H

ously described as an optically active metabolite of the slime fungus *Oudemansiella mucida*[9]. S. C, F_2, E, I, and M have dimethylallyloxy groups on their benzene rings that can be oxidatively modified. A hydroxylation of the 14-methyl group has been observed in hydroxystrobilurins A and D[10]. S. I is the simplest member of the benzodioxepin type strobilurins from which S. D (= S. G) and S. K are derived by prenylation of the hydroxy group[14]. The epoxide structures previously assigned to S. D and other compounds of this type are erroneous. 9-Methoxystrobilurins A and K represent an intermediate structure between the S. and the oudemansins[11,14]. Interestingly, S. N which occurs in fruit bodies of *Mycena crocata* shows no antifungal activity[15]. This may be due to the polar moiety attached to the benzene ring. As revealed by structure-activity relationships with synthetic analogues, the (*E*)-*β*-methoxyacrylate function is essential for the antifungal activity of the S.[12]. The corresponding (*Z*)-compounds are not active. In addition, the twisting of the methoxyacrylate unit with respect to the molecular plane caused by interactions with the planar dienyl system is also important for the activity. Accordingly, the planar (9*E*)-S. A[4] has practically no antifungal activity[12].

On account of their high specificity and efficiency towards phytopathogenic fungi in combination with their low toxicity to warm-blooded organisms, the S. are important lead structures for synthetic analogues[1,12]. Two new, broad-spectrum fungicides for plant protection: BAS 490 F (Kresoxim-methyl, Brio®, Juwel®, Discus®, Allegro®, Landmark®, Mantra®, Stroby®) from BASF AG ($C_{18}H_{19}NO_4$, M_R 313.35) and ICI A 5504 (Azoxystrobin, Amistar®, Abound®, Quadris®) from Zeneca ($C_{22}H_{17}N_3O_5$, M_R 403.39) were launched in 1992 and 1996[13]; the third product, Flint® (trifloxystrobin) from Novartis will follow in 2000. S. A itself is not suitable for practical use in plant protection on account of its instability towards sunlight (half-life: a few minutes).

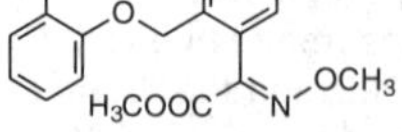

BAS 490 F

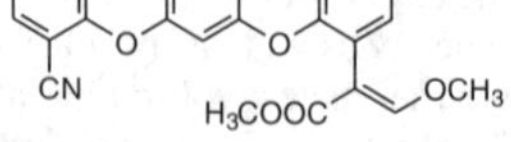

ICI A 5504

Lit.: [1]Schlunegger (ed.), Biologically Active Molecules, p. 9–25, Berlin: Springer 1989; Pestic. Sci. **31**, 499 (1991); Nat. Prod. Rep. **10**, 565 (1993); Sauter, Ammermann, & Roehl, in Copping (ed.), Crop Protection Agents from Nature: Natural Products and Analogues, Cambridge: Royal Soc. Chemis-

Table: Properties and occurrence of Strobilurins.

	molecular formula	M_R	mp. [°C]	$[\alpha]_D$	CAS	occurrence in cultures of
S. A[2,4]	$C_{16}H_{18}O_3$	258.32	oil		52110-55-1	*Strobilurus tenacellus, Agaricus, Bolinea, Cyphellopsis, Favolaschia, Hydropus, Mycena* (16 species!), *Oudemansiella, Pterula, Strobilurus, Xerula*
S. B[2,4]	$C_{17}H_{19}ClO_4$	322.79	rhomb. cryst. 98–99		65105-52-4	*Strobilurus tenacellus, Bolinea, Mycena, Strobilurus, Xerula*
S. C[5]	$C_{21}H_{26}O_4$	342.44	oil		87081-57-0	*Xerula longipes*
S. D[14] (=S. G)[7]	$C_{26}H_{34}O_6$	442.55	oil	+23.8°	129145-64-8	*Agaricus, Cyphellopsis, Bolinea*[7], *Favolaschia*
S. E[6]	$C_{26}H_{32}O_7$	456.54	oil	+78.5°	126572-78-8	*Crepidotus fulvotomentosus*
S. F_1	$C_{16}H_{18}O_4$	274.32	wax			*Agaricus, Cyphellopsis, Favolaschia*
S. F_2[7]	$C_{21}H_{26}O_4$	342.44	cryst., 77.5–78			*Bolinea lutea*
S. H[7]	$C_{17}H_{20}O_4$	288.34	oil		129145-65-9	*Bolinea lutea*
S. I[14]	$C_{21}H_{26}O_6$	374.43	oil	+20°		*Agaricus*
S. K[14]	$C_{26}H_{34}O_6$	442.55	oil	+1.6°		*Mycena tintinnabulum*
S. M[15]	$C_{26}H_{34}O_6$	442.55	oil	+65.5°		*Mycena*
S. N[16]	$C_{21}H_{26}O_7$	390.43	amorphous solid	±0°		*Mycena crocata* (fruit bodies)
S. X[8]	$C_{17}H_{20}O_4$	288.34	cryst., 77.5–78		86421-33-2	*Oudemansiella mucida*
Hydroxy-S. A[10]	$C_{16}H_{18}O_4$	274.32	oil			*Pterula*
Hydroxy-S. D[14]	$C_{26}H_{34}O_7$	448.47	oil	+22°		*Mycena sanguinolenta*
9-Methoxy-S. A	$C_{17}H_{20}O_4$	288.34	oil			*Favolaschia*
9-Methoxy-S. K[14]	$C_{27}H_{36}O_6$	456.58	oil	–8.8°		*Favolaschia*

try 1995; Chem. Br. **31**, 466 (1995) (review); Angew. Chem., Int. Ed. **38**, 1328 (1999). [2]J. Antibiot. **30**, 806 (1977). [3]Biochemistry **25**, 775 (1986); Eur. J. Biochem. **173**, 499 (1988). [4]Chem. Ber. **111**, 2779 (1978); Justus Liebigs Ann. Chem. **1984**, 1616. [5]J. Antibiot. **36**, 661 (1983). [6]J. Antibiot. **43**, 207 (1990); Tetrahedron Lett. **37**, 7955 (1996). [7]J. Antibiot. **43**, 655, 661 (1990). [8]Coll. Czech. Chem. Commun. **48**, 1508 (1983). [9]J. Antibiot. **34**, 1069 (1981). [10]J. Antibiot. **48**, 884 (1995). [11]Angew. Chem., Int. Ed. **34**, 196 (1995); Tetrahedron Lett. **38**, 7465 (1997). [12]Kleinkauf et al. (eds.), The Roots of Modern Biochemistry, p. 657–662, Berlin: De Gruyter 1988; Grabley & Thiericke (eds.), Drug Discovery from Nature, p. 320–334, Heidelberg: Springer 1998. [13]Proc. Brighton Crop Prot. Conf. Pests Dis., Vol. 1, p. 403, 435, Brighton: The British Crop Protection Council 1992. [14]Tetrahedron **55**, 10101 (1999); Tetrahedron Lett. **38**, 7465 (1997). [15]J. Antibiot. **51**, 816 (1998). [16]Z. Naturforsch. C **54**, 463 (1999).
gen.: Spec. Publ., R. Soc. Chem. **198**, 176 ff. (1997).

Strophanthins. Name for a group of *cardenolide *glycosides oxygenated at C-19, occurring mainly in African *Strophanthus* species (Apocynaceae), and used in the treatment of heart diseases. The two most important aglycons are *strophanthidin* (a) {k-strophanthidin, corchorin, 3β,5,14-trihydroxy-19-oxo-5β-card-20(22)-enolide, $C_{23}H_{32}O_6$, M_R 404.50, mp. 169–170 °C (dihydrate), 235 °C (anhydride), $[\alpha]_D$ +43.1° (CH_3OH)} and *ouabagenin* (b) {g-strophanthidin, 1β,3β,5,11α,14, 19-hexahydroxy-5β-card-20(22)-enolide, $C_{23}H_{34}O_8$, M_R 438.52, mp. 235–238 °C (monohydrate), $[\alpha]_D$ +11.3° (H_2O)}.
Strophanthidin is the aglycone of the *k-strophanthins* (the letter "k" is derived from "kombé") occurring

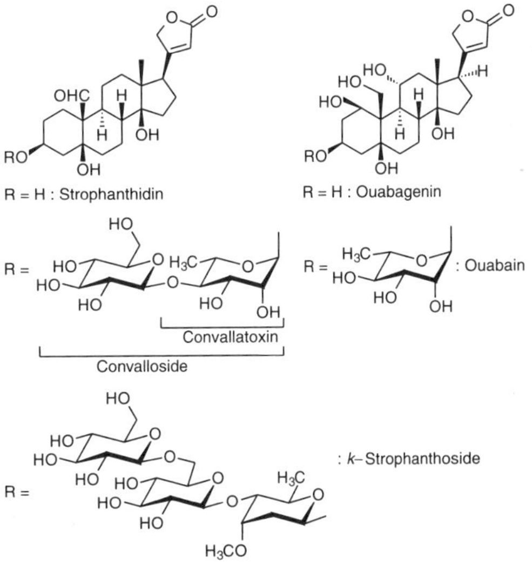

mainly in *Strophanthus kombé*. Depending on their respective carbohydrate component, distinguished are: *k-strophanthin-γ* (c) {*k*-strophanthoside, 3-*O*-[*O*-β-D-glucopyranosyl-(1→6)-*O*-β-D-glucopyranosyl-(1→4)-β-D-cymaropyranosyl]-strophanthidin, $C_{42}H_{64}O_{19}$, M_R 872.96, mp. 199–200 °C, $[\alpha]_D$ +139.9° (CH_3OH)} and *k-strophanthin-β* (d) {3-*O*-[*O*-β-D-glucopyranosyl-(1→4)-β-D-cymaropyranosyl]-strophanthidin, $C_{36}H_{54}O_{14}$, M_R 710.82, mp. 195 °C, $[\alpha]_D$ +32.6° (H_2O)}

lacking a terminal glucose unit. The glycoside containing only cymarose, namely *k-strophanthin-α* (e) {cymarin, cimarin, 3-*O*-(β-D-cymaropyranosyl)-strophanthidin, $C_{30}H_{44}O_9$, M_R 548.67, mp. 148 °C, $[\alpha]_D$ +39° ($CHCl_3$)} is also included among the k-S. Strophanthidin is also present in the glycosides *convallatoxin* (f) ($C_{29}H_{42}O_{10}$, M_R 550.65, mp. 232–240 °C) and *convalloside* (g) ($C_{35}H_{52}O_{15}$, M_R 712.79) from lily of the valley (*Convallaria majalis*). Ouabagenin is the aglycon of the glycoside *ouabain* (h) {g-strophanthin, g-strophanthoside, 3-*O*-(α-L-rhamnopyranosyl)-ouabagenin, $C_{29}H_{44}O_{12}$, M_R 584.66, mp. 190 °C, $[\alpha]_D$ –31° (CH_3OH)} isolated from the ouabaio tree (*Acokanthera ouabaio*=*A. schimperi*), from seeds of *Strophanthus gratus* and related plant species.

S. are highly toxic plant constituents that were used in Africa as arrow poisons. They belong to the group of so-called cardiac glycosides and are used (usually in the form of a mixture of g- and k-S.) in treatment of cardiac insufficiency, acute heart failure, etc. S. preparations are usually administered parenterally and, in contrast to the orally administered *Digitalis glycosides show very rapid onset of action, however, at higher doses they have cardiac paralysing effects.

Lit.: Beilstein E V **18/5**, 123–140, 625 ▪ Hager (5.) **3**, 1103 ff.; **4**, 832 ff.; 977–982; **6**, 795–820 ▪ Kleemann-Engel, p. 1756 ff. ▪ Lewin, Die Pfeilgifte (2.), Hildesheim: Gerstenberg 1984 ▪ Sax (8.), CNH 780, CQH 750, HAN 800, OKS 000, SMM 500–SMN 002 ▪ Steroids **60**, 110 (1995) ▪ Ullmann (5.) **A 5**, 274–282. – *[CAS 66-28-4 (a); 508-52-1 (b); 33279-57-1 (c); 560-53-2 (d); 508-77-0 (e); 508-75-8 (f); 13473-51-3 (g); 630-60-4 (h)]*

Strychnine (strychnidin-10-one).

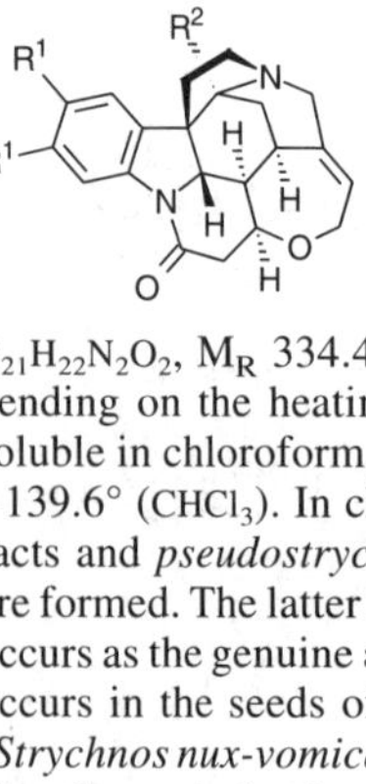

$R^1 = R^2 = H$: Strychnine
$R^1 = OCH_3$, $R^2 = H$: Brucine
$R^1 = H$, $R^2 = OH$: Pseudostrychnine

$C_{21}H_{22}N_2O_2$, M_R 334.42, cryst., mp. 268–290 °C (depending on the heating rate), bp. 270 °C (0.66 kPa), soluble in chloroform, poorly soluble in ethanol, $[\alpha]_D$ –139.6° ($CHCl_3$). In chloroform solution various artifacts and *pseudostrychnine* ($C_{21}H_{22}N_2O_3$, M_R 350.42) are formed. The latter is hydroxylated at C-16 and also occurs as the genuine alkaloid. The *indole alkaloid S. occurs in the seeds of the nux vomica or poison nut (*Strychnos nux-vomica*, Loganiaceae) together with its dimethoxy derivative *brucine, other *Strychnos* species also contain strychnine[1]. The consumption of 1–3 g of the seeds is lethal for humans.

The highest-yielding source of S. known to date is the bark of *Strychnos icaja* with a content of 6.6%. S. has an extremely bitter taste and can be detected down to a dilution of 1:130000, see also bitter principles. The poison nut tree possibly uses S. as a defence against insect attacks. S. forms salts with many acids. S. is highly toxic for all organisms (including unicellular), maximum tolerable working place concentration: 0.15 mg/m^3; LD_{50} (rat i. v.) 0.96 mg/kg.

The lethal dose for adults begins at 1 mg/kg. The extremely potent stimulating activity on the central nervous system (CNS) leads to violent convulsions under full consciousness and can lead to death through respiratory paralysis. S. thus acts similarly to tetanus toxin.

Toxicology: S. is rapidly absorbed after oral uptake, is metabolized mainly in the liver, and rapidly eliminated. Symptoms of intoxication are initial restlessness, anxiety, vomiting, then convulsions of the extensor muscles (1 or more/minute), the convulsions are again triggered by external stimuli. Therapy for acute poisoning: gastrolavage and administration of activated charcoal – for severe poisoning: barbiturate narcosis and administration of muscle relaxants.

The antagonistic action of S. towards the neurotransmitter glycine at the synapses can be used to isolate the glycine receptors since S. binds to them selectively and with high affinity[2]. Low doses of S. in humans lead to an improved sensory perception and to stimulation of the respiratory center as well as muscular and vascular tone (abuse as doping agent in sports). The former therapeutic uses of S. as analeptic agent, as an antidote for barbiturate and other centrally paralyzing sleeping agent poisonings, as well as antagonist for *curare were based on this CNS activity. S. is also occasionally used as a rodenticide (rat poison) or to trap animals without damage to their furs, in some parts of Africa *Strychnos* extracts are also used as arrow poisons.

Historical: The fruits of *Strychnos nux-vomica* (*nux vomica*, poison nut) were first described by Valerius Cordus in 1540; the alkaloid was isolated in 1818 by Caventou and Pelletier. The structure of S. was elucidated by Leuchs, Robinson, Woodward, and Prelog; Woodward achieved the synthesis (30 reaction steps) in 1954 as the first total synthesis of a structurally complicated natural product. For further syntheses, see *Lit.*[3].

Lit.: [1]Hager (5.) **6**, 817–850; **9**, 676–680. [2]Angew. Chem., Int. Ed. **24**, 365 (1985). [3]Angew. Chem., Int. Ed. **33**, 1144 (1994) (review); **38**, 395 (1999); J. Am. Chem. Soc. **115**, 8116, 9293 (1993); **117**, 5776 (1995); J. Org. Chem. **58**, 7490 (1993); **59**, 2685 (1994); **63**, 9427, 9434 (1998); Saxton (ed.), The Monoterpenoid Indole Alkaloids, Suppl. vol. 25, Heterocyclic Compounds, p. 279, Chichester: John Wiley 1994; F. A. Z., 29.7. **1992**.

gen.: Aldrichimica Acta **28**, 107–120 (1995) ▪ Beilstein E III/IV **27**, 7537 ▪ Dang. Prop. Ind. Mat. Rep. **8**, 78–83 (1988) ▪ Florey **15**, 563–646 ▪ Trends Pharmacol. Sci. **8**, 204 ff. (1987). – *Toxicology:* Biochemistry **30**, 873 (1991) ▪ Sax (8.), SMN 500–SMP 000 ▪ see also Strychnos alkaloids. – *[HS 2939 90; CAS 57-24-9 (S.), 465-62-3 (pseudo-S.)]*

Strychnos alkaloids. A group of *monoterpenoid indole alkaloids that are not exclusively isolated from *Strychnos* plants and do not exclusively belong to the strychnidine type. The most well-known members of this group of alkaloids are *strychnine, its methoxylated derivative *brucine, the *toxiferines, and *curare, as well as *vomicine which all have structures derived from the strychnidine skeleton.

Lit.: J. Am. Chem. Soc. **113**, 5085 f. (1991) ▪ J. Org. Chem. **55**, 6299–6312 (1990) ▪ Manske **34**, 211–331; **36**, 1 ▪ Nat. Prod. Rep. **9**, 425 (1992) ▪ Saxton & Husson, Indoles, Part 4: The Monoterpenoid Indole Alkaloids, Chichester: Wiley-Interscience 1983 ▪ Tetrahedron: Asymmetry **8**, 935–948 (1997). – *[HS 2939 90]*

Stubomycin see hitachimycin.

Stylopine see protoberberine alkaloids.

Styrylpyrones.

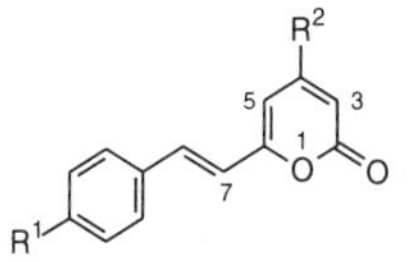

$R^1 = R^2 = H$: 6-Styryl-2-pyrone
$R^1 = R^2 = OH$: Bisnoryangonin
$R^1 = R^2 = OCH_3$: Yangonin

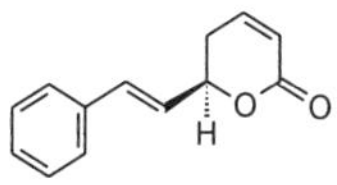

(*R*)-(+)-Goniothalamin

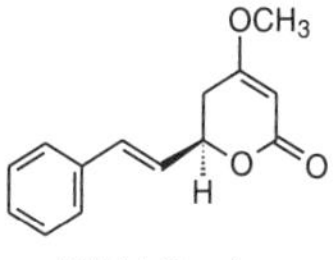

(*R*)-(+)-Kawain

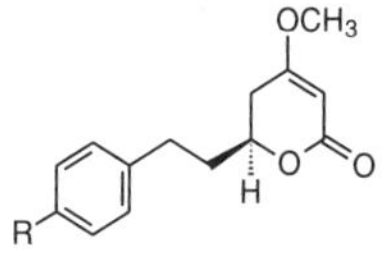

R = H : (*S*)-(+)-Marindinin
R = OCH_3 : (*S*)-(+)-5,6,7,8-Tetrahydroyangonin

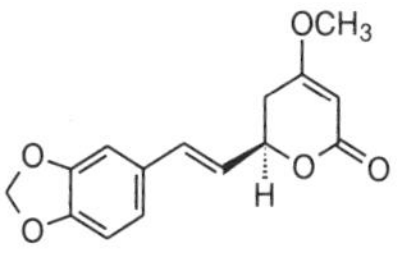

(*R*)-(+)-Methysticin

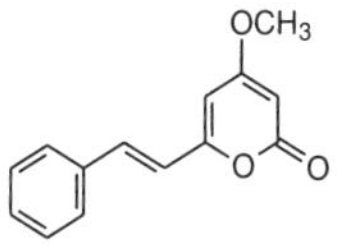

5,6-Dehydrokawain

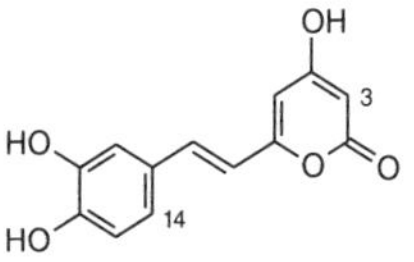

Hispidin

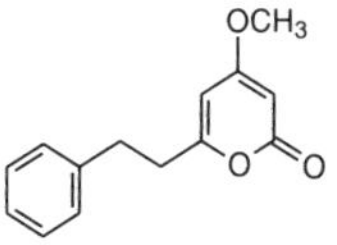

5,6-Dehydro-7,8-dihydrokawain

Large group of natural products, often with noteworthy pharmacological properties. The parent compound, 6-styryl-2-pyrone, occurs in *Goniothalamus* (Annonaceae) and *Aniba parviflora*. Goniothalamin from *Cryptocarya caloneura* and *Goniothalamus* are cytotoxic (for synthesis, see *Lit.*[1]). Goniothalamin 7,8-epoxide isolated from *G. macrophyllus* exhibits embryotoxic properties. Roots of the pepper plant *Piper methysticum* (Piperaceae, kava drug), indigenous to Oceania, contains, among others, *p*-methoxy-S. kawain, 5,6-dehydrokawain, 5,6-dehydro-7,8-dihydrokawain, yangonin, 5,6,7,8-tetrahydroyangonin, marindinin, the methylenedioxy derivative methysticin, and its derivatives. A narcotic drink (kava-kava) prepared from extracts of this plant has vasodilatory and sedative properties. Kawain and methysticin exhibit analgesic activity[3], the psychotropic properties of the drug are attributed to kawain and some other components (see *Lit.*[4]). Suggestions that these S. bind to GABA or benzodiazepine receptors have been invalidated[5]. *Hispidin*[6] is the yellow pigment of several polypores such as *Inonotus hispidus* (Aphyllophorales) and *Phaeolus schweinitzii*. Hispidin is a lead pigment in the basidiomycetes genera *Phellinus*, *Gymnopilus*, *Hypholoma*, and *Pholiota*, in which it is often accompanied by *bisnoryangonin*. When the fruit bodies ripen it is enzymatically oxidized to brown-black polymers ("fungal lignin"). Dimeric derivatives of S. include the *hypholomins and 3,14′-bihispidinyl that occurs in polypores. Hispidin has antibiotic activity against Gram-positive bacteria.

Biosynthesis: From (*p*-hydroxy)-cinnamoyl-CoA+2 malonyl-CoA.

Table: Data of Styrylpyrones.

	molecular formula	M_R	mp. [°C] [bp.]	optical activity	UV λ_{max} [nm] (log ε)	CAS
6-Styryl-2-pyrone	$C_{13}H_{10}O_2$	198.22	115–116 (yellow cryst.)			32065-58-0 [(*E*)-form]
Goniothalamin	$C_{13}H_{12}O_2$	200.24	85	$[\alpha]_D$ +170.3° ($CHCl_3$) $[\alpha]_D^{20}$ +135° (CH_3OH)		17303-67-2 [(*R*)-(+)-form]
Goniothalamin-7,8-epoxide	$C_{13}H_{12}O_3$	216.24	90–94	$[\alpha]_D$ +100.7° ($CHCl_3$)		110011-51-3
Kawain	$C_{14}H_{14}O_3$	230.26	106 (prisms) [195–197 (13 Pa)]	$[\alpha]_D^{28}$ +105° (C_2H_5OH)	CH_3OH: 210 (4.38) 245 (4.44), 282 (2.81)	500-64-1 [(*R*)-(+)-form]
5,6-Dehydro-kawain	$C_{14}H_{12}O_3$	228.25	138–140 (yellow cryst.)			1952-41-6, 15345-89-9 [(*E*)-form]
7,8-Dihydro-kawain (Marindinin)	$C_{14}H_{16}O_3$	232.28	56–60 (prisms)	$[\alpha]_D^{19}$ +30° (C_2H_5OH)	CH_3OH: 236 (4.14)	587-63-3 [(*S*)-(+)-form]
Yangonin	$C_{15}H_{14}O_4$	258.27	153–154 (green-yellow cryst.)		C_2H_5OH: 360 (4.33)	500-62-9 [(*E*)-form]
5,6,7,8-Tetra-hydroyangonin	$C_{15}H_{18}O_4$	262.31	89–90 (needles)	$[\alpha]_D^{20}$ +20° ($CHCl_3$)		49776-58-1 [(*S*)-(+)-form]
Methysticin	$C_{15}H_{14}O_5$	274.27	139–140 (132–134)	$[\alpha]_D^{28}$ +95° (acetone)	CH_3OH: 226 (4.40), 267 (4.14), 306	495-85-2 [(*R*)-(*E*)-(+)-form]
Hispidin	$C_{13}H_{10}O_5$	246.22	259 (yellow needles)	(3.93)	555-55-5	
Bisnoryangonin	$C_{13}H_{10}O_4$	230.22	240–244 (yellow cryst.)			13709-27-8

Lit.: [1]J. Chem. Soc., Perkin Trans. 1 **1991**, 519–523; Heterocycles **35**, 281–288 (1993); Justus Liebigs Ann. Chem. **1992**, 809 ff.; Tetrahedron Asymmetry **3**, 853–856 (1992). [2]J. Chem. Soc., Chem. Commun. **1992**, 1640 f.; J. Chem. Soc., Perkin Trans. 1 **1992**, 1907–1910; J. Nat. Prod. (Lloydia) **54**, 1034–1043, 1077–1081 (1991); Phytochemistry **31**, 2851–2853 (1992); Tetrahedron Lett. **33**, 3333 f. (1992). [3]Clin. Exp. Pharmacol. Physiol. **17**, 495–507 (1990). [4]Acta Ther. **14**, 249–256 (1988); Arzneim.-Forsch. **41**, 673–683 (1991); Eur. J. Pharmacol. **215**, 265–269 (1992); Hum. Psychopharmacol. **4**, 169–190 (1989); Planta Med. **60**, 88–90 (1994). [5]Pharmacol. Toxicol. (Copenhagen) **71**, 120–126 (1992). [6]Zechmeister **51**, 88–92.

gen.: Beilstein E V **17/10**, 317, 370; **18/1**, 495; **18/2**, 23, 108; **18/3**, 273, 534; **18/4** 414, 453; **19/10**, 456 ▪ J. Chem. Soc., Perkin Trans. 1 **1989**, 1793–1798 (synthesis) ▪ Luckner (3.), p. 404 ▪ Nat. Prod. Rep. **12**, 101–133 (1995) ▪ Phytochemistry **36**, 23–27 (1994) ▪ Rapid Commun. Mass Spectrom. **7**, 219–224 (1993) (MS) ▪ Sax (8.), DLR 000, GJI 250 (toxicology) ▪ Synthesis **1994**, 1133 ▪ Zechmeister **54**, 1–35.

Suberin, suberic acid. Suberin (Latin: suber = cork) in plants forms the lipophilic protective layer on the outer dermal tissue (cork) of trunks and branches as well as roots and other underground plant parts (see cutin). S. is also formed in plant tissues during wound healing. S. is a mixture of polyesters, the main components are α,ω-dicarboxylic acids, e. g., docosanedioic acid (see Japan wax) and tetracosanedioic acid ($C_{24}H_{46}O_4$, M_R 398.63, mp. 127.1 °C), ω-*hydroxy fatty acids, *fatty acids, and fatty alcohols as well as phenolic compounds.

Suberic acid (cork acid, octanedioic acid $C_8H_{14}O_4$, M_R 174.20, mp. 144 °C) is formed from cork on heating with nitric acid.

Lit.: Beilstein E IV **2**, 2028 (suberic acid) ▪ Holzforschung **51**, 219, 225 (1997) (review) ▪ J. Chem. Soc., Perkin Trans. 1 **1984**, 1061 (synthesis) ▪ see cutin. – *[HS 2917 19; CAS 505-48-6 (suberic acid); 2450-31-9 (tetracosanedioic acid)]*

Substance F. $C_{33}H_{38}N_4Na_2O_8$, M_R 664.67, uv_{max} 476 nm. S. belongs to the *bile pigments.

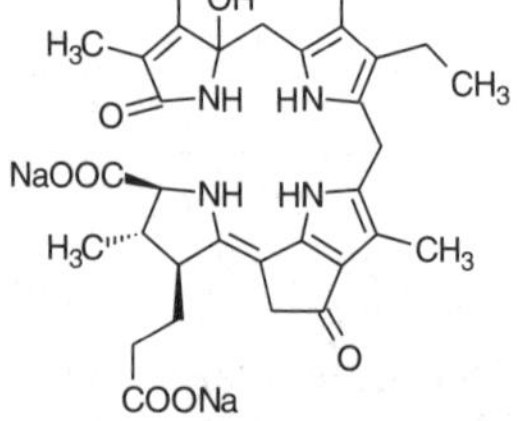

Occurrence: S. is the fluorescing substance of krill crayfish (*Euphasia pacifica*), it is structurally and biologically closely related to the *luciferins from dinoflagellates. On the basis of its substitution pattern it can be assumed that S. and the dinoflagellate luciferins are formed by cleavage of *chlorophyll.

Lit.: J. Am. Chem. Soc. **110**, 2683 (1988).

Substance P (abbreviation: SP or PPS, for pain producing substance, neurokinin P, abbreviation: NKP).

Arg–Pro–Lys–Pro–Gln–Gln–Phe–Phe–Gly–Leu–Met–NH_2

$C_{63}H_{98}N_{18}O_{13}S$, M_R 1347.64, crystals from aqueous acetic acid as the triacetate tetrahydrate, $[\alpha]_D^{27}$ −76° (5% acetic acid), a peptide hormone produced in APUD cells (from amine precursor uptake and decarboxylation) of the intestines and in the hypothalamus; these cells also form biogenic amines. SP can also be isolated from the central nervous system and gastrointestinal tract (especially the duodenum and small bowel) [1]. SP, which can be inactivated by trypsin, chymotrypsin, oxidation agents, air, and high pH values, belongs to the tachykinines; the latter are formed from various polyproteins [2] (e. g., β-preprotachykinin) and bind especially to the NK-1 receptors (neurokinin-1 receptor). SP has a stimulating effect on intestinal peristalsis, has vasodilatory and blood pressure lowering properties, and stimulates salivation. As a neurotransmitter SP is presumed to participate in the transmission of pain signals whereby *enkephalins and *endorphins can have an inhibitory effect; in addition SP interacts with the hormones *serotonin, thyroliberin, and pancreozymin as well as the neurotransmitter *acetylcholine. For the role of SP in inflammations, see *Lit.*[3]. The fragment with the amino acid sequence 4-11 shows the same biological efficacy as SP[4]. Synthetic modulators of the SP receptor are being investigated as analgesics with novel mechanisms of action[5].

Lit.: [1]J. Physiol. **72**, 74–87 (1931). [2]Bioessays **10**, 62–69 (1989). [3]Annu. Rev. Med. **40**, 341–352 (1989). [4]Chem. Ind. (London) **1991**, 792 ff. [5]Actual. Chim. Ther. **22**, 83 (1996); Science **251**, 435, 437 (1991).

gen.: Ann. N. Y. Acad. Sci. **512**, 465–475 (1987) (review) ▪ Chem. Ind. (London) **1991**, 792 (review) ▪ Dockray, Gut Pept. 1994 (Walsh, ed.), p. 401, New York: Raven Press 1994 ▪ Henry et al., Substance P and the Neurokinins, New York: Springer 1987 ▪ Jordan, Substance P: Metabolism and Biological Action, London: Taylor and Francis 1985 ▪ J. Org. Chem. **48**, 1757 ff. (1983) (synthesis) ▪ Merck-Index (12.), No. 9032 ▪ Pharmacol. Rev. **46**, 551 (1994) ▪ Proc. Natl. Acad. Sci. USA **88**, 10 042 ff. (1991) ▪ Regul. Pept. **46**, 1 (1993). – *[HS 2937 90; CAS 33507-63-0]*

Subtilin. A *peptide antibiotic of the *lantibiotics group isolated from cultures of *Bacillus subtilis* [$C_{148}H_{227}N_{39}O_{38}S_5$, M_R 3320.96, $[\alpha]_D^{23}$ −29 → −35° (structure, see *Lit.*[1])] produced, like *nisin by fermentation[2]. It consists of 32 amino acids (including eight rare acids such as *lanthionine, β-methyllanthionine, and α,β-unsaturated amino acids) and contains 5 thioether rings. S. acts on bacteria undergoing cell division and germinating bacterial spores by forming pores

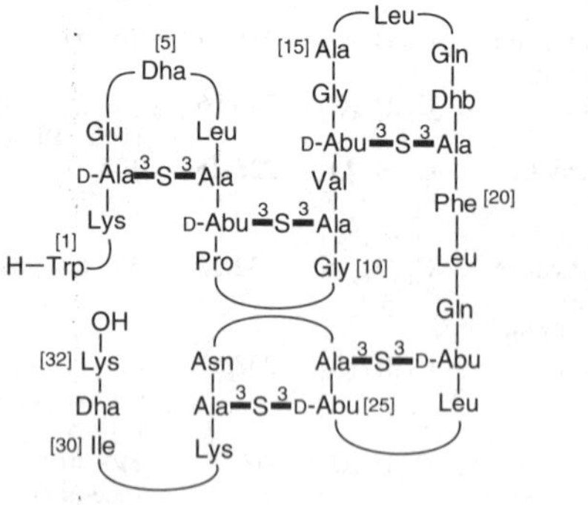

Abu: 2-Aminobutyric acid
Dha: Dehydroalanine (2-Aminoacrylic acid)
Dhb: (Z)-Dehydrobutyrine (2-Amino-2-butenoic acid)
D-Ala–S–Ala: *meso*-Lanthionine
D-Abu–S–Ala: (2*S*,3*S*,6*R*)-3-Methyllanthionine

in the bacterial membranes[3,4], thus reducing the heat resistance of the bacterial spores. S. is used in some countries as a food preservation agent.

Lit.: [1] J. Am. Chem. Soc. **75**, 2391 (1953); FEBS Lett. **300**, 56 (1992). [2] Biochem. Biophys. Res. Commun. **116**, 751 (1983). [3] Eur. J. Biochem. **182**, 181 (1989). [4] FEBS Lett. **244**, 99 (1989). *gen.:* Angew. Chem., Int. Ed. **30**, 1051 (1991) ▪ J. Food Saf. **10**, 1119 (1990) ▪ Kirk-Othmer (4.) **3**, 268. – *[HS 2941 90; CAS 1393-38-0]*

Succinic acid (butanedioic acid).
HOOC–CH_2–CH_2–COOH, $C_4H_6O_4$, M_R 118.09. Crystals with a strong sour taste, D. 1.56, mp. 185–187 °C, bp. 235 °C with formation of the anhydride, pK_{a1} 4.22, pK_{a2} 5.70 (25 °C/water); LD_{50} (rat p. o.) 2260 mg/kg. S. is a strong eye irritant. The salts are called succinates. S. was first isolated in 1546 by Agricola through dry distillation of amber. It occurs as intermediate in the citric acid cycle and is present in numerous plants and microorganisms.
Use: As flavor substance for foods; the K, Ca, and Mg salts of S. are registered under German food law for use as sodium chloride substitute in dietary foods.
Lit.: Beilstein E IV **2**, 1908–1913 ▪ Bergmeyer, Methods of Enzymatic Analysis (3.), vol. 7, p. 25–33, Weinheim: Verl. Chemie 1985 (detection) ▪ Karrer, No. 837 ▪ Merck-Index (12.), No. 9038 ▪ Ullmann (5.) **A 8**, 530; **A9**, 450. – *Toxicology:* Sax (8.), SMY 000. – *[HS 2917 19; CAS 110-15-6]*

***o*-Succinylbenzoic acid.**

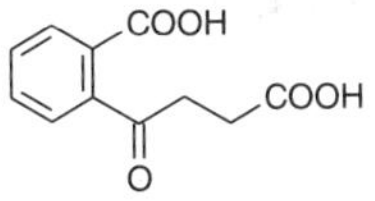

$C_{11}H_{10}O_5$, M_R 222.20, mp. 135–137 °C. An intermediate in the biosynthesis of 1,4-dihydroxynaphthalene-2-carboxylic acid, which in turn is a precursor of *alizarin(e), menaquinone (vitamin K_3), and *vitamin K_1 (phylloquinone). The biosynthesis proceeds from *isochorismic acid and 2-oxoglutaric acid.
Lit.: Eur. J. Biochem. **192**, 441–449 (1990) ▪ Haslam, Shikimic Acid: Metabolism and Metabolites, p. 220 ff., Chichester: Wiley 1993 ▪ Luckner (3.), p. 329 ff. ▪ Nat. Prod. Rep. **12**, 123 f., 101–133 (1995). – *[CAS 27415-09-41]*

Sucrose (cane sugar, beet sugar, saccharose).

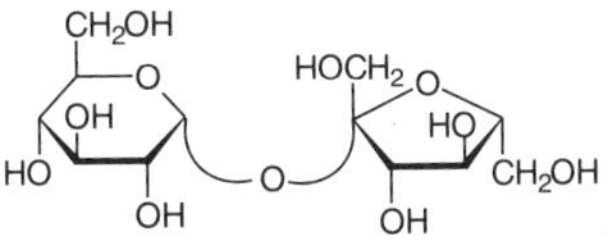

$C_{12}H_{22}O_{11}$, M_R 342.30. Sweet tasting prisms, mp. 185–186 °C (decomp., caramelization), $[\alpha]_D^{20}$ +66.4° (H_2O); the crystals exhibit triboluminescence. S. is a *disaccharide, *β-D-fructofuranosyl-α-D-glucopyranoside* = α-D-glucopyranosyl-β-D-fructofuranoside (Glcα1-2βFru) made up of α-glucose and β-fructose.
Occurrence: S. is widely distributed in the plant kingdom, present in larger amounts in many fruits and tree juices (including maple syrup). Some plants are extremely rich in S., e. g., sugar cane (*Saccharum officinarum*, Poaceae, 15–20% S.) and sugar beet (*Beta vulgaris saccharifera*, Chenopodiaceae, 5–6% S., increased to 17–20% by selective cultivation). S. is not found in the animal kingdom.
Use: S. is the economically most important sugar compound and is produced in larger amounts; for further use as an industrial feedstock, see also *Lit.*[1]. Besides its main use in the food industry and for conservation purposes (jams, etc.), S. can also be used in plastics, varnishes, as fermentation additive for the production of proteins, amino acids, yeasts, enzymes, antibiotics, steroids, fats, ethanol, ethylene, glycerol, sorbitol, sugar acids, explosives, tartaric acid, softeners, saccharose esters for detergents, etc. S.-propylene oxide adducts are used in hard PUR expanded plastics.
Lit.: [1] Pure Appl. Chem. **56**, 833–844 (1984).
gen.: Chen, Cane Sugar Handbook, New York: Wiley 1985 ▪ Kirk-Othmer (4.) **4**, 922; **10**, 323; **23**, 1–87 (review) ▪ Mathlouthi, Sucrose: Properties and Applications, London: Blackie 1995 ▪ Pure Appl. Chem. **56**, 833 (1984) ▪ Ullmann (5.) **A 4**, 65 f.; **A5**, 83 ▪ Zechmeister **55**, 114–184 ▪ Zuckerindustrie **115**, 555–565 (1990). – *[HS 1701 11, 1701 12, 1701 99; CAS 57-50-1]*

Sugar alcohols see hexitols.

Suillin.

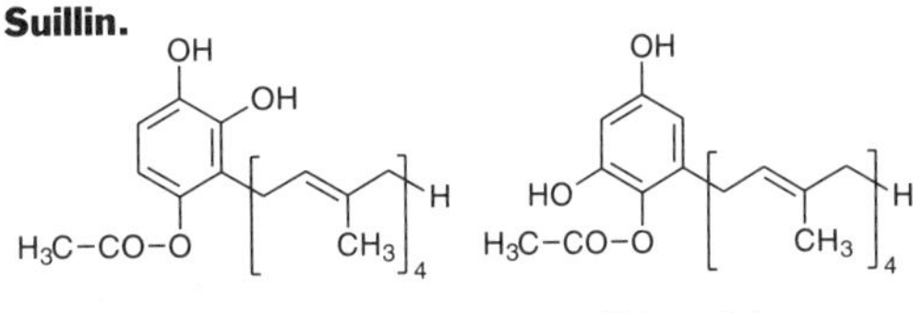

Suillin   Bolegrevilol

$C_{28}H_{40}O_4$, M_R 440.62, needles, mp. 108 °C. A polyprenylated phenol from the mushrooms *Suillus variegatus*, *S. granulatus*, *S. luteus*, and other *Suillus* species (Basidiomycetes)[1]. S. is particularly abundant in the cap skin of younger, less colored fruit bodies. S. is sensitive to oxidation and takes on a darker color with time; this may be the reason for the darker cap skin of older fruit bodies. S. is cytotoxic and shows considerable antitumor activity[2]. The isomeric *bolegrevilol* (colorless oil) can be isolated from *Suillus grevillei*[4] and, together with related compounds, from *S. granulatus*[3]. It inhibits peroxidation of rat hepatocytes (IC_{50} 1 μg/mL)[4] and has antibacterial activity[3]. Related polyprenylphenols are found in *Chroogomphus rutilus* (Basidiomycetes)[5].
Lit.: [1] Z. Naturforsch. B **41**, 645 (1986). [2] J. Nat. Prod. **52**, 844 (1989). [3] J. Nat. Prod. **52**, 941 (1989). [4] Chem. Pharm. Bull. **37**, 1424 (1989). [5] Planta Med. **58**, 383 (1992). – *[CAS 103538-03-0 (S.); 123941-64-0 (bolegrevilol)]*

Sulcatins.

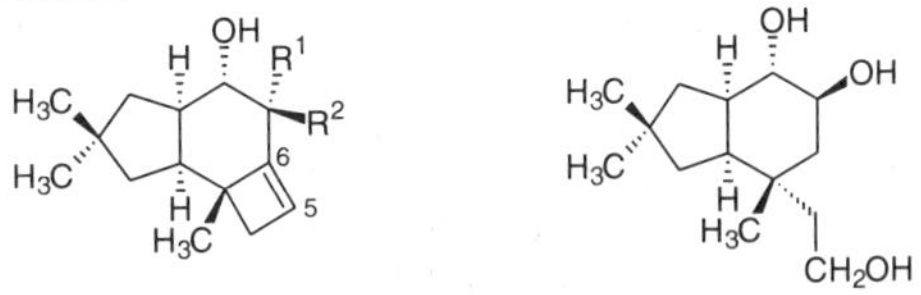

R^1 = H, R^2 = OH : Sulcatin A
R^1 = OH, R^2 = CH_2OH : Sulcatin B

Sulcatin D

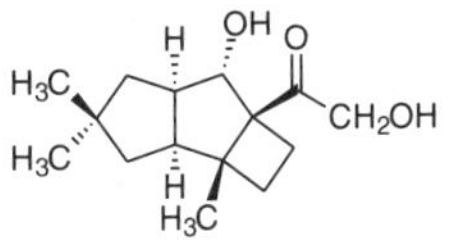

Sulcatin G

Sesquiterpenoids of various types derived from *protoilludane isolated from cultures of the taiga fungus (*Laurilia sulcata*, Basidiomycetes). *S. A* {$C_{14}H_{22}O_2$, M_R 222.33, needles, mp. 155–160 °C, $[\alpha]_D$ −61.3° ($CHCl_3$)} and *S. B.* {$C_{15}H_{24}O_3$, M_R 252.35, cryst., mp. 90–92 °C, $[\alpha]_D$ −43.4° (CH_3OH)} possess an unusual 5,6-double bond (cyclobutene). The absolute configuration of S. A dibenzoate was determined by exciton chirality methods. In some sulcatins such as *S. D* {$C_{14}H_{26}O_3$, M_R 242.36, cryst., mp. 116–118 °C, $[\alpha]_D$ −29.4° ($CHCl_3$)} the cyclobutane ring has been cleaved to furnish a new carbon skeleton (norisoilludalane). *S. G* {$C_{15}H_{24}O_3$, M_R 252.35, oil, $[\alpha]_D$ +44.5° ($CHCl_3$)} is derived from S. B by rearrangement with contraction of the cyclohexane ring. Most S. have weak antibiotic activity.

Lit.: J. Chem. Soc., Perkin Trans. 1 **1992**, 615; **1993**, 2723 ▪ Phytochemistry **26**, 1739 (1987); **31**, 2047 (1992). – *[CAS 110219-85-7 (S. A); 143049-04-1 (S. B); 141358-26-1 (S. D); 153415-33-9 (S. G)]*

Sulcatol (6-methyl-5-hepten-2-ol).

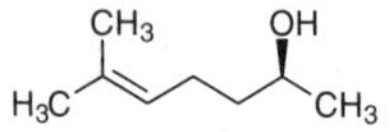

(+)-(*S*)-form

$C_8H_{16}O$, M_R 128.21, liquid, bp. 85–86 °C (2.66 kPa), (*S*)-enantiomer: $[\alpha]_D^{23}$ +17.4° (neat). Aggregation pheromone of *bark beetles of the genus *Gnathotrichus*. It occurs in various species in differing enantiomeric ratios. S. is a widely distributed natural product, it is found in the urine of vertebrates, in microorganisms, in plant fragrance substances, etc. and is possibly a degradation product of open-chain terpenes; see also pityol.

Lit.: Agric. Biol. Chem. **54**, 1577 (1990) ▪ ApSimon **9**, 280–285 (synthesis) ▪ Eur. J. Org. Chem. **1999**, 117–127 (synthesis) ▪ Indian J. Chem., Sect. B **32**, 615 (1993) (synthesis) ▪ J. Chem. Ecol. **5**, 79–88 (1979). – *[CAS 1569-60-4 ((+)-S.); 4630-06-2 ((±)-S.)]*

Sulcatoxanthin see peridinin.

Sulfazecin (SQ 26445).

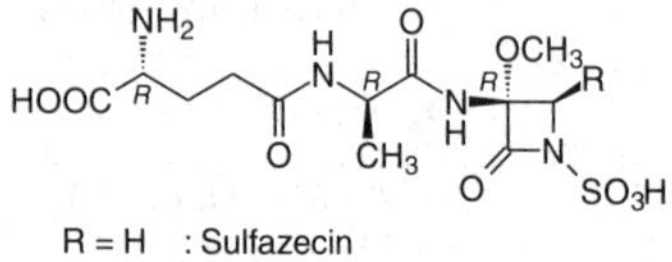

R = H : Sulfazecin
R = CH_3: MM 42842

β-Lactam antibiotic isolated from fermentation cultures of *Pseudomonas acidophila* and *Gluconobacter* sp.[1]. $C_{12}H_{20}N_4O_9S$, M_R 396.37, mp. 168–170 °C, needles, $[\alpha]_D$ +82° (H_2O). S. is readily soluble in water, insoluble in ethanol and chloroform. A noteworthy feature of this monocyclic *β*-lactam is the rare sulfamate group. S. exhibits acidic properties and is unstable in alkaline and strongly acidic media. S. has good activity against Gram-negative bacteria but only weak activity against Gram-positive species. *Isosulfazecin* and the antibiotic *MM 42842* are structurally closely related to sulfazecin. Isosulfazecin[2] is isolated from *P. mesoacidophila* sp. nov. and is the L-alanine epimer of sulfazecin. The physicochemical properties are practically identical to those of S., except $[\alpha]_D^{25}$ +4.5° (H_2O). The antibiotic MM 42842 [$C_{13}H_{22}N_4O_9S$, M_R 410.40, $[\alpha]_D^{20}$ +34° (H_2O as Na salt)][3], isolated from cultures of *P. cocovenenans*, differs from S. by a methyl group at C_4 and by its (albeit weak) activity against Gram-positive bacteria. The *β*-lactam ring of S. is formed biosynthetically from serine. The *O*-methyl group originates from methionine, and the sulfur atom from cystine[4].

Lit.: [1]J. Antibiot. **34**, 621–627 (1981). [2]J. Antibiot. **34**, 1081–1089 (1981). [3]J. Antibiot. **41**, 1, 7 (1988). [4]Antimicrob. Agents Chemother. **23**, 598–602 (1983).

gen.: Chem. Pharm. Bull. **31**, 2200 (1983) (synthesis). – *[HS 2941 90; CAS 77912-79-9 (S.); 80734-22-1 (Na salt); 77900-75-5 (isosulfazecin); 113784-28-4 (MM 42842)]*

Sulfinemycin.

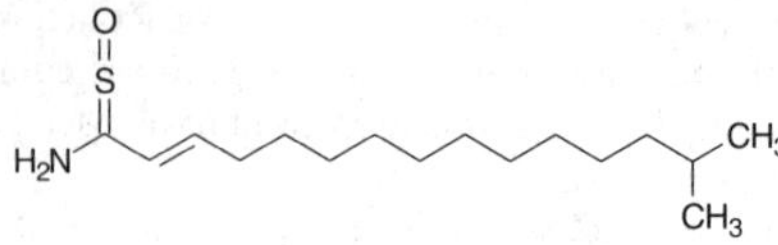

$C_{16}H_{31}NOS$, M_R 285.49, yellow amorphous powder, mp. 84 °C. An unusual thioamide *S*-oxide from cultures of *Streptomyces albus*. On heating to >105 °C it loses HSOH and is transformed to the *α,β*-unsaturated nitrile. S. has weak anthelmintic activity.

Lit.: J. Antibiot. **48**, 282 (1995). – *[HS 2941 90; CAS 162857-73-0]*

Sulforaphane, sulforaphene see mustard oils.

Sulphurenic acid.

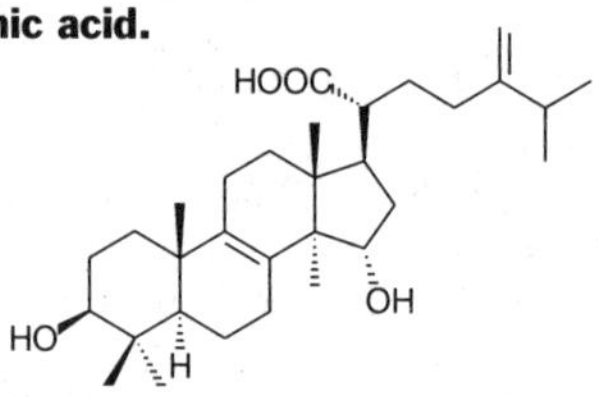

$C_{31}H_{50}O_4$, M_R 486.74, cryst., mp. 252–254 °C, $[\alpha]_D$ +42° (pyridine). S. was first isolated in ca. 7% yield from dried mycelia of the polypore fungus *Laetiporus sulphureus* (Aphyllophorales). The C_{31}-triterpene acid with the eburicane skeleton also occurs in the fruit bodies of other polypores.

Lit.: Turner **2**, 324 (1983) (occurrence). – *[CAS 1260-08-8]*

Sumatrol see rotenoids.

Supellapyrone see cockroaches.

(−)-Supinidine see necine(s).

Surugatoxin see neosurugatoxin.

Suspensolide [(3*E*,8*E*)-4,8-dimethyl-3,8-decadien-10-olide].

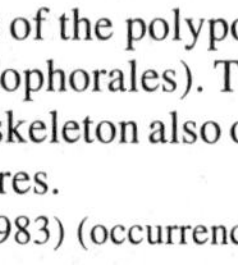

$C_{12}H_{18}O_2$, M_R 194.27. Besides *anastrephin, S. is an important component in the *pheromone systems of tropical fruit-*flies of the genus *Anastrepha*[1]. The biogenesis of S., which is possibly a precursor of anastrephin, presumably involves the sesquiterpene alcohol *farnesol[2].

Lit.: [1] Tetrahedron Lett. **29**, 6561–6564 (1988). [2] J. Chem. Ecol. **18**, 223–244 (1992).
gen.: Synthesis: ApSimon **9**, 362–363 ▪ Chem. Rev. **95**, 789-828 (1995). – *[CAS 111351-08-7]*

Swainsonine [(1*S*)-8a*β*-octahydro-1*α*,2*α*,8*β*-indolizinetriol].

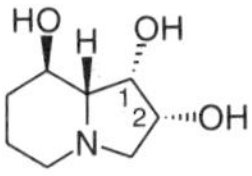

$C_8H_{15}NO_3$, M_R 173.21, mp. 144–145 °C, $[\alpha]_D^{20}$ –85.1° (CH_3OH). An *indolizidine alkaloid from the Fabaceae *Swainsona canescens*, *Astragalus lentiginosus*, and phytopathogenic fungi of the genus *Rhizoctonia*. In grazing animals S. causes the so-called "locoism", a condition with symptoms resembling those of the genetic disease mannosidosis. As an *α*-mannosidase inhibitor, S. acts on the synthesis of *N*-glycosidic oligosaccharide side chains in the Golgi apparatus. This is the basis for its inhibitory effects on growth and metastasis formation in some types of cancer. In addition, S. promotes the production of bone marrow cells and in some cases strengthens the endogenous defence against tumors. S. is in clinical phase II studies against breast cancer[1].
Lit.: [1] R&D Focus, Drug News (1. 3.) 1999, 7.
gen.: Beilstein E V **21/5**, 483 ▪ Manske **28**, 287 ▪ Pelletier **5**, 1. – *Synthesis:* Bioorg. Med. Chem. **7**, 831 (1999) ▪ Bull. Soc. Chim. Fr. **134**, 615 (1997) ▪ Chem. Pharm. Bull. **38**, 2712–2718 (1990); **41**, 1717 (1993) ▪ J. Am. Chem. Soc. **112**, 8100–8112 (1990) ▪ J. Org. Chem. **53**, 5209, 6022–6030 (1988); **61**, 7217 (1996); **62**, 1112–1124 (1997); **63**, 6281 (1998) ▪ Synlett **1995**, 404 ff.; **2000**, 53 ▪ Tetrahedron **43**, 3083–3093 (1987) ▪ Tetrahedron Lett. **31**, 7571 (1990); **36**, 5049 (1995). – *[HS 2939 90; CAS 72741-87-8]*

Sweet basil oil. Pale yellow to yellowish-green oil with a fresh, greeny, sweet, spicy more or less anise-like odor and taste[1].
Production: From the herb: sweet or common basil, *Ocimum basilicum*, by steam distillation. The "European-Mediterranean" type is cultivated mainly in the Mediterranean region, the "exotic" type in Madagascar, Réunion, Cameroon, and India.
Composition[2]***:*** The main constituent of exotic B. is **estragole* (70–90%); the Mediterranean B. contains less estragole (2–30%), with **linalool* being the main component (40–60%) and usually also large amounts of **eugenol* (up to 10%).
Use: Mainly in the manufacture of perfumes with spicy (for men) and green notes, and as an aromatic spice for foods. Indian B. is also used for isolation of pure estragole, from which *anethole can be obtained by double bond isomerization.
Lit.: [1] Fenaroli (2.) **1**, 284. [2] Perfum. Flavor. **11** (3), 49 (1986); **14** (5), 45 (1989); **15** (5), 65 (1990); **17** (4), 47 (1992); **20** (4), 35 (1995).
gen.: Arctander, p. 81 ▪ Bauer et al. (2.), p. 138 – *Toxicology:* Food Cosmet. Toxicol. **11**, 867 (1973). – *[HS 3301 29; CAS 8015-73-4]*

Sweeteners, natural. Noncaloric *sucrose substitutes for use in the sweetening of foods, beverages and medicines may be either synthetic compounds or natural products. Highly sweet, potentially noncariogenic sweeteners from plants are used in Japan and some other countries, including the diterpene glycoside *stevioside and the protein *thaumatin. Another S. of commercial importance in Japan is the triterpenoid glycoside *glycyrrhizin, of minor importance are mogroside A (a cucurbitane triterpene glycoside), phyllodulcin (a dihydroisocoumarin), perillartine, based on the monoterpene *perillaldehyde, the bisabolane sesquiterpenoid *hernandulcin and neohesperidin dihydrochalcone.
Lit.: Med. Res. Rev. **18**, 347–360 (1998) (review).

Sweroside, swertiamarin see secoiridoids.

Swinholides.

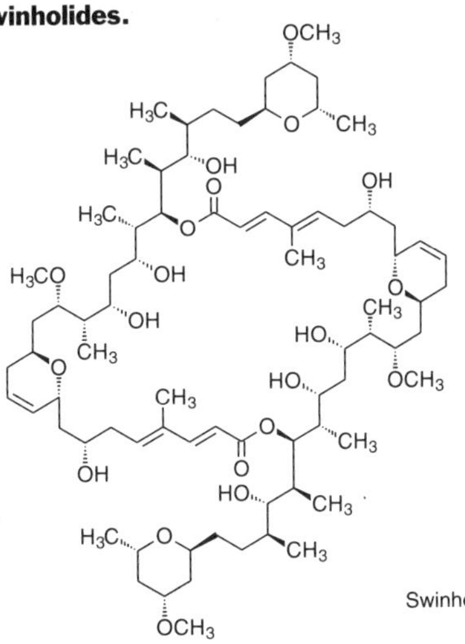

Swinholide A

Macrolides from the sponge *Theonella swinhoei* with antifungal and antineoplastic activity; e.g., *S. A:* $C_{78}H_{132}O_{20}$, M_R 1389.89, mp. 102 °C (hydrate), $[\alpha]_D^{20}$ +16° ($CHCl_3$).
Lit.: Angew. Chem., Int. Ed. **37**, 2162 (1998) (review) ▪ Chem. Eur. J. **2**, 847–868 (1996) ▪ Chem. Rev. **95**, 2041 (1995) (review) ▪ J. Am. Chem. Soc. **116**, 9391 (1994); **118**, 3059 (1996) (synthesis) ▪ J. Org. Chem. **56**, 3629 (1991); **62**, 2636 (1997); **64**, 4482 (1999) (synthesis) ▪ Tetrahedron **51**, 9393, 9413, 9437, 9467 (1995) (synthesis). – *[CAS 95927-67-6 (S. A)]*

Sylvestrosides. The S. are *iridoid/*secoiridoid glucosides in which the two molecules are joined by an ester bond. *Loganin is the alcohol component and a secologanic acid derivative (see secoiridoids) is the acid component. S. II is the acetyl derivative of S. I, while in S. III the glucose unit at the loganin part is missing.
At the beginning of the 20th century it was observed that plants of the Dipsacaceae family form intense blue pigments on slowly drying them at 35–60 °C. It was

Table: Data of Sylvestrosides.

compound	S. I	S. II	S. III
molecular formula	$C_{33}H_{48}O_{19}$	$C_{35}H_{50}O_{20}$	$C_{27}H_{36}O_{14}$
M_R	748.73	790.77	584.57
mp. [°C]	amorphous	amorphous	amorphous
$[\alpha]_D$	–106° (C_2H_5OH)	–99° (CH_3OH)	–85° (CH_3OH)
CAS	71431-22-6	71431-23-7	71431-24-8

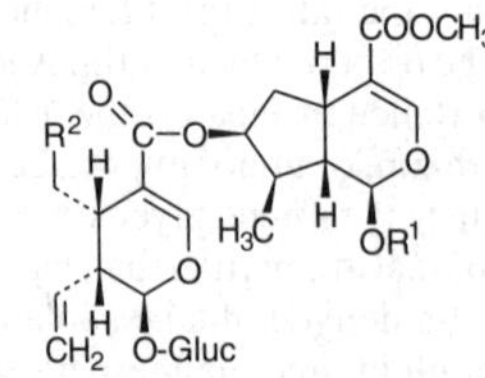

R^1 = β-D-Glucosyl, $R^2 = CH_2OH$: Sylvestroside I
R^1 = β-D-Glucosyl, $R^2 = CH_2{-}O{-}CO{-}CH_3$: Sylvestroside II
R^1 = H, R^2 = CHO : Sylvestroside III

assumed that these were iridiods [1]. However, the S. were first isolated and structurally elucidated in 1979 from leaves of *Dipsacus sylvestris* [2].

Lit.: [1] Hegnauer **IV**, 24. [2] Phytochemistry **18**, 273 (1979).
gen.: Beilstein E V **18/7**, 479; **18/9**, 31 ▪ J. Nat. Prod. **43**, 649, No. 163, 166, 167 (1980); **56**, 1486 (1993).

Symplostatin see dolastatins.

Synephrin see β-phenylethylamine alkaloids.

Synigrin(e) see glucosinolates.

Synomones. *Semiochemicals that act between organisms of different species and produce advantageous effects both for the sender (producer) and the recipient. S. play an important role in symbiosis. The homoterpene *(3E)-4,8-dimethyl-1,3,7-nonatriene* (see homoterpenes) is a typical S., it is secreted into the surroundings in increased amounts by many plants after injury by herbivorous insects and in some cases attracts the respective parasites of the plant-eating insect causing the damage [1].

Lit.: [1] Science **250**, 1251 ff. (1990); J. Chem. Ecol **19**, 411–425 (1993); **20**, 373–386, 1329–1354 (1994).
gen.: Agosta, Bombardier Beetles and Fever Trees, New York: Addison-Wesley 1995 ▪ Nordlund, Jones, & Lewis, Semiochemicals, their Role in Pest Control, New York: Wiley 1981.

Syringa aldehyde see rum flavor.

Syringenin, syringin see sinapyl alcohol.

Syringolides.

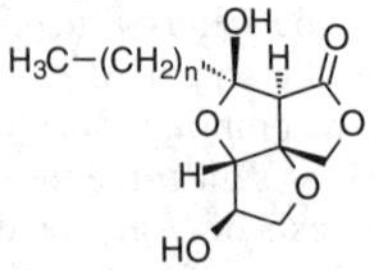

n = 4 : S. 1
n = 6 : S. 2

Bacterial signal substances of *Pseudomonas syringae*, that act on plants (e. g., tomatoes) as *phytoalexins. *S. 1:* $C_{13}H_{20}O_6$, M_R 272.30, wax, mp. 114 °C, $[\alpha]_D^{24}$ −84° ($CHCl_3$); *S. 2:* $C_{15}H_{24}O_6$, M_R 300.35, cryst., mp. 123–124 °C, $[\alpha]_D^{24}$ −76° ($CHCl_3$).

Lit.: J. Org. Chem. **58**, 2940 (1993); **60**, 286, 6641 (1995); **61**, 9374 (1996); **62**, 4780 (1997) ▪ Synlett **1996**, 335 ▪ Tetrahedron **51**, 8809 (1995) ▪ Tetrahedron Lett. **34**, 223 (1993). – *[CAS 147716-82-3 (S. 1); 147782-25-0 (S. 2)]*

T

Tabernanthine see Iboga alkaloids.

Tabersonine.

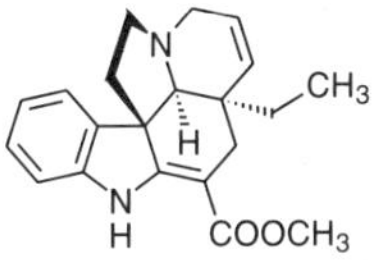

$C_{21}H_{24}N_2O_2$, M_R 336.43, mp. 196 °C (as hydrochloride), $[\alpha]_D$ +310° (CH_3OH). T. is a member of the *Aspidosperma alkaloids and occurs in numerous genera and species of the Apocynaceae. For biosynthesis, see monoterpenoid indole alkaloids. T. is a precursor of *vindoline [1,2], a building block of the dimeric alkaloids *vinblastine and *vincristine. T. is also formed in plant cell cultures but under these conditions cannot be completely transformed to vindoline. The biosynthesis of the dimeric alkaloids is not possible in cell cultures. For synthesis, see *Lit.* [3].

Lit.: [1] Plant Cell Rep. **4**, 337 (1985). [2] Plant Physiol. **86**, 447 (1988). [3] J. Am. Chem. Soc. **120**, 13523 (1998); J. Org. Chem. **58**, 1434–1442 (1993); Tetrahedron Lett. **40**, 1519 (1999).
gen.: Hager (5.) **6**, 1125. – *[HS 2939 90; CAS 4429-63-4]*

Tabtoxin (wildfire toxin).

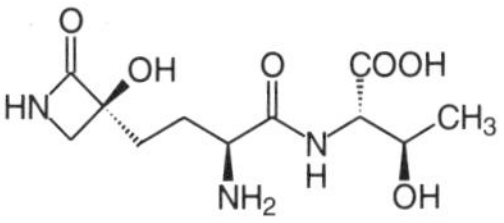

T.

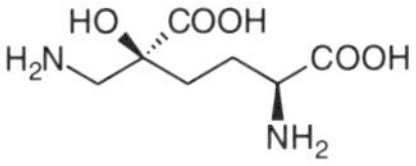

Tabtoxinin

$C_{11}H_{19}N_3O_6$, M_R 289.29, hygroscopic solid, readily soluble in water, insoluble in non-polar solvents. T. is a β-lactam phytotoxin from the bacterium *Pseudomonas tabaci* that induces the wildfire disease in tobacco plants. When not controlled this highly infectious leaf disease can rapidly cause enormous harvest losses. The isolation of T. proved to be difficult since the β-lactam ring undergoes rapid isomerization to the biologically inactive δ-lactam isotabtoxin. T. is highly toxic to a series of organisms such as bacteria, algae, higher plants, and mammals. In the host plant T. blocks the glutamine synthetase of the photorespiratory nitrogen cycle. This results in an accumulation of ammonia and the plant dies. The toxic effect is not caused by T. but rather by tabtoxinin [(2*S*,5*S*)-5-amino-2-(aminomethyl)-2-hydroxyadipic acid, $C_7H_{14}N_2O_5$, M_R 206.20], generated by hydrolysis of T. under the action of an aminopeptidase. The biosynthesis of the side chain proceeds from *L-threonine and L-aspartic acid, that of the β-lactam ring from pyruvate and a C_1 building block.

Lit.: Helv. Chim. Acta **70**, 412 (1987); **73**, 476 (1990) ▪ J. Chem. Soc., Chem. Commun. **1983**, 1049; **1985**, 1549 ▪ J. Med. Chem. **32**, 165–170 (1989) ▪ Merck-Index (12.), No. 10183 ▪ Pure Appl. Chem. **65**, 1309–1318 (1993) ▪ Stud. Org. Chem. **25**, 219–242 (1986) ▪ Tetrahedron **40**, 3695 (1984); **42**, 3097–3110 (1986). – *[CAS 40957-90-2 (T.); 40957-88-9 (tabtoxinin)]*

(6*Z*)-Tacalciol, tachysterol see calcitriol.

Tacrolimus see FK-506.

Tadeonal see polygodial.

D-Tagatose (D-*lyxo*-hexulose).

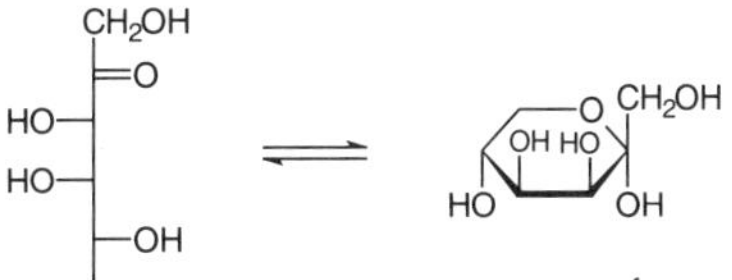

α-pyranose form

$C_6H_{12}O_6$, M_R 180.16, mp. 134–136 °C, $[\alpha]_D^{20}$ –260° (H_2O). A rare ketohexose epimer of *sorbose from plant rubbers, especially of the tropical tree *Sterculia setigera*. T. can also be formed from *lactose in strongly heated milk. The synthesis proceeds, e.g., through oxidation of D-talitol by *Acetobacter suboxydans*. For stereoselective synthesis, see *Lit.* [1].

Lit.: [1] Chem. Lett. **1982**, 1169; J. Carbohydr. Chem. **15**, 115–120 (1996).
gen.: Angew. Chem. Int. Ed. Engl. **31**, 56 (1992) (synthesis) ▪ Beilstein EIV **1**, 4414 ▪ Chem.-Ztg. **103**, 232 f. (1979) ▪ Karrer, No. 620 ▪ Merck Index (12.), No. 9202. – *[HS 2940 00; CAS 87-81-0 (D-T.); 17598-82-2 (L-T.)]*

Tagetes oil. Dark yellow to orange-colored, extremely oxidation-sensitive oil with a typical, terpene-aromatic, bitter-fruity odor and taste.

Production: By steam distillation of the flowering herbage of *Tagetes minuta* from, e. g., South Africa.

Composition [1]***:*** The main components are *(Z)-β-*ocimene* (40–50%), *dihydrotagetone* {ca. 5–10%, $C_{10}H_{18}O$, M_R 154.25, bp. 97–98 °C (0.5 hPa), $[\alpha]_D$ +1.5°}, *(Z)-tagetone* (ca. 5–10%; see ocimenone), (*Z*)- (ca. 10–20%), and (*E*)-**ocimenone* (ca. 5–10%).

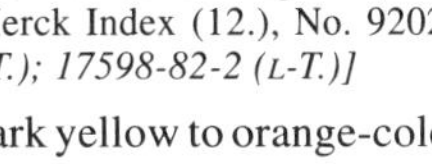

Dihydrotagetone

Use: Predominantly in the perfume industry where it is widely used, albeit in very low doses, for pronounced flowery-fruity compositions; occasionally in fruit and citrus flavors.

Lit.: [1] Perfum. Flavor. **10** (5), 73; (6), 56 (1985); **17** (5), 131 (1992).
gen.: Arctander, p. 606. – *Toxicology:* Food Cosmet. Toxicol. **20**, 829 (1982). – *[HS 3301 29; CAS 8016-84-0 (T.); 3338-55-4 (dihydrotagetone)]*

Tagetitoxin.

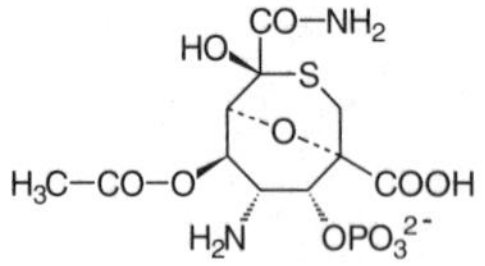

$C_{11}H_{17}N_2O_{11}PS$, M_R 416.30, isolated from liquid cultures of the plant pathogenic bacterium *Pseudomonas syringae* pv. *tagetis*. T. is phytotoxic even at a dose of 20 ng T./plant. T. induces the development of a chlorosis (bleaching) in the shoot tips. The activity is based on an impairment of chloroplast development.

Lit.: Phytochemistry **22**, 1425 (1983) ▪ Tetrahedron Lett. **30**, 501 (1989). – *[CAS 87913-21-1]*

Tagetone see ocimenone.

Talaromycins. A group of epimeric spiroketals from cultures of *Talaromyces stipitatus* (*Penicillium stipitatum*). $C_{12}H_{22}O_4$, M_R 230.30, oil. T. belong to the *mycotoxins and possess antibiotic properties.

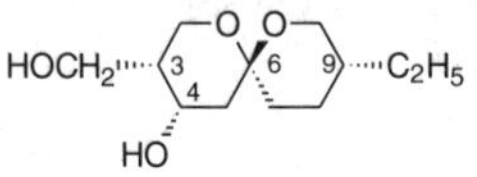

T. A

Talaromycins		CAS
T. A		83720-10-9
T. B	3-epimer	83780-27-2
T. C	4-epimer	89885-86-9
T. D	3,6-diepimer	111465-43-1
T. E	3,4-diepimer	111465-42-0
T. F	6-epimer	111465-44-2

On account of their interesting structure and the analogies to the spiroketal partial structures of, e.g., the *milbemycins and *avermectins, T. have been the target of numerous enantioselective syntheses.

Lit.: Agric. Biol. Chem. **51**, 565–571 (1987) ▪ ApSimon **8**, 569 ▪ Carbohydr. Res. **205**, 293–304 (1990) ▪ Chem. Pharm. Bull. **41**, 946 (1993) ▪ J. Chem. Soc., Chem. Commun. **1987**, 906 ▪ J. Chem. Soc., Perkin Trans. 1 **1990**, 1415–1421 ▪ J. Org. Chem. **54**, 1157–1161 (1989); **56**, 2476–2481 (1991) ▪ Synthesis **1992**, 1112 ▪ Tetrahedron **43**, 45–58 (1987) ▪ Tetrahedron Lett. **30**, 985–988 (1989).

Taleranol see zearalenone.

D-Talose (D-*talo*-hexose).

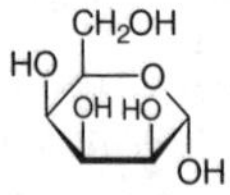

α-pyranose form

$C_6H_{12}O_6$, M_R 180.16. T. exists in the α-D-pyranose form (a) {mp. 133–134 °C, $[\alpha]_D^{20}$ +68 → +21° (H_2O)}, the β-D-pyranose form (b) {mp. 120–121 °C, $[\alpha]_D^{20}$ +13 → +21° (H_2O)}, the α-D-furanose form (c) (syrup), and the β-D-furanose form (d) (syrup). *Glycosides of all forms have been described. D-T. is a component of the *hygromycins used as antibiotics.

Lit.: Beilstein EIV **1**, 4344 ▪ Bull. Chem. Soc. Jpn. **62**, 2701 (1989) ▪ Karrer, No. 3238 ▪ Proc. Natl. Acad. Sci. USA **87**, 6684 (1990) ▪ Science **220**, 749 (1983) (synthesis). – *[CAS 7282-81-7 (a); 7283-11-6 (b); 51076-04-1 (c); 41847-63-6 (d)]*

TAN (TAN 749, TAN 1057) see sperabillins.

Tanning agents. The name T. a. is used for all chemical compounds that are able to tan animal skins, i.e., to produce leather. This means that the native skin proteins are precipitated and denaturized. The collagen chains of the skin are cross-linked in this process [1] and the durability and strength of the material increases. Distinctions are made between inorganic, mineral [2], and organic chemical tanning agents of synthetic or natural origin.

T. a. of plant origin are widely distributed and occur in various plant parts, e. g., in leaves (tea), seeds (coffee), berries, galls, woods, etc. In the stricter sense they are *tannic acids* or *tannins, they protect the plant from rot, pests, or predatory animals. They dissolve in water to form colloids or semicolloids, have an adstringent taste, and precipitate protein solutions. They form dark blue or green complexes with iron salts, and usually insoluble precipitates with lead salts and alkaloids. The colloidal particles are sufficiently small to penetrate the entire thickness of the skin. Plant T. a. are divided into two groups according to Freudenberg: the condensed T. a. or *catechin T. a.* [3] (a) which are derived from *catechin and the hydrolysable T. a. or *gallotannins* (b) derived from gallic acid. Members of both groups contain numerous phenolic hydroxy groups.

a) The monomeric catechins (flavan-3-ols) and *leucoanthocyanidins (flavan-3,4-diols) from which these T. a. are derived preferentially carry phenolic hydroxy groups in the 5,7,3′,4′,5′ positions, but do not yet have any tanning activity. Effective T. a. are first formed after self-condensation (acid-catalyzed or enzymatic). The color of fermented tea is, e. g., due to *theaflavin, a dimer of epicatechin and *epigallocatechin-3-gallate. *Flavonoids, *caffeic, *chlorogenic and ferulic acid are also involved in the formation of condensed T. a.

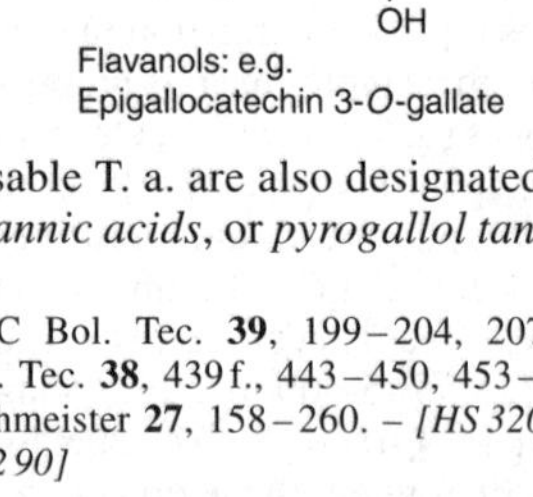

Flavanols: e.g.
Epigallocatechin 3-*O*-gallate

b) Hydrolysable T. a. are also designated as *gallotannins*, *gallotannic acids*, or *pyrogallol tannins*, see tannins.

Lit.: [1] AQEIC Bol. Tec. **39**, 199–204, 207–213 (1988). [2] AQEIC Bol. Tec. **38**, 439 f., 443–450, 453–460, 463–469 (1987). [3] Zechmeister **27**, 158–260. – *[HS 3201 10–3201 90, 3202 10, 3202 90]*

Tannins. Collective name, derived from French tanin = to tan hides, for a group of natural polyphenols with widely varying composition, more frequently known as *gallotannins* on account of the relationship with *gallic acid. The so-called *tannin* (gallotannic acid, tannic acid) is readily soluble in water (colloidal solution, acidic reaction) or alcohol. On heating it turns brown and decomposes at ca. 210 °C to pyrogallol and carbon dioxide. Aqueous solutions of T. give a blue-black, ink-like precipitate with iron chloride solution that turns yellowish brown on addition of dilute sulfuric acid. Other heavy metals are also precipitated in salt-like form. The chemical composition of the T. varies widely depending on their origin (oak or chestnut wood, divi-divi, sumac, myrobalans, trillo, valonea, or plant galls). According to E. Fischer and Freudenberg the T. are mixtures of substances of the *pentadigalloylglucose* ($C_{76}H_{52}O_{46}$, M_R 1701.22) type. The digallic acid can be replaced by gallic acid or even digalloylgallic acid.

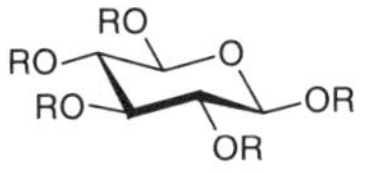

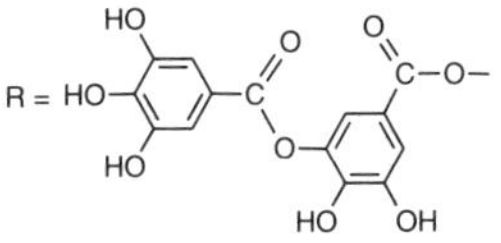

Pentadigalloylglucose

Oxidative coupling of the galloyl groups of β-1,2,3,4,6-pentagalloyl-D-glucose gives rise to T. with hexahydroxydiphenoyl residues and their subsequent products in a large diversity of structures. The *depside and ester bonds in T. are hydrolysed on boiling in dilute sulfuric acid to furnish glucose and gallic acid. The same cleavages are effected even at 20 °C by the enzyme *tannase* expressed by the mold fungi *Aspergillus niger* and *Penicillium glaucum*. For microbial degradation, see *Lit.*[1]; for biosynthesis, see *Lit.*[2].

Activity & use: T. have potent tanning and astringent properties which explains their long history of use to treat leather. T. not only tan animal skins but can also cause mild, mostly favorable "tanning effects" in living organisms: wound surfaces and catarrhally inflamed skins of the digestive organs shrink under the action of T. with a reduction in the secretion of exudates (antiseptics, hemostyptics). The permeability of skin capillaries is reduced by T. Even as 0.5% solutions T. precipitate animal gluten (gelatin solution); many heavy metals and alkaloids, with the exception of morphine, are also precipitated by T. solutions, this property can be exploited as therapy for the respective poisonings. T. from plant galls are used as antidiarrhoics. T. also occur in beer and red wine. The addition of milk to black tea reduces its bitter taste because the T. form insoluble compounds with milk proteins. The high content of T. in, e. g., *sorghum* renders it less suitable as animal fodder on account of protein precipitation.

Lit.: [1]ACS Symp. Ser. **399**, 559 (1989); J. Sci. Ind. Res. **45**, 232–243 (1986). [2]Barton-Nakanishi **3**, 747–827.
gen.: Chem. Pharm. Bull. **38**, 861–865, 2424–2428 (1990) ▪ Chem. Soc. Rev. **26**, 111–126 (1997) ▪ Hemingway & Karchesy (eds.), Chemistry and Significance of Condensed Tannins, New York: Plenum 1989 ▪ Heterocycles **30**, 1195–1218 (1990) ▪ J. Nat. Prod. **52**, 1–31, 665 (1989); **59**, 205–215 (1996) ▪ Nat. Prod. Rep. **11**, 41–66 (1994) ▪ Progr. Clin. Biol. Res. **280**, 123–134 (1988) (wine) ▪ Pure Appl. Chem. **61**, 357–360 (1989) ▪ Tetrahedron **53**, 5951 (1997) ▪ Zechmeister **13**, 70–136; **41**, 1–46, 47–76; **66**, 1–118; **77**, 21–67. – *[HS 3201 90; CAS 1401-55-4]*

Tansy (*Tanacetum vulgare*, syn. *Chrysanthemum vulgare*). A perennial, 40–120 cm high, of the Asteraceae family with fern-like leaves and golden-yellow flowers that grows preferentially on the edges of paths, deforestations, and land-fillings.

Components: In addition to polyacetylenes, flavonoids, and coumarins, the shoots of T. contain terpenes (essential oil). The latter can be subdivided into monoterpene hydrocarbons, (α- and β-sabinenes, see thujenes), monoterpene ketones [*artemisia ketone, *camphor, α- and β-thujones (see thujan-3-ones), umbellulones], monoterpene alcohols (*borneols, chrysanthenol, lyratol, 4-thujen-2α-ol), and *sesquiterpenes (germacrene D, davanone, see davana oil). Among the sesquiterpenes the *germacranolide *parthenolide* [$C_{15}H_{20}O_3$, M_R 248.32, mp. 116.5–117 °C, $[\alpha]_D$ –81.4° ($CHCl_3$)], which is presumably responsible for the prophylaxis of migraine, and the sesquiterpene lactone *crispolide* [$C_{15}H_{20}O_5$, M_R 280.32, $[\alpha]_D$ –20° (pyridine)], which shows antimalarial activity, are of particular interest.

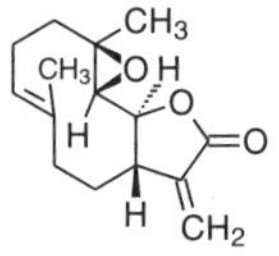

Parthenolide

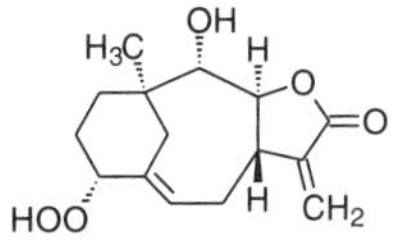

Crispolide

Use: The flowering plant is used in traditional medicine as an anthelmintic (worm remedy), antispasmodic, as remedy for stomach cramps, gout, vertigo; it was also abused for abortive purposes. T. preparations are no longer used on account of their toxicity which is mainly due to the thujone content. Recently, fungicidal[1], insect repellent[2] properties against Colorado beetles, and antimicrobial activities[3] have been reported for the essential oil.

Lit.: [1]Herba Hung. **20**, 57 (1981). [2]J. Nat. Prod. **47**, 964 (1984). [3]Planta Med. **55**, 102 (1989).
gen.: Br. J. Pharmacol. **112**, 9 (1994) ▪ Hager (5.) **3**, 1173 ▪ Phytochemistry **21**, 1099 (1982); **39**, 839 (1995) ▪ Z. Phytother. **14**, 333 (1993). – *[HS 1211 90; CAS 20554-84-1 (parthenolide); 83217-86-1 (crispolide)]*

Tantazoles.

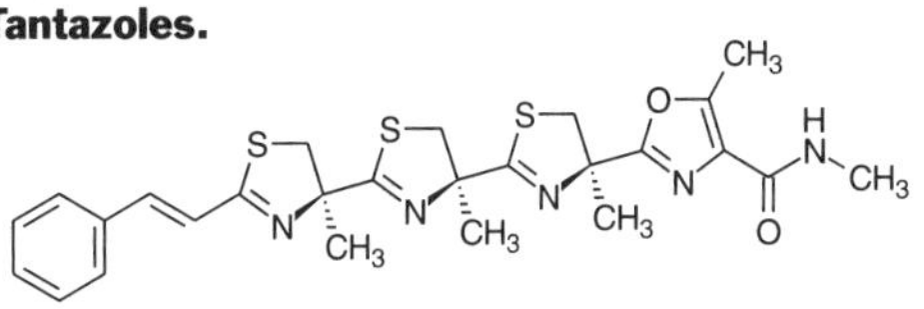

Thiangazole A

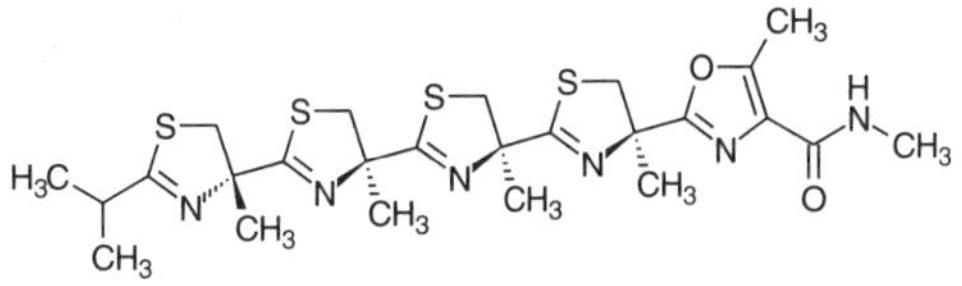

Tantazole B

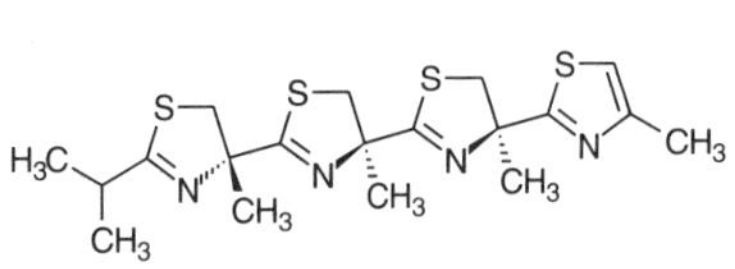

Mirabazole B

Table: Data of Tantazoles.

	molecular formula	M_R	$[\alpha]_D$	mp. [°C]	CAS
Tantazole B	$C_{25}H_{34}N_6O_2S_4$	578.85	−94°	–	129895-76-7
Mirabazole B	$C_{19}H_{26}N_4S_4$	438.71	−166°	–	135459-34-6
Thiangazole A	$C_{26}H_{29}N_5O_2S_3$	539.75	−287°	140	138667-71-7

A group of unusual oligothiazolyloxazole alkaloids from *Scytonema mirabile* (Cyanobacteria) to which the *mirabazoles* and *thiangazoles* from a *Polyangium* sp. (Myxobacteria) also belong. The thiazolines in T. are sensitive to oxidation (in air → thiazoles) and have cytotoxic activity, the thiangazoles have antifungal and insecticidal properties due to an inhibition of the respiratory chain in the NADH:ubiquinone oxidoreductase (complex I). The observed inhibition of HIV-1 replication is probably due to an energy deficit in the host cell. IC_{100} (HIV-1): 4.7 pM, has been determined for thiangazole A; no cytotoxicity was observed even at 4.7 mM (selectivity index >10^6).

Lit.: Antivir. Chem. & Chemother. **3**, 189 (1992) ▪ J. Am. Chem. Soc. **112**, 8195 (1990); **115**, 8449 (1993) ▪ J. Antibiot. **46**, 1752 (1993) ▪ Justus Liebigs Ann. Chem. **1992**, 357; **1993**, 701 ▪ Nachr. Chem. Tech. Lab. **43**, 347 (1995) ▪ Nat. Prod. Rep. **16**, 249 (1999) (review) ▪ Synlett **1996**, 1168 [synthesis of (−)-T.] ▪ Tetrahedron Lett. **32**, 2593 (1991); **34**, 6681 (1993). – *Thiangazole:* J. Org. Chem. **60**, 7224 (1995) ▪ Synlett **1994**, 702 ▪ Tetrahedron **51**, 7321 (1995); **55**, 10685 (1999) ▪ Tetrahedron Lett. **35**, 5705 (1994). – *Mirabazole B:* J. Org. Chem. **57**, 5566 (1992) ▪ Pept. Chem. **1996**, 53–56 ▪ Synlett **1994**, 587 ▪ Tetrahedron **53**, 8323 (1997).

Taraxastane(s) see triterpenes and taraxasterol.

Taraxasterol [20(30)-taraxasten-3*β*-ol]. Formula, see triterpenes. $C_{30}H_{50}O$, M_R 426.73, mp. 225–226 °C, $[\alpha]_D$ +95.9° ($CHCl_3$). T. is a pentacyclic *triterpene (taraxastane type), widely distributed in the plant kingdom, e. g., in the roots of *dandelion (*Taraxacum officinale*).

Lit.: Karrer, No. 2012 ▪ Phytochemistry **28**, 1471 (1989) (isolation). – *Biosynthesis:* Phytochemistry **11**, 1961–1966 (1972). – *[CAS 1059-14-9]*

Taraxerane(s) see triterpenes and taraxerol.

Taraxerol (14-taraxeren-3*β*-ol). Formula, see triterpenes. $C_{30}H_{50}O$, M_R 426.73, mp. 282–283 °C, $[\alpha]_D$ +0.7° ($CHCl_3$). T. is a *triterpene widely distributed in the plant kingdom, isolated for the first time in 1923 from the bark of gray alder (*Alnus incana*). T. also occurs in the roots of dandelion (*Taraxacum officinale*), in various alder species, rhododendron species, and Euphorbiaceae, as well as in some lichens. T. is derived from the taraxerane skeleton and has antigastritic properties.

Lit.: Karrer, No. 1969 ▪ Phytochemistry **25**, 453 (1986). – *[CAS 127-22-0]*

Tarichatoxin see tetrodotoxin.

Tarragon oil. Light to dark yellow oil with a strong, green-spicy, anise- and celery-like odor and a sweet-spicy, aromatic taste resembling that of anise and celery.

Prod.: By steam distillation of cultivated tarragon (*Artemisia dracunculus*) grown mostly in France, Italy, and Spain. The main component is *estragole (usually about 80%) [1].

Use: For perfume production, fresh-spicy compositions; to aromatize foods (pickles, sauces, etc.).

Lit.: [1] Perfum. Flavor. **13** (1), 49 (1988); **15** (2), 75 (1990); **20** (4), 38 (1995).
gen.: Arctander, p. 224 ▪ Bauer et al. (2.), p. 177 ▪ ISO 10115. – *Toxicology:* Food Cosmet. Toxicol. **12**, 709 (1974). – *[HS 3301 29; CAS 8016-88-4]*

Tartaric acid (2,3-dihydroxybutanedioic acid, grape acid).

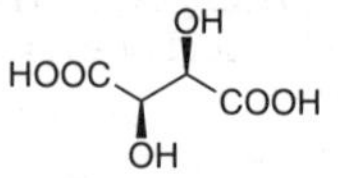

$C_4H_6O_6$, M_R 150.09, (−)-(*S,S*)-T.: monoclinic prisms, mp. 168–170 °C, $[\alpha]_D^{20}$ −12° (H_2O), pK_{a1} 2.93, pK_{a2} 4.23 (25 °C), occurs only rarely in nature, e. g., in the leaves and fruits of the West African tree *Bauhinia reticulata*. – (+)-(*R,R*)-T.: monoclinic prisms, mp. 168–170 °C, $[\alpha]_D^{20}$ +12° (H_2O), pK_{a1} 2.93, pK_{a2} 4.23 (25 °C), the pH value of a 0.1 M aqueous solution is 2.2, is widely distributed in plants and their fruits. – (±)-T. (grape acid): mp. 206 °C (anhydrous); the racemate is less soluble in water than the pure enantiomers; it does not occur in nature. – *meso*-T.: rectangular plates, monohydrate, mp. 140 °C (159–160 °C). The diesters of (+)-T. with *caffeic acid contained in chicory and endives are known as chicoric acids, e. g., *(R,R)-T. dicaffeoyl ester*, $C_{22}H_{18}O_{12}$, M_R 474.38, mp. 206 °C, $[\alpha]_D$ +384° (CH_3OH).

Toxicology: LD_{50} (mouse i. v., (+)-W.) 485 mg/kg.

Use: In the food industry as an acidify agent for soft drinks, bakery products, and puddings; in the pharmaceutical industry as a buffer; in the paint and textiles industries as additive; as a chiral auxiliary for asymmetric synthesis, for the technical production of pyruvic acid.

Lit.: Angew. Chem. Int. Ed. Engl. **24**, 513 (1985) ▪ Beilstein EIV **3**, 1219 ▪ Karrer, No. 859–861 ▪ Kirk-Othmer (4.) **13**, 1071 ▪ Luckner (3.), p. 112, 136, 451 ▪ Planta Med. **54**, 175 f. (1988) (chicoric acid) ▪ Ullmann (4.) **24**, 431–438; (5.) **A13**, 508; **A18**, 180 ▪ Z. Naturforsch. C **40**, 585 ff. (1985). – *Toxicology:* Sax (8.), TAF 750. – *[HS 2918 12; CAS 147-71-7 (S,S)-(−); 87-69-4 (R,R)-(+); 133-37-9 (±); 147-73-9 (meso-T.); 6537-80-0 ((R,R)-T. dicaffeoyl ester)]*

Tartrolones (tartrolides).

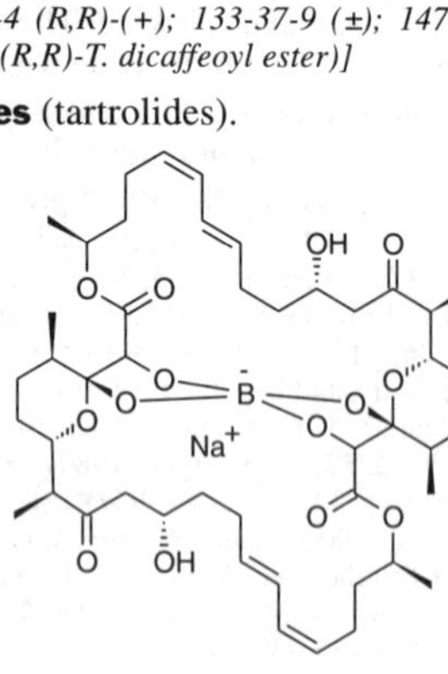

Tartrolone B

T. A is a macrodiolide, made up of two identical halves from *Sorangium cellulosum* (Myxobacteria), $C_{46}H_{72}O_{14}$, M_R 849.07, mp. 160–162 °C, $[\alpha]_D^{22}$ +252.0°, it readily takes up boron to form a Böseken complex, T. B. According to an X-ray structure analysis of the Na salt of T. B the Na and B atoms are shielded from the surroundings by the aliphatic ring segments, this explains the high lipophilicity and solubility of the complex in petroleum ether. *Boromycin and *aplasmomycin from streptomycetes and borophycin from cyanobacteria are biogenetically closely related to T. and also form Böseken complexes.
Lit.: J. Antibiot. **1995**, 26 ▪ J. Am. Chem. Soc. **121**, 8393 (1999) (synthesis) ▪ J. Heterocycl. Chem. **36**, 1421–1436 (1999) ▪ Justus Liebigs Ann. Chem. **1994**, 283; **1996**, 965 (abs. configuration, biosynthesis) ▪ Nachr. Chem. Tech. Lab. **42**, 498 (1994) (review). – *[HS 2941 90; CAS 156312-16-2 (T. A); 156312-17-3 (T. B)]*

Taurin [(11*S*)-1-oxo-4-eudesmen-12,6α-olide].

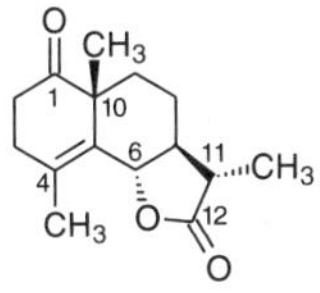

$C_{15}H_{20}O_3$, M_R 248.32, cryst., mp. 118–119 °C (109–110 °C), $[\alpha]_D^{19}$ –120° (C_2H_5OH). A sesquiterpene lactone from *Artemisia* species. T. has spasmolytic activity [1] and is used in the form of a 0.6 mM aqueous solution for the fluorimetric determination of cyanide [2] (uv_{max} 460 nm).
Lit.: [1] Plant Med. Phytother. **15**, 144–148 (1981). [2] Anal. Sci. **2**, 491 f. (1986).
gen.: Beilstein EV **17/11**, 228. – *[CAS 23522-05-6]*

Taurine (2-aminoethanesulfonic acid).

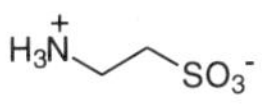

$C_2H_7NO_3S$, M_R 125.14, monoclinic prisms, mp. 328 °C (300 °C) (decomp.), soluble in water, pK_{a1} 1.5, pK_{a2} 8.74, highly toxic vapors develop on heating. T. occurs in the seeds of Fabaceae, in human and animal tissues, in bile both in the free form and as the amide bound to cholic acid (taurocholic acid). High concentrations of T. are found in the central nervous system (CNS) [1], in the retina [2], and in the heart [3]. T. plays an important role in the development of the brain; in the CNS T. influences the transport processes of divalent metal ions, e. g., as calcium, magnesium, and zinc modulator. A deficiency of T. is possibly involved in epilepsy, mongolism, and arrhythmias or leads to abnormal brain development [4]. Besides T., bacteria, giant siliceous sponges, and red algae also contain *N-monomethyl-T.*, $C_3H_9NO_3S$, M_R 139.17, prisms, mp. 241–242 °C, and *N,N-dimethyl-T.*, $C_4H_{11}NO_3S$, M_R 153.20, mp. 315–316 °C.
Use: As an intermediate product in the manufacture of dyes, drugs, cleaning agents, etc., it is active against molds, in therapy for gallstones and as adjuvant in the treatment of hypercholesterolemia. The biosynthesis proceeds from *cysteine.
Lit.: [1] Brain Res. Rev. **12**, 167–201 (1987); Prog. Neurobiol. **32**, 471–533 (1989). [2] Neurol. Neurobiol. **49**, 61–86 (1989). [3] Adv. Exp. Med. Biol. **217**, 371–387 (1987); Heitaroh, Lombardini, &. Segawa (eds.), Taurine and the Heart, Boston: Kluwer Academic 1989. [4] Nutr. Res. **8**, 955–968 (1988).
gen.: Beilstein E IV **4**, 3289 ▪ Curr. Top. Nutr. Dis. **16**, 217–243 (1987) ▪ Farm. Clin. **7**, 580–600 (1990) ▪ Huxtable, The Biology of Taurine, New York: Plenum 1987 ▪ Lombardini, Taurine: Nutritional Value and Mechanisms of Action, Plenum Press 1992 ▪ Luckner (3.) p. 278, 450, 460 ▪ Mendel & Friedman (eds.), Absorption and Utilization of Amino Acids, vol. 1, p. 199–227, Boca Raton: CRC Press 1989 ▪ Nutr. Res. Rev. **1**, 99–113 (1988) ▪ Prev. Med. **18**, 79–100 (1988) ▪ Prog. Clin. Biol. Res. **351** (1990). – *[HS 2921 19; CAS 107-35-7 (T.); 107-68-6 (N-monomethyl-T.); 637-95-6 (N,N-dimethyl-T.)]*

Taurocholic acids see bile acids.

Tautomycetin see tautomycin.

Tautomycin.

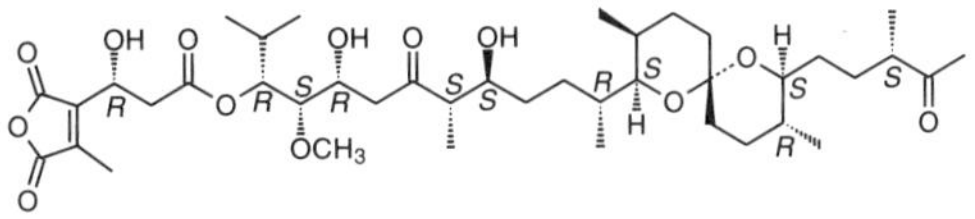

$C_{41}H_{66}O_{13}$, M_R 766.97, amorphous, yellowish powder, mp. 42–43 °C, $[\alpha]_D^{20}$ +3.4° ($CHCl_3$), a spiroketal ester and polyether antibiotic isolated from *Streptomyces griseochromogenes*, in which, in aqueous solution, the maleic acid anhydride terminal unit exists in equilibrium with the free maleic acid. T. is active against fungi, yeasts, and some Gram-negative bacteria. It is an activator of protein kinase C and protein phosphatases. Structurally related to T. is the antifungally active *tautomycetin* [$C_{33}H_{50}O_{10}$, M_R 606.75, yellow gum, $[\alpha]_D^{20}$ +19.4° ($CHCl_3$)].
Lit.: Atta-ur-Rahman **18**, 269–314 ▪ J. Antibiot. **41**, 932 (1988); **43**, 809–819 (1990); **45**, 246–257 (1992) ▪ J. Chem. Soc., Chem. Commun. **1990**, 244 ▪ J. Chem. Soc., Perkin Trans. 1 **1993**, 617–624; **1995**, 2399 (structure) ▪ J. Org. Chem. **60**, 5048–5068 (1995); **62**, 387 (1997) (synthesis) ▪ Stud. Nat. Prod. Chem. **18**, 269 (1996) ▪ Tetrahedron **52**, 13363 (1996); **53**, 5083, 5103, 5123–5143 (1997) (synthesis) ▪ Tetrahedron Lett. **36**, 7101 (1995) (synthesis). – *[CAS 109946-35-2 (T.); 119757-73-2 (tautomycetin)]*

Taxanes.

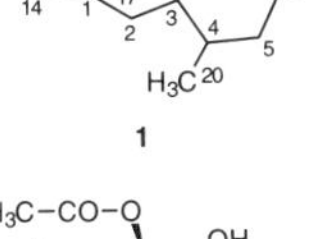

1

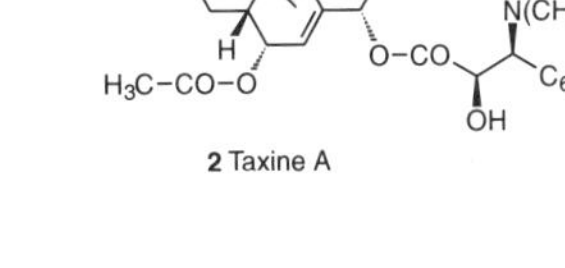

2 Taxine A

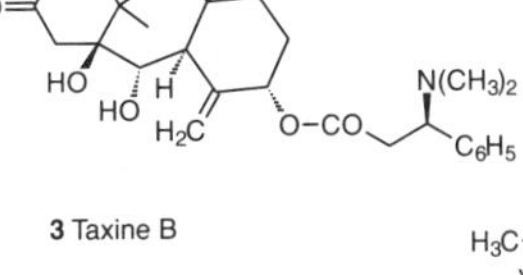

3 Taxine B

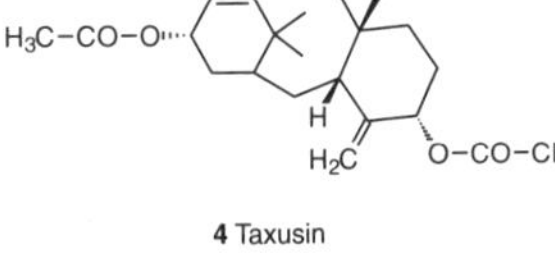

4 Taxusin

A group of *diterpenes with the skeleton **1**, isolated from yew species (*Taxus* spp.), and responsible for the toxicity of the plants. All parts of the plants with the

Table: Data of Taxanes.

	Taxine A **2**	Taxine B **3**	Taxusine **4**
molecular formula	$C_{35}H_{47}NO_{10}$	$C_{33}H_{45}NO_8$	$C_{28}H_{40}O_8$
M_R	641.76	583.72	504.62
shape	cryst.	amorphous	cryst.
mp. [°C]	204–206	115	126
$[\alpha]_D^{20}$ ($CHCl_3$)	–140°	+116.9°	+93.6°
CAS	1361-49-5	1361-51-9	19605-80-2

exception of the red seed coat contain toxic taxanes. The T. are mostly esterified with 3-amino-3-phenylpropanoic acid and thus react like alkaloids. A total fraction of this type has been named as *taxine*. Its cardiotoxic activity has been attributed in part to taxine B, comprising ca. 30–40% of taxine, LD_{50} taxine: 20 mg/kg (mouse). In addition, some other compounds have been found that have generated great excitement on account of their excellent activities against various forms of cancer, e. g., *taxol®. Taxine A is also mentioned in the literature under the (incorrect) name toxin A. Taxanes have strong antifeedant activities on insects [1]. For synthesis of taxusin, see *Lit.* [2].

Lit.: [1] Phytochemistry **49**, 1279 (1998). [2] J. Am. Chem. Soc. **118**, 9186 (1996); **120**, 5203, 5213 (1998); **121**, 3072 (1999). *gen.:* ACS Symp. Ser. **583**: Taxane Anticancer Agents (1995) ▪ Angew. Chem. Int. Ed. Engl. **34**, 1723 (1995) ▪ Atta-ur-Rahman **11**, 3–69; **20**, 79–134 ▪ Chemtracts **11**, 16 (1998) (taxusin) ▪ Contemp. Org. Synth. **1**, 47–75 (1994) ▪ Exp. Opin. Ther. Pat. **8**, 571–586 (1998) (review patents) ▪ Heterocycles **46**, 241–258, 727–764 (1997) (synthesis) ▪ J. Am. Chem. Soc. **116**, 1597 (1994) ▪ J. Nat. Prod. **59**, 117 (1996); **62**, 1448 (1999) (review) ▪ J. Org. Chem. **60**, 7215 (1995) ▪ Nachr. Chem. Tech. Lab. **37**, 172 ff. (1989) ▪ Nature (London) **367**, 630 (1994) ▪ Org. Prep. Proc. Int. **23**, 465–543 (1991) ▪ Pharmacochem. Libr. **22**, 55–101 (1995) ▪ Phytochemistry **27**, 1534 (1988); **50**, 1267–1304 (1999) (review); 633 (1999) ▪ Tetrahedron **46**, 4907–4924 (1990) ▪ Tetrahedron Lett. **38**, 7587 (1997) ▪ Zechmeister **61**, 1–206.

Taxifolin see silybin.

Taxines see taxanes.

Taxiphyllin see cyanogenic glycosides.

Taxodione.

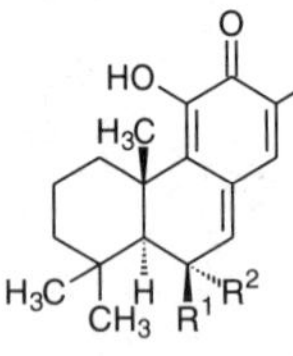

$R^1, R^2 = O$: Taxodione
$R^1 = H, R^2 = OH$: Taxodone

$C_{20}H_{26}O_3$, M_R 314.42, golden platelets, $[\alpha]_D^{25}$ +56° ($CHCl_3$), mp. 115–116 °C. An *abietane diterpene with the quinonemethide structure, occurs together with *taxodone* {$C_{20}H_{28}O_3$, M_R 316.44, yellow platelets, mp. 164–165 °C, $[\alpha]_D^{25}$ +50° ($CHCl_3$)} in *Taxodium distichum* (Taxodiaceae), *Salvia phlomoides*, *S. munzii*, and *S. moorcroftiana* (Lamiaceae). 500 g *Taxodium* leaves furnish 1 g T. and 3 g taxodone. Like taxodone, T. is a tumor inhibitor effective *in vivo* in mice and *in vitro* in humans (nasopharyngeal carcinoma) (ED_{50}: 3 µg/mL, for taxodone 0.6 µg/mL).

Lit.: Hager (5.) **6**, 490 ▪ Merck-Index (12.), No. 9247 ▪ Phytochemistry **25**, 755 (1986) ▪ Science **168**, 378 (1970). – *Synthesis:* J. Org. Chem. **47**, 2396 (1982); **59**, 5445 (1994) ▪ Justus Liebigs Ann. Chem. **1989**, 677 ▪ Synlett **1994**, 335 ▪ Synth. Commun. **13**, 201, 281 (1983) ▪ Tetrahedron **49**, 6277 (1993); **50**; 9229 (1994). – *[CAS 19026-31-4 (T.); 19039-02-2 (taxodone)]*

Taxoids. Synonymously used term for certain tetracyclic diterpenes (*taxanes); generally all taxanes, which bear oxygen functional groups.

Lit.: Chem. Pharm. Bull. **45**, 1205 (1997) ▪ Heterocycles **46**, 241–258; 727–764 (1997) ▪ Tetrahedron Lett. **38**, 7587 (1997) ▪ Zechmeister **61**, 1–206 (1993).

Taxol® (Paclitaxel).

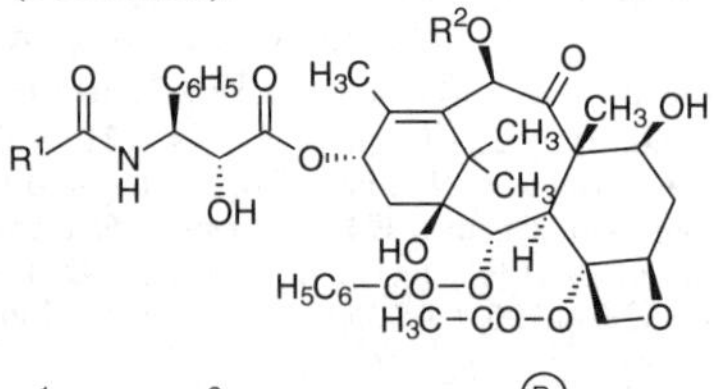

$R^1 = C_6H_5$, $R^2 = CO{-}CH_3$: Taxol®
$R^1 = O{-}C(CH_3)_3$, $R^2 = H$: Taxotere

$C_{47}H_{51}NO_{14}$, M_R 853.92, colorless needles, $[\alpha]_D^{20}$ –49° (CH_3OH), mp. 158–160 °C or 213–216 °C, soluble in CH_3OH, very poorly soluble in water, LD_{50} (dog p. o.) ca. 9 mg/kg. A *diterpene of the taxane type, isolated for the first time at the beginning of the 1960's from the bark of the Pacific yew, *Taxus brevifolia* (Taxaceae) (T. contents: 0.01–0.033%), in pure form, it was first isolated in 1971. T. and related compounds such as the synthetic derivative *taxotere* [1] {Docetaxel®, $C_{43}H_{53}NO_{14}$, M_R 807.89, cryst., mp. 232 °C, $[\alpha]_D^{20}$ –36° (C_2H_5OH)} possess very good activities against various types of tumors, e. g., malignant melanoma, non-small-cell lung cancer and metastatic or early breast cancer. The highest response rate known for a chemotherapeutic agent was observed for T. in ovarian cancer. T. influences cell division. The activity is based on a dual mechanism: 1. stabilization of the microtubuli, thus preventing the formation of the mitosis spindle and 2. binding not only to tubulin but also to Bcl-2, a human protein, that prevents cells from undergoing apoptosis, which causes cell death [8]. Thus, T. has cytotoxic effects, especially against rapidly reproducing tissues. Furthermore, cell motility, the slow transport of tubulin, actin, and polypeptides in axons, the intracellular transport of steroids, the immunoactivity of τ-MAPs in neurons, and the protein secretion from several cell types are inhibited. T. has been authorized for use as a drug in the treatment of advanced stages of cancer (especially ovarian carcinoma) since 1992 in the USA and since 1995 in Germany. Other indications for T. under clinical investigation are restenosis after angioplasty and multiple sclerosis. The high clinical demands (1997: ca. 200–300 kg) and the poor availability (3000 Pacific yew trees must be cut to obtain 1 kg of T.) triggered a 22-year race lasting to 1994 to achieve the first total synthesis. Several total syntheses have now been reported [2] as well as some partial syntheses [3] starting from baccatin III or 10-deacetylbaccatin III (from needles of the European yew *Taxus baccata* that can be cultivated) (content 0.2–0.5%). To meet the demands, *T. brevifolia* is also being grown in field culture or T. is being obtained from plant cell cultures [4]. T. is also produced in small amounts (20–1000 ng/L) by endophytic fungi such as *Taxomyces andrea-*

nae (20–50 ng/L fermentation) and *Pestalotiopsis microspora* (isolated from *Taxus wallachiana*, 60–70 µg/L fermentation) associated with *Taxus* and other conifer species[5]. On account of the poor solubility of T. in water, some very effective analogues of T. with more suitable properties for administration and liposomal encapsulated T. have been developed[6]. The biosynthesis of T. proceeds from geranylgeranyl diphosphate via taxa-4(5),11(12)-diene (ring closure to the basic tricyclic system, taxadiene synthase) and taxa-4(20),11(12)-dien-5-ol[7]. Conjugates of T. with monoclonal antibodies are developed against ovarian and breast cancer. With sales of more than 1.3 billion US $ in 1999, T. has become one of the commercially most successful secondary metabolites.

Lit.: [1] Drug News **5**, No. 4, 1 (29.1.1996); New Drug Commentary **23**, No. 11/12, 52 (1996); No. 6, 24 (1997); Tetrahedron Lett. **35**, 105, 2349, 3063 (1994). [2] Chem. Lett. **1998**, 1; Chemtracts **11**, 16, 23 (1998); Chem. Eur. J. **5**, 121–161 (1999); Enantioselective Synth. Beta Aminoacids **1997**, 1–43; J. Am. Chem. Soc. **116**, 1597, 1599 (1994); **117**, 624–659 (1995); **118**, 2843–2859, 9186, 10752 (1996); **119**, 2757 (1997); **120**, 5203, 5213, 12980 (1998); J. Org. Chem. **60**, 7209–7223 (1995); Angew. Chem. Int Ed. Engl. **34**, 1723, 2079–2090 (1995); **39**, 44–122 (2000); Nature (London) **367**, 630 (1994). [3] J. Am. Chem. Soc. **110**, 5917, 6558 (1988); J. Chem. Soc., Perkin Trans. 1 **1997**, 3501; J. Med. Chem. **40**, 236 (1997); Chem. Lett. **1998**, 1; J. Org. Chem. **56**, 1681 (1991); Tetrahedron **48**, 6985 (1992); Bioorg. Med. Chem. Lett. **2**, 295 (1992); Tetrahedron Lett. **33**, 5185 (1992). [4] J. Chem. Soc., Perkin Trans. 1 **1996**, 845. [5] Science **260**, 154, 214 (1993); Microb. **142**, 435 (1996). [6] J. Med. Chem. **32**, 788 (1989); **40**, 236 (1997); **41**, 3715–3726 (1998); **42**, 4919 (1999); J. Org. Chem. **64**, 1814 (1999). [7] Angew. Chem. Int. Ed. Engl. **36**, 2190 (1997); Biochemistry **35**, 2968 (1996); Chem. Biol. **3**, 479 (1996); Chem. Eng. News (1.7.) **1996**, 27 ff.; J. Biol. Chem. **270**, 8686 (1995); **271**, 9201 (1996); Planta Med. **63**, 291 (1997); Dewick, Medicinal Natural Products, p. 186 f., Chichester: Wiley 1997. [8] J. Molec. Biol. **285**, 197 (1999).
gen.: Acc. Chem. Res. **26**, 160 (1993) ▪ Beilstein E V **18/5**, 622 ▪ Manske **39**, 195; **50**, 509–536. – *Isolation:* J. Chromatogr. **813**, 201 (1998) (technical scale purification) ▪ J. Nat. Prod. **49**, 665 (1986); **50**, 9 (1987); **53**, 1 (1990). – *Pharmacological activity:* Alkaloids (N. Y.) **25**, 10 (1985) ▪ Angew. Chem. Int. Ed. Engl. **34**, 2710 (1995) ▪ Drugs **55**, 5 (1998) ▪ Exp. Opin. Ther. Pat. **8**, 571–586 (1998) ▪ Sax (8.), TAH 775. – *Reviews:* ACS Symp. Ser. **583**, Taxane Anticancer Agents (1995) ▪ Angew. Chem. Int. Ed. Engl. **30**, 412 (1991); **33**, 15–44, 959 (1994) ▪ Atta-ur-Rahman **20**, 79–134 ▪ Farina, The Chemistry and Pharmacology of Taxol and its Derivatives, Amsterdam: Elsevier 1995 ▪ Med. Res. Rev. **18**, 299–314, 315–331 (1998) (review development) ▪ Kleemann-Engel, p. 1432–1437 ▪ Nachr. Chem. Tech. Lab. **37**, 172 (1989) ▪ Org. Prep. Proced. Int. **23**, 465–543 (1991) ▪ Suffness (ed.), Taxol: Science and Applications, Boca Raton: CRC Press 1995. – *[CAS 33069-62-4 (T.); 114915-20-7, 148408-66-6 (taxotere)]*

Taxusin see taxanes.

Taxuspines (T. A–Z). Taxoidal diterpenoids from yew species (*Taxus* sp.) with different ring structures: The T. A, J and M are tricyclic with benzo[*f*]azulene ring structure, T. B has a central 10-membered ring system (isolated from *Taxus cuspidata*, *T. chinensis*, *T. spicata*). The *taxane skeleton of *Taxol® (6,10-methanobenzocyclodecene) is found, e.g., in T. D, P, S and T and, with an additional 3,11 ring closure, in T. H. The T. E and N belong to the taxoidal group of the baccatins (taxane with fused oxetane ring at positions 4 and 5) which can be isolated from various yew species (*T. canadensis*, *T. baccata*, *T. cuspidata*, *T. chinensis*, *T. mairei*, *T. brevifolia*, *T. media*, *T. wallichiana*, *T. yunnanensis*). More than 30 T. are known to date.

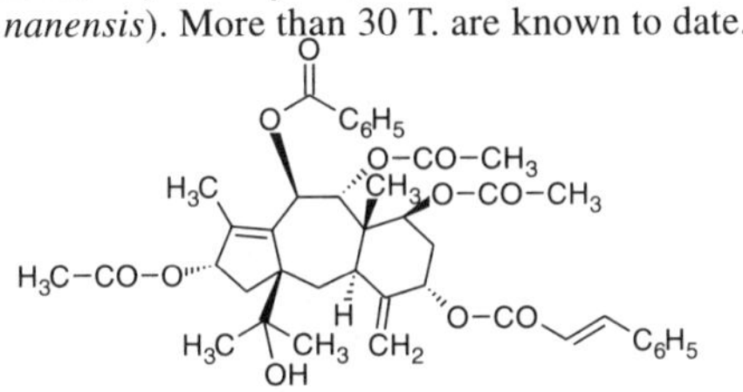

T. A

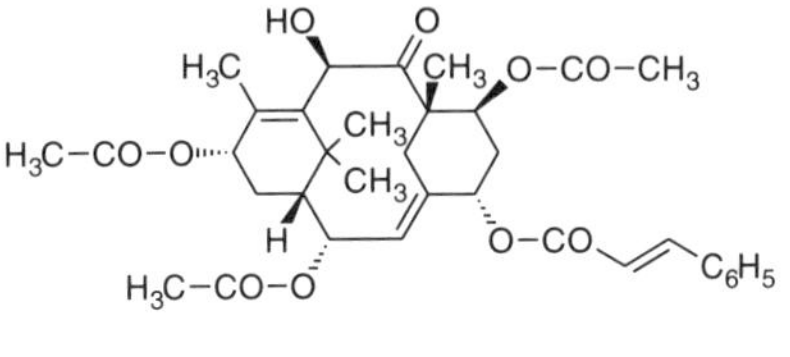

T. B

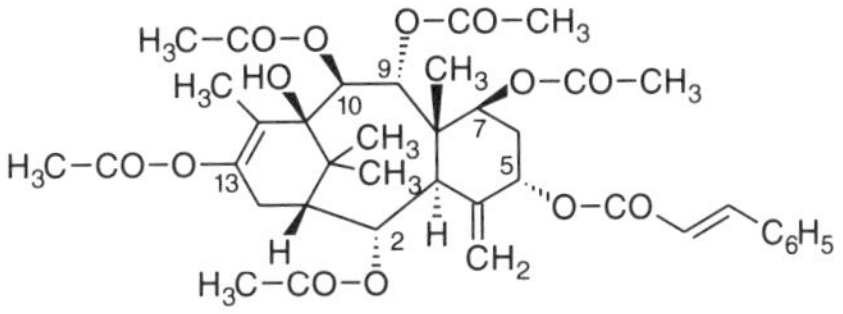

T. D

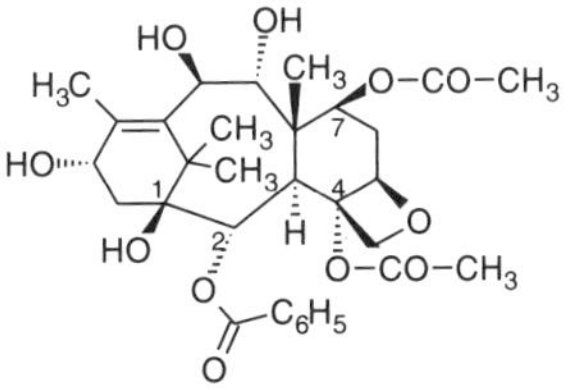

T. E

Table: Data of selected Taxuspines.

T.	molecular formula	M_R	mp.	$[\alpha]_D^{20}$	CAS
A	$C_{42}H_{48}O_{11}$	728.84	amorphous powder	–3° ($CHCl_3$)	157374-28-2
B	$C_{35}H_{42}O_{10}$	622.71	amorphous powder	–41° ($CHCl_3$)	157414-05-6
D	$C_{39}H_{48}O_{13}$	724.79	amorphous powder		166990-12-1
E	$C_{31}H_{40}O_{11}$	588.65	amorphous powder	–17° ($CHCl_3$)	165074-73-7

Lit.: *Isolation:* Chem. Pharm. Bull. **45**, 1205 (1997) (T. X–Z) ▪ Experientia **51**, 592 (1995) ▪ J. Nat. Prod. **58**, 934 (1995) ▪ Phytochemistry **48**, 857 (1998) ▪ Tetrahedron **50**, 7401 (1994); **51**, 5971 (1995); **52**, 2337, 5391, 12159 (1996). – *Synthesis:* Chem. Pharm. Bull. **45**, 1205 (1997) ▪ J. Am. Chem. Soc. **118**, 2843 (1996).

Tazettine see Amaryllidaceae alkaloids.

Tea flavor. Primary and secondary *flavor compounds (more than 500) contribute to the flavor of *black tea,* which is otherwise strongly influenced by its origin or production procedures (withering, fermentation, roasting). A major contribution is made by components with flowery notes such as *geraniol, *linalool, *citronellol, *2-phenylethanol, and *α*-/*β*-*ionones, as well as *(*Z*)-jasmone, *methyl jasmonate, *δ*-*jasmin(e) lactone, and *methyl anthranilate as trace components. Green notes such as 2- and 3-*hexen-1-ols, (*E*)-2-

*hexenals, hexanal (see alkanals), and *phenylacetaldehyde as well as the fruity and hay-like notes of the C_{13}-norisoprenoids *α-/β*-*damascones, **β*-damascenone, *theaspiranes, theaspirone (see theaspiranes), and *dihydroactinidiolide are also important. According to aroma extract dilution analysis of *green tea (Z)-1,5-octadien-3-one* ($C_8H_{12}O$, M_R 124.18, odor: metallic, geranium-like), sotolone (see hydroxyfuranones), *3-methyl-2,4-nonanedione* ($C_{10}H_{18}O_2$, M_R 170.25), (*Z*)-3-*hexenals, *β*-*damascenone, (*Z*)-4-heptenal (see alkenals), and (*E,Z*)-2,6-*nonadienals are to be considered as the most effective aroma substances[1].

Lit.: [1] Guth et al., in Schreier & Winterhalter (eds.), Progress in Flavor Precursor Studies, p. 401–407, Carol Stream (USA): Allured 1993.
gen.: Maarse, p. 617–669 ▪ TNO list (6.), Suppl. 5, p. 174–213. – *[CAS 65767-22-8 ((Z)-1,5-octadien-3-one); 113486-29-6 (3-methyl-2,4-nonanedione)]*

Teakwood see tectoquinone.

Tea polyphenols. Green tea and black tea contain antioxidative constituents with favourable pharmacological properties (*theaflavin), especially catechols and polyphenols, which show anticancer activities in animal experiments[1].

Lit.: [1] Med. Res. Rev. **17**, 327–365 (1997); Phytochemistry **46**, 1397 (1997).

Teatannin II see epigallocatechin 3-gallate.

Tecomanine (tecomine).

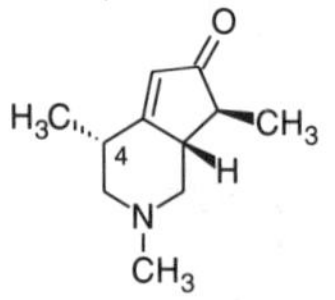

$C_{11}H_{17}NO$, M_R 179.26, bp. (13.3 Pa) 125 °C, $[\alpha]_D^{24}$ –175° ($CHCl_3$). Major alkaloid from *Tecoma stans* and *T. fulva* (Bignoniaceae). T. shows hypoglycaemic activity. *Tecoma* extracts have been used in traditional Mexican medicine to treat diabetes.

Lit.: Chem. Pharm. Bull. **40**, 1614 (1992) ▪ Phytochemistry **34**, 876 (1993) ▪ Synthesis **1998**, 552 (synthesis). – *[CAS 6878-83-7]*

Tecoside see harpagide.

Tectoquinone (2-methylanthraquinone).

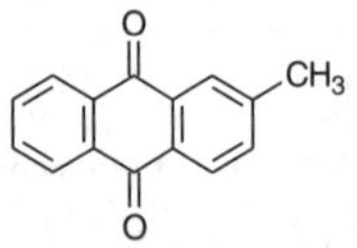

$C_{15}H_{10}O_2$, M_R 222.24, yellow needles, mp. 177–179 °C, uv_{max} 324 nm (C_2H_5OH). T. belongs to the rare class of non-hydroxylated *anthranoids and represents the yellow pigment of teak wood (*Tectona grandis*, Verbenaceae). T. serves to protect teak wood from attack by fungi and termites. T. also occurs in the roots and stems of *Morinda umbellata* (Rubiaceae), in the bark of *Clausena heptaphylla* (Rutaceae), and in the fern *Lygodium flexuosum* (Lygodiaceae).

Lit.: Acta Crystallogr., Sect. C **51**, 2094 (1995) ▪ Aust. J. Chem. **43**, 593 (1990) ▪ Beilstein E IV **7**, 2574 ▪ Planta Med. **52**, 330 (1986) ▪ Rowe, p. 849 ▪ Synth. Commun. **26**, 4507 (1996) (synthesis) ▪ Ullmann (4.) **7**, 588. – *[HS 291469; CAS 84-54-8]*

Teichoic acids. Name derived from Greek: teichos = wall, for polymeric phosphoric acid esters serving as membrane and cell wall components of Gram-positive bacteria; they may constitute up to 40 to 60% of the cell wall. T. a. are chain-like molecules in which the phosphoric acid molecules serve as linkers, either esterified with 8–50 molecules of glycerol (the free OH-groups of which may be esterified with D-*alanine) or with ribitol, the hydroxy groups in the 2,3,4-positions of which may be glycosidically linked with *N*-acetylglucosamine residues and by ester-like bonds to D-alanine.

Lit.: Adv. Polymer Sci. **79**, 139 (1986) ▪ Kirk-Othmer (4.) **4**, 922 ▪ New Compr. Biochem. **27**, 187 (1994) (review) ▪ Sutherland (ed.), Surface Carbohydrates of the Procaryotic Cell, p. 177–208, London: Academic Press 1979. – *[CAS 9041-38-7]*

Teichomycin see teicoplanins.

Teichuronic acids. *Polysaccharides, which may contain uronic acids (e. g., D-*glucuronic acid or *N*-acetyl-D-mannosaminuronic acid), isolated from the cell walls (hence the name, from Greek: teichos = wall) of Gram-positive bacteria. T. a. are either linked directly or via *oligosaccharides to the peptidoglycan (murein) of the cell wall.

Lit.: Crit. Rev. Microbiol. **17**, 129–132 (1989).

Teicoplanins (teichomycin).

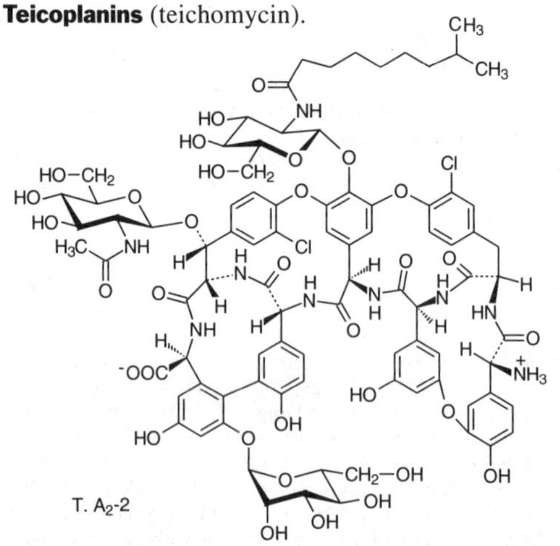

T. A_2-2

T. isolated from the bacterium *Actinoplanes teichomyceticus* proved to be a mixture of various glycopeptide antibiotics of the *vancomycin group. T. A_2 (Targocid®), obtained form the mixture, consists of five components with T. A_2-2 as the main one ($C_{88}H_{97}Cl_2N_9O_{33}$, M_R 1879.68). The T. are built up from a heptapeptide of seven unusual amino acids. The hydroxylated and, in part, chlorinated benzene rings mutually form one biphenyl ether and three diphenyl ether structures in such a way that 4 macrocyclic rings result. The periphery of the tetracyclic skeleton of T. A_2 carries the *O*-glycosidically bound sugar building blocks D-mannose, *N*-acetyl-D-glucosamine, and an *N*-acyl-D-glucosamine. The T. A_2 differ in the length, branching, and degree of saturation of the acyl group which influences the pharmacological properties. T. A_2 has bactericidal effects on Gram-positive bacteria by inhibition of cell wall biosynthesis; it is used clinically for infections with β-lactam-resistant staphylococci or streptococci and in cases of β-lactam allergies. On the whole, T. A_2-2 is more active than vancomycin. For biosynthesis and mode of action, see vancomycin.

Lit.: J. Antibiot. **31**, 170, 276 (1978); **37**, 615, 988 (1984); **42**, 361 (1989); **50**, 180 (1997). – *Biosynthesis:* J. Antibiot. **44**, 1444–1451 (1991); **48**, 1292 (1995) ▪ Zmijewski & Fayerman, in Vining & Stuttard (eds.), Genetics and Biochemistry of Antibiotic Production, p. 269–281, Boston: Butterworth-Heinemann 1995. – *Reviews:* Angew. Chem. Int. Ed. Engl. **38**, 2096–2152 (1999) ▪ Arzneimitteltherapie **9**, 331 ff. (1991) ▪ Drugs **40**, 449–486 (1990); **47**, 823 (1994) ▪ Drugs of the Future **15**, 57–72 (1990) ▪ Infektion (Munich) **22**, 430 (1994) ▪ J. Gen. Microbiol. **137**, 587–592 (1991) ▪ Kleemann-Engel, p. 1820 ff. ▪ Merck Index (12.), No. 9269 ▪ Nagarajan (ed.), Glycopept. Antibiot., p. 273, New York: Dekker 1994 ▪ Sax (8.), TAI 400. – *[CAS 61036-64-4 (T. A_2); 91032-26-7 (T. A_2-2); 61036-62-2 (T.-Komplex)]*

Teleocidins (olivoretins).

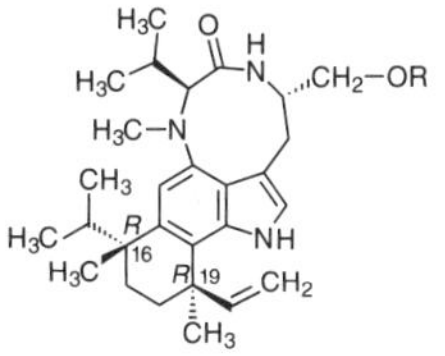

Name	R	Epimer
T. B-1	H	16*S*,19*R*
T. B-2	H	16*R*,19*S*
T. B-3	H	16*S*,19*S*
T. B-4 (Olivoretin D)	H	16*R*,19*R*
Olivoretin A	CH_3	16*R*,19*R*
Olivoretin B	CH_3	16*R*,19*S*

T. are *indole alkaloids from bacteria (actinomycetes, Eubacteriaceae) comprising a small group of toxic natural products. T. A-1, A-2, B-1, B-2, B-3, B-4 have been isolated from *Streptomyces mediocidicus*. The letter A is used to designate T. with a terpenoid, open side chain, while the B indicates the presence of a ring. Except for the terpenoid part, the T. are structurally identical to (–)-*indolactam from the cyanobacterium *Hapalosiphon*. This lactam may formally represent a biosynthetic precursor of the T. The most important member is T. A-1 (lyngbyatoxin A, see majusculamides). It induces contact dermatitis and is highly toxic. T. A-1 exhibits tumor promoting properties even at low concentrations [1]. The T. are cocarcinogens and can easily penetrate through membranes on account of their hydrophobic properties.

Table: Data of Teleocidins.

name	molecular formula	M_R	mp. [°C]	CAS
T. B-1	$C_{28}H_{41}N_3O_2$	451.65	153–155	95044-71-6
T. B-2	$C_{28}H_{41}N_3O_2$	451.65	203–204	95189-05-2
T. B-3	$C_{28}H_{41}N_3O_2$	451.65	160–162	95189-06-3
T. B-4	$C_{28}H_{41}N_3O_2$	451.65	230–232	11032-05-6
Olivoretin A	$C_{29}H_{43}N_3O_2$	465.68	251–253	90297-52-2
Olivoretin B	$C_{29}H_{43}N_3O_2$	465.68		90599-27-2

Lit.: [1] Int. J. Cancer **43**, 513 (1989).
gen.: Chem. Pharm. Bull. **34**, 4883 (1986); **37**, 563 (1989) ▪ Pure Appl. Chem. **54**, 1919 (1982) ▪ Sax (8.), TAK 750. – *Biosynthesis:* Tetrahedron **46**, 2773–2788 (1990) ▪ Tetrahedron Lett. **39**, 7929 (1998). – *Synthesis:* J. Org. Chem. **56**, 4706 (1991) ▪ Tetrahedron **47**, 8535, 8545, 8559 (1991).

Tendamistat see glycosidase inhibitors.

Tenellin. $C_{21}H_{23}NO_5$, M_R 369.42, greenish-yellow platelets, $[\alpha]_D^{24}$ –44° (acetone); pigment from *Beauveria bassiana* and *B. tenella*. Fungal metabolite with a tetrassubstituted 2-pyridone ring system of mixed

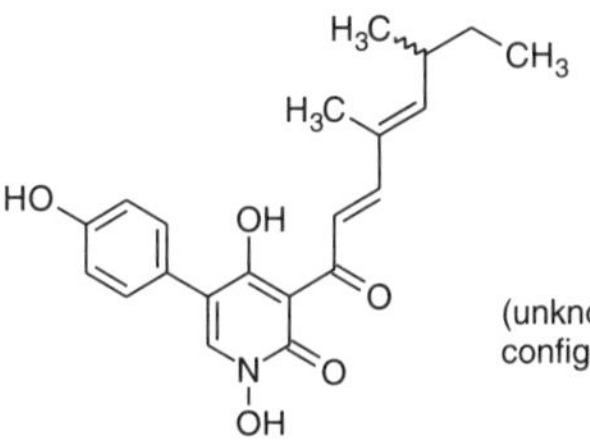

(unknown absolute configuration)

biosynthetic origin being derived from a polyketide chain and an aromatic amino acid, see sambutoxin.

Lit.: J. Chem. Soc., Perkin Trans. 1 **1991**, 2537; **1992**, 67 ▪ J. Org. Chem. **54**, 5852 (1989) ▪ Tetrahedron **54**, 9195 (1998). – *[CAS 53823-15-7]*

Tentoxin.

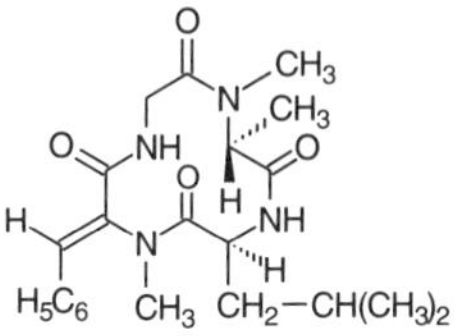

$C_{22}H_{30}N_4O_4$, M_R 414.50. A cyclic peptide antibiotic from phytopathogenic fungi of the genus *Alternaria*, e.g., *A. tenuis*, *A. alternata*, *A. solani*, *A. porri*, *A. citri*, and *A. mali* (*E*-form, isotentoxin). The *Z*-form is strongly phytotoxic. It causes chlorosis in numerous wild herbs by inhibition of protein biosynthesis through blockage of phosphorylation (ATP synthase) in chloroplasts. Sensitive spinach chloroplasts are inhibited above concentrations of 1 μM, resistant chloroplasts, e.g., of *Nicotiana tabacum* only at above 150 μM. T. could be used as a herbicide (e.g., for soy bean and corn cultivation) but the amounts produced by the fungi are too small for commercial exploitation. T. thus serves as a lead structure for synthetic work. The biosynthesis proceeds non-ribosomally at an enzyme complex, similar to other cyclopeptides.

Lit.: ACS Symp. Ser. **380**, 24–56 (1988) ▪ Agric. Biol. Chem. **54**, 2449 (1990); **56**, 139 (1992) (isolation) ▪ Böger & Sandmann (eds.), Target Sites of Herbicide Action, p. 247–282, Boca Raton: CRC Press 1989 ▪ Physiol. Plant **74**, 575–582 (1988) ▪ Phytochemistry **42**, 1537 (1996) ▪ Tetrahedron Lett. **25**, 2775 (1984); **36**, 4425 (1995) (synthesis). – *[CAS 28540-82-1 (T.); 65452-16-6 (iso-T.)]*

Tenuazonic acid [(*S*)-3-acetyl-5-(*S*)-*sec*-butyltetramic acid].

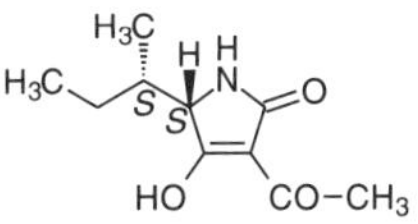

$C_{10}H_{15}NO_3$, M_R 197.23, viscous oil or pale brown cryst., mp. 74–75.5 °C, $[\alpha]_D^{20}$ –121° ($CHCl_3$); Cu salt green needles, mp. 175 °C. A phytotoxic and tremorigenic component of phytopathogenic fungi such as *Pyricularia oryzae*, *Alternaria alternata*, *A. mali*, and *A. tenuis*.

T. causes diseases in plants by inhibition of protein and nucleic acid synthesis, e.g., leaf rot and brown spot disease in rice and tobacco. T. has antiviral and insecticidal activities and inhibits the growth of mammalian cell cultures. T. can occur in moldy tomatoes and apples as well as products produced therefrom. For structure activity relationships, see *Lit.* [1].

Lit.: [1] Phytochemistry **27**, 77–84 (1988).
gen.: Agric. Biol. Chem. **38**, 791 (1974); **50**, 2401 (1986) ▪ Beilstein EV **21/11**, 555 ▪ Cole-Cox, p. 488 ▪ Turner **1**, 286; **2**, 368 ▪ Sax (8.), VTA 750. – *[CAS 27778-66-1, 75652-74-3, 610-88-8 (keto-enol tautomerism)]*

Tephrosin. A *rotenoid from *Derris* roots, *Tephrosia*, *Lonchocarpus*, and *Millettia* species.

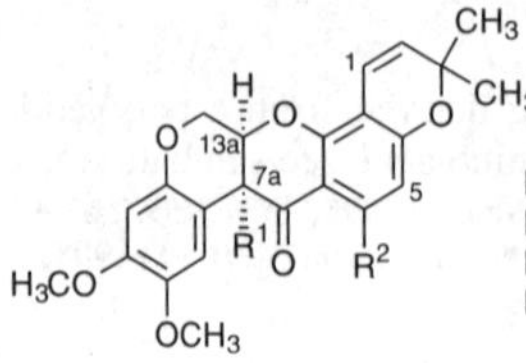

R^1 = OH, R^2 = H : Tephrosin
R^1 = H, R^2 = OH : α-Toxicarol
R^1 = R^2 = H : Deguelin
R^1 = R^2 = OH : 6-Hydroxytephrosin

Table: Data of Tephrosin and derivatives.

	configuration	molecular formula	M_R	CAS
Tephrosin	(7a*R*)-*cis*	$C_{23}H_{22}O_7$	410.42	76-80-2
	(7a*S*)-*cis*			110549-03-6
α-Toxicarol	(7a*S*)-*cis*	$C_{23}H_{22}O_7$	410.42	82-09-7
	(7a*R*)-*cis*			111057-92-2
	(±)-*trans*			123000-22-6
Deguelin	(7a*S*)-*cis*	$C_{23}H_{22}O_6$	394.42	522-17-8
	(7a*R*)-*cis*			110508-99-1
	(±)-*trans*			123000-18-0
6-Hydroxy-tephrosin	(7a*R*)-*cis*	$C_{23}H_{22}O_8$	426.42	72458-85-6
	(7a*S*)-*cis*			111058-84-5

Tephrosin: [mp. 197.5–199.5 °C (racemate)]; (7a*R*)-*cis*-form: $[\alpha]_D^{23}$ –118° (benzene). T. occurs especially in the roots of *Tephrosia purpurea* used in the traditional medicine of Sri Lanka. It is a potent fish poison and is also active against nematodes[1] and insects[2]. – *(7aS)-cis-α-Toxicarol:* greenish-yellow platelets or needles, mp. 125–127 °C, $[\alpha]_D^{20}$ –66° (benzene), $[\alpha]_D$ +58° (acetone), potent fish poison. – *(7aS)-cis-Deguelin:* yellow oil, $[\alpha]_D^{20}$ –101° (benzene), irritant, toxic on inhalation[4]. *6-Hydroxytephrosin* has potent antibacterial activity[3].

Toxicology: *Derris* roots: LD_{50} (mouse p. o.) 350 mg/kg.

Lit.: [1] Shoya Kugaku Zasshi **43**, 42–49 (1989) [Chem. Abstr. **111**, 126508 (1989)]. [2] Insect. Sci. Inst. Appl. **8**, 85 ff. (1987). [3] Heterocycles **12**, 1033 ff. (1979). [4] J. Chem. Soc., Chem. Commun. **1986**, 352 ff. (biosynthesis); Planta Med. **60**, 602 (1994) (isolation).
gen.: Agric. Biol. Chem. **38**, 1731 (1974) (synthesis) ▪ Beilstein EV **19/10**, 675 ▪ Fitoterapia **61**, 372 (1990) ▪ Harborne (1994), p. 159–166 ▪ J. Chem. Soc., Perkin Trans. 1 **1989**, 204 f. (synthesis); **1992**, 1685–1697 (biosynthesis) ▪ J. Heterocycl. Chem. **24**, 845–852 (1987) ▪ Nat. Prod. Rep. **11**, 173–203 (1994) ▪ Planta Med. **55**, 207 f. (1989); **60**, 602 (1994) ▪ Phytochemistry **28**, 3207 ff. (1989); **36**, 1523 (1997) ▪ Zechmeister **43**, 98–109 ▪ Z. Naturforsch. C **45**, 154–160 (1990). – *Pharmacology:* Chem. Pharm. Bull. **41**, 1456 (1993) ▪ J. Nat. Prod. **56**, 690–698, 843–848 (1993).

Teresantalal, teresantalol, teresantalic acid see tricyclic monoterpenes.

Termites. Mostly wingless, blind insects belonging to the Isoptera. The animals which feed preferentially on wood are often erroneously named as "white ants". They fulfill their nitrogen requirements by symbiosis with nitrogen-fixing intestinal bacteria. Methane-producing bacteria (it is estimated that more than 40% of the terrestrial methane originates from T.) and flagellates algae also participate in this symbiosis. They build very stable constructions from wood, soil, and feces in which some species even maintain fungus gardens. The T. are social insects and exhibit a pronounced class differentiation with (mostly) one conspicuously large queen, male sex partners (kings), "workers", and "soldiers".

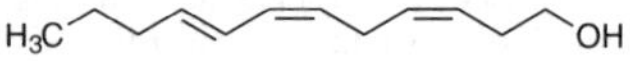

(3*Z*,6*Z*,8*E*)-3,6,8-Dodecatrien-1-ol (**1**)

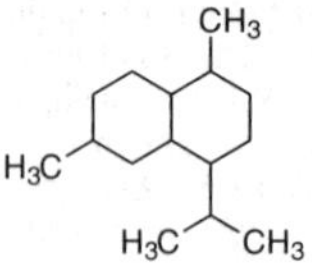

Cadinane (**2**)

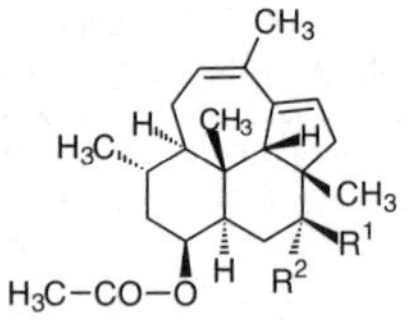

R^1 = O–CO–CH_3, R^2 = H : Kempene 1 (**3**)
R^1, R^2 = O : Kempene 2 (**4**)

Similar to the *ants T. use chemical substances for intraspecies and interspecies communication. Besides the trail pheromones (see pheromones) such as *(3Z,6Z,8E)-3,6,8-dodecatrien-1-ol* ($C_{12}H_{20}O$, M_R 180.29)[1], the *defensive secretions of the soldiers have been investigated in detail. These substances are secreted from the sometimes very large glands (frontal glands) on the head. Known components of the defensive secretions of T. include long-chain, unbranched vinyl ketones, β-ketoaldehydes, and macrocyclic lactones as well as monoterpenes (especially α-*pinenes and derivatives of *p*-menthane), sesquiterpenes (especially derivatives of *cadinane* and the *eudesmanes), and diterpenes[2] such as, e.g., *kempene 1* ($C_{24}H_{34}O_4$, M_R 386.53) and *kempene 2* ($C_{22}H_{30}O_3$, M_R 342.48, mp. 120.5–122.5 °C).

Lit.: [1] Nature (London) **219**, 963 (1969). [2] J. Am. Chem. Soc. **99**, 8082 (1977); **113**, 5883 (1991); **119**, 9584 (1997).
gen.: Ann. Rev. Entomol. **29**, 201–232 (1989) ▪ Bell & Cardé (eds.), Chemical Ecology of Insects, p. 475–519, London: Chapman & Hall 1984 ▪ Krishna & Weesner (eds.), Biology of Termites, vols. I & II, New York: Academic Press 1969, 1970 ▪ Morgan & Mandava (eds.), CRC Handbook of Natural Pesticides IVB Pheromones, p. 59–206, Boca Raton: CRC Press 1988. – *[CAS 199926-64-8 (1); 65118-73-2 (3); 65118-74-3 (4)]*

Terpenes, terpenoids. T. can formally be considered as polymerization products of the hydrocarbon isoprene (see isoprenoids). Depending on the number of isoprene units the compounds are classified as *monoterpenes (C_{10}), *sesquiterpenes (C_{15}), *diterpenes (C_{20}), *sesterterpene (C_{25}), *triterpenes (C_{30}), *tetraterpenes (C_{40}), and *polyterpenes. The *steroids are derived from a group of triterpenes, the methylsterols.
The biosynthesis of the T. proceeds according to the *isoprene rule[1]. Acyclic hydrocarbons formed in this way can be converted to a multitude of compounds by substitution, oxidation, cyclization, rearrangement, etc. reactions; accordingly a large number of T. (>40000 have been described to date) occurs in nature. Included among the T. are not only the hydrocarbons but also the alcohols, ketones, aldehydes, and esters (alternative name: terpenoids) derived from them. For nomenclature and technical use of T., see the individual entries and *Lit.*[1].

Occurrence: T. are widely distributed in plants, especially as components of the essential oils obtainable from the flowers, leaves, fruits, barks, and roots.
The amount of T. emitted from conifers (see conifer- and pine needle oils) in the hotter climatic zones has been estimated as ca. 1 000 000 000 t/year. Only small amounts of T. are found in animals, e.g., as *pheromones or repellents. Numerous halogenated T. are formed by marine organisms (see marine natural products).
Lit.: [1] Barton-Nakanishi **2**, 1–400; Nat. Prod. Rep. **16**, 97–130 (1999).
gen.: Dictionary of Terpenoids **1–3** ▪ Dev, CRC Handbook of Terpenoids: Monoterpenoids (2 vols.), Diterpenoids (4 vols.), Boca Raton: CRC Press 1982, 1985 ▪ Glasby, Encyclopedia of the Terpenoids, New York: Wiley 1982 ▪ Joulain, Progress in Terpene Chemistry, Gif-sur-Yvette: Edit. Frontières 1987 ▪ Kirk-Othmer (3.) **22**, 709–762 ▪ Thomson (2.), p. 106–136 ▪ Torssell, Natural Products Chemistry (2.), p. 251–312, London: Pergamon 1997 ▪ Ullmann (5.) **A 11**, 154–180 ▪ Verghese, Terpene Chemistry, New York: McGraw-Hill 1982. – *Biosynthesis:* Acc. Chem. Res. **23**, 70 (1990) ▪ Nat. Prod. Rep. **1**, 319, 343–450 (1984); **2**, 513 (1985); **3**, 251 (1986); **4**, 157, 377 (1987); **5**, 247, 419 (1988); **7**, 25, 387 (1990) ▪ Science **227**, 1811, 1815, 1820 (1997) ▪ Stud. Nat. Prod. Chem. **7**, 87–129 (1990). – *Synthesis:* ApSimon **4**, 451–593; **7**, 275–454 ▪ Tse-Lok Ho, Carbocycle Construction in Terpene Synthesis, Weinheim: VCH Verlagsges. 1988.

Terpestacin. Metabolite from *Arthrinium* sp.

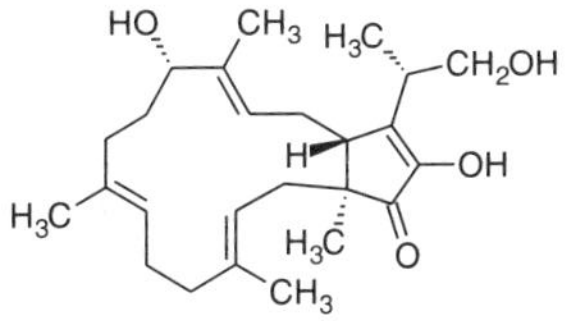

$C_{25}H_{38}O_4$, M_R 402.57, cryst., mp. 172–173 °C, $[\alpha]_D^{22}$ +26° ($CHCl_3$). T. shows activity against AIDS viruses (inhibition of syncytia formation) as well as against tumor-inducing viruses.
Lit.: *Isolation:* J. Antibiot. **46**, 367 (1993); **51**, 602 (1998) ▪ Tetrahedron Lett. **34**, 493 (1993). – *Abs. configuration, biosynthesis:* J. Org. Chem. **58**, 1875 (1993). – *Synthesis:* Tetrahedron Lett. **39**, 83 (1998); **40**, 843 (1999). – *[CAS 146436-22-8]*

Terphenylquinones.

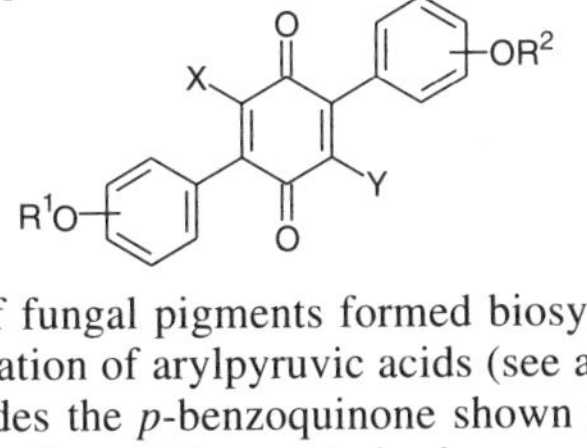

A group of fungal pigments formed biosynthetically by dimerization of arylpyruvic acids (see also grevillins). Besides the *p*-benzoquinone shown in the formula, derivatives with a central *o*-benzoquinone unit (*phlebiarubrone) are also known. T. are typical components of the Boletales (basidiomycetes) and are converted in the fungi to further pigments, e.g., to *badiones and *pulvinic acid. *Examples* of T. are *atromentin, *aurantiacin, *cycloleucomelone, *flavomentins, *leucomentins, *polyporic acid, *spiromentins, and *thelephoric acid. A general synthesis of the T. starts from grevillin derivatives which undergo rearrangement on treatment with methanolic alkali. Some T. exhibit interesting biological activities.

Lit.: Justus Liebigs Ann. Chem. **1986**, 195; **1989**, 803 (synthesis) ▪ Tetrahedron Lett. **28**, 4749 (1987) (biosynthesis). – *Reviews:* Thomson **2**; **3** ▪ Zechmeister **51**, 4–32.

Terpin (*p*-menthane-1,8-diol).

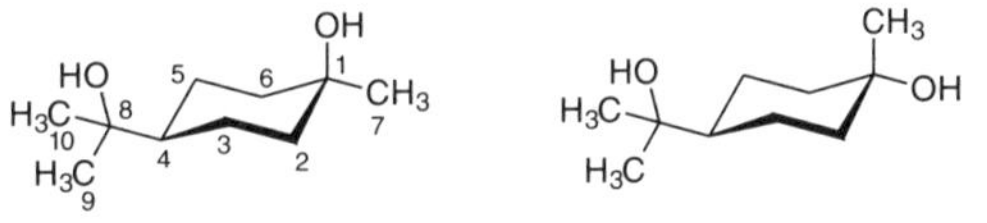

trans-p-Menthane-1,8-diol „*cis*-Terpin" *cis-p*-Menthane-1,8-diol „*trans*-Terpin"

$C_{10}H_{20}O_2$, M_R 172.27. A monocyclic *monoterpene dialcohol with the *p*-menthane structure occurring in nature in both the *trans*- and the *cis*-forms, "*trans*-T.", mp. 158–159 °C, "*cis*-T.", mp. 104–105 °C. The "*cis*-T." exists mostly as the hydrate ("*cis*-terpin hydrate", mp. 118–119 °C) which is used as an expectorant. The occurrence of T. is probably the result of secondary formation from γ-terpineol [*p*-menth-4(8)-en-1-ol]. T. occurs in Chinese star anise oil (*Illicium verum*, Illiciaceae) [1]. For synthesis, see *Lit.*[2].
Lit.: [1] Bull. Soc. Chim. Fr. **1969**, 339. [2] J. Chem Soc., Perkin Trans. **1**, 1971 (1972).
gen.: Beilstein EIV **6**, 251 ▪ Florey **14**, 273 ▪ Karrer, No. 321–323 ▪ Merck-Index (12.), No. 9314. – *[HS 2906 19; CAS 80-53-5; 565-50-4 (trans-T.); 565-48-0 (cis-T.); 2451-01-6 (cis-T. hydrate)]*

Terpinenes, terpinols see *p*-menthadienes.

Terpinenols, α-terpineol see *p*-menthenols.

α-Terpineol acetate see oil of cardamom.

Terprenin.

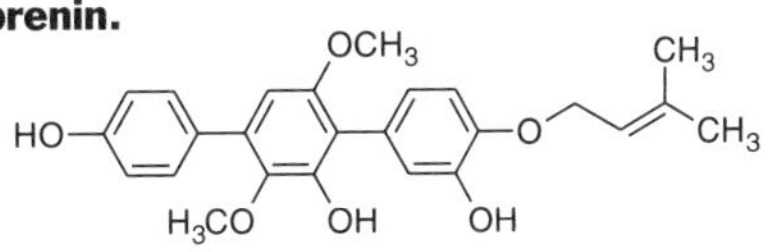

$C_{25}H_{26}O_6$, M_R 422.46, prismatic cryst., mp. 156 °C. *p*-Terphenyl derivative from cultures of *Aspergillus candidus*. T. *in vitro* inhibits the formation of immunoglobulin E in human lymphocytes (IC_{50} 0.18 nM). It is investigated as lead structure for the synthesis of therapeutics against asthma, atopic dermatitis and other allergic diseases.
Lit.: Angew. Chem. Int. Ed. Engl. **37**, 973 (1998) (synthesis) ▪ J. Antibiot. **51**, 445 (1998) (isolation) ▪ J. Org. Chem. **63**, 5831 (1998). – *[CAS 197899-11-9]*

Terramycin see tetracyclines.

Terrein [(4*S*,5*R*)-4,5-dihydroxy-3-(*E*)-propenyl-2-cyclopenten-1-one].

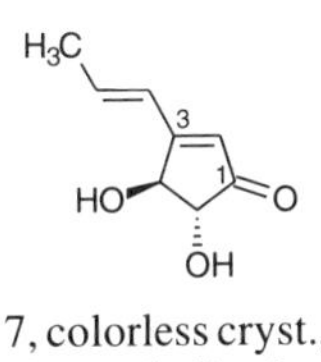

$C_8H_{10}O_3$, M_R 154.17, colorless cryst., mp. 121–122 °C, $[\alpha]_D$ +155° (H_2O). A metabolite from *Aspergillus terreus* and other fungi imperfecti. The pentaketide T. is formed biosynthetically via an aromatic intermediate by oxidative cleavage [1].

Lit.: [1] J. Chem. Soc., Perkin Trans. 1 **1981**, 2570.
gen.: Angew. Chem. Int. Ed. Engl. **29**, 67; **35**, 1560–1562 (1996) ▪ Biosci., Biotechnol., Biochem. **57**, 1208 (1993) ▪ Cole-Cox, p. 769 ▪ Tetrahedron: Asymmetry **1**, 237 (1990) ▪ Tetrahedron Lett. **22**, 4557 (1981). – *[CAS 582-46-7]*

Terrestrol. [(3*S*,6*E*)-3,7,11-trimethyl-6,10-dodecadien-1-ol, (*S*)-2,3-dihydrofarnesol].

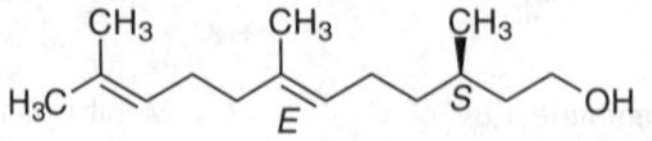

$C_{15}H_{28}O$, M_R 224.39, $[\alpha]_{546}^{23}$ −4.4° ($CHCl_3$). A territory-marking pheromone of male bumblebees of the species *Bombus terrestris* [1]. T. or its esters with short- and medium-chain fatty acids have also been identified in other hymenoptera (bees and ants).

Lit.: [1] Acta Chim. Scand. **25**, 1685–1694 (1971).
gen.: J. Chem. Ecol. **17**, 1633–1639 (1991) ▪ Physiol. Entomol. **19**, 115–123 (1994). – *[CAS 27745-36-4]*

Territrems. A group of tremorigenic *mycotoxins formed by *Aspergillus terreus*. In contrast to the *penitrems, *aflatrem, and other tremorigens, the T. do not contain nitrogen.

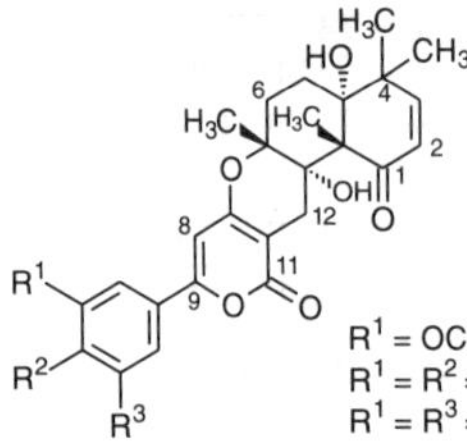

$R^1 = OCH_3$, $R^2, R^3 = –O–CH_2–O–$: T. A
$R^1 = R^2 = R^3 = OCH_3$: T. B
$R^1 = R^3 = OCH_3$, $R^2 = OH$: T. C

Table: Data of Territrems.

Territrem	molecular formula	M_R	mp. [°C]	CAS
A	$C_{28}H_{30}O_9$	510.54	288–290	70407-19-1
B	$C_{29}H_{34}O_9$	526.58	200–203	70407-20-4
C	$C_{28}H_{32}O_9$	512.56	172–173	89020-33-7

Lit.: Appl. Environ. Microbiol. **37**, 355, 358 (1979); **47**, 98 (1984); **54**, 585 (1988) ▪ J. Antibiot. **48**, 745 (1995) ▪ J. Nat. Prod. **55**, 251 (1992) ▪ Zechmeister **48**, 45–48.

Terthienyls.

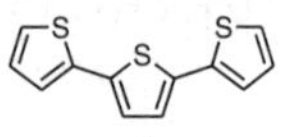

1

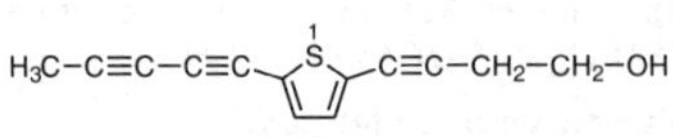

2

A group of compounds the most important of which, *α-T.* (2,2′:5′,2″-terthiophene; **1**), has been isolated from *Tagetes* species and other Asteraceae [1,2] ($C_{12}H_8S_3$, M_R 248.37, yellow orange cryst., mp. 93–94 °C). α-T. and thiophene-containing acetylenes, e.g., *4-(5-penta-1,3-diynyl-2-thienyl)-3-butyn-1-ol* [3,4] (**2**; $C_{13}H_{10}OS$, M_R 214.28) act as photoactivated insecticides and nematicides and exhibit phytotoxicty. The mechanism of action proceeds through the formation of singlet oxygen. Several syntheses have been reported [5]; for biosynthesis, see *Lit.* [6].

Lit.: [1] Bioact. Mol. 7, 21 (1988). [2] Synthesis **1991**, 462. [3] Chem. Ber. **101**, 4163–4169 (1968). [4] Phytochemistry **30**, 879 ff., 1157 ff. (1991). [5] Heterocycles **24**, 637 (1986); J. Chem. Soc., Chem. Commun. **1987**, 1021; Bull. Chem. Soc. Jpn. **62**, 1539 (1989); Chem. Rev. **92**, 711–738 (1992); Synthesis **1993**, 1099; Synth. Commun. **19**, 307 (1989); Heterocycles **30**, 645 (1990); ACS Symp. Ser. **443** (Synthesis and Chemistry of Agrochemicals II), 352–370 (1991); J. Heterocycl. Chem. **28**, 411–416 (1991). [6] Tetrahedron Lett. **1966**, 773.
gen.: ACS Symp. Ser. **387** (Insecticides of Plant Origin), 164–172 (1989) ▪ Beilstein E V **17/4**, 499 ▪ Recl. Trav. Chim. Pays-Bas **115**, 119 (1996) ▪ Synthesis **1991**, 462 ▪ Zechmeister **56**, 87–190. – *[CAS 1081-34-1 (α-T.); 1209-21-8 (4-(5-penta-1,3-diynyl-2-thienyl)-3-butyn-1-ol)]*

Testosterone see androgens.

Tetracenomycins.

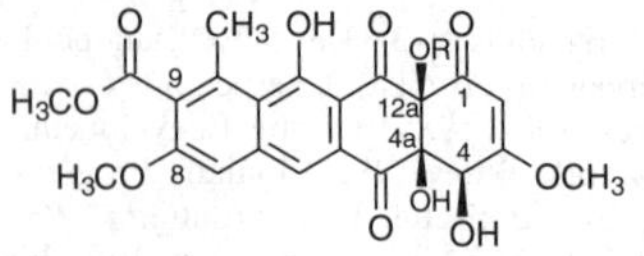

R = H : T. C
R = CH_3 : T. X

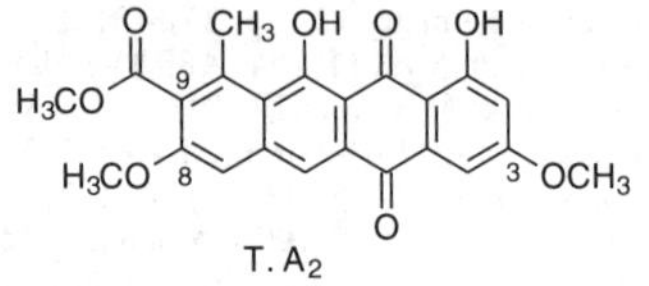

T. A_2

Tetracyclic *polyketide antibiotics from *Streptomyces glaucescens* (T. C) and *Nocardia mediterranea* (T. X). *T. C:* $C_{23}H_{20}O_{11}$, M_R 472.40, yellow cryst., decomp. above 215 °C, $[\alpha]_D$ +22° (dioxan), soluble in acetone or ethyl acetate. T. C selectively inhibits the growth of actinomycetes and related Gram-positive bacteria, it is also cytotoxic to L1210 leukemia cells. The biosynthesis of T. C and the biosynthesis gene of *S. glaucescens* have been investigated in detail. The direct precursors of T. C are the red *tetracenequinones*, e.g., T. A_2 ($C_{23}H_{18}O_8$, M_R 422.39). Attempts have been made to prepare *hybrid antibiotics by transfer of the T. C biosynthesis gene to other producers. The *elloramycins* in which the skeleton is glycosylated in position 8 are related to T. C.

Lit.: Angew. Chem. Int. Ed. Engl. **34**, 565 ff. (biosynthesis), 1107 ff. (genetics) (1995) ▪ Arch. Microbiol. **121**, 111–116 (1979) ▪ Hutchinson, in Vining & Stuttard (eds.), Genetics and Biochemistry of Antibiotic Production, p. 331–357, Boston: Butterworth-Heinemann 1995 ▪ J. Antibiot. **41**, 1066–1073 (1988); **42**, 640 ff. (1989); **43**, 1169–1178 (1990) (derivatives); **45**, 1190 ff. (1992) ▪ J. Nat. Prod. **56**, 1288 (1993) (biosynthesis). – *[HS 2941 30; CAS 71135-22-3 (T. C); 121245-07-6 (T. X); 82277-62-1 (T. A_2)]*

(*Z*)-15-Tetracosenoic acid see erucic acid.

Tetracyclines. Collective name for tetracyclic *polyketide antibiotics produced by streptomycetes (e.g., *Streptomyces aureofaciens*, *S. rimosus*) in yields of up to 20 g per L in optimized culture broths; the natural products or their semisynthetic derivatives are used for treatment of bacterial infections.

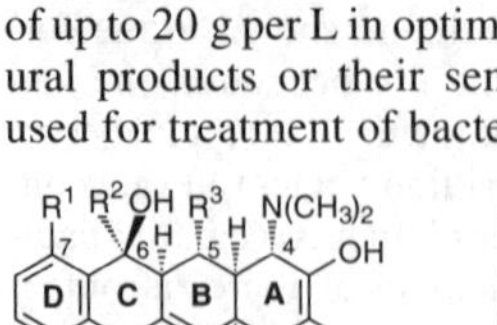

	R^1	R^2	R^3
Tetracycline (**1**)	H	CH_3	H
Chlorotetracycline (**2**)	Cl	CH_3	H
Oxytetracycline (**3**)	H	CH_3	OH
6-Demethyltetracycline (**4**)	H	H	H

The parent compound is *tetracycline* {international free name, yellow, light-sensitive cryst., mp. 170–175 °C, $[\alpha]_D$ −239° (CH_3OH), soluble in ethanol,

Table: Selection of naturally occurring Tetracyclines.

	molecular formula	M_R	CAS
1	$C_{22}H_{24}N_2O_8$	444.44	60-54-8
2	$C_{22}H_{23}ClN_2O_8$	478.89	57-62-5
3	$C_{22}H_{24}N_2O_9$	460.44	79-57-2
4	$C_{21}H_{22}N_2O_8$	430.41	987-02-0

acetone, chloroform, or ethyl acetate, poorly soluble in water and petroleum ether, amphoteric, forms a hydrochloride}. A biosynthetic precursor of T. is chlorinated at C-7 by a chloroperoxidase in *S. aureofaciens* to furnish as final product *chlorotetracycline* {brand name aureomycin, golden yellow cryst., mp. 168–169 °C, $[\alpha]_D$ −275° (CH_3OH)}. Oxygenation at C-5 which, in the case of *S. rimosus* can be controlled via the oxygen partial pressure during fermentation, affords *oxytetracycline* (terramycin, pale yellow cryst., mp. 181–182 °C). A conspicuous feature of the structure of the T. is the accumulation of oxygen atoms in the lower half of the molecule, in addition the rings A and B are twisted as a result of their *cis*-linkage. Both factors favor the formation of complexes with di- and trivalent cations (e.g., Mg^{2+}, Ca^{2+}, Al^{3+}).

Biosynthesis: Starts with malonamyl-CoA to which eight acetate units are added, methylation at C-6 (CH_3 from methionine) occurs during the cyclization. The first detectable intermediate after separation from the polyketide synthase is already a substituted naphthacene (tetracene) which is converted to T. in the later stages of the biosynthesis. Most participating enzymes and the respective biosynthetic genes have been elucidated.

Activity: The T. are broad spectrum antibiotics. They inhibit the growth of Gram-positive and Gram-negative bacteria, mycoplasma, rickettsia, some viruses and protozoa. The differentiated use of T. in human medicine, veterinary medicine, and animal breeding is based less on differences in the activity spectra but rather on differences in pharmacological behavior when, e.g., infections of the skin or the urogenital system are to be treated. The T. inhibit protein biosynthesis of pro- and eukaryotes as soon as they penetrate the cell membrane. However, only prokaryotes possess an efficient transport system (active transport). The 70S-ribosomes of bacterial cells have a preferential binding site at protein S7 between the 30S- and 50S-subunits. The T. prevent the binding of aminoacyl-tRNA to the A-side of the intact ribosomes. T.-resistant bacteria strains are easily generated because of changes at the ribosomal binding site (site of action resistance) or the activation of transport systems which carry the T. out of the cell again (export carrier). On account of the dramatic development of resistance, T. should be reserved for use in human medicine. The T. are contraindicated for pregnant women and children on account of their complex formation with Ca^{2+} (bone and teeth development).

Investigations on structure-activity relationships have shown that the naturally occurring T. are more or less optimized. There are only few successful semisynthetic T. (e.g., minocycline, doxycycline, glycocycline). Very recently, natural T. derivatives with a sugar residue (dactylocyclines) and some with inhibitory activity towards DNA topoisomerase I (TAN-1518) have been found.

Lit.: *Biosynthesis:* Appl. Microbiol. Biotechnol. **32**, 674–679 (1990) ▪ Behal, Hunter, in Vining & Stuttard (eds.), Genetics and Biochemistry of Antibiotic Production, p. 359–384, Boston: Butterworth-Heinemann 1995. – *Chloro-T.:* Beilstein E IV **14**, 2629 ▪ Florey **8**, 101–137 ▪ Hager (5.) **7**, 915 ff. ▪ Kirk-Othmer (4.) **3**, 331 ▪ Ullmann (5.) **A 2**, 516 ▪ Zechmeister **21**, 80–120. – *New T.:* Antimicrob. Agents Chemother. **38**, 637 ff. (1994) ▪ J. Antibiot. **45**, 1892–1913 (1992); **47**, 545–555 (1994) ▪ J. Med. Chem. **37**, 184–188 (1994). – *Synthesis:* J. Am. Chem. Soc. **101**, 689–701 (1979). – *Reviews:* Curr. Pharm. Des. **4**, 119–132 (1998) ▪ Hlavka & Boothe, The Tetracyclines, Heidelberg: Springer 1985 ▪ Hostalek & Vanek, in Pape & Rehm (eds.), Biotechnology, vol. 4, p. 393–429, Weinheim: VCH Verlagsges. 1986 ▪ Kleemann-Engel, p. 1844.

Tetradecanoic acid see myristic acid.

7-Tetradecen-2-one see scarabs.

Tetradecenyl acetate.

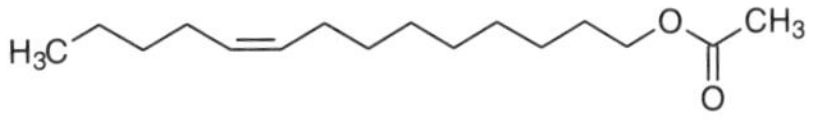

(*Z*)-9-T.

$C_{16}H_{30}O_2$, M_R 254.41, oil. The typical basic structure of many sexual pheromones of *butterflies. Particularly widely distributed are 6-, 8- and 12-, as well as (*Z*)-9-T. and (*E*)-11-tetradecenyl acetate. The corresponding free alcohols and aldehydes are also known as *pheromones. *Dodecenyl acetates* ($C_{14}H_{26}O_2$, M_R 226.36, oil) and dodecyl acetates are also widely distributed. (*Z*)-7-Dodecenyl acetate is particularly abundant[1] and is also known as the sexual pheromone of elephants[2], see elephant pheromone.

Lit.: [1] Arn, Toth, & Priesner, List of Sex Pheromones of Lepidoptera and Related Attractants (2.), Montfavet: Int. Org. for Biological Control 1992. [2] Nature (London) **379**, 684 (1996). *gen.:* Physiol. Entomol. **15**, 47 (1990) ▪ Synth. Commun. **24**, 1265 (1994); **25**, 3457 (1995). – *[CAS 16725-53-4 ((Z)-9-T).; 23192-82-7 ((E)-11-T.); 16974-10-0 ((Z)-7-T.); 14959-86-5 ((Z)-7-dodecenyl acetate); 39650-10-7 (6-T.); 35835-80-4 (8-T. a.); 35153-21-0 (12-T.)]*

Tetrahydrobiopterin [(6*R*)-2-amino-6-((1*R*,2*S*)-1,2-dihydroxypropyl)-5,6,7,8-tetrahydro-4(1*H*)-pteridinone, H_4B].

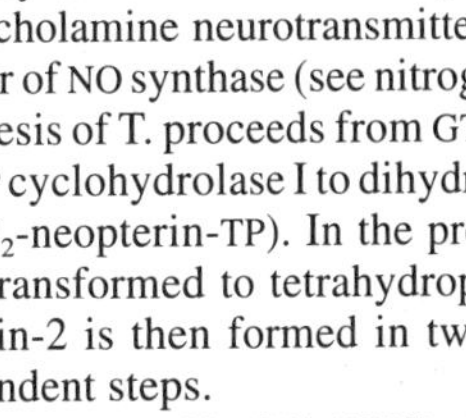

$C_9H_{15}N_5O_3$, M_R 241.25; hydrochloride: mp. 245–246 °C, $[\alpha]_D^{25}$ −6.81° (0.1 m HCl). T. plays a central role as cofactor of phenylalanine, tyrosine, and tryptophan hydroxylases that are necessary for the biosynthesis of catecholamine neurotransmitters. In addition, T. is a cofactor of NO synthase (see nitrogen monoxide). The biosynthesis of T. proceeds from GTP and leads by means of GTP cyclohydrolase I to dihydroneopterin triphosphate (H_2-neopterin-TP). In the presence of Mg^{2+} the latter is transformed to tetrahydropterin-1. Tetrahydrobiopterin-2 is then formed in two consecutive, NADPH-dependent steps.

Lit.: J. Biochem. (Tokyo) **98**, 1341 (1985) ▪ Heterocycles **23**, 3115 (1985) (synthesis) ▪ J. Nutr. Biochem. **2**, 411 (1991) (biosynthesis) ▪ Proc. Soc. Exp. Biol. Med. **203**, 1 (1993) (review). – *[CAS 17528-72-2; 99630-29-2]*

Tetrahydrocannabinols (abbr.: THC).

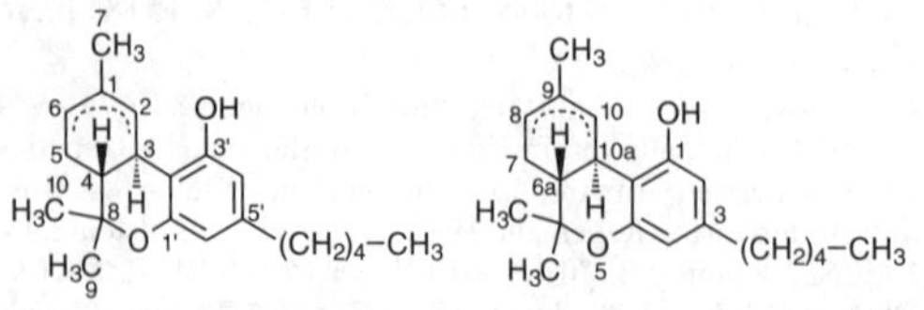

Δ^1-THC = Δ^9-THC
Δ^6-THC = $\Delta^{1(6)}$-THC = Δ^8-THC

$C_{21}H_{30}O_2$, M_R 314.47. Colorless, hallucinogenic oils from *hashish and *marihuana (*Cannabis sativa*). T. are found especially in the flower tips of the female plants. The main drug is *Δ^9-T.* which, in the (*6aR,10aR*)-form [bp. 155–157 °C (6.65 Pa), $[\alpha]_D^{20}$ –156° (C_2H_5OH)], is responsible for the euphoric and hallucinogenic effects of hashish. The enantiomer, the racemate, and the diastereomers have been prepared from Δ^9-T. Besides the name Δ^9-T., the designation Δ^1-T. according to the 8-phenoxymenthane numbering is also found in the literature. *Δ^8-T.* (Δ^6-T.) [bp. 175–178 °C (13.3 Pa), $[\alpha]_D^{27}$ –260° (C_2H_5OH)] has a similar activity.

Activity: As a lipophilic molecule Δ^9-T. is very rapidly deposited in adipose tissue. Its biological half-life is 1 week so that elimination requires at least 1 month [1]. It also has relaxing, pain-killing, and antiemetic properties and is thus used as an auxiliary agent in the chemotherapy of cancer patients (dronabinol, marinol), it is also a very efficient antiasthmatic drug and reduces intraocular pressure (glaucoma therapy). At very high doses the relaxing, pleasant properties are lost and panic, anxiety and paranoid effects dominate. Chronic use of T. leads to demotivation accompanied by personality changes.

The T. are of mixed *biogenetic* origin. They are formed from geranyl diphosphate and a polyketide precursor. The T. receptor (a polypeptide of 473 amino acids) is known [2]; see also cannabinoids. Binding to this receptor, THC act against pain (analgetically) [3].

Lit.: [1] Med. J. Aust. **145**, 82 (1986). [2] Chem. Eng. News (13.8.) **1990**, 5; Nachr. Chem. Tech. Lab. **38**, 1248 (1990). [3] Nature (London) **395**, 381 (1998).

gen.: Beilstein EV **17/4**, 421–426 ▪ J. Chem. Soc., Chem. Commun. **1989**, 1171 ff. ▪ J. Org. Chem. **56**, 6865–6872 (1991) ▪ Life Sci. **56**, 2097 (1995) (use) ▪ Merck-Index (12.), No. 9349 ▪ Sax (8.), TCM 000–TCM 250 ▪ Synlett **1991**, 553 f. – *[CAS 1972-08-3 (Δ^9-T.); 5957-75-5 (Δ^8-T.)]*

Tetrahydrofolic acid [*N*-((6*S*)-5,6,7,8-tetrahydropteroyl)-L-glutamic acid, H_4-folate, obsolete abbr.: FH_4, THF].

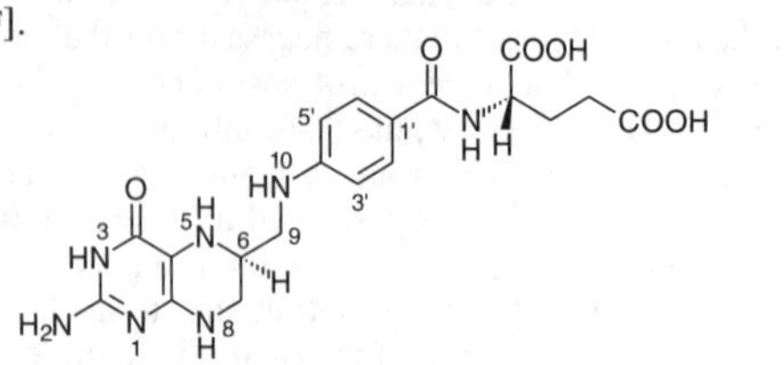

$C_{19}H_{23}N_7O_6$, M_R 445.43, $[\alpha]_D^{27}$ +14.9° (0.1 M NaOH). A *coenzyme, unstable in solution and in the air, that transfers a one-carbon unit in metabolic processes (C_1; 'activated formaldehyde, activated formic acid') (see folic acid). T. is formed by reduction of 7,8-dihydrofolic acid under the action of dihydrofolate reductase (EC 1.5.1.3) [1]. Production by chemical [2] or enzymatic [3] reduction of *folic acid.

Lit.: [1] Pteridines **1**, 37–43 (1989). [2] Chem. Ber. **125**, 2085–2095 (1992). [3] Biosci. Biotechnol. Biochem. **56**, 437 ff., 701 ff. (1992); Methods Enzymol. **281**, 3–16 (1997).

gen.: Beilstein EV **26/18**, 9 f. ▪ Isler et al., Vitamine II, p. 264–308, Stuttgart: Thieme 1988 ▪ J. Biol. Chem. **264**, 328–333 (1989) ▪ J. Org. Chem. **57**, 4470–4477 (1992) (synthesis) ▪ Neurochem. Res. **16**, 1031–1036 (1991) (review, pharmacology) ▪ see also folic acid. – *[CAS 135-16-0; 74708-38-6 ((R)-form); 71963-69-4 ((S)-form)]*

Tetrahydrofuran fatty acids see F-acids.

Tetrahydrolipstatin (THL) see lipstatin.

Tetrahydromethanopterin (5,6,7,8-tetrahydromethanopterin, H_4-MPT).

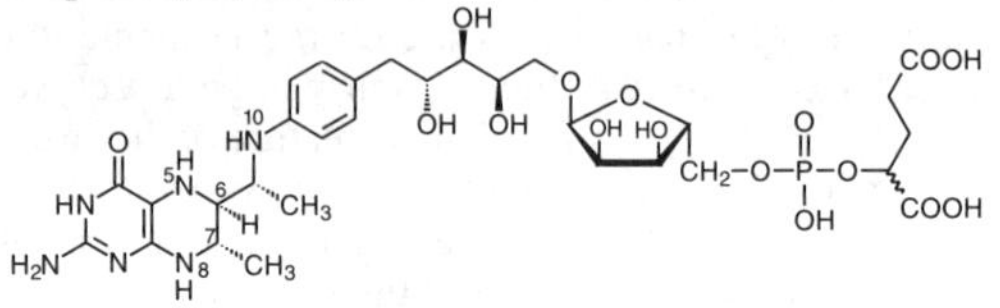

$C_{30}H_{45}N_6O_{16}P$, M_R 776.69. An oxygen-sensitive pterin derivative isolated from methanobacteria [1] that participates as a *coenzyme in the reduction of carbon dioxide to methane by accepting at its 5-position the formyl group from *N*-formylmethanofuran (catalysed by formylmethanofuran T. formyl transferase [2], EC 2.3.1.101). 5-Formyl-T. is then cyclized to 5,10-methenyl-T., reduced to 5,10-methylene-T. (methylene-T. dehydrogenase [3], EC 1.5.99.9; coenzyme F_{420} as electron donor), and reduced further to 5-methyl-T. (methylene-T. reductase [4], also with the aid of coenzyme F_{420}), finally, the methyl group is passed on to coenzyme M (enzyme: T. methyltransferase [5], EC 2.1.1.86).

Lit.: [1] Proc. Natl. Acad. Sci. USA **81**, 1976–1980 (1984). [2] FEBS Lett. **275**, 226–230 (1990); J. Biol. Chem. **261**, 16653–16659 (1986). [3] Biochem. Biophys. Res. Commun. **133**, 884–890 (1985); Angew. Chem. Int. Ed. Engl. **34**, 2247 (1995). [4] Eur. J. Biochem. **191**, 187–193 (1990). [5] Biochem. Biophys. Res. Commun. **136**, 542–547 (1986).

gen.: Annu. Rev. Biochem. **59**, 361–370 (1990) ▪ FEMS Microbiol. Rev. **87**, 327–333 (1990). – *[CAS 92481-94-2]*

Tetrahydropalmatine (Hindarine, Gindarine, Rotundine).

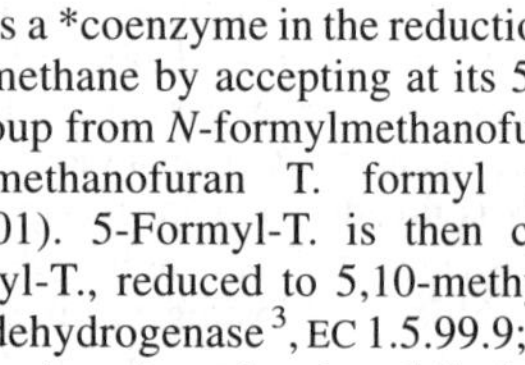

(-)(*S*)-form

$C_{21}H_{25}NO_4$, M_R 355.43, cryst., mp. 141–142 °C, pharmacologically active (*S*) enantiomer: $[\alpha]_D^{21}$ –291° (ethanol). Both enantiomers of the berberine alkaloid (see protoberberine alkaloids) T. occur in *Corydalis-*, *Stephania*, *Berberis*, *Coptis* and *Pachypodanthium* species.

The (*S*) enantiomer is marketed as sedative and analgesic with papaverine-like activity in several Asian countries and the U. S. A.

Lit.: *Structure:* Acta Crystallogr., Sect. C **49**, 1691 (1993). – *Synthesis:* J. Chem. Soc., Perkin Trans. 1 **1997**, 2291 ▪ J. Org. Chem. **55**, 1932 (1990); **57**, 6716 (1992) ▪ Org. Prep. Proced. Int. **21**, 309 (1989). – *[CAS 10097-84-4 (T., general); 2934-97-6 ((±)-T.); 483-14-7 ((S)-T.); 3520-14-7 ((R)-T.)]*

Tetramethylarsonium chloride see arsenic in natural products.

Tetramic acid (2,4-pyrrolidinedione).

$C_4H_5NO_2$, M_R 99.09, pale yellow solid, mp. >360 °C. T. is the parent compound of various T. derivatives occurring especially in microorganisms and marine invertebrates, and characterized by a wide diversity of biological activities. T. derivatives are also responsible for the colors of certain slime molds and sponges (e. g., *fuligorubin A). Most natural derivatives of T. such as, e. g., *tenuazonic acid, α- and β-*cyclopiazonic acid, or *magnesidin are acylated at C-3. The pigments *aurantoside A* ($C_{35}H_{44}Cl_2N_2O_{15}$, M_R 803.64) and *aurantoside B* ($C_{36}H_{46}Cl_2N_2O_{15}$, M_R 817.66) isolated from *Theonella* sp. show high activities against certain leukemia cell lines. *Discodermide* {$C_{27}H_{34}N_2O_6$, M_R 482.58, powder, mp. ~200 °C, $[\alpha]_D$ +97.5°} isolated from the sponge *Discodermia dissoluta* was the first T.-containing macrolactam.

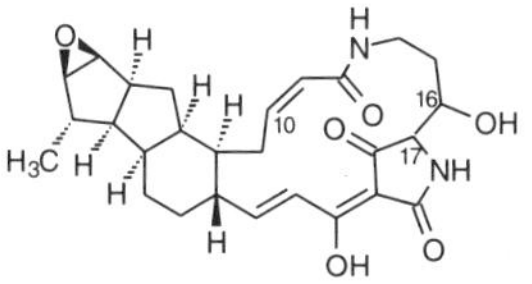

Discodermide (**1**)

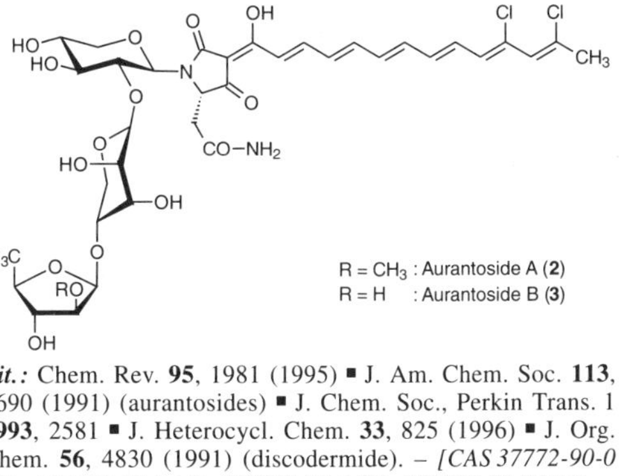

R = CH_3 : Aurantoside A (**2**)
R = H : Aurantoside B (**3**)

Lit.: Chem. Rev. **95**, 1981 (1995) ▪ J. Am. Chem. Soc. **113**, 9690 (1991) (aurantosides) ▪ J. Chem. Soc., Perkin Trans. 1 **1993**, 2581 ▪ J. Heterocycl. Chem. **33**, 825 (1996) ▪ J. Org. Chem. **56**, 4830 (1991) (discodermide). – *[CAS 37772-90-0 (T.); 134458-00-7 (1); 137895-70-6 (2); 137895-71-7 (3)]*

Tetranactin see nonactin.

Tetrandrine see bisbenzylisoquinoline alkaloids.

Tetraponerines. A group of alkaloids from the venom of ants of the genus *Tetraponera*. The currently known members (T.-1 to T.-8) have a perhydrated pyridopyrrolopyrimidine or dipyrrolopyrimidine skeleton. Some of the originally proposed structures have been revised [1]. T.-8 is formed in the ants from glutamic acid and six acetate units [2]. T.-8 is highly toxic and exhibits paralytic properties [LD_{50} 0.5 µg/mg (*Tetraponera* against the ant *Myrmica rubra*)].

R^1 = H , R^2 = C_3H_7 : (+)-T.-3
R^1 = C_3H_7 , R^2 = H : (+)-T.-4
R^1 = H , R^2 = C_5H_{11} : (+)-T.-7
R^1 = C_5H_{11} , R^2 = H : (+)-T.-8

R^1 = H , R^2 = C_3H_7* : T.-1
R^1 = C_3H_7 , R^2 = H* : T.-2
R^1 = H , R^2 = C_5H_{11} : (+)-T.-5
R^1 = C_5H_{11} , R^2 = H : (+)-T.-6
*relative configuration

Table: Data of Tetraponerines.

	molecular formula	M_R	$[\alpha]_D^{20}$ ($CHCl_3$)	CAS
T.-1	$C_{13}H_{24}N_2$	208.35		
T.-2	$C_{13}H_{24}N_2$	208.35		
T.-3	$C_{14}H_{26}N_2$	222.37	+27°	114475-95-5
T.-4	$C_{14}H_{26}N_2$	222.37	+94°	114530-37-9
T.-5	$C_{15}H_{28}N_2$	236.40	+10°	114475-96-6
T.-6	$C_{15}H_{28}N_2$	236.40	+35°	114530-38-0
T.-7	$C_{16}H_{30}N_2$	250.43	+30°	114530-39-1
T.-8	$C_{16}H_{30}N_2$	250.43	+102°	109269-89-8

Lit.: [1] Tetrahedron **51**, 1415–1428; 10913–10922 (1995). [2] Can. J. Chem. **72**, 105–109 (1994); J. Org. Chem. **59**, 3699 (1994).
gen.: J. Chem. Ecol. **14**, 517–527 (1988) ▪ J. Org. Chem. **61**, 4949 (1996) (synthesis) ▪ J. Toxicol., Toxin Rev. **11**, 115–164 (1992) ▪ Tetrahedron **47**, 3805–3816 (1991) ▪ Z. Naturforsch., C **42**, 627 ff. (1987). – *[HS 2939 90]*

Tetrapyrroles. Name for natural products, most of which are colored, containing four pyrrole units in a cyclic or linear basic skeleton, e. g., *chlorins, *corrins, *corphins, *bile pigments, *isobacteriochlorins, *phycobilins, *porphyrins, *porphyrinogens. For nomenclature of the T., see *Lit.* [1].
Lit.: [1] Pure Appl. Chem. **59**, 779 (1987).
gen.: Arigoni, The Biosynthesis of the Tetrapyrrole Pigments, Ciba Found. Symp. **180**, New York: Wiley 1993 ▪ Buchler, Metal Complexes with Tetrapyrrole Ligands, Berlin: Springer 1987, 1989 ▪ Dolphin ▪ Jordan, Biosynthesis of Tetrapyrroles, Amsterdam: Elsevier 1991.

Tetraquinanes. Diterpenoids with the 12-isopropyl-4,8,11-trimethyltetracyclo[6.6.0.0^{1,11}.0^{3,7}]tetradecane skeleton, which was first found in the *crinipellins from the fungus *Crinipellis stipitaria* (Agaricales).

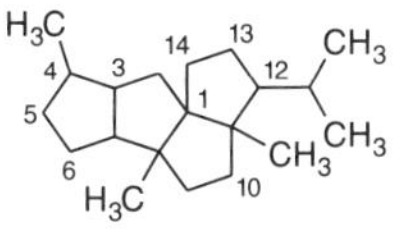

Lit.: Org. Lett. **1**, 375 (1999) (synthesis waihoensenes).

Tetrasaccharides see oligosaccharides.

Tetraterpenes. Collective name for natural products containing 40 carbon atoms (*polyterpenes) mostly formed according to the *isoprene rule. The most important group of the T. are the *carotinoids.
Lit.: Dictionary of Terpenoids **2**.

2,4,5,7-Tetrathiaoctane 2,2-dioxide see dysoxysulfone.

Tetrodotoxin (TTX, tarichatoxin, spheroidin, fugu toxin, maculotoxin).

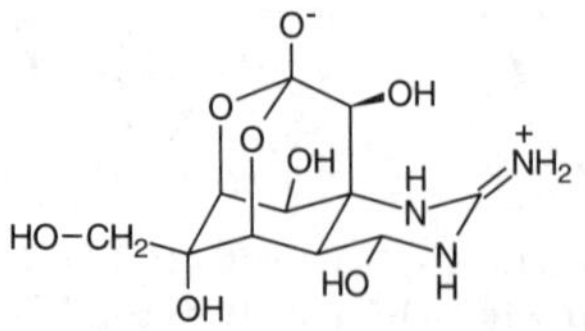

$C_{11}H_{17}N_3O_8$, M_R 319.27, cryst., dark coloration without decomp. >200 °C, soluble in dilute acids, poorly soluble in water, ether. A basic neurotoxin (pK_a 8.76) principally from the ovaries and liver of Japanese trunkfish, boxfish, cowfish, puffer fish (*Spheroides* species, Tetraodontidae), from the skin, eggs, muscles, and blood of the Californian salamander *Taricha torosa* as well as from other salamanders, frogs (e. g., *Atelopus* species), the Australian octopus *Octopus maculosus*, and the Japanese ivory snail *Babylonia japonica* [1]. T.[2] is acquired by the mentioned organisms via the food chain or via symbiotic routes as a metabolic product of the bacterium *Shewanella alga* [3] and is not formed by the higher organisms themselves. T. is extremely toxic [LD_{50} (human p.o.) ~10–15 µg/kg], thus, the much valued and appreciated Tetraodontidae fish in Japan may only be prepared for consumption by specially trained (licensed) fugu cooks. Even so frequently fatal poisonings occur. It is claimed that the low T. content of prepared fugu causes a pleasant tickling sensation in the limbs in combination with a feeling of warmth and euphoria. These pleasant effects disappear above the lethal dose. Death usually occurs by respiratory paralysis. T. blocks the sodium channels of nerve cell membranes and is toxic not only for mammals but also for reptiles, amphibians, birds, and fish.

Lit.: [1] Pure Appl. Chem. **61**, 505 (1989). [2] J. Agric. Biol. Chem. **50**, 793 (1986); **53**, 893 (1989); J. Am. Chem. Soc. **110**, 2344 (1989); Biosci. Biotech. Biochem. **56**, 370 (1992). [3] Int. J. System. Bacteriol. **40**, 331 (1990).
gen.: ACS Symp. Ser. **262**, Seafood Toxins, 333–416 (1984); **418**, Marine Toxins, 78–86 (1990) ▪ Ann. N. Y. Acad. Sci. **479**, 1, 24, 32, 52, 385 (1986) ▪ Beilstein E III/IV **27**, 8206 ▪ Chem. Rev. **93**, 1897 (1993) ▪ Kao et al., Tetrodotoxin, Saxitoxin and the Molecular Biology of the Sodium Channel, N. Y.: New York Academy of Sciences 1986 ▪ Merck Index (12.), No. 9832 ▪ Trends Biochem. Sci. **13**, 76 (1988) ▪ Zechmeister **27**, 322–339; **39**, 1–62; **46**, 159–230. – *Pharmacology:* Bioact. Mol. **10** (Mycotoxins Phycotoxins 88), 417 (1989) ▪ J. Gen. Physiol. **100**, 609 (1992) ▪ Sax (8.), FOQ 000. – *Synthesis:* Angew. Chem., Int. Ed. Engl. **38**, 3081 (1999); **39**, 61 (2000) ▪ Pure Appl. Chem. **59**, 399 (1987) ▪ Tetrahedron **55**, 4325 (1999) ▪ Tetrahedron Lett. **28**, 6485 (1987); **29**, 4127 (1988). – *[HS 3002 90; CAS 4368-28-9]*

Tetronasin.

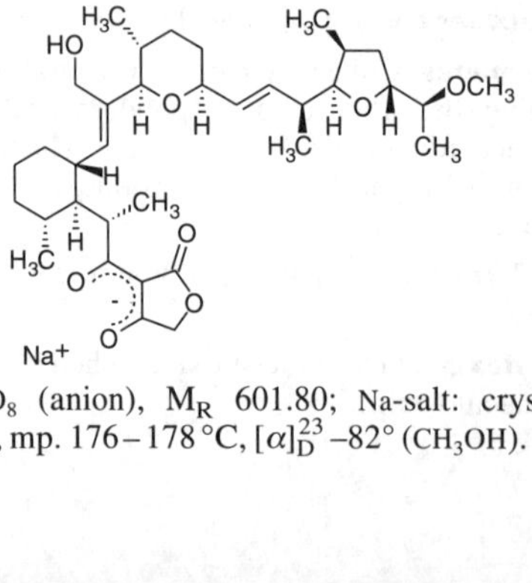

$C_{35}H_{53}O_8$ (anion), M_R 601.80; Na-salt: cryst., M_R 624.80, mp. 176–178 °C, $[\alpha]_D^{23}$ –82° (CH_3OH). Ca-ionophoric antibiotic from *Streptomyces longisporoflavus*, being tested as fodder additive in feedstock nutrition.

Lit.: *Isolation, Biosynthesis:* Tetrahedron Lett. **35**, 307, 311, 315, 319, 323 (1994); **37**, 3511, 3515, 3519 (1996). – *Synthesis:* Farmaco **51**, 147–157 (1996) ▪ J. Heterocycl. Chem. **33**, 1533 (1996) ▪ J. Chem. Soc., Perkin Trans. 1 **1998**, 2259–2276. – *Structure:* J. Chem. Soc., Chem. Commun. **1998**, 1893. – *[CAS 75139-06-9 (T.); 75139-05-8 (Na salt)]*

Tetronic acid [2,4(3*H*,5*H*)-furandione, 3-oxobutyrolactone).

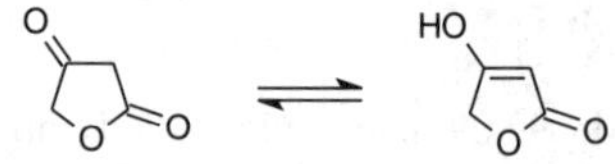

$C_4H_4O_3$, M_R 100.07, mp. 141 °C, readily soluble in warm water or alcohol. T. is a monoprotic acid, gives a dark red color with $FeCl_3$, and forms an oxime or phenylhydrazone. T. is the parent compound of *ascorbic acid. 3-Acyl-T. occur as metabolites of actinomycetes (e. g., tetronomycin, tetracarcin) or in plants. Derivatives of T. have numerous biological activities.

Lit.: Chem. Rev. **76**, 625 (1976) ▪ Chimia **41**, 73 (1987) ▪ Helv. Chim. Acta **64**, 775–786 (1981) ▪ Heterocycles **14**, 661 (1980) ▪ J. Antibiot. **47**, 143–147 (1994) ▪ Zechmeister **35**, 133–198. – *[HS 2932 29; CAS 4971-56-6]*

Teucrium lactones. C_{10}-*Iridoids occurring in the germander species *Teucrium marum* (Lamiaceae), a plant indigenous to the Mediterranean region, the essential oil of which has strong lacrimatory properties [1] (e. g., T. l. A–D, $C_{10}H_{14}O_2$, M_R 166.22).

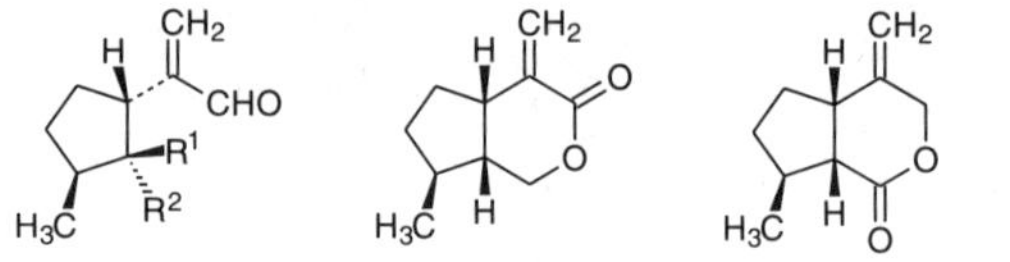

R^1 = H, R^2 = CHO : *cis*-(-)-Dolichodial (**1**)
R^1 = CHO, R^2 = H : *trans*-(+)-Dolichodial (**2**)

Teucrium lactone C (Allodolicholactone) (**3**)

Teucrium lactone D (Dolicholactone) (**4**)

Table: Teucrium lactones.

	bp. [°]	$[\alpha]_D$ (benzene)	CAS
1	96 (2 hPa)	–72°	60478-52-6
2	–	+3.5°	3671-76-9
3	110–113 (0.5 hPa)	+145.3°	60419-55-8
4	–	+31.7°	60419-56-9

T. l. A was first isolated in 1960 from the Australian Dolichoderus ants (*Dolichoderus acanthochinea clarki* and *D. acanthochinea dentata*) and thus named as *cis*-(–)-dolichodial [2]. The C-2-epimeric mixture *cis/trans*-dolichodial and T. l. D were prepared from labelled *citronellol and *iridodial, respectively, in *Teucrium marum* (biosynthesis) [3].

Lit.: [1] Aust. J. Chem. **29**, 1375 (1976). [2] Aust. J. Chem. **13**, 514 (1960). [3] Phytochemistry **22**, 2197, 2723 (1983).
gen.: J. Nat. Prod. **43**, 646, No. 253–256 (1980) ▪ Tetrahedron **52**, 3435 (1996) (synthesis T. l. C) ▪ Tetrahedron Lett. **24**, 5797 (1983) (synthesis T. l. C).

Teurilene.

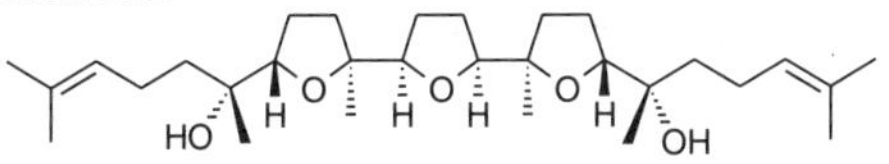

Red algae of the genus *Laurencia* contain cyclic polyethers with antiviral and cytotoxic activity derived from *squalene. T. is a *meso*-triterpene ether from *L. obtusa*, $C_{30}H_{52}O_5$, M_R 492.73, mp. 84–85 °C, $[\alpha]_D$ 0°.
Lit.: J. Am. Chem. Soc. **121**, 6792 (1999) (synthesis) ▪ J. Org. Chem. **56**, 2299 (1991) (synthesis) ▪ Tetrahedron Lett. **26**, 1329 (1985); **29**, 5947 (1988). – *[CAS 96304-92-6]*

Thaumatin (Talin®). Protein mixture isolated from the shrub *Thaumatococcus daniellii* (Marantaceae) indigenous to the African rain forests and used in Japan to sweeten chewing gum, desserts, soups, etc. The sweetness of T. referred to M_R is ca. 2000-fold greater than that of *sucrose (ca. 100000-fold higher referred to molar amounts). Wider application of T. is hindered by its sensitivity to heat. T. consists of 2 components, T. I and T. II, with M_R ca. 21000 and isoelectric point at 12 (separable on ion exchange resins). T. I is composed of 207 amino acid residues, five tripeptide sequences are identical with those of *monellin. The sequence 57–59 localized in a β-turn is considered to be the part of the structure coming into contact with the taste receptor for "sweet". The T. gene has been cloned and introduced into fruit-carrying plants, bacteria, and yeasts [1].
Lit.: [1] Biotechnol. Ser. **13**, 305–318 (1989); Biochemistry **27**, 5101 (1988).
gen.: ACS Symp. Ser. **450**, Sweeteners, 28–40 (1991) ▪ Chem. Ind. (London) **1983**, 19 ▪ J. Nutr. Biochem. **2**, 236–244 (1991) ▪ Merck-Index (12.), No. 9408. – *[CAS 53850-34-3]*

THC. Abbreviation for *tetrahydrocannabinols, see also cannabinoids.

Theaflavin.

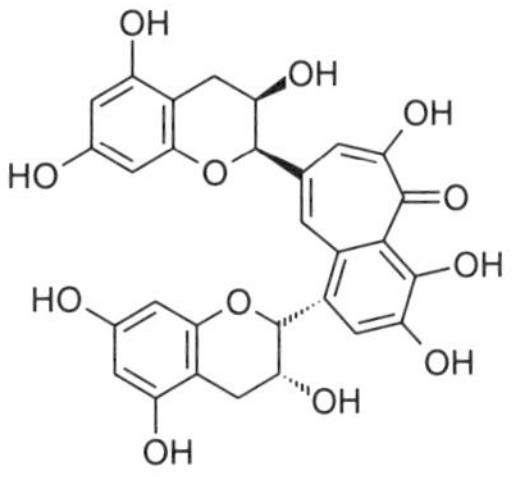

$C_{29}H_{24}O_{12}$, M_R 564.50, cryst., mp. 237–240 °C. The benzotropolone derivative T., essential for the red color and taste of black tea (*Thea sinensis*), is formed after rolling of the leaves by enzymatic oxidation from epicatechin via intermediate *o*-quinones. T. is the parent substance of a whole series of compounds (theaflavin 3-gallate, theaflavin 3,3′-digallate, epitheaflavin, etc.) detected in the ethyl acetate extract of black tea. The acidic, brownish pigments insoluble in ethyl acetate are known as *thearubigens*. Their exact composition is not yet known but their formation from catechins or gallocatechins has been demonstrated [1]. The theaflavins are claimed to have antioxidative and antimutagenic activities [2]. In tea they have astringent effects and are components of the precipitates ("tea cream") forming in cold tea [3].
Lit.: [1] J. Sci. Food Agric. **67**, 501 (1995). [2] Mutat. Res. **150**, 127 (1985); Crit. Rev. Food Sci. Nutr. **29**, 273 (1990); Mutat. Res. **323**, 29–34 (1994). [3] J. Sci. Food. Agric. **63**, 77 (1993).
gen.: Beilstein EV **19/7**, 216 ▪ Chem. Pharm. Bull. **34**, 61 (1986). – *[CAS 4670-05-7]*

Theanine (N^5-ethyl-L-glutamine).

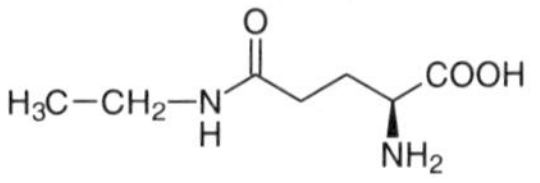

$C_7H_{14}N_2O_3$, M_R 174.20, mp. 217–218 °C (decomp.), $[\alpha]_D$ +8.7° (H_2O). A component of leaves of the tea plant (*Camellia sinensis*), also occurring in the fruit bodies of the fungus *Xerocomus badius*.
Biosynthesis: From Glu, ethylamine, and ATP. Ethylamine in tea shoots probably originates from L-Ala by the action of alanine decarboxylase [1–3]. In green tea T. constitutes ca. 0.5% of the dry weight and is decisive for the quality. Content in black tea up to 3.6%.
Lit.: [1] Phytochemistry **13**, 1401–1406 (1974). [2] Phytochemistry **17**, 313 (1978). [3] Agric. Biol. Chem. **54**, 2283–2286 (1990); J. Sci. Food Agric. **37**, 527–534 (1986).
gen.: Beilstein EIV **4**, 3038 ▪ Biosci. Biotechnol. Biochem. **56**, 689 (1992) (synthesis). – *[CAS 3081-61-6]*

Thearubigens see theaflavin.

Theaspiranes.

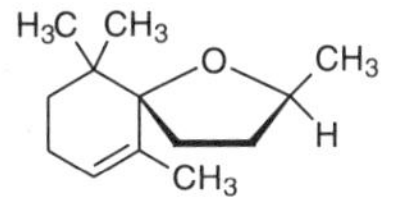

(+)-Theaspirane A

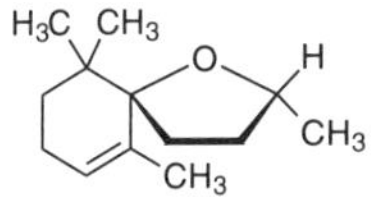

(+)-Theaspirane B

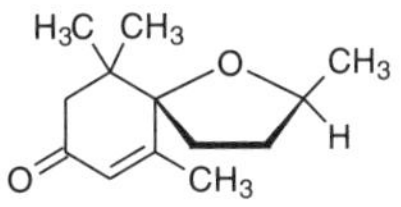

(-)-Theaspirone A

Collective name for a group of oxidized degradation products of *β-carotene. $C_{13}H_{22}O$, M_R 194.32, *T. A*: natural (+)-form, $[\alpha]_D^{22}$ +57.4° ($CHCl_3$); *(+)-T. B*, $[\alpha]_D^{22}$ +113.7° ($CHCl_3$). The T. were originally discovered in tea and belong, together with the corresponding *(–)-theaspirones A and B* {$C_{13}H_{20}O_2$, M_R 208.30, $[\alpha]_D^{22}$ –4.9° ($CHCl_3$)}, to its flavor principles [1]. The volatile components of the flower oil of *Osmanthus fragrans*, an Oleaceae species indigenous to East Asia, represent an especially high-yielding source of these organoleptically interesting cyclic ethers (osmanthus oil). The flowers are used in China to perfume black tea. The flower absolute is a valuable starting material for the fragrance and aroma industries. T. also occur in *geranium oil (*Pelargonium graveolens*), in raspberry flavor, and in the yellow passion fruit as well as in wine, quinces, and tobacco. The olfactory impression is described as hay-like, earthy, and green. The olfactory thresholds (ca. 0.2 ppb) of the hydroxy and epoxy derivatives are particularly low, the alcohol exhibits an odor reminiscent of patchouli (see patchouli oil).
Lit.: [1] Zechmeister **35**, 431.
gen.: Chem. Mikrobiol. Technol. Lebensm. **13**, 129 (1991) ▪ Morton et al. (eds.), Food Flavours, Part B The Flavour of Bev-

erages, p. 49, Amsterdam: Elsevier 1986 ▪ Ohloff, p. 165 ▪ Schreier (ed.), Bioflavour '87, p. 255, Berlin: de Gruyter 1988. – *Synthesis:* Agric. Biol. Chem. **49**, 861 (1985) ▪ Tetrahedron Lett. **22**, 2291 (1981); **26**, 4637 (1985). – *[CAS 66537-39-1 (T. A); 66537-40-4 (T. B); 24399-19-7 (theaspirone A); 50763-74-1 (theaspirone B)]*

Thebaine see morphinan alkaloids.

Theine. Synonym for *caffein(e).

Thelephoric acid.

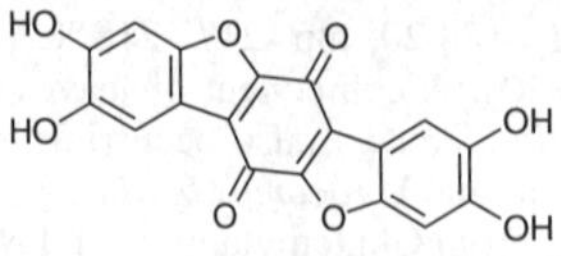

$C_{18}H_8O_8$, M_R 352.26, black powder with metallic glance, mp. >300 °C, insoluble in most organic solvents except pyridine. A characteristic pigment of the Thelephoraceae and Bankeraceae (Basidiomycetes). The *terphenylquinone T. is responsible for the dark violet color tones of the fruit bodies and also occurs occasionally in boletes such as the North American *Tylopilus plumboviolaceus* and in the cap skin of *Suillus grevillei*. T. is readily formed by oxidative cyclization of cyclovariegatin (3'-hydroxycycloleucomelone). The leucohexaacetate occurs together with *cycloleucomelone in the fruit bodies of *Boletopsis leucomelaena*, the leucohexamethyl ether is found in the deep blue mycelia of *Pulcherricium caeruleum*. T. changes color to deep green on treatment with alkali.

Lit.: Beilstein E IV **19**, 3177; E V **19/7**, 125. – *Reviews:* Thomson **2**, 161 f.; **3**, 100–101 ▪ Zechmeister **51**, 31–32. – *[CAS 479-64-1]*

Theobromine (3,7-dimethylxanthine, 3,7-dihydro-3,7-dimethyl-1*H*-purine-2,6-dione). $C_7H_8N_4O_2$, M_R 180.17; formula, see under theophylline. Monoclinic, bitter tasting needles, mp. 357 °C, sublimes at 290–295 °C, soluble in hot water, alkali hydroxides, concentrated acids, moderately soluble in ammonia, poorly soluble in cold water and alcohol. With acids T. forms salts which decompose in water; detection by the murexid reaction. T. is the main alkaloid of cocoa (*Theobroma cacao*, 1.5–3 wt.-%), from which it is obtained – especially from the husks in which it accumulates during fermentation. The typical bitter taste of cocoa is the result of interactions between T. and the piperazinediones formed in the roasting process [1]. T. has diuretic, vasodilatory, and stimulating effects on cardiac muscle. The activities are weaker than those of the structurally related *caffeine (a methylation product of T.) with which it co-occurs in cola nuts. For further pharmacological properties, see table under theophylline.

Lit.: [1] Helv. Chim. Acta **58**, 1078–1086 (1975).

gen.: Actual. Pharm. **191**, 36 (1982) ▪ Beilstein E III/IV **26**, 2336 f.; E V **26/13**, 553 f. ▪ Chem. Pharm. Bull. **39**, 2833 (1991) (synthesis) ▪ Hager (5.) **4**, 631 f.; 940 f.; **6**, 54 f.; **9**, 847 ▪ IARC Monogr. **51**, 421 (1991) ▪ Karrer, No. 2564 ▪ Martindale (29.), p. 1526 ▪ Merck-Index (12.), No. 9418 ▪ Negwer (6.), p. 688 ▪ Sax (8.), TEO 500. – *[HS 2939 90; CAS 83-67-0]*

Theonella. Sponges of the order Lithistida, family Theonellidae contain secondary metabolites from various structural classes, e.g., the genera *Theonella*: theonellasterols, *theonellamides and *swinholides; *Discodermia*: *discodermolides.

Lit.: Angew. Chem. Int. Ed. Engl. **37**, 2162–2178 (1998) (review).

Theonellamides.

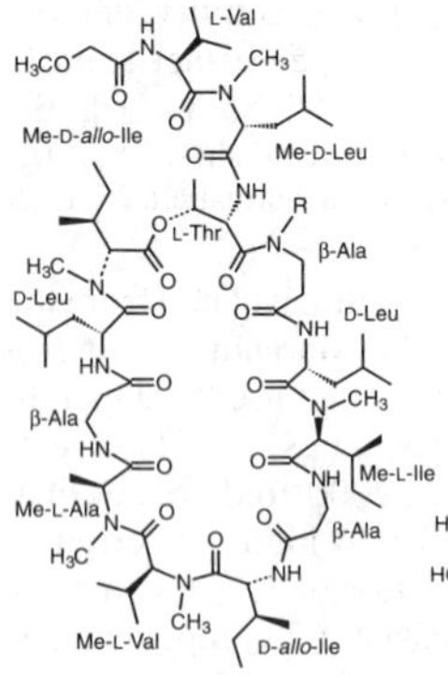

R = H : Theonellapeptolide Id (**1**)
R = CH_3 : Theonellapeptolide Ie (**2**)

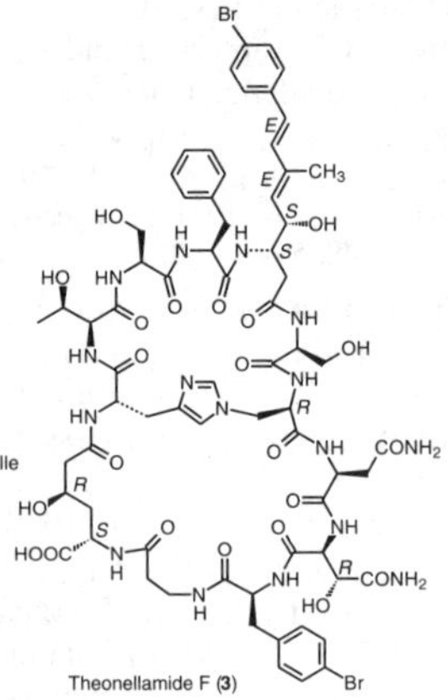

Theonellamide F (**3**)

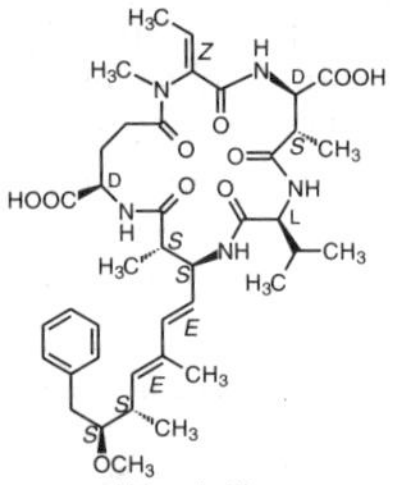

Motuporin (**4**)

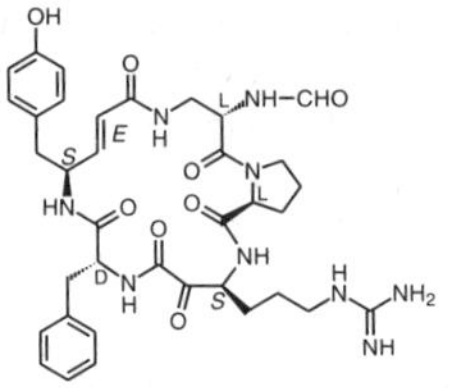

Cyclotheonamide A (**5**)

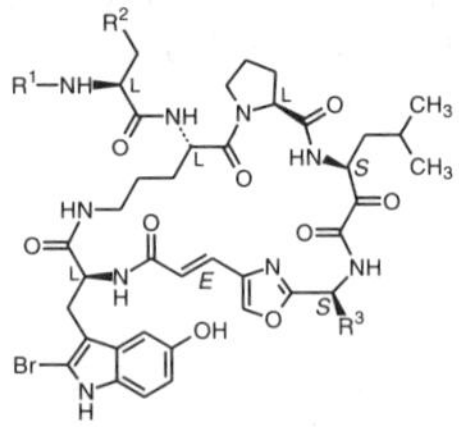

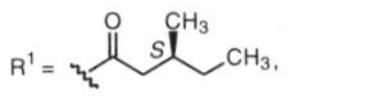

$R^2 = R^3 = CH_3$: Orbiculamide A (**6**)

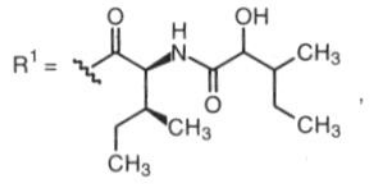

$R^2 = R^3 = C_2H_5$: Keramamide B (**7**)

Cyclic depsipeptide antibiotics from Pacific marine sponges of the genus *Theonella*. On account of their macrolide structures, they are also known as *Theonella* peptolides.

Examples: *Theonellamine B* [1] (a tridecadepsipeptide, $C_{70}H_{125}N_{13}O_{16}$, M_R 1404.84, amorphous, mp. 149–151 °C, $[\alpha]_D$ −572° (CH_3OH), inhibits Na^+/K^+-ATPase), *theonellamide* F [2] (a dodecapeptide, $C_{69}H_{86}Br_2N_{16}O_{22}$, M_R 1651.34), *theonellapeptolide Id* [3] (theonellamine B, $C_{70}H_{125}N_{13}O_{16}$, M_R 1404.84, cryst., mp. 168–169 °C), *theonellapeptolide Ie* [4] (a tridecapeptide, $C_{71}H_{127}N_{13}O_{16}$, M_R 1418.87, cryst., mp. 153–155 °C), etc. As a consequence of their antineoplastic, antiviral, and fungicidal properties, the T. are of pharmacological interest. A series of further interesting peptides has been isolated from *Theonella swinhoei*: *motuporin* [5] ($C_{40}H_{57}N_5O_{10}$, M_R 767.92, amorphous, $[\alpha]_D$ −83.8°), a cyclic pentapeptide, inhibits protein phosphatase 1 (IC_{50}<1 nM); the cyclotheonamides [6] (*cyclotheonamide A:* $C_{36}H_{45}N_9O_8$, M_R 731.81) inhibit thrombin, trypsin, and plasmin. The orbiculamides [7] (*orbiculamide A:* $C_{46}H_{62}BrN_9O_{10}$, M_R 980.96, amorphous, $[\alpha]_D$ –60°) and the keramamides[8] (*keramamide B:* $C_{54}H_{77}BrN_{10}O_{12}$, M_R 1138.17) are cyclopeptides containing unusual amino acids.

Lit.: [1]Tetrahedron Lett. **27**, 4319 (1986). [2]J. Am. Chem. Soc. **111**, 2582 (1989); Synlett **1994**, 247; Pept. Chem. **31**, 37 (1993). [3]Chem. Pharm. Bull. **39**, 1177 (1991); Tetrahedron **47**, 2169 (1991); **55**, 10305 (1999). [4]Chem. Pharm. Bull. **35**, 2129 (1987); J. Nat. Prod. **61**, 724 (1998) (Th. III E). [5]J. Org. Chem. **64**, 2711, 3000 (1999); J. Am. Chem. Soc. **121**, 6355 (1999); Tetrahedron Lett. **33**, 1561 (1992); Org. Lett. **1**, 1471 (1999). [6]J. Am. Chem. Soc. **112**, 7053 (1990); **114**, 6570 (1992); J. Org. Chem. **62**, 3880 (1997) (synthesis); Proc. Natl. Acad. Sci. USA **90**, 8048 (1993) (structure); Chem. Eng. News (20.9.) **1993** (activity). [7]J. Am. Chem. Soc. **113**, 7811 (1991). [8]J. Chem. Soc., Chem. Commun. **1999**, 981; J. Am. Chem. Soc. **113**, 7812 (1991); Tetrahedron **54**, 6719 (1998).
gen.: Chem. Rev. **93**, 1793 (1993) ▪ J. Org. Chem. **56**, 4545 (1991); **58**, 5592 (1993). – *[HS 2941 90; CAS 105115-91-1 (T. B); 105091-14-3 (1); 109767-22-8 (2); 119455-31-1 (3); 141672-08-4 (4); 129033-04-1 (5); 137041-28-2 (6); 137041-25-9 (7)]*

Theophylline (1,3-dimethylxanthine, 3,7-dihydro-1,3-dimethyl-1*H*-purine-2,6-dione).

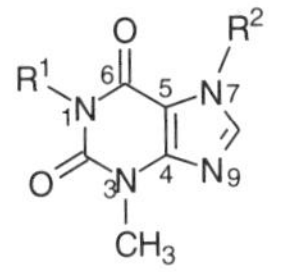

R^1 = H , R^2 = CH_3 : Theobromine
R^1 = CH_3 , R^2 = H : Theophylline
R^1 = R^2 = CH_3 : Caffeine

$C_7H_8N_4O_2$, M_R 180.17, thin, bitter-tasting platelets, mp. 270–274 °C (monohydrate), moderately soluble in water, very soluble in hot water, alkali hydroxides, ammonia, dilute acids. A *purine alkaloid from the leaves of the tea plant, also contained in maté (*Ilex paraguariensis* and *Paullinia cupana*). The methylxanthines T., *caffein(e), and *theobromine are among the oldest known stimulants and drugs. Their pharmacological properties are compared in the table. In high doses T. can cause epilepsy-like convulsions. T. also exhibits antiasthmatic activity[1].

Table: Pharmacological properties of Theophylline, compared to Caffeine and Theobromine.

	Caffeine	Theo-phylline	Theo-bromine
stimulation of CNS	+++	+++	–
cardioactivity	+	+++	++
broncho- and vasodilatation	+	+++	++
stimulation of skeletal muscles	+++	++	+
diuresis	+	+++	++

Lit.: [1]Andersson & Persson, Anti-Asthma Xanthines and Adenosine, Amsterdam: Excerpta Medica 1985.
gen.: Ann. Clin. Biochem. **25**, 4 (1988) ▪ Ann. Intern. Med. **101**, 63 (1984) ▪ Beilstein EV **26/13**, 529 ff. ▪ Chem. Eng. World **33**, 110 (1998) ▪ Florey **4**, 466–493 ▪ Hager (5.) **4**, 631; **6**, 54 f.; **9**, 853 ▪ Karrer, No. 2563 ▪ Kirk-Othmer (4.) **6**, 199 ▪ Kleemann-Engel, p. 1852 ▪ Maes et al., Theophylline Profile (DFG Komm. Klin.-toxikolog. Analytik Mitt. 5), Weinheim: Verl. Chemie 1985 ▪ Martindale (30.), p. 1037, 1317 ▪ Merck-Index (12.), No. 9421 ▪ Sax (8.), TEP 000. – *[HS 2939 50; CAS 58-55-9]*

Thermopsine see anagyrine.

Thermozymocidin see myriocin.

Thesine see truxillic and truxinic acids.

Thespesin see gossypol.

Thiactin see thiostrepton.

Thialdine see dihydro-4*H*-1,3,5-dithiazines.

Thiamin (international free name for vitamin B_1, aneurine).

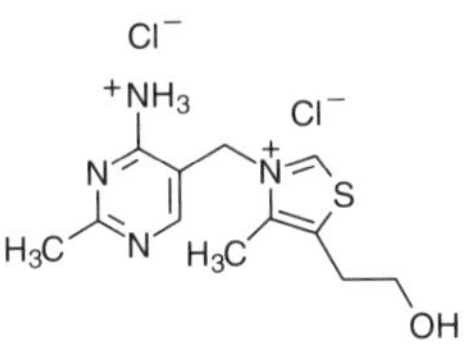

Thiamin hydrochloride

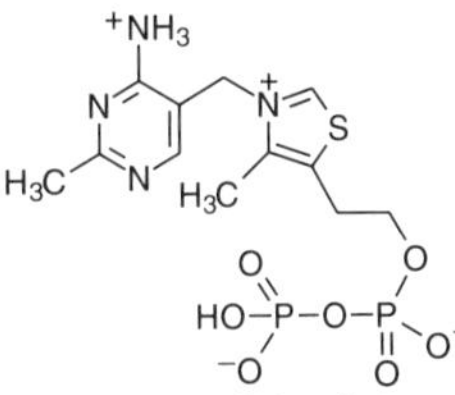

Thiamin diphosphate

Monochloride: $C_{12}H_{17}ClN_4OS$, M_R 300.81, cryst. · H_2O, mp. 120–122 °C (decomp.); anhydrous, mp. 163–165 °C (decomp.); hydrochloride: $C_{12}H_{18}Cl_2N_4OS$, M_R 337.27, monoclinic platelets or crystals in rosette-like aggregations with a weak moldy odor and bitter taste, mp. 248 °C (decomp.). T. occurs in plant and animal tissue, especially in the husks of rice, cereal grains, yeast, liver, eggs, milk, green leaves, roots, and tubers. T. is a metabolic product of many bacteria and functions in the form of the diphosphate (*thiamin diphosphate*, thiamin pyrophosphate, TPP) as a *coenzyme ("cocarboxylase"). A nutritional deficiency of T. leads to serious clinical symptoms such as beri-beri, a disease occurring especially in East Asia with severe neurological impairments.

Toxicology: T. monochloride: LD_{50} (rat i. v.) 118 mg/kg; T. dichloride: LD_{50} (mouse i. v.) 89 mg/kg.

Use: As vitamin additive in food.

Biosynthesis: From 4-amino-2-methyl-5-pyrimidinylmethyl diphosphate and 2-(4-methyl-5-thiazolyl)-ethyl phosphate[1].

Detection: By oxidation in alkaline medium to thiochrome and fluorescence measurements of the latter in the UV region.

Lit.: [1]Angew. Chem. Int. Ed. Engl. **36**, 1032–1046 (1997).
gen.: Adv. Clin. Chem. **23**, 93–140 (1983) ▪ Adv. Enzymol. **53**, 345–481 (1982) (biosynthesis, review) ▪ Beilstein E III/IV **27**, 1764–1778 ▪ Food Sci. Technol. **13**, 245–297 (1984) (review) ▪ Hager (5.) **9**, 864–870 ▪ Helv. Chim. Acta **73**, 1300–1305 (1990) (synthesis) ▪ Kleemann-Engel, p. 1855 ▪ Luckner (3.), p. 439, 453 ▪ Martindale (29.), p. 1277 f. ▪ Merck-Index (12.), No. 9430–9433 ▪ Nat. Prod. Rep. **3**, 395–419 (1986) ▪ Negwer (6.), No. 2515, 2540, 4574 ▪ New J. Chem. **20**, 607–629 (1996) (biosynthesis) ▪ Trends Biochem. Sci. **9**, 536 f. (1984) (review). – *Toxicology:* Sax (8.), TES 750, TET 300. – *[HS 2936 22; CAS 59-43-8 (T. monochloride); 67-03-8 (T. hydrochloride)]*

Thiamulin see pleuromutilin.

Thiangazoles see tantazoles.

Thiarubrins. The light-sensitive, red *T. A* ($C_{13}H_8S_2$, M_R 228.33, oil) and *B* contain the anti-aromatic 1,2-dithiin system. They occur in various taxa of the Asteraceae and exhibit antibacterial, antifungal, nematicidic, and antiviral properties. Leaves of *Aspilia africana* con-

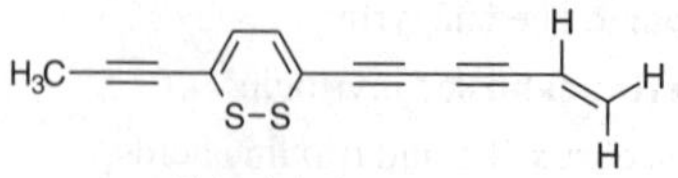

Thiarubrin A

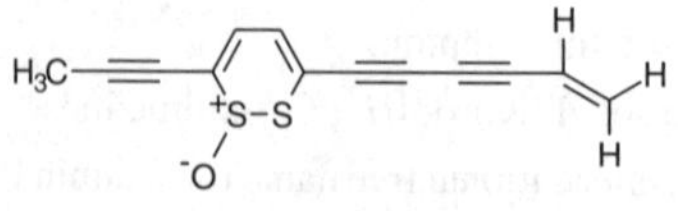

Thiarubrin 1-oxide

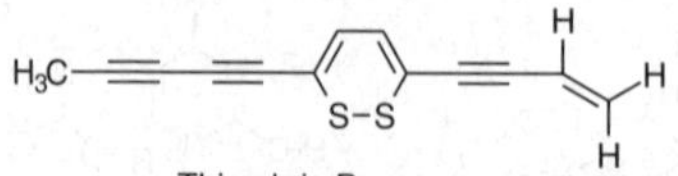

Thiarubrin B

taining T. are used in Africa as a remedy for skin infections and stomach pains. It is interesting to note that chimpanzees specifically pluck the leaves of such plants and swallow them unchewed, this is suggestive of a "medicinal" use. Various T. as well as the corresponding thiophenes and also *thiarubrin 1-oxide* ($C_{13}H_8OS_2$, M_R 244.33) have been isolated from *Ambrosia chamonissonis*. The T. are probably formed biogenetically by sulfuration of *polyyns precursors. T. have DNA-cleaving properties.

Lit.: Hager (5.) **6**, 504 f. ▪ J. Am. Chem. Soc. **116**, 9403, 10793 (1994) (synthesis, bibliography) ▪ J. Med. Chem. **38**, 2628–2648 (1995) (activity) ▪ J. Org. Chem. **63**, 8644 (1998) (T. C) ▪ Merck-Index (12.), No. 9439 ▪ Phytochemistry **28**, 3523 (1989); **29**, 2901 (1990); **31**, 2703 (1992); **33**, 224 (1993) (isolation, occurrence) ▪ Primates **24**, 276 (1983) (activity). – *[CAS 63543-09-9 (T. A); 71539-72-5 (T. B); 131836-16-3 (thiarubrin 1-oxide)]*

Thienamycins. Collective name for a class of *β-lactam antibiotics. The T. have the carbapenem structure in common with the olivanic acids and the epithienamycins.

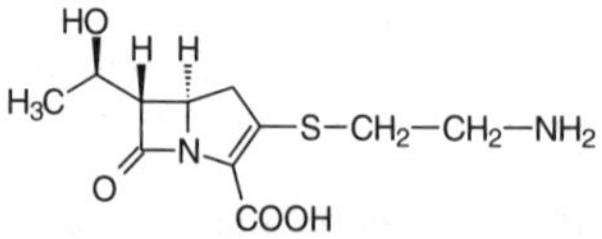

The parent compound of the T. ($C_{11}H_{16}N_2O_4S$, M_R 272.33, mp. 205 °C) was first isolated in 1976 from cultures of *Streptomyces cattleya* during a screening for β-lactamase inhibitors, but is now mainly produced synthetically [1]. As a β-lactamase inhibitor and broad spectrum *antibiotic T. has potential applications against Gram-positive and Gram-negative pathogens but is extremely unstable. It is used in the form of its chemically more stable *N*-formimidoyl derivative [2] in combination with dehydropeptidase I inhibitors such as, e. g., cilastatin in clinical practice.

Lit.: [1] Morin & Gorman (eds.), Chemistry and Biology of β-Lactam-Antibiotics, vol. 2, p. 227–313, New York: Academic Press 1982. [2] J. Med. Chem. **22**, 1435 (1979).

gen.: Chem. Pharm. Bull. **56**, 12 (1992) ▪ Demain & Solomon (eds.), Antibiotics Containing the β-Lactam Structure 2, p. 119–245, New York: Springer 1983 ▪ Heterocycles **38**, 1533 (1994) ▪ J. Am. Chem. Soc. **109**, 1129 (1987) ▪ J. Chem. Soc., Perkin Trans. 1 **1994**, 379 ▪ J. Org. Chem. **55**, 3098–3103 (1990); **61** , 2413–2427 (1996). – *[HS 2941 90; CAS 59995-64-1]*

Thiobinupharidine.

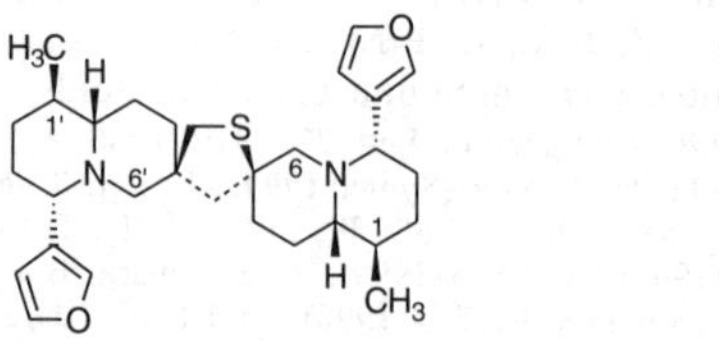

$C_{30}H_{42}N_2O_2S$, M_R 494.74, mp. 129–130 °C, $[\alpha]_D^{25}$ +7.8° (CH_3OH). T. is a typical representative of the group of sulfur-containing dimers of the *Nuphar alkaloids. Further structurally closely related alkaloids are 1-epi-T., 1′-epi-T., as well as 6α-hydroxy-T. (thionupharoline) and 6′-hydroxy-T. {$[\alpha]_D^{25}$ +34° (C_2H_5OH)}. About 30 alkaloids of this type, including some *S*-oxides, are known.

Lit.: Manske **9**, 441–465; **16**, 181–214; **35**, 215–257 ▪ Phytochemistry **25**, 2227 (1986). – *[HS 2939 90; CAS 30343-72-7 (T.); 58845-61-7 (1-epi-T.); 58845-62-8 (1′-epi-T.); 50478-55-2 (6α-hydroxy-T.); 52002-85-4 (6′-hydroxy-T.)]*

Thiocillins. Sulfur-rich polypeptides from cultures of *Bacillus badicus* and *B. cereus*. Three different T. (I, II, III) are distinguished and are related to *micrococcin P_1. Like the latter, The T. are active against Gram-

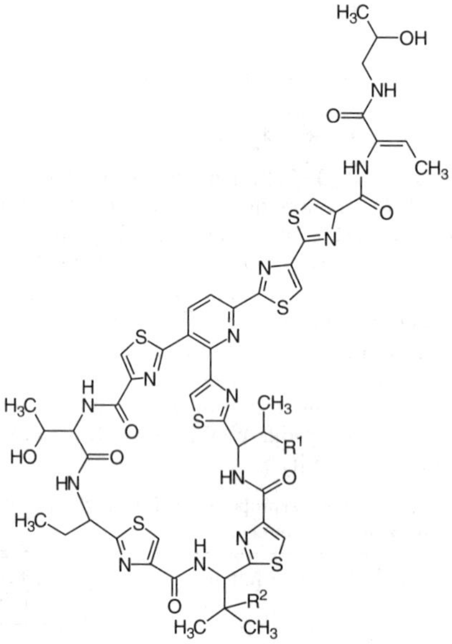

	R^1	R^2
*Micrococcin P_1	OH	H
Thiocillin I	OH	OH
Thiocillin II	OCH_3	OH
Thiocillin III	OH	OCH_3

Table: Data of Thiocillins.

Thiocillin	molecular formula	M_R	$[\alpha]_D^{25}$ (C_2H_5OH)	CAS
T. I	$C_{48}H_{49}N_{13}O_{10}S_6$	1160.4	+97.7 (*c* 2.028)	59979-01-0
T. II	$C_{49}H_{51}N_{13}O_{10}S_6$	1174.4	+93.4 (*c* 0.59)	59979-02-1
T. III	$C_{49}H_{51}N_{13}O_9S_6$	1174.4	+88 (*c* 0.85)	59979-03-2

positive bacteria [MIC (*Staphylococcus aureus*) 0.2 mg/L]. T. are obtained as colorless powders, soluble in DMSO, DMF, and THF, insoluble in ethyl acetate and water, uv_{max}: 217, 348 nm. Thiocillin is also the international free name of the semisynthetic penicillin derivative [(*Z*)-2-(2-chlorophenyl)-2-(methylthio)vinyl]-penicillin.

Lit.: J. Antibiot. **29**, 366–374 (1976); **34**, 1126–1136 (1981) ▪ J. Chem. Soc., Chem. Commun. **1978**, 256 ▪ Tetrahedron Lett. **32**, 4263 (1991) (configuration) ▪ see also micrococcin P_1.

Thioctic acid see lipoic acid.

Thiocyanates, 2-thiocyanatopupukeanane see di-isocyanoadociane.

Thioformins see fluopsins.

Thiolactomycin.

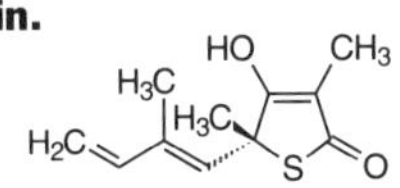

$C_{11}H_{14}O_2S$, M_R 210.30, yellowish cryst., mp. 120 °C, $[\alpha]_D^{20}$ +176° (CH_3OH), a thiobutyrolactone from *Nocardia* and *Streptomyces* cultures. T. is a *β*-lactamase inhibitor and has antibacterial activity with low toxicity.

Lit.: J. Antibiot. **42**, 391, 396, 890–896 (1989) ▪ J. Chem. Soc., Chem. Commun. **1989**, 23 f. ▪ J. Chem. Soc., Perkin Trans. 1 **1997**, 417–431 (synthesis, absolute configuration). – *[HS 2941 90; CAS 82079-32-1]*

Thiolutin see holomycin.

Thiomarinols. With *pseudomonic acid structurally related antibacterially active substances from the marine bacterium *Alteromonas rava*, e.g., T. A: $C_{30}H_{44}N_2O_9S_2$, M_R 640.82, orange cryst., mp. 106–110 °C, $[\alpha]_D^{25}$ +4.3° (CH_3OH).

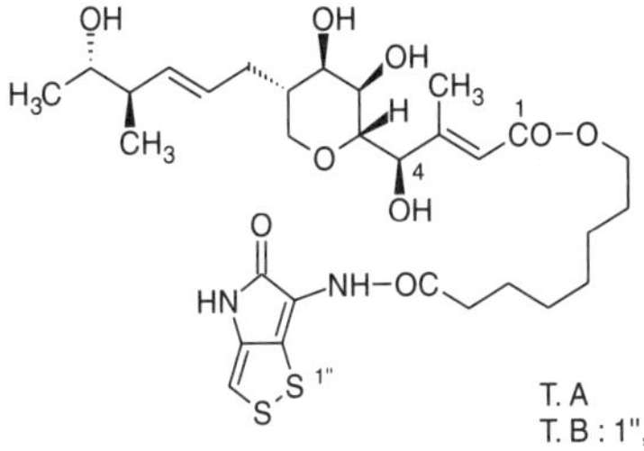

T. A
T. B : 1'',1''-dioxide of T. A
T. C : 4-deoxy-T. A

Lit.: J. Antibiot. **46**, 1834 (1993); **47**, 851 (1994); **48**, 907 (1995); **50**, 449 (1997). – *[CAS 146697-04-3 (T. A)]*

Thiophanic acid see lichen xanthones.

Thiostrepton (bryamycin, thiactin). $C_{72}H_{85}N_{19}O_{18}S_5$, M_R 1664.91, mp. 246–255 °C, $[\alpha]_D^{20}$ −98.5° (CH_3COOH), a bicyclic thiopeptide produced by *Streptomyces aureus* with activity against Gram-positive bacteria and used as a fodder additive in animal breeding. T. inhibits protein biosynthesis, the mechanism of action corresponds to that of the structurally related *nosiheptide. The biosynthesis of T. is the subject of intense investigation, it is assumed that the core peptide chain is formed non-ribosomally via a multi-enzyme complex.

Lit.: Appl. Microbiol. Biotechnol. **25**, 526 (1987) ▪ J. Antibiot. **36**, 799, 814, 831 (1983); **42**, 1649 ff. (1989). – *Biosynthesis:*

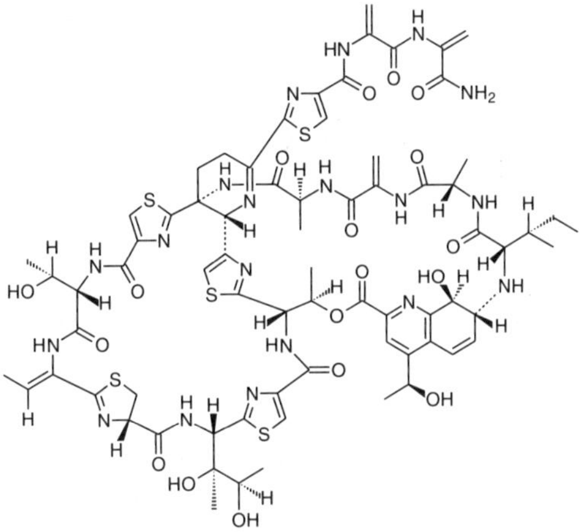

J. Am. Chem. Soc. **115**, 7992 (1993) ▪ Strohl & Floss, in Vining & Stuttard (eds.), Genetics and Biochemistry of Antibiotics Production, p. 223–238, Boston: Butterworths 1995. – *[HS 2941 90; CAS 1393-48-2]*

Thiotropocin.

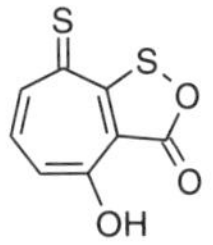

$C_8H_4O_3S_2$, M_R 212.24, orange needles, mp. 222–225 °C (decomp.). A thiotropolone derivative with broad spectrum antibiotic activities from cultures of *Pseudomonas* sp. CB-104.

Lit.: J. Am. Chem. Soc. **114**, 8479 (1992) (biosynthesis) ▪ J. Antibiot. **37**, 1294 (1984) ▪ Tetrahedron Lett. **25**, 419 (1984) (isolation). – *[CAS 89550-93-6]*

THL see lipstatin.

Threitol (*threo*-1,2,3,4-butanetetrol).

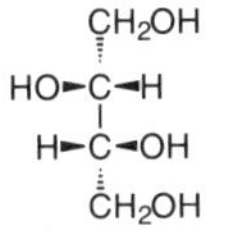

(2*R*,3*R*)-, D-form

$C_4H_{10}O_4$, M_R 122.12, a sugar alcohol occurring in the (2*R*,3*R*)-form in the edible honey mushroom (*Armillaria mellea*) and some plants. The 1,4-*O*-bismethanesulfonate of the (2*S*,3*S*)-form is used as a cytostatic drug. Both enantiomers of T. melt between 88–89 °C, the racemate at 72 °C. D-T.: $[\alpha]_D^{20}$ +4.3° (H_2O). The *meso*-form is known as erythritol.

Lit.: Beilstein EIV **1**, 2807 ▪ Br. J. Cancer **48**, 739 (1983) ▪ Carbohydr. Res. **223**, 11 (1992); **247**, 119 (1993) ▪ J. Org. Chem. **54**, 5257 (1989) (synthesis) ▪ Karrer, No. 145 ▪ Phytochemistry **34**, 715 (1993) (isolation). – *[CAS 7493-90-5 (T.); 2418-52-2 (2R,3R); 2319-57-5 (2S,3S)]*

L-Threonine [(2*S*,3*R*)-2-amino-3-hydroxybutanoic acid, abbreviation: Thr or T].

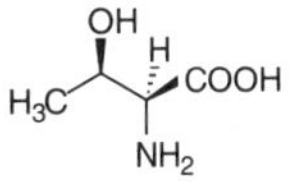

(2*S*, 3*R*)-, L-form

$C_4H_9NO_3$, M_R 119.12, mp. 255–257 °C (decomp.), $[\alpha]_D^{26}$ −28.3° (H_2O), pK_a 2.15, 9.12, pI 5.64, essential amino acid. Genetic code: ACU, ACC, ACA, ACG. Average content in proteins 6.0% [1], e.g. in eggs 5.3%, casein 4%, and gelatin 1.4%.

Biosynthesis: Thr belongs biogenetically to the Asp group and is formed from Asp. The direct precursor is L-*homoserine, which also forms Met via cystathionine and homocysteine. Homoserine is first converted to *O*-phosphohomoserine by ATP under the action of homoserine kinase (EC 2.7.1.39) and then by threonine synthase (EC 4.2.99.2) to Thr. Thr is a component of glycoproteins. It frequently occurs in the free form, see also L-serine.

Lit.: [1] Biochem. Biophys. Res. Commun. **78**, 1018–1024 (1977).

gen.: Adv. Carbohydr. Chem. Biochem. **43**, 135–202 (1985) ▪ Beilstein EIV **4**, 3171 f. ▪ Bull. Chem. Soc. Jpn. **67**, 1899 (1994) (resolution of racemate) ▪ Helv. Chim. Acta **70**, 232, 237 (1987) ▪ Karrer, No. 2357 ▪ Merck-Index (12.), No. 9522 ▪ Synthesis **1985**, 55 f. ▪ Tetrahedron Lett. **30**, 6637 (1989). – *[HS 2922 50; CAS 72-19-5 (L-T.); 36676-50-3 (racem.); 632-20-2 (D-T.)]*

Thromboxanes (TX). Cyclic *eicosanoids occurring in all body tissues. The compounds are classified into a *TXA*- (bicyclic structure with an oxetane ring) and a *TXB*-series (cyclic structure in which the oxetane ring has been hydrolytically cleaved); an index number indicates the number of C,C-double bonds in the molecule. T. are formed in the cyclooxygenase biosynthesis pathway from *arachidonic acid and other multiply unsaturated fatty acids, the first biosynthetic steps on the way to the T. and to the *prostaglandins are identical (see figure). The multiply unsaturated fatty acids are first released from phosphatidylcholines and other phospholipids of the cell membrane by phospholipase A_2.

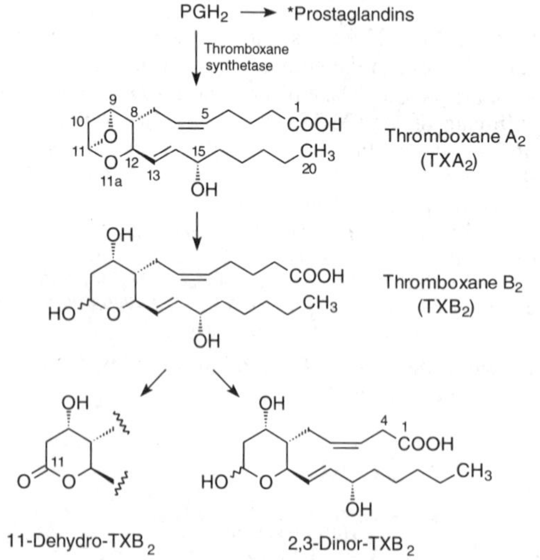

Figure: Biosynthesis of Thromboxanes (PGG_2, PGH_2 = prostaglandin 9,11-endoperoxides).

The physiologically highly active *TXA_2* ($C_{20}H_{32}O_5$, M_R 352.47) promotes thrombocyte aggregation as well as vaso- and bronchoconstriction. The biological activities of TXA_2 are associated with the development of arteriosclerosis, thrombosis, and coronary heart disease. Thus, synthetic antagonists of T. biosynthesis have been developed for therapeutic use in the treatment of these conditions. One of the physiological antagonists of TXA_2 is *prostacyclin. Under physiological conditions the oxetane ring of TXA_2 is rapidly hydrolysed (biological half-life ca. 30 s at 37 °C) to furnish the stable *TXB_2* [9α,11,15*S*-trihydroxythromboxa-5*Z*,13*E*-dienoic acid, $C_{20}H_{34}O_6$, M_R 370.49, mp. 95–96 °C, $[\alpha]_D^{25}$ +57.4° (ethyl acetate)] with weak biological activity. TXB_2 and its catabolic subsequent products such as, e.g., *11-dehydro-TXB_2* ($C_{20}H_{32}O_6$, M_R 368.47) and *2,3-dinor-TXB_2* ($C_{18}H_{30}O_6$, M_R 342.43) are excreted in the urine. Analogous T. with three C,C-double bonds are formed from eicosapentaenoic acid, e.g., *TXA_3*, which acts as an antagonist of *TXA_2* on account of its non-aggregatory and non-vasoconstrictory properties.

Lit.: Adv. Lipid Res. **22**, 63 (1987); **23**, 169 (1989) ▪ Biochim. Biophys. Acta **754**, 190 (1983) (biosynthesis) ▪ Merck-Index (12.), No. 9530 ▪ Prog. Lipid Res. **28**, 273–301 (1989); **33**, 403 ff. (1994) ▪ Synth. Commun. **25**, 1461 (1995) (synthesis). – *[CAS 54397-85-2 (TXB_2); 57576-52-0 (TXA_2); 67910-12-7 (11-dehydro-TXB_2); 63250-09-9 (2,3-dinor-TXB_2)]*

Thujane {sabinane, 4-methyl-1-(1-methylethyl)bicyclo[3.1.0]hexane}.

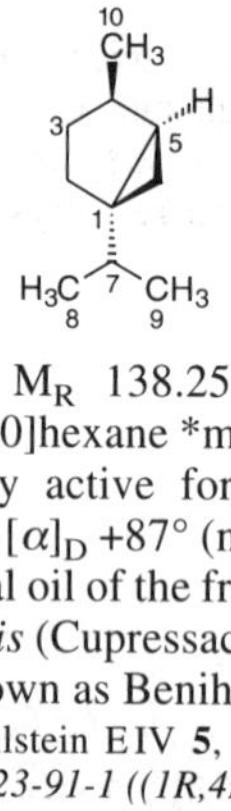

(+)-(1*R*, 4*R*, 5*R*)-Thujane

$C_{10}H_{18}$, M_R 138.25. Parent skeleton of the bicyclo[3.1.0]hexane *monoterpenes. Of the four possible optically active forms only (+)-(1*R*,4*R*,5*R*)-T., bp. 151 °C, $[\alpha]_D$ +87° (neat), occurs in nature, e.g., in the essential oil of the fresh leaves of *Chamaecyparis formosensis* (Cupressaceae), a tree indigenous to Taiwan and known as Benihio or Beniki.

Lit.: Beilstein EIV **5**, 317 ▪ Chem. Rev. **72**, 305 (1972). – *[CAS 5523-91-1 ((1R,4R,5R)-T.)]*

Thujanols. $C_{10}H_{18}O$, M_R 154.24. Bicyclic *monoterpene alcohols with the thujane structure.

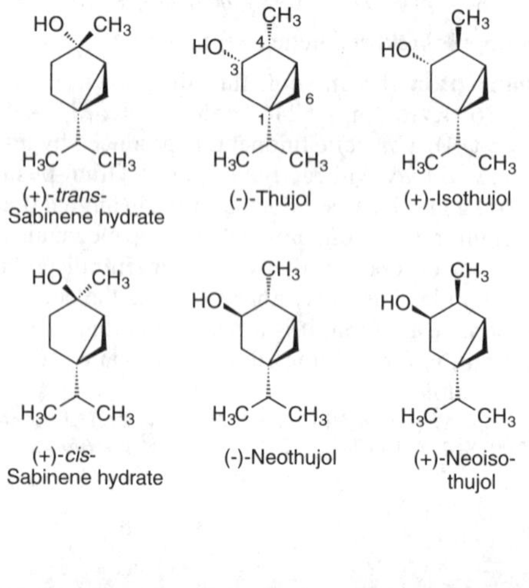

Table: Physical data and occurrence of Thujanols.

configuration* {C. A. name}	trivial name	mp. [°C]	$[\alpha]_D$	CAS	occurrence	yield %
(1*R*,4*S*,5*S*)(+)-*trans*-Sabinene hydrate		60–61	+32.2° (C_2H_5OH)	15537-55-0	peppermint oil (*Mentha piperita*, Lamiaceae)	0.8
(1*R*,4*R*,5*S*)(+)-*cis*-Sabinene hydrate {2-Methyl-5-(1-methylethyl)-bicyclo[3.1.0]hexan-2-ol}		36.5–37.2	+47.2° (C_2H_5OH)	17699-16-0	oil of marjoram (*Origanum majorana*, Lamiaceae)	3.2
(1*S*,3*S*,4*R*,5*R*)(–)-Thujol		66–67	–22.5° (C_2H_5OH)	21653-20-3	vermouth oil (*Artemisia absinthium*, Asteraceae)	10.0
(1*S*,3*R*,4*R*,5*R*)(–)-Neothujol		22–23	–8.8° (C_2H_5OH)	21653-18-9	essential oil from *Artemisia serotina* (Asteraceae)	2–3
(1*S*,3*S*,4*S*,5*R*)(+)-Isothujol		bp. 91 (9.5 mbar)	+117.2° (undiluted)	7712-79-0	tansy oil (*Tanacetum vulgare*, Asteraceae)	58
(1*S*,3*R*,4*S*,5*R*)(+)-Neoisothujol {4-Methyl-1-(1-methylethyl)-bicyclo[3.1.0]hexan-3-ol}		liquid	+42° ($CHCl_3$)	21653-19-0	vermouth oil together with (–)-T.	0.2

* numbering according to Thujane

T. are used for the production of other monoterpenes used in the perfume and soap industries.

Lit.: Beilstein EIII **6**, 279; EIV **6**, 275 ▪ Chem. Rev. **72**, 305 (1972) ▪ J. Org. Chem. **33**, 1656 (1968); **42**, 1616 (1977) ▪ Karrer, No. 308, 3539, 5645a.

Thujan-3-ones {thujone, 4-methyl-1-(1-methylethyl)bicyclo[3.1.0]hexan-3-one}.

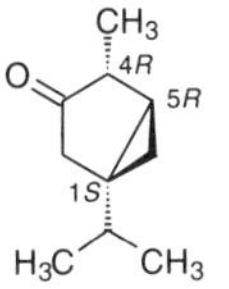

(-)-α-Thujone (Thujone)

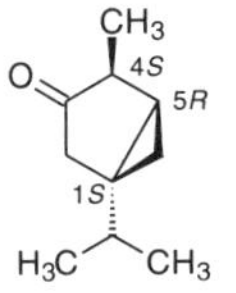

(+)-β-Thujone (Isothujone)

$C_{10}H_{16}O$, M_R 152.24, oil with a menthol-like odor. Bicyclic *monoterpene ketones with the thujane structure occurring in two C-4-epimeric forms in nature: (1*S*,4*R*,5*R*)(–)-α-T., bp. 83.8–84.1 °C (17 hPa), $[\alpha]_D$ –19.2° (neat), (1*S*, 4*S*,5*R*)(+)-β-T., bp. 85.7–86.2 °C (17 hPa), $[\alpha]_D$ +72.5° (neat). Not only the names α- and β-thujone but also *thujone* and *isothujone* are sometimes used incorrectly in the literature. T. are potent neurotoxins, cause epileptic seizures, and can lead to severe psychic damage. It is not yet known if both epimers exhibit the same biological effects.

Occurrence: Very widely distributed in the essential oils of Asteraceae, Cupressaceae, Lamiaceae, Pinaceae species. Thuja oil (*Thuja occidentalis*, Cupressaceae) contains 40% (–)-α-T. and *tansy oil (*Tanacetum vulgare*, Asteraceae) 58% (+)-β-T.

Use: Tansy oil tastes bitter and is used in the production of some bitter liquors.

Lit.: Beilstein EIV **7**, 207 ▪ Helv. Chim. Acta **80**, 623–639 (1997) (synthesis) ▪ Karrer, No. 563 ▪ Merck-Index (12.), No. 9533 ▪ Sax (8.), TFW 000 (toxicology). – *[HS 2914 29; CAS 546-80-5 ((–)-α-T.); 471-15-8 ((+)-β-T.)]*

Thujaplicins. $C_{10}H_{12}O_2$, M_R 164.20, colorless crystals. Monocyclic *monoterpenes with the tropolone structure (2-hydroxy-2,4,6-cycloheptatrien-1-one) that differ from each other in the position of the isopropyl group: α-T., mp. 34 °C, β-T., mp. 52–52.5 °C, γ-T., mp. 82 °C.

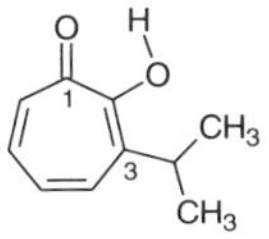

α

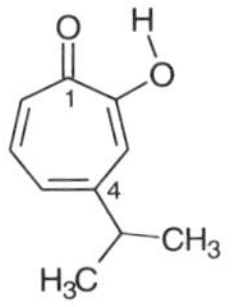

β

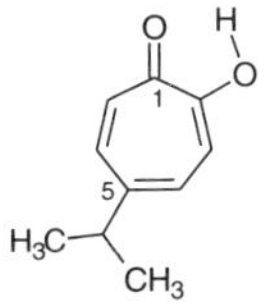

γ

Occurrence: Mainly in the wood from trees of the cypress family (Cupressaceae). The heartwood of the giant western arbor vitae (*Thuja plicata*, Cupressaceae) is particularly rich in T. For synthesis, see *Lit.*[1]. T. have nerve paralysing and spasmodic activities. They exhibit antifungal activity and thus protect cypress species from attack by fungi. However, *Sporothrix* fungal species can degrade T., trees "detoxified" in this way can then be attacked by other fungi.

Lit.: [1] Aust. J. Chem. **44**, 1–36, 705 (1991); J. Org. Chem. **52**, 2602 (1987); Tetrahedron Lett. **32**, 1051 (1991); Zechmeister **13**, 232–289; **24**, 216–229, 261 ff.

gen.: Beilstein EIV **8**, 486, 488, 495 ▪ Justus Liebigs Ann. Chem. **1996**, 99 (isolation) ▪ Karrer, No. 572–574 ▪ Sax (8.), IRR 000, TFV 750. – *[HS 2914 49; CAS 1946-74-3 (α-T.); 499-44-5 (β-T.); 672-76-4 (γ-T.)]*

Thujenes.

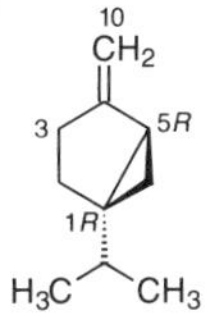

(+)-4(10)-Thujene [(+)-Sabinene]

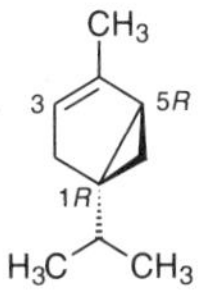

(-)-3-Thujene [(-)-α-Thujene]

$C_{10}H_{16}$, M_R 136.24, oil. Unsaturated bicyclic *monoterpenes with the thujane structure occurring in both enantiomeric forms.

Occurrence: 30% (+)-S. in the Russian savin(e) or creeping juniper oil (*Juniperus sabina*, Cupressa-

Table: Data of Thujenes.

trivial name {C. A. name}	bp. [°C]	$[\alpha]_D$	CAS
(+)-Sabinene	69	+107°	2009-00-9
(−)-Sabinene	(30 hPa)	−107°	10408-16-9
{4-Methylene-1-(1-methylethyl)bicyclo-[3.1.0]hexane}		(undiluted)	3387-41-5
(+)-3-Thujene	152	+38.8°	563-34-9
(−)-3-Thujene	(699 hPa)	−38.8°	3917-48-4
{2-Methyl-5-(1-methylethyl)bicyclo-[3.1.0]hex-2-ene}		(undiluted)	2867-05-2

ceae)[1]. (−)-S. in the Indian medicinal plant *Zanthoxylum rhetsa* (Rutaceae)[2]. (+)-T. in rubber resin of *Boswellia serrata* (Burseraceae). (−)-T. in mango leaf oil (*Mangifera indica*, Anacardiaceae) and in thuja oil (*Thuja occidentalis*, Cupressaceae).

Lit.: [1]Hegnauer **I**, 370. [2]Hegnauer **VI**, 264.
gen.: Beilstein EIV **5**, 451, 452 ▪ Chem. Rev. **72**, 305 (1972) ▪ Karrer, No. 59, 60. – *[HS 2902 19]*

(−)-Thujol see thujanols.

Thujones see thujan-3-ones.

Thujopsene.

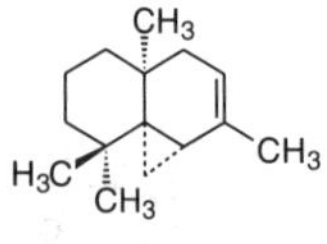

Thujopsene

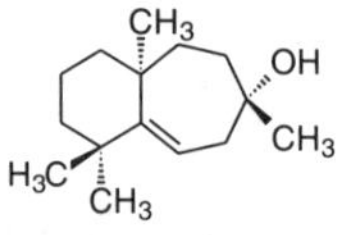

Widdrol

(−)-T. (widdrene), $C_{15}H_{24}$, M_R 204.36, oil, bp. 121–122 °C (1.6 kPa), $[\alpha]_D^{25}$ −110° ($CHCl_3$), was originally found in Japanese hiba oil (*Thujopsis dolabrata*) and in South African *Widdringtonia* species (both Cupressaceae). (−)-T. and the tricyclic (−)-α-*cedrene comprise in about equal amounts the main portion (ca. 50%) of cedar wood oil(*Juniperus* spp.). The diastereomeric (−)-3-thujopsanones with valuable ambergris note are obtained from the only weakly smelling hydrocarbon mixture of Hiba or cedar wood oils by epoxidation. Friedel-Crafts acylation of the essential oil furnishes a complex mixture with a woody odor that is widely used in the perfume industry. Under weakly acidic conditions T. undergoes rearrangement with opening of the cyclopropane ring to furnish the bicyclic *widdrol* ($C_{15}H_{26}O$, M_R 222.37).

Lit.: Agric. Biol. Chem. **50**, 2915 (1986) (biosynthesis) ▪ J. Am. Chem. Soc. **104**, 4290, (1982) (synthesis) ▪ Ohloff, p. 170 ff. ▪ Tetrahedron **43**, 5605 (1987) (synthesis widdrol). – *[CAS 470-40-6 (T.); 6892-80-4 (widdrol)]*

Thuringiensin {2-*O*-[4-*O*-(5′-deoxyadenosin-5′-yl)-β-D-glucopyranosyl]-D-allaric acid 4-phosphate}.

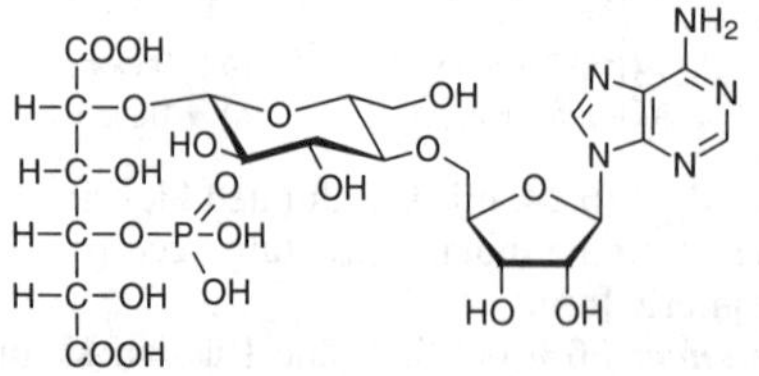

$C_{22}H_{32}N_5O_{19}P$, M_R 701.49, $[\alpha]_D^{25}$ +30.9° (H_2O), a thermostable nucleotide toxin from cultures of *Bacillus thuringiensis* var. *gelechiae*. In contrast to the *Thuringiensis* *endotoxins (proteins), T. are only formed and excreted by certain serotypes. The insecticidal and cytotoxic activity is based on the inhibition of DNA-dependent RNA polymerase.

Lit.: J. Antibiot. **41**, 1711 (1988). – *[CAS 23526-02-5]*

Thuringione see lichen xanthones.

Thymelaeaceae toxins see Euphorbiaceae and Thymelaeaceae poisons.

Thyme oil. Reddish-brown oil with a strong, sweet-herby, spicy-aromatic, somewhat medicinally pungent odor and a warm herby-spicy, slightly pungent taste.

Production: By steam distillation from the flowering herb of the thyme species *Thymus vulgaris* and *T. zygis*. Origin: Spain, France, North Africa.

Composition[1]***:*** The main component and decisive for the organoleptic impression is *thymol* (see cymenols) (over 50%), followed by *p-cymol* (see cymenes) (ca. 20%).

Use: In the perfume industry; it is present in many perfumes with a pronounced herby note (masculine tones). To improve the aromatic character of foods such as sauces, dressings, meat products, pickles, etc., in products for oral hygiene and pharmaceutical preparations such as bronchologic agents, expectorants, balneotherapeutics.

Lit.: [1]Perfum. Flavor. **7** (2), 39 (1982); **9** (2), 28 (1984); **13** (6), 60 (1988); **15** (5), 59 (1990); **17** (5), 140 (1992).
gen.: Arctander, p. 614 ▪ Bauer et al. (2.), p. 178. – *Toxicology:* Food Cosmet. Toxicol. **12**, 1003 (1974). – *[HS 3301 29; CAS 8007-46-3]*

Thymine see pyrimidines.

Thymol see cymenols.

Thyroid hormone see iodoamino acids.

L-Thyronine {(*S*)-2-amino-3-[4-(4-hydroxyphenoxy)-phenyl]-propanoic acid, *O*-(4-hydroxyphenyl)-L-tyrosine}.

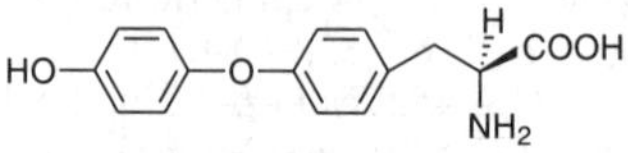

$C_{15}H_{15}NO_4$, M_R 273.29, mp. 255–260 °C (decomp.), $[\alpha]_D^{18}$ +15.0° (1 m HCl+C_2H_5OH, 1:1). Iodine-free form of the *iodoamino acids occurring in the thyroid gland.

Lit.: Beilstein EIV **14**, 2269 f. – *[HS 2922 50; CAS 1596-67-4]*

L-Thyroxine see iodoamino acids.

Thyrsiferol.

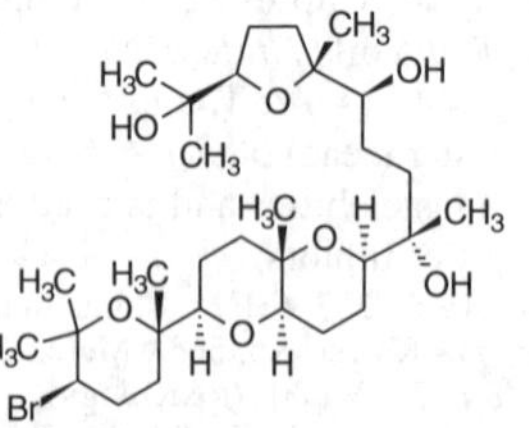

$C_{30}H_{53}BrO_7$, M_R 605.65. A bromine-containing triterpenoid from the red alga *Laurencia thyrsifera* (*L. obtusa*) with the *squalene skeleton. T. and especially *T.*

23-acetate ($C_{32}H_{55}BrO_8$, M_R 647.69, mp. 118–119 °C) have strong cytotoxic activity and are effective against P388 leukemia cells.

Lit.: J. Org. Chem. **55**, 5088 (1990) ▪ Scheuer II **1**, 131; **5**, 256 (synthesis) ▪ Tetrahedron Lett. **26**, 1329 (1985); **27**, 4287 (1986); **29**, 1143 (1987). – *[CAS 66873-39-0 (T.); 96304-95-9 (T. 23-acetate)]*

Tiamutilin see pleuromutilin.

Ticks. A family of parasitic *mites – blood-sucking on warm-blooded animals and reptiles – belonging to the arachnids, with flattened bodies that swell enormously when blood is taken up. They are carriers of different, sometimes dangerous diseases. The European species *Ixodes ricinus* (wood or castor-bean tick) can transmit not only cerebral encephalitis (meningitis) but also Lyme disease, named after the place in the north east of the USA where it was first recognized (borreliosis, the pathogen is the spirochete bacterium *Borrelia burgdorferi*). Substituted aromatic compounds such as *2-nitrophenol* ($C_6H_5NO_3$, M_R 139.11) or *2,6-dichlorophenol* ($C_6H_4Cl_2O$, M_R 163.0) are known as the *pheromones of some tick species.

Lit.: J. Chem. Ecol. **17**, 833–847 (1991) (pheromones) ▪ Tu, Handbook of Natural Toxins, vol. 2, p. 371–396, New York: Dekker 1984 ▪ Wehner & Gehring, Zoologie, p. 525, 665, 669, Stuttgart: Thieme 1990. – *[CAS 88-75-5 (2-nitrophenol); 87-65-0 (2,6-dichlorophenol)]*

Tiglianes.

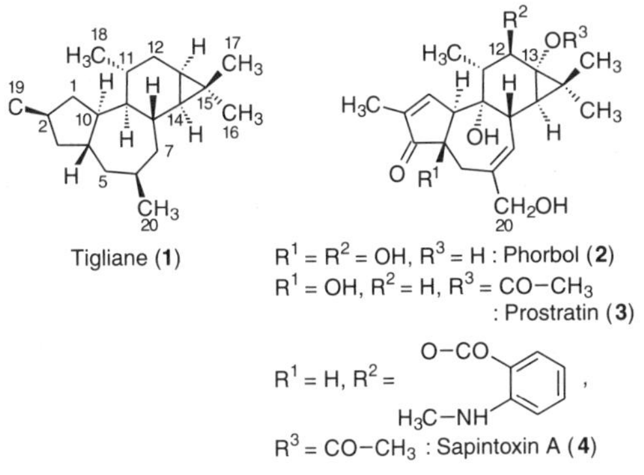

Table: Data of Tiglianes.

	molecular formula	M_R	mp. [°C]	$[\alpha]_D$	CAS
2	$C_{20}H_{28}O_6$	364.44	250–251	+118° (dioxane)	17673-25-5
3	$C_{22}H_{30}O_6$	390.48	217–218	+64° (CH_3OH)	60857-08-1
4	$C_{30}H_{37}NO_7$	523.63	274–275		79083-69-5

A class of *diterpenes with the parent skeleton **1**, the members of which mostly occur in Euphorbiaceae species such as, e. g., *Croton tiglium*, *C. sparsiflorus*, *Sapium japonicum*, *S. indicum*, *Euphorbia franckiana*, *E. tirucalli*, and *E. coerulescens*. Together with the ingenanes (see also Euphorbiaceae and Thymelaeaceae poisons) T. are responsible for the toxicity of the plants; accidental consumption by animals and humans can be fatal. Although phorbol[1] itself is neither an irritant for the skin nor cocarcinogenic (tumor-promoting), 12-tetradecanoylphorbol 13-acetate (TPA) is a highly effective cocarcinogen[2] of the promotor-type (tumor incidence after 24 weeks: 66.7) and has become a standard promotor in experimental cancer research. Other diesters from Euphorbiaceae are also effective promotors (e. g., ID_{50} for 13-*O*-phenyl-12-deoxyphorbol 20-acetate: 2.7 pmol/mouse ear). They are not mutagenic but rather influence cell growth and proliferation by activation of protein kinase C[2]. Thus, for example, mouse epidermal cells switch directly from the G1 phase to DNS synthesis and mitosis. In addition to these promotors, the Euphorbiaceae also contain so-called "cryptic" promotors, characterized by a third ester group at position 20 of the diterpene skeleton, which are biologically inactive until selective, enzymatic cleavage of the ester group at C-20 releases the diester as the active promotor. Cryptic promotors may be even more dangerous than the promotors themselves since it is difficult to detect their activity. For the synthesis of phorbol, see *Lit.*[3]. The T. skeleton is formed biosynthetically by cyclization of a *casbene derivative.

Lit.: [1]Z. Naturforsch., B **21**, 1204 (1966). [2]Chem. Eng. News (23.10.) **1995**, 21 ff.; J. Med. Chem. **39**, 1005 (1996); Nature (London) **306**, 487 (1983). [3]J. Am. Chem. Soc. **119**, 7897 (1997).

gen.: Arch. Toxikol. **44**, 279 (1980) ▪ Experientia **30**, 1438 (1974); **34**, 1595 (1978); **37**, 681 (1981) ▪ Evans, Naturally Occurring Phorbol Esters, Boca Raton: CRC Press 1986 ▪ J. Am. Chem. Soc. **112**, 4956 (1990) ▪ J. Nat. Prod. **42**, 112 (1979) ▪ Merck-Index (12.), No. 7487 ▪ Nature (London) **301**, 621 (1983) ▪ Sax (8.), p. 2779 ff. ▪ Science **213**, 655 (1981); **222**, 1036 (1983) ▪ Synlett **5**, 325 (1994) ▪ Synthesis **1999**, 1401 ▪ Tetrahedron Lett. **1969**, 3509; **1975**, 1587; **1976**, 1735; **1977**, 925 ▪ Zechmeister **31**, 377–467; **44**, 1–100. – *[CAS 16561-29-8 (TPA)]*

Tigogenin see steroid sapogenins.

Tilivalline.

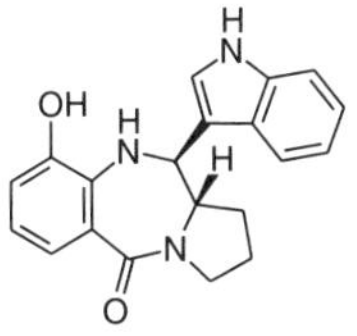

$C_{20}H_{19}N_3O_2$, M_R 333.39, mp. 168 °C, $[\alpha]_D^{25}$ +126.8° (CH_3OH). The *indole alkaloid T. was discovered during investigations on the metabolic products of *Klebsiella pneumoniae* var. *oxytoca*. T. resembles the *anthramycin antibiotics.

Lit.: J. Org. Chem. **63**, 6797 (1998) [synthesis of (+)-T.] ▪ Tetrahedron **38**, 147–152 (1982) ▪ Tetrahedron Lett. **27**, 6111–6114 (1986); **30**, 1871 f. (1989) (synthesis). – *[CAS 80279-24-9]*

Tirandamycins.

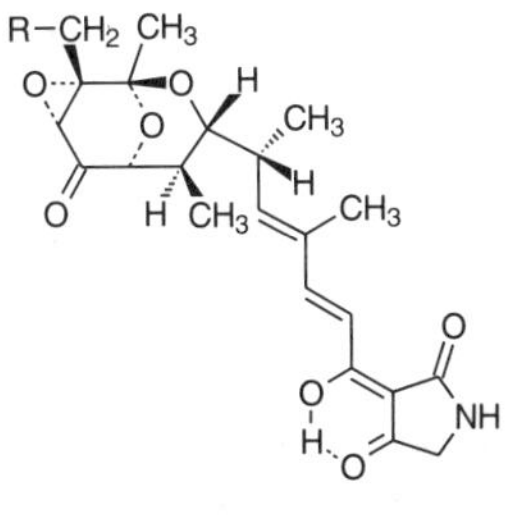

R = H : T. A
R = OH : T. B

*Tetramic acid antibiotics with a dienoyl side chain containing a terminal bicyclic ketal unit isolated from streptomycetes (e.g., *Streptomyces tirandis*, *S. flaveolus*). *T. A* [$C_{22}H_{27}NO_7$, M_R 417.46, cryst., mp. 98–102 °C, $[\alpha]_D$ +51° (C_2H_5OH, sodium salt)] and *T. B* ($C_{22}H_{27}NO_8$, M_R 433.46) differ by one hydroxy group. The T. are active against Gram-positive bacteria, especially also *Clostridia*. They are inhibitors of terminal DNA transferase and bacterial RNA polymerase.

Lit.: *Structure:* Arch. Microbiol. **109**, 65 (1976) ▪ J. Chem. Soc., Perkin Trans. 1 **1980**, 1057. – *Synthesis:* J. Am. Chem. Soc. **107**, 1777, 5219 (1985); **113**, 8791 (1991) ▪ J. Chem. Soc., Chem. Commun. **1996**, 21 f. (T. B) ▪ Tetrahedron **44**, 3171–3180 (1988). – *Review:* Chem. Rev. **95**, 1981–2001 (1995). – *[CAS 34429-70-4 (T. A); 60587-14-6 (T. B)]*

Tirucalicin see lathyranes.

Tirucallol see triterpenes.

Toad poisons (toad toxins, toad venoms). General name for the *toxins secreted by genuine toads (Bufonidae) belonging to the salientia (Anura); together with frog and salamander toxins, toad poisons represent the major group of *amphibian venoms. The toads have protruding skin glands, especially near the ears, that secrete a toxic liquid when pressed; the secretion dries on the skin to leave a residue. Toad poisons are extracted from this residue or the skin. They consist not only of biogenic amines such as the catecholamines adrenaline, noradrenaline (see β-phenylethylamine alkaloids), *dopamine and its *N*-methyl derivative epinine, tryptamine derivatives such as, e. g., bufotenine, and peptides (see indolylalkylamines, magainins) but also of the bufogenins, some of which are first formed from bufotoxins (suberoylarginine esters of bufogenins) on drying. The bufogenins have cardiac activities similar to those of digitalis or strophanthine; they have a steroid skeleton. The glycosides of the *cardenolides (aglycones of the Digitalis series) and *bufadienolides are designated as cardiac glycosides; they differ mainly through the ring expansion of the lactone ring from a five- to a six-membered ring in the bufadienolides. Toads use these toxins not only as a defence against natural predators but also and mainly as protection against attack by microorganisms [1] (see also magainins).

Use: In modern medicine toad toxins have been replaced by the more easily accessible *Digitalis glycosides.

Other animals also make use of toad toxins, for example, after consumption of toads, hedgehogs spread saliva over their spines [2]. Like the toads, some frogs also secrete active substances from their skin glands. These highly poisonous toxins (see frog toxins) are mostly alkaloids, some of which have also been identified in toads.

Lit.: [1] Naturwissenschaften **62**, 15–21 (1975). [2] Nature (London) **268**, 628 (1977).

gen.: J. Nat. Prod. **56**, 357–373 (1993) ▪ Manske **43**, 185–288 ▪ Zechmeister **41**, 205–340.

Toadstool (poisonous). Name for higher fungi from the group of basidiomycetes and ascomycetes that are toxic in both raw and cooked form. The toxins of the T. possess widely differing structures and activities. Thus, for example, the deadly amanita and the destroying angel (*Amanita phalloides*, *A. verna*) as well as *Galerina marginata* (Deadly Galerina) are lethal on account of their contents of *amanitins which inhibit enzyme activities (RNA polymerase II) in the cell nucleus [1]. The *phallotoxins and *virotoxins also present in amanitas are less toxic. *Silybin [2] and *antamanide have been discussed as specific antidotes. The toxic activities of the related toadstools *Amanita pantherina* (panther amanita) and *A. muscaria* (fly agaric) are based on their contents of ibotenic acid and muscimol (see fly agaric constituents). Ibotenic acid also occurs in various *Inocybe* and *Clitocybe* species. The toxins *clitidine and *acromelic acid have been detected in the Japanese toadstool *C. acromelalga*. The false morel (*Gyromitra esculenta*) (toxin: *gyromitrin) and some *Entoloma* species [3] as well as the yellow-foot (*Agaricus xanthoderma*) (constituents: agaricone, *xanthodermin) [4] are also poisonous. *Cortinarius orellanus* and related toadstools may also have lethal effects on account of the content of *orellanine, *Coprinus* species (inky caps) in combination with alcohol may lead to disulfiram-like intoxications (toxin: *coprin). Some fungi contain substances with mutagenic properties: e. g., milkcaps (*velleral), *Lyophyllum connatum* (*lyophyllin), *Lactarius necator* (*necatorone), or *Paxillus atrotomentosus* (*leucomentins) so that their consumption is not recommended. Some fungi are poisonous in the raw state (e. g., honey agaric, *Amanita rubescens*, or some blueing boletes) but edible when cooked, others are not actually poisonous but inedible on account of their extremely bitter tastes. The consumption of some T. results in gastrointestinal or allergic intoxications, sometimes with severe symptoms. Very little is known about the structures of the constituents, e. g., of Satan's bolete, common paxillus, etc. In the wider sense, the metabolic products *mycotoxins, secreted mostly by small fungi must also be included among the fungal toxins since they make foodstuffs and fodder attacked in particular by mold fungi and yeasts unsuitable for consumption or cause food poisoning.

Lit.: [1] Naturwiss. Rundsch. **33**, 370–378 (1980). [2] Vogel, in Faulstich et al., Amanita Toxins and Poisoning, p. 180–189, Baden-Baden: Witzstrock 1980. [3] Chem. Labor Betr. **20**, 255–260 (1969). [4] Angew. Chem. Int. Ed. Engl. **24**, 1063 (1985).

gen.: Audubon Society Field Guide to North American Mushrooms, New York: Chanticleer Press 1981 ▪ Bresinsky & Besl, Colour Atlas of Poisonous Fungi, Boca Raton: CRC Press 1989 ▪ Dähncke & Dähncke, 700 Pilze in Farbfotos, Aarau: AT Verlag 1980 ▪ Faulstich et al., Amanita Toxins and Poisoning, Baden-Baden: Witzstrock 1980 ▪ Michael, Henning & Kreisel, Handbuch für Pilzfreunde (2.), vol. 1–6, Stuttgart: Fischer 1988 ▪ Miller, Mushrooms of North America, New York: E. P. Dutton 1981.

Tobacco alkaloids. Collective name for alkaloids of tobacco plants (genus *Nicotiana*, Solanaceae).

Structure & occurrence: All T. a. contain a pyridine ring; they are therefore also occasionally known as *pyridine alkaloids*. Almost all T. a. are 3-pyridyl derivatives, the main alkaloid is *nicotine which is accompanied by a series of minor alkaloids such as *nornicotine, *anabasine, *cotinine, *nicotyrine, and *nicotelline* ($C_{15}H_{11}N_3$, M_R 233.27, mp. 147–148 °C, see figure). Their occurrence is typical for the genus *Ni-*

cotiana, comprising about 60, mostly annual plants, indigenous to the temperate and subtropical regions of North and South America as well as some Pacific islands and Australia. Nowadays, tobacco plants are cultivated on all continents. The tobacco plants grown for tobacco production are cultivated types and crossbreeds (*Nicotiana tabacum*, *N. rustica*, *N. robusta*, etc.), the wild types of which are often extinct.

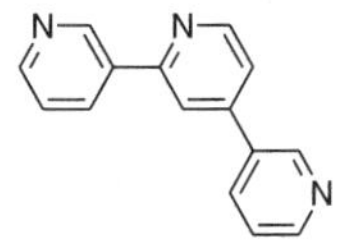

Nicotelline

Physiology & biosynthesis: Like the *tropane alkaloids, T. a. are formed in the roots and transported to the aboveground parts for storage by the plant's phloem system. In some sorts of tobacco plants a part of the nicotine is demethylated to nornicotine during transport to the shoot. Nornicotine and anabasine are often the main alkaloids in the so-called nicotine-poor tobacco plant types. The T. a. are formed biogenetically from nicotinic acid, made available via the pyridine nucleotide cycle (see nicotinamide), and a pyrrolidine or piperidine building block (figure). In the case of nicotine, like for the tropane alkaloids, *N*-methylpyrroline is an intermediate, in the biosynthesis of anabasine the intermediate is a piperidine derived from the amino acid lysine (see piperidine alkaloids).

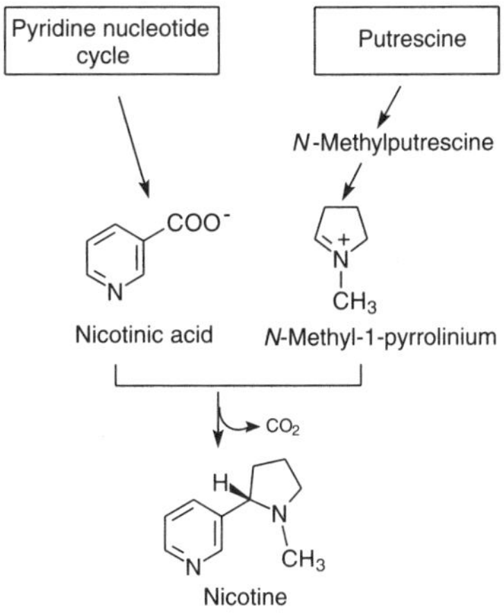

Figure: Biosynthesis of Nicotine.

Chem. ecology: The T. a. show powerful insecticidal activity. They were first used against plant pests in the form of tobacco broth in the middle of the 19th century (defined as the starting point of chemical crop protection). However, this early form of "biological pest control" was later abandoned on account of the high toxicity of the T. a. In the natural surroundings of *Nicotiana* plants, the T. a. also play an important role as chemical protection against herbivores. It was demonstrated unambiguously for the first time in T. a., that injury by an insect bite induces the synthesis of alkaloids and thus increases their content in the plant's leaves. The result is improved protection against herbivores[1]. For the chemistry, pharmacology, and toxicology of T. a., see nicotine.

Lit.: [1] Bernays (ed.), Insect-Plant Interactions, p. 1–23, Boca Raton, Florida: CRC Press 1993.
gen.: Mothes et al., p. 107 ▪ Pelletier **1**, 85–152 ▪ Ullmann (5.) **A 1**, 360 ▪ Zechmeister **34**, 1–80. – *[HS 2939 70, 2939 90; CAS 494-04-2 (nicotelline)]*

Tobacco flavor.

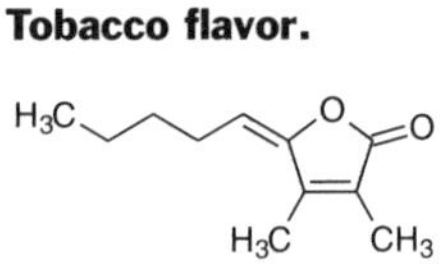

Bovolide (**1**)

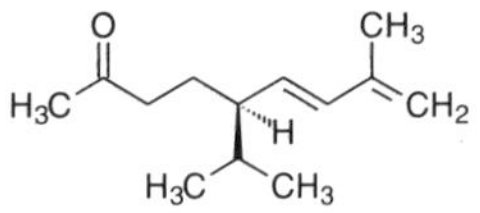

Solanone (**2**)

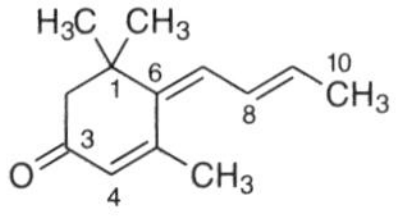

Megastigmatrienone (**3**)

The characteristic flavor of the respective tobacco types (Burley, Virginia, Oriental, etc.) depends on their origin and results from fermentative, oxidative, and photochemical degradation reactions of terpenoid precursors such as *carotinoids, *cembranoids, and labdanoids. Numerous heterocyclic compounds are formed by the Maillard reaction. In total more than 3000 components of tobacco are known[1]. The nonvolatile, e. g., glycosidically bound *flavor compounds are also of significance for the smoke flavor. No *impact compound is known. Some typical tobacco substances are: *β*-*damascones, **β*-damascenone, *megastigma-4,6,8-trien-3-one* ($C_{13}H_{18}O$, M_R 190.29), *dihydroactinidiolide, *bovolide* ($C_{11}H_{16}O_2$, M_R 180.25), *solanone* ($C_{13}H_{22}O$, M_R 194.32, $[\alpha]_D^{23}$ +13.6° {pure}), sclareolides, and ambra oxide[2,3] (see clary sage oil).

Lit.: [1] Perfum. Flavor. **15**, 27–49 (1990). [2] Enzell, in Schreier (ed.), Flavour '81, p. 449–478, Berlin, New York: de Gruyter 1981. [3] J. Agric. Food Chem. **41**, 2097–2103 (1993). – *[CAS 774-64-1 (1); 1937-45-8 (2); 13215-88-8 (3)]*

Tobramycin see kanamycins.

Tocopherols. Name for 3,4-dihydro-2*H*-1-benzopyran-6-ols (6-chromanols) substituted in the 2-position by a saturated or unsaturated 4,8,12-trimethyltridecyl residue. Oxidation of T. furnishes the tocoquinones. The latter, like the *plastoquinones, *ubiquinones, *boviquinones, and K vitamins (see vitamin K_1) belong to the polyprenylated *p*-benzo- or *p*-naphthoquinones, which fullfil important biological functions as redox pairs with the corresponding hydroquinones (so-called bioquinones).

The physiological activity of the T. is based on their radical scavenging properties[1], they are of major importance in cell membranes and lipoproteins as antioxidants. Long-term supplementation with *α*-T. can reduce the incidence of certain forms of cancer. T. are pale yellow, viscous oils. They are oxidized by atmospheric oxygen in the presence of metal ions or under the action of light. T. occur in plant oils; seed oils of soya, wheat, corn, rice, cotton, alfalfa, and nuts are particularly rich in T. Fruits and vegetables such as raspberries, legumes, fennel, paprika and celery also con-

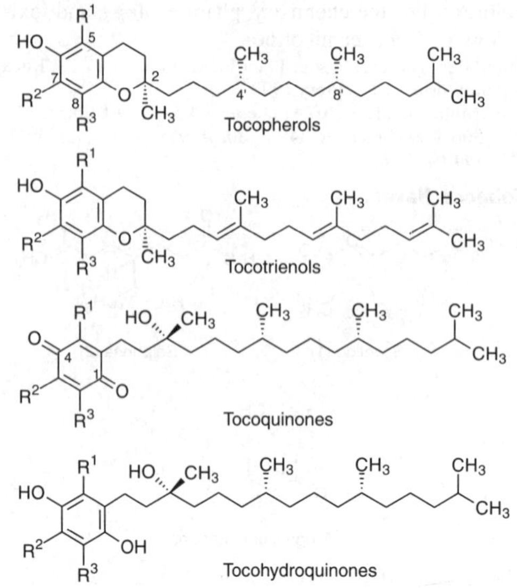

tain T.; furthermore, low concentrations of T. are found in animal products such as giblets, eggs, and fish. *α*-T. (vitamin E, antisterility factor) is the most active and most widely distributed natural T.; *β*-T. (cumo-T., neo-T.) has been isolated, among others, from spinach chloroplasts (*Spinacea oleracea*), ζ_2-T. from rice oil. *α*-Tocotrienol is present in wheat, barley, rye, and palm oil; *γ*-tocotrienol occurs in human bone marrow and in palm oil. For synthesis, see *Lit.*[2].

Biosynthesis: T. are formed in higher plants, algae, and bacteria from phytyl pyrophosphate and *homogentisic acid[3], animal organisms are not able to synthesize T.

Lit.: [1]J. Am. Chem. Soc. **103**, 6472–6477 (1981). [2]Helv. Chim. Acta **73**, 1068–1086; 1087–1107 (1990); J. Org. Chem. **57**, 5783 ff. (1992); Synlett **1996**, 1041, 1045 (synthesis of *α*-T.); **1997**, 208 (*γ*-T.). [3]Fett Wiss. Technol. **92**, 86–91 (1990); Phytochemistry **26**, 2741–2747 (1987).
gen.: Adv. Food Nutr. Res. **33**, 157–232 (1989) ▪ Angew. Chem., Int. Ed. Engl. **38**, 2715 (1999) ▪ Bourgeois, Determination of Vitamin E, Tocopherols and Tocotrienols, Amsterdam: Elsevier 1992 ▪ Eur. J. Biochem. **46**, 217 f. (1974) (nomenclature) ▪ Kleemann-Engel, p. 1901 ▪ Lipid Synth. Manuf. **1999**, 250–267 (manufacture) ▪ Luckner (3.) p. 415 ff. ▪ Nat. Prod. Rep. **12**, 101–133 (1995) ▪ N. Engl. J. Med. **308**, 1063–1071 (1983) (review) ▪ Negwer (6.), No. 7835, 7936, 8009 ▪ Stryer 1995, p. 454. – *[HS 2936 28]*

Todomatsu acid (methyl ester) see juvabione.

Toluquinone see methyl-1,4-benzoquinone.

Tolyporphin A.

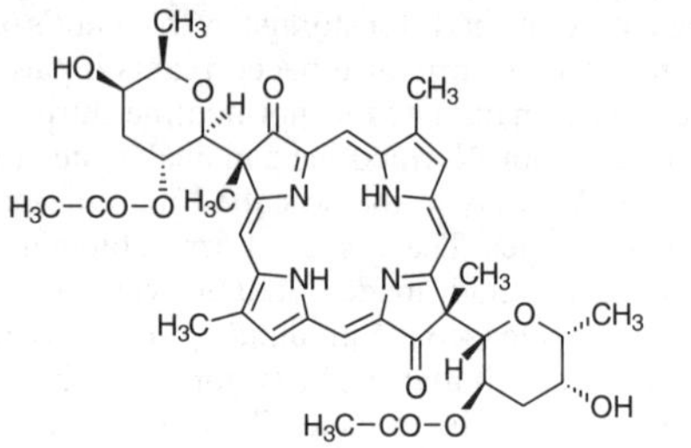

$C_{40}H_{46}N_4O_{10}$, M_R 742.83; uv$_{max}$ (CH_3OH) 276, 294, 401, 479, 504, 547, 611, 619, 642, 675 nm. T. A belongs to the structural class of the *bacteriochlorins, unusual substitution pattern.

Occurrence: T. A and several closely related T. were isolated from the Pacific alga *Tolypothrix nodosa*. T. A is able to break the resistance of cancer cells, directed concomitantly against several anticancer drugs [mul-

Table: Structure and data of Tocopherols.

	R[1]	R[2]	R[3]	configuration	molecular formula	M_R	mp. [°C]	bp. [°C]	optical activity	CAS
Tocopherols										
α-T.	CH_3	CH_3	CH_3		$C_{29}H_{50}O_2$	430.71	2.5–3.5	140	$[\alpha]_D^{25}$ +0.65°	10191-41-0
				2*R*,4′*R*,8′*R*				(0.13 mPa)	(C_2H_5OH)	59-02-9
β-T.	CH_3	H	CH_3		$C_{28}H_{48}O_2$	416.69		200–210	$[\alpha]_D^{20}$ +6.37°	148-03-8
				2*R*,4′*R*,8′*R*				(13 Pa)		16698-35-4
γ-T.	H	CH_3	CH_3		$C_{28}H_{48}O_2$	416.69	–2 to –3	200–210	$[\alpha]_{456}^{25}$ –2.4°	7616-22-0
				2*R*,4′*R*,8′*R*				(13 Pa)	(C_2H_5OH)	54-28-4
				(±)-form						7540-59-2
δ-T.	H	H	CH_3		$C_{27}H_{46}O_2$	402.66		250		119-13-1
				2*R*,4′*R*,8′*R*				(1.3 Pa)		5488-58-4
ζ_2-T.	CH_3	CH_3	H		$C_{28}H_{48}O_2$	416.69	–4			493-35-6
				2*R*,4′*R*,8′*R*						17976-95-3
η-T.	H	CH_3	H		$C_{27}H_{46}O_2$	402.64				91-86-1
				2*R*,4′*R*,8′*R*						78656-16-3
Tocotrienols										
α	CH_3	CH_3	CH_3		$C_{29}H_{44}O_2$	424.67	30–31			1721-51-3
				(*E*,*E*)					$[\alpha]_D^{25}$ –5.7°	47686-42-0
				R-(*E*,*E*)					($CHCl_3$)	58864-81-6
β (*ε*-T.)	CH_3	H	CH_3	*R*-(*E*,*E*)	$C_{28}H_{42}O_2$	410.64		140		490-23-3
								(6.5 mPa)		
γ	H	CH_3	CH_3	*R*-(*E*,*E*)	$C_{28}H_{42}O_2$	410.64				14101-61-2
				(*E*,*E*)						135970-12-6
Tocoquinones										
α	CH_3	CH_3	CH_3	3′*R*,7′*R*,11′*R*	$C_{29}H_{50}O_3$	446.71		120		7559-04-8
								(2.6 Pa)		
Tocohydroquinones										
α	CH_3	CH_3	CH_3	3′*R*,7′*R*,11′*R*	$C_{29}H_{52}O_3$	448.73				14745-36-9

tidrug resistance (MDR)], in vinblastine-resistant cell lines of human ovarian adenocarcinomas.

Lit.: Angew. Chem., Int. Ed. Engl. **38**, 923, 926 (1999) (absolute configuration) ▪ Chem. Rev. **94**, 327 (1994) ▪ J. Am. Chem. Soc. **114**, 385 (1992) ▪ Org. Lett. **1**, 1129 (1999) (synthesis). – *[CAS 138630-63-4]*

Tomamycin see anthramycins.

Tomatidenol, tomatidine see Solanum steroid alkaloids.

Tomatine see Solanum steroid alkaloid glycosides.

Tomato flavor see vegetable flavors.

Tonghaosu see polyynes.

Topostatin.

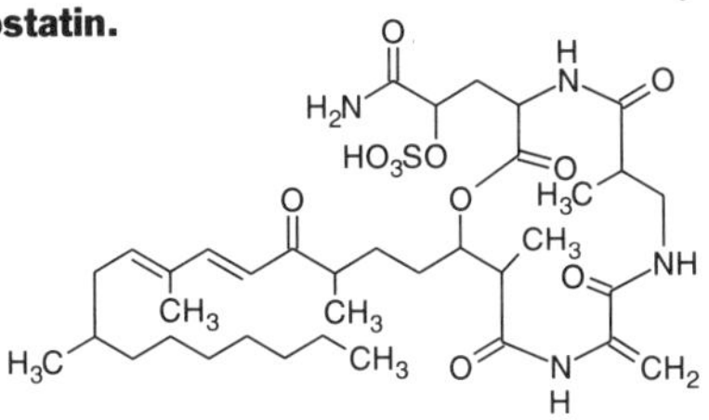

$C_{36}H_{58}N_4O_{11}S$, M_R 754.93, mp. 179–186 °C (decomp.), $[\alpha]_D^{28}$ +18.3° (CH_3OH), soluble in pyridine, partially soluble in H_2O, CH_3OH, insoluble in acetone, $CHCl_3$. Potent inhibitor of topoisomerases I and II from the culture filtrate of *Thermomonospora alba*.

Lit.: J. Antibiot. **51**, 991, 999 (1998); **52**, 460 (1999). – *[CAS 219774-62-6]*

Topotecan see camptothecin.

Topsentins.

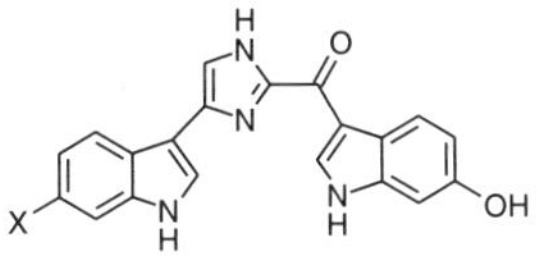

X = H : Topsentin B_1
X = Br : Topsentin B_2

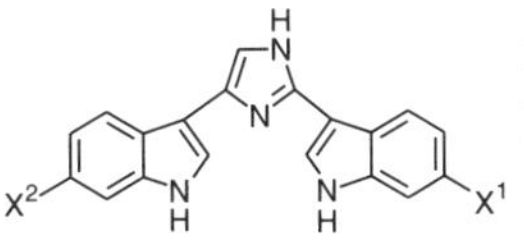

$X^1 = X^2$ = Br : Nortopsentin A
X^1 = Br , X^2 = H : Nortopsentin B
X^1 = H , X^2 = Br : Nortopsentin C

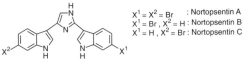

Table: Data of Topsentins.

	molecular formula	M_R	CAS
Topsentin B_1*	$C_{20}H_{14}N_4O_2$	342.36	112515-43-2
Topsentin B_2	$C_{20}H_{13}BrN_4O_2$	421.25	112515-44-3
Nortopsentin A	$C_{19}H_{12}Br_2N_4$	456.14	134029-43-9
Nortopsentin B	$C_{19}H_{13}BrN_4$	377.24	134029-44-0
Nortopsentin C	$C_{19}H_{13}BrN_4$	377.24	134029-45-1

* mp. 265–268 °C

The cytotoxic T.[1] and nortopsentins[2] are isolated from the Mediterranean sponge *Topsentia genitrix* as well as from deep-sea sponges of the genus *Spongosorites*[2]. The T. are active against P388 leukemia cells and B16 melanoma cells.

Lit.: [1]Can. J. Chem. **65**, 2118 (1987); **67**, 677 (1989); Chem. Pharm. Bull. **44**, 1831 (1996); Heterocycles **48**, 1887 (1998); J. Org. Chem. **53**, 5446 (1988); Tetrahedron Lett. **37**, 5503 (1996) (synthesis). [2]Bioorg. Med. Chem. Lett. **6**, 2103 (1996); Heterocycles **48**, 1887 (1998); J. Nat. Prod. **62**, 647 (1999); J. Chem. Soc., Chem. Commun. 1994, 2085; J. Org. Chem. **56**, 4304 (1991).

Torachrysone.

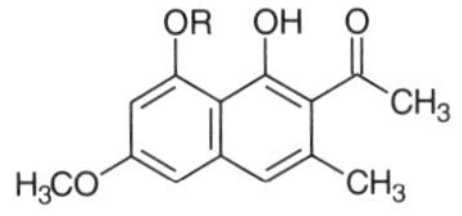

R = H : Torachrysone
R = CH_3 : Torachrysone 8-*O*-methyl ether

$C_{14}H_{14}O_4$, M_R 246.26, yellow needles, mp. 214–215 °C (decomp.), changes color to orange on treatment with NaOH. T. and its 8-*O*-methyl ether are isolated from seeds of the Asian Fabaceae *Cassia tora*[1]. T. 8-*O*-methyl ether ($C_{15}H_{16}O_4$, M_R 260.29, colorless needles, mp. 98–99 °C) also occurs in the fruit bodies of the agaric *Cortinarius rufoolivaceus* and in cultures of *C. orichalceus*[2]. T. is linked to a 1,2- or, respectively, 1,4-anthracenequinone unit in the red-violet pigments of this fungus (*rufoolivacins).

Lit.: [1]Chem. Pharm. Bull. **17**, 454 (1969); J. Am. Chem. Soc. **97**, 3270 (1975) (synthesis). [2]Zechmeister **51**, 121f. – *[CAS 22649-04-3]*

Toromycin B see gilvocarcins.

Torosachrysone.

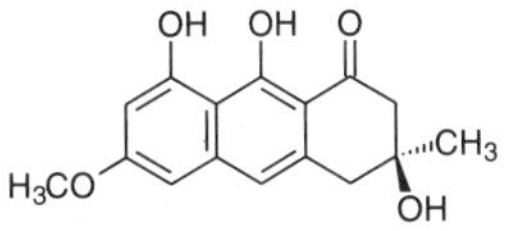

(-)-(*S*)-form

$C_{16}H_{16}O_5$, M_R 288.30, lemon yellow needles, mp. 191–194 °C, $[\alpha]_D$ −6° (CH_3OH). A monomeric *preanthraquinone occurring as building block in various dimeric fungal pigments (*flavomannins, *phlegmacins)[4] and as (−)-(3*S*)-T. in the seeds of the medicinal plant *Cassia torosa* (Fabaceae)[2]. (−)-(3*S*)-T. also exists together with related preanthraquinones in the fruit bodies of *Dermocybe splendida* and other Australian *D.* species (Basidiomycetes), sometimes as the strongly fluorescing 8-*O*-*β*-D-gentiobioside[3,5]. The (+)-(3*R*)-enantiomer occurs in Australian *Dermocybe* species[5] and in the bark of *Araliorhamnus vaginata*[4], an Araliaceae species endemic to Madagascar. (−)-(3*S*)-T. 8-*O*-methyl ether {$C_{17}H_{18}O_5$, M_R 302.33, light yellow cryst., mp. 157 °C, $[\alpha]_{546}$ ca. −15°} occurs in various European *Cortinarius* and *Tricholoma* species[6]. For determination of the absolute configuration of T. by synthesis, see *Lit.*[7]. Asperflavin from cultures of *Aspergillus flavus* is (+)-T. 8-*O*-methyl ether[1].

Lit.: [1]J. Chem. Soc., Perkin Trans. 1 **1972**, 2406. [2]J. Nat. Prod. **40**, 191 (1977). [3]Phytochemistry **28**, 2647 (1989). [4]Phytochemistry **31**, 3577 (1992). [5]Nat. Prod. Lett. **2**, 151 (1993). [6]Zechmeister **51**, 143ff.; Nat. Prod. Rep. **11**, 76ff. (1994). [7]Steglich, in Atta-ur-Rahman & Le Quesne (eds.), Natural Products Chemistry III, p. 305–315, Berlin: Springer 1988 (synthesis). – *[CAS 61419-07-6 ((−)-(S)-T.)]*

Totomycin see hygromycins.

***α*-Toxicarol** see tephrosin.

Toxicodendrol see urushiol(s).

Toxicols. Heptacyclic meroterpenoids from the sponge *Toxiclona toxius* indigenous to the Red Sea, e.g., *T. B:* $C_{36}H_{54}O_3$, M_R 534.82, powder, $[\alpha]_D$ −16° (CH_3OH/CH_2Cl_2). T. B. occurs together with the mono- and di-*O*-sulfates T. C and T. A.

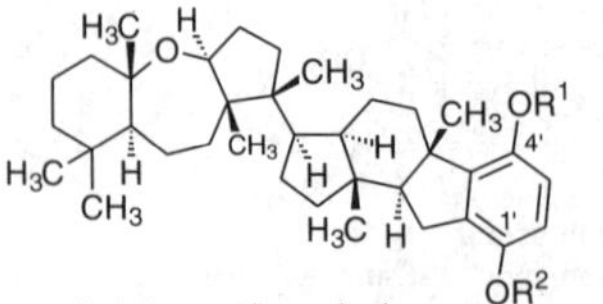

(relative configuration)

$R^1, R^2 = SO_3H$: T. A
$R^1, R^2 = H$: T. B
$R^1 = SO_3H, R^2 = H$: T. C

Lit.: J. Nat. Prod. **56**, 2120 (1993) ▪ Tetrahedron **49**, 4275–4282 (1993). – *[CAS 149764-32-9 (T. B); 149764-31-8 (T. A); 149764-33-0 (T. C)]*

C-Toxiferine I (toxiferine I).

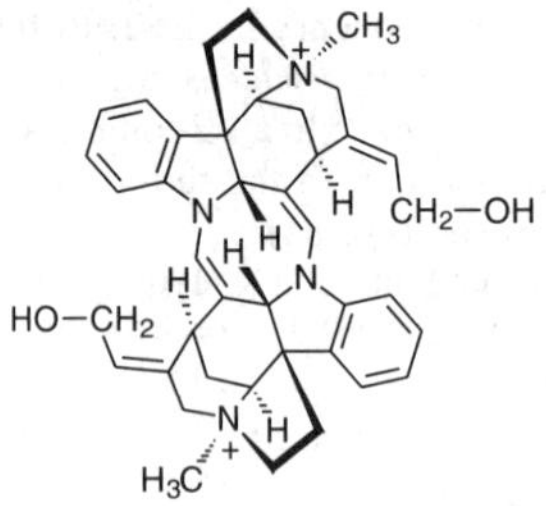

$C_{40}H_{46}N_4O_2^{2+}$, M_R 614.83; *C*-T. I, belonging to the group of *toxiferines, forms a crystalline dichloride readily soluble in water. It is the most important of the more than 40 alkaloids occurring in calabash *curare (*Strychnos toxifera, S. froesii*, Strychnaceae, Loganiaceae), $C_{40}H_{46}Cl_2N_4O_2$, M_R 685.74, $[\alpha]_D$ −511° (CH_3OH), it gives a red-violet color with cerium sulfate. *C*-T. I belongs to the dimeric C_{40}-alkaloids that exist as quaternary salts and can be precipitated (enriched) as Reinecke salts.

Activity: *C*-T. I is probably the most effective peripheral muscle relaxant. It is hardly used in current therapy because the corresponding diallylnortoxiferinium chloride (alcuronium chloride, Alloferin®) has a shorter duration of action. The muscle relaxant effect results from a (competitive) expulsion of acetylcholine from the receptors of the motoric endplates with paralysis of the transversely striated musculature (overdoses cause death by asphyxiation).

Toxicity: Orally administered *C*-T. I is only poorly absorbed and rapidly eliminated. On parenteral administration it is highly toxic, like other structurally related bisindole alkaloids. LD_{50} (mouse i.v.) 0.01 to 0.06 mg/kg. Overdoses result in muscle paralysis that starts in the face and spreads through the neck, back, and stomach, followed by visual impairments, swallowing difficulties, disorders of circulation and respiration. The main therapeutic interventions are treatment with cholinesterase inhibitors and mechanical ventilation.

Biosynthesis: "Dimerization" of two caracurine-VII molecules, monomeric indoles with a C_9-secoiridoid unit from *Strychnos* species, followed by methylation and thus quaternization at the non-indole nitrogen atoms.

Lit.: Beilstein EV **26/13**, 63 ▪ Hager (5.) **6**, 818–823, 842 ▪ Sax (8.), THI 000. – *[CAS 6888-23-9]*

Toxiferines.

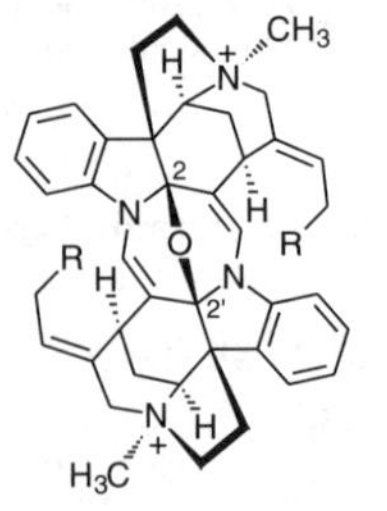

R = H : *C*-Curarine I = Toxiferine
R = OH: *C*-Toxiferine I, ether bridge 2,2' missing
R = H : *C*-Dihydrotoxiferine, ether bridge 2,2' missing

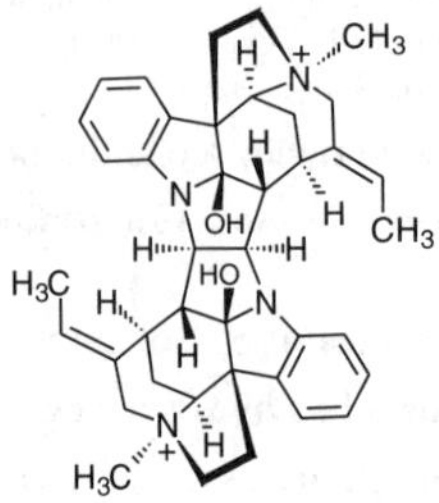

C-Calebassine = *C*-Toxiferine II

A group of *curare alkaloids from *Strychnos* species (calabash curare) consisting principally of bis-quaternary bis-indole alkaloids. The most important T. is **C*-toxiferine I; others are *toxiferine* (*C*-curarine I, $C_{40}H_{44}N_4O^{2+}$, M_R 596.82; dichloride $C_{40}H_{44}Cl_2N_4O$, M_R 667.72, $[\alpha]_D^{20}$ +73.6°), *C-toxiferine II* (*C*-calebassine, $C_{40}H_{48}N_4O_2^{2+}$, M_R 616.85; dichloride $C_{40}H_{48}Cl_2N_2O_2$, M_R 659.74, $[\alpha]_D^{20}$ +71.1°). They exhibit muscle relaxing activities; crude extracts of the stem and root bark of not only South American but also African *Strychnos* species were used by the native populations to prepare arrow poisons. For activity, see *C*-toxiferine I.

Lit.: Hager (5.) **6**, 818–825, 842 ▪ J. Org. Chem. **48**, 1869 (1983) ▪ Phytologia **51**, 433 (1982) ▪ Sax (8.), THI 000. – *[HS 2939 90; CAS 7168-64-1 (T.); 7257-29-6 (C-calebassine)]*

Toxins. Metabolic products from microorganisms, plants, or animals with a poisonous effect on mammals, especially, humans. T. are mostly immunogenic, i.e., they can induce the formation of specific antibodies (antitoxins) as a result of their antigen character. They belong to widely differing classes of compounds such as proteins, lipopolysaccharides, alkaloids, terpenoids, steroids, aliphatic acids, biogenic amines, and guanine derivatives.

T. from microorganisms are principally the poisons produced by bacteria. Distinctions are made between thermolabile, protein-like *exotoxins* secreted by living bacteria (e.g., *cholera toxin, *diphtheria toxin, tetanus, *botulism toxin, gas gangrene toxin) and the thermostable, glycolipid-like *endotoxins as components of the bacterial organism released after its destruction. Other toxins of microbial origin are the *mycotoxins (*ergot alkaloids, *aflatoxins, *citrinin, *ochratoxins, *patulin) from yeasts, mold fungi, and microfungi. The higher fungi include numerous poisonous species containing T. with oligopeptide structures (*amanitins of the death cap) or low-molecular weight, alkaloid-like compounds (*fly agaric constituents).

T. from plants (phytotoxins) include both high- and low-molecular weight substances. Examples of the former are the lectins distributed in legumes (abrin, *concanavalin A, crotin, phasin, *ricin). The low-molecular weight plant T. are more numerous and of greater diversity, including the cardiac glycosides, *saponins,

*glucosinolates, *mustard oils and, in particular, the *alkaloids.
T. from animals (zoo toxins) see animal toxins. Bacterial T. and the poisonous lectins contain a so-called *haptophoric group (haptomer)* for specific binding of the T. to receptors on the cell surface (often gangliosides) and a *toxophoric group (effectomer)* which is the carrier of the actual toxic activity. If the toxophoric group is destroyed (e. g., with formaldehyde), a non-poisonous *toxoid* (former name anatoxin) is obtained which retains the antigen activity. Toxoids of this type are used for active immunization, e. g., against diphtheria and tetanus.
Many T. are important tools in molecular biology and neurochemistry on account of their high specificities for certain metabolic processes (e. g., *tetrodotoxin, *saxitoxin, *bungarotoxin). A number of T. or their derivatives have attained significance as pharmaceutical agents (opiates, *ergot alkaloids, *Digitalis glycosides, *curare), others serve as stimulants or are abused as narcotics (*cannabinoids, *morphine, *caffein(e), *nicotine).
Lit.: Annu. Rev. Biochem. **55**, 195–224 (1986) ▪ Annu. Rev. Med. **38**, 417–432 (1987) ▪ Harborne, Dictionary of Plant Toxins, New York: Wiley 1996 ▪ Ownby & O'Dell (eds.), Natural Toxins, Elmsford, N. Y.: Pergamon 1989 ▪ Pierson & Stern, Foodborne Microorganisms and Their Toxins, New York: Dekker 1986 ▪ Science **259**, 1394 (1993) ▪ Steyn & Vleggaar, Mycotoxins and Phycotoxins, Amsterdam: Elsevier 1986 ▪ Tu, Handbook of Natural Toxins (4 vols.), New York: Dekker 1983, 1984, 1988 ▪ Tyson & Frazier, In vitro Toxicity Indicators, San Diego: Academic Press 1994.

Toxoflavin {xanthothricin, 1,6-dimethylpyrimido-[5,4-*e*]-1,2,4-triazine-5,7(1*H*,6*H*)-dione}.

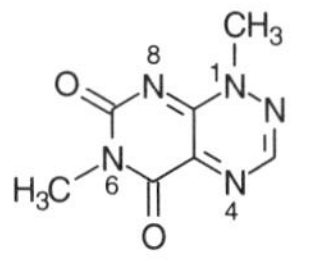

$C_7H_7N_5O_2$, M_R 193.17, yellow needles, mp. 172 °C, soluble in water, ethanol, ethyl acetate, and chloroform. A highly toxic [LD_{50} (mouse p. o.) 8.4 mg/kg] antibiotic obtained together with *bongkrekic acid from cultures of *Pseudomonas cocovenenans*. T. inhibits xanthine oxidase and exhibits antimicrobial as well as many other pharmacological activities [1]. The biosynthesis [2] proceeds from a purine derivative with C-8 of the imidazole ring being replaced by an aminomethyl unit from glycine. For synthesis, see *Lit.* [3].
Lit.: [1] Eur. J. Med. Chem. **29**, 245 (1994). [2] J. Biol. Chem. **241**, 846 (1966). [3] J. Chem. Soc., Perkin Trans. 1 **1976**, 713.
gen.: Beilstein E V **26/19**, 135 ▪ Merck-Index (12.), No. 9695. – *[HS 2941 90; CAS 84-82-2]*

Toxoid see toxins.

Toyocamycin (toyokamycin).

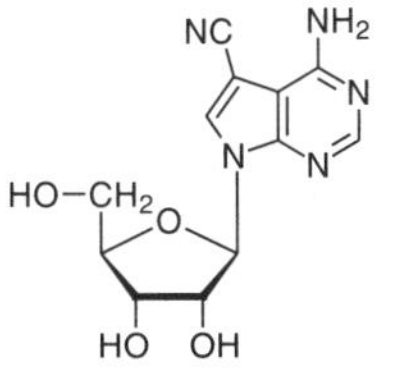

$C_{12}H_{13}N_5O_4$, M_R 291.27, needles, mp. 243 °C, $[\alpha]_D$ –55.6° (0.1 m HCl), LD_{50} (mouse p. o.) 8 mg/kg. A *nucleoside antibiotic from *Streptomyces toyocaensis* related to *tubercidin.
Lit.: J. Antibiot. **41**, 1711 (1988) ▪ J. Chem. Soc., Perkin Trans. 1 **1995**, 1543 (synthesis) ▪ Merck-Index (12.), No. 9698 ▪ Nucleosides Nucleotides **12**, 643 (1993); **18**, 153 (1999) (synthesis) ▪ Sax (8.), VGZ 000. – *[CAS 606-58-6]*

TPA see tiglianes.

TPP see thiamin(e).

TR-2 see fumitremorgins.

(+)-Trachelanthamidine see necine(s).

Trachylobanes.

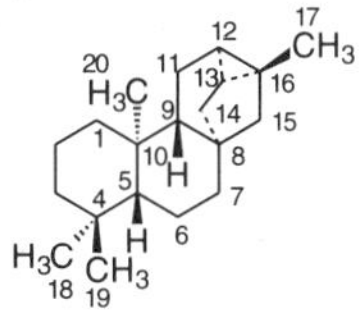

1

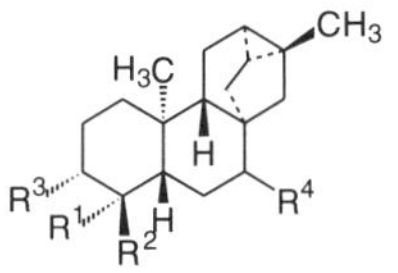

2-7

Nr.	R¹	R²	R³	R⁴	
2	CH_3	CH_3	H	H	: Trachylobane
3	CH_3	COOH	H	H	: Trachylobanic acid
4	COOH	CH_3	H	α-OH	: Ciliaric acid
5	COOH	CH_3	OH	H	: 3-Hydroxy-18-trachylobanic acid
6	CH_3	CH_3	H	β-OH	: Trachinol
7	CH_3	CH_2OH	H	β-OH	: Trachinodiol

Table: Data of Trachylobanes.

	molecular formula	M_R	mp. [°C]	$[\alpha]_D$	CAS
2	$C_{20}H_{32}$	272.47	oil –5.5		5282-35-9
3	$C_{20}H_{30}O_2$	302.46	110–112, methyl ester	–41° ($CHCl_3$) (methyl ester)	26263-39-8
4	$C_{20}H_{30}O_3$	318.46	304–405		35682-59-8
5	$C_{20}H_{30}O_3$	318.46	268	–72° (C_2H_5OH/$CHCl_3$)	
6	$C_{20}H_{32}O$	288.47	105–106	+4° ($CHCl_3$)	34144-23-5
7	$C_{20}H_{32}O_2$	304.47	175–176	–10° ($CHCl_3$)	34144-22-4

A class of *diterpenes with the trachylobane skeleton **1**, members of which have been isolated from various higher plants including *Trachylobium*, *Helianthus*, *Sideritis*, *Araucaria* and bryophytes of the genus *Jungermannia*. Native T. (**2** = **1**) has been found in the latter two genera. The T. possess an interesting tricyclic subunit, the cation of which had been postulated as an intermediate in the biosynthesis of several classes of diterpenes prior to the discovery of the T. Trachylobanic acid [2] (**3**) exhibits inhibitory activity on the growth of larvae of various butterfly species, such as the significant pest *Homeosoma electellum* (sunflower moth).
Lit.: [1] Can. J. Chem. **51**, 2534 (1973); Tetrahedron **31**, 1311 (1975). [2] Naturwissenschaften **64**, 341 (1977); Experientia **32**, 1364 (1976); J. Org. Chem. **45**, 1852 (1980).

gen.: J. Chem. Res. (S) **1982**, 200 ▪ Justus Liebigs Ann. Chem. **1984**, 495 ▪ Phytochemistry **28**, 1533 (1989); **30**, 913 (trachinodiol), 3361 (1991) (trachinol).

Trail pheromones see pheromones.

Trametin.

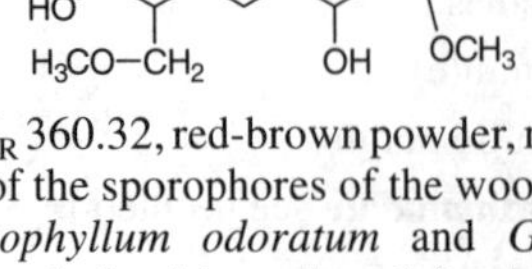

$C_{18}H_{16}O_8$, M_R 360.32, red-brown powder, mp. > 300 °C. Extraction of the sporophores of the wood-destroying fungi *Gloeophyllum odoratum* and *G. sepiarium* (Aphyllophorales) with methanol furnishes the red-brown pigment trametin. T. is formed by methanolysis of a precursor contained in the fungi, thus the methoxy groups originate from the solvent. T. can be detected with aluminum oxide which, when added to the ethanolic extract of the fungus, causes a green color.
Lit.: Zechmeister **51**, 112–116.

Trapoxin A.

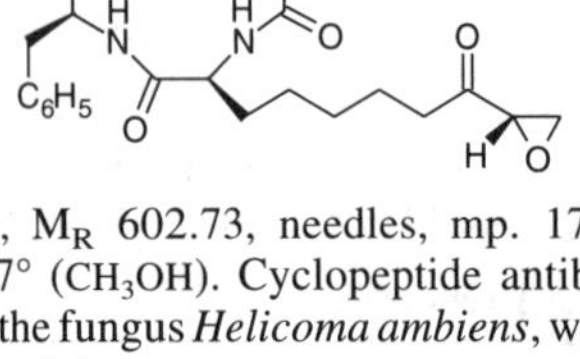

$C_{34}H_{42}N_4O_6$, M_R 602.73, needles, mp. 173–174 °C, $[\alpha]_D^{25}$ −63.7° (CH_3OH). Cyclopeptide antibiotic from cultures of the fungus *Helicoma ambiens*, which shows antitumor activity.
Lit.: Acta Crystallogr., Sect. C **47**, 1496 (1991) ▪ J. Am. Chem. Soc. **118**, 10412 (1996) (synthesis) ▪ J. Antibiot. **43**, 1524 (1990). – *[CAS 133155-89-2]*

Traumatic acid [(*E*)-2-dodecenedioic acid].

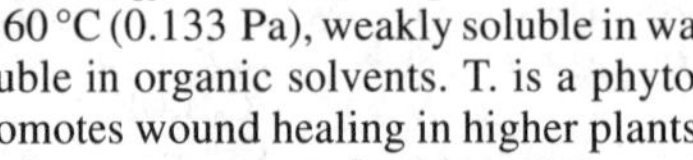

$C_{12}H_{20}O_4$, M_R 228.29, mp. 166–167 °C, bp. 150–160 °C (0.133 Pa), weakly soluble in water, readily soluble in organic solvents. T. is a phytohormone that promotes wound healing in higher plants by stimulating the regeneration of epidermal tissue. T. occurs in green beans among others.
Lit.: Beilstein E IV **2**, 2279 ▪ Chem. Ind. (London) **1986**, 36 ▪ Merck-Index (12.), No. 9710 ▪ Org. Prep. Proced. Int. **19**, 461 (1987) ▪ Synth. Commun. **21**, 183–190 (1991). – *[HS 2917 19; CAS 6402-36-4]*

Trehalose [α-D-glucopyranosyl-α-D-glucopyranoside]

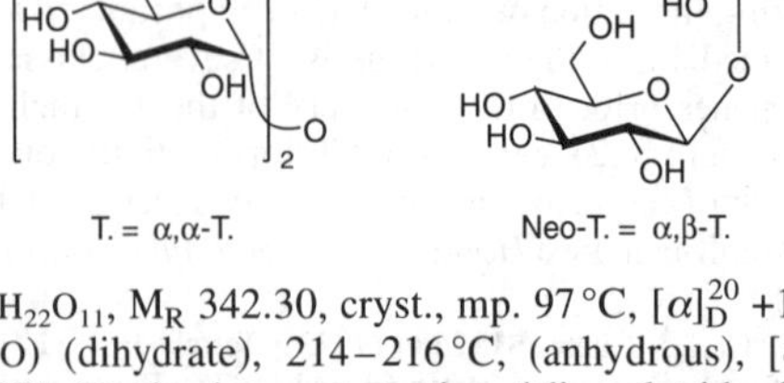

T. = α,α-T. Neo-T. = α,β-T.

$C_{12}H_{22}O_{11}$, M_R 342.30, cryst., mp. 97 °C, $[\alpha]_D^{20}$ +178° (H_2O) (dihydrate), 214–216 °C, (anhydrous), $[\alpha]_D^{20}$ +197° (H_2O). A sweet tasting *disaccharide from (mold) fungi, e. g. *ergot, as well as yeasts, lichens, algae, bacteria, mosses; it also occurs in the hemolymph of insects. α,β-T. {*neotrehalose*, mp. 210–220 °C, $[\alpha]_D^{20}$ +95° (H_2O)} occurs in honey.
Use: T. obtained from compressed yeast is used as a nutrient substrate for bacteria. Diesters of T. have immunostimulating activity[1].
Lit.: [1] Med. Res. Rev. **6**, 243–274 (1986).
gen.: Acta Crystallogr., Sect. C **53**, 234 (1997) ▪ Annu. Rev. Entomol. **23**, 389–408 (1978) ▪ Beilstein E V **17/8**, 3 ▪ Carbohydr. Res. **261**, 25 (1994); **266**, 147 (1995) ▪ J. Biotechnol. **3**, 121–130 (1985) ▪ Karrer, No. 644 ▪ Merck-Index (12.), No. 9713 ▪ Plant Sci. (Limerick, Irel.) **112**, 1 (1995) (review). – *[HS 2940 00; CAS 99-20-7 (α,α-T.); 585-91-1 (α,β-T.)]*

Trehalostatin (trehazolin).

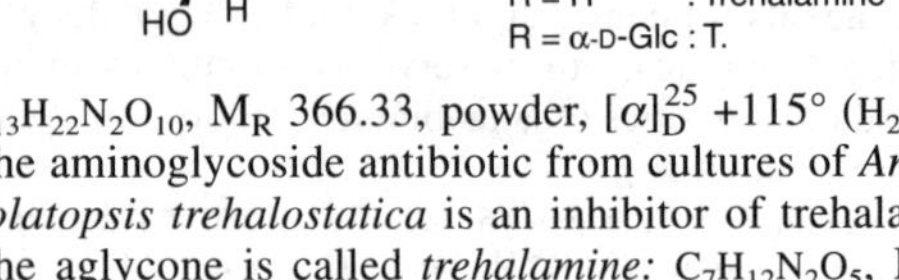

$C_{13}H_{22}N_2O_{10}$, M_R 366.33, powder, $[\alpha]_D^{25}$ +115° (H_2O). The aminoglycoside antibiotic from cultures of *Amycolatopsis trehalostatica* is an inhibitor of trehalase. The aglycone is called *trehalamine:* $C_7H_{12}N_2O_5$, M_R 204.18, mp. 163–165 °C, $[\alpha]_D^{25}$ +13.5° (H_2O).
Lit.: *Isolation:* J. Antibiot. **44**, 1165 (1991); **46**, 1116 (1993); **47**, 932 (1994). – *Synthesis:* Carbohydr. Res. **288**, 99 (1996) (analogues); **315**, 1–15 (1999) ▪ Chem. Lett. **1993**, 173, 971 ▪ J. Org. Chem. **59**, 813, 4450 (1994); **63**, 3403, 5877 (1998) ▪ J. Chem. Soc., Perkin Trans. 1 **1994**, 589 ▪ J. Am. Chem. Soc. **117**, 11811 (1995) ▪ Org. Lett. **1**, 1705 (1999). – *[CAS 132729-37-4 (T.); 144811-33-6 (trehalamine)]*

Tremetriol see sterpuranes.

Tremortins see penitrems.

Tremulanes.

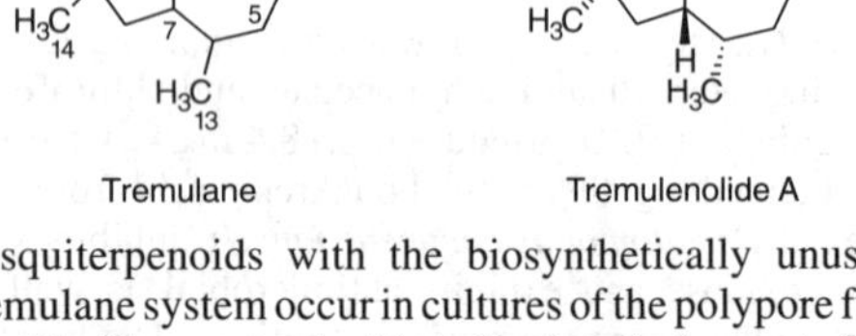

Tremulane Tremulenolide A

Sesquiterpenoids with the biosynthetically unusual tremulane system occur in cultures of the polypore fungus *Phellinus tremulae* (Aphyllophorales), a pest of the trembling poplar, e. g., *tremulenolide A* {$C_{15}H_{22}O_2$, M_R 234.34, needles, mp. 110–112 °C, $[\alpha]_D$ +110.7°}.
Lit.: Can. J. Chem. **75**, 834 (1997) (biosynthesis) ▪ J. Org. Chem. **58**, 7529 (1993). – *[CAS 152075-90-6 (tremulenolide A)]*

Tremuloidin see salicyl alcohol.

Trestatins see glycosidase inhibitors.

Tretinoin (retinoic acid) see retinol.

Triacontanoic acid see melissic acid.

1-Triacontanol (melissyl alcohol, myricyl alcohol). $H_3C–(CH_2)_{28}–CH_2OH$, $C_{30}H_{62}O$, M_R 438.82, mp. 86.5 °C, soluble in toluene, insoluble in water. T. is a component of many plant waxes and occurs as the palmitate in bees wax. *Triacontanal* ($C_{30}H_{60}O$, M_R 436.80) has also been isolated from plant waxes[1]. T. acts as a plant growth factor[2] that elicits (+)-β-L-adenosine as a second messenger[3]. T. is used to improve plant growth and increase harvests[4]. For synthesis, see *Lit.*[5]. The biosynthesis of plant waxes has been investigated with the help of genetechnological methods in *Arabidopsis thaliana*. Defined genes have been assigned to individual steps of the biosynthesis[6].

Lit.: [1]Hoppe Seyler's Z. Physiol. Chem. **350**, 462 (1969). [2]Science **195**, 1339 (1977). [3]Plant Physiol. **95**, 986–989 (1991). [4]Science **219**, 1219 (1983). [5]Monatsh. Chem. **126**, 565 (1995); Justus Liebigs Ann. Chem. **1985**, 214; Chem. Ind. (London) **1986**, 398; Tetrahedron **48**, 9187–9194 (1992). [6]Genome **36**, 610–618 (1993); Phytochemistry **33**, 851–855 (1993).

gen.: Beilstein EIV **1**, 1918 ▪ Karrer, No. 100, 5055 ▪ Z. Chem. **20**, 397–406 (1980). – *Plant physiology:* Angew. Bot. **66**, 109–114 (1992) ▪ J. Agric. Food Chem. **41**, 814–818 (1993). – *Pharmacology:* J. Ethnopharmacol. **14**, 41 ff. (1985) ▪ Science **195**, 1339 (1977). – *[CAS 593-50-0 (T.); 22725-63-9 (triacontanal)]*

Triacsins.

$H_3C–CH_2–(CH=CH)_4–CH=N–N=N–OH$

T. B

T. A: 6,7,8,9-tetrahydro-T. B, T. C: 8,9-dihydro-T. B, T. D: 6,9-dihydro-T. B

Nitrosohydrazones from culture filtrates of *Streptomyces* sp. SK 1894, e.g., *T. B:* $C_{11}H_{15}N_3O$, M_R 205.26, yellow powder, mp. 144–147 °C; *T. A:* $C_{11}H_{19}N_3O$, M_R 209.29, yellow powder, mp. 116–118 °C. T. inhibit various acyl-CoA synthetases, e.g., arachidonoyl-CoA synthetase and the synthetases of longer-chain fatty acids; the biosynthesis of acetyl-CoA remains unaffected. Common structural feature of the T. is the 1-*N*-hydroxy-1-triazene unit, bound to an unsaturated aliphatic chain.

Lit.: Biochim. Biophys. Acta **921**, 595 (1987) ▪ J. Antibiot. **39**, 1211 (1986) ▪ J. Biol. Chem. **266**, 4214 (1991) ▪ Prostaglandins **37**, 655–671 (1989). – *[CAS 105201-46-5 (T. A); 105201-47-6 (T. B)]*

Triandrin see 4-hydroxycinnamyl alcohol.

Triangularine see pyrrolizidine alkaloids.

Trichaurantins (trichoaurantianolides).

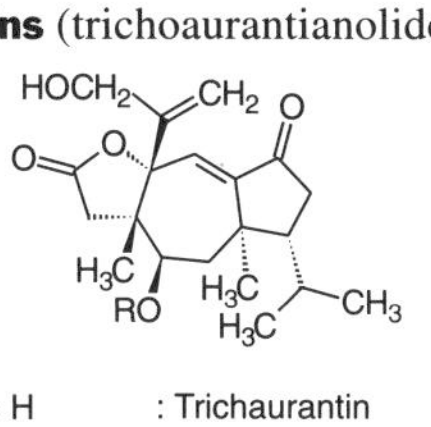

R = H : Trichaurantin
R = CO—CH$_3$: 6-*O*-Acetyltrichaurantin

Diterpenoids with the unusual 2,3-seconeodolastane skeleton from the toadstools *Tricholoma aurantium* and *T. fracticum* (Agaricales). The structure of *trichaurantin* ($C_{20}H_{28}O_5$, M_R 348.44, cryst., mp. 185 °C, $[\alpha]_D$ +33°) was confirmed by X-ray crystallography. In addition to *6-O-acetyltrichaurantin* ($C_{22}H_{30}O_6$, M_R 390.48, oil, $[\alpha]_D$ +3°) the corresponding aldehyde and two derivatives formed by intramolecular Michael addition of the allylic hydroxy group to the α,β-unsaturated double bond also occur in the fungi.

Lit.: Justus Liebigs Ann. Chem. **1995**, 77 ▪ Tetrahedron Lett. **36**, 1905, 3035 (1995). – *[CAS 160525-60-0 (T.); 160489-11-2 (6-O-acetyl-T.)]*

Trichilins see meliatoxins.

Trichione.

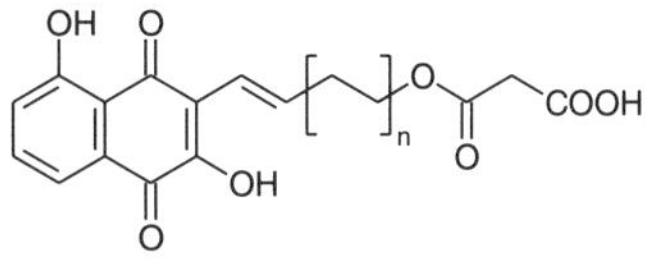

n = 1 : Trichione
n = 2 : Homotrichione

$C_{17}H_{14}O_8$, M_R 346.29, red needles, mp. 157–159 °C. A naphthoquinone pigment from sporangia of the slime molds *Trichia floriformis* and *Metatrichia vesparium* (Myxomycetes). *Homotrichione* ($C_{19}H_{18}O_8$, M_R 374.35, red needles, mp. 128–129 °C) differs from T. only in having a side chain extended by two methylene groups. The tetracyclic *vesparione is biosynthetically closely related to homotrichione.

Lit.: Justus Liebigs Ann. Chem. **1982**, 1722 ▪ Zechmeister **51**, 122, 170 (review). – *[CAS 83447-92-1 (T.); 83447-93-2 (homo-T.)]*

2,4,6-Trichloroanisol see wine flavor.

1,2,4-Trichloro-3,6-dimethoxy-5-nitrobenzene see drosophilin A.

Trichlorophenols. In fermentation cultures, the basidiomycete *Hypholoma elongatum* produces trichlorinated phenols, e.g., 2,4,6-trichloro-3-methoxyphenol (**1**), $C_7H_5Cl_3O_2$, M_R 227.47, yellow crystals, mp. 62–66 °C; 2,4,5-trichloro-3,6-dimethoxyphenol (**2**), $C_8H_7Cl_3O_3$, M_R 257.50; 2,3,5-trichloro-4,6-dimethoxyphenol (**3**), $C_8H_7Cl_3O_3$, M_R 257.50.

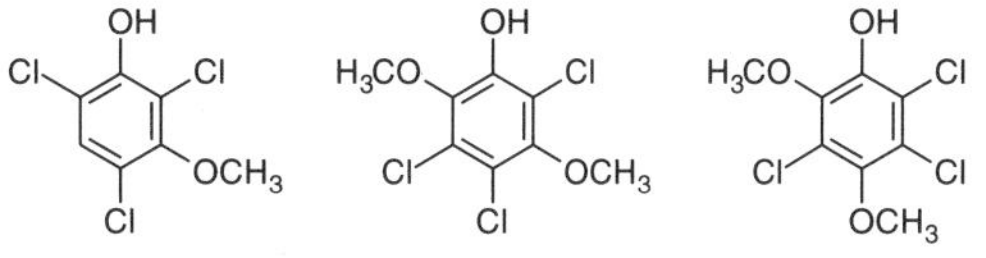

1 **2** **3**

Lit.: J. Nat. Prod. **61**, 913 (1998) ▪ Phytochemistry **49**, 203 (1998). – *[CAS 26378-18-7 (1); 101421-06-1 (2); 215176-75-3 (3)]*

Trichocarpin see gentisic acid.

Trichochromes (T. A–F). Pigments of the pheomelanin group (see melanins) isolated for the first time in 1878 from red hair. The T., formerly also known as *trichosiderins*, are pigments of human red hair and red animal feathers. The chromophore of T. is the protonated 2,2′-bi[1,4]benzothiazinylidene skeleton (see figure), the mesomeric cation of which is formed by protonation in acidic solution. Various yellow and violet pigments are also known; see also melanins.

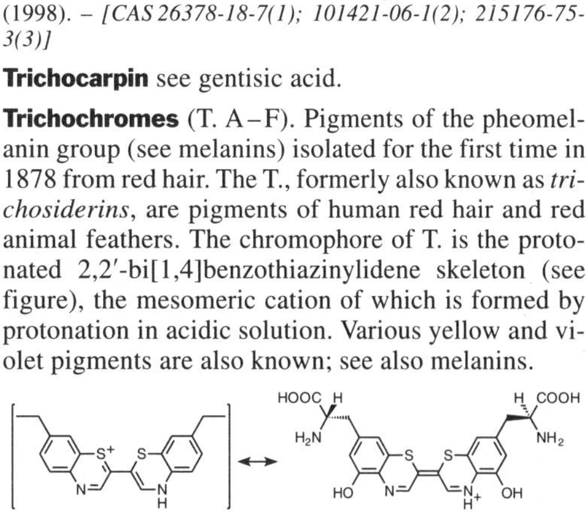

Chromophore of T. Trichochrome F

Lit.: Angew. Chem. Int. Ed. Engl. **13**, 305–312 (1974) ▪ J. Invest. Dermatol. **75**, 122 (1980) ▪ Tetrahedron **30**, 2781 ff. (1974); **46**, 6831–6838 (1990). – *[CAS 25942-17-0 (T. F)]*

Trichodermin, trichodermol see scirpenols.

Trichodiene [(*S,S*)-1,4-dimethyl-4-(1-methyl-2-methylenecyclopentyl)-cyclohexene].

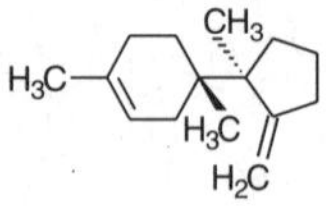

$C_{15}H_{24}$, M_R 204.36, oil, $[\alpha]_D$ +21° ($CHCl_3$). A *sesquiterpene from cultures of *Trichothecium roseum* constituting the biogenetic precursor of all *tric(h)othecenes. T. is formed from *trans,trans*-farnesyl pyrophosphate. The cyclizing enzyme, trichodiene synthase, has been isolated and characterized from *T. roseum* and *Fusarium sporotrichioides*. The corresponding gene (Tox5) has been cloned and expressed in *Escherichia coli* and tobacco plants.

Lit.: Appl. Environ. Microbiol. **59**, 2359 (1993) (Tox5 gene) ▪ Arch. Biochem. Biophys. **251**, 756 (1986) (T. synthase) ▪ Chem. Rev. **90**, 1089-1103 (1990) ▪ J. Am. Chem. Soc. **120**, 5453 (1998) [synthesis of (–)-T.] ▪ J. Chem. Soc., Chem. Commun. **1991**, 1033 (12,13-diol) ▪ J. Chem. Soc., Perkin Trans. 1 **1996**, 829 (synthesis) ▪ J. Org. Chem. **55**, 4403 (1990); **58**, 6255–6265 (1993) (synthesis) ▪ Microbiol. Rev. **57**, 595 (1993) ▪ Tetrahedron Lett. **39**, 5489 (1998) [synthesis of (±)-T.] ▪ Phytochemistry **43**, 1235 (1996). – *[CAS 28624-60-4 (S-form)]*

Tricholomenyns.

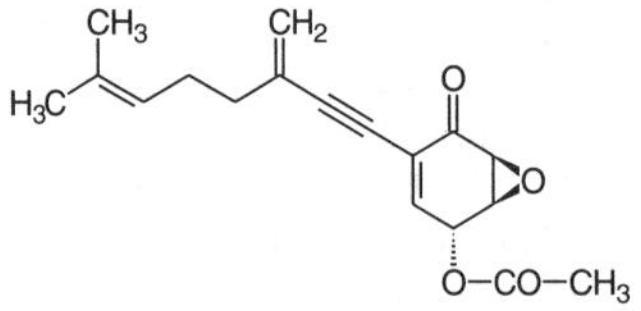

Tricholomenyn A

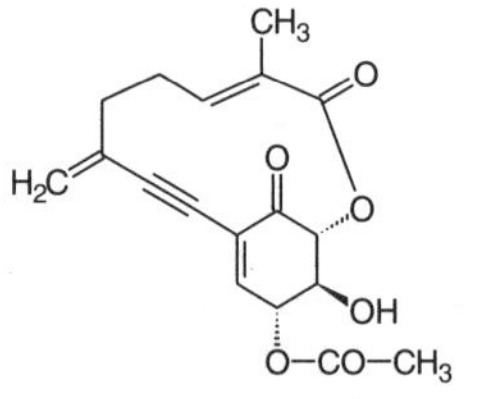

Tricholomenyn B

Fresh fruit bodies of the toadstool *Tricholoma acerbum* (Agaricales) contain *T. A* ($C_{18}H_{20}O_4$, M_R 300.35, oil) and *T. B* ($C_{18}H_{18}O_6$, M_R 330.34, oil) which prove to be effective inhibitors of mitosis in T lymphocyte cultures. The T. can be considered as geranyl analogues of the *siccayne group.

Lit.: *Synthesis:* J. Chem. Soc., Chem. Commun. **1996**, 1679 f.; J. Chem. Soc., Perkin Trans. 1 **1997**, 1087 ▪ J. Org. Chem. **62**, 1582 (1997) ▪ Tetrahedron Lett. **36**, 5633 (1995) (isolation); **37**, 6223 (1996) (synthesis).

Tricholomic acid [(*S,S*)-α-amino-3-oxo-5-isoxazolidinylacetic acid].

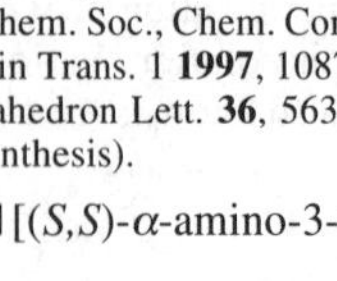

$C_5H_8N_2O_4$, M_R 160.13, mp. 207 °C (decomp.), $[\alpha]_D^{20}$ +80.0° (H_2O), pK_a ca. 2, 6.0, 8.6. A non-proteinogenic, neurotoxic amino acid in fruit bodies of the Japanese toadstool *Tricholoma muscarium* (Basidiomycetes). It is a lethal poison for houseflies (*Musca domestica*) and the dried fungus was used in the past for this purpose; it has a taste similar to that of glutamic acid. An unsaturated analogue is ibotenic acid (see fly agaric constituents).

Lit.: Can. J. Chem. **65**, 195–199 (1987). – *[CAS 2644-49-7]*

Trichorovins. Peptide antibiotics complex (peptaibols) from fermentation cultures of *Trichoderma viride*. T. contain up to 20 amino acid moieties. Ca. 30 T. are known to date, e.g. *T. XII*: $C_{58}H_{102}N_{12}O_{13}$, M_R 1175.50, amorphous solid.

Ac–Aib–Asn–Ile–Ile–Aib–Pro–Leu–Leu–Aib–Pro–Isoleucinol
T. XII (Aib = α-Aminoisobutyric acid)

Lit.: Chem. Pharm. Bull. **43**, 910 (1995). – *[CAS 156986-02-6 (T. XII)]*

Trichosanthin. The active principle of the traditional Chinese drug Tian Hua Fen (roots of *Trichosanthes kirilowii*, Cucurbitaceae). The drug has been used in China for over 2000 years to induce abortions and for treatment of menstrual complaints. T. is a basic polypeptide of 234 amino acids, M_R 25682, monoclinic crystals. T. inhibits ribosomal protein biosynthesis. It is being clinically tested in China against AIDS [1].

Lit.: [1] AIDS **4**, 1189 (1990); AIDS Res. Hum. Retroviruses **10**, 413 (1994).

gen.: Life Sci. **55**, 253 (1994) ▪ Merck-Index (12.), No. 9780 ▪ Proc. Natl. Acad. Sci. USA **86**, 2844 (1989) ▪ Proteins **19**, 4 (1994) ▪ Sax (8.), TJF 350 ▪ Tang u. Eisenbrand, Chinese Drugs of Plant Origin, p. 983–988, Berlin: Springer 1992. – *[CAS 60318-52-7]*

Trichosiderins see trichochromes.

Trichostatic acid see trichostatins.

Trichostatins.

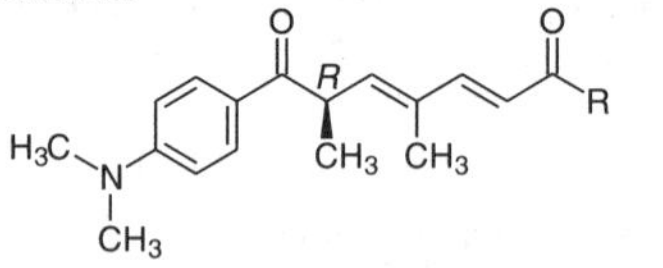

R = NHOH : T. A
R = NH–O-β-D-Glucose : T. C
R = OH : Trichostatic acid

Table: Data of Trichostatins.

	molecular formula	M_R	mp. [°C]	CAS
T. A	$C_{17}H_{22}N_2O_3$	302.37	150	58880-19-6
T. B[a]	$C_{51}H_{63}FeN_6O_9$	959.94	192 (decomp.)	58895-00-4
T. C	$C_{23}H_{32}N_2O_8$	464.52	171–173	68676-88-0
Trichostatic acid	$C_{17}H_{21}NO_3$	287.36	138–140	68690-19-7

[a] iron(III)-complex of T. A; composition 3 : 1 (T. A : Fe)

A group of *antibiotics from *Streptomyces hygroscopicus*, *S. sioyaensis*, and *S. platensis* with biological activities against fungi and certain tumors; T. A also has immunosuppressive activity. T. A is being tested against liver fibrosis.

Lit.: Agric. Biol., Chem. **49**, 563.1365 (1985); **52**, 251 (1988) ▪ Merck-Index (12.), No. 9781 ▪ Tetrahedron **39**, 841 (1983); **44**, 6013 (1988) (synthesis). – *Activity:* Cancer Research **52**,

168–172 (1992) ▪ J. Antibiot. **49**, 453–457 (1996) ▪ J. Biol. Chem. **265**, 17174 (1990) ▪ J. Med. Chem. **42**, 4669 (1999) (analogues). – *[HS 2941 90]*

Tric(h)othecenes. A group of more than 100 sesquiterpene *mycotoxins from various ascomycetes and fungi imperfecti growing on grain and grain products (*Gibberella, Calonectria, Fusarium, Myrothecium, Trichothecium, Trichoderma, Stachybotrys, Cephalosporium, Cylindrocarpon, Verticimonosporium* species) with the 12,13-epoxy-9-trichothecene ring system as common structural feature. They are classified into four subgroups. Group A: T. without oxygen functions at C-8 [e. g., trichodermol, calonectrin, scirpenetriol (see scirpenols), *trichoverroids]; – group B: T. with oxygen functions other than oxo groups at C-8 (e. g., *T-2 toxin, *HT-2 toxin, *neosolaniol, *crotocin*); – group C: T. with oxo group at C-8 (e. g., *nivalenol, *fusarenone X*, *deoxynivalenol). The 4th group includes macrocyclic T. containing di- or triester bridges between C-4 and C-15. Distinctions are made between ether-diesters, *roridins, and triesters (*verrucarins) of verrucarol. The *baccharinoids occurring in plants and miotoxins also belong to this group. T. with unsaturated ester side chains at C-4 belong to the *trichoverroids. The hydroxy groups at C-3, C-4, C-7, C-8, or C-15 are often esterified, mostly by acetic acid.

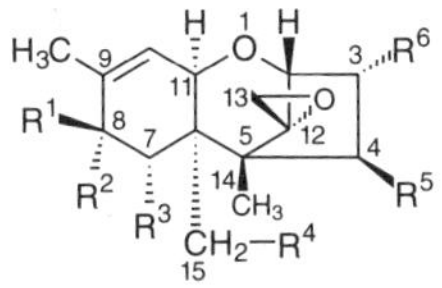

The T. possess numerous biological properties such as phytotoxic, insecticidal, antifungal, antiviral, and cytotoxic activities. All T. are contact poisons (biological weapons!) and are among the most important mycotoxins. T. inhibit protein biosynthesis in mammalian cells, sometimes even at concentrations as low as 1 ng/mL (e. g., vertisporin in HeLa cells). *Verrucarin A inhibits the growth of P-815 mouse tumor cells at 0.6 ng/mL. Examples of poisonings by T. have been observed in the USA in cattle, pigs, and poultry fed with moldy corn ("moldy corn toxicose"), as well as the Akakabi-Byo disease or "bean hull toxicose" in Japan, "alimentary toxic aleukia" in the former USSR, and a *Stachybotrys* poisoning of epidemic extent in 1994 in the southern Urals (Russia). Symptoms of T. poisonings include vomiting, diarrhea, anorexia, ataxia, leukocytosis and subsequent severe leukopenia, inflammations of the gastrointestinal tract, destruction of nerve cells of the central nervous system, hemorrhagia of cardiac muscles, lesions of the lymph nodes, testes, and thymus as well as tissue necrosis. The LD_{50} values (mice i. p.) vary between 0.5 mg/kg (verrucarin) and 23 mg/kg (diacetylscirpenetriol). The biosynthesis of the T. proceeds through *trichodiene.

Lit.: Adv. Environ. Sci. Technol. **23**, 613–638 (1990) ▪ Appl. Environ. Microbiol. **57**, 2101 (1991) ▪ Biochem. Biotech. Biosci. **56**, 523 (1992) ▪ Cole-Cox, p. 152–263 ▪ J. Biol. Chem. **271**, 27353 (1996) (biosynthesis) ▪ Lacey, Trichothecenes and Other Mycotoxins, New York: Wiley 1986 ▪ Merck-Index (12.), No. 9782 ▪ Microbiol. Rev. **57**, 595–604 (1993) (biosynthesis) ▪ Nat. Prod. Rep. **5**, 187–209 (1988) ▪ Tetrahedron **45**, 2277–2305 (1989) ▪ Zechmeister **47**, 153–219; **69**, 1–70. – *Isolation of new T.:* J. Nat. Prod. **54**, 1303 (1991); **55**, 1441

Table: Structures and data of selected Trichothecenes.

Trichothecene	R^1	R^2	R^3	R^4	R^5	R^6	molecular formula	M_R	mp. [°C]	CAS
Trichothecene (3-Scirpene)	H	H	H	H	H	H	$C_{15}H_{22}O_2$	234.34		114882-97-2
Trichodermol	H	H	H	H	OH	H	$C_{15}H_{22}O_3$	250.34	116–119	2198-93-8
Verrucarol	H	H	H	OH	OH	H	$C_{15}H_{22}O_4$	266.34	158–159	2198-92-7
3,4,15-Scirpenetriol	H	H	H	OH	OH	OH	$C_{15}H_{22}O_5$	282.34	189–191	2270-41-9
T-2-Tetraol	H	OH	H	OH	OH	OH	$C_{15}H_{22}O_6$	298.34		34114-99-3
Trichodermin	H	H	H	H	$OCOCH_3$	H	$C_{17}H_{24}O_4$	292.39	45–46	4682-50-2
Monoacetoxyscirpenol	H	H	H	$OCOCH_3$	OH	OH	$C_{17}H_{24}O_6$	324.38	172–173	2623-22-5
Diacetoxyscirpenol	H	H	H	$OCOCH_3$	$OCOCH_3$	OH	$C_{19}H_{26}O_7$	366.41	160–164	2270-40-8
Neosolaniol	H	OH	H	$OCOCH_3$	$OCOCH_3$	OH	$C_{19}H_{26}O_8$	382.41	171–172	36519-25-2
HT-2	H	A	H	$OCOCH_3$	OH	OH	$C_{22}H_{32}O_8$	424.49	oil	26934-87-2
T-2 Toxin	H	A	H	$OCOCH_3$	$OCOCH_3$	OH	$C_{24}H_{34}O_9$	466.51	151–152	21259-20-1
Calonectrin	H	H	H	$OCOCH_3$	OH	$OCOCH_3$	$C_{19}H_{26}O_6$	350.41	83–85	38818-51-8
Crotocin	H	β-O–		H	B	H	$C_{19}H_{24}O_5$	332.40	126–128	21284-11-7
Deoxynivalenol	=O		OH	OH	H	OH	$C_{15}H_{20}O_6$	296.32	151–153	51481-10-8
Nivalenol	=O		OH	OH	OH	OH	$C_{15}H_{20}O_7$	312.32	222–225	23282-20-4
Fusarenon-X	=O		OH	OH	$OCOCH_3$	OH	$C_{17}H_{22}O_8$	354.36	91–92	23255-69-8
Trichothecolone	=O		H	H	OH	H	$C_{15}H_{20}O_4$	264.32	183–184	2199-06-6
Trichothecin	=O		H	H	B	H	$C_{19}H_{24}O_5$	332.38	118	6379-69-7
NT-1	H	$OCOCH_3$	H	OH	$OCOCH_3$	OH	$C_{19}H_{26}O_8$	382.41	173	65180-29-2
NT-2	H	OH	H	OH	$OCOCH_3$	OH	$C_{17}H_{24}O_7$	340.37	172	76348-84-0
Sporotrichiol	H	A	H	OH	H	OH	$C_{20}H_{30}O_6$	366.45		101401-89-2
CBD_2	=O		OH	OH	H	C	$C_{18}H_{24}O_8$	368.38		81662-72-8
Trichoverrol A	see formula						$C_{23}H_{32}O_7$	420.50	178	76739-71-4
Trichoverrol B	see formula						$C_{23}H_{32}O_7$	420.50	oil	76685-83-1
Trichoverritone	see formula						$C_{35}H_{46}O_{11}$	642.74	oil	87292-19-1

A = O–CO–CH_2–$CH(CH_3)_2$
B = O–CO–CH $\overset{Z}{=}$ CH–CH_3
C = O–CO–CH(OH)–CH_3

(1992); **56**, 1890 (1993); **57**, 422 (1994); **59**, 25, 254 (1996) ▪ Tetrahedron Lett. **51**, 9219 (1996). – *Pharmacology:* Bull. Inst. Pasteur (Paris) **88**, 159–192 (1990) ▪ J. Pathol. **154**, 301–311 (1988). – *Synthesis:* J. Am. Chem. Soc. **112**, 2749 (1990) ▪ J. Org. Chem. **54**, 4663 (1989); **55**, 4403 (1990) ▪ J. Prakt. Chem. **334**, 53 (1992) ▪ Tetrahedron **45**, 2385–2415 (1989). – *Occurrence:* Appl. Environ. Microbiol. **59**, 3798 (1993) ▪ Cereal Foods World **35**, 661–666 (1990) ▪ Krogh (ed.), Mycotoxins in Food, p. 123–147, 217–249, London: Academic Press 1987 ▪ Sax (8.), TJE 750.

Trichoverroids. T. are open chain ester derivatives of verrucarol (see scirpenols) and some are biogenetic precursors of the macrocyclic *tric(h)othecenes (*baccharinoids, *roridins, *verrucarins); they are isolated from cultures of *Myrothecium verrucaria*, *M. roridum*, and *Fusarium graminearum*. The T., of which more than 20 are now known, exhibit weak or no cytotoxic properties and weak antifungal activities. Trichoverritone also has antibacterial activity.

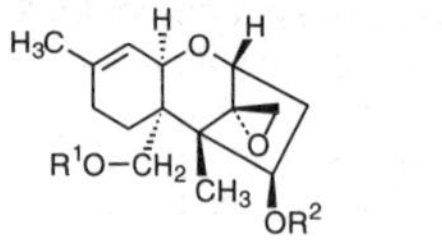

$R^1 = R^2 = H$: *Verrucarol
$R^1 = $ **A**, $R^2 = H$: Verrol (**1**)
$R^1 = $ **A**, $R^2 = $ **B** : Trichoverrin A (**2**)
(7'*R*)-Epimer : Trichoverrin B (**3**)
$R^1 = H$, $R^2 = $ **B** : Trichoverrol A (**4**)
$R^1 = $ **A**, $R^2 = $ **C** : Trichoverritone (**5**)

Acyl substituents:

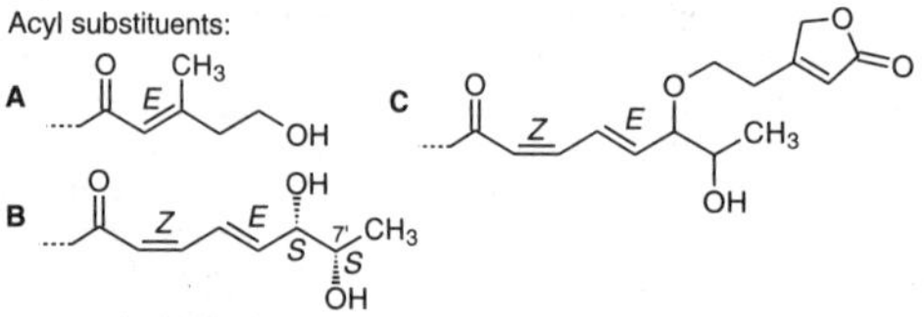

Table: Data of Trichoverroids.

	molecular formula	M_R	mp. [°C]	CAS
1	$C_{21}H_{30}O_6$	378.47	oil	84412-91-9
2	$C_{29}H_{40}O_9$	532.63	78–79	76739-70-3
3	$C_{29}H_{40}O_9$	532.63		76685-82-0
4	$C_{23}H_{32}O_7$	420.50	177–179	76739-71-4
5	$C_{35}H_{46}O_{11}$	642.74		87292-19-1

Lit.: J. Antibiot. **36**, 459 (1983) ▪ J. Nat. Prod. **52**, 663 (1989); **55**, 1441 (1992); **57**, 422 (1994); **59**, 254 (1996) ▪ J. Org. Chem. **47**, 1117 (1982) ▪ Tetrahedron Lett. **24**, 3539 (1983) ▪ Zechmeister **47**, 153–219.

Trichoviridin see isocyanides.

Tricolorin A see Ipomoea resins.

Tricosanedioic acid see Japan wax.

(*Z*)-9-Tricosene see muscalure.

Tricyclene see tricyclic monoterpenes.

Tricyclic monoterpenes.

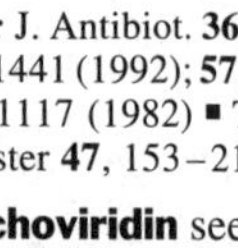

(numbers valid for systematic names)

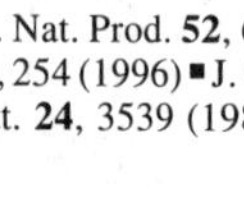

Tricyclene (**1**)
(1,7,7-Trimethyl-tricyclo[2.2.1.$0^{2,6}$]heptane)

R = CH_2OH : Teresantalol (**2**)
R = CHO : Teresantalal (**3**)
R = COOH : Teresantalic acid (**4**)

All natural T. m. have the tricyclo[2.2.1.$0^{2,6}$]heptane structure, the parent C_{10}-skeleton is called tricyclene. Although the ring system is strained it seems to be formed relatively easily. The basic skeletons of the two forms shown are identical but are numbered differently in the systematic nomenclature.

Table: Tricyclic monoterpenes.

	molecular formula	M_R	mp. [°C]	$[\alpha]_D$	CAS
1	$C_{10}H_{16}$	136.24	69–70	–	508-32-7
2	$C_{10}H_{16}O$	152.24	113	+11.9°	29550-55-8
3	$C_{10}H_{14}O$	150.22			59300-39-9
4	$C_{10}H_{14}O_2$	166.22	158	–76.6°	562-66-3

Occurrence: Tricyclene in juniper oil (*Juniperus communis*, Cupressaceae) and in the essential oil of the fir *Abies balsamea* (Pinaceae); ca. 1.3%. Teresantalol, tersantalal, and teresantalic acid are isolated from East Indian sandalwood oil. *sandalwood oil (*Santalum album*, Santalaceae).

Lit.: Beilstein E II **7**, 134 (teresantalal); E III **6**, 393 (teresantalol); E IV **5**, 468 (tricyclene); E IV **9**, 247 (teresantalic acid) ▪ Helv. Chim. Acta **59**, 737 (1976) ▪ J. Org. Chem. **43**, 2282 (1978) (synthesis) ▪ Karrer, No. 320, 998, 5622.

(*S*)-2-Tridecyl acetate see flies.

Tridentatols. Sulfur-containing metabolites from the marine hydroid *Tridentata marginata* which grows on sargasso seaweed. The T. constitute about 10% of the dry weight of *T. marginata*. T. A shows antifeedant activity on fish.

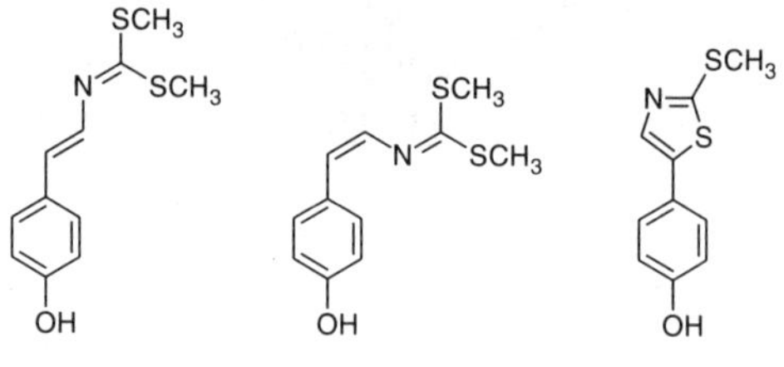

T. A T. B T. C

Table: Data of Tridentatols.

T.	molecular formula	M_R	description	CAS
A	$C_{11}H_{13}NOS_2$	239.36	amorphous solid	185548-48-5
B	$C_{11}H_{13}NOS_2$	239.36	amorphous solid	185548-49-6
C	$C_{10}H_9NOS_2$	223.31	cryst. plates	185548-50-9

Lit.: Org. Lett. **1**, 661 (1999) (synthesis) ▪ Tetrahedron Lett. **37**, 9131 (1996).

Tridentoquinone. $C_{26}H_{34}O_4$, M_R 410.55, ruby-red cryst., mp. 78–80 °C, $[\alpha]_{578}$ +1165° ($CHCl_3$). A macrocarbocyclic pigment from the mushroom (bolete) *Suillus tridentinus* (Basidiomycetes), structurally related to *suillin and *boviquinone-4. T. and tridentorubin ($C_{52}H_{70}O_7$, M_R 807.13, red cryst.) from *S. tridentinus* besides *rhizopogone* {$C_{26}H_{34}O_4$, M_R 410.55, orange-red needles, mp. 110 °C, $[\alpha]_{578}$ +28° ($CHCl_3$)} from the gasteromycete *Rhizopogon pumilionus*, are the only carbo- and ansa-cyclic terpenoids to be found in fungi as yet. T. is formed biosynthetically from *4-hydroxybenzoic acid and geranylgeranyl diphosphate. The formula of T. shows the absolute configuration

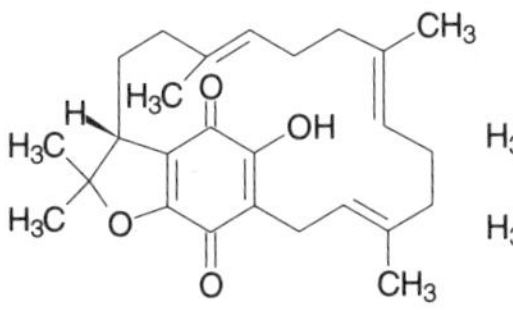

Tridentoquinone

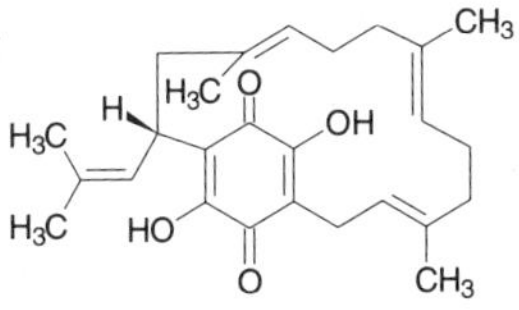

Rhizopogone

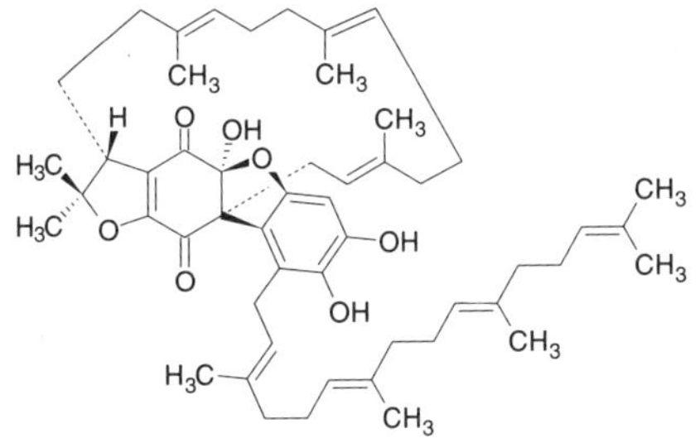

Tridentorubin

confirmed by X-ray crystallography of the camphorsulfonic acid ester.
Lit.: Aust. J. Chem. **48**, 515–529 (1995) (synthetic investigations) ▪ Chem. Ber. **108**, 3675 (1975) ▪ Z. Mykologie **55**, 169 (1979) ▪ Zechmeister **51**, 100 ff. (review). – *[CAS 58262-68-3]*

Triglycerides (triacylglycerols). Collective name for triesters of glycerol with carboxylic acids. T. ("neutral fats") are the main components of *fats and oils. In nature T. occur principally as esters of various fatty acids. T. containing only one *type* of fatty acid are preferred for oleochemical purposes. Plants whose seeds contain such uniform ("one-acid") T. occur in nature and have recently been selectively cultivated, e.g., so-called *high-oleic* species that contain 80–90% oleic acid [1]. Medium-chain T. (MCT) [2] are prepared from the mixtures of medium-chain fatty acids with 6 to 12 C atoms obtained from coconut oil or *palm kernel oil. They differ from the normal fats and oils in their particular physical properties, especially their low mp. of ca. 17–19 °C and their solubility in water. Biologically the MCT differ from fats and oils in that they are absorbed in the stomach as intact molecules. They are then transported through the portal vein directly to the liver where they serve preferentially for energy production. MCT are used in the treatment of disorders of fat absorption.
Lit.: [1] Fat Sci. Technol. **96**, 23 (1994). [2] Applewhite (ed.), Proceedings of the World Conference on Lauric Oils: Sources, Processing, and Applications, Champaign: AOCS Press 1994; Lipid Technol. **1994** (3), 61.
gen.: see fats and oils.

Trigonelline (1-methylpyridinium 3-carboxylate, *N*-methylnicotinic acid betaine).

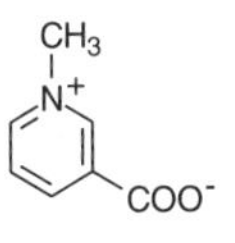

$C_7H_7NO_2$, M_R 137.14; hydrochloride: mp. 258–259 °C, as hydrate 218 °C. A widely distributed alkaloid in plants and animals, e.g., in seeds of fenugreek (*Trigonella foenum-graecum*, Fabaceae) and in coffee beans. T. regulates cell growth in the roots of Fabaceae [1].
Lit.: [1] Phytochemistry **26**, 2891 ff. (1987).
gen.: Beilstein E V **22/2**, 143 ▪ Karrer, No. 2437 ▪ Merck Index (12.), No. 9822 ▪ Planta Med. **53**, 262 (1987). – *[HS 2933 39; CAS 535-81-1]*

Trihydroxybenzenes see benzene-1,2,4-triol and phloroglucinol.

Triiodothyronine see iodoamino acids.

Trilobine see bisbenzylisoquinoline alkaloids.

Trimethylarsine oxide see arsenic in natural products.

15,19,23-Trimethylheptatriacontane see flies.

6,10,13-Trimethyl-1-tetradecanol see bugs.

4,8,12-Trimethyltrideca-1,3,7,11-tetraene see homoterpenes.

Trinactin see nonactin.

Trioses see monosaccharides.

Triptolide.

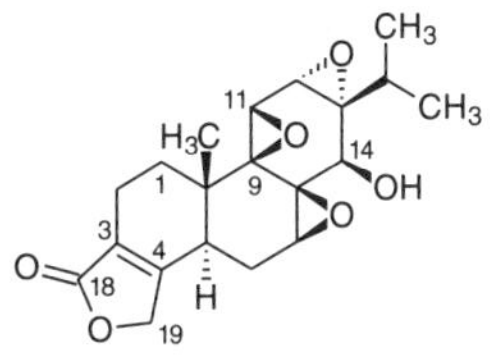

$C_{20}H_{24}O_6$, M_R 360.41, cryst., mp. 226–227 °C, $[\alpha]_D$ −154° (CH_2Cl_2). A diterpene of the 18-(4→3)-*abeo*-*abietane type, isolated from *Tripterygium wilfordii* (Celastraceae). T. exhibits *in vitro* antileukemic activity in mice and dogs as well as in human cells [1]. The activity is based mainly on the nucleophilic ring opening of the 9,11-epoxide. From *T. wilfordii* plants, and plant cell cultures, besides T. and tripdiolide (2-OH), more than 50 other tricyclic diterpenes have been isolated. As long as 2000 years ago, *T. wilfordii* was used in trad. Chinese medicine against rheumatic diseases, skin disorders and hepatitis. T. shows picomolar immunosuppressive and male contraceptive activity. The biosynthesis of *abeo*-abietanes proceeds by cyclization of *labdanes and migration of two methyl groups [2].
Lit.: [1] Science **185**, 791 (1974). [2] Can. J. Chem. **59**, 2677 (1981).
gen.: Synthesis: J. Am. Chem. Soc. **104**, 867, 1785 (1982) ▪ J. Org. Chem. **47**, 2364 (1982); **63**, 6446 (1998) [(±)-T.] ▪ Tetrahedron **38**, 2559 (1982). – *[CAS 38748-32-2]*

Triquinanes. Group of natural products with parent skeletons containing three fused cyclopentane rings. Linear T. from fungi are e.g. hirsutene, *hirsutic acid and *complicatic acid, the *coriolins, hypnophilin and pleurotellol. The *crinipellins consist of four annelated cyclopentane rings, belonging to the *tetraquinanes. *Capnellene is the metabolite of a soft coral. *Modhephene is a [3.3.3]-propellane derivative. Examples of angularly annellated T. are the hydrocarbons silphinene (from roots of *Silphium perfoliatum*), α-isocomene (from *Isocoma wrightii*), *pentalenene and pentalenic acid.
The biosynthesis of T. can proceed directly from farnesyl diphosphate to the tricyclic skeleton, e.g., to pentalenene under the influence of pentalenene synthase [1].

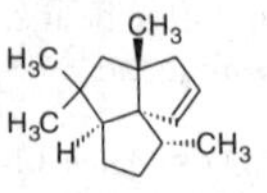

Silphinene (**1**)

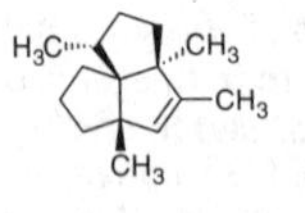

α-Isocomene (**2**)

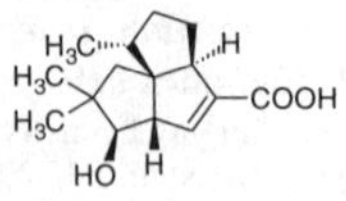

Pentalenic acid (**3**)

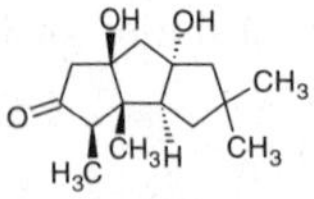

Arthrosporone (**4**)

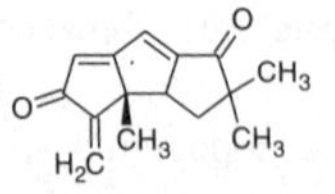

Cucumin A (**5**)

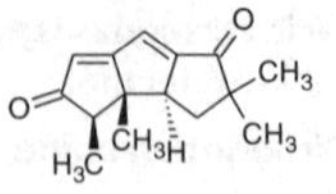

Cucumin B (**6**)

Table: Data of Triquinanes.

no.	molecular formula	M_R	mp.	$[\alpha]_D$ ($CHCl_3$)	CAS
1	$C_{15}H_{24}$	204.36	oil	$[\alpha]_D^{24}$ −21.3°	74284-57-4
2	$C_{15}H_{24}$	204.36	oil	$[\alpha]_D^{24}$ −85°	65372-78-3
3	$C_{15}H_{22}O_3$	250.34	oil		69394-19-0
4	$C_{15}H_{24}O_3$	252.35	140 °C	$[\alpha]_D^{20}$ −141°	124724-98-7
5	$C_{15}H_{16}O_2$	228.29	yellow oil	$[\alpha]_D^{18}$ −101°	202850-71-3
6	$C_{15}H_{18}O_2$	230.30	yellow oil	$[\alpha]_D^{18}$ −136°	202851-13-6

Other cyclisations seem to proceed via *humulene. From the basidiomycete *Macrocystidia cucumis* the linear T. arthrosporone (hirsutane type) and novel cucumane derivatives, the cucumins A–H have been isolated[2], e.g. cucumins A and B.

T. have been the target of several total syntheses.

Lit.: [1]Science **277**, 1811, 1815, 1820 (1997). [2]Eur. J. Org. Chem. **1**, 73 (1998).

gen.: Biosynthesis: Bioorg. Chem. **12**, 312 (1984) ▪ Can. J. Chem. **72**, 118 (1994). – *Synthesis:* Adv. Free Radical Chem. **1**, 121–157 (1990) ▪ Angew. Chem. Int. Ed. Engl. **28**, 447 (1989) ▪ Chem. Commun. **1994**, 1797 ▪ Helv. Chim. Acta **79**, 1026 (1996) (isocomene) ▪ J. Org. Chem. **62**, 4554, 4851 (1997); **63**, 2699, 5302, 6764 (1998); **64**, 6060 (1999) ▪ J. Chem. Soc., Perkin Trans. 1 **1988**, 2963 (pentalenic acid) ▪ Lindberg **2**, 153–158, 415–438 ▪ Nicolaou, p. 221 ▪ Pure Appl. Chem. **68**, 675 (1996) ▪ Synthesis **1998**, 1559–1583 (review) ▪ Stud. Nat. Prod. Chem. **3**, 3–72 (1989) (review) ▪ Tetrahedron **53**, 8975 (1997) (α-isocomene); **53**, 2103 (1997) (pentalenene); **53**, 2111 (1997) (silphinene); **54**, 6539, 14803, 15701 (1998).

Trisaccharides see oligosaccharides.

Triterpenes. A group of natural products currently encompassing over 4000 compounds. T. are formed biosynthetically from 6 isoprene units (*isoprene rule). This proceeds through head-to-head addition of 2 molecules of farnesyl diphosphate to *squalene, the parent precursor of all T. and *steroids. The various polycyclic ring systems of the T. are the result of the different folding possibilities for squalene on the enzyme surface. In eukaryotes the folding of squalene usually proceeds via (3*S*)-squalene 2,3-epoxide. The site of the biosynthesis in plants is the cytoplasm, in animals the hepatic microsomes. In prokaryotes squalene is cyclized directly by protonation. The epoxide oxygen atom furnishes the β-equatorial hydroxy group in the cyclization product. Hydroxy groups with the 3α configuration are rare and arise by subsequent epimerization. Cyclization of squalene epoxide preferentially furnishes 6-membered rings as well as 5-membered rings in most (e.g., *cucurbitacins) and in some pentacyclic (e.g., *lupanes*) T. Since the 6-membered rings can exist in the boat and chair forms and the 5-membered rings can be planar or bent, numerous different skeletons are possible. Currently, ca. 40 different T. skeletons are known. Substitution creates centers of asymmetry (e.g. in *lanosterol 7 and in α-*amyrins 10 asymmetric centers). During the folding of squalene epoxide the double bonds arrange themselves in such a way that a *trans*-addition to a neighboring double bond is always possible. In the formation of *tetracyclic T.* the folding only extends over the region to C-19 of the squalene chain, the rest remains unfolded and so is not cyclized. Several proton and methyl group rearrangements are then required to reach the lanosterol stage. In plants, numerous variations in the folding occur. Most plant tetracyclic T. are derived from *cycloartenol. *Euphol* {$C_{30}H_{50}O$, M_R 426.73, mp. 116 °C} and its (20*S*)-epimer *tirucallol* (mp. 133–134 °C) correspond to lanosterol with the opposite stereochemistry at the C atoms 13 (α), 14 (β), and 17 (α). In the *dammarenes the configuration of the methyl groups at the C atoms 8 and 14 are not changed. Addition of water to the Δ^{20}-double bond leads to dammarenediol, the (20*S*)-epimer of which represents the skeleton of the ginsenosides (see ginseng). Some tetracyclic T. possess 31 C atoms (e.g., 24-methylenecycloartenol) as a result of methyl group transfer from *S*-adenosylmethionine.

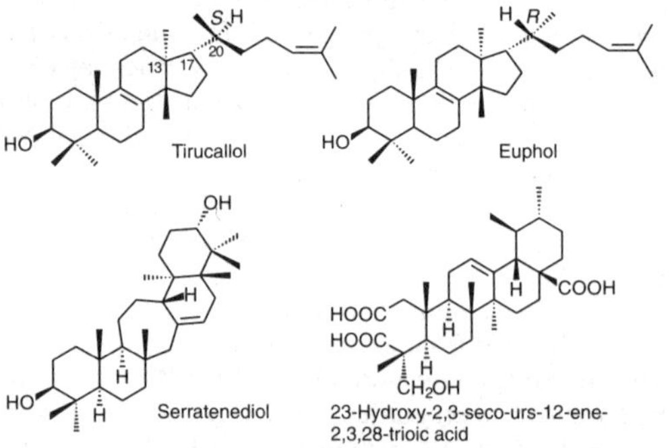

Figure: Some Triterpenes.

The biosynthesis of the tetracyclic *α-*onocerin* can be rationalized by folding of the 2,3;22,23-bisepoxide of squalene, the methyl and methylene branchings correspond to those of squalene without any changes. Tetracyclic T. with four 6-membered rings are rare (e.g., *baccharane* type and shionanes). The *pentacyclic T.* with five 6-membered rings are divided into the *oleanane type most frequently occurring in nature (e.g., β-*amyrins), the *friedelanes (formula, see there), glutanes and the *taraxerane* type with geminal methyl groups at C-20, the *taraxastane* type and the *ursanes* (e.g., α-*amyrins) with methyl groups at C-19 and C-20 as well as the gammaceranes with geminal methyl groups at C-22. Taraxastanes and ursanes differ from each other in the configuration at the C-atoms 18, 19, and 20. *Serratenediol*, (3β,21α-form, $C_{30}H_{50}O_2$, M_R 442.73, mp. 302.5–304.5 °C), the biosynthesis of which proceeds via α-onocerine, has a 7-membered ring as central structure. The *hopanes (see also hopanoids), formed by prokaryotes, possess a cyclopen-

tane ring as ring E. Since the hopanoids are generally not formed from squalene 2,3-epoxide, but rather by protonation of squalene, they usually lack the hydroxy group at C-3.
The methyl groups of T. are oxidized by oxidases and peroxidases. Water can be added to the double bonds. In some cases ring opening occurs, most often at ring A. Those T. occurring predominantly in the above-ground parts of plants are given the prefix *"seco"*, e. g., 23-hydroxy-2,3-seco-urs-12-ene-2,3,28-trioic acid (see figure). T., in which one or more methyl groups are removed are indicated by the prefixes *nor*, *dinor*, *trinor*, etc. (example for tetranortriterpenes, see limonin). The nomenclature is defined in the IUPAC Rules Parts F (natural products) and S (steroids).

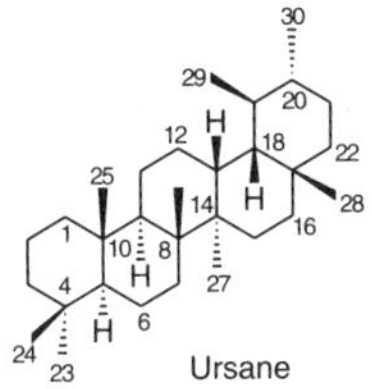

Ursane

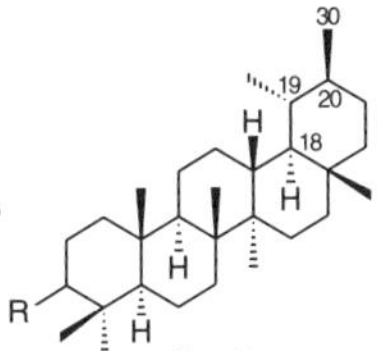

R = H : Taraxastane
R = β-OH, $\Delta^{20(30)}$: Taraxasterol

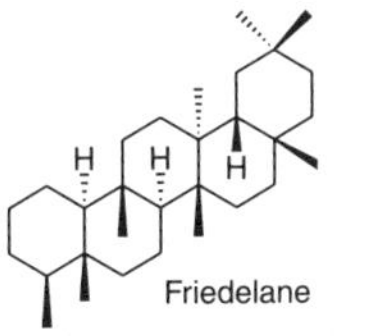
Friedelane

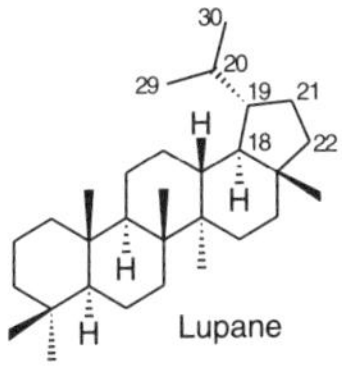

Lupane

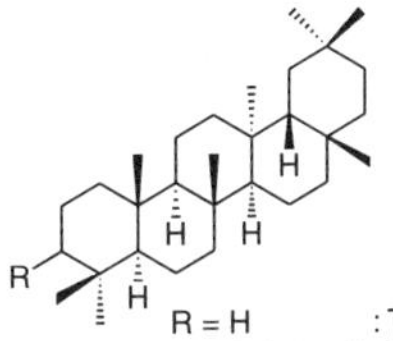
R = H : Taraxerane
R = β-OH, Δ^{14} : Taraxerol

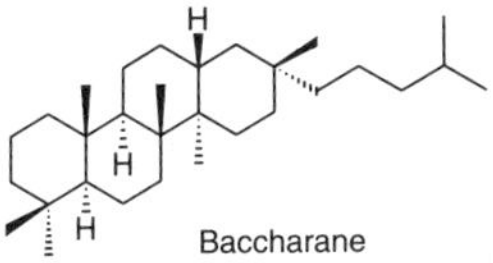
Baccharane

Some T. have major physiological significance. Thus, *lanosterol is converted biosynthetically to *cholesterol, the precursor of all steroid hormones, bile acids, and vitamin D_3. In fungi, lanosterol is converted to ergosterol (see sterols), an essential component of the fungal cell membrane. Plant cell membranes also incorporate steroids (phytosterols). In prokaryotes, the hopanoids take over the functions of steroids in the cell membranes. As a component of animal and plant waxes T. strengthen the structures. They protect the plant surface from desiccation and attack by microorganisms (e. g., *betulin, *lupeol, *oleanolic acid, and *ursolic acid).
The medical uses of the T. are as yet very limited. For example, antitumor activities have been described for *glycyrrhet(in)ic acid and *betulinic acid (3β-hydroxy-lup-20(29)-ene-28-oic acid), various lanostane derivatives exhibit antimicrobial and cholesterol-lowering activities (e. g., lanost-8-ene-3,7-dione), and *dammar resin T. have antiviral properties. Some *cucurbitacins show anti-HIV activity.

Lit.: Nat. Prod. Rep. **16**, 221–240 (1990) ▪ Phytochemistry **22**, 1071–1095 (1983); **31**, 2199–2249 (1992); **44**, 1185–1236 (1996). – *Biosynthesis:* Barton-Nakanishi **2**, 1–400 ▪ Nat. Prod. Rep. **2**, 525–560 (1985); **5**, 387–415 (1988); **7**, 459–484 (1990); **14**, 661–679 (1997) ▪ Zechmeister **29**, 307–362; **45**, 1–102; **67**, 1–123. – *[CAS 514-47-6 (euphol); 514-46-5 (tirucallol); 2239-24-9 (serratenediol)]*

Triterpene saponins. T. s. are composed of a tetra- or pentacyclic *triterpene aglycone (*sapogenins) and one (monodesmosides), two (bisdesmosides), or, in rare cases three, glycosidically bound sugar chain(s). They mostly contain 2 to 5 sugar molecules, the oligosaccharide residues are linear or branched, and are frequently bonded to the hydroxy group at C-3. The sugar residues may also by acylated with carboxylic acids (ester saponins). With the exception of a few C_{29}- and C_{31}-compounds, the aglycone is made up of 30 C atoms. T. s. occur in numerous plant families, especially in dicotyledons. In the animal kingdom their occurrence is apparently limited to sea cucumbers and other holothurians (*tunicates). More than half of the as yet known >1500 substances belong to the *oleanane type. The lanostane and dammarane types (see lanosterol and dammarenes), to which the ginsenosides belong, are also important.
Many herbal drugs contain complex mixtures of T. s. that differ only slightly in their sugar residues. In spite of the enormous advances in chromatographic separation methods it is still difficult to isolate pure T. s. on account of the often only slight differences in their structures, thus standardized plant extracts are mostly used for pharmacological purposes. Examples of the structures can be found under *escin, *glycyrrhizin, *α-hederin, *medicagenic acid.
T. show a broad spectrum of activities. Saponin-containing drugs have played a major role in traditional medicine for over 2000 years (China). T. s. are used mainly for their expectorant and secretolytic activities (e. g., *Primula*, etc.). They are characterized by hemolytic, mucus-irritating, and surfactant activities. Some T. s. also exhibit spasmolytic, sedative, hepatoprotective, anticoagulative, spermicidal, antibacterial, and molluscicidal activities. Antiviral and antifungal properties of *escin and *α-hederin have been reported. Some T. s. have antitumor (*in vivo*) and cytotoxic (*in vitro*) properties such as, e. g. ginsenosides, *glycyrrhizin, *escin, saikosaponins, and tubeimoside I. Immunostimulating effects have been observed for, e. g., ginsenosides and *quillaja saponin. Cholesterol-lowering effects have been demonstrated in animal experiments (e. g., alfalfa saponins, see medicagenic acid). This activity is based on the ability of T. s. to form insoluble complexes with *cholesterol, which are excreted.

Lit.: Phytochemical Dictionary, p. 670–688 ▪ Pharmazie **45**, 313–342 (1990); **49**, 391–400 (1994) ▪ Phytochemistry **27**, 3037–3067 (1988); **30**, 1357–1390 (1991) ▪ Zechmeister **74**, 1–196 (review).

Triticenes.

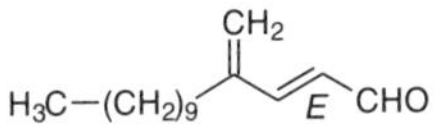

α-Triticene

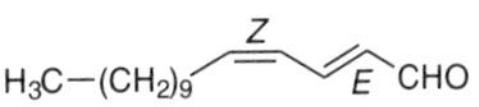

β-Triticene

$C_{15}H_{26}O$, M_R 222.37. *α-T.* [(*E*)-4-decyl-2,4-pentadienal] and *β-T.* [(2*E*,4*Z*)-2,4-pentadecadienal] are isolated from wheat (*Triticum aestivum*)[1]. They are colorless oils with herbicidal and fungicidal[2] activities.

Lit.: [1]Phytochemistry **21**, 2403–2404 (1982). [2]Trans. Br. Mycol. Soc. **82**, 282–288 (1984). – *[CAS 85769-28-4 (α-T.); 85769-29-5 (β-T.)]*

Triticones (T. A–F). Spirocyclic phytotoxins from the deuteromycetes *Drechslera tritici-repentis* and *Staphylotrichum coccosporum*. *T. A:* $C_{14}H_{15}NO_5$, M_R 277.28, cryst., mp. 133–138 °C. T. B is the 6-epimer of T. A. *T. E:* $C_{14}H_{19}NO_6$, M_R 297.31, oil, $[\alpha]_D^{25}$ –70° (CH_3OH).

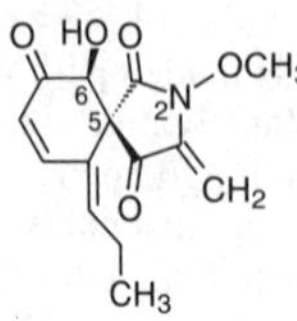

(relative configuration)

T. A

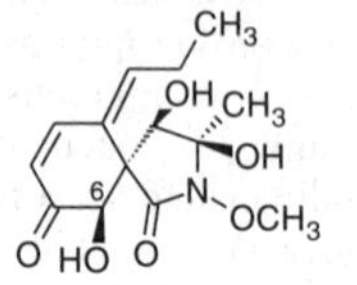

T. E (6-epimer = T. F)

Lit.: Helv. Chim. Acta **72**, 774, 784 (1989) (biosynthesis) ▪ J. Am. Chem. Soc. **110**, 4086 (1988) (isolation) ▪ J. Nat. Prod. **56**, 747 (1993) ▪ Tetrahedron **51**, 4947 (1995) ▪ Zechmeister **67**, 159–165. – *[CAS 114582-74-0 (T. A); 114613-32-0 (T. B); 149564-25-0 (T. E); 149654-98-8 (T. F)]*

Trogodermal [(*R*)-14-methyl-8-hexadecenal].

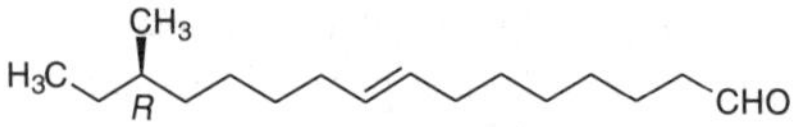

(*E*)-Trogodermal

$C_{17}H_{32}O$, M_R 252.4, liquid. Sexual pheromone of female khapra beetles (*Trogoderma granarium*) and other *Trogoderma* species belonging to the Dermestidae. Both the (*Z*)-form {bp. 127–128 °C (84.5 Pa), $[\alpha]_D^{21}$ –5.94° ($CHCl_3$)} and the (*E*)-form {bp. 127–128 °C (108 Pa), $[\alpha]_D^{22}$ –4.72° (diethyl ether)} are used by the various species.

Lit.: ApSimon **9**, 152–157 (1992) (synthesis) ▪ J. Chem. Ecol. **6**, 911–917 (1980) ▪ Justus Liebigs Ann. Chem. **1993**, 445 (synthesis). – *[CAS 70224-30-5 (Z-isomer); 70144-70-6 (E-isomer)]*

Troleandomycin see oleandomycin.

Tropane alkaloids. Collective name for alkaloids with the tropane ring system.

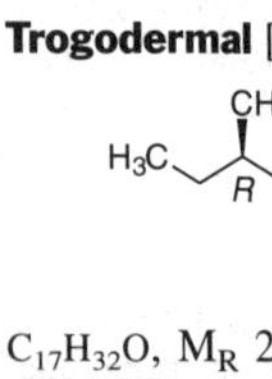

$R^1 = R^2 = H$: Tropane
$R^1 = OH$, $R^2 = H$: 3α-Tropanol (Tropine)
$R^1 = H$, $R^2 = OH$: 3β-Tropanol (Pseudotropine)

Structure & occurrence: The T. a. occur principally in plants of the Solanaceae, Convolvulaceae, Erythroxylaceae, Proteaceae, and Rhizophoraceae families as well as in isolated species of the Euphorbiaceae and Brassicaceae. The most important of the ca. 140 known T. a. are either esters of 3α-tropanol (tropine) or, less commonly, of 3β-tropanol (pseudotropine). Prominent examples are the T. a. of the Solanaceae: *hyoscyamine [racemate (±)-hyoscyamine = *atropine], *scopolamine, and the T. a. of the Erythroxylaceae (*Coca alkaloids*) including the addictive *cocaine. The T. a. are often accompanied by simple, non-esterified *pyrrolidine alkaloids*, such as *hygrine and *cuscohygrine. Most recently discovered were the calystegines as polyhydroxylated derivatives of 3β-tropanol with pronounced glycosidase inhibitory activities.

Biosynthesis & physiology: The T. a. of the Solanaceae are formed in the roots and transported in the phloem to the aboveground parts of the plants for storage (e. g., in the genera *Atropa, Hyoscyamus, Datura*). The biosynthesis of the T. a. starts from ornithine and arginine. *N*-Methylpyrrolidine, formed from *N*-methylputrescine, is an important branching point: linkage with 2 acetate units furnishes cocaine via methylecgonine or decarboxylation leads to 3α- and 3β-tropanol via tropinone. The ester composed of 3α-tropanol and tropic acid derived, in turn, from the amino acid phenylalanine, is hyoscyamine, formed by intramolecular C–C-bond shift from littorine, the 3α-tropanyl ester of (*R*)-3-phenyllactic acid[1]. Hygrine is formed as a byproduct from *N*-methylpyrroline and can be transformed to cuscohygrine by coupling with a further molecule of *N*-methylpyrroline. The *calystegines are probably derived from 3β-tropanol[2].

The medically valued *scopolamine is formed enzymatically from (–)-hyoscyamine by the action of hyoscyamine 6β-hydroxylase, which closes the epoxide ring in a two-step process via 6-hydroxyhyoscyamine. Plants of *Atropa belladonna* have been successfully transformed with cDNA coding for hydroxylase. The resulting transgenic *A. belladonna* plants produce scopolamine in increased concentrations. This was the first concrete example for a successful alteration of a medicinal plant by genetic engineering[3]. T. a. can also be formed in cell cultures[4]. For chemistry, pharmacology, and toxicology of the T. a., see atropine and scopolamine.

Lit.: [1]J. Chem. Soc., Perkin Trans. 1 **1994**, 615–619. [2]Plant Cell, Tissue Organ. Culture **38**, 235–240 (1994). [3]J. Biol. Chem. **266**, 4648–4653, 9460–9464 (1991); Proc. Natl. Acad. Sci. USA **89**, 11799–11802 (1992). [4]Atta-ur-Rahman **17**, 395–420.

gen.: Manske **16**, 84–180; **33**, 1–81; **44**, 1–114, 115–187 ▪ Nat. Prod. Rep. **14**, 637–651 (1997) ▪ Rodd's Chem. Carbon Compds. (2.) **4 B**, 251–276 (1997) ▪ Ullmann (5.) **A 1**, 360f. – *Biosynthesis:* J. Am. Chem. Soc. **117**, 8100 (1995) ▪ J. Chem. Soc., Chem. Commun. **1998**, 1045 ▪ Phytochemistry **41**, 767 (1996) ▪ Planta Med. **36**, 97–112 (1979); **56**, 339–352 (1990). – *Synthesis:* Nachr. Chem. Tech. Lab. **26**, 369f. (1978). – *Tropanol:* Hager (5.) **3**, 112; **4**, 425–433, 923, 1140; **5**, 89, 461, 465 ▪ Manske **33**, 1 ▪ Merck Index (12.), No. 9912. – *Tropinone:* Angew. Chem., Int. Ed. Engl. **39**, 52 (2000) ▪ J. Org. Chem. **62**, 7888 (1997). – *[HS 2939 90]*

Tropic acid (α-hydroxymethylphenylacetic acid).

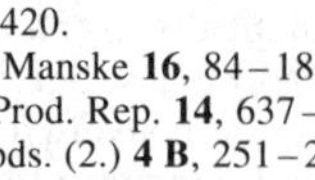

(S)-Tropic acid

Atropic acid

$C_9H_{10}O_3$, M_R 166.18, mp. 129–130 °C, $[\alpha]_D^{15}$ –72.5° (C_2H_5OH). Constituent of *atropine, *scopolamine and other Solanum alkaloids. On heating or enzymatic conversion, *atropic acid* forms by dehydration: $C_9H_8O_2$, M_R 148.16, mp. 106–107 °C, bp. 267 °C (decomp.).

Lit.: *Synthesis:* J. Chem. Soc., Perkin Trans. 1, **1996**, 2895 ▪ Tetrahedron Lett. **1971**, 2283. – *Biosynthesis:* Chem. Commun. **1995**, 127, 129 ▪ J. Chem. Soc., Perkin Trans. 1 **1994**, 1159 ▪ Phytochemistry **35**, 935 (1994).
gen.: Merck Index (12.), No. 906, No. 9910. – *[CAS 552-63-6; 529-64-6 (rac. T.); 16202-15-6 (S-T.)]*

Truxillic and truxinic acids.

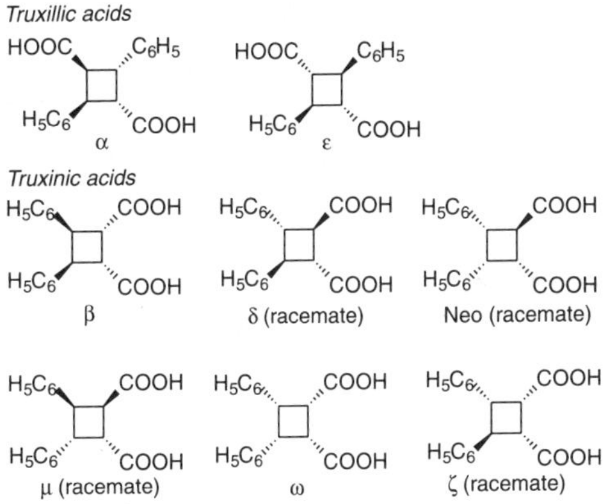

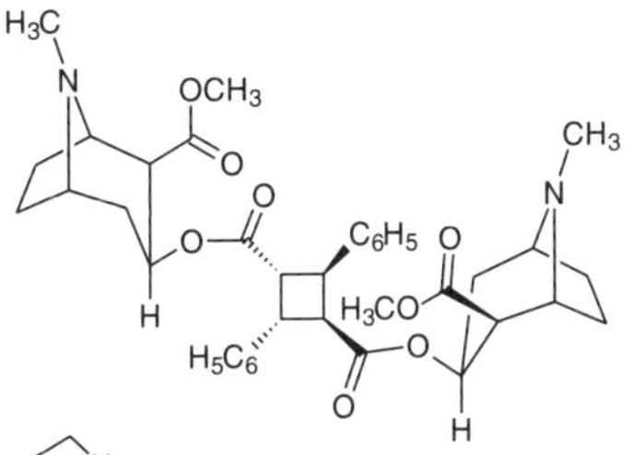

α-Truxilline

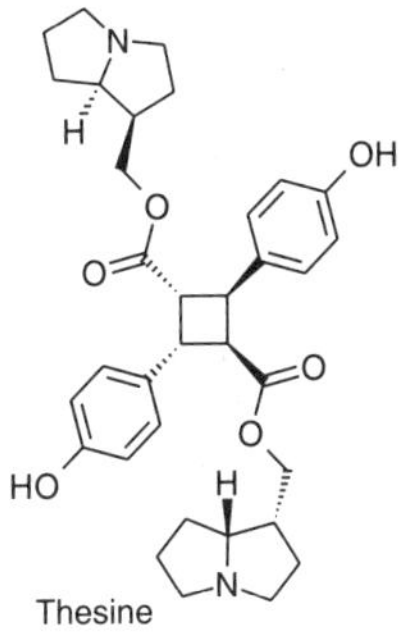

Thesine

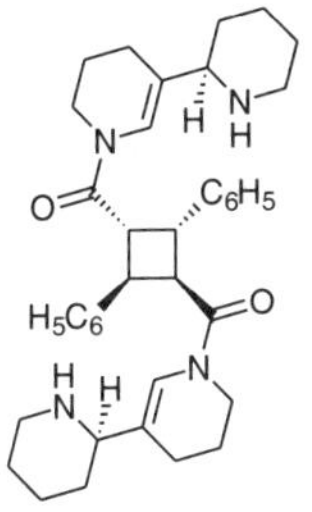

(2*R*,2'*R*)-Santiaguine

Table: Data of Truxillic and Truxinic acids.

	mp. [°C]	CAS
Truxillic acids		
α-T. (γ-isatropic acid, cocaic acid)	274–275	490-20-0
ε-T. (β-cocaic acid)	192	528-38-1
Truxinic acids		
β-T. (isococaic acid)	209–210	528-34-7
δ-T.	209–210	528-33-6
Neo-T.	209	490-16-4
μ-T.		528-35-8
ω-T.		528-36-9
ξ-T.		528-32-5

$C_{18}H_{16}O_4$, M_R 296.32. These acids are formed as photodimerization products of *cinnamic acid. The α- and ε-truxillic acids as well as the β-, δ-, and neo-truxinic acids are isolated from minor alkaloid fractions of the *coca plant (*Erythroxylum coca*). They exist in the plant esterified with the alkaloid *ecgonine methyl ester as truxillines, e.g., *α-truxilline*, $C_{38}H_{46}N_2O_8$, M_R 658.79, mp. 274 °C (hydrochloride). Other alkaloid derivatives of α-truxillic acid are *thesine*, $C_{34}H_{42}N_2O_6$, M_R 574.72, mp. 254–256 °C, the diester of *p,p'*-dihydroxy-α-truxillic acid with the *pyrrolizidine alkaloid (+)-isoretronecanol [see necine(s)] from *Thesium minkwitzianum* and *santiaguine*, $C_{38}H_{48}N_4O_2$, M_R 592.82, the diamide with Δ^2-tetrahydro-*anabasine from Fabaceae (including *Adenocarpus grandiflorus*) occurring as four stereoisomers. Esters of the α-truxillic acid type from hydroxycinnamic acids [*(*E*)-*p*-c(o)umaric acid and ferulic acid, see caffeic acid] with arabinoxylans and other oligosaccharides are present in the cell walls of sugar beet, corn, and barley [1], where they increase the stability of the cell walls [2].

Lit.: [1]Carbohydr. Res. **137**, 139–150 (1985); **148**, 71–85 (1986). [2]Phytochemistry **29**, 3705–3709 (1990).
gen.: J. Am. Chem. Soc. **103**, 3859–3863 (1981) (photochemistry) ▪ Karrer, No. 947 ▪ Merck-Index (12.), No. 9922 ▪ Nat. Prod. Rep. **12**, 101–133 (1995) ▪ Phytochemistry **27**, 349 ff. (1988); **29**, 3699–3703 (1990); **36**, 529 (1994). – *[CAS 4462-95-7 (truxillic acid); 4482-52-4 (truxinic acid); 528-37-0 (thesine); 528-31-4 (santiaguine)]*

Trypethelone.

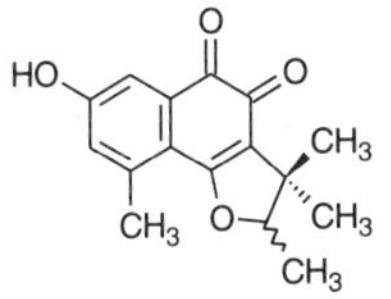

$C_{16}H_{16}O_4$, M_R 272.30, black-violet prisms, mp. 264–266 °C, $[\alpha]_D^{25}$ +365° (CH_3OH). An *o*-*naphthoquinone from the mycosymbiont of the lichen *Trypethelium eluteriae*.

Lit.: Can. J. Chem. **67**, 2089 (1989) ▪ J. Org. Chem. **49**, 1853 (1984) ▪ Justus Liebigs Ann. Chem. **1980**, 779–785. – *[CAS 74513-59-0]*

Tryprostatins.

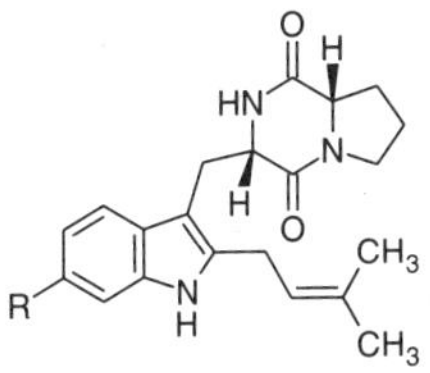

R = OCH_3 : T. A
R = H : T. B

Fermentation products from cultures of *Aspergillus fumigatus* with cytostatic activity (inhibition of the cell cycle); *T. A*: $C_{22}H_{27}N_3O_3$, M_R 381.47, yellow cryst., mp. 120–123 °C, $[\alpha]_D^{27}$ –70° ($CHCl_3$); *T. B*: $C_{21}H_{25}N_3O_2$, M_R 351.45, yellow cryst., mp. 102–105 °C, $[\alpha]_D^{27}$ –71° ($CHCl_3$); see also spirotryprostatins.

Lit.: J. Am. Chem. Soc. **121**, 2147, 11964–11975 (1999) ▪ J. Antibiotics **48**, 1382 (1995); **49**, 527, 534 (1996) (isolation) ▪ J. Org. Chem. **62**, 9298 (1997) (synthesis T. A) ▪ Tetrahedron Lett. **38**, 1301 (1997); **39**, 7009 (1998) (synthesis). – *[CAS 171864-80-5 (T. A); 179936-52-8 (T. B)]*

Tryptamine see indolylalkylamines.

L-Tryptophan [(*S*)-2-amino-3-(3-indolyl)propanoic acid, β-(3-indolyl)-L-alanine, abbreviation: Trp or W].

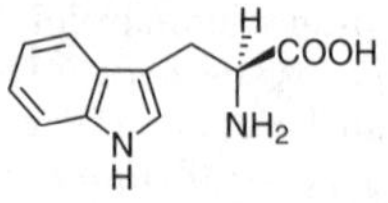

$C_{11}H_{12}N_2O_2$, M_R 204.23, mp. 289 °C (decomp.); monohydrochloride, mp. 257 °C (decomp.); $[\alpha]_D^{23}$ −31.5° (H_2O), $[\alpha]_D^{20}$ +2.4° (0.5 N HCl), $[\alpha]_D^{20}$ +0.15° (0.5 m HCl); pK_a 2.38, 9.39, pI 5.89. An essential proteinogenic amino acid. Genetic code: UGG. Of all proteinogenic amino acids, Trp has the lowest average content in proteins (1.1%)[1]. L-Trp is tasteless, D-Trp is sweet. In order to isolate Trp from proteins they must by hydrolysed with proteinases because Trp is labile in hot, strong acids.

Biosynthesis: Like other aromatic amino acids, e. g., Phe and Tyr, Trp is formed on the *shikimic acid pathway. There is a branching point at *chorismic acid: one branch leads to Phe and Tyr, the other to Trp: chorismic acid → anthranilic acid (anthranilic acid synthase, EC 4.1.3.27) → *N*-(5′-*O*-phosphoribosyl)-anthranilic acid (anthranilic acid phosphoribosyl transferase, EC 2.4.2.18) → 1-*o*-carboxyphenylamino-1-deoxyribulose 5-phosphate [*N*-(5′-phosphoribosyl)anthranilic acid isomerase] → indole-3-glycerol phosphate (indole-3-glycerol phosphate synthase, EC 4.1.1.48) → indole (tryptophan synthase, EC 4.2.1.20)+serine → Trp. Many biologically active indole compounds are derived from Trp, e. g., 5-hydroxytryptophan, 5-hydroxytryptamine (*serotonin), and *melatonin as well as many *indole alkaloids.

Use: In diagnostics for vitamin B_6-deficiency conditions, in infusion solutions for parenteral nutrition, as additive to protein hydrolysates, and in microbiology as well as an additive in animal fodder.

Lit.: [1]Biochem. Biophys. Res. Commun. **78**, 1018–1024 (1977).

gen.: Beilstein E V **22/14**, 14–29 ▪ Hager (5.) **9**, 1112–1118 ▪ J. Heterocycl. Chem. **29**, 953, 1181 (1992) (synthesis) ▪ Karrer, No. 2408 ▪ Merck-Index (12.), No. 9929 ▪ Schwarzc et al., Progress in Tryptophan Research, New York: Plenum Press 1991 ▪ Stryer 1995, p. 725 f., 964–967 ▪ Ullmann (5.) **A 2**, 63, 73, 86. – *[HS 2933 90; CAS 73-22-3]*

Tsugalactone, tsugaresinol see conidendrin.

Tsukubaenolide see FK-506.

T-2-Tetraol see tric(h)othecenes.

T-2 Toxin [8-*O*-(3-methylbutyryl)-neosolaniol, fusariotoxin T-2].

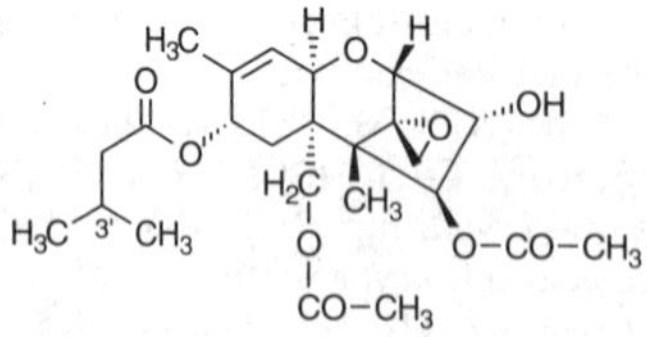

$C_{24}H_{34}O_9$, M_R 466.53, mp. 151–152 °C, needles, $[\alpha]_D^{21}$ −15.5° ($CHCl_3$), a *mycotoxin of the *tric(h)othecene type formed by *Gibberella zeae*, various *Fusarium* species, and *Trichoderma lignorum*. T. exhibits the typical properties of a trichothecene compound, such as emetic, skin irritating, cytotoxic activities and skin necrosis. LD_{50} (mouse i. p.) 3 mg/kg. The previously suggested cancerogenic activity was not confirmed. As a result of its inhibition of protein synthesis, T. also exhibits a useful antiviral effect, e. g., against *Herpes simplex*. T. is metabolized in the liver or by soil bacteria via *HT-2 toxin and *neosolaniol to T-2-tetraol (see tric(h)othecenes). The latter has a 20-fold lower toxicity than T-2 toxin. T. occurs in moldy barley, corn, wheat, oats, and rye.

Lit.: Appl. Environ. Microbiol. **55**, 190 (1989) (metabolism) ▪ Biosci. Biotech. Biochem. **56**, 523 (1992) (antiviral activity) ▪ Chelkowski (ed.), Fusarium – Mycotoxins, Taxonomy and Pathogenicity, p. 441–472, Amsterdam: Elsevier 1989 (occurrence) ▪ Cole-Cox, p. 185–188 ▪ J. Nat. Prod. **54**, 1303 (1991) (neosolaniol) ▪ J. Org. Chem. **55**, 1237 (1990) (biosynthesis) ▪ J. Prakt. Chem. **334**, 53–59 (1992) ▪ Merck-Index (12.), No. 9933 ▪ Sax (8.), No. FQS 000 ▪ Trends Biochem. Sci. **14**, 204 (1989). – *[CAS 21259-20-1]*

Tuberactinomycin B see viomycin.

Tubercidin (7-*β*-D-ribofuranosyl-7*H*-pyrrolo[2,3-*d*]-pyrimidin-4-amine).

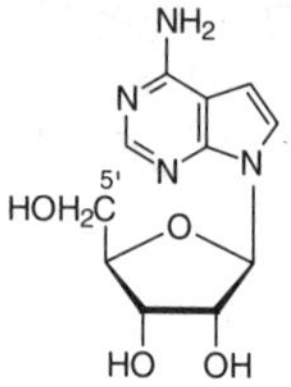

$C_{11}H_{14}N_4O_4$, M_R 266.26, cryst., mp. 247 °C (decomp.), cytotoxic *nucleoside antibiotic from *Streptomyces tubercidicus*[1], LD_{50} (rat p. o.) 16 mg/kg, with antitumor, antifungal, and antiviral activities. T. was also isolated as one of the main cytotoxins from *Tolypothrix byssoidea*[2]. Known derivatives of T. are the *5′-O-sulfamoyl* compound[3] ($C_{11}H_{15}N_5O_6S$, M_R 345.34, herbicide) isolated from *Streptomyces mirabilis* as well as the cytotoxic and antifungal *5′-O-α-D-glucopyranosyl* compound[4] {$C_{17}H_{24}N_4O_9$, M_R 428.40, $[\alpha]_D^{20}$ +10° (H_2O)} isolated from *Plectonema radiosum* and *Tolypothrix distorta*.

Lit.: [1]J. Antibiot. **42**, 1711 (1988). [2]Phytochemistry **22**, 2851 (1983). [3]Chem. Abstr. **109**, 3509 (1988). [4]J. Antibiot. **41**, 1048 (1988).

gen.: Tetrahedron Lett. **28**, 5107 (1987) (synthesis). – *[HS 2941 90; CAS 69-33-0 (T.); 117446-65-5 (sulfamoyl-T.); 117456-78-7 (glucopyranosyl-T.)]*

Tuberin. Name for 4 different compounds:

1. Globulin from potatoes. Another protein from potatoes is called tuberinin.

2. *N*-[2-(4-methoxystyryl)]formamide

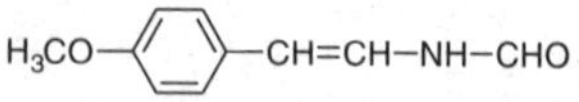

$C_{10}H_{11}NO_2$, M_R 177.20, prisms, mp. 132–133 °C, antibiotic from *Streptomyces amakusaensis*, active against mycobacteria.

3. Anthocyanidine glucoside from potatoes with violet colouring: $C_{22}H_{23}ClO_{12}$, M_R 479.42 with the redbrown aglycone tuberidin chloride ($C_{16}H_{13}ClO_7$, M_R 352.71, amorphous).

4. Protein, acting as a tumor-suppressor. Defect of the corresponding gene causes tuberous sclerosis, a severe human hereditary disease.

Lit. (to 2.): J. Antibiot. **37**, 469 (1984) ▪ J. Med. Chem. **21**, 588 (1978) ▪ Merck Index (12.), No. 9938 ▪ Tetrahedron Lett. **25**,

4263 (1984). – *[CAS 53643-53-1 (2.); 69915-09-9 (3.); 169027-60-5 (4.)]*

Tuberine. Metabolite from *Haplophyllum tuberculatum* (Rutaceae) (aerial parts) with antibiotic activity against *Staphylococcus aureus* and *Escherichia coli*:

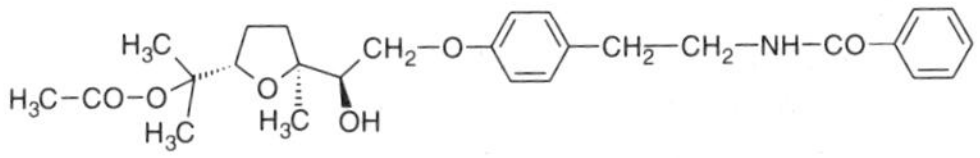

$C_{27}H_{35}NO_6$, M_R 469.58, mp. 150–152 °C, $[\alpha]_D$ +12.5° ($CHCl_3$).
Lit.: Phytochemistry **24**, 884 (1985); **29**, 3055 (1990). – *[CAS 97400-75-4]*

Tuberose absolute. Yellowish-brown to reddish-brown, slightly viscous oil with a heavy, honey-like sweet, anesthetizing, flowery odor that is often found to be pleasant on dilution.
Production: By solvent extraction from flowers of the tuberose (*Polianthes tuberosa*, Agavaceae). Yield ca. 0.1% of T. concrete from which the oil is obtained in ca. 30% yield. The tuberose is cultivated in southern France, Egypt, and India for production of the absolute.
Composition [1]***:*** Main components of the volatile portion include *methyl benzoate* (see ylang-ylang oil) and *benzyl benzoate* (see cinnamon leaf (bark) oil), content ca. 10%. Characteristic for the tuberose odor is the combination of *methyl salicylate*, **methyl anthranilate*, *α-terpineol* (see *p*-menthenols), and *methylisoeugenol* (see isoeugenol) (ca. 5%) with small amounts of *4-decanolide* (see alkanolides) and *δ-*jasmin(e) lactone*.
Use: T. is a very expensive raw material for perfumes and is thus used in very small amounts but even then influences the odor character of many modern flowery perfumes on account of its high odor yield.
Lit.: [1] Müller-Lamparsky, p. 511.
gen.: Arctander, p. 630. – *[HS 3301 29; CAS 8024-05-3]*

Tubocurarine see bisbenzylisoquinoline alkaloids and curare.

Tubulosine see ipecac alkaloids.

Tubulysins. Tetrapeptides from fermentation cultures of *Archangium gephyra* with antimycotic and cytostatic activity. T. A–D inhibit the growth of certain tumor cell lines very potently (e.g., KB 3.1 cervix, PC-3 prostata, SK-OV3 ovarials, K-562 leukemia) with IC_{50} values between 1 and 200 pg/mL (mechanism: induction of apoptosis).

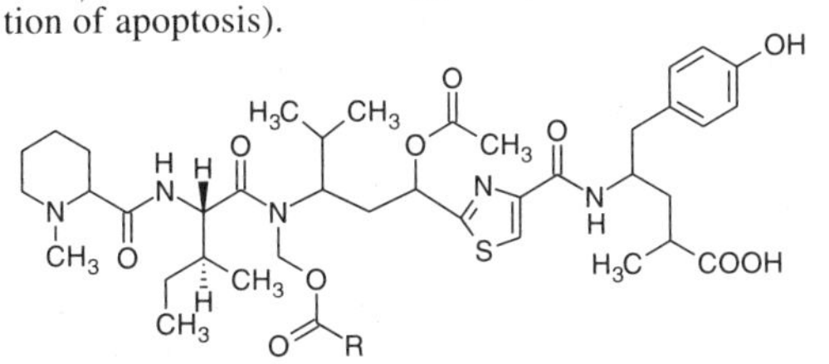

T. A : R = CH_2–$CH(CH_3)_2$
T. B : R = $(CH_2)_2$–CH_3
T. C : R = C_2H_5

T. A: $C_{43}H_{65}N_5O_{10}S$, M_R 844.05; *T. B*: $C_{42}H_{63}N_5O_{10}S$, M_R 830.03; *T. C*: $C_{41}H_{61}N_5O_{10}S$, M_R 816.01, amorphous solids.
Lit.: Ger. Pat. Appl. DE 19638870 (publ. 26.3.1998). – *[CAS 205304-86-5 (T. A); 205304-87-6 (T. B); 205304-88-7 (T. C)]*

Tuckolide.

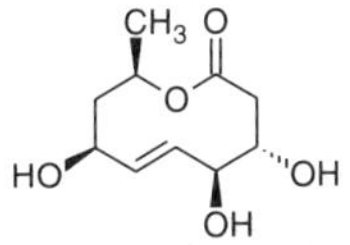

$C_{10}H_{16}O_5$, M_R 216.23, needles, mp. 114–115 °C, $[\alpha]_D^{25}$ –31° ($CHCl_3$). T. (decarestrictin D) inhibits the cholesterol biosynthesis in liver cells (IC_{50} 0.1 µg/mL). T. is isolated from fermentation cultures of *Penicillium corylophilum* and *Polyporus tuberaster*.
Lit.: J. Nat. Prod. **55**, 649 (1992) (isolation) ▪ J. Org. Chem. **61**, 8780 (1996). – *[CAS 127393-89-9]*

Tuftsin (L-Threonyl L-lysyl→L-prolyl→L-arginine). Thr-Lys-Pro-Arg, $C_{21}H_{40}N_8O_6$, M_R 500.60. Tetrapeptide from the γ-globulin fraction of blood and part of the protein leucokinin. T. stimulates the human immune system, especially phagocytosis and the activity of monocytes and granulocytes by different mechanisms, e.g., by induction of *nitrogen monoxide synthesis. T. shows antitumor and antibacterial properties.
Lit.: Biochemistry **31**, 9581 (1992) (structure) ▪ Crit. Rev. Biochem. Mol. Biol. **24**, 1–40 (1989) ▪ Immunopharmacol. **25**, 261 (1993) ▪ Indian J. Chem. B. **25**, 1138 (1986) (synthesis) ▪ Infect. Immun. **62**, 2649 (1994) ▪ Int. J. Pept. Protein Res. **37**, 112–121 (1991) ▪ J. Med. Chem. **29**, 1961 (1986) ▪ Martindale (30.), p. 1424 ▪ Merck Index (12.), No. 9941 ▪ New Methods Drug Res. **2**, 215–231 (1988) ▪ Tetrahedron Lett. **27**, 4841 (1986) (synthesis). – *[CAS 9063-57-4]*

Tulipalin A (tulipalactone, α-methylenebutyrolactone).

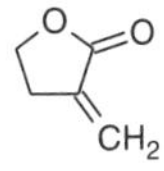

$C_5H_6O_2$, M_R 98.10, liquid, bp. (0.27 kPa) 56–57 °C. Skin-irritating, weakly antibacterially active compound from tulip bulbs (*Tulipa gesneriana*, *Erythronium americanum*).
Lit.: J. Med. Chem. **35**, 2392 (1992) ▪ Tetrahedron Lett. **28**, 2111 (1987) (synthesis). – *[CAS 547-65-9]*

Tumulosic acid.

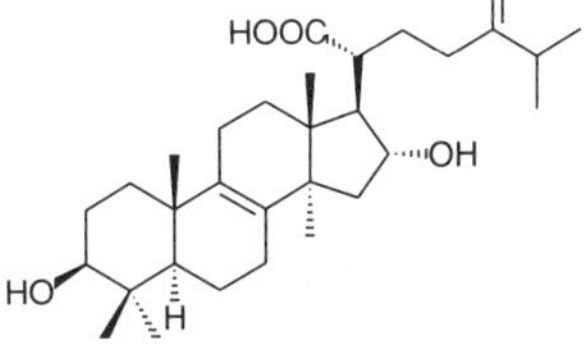

$C_{31}H_{50}O_4$, M_R 486.74, needles, mp. 306 °C (decomp.), $[\alpha]_D$ +8.1° ($CHCl_3$). A C_{31}-triterpene acid from various polypore fungi of the genera *Melanoporia*, *Polyporus*, *Poria*, and *Trametes* (Aphyllophorales). The 3-*O*-acetate (*pachymic acid*, $C_{33}H_{52}O_5$, M_R 528.77) occurs in *Poria cocos* and other wood-degrading fungi.
Lit.: Phytochemistry **31**, 2548 (1992); **40**, 225 (1995); **41**, 1041 (1996) ▪ Turner **1**, 260; **2**, 323–324 (occurrence) – *[CAS 508-24-7 (T.); 29070-92-6 (pachymic acid)]*

Tunicamycins (streptovirudins, corynetoxins). A complex of closely related *nucleoside antibiotics

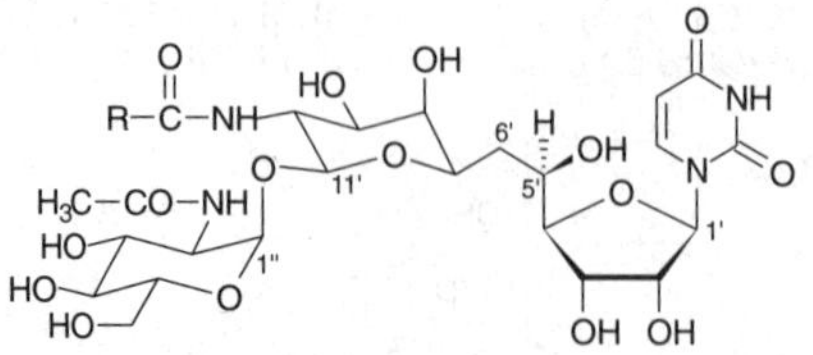

R = CH=CH—$(CH_2)_7$—$CH(CH_3)_2$: T. I (T. A_0)
R = CH=CH—$(CH_2)_8$—$CH(CH_3)_2$: T. II (T. A_1)
R = CH=CH—$(CH_2)_{10}$—CH_3 : T. III (T. A_2)
R = CH=CH—$(CH_2)_{11}$—CH_3 : T. IV (T. B_2)
R = CH=CH—$(CH_2)_9$—$CH(CH_3)_2$: T. V (T. B_1)
R = $(CH_2)_{11}$—$CH(CH_3)_2$: T. VI
R = CH=CH—$(CH_2)_{10}$—$CH(CH_3)_2$: T. VII (T. B, T. C_1)
R = CH=CH—$(CH_2)_{12}$—CH_3 : T. VIII (T. C_2)
R = CH=CH—$(CH_2)_{13}$—CH_3 : T. IX (T. D_1)
R = CH=CH—$(CH_2)_{11}$—$CH(CH_3)_2$: T. X (T. D, T. D_2)

from *Streptomyces* species, M_R between 802.87 {*T. I:* $C_{36}H_{58}N_4O_{16}$, mp. 215–220 °C (decomp.), $[\alpha]_D$ +44° (CH_3OH)} and 858.98 {*T. X:* $C_{40}H_{66}N_4O_{16}$, mp. 239–254 °C (decomp.), $[\alpha]_D$ +37° (CH_3OH)}. The T. I–X, formerly known as T. A–D, are distinguished, in total over 30 components are currently known, many of which are named as *streptovirudins* and *corynetoxins*. The T. all possess the same parent skeleton and differ in the chain length (C_{12-19}), the branching, and/or the presence of a (2*E*)-double bond in the *N*-acyl group. The parent skeleton consists of uracil, an *N*-glycosidically linked pseudodisaccharide (*tunicamine*: undecose formally composed of D-ribose and D-galactosamine), and an *O*-glycosidically (*α,β*-trehalose-type) linked *N*-acetyl-D-glucosamine. T. inhibit glycosyl transferase processes on phospholipid intermediates and influence the action of interferon, they have broad activities against Gram-positive bacteria, yeasts, fungi, viruses, and tumor cells. In prokaryotes T. inhibit cell wall biosynthesis by blockade of the phospho-*N*-acetylmuramyl pentapeptide translocase and inhibition of the incorporation of undecaprenyl phosphate, in eukaryotes they inhibit oligosaccharide synthesis, and in fungi, e. g., 1,3-*β*-glucan synthase. T. are used in investigations of the formation and secretion of immunoglobulins. They cannot be used in therapy on account of their toxicity.

Lit.: Synthesis: J. Am. Chem. Soc. **115**, 2036 (1993); **116**, 4697 (1994). – *Review:* J. Antibiot. **41**, 1711–1739 (1988) ▪ J. Nat. Prod. **46**, 544 (1983) ▪ Merck-Index (12.), No. 9949 ▪ Tamura, Tunicamycin, Tokyo: Jap. Sci. Soc. Press 1982. – *[CAS 73942-10-6 (T. I); 66081-38-7 (T. X)]*

Tunicates. Marine organisms that belong to the subphylum *Urochorda* and represent the simplest members of the chordates (*Chordata*). Characteristic features of the T. are an outer layer or sheath composed of a mixture of polymeric carbohydrates and polypeptides, gill clefts (for respiration and uptake of food by filtration), a lacuna-like blood vessel system, a heart that pumps the colorless blood first in one direction and then in the other, stomach, kidneys, and nervous system; the *chorda dorsalis* is only present in the larvae that undergo a remarkable metamorphosis to the adult form; about 2000 exclusively marine species are known, occurring either as isolated organisms or in groups or colonies of many individuals (zooids) in depths of 1–1500 m.

Class Appendicularia: free-swimming, somewhat fish-shaped tunicates, with chorda and swimming tails throughout their life, about 2–10 mm long, e. g., *Oikopleura dioica* and *Fritillaria borealis*.
Class Ascidiacea (see squirts): about 0.5–20 cm long, sessile and sometimes intensively colored;
Order Aplousobranchia, families:
a) Clavelinidae, e. g., *Clavelina lepadiformis;*
b) *Aplidium conicum*, *Sydnyum turbinatum*;
c) *Trididemnum solidum* (see didemnins), *Lissoclinum patella* (see patellamides), *L. vareau* (see varacin);
d) *Eudistoma olivaceum* (see eudistomins), *Cystodytes dellechiajei* (see shermilamines);
Order Phlebobranchiata, families:
a) Cionidae, e. g., *Ciona intestinalis;*
b) *Ascidia nigra*, *Phallusia mammilata* (see tunichromes);
c) *Ecteinascidia turbinata*, mangrove ascidians (see ecteinascidines);
Order Stolidobranchiata, families:
a) *Styela plicata*, *Dendrodoa grossularia* (see grossularines);
b) *Botryllus schlosseri*, *Distomus variolosus*;
c) *Halocynthia papillosa* (see tunichromes), *Microcosmus sulcatus*.

Lit.: Bone, The Biology of Pelagic Tunicates, Oxford: Oxford Univ. Press 1998 ▪ Med. Res. Rev. **20**, 1–27 (2000).

Tunichlorin [Ni(II)-2-devinyl-2-(hydroxymethyl)pyropheophorbide a].

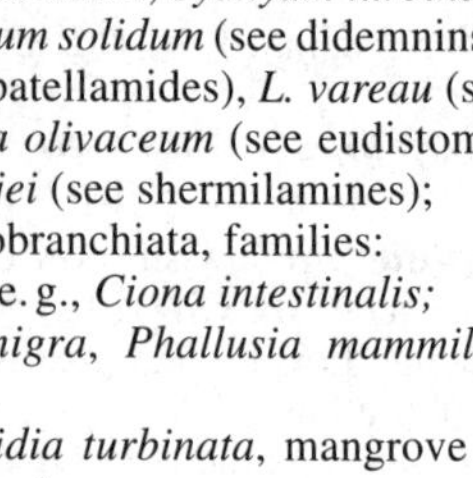

$C_{32}H_{32}N_4NiO_4$, M_R 595.32, uv_{max} (CH_2Cl_2) 389, 416, 641 nm. A blue-green solid, nickel *chlorin from the Caribbean tunicate *Trididemnum solidum* and thus, together with the *tunichromes, a further example of transition metal-containing pigments from *tunicates. Since *T. solidum* lives in symbiosis with a blue alga, T. may also be an algal pigment, however as yet no nickel porphyrinoids are known from algae. As a nickel chlorin T. is an unusual natural product. From *T. solidum* also acylated tunichlorins have been isolated [1].

Lit.: [1] Proc. Natl. Acad. Sci. USA **93**, 10560 (1996).
gen.: Chem. Rev. **94**, 327 (1994) ▪ J. Nat. Prod. **51**, 1 (1988); **56**, 1981 (1993) ▪ Naturwiss. Rundsch. **42**, 333 f. (1989) ▪ Proc. Natl. Acad. Sci. USA **85**, 4582 (1988) ▪ Pure Appl. Chem. **61**, 525 (1989). – *[CAS 114571-91-4]*

Tunichromes. Polyphenolic yellow blood pigments from the *tunicates *Ascidia nigra* and *Phallusia mammilata*. They are powerful biological reducing agents that decompose on warming or on exposure to air.
The T. are hydroxy-DOPA-containing di- and tripeptides that form complexes with oxovanadium(IV) ions (VO^{2+}) (*T. B_1*, yellow solid, $C_{26}H_{25}N_3O_{11}$, M_R 555.50) [1]. Vanadium exists in blood cells (vanadocytes) at pH ~ 7 predominantly in the oxidation states +3 and +4. Tu-

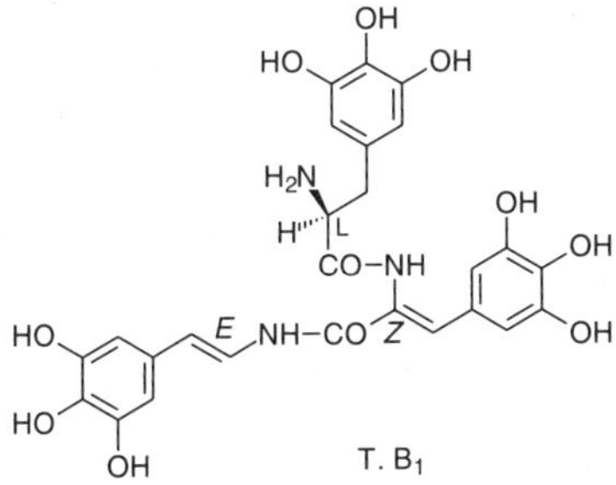

T. B_1

nicates selectively accumulate vanadium from sea water. T. can comprise up to 20% of the dry weight of the tunicate *Ascidia nigra*. They are probably of decisive importance for the construction of the membrane (Latin: tunica) of the tunicates.

The tunicate *Halocynthia roretzii* contains similar compounds, the halocyamines[2]. In addition to T. the nickel chlorin *tunichlorin is also found in tunicates.

Lit.: [1] J. Am. Chem. Soc. **110**, 6162 (1988); Angew. Chem., Int. Ed. Engl. **31**, 52 (1992); Biochim. Biophys. Acta **1033**, 311 (1990). [2] Biochemistry **29**, 159 (1990).

gen.: Acc. Chem. Res. **24**, 117 (1991) ▪ Attaway & Zaborsky, Marine Biotechnology, vol. 1, p. 66–76, New York: Plenum 1993 ▪ Chem. Rev. **97**, 333–346 (1997) (review) ▪ Coord. Chem. Rev. **109**, 61–105 (1991) ▪ J. Nat. Prod. **49**, 193 (1986) ▪ Merck-Index (12.), No. 9950. – *Biosynthesis:* Experientia **48**, 367 (1992) ▪ J. Nat. Prod. **49**, 193 (1986); **54**, 918 (1991). – *Synthesis:* J. Am. Chem. Soc. **111**, 6242 (1989); **112**, 2627 (1990) ▪ Tetrahedron Lett. **31**, 7119 (1990). – *Isolation:* J. Chem. Soc., Chem. Commun. **1991**, 9. – *[CAS 97689-87-7 (T. B_1)]*

Turacin see uroporphyrins.

Turanose (3-*O*-α-D-glucopyranosyl-D-fructose).

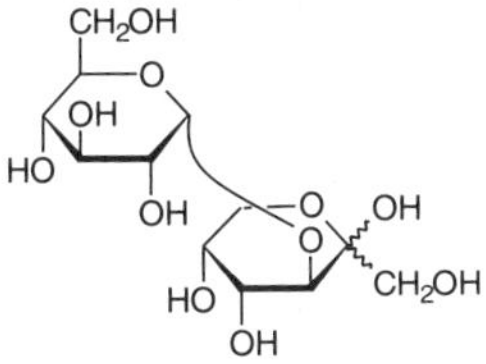

$C_{12}H_{22}O_{11}$, M_R 342.30. Sweet tasting cryst., mp. 157 °C (decomp.), $[\alpha]_D^{20}$ +22 → +75° (H_2O). The disaccharide T. is isomeric with *sucrose and is cleaved by acids and glucosidases to a mixture of its two components D-*glucose and D-*fructose. It can be detected in pollen.

Lit.: Beilstein E V **17/7**, 213 ▪ J. Chem. Soc., Perkin Trans. 2 **1990**, 1489 (mutarotation) ▪ Karrer, No. 655 ▪ Merck-Index (12.), No. 9951. – *[HS 2940 00; CAS 547-25-1]*

Turbotoxins. Diiodotyramine derivatives from the Japanese gastropod *Turbo marmorata*, e.g., T. A and T. B, which were isolated as the trifluoroacetates.

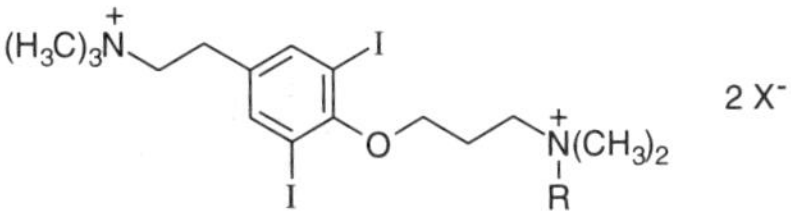

T. A : R = CH_3; T. B : R = H; X = CF_3COO^-

T. exhibit acute toxicity in mice: LD_{100} (T. A): 1 mg/kg; (T. B): 4 mg/kg. T. contribute to shellfish poisonings in Japan.

Lit.: Tetrahedron Lett. **40**, 5745 (1999). – *[CAS 245128-77-2 (T. A dication); 245128-79-4 (T. B dication)]*

Turgorins see leaf-movement factors.

Turmeric see curcuminoids.

Turpentine oil. A clear, readily mobile oil with a unique, penetrating, fresh, warm-balsamy odor (of wet paint) and a bitter-hot taste. With a yearly production of ca. 300 000 t, T. is the quantitatively most important essential oil. The main producing country are the USA with about 50% of the world production, followed by countries of the former USSR, Scandinavia, China, and Portugal/Spain. T. is obtained from the wood of various conifer species, e.g., *Pinus palustris* ("long-leaved pine"), *P. caribea* ("slash pine"), and *P. pinaster*. The former two species are used in the southern states of the USA, mainly Georgia, for the production of T., the latter in Portugal. Depending on the processing three types of oil are distinguished:

1. *Balsam T.:* Produced by steam distillation from the oil-containing resin (balsamic resin, turpentine), secreted from the trunks of pine trees as pathological symptom (wound balsam) when the outer wood layer is injured. The residue is colophony consisting of diterpene resin acids.
2. *Wood or root T.*: By extraction of pine cuttings and roots with petroleum ether or a similar solvent and rectification of the obtained extract. The concomitantly obtained fraction containing mostly monoterpene alcohols is the so-called *pine oil. The residue consists of resin acids (see above).
3. *Sulfate T.* is a by-product of cellulose production. It forms during extraction of the wood according to the sulfate cellulose process; about 10 kg/t cellulose are obtained.

Balsam and wood T. play a quantitatively minor role in comparison to sulfate T.

Composition: All oils consist mainly of α- (70% or more) and β-**pinenes* (up to 20%). Some qualities also contain 10–20% 3-**carene*. Sulfate T. is also a source of **estragole* or, respectively, **anethole* that are present in smaller amounts and can be obtained in the pure form by rectification.

Use: Balsam T. is used in the perfume industry mainly for technical perfuming (e.g., for cleaning agents, etc.). In medicine it is used, e.g., in balneotherapeutics. Sulfate T., or the pinene-rich fractions obtained from it, are used as solvents and basic chemicals for polymers, paint resins, thermoplastic adhesives, etc. A large portion is hydrated to furnish synthetic *pine oil. Rectification yields pure α- and β-pinenes that are of importance as starting materials for the synthesis of a series of fragrance and flavor compounds[1,2].

Lit.: [1] Parfüm. Kosmet. **61**, 457 (1980). [2] Ohloff, p. 98.

gen.: Arctander, p. 633 ▪ Bauer et al. (2.), p. 179 ▪ Chem. Technol. **5**, 235 (1975) ▪ Ullmann (5.) **A11**, 242 f.; **A24**, 469 ff. – *[HS 3805 10; CAS 8006-64-2]*

Tutin see coriamyrtin.

Tylonolide see tylosin.

Tylophora alkaloids. The T. a. belong to the phenanthroindolizidine alkaloids and constitute a small group of compounds isolated solely from genera of the As-

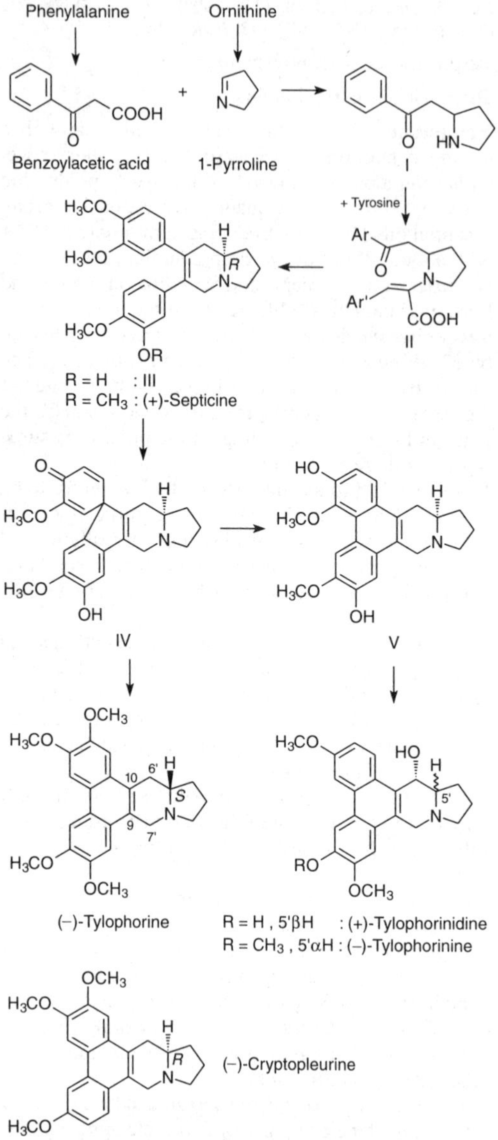

Figure: Biosynthesis of the Tylophora alkaloids.

Table: Data of Tylophora alkaloids.

name	molecular formula	M_R	mp. [°C]	$[\alpha]_D$	CAS
(–)-Cryptopleurine	$C_{24}H_{27}NO_3$	377.48	215–217	–74° (CH_3OH)	482-22-4
(+)-Septicine	$C_{24}H_{29}NO_4$	395.50	136	+38.8° (CH_3OH)	42922-10-1
(–)-Tylophorine	$C_{24}H_{27}NO_4$	393.48	286–287	–11.6° ($CHCl_3$)	482-20-2
(+)-Tylophorinidine	$C_{22}H_{23}NO_4$	365.43	216–218	+105° ($CHCl_3$)	32523-69-6
(–)-Tylophorinine	$C_{23}H_{25}NO_4$	379.46	248–249	–14.2° ($CHCl_3$)	571-70-0

clepiadaceae family, *Tylophora*, *Vincetoxicum*, *Pergularia*, and *Cynanchum*, as well as *Ficus septica* (Moraceae). On the other hand, only few biosynthetically related phenanthroquinolizidine alkaloids are known. The most important example of the latter is cryptopleurine from *Cryptocarya pleurosperma*.

Both groups of alkaloids are built up from a phenylalanine unit and a tyrosine unit as well as ornithine or lysine. Phenylalanine provides ring A, C-10, and C-6′ while tyrosine furnishes ring B, C-9, and C-7′. Ornithine reacts with benzoylacetic acid to give compound I, which reacts further via II to form *O*-demethylsepticine III; the latter undergoes phenolic oxidation to yield the dienone IV which can be transformed either by dienone-phenol rearrangement to tylophorine or after reduction to the dienol and subsequent dienol-benzene rearrangement to tylophorinine and tylophorinidine.

Lit.: Alkaloids: Chem. Biol. Perspectives **3**, 241–273 (1985) ▪ Angew. Chem., Int. Ed. Engl. **32**, 1010 (1993) (abs. configuration) ▪ Chem. Pharm. Bull. **46**, 767 (1998) (isolation) ▪ Heterocycles **28**, 63 (1989) ▪ J. Am. Chem. Soc. **118**, 12082 (1996) (synthesis) ▪ J. Chem. Soc., Perkin Trans. 1 **1990**, 2287 ▪ J. Org. Chem. **62**, 7435 (1997) (synthesis septicine, tylophorine) ▪ Manske **9**, 517–528; **19**, 193–218; **25**, 156–163 ▪ Merck-Index (12.), No. 9961 f. ▪ Mothes et al., p. 250–251 ▪ Pelletier **3**, 241–273 ▪ Tetrahedron **50**, 12293 (1994); **55**, 2659 (1999) (cryptopleurine) ▪ Tetrahedron Lett. **32**, 5919 (1991) (synthesis). – *[HS 2939 90]*

Tylosin.

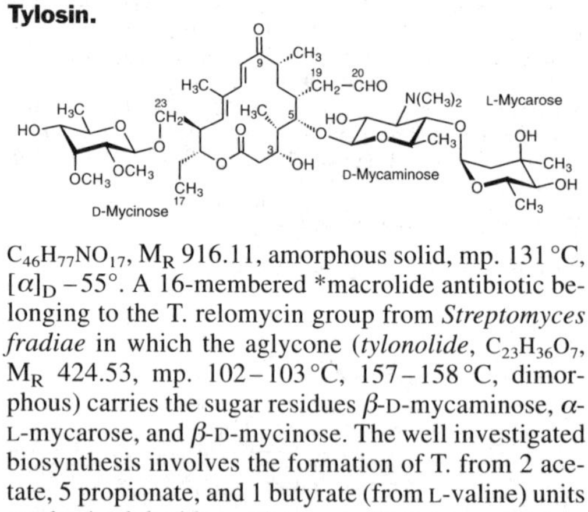

$C_{46}H_{77}NO_{17}$, M_R 916.11, amorphous solid, mp. 131 °C, $[\alpha]_D$ –55°. A 16-membered *macrolide antibiotic belonging to the T. relomycin group from *Streptomyces fradiae* in which the aglycone (*tylonolide*, $C_{23}H_{36}O_7$, M_R 424.53, mp. 102–103 °C, 157–158 °C, dimorphous) carries the sugar residues *β*-D-mycaminose, *α*-L-mycarose, and *β*-D-mycinose. The well investigated biosynthesis involves the formation of T. from 2 acetate, 5 propionate, and 1 butyrate (from L-valine) units on the *polyketides pathway via the precursor tylactone (orotylonolide) and subsequent glycosylation and oxidation or methylation, respectively. The biosynthesis genes of T. have been cloned and used for the preparation of hybrid macrolides (see hybrid antibiotics). T. is active against Gram-positive bacteria, mycoplasma, and spirochetes, it is used as the free base or as the tartrate in veterinary medicine and animal breeding. Further derivatives are T. B (*demycarosyltylosin*, $C_{39}H_{65}NO_{14}$, M_R 771.94, mp. 114–116 °C), a hydrolysis product of T., and T. D (*relomycin*, $C_{46}H_{79}NO_{17}$, M_R 918.13, mp. 172–175 °C, in which the 20-oxo group of T. is reduced to a 20-hydroxy group), both are less active than tylosin.

Lit.: Merck-Index (12.), No. 9963 ▪ Omura (eds.), Macrolide Antibiotics, New York: Academic Press 1984 ▪ Omura & Tanaka, in Pape & Rehm (eds.), Biotechnology, vol. 4, p. 359–391, Weinheim: VCH Verlagsges. 1986 ▪ Sax (8.), TOE 600, TOE 750. – *Biosynthesis:* Annu. Rev. Microbiol. **42**, 547–574 (1988); **47**, 875–912 (1993); **49**, 201–238 (1995) ▪ Antimicrob. Agents Chemother. **21**, 758 (1982) ▪ Katz & Donadio, in Genetics and Biochemistry of Antibiotic Production, p. 385–420, Boston: Butterworth-Heinemann, 1995 ▪ Ullmann

(5.) **A 2**, 496, 525. – *Synthesis:* Ann. Pharm. Fr. **56**, 152 (1998) ▪ Chem. Pharm. Bull. **30**, 97–110 (1982) ▪ Drugs Pharm. Sci. **22**, 743 (1984) ▪ J. Am. Chem. Soc. **104** 5523–5528 (1982) ▪ J. Antibiot. **40**, 1123–1130 (1987) ▪ J. Chem. Soc., Chem. Commun. **1982**, 855 ▪ J. Org. Chem. **48**, 892–895, (1983) ▪ Lindberg **1**, 123–152 ▪ Pol. J. Chem. **73**, 77–87 (1999) ▪ Tetrahedron Lett. **1982**, 3375. – *[HS 2941 90; CAS 1401-69-0 (T.); 61219-82-7 (tylonolide); 11032-98-7 (T. B); 1404-48-4 (T. D)]*

Tyramine see *β*-phenylethylamine and *β*-phenylethylamine alkaloids.

Tyrocidines.

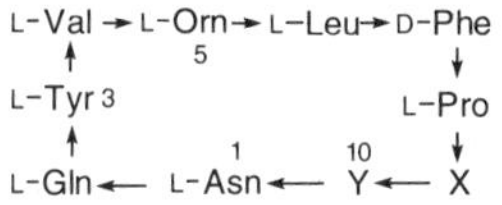

X = L-Phe, Y = D-Phe : T. A
X = L-Trp, Y = D-Phe : T. B
X = L-Trp, Y = D-Trp : T. C
X = L-Trp, Y = L-Trp, with 3-L-Trp instead L-Tyr : T. D
X = L-Phe, Y = D-Phe, with 3-L-Phe instead L-Tyr : T. E

A group of *peptide antibiotics formed by cultures of *Bacillus brevis*, the main component is tyrothricin (80%, the rest consists of the *gramicidins A–D). T. exhibit bactericidal activity against Gram-positive pathogens through damage to the cell membrane. The initially obtained mixture was first separated into 3 cyclic decapeptides (T. A, B, C) differing at 2 positions on the ring, T. D and E were discovered later. The hydrochloride of the mixture of *T. A* ($C_{66}H_{87}N_{13}O_{13}$, M_R 1270.50, mp. 240–242 °C), *T. B* ($C_{68}H_{88}N_{14}O_{13}$, M_R 1309.53, mp. 236–237 °C) and *T. C* ($C_{70}H_{89}N_{15}O_{13}$, M_R 1348.57) forms needles, soluble in 95% ethanol, poorly soluble in water. Although the individual T. can be synthesized, in general the T. mixture produced on a technical scale by fermentation is used.

Lit.: J. Mol. Biol. **135**, 757 (1979) ▪ Kirk-Othmer (4.) **3**, 268–276 ▪ Merck-Index (12.), No. 9967, 9972 ▪ Tetrahedron Lett. **31**, 6121 (1990) (synthesis). – *[HS 2941 90; CAS 1481-70-5 (T. A); 865-28-1 (T. B); 3252-29-7 (T. C); 8011-61-8 (T. complex)]*

Tyromycin A.

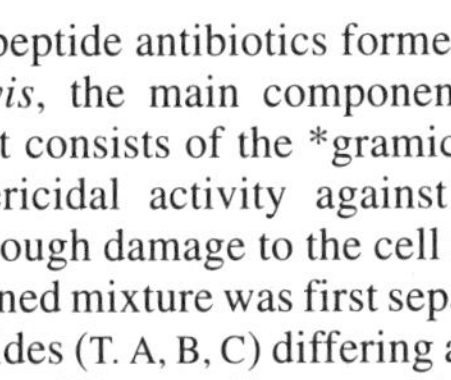

$C_{26}H_{38}O_6$, M_R 446.58, cryst., mp. 59–60 °C. Bis-citraconic acid anhydride derivative and an inhibitor of leucine and cysteine aminopeptidases from cultures of the wood-rotting fungus *Tyromyces lacteus* (Basidiomycetes). T. shows cytostatic activity.

Lit.: J. Org. Chem. **63**, 1342 (1998) (synthesis) ▪ Planta Med. **58**, 56 (1992) (isolation). – *[CAS 141364-77-4]*

L-Tyrosine [(*S*)-2-amino-3-(4-hydroxyphenyl)-propanoic acid, abbr.: Tyr or Y].

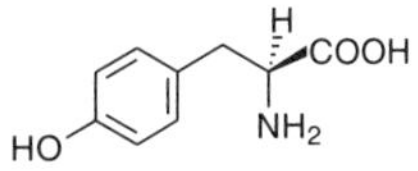

$C_9H_{11}NO_3$, M_R 181.19, mp. 342–344 °C (decomp., in sealed capillary), $[\alpha]_D^{22}$ –10.6° (1 m HCl), pK_a 2.20, 9.11, 10.07 (OH), pI 5.66, poorly soluble in water. For humans a non-essential proteinogenic amino acid. Genetic code: UAU, UAC. Average content in many proteins 3.5% [1]. L-Tyr is almost tasteless or somewhat bitter, D-Tyr is sweet. Papain and coral proteins are rich in Tyr. (Modified) Tyr residues occur in enzymes and determine their catalytic activity, e. g., the aminoxidase from bovine plasma[2] and galactose oxidase (EC 1.1.3.9) as well as the activity of redox enzymes. Tyrosinase is responsible for the formation of *melanins.
Biosynthesis: Prephenic acid is oxidized by prephenic acid dehydrogenase (EC 1.3.1.12) to 4-hydroxyphenylpyruvate and then aminated by T. aminotransferase (EC 2.6.1.5; coenzyme: *vitamin B_6). Tyr is hydroxylated by T. monooxygenase to DOPA and transformed in consecutive steps to *dopamine, *noradrenaline, and *adrenaline, all of which exhibit specific biological functions as neurotransmitters. In animals T. is converted to various iodotyrosines (see iodoamino acids). T.-bound proteins have important functions in phosphorylation, blood coagulation, and other biological processes. For metabolic diseases, see also phenylalanine.

Lit.: [1]Biochem. Biophys. Res. Commun. **78**, 1018–1024 (1977). [2]Science **248**, 981–987 (1990); Eur. J. Biochem. **200**, 271–284 (1991).
gen.: Beilstein EIV **14**, 2264–2267 ▪ Hager (5.) **9**, 952 f., 1126 ff. ▪ Karrer, No. 2363 ▪ Merck-Index (12.), No. 9970 f. ▪ Stryer 1990, p. 533 f., 604 f., 610 f. ▪ Ullmann (5.) **A 2**, 58, 63, 86. – *[HS 2922 50; CAS 60-18-4]*

U

U 106305.

$C_{28}H_{41}NO$, M_R 407.61, oil, $[\alpha]_D$ –270° ($CHCl_3$). Inhibitor of cholesteryl ester transfer protein (CETP), a key factor of lipid metabolism, that controls the transformation of HDL- into LDL-lipids, from cultures of a *Streptoverticillium* strain. The methylene groups of the quinquecyclopropane units originate biosynthetically from methionine. U. is structurally closely related to *FR 900848.

Lit.: J. Am. Chem. Soc. **117**, 10629 (1995); **118**, 7863, 10327 (1996); **119**, 8608 (1997) (synthesis) ▪ J. Org. Chem. **62**, 1215 (1997) ▪ J. Chem. Soc., Chem. Commun. **1997**, 1693 (review). – *[CAS 170591-54-5]*

Ubenimex see bestatin.

Ubichromenols see ubiquinones.

Ubiquinones (coenzymes Q, mitoquinones). Widely distributed bioquinones in animals and microorganisms (but not in Gram-positive bacteria and cyanobacteria). The nomenclature is based on the number of isopropene units in the side chain as coenzyme Q_n (n=1 to 12) or on the number of carbon atoms in the side chain as U.-5 to U.-60. Most naturally occurring U. possess side chains with 30–50 C atoms and are thus named U.-30 to U.-50 (or coenzymes Q_{6-10}). Q_6 is mainly found in microorganisms and bacteria, Q_9 in fish, and Q_{10} in humans. The manufacturing costs of Q_{10} are about 1000 US $/kg (Japan). 38′,39′-Dihydro-Q_{10}, $C_{59}H_{92}O_4$, M_R 865.38, mp. 29 °C, is isolated from *Penicillium stipitatum*, *Gliocladium roseum*, and *Gibberella fujikuroi*. The U. occur as yellow-orange crystals and are soluble in organic solvents and fats. U. cyclize readily to *ubichromenols* under the action of light.

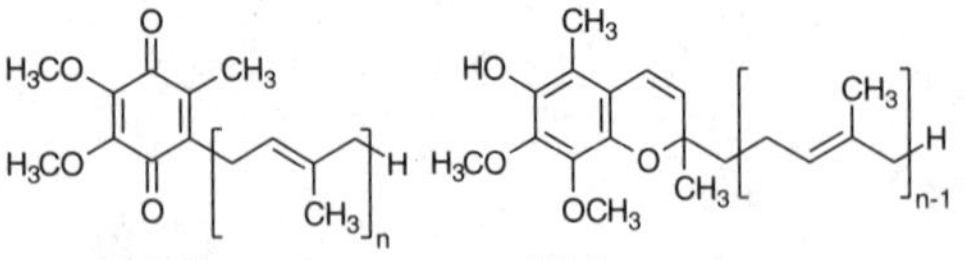

Ubiquinones Ubichromenols

2,3-Dimethoxy-5-methyl-1,4-benzoquinone (U.-0, Q_0, *fumigatin methyl ether*, $C_9H_{10}O_4$, M_R 182.18, mp. 58–59 °C) is also classified as a U., it occurs in the defensive secretions of some millipedes, such as *Rhapidostreptus innominatus*, *Metiche tanganyicense*.

Biological activity: U. serve as essential electron-transfer agents in the respiratory chain. They are present in mitochondria and enable the cyclic oxidation and reduction of substrates in the citric acid cycle. In phototrophic bacteria U. fulfill the functions of the *plastoquinones in higher plants.

Table: Structure and Data of Ubiquinones.

Ubiquinone	molecular formula	M_R	mp. [°C]	CAS
U.-30 (Q_6)	$C_{39}H_{58}O_4$	590.89	19–20 (16)	1065-31-2
U.-35 (Q_7)	$C_{44}H_{66}O_4$	659.01	31–32	303-95-7
U.-40 (Q_8)	$C_{49}H_{74}O_4$	727.12	37–38	2394-68-5
U.-45 (Q_9)	$C_{54}H_{82}O_4$	795.24	45	303-97-9
U.-50 (Q_{10}, Ubidecarenone)	$C_{59}H_{90}O_4$	863.36	50	303-98-0

Use: Various Q. are used as "neutraceuticals" for the treatment of heart disease, (the most important) Q_{10} as inhibitor of myocardiosis, Q_{11} against diabetes, obesity, periodental disease and to improve athletic performance.

Lit.: Acta Chem. Scand., Ser. B **41**, 70 ff. (1987) (biosynthesis) ▪ Angew. Chem. Int. Ed. Engl. **19**, 659–675 (1980) ▪ Chem. Pharm. Bull. **32**, 3952–3958 (1984) (synthesis); **35**, 1531–1537 (1987) (biochemistry) ▪ Helv. Chim. Acta **73**, 790–796 (1990) (synthesis) ▪ J. Am. Chem. Soc. **110**, 4356–4362 (1988); **118**, 5512 (1996) (synthesis) ▪ Lee (ed.), Structure, Biogenesis and Assembly of Energy Transducing Enzyme Systems, Orlando: Academic Press 1987 ▪ Lenaz (ed.), Coenzyme Q, Biochemistry, Bioenergetics and Clinical Application of Ubiquinone, Chichester: Wiley 1985 ▪ Luckner (3.), p. 331, 454 ▪ Merck-Index (12.), No. 9974 ▪ Nat. Prod. Rep. **12**, 101–133 (1995) ▪ Negwer (6.), No. 8345 ▪ Stryer 1995, p. 535 f., 708 ▪ Synthesis **1991**, 1130; **1993**, 797 (synthesis Q2) ▪ Tetrahedron **53**, 2505 (synthesis Q3) ▪ Tetrahedron Lett. **38**, 6181 (1997) (biosynthesis). – *Pharmacology:* Med. Clin. North Am. **72**, 243–258 (1988). – *[CAS 992-78-9 (ubiquinol 50); 605-94-7 (fumigatin methyl ether)]*

UK-1.

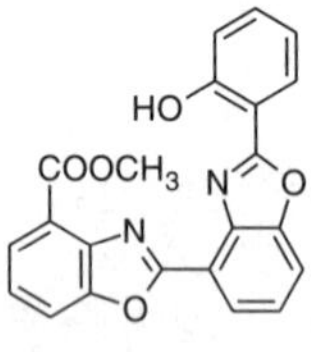

Bibenzoxazole derivative from cultures of *Streptomyces morookaensis*, $C_{22}H_{14}N_2O_5$, M_R 386.36, needles, mp. 217–219 °C. UK-1 is cytotoxic and shows activity against certain tumor cell lines.

Lit.: J. Antibiot. **46**, 1089, 1095 (1993) (isolation) ▪ Tetrahedron Lett. **38**, 199 (1997) (synthesis). – *[CAS 151271-53-3]*

Ulapualides. U. are macrolides containing 3 linearly linked oxazoles as part of the macrocyclic ring. U. are isolated from eggs of nudibranchia of the species *Hexabranchus sanguineus*[1]. Two members are known: *U. A* ($C_{46}H_{64}N_4O_{13}$, M_R 881.03, oil) and *U. B* ($C_{51}H_{74}N_4O_{16}$,

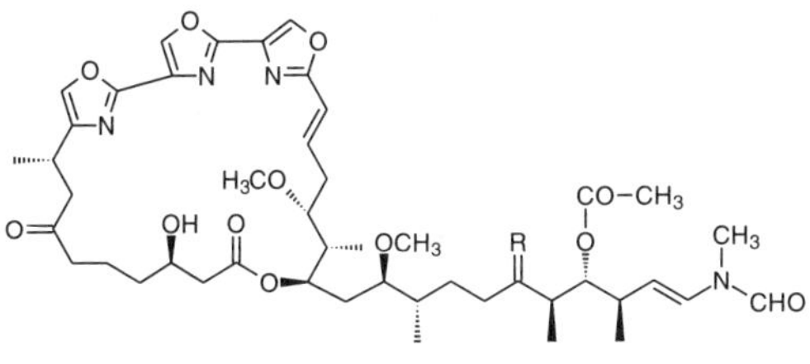

R = O : Ulapualide A
R = H , O–CO–CH(OCH3)–CH2–OCH3 : Ulapualide B

M_R 999.17, oil). The compounds have antimicrobial and cytotoxic activity against L1210 cells (ED_{50} 0.01–0.03 μg/mL). A similar macrolide, *kabiramide C* ($C_{48}H_{71}N_5O_{14}$, M_R 942.12, amorphous, $[\alpha]_D$ +20°), with antifungal activity is isolated from the egg mass of the nudibranch *Dendrodoris nigra*[2].

Lit.: [1] J. Am. Chem. Soc. **108**, 846 (1986). [2] J. Am. Chem. Soc. **108**, 847 (1986).
gen.: Synlett **1990**, 36; **1997**, 1345 (synthesis) ▪ Tetrahedron Lett. **39**, 6095 (1998) (synthesis). – *[CAS 10045-73-6 (U. A); 100045-74-7 (U. B); 100045-78-1 (kabiramide C)]*

Ulicyclamide, ulithiacyclamides see patellamides.

Umbelliferone (7-hydroxy-2*H*-1-benzopyran-2-one, 7-hydroxycoumarin).

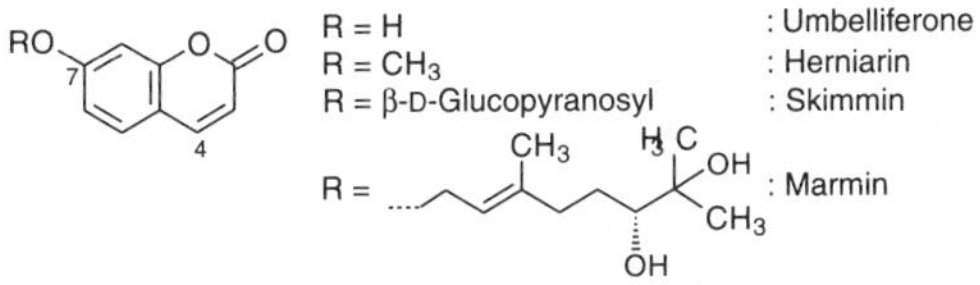

$C_9H_6O_3$, M_R 162.14, yellow crystals or needles, mp. 230–232 °C (223–224 °C, 227–228 °C), pK_a 7.7, readily soluble in alcohol, chloroform, and glacial acetic acid, the solutions exhibit fluorescence and on heating an odor of *coumarin. U. occurs in sweet potatoes as a *phytoalexin. U. is widely distributed in nature in the free form and as ether and in glycoside derivatives, e. g.: in *Angelica* (Apiaceae), *Artemisia* (mugwort), *Coronilla* (crown vetch), and *Ruta* species (Rutaceae), in the bark of *Daphne* species, fruits of coriander, roots of *Atropa* species (deadly nightshade) as well as roots of carrots and chamomile. Derivatives of U.: 7-*O*-*β*-D-glucopyranosyl-U. (*skimmin*), $C_{15}H_{16}O_8$, M_R 324.29, mp. 215–221 °C, $[\alpha]_D^{18}$ −79.8° (pyridine); *herniarin* (umbelliferone methyl ether), $C_{10}H_8O_3$, M_R 176.17, shining platelets, mp. 117 °C, component of the leaves of various plants such as rupturewort (*Herniaria hirsuta*) and water hemp (*Eupatorium ayapana*). Isoprenylated derivatives of U. are also widely distributed, e. g., (*R*)-*marmin*, $C_{19}H_{24}O_5$, M_R 332.40, cryst., mp. 123–124 °C, $[\alpha]_D$ +25° (C_2H_5OH), $[\alpha]_D$ +18.2° ($CHCl_3$) from the trunk bark of *Aegle marmelos* and grapefruit skin. *4-Methyl-U.*, $C_{10}H_8O_3$, M_R 176.17, needles, mp. 194–195 °C, pK_a 7.80 (25 °C), occurs in young twigs of *Dalbergia volubilis* and in *Eupatorium pauciflorum*.
Use: U. serves as a pH indicator in the range 6.5–8.9; 0.1% aqueous solutions are used as a fluorescence indicator to detect calcium and copper ions. U. is used in sun blockers and as an optical brightener for textiles. It also serves as a laser dye while the synthetic 7-*O*-glycosides are used for fluorimetric determination of glycosidases.
Biosynth.: *Coumarinic acid → U. or hydroxylation of *coumarin in position 7; methylation with *S-adenosylmethionine.

Lit.: Beilstein EV **18/1**, 386 ▪ Chem. Pharm. Bull. **39**, 3100 (1991) (synthesis) ▪ Hager (5.) **5**, 665 f. ▪ Luckner (3.), p. 389 ▪ Nat. Prod. Rep. **6**, 591–624 (1989) ▪ Negwer (6.), No. 1420 ▪ Phytochemistry **26**, 257–260 (1987); **29**, 1137–1142 (1990) (biosynthesis) ▪ Ullmann (5.) **A 12**, 147 (4-methyl-U.) ▪ Zechmeister **35**, 199–210. – *Pharmacology:* Pharm. Ind. **53**, 83 ff. (1991) ▪ see also coumarin. – *[HS 2932 29; CAS 93-35-6 (U.); 93-39-0 (skimmin); 531-59-9 (herniarin); 14957-38-1 ((R)-marmin); 90-33-5 (4-methyl-U.)]*

Undecanal see alkanals.

Undecane. $H_3C–(CH_2)_9–CH_3$, $C_{11}H_{24}$, M_R 156.31, inflammable liquid; mp. −26 °C, bp. 196 °C; occurs in some turpentine oils and in crude oil; it is also known as an *alarm substance of red ants[1]. U. acts as a sexual pheromone for the alpine ant *Formica lugubris*[2].

Lit.: [1] J. Insect Physiol. **22**, 1331–1348 (1976). [2] Naturwissenschaften **80**, 30–34 (1993).
gen.: Beilstein EIV **1**, 487. – *[HS 2901 10; CAS 1120-21-4]*

4-Undecanolide see alkanolides.

(3*E*,5*Z*)-1,3,5-Undecatriene see galbanum oil.

Unedoside.

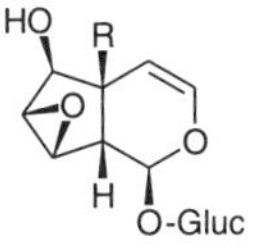

R = H : Unedoside
R = OH : Stilbericoside

U. and stilbericoside are as yet the only *iridoid glucosides with a C_8-aglycone. *U.*, $C_{14}H_{20}O_9$, M_R 332.31, mp. 232–234 °C, $[\alpha]_D$ −112.4° (H_2O); *stilbericoside*, $C_{14}H_{20}O_{10}$, M_R 348.31, amorphous, $[\alpha]_D$ −61.5° (H_2O). Both compounds are isolated from the herbage of the South African plants *Arbutus unedo* (Ericaceae) and *Stilbe ericoides* (Verbenaceae)[1]. They are of importance in phytomedicine on account of their astringent, diuretic, and antiseptic properties[2].

Lit.: [1] Fitoterapia **62**, 176 (1991); Phytochemistry **28**, 3059 (1989). [2] Planta Med. **53**, 223 (1987).
gen.: Justus Liebigs Ann. Chem. **1982**, 872 ▪ Phytochemistry **37**, 1599 (1994); **47**, 1007 (1998) (biosynthesis). – *[HS 2938 90; CAS 38965-81-0 (U.); 53852-43-0 (stilbericoside)]*

Uracil see pyrimidines.

Urdamycins. Colored (yellow, orange, and red) *angucycline antibiotics from *Streptomyces fradiae*, they are related to the *saquayamycins and *vineomycins. Acid hydrolysis furnishes the aglycones of the U., from U. A (=kerriamycin B) *urdamycinone A* (=*aquayamycin) together with 2 mol. L-rhodinose (sugar A and C) and 1 mol. D-olivose (sugar B), the 9 *C*-glycosidically bound D-olivose at position 9 is not cleaved. In the later stages of the biosynthesis, amino acids can be connected with the positions 11 and 12 of the tetracyclic angular framework of U. A, a *polyketide, to afford U. C, D, and H. The U. have antibacterial and antineoplastic properties.

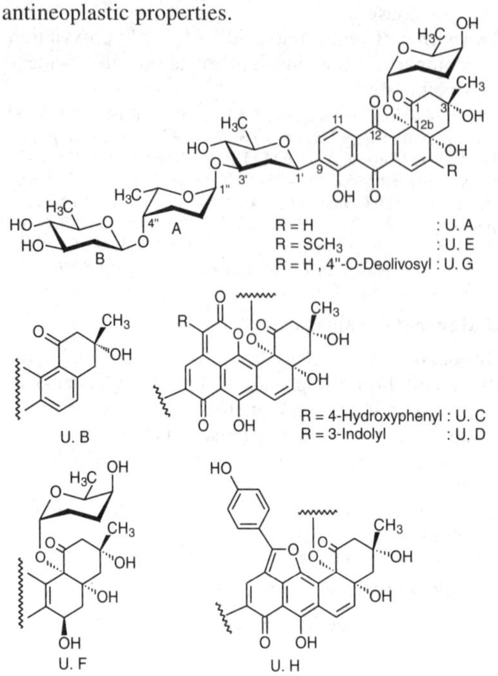

Table: Data of Urdamycins.

U.	molecular formula	M_R	mp. [°C]	CAS
A	$C_{43}H_{56}O_{17}$	844.91	160 (decomp.)	98474-21-6
B	$C_{37}H_{44}O_{13}$	696.75		104542-46-3
C	$C_{51}H_{60}O_{19}$	977.03	200 (decomp.)	104443-43-8
D	$C_{53}H_{61}NO_{18}$	1000.06	220 (decomp.)	104443-44-9
E	$C_{44}H_{58}O_{17}S$	890.99		104542-47-4
F	$C_{43}H_{58}O_{18}$	862.92		104562-12-1
G	$C_{37}H_{46}O_{14}$	714.76	141	115626-67-0
H	$C_{50}H_{60}O_{18}$	949.02		126121-78-6

Lit.: J. Antibiot. **41**, 126, 812 (1988); **42**, 299–311, 1482–1488 (1989). – *Biosynthesis:* J. Antibiot. **42**, 1151–1157 (1989) ▪ Angew. Chem. Int. Ed. Engl. **29**, 1051–1053 (1990) ▪ J. Org. Chem. **58**, 2547–2551 (1993). – *Review:* Nat. Prod. Rep. **9**, 103–137 (1992). – *Synthesis:* J. Chem. Soc., Chem. Commun. **1996**, 225 ▪ J. Am. Chem. Soc. **117**, 8472 (1995).

Uric acid [7,9-dihydro-1*H*-purine-2,6,8(3*H*)-trione and tautomeric forms].

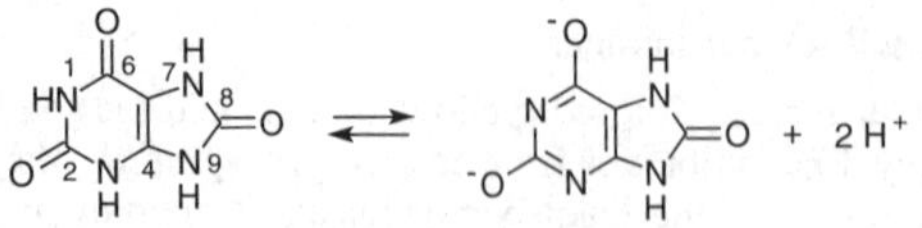

$C_5H_4N_4O_3$, M_R 168.11, cryst., D. 1.893, mp. >300 °C, decomposes with evolution of hydrogen cyanide on heating; pK_{a1} 5.75, pK_{a2} 10.6. On account of its tautomerism U. a. forms various different derivatives. One or two of the four hydrogen atoms can undergo ionic dissociation (see figure); two series of uric acid salts or *urates* are formed. U. a. can be detected in the murexide reaction.
Uric acid occurs in urine, in the kidneys and urinary calculi as well as in smaller amounts in higher plants, especially in seeds. In humans and primates U. a. is the final product of the *purine metabolism whereas, in other mammals and some reptiles, it can be degraded to *allantoin by the action of the enzyme uricase, and in amphibians and fish to urea and glyoxylic acid. In birds and reptiles U. a. is also formed by detoxification of ammonia resulting from the degradation of proteins; thus, for example, the solid excretion products of reptiles and birds (guano) consist of up to about 90% of the ammonium salt of uric acid.

Lit.: Beilstein E IV **26**, 2619 ▪ Kirk-Othmer (3.) **23**, 608 ▪ Luckner (3.), p. 437–439 ▪ Stryer 1995, p. 756. – *Toxicology:* Sax (8.), UVA 400. – *[HS 2933 59; CAS 69-93-2]*

Urobilin.

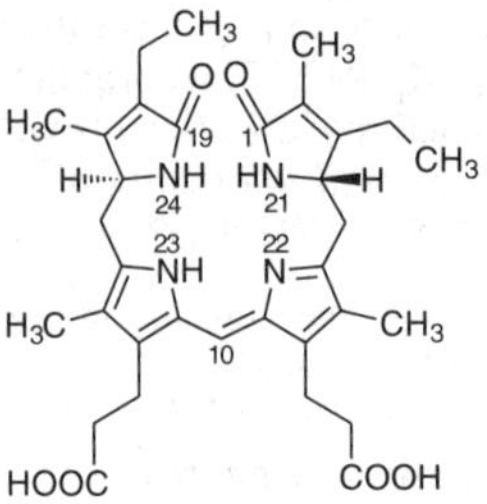

$C_{33}H_{42}N_4O_6$, M_R 590.72. According to IUPAC/IUB rule TP-6.4 (see *Lit.*[1]) the name U. is applied to what was formerly known as U. IXα. Racemate, orange yellow needles, mp. (hydrochloride): 174–175 °C, $[\alpha]_D^{20}$ +5000°! ($CHCl_3$), free base: mp. 162–165 °C, uv_{max} 460 nm. U. is formed in the intestines of warm-blooded organisms by non-enzymatic oxidation of *urobilinogen which, in turn, is formed from *bilirubin during degradation of the red blood pigment *heme (see also bile pigments). The color of feces depends on the content of U.

Lit.: [1] Pure Appl. Chem. **59**, 811 (1987).
gen.: Beilstein E V **26/15**, 515–519 ▪ Dolphin **VI**, 524–545 ▪ Falk, The Chemistry of Linear Oligopyrroles and Bile Pigments, p. 46–50, 336 f., Wien: Springer 1989 ▪ Helv. Chim. Acta **70**, 2098 (1987) (abs. configuration) ▪ Merck-Index (12.), No. 10022 ▪ Zechmeister **29**, 94–99. – *[CAS 1856-98-0]*

Urobilinogen (10,22-dihydrourobilin). Formula see urobilin. $C_{33}H_{44}N_4O_6$, M_R 592.74. U. and its 14,16-diepimer are formed from *bilirubin during the degradation of *heme in warm-blooded organisms (see also bile pigments). In the intestines a small part of U. is non-enzymatically oxidized to *urobilin while the main part is reduced by intestinal bacteria to *stercobilinogen.

Lit.: see urobilin. – *[CAS 14684-37-8 (U.); 74778-38-4 (diepimer)]*

Urocan(in)ic acid [(*E*)-3-(1*H*-imidazol-4-yl)-2-propenoic acid].

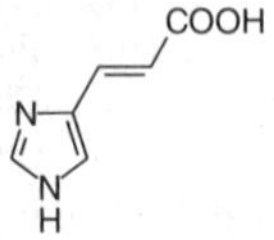

$C_6H_6N_2O_2$, M_R 138.13, mp. 243–245 °C. A naturally occurring metabolite originating from the degradation of *histidine by histidine ammonia lyase (EC 4.3.1.3). It is formed by fermentation, e. g., of *Bacillus* spp. or by fungi, it is produced technically by biotransformation of histidine with *Achromobacter liquidum*. It exhibits anti-tumor activity and is used as a sun blocking agent in cosmetics. The (*Z*)-form, which is generated on irradiation with UV light, is inactive.

Lit.: Appl. Microbiol. **27**, 688–694 (1974) ▪ Beilstein E V **25/4**, 204 f. ▪ J. Am. Chem. Soc. **106**, 4136–4144 (1984); **119**, 2715 (1997). – *[HS 2941 90; CAS 104-08-3 (U.); 3465-72-3 ((Z)-U.)]*

Uroporphyrins. The U. comprise 4 isomeric compounds; $C_{40}H_{38}N_4O_{16}$, M_R 830.76.

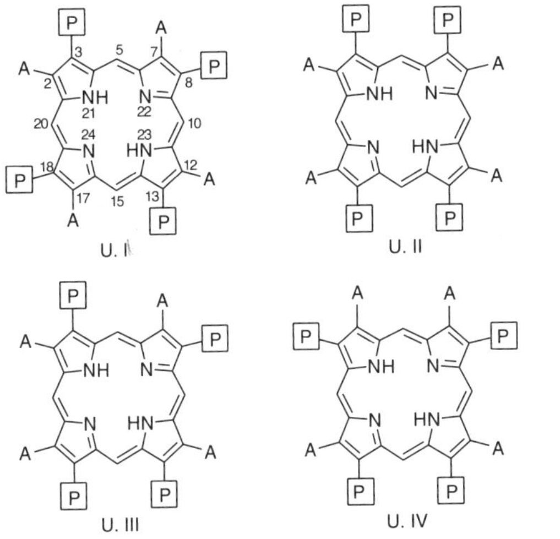

A = CH_2—COOH [P] = CH_2—CH_2—COOH

Figure: Structures of the Uroporphyrins.

U. III is the longest known, natural porphyrin (see porphine) (Hammarsten, 1892). U. II and U. IV have no physiological significance. They are formed together with the other U. by acid-catalyzed polymerization and oxidation of *porphobilinogen.

Table: Data of Uroporphyrins and Uroporphyrinogens.

	mp. [°C]	CAS
U. I	295*	607-14-7
U. II	>300	531-42-0
U. III	255–259*	18273-06-8
U. IV	>300	613-02-5
Uroporphyrinogen I		1867-62-5
Uroporphyrinogen II		53790-13-9
Uroporphyrinogen III		1976-85-8
Uroporphyrinogen IV		53790-14-0

* mp. of octamethyl esters

Patients with porphyria excrete I in their urine. The copper(II) complex of U. III, *turacin* ($C_{40}H_{36}CuN_4O_{16}$, M_R 829.29), occurs as red pigment in the feathers of African touracos (banana eater) and related birds. The genuine natural products are the *uroporphyrinogens* III and I, from which U. III and I are formed as artifacts during isolation. The uroporphyrinogens I–IV are the 5,10,15,20,22,24-hexahydroporphyrin derivatives, the *porphyrinogens, of the corresponding U. I–IV with 4 isolated pyrrole rings ($C_{40}H_{44}N_4O_{16}$, M_R 836.81). Like all porphyrinogens, the uroporphyrinogens are extremely sensitive to oxidation and are thus readily converted to U. Uroporphyrinogen III is the key building block of *porphyrin biosynthesis, from which all known *tetrapyrroles are formed. Interactions between the enzyme porphobilinogen ammonia lyase (EC 4.3.1.8) and uroporphyrinogen III cosynthetase lead to uroporphyrinogen III from *porphobilinogen. When the cosynthetase is absent uroporphyrinogen I is formed. One of the fascinating aspects of porphyrin biosynthesis is the question why and how nature did construct uroporphyrinogen III with an inverted arrangement of the acetic acid and propanoic acid side chains in ring D and not the simpler uroporphyrinogen I with serial arrangement of the side chains. Research on the prebiological formation [1] has shown that uroporphyrinogen III is statistically preferred and thus provides a plausible explanation.

Lit.: [1] Angew. Chem. Int. Ed. Engl. **27**, 5–39 (1988).
gen.: Angew. Chem. Int. Ed. Engl. **32**, 1040 (1993) (biosynthesis) ▪ Beilstein E V **26/15**, 354 ▪ Chem. Rev. **90**, 1261–1274 (1990) ▪ Dolphin **I**, 316–319; **II**, 91–101; **VI**, 46, 67–91, 144, 243–248 ▪ Dolphin B_{12} **I**, 110–117 ▪ J. Chem. Soc., Perkin Trans. 1 **1983**, 3031, 3041 (biosynthesis) ▪ Jordan, p. 30–59, 67–77 ▪ Smith, p. 75–83.

Urothion(e).

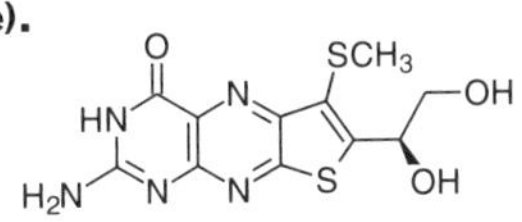

$C_{11}H_{11}N_5O_3S_2$, M_R 325.37, orange cryst., mp. >360 °C (decomp.), $[\alpha]_D^{20}$ −20° (0.1 N NaOH), poorly soluble in H_2O. U. is a metabolic component of human urine.

Lit.: Merck-Index (12.), No. 10025 ▪ Tetrahedron Lett. **1967**, 4507; **1968**, 2941; **36**, 2631 (1995) (absolute configuration). – *[CAS 17801-77-3]*

Ursane. $C_{30}H_{52}$, M_R 412.74, see figure under triterpenes. The pentacyclic, saturated triterpene skeleton, substituted by 8 methyl groups, of *ursolic acid and α-*amyrin.

Ursodeoxycholic acid see bile acids.

Ursolic acid (3β-hydroxy-12-ursen-28-oic acid).

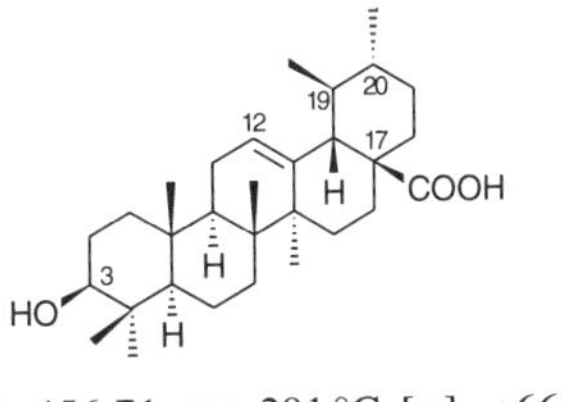

$C_{30}H_{48}O_3$, M_R 456.71, mp. 291 °C, $[\alpha]_D$ +66° (C_2H_5OH), somewhat soluble in methanol and ether. U. is a *triterpene carboxylic acid with the ursane skeleton (see triterpenes) isolated for the first time in 1854 by Trommsdorf from the leaves of the bearberry (*Arctostaphylos uva-ursi*). U. occurs in many plants such as, e.g., in leaves and berries of *Vaccinium*, *Arbutus*, *Erica*, and *Rhododendron* species, in wax coatings of apples and pears, in the fruit skins of cranberries and bilberries, in thyme, marjoram, rosemary, sage. U. exhibits cytotoxic and antileukemic activities.

Lit.: Beilstein E IV **10**, 1160 ▪ Karrer, No. 2015 ▪ Merck-Index (12.), No. 10027 ▪ Pharm. Pharmacol. Lett. **1**, 115 (1992) (isolation) ▪ Phytochemical Dictionary, p. 687 ▪ Phytochemistry **35**, 121 (1994) (biosynthesis). – *[CAS 77-52-1]*

Table: Structure and data of Urushiols.

Urushiol	R	molecular formula	M_R	mp. [°C]	bp. [°C]	CAS
U.-I	pentadecyl	$C_{21}H_{36}O_2$	320.52	59–60		492-89-7
U.-II	8-pentadecenyl	$C_{21}H_{34}O_2$	318.50	33		2764-91-2 (*Z*): 35237-02-6
Renghol	10-pentadecenyl	$C_{21}H_{34}O_2$	318.50	14–15	170–172 (0.13 Pa)	492-90-0 (*Z*): 83532-37-0
U.-III (Urushin)	8,11-pentadecadienyl	$C_{21}H_{32}O_2$	316.48			(*Z,Z*): 492-91-1 (*E,Z*): 83532-38-1
U.-IV	8,11,13-pentadecatrienyl	$C_{21}H_{30}O_2$	314.47			(*Z,Z,Z*): 21104-16-5 (*Z,E,Z*): 83532-40-5 (*E,E,Z*): 83532-39-2
U.-V	8,11,14-pentadecatrienyl	$C_{21}H_{30}O_2$	314.47			(*Z,Z,Z*): 2790-58-1
Hydrolaccol	heptadecyl	$C_{23}H_{40}O_2$	348.57	63.5–65	209–217 (3 Pa)	5862-27-1
	8-heptadecenyl	$C_{23}H_{38}O_2$	346.55	33–36		(*E*): 122850-19-5 (*Z*): 54954-20-0
	10-heptadecenyl					89838-23-3
	11-heptadecenyl					(*Z*): 83532-41-6
	8,11,14-heptadecatrienyl	$C_{23}H_{34}O_2$	342.52		209–217 (3 Pa)	(*Z,Z,Z*): 4807-34-5

Urushiol(s) [toxicodendrol(s)].

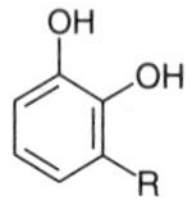

R = $(CH_2)_{14}-CH_3$: U.-I
R = $(CH_2)_7-CH{=}CH-(CH_2)_5-CH_3$: U.-II
R = $(CH_2)_7-CH{=}CH-CH_2-CH{=}CH-(CH_2)_2-CH_3$: U.-III
R = $(CH_2)_7-CH{=}CH-CH_2-CH{=}CH-CH{=}CH-CH_3$: U.-IV
R = $(CH_2)_7-CH{=}CH-CH_2-CH{=}CH-CH_2-CH{=}CH_2$: U.-V

Name for the allergenic components of poison ivy (poison vine, *Rhus radicans = Toxicodendron radicans*, Anacardiaceae), poison oak (*T. diversilobum*), and the Japanese lacquer tree (*Rhus verniciflua*), colorless to pale yellow oil, bp. 200–210 °C, soluble in alcohol, ether, and benzene. The name U. also includes mixtures of catechols substituted in the 3 position with saturated or unsaturated C_{15} or C_{17} side chains as well as some of the pure compounds (see table). Higher homologues, e. g., hydrolaccol and 3-(8,11,14-heptadecatrienyl)-1,2-benzenediol, occur together with long-chain 4-alkyl-1,2-benzenediols in the Burmese lacquer tree (*Melanorrhoea usitata*).

Activity: U. induce very severe allergic skin reactions. They are used therapeutically as antiallergenic agents for hypersensitization. U.-II has antibacterial activity.

Biosynthesis: U. are formed in the polyketide pathway.

Lit.: Hager (5.) **3**, 1232 ▪ J. Nat. Prod. **45**, 532–538 (1982) (^{1}H- & ^{13}C-NMR) ▪ Merck-Index (12.), No. 10028 ▪ J. Pharm. Sci. **70**, 785 ff. (1981) ▪ Luckner (3.), p. 401, 476, 478 ▪ Planta Med. **1986**, 20 ff. (renghol). – *Pharmacology:* J. Invest. Dermatol. **76**, 164–170 (1981); **80**, 149–155 (1983). – *Synthesis:* Bull. Chem. Soc. Jpn. **64**, 1054 ff., 2560 ff. (1990) ▪ Chem. Soc. Rev. **8**, 499–537 (1979) ▪ Synlett **9**, 555 (1990) ▪ Synthesis **1990**, 407 ff.

Uscharidin. $C_{29}H_{38}O_9$, M_R 530.62, rhombic plates, mp. 290 °C (decomp.), $[\alpha]_D^{20}$ +38° (CH_3OH). A cardiac glycoside (see Digitalis glycosides and strophanthins) from *Calotropis procera* (Asclepiadaceae), used in Africa as an *arrow poison.

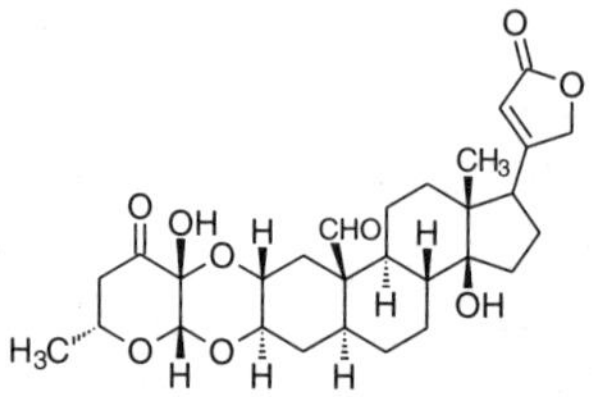

Lit.: J. Chem. Soc., Perkin Trans. 1 **1980**, 2162 ▪ Merck-Index (12.), No. 10030 ▪ Phytochemistry **37**, 217, 801, 1605 (1994). – *[CAS 20304-48-7]*

Usnic acid.

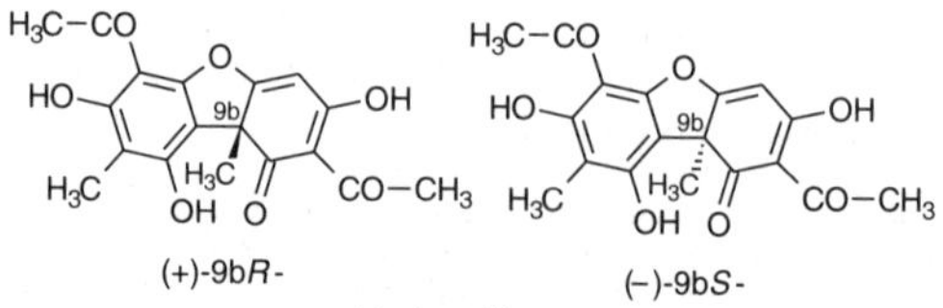

Usnic acid

$C_{18}H_{16}O_7$, M_R 344.32, yellow prisms, mp. 203 °C, $[\alpha]_D^{25}$ +495° and –495° ($CHCl_3$), respectively. A biologically highly active dibenzofuran derivative, occurring both as the (+)- and the (–)-forms in numerous lichens (especially in *Usnea* species): antibiotic, antileukemic, growth regulatory (in plants), and antifeedant (in insects) activities have been described. It is used as the free acid and as the sodium salt in pharmaceutical preparations. The biosynthesis presumably proceeds via phenolic oxidation. Other dibenzofuran derivatives from lichens are didymic acid, *pannaric acid, *placodiolic acid, and schizopeltic acid.

Lit.: Asahina-Shibata, p. 171–198 ▪ Beilstein E V **18/5**, 586 ▪ J. Crystallogr. Spectrosc. Res. **23**, 107 (1993) (structure) ▪ Karrer, No. 1802 ▪ Merck-Index (12.), No. 10031 ▪ Pharmazie **51**, 195 f. (1996) ▪ Tetrahedron Lett. **22**, 351 f. (1981) ▪ Turner **1**, 110 f. ▪ Zechmeister **45**, 204–207. – *[HS 2932 90; CAS 7562-61-0 (R); 125-46-2 (racemate)]*

Uvaricin see annonaceous acetogenins.

Uvidins.

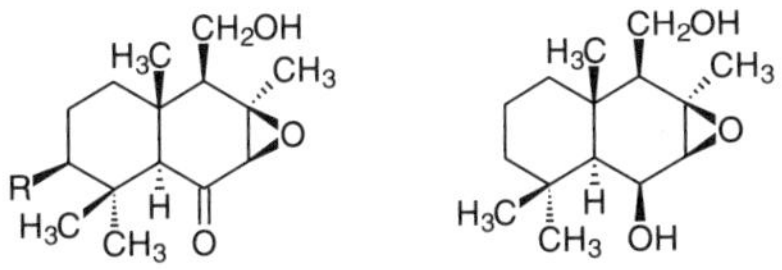

R = H : Uvidin A
R = OH : Uvidin B

Dihydrouvidin A
= Uvidin C

Sesquiterpenoids of the *drimane type from the agaric *Lactarius uvidus* (Basidiomycetes). *U. A* {$C_{15}H_{24}O_3$, M_R 252.35, cryst., mp. 123–124 °C, $[\alpha]_D$ +171°} is accompanied by its hydroxy derivative *U. B* {$C_{15}H_{24}O_4$, M_R 268.35, cryst., mp. 180–181 °C, $[\alpha]_D$ +171° (acetone)}, *dihydrouvidin A* (U. C, $C_{15}H_{26}O_3$, M_R 254.37, cryst., mp. 110–112 °C), and *(–)-drimenol* (see drimanes).

Lit.: Chem. Lett. **1986**, 2073. – *Review:* J. Chem. Soc., Perkin Trans. 1 **1980**, 221 ▪ J. Nat. Prod. **57**, 905 (1994) ▪ Nat. Prod. Rep. **8**, 309 (1991) ▪ Turner **2**, 256. – *[CAS 74636-06-9 (U. A); 74636-05-8 (U. B); 74635-85-1 (U. C)]*

V

Vaccinoside see monotropein.

Valanimycin.

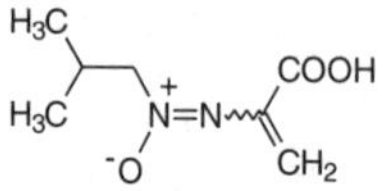

$C_7H_{12}N_2O_3$, M_R 172.18, unstable oil, soluble in water, methanol, ethanol, ethyl acetate; readily adds ammonia and amines. Like *cycasin, *elaiomycin, and *lyophyllin, V. is one of the few *azoxy compounds (natural) occurring in nature. V. is derived biosynthetically from L-*valine and *alanine. V. is active against some Gram-positive and Gram-negative bacteria, *in vitro* against leukemia and Ehrlich carcinoma. The mechanism of action seems to be similar to that of *cycasin.

Lit.: J. Antibiot. (Tokyo) **39**, 184–191, 1263–1269 (1986) (biosynthesis). – *[CAS 101961-60-8]*

Valencene [(4α,5α)-1(10),11-eremophiladiene].

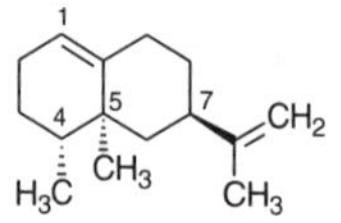

(+)-Valencene

$C_{15}H_{24}$, M_R 204.36, oil, bp. 94 °C (2.66 kPa), $[\alpha]_D^{20}$ +79.2° ($CHCl_3$); together with limonene (see *p*-menthadienes), *myrcene, and *linalool V. is a component[1] (up to ca. 1.5% in Valencia oranges) of *orange oils that are quantitatively very important for the fragrance industry. The presence of (+)-V. distinguishes orange and grapefruit oils from other *citrus oils. The C-4 and C-5 epimeric *eremophilene*, oil, bp. 129.5 °C (1.73 kPa), $[\alpha]_D^{24}$ –142.5° (neat), occurs in essential oils of *Petasites* spp. Further isomers of the eremophiladiene group are present in other essential oils. V. is oxidized by chromates to nootkatone (see eremophilanes), the typical impact component of grapefruit flavor[2]. Metal-catalyzed oxidation processes have been responsible for flavor falsifications (grapefruit taste) of orange juices that have been stored in tinplate containers.

Lit.: [1]Chem. Ind. (London) **1970**, 312; Ohloff, p. 132. [2]Tetrahedron Lett. **31**, 1943 (1990).

gen.: Beilstein EIV **5**, 1186. – *Biosynthesis:* J. Chem. Soc. Chem. Commun. **1970**, 230. – *Synthesis:* Bull. Chem. Soc. Jpn. **55**, 887 (1982) ▪ Helv. Chim Acta **65**, 2212 (1982) ▪ J. Org. Chem. **50**, 3615 (1985). – *[CAS 4630-07-3 (V.); 10219-75-7 ((–)-V.); 10219-75-7 (eremophilene)]*

Valepotriates. V. are *iridoids in which the OH groups have been esterified with isovaleric acid and acetic acid, they also possess an exocyclic epoxide ring. They occur in up to 5% in the herbage of various Valerianaceae species, e.g., *Valeriana officinalis* (common valerian), *V. wallichii*, and *V. mexicana*, as well as in the red valerian (*Centranthus ruber*, Valerianaceae)[1]. The name is derived from *Val*eriana-*epo*xy-*tri*esters.

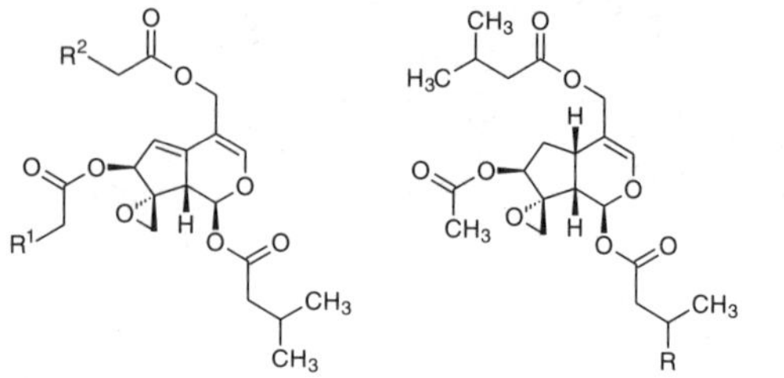

$R^1 = CH(CH_3)_2$, $R^2 = H$: Valtrate
$R^1 = H$, $R^2 = CH(CH_3)_2$: Isovaltrate
$R = CH_3$: Dihydrovaltrate
$R = C_2H_5$: Homodihydrovaltrate

Table: Data of Valepotriates.

trivial name	Valtrate	Isovaltrate[5]	Dihydrovaltrate	Homodihydrovaltrate
molecular formula	$C_{22}H_{30}O_8$	$C_{22}H_{30}O_8$	$C_{22}H_{32}O_8$	$C_{23}H_{34}O_8$
M_R	422.48	422.48	424.49	438.52
mp. [°C]	oil	–	64–65	50–51
$[\alpha]_D$ (CH_3OH)	+172.7°	+151°	–80.8°	–72°
CAS	18296-44-1	31078-10-1	18296-45-2	18361-41-6

Biological activity: V. have an equilibrating effect, i.e., sedating in states of excitement and stimulating in states of tiredness. Valerian teas and tinctures do not contain V. but rather valerenic, valerenolic, and acetylvalerenolic acids[2], that are purported to be responsible for the sedative activity of valerian preparations[3]. V. or their metabolites exhibit *in vivo* alkylating, cell toxic, and mutagenic activities[4].

Lit.: [1]Tetrahedron **24**, 313 (1968). [2]J. Am. Chem. Soc. **82**, 2962 (1960). [3]Dragoco-Rep. **31**, 3 (1984); Planta Med. **42**, 62 (1981). [4]Planta Med. **52**, 446 (1986). [5]Phytochemistry **13**, 2815 (1974).

gen.: Arzneim.-Forsch. **34**, 170 (1984) ▪ Beilstein EV **19/3**, 577 ▪ Hager (5.) **6**, 1068–1086 ▪ J. Nat. Prod. **43**, 649, No. 201, 203, 206, 207 (1980).

Valienamine.

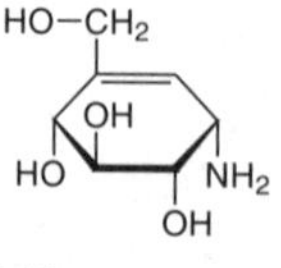

$C_7H_{13}NO_4$, M_R 175.18, mp. (pentaacetate) 95 °C. An aminocyclitol *antibiotic from *Streptomyces hygroscopicus* var. *limoneus*, it is active against *Bacillus* species and has hypoglycemic activity as an α-glucosidase inhibitor. V. is a component of the validamycins from which it is released by soil bacteria, it is also present in a series of pseudooligosaccharide α-glucosidase inhibitors, e.g., the orally active antidiabetic agent *acarbose.

Lit.: J. Antibiot. **33**, 1575 (1980); **37**, 1301 (1984). – *Synthesis:* Angew. Chem. Int. Ed. Engl. **26**, 482 (1987) ▪ Aust. J. Chem. **50**, 193 (1997) ▪ Chem. Pharm. Bull. **36**, 4236 ff. (1988) ▪ Gazz. Chim. Ital. **119**, 577 ff. (1989) ▪ J. Am. Chem. Soc. **120**, 1732 (1998) (synthesis) ▪ J. Org. Chem. **64**, 1941 (1999) (synthesis) ▪ Kleemann-Engel, p. 2008. – *[HS 2941 90; CAS 38231-86-6]*

L-Valine [(*S*)-2-amino-3-methylbutanoic acid, (*S*)-α-aminoisovalerianic acid, abbr.: Val or V].

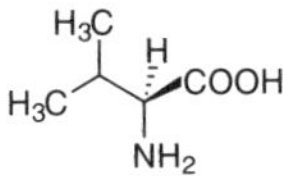

$C_5H_{11}NO_2$, M_R 117.15, mp. 315 °C (decomp., in a sealed capillary), $[\alpha]_D^{20}$ +6.3° (H_2O), +23.4° (1 m HCl), pK_a 2.32, 9.62, pI 5.96. An essential proteinogenic amino acid. Genetic code: GUU, GUC, GUA, GUG. Component of almost all proteins, 5–8% in grain, meat, and eggs, over 15% in elastin. Average content in proteins: 6.9% [1].

Biosynthesis: In microorganisms and plants from pyruvic acid: 2 pyruvate → 2-acetolactic acid (acetolactate synthase, EC 4.1.3.18; coenzyme: *thiamin(e) diphosphate) → 2,3-dihydroxyisovaleric acid (2-acetolactate mutase, EC 5.4.99.3) → 2-oxoisovaleric acid (dihydroxy acid dehydratase, EC 4.2.1.9). This is finally aminated by branched chain amino acid aminotransferase (EC 2.6.1.42). 2-Oxoisovaleric acid is also a precursor of Leu.

Adults require about 1.6 g Val per day since the amino acid is important for the normal function of the nerve/muscle apparatus; in cases of an insufficient supply of Val, hypersensitivity, motoric disorders, rotatory spasms, and degeneration of muscle cells are observed. D-Valine occurs in some peptide antibiotics such as *gramicidins, *actinomycins, amidomycin, and *valinomycin.

Lit.: [1] Biochem. Biophys. Res. Commun. **78**, 1018–1024 (1977).
gen.: Beilstein E IV **4**, 2659 ff. ▪ Hager (5.) **9**, 1149 ff. ▪ Karrer, No. 2348 ▪ Merck-Index (12.), No. 10046 ▪ Ullmann (5.) **A 2**, 58, 63, 74, 86. – *[HS 2922 49; CAS 72-18-4]*

Valinomycin.

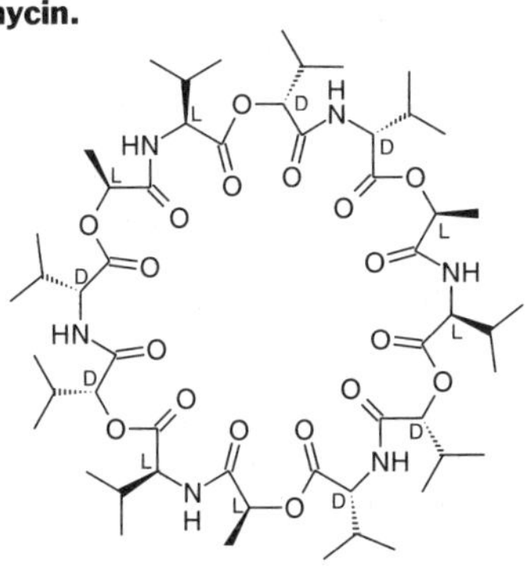

$C_{54}H_{90}N_6O_{18}$, M_R 1111.34, cryst., mp. 187 °C, $[\alpha]_D^{20}$ +31° (benzene), soluble in organic solvents. Cyclodepsipeptide *antibiotic from *Streptomyces fulvissimus* structurally related to the *enniatins. V. is composed of three residues each of L-*valine, D-valine, D-α-hydroxyisovaleric acid, and L-*lactic acid that form a 36-membered ring. V. is active against the tuberculosis pathogen (*Mycobacterium tuberculosis*). LD_{50} (rat p.o.) 4 mg/kg, it is highly toxic when swallowed or on contact with the skin.

Use: In experimental biochemistry as a decoupler of oxidative phosphorylation, also as a potassium carrier or *ionophore in studies of ion transport in biomembranes, and for ion-selective electrodes. V. is used as an insecticide and nematicide. For synthesis of V., see *Lit.* [1].

Lit.: [1] J. Org. Chem. **64**, 8063–8075 (1999) (analogues); Tetrahedron **45**, 5039–5050 (1989).
gen.: Beilstein E III/IV **27**, 9728 ▪ Int. J. Pept. Prot. Res. **39**, 291 (1992) ▪ Merck-Index (12.), No. 10047 ▪ Neurath & Hilmer (eds.), The Proteins, vol. 5, p. 563–573, New York: Academic Press 1982 ▪ Sax (8.), VBZ 000 ▪ Tetrahedron Lett. **30**, 1695 (1989) ▪ Top. Curr. Chem. **128**, 175–218 (1985). – *[HS 2941 90; CAS 2001-95-8]*

Valiolamine see glycosidase inhibitors.

Valtrate see valepotriates.

Vancomycin (vancocin).

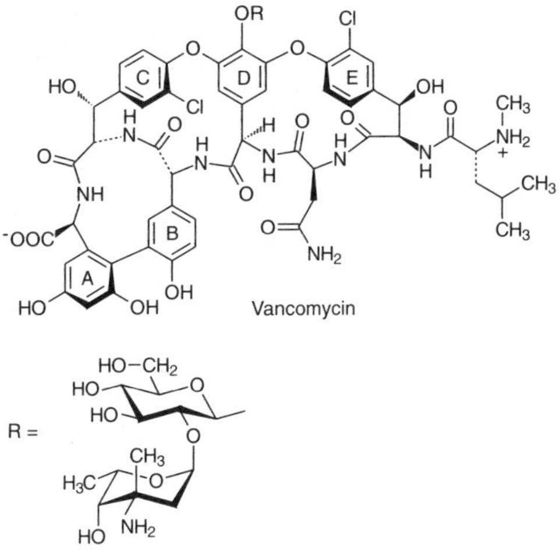

The parent compound, discovered in 1956, of a group of microbial glycopeptide antibiotics (dalbaheptides), composed of a heptapeptide with unusual aromatic amino acids and sugar residues bound *O*-glycosidically on the periphery. V. ($C_{66}H_{75}Cl_2N_9O_{24}$, M_R 1449.27, soluble in water and polar organic solvents, amphoteric, forms a hydrochloride) is produced by the rare *actinomycete *Amycolatopsis orientalis* (formerly incorrectly assigned to the genus *Streptomyces* or *Nocardia*), contains five aromatic amino acids, L-asparagine and *N*-methyl-D-leucine as well as the sugar building blocks D-glucose and vancosamine (3-amino-2,3,6-trideoxy-3-*C*-methyl-L-*lyxo*-hexopyranose). In the biosynthesis, the aromatic amino acid A is derived from acetate while the others (B to E) originate from tyrosine which, in the cases of B and D is degraded to *p*-hydroxyphenylglycine. The heptapeptide is formed non-ribosomally by a peptide synthetase and is presumably modified subsequently. V. and other members of the vancomycin group (e. g., *teicoplanins, ristocetin) have bactericidal effects on Gram-positive bacteria and are used clinically. V. is the agent of choice for infections by *β*-lactam-resistant staphylococci or streptococci. The activity is based on an inhibition of cell wall synthesis in sensitive bacteria. At least one other mechanism contributes to V. activity [1]. V. binds to the D-alanyl-D-alanine end of UDP-muramyl pentapeptide and thus prevents elongation and cross-linking of the peptidoglycan chain. The tricyclic peptide in V. is responsible for this binding, it forms a binding pocket (receptor) for D-Ala-D-Ala, modifications of the periphery influence strength of action and pharmacokinetics. The recently observed V.-resistance in enterococci is currently under investigation. Novel derivatives like eremomycin or Ly 307599 show no resistance yet [2]. The total synthesis of V. has recently been accomplished (1998/1999).

Lit.: [1] Angew. Chem. Int. Ed. Engl. **38**, 1172–1193 (1999). [2] Science **284**, 507 (1999); J. Chem. Soc., Chem. Commun. **1999**, 1361; J. Chem. Soc., Perkin Trans. 1 **1999**, 2267.
gen.: Biosynthesis: Angew. Chem. Int. Ed. Engl. **38**, 1976 (1999) ▪ FEMS Microbiol. Lett. **59**, 129–134 (1989) ▪ Nature (London) **397**, 567 (1999) ▪ Zmijewski & Fayerman, in Vining & Stuttard (eds.), Genetics and Biochemistry of Antibiotic Production, p. 269–281, Boston: Butterworth-Heinemann 1995. – *Total synthesis:* Angew. Chem. Int. Ed. Engl. **37**, 2700, 2704, 2708, 2714, 2717 (1998); **38**, 240, 634 (1999); **39**, 44–122, 1084 (2000) ▪ Bioorg. Med. Chem. Lett. **8**, 721 (1998) ▪ Chem.-Eur. J. **5**, 2584, 2602, 2622, 2648 (1999) ▪ J. Am. Chem. Soc. **120**, 11014 (1998); **121**, 1237, 3226, 10004 (1999) ▪ Nachr. Chem. Tech. Lab. **46**, 1182 (1998) – *Reviews:* Angew. Chem. Int. Ed. Engl. **38**, 2096–2152 (1999) ▪ Drugs of the Future **15**, 57–72 (1990) ▪ J. Antibiot. **42**, 1882 f. (1989) ▪ Lancini & Cavalleri, in Kleinkauf & von Döhren (eds.), Biochemistry of Peptide Antibiotics, p. 159–178, Berlin: De Gruyter 1990 ▪ Merck-Index (12.), No. 10066. – *Activity & resistance:* Antimicrob. Agents Chemother. **35**, 605–609 (1991) ▪ Chem. Biol. **3**, 21–28 (1996) ▪ Helv. Chim. Acta **79**, 942–960 (1996) ▪ J. Am. Chem. Soc. **113**, 2264–2270 (1991) ▪ J. Antibiot. **46**, 1181 (1989); **48**, 805 (1995); **49**, 575–581 (1996) ▪ J. Chem. Soc., Perkin Trans. 2 **1995**, 159 ▪ J. Med. Chem. **41**, 2090 (1998); **42**, 4714 (1999) ▪ Kleemann-Engel, p. 1985. – *[CAS 1404-90-6]*

Vanilla. V. is the dried and fermented pod fruit of the vanilla plant, *Vanilla planifolia*, syn. *V. fragrans*, a tropical creeper belonging to the Orchidaceae. V. pods contain a dark, paste-like mass with many small black grains.

Originally indigenous to Mexico, V. is now cultivated in many tropical regions. Main cultivating countries are Madagascar, Indonesia, Comore Islands, Mexico, Réunion ("Bourbon-Vanilla"). In Tahiti smaller amounts of V. pods are obtained from a different V. species, *Vanilla tahitensis*. The annual world-wide production amounts to about 2000 t. The USA alone imported ca. 1100 t of V. pods in 1990, corresponding to a monetary value of about US $ 48 000 000.

After harvest, the V. pods are subjected to a fermentation process (supported by heating) and subsequently to drying and conditioning processes which can last for several months. During the fermentation, the odorless glucoside glucovanillin is converted to free *vanillin* (see 3,4-dihydroxybenzaldehydes), the main contributor to V. flavor. A V. pod contains about 2 wt.-% vanillin. In addition, many other substances, some of which are only present in trace amounts, contribute to the organoleptic impression [1]. The odor and taste of V. pods is of a heavy sweet, somewhat phenolic, creamy, spicy-balsamy, tobacco-like nature. V. from Tahiti has a markedly different organoleptic profile; it is flowery, perfume-like on account of its appreciable contents of **anethole* and **p*-anisaldehyde.

Use: In the food and perfume industries, V. extracts are used preferentially for technical reasons (e. g., V. tincture, V. oleoresin, V. absolute). The main use of these products is to improve the aromatic character of foods. In the perfume industry, V. extracts are only used in high-quality, expensive perfume compositions. For example, they add a unique fullness and brilliance to sweet-flowery and oriental notes.

Lit.: [1] Perfum. Flavor. **18** (2), 25 (1993).
gen.: Arctander, p. 638, 641, 642 ▪ Dev. Food. Sci. **34**, 517–577 (1994). – *Toxicology:* Food Cosmet. Toxicol. **20**, 849 (1982). – *[HS 0905 00; CAS 8023-78-7 (V. oleoresin); 8024-06-4 (V. absolute)]*

Vanillic acid see dihydroxybenzoic acids.

Vanillin see 3,4-dihydroxybenzaldehydes.

Varacin.

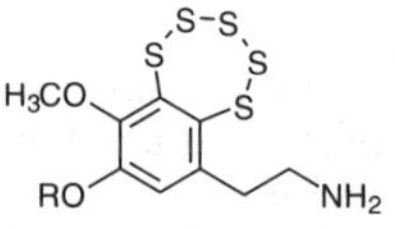

R = CH_3 : Varacin (**1**)
R = H : Lissoclinotoxin A (**2**)

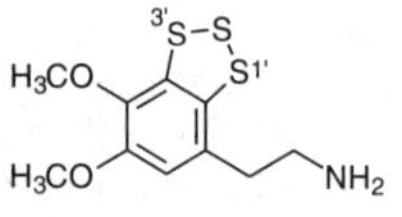

Trithiolane derivative of Varacin (**3**)
(Varacin A)

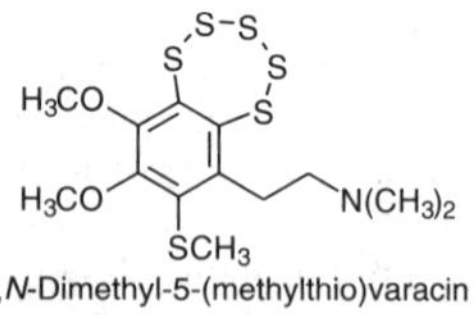

N,N-Dimethyl-5-(methylthio)varacin (**4**)

$C_{10}H_{13}NO_2S_5$, M_R 339.52, light yellow powder, mp. 258–260 °C (decomp.). Ascidia of the genera *Lissoclinum* and *Polycitor* produce unusual cyclic polysulfides of the benzopentathiepin and benzotrithiolane types. V. from *L. vareau* has strong antifungal and cytotoxic activities and proved to be 100-fold more active than 5-fluorouracil against the human colon cancer cell line HCT 116 [1]. The cytotoxic activity is probably attributable to damage to DNA. The corresponding trithiolane derivative (*V. A*, $C_{10}H_{13}NO_2S_3$, M_R 275.40) and its 1′- and 3′-*S*-oxides have been detected in a Far Eastern *Polycitor* species [2]. *Lissoclinotoxin A* ($C_9H_{11}NO_2S_5$, M_R 325.49, beige amorphous solid, mp. 245–250 °C) from *L. perforatum* is active against fungi and bacteria and was originally but incorrectly considered to be a benzotrithiolane [3]. *L. japonicum* contains other derivatives of V. as well as *N,N-dimethyl-5-(methylthio)varacin* ($C_{13}H_{19}NO_2S_6$, M_R 413.66, trifluoroacetate: pale yellow oil) and the corresponding trithiolane ($C_{13}H_{19}NO_2S_4$, M_R 349.54, trifluoroacetate: pale yellow oil), both are inhibitors of protein kinase C [4].

Lit.: [1] J. Am. Chem. Soc. **113**, 4709 (1991); **115**, 7017 (1993); **117**, 7261 (1995); J. Org. Chem. **58**, 4522 (1993) (synthesis). [2] J. Nat. Prod. **58**, 254 (1995). [3] Tetrahedron Lett. **32**, 911 (1991); Tetrahedron **50**, 5323 (1994). [4] Tetrahedron **50**, 12785 (1994).
gen.: Nachr. Chem. Tech. Lab. **40**, 165 (1992). – *[CAS 134029-48-4 (1); 133883-05-3 (2); 159645-85-9 (3); 159645-83-7 (4)]*

Variabilin.

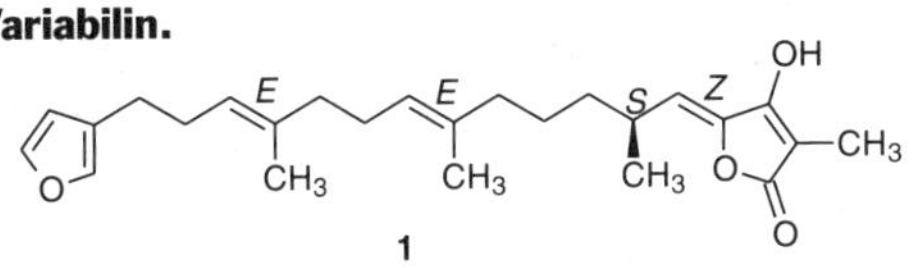

1

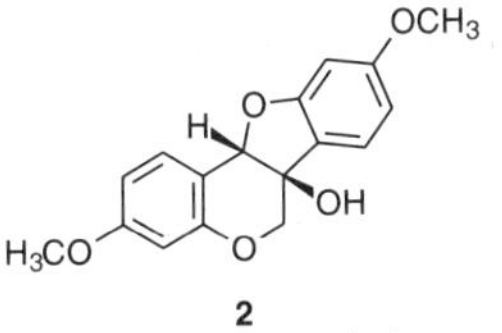

2
(relative configuration)

1. A *sesterterpenoid from marine sponges of the order Dictyoceratida: *Sarcotragus* spp., *Ircinia* spp., and *Psammocinia* spp., $C_{25}H_{34}O_4$, M_R 398.54, (*S*)-(4*Z*,10*E*,14*E*)-(–)-form, oil, $[\alpha]_D$ –4° ($CHCl_3$). V. has antiviral activity against *Herpes simplex*, but is concomitantly highly cytotoxic which prevents its clinical use.
2. 6a-Hydroxy-3,9-dimethoxypterocarpan (homo-*pisatin), $C_{17}H_{16}O_5$, M_R 300.31, a *phytoalexin from the cotyledons of Fabaceae plants such as lentils (*Lens culinaris* and *L. nigricans*) and clover (*Trifolium pratense*), also present in the wood of palisander (*Dalbergia variabilis*), exists as both antipodes.

Lit. (to 1): Biochem. Syst. Ecol. **15**, 373 ff. (1987) ▪ J. Nat. Prod. **51**, 275–281, 1294–1298 (1988); **52**, 346–359 (1989) ▪ Nat. Prod. Lett. **3**, 189 ff. (1993); **4**, 51–56 (1994); **6**, 281 (1995) (abs. configuration). – *(to 2):* Biochem. Syst. Ecol. **18**, 329–343 (1990); **19**, 497–506 (1991) ▪ Phytochemistry **36**, 189–194 (1994) ▪ Zechmeister **43**, 147–155. – *[CAS 51847-87-1 ((S)-(4Z,10E,14E)-(–)-form of 1); 3187-52-8 ((6aR,11aR)-form of 2)]*

Variabiline.

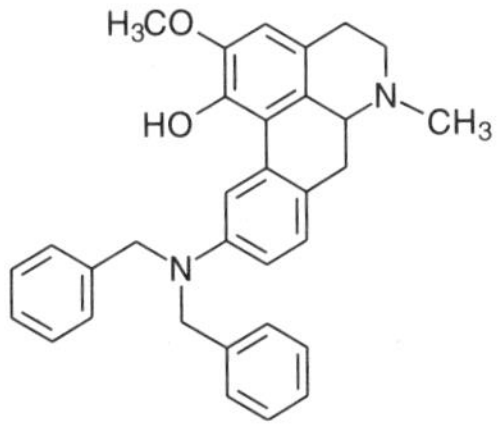

An *aporphine alkaloid from *Ocotea variabilis* (Lauraceae), $C_{32}H_{32}N_2O_2$, M_R 476.62, cryst., mp. 116–117 °C, as hydrochloride mp. 230–232 °C (decomp.).

Lit.: Tetrahedron Lett. **1972**, 4647 ff. – *[CAS 40374-54-7]*

Variegatic acid (3,3′,4,4′-tetrahydroxypulvinic acid). $C_{18}H_{12}O_9$, M_R 372.29, red needles, mp. 235 °C (decomp.). A hydroxylated *pulvinic acid derivative from the mushroom *Suillus variegatus* and many other boletes of the genera *Boletellus*, *Boletus*, *Chalciporus*, *Suillus*, *Xerocomus*. V. may be considered as a characteristic pigment of the order Boletales and has also been

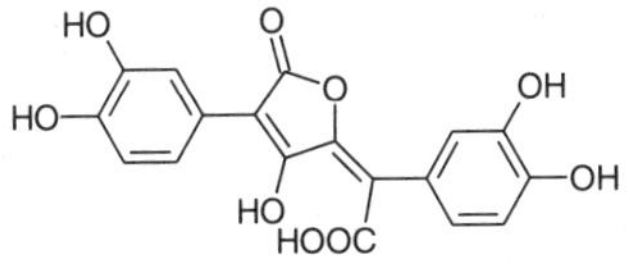

Variegatic acid

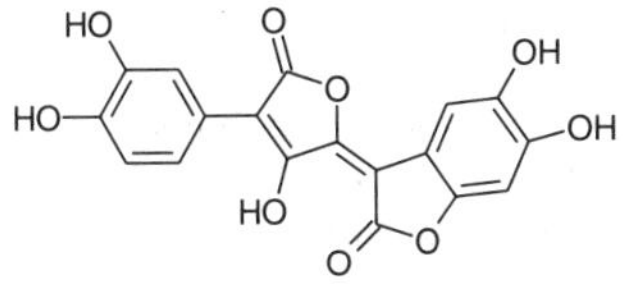

Variegatorubin

detected in, e.g., the dry-rot fungus (*Serpula lacrimans*) and in the gasteromycete *Rhizopogon luteolus*. Like *xerocomic acid, V. is responsible for the blue staining of fungal tissue upon injury. This blue color results from the formation of hydroxyquinone methide anions by oxidases and atmospheric oxygen. Oxidation of V. (in solution on contact with air or on a preparative scale with hydrogen peroxide) affords the red pigment *variegatorubin* [$C_{18}H_{10}O_9$, M_R 370.27, brown-violet needles, mp. >320 °C (decomp.)] often present in the cap skins of boletes, e.g., *Boletus frostii*.

Lit.: Beilstein E V **18/9**, 411 (V.); E V **19/7**, 147 (variegatorubin) ▪ J. Chem. Soc., Perkin Trans. 1 **1973**, 1529 (synthesis) ▪ Zechmeister **51**, 32–51 (review). – *[CAS 20988-30-1 (V.); 27286-59-5 (variegatorubin)]*

Variotin (pecilocin).

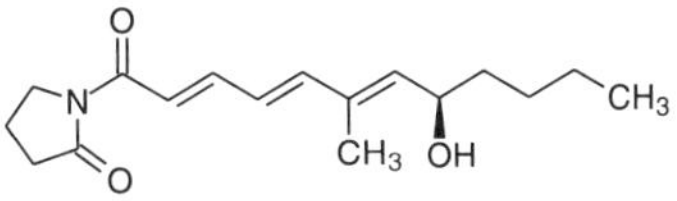

$C_{17}H_{25}NO_3$, M_R 291.38, mp. 41.5–42.5 °C, $[\alpha]_D$ –5.7° (CH_3OH). International free name for the oxidation-sensitive trienecarboxamide 1-[(2*E*,4*E*,6*E*,8*R*)-8-hydroxy-6-methyl-2,4,6-dodecatrienoyl]-2-pyrrolidinone in the all-*E* configuration isolated from cultures of the fungus *Paecilomyces variotii*. V. is active against a series of pathogenic fungi and is used topically in the treatment of dermatomycoses.

Lit.: Agric. Biol. Chem. **37**, 903, 911, 1131 (1973) ▪ J. Antibiot. **A12**, 109, 195 (1959); **22**, 179 f. (1980). – *Synthesis:* Bull. Chem. Soc. Jpn. **51**, 2077–2081 (1978). – *[HS 2941 90; CAS 19504-77-9]*

Vasicin(one) see peganine.

Vatine, vatamine, vatamidine see pyrroloindole alkaloids.

Veatchine.

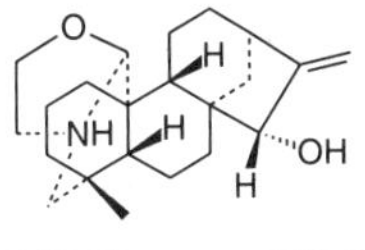

$C_{22}H_{33}NO_2$, M_R 343.51, mp. 128.5 °C, $[\alpha]_D^{30}$ –67.5° ($CHCl_3$). A diterpene alkaloid from the bark of *Garrya veatchii*.

Lit.: J. Nat. Prod. **43**, 41–71 (1980). – *[HS 2939 90; CAS 76-53-9]*

Vegetable flavors. Only about 40 of the 80 known sorts of vegetable[1], including the edible fungi (see mushrooms aroma constituents), have as yet been studied for their volatile constituents and more than 1000, sometimes species-specific, *flavor compounds have been identified[2]. Often the typical flavor is only released after destruction of the cellular structure by enzymatic processes, e.g., isothiocyanates (*mustard oils) from *glucosinolates, in cabbage and radish species (Brassicaceae), or the onion and garlic constituents. The characteristic flavor compounds of some selected vegetables are given below:

Asparagus: The main volatile components of raw asparagus are *1,2-dithiolane-4-carboxylic acid* (asparagus acid) ($C_4H_6O_2S_2$, M_R 150.21, CAS [2224-02-4]) and its *methyl ester* ($C_5H_8O_2S_2$, M_R 164.24, CAS [7389-17-5]). The key compounds of boiled asparagus flavor are *1,2-dithiacyclopentene* (3*H*-1,2-dithiole) ($C_3H_4S_2$, M_R 104.18, CAS [288-26-6]), dimethyl sulfide and 2-acetylthiazole (see meat flavor). The strong smell in urine after consumption of asparagus is due to *S-methyl 2-propenethioate* (C_4H_6OS, M_R 102.15, CAS [5883-16-9]) and *S-methyl 3-(methylthio)thiopropanoate* ($C_5H_{10}OS_2$, M_R 150.25, CAS [42919-65-3]).

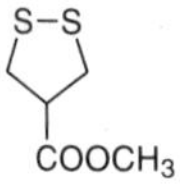

1,2-Dithiolane-4-carboxylic acid methyl ester

1,2-Dithiacyclopentene

Beetroots: The flavor is decisively characterized by the earthy note of *geosmin

Carrots: The impact compound in fresh carrots is 2-*sec*-butyl-3-methoxypyrazine together with numerous terpenes, sesquiterpenes, and *β*-*ionone, which also contribute to the flavor of boiled carrots; the latter contain additionally (*E*)-2-nonenal, (*E*)-2-decenal (see alkenals), and (*E,E*)-*2,4-decadienal as active principles.

Cucumbers: The impact compound is (*E,Z*)-2,6-*nonadienal, together with (*E*)- and (*Z*)-2-nonenal (see alkenals), hexanal (see alkanals), and (*Z*)-1,5-octadien-3-one (see tea flavor).

Green peas (raw): 2-alkyl-3-methoxypyrazines (2-isopropyl-, 2-isobutyl-, and 2-*sec*-butyl-, see pyrazines) together with (*E*)-2-nonenal (see alkenals) (*E,Z*)-2,6-*nonadienal, (*E,E*)-*2,4-decadienal, hexan-1-ol, and (*Z*)-3-*hexen-1-ol.

Leeks: Leek oil contains numerous thiols, sulfides, di-, tri-, tetra-, and pentasulfides, trithiolanes and thioesters[3]. One isomer of *3-ethyl-5-methyl-1,2,4-trithiolane* ($C_5H_{10}S_3$, M_R 166.31, CAS [116505-59-0]) has a mild leek odor, the other resembles the odor of onions.

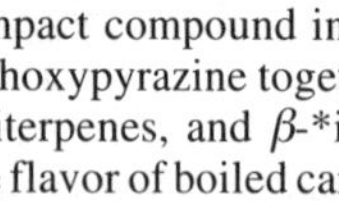

3-Ethyl-5-methyl-1,2,4-trithiolane

Paprikas: The main flavor components of raw pepper pods are 2-alkyl-3-methoxypyrazines (especially 2-isobutyl-, see pyrazines), as well as (*E,Z*)-2,6-*nonadienal and (*E,E*)-*2,4-decadienal.

Potatoes: 2-Isopropyl-3-methoxypyrazine together with 2,5-dimethylpyrazine (see pyrazines) dominate in raw potatoes. Key compounds in boiled potatoes are *methional and 2-ethyl-3,6-dimethylpyrazine.

Tomatoes: The highest flavor values in fresh tomatoes are exhibited by (*Z*)-3- and (*E*)-2-hexenals (see alkenals), *β*-*ionone, hexanal (see alkanals), **β*-damascenone, *1-penten-3-one* (C_5H_8O, M_R 84.12, CAS [1629-58-9], 3-methylbutanal (see alkanals), and *2-isobutylthiazole[1]. The most important flavor components in tomato pulp are dimethyl sulfide, *β*-damascenone, *β*-ionone, 3-methylbutanal, *1-nitro-2-phenylethane* ($C_8H_9NO_2$, M_R 151.16, CAS [24330-46-9]), *eugenol, *methional, and 3-methylbutanoic acid[4].

Lit.: [1]Z. Naturforsch. C **43**, 731 (1988). [2]TNO list (6.), Suppl. 1, p. 1–256; Maarse, p. 203–281. [3]Justus Liebig Ann. Chem. **1993**, 871. [4]J. Agric. Food Chem. **38**, 336–340, 792–795 (1990).

Velban, Velbe see vinblastine.

Velleral.

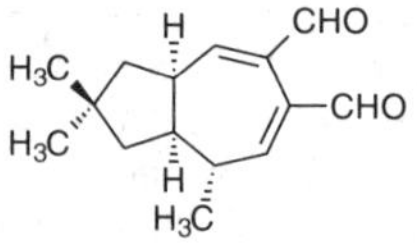

Velleral

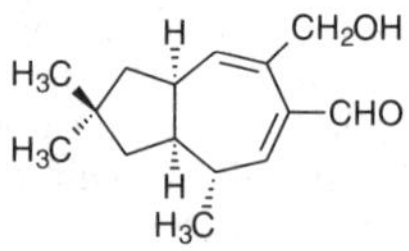

Vellerol

$C_{15}H_{20}O_2$, M_R 232.32, cryst., mp. 86.5–87.5 °C, $[\alpha]_D$ –25° ($CHCl_3$). A *lactarane derivative with a pungent taste and mutagenic and antimicrobial activities from toadstools of the genera *Lactarius* (milk caps), *Russula*, and other Russulaceae; like the isomeric *isovelleral V. is formed enzymatically from *stearoylvelutinal within a few seconds after injury of the fruit body. This is followed in fungi such as *L. vellereus* by an enzymatic "detoxification" in which V. is reduced within a few minutes to *vellerol* ($C_{15}H_{22}O_2$, M_R 234.34). Velleral and isovelleral have antifeedant effects on insects and mammals (opossum).

Lit.: Acta Chem. Scand., Ser. B **43**, 694 (1989) ▪ J. Am. Chem. Soc. **100**, 6728 (1978) ▪ J. Chem. Ecol. **10**, 1439 (1983) ▪ J. Nat. Prod. **48**, 279 (1985) ▪ Mutat. Res. **188**, 169 (1987) ▪ Toxicol. in Vitro **5**, 9 (1991). – *[CAS 50656-61-6 (V.); 96910-72-4 (vellerol)]*

Velutinal see stearoylvelutinal.

Vera... see Veratrum steroid alkaloids.

Veratraldehyde, Veratrumaldehyde see 3,4-dihydroxybenzaldehydes.

Veratrum steroid alkaloids. These include *steroid alkaloids with the characteristically changed (rearranged) cholestane C_{27} skeleton occurring in the plant family Liliaceae, in particular the genus *Veratrum*, as well as the genera *Schoenocaulon*, *Amianthium*, *Fritillaria*, *Rhinopetalum*, and *Zygadenus*. These alkaloids can be divided into three groups characterized by the following structural features:

1) Compounds with a 22,26-epimino-14(13→12)-*abeo*-cholestane ring system (*veratraman(e)* type, lead compound veratramine, see figure 1),

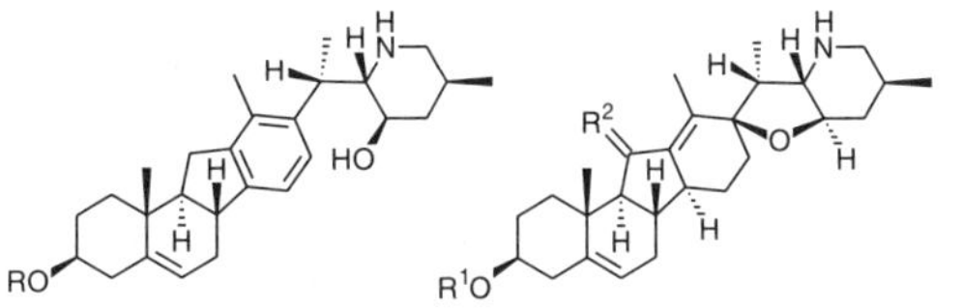

Figure 1: Alkaloids of the Veratramane type.

2) Compounds with the 18,22,26-nitrilo-14(13→12)-*abeo*-5α-cholestane ring system (*5α-cevane* type, e. g., veracevine, see figure 2), and finally

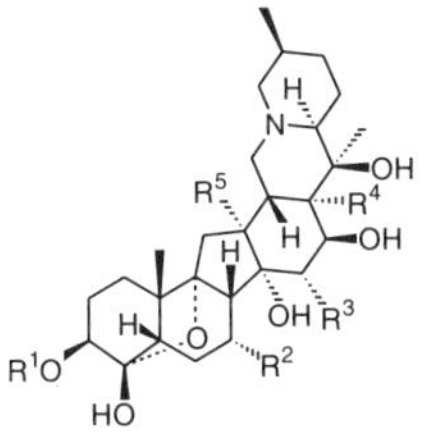

R^1	R^2	R^3	R^4	R^5	
H	H	H	OH	OH	: Veracevine
An	H	H	OH	OH	: Cevadine
Ve	H	H	OH	OH	: Veratridine
H	OH	OH	H	H	: Germine
HMB	OAc	OMB	H	H	: Germitrine

Ac = acetyl, An = angeloyl,
Ve = veratroyl,
HMB = (*S*)-2-hydroxy-2-methylbutyryl,
MB = (*R*)-2-methylbutyryl

Figure 2: Alkaloids of the Cevane type.

3) Compounds with the 22,26-epimino-18(13→17)-*abeo*-5α-cholestane ring system (*5α-veralinane* type, lead compound veralinine, see figure 3).

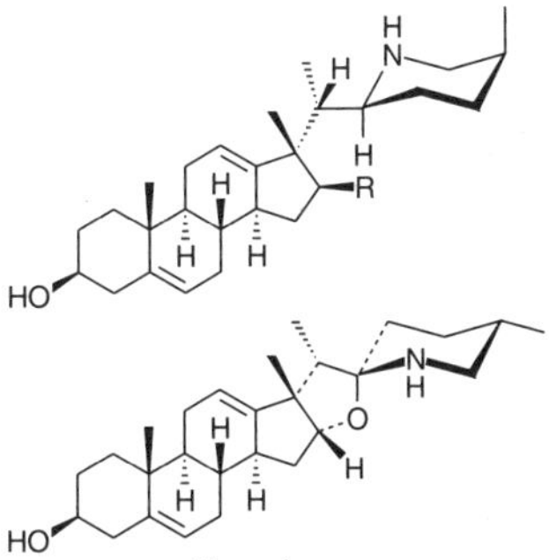

Figure 3: Alkaloids of the Veralinane type.

The Veratrum steroid alkaloids are often also divided into the groups of the *jerveratrum alkaloids* (with 1–3 oxygen atoms) and *ceveratrum alkaloids* (with 7–9 oxygen atoms). Glycosides have to date only been found among the jerveratrum alkaloids. The oxygen-rich ceveratrum alkaloids are mainly found in the form of various ester derivatives, the acid components include principally acetic, angelic, tiglic, veratric, and vanillic acids as well as (less common) (*R*)-2-methyl-, (*S*)-2-hydroxy-2-methyl-, and D-*threo*- or L-*erythro*-2,3-dihydroxy-2-methylbutanoic acids. In addition to these alkaloids, the above-mentioned Liliaceae genera also contain numerous *Solanum steroid alkaloids, which play a role as biosynthetic intermediates of the Veratrum steroid alkaloids of groups 1–3. Of the well over 100 members, a few characteristic examples are listed in the table, p. 688.

Veratrine, used officinally in the past and obtainable from sabadilla seeds (*S. officinale* = *V. sabadilla*), is a mixture of several cevane alkaloids, including cevadine (ca. 70%) and veratridine (ca. 25%) as well as cevine and sabadine.

Biosynthesis: These alkaloids are formed by the usual plant steroid biosynthetic pathway via *cycloartenol and *cholesterol. The Solanum steroid alkaloids and *procevine* {(22*R*,25*S*)-18,22,26-nitrilocholest-5-en-3β-ol, CAS [468-24-6]} also present in Liliaceae are important intermediates in these processes and furnish all the other types of compounds by carbon skeleton rearrangement and ring closure reactions [1].

Activity: Many of these alkaloids are highly toxic [2]. Veratrine was formerly used externally as an antineuralgic agent but also has cardiotoxic properties. Veracevine and germine are supposed to have blood pressure lowering effects, and veratramine cardiostimulating activity. The jerveratrum alkaloids cyclopamine, cycloposine, jervine, and pseudojervine have strong teratogenic activities. It has been found that these alkaloids cause fetal malformations in pregnant sheep, e. g., cyclopia (one eye only) [2,3]. Veratridine is used in experimental pharmacology because it binds to sodium channels, keeps them open and permanently active. Cevane alkaloids are occasionally used as insect poisons [2].

Lit.: [1] Mothes et al., p. 363–384. [2] Manske **43**, 1–118. [3] J. Nat. Prod. **38**, 56–86 (1975); Pelletier **4**, 389–425.
gen.: Fieser & Fieser, Steroide, p. 957–988, Weinheim: Verl. Chemie 1961 ▪ Manske **3**, 247–312; **7**, 363–417; **10**, 193–285; **14**, 1–82; **41**, 177–237 ▪ Merck-Index (12.), No. 10082, 10085, 10087 f. ▪ Phytochemistry **43**, 907 (1996) (pseudojervine); **44**, 1257 (1997) ▪ Ullmann (5.) **A 1**, 400. – *[HS 2939 90]*

Verazine see Solanum steroid alkaloids.

Verbascine, Verballocine see Verbascum alkaloids.

Verbascum alkaloids. Group of macrocyclic *spermine alkaloids from the leaves of *Verbascum pseudonobile* (Scrophulariaceae), e. g. the main alkaloids (–)-(*S*)-verbascine and (–)-(*S*)-verballocine; both are glass-like solids, $C_{28}H_{38}N_4O_2$, M_R 462.63.

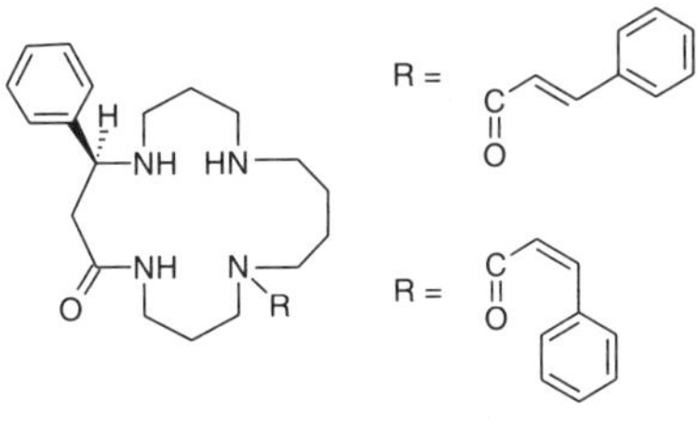

Lit.: Helv. Chim. Acta **82**, 229 (1999). – *[CAS 164230-52-8 (verbascine); 164176-10-7 (verballocine)]*

Verbenols see pinenols.

(+)-Verbenone see pinenones.

Vermeerin see hymenoxone.

Vernolepin.

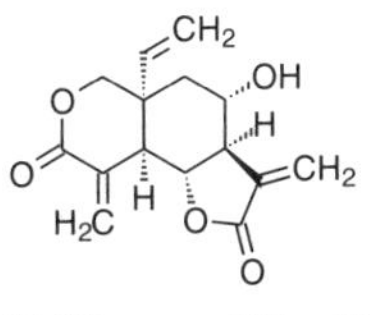

$C_{15}H_{16}O_5$, M_R 276.29, mp. 181–182 °C, $[\alpha]_D^{28}$ +72° (acetone). A sesquiterpene lactone of the elemanolide

Table: Data of Veratrum steroid alkaloids.

alkaloid	molecular formula	M_R	mp. [°C]	$[\alpha]_D$	occurrence	CAS
Veratramine	$C_{27}H_{39}NO_2$	409.61	209.5–210.5	–70° (CH_3OH)	*Veratrum album*, var. *grandiflorum* and other *V.* species	60-70-8
Veratrosine [3-*O*-(*β*-D-gluco-pyranosyl)-veratramine]	$C_{33}H_{49}NO_7$	571.75	242–243 (decomp.)	–53° ($CHCl_3/C_2H_5OH$)	*V. viride*, *V. eschscholtzii*	475-00-3
Jervine	$C_{27}H_{39}NO_3$	425.61	243–244	–147° (C_2H_5OH)	*V. album* var. *grandiflorum*, *V. californicum* and other *V.* species, *Amianthium muscaetoxicum*	469-59-0
Pseudojervine [3-*O*-(*β*-D-gluco-pyranosyl)-jervine]	$C_{33}H_{49}NO_8$	587.75	300–301 (decomp.)	–131° ($CHCl_3/C_2H_5OH$)	*V. album* and other *V.* species	36069-05-3
Cyclopamine (11-deoxojervine)	$C_{27}H_{41}NO_2$	411.63	237–238	–48° ($CHCl_3/CH_3OH$)	*V. californicum*, *V. album*	4449-51-8
Cycloposine [3-*O*-(*β*-D-gluco-pyranosyl)-cyclopamine]	$C_{33}H_{51}NO_7$	573.77	267–269	–51° (C_2H_5OH)	*V. californicum*	23185-94-6
Veracevine (protocevine)	$C_{27}H_{43}NO_8$	509.64	220–225	–26° (C_2H_5OH)	*Schoenocaulon officinale*	5876-23-3
Cevadine (3-*O*-angeloyl-veracevine)	$C_{32}H_{49}NO_9$	591.74	209–211	+11° (C_2H_5OH)	*S. officinale*	62-59-9
Veratridine (3-*O*-veratroyl-veracevine)	$C_{36}H_{51}NO_{11}$	673.80	170–178	+8.2° (C_2H_5OH)	*S. officinale*	71-62-5
Germine	$C_{27}H_{43}NO_8$	509.64	220–225	+4° (C_2H_5OH)	*V. album*, *V. viride* and other *V.* and *Zygadenus* species	508-65-6
Germitrine	$C_{39}H_{61}NO_{12}$	735.91	216–219	–4° ($CHCl_3$)	*V. album*, *V. viride* and other *V.* and *Z.* species	560-48-5
Veralinine	$C_{27}H_{43}NO$	397.64	124–126	–80° ($CHCl_3$)	*V. album* ssp. *lobelianum*	212333-19-2
Veralkamine (16*β*-hydroxy-veralinine)	$C_{27}H_{43}NO_2$	413.64	165–169	–84.1° ($CHCl_3$)	*V. album* ssp. *lobelianum*, *Fritillaria camtschatcensis*	17155-31-6
Veramine	$C_{27}H_{41}NO_2$	411.63	amorphous	–93.9° ($CHCl_3$)	*V. album* ssp. *lobelianum*, *V. nigrum*, *V. oxysepalum*	21059-48-3

dilactone type from *Vernonia hymenolepis* (Asteraceae) with antitumor activity.

Lit.: J. Org. Chem. **59**, 6395 (1994) ▪ Merck-Index (12.), No. 10098 ▪ Phytochemistry **37**, 191 (1994) ▪ Tetrahedron Lett. **34**, 4219 (1993). – *[CAS 18542-37-5]*

Vernolic acid [(9*Z*,12*S*,13*R*)-12,13-epoxy-9-octadecenoic acid, leukotoxin B].

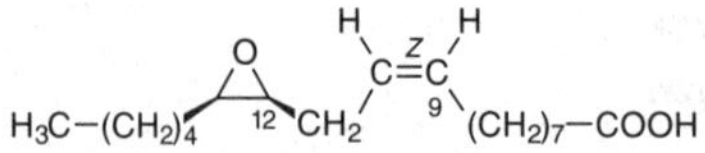

$C_{18}H_{32}O_3$, M_R 296.45, mp. 32.5 °C, $[\alpha]_D^{20}$ +2.03° (hexane), soluble in organic solvents. V. occurs in large amounts (60–80%) as glycerol esters in the seed oils of *Vernonia anthelmintica* and *V. galamensis* (Asteraceae) as well as *Euphorbia lagascae* (Euphorbiaceae). V. and other epoxy fatty acids are toxic compounds; they are used for technical purposes, especially in the manufacture of plastics and paints.

Lit.: Beilstein E V **18/6**, 99 ▪ Fat Sci. Technol. **91**, 488 (1989) ▪ Lipids **33**, 1217 (1998) (biosynthesis) ▪ Merck-Index (12.), No. 10099 ▪ Murphy (ed.), Designer Oil Crops, p. 119 f., 257, Weinheim: VCH Verlagsges. 1994 ▪ Tetrahedron Lett. **24**, 4715 (1983); **27**, 303 (1986) (synthesis) ▪ Ullmann (5.) **A 10**, 233. – *[CAS 503-07-1]*

Verpacrocin.

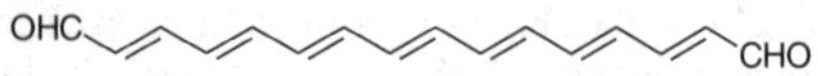

$C_{16}H_{16}O_2$, M_R 240.30, orange-colored cryst., mp. 208 °C (decomp.), very poorly soluble in organic solvents. A polyene dialdehyde from mycelium cultures of the morel *Verpa digitaliformis* (Ascomycetes).

Lit.: Z. Naturforsch. C **38**, 492 (1983). – *[CAS 86827-80-7]*

Verrol see trichoverroids.

Verrucarins. *Mycotoxins of the *tric(h)othecene group. They are macrocyclic triesters of the parent compound verrucarol (see scirpenols). Compounds with two separated side chains belong to the *trichoverroids. Examples of the V. are shown in the table. The V. decompose at >315–360 °C. They are formed by deuteromycetes, e. g., *Myrothecium verrucaria, M. roridum, Stachybotrys chartarum*, and *Dendrodochium toxicum*. V. exhibit toxicities typical for the trichothecene type: they are skin irritants, cytotoxic, and potent poisons for vertebrates (e. g., LD_{50} mouse p. o. 7 mg/kg; i. p. 0.5 mg/kg or i. v. 1.5 mg/kg). A pyrrole derivative also obtained from *M. verrucaria* is known as *verrucarin E* {1-[4-(hydroxymethyl)-3-pyrrolyl]-ethanone, $C_7H_9NO_2$, M_R 139.15, mp. 90–91 °C}.

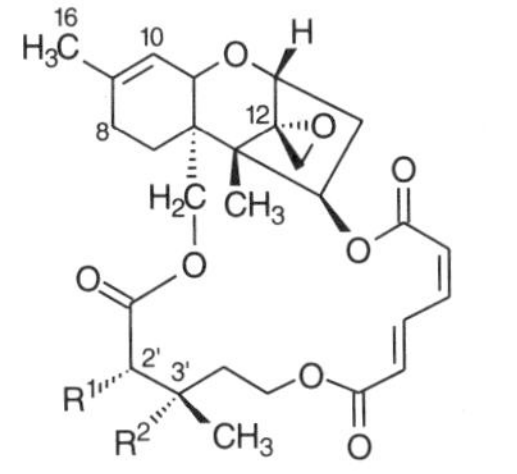

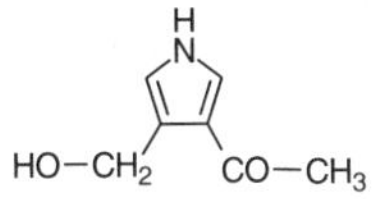
V. E

Table: Structure and data of Verrucarins.

Verrucarin (double bond in side chain)	R¹	R²	molecular formula	M_R	CAS
V. A	OH	H	$C_{27}H_{34}O_9$	502.56	3148-09-2
V. B	–O–		$C_{27}H_{32}O_9$	500.55	2290-11-1
V. J ($\Delta^{2'E}$)	H	H	$C_{27}H_{32}O_8$	484.55	4643-58-7
V. K ($\Delta^{12,13}$, deepoxy)	OH	H	$C_{27}H_{34}O_8$	486.56	63739-93-5
V. L ($\Delta^{2'E}$, 8α-OH)	H	H	$C_{27}H_{32}O_9$	500.55	77101-87-2

Lit.: Bioact. Mol. **10**, 197–204 (1989) ▪ Cole-Cox, p. 247–260 ▪ Helv. Chim. Acta **67**, 1168 (1984) (V. E) ▪ J. Org. Chem. **49**, 4332 (1984) (synthesis) ▪ Merck-Index (12.), No. 10100 ▪ Zechmeister **47**, 153–220. – *Toxicology:* Betina, p. 212–215 ▪ Food Chem. Toxicol. **25**, 379 (1987) ▪ Sax (8.), MRV 500. – *[CAS 24445-13-4 (V. E)]*

Verrucarol see scirpenols.

Verrucosidin.

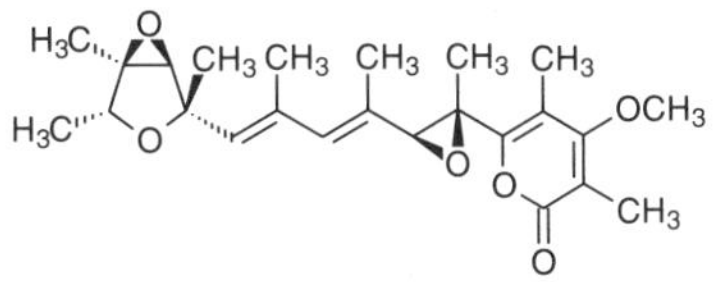

$C_{24}H_{32}O_6$, M_R 416.51, mp. 90–91 °C , $[\alpha]_D$ +92.4° (CH_3OH); a tremorigenic *mycotoxin formed by *Penicillium verrucosum* var. *cyclopium* and other *Penicillium* species. V. is structurally related with the *citreoviridins and *asteltoxin.

Lit.: J. Am. Chem. Soc. **112**, 8985 (1990) ▪ J. Org. Chem. **59**, 3762 (1994) ▪ Zechmeister **48**, 49 f. – *[CAS 88389-71-3]*

Verruculogen see fumitremorgins.

Verruculotoxin. $C_{15}H_{20}N_2O$, M_R 244.34, mp. 152 °C, $[\alpha]_D$ –56° (CH_3OH); a tremorigenic *mycotoxin from

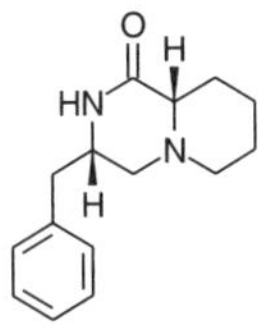

green peanuts infected with *Penicillium verruculosum*. The LD_{50} values for chicks are around 20 mg/kg, the animals die within 4–6 h after administration.

Lit.: Cole-Cox, p. 838 ▪ Tetrahedron Lett. **32**, 1417 (1991) (synthesis) ▪ Toxicol. Appl. Pharmacol. **31**, 465 (1975); **46**, 529 (1978) ▪ Zechmeister **48**, 51. – *[CAS 56092-63-8]*

Verticillol.

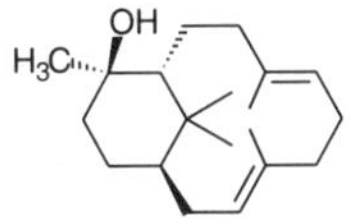

$C_{20}H_{34}O$, M_R 290.49, cryst., mp. 104–105 °C, $[\alpha]_D$ +168° ($CHCl_3$). A *diterpene of the verticillane type, isolated from the wood of *Sciadopitys verticillata* (Taxodiaceae). The verticillanes may be biosynthetically derived from formal cyclization of a cembrane precursor.

Lit.: Phytochemistry **46**, 1203 (1997) ▪ Tetrahedron **34**, 2349 (1978). – *[CAS 70000-19-0]*

Vertine see Lythraceae alkaloids.

Vesparione.

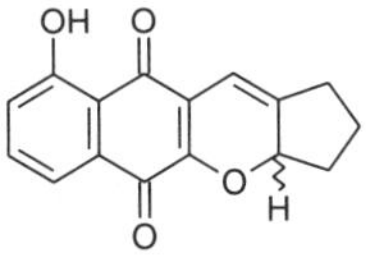
(racemate)

$C_{16}H_{12}O_4$, M_R 268.27, red needles, mp. 145–147 °C. A pigment from the sporangia of the slime fungus *Metatrichia vesparium* (Myxomycetes), biosynthetically, it is an octaketide.

Lit.: Justus Liebigs Ann. Chem. **1987**, 793 ▪ Zechmeister **51**, 171. – *[CAS 108834-58-8]*

Vetispiranes (spirovetivanes). A group of *sesquiterpenes with the skeleton (**1**). Important sources include *vetiver oil [e. g., β-*vetivone and α-vetispirene (**2**)], a valuable fragrance substance for the perfume industry, and potatoes (*Solanum tuberosum*) [e. g., solavetivone (**3**), isolubimin (**4**), lubimin [1] (**5**), and hydroxylubimin (**6**)].

Table: Data of Vetispiranes.

	molecular formula	M_R	shape (mp. [°C])	$[\alpha]_D$	CAS
2	$C_{15}H_{22}$	202.34	oil	+220° (C_2H_5OH)	28908-28-3
3	$C_{15}H_{22}O$	218.34	oil	–119° (C_2H_5OH)	54878-25-0
4	$C_{15}H_{24}O_2$	236.35	oil	+34.4° ($CHCl_3$)	60077-68-1
5	$C_{15}H_{24}O_2$	236.35	oil	+39° (C_2H_5OH)	35951-50-9
6	$C_{15}H_{24}O_3$	252.35	cryst. (96–98)	+55° (C_2H_5OH)	55784-90-2

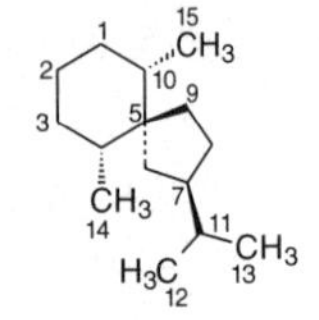

1

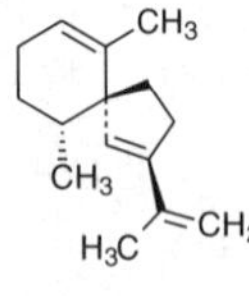

2

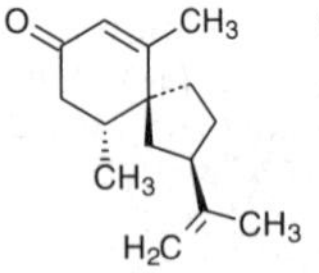

3

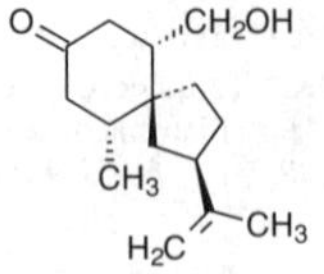

4

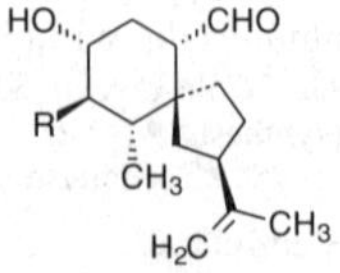

5 R = H
6 R = OH

The latter are formed as stress metabolites after infection by fungi or bacteria such as *Erwinia carotovora* and *Phytophthora infestans* together with other compounds such as *phytuberin. Compounds (**3**), (**4**), (**5**), and (**6**) in the given order are precursors in the biosynthesis of the *phytoalexin *rishitin. The V. are formed biosynthetically from *eudesmane type sesquiterpenes.

Lit.: [1] J. Am. Chem. Soc. **120**, 1747 (1998) (synthesis); Tetrahedron Lett. **37**, 8703 (1996).

gen.: Can. J. Chem. **61**, 1766 (1983) ▪ Chem. Pharm. Bull. **38**, 361 (1990) (hydroxylubimin) ▪ J. Org. Chem. **57**, 922 (1992) (solavetivone) ▪ Phytochemistry **24**, 1963 (1985) ▪ Zechmeister **31**, 283. – *Synthesis:* Bull. Chem. Soc. Jpn. **55**, 2424 (1982); **57**, 2286, 2291 (1984) (lubimin) ▪ J. Am. Chem. Soc. **120**, 1747 (1998) ((±)-lubiminol) ▪ Synth. Commun. **14**, 445 (1984) ▪ Tetrahedron **48**, 4677 (1992) (*β*-vetivone) ▪ Tetrahedron Lett. **26**, 2309 (1985) (*α*-vetispirene).

Vetiver oil. Viscous, brown to reddish brown oil with a heavy, very adherent, earthy, slightly acidic, woody, balsamy odor, often with a fatty-moldy head note.

Production: By steam distillation from the roots of the tropical grass species *Vetiveria zizanioides* (Poaceae).

Origins: Réunion ("Bourbon"), Haiti, Indonesia, Brazil, China, India. The yearly world-wide production is probably about 200 t.

Composition [1]***:*** V. is an extremely complex mixture of sesquiterpenoids (hydrocarbons, alcohols, ketones, aldehydes, acids). Lead substances are considered to be *α*- (5–10%) and *β*-**vetivone* (3–5%), *khusimol* (10–15%, $C_{15}H_{24}O$, M_R 220.35), and *isovalencenol* (5–10%, $C_{15}H_{24}O$, M_R 220.35).

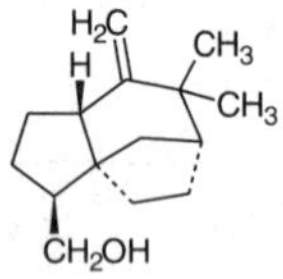

Khusimol

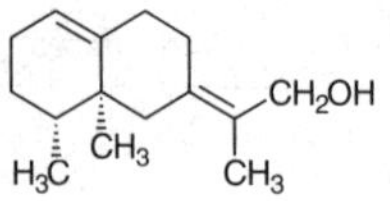

Isovalencenol

Use: In the perfume industry, mainly for masculine compositions. A large portion of V. is converted to the so-called vetiveryl acetate which has an appreciably finer odor and contains principally the acetates of the esterifiable alcohols of V.; it is used in more luxurious perfumes for noble and elegant woody tones.

Lit.: [1] Parfums Cosmet. Savons Fr. **1**, 569 (1971); Helv. Chim. Acta **55**, 2371 (1972); **80**, 139, 1857 (1997); **81**, 40 (1998); Parfum. Cosmet. Arom. **84**, 61 (1988); J. Chromatogr. **557**, 451 (1991).

gen.: Arctander, p. 649 ▪ Bauer et al. (2.), p. 180 ▪ ISO 4716 (1987) ▪ Justus Liebigs Ann. Chem. **1996**, 1195 ▪ Ohloff, p. 169. – *Toxicology:* Food Cosmet. Toxicol. **12**, 1013 (1974). – *[HS 3301 26; CAS 8016-94-4 (V.); 16223-63-5 (khusimol); 22387-74-2 (isovalencenol)]*

Vetivone.

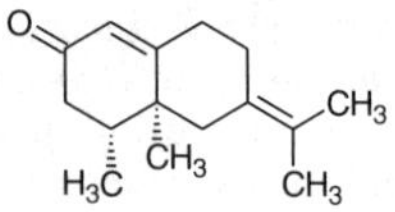

α - Vetivone

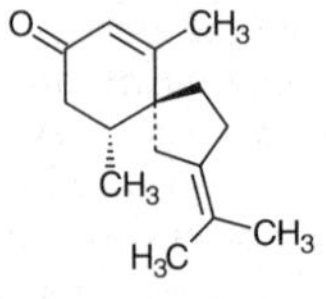

β - Vetivone

$C_{15}H_{22}O$, M_R 218.34. The structural isomers *α-V.* {isonootkatone, *α*-vetiverone, cryst., mp. 51–51.5 °C, bp. 152–153 °C (5.32·10^2 Pa), soluble in all common organic solvents, $[\alpha]_D$ +225° (C_2H_5OH)} and *β-V.* [cryst., mp. 44–45 °C, $[\alpha]_D$ −38.9° (C_2H_5OH)] are known. Both isomers occur in *vetiver oil. The biosynthesis proceeds from *eudesmane sesquiterpenes. For some time *α*-V. was erroneously considered to be a hydroazulenone derivative.

Lit.: Merck-Index (12.), No. 10106 ▪ Ullmann (5.) **A 11**, 244. – *α-V. synthesis:* J. Am. Chem. Soc. **96**, 2605 (1974). – *β-V., synthesis:* Chirality **11**, 133 (1999) ▪ J. Org. Chem. **53**, 6031 (1988) ▪ Tetrahedron **48**, 4677 (1992). – *[CAS 15764-04-2 (α-V.); 18444-79-6 (β-V.)]*

Vibriobactin see agrobactin.

Viburnols. Tetra- and pentacyclic *triterpenes from leaves of the Japanese tree *Viburnum dilatatum* (dammarane type). More than 15 V. have been described.

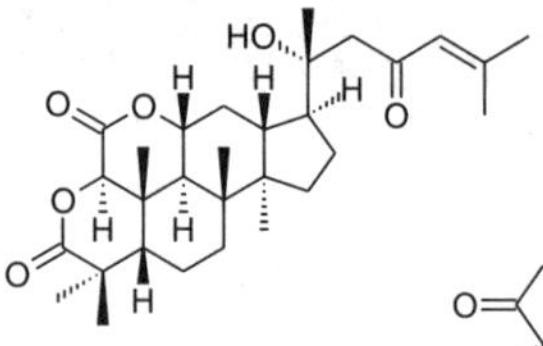

Viburnol A

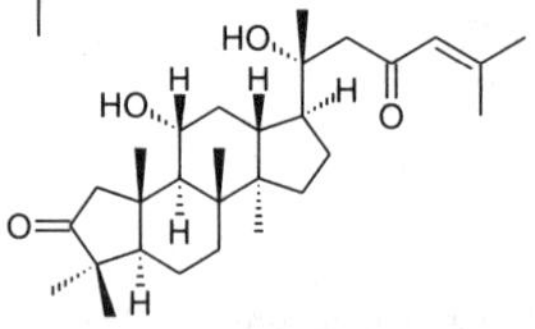

Viburnol E

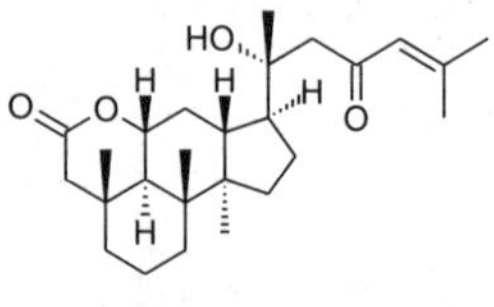

Viburnol K

Table: Data of Viburnols.

V.	molecular formula	M_R	$[\alpha]_D^{20}$ ($CHCl_3$)	CAS
A	$C_{30}H_{44}O_6$	500.67	+26.4°	179088-83-6
E	$C_{29}H_{46}O_4$	458.68	+76.4°	179088-87-0
K	$C_{26}H_{40}O_4$	416.61	+25.5°	188554-06-5

All V. are amorphous solids.

Lit.: Chem. Pharm. Bull. **45**, 1928–1931 (1997) ▪ Tetrahedron Lett. **37**, 4157–4160 (1996); **38**, 571–574 (1997).

Vicine (divicine 5-β-D-glucoside).

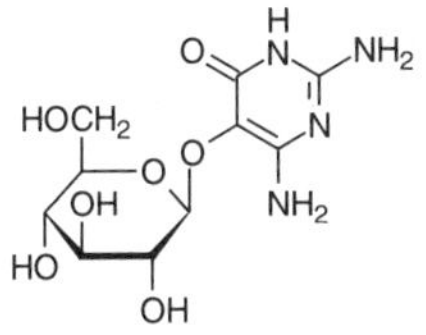

$C_{10}H_{16}N_4O_7$, M_R 304.26, needles, mp. 243–244 °C (decomp.). V. occurs in broad beans and vetches (*Vicia* sp.); the aglycone *divicine* can be released into the gastrointestinal tract by β-glycosidases. A congenital disease associated with a deficit of glucose 6-phosphate dehydrogenase in erythrocytes is endemic in some parts of the Mediterranean region and the Middle East. Divicine can then cause a severe hemolytic anemia in the metabolism of afflicted people, the so-called *favism*.

Lit.: Beilstein EIII/IV **25**, 4285 ▪ Hager (5.) **3**, 1239 ▪ Justus Liebigs Ann. Chem. **1994**, 1059 (synthesis) ▪ Karrer, No. 2546, 2547 ▪ Merck-Index (12.), No. 10112. – *[HS 2933 59; CAS 152-93-2]*

Vidarabin see mycalamides.

Vinacin A see viomycin.

Vinblastine (VLB, vincaleukoblastine).

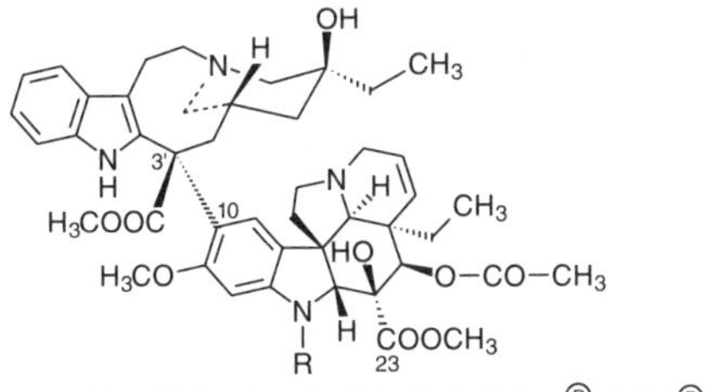

R = CH_3 : Vinblastine; Sulfate (1:1): Velban®, Velbe®
R = CHO : Vincristine

$C_{46}H_{58}N_4O_9$, M_R 810.99, cryst., mp. 216 °C, $[\alpha]_D$ +42° ($CHCl_3$); a dimeric *indole alkaloid from *Catharanthus roseus* (see Vinca alkaloids), composed of two parts: *vindoline and *catharanthine with a 10,3′-linkage. V. is a cytostatically active trace alkaloid and is accompanied by demethylated, deformylated, and deacetylated alkaloids. V. also occurs in other *Catharanthus* species such as *C. ovalis*, *C. longifolius*, *C. trichophyllus*. It is one of the most important *monoterpenoid indole alkaloids.

Pharmacology: V., together with *vincristine and the semisynthetic vindesine (deacetyl-V.-23-amide), is now established in modern cytostatic therapy (i. v.). It acts in the same was as colchicine (see phenethyltetrahydroisoquinoline alkaloids) as an inhibitor of mitosis and arrests the metaphase by binding to microtubular proteins. In addition, V. inhibits DNA polymerase and thus also DNA biosynthesis. In combination with other cytostatic agents it is used in therapy for (non)-Hodgkin's lymphomas, testicular, breast, and lung cancer [1].

Toxicity: The pronounced toxic activity affects bone marrow cells, leukocytes and granulocytes are damaged dose-dependently. Granulocytopenia and leukopenia occur much more frequently than under therapy with the structurally related vincristine.

Metabolism: V. is excreted via the bile in unchanged form or, respectively, via as yet unidentified metabolites. LD_{50} rat 7 mg/kg on oral administration of Velbe®, intravenously 2.9 ± 1.5 mg/kg. For synthesis, see *Lit.* [2–4].

Biosynthesis: Much work has been done to clarify the linkage of the two halves of the alkaloid (also with cell-free extracts) [5–7].

Lit.: [1] Kleemann-Engel, p. 1996; Manske **37**, 229–240; Sax (8.), VGU 750, VKZ 000, VLA 000. [2] Manske **37**, 77–131. [3] J. Am. Chem. Soc. **112**, 8210 (1990); **114**, 10232 (1992). [4] J. Org. Chem. **56**, 513–528 (1991); **57**, 1752 (1992). [5] J. Chem. Soc., Chem. Commun. **1982**, 791. [6] Tetrahedron **39**, 3777 (1983). [7] Nat. Prod. Rep. **7**, 85–103 (1990).
gen.: Beilstein E V **26/15**, 418 ▪ Hager (5.) **3**, 259; **9**, 1176–1185 ▪ Merck-Index (12.), No. 10119 ▪ Tetrahedron **54**, 15739–15758 (1998) (CD-spectra). – *[HS 2939 90; CAS 865-21-4]*

Vinca alkaloids. A group of alkaloids occurring principally in 3 species of the genus *Vinca* that are derived from *tryptophan and secologanin (see secoiridoids) and thus belong to the *monoterpenoid indole alkaloids. Since the plant genus *Catharanthus* was formerly known as *Vinca*, the V. a. grouped together in the older literature are in part *Catharanthus alkaloids. Well over 100 different V. a. have been isolated from *Vinca* species, however, these compounds are often not typical of the genus *Vinca* alone but rather also of, e. g., *Amsonia, Aspidosperma, Catharanthus, Rhazya, Kopsia, Rauvolfia, Tabernaemontana*. The most important V. a. are *vincamine and *eburnamonine.

Lit.: Hager (5.) **6**, 1123–1135 ▪ J. Chem. Res. (S) **1996**, 344 (synthesis) ▪ Taylor & Farnsworth, The Vinca Alkaloids: Botany, Chemistry, and Pharmacology, New York: Dekker 1973. – *[HS 2939 90]*

Vincaleukoblastine see vinblastine.

Vincamine.

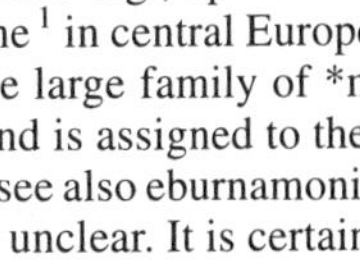

$C_{21}H_{26}N_2O_3$, M_R 354.45, yellowish cryst., mp. 232–233 °C. V. is a typical *Vinca alkaloid, the main component, besides e. g., apovincamine, vincadifformine, and vincine [1] in central European *Vinca* species. V. belongs to the large family of *monoterpenoid indole alkaloids and is assigned to the group of eburnamine alkaloids (see also eburnamonine). Its biosynthesis is still partly unclear. It is certain that V. is formed from *tryptophan via tryptamine and the iridoid system secologanin (see secoiridoids) which undergoes rearrangement during the biosynthesis. The mechanism of the rearrangement and the responsible enzymes are unknown.

Activity: V. reduces peripheral vascular resistance (blood pressure lowering activity), it promotes cranial perfusion and is purported to improve cerebral metabolism. It is thus recommended as treatment for cerebrovascular disorders. V. is also used in the treatment of headaches and migraine. The pharmacokinetics and metabolism remain practically unknown. Polar metabolites are excreted in the urine. Other Vinca alkaloids such as vincadifformine and vincaminorine inhibit the biosynthesis of nucleic acids and proteins as well as having a proliferation inhibitory effect on P 388 leukemia cells[2].

Lit.: [1]Khim. Prit. Soedin **1980**, 735. [2]Pharmazie **41**, 270 (1986).

gen.: Beilstein E V **25/6**, 373 f. ▪ Merck-Index (12.), No. 10120 ▪ Sax (8.), VLF 000 ▪ Szántay, in Saxton (ed.), The Monoterpenoid Indole Alkaloids, Suppl. Vol. 25, Heterocyclic Compounds, p. 478 ff., 733 ff., Chichester: Wiley 1994. – *Pharmacology:* Chemtracts: Org. Chem. **1**, 118–122 (1988) ▪ Eur. J. Drug. Metab. Pharmakokinet. **10**, 89 (1985) ▪ Hager (5.) **6**, 1124–1130 ▪ Kleemann-Engel, p. 1998. – *Synthesis:* Heterocycles **24**, 1663 (1986); **45**, 2007 (1997) ▪ J. Chem. Soc., Chem. Commun. **1986**, 156 ▪ J. Org. Chem. **55**, 3068–3074 (1990); **61**, 6001 (1996); **62**, 1223, 3890 (1997) ▪ Tetrahedron: Asymmetry **8**, 1963 (1997); **10**, 297 (1999). – *[HS 2939 90; CAS 1617-90-9]*

Vincamone, vincanorine see eburnamonine.

Vincarubine.

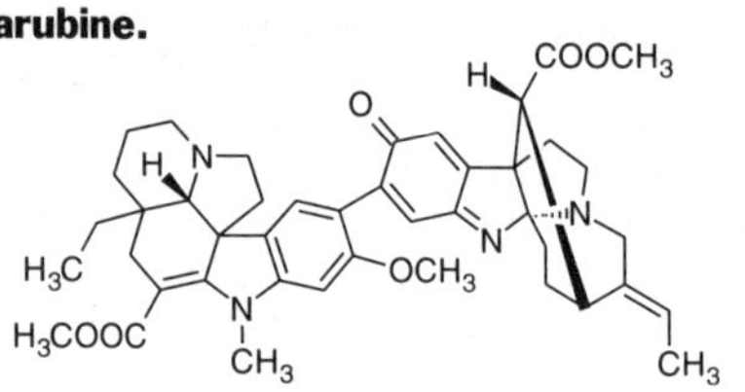

$C_{43}H_{50}N_4O_6$, M_R 718.89, dark red cryst., mp. 179 °C (decomp.), $[\alpha]_D^{23}$ −550° (C_2H_5OH). Cytotoxic alkaloid from the European periwinkle, *Vinca minor* (Apocynaceae). Like the *Vinca alkaloids, *vinblastine and *vincristine, V. consists of a *catharanthine and a *vindoline part, but being connected in the 10-position.

Lit.: Planta Med. **54**, 214 (1988). – *[CAS 107290-03-9]*

Vincoside.

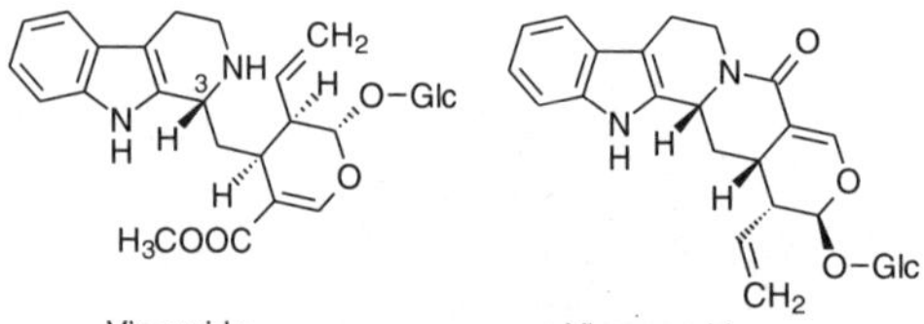

Vincoside Vincosamide

$C_{27}H_{34}N_2O_9$, M_R 530.57. V. is a very unstable glucoalkaloid that is easily transformed biosynthetically into its lactam (vincoside lactam, *vincosamide*) [$C_{26}H_{30}N_2O_8$, M_R 498.53, mp. 201–202 °C, $[\alpha]_D$ −118° (CH_3OH)]. The originally assumed role of glucoalkaloid V. as the first and essential biosynthetic precursor of *monoterpenoid indole alkaloids[1,2] could not be verified since, in contrast to V., its (3*S*)-isomer (*strictosidine) is incorporated under *in vivo* conditions into the alkaloids *ajmalicine, serpentine, *vindoline. and *catharanthine[2,3].

Lit.: [1]J. Chem. Soc., Chem. Commun. **1968**, 1582. [2]J. Chem. Soc. C **1969**, 1193. [3]J. Chem. Soc., Chem. Commun. **1977**, 646.

gen.: Hager (5.) **6**, 1129 ▪ J. Am. Chem. Soc. **97**, 6270 (1975). – *[HS 2939 90; CAS 19456-89-4 (V.); 23141-27-7 (vincosamide)]*

Vincristine (VCR, LCR, 22-oxovinblastine, leurocristine). Formula, see vinblastine. $C_{46}H_{56}N_4O_{10}$, M_R 824.97, cryst., mp. 218–220 °C, $[\alpha]_D$ +26° (CH_2Cl_2); a *monoterpenoid indole alkaloid from *Catharanthus roseus* (Apocynaceae). V. belongs to a group of ca. 25 dimeric indole alkaloids (see also vinblastine) that have been isolated from *C. roseus*. The content of V. in *C. roseus* is extremely low, $<3\times10^{-4}$% of the plant's dry weight. V. is the only natural product that is still isolated commercially from a natural source in spite of its trace occurrence and used in therapy. Like *vinblastine, V. is made up of a vindoline half and a catharanthine half; it differs structurally from vinblastine in having an N^1-formyl group instead of an N^1-methyl group. V. acts in the same way as vinblastine and is used in the chemotherapy of various types of cancer; the doses required are 3–5 times smaller than those of vinblastine[1]. V. is excreted via the liver unidentified metabolites. Overdoses of V. can be lethal, LD_{50} (rat i. v.) 1.0±0.1 mg/kg[1].

Lit.: [1]J. Clin. Oncol. **9**, 877–887 (1991); Kleemann-Engel, p. 2002.

gen.: Beilstein E V **26/15**, 417 f. ▪ Florey **1**, 463–480 ▪ Hager (5.) **3**, 1241 ff.; **9**, 1178–1185 ▪ Manske **37**, 205–226, 229–239 ▪ Merck-Index (12.), No. 10124 ▪ Sax (8.), LEY 000, LEZ 000. – *[HS 2939 90; CAS 57-22-7]*

Vindoline.

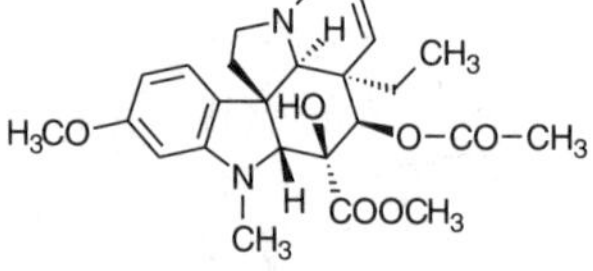

absolute configuration

$C_{25}H_{32}N_2O_6$, M_R 456.54, cryst., dimorphous, mp. 174–175 and 164–165 °C, $[\alpha]_D^{27}$ +42° ($CHCl_3$). V. is an *indole alkaloid of monoterpenoid origin belonging to the Aspidosperma type (see Aspidosperma alkaloids) and a characteristic main alkaloid of the plants *Catharanthus roseus* (formerly *Vinca rosea*), *C. pusillus* (*V. pusilla*, Apocynaceae). It occurs together with its *O*-deacetyl derivative, *deacetylvindoline* ($C_{23}H_{30}N_2O_5$, M_R 414.50, mp. 163–165 °C) as well as traces of *4-deacetoxyvindoline* ($C_{23}H_{30}N_2O_4$, M_R 398.50), which are both important as biogenetic precursors of V. V. is of major biogenetic significance since it constitutes one half of therapeutically important "dimers" such as *vinblastine and *vincristine (also vindesine, etc.). In the biosynthesis of these "dimers" V. is linked with *catharanthine. For synthesis, see *Lit.*[1,2].

Biosynthesis: The biosynthesis of V., and especially the final stage of a ca. 15-step chain of reactions, has been studied on the enzymatic level[3,4]. The last 6 reactions, beginning at the level of the *Catharanthus* alkaloid *tabersonine are accordingly hydroxylations, methylations, and acetylation. The final reaction in the biosynthesis of V. is catalyzed by a highly specific acetyl-CoA-dependent acetyltransferase which converts deacetylvindoline to V. Although catharanthine and tabersonine can be produced in cell cultures of *C. roseus*, the biosynthesis of V. in these cell systems is interrupted at the stage of tabersonine; thus the dimeric indole alkaloids with a vindoline half cannot be obtained biosynthetically from such *in vitro* systems.

Lit.: [1] J. Am. Chem. Soc. **97**, 6880 (1975). [2] J. Am. Chem. Soc. **100**, 4220 (1978). [3] Manske **37**, 61 – 63. [4] Helv. Chim. Acta **65**, 2088 (1982).
gen.: Beilstein E V **25/7**, 127 ▪ Hager (5.) **3**, 259 ▪ Merck-Index (12.), No. 10126 ▪ Phytochemistry **32**, 493 – 506 (1993). – *[HS 2939 90; CAS 2182-14-1 (V.); 3633-92-9 (deacetyl-V.)]*

Vineomycins.

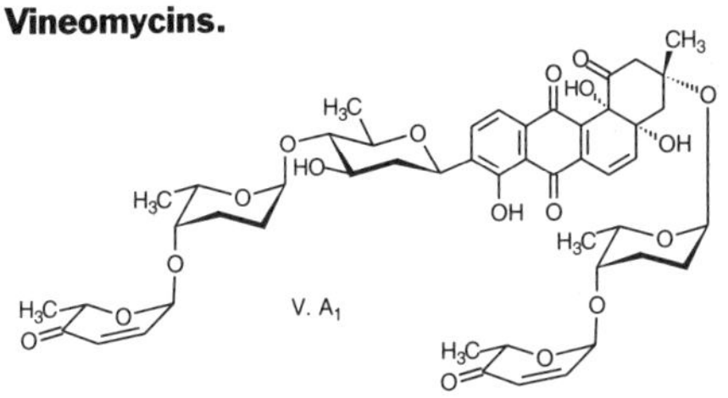

V. A_1

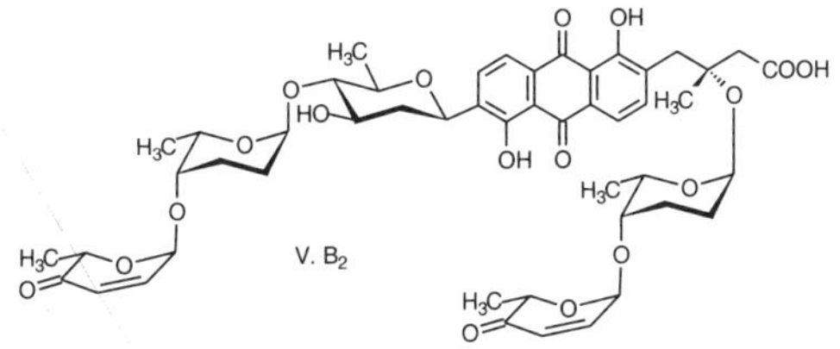

V. B_2

Table: Data of Vineomycins.

V.	molecular formula	M_R	mp. [°C]	$[\alpha]_D^{26}$ (CHCl$_3$)	CAS
A_1	$C_{49}H_{58}O_{18}$	934.99	164 – 166	+92°	78164-00-8
A_2	–	–	173 – 176	+3°	66198-31-0
B_1	–	–	–	–	66198-32-1
B_2	$C_{49}H_{58}O_{18}$	934.99	128 – 131	+31°	80928-52-5

Quinonoid antibiotics of the *angucycline type from *streptomycetes (e. g., *Streptomyces matensis vineus*). Acid hydrolysis of V. A_1 furnishes the aglycone *aquayamycin together with 2 mol. L-aculose and 2 mol. L-rhodinose. The aglycone of V. B_2, the anthraquinone vineomycinone B_2, is formed biosynthetically from aquayamycin by a Baeyer-Villiger oxidation. The V. are yellow-red compounds with activities against Gram-positive bacteria and tumor cells (*in vitro*). V. A_1 also inhibits prolyl hydroxylase and thus collagen biosynthesis. The structures of V. A_2 and B_1 are still unknown.

Lit.: Angew. Chem. Int. Ed. Engl. **26**, 15 – 23 (1987) ▪ Biochem. Pharmacol. **31**, 915 (1984) ▪ Chem. Pharm. Bull. **30**, 762 (1982) ▪ J. Am. Chem. Soc. **113**, 5775, 6320, 6982 (1991) (synthesis) ▪ J. Antibiot. **34**, 1355 (1981); **35**, 602 (1982) (biosynthesis) ▪ Prog. Med. Chem. **22**, 1 – 65 (1985) (pharmacology).

Viniferins see wine, phenolic constituents of.

Vinigrol.

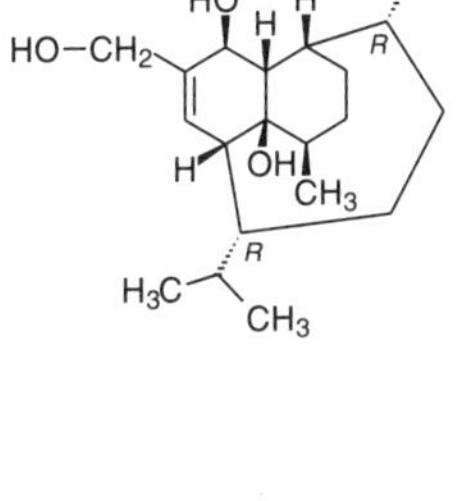

$C_{20}H_{34}O_3$, M_R 322.49, cryst., mp. 108 °C, $[\alpha]_D$ –96.2° (CHCl$_3$). Antihypertensive substance from the soil fungus *Virgaria nigra* (Hyphomycetes).

Lit.: J. Antibiot. **41**, 25, 31 (1988) ▪ J. Org. Chem. **52**, 5292 (1987); **58**, 2349 (1993); **62**, 5062 (1997) (synthesis). – *[CAS 111025-83-3]*

4-Vinylguaiacol.

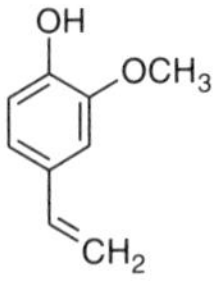

$C_9H_{10}O_2$, M_R 150.18, bp. 224 °C, mp. 9 – 10 °C (57 °C). Yellowish oil or crystals [1] with a smell like cloves, olfactory threshold: 10 ppb [2]. V. forms by fermentation or thermally from ferulic acid and other phenol carboxylic acids and is a constituent of the flavors of beer, bread, coffee, popcorn, wine, tar and smoked food.

Lit.: [1] Arctander, No. 1891; Biochim. Biophys. Acta **789**, 111 (1984). [2] Maarse, p. 515. – *[CAS 7786-61-0]*

Viocin see viomycin.

Violacein.

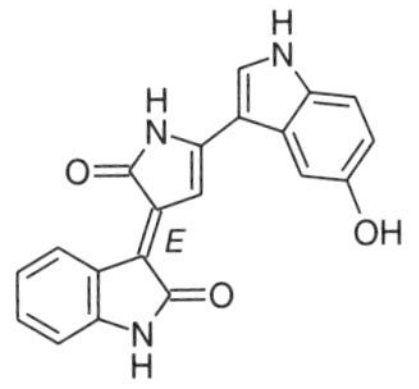

$C_{20}H_{13}N_3O_3$, M_R 343.34, black-violet prisms, mp. >350 °C (decomp.). Representative of a group of pigments from *Chromobacterium violaceum* structurally derived from isatin (2,3-indoledione) with weak activities against Gram-positive and Gram-negative bacteria.

Lit.: Agric. Biol. Chem. **51**, 965, 2733 (1987); **54**, 2339 (1990) (biosynthesis) ▪ J. Chem. Soc., Perkin Trans. 1 **1995**, 1565 – 1571; **1998**, 3087 (biosynthesis) ▪ Merck-Index (12.), No. 10135. – *[HS 3203 00; CAS 548-54-9]*

Violanin [delphinidin 3-*O*-(4′-*O*-*p*-cumaroylrutinoside)-5-*O*-glucoside].

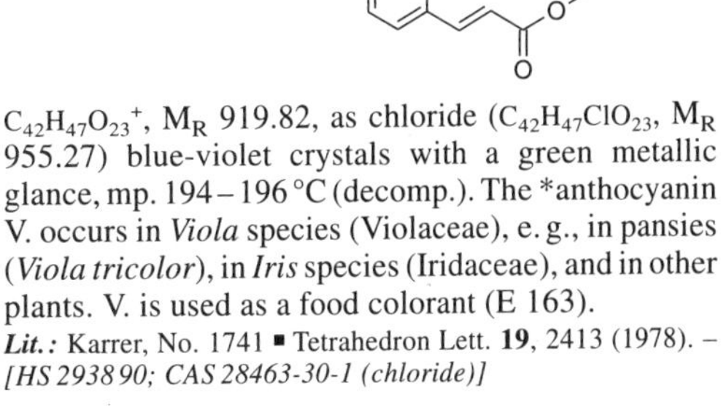

$C_{42}H_{47}O_{23}^+$, M_R 919.82, as chloride ($C_{42}H_{47}ClO_{23}$, M_R 955.27) blue-violet crystals with a green metallic glance, mp. 194 – 196 °C (decomp.). The *anthocyanin V. occurs in *Viola* species (Violaceae), e. g., in pansies (*Viola tricolor*), in *Iris* species (Iridaceae), and in other plants. V. is used as a food colorant (E 163).

Lit.: Karrer, No. 1741 ▪ Tetrahedron Lett. **19**, 2413 (1978). – *[HS 2938 90; CAS 28463-30-1 (chloride)]*

Violaxanthin (zeaxanthin 5*R*,6*S*:5′*R*,6′*S*-diepoxide, C. I. 75 138).

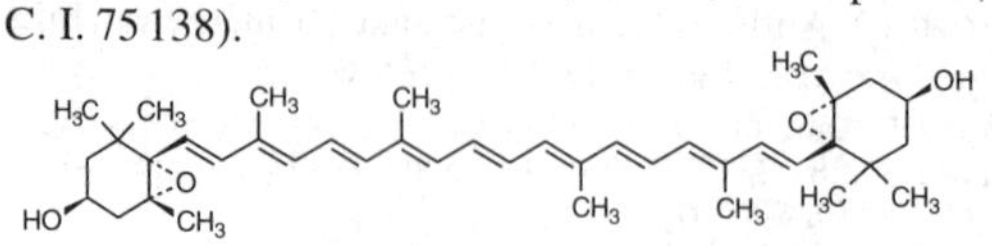

$C_{40}H_{56}O_4$, M_R 600.88, red cryst., mp. 208 °C, uv_{max} 418, 443, 472 nm (C_2H_5OH); $[\alpha]_D^{20}$ +35° ($CHCl_3$); soluble in ethanol, methanol, ether. The xanthophyll V. is a widely distributed *carotinoid in plants, e. g., in *Viola tricolor* (Violaceae). It is formed biosynthetically from *zeaxanthin and is degraded by photooxidation to *xanthoxin.

Lit.: Beilstein E V **19/3**, 463 ▪ Helv. Chim. Acta **71**, 931–956 (1988) (synthesis) ▪ J. Chem. Soc. Perkin Trans. 1 **1972**, 1769 (biosynthesis) ▪ Karrer, No. 1482 ▪ Merck-Index (12.), No. 10136 ▪ Pure Appl. Chem. **35**, 47 (1973) (review) ▪ see also zeaxanthin. – *[CAS 126-29-4]*

Viomycin (viocin, vinacin A, tuberactinomycin B).

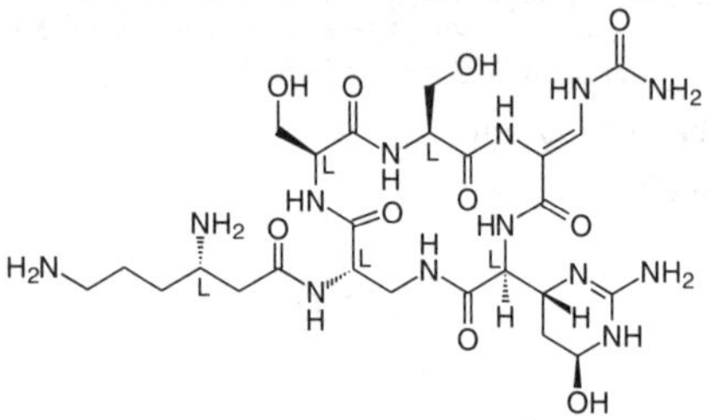

$C_{25}H_{43}N_{13}O_{10}$, M_R 685.70, red-violet crystals, mp. 226 °C (hydrochloride), readily soluble in water, poorly soluble in organic solvents. International free name for a strongly basic, cyclic *peptide antibiotic from cultures of *Streptomyces puniceus* and *S. floridae*. V. is built up from 5 amino acids with an (*S*)-3,6-diaminohexanoyl residue (L-*β*-lysyl) as side chain; of these amino acids only L-serine is of proteinogenic origin. V. is active against Gram-positive and Gram-negative bacteria, inhibits protein biosynthesis, and is an effective tuberculostatic agent. However, therapeutic use of V. has been discontinued on account of side effects leading to hearing impairments. LD_{50} (rat i. m.) 1300 mg/kg, it is teratogenic in higher doses.

Lit.: Beilstein E V **16/19**, 573 ▪ Chem. Pharm. Bull. **27**, 2551–2556 (1979) ▪ Heterocycles, **15**, 999–1005 (1981) ▪ Kleemann-Engel, p. 2007 ▪ Sax (8.), VQZ 000 ▪ Zechmeister **66**, 134–142. – *[HS 2941 90; CAS 32988-50-4]*

Virantmycin.

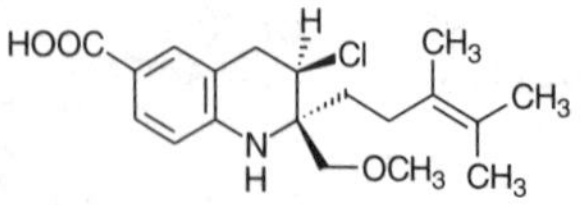

$C_{19}H_{26}ClNO_3$, M_R 351.87, cryst., mp. 59 °C, $[\alpha]_D$ –11.1° ($CHCl_3$), an antiviral antibiotic isolated from *Streptomyces nitrosporeus*. V. inhibits the growth of RNA and DNA viruses (e. g., *Herpes simplex* 1 and 2) and has weak antifungal activity. The stereochemistry was revised on the basis of NOE experiments [1].

Lit.: [1] J. Chem. Soc., Perkin Trans. 1 **1990**, 409–411.
gen.: J. Antibiot. **33**, 1395–1397 (1980); **34**, 1408–1415 (1981). – *Synthesis:* Org. Lett. **1**, 823 (1999) ▪ Synlett **1991**, 202–203 ▪ Tetrahedron **46**, 4587–4594 (1990); **52**, 10609, 10631–10652 (1996). – *[HS 2941 90; CAS 76417-04-4]*

Virginiamycin (staphylomycin, pristinamycin II).

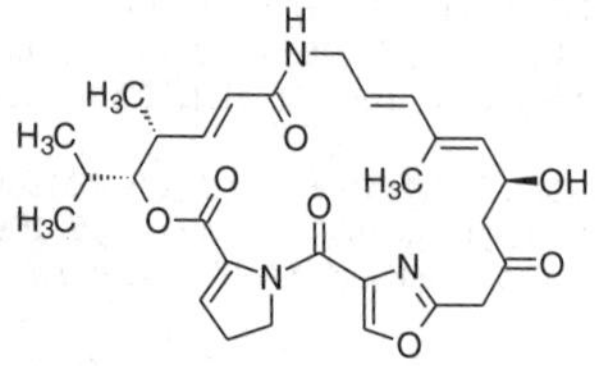

V. M_1

An antibiotic complex from *Streptomyces virginiae* cultures, e. g., *V. M_1*: $C_{28}H_{35}N_3O_7$, M_R 525.60, mp. 165–167 °C (decomp.), $[\alpha]_D$ –190° (C_2H_5OH), soluble in DMF. V. is used as a fodder additive in veterinary medicine.

Lit.: Curr. Pharm. Des. **4**, 155 (1998) (activity) ▪ Environ. Sci. Res. **44**, 105 (1992) (biosynthesis) ▪ J. Am. Chem. Soc. **102**, 5964 (1980); **118**, 3301 (1996) ▪ Merck-Index (12.), No. 10142 ▪ Synthesis **1998**, 603; **1999**, 1510 ▪ Tetrahedron Lett. **39**, 3145, 3149 (1998) (pristinamycin II_B). – *[CAS 11006-76-1 (V.); 21411-53-0 (V. M_1)]*

Viridicatumtoxin.

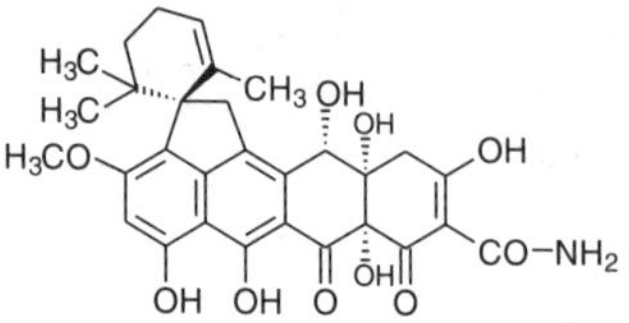

(relative configuration)

$C_{30}H_{31}NO_{10}$, M_R 565.58, light yellow cryst., mp. 235 °C (decomp.), $[\alpha]_D^{23}$ –12° (C_2H_5OH). *Mycotoxin from *Penicillium viridicatum*. Biosynthetically, it is a nonaketide with a monoterpene unit at C-6. V. is toxic (LD_{50} rat p. o. 122 mg/kg) and can be ingested with moldy food.

Lit.: Cole-Cox, p. 882 ▪ J. Chem. Soc., Chem. Commun. **1976**, 728; **1982**, 902; **1988**, 1562 (biosynthesis) ▪ Sax (8.), VRP 000 (toxicology). – *[CAS 39277-41-3]*

Viridiene see algal pheromones.

Viridin see wortmannin.

Viriplanins. An antitumor antibiotic complex of the *nogalamycin type, related to *anthracycline, isolated from *Ampullariella regularis*. It has a high toxicity [LD_{50} (mouse i. v.) 1–2 mg/kg] and also has antibacterial and antiviral activities. The main component *V. A* [$C_{79}H_{109}N_3O_{36}$, M_R 1676.73, red amorphous powder, decomposes above 210 °C, $[\alpha]_D$ +144° ($CHCl_3/CH_3OH$, 9:1)] contains the aglycone *viriplanol*, to which two oligosaccharide units are bound *O*-glycosidically. A conspicuous sugar building block is D-virinose (2×), the others are 2-deoxy-L-fucose (L-lyxo-2,6-dideoxyhexose) (4′) and L-diginose (1′) which is esterified with mesaconic acid. When dissolved in methanol and irradiated V. A is converted to *V. D* ($C_{79}H_{109}N_3O_{38}$, M_R 1708.73), this involves a photooxidation of the nitroso to a nitro group. After hydrolysis of V. D the nitro sugar D-decilonitrose is obtained, contradictory results about the absolute configuration of the latter have been published.The antibiotics decilorubicin and arugomycin are related to V. D. The activity is based on intercalation into DNA. (see formula, p. 695)

Lit.: Anti-Cancer Drug Design **3**, 157–168 (1988) ▪ J. Antibiot. **39**, 1193–1204 (1986); **42**, 54–61 (1990); **45**, 1313 (1992). – *[CAS 106153-36-0 (V. A), 121276-86-6 (V. D)]*

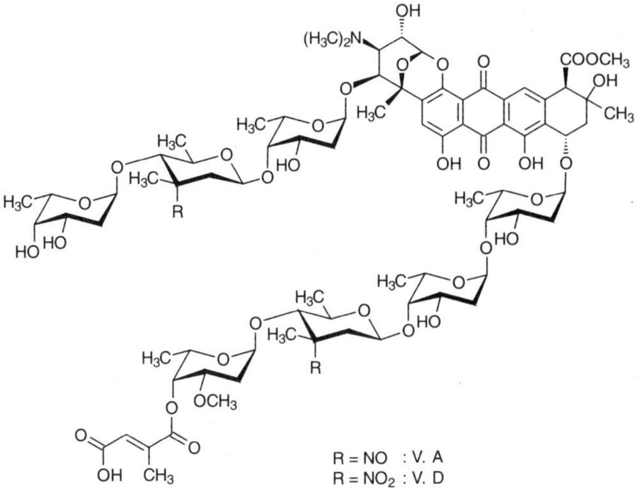

Viroidin, viroisin see virotoxins.

Virosecurinine see Securinega alkaloids.

Virotoxins. A group of monocyclic heptapeptides from the toadstool *Amanita virosa* (destroying angel). Together with the *amanitins and *phallotoxins, they represent the third class of poisons in death caps. Six compounds are known; viroisin is the main toxin.

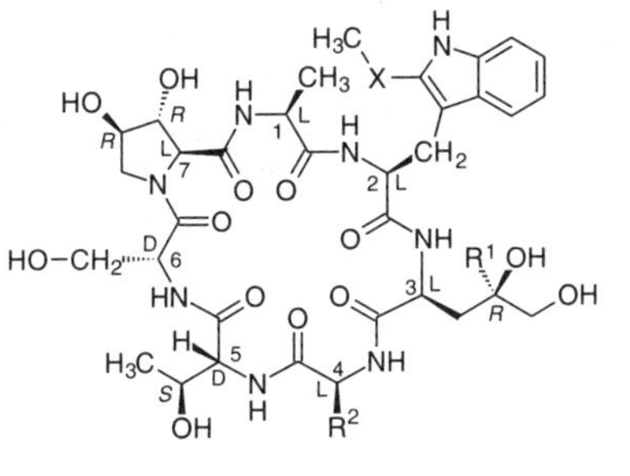

Virotoxin	X	R^1	R^2
Viroidin	SO_2	CH_3	$CH(CH_3)_2$
Deoxyviroidin	SO	CH_3	$CH(CH_3)_2$
[Ala[4]]Viroidin	SO_2	CH_3	CH_3
[Ala4]Deoxyviroidin	SO	CH_3	CH_3
Viroisin	SO_2	CH_2OH	$CH(CH_3)_2$
Deoxyviroisin	SO	CH_2OH	$CH(CH_3)_2$

Table: Data of Virotoxins.

Virotoxin	molecular formula	M_R	CAS
Viroidin	$C_{38}H_{56}N_8O_{15}S$	896.97	53568-33-5
Deoxyviroidin	$C_{38}H_{56}N_8O_{14}S$	880.97	74125-14-7
[Ala[4]]Viroidin	$C_{36}H_{52}N_8O_{15}S$	868.91	74125-15-8
[Ala[4]]Deoxyviroidin	$C_{36}H_{52}N_8O_{14}S$	852.91	74125-16-9
Viroisin	$C_{38}H_{56}N_8O_{16}S$	912.97	74113-57-8
Deoxyviroisin	$C_{38}H_{56}N_8O_{15}S$	896.97	74125-17-0

Lit.: Biochemistry **19**, 3334 (1980).

Visamminol.

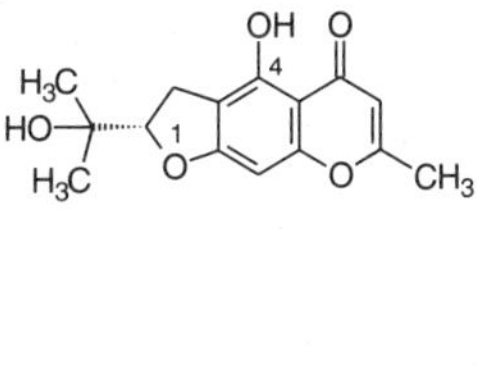

$C_{15}H_{16}O_5$, M_R 276.29, needles, mp. 160 °C, bp. 140 °C (0.26 Pa), $[\alpha]_D^{18}$ +92° ($CHCl_3$); a furochromone from the fruit of *Ammi visnaga* (see khellin). The 4-methyl ether, $C_{16}H_{18}O_5$, M_R 290.32, needles, mp. 141–142 °C, $[\alpha]_D^{25}$ +91.8° ($CHCl_3$) and some glycosides of V. occur in the roots of *Ledebouriella seseloides*, *Libanotis laticalycina* (Chinese drug Shuifangfeng) and *Saposhnikovia divaricata* (Chinese drug Quangangfeng).

Activity: The drugs have analgesic and antipyretic properties; derivatives of V. have blood pressure lowering effects.

Lit.: Chem. Pharm. Bull. **29**, 2565 (1981); **30**, 3555 (1982) ▪ J. Chem. Soc., Perkin Trans. 1 **1975**, 150–154 (synthesis) ▪ Phytochemistry **38**, 427 (1995) ▪ Planta Med. **56**, 134 (1990). – *[CAS 492-52-4 (unspecified V.); 173293-90-8 ((+)-(S)-form); 80681-42-1 ((+)-(S)-4-O-methyl V.)]*

Visnadin (visnacorin, carduben®).

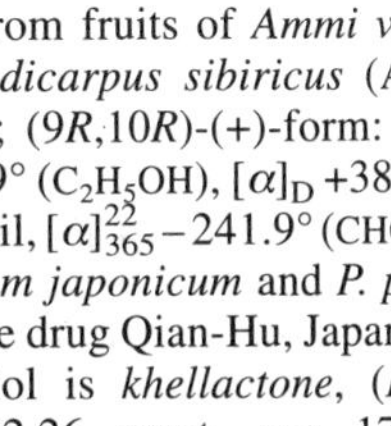

(9*S*,10*S*)-(-)-form

International free name for a coronary vasodilator isolated from fruits of *Ammi visnaga* (see khellin) and *Phlododicarpus sibiricus* (Apiaceae), $C_{21}H_{24}O_7$, M_R 388.42; (9*R*,10*R*)-(+)-form: needles, mp. 85–88 °C, $[\alpha]_D$ +9° (C_2H_5OH), $[\alpha]_D$ +38° (dioxan); (9*S*,10*S*)-(–)-form: oil, $[\alpha]_{365}^{22}$ –241.9° ($CHCl_3$) from the roots of *Peucedanum japonicum* and *P. praeruptorum* (Apiaceae, Chinese drug Qian-Hu, Japanese Shoku-Bohfuu). The free diol is *khellactone*, (*R,R*)-(+)-form, $C_{14}H_{14}O_5$, M_R 262.26, cryst., mp. 174–175 °C, $[\alpha]_D^{22}$ +104° (C_2H_5OH), it occurs in *Seseli gummiferum*.

Activity: (+)-V. blocks calcium channels[1]; the drugs are used in traditional Chinese and Japanese medicine for treatment of coughs, headaches, and as a tonic.

Lit.: [1] Planta Med. **60**, 101–105 (1994).
gen.: Arzneim. Forsch. **17**, 283, 288, 290 (1967); **23**, 201 (1973) ▪ Beilstein E V **19/6**, 272 (V.), 30 (visnagin) ▪ Hager (5.) **9**, 1187 ▪ J. Drug Res. **7**, 1–7, 9–17, 45–57 (1975) (pharmacology) ▪ Kleemann-Engel, p. 2007 ▪ Merck-Index (12.), No.

10147 ▪ Phytochemistry **31**, 4303–4306 (1992). – *[HS 2932 29; CAS 477-32-7 ((9R,10R)-(+)-form); 146074-44-4 ((9S,10S)-(–)-form); 24144-61-4 ((R,R)-(+)-khellactone)]*

Visnagin see khellin.

Visual purple see rhodopsin.

Vitamin A_1 see retinol.

Vitamin A_1 aldehyde see retinal.

Vitamin A_2, vitamin A_2 aldehyde see 3,4-didehydroretinal.

Vitamin B_1 see thiamin(e).

Vitamin B_2 (acid) see riboflavin(e).

Vitamin B_3. Obsolete name for nicotinic acid, *nicotinamide (see also nicotinamide coenzymes), and *pantothenic acid.

Vitamin B_5. Obsolete name for *pantothenic acid.

Vitamin B_6. Collective name for pyridoxal, pyridoxamine, and pyridoxine.

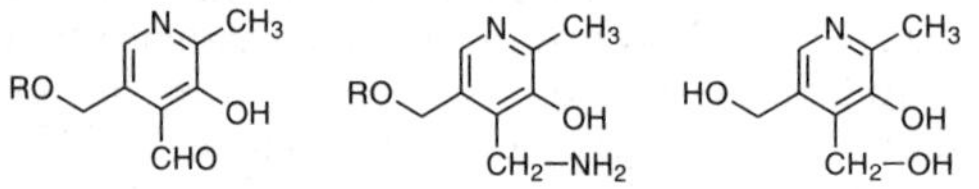

R = H : Pyridoxal (PL, **1**)
R = PO_3H_2: Pyridoxal 5'-phosphat (PLP, **2**)

R = H : Pyridoxamin (PM, **3**)
R = PO_3H_2: Pyridoxamine 5'-phosphat (PMP, **4**)

Pyridoxin (**5**)

Table: Data of Vitamin B_6 and derivatives.

	molecular formula	M_R	mp. [°C]	CAS
1	$C_8H_9NO_3$	167.16	165 (decomp.)	66-72-8 65-22-5 (hydrochloride)
2	$C_8H_{10}NO_6P$	247.14	139–142 (monohydrate, decomp.)	54-47-7
3	$C_8H_{12}N_2O_2$	168.20	193	85-87-0
4	$C_8H_{13}N_2O_5P$	248.18		529-96-4
5	$C_8H_{11}NO_3$	169.18	160	65-23-6

Pyridoxal (3-hydroxy-5-hydroxymethyl-2-methylpyridine-4-carbaldehyde); the aldehyde form exists in equilibrium with a hemiacetal furopyridine. PL can be used for a sensitive detection of amino acids[1]. *Pyridoxal 5'-phosphate* (PLP), as free acid yellow, monoclinic crystals. *Pyridoxamine* (4-aminomethyl-5-hydroxymethyl-2-methylpyridin-3-ol), as dihydrochloride mp. 228 °C (decomp.). *Pyridoxine* (pyridoxol, 4,5-bis-hydroxymethyl-2-methylpyridin-3-ol), as hydrochloride mp. 206–208 °C. The mentioned compounds are sensitive to light. PL hydrochloride: LD_{50} (mouse p.o.) 1800 mg/kg. For the detection of V. B_6, see *Lit.*[2]. PLP is a *coenzyme for transaminases, decarboxylases (hence the formerly used name codecarboxylase), and racemases, which play central roles in the metabolism of amino acids. In transamination processes, an amino group of an α-L-amino acid is transferred to a 2-oxo acid with the Schiff base (phosphopyridoxylidene imine) and *pyridoxamine 5'-phosphate* occurring as intermediates. In the absence of substrates PLP forms a Schiff base with the ε-amino group on a lysine residue of the enzyme. PLP is also a component of glycogen phosphorylase (see glycogen). In this case the phosphate group of PLP functions as an acid-base catalyst[3]. PLP can also react non-enzymatically with free amines or amino acids to furnish endogenous alkaloids[4].

Biosynthesis: In plants and most microorganisms PL is formed from acetaldehyde (or its biological equivalent glycine), dihydroxyacetone phosphate, and glyceraldehyde phosphate. Phosphorylation by means of ATP and pyridoxal kinase (EC 2.7.1.35) affords PLP. The latter can also be obtained from pyridoxine 5'-phosphate or PMP under the action of pyridoxine 4-oxidase (EC 1.1.3.12). V. B_6 taken up in food – especially from liver, yeast, bananas, grain, vegetables – is absorbed in the upper part of the small intestine and stored in tissue, in particular in the brain, liver, and kidneys, in the form of PLP and PM. Degradation products appearing in the urine are, besides unchanged pyridoxines, 4-*pyridoxic acids.

Use: As fodder additive, in medicine for treatment of vitamin deficiency conditions and for travel sickness, vomiting in pregnancy, radiation sickness, as glyoxylate (pyridoxilate), and also as a vasodilator.

Historical: Feeding experiments performed in 1934 by György in rats fed on a diet free of vitamin B complex resulted in reduced growth and a Pellagra-like dermatitis on the extremities; thus the presumed active principle was first named as antidermatitis factor or adermine. Pyridoxine was first obtained in crystalline form in 1938, the constitution was elucidated in 1939 independently by Folkers and R. Kuhn. PL and PM were discovered in 1942 by E. E. Snell. In 1996, the world production of V. B_6 was 15 000 tons, valued at 125 million U.S. $.

Lit.: [1]Angew. Chem. Int. Ed. Engl. **11**, 227 ff. (1972). [2]Int. J. Vitam. Nutrit. Res. **59**, 338–343 (1989). [3]Biochemistry (USA) **29**, 1099–1107 (1990); Coord. Chem. Rev. **114**, 147–197 (1995). [4]Angew. Chem. Int. Ed. Engl. **25**, 177 f., (1986).
gen.: Ann. N. Y. Acad. Sci. **585**, 1 (1990) ▪ Annu. Rev. Biochem. **59**, 87–110 (1990) ▪ Beilstein E V **21/5**, 492 f. (pyridoxine); **21/13**, 44 f. (pyridoxal); **22/12**, 324 f. (pyridoxamine) ▪ Dolphin (ed.), Vitamin B_6 Pyridoxal Phosphate, New York: Wiley 1986 ▪ Isler et al., Vitamine II, p. 193–229, Stuttgart: Thieme 1988 ▪ Nat. Prod. Rep. **12**, 555 (1995) (biosynthesis pyridoxine) ▪ Physiol. Rev. **69**, 1170–1198 (1989) ▪ Reynolds & Leklem, Vitamin B_6: Its Role in Health and Disease, New York: Alan Liss 1986 ▪ Sax (8.), PII 100, VSU 000 (toxicology). – *[HS 2936 25]*

Vitamin B_7 see biotin.

Vitamin B_9 (B_c). Obsolete name for *folic acid.

Vitamin B_{12} {cyanocobalamin, coβ-cyano-Coα-[α-(5,6-dimethylbenzimidazolyl)]-cobamide}. $C_{63}H_{88}CoN_{14}O_{14}P$, M_R 1355.38. uv_{max} (in H_2O) 548, 361, 278 nm; dark red, needle-like, hygroscopic crystals, which in hydrated form are stable in the air, become dark on heating at 210–220 °C, are sparingly soluble in alcohol, very poorly soluble in water, and insoluble in acetone, ether, and chloroform. V. B_{12} decomposes in strongly acidic or alkaline aqueous solutions under the action of light and reducing agents such as *ascorbic acid, vanillin, iron(II) sulfate. The term V. B_{12} is

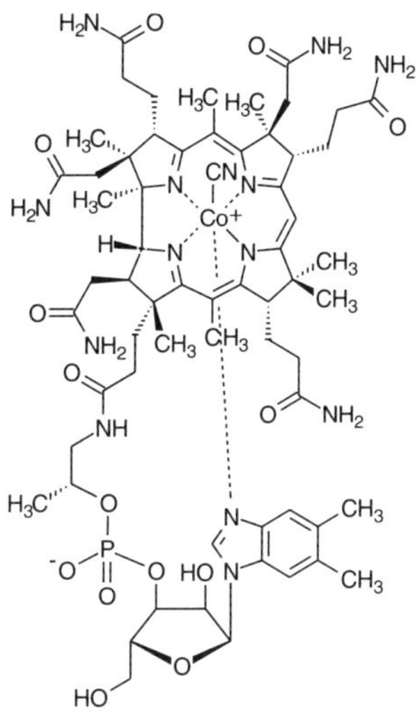

generally understood to mean *cyanocobalamin*, which occurs as an artifact on isolation from naturally occurring *coenzyme B_{12} (the physiologically active form of V. B_{12}) in the presence of cyanide ions or, in cases of cyanide poisoning, from hydroxycobalamin injected as an antidote. V. B_{12} has a *corrin skeleton with a central cobalt(III) atom. The fifth and sixth coordination sites at cobalt are occupied by a nitrogen atom of an imidazole ring and a cyanide ion. V. B_{12} can take up one or two electrons to form the Co(II) complex (V. B_{12r}) or the Co(I) complex (V. B_{12s}), respectively.

Physiological activity: A sufficient supply of V. B_{12} is necessary, above all, for the growth of red blood cells (erythrocytes), as well as for the function of nerve cells and – especially in animal fodder – for growth of livestock. In cases of V. B_{12} deficiency the development of erythrocytes in bone marrow is impaired, they are arrested in an embryonic stage with formation of so-called megaloblasts, i. e., particularly large cells with cell nuclei containing *h(a)emoglobin. In the resultant clinical situations of *megaloblastic* or *pernicious anemia* or *perniciosa*, hyperchromia is observed in the blood picture together with inflammations of the tongue, deficient formation of gastric juices, and changes in the spinal marrow which can result in severe paralysis. V. B_{12} deficiency symptoms can occur in vegetarians since plants contain practically no V. B_{12}. Two V. B_{12} derivatives participate as coenzymes of V. B_{12} in human metabolic processes: methylcobalamin (mecobalamin) and 5′-deoxyadenosylcobalamin (cobamamide, coenzyme B_{12}).

V. B_{12} also functions as an important biochemical catalyst in alkylation and isomerization processes. All reactions are based on the homolytic cleavage of the unique cobalt-carbon bond incorporated in a water-soluble structure [Co(III)→ Co(II)].

Use: V. B_{12} is administered in the treatment of various forms of anemia, especially those with hyperchromia as well as for neuritides (nerve inflammations). In young animals the use of V. B_{12} as an additive results in a better utilization of the fodder.

Occurrence: V. B_{12} is formed during work-up from coenzyme B_{12} by replacement of the 5′-deoxyadenosyl ligand by cyanide.

Production: V. B_{12} is currently mainly produced by microbial fermentation in submersion cultures or by isolation from sludge or activated sludge. In 1996, the world production of V. B_{12} was 8.7 tons, valued at 70 million U.S. $.

Biosynthesis: Only microorganisms such as, e. g., *Propionibacterium shermanii* are able to biosynthesize V. B_{12} [1,2,3]. ^{13}C-FT-NMR spectroscopy played a crucial role in the elucidation of the biosynthetic pathways of V. B_{12} [2] since, in contrast to laborious labelling experiments with ^{14}C and ^{3}H, the biosynthetic processes can be observed directly in NMR tubes by this method. The biosynthesis of V. B_{12} starts from *5-amino-4-oxovaleric acid (δ-ALA) and proceeds via *porphobilinogen to uroporphyrinogen III (see uroporphyrins). The biosynthesis of V. B_{12} involves ca. 20 enzymes and can be divided into three main sections:

1) Synthesis of uroporphyrinogen III from δ-ALA via porphobilinogen. – 2) *C*-Methylation of uroporphyrinogen III by a series of different methyltransferases with use of **S*-adenosylmethionine (SAM), decarboxylation, ring contraction, and incorporation of cobalt with formation of cobyrinic acid, the simplest representative of the naturally occurring corrins. – 3) Conversion of cobyrinic acid to cobalamin and then to the coenzyme form of V. B_{12}.

The stage of cobalt incorporation differs for aerobic and anaerobic organisms producing V. B_{12}. Thus, in *P. shermanii* incorporation of cobalt occurs at the stage of precorrin 2, while in *P. denitrificans* it occurs later at the stage of hydrogenobyrinic acid. Today the biosynthesis of V. B_{12} is considered to be clarified except for a few details.

Detection: By UV spectroscopy, by colorimetric determination of the hydrogen cyanide released under irradiation, biologically via lactic acid formation with *Lactobacillus* species or by nephelometric measurement of their multiplication.

Historical: The curative effects of raw liver in cases of pernicious anemia were discovered in 1926 by Minot and Murphy. In 1948 Folkers and E. L. Smith independently isolated the active principle against pernicious anemia in red crystalline form from liver, milk, and cultures of *Streptomyces griseus*. The constitution was elucidated in 1955 by the research groups of Todd, E. L. Smith, and Crowfoot-Hodgkin (X-ray crystal structure analysis, 1956), for structure, see also *Lit.* [7]. The total synthesis was completed in 1971 by the groups of Woodward [4] and Eschenmoser [5]. Eschenmoser's simulation of the prebiotic synthesis of uroporphyrinogen III [6] and further transformations of uroporphyrinogen III indicate that the corrins may also have a prebiotic origin.

Lit.: [1] Chem. Ber. **1994**, 923–927. [2] Angew. Chem. Int. Ed. Engl. **32**, 1223–1243 (1993); **34**, 383–411 (1995). [3] Chem. Biol. **1**, 119 (1994); Acc. Chem. Res. **26**, 15–21 (1993); Heterocycles **47**, 1051–1066 (1998); Science **264**, 1551–1557 (1994). [4] Pure Appl. Chem. **33**, 145–177 (1973). [5] Angew. Chem. Int. Ed. Engl. **39**, 61 (2000); Naturwissenschaften **61**, 513–525 (1974); Nicolaou, p. 99–136; Science **196**,

1410–1420 (1977). [6]Helv. Chim. Acta **70**, 1115 (1987). [7]Angew. Chem. Int. Ed. Engl. **35**, 167–170 (1996).
gen.: Beilstein E V **26/15**, 338–342 ▪ Dolphin B_{12} ▪ Hager (5.) **7**, 1060, 1117–1120 ▪ Jordan, p. 112–133, 279 ▪ Kräutler, Golding & Arigoni (eds.), Vitamin B_{12} and B_{12}-Proteins, Weinheim: Wiley VCH 1998 ▪ Merck-Index (12.), No. 10152 ff. – *[HS 2936 26; CAS 68-19-9]*

Vitamin C see ascorbic acid.

Vitamin D. Collective name for the fat-soluble steroids with more or less pronounced rickets preventing activity obtained biosynthetically from $\Delta^{5,7}$-*sterols; they are often summarized under the group name *calciferols* (see calcitriol). To date 7 vitamins of the D series have been isolated or synthesized; however, only V. D_2 and V. D_3 have high activities. V. D_2 (ercalciol, ergocalciferol) is formed by irradiation of ergosterol, V. D_3 (cholecalciferol) is a component of fish liver oils. In the strictest sense, V. D is not a vitamin since the human organism is itself able to synthesize the steroid hormone *calcitriol, 1α,25-dihydroxycholecalciferol (1α,25-dihydroxyvitamin D_3), and the precursor cholecalciferol (vitamin D_3). The historical name vitamin D thus does not fit with the current state of knowledge.
Lit.: Adv. Drug Res. **28**, 269–312 (1996) (pharmacology) ▪ Merck-Index (12.), No. 10155–10158 ▪ Ullmann (5.) **A 13**, 154–163. – *Synthesis:* Atta-ur-Rahman **10**, 43–75 ▪ Chem. Rev. **95**, 1877–1952 (1995) ▪ J. Org. Chem. **63**, 2325 (1998) (V. D_3); **64**, 3196 (1999) ▪ Pol. J. Chem. **73**, 1209 (1999) ▪ Synthesis **1994**, 1383–1398; **1999**, 1209.

Vitamin E see tocopherols.

Vitamin H. Synonym for *biotin.

Vitamin H′. Obsolete name for *4-aminobenzoic acid.

Vitamin K₁ [vitamin K_1(20), phytomenadione (INN), phylloquinone].

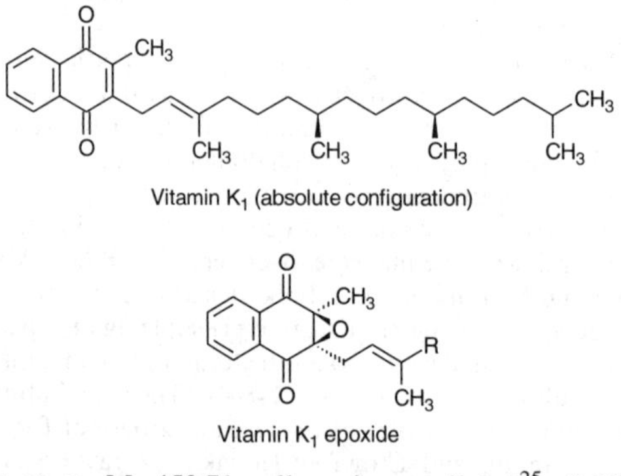

Vitamin K_1 (absolute configuration)

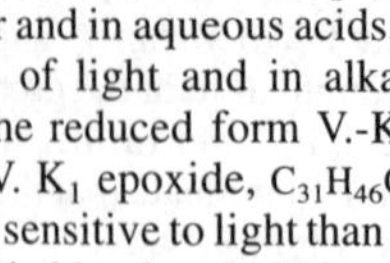

Vitamin K_1 epoxide

$C_{31}H_{46}O_2$, M_R 450.71, yellow viscous oil, $[\alpha]_D^{25}$ –0.28° (dioxan), crystallizes at –20 °C, mp. –4 °C. V. K_1 is stable in moist air and in aqueous acids, it decomposes under the action of light and in alkaline solutions. Besides V. K_1, the reduced form V.-K_1H_2, $C_{31}H_{48}O_2$, M_R 452.72 and V. K_1 epoxide, $C_{31}H_{46}O_3$, M_R 466.70, colorless oil, less sensitive to light than V. K_1, also play an important role in blood coagulation. The mixture of homologues present in many plants is also known as V.-K_2, but distinctions are made according to the number of isoprene residues: V.-K_2(30) (=menaquinone K_6), V.-K_2(35) (=menaquinone K_7), etc. The non-isoprenylated 2-methyl-1,4-naphthoquinone is called menaquinone, menadione, or vitamin K_3.

Use: As an antidote for *dic(o)umarol and warfarin poisonings.
Lit.: Beilstein E IV **7**, 2430, 2496 f., 2681 ▪ Florey **17**, 449–531 (1988) (review) ▪ Helv. Chim. Acta **73**, 1087–1107, 1276–1299 (1990) (synthesis) ▪ J. Chem. Res. (S) **1995**, 104 (synthesis) ▪ Merck-Index (12.), No. 10161 f. ▪ Nat. Prod. Rep. **11**, 251–264 (1994) ▪ Negwer (6.), No. 7926 ▪ Phytochemistry **30**, 3585–3589 (1991) ▪ Rec. Adv. Phytochem. **20**, 243–261 (1986) (biosynthesis). – *[HS 2936 29; CAS 84-80-0 (R,R)-(E)-(–)]*

Vitamin M. Obsolete name for *folic acid.

Vitamin PP. Obsolete name for nicotinic acid and *nicotinamide (see nicotinamide coenzymes).

Vitisins see wine, phenolic constituents of.

Vitispirane [6,9-epoxy-3,5(13)-megastigmadiene].

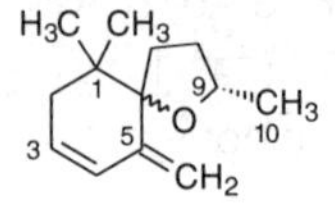

$C_{13}H_{20}O$, M_R 192.30, oil, bp. 57–59 °C (26.6 Pa). A megastigmane sesquiterpene. Natural V. is a mixture of the (6*R*)- and (6*S*)-epimers. It occurs in grape juice [1], vanilla flavor [2], honey, quinces [3], grapes [4], grapefruit juice, and geranium oil. It is presumably formed biosynthetically from *abscisic acid.
Lit.: [1]Chem. Ind. (London) **1977**, 663. [2]Helv. Chim. Acta **61**, 1125 (1978). [3]J. Agric. Food Chem. **36**, 1251 (1988). [4]J. Agric. Food Chem. **39**, 1825 (1991).
gen.: J. Org. Chem. **61**, 1825 (1996) (synthesis) ▪ Ohloff, p. 137, 157, 168 f. – *[CAS 65416-59-3]*

Vogeloside see secoiridoids.

Volicitin.

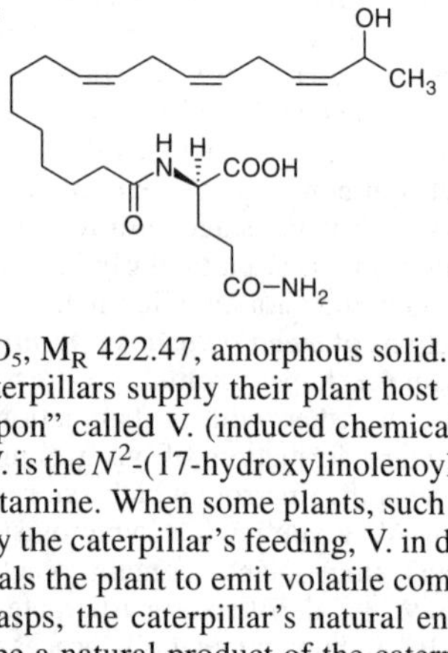

$C_{23}H_{38}N_2O_5$, M_R 422.47, amorphous solid. Beet armyworm caterpillars supply their plant host with a "suicide weapon" called V. (induced chemical defense in plants). V. is the N^2-(17-hydroxylinolenoyl) derivative of L-*glutamine. When some plants, such as corn, are injured by the caterpillar's feeding, V. in damaged tissues signals the plant to emit volatile compounds that attract wasps, the caterpillar's natural enemy. V. appears to be a natural product of the caterpillar, rather than of the plants.
Lit.: Angew. Chem. Int. Ed. Engl. **37**, 1213–1216 (1998) ▪ Proc. Natl. Acad. Sci. USA **95**, 13971 (1998). – *[CAS 191548-06-8]*

Vomicine (12-hydroxy-*N*-methylpseudostrychnine, strychnicine, struxine). $C_{22}H_{24}N_2O_4$, M_R 380.44, $[\alpha]_D^{22}$ +80° (C_2H_5OH), +100° ($CHCl_3$), mp. 284 °C, hexagonal prisms (acetone). An *indole alkaloid isolated, besides *strychnine from seeds of the poison nut tree *Strychnos nux-vomica* (Loganiaceae). V. has an extremely strong nauseous effect (induces vomiting, hence the name).

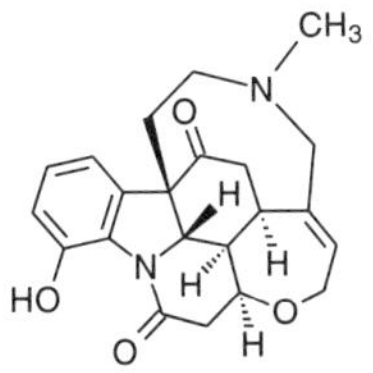

Vomicine

Lit.: Hager (5.) **6**, 817–830 ▪ Manske **6**, 195–204 ▪ Merck-Index (12.), No. 10170 ▪ Phytochemistry **15**, 1973 (1976). – *[CAS 125-15-5]*

Vomitoxin see deoxynivalenol.

Vulgamycin (enterocin).

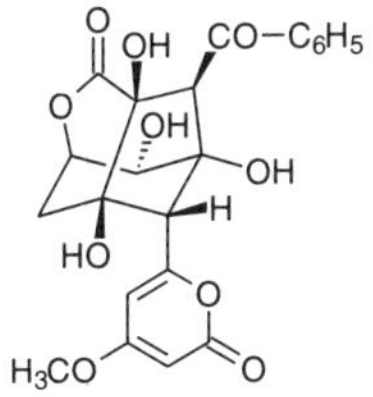

$C_{22}H_{20}O_{10}$, M_R 444.39, mp. 171–173 °C (other reports 163–167 °C), $[\alpha]_D^{20}$ –10.5° (CH_3OH). V. is an antibiotic from *Streptomyces* species with activities against Gram-positive and Gram-negative bacteria, especially intestinal bacteria. V. exhibits synergistic effects with *streptomycin and *chloramphenicol; it also has herbicidal properties and influences the isoleucine metabolism.

Lit.: J. Antibiot. **38**, 1499 (1985) ▪ J. Org. Chem. **61**, 1543 (1996) (isolation) ▪ Pestic. Sci. **33**, 439 (1991) ▪ Tetrahedron **52**, 8073 (1996) (isolation) ▪ Tetrahedron Lett. **48**, 4367 (1976) (biosynthesis). – *[HS 2941 90; CAS 59678-46-5]*

Vulgaxanthins. V. belong to the betaxanthins (see betalains). They are immonium conjugates of *betalamic acid with glutamine (V. I, $C_{14}H_{17}N_3O_7$, M_R 339.31) or

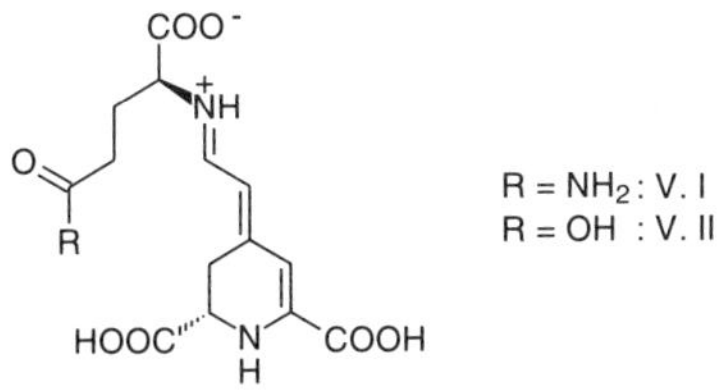

glutamate (V. II, $C_{14}H_{16}N_2O_8$, M_R 340.29). They occur as yellow pigments in *Beta vulgaris*, *Chenopodium rubrum* (Chenopodiaceae), and *Portulaca grandiflora* (Portulacaceae) as well as in other species of the Aizoaceae, Cactaceae, Chenopodiaceae, and Portulacaceae families.

Lit.: J. Food Sci. **45**, 489 (1980) ▪ Manske **39**, 22, 23, 38–51. – *[CAS 904-62-1 (V. I); 1047-87-6 (V. II)]*

Vulp(in)ic acid (methylpulvinic acid).

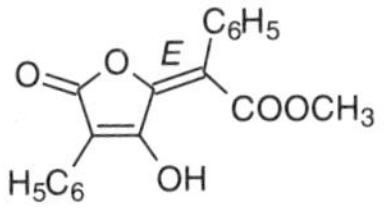

$C_{19}H_{14}O_5$, M_R 322.32, yellow cryst., mp. 148–149 °C (other reports 145–146 °C), soluble in chloroform. A typical lichen acid, e.g., in *Letharia vulpina*. V. a. also occurs in some mushrooms of the order Boletales. Saponification of V. a. affords *pulvinic acid (see also gomphidic acid, xerocomic acid). V. a. has antibacterial as well as antiviral activities[1] and inhibits the growth of plants. V. a. is used in some countries as a natural dye for wool[2].

Lit.: [1] World J. Microbiol. Biotechnol. **6**, 155–158 (1990). [2] Sundström & Sundström, Mit Pilzen Färben, Zürich: Orell Füssli 1984.

gen.: Beilstein E V **18/8**, 611 ▪ J. Chem. Soc., Perkin Trans. 1 **1984**, 1547 ▪ Karrer, No. 1118 ▪ Zechmeister **35**, 177 ff.; **51**, 40 f. ▪ Z. Naturforsch., C **33**, 449 (1978). – *[HS 2932 19; CAS 521-52-8]*

W

Wairol (3-hydroxy-7,9-dimethoxycoumestan).

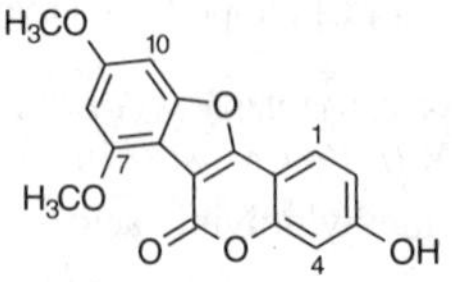

$C_{17}H_{12}O_6$, M_R 312.28, mp. 292–294 °C. W. is a *phytoalexin isolated from alfalfa infected with *Ascochyta imperfecta* [1]. For synthesis, see *Lit.* [2].
Lit.: [1] Phytochemistry **19**, 2801 (1980). [2] Phytochemistry **21**, 249 (1982). – *[CAS 77331-73-8]*

Warburganal (9-hydroxy-7-drimene-11,12-dial).

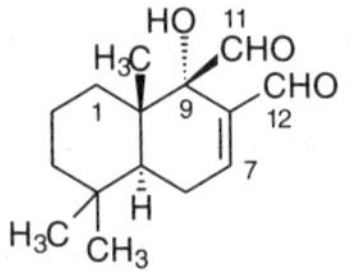

$C_{15}H_{22}O_3$, M_R 250.34, mp. 106–107 °C, $[\alpha]_D^{22}$ +260° ($CDCl_3$). A sesquiterpenoid (drimane) dialdehyde occurring in *Polygonum hydropiper* as well as in East African trees of the genus *Warburgia* (Cannelaceae). W. is an efficient inhibitor of fungal growth [1] and an antifeedant for many insects [2], e.g., the significant pest *Spodoptera*. The biosynthesis of the drimane proceeds through direct cyclization of farnesane derivatives.
Lit.: [1] Isr. J. Chem. **16**, 28 (1977). [2] Agric. Biol. Chem. **48**, 73 (1984); Phytochemistry **21**, 2895 (1982).
gen.: Merck-Index (12.), No. 10173. – *Synthesis:* Atta-ur-Rahman **17**, 233–250 (1995) ▪ Can. J. Chem. **66**, 1675 (1988) ▪ J. Org. Chem. **53**, 855 (1988) ▪ Tetrahedron **50**, 10995 (1994); **51**, 1845–1860 (1995); **53**, 6313 (1997) ▪ Tetrahedron Lett. **35**, 3781 (1994). – *[CAS 62994-47-2]*

Warning substances (odors). W. s. are used by animals, similar to warning colors, to intimidate opponents. Particularly strong-smelling, alkylated methoxypyrazines have been discussed as the W. s. of certain insects [1]; see also alarm substances, defensive secretions.
Lit.: [1] Insect Biochem. **11**, 493–499 (1981).

Wasps. Insects belonging, like *bees, to the order *Hymenoptera* (more than 100000 species) that feed especially on plants but also on meat and other insects. Gall-flies and some hymenopters live as parasites; the latter in particular have been employed for biological pest control since their larvae, after hatching, consume the host insect in which the egg has been placed from within. The best known are the social, yellow jacketed wasps of the family Vespidae (genera *Vespa, Polistes, Vespula, Dolichovespula*) to which the hornets (*Vespa crabro*) also belong, with their mostly characteristic yellow (xanthopterin) and black stripes and poisonous sting. The so-called paper envelopes of the nests are made of a chitin-containing salivary secretion that hardens to a horn-like mass in the air. W. are harmful to man on account of their poisonous sting (see wasp venom and hornet venom) and the transmission of pathogens to food. Control methods include insecticides and fumigating the nests. Many wasp species are now protected on account of their rare occurrence.
Lit.: Annu. Rev. Entomol. **23**, 215–238 (1978) ▪ Jacobs & Renner, Biologie u. Ökologie der Insekten, Stuttgart: Fischer 1988.

Wasp venom. Secretion from the poison bladder of *wasps (similar to *hornet venom) consisting of the biogenic amines *histamine and *serotonin, peptides such as *mastoparan, and wasp kinins (similar to bradykinin) as well as enzymes such as hyaluronidases and phospholipases. Just recently, low-molecular weight polyacylamines such as, e.g., *δ-philanthotoxin have been identified in wasp venoms as well as in *spider venoms. Wasp venoms are hazardous to humans on account of allergies induced by the antigenic action of the venom as well as the possible anaphylactic shock conditions following a sting.
Lit.: Toxikon **29**, 139–149 (1991).

Waxes. Phenomenological or commercial name for a series of natural or artificial substances that generally possess the following properties: kneadable at 20 °C, rigid to brittle hardness, coarse to fine crystalline, translucent to opaque but not glassy, melting at above 40 °C without decomposition, low viscosity and not threading at only a little above the melting point, able to be polished under slight pressure, generally burning with a smoky flame and without formation of ash. Natural W. are classified as plant W. [e.g., candelilla wax, *carnauba wax, *Japan wax, esparto (grass) wax, cork wax, guaruma wax, rice seed oil wax, sugar beet wax, ouricuri wax, montan wax] and animal W. [e.g., *beeswax, *shellac (shellac wax), cetaceum, wool wax (lanolin), rump wax].
Lit.: Sayers, Wax, an Introduction, European Wax Federation, London: Gentry Books 1983. ▪ SÖFW J. **126**, 40 (2000) (review).

Wedelolactone (1,8,9-trihydroxy-3-methoxycoumestan).

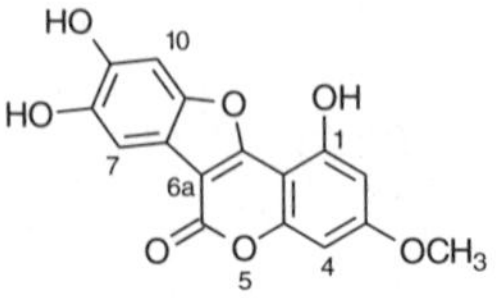

$C_{16}H_{10}O_7$, M_R 314.25. Green-yellow needles, mp. 327–330 °C (decomp.). A coumestan derivative from

Eclipta prostrata, *E. alba*, and *Wedelia calendulacea* (Asteraceae).
In vitro W. protects liver cells from the toxic effects of tetrachloromethane, *galactosamine, and phalloidin (see phallotoxins). The hepatoprotective effect is not present in derivatives lacking a substituent at C-8. The free hydroxy group in the 1 position is also important[1]. In animal experiments W. has a life-saving effect against the venom of the Brazilian rattlesnake (*Bothrops jararaca*)[2]. W. inhibits 5-lipoxygenase and thus acts as an anti-inflammatory agent[1,3]. The biosynthesis proceeds through oxidative cyclization of the corresponding *isoflavone.

Lit.: [1]Arzneim. Forsch. **38**, 661–665 (1988). [2]Toxicon **27**, 1003–1009 (1989). [3]Planta Med. **1986**, 374–377.
gen.: Beilstein EV **19/7**, 60 ▪ J. Heterocycl. Chem. **20**, 635 (1983) (synthesis) ▪ Harborne (1994), p. 189–193. – *[CAS 524-12-9]*

Weevils (Curculionidae). A species-rich beetle family having a trunk-like extended head with the chewing mouthparts at its conspicuous tip. For some of these insects, which include species that cause serious economic damage, their *pheromones are known. Propionate units are mostly involved in the biogenesis of these substances[1]. Similar structures are found in the pheromones of Anobiidae (see serricornin, stegobinone, invictolide, cockroaches, and ants). The crotonate of (*Z*)-3-dodecen-1-ol ($C_{16}H_{28}O_2$, M_R 252.4) is the sexual pheromone of the female sweet potato beetle *Cylas formicarius*[2]. For further pheromone structures, see sitophilate, sitophilure, grandisol, rhynchophorol.

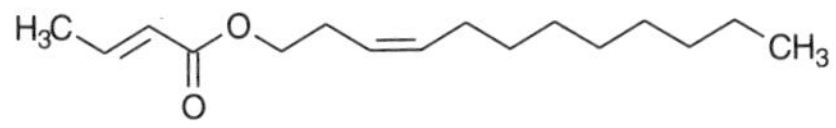

(*Z*)-3-Dodecenyl-crotonate

Lit.: [1]Naturwissenschaften **80**, 90 f., 328 ff. (1993); Justus Liebigs Ann. Chem. **1993**, 865–870, 1201 ff.; J. Chem. Ecol. **20**, 505–515, 2653–2671 (1994). [2]J. Chem. Ecol. **12**, 1489–1503 (1986).

Welwitindolinone A isonitrile see indole alkaloids from cyanobacteria.

WF-3681.

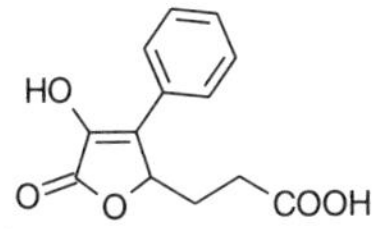

$C_{13}H_{12}O_5$, M_R 248.24, needles, mp. 179–180 °C, (+)-form: $[\alpha]_D$ +130.4° (CH_3OH), an aldose reductase inhibitor from *Chaetomella raphigera*. Various derivatives have been prepared and some of them have more advantageous activities. The enzyme inhibition with the substrate glyceraldehyde is non-competitive.

Lit.: J. Antibiot. **40**, 1394–1399, 1400–1407 (1987) ▪ Tetrahedron Lett. **27**, 2015–2028 (1986). – *Synthesis & derivatives:* Chem. Pharm. Bull. **35**, 2594 (1987); **36**, 1404–1414 (1988) ▪ J. Antibiot. **43**, 1186–1188 (1990); **44**, 441–444 (1991). – *[CAS 90055-27-9 ((+)-form)]*

Whisk(e)y flavor. The typical W. flavor develops on maturation in charred (toasted) wooden barrels (Bourbon) or old sherry barrels (Scotch). Components relevant for the sensory impression are the fusel alcohols (ca. 2000–4000 ppm) and their acetates, ethyl acetate (see fruit esters), and the ethyl esters of saturated and unsaturated fatty acids, acetaldehyde, 3-methylbutanal (see alkanals) and their acetals, as well as the smoky and leathery notes contributed by alkyl- and methoxyphenols [e.g., *guaiacol, 4-ethylguaiacol (see coffee flavor), *eugenol] and the rounding off by the substances extracted from the wood, including vanillin (see 3,4-dihydroxybenzaldehydes), syringa aldehyde (see rum flavor), *sinapaldehyde* ($C_{11}H_{12}O_4$, M_R 208.21), and *whisk(e)y lactone.

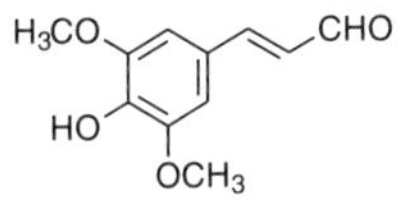

Sinapaldehyde

Lit.: Chem. Ind. (London) **1993**, 827 ▪ Dev. Food. Sci. **37B**, 1767–1778 (1995) ▪ Helv. Chim. Acta **72**, 1362 (1989) ▪ Heterocycles **31**, 1585 ff. (1990) ▪ Int. J. Food Sci. Technol. **28**, 308–318 (1993) ▪ J. Org. Chem. **60**, 5628 ff. (1995) ▪ Maarse, p. 547–580 ▪ Merck-Index (12.), No. 10179 ▪ TNO list (6.), p. 549–576. – *[CAS 4206-58-0 (sinapaldehyde)]*

Whisk(e)y lactone, oak lactone (3-methyl-4-octanolide).

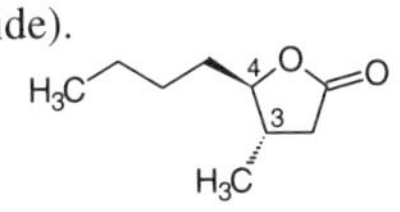

(+)-(4*R*)-*trans*-W.

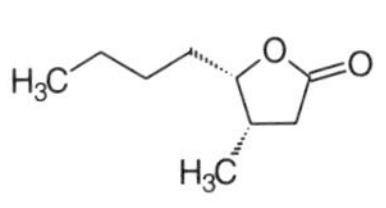

(-)-(4*S*)-*cis*-W.

$C_9H_{16}O_2$, M_R 156.22. Liquid with an odor like celery and coconut, bp. 93–94 °C (0.7 kPa). (+)-(4*R*)-*trans*- and (–)-(4*S*)-*cis*-W. {$[\alpha]_D^{20}$ +96° and –78° (CH_3OH), respectively} occur as oak wood extractives in barrel-matured wines and brandies[1], the *trans*-compound has the stronger aroma (olfactory threshold in 30% alcohol solution 67 ppb, *cis*-W. l. 790 ppb)[2]. For synthesis, see *Lit.*[3].

Lit.: [1]Dev. Food Sci. **29**, 469 (1992); **37B**, 1767–1778 (1995). [2]Maarse, p. 501–504. [3]Chem. Lett. **1995**, 1027; J. Org. Chem. **60**, 5628–5633 (1995); **63**, 1102 (1998); Heterocycles **36**, 1017–1026 (1993); Tetrahedron **52**, 3905–3920 (1996); Tetrahedron: Asymmetry **9**, 1165 (1998). – *[HS 2932 29; CAS 80041-01-6 (+)-(4R); 80041-00-5 (–)-(4S)]*

Whooping cough toxin see pertussis toxin.

Widdrene, widdrol see thujopsene.

Wiedendiols.

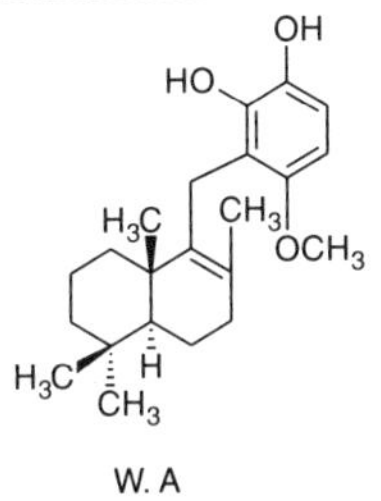

W. A

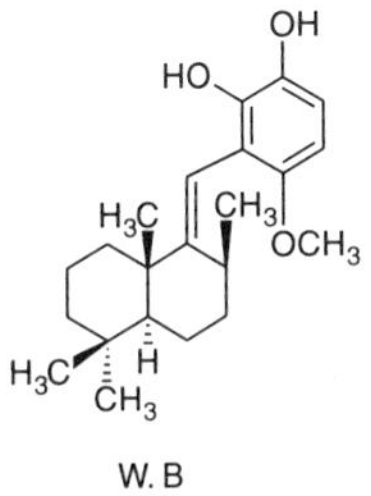

W. B

$C_{22}H_{32}O_3$, M_R 344.48, amorphous solids. Merosesquiterpenes from the sponge *Xestospongia wiedenmayeri*. W. inhibit cholesterylester-transferprotein (CETP), having anti-atherosclerotic potential, see avarol.

Lit.: Tetrahedron **54**, 5635–5650 (1998) (synthesis) ▪ Tetrahedron Lett. **38**, 8101 (1997). – *[CAS 162341-30-2 (W. A); 162341-31-3 (W. B)]*

Wieland-Gumlich aldehyde [(17*R*)-19,20-didehydro-17,18-epoxycuran-17-ol, deacetyldiaboline, caracurine VII].

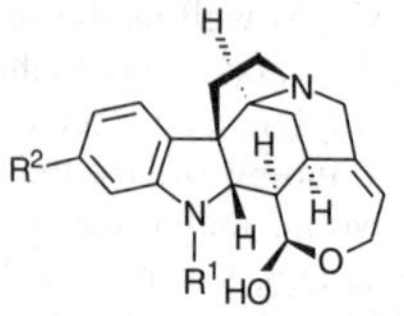

$R^1 = R^2 = H$: Wieland-Gumlich aldehyde
$R^1 = CO-CH_3$, $R^2 = H$: Diaboline
$R^1 = CO-CH_3$, $R^2 = OCH_3$: 11-Methoxydiaboline

$C_{19}H_{22}N_2O_2$, M_R 310.40, $[\alpha]_D^{22}$ –133.8° (CH_3OH); mp. 213–214 °C (decomp.). W.-G. a. is a degradation product[1] of *strychnine (Wieland 1933) and was of decisive significance for the structural elucidation of that alkaloid. The aldehyde also occurs as a natural product and is isolated together with *diaboline* {$C_{21}H_{24}N_2O_3$, M_R 352.43, mp. 187–189 °C, $[\alpha]_D^{27}$ +44.8° (C_2H_5OH)} and *11-methoxydiaboline* {$C_{22}H_{26}N_2O_4$, M_R 382.46, mp. 214–216 °C, $[\alpha]_D^{20}$ +20° ($CHCl_3$)} as the main monomeric products from numerous African *Strychnos* species.

Lit.: [1]Justus Liebigs Ann. Chem. **506**, 60 (1933).
gen.: J. Am. Chem. Soc. **117**, 5776–5788 (1995) (asymmetric synthesis) ▪ Merck-Index (12.), No. 10181 ▪ Saxton (ed.), The Monoterpenoid Indole Alkaloids, Suppl. Vol. 25, Heterocyclic Compounds, p. 279, Chichester: Wiley 1994 ▪ Ullmann (5.) **A 1**, 390. – *[CAS 466-85-3 (W.-G. a.); 509-40-0 (diaboline)]*

Wikstromol [(3*R*)-*cis*-(+)-dihydro-3-hydroxy-3,4-bis(4-hydroxy-3-methoxybenzyl)-2(3*H*)-furanone].

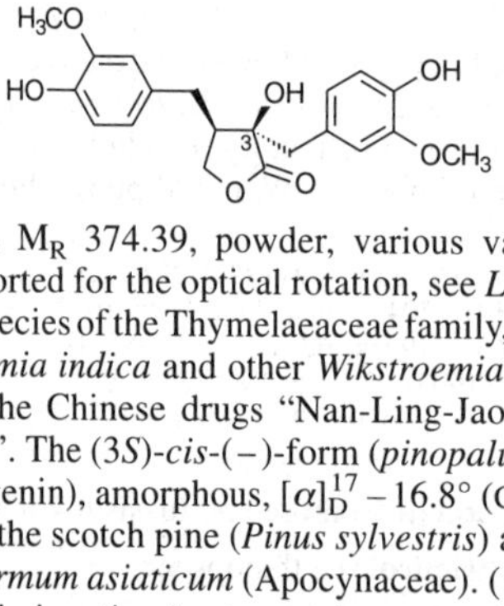

$C_{20}H_{22}O_7$, M_R 374.39, powder, various values have been reported for the optical rotation, see *Lit.*[1]. W. occurs in species of the Thymelaeaceae family, especially *Wikstroemia indica* and other *Wikstroemia* species as well as the Chinese drugs "Nan-Ling-Jao-Hua" and "Po-Lun". The (3*S*)-*cis*-(–)-form (*pinopalustrin*, nortrachelogenin), amorphous, $[\alpha]_D^{17}$ –16.8° (C_2H_5OH), is found in the scotch pine (*Pinus sylvestris*) and in *Trachelospermum asiaticum* (Apocynaceae). (–)-Nortrachelogenin is active *in vitro* against HIV and leukemia P 388. W. is formed as a *lignan from *coniferyl alcohol.

Lit.: [1]Tetrahedron **48**, 10 115–10 126 (1992) (synthesis).
gen.: Beilstein E V **18/5**, 463 ▪ J. Nat. Prod. **42**, 159 ff. (1979); **53**, 1587–1592 (1990) ▪ J. Org. Chem. **53**, 4724–4729 (1988) ▪ Nat. Prod. Rep. **12**, 183–205 (1995). – *[CAS 61521-74-2 (3R-cis); 34444-37-6 (3S-cis)]*

Wildfire toxin see tabtoxin.

Willardiine (*β*-uracil-1-yl-L-alanine).

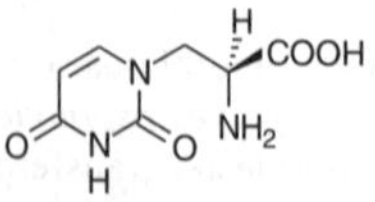

$C_7H_9N_3O_4$, M_R 199.17, mp. 206–211 °C (decomp.), $[\alpha]_D$ –20° (1 m HCl). A non-proteinogenic amino acid in *Acacia willardiana*[1] and other *A.* spp[2], as well as in other Mimosaceae, see isowillardiine.

Lit.: [1]Z. Physiol. Chem. **316**, 164 (1959). [2]Phytochemistry **16**, 565–570 (1977); **17**, 1571 (1978) (biosynthesis).
gen.: Beilstein E V **24/6**, 87 ▪ Org. Magn. Reson. **29**, 641 (1991). – *[CAS 21416-43-3]*

Wilting (withering) agents. Collective name for all substances which can cause a non-physiological wilting of plants – in the broadest sense this can also include defoliants. For example, phytopathogenic microfungi living as parasites or symbionts on plants (e.g., *Fusarium* species) produce so-called *wilting toxins*[1] (also known as *marasmins*, from Greek: marasmos = to wilt away) that can act in higher plants both directly at the site of attack (near symptoms) and at leaves, stems, etc. more distant from the hyphae of the fungus. The W. a. impair respiration and permeability. The permeability of the cell plasma for water and dissolved substances is increased. The constitution of some of these W. a. has been elucidated, e.g., *fusaric acid, as well as various peptide W. a. (marasmin; L-*γ*-glutamyl-L-marasmin[2,3]) and amino acids (*lycomarasmin, aspergillomarasmins A and B[4]) derived from marasmins. An aspergillomarasmin A/Fe^{3+} chelate catalyzes the photooxidation in plant leaves (Fenton reaction) and thus leads to phytotoxicity[5]. The W. a. probably act in a similar manner to *plant growth substances also in the natural processes of wilting and winter rest. Signs of wilting may also occur as deficiency symptoms (e.g., in cotton under vanadium-deficient conditions) or under the influence of competing plants (*example:* *canavanine). Some plants are able to detoxify W. a., thus, for example, fusaric acid is detoxified by some tomato species through methylation. In the case of wilting diseases of fungal origin, fungicides are used to control the pathogenic fungus.

Lit.: [1]Helv. Chim. Acta **54**, 845–851 (1971); Physiol. Plant. Pathol. **1** 495–514 (1971). [2]Phytochemistry **15**, 1717 (1976). [3]J. Org. Chem. **52**, 1511 (1987). [4]Helv. Chim. Acta **48**, 729 (1967); Annu. Phytopathol. **3**, 407 f. (1971); Agric. Biol. Chem. **47**, 2693 (1983). [5]Photobiochem. Photobiophys. **7**, 53–58 (1984).
gen.: Addicott (ed.), Abscisic Acid in Abscission, p. 269–300, New York: Praeger 1983 ▪ Naturwissenschaften **71**, 210 f. (1984).

Wine, phenolic constituents of. Besides the volatile constituents of *wine flavor, alcohols, *tartaric acid and tartrates, especially red wines contain phenolic constituents: stilbenes, e.g. resveratrol, the vitisins[1], *tannins and other polyphenols like *ellagic acid and the viniferins[2], and other oligomeric stilbenes, e.g. *ε*-viniferin. Mainly to these phenolic compounds, the anti-atherosclerotic and infarction-prophylactic activity is attributed, explaining the so-called "French Para-

Table: Data of Wine phenolic compounds.

	molecular formula	M_R	CAS
Vitisin D	$C_{56}H_{42}O_{12}$	906.93	206192-29-2
Vitisin E	$C_{42}H_{32}O_9$	680.71	206192-30-5
Castavinol	$C_{26}H_{30}O_{13}$	550.52	183607-09-2
5′-Methoxy-castavinol	$C_{27}H_{32}O_{14}$	580.54	183607-16-1
5′-Hydroxy-castavinol	$C_{26}H_{30}O_{14}$	566.52	183607-17-2
ε-Viniferin	$C_{28}H_{22}O_6$	454.48	62218-08-0

doxon" of red wines[3]. Polyphenols and tannins often originate from the storage of young wines in oakwood barrels (barriques); but the vitisins, the viniferins or castavinols[4] originate from the grape skins and wooden parts of the grapewines.

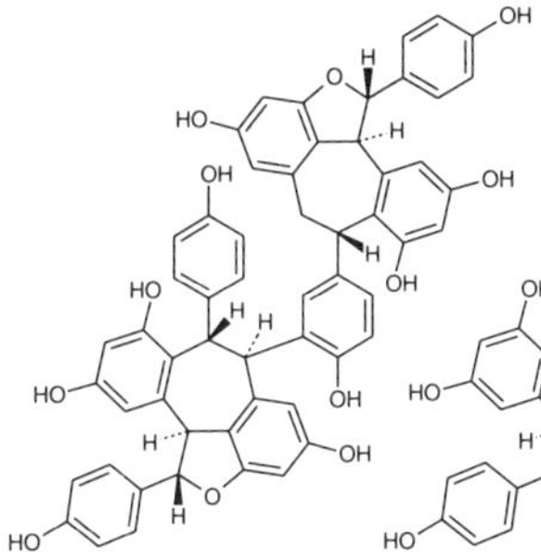
Vitisin D

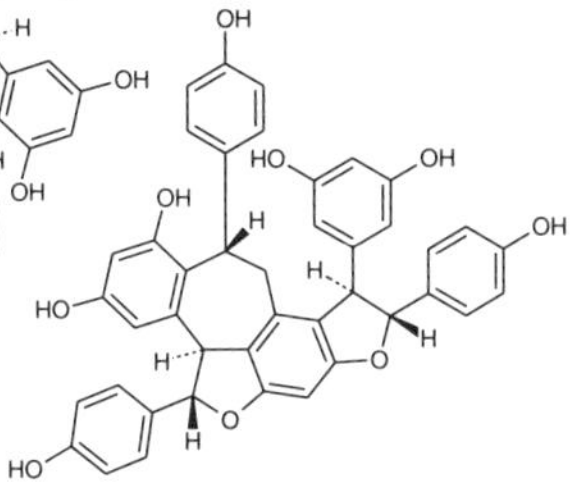
Vitisin E

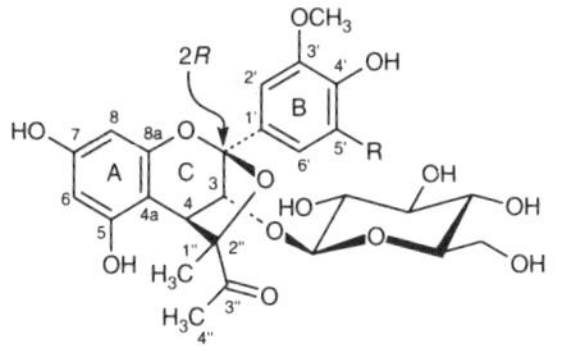
R = H : Castavinol
R = OCH_3 : 5'-Methoxycastavinol
R = OH : 5'-Hydroxycastavinol

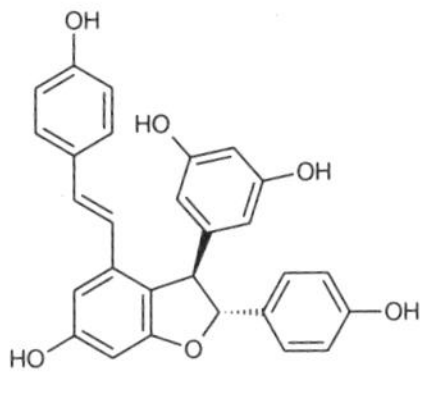
ε-Viniferin

Also *flavonoid constituents of red wine contribute to the antioxidative activity of wine and – like the polyphenols – are altered by microbial processes and ageing (air exchange through the cork stopper). Phenolic constituents inhibit the oxidation of human low density lipoprotein (LDL) *in vitro*[5].

Lit.: [1] J. Org. Chem. **58**, 850 (1993); Heterocycles **46**, 169 (1997). [2] Chem. Pharm. Bull. **46**, 655 (1998); Phytochemistry **16**, 1193 (1977) (biosynthesis); **32**, 1083 (1993); **37**, 1157 (1994); **38**, 1501 (1995). [3] J. Agric. Food Chem. **45**, 3995–4003 (1997). [4] Bull. Soc. Pharm. Bordeaux **136**, 19–36 (1997); Tetrahedron Lett. **37**, 7739 (1996). [5] J. Agric. Food Chem. **45**, 1638 (1997); Lancet **341**, 454, 1103 (1993).

gen.: ACS Symp. Ser. **661**: Wine, Nutritional and Therapeutic Benefits (Watkins, ed.), Washington, DC: ACS 1997 ▪ Dubourdieu et al., Handbook of Enology, Weinheim: Wiley-VCH 2000 ▪ GIT Fachz. Lab. **41**, 882–886, 890 (1997) ▪ J. Wine Res. **4**, 79 (1993) ▪ Science **275**, 218 (1997).

Wine flavor. Currently ca. 800 components of wine have been identified but very little research has been directed at their contributions to the flavor and bouquet of wines. Quantitatively and qualitatively, ethyl esters and acetates dominate and, of these, the ethyl, 3-methylbutyl, and hexyl acetates as well as ethyl hexanoates and octanoates (see fruit esters) exhibit higher aroma values. Of the lactones, *wine lactone*[1] (*P*-menth-1-ene-9,3-olide, $C_{10}H_{14}O_2$, M_R 166.22, *4-nonanolide* [$C_9H_{16}O_2$, M_R 156.22, bp. 136 °C (1.7 kPa), $[\alpha]_D^{20}$ +51.8° (*R*)] is of some significance in Pinot Noir and Merlot, *whisk(e)y lactone in wines kept in barrels, and sotolone (see hydroxyfuranones) in *Botrytis*-infected wines (e.g. sweet wines from region Sauternes). Ethyl cinnamate, *linalool, *geraniol, and nerol (see geraniol) are characteristic for specific types of wine, as are *β*-*ionones in muscatel, *β*-

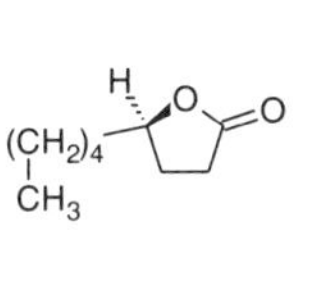
(*R*)-4-Nonanolide

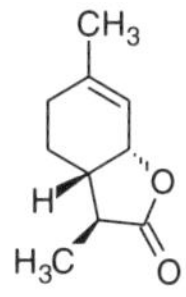
Wine lactone

*damascenone in Riesling and Chardonnay; *4-vinylguaiacol in Gewurztraminer, and 2-alkyl-3-methoxypyrazines (see pyrazines) in red and white Sauvignon wines (Cabernet Sauvignon, Sauvignon Blanc). The distribution pattern of the monoterpene compounds provides a means to distinguish between various sorts of white grapes[2]. The most important flavor compound in Semillon wines is (3*S*,5*R*,6*S*,9ζ)-megastigm-7-ene-3,6,9-triol[3]. *Vitispirane and dimethyl sulfide are found in mature white wines, phenols, e.g., *4-ethylphenol* ($C_8H_{10}O$, M_R 122.17) and ethylguaiacol (see coffee flavor), especially in red wines. A frequent complaint, the taste of cork, is due to a concentration of >20 ppt of *2,4,6-trichloroanisole* ($C_7H_5Cl_3O$, M_R 211.48) which can be formed microbially in chlorine-bleached corks. Another explanation for the taste of cork are secondary metabolites from the mycelium of *Armillaria mellea*, a basidiomycete, that invades the bark of the cork tree (*Quercus suber*[4]).

Lit.: [1] Eur. J. Org Chem. **2000**, 419; Helv. Chim. Acta **79**, 1559–1571 (1996). [2] Fresenius J. Anal. Chem. **337**, 777–785 (1990). [3] Aust. J. Grape Wine Res. **2**, 179 (1996). [4] Chem. Eng. News (11.9.) **1995**, 39 f.

gen.: Chem. Ind. (London) **1995**, 338–341 ▪ Chem. Unserer Zeit **32**, 87–93 (1998) (review) ▪ J. Agric. Food Chem. **39**, 1833 ff. (1991); **46**, 1474 (1998) (Riesling) ▪ Maarse, p. 483–546 ▪ Spec. Publ.-Royal Soc. Chem. **197**, 163–167 (1996) ▪ TNO-list (6.), p. 593–627. – *[CAS 123-07-9 (4-ethylphenol); 87-40-1 (2,4,6-trichloroanisole); 104-61-0 (4-nonanolide); 63357-96-0 (R-4-nonanolide); 182699-77-0 (3R,4S,8S-wine lactone)]*

Winterin.

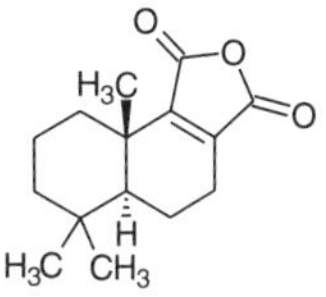

$C_{15}H_{20}O_3$, M_R 248.32, mp. 158 °C, $[\alpha]_D$ +109° ($CHCl_3$). A *sesquiterpene of the *drimane group, isolated from *Drimys winteri* (Winteraceae), a South American tree or bush now grown as an ornamental tree in Ireland and the Mediterranean region. The biosynthesis of the drimanes proceeds by direct cyclization of farnesane derivatives.

Lit.: Aust. J. Chem. **45**, 969 (1992) ▪ Beilstein E V **17/11**, 229 ▪ Chem. Pharm. Bull. **29**, 1580, 1586 (1981) ▪ J. Chem. Res. (*S*) **1998**, 560 (synthesis) ▪ Tetrahedron **45**, 1567 (1989). – *[CAS 6754-56-9]*

Winterstein's acid [(*R*)-3-(dimethylamino)-3-phenylpropanoic acid].

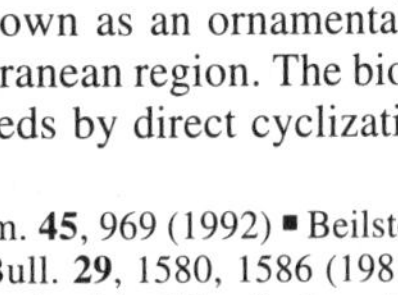

$C_{11}H_{15}NO_2$, M_R 193.24. A non-proteinogenic amino acid. Component of the *Taxus* alkaloids.

Lit.: Beilstein E IV **4**, 1538 ▪ Tetrahedron: Asymmetry **1**, 279 (1990). – *[HS 2916 39; CAS 94099-69-1]*

Wistarin.

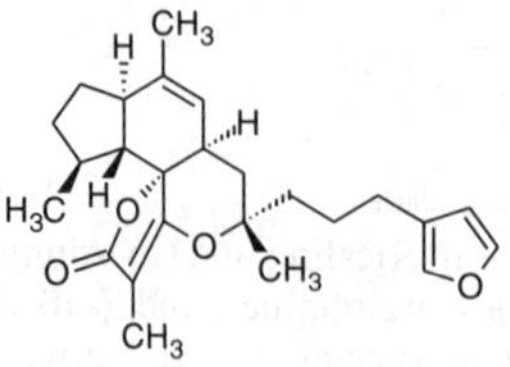

$C_{25}H_{32}O_4$, M_R 396.53, oil, $[\alpha]_D^{20}$ +130° (CH_2Cl_2). Diterpenoid constituent of *Ircinia wistarii.*

Lit.: J. Nat. Prod. **45**, 412 (1982) (isolation) ▪ J. Org. Chem. **62**, 1691 (1997) (synthesis) ▪ Tetrahedron: Asymmetry **10**, 3869 (1999) (structure). – *[CAS 83995-04-4]*

Withanolides.

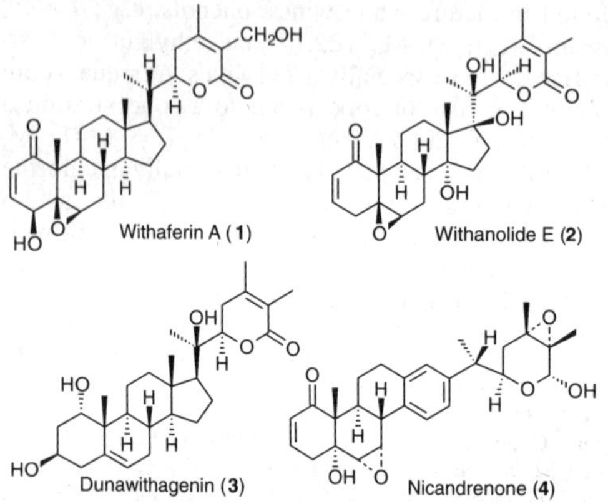

Table: Data of Withanolides.

	molecular formula	M_R	mp. [°C]	$[\alpha]_D$ ($CHCl_3$)	CAS
1	$C_{28}H_{38}O_6$	470.61	252–253	+125°	5119-48-2
2	$C_{28}H_{38}O_7$	486.61	167–168	+103.5°	38254-15-8
3	$C_{28}H_{42}O_5$	458.64	273	+19.8°	79240-65-6
4	$C_{28}H_{34}O_6$	466.57	117		40071-64-5

Highly substituted C_{28}-steroids in plants that are derived from ergostane (see sterols) and possess a characteristic δ-lactone side chain at position 17 and sometimes also a γ-lactone side chain. Today over 100 W. are known from the Solanaceae genera *Withania*, *Acnistus*, *Datura*, *Dunalia*, *Lycium*, *Nicandra*, and *Physalis*. They usually occur in the free form, in a few cases also as glycosides. The most important and longest known W. is *withaferin A* from *Withania somnifera* and *Acnistus arborescens* and is purported to have antitumor activity. Further representatives are, e.g., *withanolide E* and *dunawithagenin* (also from *W. somnifera*) as well as *nicandrenone* (Nic 1) with the 17 (13→18)*abeo*-structure, an aromatic ring D, and a cyclohemiacetal function instead of a lactone function (an insect antifeedant compound from *Nicandra physaloides*).

Lit.: Atta-ur-Rahman **20**, 135–262 (1998) (review) ▪ D'Arcy (ed.), Solanaceae, Biology and Systematic, p. 187–200, New York: Columbia Univ. Press 1986 ▪ Fitoterapia **69**, 433 (1998) ▪ Indian J. Chem. Sect. B **35**, 1311 (1996) ▪ J. Am. Chem. Soc. **113**, 1057 ff. (1991) ▪ J. Chem. Soc., Perkin Trans. 1 **1991**, 739 ▪ J. Indian Chem. Soc. **75**, 672 (1998) ▪ J. Nat. Prod. **54**, 554–563, 599 ff. (1991); **59**, 717 (1996); **62**, 1445 (1999) ▪ Merck-Index (12.), No. 10184 (withaferin A) ▪ Phytochemistry **35**, 927 (1994) (biosynthesis); **44**, 1163 (1997); **47**, 1427 (1998) ▪ Zechmeister **63**, 1–106.

Wood sugar see D-xylose.

Wormseed oil see oil of chenopodium.

Wortmannin.

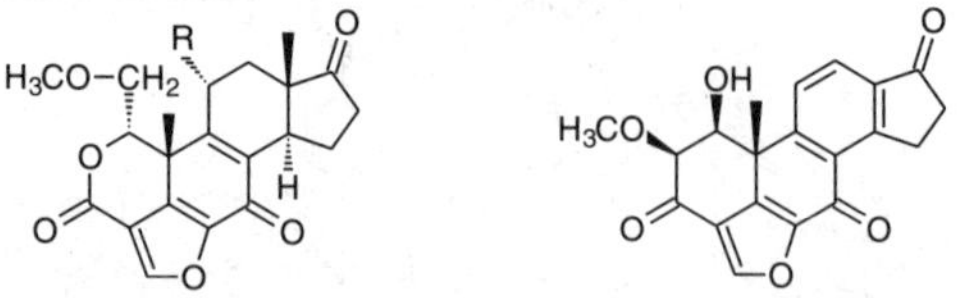

R = O–CO–CH_3 : Wortmannin
R = H : 11-Deacetoxyw.

Viridin

$C_{23}H_{24}O_8$, M_R 428.44, needles, mp. 240 °C, $[\alpha]_D$ +89° ($CHCl_3$). A steroid derivative from cultures of *Penicillium wortmanni* and *Myrothecium roridum.*[1] W. has antifungal and anti-inflammatory activities and shows cytotoxicity. Its *11-deacetoxy* derivative [$C_{21}H_{22}O_6$, M_R 370.40, cryst., mp. 178–180 °C, $[\alpha]_D$ +104° ($CHCl_3$)] is used in the treatment of rheumatism, other derivatives show antineoplastic activities (inhibition of phosphatidylinositol-3-kinase)[2]. A close relative is *viridin* [$C_{20}H_{16}O_6$, M_R 352.34, prisms, mp. 208–217 °C (245 °C decomp.), $[\alpha]_D$ –224° ($CHCl_3$)][3] from cultures of *Trichoderma viride* and *Gliogladium virens* with potent antifungal activity. The biosynthesis of this fungal metabolite proceeds similarly to that of steroids via squalene and lanosterol[1,4].

Lit.: [1]Merck-Index (12.), No. 10188. [2]Helv. Chim. Acta **56**, 2901 (1973); J. Med. Chem. **39**, 5021 (1996). [3]Chem. Commun. **1999**, 1947 (review); J. Chem. Soc., Perkin Trans. 2 **1972**, 760; Phytochemistry **27**, 387 (1988). [4]J. Chem. Soc., Perkin Trans. 1 **1980**, 422; Can. J. Microbiol. **33**, 963 (1987).
gen.: J. Org. Chem. **48**, 741 (1983) ▪ Nat. Prod. Rep. **12**, 381 (1995) (review) ▪ Tetrahedron Lett. **37**, 6141 (1996); **40**, 311 (1999) (synthesis). – *[CAS 19545-26-7 (W.); 31652-69-4 (11-deacetoxy-W.); 3306-52-3 (viridin)]*

WS75624-A and B. Endothelin-converting enzyme (ECE) inhibitors from fermentation broths of *Saccharothrix* sp. no. 75624. ECE inhibitors are a new therapeutic principle for therapy of cardiac insufficiency, kidney failure and other circulatory diseases. W. belong to the few non-peptidic small molecules with this effect, IC_{50} (W. B): 30 ng/mL.

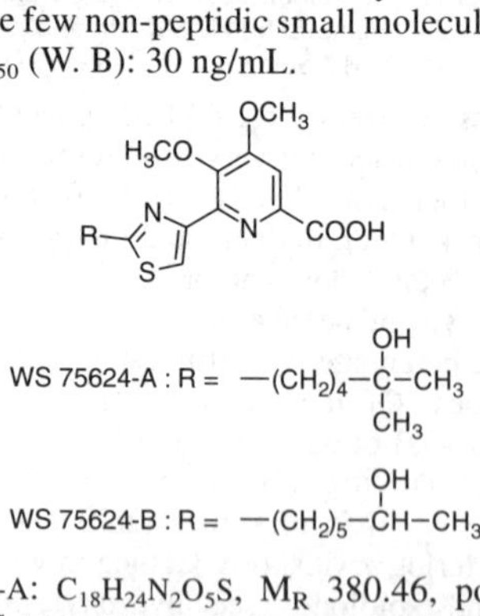

WS 75624-A: $C_{18}H_{24}N_2O_5S$, M_R 380.46, powder, mp. 53–54 °C; WS 75624-B: $C_{18}H_{24}N_2O_5S$, M_R 380.46, powder, mp. 139–140 °C, $[\alpha]_D$ +3° (CH_3OH).

Lit.: J. Antibiot. **48**, 1066, 1073 (1995) ▪ Tetrahedron Lett. **38**, 1297 (1997). – *[CAS 170663-44-2 (A); 157242-75-6 (B)]*

Wyerone see polyynes.

X

Xanthine [2,6-purinediol, 3,7-dihydro-1*H*-purine-2,6-dione].

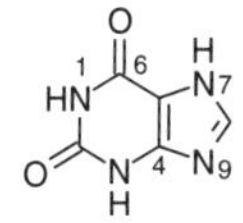

7*H*-form

$C_5H_4N_4O_2$, M_R 152.11, cryst., mp. >350 °C, decomposition to carbon dioxide, ammonia, and hydrogen cyanide on further heating. X. can be detected by the murexid or Weidel-Kossel reaction.

Occurrence: In small amounts in muscles, liver, urinary calculi, beet juice, barley shoots, fly agarics, peanut kernels, potatoes, yeasts, coffee beans, tea leaves. X. is formed in the metabolism of higher animals by deamination of guanine (component of nucleic acid) or oxidation of hypoxanthine by xanthine oxidase present in muscles which then also oxidizes X. further to uric acid, the final product of the *purine metabolism in humans. The nucleoside derived from X. is *xanthosine. Although X. is chemically closely related to *caffein(e) and other methylxanthines, its activity is different. Stimulation of the central nervous system is less pronounced, the paralyzing effects dominate, and cardiac muscles are severely damaged.

Lit.: Adv. Chromatogr. **25**, 245–278 (1986) ▪ Beilstein EV **26/13**, 527 f. ▪ Chest **106**, 1407 (1994) (activity) ▪ Karrer, No. 2554 ▪ Merck-Index (12.), No. 10193 ▪ Neish, Xanthines and Cancer, Oxford: Pergamon Press 1988 ▪ Prog. Clin. Biol. Res. **158** (The Methylxanthine Beverages and Foods) (1984) ▪ Sax (8.), XCA 000 ▪ Ullmann (5.) **A 2**, 461 ff. – *[HS 2933 59; CAS 69-89-6]*

Xanthobaccin A.

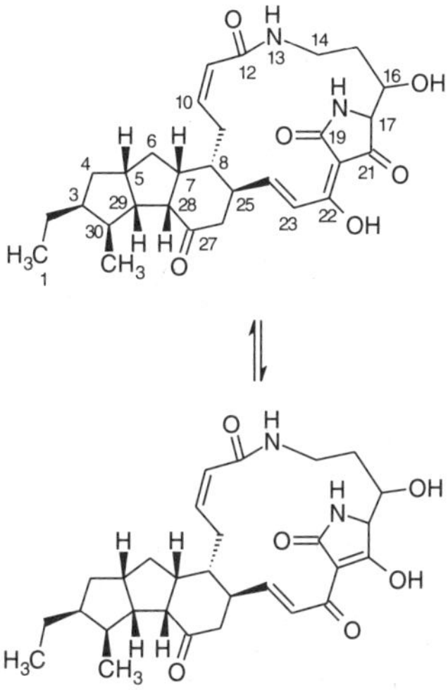

Antibiotic (antifungal activity) of the *ikarugamycin type from broth cultures of *Stenotrophomonas* sp., tautomeric equilibrium (see formula): $C_{29}H_{38}N_2O_6$, M_R 510.61, colourless, amorphous powder.

Lit.: Tetrahedron Lett. **40**, 2957 (1999). – *[CAS 184877-66-5]*

Xanthocillin (xantocillin).

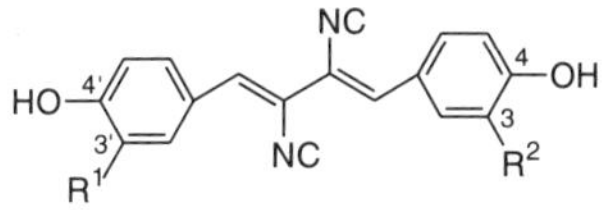

R^1	R^2	
H	H	X. X
OH	H	X. Y_1
OH	OH	X. Y_2

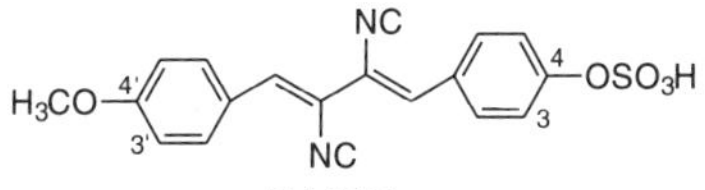

BU 4704

Table: Data of Xanthocillin.

	molecular formula	M_R	mp. [°C]	CAS
X. X	$C_{18}H_{12}N_2O_2$	288.31	200	580-74-5
X. Y_1	$C_{18}H_{12}N_2O_3$	304.30		38965-69-4
X. Y_2	$C_{18}H_{12}N_2O_4$	320.30		38965-70-7
BU 4704	$C_{19}H_{14}N_2O_5S$	382.40	165–170 (decomp.)	149997-61-5

An *antibiotic mixture from cultures of *Penicillium notatum* and *Eupenicillium egyptiacum* consisting of at least 3 components with central 2,3-diisocyano-1,3-butadiene units. The cytotoxic *X. X* (yellow crystals) can be isolated as the main component [1] and is also accessible by total synthesis [2]. X. X has bacteriostatic activity against Gram-positive and Gram-negative pathogens; it is used in dermatology and against infections of the pharynx [3]. The 4-monomethyl ether of X. X (from *Dichotomomyces albus* and *D. cejpii*, $C_{19}H_{14}N_2O_2$, M_R 302.33, yellow cryst.), represents a potent inhibitor of *prostaglandin biosynthesis and is active against some Gram-positive and Gram-negative bacteria. *X. Y_1* (yellow needles/rhombic crystals) and *X. Y_2* have been described as further components of the mixture [4]. Another member of the X. family is the antibiotic *BU 4704* (white needles, K salt) first isolated from cultures of *Aspergillus* sp. in 1993 [5].

Lit.: [1] Pharmazie **26**, 503 (1971); Chem. Ber. **105**, 784 (1972); Chem. Pharm. Bull. **24**, 2317 (1976); Cryst. Struct. Commun. **10**, 1497 (1981); J. Chem. Soc., Perkin Trans. 1 **1991**, 595 (biosynthesis). [2] Angew. Chem. Int. Ed. Engl. **1**, 212 (1962). [3] Pharmazie **26**, 503 (1971). [4] J. Org. Chem. **42**, 244 (1977). [5] J. Antibiot. **46**, 687 (1993).

gen.: Merck-Index (12.), No. 10195 ▪ Sax (8.), No. DVU 000, XCS 680. – *[HS 2941 10; CAS 19559-24-1 (4-O-methyl-X. X)]*

Xanthodermin {L-glutamic acid 5-[N'-(4-hydroxyphenyl)hydrazide}.

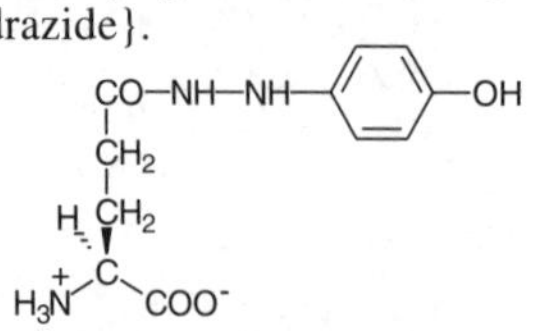

$C_{11}H_{15}N_3O_4$, M_R 253.26, cryst., mp. 164–166 °C, $[\alpha]_D$ +4.5° (1 N HCl). X. exhibits an intense yellow coloration with potassium hexacyanoferrate(III) in weakly alkaline solution. X. occurs in the toadstool *Agaricus xanthoderma* (Basidiomycetes) and is responsible for the chrome-yellow color that develops when the fungus is injured. Besides X., the fungus contains a further chromogen, leukoagaricone, which on contact with atmospheric oxygen (oxidases) is oxidized to the orange-yellow *agaricone.

Lit.: Angew. Chem. Int Ed. Engl. **24**, 709 (1985) ▪ Zechmeister **51**, 238–242 (review). – *[CAS 99280-76-9]*

Xanthokermic acid see laccaic acids.

Xanthomegnin.

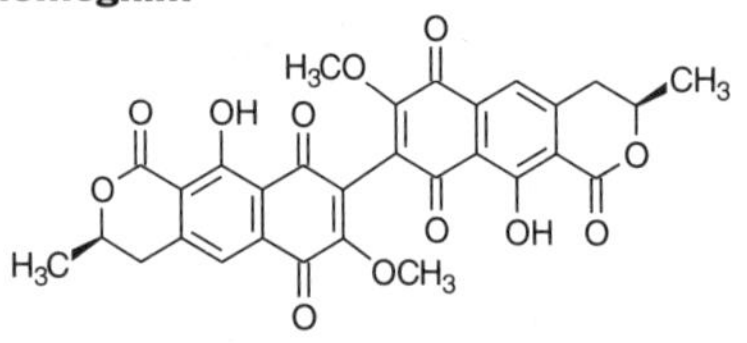

$C_{30}H_{22}O_{12}$, M_R 574.50, yellow-red crystals, decomp. at 264 °C. A hydroxynaphthoquinone produced by various species of fungi (e.g., *Trichophyton, Penicillium, Aspergillus*). X. is formed biosynthetically on the polyketide pathway and often occurs together with its precursors (e.g., *semi*-vioxanthin, vioxanthin, viomellein). X. is a *mycotoxin, it has antibacterial and insecticidal activities and influences oxidative phosphorylation in rat liver mitochondria.

Lit.: Appl. Environ. Microbiol. **45**, 1937 (1983) ▪ Chem. Ber. **112**, 957 (1979) ▪ Mycopathologia **87**, 43 (1984). – *[CAS 1685-91-2]*

Xanthommatin see ommochromes.

Xanthophylls see carotinoids, lutein, and zeaxanthin.

Xanthopterin see pteridines.

Xanthosine (ribosylxanthine, 3,9-dihydro-9-β-D-ribofuranosyl-1*H*-purine-2,6-dione).

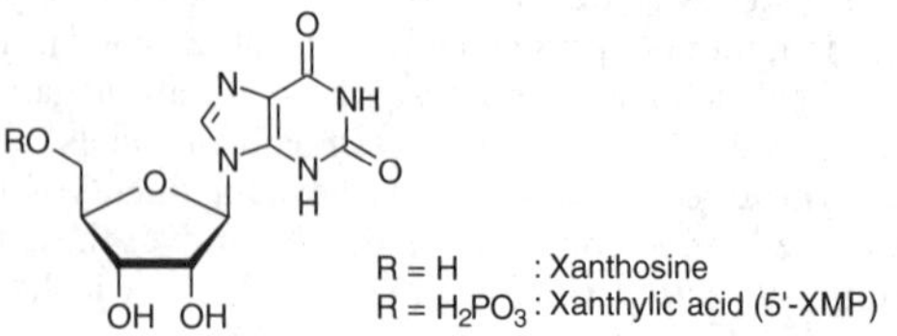

$C_{10}H_{12}N_4O_6$, M_R 284.23, prisms. The dihydrate decomposes on heating without a discrete mp., $[\alpha]_D^{30}$ –51.2° (0.3 N NaOH), poorly soluble in cold, readily soluble in hot water and hot alcohol, it is easily hydrolyzed by acids. The nucleoside from *xanthine and D-*ribose is formed by guanine-deficient mutants of some bacterial strains (e.g., *Bacillus subtilis* or *Aerobacter aerogenes*) and also occurs in the seeds of *Trifolium alexandrinum. Xanthylic acid* (5'-XMP), $C_{10}H_{13}N_4O_9P$, M_R 364.21, which does not occur in nucleic acids, is formed in the purine nucleotide metabolism from inosine 5'-monophosphate and is the precursor of guanosine 5'-monophosphate.

Activity: Similar to inosine and guanosine 5'-monophosphates, XMP also has a taste-intensifying effect, in this case for the taste "salty". It is produced technically from yeast or by bacterial fermentation.

Lit.: Beilstein E V **26/13**, 538–541 ▪ J. Org. Chem. **32**, 3258 ff. (1967) (synthesis) ▪ Karrer, No. 7644, 7720 ▪ Luckner (3.) p. 427. – *[HS 2934 90; CAS 146-80-5 (X.); 523-98-8 (xanthylic acid)]*

Xanthothricin see toxoflavin.

Xanthotoxin see bergapten.

Xanthoxin (Xanthoxal).

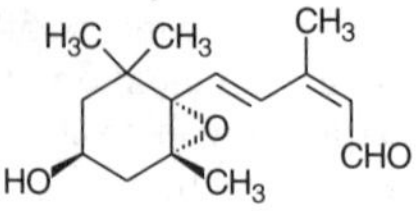

$C_{15}H_{22}O_3$, M_R 250.34, cryst., mp. 85–86 °C, $[\alpha]_D$ –56° ($CHCl_3$), a cyclofarnesane sesquiterpene formed by photochemical oxidation of *violaxanthin. It has been detected in several marine and higher plants. Naturally occurring X. exists as a mixture of *E/Z*-isomers of the 2,3-double bond of the 2,4-pentadienal side chain. It is an inhibitor of plant growth [1] and is involved in the winter rest phase of plants (senescence) [2], in some cases it is also involved in the damage to trees [3]. X. is closely related structurally and in its biological activity to *abscisic acid and presumably functions as an intermediate in the biosynthesis of it.

Lit.: [1] Phytochemistry **9**, 2217 (1970). [2] Leshem et al., Processes and Control of Plant Senescence, Amsterdam: Elsevier 1986. [3] Naturwiss. Rundsch. **37**, 52–61 (1984); Agric. Biol. Chem. **54**, 2723 (1990).

gen.: Beilstein E V **18/1**, 219. – *Synthesis:* Helv. Chim. Acta **61**, 2616 (1978) ▪ Phytochemistry **30**, 815 (1991); **44**, 977 (1997) ▪ Tetrahedron **48**, 8229–8238 (1992). – *[HS 2934 90; CAS 8066-07-7]*

Xanthylic acid see xanthosine.

Xantocillin see xanthocillin.

Xenorhabdin 1 see holomycin.

Xerocomic acid (3,4,4'-trihydroxypulvinic acid).

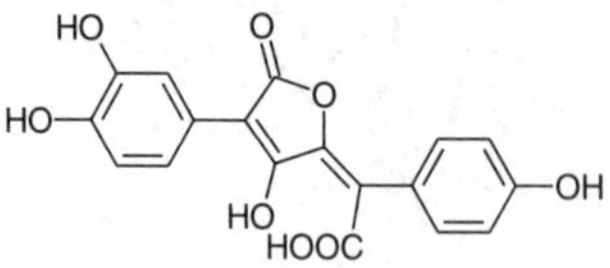

$C_{18}H_{12}O_8$, M_R 356.29, orange cryst., mp. 295 °C. X. a. and *variegatic acid are yellow *pulvinic acid derivatives that cause the characteristic blueing of the flesh which occurs when the fruit bodies of many fungi of the order Boletales are cut or bruised (see also gyrocyanin). Oxygen and the presence of oxidases are necessary for the color reaction. As a consequence of the unsymmetrical substitution of the phenyl ring with hydroxy groups there are two isomeric forms, X. and iso-X. (3',4,4'-trihydroxypulvinic acid). X. inhibits phytopathogenic bacteria.

Lit.: J. Chem. Soc., Perkin Trans. 1 **1984**, 1547 (synthesis) ▪ Zechmeister **51**, 32–51 (review). – *[CAS 25287-88-1 (X.); 27711-61-1 (Iso-X.)]*

Xerulin.

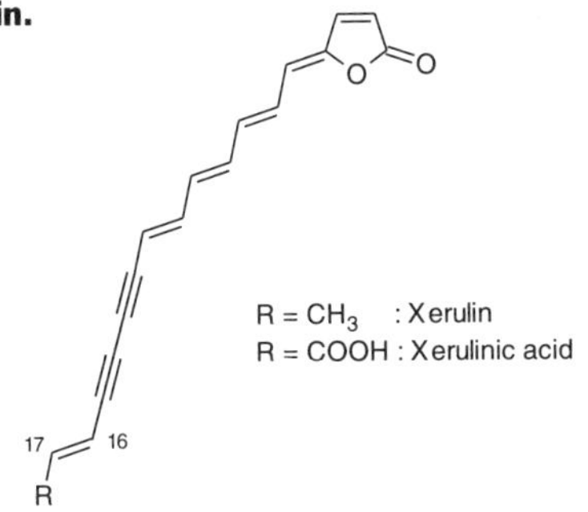

$C_{18}H_{14}O_2$, M_R 262.31, orange-yellow cryst., mp. 143–153 °C. Highly conjugated eneyne-furanone from cultures of the toadstool *Xerula melanotricha* (Basidiomycetes). X. co-occurs with the 16,17-dihydrocompound ($C_{18}H_{16}O_2$, M_R 264.32) and xerulinic acid ($C_{18}H_{12}O_4$, M_R 292.29, orange-yellow cryst., mp. 250 °C – decomp.). All three compounds inhibit cholesterol biosynthesis.

Lit.: J. Antibiot. **43**, 1413 (1990) (isolation). – *Synthesis:* Org. Lett. **2**, 65 (2000) ▪ Synlett **1999**, 1227 ▪ Tetrahedron Lett. **56**, 479 (2000). – *[CAS 132971-61-0 (X.); 132971-60-9 (dihydro-X.); 132971-62-1 (xerulinic acid)]*

Xestoquinone see halenaquinone.

Xestospongins. Epimeric alkaloids (X. A–D) from the marine sponge *Xestospongia exigua*, related to *petrosin, e.g.: *X. A* (araguspongin D): $C_{28}H_{50}N_2O_2$, M_R 446.72, cryst., mp. 135–136 °C, $[\alpha]_D$ +7° ($CHCl_3$). X. show vasodilatory and antimicrobial activity and bind to the IP_3-receptor[1].

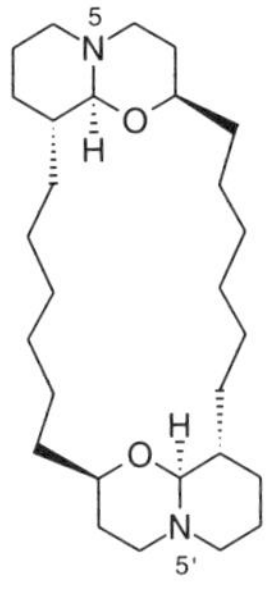

(+)
X. A
with corrected abs. configuration

Lit.: [1] Neuron **19**, 723–733 (1997).
gen.: Bioorg. Med. Chem. Lett. **6**, 1313 (1996) ▪ Chem. Pharm. Bull. **37**, 1676 (1989) ▪ J. Am. Chem. Soc. **116**, 2617 (1994); **120**, 8559 (1998) (synthesis, abs. configuration) ▪ Tetrahedron Lett. **25**, 3227 (1984). – *[CAS 123048-14-6 (X. A)]*

Xylans. Polysaccharides composed of D-xylose units. Pure X. are very rare. The wood of many terrestrial plants contains X. in the form of heteroxylans, e.g., deciduous wood (hardwood) contains 10–35% *O-acetyl-(4-O-methylglucuronosyl)-xylan*[1], which is composed of 1→4-glycosidically linked β-D-xylopyranose and 1→2-glycosidically linked 4-*O*-methyl-α-D-glucopyranuronic acid with some of the oxygen atoms being acetylated. Coniferous wood (softwood) contains ca. 10–15% *arabino-(4-O-methylglucuronosyl)-xylan*[1,2] which does not have any acetyl groups while the oxygen in the 3 position is linked to an L-arabinofuranose unit (1→3). For the biosynthesis of X., see *Lit.*[1]. X. are hydrolyzed to xylose and other pentoses by *xylanases*, certain hemicellulases. Xylanases of this type are formed by fungi[2]. Genetically engineered *Zymomonas mobilis* bacteria are able to ferment X. to ethanol.

Lit.: [1] Phytochemistry **11**, 2489–2495 (1972). [2] ACS Symp. Ser. **399** (Plant Cell Wall Polym.), 619–629 (1989); Appl. Microbiol. Biotechnol. **33**, 506–510 (1990).
gen.: Infect Agents Dis. **3**, 54 (1994) ▪ Microbiol. Rev. **52**, 305–317 (1988) ▪ Thromb. Res. **64**, 301 (1991) ▪ Topics Enzyme Ferment. Biotechnol. **8**, 9 ff. (1984) ▪ Ullmann (5.) **A 12**, 122 ▪ see also polyoses. – *[CAS 9014-63-5]*

Xylaral.

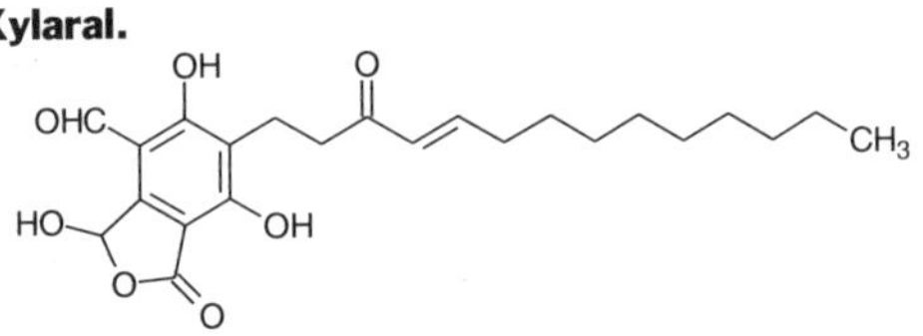

$C_{23}H_{30}O_7$, M_R 418.49, colorless cryst., mp. 180–183 °C. Hydroxyphthalide derivative from fruit bodies of the fungus *Xylaria polymorpha* (Ascomycetes); X. gives a violet color with ammonia and a yellow color with dilute alkali.

Lit.: Justus Liebigs Ann. Chem. **1990**, 825. – *[CAS 127913-82-0]*

Xylerythrin.

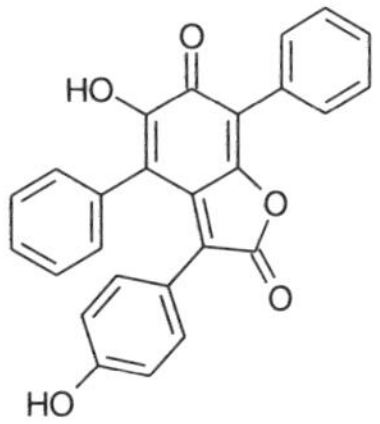

$C_{26}H_{16}O_5$, M_R 408.41, shining black cryst., mp. 265–268 °C. Wood attacked by the fungus *Phanerochaete (= Peniophora) sanguinea* takes on a vermilion color ("red rot"). A series of closely related quinone methide pigments with X. as the main component has been isolated from such substrates.

Lit.: Beilstein E V **18/4**, 696 ▪ Chem. Ber. **101**, 3753 (1968) (synthesis) ▪ Phytochemistry **16**, 1695 (1977) (biosynthesis) ▪ Zechmeister **51**, 70–75 (review). – *[CAS 7611-34-9]*

Xylindein.

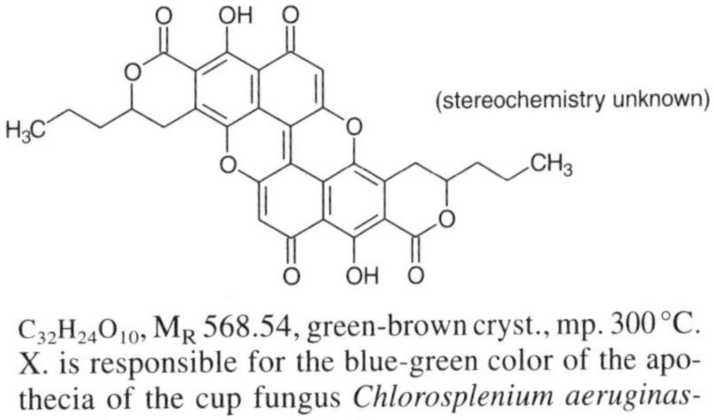

$C_{32}H_{24}O_{10}$, M_R 568.54, green-brown cryst., mp. 300 °C. X. is responsible for the blue-green color of the apothecia of the cup fungus *Chlorosplenium aeruginascens* ("green rot" on twigs).

Lit.: Beilstein E IV **19**, 5997 ▪ Proc. Chem. Soc. **1962**, 327 ▪ Tetrahedron **21**, 2095 (1965) (synthesis) ▪ Thomson **2**, 623 (review).

D-Xylose (wood sugar).

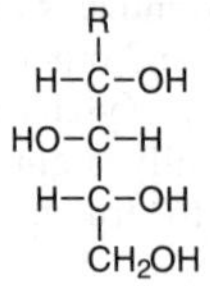

R = CHO : D-Xylose
R = CH_2OH : Xylitol

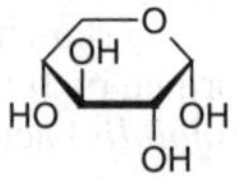

α-D-Xylopyranose

α-D-Xylopyranose: $C_5H_{10}O_5$, M_R 150.13, mp. 153 °C, $[\alpha]_D^{20}$ +94→+19° (H_2O). A widely distributed pentose in plants, especially in wood; however it does not occur in the free form but rather as xylans and in *glycosides (e.g., *primeverose), together with *cellulose in numerous deciduous and coniferous trees, straw, brans, plant rubbers, peanut and cotton seeds, in the husks of apricot stones, etc. X. exists in the pyranose and furanose forms as well as the free aldehyde. *L-Xylofuranose:* mp. 144 °C; *α-L-xylopyranose*: mp. 146–150 °C, $[\alpha]_D^{20}$ –75°→–19° (H_2O); *β-DL-xylopyranose*, DL-form, mp. 129–131 °C. The L-forms are not natural. An aqueous solution of X. at, e.g., 31 °C, contains 36.5% α-pyranose, 63% β-pyranose, <1% α- and β-furanose, and 0.02% aldehyde. X. is obtained technically from the saccharification of wood and from the residues of cellulose production or, respectively, by isolation from *xylans or from ears of corn by hydrolysis with dilute acids. Reduction of X. furnishes xylitol which can be used as a sugar substitute. X. itself would also be suitable for use as a sweetener but has only half the sweetness of saccharose and also shows laxative effects.

Lit.: Adv. Biochem. Eng./Biotechnol. **27**, 1–83 (1983) ▪ Beilstein EIV **1**, 4223–4229 ▪ Collins-Ferrier, p. 47, 54 (synthesis, biosynthesis) ▪ Justus Liebigs Ann. Chem. **1992**, 95 (synthesis) ▪ Karrer, No. 585 ▪ Merck-Index (12.), No. 10220. – *[HS 1702 90, 2940 00; CAS 6763-34-4 (α-D-pyranose); 2460-44-8 (β-D-pyranose); 37110-85-3 (β-D-furanose); 58-86-6 (D-X.); 609-06-3 (L-X.); 7296-58-4 (α-L-pyranose)]*

Y

Yamogenin see steroid sapogenins.

Yangonin see styrylpyrones.

Yazumycins see streptothricins.

Yemuoside YM$_2$ see olivil.

Yessotoxin (YTX).

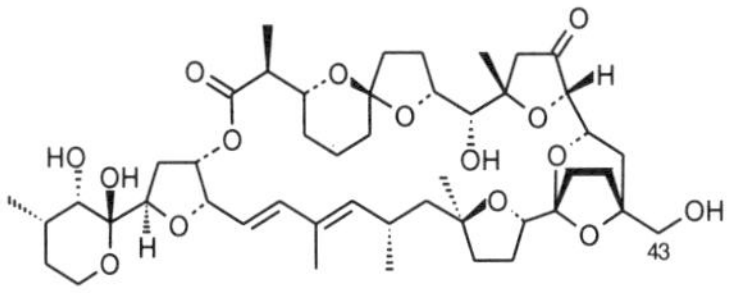

Pectenotoxin-1

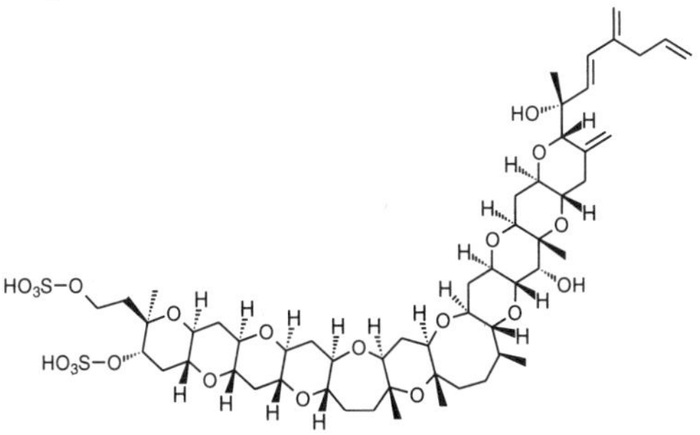

Yessotoxin

$C_{55}H_{82}O_{21}S_2$, M_R 1143.38, $[\alpha]_D$ +3° (CH_3OH), amorphous solid as disodium salt[1]. Y. and *pectenotoxin* (PTX, pectenotoxin 1: $C_{47}H_{70}O_{15}$, M_R 875.06, cryst., $[\alpha]_D$ +17.1°)[2] are isolated as DSP toxins from the digestive glands of the scallop *Patinopecten yessoensis*; the factual producer is the dinoflagellate *Dinophysis fortii* that is taken up with food and accumulated in the scallop; the toxins cause poisonings in humans. Pectenotoxin is hepatotoxic and induces a rapid necrosis of hepatocytes[3]. Its behavior is similar to that of phalloidin (see phallotoxins) while the structure rather resembles those of the *brevetoxins.

Lit.: [1]Tetrahedron Lett. **28**, 5869 (1987); **37**, 5955 (1996) (relative configuration); **37**, 7087 (1996) (abs. configuration); **39**, 8897 (1998) (analogues); Rapid Commun. Mass Spectrom. **7**, 179 (1993). [2]Tetrahedron **41**, 1019 (1985); J. Appl. Phycol. **1**, 147 (1989). [3]Toxicon **28**, 1095 (1990). – *[CAS 112514-54-2 (Y.); 97564-90-4 (pectenotoxin-1)]*

Yingzhaosu.

(natural product almost racemic)

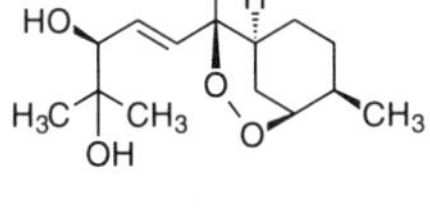

Y. A

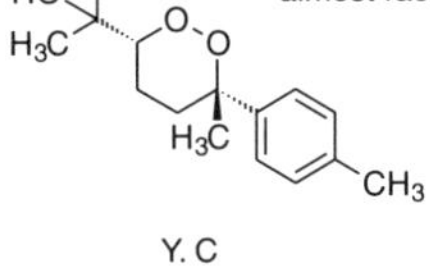

Y. C

Y. A {$C_{15}H_{26}O_4$, M_R 270.37, cryst., mp. 95–96 °C, $[\alpha]_D^{25}$ +226° ($CHCl_3$)} and *Y. C* {$C_{15}H_{22}O_3$, M_R 250.34, oil, $[\alpha]_D^{14}$ +2.9° (CH_3OH)} from the Chinese plant Yingzhao (*Artabotrys uncinatus*, Annonaceae) with activity against *Plasmodium falciparum* (malaria) served as lead structure for the synthesis of more active compounds. These are, like *qinghaosu (artemisinin), cyclic peroxides, however, their *sesquiterpene skeleton is of the *bisabolene type.

Lit.: Acc. Chem. Res. **27**, 211 (1994) ▪ J. Org. Chem. **60**, 3039 (1995) ▪ Tetrahedron Lett. **32**, 5785 (1991); **35**, 9429 (1994); **36**, 4167 (1995) ▪ Trop. Med. Parasitol. **45**, 261 (1994). – *[CAS 73301-54-9 (Y. A); 121067-52-5 (Y. C)]*

Ylang-ylang oils, cananga oils. The flowers of the cananga tree furnish essential oils with differing compositions depending on the flower variety and the distillation technique.

1) ***Ylang-ylang oils:*** Yellow to dark yellow oils with a typical radiant, narcotic-sweet, jasmine-like odor.

Prod.: By steam distillation of flowers of the ylang-ylang tree (*Cananga odorata* subsp. *genuina*, Annonaceae) cultivated in Madagascar, the Comoro Islands, and Nosy Bé. The annual production world-wide is probably somewhat more than 50 t.

Composition[1]: The components responsible for the typical ylang odor are present in the largest amounts in the first fractions, e. g.: **linalool* (up to ca. 10%), *p-cresol methyl ether* (up to ca. 10%, $C_8H_{10}O$, M_R 122.17), *methyl benzoate* (up to ca. 10%, $C_8H_8O_2$, M_R 136.15), and **benzyl acetate* (up to ca. 25%). The share of these components decreases during the distillation and increasing amounts of sesquiterpene hydrocarbons such as *caryophyllenes, *germacrene D* (see germacrenes), *trans,trans-α-**farnesene, etc. are formed.

Use: A building block for perfumes with a wide scope of application, mainly used in compositions with pronounced warm-flowery notes.

2) ***Cananga oil:*** Light yellow to dark yellow, slightly viscous oil with a sweet-flowery, somewhat woody odor that is heavier and more adherent than that of Y.-Y. oil.

Production: By steam distillation of the flowers of a subspecies of the ylang tree growing mainly in Java, *Cananga odorata* subsp. *macrophylla*. The annual production world-wide is between 50 and 100 t.

Composition[1]: Qualitatively, cananga oil resembles the later fractions of Y.-Y. oil but contains even larger amounts of the sesquiterpene hydrocarbons and only very little *methyl benzoate* and *benzyl acetate* (see benzoic acid).

Use: Similar to Y.-Y. oil. On account of the lower ester content, cananga oil is more suitable for use as a soap perfume.

Lit.: [1]Perfum. Flavor. **1** (1), 4 (1976); **11** (5), 118 (1986); **14** (3), 79 (1989); **20** (2), 57 (1995).
gen.: Arctander, p. 668, 117 ▪ Bauer et al. (2.), p. 181 ▪ ISO 3063 (1983), 3523 (1976). – *Toxicology:* Food Cosmet. Toxicol. **11**, 1049 (1973); **12**, 1015 (1974). – *[HS 3301 29; CAS 8006-81-3 (1.); 68608-83-7 (2.); 104-93-8 (p-cresol methyl ether); 93-58-3 (methyl benzoate)]*

Yohimbine (obsolete synonyms: johimbine, quebrachine, aphrodine). $C_{21}H_{26}N_2O_3$, M_R 354.45. The parent *indole alkaloid of the yohimbine type occurring in nature as the (+)-form, the racemate is accessible by synthesis [1]. Other, naturally occurring stereoisomers of Y. have been isolated: α-Y. (rauwolscine, corynanthidine, isoyohimbine, mesoyohimbine, 16β,17α, 20α-Y.), *allo*-Y. (16α,17α,20α-Y.), 3-*epi*-α-Y. (isorauhimbine), β-Y. (amsonine), and corynanthine (rauhimbine, 16β,17α-Y.).

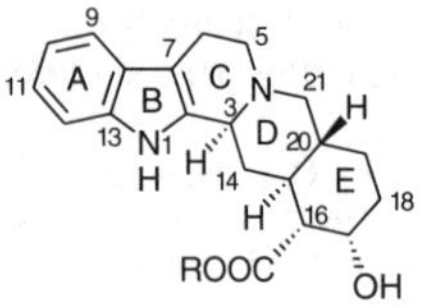

R = H : Yohimb(in)ic acid
R = CH_3 : Yohimbine

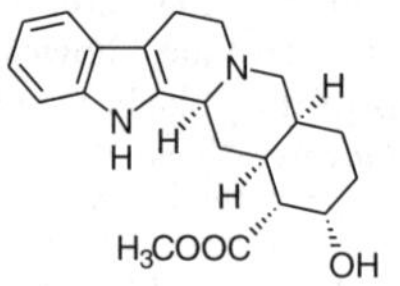

allo-Yohimbine

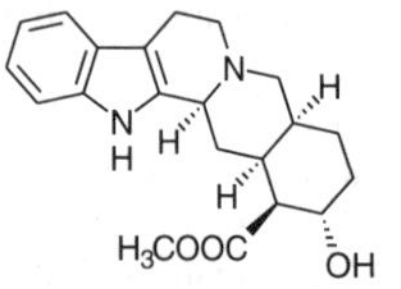

α-Yohimbine

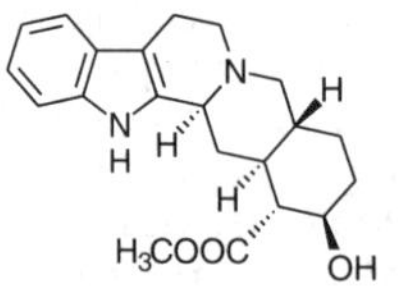

β-Yohimbine

Table: Stereoisomers of Yohimbine.

Stereoisomer		mp. [°C]	CAS
Yohimbine	(+)-form	241 (235–236)	146-48-5
	(±)-form	218–220	24252-70-8
α-Yohimbine	(–)-form	243	131-03-3
3-*epi*-α-Yohimbine	(–)-form	225	483-09-0
β-Yohimbine	(–)-form	236	549-84-8
	(±)-form	232–236	

Y. is the main alkaloid of many *Aspidosperma* species, it also occurs in Rubiaceae such as *Corynanthe johimbe* (=*Pausinystalia*) and related trees as well as in some Apocynaceae (e.g., *Rauvolfia* and *Vinca* species). Y. is structurally related to the *Rauvolfia alkaloids, e.g., *reserpine. Y. is the methyl ester of *yohimb(in)ic acid* ($C_{20}H_{24}N_2O_3$, M_R 340.42, mp. 280–300 °C). In medicine Y. is used especially in the form of its more water-soluble hydrochloride (decomp. at 302 °C) as a sympathicolytic agent, α_2-adrenoceptor blocker, antihypertensive agent, and – but only in high doses – as an aphrodisiac (especially in veterinary medicine) on account of its ability to dilate penile vessels and its specific stimulating activity on genital centers in the spinal cord [2].

Lit.: [1]Atta-ur-Rahman **3**, 399–416; J. Chem. Soc., Chem. Commun. **1983**, 1231 f.; J. Org. Chem. **56**, 2701–2712, 2947 ff., 2960–2964 (1991); Pure Appl. Chem. **58**, 685–692 (1986). [2]J. Psychoact. Drugs **17**, 131 f. (1985).
gen.: Adv. Heterocycl. Nat. Prod. Synth. **1996**, 99–150 (synthesis) ▪ Atta-ur-Rahman & Basha, Biosynthesis of Indole Alkaloids, Oxford: Clarendon Press 1983 ▪ Beilstein E V **25/6**, 358–362 ▪ Bioorg. Med. Chem. **4**, 1097 (1996) (yohimbic acid in combinatorial synthesis) ▪ Chem. Heterocycl. Compds. **25**, 147–199 (1983) ▪ Florey **16**, 731–768 ▪ Hager (5.) **4**, 402 f., 1030; **6**, 361–380; **9**, 1221–1225 ▪ J. Clin. Pharmacol. **34**, 418 (1994) ▪ J. Org. Chem. **64**, 3237 (1999) (synthesis yohimbic acids) ▪ Manske **27**, 131–268, 407 ff. ▪ Merck-Index (12.), No. 10236–10238 ▪ Phillipson & Zenk (eds.), Indole and Biogenetically Related Alkaloids, p. 113–141, 142–184, 185–200, London: Academic Press 1980. – *[HS 2939 90; CAS 522-87-2 (yohimbic acid)]*

Yuehchukene.

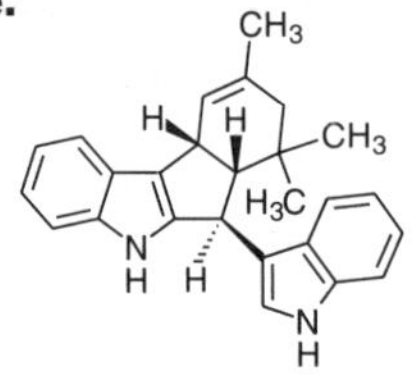

$C_{26}H_{26}N_2$, M_R 366.51. Y. is a racemic bisindole alkaloid, readily soluble in chloroform, methanol; it is isolated from the roots of *Micromelum* species (Rutaceae) and *Murraya paniculata* [1] as well as other *Murraya* species [2], it also occurs in the trunk and root barks of *Merrillia caloxylon* [3].

Biological activity: Tests in rats and mice have shown that Y. exhibits anti-implantation activity (termination of pregnancy) [4]. Its derivatives are being studied for their suitability in the development of substances for the specific termination of pregnancies on account of their *estrogen activity. Several syntheses have been described [5].

Biosynthesis: Unknown, the racemic form of Y., however, is indicative of a non-enzymatic formation through condensation of the corresponding prenylated indoles.

Lit.: [1]J. Chem. Soc., Chem. Commun. **1985**, 47. [2]Biochem. Syst. Ecol. **14**, 491 (1987); **15**, 195 (1987). [3]Biochem. Syst. Ecol. **16**, 47 (1988). [4]Planta Med. **50**, 304 (1985). [5]Heterocycles **41**, 1385 ff. (1995); J. Chem. Soc., Perkin Trans. 1 **1996**, 1213; **1998**, 1959; J. Org. Chem. **58**, 5784 (1993); Nat. Prod. Lett. **5**, 225 (1994); Tetrahedron **48**, 759 (1992). – *[HS 2939 90; CAS 96624-37-2]*

Z

Zaragozic acids (squalestatins). Highly active, competitive inhibitors of squalene synthase (SQS) and thus of cholesterol biosynthesis. The different names are a result of the independent, simultaneous discovery of this class of natural products in various ascomycete strains (e. g., *Phoma* sp., *Setosphaeria khartoumensis*) by three research groups. All Z. possess a 4,6,7-trihydroxy-2,8-dioxabicyclo[3.2.1]octane-3,4,5-tricarboxylic acid skeleton carrying acyl groups on the C-6 oxygen atom as lipophilic side chains and alkyl groups on the C-1 carbon atom.

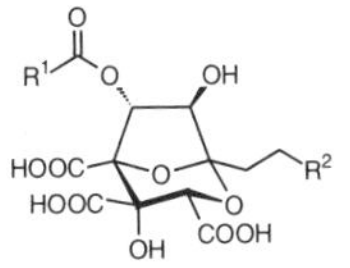

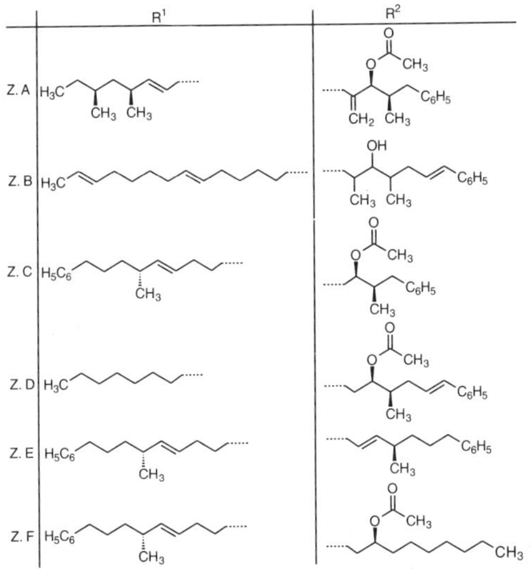

Table: Data of Zaragozic acids.

	molecular formula	M_R	$[\alpha]_D^{25}$	CAS
Z. A (= Squalestatin S1)	$C_{35}H_{46}O_{14}$	690.72	+37° (CH_3OH)	142561-96-4
Z. B	$C_{39}H_{54}O_{13}$	730.82	+5.9° (CH_3OH)	146389-61-9
Z. C	$C_{40}H_{50}O_{14}$	754.80	+9.6° (C_2H_5OH)	146389-62-0
Z. D	$C_{34}H_{46}O_{14}$	678.71	+13.7° (CH_3OH)	155179-14-9
Z. E	$C_{40}H_{50}O_{12}$	722.80	–	151990-70-4
Z. F	$C_{38}H_{54}O_{14}$	734.81	–	151990-71-5

Data for Z. A: IC_{50} (rat SQS) 15 nMol; lowering of the cholesterol level in marmosets by 50% after 7 daily doses of 10 mg/kg. SQS catalyzes the reductive dimerization of farnesyl pyrophosphate to *squalene. All Z. thus show pronounced antifungal activities (IC_{50} in nMol range). As inhibitors of farnesyl protein transferase the Z. are potential cytostatic agents. Several total syntheses of the Z. have been reported [1]. Z. are structurally related to the CP compounds CP-225,917 and CP-263,114 [2].

Lit.: [1] Angew. Chem., Int. Ed. **33**, 2184, 2187, 2190 (1994); **34**, 773 (1995) (review); Chem. Pharm. Bull. **47**, 330 (1999); Nicolaou, p. 673; J. Org. Chem. **61**, 9115–9134 (1996) (Z. A); **63**, 7592 (1998) (Z. A); Synlett **1997**, 451 (Z. C); Tetrahedron Lett. **39**, 3337 (1998). [2] Angew. Chem., Int. Ed. **36**, 1194 (1997).

gen.: Angew. Chem., Int. Ed. Engl. **35**, 1622–1656 (1996) (review) ▪ Annu. Rev. Microbiol. **49**, 607 (1995) ▪ Biol. Med. Chem. Lett. **3**, 2527, 2541 (1993) ▪ Chem. Eur. J. **1**, 467–494 (1995) ▪ Exp. Opin. Ther. Pat. **8**, 521–530 (1998) (review activities) ▪ J. Am. Chem. Soc. **117**, 8106 (1995) ▪ J. Antibiot. **44**, 614, 623, 639 (1991); **46**, 648, 689 (1993); **47**, 741 (1994); **48**, 311 (1995); **51**, 428 (1998) ▪ J. Biol. Chem. **267**, 11705 (1992) ▪ J. Med. Chem. **37**, 3274 (1994); **39**, 207–216, 1413–1422 (1996) ▪ J. Nat. Prod. **56**, 1923 (1993) ▪ Nat. Prod. Rep. **11**, 279 (1994) ▪ Proc. Natl. Acad. USA **90**, 80 (1993) ▪ Prog. Med. Chem. **33**, 331 (1996) ▪ Tetrahedron **48**, 1022 (1992) ▪ Tetrahedron Lett. **34**, 399 (1993).

Zarzissine.

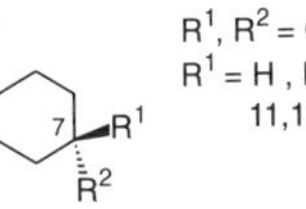

$C_5H_5N_5$, M_R 135.13, powder, subl. (270–271 °C). A cytotoxic pyridazine from the Mediterranean sponge *Anchinoe paupertas*.

Lit.: J. Nat. Prod. **57**, 1455 (1994). – *[CAS 160568-14-9]*

Zearalenone.

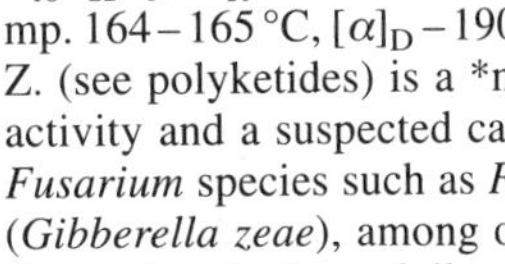

$R^1, R^2 = O$: Zearalenone
$R^1 = H$, $R^2 = OH$; 11,12-dihydro : Zeranol

$C_{18}H_{22}O_5$, M_R 318.37. Colorless, skin-irritating cryst., mp. 164–165 °C, $[\alpha]_D$ –190° ($CHCl_3$). The nonaketide Z. (see polyketides) is a *mycotoxin with estrogenic activity and a suspected carcinogen. Z. is formed by *Fusarium* species such as *F. roseum*, *F. graminearum* (*Gibberella zeae*), among others, in hay, animal fodder, and grain (especially corn = *Zea mays*) preferentially at lower temperatures (12–14 °C) and can cause fertility disorders as well as premature births and stillbirths in grazing animals, but does not exert acute toxicity. In *F. roseum* Z. is assumed to induce the formation of perithecia with asci and ascospores and thus acts as a sexual hormone; however, recent results contradict this assumption[1]. For biosynthesis of Z., see *Lit.*[2], and for analysis, see *Lit.*[3]. Z. is produced in submersion cultures and serves as an intermediate in the production of α-zearalanol (international free name: zeranol). *Zeranol* {$C_{18}H_{26}O_5$, M_R 322.40, (3*S*,7*R*)-form: mp. 182–184 °C, $[\alpha]_D$ +46.3° ($CHCl_3$)}, is formed together with other diastereoisomers by catalytic reduction and is also known as a metabolite of *Fusarium* cultures; it is used in veterinary medicine as an anabolic agent (Brand names: Ralgro®; Ralabol®, etc.). In sensitive animals and mammalian cell cultures zeranol binds to the estrogen receptors and thus stimulates the syntheses of proteins, DNA, and RNA. β-Zearalanol, the C-7 epimeric reduction product from fermentation cultures (international free name: taleranol), is a very effective inhibitor of gonadotropin. Besides Z., 14-*O*-acetyl-Z., 5-hydroxy-Z., and 13-formyl-Z., Z. 14-sulfate is also formed as a natural fungal metabolite.

Lit.: [1]Appl. Environ. Microbiol. **33**, 546 (1977); Mycopathologia **81**, 272 (1989). [2]Pure Appl. Chem. **52**, 189–204 (1980). [3]IARC Monogr. **31**, 279–291 (1983); Williams (ed.), Official Methods of Analysis of the Assoc. Off. Anal. Chem., p. 477–500, Arlington: AOAC Inc. 1984; J. Assoc. Off. Anal. Chem. **68**, 958 (1985); Appl. Environ. Microbiol. **45**, 15 (1983); **50**, 332 (1985).

gen.: Acta Crystallogr. Sect C **52**, 1995, 3095 (1996) ▪ Adv. Cancer Res. **45**, 217–290 (1985) ▪ Betina, chap. 12 ▪ Hager (5.) **9**, 1228 ▪ Naturwissenschaften **75**, 309 f. (1988) ▪ Sax (8.), RBF 100, ZAT 000. – *Synthesis:* Angew. Chem. Int. Ed. Engl. **37**, 2534 (1998) ▪ J. Chem. Soc., Perkin Trans. 1 **1992**, 1323–1328. – *Occurrence:* Chelkowski (ed.), Fusarium – Mycotoxins, Taxonomy and Pathogenicity, p. 441–472, Amsterdam: Elsevier 1989 ▪ J. Agric. Food Chem. **36**, 979 (1988). – *[CAS 17924-92-4 (Z.); 26538-44-3 (zeranol); 42422-68-4 (β-zeranol)]*

Zeatin. *trans*-Z. [(*E*)-2-methyl-4-(9*H*-purin-6-yl-amino)-2-buten-1-ol, $C_{10}H_{13}N_5O$, M_R 219.25, mp. 212–213 °C (207–208°)], like *kinetin, is a *cytokinin and was originally isolated from corn ears (*Zea mays*)[1] as well as from many other plants. In higher plants, *trans*-Z. acts as a promoter of cell division and, together with *plant growth substances, has a regulating effect on various processes of plant growth. It is the most effective, naturally occurring cytokinin and at least 50-fold more active than the *cis*-form.

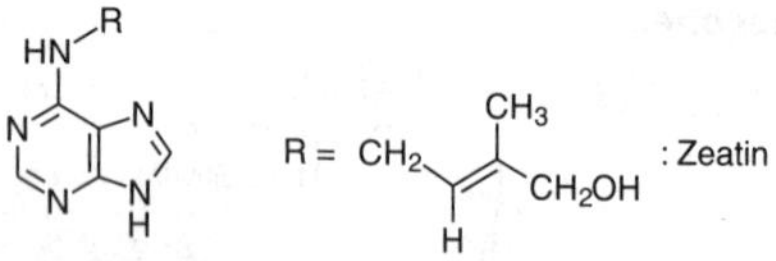

Z. is used in the commercial cultivation of plants, e. g., for *in vitro* regeneration[2]. Several syntheses of *trans*-Z. have been reported[3]. The biosynthesis proceeds from adenine[4].

Lit.: [1]Merck-Index (12.), No. 10247. [2]Plant Cell, Tissue Organ Cult. **28**, 45–51 (1992). [3]Collect. Czech. Chem. Commun. **64**, 696 (1999); Synthesis **1983**, 488–491; J. Chem. Soc., Perkin Trans. 1 **1976**, 1446; Chem. Pharm. Bull. **37**, 3119 ff. (1989); **40**, 1937 (1992). [4]J. Exp. Bot. **44**, 1411–1414 (1993).

gen.: Beilstein E V **26/16**, 193 f., 497 f. ▪ Karrer, No. 7099; 7100 ▪ Org. Prep. Proced. Int. **29**, 549 (1997). – *Plant physiology:* Annu. Rev. Plant Physiol. **34**, 163–197 (1983). – *[CAS 1637-39-4]*

Zeaxanthin [(3*R*,3′*R*)-β,β-carotene-3,3′-diol; C. I. 75 137].

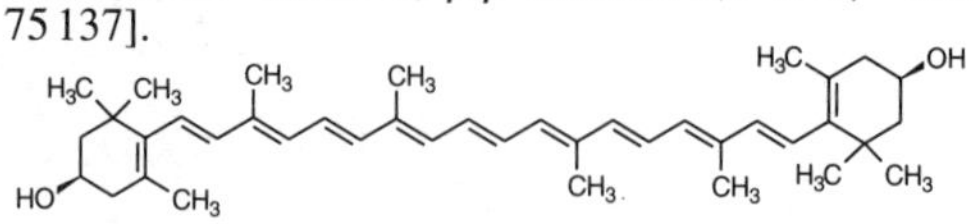

$C_{40}H_{56}O_2$, M_R 568.88, orange-yellow crystals with a steel blue metallic luster, mp. 215 °C (other reports 207 °C), soluble in petroleum ether, methanol; uv_{max} 452, 483 nm, optically inactive. Z. is a widely distributed, natural xanthophyll (see carotinoids). The yellow pigment occurs in e. g., corn (*Zea mays*, Poaceae), barley, egg yolk, crustaceans, fish, birds, fungi, and bacteria[1], as well as in partially or completely esterified form in many flowers and fruits, e. g., crocus, saffron, hips, peppers, oranges, etc. The dipalmitate of Z., *physalien*, $C_{72}H_{116}O_4$, M_R 1045.71, red cryst., mp. 98–99 °C, is the red pigment of *Physalis* species, spindle tree, sea buckthorn, ground-cherry, etc. see helenien. It also occurs in *Lycium* species (Solanaceae), *Asparagus officinalis* (Liliaceae), and other plants, often as glycosides[2]. Red peppers (*Capsicum annuum*) contain 5,6-dihydro-5β,6α-dihydroxyzeaxanthin ($C_{40}H_{58}O_4$, M_R 602.90, yellow cryst., mp. 174 °C). Z. is converted by photochemical oxidation to *violaxanthin, *xanthoxin, *capsorubin, etc. Z. is used as a food colorant (E 141).

Lit.: [1]J. Appl. Bacteriol. **70**, 181–191 (1991). [2]J. Nat. Prod. **60**, 371 (1997).

gen.: Beilstein E IV **6**, 7017 ▪ J. Org. Chem. **63**, 35–44 (1991) ▪ Karrer, No. 1839, 1842 ▪ Merck-Index (12.), No. 10248 ▪ Tetrahedron Lett. **27**, 2535 (1986). – *Biosynthesis:* J. Chem. Soc., Chem. Commun. **1977**, 655 ▪ Phytochemistry **30**, 2983 (1991). – *Synthesis:* Acta Chem. Scand. **50**, 637 (1996) ▪ Helv. Chim. Acta **73**, 861–873 (1990); **74**, 969–982 (1991). – *[CAS 144-68-3 (Z.); 144-67-2 (physalien)]*

Zeorin (hopane-6α,22-diol).

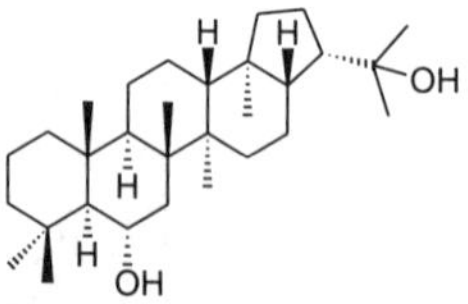

$C_{30}H_{52}O_2$, M_R 444.74, mp. 236–242 °C, $[\alpha]_D^{22}$ +63.3°. A pentacyclic triterpenediol of the *hopane series occurring in numerous lichens, e. g., the widely distributed wall lichen *Lecanora muralis*.

Lit.: Aust. J. Chem. **35**, 641 (1982); **42**, 1415 (1989) ▪ Culberson, p. 206 ff. – *[CAS 22570-53-2]*

Zeorinane see hopanes.

Zeranol see zearalenone.

Zierin see cyanogenic glycosides.

Zincophorin (griseochelin).

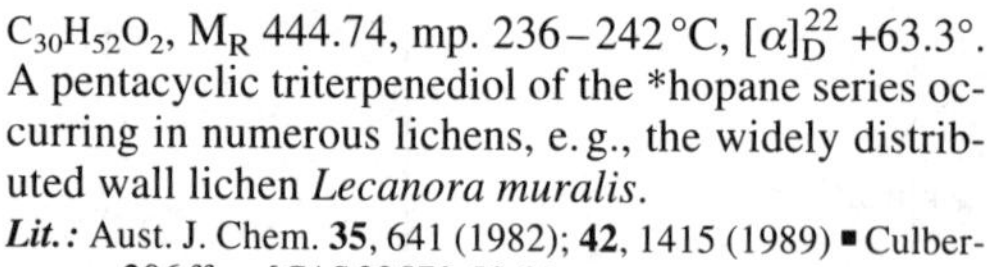

$C_{33}H_{60}O_7$, M_R 568.84, cryst., mp. 72–74 °C. Zn/Mg salt: cryst., mp. 221–222 °C, $[\alpha]_D$ +17.9 ° ($CHCl_3$). An ionophoric antibiotic produced by *Streptomyces griseus*

with a high affinity for divalent cations, especially zinc. It is active against Gram-positive bacteria, especially *Clostridia*. Its use against coccidiosis (poultry breeding) is possible. Z. methyl ester ($C_{34}H_{62}O_7$, M_R 582.86) is active against influenza viruses.

Lit.: J. Antibiot. **37**, 836, 1501 (1984); **42**, 1213 (1989) ▪ Pharmazie **46**, 781 (1991). – *Synthesis:* J. Am. Chem. Soc. **110**, 4368 (1988) ▪ Tetrahedron Lett. **29**, 6905 (1988). – *[HS 2941 90; CAS 95673-10-2 (Z. Zn salt); 91920-88-6 (Z.)]*

Zingerone [4-(4-hydroxy-3-methoxyphenyl)-2-butanone].

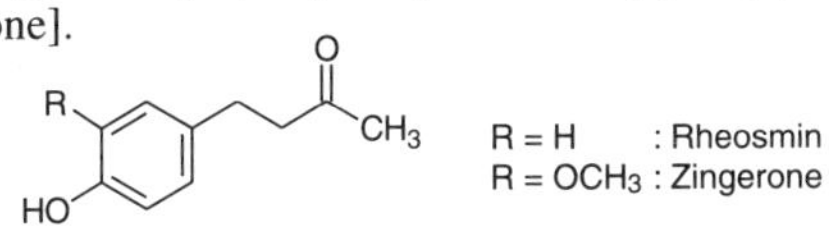

$C_{11}H_{14}O_3$, M_R 194.23, hot-tasting crystals, mp. 40 °C, bp. 187–188 °C (1.9 kPa). Z. is the *pungent (hot) flavor principle in the oleoresin of ginger. 4-(4-hydroxyphenyl)-2-butanone (*rheosmin*, raspberry ketone, frambinone), $C_{10}H_{12}O_2$, M_R 164.20, needles, mp. 82 °C, is a component of raspberry flavor. It acts as a *kairomone on the melon fly *Dacus cucurbitae* and related species. It is used as a flavor substance.

Lit.: Beilstein EIV **8**, 1866 ▪ Chem. Ber. **112**, 3703–3714 (1979) ▪ J. Agric. Food Chem. **40**, 263 ff. (1992); **41**, 1452 (1993) ▪ Karrer, No. 512. – *Toxicology:* Sax (8.), VFP 100, AAR 500, RBU 000. – *[CAS 122-48-5 (Z.); 5471-51-2 (rheosmin)]*

Zingiberene [(4*R*,7*S*)-1,5,10-bisabolatriene].

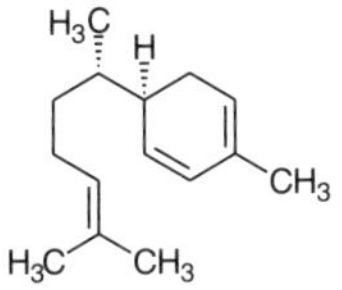

$C_{15}H_{24}$, M_R 204.36, liquid, bp. 128–130 °C (1.46 kPa), $[\alpha]_D$ −111.6° (neat). A *sesquiterpene of the bisabolane type. A component of ginger oil (ca. 70%), curcuma oil (up to 25%), violet flower oil (ca. 17%), and zedoary oil (essential oil of sweet flag) which are used in the food and perfume industries. Z. is often accompanied by *bisabolene.

Lit.: Beilstein EIII **5**, 1078 ▪ Karrer, N. 1869 ▪ Riechst., Aromen, Körperpflegem. **25**, 15 (1975). – *Isolation:* Tetrahedron **50**, 11123 (1994). – *Synthesis:* Ind. J. Chem. Sect. B **33**, 313 (1994). – *[HS 2902 19; CAS 495-60-3]*

Zinostatin see neocarzinostatins.

Ziracin (everninomicin). Antibiotic from *Micromonospora carbonacea* var. *africana* with activity against methicillin-resistant *Staphylococci* and *vancomycin-resistant *Streptococci* and *Enterococci*. It belongs to the orthosomycin (see avilamycin) group and is investigated in clinical trials: $C_{70}H_{97}Cl_2NO_{38}$, M_R 1631.43, amorphous, $[\alpha]_D^{26}$ −47.2° (CH_3OH).

Lit.: Angew. Chem. Int. Ed. Engl. **38**, 3334, 3340, 3345 (1999); **39**, 1089 (2000) (synthesis) ▪ Chem. Eng. News (22.11.) **1999**, 67 (review) ▪ Drugs **1999**, 446–453 ▪ Heterocycles **28**, 83 (1989) ▪ Tetrahedron Lett. **38**, 7989 (1997) (abs. configuration). – *[CAS 109545-84-8]*

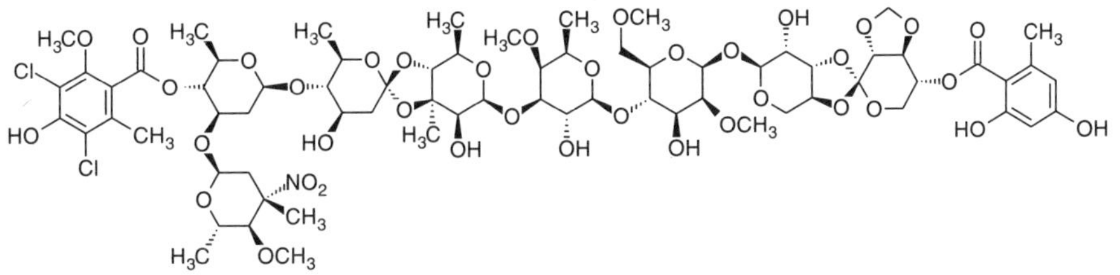

Zizanin see ophiobolins.

Zizyphine A see cyclopeptide alkaloids.

Zoanthoxanthin.

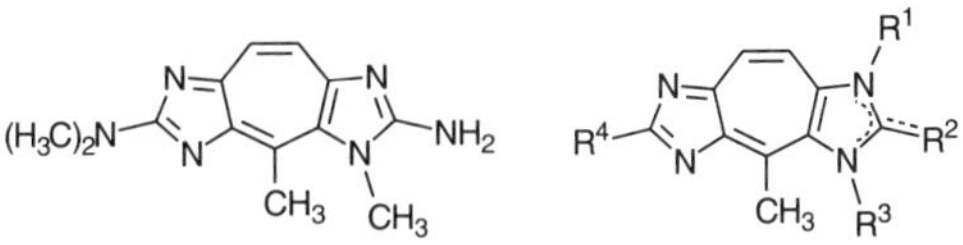

Zoanthoxanthin (Parazoanthoxanthin E)

Parazoanthoxanthin

Parazoantho-xanthin	R^1	R^2	R^3	R^4
A	–	NH_2	H	NH_2
B	–	NH_2	CH_3	NH_2
D	–	NH_2	H	$N(CH_3)_2$
E (= Z.)	–	NH_2	CH_3	$N(CH_3)_2$
F	CH_3	=NH	CH_3	$N(CH_3)_2$
G	–	NH_2	CH_3	$NH-CH_3$

Table: Data of Parazoanthoxanthins.

Parazoantho-xanthin	molecular formula	M_R	CAS
A	$C_{10}H_{10}N_6$	214.23	53823-11-3
B	$C_{11}H_{12}N_6$	228.26	53571-91-8
D	$C_{12}H_{14}N_6$	242.28	53941-25-6
E (=Z.)	$C_{13}H_{16}N_6$	256.31	40451-47-6
F	$C_{14}H_{14}N_6$	266.31	55084-58-7
G	$C_{12}H_{14}N_6$	242.28	71827-19-5

$C_{13}H_{16}N_6$, M_R 256.31, strongly fluorescent yellow needles, mp. 275–277 °C (decomp.). The main fluorescent component of crustaceous anemones (*Parazoanthus* species) of the Mediterranean region and Indonesian *Zoanthus* and *Palythoa* species, further similar pigments are the *parazoanthoxanthins* A to G (see table), all with mp. above 300 °C.

Lit.: Comp. Biochem. Physiol. B **63**, 77 (1979) ▪ Tetrahedron Lett. **38**, 717 (1997). – *Synthesis:* J. Am. Chem. Soc. **98**, 3049 (1976); **100**, 4208 (1978) ▪ Tetrahedron **30**, 3611 (1974) ▪ Tetrahedron Lett. **33**, 4385 (1992).

Zoapatanol [(2*E*,6*R*,7*S*,11*R*)-7,20-epoxy-1,6-dihydroxy-2,14-phytadien-12-one).

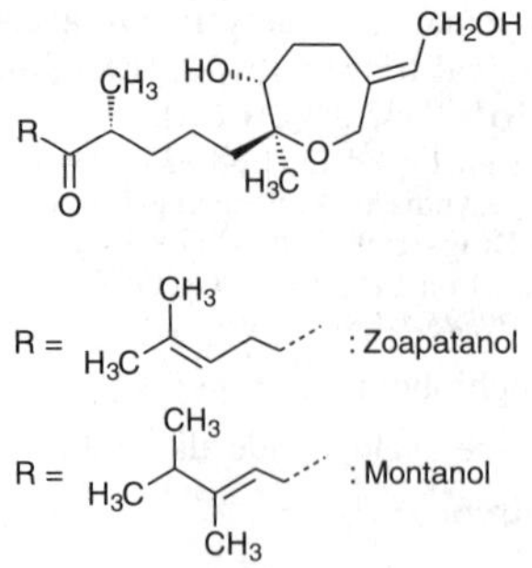

$C_{20}H_{34}O_4$, M_R 338.49, yellowish oil. A *diterpene found together with *montanol* ($C_{21}H_{36}O_4$, M_R 352.51, oil) in *Montanoa tomentosa* (Asteraceae). 10 kg plant material furnish 1.22 g Z. The name is derived from the Mexican name of the plant, Zoapatl. Natives of Mexico for centuries have used a tea made from the plant for abortive purposes. Z. was isolated as the active principle and induces menstruation and rejection of the fetus in pregnant hamsters.

Lit.: J. Am. Chem. Soc. **101**, 3404 (1979) ▪ J. Org. Chem. **47**, 1310 (1982). ▪ J. Reprod. Med. **1981**, 524. – *Synthesis:* J. Am. Chem. Soc. **102**, 6609, 6611 (1980) ▪ J. Chem. Soc., Perkin Trans. 1 **1985**, 1589. – *[CAS 71117-51-6 (Z.); 71117-50-5 (montanol)]*

Zoosterols see sterols.

Zootoxins see animal toxins.

Zygosporins.

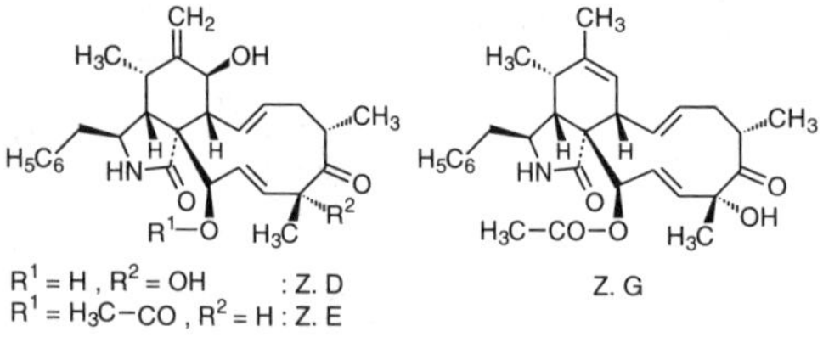

Antibiotics from *Zygosporium mansonii*, structurally related to the *cytochalasins but possessing a carbocyclic ring in place of the macrocyclic lactone ring, e.g., *Z. D* {$C_{28}H_{35}NO_5$, M_R 465.59, platelets, mp. 180–190 °C, $[\alpha]_D^{23}$ – 14.9° (dioxan)}, *Z. E* {$C_{30}H_{37}NO_5$, M_R 491.63, mp. 218–223.5 °C, $[\alpha]_D^{24}$+6.2° (dioxan)}, *Z. F* {7,21-di-*O*-acetyl-Z. D, $C_{32}H_{39}NO_7$, M_R 549.66, mp. 126–129 °C, $[\alpha]_D^{24}$ –12.0° (dioxan)} u. *Z. G* {$C_{30}H_{37}NO_5$, M_R 491.62, mp. 115–125 °C, $[\alpha]_D^{24}$ –82° (dioxan)}.

Lit.: Cole-Cox, p. 338 ff. ▪ Helv. Chim. Acta **58**, 1986 (1975) (biosynthesis); **65**, 521 (1982) (synthesis) ▪ J. Am. Chem. Soc. **112**, 4357–4364 (1990) ▪ J. Chem. Soc., Chem. Commun. **1990**, 464, 467 (synthesis) ▪ Sax (8.), ZUS 000 (toxicology). – *[CAS 25374-67-8 (Z. D); 26399-27-9 (Z. E); 25374-68-9 (Z. F); 25374-69-0 (Z. G)]*

Zyzzin.

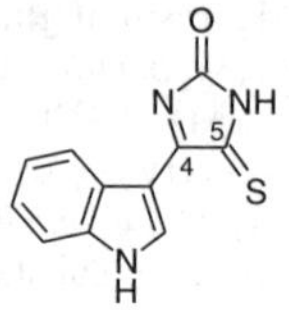

$C_{11}H_7N_3OS$, M_R 229.26, red-orange, amorphous solid. Metabolite from the sponge *Zyzzya massalis*.

Lit.: Helv. Chim. Acta **77**, 1886 (1994). – *[CAS 159509-39-4]*

Index of Molecular Formulae

CH_2O_2=Formic acid
CO=Carbon monoxide

$C_2H_3FO_2$=Fluoroacetic acid, Fluoroorganic Natural Products
C_2H_4=Plant growth substances
$C_2H_4ClO_4P$=Fosfonochlorin
C_2H_4O=Alkanals
$C_2H_4S_5$=Lenthionine
$C_2H_5NO_2$=Glycine
$C_2H_6N_2O_2$=Cycasin
$C_2H_6S_3$=Alarm substances
$C_2H_7NO_3S$=Taurine
$C_2H_8NO_3P$=Ciliatine

C_3Cl_6O=Halogen compounds
$C_3H_2Br_2O$=Halogen compounds
$C_3H_4S_2$=Vegetable flavors
C_3H_5FO=Fluoroorganic Natural Products
$C_3H_5NO_4$=3-Nitropropanoic acid
$C_3H_5O_6P$=Phosphoenolpyruvic acid
$C_3H_6N_2O_2$=Cycloserine
$C_3H_6O_2$=Fruit esters
$C_3H_6O_3$=Lactic acid
$C_3H_7NO_2$=Alanine, *β*-Alanine, Sarcosine
$C_3H_7NO_2S$=Cysteine
$C_3H_7NO_2Se$=L-Selenocysteine
$C_3H_7NO_3$=3-Nitropropanoic acid, L-Serine
$C_3H_7NO_5S$=Cysteic acid
$C_3H_7N_3O_4$=L-Alanosine, 3-(Nitramino)-L-alanine
$C_3H_7O_4P$=Fosfomycin
$C_3H_8N_2O_2$=Connatin
C_3H_9AsO=Arsenic in natural products
$C_3H_9NO_3S$=Taurine

C_4HKO_3=Moniliformin
$C_4H_2Br_2Cl_2O$=Halogen compounds
$C_4H_2Cl_2N_2O_2$=Pyrrolomycins
$C_4H_4N_2O_2$=Pyrimidines
$C_4H_4O_3$=Tetronic acid
$C_4H_5NO_2$=Azirinomycin, Tetramic acid
C_4H_5NS=Mustard oils
$C_4H_5N_3O$=Pyrimidines
$C_4H_6N_2O_2$=3-Cyano-L-alanine, Fly agaric constituents
$C_4H_6N_4O_3$=Allantoin
C_4H_6OS=Vegetable flavors
$C_4H_6O_2$=2,3-Butanedione
$C_4H_6O_2S_2$=Vegetable flavors
$C_4H_6O_4$=Succinic acid
$C_4H_6O_5$=Malic acid
$C_4H_6O_6$=Tartaric acid
$C_4H_7FO_2$=Fluoroorganic Natural Products
$C_4H_7NO_2$=1-Aminocyclopropanecarboxylic acid, Azetidine-2-carboxylic acid
$C_4H_7NO_3$=Oxetin
$C_4H_7NO_4$=Aspartic acid
C_4H_7NS=Mustard oils
$C_4H_7N_3O$=Creatine
$C_4H_7O_7P$=D-Erythrose 4-phosphate
$C_4H_8CuN_2O_2S_2$=Fluopsins
$C_4H_8FNO_3$=Fluoroorganic Natural Products
$C_4H_8N_2O$=Gyromitrin
$C_4H_8N_2O_3$=Asparagine
$C_4H_8N_4O_4$=Allantoic acid
C_4H_8OS=Methional
$C_4H_8O_2$=Acetoin, Fruit esters
$C_4H_8S_3$=Charamin
$C_4H_8S_4$=Charamin
$C_4H_9NO_2$=4-Aminobutanoic acid
$C_4H_9NO_2S$=Cystathionine, *S*-Methyl-L-cysteine (*S*-oxide)
$C_4H_9NO_3$=Homoserine, L-Threonine
$C_4H_9NO_3S$=*S*-Methyl-L-cysteine (*S*-oxide)
$C_4H_9N_3O_2$=Creatine, Lyophyllin
$C_4H_9N_3O_3$=Albizziine
$C_4H_{10}N_3O_5P$=Creatine
$C_4H_{10}O_2S_4$=Dysoxysulfone
$C_4H_{10}O_4$=Threitol
$C_4H_{11}NO_3S$=Taurine
$C_4H_{12}AsCl$=Arsenic in natural products
$C_4H_{12}N_2$=Putrescine

$C_5H_4N_2O_4$=Orotic acid
$C_5H_4N_4$=Purines
$C_5H_4N_4O_2$=Xanthine
$C_5H_4N_4O_3$=Uric acid
$C_5H_4O_2$=Anemonin, Pyrones
$C_5H_4O_3$=Citric acid
C_5H_5NOS=Meat flavor
$C_5H_5N_5$=Zarzissine
$C_5H_5N_5O$=Nostocine A
$C_5H_6N_2$=Pyrazines
$C_5H_6N_2O_2$=Pyrimidines
$C_5H_6N_2O_4$=Fly agaric constituents
C_5H_6OS=2-Furylmethanethiol, 2-Methyl-3-furanthiol
$C_5H_6O_2$=Tulipalin A
$C_5H_6O_3$=Hydroxyfuranones
$C_5H_7ClN_2O_3$=Acivicin
C_5H_7NOS=Goitrin, Meat flavor
$C_5H_7NO_2$=(*S*)-2-Amino-4-pentynoic acid, Jatropham
$C_5H_7NO_3$=L-Pyroglutamic acid
C_5H_7NS=Mustard oils
$C_5H_7N_3O_4$=Azaserine
$C_5H_7N_3O_5$=Quisqualic acid
$C_5H_8ClNO_2$=2-Amino-4-pentenoic acid
$C_5H_8N_2O_4$=Tricholomic acid
$C_5H_8N_2O_5$=N^2-, N^3-Oxalyl-2,3-diaminopropanoic acid
C_5H_8O=Vegetable flavors
$C_5H_8O_2$=Butter flavor
$C_5H_8O_2S_2$=Vegetable flavors
$C_5H_9NO_2$=2-Amino-4-pentenoic acid, L-Proline
$C_5H_9NO_3$=5-Amino-4-oxovaleric acid, 4-/3-Hydroxyproline
$C_5H_9NO_3S$=Chondrine
$C_5H_9NO_4$=Glutamic acid, 5-Hydroxy-4-oxo-L-norvaline
$C_5H_9NO_5$=4-Hydroxyglutamic acid
C_5H_9NS=Mustard oils
$C_5H_9N_3$=Histamine
$C_5H_{10}N_2O_2$=3-Aminoproline
$C_5H_{10}N_2O_3$=Glutamine
$C_5H_{10}O$=Alkanals
$C_5H_{10}OS_2$=Vegetable flavors
$C_5H_{10}O_2$=Fruit esters, Fruit flavors
$C_5H_{10}O_2S$=Fruit flavors
$C_5H_{10}O_4$=2-Deoxy-D-ribose
$C_5H_{10}O_5$=Arabinose, D-Ribose, D-Xylose
$C_5H_{10}S_3$=Vegetable flavors
$C_5H_{11}AsO_2$=Arsenic in natural products
$C_5H_{11}NO_2$=L-Valine
$C_5H_{11}NO_2S$=L-Methionine
$C_5H_{11}NS_2$=Nereistoxin
$C_5H_{11}N_3O_3$=Azoxybacilin
$C_5H_{11}O_6P$=Phosphonothrixin
$C_5H_{12}NO_4P$=Bialaphos
$C_5H_{12}N_2O_2$=L-Ornithine
$C_5H_{12}N_4O_3$=Canavanine
$C_5H_{12}O_4S_5$=Dysoxysulfone
$C_5H_{12}O_5$=Arabitol (arabinitol)
$C_5H_{14}ClNO$=Choline
$C_5H_{14}NO^+$=Choline
$C_5H_{14}N_2$=Cadaverine
$C_5H_{14}N_4$=Agmatine
$C_5H_{15}NO_2$=Choline
$C_5H_{15}N_4O_6PS$=Phaseolotoxin

$C_6H_2Br_4O_2$=Halogen compounds
$C_6H_2Cl_4$=Halogen compounds
$C_6H_4ClNO_3$=Drosophilin A
$C_6H_4Cl_2O$=Ticks
$C_6H_4N_2O$=4-Diazo-2,5-cyclohexadien-1-one
$C_6H_4N_4$=Pteridines
$C_6H_4N_4O_2$=Pteridines
$C_6H_5NO_2$=2-Hydroxy-4-imino-2,5-cyclohexadienone
$C_6H_5NO_3$=Basidalin, Ticks
$C_6H_5N_2NaO_4S$=4-Diazo-2,5-cyclohexadien-1-one
$C_6H_5N_5O$=Pteridines
$C_6H_5N_5O_2$=Pteridines
$C_6H_5N_5O_3$=Pteridines
$C_6H_6N_2O$=Nicotinamide, Pyrazines
$C_6H_6N_2O_2$=Nicotinamide, Urocan(in)ic acid
$C_6H_6N_2O_3$=Incaflavin
$C_6H_6N_4$=6-Methylpurine
$C_6H_6O_2$=Hydroquinone
$C_6H_6O_3$=Benzene-1,2,4-triol, Maltol, Phloroglucinol
$C_6H_6O_4$=Kojic acid
$C_6H_6O_6$=Aconitic acid
$C_6H_7ClO_3$=Lepiochlorin
$C_6H_7FO_7$=Fluoroacetic acid, Fluoroorganic Natural Products
C_6H_7NO=2-Aminophenol
$C_6H_7NO_3$=Butenolide(s)
$C_6H_7NO_4$=Clavams
C_6H_7NS=Cocoa flavor
$C_6H_8N_2$=Pyrazines
$C_6H_8N_2O_5$=Enteromycin
$C_6H_8O_2$=Bugs, 2-Hydroxy-3-methyl-2-cyclopenten-1-one, Sorbic acid
$C_6H_8O_3$=Furaneol®, Hydroxyfuranones, Osmundalactone, Sherry flavor
$C_6H_8O_6$=Ascorbic acid, D-Glucurono-6,3-lactone, 1,2,3-Propanetricarboxylic acid
$C_6H_8O_7$=Citric acid
C_6H_9NO=2-Acetyl-1-pyrroline
$C_6H_9NOS_2$=Mustard oils
$C_6H_9NO_2$=2-Amino-4,5-hexadienoic acid, 2-Amino-4-hexynoic acid, Areca alkaloids, Baikiain, Pheromones, L-Proline
$C_6H_9NO_3$=Baikiain
$C_6H_9NO_4$=4-Methyleneglutamic acid
$C_6H_9N_3O_2$=Bacimethrin, Histidine
$C_6H_9N_3O_3$=2-Amino-6-diazo-5-oxohexanoic acid
$C_6H_{10}N_2O_3$=4-Methyleneglutamic acid
$C_6H_{10}O$=Hexenals
$C_6H_{10}OS_2$=Allium
$C_6H_{10}O_2$=Bugs
$C_6H_{10}O_2S$=Gonyauline

$C_6H_{10}O_3$=(*R*)-Mevalonolactone, Sherry flavor
$C_6H_{10}O_4$=Conduritols
$C_6H_{10}O_7$=D-Glucuronic acid, L-Iduronic acid
$C_6H_{11}ClN_4O$=Girolline
$C_6H_{11}NOS_2$=Mustard oils
$C_6H_{11}NO_2$=(*S*)-2-Amino-3-cyclopropylpropanoic acid, Cispentacin, L-Pipecolic acid, L-Proline
$C_6H_{11}NO_2S_2$=Mustard oils
$C_6H_{11}NO_3$=4-/5-Hydroxypipecolic acid
$C_6H_{11}NO_3S$=Allium, Cycloalliin
$C_6H_{11}NO_4$=2-Aminoadipic acid
$C_6H_{11}NS_2$=Mustard oils
$C_6H_{12}ClNO_2$=Charamin
$C_6H_{12}FeN_3O_3S_3$=Fluopsins
$C_6H_{12}N_2O_4S$=Lanthionine
$C_6H_{12}N_2O_4S_2$=Cystine
$C_6H_{12}O$=Alkanals, Hexen-1-ols
$C_6H_{12}O_2$=Fruit esters
$C_6H_{12}O_2S$=Coffee flavor, Fruit flavors
$C_6H_{12}O_4$=Digitoxose, Mevalonic acid, Pantothenic acid
$C_6H_{12}O_5$=Fucose, L-Rhamnose
$C_6H_{12}O_6$=D-Fructose, Galactose, D-Glucose, Inositols, Mannose, Sorbose, D-Tagatose, D-Talose
$C_6H_{13}NO_2$=L-Isoleucine, L-Leucine
$C_6H_{13}NO_3S$=Glycosidase inhibitors
$C_6H_{13}NO_4$=Deoxymannojirimycin, Deoxynojirimycin, (2*R*)-3β,4α-Dihydroxy-2α,5β-pyrrolidinedimethanol
$C_6H_{13}NO_4S$=Glycosidase inhibitors
$C_6H_{13}NO_5$=D-Galactosamine, Galactostatin, D-Glucosamine, Mannojirimycin, Mannosamine, Nojirimycin
$C_6H_{13}NS_2$=Dihydro-4*H*-1,3,5-dithiazines
$C_6H_{13}N_3$=Galegine
$C_6H_{13}N_3O_3$=Citrulline
$C_6H_{14}ClNO_2S$=*S*-Methyl-L-methioninesulfonium chloride
$C_6H_{14}INO_2S$=*S*-Methyl-L-methioninesulfonium chloride
$C_6H_{14}N_2O_2$=L-Lysine
$C_6H_{14}N_2O_3$=5-Hydroxylysine
$C_6H_{14}N_4O_2$=L-Arginine
$C_6H_{14}O$=Alarm substances
$C_6H_{14}OS$=Fruit flavors
$C_6H_{14}O_6$=Galactitol, D-Glucitol, Iditol, Mannitol
$C_6H_{16}N_6$=Arcaine
$C_6H_{18}O_{24}P_6$=Inositols

$C_7H_4Cl_4O_2$=Drosophilin A
$C_7H_4O_7$=Poppy acid
$C_7H_5Cl_3O$=Wine flavor
$C_7H_5Cl_3O_2$=Trichlorophenols
$C_7H_5NO_4$=Dipicolinic acid
$C_7H_6Cl_2O_2$=Russuphelins
$C_7H_6N_2O_2$=4-Diazo-3-methoxy-2,5-cyclohexadien-1-one
$C_7H_6N_2O_2S_2$=Holomycin
$C_7H_6N_4O_3$=Luciferins
C_7H_6O=Benzaldehyde
$C_7H_6O_2$=Benzoic acid, Cassia oil, Methyl-1,4-benzoquinone
$C_7H_6O_3$=3,4-Dihydroxybenzaldehydes, 4-Hydroxybenzoic acid, Salicylic acid, Sesamol
$C_7H_6O_4$=Dihydroxybenzoic acids, Gentisic acid, Patulin
$C_7H_6O_5$=Gallic acid
$C_7H_7ClN_2O$=4-(Hydroxymethyl)benzenediazonium ion
C_7H_7NO=Cocoa flavor
$C_7H_7NO_2$=4-Aminobenzoic acid, Anthranilic acid, Trigonelline
$C_7H_7NO_3$=3-Hydroxyanthranilic acid
$C_7H_7N_2O^+$=4-(Hydroxymethyl)benzenediazonium ion
$C_7H_7N_5O_2$=Fervenulin, Toxoflavin
$C_7H_8N_2O_2$=Nicotinamide
$C_7H_8N_2O_3$=Dephostatin
$C_7H_8N_4O_2$=Theobromine, Theophylline
C_7H_8O=Benzyl alcohol, Jasmin(e) absolute
C_7H_8OS=Coffee flavor
$C_7H_8O_2$=Guaiacol, Hydroquinone, Orcinol, Salicyl alcohol
$C_7H_8O_3$=Sarkomycin A
$C_7H_8O_5$=3-Dehydroshikimic acid
$C_7H_9NO_2$=Gabaculine, Verrucarins
$C_7H_9N_3O_4$=Isowillardiine, Willardiine
$C_7H_{10}N_2O_6$=Hydantocidin
$C_7H_{10}N_4O_2$=Lathyrine
$C_7H_{10}O_3$=Furaneol®, Hydroxyfuranones
$C_7H_{10}O_4$=Gabosins
$C_7H_{10}O_5$=Gabosins, Shikimic acid
$C_7H_{10}O_6$=3-Dehydroquinic acid
$C_7H_{11}Br_3O$=Halogen compounds
$C_7H_{11}NO$=2-Acetyl-1-pyrroline
$C_7H_{11}NO_2$=Areca alkaloids, Hypoglycines
$C_7H_{11}NO_3$=Clavams
$C_7H_{11}NO_4$=4-Methyleneglutamic acid
$C_7H_{11}NS$=2-Isobutylthiazole, Meat flavor
$C_7H_{11}N_3O_2S$=Ovothiols
$C_7H_{11}O_{10}P$=3-Deoxy-D-*arabino*-2-heptulosonic acid 7-phosphate
$C_7H_{12}N_2O$=Lolines
$C_7H_{12}N_2O_3$=Valanimycin
$C_7H_{12}N_2O_5$=Trehalostatin
$C_7H_{12}O$=Alkenals, Fruit flavors, Seudenol
$C_7H_{12}O_2$=Alarm substances
$C_7H_{12}O_5$=Cyclophellitol
$C_7H_{12}O_6$=Quinic acid
$C_7H_{13}NO_3$=Calystegines
$C_7H_{13}NO_4$=2-Aminopimelic acid, Calystegines, Valienamine
$C_7H_{14}N_2O_3$=Theanine
$C_7H_{14}N_2O_4$=Diaminopimelic acid, Rhizobitoxin
$C_7H_{14}N_2O_4S$=Cystathionine
$C_7H_{14}N_2O_4S_2$=Djenkolic acid
$C_7H_{14}N_2O_5$=Tabtoxin
$C_7H_{14}O$=Ants, Fruit flavors
$C_7H_{14}O_2$=Bees, Fruit esters, (O)enanthic acid
$C_7H_{14}O_3$=Fruit flavors
$C_7H_{14}O_4$=Mycarose, L-Oleandrose
$C_7H_{14}O_7$=Sedoheptulose
$C_7H_{15}NO_2$=Scarabs
$C_7H_{15}NO_3$=Carnitine
$C_7H_{15}NO_5$=Glycosidase inhibitors
$C_7H_{15}N_3O_2$=Indospicine
$C_7H_{15}N_5O_3$=Gigartinine
$C_7H_{16}ClNO_2$=Acetylcholine
$C_7H_{16}N_4O_2$=L-Homoarginine
$C_7H_{17}NO_3$=Acetylcholine
$C_7H_{17}N_4O_4P$=Anatoxins
$C_7H_{19}N_3$=Spermidine

$C_8H_4O_3S_2$=Thiotropocin
C_8H_5NO=3*H*-Indol-3-one
$C_8H_5N_3O_3$=Calvatic acid
$C_8H_6Cl_3NO_4$=Drosophilin A
$C_8H_6Cl_4O_2$=Drosophilin A
$C_8H_6N_2O_4$=Cremeomycin
$C_8H_6O_3$=Piperonal
$C_8H_6O_5$=Stipitatic acid
$C_8H_7Br_2NO_3$=Homogentisic acid
$C_8H_7Cl_3O_3$=Trichlorophenols
C_8H_7N=Indole
C_8H_7NO=Isocyanides, Locustol
$C_8H_7NO_4$=DIBOA
C_8H_8=Storax, styrax oil (resin)
$C_8H_8N_2O_2$=4-(Methylnitrosamino)-benzaldehyde, Ricinine
$C_8H_8N_2O_2S_2$=Holomycin
C_8H_8O=Phenylacetaldehyde
$C_8H_8O_2$=*p*-Anisaldehyde, Ants, Ethyl-1,4-benzoquinone, Honey flavor, Picein, Ylang-ylang oils, cananga oils
$C_8H_8O_3$=3,4-Dihydroxybenzaldehydes, Methyl oxepine-2-carboxylate, Queen substance
$C_8H_8O_4$=Dihydroxybenzoic acids, Fumigatin, Homogentisic acid, Orsellinic acid
$C_8H_8O_5$=Spinulosin
C_8H_9NO=Danaidal, Fruit flavors
$C_8H_9NO_2$=Danaidal, Methyl anthranilate, Vegetable flavors
$C_8H_9NO_3$=Vitamin B_6
$C_8H_9NO_4$=Isocyanides, Pyridoxic acids
$C_8H_9NO_5$=Clavulanic acid
$C_8H_{10}NO_6P$=Vitamin B_6
$C_8H_{10}N_2O$=Sibyllimycin
$C_8H_{10}N_2O_4$=Mimosine
$C_8H_{10}N_4O_2$=Caffeine, coffein(e)
$C_8H_{10}O$=Methylenomycins, 2-Phenylethanol, Wine flavor, Ylang-ylang oils, cananga oils
$C_8H_{10}O_2$=Oak moss absolute, Orcinol
$C_8H_{10}O_3$=Echinacoside, Terrein
$C_8H_{10}O_4$=Penicillic acid
$C_8H_{10}S$=Sesame seed (roasted) flavor
$C_8H_{11}N$=β-Phenylethylamine alkaloids
$C_8H_{11}NO$=β-Phenylethylamine alkaloids
$C_8H_{11}NO_2$=Dopamine
$C_8H_{11}NO_3$=Noradrenaline, β-Phenylethylamine alkaloids, Vitamin B_6
$C_8H_{11}NO_6$=Lycoperdic acid
$C_8H_{11}N_3O_4$=2-Amino-6-diazo-5-oxohexanoic acid
$C_8H_{11}O_6P$=Cyclophostin
C_8H_{12}=Algal pheromones
$C_8H_{12}ClNO_3$=Noradrenaline
$C_8H_{12}N_2$=Ligustrazine, Pheromones, Pyrazines
$C_8H_{12}N_2O$=Pyrazines
$C_8H_{12}N_2O_2$=Lolines, Vitamin B_6
$C_8H_{12}N_2O_3S$=6-Aminopenicillanic acid
$C_8H_{12}N_2O_4$=Clavams
$C_8H_{12}N_2O_6$=Kifunensine
$C_8H_{12}O$=Tea flavor
$C_8H_{12}O_2$=Hepialone
$C_8H_{13}NO$=FK-409, Necine(s)
$C_8H_{13}NO_2$=Areca alkaloids, Necine(s)
$C_8H_{13}NO_4$=4-Methyleneglutamic acid
$C_8H_{13}NS$=Peanut flavor
$C_8H_{13}N_2O_5P$=Vitamin B_6
$C_8H_{13}N_3O_2S$=Ovothiols
$C_8H_{13}N_3O_4$=FK-409
$C_8H_{14}N_2O$=Lolines
$C_8H_{14}N_2O_4$=Coprine
$C_8H_{14}N_2O_5$=Siastatin B
$C_8H_{14}O$=Alkenals, Ants, Filbertone, 1-Octen-3-one
$C_8H_{14}O_2$=Alkanolides, Frontalin, Fruit esters
$C_8H_{14}O_2S_2$=Lipoic acid
$C_8H_{14}O_3S_2$=Lipoic acid
$C_8H_{14}O_4$=Suberin, suberic acid
$C_8H_{14}O_8$=KDO
$C_8H_{15}N$=Hemlock alkaloids
$C_8H_{15}NO$=Hygrine, Necine(s), Pelletierine
$C_8H_{15}NOS_2$=Guinesines
$C_8H_{15}NO_2$=Necine(s)
$C_8H_{15}NO_3$=Swainsonine
$C_8H_{15}NO_4$=Alexine, Castanospermine
$C_8H_{15}NO_9S_2$=Glucosinolates
$C_8H_{15}NS_2$=Dihydro-4*H*-1,3,5-dithiazines
$C_8H_{15}N_3O_7$=Streptozocin
$C_8H_{16}N_2O_7$=Cycasin
$C_8H_{16}O$=Alkanals, 4-Methyl-3-heptanone, (–)-(*R*)-1-Octen-3-ol, Rhynchophorol, Sulcatol
$C_8H_{16}OS$=2-Methyl-4-propyl-1,3-oxathian(e)
$C_8H_{16}O_2$=Fruit esters, Pityol, Sitophilure
$C_8H_{16}O_2S_2$=Lipoic acid
$C_8H_{17}N$=Hemlock alkaloids
$C_8H_{17}NO$=Hemlock alkaloids
$C_8H_{17}NO_3$=Statine
$C_8H_{17}N_3O_4$=Connatin
$C_8H_{18}O$=4-Methyl-3-heptanol

$C_9H_6N_2S$=Brassinin
$C_9H_6O_2$=Coumarin
$C_9H_6O_3$=4-Hydroxycoumarin, Umbelliferone
$C_9H_6O_4$=Caffeic acid, Daphnetin, 6,7-Dihydroxycoumarins

$C_9H_7NO_3$=Isocyanides
$C_9H_7N_5O_5$=Erythropterin
C_9H_8O=Cinnamaldehyde
$C_9H_8O_2$=Cinnamic acid, Melilotic acid, Tropic acid
$C_9H_8O_3$=(*E*)-*p*-Coumaric acid, Coumarinic acid, Phenylpyruvic acid
$C_9H_8O_4$=Caffeic acid, Phenylpyruvic acid
$C_9H_8O_5$=Phenylpyruvic acid
$C_9H_8O_6S$=*p*-Coumaric acid 4-*O*-(hydrogen sulfate)
$C_9H_9I_2NO_3$=Iodoamino acids
C_9H_9N=Skatole
C_9H_9NO=Cinnamic acid
$C_9H_9NO_3$=Erbstatin, Hippuric acid
$C_9H_9NO_5$=Betalamic acid, DIBOA, Fly agaric pigments
$C_9H_9NO_6$=Stizolobic acid
C_9H_9NS=Mustard oils
$C_9H_9N_3O_2$=Pyridopyrazines
$C_9H_{10}N_2$=Myosmine
$C_9H_{10}N_2O_2S_2$=Holomycin
$C_9H_{10}O$=Cinnamyl alcohol, Estragole, Hydrocinnamaldehyde
$C_9H_{10}O_2$=Benzyl acetate, Fruit flavors, Hydrocinnamic acid, 4-Hydroxycinnamyl alcohol, 4-Vinylguaiacol
$C_9H_{10}O_3$=3,4-Dihydroxybenzaldehydes, Ipomeanine, Melilotic acid, Tropic acid
$C_9H_{10}O_4$=Avenaciolide, Dihydrocaffeic acid, Dihydroxybenzoic acids, Methylenomycins, Rum flavor, Ubiquinones
$C_9H_{10}O_5$=3-(3,4-Dihydroxyphenyl)lactic acid
$C_9H_{11}NO$=*β*-Phenylethylamine alkaloids
$C_9H_{11}NO_2$=Methyl *N*-methylanthranilate, L-Phenylalanine
$C_9H_{11}NO_2S_5$=Varacin
$C_9H_{11}NO_3$=L-Tyrosine
$C_9H_{11}NO_4$=*α*-3,4-Dihydroxyphenylalanine, *β*-3,4-Dihydroxyphenylalanine
$C_9H_{11}NO_6$=Showdomycin
$C_9H_{11}N_3O_3S_2$=Rubroflavin
$C_9H_{11}N_5O_2$=Isosepiapterin
$C_9H_{11}N_5O_3$=Biopterin
$C_9H_{11}N_5O_4$=Eritadenine, Ichthyopterin, Neopterin
$C_9H_{12}N_2$=Nornicotine
$C_9H_{12}O$=Hydrocinnamyl alcohol
$C_9H_{12}O_2$=Coffee flavor, Locustol, Rotundial
$C_9H_{12}O_3$=Dihydroconiferyl alcohol, Ipomeanine, Queen substance
$C_9H_{13}NO$=*β*-Phenylethylamine alkaloids
$C_9H_{13}NO_2$=*β*-Phenylethylamine alkaloids
$C_9H_{13}NO_3$=(*R*)-Adrenaline, *β*-Phenylethylamine alkaloids
$C_9H_{13}NO_4$=Chlorotetaine
$C_9H_{13}N_5O_6$=Clitocine
$C_9H_{14}N_2O$=Galbanum oil or resin, Pyrazines
$C_9H_{14}N_2O_2$=Lolines
$C_9H_{14}N_4O_3$=Carnosine
$C_9H_{14}O$=Nonadienals
$C_9H_{14}OS_3$=Allium
$C_9H_{14}O_2$=Brevicomin, Chamomile oils, Hepialone
$C_9H_{14}O_3$=Ipomeanine
$C_9H_{15}Cl_2N$=Oxypterine
$C_9H_{15}NO$=Pseudopelletierine
$C_9H_{15}NO_3$=Ecgonine, Necine(s)
$C_9H_{15}N_3O_2$=Hercynin
$C_9H_{15}N_3O_2S$=Ergothioneine, Ovothiols
$C_9H_{15}N_3O_7$=Lycomarasmine
$C_9H_{15}N_5O_3$=Sapropterin, Tetrahydrobiopterin
$C_9H_{16}NNaO_5$=Pantothenic acid
$C_9H_{16}N_2O$=Lolines
$C_9H_{16}O$=Alkenals, Brevicomin, Fruit flavors, Manicone
$C_9H_{16}OS$=Galbanum oil or resin
$C_9H_{16}O_2$=Alkanolides, Brevicomin, Chalcogran, Chamomile oils, Conophthorin, Olean, Whisk(e)y lactone, oak lactone, Wine flavor
$C_9H_{16}O_4$=Eucommiol
$C_9H_{17}N$=Pinus alkaloids
$C_9H_{17}NO_5$=Pantothenic acid
$C_9H_{17}NO_7$=Muramic acid
$C_9H_{17}NO_8$=Neuraminic acid, 3-Nitropropanoic acid
$C_9H_{17}N_3O_3$=Dopastin
$C_9H_{18}N_4O_4$=Octopine
$C_9H_{18}O$=Alkanals, Scarabs
$C_9H_{18}O_2$=Fruit esters
$C_9H_{20}ClNO_2$=Fly agaric constituents

$\mathbf{C_{10}}H_4Cl_5NO$=Neopyrrolomycin
$C_{10}H_5Cl_3N_2O_3$=Pyrrolomycins
$C_{10}H_6Cl_2N_2O_2$=Pyrrolnitrin
$C_{10}H_6O_3$=Juglone, Lawsone
$C_{10}H_6O_6$=Mompain
$C_{10}H_8$=Azulenes
$C_{10}H_8N_2OS$=Sinalexin
$C_{10}H_8N_2O_4$=Orellanine
$C_{10}H_8N_2O_5$=Orellanine
$C_{10}H_8N_2O_6$=Orellanine
$C_{10}H_8N_4O_4$=Indigoidin
$C_{10}H_8O_2$=1,8-Naphthalenediol
$C_{10}H_8O_3$=4-Hydroxycoumarin, 4-Hydroxy-5-methylcoumarin, Umbelliferone
$C_{10}H_8O_4$=Anemonin, 6,7-Dihydroxycoumarins
$C_{10}H_9NOS$=Brassinin
$C_{10}H_9NOS_2$=Tridentatols
$C_{10}H_9NO_2$=3-Indolylacetic acid
$C_{10}H_9N_5O$=Kinetin
$C_{10}H_{10}Cl_2O_4$=Cryptosporiopsin, Lachnumone
$C_{10}H_{10}N_2$=Nicotyrine
$C_{10}H_{10}N_2O$=Glomerine
$C_{10}H_{10}N_2O_3S$=Ferrithiocin
$C_{10}H_{10}N_6$=Zoanthoxanthin
$C_{10}H_{10}O_2$=Cassia oil, Cinnamic acid, Safrole
$C_{10}H_{10}O_2S_2$=2-Methyl-3-furanthiol
$C_{10}H_{10}O_3$=Mellein
$C_{10}H_{10}O_4$=Caffeic acid, Repandiol
$C_{10}H_{10}O_6$=Chorismic acid, Isochorismic acid, Prephenic acid
$C_{10}H_{11}N$=Lascivol
$C_{10}H_{11}NO_2$=Tuberin
$C_{10}H_{11}N_3O$=Agaricone
$C_{10}H_{12}N_2$=Indolylalkylamines
$C_{10}H_{12}N_2O$=Cotinine, Indolylalkylamines, Serotonin
$C_{10}H_{12}N_2O_3$=L-Kynurenine
$C_{10}H_{12}N_2O_4$=Nicotianine
$C_{10}H_{12}N_2O_7$=Lactivicin
$C_{10}H_{12}N_2O_8$=Orotic acid
$C_{10}H_{12}N_4O_4$=Nebularine
$C_{10}H_{12}N_4O_6$=Xanthosine
$C_{10}H_{12}O$=Anethole, Estragole
$C_{10}H_{12}O_2$=Egomaketone, Eugenol, Insect attractants, Invictolide, Isoeugenol, Mites, Thujaplicins, Zingerone
$C_{10}H_{12}O_3$=Coniferyl alcohol, 3,4-Dihydroxybenzaldehydes
$C_{10}H_{12}O_4$=Acetosyringone, Cantharidin, Oak moss absolute
$C_{10}H_{13}ClN_2O$=Serotonin
$C_{10}H_{13}FN_6O_6S$=Nucleocidin
$C_{10}H_{13}N$=Actinidine
$C_{10}H_{13}NO_2$=Fusaric acid
$C_{10}H_{13}NO_2S_3$=Varacin
$C_{10}H_{13}NO_2S_5$=Varacin
$C_{10}H_{13}NO_3$=*N*-Acylcatecholamines, 3-Hydroxyanthranilic acid
$C_{10}H_{13}NO_4$=*N*-Acylcatecholamines
$C_{10}H_{13}NO_5$=L-Pretyrosine
$C_{10}H_{13}N_2O_4P$=Psilocybin
$C_{10}H_{13}N_3O$=Agaricone
$C_{10}H_{13}N_4O_9P$=Xanthosine
$C_{10}H_{13}N_5O$=Zeatin
$C_{10}H_{13}N_5O_3$=Cordycepin, Oxetanocin
$C_{10}H_{13}N_5O_4$=Mycalamides, Oxetanocin
$C_{10}H_{14}$=Cosmene, Cymenes, Parsley leaf/seed oil
$C_{10}H_{14}Cl_4$=Halogen compounds
$C_{10}H_{14}N_2$=Anabasine, Nicotine
$C_{10}H_{14}N_2O_2$=Laccarin, Pilocarpine
$C_{10}H_{14}N_2O_4$=Porphobilinogen
$C_{10}H_{14}O$=Carvone, Cymenols, Eucarvone, Filifolone, Honey flavor, Mites, Ocimenone, Peppermint oils, Perillaldehyde, Pinenones, Safranal, Tricyclic monoterpenes
$C_{10}H_{14}O_2$=Egomaketone, Fruit flavors, Iridodial, Mintlactone, Mites, Pyrethrum, Teucrium lactones, Tricyclic monoterpenes
$C_{10}H_{14}O_3$=Dihydroconiferyl alcohol, Germicidins
$C_{10}H_{14}O_5$=Acetomycin, Decarestrictins
$C_{10}H_{15}Br_2Cl_3$=Halomon
$C_{10}H_{15}ClN_4O_3$=Pyridazomycin
$C_{10}H_{15}N$=Sesame seed (roasted) flavor
$C_{10}H_{15}NO$=Anatoxins, Ferruginine, *β*-Phenylethylamine alkaloids
$C_{10}H_{15}NO_3$=*β*-Phenylethylamine alkaloids, Tenuazonic acid
$C_{10}H_{15}NO_4$=Domoic acid
$C_{10}H_{15}NO_6$=Mycosporins
$C_{10}H_{16}$=Camphene, Carenes, Fenchenes, *p*-Menthadienes, Myrcene, Ocimene, Pinenes, Thujenes, Tricyclic monoterpenes
$C_{10}H_{16}Br_3ClO$=Halogen compounds
$C_{10}H_{16}KNO_9S_2$=Glucosinolates
$C_{10}H_{16}N_2O_2$=Lepistine, Lolines
$C_{10}H_{16}N_2O_3S$=Biotin
$C_{10}H_{16}N_4O_3$=Anserine
$C_{10}H_{16}N_4O_7$=Vicine
$C_{10}H_{16}O$=Artemisia ketone, Calamus oil, Camphor, Citrals, 2,4-Decadienals, Dill (weed) oil, Fenchone, Fruit flavors, Grandisol, Ipsdienol, Matatabiether, *p*-Menthenones, Ocimenone, Perilla alcohol, Pinenols, Thujan-3-ones, Tricyclic monoterpenes
$C_{10}H_{16}O_2$=Ascaridole, Buchu leaf oil, Geranic acid, Iridodial, Iridomyrmecin, Jasmin(e) lactone, Lineatin, Mites, Pyrethrum
$C_{10}H_{16}O_3$=Queen substance
$C_{10}H_{16}O_5$=Decarestrictins, Tuckolide
$C_{10}H_{17}N$=Polyzonimine
$C_{10}H_{17}NO_6$=Cyanogenic Glycosides
$C_{10}H_{17}NO_9S_2$=Glucosinolates
$C_{10}H_{17}N_3O_5$=Linatine
$C_{10}H_{17}N_3O_6S$=Glutathione
$C_{10}H_{17}N_7O_4$=Saxitoxin
$C_{10}H_{17}N_7O_8S$=Saxitoxin
$C_{10}H_{18}$=Thujane
$C_{10}H_{18}N_2O_2$=Maniwamycins, Slaframine
$C_{10}H_{18}N_2O_6S_2$=*γ*-Glutamylmarasmine
$C_{10}H_{18}O$=Alkenals, Borneols, Cineoles, Citronellals, Fenchol, Fragranol, Geraniol, Grandisol, Ipsdienol, Lavandin [oil], Linalool, Manicone, *p*-Menthanones, *p*-Menthenols, Necrodols, Rose oxide, Tagetes oil, Thujanols
$C_{10}H_{18}OS$=Buchu leaf oil
$C_{10}H_{18}O_2$=Alkanolides, Chamomile oils, Cognaclactone, Longhorn beetles, Multistriatin, Tea flavor
$C_{10}H_{18}O_3$=Queen substance
$C_{10}H_{18}O_4$=Nonactin
$C_{10}H_{18}S$=*p*-Menth-1-ene-8-thiol
$C_{10}H_{19}FO_2$=Fluoroorganic Natural Products
$C_{10}H_{19}NO$=Lupinine, Nitramine
$C_{10}H_{19}NOS_2$=Mustard oils
$C_{10}H_{20}$=*p*-Menthane
$C_{10}H_{20}N_2O_2$=Maniwamycins
$C_{10}H_{20}O$=Alkanals, Citronellols, 3-*p*-Menthanols
$C_{10}H_{20}O_2$=Fruit esters, Terpin
$C_{10}H_{20}O_3$=Myrmicacin
$C_{10}H_{21}NO$=Longhorn beetles
$C_{10}H_{23}N_6O_6P$=Fosfazinomycin A
$C_{10}H_{26}N_4$=Spermine

$\mathbf{C_{11}}H_4Cl_5NO_2$=Pyrrolomycins
$C_{11}H_5Br_2Cl_2NO_2$=Pyrrolomycins
$C_{11}H_5Br_3ClNO_2$=Pyrrolomycins

$C_{11}H_5Br_4NO_2$=Pyrrolomycins
$C_{11}H_5Cl_3O_3$=Mycenone
$C_{11}H_5Cl_4NO_2$=Pyrrolomycins
$C_{11}H_6Cl_4N_2O_3$=Pyrrolomycins
$C_{11}H_6O_3$=Furocoumarins
$C_{11}H_6O_4$=Bergapten
$C_{11}H_7Cl_2NO_3$=Pyoluteorin
$C_{11}H_7N_3OS$=Zyzzin
$C_{11}H_8N_2$=β-Carboline, 3-((*Z*)-2-Isocyanovinyl)indole
$C_{11}H_8N_2O_3S_2$=Luciferins
$C_{11}H_8O_2$=Polyynes
$C_{11}H_8O_3$=Lawsone, Omphalone, Phthiocol, Plumbagin
$C_{11}H_9Br_2N_5O$=Imidazole alkaloids
$C_{11}H_9ClO_4$=Sordidone
$C_{11}H_{10}BrN_5O_2$=Oroidin
$C_{11}H_{10}N_2O$=Peganine
$C_{11}H_{10}N_2O_2$=Peganine
$C_{11}H_{10}N_2O_2S_2$=Oxindole alkaloids
$C_{11}H_{10}N_2S_2$=Brassinin
$C_{11}H_{10}O_2$=1,8-Naphthalenediol, Siccayne
$C_{11}H_{10}O_3$=Harveynone, Psilotin
$C_{11}H_{10}O_4$=Caffeic acid, 6,7-Dihydroxycoumarins
$C_{11}H_{10}O_5$=Sphagnum acid, *o*-Succinylbenzoic acid
$C_{11}H_{10}O_6$=Dihydroxybenzoic acids
$C_{11}H_{11}Br_2N_5O$=Oroidin
$C_{11}H_{11}ClO_5$=Oak moss absolute
$C_{11}H_{11}N_5O_3S_2$=Urothione
$C_{11}H_{12}Cl_2N_2O_5$=Chloramphenicol
$C_{11}H_{12}N_2O$=Glomerine, Peganine
$C_{11}H_{12}N_2OS_2$=Oxindole alkaloids
$C_{11}H_{12}N_2O_2$=L-Tryptophan
$C_{11}H_{12}N_2O_4$=*N*-Formyl-L-kynurenine
$C_{11}H_{12}N_2O_5$=Agaritine
$C_{11}H_{12}N_2S_2$=Brassinin
$C_{11}H_{12}N_6$=Zoanthoxanthin
$C_{11}H_{12}O_2$=Cassia oil, Cinnamic acid
$C_{11}H_{12}O_3$=Safrole, Siccayne
$C_{11}H_{12}O_4$=Whisk(e)y flavor
$C_{11}H_{12}O_5$=Sepedonin, Sinapine
$C_{11}H_{13}ClN_2$=Epibatidine
$C_{11}H_{13}NOS_2$=Tridentatols
$C_{11}H_{13}NO_3$=Streptazolin
$C_{11}H_{13}N_3O$=Pyridopyrazines
$C_{11}H_{13}N_5O_3$=Neplanocins
$C_{11}H_{13}N_5O_4$=Neplanocins
$C_{11}H_{13}N_5O_5$=Ichthyopterin
$C_{11}H_{14}$=Algal pheromones
$C_{11}H_{14}N_2$=Gramine
$C_{11}H_{14}N_2O$=Cytisine
$C_{11}H_{14}N_2O_2S_2$=Holomycin
$C_{11}H_{14}N_2O_4$=Agaritine
$C_{11}H_{14}N_2O_5$=Agaritine
$C_{11}H_{14}N_2O_6$=Clitidine
$C_{11}H_{14}N_4O_4$=6-Methylpurine, Tubercidin
$C_{11}H_{14}N_4O_5$=6-Methylpurine
$C_{11}H_{14}O$=Algal pheromones
$C_{11}H_{14}O_2$=Eugenol, Isoeugenol, Pyrethrum
$C_{11}H_{14}O_2S$=Thiolactomycin
$C_{11}H_{14}O_3$=Zingerone
$C_{11}H_{14}O_4$=Canadensolide, Oak moss absolute, Panepoxydone, Sinapyl alcohol
$C_{11}H_{14}O_5$=Genipin, Papyracillic Acid
$C_{11}H_{14}O_6$=Gabosins
$C_{11}H_{15}NO_2$=Winterstein's acid
$C_{11}H_{15}NO_3$=Anhalonium alkaloids
$C_{11}H_{15}N_2O_4P$=Psilocybin
$C_{11}H_{15}N_3O$=Triacsins
$C_{11}H_{15}N_3O_4$=Xanthodermin
$C_{11}H_{15}N_3O_8$=Polyoxins
$C_{11}H_{15}N_5O_3$=Neplanocins
$C_{11}H_{15}N_5O_6S$=Tubercidin
$C_{11}H_{16}$=Algal pheromones
$C_{11}H_{16}N_2O_2$=Pilocarpine
$C_{11}H_{16}N_2O_3$=*N*-Acylcatecholamines
$C_{11}H_{16}N_2O_4$=*N*-Acylcatecholamines
$C_{11}H_{16}N_2O_4S$=Thienamycins
$C_{11}H_{16}N_4O_4$=Pentostatin
$C_{11}H_{16}O$=(*Z*)-Jasmone
$C_{11}H_{16}O_2$=Dihydroactinidiolide, Invictolide, Pyrethrum, Tobacco flavor
$C_{11}H_{16}O_3$=Germicidins, Hepialone
$C_{11}H_{16}O_4$=Depudecin, Pyrethrum
$C_{11}H_{17}NO$=Tecomanine
$C_{11}H_{17}NO_3$=β-Phenylethylamine alkaloids
$C_{11}H_{17}NO_6$=Multifidin
$C_{11}H_{17}NO_7$=Jatropham
$C_{11}H_{17}N_2O_{11}PS$=Tagetitoxin
$C_{11}H_{17}N_3O$=Imidazole alkaloids
$C_{11}H_{17}N_3O_8$=Tetrodotoxin
$C_{11}H_{18}$=Galbanum oil or resin, Homoterpenes
$C_{11}H_{18}N_2$=Pyrazines
$C_{11}H_{18}N_2O_2$=Lolines
$C_{11}H_{18}O_2$=Chalcogran, Fruit flavors, Geranium oil, Hepialone
$C_{11}H_{18}O_4$=Pestalotin
$C_{11}H_{19}NO_6$=Cyanogenic Glycosides, Mycosporins
$C_{11}H_{19}NO_9$=Neuraminic acid
$C_{11}H_{19}NO_{10}S_2$=Glucosinolates
$C_{11}H_{19}N_3O$=Triacsins
$C_{11}H_{19}N_3O_3$=Leuhistin
$C_{11}H_{19}N_3O_6$=Tabtoxin
$C_{11}H_{20}N_2O_6$=Enteromycin, Saccharopine
$C_{11}H_{20}N_4O_6$=Nopaline
$C_{11}H_{20}O$=Manicone
$C_{11}H_{20}O_2$=Alkanolides, Dominicalure, Geranium oil, Sordidin
$C_{11}H_{20}O_{10}$=Primeverose
$C_{11}H_{21}N$=Skytanthus alkaloids
$C_{11}H_{21}NO$=Nitramine
$C_{11}H_{22}N_2O_4S$=Pantothenic acid
$C_{11}H_{22}N_3O_6P$=Bialaphos
$C_{11}H_{22}N_5O_6P$=Rhizocticins
$C_{11}H_{22}O$=Alkanals
$C_{11}H_{22}O_2$=Serricornin
$C_{11}H_{22}O_3$=Sitophilate
$C_{11}H_{24}$=Undecane

$C_{12}H_6Cl_4N_2O_4$=Pyrrolomycins
$C_{12}H_6O_6$=Benzene-1,2,4-triol
$C_{12}H_7NO_2$=Cherimoline, Cinnabarin
$C_{12}H_8N_2O_4$=Iodinin
$C_{12}H_8O_4$=Bergapten, Nitidone
$C_{12}H_8S_3$=Terthienyls
$C_{12}H_{10}N_2$=Harmans
$C_{12}H_{10}O_2$=Siccayne
$C_{12}H_{10}O_3$=4-Hydroxycoumarin
$C_{12}H_{10}O_7$=Echinochromes
$C_{12}H_{11}ClN_2O_2S$=Batzellines
$C_{12}H_{11}NO_2S$=Chuangxinmycin
$C_{12}H_{11}NO_4$=Mimosamycin, Renierol
$C_{12}H_{11}N_3O_2$=Caerulomycin A
$C_{12}H_{12}$=Azulenes
$C_{12}H_{12}ClN_3OS$=Batzellines
$C_{12}H_{12}N_2O$=Alkaloidal dyes
$C_{12}H_{12}N_2O_3$=Sesbanimides
$C_{12}H_{12}O_2$=Fomannoxin, 1,8-Naphthalenediol
$C_{12}H_{13}BrN_4O_3$=Agelastatines, Oroidin
$C_{12}H_{13}N_5O_4$=Toyocamycin
$C_{12}H_{14}N_2OS_2$=Brassinin
$C_{12}H_{14}N_2O_2$=Abrine
$C_{12}H_{14}N_6$=Zoanthoxanthin
$C_{12}H_{14}O_2$=Celery seed oil, Precocenes
$C_{12}H_{14}O_3$=Cinnamon leaf (bark) oil, Eugenol
$C_{12}H_{14}O_4$=Apiol, Dill apiol
$C_{12}H_{15}NO_3$=Anhalonium alkaloids
$C_{12}H_{16}N_2O$=Cytisine, Indolylalkylamines, Psilocybin
$C_{12}H_{16}N_2O_4S$=Indolylalkylamines
$C_{12}H_{16}N_4O_7$=Russupteridines
$C_{12}H_{16}N_6O_7$=Russupteridines
$C_{12}H_{16}O_2$=Celery seed oil
$C_{12}H_{16}O_3$=Asarones, Oudenone, Primin, Safrole
$C_{12}H_{16}O_7$=Arbutin, Erythrin
$C_{12}H_{17}ClN_2O_4$=Chlorotetaine
$C_{12}H_{17}ClN_4OS$=Thiamin
$C_{12}H_{17}NO_2$=Anhalonium alkaloids
$C_{12}H_{17}NO_3$=Anhalonium alkaloids, Cerulenin, Stimona alkaloids
$C_{12}H_{17}N_2O_4P$=Psilocybin
$C_{12}H_{17}N_3O_4$=Agaritine
$C_{12}H_{17}N_3O_6$=Nagstatin
$C_{12}H_{17}N_5O$=Peramine
$C_{12}H_{18}Cl_2N_4OS$=Thiamin
$C_{12}H_{18}N_2O_5$=Chlorotetaine
$C_{12}H_{18}N_2O_7$=Bicyclomycin
$C_{12}H_{18}N_2O_{10}V$=Amavadin
$C_{12}H_{18}O$=Fruit flavors
$C_{12}H_{18}O_2$=Anastrephin, Suspensolide
$C_{12}H_{18}O_3$=Jasmonic acid, Methyl jasmonate, Primin
$C_{12}H_{18}O_4$=Allixin
$C_{12}H_{18}O_8$=Osmundalactone
$C_{12}H_{19}NO$=Elaeocarpus alkaloids
$C_{12}H_{19}NO_3$=β-Phenylethylamine alkaloids
$C_{12}H_{20}N_2O$=Ammodendrine
$C_{12}H_{20}N_2O_2$=Aspergillic acid, Lolines
$C_{12}H_{20}N_2O_8$=Nicotianamine
$C_{12}H_{20}N_4O_9S$=Sulfazecin
$C_{12}H_{20}O$=Dehydrogeosmin, Fruit flavors, Termites
$C_{12}H_{20}O_2$=Bugs, Ferrulactone II, Fir and pine needle oils, Fruit esters, Juniper berry oil, Lavandin [oil], Linalool, Oil of cardamom, Palmarosa oil
$C_{12}H_{20}O_4$=Traumatic acid
$C_{12}H_{21}N_3O_6$=Nicotianamine
$C_{12}H_{22}N_2O_6$=Enteromycin
$C_{12}H_{22}N_2O_{10}S_4$=Lentinic acid
$C_{12}H_{22}O$=Alkenals, Codlemone, Geosmin
$C_{12}H_{22}O_2$=Alkanolides, Dominicalure, Invictolide, Peppermint oils, Quadrilure
$C_{12}H_{22}O_4$=Butenolide(s), Talaromycins
$C_{12}H_{22}O_{10}$=Rutinose
$C_{12}H_{22}O_{11}$=Gentiobiose, Lactose, Palatinose, Sucrose, Trehalose, Turanose
$C_{12}H_{24}O$=Alkanals
$C_{12}H_{24}O_3$=Hydroxyfatty acids
$C_{12}H_{25}N$=Stenusine
$C_{12}H_{26}N_4O_6$=Neomycin

$C_{13}H_8OS_2$=Thiarubrins
$C_{13}H_8O_4$=Indian yellow
$C_{13}H_8O_5$=Gentisein
$C_{13}H_8O_6$=Hymenoquinone
$C_{13}H_8S_2$=Thiarubrins
$C_{13}H_9NO_2S$=Cinnabarin
$C_{13}H_{10}BrNO_3$=Dihydroxybenzoic acids
$C_{13}H_{10}N_2O$=Pyocyanine
$C_{13}H_{10}N_2O_4$=Iodinin
$C_{13}H_{10}N_2O_9$=Rubrifacine
$C_{13}H_{10}OS$=Terthienyls
$C_{13}H_{10}O_2$=Polyynes, Styrylpyrones
$C_{13}H_{10}O_4$=Khellin, Styrylpyrones
$C_{13}H_{10}O_5$=Khellin, Pimpinellin, Styrylpyrones
$C_{13}H_{10}O_6$=Maclurin
$C_{13}H_{11}ClO_2$=Pterulinic acid and pterulones
$C_{13}H_{12}N_2O$=Harmans
$C_{13}H_{12}N_2O_2$=Aaptamine
$C_{13}H_{12}N_2O_3$=Haematopodin
$C_{13}H_{12}N_2O_{10}$=Octosyl acids
$C_{13}H_{12}N_4OS$=Grossularines
$C_{13}H_{12}N_4O_2$=Alboinone
$C_{13}H_{12}O_2$=Azulenes, Chamomile oils, Polyynes, Styrylpyrones
$C_{13}H_{12}O_3$=Styrylpyrones
$C_{13}H_{12}O_5$=WF-3681
$C_{13}H_{13}N_3O_2S$=Caerulomycin A
$C_{13}H_{14}N_2O$=Harmans
$C_{13}H_{14}N_2O_3$=Nigellidine
$C_{13}H_{14}N_2O_4S_2$=Gliotoxin
$C_{13}H_{14}N_2O_4S_3$=Gliotoxin
$C_{13}H_{14}N_2O_4S_4$=Gliotoxin
$C_{13}H_{14}N_2O_7$=Acromelic acids
$C_{13}H_{14}N_2O_8$=Miraxanthins
$C_{13}H_{14}N_2O_{10}$=Octosyl acids
$C_{13}H_{14}O_3$=Asperfuran
$C_{13}H_{14}O_5$=Citrinin, Papulinone
$C_{13}H_{15}Cl_3O_2$=Neocarzilins
$C_{13}H_{15}KO_9$=Leaf-Movement Factors
$C_{13}H_{15}NO_2$=Securinega alkaloids

$C_{13}H_{16}Cl_2O_4$=Halogen compounds
$C_{13}H_{16}N_2OS_2$=Brassinin
$C_{13}H_{16}N_2O_2$=Melatonin, Oxindole alkaloids
$C_{13}H_{16}N_2O_3$=Anthramycins
$C_{13}H_{16}N_2O_9$=Octosyl acids
$C_{13}H_{16}N_6$=Zoanthoxanthin
$C_{13}H_{16}O$=Cocoa flavor
$C_{13}H_{16}O_3$=Precocenes
$C_{13}H_{16}O_9$=Gentisic acid
$C_{13}H_{17}NO_3$=Anhalonium alkaloids, Securinega alkaloids
$C_{13}H_{18}N_2O$=Indolylalkylamines
$C_{13}H_{18}N_6O_3$=Lupinic acid
$C_{13}H_{18}O$=β-Damascenone, Tobacco flavor
$C_{13}H_{18}O_3$=Honey flavor
$C_{13}H_{18}O_4$=Parsley leaf/seed oil
$C_{13}H_{18}O_7$=Salicyl alcohol
$C_{13}H_{19}NO_2$=Anhalonium alkaloids, Dioscorine
$C_{13}H_{19}NO_2S_4$=Varacin
$C_{13}H_{19}NO_2S_6$=Varacin
$C_{13}H_{19}NO_3$=Anhalonium alkaloids
$C_{13}H_{19}NO_6$=Mycosporins
$C_{13}H_{19}N_3O_5S_2$=Sparsomycin
$C_{13}H_{19}N_3O_6S_2$=Sparsomycin
$C_{13}H_{20}ClNO_4$=Sticticin
$C_{13}H_{20}N_2O_6$=Bactobolin
$C_{13}H_{20}O$=Damascones, Ionones, Vitispirane
$C_{13}H_{20}O_2$=Theaspiranes
$C_{13}H_{20}O_3$=Methyl jasmonate, Stegobinone
$C_{13}H_{20}O_6$=Syringolides
$C_{13}H_{21}N$=Hippocasine
$C_{13}H_{22}N_2O$=Polyzonimine
$C_{13}H_{22}N_2O_2$=Polyzonimine
$C_{13}H_{22}N_2O_{10}$=Trehalostatin
$C_{13}H_{22}N_4O_5S$=Clithioneine
$C_{13}H_{22}N_4O_9S$=Sulfazecin
$C_{13}H_{22}O$=Theaspiranes, Tobacco flavor
$C_{13}H_{22}O_2$=Scales
$C_{13}H_{22}O_3$=Methyl jasmonate, Stegobinone
$C_{13}H_{22}O_4$=A-factor
$C_{13}H_{23}N$=Coccinelline
$C_{13}H_{23}NO$=Adaline, Coccinelline
$C_{13}H_{23}NO_8$=Salbostatin
$C_{13}H_{24}N_2$=Tetraponerines
$C_{13}H_{24}N_2O$=Cuscohygrine
$C_{13}H_{24}N_2O_{11}$=Cycasin
$C_{13}H_{24}N_6O_3$=Ficellomycin
$C_{13}H_{25}N$=Monomorines, Pumiliotoxins
$C_{13}H_{25}N_9O_3$=Sperabillins
$C_{13}H_{26}N_2O_3$=Elaiomycin

$\mathbf{C_{14}}H_6Cl_4O_5$=Lichen xanthones
$C_{14}H_6N_2O_8$=Pyrroloquinoline quinone
$C_{14}H_6O_6$=Ellagic acid
$C_{14}H_6O_8$=Ellagic acid
$C_{14}H_8N_2O$=Canthin-6-one
$C_{14}H_8N_2O_2$=Canthin-6-one
$C_{14}H_8O_4$=Alizarin(e)
$C_{14}H_8O_5$=Purpurin
$C_{14}H_8O_6$=Canarione
$C_{14}H_9NO_3$=Dianthalexine
$C_{14}H_{10}N_2O_5$=Azoxybenzene-4,4′-dicarboxylic acid, Cinnabarin
$C_{14}H_{10}O_5$=Alternariol, Gentisin, Lichen xanthones
$C_{14}H_{10}O_6$=Juglomycins, Oosporein, Simonyellin
$C_{14}H_{10}O_8$=Oosporein
$C_{14}H_{11}Cl_3O_4$=Russuphelins
$C_{14}H_{12}Br_2O_4$=Halogen compounds
$C_{14}H_{12}ClNO_4$=Streptopyrrole
$C_{14}H_{12}N_2O$=4-(Hydroxymethyl)benzenediazonium ion
$C_{14}H_{12}N_2O_2$=4-(Hydroxymethyl)benzenediazonium ion, Infractines
$C_{14}H_{12}N_2O_4$=Azoxybenzene-4,4′-dicarboxylic acid
$C_{14}H_{12}N_4O_4$=Boxazomycins
$C_{14}H_{12}N_4O_5$=Boxazomycins
$C_{14}H_{12}O_2$=Benzoic acid, Hydroxystilbenes
$C_{14}H_{12}O_3$=Hydroxystilbenes, Styrylpyrones
$C_{14}H_{12}O_4$=Hydroxystilbenes
$C_{14}H_{12}O_5$=Khellin
$C_{14}H_{13}NO_7$=Narciclasine
$C_{14}H_{13}N_3O_8$=Fly agaric pigments
$C_{14}H_{13}N_5O_5$=Phidolopin
$C_{14}H_{13}N_5O_7$=Griseolic acids
$C_{14}H_{13}N_5O_8$=Griseolic acids
$C_{14}H_{14}N_4O$=Aplysi…
$C_{14}H_{14}N_6$=Zoanthoxanthin
$C_{14}H_{14}O_3$=Hydroxystilbenes, Senoxepin, Styrylpyrones
$C_{14}H_{14}O_4$=Corticrocin, Marmesin, Torachrysone
$C_{14}H_{14}O_5$=Visnadin
$C_{14}H_{15}ClO_4$=Mycorrhizin
$C_{14}H_{15}ClO_5$=Mycorrhizin
$C_{14}H_{15}NO_5$=Triticones
$C_{14}H_{15}NO_7$=Isatan B
$C_{14}H_{15}NO_8$=Pancratistatin
$C_{14}H_{15}N_3O_2$=Indolmycin
$C_{14}H_{15}N_5O_7$=Griseolic acids
$C_{14}H_{16}$=Azulenes
$C_{14}H_{16}BrN_3OS$=Eudistomins
$C_{14}H_{16}N_2O_5$=Monatin
$C_{14}H_{16}N_2O_6$=Indicaxanthin
$C_{14}H_{16}N_2O_8$=Vulgaxanthins
$C_{14}H_{16}O_3$=Styrylpyrones
$C_{14}H_{17}ClO$=Pterosins
$C_{14}H_{17}Cl_3O_2$=Neocarzilins
$C_{14}H_{17}NO_6$=Cyanogenic Glycosides, Indican
$C_{14}H_{17}NO_7$=Cyanogenic Glycosides
$C_{14}H_{17}N_3O_7$=Vulgaxanthins
$C_{14}H_{18}N_2O$=Camoensine
$C_{14}H_{18}N_2O_7S$=Miraxanthins
$C_{14}H_{18}O$=Eudesmanes
$C_{14}H_{18}O_2$=Pterosins
$C_{14}H_{18}O_3$=Gyrinidal, Precocenes
$C_{14}H_{18}O_4$=Mites
$C_{14}H_{18}O_5$=Mycorrhizin
$C_{14}H_{18}O_7$=Picein
$C_{14}H_{18}O_8$=3,4-Dihydroxybenzaldehydes, Leaf-Movement Factors
$C_{14}H_{18}O_9$=Homogentisic acid
$C_{14}H_{19}NO_4$=Anisomycin
$C_{14}H_{19}NO_6$=Triticones
$C_{14}H_{19}NO_{10}S_2$=Glucosinolates
$C_{14}H_{20}Cl_2N_2O_6$=Bactobolin
$C_{14}H_{20}N_5O_5$=Leaf-Movement Factors
$C_{14}H_{20}O_3$=Gyrinidal
$C_{14}H_{20}O_9$=Unedoside
$C_{14}H_{20}O_{10}$=Unedoside
$C_{14}H_{21}NO$=Sedum alkaloids
$C_{14}H_{22}N_2O$=Angustifoline, Camoensine
$C_{14}H_{22}N_5O_7P$=B-factor
$C_{14}H_{22}O$=Irones, Patchouli oil
$C_{14}H_{22}O_2$=Rishitin, Sulcatins
$C_{14}H_{22}O_3$=Serricorole
$C_{14}H_{24}N_2O$=Polyzonimine
$C_{14}H_{24}N_2O_7$=Spectinomycin
$C_{14}H_{24}O_2$=Japonilure
$C_{14}H_{24}O_3$=Serricorole
$C_{14}H_{25}N_3O_9$=Kasugamycin
$C_{14}H_{26}N_2$=Tetraponerines
$C_{14}H_{26}O$=Cockroaches, Scarabs
$C_{14}H_{26}O_2$=Elephant pheromone, Tetradecenyl acetate
$C_{14}H_{26}O_3$=Sulcatins
$C_{14}H_{27}FO_2$=Fluoroorganic Natural Products
$C_{14}H_{28}O$=Leaf beetles, Meat flavor
$C_{14}H_{28}O_2$=Leaf beetles, Myristic acid
$C_{14}H_{30}N_2O_2$=Nitrosoxacin

$\mathbf{C_{15}}H_2O_2$=Lactardial
$C_{15}H_8N_2O$=Sampangine
$C_{15}H_8N_2O_2$=Monomargine
$C_{15}H_8N_2O_3$=Necatorone
$C_{15}H_8O_5$=Coumestrol
$C_{15}H_8O_6$=Rhein
$C_{15}H_9Cl_3O_5$=Lichen xanthones
$C_{15}H_9O_6^+$=Riccionidins
$C_{15}H_{10}BrN_3O$=Eudistomins
$C_{15}H_{10}N_2OS$=Curtisine A
$C_{15}H_{10}N_2O_3S$=Curtisine A
$C_{15}H_{10}N_2O_4S$=Curtisine A
$C_{15}H_{10}N_4O_2$=Didemnimides
$C_{15}H_{10}O_2$=Isoflavones, Tectoquinone
$C_{15}H_{10}O_4$=Chrysin, Chrysophanol, Isoflavones, Marginalin, Primetin
$C_{15}H_{10}O_5$=Aloe-emodin, Apigenin, Emodin, Galangin, Islandicin, Isoflavones, Morindone
$C_{15}H_{10}O_6$=Fisetin, Islandicin, Isoflavones, Kaempferol, Luteolin, Maritimetin
$C_{15}H_{10}O_7$=Morin, Quercetin, Robinetin
$C_{15}H_{10}O_8$=Gossypetin, Myricetin, Rhodocladonic acid
$C_{15}H_{11}ClO_4$=Pterulinic acid and pterulones
$C_{15}H_{11}ClO_5$=Pelargonidin
$C_{15}H_{11}ClO_6$=Cyanidin
$C_{15}H_{11}ClO_7$=Delphinidin
$C_{15}H_{11}I_4NO_4$=Iodoamino acids
$C_{15}H_{11}N_3$=Tobacco alkaloids
$C_{15}H_{11}O_4^+$=Apigeninidin
$C_{15}H_{11}O_5^+$=Pelargonidin
$C_{15}H_{11}O_6^+$=Cyanidin
$C_{15}H_{11}O_7^+$=Delphinidin
$C_{15}H_{12}I_3NO_4$=Iodoamino acids
$C_{15}H_{12}O_3$=Chrysarobin
$C_{15}H_{12}O_4$=Chalcones, Garcifurans
$C_{15}H_{12}O_5$=Butein, Naringin
$C_{15}H_{12}O_6$=Chalcones
$C_{15}H_{12}O_7$=Haemoventosin, Silybin
$C_{15}H_{13}I_2NO_4$=Iodoamino acids
$C_{15}H_{14}N_2O_2$=Infractines
$C_{15}H_{14}O$=Azulenes, Lactarius pigments
$C_{15}H_{14}O_2$=Hydroxystilbenes
$C_{15}H_{14}O_3$=Lapachol
$C_{15}H_{14}O_4$=Lomatiol, Lunularic acid, Polyynes, Styrylpyrones
$C_{15}H_{14}O_5$=Atrochrysone, Phloretin, Styrylpyrones
$C_{15}H_{14}O_6$=Catechins, Plumeria iridoids
$C_{15}H_{14}O_7$=Plagiogyrins
$C_{15}H_{15}BrN_4$=Lissoclinum Alkaloids
$C_{15}H_{15}BrO_2$=Halogen compounds
$C_{15}H_{15}NO$=Navenone
$C_{15}H_{15}NO_4$=L-Thyronine
$C_{15}H_{16}Br_2O_3$=Okamurallene
$C_{15}H_{16}NO^+$=Ipom(o)ea alkaloids
$C_{15}H_{16}N_4O_4S$=Anguibactin
$C_{15}H_{16}N_{10}O_2$=Pteridines
$C_{15}H_{16}O$=Lactarius pigments
$C_{15}H_{16}O_2$=Triquinanes
$C_{15}H_{16}O_3$=Germacranolides
$C_{15}H_{16}O_4$=Pleurotellic acid, Sterepolide, Torachrysone
$C_{15}H_{16}O_5$=Ascochitin, Collybolide, Illudinine, Lactucin, Nidulal, Pentalenolactones, Vernolepin, Visamminol
$C_{15}H_{16}O_6$=Picrotoxin
$C_{15}H_{16}O_7$=Allamandin
$C_{15}H_{16}O_8$=Plagiogyrins, Umbelliferone
$C_{15}H_{16}O_9$=Daphnetin, 6,7-Dihydroxycoumarins
$C_{15}H_{17}Br$=Azulenes
$C_{15}H_{17}BrO_2$=Aplysi…
$C_{15}H_{17}Br_2ClO_2$=Obtusallene I
$C_{15}H_{18}$=Azulenes
$C_{15}H_{18}N_2O$=Huperzine A
$C_{15}H_{18}N_4O_5$=Mitomycins
$C_{15}H_{18}O_2$=Chrysorrhealactone, Herbertenes, Triquinanes
$C_{15}H_{18}O_3$=Fulvoferruginin, Lagopodins, Pleurotellic acid, Santonin
$C_{15}H_{18}O_4$=Artemisin, Clavicoronic acid, Complicatic acid, Fomannosin, Helenalin, Lagopodins, Leaianafulvene, Marasmic acid, Pentalenolactones, Styrylpyrones
$C_{15}H_{18}O_5$=Coriamyrtin, Lagopodins, Marasmenes, Pentalenolactones
$C_{15}H_{18}O_6$=Coriamyrtin, Pentalenolactones
$C_{15}H_{18}O_7$=Coriamyrtin, Picrotoxin
$C_{15}H_{18}O_8$=Bilobalide, Coumarinic acid
$C_{15}H_{19}BrO$=Aplysi…
$C_{15}H_{19}BrO_2$=Aplysi…

$C_{15}H_{19}ClO$=Pterosins
$C_{15}H_{19}N$=Myrmicarins
$C_{15}H_{19}NO$=Ipom(o)ea alkaloids
$C_{15}H_{19}NO_2$=Mesembrine alkaloids
$C_{15}H_{20}$=Laurene
$C_{15}H_{20}N_2O$=Anagyrine, Verruculotoxin
$C_{15}H_{20}N_2O_2$=Baptifoline
$C_{15}H_{20}N_2O_4$=Epiderstatin, Fusarochromanones
$C_{15}H_{20}O$=Cuparenone, Nuciferal, Periplanones
$C_{15}H_{20}O_2$=Aristolactone, Chrysorrhealactone, Frullanolide, Germacranolides, Helenine, Isovelleral, Lactaranes, Periplanones, Pterosins, Velleral
$C_{15}H_{20}O_3$=Arborescin, Helicobasidin, Illudanes, Merulidial, Perezone, Periplanones, Pleurotellic acid, Proazulenes, PR toxin, Quadrone, Ricciocarpins, Tansy, Taurin, Winterin
$C_{15}H_{20}O_4$=Abscisic acid, Alliacol(ide), Helicobasidin, Hirsutic acid, Hymenoxon(e), Illudanes, Pentalenolactones, Ricciocarpins, Tric(h)othecenes
$C_{15}H_{20}O_5$=Abscisic acid, Coriolin, Tansy
$C_{15}H_{20}O_6$=Deoxynivalenol, Marasmenes, Shellac (shellac wax), Tric(h)othecenes
$C_{15}H_{20}O_7$=Anisatin, 4-Hydroxycinnamyl alcohol, Nivalenol, Tric(h)othecenes
$C_{15}H_{20}O_8$=Anisatin, Melilotic acid, Salicyl alcohol
$C_{15}H_{20}O_{10}$=3-(3,4-Dihydroxyphenyl)lactic acid
$C_{15}H_{21}BrO_3$=Aplysi...
$C_{15}H_{21}Br_2ClO_2$=Halogen compounds
$C_{15}H_{21}NO_3$=Pyridoxatin
$C_{15}H_{21}NO_6$=Domoic acid
$C_{15}H_{21}NO_6S$=Niazinins
$C_{15}H_{21}NO_7$=Sesbanimides
$C_{15}H_{21}NO_9S_2$=Glucosinolates
$C_{15}H_{21}NO_{10}S_2$=Glucosinolates
$C_{15}H_{21}N_3O_2$=Physostigmine
$C_{15}H_{21}N_3O_{15}$=3-Nitropropanoic acid
$C_{15}H_{21}N_5O_7S$=Cylindrospermopsin
$C_{15}H_{22}$=Curcumenes, Herbertenes, Vetispiranes
$C_{15}H_{22}BrClO$=Elatol
$C_{15}H_{22}N_2O$=Multiflorine
$C_{15}H_{22}N_2O_2$=Multiflorine
$C_{15}H_{22}N_2O_3$=Brevioxime
$C_{15}H_{22}N_6O_5S$=*S*-Adenosylmethionine
$C_{15}H_{22}O$=Cyperones, Dendrolasin, Eremophilanes, Germacranes, Herbertenes, (–)-Isobicyclogermacrenal, Nuciferal, Patchouli oil, Periplanones, Sinensals, Vetispiranes, Vetivone
$C_{15}H_{22}O_2$=Ancistrodial, Eremophilanes, Helminthosporal, Herbertenes, Lepistirone, Marasmenes, Polygodial, Rishitin, Tremulanes, Tric(h)othecenes, Velleral
$C_{15}H_{22}O_3$=Fecapentaenes, Illudanes, Ipomeamarone, Ipomeanine, Lactaranes, Lactardial, Myoporone, Piperdial, Scirpenols, Tric(h)othecenes, Triquinanes, Warburganal, Xanthoxin, Yingzhaosu
$C_{15}H_{22}O_4$=Alliacol(ide), Avenaciolide, Ipomeanine, Isolactaranes, Mniopetals, Scirpenols, Tric(h)othecenes
$C_{15}H_{22}O_5$=Hymenoxon(e), Qinghaosu, Scirpenols, Tric(h)othecenes
$C_{15}H_{22}O_6$=Tric(h)othecenes
$C_{15}H_{22}O_9$=Aucubin, Deutzioside
$C_{15}H_{22}O_{10}$=Antirrhinoside, Catalpol, Monomelittoside
$C_{15}H_{23}NO$=Nuphar alkaloids
$C_{15}H_{23}NO_2$=Castoramine
$C_{15}H_{23}NO_5$=Callimorphine
$C_{15}H_{23}N_7O_5$=Sinefungin
$C_{15}H_{23}O_2$=Sirenin
$C_{15}H_{24}$=Africanol, β-Bergamotene, Bisabolenes, Cadinene, Capnellene, Caryophyllenes, Cedrene, Chamigrenes, Curcumenes, Elemenes, Farnesene, Geranium oil, Germacrenes, Ginger oil, Gymnomitranes, Hirsutanes, Humulene, Ishwarane, Isocomene, Longifolene, Modhephene, Patchouli oil, Pentalenene, Protoilludanes, Santalenes, Sativene, Selinenes, Seychellene, Sterpuranes, Thujopsene, Trichodiene, Triquinanes, Valencene, Zingiberene
$C_{15}H_{24}N_2O$=Lupanine, Matrine
$C_{15}H_{24}N_2O_2$=13-Hydroxylupanine
$C_{15}H_{24}N_2O_7S$=Lactacystin
$C_{15}H_{24}O$=Bugs, Calamus oil, Elemenes, Eudesmanes, Gymnomitranes, Ishwarane, Patchouli oil, (+)-2-Pupukeanone, Santalols, Vetiver oil
$C_{15}H_{24}O_2$=Capsidiol, Cockroaches, Davana oil, Hernandulcin, Luciferins, Rishitin, Sativene, Vetispiranes
$C_{15}H_{24}O_3$=Fomannosin, Myoporone, Phytuberin, Piperdial, Protoilludanes, Punctaporonins, Sterpuranes, Sulcatins, Triquinanes, Uvidins, Vetispiranes
$C_{15}H_{24}O_4$=Uvidins
$C_{15}H_{24}O_5$=Qinghaosu
$C_{15}H_{24}O_6$=Syringolides
$C_{15}H_{24}O_9$=Deutzioside
$C_{15}H_{24}O_{10}$=Harpagide
$C_{15}H_{25}NO$=Histrionicotoxins
$C_{15}H_{25}NO_2$=Nuphar alkaloids
$C_{15}H_{25}NO_4$=Aerocyanidin
$C_{15}H_{25}NO_5$=Indicine
$C_{15}H_{25}N_3$=Ptilomycalins
$C_{15}H_{26}N_2$=Sparteine
$C_{15}H_{26}N_2O$=Retamine
$C_{15}H_{26}O$=Africanol, (–)-α-Bisabolol, Cedrol, Drimanes, Elemenes, Eudesmanes, Farnesol, Geranium oil, Guaiac(um), Ledol, Orange flower absolute (oil), Patchouli oil, Thujopsene, Triticenes
$C_{15}H_{26}O_3$=Uvidins
$C_{15}H_{26}O_4$=Agarofurans, Yingzhaosu
$C_{15}H_{27}NO_{10}$=Pantothenic acid
$C_{15}H_{27}N_5O_3$=Sperabillins
$C_{15}H_{28}N_2$=Tetraponerines
$C_{15}H_{28}O$=Terrestrol
$C_{15}H_{28}O_2$=15-Pentadecanolide, Scarabs
$C_{15}H_{29}NO_2$=Nitro compounds
$C_{15}H_{30}$=Mites
$C_{15}H_{30}O_2$=Flies, Lardolure
$C_{15}H_{32}N_7O_7P$=Fosfazinomycin A
$C_{15}H_{34}N_9O_8PS$=Phaseolotoxin

$C_{16}H_7N_3OS$=Kuanoniamines
$C_{16}H_8Br_2N_2O_2$=Purple, Tyrian purple
$C_{16}H_9NO_7$=Aristolochic acids
$C_{16}H_{10}Cl_4O_5$=Diploicin
$C_{16}H_{10}N_2O_2$=Indigo red, Indigotin, indigo
$C_{16}H_{10}O_6$=Aflatoxins
$C_{16}H_{10}O_7$=Endocrocin, Laccaic acids, Wedelolactone
$C_{16}H_{10}O_8$=Ellagic acid, Kermesic acid
$C_{16}H_{11}NO_4$=Avenalumin
$C_{16}H_{12}N_2$=Cryptolepine
$C_{16}H_{12}N_2O_2$=Perloline
$C_{16}H_{12}O_4$=Hallachrome, Vesparione
$C_{16}H_{12}O_5$=Brazilin, Isoflavones, Physcion, Pterocarpans
$C_{16}H_{12}O_6$=Dermocybe pigments, Fusidienol A, Haematoxylin, Isoflavones, Kalafungin, Maritimetin, Nanaomycins
$C_{16}H_{12}O_7$=Dermocybe pigments, Pannaric acid, Selgin
$C_{16}H_{13}ClO_6$=Peonidin
$C_{16}H_{13}ClO_7$=Petunidin, Tuberin
$C_{16}H_{13}NO_5$=Avenalumin
$C_{16}H_{13}NO_7$=Cercosporamide
$C_{16}H_{13}NO_8$=Oviposition deterring pheromones (host marking pheromones)
$C_{16}H_{13}O_6^+$=Peonidin
$C_{16}H_{13}O_7^+$=Petunidin
$C_{16}H_{14}N_4O_3$=Lymphostin
$C_{16}H_{14}O_2$=Peru balsam, Peru oil
$C_{16}H_{14}O_4$=Imperatorin, Parviflorin
$C_{16}H_{14}O_5$=Brazilin, Clavilactones, Garcifurans, Moracin(s)
$C_{16}H_{14}O_6$=Aflatoxins, Haematoxylin, Hesperetin, Nanaomycins
$C_{16}H_{14}O_7$=Lecanoric acid, Nanaomycins
$C_{16}H_{15}NO_5$=Nanaomycins
$C_{16}H_{15}N_5O_4$=Clathridine
$C_{16}H_{16}N_2O_2$=Lysergic acid
$C_{16}H_{16}O_2$=Verpacrocin
$C_{16}H_{16}O_4$=Trypethelone
$C_{16}H_{16}O_5$=Alkannin, Clavilactones, Shikonin, Torosachrysone
$C_{16}H_{16}O_6$=Austrocortilutein, Clavilactones
$C_{16}H_{16}O_7$=Austrocortilutein, Nanaomycins
$C_{16}H_{17}NO_2$=Carbazomycins
$C_{16}H_{17}NO_3$=Amaryllidaceae alkaloids, Illudinine
$C_{16}H_{17}NO_4$=Amaryllidaceae alkaloids
$C_{16}H_{17}N_3O_4$=Anthramycins
$C_{16}H_{18}N_2$=Agroclavine
$C_{16}H_{18}N_2O_2$=Clavicipitic acid
$C_{16}H_{18}N_2O_3$=Pilocarpine
$C_{16}H_{18}O_3$=Strobilurins
$C_{16}H_{18}O_4$=Strobilurins
$C_{16}H_{18}O_5$=Curvularin, Visamminol
$C_{16}H_{18}O_9$=Chlorogenic acid, 6,7-Dihydroxycoumarins
$C_{16}H_{19}NO_3$=Amaryllidaceae alkaloids, Erythrina alkaloids
$C_{16}H_{19}NO_4$=Scopolamine
$C_{16}H_{19}NO_5$=Rohitukine
$C_{16}H_{19}N_3O_2$=Imidazole alkaloids
$C_{16}H_{19}N_3O_3$=Febrifugine
$C_{16}H_{19}N_3O_5S$=Karnamicins
$C_{16}H_{19}N_3O_6$=Mitomycins
$C_{16}H_{20}N_2O$=Huperzine A, Roquefortines
$C_{16}H_{20}N_2O_2$=Nematophin
$C_{16}H_{20}N_2O_4$=Anthramycins
$C_{16}H_{20}N_2O_9S_2$=Glucosinolates
$C_{16}H_{20}N_4O_5$=Mitomycins
$C_{16}H_{20}O_5$=Cladosporin, Curvularin
$C_{16}H_{20}O_6$=Curvularin
$C_{16}H_{20}O_8$=(*E*)-*p*-Coumaric acid
$C_{16}H_{20}O_9$=Gentiopicroside
$C_{16}H_{20}O_{10}$=Asperuloside
$C_{16}H_{21}NO$=Chalciporone, 2-Heptyl-1-hydroxy-4(1*H*)-quinolinone
$C_{16}H_{21}NO_2$=2-Heptyl-1-hydroxy-4(1*H*)-quinolinone
$C_{16}H_{21}NO_3$=Annotinine
$C_{16}H_{21}N_3O_6S$=Karnamicins
$C_{16}H_{21}N_3O_8S$=Cephalosporins
$C_{16}H_{22}N_2$=Lycodine
$C_{16}H_{22}N_2O_3$=Phthoxazolin A
$C_{16}H_{22}O_7$=Eugenol
$C_{16}H_{22}O_8$=Coniferin
$C_{16}H_{22}O_9$=Acetosyringone, Secoiridoids
$C_{16}H_{22}O_{10}$=Genipin, Secoiridoids
$C_{16}H_{22}O_{11}$=Asperuloside, Monotropein, Secoiridoids
$C_{16}H_{23}N$=Isocyanides
$C_{16}H_{23}NO_6$=Monocrotaline
$C_{16}H_{23}N_3O_7$=Siderophores
$C_{16}H_{24}INO_5$=Sinapine
$C_{16}H_{24}NO_5^+$=Sinapine
$C_{16}H_{24}N_2O_4$=Bestatin
$C_{16}H_{24}N_2O_{12}$=Siderophores
$C_{16}H_{24}O_3$=Brefeldins
$C_{16}H_{24}O_4$=Brefeldins, Gingerol
$C_{16}H_{24}O_5$=Ovalicin
$C_{16}H_{24}O_8$=Boschnaloside
$C_{16}H_{24}O_{10}$=Loganin
$C_{16}H_{24}O_{12}$=Secoiridoids
$C_{16}H_{25}N$=Diisocyanoadociane, Isocyanides
$C_{16}H_{25}NO$=Lycopodine
$C_{16}H_{25}NO_2$=Dendrobium alkaloids, Lycopodium alkaloids
$C_{16}H_{25}NO_9S$=Sinapine
$C_{16}H_{25}NS$=Diisocyanoadociane
$C_{16}H_{25}N_3O_3$=Amphikuemine
$C_{16}H_{26}$=Homoterpenes

$C_{16}H_{26}N_2O$=Cernuine
$C_{16}H_{26}O_2$=Clary sage oil, Scales
$C_{16}H_{26}O_3$=Juvabione, Juvenile hormones
$C_{16}H_{26}O_4$=Fumagillin, Podoscyphic acid
$C_{16}H_{26}O_5$=Qinghaosu
$C_{16}H_{26}O_7$=Safranal
$C_{16}H_{27}NO_{11}$=Cyanogenic Glycosides
$C_{16}H_{28}N_4O_9$=Aerobactin
$C_{16}H_{28}O$=Ambrox, Bombykol, Clary sage oil
$C_{16}H_{28}O_2$=Ambrettolide, Chaulmoogric acid, Prodlure, Scales, Weevils
$C_{16}H_{29}FO_2$=Fluoroorganic Natural Products
$C_{16}H_{29}NO_2$=Epilachnene
$C_{16}H_{30}N_2$=Tetraponerines
$C_{16}H_{30}O$=Bombykol, Muscone
$C_{16}H_{30}O_2$=Tetradecenyl acetate
$C_{16}H_{30}O_3$=Majusculamides
$C_{16}H_{30}O_4$=F-acids
$C_{16}H_{31}FO_2$=Fluoroorganic Natural Products
$C_{16}H_{31}NOS$=Sulfinemycin
$C_{16}H_{31}N_6O_7P$=Rhizocticins
$C_{16}H_{32}O_3$=Hydroxyfatty acids
$C_{16}H_{32}O_5$=Hydroxyfatty acids
$C_{16}H_{34}N_2O_2$=Nitrosoxacin

$\mathbf{C_{17}}H_{10}O_6$=Austocystins
$C_{17}H_{10}O_7$=Austocystins
$C_{17}H_{11}NO_7$=Aristolochic acids
$C_{17}H_{11}NO_8$=Aristolochic acids
$C_{17}H_{12}O_5$=Gyrocyanin
$C_{17}H_{12}O_6$=Aflatoxins, Gyrocyanin, Wairol
$C_{17}H_{12}O_7$=Aflatoxins, Cinnalutein, Dermocybe pigments
$C_{17}H_{12}O_8$=Aflatoxins, Cinnalutein, Dermocybe pigments, Ellagic acid
$C_{17}H_{13}ClN_2O$=Infractines
$C_{17}H_{13}NO_4$=Oxindole alkaloids
$C_{17}H_{14}N_2$=Ellipticine
$C_{17}H_{14}N_2O_3$=Cyclopenine
$C_{17}H_{14}N_2O_7$=Obafluorin
$C_{17}H_{14}O_5$=Chamonixin, Pterocarpans
$C_{17}H_{14}O_6$=Aflatoxins, Involutin, Pisatin
$C_{17}H_{14}O_7$=Aflatoxins
$C_{17}H_{14}O_8$=Aflatoxins, Trichione
$C_{17}H_{15}ClO_7$=Malvidin
$C_{17}H_{15}NO_2$=Aporphine alkaloids
$C_{17}H_{15}O_7^+$=Malvidin
$C_{17}H_{16}O_5$=Combretastatins, Parviflorin, Variabilin
$C_{17}H_{16}O_7$=Fomentariol
$C_{17}H_{17}ClO_2$=Pterulinic acid and pterulones
$C_{17}H_{17}ClO_6$=Griseofulvin
$C_{17}H_{17}NO_2$=Aporphine alkaloids
$C_{17}H_{17}NO_3$=Proaporphine alkaloids
$C_{17}H_{17}NO_5$=Amaryllidaceae alkaloids
$C_{17}H_{18}N_2O_5$=Miraxanthins
$C_{17}H_{18}N_4O_7$=Riboflavin(e)
$C_{17}H_{18}O_5$=Notholaenic acid, Torosachrysone
$C_{17}H_{19}ClO_4$=Strobilurins
$C_{17}H_{19}NO_2$=Elaeocarpus alkaloids
$C_{17}H_{19}NO_3$=Amaryllidaceae alkaloids, Benzyl(tetrahydro)isoquinoline alkaloids, Erythrina alkaloids, Morphinan alkaloids, Piperine, Proaporphine alkaloids, Scopolamine
$C_{17}H_{19}NO_4$=Amaryllidaceae alkaloids, Cephalotaxus alkaloids
$C_{17}H_{19}NO_5$=Amaryllidaceae alkaloids
$C_{17}H_{20}N_4O_6$=Riboflavin(e)
$C_{17}H_{20}O_4$=Strobilurins
$C_{17}H_{20}O_5$=Proazulenes
$C_{17}H_{20}O_6$=Mycophenolic acid, PR toxin
$C_{17}H_{20}O_8$=Psilotin
$C_{17}H_{21}NO_2$=Apoatropine, Erythrina alkaloids
$C_{17}H_{21}NO_3$=Amaryllidaceae alkaloids, Trichostatins
$C_{17}H_{21}NO_4$=Cocaine, Scopolamine
$C_{17}H_{21}NO_5$=PR toxin, Scopolamine
$C_{17}H_{21}NO_6$=PR toxin
$C_{17}H_{21}N_4O_9P$=Riboflavin(e)
$C_{17}H_{22}ClNO_4$=Cocaine
$C_{17}H_{22}ClN_9O_2$=Palauamine
$C_{17}H_{22}N_2O_3$=Trichostatins
$C_{17}H_{22}N_2O_{10}S_2$=Glucosinolates
$C_{17}H_{22}O_2$=Cicu(to)toxin, Flavidulols
$C_{17}H_{22}O_3$=Podocarpic acid
$C_{17}H_{22}O_4$=Oudemansins
$C_{17}H_{22}O_5$=Germacranolides, Matricin, Naematolin, PR toxin
$C_{17}H_{22}O_6$=PR toxin
$C_{17}H_{22}O_8$=Tric(h)othecenes
$C_{17}H_{23}BrO_3$=Halogen compounds
$C_{17}H_{23}NO$=Navenone
$C_{17}H_{23}NO_3$=Atropine, Hyoscyamine, Mesembrine alkaloids
$C_{17}H_{23}N_3O_2$=Indolactams
$C_{17}H_{24}N_4O_9$=Tubercidin
$C_{17}H_{24}O$=Falcarinol
$C_{17}H_{24}O_2$=Falcarinol
$C_{17}H_{24}O_4$=Scirpenols, Tric(h)othecenes
$C_{17}H_{24}O_5$=Naematolin, Scirpenols
$C_{17}H_{24}O_6$=Scirpenols, Tric(h)othecenes
$C_{17}H_{24}O_7$=Tric(h)othecenes
$C_{17}H_{24}O_9$=Dihydrocaffeic acid, Sinapyl alcohol
$C_{17}H_{24}O_{10}$=Genipin, Loganin, Secoiridoids
$C_{17}H_{24}O_{11}$=Monotropein, Secoiridoids
$C_{17}H_{25}FN_8O_5$=Fluoroorganic Natural Products
$C_{17}H_{25}NO$=Histrionicotoxins
$C_{17}H_{25}NO_3$=Lythraceae alkaloids, Variotin
$C_{17}H_{26}N_8O_5$=Blasticidins
$C_{17}H_{26}O_2$=Sterpuranes
$C_{17}H_{26}O_3$=Falcarinol, Fecapentaenes
$C_{17}H_{26}O_4$=Gingerol, Phytube-rin
$C_{17}H_{26}O_5$=Botrydial
$C_{17}H_{26}O_6$=Naematolin
$C_{17}H_{26}O_{10}$=Loganin
$C_{17}H_{26}O_{11}$=Harpagide
$C_{17}H_{26}O_{12}$=Lamiide
$C_{17}H_{28}N_2O_7$=Lascivol
$C_{17}H_{28}N_8O_6$=Blasticidins
$C_{17}H_{28}O_2$=Ambrette seed oil
$C_{17}H_{28}O_3$=Polygodial
$C_{17}H_{28}O_5$=Qinghaosu
$C_{17}H_{29}N_3O_6$=Propioxatins
$C_{17}H_{30}O$=Civet(t)one, Faranal, Matsuone
$C_{17}H_{30}O_2$=Scales
$C_{17}H_{31}N_3O_2$=Palustrine
$C_{17}H_{32}N_2O_6S$=Lincomycins
$C_{17}H_{32}O$=Flies, Trogodermal
$C_{17}H_{32}O_4$=Roccellic acid
$C_{17}H_{34}O$=Leaf beetles
$C_{17}H_{34}O_2$=Bugs, Flies
$C_{17}H_{36}O$=Bugs, Diprionol
$C_{17}H_{37}N_7O_3$=15-Deoxyspergualin
$C_{17}H_{37}N_7O_4$=Spergualin

$\mathbf{C_{18}}H_6Br_8O_6$=Halogen compounds
$C_{18}H_8O_8$=Thelephoric acid
$C_{18}H_9NO_8$=Lamellarins
$C_{18}H_9N_3O$=Shermilamines
$C_{18}H_9N_3O_2$=Meridine
$C_{18}H_{10}N_2O_4$=Kinamycins
$C_{18}H_{10}O_7$=Cycloleucomelone
$C_{18}H_{10}O_9$=Hydnuferrugin, hydnuferruginin, Variegatic acid
$C_{18}H_{11}ClO_7$=Austocystins
$C_{18}H_{11}NO_3$=Nostodione A
$C_{18}H_{12}BrN_3O_2S$=Discorhabdins
$C_{18}H_{12}N_2O_2$=Xanthocillin
$C_{18}H_{12}N_2O_3$=Xanthocillin
$C_{18}H_{12}N_2O_4$=Xanthocillin
$C_{18}H_{12}O_4$=Polyporic acid, Xerulin
$C_{18}H_{12}O_5$=Pulvinic acid
$C_{18}H_{12}O_6$=Atromentin, Grevillins, Sterigmatocystin
$C_{18}H_{12}O_7$=Austocystins, Grevillins
$C_{18}H_{12}O_8$=Grevillins, Leprocybe pigments, Xerocomic acid
$C_{18}H_{12}O_9$=Gomphidic acid, Leprocybe pigments, Norstictic acid, Variegatic acid
$C_{18}H_{12}O_{10}$=Hydnuferrugin, hydnuferruginin, Salazinic acid
$C_{18}H_{13}Br_2N_3O_2$=Discorhabdins
$C_{18}H_{13}NO_8$=Aristolochic acids
$C_{18}H_{14}BrN_3O_2$=Discorhabdins
$C_{18}H_{14}BrN_3O_2S$=Discorhabdins
$C_{18}H_{14}ClN_3O_2S$=Discorhabdins
$C_{18}H_{14}Cl_2O_6$=Argopsin
$C_{18}H_{14}O_2$=Xerulin
$C_{18}H_{14}O_6$=Sterigmatocystin
$C_{18}H_{14}O_8$=Psoromic acid
$C_{18}H_{15}ClO_6$=Pannarin
$C_{18}H_{16}N_2O_8$=Betanidin
$C_{18}H_{16}N_2O_{10}$=Fly agaric pigments
$C_{18}H_{16}O_2$=Cinnamic acid, Xerulin
$C_{18}H_{16}O_4$=Combretastatins, Truxillic and truxinic acids
$C_{18}H_{16}O_6$=Americanin A
$C_{18}H_{16}O_7$=Isousnic acid, Usnic acid
$C_{18}H_{16}O_8$=Irigenin, Rosmarinic acid, Trametin
$C_{18}H_{17}ClO_6$=Radicicol
$C_{18}H_{17}NO_2$=Aporphine alkaloids, Clausenamide
$C_{18}H_{17}NO_3$=Proaporphine alkaloids
$C_{18}H_{17}NO_4$=Cularine alkaloids
$C_{18}H_{18}N_2O_2$=Nigellidine
$C_{18}H_{18}N_2O_9$=Fly agaric pigments
$C_{18}H_{18}O_2$=Estrogens, Storax, styrax oil (resin)
$C_{18}H_{18}O_4$=Lignans
$C_{18}H_{18}O_6$=Dermochrysone, Tricholomenyns
$C_{18}H_{18}O_8$=Lepraric acid
$C_{18}H_{19}NO_2$=Aporphine alkaloids
$C_{18}H_{19}NO_3$=Clausenamide, Erythrina alkaloids, Morphinan alkaloids, Proaporphine alkaloids
$C_{18}H_{19}NO_4$=Cephalotaxus alkaloids, Cularine alkaloids, Strobilurins
$C_{18}H_{19}NO_5$=Amaryllidaceae alkaloids
$C_{18}H_{20}ClN_3O_5S_2$=Sporidesmins
$C_{18}H_{20}ClN_3O_6S_2$=Sporidesmins
$C_{18}H_{20}ClN_3O_6S_3$=Sporidesmins
$C_{18}H_{20}ClN_3O_6S_4$=Sporidesmins
$C_{18}H_{20}N_6O_6S$=Clathridine
$C_{18}H_{20}O_2$=Estrogens
$C_{18}H_{20}O_4$=Tricholomenyns
$C_{18}H_{20}O_5$=Combretastatins, Lentinellic acid
$C_{18}H_{20}O_6$=Byssochlamic acid, Combretastatins
$C_{18}H_{20}O_8$=Nagilactone C
$C_{18}H_{21}NO_3$=Erythrina alkaloids, Morphinan alkaloids
$C_{18}H_{21}NO_4$=Amaryllidaceae alkaloids, Cephalotaxus alkaloids, Erythrina alkaloids
$C_{18}H_{21}NO_5$=Amaryllidaceae alkaloids, Stemona alkaloids
$C_{18}H_{21}N_3$=Elaeocarpus alkaloids
$C_{18}H_{22}N_2O_2$=Roquefortines
$C_{18}H_{22}N_2O_4$=Quinocarcin
$C_{18}H_{22}O_2$=Estrogens
$C_{18}H_{22}O_4$=Nordihydroguaiaretic acid
$C_{18}H_{22}O_5$=Zearalenone
$C_{18}H_{22}O_{11}$=Asperuloside
$C_{18}H_{22}O_{11}S$=Asperuloside
$C_{18}H_{23}ClO_5$=Oudemansins
$C_{18}H_{23}NO_2$=Erythrina alkaloids
$C_{18}H_{23}NO_4$=Amaryllidaceae alkaloids
$C_{18}H_{23}NO_5$=Amaryllidaceae alkaloids, Senecionine
$C_{18}H_{23}NO_6$=Jacobine
$C_{18}H_{23}N_5O_9$=Herbicidins
$C_{18}H_{24}N_2O_5S$=WS75624-A and B
$C_{18}H_{24}N_4O_{18}$=3-Nitropropanoic acid
$C_{18}H_{24}O_2$=Estrogens, Neem tree
$C_{18}H_{24}O_3$=Estrogens
$C_{18}H_{24}O_5$=Oudemansins
$C_{18}H_{24}O_8$=Tric(h)othecenes
$C_{18}H_{24}O_{12}$=Asperuloside
$C_{18}H_{24}O_{12}S$=Asperuloside
$C_{18}H_{25}NO_2$=Lobelia alkaloids
$C_{18}H_{25}NO_4$=Coronatine
$C_{18}H_{25}NO_5$=Senecionine
$C_{18}H_{25}NO_6$=Jacobine, Retrorsine
$C_{18}H_{25}NO_7$=Monocrotaline

$C_{18}H_{25}N_3O_2$=Imidazole alkaloids
$C_{18}H_{26}ClNO_6$=Jacobine
$C_{18}H_{26}O_5$=Zearalenone
$C_{18}H_{27}NO_2$=Lobelia alkaloids
$C_{18}H_{27}NO_3$=Capsaicin
$C_{18}H_{27}NO_5$=Platyphylline
$C_{18}H_{27}NO_7$=Jacobine
$C_{18}H_{28}O_5$=Cibaric acid
$C_{18}H_{29}NO_2$=Lobelia alkaloids
$C_{18}H_{30}O_2$=Calendula oil, Chaulmoogric acid, Linolenic acid, Polyynes, Scales
$C_{18}H_{30}O_3$=F-acids, Juvenile hormones
$C_{18}H_{30}O_5$=Gloeosporone
$C_{18}H_{30}O_6$=Thromboxanes
$C_{18}H_{31}FO_2$=Fluoroorganic Natural Products
$C_{18}H_{31}N_3O_3$=Palustrine
$C_{18}H_{31}N_3O_6$=Propioxatins
$C_{18}H_{32}CaN_2O_{10}$=Pantothenic acid
$C_{18}H_{32}O_2$=Chaulmoogric acid, Linoleic acid, Malvalic acid
$C_{18}H_{32}O_3$=Vernolic acid
$C_{18}H_{32}O_5$=Aspicilin
$C_{18}H_{32}O_{16}$=Gentianose, Kestoses, Raffinose
$C_{18}H_{33}FO_2$=Fluoroacetic acid, Fluoroorganic Natural Products
$C_{18}H_{33}NO_6$=Broussonetine G
$C_{18}H_{34}N_2O_2$=Jietacin A
$C_{18}H_{34}N_2O_6S$=Lincomycins
$C_{18}H_{34}O_2$=Petroselinic acid
$C_{18}H_{34}O_3$=Ricinoleic acid
$C_{18}H_{35}FO_2$=Fluoroorganic Natural Products
$C_{18}H_{35}FO_4$=Fluoroorganic Natural Products
$C_{18}H_{35}NO$=Cerebrodienes, Oleamide
$C_{18}H_{36}N_4O_{11}$=Kanamycins
$C_{18}H_{36}O_4$=Hydroxyfatty acids
$C_{18}H_{37}NO_2$=Sphingosine
$C_{18}H_{37}N_5O_9$=Kanamycins
$C_{18}H_{37}N_5O_{10}$=Kanamycins
$C_{18}H_{38}N_2$=Harmonine
$C_{18}H_{39}NO_2$=Sphingosine
$C_{18}H_{39}NO_3$=Sphingosine

$C_{19}H_{10}O_5$=Phenalenones
$C_{19}H_{12}Br_2N_4$=Topsentins
$C_{19}H_{12}O_2$=Phenalenones
$C_{19}H_{12}O_3$=Phenalenones
$C_{19}H_{12}O_4$=Phlebiarubrone
$C_{19}H_{12}O_6$=Dicoumarol
$C_{19}H_{13}BrN_4$=Topsentins
$C_{19}H_{13}ClO_6$=Austocystins
$C_{19}H_{14}NO_4^+$=Protoberberine alkaloids
$C_{19}H_{14}N_2O_2$=Xanthocillin
$C_{19}H_{14}N_2O_2S$=Xanthocillin
$C_{19}H_{14}N_4$=Corrin(s)
$C_{19}H_{14}O_5$=Vulp(in)ic acid
$C_{19}H_{14}O_6$=Sterigmatocystin
$C_{19}H_{14}O_7$=Austrocorticin, Dermochrysone
$C_{19}H_{14}O_8$=Leprocybe pigments
$C_{19}H_{14}O_9$=Stictic acid
$C_{19}H_{16}N_2$=Sempervirine
$C_{19}H_{16}N_2O_3$=Darlucins
$C_{19}H_{16}O_4$=Curcuminoids, Moracin(s)
$C_{19}H_{16}O_6$=Sterigmatocystin
$C_{19}H_{16}O_8$=Sakyomycins
$C_{19}H_{16}O_{10}$=Indian yellow
$C_{19}H_{17}ClO_8$=Chloroatranorin
$C_{19}H_{17}NO_3$=Acridones
$C_{19}H_{17}NO_4$=Pavine and isopavine alkaloids, Protoberberine alkaloids
$C_{19}H_{18}N_2O_5$=Anthocerodiazonin
$C_{19}H_{18}O_5$=Geiparvarin
$C_{19}H_{18}O_6$=Atrovenetin
$C_{19}H_{18}O_8$=Atranorin, Sakyomycins, Trichione
$C_{19}H_{19}NO_4$=Aporphine alkaloids, Morphinan alkaloids, Pavine and isopavine alkaloids, Protoberberine alkaloids
$C_{19}H_{19}N_7O_6$=Folic acid
$C_{19}H_{20}NO_4^+$=Dibenzo[*b,g*]pyrrocoline alkaloids
$C_{19}H_{20}N_2O_3$=Darlucins
$C_{19}H_{20}N_2O_6$=Cinnabarin
$C_{19}H_{20}O_5$=Geiparvarin
$C_{19}H_{20}O_6$=Coleons, Cynaropicrin
$C_{19}H_{20}O_8$=Placodiolic acid
$C_{19}H_{20}O_{10}$=Khellin
$C_{19}H_{21}NO_2$=Aporphine alkaloids
$C_{19}H_{21}NO_3$=Aporphine alkaloids, Morphinan alkaloids, Proaporphine alkaloids
$C_{19}H_{21}NO_4$=Aporphine alkaloids, Cularine alkaloids, Morphinan alkaloids, Pavine and isopavine alkaloids, 1-Phenyltetrahydroisoquinoline alkaloids, Proaporphine alkaloids, Protoberberine alkaloids
$C_{19}H_{22}ClN_3O_6S$=Sporidesmins
$C_{19}H_{22}N_2O$=Cinchonaminal, Cinchonidine, Cinchonine, Corynantheine alkaloids, Eburnamonine
$C_{19}H_{22}N_2O_2$=Sarpagine, Wieland-Gumlich aldehyde
$C_{19}H_{22}N_2O_3$=Gelsemine
$C_{19}H_{22}N_4$=Corrin(s)
$C_{19}H_{22}O_2$=Cannabinoids
$C_{19}H_{22}O_5$=Acanthostral, Hericenones
$C_{19}H_{22}O_6$=Cynaropicrin, Gibberellic acid, Nagilactone C, Strigol
$C_{19}H_{22}O_7$=Nagilactone C
$C_{19}H_{23}NO_3$=Cephalotaxus alkaloids, Erythrina alkaloids, Morphinan alkaloids
$C_{19}H_{23}NO_4$=Benzyl(tetrahydro)isoquinoline alkaloids, Cephalotaxus alkaloids, Hirsutin(e), Morphinan alkaloids
$C_{19}H_{23}N_3O_2$=Ergot alkaloids
$C_{19}H_{23}N_7O_6$=Tetrahydrofolic acid
$C_{19}H_{24}N_2$=Iboga alkaloids
$C_{19}H_{24}N_2O$=Cinchonamine
$C_{19}H_{24}O_3$=Maracenins
$C_{19}H_{24}O_5$=Tric(h)othecenes, Umbelliferone
$C_{19}H_{24}O_6$=Nagilactones
$C_{19}H_{24}O_7$=Nagilactones
$C_{19}H_{24}O_9$=Nivalenol
$C_{19}H_{25}ClO_3$=Maracenins
$C_{19}H_{25}NO$=Histrionicotoxins
$C_{19}H_{26}ClNO_3$=Virantmycin
$C_{19}H_{26}N_2$=Quebrachamine
$C_{19}H_{26}N_4S_4$=Tantazoles
$C_{19}H_{26}O_2$=Androgens, Cannabinoids
$C_{19}H_{26}O_5$=Humulone
$C_{19}H_{26}O_6$=Plagiochilins, Scirpenols, Tric(h)othecenes
$C_{19}H_{26}O_7$=Scirpenols, Tric(h)othecenes
$C_{19}H_{26}O_8$=Neosolaniol, Scirpenols, Tric(h)othecenes
$C_{19}H_{27}NO$=2-Heptyl-1-hydroxy-4(1*H*)-quinolinone, Histrionicotoxins
$C_{19}H_{27}NO_3$=Ipecac alkaloids
$C_{19}H_{27}NO_6$=Senkirkine
$C_{19}H_{27}N_5O_3$=Argiopinins
$C_{19}H_{27}O_9P$=Fostriecin
$C_{19}H_{28}O$=Steroid odorants
$C_{19}H_{28}O_2$=Androgens
$C_{19}H_{28}O_4$=Atractyligenin, Flexilin
$C_{19}H_{28}O_7$=Punctaporonins
$C_{19}H_{29}NO$=Gephyrotoxin
$C_{19}H_{29}NO_2$=Salamander steroid alkaloids
$C_{19}H_{30}N_8O_9$=Mildiomycin
$C_{19}H_{30}O$=Steroid odorants
$C_{19}H_{30}O_2$=Androgens
$C_{19}H_{31}NO_2$=Salamander steroid alkaloids
$C_{19}H_{32}O_4$=Lichesterinic acid, Pironetin, Protolichester(in)ic acid
$C_{19}H_{32}O_6$=Fungi
$C_{19}H_{33}NO$=Histrionicotoxins, Salamander steroid alkaloids
$C_{19}H_{33}NO_2$=Dysidazirine, Pumiliotoxins
$C_{19}H_{33}NO_3$=Pumiliotoxins
$C_{19}H_{34}N_8O_8$=Streptothricins
$C_{19}H_{34}O_2$=Malvalic acid
$C_{19}H_{34}O_4$=Roccellaric acid
$C_{19}H_{36}N_2O_6S$=Lincomycins
$C_{19}H_{36}O_2$=Lactobacillic acid
$C_{19}H_{37}N_5O_7$=Sisomicin
$C_{19}H_{38}N_2O_4$=Poecillanosine
$C_{19}H_{38}O$=Disparlure
$C_{19}H_{38}O_2$=Pristane
$C_{19}H_{39}NO_3$=Penaresidins
$C_{19}H_{40}$=Butterflies, Pristane
$C_{19}H_{40}O_3$=Batyl alcohol

$C_{20}H_9ClO_8$=Spiroxins
$C_{20}H_{10}O_4$=4,9-Dihydroxy-3,10-perylendione
$C_{20}H_{10}O_5$=Bulgarein
$C_{20}H_{10}O_6$=Bulgarein
$C_{20}H_{11}N_3O_2$=Arcyria pigments
$C_{20}H_{11}N_3O_3$=Arcyria pigments
$C_{20}H_{11}N_3O_4$=Arcyria pigments
$C_{20}H_{11}N_3O_6$=Arcyria pigments
$C_{20}H_{12}O_5$=Halenaquinone, Hypoxylone
$C_{20}H_{13}BrN_4O_2$=Topsentins
$C_{20}H_{13}Br_4ClN_4O$=Chartelline A
$C_{20}H_{13}N_3$=Nitramarine
$C_{20}H_{13}N_3O_2$=Arcyria pigments
$C_{20}H_{13}N_3O_3$=Arcyria pigments, Violacein
$C_{20}H_{13}N_3O_4$=Arcyria pigments
$C_{20}H_{13}N_3O_8$=Ommochromes
$C_{20}H_{14}Cl_4O_6$=Russuphelins
$C_{20}H_{14}NO_4^+$=Benzo[*c*]phenanthridine alkaloids
$C_{20}H_{14}N_4$=Porphine, Quaterpyridine
$C_{20}H_{14}N_4O_2$=Topsentins
$C_{20}H_{14}O_4$=Halenaquinone, Phenalenones
$C_{20}H_{14}O_5$=Coumestrol, Halenaquinone
$C_{20}H_{14}O_7$=Preussomerins
$C_{20}H_{15}N_3O_{11}S$=Ommochromes
$C_{20}H_{16}FeN_4O_6S_2$=Ferrithiocin
$C_{20}H_{16}N_2O_2$=Emerin
$C_{20}H_{16}N_2O_4$=Camptothecin
$C_{20}H_{16}O_6$=Rotenoids, Wortmannin
$C_{20}H_{16}O_7$=Averufin
$C_{20}H_{16}O_8$=Sterigmatocystin
$C_{20}H_{17}N_2O_3^+$=Perloline
$C_{20}H_{17}N_3O_2S$=Lissoclinum Alkaloids
$C_{20}H_{18}BrClN_4O_2$=Securamines
$C_{20}H_{18}ClNO_6$=Ochratoxins
$C_{20}H_{18}ClNO_7$=Ochratoxins
$C_{20}H_{18}NO_4^+$=Protoberberine alkaloids
$C_{20}H_{18}N_2O_2$=Streptindole
$C_{20}H_{18}N_2O_4$=Perloline
$C_{20}H_{18}O_3$=Cortisalin
$C_{20}H_{18}O_4$=Phaseoline
$C_{20}H_{18}O_5$=Curcuminoids, Glyceollins
$C_{20}H_{18}O_6$=Sesamol
$C_{20}H_{18}O_7$=Sesamol
$C_{20}H_{18}O_9$=Frangulins
$C_{20}H_{19}Br_2ClN_4O$=Securamines
$C_{20}H_{19}ClN_4O_2$=Securamines
$C_{20}H_{19}NO_3$=Acridones
$C_{20}H_{19}NO_4$=Ficine
$C_{20}H_{19}NO_5$=Alkaloidal dyes, Benzo[*c*]phenanthridine alkaloids, Protopine alkaloids
$C_{20}H_{19}NO_6$=Ochratoxins, Rhoeadine alkaloids
$C_{20}H_{19}N_3O_2$=Tilivalline
$C_{20}H_{20}BrClN_4O$=Securamines
$C_{20}H_{20}NO_4^+$=Protoberberine alkaloids
$C_{20}H_{20}N_2O_2$=Peronatins
$C_{20}H_{20}N_2O_3$=Cyclopiazonic acid
$C_{20}H_{20}N_2O_4$=Alstonia alkaloids
$C_{20}H_{20}O_2$=Serpentene
$C_{20}H_{20}O_6$=Conidendrin, Lignans
$C_{20}H_{21}NO_4$=Aporphine alkaloids, Benzyl(tetrahydro)isoquinoline alkaloids, Protoberberine alkaloids
$C_{20}H_{21}NO_5$=Protoberberine alkaloids
$C_{20}H_{22}N_2O_2$=Alstonia alkaloids, Gelsemine
$C_{20}H_{22}N_2O_3$=Alstonia alkaloids
$C_{20}H_{22}N_2O_4$=Picroroccellin
$C_{20}H_{22}N_2O_8$=Azotochelin
$C_{20}H_{22}N_4O_9$=Riboflavin(e)
$C_{20}H_{22}N_6O_4S_2$=Nostocyclamide
$C_{20}H_{22}O_6$=Coleons, Columbin, Lignans, Miroestrol
$C_{20}H_{22}O_7$=Saudin, Wikstromol
$C_{20}H_{22}O_8$=Salicyl alcohol
$C_{20}H_{22}O_9$=Gentisic acid
$C_{20}H_{22}O_{10}$=Erythrin
$C_{20}H_{23}Cl_3N_2O_2S$=Barbamide

$C_{20}H_{23}NO_4$=Aporphine alkaloids, Cularine alkaloids, Morphinan alkaloids, Pavine and isopavine alkaloids, Phenethyltetrahydroisoquinoline alkaloids, Protoberberine alkaloids
$C_{20}H_{23}NO_5$=Fuligorubin A
$C_{20}H_{23}N_7O_7$=Folinic acid
$C_{20}H_{24}NO_4{}^+$=Aporphine alkaloids, Dibenzo[*b,g*]pyrrocoline alkaloids
$C_{20}H_{24}N_2$=Aristotelia alkaloids
$C_{20}H_{24}N_2O$=Kopsine
$C_{20}H_{24}N_2O_2$=Alstonia alkaloids, Quinidine, Quinine, Sarpagine
$C_{20}H_{24}N_2O_3$=Yohimbine
$C_{20}H_{24}N_2O_7$=Azotochelin, 1-Nitroaknadinine
$C_{20}H_{24}O_3$=Jatrophone
$C_{20}H_{24}O_4$=Crocetin, Jatrophone, Lignans, Nordihydroguaiaretic acid, Pinnatins A–E
$C_{20}H_{24}O_6$=Triptolide
$C_{20}H_{24}O_7$=Olivil
$C_{20}H_{24}O_9$=Ginkgolides, Marmesin
$C_{20}H_{24}O_{10}$=Galapagin, Ginkgolides, Salicortin
$C_{20}H_{24}O_{11}$=Ginkgolides
$C_{20}H_{25}NO_4$=Benzyl(tetrahydro)isoquinoline alkaloids
$C_{20}H_{25}NO_5$=1-Nitroaknadinine
$C_{20}H_{25}N_3O$=Prodigiosin
$C_{20}H_{25}N_5O_{10}$=Nikkomycins
$C_{20}H_{26}ClN_3O_6S_2$=Sporidesmins
$C_{20}H_{26}NO_3P$=PB toxin
$C_{20}H_{26}N_2$=Aristotelia alkaloids
$C_{20}H_{26}N_2O$=Iboga alkaloids
$C_{20}H_{26}N_2O_2$=Ajmaline, Cinchonamine
$C_{20}H_{26}N_2O_4$=Gelsemine
$C_{20}H_{26}N_2O_7$=Amicoumacins
$C_{20}H_{26}O$=3,4-Didehydroretinal
$C_{20}H_{26}O_3$=Jatrophatrione, Pimaranes, Taxodione
$C_{20}H_{26}O_4$=Cannabinoids, Carnosol, Crinipellins, Pimaranes
$C_{20}H_{26}O_6$=Coleons, Enmein, Pimaranes
$C_{20}H_{26}O_7$=Cnicin
$C_{20}H_{26}O_8$=Specionin
$C_{20}H_{27}NO_3$=Hetisine
$C_{20}H_{27}NO_{11}$=Cyanogenic Glycosides
$C_{20}H_{28}N_2O_8$=Amicoumacins
$C_{20}H_{28}O$=3,4-Didehydroretinal, Retinal
$C_{20}H_{28}O_2$=Cembranolides, Cyathanes, Spruceanol
$C_{20}H_{28}O_3$=Cafestol, Cinerins, Coffee flavor, Hardwickiic acid, Lathyranes, Petasin, Retinol, Rosanes, Sarcodonin A, Taxodione
$C_{20}H_{28}O_4$=Callicarpone, Cannabinoids, Lathyranes, Marrubiin, Prostaglandins, Sordarin, Spongianes
$C_{20}H_{28}O_5$=Humulone, Ingenol, Trichaurantins
$C_{20}H_{28}O_6$=Resiniferonol, Tiglianes
$C_{20}H_{28}O_{13}$=Primeverin
$C_{20}H_{29}NO_4$=Saussureamines
$C_{20}H_{29}N_3O_7$=Amicoumacins
$C_{20}H_{30}$=Cembranoids
$C_{20}H_{30}N_4O_6$=HC toxins
$C_{20}H_{30}O$=Retinol
$C_{20}H_{30}O_2$=Abietic acid, Eicosapentaenoic acid, Palustric acid, Sacculatanes, Sarcophytol A, Trachylobanes
$C_{20}H_{30}O_3$=Cyathanes, Leukotrienes, Rosanes, Spatol, Steviol, Trachylobanes
$C_{20}H_{30}O_4$=Fungi, Lathyranes, Prostaglandins, Schizostatin, Spongianes
$C_{20}H_{30}O_5$=Prostacyclins, Prostaglandins
$C_{20}H_{30}O_6$=Tric(h)othecenes
$C_{20}H_{30}O_8$=Pterosins
$C_{20}H_{32}$=Abieta-7,13-diene, Casbene, Cembrenes, Kauranes, Pimaranes, Rosanes, Trachylobanes
$C_{20}H_{32}N_6O_{12}S_2$=Glutathione
$C_{20}H_{32}O$=Dolabellanes, Rippertenol, Sarcophytol A, Trachylobanes
$C_{20}H_{32}O_2$=Arachidonic acid, Trachylobanes
$C_{20}H_{32}O_3$=Aplysi…, Asperdiol, Cyathanes, Hydroxyfatty acids, Labdanes, Pleuromutilin
$C_{20}H_{32}O_4$=Hepoxilins, Leukotrienes, Portulal, Prostaglandins
$C_{20}H_{32}O_5$=Grayanotoxins, Lipoxins, Portulal, Prostacyclins, Thromboxanes
$C_{20}H_{32}O_6$=Thromboxanes
$C_{20}H_{33}N_3$=Panamine
$C_{20}H_{34}$=Gibberellins, Kauranes
$C_{20}H_{34}O$=Geranylgeraniol, Verticillol
$C_{20}H_{34}O_2$=Dihomo-γ-linolenic acid, Labdanes, Plaunotol
$C_{20}H_{34}O_3$=Aplysi…, Labdanes, Pimaranes, Vinigrol
$C_{20}H_{34}O_4$=Aphidicolin, Ebelactones, Zoapatanol
$C_{20}H_{34}O_5$=Prostaglandins
$C_{20}H_{34}O_6$=Grayanotoxins, Thromboxanes
$C_{20}H_{35}BrO_2$=Aplysi…
$C_{20}H_{35}N_3$=Ormosanine
$C_{20}H_{36}N_4O_9$=Aerobactin
$C_{20}H_{36}O_2$=Labdanes
$C_{20}H_{36}O_3$=Labdanum (absolute/oil/resinoid)
$C_{20}H_{36}O_5$=Prostaglandins
$C_{20}H_{37}N_3O_{13}$=Hygromycins
$C_{20}H_{38}N_2O_6S$=Lincomycins
$C_{20}H_{38}O$=Phytol
$C_{20}H_{38}O_2$=Labdanum (absolute/oil/resinoid), Rapeseed oil
$C_{20}H_{39}FO_2$=Fluoroorganic Natural Products
$C_{20}H_{39}NO_6$=Sphingofungins
$C_{20}H_{40}O$=Phytol
$C_{20}H_{40}O_2$=Arach(id)ic acid

$\mathbf{C_{21}}H_{15}N_3O_5$=Halenaquinone
$C_{21}H_{16}O_9$=Mitorubrin
$C_{21}H_{17}N_5O_2$=Grossularines
$C_{21}H_{18}Cl_2O_3$=Nostoclides
$C_{21}H_{18}NO_4{}^+$=Benzo[*c*]phenanthridine alkaloids
$C_{21}H_{18}N_4O_2S$=Shermilamines
$C_{21}H_{18}O_7$=Mitorubrin
$C_{21}H_{18}O_8$=Daunorubicin, Mitorubrin
$C_{21}H_{18}O_9$=Mitorubrin
$C_{21}H_{19}ClO_3$=Nostoclides
$C_{21}H_{20}N_4O_2$=4-(Hydroxymethyl)benzenediazonium ion
$C_{21}H_{20}O_3$=Cortisalin
$C_{21}H_{20}O_6$=Cabenegrins, Curcuminoids
$C_{21}H_{20}O_7$=Solorinic acid
$C_{21}H_{20}O_8$=Peltatin
$C_{21}H_{20}O_9$=Daidzin, Frangulins
$C_{21}H_{20}O_{10}$=Galangin
$C_{21}H_{20}O_{11}$=Maritimetin, Quercetin
$C_{21}H_{20}O_{12}$=Quercetin
$C_{21}H_{20}O_{13}$=Gossypetin
$C_{21}H_{21}ClN_2O$=Indole alkaloids from cyanobacteria
$C_{21}H_{21}ClO_{11}$=Id(a)ein
$C_{21}H_{21}NO_4$=Spirobenzylisoquinoline alkaloids
$C_{21}H_{21}NO_6$=Phthalide-isoquinoline alkaloids, Rhoeadine alkaloids
$C_{21}H_{21}O_{10}{}^+$=Callistephin, Fragarin
$C_{21}H_{21}O_{11}{}^+$=Chrysanthemin, Id(a)ein
$C_{21}H_{22}NO_4{}^+$=Protoberberine alkaloids
$C_{21}H_{22}N_2O_2$=Strychnine
$C_{21}H_{22}N_2O_3$=Cathenamine, Strychnine
$C_{21}H_{22}N_2O_4$=Alstonia alkaloids
$C_{21}H_{22}N_2O_8$=Tetracyclines
$C_{21}H_{22}O_2$=Endiandric acids
$C_{21}H_{22}O_4$=Licochalcones
$C_{21}H_{22}O_5$=Fungi, Pleurotin
$C_{21}H_{22}O_6$=Wortmannin
$C_{21}H_{22}O_8$=Pseudocyphellarins
$C_{21}H_{23}Br_4N_3O_6$=Aerothionin
$C_{21}H_{23}ClN_2$=Hapalindoles, Indole alkaloids from cyanobacteria
$C_{21}H_{23}NO_5$=Morphinan alkaloids, Protopine alkaloids, Tenellin
$C_{21}H_{24}N_2O_2$=Apovincamine, Catharanthine, Tabersonine
$C_{21}H_{24}N_2O_3$=Ajmalicine, Alstonia alkaloids, Geissoschizine, Wieland-Gumlich aldehyde
$C_{21}H_{24}O_3$=Piptoporic acid
$C_{21}H_{24}O_5$=Kadsurenone
$C_{21}H_{24}O_6$=Purpactins
$C_{21}H_{24}O_7$=Visnadin
$C_{21}H_{24}O_8$=Pseudocyphellarins
$C_{21}H_{24}O_{10}$=Phloretin
$C_{21}H_{25}NO_4$=Aporphine alkaloids, Pavine and isopavine alkaloids, Protoberberine alkaloids, Tetrahydropalmatine
$C_{21}H_{25}NO_5$=Phenethyltetrahydroisoquinoline alkaloids
$C_{21}H_{25}N_3O_2$=Tryprostatins
$C_{21}H_{26}N_2O_2$=Iboga alkaloids, Kopsine
$C_{21}H_{26}N_2O_3$=Iboga alkaloids, Vincamine, Yohimbine
$C_{21}H_{26}N_4O_6$=Papiliochrome II
$C_{21}H_{26}N_4O_{10}$=Nikkomycins
$C_{21}H_{26}O_2$=Cannabinoids, Longithorones
$C_{21}H_{26}O_4$=Crocetin, Strobilurins
$C_{21}H_{26}O_6$=Caulerpenyne, Strobilurins
$C_{21}H_{26}O_7$=Clementeins, Strobilurins
$C_{21}H_{26}O_{12}$=Plumeria iridoids
$C_{21}H_{27}NO_4$=Benzyl(tetrahydro)isoquinoline alkaloids, Morphinan alkaloids
$C_{21}H_{27}NO_5$=Morphinan alkaloids, Phenethyltetrahydroisoquinoline alkaloids
$C_{21}H_{27}N_2O_2{}^+$=Sarpagine
$C_{21}H_{27}N_3O_6$=Naphthyridinomycin A, Pistillarine
$C_{21}H_{27}N_7O_{14}P_2$=Nicotinamide coenzymes
$C_{21}H_{28}N_7O_{17}P_3$=Nicotinamide coenzymes
$C_{21}H_{28}O_2$=Avarol, Frondosins
$C_{21}H_{28}O_3$=Cannabinoids
$C_{21}H_{28}O_4$=Boviquinones, Corticosteroids
$C_{21}H_{28}O_5$=Aldosterone, Cinerins, Corticosteroids
$C_{21}H_{29}BrN_2$=Flustramines
$C_{21}H_{29}NO_6$=Ipom(o)ea alkaloids
$C_{21}H_{29}N_7O_{14}P_2$=Nicotinamide coenzymes
$C_{21}H_{30}N_7O_{17}P_3$=Nicotinamide coenzymes
$C_{21}H_{30}O_2$=Avarol, Cannabinoids, Progestogens, Tetrahydrocannabinols, Urushiol(s)
$C_{21}H_{30}O_3$=Ceratiopyrones, Corticosteroids
$C_{21}H_{30}O_4$=Cannabinoids, Corticosteroids
$C_{21}H_{30}O_5$=Corticosteroids, Humulone
$C_{21}H_{30}O_6$=Trichoverroids
$C_{21}H_{30}O_9$=Abscisic acid
$C_{21}H_{31}NO_7$=Monocrotaline
$C_{21}H_{32}N_2$=Apocynaceae steroid alkaloids
$C_{21}H_{32}N_4O_6$=HC toxins
$C_{21}H_{32}N_4O_7$=HC toxins
$C_{21}H_{32}O_2$=Cardol, Urushiol(s)
$C_{21}H_{32}O_3$=Asterosaponins
$C_{21}H_{32}O_4$=Chatancin
$C_{21}H_{32}O_6$=Galbonolides
$C_{21}H_{32}O_{15}$=Monomelittoside
$C_{21}H_{33}NO_3$=Salamander steroid alkaloids
$C_{21}H_{34}O$=Ginkgo extract
$C_{21}H_{34}O_2$=Urushiol(s)
$C_{21}H_{34}O_3$=Asterosaponins
$C_{21}H_{34}O_7$=Phlebiakauranol
$C_{21}H_{35}NO$=Apocynaceae steroid alkaloids, Preussine
$C_{21}H_{35}N_5O_2$=Hebestatis toxins
$C_{21}H_{36}N_2O$=Cribrochalinamine oxide A
$C_{21}H_{36}O_2$=Urushiol(s)
$C_{21}H_{36}O_4$=Ebelactones, Zoapatanol
$C_{21}H_{38}O_6$=Erythronolide B, Rangiformic acid
$C_{21}H_{38}O_7$=Erythronolide B
$C_{21}H_{39}NO_6$=Myriocin
$C_{21}H_{39}N_7O_{12}$=Streptomycin
$C_{21}H_{40}N_8O_6$=Tuftsin
$C_{21}H_{40}O$=Butterflies
$C_{21}H_{40}O_4$=Japan wax
$C_{21}H_{42}O_3$=Batyl alcohol
$C_{21}H_{44}O_3$=Batyl alcohol

$\mathbf{C_{22}}H_{14}N_2O_5$=UK-1
$C_{22}H_{14}N_4O_4$=Lavendamycin

$C_{22}H_{14}O_7$=Hypoxylone
$C_{22}H_{16}O_6$=Resistomycin
$C_{22}H_{16}O_7$=Resistomycin
$C_{22}H_{16}O_{12}$=Fumarprotocetraric acid
$C_{22}H_{17}N_3O_5$=Strobilurins
$C_{22}H_{18}NO_6^+$=Benzo[*c*]phenanthridine alkaloids
$C_{22}H_{18}N_2O_9$=Kinamycins
$C_{22}H_{18}O_7$=Austocystins
$C_{22}H_{18}O_8$=Podophyllotoxin
$C_{22}H_{18}O_{11}$=Epigallocatechin 3-gallate
$C_{22}H_{18}O_{12}$=Tartaric acid
$C_{22}H_{19}N_3O_2$=Shermilamines
$C_{22}H_{20}O_7$=Austocystins, Collybolide
$C_{22}H_{20}O_8$=Austocystins
$C_{22}H_{20}O_{10}$=Granaticins, Vulgamycin
$C_{22}H_{20}O_{13}$=Carminic acid
$C_{22}H_{21}ClN_2O_2S$=Indole alkaloids from cyanobacteria
$C_{22}H_{22}Br_4N_{10}O_2$=Oroidin
$C_{22}H_{22}ClNO_6$=Ochratoxins
$C_{22}H_{22}O_7$=Podophyllotoxin
$C_{22}H_{22}O_8$=Peltatin, Podophyllotoxin
$C_{22}H_{22}O_{11}$=Maritimetin
$C_{22}H_{23}ClN_2O_8$=Tetracyclines
$C_{22}H_{23}ClO_{12}$=Tuberin
$C_{22}H_{23}NO_4$=Spirobenzylisoquinoline alkaloids, Tylophora alkaloids
$C_{22}H_{23}NO_6$=Aureothin
$C_{22}H_{23}NO_7$=Phthalide-isoquinoline alkaloids
$C_{22}H_{23}NO_9$=Azicemicins
$C_{22}H_{23}N_5O_2$=Roquefortines
$C_{22}H_{24}N_2O_4$=Kopsine, Vomicine
$C_{22}H_{24}N_2O_8$=Tetracyclines
$C_{22}H_{24}N_2O_9$=Tetracyclines
$C_{22}H_{24}O_8$=Lophotoxin
$C_{22}H_{24}O_{11}$=Hesperetin
$C_{22}H_{25}NO_6$=Phenethyltetrahydroisoquinoline alkaloids, Rhoeadine alkaloids
$C_{22}H_{25}NO_8$=Pseurotin A
$C_{22}H_{25}N_3O_3$=Fumitremorgins
$C_{22}H_{25}N_3O_4$=Spirotryprostatins
$C_{22}H_{25}N_3O_5$=Fumitremorgins
$C_{22}H_{25}N_5O_2$=Roquefortines
$C_{22}H_{26}Br_2Cl_2N_{10}O_2$=Sceptrin
$C_{22}H_{26}Br_2N_{10}O_2$=Sceptrin
$C_{22}H_{26}N_2O_3$=Corynantheine alkaloids, Geissoschizine
$C_{22}H_{26}N_2O_4$=Iboga alkaloids, Wieland-Gumlich aldehyde
$C_{22}H_{26}N_2O_6S_3$=Leinamycin
$C_{22}H_{26}N_4$=Calycanthine, Pyrroloindole alkaloids
$C_{22}H_{26}O_4$=Cannabinoids
$C_{22}H_{26}O_5$=Calanolides
$C_{22}H_{26}O_6$=Lignans
$C_{22}H_{26}O_8$=Contortin
$C_{22}H_{26}O_{10}$=Aucubin
$C_{22}H_{26}O_{11}$=Aucubin, Monomelittoside
$C_{22}H_{26}O_{12}$=Catalpol
$C_{22}H_{27}NO_2$=Lobelia alkaloids
$C_{22}H_{27}NO_4$=Madindolines, Protoberberine alkaloids
$C_{22}H_{27}NO_5$=Asparagamine A, Phenethyltetrahydroisoquinoline alkaloids, Protopine alkaloids
$C_{22}H_{27}NO_6$=Rhoeadine alkaloids
$C_{22}H_{27}NO_7$=Tirandamycins
$C_{22}H_{27}NO_8$=Tirandamycins
$C_{22}H_{27}N_3O_3$=Tryprostatins
$C_{22}H_{27}N_3O_6$=Fumitremorgins
$C_{22}H_{27}N_7O$=Luciferins
$C_{22}H_{27}N_9O_4$=Distamycins
$C_{22}H_{28}N_2O_3$=Corynantheine alkaloids, Hirsutin(e)
$C_{22}H_{28}N_4O_{10}$=Lampteroflavin
$C_{22}H_{28}O_4$=Crocetin
$C_{22}H_{28}O_5$=Crinipellins
$C_{22}H_{28}O_6$=Gymnocolin, Quassin(oids)
$C_{22}H_{28}O_8$=Gracilins
$C_{22}H_{29}NO_2$=Lobelia alkaloids
$C_{22}H_{29}N_7O_5$=Puromycin
$C_{22}H_{30}N_2O_4$=Alstonia alkaloids
$C_{22}H_{30}N_4O_4$=Tentoxin
$C_{22}H_{30}O_3$=Siccanin, Termites
$C_{22}H_{30}O_4$=Ilimaquinone
$C_{22}H_{30}O_6$=Tiglianes, Trichaurantins
$C_{22}H_{30}O_8$=Valepotriates
$C_{22}H_{31}NO$=Apocynaceae steroid alkaloids
$C_{22}H_{31}NO_2$=Daphniphyllum alkaloids
$C_{22}H_{32}N_2$=Diisocyanoadociane
$C_{22}H_{32}N_4O_5$=Radiosumin
$C_{22}H_{32}N_5O_{19}P$=Thuringiensin
$C_{22}H_{32}O_2$=Grifolin
$C_{22}H_{32}O_3$=Anacardic acid, Wiedendiols
$C_{22}H_{32}O_5$=Ajugarins, Humulone, Sarcodontic acid
$C_{22}H_{32}O_6$=Ajugarins, Spongianes
$C_{22}H_{32}O_7$=Cascarillin
$C_{22}H_{32}O_8$=HT-2 toxin, Tric(h)othecenes, Valepotriates
$C_{22}H_{32}O_{14}$=Dihydrocaffeic acid
$C_{22}H_{33}NO_2$=Apocynaceae steroid alkaloids, Garryine, Veatchine
$C_{22}H_{33}NO_3$=Napelline
$C_{22}H_{33}NO_{20}$=Atisine
$C_{22}H_{34}O_2$=Clupa(no)donic acid
$C_{22}H_{34}O_3$=Ginkgo extract
$C_{22}H_{34}O_4$=Sphenolobanes
$C_{22}H_{34}O_5$=Pleuromutilin
$C_{22}H_{34}O_7$=Forskolin
$C_{22}H_{35}NO_2$=Himbacine
$C_{22}H_{36}N_4O_{13}$=Aerobactin
$C_{22}H_{36}N_6O_{16}P_2$=Agrocin 84
$C_{22}H_{36}O_7$=Grayanotoxins
$C_{22}H_{37}NO_2$=Anandamide
$C_{22}H_{37}N_5O_2$=Hebestatis toxins
$C_{22}H_{37}N_5O_5S_2$=Malformins
$C_{22}H_{38}N_6O_5$=JSTX
$C_{22}H_{38}O_{10}S$=Oreganic acid
$C_{22}H_{42}O_2$=Erucic acid, Flies
$C_{22}H_{42}O_4$=Japan wax
$C_{22}H_{43}NO$=Oleamide
$C_{22}H_{44}O_2$=Behenic acid

$\mathbf{C_{23}}H_{14}O_6$=Sphagnorubins
$C_{23}H_{18}N_6O$=Grossularines
$C_{23}H_{18}O_8$=Tetracenomycins
$C_{23}H_{20}O_{11}$=Tetracenomycins
$C_{23}H_{21}NO_{13}$=Aristolochic acids
$C_{23}H_{22}N_4OS$=Kuanoniamines
$C_{23}H_{22}O_6$=Rotenone, Tephrosin
$C_{23}H_{22}O_7$=Austocystins, Lactucin, Rotenoids, Tephrosin
$C_{23}H_{22}O_8$=Austocystins, Tephrosin
$C_{23}H_{22}O_{10}$=Phrymarolins
$C_{23}H_{23}N_3O_3$=Imidazole alkaloids
$C_{23}H_{24}ClNO_6$=Ochratoxins
$C_{23}H_{24}Cl_2O_7$=Falconensins
$C_{23}H_{24}N_4O_9$=Nocardicins
$C_{23}H_{24}O_2$=Endiandric acids
$C_{23}H_{24}O_7$=Purpactins
$C_{23}H_{24}O_8$=Peltatin, Wortmannin
$C_{23}H_{25}NO_3$=Naphthyl-(tetrahydro)isoquinoline alkaloids
$C_{23}H_{25}NO_4$=Naphthyl-(tetrahydro)isoquinoline alkaloids, Phenoxan, Tylophora alkaloids
$C_{23}H_{25}NO_9$=Azicemicins
$C_{23}H_{25}O_{12}^+$=Oenin
$C_{23}H_{26}N_2O_4$=Brucine
$C_{23}H_{26}O_3$=Ceratiopyrones
$C_{23}H_{26}O_7$=Purpactins
$C_{23}H_{28}O_5$=Cristatic acid
$C_{23}H_{28}O_8$=Armillaria sesquiterpenoids, Pseudolaric acids
$C_{23}H_{29}ClO_4$=Ascochlorin
$C_{23}H_{29}NO_7$=Fusarin C
$C_{23}H_{29}NO_{12}$=Hygromycins
$C_{23}H_{29}N_3$=Manzamines
$C_{23}H_{30}N_2O_4$=Vindoline
$C_{23}H_{30}N_2O_5$=Vindoline
$C_{23}H_{30}O_6$=Armillaria sesquiterpenoids, Citreoviridins
$C_{23}H_{30}O_7$=Asteltoxin, Xylaral
$C_{23}H_{31}NO_6$=Myxopyronins, Stemona alkaloids
$C_{23}H_{31}NO_7$=Mycophenolic acid
$C_{23}H_{32}N_2O_8$=Trichostatins
$C_{23}H_{32}N_6O_{14}$=Polyoxins
$C_{23}H_{32}O_4$=Scutigeral
$C_{23}H_{32}O_6$=Phellodonic acid, Strophanthins
$C_{23}H_{32}O_7$=Tric(h)othecenes, Trichoverroids
$C_{23}H_{33}N_3O_2$=Isocyanides
$C_{23}H_{34}N_3O_{10}P$=Phosphoramidon
$C_{23}H_{34}O_2$=Urushiol(s)
$C_{23}H_{34}O_4$=Digitalis glycosides
$C_{23}H_{34}O_5$=Compactin, Digitalis glycosides
$C_{23}H_{34}O_6$=Ajugarins, Coriolin
$C_{23}H_{34}O_7$=Coriolin
$C_{23}H_{34}O_8$=Strophanthins, Valepotriates
$C_{23}H_{35}NOS$=Curacin A
$C_{23}H_{35}NO_2$=Daphniphyllum alkaloids
$C_{23}H_{36}O_7$=Cryptoporic acids, Pravastatin, Tylosin
$C_{23}H_{37}NO_5S$=Leukotrienes
$C_{23}H_{38}N_2O_5$=Volicitin
$C_{23}H_{38}N_7O_{17}P_3S$=Acetyl-CoA
$C_{23}H_{38}O_2$=Urushiol(s)
$C_{23}H_{39}N_3O_8$=Roseotoxins
$C_{23}H_{39}N_5O_5S_2$=Malformins
$C_{23}H_{40}O_2$=Urushiol(s)
$C_{23}H_{40}O_5$=Denticulatins
$C_{23}H_{41}N_5O_3$=δ-Philanthotoxin
$C_{23}H_{42}N_2$=Apocynaceae steroid alkaloids
$C_{23}H_{44}O_4$=Japan wax
$C_{23}H_{45}N_3O$=Inandenines
$C_{23}H_{45}N_3O_2$=Inandenines
$C_{23}H_{45}N_5O_{14}$=Neomycin
$C_{23}H_{46}$=Muscalure
$C_{23}H_{46}ClNO_4$=Pahutoxin
$C_{23}H_{46}N_6O_{13}$=Neomycin

$\mathbf{C_{24}}H_{16}O_6$=Sphagnorubins
$C_{24}H_{16}O_9$=Fasciculins
$C_{24}H_{16}O_{10}$=Fasciculins
$C_{24}H_{16}O_{12}$=Laccaic acids
$C_{24}H_{17}NO_{11}$=Laccaic acids
$C_{24}H_{17}N_3O_2$=Benzomalvines
$C_{24}H_{18}N_2O_4$=Caulerpin
$C_{24}H_{18}O_8$=Flavomentins, Knipholone, Spiromentins
$C_{24}H_{20}O_{10}$=Gyrophoric acid
$C_{24}H_{20}O_{11}$=Crustinic acid
$C_{24}H_{20}O_{13}$=Leprocybe pigments
$C_{24}H_{23}N_5O_4$=Fungi
$C_{24}H_{24}O_9$=Lignans
$C_{24}H_{24}O_{11}$=Phrymarolins
$C_{24}H_{26}Br_4N_4O_8$=Aerothionin
$C_{24}H_{26}N_2O_{13}$=Betanidin
$C_{24}H_{26}N_4O_7S$=Imbricatine
$C_{24}H_{26}O_7$=Siphulin
$C_{24}H_{26}O_{13}$=Irigenin
$C_{24}H_{27}NO_3$=Tylophora alkaloids
$C_{24}H_{27}NO_4$=Tylophora alkaloids
$C_{24}H_{27}N_3O_6$=Physarochrome
$C_{24}H_{28}N_2O_7$=Kopsine
$C_{24}H_{28}O_4$=Norbixin
$C_{24}H_{28}O_{11}$=Catalpol
$C_{24}H_{29}NO_4$=Tylophora alkaloids
$C_{24}H_{30}N_2$=*N,N'*-Bis(2,6-dimethylphenyl)-2,2,4,4-tetramethyl-1,3-cyclobutanediimine
$C_{24}H_{30}N_4$=Pyrroloindole alkaloids
$C_{24}H_{30}O_3$=Ferulenol
$C_{24}H_{30}O_{11}$=Harpagide
$C_{24}H_{30}O_{12}$=Globularimin
$C_{24}H_{31}Cl_2N_5O_7$=Cyclochlorotine
$C_{24}H_{32}O_4$=Bufadienolides
$C_{24}H_{32}O_5$=Marinobufagin
$C_{24}H_{32}O_6$=Verrucosidin
$C_{24}H_{32}O_{13}$=Harpagide
$C_{24}H_{33}NO_4$=Aspochalasins
$C_{24}H_{33}NO_6$=Myxopyronins
$C_{24}H_{33}N_3O_7$=Anthramycins
$C_{24}H_{34}N_2O_6$=Arthogalin
$C_{24}H_{34}O_4$=Plakinic acids, Polyfibrospongols, Termites
$C_{24}H_{34}O_5$=Macrolactins, Polyfibrospongols

$C_{24}H_{34}O_7$=Ajugarins, Clerodin
$C_{24}H_{34}O_9$=Tric(h)othecenes, T-2 Toxin
$C_{24}H_{35}Cl_2NO_5$=Gymnastatins
$C_{24}H_{35}NO_5$=Batrachotoxins
$C_{24}H_{36}O_5$=Mevinolin
$C_{24}H_{36}O_6$=Macrolactins
$C_{24}H_{36}O_8$=Ajugarins
$C_{24}H_{39}NO_4$=Erythrophleum alkaloids
$C_{24}H_{39}NO_5$=Erythrophleum alkaloids
$C_{24}H_{39}N_9O_7$=Blasticidins
$C_{24}H_{40}N_2$=Apocynaceae steroid alkaloids
$C_{24}H_{40}O_3$=Bile acids
$C_{24}H_{40}O_4$=Bile acids
$C_{24}H_{40}O_5$=Bile acids
$C_{24}H_{41}NO_4$=Erythrophleum alkaloids
$C_{24}H_{41}NO_{10}$=Mycalamides
$C_{24}H_{42}N_6O_5$=Argiotoxins
$C_{24}H_{42}O_{21}$=Stachyose
$C_{24}H_{44}O_5$=Plakinic acids
$C_{24}H_{46}O_2$=Erucic acid
$C_{24}H_{46}O_4$=Suberin, suberic acid
$C_{24}H_{47}NO_7$=Sphingosine
$C_{24}H_{47}NO_{11}S$=Oviposition deterring pheromones (host marking pheromones)

C_{25} $H_{15}NO_8$=Lamellarines
$C_{25}H_{17}NO_{13}$=Laccaic acids
$C_{25}H_{18}O_6$=Sphagnorubins
$C_{25}H_{21}N_3O_3$=Coelenterazine
$C_{25}H_{22}N_4O_8$=Streptonigrin
$C_{25}H_{22}O_{10}$=Silybin
$C_{25}H_{23}N_3O_7$=Duocarmycins
$C_{25}H_{24}O_{12}$=Cynarin(e)
$C_{25}H_{26}BrN_5O_{11}$=Neosurugatoxin
$C_{25}H_{26}BrN_5O_{13}$=Neosurugatoxin
$C_{25}H_{26}O_6$=Terprenin
$C_{25}H_{26}O_9$=Sakyomycins
$C_{25}H_{26}O_{10}$=Aquayamycin, Sakyomycins
$C_{25}H_{26}O_{13}$=Ruberythric acid
$C_{25}H_{27}NO_5$=Lythraceae alkaloids
$C_{25}H_{28}O_8$=Lobaric acid
$C_{25}H_{28}O_{13}$=Monotropein
$C_{25}H_{29}NO_4$=Naphthyl-(tetrahydro)isoquinoline alkaloids
$C_{25}H_{29}NO_5$=Lythraceae alkaloids
$C_{25}H_{30}O_4$=Bixin
$C_{25}H_{30}O_7$=Phyllanthocin, Picrolichenic acid
$C_{25}H_{30}O_8$=Miriquidic acid
$C_{25}H_{31}NO_{10}$=Ipecac alkaloids
$C_{25}H_{31}NO_{12}$=Cripowellines
$C_{25}H_{31}N_3O_4$=Lunarine
$C_{25}H_{32}N_2O_4$=Arthonin
$C_{25}H_{32}N_2O_6$=Vindoline
$C_{25}H_{32}O_4$=Wistarin
$C_{25}H_{32}O_8$=Kosins
$C_{25}H_{32}O_{12}$=Ligustroside
$C_{25}H_{32}O_{13}$=Ligustroside
$C_{25}H_{33}NO$=Aurachins
$C_{25}H_{33}NO_2$=Aurachins
$C_{25}H_{33}NO_3$=Aurachins
$C_{25}H_{33}NO_6$=Ilimaquinone
$C_{25}H_{33}NO_{11}$=Cripowellines
$C_{25}H_{33}N_3O_3S_2$=Myxothiazol
$C_{25}H_{34}N_2O_9$=Antimycins
$C_{25}H_{34}N_6O_2S_4$=Tantazoles
$C_{25}H_{34}O_4$=Variabilin
$C_{25}H_{35}NO_9$=Ryanodine
$C_{25}H_{36}O_3$=Ophiobolins
$C_{25}H_{36}O_4$=Ophiobolins
$C_{25}H_{36}O_5$=Manoalide
$C_{25}H_{36}O_6$=Herical, Pseudopterosins
$C_{25}H_{38}O_2$=Retigeranic acid
$C_{25}H_{38}O_3$=Ophiobolins
$C_{25}H_{38}O_4$=Dysidiolide, Terpestacin
$C_{25}H_{39}NO_5$=Erythrophleum alkaloids
$C_{25}H_{39}NO_6$=Erythrophleum alkaloids
$C_{25}H_{40}N_2$=Papuamine
$C_{25}H_{40}N_2O_6S$=Leukotrienes
$C_{25}H_{41}NO_7$=Browniine
$C_{25}H_{41}NO_9$=Aconine
$C_{25}H_{42}N_2O$=Buxus steroid alkaloids
$C_{25}H_{42}N_4O_{14}$=Allosamidin
$C_{25}H_{42}O_6$=Herboxidiene
$C_{25}H_{43}NO_{10}$=Mycalamides
$C_{25}H_{43}NO_{18}$=Acarbose
$C_{25}H_{43}N_{13}O_{10}$=Viomycin
$C_{25}H_{45}FeN_6O_8$=Ferrioxamines
$C_{25}H_{45}NO_9$=Pederin
$C_{25}H_{45}NO_{10}$=Fumonisins
$C_{25}H_{46}$=Ophiobolins
$C_{25}H_{46}N_{10}O_9$=Streptothricins
$C_{25}H_{48}N_6O_8$=Ferrioxamines

C_{26} $H_{16}O_5$=Xylerythrin
$C_{26}H_{18}O_9$=Hypholomins
$C_{26}H_{18}O_{10}$=Hypholomins
$C_{26}H_{18}O_{12}$=Rubromycins
$C_{26}H_{19}NO_{12}$=Laccaic acids
$C_{26}H_{20}O_8$=Retipolide A
$C_{26}H_{20}O_9$=Retipolide A
$C_{26}H_{21}N_3O_3$=Coelenterazine
$C_{26}H_{25}N_3O_8$=Duocarmycins
$C_{26}H_{25}N_3O_{11}$=Tunichromes
$C_{26}H_{26}ClN_3O_8$=Duocarmycins
$C_{26}H_{26}N_2$=Yuehchukene
$C_{26}H_{26}O_6$=Rocaglamide
$C_{26}H_{26}O_9$=Gilvocarcins
$C_{26}H_{28}N_2O_9$=Ophiorrhiza alkaloids
$C_{26}H_{28}O_{10}S$=Sakyomycins
$C_{26}H_{28}O_{14}$=Apiin, Frangulins
$C_{26}H_{29}NO_4$=Naphthyl-(tetrahydro)isoquinoline alkaloids
$C_{26}H_{29}NO_5$=Lythraceae alkaloids
$C_{26}H_{29}N_5O_2S_3$=Tantazoles
$C_{26}H_{30}ClN_2O_3$=Hapalindoles
$C_{26}H_{30}N_2O_8$=Vincoside
$C_{26}H_{30}O_8$=Limonin, Physodic acid
$C_{26}H_{30}O_{11}$=Rubratoxin B
$C_{26}H_{30}O_{13}$=Wine, phenolic constituents of
$C_{26}H_{30}O_{14}$=Wine, phenolic constituents of
$C_{26}H_{31}ClO_{10}$=Erythrolides
$C_{26}H_{31}NO_4$=Naphthyl-(tetrahydro)isoquinoline alkaloids
$C_{26}H_{31}NO_5$=Lythraceae alkaloids
$C_{26}H_{32}O_7$=Strobilurins
$C_{26}H_{32}O_{11}$=Rubratoxin B
$C_{26}H_{32}O_{14}$=Lamiide
$C_{26}H_{34}N_2O$=Coccinelline
$C_{26}H_{34}O_4$=Tridentoquinone
$C_{26}H_{34}O_6$=Bufadienolides, Strobilurins
$C_{26}H_{34}O_7$=Fumagillin, Strobilurins
$C_{26}H_{35}N_5O_{12}S$=Cyclothialidine
$C_{26}H_{36}N_2O_2$=Fischerellins
$C_{26}H_{36}N_2O_9$=Antimycins
$C_{26}H_{36}O_4$=Boviquinones
$C_{26}H_{36}O_6$=Bufadienolides
$C_{26}H_{36}O_{12}$=Foliamenthin
$C_{26}H_{38}N_2O_2$=Manzamines
$C_{26}H_{38}O_4$=Lupulone, Plakinic acids
$C_{26}H_{38}O_6$=Tyromycin A
$C_{26}H_{38}O_8$=Striatals
$C_{26}H_{39}NO_2$=Buxus steroid alkaloids
$C_{26}H_{39}NO_6S$=Epothilones
$C_{26}H_{40}O_4$=Viburnoles
$C_{26}H_{40}O_7$=Spongianes
$C_{26}H_{41}NO_3$=Myxalamides
$C_{26}H_{41}N_7O_4$=JSTX
$C_{26}H_{42}O_3$=Eugenol
$C_{26}H_{42}O_8$=Fusicoccin H
$C_{26}H_{42}O_9$=Pseudomonic acids
$C_{26}H_{43}NO_5$=Bile acids
$C_{26}H_{43}NO_6$=Bile acids
$C_{26}H_{44}N_2$=Buxus steroid alkaloids
$C_{26}H_{44}O_8$=Pseudomonic acids
$C_{26}H_{44}O_9$=Pseudomonic acids
$C_{26}H_{44}O_{10}$=Pseudomonic acids
$C_{26}H_{45}NO_6S$=Bile acids
$C_{26}H_{45}NO_7S$=Bile acids
$C_{26}H_{46}N_2$=Buxus steroid alkaloids
$C_{26}H_{49}Cl_2N_5O$=Phloeodictines
$C_{26}H_{52}O_2$=Cerotic acid

C_{27} $H_{19}N_5O$=Shermilamines
$C_{27}H_{20}O_{12}$=Rubromycins
$C_{27}H_{21}Cl_2N_3O_7$=Rebeccamycin
$C_{27}H_{21}N_3O_5$=Staurosporine
$C_{27}H_{22}N_2O_{10}S_2$=Emestrin
$C_{27}H_{22}N_2O_{10}S_3$=Emestrin
$C_{27}H_{24}O_9$=Lignans
$C_{27}H_{29}NO_{10}$=Daunorubicin
$C_{27}H_{29}NO_{11}$=Adriamycin
$C_{27}H_{30}N_2O_9$=Ophiorrhiza alkaloids
$C_{27}H_{30}O_8$=Daphnetoxin
$C_{27}H_{30}O_{13}$=Peltatin
$C_{27}H_{30}O_{14}$=Frangulins
$C_{27}H_{30}O_{16}$=Rutin
$C_{27}H_{31}ClO_{15}$=Keracyanin
$C_{27}H_{31}ClO_{17}$=Delphin
$C_{27}H_{31}NO_3$=Paspalitrems
$C_{27}H_{31}NO_4$=Hericenones
$C_{27}H_{31}O_{15}^+$=Keracyanin, Pelargonin
$C_{27}H_{31}O_{16}^+$=Cyanin, Mecocyanin
$C_{27}H_{31}O_{17}^+$=Delphin
$C_{27}H_{32}N_2O$=Phenazinomycin
$C_{27}H_{32}N_2O_8$=Raucaffricine
$C_{27}H_{32}O_8$=Verrucarins
$C_{27}H_{32}O_9$=Austin, Verrucarins
$C_{27}H_{32}O_{14}$=Naringin, Wine, phenolic constituents of
$C_{27}H_{33}ClN_2O_2$=Roseophilin
$C_{27}H_{33}NO_3$=Hericenones
$C_{27}H_{33}N_3O_3$=Marcfortines
$C_{27}H_{33}N_3O_4$=Marcfortines
$C_{27}H_{33}N_3O_5$=Andrimid, Fumitremorgins
$C_{27}H_{33}N_3O_7$=Fumitremorgins
$C_{27}H_{33}N_9O_{15}P_2$=Riboflavin(e)
$C_{27}H_{34}N_2O_6$=Tetramic acid
$C_{27}H_{34}N_2O_9$=Strictosidine, Vincoside
$C_{27}H_{34}O_8$=Verrucarins
$C_{27}H_{34}O_9$=Verrucarins
$C_{27}H_{35}NO_5$=Lythraceae alkaloids
$C_{27}H_{35}NO_6$=Tuberine
$C_{27}H_{35}NO_{12}$=Ipecac alkaloids
$C_{27}H_{36}O_6$=Strobilurins
$C_{27}H_{36}O_{14}$=Sylvestrosides
$C_{27}H_{38}N_2O_9$=Antimycins
$C_{27}H_{38}O_7$=Pseudopterosins
$C_{27}H_{38}O_9$=Lathyranes
$C_{27}H_{39}NO_2$=Veratrum steroid alkaloids
$C_{27}H_{39}NO_3$=Veratrum steroid alkaloids
$C_{27}H_{39}N_3O_2$=Majusculamides
$C_{27}H_{40}O_8$=Herical, Sordarin
$C_{27}H_{40}O_9$=Mniopetals
$C_{27}H_{41}NO_2$=Veratrum steroid alkaloids
$C_{27}H_{41}NO_6S$=Epothilones
$C_{27}H_{42}FeN_9O_{12}$=Ferrichromes
$C_{27}H_{42}O_3$=Steroid sapogenins
$C_{27}H_{42}O_4$=Steroid sapogenins
$C_{27}H_{43}NO$=Solanum steroid alkaloids, Veratrum steroid alkaloids
$C_{27}H_{43}NO_2$=Solanum steroid alkaloids, Veratrum steroid alkaloids
$C_{27}H_{43}NO_3$=Solanum steroid alkaloids
$C_{27}H_{43}NO_8$=Veratrum steroid alkaloids
$C_{27}H_{44}N_2O_8$=Bengamides
$C_{27}H_{44}N_4O_6$=Glidobactins
$C_{27}H_{44}O$=Calcitriol, Sterols
$C_{27}H_{44}O_2$=Calcitriol
$C_{27}H_{44}O_3$=Calcitriol, Steroid sapogenins
$C_{27}H_{44}O_5$=Steroid sapogenins
$C_{27}H_{44}O_6$=Ecdysteroids
$C_{27}H_{44}O_7$=Ecdysteroids
$C_{27}H_{45}NO$=Solanum steroid alkaloids
$C_{27}H_{45}NO_2$=Solanum steroid alkaloids
$C_{27}H_{45}NO_3$=Solanum steroid alkaloids
$C_{27}H_{46}N_2O$=Solanum steroid alkaloids
$C_{27}H_{46}N_2O_2$=Buxus steroid alkaloids, Solanum steroid alkaloids
$C_{27}H_{46}O$=Cholesterol
$C_{27}H_{46}O_2$=Tocopherols
$C_{27}H_{47}N_7O_6$=JSTX
$C_{27}H_{48}O$=Sterols
$C_{27}H_{49}N_5O_7$=Fungi
$C_{27}H_{49}N_5O_8$=Fungi
$C_{27}H_{49}N_7O_{17}$=Streptomycin
$C_{27}H_{52}$=Flies
$C_{27}H_{53}Cl_3N_8OS$=Phloeodictines

C_{28} $H_{21}NO_8$=Lamellarins
$C_{28}H_{22}O_6$=Wine, phenolic constituents of

$C_{28}H_{23}NO_8$=Lamellarins
$C_{28}H_{24}O_4$=Riccardins
$C_{28}H_{24}O_5$=Marchantins
$C_{28}H_{24}O_6$=Marchantins
$C_{28}H_{24}O_{10}$=Maduramicin α
$C_{28}H_{26}N_2O_{10}$=Balanol
$C_{28}H_{26}N_4O_3$=Staurosporine
$C_{28}H_{26}O_4$=Perrottetins
$C_{28}H_{26}O_5$=Perrottetins
$C_{28}H_{30}O_9$=Territrems
$C_{28}H_{30}O_{12}$=Granaticins
$C_{28}H_{30}O_{13}$=Steffimycin
$C_{28}H_{31}N_3O_8$=Saframycins
$C_{28}H_{32}N_4O_4$=Roccanin
$C_{28}H_{32}O_9$=Territrems
$C_{28}H_{32}O_{12}$=Steffimycin
$C_{28}H_{32}O_{13}$=Peltatin, Podophyllotoxin
$C_{28}H_{33}O_{16}^+$=Peonin
$C_{28}H_{33}O_{17}^+$=Petunin
$C_{28}H_{34}O_6$=Withanolides
$C_{28}H_{34}O_{15}$=Hesperetin
$C_{28}H_{35}NO_5$=Zygosporins
$C_{28}H_{35}N_3O_4$=Marcfortines
$C_{28}H_{35}N_3O_5$=Marcfortines
$C_{28}H_{35}N_3O_7$=Streptogramins, Virginiamycin
$C_{28}H_{36}N_2O_4$=Ipecac alkaloids
$C_{28}H_{36}N_2O_6$=Sarcodictyins
$C_{28}H_{36}N_2O_7$=Sarcodictyins
$C_{28}H_{36}O_3$=Russulaflavidin
$C_{28}H_{36}O_9$=Meliatoxins
$C_{28}H_{36}O_{11}$=Bruceantin
$C_{28}H_{37}ClN_2O_8$=Maytansinoids
$C_{28}H_{37}NO_8$=Cephalotaxus alkaloids
$C_{28}H_{37}NO_9$=Cephalotaxus alkaloids
$C_{28}H_{38}N_2O_4$=Ipecac alkaloids
$C_{28}H_{38}N_4O_2$=Verbascum alkaloids
$C_{28}H_{38}O_6$=Withanolides
$C_{28}H_{38}O_7$=Bongkrekic acid, Withanolides
$C_{28}H_{39}NO_2$=Paspalitrems
$C_{28}H_{39}NO_4$=Sambutoxin
$C_{28}H_{40}N_2O_9$=Antimycins
$C_{28}H_{40}O_4$=Suillin
$C_{28}H_{40}O_8$=Striatals, Taxanes
$C_{28}H_{40}O_9$=Striatals
$C_{28}H_{41}NO$=U 106305
$C_{28}H_{41}N_3O_2$=Teleocidins
$C_{28}H_{42}O_2$=Tocopherols
$C_{28}H_{42}O_5$=Withanolides
$C_{28}H_{42}O_6$=Ambruticins
$C_{28}H_{42}O_8$=Denticulatins
$C_{28}H_{43}NO_6$=Hapalosin
$C_{28}H_{44}FeN_9O_{12}$=Ferrichromes
$C_{28}H_{44}N_4O_4$=Cyclopeptide alkaloids
$C_{28}H_{44}O$=Calcitriol, Lichesterol, Sterols
$C_{28}H_{44}O_2$=Calcitriol
$C_{28}H_{44}O_3$=Calcitriol
$C_{28}H_{45}N_3O_5$=Majusculamides
$C_{28}H_{46}N_2O_6$=Esterastin
$C_{28}H_{46}N_2O_8$=Bengamides
$C_{28}H_{46}O$=Sterols
$C_{28}H_{46}O_7$=Ecdysteroids
$C_{28}H_{47}NO_8$=Picromycin
$C_{28}H_{48}O$=Sterols
$C_{28}H_{48}O_2$=Tocopherols
$C_{28}H_{48}O_5$=Brassinosteroids
$C_{28}H_{48}O_6$=Brassinosteroids
$C_{28}H_{50}N_2O_2$=Xestospongines
$C_{28}H_{50}N_2O_4$=Carpaine
$C_{28}H_{50}O_4$=α-Onocerin

$\mathbf{C_{29}}H_{23}NO_9$=Dynemicins, Lamellarins
$C_{29}H_{24}O_6$=Marchantins
$C_{29}H_{24}O_{12}$=Theaflavin
$C_{29}H_{25}NO_9$=Lamellarins
$C_{29}H_{26}O_4$=Riccardins
$C_{29}H_{26}O_6$=Marchantins
$C_{29}H_{28}O_5$=Perrottetins
$C_{29}H_{30}O_{13}$=Amarogentin
$C_{29}H_{31}NO_7$=Rocaglamide
$C_{29}H_{32}N_2O_4$=Hypothallin
$C_{29}H_{32}O_{13}$=Steffimycin
$C_{29}H_{34}N_2O_4$=Cytochalasins
$C_{29}H_{34}O_4$=Mulberrofurans
$C_{29}H_{34}O_6$=Gamboge
$C_{29}H_{34}O_9$=Territrems
$C_{29}H_{34}O_{10}$=Satratoxins
$C_{29}H_{34}O_{12}$=Steffimycin
$C_{29}H_{35}NO_5$=Cytochalasins, Hitachimycin
$C_{29}H_{35}O_{17}^+$=Malvin
$C_{29}H_{36}O_8$=Roridins
$C_{29}H_{36}O_9$=Roridins, Satratoxins
$C_{29}H_{36}O_{10}$=Satratoxins
$C_{29}H_{37}NO_5$=Cytochalasins
$C_{29}H_{37}NO_6$=Lythraceae alkaloids
$C_{29}H_{37}NO_{12}$=Agarofurans
$C_{29}H_{37}NO_{13}$=Agarofurans
$C_{29}H_{37}N_3O_3$=Ipecac alkaloids
$C_{29}H_{37}N_3O_6$=Calcimycin
$C_{29}H_{38}N_2O_4$=Ikarugamycin
$C_{29}H_{38}N_2O_6$=Xanthobaccin A
$C_{29}H_{38}O_8$=Roridins
$C_{29}H_{38}O_9$=Baccharinoids, Roridins, Uscharidin
$C_{29}H_{38}O_{11}$=Baccharinoids
$C_{29}H_{39}NO_9$=Cephalotaxus alkaloids
$C_{29}H_{40}N_2O_4$=Ipecac alkaloids
$C_{29}H_{40}N_8O_6S_2$=Dolastatins
$C_{29}H_{40}O_7$=Petunia steroids
$C_{29}H_{40}O_9$=Roridins, Trichoverroids
$C_{29}H_{40}O_{10}$=Baccharinoids
$C_{29}H_{42}N_2O_6$=Hennoxazoles
$C_{29}H_{42}O_5$=Antheridiol
$C_{29}H_{42}O_{10}$=Strophanthins
$C_{29}H_{43}N_3O_2$=Teleocidins
$C_{29}H_{44}O_2$=Tocopherols
$C_{29}H_{44}O_8$=Ecdysteroids, Soraphens
$C_{29}H_{44}O_{12}$=Strophanthins
$C_{29}H_{45}ClN_2O_6$=Majusculamides
$C_{29}H_{46}N_4O_6$=Glidobactins
$C_{29}H_{46}O$=Calysterol
$C_{29}H_{46}O_4$=Betulinic acid, Viburnoles
$C_{29}H_{47}NO_6$=Erythrophleum alkaloids
$C_{29}H_{47}N_5O_7$=Destruxins
$C_{29}H_{47}N_5O_8$=Destruxins
$C_{29}H_{48}N_4O_6$=Glidobactins
$C_{29}H_{48}N_8O_5$=JSTX
$C_{29}H_{48}O$=Sterols
$C_{29}H_{48}O_5$=Antheridiol
$C_{29}H_{48}O_7$=Ecdysteroids
$C_{29}H_{49}NO_5$=Lipstatin
$C_{29}H_{49}N_5O_8$=Destruxins
$C_{29}H_{50}O$=Aplysi..., Sterols
$C_{29}H_{50}O_2$=Tocopherols
$C_{29}H_{50}O_3$=Tocopherols
$C_{29}H_{52}N_{10}O_6$=Argiotoxins
$C_{29}H_{52}O_3$=Hierridin, Tocopherols
$C_{29}H_{53}NO_5$=Lipstatin

$\mathbf{C_{30}}H_{14}N_4O_4$=Necatorone
$C_{30}H_{14}N_4O_6$=Necatorone
$C_{30}H_{16}O_8$=Hypericin
$C_{30}H_{16}O_{12}$=Bartramiaflavone
$C_{30}H_{17}O_{12}^+$=Riccionidins
$C_{30}H_{18}O_{10}$=Amentoflavones, Hinokiflavone, Robustaflavone, Skyrin
$C_{30}H_{18}O_{12}$=Bryoflavone
$C_{30}H_{18}O_{13}$=Bartramiaflavone
$C_{30}H_{19}NO_9$=Dynemicins
$C_{30}H_{20}O_{11}$=Hegoflavones
$C_{30}H_{20}O_{12}$=Hegoflavones
$C_{30}H_{21}NO_9$=Fredericamycin A
$C_{30}H_{22}ClNO_9$=Dynemicins
$C_{30}H_{22}O_{10}$=Rugulosin
$C_{30}H_{22}O_{11}$=Hypnogenols
$C_{30}H_{22}O_{12}$=Hypnogenols, Luteoskyrin, Xanthomegnin
$C_{30}H_{24}O_9$=Atrovirin
$C_{30}H_{24}O_{10}$=Elsinochromes
$C_{30}H_{25}NO_9$=Lamellarins
$C_{30}H_{26}O_{10}$=Atrovirin, Elsinochromes, Flavomannins, Hypocrellin, Leucomentins
$C_{30}H_{26}O_{14}$=Ergochromes
$C_{30}H_{27}NO_9$=Lamellarins
$C_{30}H_{27}NO_{10}$=Lamellarins
$C_{30}H_{27}N_3O_{15}$=Enterobactin
$C_{30}H_{27}O_{14}^+$=Gentianin
$C_{30}H_{28}O_8$=Rottlerin
$C_{30}H_{30}O_8$=Gossypol
$C_{30}H_{31}NO_{10}$=Viridicatumtoxin
$C_{30}H_{34}BrN_5O_{15}$=Neosurugatoxin
$C_{30}H_{34}N_2O_9$=Ecteinascidins
$C_{30}H_{34}N_2O_{19}$=Amarantin
$C_{30}H_{34}O_{13}$=Picrotoxin
$C_{30}H_{35}NO_8$=Ipom(o)ea alkaloids
$C_{30}H_{36}O_9$=Neem tree
$C_{30}H_{37}NO_5$=Zygosporins
$C_{30}H_{37}NO_7$=Tiglianes
$C_{30}H_{37}N_5O_5$=Ergot alkaloids
$C_{30}H_{37}O_{17}^+$=Hirsutin(e)
$C_{30}H_{39}ClN_2O_9$=Maytansinoids
$C_{30}H_{39}ClO_{13}$=Erythrolides
$C_{30}H_{39}N_7O_6$=Chymostatins
$C_{30}H_{40}O_6$=Absinthin, Proazulenes
$C_{30}H_{41}NO_7$=Myxopyronins
$C_{30}H_{42}N_2$=Myrmicarins
$C_{30}H_{42}N_2O_2S$=Nuphar alkaloids, Thiobinupharidine
$C_{30}H_{42}N_2O_8$=Macbecins
$C_{30}H_{42}N_2O_9$=Herbimycin
$C_{30}H_{42}N_2O_{15}S_2$=Glucosinolates
$C_{30}H_{42}N_4O_2$=Homaline
$C_{30}H_{42}O_3$=Pomacerone
$C_{30}H_{42}O_4$=Mastigophorenes
$C_{30}H_{42}O_7$=Stigmatellin
$C_{30}H_{42}O_8$=Bufadienolides
$C_{30}H_{44}N_2O_9S_2$=Thiomarinols
$C_{30}H_{44}O_5$=Holothurins
$C_{30}H_{44}O_6$=Viburnoles
$C_{30}H_{44}O_7$=Ganoderic acids
$C_{30}H_{44}O_9$=Strophanthins
$C_{30}H_{44}O_{10}$=Macrolactins
$C_{30}H_{45}N_6O_{16}P$=Tetrahydromethanopterin
$C_{30}H_{46}O_3$=Mastic
$C_{30}H_{46}O_4$=Glycyrrhet(in)ic acid, Holothurins, Melianol
$C_{30}H_{46}O_5$=Quillaja saponin
$C_{30}H_{46}O_6$=Medicagenic acid, Saponaceolides
$C_{30}H_{46}O_7$=Saponaceolides
$C_{30}H_{46}O_{16}S_2$=Atractyloside
$C_{30}H_{47}NO_4$=Daphniphyllum alkaloids
$C_{30}H_{47}NO_5$=Ambruticins
$C_{30}H_{47}N_3O_9S$=Leukotrienes
$C_{30}H_{48}$=Fernanes
$C_{30}H_{48}O$=Fernanes, Oleanane
$C_{30}H_{48}O_3$=Betulinic acid, Boswellic acids, Oleanolic acid, Ursolic acid
$C_{30}H_{48}O_4$=α-Hederin, Melianol, Retigeric acid A
$C_{30}H_{49}N_3O$=Lycopodium alkaloids
$C_{30}H_{49}N_5O_7$=Roseotoxins
$C_{30}H_{49}N_5O_9$=Destruxins
$C_{30}H_{50}$=Dammarenes, Fernanes, Oleanane, Squalene
$C_{30}H_{50}N_2O_2$=Petrosin
$C_{30}H_{50}O$=Amyrins, Cycloartenol, Fernanes, Friedelin, Lanosterol, Lupeol, Sterols, Taraxasterol, Taraxerol, Triterpenes
$C_{30}H_{50}O_2$=Betulin, Oleanane, α-Onocerin, Triterpenes
$C_{30}H_{50}O_3$=4-Methoxy-2-(3-methyl-2-butenyl)-phenyl stearate, Oleanane
$C_{30}H_{50}O_4$=Iridals, Oleanane
$C_{30}H_{50}O_5$=Melianol, Oleanane
$C_{30}H_{50}O_6$=Gymnemic acids, Oleanane
$C_{30}H_{51}N_5O_7$=Destruxins
$C_{30}H_{51}N_5O_8$=Destruxins
$C_{30}H_{52}$=Hopanes, Oleanane, Ursane
$C_{30}H_{52}O$=Ambergris
$C_{30}H_{52}O_2$=Dammarenes, Friedelanes, Stictane-3β,22α-diol, Zeorin
$C_{30}H_{52}O_3$=Dammarenes, Ginseng
$C_{30}H_{52}O_4$=Ginseng, Pyxinol
$C_{30}H_{52}O_5$=Teurilene
$C_{30}H_{53}BrO_7$=Thyrsiferol
$C_{30}H_{54}N_{10}O_6$=Argiotoxins
$C_{30}H_{54}O_5$=Iridals
$C_{30}H_{54}O_6$=Iridals
$C_{30}H_{58}N_4O_2$=Homaline
$C_{30}H_{58}N_4O_3$=Homaline
$C_{30}H_{60}O$=1-Triacontanol
$C_{30}H_{60}O_2$=Melissic acid
$C_{30}H_{62}O$=1-Triacontanol

$\mathbf{C_{31}}H_{28}O_{14}$=Ergochromes
$C_{31}H_{29}NO_9$=Lamellarins
$C_{31}H_{29}N_5O_4$=Asperlicin
$C_{31}H_{29}N_5O_5$=Asperlicin
$C_{31}H_{30}N_2O_6$=Aglaiastatine
$C_{31}H_{32}O_{12}$=Saquayamycins
$C_{31}H_{33}N_3O_{11}$=Carzinophilin
$C_{31}H_{34}N_4O_4$=Bonellin
$C_{31}H_{34}O_{12}$=Saquayamycins
$C_{31}H_{36}N_2O_{11}$=Novobiocin
$C_{31}H_{36}O_9$=CP-263,114
$C_{31}H_{38}N_2O_7$=Manumycins
$C_{31}H_{39}NO_7$=Lythraceae alkaloids
$C_{31}H_{39}N_5O_5$=Ergot alkaloids
$C_{31}H_{40}O_8$=Calbistrin A
$C_{31}H_{40}O_{11}$=Taxuspines (T. A–Z)
$C_{31}H_{41}N_7O_6$=Chymostatins
$C_{31}H_{42}N_2O_6$=Batrachotoxins
$C_{31}H_{42}N_4O_4$=Frangufoline
$C_{31}H_{42}O_5$=Milbemycins
$C_{31}H_{42}O_{10}$=Papulacandins
$C_{31}H_{42}O_{16}$=Olivil
$C_{31}H_{43}NO_4$=Indanomycin
$C_{31}H_{46}O_2$=Vitamin K_1
$C_{31}H_{46}O_3$=Vitamin K_1
$C_{31}H_{46}O_4$=Polyporenic acids
$C_{31}H_{46}O_7$=Amphidinolides
$C_{31}H_{46}O_{10}$=Majusculamides
$C_{31}H_{48}O_2$=Vitamin K_1
$C_{31}H_{48}O_4$=Abietospiran
$C_{31}H_{48}O_6$=Fusidic acid
$C_{31}H_{50}O_3$=Eburicoic acid
$C_{31}H_{50}O_4$=Polyporenic acids, Sulphurenic acid, Tumulosic acid
$C_{31}H_{52}N_2O_{23}$=Sialyl-Lewis x Tetrasaccharide (S Le^x)
$C_{31}H_{52}N_4O_3$=Homaline
$C_{31}H_{52}O$=Eburicol
$C_{31}H_{53}N_{11}O_5$=Argiopinins
$C_{31}H_{54}N_{10}O_4$=Argiopinins
$C_{31}H_{56}N_2O_8$=Bengamides
$C_{31}H_{58}N_{12}O_{10}$=Streptothricins
$C_{31}H_{62}O$=Cockroaches
$C_{31}H_{62}O_2$=Cockroaches, Hentriacontanoic acid
$C_{31}H_{64}O$=1-Hentriacontanol

$\mathbf{C_{32}}H_{20}O_8$=Aurantiacin
$C_{32}H_{23}NO_{10}$=Purpurone
$C_{32}H_{24}O_{10}$=Xylindein
$C_{32}H_{26}O_{10}$=Phlegmacins
$C_{32}H_{26}O_{14}$=Actinorhodin
$C_{32}H_{28}O_9$=Rufoolivacins
$C_{32}H_{30}N_2O_4$=L-783281
$C_{32}H_{30}O_{10}$=Flavomannins, Phlegmacins
$C_{32}H_{30}O_{14}$=Ergochromes
$C_{32}H_{32}N_2O_2$=Variabiline
$C_{32}H_{32}N_4NiO_4$=Tunichlorin
$C_{32}H_{32}O_{13}S$=Podophyllotoxin
$C_{32}H_{32}O_{14}$=Chartreusin
$C_{32}H_{34}N_4O_4$=Corallistin A
$C_{32}H_{34}O_{10}$=Cardinalins
$C_{32}H_{36}N_2O_5$=Chaetoglobosins
$C_{32}H_{36}N_4O_9$=Agrobactin
$C_{32}H_{36}N_4O_{10}$=Agrobactin
$C_{32}H_{36}O_{10}$=Cardinalins
$C_{32}H_{37}NO_7$=Cytochalasins
$C_{32}H_{38}N_2O_8$=Reserpine
$C_{32}H_{39}NO_4$=Aflatrem, Paspalitrems
$C_{32}H_{39}NO_7$=Zygosporins
$C_{32}H_{40}O_{11}$=Azadirachtin(s), Meliatoxins
$C_{32}H_{40}O_{12}$=Meliatoxins
$C_{32}H_{41}N_3O_7$=Fumitremorgins
$C_{32}H_{41}N_5O_5$=Ergot alkaloids
$C_{32}H_{42}N_8O_6S_4$=Patellamides
$C_{32}H_{43}ClN_2O_9$=Maytansinoids
$C_{32}H_{43}N_3O_6$=FR 900848
$C_{32}H_{44}MgN_2O_8$=Magnesidin
$C_{32}H_{44}N_2O_6$=Batrachotoxins
$C_{32}H_{44}O_8$=Cucurbitacins
$C_{32}H_{45}NO_3$=Myxalamides
$C_{32}H_{45}NO_4$=Gymnodimine
$C_{32}H_{46}O_6S$=Petunia steroids
$C_{32}H_{46}O_8$=Cucurbitacins
$C_{32}H_{46}O_9$=Cucurbitacins
$C_{32}H_{47}BrO_{10}$=Aplysi…
$C_{32}H_{48}O_4$=Echinodol
$C_{32}H_{48}O_8$=Cucurbitacins
$C_{32}H_{48}O_{10}$=Majusculamides
$C_{32}H_{49}NO_5$=Daphniphyllum alkaloids
$C_{32}H_{49}NO_9$=Veratrum steroid alkaloids
$C_{32}H_{50}N_2O_3$=Sarains
$C_{32}H_{50}O_4$=Echinodol
$C_{32}H_{52}O_{14}$=Lepranthin
$C_{32}H_{55}BrO_8$=Thyrsiferol
$C_{32}H_{56}N_{10}O_4$=Argiopinins
$C_{32}H_{58}N_2O_8$=Bengamides
$C_{32}H_{64}O_2$=Shellac (shellac wax)

$\mathbf{C_{33}}H_{26}O_{10}$=Santalins
$C_{33}H_{28}O_{10}$=Dermocanarin I
$C_{33}H_{30}FeN_4O_5$=Chlorocruoroporphyrin
$C_{33}H_{30}O_{10}$=Santalins
$C_{33}H_{32}N_4O_3$=Chlorophyllone a
$C_{33}H_{32}N_4O_5$=Chlorocruoroporphyrin
$C_{33}H_{34}N_4O_6$=Biliverdin
$C_{33}H_{34}O_{14}$=Haedoxans
$C_{33}H_{35}N_5O_5$=Ergot alkaloids
$C_{33}H_{36}N_4O_6$=Bilirubin, Phytochrome, phytochromobilin
$C_{33}H_{38}N_4Na_2O_8$=Substance F
$C_{33}H_{39}N_7O_5S_2$=Patellamides
$C_{33}H_{40}N_2O_9$=Reserpine
$C_{33}H_{40}N_4O_6$=Mesobilirubin
$C_{33}H_{40}O_{19}$=Robinin
$C_{33}H_{41}NO_8$=Lythraceae alkaloids
$C_{33}H_{42}N_4O_6$=Urobilin
$C_{33}H_{43}N_5O_4$=Cyclopeptide alkaloids
$C_{33}H_{44}N_4O_6$=Urobilinogen
$C_{33}H_{44}O$=Citranaxanthin
$C_{33}H_{45}ClN_2O_9$=Maytansinoids
$C_{33}H_{45}NO_8$=Taxanes
$C_{33}H_{46}N_4O_6$=Stercobilin
$C_{33}H_{47}NO_{13}$=Natamycin
$C_{33}H_{48}N_2O_3$=Buxus steroid alkaloids
$C_{33}H_{48}N_4O_6$=Stercobilinogen
$C_{33}H_{48}O_7$=Milbemycins
$C_{33}H_{48}O_{19}$=Sylvestrosides
$C_{33}H_{49}NO_7$=Veratrum steroid alkaloids
$C_{33}H_{49}NO_8$=Veratrum steroid alkaloids
$C_{33}H_{49}N_5O_6$=Cyclopeptide alkaloids
$C_{33}H_{50}O_{10}$=Tautomycin
$C_{33}H_{51}NO_7$=Veratrum steroid alkaloids
$C_{33}H_{52}O_2$=Lactarius pigments
$C_{33}H_{52}O_5$=Tumulosic acid
$C_{33}H_{54}O_5$=*O*-Stearoylvelutinal
$C_{33}H_{54}O_{11}$=Cotylenin F
$C_{33}H_{55}NO_8$=Discodermin A
$C_{33}H_{56}O_4$=*O*-Stearoylvelutinal
$C_{33}H_{57}NO_8$=Solanum steroid alkaloid glycosides
$C_{33}H_{57}NO_{15}$=Fumonisins
$C_{33}H_{57}N_3O_9$=Enniatins
$C_{33}H_{58}N_{10}O_{10}$=Edeines
$C_{33}H_{60}O_7$=Zincophorin

$\mathbf{C_{34}}H_{24}O_{10}$=Stentorin
$C_{34}H_{25}Br_5N_4O_8$=Bastadins
$C_{34}H_{26}O_8$=Mulberrofurans
$C_{34}H_{28}O_{10}$=Santalins
$C_{34}H_{30}FeN_4O_{10}$=Heme d_1
$C_{34}H_{32}ClFeN_4O_4$=Hemin
$C_{34}H_{32}FeN_4O_4$=Heme
$C_{34}H_{32}FeN_4O_5$=Heme d
$C_{34}H_{32}N_2O_5$=Bisbenzylisoquinoline alkaloids
$C_{34}H_{34}N_4O_4$=Protoporphyrins
$C_{34}H_{37}N_5O_5$=Ergot alkaloids
$C_{34}H_{38}N_4O_5$=Chlorophylls
$C_{34}H_{38}N_4O_6$=Hematoporphyrin
$C_{34}H_{40}N_2O_5$=Chaetoglobosins
$C_{34}H_{40}N_4O_6$=Aplysi…
$C_{34}H_{40}O_{12}$=Filixic acids
$C_{34}H_{42}N_2O_6$=Truxillic and truxinic acids
$C_{34}H_{42}N_4O_6$=Trapoxin A
$C_{34}H_{42}O_4$=Flavidulols
$C_{34}H_{42}O_{13}$=Rohitukin
$C_{34}H_{44}N_4O_{15}$=Methanofuran
$C_{34}H_{44}O_{12}$=Meliatoxins
$C_{34}H_{46}ClN_3O_{10}$=Maytansinoids
$C_{34}H_{46}O_{14}$=Zaragozic acids
$C_{34}H_{47}NO_{11}$=Aconitine
$C_{34}H_{48}O_8$=Huratoxin
$C_{34}H_{52}O_7$=Milbemycins
$C_{34}H_{54}O_8$=Lasalocids
$C_{34}H_{59}NO_{15}$=Fumonisins
$C_{34}H_{59}N_3O_9$=Enniatins
$C_{34}H_{60}N_4O_2$=Rhapsamine
$C_{34}H_{60}N_{12}O_{10}$=Edeines
$C_{34}H_{62}O_7$=Zincophorin
$C_{34}H_{63}N_5O_9$=Pepstatin A
$C_{34}H_{65}N_3O_5S$=Squalamine
$C_{34}H_{66}NO_{12}P$=Lipid X

$\mathbf{C_{35}}H_{18}O_{15}$=Badiones
$C_{35}H_{26}BrCl_2N_5O_6$=Diazonamides
$C_{35}H_{30}MgN_4O_5$=Chlorophylls
$C_{35}H_{33}NO_{12}$=Neocarzinostatins
$C_{35}H_{34}N_2O_5$=Bisbenzylisoquinoline alkaloids
$C_{35}H_{39}N_5O_5$=Ergot alkaloids
$C_{35}H_{39}N_5O_{11}$=Agrobactin
$C_{35}H_{40}N_8O_6S_4$=Patellamides
$C_{35}H_{41}KN_4O_{10}$=Chlorophylls
$C_{35}H_{42}O_{10}$=Taxuspines (T. A–Z)
$C_{35}H_{42}O_{12}$=Filixic acids
$C_{35}H_{44}Cl_2N_2O_{15}$=Tetramic acid
$C_{35}H_{44}O_{16}$=Azadirachtin(s)
$C_{35}H_{46}O_{11}$=Tric(h)othecenes, Trichoverroids
$C_{35}H_{46}O_{12}$=Meliatoxins
$C_{35}H_{46}O_{13}$=Meliatoxins
$C_{35}H_{46}O_{14}$=Zaragozic acids
$C_{35}H_{46}O_{20}$=Echinacoside
$C_{35}H_{47}NO_9$=Rhizoxin
$C_{35}H_{47}NO_{10}$=Taxanes
$C_{35}H_{48}N_2O_{10}$=Eleuthosides
$C_{35}H_{48}N_8O_9S$=Phallotoxins
$C_{35}H_{48}N_8O_{10}S$=Phallotoxins
$C_{35}H_{48}N_8O_{11}S$=Phallotoxins
$C_{35}H_{48}N_8O_{12}S$=Phallotoxins
$C_{35}H_{50}N_8O_6S_2$=Patellamides
$C_{35}H_{50}O_{20}$=Sylvestrosides
$C_{35}H_{52}Cl_2O_6$=Nostocyclophanes
$C_{35}H_{52}O_4$=Hyperforin
$C_{35}H_{52}O_{15}$=Strophanthins
$C_{35}H_{53}O_8$=Tetronasine
$C_{35}H_{56}N_6O_5$=Dolastatins
$C_{35}H_{56}O_3$=Flavidulols
$C_{35}H_{56}O_8$=Lasalocids
$C_{35}H_{57}N_6O_3^+$=Batzelladines
$C_{35}H_{58}O_9$=Bafilomycins, Filipin
$C_{35}H_{58}O_{10}$=Filipin
$C_{35}H_{58}O_{11}$=Filipin
$C_{35}H_{59}ClO_7$=Cockroaches
$C_{35}H_{59}NO_8$=Myxovirescins
$C_{35}H_{59}NO_{13}$=Leucomycin
$C_{35}H_{60}N_{12}O_5$=Argiopinins
$C_{35}H_{60}N_{12}O_6$=Argiopinins
$C_{35}H_{60}N_{12}O_9$=Clavamine
$C_{35}H_{60}O_{11}$=Monensin
$C_{35}H_{61}ClO_7$=Cockroaches
$C_{35}H_{61}NO_7$=Pamamycins
$C_{35}H_{61}NO_8$=Myxovirescins
$C_{35}H_{61}NO_{12}$=Oleandomycin
$C_{35}H_{61}N_3O_9$=Enniatins
$C_{35}H_{62}O_4$=Hopanoids
$C_{35}H_{63}NO_7$=Myxovirescins
$C_{35}H_{64}O_7$=Annonaceous acetogenins
$C_{35}H_{64}O_8$=Annonaceous acetogenins
$C_{35}H_{72}$=Flies

$\mathbf{C_{36}}H_{18}O_{16}$=Badiones
$C_{36}H_{18}O_{18}$=Badiones
$C_{36}H_{32}O_{12}$=Leucomentins
$C_{36}H_{36}N_2O_5$=Bisbenzylisoquinoline alkaloids
$C_{36}H_{36}O_{10}$=Gnidicin
$C_{36}H_{44}N_4O$=Manzamines
$C_{36}H_{44}O_{12}$=Filixic acids
$C_{36}H_{45}BrN_4O_6$=Jaspamide
$C_{36}H_{45}ClN_6O_{12}$=Phomopsins
$C_{36}H_{45}N_9O_8$=Theonellamides
$C_{36}H_{46}Cl_2N_2O_{15}$=Tetramic acid

$C_{36}H_{46}N_4O_7$=Chondramides
$C_{36}H_{46}N_6O_{12}$=Phomopsins
$C_{36}H_{48}N_2O_8$=Ansatrienins
$C_{36}H_{48}N_2O_{12}$=Rhodomycins
$C_{36}H_{50}N_2O_8$=Ansatrienins
$C_{36}H_{50}O_{10}$=Sodagnitins
$C_{36}H_{51}NO_{11}$=Veratrum steroid alkaloids
$C_{36}H_{51}NO_{12}$=Aconitine
$C_{36}H_{52}N_8O_{14}S$=Virotoxins
$C_{36}H_{52}N_8O_{15}S$=Virotoxins
$C_{36}H_{52}O_8$=Ganoderic acids, Mancinellin
$C_{36}H_{52}O_{11}$=Reveromycins
$C_{36}H_{54}Cl_2O_6$=Nostocyclophanes
$C_{36}H_{54}O_3$=Toxicols
$C_{36}H_{54}O_{14}$=Strophanthins
$C_{36}H_{55}FeN_{10}O_{18}S$=Albomycins
$C_{36}H_{56}FeN_{11}O_{17}S$=Albomycins
$C_{36}H_{56}O_{11}$=Medicagenic acid
$C_{36}H_{58}N_4O_{11}S$=Topostatin
$C_{36}H_{58}N_4O_{16}$=Tunicamycins
$C_{36}H_{60}O_8$=Fasciculic acids
$C_{36}H_{60}O_9$=Fasciculic acids
$C_{36}H_{60}O_{30}$=Cyclodextrins
$C_{36}H_{61}NO_{16}$=Fumonisins
$C_{36}H_{62}O_{11}$=Monensin
$C_{36}H_{63}NO_7$=Pamamycins
$C_{36}H_{63}N_3O_9$=Enniatins
$C_{36}H_{63}N_{12}O_5^+$=Argiopinins
$C_{36}H_{63}N_{12}O_6^+$=Argiopinins
$C_{36}H_{64}N_{10}O_8$=JSTX
$C_{36}H_{71}NO_3$=Ceramides, Sphingosine

$C_{37}H_{33}N_7O_8$=CC-1065
$C_{37}H_{38}N_2O_6$=Bisbenzylisoquinoline alkaloids
$C_{37}H_{40}N_2O_6$=Bisbenzylisoquinoline alkaloids
$C_{37}H_{40}O_9$=Resiniferatoxin
$C_{37}H_{41}N_2O_6^+$=Bisbenzylisoquinoline alkaloids
$C_{37}H_{44}ClNO_4$=Penitrems
$C_{37}H_{44}ClNO_5$=Penitrems
$C_{37}H_{44}ClNO_6$=Penitrems
$C_{37}H_{44}O_{13}$=Urdamycins
$C_{37}H_{45}NO_4$=Penitrems
$C_{37}H_{45}NO_5$=Penitrems
$C_{37}H_{45}NO_6$=Penitrems
$C_{37}H_{46}N_8O_6S_2$=Patellamides
$C_{37}H_{46}O_{14}$=Urdamycins
$C_{37}H_{47}NO_{12}$=Rifamycins
$C_{37}H_{48}O_6$=Peridinin
$C_{37}H_{49}NO_6$=Janthitrems
$C_{37}H_{50}N_8O_{12}S$=Phallotoxins
$C_{37}H_{50}N_8O_{13}S$=Phallotoxins
$C_{37}H_{50}N_8O_{14}S$=Phallotoxins
$C_{37}H_{57}FeN_{12}O_{18}S$=Albomycins
$C_{37}H_{60}N_6O_7S$=Patellamides
$C_{37}H_{64}N_6O_2$=Batzelladines
$C_{37}H_{64}O_{11}$=Monensin
$C_{37}H_{66}FNO_{13}$=Fluoroorganic Natural Products
$C_{37}H_{66}O_7$=Annonaceous acetogenins
$C_{37}H_{66}O_8$=Annonaceous acetogenins, Mucocin
$C_{37}H_{67}NO_{13}$=Erythromycin
$C_{37}H_{68}O_7$=Annonaceous acetogenins
$C_{37}H_{70}N_{14}O_{11}$=Streptothricins

$C_{38}H_{34}N_6O_8$=Gilvusmycin
$C_{38}H_{38}O_{10}$=Mezerein
$C_{38}H_{42}N_2O_6$=Bisbenzylisoquinoline alkaloids, Phenethyltetrahydroisoquinoline alkaloids
$C_{38}H_{44}O_8$=Gamboge
$C_{38}H_{46}N_2O_8$=Truxillic and truxinic acids
$C_{38}H_{48}N_4O_2$=Truxillic and truxinic acids
$C_{38}H_{48}N_4O_{13}$=Aerobactin
$C_{38}H_{48}N_8O_6S_2$=Patellamides
$C_{38}H_{50}O_6$=Garcinol
$C_{38}H_{54}O_{14}$=Zaragozic acids
$C_{38}H_{55}NO_9$=Lituarines
$C_{38}H_{55}NO_{11}$=Lituarines
$C_{38}H_{56}N_8O_{14}S$=Virotoxins
$C_{38}H_{56}N_8O_{15}S$=Virotoxins
$C_{38}H_{56}N_8O_{16}S$=Virotoxins
$C_{38}H_{60}O_{11}$=Fasciculic acids
$C_{38}H_{60}O_{18}$=Stevioside
$C_{38}H_{65}N_{11}O_7$=JSTX
$C_{38}H_{70}O_6$=Annonaceous acetogenins
$C_{38}H_{75}NO_3$=Sphingosine

$C_{39}H_{43}N_3O_{11}S$=Ecteinascidins
$C_{39}H_{44}ClNO_9$=Naphthomycins
$C_{39}H_{45}NO_9$=Naphthomycins
$C_{39}H_{48}O_{13}$=Taxuspines (T. A–Z)
$C_{39}H_{49}NO_{14}$=Rifamycins
$C_{39}H_{49}NO_{16}$=Nogalamycin
$C_{39}H_{50}O_7$=Peridinin
$C_{39}H_{51}NO_6$=Janthitrems
$C_{39}H_{51}NO_7$=Janthitrems
$C_{39}H_{53}N_9O_{13}S$=Amanitins
$C_{39}H_{53}N_9O_{14}S$=Amanitins
$C_{39}H_{53}N_9O_{15}S$=Amanitins
$C_{39}H_{54}N_{10}O_{11}S$=Amanitins
$C_{39}H_{54}N_{10}O_{12}S$=Amanitins
$C_{39}H_{54}N_{10}O_{13}S$=Amanitins
$C_{39}H_{54}N_{10}O_{14}S$=Amanitins
$C_{39}H_{54}O_{13}$=Zaragozic acids
$C_{39}H_{58}O_4$=Ubiquinones
$C_{39}H_{61}NO_{12}$=Veratrum steroid alkaloids
$C_{39}H_{63}N_5O_{12}$=Mycalamides
$C_{39}H_{64}O_{10}$=Concanamycins
$C_{39}H_{65}NO_{14}$=Leucomycin, Tylosin
$C_{39}H_{65}N_5O_9$=Microcolin A
$C_{39}H_{66}O_2S_2$=Caldariellaquinone
$C_{39}H_{68}O_7$=Annonaceous acetogenins
$C_{39}H_{80}$=Flies

$C_{40}H_{27}NO_{11}$=Purpurone
$C_{40}H_{27}NO_{12}$=Purpurone
$C_{40}H_{34}N_2O_8$=Hypericin
$C_{40}H_{36}Cl_2N_6O_7$=Diazonamides
$C_{40}H_{36}CuN_4O_{16}$=Uroporphyrins
$C_{40}H_{36}O_{12}$=Sanggenon D
$C_{40}H_{38}N_4O_{16}$=Uroporphyrins
$C_{40}H_{44}Cl_2N_4O$=Toxiferines
$C_{40}H_{44}N_2O_{18}$=Pradimicins
$C_{40}H_{44}N_4O^{2+}$=Toxiferines
$C_{40}H_{44}N_4O_{16}$=Uroporphyrins
$C_{40}H_{46}ClNO_9$=Naphthomycins
$C_{40}H_{46}Cl_2N_4O_2$=*C*-Toxiferine I
$C_{40}H_{46}N_4O_2^{2+}$=*C*-Toxiferine I
$C_{40}H_{46}N_4O_8$=Coproporphyrins
$C_{40}H_{46}N_4O_{10}$=Tolyporphin A
$C_{40}H_{47}NO_9$=Naphthomycins
$C_{40}H_{47}NO_{10}$=Naphthomycins
$C_{40}H_{48}Cl_2N_2O_2$=Toxiferines
$C_{40}H_{48}N_4O_2^{2+}$=Toxiferines
$C_{40}H_{50}O_2$=Rhodoxanthin
$C_{40}H_{50}O_{12}$=Zaragozic acids
$C_{40}H_{50}O_{14}$=Zaragozic acids
$C_{40}H_{51}NO_{14}$=Streptovaricin
$C_{40}H_{52}O_2$=Canthaxanthin
$C_{40}H_{52}O_4$=Astaxanthin
$C_{40}H_{52}O_{17}$=Phyllanthocin
$C_{40}H_{56}$=β-Carotene, Carotenes, Lycopene
$C_{40}H_{56}O$=Aleuriaxanthin, Cryptoxanthin, Rubixanthin
$C_{40}H_{56}O_2$=Aleuriaxanthin, Lutein, Zeaxanthin
$C_{40}H_{56}O_3$=Antheraxanthin, Capsanthin, Flavoxanthin
$C_{40}H_{56}O_4$=Capsorubin, Neoxanthin, Violaxanthin
$C_{40}H_{56}O_{23}S$=Breynins
$C_{40}H_{56}O_{24}S$=Breynins
$C_{40}H_{57}NO_{12}$=Lituarines
$C_{40}H_{57}N_5O_{10}$=Theonellamides
$C_{40}H_{58}$=Neurosporene
$C_{40}H_{58}O_4$=Zeaxanthin
$C_{40}H_{60}BNaO_{14}$=Aplasmomycins
$C_{40}H_{60}N_4O_{10}$=Bufadienolides
$C_{40}H_{60}O_2$=Retinol
$C_{40}H_{60}O_6$=Mactraxanthin
$C_{40}H_{62}$=Phytofluene
$C_{40}H_{64}$=Phytoene
$C_{40}H_{64}O_{12}$=Nonactin
$C_{40}H_{66}$=Lycopene
$C_{40}H_{66}N_4O_{16}$=Tunicamycins
$C_{40}H_{67}NO_{14}$=Leucomycin
$C_{40}H_{68}O_{11}$=Nigericin
$C_{40}H_{77}NO_3$=Sphingosine
$C_{40}H_{82}$=Flies

$C_{41}H_{30}O_{11}$=Blepharismins
$C_{41}H_{42}N_4O_{16}$=Factor I
$C_{41}H_{44}O_7$=Lathyranes
$C_{41}H_{58}FeN_9O_{20}$=Ferrichromes
$C_{41}H_{58}O_9$=Melianol
$C_{41}H_{60}O_{15}$=Bryostatins
$C_{41}H_{61}NO_9$=Pinnatoxins
$C_{41}H_{61}N_5O_{10}S$=Tubulysines
$C_{41}H_{64}FeN_9O_{17}$=Siderophores
$C_{41}H_{64}O_{13}$=Digitalis glycosides, Gymnemic acids
$C_{41}H_{64}O_{14}$=Digitalis glycosides
$C_{41}H_{64}O_{17}S$=Holothurins
$C_{41}H_{65}FeN_{10}O_{14}$=Ferrimycins
$C_{41}H_{65}NO_{10}$=Spinosyns (Lepicidins)
$C_{41}H_{66}O_{12}$=α-Hederin, Nonactin
$C_{41}H_{66}O_{13}$=Gymnemic acids, Tautomycin
$C_{41}H_{69}BrO_{11}$=Oscillariolide
$C_{41}H_{72}O_9$=Ionomycin

$C_{42}H_{32}O_9$=Wine, phenolic constituents of
$C_{42}H_{35}N_3O_{11}$=Storniamides
$C_{42}H_{38}O_{14}$=Leucomentins
$C_{42}H_{38}O_{20}$=Sennosides
$C_{42}H_{40}O_{19}$=Sennosides
$C_{42}H_{44}FeN_4O_{16}$=Siroheme
$C_{42}H_{46}N_4O_{16}$=Siroheme
$C_{42}H_{46}O_5$=Longithorones
$C_{42}H_{47}ClO_{23}$=Violanin
$C_{42}H_{47}O_{23}^+$=Violanin
$C_{42}H_{48}O_{11}$=Taxuspines (T. A–Z)
$C_{42}H_{51}ClN_6NiO_{17}$=Factor F 430
$C_{42}H_{53}NO_{15}$=Aclarubicin
$C_{42}H_{57}FeN_{12}O_{16}$=Pseudobactin
$C_{42}H_{58}O_6$=Peridinin
$C_{42}H_{62}BNaO_{15}$=Aplasmomycins
$C_{42}H_{62}O_{16}$=Glycyrrhizin
$C_{42}H_{63}N_5O_{10}S$=Tubulysines
$C_{42}H_{64}Cl_2O_{11}$=Nostocyclophanes
$C_{42}H_{64}O_{19}$=Strophanthins
$C_{42}H_{67}NO_{10}$=Spinosyns (Lepicidins)
$C_{42}H_{67}NO_{15}$=Carbomycin
$C_{42}H_{67}NO_{16}$=Carbomycin
$C_{42}H_{68}N_6O_6S$=Dolastatins
$C_{42}H_{68}O_{12}$=Elaiophylin, Nonactin
$C_{42}H_{69}NO_{15}$=Leucomycin
$C_{42}H_{70}O_{11}$=Salinomycin
$C_{42}H_{73}N_9O_4$=Batzelladines
$C_{43}H_{42}O_{22}$=Carthamin
$C_{43}H_{48}O_{16}$=Saquayamycins
$C_{43}H_{50}N_4O_6$=Vincarubine
$C_{43}H_{50}O_{16}$=Saquayamycins
$C_{43}H_{52}O_{16}$=Saquayamycins
$C_{43}H_{53}NO_{14}$=Taxol®
$C_{43}H_{54}O_4$=Flexirubin
$C_{43}H_{56}O_{17}$=Urdamycins
$C_{43}H_{58}N_4O_{12}$=Rifamycins
$C_{43}H_{58}O_{18}$=Urdamycins
$C_{43}H_{60}N_2O_{12}$=Kirromycin
$C_{43}H_{62}N_2O_{14}S_5$=Namenamicin
$C_{43}H_{64}O_{11}$=Gambierol
$C_{43}H_{65}N_5O_{10}S$=Tubulysines
$C_{43}H_{66}O_{14}$=Gymnemic acids
$C_{43}H_{68}O_{14}$=Gymnemic acids
$C_{43}H_{69}NO_{12}$=Ascomycin
$C_{43}H_{70}O_{12}$=Nonactin
$C_{43}H_{71}N_6O_6S$=Dolastatins
$C_{43}H_{72}O_{11}$=Narasin
$C_{43}H_{74}N_2O_{14}$=Spiramycin
$C_{43}H_{82}N_{16}O_{12}$=Streptothricins

$C_{44}H_{58}O_{17}S$=Urdamycins
$C_{44}H_{60}N_4O_{12}$=Halichondramide
$C_{44}H_{62}N_2O_{12}$=Efrotomycin
$C_{44}H_{64}BNaO_{16}$=Aplasmomycins
$C_{44}H_{64}O_{14}$=Cryptoporic acids
$C_{44}H_{64}O_{24}$=Crocetin, Crocin
$C_{44}H_{65}NO_{13}$=Bafilomycins
$C_{44}H_{66}O_4$=Ubiquinones
$C_{44}H_{68}O_{13}$=Okadaic acid

$C_{44}H_{68}O_{13}S$=Okadaic acid
$C_{44}H_{69}NO_{12}$=FK-506, tsukubaenolide
$C_{44}H_{69}N_5O_{10}$=Mycobactins
$C_{44}H_{72}O_{11}$=Oligomycins
$C_{44}H_{72}O_{12}$=Nonactin

$\mathbf{C_{45}}H_{54}N_2O_{12}S$=Naphthomycins
$C_{45}H_{65}N_3O$=Myrmicarins
$C_{45}H_{68}O_{15}$=Cryptoporic acids
$C_{45}H_{70}O_{13}$=Okadaic acid
$C_{45}H_{70}O_{19}$=Echinoderm saponins
$C_{45}H_{72}O_{12}$=Oligomycins
$C_{45}H_{72}O_{13}$=Oligomycins
$C_{45}H_{72}O_{14}$=Hebevinosides
$C_{45}H_{72}O_{16}$=Steroid saponins
$C_{45}H_{73}NO_{14}$=Solanum steroid alkaloid glycosides
$C_{45}H_{73}NO_{15}$=Solanum steroid alkaloid glycosides
$C_{45}H_{73}NO_{16}$=Solanum steroid alkaloid glycosides
$C_{45}H_{74}BNO_{15}$=Boromycin
$C_{45}H_{74}O_{10}$=Oligomycins
$C_{45}H_{74}O_{11}$=Oligomycins
$C_{45}H_{74}O_{13}$=Concanamycins
$C_{45}H_{74}O_{17}$=Osladin
$C_{45}H_{76}N_2O_{15}$=Spiramycin
$C_{45}H_{80}N_6O_5$=Ptilomycalins
$C_{45}H_{80}N_6O_7$=Ptilomycalins

$\mathbf{C_{46}}H_{30}O_{10}$=Aurantiacin
$C_{46}H_{48}BrN_5O_{14}$=Celenamides
$C_{46}H_{48}N_2O_8$=Michellamines
$C_{46}H_{56}N_2O_{12}S$=Naphthomycins
$C_{46}H_{56}N_4O_8$=15′,20′-Anhydrovinblastine
$C_{46}H_{56}N_4O_{10}$=Vincristine
$C_{46}H_{58}N_4O_9$=Vinblastine
$C_{46}H_{62}BrN_9O_{10}$=Theonellamides
$C_{46}H_{64}N_4O_{13}$=Ulapualides
$C_{46}H_{72}O_{14}$=Tartrolones
$C_{46}H_{75}NO_{14}$=Concanamycins
$C_{46}H_{76}O_{11}$=Oligomycins
$C_{46}H_{77}NO_{17}$=Tylosin
$C_{46}H_{78}N_2O_{15}$=Spiramycin
$C_{46}H_{79}NO_{17}$=Tylosin

$\mathbf{C_{47}}H_{51}NO_{14}$=Taxol®
$C_{47}H_{64}O_{17}$=Papulacandins
$C_{47}H_{66}O_{10}$=Sorangicins
$C_{47}H_{66}O_{11}$=Sorangicins
$C_{47}H_{66}O_{16}$=Papulacandins
$C_{47}H_{68}O_{17}$=Bryostatins
$C_{47}H_{70}O_{14}$=Avermectins, Pectenotoxins
$C_{47}H_{70}O_{15}$=Yessotoxin
$C_{47}H_{71}NO_{12}$=Azaspiracid
$C_{47}H_{71}NO_{17}$=Candidin
$C_{47}H_{72}O_{14}$=Ivermectin
$C_{47}H_{72}O_{15}$=Avermectins
$C_{47}H_{73}NO_{17}$=Amphotericin B
$C_{47}H_{74}O_{18}$=Digitalis glycosides
$C_{47}H_{74}O_{19}$=Digitalis glycosides
$C_{47}H_{75}NO_{16}$=Solanum steroid alkaloid glycosides
$C_{47}H_{75}NO_{17}$=Nystatin
$C_{47}H_{80}O_{17}$=Maduramicin α

$\mathbf{C_{48}}H_{46}N_4O_{19}$=Protorubradirin
$C_{48}H_{46}N_4O_{20}$=Protorubradirin
$C_{48}H_{49}N_{13}O_9S_6$=Micrococcin P_1
$C_{48}H_{49}N_{13}O_{10}S_6$=Thiocillins
$C_{48}H_{71}N_5O_{14}$=Ulapualides
$C_{48}H_{72}O_{14}$=Avermectins
$C_{48}H_{74}Cl_2O_{16}$=Nostocyclophanes
$C_{48}H_{74}O_{14}$=Ivermectin
$C_{48}H_{74}O_{15}$=Avermectins
$C_{48}H_{78}O_{19}$=Asiaticoside

$\mathbf{C_{49}}H_{51}N_{13}O_9S_6$=Thiocillins
$C_{49}H_{51}N_{13}O_{10}S_6$=Thiocillins
$C_{49}H_{54}ClN_5O_{10}$=Pepticinnamin E
$C_{49}H_{56}ClFeN_4O_6$=Cytohemin
$C_{49}H_{58}N_4O_6$=Cytohemin
$C_{49}H_{58}O_{18}$=Vineomycins
$C_{49}H_{69}ClO_{14}$=Brevetoxins
$C_{49}H_{70}O_{13}$=Brevetoxins
$C_{49}H_{74}O_4$=Ubiquinones
$C_{49}H_{74}O_{14}$=Avermectins
$C_{49}H_{75}N_{13}O_{12}$=Microcystins
$C_{49}H_{76}O_{15}$=Avermectins
$C_{49}H_{76}O_{19}$=Digitalis glycosides
$C_{49}H_{76}O_{20}$=Digitalis glycosides
$C_{49}H_{78}N_6O_{12}$=Didemnins
$C_{49}H_{94}N_{18}O_{13}$=Streptothricins

$\mathbf{C_{50}}H_{60}O_{18}$=Urdamycins
$C_{50}H_{63}ClO_{16}$=Chlorothricin
$C_{50}H_{70}O_{14}$=Brevetoxins
$C_{50}H_{80}N_8O_{12}$=Majusculamides
$C_{50}H_{81}N_4O_{15}P$=Calyculins
$C_{50}H_{83}NO_{20}$=Solanum steroid alkaloid glycosides
$C_{50}H_{83}NO_{21}$=Solanum steroid alkaloid glycosides
$C_{50}H_{86}O_{21}$=Ipomoea resins

$\mathbf{C_{51}}H_{43}N_{13}O_{12}S_6$=Nosiheptide
$C_{51}H_{60}O_{19}$=Urdamycins
$C_{51}H_{63}FeN_6O_9$=Trichostatins
$C_{51}H_{74}N_4O_{16}$=Ulapualides
$C_{51}H_{79}NO_{13}$=Rapamycin
$C_{51}H_{82}N_8O_{17}$=Pneumocandins
$C_{51}H_{83}N_4O_{15}P$=Calyculins
$C_{51}H_{84}O_{22}$=Steroid saponins
$C_{51}H_{85}NO_{21}$=Solanum steroid alkaloid glycosides
$C_{51}H_{98}O$=Cockroaches

$\mathbf{C_{52}}H_{70}O_7$=Tridentoquinone
$C_{52}H_{76}O_{24}$=Aureolic acid
$C_{52}H_{81}N_7O_{16}$=Echinocandin B
$C_{52}H_{82}N_6O_{14}$=Didemnins
$C_{52}H_{91}N_{13}O_{15}$=Herbicolins

$\mathbf{C_{53}}H_{60}ClN_3O_{16}$=Kedarcidin
$C_{53}H_{61}NO_{18}$=Urdamycins
$C_{53}H_{71}BrN_2O_{13}$=Phorboxazoles
$C_{53}H_{76}O_{15}$=Sorangicins
$C_{53}H_{76}O_{16}$=Sorangicins
$C_{53}H_{80}O_2$=Plastoquinones

$\mathbf{C_{54}}H_{74}N_2O_{10}$=Cephalostatins
$C_{54}H_{76}N_2O_{10}$=Ritterazines
$C_{54}H_{77}BrN_{10}O_{12}$=Theonellamides
$C_{54}H_{82}O_4$=Ubiquinones
$C_{54}H_{85}NaO_{27}S$=Holothurins
$C_{54}H_{88}O_{18}$=Elaiophylin
$C_{54}H_{90}N_6O_{18}$=Valinomycin
$C_{54}H_{96}O_{21}$=Arthrobacilins

$\mathbf{C_{55}}H_{69}ClN_{10}O_{14}$=Hormaomycin
$C_{55}H_{70}MgN_4O_6$=Chlorophylls
$C_{55}H_{72}MgN_4O_5$=Chlorophylls
$C_{55}H_{74}IN_3O_{21}S_4$=Calicheamicins
$C_{55}H_{75}FeN_{16}O_{28}$=Siderophores
$C_{55}H_{76}N_2O_{11}$=Cephalostatins
$C_{55}H_{82}O_{21}S_2$=Yessotoxin
$C_{55}H_{83}N_{17}O_{22}$=Pyoverdins
$C_{55}H_{84}N_{17}O_{21}S_3^+$=Bleomycins
$C_{55}H_{84}N_{18}O_{21}$=Pyoverdins
$C_{55}H_{85}N_{20}O_{21}S_2^+$=Bleomycins
$C_{55}H_{86}O_{24}$=Escin
$C_{55}H_{94}O_{17}$=Oasomycins
$C_{55}H_{96}N_{16}O_{13}$=Polymyxins
$C_{55}H_{106}N_{20}O_{14}$=Streptothricins

$\mathbf{C_{56}}H_{42}O_{12}$=Wine, phenolic constituents of
$C_{56}H_{76}N_8O_{16}$=Nostocyclin
$C_{56}H_{83}N_{17}O_{23}$=Pyoverdins
$C_{56}H_{92}O_{29}$=Steroid saponins
$C_{56}H_{95}ClN_{12}O_{16}$=Puwainaphycins
$C_{56}H_{98}N_{16}O_{13}$=Polymyxins
$C_{56}H_{100}O_{21}$=Arthrobacilins

$\mathbf{C_{57}}H_{82}O_{26}$=Chromomycins
$C_{57}H_{89}N_7O_{15}$=Didemnins
$C_{57}H_{96}O_{28}$=Steroid saponins

$\mathbf{C_{58}}H_{84}N_{14}O_{16}S$=Physal(a)emin
$C_{58}H_{84}O_{26}$=Olivomycins
$C_{58}H_{101}N_{13}O_{20}$=Herbicolins
$C_{58}H_{102}N_{12}O_{13}$=Trichorovins
$C_{58}H_{104}O_{21}$=Arthrobacilins

$\mathbf{C_{59}}H_{80}N_4O_{22}S_4$=Esperamycins
$C_{59}H_{80}N_4O_{22}S_5$=Esperamycins
$C_{59}H_{84}Cl_2O_{32}$=Avilamycins
$C_{59}H_{84}N_2O_{18}$=Candicidin
$C_{59}H_{88}N_2O_{20}$=Efrotomycin
$C_{59}H_{90}O_4$=Ubiquinones
$C_{59}H_{92}O_4$=Ubiquinones

$\mathbf{C_{60}}H_{86}O_{19}$=Ciguatoxin, Halistatins
$C_{60}H_{86}O_{20}$=Halistatins
$C_{60}H_{91}N_5O_{13}$=Sanglifehrins
$C_{60}H_{92}N_8O_{11}$=Aureobasidins
$C_{60}H_{92}N_{12}O_{10}$=Gramicidins

$\mathbf{C_{61}}H_{88}Cl_2O_{32}$=Avilamycins
$C_{61}H_{90}Cl_2O_{32}$=Avilamycins
$C_{61}H_{104}O_{22}$=Oasomycins
$C_{61}H_{109}NO_{21}$=Desertomycins

$\mathbf{C_{62}}H_{86}N_{12}O_{16}$=Actinomycins
$C_{62}H_{96}N_5O_{28}P$=Moenomycin
$C_{62}H_{111}N_{11}O_{12}$=Cyclosporins
$C_{62}H_{115}NO_{24}$=Aflastatin A

$\mathbf{C_{63}}H_{88}CoN_{14}O_{14}P$=Vitamin B_{12}
$C_{63}H_{95}ClO_{21}$=Spongistatins
$C_{63}H_{96}O_{21}$=Spongistatins
$C_{63}H_{98}N_5O_{28}P$=Moenomycin
$C_{63}H_{98}N_5O_{29}P$=Moenomycin
$C_{63}H_{98}N_{18}O_{13}S$=Substance P

$\mathbf{C_{64}}H_{78}N_{10}O_{10}$=Antamanide

$\mathbf{C_{65}}H_{85}N_{13}O_{30}$=Mycobacillin

$\mathbf{C_{66}}H_{75}Cl_2N_9O_{24}$=Vancomycin
$C_{66}H_{87}N_{13}O_{13}$=Tyrocidines
$C_{66}H_{103}N_{17}O_{16}S$=Bacitracin

$\mathbf{C_{68}}H_{88}N_{14}O_{13}$=Tyrocidines
$C_{68}H_{130}N_2O_{23}P_2$=Lipid A

$\mathbf{C_{69}}H_{86}Br_2N_{16}O_{22}$=Theonellamides
$C_{69}H_{108}N_5O_{34}P$=Moenomycin
$C_{69}H_{112}O_{21}$=Ipomoea resins

$\mathbf{C_{70}}H_{89}N_{15}O_{13}$=Tyrocidines
$C_{70}H_{97}Cl_2NO_{38}$=Ziracin
$C_{70}H_{125}N_{13}O_{16}$=Theonellamides
$C_{70}H_{131}N_{19}O_{15}$=Mastoparans

$\mathbf{C_{71}}H_{110}N_{24}O_{18}S$=Bombesin
$C_{71}H_{127}N_{13}O_{16}$=Theonellamides

$\mathbf{C_{72}}H_{85}N_{19}O_{18}S_5$=Thiostrepton
$C_{72}H_{104}N_{14}O_{20}$=Himastatin
$C_{72}H_{110}N_{12}O_{20}$=Plipastatins
$C_{72}H_{116}O_4$=Helenien, Zeaxanthin

$\mathbf{C_{73}}H_{112}N_{12}O_{20}$=Plipastatins
$C_{73}H_{126}N_{20}O_{15}S$=Mastoparans
$\mathbf{C_{74}}H_{114}N_{12}O_{20}$=Plipastatins
$C_{74}H_{123}N_{13}O_{18}$=Omphalotins

$\mathbf{C_{75}}H_{116}N_{12}O_{20}$=Plipastatins

$\mathbf{C_{76}}H_{52}O_{46}$=Tannins

$\mathbf{C_{77}}H_{116}N_{20}O_{22}S$=Discodermin A

$\mathbf{C_{78}}H_{132}O_{20}$=Swinholides

$\mathbf{C_{79}}H_{109}N_3O_{36}$=Viriplanins
$C_{79}H_{109}N_3O_{38}$=Viriplanins
$C_{79}H_{131}N_{31}O_{24}S_4$=Apamin

$\mathbf{C_{88}}H_{97}Cl_2N_9O_{33}$=Teicoplanins

$\mathbf{C_{98}}H_{141}N_{25}O_{23}S_4$=Epidermin

$\mathbf{C_{99}}H_{140}N_{20}O_{17}$=Gramicidins

$\mathbf{C_{100}}H_{164}O$=Dolichols

$\mathbf{C_{110}}H_{192}N_{40}O_{24}S_4$=Mast cell degranulating peptide

$\mathbf{C_{129}}H_{223}N_3O_{54}$=Palytoxin

$\mathbf{C_{131}}H_{229}N_{39}O_{31}$=Melittin

$\mathbf{C_{143}}H_{230}N_{42}O_{37}S_7$=Nisin

$\mathbf{C_{148}}H_{227}N_{39}O_{38}S_5$=Subtilin

$\mathbf{C_{153}}H_{256}N_{48}O_{39}S_3$=Epidermin

$\mathbf{C_{164}}H_{256}Na_2O_{68}S_2$=Maitotoxin

$\mathbf{C_{176}}H_{277}N_{57}O_{55}S_7$=Charybdotoxin

$\mathbf{C_{345}}H_{523}N_{93}O_{116}S_4$=Glycosidase inhibitors

NO=Nitrogen monoxide

Index of Latin species names

A

E

I

J

N

O

P

Stephania sasakii [Bisbenzylisoquinoline alkaloids]
Stephania sutchuenensis [1-Nitroaknadinine]
Stephania tetrandra [Bisbenzylisoquinoline alkaloids]
Stephanospora caroticolor [Drosophilin A]
Sterculia setigera [D-Tagatose]
Stereum complicatum [Complicatic acid]
Stereum hirsutum [Hirsutic acid, Mycosporins]
Stereum purpureum [Isolactaranes, Sterepolide, Sterpuranes]
Stevia rebaudiana [Steviol, Stevioside]
Sticta [Sticticin]
Sticta coronata [Polyporic acid]
Stigmatella aurantiaca [Aurachins, Stigmatellin]
Stilbe ericoides [Unedoside]
Stiretrus anchorago [Bugs]
Stizolobium [α-3,4-Dihydroxyphenylalanine]
Stizolobium hassjoo [Stizolobic acid]
Stomoxys calcitrans [Flies]
Streptococcus [Lactobacillic acid]
Streptococcus cremoris [Nisin]
Streptococcus faecalis [Enterotoxins]
Streptococcus faecium [Streptindole]
Streptococcus lactis [Nisin]
Streptomyces [Allosamidin, (*S*)-2-Amino-4-pentynoic acid, 2-Aminophenol, Anthramycins, Azaserine, Blasticidins, Butenolide(s), Cephamycins, Chloramphenicol, Clavams, Deoxynojirimycin, Efrotomycin, Erbstatin, Fungi, Geosmin, Gilvusmycin, Holomycin, 3-Hydroxyanthranilic acid, Juglomycins, Lymphostin, Madindolines, Majusculamides, Narasin, Nebularine, Neopyrrolomycin, Oxetin, Panclicins, Pepticinnamin E, Phenazinomycin, Phthoxazolin A, Polyoxins, Reveromycins, Rifamycins, Sakyomycins, Serpentene, Statine, Staurosporine, Steroids, Thiolactomycin, Triacsins, Tunicamycins, Vulgamycin]
Streptomyces sp. [Steffimycin]
Streptomyces aburaviensis [Chromomycins, Ebelactones]
Streptomyces achromogenes [Nitroso compounds, Protorubradirin, Streptozocin]
Streptomyces actuosus [Nosiheptide]
Streptomyces aizunensis [Bicyclomycin]
Streptomyces akiyoshiensis [5-Hydroxy-4-oxo-L-norvaline]
Streptomyces alanosinicus [L-Alanosine]
Streptomyces albireticuli [Enteromycin]
Streptomyces alboniger [Pamamycins, Puromycin]
Streptomyces albus [Bacimethrin, Holomycin, Indolmycin, Ionophores, Salbostatin, Salinomycin, Sulfinemycin]
Streptomyces amakusaensis [Nagstatin, Tuberin]
Streptomyces ambofaciens [2-Amino-6-diazo-5-oxohexanoic acid, Spiramycin]
Streptomyces antibioticus [Boromycin, Chlorothricin, Clavams, Cyclophostin, Ferrithiocin, Indanomycin, Ionophores, Oleandomycin, Pentostatin, Rubromycins]
Streptomyces arenae [Pentalenolactones]
Streptomyces argenteolus [Pepstatin A]
Streptomyces argillaceus [Aureolic acid]
Streptomyces armeniacus [Streptopyrrole]
Streptomyces aurantiacus [Pamamycins]
Streptomyces aureofaciens [Tetracyclines]
Streptomyces aureus [Azirinomycin, Thiostrepton]
Streptomyces avermitilis [Avermectins]
Streptomyces avidinii [Streptavidin]
Streptomyces bambergiensis [Moenomycin]
Streptomyces bobiliae [Rhodomycins]
Streptomyces cacaoi [Polyoxins]
Streptomyces cacaoi var. *asoensis* [Octosyl acids]
Streptomyces caeruleus [Caerulomycin A]
Streptomyces caespitosus [Mitomycins]
Streptomyces calvus [Nucleocidin]
Streptomyces carzinostaticus [Neocarzilins, Neocarzinostatins]
Streptomyces cattleya [Carbapenems, Fluoroorganic Natural Products, Thienamycins]
Streptomyces cerevisiae [Fluoroorganic Natural Products]
Streptomyces chartreusis [Calcimycin, Chartreusin, Ionophores]
Streptomyces chromofuscus [Herboxidiene, Pentalenolactones]
Streptomyces cinnamonensis [Ionophores, Monensin]
Streptomyces clavuligerus [Clavams, Clavulanic acid]
Streptomyces coelicolor [Actinorhodin, Methylenomycins]
Streptomyces coerulescens [Gilvocarcins]
Streptomyces collinus [Ansatrienins, Kirromycin, Rubromycins]
Streptomyces conglobatus [Ionomycin, Ionophores]
Streptomyces cremeus [Cremeomycin]
Streptomyces cyanoflavus [Pyocyanine]
Streptomyces diastatochromogenes [Concanamycins, Oligomycins]
Streptomyces diastatochromogenes var. *juglonensis* [Juglomycins]
Streptomyces distallicus [Distamycins]
Streptomyces durhamensis [Filipin]
Streptomyces elgreteus [Steffimycin]
Streptomyces erythraeus [Erythromycin, Fluoroorganic Natural Products]
Streptomyces erythrochromogenes [Sarkomycin A]
Streptomyces felleus [Picromycin]
Streptomyces fervens [Fervenulin]
Streptomyces ficellus [Ficellomycin]
Streptomyces filipinensis [Cyclothialidine, Filipin]
Streptomyces flaveolus [Sanglifehrins, Tirandamycins]
Streptomyces flavofungini [Desertomycins]
Streptomyces flavopersicus [Spectinomycin]
Streptomyces flocculus [Streptonigrin]
Streptomyces floridae [Viomycin]
Streptomyces fradiae [Neomycin, Tylosin, Urdamycins]
Streptomyces fulvissimus [Ionophores, Valinomycin]
Streptomyces fumanus [Pyrrolomycins]
Streptomyces galbus [Galbonolides]
Streptomyces galilaeus [Aclarubicin]
Streptomyces gilvotanareus [Gilvocarcins]
Streptomyces glaucescens [Tetracenomycins]
Streptomyces goldiniensis [Efrotomycin]
Streptomyces griseoaurantiacus [Griseolic acids]
Streptomyces griseocarneus [Streptomycin]
Streptomyces griseochromogenes [Fluoroorganic Natural Products, Pentalenene, Tautomycin]
Streptomyces griseoflavus [Ferrimycins, Hormaomycin]
Streptomyces griseofuscus [Carzinophilin]
Streptomyces griseolus [Anisomycin, Sinefungin]
Streptomyces griseoplanus [Chlorotetaine]
Streptomyces griseosporeus [FK-409]
Streptomyces griseoviridis [Bactobolin, Roseophilin]
Streptomyces griseus [A-factor, Aplasmomycins, Bafilomycins, Candicidin, Chromomycins, Fredericamycin A, Holomycin, Indolmycin, Streptomycin, Vitamin B_{12}, Zincophorin]
Streptomyces halstedii [Carbomycin]
Streptomyces hepaticus [Elaiomycin]
Streptomyces hygroscopicus [Ascomycin, Bialaphos, Clavams, Glycosidase inhibitors, Herbimycin, Himastatin, Hydantocidin, Hygromycins, Ionophores, Milbemycins, Nigericin, Rapamycin, Trichostatins, Valienamine]
Streptomyces incarnatus [Sinefungin]
Streptomyces jumonjinensis [Clavulanic acid]
Streptomyces kanamyceticus [Kanamycins]
Streptomyces kasugaensis [Kasugamycin]
Streptomyces katsurahamanus [Clavulanic acid]
Streptomyces kitasatoensis [Leucomycin]
Streptomyces lasaliensis [Lasalocids]
Streptomyces lavendofoliae [Fosfazinomycin A]
Streptomyces lavendulae [Cyclophostin, Deoxymannojirimycin, Esterastin, Lavendamycin, Mannojirimycin, Mimosamycin, Mitomycins, Nojirimycin, Saframycins]
Streptomyces lincolnensis [Lincomycins]
Streptomyces longisporoflavus [Tetronasine]
Streptomyces lusitanus [Naphthyridinomycin A]
Streptomyces lydicus [Galactostatin]
Streptomyces matensis vineus [Vineomycins]
Streptomyces mediocidicus [Teleocidins]
Streptomyces melanosporofaciens [Elaiophylin]
Streptomyces melanovinaceus [Quinocarcin]
Streptomyces mirabilis [Tubercidin]
Streptomyces misawanensis [Aquayamycin]
Streptomyces morookaense [UK-1]
Streptomyces murayamaensis [Kinamycins]
Streptomyces natalensis [Natamycin]
Streptomyces nitrosporeus [Virantmycin]
Streptomyces noboritoensis [Hygromycins]
Streptomyces nodosus [Amphotericin B, Saquayamycins]
Streptomyces nogalater [Nogalamycin]
Streptomyces noursei [Nystatin]
Streptomyces olivaceus [Pyrrolizidine alkaloids, Streptogramins]
Streptomyces olivoreticuli [Bestatin, Olivomycins]
Streptomyces omiyaensis [Pentalenolactones]
Streptomyces orchidaceus [Cycloserine]
Streptomyces parvulus [Manumycins]
Streptomyces peucetius [Adriamycin, Daunorubicin]
Streptomyces phaeochromogenes [Ikarugamycin]
Streptomyces pilosus [Ferrioxamines]
Streptomyces pimprina [Holomycin]
Streptomyces platensis [Trichostatins]
Streptomyces prasinopilosus [Maniwamycins]
Streptomyces pristinaespiralis [Streptogramins]
Streptomyces prunicolor [Pironetin]
Streptomyces pulveraceus [Epiderstatin, Fostriecin]
Streptomyces puniceus [Viomycin]
Streptomyces purpurascens [Rhodomycins]
Streptomyces ramulosus [Acetomycin]
Streptomyces refuineus [Anthramycins]
Streptomyces resistomycificus [Resistomycin]
Streptomyces rimosus [Tetracyclines]
Streptomyces rosa notoensis [Nanaomycins]
Streptomyces rubrireticuli [Fervenulin]
Streptomyces saganonensis [Herbicidins]
Streptomyces sahachiroi [Carzinophilin]
Streptomyces sapporonensis [Bicyclomycin]
Streptomyces setonii [Cispentacin]
Streptomyces showdoensis [Showdomycin]
Streptomyces sioyaensis [Trichostatins]
Streptomyces spadicogriseus [Anthramycins]
Streptomyces sparsogenes [Sparsomycin]
Streptomyces spectabilis [Spectinomycin, Streptovaricin]

T